Y0-ABY-897

IES
LIGHTING
HANDBOOK

Fifth Edition

ILLUMINATING
ENGINEERING
SOCIETY

Founded 1906

"Its object shall be . . . the advancement of the theory and practice of illuminating engineering and the dissemination of knowledge relating thereto."

IES LIGHTING HANDBOOK

The Standard Lighting Guide

Fifth Edition
1972

JOHN E. KAUFMAN
Editor

JACK F. CHRISTENSEN
Associate Editor

Published by the
ILLUMINATING ENGINEERING SOCIETY
345 East 47th Street, New York, N.Y. 10017

ISBN 0-87995-000-5
Library of Congress Catalog Card Number: 77-186864

Composed and Printed by
WAVERLY PRESS, INC.
Baltimore, Maryland
1972

PRINTED IN THE UNITED STATES OF AMERICA

PREFACE

This fifth edition of the *IES Lighting Handbook* is an expanded and up-to-date version of the 1966 fourth edition (first edition 1949, second edition 1952, third edition 1959)—with the continued objective of providing essential information on light and lighting in a simple, condensed style.

Each Handbook section has been reviewed for content and accuracy either by the Society's technical committees, under the counsel of the Handbook Editor as Technical Director, or where outside of the scope of the Society's committee structure by other organizations and individuals. The result of this reviewing is a revised and expanded edition with 26 sections and full color illustrations. Both the first and last sections are new—the first is a dictionary with over 560 lighting terms, including those related to new sources and concepts, such as high-intensity discharge sources and equivalent sphere illumination; the last section covers underwater lighting principles, sources, and applications. Some of the new or updated material in the sections between include: a new IES method for prescribing illumination; new data and testing procedures for high-intensity discharge light sources, more on color and color rendering index; the latest industry tabulation of lamp data; additional point illumination and luminance calculation procedures, more on the over-all designing of the luminous environment; the latest information on application ranging from offices, schools, hospitals, libraries, industrial processes to aviation, roadway, sports, and public conveyance; and, of course, the Society's latest recommended levels of illumination presented in both footcandles and dekalux.

For this revision and expansion, we wish to acknowledge sincere appreciation to the Society's technical committees (over 800 total members), and to individuals and organizations listed on the following page.

The Handbook Committee (listed below) developed policy and plans for operation, and established the format of this edition—to retain same size as fourth edition, to add full color illustrations, and to remove the Manufacturers' Reference Data section. The Publications Committee (also listed below) was responsible for sales and distribution of the book.

Handbook Committee

M. N. Waterman, *Chairman*
C. L. Amick
G. G. Bonvallet
S. Bruun
B. C. Cooper
J. R. Knisley
R. W. Morris
J. J. Neidhart

Publications Committee
(1969–1971)

R. S. Wissoker, *Chairman*
H. A. Anderson
R. H. Atkinson
B. S. Benson, Jr.
W. S. Cahill
F. Clark
M. E. Keck
G. Lanson
E. Meehan
R. R. Viehman
H. J. Wald

In addition to those acknowledged above, appreciation is also given to the officers of the Illuminating Engineering Society, during the Society's years 1969 to 1972, under whose administration and stimulation this edition was prepared, and to Paul C. Ringgold, Executive Vice President, who was helpful in consultation and budgetary operations.

The promotion and sales of this edition were handled by the Advertising Manager's Office, under the direction of Clayton E. Ellis. Editing, coordination, artwork, and production of the Handbook were carried out through the Technical Office.

With the extent of revisions and additions of new subjects and material it is evident that errors or omissions will occur. Whenever such errors or omissions are found, your comments will be appreciated.

John E. Kaufman
Editor

Jack F. Christensen
Associate Editor

Fifth Edition Handbook Contributors

The numbers in parentheses following each name refer to the sections where contributions were made.

Contributing IES Technical Committees

Aviation (8) (21)
Color (5)
Daylighting (7) (9)
Design Practice (9)
IES/ASAE Joint Farm Lighting (14)
Industrial (14)
Institutions (12)
Light Control and Equipment Design (6)
Light Projection (18)
Light Sources (2) (8)
Lighting and Air Conditioning (16)
Maintenance (16)
Merchandising (13)
Nomenclature (1)
Office (11)
Public Conveyance (22)
Recommendations on Quality and Quantity of Illumination (3) (9)
Residence (15)
Roadway (9) (20)
School and College (11)
Sports and Recreational Areas (19)
Testing Procedures (4)
Theatre, Television, and Film (8) (24)
Underwater (26)

Contributing Individuals and Organizations

H. R. Blackwell (3)
O. M. Blackwell (3)
L. J. Buttolph (25)
B. C. Cooper (10) (16)
C. L. Crouch (3)
R. T. Dorsey (10)
C. A. Douglas (3)
J. R. Fairweather (17)
J. E. Flynn (10)
P. H. Goodell (25)
R. H. Goodman (23)
A. L. Hart (17) (23)
J. P. Hoxie (22)
C. C. Mpelkas (25)
J. J. Neidhart (10)
R. M. Rinalducci (3)
H. E. Sereika (25)
R. P. Teele (22)
H. A. Williams (23)
R. L. Zahour (25)
Society of Motion Picture and Television Engineers (24)

Illuminating Engineering Society

... Programs

... Publications

... Services

IES is the recognized spokesman for the illumination field. For over 65 years, its objective has been to communicate information about all aspects of good lighting practice to individual members, the lighting industry, and consumers through a variety of programs, publications, and services. The strength of IES is in its diversified membership: engineers, architects, designers, educators, students, contractors, distributors, utility personnel, manufacturers, and scientists, all *contributing to* and *benefiting from* the Society.

Programs

IES sponsored local, regional, and transnational meetings, and conferences, symposiums, seminars, designers' forums, roadway forums, workshops, and lighting exhibitions provide an access to latest developments in the field through slide presentations, films, expert speakers, and training aids. Basic and advanced IES lighting courses are offered annually by local IES Sections and Chapters and in cooperation with other organizations. Other Society programs include liaison with school and colleges and career information for students and counsellors.

Publications

Lighting Design and Application (*LD & A*), and the *Journal of the Illuminating Engineering Society* are the official magazines of the Society. *LD & A* is an application-oriented monthly specializing in practical features on new ideas, lighting layouts, equipment, economics, and news of the industry and its people. The *Journal*, a technical quarterly, contains official transactions, approved texts of American National Standards and IES recommended practices, technical committee reports, conference papers, research reports, and technically-oriented features.

In addition to the Handbook, the Society publishes in separate booklet form, a *Lighting Fundamentals Course*, an *Advanced Lighting Problems Course*, IES Technical Committee reports covering many specific lighting tasks; various guides for measuring and reporting lighting values, lighting calculations, and performance of light sources and luminaires, etc.

Complete lists of current and available IES publications are published periodically in its official magazines or may be obtained by writing to Publications Office, Illuminating Engineering Society, 345 E. 47th St., New York, N.Y. 10017.

Services

IES provides professional staff assistance with technical problems, reference help, and inter-professional liaison with AIA, AID, IEEE, NAED, NECA, NEMA, NSID and other groups. IES is a forum for exchange, professional development, and recognition. It correlates the vast amount of research, investigation, and discussion through hundreds of qualified members of its technical committees to guide lighting experts and laymen on research-based lighting recommendations.

CONTENTS

Section		Page
1	**Dictionary of Lighting Terms**	1-1
	Definitions	1-1
	Abbreviations and Conversion Factors	1-23
2	**Physics of Light**	2-1
	Concepts	2-1
	Generation of Light	2-6
	Incandescence	2-7
	Luminescence	2-7
	Light Detection	2-12
3	**Light and Vision**	3-1
	Visual Sensory Processes	3-1
	Visual Ability	3-6
	IES Method for Prescribing Illumination	3-14
	Visual Range of Luminous Signals	3-33
4	**Measurement of Light and Other Radiant Energy**	4-1
	Basis of Photometry	4-1
	Photometers	4-4
	Laboratory Measurements	4-10
	Field Measurements	4-26
5	**Color**	5-1
	Basic Concepts	5-1
	Colorimetry	5-14
	Use of Color	5-16
	Color Rendering	5-18
	Applications	5-21
6	**Light Control and Luminaire Design**	6-1
	Physical Principles	6-1
	Shielding	6-9
	Reflector Design	6-10
	Refractor Design	6-13
	Luminaire Materials	6-14
	Luminaire Finishes	6-19
	Equipment Design	6-21
7	**Daylighting**	7-1
	Daylight	7-1
	Architectural Design	7-5
	Building Types	7-9
	Maintenance	7-12
	Measurements	7-12
8	**Light Sources**	8-1
	Incandescent Filament	8-1
	Fluorescent	8-17
	High-Intensity Discharge	8-33
	Short-Arc	8-40
	Miscellaneous Discharge	8-43
	Electroluminescent	8-47
	Light Emitting Diodes	8-48
	Nuclear	8-49
	Carbon-Arc	8-50
	Gaslights	8-52
9	**Lighting Calculations**	9-1
	Calculation Procedure	9-1
	Average Illumination	9-6
	Average Luminance	9-33
	Luminaire Spacing	9-33
	Flux Transfer	9-36
	Point Illumination	9-44
	Specific Application Methods	9-63
	Floodlighting	9-68
	Searchlighting	9-69
	Roadway Illumination	9-71
	Daylighting	9-74
	Recommended Illumination Levels	9-81
10	**Interior Lighting Design Approaches**	10-1
	Luminous Environment Design Approach	10-3
	Visual Task Oriented Design Approach	10-10
	Types of Lighting Systems	10-11
11	**Offices and Schools**	11-1
	Office Lighting	11-1
	School Lighting	11-10
12	**Institutions and Public Buildings**	12-1
	Banks	12-1
	Churches and Synagogues	12-2
	Foodservice Facilities	12-4
	Health Care Facilities	12-6
	Hotels and Motels	12-13
	Libraries	12-15
	Museums and Art Galleries	12-18
13	**Lighting for Merchandising**	13-1
	Merchandising and Architectural Trends	13-1
	Lighting and Merchandising Relationships	13-2
	Merchandising Emphasis	13-4
	Show Windows	13-7
	Exteriors	13-9
	Parking Areas	13-12
14	**Industries**	14-1
	General Lighting	14-1
	Supplementary Lighting	14-6
	Emergency Lighting	14-10
	Steel Mills	14-12
	Foundries	14-15
	Machining Metal Parts	14-17
	Sheet Metal Shops	14-19
	Automotive Assembly	14-20
	Rubber Tires	14-22
	Graphic Arts—Printing	14-24
	Cotton Gins	14-26
	Textiles	14-27

Section	Page

Cleaning and Pressing 14-31
Shoe Manufacturing 14-33
Men's Clothing 14-34
Fluid Milk Industry 14-35
Flour Mills 14-36
Bakeries 14-37
Candy Manufacturing 14-39
Fruit and Vegetable Packaging 14-41
Petrochemicals 14-43
Electrical Generating Stations 14-43
Sawmills 14-46
Railroad Yards 14-47
Farm Lighting 14-49

15 Residential Lighting **15-1**
Light as an Element of Design 15-1
Environment Objectives 15-1
Lighting Equipment 15-3
Design Considerations for Specific Visual Tasks 15-12
Decorative Accents 15-26
Exteriors and Grounds 15-26

16 Lighting System Design Factors **16-1**
Lighting and Air Conditioning 16-1
Wiring for Lighting 16-7
Dimming Devices 16-9
Maintenance of Lighting 16-11
Economic Analysis 16-20

17 Outdoor Lighting Applications **17-1**
Design Procedure 17-1
Building Floodlighting 17-1
Monuments, Expositions, and Public Gardens 17-6
Fountains 17-7
Christmas Lighting 17-8
Service Stations 17-8

18 Light Projection Equipment and Protective Lighting **18-1**
Light Projection Equipment 18-1
Protective Lighting 18-5

19 Sports Lighting **19-1**
Seeing Problems 19-1
Quality and Quantity of Illumination 19-2
Light Sources 19-3
Indoor Sports 19-4
Outdoor Sports 19-10

20 Roadway Lighting **20-1**
Visibility, Discernment, Glare 20-1
Roadway, Walkway, and Area Classifications 20-2
Luminaire Distribution Classifications 20-3
Design of Roadway and Walkway Lighting 20-5

21 Aviation Lighting **21-1**
Aircraft Lighting 21-1
Ground Lighting to Aid Navigation 21-6
Heliport and VTOL Lighting 21-12
STOL Port Lighting 21-12
Obstruction Lights 21-13
Ground Operational Lighting 21-13

22 Transportation Lighting **22-1**
Automobile Lighting 22-1
Specifications 22-3
Reflex Devices 22-3
Public Conveyance Lighting—Road and Rail 22-5
Lamps and Power Systems 22-11
Marine Lighting 22-12
Railway Guidance Systems 22-15

23 Lighting for Advertising **23-1**
Exposed Lamp Signs 23-1
Luminous Element Signs 23-4
Floodlighted Signs 23-7

24 Theatre, Television and Photographic Lighting **24-1**
Luminaires, Lamps, and Control Systems 24-1
Lighting for Theatres 24-6
Lighting for Television 24-12
Photographic Lighting 24-19
Picture Projection Lighting 24-25
Photochemical Reproduction Processes 24-35

25 Miscellaneous Applications of Radiant Energy **25-1**
Light 25-2
Ultraviolet Energy 25-14
Infrared Energy 25-19

26 Underwater Lighting **26-1**
Terms and Definitions 26-1
Filtering Properties of Sea Water 26-2
Seeing Distance in Water 26-2
Light Sources 26-4
Sensor Characteristics 26-4
Underwater Lighting Calculations 26-5
Measurement Techniques and Instrumentation 26-7

C Credits for Illustrations and Tables **C-1**

I Index **I-1**

SECTION
1

DICTIONARY OF LIGHTING TERMS

Definitions 1–1 *Abbreviations and Conversion Factors 1–23*

As the title implies this first section contains a selection of terms directly related to light and lighting practice. Many of these terms may be found in the *American National Standard Nomenclature and Definitions for Illuminating Engineering,* ANSI Z7.1-1967.[1] Others have appeared in more recent publications. Definitions of electrical terms common to lighting and to other fields are available in American National Standard Definitions of Electrical Terms, C42 (Series).

All terms are presented in alphabetical order and are followed by their standard symbols and defining equations where applicable, by their definitions and by other related terms of interest. No attempt has been made to provide information on pronunciations or etymologies.

Any of the following radiometric and photometric quantities may be restricted to a narrow wavelength interval $\Delta\lambda$ by the addition of the word spectral and the specification of the wavelength λ. The corresponding symbols are changed by adding a subscript λ, *i.e.,* Q_λ, for a spectral concentration, or a λ in parentheses, *i.e.,* $K(\lambda)$ for a function of wavelength. Fig. 1–1 is a tabulated summary of standard symbols, defining equations, units and symbolic abbreviations for the fundamental photometric and radiometric quantities. Other abbreviations and conversion factors are given in Figs. 1–4 to 1–11 at the end of this section.

DEFINITIONS

A

absolute threshold: the luminance threshold or minimum perceptible luminance (photometric brightness) when the eye is completely dark adapted.

absorptance, $\alpha = \Phi_a/\Phi_i$**:** the ratio of the flux absorbed by a medium to the incident flux. See *absorption.*

NOTE: The sum of the hemispherical reflectance, the hemispherical transmittance, and the absorptance is one.

absorption: a general term for the process by which incident flux is dissipated.

NOTE: All of the incident flux is accounted for by the processes of reflection, transmission and absorption.

NOTE: References are listed at the end of each section.

accent lighting: directional lighting to emphasize a particular object or draw attention to a part of the field of view. See *directional lighting.*

accommodation: the process by which the eye changes focus from one distance to another.

adaptation: the process by which the retina becomes accustomed to more or less light than it was exposed to during an immediately preceding period. It results in a change in the sensitivity of the eye to light. See *scotopic vision* and *photopic vision.*

NOTE: Adaptation is also used to refer to the final state of the process, as reaching a condition of dark adaptation or light adaptation.

adaptive color shift: the change in the perceived object color caused solely by change of chromatic adaptation.

aerodrome beacon: an aeronautical beacon used to indicate the location of an aerodrome.

NOTE: An aerodrome is any defined area on land or water—including any buildings, installations, and equipment—intended to be used either wholly or in part for the arrival, departure and movement of aircraft.

aeronautical beacon: an aeronautical ground light visible at all azimuths, either continuously or intermittently, to designate a particular location on the surface of the earth. See *aerodrome beacon, airway beacon, hazard or obstruction beacon, landmark beacon.*

aeronautical ground light: any light specially provided as an aid to air navigation, other than a light displayed on an aircraft. See *aeronautical beacon, angle-of-approach lights, approach lights, approach-light beacon, bar (of lights), boundary lights, circling guidance lights, course light, channel lights, obstruction lights, runway alignment indicator, runway-end identification light, perimeter lights, runway lights, taxi-channel lights, taxiway lights.*

aeronautical light: any luminous sign or signal specially provided as an aid to air navigation.

after image: a visual response that occurs after the stimulus causing it has ceased.

aircraft aeronautical light: any aeronautical light specially provided on an aircraft. See *navigation light system, anti-collision light, ice detection light, fuselage lights, landing light, position lights, taxi light.*

airway beacon: an aeronautical beacon used to indicate a point on the airway.

altitude (in daylighting): the angular distance of a heavenly body measured on the great circle that passes perpendicular to the plane of the horizon through the body and through the zenith. It is measured positively from the horizon to the zenith, from 0 to 90 degrees.

Fig. 1–1. Standard Units, Symbols, and Defining Equations for Fundamental Photometric and Radiometric Quantities*

Quantity†	Symbol†	Defining Equation	Unit	Symbolic Abbreviation
Radiant energy	Q		erg joule calorie kilowatt-hour	erg J cal kWh
Radiant density	w	$w = dQ/dV$	joule per cubic meter‡ erg per cubic centimeter	J/m^3 erg/cm^3
Radiant flux	Φ	$\Phi = dQ/dt$	erg per second watt‡	erg/s W
Radiant flux density at a surface Radiant exitance (Radiant emittance§) Irradiance	 M E	 $M = d\Phi/dA$ $E = d\Phi/dA$	watt per square centimeter, watt per square meter,‡ etc.	W/cm^2 W/m^2
Radiant intensity	I	$I = d\Phi/d\omega$ (ω = solid angle through which flux from point source is radiated)	watt per steradian‡	W/sr
Radiance	L	$L = d^2\Phi/d\omega(dA \cos\theta)$ $= dI/(dA \cos\theta)$ (θ = angle between line of sight and normal to surface considered)	watt per steradian and square centimeter watt per steradian and square meter‡	$W/sr \cdot cm^2$ $W/sr \cdot m^2$
Emissivity	ϵ	$\epsilon = M/M_{blackbody}$ (M and $M_{blackbody}$ are respectively the radiant exitance of the measured specimen and that of a blackbody at the same temperature as the specimen)	one (numeric)	
Absorptance	α	$\alpha = \Phi_a/\Phi_i$‖	one (numeric)	
Reflectance	ρ	$\rho = \Phi_r/\Phi_i$‖	one (numeric)	
Transmittance	τ	$\tau = \Phi_t/\Phi_i$‖	one (numeric)	

anchor light (aircraft): a light designed for use on a seaplane or amphibian to indicate its position when at anchor or moored.

angle-of-approach lights: aeronautical ground lights arranged so as to indicate a desired angle of descent during an approach to an aerodrome runway. (Also called optical glide path lights.)

angle of collimation: the angle subtended by a light source at a point on an irradiated surface.

angstrom, Å: unit of wavelength equal to 10^{-10} (one ten-billionth) meter.

anti-collision light: a flashing aircraft aeronautical light or system of lights designed to provide a red signal throughout 360 degrees of azimuth for the purpose of giving long-range indication of an aircraft's location to pilots of other aircraft.

apostilb, *asb*: a unit of luminance equal to $1/\pi$ candela per square meter.

apparent candlepower of an extended source at a specific distance: the candlepower of a point source that would produce the same illumination at that distance.

approach-light beacon: an aeronautical ground light placed on the extended centerline of the runway at a fixed distance from the runway threshold to provide an early indication of position during an approach to a runway.

NOTE: The runway threshold is the beginning of the runway usable for landing.

approach lights: a configuration of aeronautical ground lights located in extension of a runway or channel before the threshold to provide visual approach and landing guidance to pilots. See *angle-of-approach lights, approach-light beacon, VASIS.*

arc discharge: an electric discharge characterized by high cathode current densities and a low voltage drop at the cathode.

Fig. 1-1. *Continued*

Quantity†	Symbol†	Defining Equation	Unit	Symbolic Abbreviation
Luminous energy (quantity of light)	Q	$Q_v = \int_{380}^{780} K(\lambda) Q_{e\lambda} d\lambda$	lumen-hour lumen-second‡ (talbot)	lm·h lm·s
Luminous density	w	$w = dQ/dV$	lumen-hour per cubic centimeter‡	lm·h/cm^3
Luminous flux	Φ	$\Phi = dQ/dt$	lumen‡	lm
Luminous flux density at a surface				
Luminous exitance (Luminous emittance)§	M	$M = d\Phi/dA$	lumen per square foot	lm/ft^2
Illumination (Illuminance)	E	$E = d\Phi/dA$	footcandle (lumen per square foot) lux (lm/m^2)‡ phot (lm/cm^2)	fc lx ph
Luminous intensity (candlepower)	I	$I = d\Phi/d\omega$ (ω = solid angle through which flux from point source is radiated)	candela‡ (lumen per steradian)	cd
Luminance (photometric brightness)	L	$L = d^2\Phi/d\omega(dA \cos\theta)$ $= dI/(dA \cos\theta)$ (θ = angle between line of sight and normal to surface considered)	candela per unit area stilb (cd/cm^2) nit $(\text{cd/m}^2$‡$)$ footlambert $(\text{cd}/\pi\text{ft}^2)$ lambert $(\text{cd}/\pi\text{cm}^2)$ apostilb $(\text{cd}/\pi\text{m}^2)$	cd/in^2, etc. sb nt fL L asb
Luminous efficacy	K	$K = \Phi_v/\Phi_e$	lumen per watt‡	lm/W
Luminous efficiency	V	$V = K/K_{\text{maximum}}$ (K_{maximum} = maximum value of $K(\lambda)$ function)	one (numeric)	

* The symbols for photometric quantities are the same as those for the corresponding radiometric quantities. When it is necessary to differentiate them the subscripts v and e respectively should be used, *e.g.*, Q_v and Q_e .

† Quantities may be restricted to a narrow wavelength band by adding the word spectral and indicating the wavelength. The corresponding symbols are changed by adding a subscript λ, *e.g.*, Q_λ, for a spectral concentration or a λ in parentheses, *e.g.*, $K(\lambda)$, for a function of wavelength.

‡ International System (SI) unit

§ To be deprecated

‖ Φ_i = incident flux, Φ_a = absorbed flux, Φ_r = reflected flux, Φ_t = transmitted flux

artificial pupil: a diaphragm or other limitation confining the light entering the eye to an aperture smaller than the natural pupil.

atmospheric transmissivity: the ratio of the directly transmitted flux incident on a surface after passing through unit thickness of the atmosphere to the flux that would be incident on the same surface if the flux had passed through a vacuum.

average luminance (of a surface): the average luminance (average photometric brightness) of a surface may be expressed in terms of the total luminous flux (lumens) actually leaving the surface per unit area. Average luminance specified in this way is identical in magnitude with luminous exitance, which is the preferred term. See note under *luminous flux density at a surface.*

In general, the concept of average luminance is useful only when the luminance is reasonably uniform throughout a very wide angle of observation and over a large area of the surface considered. It has the advantage that it can be readily computed for reflecting surfaces by multiplying the incident luminous flux density (illumination) by the luminous reflectance of the surface. For a transmitting body it can be computed by multiplying the incident luminous flux density by the luminous transmittance of the body.

average luminance (of a luminaire): the luminous intensity at a given angle divided by the projected area of the luminaire at that angle.

azimuth: the angular distance between the vertical plane containing a given line or celestial body and the plane of the meridian.

B

back light: illumination from behind a subject directed subtantially parallel to a vertical plane through the optical axis of the camera. See *side back light.*

backup lamp: a lighting device mounted on the rear of a vehicle for illuminating the region near the rear of the vehicle while moving or about to move in reverse. It normally can be used only while backing up.

bactericidal (germicidal) effectiveness: the capacity of various portions of the ultraviolet spectrum to destroy bacteria, fungi and viruses.

bactericidal (germicidal) efficiency of radiant flux (for a particular wavelength): the ratio of the bactericidal effectiveness at a particular wavelength to that at wavelength 265.0 nanometers, which is rated as unity.

NOTE: Tentative bactericidal efficiency of various wavelengths of radiant flux is given in Fig. 25–25.

bactericidal (germicidal) exposure: the product of bactericidal flux density on a surface and time. It usually is measured in bactericidal microwatt-minutes per square centimeter or bactericidal watt-minutes per square foot.

bactericidal (germicidal) flux: radiant flux evaluated according to its capacity to produce bactericidal effects. It usually is measured in microwatts of ultraviolet radiation weighted in accordance with its bactericidal efficiency. Such quantities of bactericidal flux would be in bactericidal microwatts.

NOTE: Ultraviolet radiation of wavelength 253.7 nanometers usually is referred to as "ultraviolet microwatts" or "UV watts." These terms should not be confused with "bactericidal microwatts" because the radiation has not been weighted in accordance with the values given in Fig. 25–25.

bactericidal (germicidal) flux density: the bactericidal flux per unit area of the surface being irradiated. It is equal to the quotient of the incident bactericidal flux divided by the area of the surface when the flux is uniformly distributed. It usually is measured in microwatts per square centimeter or watts per square foot of bactericidally weighted ultraviolet radiation (bactericidal microwatts per square centimeter or bactericidal watts per square foot).

baffle: a single opaque or translucent element to shield a source from direct view at certain angles, or to absorb unwanted light.

ballast: a device used with an electric-discharge lamp to obtain the necessary circuit conditions (voltage, current and wave form) for starting and operating. See *reference ballast.*

bar (of lights): a group of three or more aeronautical ground lights placed in a line transverse to the axis, or extended axis, of the runway. See *barette.*

bare (exposed) lamp: a light source with no shielding.

barette (in aviation): a short bar in which the lights are closely spaced so that from a distance they appear to be a linear light.

NOTE: Barettes are usually less than 15 feet in length.

base light: uniform, diffuse illumination approaching a shadowless condition, which is sufficient for a television picture of technical acceptability, and which may be supplemented by other lighting.

beacon: a light (or mark) used to indicate a geographic location. See *aerodrome beacon, aeronautical beacon, airway beacon, approach-light beacon, hazard or obstruction beacon, identification beacon, landmark beacon.*

beam axis of a projector: a line midway between two lines that intersect the candlepower distribution curve at points equal to a stated per cent of its maximum (usually 50 per cent).

beam spread: (in any plane) the angle between the two directions in the plane in which the candlepower is equal to a stated per cent (usually 10 per cent) of the maximum candlepower in the beam.

binocular visual field: that portion of the visual field where the fields of the two eyes overlap. It has a half angle of roughly 60 degrees.

blackbody: a temperature radiator of uniform temperature whose radiant exitance in all parts of the spectrum is the maximum obtainable from any temperature radiator at the same temperature.

Such a radiator is called a blackbody because it will absorb all the radiant energy that falls upon it. All other temperature radiators may be classed as non-blackbodies. They radiate less in some or all wavelength intervals than a blackbody of the same size and the same temperature.

NOTE: the blackbody is practically realized in the form of a cavity with opaque walls at a uniform temperature and with a small opening for observation purposes. It also is called a full radiator, standard radiator, complete radiator or ideal radiator.

blackbody (Planckian) locus: the locus of points on a chromaticity diagram representing the chromaticities of blackbodies having various (color) temperatures.

"black light:" the popular term for ultraviolet energy near the visible spectrum.

NOTE: For engineering purposes the wavelength range 320–400 nanometers has been found useful for rating lamps and their effectiveness upon fluorescent materials (excluding phosphors used in fluorescent lamps). By confining "black light" applications to this region, germicidal and erythemal effects are, for practical purposes, eliminated.

"black light" flux: radiant flux within the wavelength range 320 to 400 nanometers. It is usually measured in milliwatts. See *fluoren*

NOTE: Because of the variability of the spectral sensitivity of materials irradiated by "black light" in practice, no attempt is made to evaluate "black light" flux according to its capacity to produce effects.

"black light" flux density: "black light" flux per unit area of the surface being irradiated. It is equal to the incident "black light" flux divided by the area of the surface when the flux is uniformly distributed. It usually is measured in milliwatts per square foot of "black light" flux.

blinding glare: glare that is so intense that for an appreciable length of time no object can be seen.

boundary lights: aeronautical ground lights delimiting the boundary of a land aerodrome without runways. See *range lights.*

bowl: an open top diffusing glass or plastic enclosure used to shield a light source from direct view and to redirect or scatter the light.

bracket (mast arm): an attachment to a lamp post or pole from which a luminaire is suspended.

brightness: see *subjective brightness, luminance, veiling brightness, brightness of a perceived light-source color.*

brightness of a perceived light-source color: the attribute in accordance with which the source seems to emit more or less luminous flux per unit area.

C

candela, cd: (formerly candle) the unit of luminous intensity. One candela is defined as the luminous intensity of $\frac{1}{600,000}$ square meter of projected area of a blackbody radiator operating at the temperature of

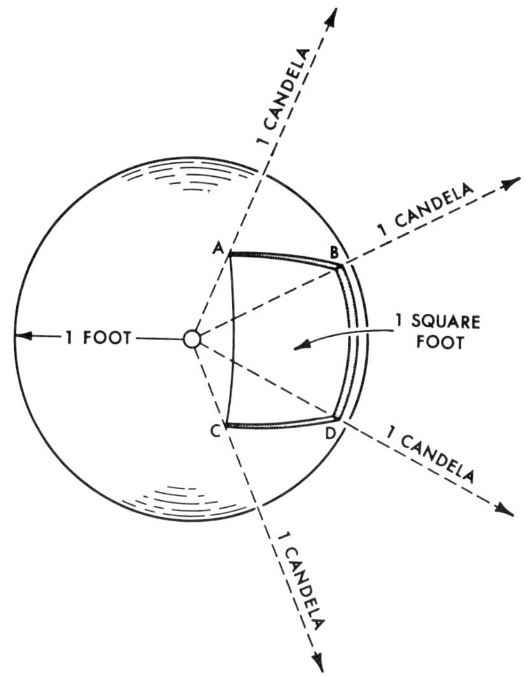

Fig. 1–2. Relationship between candelas, lumens and footcandles. A uniform point source (luminous intensity or candlepower = 1 candela) is shown at the center of a sphere of 1-foot radius. It is assumed that the sphere surface has zero reflectance.

The illumination at any point on the sphere is 1 footcandle (1 lumen per square foot).

The solid angle subtended by the area, A, B, C, D is 1 steradian. The flux density is therefore 1 lumen per steradian, which corresponds to a luminous intensity of 1 candela, as originally assumed.

The sphere has a total area of 12.57 (4π) square feet, and there is a luminous flux of 1 lumen falling on each square foot. Thus the source provides a total of 12.57 lumens.

solidification of platinum under a pressure of 101,325 newtons per square meter. Values for standards having other spectral distributions are derived by the use of accepted spectral luminous efficiency data for photopic vision. See Fig. 1–2 and *spectral luminous efficiency of radiant flux* and *values of spectral luminous efficiency for photopic vision.*

NOTE: From 1909 until the introduction of the present photometric system on January 1, 1948, the unit of luminous intensity in the United States, as well as in France and Great Britain, was the "international candle" which was maintained by a group of carbon-filament vacuum lamps. For the present unit as defined above, the internationally accepted term is *candela.* The difference between the candela and the old international candle is so small that only measurements of high precision are affected.

candlepower, $I = d\Phi/d\omega$; abbreviation, *cp.*: luminous intensity expressed in candelas.

candlepower distribution curve: a curve, generally polar, representing the variation of luminous intensity of a lamp or luminaire in a plane through the light center.

NOTE: A vertical candlepower distribution curve is obtained by taking measurements at various angles of elevation in a vertical plane through the light center; unless the plane is specified, the vertical curve is assumed to represent an average such as would be obtained by rotating the lamp or luminaire about its vertical axis. A horizontal candlepower distribution curve represents measurements made at various angles of azimuth in a horizontal plane through the light center.

carbon-arc lamp: an electric-discharge lamp employing an arc discharge between carbon electrodes. One or more of these electrodes may have cores of special chemicals that contribute importantly to the radiation.

cavity ratio: a number indicating cavity proportions calculated from length, width and height. See *ceiling cavity ratio, floor cavity ratio* and *room cavity ratio.*

ceiling area lighting: a general lighting system in which the entire ceiling is, in effect, one large luminaire.

NOTE: Ceiling area lighting includes *luminous ceilings* and *louvered ceilings.*

ceiling cavity: the cavity formed by the ceiling, the plane of the luminaires, and the wall surfaces between these two planes.

ceiling cavity ratio, *CCR*: a number indicating ceiling cavity proportions calculated from length, width and height. See page 9–8.

ceiling projector: a device designed to produce a well-defined illuminated spot on the lower portion of a cloud for the purpose of providing a reference mark for the determination of the height of that part of the cloud.

ceiling ratio: the ratio of the luminous flux reaching the ceiling directly to the upward component from the luminaire.

central (foveal) vision: the seeing of objects in the central or foveal part of the visual field, approximately two degrees in diameter. It permits seeing much finer detail than does peripheral vision.

central visual field: that region of the visual field corresponding to the foveal portion of the retina.

channel: an enclosure containing the ballast, starter, lamp holders and wiring for a fluorescent lamp, or a similar enclosure on which filament lamps (usually tubular) are mounted.

channel lights: aeronautical ground lights arranged along the sides of a channel of a water aerodrome. See *taxi-channel lights.*

characteristic curve: a curve expressing the relationship between two variable properties of a light source, such as candlepower and voltage, flux and voltage, etc.

chromatic adaptation: see *state of chromatic adaptation.*

chromaticity coordinates (of a light), x, y, z: the ratios of each of the tristimulus values of the light to the sum of the three tristimulus values.

chromaticity diagram: a plane diagram formed by plotting one of the three chromaticity coordinates against another.

chromaticity of a color: consists of the dominant or complementary wavelength and purity aspects of the color taken together, or of the aspects specified by the chromaticity coordinates of the color taken together.

CIE standard chromaticity diagram: a diagram in which the x and y chromaticity coordinates are plotted in rectangular coordinates.

circling guidance lights: aeronautical ground lights provided to supply additional guidance during a circling approach when the circling guidance furnished by the approach and runway lights is inadequate.

clear sky: a sky that has less than 30 per cent cloud cover.

clearance lamp: a lighting device mounted on a vehicle for the purpose of indicating the overall width and height of the vehicle.

clerestory: that part of a building rising clear of the roofs or other parts and whose walls contain windows for lighting the interior.

cloudy sky: a sky having more than 70 per cent cloud cover.

coefficient of attenuation, μ: the decrement in flux per unit distance in a given direction within a medium and defined by the relation: $\Phi_x = \Phi_o e^{-\mu x}$ where Φ_x is the flux at any distance x from a reference point having flux Φ_o.

coefficient of beam utilization, *CBU:* the ratio of the luminous flux (lumens) reaching a specified area directly from a floodlight or projector to the total beam luminous flux.

coefficient of utilization, *CU:* the ratio of the luminous flux (lumens) from a luminaire received on the work-plane to the lumens emitted by the luminaire's lamps alone.

coffer: a recessed panel or dome in the ceiling.

cold-cathode lamp: an electric-discharge lamp whose mode of operation is that of a glow discharge, and having electrodes so spaced that most of the light comes from the positive column between them.

color: the characteristics of light by which a human observer may distinguish between two structure-free patches of light of the same size and shape. See *light-source color* and *object color*.

color comparison, or color grading (*CIE*, object color inspection): the judgment of equality, or of the amount and character of difference, of the color of two objects viewed under identical illumination.

color correction (of a photograph or printed picture): the adjustment of a color reproduction process to improve the color conformity of the reproduction to the original.

color discrimination: the perception of differences between two or more colors.

color matching: the process of adjusting the color of one area so that it is the same color as another.

color rendering: general expression for the effect of a light source on the color appearance of objects in conscious or subconscious comparison with their color appearance under a reference light source.

color rendering improvement (of a light source): the adjustment of spectral composition to improve color rendering.

color rendering index (of a light source): measure of the degree to which the perceived colors of objects illuminated by the source conform to those of the same objects illuminated by a reference source for specified conditions.

color temperature (of a light source): the absolute temperature at which a blackbody radiator must be operated to have a chromaticity equal to that of the light source. See also *correlated color temperature* and *distribution temperature*.

colorimetric purity (of a light), p_c: the ratio L_1/L_2 where L_1 is the luminance of the single frequency component that must be mixed with a reference standard to match the color of the light and L_2 is the luminance (photometric brightness) of the light. See *excitation purity*.

colorimetric shift: the change of chromaticity and luminance factor of an object color due to change of the light source. See *adaptive color shift* and *resultant color shift*.

colorimetry: the measurement of color.

comparison lamp: a light source having a constant, but not necessarily known, luminous intensity with which standard and test lamps are compared successively.

complementary wavelength (of a light), λ_c: the wavelength of radiant energy of a single frequency that, when combined in suitable proportion with the light, matches the color of a reference standard. See *dominant wavelength*.

complete diffusion: diffusion in which the diffusing medium redirects the flux incident upon it so that none is in an image-forming state.

cone: a retinal element primarily concerned with the perception of detail and color by the light-adapted eye.

conspicuity: the capacity of a signal to stand out in relation to its background so as to be readily discovered by the eye.

contrast: see *luminance contrast*.

contrast rendition factor, *CRF:* the ratio of visual task contrast with a given lighting environment to the contrast with sphere illumination.

contrast sensitivity: the ability to detect the presence of luminance differences. Quantitatively, it is equal to the reciprocal of the contrast threshold.

contrast threshold: the minimal perceptible contrast for a given state of adaptation of the eye. It also is defined as the luminance contrast detectable during some specific fraction of the times it is presented to an observer, usually 50 per cent.

cornice lighting: lighting by means of light sources shielded by a panel parallel to the wall and attached to the ceiling, and distributing light over the wall.

correlated color temperature (of a light source): the absolute temperature of a blackbody whose chromaticity most nearly resembles that of the light source.

cosine-cubed law: an extension of the cosine law in which the distance d between the source and surface is replaced by $h/\cos\theta$, where h is the perpendicular distance of the source from the plane in which the point is located. It is expressed by $E = (I \cos^3\theta)/h^2$.

cosine law: the law that the illumination on any surface varies as the cosine of the angle of incidence. The angle of incidence θ is the angle between the normal to the surface and the direction of the incident light. The inverse-square law and the cosine law can be combined as $E = (I\cos\theta)/d^2$. See *cosine-cubed law* and *inverse-square law*.

country beam: see *upper (driving) beams*.

course light: an aeronautical ground light, supplementing an airway beacon, for indicating the direction of the airway and to identify by a coded signal the location of the airway beacon with which it is associated.

cove lighting: lighting by means of sources shielded by a ledge or horizontal recess, and distributing light over the ceiling and upper wall.

critical fusion frequency, *cff:* see *flicker fusion frequency*.

critical flicker frequency, *cff:* see *flicker fusion frequency*.

cross light: equal illumination in front of the subject from two directions at substantially equal and opposite angles with the optical axis of the camera and a horizontal plane.

cut-off angle (of a luminaire): the angle, measured up from nadir, between the vertical axis and the first line of sight at which the bare source is not visible.

D

dark adaptation: the process by which the retina becomes adapted to a luminance less than about 0.01 footlambert.

daylight factor: a measure of daylight illumination at a point on a given plane expressed as a ratio of the illumination on the given plane at that point to the simultaneous exterior illumination on a horizontal plane from the whole of an unobstructed sky of assumed or known luminance (photometric brightness) distribution. Direct sunlight is excluded from both interior and exterior values of illumination.

densitometer: a photometer for measuring the optical density (common logarithm of the reciprocal of the transmittance) of materials.

diffuse reflectance: the ratio of the flux leaving a surface or medium by diffuse reflection to the incident flux.

diffuse reflection: the process by which incident flux is re-directed over a range of angles.

diffuse transmission: the process by which the incident flux passing through a surface or medium is scattered.

diffuse transmittance: the ratio of the diffusely transmitted flux leaving a surface or medium to the incident flux.

diffused lighting: light that is not predominantly incident from any particular direction.

diffuser: a device to redirect or scatter the light from a source, primarily by the process of diffuse transmission.

diffusing panel: a translucent material covering the lamps in a luminaire to reduce the luminance by distributing the flux over an extended area.

diffusing surfaces and media: those that redistribute some of the incident flux by scattering in all directions. See *complete diffusion, incomplete diffusion, perfect diffusion, narrow-angle diffusion, wide-angle diffusion.*

direct component: that portion of the light from a luminaire which arrives at the work-plane without being reflected by room surfaces. See *indirect component.*

direct glare: glare resulting from high luminances or insufficiently shielded light sources in the field of view or from reflecting areas of high luminance. It usually is associated with bright areas, such as luminaires, ceilings and windows that are outside the visual task or region being viewed.

direct-indirect lighting: a variant of general diffuse lighting in which the luminaires emit little or no light at angles near the horizontal.

direct lighting: lighting by luminaires distributing 90 to 100 per cent of the emitted light in the general direction of the surface to be illuminated. The term usually refers to light emitted in a downward direction.

direct ratio: the ratio of the luminous flux reaching the work-plane directly to the downward component from the luminaire.

directional lighting: illumination on the work plane or on an object that is predominantly from a single direction. See *accent lighting, key light, cross light.*

disability glare: glare resulting in reduced visual performance and visibility. It often is accompanied by discomfort. See *veiling brightness.*

discomfort glare: glare producing discomfort. It does not necessarily interfere with visual performance or visibility.

discomfort glare factor, *DGF:* the numerical assessment of the capacity of a single source of brightness, such as a luminaire, in a given visual environment for producing discomfort. See *glare* and *discomfort glare.*

discomfort glare rating, *DGR:* a numerical assessment of the capacity of a number of sources of brightness, such as luminaires, in a given visual environment for producing discomfort. It usually is derived from the discomfort glare factors of the individual sources. See *discomfort glare factor.*

distribution temperature (of a light source): the absolute temperature of a blackbody whose relative spectral distribution is the same (or nearly so) in the visible region of the spectrum as that of the light source.

dominant wavelength (of a light), λ_d: the wavelength of radiant energy of a single frequency that, when combined in suitable proportion with the radiant energy of a reference standard, matches the color of the light. See *complementary wavelength.*

downlight: a small direct lighting unit that can be recessed, surface mounted or suspended.

downward component: that portion of the luminous flux from a luminaire emitted at angles below the horizontal. See *upward component.*

dual headlighting system: headlighting by means of two double units, one mounted on each side of the front end of a vehicle. Each unit consists of two lamps mounted in a single housing. The upper or outer lamps may have two filaments supplying the lower beam and part of the upper beam, respectively. The lower or inner lamps have one filament providing the primary source of light for the upper beam.

dust-proof luminaire: a luminaire so constructed or protected that dust will not interfere with its successful operation.

dust-tight luminaire: a luminaire so constructed that dust will not enter the enclosing case.

E

effective ceiling cavity reflectance, ρ_{CC}: the effective reflectance of all the area above the luminaire plane as seen from the room cavity.

effective floor cavity reflectance, ρ_{FC}: the effective reflectance of all the area below the work-plane as seen from the room cavity.

efficacy: see *luminous efficacy of a source of light* and *spectral luminous efficacy of radiant flux.*

efficiency: see *luminaire efficiency, luminous efficacy of a source of light* and *spectral luminous efficiency of radiant flux.*

electric discharge: see *arc discharge, gaseous discharge* and *glow discharge.*

electric-discharge lamp: a lamp in which light (or radiant energy near the visible spectrum) is produced by the passage of an electric current through a vapor or a gas. See *fluorescent-mercury lamp, cold-cathode lamp, hot-cathode lamp, carbon-arc lamp, glow lamp, fluorescent lamp, mercury lamp.*

NOTE: Electric-discharge lamps may be named after the filling gas or vapor that is responsible for the major portion of the radiation; *e.g.* mercury lamps, sodium lamps, neon lamps, argon lamps, etc.

A second method of designating electric-discharge lamps is by psysical dimensions or operating parameters; *e.g.* short-arc lamps, high-pressure lamps, low-pressure lamps, etc.

A third method of designating electric-discharge lamps is by their application; in addition to lamps for illumination there are photochemical lamps, bactericidal lamps, blacklight lamps, sun lamps, etc.

electroluminescence: the emission of light from a phosphor excited by an electromagnetic field.

electromagnetic spectrum: a continuum of electric and magnetic radiation encompassing all wavelengths. See *regions of electromagnetic spectrum.*

elevation: the angle between the axis of a searchlight drum and the horizontal. For angles above the horizontal, elevation is positive, and below the horizontal negative.

emergency lighting: lighting designed to supply illumination essential to safety of life and property in the event of failure of the normal lighting.

emissivity, ϵ: the ratio of the radiance (for directional emissivity) or radiant exitance (for hemispherical emissivity) of an element of surface of a temperature radiator to that of a blackbody at the same temperature.

enclosed and gasketed: see *vapor-tight.*

equal interval (isophase) light: a rhythmic light in which the light and dark periods are equal.

equivalent sphere illumination: the level of sphere illumination which would produce task visibility equivalent to that produced by a specific lighting environment.

erythema: the temporary reddening of the skin produced by exposure to ultraviolet energy.

NOTE: The degree of erythema is used as a guide to dosages applied in ultraviolet therapy.

erythemal effectiveness: the capacity of various portions of the ultraviolet spectrum to produce erythema.

erythemal efficiency of radiant flux: the ratio of the erythemal effectiveness of a particular wavelength to that of wavelength 296.7 nanometers, which is rated as unity.

NOTE: This term formerly was called "relative erythemal factor."

The erythemal efficiency of radiant flux of various wavelengths for producing a minimum perceptible erythema (MPE) is given in Fig. 25–25. These values have been accepted for evaluating the erythemal effectiveness of sun lamps.

erythemal exposure: the product of erythemal flux density on a surface and time. It usually is measured in erythemal microwatt-minutes per square centimeter.

NOTE: For average untanned skin a minimum perceptible erythema requires about 300 microwatt-minutes per square centimeter of radiation at 296.7 nanometers.

erythemal flux: radiant flux evaluated according to its capacity to produce erythema of the untanned human skin. It usually is measured in microwatts of ultraviolet radiation weighted in accordance with its erythemal efficiency. Such quantities of erythemal flux would be in erythemal microwatts. See *erythemal efficiency of radiant flux.*

NOTE: A commonly used practical unit of erythemal flux is the erythemal unit (EU) or E-viton (erytheme) which is equal to the amount of radiant flux that will produce the same erythemal effect as 10 microwatts of radiant flux at wavelength 296.7 nanometers. See also *erythemal unit or E-viton.*

erythemal flux density: the erythemal flux per unit area of the surface being irradiated. It is equal to the quotient of the incident erythemal flux divided by the area of the surface when the flux is uniformly distributed. It usually is measured in microwatts per square centimeter of erythemally weighted ultraviolet radiation (erythemal microwatts per square centimeter). See *Finsen.*

erythemal unit, *EU*: a unit of erythemal flux that is equal to the amount of radiant flux that will produce the same erythemal effect as 10 microwatts of radiant flux at wavelength 296.7 nanometers. Also called E-viton.

E-viton (erytheme): See *erythemal unit.*

exitance: see *luminous exitance* and *radiant exitance.*

excitation purity (of a light), p_e: the ratio of the distance on the CIE (x,y) chromaticity diagram between the reference point and the light point to the distance in the same direction between the reference point and the spectrum locus or the purple boundary. See *colorimetric purity.*

explosion-proof luminaire: a completely enclosed luminaire capable of withstanding an explosion of a specific gas or vapor that may occur within it, and preventing the ignition of a specific gas or vapor surrounding the enclosure by sparks, flashes or explosion of the gas or vapor within. It must operate at such an external temperature that a surrounding flammable atmosphere will not be ignited thereby.

eye light: illumination on a person to provide a specular reflection from the eyes (and teeth) without adding a significant increase in light on the subject.

F

fenestra method: a procedure for predicting the interior illumination received from daylight through windows.

fenestration: any opening or arrangement of openings (normally filled with media for control) for the admission of daylight.

fill light: supplementary illumination to reduce shadow or contrast range.

filter: a device for changing, by transmission, the magnitude and/or the spectral composition of the flux incident on it. Filters are called *selective* (or *colored*) or *neutral,* according to whether or not they alter the spectral distribution of the incident flux.

filter factor: the transmittance of "black light" by a filter.

NOTE: The application of this term is illustrated by the following formula for determining the luminance of fluorescent materials exposed to "black light":

$$\text{Footlamberts} = \frac{\text{Fluorens}}{\text{square feet}} \times \text{glow factor} \times \text{filter factor}$$

When integral-filter "black light" lamps are used, the filter factor is dropped from the formula because it already has been applied in assigning fluoren ratings to these lamps. See *fluoren* and *glow factor.*

Finsen: a suggested practical unit of erythemal flux density equal to one E-viton per square centimeter.

fixed light: a light having a constant luminous intensity when observed from a fixed point.

fixture: see *luminaire.*

flashing light: a rhythmic light in which the periods of light are of equal duration and are clearly shorter than the periods of darkness. See *group flashing light, interrupted quick-flashing light, quick-flashing light.*

flicker fusion frequency, *fff*: the frequency of intermittent stimulation of the eye at which flicker disappears. It also is called critical fusion frequency (cff) or critical flicker frequency (cff).

floodlight: a projector designed for lighting a scene or object to a luminance considerably greater than its surroundings. It usually is capable of being pointed in any direction and is of weatherproof construction. See *heavy duty floodlight, general purpose floodlight, ground-area open floodlight, ground-area floodlight with reflector insert.*

floodlighting: a system designed for lighting a scene or object to a luminance greater than its surroundings. It may be for utility, advertising or decorative purposes.

floor cavity: the cavity formed by the work-plane, the floor, and the wall surfaces between these two planes.

floor cavity ratio, *FCR:* a number indicating floor cavity proportions calculated from length, width and height. See page 9–8.

floor lamp: a portable luminaire on a high stand suitable for standing on the floor. See *torchere*.

fluoren: a unit of "black light" flux equal to one milliwatt of radiant flux in the wavelength range 320 to 400 nanometers.

fluorescence: the emission of light (luminescence) as the result of, and only during, the absorption of radiation of other (mostly shorter) wavelengths.

fluorescent lamp: a low-pressure mercury electric-discharge lamp in which a fluorescing coating (phosphor) transforms some of the ultraviolet energy generated by the discharge into light. See *instant start fluorescent lamp, preheat (switch start) fluorescent lamp, rapid start fluorescent lamp.*

fluorescent-mercury lamp: an electric-discharge lamp having a high-pressure mercury arc in an arc tube, and an outer envelope coated with a fluorescing substance (phosphor) that transforms some of the ultraviolet energy generated by the arc into light.

flush mounted or recessed: a luminaire mounted above the ceiling (or behind a wall or other surface) with the opening of the luminaire level with the surface.

fog (adverse-weather) lamps: lamps that may be used with, or in lieu of, lower beam headlamps to provide road illumination under conditions of rain, snow, dust or fog.

footcandle, *fc:* the unit of illumination when the foot is taken as the unit of length. It is the illumination on a surface one square foot in area on which there is a uniformly distributed flux of one lumen, or the illumination produced on a surface all points of which are at a distance of one foot from a directionally uniform point source of one candela. See Fig. 1–2.

footlambert, *fL:* a unit of luminance (photometric brightness) equal to $1/\pi$ candela per square foot, or to the uniform luminance of a perfectly diffusing surface emitting or reflecting light at the rate of one lumen per square foot, or to the average luminance of any surface emitting or reflecting light at that rate. See *units of luminance*.

Note: The average luminance of any reflecting surface in footlamberts is, therefore, the product of the illumination in footcandles by the luminous reflectance of the surface.

formation light: a navigation light especially provided to facilitate formation flying.

fovea: a small region at the center of the retina, subtending about two degrees, containing only cones and forming the site of most distinct vision.

fuselage lights: aircraft aeronautical lights, mounted on the top and bottom of the fuselage, used to supplement the navigation light.

G

gaseous discharge: the emission of light from gas atoms excited by an electric current.

general diffuse lighting: lighting by luminaires distributing 40 to 60 per cent of the emitted light downward and the balance upward, sometimes with a strong component at 90 degrees (horizontal). See *direct-indirect lighting*.

general lighting: lighting designed to provide a substantially uniform level of illumination throughout an area, exclusive of any provision for special local requirements. See *direct lighting, semi-direct lighting, general diffuse lighting, direct-indirect lighting, semi-indirect lighting, indirect lighting, ceiling area lighting, localized general lighting.*

general purpose floodlight (*GP*): a weatherproof unit so constructed that the housing forms the reflecting surface. The assembly is enclosed by a cover glass.

glare: the sensation produced by luminance within the visual field that are sufficiently greater than the luminance to which the eyes are adapted to cause annoyance, discomfort, or loss in visual performance and visibility. See *blinding glare, direct glare, disability glare, discomfort glare.*

Note: The magnitude of the sensation of glare depends upon such factors as the size, position and luminance of a source, the number of sources and the luminance to which the eyes are adapted.

globe: a transparent or diffusing enclosure intended to protect a lamp, to diffuse and redirect its light, or to change the color of the light.

glow discharge: an electric discharge characterized by a low, approximately constant, current density at the cathode, low cathode temperature, and a high, approximately constant, voltage drop.

glow factor: a measure of the visible light response of a fluorescent material to "black light." It is equal to the luminance in footlamberts produced on the material divided by the incident "black light" flux density in milliwatts per square foot. It may be measured in lumens per milliwatt.

Note: See note after *filter factor.*

glow lamp: an electric-discharge lamp whose mode of operation is that of a glow discharge, and in which light is generated in the space close to the electrodes.

goniophotometer: a photometer for measuring the directional light distribution characteristics of sources, luminaires, media and surfaces.

graybody: a temperature radiator whose spectral emissivity is less than unity and the same at all wavelengths.

ground-area open floodlight (*O*): a unit providing a weatherproof enclosure for the lamp socket and housing. No cover glass is required.

ground-area open floodlight with reflector insert (*OI*): a weatherproof unit so constructed that the housing forms only part of the reflecting surface. An auxiliary reflector is used to modify the distribution of light. No cover glass is required.

ground light: visible radiation from the sun and sky reflected by surfaces below the plane of the horizon.

group flashing light: a flashing light in which the flashes are combined in groups, each including the same number of flashes, and in which the groups are repeated at regular intervals. The duration of each flash is clearly less than the duration of the dark periods between flashes, and the duration of the dark periods between flashes is clearly less than the duration of the dark periods between groups.

H

hazard or obstruction beacon: an aeronautical beacon used to designate a danger to air navigation.

hazardous location: an area where ignitable vapors or dust may cause a fire or explosion created by energy emitted from lighting or other electrical equipment or by electrostatic generation.

headlamp: a major lighting device mounted on a vehicle and used to provide illumination ahead of it. Also called headlight. See *multiple-beam headlamp* and *sealed beam headlamp*.

headlight: an alternate term for *headlamp*.

heavy duty floodlight (*HD*): a weatherproof unit having a substantially constructed metal housing into which is placed a separate and removable reflector. A weatherproof hinged door with cover glass encloses the assembly but provides an unobstructed light opening at least equal to the effective diameter of the reflector.

hemispherical reflectance: the ratio of all of the flux leaving a surface or medium by reflection to the incident flux. See *hemispherical transmittance*.

Note: If reflectance is not preceded by an adjective descriptive of the angles of view, hemispherical reflectance is implied.

hemispherical transmittance: the ratio of the transmitted flux leaving a surface or medium to the incident flux.

Note: If transmittance is not preceded by an adjective, descriptive of the angles of view, hemispherical reflectance is implied.

high-intensity discharge lamps: a general group of lamps consisting of mercury, metal halide and high pressure sodium lamps.

high-key lighting: a type of lighting that, applied to a scene, results in a picture having gradations falling primarily between gray and white; dark grays or blacks are present, but in very limited areas. See *low-key lighting*.

high pressure sodium lamp: a sodium vapor lamp in which the partial pressure of the vapor during operation is of the order of 10^4 $N \cdot m^{-2}$ (0.1 atm).

horizontal plane of a searchlight: the plane perpendicular to the elevation axis and in which the train lies. See *vertical plane of a searchlight*.

hot-cathode lamp: an electric-discharge lamp whose mode of operation is that of an arc discharge. The cathodes may be heated by the discharge or by external means.

hue of a perceived light-source color: the attribute that determines whether the color is red, yellow, green, blue, or the like. See *hue of a perceived object color*.

hue of a perceived object color: the attribute that determines whether the color is red, yellow, green, blue, or the like.

I

ice detection light: an inspection light designed to illuminate the leading edge of an aircraft wing to check for ice formation.

identification beacon: an aeronautical beacon emitting a coded signal by means of which a particular point of reference can be identified.

illuminance, *E*: an alternate term for *illumination* that is receiving growing worldwide acceptance, especially in science. The term is subject to confusion with luminance and illuminants, especially when not clearly pronounced, and is therefore not yet generally used by illuminating engineers.

illumination, $E = d\Phi/dA$: the density of the luminous flux incident on a surface; it is the quotient of the luminous flux by the area of the surface when the latter is uniformly illuminated. See Fig. 1–9 for units and conversion factors.

Note: The term illumination also is commonly used in a qualitative or general sense to designate the act of illuminating or the state of being illuminated. Usually the context will indicate which meaning is intended, but occasionally it is desirable to use the expression *level of illumination* to indicate that the quantitative meaning is intended.

illumination (footcandle) meter: an instrument for measuring the illumination on a surface. Most such instruments consist of one or more barrier-layer cells connected to a meter calibrated in footcandles.

incandescence: the self-emission of radiant energy in the visible spectrum due to the thermal excitation of atoms or molecules.

incandescent filament lamp: a lamp in which light is produced by a filament heated to incandescence by an electric current.

incomplete diffusion: that in which the diffusing medium re-directs some of the flux incident upon it in an image-forming state and another portion in a non-image-forming state.

indirect component: the portion of the luminous flux from a luminaire arriving at the work-plane after being reflected by room surfaces. See *direct component*.

indirect lighting: lighting by luminaires distributing 90 to 100 per cent of the emitted light upward.

infrared radiation: for practical purposes any radiant energy within the wavelength range 780 to 10^5 nanometers.

Note: In general, unlike ultraviolet energy, infrared energy is not evaluated on a wavelength basis but rather in terms of all of such energy incident upon a surface. Examples of these applications are industrial heating, drying, baking and photoreproduction. However, some applications, such as infrared viewing devices, involve detectors sensitive to a restricted range of wavelengths; in such cases the spectral characteristics of the source and receiver are of importance.

initial luminous exitance: the density of luminous flux leaving a surface within an enclosure before interreflections occur.

Note: For light sources this is the luminous exitance as defined in *luminous flux density at a surface*. For non-self-luminous surfaces it is the reflected luminous exitance of the flux received directly from sources within the enclosure or from daylight.

instant start fluorescent lamp: a fluorescent lamp designed for starting by a high voltage without pre-heating of the electrodes.

integrating photometer: a photometer that enables total luminous flux to be determined by a single measurement. The usual type is the Ulbricht sphere with associated photometric equipment for measuring the luminance of the inner surface of the sphere.

intensity: a shortening of the terms *luminous intensity* and *radiant intensity*. Often misused for level of illumination.

interflectance: an alternate term for *room utilization factor*.

interflectance method: a lighting design procedure for predetermining the luminances (photometric brightnesses) of walls, ceiling and floor and the average illumination on the work-plane. It takes into acount both direct and reflected flux.

interflected component: that portion of the luminous

flux from a luminaire arriving at the work-plane after being reflected one or more times from room surfaces.

interflection: the multiple reflection of light by the various room surfaces before it reaches the work-plane or other specified surface of a room.

inter-reflectance: the portion of the luminous flux (lumens) reaching the work-plane that has been reflected one or more times as determined by the flux transfer theory. See page 9–36.

interrupted quick-flashing light: a quick flashing light in which the rapid alternations are interrupted by periods of darkness at regular intervals.

inverse-square law: the law stating that the illumination E at a point on a surface varies directly with the candlepower I of a point source, and inversely as the square of the distance d between the source and the point. If the surface at the point is normal to the direction of the incident light, the law is expressed by $E = I/d^2$.

NOTE: For sources of finite size this gives results that are accurate within one-half per cent when d is at least five times the maximum dimension of the source (or luminaire) as viewed from the point on the surface.

irradiance, *E*: the density of radiant flux incident on a surface.

isocandela line: one plotted on any appropriate coordinates to show directions in space, about a source of light, in which the candlepower is the same. For a complete exploration the line always is a closed curve. A series of such curves, usually for equal increments of candlepower, is called an isocandela diagram. See *isolux line.*

isolux (isofootcandle) line: one plotted on any appropriate coordinates to show all the points on a surface where the illumination is the same. For a complete exploration the line is a closed curve. A series of such lines for various illumination values is called an isolux (isofootcandle) diagram. See *isocandela line.*

K

key light: the apparent principal source of directional illumination falling upon a subject or area.

L

laboratory reference standards: the highest ranking order of standards at each laboratory.

lambert, *L*: a unit of luminance (photometric brightness) equal to $1/\pi$ candela per square centimeter, and, therefore, equal to the uniform luminance of a prefectly diffusing surface emitting or reflecting light at the rate of one lumen per square centimeter.

NOTE: The lambert also is the average luminance of any surface emitting or reflecting light at the rate of one lumen per square centimeter. For the general case, the average must take account of variation of luminance with angle of observation; also of its variation from point to point on the surface considered.

Lambert's cosine law, $I_\theta = I_0 \cos \theta$: the law stating that the flux per solid angle in any direction from a perfectly diffusing surface varies as the cosine of the angle between that direction and the perpendicular to the surface.

lambert surface: a surface that emits or reflects light in accordance with Lambert's cosine law. A lambert surface has the same luminance regardless of viewing angle.

lamp: a generic term for a man-made source of light. By extension, the term is also used to denote sources that radiate in regions of the spectrum adjacent to the visible.

NOTE: A lighting unit consisting of a lamp with shade, reflector, enclosing globe, housing, or other accessories is also called a "lamp." In such cases, in order to distinguish between the assembled unit and the light source within it, the latter is often called a "bulb" or "tube," if it is electrically powered. See also *luminaire*

lamp lumen depreciation factor, *LLD*: the multiplier to be used in illumination calculations to relate the initial rated output of light sources to the anticipated minimum rated output based on the relamping program to be used.

lamp post: a standard support provided with the necessary internal attachments for wiring and the external attachments for the bracket and luminaire.

lamp shielding angle: the angle between the plane of the baffles or louver grid and the plane most nearly horizontal that is tangent to both the lamps and the louver blades.

NOTE: The lamp shielding angle frequently is larger than the louver shielding angle, but never smaller. See *louver shielding angle.*

landing direction indicator: a device to indicate visually the direction currently designated for landing and take-off.

landing light: an aircraft aeronautical light designed to illuminate a ground area from the aircraft.

landmark beacon: an aeronautical beacon used to indicate the location of a landmark used by pilots as an aid to enroute navigation.

laser: an acronym for *L*ight *A*mplification by *S*timulated *E*mission of *R*adiation. The Laser produces a highly monochromatic and coherent beam of radiation. A steady oscillation of nearly a single electromagnetic mode is maintained in a volume of an active material bounded by highly reflecting surfaces, called a resonator. The frequency of oscillation varies according to the material used and by the methods of initially exciting or pumping the material.

lateral width of a light distribution: (in roadway lighting) the lateral angle between the reference line and the width line, measured in the cone of maximum candlepower. This angular width includes the line of maximum candlepower. See *reference line* and *width line.*

level of illumination: see *illumination.*

life performance curve: one which represents the variation of a particular characteristic of a light source (luminous flux, candlepower, etc.) throughout the life of the source.

NOTE: Life performance curves sometimes are called maintenance curves as, for example, lumen maintenance curves.

life test of lamps: one in which lamps are operated under specified conditions for a specified length of time, for the purpose of obtaining information on lamp life. Measurements of photometric and electrical characteristics may be made at specified intervals of time during this test.

light: for the purposes of illuminating engineering, visually evaluated radiant energy.

NOTE: Light is psychophysical, neither purely physical nor purely psychological. Light is not synonymous with radiant energy, however restricted, nor is it merely sensation. In a general non-specialized sense, light is the aspect of radiant energy of which a human observer is aware through the stimulation of the retina of the eye.

The present basis for the engineering evaluation of light consists of the color-mixture data $\bar{x}$, $\bar{y}$, $\bar{z}$, (see *spectral tristimulus values*) adopted in 1931 by the International Commission of Illumination (CIE). These data include the spectral luminous efficiency data for photopic vision (see *values of spectral luminous efficiency for photopic vision*) adopted in 1924 by the CIE.

light adaptation: the process by which the retina becomes adapted to a luminance greater than about one footlambert. See also *dark adaptation*.

light center (of a lamp): the center of the smallest sphere that would completely contain the light-emitting element of the lamp.

light center length (of a lamp): the distance from the light center to a specified reference point on the lamp.

light loss factor: a factor used in calculating the level of illumination after a given period of time and under given conditions. It takes into account temperature and voltage variations, dirt accumulation on luminaire and room surfaces, lamp depreciation, maintenance procedures and atmosphere conditions.

light-source color: the color of the light emitted by the source.

NOTE: The color of a point source may be defined by its luminous intensity and chromaticity coordinates; the color of an extended source may be defined by its luminance and chromaticity coordinates. See *perceived light-source color*.

lighting effectiveness factor, LEF_V: the ratio of equivalent sphere illumination to ordinary measured or calculated illumination.

lightness of a perceived object color: the attribute by which it seems to transmit or reflect a greater or lesser fraction of the incident light.

linear light: a luminous signal having a perceptible physical length.

local lighting: lighting designed to provide illumination over a relatively small area or confined space without providing any significant general surrounding lighting.

localized general lighting: lighting that utilizes luminaires above the visual task and contributes also to the illumination of the surround.

louver: a series of baffles used to shield a source from view at certain angles or to absorb unwanted light. The baffles usually are arranged in a geometric pattern.

louver shielding angle, θ: the angle between the horizontal plane of the baffles or louver grid and the plane at which the louver conceals all objects above. See *lamp shielding angle*.

NOTE: The planes usually are so chosen that their intersection is parallel with the louvered blade.

louvered ceiling: a ceiling area lighting system comprising a wall-to-wall installation of multicell louvers shielding the light sources mounted above it. See *luminous ceiling*.

lower (passing) beams: one or more beams directed low enough on the left to avoid glare in the eyes of oncoming drivers, and intended for use in congested areas and on highways when meeting other vehicles within a distance of 1000 feet. Formerly "traffic beam."

low-key lighting: a type of lighting that, applied to a scene, results in a picture having gradations falling primarily between middle gray and black, with comparatively limited areas of light grays and whites. See *high-key lighting*.

lumen, lm: the unit of luminous flux. It is equal to the flux through a unit solid angle (steradian), from a uniform point source of one candela (candle), or to the flux on a unit surface all points of which are at unit distance from a uniform point source of one candela. See Fig. 1–2 for a diagrammatic representation.

NOTE: For some purposes, the kilolumen, equal to 1000 lumens, is a convenient unit.

lumen-hour, *lm-h*: a unit of quantity of light (luminous energy). It is the quantity of light delivered in one hour by a flux of one lumen.

lumen (or flux) method: a lighting design procedure used for predetermining the number and types of lamps or luminaires that will provide a desired average level of illumination on the work-plane. It takes into account both direct and reflected flux.

luminaire: a complete lighting unit consisting of a lamp or lamps together with the parts designed to distribute the light, to position and protect the lamps and to connect the lamps to the power supply.

luminaire dirt depreciation factor, *LDD*: the multiplier to be used in illumination calculations to relate the initial illumination provided by clean, new luminaires to the reduced illumination that they will provide due to dirt collection on the luminaires at the time at which it is anticipated that cleaning procedures will be instituted.

luminaire efficiency: the ratio of luminous flux (lumens) emitted by a luminaire to that emitted by the lamp or lamps used therein.

luminance (photometric brightness), $L = d^2\Phi/d\omega\,(dA \cos\theta) = dI/(dA \cos\theta)$: Luminance (photometric brightness) in a direction, at a point on the surface of a source, of a receiver, or of any other real or virtual surface is the quotient of the luminous flux leaving, passing through, or arriving at an element of the surface surrounding the point, and propagated in directions defined by an elementary cone containing the given direction, by the product of the solid angle of the cone and the area of the orthogonal projection of the element of the surface on a plane perpendicular to the given direction; or it is the luminous intensity of any surface in a given direction per unit of projected area of the surface as viewed from that direction. See *Units of Luminance* and Fig. 1–7.

NOTE: In the defining equation θ is the angle between the direction of observation and the normal to the surface.

In common usage the term *brightness* usually refers to the intensity of *sensation* resulting from viewing surfaces or spaces from which light comes to the eye. This sensation is determined in part by the definitely measurable luminance defined above and in part by conditions of observation such as the state of adaptation of the eye. (See *subjective brightness*).

In much of the literature the term brightness, used alone, refers to both luminance and sensation. The context usually indicates which meaning is intended.

luminance contrast: the relationship between the luminances of an object and its immediate background. It is equal to $(L_1 - L_2)/L_1$ or $(L_2 - L_1)/L_1$, where L_1 and L_2 are the luminances of the background and object, respectively. The form of the equation must be specified.

NOTE: See last paragraph of the note under *luminance*. Because of the relationship among luminance, illumination and reflectance, contrast often is expressed in terms of reflectance when only reflecting surfaces are involved. Thus, contrast is equal to $(\rho_1 - \rho_2)/\rho_1$ or $(\rho_2 - \rho_1)/\rho_1$ where ρ_1 and ρ_2 are the reflectances of the background and object, respectively. This method of computing contrast holds only for perfectly diffusing surfaces; for other surfaces it is only an approximation unless the angles of incidence and view are taken into consideration. (See *reflectance*.)

luminance (brightness) coefficient: a coefficient similar to the coefficient of utilization used to determine wall and ceiling luminances.

luminance difference: the difference in luminance between two areas. It usually is applied to contiguous areas, such as the detail of a visual task and its immediate background, in which case it is quantitatively equal to the numerator in the formula for luminance contrast.

NOTE: See last paragraph of the note under *luminance.*

luminance factor, β: the ratio of the luminance of a surface or medium under specified conditions of incidence, observation, and light source, to the luminance of a perfectly reflecting or transmitting, perfectly diffusing surface or medium under the same conditions.

NOTE: Reflectance or transmittance cannot exceed unity, but luminance factor may have any value from zero to values approaching infinity.

luminance factor of room surfaces: factors by which the average work-plane illumination is multiplied to obtain the average luminances of walls, ceilings and floors.

luminance ratio: the ratio between the luminances (photometric brightnesses) of any two areas in the visual field.

NOTE: See last paragraph of the note under *luminance.*

luminance threshold: the minimum perceptible difference in luminance for a given state of adaptation of the eye.

luminescence: any emission of light not ascribable directly to incandescence. See *electroluminescence, fluorescence, phosphorescence.*

luminosity factor: previously used term for *spectral luminous efficacy of radiant flux.*

luminous ceiling: a ceiling area lighting system comprising a continuous surface of transmitting material of a diffusing or light-controlling character with light sources mounted above it. See *louvered ceiling.*

luminous density, $w = dQ/dV$: quantity of light (luminous energy) per unit volume.

luminous efficacy of radiant flux: the quotient of the total luminous flux by the total radiant flux. It is expressed in lumens per watt.

luminous efficacy of a source of light: the quotient of the total luminous flux emitted by the total lamp power input. It is expressed in lumens per watt.

NOTE: The term luminous efficiency has in the past been extensively used for this concept.

luminous efficiency: see *spectral luminous efficiency of radiant flux.*

luminous exitance, M: density of luminous flux leaving a surface. It is expressed in lumens per unit area of the surface.

luminous flux, Φ: the time rate of flow of light.

luminous flux density at a surface, $d\Phi/dA$: luminous flux per unit area of the surface.

NOTE: When referring to luminous flux *emitted* from a surface, this has been called *luminous emittance* (symbol M). The preferred term for luminous flux *leaving* a surface is *luminous exitance* (symbol M). When referring to flux incident on a surface, it is identical with *illumination* (symbol E).

luminous intensity, $I = d\Phi/d\omega$: the luminous flux per unit solid angle in a specific direction. Hence, it is the luminous flux on a small surface normal to that direction, divided by the solid angle (in steradians) that the surface subtends at the source.

NOTE: Mathematically a solid angle must have a point as its apex; the definition of luminous intensity, therefore, applies strictly only to a point source. In practice, however, light emanating from a source whose dimensions are negligible in comparison with the distance from which it is observed may be considered as coming from a point. For extended sources see *apparent candlepower of an extended source.*

lux, *lx*: the International System (SI) unit of illumination. It is the illumination on a surface one square meter in area on which there is a uniformly distributed flux of one lumen, or the illumination produced at a surface all points of which are at a distance of one meter from a uniform point source of one candela. See Fig. 1–9.

M

maintenance factor, *MF*: a factor formerly used to denote the ratio of the illumination on a given area after a period of time to the initial illumination on the same area. See *light loss factor.*

matte surface: one from which the reflection is predominantly diffuse, with or without a negligible specular component. See *diffuse reflection.*

mercury lamp: an electric discharge lamp in which the major portion of the radiation is produced by the excitation of mercury atoms.

mesopic vision: vision with luminance conditions between those of photopic and scotopic vision, that is, between about one footlambert and 0.01 footlambert. See *photopic* and *scotopic vision.*

metal halide lamp: a discharge lamp in which the light is produced by the radiation from a mixture of a metallic vapor (for example, mercury) and the products of the disassociation of halides (for example, halides of thallium, indium or sodium.)

metamers: lights of the same color but of different spectral energy distribution.

minimal perceptible erythema, *MPE*: the erythemal threshold.

monocular visual field: the visual field of a single eye. See *binocular visual field.*

mounting height (roadway): the vertical distance between the roadway surface and the center of the apparent light source of the luminaire.

mounting height above the floor, MH_f: the distance from the floor to the light center of the luminaire or to the plane of the ceiling for recessed equipment.

mounting height above the work-plane, MH_{wp}: the distance from the work-plane to the light center of the luminaire or to the plane of the ceiling for recessed equipment.

multiple-beam headlamp: a headlamp designed to permit the driver of a vehicle to use any one of two or more distributions of light on the road.

Munsell chroma, C: the index of saturation of the perceived object color defined in terms of the Y-value and chromaticity coordinates (x, y) of the color of light reflected or transmitted by the object.

Munsell color system: a system of surface-color specification based on perceptually uniform color scales for the three variables: Munsell hue, Munsell value, and Munsell chroma. For an observer of normal color vision, adapted to daylight, and viewing a specimen when illuminated by daylight and surrounded with a middle gray to white background, the Munsell hue, value and chroma of the color correlate well with the hue, lightness and saturation of the perceived color.

Munsell hue, *H*: the index of the hue of the perceived object color defined in terms of the Y-value and chromaticity coordinates (x, y) of the color of the light reflected or transmitted by the object.

Munsell value, *V*: the index of the lightness of the perceived object color defined in terms of the Y-value.

NOTE: Munsell value is approximately equal to the square root of the reflectance expressed in per cent.

N

nanometer: unit of wavelength equal to 10^{-9} meter. See Fig. 1–8.

narrow-angle diffusion: that in which flux is scattered at angles near the direction that the flux would take by regular reflection or transmission. See *wide-angle-diffusion.*

narrow angle luminaire: a luminaire that concentrates the light within a cone of comparatively small solid angle. See *wide angle luminaire.*

national standard of length: 1,650,763.73 wavelengths of the orange-red radiation of krypton 86 under specified conditions equal a meter.

national standard: an alternate term for *primary standard.*

navigation lights: an alternate term for *position lights.*

navigation light system: a set of aircraft aeronautical lights provided to indicate the position and direction of motion of an aircraft to pilots of other aircraft or to ground observers.

night: the hours between the end of evening civil twilight and the beginning of morning civil twilight.

NOTE: Civil twilight ends in the evening when the center of the sun's disk is six degrees below the horizon and begins in the morning when the center of the sun's disk is six degrees below the horizon.

nit, *nt*: a unit of luminance (photometric brightness) equal to one candela per square meter.

NOTE: Candela per square meter is the International Standard (SI) unit of luminance.

O

object color: the color of the light reflected or transmitted by the object when illuminated by a standard light source, such as CIE source A, B, C or D_{65}. See *standard source* and *perceived object color.*

obstruction lights: aeronautical ground lights provided to indicate obstructions.

occulting light: a rhythmic light in which the periods of light are clearly longer than the periods of darkness.

orientation: the position of a building with respect to compass directions.

overcast sky: one that has 100 per cent cloud cover; the sun is not visible.

overhang: the distance between a vertical line passing through the luminaire and the curb or edge of the roadway.

ozone-producing radiation: ultraviolet energy shorter than about 220 nanometers that decomposes oxygen O_2 thereby producing ozone O_3. Some ultraviolet sources generate energy at 184.9 nanometers that is particularly effective in producing ozone.

P

parking lamp: one placed on a vehicle to indicate its presence when parked.

partly cloudy sky: one that has 30 to 70 per cent cloud cover.

perceived light-source color: the color perceived to belong to a light source.

perceived object color: the color perceived to belong to an object, resulting from characteristics of the object, of the incident light, and of the surround, the viewing direction and observer adaptation. See *object color.*

perfect diffusion: that in which flux is scattered in accord with Lambert's cosine law. See *Lambert's cosine law.*

perimeter lights: aeronautical ground lights provided to indicate the perimeter of a landing pad for helicopters.

peripheral vision: the seeing of objects displaced from the primary line of sight and outside the central visual field.

peripheral visual field: that portion of the visual field that falls outside the region corresponding to the foveal portion of the retina.

phosphorescence: the emission of light (luminescence) as the result of the absorption of radiation, and continuing for a noticeable length of time after excitation.

phot, *ph*: the unit of illumination when the centimeter is taken as the unit of length; it is equal to one lumen per square centimeter.

photochemical radiation: energy in the ultraviolet, visible and infrared regions capable of producing chemical changes in materials.

NOTE: Examples of photochemical processes are accelerated fading tests, photography, photoreproduction and chemical manufacturing. In many such applications a specific spectral region is of importance.

photolectric receiver: a device that reacts electrically in a measurable manner in response to incident radiant energy.

photoflash lamp: a lamp in which combustible metal or other solid material is burned in an oxidizing atmosphere to produce light of high intensity and short duration for photographic purposes.

photometer: an instrument for measuring photometric quantities such as luminance (photometric brightness), luminous intensity, luminous flux and illumination. See *densitometer, goniophotometer, illumination (footcandle) meter, integrating photometer, reflectometer, spectrophotometer, transmissometer.*

photometry: the measurement of quantities associated with light.

NOTE: Photometry may be visual in which the eye is used to make a comparison, or physical in which measurements are made by means of physical receptors.

photopic vision: vision mediated essentially or exclusively by the cones. It is generally associated with adaptation to a luminance of at least one footlambert. See *scotopic vision.*

physical photometer: an instrument containing a physical receptor (photoemissive cell, barrier-layer cell, thermopile, etc.) and associated filters, that is calibrated so as to read photometric quantities directly. See *visual photometer.*

pilot house control: a mechanical means for controlling the elevation and train of a searchlight from a position on the other side of the bulkhead or deck on which it is mounted.

Planck radiation law: an expression representing the spectral radiance of a blackbody as a function of the wavelength and temperature. This law commonly is expressed by the formula

$$L_\lambda = I_\lambda/A' = c_{1L}\lambda^{-5}[e^{(c_2/\lambda T)} - 1]^{-1}$$

in which L_λ is the spectral radiance, I_λ is the spectral radiant intensity, A' is the projected area (A cos θ)

of the aperture of the blackbody, e is the base of natural logarithms (2.718+), T is absolute temperature, c_{1L} and c_2 are constants designated as the first and second radiation constants.

NOTE: The designation c_{1L} is used to indicate that the equation in the form given here refers to the radiance L, or to the intensity I per unit projected area A', of the source. Numerical values are commonly given not for c_{1L} but for c_1 which applies to the total flux radiated from a blackbody aperture, that is, in a hemisphere (2 π steradians), so that, with the Lambert cosine law taken into account, $c_1 = \pi c_{1L}$. The currently recommended value of c_1 is 3.7415 × 10^{-16} $W\ m^2$ or 3.7415 × 10^{-12} $W\ cm^2$. Then c_{1L} is 1.1910 × 10^{-16} $W\ m^2\ sr^{-1}$ or 1.1910 × 10^{-12} $W\ cm^2\ sr^{-1}$. If, as is more convenient, wavelengths are expressed in micrometers and area in square centimeters. c_{1L} = 1.1910 × 10^4 $W\ \mu m^4\ cm^{-2}\ sr^{-1}$, L_λ being given in $W\ cm^{-2}\ sr^{-1}\ \mu m^{-1}$. The currently recommended value of c_2 is 1.4388 cm K.

The Planck law in the following form gives the energy radiated from the blackbody in a given wavelength interval ($\lambda_1 - \lambda_2$):

$$Q = \int_{\lambda_1}^{\lambda_2} Q_\lambda\, d\lambda = Atc_1 \int_{\lambda_1}^{\lambda_2} \lambda^{-5}(e^{(c_2/\lambda T)} - 1)^{-1}\, d\lambda$$

If A is the area of the radiation aperture or surface in square centimeters, t is time in seconds, λ is wavelength in micrometers, and c_1 = 3.7415 × 10^4 $W\ \mu m^4\ cm^{-2}$, then Q is the total energy in watt seconds emitted from this area (that is, in the solid angle 2 π), in time t, within the wavelength interval ($\lambda_1 - \lambda_2$).

NOTE: It often is convenient, as is done here, to use different units of length in specifying wavelengths and areas, respectively. If both quantities are expressed in centimeters and the corresponding value for c_1 (3.7415 × 10^{-5} erg cm^2 sec^{-1}) is used, this equation gives the total emission of energy in ergs from area A (that is in the solid angle 2 π), for time t, and for the interval $\lambda_1 - \lambda_2$ in centimeters.

point of fixation: a point or object in the visual field at which the eyes look and upon which they are focused.

point of observation: the midpoint of the base line connecting the centers of rotation of the two eyes. For practical purposes, the center of the pupil of the eye often is taken as the point of observation.

point-by-point method: a lighting design procedure for predetermining the illumination at various locations in lighting installations, by use of luminaire photometric data.

NOTE: Since interreflections are not taken into account, the point-by-point method may indicate lower levels of illumination than are actually realized.

polarization: the process by which the transverse vibrations of light waves are oriented in a specific plane.

pole (roadway lighting): a standard support generally used where overhead lighting distribution circuits are employed.

portable lighting: lighting by means of equipment designed for manual portability.

portable luminaire: a lighting unit that is not permanently fixed in place. See *table lamp* and *floor lamp*.

portable traffic control light: a signalling light producing a controllable distinctive signal for purposes of directing aircraft operations in the vicinity of an aerodrome.

position index, P: a factor representing the relative average luminance, for a source located anywhere within the visual field producing a sensation at the borderline between comfort and discomfort (BCD).

position lights: aircraft aeronautical lights forming the basic or internationally recognized navigation light system.

NOTE: The system is composed of a red light showing from dead ahead to 110° to the left, a green light showing from dead ahead to 110° to the right, and a white light showing to the rear through 140°. Position lights are also called navigation lights.

preheat (switch start) fluorescent lamp: a fluorescent lamp designed for operation in a circuit requiring a manual or automatic starting switch to preheat the electrodes in order to start the arc.

primary: any one of three lights in terms of which a color is specified by giving the amounts required to duplicate it by additive combination. See *tristimulus values of a light, X, Y, Z.*

primary line of sight: the line connecting the point of observation and the point of fixation.

primary standard: a light source by which the unit of light is established and from which the values of other standards are derived. This order of standard also is designated as the national standard.

NOTE: A satisfactory primary (national) standard must be reproducible from specifications (see *candela*). Primary (national) standards usually are found in national physical laboratories.

projector: a lighting unit that by means of mirrors and lenses, concentrates the light to a limited solid angle so as to obtain a high value of luminous intensity. See *floodlight, searchlight, signalling light.*

protective lighting: a system intended to facilitate the nighttime policing of industrial and other properties.

pupil (pupillary aperture): the opening in the iris that admits light into the eye. See *artificial pupil.*

Purkinje phenomenon: the reduction in subjective brightness of a red light relative to that of a blue light when the luminances are reduced in the same proportion without changing the respective spectral distributions. In passing from photopic to scotopic vision, the curve of spectral luminous efficiency changes, the wavelength of maximum efficiency being displaced toward the shorter wavelengths.

purple boundary: the straight line drawn between the ends of the spectrum locus on a chromaticity diagram.

Q

quality of lighting: pertains to the distribution of luminance in a visual environment. The term is used in a positive sense and implies that all luminances contribute favorably to visual performance, visual comfort, ease of seeing, safety, and esthetics for the specific visual tasks involved.

quantity of light (luminous energy, $Q = \int \Phi dt$: the product of the luminous flux by the time it is maintained. It is the time integral of luminous flux (compare *light* and *luminous flux.*)

quick-flashing light: a rhythmic light exhibiting very rapid regular alternations of light and darkness. There is no restriction on the ratio of the durations of the light to the dark periods.

R

radiance, $L = d^2\Phi/d\omega(dA\cos\theta) = dI/(dA\cos\theta)$: radiance in a direction, at a point of the surface of a source, of a receiver, or of any other real or virtual surface, is the quotient of the radiant flux leaving, passing through, or arriving at an element of the surface surrounding the point, and propagated in directions defined by an elementary cone containing the given direction, by the product of the solid angle of the cone and the area of the orthogonal projection of the element of the surface on a plane perpendicular to the given direction.

Note: In the defining equation θ is the angle between the normal to the element of the source and the direction of observation.

radiant density, $w = dQ/dV$: radiant energy per unit volume; *e.g.*, joules/m^3.

radiant energy, Q: energy traveling in the form of electromagnetic waves. It is measured in units of energy such as joules, ergs, or kilowatt-hours. See *spectral radiant energy*.

radiant exitance, *M*: the density of radiant flux leaving a surface. It is expressed in watts per unit area of the surface.

radiant flux, $\Phi = dQ/dt$: the time rate of flow of radiant energy. It is expressed preferably in watts, or in joules per second. See *spectral radiant flux*.

radiant flux density at an element of a surface, $d\Phi/dA$: the quotient of radiant flux at that element of surface to the area of that element; *e.g.*, watts/cm^2. When referring to radiant flux *emitted* from a surface, this has been called *radiant emittance* (symbol: *M*). The preferred term for radiant flux *leaving* a surface is *radiant exitance* (*M*). When referring to radiant flux incident on a surface, it is called *irradiance* (*E*).

radiant intensity, $I = d\Phi/d\omega$: the radiant flux proceeding from the source per unit solid angle in the direction considered; *e.g.*, watts/steradian. See *spectral radiant intensity*.

radiator: an emitter of radiant energy.

radiometry: the measurement of quantities associated with radiant energy.

range lights: groups of color-coded boundary lights provided to indicate the direction and limits of a preferred landing path normally on an aerodrome without runways but exceptionally on an aerodrome with runways.

rapid start fluorescent lamp: a fluorescent lamp designed for operation with a ballast that provides a low-voltage winding for preheating the electrodes and initiating the arc without a starting switch or the application of high voltage.

reaction time: the interval between the beginning of a stimulus and the beginning of the response of an observer.

redirecting surfaces and media: those surfaces and media that change the direction of the flux without scattering the redirected flux.

reference ballast: a ballast specially constructed to have certain prescribed characteristics for use in testing electric discharge lamps and other ballasts.

reference line (roadway lighting): either of two radial lines where the surface of the cone of maximum candlepower is intersected by a vertical plane parallel to the curb line and passing through the light-center of the luminaire.

reference standard: an alternate term for *secondary standard*.

reflectance of a surface or medium, $\rho = \Phi_r/\Phi_i$: the ratio of the reflected flux to the incident flux. See *diffuse reflectance, hemispherical reflectance, regular reflectance, spectral reflectance*.

Note: Measured values of reflectance depend upon the angles of incidence and view and on the spectral character of the incident flux. Because of this dependence, the angles of incidence and view and the spectral characteristics of the source should be specified. See *reflection*.

reflectance factor (at a point of a surface, for the part of the reflected radiation contained in a cone with apex at the point of the surface, and for incident radiation of a given spectral composition and geometrical distribution): ratio of the radiant (or luminous) flux reflected in the directions delimited by the cone to that reflected in the same directions by a perfect reflecting diffuser identically irradiated (or illuminated).

reflected glare: glare resulting from specular reflections of high luminances in polished or glossy surfaces in the field of view. It usually is associated with reflections from within a visual task or areas in close proximity to the region being viewed. See *veiling reflection*.

reflection: a general term for the process by which the incident flux leaves a surface or medium from the incident side.

Note: Reflection is usually a combination of regular and diffuse reflection. See *regular (specular) reflection* and *diffuse reflection*.

reflectometer: a photometer for measuring reflectance.

Note: Reflectometers may be visual or physical instruments.

reflector: a device used to redirect the luminous flux from a source by the process of reflection. See *retro-reflector*.

refraction: the process by which the direction of a ray of light changes as it passes obliquely from one medium to another in which its speed is different.

refractor: a device used to redirect the luminous flux from a source, primarily by the process of refraction.

regions of electromagnetic spectrum: for convenience of reference the electromagnetic spectrum is divided as follows:

Ultraviolet Spectrum:	
Far ultraviolet	10–280 nm
Middle ultraviolet	280–315 nm
Near ultraviolet	315–380 nm
Visible Spectrum	380–780 nm
Infrared Spectrum	780–10^5 nm

Note: The spectral limits indicated above should not be construed to represent sharp delineations between the various regions. There is a gradual transition from region to region. The above ranges have been established for practical purposes.

regressed luminaire: a luminaire mounted above the ceiling with the opening of the luminaire above the ceiling line. See *flush-mounted, surface-mounted, suspended* and *troffer*.

regular (specular) reflectance: the ratio of the flux leaving a surface or medium by regular (specular) reflection to the incident flux. See *regular (specular) reflection*.

regular (specular) reflection: that process by which incident flux is re-directed at the specular angle. See *specular angle*.

regular transmission: that process by which incident flux passes through a surface or medium without scattering. See *regular transmittance*.

regular transmittance: the ratio of the regularly transmitted flux leaving a surface or medium to the incident flux.

relative luminosity: previously used term for *spectral luminous efficiency of radiant flux*.

relative luminosity factor: previously used term for *spectral luminous efficiency of radiant flux*.

resultant color shift: the difference between the perceived color of an object illuminated by a test source and that of the same object illuminated by the reference source, taking account of the state of chromatic adaptation in each case; *i.e.*, the resultant of colorimetric shift and adaptive color shift.

retina: a membrane lining the more posterior part of the inside of the eye. It comprises photoreceptors (cones and rods) that are sensitive to light and nerve cells that transmit to the optic nerve the stimulation of the receptor elements.

retro-reflector (reflex reflector): a device designed to reflect light in a direction close to that at which it is incident, whatever the angle of incidence.

rhythmic light: a light, when observed from a fixed point, having a luminous intensity that changes periodically. See *equal interval light, flashing light, group flashing light, interrupted quick-flashing light, quick flashing light, occulting light*.

rod: a retinal element that is primarily concerned with the perception of light. Rods play little or no part in color discrimination or the discrimination of fine detail.

room cavity: the cavity formed by the plane of the luminaires, the work-plane, and the wall surfaces between these two planes.

room cavity ratio, *RCR:* a number indicating room cavity proportions calculated from length, width and height. See page 9–8.

room index: formerly a letter designation for a range of room ratios.

room ratio: formerly a number indicating room proportions, calculated from the length, width and ceiling height (or luminaire mounting height) above the work-plane.

room utilization factor (utilance): the ratio of the luminous flux (lumens) received on the work-plane to that emitted by the luminaire.

Note: This ratio sometimes is called interflectance. Room utilization factor is based on the flux emitted by a complete luminaire, whereas coefficient of utilization is based on the flux generated by the bare lamps in a luminaire.

runway alignment indicator: a group of aeronautical ground lights arranged and located to provide early direction and roll guidance on the approach to a runway.

runway centerline lights: lights installed in the surface of the runway along the centerline indicating the location and direction of the runway centerline; of particular value in conditions of very poor visibility.

runway-edge lights: lights installed along the edges of a runway marking its lateral limits and indicating its direction.

runway-end identification light: a pair of flashing aeronautical ground lights symmetrically disposed on each side of the runway at the threshold to provide additional threshold conspicuity.

runway exit lights: lights placed on the surface of a runway to indicate a path to the taxiway centerline.

runway lights: aeronautical ground lights arranged along or on a runway. See *runway centerline lights, runway-edge lights, runway-end identification light, runway-exit lights*.

runway threshold: the beginning of the runway usable for landing.

runway visibility: the meteorological visibility along an identified runway. Where a transmissometer is used for measurement, the instrument is calibrated in terms of a human observer; *e.g.*, the sighting of dark objects against the horizon sky during daylight and the sighting of moderately intense unfocused lights of the order of 25 candelas at night.

runway visual range, *RVR:* In the United States, an instrumentally derived value, based on standard calibrations, representing the horizontal distance a pilot will see down the runway from the approach end; it is based on the sighting of either high intensity runway lights or on the visual contrast of other targets —whichever yields the greater visual range.

S

saturation of a perceived light source color: the attribute used to describe the departure of a perceived light-source color from a light-source color of the same brightness perceived to have no hue. See *saturation of a perceived object color*.

saturation of a perceived object color: the attribute used to describe the departure of a perceived light-source color from gray of the same lightness.

scotopic vision: vision mediated essentially or exclusively by the rods. It is generally associated with adaptation to a luminance below about 0.01 footlambert. See *photopic vision*.

sealed beam headlamp: an integral optical assembly designed for headlighting purposes, identified by the name "Sealed Beam" branded on the lens.

searchlight: a projector designed to produce an approximately parallel beam of light, and having an optical system with an aperture of eight inches or more.

secondary standard: a constant and reproducible light source calibrated directly or indirectly by comparison with a primary standard. This order of standard also is designated as a reference standard.

Note: National secondary (reference) standards are maintained at national physical laboratories; laboratory secondary (reference) standards are maintained at other photometric laboratories.

self-ballasted lamp: any arc discharge lamp of which the current-limiting device is an integral part.

semi-direct lighting: lighting by luminaires distributing 60 to 90 per cent of their emitted light downward and the balance upward.

semi-indirect lighting: lighting by luminaires distributing 60 to 90 per cent of their emitted light upward and the balance downward.

service period: the number of hours per day for which daylighting provides a specified illumination level. It often is stated as a monthly average.

set light: the separate illumination of the background or set, other than that provided for principal subjects or areas.

shade: a screen made of opaque or diffusing material designed to prevent a light source from being directly visible at normal angles of view.

shielding angle (of a luminaire): the angle between a horizontal line through the light center and the line of sight at which the bare source first becomes visible. See *cut-off angle (of a luminaire)*.

side back light: illumination from behind the subject in a direction not parallel to a vertical plane through the optical axis of the camera. See *back light*.

side marker lamp: lights indicating the presence and overall length of a vehicle when seen from the side.

signal shutter: a device that modulates a beam of light by mechanical means for the purpose of transmitting intelligence.

signalling light: a projector used for directing light signals toward a designated target zone.

size threshold: the minimum perceptible size of an object. It also is defined as the size that can be detected some specific fraction of the times it is presented to an observer, usually 50 per cent. It usually is measured in minutes of arc. See *visual acuity*.

sky factor: the ratio of the illumination on a horizontal plane at a given point inside a building due to the light received directly from the sky, to the illumination due to an unobstructed hemisphere of sky of uniform luminance equal to that of the visible sky.

sky light: visible radiation from the sun redirected by the atmosphere.

solar constant: the irradiance (averaging 1,393 W/m^2) from the sun at its mean distance from the earth (92.9 $\times$ 10^6 miles), before modification by the earth's atmosphere.

solar radiation simulator: a device designed to produce a beam of collimated radiation having a spectrum, flux density, and geometric characteristics similar to those of the sun outside the earth's atmosphere.

solid angle, ω: a ratio of the area on the surface of a sphere to the square of the radius of the sphere. It is expressed in steradians.

NOTE: Solid angle is a convenient way of expressing the area of light sources and luminaires for computations of discomfort glare factors. It combines into a single number the projected area A_p of the luminaire and the distance D between the luminaire and the eye. It usually is computed by means of the approximate formula

$$\omega = \frac{A_p}{D^2}$$

in which A_p and D^2 are expressed in the same units. This formula is satisfactory when the distance D is greater than about three times the maximum linear dimension of the projected area of the source. Larger projected areas should be sub-divided into several elements.

spacing: for roadway lighting the distance between successive lighting units, measured along the center line of the street. For interior applications see page 9–33.

spacing-to-mounting height ratio, S/MH_{wp}: the ratio of the distance between luminaire centers to the mounting height above the work-plane.

spectral emissivity, ϵ (λ): (of an element of surface of a temperature radiator at any wavelength) the ratio of its radiant flux density per unit wavelength interval (spectral radiant exitance) at that wavelength to that of a blackbody at the same temperature. See *emissivity* and *total emissivity*.

spectral luminous efficacy of radiant flux, $K(\lambda) = \Phi_{v\lambda}/\Phi_{e\lambda}$: the quotient of the luminous flux at a given wavelength by the radiant flux at that wavelength. It is expressed in lumens per watt. See *spectral luminous efficiency of radiant flux* and *luminous efficacy of radiant flux*.

NOTE: This term formerly was called "luminosity factor".

The reciprocal of the maximum luminous efficacy of radiant flux is sometimes called "mechanical equivalent of light"; that is, the watts per lumen at the wavelength of maximum luminous efficacy. The most probable value is 0.00149 watt per lumen, corresponding to 673 lumens per watt as the maximum possible luminous efficacy. These values are based on the definition of *candela*, on the 1968 International Temperature Scale by using Planck's equation with c_2 = 1.4388 cm deg K, c_1 = 3.741 $\times$ 10^{-12} watt cm^2 and 2045 K as the temperature of freezing platinum, and on the standard CIE photopic spectral luminous efficiency values given in Fig. 3–7. If the standard CIE scotopic spectral luminous efficiency values (Fig. 3–7) are used, the maximum luminous efficacy is 1725 "scotopic" lumens per watt.

spectral luminous efficiency of radiant flux: the ratio of the luminous efficacy for a given wavelength to the value of the wavelength of maximum luminous efficacy. It is dimensionless.

NOTE: The term *spectral luminous efficiency* replaces the previously used terms *relative luminosity* and *relative luminosity factor*.

spectral radiant energy, $Q_\lambda = dQ/d\lambda$: radiant energy per unit wavelength interval at wavelength λ; *e.g.*, joules/nanometer.

spectral radiant flux, $\Phi_\lambda = d\Phi/d\lambda$: radiant flux per unit wavelength interval at wavelength λ; *e.g.*, watts/nanometer.

spectral radiant intensity, $I_\lambda = dI/d\lambda$: radiant intensity per unit wavelength interval; *e.g.*, watts/(steradian-nanometer).

spectral reflectance of a surface or medium, $\rho(\lambda) = \Phi_{r\lambda}/\Phi_{i\lambda}$: the ratio of the reflected flux to the incident flux at a particular wavelength, λ, or within a small band of wavelengths, $\Delta\lambda$, about λ.

NOTE: The terms *hemispherical, regular,* or *diffuse reflectance* may each be considered restricted to a specific region of the spectrum and may be so designated by the addition of the adjective "spectral."

spectral transmittance of a medium, $\tau(\lambda) = \Phi_{t\lambda}/\Phi_{i\lambda}$: the ratio of the transmitted flux to the incident flux at a particular wavelength, λ or within a small band of wavelengths, $\Delta\lambda$, about λ.

NOTE: The terms *hemispherical, regular,* or *diffuse transmittance* may each be considered restricted to a specific region of the spectrum and may be so designated by the addition of the adjective "spectral."

spectral tristimulus values: $\bar{x}(\lambda) = X_\lambda/\Phi_{e\lambda}$; $\bar{y}(\lambda) = Y_\lambda/\Phi_{e\lambda}$; $\bar{z}(\lambda) = Z_\lambda/\Phi_{e\lambda}$: the tristimulus values per unit wavelength interval and unit spectral radiant flux.

NOTE: Spectral tristimulus values have been adopted by the International Commission on Illumination. They are tabulated as functions of wavelength throughout the spectrum and are the basis for the evaluation of radiant energy as light. The standard values adopted by the CIE in 1931 are given in Fig. 5–1. The $\bar{y}$ values are identical with the values of spectral luminous efficiency for photopic vision (See *values of spectral luminous efficiency for photopic vision* and Fig. 3–7.)

Tristimulus computational data for several CIE standard sources are given in Fig. 5–4.

spectrophotometer: an instrument for measuring the transmittance and reflectance of surfaces and media as a function of wavelength.

spectroradiometer: an instrument for measuring radiant flux as a function of wavelength.

spectrum locus: the locus of points representing the colors of the visible spectrum in a chromaticity diagram.

specular angle: the angle between the perpendicular to the surface and the reflected ray that is numerically equal to the angle of incidence and that lies in the same plane as the incident ray and the perpendicular but on the opposite side of the perpendicular to the surface.

specular surface: one from which the reflection is

Fig. 1–3. Relative Energy Distribution of Sources A, B, and C

Wavelength (nanometers)	Relative Energy A	Relative Energy B	Relative Energy C	Wavelength (nanometers)	Relative Energy A	Relative Energy B	Relative Energy C	Wavelength (nanometers)	Relative Energy A	Relative Energy B	Relative Energy C
380	9.79	22.40	33.00	510	66.06	90.70	102.30	640	157.98	102.20	87.80
390	12.09	31.30	47.40	520	72.50	89.50	96.90	650	165.03	103.90	88.20
400	14.71	41.30	63.30	530	79.13	92.20	98.00	660	171.96	105.00	87.90
410	17.68	52.10	80.60	540	85.95	96.90	102.10	670	178.77	104.90	86.30
420	21.00	63.20	98.10	550	92.91	101.00	105.20	680	185.43	103.90	84.00
430	24.67	73.10	112.40	560	100.00	102.80	105.30	690	191.93	101.60	80.20
440	28.70	80.80	121.50	570	107.18	102.60	102.30	700	198.26	99.10	76.30
450	33.09	85.40	124.00	580	114.44	101.00	97.80	710	204.41	96.20	72.40
460	37.82	88.30	123.10	590	121.73	99.20	93.20	720	210.36	92.90	68.30
470	42.87	92.00	123.80	600	129.04	98.00	89.70	730	216.12	89.40	64.40
480	48.25	95.20	123.90	610	136.34	98.50	88.40	740	221.66	86.90	61.50
490	53.91	96.50	120.70	620	143.62	99.70	88.10	750	227.00	85.20	59.20
500	59.86	94.20	112.10	630	150.83	101.00	88.00	760	232.11	84.70	58.10

predominantly regular. See *regular (specular) reflection.*

speed of light: the speed of all radiant energy, including light, is 2.997925 × 10^8 meters per second in vacuum (approximately 186,000 miles per second). In all material media the speed is less and varies with the material's index of refraction, which itself varies with wavelength.

speed of vision: the reciprocal of the duration of the exposure time required for something to be seen.

sphere illumination: the illumination on a task from a source providing equal luminous intensity (candelas) from all directions, such as an illuminated sphere with the task located at the center.

standard source: in colorimetry, a source that has a specified spectral distribution and is used as a standard.

NOTE: In 1931 the International Commission on Illumination specified the spectral energy distributions for three standard sources, *A*, *B* and *C*. See Fig. 1–3. Recently a standard source D_{65} has been added.

standard source *A*: a tungsten filament lamp operated at a color temperature of 2856 K, and approximating a blackbody operating at that temperature.

standard source *B*: an approximation of noon sunlight having a correlated color temperature of approximately 4874 K. It is obtained by a combination of Source *A* and a special filter consisting of a layer one centimeter thick of each of the following solutions, contained in a double cell constructed of non-selective optical glass:

No. 1

Copper sulphate ($CuSO_4 \cdot 5H_2O$)	2.452 g
Mannite ($C_6H_8(OH)_6$)	2.452 g
Pyridine (C_5H_5N)	30.0 cm³
Distilled water to make	1000.0 cm³

No. 2

Cobalt-ammonium sulphate ($CoSO_4 \cdot (NH_4)_2SO_4 \cdot 6H_2O$)	21.71 g
Copper sulphate ($CuSO_4 \cdot 5H_2O$)	16.11 g
Sulphuric acid (density 1.835)	10.0 cm³
Distilled water to make	1000.0 cm³

standard source *C*: an approximation of daylight provided by a combination of direct sunlight and clear sky having a correlated color temperature of approximately 6774 K. It is obtained by a combination of Source *A* plus a cell identical with that used for Source B, except that the solutions are:

No. 1

Copper sulphate ($CuSO_4 \cdot 5H_2O$)	3.412 g
Mannite ($C_6H_8(OH)_6$)	3.412 g
Pyridine (C_5H_5N)	30.0 cm³
Distilled water to make	1000.0 cm³

No. 2

Cobalt-ammonium sulphate ($CoSO_4 \cdot (NH_4)_2SO_4 \cdot 6H_2O$)	30.580 g
Copper sulphate ($CuSO_4 \cdot 5H_2O$)	22.520 g
Sulphuric acid (density 1.835)	10.0 cm³
Distilled water to make	1000.0 cm³

standard source D_{65}: an approximation of daylight at a correlated color temperature of 6500 K. It is defined by its relative spectral power distribution over the range from 300 to 830 nanometers.

starter: a device used in conjunction with a ballast for the purpose of starting an electric-discharge lamp.

state of chromatic adaptation: the condition of the eye in equilibrium with the average color of the visual field.

Stefan-Boltzmann law: the statement that the radiant exitance of a blackbody is proportional to the fourth power of its absolute temperature; that is,

$$M = \sigma T^4$$

NOTE: The currently recommended value of the Stefan-Boltzmann constant σ is 5.6697 × 10^{-8} $W\ m^{-2}\ K^{-4}$ or 5.6697 × 10^{-12} $W\ cm^{-2}\ K^{-4}$.

steradian, *sr* (unit solid angle): a solid angle subtending an area on the surface of a sphere equal to the square of the sphere radius.

stilb, *sb*: a unit of luminance (photometric brightness) equal to one candela per square centimeter.

NOTE: The name stilb has been adopted by the International Commission on Illumination and is commonly used in European publications. In the United States and Canada the preferred practice is to use self-explanatory terms such as candela per square inch and candela per square centimeter.

Stiles-Crawford effect: the reduced luminous efficiency of rays entering the peripheral portion of the pupil of the eye.

stop lamp: a device giving a steady warning light to the rear of a vehicle or train of vehicles, to indicate the intention of the operator to diminish speed or to stop.

stray light: light from a source that is scattered onto parts of the retina lying outside the retinal image of the source.

street lighting luminaire: a complete lighting device consisting of a light source together with its direct appurtenances such as globe, reflector, refractor, housing, and such support as is integral with the housing. The pole, post or bracket is not considered part of the luminaire.

street lighting unit: the assembly of a pole or lamp post with a bracket and a luminaire.

subjective brightness: the subjective attribute of any light sensation giving rise to the percept of luminous intensity, including the whole scale of qualities of being bright, light, brilliant, dim or dark.

NOTE: The term brightness often is used when referring to the measurable "photometric brightness." While the context usually makes it clear as to which meaning is intended, the preferable term for the photometric quantity is *luminance*, thus reserving *brightness* for the subjective sensation.

sun bearing: the angle measured in the plane of the horizon between a vertical plane at a right angle to the window wall and the position of this plane after it has been rotated to contain the sun.

sun light: direct visible radiation from the sun.

supplementary lighting: lighting used to provide an additional quantity and quality of illumination that cannot readily be obtained by a general lighting system and that supplements the general lighting level, usually for specific work requirements.

surface mounted luminaire: a luminaire mounted directly on the ceiling.

suspended (pendant) luminaire: a luminaire hung from a ceiling by supports.

T

table lamp: a portable luminaire with a short stand suitable for standing on furniture.

tail lamp: a lamp used to designate the rear of a vehicle by a warning light.

talbot, T.: a unit of light; equal to one lumen-second.

taxi-channel lights: aeronautical ground lights arranged along a taxi-channel of a water aerodrome to indicate the route to be followed by taxiing aircraft.

taxi light: an aircraft aeronautical light designed to provide necessary illumination for taxiing.

taxiway lights: aeronautical ground lights provided to indicate the route to be followed by taxiing aircraft. See *taxiway-centerline lights, taxiway-edge lights, taxiway holding-post light*.

taxiway-centerline lights: taxiway lights placed along the centerline of a taxiway except that on curves or corners having fillets, these lights are placed a distance equal to half the normal width of the taxiway from the outside edge of the curve or corner.

taxiway-edge lights: taxiway lights placed along or near the edges of a taxiway.

taxiway holding-post light: a light or group of lights installed at the edge of a taxiway near an entrance to a runway, or to another taxiway, to indicate the position at which the aircraft should stop and obtain clearance to proceed.

temperature radiator: a radiator whose radiant exitance is determined by its temperature and the material and character of its surface, and is independent of its previous history. See *blackbody* and *graybody*.

thermopile: a number of interconnected thermocouples in order to increase the sensitivity to incident radiant flux.

threshold: the value of a physical stimulus (such as size, luminance, contrast or time) that permits an object to be seen a specific percentage of the time or at a specific accuracy level. In many psychophysical experiments, thresholds are presented in terms of 50 per cent accuracy or accurately 50 per cent of the time. However, the threshold also is expressed as the value of the physical variable that permits the object to be just barely seen. The threshold may be determined by merely detecting the presence of an object or it may be determined by discriminating certain details of the object. See *absolute threshold, contrast threshold, luminance threshold, size threshold*.

threshold lights: runway lights placed to indicate the longitudinal limits of that portion of a runway, channel or landing path usable for landing.

torchere: an indirect floor lamp sending all or nearly all of its light upward.

total emissivity of an element of surface of a temperature radiator, ϵ: the ratio of its radiant exitance to that of a blackbody at the same temperature. See *spectral emissivity*.

touchdown zone lights: barettes of runway lights installed in the surface of the runway between the runway edge lights and the runway centerline lights to provide additional guidance during the touchdown phase of a landing in conditions of very poor visibility.

traffic beam: see *lower (passing) beams*.

train: the angle between the vertical plane through the axis of the searchlight drum and the plane in which this plane lies when the searchlight is in a position designated as having zero train.

transmission: a general term for the process by which incident flux leaves a surface or medium on a side other than the incident side.

NOTE: Transmission through a medium is often a combination of regular and diffuse transmission. See *regular transmission, diffuse transmission*, and *transmittance*.

transmissometer: a photometer for measuring transmittance.

NOTE: Transmissometers may be visual or physical instruments.

transmittance, $\tau = \Phi_t/\Phi_i$: the ratio of the transmitted flux to the incident flux.

NOTE: Measured values of transmittance depend upon the angle of incidence, the method of measurement of the transmitted flux, and the spectral character of the incident flux. Because of this dependence complete information on the technique and conditions of measurement should be specified.

It should be noted that transmittance refers to the ratio of flux emerging to flux incident; therefore, reflections at the surface as well as absorption within the material operate to reduce the transmittance.

tristimulus values of a light, *X, Y, Z:* the amounts of each of three primaries required to match the color of the light.

troffer: a recessed lighting unit, usually long and installed with the opening flush with the ceiling. The term is derived from "trough" and "coffer."

troland: a unit used for expressing the magnitude of the external light stimulus applied to the eye. When the eye is viewing a surface of uniform luminance, the number of trolands is equal to the product of the area in square millimeters of the limiting pupil (natural or artificial) and the luminance of the surface in candelas per square meter.

tungsten-halogen lamp: a gas filled tungsten incandescent lamp containing a certain proportion of halogens.

NOTE: The tungsten-iodine lamp (UK) and quartz-iodine lamp (USA) belong to this category.

turn signal operating unit: that part of a signal system by which the operator of a vehicle indicates the direction a turn will be made, usually by a flashing light.

U

ultraviolet radiation: for practical purposes any radiant energy within the wavelength range 10 to 380 nanometers. See *units of wavelength:*

NOTE: On the basis of practical applications and the effect obtained, the ultraviolet region often is divided into the following bands:

Ozone-producing	180–220 nanometers
Bactericidal (germicidal)	220–300 nanometers
Erythemal	280–320 nanometers
"Black light"	320–400 nanometers

There are no sharp demarcations between these bands, the indicated effects usually being produced to a lesser extent by longer and shorter wavelengths. For engineering purposes, the "black light" region extends slightly into the visible portion of the spectrum.

units of luminance (photometric brightness): the luminance (photometric brightness) of a surface in a specified direction may be expressed in luminous intensity per unit of projected area of surface.

NOTE: Typical units in this system are the candela per square meter and the candela per square inch. See Fig. 1-7.

The luminance of a surface in a specified direction also may be expressed in terms of the number of lumens per unit area that would leave the surface *if the luminances in all directions within the hemisphere on the side of the surface being considered were the same as the luminance in the specified direction.*

NOTE: A typical unit in this system is the footlambert, equal to one lumen per square foot.

This method of specifying luminance is equivalent to stating the number of lumens that would leave the surface *if the surface were replaced by a perfectly diffusing surface with a luminance in all directions within the hemisphere equal to the luminance of the actual surface in the direction specified.* In practice no surface follows exactly the cosine formula of emission or reflection; hence the luminance is not uniform but varies with the angle from which it is viewed.

upper (driving) beams: one or more beams intended for distant illumination and for use on the open highway when not meeting other vehicles. Formerly "country beam." See *lower (passing) beams.*

upward component: that portion of the luminous flux from a luminaire emitted at angles above the horizontal. See *downward component.*

utilance: See *room utilization factor.*

V

valance: a longitudinal shielding member mounted across the top of a window or along a wall and usually parallel to the wall, to conceal light sources giving both upward and downward distributions.

values of spectral luminous efficiency for photopic vision, $V(\lambda)$: values for spectral luminous efficiency at 10-nanometer intervals (see Fig. 3-7) were provisionally adopted by the International Commission on Illumination in 1924 and were adopted in 1933 by the International Committee on Weights and Measures as a basis for the establishment of photometric standards of types of sources differing from the primary standard in spectral distribution of radiant flux. These values are given in the second column of Fig. 3-7; values given in the other column are for spectral luminous efficiency for scotopic vision.

NOTE: These standard values of spectral luminous efficiency were determined by observations with a two-degree photometric field having a moderately high luminance (photometric brightness), and photometric evaluations based upon them consequently do not apply exactly to other conditions of observation. Power in watts weighted in accord with these standard values are often referred to as *light-watts.*

values of spectral luminous efficiency for scotopic vision $V'(\lambda)$: values of spectral luminous efficiency at 10-nanometer intervals (see Fig. 3-7) were provisionally adopted by the International Commission on Illumination in 1951.

NOTE: These values of spectral luminous efficiency were determined by observation by young dark-adapted observers using extra-foveal vision at near-threshold luminance.

vapor-tight luminaire: a luminaire designed and approved for installation in damp or wet locations. It also is described as "enclosed and gasketed."

VASIS (Visual Approach Slope Indicator System): the system of angle-of-approach lights accepted as a standard by the International Civil Aviation Organization, comprising two bars of lights located at each side of the runway near the threshold and showing red or white or a combination of both (pink) to the approaching pilot depending upon his position with respect to the glide path.

veiling brightness: a brightness superimposed on the retinal image which reduces its contrast. It is this veiling effect produced by bright sources or areas in the visual field that results in decreased visual performance and visibility.

veiling reflection: regular reflections superimposed upon diffuse reflections from an object that partially or totally obscure the details to be seen by reducing the contrast. This sometimes is called reflected glare.

vertical plane of a searchlight: the plane perpendicular to the train axis and in which the elevation axis lies. See *horizontal plane of a searchlight.*

visibility: the quality or state of being perceivable by the eye. In many outdoor applications, visibility is defined in terms of the distance at which an object can be just perceived by the eye. In indoor applications it usually is defined in terms of the contrast or size of a standard test object, observed under standardized view-conditions, having the same threshold as the given object. See *visibility (meteorological).*

visibility (meteorological): a term that denotes the greatest distance, expressed in miles, that selected objects (visibility markers) or lights of moderate intensity (25 candelas) can be seen and identified under specified conditions of observation.

vision: see *central vision, foveal vision, mesopic vision, peripheral vision, photopic vision and scotopic vision.*

visual acuity: a measure of the ability to distinguish fine details. Quantitatively, it is the reciprocal of the angular size in minutes of the critical detail that is just large enough to be seen. See *size threshold*.

visual angle: the angle subtended by an object or detail at the point of observation. It usually is measured in minutes of arc.

visual field: the locus of objects or points in space that can be perceived when the head and eyes are kept fixed. The field may be monocular or binocular. See *monocular visual field, binocular visual field, central visual field, peripheral visual field*.

visual perception: the interpretation of impressions transmitted from the retina to the brain in terms of information about a physical world displayed before the eye.

NOTE: Visual perception involves any one or more of the following: recognition of the presence of something (object, aperture or medium); identifying it; locating it in space; noting its relation to other things; identifying its movement, color, brightness or form.

visual performance: the quantitative assessment of the performance of a task taking into consideration speed and accuracy.

visual photometer: a photometer in which the equality of brightness of two surfaces is established visually. See *physical photometer*.

NOTE: The two surfaces usually are viewed simultaneously side by side. This method is used in portable visual luminance (photometric brightness) meters. This is satisfactory when the color difference between the test source and comparison source is small. However, when there is a color difference, a flicker photometer provides more precise measurements. In this type of photometer the two surfaces are viewed alternately at such a rate that the color sensations either nearly or completely blend and the flicker due to brightness difference is balanced by adjusting the comparison source.

visual range: the maximum distance at which a particular light (or object) can be seen and identified.

visual surround: includes all portions of the visual field except the visual task.

visual task: conventionally designates those details and objects that must be seen for the performance of a given activity, and includes the immediate background of the details or objects.

W

wavelength: wavelength is the distance between two successive points of a periodic wave in the direction of propagation, in which the oscillation has the same phase. The three commonly used units are listed in the following table:

Name	Symbol	Value
Micrometer	μm	1 μm = 10^{-6} m
Nanometer	nm	1 nm = 10^{-9} m
Angstrom	Å	1 Å = 10^{-10} m

wide-angle diffusion: diffusion in which flux is scattered at angles far from the direction that the flux would take by regular reflection or transmission. See *narrow-angle diffusion*.

wide angle luminaire: a luminaire distributing the light through a comparatively wide solid angle. See also *narrow angle luminaire*.

width line: the radial line (the one that makes the larger angle with the reference line) that passes through the point of one-half maximum candlepower on the lateral candlepower distribution curve plotted on the surface of the cone of maximum candlepower.

Wien displacement law: an expression representing, in a functional form, the spectral radiance of a blackbody as a function of the wavelength and the temperature.

$$L_\lambda = I_\lambda/A' = c_1\lambda^{-5}f(\lambda T)$$

The two principal corollaries of this law are:

$$\lambda_m T = b$$
$$L_m/T^5 = b'$$

which show how the maximum spectral radiance L_m and the wavelength λ_m at which it occurs are related to the absolute temperature T. See *Wien radiation law*.

NOTE: The currently recommended value of b is 2.8978 $\times 10^{-3}$ m K or 2.8978 $\times 10^{-1}$ cm K. From the Planck radiation law, and with the use of the value of b, c_1, and c_2 as given above, b' is found to be 4.10 $\times 10^{-12}$ W cm^{-3} K^{-5} sr^{-1}.

Wien radiation law: an expression representing approximately the spectral radiance of a blackbody as a function of its wavelength and temperature. It commonly is expressed by the formula

$$L_\lambda = I_\lambda/A' = c_{1L}\lambda^{-5}e^{-(c_2/\lambda T)}$$

This formula is accurate to one per cent or better for values of λT less than 3000 micrometer kelvin.

wing clearance lights: aircraft lights provided at the wing tips to indicate the extent of the wing span when the navigation lights are located an appreciable distance inboard of the wing tips.

work-plane: the plane at which work usually is done, and at which the illumination is specified and measured. Unless otherwise indicated, this is assumed to be a horizontal plane 30 inches above the floor.

working standard: a standardized light source for regular use in photometry.

Z

zonal-cavity inter-reflectance method: a procedure for calculating coefficients of utilization taking into consideration the luminaire intensity distribution, room size (cavity ratio concepts) and room reflectances.

zonal constant: a factor by which the mean candlepower emitted by a source of light in a given angular zone is multiplied to obtain the lumens in the zone. See Fig. 4–13.

zonal factor interflection method: a former procedure for calculating coefficients of utilization.

Fig. 1–4. Abbreviations for Use in Text*

A	
absolute	abs
alternating current	ac
ampere	A or amp
ampere-hour	Ah or amp h
Angstrom unit	Å
antilogarithm	antilog
atmosphere	atm
avoirdupois	avdp
azimuth	az
B	
boiling point	bp
British thermal unit	Btu
C	
calorie	cal
candela	cd
candlepower	cp
centimeter	cm
centimeter-gram-second (system)	cgs
chemically pure	cp
coefficient	coef
cologarithm	colog
conductivity	cndct
constant	const
cosecant	csc
cosine	cos
cotangent	cot
coulomb	spell out or C
counter electromotive force	cemf
D	
decibel	dB
degree	deg or °
degree Celsius	°C
degree Fahrenheit	°F
degree Kelvin	K
diameter	dia
direct current	dc
E	
electric	elec
electromotive force	emf
F	
farad	spell out or F
foot per minute	fpm
foot per second	fps
foot	ft
footcandle	fc
footlambert	fL
foot-pound-second (system)	fps
freezing point	fp
frequency	freq
fusion point	fnpt
G	
greatest common divisor	gcd
H	
henry	H
hertz	Hz
horsepower	hp
horsepower-hour	hph
hour	h
I	
inch	in
inch per second	ips
inside diameter	id
J	
joule	J
K	
kilocalorie	kcal
kilohertz per second	kHz/s
kilogram	kg
kilometer	km
kilometer per second	km/s
kilovolt	kV
kilovolt-ampere	kVA
kilowatt	kW
kilowatthour	kWh
L	
lambert	L
least common multiple	lcm
logarithm (common)	log
logarithm (natural)	ln
lumen	lm
M	
mass	spell out
maximum	max
mean horizontal candlepower	mhcp
megahertz per second	MHz/s
megohm	MΩ
melting point	mp
mho	spell out
mocroampere	μA
microfarad	μF
microvolt	μV
microwatt	μW
mile	mi or spell out
mile per hour	mph
milliampere	mA
milligram	mg
millihenry	mH
millimeter	mm
minimum	min
minute (time)	min
minute (angular measure)	′
molecular weight	mol wt
N	
nanometer	nm
National Electrical Code	NEC
O	
ohm	spell out or Ω
ohm-centimeter	ohm-cm
outside diameter	od
P	
parts per million	ppm
power factor	spell out or pf
R	
radian	rad
reactive kilovolt-ampere	kVAr
reactive volt-ampere	VAr
revolution per minute	rpm
revolutions per second	rps
root mean square	rms
S	
secant	sec
second (time)	s
second (angular measure)	″
sine	sin
specific gravity	sp gr
specific heat	sp ht
square	sq
square centimeter	cm^2
square foot	ft^2
square inch	in^2
square kilometer	km^2
square meter	m^2
square millimeter	mm^2
steradian	sr
T	
tangent	tan
temperature	temp
V	
volt	V
volt-ampere	VA
volt-coulomb	spell out
W	
watt	W
watthour	Wh
weight	wt
Y	
yard	yd
year	yr

* Essentially in accord with latest available proposed draft of American National Standard Y1.

Fig. 1–5. Conversion from Values in Customary IES Units to Values in SI Units[13]

(fc* → lx; fL → cd/m²; cd/in² → kcd/m²; in → cm; ft → m)

ft / in / cd/in² / fL / fc*	lx	cd/m²	kcd/m²	cm	m	ft / in / cd/in² / fL / fc*	lx	cd/m²	kcd/m²	cm	m
1	10.76	3.4	1.55	2.54	.30	500	5380	1713	775.0	1270	152.4
2	21.5	6.9	3.00	5.08	.61	510	5488	1747	790.5	1295	155.4
3	33.3	10.3	4.65	7.62	.91	520	5595	1782	806.0	1321	158.5
4	43.0	13.7	6.20	10.16	1.22	530	5703	1816	821.6	1346	161.5
5	53.8	17.1	7.75	12.70	1.52	540	5810	1850	837.0	1372	164.6
6	64.6	20.6	9.30	15.24	1.83	550	5918	1884	852.5	1397	167.6
7	75.3	24.0	10.85	17.78	2.13	560	6026	1919	868.0	1422	170.7
8	86.1	27.4	12.40	20.32	2.44	570	6133	1953	883.5	1448	173.7
9	96.8	30.8	13.95	22.86	2.74	580	6241	1987	899.0	1473	176.8
						590	6348	2021	914.5	1499	179.8
100	1076	343	155.0	254	30.5	600	6456	2056	930.0	1524	182.9
110	1184	377	170.5	279	33.5	610	6564	2090	945.5	1549	185.9
120	1291	411	186.0	305	36.6	620	6671	2124	961.0	1575	189.0
130	1399	445	201.5	330	39.6	630	6779	2158	976.5	1600	192.0
140	1506	480	217.0	356	42.7	640	6886	2193	992.0	1626	195.1
150	1614	514	232.5	381	45.7	650	6994	2227	1007.5	1651	198.1
160	1722	548	248.0	406	48.8	660	7102	2261	1023.0	1676	201.2
170	1829	582	263.5	432	51.8	670	7209	2295	1038.5	1702	204.2
180	1937	617	279.0	457	54.9	680	7317	2330	1054.0	1727	207.3
190	2044	651	294.5	483	57.9	690	7424	2364	1069.5	1753	210.3
200	2152	685	310.0	508	61.0	700	7532	2398	1085.0	1778	213.4
210	2260	719	325.5	533	64.0	710	7640	2432	1100.5	1803	216.4
220	2367	754	341.0	559	67.1	720	7747	2467	1116.0	1829	219.5
230	2475	788	356.5	584	70.1	730	7855	2501	1131.5	1854	222.5
240	2582	822	372.0	610	73.2	740	7962	2535	1147.0	1880	225.6
250	2690	857	387.5	635	76.2	750	8070	2570	1162.5	1905	228.6
260	2798	891	403.0	660	79.2	760	8178	2604	1178.0	1930	231.6
270	2905	925	418.5	686	82.3	770	8285	2638	1193.5	1956	234.7
280	3013	959	434.0	711	85.3	780	8393	2672	1209.0	1981	237.7
290	3120	994	449.5	737	88.4	790	8500	2702	1224.5	2007	240.8
300	3228	1028	465.0	762	91.4	800	8608	2741	1240.0	2032	243.8
310	3336	1062	480.5	787	94.5	810	8716	2775	1255.5	2057	246.9
320	3443	1096	496.0	813	97.5	820	8823	2809	1271.0	2083	249.9
330	3551	1131	511.5	838	100.6	830	8931	2844	1286.5	2108	253.0
340	3658	1165	527.0	864	103.6	840	9038	2878	1302.0	2134	256.0
350	3766	1199	542.5	889	106.7	850	9146	2912	1317.5	2159	259.1
360	3874	1233	558.0	914	109.7	860	9254	2946	1333.0	2184	262.1
370	3981	1268	573.5	940	112.8	870	9361	2981	1348.5	2210	265.2
380	4089	1302	589.0	965	115.8	880	9469	3015	1364.0	2235	268.2
390	4196	1336	604.5	991	118.9	890	9576	3049	1379.5	2261	271.3
400	4304	1370	620.0	1016	121.9	900	9684	3083	1395.0	2286	274.3
410	4412	1405	635.5	1041	125.0	910	9792	3118	1410.5	2311	277.4
420	4519	1439	651.0	1067	128.0	920	9899	3152	1426.0	2337	280.4
430	4627	1473	666.5	1092	131.1	930	10010	3186	1441.5	2362	283.5
440	4734	1507	682.0	1118	134.1	940	10110	3220	1457.0	2388	286.5
450	4842	1542	697.5	1143	137.2	950	10220	3255	1472.5	2413	289.6
460	4950	1576	713.0	1168	140.2	960	10330	3289	1488.0	2438	292.6
470	5057	1610	728.5	1194	143.3	970	10440	3323	1503.5	2464	295.7
480	5165	1644	744.0	1219	146.3	980	10540	3357	1519.0	2489	298.7
490	5272	1679	759.5	1245	149.4	990	10650	3392	1534.5	2515	301.8

*Also useful for converting from m² to ft²

Fig. 1–6. Conversion from Values in SI Units to Values in Customary IES Units[13]

m → ft; cm → in; kcd/m² → cd/in²; cd/m² → fL; lx* → fc	fc	fL	cd/in²	in	ft	m → ft; cm → in; kcd/m² → cd/in²; cd/m² → fL; lx* → fc	fc	fL	cd/in²	in	ft
1	.09	.29	.65	.39	3.3	500	46.5	146.0	322.5	196.9	1641
2	.19	.58	1.29	.79	6.6	510	47.4	148.9	329.0	200.8	1673
3	.28	.88	1.94	1.18	9.8	520	48.3	151.8	335.4	204.7	1706
4	.37	1.17	2.58	1.57	13.1	530	49.2	154.7	341.9	208.7	1739
5	.47	1.46	3.23	1.97	16.4	540	50.2	157.6	348.3	212.6	1772
6	.56	1.75	3.87	2.36	19.7	550	51.1	160.5	354.8	216.5	1805
7	.65	2.04	4.52	2.76	23.0	560	52.0	163.5	361.2	220.5	1837
8	.74	2.34	5.16	3.15	26.2	570	53.0	166.4	367.7	224.4	1870
9	.84	2.63	5.81	3.54	29.5	580	53.9	169.3	374.1	228.3	1903
						590	54.8	172.2	380.6	232.3	1936
100	9.3	29.2	64.5	39.4	328	600	55.7	175.1	387.0	236.2	1969
110	10.2	32.1	71.0	43.3	361	610	56.7	178.1	393.5	240.2	2001
120	11.1	35.0	77.4	47.2	394	620	57.6	181.0	399.9	244.1	2034
130	12.1	37.9	83.9	51.2	427	630	58.5	183.9	406.4	248.0	2067
140	13.0	40.9	90.3	55.1	459	640	59.5	186.8	412.8	252.0	2100
150	13.9	43.8	96.8	59.1	492	650	60.4	189.7	419.3	255.9	2133
160	14.9	46.7	103.2	63.0	525	660	61.3	192.7	425.7	259.8	2165
170	15.8	49.6	109.7	66.9	558	670	62.2	195.6	432.2	263.8	2198
180	16.7	52.5	116.1	70.9	591	680	63.2	198.5	438.6	267.7	2231
190	17.7	55.5	122.6	74.8	623	690	64.1	201.4	445.1	271.7	2264
200	18.6	58.4	129.0	78.7	656	700	65.0	204.3	451.5	275.6	2297
210	19.5	61.3	135.5	82.7	689	710	66.0	207.2	458.0	279.5	2330
220	20.4	64.2	141.9	86.6	722	720	66.9	210.2	464.4	283.5	2362
230	21.4	67.1	148.4	90.6	755	730	67.8	213.1	470.9	287.4	2395
240	22.3	70.1	154.8	94.5	787	740	68.7	216.0	477.3	291.3	2428
250	23.2	73.0	161.3	98.4	820	750	69.7	218.9	483.8	295.3	2461
260	24.2	75.9	167.7	102.4	853	760	70.6	221.8	490.2	299.2	2494
270	25.1	78.8	174.2	106.3	886	770	71.5	224.8	496.7	303.1	2526
280	26.0	81.7	180.6	110.2	919	780	72.5	227.7	503.1	307.1	2559
290	26.9	84.7	187.1	114.2	951	790	73.4	230.6	509.6	311.0	2592
300	27.9	87.6	193.5	118.1	984	800	74.3	233.5	516.0	315.0	2625
310	28.8	90.5	200.0	122.0	1017	810	75.2	236.4	522.5	318.9	2658
320	29.7	93.4	206.4	126.0	1050	820	76.2	239.4	528.9	322.8	2690
330	30.7	96.3	212.9	130.0	1083	830	77.1	242.3	535.4	326.8	2723
340	31.6	99.2	219.3	133.9	1116	840	78.0	245.2	541.8	330.7	2756
350	32.5	102.2	225.8	137.8	1148	850	79.0	248.1	548.3	334.6	2789
360	33.4	105.8	232.2	141.7	1181	860	79.9	251.0	554.7	338.6	2822
370	34.4	108.0	238.7	145.7	1214	870	80.8	254.0	561.2	342.5	2854
380	35.3	110.9	245.1	149.6	1247	880	81.8	256.9	567.6	346.5	2887
390	36.2	113.8	251.6	153.5	1280	890	82.7	259.8	574.1	350.4	2920
400	37.2	116.8	258.0	157.5	1312	900	83.6	262.7	580.5	354.3	2953
410	38.1	119.7	264.5	161.4	1345	910	84.5	265.6	587.0	358.3	2986
420	39.0	122.6	270.9	165.4	1378	920	85.5	268.5	593.4	362.2	3019
430	39.9	125.5	277.4	169.3	1411	930	86.4	271.5	600.0	366.1	3051
440	40.9	128.4	283.8	173.2	1444	940	87.3	274.4	606.3	370.1	3084
450	41.8	131.4	290.3	177.2	1476	950	88.3	277.3	612.8	374.0	3117
460	42.7	134.3	296.7	181.1	1509	960	89.2	280.2	619.2	378.0	3150
470	43.7	137.2	303.2	185.0	1542	970	90.1	283.1	625.7	381.9	3183
480	44.6	140.1	309.6	189.0	1575	980	91.0	286.1	632.1	385.8	3215
490	45.5	143.0	316.1	192.9	1608	990	92.0	289.0	638.6	389.8	3248

*Also useful for converting from ft² to m².

Fig. 1-7. Luminance (Photometric Brightness) Conversion Factors

1 nit = 1 candela/m^2
1 stilb = 1 candela/cm^2
1 apostilb (international) = 0.1 millilambert = 1 blondel
1 apostilb (German Hefner) = 0.09 millilambert
1 lambert = 1000 millilamberts

Multiply Number of → / To Obtain Number of ↓ By ↘	Footlambert	Nit	Millilambert	Candela/in^2	Candela/ft^2	Stilb
Footlambert	1	0.2919	0.929	452	3.142	2,919
Nit	3.426	1	3.183	1,550	10.76	10,000
Millilambert	1.076	0.3142	1	487	3.382	3,142
Candela/in^2	0.00221	0.000645	0.00205	1	0.00694	6.45
Candela/ft^2	0.3183	0.0929	0.2957	144	1	929
Stilb	0.00034	0.0001	0.00032	0.155	0.00108	1

Fig. 1-8. Conversion Factors for Units of Length

Multiply Number of → / To Obtain Number of ↓ By ↘	Angstroms	Nanometers	Microns	Millimeters	Centimeters	Meters	Kilometers	Mils	Inches	Feet	Miles
Angstroms	1	10	10^4	10^7	10^8	10^{10}	10^{13}	2.540×10^5	2.540×10^8	3.048×10^9	1.609×10^{13}
Nanometers	10^{-1}	1	10^3	10^6	10^7	10^9	10^{12}	2.540×10^4	2.540×10^7	3.048×10^8	1.609×10^{12}
Microns	10^{-4}	10^{-3}	1	10^3	10^4	10^6	10^9	2.540×10	2.540×10^4	3.048×10^5	1.609×10^9
Millimeters	10^{-7}	10^{-6}	10^{-3}	1	10	10^3	10^6	2.540×10^{-2}	2.540×10	3.048×10^2	1.609×10^6
Centimeters	10^{-8}	10^{-7}	10^{-4}	10^{-1}	1	10^2	10^5	2.540×10^{-3}	2.540	3.048×10	1.609×10^5
Meters	10^{-10}	10^{-9}	10^{-6}	10^{-3}	10^{-2}	1	10^3	2.540×10^{-5}	2.540×10^{-2}	3.048×10^{-1}	1.609×10^3
Kilometers	10^{-13}	10^{-12}	10^{-9}	10^{-6}	10^{-5}	10^{-3}	1	2.540×10^{-8}	3.048×10^{-5}	3.048×10^{-4}	1.609
Mils	3.937×10^{-6}	3.937×10^{-5}	3.937×10^{-2}	3.937×10	3.937×10^2	3.937×10^4	3.937×10^7	1	10^3	1.2×10^4	6.336×10^7
Inches	3.937×10^{-9}	3.937×10^{-8}	3.937×10^{-5}	3.937×10^{-2}	3.937×10^{-1}	3.937×10	3.937×10^4	10^{-3}	1	12	6.336×10^4
Feet	3.281×10^{-10}	3.281×10^{-9}	3.281×10^{-6}	3.281×10^{-3}	3.281×10^{-2}	3.281	3.281×10^3	8.333×10^{-5}	8.333×10^{-2}	1	5.280×10^3
Miles	6.214×10^{-14}	6.214×10^{-13}	6.214×10^{-10}	6.214×10^{-7}	6.214×10^{-6}	6.214×10^{-4}	6.214×10^{-1}	1.578×10^{-8}	1.578×10^{-5}	1.894×10^{-4}	1

Fig. 1-9. Illumination Conversion Factors

1 lumen = 1/673 lightwatt	1 watt-second = 10^7 ergs
1 lumen-hour = 60 lumen-minutes	1 phot = 1 lumen/cm^2
1 footcandle = 1 lumen/ft^2	1 lux = 1 lumen/m^2 = 1 metercandle

Multiply Number of → To Obtain Number of ↓ By ↘	Footcandles	Lux	Phot	Milliphot
Footcandles	1	0.0929	929	0.929
Lux	10.76	1	10,000	10
Phot	0.00108	0.0001	1	0.001
Milliphot	1.076	0.1	1,000	1

Fig. 1-10. Angular Measure, Temperature, Power and Pressure Conversion Equations

Angle
1 radian = 57.29578 degrees
Temperature
(F to C) C = 5/9 (F − 32)
(C to F) F = 9/5 C + 32
(C to K) K = C + 273
Power
1 kilowatt = 1.341 horsepower
= 56.89 Btu per minute
1 Btu = 0.2928 watt-hour
Pressure
1 atmosphere = 760 millimeters of mercury at 0° C
= 29.92 inches of mercury at 0° C
= 14.7 pounds per square inch

Fig. 1-11. Greek Alphabet (Capital and Lower Case)

Capital	Lower Case	Greek Name
Α	α	Alpha
Β	β	Beta
Γ	γ	Gamma
Δ	δ	Delta
Ε	ϵ	Epsilon
Ζ	ζ	Zeta
Η	η	Eta
Θ	θ	Theta
Ι	ι	Iota
Κ	κ	Kappa
Λ	λ	Lambda
Μ	μ	Mu
Ν	ν	Nu
Ξ	ξ	Xi
Ο	ο	Omicron
Π	π	Pi
Ρ	ρ	Rho
Σ	σ, ς	Sigma
Τ	τ	Tau
Υ	υ	Upsilon
Φ	φ, ϕ	Phi
Χ	χ	Chi
Ψ	ψ	Psi
Ω	ω	Omega

Fig. 1-12. Unit Prefixes[12]

Prefix	Symbol	Factor by Which the Unit is Multiplied
tera	T	1,000,000,000,000 = 10^{12}
giga	G	1,000,000,000 = 10^9
mega	M	1,000,000 = 10^6
kilo	k	1,000 = 10^3
hecto	h	100 = 10^2
deka	da	10 = 10^1
deci	d	0.1 = 10^{-1}
centi	c	0.01 = 10^{-2}
milli	m	0.001 = 10^{-3}
micro	μ	0.000,001 = 10^{-6}
nano	n	0.000,000,001 = 10^{-9}
pico	p	0.000,000,000,001 = 10^{-12}
atto	a	(0.000,000,000,000,001 = 10^{-15}
femto	f	(0.000,000,000,000,000,001 = 10^{-18}

REFERENCES

1. *American National Standard Nomenclature and Definitions for Illuminating Engineering*, ANSI Z7.1-1967, American National Standards Institute, New York, 1967 (Sponsored by the Illuminating Engineering Society).
2. Nomenclature and Light Sources Committee of the IES: "Nomenclature and Definitions for Newer Lamps," *Trans. Illum. Eng. Soc.*, Vol. 63, p. 581, November, 1968.
3. "Nomenclature and Definitions Applicable to Radiometric and Photometric Characteristics of Matter," *ASTM Spec. Tech. Publ. 475*, American Society for Testing and Materials, Philadelphia, Pennsylvania, 1970.
4. "Announcement of Changes in Electrical and Photometric Units," *Nat. Bur. Stand. Circ. 459*, U.S. Department of Commerce, Washington, D.C., May 15, 1947.
5. Committee on Colorimetry of the Optical Society of America: *The Science of Color*, Thomas Y. Cromwell Company, New York, 1953.
6. Council on Physical Medicine and Rehabilitation: "Minimal Requirements for Acceptance of Sun Lamps," *J. Amer. Med. Ass.*, Vol. 144, p. 625, October 21, 1950.
7. Stickney, G. H. and Crittenden, E. C.: "Comments on Proper Use of Illuminating Terms," *Trans. Illum. Eng. Soc.*, Vol. 33, p. 193, February, 1938.
8. Newhall, S. M., Nickerson, D. and Judd, D. B.: "Final Report of the O.S.A. Subcommittee on the Spacing of the Munsell Colors," *J. Opt. Soc. Amer.*, Vol. 33, p. 385, July, 1943.
9. Committee on Illumination Performance Recommendations of the IES: "Classification of Luminaires by Distribution," *Illum. Eng.*, Vol. 49, p. 552, November, 1954.
10. Hardy, A. C.: *Handbook of Colorimetry*, Technology Press, Cambridge, Massachusetts, 1936.
11. *NBS Technical News Bulletin:* October, 1963.
12. Barbow, L. E.: "The Metric System in Illuminating Engineering," *Illum. Eng.*, Vol. 62, p. 638, November, 1967.
13. Kaufman, J. E.: "Introducing SI Units," *Illum. Eng.*, Vol. 63, p. 537, October, 1968.
14. Schapero, M., Cline, D. and Hofstetter, H. W.: *Dictionary of Visual Science*, Chilton Company, Philadelphia and New York, 1960.
15. Hechtinger, A., *Modern Science Dictionary*, Franklin Publishing Company, Inc., New Jersey, 1959.
16. McNish, A. G.: "Classification and Nomenclature for Standards of Measurement," *IRE (Inst. Radio Eng.) Trans. on Instrum. 371*, December, 1958.
17. "Steps to a New International Lighting Vocabulary," *Illum. Eng.*, Vol. 59, p. 228, April, 1964.

SECTION

2

PHYSICS OF LIGHT

Concepts 2–1 *Generation of Light 2–6* *Incandescence 2–7* *Luminescence 2–7*
Light Detection 2–12

CONCEPTS

For illuminating engineering purposes, the Illuminating Engineering Society has defined light as *visually evaluated radiant energy*.

From a physical viewpoint, light can be regarded as that portion of the electromagnetic spectrum which lies between the wavelength limits of 380 nanometers and 780 nanometers. Visually there is some individual variation in these limits.

Radiant energy of the proper wavelength makes visible anything from which it is emitted or reflected in sufficient quantity to activate the receptors in the eye.

Radiant energy may be evaluated in a number of different ways. Two are listed below.

Radiant Flux. The time rate of the flow of any part of the radiant energy spectrum measured in joules per second or in watts.

Luminous Flux. The time rate of the flow of the luminous parts of the radiant energy spectrum measured in lumens.

For further information on these terms see Section 1.

Theories

Several theories describing radiant energy have been advanced.[1] They are briefly discussed below.

Corpuscular Theory. The theory advocated by Newton, based on these premises:

1. That luminous bodies emit radiant energy in particles.
2. That these particles are intermittently ejected in straight lines.
3. That the particles act on the retina of the eye stimulating the optic nerves to produce the sensation of light.

Wave Theory. The theory advocated by Huygens, based on these premises:

1. That light is the resultant of molecular vibration in the luminous material.
2. That vibrations are transmitted through the ether as wavelike movements (comparable to ripples in water).
3. That the vibrations thus transmitted act on the retina of the eye stimulating the optic nerves to produce visual sensation.

Electromagnetic Theory.[2] The theory advanced by Maxwell, based on these premises:

1. That luminous bodies emit light in the form of radiant energy.
2. That this radiant energy is propagated in the form of electromagnetic waves.
3. That the electromagnetic waves act upon the retina of the eye thus stimulating the optic nerves to produce the sensation of light.

Quantum Theory. A modern form of the corpuscular theory advanced by Planck and others and based on these premises:

1. That energy is *emitted* and *absorbed* in discrete quanta (photons).
2. That the magnitude of each quantum is $h\nu$,
where $h = 6.6256 \times 10^{-27}$ erg-sec (Planck's constant),
and
ν = frequency in hertz.

Unified Theory. The theory proposed by De Broglie and Heisenberg and based on these premises:

1. Every moving element of mass has associated with it a wave whose length is given by $\lambda = h/mv$
where λ = wavelength of the wave motion
h = Planck's constant
m = mass of the particle
v = velocity of the particle
2. It is impossible to simultaneously determine all of the properties that are distinctive of a wave or a corpuscle.

The quantum and electromagnetic wave theories provide an explanation of those characteristics of radiant energy of concern to the illuminating engineer. Whether light is thought of as wave-like or

NOTE: References are listed at the end of each section.

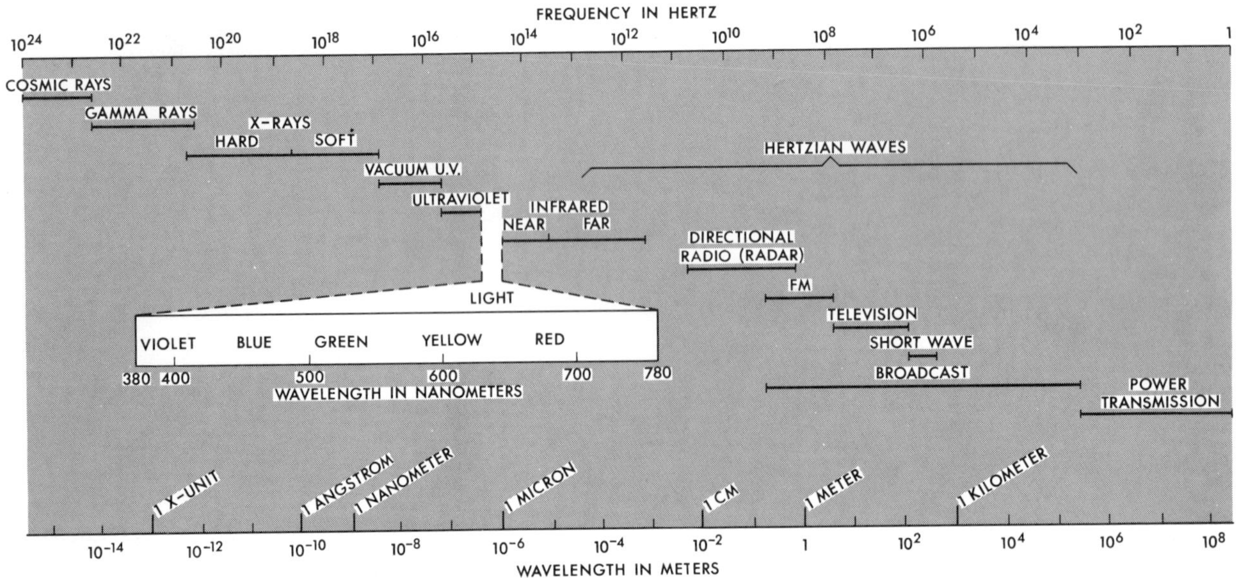

Fig. 2–1. The radiant energy (electromagnetic) spectrum.

photon-like in nature, it is radiation that is produced by electronic processes in the most exact sense of the term. It is produced in an incandescent body, a luminous discharge or a solid-state device, by excited electrons just having reverted to more stable positions in their respective atoms, releasing energy.

Light and the Energy Spectrum

The wave theory permits a convenient graphical representation of radiant energy in an orderly arrangement according to its wavelength or frequency. This arrangement is called a *spectrum* (see Fig. 2–1). It is useful in indicating the relationship between various radiant energy wavelength regions. Such a graphical representation should not be construed to indicate that each region of the spectrum is divided from the others in any physical way whatsoever. Actually there is a gradual transition from one region to another.

The practical limits of the radiant energy spectrum extend over a range of wavelengths varying from a few picometers (10^{-12} meters) to one hundred thousand miles (1.6×10^8 meters). Radiant energy in the visible spectrum has wavelengths between 380×10^{-9} and 780×10^{-9} meters (380 and 780 nanometers).

The Angstrom unit (Å), the micron (μ), and the nanometer* (nm), which are respectively 10^{-10}, 10^{-6}, and 10^{-9} meter, are commonly used units of length in the visible spectrum region. The relationship of several units for measuring wavelength is given in Fig. 1–8 in Section 1.

All forms of radiant energy are transmitted at the same speed in a vacuum (186,300 miles per second). However, each form differs in wavelength and thus in frequency. The wavelength and velocity may be altered by the medium through which it passes, but frequency is fixed independently of the medium. Thus, through the equation:

$$v = \frac{\lambda \nu}{n}$$

where v = velocity of waves in the medium, *e.g.*, cm/s
n = index of refraction of the medium
λ = wavelength in a vacuum, *e.g.*, cm
ν = frequency, *e.g.*, hertz

it is possible to determine the velocity of radiant energy and also to indicate the relationship between frequency and wavelength.

Fig. 2–2 gives the speed of light in different media for a frequency corresponding to a wavelength of 589 nanometers in air.

Blackbody Radiation

The light from practical light sources, particularly that from incandescent sources, is often described by comparison with that from a blackbody or com-

Fig. 2–2. Speed of Light for a Wavelength of 589 Nanometers (Sodium D-Lines)

Medium	Speed (centimeters per second)
Vacuum	2.997925×10^{10}
Air (760 mm at 0°C)	2.99724×10^{10}
Crown Glass	1.98223×10^{10}
Water	2.24915×10^{10}

* Formerly millimicron.

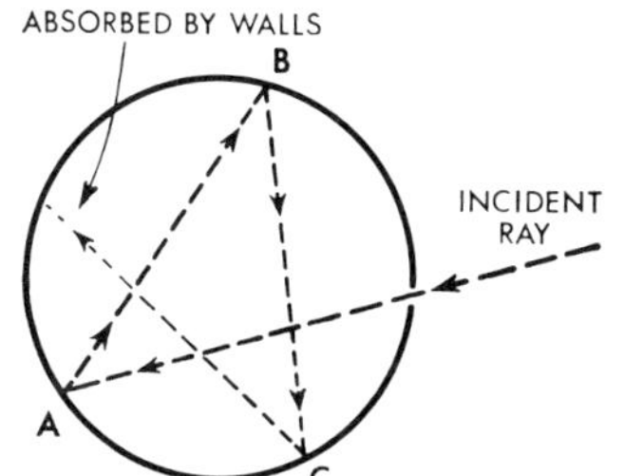

Fig. 2-3. Small aperture in an enclosure exhibits blackbody characteristics.

plete radiator. A blackbody is defined as a body which absorbs all radiation incident upon it, transmitting none and reflecting none. A blackbody will for equal area radiate more total power and more power at any given wavelength than any other source operating at the same temperature.

For experimental purposes, laboratory sources have been devised which approach very closely the ideal blackbody. Designs of the sources are based on the fact that a hole in the wall of a closed chamber, small in size as compared with the size of the enclosure exhibits blackbody characteristics. This is understood if one considers what happens to a ray of light entering such an enclosure. See Fig. 2-3. At reflection (A, B, C, etc.) some energy is absorbed. In time, all incoming energy will be absorbed by the walls.

Since 1948 the luminance of a blackbody operated at the temperature of freezing platinum has been used as an international reference standard for maintaining the unit of candlepower. (See Section 1, definition of *candela.*)

Planck's Equation. Data describing blackbody radiation curves were obtained by Lummer and Pringsheim using a specially constructed and uniformly heated tube as the source. Planck, introducing the concept of discrete quanta of energy, developed the following equation depicting these curves. It has the form:

$$M_\lambda = c_1\lambda^{-5}(e^{c_2/\lambda T} - 1)^{-1}$$

where M_λ = watts radiated by a blackbody (per cm^2 of surface) in each wavelength band one micron wide, at wavelength λ

λ = wavelength in microns (μ)

T = absolute temperature of the blackbody (degree Kelvin)

c_1 = 37,415*

c_2 = 14,388 micron-degrees†

e = 2.718 +

The curves for several values of T are plotted on a logarithmic scale in Fig. 2-4.

Wien Radiation Law. In the temperature range of incandescent filament lamps (2000 K to 3400 K) and in the visible wavelength region (380 to 780 nm), the following simplification of the Planck equation known as the Wien Radiation Law gives a good representation of the blackbody distribution:

$$M_\lambda = c_1\lambda^{-5}e^{-c_2/\lambda T}$$

Wien Displacement Law. This gives the relation between blackbody distributions for various temperatures (see line AB, Fig. 2-4):

$$\lambda_{max}\, T = b \text{ (2898 micron-degrees)}$$

where λ_{max} is the wavelength, in microns, at which blackbody radiation is a maximum, found by setting the differential of Planck's equation equal to zero.

$$M_{max}\, T^{-5} = b_1 = 1.3 \times 10^{-11} \text{ watt cm}^{-3} \text{ deg}^{-5}$$

Stefan-Boltzmann Law. This law, obtained by integrating Planck's expression for M_λ from zero to infinity, states that the total radiant power per unit area of a blackbody varies as the fourth power of the absolute temperature:

$$M = \sigma T^4 \text{ watt cm}^{-2}$$

where M = summation of power per unit area radiated by a blackbody at all wavelengths

σ = 5.6697×10^{-12} watt cm^{-2} deg^{-4}

T = temperature of the radiator (deg. K)

It should be noted that this equation applies to the total power, that is, the whole spectrum. It cannot be used to estimate the power in the visible portion of the spectrum alone.

Spectral Emissivity

No known radiator has the same emissive power as a blackbody. The ratio of the output of a radi-

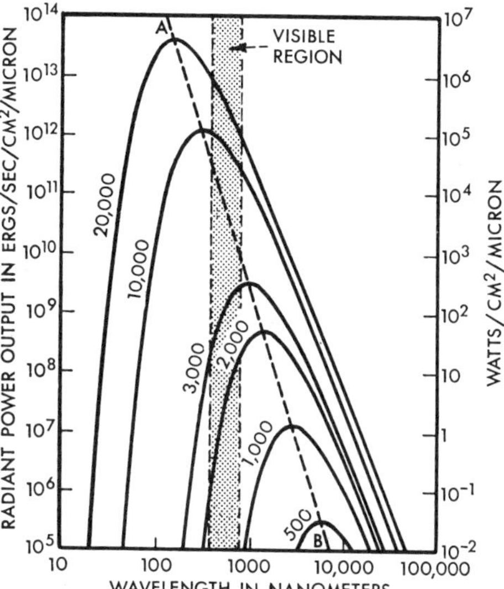

Fig. 2-4. Blackbody radiation curves for operating temperatures between 500 K and 20,000 K showing Wien displacement of peaks. Shaded area is region of visible wavelengths.

* National Bureau of Standards Technical News Bulletin 47, October 1963.

† This value is based on the 1968 International Practical Temperature Scale.

ator at any wavelength to that of a blackbody at the same temperature and the same wavelength is known as the spectral emissivity, $\epsilon(\lambda)$, of the radiator.

Graybody Radiation

When the spectral emissivity is constant for all wavelengths, the radiator is known as a graybody. No known radiator has a constant spectral emissivity for all visible, infrared, and ultraviolet wavelengths, but in the visible region a carbon filament exhibits very nearly uniform emissivity, that is, nearly a graybody.

Selective Radiators

The emissivity of all known material varies with wavelength. Therefore, they are called selective radiators. In Fig. 2–5 the radiation curves for a blackbody, a graybody, and a selective radiator (tungsten), all operating at 3000 K, are plotted on the same logarithmic scale to show the characteristic differences in output.

Radiation Equations. These equations into which the spectral emissivity factor has been introduced, are applicable to any incandescent source:

Planck's equation:

$$M_\lambda = 37{,}415\epsilon_\lambda\lambda^{-5}(10^{6245/\lambda T} - 1)^{-1}$$

Wien radiation law:

$$M_\lambda = 37{,}415\epsilon_\lambda\lambda^{-5}10^{-6245/\lambda T}$$

Stefan-Boltzmann law:

$$M = 5.6697 \times 10^{-12}\epsilon_t T^4$$

where

$$\epsilon_t = \frac{\int_0^\infty \epsilon_\lambda M_\lambda d\lambda}{\int_0^\infty M_\lambda d\lambda} \quad \text{(total emissivity)}$$

Color Temperature

The radiation characteristics of a blackbody of unknown area may be specified with the aid of the above equations by fixing only two quantities; the magnitude of the radiation at any given wavelength and the absolute temperature. The same type of specification may be used with reasonable accuracy in the visible region of the spectrum for tungsten filaments and other incandescent sources. However, the temperature used in the case of selective radiators is not that of the filament but a value called the color temperature.

The color temperature of a selective radiator is equal to that temperature at which a blackbody must be operated in order that its output be the

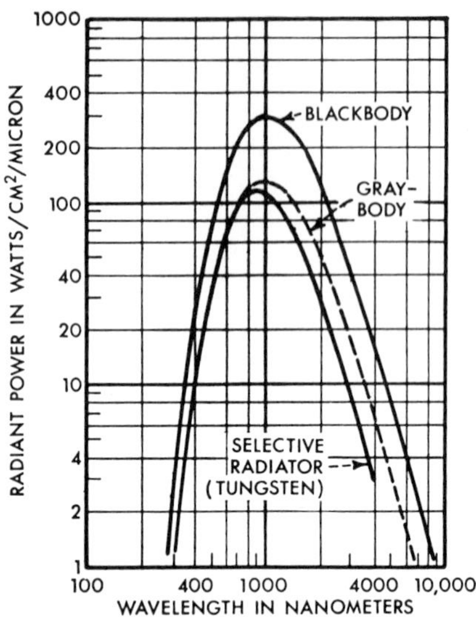

Fig. 2–5. Radiation curves for blackbody, graybody, and selective radiators operating at 3000 K.

closest possible approximation to a perfect color match with the output of the selective radiator (see Section 5). While the match is never perfect the small deviations that occur in the case of incandescent filament lamps are not of practical importance.

The apertures between coils of the filaments used in many tungsten lamps act somewhat as a blackbody because of the interreflections which occur at the inner surfaces of the helix formed by the coil. For this reason the distribution from coiled filaments exhibits a combination of the characteristics of the straight filament and of a blackbody operating at the same temperature.

The application of the color temperature method to deduce the spectral distribution from other than incandescent sources even in the visible region of the spectrum will usually result in appreciable error. Color temperature values associated with light sources other than incandescent are correlated color temperature and not true color temperatures.

Atomic Structure

The atomic theories first proposed by Rutherford and Bohr in 1913 have since been expanded upon and verified repeatedly by careful experiment. They propose that each atom is in reality a minute solar system, such as that shown in Fig. 2–6.

The *atom* consists of a central nucleus possessing a positive charge n about which rotate n negatively charged electrons. In the normal state these electrons remain in particular orbits or energy levels and radiation is not emitted by the atom.

The *orbit* described by a particular electron rotating about the nucleus is determined by the energy of that electron. That is to say, there is a particular energy associated with each orbit. The system of orbits or energy levels is characteristic of each element and remains stable until disturbed by external forces.

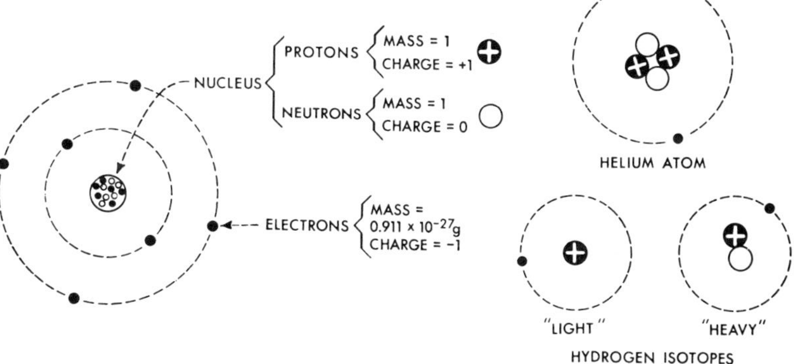

Fig. 2–6. Structure of the atom showing electron orbits around central nucleus. Hydrogen isotopes and helium atom are simplest of all atomic structures.

The *electrons* of an atom can be divided into two classes. The first includes the inner shell electrons which are not readily removed or excited except by high energy radiation. The second class includes the outer shell valence electrons which cause chemical bonding into molecules. They are readily excited by ultraviolet radiation or by electron impact and can be removed completely with relative ease. The valence electrons of an atom in a solid when removed from their associated nucleus enter the conduction band and confer on the solid, the property of electrical conductivity.

Upon the absorption of sufficient energy by an atom in the gaseous state, the valence electron is pushed to a higher energy level further from the nucleus. Eventually, the electron returns to the normal orbit, or an intermediate one, and in so doing the energy that the atom loses is emitted as a quantum of radiation. The wavelength of the radiation is determined by Planck's formula:

$$E_1 - E_2 = h\nu$$

where E_1 = energy associated with the excited orbit
E_2 = energy associated with the normal orbit
h = Planck's constant
ν = frequency of the emitted radiation.

This formula can be converted to a more usable form:

$$\text{wavelength} = \frac{1239.76}{V_d} \text{ nanometers}$$

where V_d is the potential difference (volts) between two energy levels through which the displaced electron has fallen in one transition.

Luminous Efficiency of Radiant Energy

Many apparent differences in intensity between radiant energy of various wavelengths are in reality differences in ability of various receiving and sensing devices to detect them uniformly.[3]

The reception characteristics of the human eye have been subject to extensive investigations. The results may be summarized as follows:

1. The spectral response characteristic of the human eye varies between individuals, with time, and with the age and the state of health of any individual, to the extent that the selection of any individual to act as a standard observer is not scientifically feasible.

2. However, from the wealth of data available, a luminous efficiency curve has been selected for engineering purposes to represent a typical human observer. This curve may be applied mathematically to the solution of photometric problems. (See also Section 3.)

Recognizing these facts, the Illuminating Engineering Society in 1923 and the International Commission on Illumination (C. I. E.) in 1924 adopted the values of photopic spectral luminous efficiency of Fig. 3–7 from which the spectral luminous efficiency curve of Fig. 2–7 was plotted.

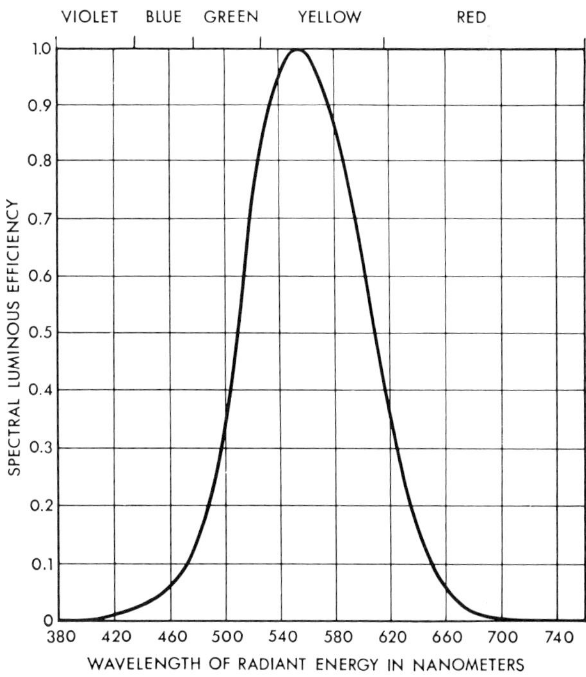

Fig. 2–7. The standard (CIE) spectral luminous efficiency curve (for photopic vision) showing the relative capacity of radiant energy of various wavelengths to produce visual sensation.

The standard spectral luminous efficiency curve for photopic vision represents a typical characteristic, adopted arbitrarily to give unique solutions to photometric problems, from which the characteristics of any individual may be expected to vary. Goodeve's data (See Section 3) indicate that most human observers are capable of experiencing a visual sensation upon exposure to radiant energy of wavelengths longer than 780 nanometers, usually termed "infrared," provided the radiant energy reached the eye at a sufficiently high rate. It also is known that "ultraviolet" radiant energy (wavelength less than 380 nanometers) can easily be seen if it reached the retina at even a moderate rate. Most observers, however, yield only a slight response to ultraviolet radiant energy because the crystalline lens of the eye absorbs nearly all of it.

Luminous Efficacy

The luminous efficacy of a light source is defined as the ratio of the total luminous flux (lumens) to the total power input (watts or equivalent).

The maximum luminous efficacy of an "ideal" white source, defined as a radiator with constant output over the visible spectrum and no radiation in other parts of the spectrum is approximately 220 lumens per watt.

GENERATION OF LIGHT

Natural Phenomena

Sunlight. Energy of color temperature about 6500 K is received from the sun at the outside of the earth's atmosphere at an average rate of about 0.135 watt per square centimeter. About 75 per cent of this energy reaches the earth's surface at sea level (equator) on a clear day.

The average luminance of the sun is approximately 160,000 candelas per square centimeter viewed from sea level. Illumination of the earth's surface by the sun may exceed 10,000 footcandles; on cloudy days the illumination drops to less than 1000 footcandles. See Section 7.

Sky Light. A considerable amount of light is scattered in all directions by the earth's atmosphere. The investigations of Rayleigh first showed that this was a true scattering effect. On theoretical grounds the scattering should vary inversely as the fourth power of the wavelength when the size of the scattering particles is small compared to the wavelength of light, as in the case of the air molecules themselves. The blue color of a clear sky and the reddish appearance of the rising or setting sun are common examples of this scattering effect. If the scattering particles are of appreciable size (the water droplets in a cloud, for example), scattering is essentially the same for all wavelengths. (Clouds appear white.) Polarization in parts of the sky may be 50 per cent complete.

Moonlight. The moon shines purely by virtue of its ability to reflect sunlight. Since the reflectance of its surface is rather low, its luminance is approximately 0.25 candelas per square centimeter. Illumination of the earth's surface by the moon may be as high as 0.01 footcandle.

Lightning. Lightning is a meterological phenomenon arising from the accumulation, in the formation of clouds, of tremendous electrical charges, usually positive, which are suddenly released in a spark type of discharge. The lightning spectrum corresponds closely to that of an ordinary spark in air, consisting principally of nitrogen bands, though hydrogen lines may sometimes appear owing to dissociation of water vapor.

Aurora Borealis (Northern Lights) and Aurora Australis (Southern Lights). These hazy horizontal patches or bands of greenish light on which white, pink, or red streamers sometimes are superposed appear 60 to 120 miles above the earth. They are caused by electron streams spiraling into the atmosphere, primarily at polar latitudes. Some of their spectrum lines have been identified with transitions of valence electrons from metastable states of oxygen and nitrogen atoms.

Bioluminescence. Living light is a form of chemiluminescence (see page 2–12) in which special compounds manufactured by plants and animals are oxidized, producing light. Although it has been proven that oxygen is required to produce light, there is no evidence that the light producing compound is a "living" material. The light producing compound can be dried and stored many years and upon exposure to oxygen emit light. Bioluminescent sources can have very high lumen per watt ratings.

"Man-Made" Sources

Historically light sources have been divided into two types—incandescent and luminescent. Fundamentally, the cause of light emission is the same, *i.e.*, electronic transitions from higher to lower energy states. The mode of electron excitation is different, however, as well as the spectral distribution of radiation, although even here there is some overlapping as with the lines of radiation from incandescent rare earth elements and the continuum from high pressure discharges.

The two classical types, with subdivisions showing associated devices or processes, are listed as follows (see Section 8):

1. Incandescence
   - Filament
   - Pyroluminescence (flame)
   - Candoluminescence (gas mantle)
   - Arc Crater
2. Luminescence
   - Gaseous Discharge
   - Photoluminescence (light)

a) Fluorescence
b) Phosphorescence
c) Solid laser

Electroluminescence (electromagnetic)
a) Cathodoluminescence (electron)
b) Electroluminescent Lamp (ac capacitive)
c) Light Emitting Diode
d) Gas Laser

Galvanoluminescence (chemical)
Crystalloluminescence (crystallization)
Chemiluminescence (oxidation)
Thermoluminescence (heat)
Triboluminescence (friction/fracture)
Sonoluminescence (ultrasonics)
Radioluminescence (α, β, γ, X-rays)
Ionoluminescence (ions)

INCANDESCENCE

Filament

Familiar physical objects are simple or complex combinations of chemically identifiable molecules, which in turn are made up of atoms. In solid materials the molecules are packed together and the substances hold their shape over a wide range of physical conditions. In contrast, the molecules of a gas are highly mobile and occupy only a small part of the space filled by the gas.

Molecules of both gases and solids are constantly in motion at temperatures above absolute zero and their movement is a function of temperature. If the solid or gas is hot, the molecules move rapidly; if it is cold, they move more slowly.

At temperatures below about 873 K (600°C) only invisible energy of the longer infrared (heat) wavelengths is emitted by any body; a coal stove or an electric iron, for example. Electronic transitions in atoms and molecules at temperatures of about 600° C result in the release of visible radiation along with the heat.

The incandescence of a filament lamp is caused by the heating action of an electric current. This heating action raises the filament temperature above 600°C, producing light.

Pyroluminescence (Flame Luminescence)

A flame is the most often "noted visible" evidence of combustion. Flame luminescent light may be due to recombination of ions to form molecules, reflection from solid particles in the flame, incandescence of carbon or other solid particles, and any combination of these.

The combustion process is a high temperature energy exchange between highly excited molecules and atoms. The process releases and radiates energy, some of which is in that portion of the electromagnetic spectrum called light. The quality and the amount of light generated depend upon the material undergoing combustion. For example, a flashbulb containing zirconium yields the equivalent of 56 lumens per watt where as an acetylene flame yields ⅕ lumen per watt.

Candoluminescence (Gas Mantle)

Incandescence is exhibited by heated bodies which give off shorter wavelength radiation than would be expected by radiation laws—the effect due to fluorescence excited by incandescent radiation. Such materials are zinc oxide and rare earths (cerium, thorium) used in a Welsbach mantle.

Arc Crater

A carbon arc source radiates because of incandescence of the electrodes (crater) and because of luminescence of vaporized electrode material and other constituents of the surrounding gaseous atmosphere. Considerable spread in the luminance, total radiation, and spectral energy distribution may be achieved by varying the electrode materials.

LUMINESCENCE[6, 7]

Radiation from luminescent sources results from the excitation of single valence electrons of an atom, either in a gaseous state, where each atom is free from interference from its neighbors, or in a crystalline solid or organic molecule, where the action of its neighbors exerts a marked effect. In the first case line spectra, such as those of mercury or sodium arcs, result. In the second case relatively narrow emission bands, which cover a portion of the spectrum (usually in the visible region) result. Both of these cases contrast with the radiation from incandescent sources, where the irregular excitation at high temperature of the free electrons of innumerable atoms gives rise to all wavelengths of radiation to form a continuous spectrum of radiation as discussed in the section on blackbody radiation.

Gaseous Discharge

Radiation, including light can be produced by gaseous discharge previously discussed under *Atomic Structure*.

Photoluminescence

Photoluminescence results if the existing radiation is visible or ultraviolet light (photons). A typical mechanism for generating photons from a gaseous discharge (such as in a fluorescent lamp) is described below. See Fig. 2-8.

1. A free electron emitted from the cathode collides with one of the two valence electrons of a mercury atom and excites it by imparting to it part of the kinetic energy of the moving electron, thus raising the valence electron from its normal energy level to a higher one.

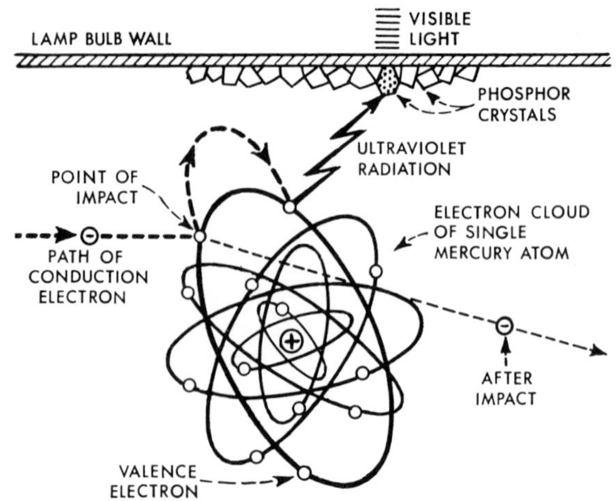

Fig. 2-8. Magnified cross-section of fluorescent lamp showing progressive steps in luminescent process which finally result in the release of visible light.

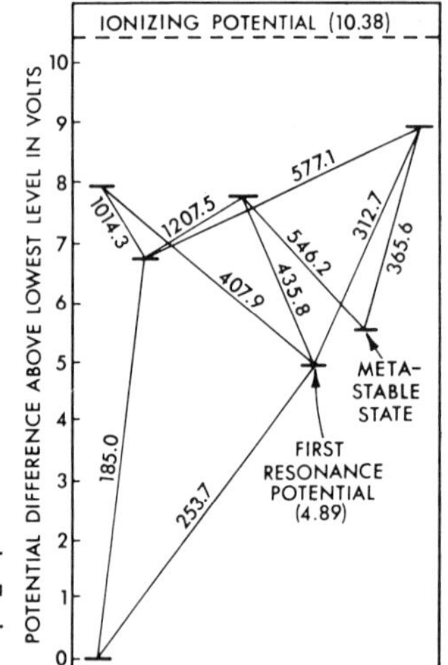

Fig. 2-9. Simplified energy diagram for mercury showing a few of the characteristic spectral lines.

2. The conduction electron loses speed at the impact and changes direction, but continues along the tube to excite or ionize one or more additional atoms before losing its energy stepwise and completing its path. It generally ends at the wall of the tube where it recombines with an ionized atom. A part of the electron current is collected at the anode.

3. Conduction electrons, either from the cathode or formed by collision processes, gain energy from the electric field thus maintaining the discharge along the length of the tube.

4. After a short delay the valence electron returns to its normal energy level either in a single transition or by a series of steps from one excited level to a lower level. At each of these steps a photon (quantum of radiant energy) is emitted. See Fig. 2-9.

Fluorescence. In the fluorescent lamp and in the fluorescent mercury lamp ultraviolet radiation resulting from luminescence of the mercury vapor due to a gas discharge is converted into visible light by a phosphor coating on the inside of the tube or outer jacket. If this emission continues only during the excitation it is called "fluorescence."

Fig. 2-8 shows schematically a greatly magnified section of a part of a fluorescent lamp.

Ultraviolet photons generated in an arc discharge such as in a fluorescent or fluorescent-mercury lamp eventually strike one of the phosphor crystals or the surface of the tube.

The phosphor will transmit this energy through the crystal until it reaches an activator ion. At this point, if the photon has a wavelength within the excitation band of the phosphor, it will be absorbed and converted into a photon of longer wavelength usually in the near ultraviolet or in the visible spectrum (see Fig. 2-10).

The phosphors used in fluorescent lamps are crystalline inorganic compounds, of exceptionally high chemical purity and of controlled composition to which small percentages of intentional impurities (the activators) have been added to convert them into efficient fluorescent materials. By choice of the right combination of activator and inorganic compound the color of emission can be controlled. A typical schematic model for a phosphor is given in Fig. 2-11. In the normal state the electron oscillates about position A on the energy curve, as the lattice expands and contracts due to thermal vibration. For the phosphor to emit light it must first absorb radiation. In the fluorescent lamp this is chiefly that at 253.7 nanometers, while in the mercury lamp it may be ultraviolet of this and longer wavelengths generated in the arc. The absorbed energy transfers the electron to an excited state at position B. After loss of excess energy to the lattice as vibrational (heat) energy the electron again oscillates around a stable position, C, for a very short time after which it return to position D on the normal energy curve with simultaneous emission of a photon of

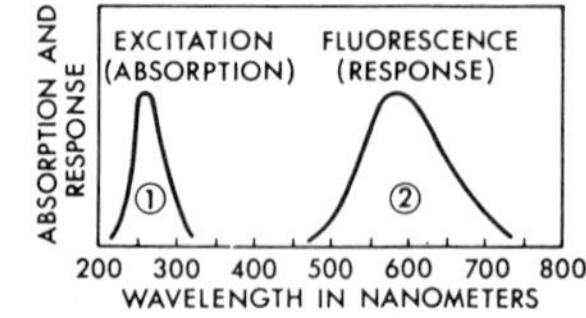

Fig. 2-10. Fluorescence curve of typical phosphor showing initial excitation by ultraviolet rays and subsequent release of visible radiation.

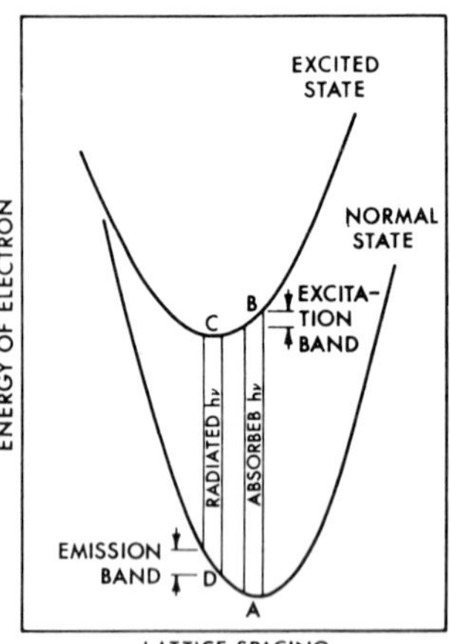

Fig. 2-11. Simplified energy diagram for a typical phosphor.

Fig. 2–12. Color Characteristic of Important Fluorescent Lamp Phosphors

Material	Activator	Peak of Fluorescent* Band and Color	Lamp Use
Barium disilicate	Lead	355 Pale blue	Blacklight
Barium strontium magnesium silicate	Lead	370 Pale blue	Blacklight
Calcium tungstate	Lead	440 Blue	Blue
Strontium pyrophosphate	Tin	470 Blue	Various whites
Magnesium tunsgtate	None	480 Blue white }	Daylight blends
Calcium halophosphate	Antimony	480 Blue white }	
Barium titanium phosphate	Titanium	490 Blue-green	Various whites
Zinc silicate	Manganese	520 Green	Green
Calcium halophosphate	Antimony and manganese	590 White to yellow	Cool white, white, warm white, daylight
Calcium silicate	Lead and manganese	610 Pink }	{ Improved-color cool and warm white
Strontium magnesium phosphate	Tin	620 Pink }	
Calcium-strontium phosphate	Tin	640 White }	
Cadmium borate	Manganese	615 Pink	Pink and red
Magnesium fluorogermanate	Manganese	660 Red	Improved-color whites

* Wavelengths in nanometers.

radiation. Stokes Law, stating that the radiation emitted must be of longer wavelength than that absorbed, is readily explained by this model. It then returns to A with a further loss of energy as heat and is then ready for another cycle of excitation and emission.

Because of the oscillation around both stable positions A and C the excitation and emission processes cover a range of wavelength commonly referred to as bands.

In some phosphors two activators are present. One of these, the primary activator, determines the absorption characteristics and can be used alone as it also gives emission. The other, or secondary activator, does not enter into the absorption mechanism but receives its energy by transfer within the crystal from a neighboring primary activator. The emitted light from the secondary activator is longer in wavelength than that from the primary activator. The relative amount of emission from the two activators is determined by the concentration of the secondary activator. The phosphors now used in most "white" fluorescent lamps are doubly activated calcium halophosphate phosphors.

Fig. 2–12 shows the characteristic colors and uses of phosphors currently employed in manufacture of fluorescent lamps. Fig. 2–13 gives the characteristics of some phosphors useful in fluorescent-mercury lamps.

Impurities other than activators and excessive amount of activators have a serious deleterious effect on the efficiency of the phosphor.[8]

Phosphorescence.[6] In some fluorescent materials there are metastable excited states in which the electron can be trapped, for a time varying from milliseconds to days. After release from these traps they emit light. This phenomenon is called "phosphorescence." These metastable states lie slightly below the usual excited states responsible for fluorescence, and energy usually derived from heat is required to transfer the electron from the metastable state to the emitting state. Since the same emitting state is usually involved, the color of fluorescence and phosphorescence is generally the same for a given phosphor. In doubly activated phosphors the secondary activator has longer phosphorescence than the primary activator so the color changes with time.

Short duration phosphorescence is important in fluorescent lamps in reducing the stroboscopic effect with alternating current operation.

The infrared stimulable phosphors have an unusual type of phosphorescence. After excitation they show phosphorescence, but decay to invisibility in a few seconds. However, they retain a considerable amount of energy trapped in metastable states which can be released as visible light by infrared radiation, when a wavelength within the stimulation band is allowed to fall on the phosphor.

Fig. 2–13. Color Characteristics of Some Mercury Lamp Phosphors

Material	Activator	Peak of Fluorescent Band*	Color of Fluorescence
Magnesium arsenate	Manganese	660	Deep red
Magnesium fluorogermanate	Manganese	660	Deep red
Strontium-zinc phosphate	Tin	610	Orange red
Strontium-magnesium phosphate	Tin	610	Orange red
Yttrium-vanadate	Europium	612	Orange red

* Wavelengths in nanometers.

Solid Laser.[9] A laser is defined by its component letters as a light source in which there is *L*ight *A*mplification by *S*timulated *E*mission of *R*adiation. However, there are other characteristics of laser light which are of major interest to illuminating engineers. In addition to amplifying light, lasers produce intense, highly monochromatic, well collimated, coherent light.

Coherent light consists of radiation whose waves are in phase with regard to time and space. Ordinary light, although it may contain a finite proportion of coherent light, is basically incoherent because light emission is occurring in a random fashion.

In a laser, however, electronic transitions are triggered (stimulated) by a wave of the same frequency as the emitted light instead of occurring at random. As a consequence, a beam of light is emitted, all of whose waves are in phase and of the same frequency.

A prerequisite to laser action is a pumping process whereby an upper and a lower electron level in the active material undergo a population inversion. The pumping source may be a light as in a ruby laser or electronic excitation as in a gas laser.

The choice of laser materials is quite limited. First, it must be possible to highly populate an upper electronic level; second, there must be a light emitting transition from this upper level with a long lifetime; and third, a lower level must exist which can be depopulated either spontaneously or through pumping.

Laser construction is as important to laser action as is the source material. Since light wavelengths are too short to allow building a resonant cavity, long multi-nodal chambers are made with parallel reflectors at each end to feed back radiation until lasing takes place. The effect is to produce well collimated light that is highly directional.

An example of a laser is the pink ruby device whose electronic transitions are shown in Fig. 2–14 and whose mechanical construction is shown in Fig. 2–15. This laser is pumped by a flash lamp (a) and electrons in the ruby (b) are raised from level E_1 to E_3. The electrons decay rapidly and spontaneously from E_3 to E_2. They can then move from E_2 to E_1 spontaneously and slowly emit fluorescent light, $h\nu_{21}$, or they can be stimulated to emit coherent light, $h\nu_{21}$. The full reflector (c) and the partial reflector (d) channel the coherent radiation, $h\nu_{21}$, until it has built up enough to emit coherent light $h\nu_{21}$ through (d). The fact that this light has been reflected many times by parallel mirrors ensures that it is well collimated. The electrons are then available for further pumping. See Fig. 2–16.

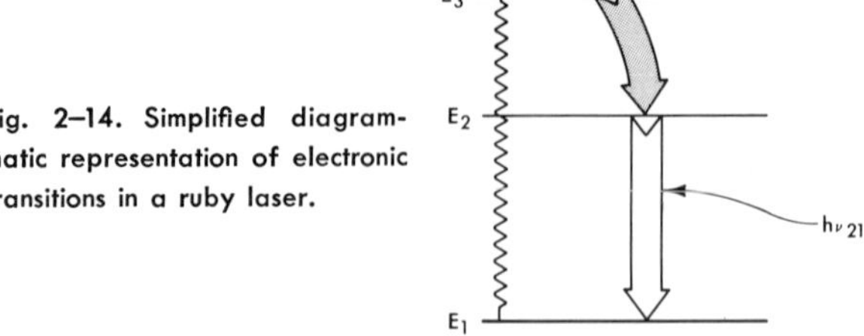

Fig. 2–14. Simplified diagrammatic representation of electronic transitions in a ruby laser.

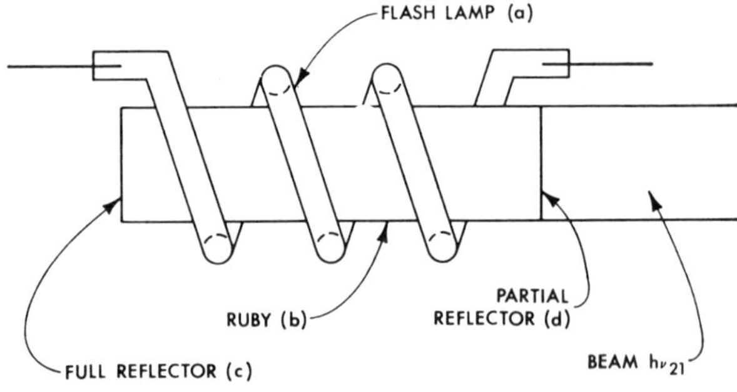

Fig. 2–15. Simplified diagram of a ruby laser.

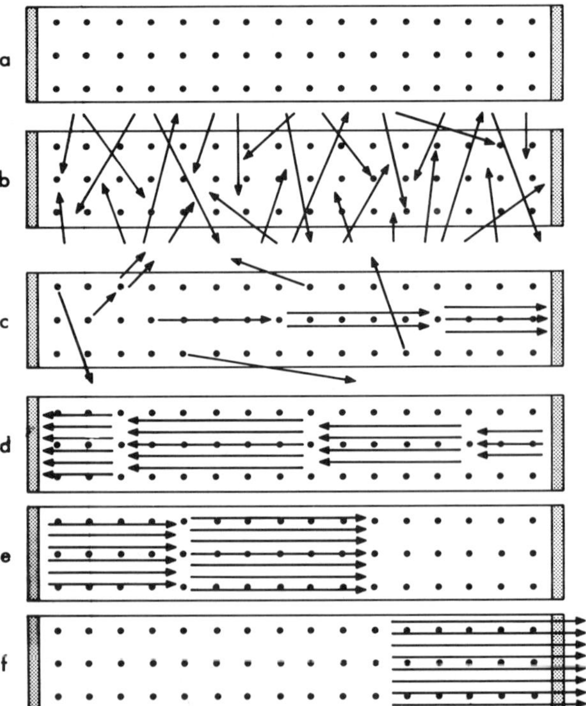

Fig. 2–16. Photon cascade in a solid laser.[12] Before the build-up begins, atoms in the laser crystal are in the ground state (a). Pumping light (arrows in b) raises most of the atoms to the excited state. The cascade (c) begins when an excited atom spontaneously emits a photon parallel to the axis of the crystal (photons emitted in other directions pass out of the crystal). The build-up continues (d) and (e) through thousands of reflections back and forth from the silvered surfaces at the ends of the crystal. When amplification is great enough, light passes out (f).*

* Adapted from *Scientific American*.

Electroluminescence

In fluorescent and fluorescent-mercury lamps, generation of ultraviolet radiation in the arc discharge is one step in the conversion of electrical energy into visible light by the use of phosphors. With certain special phosphors it is possible to convert alternating current energy directly into light without using this intermediate step by utilizing the phenomenon of electroluminescence.

Cathodoluminescence. Cathodoluminescence is light emitted when a substance is bombarded by cathode rays. If the energy is from the anode, the phenomenon is called anodoluminescence.

Electroluminescent Lamp (AC Capacitive). An electroluminescent lamp is composed of a conductor (transparent or opaque) unto which a dielectric-phosphor mixture is laid. A second conductor, of transparent material, is deposited over the dielectric-phosphor mixture. An electric field is now capable of being generated between the two conductors.

The electrons in electroluminescent phosphors are excited to higher energy levels by the influence of rapidly fluctuating, high potential radiant fields. Their spontaneous return to lower levels results in the emission of visible light. The light is in the form of broad bands just as in the case of fluorescent phosphors.

Fig. 2–17 shows the cross-section of a lamp diagrammatically, while Fig. 2–18 gives the properties of some electroluminescent phosphors.

The color of light emitted by the electroluminescent lamp is somewhat dependent on frequency, while the luminance is strongly dependent on both voltage and frequency. The effects of both voltage and frequency change with the specific phosphors. The present efficacy is low compared even to incandescent lamps, though comparable to colored incandescent with filter coats giving blue or green light.

Light Emitting Diodes. Light emitting diodes (LED), also called solid state lamps (SSL), produce light by electroluminescence when low-voltage direct current is applied to a suitably doped crystal containing a p-n junction. The phenomenon has been observed as early as 1923 in naturally occurring junctions but was not considered practical due to the low luminous efficacy in converting electric energy to light. Recently it was discovered that under certain conditions the conversion was significant.

The efficacy is dependent upon the visible energy generated at the junction and losses due to reabsorption when light tries to escape through the crystal. Due to the high index of refraction of most semiconductors, light is re-reflected back from the surface into the crystal and highly attenuated before finally exiting. The term used to express this ultimate measurable visible energy is "external" efficacy. While external efficacies are moderate, internal efficacies are calculated to be very high.

Gas Laser. In a solid laser there are three requirements—a material which reacts energetically to light, a population inversion generated by pumping in energy at the proper energy level, and a growth of the internal energy caused by the reflection of photons within the solid. While the same requirements are met in a gas laser, two other characteristics are available—strong, narrow spectral lines and unequal emission at different energy levels. An example of such a gas laser is that containing a mixture of helium and neon. See Fig. 2–19. Helium is used as the energizing gas since it has a level of energy at which it can lose energy only by collision. This level corresponds to the level at which neon radiates energy in the form of red light. By energizing helium in a gas discharge inside a cavity whose ends are reflecting and containing both helium and

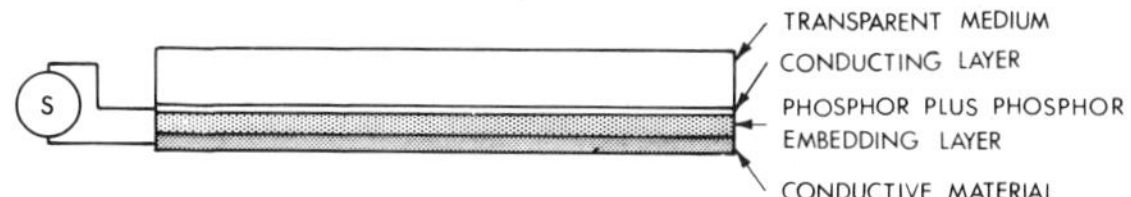

Fig. 2–17. Diagrammatic cross-section of an electroluminescent lamp.

Fig. 2–18. Properties of Some Electroluminescent Phosphors

Material	Activators	Color of Light
Cubic zinc sulfide	Copper (low), lead	Blue
Cubic zinc sulfide	Copper (high), lead	Green
Cubic zinc sulfide	Copper (high), lead, manganese	Yellow
Hexagonal zinc sulfide	Copper (very high)	Green
Hexagonal zinc sulfide	Copper (very high), manganese	Yellow
Zinc sulfo selenide	Copper	Green to yellow
Zinc cadmium sulfo selenide	Copper	Yellow to pink

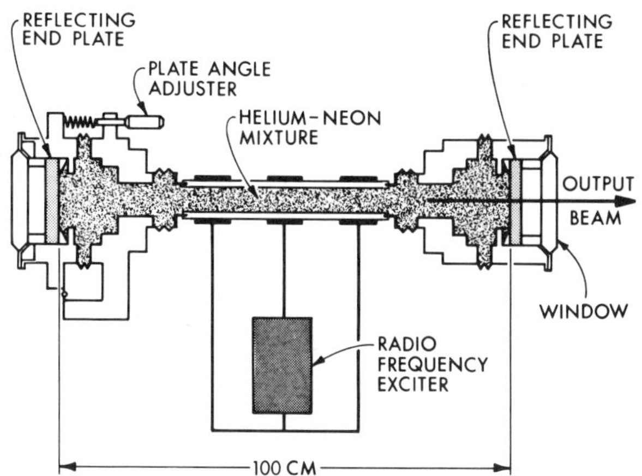

Fig. 2–19. Structure of CW helium-neon gas laser, showing essential parts.* Operation of a gas laser depends on the right mixture of helium and neon gases to provide an active medium. Radio frequency exciter puts energy into the medium. The output beam is built up by repeated passes back and forth between reflecting end plates.[12]

* Adapted from *Scientific American*.

neon, helium transfers energy by collision with neon. The excited neon emits photons which begin to amplify by cascading between the two reflecting surfaces until the internal energy is so large that the losses through the partially transmitting mirror become equal to the internal gains and the laser becomes saturated.

Galvanoluminescence[11]

Galvanoluminescence is light which appears at either the anode or cathode when solutions are electrolyzed.

Crystalloluminescence[11]

Crystalloluminescence (lyoluminescence) is observed when solutions crystallize and is believed to be due to the rapid reformation of molecules from ions. The light intensity increases upon stirring which might be due to some *triboluminescence.*

Chemiluminescence[11]

Chemiluminescence (oxyluminescence) is the production of light during a chemical reaction at room temperatures. True chemiluminescences are oxidation reactions involving valence changes.

Thermoluminescence[11]

Thermoluminescence (heat) is exhibited by some materials when slightly heated. In all cases of thermoluminescence the effect is dependent upon some previous illumination or radiation of the crystal. Diamonds, marble apatite, quartz, and fluorspar are thermoluminescent.

Triboluminescence[11]

Triboluminescence (piezoluminescence) is light produced by shaking, rubbing, or crushing crystals. Triboluminescent light may result from unstable light centers previously exposed to some source or radiation, such as, light, X rays, radium, and cathode rays; centers not exposed to previous radiation but characteristic of the crystal itself; or electrical discharges from fracturing crystals.

Sonoluminescence[11]

Sonoluminescence is the light which is observed when sound waves are passed through fluids. Luminescence occurs when fluids are completely shielded from an electrical field and is always connected with cavitation (the formation of gas or vapor cavities in a liquid). It is believed the minute gas bubbles of cavitated gas develop a considerable charge as their surface increases. When they collapse their capacity decreases and their voltage rises until a discharge takes place in the gas causing a faint luminescence.

Radioluminescence

Radioluminescence is light emitted from a material under bombardment from alpha rays, beta rays, gamma rays or X rays.

LIGHT DETECTION

The most universally used detector of light is the human eye (see Section 3). Other common detectors are photoelectric tubes, photovoltaic cells, photoconductor cells and photographic plates.

Photoelectric Tubes

Photoelectric effect is the emission of an electron from a surface when bombarded by sufficiently energetic photons. If the surface is connected as a cathode in an electric field (see Fig. 2–20) the liberated electrons will flow to the anode creating a photoelectric current. An arrangement of this sort may be used as an illumination meter and can be calibrated in footcandles.

Effects of Illumination. It has been found that the photoelectric current in vacuum varies directly with the level of illumination over a very wide range (spectral distribution, polarization, and cathode potential remaining the same). In gas-filled tubes the response is linear over only a limited range.

Effect of Polarization. If the radiant energy is polarized, the photoelectric current will vary as the orientation of the polarization is changed (except at normal incidence).

Effect of Wavelength. The more electropositive the metal the longer the wavelength of its maximum photoelectric emission and the lower the frequency threshold below which electrons are not liberated. See Fig. 2–21.

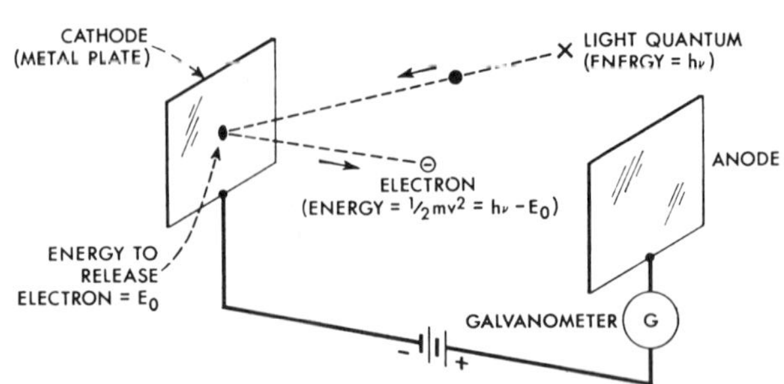

Fig. 2–20. By the photoelectric effect, electrons may be liberated from illuminated metal surfaces. In an electric field these will flow to an anode and create an electric current which may be detected by means of a galvanometer.

Fig. 2-21. The Electrode Potential Series

Li	Rb	K	Cs	Na	Ba	Sr	Ca	Mg	Mn	Zn	Cr	Fe*	Cd	Tl	Co	Ni	Sn	Pb	Fe†	Sb	Bi	As	Cu	Ti	Pt	Hg	Ag	Au
High																												Low

* ferrous † ferric

The maximum value of the initial velocity of a photoelectron, and therefore its maximum kinetic energy, decreases as the wavelength of the radiant energy increases.

The quantum theory provides the energy relationships which explain this phenomenon. The energy E of a light quantum equals the product of Planck's constant h by the frequency ν:

$$E = h\nu$$

It is known that an amount of energy E_0 (different from each metal) is required to separate an electron from the atom with which it is associated. Therefore, the energy of the liberated electron ($\frac{1}{2}mv^2$) is equal to that of the incident quantum $h\nu$ less E_0, that required to free it from the metal:

$$\tfrac{1}{2}mv^2 = h\nu - E_0$$

where m = mass of electron
v = velocity of electron

Photovoltaic Cells

Photovoltaic cells depend upon the generation of a voltage as a result of the absorption of a photon. The cell is comprised of a metal plate coated with a semiconductor material, *e.g.*, selenium on iron or cuprous oxide on copper. Upon exposure to light, electrons liberated from the metal surface are trapped at the interface unless there is an external circuit provided through which they may escape.

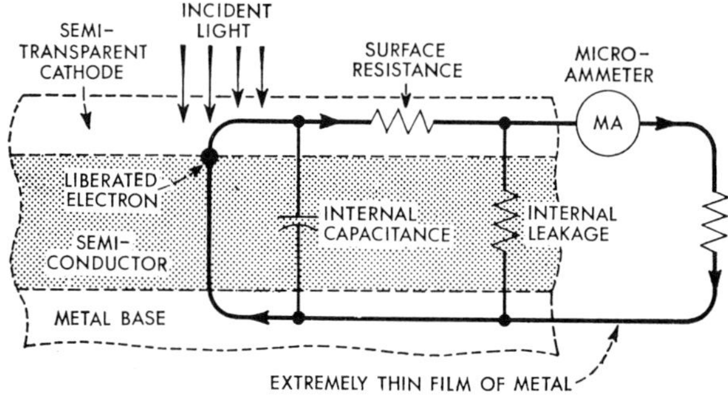

Fig. 2-22. Cross section of barrier-layer or photovoltaic cell showing motion of photoelectrons through microameter circuit.

In photographic and illumination meters, this circuit includes a small microammeter calibrated in units of illumination. See Fig. 2-22. This type is commonly used in photographic exposure meters and portable illumination meters.

Photoconductor Cell

Photoconductor cells depend upon the resistance of the cell changing directly as a result of photon absorption. These detectors use materials, such as, cadmium sulfide, cadmium selenide, and selenium. Cadmium sulfide and cadmium selenide are available in transparent resin or glass envelopes and are suitable for low levels of illumination ($<10^{-5}$ footcandles).

REFERENCES

1. Richtmeyer, F. K., Kennard, E. H., and Lauritsen, T. *Introduction to Atomic Physics*, McGraw-Hill Book Company, New York, 1955. Born, Max: *Atomic Physics*, Hafner Publishing Company, New York, 1962.
2. Maxwell, J. C. *A Treatise on Electricity & Magnetism*, Dover Publications, New York, 1960.
3. Forsythe, W. E. *Measurement of Radiant Energy*, McGraw-Hill Book Company, New York, 1937.
4. Elenbaas, W. *Fluorescent Lamps and Lighting*, The Macmillan Company, New York, 1959.
5. Funke, J. and Oranje, P. J. *Gas Discharge Lamps*, N. V. Philips, Eindhoven, 1951.
6. Fonda, G. R. and Seitz, F. *Solid Luminescent Materials*, John Wiley & Sons, New York, 1948. Leverenz, H. W. *Luminescence of Solids*, John Wiley & Sons, New York, 1950. Pringsheim, P. *Fluorescence and Phosphorescence*, Interscience Publishers, New York, 1949. Garlick, G. F. *Luminescent Materials*, Clarendon Press, Oxford, 1949. Kroger, F. A. *Aspects of the Luminescence of Solids*, Elsevier Publishing Company, New York, 1948.
7. Harvey, E. N. *A History of Luminescence*, American Philosophical Society, Philadelphia, 1957.
8. Wachtel, A. "The Effect of Impurities on the Plaque Brightness of a 3000° K Calcium Halosphosphate Phosphor", *J. of Electro-Chem. Soc.*, Vol. 105, p. 256-60, May, 1958.
9. Lengyel, B. A. *Lasers*, John Wiley & Sons, New York, 1962.
10. Ditchburn, R. W. *Light*, Second Edition, John Wiley & Sons, New York, 1963.
11. Newton, H. E. *Living Light*, Hafner Publishing Company, New York, 1965.
12. Sucsy, L. G. *Masers and Lasers*, Maser/Laser Associates, Cambridge, 1962.
13. Lorenz, M. R. and Pilkuhn, M. H. "Semiconductor-Diode Light Sources", *IEEE Spectrum*, Vol. 4, No. 4, p. 87, April, 1967.

SECTION

3

LIGHT AND VISION

Visual Sensory Processes 3–1 Visual Ability 3–6 IES Method for Prescribing Illumination 3–14 Visual Range of Luminous Signals 3–33

Those who are concerned with developing or applying lighting recommendations should be familiar with certain fundamental relationships among light, vision, and seeing. The eye is our primary gateway to the world about us. Without light we cannot see; with inadequate light or the wrong kind of lighting, seeing may be inefficient, uncomfortable, or hazardous. No single simple formula or procedure will solve all lighting problems. On the contrary, many factors need to be taken into consideration, and their relative importance can vary widely depending upon the seeing requirements that exist or are imposed upon people. The visual demands may range from the difficult ones involving prolonged critical seeing of very fine assembly or precision machine work to what might be termed casual seeing of large objects. The time available for seeing may be short or long. The things to be seen may require color discrimination or merely the differentiation of black, gray, and white.

It is evident that the illuminating engineer has a great responsibility to:

1. provide adequate visibility so that tasks can be performed with required standards of speed and accuracy,
2. provide lighting levels that will permit one to work with minimum effort, and
3. provide lighting conditions that will result in maximum safety and absence of visual disability and visual discomfort.

But this is a shared responsibility with other professions which are concerned with providing the public with the means for achieving and maintaining the best vision attainable within the limits of scientific and engineering development and of economic feasibility. Ophthalmologists and optometrists are responsible for the care of the eyes.[1] The coordinated skills of architects, designers, and illuminating engineers are required for the development of the visual environment. All are greatly dependent upon the fundamental data obtained by the significant researches of many disciplines—physiology, psychology, physics, engineering, etc.

Extensive investigations of light and vision and associated subjects are being carried on in many university, industrial, and government laboratories throughout the world. Agencies such as the Illuminating Engineering Research Institute, the Armed Forces-National Research Council Committee on Vision, The National Institutes of Health, and others are sponsoring or actively stimulating many of these researches. All such activities are making material contributions to a better knowledge of the visual process. More significantly, they are enabling the development of a firmer basis upon which to build a sound lighting practice.

VISUAL SENSORY PROCESSES

The Eye

A simplified diagram illustrating the camera-like structure of the eye is shown in Fig. 3–1. Its spherical shape is maintained by the *sclera,* a tough outer coat that also protects the eye, and the *cornea,* the clear front covering of the eye. The *iris,* lens, and associated muscles divide the interior into two chambers, each filled with a transparent fluid. The optical system of the eye consists of three major components: the *cornea,* which supplies about 75 per cent of the total dioptric power; the *pupil,* which controls the amount of light entering the eye; and the *lens,* which is the adjustable focusing element.

The *ciliary muscle* delicately controls the curvature of the lens so that the eye can focus accurately on an object at one distance and then change its focus to another object at a different distance. This process of being able to change the focus of the eye is called *accommodation.* It is accompanied by a reduction in pupil diameter when nearby objects are viewed. This pupillary constriction increases the sharpness with which these nearby objects are seen.

Note: References are listed at the end of each section.

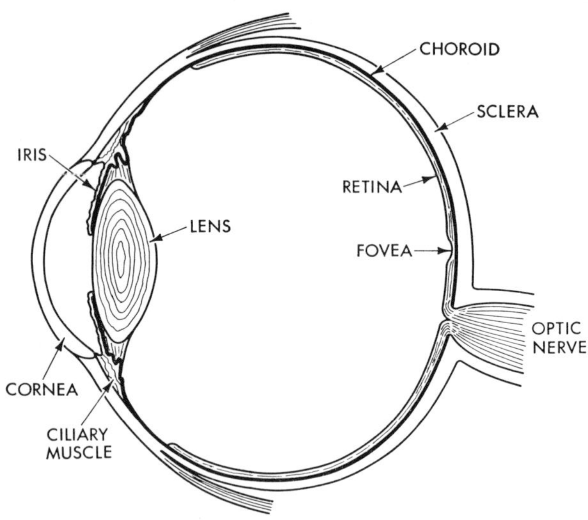

Fig. 3-1. A horizontal cross-section of the human eye. Approximate length from cornea to cone layer of retina is 24 mm. Thickness of choroid is about 0.05 mm and the sclera 1.0 mm.

The *retina* consists of a mosaic of photoreceptors and several layers of connecting nerve cells as well as supporting cells which line the rear part of the inside of the eye. Fig. 3-2 shows an enlarged and simplified diagram of the retina. Light reaches the photosensitive receptor cells only after passing through the various retinal layers. The outer segments of the receptor cells which contain molecules of photopigment are oriented toward the back of the eye. This arrangement results from the embryonic development of the retina as an outgrowth of the brain. The human eye is, therefore, described as an "inverted retina". The photosensitive receptor cells are of two main types called *rods* and *cones.* They are designated as such primarily on the basis of their shape, but they differ in other structural characteristics as well. The distribution of the receptors varies across the retina; the *fovea,* which is located in the central region of the retina, contains only cones and appears as a small pit of about 0.5 mm in diameter. From the fovea to the periphery of the eye, the density of cones decreases while that of rods increases. Due to the absence of several retinal layers and blood vessels in the fovea, as compared to the more peripheral regions of the retina, there is less light interference present. This along with the suggestion of a greater number of neural connections from the optic nerve are two reasons for the excellent acuity or ability to resolve fine detail found in the fovea. The mechanisms of accommodation and fixation attempt to produce a retinal image in sharp focus on the fovea.

Photochemical Process

It is generally accepted that a photochemical mechanism in the retina makes it possible for the photoreceptors to generate nerve impulses in response to light. That is, a photochemical substance present in the rods and cones forms several products when exposed to light. Though the exact nature of these products is not known, there is considerable support[2] for the existence of the photochemical cycle shown in Fig. 3-3. In the rods there is a rose colored chemical known as *visual purple* or *rhodopsin.* Upon exposure to light, this breaks down into *retinene* plus *opsin* (a protein) and finally into Vitamin A. It has been hypothesized[3] that the only action of light on vision consists of the absorption of a quantum of light by the rhodopsin molecule effecting a change in the molecular configuration of retinene (or *retinal* as it also has been designated[3]) which is the photosensitive component of the visual pigment. This change leads to the progressive opening up of the protein base opsin (also designated *retinol*[3]) which by an as yet unknown amplification process results in a change in the electrical energy of the receptor. The primary reaction leads to a series of electrical changes eventually resulting in the development of a nerve impulse in the ganglion cell. The nerve impulse is conducted to the higher visual centers of the brain via the optic nerve. It is

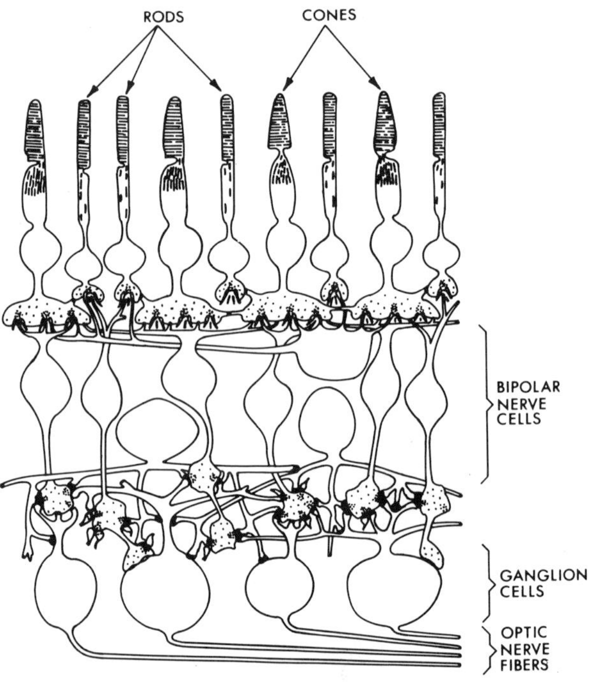

Fig. 3-2. Simplified diagram of the connections among neural elements in the retina. The regions where the cells are contiguous are synapses. The direction of light is from the bottom up in this diagram.

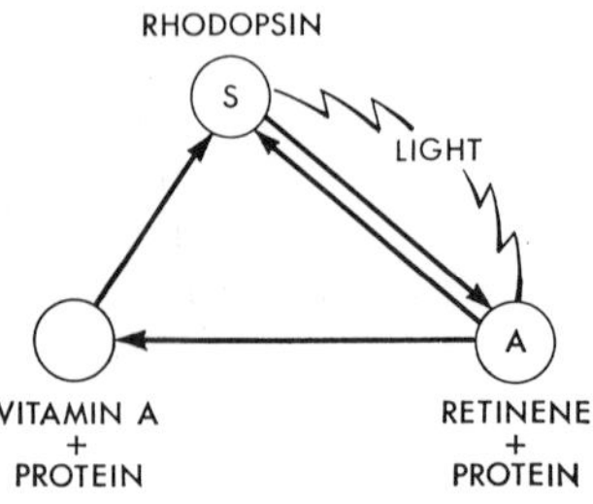

Fig. 3-3. In the rods, both chemical and photochemical activity have been observed involving rhodopsin, retinene, Vitamin A, and protein.

generally believed that similar reactions take place in the cone receptors. The intensity of the light stimulus is encoded in terms of the frequency of nerve impulses. Empirical evidence indicates that it takes only one quantum absorption by a molecule of rhodopsin to stimulate a rod.

Accommodation

The process of focusing the eye on an object at a given distance is called accommodation. The lens is an elastic structure, its form and position being determined by the tension of the suspensary ligaments. During accommodation the ciliary muscle contracts and relaxes the suspensary ligaments; this permits the inherent elasticity of the lens to assume a more spherical shape, thereby increasing its power in order to focus on near objects. When viewing distant objects, the ciliary muscle relaxes and tension is exerted on the lens, which becomes more flattened and has less power. These changes in the form, and to a slight extent, the position of the lens inside the eye, are illustrated in Fig. 3-4. The amplitude of accommodation is expressed in *diopters* and is numerically equal to the reciprocal of the distance in meters from the spectacle plane (1.4 cm in front of the cornea) to the nearest point on which the eye can focus when a lens is worn which permits the eye to see a distant object clearly with relaxed accommodation.

Adaptation

The response of any portion of the retina at a given moment depends upon the light that fell on that region during the just preceding period of time. Under steady retinal illumination the concentration of the photochemical substances reaches an equilibrium level and the area of the retina exposed is described as being adapted to this level. Any change in the adapting illumination produces a change in the photochemical substances toward a new equilibrium level. Since the photochemical reactions take a finite time for their completion, it is evident that there will be a time delay before the retina reaches maximum sensitivity for a given adaptation level. When the eye is in total darkness for 30 to 40 minutes the concentration of rhodopsin in the rods reaches its maximum. On the other hand, cones will achieve fairly complete dark adaptation in less than two minutes. The process of light adaptation is much faster in both types of photoreceptors, being rapid in the cones.

Using fundus reflectometry* there has been an attempt to establish a causal relationship between photopigment concentration in the receptors and visual adaptation.[4] The experiments suggest that the visual threshold is not directly proportional to pigment concentration but is logarithmically related.

* Fundus reflectometry refers to quantitative changes in the amount of light reflected from the back of the eye after the photopigment is bleached by exposure to a strong light.

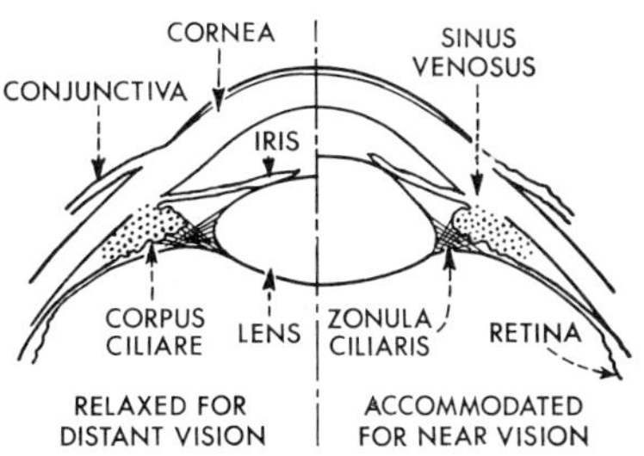

Fig. 3-4. Change in the form and position of the lens during accommodation.

Adaptation in general appears to be of two basic types. One seems to be primarily neural in nature and has a very short time course. It has been called transient or early light and dark adaptation.[5] Transient adaptation involves changes in the eye's sensitivity as a result of sudden increases or decreases in the prevailing level of illumination. A loss in visibility is often produced subsequent to a change in luminance level occuring when an individual shifts his point of fixation to surfaces having different luminances, when viewing a variegated surface, or simply as the result of illumination changes occurring naturally in the visual environment. Typical results show an increase in threshold (corresponding to a decrease in sensitivity) just prior to an increment or a decrement in the prevailing level of illumination. This is followed by an abrupt fall in threshold (or increase in sensitivity) which turns into a more gradual decline. The initial loss in sensitivity is believed to be produced by a neural inhibitory effect of the changing background field acting back upon the neural response produced by the target used to measure sensitivity. The course of early light and dark adaptation is usually complete within a few hundred milliseconds. Assuming that some variation in the visual environment is necessary (a complete lack of change produced by stopping the visual image on the retina or a complete uniformity of the visual field leads to a loss in visibility),[6] research on transient adaptation has adressed itself to the question of how much variation in luminance should be permitted while still maintaining adequate visibility.[7]

The second type of adaptation process is believed to be more closely linked to the concentration of photopigment in the receptor cells although a neural component is also indicated.[8] It is well known that the human dark-adaptation curve under the appropriate experimental conditions consists of two branches. The earlier or photopic branch is attributed to the increasing sensitivity of the cone receptors with increasing time in the dark, whereas the later scotopic branch is attributed to the increasing sensitivity of the rod receptors.

Pupil Size

The size of the pupil is affected by a number of factors. Stimulating the retina with light produces a constriction of the pupil. Conversely, when the eye is in the dark, the pupil dilates. The pupil diameter may range from approximately 8 mm to as small as 2 mm, depending upon the luminance to which the eye is exposed. Fig. 3-5 illustrates the effect of variations of the luminance of a large uniform field upon pupil diameter.[9] An important optical function of the pupil, in addition to controlling the retinal illumination, is to confine light rays entering the eye to a limited central area of the lens. This minimizes the aberrations of the eye by eliminating the outer edges of the cornea and lens. The pupillary constriction during accommodation and convergence for near vision is a similar aid to seeing.

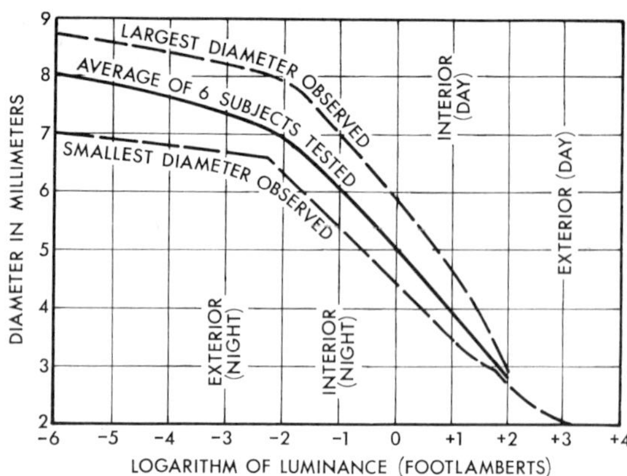

Fig. 3-5. The effect upon diameter when the eyes are exposed to a large uniform field of fixed luminance.

Central and Peripheral Vision

When a person with a normal seeing apparatus looks at an object with both eyes at the same time, the visual fields of the two eyes intermesh as is illustrated in Fig. 3-6. The areas seen by the two eyes are not coextensive because a portion of the field of each eye is blocked off by the nose, eyebrow, and cheek. Thus the visual field for both eyes includes more space than the field of either eye. The binocular field is that seen by both eyes simultaneously and is approximately 60 degrees in radius.

Central vision differs from peripheral vision in several respects. The cones in the fovea, because of their small diameter, close packing (about 150,000 per square millimeter), and individual connections with the brain, transmit a very sharp image showing the greatest detail of which the eye is capable. Outside the fovea, the density of cones drops rapidly, while that of the rods at first increases, reaching a maximum at about 20 degrees from the fovea, and then falls off toward the periphery of the retina. Because of the sparse distribution of photoreceptors in the periphery and because in this region of the retina each nerve cell branches to provide connections with many photoreceptors, sharp images are not transmitted to the brain and objects appear as fuzzy silhouettes.

The rods have a lower threshold; that is, they are more sensitive to light than the cones. When the luminance of the visual field is less than about 0.01 footlambert, as at night, the cones do not respond; the rod-free area is blind and vision is limited to the periphery and is due entirely to activity of the rods. An example of this is that less bright stars cannot be seen when looked at directly, but are readily visible when viewed with slightly averted vision. Vision at these low levels of field luminance is called *scotopic vision.* When primarily cones are involved, as at higher luminance levels, this is called *photopic vision.* Vision at luminance levels intermediate to the photopic and scotopic levels is referred to as *mesopic vision.* This roughly corresponds to a twilight range and probably refers to combinations of both rod and cone functioning.

Rods and cones differ in the time factors associated with their activities. The rods are much slower in action than the cones. As a result cones show a greater temporal resolution or a higher frequency at which a flickering light appears constant in brightness than do the rods. The rods also are notably slow in recovery from light adaptation.

Color Vision

A great deal is known about color vision, but the basic physiological processes are still subject to considerable speculation.[10] Two basic theories[11] of color vision—the Young-Helmholtz three-component theory and the Hering opponent-color theory—have been developed. The Young-Helmholtz theory is derived from the fact that any color can be represented by a unique mixture of three differently colored lights and assumes that there are three kinds of receptors in the eye that react selectively to light according to its wavelength. In effect, this theory attempts to explain color vision in terms of the stimulus. On the other hand, the Hering theory is based upon visual responses and assumes that

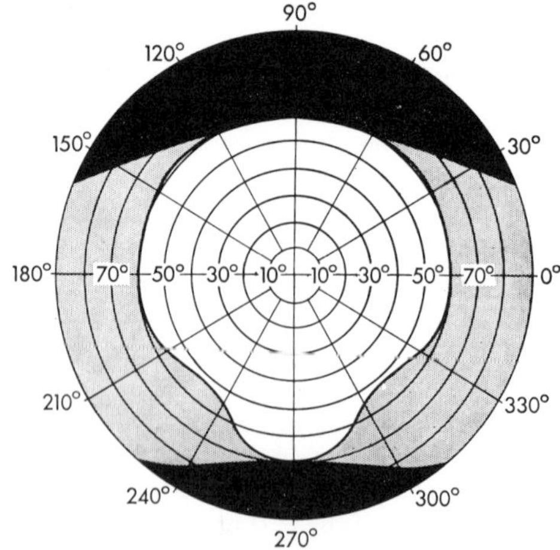

Fig. 3-6. The normal field of view of a pair of human eyes. The white central portion represents the region seen by both eyes. The gray portions, right and left, represent the regions seen by the respective eyes alone. The cut-off by the eyebrows, cheeks, and nose is shown by the dark areas.

there are six basic unitary colors (red, yellow, green, blue, white, and black). It also is assumed that visual activity is found in three pairs of processes with two members of each pair being opposites, *i.e.*, there are blue-yellow, green-red, and white-black pairs of processes.

In recent years, considerable evidence has been gathered on the response of the retina to color with particular regard to photochemistry and electrophysiology by a number of investigators using a variety of techniques. It has been determined that there are at least three cone photopigments. These pigments show a maximum absorption in the visible spectrum in either the short-, medium-, or long-wavelength regions.[12] In other words, there are pigments which maximally absorb light in the violet, green, and yellow to red regions of the spectrum. In addition, electrical response records of single receptor cells from at least one vertebrate with color vision indicate that there are three different cone types showing maximal sensitivity in one of three regions of the visible spectrum.[13] These results provide an impressive argument for the Young-Helmholtz trichromatic theory, at least at the level of the photoreceptors. Beyond the photoreceptor level, however, empirical evidence gathered from psychophysical research with human subjects and electrophysiological research with animals indicates the functioning of a Hering opponent-colors neural mechanism.[14] Both theories are, therefore, necessary to provide an adequate explanation of color vision phenomena. Both also recognize the need for a set of three functions whether they are based on cone photopigments or on neural responses in order to treat the facts of color vision. Both theories have been described as being trichromatic in nature. See Section 5.

Rods acting alone are insensitive to color as such, and objects appear gray when these receptors alone are stimulated as under certain nighttime conditions. When rods and cones are stimulated simultaneously, as under conditions of moderate illumination, the periphery of the retinal field yields a gray response due to rod stimulation which has to be integrated with the color response of the cones that are located in this region. In the central or rod-free portion of the retina, the color response of the cones is not diluted with the gray response of the rods.

The *spectral luminous efficiency curves* for cones[15] and rods[16] at 10 nanometer intervals are given in Fig. 3-7 and plotted in Fig. 3-8a. The ordinates of the curves represent the reciprocals of the relative amounts of energy required to produce equal perceived luminances. These equivalent luminances were determined for a number of different wavelengths of light as seen (1) at the fovea and (2) toward the periphery of the retina. As the luminance level is gradually increased and exceeds the threshold of cone vision, the spectral luminous efficiency curve for the peripheral retina undergoes a gradual transition until it becomes indentical with the curve for the fovea. This means that at higher luminance levels the cones in the periphery almost completely dominate the rods. Goodeve's data[17] indicate that most human observers are capable of experiencing a visual sensation upon exposure to radiant energy of wavelengths usually termed "infrared" (longer than 780 nanometers) provided the energy reaches the eye at a sufficiently high rate per unit area. See Fig. 3-8b and page 2-5.

The *macular pigment* which covers the fovea and the area immediately surrounding it acts as a yellow filter. The spectral composition of light passing through this pigment layer is modified before the cones are stimulated. This influences the judgment of an observer making either a color match or a

Fig. 3-7. Photopic, $V(\lambda)$, and Scotopic, $V'(\lambda)$, Spectral Luminous Efficiency Values. (Unity at Wavelength of Maximum Luminous Efficacy)

Wavelength λ (nanometers)	Values	
	Photopic $V(\lambda)$	Scotopic $V'(\lambda)$
380	0.00004	0.000589
390	.00012	.002209
400	.0004	.00929
410	.0012	.03484
420	.0040	.0966
430	.0116	.1998
440	.023	.3281
450	.038	.455
460	.060	.567
470	.091	.676
480	.139	.793
490	.208	.904
500	.323	.982
510	.503	.997
520	.710	.935
530	.862	.811
540	.954	.650
550	.995	.481
560	.995	.3288
570	.952	.2076
580	.870	.1212
590	.757	.0655
600	.631	.03315
610	.503	.01593
620	.381	.00737
630	.265	.003335
640	.175	.001497
650	.107	.000677
660	.061	.0003129
670	.032	.0001480
680	.017	.0000715
690	.0082	.00003533
700	.0041	.00001780
710	.0021	.00000914
720	.00105	.00000478
730	.00052	.000002546
740	.00025	.000001379
750	.00012	.000000760
760	.00006	.000000425
770	.00003	.0000002413
780	.000015	.0000001390

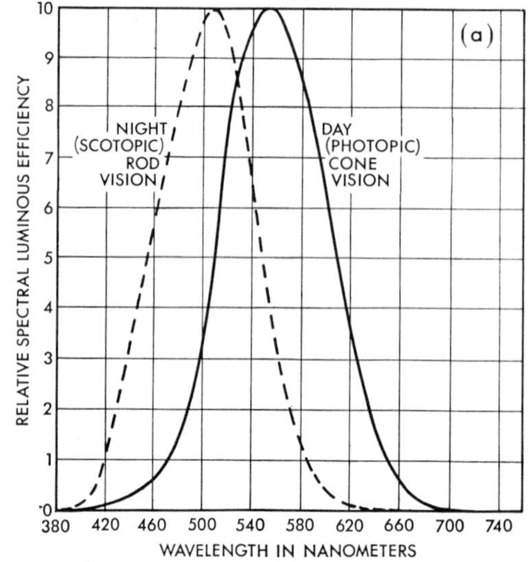

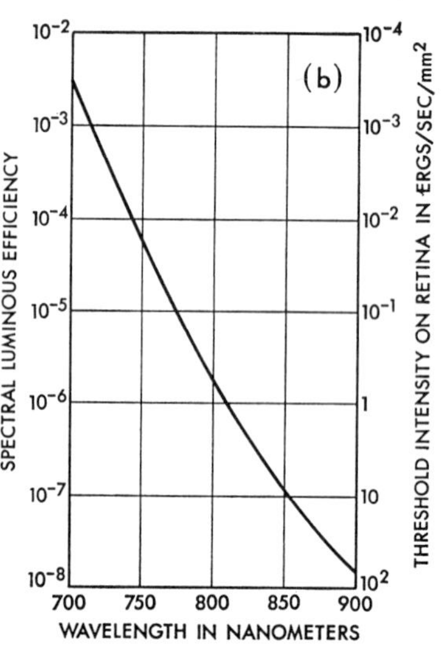

Fig. 3-8. a. Relative spectral luminous efficiency curves for photopic and scotopic vision, showing the Purkinje shift on the wavelength of maximum efficiency. b. Goodeve's investigations reveal that radiant energy of wavelength outside the "visible region" (380 to 780 nanometers) is capable of producing visual sensations provided the energy is incident on the retina at a high rate per unit area.

color evaluation. The pigment differs in density and composition in different individuals and gets darker with age. This variation is believed to be one cause of the difficulties experienced in obtaining identical color matches from different observers.

There are a number of illusions or perceptions which can be attributed to structures which lie entirely within the eye. These effects are called entopic phenomena. Many believe that the entopic phenomenon known as Maxwell's spot is mainly the result of a filtering effect on the visual receptors by the yellow pigment of the macula. The effect is made observable in the direction of fixation as a dark spot and surrounding ring after alternate viewing through yellow and blue filters of a brightly illuminated white paper. This phenomenon has also been described as influencing color matches of small fields.[8]

VISUAL ABILITY

Many factors influence visual ability. They can be divided into two separate but related categories: physiological factors which depend upon adequate functioning of the visual sensory processes such as accommodation, adaptation, photochemical mechanism of the eyes, and the relative sensitivity of various retinal areas; and physical factors which describe certain characteristics of the visual task and the environment in which it is seen.

Physiological Factors

When a person looks at a point in space he brings his eyes into a definite position with respect to the point and will hold his eyes in that position. The line connecting the point of fixation and the center of the pupil of the eye is known as the *primary line of sight.* For practical purposes, with binocular vision, the point of observation often is taken as the midpoint between the eyes. In a normal eye the image of the point of fixation falls at or near the center of the fovea. It is this portion of the retina—an area about two degrees in diameter—which is especially adapted for the perception of fine detail. When performing most tasks, the eyes jump from one fixation point to another so that those things which require critcial seeing are brought within the central or foveal portion of the visual field.

For a task such as threading a needle, foveal visual ability obviously is of paramount importance; peripheral vision plays a very subordinate role. However, when the task, for example, involves stacking cartons in a warehouse, peripheral vision is just about as important as central vision. In this case the worker must be aware of the spatial relations among objects which are widely distributed in the visual field.

Peripheral vision gives an over-all picture of the location of various objects in the field of view and thereby provides information which guides foveal vision from one point to another. It also aids in maintaining the eyes converged and accommodated in the proper amount to permit clear binocular vision. Peripheral vision helps to give a person an awareness of his position and orientation. It provides a feeling of security in the spatial framework in which he must exist. It is useful in the detection of potential hazards and thus plays a role in accident prevention. This makes it necessary to draw a distinction between the lighting requirements for objects that have to be seen with foveal vision and those which are seen peripherally.

When performing a visual task a person moves his eyes from one fixation point to another. In association with this eye movement there may be movements of the head and even of the whole body. He may move from one part of a room to another, or to another room. All of these actions produce a redistribution of the luminances in the field of view and thereby influence both local and general adaptation of the retinas of the two eyes. The effects of such luminance changes upon ability to see and visual discomfort can be evaluated.

Physical Factors

The Visual Task. When one looks at an object at any moment during the performance of a task, the head takes a certain position with the lines of sight of the two eyes converging to a given point. This results in a subjective visual sensation which represents the completion of the ocular part of the visual activity of the moment. The seeing of all the things that have to be seen at that moment constitutes the visual part of any task, and the term visual task conventionally designates the sum total of all the things that have to be seen at the given moment. The character of this visual task may change from moment to moment. Hence, in determining whether the illumination is adequate for a given task, it is necessary to consider the nature of the visual task during each phase—ideally each instant—of any occupation.

Fundamental Factors. A visual task can be described in terms of its size, contrast, luminance, and color. It also is often necessary to include the factor of the time available to see the task, especially when a series of tasks must be seen sequentially or if the worker or the task is in motion. The following discussions are limited to the consideration of seeing of achromatic objects illuminated by "white" light. For color, see Section 5.

The luminance (L) of a non-luminous perfectly diffusely reflecting surface is dependent upon the illumination (E) incident on the surface and its reflectance (ρ) and is represented by:

$$L = E \times \rho \qquad (3\text{-}1)$$

When the illumination is expressed in footcandles, the luminance will be in footlamberts.

Luminance variations in the visual task occur in the form of gradients or abrupt transitions which are called *contrast borders*. Points and lines do not constitute exceptions; a point is treated as a finite area surrounded by a border, and a line as a strip bounded by parallel borders. In the case of a luminance contrast border, the contrast (C) between the two adjoining areas is defined as:

$$C = \frac{L_b - L_o}{L_b} \qquad (3\text{-}2)$$

where L_b and L_o are the luminances of the two contrasting areas, the background and the object, respectively. In most instances, the larger of the two areas constitutes the background, but in some cases, such as an evenly divided photometric field or a grating consisting of bright and dark bars of equal width, it is necessary to be arbitrary in defining which represents the background and the object. It is often convenient to write Equation 3-2 as:

$$C = \frac{\Delta L}{L_b} \qquad (3\text{-}3)$$

in which the numerator is the luminance difference between the object and its background. This form has the convenience that it can be used for objects either darker or brighter than the background because ΔL represents the numerical luminance difference regardless of whether L_b or L_o is the larger.

The minimum perceptible contrast is a function of the luminance of the background against which the object is seen. The general relationship is illustrated in Fig. 3-9 for a small circular luminous spot on a less bright background.[19] This attribute of visual ability also may be expressed in terms of contrast sensitivity which is quantitatively defined as the reciprocal of the minimum perceptible contrast. Fig. 3-10, which represents the relationship between contrast sensitivity and background luminance, has been derived from the same data as Fig. 3-9. It sometimes is convenient to express such relationships in terms of luminance differences, *i.e.*, the numerator of Equation 3-3. Using the data of Fig. 3-9, luminance difference varies with background luminance as is shown in Fig. 3-11. Each method of presenting the data has certain advantages and disadvantages. Fig. 3-9 illustrates that smaller contrasts can be seen with higher background lumi-

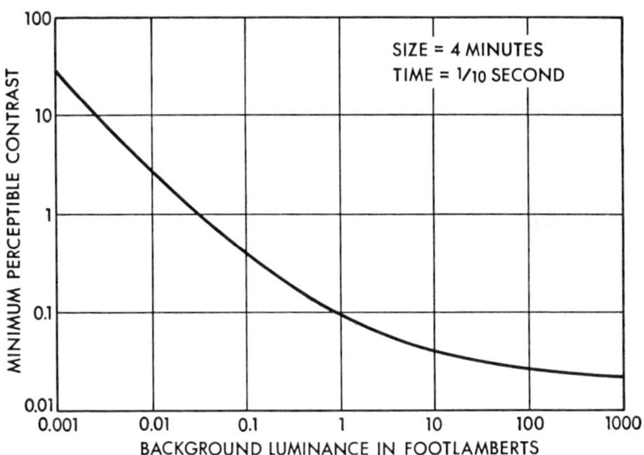

Fig. 3-9. Variation of minimum perceptible contrast with background luminance.

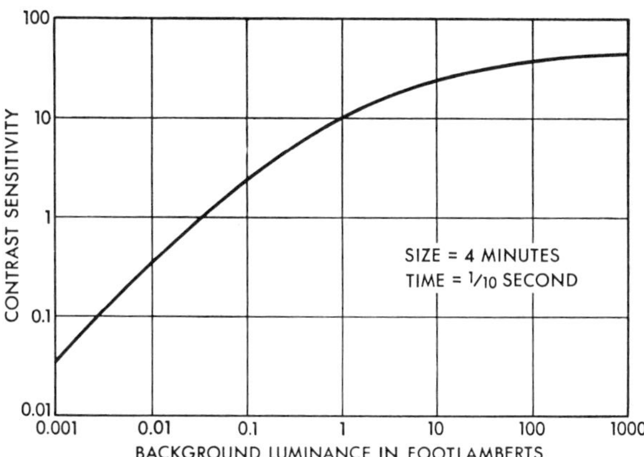

Fig. 3-10. Variation of contrast sensitivity with background luminance.

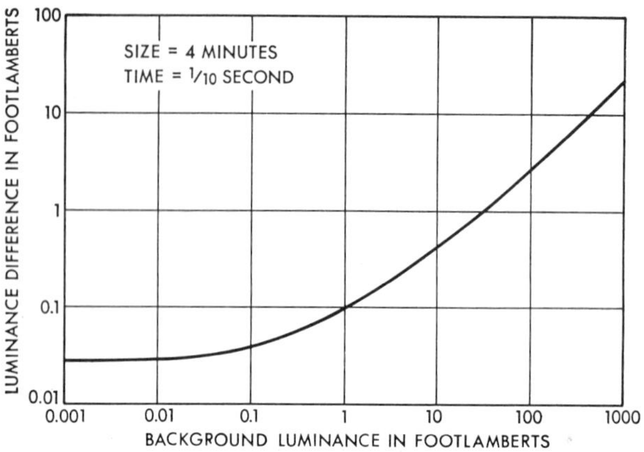

Fig. 3–11. Variation of luminance difference with background luminance.

nances, but the improvement of visual sensitivity is more readily apparent in Fig. 3–10. The curves of Figs. 3–9 and 3–10 become asymptotic at higher values of background luminance, making it difficult to determine precise values of contrast in this region. When presented in terms of luminance differences, as in Fig. 3–11, this portion of the curve approaches linearity with a positive slope and makes it easier to determine the effects of higher background luminances.

Visual data may be expressed most usefully in terms of the luminance difference when considering problems related to the visibility of self-luminous signals. The curve shows that such signals require a higher intensity the higher the background luminance against which the signal is viewed. The fundamental relationships expressed in terms of contrast have been found most useful for prescribing quantity of illumination. However, there are special problems involving interior illumination in which the luminance difference approach is useful.[20] For example, the visibility of objects of the same contrast but different reflectances may be evaluated conveniently in terms of the luminance difference when the objects are illuminated to the same level.

The visual size of any detail that needs to be seen is a function of its physical size and its distance from the point of observation. By combining these two dimensions one can express the size as a *visual angle* which usually is measured in minutes of arc. Thus, the farther a given object is from the eyes, the smaller its visual size becomes. Various types of test objects, four of which are shown in Fig. 3–12, have been used for evaluating the size discriminatory ability of the eye. In each case the critical detail which must be discriminated is indicated by the dimension *d;* for a constant viewing distance, the visual angle subtended at the eye by *d* is the same for the four objects, even though the maximum dimensions are different. Another often used way of expressing the size threshold of the eye is in terms of *visual acuity*. Quantitatively, this is the reciprocal of the visual angle. A typical curve[21] relating visual acuity and background luminance for a high contrast object is shown in Fig. 3–13. The scale on the right-hand side of Fig. 3–13 illustrates the variation in visual size at the threshold. The form of the test object used has a considerable effect upon the absolute values of the size threshold and this must be taken into consideration when comparing the results of various investigators.

It takes time to see. As is shown in Fig. 3–14, higher background luminances make it possible to see a specific test object in progressively shorter time intervals.[19] This visual ability sometimes is expressed in terms of speed of vision which is equal to the reciprocal of the time (seconds) required to just barely see the object. The right-hand scale of Fig. 3–14 illustrates the relationship for speed of vision.

Since size, contrast, and time are functions of

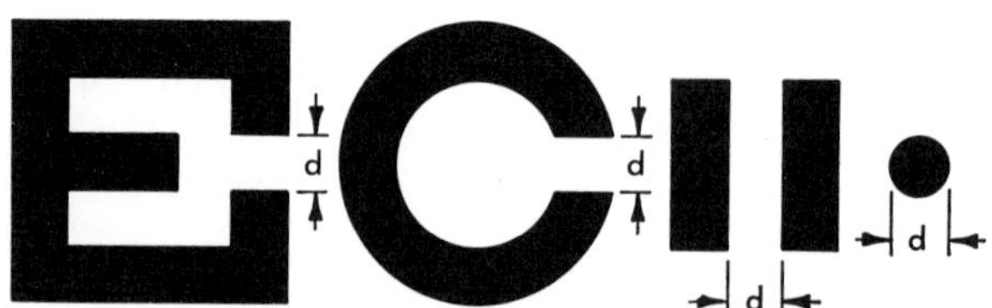

Fig. 3–12. Commonly used test objects for determining size discrimination and visual acuity.

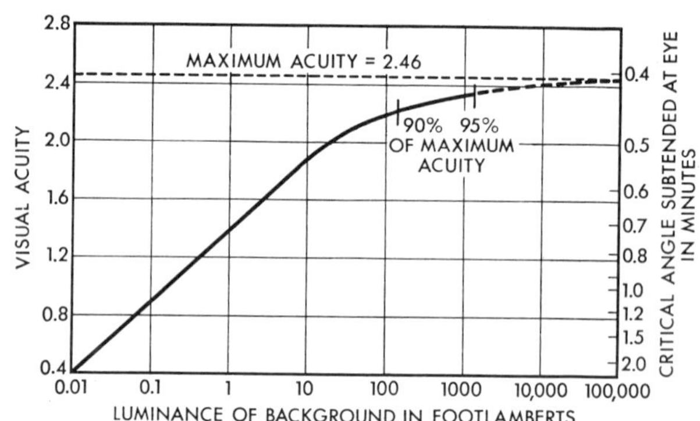

Fig. 3–13. The variation in visual acuity and visual size with background luminance for a black object on a white background.

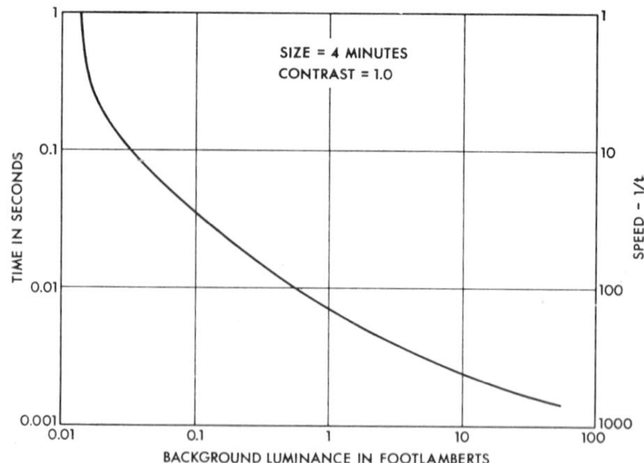

Fig. 3–14. The effect of background luminance on the time required to see a high contrast test object. The right hand scale represents speed of vision.

the background luminance, it is evident that these factors are interrelated, *i.e.*, a change in any one will influence the threshold values of others. In Figs. 3–15A and 3–15B are shown the effects upon threshold contrast when size and exposure time are varied.[19] It is emphasized that the absolute value of threshold contrast also depends upon the type and shape of test objects used and the criterion—detection of presence or discrimination of detail—employed. However, all exhibit very nearly the general relationships shown in Fig. 3–15. It has been shown that most practical tasks can be related to any of the laboratory test objects for which a great deal of information has been obtained. This is discussed later in this section.

The differential effects of high (white) and low (gray) reflectance backgrounds against which a dark object is viewed are shown in Fig. 3–16.[23] In this case speed of seeing is represented as relative visual performance. With a white background maximum performance is obtained with about 50 footcandles, and the loss at lower levels of illumination is relatively small. However, when the same object is viewed against a gray background, it is seen that the improvement in visual performance is very marked as the illumination is increased; to achieve a relative performance of 90, this task will require about 1000 footcandles.

Effects of Surroundings

Luminances in the visual field which surrounds the task can have different effects depending upon the areas involved, their location with respect to the line of sight, and their actual luminances as compared with that of the task. These luminances may produce a decrement in visual ability, visual comfort, or both. Researches indicate that best seeing conditions are achieved when the whole field of view is as uniform as possible.[24] In the case of a visual task which involves a simple object on a background this means that the entire surround should have the same luminance as the immediate background of the object. Where the task involves a more complex pattern, the field surrounding the task should be uniform and have a luminance equal to the average of the central portion of the field in which the task is located.[25]

Lines, points, and low contrast borders which demarcate the field are not to be construed as defeating this objective. Some demarcation is necessary for the perception of objects in the periphery of the visual field. It helps to stabilize the perception of objects in space and the observer's perception of his own position and orientation in space. It also helps to stabilize the motor adjustments of the eyes such as accommodation[26] and convergence.[27]

A simplified diagrammatic sketch of a task and surrounding areas is illustrated in Fig. 3–17. The disk at the center is a low contrast object, the detection of which constitutes the visual task. About it are the immediate background and the

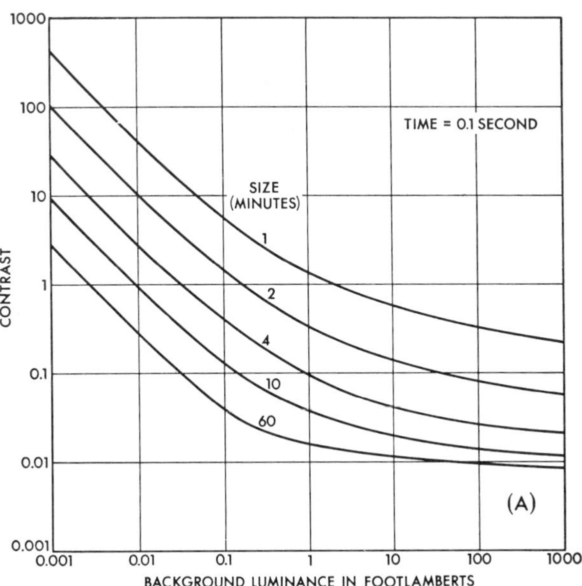

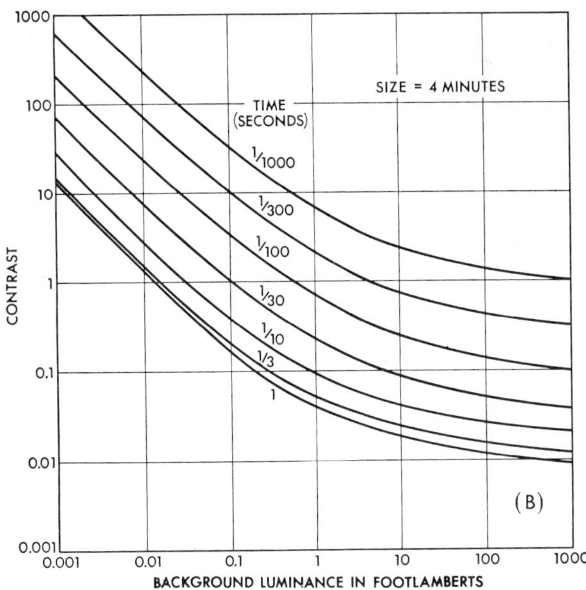

Fig. 3–15. The relationship between threshold contrast and background luminance for (A) various sizes of object and (B) various exposure times.

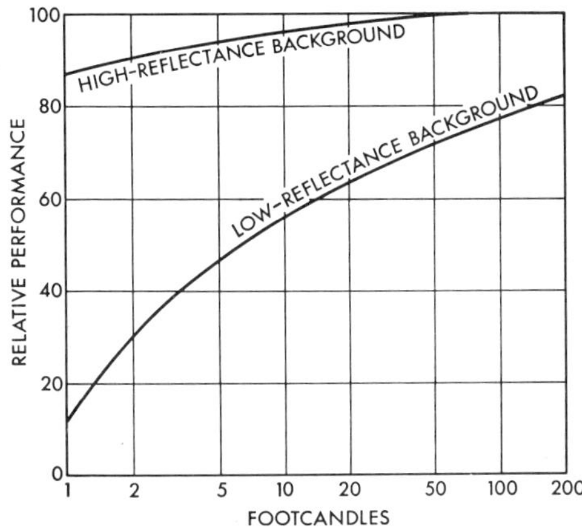

Fig. 3–16. The effect of level of illumination on relative visual performance when the task involves high contrast (upper curve) and low contrast (lower curve).

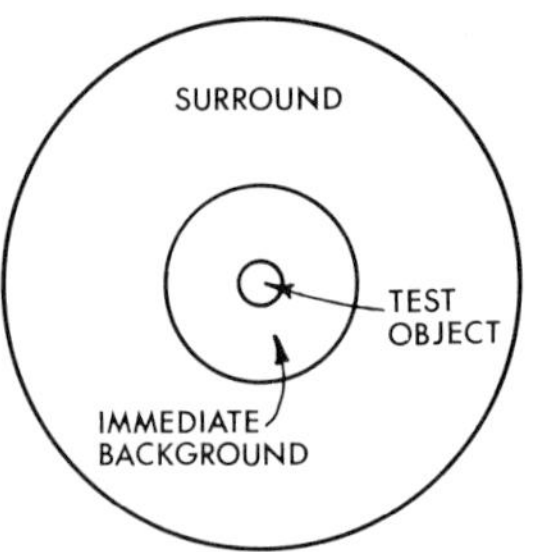

Fig. 3-17. A simplified representation of a task, its immediate background, and the surround.

general surround. This type of arrangement relates to the problem of general versus local lighting. Many investigations have been made of such conditions and have included the variables of the size of the immediate background and various luminances of the background and surround.[21, 28, 29] The results of such experiments emphasize that the luminance of the surround should not be excessively brighter or darker than that of the background to prevent significant losses in contrast sensitivity. Furthermore, when there is a high contrast border between the background and the surround, the diameter of the former should be at least 5 degrees so that the border has no effect on the visibility of the task.[28, 30]

A special case of non-uniform surround is the superposition of a glare source or a number of such sources on the otherwise approximately uniform luminance. This sort of luminance distribution may reduce visual ability and also produce visual discomfort.

As has been shown, pupil diameter is dependent upon the field luminance. Even though a great deal of information is available relative to the effects of uniform and various types of non-uniform fields,[31, 32] it is not possible to predict the size of the pupil for such conditions. It would be worthwhile to be able to make such predictions because pupil size affects foveal vision and may have some bearing on discomfort.

Disability Glare. In most interior lighting conditions stray light produced within the eye by glare sources does not present a serious problem. However, in the case of roadway lighting, vehicle headlights, and certain types of industrial lighting, the effects of stray light may have to be taken into consideration.

Stray light within the eye produces a veiling luminance which is superimposed upon the retinal image of the object to be seen. This alters the luminances of the image and its background and hence the contrast. The apparent luminances of the contrasting areas are given by:

$$L'_b = L_b + L_v \qquad (3\text{–}4)$$

$$L'_o = L_o + L_v \qquad (3\text{–}5)$$

where L_v is the added luminance due to stray light. Substituting these in Equation 3–2 the resulting apparent contrast is:

$$C' = \frac{L_b - L_o}{L_b + L_v} \qquad (3\text{–}6)$$

It has been shown[31, 33] that a small source located at an angular distance θ degrees from the line of sight produces an effect at the fovea which is equivalent to a veiling luminance L_v given by the following formula:

$$L_v = kE\theta^{-n} \qquad (3\text{–}7)$$

in which E represents the illumination produced by the glare source in the plane of the pupil, and k and n are constants. Even though the smallest angle investigated was 1 degree and according to the equation, L_v approaches infinity as θ approaches zero, these are not serious drawbacks. In practical situations, at very small angles the glare source will physically obstruct the view of the object being viewed. This equation was found to hold for glare sources in any meridian of the field of view.

The facts that (1) the equivalent veiling luminance is directly proportional to the intensity of the glare source and (2) the effects of two or more glare sources are strictly additive, conform to the stray light hypothesis. Therefore, it has been assumed[34] that the entire effect is due to stray light and that Equation 3–7 is used for computing the combined effects of a number of glare sources. For this purpose values of 10 and 2 for the constant k and n are used. The general equation for a number of glare sources is:

$$L_v = 10\pi\,[E_1\theta_1^{-2} + E_2\theta_2^{-2} + \ldots E_n\theta_n^{-2}] \qquad (3\text{–}8)$$

where L_v = the equivalent veiling luminance (footlamberts),
E = illumination on a plane through the center of the pupil, perpendicular to the line of sight, contributed by each glare source (footcandles), and
θ = angular displacement between the line of sight and each glare source (degrees).

Substituting the computed L_v in Equation 3–6 one can determine the apparent contrast of the object being viewed.

Following the same principles, the equivalent veiling luminance produced by the peripheral portion (excluding a 2-degree central area) of a uniform field has been computed[34] and was found to be about 7 per cent of the field luminance producing it. Consequently, a uniform surround equal to the luminance of the immediate background of an object cannot be objected to on the ground that it produces harmful stray light. The advantages far outweigh the very small loss in contrast resulting from the veiling luminance produced by the stray light.

Transient Adapation. Many basic studies of ocular physiology have shown that the performance of visual sensory functions is influenced by sudden changes in the level of luminance to which the eye is adapted. This implies that glances at different portions of the surround of a visual task will produce transient adaptive effects except when all luminances in the visual field equal the luminance of the task. The classical studies of transient adaptation do not provide directly useful information for illuminating engineering practice, however, since these studies involved fixing the size of the ocular pupils in order to isolate the sensory effects from effects due to changes in the pupillary apertures. The practical situation involves the compound effect of changes in the performance of visual sensory functions due to ocular physiology, together with changes in the pupil size which influence the level of illumination reaching the photoreceptors in the retina from a given luminance in the visual field. The practical situation will also usually involve complex patterns of luminance in the visual field and complex patterns of eye movements. Thus, a complete understanding of the transient adaptive effects due to realistic fields of non-uniform luminance must await the collection of a considerable body of new knowledge.

In recent years, a series of studies of the transient effect due to a simple step change in the luminance to which the eyes are exposed, with naturally-functioning ocular pupils, have been carried out.[7] The visual task involved recognizing letters first under conditions of steady adaptation to a single luminance and then under transient adaptive conditions of two types. In one situation, the eyes were adapted to a second luminance, then letter recognition was tested after reexposure to a first luminance which had previously been used for steady adaptation. In the second, the eyes were steadily adapted to a first luminance, then exposed momentarily to a second luminance, with letter recognition tested after reexposure to the first luminance. Results of these studies were as follows:

1. Transient adaptive effects occur immediately after sudden changes in luminance, their magnitude being dissipated rather quickly. For this reason, transient adaptive effects are to be studied primarily by measuring visual sensory performance at a fixed interval after an adaptive change. An interval of 0.3 second seems most satisfactory.

2. Transient adaptive effects can be described in terms of the increase in luminance contrast required to compensate for the effect by restoring visual sensory performance to the level achieved under steady adaptation to a single luminance level.

3. So long as luminances do not reach about 4000 footlamberts, transient adaptive effects depend only upon the ratio of luminances involved in the change, and not upon the absolute values of the luminances.

Further work will be required to characterize the transient adaptive effects which occur when the eyes are exposed to a series of different luminances as will usually be the case in a realistic environment when normal patterns of eye movements are used to glance around the visual field.

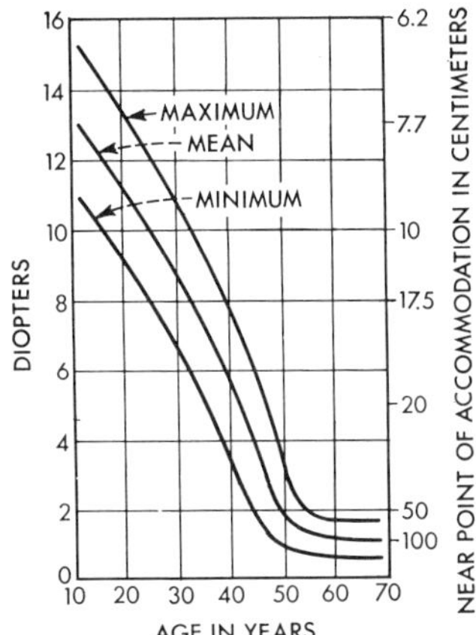

Fig. 3-18. The decrease of the amplitude of accommodation with age.

Effects of Age

If the characteristics of the eye and visual system of the average young adult are defined as normal, then all differences from this condition may be considered abnormalities. From this point of view, increasing age is accompanied by abnormalities in the physiology of the eye and visual system which reduce visual ability. Some of these abnormalities exist to some degree in members of the young adult population and become more pronounced with age. Others appear only with advancing age. Thus the population as a whole can be conceived of as possessing a distribution of abnormalities differing in kind and amount, all of which influence visual ability to some extent. These abnormalities include decrease in amplitude of accommodation, reduction in pupil size, decrease in rate and amount of dark and light adaptation, reduction in sensitivity especially at low luminance levels, loss of transmission of light due to increased opacity of the eye media, and degenerative changes in the various parts of the visual system including the retina.

Age has a marked effect upon the accommodative ability of the eyes. As a person becomes older the lens hardens and loses its elasticity and no longer can be changed to the more spherical shape that is needed to focus on near objects. This results in the decrease in the amplitude of accommodation[35] shown in Fig. 3-18. The right-hand scale indicates how the near point progressively recedes. When a person finds it necessary to wear bifocal lenses in order to be able to focus nearby and at a distance, he is said to be presbyopic.

Data,[36] plotted in Fig. 3-19, is illustrative of the change in pupil size with age as measured in the dark (upper curve) and at one footlambert (lower curve) for 222 observers of age 20 to 89 years. The mean difference in pupil size between ages 20 to 29 and 80 to 89 was 2.6 mm in the dark and 1.7 mm at a level of one footlambert. The reduction in pupillary area with increasing age is equivalent to approximately a 0.3 log unit reduction in luminance. This decrease in the amount of light reaching the retina for image-forming purposes is one of the underlying causes of loss in visual sensitivity with age, particularly at low luminance levels, but the results of many studies indicate that the magnitude of this

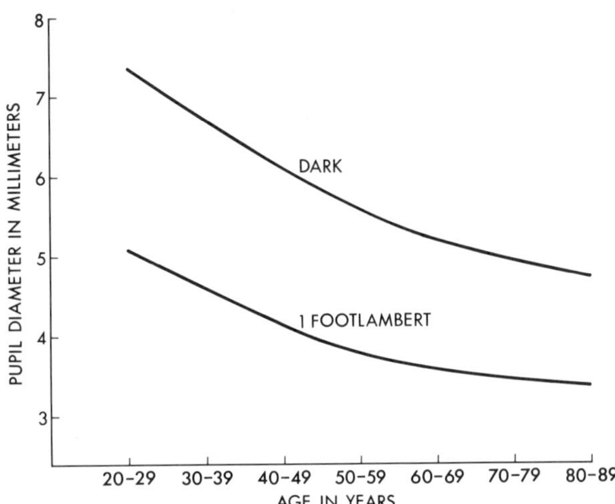

Fig. 3-19. The change in pupil diameter with age as measured in the dark and at one footlambert for 222 observers.[36]

effect is not sufficient to explain the demonstrated reduction in the dark adapted threshold as a function of age. One extensive study[37] has shown a marked reduction in sensitivity with age for 241 observers at all stages of dark adaptation from 59 seconds up to 40 minutes after exposure to light. The loss of sensitivity increased with increasing time in the dark. However, it has been demonstrated[38] that the magnitude of the decrease in dark adapted sensitivity with age depends markedly on the spectral composition of the light used in testing it, being greatest for violet test light such as that used by others.[37] It was also shown,[38] by comparative studies with observers of different ages whose lenses had been removed, that this greater decrement in the violet to blue spectral range is due to lens yellowing with age. Thus, the data of the others,[37] cannot be directly applied to practical problems since violet light is seldom used. It should be noted that lens yellowing also affects color discrimination at the blue end of the spectrum.

Another important loss in visual ability with age is the reduction in visual acuity even with the best possible refractive correction. This is illustrated in Fig. 3-20.[39] It is interesting to note the difference

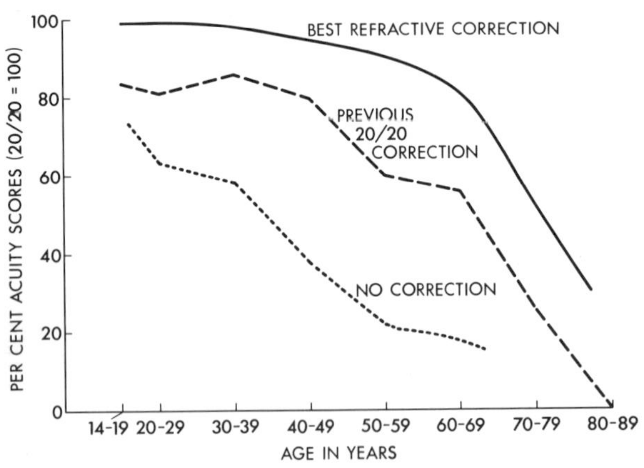

Fig. 3-20. The reduction in visual acuity with age.[39]

between the percentage of 696 observers of different ages achieving 20/20 visual acuity when corrected as well as possible, compared to the percentage who were actually corrected to 20/20 before testing. These latter data may be most relevant for practical application. The lower curve is based on a compilation of data from insurance policy physical examinations and vision tests of school children in four states, and shows how misleading statistics from the uncorrected eye can be.

There is also marked increase in sensitivity to disability glare as a function of age. There has been a beginning at quantifying this effect showing that the constant k in equation (3-7) varies threefold with increasing age.[40] Also, it has been found that there is a reduction in visual ability for 112 observers from age 10-80 as measured by the contrast required to see the break in a Landolt C presented at different distances from a glare source.[41] However, these results reflect a combination of the loss of contrast sensitivity with age *per se* with a loss of contrast in the retinal image due to the scattered light in the eye media introduced by the glare source. A study[42] has been conducted to find out if light scatter in the anterior eye media was responsible for the loss in contrast sensitivity with increasing age in the absence of a glare source. In 62 observers of age 10-80 it was found that the contrast required to see a break in a Landolt C had to be increased with increasing age. In the same observers, a four-fold increase in ocular light scatter in the anterior eye media was found as age increased from 10 to 80. However, since there was no correlation between the per cent contrast needed to resolve a Landolt C and the per cent light scatter measured by slit lamp photometry for different individuals, it was concluded that additional factors must be responsible for the reduction in contrast sensitivity with age.

Since contrast has been adopted as the basic metric for evaluating visual performance in the IES method of prescribing illumination (see page 3-14), the loss of contrast sensitivity with age is of special interest. A study[43] of contrast sensitivity as a function of age deliberately designed to permit direct comparison with the standard performance curve used in the IES method to represent the average young adult has been conducted. For this purpose a 4-minute disk exposed for ⅕ second on an extended uniform background was used as the task. One hundred workers with no known pathology were tested at two luminance levels, one and 100 footlamberts and it was found that the contrast required to see the disk increased with increasing age from 18 to 63 years. Recent research[44] has been reported on a long-range study with the same goal where a contrast sensitivity function of luminance was obtained using a standard 4-minute disk exposed for ⅕ second. Uniform background luminances were studied over the range from $10^{2.7}$ to 10^{-3} footlambert for 156 observers from age 20 to 70, all of whom were free from ocular pathology discernible under the usual clinical examination. All observers had visual acuity of not less than 20/30 when corrected. The results represent a conservative estimate of the losses in contrast sensitivity to be expected in a real population, especially in the oldest age groups. The results for 10-year arbitrary age spans can be seen in Fig. 3-21. The solid curve represents data for the 20 to 30 year age group obtained with the psychophysical method of adjustment used throughout the study (see page 3-14). The points in the figure are the average data for each 10-year span. It can be seen that the contrast required to perform the task not only increases with increasing age but also increases relatively more at the lower luminances. Evidence presented indicates that it is reasonably adequate to use a single contrast multiplier to describe the increase in contrast required for each older age group to perform as well as the 20 to 30 year olds, provided that the luminances are restricted to 0.29 footlamberts and above and the age groups to those under 60 years.

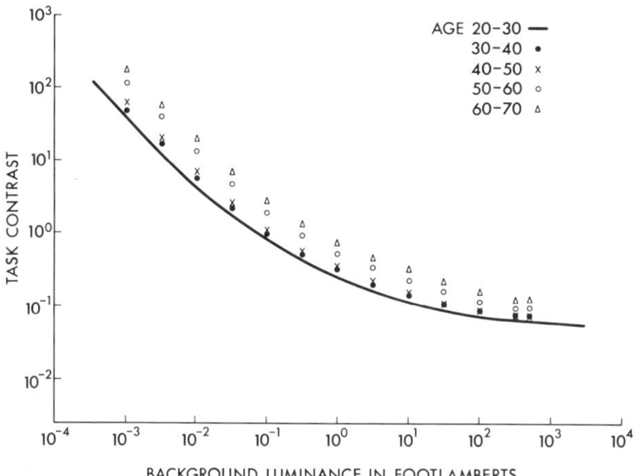

Fig. 3-21. Effect of age on required task contrast for task performance.[44]

Effects of Shadows

The direction of illumination is especially important when viewing three dimensional objects. As is illustrated in Figs. 3-22 and 3-23, shadows can aid or hinder the seeing of details.[45, 46] In the case of curved and faceted surfaces which are polished or semi-polished, the direction of the lighting is important in controlling highlights. Some shadow contributes to the identification of form.

Even in the case of apparently flat surfaces, the direction of illumination may be important if gloss

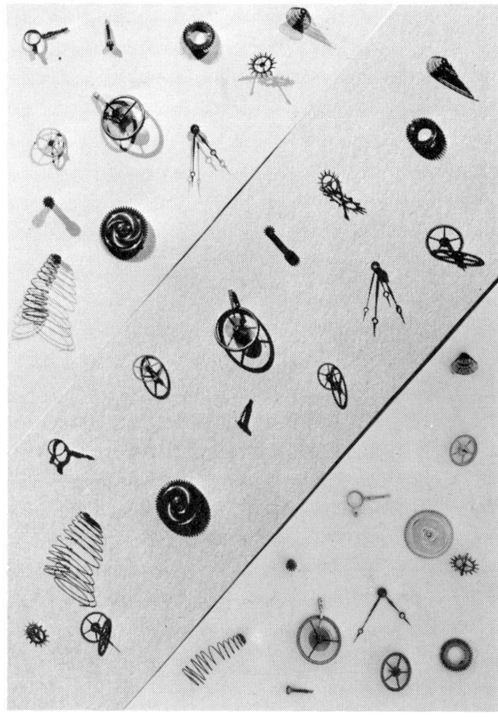

Fig. 3-23. Multiple shadows (upper left) are confusing; single shadows (center) may confuse but can help; diffused light (lower right) erases the shadows.

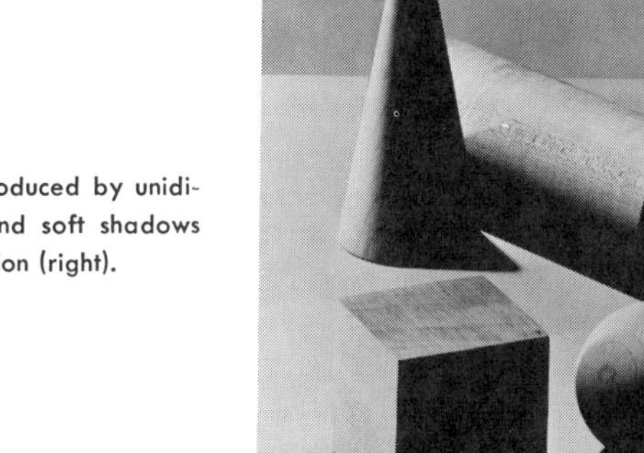

Fig. 3-22. Harsh shadows produced by unidirectional illumination (left) and soft shadows produced by diffuse illumination (right).

is present. The gloss can produce veiling reflections which alter the apparent contrast in a way similar to what stray light does to retinal images in the eye. An example of this is printing or writing on the usual types of paper. For a more detailed discussion of veiling reflections, see page 3-28.

IES METHOD FOR PRESCRIBING ILLUMINATION

The goal of illumination prescription is to insure adequate performance of tasks with significant visual components, without introducing discomfort. Accordingly, the IES method for prescribing illumination is based upon both visual performance and visual comfort criteria.

Visual Performance Criteria

The term "visual performance" is a short-hand phrase to describe the visual components of total human performance of tasks having significant visual components. Visual performance may be measured in the full context of human performance or abstracted from total human performance to a variable extent. The most abstracted form of visual performance may be designated *visual performance potential.* This quantity represents the visibility of a task during a series of controlled ocular fixations. Visual performance can never exceed the level of visual performance potential defined by task visibility but it can fall short of the potential level. *Visual performance capability* represents the degree to which visual work may be accomplished with a task of known visibility under optimal conditions of observer training, motivation, and fatigue. Visual performance capability will depend upon the conditions of observation and response defined by the visual work situation. The actual visual performance to be expected in a real visual work situation will depend upon the visual performance capability and the extent to which the worker operates under optimal conditions of training, motivation, and fatigue.

The visual performance criterion used in the IES method for prescribing illumination is expressed in terms of visual performance potential, measured by task visibility, since knowledge of this aspect of visual performance is most complete. Indications are given as to corresponding levels of visual performance capability. Worker training, motivation, and fatigue are not considered.

Task visibility will depend upon two classes of variables, the intrinsic characteristics of the task and aspects of the task altered by characteristics of the luminous environment. Intrinsic task characteristics will include size and configuration of task detail, and spatial patterns of luminous reflectance and chromaticity of task detail and task background. Physical characteristics of the luminous environment which will influence task visibility may be listed as follows:

1. Level of illumination, which will interact with task background luminance factors to determine task background luminance.
2. Spatial pattern of illumination reaching the task, plus the plane of polarization and the spectral composition of the illumination, which will interact with task luminance factors and spectrophotometric characteristics to determine luminance and chromatic contrast of the task detail with respect to the task background.
3. Spatial pattern of luminances within the environment relative to task background luminance, which will determine the disability glare and transient adaptation effects.

In assessing visual performance potential the IES method takes quantitative account of task background luminance, the luminance contrast of the task, and the disability glare effect. No account is taken of the chromatic contrast of the task, since insufficient data are available at this time. Simplified indices of the transient adaptation effect are used to provide guidance as to the quantitative aspects of this effect.

Selection of a Criterion Level of Visual Performance Potential

The visual performance criterion of the IES method is described in terms of Visibility Level (VL). Different values of VL are defined relative to the Visibility Reference Function shown as the solid curve in Fig. 3-24. This function represents the luminance contrast required at different levels of task background luminance to achieve visibility threshold for the visibility reference task consisting of a 4-minute luminous disk exposed for ⅕ second. The data points were obtained on 68 normal observers in the 20-30 year age group.[44] The observers adjusted the contrast to the borderline between visi-

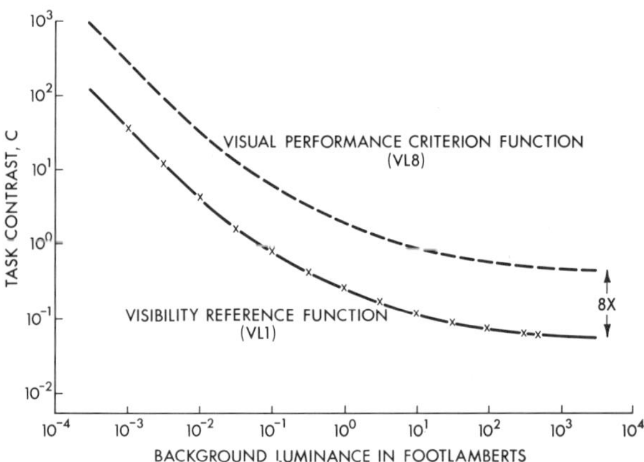

Fig. 3-24. Plot of the Visibility Reference Function (solid curve) representing task contrast required at different levels of task background luminance to achieve threshold visibility for a 4-minute luminous disk exposed for ⅕ second. The dashed curve represents the Visual Performance Criterion Function (values of the solid curve multiplied by 8).

bility and invisibility. The solid curve was derived from data[48] representing nearly 500,000 observations by 35 normal observers in the 20 to 30 year age group. The curve has been fitted to the data points by adjustment up the vertical scale of task contrast. Since the scale is logarithmic, this adjustment amounts to multiplying each value on the curve by a constant. The constant used to adjust the curve to fit the data was 2.51.

In the latter experiments,[48] observers were required to prove they could detect the presence of the target by correctly identifying the time interval in a sequence of four during which the target appeared. The data represents a compilation of data from a total of eight individual studies, including the studies[19, 49] reported in earlier editions of this Handbook.[50, 51] The psychophysical method used in these studies is a laboratory technique of high precision, well suited for determination of the shape of the curve relating threshold contrast to task background luminance for the visibility reference task. However, the psychophysical criterion used in more recent studies[44] corresponds more nearly to a common-sense criterion of the visibility threshold. Hence these data are considered to represent a more reasonable reference base for use in defining visual performance criteria. The agreement between the data points and the solid curve in Fig. 3–24 indicates the adequacy of the curve for use as the standard Visibility Reference Function.

It may be of interest to note that the factor 2.51 used to adjust the curve to the data provides a direct assessment of the factor for "common sense" seeing described in earlier editions of this Handbook.[50, 51] The value taken in these earlier references was 2.4 which agrees well with the value 2.51 derived from the more recent and more complete studies.

The Visibility Level corresponding to the Visibility Reference Function is designated VL1. The entire curve shown in Fig. 3–24 bears this designation, representing its use as a reference curve. Other Visibility Levels are represented by curves constructed parallel to the reference curve by displacement up along the vertical scale. The Visibility Level numeral identifies the contrast multiplier used to construct a given curve. For example, VL2 represents a curve constructed by moving up the vertical scale of the figure by a factor of 2. All points in the VL2 curve represent instances in which task visibility exceeds threshold visibility by a factor of 2.

In selecting a visual performance criterion for the IES method an analysis has been made of the factors which influence visual performance potential under conditions of realistic visual work. This procedure is very similar to the selection of a Field Factor described in the earlier Handbooks.[50, 51] The new data represent visibility thresholds obtained with a commonsense psychophysical criterion under conditions in which the observers knew where and when the task would appear, and in which they were not required to undertake ocular search or scanning movements. It has been shown[49] that a contrast multiplier of 2.78 is required to compensate for the differences between static and dynamic presentations of the same task. It has also been shown[19] that a contrast multiplier of 1.5 is required to compensate for the difference between not knowing and knowing where and when a task will appear. It has been shown[44] that a contrast multiplier of 1.9 is required to correct from the visibility threshold obtained by the common-sense psychophysical method used to a common-sense level of 99 per cent certainty that the task was barely visible. These three factors may be multiplied together to obtain a suitable visual performance criterion,[19] expressed in the language of visual performance potential as VL8. The dashed curve in Fig. 3–24 represents the visual performance criterion corresponding to VL8. It will be found that this visual performance criterion is virtually identical to that used in earlier editions of this Handbook.[50, 51] The dashed curve in Fig. 3–24 is designated the Visual Performance Criterion Function. It represents values of luminance contrast of the visibility reference task required at different levels of task background luminance to satisfy the criterion value of VL8.

It is possible to suggest the approximate level of visual performance capability to be expected from the IES visual performance criterion level of VL8 by analyzing data obtained under different conditions of visual work.

Visual performance capability is characterized in terms of realistic measures of visual work such as speed and accuracy. Visual performance potential is described in terms of Visibility Level. Fig. 3–25 shows data from four studies of visual work situations. In each case, the vertical scale presents the measure of visual work used by each of the investigators, representing visual performance capability. The measures used in the different studies represent either speed alone,[52] accuracy alone[49, 53, 54] or a combination of speed and accuracy.[55] The horizontal scales each present values of the Visibility Level, representing visual performance potential.

Values of the Visibility Level were obtained with a contrast-reducing visibility meter in the following way. The tasks used by the different investigators of visual work situations were reconstructed, and the visibility of each task was found by reducing it to threshold using the variable contrast transmittance of the visibility meter. Then, a value of the luminance contrast of the standard visibility reference task was found which was at visibility threshold at the same setting of the visibility meter when background luminances were the same for the work task and the reference task. This value of luminance contrast is designated the Equivalent Contrast, $\tilde{C}$, and is used as a numerical description of the relative visibility of a task. The value of VL equals the value of the Equivalent Contrast, $\tilde{C}$, for a task, divided by the value of the Visibility Reference Function at the value of task background luminance used in the visual work study. To per-

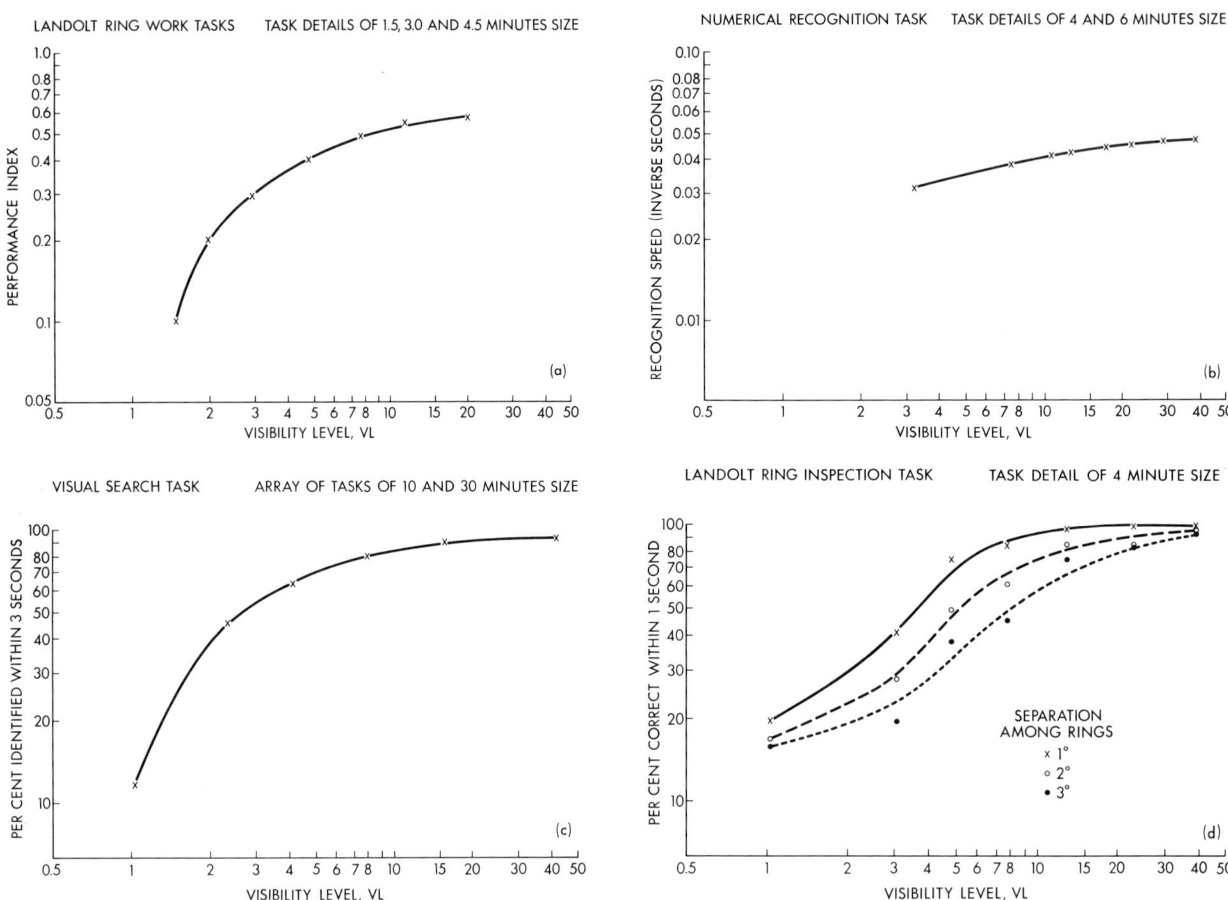

Fig. 3–25. Curves from four studies showing that visual performance capabilities increase as visual performance potential increases.

form such calculations, numerical values of the Visibility Reference Function may be read from Fig. 3–24, or computed in a manner to be described below.

The four graphs of Fig. 3–25 illustrate that visual performance capability increases as visual performance potential increases, the curve being generally steep at first and then increasingly flat. There is a general similarity in the shapes of the four curves in Fig. 3–25, which have been plotted on double logarithmic grids to facilitate this comparison. However, there are clearly differences in the curves representing the results of the four studies, suggesting that visual performance capability is related to visual performance potential in a manner which depends upon the precise conditions of observation and response defined by a given work situation. This point is emphasized by the diverse curves in Fig. 3–25(d) which were obtained with the same task and the same observers under only slightly different work conditions.

It has been suggested[56] that all visual work data may be considered to fall upon a single curve when relative measures of visual performance capability are plotted against values of Visibility Level, and that the average curve for all available visual work data may be used to establish a visual performance criterion expressed in terms of the percentage of maximum visual performance capability. The simplicity of this criterion is appealing. However, various data presented in Fig. 3–25(d) show that such a simplification of the problem of describing visual performances capability is not warranted. For this reason, the IES method does not assign a fixed value of visual performance capability to correspond to the criterion level of visual performance potential, preferring to exhibit the data of Fig. 3–25 to illustrate the approximate levels to be expected under different specified conditions of visual work. Further study needs to be made of the shape of the curve relating visual performance capability to visual performance potential as a function of such variables as extent of search and patterns of eye movements, and the extent of response limitations. It may ultimately be possible to select different levels of visual performance capability for different tasks or classes of tasks in place of a fixed value of visual performance potential such as the criterion presently used in the IES method.

Determination of Values of Required Illumination

The visual data presented in both Figs. 3–24 and 3–25 were obtained under what is designated as reference lighting conditions, characterized by diffuse illumination of the task and uniform luminance of the surround equal in magnitude to the task background luminance. Thus, the IES visual performance criterion, VL8, refers to this reference lighting condition. The dashed curve in Fig. 3-24 may be used to determine values of required illumination (E_r) for different tasks to meet the criterion level of visual performance potential as follows. A value of equivalent contrast, ($\tilde{C}$) for each task is determined using a contrast-reducing visibility meter. As noted above, $\tilde{C}$ equals the luminance contrast of the 4-minute disk standard visibility reference task found equal in visibility to the task at visibility threshold. To determine values of E_r, each task of interest is placed under the diffuse illumination of a photometric sphere, and viewed in a uniform surround of luminance matched to the task background luminance.

The concept of visual performance potential depends upon assessment of task visibility during a controlled exposure so short that eye movements cannot occur. The use of ⅕ second exposures is recommended, since this exposure duration was involved in collection of the data for the Visibility Reference Function shown in Fig. 3-24. The ⅕ second exposures are to be used for each task of interest and for the visibility reference task. In many cases, it will be found equivalent to assess values of $\tilde{C}$ with uncontrolled long exposures of both the tasks of interest and the visibility reference task.

Contrast reducing visibility meters have been developed in recent years.[19, 57-62] In principle, any of these instruments may be used to establish values of $\tilde{C}$ under the recommended conditions. Currently, only the Blackwell Visual Task Evaluator (VTE) Model No. 3 shown in Fig. 3-26 permits use of ⅕ second exposures. This instrument is more sophisticated in comparison with the models described in the earlier Handbooks.[50, 51] It contains a substitution field whose luminance is matched to the task background luminance. This field appears whenever the task is not in view. The instrument also features the use of two pairs of light attenuating wedges which are mechanically linked with an antilogarithmic cam. One pair of wedges reduces focused light from the task as the other increases unfocused light from a luminance veil produced within the instrument. A single control operates the two pairs of wedges through the non-linear cam. The result is that the total luminance of the photometric field produced by both focused and unfocused light remains very nearly constant as contrast is reduced by more than one thousand to one. This feature of the device is desirable to maintain close control over the visual adaptation of the instrument operator.

In determining values of $\tilde{C}$ for tasks of interest, care must be taken to select a suitable informational requirement for each task. Regardless of what requirement is used for a given task, the informational requirement for the standard visibility reference task is always detection of presence. It will also be important to determine values of $\tilde{C}$ for each viewing angle of interest, since it has shown[63] that the value of $\tilde{C}$ depends upon complex interplay between the viewing angle and the informational requirements involved for a given task.

Values of $\tilde{C}$ may be used to establish values of required luminance (L_r) by reading from the dashed curve in Fig. 3-24 or by computation following a method to be described below. Values of required illumination, E_r equal values of L_r divided by ρ, the reflectance of the task background under the reference lighting conditions of the photometric sphere.

Determination of Values of the Design Illumination in Real Luminous Environments

Allowance for Veiling Reflections. The spatial pattern and the plane of polarization of illumination striking the task affect task visibility due to the veiling reflection effect. The effect comes about from light rays which are reflected from the surface of tasks rather than absorbed and re-radiated diffusely in the direction of the observer's eye. The result is a reflected luminance veil which reduces task contrast and visibility. Reference lighting conditions involve the diffuse illumination of a photometric sphere. Thus, the extent to which veiling reflections

Fig. 3-26. The Blackwell Visual Task Evaluator. The view on the left shows the instrument with its power supply in the background. The view on the right shows the other side of the instrument with a calibrator unit installed.

exist is referenced to this condition. The result is that some luminous environments produce less veiling reflections than the reference condition while others produce more.

The veiling reflection effect is described by a measure of task visibility in a real luminous environment relative to task visibility in the photometric sphere. The measure used is the Contrast Rendition Factor (CRF). CRF is basically defined by measurements of visibility obtained with a contrast-reducing visibility meter. However, it has been shown[63] that physical measurements of luminance contrast will yield equivalent values under many conditions.

Task difficulty in a real luminous environment producing non-diffuse illumination is measured by the Effective $\tilde{C}$, $\tilde{C}_e = \tilde{C} \times$ CRF. Once a value of $\tilde{C}_e$ is known, it may be used in place of the original value of $\tilde{C}$ measured in a photometric sphere in deriving values of design luminance (L_d) and design illumination (E_d) from the standard curve of the Visual Performance Criteria Function.

Fig. 3–27 illustrates the effect of CRF upon L_d and E_d with respect to L_r and E_r. Because of the relative flatness of the Visual Performance Criterion Function, a small percentage difference in CRF from unity produces a large percentage change in either L or E. This emphasizes the great importance of illumination quality as opposed to quantity even when a visual performance criterion is used.

Allowance for Disability Glare. Luminances in the surround near the task will influence task visibility due to the disability glare effect.[64, 65] It is now generally agreed that the disability glare effect is due to stray light produced within the eye. In any case, disability glare operates as though each glare source produced a light veil over the image of the task on the retina. As shown in equations (3–4) through (3–6), the equivalent veiling luminance, L_v, both increases task background luminance and

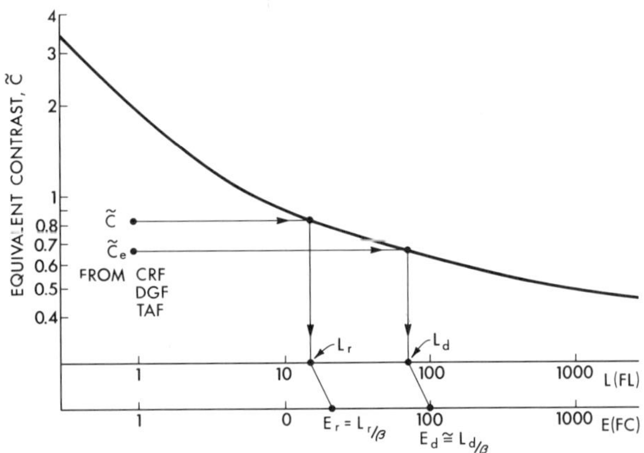

Fig. 3–27. Illustration of the determination of design illumination, E_d. The known task with a $\tilde{C}$ as shown requires an L_r and E_r as shown (under reference lighting conditions—sphere illumination). Under the real luminous environment values of CRF, DGF, and TAF have been used to determine $\tilde{C}_e$ (where $\tilde{C}_e = \tilde{C} \times$ CRF $\times$ DGF $\times$ TAF). $\tilde{C}_e$ is then used to determine L_d and finally E_d.

decreases task contrast. Since reference lighting conditions consist of a field of luminance equal to task background luminance, allowance for the disability glare effect in a real environment must take account of the difference in the effect between the real environment and the reference sphere. Using the constants of 10 for k and 2 for n in Equation (3-8), it was found that $L_v = 0.074\,L$ under reference lighting conditions. Thus, all values of L_v are referred to this value in taking account of the disability glare effect of a real environment. The value of task background luminance in a real luminous environment, Equivalent L, equals $(L + L_v)/1.074$. The Disability Glare Factor (DGF) may be computed by allowing first for the change in task background luminance, then for the change in task contrast.[66] The first effect is computed by determining values of $\tilde{C}$ from the Visual Performance Criterion Function, first at L and then at Equivalent L. The value of $\tilde{C}$ at L is divided by the value at Equivalent L to represent the reduction in equivalent contrast required for equal visual performance potential due to the increase in task background luminance. The second effect is computed by dividing L by Equivalent L, representing the loss in task contrast due to the addition of a luminous veil. The value of DGF equals the product of the two quantities. At the values of task background luminance encountered in interior illumination, the second term is large in comparison to the first, but the two terms are of more nearly equal magnitude at the levels of roadway lighting.

Allowance for the disability glare effect is made by computing Effective $\tilde{C}$, $\tilde{C}_e = \tilde{C} \times$ CRF $\times$ DGF. The value of $\tilde{C}_e$ is used in place of the original value of $\tilde{C}$ in deriving values of L_d and E_d for a given task when it is viewed in a particular luminous environment, as shown in Fig. 3–27. It is necessary to indicate by notation whether values of $\tilde{C}$, L_d, and E_d are based upon CRF $\times$ DGF or upon CRF alone.

In many real environments involving only ceiling-mounted luminaires in which tasks are located on horizontal surfaces, DGF will equal or exceed unity. The value of DGF will become significantly less than unity for tasks located on vertical surfaces, particularly if there are windows exposing a view of outside areas of high luminance.

The value of L_v may be computed from equation (3-8). Alternatively, it may be measured with an optical analog device.[67]

Estimation of the Transient Adaptation Effect. Allowance for the disability glare effect will express the entire influence of the pattern of luminances in the environment so long as the task is steadily fixated. However, whenever the observer glances about the environment, his foveal retina will be exposed to whatever luminances exist at points in the environment which are fixated. Unless all luminances equal the task background luminance, such eye movements will produce transient adaptive effects, resulting in a reduction in task visibility, as described on page 3–10. Although the effects have been shown to be sizable, they are of brief duration.[7]

What is needed is a method for describing the loss in sensitivity to be expected from realistic patterns of eye movements involving continuously varying luminances. Current state of our knowledge does not permit this precise an allowance for the transient adaptive effects to be expected in different real environments. However, data now exist on the basis of which two simple types of transient adaptive effects may be evaluated. In each case, the transient adaptive effects are expressed in terms of a Transient Adaptation Factor (TAF). The value of TAF represents a measure of task visibility under transient conditions divided by task visibility under conditions of steady fixation and equilibrium adaptation.

The first estimate of TAF represents the readaptation process involved when the observer has two or more visual tasks to perform in alternation. Fig. 3–28 is based upon experimental data relating the value of TAF to the ratio of luminances in the two tasks (R), the larger always being divided by the smaller. The loss in task visibility was assessed 0.3 seconds after a change from one task to the other.[7] The value of this estimate of TAF, designated TAF_1, may be computed from the equation $TAF_1 = 1 - .315 \log R$, so long as no luminance is in excess of 372 footlamberts.

The second estimate of TAF represents the glance adaptation process involved when the observer looks momentarily at a luminance different from the task background luminance. Values of TAF obtained[7] under these conditions are shown in Fig. 3–29. In this case, the observer was assumed to look for a full second at the different luminance, and task visibility was measured 0.3 second after re-fixation of the visual task. The value of this estimate of TAF, designated TAF_2, may be calculated from the equation $TAF_2 = 1.009 - .0092R$, so long as the luminance ratio does not exceed 20 as will usually be the case with interior lighting.

It is possible to calculate values of $\tilde{C}_e = \tilde{C} \times CRF \times DGF \times TAF$ using values of TAF_1 or TAF_2, and to determine corresponding values of L_d and E_d. However, it must be emphasized that these values characterize task visibility only for a brief period following an eye movement producing a transient adaptive effect, and do not apply at all to the considerably greater periods during which the task is fixated steadily. The values representing conditions of transient adaptation should be given weight in accordance with the frequency and significance of eye movements in comparison with steady fixations of the task.

Procedures for Achieving Values of Design Illumination. The basic procedure involved in designing illumination to satisfy the IES visual performance criterion involves iteration. It is obvious from the construction of Fig. 3–27 that an installation will provide the criterion level of visual performance potential only if the design illumination, E_d, is achieved without any alteration in the value of ρ, CRF, DGF and perhaps TAF used in its determination. There will almost always be some error involved, since the luminance factor, β, will rarely be equal to the reflectance, ρ, obtained under conditions of diffuse illumination in the photometric sphere. Values of CRF, DGF, and TAF will usually vary with different selections of layout or lighting materials which may be used to produce E_d.

Pre-determination programs will ultimately make it possible for the designer to compute values of E, β, CRF, DGF, and TAF for different lighting installations under consideration. These may be used to pre-determine values of the Equivalent Sphere Illumination, E_s,* as described below. The value of E_s may be compared with the value of the Required Illumination, E_r. Design must continue until the value of E_s equals or exceeds E_r.

Experience with pre-determination programs and with measurements of E_s in actual installations may make possible an approximate design method based upon the Lighting Effectiveness Factor, $LEF = E_s/E$ where E_s is the Equivalent Sphere Illumination and E is the actual task illumination. The value of LEF depends upon the values of CRF, DGF, and TAF if used, and the level of task background lumi-

* Also abbreviated ESI.

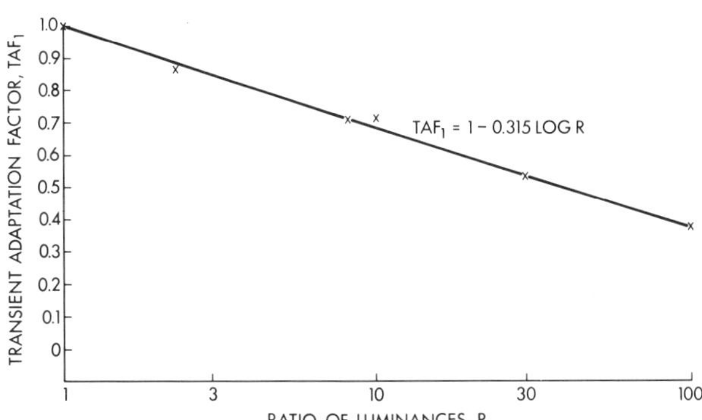

Fig. 3–28. Plot of experimental data relating TAF to the ratio R between task luminances. TAF represents the readaptation process involved when an observer has two or more visual tasks to perform in alternation. Luminances range from 0.02 to 372 footlamberts.

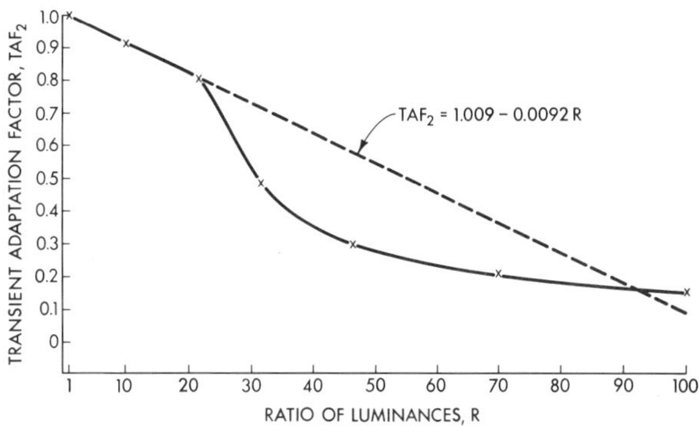

Fig. 3–29. Plot of experimental data relating TAF to the ratio R between task background luminance and the luminance of another surface viewed, when TAF represents the glance adaptation process. Luminances range from 4 to 4000 footlamberts.

nance which determines the slope of the Visual Performance Criterion Function at which these factors are assessed. After experience, the designer may develop a catalog of approximate values of LEF for different types of lighting installations. Values of E_d may be estimated from E_r/LEF.

Evaluation of Equivalent Sphere Illumination

As noted above, the extent to which a given lighting installation will satisfy the IES visual performance criterion requires calculation of values of the Equivalent Sphere Illumination, E_s. Calculation of E_s involves describing the visual performance potential of a real environment of known actual illumination in terms of the illumination under the reference conditions of a photometric sphere providing equivalent performance potential. The value of E_s is computed from the actual task illumination after allowance for the veiling reflection and disability glare effects, and the transient adaptive effects if desired. This is accomplished in terms of a standard curve relating contrast sensitivity to illumination.

Standard RCS Function of Luminance. It has been traditional to describe data involving contrast discrimination either in terms of the contrast required for detection at different levels of luminance or in terms of the corresponding level of contrast sensitivity, defined as the inverse of the contrast (see, for example, Figs. 3-9 and 3-10). It also has been traditional to describe data involving spatial resolution obtained at different levels of luminance in terms of either the angular size of the detail, or its inverse, the visual acuity. However, it has been shown[68] that data from both types of visual task can be brought into agreement by means of a broadened conception of contrast sensitivity. It also has been suggested[68] that data for any task, whether believed to involve contrast discrimination or spatial resolution, may be described in terms of the inverse of the luminance contrast required for a given level of performance, and that this quantity is to be designated Contrast Sensitivity. Relative Contrast Sensitivity (RCS) is simply the value of Contrast Sensitivity expressed as a percentage of the value obtained under a very high level of diffuse task illumination. It was shown that contrast discrimination and visual acuity data agreed when expressed in terms of RCS, suggesting that a standard function can be developed describing the relation between RCS and task luminance under reference lighting conditions. The solid curve of Fig. 3-24 is suitable for this purpose and its reciprocal is replotted as the RCS function of luminance in Fig. 3-30. Since logarithmic scales are used in Figs. 3-24 and 3-30, the curves have the same shape. Fig. 3-30 may be produced by inverting Fig. 3-24 and shifting the curve along the vertical scale so that the lowest value of threshold contrast in Fig. 3-24 becomes the 100 per cent RCS in Fig. 3-30. Sample values of RCS are presented in Fig. 3-31 covering the entire range of luminances. More complete values are presented in Fig. 3-32 for the range of luminances of interest in connection with interior lighting.

Values of the Visibility Reference Function, C_1, at various values of luminance may be computed from the relation $C_1 = 5.74/\text{RCS}$ where RCS is read from either Figs. 3-31 or 3-32. Values of L_r corresponding to values of $\tilde{C}$ may be computed as follows. First compute a value of RCS_r, the required value of RCS for the task to meet the IES visual performance criterion. For the Visual Performance Criterion Function, $\text{RCS} = 45.9/\tilde{C}$. Then, read a value of L_r corresponding to each value of RCS_r from either Figs. 3-31 or 3-32.

Determination of RCS for Different Lighting Installations. Values of RCS may be read from Figs. 3-30, 3-31, or 3-32 corresponding to any value of task background luminance (L_t) obtained in a lighting installation. Values of L_t may be measured with a non-polarizing luminance photometer, or calculated from the relation $L_t = E_t\beta$ where E_t is

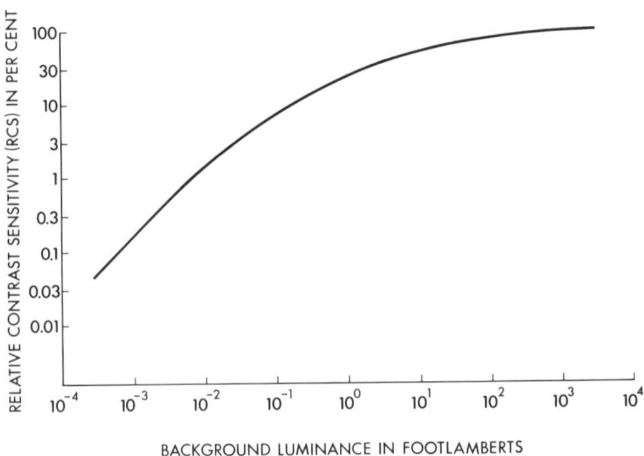

Fig. 3-30. Standard RCS function of luminance.

Fig. 3-31. Complete Range of Values of Relative Contrast Sensitivity, RCS, as a Function of Task Luminance L

RCS (per cent)	Task Luminance (footlamberts)
100.	2,920.
93.1	875.
83.3	292.
72.4	87.5
62.2	29.2
49.8	8.75
36.2	2.92
22.4	0.875
13.5	0.292
6.9	0.0875
3.3	0.0292
1.3	0.00875
0.46	0.00292
0.14	0.000875
0.047	0.000292

Fig. 3-32. Tabular Values of Relative Contrast Sensitivity, RCS, in Per Cent as a Function of Luminance, L, in Footlamberts

RCS	L	RCS	L	RCS	L	RCS	L	RCS	L
100.	2,920.	93.7	944.	87.4	461.	81.1	229.	74.8	115.9
99.9	2,767.	93.6	930.	87.3	456.	81.0	226.	74.7	114.7
99.8	2,606.	93.5	920.	87.2	452.	80.9	224.	74.6	113.5
99.7	2,550.	93.4	910.	87.1	447.	80.8	222.	74.5	112.3
99.6	2,458.	93.3	900.	87.0	442.	80.7	220.	74.4	111.0
99.5	2,390.	93.2	888.	86.9	437.	80.6	218.	74.3	109.8
99.4	2,327.	93.1	878.	86.8	433.	80.5	215.	74.2	108.4
99.3	2,264.	93.0	865.	86.7	428.	80.4	213.	74.1	107.2
99.2	2,206.	92.9	856.	86.6	423.	80.3	210.	74.0	106.0
99.1	2,153.	92.8	844.	86.5	419.	80.2	208.	73.9	105.0
99.0	2,110.	92.7	834.	86.4	414.	80.1	206.	73.8	103.9
98.9	2,062.	92.6	824.	86.3	410.	80.0	204.	73.7	102.5
98.8	2,022.	92.5	815.	86.2	405.	79.9	201.	73.6	101.5
98.7	1,979.	92.4	805.	86.1	401.	79.8	199.0	73.5	100.2
98.6	1,938.	92.3	795.	86.0	397.	79.7	196.3	73.4	99.0
98.5	1,900.	92.2	785.	85.9	392.	79.6	194.2	73.3	97.8
98.4	1,868.	92.1	776.	85.8	388.	79.5	192.2	73.2	96.6
98.3	1,835.	92.0	767.	85.7	384.	79.4	190.0	73.1	95.4
98.2	1,810.	91.9	758.	85.6	379.	79.3	188.0	73.0	94.2
98.1	1,780.	91.8	749.	85.5	375.	79.2	186.0	72.9	93.0
98.0	1,750.	91.7	741.	85.4	371.	79.1	183.5	72.8	92.0
97.9	1,721.	91.6	732.	85.3	367.	79.0	181.5	72.7	91.1
97.8	1,692.	91.5	724.	85.2	363.	78.9	179.5	72.6	90.0
97.7	1,664.	91.4	717.	85.1	358.	78.8	177.5	72.5	88.9
97.6	1,635.	91.3	708.	85.0	354.	78.7	175.5	72.4	87.8
97.5	1,608.	91.2	700.	84.9	350.	78.6	173.6	72.3	86.7
97.4	1,581.	91.1	692.	84.8	346.	78.5	171.5	72.2	85.8
97.3	1,559.	91.0	682.	84.7	342.	78.4	169.8	72.1	84.8
97.2	1,534.	90.9	676.	84.6	339.	78.3	167.8	72.0	83.7
97.1	1,510.	90.8	670.	84.5	335.	78.2	166.1	71.9	82.8
97.0	1,484.	90.7	662.	84.4	330.	78.1	164.6	71.8	82.2
96.9	1,460.	90.6	655.	84.3	327.	78.0	162.9	71.7	81.3
96.8	1,440.	90.5	648.	84.2	323.	77.9	161.0	71.6	80.5
96.7	1,420.	90.4	640.	84.1	320.	77.8	159.0	71.5	79.5
96.6	1,400.	90.3	634.	84.0	316.	77.7	157.3	71.4	78.7
96.5	1,380.	90.2	626.	83.9	312.	77.6	155.9	71.3	77.8
96.4	1,363.	90.1	620.	83.8	309.	77.5	154.1	71.2	77.0
96.3	1,340.	90.0	614.	83.7	306.	77.4	152.4	71.1	76.2
96.2	1,320.	89.9	608.	83.6	302.	77.3	151.0	71.0	75.4
96.1	1,300.	89.8	602.	83.5	299.	77.2	149.5	70.9	74.6
96.0	1,280.	89.7	596.	83.4	295.	77.1	148.0	70.8	74.0
95.9	1,262.	89.6	590.	83.3	292.	77.0	146.5	70.7	73.2
95.8	1,245.	89.5	584.	83.2	289.	76.9	145.1	70.6	72.4
95.7	1,230.	89.4	577.	83.1	286.	76.8	143.6	70.5	71.8
95.6	1,212.	89.3	571.	83.0	283.	76.7	142.1	70.4	71.1
95.5	1,196.	89.2	565.	82.9	280.	76.6	140.7	70.3	70.4
95.4	1,179.	89.1	559.	82.8	276.	76.5	139.3	70.2	69.8
95.3	1,165.	89.0	552.	82.7	274.	76.4	137.8	70.1	69.1
95.2	1,149.	88.9	546.	82.6	271.	76.3	136.4	70.0	68.4
95.1	1,134.	88.8	540.	82.5	268.	76.2	134.9	69.9	67.7
95.0	1,118.	88.7	534.	82.4	265.	76.1	133.4	69.8	67.0
94.9	1,105.	88.6	528.	82.3	262.	76.0	132.0	69.7	66.2
94.8	1,091.	88.5	522.	82.2	259.	75.9	130.5	69.6	65.6
94.7	1,078.	88.4	516.	82.1	256.	75.8	129.0	69.5	65.0
94.6	1,063.	88.3	510.	82.0	253.	75.7	127.5	69.4	64.3
94.5	1,048.	88.2	504.	81.9	250.	75.6	126.1	69.3	63.7
94.4	1,034.	88.1	500.	81.8	248.	75.5	125.0	69.2	63.0
94.3	1,019.	88.0	493.	81.7	245.	75.4	123.5	69.1	62.4
94.2	1,007.	87.9	487.	81.6	242.	75.3	122.3	69.0	61.8
94.1	995.	87.8	482.	81.5	240.	75.2	120.9	68.9	61.1
94.0	982.	87.7	477.	81.4	239.	75.1	119.6	68.8	60.4
93.9	970.	87.6	471.	81.3	235.	75.0	118.5	68.7	59.8
93.8	956.	87.5	467.	81.2	232.	74.9	117.0	68.6	59.2

Fig. 3-32. Continued

RCS	L	RCS	L	RCS	L	RCS	L	RCS	L
68.5	58.5	62.0	28.5	55.5	14.88	49.0	8.23	42.6	4.80
68.4	57.9	61.9	28.2	55.4	14.73	48.9	8.17	42.5	4.76
68.3	57.2	61.8	28.0	55.3	14.58	48.8	8.11	42.4	4.72
68.2	56.6	61.7	27.7	55.2	14.42	48.7	8.03	42.3	4.68
68.1	56.0	61.6	27.4	55.1	14.27	48.6	7.96	42.2	4.64
68.0	55.4	61.5	27.1	55.0	14.15	48.5	7.90	42.1	4.60
67.9	54.8	61.4	26.8	54.9	14.01	48.4	7.83	42.0	4.56
67.8	54.1	61.3	26.6	54.8	13.86	48.3	7.77	41.9	4.52
67.7	53.5	61.2	26.3	54.7	13.72	48.2	7.71	41.8	4.48
67.6	52.9	61.1	26.0	54.6	13.60	48.1	7.65	41.7	4.44
67.5	52.3	61.0	25.8	54.5	13.45	48.0	7.59	41.6	4.40
67.4	51.6	60.9	25.6	54.4	13.32	47.9	7.53	41.5	4.36
67.3	51.0	60.8	25.3	54.3	13.19	47.8	7.47	41.4	4.32
67.2	50.4	60.7	25.0	54.2	13.08	47.7	7.41	41.3	4.29
67.1	50.0	60.6	24.79	54.1	12.93	47.6	7.35	41.2	4.25
67.0	49.3	60.5	24.54	54.0	12.81	47.5	7.29	41.1	4.21
66.9	48.7	60.4	24.32	53.9	12.69	47.4	7.23	41.0	4.17
66.8	48.1	60.3	24.04	53.8	12.58	47.3	7.17	40.9	4.14
66.7	47.6	60.2	23.82	53.7	12.43	47.2	7.12	40.8	4.11
66.6	47.0	60.1	23.53	53.6	12.31	47.1	7.05	40.7	4.08
66.5	46.4	60.0	23.32	53.5	12.18	47.0	6.99	40.6	4.04
66.4	45.8	59.9	23.08	53.4	12.08	46.9	6.93	40.5	4.01
66.3	45.3	59.8	22.88	53.3	11.94	46.8	6.87	40.4	3.98
66.2	44.8	59.7	22.65	53.2	11.82	46.7	6.82	40.3	3.95
66.1	44.2	59.6	22.41	53.1	11.70	46.6	6.76	40.2	3.91
66.0	43.7	59.5	22.20	53.0	11.59	46.5	6.70	40.1	3.88
65.9	43.2	59.4	21.96	52.9	11.48	46.4	6.65	40.0	3.85
65.8	42.8	59.3	21.76	52.8	11.37	46.3	6.59	39.9	3.82
65.7	42.3	59.2	21.54	52.7	11.25	46.2	6.54	39.8	3.79
65.6	41.8	59.1	21.32	52.6	11.13	46.1	6.48	39.7	3.76
65.5	41.3	59.0	21.10	52.5	11.04	46.0	6.42	39.6	3.73
65.4	40.8	58.9	20.86	52.4	10.92	45.9	6.36	39.5	3.70
65.3	40.4	58.8	20.66	52.3	10.83	45.8	6.31	39.4	3.67
65.2	40.0	58.7	20.45	52.2	10.72	45.7	6.26	39.3	3.64
65.1	39.5	58.6	20.22	52.1	10.62	45.6	6.21	39.2	3.61
65.0	39.1	58.5	20.04	52.0	10.52	45.5	6.16	39.1	3.59
64.9	38.7	58.4	19.85	51.9	10.42	45.4	6.10	39.0	3.56
64.8	38.3	58.3	19.66	51.8	10.34	45.3	6.05	38.9	3.53
64.7	37.9	58.2	19.45	51.7	10.22	45.2	6.00	38.8	3.50
64.6	37.4	58.1	19.28	51.6	10.13	45.1	5.95	38.7	3.48
64.5	37.0	58.0	19.09	51.5	10.04	45.0	5.89	38.6	3.45
64.4	36.6	57.9	18.90	51.4	9.96	44.9	5.84	38.5	3.43
64.3	36.2	57.8	18.70	51.3	9.87	44.8	5.80	38.4	3.40
64.2	35.8	57.7	18.52	51.2	9.79	44.7	5.74	38.3	3.37
64.1	35.4	57.6	18.35	51.1	9.70	44.6	5.69	38.2	3.35
64.0	35.0	57.5	18.16	51.0	9.62	44.5	5.65	38.1	3.33
63.9	34.7	57.4	17.98	50.9	9.55	44.4	5.60	38.0	3.30
63.8	34.3	57.3	17.80	50.8	9.48	44.3	5.55	37.9	3.28
63.7	33.9	57.2	17.63	50.7	9.40	44.2	5.50	37.8	3.26
63.6	33.6	57.1	17.45	50.6	9.34	44.1	5.46	37.7	3.23
63.5	33.2	57.0	17.30	50.5	9.26	44.0	5.41	37.6	3.20
63.4	32.9	56.9	17.09	50.4	9.18	43.9	5.36	37.5	3.18
63.3	32.5	56.8	16.94	50.3	9.11	43.8	5.32	37.4	3.16
63.2	32.2	56.7	16.78	50.2	9.04	43.7	5.28	37.3	3.14
63.1	31.9	56.6	16.60	50.1	8.97	43.6	5.23	37.2	3.12
63.0	31.6	56.5	16.43	50.0	8.90	43.5	5.18	37.1	3.10
62.9	31.2	56.4	16.30	49.9	8.84	43.4	5.13	37.0	3.08
62.8	30.9	56.3	16.11	49.8	8.77	43.3	5.09	36.9	3.06
62.7	30.6	56.2	15.96	49.7	8.70	43.2	5.05	36.8	3.04
62.6	30.3	56.1	15.80	49.6	8.63	43.1	5.01	36.7	3.02
62.5	30.0	56.0	15.63	49.5	8.57	43.0	4.96	36.6	3.00
62.4	29.7	55.9	15.49	49.4	8.49	42.9	4.92	36.5	2.98
62.3	29.4	55.8	15.33	49.3	8.43	42.8	4.88	36.4	2.96
62.2	29.1	55.7	15.17	49.2	8.35	42.7	4.84	36.3	2.94
62.1	28.8	55.6	15.03	49.1	8.29			36.2	2.92

task illumination and β is the luminance factor under appropriate conditions. In any event, the value of RCS so determined represents a measure of the generalized visual sensitivity provided by the level of task luminance, disregarding effects due to veiling reflections, disability glare, and transient adaptation.

Allowance for Veiling Reflections.[69] Differences between the spatial pattern and polarization of the illumination striking a task in a real environment and the reference lighting conditions of a photometric sphere are evaluated in terms of CRF as before. The Effective RCS, RCS_e = RCS × CRF. The value of RCS_e reflects the level of visual sensitivity provided by the task background luminance, modified to take account of gains or losses in task visibility due to the veiling reflection effect.

The value of luminance corresponding to RCS_e is designated the Effective Luminance, L_e. It represents the actual task luminance in the reference sphere providing the same visual performance potential as the real installation, account having been taken of the level of actual task luminance and the veiling reflections effect. The value of Equivalent Sphere Illumination, E_s, equals L_e/β. This procedure is shown schematically in Fig. 3–33. The value of the Lighting Effectiveness Factor, LEF equals E_s/E_t.

Allowance for Disability Glare. Differences between the pattern of luminances in the surround near the task in a real environment and under the reference lighting conditions of a photometric sphere are evaluated in terms of DGF as before. A new value of Effective RCS, RCS_e is defined as RCS × CRF × DGF. The value of luminance corresponding to this second value of RCS_e is again designated Effective Luminance, L_e, and there are also second values of Equivalent Sphere Illumination, and Lighting Effectiveness Factor.

Estimation of the Transient Adaptation Effect. Differences between the pattern of luminances of the entire surround in a real environment and under the reference lighting conditions of a photometric sphere may be evaluated approximately as before in terms of TAF_1 and TAF_2.

Procedures for Utilizing Values of Equivalent Sphere Illumination in Illumination Design. As noted above, required levels of illumination for selected tasks are expressed in terms of E_r, the illumination under the reference lighting conditions of a photometric sphere needed to satisfy the IES visual performance criterion. Procedures described above lead to a description of real luminous environments in terms of E_s, the level of illumination within the reference sphere yielding a level of visual performance potential equal to that provided by the real environment. Design proceeds by attempting to match E_s to E_r. As before, this will be an iterative procedure, but values of LEF may be useful for approximate calculations.

Evaluation of the Adequacy of Illumination for Tasks not Studied under Reference Lighting Conditions

It will be found impractical to study some tasks, such as those encountered in roadway lighting, under the reference lighting conditions of a photometric sphere. It will prove convenient to study other tasks such as highly specular industrial tasks under more realistic types of lighting than the diffuse lighting of a photometric sphere. In these cases, the adequacy of illumination may be established in terms of the Effective Visibility Level (VL_e).

Determination of VL. The value of Equivalent Contrast ($\tilde{C}$) is determined with the task illuminated realistically. The field of view of the visibility meter is restricted to the task background so that disability glare and transient adaptation effects are excluded. Values of VL are computed from the relation, VL = $\tilde{C}/C_1$, where values of C_1 are computed from the standard RCS function of L in the manner described above.

Allowance for Disability Glare. The effect of the luminance pattern in the surround near the task is evaluated as before in terms of DGF. The Effective Visibility Level, VL_e equals VL × DGF. The adequacy of the lighting system while the observer steadily fixates the task may be evaluated by comparing the value of VL_e with 8, the level of VL corresponding to the IES visual performance criterion.

Estimation of the Transient Adaptation Effect. The effect of the luminance pattern of the entire surround may be evaluated during eye movements away from the task in terms of TAF_1 and TAF_2 as before. New values of the Effective Visibility Level, VL_e equal VL × DGF × TAF_1 or TAF_2. The adequacy of the lighting system for brief periods following eye movements may be evaluated by comparing these values of VL_e with 8.

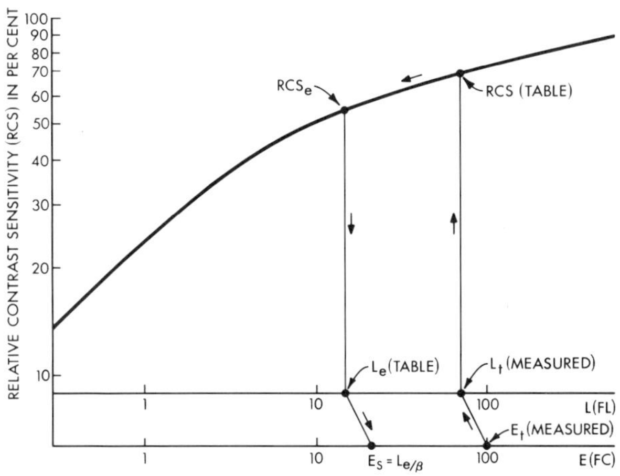

Fig. 3–33. Determination of Equivalent Sphere Illumination, E_s.

Visual Comfort Criteria

When luminances and their relationships in the field of view cause visual discomfort but do not necessarily interfere with seeing, the sensation experienced by an observer is termed discomfort glare. It usually is produced by direct glare from light sources or luminaries which are too bright, inadequately shielded, or of two great an area. Discomfort glare also can be caused by annoying reflection of bright areas in specular surfaces (known as reflected glare). The latter should not be confused with veiling reflections which impair visual performance rather than cause discomfort.

Basis for Evaluating Visual Comfort. The visual comfort of a lighting system depends upon many factors: room size and shape; room surface reflectances; illumination level; luminaire type, size and light distribution; number and location of luminaires; luminance of entire field of view; observer location and line of sight; and differences in individual glare sensitivity. Extensive investigations[70, 71, 72] and analyses[73, 74] have resulted in a comprehensive standard discomfort glare evaluation procedure[75] which takes these factors into account. The final product of the evaluation procedure is a Visual Comfort Probability (VCP) rating of the lighting system expressed as a per cent of people who, if seated in the most undesirable location, will be expected to find it acceptable.

Comprehensive Procedure. The fundamental glare formula[74] expresses the index of sensation (M)

$$M = \frac{L_s Q}{P F^{0.44}} \qquad (3\text{–}9)$$

of a source as functions of its luminance (L_s), solid angle factor (Q), and position index (P), and the field luminance (F).

In order to systematize the calculations and to aid in the development of VCP tables, standard conditions have been adopted[75]:

1. An initial level of 100 footcandles.
2. Room surfaces of 80 per cent for the effective ceiling cavity reflectance, 50 per cent for the wall reflectance and 20 per cent for the effective floor cavity reflectance.
3. Mounting heights above the floor of 8.5, 10, 13, and 16 feet.
4. A range of room dimensions to include square, long-narrow, and short-wide rooms.
5. A standard layout involving luminaires uniformly distributed throughout the space.
6. An observation point four feet in front of the center of the rear wall and four feet above the floor.
7. A horizontal line of sight directly forward.
8. A limit to the field of view corresponding to an angle of 53 degrees above and directly forward from the observer.

These have permitted preparation of tables of constants of certain factors used for determining the quantities in Equation 3-9. While the basic procedure is directed toward the preparation of VCP tables, it also can be used for obtaining comfort ratings of specific lighting systems[76] and for other illumination levels and room surface reflectances.[77]

The standard layout shown in Fig. 3-34 illustrates the left half of rooms up to 100 by 100 feet in size and a 10-foot mounting height above the floor. The number of luminaires required for 100 footcandles is assumed to be uniformly divided among the 5 by 5-foot modules, some of which are combined to form the numbered ceiling elements.

The geometry involved in specifying the location of a luminaire at any position (S) in a room and for determining its projected area in terms of the directly crosswise projected area (at S_0) is illustrated in Fig. 3-35. The directly crosswise projected area (A_{po}) may be determined graphically[78] as shown in Fig. 3-36. This area is, in effect, moved along the dashed line from S_0 to S (Fig. 3-35) and the lengthwise foreshortening is accounted for by a factor cos θ.

The average luminance of the luminaires (L_s) in any of the ceiling elements is obtained from

$$L_s = \frac{\pi I \cos \theta_0}{A_{po} \cos \theta} \qquad (3\text{–}10)$$

where I = luminous intensity in the direction of the reference observer position
θ_0 = directly crosswise angle from nadir to center line of each row
A_{po} = directly crosswise projected area
θ = angle from nadir of each ceiling element.

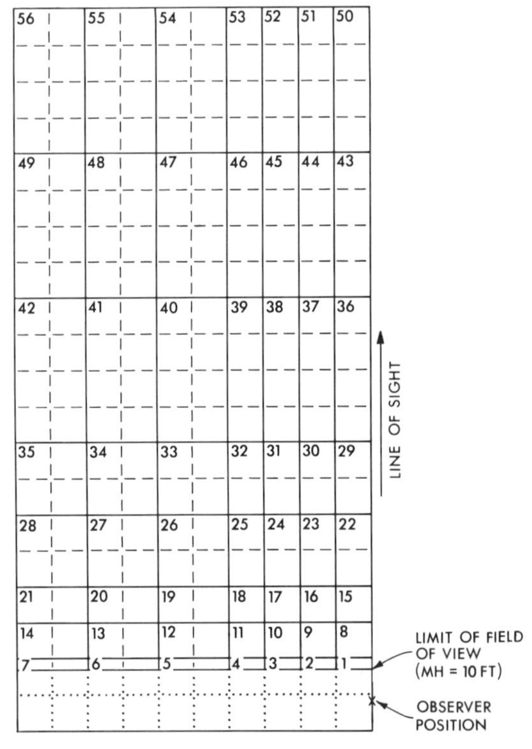

Fig. 3-34. The standard layout used as a basis for VCP tables. This illustrates the left half of rooms up to 100 by 100 feet in size and a 10-foot mounting height above the floor.

The quantity $A_{po}/\cos\theta_0$ can be considered the effective horizontal plane area[77] for a luminaire located in each crosswise or lengthwise row of the standard layout. This simplifies the determination of projected areas because they are based upon directly crosswise projections of a luminaire. The effective horizontal plane area converts all types of luminaires into an area corresponding to the bottom panel of a troffer. For regressed luminaires it is necessary to combine the lengthwise and crosswise projections.

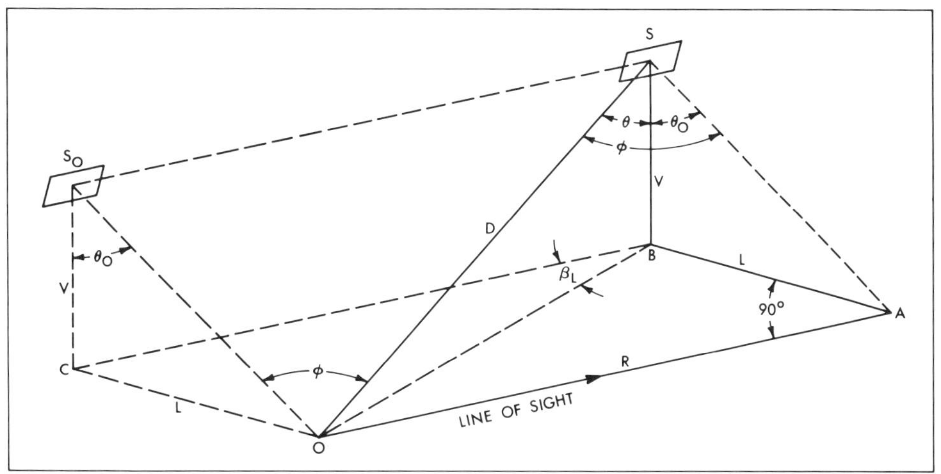

Fig. 3–35. Illustrating the geometrical relationship for determining the projected area of a luminaire at any location in a room in terms of the directly crosswise projected area.

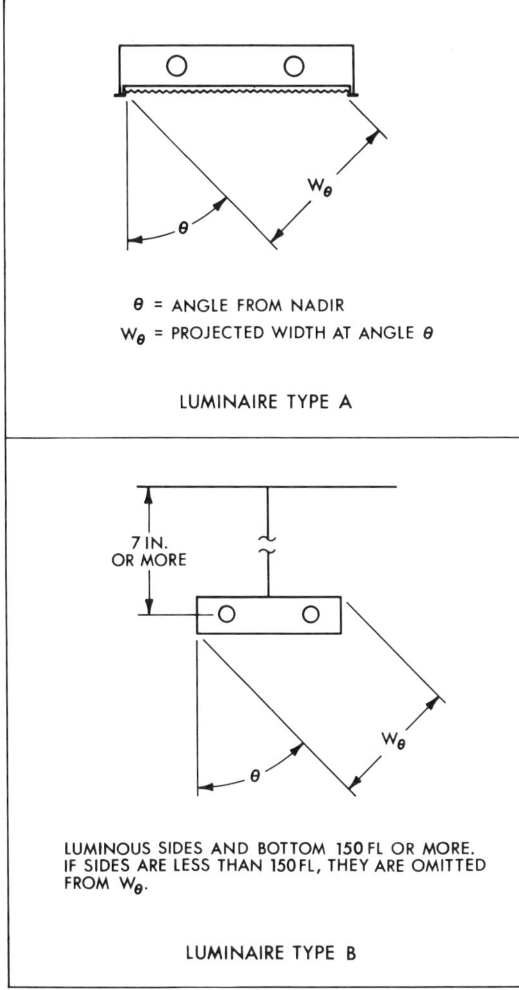

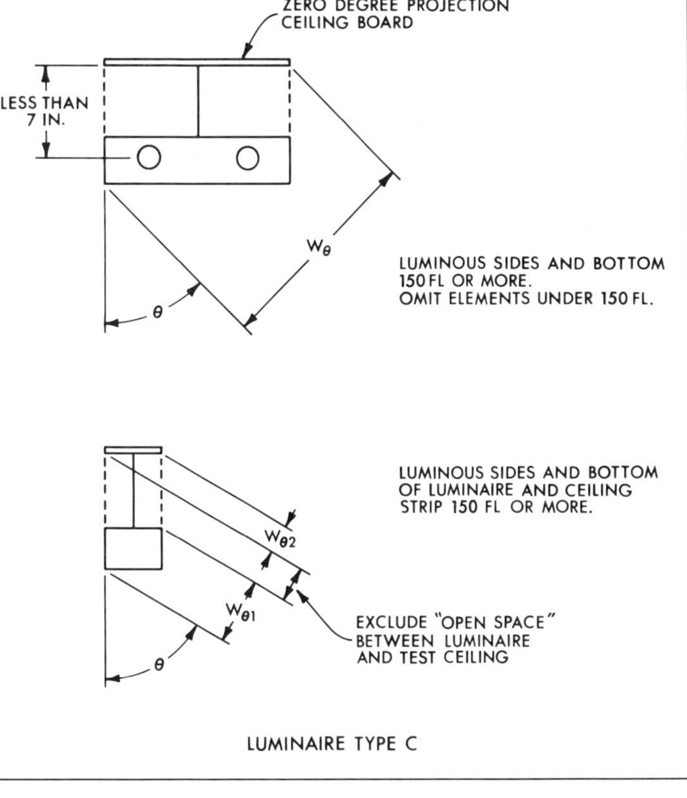

Fig. 3–36. Illustrating the graphical procedure for determining the projected width of the luminous area of three basic types of luminaires. Only luminous elements with a luminance of 150 footlamberts or more are included. Projected area is obtained by multiplying this projected width by the luminous length of the luminaire.

In a restated version[77] of Equation 3–9, the quantity Q is replaced by $\bar{K}$ which is a variable function of solid angle:

$$\bar{K} = 20.4 + 1.52\omega^{-0.8} - 0.075\omega^{-1} \quad (3\text{–}11)$$

The formula for solid angle is

$$\omega = \frac{2500c \cos \theta A_{po}}{(\text{lamp lumens/luminaire})\, D^2\, \text{CU} \cos \theta_0} \quad (3\text{–}12)$$

where c = factor for number of 5 by 5-foot modules in each ceiling element of the standard layout.
D = direct distance from observation point to center of each ceiling element.
CU = coefficient of utilization.
2500 = product of area of module (25 square feet) by illumination level (100 footcandles).

Since certain quantities in Equations 3–10 and 3–12 are fixed for each ceiling element, they can be combined into a series of tabulated constants.[79] Thus,

$$K_1 = \frac{\pi}{\cos \theta} \quad (3\text{–}13)$$

$$K_2 = \frac{2500c \cos \theta}{D^2} \quad (3\text{–}14)$$

and the restated discomfort glare formula becomes:

$$M = \left[\frac{IK_1 \cos \theta_0}{A_{po}}\right]\left[\frac{\bar{K}}{PF^{0.44}}\right] \times \left[\frac{A_{po}K_2}{(\text{lamp lumens/luminaire})\, \text{CU} \cos \theta_0}\right] \quad (3\text{–}15)$$

which has been found to be more suitable, especially for preparing computer programs.[77] The position indices (P) also have been tabulated.[79]

The average field luminance (F) is computed by

$$F = \frac{L_w\omega_w + L_{fc}\omega_f + L_{cc}(\omega_c - \Sigma\omega_s) + \Sigma L_s\omega_s}{5} \quad (3\text{–}16)$$

This takes into account the luminances of the walls (L_w) floor cavity (L_{fc}), ceiling cavity (L_{cc}) and the luminaires (L_s); these are weighted in accordance with the solid angle subtended by each. While room surface solid angles have been tabulated,[79] they may be calculated.[80] Room surface luminances are computed from luminance coefficients obtained by standard procedures (see Section 9).

Equation 3–15 as written, is to be used with English units (feet, footcandles, footlamberts). When SI units (meters, lux, cd/m^2) are employed it must be prefixed by the constant 0.16 in order to obtain appropriate values of the index of sensation (M).

The numerical discomfort glare rating (DGR) is computed from the sum of the individual values of M for all luminous areas:

$$\text{DGR} = (\Sigma M)^a \quad (3\text{–}17)$$

where $a = n^{-0.0914}$ (3–18)
n = number of sources (ceiling elements) in the field of view.

The variable exponent a provides a direct relationship between subjective and computed ratings.[74] As a final step, using Fig. 3–37 the DGR is converted into a Visual Comfort Rating (VCP). A typical VCP table is shown in Fig. 3–38.

Direct glare will not be a problem in lighting installations if all three of the following conditions are satisfied:[75]

1. The VCP is 70 or more;
2. The ratio of maximum-to-average luminaire luminance does not exceed five to one (preferably three to one) at 45, 55, 65, 75 and 85 degrees from nadir crosswise and lengthwise; and
3. Maximum luminaire luminances crosswise and lengthwise do not exceed the following values:

Angle Above Nadir (degrees)	Maximum Luminance (footlamberts)
45	2250
55	1605
65	1125
75	750
85	495

Simplified Procedures. An approximate assessment can be obtained by an alternate simplified procedure.[81, 82] While this simplified procedure is applicable only to flat-bottomed, nonluminous sided equipment in a 60 by 60-foot room with a 10-foot mounting height, its extension to other types is being studied. Although the method has been derived from Equation 3–9, it involves several assumptions:

1. Discomfort glare is independent of the number of luminaires used to provide 100 footcandles.

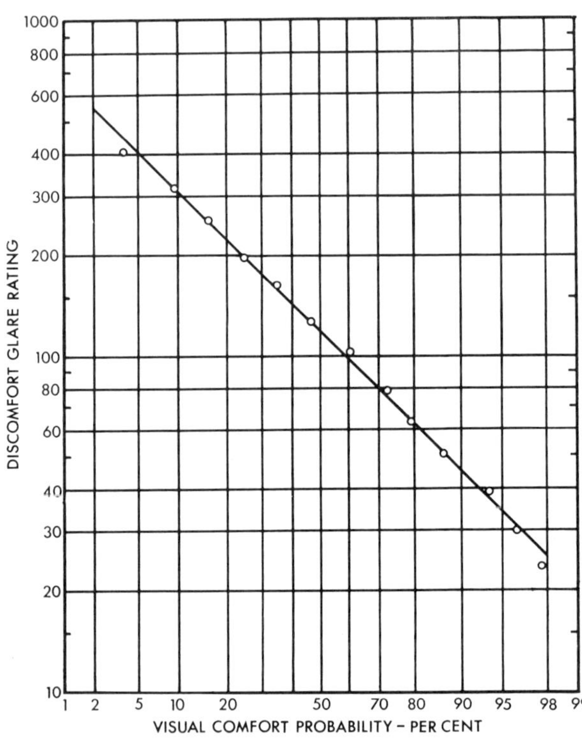

Fig. 3–37. A chart for converting discomfort glare ratings to visual comfort probabilities (VCP) (the per cent of observers who would be expected to judge a lighting condition to be at or more comfortable than the borderline between comfort and discomfort).

Fig. 3-38. Examples of a Typical Tabulation of Visual Comfort Probability Values

WALL REFL 50%, EFF CEILING CAV REFL 80%, EFF FLOOR CAV REFL 20%
LUMINAIRE NO. 000
WORK-PLANE ILLUMINATION 100 FOOTCANDLES

Room		Luminaires Lengthwise				Luminaires Crosswise			
W	L	8.5	10.0	13.0	16.0	8.5	10.0	13.0	16.0
20	20	78	82	90	94	77	81	89	93
20	30	73	76	82	88	72	75	81	86
20	40	71	73	78	82	70	72	76	80
20	60	69	71	74	78	68	70	73	76
30	20	78	82	88	92	77	81	87	92
30	30	73	75	80	85	72	74	79	84
30	40	70	72	75	78	69	71	74	77
30	60	68	69	71	74	67	69	70	73
30	80	67	69	69	72	67	68	68	71
40	20	79	82	87	92	79	82	87	91
40	30	74	76	79	84	73	75	78	83
40	40	71	72	74	77	70	71	73	76
40	60	68	69	70	72	68	69	69	71
40	80	67	68	68	70	67	68	67	69
40	100	67	68	67	69	67	67	66	68
60	30	75	76	79	83	74	76	78	82
60	40	71	72	74	76	71	72	73	76
60	60	69	69	69	71	68	69	68	70
60	80	68	68	67	69	67	68	66	68
60	100	67	67	66	67	67	67	65	66
100	40	74	75	75	78	74	74	75	77
100	60	71	71	71	72	71	71	70	72
100	80	70	70	68	69	70	69	67	69
100	100	69	68	66	67	69	68	66	67

2. The average luminance of the field of view is proportional to the number of luminaires in the field of view.

3. VCP may be regarded as a function of $\bar{L}$ which is the summation of the product LT for the 24 segments into which the ceiling is divided (see Fig. 3-39). L is equal to the average luminance of a luminaire in the lengthwise, diagonal, and crosswise vertical planes at the midzone angles above nadir listed in Fig. 3-40; T is a factor which takes into account the position index and angular size of the segment.

4. The effect of the luminous area of the luminaires is included in the quantity $\bar{L}$.

5. If the value of $\bar{L}$ is as low as, or lower than 320, this may be taken as an indication that the VCP is 70 or higher.

A work sheet for computing $\bar{L}$ is shown in Fig. 3-40 on which are included the values of T for the different ceiling segments. When luminaires are viewed crosswise, crosswise values of luminance L are entered in the upper section and the lengthwise values in the lower section. For lengthwise viewing, lengthwise values are entered in the upper section and crosswise in the lower section. The sum of the LT values represents the rating. It is emphasized that if VCP ratings using the comprehensive and simplified procedures are different, the former takes precedence provided both are calculated correctly.

A previously described system based upon a scissors curve criterion[82] may be employed to determine whether a luminaire complies with lengthwise and crosswise luminance limitations. This will serve as an approximate guide to estimate compliance with a VCP rating of 70.

Reflected Glare

Glare resulting from specular reflection of high luminance in polished or glossy surfaces is known as reflected glare and, as mentioned above under Visual Comfort Criteria, it may cause discomfort. If the reflections are of high luminance and of significant area, they also may produce disability glare if near the line of sight and may prove distracting to workers.[84] Bright reflections of small size often do not cause reflected glare but provide desired sparkle, such as in jewelry, glassware, or tableware on display. When specular reflections of high or low luminance veil the visual task, this effect is known as veiling reflections. (See below.)

There is little in the way of research to quantify the visual impact of reflections in specular surfaces. Generally the conditions which cause reflected glare are sufficiently obvious as to indicate the lighting solution. In general, large area low luminance sources are preferred where the task has even a small degree of specularity. Additional considerations in this regard are discussed in the various application sections. For example, finishes on office furniture and equipment should be matte. See Section 11.

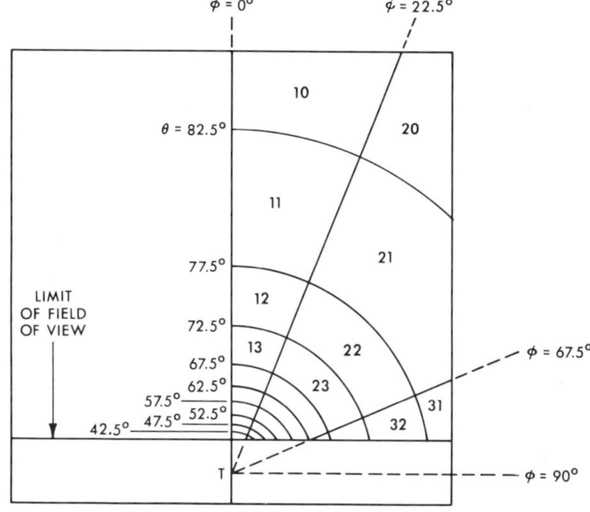

Fig. 3-39. Division of the ceiling into zones and sectors for use with the simplified procedure for assessment of visual comfort.

Fig. 3–40. Work Sheet for Computing $\bar{L}$ for Determining the Acceptability of a Luminaire from a VCP Standpoint for Use in Large Rooms

Luminaire		Viewing Direction			
	Angles Above Nadir (degrees)	L Average Luminance (foot-lamberts)	T Multiplier	L × T	Totals
Vewing Direction	85		.0375		
	80		.1080		
	75		.0884		
	70		.0703		
	65		.0543		
	60		.0406		
	55		.0312		
	50		.0229		
	45		.0159		
	40		.0102		
Diagonal (45°)	85		.0203		
	80		.1065		
	75		.1022		
	70		.0841		
	65		.0681		
	60		.0507		
	55		.0333		
	50		.0214		
	45		.0109		
	40		.0021		
90° to Viewing Direction	80		.0046		
	75		.0096		
	70		.0052		
	65		.0017		

$\bar{L} = \sum L \times T =$

To be acceptable $\bar{L}$ should be equal to or less than 320.*

* When the ratio $B = \frac{\text{Flux in 0–60° zone}}{\text{Flux in 60–90° zone}}$

as calculated from available luminaire zonal flux data, is greater than 10 or less than 4 it becomes necessary to correct $\bar{L}$ by multiplying it by a correction factor $\phi = 2.34\ (B^{-0.48})$. Then the value of 320 for $\phi\bar{L}$ is used as the cut off level for regarding a luminaire as acceptable. If B falls between 4 and 10 no correction (ϕ) is necessary but may be used if desired.

A table relating values of ϕ for values of B is shown below.

B	ϕ	B	ϕ
1	2.34	22	.531
2	1.68	24	.509
3	1.38	26	.490
4	1.20	28	.473
5	1.08	30	.457
6	.990		
7	.920	35	.425
8	.862	40	.398
9	.815	45	.376
10	.775	50	.358
11	.740	55	.342
12	.710	60	.328
13	.683	65	.316
14	.659	70	.304
15	.638	75	.295
16	.613	80	.286
17	.601	85	.277
18	.584	90	.270
19	.569	95	.263
20	.556	100	.257

Veiling Reflections

For years lighting engineers and those engaged in vision research have recognized that substantial losses in contrast, hence, in visibility and visual performance, can result when light sources are reflected in specular or semi-specular visual tasks.[85, 86] This has been known as the general subject of reflected glare. The effects vary from the reflection of an incandescent filament in a polished metal surface at one extreme, in which case the result is annoying, distracting, and disabling from the visual standpoint, to the other extreme where there is a reflection of a large luminous area in the surface of a magazine printed on dull or matte paper. With the latter, the effect may be undetectable by the naked eye and may be almost unmeasurable by instruments. The term "reflected glare" is reserved today for effects toward the first extreme and the term "veiling reflection" for effects near the second. Perhaps the greatest and certainly the most insidious problem is the reflection of luminaires or skylights in semi-specular and semi-matte surfaces such as a printed page and pencil writing on paper.

There are a great many factors that contribute to veiling reflections and each of them individually has long been known. The problem is to integrate the effects of these interrelated factors. Much has been learned in the last few years, particularly the contributions made by Illuminating Engineering Research Institute (IERI) research.[63, 87]

Factors Causing Contrast Loss. In the study of contrast losses due to veiling reflections, the visual task (printing or handwriting on paper are usually considered as the visual task in the following discussions), the workers' orientation and viewing angle, and the lighting system all must be analyzed.

The Visual Task. The specularity of paper covers a wide range of degrees and modes of appearance. Most papers consist of rough fibers that have been matted together. Generally the fibers are somewhat shiny in themselves, but because of their random orientation they reflect light more or less equally in all directions. The harder the paper is pressed the more specular it becomes. Some papers are filled with clay or other coating so that the surface is very smooth. Some papers are actually glazed. The luminance of the paper depends both on the amount of light being diffusely reflected from it and whatever bright surface may be reflected in it. This reflection may be discernible as in the case of coated papers or glossy photographs, but frequently it is so indistinct as to go undetected even though serious losses in visibility occur.

The specularity of the graphic medium—pencil, pen, ink, carbon, etc.—again covers a very wide range. The degree of specularity depends on how the medium is deposited. For example, a very soft pencil brushed lightly across rough paper would leave a very diffuse mark. On the other hand, a hard pencil applied with pressure on a smooth surface can be very shiny. The luminance of the mark

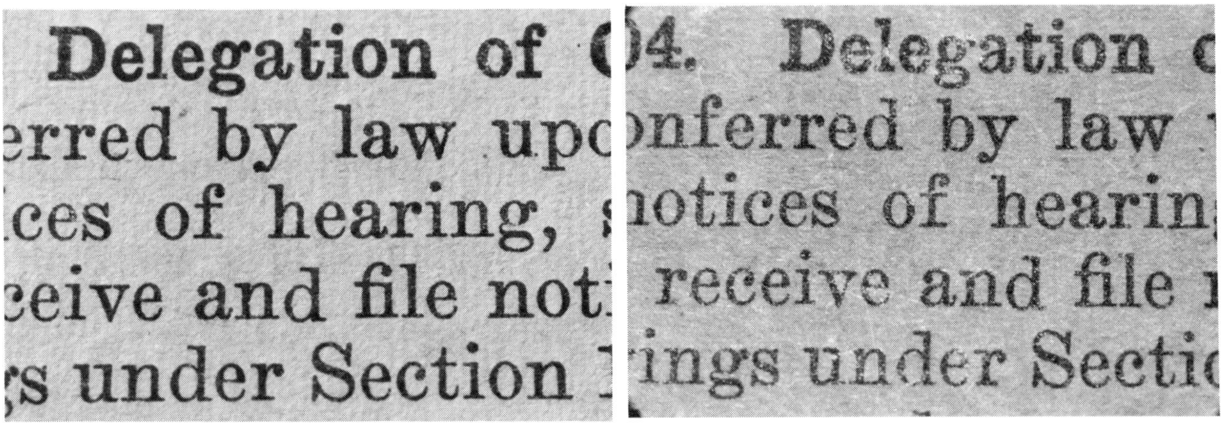

Fig. 3–41. (Left) An enlarged example of printing on a soft paper with matte black ink. Illumination is provided by a spotlight behind the camera. Note that there are very few highlights in the black letter strokes and that the texture of the paper can be observed. (Right) The same soft paper with matte black ink. Lighting in this case is with a spotlight in front of and above the task where it is in a position to cause the maximum veiling reflection effect. Note the many highlights in the black stroke and on the paper as well. These highlights in both areas are specular reflections which tend to veil the task and reduce contrast.

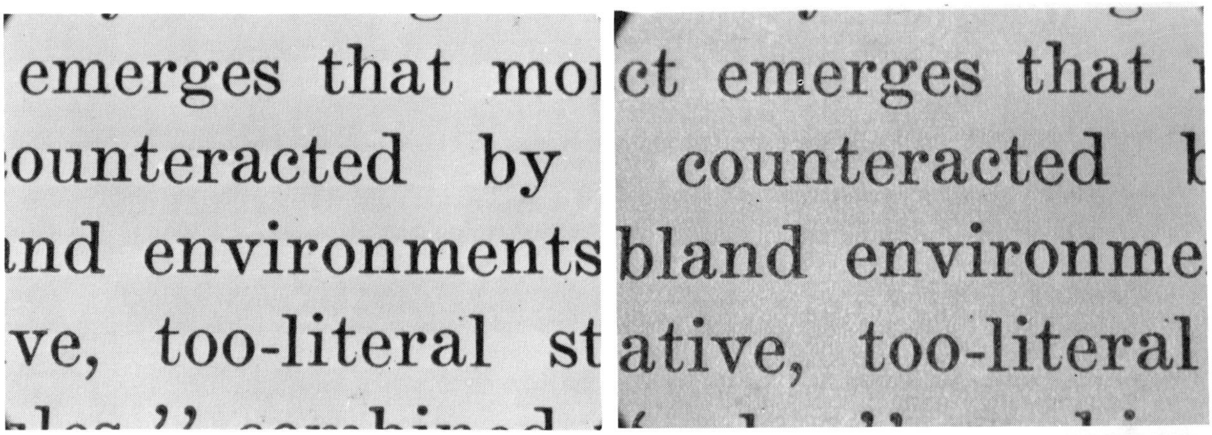

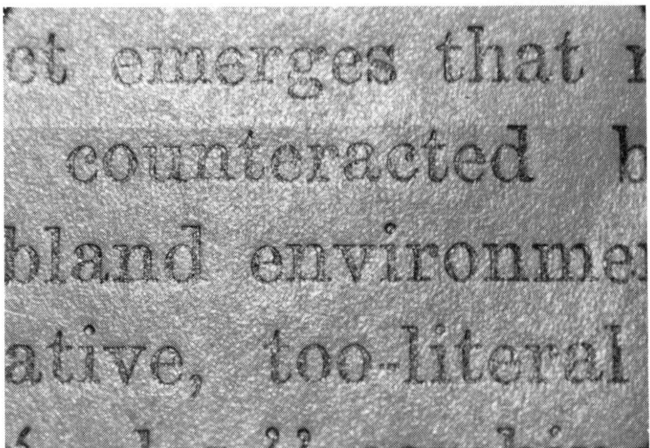

Fig. 3–42. An example of black printing on glossy paper. In the photograph above, a spotlight is located behind the camera in a position to cause minimum veiling reflections. The upper right photograph shows the same task lighted by a troffer that is positioned above and in front of the task in the offending zone. Note the subtle character of the specular reflections so that the loss of contrast is not immediately obvious. Note also that the specular reflections occur in different positions on different letters. The photograph at the right shows the same paper with a spotlight in the offending zone which creates the maximum veiling reflection effect. This condition might be called reflected glare. The image on the page is apparent as are the highlights in the letter stroke. Note how highlights tend to occur along the edges of the letter indicating that the type has embossed the paper.

again depends on the amount of light being diffusely reflected from it and the reflection in it of luminous areas. Therefore, when considering task contrast, both diffuse and specular reflectance of both paper and graphic medium should be considered as well as the reflection of light sources in relation to the level of illumination. Figs. 3–41 through 3–43 illustrate a range of conditions of veiling reflections.

If the paper and the graphic medium could be considered as being perfect planes, the problem would be simpler than is the actual case. The pressure applied by the pencil, pen, typewriter key, or printing type actually embosses the paper. The groove thus created causes the reflection of the light source to occur from positions on the ceiling other than the normal angle of reflection from the plane of the paper. Thus the part of the ceiling that is causing the problem may not be immediately obvious.

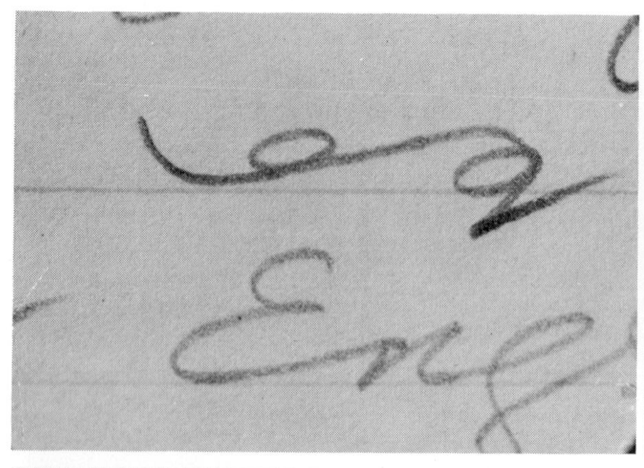

Fig. 3–43. The upper left photograph shows pencil stenographic notes lighted by a spotlight located behind the camera. It can be seen that in this case the pencil stroke is relatively light because even in the darker parts of the stroke, areas of white paper show through. The photograph directly above shows the same task with the spotlight in the offending zone (above and in front of the task). As is frequently the case with this type of task, negative contrast or contrast reversal occurs where the pencil stroke can actually become brighter than the paper. In the photograph at the left the same task is lighted with an indirect lighting system.

Consideration should be given to whether the task is lying in a horizontal plane or is slanted as is the case of a letter being hand-held, a book on a slanted school desk top, or a pencil drawing on a vertical drafting board. Chalkboards in classrooms, merchandise in stores, and signs are tasks of prime importance that lie in a vertical plane. The relative importance of various tasks and the planes in which they occur also should be considered. For example, office lighting should really be designed not just in terms of a task lying flat on a desk but also for one hand-held at about 45 degrees.

The Worker. The orientation of the worker with respect to the task greatly influences the magnitude of the effect of veiling reflections. First, for one eye position and one point of regard, a simplified relationship between the eye, the task, the perpendicular to the task, and an "offending zone" can be established. See Fig. 3–44. If the task were perfectly specular and flat, the offending zone would merely be a point. However, since the types of tasks involved here are more or less diffusing, the theoretical offending point becomes enlarged to an offending zone. Now if the eye is in such a position that the rays of light from the offending zone are reflected toward it, veiling reflections will occur. The angle of reflection is considered as the viewing angle. As the viewing angle increases, effects of the specular characteristics of the paper and the ink or pencil increase.

It has been found that people work throughout a range of viewing angles[88, 89] with a peak at about 25 degrees as indicated by the approximate frequency distribution curves for office workers and for school children shown in Fig. 3–45a. Fig. 3–45b shows that 85 per cent of seeing occurs within 0 to 40-degree viewing angles, with higher angles used for only occasional glances. This is due to foreshort-

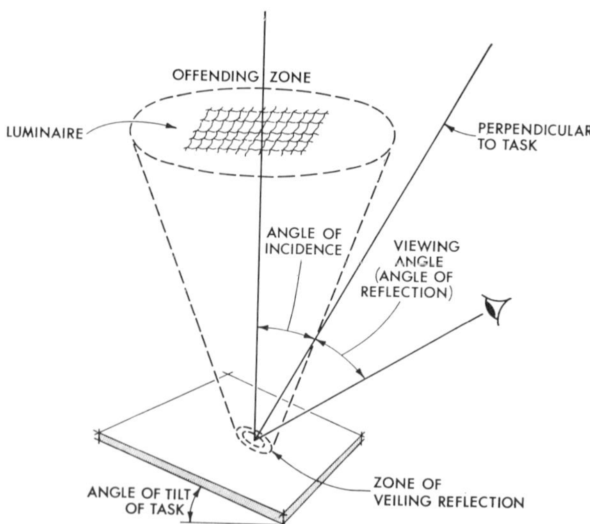

Fig. 3–44. Generalized description of angular relationships in analyzing veiling reflections.

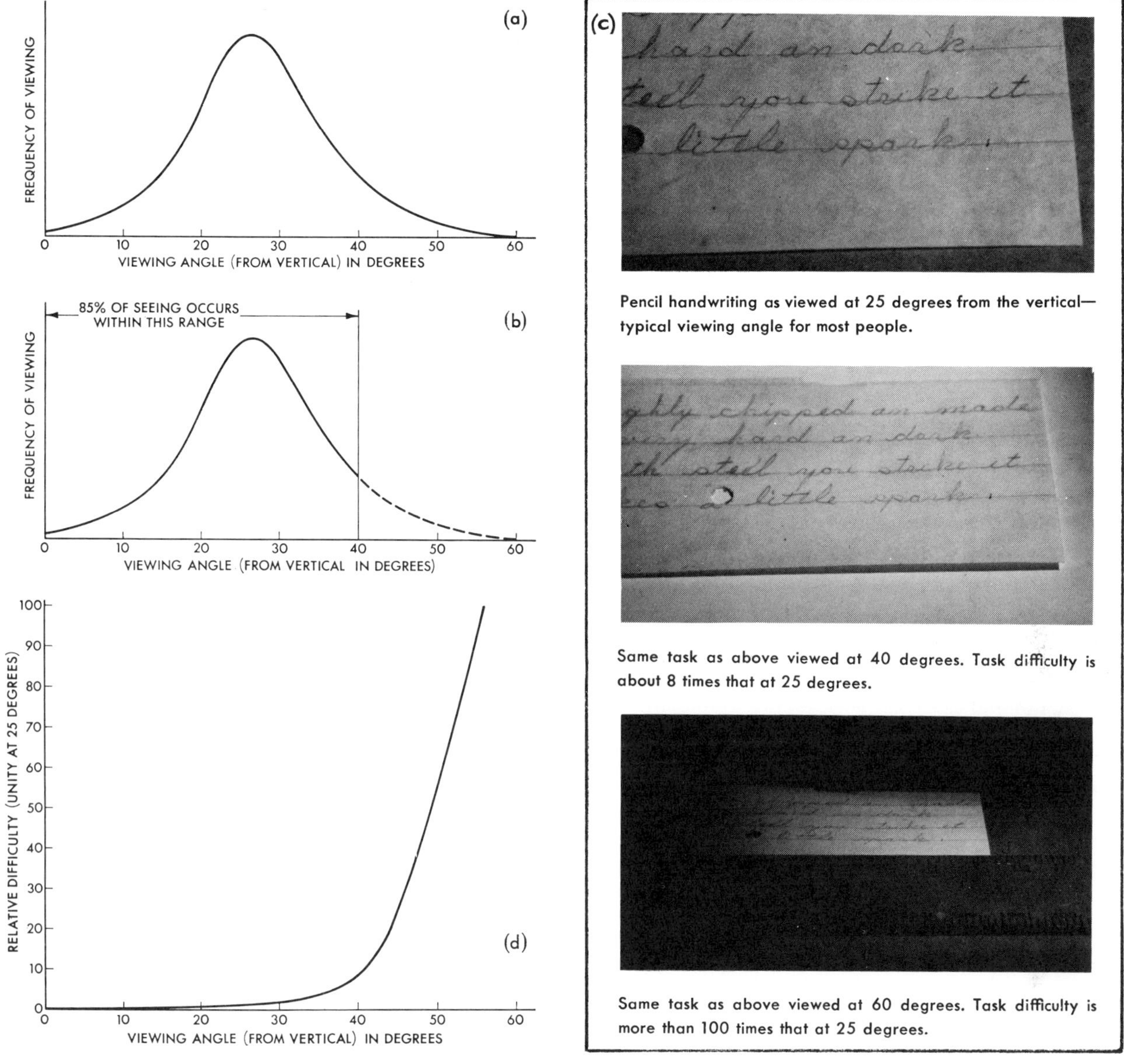

Pencil handwriting as viewed at 25 degrees from the vertical—typical viewing angle for most people.

Same task as above viewed at 40 degrees. Task difficulty is about 8 times that at 25 degrees.

Same task as above viewed at 60 degrees. Task difficulty is more than 100 times that at 25 degrees.

Fig. 3–45. Task viewing angles. *a.* People use a range of viewing angles in their work but the peak is about 25 degrees. *b.* Eighty-five per cent of seeing occurs within a range of 0 to 40 degrees, with seeing at larger angles limited to occasional glances due to foreshortening and increased viewing distance. *c.* Photographs of an actual pencil handwriting sample (handwriting used in the original research to determine levels of illumination) as seen at viewing angles of 25, 40, and 60 degrees. *d.* Curve showing the relative difficulty of the pencil handwriting shown in c as measured with the Visual Task Evaluator.

ening of the task and to the increased viewing distance (see Fig. 3–45c) and to the resulting increase of task difficulty as shown in Fig. 3–45d. On this basis it seems reasonable to use 0 to 40 degrees as the practical range of viewing angles for design purposes.

Next the location and orientation of the worker and the task in the room must be evaluated. A worker with his back to a wall and facing out toward the center of the room has the maximum ceiling area as a potential offending zone. Furthermore, he has relatively little light coming from behind him which would cause little or no veiling reflections. A worker facing the wall would have minimum veiling reflections (not an appealing position psychologically). See Fig. 3–46. A person in the center of the room has light falling on the task from all directions. The ceiling may or may not constitute the offending zone, depending on the viewing angle.

A person sitting beside a window would have greatly reduced veiling reflections because the major source of illumination is outside the offending zone. It is also true that if a person were seated facing a window he would have potentially serious veiling reflection conditions.

The Lighting System. The worst condition is represented by a highly concentrated, very bright source with maximum candlepower directed toward the task. It is also likely to be uncomfortable as one looks around the room, and it may create shadows that interfere with writing, etc. Paradoxically, it is also the condition under which the worker can most easily escape veiling reflections by tilting the task so that the reflected rays do not reach his eye. When the task is truly a flat surface this is an effective solution. Because of embossing and curvature of many tasks such as books and magazines, reflections are not so easily eliminated.

At the other extreme would be a luminous dome placed over the worker and the task. Here the effect of veiling reflections would be minimized but could not be escaped since there would always be a luminous area in the offending zone. Furthermore, through a rather wide range of levels of illumination the lighting system would generally be considered comfortable.

Between these two extremes lies the full range of luminaires of various sizes spaced so as to occupy various proportions of the ceiling and employing materials that produce varying candlepower distribution.

Lighting materials with reference to horizontal tasks may be compared as follows:

Diffusing. At small viewing angles (looking nearly straight down at the task) this material has the minimum luminance exposed to be reflected in the task. At large viewing angles it tends to have the highest luminance. This helps reduce veiling reflections but tends to reduce visual comfort per unit area.

Prismatic. A wide range of materials comes under this category, but generally they tend to expose greater flux toward the work for small viewing angles. Many of them have good light control at large viewing angles. Prismatic materials have been designed to produce special flux distributions referred to below.

Louvers. These materials expose the maximum luminance to the work at small viewing angles. They generally have lower luminance at large viewing angles. Translucent louvers have the least control; opaque louvers next; and specular parabolic wedge louvers have the maximum.

Polarizing. Available materials of the flake or layer type have a degree of diffusion that exposes less luminance to the work than prismatic or louver materials. They have less luminance at large angles than diffusing, about the same as prismatic, and more than opaque louvers.

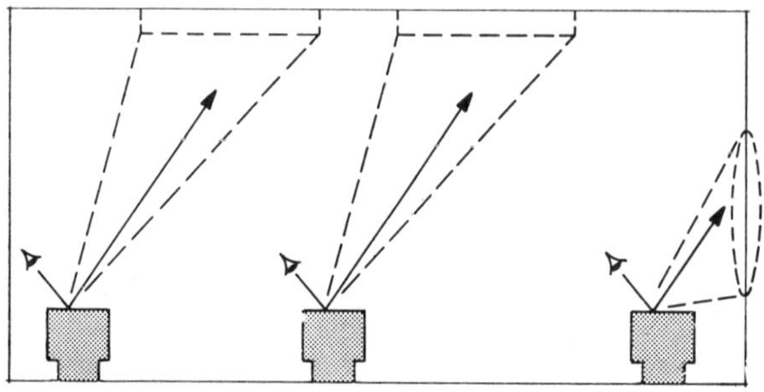

Fig. 3-46. Generally the position at the rear of the room is least favorable from the standpoint of veiling reflections. The center of the room has the advantage of light coming from the rear where it cannot cause veiling reflections. If a worker were in a position facing the wall, the ceiling may not come into the offending zone.

Polarization can reduce veiling reflections. The effect is greatest at large viewing angles and least at small viewing angles. For any single ray of light, polarization in a plane perpendicular to the task always tends to reduce veiling reflections.

These have been termed "radial" polarizers because in azimuth they produce the same degree of polarization in all directions. "Linear" and "dichroic" polarizers can also be useful—particularly for specialized application.

Special. New optical designs of luminaires and materials have been produced with candlepower distributions that reduce the flux coming from the offending zone and minimize the luminance directed to the eye of the worker. While there are distinct variations in effective illumination and visual comfort for various orientations and positions of the worker, very significant improvements are provided in controlling veiling reflections.

Methods of Evaluation. It can be seen that the effect of veiling reflections must be evaluated in terms of the amount of light reaching the task from each direction, the polarization of the light in that direction, the layout of the lighting system as well as the position, orientation, and viewing angle of the worker.[69] See page 3–20, Evaluation of Equivalent Sphere Illumination, for the method used in evaluating the effect of veiling reflections on task visibility in real luminous environments.

Visual Significance. The visual significance of contrast losses caused by veiling reflections has been shown in theory by Figs. 3–27 and 3–31 and their associated texts. Recent laboratory and field evaluations[90-93] have illustrated these losses in terms of real lighting systems in actual environments.

Guides for Reducing Veiling Reflections.

The Task. Where possible the written or printed task should be on matte paper using non-glossy inks. The use of glossy paper stock and hard pencils should be minimized.

The Worker. The orientation and the position of the worker is very important. The desirable position only can be determined by actually determining CRF values. It is also true that various orientations will produce varying degrees of visual comfort.

The Lighting System. In smaller offices where desk positions can be determined, substantial gains can be made by not positioning lighting equipment in the general area above and forward of the desk. Positions on either side and behind the worker are preferred. Where desk positions are random, as in large general offices, it is desirable to have as much light as possible reach the task from sources outside the offending zone. Taken to the extreme this will suggest the utilization of over-all ceiling treatments.

Any decision on a lighting installation should be

made on an over-all basis rather than on any one factor. Thus in addition to considering the level of illumination and the effect of veiling reflections produced by a lighting system and the material, the efficiency of the system and the visual comfort in the space should be considered as prime factors.

Practical Guides for Design of Lighting Installations to Satisfy the IES Method for Prescribing Illumination

It is evident that seeing is a complicated human activity which requires consideration of many factors in order to provide optimal visual ability. When working out a specific lighting design or developing generalized recommendations it is necessary to single out typical visual situations and determine for each what constitutes adequate illumination for the objects to be seen with central and peripheral vision; *i.e.*, adequate to insure the desired level of speed and accuracy in performing the tasks and also to provide for the safety of the worker. It is necessary to consider the total effect of veiling reflections, reflected glare, and transient adaptation on the worker's visual performance as discussed above. Since the worker may perform many tasks in the same room or several people may occupy the same room at the same time, final lighting design should satisfy the needs for each phase of the task and the many that may have to be performed. Last, but not least, the cost and the relative importance of the task for which the light is needed should also be considered.

As pointed out in Section 10, page 10–10, when the Visual Task Oriented Design Approach is used in lighting design, the starting point is to determine what visual tasks are to be lighted and what levels of illumination are required. In using Fig. 9–80, Levels of Illumination Currently Recommended, the designer will find that for certain tasks the recommendations call for equivalent sphere illumination, E_s. This will require the selection of equipment and layouts to take into consideration effects of veiling reflections. Regardless of illumination level recommended, and equipment and layout selection, the designer also should consider the visual comfort aspects and the possibility of reflected glare, disability glare and transient adaptation, as discussed above.

In performing calculations for preliminary layout purposes where exact task locations are not known the lumen method (see page 9–6) is used to determine the number of luminaires required. After a layout is made illumination uniformity and levels at specific locations can be checked using the methods of point calculation covered in Section 9. These present methods do not cover a procedure for checking for equivalent sphere illumination, E_s; however, methods have been developed that provide systems for determining CRF and E_s values for the design process.[94-96] Until a system has been officially adopted, the designer should use the current illumination calculation procedures, carefully considering the guidelines for reducing veiling reflections on page 3–32 and data available on CRF from evaluations of actual installations.

In using the latter data, it must be emphasized that when CRF is less than unity, E_s will be less than the illumination E_d calculated by the lumen or point methods. Generally speaking when luminaires that are not designed to minimize veiling reflections are in the offending zone, CRF is less than unity no matter what their flux distribution may be. Since, when one must assume random orientation of people, one cannot escape the offending zone, it seldom happens that CRF will be greater than 1 for all viewing positions and directions in a room. This means that for the general case E_s will be less than the calculated illumination.

On the other hand, there are circumstances (where sources of illumination are far removed from the offending zone) under which contrast can be greater than under reference conditions. It follows then that in some cases E_s *can* be greater than the calculated. One such case is a worker location with a window or luminous panel on his left side and his angle of view in azimuth is parallel or away from the window. Here, E_s may be 1.5 times E_d. This can be accomplished also by special orientations in private offices and study carrels. For other orientations, this relationship does not necessarily hold.

VISUAL RANGE OF LUMINOUS SIGNALS

Basic Principles

It is the illumination at an observer's eye, produced by a light, that determines if the light will be seen. The illumination E produced at a distance x by a source of luminous intensity I in an atmosphere having a transmissivity (transmittance per unit distance) T is:

$$E = IT^x/x^2 \qquad (3\text{–}19)$$

If the illumination E, at the eye, is greater than E_m, the minimum perceptible (or threshold) illumination, the light will be visible. The distance at which E is equal to E_m is designated as V, the visual range of the light. Then[97]:

$$E_m = IT^V/V^2 \qquad (3\text{–}20)$$

Equation 3–20 is generally known as *Allard's law*.[98]

Equations 3–19 and 3–20 are strictly applicable only when the luminance of the background is small compared to the average luminance of the light.[99] Otherwise Equation 3–19 becomes:

$$E = [I - (L - L')A]\ T^x/x^2 \qquad (3\text{–}21)$$

where L is the luminance of the background of the light, L' is the average luminance of the unlighted projector, both in candelas per unit area, and A is the area of the entire projector projected on a plane normal to the line of sight.

Both L and L' are measured in the direction of the line of sight from a position near the light.

The quantity $(L - L')A$ is the intensity required of the light to make its average luminance equal to that of the background. The visual range of the light is determined by the net intensity, that is, the difference between the measured intensity of the light and this intensity. Typically, the term $(L - L')A$ has a significant effect on the visual range of a signal light only under daylight conditions when the light is dimmed or when the light has a low average luminance in the direction of view.

Effects of the Atmosphere. The atmosphere is never perfectly transparent. Hence, unless the viewing distance is short, atmospheric losses may have a significant effect upon the illumination at the observer's eye. In fog the law of diminishing returns takes effect at relatively short distances. For example, if the transmissivity is 0.01 per mile (light fog), a light with an intensity of 100 candelas will produce an illumination of one milecandle at a distance of 1 mile; an intensity of 40,000 candelas is required to produce that illumination at 2 miles; and 9,000,000 candelas is required at 3 miles.

Equations 3–19, 3–20, and 3–21 can be best solved graphically. In Fig. 3–47 curves relating distance and the ratio of intensity to illumination are shown for several values of atmospheric transmissivity. These curves may be used with any consistent set of units; for example, I in candelas, D in miles, E in lumens per square mile, and T per mile.

The clarity of the atmosphere may be conveniently expressed by a distance defined as the meteorological optical range, the distance at which the transmittance of the light path for light of color temperature of 2700 K is equal to 0.05. This distance corresponds closely with the maximum distance at which dark objects may be observed through such an atmosphere in the daytime.[98, 100]

The distance V_o at which a large black object can be seen against the horizon sky is given by the relation:

$$\epsilon = T^{V_o} \qquad (3\text{–}22)$$

where ϵ is the minimum perceptible contrast, or threshold contrast, of the observer. A value of 0.05 is considered representative of the daylight contrast threshold of a meteorological observer.[100]

Values of transmissivity for various visibility descriptions are given in Fig. 3–48.

Equation 3–22 is a particular case of Koschmieder's law.[98, 101, 102] If the object is not black, equation 3–22 becomes

$$\epsilon = [(L_o - L_H)/L_H]\, T^V \qquad (3\text{–}23a)$$

or

$$\epsilon = C_o T^V \qquad (3\text{–}23b)$$

where L_o is the luminance of the object, L_H is the luminance of the horizon sky, V is the visual range of the object, and C_o is the inherent contrast between the object and the sky. (Note that ϵ may be either positive or negative, having the same sign of $L_o - L_H$.) Equation 3–23 applies to artificially lighted as well as naturally lighted objects.

If the object, or area, is viewed against a background other than the horizon sky equation 3–23 becomes[98]

$$\epsilon = \frac{[L_o - L_H - (L_b - L_H)T^d]T^V}{[(L_b - L_H)T^{(d+V)} + L_H]} \qquad (3\text{–}24)$$

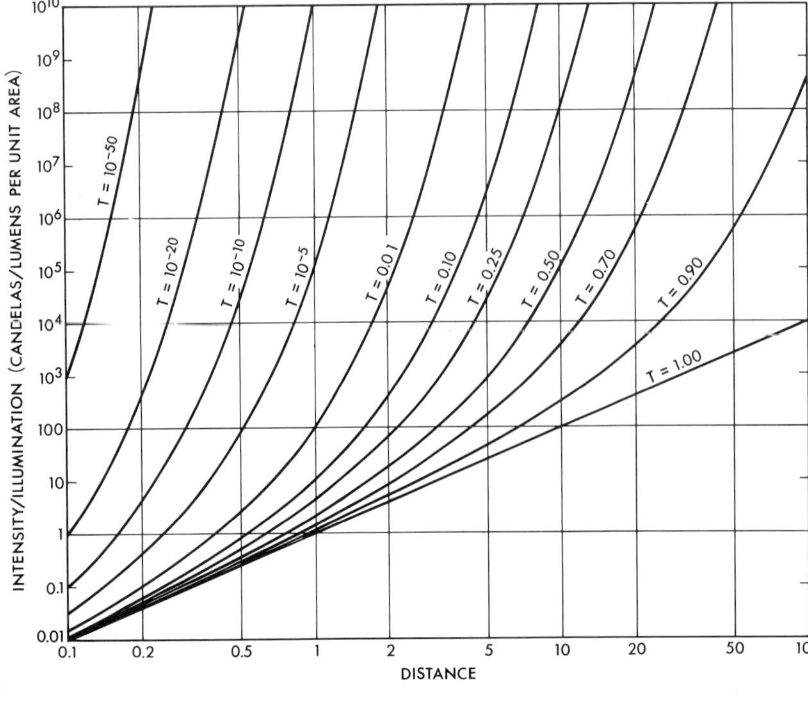

Fig. 3–47. Curves for solving Equations 3–19 and 3–20 (Allard's law) for several values of atmospheric transmissivity. Illumination, distance, and transmissivity (T) must be in consistent units.

where L_b is the inherent luminance of the background and d is the distance between the object and its background.

Equations 3–22 and 3–23 may be used without significant error in computing the visual range of objects, or area sources, viewed against a terrestrial background if the distance, d, between the object and its background exceeds one half of V_o of equation 3–22.

Equations 3–19 through 3–24 are based upon the assumption that the transmittance of the atmosphere is independent of wavelength throughout the visible portion of the spectrum. In clean fogs and in rain this assumption is usually valid. However, in smoke or dust there may, on occasion, be significant differences with the transmittance of red light being greater than the transmittance of blue light, and these equations must be applied wavelength by wavelength.

For example, equation 3–19 takes the form

$$E_m = \sum_{380}^{830} I_\lambda \, V(\lambda)[T(\lambda)]^x \Delta\lambda / x^2 \qquad (3\text{-}25)$$

where I_λ is the spectral radiant intensity of the source.

Threshold Illumination. The threshold illumination is not a constant. It is a function of the luminance of the background of the light, the position of the light in the field of view, the angular size and shape of the light, its color, and, if it is not steady burning, its flash characteristics. In addition, the observer's knowledge of the position of the light and his time for search have a significant influence on the threshold.

Fig. 3–49 shows the relation between threshold illumination and background luminance for steady-burning, white, point sources for about 98 per cent probability of detection. The threshold illumination values shown are applicable only when the observer knows precisely where to look for the light. Even if

Fig. 3–48. Transmissivities and Meteorological Optical Ranges for Various Visibility Descriptions

Visibility Description	Meteorological Optical Range, V_0 (miles)	Transmissivity (T per mile)
Exceptionally clear	30+	>0.90
Very clear	30	.90
Clear	10	.74
Light Haze	5	.55
Haze	2	.22
Thin fog	1	.05
Light fog	½	.0025
Moderate fog	¼	$10^{-5.2}$
Thick fog	⅛	$10^{-10.4}$
Dense fog	1/16	$10^{-20.8}$
Very dense fog	100 ft.	10^{-69}
Exceptionally dense fog	50 ft.	10^{-137}

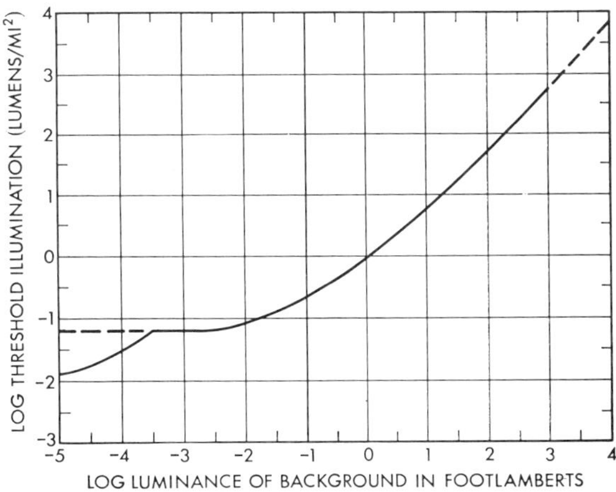

Fig. 3–49. Threshold illumination at the eye from an achromatic (white) point source for about 98 per cent probability of detection as a function of the background luminance.

the illumination is twice the values shown the light will be hard to find. The illumination values must be increased by a factor of 5 to 10 if the light is to be easy to find.[104]

These increases in illumination are applicable only when the observer is looking for the light signal. Much greater increases are needed if the light signal is to attract the attention of an observer who is not searching for it. Factors of 100 to 1000 are not excessive.[105]

The break in the curve represents the changes from cone to rod vision. At low background luminances, the threshold illumination for cone vision remains essentially unchanged as indicated by the broken line at the left. The horizontal portion of the curve represents most night seeing conditions since a light used as a signal is usually observed by looking directly at it; hence cone, not rod, vision is used. Moreover, it is doubtful whether those engaged in transport, with the possible exception of lookouts on ships, ever reach the state of dark adaptation required for rod vision.

Representative background luminances are in Fig. 3–50. It should be noted that the luminance of the night sky in the vicinity of cities and airports seldom falls below 0.001 footlambert because of the effects of man-made sources. Note also that, unless there are glare sources in the field of view, it is probably necessary to consider only the background in the immediate vicinity of the light.[106]

Source Size. The threshold illumination values shown in Fig. 3–49 are applicable to sources which are in effect point sources. Fig. 3–51 shows the maximum diameter of a source which may be considered a point source.[47] Most signal lights behave as point sources. Approximate thresholds for sources which are too large to be considered point sources may be obtained by multiplying the thresholds obtained from Fig. 3–49 by the size factors given in Fig. 3–52.[107]

Figs. 3–51 and 3–52 apply only to threshold and near threshold viewing. Recent work has shown that the intensity of a red traffic signal light required to produce optimum recognition under bright daylight conditions is independent of source size for sources subtending up to 16.5 minutes of arc.[108]

Fig 3-50. Luminance Values of Various Backgrounds Against Which Luminous Signals Are Viewed

Background	Background Luminance (footlamberts)
Horizon sky	
Overcast, no moon	0.00001
Clear, no moon	.0001
Overcast, moon	.001
Clear, moonlight	.01
Deep twilight	.1
Twilight	1
Very dark day	10
Overcast day	100
Clear day	1000
Clouds, sun-lighted	10000
Daylight fog	
Dull	100-300
Typical	300-1000
Bright	1000-5000
Ground	
On sunny day	100
On overcast day	10-30
Snow, full sunlight	5000

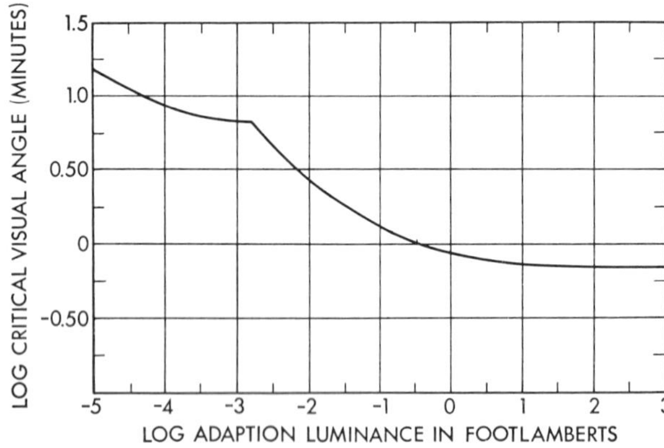

Fig. 3-51. Maximum angular diameter of a circular source which can be considered strictly as a point source.

Fig. 3-52. Size Factors for Sources Other Than Point Sources

Ratio of Source Diameter to Viewing Distance	Size Factor	
	Night	Day
0.0005	1.0	1.0
0.001	1.0	1.2
0.003	1.1	2.5
0.005	1.4	4.9
0.01	2.5	20.0

Colored Light Signals

The threshold illumination for colored light signals is about the same as that for white light, slightly less for red, somewhat more for green. However, 5 to 20 times as much light is required to permit identification of the color.[109] This increase is of the same order as that required to make a light easy to find. Thus, in service, the color of a light signal will usually be recognizable when the light is first seen. The apparent color of distant lights may be changed, sometimes very considerably, by the selective attenuation of the atmosphere. This effect is most marked when the restriction to vision is due primarily to smoke.

Flashing Light Signals

When a light signal consists of separate flashes, the instantaneous intensity during the flashes must be greater than the intensity required of a fixed (steady-burning) light in order that the light may be seen.[110] It is convenient to evaluate flashing lights in terms of their effective intensity, *i.e.*, the intensity of a fixed light of the same color, size, and shape which will produce the same visual range under identical conditions of observation. The effective intensity is computed from the relation:[111]

$$I_e = \left(\int_{t_1}^{t_2} I dt\right) / (a + t_2 - t_1) \qquad (3\text{-}26)$$

where I_e is the effective intensity, I is the instantaneous intensity, and t_1 and t_2 are the times in seconds of the beginning and end of that part of the flash when the value of I exceeds I_e. This choice of times maximizes the value of I_e.

Methods have been developed whereby the maximum value of I_e can be obtained in two or three steps. "Cut and try" methods are not necessary.

The term a is a function of the effective illumination produced by the flashing source at the eye. The value of a decreases as this illumination increases above threshold. Various values of a have been suggested from time to time as the most applicable to a particular signal problem. In the United States a value of 0.2 is usually used for specification purposes and, to promote uniformity, as the nighttime threshold value.[111] This value was chosen as an interim value until a value appropriate to the search situation is determined by definitive experimental work. Thus, by definition

$$I_e = \left(\int_{t_1}^{t_2} I dt\right) / (0.2 + t_2 - t_1) \qquad (3\text{-}27)$$

If the duration of the flash is less than about one millisecond, the effective intensity is given by:

$$I_e = 5 \int I dt \qquad (3\text{-}28)$$

where the integration is performed over the entire flash cycle. The effective intensity can then be measured directly using simple electronic integration.

Flashing lights are often used where increased conspicuity is desired. In addition the use of a flashing instead of a steady signal will provide an increase in effectiveness.[112]

REFERENCES

1. Kuhn, H. S.: *Eyes and Industry*, C. V. Mosby Company, St. Louis, 1950. Reznick, L.: *Eye Hazards in Industry*, Columbia University Press, New York, 1941. *Proceedings of a Conference on Industrial Ophthalmology*, Columbia University Press, New York, 1947. Hofstetter, H. W.: *Industrial Vision*, Chilton Company, Philadelphia, 1956. *Industrial and Traumatic Ophthalmology*, C. V. Mosby Company, St. Louis, 1964.
2. Adler, F. H.: *Physiology of the Eye*, C. V. Mosby Company, St. Louis, 1965.
3. Wald, G., Brown, P. K., and Gibbons, I. R.: "The Problem of Visual Excitation," *J. Opt. Soc. Amer.*, Vol. 53, p. 20, 1963. Wald, G.: "Molecular Basis of Visual Excitation," *Sci.*, Vol. 162, p. 230, 1968.
4. Rushton, W. A. H.: "Dark-Adaptation and the Regeneration of Rhodopsin," *J. Physiol.*, Vol. 156, p. 166, 1961. Rushton, W. A. H.: "Bleached Rhodopsin and Visual Adaptation," *J. Physiol.*, Vol. 181, p. 645, 1965.
5. Crawford, B. H.: "Visual Adaptation in Relation to Brief Conditioning Stimuli," *Proc. Roy. Soc.*, (London), Vol. 134B, p. 283, 1947. Bush, W. R.: "Foveal Light Adaptation as Affected by the Spectral Composition of the Test and Adapting Stimuli," *J. Opt. Soc. Amer.*, Vol. 45, p. 1047, 1955. Boynton, R. M., and Kandel, G.: "On-Responses in the Human Visual System as a Function of Adaptation Level," *J. Opt. Soc. Amer.*, Vol. 47, pp. 275, 1957. Battersby, W. S. and Wagman, I. W.: "Neural Limitations of Visual Exitability, I.: The Time Course of Monocular Light Adaptation," *J. Opt. Soc. Amer.*, Vol. 49, p. 752, 1959. Wagman, I. W. and Battersby, W. S.: "Neural Limitations of Visual Excitability, II: Retrochiasmal Interaction," *Amer. J. Physiol.*, Vol. 197, p. 1237, 1959. Boynton, R. M.: "Some Temporal Factors in Vision," *Sensory Communication*, Rosenblith, W. A. (Editor), MIT Press and Wiley, Cambridge and New York, 1961. Baker, H. D.: "Initial Stages of Light and Dark Adaptation," *J. Opt. Soc. Amer.*, Vol. 53, p. 98, 1963. Rinalducci, E. J.: "Early Dark Adaptation as a Function of Wavelength and Preadapting Level," *J. Opt. Soc. Amer.*, Vol. 57, p. 1270, 1967.
6. Riggs, L. A., Ratliff, F., Cornsweet, J. C., and Cornsweet, T. N.: "The Disappearance of Steadily Fixated Test Objects," *J. Opt. Soc. Amer.*, Vol. 43, p. 495, 1953. Ditchburn, R. W. and Fender, D. H.: "The Stabilized Retinal Image," *Optica Acta*, Vol. 2, p. 128, 1955. Heckenmueller, E. G.: "Stabilization of the Retinal Image—A Review of Method, Effects, and Theory." *Psychol. Bull.*, Vol. 63, p. 157, 1965. Cohen, W.: "Spatial and Textural Characteristics of the Ganzfeld," *Amer. J. Psychol.*, Vol. 70, p. 403, 1957. Avant, L. L.: "Vision in the Ganzfeld," *Psychol. Bull.*, Vol. 64, p. 246, 1965. Clarke, F. J. J.: "Visual Recovery Following Local Adaptation of the Peripheral Retina (Troxler's Effect)," *Optica Acta*, Vol. 8, p. 121, 1961.
7. Boynton, R. M. and Miller, N. D.: "Visual Performance Under Conditions of Transient Adaptation," *Illum. Eng.*, Vol. LVII, p. 541, August, 1963. Boynton, R. M., Rinalducci, E. J., and Sternheim, C.: "Visibility Losses Produced by Transient Adaptational Changes in the Range from 0.4 to 4000 Footlamberts," *Illum. Eng.*, Vol. 64, p. 217, April, 1969. Boynton, R. M., Corwin, T. R., and Sternheim, C.: "Visibility Losses Produced by Flash Adaptation," *Illum. Eng.*, Vol. 65, p. 259, April, 1970.
8. Hecht, S.: "Rods, Cones, and the Chemical Basis of Vision," *Physiol. Rev.*, Vol. 125, p. 239, 1937. Lythgoe, R. J.: "The Mechanisms of Dark Adaptation—A Critical Resume," *Brit. J. Ophthalmol.*, Vol. 24, p. 21, 1940. Craik, J. W., and Vernon, M. D.: "The Nature of Dark Adaptation," *Brit. J. Psychol.*, Vol. 32, p. 62, 1941. Arden, G. B., and Weale, R. A.: "Nervous Mechanisms and Dark Adaptation," *J. Physiol.*, Vol. 125, p. 417, 1954. Barlow, H. B.: "Dark-Adaptation: A New Hypothesis," *Vis. Res.*, Vol. 4, p. 47, 1964. Barlow, H. B., and Sparrock, J. M. B.: "The Role of Afterimages in Dark Adaptation," *Sci.*, Vol. 44, p. 1309, 1964. Rushton, W. A. H.: "The Ferrier Lecturer: Visual Adaptation," *Proc. Roy. Soc.*, (London), Vol. 162B, p. 20, 1965. Rushton, W. A. H.: "Light and Dark Adaptation of the Retina," *The Retina: Morphology, Function, and Clinical Characteristics*, Straatsma, B. R., Hall, M. O., Allen, R. A., and Crescitelli, F. (Editors), Berkeley and Los Angeles, University of California Press, 1969.
9. Reeves, P.: "The Response of the Average Pupil to Various Intensities of Light," *J. Opt. Soc. Amer.*, Vol. 4, p. 35, January, 1920.
10. Graham, C. H.: "Color Theory," in S. Koch, Ed., *Psychology: A Study of a Science*, Study I. Vol. I, McGraw-Hill Book Co., Inc., New York, 1959. LeGrand, Y.: *Light, Colour and Vision*, John Wiley & Sons, Inc., New York, 1957.
11. Burnham, R. W., Hanes, R. M., and Bartleson, C. J.: *Color: A Guide to Basic Facts and Concepts*, John Wiley & Sons, Inc., New York, 1963.
12. Ripps, H., and Weale, R. A.: "Cone Pigments in the Normal Human Fovea," *Vis. Res.*, Vol. 3, p. 531, 1963. Weale, R. A.: "Vision and Fundus Reflectometry—A Review," *Vis. Res.*, Vol. 4, p. 67, 1965. Rushton, W. A. H.: "The Density of Chlorolabe in the Foveal Cones of the Protanope," *J. Physiol.*, Vol. 168, p. 360, 1963. Ruston, W. A. H.: "A Foveal Pigment in the Deuteranope," *J. Physiol.*, Vol. 176, p. 24, 1965. Rushton, W. A. H.: "Cone Pigment Kinetics in the Deuteranope," *J. Physiol.*, Vol. 176, p. 38, 1965. Marks, W. B., Dobelle, W. H., and MacNichol, E. F.: "Visual Pigments of Single Primate Cones," *Sci.*, Vol. 143, p. 1181, 1964. Brown, P. K., and Wald, G.: "Visual Pigments in Single Rods and Cones of the Human Retina," *Sci.*, Vol. 144, p. 145, 1964.
13. Tomita, T., Kaneko, A., Murakami, M., and Pautler, E. L.: "Spectral Response Curves of Single Cones in the Carp," *Vis. Res.*, Vol. 7, p. 519, 1967.
14. Wagner, H. G., MacNichol, E. F., Jr., and Wolbarsht, M. L.: "Functional Basis for 'On'-Center and 'Off'-Center Receptive Fields in the Retina," *J. Opt. Soc. Amer.*, Vol. 53, p. 66, 1963. MacNichol, E. F., Jr.: "Retinal Mechanisms of Color Vision," *Vis. Res.*, Vol. 4, p. 119, 1964. Wiesel, T. N., and Hubel, D. H.: "Spatial and Chromatic Interactions in the Lateral Geniculate Body of the Rhesus Monkey," *J. Neurophysiol.*, Vol. 29, p. 1115, 1966. DeValois, R. L., Abramov, I., and Mead, W. R.: "Single Cell Analysis of Wavelength Discrimination at the Lateral Geniculate Nucleus in the Macaque," *J. Neurophysiol.*, Vol. 30, p. 415, 1967. Hubel, D. H., and Wiesel, T. N.: "Receptive Fields and Functional Architecture of Monkey Striate Cortex," *J. Physiol.*, Vol. 195, p. 215, 1968. Michael, C. R.: "Receptive Fields of Single Optic Nerve Fibers in a Mammal With an All-Cone Retina," *J. Neurophysiol.*, Vol. 31, p. 249, 1968.
15. Commission Internationale de l'Eclairage, *Compte Rendu*, Geneva, 1924.
16. Commission Internationale de l'Eclairage, *Compte Rendu*, Stockholm, 1951.
17. Goodeve, C. F.: "Relative Luminosity in the Extreme Red," *Proc. Roy. Soc.* (London), Vol. A155, p. 664, July, 1936.
18. Maxwell, J. C.: "On the Unequal Sensibility of the Foramen Centrale to Light of Different Colors," *Rep. Brit. Ass.*, Vol. 12, 1956. Brindley, G. S., and Wilmer, E. N.: "The Reflexion of Light from the Macular and Peripheral Fundus Oculi in Man," *J. Physiol.*, Vol. 116, p. 350, 1952. Brindley, G. S.: *Physiology of the Retina and the Visual Pathway*, London, Edward Arnold Ltd., 1960. Marriott, F. H. C.: "Colour Vision," *The Eye, Vol. II: The Visual Process*, H. Davson (Editor), New York, Academic Press, 1962.
19. Blackwell, H. R.: "Development and Use of a Quantitative Method for Specification of Interior Illumination Levels on the Basis of Performance Data," *Illum. Eng.*, Vol. LIV, p. 317, June, 1959.
20. Guth, S. K., Eastman, A. A., and Rodgers, R. C.: "Brightness Difference a Basic Factor in Supathreshold Seeing," *Illum. Eng.*, Vol. LVIII, p. 233, May, 1953.
21. Lythgoe, R. J.: "Measurement of Visual Acuity," Special Report No. 173, Medical Research Council, H. M. Stationery Office, London, 1932.
22. Guth, S. K. and McNelis, J. F.: "Visual Performance: A Comparison in Terms of Detection of Presence and Discrimination of Detail," *Illum. Eng.*, Vol. 63, p. 32, January, 1968. McNelis, J. F. and Guth, S. K.: "Visual Performance—Further Data on Complex Test Objects," *Illum. Eng.*, Vol. 64, p. 99, February, 1969.
23. Guth, S. K.: "Lighting Research," *Amer. Ind. Hyg. Ass. J.*, Vol. 23, p. 359, 1962. Luckiesh, M. and Moss, F. K., *The Science of Seeing*, D. Van Nostrand Company, Inc., New York, 1937.
24. Committee on Standards of Quality and Quantity for Interior Illumination of the IES: "Brightness and Brightness Ratios," Report No. 1, *Illum. Eng.*, Vol. XXXIX, p. 713, December, 1944.
25. Moon, P. and Spencer, D. E.: "The Visual Effects of Non-Uniform Surrounds," *J. Opt. Soc. Amer.*, Vol. 35, p. 233, March, 1945.
26. McFadden, H. B.: *Three Studies in Psychological Optics*, Optometric Extension Program, Duncan, Oklahoma, 1941.
27. Burian, H. M.: "Fusion Movements, Role of Peripheral Retinal Stimuli," *Arch. Ophthalmol.*, Vol. 21, p. 486, March, 1939.
28. Crawford, B. H.: "The Effect of Field Size and Pattern on the Change of Visual Sensitivity with Time," *Proc. Roy. Soc.* (London), Series B, Vol. 129, p. 94, March, 1940.
29. Fisher, M. B.: "The Relationship of the Size of the Surrounding Field to Visual Acuity in the Fovea," *J. Exp. Psychol.*, Vol. 23, p.

215, September, 1938. Luckiesh, M. and Moss, F. K.: *Seeing*, Williams and Wilkins Company, Baltimore, 1931. Cobb, P. W. and Moss, F. K.: "The Effect of Brightness on the Precision of Visually Controlled Operations," *J. Franklin Inst.*, Vol. 199, p. 507, April, 1925.

30. Fry, G. A. and Bartley, S. H.: "The Effect of One Border Upon the Threshold of Another," *Amer. J. Physiol.*, Vol. 112, p. 414, June, 1935. Fry, G. A.: "Effects of Uniform and Non-Uniform Surrounds on Foveal Vision," *Amer. J. Optm.*, Vol. 27, p. 423, September, 1950.
31. Holladay, L. L.: "The Fundamentals of Glare and Visibility," J. Opt. Soc. Amer., Vol. 12, p. 271, April, 1926.
32. Stiles, W. S.: "The Scattering Theory of the Effect of Glare on the Brightness Difference Theshold," Technical Paper No. 8 Illumination Research, Dept. of Scientific and Industrial Research, H. M. Stationery Office, London, 1929. Luckiesh, M. and Moss, F. K.: "Area and Brightness of Stimulus Related to Pupillary Light Reflex," *J. Opt. Soc. Amer.*, Vol. 24, p. 130, May, 1934. Ferree, G. E., Rand, G., and Harris, E. T.: "Intensity of Light and Area of Illuminated Field as Interacting Factors," *J. Exp. Psychol.*, Vol. 16, p. 408, 1933.
33. Stiles, W. S.: "Recent Measurements of the Effect of Glare on the Brightness Difference Theshold," *Compte Rendu*, Commission Internationale de l'Eclairage, Saranac Inn, New York, 1928.
34. Moon, P. and Spencer, D. E.: "The Specification of Foveal Adaptation," *J. Opt. Soc. Amer.*, Vol. 33, p. 444, August, 1943.
35. Duane, A.: "Normal Values of the Accommodation at All Ages," *J. Amer. Med. Ass.*, Vol. 59, p. 1010, September, 1912. Hofstetter, H. W.: "A Comparison of Duane's and Donder's Tables of the Amplitude of Accommodation," *Amer. J. Optm.*, Vol. 21, p. 345, September, 1944.
36. Birren, J. E., Casperson, R. C., and Botwinick, J.: "Age Changes in Pupil Size," *J. Gerontol.*, Vol. 5, p. 216, July, 1950.
37. Domey, R. G. and McFarland, R. A.: "Dark Adaptation and Age," *Amer. J. Ophthal.*, Vol. 51, p. 1262, 1961.
38. Gunkel, R. D. and Gouras, P.: "Changes in Scotopic Visibility Thresholds with Age," *Arch. Ophthal.*, Vol. 69, p. 4, 1963.
39. Zerbe, L. B. and Hofstetter, H. W.: "Prevalence of 20/20 with Best Previous and No Lens Correction," *J. Amer. Optom. Ass.*, Vol. 29, p. 772, 1957-58.
40. Fisher, A. J. and Christie, A. W.: "A Note on Disability Glare," *Vis. Res.*, Vol. 5, p. 565, 1965.
41. Wolf, E.: "Glare and Age," *Arch. Ophthal.*, Vol. 64, p. 502, 1960.
42. Allen, M. J. and Vos, J. J.: "Ocular Scattered Light and Visual Performance as a Function of Age," *Amer. J. Optom.*, Vol. 44, p. 717, 1967.
43. Guth, S. K. and McNelis, J. F.: "Visual Performance—Subjective Differences," *Illum. Eng.*, Vol. 64, p. 723, December, 1969.
44. Blackwell, O. M. and Blackwell, H. R.: "Visual Performance Data for 156 Normal Observers of Various Ages," *J. Illum. Eng. Soc.*, Vol. 1, October, 1971.
45. Luckiesh, M. and Moss, F. K.: *Light, Vision and Seeing*, D. Van Nostrand Company, New York, 1944.
46. Harmon, D. B.: *The Coordinated Classroom*, American Seating Company, Grand Rapids, 1949.
47. Blackwell, H. R.: "Contrast Thresholds of the Human Eye," *J. Opt. Soc. Amer.*, Vol. 36, p. 624, November, 1946.
48. Blackwell, H. R. and Taylor, J. H.: "A Consolidated Set of Foveal Contrast Thresholds for Normal Human Binocular Vision," Ohio State University and University of California at San Diego Report, 1970.
49. Blackwell, H. R. and Smith, S. W.: "Additional Visual Performance Data for Use in Illumination Specification Systems," *Illum. Eng.*, Vol. 65, p. 389, June, 1970.
50. *IES Lighting Handbook*, Third Edition, p. 2-22, Illuminating Engineering Society, 345 E. 47th Street, New York, N. Y., 1959.
51. *IES Lighting Handbook*, Fourth Edition, p. 2-12, Illuminating Engineering Society, 345 E. 47th Street, New York, N. Y., 1966.
52. Bodmann, H. W.: "Illumination Levels and Visual Performance," *Int. Light. Rev.*, Vol. 13, p. 41, 1962.
53. Boynton, R. M. and Boss, D. E.: "The Effect of Background Luminance Upon Visual Search Performance," *Illum. Eng.*, Vol. 66, p. 173, April, 1971.
54. Scott, D. E. and Blackwell, H. R.: "Visual Performance in a Landlot Ring Task Without Response Limitation," Ohio State University Report to the Illuminating Engineering Research Institute, 345 E. 47th Street, New York, N. Y., 1970.
55. Weston, H. C.: "The Relation Between Illumination and Visual Efficiency—The Effect of Brightness Contrast," Industrial Health Research Board Report No. 87, Great Britain Medical Research Council, p. 35, 1945.
56. CIE Committee E-1.4.2: "Recommended Method for Evaluating Visual Performance Aspects of Lighting," CIE Report No. 19, January, 1971.
57. Cottrell, C. L.: "Measurement of Visibility," *Illum. Eng.*, Vol. XLVI, p. 95, February, 1951.
58. Simmons, A. E. and Finch, D. M.: "An Instrument for the Evaluation of Night Visibility on Highways," *Illum. Eng.*, Vol. XLVIII, p. 517, October, 1953.
59. Schwab, R. N.: "Use of Veiling Luminance for Evaluating the Visibility of Complex Visual Tasks," Master's Dissertation, The Ohio State University, p. 47, 1962.
60. Popp, H. P.: "Die Bewertung von Sehaufgaben mit dem Geraet nach Blackwell," Doctoral Dissertation, University of Munich, Germany, p. 50, 1962.
61. Eastman, A. A.: "A New Contrast Threshold Visibility Meter," *Illum. Eng.*, Vol. 63, p. 37, January, 1968.
62. Blackwell, H. Richard: "Development of Procedures and Instruments for Visual Task Evaluation," *Illum. Eng.*, Vol. 65, p. 267, April, 1970.
63. Blackwell, H. R.: "A More Complete Quantitative Method for Specification of Interior Illumination Levels on the Basis of Performance Data," *Illum. Eng.*, Vol. 64, p. 289, April, 1969.
64. Holladay, L. L.: "Action of a Light Source in the Field of View in Lowering Visibility," *J. Opt. Soc. Amer.*, Vol. 14, p. 1, 1927.
65. Stiles, W. S.: "The Effect of Glare on the Brightness Difference Threshold," *Proc. Roy. Soc.*, London, Vol. B104, p. 322, 1929.
66. Blackwell, H. R.: "Use of Performance Data to Specify Quantity and Quality of Interior Illumination," *Illum. Eng.*, Vol. L, p. 286, June, 1955.
67. Fry, G. A., Pritchard, B. S., and Blackwell, H. R.: "Design and Calibration of a Disability Glare Lens," *Illum. Eng.*, Vol. LVIII, p. 120, March, 1963.
68. Blackwell, H. R. and Blackwell, O. M.: "The Effect of Illumination Quantity upon the Performance of Different Visual Tasks," *Illum. Eng.*, Vol. 63, p. 143, April, 1968.
69. Committee on Recommendations for Quality and Quantity of Illumination of the IES: "RQQ Report No. 4—A Method of Evaluating the Visual Effectiveness of Lighting Systems," *Illum. Eng.*, Vol. 65, p. 505, August, 1970.
70. Luckliesh, M. and Guth, S. K.: "Brightnesses in Visual Field at Borderline Between Comfort and Discomfort (BCD)," *Illum. Eng.*, Vol. XLIV, p. 650, November, 1949.
71. Guth, S. K. and McNelis, J. F.: "A Discomfort Glare Evaluator," *Illum. Eng.*, Vol. LIV, p. 398, June, 1959.
72. Guth, S. K. and McNelis, J. F.: "Further Data on Discomfort Glare From Multiple Sources," *Illum. Eng.*, Vol. LVI, p. 46, January, 1961.
73. Bradley, R. D .and Logan, H. L.: "A Uniform Method for Computing the Probability of Comfort Responses in a Visual Field," *Illum. Eng.*, Vol. LIX, p. 189, March, 1964.
74. Guth, S. K.: "A Method for the Evaluation of Discomfort Glare," *Illum. Eng.*, Vol. LVIII, p. 351, May, 1963.
75. Committee on Recommendations of Quality and Quantity of Illumination of the IES: "Outline of a Standard Procedure for Computing Visual Comfort Ratings for Interior Lighting—Report No. 2," *Illum. Eng.*, Vol. LXI, p. 643, October, 1966.
76. Guth, S. K.: "Computing Visual Comfort Ratings for a Specific Interior Lighting System," *Illum. Eng.*, Vol. LXI, p. 634, October, 1966.
77. McGowan, T. K. and Guth, S. K.: "Extending and Applying the IES Visual Comfort Rating Procedure," *Illum. Eng.*, Vol. 64, p. 253, April, 1969.
78. Committee on Testing Procedures of the IES: "Determination of Average Luminance of Luminaires," *J. Illum. Eng. Soc.*, Vol. 1, January, 1972.
79. Committee on Recommendations of Quality and Quantity of Illumination of the IES: "Extending the IES Visual Comfort Rating System—Report No. 5," in preparation.
80. Fry, G. A.: "Limits of the Field of View," *Illum. Eng.*, Vol. 64, p. 403, May, 1969. Lighting Design Practice Committee of the IES: "Calculation of Luminance Coefficients Based Upon the Zonal-Cavity Method," *Illum. Eng.*, Vol. 63, p. 423, August, 1968.
81. Committee on Recommendations of Quality and Quantity of Illumination of the IES: "An Alternate Simplified Method for Determining the Acceptability of a Luminaire, from the VCP Standpoint, for use in Large Rooms—Report No. 3," *J. Illum. Eng. Soc.*, to be published.
82. Fry, G. A.: "An Approximate Formula for Discomfort Glare," to be published.
83. Crouch, C. L.: "Derivation, Background and Use of the Scissors Curve," *Illum. Eng.*, Vol. LX, p. 399, June, 1965.
84. Petherbridge, P. and Hopkinson, R. G.: "A Preliminary Study of Reflected Glare," *Trans. Illum. Eng. Soc.* (London), Vol. XX, No. 8, 1955.
85. Finch, D. M.: "The Effect of Specular Reflection on Visibility, Part I—Physical Measurements for the Determination of Brightness and Contrast," *Illum. Eng.*, Vol. LIV, p. 474, August, 1959.

86. Chorlton, J. F. and Davidson, H. F.: "The Effect of Specular Reflection on Visibility, Part II—Field Measurements of Loss of Contrast," *Illum. Eng.*, Vol. LIV, p. 482, August, 1959.

87. Blackwell, H. R.: "A General Quantitative Method for Evaluating the Visual Significance of Reflected Glare, Utilizing Visual Performance Data," *Illum. Eng.*, Vol. LVIII, p. 161, April, 1963.

88. Allphin, W.: "Sight Lines to Desk Tasks in Schools and Offices," *Illum. Eng.*, Vol. LVIII, p. 244, April, 1963.

89. Crouch, C. L. and Kaufman, J. E.: "Practical Application of Polarization and Light Control for Reduction of Reflected Glare," *Illum. Eng.*, Vol. LVIII, p. 277, April, 1963.

90. Sampson, F. K.: "Contrast Rendition in School Lighting," EFL Technical Report No. 4, Educational Facilities Laboratories, 477 Madison Avenue, New York, January, 1970.

91. Lewin, I. and Griffith, J. W.: "Veiling Reflection Control by Candlepower Distribution," *Illum. Eng.*, Vol. 65, p. 594, October, 1970.

92. Blackwell, H. R.: "Measurement of the Contrast Rendition Factor for Pencil Handwritten Tasks," *Illum. Eng.*, Vol. 65, p. 199, April, 1970.

93. Florence, N. S. and Glickman, S. B.: "An Evaluation of Troffer Lighting Systems with Respect to Veiling Reflections," *Illum. Eng.*, Vol. 66, p. 149, March, 1971.

94. Jones, B. F. and Sampson, F. K.: "Contrast Rendition Factor Calculation and Measurement for the Standard Classroom Task," a paper presented at the Annual IES Conference, Chicago, Illnois, August, 1971.

95. Squillace, S. S. and DiLaura, D.: "Comprehensive Prediction of Contrast Losses," a paper presented at the Annual IES Conference, Chicago, Illinois, 1971.

96. Goodbar, I., "Simplified Prediction of Contrast Losses," a paper presented at the Annual IES Conference, Chicago, Illinois, August, 1971.

97. Reynaud, L.: "Memoir upon the Illumination and Beaconage of the Coasts of France," Paris, 1864 (Translation—Hains, Peter C., Government Printing Office, Washington, 1876).

98. Middleton, W. E. K.: *Vision Through the Atmosphere*, University of Toronto Press, Toronto, 1952.

99. Kevern, G. M.: "Effect of Source Size upon Approach Light Performance," *Illum. Eng.*, Vol. XLV, p. 96, February, 1950.

100. Douglas, C. A. and Young, L. L.: "Development of a Transmissimeter for Determining Visual Range," Civil Aeronautics Administration Technical Development Report No. 47, Washington, 1945.

101. Koschmieder, H.: "Theorie der Horizontalen Sichtweite," *Beitr. Phys. Freien Atm.* 12, 33, and 171, 1924. Duntley, S. Q.: "The Reduction of Apparent Contrast by the Atmosphere," *J. Opt. Soc. Amer.*, Vol. 38, p. 179, February, 1948.

102. Duntley, S. Q., Gordon, J. I., Taylor, J. H., White, C. T., Boileau, A. R., Tyler, J. E., Austin, R. W., and Harris, J. L.: "Visibility," *Appl. Opt.*, Vol. 3, p. 549, May, 1964.

103. Zuev, V. E.: "Atmospheric Transparency in the Visible and the Infrared, (*Prozrachnost' atmosfery dlya vidimykh i infrakrasykh lnchei*)," *Izdatel' stv "Sovetskoe Radio,"* Moskva, 1966,—Translated into English by Z. Lerman, Israel Program for Scientific Translations, Jerusalem, 1970. Available from National Technical Information Service, Springfield, Virginia 22151 as TT69-55102.

104. Tousey, R. and Koomen, M. J.: "The Visibility of Stars and Planets during Twilight," *J. Opt. Soc. Amer.*, Vol. 43, p. 177, March, 1953.

105. Breckenridge, F. C. and Douglas, C. A.: "Development of Approach- and Contact-Light Systems," *Illum. Eng.*, Vol. XL, p. 785, November, 1945.

106. Knoll, H. A., Tousey, R., and Hulbert, E. O.: "Visual Thresholds of Steady Point Sources of Light in Fields of Brightness from Dark to Daylight," *J. Opt. Soc. Amer.*, Vol. 36, p. 480, August, 1946.

107. de Boer, J. B.: "Visibility of Approach and Runway Lights," *Philips Res. Rep.*, Vol. 6, p. 224, 1951.

108. Cole, B. L. and Brown, B.: "Specifications of Road Traffic Signal Light Intensity," *Hum. Factors*, Vol. 10, p. 245, June, 1968.

109. Stiles, W. S., Bennett, M. G., and Green, H. N.: "Visibility of Light Signals with Special Reference to Aviation Lights," A. R. C. Reports and Memoranda No. 1793, H. M. Stationery Office, London, 1837. Hill, N. E. G.: "The Recognition of Colored Light Signals which are Near the Limit of Visibility," *Proc. Phys. Soc.*, Vol. 59, p. 560, 1947. Hill, N. E. G.: "The Measurement of the Chromatic and Achromatic Thresholds of Colored Point Sources against a White Background," *Proc. Phys. Soc.*, Vol. 59, p. 574, 1947.

110. Blondel, A. and Rey, J.: "The Perception of Lights of Short Duration at Their Range Limits," *Trans. Illum. Eng. Soc.*, Vol. VII, p. 625, November, 1912. The paper is a complete account in English of the material contained in two papers published in the *J. de Phy. et Radium*, Vol. 1, Series 5, pp. 530, 643, 1911. Projector, T. H.: "Effective Intensity of Flashing Lights," *Illum. Eng.*, Vol. LII, p. 630, December, 1957.

111. Committee on Aviation of the IES: "IES Guide for Calculating the Effective Intensity of Flashing Signal Lights," *Illum. Eng.*, Vol. LIV, p. 747, November, 1964. Douglas, C. A.: "Computation of the Effective Intensity of Flashing Lights," *Illum. Eng.*, Vol. LII, p. 641, December, 1957.

112. Projector, T. H.: "Efficiency of Flashing Lights," *Illum. Eng.*, Vol. LIII, p. 600, November, 1958.

SECTION

4

MEASUREMENT OF LIGHT AND OTHER RADIANT ENERGY

Basis of Photometry *4–1* *Photometers* *4–4* *Laboratory Measurements* *4–10*
Field Measurements *4–26*

Progress in a branch of science or engineering is dependent, to a very large extent, on the ability to measure the quantities associated with that subject. Thus, each advance in measuring technique means a broadening of knowledge.

The measurement of light is called photometry, and the basic instrument employed is known as a photometer. The earlier photometers depended on visual appraisal as the means of measurement. The eye can judge only brightness; visual photometry usually requires that light first strike a diffusing surface from which it is directed toward the eye. Further, since the eye is an accurate instrument only when deciding that two adjacent brightnesses are equal, visual photometers are always devices for comparing two brightnesses; one being unknown, and the other having at least a relation to a known value.

"Physical" photometers differ in performance from the human eye because they respond to illumination, or concentration of radiant energy. Radiant energy incident upon physical receivers produces a change in electrical quantities which can be measured.

Spectral Luminous Efficiency Curve. In general, light measurements with physical instruments are useful only if they indicate reliably how the eye would react to a certain stimulus. In other words, such an instrument should be sensitive to the spectral energy distribution of light in the same way as the eye. Because of very substantial differences between individual pairs of eyes in the latter respect, the CIE standard observer response curve has been established (also known as the eye sensitivity curve.) See Figs. 3–7, 3–8 and 4–5. Therefore, the sensitivity characteristics of a physical receiver should be the equivalent of this standard observer. The required match is usually accomplished by adding filters between the light-sensitive element and the light source. See page 4–6.

Photopic and Scotopic Vision. All ordinary photometric measurements are assumed to be made with the observer's eye in the photopic or light-adapted state. When the luminance (photometric brightness) of the surface to be measured is much below 0.01 footlambert, the eye can no longer be regarded as light-adapted. Most measurements of fluorescent and phosphorescent materials are made in the scotopic (dark-adapted) range, or the region between the scotopic and photopic. Because of the change in the spectral response of the eye at these low levels (see Fig. 3–8), it is necessary to take special measures in evaluating the results of such measurements.[1]

NOTE: References are listed at the end of each section.

Measurable Quantities

As indicated in Fig. 4–1, many characteristics of light, light sources, lighting materials, and lighting installations, may be measured. In addition to the characteristics covered in Fig. 4–1, most of the standard photometric terms that apply to the material in this section are also defined in Section 1. Further information on ultraviolet, visible light, and infrared are covered in Section 25.

The measurements of most general interest are:

1. Illumination.
2. Luminance (photometric brightness).
3. Luminous intensity.
4. Luminous flux.
5. Contrast.
6. Color (appearance and rendering).
7. Spectral distribution.
8. Electrical characteristics.
9. Radiant energy.

BASIS OF PHOTOMETRY[2]

Principles

Photometric measurements frequently involve a consideration of the inverse-square law (which is strictly applicable only for point sources) and the cosine law.

Fig. 4—1. Some Measurable Characteristics of Light, Light Sources, and Lighting Materials

Characteristic	Dimensional Unit	Equipment	Technique
Light			
Wavelength	Nanometer†	Interference grating	Laboratory
Color	None	Spectrophotometer and colorimeter	Laboratory
Flux density (illumination)	Lumen per square foot (footcandle)‡	Photometer	Laboratory or field
Orientation of polarization	Degree (angle)	Analyzing Nicol prism	Laboratory
Degree of polarization*	Per cent (dimensionless ratio)	Polarization photometer	Laboratory
Light Sources			
Energy radiated	Joule per square meter	Calibrated radiometer	Laboratory
Color temperature	Kelvin (K)	Colorimeter or filtered photometer	Laboratory or field
Luminous intensity	Candela§	Photometer	Laboratory or field
Luminance (photometric brightness)	Candela per unit area§	Photometer or luminance meter	Laboratory or field
Spectral energy distribution	Joule per nanometer	Spectrometer	Laboratory
Power consumption	Watt	Wattmeter, or voltmeter and ammeter for dc, and unity power factor ac circuits	Laboratory or field
Light output (total flux)	Lumen	Integrating sphere photometer	Laboratory
Zonal distribution	Lumen or candelas§	Distribution or goniophotometer	Laboratory
Lighting Materials			
Reflectance	Per cent (dimensionless ratio)§	Reflectometer	Laboratory or field
Transmittance	Per cent (dimensionless ratio)§	Photometer	Laboratory or field
Spectral reflectance and transmittance	Per cent (at specific wavelengths)§	Spectrophotometer	Laboratory
Optical density	Dimensionless number§	Densitometer	Laboratory

* Committee on Testing Procedures for Illumination Characteristics of the IES: "Resolution on Reporting Polarization," *Illum. Eng.*, Vol. LVIII, p. 386, May 1963.
† See Fig. 1-8 for conversion factors for units of length.
‡ See Fig. 1-9 for illumination conversion factors.
§ See Fig. 1-1 for standard units, symbols and defining equations for fundamental photometric and radiometric quantities.

Inverse-Square Law. The inverse-square law (see Fig. 4–2a) states that the illumination E at a point on a surface varies directly with the candlepower I of the source, and inversely as the square of the distance d between the source and the point. If the surface at the point is normal to the direction of the incident light, the law may be expressed as follows:

$$E = \frac{I}{d^2}$$

This equation holds true within one-half per cent when d is at least five times the maximum dimension of the source (or luminaire) as viewed from the point on the surface.

Cosine Law. The Lambert cosine law (see Fig. 4–2b) states that the illumination of any surface varies as the cosine of the angle of incidence. The angle of incidence θ is the angle between the normal to the surface and the direction of the incident light. The inverse-square law, and the cosine law, can be combined as follows:

$$E = \frac{I}{d^2} \cos \theta$$

Cosine-Cubed Law. A useful extension of the cosine law is the "cosine-cubed" equation (see Fig. 4–2c). By substituting $h/\cos\theta$ for d, the above equation may be written:

$$E = \frac{I \cos^3 \theta}{h^2}$$

Standards[3]

Standards of candlepower, luminous flux, and color are established by national physical laboratories. Various types of standards may be used in photometric laboratories.

Primary Standards. A primary standard, reproducible from specifications, is usually found only in a national physical laboratory. See Section 1.

Secondary Standards. Secondary standards are usually derived directly from primary standards. A secondary standard must be prepared using precise electrical and photometric equipment, and is usually maintained at a national physical laboratory. See Section 1.

Reference Standards. Reference standards are lamps standardized against others of known preci-

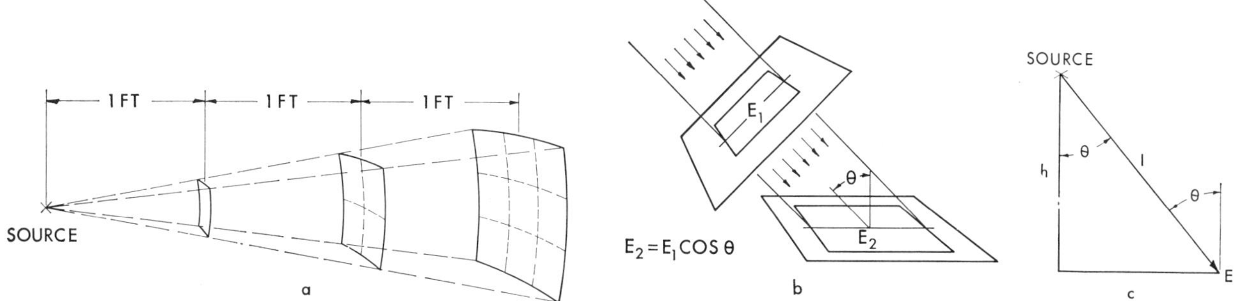

Fig. 4–2. *a.* The inverse-square law; illustrating how the same quantity of light flux is distributed over a greater area, as the distance from source to surface is increased. *b.* The Lambert cosine law; showing that light flux striking a surface at angles other than normal is distributed over a greater area. c. The cosine-cubed law; explaining the transformation of the formula.

sion. They are used in photometric laboratories throughout the industry. There are various grades of such standards, depending on the number of steps removed from secondary standards. There is no established terminology to distinguish between these grades. Preservation of the rating is of prime importance; accordingly, a standard will be used as seldom as possible. Such a standard is used, handled, and stored with the greatest care.

The highest precision results from standardization against a group of secondary standards at a national physical laboratory. Such lamps will be found wherever photometric accuracy warrants the highest attainable precision.

Standards may be prepared from the above by any well equipped laboratory, for broader distribution.

Working Standards. Working standards may be prepared for everyday use, because of the cost of the reference standards. A laboratory prepares its own working standards for use in calibrating a photometer. Even a working standard, however, is not used during the conduct of a test, except where a direct comparison is necessary.

General Methods

Photometric measurements, in general, make use of the basic laws of photometry mentioned on page 4–2, applied to readings from visual photometric comparisons, or photoelectric instruments. Various types of procedures are designated as follows:

Direct Photometry. Direct photometry consists of the simultaneous comparison of a standard and an unknown light source.

Substitution Photometry. Substitution photometry consists of the sequential evaluation of the desired photometric characteristics of a standard and an unknown light source in terms of an arbitrary reference.

Relative Photometry. To avoid the use of standard lamps, the relative method is widely applied. Relative photometry consists of the evaluation of the desired photometric characteristic of one uncalibrated light source in terms of another uncalibrated light source.

Photometry of Non-Steady-State Sources. It is sometimes necessary to measure the output of sources which are non-steady or cyclic, in which case extreme care should be taken.[4]

Means of Attenuation. In making photometric measurements, it often becomes necessary to reduce the luminous intensity of a source in a known ratio, to bring it within the range of the measuring instrument. A rotating sector disk with one or more angular apertures is one means of doing this. If such a disk is placed between a source and a surface, and rotated at such speed that the eye perceives no flicker, the effective luminance of the surface is reduced in the ratio of the time of exposure to the total time (Talbot's law). The reduction is by the factor $\theta/360$ degrees, where θ is the total angular aperture in degrees. The sector disk has advantages over many filters. It is not affected by a change of characteristics over a period of time. It reduces luminous flux without changing its spectral composition. Sector disks should not be used with light sources having cyclical variation in output.[5]

Various types of neutral filters of known transmittance are also used for attenuation. Wire mesh or perforated metal filters are perfectly neutral, but have a limited range. Partially silvered mirrors have high reflectance, and the reflected light must be controlled to avoid errors in the photometer. When this type of filter is perpendicular to the light source photometer axis, serious errors may be caused by multiple reflections between the filter and receiver surface. These may be avoided by mounting the filter at a small angle (not over 3 degrees) from the perpendicular, and at a sufficient distance from the receiver surface to throw reflections away from the photometric axis. In this canted position, care must be taken not to reflect light from adjacent surfaces on to the receiver. Also, it is difficult to secure completely uniform transmission over all parts of the surface.

So-called "neutral" glass filters are seldom neutral. In general, they have a characteristic high transmittance in the red region, and low in the blue. This may be reasonably well corrected by the use of a second filter of yellow-green glass which absorbs in the extreme red region. However, this type of filter varies in transmittance with ambient temperature, as do many other optical filters.

"Neutral" gelatin filters are quite satisfactory, although not entirely neutral. Some have a small seasoning effect, losing neutrality over a period of time. They must be protected by being cemented between two pieces of glass and watched carefully for loss of contact between the glass and gelatin.

PHOTOMETERS

A photometer is a device for measuring radiant energy in the visible spectrum. Various types of physical instruments consisting of a radiant energy sensitive element and appropriate measuring equipment are used to measure ultraviolet and infrared radiant energy. When used with a filter to correct their response to the CIE standard observer, they measure visible light and are usually called physical photometers.

In general, photometers may be divided into two classifications: (1) laboratory photometers which are usually fixed in position and yield results of highest accuracy, and (2) portable photometers of lower accuracy for making photometric measurements in the field or outside the laboratory. Each of these two classes may be subdivided into visual (subjective) photometers, and photoelectric (objective or physical) photometers. These in turn may be grouped according to function, such as photometers to measure luminous intensity (candlepower), luminous flux, illumination (illuminance), luminance (photometric brightness), light distribution, light reflectance and transmittance, color, spectral distribution, and visibility.

In recent years, visual photometric methods have largely been supplanted commercially by physical methods; however, visual methods, because of their simplicity, are still used in educational laboratories for demonstrating photometric principles, and the less routine types of photometric measurements. Even in physical photometry, the aim is to obtain an instrument which will give results in accordance with the photometric standard observer for scotopic vision so that the eye remains the ultimate judge of light, both qualitatively and quantitatively.

Visual Photometers[6]

Visual photometers are luminance comparison devices in which the eye is used to compare the brightness of two surfaces, and usually to adjust them to equality.

Bunsen "Grease Spot" Photometer. One of the oldest visual photometers, still in limited use, is the Bunsen "grease spot" disk. This consists of a translucent paraffin spot in the center of a thin opaque white paper, flanked by two mirrors forming an angle of about 135 degrees. This angle is bisected by the paper. From a point in the plane of the paper, images of both sides of the paper, formed by the mirrors, can be seen in close proximity. The photometer is mounted between two light sources to be compared, and a photometric balance is made by adjusting the distance of one or both sources until the two reflected images appear equally bright. The luminous intensity of one source may then be calculated in terms of the other, by means of the inverse-square law.

Lummer-Brodhun Photometer.[6] The Lummer-Brodhun photometer is an optical device for viewing simultaneously two sides of an opaque diffuse white plaster screen illuminated by the light sources being compared. The more precise *equality-of-contrast* sight-box is most commonly used for accurate visual photometry. The cube consists of two identical 45- by 90-degree prisms with a pattern etched in the hypotenuse face of one. The two hypotenuse faces are assembled together to make optical contact. Where the surfaces are in optical contact, light is transmitted, and the luminance of the two fields is compared through a viewing telescope. In the contrast cube, a balance is secured by matching one contrast field centered in an outer field, with another contrast and outer field. In the second, or simple version of the Lummer-Brodhun cube, a photometric balance is secured when the two fields, illuminated by the standard and test sources, respectively, merge and the lines of demarcation between the two fields disappear. This *equality-of-brightness* sight-box is less sensitive than the *equality-of-contrast* sight-box. If there should happen to be a color difference in the two light sources being compared, difficulty may be experienced in making a photometric balance. In this case, methods of heterochromatic photometry, such as one involving the flicker photometer may be used.

Flicker Photometer. The flicker photometer was developed as a means of comparing two sources of different colors.[7] A single field is alternately illuminated by two sources so that the observer sees a flicker, which may be due to color difference, difference in luminance, or both. Above a certain rate of alternation the color differences disappear while the photometric balance is obtained by moving the sight-box back and forth, the criterion of equality of luminance of the two sources being the disappearance of flicker. In the period from 1948 to 1952, the *Bureau of International des Poids et Mesures* organized measurements and concluded from the results that measurements by flicker photometry were not in agreement with results obtained from spectrophotometric data. Other systems of heterochromatic photometry are the cascade method, the composition or mixture method and the use of color matching or color equalizing optical filters.[8]

Macbeth Illuminometer.[9] One of the more widely used portable visual photometers is the Macbeth illuminometer (Fig. 4-3a). It contains a Lummer-Brodhun cube of the *equality of brightness* type of which one field is a view of the test surface to be measured. The other field is lighted by a small com-

Fig. 4-3. a. The Macbeth illuminometer. b. The Luckiesh-Taylor luminance meter. c. The Freund brightness spot meter. d. The Pritchard photometer.

parison lamp mounted on a rack and pinion in a tube. The two fields are balanced by moving the comparison lamp back and forth in the tube. The inverse-square principle determines the result. The illuminometer may be equipped with a lens to restrict the test field and bring it into focus. The range can be extended over very wide limits by inserting calibrated neutral filters in either the test or the comparison lamp side. Colored filters may also be used to correct the color of the comparison lamp to that of a test surface. The measurement of illumination is generally accomplished by measuring the luminance of a small circular test plate of known reflectance furnished with the instrument. Illuminometers of the visual type may also be used for measuring the luminance of any surface which is large enough to fill the instrument's field of view.

Luckiesh-Taylor Luminance Meter.[10] The Luckiesh-Taylor meter (Fig. 4-3b) is a small, completely self-contained instrument for measurement of luminance or illumination. A lens in the eyepiece brings the external test field into focus in the same plane as the comparison field, which is seen as two small trapezoids against the circular test field. Light for the comparison field is supplied by a small lamp. The luminance of the comparison field is adjusted to match the test field by rotating a photographic film gradient. An illuminated scale, calibrated in footlamberts and candelas per square inch, is seen through a second eyepiece. Neutral filters greatly extend the range of values that can be measured.

Taylor Reflectometer. The Taylor reflectometer is a forerunner of the Baumgartner reflectometer (see page 4-7) and is a considerably smaller instrument adapted to employ the Macbeth illuminometer as the measuring element. With this instrument, the luminance of a very small spot on the sphere wall is measured; the openings in the sphere are smaller in proportion, and smaller baffles are used. The Taylor instrument is very accurate and has supplied the means for an intensive study of the theory of reflectometry. A modern version of this instrument which substitutes electronic for visual measurements has been described in available literature.[10]

Photoelectric Photometers[11]

Photoelectric photometers are convenient to use, eliminate personal judgment and variations between individual observers, and, under the best conditions, give very accurate results. Hence, these receivers have largely replaced visual devices for most routine measuring procedures. Photoelectric photometers may be divided into two classes: those employing solid state devices such as photovoltaic and photoconductive cells; and those employing photoemissive tubes, which require considerable additional equipment for operation.

Photovoltaic Cell Meters.[12] A photovoltaic cell is one that directly converts radiant energy into electrical energy. It not only provides a small current which is approximately proportional to the inci-

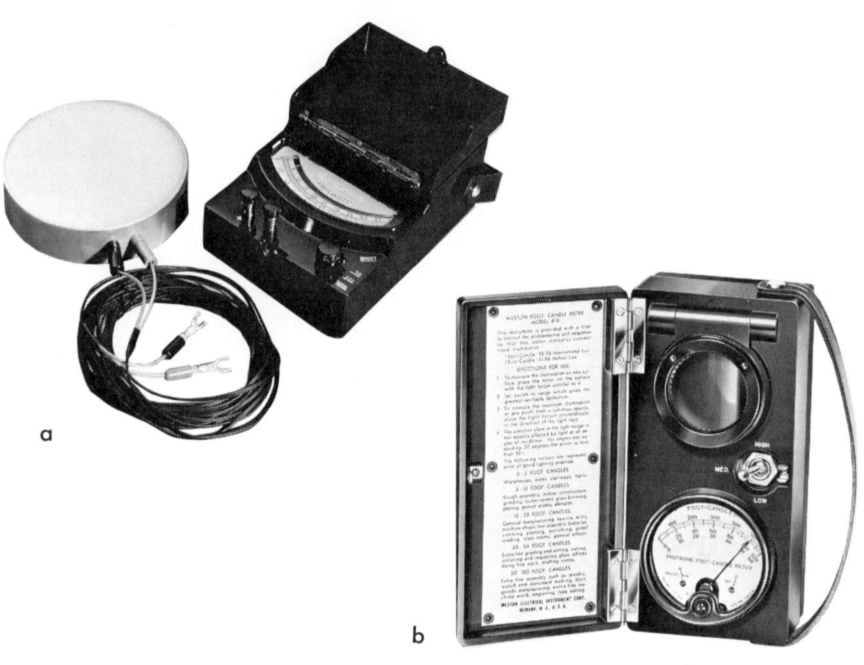

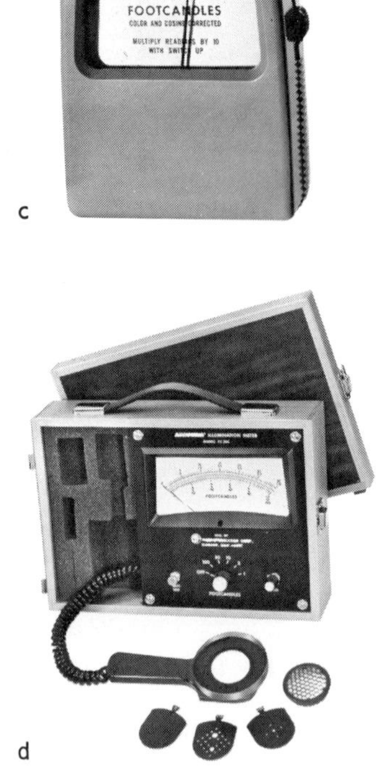

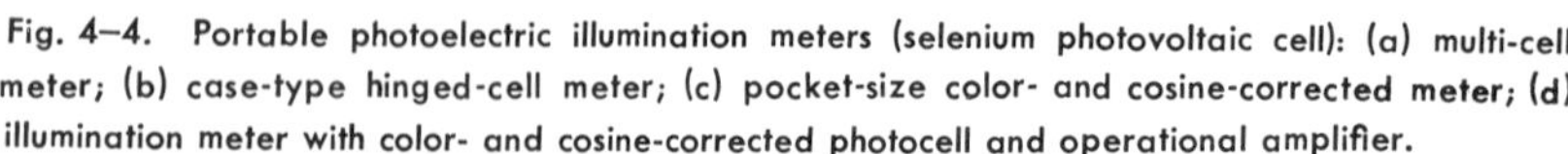
Fig. 4–4. Portable photoelectric illumination meters (selenium photovoltaic cell): (a) multi-cell meter; (b) case-type hinged-cell meter; (c) pocket-size color- and cosine-corrected meter; (d) illumination meter with color- and cosine-corrected photocell and operational amplifier.

dent illumination, but also produces a small electromotive force capable of forcing this current through a low resistance circuit. Photovoltaic cells provide much larger currents than do photoemissive cells, and can directly operate a sensitive instrument such as a microammeter or galvanometer. However, photovoltaic cells depart from linearity of response at higher levels of incident illumination as the resistance of circuit to which they are connected increases; therefore, for precise results, a current-balancing measuring instrument giving zero external resistance is necessary.[13]

Some of the portable illumination meters in use today consist of a photovoltaic cell, or cells, connected to a meter calibrateed directly in footcandles. See Fig. 4–4. However, with modern solid state electronic devices, operational amplifiers have been used successfully to amplify the output of photovoltaic cells, and the condition which produces most favorable linearity between cell output and incident light is automatically achieved, namely, by reducing the potential difference across the cell to zero (see Fig. 4–4d). The amplifier power requirements are small, and easily supplied by small batteries. In addition, digital readouts may be conveniently used to eliminate the ambiguities inherent in deflection type instruments.[14]

A. ***Spectral Response.*** The spectral response of photovoltaic cells is quite different from that of the human eye, and color-correcting filters are usually employed.[15] Fig. 4–5a illustrates the degree to which a typical commercially corrected selenium photovoltaic cell, as commonly used in footcandle meters, approximates the standard spectral luminous efficiency curve. Cells of a given type vary considerably in this respect, and for precision laboratory photometry each cell should be individually color-corrected. A spectrally uncorrected cell can of course be used to compare sources having identical spectral distributions, or to measure illumination of the same spectral distribution as that of the source with which it was calibrated.

B. ***Temporary Fatigue.*** The output of photovoltaic cells, when exposed to a constant illumination, may decrease slightly over a period of time before reaching its final value.[16] The cell should be well fatigued before beginning a series of measurements. This is accomplished by exposing it to illumination at the approximate level at which the measurements will be made. When the output (meter reading) stops drifting, the cell is adequately fatigued.

C. ***Effect of Angle of Incidence (Cosine Effect).*** Part of the light reaching a photovoltaic cell at high angles of incidence is reflected by the cell surface and the cover glass, and some may be obstructed by the rim of the case. The resultant error increases with angle of incidence, and in measuring illumination where an appreciable portion of the flux comes at wide angles, an uncorrected meter may read as much as 25 per cent below the true value. The cells used in most footcandle meters are now provided with diffusing covers or some other means of correcting the light-sensitive surface to approximate the true cosine response. The component of illumination contributed by single sources at wide angles of incidence may be determined by orienting the cell perpendicular to the direction of the light, and multiplying the reading thus obtained by the cosine of the angle of incidence. Other methods have been proposed. The possibility of cosine error must be taken into consideration in some laboratory applications of photovoltaic cells. One satisfactory solution to the problem consists in placing an opal diffusing acrylic plastic disk with a matte surface over the cell. At high angles of incidence, however, it reflects the light specularly so that the readings are too low. This can be compensated by allowing light to reach the cell through the edges of the plastic. The readings at very high angles will then be too high but can be corrected by using a screening ring.[17]

D. ***Effect of Temperature.***[11, 18] Wide temperature variations, either high or low, affect the performance of photovoltaic cells, particularly where the external resistance of the circuit is high. Prolonged exposure to temperatures above 120° F will permanently damage selenium cells. Measurements at high temperatures and at high illumination levels should, therefore, be made rapidly to avoid overheating the cell. Hermetically sealed cells provide greater protection from the effects of temperature and humidity. When using photovoltaic cells at other than their calibrated temperature, conversion factors may be employed, or means provided to maintain cell temperature in the vicinity of 77° F.

E. ***Effect of Cyclical Variation of Light.*** When electric discharge sources are operated on alternating current power supplies, precautions should be taken with regard to the effect of frequency on photocell response.[19] In some cases, these light sources may be modulated at several thousand Hertz. Consideration should then be given to whether the response of the cell is exactly equivalent to the Talbot's Law response of the eye for cyclic varying light. Due to the internal capacitance of the cell, it cannot always be assumed that its dynamic response exactly corresponds to the mean value of the illumination. It has been found that a low or zero-resistance circuit is the most satisfactory for determining the average intensity of modulated or steady state light sources with which photovoltaic cell instruments are generally calibrated. Although a microammeter or galvanometer appears to register a steady photocell current, it may not be receiving such a current, and may be actually receiving a pulsating current which it integrates since its natural period of oscillation is long compared to the pulses.

F. ***Meter Accuracy.***[20] The simplest circuit comprises merely a photovoltaic cell and microammeter or galvanometer of suitable range and resistance. The limitations imposed on the meter resistance by considerations of linearity were discussed above. These limitations may be relaxed to some extent if no assumption is made as to proportionality if the light meter is empirically calibrated as a whole. Instruments of this type have no provision for field calibration other than a zero-reading correction. This instrument, therefore, together with its attendant photovoltaic cell, should be checked frequently against known illumination values.

Baumgartner Reflectometer.[21] The Baumgartner reflectometer (Fig. 4–6) is an instrument designed to measure reflectance and diffuse transmittance by means of photovoltaic cells. The upper sphere is used alone for the measurement of reflectance. The sample to be measured is placed over an opening at the bottom of the sphere, a collimated beam of light is directed on it at about 30 degrees from the normal, and the total reflected light, integrated by the sphere, is measured by two cells mounted in the sphere wall. The tube carrying the light source and the collimating lenses is then rotated so that the light is incident on the sphere wall, and a second reading taken. The test sample is

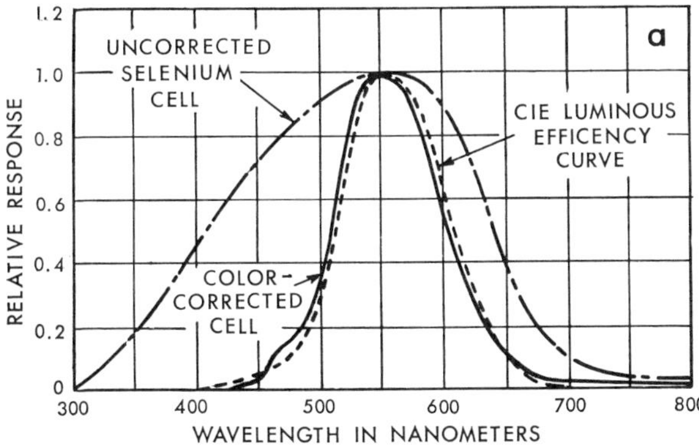

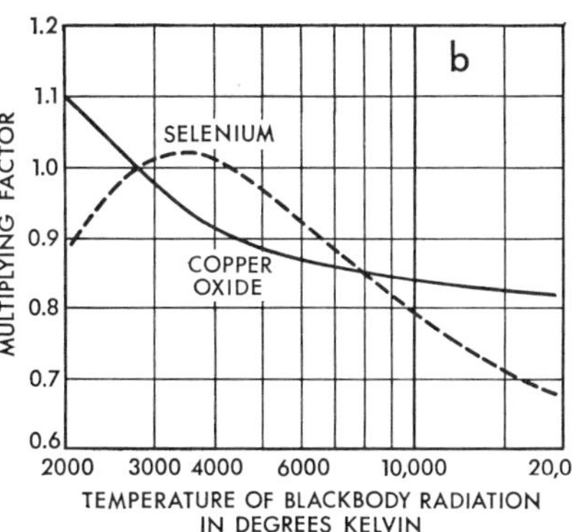

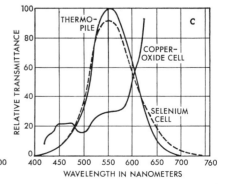

Fig. 4–5. *a.* Average spectral sensitivity characteristics of selenium photovoltaic cells, compared with CIE spectral luminous efficiency curve. *b.* Multiplying (correction) factor versus color temperature for uncorrected cells of representative spectral characteristics.[5] *c.* Spectral transmittance curves for filters designed to correct the spectral sensitivity characteristic of photovoltaic cells and thermopiles to correspond with CIE observer. "Viscor" and "Barnes" filters are designed for use with selenium photovoltaic cells.

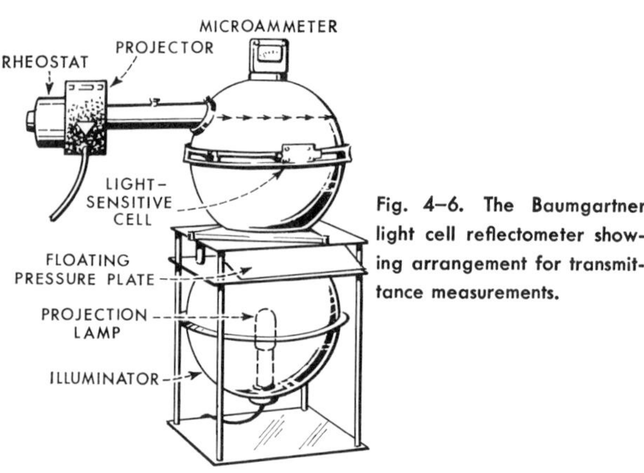

Fig. 4–6. The Baumgartner light cell reflectometer showing arrangement for transmittance measurements.

in place during both measurements, so that the effect on both readings of the small area of the sphere surface it occupies is the same.[22] The ratio of the first reading to the second is the reflectance of the sample for the conditions of the test. Test specimens of translucent materials should be backed by a non-reflecting diffuse material.

Transmittance for diffuse incident light is measured by using the light source in the lower sphere, and taking readings with and without the sample in the opening between the two spheres. The introduction of the sample will change the characteristics of the upper sphere. Correction must be made to compensate for the error thus introduced.[23, 24]

Other Types of Instruments. Various instruments are available for measuring such properties as specular and diffuse reflectance, and the gloss characteristics of materials.[25]

Photoelectric Tube Meters.[26] Photoelectric (photoemissive) tubes, like photovoltaic cells, produce current when radiant energy is received on the sensitive surface. They require a battery or external power supply for operation, and may require an amplifier to provide the necessary circuit sensitivity. The resultant photoelectric current may be measured by either a galvanometer, electrometer, oscillograph, deflection instrument, or bridge circuit. Special precautions must be taken to prevent leakage currents through moisture or dirt on the surface of the photoemissive tube or the grid lead of the amplifier tube. "Dark current" is the current passing through the photoemissive tube when it is not illuminated, and must be compensated for by circuitry, or by subtracting it from the lighted tube output.

No photoelectric tube has a spectral response similar to that of the human eye, and for photometric measurements, a filter which will correct the response to the standard spectral luminous efficiency curve is essential.[27] Photoelectric tubes are also used for measurements of other wavelength bands outside the visible region.

Freund Brightness Spot Meter.[28] This is a photoelectric photometer for measuring the luminance of small areas (Fig. 4–3c). Models available cover ¼, ½ or 1½ degrees field of view. A beam splitter allows a portion of the light from the objective lens to reach a reticule viewed by the eyepiece. The remainder of the light is reflected on to the field operative in front of the photomultiplier tube, and the output of the tube after amplification is read on a microameter with a scale calibrated in footlamberts. One of the filters provided with the instrument approximately corrects the response of the photomultiplier to the standard spectral luminous efficiency curve. Full scale deflection is produced by 10^{-2} to 10^{9} footlamberts.

Pritchard Photometer. This is a high sensitivity precision photomultiplier photometer with interchangeable field apertures covering fields from 2 arc minutes to 2 degrees in diameter (Fig. 4–3d). Full scale sensitivity ranges are from 10^{-4} to 10^{9} footlamberts. The readings of the light being measured are free from the effects of polarization since there are no internal reflections of the beam. The spectral response of each photometer is individually measured and the filters to best match it to the standard spectral luminous efficiency curve are determined and inserted. Filters are also included to permit evaluation of polarization and color factors.[29]

Visual Task Photometer.[30] This photometer incorporates a Pritchard photometer and is used to measure quantities based upon visual task contrast. The photometer incorporates a system which allows the Pritchard photometer to view the visual task, and then change rapidly to view the task background. In this way, the difference in readings provides a measure of task contrast. A mechanism allows the Pritchard photometer to be directed at the task over a range of viewing angles from 0 to 60 degrees. See Section 3.

Radiometric Photometers

Radiometric photometers are used to measure radiant flux over a wide range of wavelengths. They may employ receivers which are non-selective in wavelength response, or receivers that give adequate response in the desired wavelength band. Non-selective receivers are thermocouples, thermopiles or bolometers.[31] For measurement in the visible range, they can be color-corrected by means of a filter that duplicates the standard spectral luminous efficiency curve.[32] Another filtering method is to disperse the incident flux by means of a prism, placing in its path a template[33] which passes, at each wavelength, a percentage of flux proportional to the luminous efficacy for that wavelength. Corrections for any selectivity in the spectral transmittance of the optical system should be applied. Since radiometers respond to radiation outside the visible range as readily as to that within the visible range, the filter must be so designed that all radiation outside the visible is completely eliminated.

For short wavelengths, such as the ozone-producing region around 185 nanometers, platinum phototubes can be used. For the germicidal range around 250 nanometers, tantalum or tungsten phototubes are available. For the erythemal band, sodium cells are useful; and in the infrared, silver cesium oxide, lead sulphide, and thallous sulphide tubes can be employed.

Detectors may be galvanometers or special circuits to measure the voltage output of the receptor. Since the electrical output of a radiometer is very small, special precautions should be observed to avoid errors from such factors as ambient temperature and instrument zero shift.

In all radiometric work it is of the utmost importance to avoid stray radiation, and great care must be taken to be sure it is excluded. This is difficult, because such stray radiation is not visible to the eye, and a surface appearing black to the eye may actually be an excellent reflector of radiant energy outside of the visible spectrum.

Since radiated flux of some wavelengths is dispersed or absorbed by a layer of air between the radiator and the detector, consideration must be given to the placement of the source and detector, and to the medium surrounding them.

Basic Equipment Types

Bar Photometer. A bar photometer, most commonly used for the measurement of luminous intensity, provides a means for mounting sources and photometer in proper alignment and easily determining the distances between them. If a visual photometer is used, it is placed between two sources, a photometric balance is obtained, and the inverse-square law used to compute luminous intensity.

Goniometer. A goniometer is a device for rigidly mounting a light source, rotating it through angular traverses and measuring the angles within the traverses.

Distribution (Gonio) Photometer. Candlepower distribution measurements of non-beam producing sources and luminaires, resulting in a candlepower distribution curve, are made on a distribution (gonio) photometer of the general type shown in Fig. 4–7.[34] These instruments usually have provision for rotating the source about its vertical axis while measurements are being made; otherwise, measurements must be made in a sufficient number of vertical planes to provide a satisfactory average. For some types of equipment, vertical curves at certain specific angles, rather than an average curve, may be desired. The test distance (source to photometer) must be great enough for the inverse-square law to be valid (see page 4–2), and this often necessitates the use of one or more mirrors to direct the light through a sufficiently long path to the receiver.[35]

Because of the relatively large size of the sources to be tested, distribution photometers designed for fluorescent lamps and luminaires often have the test source mounted perpendicular to the normal position of use.[36] The axis of measurement is then horizontal, rather than vertical, and a long test distance is more easily provided. The photometer is stationary, and angular settings are obtained by turning the source rather than the receiver.

Integrating Sphere Photometer. The total luminous flux from a source (lamp or luminaire) is measured by some form of integrator, the most common one being the Ulbricht[37] sphere. See Fig. 4–8. Other geometric forms are sometimes used.[38] The theory of the integrating sphere assumes an empty sphere whose inner surface is perfectly diffusing and of uniform non-selective reflectance. Every point on the inner surface then reflects to every other point, and the illumination at any point is, therefore, made up of two components; the flux coming directly from the source, and that reflected from other parts of the sphere wall. With these assumptions, it follows that the illumination, and hence the luminance, of any part of the wall due to *reflected light only* is proportional to the total flux from the source, regardless of its distribution. The luminance

Fig. 4–7. A distribution photometer or goniophotometer used in a laboratory for obtaining data for candlepower distribution curves.

Fig. 4–8. Integrating (Ulbricht) sphere used at the Canadian Standards Association. It is used for directly obtaining total light output of lamps and luminaires.

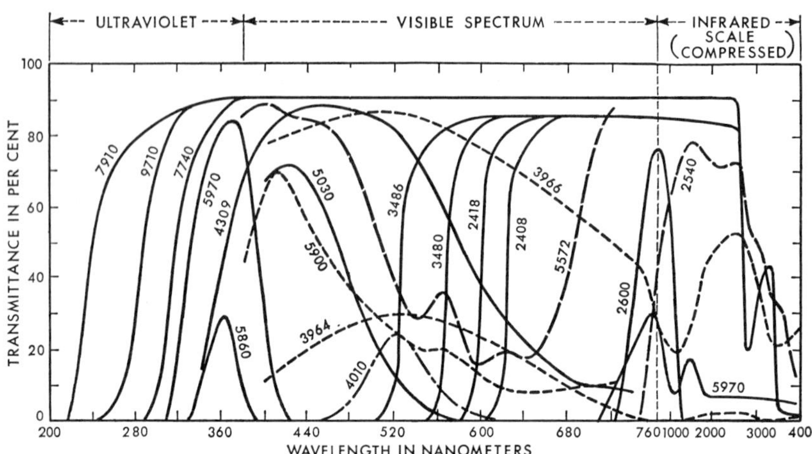

Fig. 4-9. Spectral transmittance curves for a number of ground and polished samples of various melts* of glass.

of a small area of the wall, or the luminance of the outer surface of a diffusely transmitting window in the wall, carefully screened from direct light from the source, but receiving light from other portions of the sphere, is therefore a relative measurement of the flux output of the source. The presence of a finite source, its supports, electrical connections, the necessary shield, and the aperture or window, are all obvious departures from the basic assumptions of the integrating sphere theory. The various elements entering into the considerations of a sphere, as an integrator, make it generally undesirable to use a sphere for absolute measurement of flux, but do not detract in the least from its use when a substitution method is employed.[39]

Spectrophotometers. Spectrophotometry is the term applied to the measurement of radiant energy or radiant flux as a function of wavelength. Its applications extend from precise quantitative chemical analysis to the exact determination of the physical properties of matter. In illuminating engineering, spectrophotometry is important in the determination of the spectral transmittance (see Fig. 4-9), spectral reflectance, and spectral emittance of materials, and objects related primarily to the ultraviolet, visible, and near infrared portions of the electromagnetic spectrum. The instrument used for these measurements is usually termed a spectrophotometer. However, for the measurement of spectral emittance of lamps, the instruments used are more often termed *spectroradiometers.* These instruments separate or disperse the various wavelengths of the spectrum by means of prisms or gratings as the dispersing element and have a collimating lens or mirror and a focusing lens or mirror. When the spectrum is examined by means of a lens or ocular, the instrument is called a spectroscope. When the ocular is replaced with an exit slit, it is called monochromator. A monochromator, when combined with a photometer for the measurement of spectral transmittance, spectral reflectance or spectral emittance, is called a spectrophotometer. Even when the energy detector is photoelectric and automatic the same term applies.

In the visible spectrum, the only fundamental means of examining a color for analysis, standardization and specification is by means of spectrophotometry. In addition, it is the only means of color standardization that is independent of material color standards (always of questionable permanence), and independent of the abnormalities of color vision existing among even so-called normal observers.

Colorimetric data may be computed from spectrophotometry by means of the methods of measuring and specifying color in the CIE standard observer and coordinate system for colorimetry. See also Section 5.

Commercial development of spectrophotometers has served to extend the wavelength range from about 200 to 2500 nanometers; to make them automatic recording and to add a tristimulus integrator.

LABORATORY MEASUREMENTS[40]

Precision of laboratory measurements results only from a combination of several elements; perhaps the most important is properly trained personnel. Others are control of electrical power, good electrical and light measuring instruments, and meticulous care to avoid sources of error. Errors frequently arise from insecure luminaire mounting; stray light; insufficiently frequent check of meters, cells, and standards; inadequate warm-up of discharge lamps and ballasts; and failure to recognize inherent nonsymmetry of sources and luminaires.

The availability of a room providing an adequate test distance is frequently an important problem. Long ranges are needed for floodlight and projector measurement. The use of mirrors to permit obtaining the effective test distance may help considerably. Temperature and air circulation control are serious requirements for discharge lamp photometry. Cleanliness, provision for suitable electric power, and storage space for lamps, luminaires, and delicate

* Figures on curves refer to standard melts of the Corning Glass Works. Transmittance decreases rapidly as the sample thickness increases and vice versa.

instruments are requirements that scarcely need mentioning.

Special techniques and current procedures are described in the pertinent IES and other testing and measurement guides found in the references at the end of this section.

Electrical Measurements of Incandescent Filament Lamps

It is often desirable to determine the electrical characteristics of light sources in connection with photometric measurements. Where additional information is required the reader is referred to any of the many texts or handbooks on electrical engineering.[41]

When electrical instruments are used, they should be selected to have current and voltage ratings corresponding to the circuit conditions to be encountered and to give indications at half-scale or above. The use of instruments at lower scale indications make the accuracy of measurement questionable.[42]

Corrections for instrument currents or voltage drops must be taken into account for precise photometric measurements since the light output of incandescent filament lamps is a high order exponential function of current and voltage. Preferably, instruments should be selected that have ½ per cent full scale accuracy or better to give measurements of high accuracy and precision.

Direct Current Circuits.

Power. Power is determined by observing the current and the voltage and computing the power as their product.

Current. The current taken by the light source is measured by means of an ammeter. The location of the ammeter in the circuit should be such that all of the current to the light source goes through the ammeter. When the current taken by the voltmeter is included in the ammeter indication, a correction must be made to obtain true lamp current.

Voltage. The voltage applied to the light source is measured by means of a voltmeter. To avoid correction for voltage drop in the ammeter, the voltmeter is generally connected directly across the load. Where precise measurements are desired, separate voltage leads are connected to the base of the lamp through special lampholders.

Precise Method. For the measurement of dc current and voltage, the deflection potentiometer,[43] equipped with suitable multipliers and shunts, provides a rapid and accurate method of measuring current and voltage. Adjustment of a single dial gives an approximate balance of the unknown value against a standard cell. The residual unbalance is read from the deflection of a sensitive millivoltmeter, which replaces the galvanometer of the null potentiometer.

The initial cost of a deflection potentiometer is considerably greater than a voltmeter and ammeter; but speed of operation, and increased accuracy, often justify the expenditure.

High precision can also be obtained with modern digital voltmeters (DVM) having accuracies of .01 per cent of reading and better. These meters draw very little power from the circuit being measured compared to moving element meters.

Alternating Current Circuits.

Power. In ac circuits the power is best determined by a wattmeter measurement. The current coil of the wattmeter is connected into the line in a manner similar to an ammeter and the voltage or potential circuit of the wattmeter connected directly across the load, as for a voltmeter. Under this condition, the wattmeter indication will include the power taken by the wattmeter voltage circuit and a correction should be made. The correction is determined by computation and is equal to the square of the load (lamp) voltage divided by the resistance of the potential circuit of the wattmeter. Certain wattmeters are designed to compensate automatically for the power taken by the wattmeter potential coil. In such cases the wattmeter is referred to as a "compensated wattmeter" [44] and when so arranged, no correction need be made.

Current. Ammeters for use in ac circuits may be self-contained or designed for use with a current transformer, in order to adapt the current rating of the instrument to the actual current in the lamp circuit.

Voltage. Voltmeters for use in ac circuits may be self-contained or arranged for use with either a multiplier or a potential transformer. With the voltmeter connected directly across the load, the current taken by the voltmeter is included in the ammeter indication, and the power taken by the voltmeter is included in any wattmeter indication. As in the case of direct-current measurements, the voltmeter current is computed by dividing the voltmeter indication in volts by the voltmeter resistance in ohms. The power taken by the voltmeter is computed in the same manner as that taken by the wattmeter potential circuit, and a correction made to the wattmeter indication.

Care must be exercised in selecting ac instruments compatible with ac wave shape. Non-sinusoidal ac wave shape may require true rms measuring instruments for highest accuracy.

Electrical Measurements of Electric Discharge Lamps and Circuits

All electric discharge lamps have negative volt-ampere characteristics and must therefore be operated in conjunction with current limiting devices, known as ballasts. These are described in Section 8.

Electric discharge lamp circuit measurements may involve lamps or ballasts, but in many cases the two are inseparable and the combined operation must be considered. Ballasts used commercially, because of normal manufacturing tolerances, supply lamps with some variation in voltage and current characteristics which affect the electrical input and the light output of lamps. To promote uniformity of testing, the International Electrotechnical Commission (IEC), working through the American National Standards Institute (ANSI), the Canadian Standards Association (CSA), and similar national standardizing bodies throughout the world, has or is establishing standardized testing procedures for determining the electrical characteristics for most of the common arc discharge lamps.

Where international standards have not been established, national standards are used.

Instrumention.[42]

A. ***Type.*** The voltage across an electric discharge lamp has a distorted wave shape, hence the instruments used must be of a type whose deflection depends on rms values.

B. ***Impedance.*** The combined impedance of instruments connected in parallel with the lamp should not draw more than three per cent of the rated lamp current. The voltage drop across instruments connected in series with the lamp should be less than two per cent of the rated lamp voltage.

Testing Procedures—Lamps. Lamp parameters are influenced by many factors and detailed accepted testing procedures are described in the appropriate IES guides.[45, 46] Some of the more important conditions affecting lamp and ballast testing are discussed below.

A. ***Ambient Temperature.*** If a lamp is operated within an enclosure, such as an Ulbricht sphere, the air temperature within the enclosure becomes the ambient. An ambient temperature of 25°C plus or minus 1°C should be maintained for fluorescent lamps. High intensity discharge lamps are less sensitive to temperature and for them ±5°C is satisfactory.

B. ***Drafts.*** In the testing of fluorescent lamps, extreme caution should be exercised to reduce air movement over the surface of the lamp to the lowest possible minimum.

C. ***Lamp Position.*** Unless tests are designed to meet special conditions, fluorescent lamps should be tested in a horizontal position, and high intensity discharge lamps in a vertical position. High intensity discharge lamps not operated in a vertical position should have the angular orientation from the vertical as well as the rotational orientation specified.

D. ***Lamp Connections.*** Fluorescent lamps of the hot cathode bipin type should be operated with the same two pins connected to the operating circuit for all conditions during tests.[47]

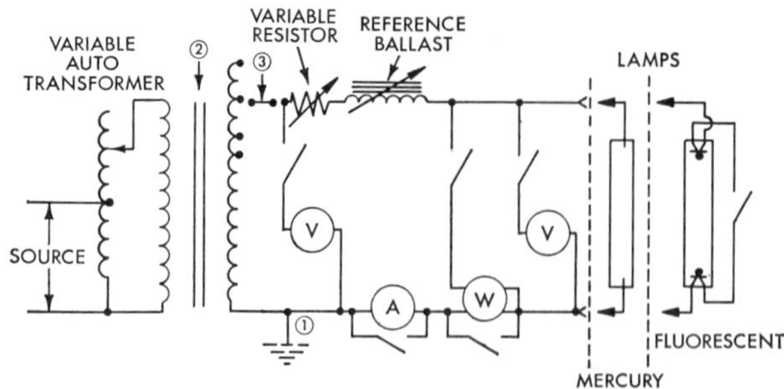

Fig. 4–10. Basic measurement circuit for electric discharge lamps and ballasts. Note:

(1) This side of circuit should be the one connected to ground if a grounded supply system is used.

(2) It may be necessary to insert the transformer shown at this point in order to obtain the open circuit voltage required by the Reference Ballast for certain sizes of lamps. Either an autotransformer or a double-wound transformer (as illustrated) may be used.

(3) Insert magnet between these points.

High intensity discharge lamp circuit connections to the center contact or shell contact may be interchanged. See Fig. 4–10.

E. ***Lamp Stabilization.*** Before any measurements are taken, the lamp should be operated until the performance characteristics are stable. If the nature of the test requires seasoned lamps, a minimum of 100 hours operation is recommended.

If lamps are warmed up on one ballast and then transferred to a different ballast for measurement, an additional period of burning in the measurement circuit is usually necessary to bring the lamp to equilibrium.

No lamp which shows swirling or other abnormal behavior should be considered to be stabilized for measurement purposes.

F. ***Power Supply.*** The ac wave shape should be such that the rms summation of the harmonic components does not exceed three per cent of the fundamental.

The input impedance, looking into the power source from the lamp and ballast, should not exceed ten per cent of the ballast impedance.

Unless automatic voltage regulation within 0.1 per cent is available, constant checking and readjustment are necessary. If a static type voltage stabilizer is used, it is particularly important to check the wave shape.

G. ***Ballasts.*** Rating tests of lamps should be conducted with the lamp operating in series with the assigned Reference Ballast.

If no Reference Ballast specification has been issued for the lamp under test, a reactor complying with the general requirements should be used.

Special tests may be made with other than specified Reference Ballasts. The results of these tests are not directly comparable with data taken on a Reference Ballast.

H. ***Rating Measurements of Fluorescent Lamps with Continuously Heated Electrodes.***[48] Lamps using the rapid start principle are designed to be operated with the electrode continuously heated by separate windings within the ballast. Because of the complexity of circuit conditions, measurements of these lamps, operated in conjunction with a series ballast, may not exactly represent the values when the lamp is operated by a rapid start ballast.

I. ***Test Circuits.*** A typical measurement circuit is shown in Fig. 4-10. Since the circuit voltage may be insufficient for lamp starting or operation, a variable tap step-up transformer has been shown.

In the case of fluorescent lamps, a preheat lamp has been shown; omission of the starting circuit provides for instant-start lamps. The procedure for testing lamps designed for operation with continuously heated electrodes requires special attention.[46]

Switches are shown for removing instruments from the circuit. Those used to short circuit current coils should have very low resistance. The Reference Ballast and the accompanying variable resistor can be set for impedance and power factor, with or without the ammeter and wattmeter current coils included. Either procedure is satisfactory, but instrument corrections must be determined for the method used.

Testing Procedures—Ballasts.[49] Ballast parameters are influenced by many factors and detailed accepted testing procedures are described in the appropriate ANSI standards.[49] Some of the more important conditions affecting ballast testing are discussed below.

A. ***Voltage Range.*** For most tests, ballasts should be operated at their rated primary voltage. If they are rated for a range, the voltage should be the center of the range; the center for 110–125 should be 118 volts.

B. ***Reference Lamps.*** Some tests on ballast specify that the ballast shall be operating a Reference Lamp. These are seasoned lamps which, when operated under stated conditions with the specified Reference Ballast, operate at values of lamp volts, amperes, and watts each within plus or minus 2.5 per cent of the values established by the appropriate existing or proposed specifications.

C. ***Open Circuit Voltage.*** This measurement is necessary only for ballasts containing a transformer.

Series-sequence fluorescent ballasts are so designed that two lamps operate in series. The open circuit voltage of each of the lamp positions should be measured with an operable lamp in the other position.

Fluorescent ballasts for lamps with continuously heated cathodes have four terminals which connect to the lamp. Open circuit voltage should be measured between the two giving the highest voltage with all cathode heating windings loaded with the appropriate dummy resistance load.

D. ***Electrode Heating Voltage.*** On ballasts for use with lamps having continuously heated electrodes, the electrode heating voltages are measured with the electrode windings loaded with the specified dummy load.

E. ***Short Circuit Current (High Intensity Discharge Ballasts).*** An ammeter is inserted in the circuit in place of the lamp, and the short circuit current of the ballast measured.

F. ***Starting Current (Fluorescent Instant Start Ballasts).*** A resistor and ammeter, in series, with a total resistance equivalent to the value specified in the appropriate standard[49] should be used instead of the lamp.

G. ***Electrode Preheating Current (Fluorescent Preheat Ballasts).*** This measurement is made with an ammeter connected in series with the lamp electrodes while the lamp is maintained in the preheat condition.

H. ***Fluorescent Ballast Output.***

Preheat and Instant Start. Specifications are in terms of the power delivered to a Reference Lamp operated by the ballast under test, as compared to the power delivered to the same lamp by the appropriate Reference Ballast.

The general circuit for comparing fluorescent lamp performance on commercial and Reference Ballasts is shown in Fig. 4-11.

Continuously Heated Electrodes. Specifications are in terms of the light output of a Reference Lamp operated by the ballast under test, as compared with the light output of the same Reference Lamp when operated by the appropriate Reference Ballast.

I. ***Fluorescent Ballast Regulation.*** This measurement involves the relative lamp power input and light output at 90 per cent and 110 per cent of rated ballast input voltage.

J. ***Fluorescent Lamp Current.***

Reference Lamps. The current of a Reference Lamp should be measured on both the ballast under test and the Reference Ballast.

Lamps with Continuously Heated Electrodes.[48] Unless the internal connections of the ballasts are

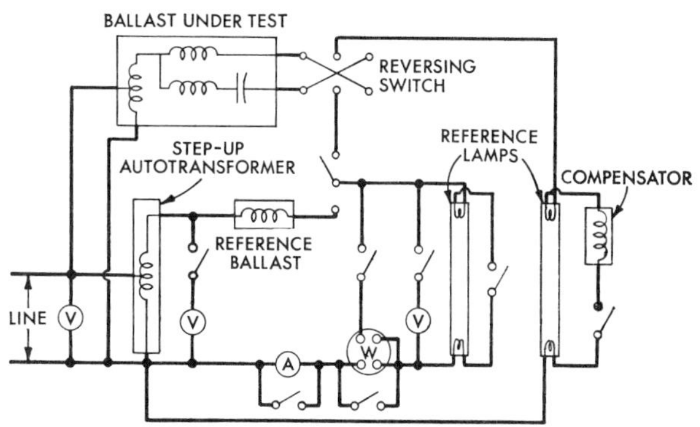

Fig. 4-11. Circuit for comparing fluorescent lamp performance with two different ballasts.

accessible, measurement of lamp current requires special instrumentation to provide the vector summation of currents in the two leads to an electrode.

Photometric Measurements of Electric Discharge Lamps[50]

The photometric characteristics of electric discharge lamps are usually determined in conjunction with electrical measurements. Lamp testing procedures, circuits and instrumentation are similar and the same general precautions should be observed during photometry as in making electrical measurements (see page 4–11). Details concerning photometric measurements will be found in the appropriate IES guides.[46, 51]

Ballasts. When a lamp is being photometered for rating purposes it should be operating on the appropriate Reference Ballast.[49] If no standard exists, the ballast used should comply with the general requirements.[49] Photometric measurements of lamps burning on commerical ballasts should be made with the ballast operating at rated input voltage and after the ballast has been operated long enough to reach thermal equilibrium. The use of commercial ballasts should conform to the procedures given in the appropriate standards.[49]

Photometers. Most measurements are made by the substitution method in photometers equipped with photoelectric receivers. Photovoltaic photocells are most commonly employed as receivers and should be selected for linearity and absence of fatigue. Spectral response should be corrected, by means of filters, to the standard spectral luminous efficiency curve (see page 4-6 and Fig. 4-5). Standards should have characteristics similar to the lamp under test with respect to light output, physical size, shape, and spectral distribution.

A. ***Integrating Sphere Photometer.*** The sphere diameter should be at least 1.2 times the length of the lamp for straight lamps; the area of the light source should not exceed 2 per cent of the interior surface of the sphere, unless strict substitution methods are employed. Precautions should be taken to prevent supports and baffles from absorbing light or affecting the lamp temperature. There should be no selective spectral absorption by the sphere, sphere window or internal supports. The lamp should be positioned in the center of the sphere, but not remain so long as to raise ambient temperature. See Fig. 4–8. The light measuring device should receive no direct light from the lamp.

B. ***Candlepower Distribution Photometer.***[34, 52] The lamp is mounted in open air with the distance between receiver and lamp at least five times lamp length, or ten feet for compact sources. When used as a normal-candlepower photometer for fluorescent lamps, the lamps are mounted in a horizontal position and measurements taken normal to the axis of the lamp. For accuracy, measurements are taken at several points about the axis of the lamp by rotating the lamp around its axis between each measurement. Total light output can be computed if the lumen-candlepower ratio is known for the lamp, or if strict substitution is practiced. For high intensity discharge lamps the operating position of the lamp should be as specified (see page 4–12) and sufficient readings should be taken to allow proper calculation of candlepower distribution and total lumen output. Candlepower values of lamps are established by multiplying the photometric calibration constant by the test readings. Total light output of the lamps is established by summing the products of the candlepower values and the appropriate zonal lumen constants. Care should be taken to exclude stray light, to control ambient temperature and drafts, and to reduce the effects of light absorbing or reflecting materials.

Color Appearance of Light Sources. For measurement of color appearance of light source, see page 5–14.

Photometric Measurements of Incandescent Filament Lamps

For photometric purposes, incandescent filament lamps can be divided into two classes; multiple and series lamps. Multiple lamps are designed and rated to be operated at rated voltage. Series lamps are designed for constant current operation.

Photometric samples should be seasoned to assure stable operation. This consists of burning multiple lamps at rated voltage for approximately one per cent of their rated life, or at over-voltage for a time that will produce equivalent seasoning. Series lamps are usually seasoned four hours at 110 per cent rated current.

Reference standard lamps[53] may be purchased from the National Bureau of Standards or established laboratories. Multiple lamp standards are usually rated for lumens and amperes at a voltage a little below their rating to improve the length of burning time without important change in the assigned values. Series lamps are standardized at current.

Working standards are usually made by comparison with reference standards. They should have the loops of filament supports closed firmly around the filament to avoid the possibility of random short circuiting of a portion of the filament by a twisted loop. They should be adequately seasoned and selected by successive comparisons with reference standards for stability. All standards should be handled carefully to avoid exposure to mechanical shocks. Exposure to current or voltage above the standardized value may alter lamp ratings.

Lamps may be measured on a bar photometer if candlepower in an oriented direction, or mean horizontal candlepower, is desired. Lamps standardized for unidirectional measurements are usually marked to indicate the orientation. A common practice is to inscribe a circle and vertical line on opposite sides of the bulb. The standardized direction is from the circle toward the line, when they are centered on each other, looking toward the receiver.

Most routine photometric measurements of incandescent filament lamps require total light output and are made in a sphere (see Fig. 4-8). Details of operation are described on page 4-9. Strict substitution procedures should be followed, or corrections made for any unavoidable departures.

A standard should be selected of about the same physical size, lumen output, and color temperature as the lamp under test. The unknown lamp should be measured in the sphere in the same position as the standard.

If photometric readings are to be taken during lamp life, some special precautions should be observed. Readings to determine lamp depreciation are usually taken at 70 per cent rated lamp life. At this time, some blackening of the bulb may have occurred, hence, a blackened unknown lamp replaces a clear standard in the sphere, violating strict substitution. The rather important errors which this may introduce can be avoided by using a third lamp shielded from the light measuring device and the test lamps. This third lamp, commonly called the "absorption," "comparison" or "sub-standard" lamp, is lighted and consecutive measurements made with the test and standard lamps in the test position but not in operation. The difference in the two readings indicates the amount of light absorbed by the blackened test lamp. The same general procedure can be followed in most cases where the characteristics of the integrator are altered during the test by the introduction of light absorbing elements.

Photometric Measurements of Reflector-Type-Lamps[54]

For purposes of identification, a reflector-type lamp is defined as a lamp having a reflective coating applied to the reflector part of a bulb, the reflector being specifically contoured for control of the luminous distribution. Included are pressed or blown lamps such as PAR and EAR lamps plus other lamps with optically contoured reflectors. Excluded are: lamps of standard bulb shape to which an integral reflector is added, such as silvered-bowl and silvered-neck lamps; lamps designed for special applications, such as automotive headlamps and picture projection lamps, for which special test procedures are already established; lamps having translucent coatings, such as partially phosphor-coated mercury lamps; and reflector fluorescent lamps.

Lamp Selection and Seasoning. Test lamps are selected according to the American Society for Testing Materials (ASTM) sampling procedures contained in Part I, Presentation of Data, *ASTM Manual on Quality Control of Materials,* Special Technical Publication 15C and also *Recommended Practice of Probability Sampling of Materials,* ASTM designation E105.

Incandescent Lamps are seasoned for a minimum of one per cent of rated life. For forced seasoning see Table I in the "IES Approved Method for Photometric Testing of Floodlights Using Incandescent Filament or Discharge Lamps."[55] Standard practice for seasoning mercury, metal halide, high pressure sodium and low pressure sodium lamps is to operate them not less than 200 hours on commercial ballasts at rated input conditions.

Intensity Distribution. Several different methods of making intensity distribution measurements may be used depending on the type of lamp and the purpose of the test. The *photometric center of lamp* normally should be taken as center of bulb face disregarding any protuberences or recesses in face center, and the *test distance* should be great enough so that the inverse-square law applies. The *receptor* should subtend an angle of one square degree.

The intensity distribution of a normally *circular beam* is commonly represented by an average curve in the plane of the *beam axis* (defined as the axis around which the average distribution is substantially symmetrical—beam axis and photometric axis are adjusted to coincide). The curve is obtained either by taking measurements with the lamp rotating about the beam axis, or by averaging a number of curves (at least three) taken in planes at equally spaced azimuthal intervals about the axis.

The intensity distribution of a lamp whose beam is nominally oval or rectangular in cross section is not adequately represented by one average curve. For some lamps two curves through the beam axis, one in the plane of each axis of symmetry, may supply sufficient information. The necessary number of traverses, their distribution within the beam, and the intervals between individual readings vary considerably with the type of lamp; sufficient measurements should be made to adequately describe the distribution pattern.

When reflector-type lamps are considered for a specific application, test results will be most readily comparable when in the same form as that for equipment used for the same application. For example, when a direct performance comparison of a reflector lamp with floodlighting luminaires is desired, the lamp should be tested according to approved floodlight testing procedures.[55] The same is true for indoor luminaire applications.

Total Flux Measurements. Total flux may be obtained by direct measurement in an integrating sphere or by calculation from intensity distribution data. Because of the high intensity spot produced by most reflector-type lamps, special precautions should be taken when using an integrating sphere. One possible position for the test lamp in the sphere is with its base close to the sphere wall and the beam aimed through the sphere center, thus distributing the flux over as large an area of the sphere as possible. An appropriate baffle should be placed between the light source and the receptor.

When reflector-type standards are available the calibration of the sphere follows the usual substitution procedure and for maximum accuracy the standard lamps should be of the same type as the test lamps which are used.

Beam and Field Flux. Beam and field flux may be calculated from an average intensity distribution curve or from an isocandela diagram. Of particular interest is the flux contained within the limits of 50 per cent and 10 per cent respectively of the maximum intensity. *Beam angle* is designated as the total angular spread of the cone intercepting the 50 per cent of maximum intensity. *Field angle* is designated as the total angular spread of the cone intercepting the 10 per cent of maximum intensity.

Life Performance Testing of Lamps

Life tests are performed on a very small portion of the product under consideration. Under such conditions test program planning, sampling techniques, and data evaluation become especially important. Helpful guidance can be found in a variety of documents.[56]

It is recognized that it is not practical to test lamps under all of the many variables that occur in service hence, specific reproducible procedures must be included in the test experiment plan.

Incandescent Lamp Life Testing.[57] Life tests of incandescent filament lamps may be divided into two classes, *rated-voltage* and *over-voltage tests.*

A. ***Rated Voltage.*** Lamps are operated in the specified burning position at labeled voltage or current held within plus or minus 0.25 per cent of rated value. Sockets should be designed to assure good contact with lamp bases, and the racks should not be subjected to excessive shocks or vibration. If lamps are removed for interim photometric readings, great care should be taken to avoid accidental filament breakage. Sockets should be slightly lubricated because the vibration of a "squeak", as the lamp is removed or replaced, may be sufficient to break a filament which has been rendered brittle by burning.

B. ***Over Voltage Tests.*** Lamp life is shortened by the application of voltage in excess of rated. The mathematical relationship between voltage and lamp life is shown in Section 8. By means of extreme over voltage life testing, sometimes called "high forced testing", lamp life may be shortened so that an evaluation of 1000-hour lamps may be obtained in an eight hour day. The exponents are empirical and require many comparison tests at rated voltages to determine them. This type of testing introduces additional uncertainties which may cause appreciable errors; hence, the results are only approximate.

Electric Discharge Lamp Life Testing.[58] No method of accelerated life testing for electric discharge lamps, which is in any degree comparable with normal testing, is known. The usual test cycle is three hours "on" and twenty minutes "off." It is well known that increasing the rapidity of this cycling will materially shorten lamp life, but exponents have not been determined which give reasonable accuracy.

Line voltage fluctuations of plus or minus 5 per cent from the rated ballast input voltage are not an important factor in fluorescent lamp life. However, many other factors make the evaluation of life characteristics of these lamps difficult.

Because of the long rated life, about 18 months are necessary for evaluation of life.

An electric discharge lamp must be operated with auxiliaries which may affect lamp life. To satisfactorily evaluate lamp life the auxiliaries must be selected to conform with the requirements of the appropriate guides and specifications.

Tests must be made on ac, and the power supply should have a voltage wave shape in which the harmonic content does not exceed 3 per cent of the fundamental.

These factors, combined with those of vibration, shock, room temperature, etc., result in wide deviations unless the results from many samples are averaged.

Photometry of Luminaires

The purpose in photometering or measuring the light distribution of a luminaire is to accurately determine and report the characteristics of the luminaire that will adequately describe its performance. Data such as luminaire efficiency, candlepower distribution, beam widths, zonal lumens, and luminances are necessary in designing, specifying, and selecting lighting equipment. Photometric data is also essential in deriving and developing additional application information for the luminaires.

The information that follows will serve as a rudimentary guide in outlining the photometry of luminaires. Specific photometric guides and practices are referenced and should be consulted to obtain the detailed testing procedure for each type of luminaire. The "IES General Guide to Photometry,"[6] and the "IES Practical Guide to Photometry"[59] provide information covering general photometric practices, equipment, and related information.

Each specific type of luminaire, e.g., fluorescent, incandescent, floodlight, etc., require different testing procedures. However, there are several general requirements that should be met in all tests. The luminaire to be tested should be: (1) typical of the unit that it represents; (2) clean and free of mechanical defects; (3) equipped with the proper fittings; (4) equipped with lamps of size and type recommended for use in service; and (5) equipped with the light source in position as recommended in service. If the source in a beam-producing luminaire is adjustable, it should be positioned to produce the spot of intended size and shape.

To provide an accurate description of the characteristics of the materials used in the manufacture of a luminaire, measurements should be made of the reflectance of the light controlling surfaces.

Luminaires should be tested in a controlled environment under controlled conditions. The photometric laboratory should be draft free. Room temperatures should be held steady. Typically, for fluorescent photometry, where lamps and luminaires are sensitive to temperature variations, the room temperature should be held to 25°C ± 1°C. Power supplies should be free of distortion and regulated to minimize any effects of voltage variations. Test rooms should be painted black and/or provided with sufficient baffling to minimize or eliminate extraneous and reflected light during testing.

Luminaires are mounted in goniometers to allow positioning to definite angles about both horizontal and vertical axes. There are many different versions of goniometers, each being related to the type of lighting equipment being measured and the facilities in which it is located. Use of computers to mathematically change coordinate systems has allowed the development of universal goniometers which meet the measuring needs of several different types of luminaires with minimum modifications. It is desirable to use a goniometer which holds the test luminaire in its normal operating position during the test. For floodlights and searchlights, the goniometer usually has axes of rotation corresponding to an isocandela web with its polar axis in a horizontal plane. See Fig. 4–12.

For accurate measurements, the distance between the luminaire and light sensing device should be great enough so that the inverse-square law applies. The minimum test distance is governed by the dimensions of the luminaire; this distance should be not less than 10 feet, and at least five times the maximum dimension of the luminaire. For maximum precision the test distance should be measured from the surface of the receiver to the photometric centroid of the luminaire. However, from a practical standpoint, the following rules should be followed: (1) for recessed and totally direct fluorescent and incandescent luminaires, the test distance should be measured to the plane of the light opening (plane of the ceiling); (2) for suspended and luminous sided luminaires, the test distance should be measured to the geometric center of the lamps; (3) for floodlights it is recommended that a photometric test distance of not less than 100 feet be used.

General Lighting Luminaires [Candlepower Distribution and Luminance (Photometric Brightness)]. For specific information on testing general lighting luminaires, the following *IES*

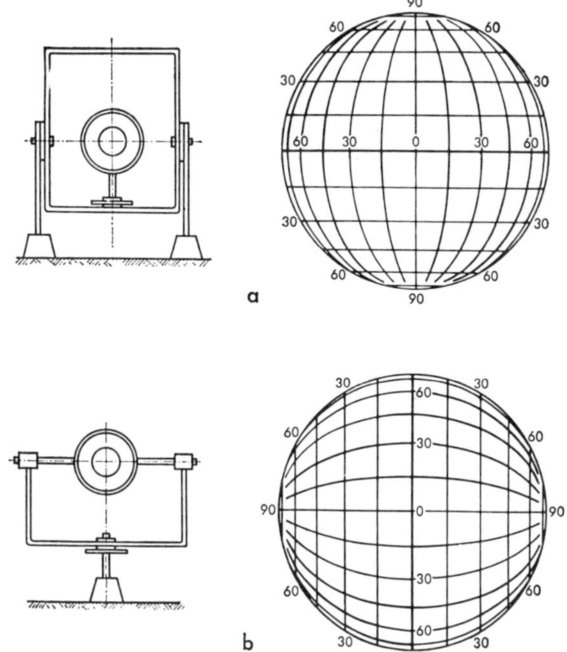

Fig. 4–12. *a.* A measuring apparatus where the projector turns about a fixed horizontal axis and also about an axis which, in the position of rest, is vertical, and upon rotation follows the movement of the horizontal axis. *b.* A measuring apparatus where the light souce turns about a fixed vertical axis and also about a horizontal axis following the movement of the vertical axis.

Guides should be consulted: *Photometric Testing of Fluorescent Luminaires,*[60] *Measurement of Luminance (Photometric Brightness),*[61] *Photometric Testing of Filament Type Luminaires for General Lighting Service,*[62] *Reporting General Lighting Equipment Engineering Data,*[63] and *Determination of Average Luminance of Luminaires.*[64]

The basic measurements made in a photometric test of a luminaire are the luminous intensity output in specified planes and angles. The resulting candlepower distribution is used to determine the luminaire efficiency, average luminance, and zonal lumens. It is therefore essential that sufficient data of the luminous intensity distribution be taken to adequately describe the luminaire's total light output.

Incandescent filament lamp luminaires and others having a symmetrical distribution are usually rotated at about 50 rpm during the test, thereby averaging the candlepowers in all planes about the luminaire. Most fluorescent luminaires are tested in three or five planes in order to evaluate the candlepower distribution. Substantially symmetrical luminaires which are too bulky or heavy to be rotated are also tested in three or more planes. It is sometimes necessary to test highly asymmetric luminaires in planes at intervals of 10 degrees or less.

The distribution in each vertical plane is determined by taking readings at 10-degree intervals. To facilitate calculation of lumens by zones, candlepower measurements are usually taken at mid-zone angles; *i.e.*, 5°, 15°, 25°, etc. If the light distribution is changing very rapidly, candlepower measurements should be taken at closer intervals. Five degree intervals are preferable when calculations of visual comfort probability are to be made.

The great majority of candlepower distribution tests on luminaires using either incandescent or electric discharge lamps are made by the relative method. The intensity of the set of lamps with which the luminaire is to be equipped is first read alone and recorded. This reading is assigned a value in candlepower equal to what the lamps would produce if they were delivering rated output. This same calibration constant is then used to determine the light output of the luminaire at every measuring station. These corrected candlepower measurements are those that would result if the lamp (or lamps) output were equal to the published rated values. No measurements of actual candlepower are required.

The candlepower distribution data are generally presented in tabular form on the test data sheet. Data for a fluorescent luminaire are given in three planes; one being for a vertical plane parallel to the lamps; one perpendicular to the lamps; and an intermediate plane at 45 degrees. These data may be given in graphical form as a polar distribution.

Either before or after the complete test, while the appropriate lamps are installed in the luminaire, the maximum luminance of the luminaire in candelas per square inch should be measured at the angles prescribed in the guides and at the shielding angles. The readings should be taken both crosswise and lengthwise in the case of fluorescent luminaires. The projected area observed should be circular and of one square inch. Care should be taken that the luminance measurement be related to the rated performance of the lamps. In other words, the luminance measuring instrument must be calibrated against the test lamps, employing the fundamentals of the relative technique described above.

If average luminance values are desired, they can be obtained by using the candlepower measurements described above. By definition luminance is the luminous intensity of any surface in a given direction per unit of projected area of the surface as viewed from that direction. Specific procedures for calculating average luminances are found in reference 64.

General Lighting Luminaires (Total Lumen Output). The total light output of the luminaire, needed to establish its efficiency in terms of the lumen output of the lamps with which it is equipped, can be determined in a spherical integrating photometer, or by computations from the candlepower distribution data.

If measured in a sphere, the efficiency can be determined very simply by the relative method. The lamps are first mounted in a large sphere and a reading taken. The luminaire is then placed in the sphere; its reading, in terms of the reading of the lamps alone, will give the efficiency.

Average candlepower distribution data permits the computation of the luminous flux in any angular zone, from nadir to 180 degrees. The product of the mid-zone candlepower value and the zonal factor, gives the number of lumens in the zone.

Calculation of Luminaire Lumens From Candlepower Data. Constants useful in calculating luminous flux from candlepower data are given in Figs. 4–13 and 4–14. For computing luminous flux, multiply the average candlepower at the center of each zone by the zonal constant (see Fig. 4–13) equal to 2π ($\cos \theta_1 - \cos \theta_2$). A nomogram is given in Fig. 4–15 for obtaining the lumens in 10-degree zones.

The constants in Fig. 4–14 are computed for candlepower measurements of projector-type luminaires made on a goniometer of the type shown in Fig. 4–12*b*. If the measurements have been made with the type as shown in Fig. 4–12*a* the constants may be used by interchanging the vertical and horizontal angular arguments, *i.e.*, by substituting the word "vertical" wherever "horizontal" appears. The zonal constants for Fig. 4–14 were computed as equal to $\phi\pi(\sin \theta_2 - \sin \theta_1)/180$, where ϕ is the vertical interval and θ_1 and θ_2 are the limits of the horizontal interval. If a goniometer of the type in Fig. 4–12*a* is used the constant is equal to $\theta\pi(\sin \phi_2 - \sin \phi_1)/180$, where θ is the horizontal interval and ϕ_1 and ϕ_2 are the limits of the vertical interval.

If a number of constants are to be calculated for the same interval, the following *shortcut method* is a time-saver but is completely accurate.

Fig. 4–13. Constants for Use in the Zonal Method of Computing Luminous Flux from Candlepower Data. (1, 2, 5, and 10 Degree Zones)

1 Degree Zones		2 Degree Zones		5 Degree Zones		10 Degree Zones	
Zone Limits (degrees)	Zonal Constant	Zone Limits (degrees)	Zonal Constant	Zone Limits (degrees)	Zonal Constant	Zone Limits (degrees)	Zonal Constant
0–1	0.0009	0–2	0.0038	0–5	0.0239	0–10	0.095
1–2	.0029	2–4	.0115	5–10	.0715	10–20	.283
2–3	.0048	4–6	.0191	10–15	.1186	20–30	.463
3–4	.0067	6–8	.0267	15–20	.1649	30–40	.628
4–5	.0086	8–10	.0343	20–25	.2097	40–50	.774
5–6	.0105	10–12	.0418	25–30	.2531	50–60	.897
6–7	.0124	12–14	.0493	30–35	.2946	60–70	.993
7–8	.0143	14–16	.0568	35–40	.3337	70–80	1.058
8–9	.0162	16–18	.0641	40–45	.3703	80–90	1.091
9–10	.0181	18–20	.0714	45–50	.4041		
				50–55	.4349		
				55–60	.4623		
				60–65	.4862		
				65–70	.5064		
				70–75	.5228		
				75–80	.5351		
				80–85	.5434		
				85–90	.5476		

Fig. 4–14. Constants for Converting Beam Candlepower of Projector-Type Luminaires (Searchlights, Floodlights, Spotlights) into Lumens
(0.1 × 0.1 to 10 × 10 degree steps)

Spacing 0.1° Vertical

0.1° Horizontal

Horizontal Angle and Setting	K	Horizontal Angle and Setting	K	Horizontal Angle and Setting	K
0.05	$0.^{5}3046$	2.75	$0.^{5}3043$	5.45	$0.^{5}3032$
0.15	3046	2.85	3042	5.55	3032
0.25	3046	2.95	3042	5.65	3031
0.35	3046	3.05	3042	5.75	3031
0.45	3046	3.15	3042	5.85	3030
0.55	3046	3.25	3041	5.95	3030
0.65	3046	3.35	3041	6.05	3029
0.75	3046	3.45	3041	6.15	3029
0.85	3046	3.55	3040	6.25	3028
0.95	3046	3.65	3040	6.35	3028
1.05	3046	3.75	3040	6.45	3027
1.15	3046	3.85	3039	6.55	3026
1.25	3045	3.95	3039	6.65	3026
1.35	3045	4.05	3039	6.75	3025
1.45	3045	4.15	3038	6.85	3024
1.55	3045	4.25	3038	6.95	3024
1.65	3045	4.35	3037	7.05	3023
1.75	3045	4.45	3037	7.15	3023
1.85	3045	4.55	3037	7.25	3022
1.95	3044	4.65	3036	7.35	3021
2.05	3044	4.75	3036	7.45	3021
2.15	3044	4.85	3035	7.55	3020
2.25	3044	4.95	3035	7.65	3019
2.35	3044	5.05	3034	7.75	3018
2.45	3043	5.15	3034	7.85	3018
2.55	3043	5.25	3033	7.95	3017
2.65	3043	5.35	3033		

0.2° Horizontal

Horizontal Angle and Setting	K	Horizontal Angle and Setting	K	Horizontal Angle and Setting	K
0.1	$0.^{5}6092$	2.9	$0.^{5}6085$	5.5	$0.^{5}6065$
0.3	6092	3.1	6084	5.7	6063
0.5	6092	3.3	6083	5.9	6061
0.7	6092	3.5	6081	6.1	6058
0.9	6091	3.7	6080	6.3	6056
1.1	6091	3.9	6079	6.5	6054
1.3	6091	4.1	6077	6.7	6051
1.5	6090	4.3	6076	6.9	6049
1.7	6089	4.5	6074	7.1	6046
1.9	6089	4.7	6073	7.3	6044
2.1	6088	4.9	6071	7.5	6041
2.3	6087	5.1	6069	7.7	6038
2.5	6087	5.3	6067	7.9	6035
2.7	6086				

0.4° Horizontal

Horizontal Angle and Setting	K	Horizontal Angle and Setting	K	Horizontal Angle and Setting	K
0.2	$0.^{4}1219$	3.0	$0.^{4}1217$	5.8	$0.^{4}1212$
0.6	1218	3.4	1216	6.2	1211
1.0	1218	3.8	1216	6.6	1210
1.4	1218	4.2	1215	7.0	1210
1.8	1218	4.6	1215	7.4	1208
2.2	1217	5.0	1214	7.8	1207
2.6	1217	5.4	1213		

0.6° Horizontal

Horizontal Angle and Setting	K	Horizontal Angle and Setting	K	Horizontal Angle and Setting	K
0.3	$0.^{4}1827$	3.3	$0.^{4}1825$	5.7	$0.^{4}1819$
0.9	1827	3.9	1824	6.3	1816
1.5	1827	4.5	1822	6.9	1814
2.1	1827	5.1	1821	7.5	1810
2.7	1826				

0.8° Horizontal

Horizontal Angle and Setting	K	Horizontal Angle and Setting	K	Horizontal Angle and Setting	K
0.4	$0.^{4}2437$	3.6	$0.^{4}2432$	6.0	$0.^{4}2424$
1.2	2436	4.4	2429	6.8	2420
2.0	2435	5.2	2427	7.6	2416
2.8	2434				

1.0° Horizontal

Horizontal Angle and Setting	K	Horizontal Angle and Setting	K	Horizontal Angle and Setting	K
0.5	$0.^{4}3046$	3.5	$0.^{4}3041$	6.5	$0.^{4}3027$
1.5	3045	4.5	3037	7.5	3020
2.5	3043	5.5	3032		

Spacing 0.2° Vertical

0.2° Horizontal

Horizontal Angle and Setting	K	Horizontal Angle and Setting	K	Horizontal Angle and Setting	K
0.1	$0.^{4}1219$	2.9	$0.^{4}1217$	5.5	$0.^{4}1213$
0.3	1219	3.1	1217	5.7	1212
0.5	1218	3.3	1217	5.9	1212
0.7	1218	3.5	1216	6.1	1212
0.9	1218	3.7	1216	6.3	1211
1.1	1218	3.9	1216	6.5	1211
1.3	1218	4.1	1215	6.7	1210
1.5	1218	4.3	1215	6.9	1210
1.7	1218	4.5	1215	7.1	1209
1.9	1218	4.7	1214	7.3	1209
2.1	1218	4.9	1214	7.5	1208
2.3	1218	5.1	1214	7.7	1208
2.5	1217	5.3	1213	7.9	1207
2.7	1217				

0.4° Horizontal

Horizontal Angle and Setting	K	Horizontal Angle and Setting	K	Horizontal Angle and Setting	K
0.2	$0.^{4}2437$	3.0	$0.^{4}2434$	5.8	$0.^{4}2425$
0.6	2437	3.4	2433	6.2	2423
1.0	2437	3.8	2432	6.6	2421
1.4	2436	4.2	2430	7.0	2419
1.8	2436	4.6	2429	7.4	2417
2.2	2435	5.0	2428	7.8	2414
2.6	2434	5.4	2426		

0.6° Horizontal

Horizontal Angle and Setting	K	Horizontal Angle and Setting	K	Horizontal Angle and Setting	K
0.3	$0.^{4}3655$	3.3	$0.^{4}3649$	5.7	$0.^{4}3637$
0.9	3655	3.9	3647	6.3	3633
1.5	3654	4.5	3644	6.9	3629
2.1	3653	5.1	3641	7.5	3624
2.7	3651				

0.8° Horizontal

Horizontal Angle and Setting	K	Horizontal Angle and Setting	K	Horizontal Angle and Setting	K
0.4	$0.^{4}4874$	3.6	$0.^{4}4864$	6.0	$0.^{4}4848$
1.2	4872	4.4	4858	6.8	4840
2.0	4870	5.2	4854	7.6	4832
2.8	4868				

1.0° Horizontal

Horizontal Angle and Setting	K	Horizontal Angle and Setting	K	Horizontal Angle and Setting	K
0.5	$0.^{4}6092$	3.5	$0.^{4}6082$	6.5	$0.^{4}6054$
1.5	6090	4.5	6074	7.5	6040
2.5	6086	5.5	6064		

Spacing 0.4° Vertical

0.2° Horizontal

Horizontal Angle and Setting	K	Horizontal Angle and Setting	K	Horizontal Angle and Setting	K
0.1	$0.^{4}2437$	2.9	$0.^{4}2434$	5.5	$0.^{4}2426$
0.3	2437	3.1	2434	5.7	2425
0.5	2437	3.3	2433	5.9	2424
0.7	2437	3.5	2432	6.1	2423
0.9	2437	3.7	2432	6.3	2422
1.1	2437	3.9	2431	6.5	2421
1.3	2436	4.1	2431	6.7	2420
1.5	2436	4.3	2430	6.9	2419
1.7	2436	4.5	2430	7.1	2418
1.9	2436	4.7	2429	7.3	2417
2.1	2435	4.9	2428	7.5	2416
2.3	2435	5.1	2427	7.7	2415
2.5	2434	5.3	2426	7.9	2414
2.7	2434				

0.4° Horizontal

Horizontal Angle and Setting	K	Horizontal Angle and Setting	K	Horizontal Angle and Setting	K
0.2	$0.^{4}4874$	3.0	$0.^{4}4867$	5.8	$0.^{4}4849$
0.6	4874	3.4	4865	6.2	4846
1.0	4874	3.8	4863	6.6	4842
1.4	4872	4.2	4861	7.0	4838
1.8	4872	4.6	4858	7.4	4834
2.2	4870	5.0	4855	7.8	4829
2.6	4869	5.4	4852		

Spacing 0.4° Vertical *Continued*

0.6° Horizontal

Horizontal Angle and Setting	K	Horizontal Angle and Setting	K	Horizontal Angle and Setting	K
0.3	$0.^{4}7310$	3.3	$0.^{4}7299$	5.7	$0.^{4}7274$
0.6	7310	3.9	7294	6.3	7267
1.5	7308	4.5	7288	6.9	7258
2.1	7306	5.1	7282	7.5	7248
2.7	7302				

0.8° Horizontal

Horizontal Angle and Setting	K	Horizontal Angle and Setting	K	Horizontal Angle and Setting	K
0.4	$0.^{4}9747$	3.6	$0.^{4}9728$	6.0	$0.^{4}9694$
1.2	9746	4.4	9719	6.8	9679
2.0	9742	5.2	9707	7.6	9662
2.8	9736				

1.0° Horizontal

Horizontal Angle and Setting	K	Horizontal Angle and Setting	K	Horizontal Angle and Setting	K
0.5	$0.^{3}1218$	3.5	$0.^{3}1216$	6.5	$0.^{3}1211$
1.5	1218	4.5	1215	7.5	1208
2.5	1217	5.5	1212		

Spacing 0.6° Vertical

0.2° Horizontal

Horizontal Angle and Setting	K	Horizontal Angle and Setting	K	Horizontal Angle and Setting	K
0.1	$0.^{4}3655$	2.9	$0.^{4}3650$	5.5	$0.^{4}3638$
0.3	3655	3.1	3650	5.7	3637
0.5	3655	3.3	3649	5.9	3636
0.7	3655	3.5	3649	6.1	3635
0.9	3655	3.7	3648	6.3	3634
1.1	3655	3.9	3647	6.5	3632
1.3	3654	4.1	3646	6.7	3631
1.5	3654	4.3	3645	6.9	3629
1.7	3654	4.5	3644	7.1	3628
1.9	3653	4.7	3643	7.3	3626
2.1	3653	4.9	3642	7.5	3625
2.3	3653	5.1	3641	7.7	3622
2.5	3652	5.3	3640	7.9	3621
2.7	3652				

0.4° Horizontal

Horizontal Angle and Setting	K	Horizontal Angle and Setting	K	Horizontal Angle and Setting	K
0.2	$0.^{4}7310$	3.0	$0.^{4}7301$	5.8	$0.^{4}7273$
0.6	7310	3.4	7299	6.2	7268
1.0	7310	3.8	7295	6.6	7262
1.4	7308	4.2	7292	7.0	7256
1.8	7307	4.6	7288	7.4	7250
2.2	7306	5.0	7283	7.8	7243
2.6	7303	5.4	7278		

0.6° Horizontal

Horizontal Angle and Setting	K	Horizontal Angle and Setting	K	Horizontal Angle and Setting	K
0.3	$0.^{3}1097$	3.3	$0.^{3}1095$	5.7	$0.^{3}1091$
0.9	1097	3.9	1094	6.3	1090
1.5	1096	4.5	1093	6.9	1089
2.1	1096	5.1	1092	7.5	1087
2.7	1095				

0.8° Horizontal

Horizontal Angle and Setting	K	Horizontal Angle and Setting	K	Horizontal Angle and Setting	K
0.4	$0.^{3}1462$	3.6	$0.^{3}1459$	6.0	$0.^{3}1454$
1.2	1462	4.4	1458	6.8	1452
2.0	1461	5.2	1456	7.6	1449
2.8	1460				

1.0° Horizontal

Horizontal Angle and Setting	K	Horizontal Angle and Setting	K	Horizontal Angle and Setting	K
0.5	$0.^{3}1828$	3.5	$0.^{3}1824$	6.5	$0.^{3}1816$
1.5	1827	4.5	1822	7.5	1812
2.5	1826	5.5	1819		

Fig. 4–14. *Continued*

Constants for Converting Beam Candlepower of Projector-Type Luminaires (Searchlights, Floodlights, Spotlights) into Lumens

(0.1 × 0.1 to 10 × 10 degree steps)

Horizontal Angle and Setting	K	Horizontal Angle and Setting	K	Horizontal Angle and Setting	K
Spacing 0.8° Vertical					
0.2° Horizontal					
0.1	$0.^{4}4874$	2.9	$0.^{4}4867$	5.5	$0.^{4}4851$
0.3	4874	3.1	4867	4.7	4850
0.5	4874	3.3	4866	5.9	4848
0.7	4874	3.5	4865	6.1	4846
0.9	4874	3.7	4864	6.3	4845
1.1	4874	3.9	4862	6.5	4842
1.3	4872	4.1	4862	6.7	4841
1.5	4872	4.3	4860	6.9	4838
1.7	4872	4.5	4859	7.1	4837
1.9	4871	4.7	4858	7.3	4834
2.1	4870	4.9	4856	7.5	4833
2.3	4870	5.1	4854	7.7	4830
2.5	4869	5.3	4853	7.9	4828
2.7	4869				
0.4° Horizontal					
0.2	$0.^{4}9747$	3.0	$0.^{4}9734$	5.8	$0.^{4}9698$
0.6	9747	3.4	9730	6.2	9691
1.0	9747	3.8	9726	6.6	9683
1.4	9744	4.2	9722	7.0	9675
1.8	9743	4.6	9717	7.4	9667
2.2	9741	5.0	9710	7.8	9658
2.6	9738	5.4	9704		
0.6° Horizontal					
0.3	$0.^{3}1462$	3.3	$0.^{3}1460$	5.7	$0.^{3}1455$
0.9	1462	3.9	1459	6.3	1453
1.5	1462	4.5	1458	6.9	1452
2.1	1461	5.1	1456	7.5	1450
2.7	1460				
0.8° Horizontal					
0.4	$0.^{3}1949$	3.6	$0.^{3}1946$	6.0	$0.^{3}1939$
1.2	1949	4.4	1944	6.8	1936
2.0	1948	5.2	1941	7.6	1932
2.8	1947				
1.0° Horizontal					
0.5	$0.^{3}2437$	3.5	$0.^{3}2432$	6.5	$0.^{3}2421$
1.5	2436	4.5	2429	7.5	2416
2.5	2435	5.5	2426		
Spacing 1.0° Vertical					
0.1° Horizontal					
0.05	$0.^{4}3046$	2.75	$0.^{4}3043$	5.45	$0.^{4}3032$
0.15	3046	2.85	3042	5.55	3032
0.25	3046	2.95	3042	5.65	3031
0.35	3046	3.05	3042	5.75	3031
0.45	3046	3.15	3042	5.85	3030
0.55	3046	3.25	3041	5.95	3030
0.65	3046	3.35	3041	6.05	3029
0.75	3046	3.45	3041	6.15	3029
0.85	3046	3.55	3040	6.25	3028
0.95	3046	3.65	3040	6.35	3028
Spacing 1.0° Vertical *Continued*					
0.1° Horizontal *Continued*					
1.05	3046	3.75	3040	6.45	3027
1.15	3046	3.85	3039	6.55	3026
1.25	3045	3.95	3039	6.65	3026
1.35	3045	4.05	3039	6.75	3025
1.45	3045	4.15	3038	6.85	3024
1.55	3045	4.25	3038	6.95	3024
1.65	3045	4.35	3037	7.05	3023
1.75	3045	4.45	3037	7.15	3023
1.85	3045	4.55	3037	7.25	3022
1.95	3044	4.65	3036	7.35	3021
2.05	3044	4.75	3036	7.45	3021
2.15	3044	4.85	3035	7.55	3020
2.25	3044	4.95	3035	7.65	3019
2.35	3044	5.05	3034	7.75	3018
2.45	3043	5.15	3034	7.85	3018
2.55	3043	5.25	3033	7.95	3017
2.65	3043	5.35	3033		
0.2° Horizontal					
0.1	$0.^{4}6092$	2.9	$0.^{4}6085$	5.5	$0.^{4}6065$
0.3	6092	3.1	6084	5.7	6063
0.5	6092	3.3	6083	5.9	6061
0.7	6092	3.5	6081	6.1	6058
0.9	6092	3.7	6080	6.3	6056
1.1	6092	3.9	6079	6.5	6054
1.3	6090	4.1	6077	6.7	6051
1.5	6090	4.3	6076	6.9	6049
1.7	6089	4.5	6074	7.1	6046
1.9	6089	4.7	6073	7.3	6044
2.1	6088	4.9	6071	7.5	6041
2.3	6087	5.1	6069	7.7	6038
2.5	6087	5.3	6067	7.9	6035
2.7	6086				
0.4° Horizontal					
0.2	$0.^{3}1219$	3.0	$0.^{3}1217$	5.8	$0.^{3}1212$
0.6	1218	3.4	1216	6.2	1211
1.0	1218	3.8	1216	6.6	1210
1.4	1218	4.2	1215	7.0	1210
1.8	1218	4.6	1215	7.4	1208
2.2	1217	5.0	1214	7.8	1207
2.6	1217	5.4	1213		
0.6° Horizontal					
0.3	$0.^{3}1828$	3.3	$0.^{3}1825$	5.7	$0.^{3}1819$
0.9	1828	3.9	1824	6.3	1816
1.5	1827	4.5	1822	6.9	1814
2.1	1827	5.1	1821	7.5	1810
2.7	1826				
0.8° Horizontal					
0.4	$0.^{3}2437$	3.6	$0.^{3}2432$	6.0	$0.^{3}2424$
1.2	2436	4.4	2430	6.8	2420
2.0	2436	5.2	2427	7.6	2416
2.8	2434				
Spacing 1.0° Vertical *Continued*					
1.0° Horizontal					
0.5	$0.^{3}3046$	3.5	$0.^{3}3041$	6.5	$0.^{3}3027$
1.5	3045	4.5	3037	7.5	3020
2.5	3043	5.5	3032		
Spacing 2° Vertical					
2° Horizontal					
1	$0.^{2}122$	31	$0.^{2}104$	61	$0.^{3}59$
3	122	33	102	63	55
5	121	35	100	65	51
7	121	37	097	67	48
9	120	39	095	69	44
11	120	41	092	71	40
13	119	43	089	73	36
15	118	45	086	75	32
17	116	47	083	77	27
19	115	49	080	79	23
21	114	51	077	81	19
23	112	53	073	83	15
25	110	55	070	85	11
27	108	57	066	87	06
29	107	59	063	89	02
5° Horizontal					
2.5	$0.^{2}3046$	32.5	$0.^{2}2570$	62.5	$0.^{2}1406$
7.5	3020	37.5	2416	67.5	1166
12.5	2970	42.5	2246	72.5	0918
17.5	2906	47.5	2060	77.5	658
22.5	2814	52.5	1856	82.5	396
27.5	2702	57.5	1638	87.5	134
10° Horizontal					
5	$0.^{2}6066$	35	$0.^{2}4986$	65	$0.^{2}2572$
15	5876	45	4306	75	1576
25	5576	55	3494	85	0530
Spacing 5° Vertical					
5° Horizontal					
2.5	$0.^{2}760$	32.5	$0.^{2}642$	62.5	$0.^{2}352$
7.5	755	37.5	604	67.5	291
12.5	744	42.5	562	72.5	229
17.5	726	47.5	514	77.5	165
22.5	704	52.5	463	82.5	099
27.5	676	57.5	409	87.5	033
10° Horizontal					
5	0.015165	35	0.012465	65	$0.^{2}6430$
15	14690	45	10765	75	3940
25	13790	55	08735	85	1325
5	0.0304	35	0.0249	65	0.0129
15	294	45	214	75	076
25	276	55	174	85	026

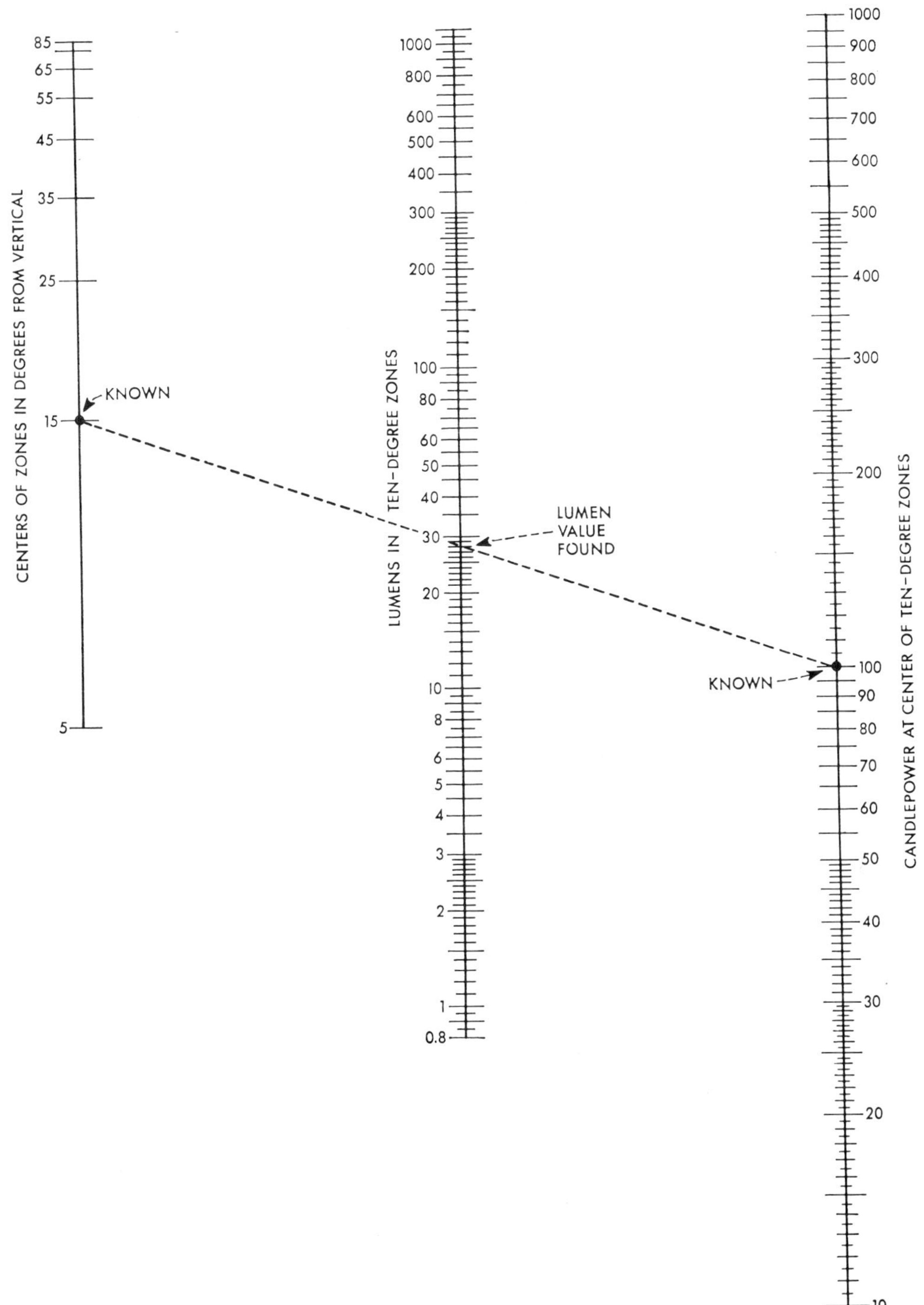

Fig. 4–15. Nomogram for obtaining zonal lumens when the average candlepower at the center of ten-degree zones is known.

For the first formula on page 4-18, let θ_m be the mid-zone angle and let P equal ½ of the zone interval. The formula then becomes:

$$4\pi \sin P\ (\sin \theta_m)$$

Since the zone width is often the same for a series of constants, the first term will simply be multiplied successively by the sines of the mid-zone angles.

For the second formula on page 4-18, let θ_m be the median angle on the horizontal and let P equal ½ of the horizontal interval. The formula then becomes:

$$2\pi \frac{\theta}{180} \sin P\ (\cos \theta_m)$$

Where the horizontal interval is the same for a number of constants, the first term will simply be multiplied successively by the cosines of the median angles.

Floodlight-Type Luminaires. The following applies to floodlighting equipment having a total beam spread (divergence) of more than 10 degrees. For specific information on testing this type of equipment, consult the "IES Approved Method for Photometric Testing of Floodlights Using Incandescent Filament or Discharge Lamps." [55] For lighting equipment having a beam spread less than 10 degrees, see the "IES Guide for Photometric Testing of Searchlights." [65]

The classification of floodlights is based upon the beam width of the floodlight on the horizontal and vertical axes of the floodlight. The classification is designated by NEMA type numbers as listed in Fig. 9-64. For symmetrical beams the floodlight type is defined as the average of the horizontal and vertical beam spreads. For asymmetrical beam floodlights the type is designated by the horizontal and vertical beam spreads in that order, *e.g.*, a floodlight with a horizontal beam spread of 75 degrees (Type 5) and a vertical beam of 35 degrees (Type 3) would be designated as a Type 5 × 3 floodlight.

Stray light may be defined as that light emitted by the floodlight which is outside the floodlight beam as defined by the beam classification. In some instances stray light may be useful in illumination or detrimental to vision depending upon its magnitude and direction. When it is desired to determine the amount and direction of stray light, it is necessary to make measurements as far horizontally and vertically as the readings have significant values in relation to the measuring system.

If the light center of the test lamp (if more than one lamp, the geometric center of lamp light centers) is not enclosed by the reflector, the floodlight should be mounted on the goniometer so that the light center of the lamp is at the goniometer center. If the lamp light center is within the reflector, the floodlight should be positioned so that the center of the reflector opening coincides with the goniometer center.

Either the direct or the relative method of photometry may be used for floodlights, but the relative method lends itself particularly well since cumulative errors may be reduced, and maintenance of standards of luminous intensity and flux are not necessary. In the latter method, relative candlepower readings of the test lamp alone made on a distribution photometer, and of the lamp-floodlight combination made on the floodlight photometer are taken with the lamp operating under identical electrical conditions in both tests.

Beams produced by projector-type luminaires utilizing specular reflectors and filament lamps are likely to be non-uniform in intensity. To average out such variations in the beams, an integrating device should be used which will integrate the illumination over one square degree (20.94 inches by 20.94 inches at 100-foot testing distance) or over a circular degree at a distance of 100 feet.

The method of taking candlepower readings is to traverse the beam with such angular spacings as to give approximately 100 reading stations uniformly spaced throughout the beam, the beam limit being as described above. By interpolating between these readings an isocandela diagram may be plotted on rectangular co-ordinates (see Fig. 9-65, page 9-69). The lumens in the beam may be computed using the constants in Fig. 4-14.

The information usually reported for floodlights includes the following: NEMA type, horizontal and vertical beam distribution curves, maximum beam luminous intensity, average maximum beam luminous intensity, beam spread in both horizontal and vertical directions, beam flux, beam efficiency, total floodlight flux and total efficiency.[66]

Roadway Luminaires. A guide has been prepared to provide test procedures and methods of reporting data to promote the uniform evaluation of the optical performance of roadway luminaires using incandescent filament and high intensity discharge lamps.[67]

Luminaires selected for test should be representative of the manufacturer's typical product. The testing distance should approximate the mounting height at which the luminaire is to be used. A distance of 25 feet should be sufficient for most beam forming luminaires. Photometric test distance is, in general, defined as the distance from the goniometer center to the surface of the test plate or photosensitive element, taking into account the distance to and from any mirror or mirrors that may be used.

The number of planes explored during photometric measurements should be determined by the symmetry or irregularity of distribution and by the end results desired from the test. The number of vertical angles at which readings are taken will depend on how the readings are to be used. If an isocandela diagram is to be plotted, readings may have to be taken at close intervals, especially if values are changing rapidly.

For luminaires having a distribution that is symmetric about a vertical axis (IES Type V),* readings may be taken in one vertical plane while the entire luminaire is being rotated below 45 rpm. The lamp may be rotated within a stationary optical assembly provided readings are taken in several vertical planes and averaged. If neither luminaire nor lamp is rotated, average of the readings taken in ten equally spaced vertical planes should be sufficient.

For luminaires having a distribution that is symmetric about a single vertical plane (IES Types II, III, IV and II 4-way) readings may be taken in vertical planes that are ten degrees apart. Due to the method used in data processing, it may be advantageous to divide laterally into 10 degree zones and measure at the midzone angle. Averages may be made of the readings taken at corresponding angles on the opposite sides of the plane of symmetry. Any computations that are to be performed may then be done on one side of the plane of symmetry.

For luminaires having a distribution that is symmetric about two vertical planes (IES Type I), readings may be taken as directly above, but computations may be performed in one quadrant of the sphere.

For luminaires having a distribution that is symmetric about four vertical planes (IES Type I 4-way), readings may be taken as above, but computations may be performed in one octant of the sphere.

For luminaires having an asymmetric type distribution (unidirectional), readings may be taken in vertical planes that are ten degrees apart. Since there is no symmetry, any computations performed should be done without averaging.

Sufficient data should be accumulated to permit: classifying the light distribution in accordance with recommended practice (see Section 20); an isofootcandle diagram; the utilization efficiency; and the total and four quadrant efficiencies.

Luminaire-Lamp-Ballast Operating Factor. A procedure has been developed[68] to provide a factor to be applied to high intensity discharge luminaire photometric data to adjust them to the specific combination of luminaire, lamp type, and ballast used in a system. By repetitive tests, this procedure may be used to determine variations of system performance exclusive of lamp variations. Such a factor can be applied specifically to lumens, candelas, and footcandles as they appear on photometric data sheets.

Two possibilities are recognized: the first, in which the lamp is used in the operating position for which it is rated; the second, in which the lamp is operated in a position other than the one for which it is rated. The factors determined relate to equipment (luminaire, lamp, ballast combination) operating factor under initial conditions unless otherwise specified.

* See Section 20 for a description of the distribution types.

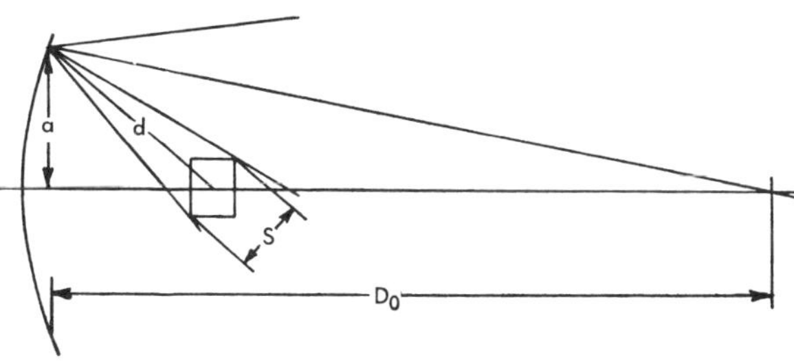

Fig. 4-16. Diagram showing distances and dimensions used to determine minimum inverse-square distance.

The procedures essentially involve a relative light output measurement using the selected lamp in the test luminaire with its test ballast operated at rated supply voltage, after operating conditions have stabilized, and a relative light output measurement with the lamp switched to a *reference ballast* without being extinguished, holding the rated input voltage specified for the *reference ballast*. The *equipment operating factor* is the ratio of the light output measurements made in the first measurement to that made in the second.

Projector Photometry[65]

The illumination from searchlights or other highly collimating luminaires, if measured at distances greater than a certain minimum, obeys the inverse-square law. At such distances, the aperture appears flashed† all over and the luminaire may be considered to behave as a point source lying in the aperture plane and having an intensity equal to the sum of the intensities of the elemental areas of the aperture.

To determine the minimum inverse-square distance for a searchlight, the distance must be determined for each small area on the searchlight optic, and the maximum distance thus obtained is taken as the required minimum inverse-square distance for the searchlight as a whole. The following general formula may be used (see Fig. 4-16):

$$D_0 = \frac{ad}{6s}$$

where D_0 is the minimum inverse-square distance for the small area of optic under consideration (feet).

a is the distance from the small area to the axis of the searchlight (inches).

d is the distance from the light source to the small area (inches).

s is the smallest projected dimension of the light source as viewed from the small area (inches). For example, s is the diameter of one coil of a multi-coil filament lamp, etc.

It is usually unnecessary to compute D_0 for more than one area of the optic, since the area that will

† When viewed through a telescope at these distances the reflector aperture appears bright all over.

Fig. 4-17. Goniometer for 100-foot indoor photometric range, National Bureau of Standards.[69] Two rotary tables of the type used in large machine tools are incorporated into the goniometer to provide two of the three rotations available. For horizontal distributions either the rotary table on which the outer frame is mounted or the inner table on which the test equipment is mounted may be used. This makes for flexibility and is especially useful for obtaining polar angle distributions. The pinion gear in the vertical drive may be disengaged readily and the inner frame with the test unit mounted can then be balanced with the adjustable counterweights. After balancing, a small constant torque is applied to the inner frame through the use of the pulley and weight at the side, thus eliminating backlash.

make D_0 a maximum can often be determined by inspection.

As an example, a carbon arc searchlight with a reflector 19 inches in diameter and 6 inches in focal length, and with a disk-shaped source 7 mm in diameter (crater of an 11 mm positive carbon), would require a minimum inverse-square distance of 240 feet.

The foregoing treatment of the determination of minimum inverse-square distance is based on ideal sources and axial measurements, and should be considered approximate. In practice, a range two or three times the minimum computed length is adequate. If it is impractical to make measurements on a range of this length, shorter ranges may be used, even less than the minimum inverse-square distance, provided that the searchlight is focused for maximum candlepower at the range distance. While a full length range is preferable for highest accuracy, the error can be kept to a minimum by this method. For additional information consult references 65 and 69.

Goniometers. The usual method for measuring angles in candlepower distribution measurements is to mount the test equipment on a goniometer, keeping the photometric equipment fixed.[69] Basically goniometers provide means for mounting the equipment, for rotating it around the two axes (horizontal and vertical) and for measuring the angles of rotation. (See Figs. 4-12 and 4-17.)

When the searchlight is of unusual size and weight it may be necessary to utilize its own mounting and goniometric facilities for the photometric work or to hold the searchlight fixed and traverse the beam by moving the photometric receptor.

Since searchlights may have a total beam spread of less than one degree and may furthermore be of unusual weight, the mechanical requirements of the goniometer are severe. Suitable rigidity, freedom from backlash, and accuracy of angular measurements are prime requirements. There should be provision for accurate angular settings of the order of 0.1 degree. In special cases higher orders of accuracy may be required.

Photometric Ranges. Indoor ranges are in general preferable, but are frequently impracticable because of the lengths required. For photometry on the relatively shorter indoor ranges, the photometric equipment and procedures are conventional. For such ranges where the cross section is small, some difficulty may arise from reflections off the walls, floor, and ceiling. This can usually be controlled by careful attention to baffling and diaphragming. Outdoor ranges suitable for daytime measurements require much more attention to methods of reducing stray light, minimizing atmospheric disturbances, and correcting for atmospheric transmission. Range sites should be selected where the terrain is flat and homogeneous. The range should be as high off the ground as practicable. Ranges should not be located where atmospheric disturbances occur regularly or where dust or moisture are prevalent. Stray light should be minimized by a suitable system of diaphragms. Any remaining stray light should be measured and subtracted from all readings. In the northern hemisphere, ranges should lie in the north-south line, with the goniometer in an enclosure at the south end. In this way, the shaded north wall of the enclosure will face the photometer.

Correction for Atmospheric Transmission. The absorption of light by moisture, smoke, and dust particles, even in an apparently clear atmosphere, may introduce considerable errors in measurements.[70] It is therefore desirable to measure the atmospheric transmittance before and after the test has been made. A standard reference projector is frequently employed, and is calibrated either by repeated observations in the clearest weather, when the atmospheric transmittance may be accurately estimated or independently measured, or by laboratory methods.

Photometers for Searchlight Measurements.[71] The equipment required for photometric measurements is basically similar in most respects to that required for other types of photometric measurements.

Automatic recording photometers are frequently employed in order to obtain rapid measurements or to obtain statistically valid quantities of data. Multi-element oscillographs are particularly valuable since they facilitate simultaneous recording of searchlight parameters, such as voltage and current, along with the basic photometric data.

Luminance (Photometric Brightness) Measurements[61]

Lamps and Luminaires. Luminance measurements of lamps and luminaires should be made by either the absolute or relative method. With the absolute method, reference standards must be available for equipment calibration. In practice the relative method is generally used.

The published luminances of non-asymmetric fluorescent lamps are computed from the rated lumen output of the lamp according to the following formula:[72]

L_{avg} (candelas per square inch) =

$$\frac{K \times \text{Total Lamp Lumens}}{\text{Lamp Diameter} \times \text{Lamp Luminous Length}}$$

where L_{avg} is the average luminance of the full width of the lamp at its center. The diameter and length are expressed in inches and K is:

$$\text{for 48-inch T-12 lamps } K = \frac{1.09}{9.25}$$

$$\text{72-inch T-12 lamps } K = \frac{1.07}{9.28}$$

$$\text{96-inch T-12 lamps } K = \frac{1.05}{9.28}$$

To compute approximate luminance L_θ of a fluorescent lamp at any angle with the lamp axis, the following formula is used:

$$L_\theta = \frac{\text{Total Lamp Lumens}}{K_\theta \times \text{Lamp Diameter} \times \text{Lamp Luminous Length} \times \sin\theta}$$

where K_θ is as shown in Fig. 4–18.

In the laboratory, an accurate means of establishing lamp luminance is secured by construction of a baffled collimating tube 2 to 3 feet long, with a receptor at one end, and at the other end a rectangular aperture having one dimension equal to the lamp diameter, and the second dimension such that the product of the two dimensions equals one square inch.[36] This is placed against the lamp near its midpoint so that the entire diameter of the lamp fills the aperture. The deflection of the indicating equipment is directly proportional to L_{avg}. The instrument is thus calibrated when a lamp of known lumen output is viewed, and direct comparisons can be made with lamps of similar diameters. If the lumen output is assumed to be the rated value, the instrument is calibrated for the relative method so that all subsequent luminance measurements are related to rated lumen output.

For measuring luminance of a luminaire, an instrument having a circular aperture is preferable, so that rotation of the instrument or angular position of luminaire will not affect readings. It is convenient, therefore, to use another collimating tube with an aperture 1⅛-inches in diameter, or luminance meters with circular fields of view, encompassing 1 square inch at the measuring distance, and calibrated against a "working standard" which can be similar to the device shown in Fig. 4–19. In use, luminance of the device's diffusing face is adjusted such that the deflection of the indicating instrument used with the collimating tube equipped with a rectangular aperture is the same as that indicated when the same aperture is placed against the reference fluorescent lamp. The collimator with circular aperture is then substituted and the deflection noted—then this will be the deflection for the lamp luminance. All luminaire luminance readings made with this circular aperture are then directly related to the lamp luminance.

Meter Calibration. Where luminance meters are calibrated by means of a diffusing test plate with a known illumination, the characteristics of the test plate for specific conditions of incidence, spectral energy distribution of the source, and angle of view must be known. For a perfectly diffusing reflecting test plate the luminance is uniform at all angles, and its value in footlamberts is the product of the reflectance and the illumination in footcandles. However, no surface is perfectly diffusing, and serious error can result from the assumption that this simple relationship holds. It is essential that the characteristics (luminance vs. angle of view) of the test plate be known, and taken into consideration in the calibration.

Fig. 4–18. Values at Various Angles of the Lamp: Candlepower Ratio K_θ for Preheat-Starting Types of Fluorescent Lamps*

Angle (degrees)	0	10	20	30	40	50	60	70	80	90
K_θ	—	172.0	46.0	24.7	17.1	13.3	11.3	10.1	9.5	9.25
Sin θ	0	.174	.342	.500	.643	.766	.866	.940	.985	1.000

* Average for 15-, 20-, 30-, 40-, and 100-watt lamps.

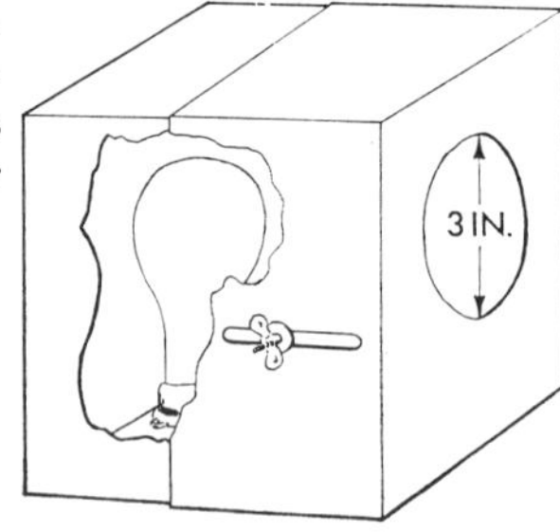

Fig. 4-19. An adjustable source of luminance, useful when calibrating luminance, measuring devices having apertures of different shapes.

A convenient method of measuring source luminance of carbon arcs,[61, 73] high-luminance discharge lamps, and other similar sources involves measurement of the illumination in a projected image of the source. Measurements are most readily made with a photovoltaic cell placed in the plane at which the image is focused. The cell may be stationary, or may be moved across the image to scan the luminance pattern. The cell aperture must of course be relatively small with respect to the size of the image, so that the area measured at any one time is virtually uniform. The cell and its measuring instrument are calibrated in footcandles. Luminance of the source is calculated from the footcandle reading in the image, the magnification of the image, the transmittance of the projected lens, the diameter of the lens aperture, and the distance between the lens aperture and the source.

Thermal Testing of Luminaires

Thermal testing methods have been developed for compiling data on air cooled heat transfer luminaires.[74] In that method, the entire laboratory room in which the calorimeter is located becomes a part of the calorimetric system. (A calorimeter is a device used to measure the thermal energy distribution, and/or dissipation, of luminaires; primarily those luminaires which handle either air, water, or both.) The room must be controlled closely with respect to temperature, air motion and relative humidity. The varying conditions can materially affect the results of the calorimetric measurements. The room conditions should be as follows: the temperature should be controlled within 0.5°F above or below the set test temperature; the velocity of the air in the space containing the calorimeter should be held constant and not exceed 30 feet per minute; the relative humidity should be held constant at any convenient value; and the room should be located in an area that will not be affected by external conditions.

The category of calorimeters to be used is determined by the specific purpose of the device and the degree to which its conditions can be controlled. Three types of calorimeters are: the *zero heat-loss calorimeter,* a calorimeter so constructed to compensate for the heat transfer through its walls; the *calibrated heat-loss calorimeter,* a box in which the heat loss can be determined by dissipating a measured quantity of energy in the plenum; and the *continuous fluid-flow calorimeter,* a modification of the zero heat-loss calorimeter consisting of a heavily insulated heat exchanger installed over the luminaire.

Instrumentation for calorimetry of luminaires must be precise to measure small variations from the total. Instrumentation is needed to measure temperature (thermometers, thermocouples, thermistors, and resistance elements), pressure (manometers, micromanometers, draft gages, and swinging vane type gages), mass flow rate of air and water, electrical quantities, and light output.

Each luminaire that is to be tested for energy distribution should first be photometered in accordance with accepted IES procedures.[6] During calorimetry, a light meter or photovoltaic cell and microammeter should be installed at luminaire nadir, not less than one foot from the bottom of the enclosure. The maximum distance at which the cell should be mounted below the luminaire should be governed by the distance required for the cell to see the full luminous area of the luminaire. It is necessary that precautions be taken to prevent the cell from seeing luminous flux other than that transmitted by the luminaire under test. This position must not be changed during test.

The data to be recorded and reported should include: description and size of luminaire; mode of operation; test conditions—space and plenum temperatures; relative light output from a 77°F base; energy to plenum (total); energy to space; exhaust air volume; energy removed by return air stream; and exhaust air temperature. Energy and relative light output values should be plotted as ordinates vs. air volumes as abscissa.

FIELD MEASUREMENTS—INTERIORS

In evaluating an actual lighting installation in the field it is necessary to measure or survey the quality and quantity of lighting in the particular environment. To help do this, the IES has developed a uniform survey method of measuring and reporting the necessary data.[75] The results of these uniform surveys can be used alone or with other surveys for comparison purposes, can be used to determine compliance with specifications or recommended practice, and can be used to reveal the need for maintenance, modification or replacement.

Field measurements apply only to the conditions that exist during the survey. Recognizing this it is very important to record a complete detailed description of the surveyed area and all factors that might affect results, such as: interior surface reflectances, lamp type and age, voltage, and instruments used in the survey.

In measuring illumination, cell type instruments should be used which are cosine and color corrected. They should be used at a temperature above 60° F and below 120° F, if possible. Before taking readings, the cells should be exposed to the approximate illumination level to be measured until they become stabilized. This usually requires 5 to 15 minutes. Casting a shadow on the light sensitive cell should be avoided while reading the instrument. A high intensity discharge or fluorescent system must be lighted for at least one hour before measurements are taken to be sure that normal operating output has been attained. In relatively new lamp installations, at least 100 hours of operation of a gaseous source should elapse before measurements are taken. With incandescent lamps, seasoning is accomplished in a shorter time (20 hours or less for common sizes).

Illumination Measurements—Average

Determination of Average Footcandles on a Horizontal Plane from General Lighting Only. The use of this method in the types of areas described should result in values of average illumination within 10 per cent of the values that would be obtained by dividing the area into 2 foot squares, taking a reading in each square and averaging.

The measuring instrument should be positioned so that when readings are taken, the surface of the light sensitive cell is in a horizontal plane and 30 inches above the floor. This can be facilitated by means of a small portable stand of wood or other material that will support the cell at the correct height and in the proper plane. Daylight should be excluded during illumination measurements. Readings should be taken at night or with shades, blinds or other opaque covering on the fenestration.

Regular Area With Symmetrically Spaced Luminaires in Two or More Rows. (See Fig. 4–20*a*.) (1) Take readings at stations r-1, r-2, r-3, and r-4 for a typical inner bay. Repeat at stations r-5, r-6, r-7 and r-8 for a typical centrally located bay. Average the 8 readings. This is R in the equation directly below. (2) Take readings at stations q-1, q-2, q-3 and q-4 in two typical half bays on each side of room. Average the 4 readings. This is Q in the equation below. (3) Take readings at stations t-1, t-2, t-3 and t-4 in two typical half bays at each end of room. Average the 4 readings. This is T in the equation. (4) Take readings at stations p-1 and p-2 in two typical corner quarter bays. Average the 2 readings. This is P in the equation. (5) Determine the average illumination in the area by solving the equation:

$$\text{Average Illumination} = \frac{R(N-1)(M-1) + Q(N-1) + T(M-1) + P}{NM}$$

where N = number of luminaires per row;
M = number of rows.

Regular Area With Symmetrically Located Single Luminaire. (See Fig. 4–20b.) (1) Take readings at stations p-1, p-2, p-3 and p-4 in all 4 quarter bays. Average the 4 readings. This is P, the average illumination in the area.

Regular Area With Single Row of Individual Luminaires. (See Fig. 4–20*c*.) (1) Take readings at stations q-1 thru q-8 in 4 typical half bays located two on each side of the area. Average the 8 readings. This is Q in the equation directly below. (2) Take readings at stations p-1 and p-2 for two typical corner quarter bays. Average the 2 readings. This is P in the equation. (3) Determine the average illumination in the area by solving the equation:

$$\text{Average Illumination} = \frac{Q(N-1) + P}{N}$$

where N = number of luminaires.

Regular Area With Two or More Continuous Rows of Luminaires. (See Fig. 4–20d.) (1) Take readings at stations r-1 thru r-4 located near the center of the area. Average the 4 readings. This is R in the equation directly below. (2) Take readings at stations q-1 and q-2 located at each midside of the room and midway between the outside row of luminaires and the wall. Average the 2 readings. This is Q in the equation below. (3) Take readings at stations t-1 thru t-4 at each end of the room. Average the 4 readings. This is T in the equation. (4) Take readings at stations p-1 and p-2 in two typical corners. Average the 2 readings. This is P in the equation. (5) Determine the average illumination in the area by solving the equation:

$$\text{Average Illumination} = \frac{RN(M-1) + QN + T(M-1) + P}{M(N+1)}$$

where N = number of luminaires per row;
M = number of rows.

Regular Area With Single Row of Continuous Luminaires. (See Fig. 4–20*e*.) (1) Take readings at stations q-1 thru q-6. Average the 6 readings. This is Q in the equation directly below. (2) Take readings at stations p-1 and p-2 in typical corners. Average the 2 readings. This is P in the equation below. (3) Determine the average illumination in the area by solving the equation:

$$\text{Average Illumination} = \frac{QN + P}{N + 1}$$

where N = number of luminaires.

Regular Area With Luminous or Louverall Ceiling. (See Fig. 4–20*f*.) (1) Take readings at stations r-1 thru r-4 located at random in the central portion of the area. Average the 4 readings. This is R in the equation directly below. (2) Take readings at stations q-1 and q-2 located 2 feet from the long walls at random lengthwise of the room. Average the 2 readings. This is Q in the equation below. (3) Take readings at stations t-1 and t-2 located 2 feet from the short walls at random crosswise of the room. Average the 2 readings. This is T in the equation below. (4) Take readings at stations p-1 and

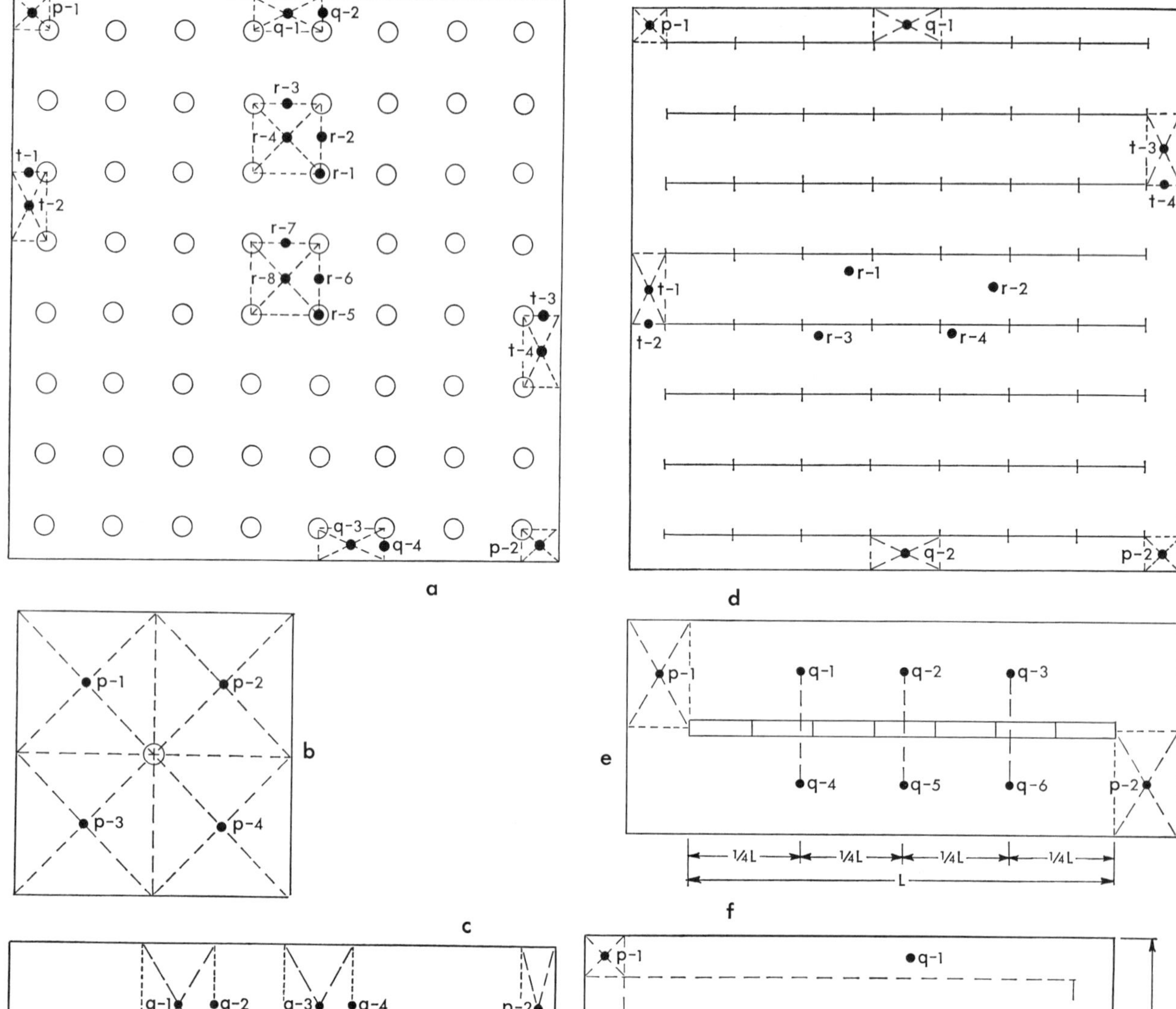

Fig. 4–20. Location of illumination measurement stations in: (*a*) regular area with symmetrically spaced luminaires in 2 or more rows: (*b*) regular area with symmetrically located single luminaire; (*c*) regular area with single row of individual luminaires; (*d*) regular area with 2 or more continuous rows of luminaires; (*e*) regular area with single row of continuous luminaires; (*f*) regular area with luminous or louverall ceiling.

p-2 located at diagonally opposite corners 2 feet from each wall. Average the 2 readings. This is P in the equation. (5) Determine the average illumination in the area by solving the equation:

$$\text{Average Illumination} = \frac{R(L-8)(W-8) + 8Q(L-8) + 8T(W-8) + 64P}{WL}$$

where W = width of room;
L = length of room.

Illumination Measurements—Spot

With general plus supplementary lighting in use, the illumination at the point of work should be measured with the worker in his normal working position. The measuring instrument should be located so that when readings are taken, the surface of the light sensitive cell is in the plane of the work or of that portion of the work on which the critical visual task is performed—horizontal, vertical or inclined. Readings should be taken at night or with fenestration covered. Readings as shown in Fig. 4-21 should be recorded.

Luminance Measurements

Footlambert surveys, unlike footcandle surveys, should be made under actual working conditions and from a specified work point location with the combinations of daylight and electric lighting facilities available. Consideration should be given to sun position and weather conditions, both of which may have marked effect on luminance distribution. All lighting in the area, both general and supplementary, should be in normal use. Work areas used only in the daytime should be surveyed in the daytime; work areas used both daytime and nighttime should preferably have two luminance surveys made under the two sets of conditions, as the luminance distribution and the possible comfort or discomfort will differ markedly at these times. Nighttime surveys should be made with shades drawn. Daytime surveys should be made with shades adjusted for best control of daylight.

On a floor plan sketch of the area, an indication should be made of which exterior wall or walls, if any, were exposed to direct sunlight during the time of the survey by writing the word "Sun" in the appropriate location. Readings should be taken, successively, from the worker's position at each work point location A, B, C, etc. and luminance readings from each location recorded as shown in Fig. 4-22.

Fig. 4-21. Form for Tabulation of Spot Illumination Measurements

Work Point	Description of Work Point	Height Above Floor (feet)	Plane (horizontal, vertical, or inclined)	Footcandles	
				Total (general + supplementary)	General Only
1—(max.)					
2—(min.)					
3—					
4—					
5—					

Fig. 4-22. Form for Tabulation of Luminance Measurements

Work Point Location*	Luminance (footlamberts)					
	A	B	C	D	E	F
Luminaire at 45° above eye level						
Luminaire at 30° above eye level						
Luminaire at 15° above eye level						
Ceiling, above luminaire						
Ceiling, between luminaires						
Upper wall or ceiling adjacent to a luminaire						
Upper wall between two luminaires						
Wall at eye level						
Dado						
Floor						
Shades and blinds						
Windows						
Task						
Immediate surroundings of task						
Peripheral surroundings of task						
Highest luminance in field of view						

* Describe locations A thru F.

FIELD MEASUREMENTS—OUTDOOR

In roadway and many floodlight installations light is projected in a direction forming a large angle of incidence with the surface to be lighted, and each unit must be adjusted carefully to produce the best utilization and quality of illumination. For an accurate evaluation of this type of installation, special care must be taken in the measurement of the resultant illumination. A summary of the IES guides for Roadway Illumination Measurements and Sports Illumination Measurements follows, but the full guides should be consulted before making an actual survey.[76]

Preparation for the Survey. (1) Inspect and record the condition of the luminaires (globes, reflectors, refractors, lamp positioning, etc.). Unless the purpose of the test is to check depreciation, all units should be cleaned and new lamps (preferably seasoned) installed. (2) Record the mounting height of the luminaires. (3) Record the location of the poles, the number of units per pole, the wattage of the lamps and other pertinent data. Check these data against the recommended layout; a small change in the location or adjustment of the luminaires may make a considerable difference in the resultant illumination. (4) Determine and record the hours of burning of the installed lamps. (5) Record the atmospheric conditions. Because of the effect of adverse atmospheric conditions, the survey should be made only when the atmosphere is clear. Extraneous light produced by a store, service station, or other lights in the vicinity, requires careful attention in street lighting tests. (6) Because of the influence of the electrical circuit operating conditions on lamp light output, it is usually necessary to know precisely the electrical circuit operating conditions at the luminaires in the system at the time the photometric measurements are being made. At night, during the hours when the luminaires will normally be used, record the voltage at the lamp socket with all of the lamps operating. The voltage at the main switch may be measured provided allowance is made for the voltage drop to the individual units. The light output of large incandescent lamps varies approximately 3.5 per cent with every 1 per cent change in voltage. If the measured voltage is not exactly the rated voltage of the lamp, this correction should be applied to the footcandle readings obtained before comparing them with calculated values. Discharge lamps should be operated at least one half hour to reach normal operating conditions before measurements are made.

Survey Procedures. Measurements should be made with a recently calibrated, color- and cosine-corrected photometer capable of being leveled for horizontal measurements or positioned accurately for other measurement planes as required. The photometer should be selected for its portability and repeatability of measurements at any point of the scale which is used. If required by the spectral characteristic of the light source in the system being measured, appropriate corrections should be made to each reading.

1. For roadways, divide the distance between poles into an even number of divisions (as near 10-foot intervals as possible) and take a reading in the center of each rectangle formed by the above divisions and the lanes of the roadway. Additional measurements may be taken at points of special significance, but these readings should not be used in calculating the average horizontal illumination. In some instances, luminance measurements may also be desirable.

2. For sports installations, the sports area, or that portion of the area under immediate consideration, should be divided into test areas of approximately 5 per cent of the total area and readings should be taken in the center of each area. Where illumination for color television is involved, multiple readings should be taken at each station: one reading with the meter cell tilted 15 degrees from vertical in the direction of each camera location; and a final reading with the cell in the horizontal position.

3. Readings should be made at each test station with repeat measurements at the first station frequently enough to assure stability of the system and repeatability of results. Readings should be reproducible within 5 per cent. Enough readings should be taken so that additional readings in similar locations will not change the average results significantly. Avoid casting shadows on the test plate during the readings.

REFERENCES

1. Wright, W. D.: *Photometry and the Eye*, Hatton Press, Ltd., London, pp. 31, 80, 123, 124, 1949.
2. Walsh, J. W. T.: *Photometry*, 3rd ed., Constable and Co., Ltd., London, pp. 120–173, 1958. Committee on Testing Procedures of the IES: "IES General Guide to Photometry," *Illum. Eng.*, Vol. L, p. 147, March, 1955.
3. Walsh, J. W. T.: *Photometry*, 3rd ed., Constable and Co., Ltd., London, pp. 174–188, 1958.
4. Walsh, J. W. T.: *Photometry*, 3rd ed., Constable and Co., Ltd., London, pp. 235–237, 1958. Projector, T. E.: "Effective Intensity of Flashing Lights," *Illum. Eng.*, Vol. LII, p. 630, December, 1957. Douglas, C. A.: "Computation of the Effective Intensity of Flashing Lights," *Illum. Eng.*, Vol. LII, p. 641, December, 1957. Lash, J. D. and Prideaux, G. F.: "Visibility of Signal Lights," *Illum. Eng.*, Vol. XXXVIII, p. 481, November, 1943. Preston, J. S.: "Note on the Photoelectric Measurement of the Average Intensity of Fluctuating Light Sources," *J. Sci. Insts.*, Vol. 18, p. 57, April, 1941. Schuil, A. E.: "The Effect of Flash Frequency on the Apparent Intensity of Flashing Lights Having Constant Flash Duration," *Trans. Illum. Eng. Soc.* (London), Vol. XXXV, p. 117, September, 1940. Neeland, G. K., Laufer, M. K., and Schaub, W. R.: "Measurement of the Equivalent Luminous Intensity of Rotating Beacons," *J. Opt. Soc. Amer.*, Vol. 28, p. 280, August, 1938. Blondel, A. and Rey, J.: "The Perception of Lights of Short Duration at Their Range Limits," *Trans. Illum. Eng. Soc.*, Vol. VII, p. 625, November, 1912.
5. Walsh, J. W. T.: *Photometry*, 3rd ed., Constable and Co., Ltd., London, p. 223, 1958.
6. Committee on Testing Procedures of the IES: "IES General Guide to Photometry," *Illum. Eng.*, Vol. L, p. 149, March, 1955.
7. Kingsbury, E. F.: "A Flicker Photometer Attachment for the Lummer-Brodhun Contrast Photometer," *J. Franklin Inst.*, August, 1915. Guild, J.: "A New Flicker Photometer for Heterochromatic Photometry," *J. Sci. Instrum.*, March, 1924. Feree, C. E. and Rand, G.: "Flicker Photometry," *Trans. Illum. Eng. Soc.*, Vol. XVIII, p. 151, February, 1923.
8. Little, W. F. and Estey, R. S.: "The Use of Color Filters in Visual Photometry," *Trans. Illum. Eng. Soc.*, Vol. XXXII, p. 628, June, 1937. Johnson, L. B.: "Photometry of Gaseous-Conductor Lamps," *Trans. Illum. Eng. Soc.*, Vol. XXXII, p. 646, June, 1937.
9. Sharp, C. H. and Little, W. F.: "Compensated Test Plate for Illumination Photometers," *Trans. Illum. Eng. Soc.*, Vol. X, p. 727, November, 1915. Little, W. F.: "Practical Hints on the Use of Portable Photometers," *Trans. Illum. Eng. Soc.*, Vol. X, p. 766, November, 1915. Morris, A., McGuire, F. L., and Van Cott, H. P.: "Accuracy of Macbeth Illuminometer as a Function of Operator Variability, Calibration, and Sensitivity," *J. Opt. Soc. Amer.*, Vol. 45, p. 525, July, 1955.
10. Taylor, A. H.: "A Portable Reflectance Meter," *Illum. Eng.*, Vol. LV, p. 614, November, 1960.
11. Preston, J. S.: "Photoelectric Photometers: Their Properties, Uses, and Maintenance," *Trans. Illum. Eng. Soc.* (London), Vol. VIII, p. 121, July, 1943. Committee on Testing Procedures of the IES: "IES Guide for the Use and Care of Electrical Instruments," *Illum. Eng.*, Vol. 63, p. 376, July, 1968.
12. "Report of Committee on Portable Photoelectric Photometers," *Trans. Illum. Eng. Soc.*, Vol. XXXII, p. 379, April, 1937. Lange, B., *Photoelements and Their Application*, Reinhold Publishing Corp., New York, May, 1938.
13. Projector, T. H., Laufer, M. K., and Douglas, C. A.: "An Improved 'Zero-Resistance' Circuit for Photo-Cell Photometry," *Rev. Sci. Instrum.*, April, 1944. Barbrow, L. E.: "A Photometric Procedure Using Barrier-Layer Photocells," *J. Res. N. B. S.*, Vol. 25, p. 703, December, 1940.
14. Horton, G. A.: "Electronic Instrumentation in Light Measurements," *Illum. Eng.*, Vol. 64, p. 701, December, 1969.
15. Fogle, M. E.: "New Color Corrected Photronic Cells for Accurate Light Measurements," *Trans. Illum. Eng. Soc.*, Vol. XXXI, p. 773, September, 1936. Parker, A. E.: "Measurement of Illumination From Gaseous Discharge Lamps," *Illum. Eng.*, Vol. XXXV, p. 883, November, 1940. Preston, J. S.: "The Relative Spectral Response of the Selenium Rectifier Photocell in Relation to Photometry and the Design of Spectral Correction Filters," *J. Sci. Instrum.*, Vol. 27, p. 135, May, 1950.
16. Elvegard, E., Lindroth, S., and Larson, E.: "The Drift Effect in Selenium Photovoltaic Cells," *J. Opt. Soc. Amer.*, Vol. 28, p. 33, February, 1938. Houston, R. A.: "The Drift of the Selenium Barrier-Layer Photo-Cell," *Phil. Mag.*, June, 1941. Preston, J. S.: "Fatigue in Selenium Rectifier Photocells," *Nature*, Vol. 153, p. 680, June, 1944.
17. "Report of the Committee on Portable Photoelectric Photometers," *Trans. Illum. Eng. Soc.*, Vol. XXXII, p. 379, April, 1937. Goodbar, I.: "New Procedure to Measure Accurately Illumination at Large Angles of Incidence, with a Barrier-Layer Cell," *Illum. Eng.*, Vol. XL, p. 830, November, 1945. Morton, C. A.: "Cosine Response of Photocells and the Photometry of Linear Light Sources," *Light and Lighting* (London), Vol. XXVIII, p. 157, November, 1945. Buck, G. B., II: "Correction of Light-Sensitive Cells for Angle of Incidence and Spec-

tral Quality of Light," *Illum. Eng.*, Vol. XLIV, p. 293, May, 1949. Dows, C. L.: "Illumination Measurements with Light Sensitive Cells," *Illum. Eng.* Vol. XXXVII, p. 103, February, 1942. Pleijel, G., and Longmore, J.: "A Method of Correcting the Cosine Error of Selenium Rectifier Photocells," *J. Sci. Instrum.*, Vol. 29, p. 137, May, 1952.

18. Atkinson, J. R., Campbell, N. R., Palmer, E. H., and Winch, G. T.: "The Accuracy of Rectifier-Photoelectric Cells," *Proc. Phys. Soc.* (London), November, 1936. Also see ref. 12.

19. MacGregor-Morris, J. T. and Billington, R. M.: "The Selenium Rectifier Photocell: Its Characteristics and Response to Intermittent Illumination," *J. Inst. Elec. Eng.* (London), Vol. 78, p. 435, October, 1936. Zworykin, V. K. and Ramberg, E. G.: *Photoelectricity and Its Application*, John Wiley & Sons, New York, p. 211, 1949. Gleason, P. R.: "Failure of Talbot's Law for Barrier-Layer Photocells," *Phys. Rev.*, Vol. 45, p. 745, 2nd ser., 1934. Walsh, J. W. T.: *Photometry*, 3rd ed., Constable and Co., Ltd., London, p. 98, 107, 1958. Lange, B., *Photoelements and Their Application*, Reinhold Publishing Co., New York, p. 151, 1938.

20. Brooks, H. B.: "The Accuracy of Commercial Electrical Measurements," *Trans. Amer. Inst. Elec. Eng.*, Vol. XXXIX, p. 495, 1920.

21. Baumgartner, G. R.: "A Light-Sensitive Cell Reflectometer," *Gen. Elec. Rev.*, Vol. 40, p. 525, November, 1937.

22. Taylor, A. H.: "Errors in Reflectometry," *J. Opt. Soc. Amer.*, Vol. 43, p. 51, February, 1953.

23. McNicholas, H. J.: "Absolute Methods in Reflectometry," Research Paper No. 3, *J. Res. Bur. Stand.*, p. 29, 1928. Hunter, R. S.: "A Multipurpose Photoelectric Reflectometer," Research Paper RP 1345, *J. Res. Nat. Bur. Stand.*, p. 581, November, 1940. McNicholas, H. J.: "Equipment for Measuring the Reflective and Transmissive Properties of Diffusing Media," Research Paper RP 704, *J. Res. Nat. Bur. Stand.*, p. 211, August, 1934. Hunter, R. S.: "A New Goniophotometer and Its Applications," *Bulletin No. 106*, Henry A. Gardner Laboratory, Inc., Bethesda 14, Maryland, December, 1951. Sharp, C. H. and Little, W. F.: "Measurement of Reflection Factors," *Trans. Illum. Eng. Soc.*, Vol. XV, p. 802, December, 1920.

24. Taylor, A. H.: "A Simple Portable Instrument for the Absolute Measurement of Reflection and Transmission Factors," Sci. Paper No. 405, *Bull. Bur. Stand.*, p. 421, July 28, 1920.

25. Dows, C. L. and Baumgartner, G. R.: "Two Photo-Voltaic Cell Photometers for Measurement of Light Distribution," *Trans. Illum. Eng. Soc.*, Vol. XXX, p. 476, June, 1935. "Permanent Gloss Standards," *Illum. Eng.* Vol. XLV, p. 101, February, 1950. Hunter, R. S.: *A New Goniophotometer and Its Applications*, Bulletin No. 106, Gardner Laboratory, Incorporated, Bethesda, Maryland, December, 1951. Spencer, D. E. and Gray, S. M.: "On the Foundations of Goniophotometry," *Illum. Eng.*, Vol. LV, p. 228–229, April, 1960. Nimeroff, I.: "Analysis of Goniophotometric Curves," *J. Res. of NBS*, RP2335, Vol. 48, p. 441–447, June, 1952.

26. *R.C.A. Photo and Image Tubes*, Booklet 10E-269, R.C.A. Electron Tube Division, Harrison, New Jersey. Sharpe, J.: *Photoelectric Cells and Photomultipliers*, Whittaker Corporation, 80 Express Street, Plainville, New York, 1961. Sharpe, J.: *Dark Current in Photomultiplier Tubes*, Document No. CP. 5475, E.M.I. Electronics, Ltd., Hayes, Middlesex, England, October, 1964.

27. Preston, J. S.: "The Specification of a Spectral Correction Filter for Photometry with Emission Photocells," *J. Sci. Instrum.*, Vol. 23, p. 211, September, 1946. Winch, G. T. and Machin, C. F.: "The Physical Realization of the CIE Average Eye," *Trans. Illum. Eng. Soc.* (London), Vol. XXXIV, p. 89, August, 1940. Winch, G. T.: "Photometry and Colorimetry of Fluorescent and other Electric Discharge Lamps," *Trans. Illum. Eng. Soc.* (London), Vol. XLI, p. 107, June, 1946. Voogd, J.: "Physical Photometry," *Philips Tech. Rev.*, Vol. 4, p. 260, September, 1939.

28. Freund, K.: "Design Characteristics of a Photoelectric Brightness Meter," *Illum. Eng.*, Vol. XLVIII, p. 524, October, 1953.

29. Horton, G. A.: "Evaluation of Capabilities and Limitations of Various Luminance Measuring Instruments," *Illum. Eng.*, Vol. LX, p. 217, April, 1965. Eastman, A. A.: "Contrast Determination with the Pritchard Telephotometer," *Illum. Eng.*, Vol. LX, p. 179, April, 1965. Spencer, D. E.: "Out of Focus Photometry," *J. Opt. Soc. Amer.*, Vol. 55, p. 396, April, 1965. Spencer, D. E. and Levin, R. E.: "On the significance of Photometric Measurements," *Illum. Eng.*, Vol. LXI, p. 196, April, 1966.

30. Blackwell, Richard H.: "A Recommended Field Test Method for Evaluating Over-All Visual Efficiency of Lighting Installations," *Illum. Eng.*, Vol. LVIII, p. 642, October, 1963. Blackwell, Richard H.: "Development of Procedures and Instruments for Visual Task Evaluation," *Illum. Eng.*, Vol. 65, p. 267, April, 1970.

31. Teele, R. P.: "A Physical Photometer," Res. Paper RP 1415, *J. Res. Nat. Bur. Stand.*, Vol. 27, p. 217, September, 1941. Teele, R. P.: "Measuring Circuit for Radiometers," *J. Opt. Soc. Amer.*, Vol. 44, p. 860, November, 1954. Forsythe, W. E.: *Measurement of Radiant Energy*, McGraw-Hill Book Co., Inc., New York, p. 207, 1937. Moon, P.: "Theory of the Alternating Current Bolometer," *J. Franklin Inst.*, p. 17, January, 1935. Moon, P. and Mills, W. R.: "Construction and Test of an Alternating Current Bolometer," *Rev. Sci. Instrum.*, p. 8, January, 1935. Moon, P. and Steinhardt, L. R.: "The Dielectric Bolometer," *J. Opt. Soc. Amer.*, Vol. 28, p. 148, May, 1938.

32. Teele, R. P. and Gibson, K. S.: "A Standard Luminosity Filter," *J. Opt. Soc. Amer.*, Vol. 38, p. 1096, December, 1948.

33. Ives, H. E.: "A Precision Artificial Eye," *Phys. Rev.*, Vol. 6, p. 334, 2nd ser., November, 1915.

34. Franck, K. and Smith, R. L.: "A Photometric Laboratory for Today's Light Sources," *Illum Eng.*, Vol. XLIX, p. 287, June, 1954. Baumgartner, G. R.: "New Semi-Automatic Distribution Photometer and Simplified Calculation of Light Flux," *Illum. Eng.*, Vol. XLV, p. 253, April, 1950.

35. Horton, G. A.: "Modern Photometry of Street Lighting Luminaires," *Illum. Eng.*, Vol. XLIII, p. 989, November, 1948. Colby, C. C., Jr., and Doolittle, C. M.: "A Distribution Photometer of New Design," *Trans. Illum. Eng. Soc.*, Vol. XVIII, p. 273, March, 1923.

36. Horton, G. A.: "Modern Photometry of Fluorescent Luminaires," *Illum. Eng.*, Vol. XLV, p. 458, July, 1950.

37. Rosa, E. B. and Taylor, A. H.: "Theory Construction and Use of the Photometric Integrating Sphere," Sci. Paper No. 447, *Bull. Bur. Stand.*, September 26, 1921. Walsh, J. W. T.: *Photometry*, 3rd ed., Constable and Co., Ltd., London, p. 257, 1958. Buckley, H.: "The Effect of Non-Uniform Reflectance of the Interior Surface of Spherical Photometric Integrators," *Trans. Illum. Eng. Soc.* (London), Vol. XLI, p. 167, July, 1946. Hardy, A. C. and Pineo, O. W.: "The Errors Due to the Finite Size of Holes and Sample in Integrating Spheres," *J. Opt. Soc. Amer.*, Vol. 21, p. 502, August, 1931. Gabriel, M. H., Koenig, C. F., and Steeb, E. S.: "Photometry, Parts I and II," *Gen. Elec. Rev.*, Vol. 54, pp. 30, 23, September and October, 1951.

38. Weaver, K. S. and Shackleford, B. E.: "The Regular Icosahedron as a Substitute for the Ulbricht Sphere," *Trans. Illum. Eng. Soc.*, Vol. XVIII, p. 290, March, 1923.

39. Walsh, J. W. T.: *Photometry*, Constable and Co., Ltd., London, 3rd ed., p. 265, 1958.

40. Committee of Testing Procedures of IES: "IES General Guide to Photometry," *Illum. Eng.*, Vol. L, p. 201, April, 1955. Stephenson, H. F.: "The Equipment and Functions of an Illumination Laboratory," *Trans. Illum. Eng. Soc.* (London), Vol. XVII, p. 1, January, 1952.

41. Knowlton, A. E.: *Standard Handbook for Electrical Engineers*, 9th ed., McGraw-Hill Book Co., Inc., New York, 1968. Pender, H. and Delmar, W. A.: *Electrical Engineers Handbook: Electric Power*, 4th ed., John Wiley and Sons, Inc., New York, 1949.

42. Committee on Testing Procedures of the IES: "IES Guide for the Use and Care of Electrical Instruments," *Illum. Eng.*, Vol. 63, p. 376, July, 1968.

43. Brooks, H. B.: "Deflection Potentiometers for Current and Voltage Measurements," *Nat. Bur. of Stand. Tech. News Bull.*, p. 395, 1912.

44. Karapetoff, V. and Dennison, B. C.: *Experimental Electric Engineering*, Vol. 1, 4th ed., John Wiley and Sons, New York, p. 132, 1933. Gabriel, M. H., Koenig, C. F., and Steeb, E. S.: "Photometry, Parts I and II," *Gen. Elec. Rev.*, Vol. 54, pp. 30, 23, September and October, 1951. John Fluke Mfg. Co., Inc., Seattle 99, Wash. "Electronic Volt-Ampere-Wattmeter with Negligible Circuit Burden."

45. "Methods of Measurement of Mercury Lamp Characteristics," ANSI C78.386-1965, American National Standards Institute, N.Y.

46. Committee on Testing Procedures of the IES: "IES Approved Method for the Electrical and Photometric Measurements of Fluorescent Lamps," *Illum. Eng.*, Vol. LXII, p. 556, September, 1967.

47. Little, W. F. and Salter, E. H.: "The Measurement of Fluorescent Lamps and Luminaires," *Illum. Eng.*, Vol. XLII, p. 217, February, 1947.

48. Miller, J. H.: "A Vector Sum Ammeter," *Weston Eng. Notes*, p. 7, December, 1954.

49. "Specifications for Fluorescent Lamp Ballasts," ANSI C82.1-1969 and supplement ANSI C82.la-1970. "Methods of Measurement of Fluorescent Lamp Ballasts," ANSI C82.2-1963. "Specifications for Fluorescent Lamp Reference Ballasts," ANSI C82.3-1962. "Specifications for Mercury Lamp Ballasts (Multiple-Supply Type)," ANSI C82.4-1968. "Specifications for Mercury Lamp Reference Ballasts," ANSI C82.5-1966. "Methods of Measurement of Mercury Lamp Ballasts and Transformers," ANSI C82.6-1965 and supplement C82.6a, American National Standards Institute, New York.

50. Winch, G. T.: "Photometry and Colorimetry of Fluorescent and Other Discharge Lamps," *Trans. Illum. Eng. Soc.* (London), Vol. XI, p. 107, June, 1946. Voogd, J.: "Physical Photometry," *Philips Tech. Rev.*, Vol. 4, p. 260, September, 1939. Winch, G. T.: "The Measurement of Light and Color," *Proc. Inst. Elec. Eng.* (London), p. 452, June, 1949.

51. Committee on Testing Procedures of the IES: "IES Guide for Photometric Measurements of Mercury Lamps," *Illum. Eng.*, Vol. 64, p. 323, April, 1969.
52. Baumgartner, G. R.: "Practical Photometry of Fluorescent Lamps and Reflectors," *Illum. Eng.*, Vol. XXXVI, p. 1340, December, 1941.
53. Teele, R. P.: "Gas-Filled Lamps as Photometric Standards," *Trans. Illum. Eng. Soc.*, Vol. XXV, p. 78, January, 1930. Knowles-Middleton, W. E. and Mayo, E. G.: "Variation in the Horizontal Distribution of Light from Candlepower Standards," *J. Opt. Soc. Amer.*, Vol. 41, p. 513, August, 1951. Winch, G. T.: "Recent Developments in Photometry and Colorimetry," *Trans. Illum. Eng. Soc.* (London), Vol. XXI, p. 91, May, 1956. Also see ref. 3.
54. Committee on Testing Procedures of the IES: "IES Guide for Photometric Measurements of Reflector-Type Lamps," *Illum. Eng.*, Vol. LVII, p. 688, October, 1962.
55. Committee on Testing Procedures of the IES: "IES Approved Method for Photometric Testing of Floodlights Using Incandescent Filament or Discharge Lamps," *Illum. Eng.*, Vol. 66, p. 107, February, 1971.
56. "Experimental Statistics," *Handbook 91*, Section 3, National Bureau of Standards. "Recommended Practice for Probability Sampling of Materials," ASTM E105-58, American Society for Testing Materials. "Evaluation of Test Data," *Handbook 91*, Section 2, National Bureau of Standards.
57. Lewinson, L. J.: "The Interpretation of Forced Life Tests of Incandescent Electric Lamps," *Trans. Illum. Eng. Soc.*, Vol. XI, p. 815, November, 1916. Lewinson, L. J. and Millar, P. S.: "The Evaluation of Lamp Life," *Trans. Illum. Eng. Soc.*, Vol. VI, p. 774, November, 1911. Purcell, W. R.: "Saving Time in Testing Life of Incandescent Lamps," *Proc. AIEE*, Vol. 68, p. 617, July, 1949.
58. Committee on Testing Procedures of the IES: "IES Guide for Life Performance Testing of Fluorescent Lamps," *Illum. Eng.*, Vol. LI, p. 595, August, 1956.
59. Committee on Testing Procedures of the IES: "IES Practical Guide to Photometry," *J. Illum. Eng. Soc.*, Vol. 1, October, 1971.
60. Committee on Testing Procedures of the IES: "IES Guide for Photometric Testing of Fluorescent Luminaires," *Illum. Eng.*, Vol. XLIV, p. 413, July, 1949. Also see ref. 34, 36, and 52.
61. Committee on Testing Procedures of the IES: "IES Guide for Measurement of Photometric Brightness (Luminance)," *Illum. Eng.*, Vol. LVI, p. 457, July, 1961.
62. Committee on Testing Procedures of the IES: "IES Approved Method for the Photometric Testing of Filament Type Luminaires for General Lighting Service," *Illum. Eng.*, Vol. LXII, p. 587, October, 1967.
63. Committee on Testing Procedures of the IES: "IES Guide for Reporting General Lighting Equipment Engineering Data," to be published.
64. Committee on Testing Procedures of the IES: "Determination of Average Luminance of Luminaires," *J. Illum. Eng. Soc.*, Vol. 1, January, 1972.
65. Committee on Testing Procedures of the IES: "IES Guide for Photometric Testing of Searchlights," *Illum. Eng.*, Vol. LIII, p. 155, March, 1958.
66. Committee on Testing Procedures of the IES: "IES Approved Method for Photometric Testing of Floodlights Using Incandescent Filament or Discharge Lamps," *Illum. Eng.*, Vol. 66, p. 107, February, 1971. Committee on Equipment Performance Ratings, Joint IES-SMPTE: "Recommended Practice for Reporting Photometric Performance of Incandescent Filament Lighting Units Used in Theatre and Television Production," *Illum. Eng.*, Vol. LIII, p. 516, September, 1958.
67. Committee on Testing Procedures of the IES: "Photometric Testing of Roadway Luminaires," *Illum. Eng.*, Vol. 63, p. 541, October, 1968.
68. Committee on Testing Procedures of the IES: "IES Approved Method for Determining Luminaire-Lamp-Ballast Combination Operating Factors for High Intensity Discharge Luminaires," *Illum. Eng.*, Vol. 65, p. 718, December, 1970.
69. Breckenridge, F. C. and Projector, T. H.: "Construction of a Goniometer for Use in Determining the Candlepower Characteristics of Beacons," CAA Technical Development Report No. 39, February, 1944. Projector, T. H.: "Versatile Goniometer for Projection Photometry," *Illum. Eng.*, Vol. XLVIII, p. 192, April, 1953.
70. Douglas, C. A. and Young, L. L.: "Development of a Transmissiometer for Determining Visual Range," *CAA Tech. Dev. Rep. No. 47*, U. S. Dept. of Commerce, Knowles-Middleton, W. E.: *Vision Through the Atmosphere*, Univ. of Toronto Press, 1952. Committee on Instruments and Measurements of the IES, Annual Report of "Part II—Description of Method for Measuring Atmospheric Transmission," *Illum. Eng.*, Vol. XXXVIII, p. 515, November, 1943.
71. Chernoff, L.: "Photometry of Projectors at the National Bureau of Standards," *Nat. Bur. of Stand. Tech. Note No. 198*, December, 1963. Johnson, J.: "Zero Length Searchlight Photometry System," *Illum. Eng.*, Vol. LVII, p. 187, March, 1962.
72. Linsday, E. A.: "Brightness of Cylindrical Fluorescent Sources," *Illum. Eng.*, Vol. XXXIX, p. 23, January, 1944.
73. Jones, M. T., Zavesky, R. J., and Lozier, W. E.: "Method for Measurement of Brightness of Carbon Arcs," *J. Soc. Motion Pict. Eng.*, Vol. 44, p. 10, July, 1945. Thomas, E. R.: "A Portable Brightness Distribution Photometer," *J. Sci. Instrum.*, August, 1946.
74. Committee on Testing Procedures of the IES: "Photometric and Thermal Testing of Air and Liquid Cooled Heat Transfer Luminaires," *Illum. Eng.*, Vol. 63, p. 485, September, 1968.
75. Lighting Survey Committee of the IES: "How to Make a Lighting Survey," *Illum. Eng.*, Vol. LVII, p. 87, February, 1963.
76. Committee on Testing Procedures of the IES: "IES Guide for Outdoor Illumination Tests," *Illum. Eng.*, Vol. XLVI, p. 425, August, 1951. Committee on Roadway Lighting of the IES: "American Standards Practice for Roadway Lighting," *Illum. Eng.*, Vol. LIX, p. 73, February, 1964 (revision to be published). Committee on Testing Procedures of the IES: "Photometric Testing of Roadway Luminaires," *Illum. Eng.*, Vol. 63, p. 541, October, 1968. Committee on Sports and Recreational Areas of the IES: "Interim Report-Design Criteria for Lighting of Sports Events for Color Television Broadcasting," *Illum. Eng.*, Vol. 64, March, 1969. Committee on Sports and Recreational Areas of the IES: "Current Recommended Practice for Sports and Recreational Areas Lighting," *Illum. Eng.*, Vol. 64, p. 457, July 1969. Committee on Testing Procedures of the IES: "IES Guide for Photometric Measurements of Area and Sports Lighting Installations," to be published.

SECTION

5

COLOR

Basic Concepts 5–1 *Colorimetry 5–14* *Use of Color 5–16*
Color Rendering 5–18 *Applications 5–21*

Those involved in illumination design need to know a great deal about color, more than just the visual sensation produced. When working with light and color, the characteristic of greatest importance is the distribution of the radiant power within the light spectrum. This distribution is something that the eye integrates, but it cannot be differentiated without the use of an instrument. Because the color of light sources, and the color rendering properties of light sources, are becoming increasingly important in the design of a luminous environment, today's illuminating engineer needs a good working knowledge of the vocabulary and practices of modern colorimetry. Consequently, in 1960 a subcommittee of the IES Committee on Testing Procedures for Illumination Characteristics prepared a "Practical Guide to Colorimetry." [1]

The esthetic use of color to produce a pleasing interior is the province of the interior designer. Nevertheless, the lighting engineer also should have a considerable knowledge of certain aspects of color if he is to make efficient use of otherwise well planned lighting installations. He needs to know how to use color to help provide required illumination levels and distributions. Essentially it is this aspect of color for which the IES Color Committee in 1961 issued its guide on "Color and the Use of Color by the Illuminating Engineer." [2]

Today's lighting engineer not only is faced with a choice of color in light sources, but with wide variations in color rendering properties of light sources that may be identical in color. It is with this aspect that the Subcommittee on Color Rendering of the IES Light Sources Committee is concerned in its 1962 report on an "Interim Method of Measuring and Specifying Color Rendering of Light Sources," [3] a report that essentially is incorporated in CIE* recommendations published in 1965: "Method of Measuring and Specifying Color Rendering Properties of Light Sources." [4, 5]

* Commission Internationale de l'Eclairage (International Commission on Illumination).

Note: References are listed at the end of each section.

These reports of Illuminating Engineering Society committees, plus definitions of color terms listed in Section 1, provide the lighting engineer with a nucleus for his studies in color. Following a consideration of basic concepts, these three reports are used in the general plan for this section, which concludes with examples in several fields of special application. A few other sections of this Handbook contain brief discussions of color, with specialized applications.

BASIC CONCEPTS OF COLOR

Color Terms

In Section 1 color terms are carefully defined to provide a way of distinguishing between several commonly confused meanings of the term *color*. Whether one makes strict use of the definitions or not, an understanding of the purpose and need for the differentiations that are made is basic to an understanding of the subject.

The *perceived object color*, the color perceived as belonging to an object, is something perceived instantaneously. It is so common an experience that many persons find it hard to understand why color is not simple to explain in a few easy lessons. But a color perception results from the interaction of many highly complex factors such as: the characteristics of the object, the incident light, and the surround; the viewing direction; observer characteristics; and the observer's adaptation. Characteristics of object, light, surround, and observer may vary both spectrally and directionally, each in a different manner. The observer may vary in regard to time of seeing, what he saw last, or how his attention was focussed in relation to time of seeing. Unless the circumstances of a former situation with which the layman, interior designer, or lighting engineer may be familiar, are similar enough in all important respects, a new situation cannot be answered by reference to past experience alone. The layman or interior designer may answer a new situation by making certain assumptions, or by limiting himself to use of conditions with which he is familiar. But the lighting engineer may not do this if he is to deal with all types of architectural situations, with all types of light sources, and with requirements that will fit new or specialized situations.

The color of an object, or *object color*, as distinct from the *color perceived as belonging to that object*, may be defined by limiting the factors to be considered to the color of light reflected or transmitted by an object when it is illuminated by a standard light source such as those adopted for colorimetry by the CIE. See Fig. 5–6. For this purpose, a CIE standard observer, using standardized conditions of observation, must also be assumed.

Color, as distinct from *color of an object*, or *color perceived as belonging to an object*, is defined as the characteristics of light by which an observer may distinguish between patches of light of the same size, shape, and structure. It reduces itself to a basic description of light in terms of relative amounts of radiant power at the different wavelengths of the visually effective spectrum, which quite generally is considered to extend from 380 to 780 nanometers. (To identify colors resulting from wide use of fluorescent dyes activated by energy in the ultraviolet, it is becoming necessary in specifying the spectral distribution of a light source to extend the wavelength range beyond that which is visually effective, down to 300 nanometers in the ultraviolet, particularly for sources that are intended to reproduce daylight.) Identical colors are produced not only by identical combinations of radiant power and spectral distribution but also by many different combinations, depending upon their relative visual effectiveness. Such different combinations cause metamerism (see page 5–13).

The word *color* often is used to cover the three quite different meanings discussed in the three preceding paragraphs. They reduce themselves to the same perceived color when, as in the previous paragraph, only a spectral distribution is required and all other conditions are assumed to be standard, or when, as in the paragraph before that, spectral distributions are required for the light source and object, with observer and observing conditions assumed to be standard. When assumed standard conditions are the same, then there is less need for distinguishing between the three meanings, the *color*, *color of object*, and *perceived color of object*. The layman makes assumptions of standard conditions all the time. Even if he knows the difference, he simply assumes the standard conditions to which his past experience has accustomed him. But the lighting engineer, if he is to handle new problems in color, including new light sources that may vary widely in spectral distributions, not only must know the basic difference of the meaning for color under the above conditions, but he must understand why it is necessary to keep this distinction in mind even when he uses the one term *color* to cover all three situations.

Color Rendering

The *color rendering* properties of a light source cannot be assessed by visual inspection, or by a knowledge of its color.[6] For this purpose, full knowledge of its spectral energy distribution is required. Viewed in succession under lamps that look quite alike but are different in spectral distribution, objects may look entirely different in color. An extreme case would be a group of objects seen under a pair of color-matched yellow sodium and yellow fluorescent lamps. Objects that in daylight look red, yellow, green, blue, and purple, would appear quite different under these two lamps. In the case of the sodium lamp all objects would lose their daylight appearance, all appearing as grays, from light to very dark, near-black. Under the yellow fluorescent lamp, more hues could be recognized, but the color of objects would still vary considerably from their daylight color.

Methods of measuring and specifying color rendering properties of light sources depend upon a comparison of the color appearance of objects under a test source with their color appearance under a reference, or standard, light source (see page 5–18).

Basis for Measurements

If *color* is the characteristic of light by which a human observer distinguishes patches of light, and if *light* is visually evaluated radiant energy, then color may be computed by combining physical measurements of radiant power, wavelength by wavelength, with an observer's ability to respond to light. The relative spectral responses of the internationally adopted CIE standard observers, defined by the tristimulus values of an equal power spectrum, are provided in Figs. 5–1 and 5–2. With data for a standard observer and the spectroradiometric measurement of a light source, the color of that light source can be calculated. Thus spectroradiometry becomes a tool for color measurement. Measurements of radiant power may be purely physical, while evaluation of radiant power by a human observer, based solely on his perceptions, may be purely psychological. Visual evaluations, tied down to numbers through measurements made for standardized conditions of test, provide psychophysical methods of measurement.

Visual evaluation of appearance of objects and light sources may be in terms derived wholly from one's perceptions. One convenient and useful set of such psychological terms involves hue, lightness, and saturation for objects, and hue, brightness, and saturation for light sources. Approximate psychophysical correlates for these terms are dominant (or com-

Fig. 5-1. Spectral Tristimulus Values for Equal Spectral Power Source

a. 1931 CIE Standard Observer

Wavelength (nanometer)	$\bar{x}(\lambda)$	$\bar{y}(\lambda)$	$\bar{z}(\lambda)$	Wavelength (nanometer)	$\bar{x}(\lambda)$	$\bar{y}(\lambda)$	$\bar{z}(\lambda)$
380	0.0014	0.0000	0.0065	580	0.9163	0.8700	0.0017
385	0.0022	0.0001	0.0105	585	0.9786	0.8163	0.0014
390	0.0042	0.0001	0.0201	590	1.0263	0.7570	0.0011
395	0.0076	0.0002	0.0362	595	1.0567	0.6949	0.0010
400	0.0143	0.0004	0.0679	600	1.0622	0.6310	0.0008
405	0.0232	0.0006	0.1102	605	1.0456	0.5668	0.0006
410	0.0435	0.0012	0.2074	610	1.0026	0.5030	0.0003
415	0.0776	0.0022	0.3713	615	0.9384	0.4412	0.0002
420	0.1344	0.0040	0.6456	620	0.8544	0.3810	0.0002
425	0.2148	0.0073	1.0391	625	0.7514	0.3210	0.0001
430	0.2839	0.0116	1.3856	630	0.6424	0.2650	0.0000
435	0.3285	0.0168	1.6230	635	0.5419	0.2170	0.0000
440	0.3483	0.0230	1.7471	640	0.4479	0.1750	0.0000
445	0.3481	0.0298	1.7826	645	0.3608	0.1382	0.0000
450	0.3362	0.0380	1.7721	650	0.2835	0.1070	0.0000
455	0.3187	0.0480	1.7441	655	0.2187	0.0816	0.0000
460	0.2908	0.0600	1.6692	660	0.1649	0.0610	0.0000
465	0.2511	0.0739	1.5281	665	0.1212	0.0446	0.0000
470	0.1954	0.0910	1.2876	670	0.0874	0.0320	0.0000
475	0.1421	0.1126	1.0419	675	0.0636	0.0232	0.0000
480	0.0956	0.1390	0.8130	680	0.0468	0.0170	0.0000
485	0.0580	0.1693	0.6162	685	0.0329	0.0119	0.0000
490	0.0320	0.2080	0.4652	690	0.0227	0.0082	0.0000
495	0.0147	0.2586	0.3533	695	0.0158	0.0057	0.0000
500	0.0049	0.3230	0.2720	700	0.0114	0.0041	0.0000
505	0.0024	0.4073	0.2123	705	0.0081	0.0029	0.0000
510	0.0093	0.5030	0.1582	710	0.0058	0.0021	0.0000
515	0.0291	0.6082	0.1117	715	0.0041	0.0015	0.0000
520	0.0633	0.7100	0.0782	720	0.0029	0.0010	0.0000
525	0.1096	0.7932	0.0573	725	0.0020	0.0007	0.0000
530	0.1655	0.8620	0.0422	730	0.0014	0.0005	0.0000
535	0.2257	0.9149	0.0298	735	0.0010	0.0004	0.0000
540	0.2904	0.9540	0.0203	740	0.0007	0.0003	0.0000
545	0.3597	0.9803	0.0134	745	0.0005	0.0002	0.0000
550	0.4334	0.9950	0.0087	750	0.0003	0.0001	0.0000
555	0.5121	1.0002	0.0057	755	0.0002	0.0001	0.0000
560	0.5945	0.9950	0.0039	660	0.0002	0.0001	0.0000
565	0.6784	0.9786	0.0027	765	0.0001	0.0000	0.0000
570	0.7621	0.9520	0.0021	770	0.0001	0.0000	0.0000
575	0.8425	0.9154	0.0018	775	0.0000	0.0000	0.0000
580	0.9163	0.8700	0.0017	780	0.0000	0.0000	0.0000
Totals					21.3713	21.3714	21.3715

b. 1964 CIE Supplementary Observer

Wavelength (nanometer)	$\bar{x}(10,\lambda)$	$\bar{y}(10,\lambda)$	$\bar{z}(10,\lambda)$	Wavelength (nanometer)	$\bar{x}(10,\lambda)$	$\bar{y}(10,\lambda)$	$\bar{z}(10,\lambda)$
380	0.0002	0.0000	0.0007	580	1.0142	0.8689	0.0000
385	0.0007	0.0001	0.0029	585	1.0743	0.8256	0.0000
390	0.0024	0.0003	0.0105	590	1.1185	0.7774	0.0000
395	0.0072	0.0008	0.0323	595	1.1343	0.7204	0.0000
400	0.0191	0.0020	0.0860	600	1.1240	0.6583	0.0000
405	0.0434	0.0045	0.1971	605	1.0891	0.5939	0.0000
410	0.0847	0.0088	0.3894	610	1.0305	0.5280	0.0000
415	0.1406	0.0145	0.6568	615	0.9507	0.4618	0.0000
420	0.2045	0.0214	0.9725	620	0.8563	0.3981	0.0000
425	0.2647	0.0295	1.2825	625	0.7549	0.3396	0.0000
430	0.3147	0.0387	1.5535	630	0.6475	0.2835	0.0000
435	0.3577	0.0496	1.7985	635	0.5351	0.2283	0.0000
440	0.3837	0.0621	1.9673	640	0.4316	0.1798	0.0000
445	0.3867	0.0747	2.0273	645	0.3437	0.1402	0.0000
450	0.3707	0.0895	1.9948	650	0.2683	0.1076	0.0000
455	0.3430	0.1063	1.9007	655	0.2043	0.0812	0.0000
460	0.3023	0.1282	1.7454	660	0.1526	0.0603	0.0000
465	0.2541	0.1528	1.5549	665	0.1122	0.0441	0.0000
470	0.1956	0.1852	1.3176	670	0.0813	0.0318	0.0000
475	0.1323	0.2199	1.0302	675	0.0579	0.0226	0.0000
480	0.0805	0.2536	0.7721	680	0.0409	0.0159	0.0000
485	0.0411	0.2977	0.5701	685	0.0286	0.0111	0.0000
490	0.0162	0.3391	0.4153	690	0.0199	0.0077	0.0000
495	0.0051	0.3954	0.3024	695	0.0138	0.0054	0.0000
500	0.0038	0.4608	0.2185	700	0.0096	0.0037	0.0000
505	0.0154	0.5314	0.1592	705	0.0066	0.0026	0.0000
510	0.0375	0.6067	0.1120	710	0.0046	0.0018	0.0000
515	0.0714	0.6857	0.0822	715	0.0031	0.0012	0.0000
520	0.1177	0.7618	0.0607	720	0.0022	0.0008	0.0000
525	0.1730	0.8233	0.0431	725	0.0015	0.0006	0.0000
530	0.2365	0.8752	0.0305	730	0.0010	0.0004	0.0000
535	0.3042	0.9238	0.0206	735	0.0007	0.0003	0.0000
540	0.3768	0.9620	0.0137	740	0.0005	0.0002	0.0000
545	0.4516	0.9822	0.0079	745	0.0004	0.0001	0.0000
550	0.5298	0.9918	0.0040	750	0.0003	0.0001	0.0000
555	0.6161	0.9991	0.0011	755	0.0002	0.0001	0.0000
560	0.7052	0.9973	0.0000	760	0.0001	0.0000	0.0000
565	0.7938	0.9824	0.0000	765	0.0001	0.0000	0.0000
570	0.8787	0.9555	0.0000	770	0.0001	0.0000	0.0000
575	0.9512	0.9152	0.0000	775	0.0000	0.0000	0.0000
580	1.0142	0.8689	0.0000	780	0.0000	0.0000	0.0000
Totals					23.3294	23.3323	23.3343

plementary) wavelength, luminance (photometric brightness), and purity. Other widely used psychophysical methods for describing and specifying color show less correlation with perceptual factors, and these often are converted to more meaningful visual terms, usually to some sort of more uniform color spacing, of which Munsell spacing[7] and a spacing based on "MacAdam ellipses"[8] are prime examples.

Systems of Color Specification

CIE System. This is a system originally recommended in 1931 by the CIE to define all metameric pairs (see page 5–13) by giving the amounts X, Y, Z, of three imaginary primary colors required by a standard observer to match the color being specified. These amounts may be calculated as an integral or suitably approximated as a summation from the spectral compositions of the radiant power of the source (and/or of the color specimen), and the spectral tristimulus values for an equal power source (see Fig. 5–1). For example:

$$X = k \sum_{\lambda=380}^{\lambda=780} S(\lambda)\rho(\lambda)\bar{x}(\lambda)\Delta\lambda$$

where

$S(\lambda)$ is the spectral irradiance distribution of the source (see Fig. 5–6, page 5–7),

$\rho(\lambda)$ is the spectral reflectance of the specimen,

k is a normalizing factor,

$\bar{x}(\lambda)$ is the spectral tristimulus value from Fig. 5–1 with similar expressions for Y and Z wherein $\bar{y}(\lambda)$ and $\bar{z}(\lambda)$ respectively are substituted for $\bar{x}(\lambda)$.

The normalizing factor, k, may be assigned any arbitrary value provided it is kept constant throughout any particular discussion. Where only the relative values of X, Y, and Z are required, the value of X is usually chosen so that Y has the value 100.0. For object colors the normalizing factor is usually given the value:

$$k = \frac{100}{\Sigma S(\lambda)\bar{y}(\lambda)\Delta\lambda}$$

According to this normalization, the value of Y of a reflecting object is its luminous reflectance expressed as a percentage, and the value of Y for a transmitting object is its percentage luminous transmittance.*

To simplify this computation, the values of $\bar{x}(\lambda)S(\lambda)$, $\bar{y}(\lambda)S(\lambda)$ and $\bar{z}(\lambda)S(\lambda)$ for each of the CIE standard sources are given for $\Delta\lambda = 10$ nm in Fig. 5–7. (Values of $S(\lambda)$ are given in Fig. 5–6.)

The fractions, $X/(X + Y + Z)$, $Y/(X + Y + Z)$, $Z/(X + Y + Z)$ are known as chromaticity coordinates, x, y, z respectively. Note that $x + y + z = 1$ whereby specification of any two fixes the third. By convention, chromaticity usually is stated in terms of x and y as illustrated in Fig. 5–3.

A sample calculation for determining the CIE coordinates x, y, and Y are shown in Fig. 5–8. For a deep red surface, specified R 4/14 in Munsell rotation (see below), when illuminated by CIE source C. In Fig. 5–8, column I is a listing of wavelengths in 10-nanometer steps, column II is a tabulation of spectral reflectance values for the deep red surface at each wavelength in column I, and column III lists the CIE tristimulus computational data from Fig. 5–7. By multiplying column II by column III and summing each tabulation in column IV, X, Y and Z values are determined. Then by using the three fractions in the previous paragraph, the chromaticities are determined. The surface reflectance can be determined by dividing the Y value by 100,000, the sum of $S(\lambda)\bar{y}(\lambda)$ in column III of Fig. 5–8.

CIE standards include data for a 1931 Standard Observer (for angular subtense between 1° and 4°) as shown in Fig. 5–1a, a 1964 Supplementary Observer (for angular subtense greater than 4°) as shown in Fig. 5–1b, four standard sources (A representing tungsten at 2856 K; B representing direct sunlight with a correlated color temperature of 4874 K; C representing average daylight with a correlated color temperature of 6774 K; and D_{65} representing a phase of natural daylight at 6504 K), and two supplementary standard sources (D_{55} representing sun plus gray sky at 5503 K, and D_{75} representing lightly overcast north sky at 7504 K). Four alternative conditions for illuminating/viewing a test sample are specified: a) 45°/normal, b) normal/45°, c) diffuse (sphere)/normal, and d) normal/diffuse (sphere). For an extended discussion of the calculation and application of CIE data, including extensive tables of quantitative data and methods of colorimetry, consult *Color Science.*[9]

In 1960 the CIE provisionally recommended that whenever a diagram is desired to yield chromaticity spacing more uniform than the CIE (x,y)-diagram, a uniform chromaticity-scale diagram based on that described in 1937 by MacAdam[10] be used. The ordinate and abscissa of this (u,v)-diagram are defined as:

$$u = \frac{4X}{(X + 15Y + 3Z)} \quad \text{or,} \quad u = \frac{4x}{(-2x + 12y + 3)}$$

$$v = \frac{6Y}{(X + 15Y + 3Z)} \quad \text{or,} \quad v = \frac{6y}{(-2x + 12y + 3)}$$

To convert this to a three-dimensional metric system that is useful in studying color differences, the CIE has added a recommendation developed for the purpose in 1963 by Wyszecki[11] that converts the luminance factor, Y, to a lightness index, W^*, by the relationship:

$$W^* = 25Y^{1/3} - 17 \qquad (1 \leqq Y \leqq 100)$$

and converts the chromaticity coordinates (u,v) to chromaticness indices U^* and V^* by the relationships:

$$U^* = 13\,W^*(u - u_0)$$

$$V^* = 13\,W^*(v - v_0)$$

The lightness index, W^*, approximates the Munsell value function in the range of Y from 1 to 98 per cent. The chromaticity coordinates (u_0, v_0) refer to the nominally achromatic color (usually that of the source) placed at the origin of the (U^*, V^*)-system.

* In the special case where the absolute values of $S(\lambda)\Delta\lambda$ are given (e.g., in watts), it is convenient to take k = Km = 680 whereby the value of Y gives the luminous flux in the stimulus in lumens. In this case the accepted symbolism for $S(\lambda)$ is $\Phi_{e,\lambda}$ (see Section 1).

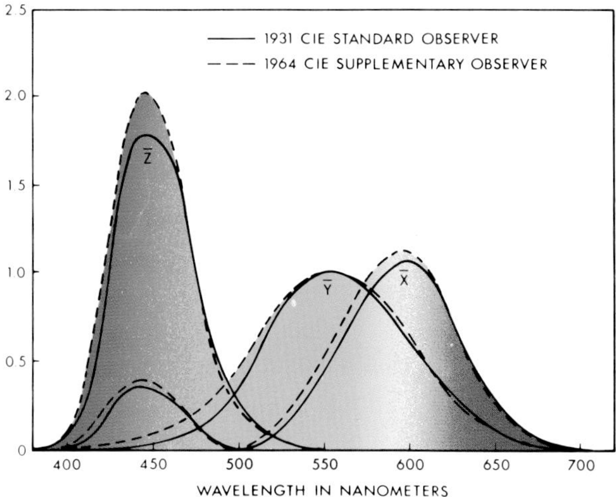

Fig. 5-2. Spectral tristimulus values for equal spectral power source. (Colors shown are approximate representations.)

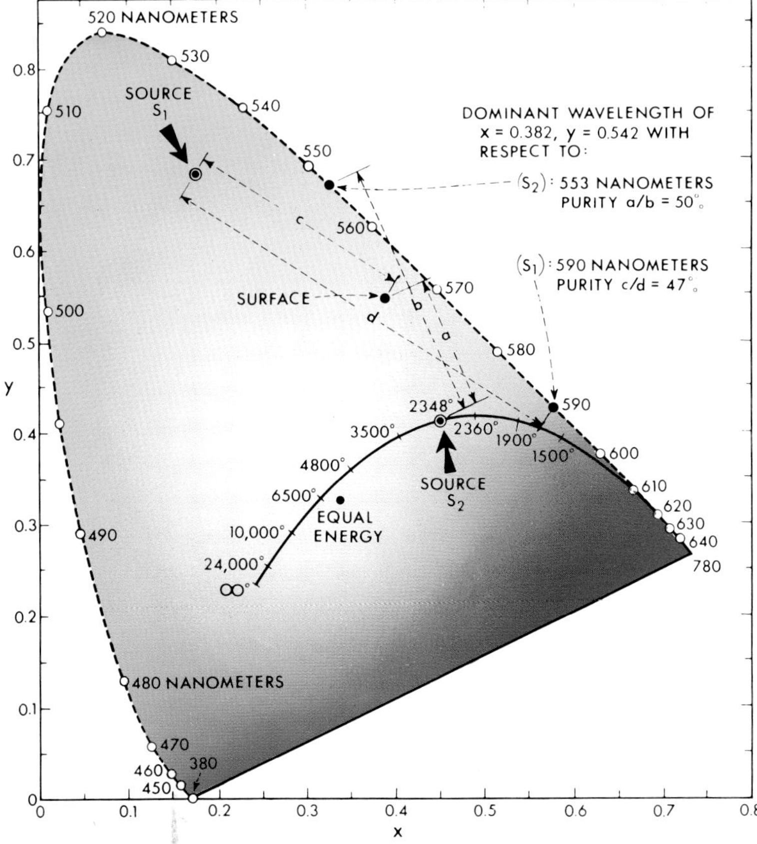

Fig. 5-3. Locus of spectrum colors plotted on 1931 CIE chromaticity diagram showing method of obtaining dominant wavelength and purity for different samples under different light sources. (Colors shown are approximate representations.)

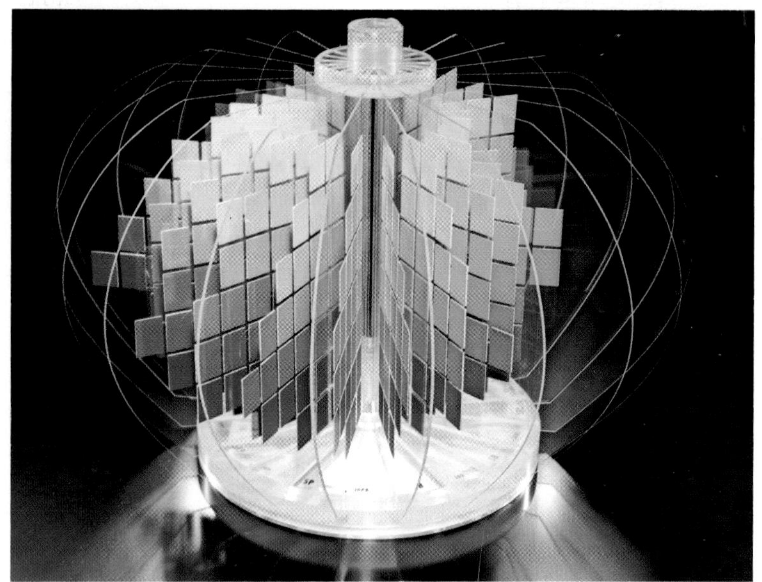

Fig. 5-4. Form of Munsell color solid is shown in the photograph at the left. Diagram below is a cut-away view showing notation scales of hue, value, and chroma (e.g., 5Y 5/4), and the relation of constant hue charts to the three-dimensional representation. See inside back cover for a set of color chips of neutral grays and chromatic colors, with reflectances, selected for use in interior applications.

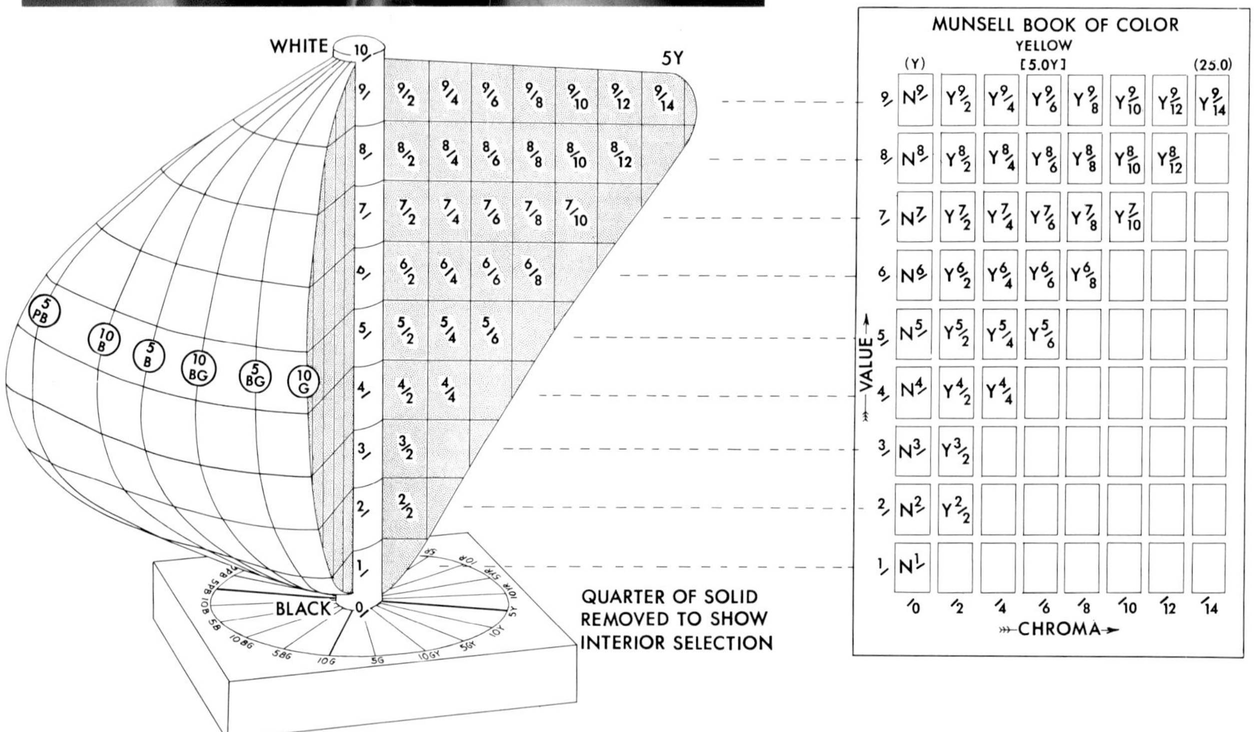

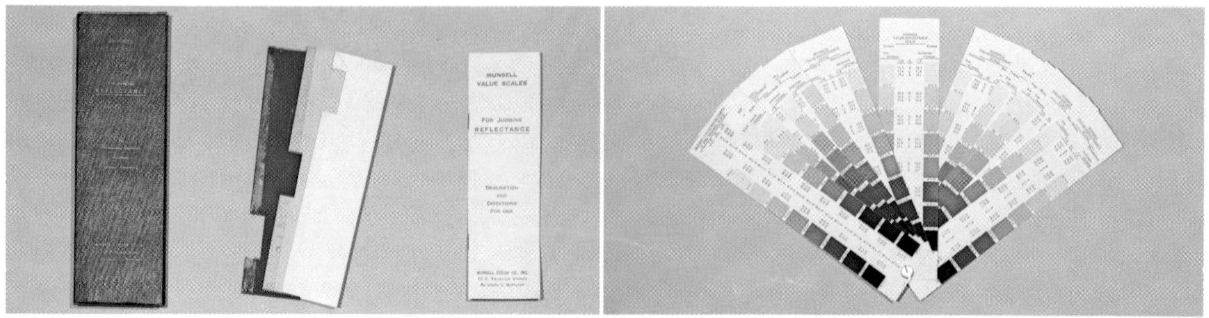

Fig. 5-5. Munsell value scales for judging reflectance.

Fig. 5-6. Spectral Power Distributions of CIE Standard Sources

Wavelength (nanometer)	Standard Source					
	A	B	C	D_{55}	D_{65}	D_{75}
300	9			0.2	0.3	0.4
310	14			21	33	52
320	19	0.2	0.1	112	202	298
330	27	5	4	207	371	550
340	36	23	26	240	400	573
350	47	54	66	279	450	627
360	62	93	123	307	467	630
370	78	148	203	344	522	703
380	98	218	313	326	500	668
390	121	304	450	382	547	700
400	147	402	601	610	828	1019
410	177	507	765	686	916	1119
420	210	615	932	716	935	1128
430	247	711	1067	679	868	1033
440	287	786	1154	856	1049	1211
450	331	831	1178	981	1171	1330
460	378	859	1169	1004	1178	1323
470	429	895	1176	999	1149	1272
480	483	926	1177	1026	1159	1269
490	539	939	1146	980	1088	1177
500	599	916	1065	1007	1094	1165
510	661	882	972	1008	1078	1137
520	725	871	920	1000	1049	1086
530	791	897	931	1042	1077	1105
540	860	943	970	1021	1044	1063
550	929	982	999	1030	1040	1049
560	1000	1000	1000	1000	1000	1000
570	1072	998	972	973	964	956
580	1144	982	929	977	957	942
590	1217	965	885	914	886	870
600	1290	953	852	944	900	873
610	1363	958	840	951	896	862
620	1436	970	837	942	876	836
630	1508	982	836	904	833	787
640	1480	994	834	923	837	785
650	1650	1011	838	889	800	748
660	1720	1021	835	903	802	745
670	1788	1020	820	940	822	755
680	1854	1011	798	900	783	717
690	1919	988	762	797	697	640
700	1983	964	725	829	716	652
710	2048	936	688	849	743	681
720	2104	904	649	702	616	565
730	2161	870	612	793	699	643
740	2217	845	584	850	751	692
750	2270	829	562	719	636	587
760	2321	824	552	528	464	427
770	2370	831	553	759	668	614
780	2417			718	634	584
790	2461			729	643	592
800	2503			674	594	548
810	2543			587	519	480
820	2581			650	574	530
830	2616			683	603	556
Chromaticity coordinates (1931 CIE Sytem)						
x	0.4465	0.3485	0.3101	0.3324	0.3127	0.2991
y	0.4075	0.3518	0.3163	0.3475	0.3291	0.3150

Dominant Wavelength, Purity, and Luminous Reflectance. Dominant wavelength and purity are quantities more suggestive of color appearance of objects than a CIE x,y-specification, and may be determined on an x,y-diagram[12] in relation to the spectrum locus and the point for the light source. See Fig. 5-3. The *dominant wavelength* of all colors whose x,y-coordinates fall on a straight line connecting the light source point with a point on the spectrum locus is the wavelength indicated at the intersection of that line with the spectrum locus. *Excitation purity* of such a color is defined as the distance of the point in the x,y-diagram from an assumed achromatic point expressed relative to the total distance in the same direction from achromatic point to the spectrum locus. If the color belongs to an object, the source point is taken as the achromatic point. *Luminous reflectance* or luminance factor equals CIE tristimulus value Y. An x,y-specification of any object color relates it only to the light source for which the object color is calculated; consequently, dominant wavelength and purity of any object depend on the spectral composition of its illumination.

Munsell System. This is a system of specifying color on scales of hue, value, and chroma. These are exemplified by a collection of color chips forming an atlas of charts that show scales for which two of the three variables are constant. See Fig. 5-4. The hue scales contain five principal and five intermediate hues. The value scale contains ten steps from black to white, 0 to 10. The chroma scales contain up to 16 or more steps from the equivalent gray. Each of the three scales is intended to represent equal visual intervals for a normal observer under daylight viewing conditions with gray to white surroundings. Under these conditions the hue, value, and chroma of a color correlate closely with hue, lightness, and saturation of color perception; under other conditions the correlation is lost. It is only for daylight conditions that Munsell samples are expected to appear equally spaced; when problems of color adaptation are fully solved, it may be possible to calculate the change in appearance that takes place when samples are viewed under a change in color of light source. Munsell specifications are useful whether or not reference is made to Munsell samples. For daylight conditions the system represents uniform color spacing as adequately as anything that is available. The Munsell notation is written in hue, value/chroma order, *e.g.*, 5R 4/10. This reads "five red, four-ten." One widely used approximation of equivalence between hue value/chroma units is 1 value step = 2 chroma steps = 3 hue steps (when the hue is at chroma 5). For use as standards or in technical color control, collections of carefully standardized color chips in matte or glossy surface may be obtained from the Munsell Color Company, Inc., 2441 North Calvert Street, Baltimore, Maryland 21218, in several different forms.

Fig. 5-7. CIE Tristimulus Computational Data for Several Sources Computed for the 1931 CIE Standard Observer

Wavelength (nanometer)	A 2854 K, c_2 = 1.438			B			C			D_{65}			D_{55}			D_{75}		
	$\bar{x}(\lambda)S(\lambda)$	$\bar{y}(\lambda)S(\lambda)$	$\bar{z}(\lambda)S(\lambda)$	$\bar{x}(\lambda)S(\lambda)$	$\bar{y}(\lambda)S(\lambda)$	$\bar{z}(\lambda)S(\lambda)$	$\bar{x}(\lambda)S(\lambda)$	$\bar{y}(\lambda)S(\lambda)$	$\bar{z}(\lambda)S(\lambda)$	$\bar{x}(\lambda)S(\lambda)$	$\bar{y}(\lambda)S(\lambda)$	$\bar{z}(\lambda)S(\lambda)$	$\bar{x}(\lambda)S(\lambda)$	$\bar{y}(\lambda)S(\lambda)$	$\bar{z}(\lambda)S(\lambda)$	$\bar{x}(\lambda)S(\lambda)$	$\bar{y}(\lambda)S(\lambda)$	$\bar{z}(\lambda)S(\lambda)$
380	0.001	0.000	0.006	0.003	0.000	0.014	0.004	0.000	0.020	0.007	0.000	0.031	0.004	0.000	0.020	0.009	0.000	0.041
390	0.005	0.000	0.023	0.013	0.000	0.060	0.019	0.000	0.089	0.022	0.001	0.104	0.015	0.000	0.073	0.028	0.001	0.132
400	0.019	0.001	0.093	0.056	0.002	0.268	0.085	0.002	0.404	0.112	0.003	0.532	0.083	0.002	0.394	0.137	0.004	0.650
410	0.071	0.002	0.340	0.217	0.006	1.033	0.329	0.009	1.570	0.377	0.010	1.796	0.284	0.008	1.354	0.457	0.013	2.179
420	0.262	0.008	1.256	0.812	0.024	3.890	1.238	0.037	5.949	1.189	0.035	5.711	0.916	0.027	4.400	1.424	0.042	6.839
430	0.649	0.027	3.167	1.983	0.081	9.678	2.997	0.122	14.628	2.330	0.095	11.370	1.836	0.075	8.959	2.749	0.112	13.414
440	0.926	0.061	4.647	2.689	0.178	13.489	3.975	0.262	19.938	3.458	0.228	17.343	2.838	0.187	14.235	3.964	0.262	19.883
450	1.031	0.117	5.435	2.744	0.310	14.462	3.915	0.443	20.638	3.724	0.421	19.627	3.136	0.354	16.528	4.199	0.475	22.133
460	1.019	0.210	5.851	2.454	0.506	14.085	3.362	0.694	19.299	3.243	0.669	18.614	2.780	0.574	15.959	3.614	0.746	20.746
470	0.776	0.362	5.116	1.718	0.800	11.319	2.272	1.058	14.972	2.124	0.989	13.998	1.858	0.865	12.243	2.336	1.088	15.395
480	0.428	0.622	3.636	0.870	1.265	7.396	1.112	1.618	9.461	1.048	1.524	8.915	0.933	1.357	7.936	1.139	1.655	9.682
490	0.160	1.039	2.324	0.295	1.918	4.290	0.363	2.358	5.274	0.330	2.142	4.791	0.299	1.941	4.341	0.354	2.301	5.146
500	0.027	1.792	1.509	0.044	2.908	2.449	0.052	3.401	2.864	0.051	3.343	2.815	0.047	3.094	2.606	0.054	3.536	2.978
510	0.057	3.080	0.969	0.081	4.360	1.371	0.089	4.833	1.520	0.095	5.132	1.614	0.089	4.819	1.516	0.099	5.371	1.689
520	0.425	4.771	0.525	0.541	6.072	0.669	0.576	6.462	0.712	0.628	7.041	0.775	0.602	6.754	0.744	0.646	7.245	0.798
530	1.214	6.322	0.309	1.458	7.594	0.372	1.523	7.934	0.338	1.687	8.785	0.430	1.641	8.546	0.418	1.717	8.941	0.438
540	2.313	7.600	0.162	2.689	8.834	0.188	2.785	9.149	0.195	2.869	9.425	0.201	2.821	9.267	0.197	2.899	9.523	0.203
550	3.732	8.568	0.075	4.183	9.603	0.084	4.282	9.832	0.086	4.267	9.796	0.086	4.245	9.747	0.085	4.270	9.803	0.086
560	5.510	9.222	0.036	5.840	9.774	0.038	5.880	9.841	0.039	5.625	9.415	0.037	5.656	9.466	0.037	5.584	9.345	0.037
570	7.571	9.457	0.021	7.472	9.334	0.021	7.322	9.147	0.020	6.947	8.678	0.019	7.048	8.804	0.019	6.844	8.550	0.019
580	9.719	9.228	0.018	8.843	8.396	0.016	8.417	7.992	0.016	8.304	7.885	0.015	8.520	8.089	0.016	8.108	7.699	0.015
590	11.579	8.540	0.012	9.728	7.176	0.010	8.984	6.627	0.010	8.612	6.352	0.009	8.927	6.584	0.010	8.387	6.186	0.009
600	12.704	7.547	0.010	9.948	5.909	0.007	8.949	5.316	0.007	9.046	5.374	0.007	9.541	5.668	0.007	8.704	5.171	0.007
610	12.669	6.356	0.004	9.436	4.734	0.003	8.325	4.176	0.002	8.499	4.264	0.003	9.074	4.553	0.003	8.114	4.071	0.002
620	11.373	5.071	0.003	8.140	3.630	0.002	7.070	3.153	0.002	7.089	3.161	0.002	7.658	3.415	0.002	6.710	2.992	0.002
630	8.980	3.704	0.000	6.200	2.558	0.000	5.309	2.190	0.000	5.062	2.088	0.000	5.528	2.280	0.000	4.754	1.961	0.000
640	6.558	2.562	0.000	4.374	1.709	0.000	3.693	1.443	0.000	3.547	1.386	0.000	3.934	1.537	0.000	3.301	1.290	0.000
650	4.336	1.637	0.000	2.815	1.062	0.000	2.349	0.886	0.000	2.147	0.810	0.000	2.397	0.905	0.000	1.993	0.752	0.000
660	2.628	0.972	0.000	1.655	0.612	0.000	1.361	0.504	0.000	1.252	0.463	0.000	1.417	0.524	0.000	1.152	0.426	0.000
670	1.448	0.530	0.000	0.876	0.321	0.000	0.708	0.259	0.000	0.680	0.249	0.000	0.781	0.286	0.000	0.620	0.227	0.000
680	0.804	0.292	0.000	0.465	0.169	0.000	0.369	0.134	0.000	0.347	0.126	0.000	0.401	0.145	0.000	0.315	0.114	0.000
690	0.404	0.146	0.000	0.220	0.080	0.000	0.171	0.062	0.000	0.150	0.054	0.000	0.172	0.062	0.000	0.136	0.049	0.000
700	0.209	0.075	0.000	0.108	0.039	0.000	0.082	0.029	0.000	0.077	0.028	0.000	0.090	0.032	0.000	0.070	0.025	0.000
710	0.110	0.040	0.000	0.053	0.019	0.000	0.039	0.014	0.000	0.041	0.015	0.000	0.047	0.017	0.000	0.037	0.013	0.000
720	0.057	0.019	0.000	0.026	0.009	0.000	0.019	0.006	0.000	0.017	0.006	0.000	0.019	0.007	0.000	0.015	0.005	0.000
730	0.028	0.010	0.000	0.012	0.004	0.000	0.008	0.003	0.000	0.009	0.003	0.000	0.011	0.004	0.000	0.008	0.003	0.000
740	0.011	0.006	0.000	0.006	0.002	0.000	0.004	0.002	0.000	0.005	0.002	0.000	0.006	0.002	0.000	0.005	0.002	0.000
750	0.006	0.002	0.000	0.002	0.001	0.000	0.002	0.001	0.000	0.002	0.001	0.000	0.002	0.001	0.000	0.002	0.001	0.000
760	0.004	0.002	0.000	0.002	0.001	0.000	0.001	0.001	0.000	0.001	0.000	0.000	0.001	0.001	0.000	0.001	0.000	0.000
770	0.002	0.000	0.000	0.001	0.000	0.000	0.001	0.000	0.000	0.001	0.000	0.000	0.001	0.000	0.000	0.001	0.001	0.000
Total	109.828	100.000	35.547	99.072	100.000	85.223	98.041	100.000	118.103	95.018	100.000	108.845	95.655	100.000	92.102	94.954	100.000	122.520

Fig. 5–8. Determination from Spectrophotometric Curve of CIE Coordinates for a Surface Illuminated by Source C

I	II	III CIE Data for Source C (from Fig. 5–7) (multiplied by 1000)			IV (II × III)		
Wavelength (nonometers)	Reflectance R 4/14 $\rho(\lambda)$	$\bar{x}(\lambda)S(\lambda)$	$\bar{y}(\lambda)S(\lambda)$	$\bar{z}(\lambda)S(\lambda)$	$\rho(\lambda)\bar{x}(\lambda)S(\lambda)$	$\rho(\lambda)\bar{y}(\lambda)S(\lambda)$	$\rho(\lambda)\bar{z}(\lambda)S(\lambda)$
380	0.051	4		20			1
390	.051	19		89	1		5
400	.051	85	2	404	4		21
410	.051	329	9	1,570	17		80
420	.050	1,238	37	5,949	62	2	297
430	.050	2,997	122	14,628	150	6	731
440	.050	3,975	262	19,938	199	13	997
450	.047	3,915	443	20,638	184	21	970
460	.045	3,362	694	19,299	151	31	868
470	.044	2,272	1,058	14,972	100	47	659
480	.043	1,112	1,618	9,461	48	70	407
490	.041	363	2,358	5,274	15	97	216
500	.041	52	3,401	2,864	2	139	117
510	.041	89	4,833	1,520	4	198	62
520	.041	576	6,462	712	24	265	29
530	.041	1,523	7,934	388	62	325	16
540	.041	2,785	9,149	195	114	375	8
550	.042	4,282	9,832	86	180	413	4
560	.043	5,880	9,841	39	253	423	2
570	.050	7,322	9,147	20	366	457	1
580	.075	8,417	7,992	16	631	599	1
590	.145	8,984	6,627	10	1,303	961	1
600	.290	8,949	5,316	7	2,595	1,542	2
610	.465	8,325	4,176	2	3,871	1,942	1
620	.575	7,070	3,153	2	4,065	1,813	1
630	.623	5,309	2,190		3,308	1,364	
640	.648	3,693	1,443		2,393	935	
650	.667	2,349	886		1,567	591	
660	.683	1,361	504		930	344	
670	.699	708	259		495	181	
680	.713	369	134		263	96	
690	.725	171	62		124	45	
700	.739	82	29		61	21	
710	.749	39	14		29	10	
720	.762	19	6		14	5	
730	.775	8	3		6	2	
740	.785	4	2		3	2	
750	.791	2	1		2	1	
760	.795	1	1		1	1	
Sums		98,040	100,000	118,103	$X = 23,597$	$Y = 13,337$	$Z = 5,497$

$$x = \frac{X}{X + Y + Z} = 0.56 \qquad y = \frac{Y}{X + Y + Z} = 0.31 \qquad z = \frac{Z}{X + Y + Z} = 0.13$$

Since 1943 the smoothed renotations for the system recommended by the Optical Society of America's Colorimetry Committee have been recognized as the primary standard for these papers. Instructions for obtaining Munsell notations by calculation, or by conversion through CIE are contained in several publications.[7, 13, 14]

ISCC-NBS Method of Designating Colors. The Inter-Society Color Council—National Bureau of Standards method of designating colors appeared in its original form in 1939 as NBS Research Paper 1239. The current version is available in book form (1955)[15] with a supplement (1965)[16] that provides useful, low cost color charts that illustrate with one-inch square samples the centroid color for as many of the 267 names in the system as can be matched in paint. Each of the names defines a block in color space. This method is distinguished from all others in that the boundaries of each name are fixed. These limits are defined in Munsell notations. The method does not provide for pinpointing colors, but provides an understandable color description. When close distinctions must be made among samples that might bear the same ISCC-NBS designation, other specifications such as CIE or Munsell should be used. The method is simple in principle: terms *light*, *medium*, and *dark* designate decreasing degrees of lightness, and the adverb *very* extends the scale to *very light* and *very dark;* adjectives *grayish*, *moderate*, *strong*, and *vivid* designate increasing degrees of saturation. These, and a series of hue names, used in both noun and adjective forms, are combined to form names for describing color in terms of its three perceptual attributes: hue, lightness, and saturation. A few adjectives are added to cover combinations of lightness and saturation: *brilliant* for *light and strong; pale* for *light and grayish;* and *deep* for *dark and strong.* Hue names and modifiers are listed in Fig. 5–9. The latest edition includes a dictionary of color names that covers and correlates names from all well known volumes on the subject. A 61-item reference list makes it as complete a volume on color names as available anywhere. With the addition of the centroid charts it can be useful in many ways. Reflectance of samples can be found by conversion from Munsell value according to tables such as are summarized in Fig. 5–10.

Fig. 5–9. ISCC—NBS Standard Hue Names and Abbreviations

Name	Abbreviation	Name	Abbreviation
red	R	purple	P
reddish orange	rO	reddish purple	rP
orange	O	purplish red	pR
orange yellow	OY	purplish pink	pPk
yellow	Y	pink	Pk
greenish yellow	gY	yellowish pink	yPk
yellow green	YG	brownish pink	brPk
yellowish green	yG	brownish orange	brO
green	G	reddish brown	rBr
bluish green	bG	brown	Br
greenish blue	gB	yellowish brown	yBr
blue	B	olive brown	OlBr
purplish blue	pB	olive	Ol
violet	V	olive green	OlG

Fig. 5–10. Relationship between Munsell Value and Luminance Factor

Munsell Value	Luminance Factor*
10.0	102.6
9.5	90.0
9.0	78.7
8.5	68.4
8.0	59.1
7.5	50.7
7.0	43.1
6.5	36.2
6.0	30.0
5.5	24.6
5.0	19.8
4.5	15.6
4.0	12.0
3.5	9.0
3.0	6.5
2.5	4.6
2.0	3.1
1.5	2.0
1.0	1.2
0	0

* Relative to smoked layer of magnesium oxide. To obtain approximate *absolute* luminance factor multiply factors given by 0.974.

ASTM and ANSI Methods of Measuring and Specifying Color. The American Society for Test-

Fig. 5–11. ASTM and ANSI Documents on Color (Numbers in Parentheses are ANSI Designations)

C609 Measurement of Small Color Differences Between Ceramic Wall or Floor Tile
D1003 (K65.5-1960) Haze and Luminous Transmittance of Transparent Plastics
D1535 Specifying Color by the Munsell System
D1684 (L14.135-1963) Lighting Cotton Classing Rooms for Color Grading
D1729 Visual Evaluation of Color Differences of Opaque Materials*
D1925 Yellowness Index of Plastics
D2244 Color Differences of Opaque Materials
D2253 (L14.187-1966) Color of Raw Cotton Using the Nickerson-Hunter Colorimeter
D2616 Evaluating Change in Color with a Gray Scale
E97 (Z172.1-1969) Test for 45°, 0° Directional Reflectance of Opaque Specimens by Filter Photometry
E166 (Z172.2-1969) Goniophotometry of Transmitting Objects and Materials
E167 (Z172.3-1969) Goniophotometry of Reflecting Objects and Materials
E179 (Z138.3-1969) Selection of Geometric Conditions for Measurement of Reflectance and Transmittance
E259 (Z138.4-1969) Preparation of Reference White Reflectance Standards
E284 (Z138.1-1969) Standard Definitions of Terms Relating to Appearance of Materials
E306 (Z172.4-1969) Absolute Calibration of Reflectance Standards
E308 (Z138.2-1969) Spectrophotometry and Description of Color in CIE 1931 System
E313 (Z172.6-1969) Indexes of Whiteness and Yellowness of Near-White Opaque Materials
(Z7.1-1967) Illuminating Engineering, Nomenclature and Definitions for
(PH 2.31-1969) Direct Viewing of Photographic Color Transparencies
(PH 2.32-1970) Viewing Conditions for the Appraisal of Color Quality and Color Uniformity in the Graphic Arts Industry

* Method of Test D1729 also includes specification of Light Sources.

ing and Materials (ASTM) and the American National Standards Institute (ANSI) have adopted a group of standards dealing with color. See Fig. 5–11 for a listing.

Ostwald System. This is a system describing colors in terms of Color Content, White Content, and Black Content. It is usually exemplified by color charts in triangular form with Full Color, White, and Black at the apexes providing a gray scale of White and Black mixtures, and parallel scales of constant White Content and Black Content as these grays are mixed with varying proportions of the Full Color. Twenty-four or more triangles form a collection of charts with color samples on each illustrating constant dominant wavelength (called hue), those colors lying parallel to the gray scale illustrating constant purity (called Shadow Series). The Ostwald system of color order is similar to another color order system represented by the *Color Harmony Manual* produced by the Container Corporation of America, 1 First National Plaza, Chicago, Illinois. CIE *x,y*-coordinates for each chip in the first three editions of the *Color Harmony Manual* have been determined, and the notations for these chips may be transposed through CIE coordinates to Munsell notations or any other notations for which CIE data are available.

Correlation Among Methods. Frequently it is desirable to convert from one system of specification to another, or to convert or identify the color of samples on a chart or color card to terms of another. If the coordinates or samples of one system are given in CIE or Munsell terms, they may be converted or compared to any other system for which a similar conversion is available. Color charts of the German standard 6164 DIN system are provided with both CIE and Munsell equivalents. The Japanese standard system of color specification, JIS Z 8721-1958, is in terms of hue, value, and chroma of the Munsell renotation system, according to the CIE *Y,x,y*-coordinates recommended by the Optical Society of America's 1943 subcommittee report. The name blocks of the ISCC-NBS method are in terms of the Munsell renotation system with samples measured in CIE terms. Having a common conversion language helps promote international cooperation and understanding of the subject. Complete sets of CIE-Munsell conversion charts are contained in the OSA subcommittee report,[7] and are available from the Munsell Color Company. Many of the available conversions are referenced in a 1957 paper by Nickerson.[17] For more detailed descriptions of color systems or conversions, consult *Color in Business, Science and Industry*.[13]

Color Temperature

This term is widely used, and often misused, in illumination work. It relates to the color of a completely radiating (blackbody) source and of light sources that color match. A completely radiating source would be red at a temperature of 800 K to 900 K, yellow at 3000 K, white (neutral) at about 5000 K, a pale bluish color at 8000 K to 10,000 K, and a more brilliant sky blue at 60,000 K to 100,000 K. Blackbody characteristics at different temperatures are defined by Planck's equation. See page 2–3.

The locus of blackbody chromaticities on the *x,y*-diagram is known as the Planckian locus. Any chromaticity represented by a point on this locus may be specified by color temperature. Strictly speaking, color temperature should not be used to specify a chromaticity that does not lie on the Planckian locus. However, what is called the *correlated* color temperature is sometimes of interest, and the loci of isotemperature lines may be used as an approximation to obtain a reading of the correlated color temperature from diagrams[18] similar to the one shown in Fig. 5–12.

Equal *color differences* on the Planckian locus are more nearly expressed by equal steps of *reciprocal* color temperature than by equal steps of color temperature itself. A difference of one microreciprocal degree $[(1/CT) \times 10^6 = 1$ mired, pronounced my′-red] indicates approximately the same color difference anywhere on the color temperature scale above 1800 K; yet 1 mired varies from about 4 degrees at 2000 K to 100 degrees at 10,000 K.

Color temperature is a specification of chromaticity only. It does not represent the power distribution of a light source. Chromaticities of many "daylight" lamps plot very close to the Planckian locus, and their color may be specified in terms of correlated *color temperature*. However, this specification gives no information about their spectral power distribution which can, and often does, depart widely from that of daylight. The addition of light from two sources each having blackbody distribution but different color temperatures does not necessarily produce a blackbody mixture. Figs. 5–13 and 5–14 show spectral curves for Planckian distributions for different color temperatures, and distributions based on daylight[19] for several correlated color temperatures.

Most tungsten filament lamps approach the power distribution of a blackbody quite closely. The color temperature of such lamps varies with the current passing through them. By varying the voltage across such a lamp a series of color temperatures can be obtained covering a very wide range up to about 3600 K.

Color Constancy and Adaptation[3, 20, 21]

A non-luminous chromatic object contributes to observed color by modifying the spectral power distribution of the light. The color of the light reflected or transmitted by the object is known as the *object color*. The color seen when the object is viewed normally in daylight is a mental phenomenon, and is referred to as the perceived color of the object. While

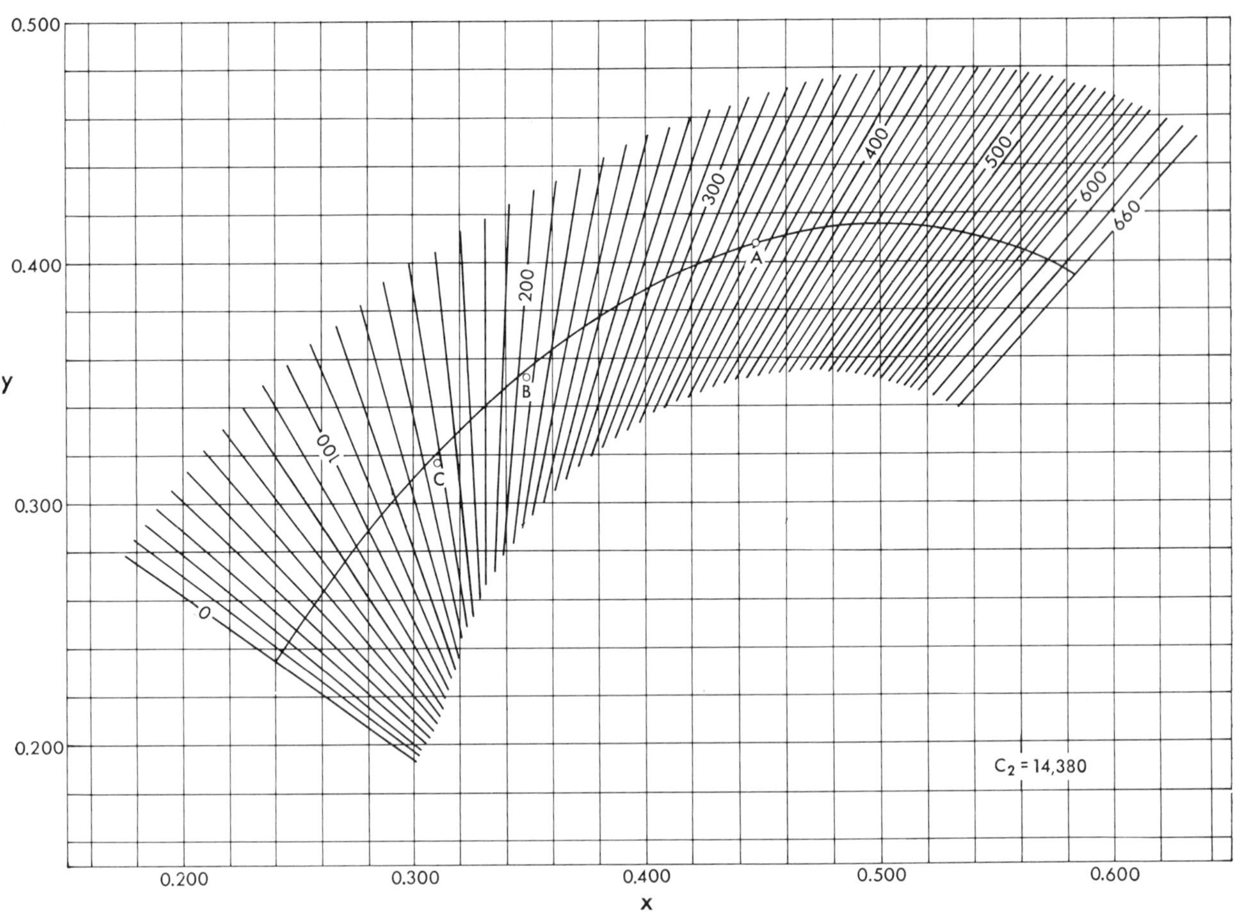

Fig. 5-12. 1931 CIE (x,y) chromaticity diagram showing isotemperature lines.[18] Lines of constant correlated color temperature are given at every 10 microreciprocal degrees.

there are exceptions, the perceived colors of objects do not vary greatly under those illuminations in which light is present in reasonable quantity throughout the spectrum, although slight shifts may be observed with almost any two light sources. Objects whose perceived colors change greatly as there is a wide change in illumination, as for example from daylight to incandescent filament light, are said to have unstable colors.

The fact that the perceived colors of most objects do not change greatly with the power distribution of the light source is due primarily to a low degree of spectral selectivity in the illumination. Color constancy requires awareness of the illumination as separate from the object. It is affected by such factors as persistence of memory colors, the prevalence of object attitude, and the color adaptation of the visual mechanism. Adaptation contributes to this relative color constancy because it tends to reduce to huelessness the chromatic response to the illumination. In other words, adaptation tends to counteract the shift of the color correlated with the shift of the source color, although there are cases where even slight residual shifts may be annoying or even intolerable. Such cases may be encountered with foodstuffs, displayed merchandise, or in the grading of various commercial products.

The facts of color constancy and adaptation are not yet well enough known to make possible the computation of color rendering properties of a lamp with sufficient accuracy except when the reference or standard lamp is limited to the same color as the test lamp. When it becomes possible to compute the effects of constancy and adaptation so that the results agree accurately with the facts of experience, it will then be possible in calculating the color rendering properties of a lamp to take into consideration differences in lamp color, as well as differences in spectral power distributions of lamps that have the same color.

Color Contrast

The appearance of chromatic areas is affected markedly by adjacent areas, particularly if one surrounds the other. Thus a central chromatic patch appears brighter or less gray if surrounded by a sufficiently large and relatively dark area, but dimmer or more gray if surrounded by a relatively light area. Juxtaposed chromatic areas also produce shifts in hue and saturation. Hues shift in opposite directions, tending to make them complementary, and saturations tend to shift away from each other, magnifying the saturation difference. In the general case

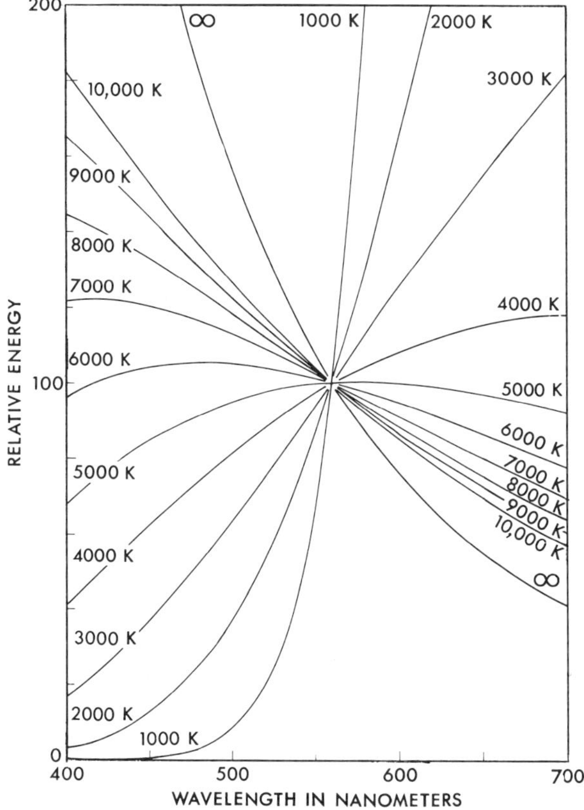

Fig. 5-13. Family of Planckian distribution curves.

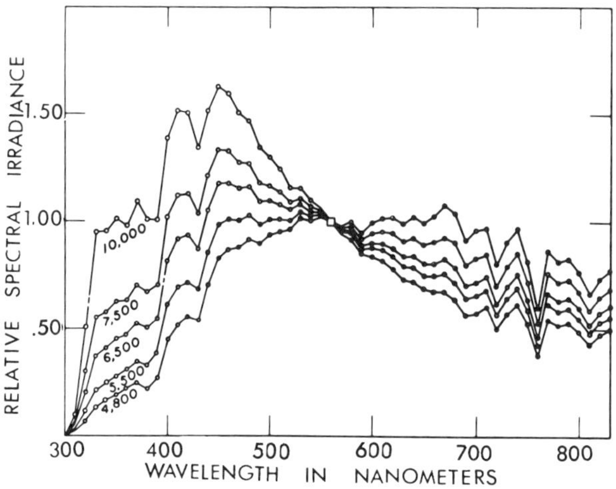

Fig. 5-14. Relative spectral distributions of irradiance of five phases of daylight, of correlated color temperatures 4800, 5500, 6500, 7500, 10,000 K.[19]

there tends to be a simultaneous complex shift with respect to all three attributes.

Metameric and Conditional Color Matches

If two beams of light are visually indistinguishable because they have the same spectral compositions, they are said to form a spectral (or nonmetameric) match. A spectral match is an unconditional match; nobody can tell them apart. Two beams of light may, however, be visually identical in spite of having quite different spectral compositions. Such a color match is said to be metameric. The match is identified by application of color-matching functions which show that the tristimulus values for one beam are identical to those for the other. If the beams are viewed by an observer characterized by different color-matching functions, the match may be upset. All metameric matches are therefore conditional matches. Even though one observer cannot, another one may be able to tell them apart.

Light sources emitting metameric beams look alike in spite of the different spectral compositions (e.g., incandescent filament compared to warm white fluorescent). The metameric character of such a match will usually be revealed by looking at a spectrally selective object and noting that the object is of different color under the two sources.

Objects, identically illuminated, that produce spectrally identical reflected beams (see samples A and B, Fig. 5-15) are said to produce an unconditional match. Nobody can tell them apart no matter what source is used. If, however, the color-matched reflected beams come from identically illuminated objects whose reflectances have different wavelength dependencies, the match is metameric. Substitution of another light source, or another ob-

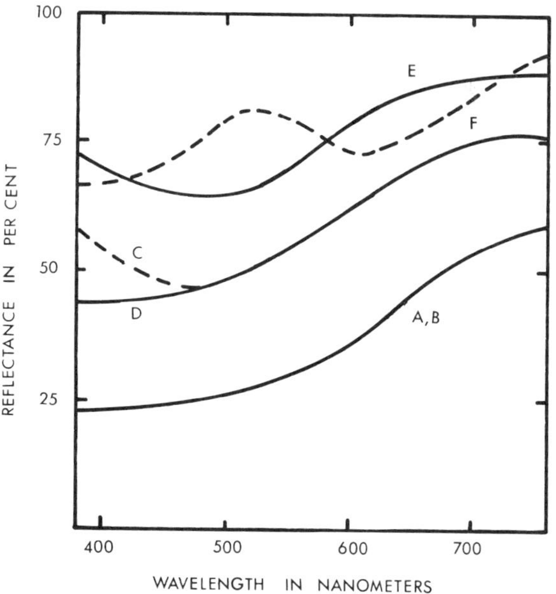

Fig. 5-15. Color matches. A and B are non-metameric matches, *i.e.*, will match for any observer under any light source. C and D will match for any observer under a source with no power of wavelength less than 500 nm, but will not match for some observers under a source that does have some power in the wavelength region 380 to 480 nm. C and D produce a source-conditional match, but they do not form a metameric pair, any more than A and B do, because the matching beams fail to have different spectral compositions. E and F may form a metameric pair for some source-observer combination.

server, may upset the match; so objects that can produce a metameric match, though identically illuminated, may be said to produce a match that is both observer-conditional and source conditional.

It has sometimes been wrongly argued that, because the presence of metamerism always corresponds to a conditional match, a metameric match is the same as a conditional match. Not all conditional matches, however, are metameric. Fig. 5–15 shows the reflectance curves of two samples C and D that have different colors if the light source contains a significant amount of power of wavelength between 380 and 480 nm, but produce a spectral match if the power of the source is confined to wavelengths greater than 500 nm. Samples C and D thus form a source-conditional match; but no metamerism is involved.

There is a necessary, though not sufficient, condition that must be satisfied by the spectral reflectances of objects that, identically illuminated, may produce metameric beams of light. First, the two reflectance curves must be different in some part of the visible spectrum, or the reflected beams will be a spectral rather than a metameric match. Second, to be a color match the two objects must reflect equal fractions of the incident light, and this means that the curves must cross at least once within the visible spectrum. Third, the two objects must not differ in the yellow-blue sense; but if the curves cross at only one wavelength within the visible spectrum, a yellow-blue difference is implied. Therefore, the curves must cross at least at two wavelengths. Finally, the two objects must not differ in the purple-green sense; but if the curves cross at only two wavelengths, a purple-green difference is implied. Therefore, the reflectance curves of objects capable of producing a metameric match for some combination of trichromatic observer and source must at least cross at three wavelengths in the visible spectrum. Samples E and F, Fig. 5–15, have this property, and so for some, source-observer combination might produce a metameric pair of reflected beams. For a discussion of the exact conditions under which a minimum of three or fewer than three intersections is required, see Stiles and Wyszecki[22].

COLORIMETRY OF LIGHT SOURCES

Color of a Light Source

The subjective color of a light source is its appearance apart from its brightness, geometry and time variations. This in turn depends upon the spectral power distribution of the radiant flux from the source and the observer adaptation. Two sources may have the same color appearance and yet have entirely different spectral distributions (*i.e.*, they may produce a metameric match). But if they have the same spectral distributions they must have the same color appearance, all other factors being constant.

In measuring the color of a light source one makes use of the 1931 or 1964 CIE standard observer functions. (See Fig. 5–1 and Fig. 5–2.) Usually the "color" is expressed in terms of tristimulus values (X,Y,Z) or chromaticity coordinates (x,y). It can also be expressed as color temperature if the chromaticity falls on the blackbody locus.

Measurement of Light Source Color

The fundamental but indirect determination of chromaticity of light sources is by means of spectroradiometry.[24, 25, 26, 27] Here the spectral power distribution of the source is measured and the chromaticity computed using the 1931 or 1964 CIE standard observer tristimulus functions. The spectroradiometric curve is fundamental to determinations of chromaticity, color temperature and color rendering attributes of light sources. Current fluorescent lamp chromaticity standards in the United States are now based on spectroradiometrically determined assignments by the National Bureau of Standards. See ANSI Standard C78.3768-1966 and Fig. 5–16. Incandescent lamp color standards on the other hand are still based on visual color temperature comparisons while no standards exist for mercury, xenon or metal halide lamps. Colorimetry of light sources can be done directly through the use of calibrated photoelectric colorimeters,[23] visual tristimulus colorimeters,[13] or by visual color temperature comparisons.

Photoelectric Colorimeters

The Barnes colorimeter[23] or variations of it, is used almost universally in the United States for measuring fluorescent lamp chromaticity. The instrument produces electrical responses to light of various wavelengths that are intended to be proportional to the computed values of the CIE system's X,Y,Z. The IES "Practical Guide to Colorimetry"[1] describes the use of this type instrument and the precautions required. For example, due to the mismatch between the spectral sensitivity of the available detectors and the 1931 CIE tristimulus functions, it is necessary to have a lamp standard with a spectral distribution very close to that of each test lamp to be measured. It is not sufficient just to have a standard of the same chromaticity—a near spectral match is also required. Through the use of operational amplifiers and logic circuitry the instrument has now been modified to read chromaticity directly although with the same fundamental limitations as before. Interlaboratory precision of the Barnes-type colorimeter has been shown to be about .003 in x and .002 in y. Incandescent and vapor type lamps are not normally read on this type of instrument but are usually specified by color temperature and spectroradiometrically based chromaticity respectively.

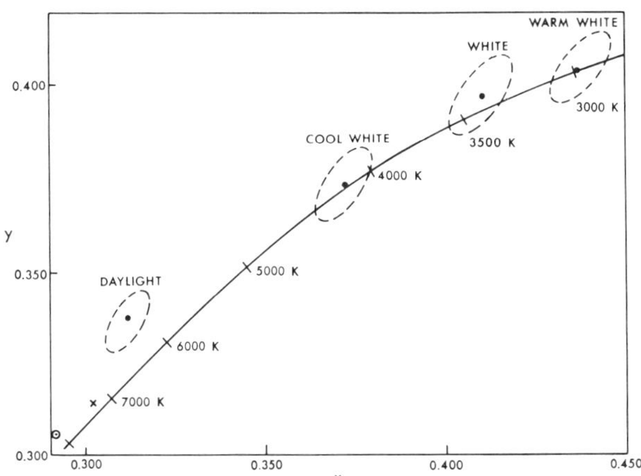

Fig. 5-16. ANSI colorimetric standards for color of fluorescent lamps.

Color Temperature Specification

For a true blackbody light source the specification of color temperature fixes the chromaticity and vice versa. Because the spectral power distribution can be represented as a function of temperature this also is fixed by either the chromaticity or color temperature specification. For reasons already discussed (see page 5-11) the incandescent lamp is the only common light source for which color appearance or chromaticity is a sufficiently accurate function of the color temperature for most practical applications. Even here, however, it is not always exact since the incandescent lamp chromaticities do not always fall precisely on the blackbody locus. Nearest correlated color temperature does not adequately specify color appearance since it is implicit that the corresponding chromaticity lies off the blackbody locus. Any light source (incandescent, fluorescent, high intensity discharge) having a chromaticity falling directly on the locus may be described as having the color temperature associated with that chromaticity without implications as to spectral characteristics.

Computer programs have been written which will operate on the basic spectroradiometric curve to give values of color temperature or nearest correlated color temperature for any light source.[28]

Visual Colorimeters

The eye is a remarkably sensitive color difference detector and can be used effectively to determine when a color match is made between a test light source and a comparison source consisting of a combination of three known "primary" sources. An instrument for facilitating such a comparison and for determining the proportions of the primaries needed to effect the match is called a visual tristimulus colorimeter.[13] The proportions of the primaries needed can be converted into CIE tristimulus values or chromaticity coordinates. Visual colorimeters thus afford an indirect measurement of CIE chromaticity coordinates of a light source.

Color from Spectroradiometry

All of the information needed to determine the chromaticity, color temperature, and color rendering characteristics of a light source is found in the spectral power distribution curve of the source.[24] Where visual phenomena such as color are involved the curve usually is plotted between the wavelengths 380 to 780 nanometers and should be plotted at sufficiently narrow wavelength intervals and be based on sufficiently narrow bandwidths to show the desired resolution. Generally, the curves are plotted at 10 nm intervals and are based on a 10 nm spectral band pass. With this convention monochromatic lines (such as in fluorescent lamps) are represented as 10 nm wide rectangles superimposed on any continuum which may be present, whereas they are actually virtual discontinuities.

Chromaticity is calculated by the methods described earlier (see page 5-4) and has been greatly simplified through the use of modern computers, which in some systems have been tied directly to the spectroradiometer output.[25] Sufficient accuracy is attainable with modern spectroradiometric systems so that they can be used for defining standards for other methods such as employed with the Barnes colorimeter for fluorescent lamps.

Direct Visual Comparisons

The purpose of evaluating the color appearance of light sources usually is to keep such products as fluorescent lamps within limits so that they look alike. Therefore, it seems reasonable to ask the opinion of qualified observers and abide by their findings. This probably is the method most used today, backed up with instrument measurements of standards and limit lamps. It does, however, have limitations.

The abilities of observers to discriminate between small color differences are by no means equal, even when persons of deficient color vision are excluded. A good test for detecting color defective observers is the Hardy-Rand-Rittler Test published by the American Optical Company. For selecting individuals with an innate ability to distinguish small saturation differences, the ISCC Color Aptitude Test, devised and evaluated by a committee of the Inter-Society Color Council,* offers considerable promise.

The technique of matching light sources is important. Lamps should be adjacent to each other, and neither standard nor sample should be the end lamp, that is, they should be surrounded by lamps that are similar in chromaticity and luminance to those being judged. The background should be spec-

* Available from Federation of Societies for Paint Technology, 121 South Broad Street, Philadelphia 7, Pennsylvania.

trally non-selective and of matte finish. Usually the center portion of the lamp is used for color judgment. The observer should inspect standard and sample from at least two directions and from grazing angles.

Skilled observers can learn to judge from a color difference the percentage change in components needed to perfect a match.

Visual judgments, combined with instrument data, can be very useful if the limitations of both are well understood. The problem of metamerism (see page 5–13) in which light sources of different spectral composition may appear to have the same color, makes it quite possible for visual and instrumental results to provide complementary information.

USE OF COLOR

Reflectance

Every object color reflects some portion of the light it receives. This quantity, often expressed as a percentage, may be total reflectance, regular (specular) reflectance, diffuse reflectance, or spectral reflectance, depending on the component considered. In the CIE system, measurements of luminance factor are in terms of the Y tristimulus value. The reflectance scale is not visually uniform between 0 and 100 per cent, black to white. An object that reflects 50 per cent does not look halfway between black and white, but looks much lighter than halfway. On the other hand, the purpose of the Munsell value scale is to illustrate equal visual steps for a given set of standard conditions. Thus, under daylight conditions, and for a light gray surround, a Munsell 5 value should look about halfway in lightness between the black and white end points of the scale. Yet the luminance factor of a 5 value sample is only about 20 per cent. In a condensed table, Fig. 5–10, reflectance and Munsell value units are related. A Munsell 7 value color is a light color, yet it reflects less than half—actually only 43 per cent—of the light it receives. This is an important point for an illuminating engineer to consider, for unless all the colors in the color scheme of a room layout are very light, he may find that well over 50 per cent of his light is absorbed. If he uses 5 value colors, as much as 80 per cent of the incident light may be absorbed. With practice in the use of a Munsell value scale, either in a neutral series of grays and a hue series in 6, 7, and 8 values as illustrated in a color chart of the IES Color Report[2] (see inside of rear cover), or in the special set of Munsell scales developed for illuminating engineers and interior designers illustrated in Fig. 5–5 (page 5–6) one may learn to estimate value rather accurately, and convert to reflectance from Fig. 5–10. Value-reflectance conversion tables for every 0.1 value are available in several publications.[7, 13, 14]

The reflectance of spectrally non-selective gray objects remains constant for all light sources, but for objects that are spectrally highly selective, and thus appear highly saturated, their reflectances will differ in accordance with the spectral selectivity of the light source. For example, with illumination from incandescent sources which are relatively rich in the red and yellow portions of the spectrum and poor in the blue end, yellow objects appear lighter and blue objects darker than they do under daylight illumination; under blue sky the reverse would be true. On the special set of Munsell scales for judging reflectance, shown in Fig. 5–5, reflectances of each sample are given for three light sources; CIE "A" at 2856 K, CIE "C" at 6774 K, and blue sky at 100,000 K.

Light walls and ceilings, whether neutral or chromatic, are much more efficient than dark walls in conserving light and distributing it uniformly. Step by step changes studied by Brainerd and Massey in 1942[29] have been reported in terms of footcandles and utilization coefficients, and are shown in Fig. 5–17. Mathematical analyses[30] of effect of wall colors on illumination and luminance ratios in cubical rooms show that an increase of wall reflectance by a factor of 9 could result in an increase of illumination by a factor of about 3. Moon has published much information concerning spectral and colorimetric characteristics of material used in room interiors.

When neutral and chromatic surfaces of equal reflectance are equally and directly illuminated, they will be equally light. But by *interreflections* in a room, the light reaching a working surface having undergone several reflections from ceiling and walls, greater illumination will result from use of chromatic surfaces than from neutral surfaces. The calculated amount of improvement is shown in Fig. 5–18 for yellow, blue-green, and pink walls, each in comparison to gray walls of the same paired reflectances. The greater the number of interreflections, the greater the advantage. Spencer,[31] O'Brien,[32] and Jones and Jones[33] have published basic studies in this field. Spencer has analytically solved the color shift due to interreflection in the infinite room and in a finite rectangular room, and O'Brien has developed and used computer methods and results to provide charts and tables to aid designers in making detailed predictions of illumination and luminance distributions in rooms, a prerequisite for solving the problem for color interreflections. In France, Barthès has published experimental measurements for a model room.[34] In Japan, Krossawa[35] has computed data on a closed surface painted with a uniform color and derived a general empirical formula on a color shift due to interreflections for different colors. Yamanaka and Nayatani[36] have compared results for computed and actual rooms, and consider the agreements to be quite satisfactory under the conditions. Gradually the data based on such studies will reach a form in which the practicing lighting engineer can use them.

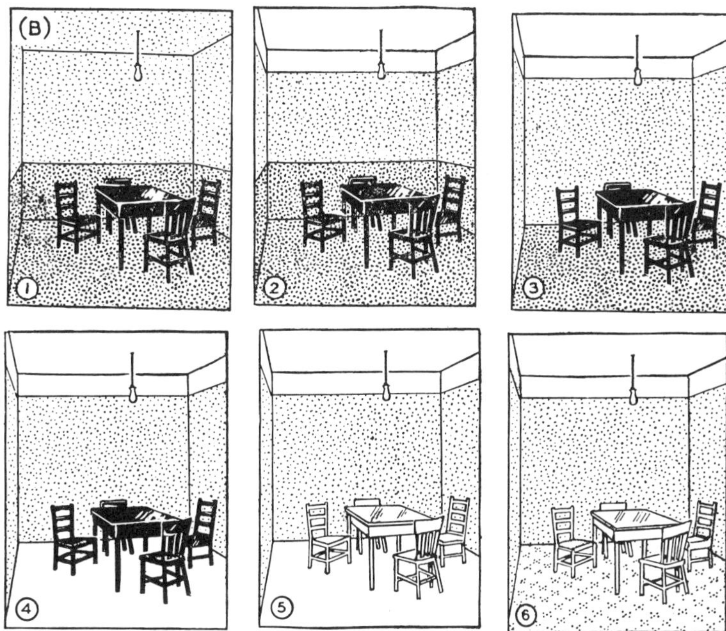

STEP	CEILING HEIGHT	CEILING COLOR	CEILING ρ*	WALLS COLOR	WALLS ρ*	FLOOR COLOR	FLOOR ρ*	FURNITURE COLOR	FURNITURE ρ*
START	12 FT	WHITE	30						
1	10 FT		65	WHITE AND GRAY	40	DARK RED	12	DARK OAK	
2	10 FT		85						28
3	10 FT	CREAM	85						
4	10 FT		85	GREEN	72	WHITE	85		
5	10 FT		85					BLOND	50
6	10 FT		85			WHITE AND RUSSET	70		

* REFLECTANCE

AVERAGE FOOTCANDLES

UTILIZATION COEFFICIENT

0 10 20 30 40 50 60

(D)

Fig. 5–17. Effect of color scheme on appearance, coefficient of utilization, and illumination level in a small room in an industrial area.[29] (A) Test room before changes in color scheme. (B) Step by step changes. (C) Test room with light walls, ceiling, floor, and furniture. (D) Variation of illumination and utilization coefficient with color scheme.

Meanwhile he should understand the general principles so that he may take the facts of color change by room interreflections into consideration in planning a lighting layout.

Color Schemes — Choosing the Right Color

No set of simple rules can allow for tastes of different people, or for different conditions and changing fashions. However, the following few simple rules[2] provide a place to start:

1. *Ceilings* are assumed to be white, or slightly tinted with wall color, except some hospital ceilings that can be treated as a fifth wall.
2. *Walls and floors* (the largest areas) and things that cannot be changed (such as existing floors) are considered first.
3. *Smaller areas,* machinery or furniture, need only blend or contrast with walls and floors.

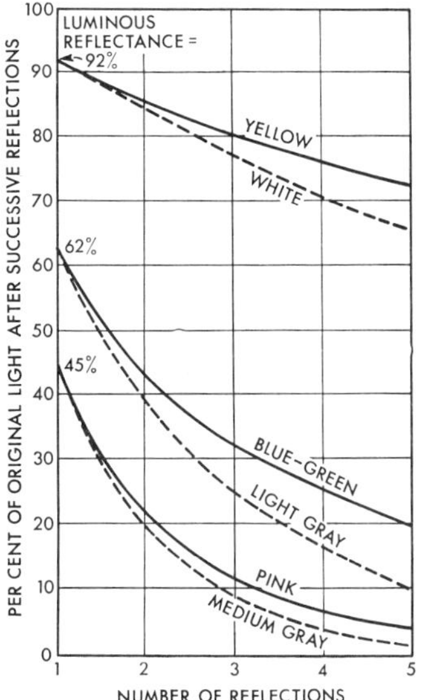

Fig. 5–18. Comparison of interreflection efficiencies of colored and neutral surfaces having the same luminous reflectance.

4. The *purpose* of the color scheme needs consideration (it may be seeing efficiency in a schoolroom, dignity in a church, a sense of well being for a factory, atmosphere of excitement for a circus, quiet for an office).

5. *Limitations* may exist in which redecoration schemes must be built around existing colors of rugs, linoleum, draperies, or furniture.

The 1961 report of the Color Committee,[2] has three useful color charts. The first provides scales of hue, value, and chroma to help in understanding color terminology used in interior design. The second (see inside of rear cover) provides a series of 66 color chips arranged to show strong versus weak chromas, and warm versus cool colors, with reflectances and Munsell notations for each sample, of colors used for interior surfaces. The third illustrates a 10-sample hue circle of typical wall colors at 60 per cent reflectance, and three sample color scheme selections.

Color schemes usually are variations of basic plans classified as monochromatic, complementary, adjacent or analogous, split complementary, or triads. The dominant character usually is determined by the largest area, and in three-hued schemes this usually is the grayest. A large pattern, strong in value contrast, makes a room seem smaller; a small pattern, in gentle contrast, can make a room appear larger. Absence of pattern can provide the illusion of maximum space. Strong contrasts of color or pattern have similar effects; they are stimulating, make people restless, and make time seem longer. They are effective for corridors, places of entertainment, entrance halls, public washrooms, quick lunch counters, and other locations where it is desired that people spend a short time. Gentle contrast is restful and makes time seem shorter. The play of molded form and texture can add interest; contrasts of natural wood, brick, stone, and woven materials add life to smooth, painted walls.

Although personal tastes in color vary with climate, nationality, age, sex, and personality, there is almost universal agreement to call yellows, yellow-reds, reds, and red-purples warm colors, and to call greens, blue-greens, blues, and purple-blues cool colors. All grayed colors approach neutrality of character, whether from the warm or cool side.

Apparent size and position of objects are affected by color. High chroma warm colors usually are the most advancing, with cool colors the most receding. Graying the chroma reduces the effect on position. Light colors make objects appear larger, conversely, dark colors make objects appear smaller.

Selection Guide

By considering factors such as warmth, spaciousness, excitement level, etc., it is possible to determine a suitable dominant color and degree of contrast. These considerations may be made in four steps[2] to help decide on values, hues, chromas, and contrasts.

Step 1: This step helps decide *value,* that is, how light or dark a color scheme should be. If a high level of illumination is necessary, colors with a high reflectance should be used. Dark color combinations tend to produce luminance ratios that are unsatisfactory for efficient seeing. For areas in which illumination levels of 70 footcandles or higher are recommended, the dominant values should be kept high, with reflectances of 40 to 60 per cent—the higher reflectance where the task is critical. Where lower footcandle levels are recommended, around 30, lower values may be introduced, at reflectances of 35 to 60 per cent. For footcandle levels recommended at 20 or under, the dominant values may be still lower, with reflectances for large areas down to 15 to 35 per cent.

Step 2: This step helps decide the "color temperature." Use warm, exciting, advancing colors where rooms have northern exposure, cool temperatures, and low noise element; where the room is too large and has smooth textures; where there is light physical exertion, time exposure is short, and stimulating atmosphere is required; and where lamps are cool fluorescent. Use cool, restful, receding colors for rooms with southern exposure, warm temperatures, and high noise element; for small rooms with rough texture; where physical exertion is heavy, time exposure is long, and a restful atmosphere is desired; and where lamps are incandescent or warm fluorescent.

Step 3: This step helps decide the *chroma.* Strong chromas are used primarily for advertising, display, accents, and food merchandising; grayed colors of low chromas are primarily used for fashion areas, general interiors, and merchandising. Use strong chromas if time exposure is short, responsibility low, lively atmosphere desired, noise level is low, and sense of taste or smell unimportant. Use grayed, low chroma colors if time exposure is long, responsibility high, atmosphere of dignity desired, noise level is high, and sense of taste or smell important.

Step 4: This step helps decide the amount of *contrast.* Contrast is obtained by using light with dark, low with high chromas, and hues that are complementary. Consider use of little or no contrast if time exposure is long, room size is small, dignified atmosphere is required, and wall surfaces are textured. Consider use of strong contrast if time exposure is short, room size is large, lively or exciting atmosphere is desired, and wall surfaces are flat.

COLOR RENDERING

As discussed previously under Basic Concepts, lamps cannot be assessed for color rendering properties by visual inspection of the lamps alone. To provide a color-rendering index that will in all cases represent the facts, it is necessary to have accurate and precise spectroradiometric measurements of light sources, and an adequate understanding of the concepts of color vision. Neither is routinely available, so that today no complete answer to the problem can be provided. But, if such facts as now can be ascertained are considered to have a sufficient degree of accuracy and precision, and the application is held within prescribed limits, it is possible to provide a practical, though not complete answer. This is what has been recommended, based on the following assumptions.

The color shift that occurs when an object is observed under different light sources may be classified in three ways—as a *colorimetric shift*, an *adaptive shift*, or a *resultant color shift* in which the first two are combined. The German language has a separate word for each concept, but there are no separate words for them in English. To understand the subject it is extremely important that the three concepts be separately understood.

1. *Colorimetric shift* is the change in color (luminance and chromaticity, *e.g.*, Y, x, y) of an object illuminated by a non-standard source and the color of the same object illuminated by the standard source, usually measured on a scale that is appropriate for color differences (*e.g.*, W^*, U^*, V^*).
2. *Adaptive color shift* is the difference in the perceived color of an object caused solely by chromatic adaptation.
3. *Resultant color shift* is the difference between the perceived color of an object illuminated by a non-standard source and that of the same object illuminated by the standard source for specified viewing conditions. The conditions usually are that the observer shall have normal color vision and be adapted to the environment illuminated by each source in turn. Resultant color shift is the resultant of colorimetric and adaptive color shift.

CIE Test Color Method

In 1965 a CIE test color method for measuring and specifying the color rendering properties of light sources[4, 5] was recommended for international use. It rates lamps in terms of a color rendering index, R, that is limited to consideration of the degree of colorimetric shift of a test object under a test lamp in comparison to its color under a standard lamp of the same chromaticity. The indexes are based on a general comparison of the lengths of chromaticity-difference vectors on the 1960 CIE uniform chromaticity u,v-diagram. The rating consists of a General Index, R_a, based on a set of eight test-color samples that have been found adequate to cover the hue circuit. The CIE General Index is identical with the rating provided by the 1962 IES method.[3] This may be supplemented by Special Indexes, based on special-purpose test samples. Unless otherwise specified the reference light source for sources with a correlated color temperature of 5000 K and below is a Planckian radiator of the nearest color temperature (see Fig. 5–13). Above 5000 K the reference source is a series of spectral energy distributions of daylight based on reconstituted daylight data[19] developed from recent daylight measurements made in Enfield, England; Rochester, N. Y.; and Ottawa, Canada, see Fig. 5–14. Tables of (u,v)-data are included in the CIE recommendations for Planckian radiators through 5000 K, and on these reconstituted daylight curves above 5000 K to infinity, for eight general and six special test-color samples. The IES subcommittee has computed similar reference data for test samples 1 to 8 at 100 K intervals for Planckian radiators from 1000 K to 8500 K, and for reconstituted spectral curves of daylight from 4800 K to 8500 K; above 8500 K, at intervals greater than 100 K for both series.

A paper by Nickerson and Jerome[5] provides a working text and formulas, discusses the meaning of the index, and shows applications to a number of lamps. The 1962 IES report[3] discusses in more detail the problems involved in its more than 10-year study of the subject. It indicates some of the problems, particularly those of chromatic adaptation, that remain to be solved before an all-purpose, completely satisfactory method can be established for rating a lamp, regardless of its color, against a single

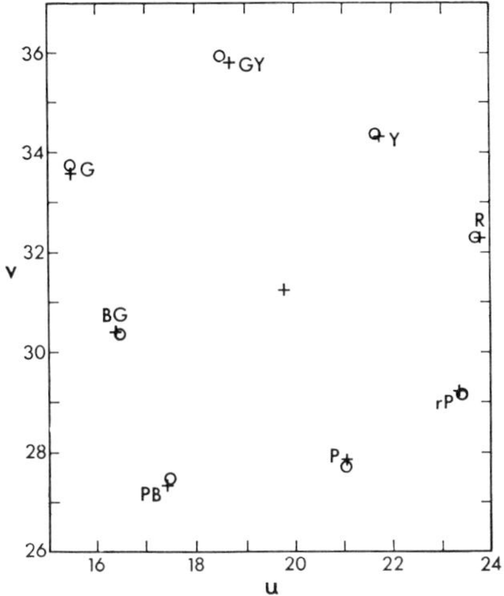

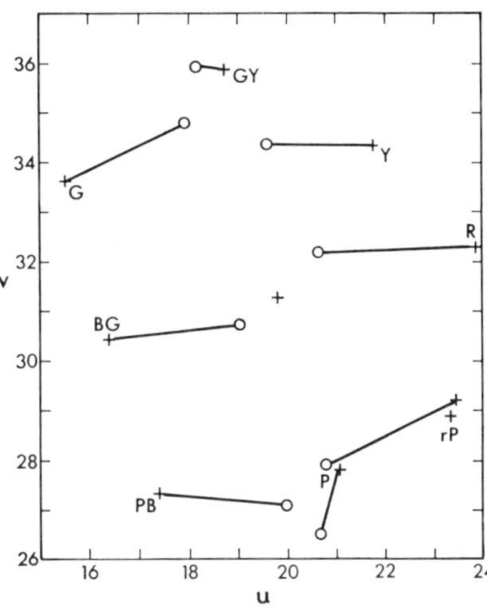

Fig. 5–19. The entire basis for CIE Color Rendering Index, R_a, is illustrated in this one figure: For test lamp A (left) the index is 95, with each of the test colors (0) almost the same as under the standard (+); for test lamp B (right) the index is 18, with each of the test colors (0) a long way from its color under the standard (+). Eight standard test colors, R, Y, GY, etc., are illustrated. When differences between test and standard lamp results are small, the index is high; when differences are large, the index becomes low.

Fig. 5-20. Color and Color Rendering Characteristics of Common Light Sources

Test Lamp Designation[a]	CIE Chromaticity	Correlated Color Temperature (Kelvins)	Reference Source Identification[b] (Kelvins)	Color Rendering Index,[g] R_a	$\Delta E_{a,i}$ = [u,v Vector × 800] (See Footnote[d])								
					Source and Reference[c]	8 CIE Test Samples							
						R	Y	GY	G	BG	PB	P	rP
Fluorescent													
Warm white	$x = .431$ $y = .406$	3105	P3100	55	1.27	12.0	6.4	0.6	13.5	12.7	9.9	6.6	16.8
Warm white deluxe	$x = .434$ $y = .400$	3020	P3000	77	0.9	4.1	3.8	4.0	5.0	5.4	6.0	4.2	6.7
White	$x = .402$ $y = .394$	3595	P3600	61	1.67	10.8	5.7	0.2	11.5	10.8	8.3	5.9	15.1
Soft white	$x = .406$ $y = .372$	(3350)[e]	P3350	62	12.2	6.9	6.7	7.7	7.9	6.9	9.6	8.6	10.7
Cool white	$x = .367$ $y = .375$	4370	P4400	67	2.54	9.3	5.0	0.3	9.4	8.8	6.7	4.9	12.9
Cool white deluxe	$x = .376$ $y = .366$	4050	P4000	85	2.9	2.4	2.9	3.7	3.3	2.8	4.4	3.5	3.3
Daylight	$x = .313$ $y = .332$	6500	D_{6500}	75	1.1	7.3	4.0	0.3	7.0	6.4	4.8	3.8	10.5
D-5000 (a)	$x = .348$ $y = .360$	4942	P4900	92	0	1.2	0.8	0.8	1.0	1.2	1.2	0.6	1.1
D-5000 (b)	$x = .342$ $y = .364$	5149	D_{5100}	90	4.5	2.4	2.2	2.1	2.1	2.6	2.8	1.5	2.4
D-5000 (d)	$x = .344$ $y = .359$	5056	D_{5056}	91	1.2	1.9	1.8	2.2	1.9	2.0	2.6	1.7	1.8
D-5500 (a)	$x = .340$ $y = .345$	5180	D_{5200}	89	3.84	1.8	2.2	3.0	2.4	1.6	3.0	2.7	2.3
D-5500 (b)	$x = .332$ $y = .347$	5517	D_{5500}	91	0.1	1.4	1.7	2.4	2.0	1.6	2.5	2.3	2.0
D-5500 (c)	$x = .333$ $y = .340$	5460	D_{5500}	87	1.91	3.0	2.6	2.7	2.9	2.5	3.3	2.8	3.6
D-6500 (d)	$x = .313$ $y = .325$	6520	D_{6520}	91	2.2	1.5	1.9	3.0	2.0	1.5	2.7	2.3	1.7
D-7000 (a)	$x = .307$ $y = .313$	7070	D_{7100}	93	4.37	0.7	1.4	1.7	2.5	0.3	1.7	1.8	1.2
D-7500 (a)	$x = .297$ $y = .312$	7732	D_{7700}	94	0	0.5	0.6	0.9	0.7	0.6	0.9	0.8	0.7
D-7500 (b)	$x = .299$ $y = .315$	7502	D_{7500}	93	0.2	1.7	1.5	1.8	1.5	1.4	1.8	1.4	1.8
D-7500 (d)[e]	$x = .310$ $y = .328$	6640	D_{6640}	91	0.63	2.1	1.8	1.9	1.9	2.0	2.5	1.6	2.2
D-7500 (d)	$x = .301$ $y = .317$	7350	D_{7350}	92	.84	1.4	2.1	3.1	1.8	1.2	2.3	2.1	1.6
High intensity discharge													
Mercury, 400 W. BT-37	$x = .331$ $y = .379$	(5690)[e]	D_{5500}	22	12.3	24.4	13.4	2.5	21.8	21.5	19.1	9.0	23.3
Mercury, color improved, 400 W. BT-37	$x = .392$ $y = .419$	(3980)[e]	P4000	51	12.0	14.1	9.9	7.4	13.8	14.8	16.5	3.5	4.6
Mercury, deluxe white, 400 W. BT-37	$x = .409$ $y = .387$	3351	P3400	47	3.62	13.9	8.6	4.4	15.6	14.9	14.6	7.5	12.6
Mercury, self ballasted, 450 W. BT-37-White	$x = .388$ $y = .398$	3940	P3900	47	5.9	15.4	9.0	3.0	17.7	17.7	17.4	3.4	8.1
Metal halide (a),[f] 400 W.	$x = .343$ $y = .382$	5200	D_{5200}	72	9.52	6.0	4.9	5.0	4.2	6.3	5.2	4.2	13.7
Metal halide (e)[f]	$x = .359$ $y = .376$	4619	P4600	53	5.26	14.2	6.8	2.0	12.9	12.2	8.3	7.4	18.6
Sodium, high pressure,[f] 400 W.	$x = .515$ $y = .435$	2250	P2200	21	5.0	19.0	8.7	9.3	25.0	20.8	11.7	13.0	28.3
Xenon, 150 W. short arc	$x = .331$ $y = .333$	5550	D_{5500}	97	5.7	0.9	1.1	1.2	0.3	0.8	0.9	0.2	0.6
Xenon, 10 kW short arc	$x = .317$ $y = .320$	6300	D_{6100}	95	5.95	0.7	1.5	2.1	0.9	0.4	1.2	0.9	0.4
Incandescent													
Illuminant "A"	$x = .448$ $y = .407$	2854	P2800	97	1.29	0.7	0.7	0.8	0.5	0.4	0.8	1.0	1.1
	$x = .448$ $y = .407$	2854	P2900	97	2.04	0.6	0.8	1.1	0.9	0.8	0.7	0.5	0.4
Tungsten halogen (NBS Standard)	$x = .430$ $y = .400$	3086	P3100	99	.69	0.2	0.3	0.4	0.2	0.2	0.3	0.2	0.2
2500 K tungsten	$x = .475$ $y = .413$	2500	P2500	92	.51	1.8	1.8	1.8	1.5	1.5	1.9	2.2	2.3
3000 K tungsten	$x = .437$ $y = .404$	3000	P3000	89	.22	2.4	2.3	2.1	1.8	1.9	2.3	2.7	2.8
3200 K tungsten	$x = .423$ $y = .399$	3200	P3200	89	.08	2.5	2.3	2.1	1.9	2.0	2.4	2.8	3.0
Filtered tungsten (d)	$x = .300$ $y = .317$	7450	D_{7500}	96	.82	1.3	.56	.79	1.2	.98	.96	.47	1.5

Fig. 5—20. Color and Color Rendering Characteristics of Common Light Sources—*Continued*

[a] Standard lamps representative of the industry are listed. Certain daylight fluorescent types as well as metal halide types are listed by manufacturer as a, b, c, d, e.

[b] Reference sources are Planckian (P) to 5000 K, and above 5000 K are standard curves for reconstituted daylight, recommended by the CIE colorimetry committee in 1965, and designated by them with the letter D.

[c] See CIE color rendering method, page 5-19, for warning regarding reduction in accuracy that is expected when suggested tolerance is exceeded.

[d] For CIE test color samples 1-8, all at Munsell value 6, $\Delta E_a = 800\ \Delta E_{u,v}$. For lighter or darker test colors the differences in W* (lightness index) must be considered, as well as an adjustment in u,v-vectors, to allow tor this value difference. The ΔE_a unit is approximately the size of 1 NBS unit of color-difference.

[e] Mixed incandescent and fluorescent luminaire.

[f] Color instability of these types introduces uncertainty in the color parameters shown.

[g] Color Rendering Index is dependent upon accurate spectral power distribution data for the light source. For the state of the art in spectroradiometry see NBS Technical Note No. 559, November 1970.

standard (probably daylight since it is the daylight-color of objects that usually is used as a memory reference standard).

CIE ratings are in terms of a single index, R_a, but to provide more information on the color rendering properties of a lamp, it is recommended that this be accompanied by a listing of the chromaticity-difference vectors on which the rating is based, with the index accompanied by a number indicating the standard deviation of the vectors. Since the vectors refer to eight test samples that cover the hue circuit, this makes it possible to obtain a record of the relative colorimetric shift of the different hues under the test lamp. Plotting the vectors provides even more information, for this indicates the direction, as well as the degree, of colorimetric shift that is involved.

The color rendering properties of a lamp with an index of 100 are identical to those of a standard reference lamp. A lamp with an index of 50 has color rendering properties that shift the colors of objects on the average about as much as they are shifted by a standard fluorescent white lamp at 3000 K (such as used in IES subcommittee studies) in comparison to the colors of the same objects under an incandescent reference lamp at 3000 K. Indexes below 50 indicate lamps under which the average colorimetric shift is relatively greater than this. Fig. 5-19 shows the entire basis for the CIE index. At the left, on a uniform chromaticity diagram, there is a plot of the center match point of reference and test lamp with a high index, $R_a = 95$, and at the right the same center match point for a lamp of very low index, $R_a = 18$. The difference in color rendering properties of the two lamps is shown by the differences in the colorimetric shift for the eight test colors, red through a reddish-purple, under each lamp in turn. The shifts under lamp *A* are very small; under lamp *B* they are very large. Yet the lamps themselves are color-matched.

Ratings, accompanied by vector lengths on which they are based, are illustrated in Fig. 5-20 for a number of typical lamps. For high-index, special-purpose lamps, it is possible for ratings to be as high as 95, thus approaching the theoretical maximum of 100. The best color rendering lamps not only must have a high index, but also have the least variation in vector lengths for the different hues that are used as test samples. The closer one comes to an optimum color-matching lamp, the tighter must be these tolerances. As shown in Fig. 5-20, several lamps already available, rate up to 94 and 95, including xenon, and filtered-tungsten. Special fluorescent lamps have been reported to have similarly high indices.

A 5-point change in the index, R_a, is roughly equal to one NBS unit of color difference. It follows that a rating of 95 indicates, in comparison to a reference standard, that a lamp will duplicate the colors of objects within an average of about one NBS unit of color difference, or an average of about 2 units for a rating of 90, while a lamp that rates 50 will duplicate the colors of objects only within an average of about 10 NBS units. Since in commercial color-matching work tolerances tighter than one NBS unit are often required, the relative importance of a difference of 2 or 5 in the color rendering index becomes apparent.

LIGHT SOURCES FOR SPECIAL PURPOSE APPLICATIONS

New Data for Daylight Standards. Standard specifications for the color and spectral quality of light sources for special applications have been adopted in several fields. Some have been adopted formally, as in the graphic arts field, by the IES; others, as for photographic prints and transparencies, by the American National Standards Institute; and still others by the American Society for Testing and Materials. Other less formal standard specifications are recommended, as those for textiles by the American Association of Textile Chemists and Colorists, and for diamond grading by the Gemological Institute of America. In many of these, the target standard is the spectral quality of daylight at around 7500 K, sometimes used with an additional lamp at lower color temperature, sometimes with the addition of ultraviolet. For many years the target standard for 7500 K daylight was the Abbot-Gibson daylight curve at 7400 K (85 per cent daylight component plus 15 per cent blue sky component). Early studies produced this as the best standard at the time for supplying a family of curves, based on daylight measurements, that would provide any color temperature from 6100 K to infinity that could be produced by mixtures of Abbot daylight-outside-the-atmosphere, at 6100 K, with various proportions of blue sky (calculated from the Abbot-Gibson 6100 K data by use of the Rayleigh equation, $1/\lambda^4$). On the CIE (x,y)-diagram, chromaticities of such mixtures lie on a straight line connecting the two end points, but from 6100 K to 15,000 K these are on the purple side of the Planckian locus instead of the green side where the majority of daylight measurements seem to lie. Information on this point is well summarized in papers by Nayatani and Wyszecki,[37] by Macbeth and Reese,[38] and by Judd, MacAdam, and Wyszecki.[19] See Fig. 5–21.

These and other daylight measurements have led to adoption by the CIE of target curves for daylight based on a family of reconstituted spectral distribution curves of typical daylight[19] such as are shown in Fig. 5–14. In the CIE method for specifying color rendering[4] the reconstituted curves for daylight are used as the reference sources for correlated color temperatures above 5000 K.

Artificial Daylighting. Specifications for the best artificial daylighting for accurate work include: a large source of relatively low luminance; duplication of color of a moderately overcast north sky; and more light for inspecting dark colored samples than light colored samples. Levels of at least 80 footcandles for white and light samples (40 per cent reflectance and over), and up to 300 for dark (6 per cent or less) samples are indicated.

The color specification for an artificial daylight source should be aimed at the best obtainable duplicate of preferred natural daylight conditions, and today this is possible with lamps that have a color rendering index above 90. Most commercial color appraisals are made under natural daylight. When artificial daylight is substituted for natural daylighting it should give the same result.

Inspection for suitability of color of materials to be used in daylight (as by a customer in a retail store) usually requires a less rigid approximate duplication of the spectral power distribution of natural daylight, because larger object-color variations are tolerable, but even in this case the target for a color-rendering index should be well up to 90 for lamp colors in the daylight range. The normal eye adapts to rather large changes in the chromaticity of a light source, and if the color rendering properties of the two lamps are sufficiently good, the apparent colors of objects often will remain approximately constant.

Preference of Textile Color Matchers

Data obtained by an Inter-Society Color Council committee indicates that the footcandle and color temperature combinations of natural daylight preferred by textile color matchers are as shown in Fig. 5–22. At 100 footcandles the minimum color temperature preferred is close to 7500 K, ranging upward to a maximum above 25,000 K. The preferred color temperature may drop to 5700 K when the illumination is 300 or more footcandles.[39]

Color Photography

Color photography[40, 41] is in many respects analogous to color vision. Light sources intended as substitutes for, or to supplement daylight for color pho-

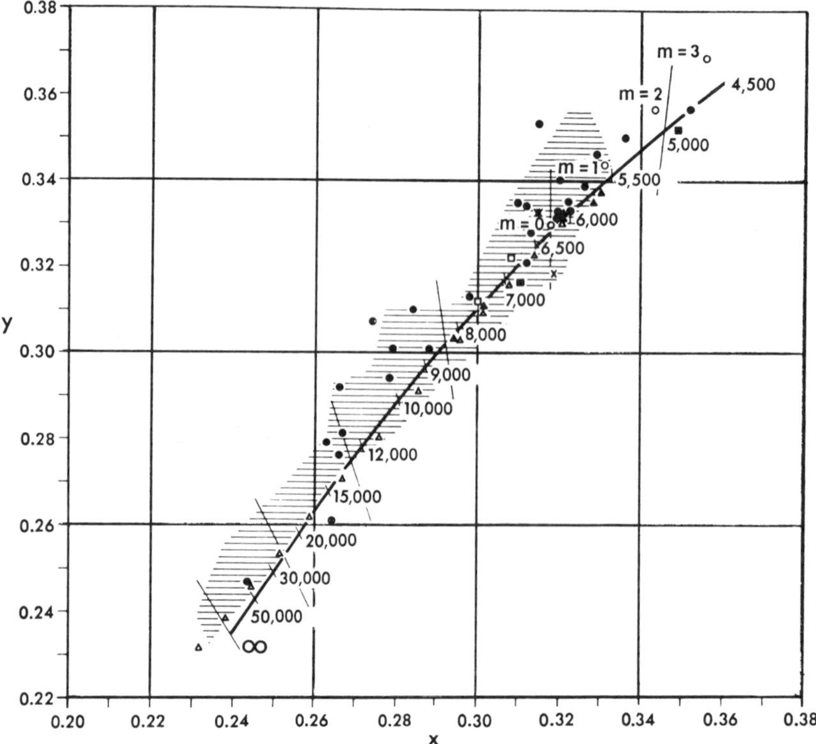

Fig. 5–21. Chromaticity points of daylight and electric sources. Hatched area is north sky as reported by Nayatani and Wyszecki.[37] Solid dots are different phases of sunlight and skylight after Taylor and Kerr. Open circles are direct sunlight at sea level for air masses m = 0, 1, 2, and 3, after Moon. Open circles with cross insert are clear and overcast sky after Masaki. Open triangles are Abbot-Gibson daylight after Nickerson. Solid triangles are overcast sky after Middleton. Solid squares are CIE standard sources B and C. Open squares are Macbeth daylight (6800 K and 7500 K). Crosses are high pressure xenon arcs after Frühlung. Double crosses are carbon arcs. Solid line is Planckian locus.

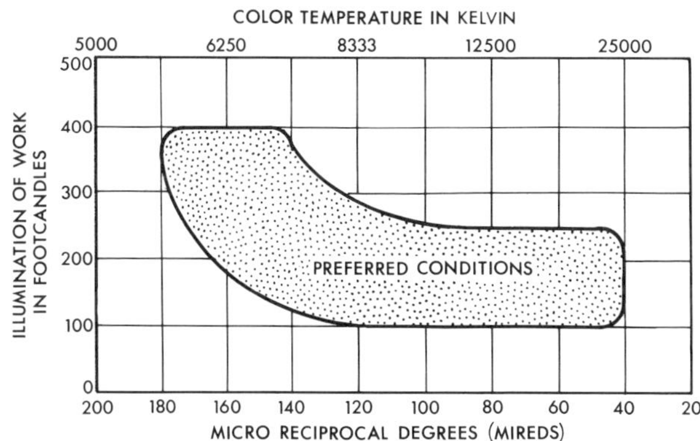

Fig. 5-22. Tests conducted under the direction of the Inter-Society Color Council show the characteristics of preferred daylight illumination conditions for color matching, grading, and classing.

tography should have very nearly the same spectral distribution as the light source replaced. Neither color temperature nor a visual color match is a sufficient specification for light sources used in color photography. See Section 24.

Lighting of Classing Rooms for Color Grading

A large scale application of artificial daylighting to replace natural daylighting in color grading is in the field of raw cotton classification. Studies made in the U. S. Department of Agriculture and reported to the IES as early as 1939 provided the basis for specifying no less than 60 to 80 footcandles on the classing table, and color and spectral quality as close as possible to that of daylight of a moderately overcast sky, about 7500 K. This specification was first met by filtered light from high wattage incandescent lamps. Early experiments included a lamp-and-filter combination close to 6774 K, the color of CIE standard source C, and several other "daylight" sources, such as high temperature carbon arc, carbon dioxide, and fluorescent lamps, but all were found less satisfactory than the lamp-and-filter at 7500 K. (Filtered xenon has now been shown to provide essentially as high color rendering indices as those of filtered incandescent. In addition, filtered tungsten lamps have been supplemented by the addition or combination of ultraviolet sources to provide systems that simulate daylight for wavelengths as low as 300 nm.) Following World War II, a combination of daylight and blue fluorescent lamps, plus incandescent, was developed to match the color of 7500 K with a spectral power as close as possible to that of daylight. Since then there have been improvements, first by combining the phosphors of the daylight and blue lamps into a single fluorescent lamp and using with it low voltage incandescent lamps to increase their life for use with the long-life fluorescent lamp, and in 1956 by the production of fluorescent lamps with much improved spectral power distributions. Mercury lines are still a factor in fluorescent lamps, but except for these lines it now seems possible to produce a fluorescent lamp with a spectral power distribution very close to that of daylight. The advantage of greater efficacy and lower cost over the filtered-tungsten lamps first used in classing rooms has made it the accepted lighting standard in practically all modern cotton classing rooms in the United States as well as in many countries where American cotton is used, for U.S. cotton standards are used around the world. See Fig. 5-23 for a typical modern cotton classing room.

In such installations as in Fig. 5-23 the geometry of the illumination, as well as the spectral power distribution, needs control in order to minimize glare, for as the amount of illumination is increased this factor becomes increasingly important. Control of surrounding conditions is very important, and for these the following instructions, as given by the United States Department of Agriculture, are included in ASTM D-1684-61:[42]

"For rooms with artificial lighting, the surroundings should be light (near-white) in color in order to conserve the lighting and to reduce brightness contrasts as much as possible.
Walls, preferably Munsell Neutral 8.5/, certainly not darker than Munsell Neutral 8.0/;
Ceilings, white or as near white as possible, certainly not darker than Munsell Neutral 8.5/;
Floors, preferably a light gray, about Munsell Neutral 7.0/ (although darker floors may be satisfactory);
Mats, on which a classer stands, should be black (so that they may be used as a background for stapling);
Tables for classing, black, although gray may be satisfactory. (When in use tables are covered with cotton which keeps the light reflectance high during periods of use.)
The grays used in the classing room should be neutral grays, showing no trace of any hue; that is, they should not be yellowish grays, greenish grays, or bluish grays."

Lighting for Graphic Arts Color Work

For recommendations for lighting for color appraisal of reflection type materials in the graphic

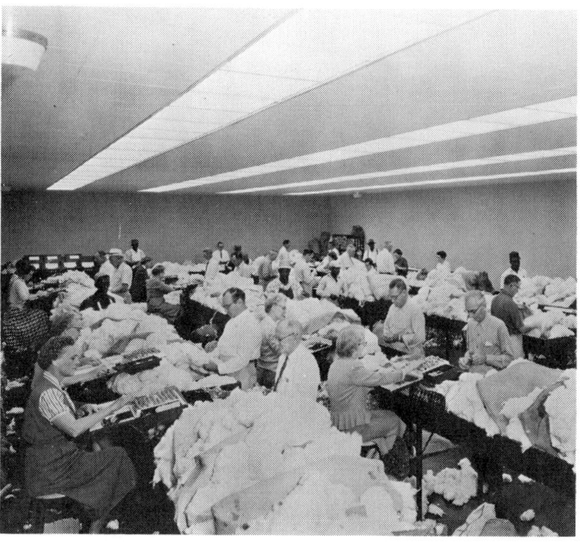

Fig. 5-23. Typical modern cotton classing room.

arts industry[43] see Section 14. The following are four basic tasks in graphic arts color work, and in these, the precision of the lighting requirements is not necessarily the same.

1. The original color selection, or artist's choice of colors.
2. Color matching of basic materials, selection of pigments.
3. Visual appraisal of color quality of printed result compared with original selection, that is, proofs versus copy.
4. Visual appraisal of color uniformity of production sheets.

The spectral power distribution of CIE sources D5000 and D7500 are recommended as the design standard for the various tasks enumerated above. These two standards are specified to serve as a basis for making or specifying color corrections and to establish a common understanding between customer and supplier.

For task 1, copy should be evaluated under commonly used light sources as well as under a Standard Source; many undesirable choices would be avoided if original color selections were checked under lighting conditions in which they will ultimately be used. For task 2, two light sources of widely different color could be used to detect metamerism, one should be CIE source D7500 (an approximation to overcast north sky daylight), the other may be an incandescent filament lamp. For task 3, appraisal of color quality, use of CIE source D5000 is specified, and for task 4, the appraisal of color uniformity, use of CIE source D7500 is specified in the United States.[43]

In each case, the standard sources are to have CIE general color rendering indices, R_a, above 90. Geometrical, background, and surround conditions are specified to avoid glare and undesirable reflections.

Control of Color

Color Selection. The problem may be one of simple selection, as for example that faced by a housewife about to choose from an assortment of meat at the meat dealer's, or of fruit or vegetables at the grocer's, or from an assortment of dress or upholstery fabrics, paints, or wallpapers at a department store; the decision will be based on the appearance of the object on display and upon the customer's estimate of its probable appearance under the conditions most likely to be encountered in use. The conditions of display and use differ more often than they coincide, something that should be corrected by the illuminating engineer as rapidly as lamps with better color rendering properties become available.

Color Matching. Visual color matches satisfactory for many purposes may usually be assured by the simple expedient of checking the match under each of two widely different light sources of continuous power distribution. A practical application of this general method is used in dye houses. When samples are to be matched they are viewed under two light sources selected at or near wide extremes of daylight color temperatures.

A perfect match under all conditions will be obtained only in materials having matching spectrophotometric curves. Even then, if their surface textures are not the same (smooth paint and rough textiles, for example), their appearance may vary slightly depending on the angle at which they are illuminated and viewed.

Color Grading. The market value of many things —raw cotton, tobacco, fruit, vegetables, furs, textiles, and so forth—varies with their colors over a very wide range. In some instances such products are accepted or rejected on the basis of color specifications or standards. They may be separated according to nearly imperceptible color differences into a number of "standard" grades, each of which may have a different market value.

Color grading of a group of materials known to have similar spectral reflectance characteristics may not always require close duplication of the spectral power distribution of daylight. This use, however, differs from the other uses listed because large departures from the spectral power distribution of daylight are allowable as long as they yield in undiminished amount the object-color differences characteristic of daylight inspection.

Whenever the electric light source magnifies the characteristic differences so that they may more easily be detected, departures from the spectral power distribution of daylight may be desirable. If yellow samples are to be examined, a light source rich in energy in the blue portion of the spectrum, where the spectral reflectances of yellow samples are apt to differ most widely, will enable the observer to discriminate differences more easily than when using a source deficient in the blue portion of the spectrum. When blue samples are to be examined, the reverse is true, *i.e.*, a light source rich in energy in the yellow portion of the spectrum, will facilitate discrimination. It should be remembered, however, that while *differences* may be revealed by such a method, the average daylight appearance of the samples may be quite distorted. To an experienced color grader, duplication of the spectral distribution of the light with which he is most familiar permits him to take full advantage of his experience and makes conversion to new conditions much less difficult.

Color Control in a Lighting Installation. The artist, architect, interior designer, and illuminating engineer, after agreeing on a design having suitable decorative qualities and which at the same time will provide the proper quantity and quality of illumination, have the problem of transferring their plans to the location in question. This must be done by specifying to the contractor and builder, as well as to the furniture, wall covering, drapery, and paint manufacturers, what the restrictions are from the standpoint of colored surfaces.

Color Preference

Work has been done under grants from IERI, that is expected to add to our knowledge of color preferences in lighting. A few Helson reports[44-47] already are published in this and related fields, and reports[48, 49] on recent and more extensive studies with fluorescent sources are now available.

In his report, Dr. Helson states that the pleasantness of object colors depends on the interaction of light source with background color and the hue, lightness, and saturation of the object color. The best background colors for enhancing the pleasantness of object colors were found to have either very high Munsell values (8/ or 9/) or very low values (2/ or 1/), and, with only one exception, very low or zero chroma. Color of background was found to be more important than spectral distribution of light source. Neutrals rank high as background colors, but almost at the bottom for object colors. High chromas are preferred to low for object colors.

The chief single factor responsible for pleasant color harmonies was found to be lightness contrast between object and background colors. The greater the lightness contrast, the greater are the chances of good color combinations. Some lightness contrast must be preserved for pattern vision to occur, and this may be why this factor is the most important single factor in good color combinations. The influences of hue and saturation difference cannot be stated simply. A certain amount of variety, change, differentiation, or contrast is pleasant; sameness, monotony, repetition tend to be unpleasant. Static configurations of colors should contain some variations in hue, lightness, and saturation, and over a period of time different configurations of colors should be employed to prevent satiation of overfamiliar patterns of stimulation.

REFERENCES

1. Subcommittee on Practical Guide to Colorimetry of Light Sources of the Committee on Testing Procedures for Illumination Characteristics of the IES: "Practical Guide to Colorimetry," *Illum. Eng.*, Vol. LV, p. 109, February, 1960.
2. Color Committee of the IES: "Color and the Use of Color by the Illuminating Engineer," *Illum. Eng.*, Vol. LVII, p. 764, December, 1962.
3. Subcommittee on Color Rendering of the Light Sources Committee of the IES: "Interim Method of Measuring and Specifying Color Rendering of Light Sources," *Illum. Eng.*, Vol. LVII, p. 471, July, 1962.
4. CIE Committee E-1.3.2: *Method of Measuring and Specifying Color Rendering Properties of Light Sources*, 1st Edition, CIE Publication No. 13, 1965.
5. Nickerson, D. and Jerome, C. W.: "Color Rendering of Light Sources: CIE Method of Specification and Its Application," *Illum. Eng.*, Vol. LX, p. 262, April, 1965.
6. Nickerson, D.: "Light Sources and Color Rendering," *J. Opt. Soc. Amer.*, Vol. 50, p. 57, January, 1960.
7. Newhall, S. M., Nickerson, D., and Judd, D. B.: "Final Report of the OSA Subcommittee on Specifying of the Munsell Colors," *J. Opt. Soc. Amer.*, Vol. 33, p. 385, July, 1943.
8. MacAdam, D. L.: "Visual Sensitivities to Color Differences in Daylight," *J. Opt. Soc. Amer.*, Vol. 32, p. 247, April, 1942.
9. Wyszecki, G. and Stiles, W. S.: "Color Science," John Wiley and Sons, Inc., New York, 1967.
10. MacAdam, D. L.: "Projective Transformations of ICI Color Specifications," *J. Opt. Soc. Amer.*, Vol. 27, p. 294, 1937.
11. Wyszecki, G.: "Proposal for a New Color-Difference Formula," *J. Opt. Soc. Amer.*, Vol. 53, p. 1318, 1963.
12. Committee on Colorimetry of Optical Society of America: *The Science of Color*, Thomas Y. Crowell Co., New York, 1953: Optical Society of America, Washington, D. C., 1963.
13. Judd, D. B. and Wyszecki, G.: *Color in Business, Science and Industry*, 2nd Edition, John Wiley & Sons, Inc., New York, 1963.
14. American Society for Testing Materials: *Method for Specifying Color by the Munsell System D1535-62*, American Society for Testing and Materials, Philadelphia, Pa., 1962.
15. Kelly, K. L. and Judd, D. B.: *The ISCC-NBS Method of Designating Colors and a Dictionary of Color Names*, National Bureau of Standards Circular No. 553, U. S. Government Printing Office, Washington, D. C., 1963.
16. *ISCC-NBS Centroid Color Charts*, Std. No. 2106, Supplement to Circular No. 553, National Bureau of Standards, Washington, D. C.
17. Nickerson, D.: "Horticultural Color Chart Names with Munsell Key," *J. Opt. Soc. Amer.*, Vol. 47, p. 619, July, 1957.
18. Kelly, K. L.: "Lines of Constant Correlated Color Temperature Based on MacAdam's (*u,v*) Uniform Chromaticity Transformation of the CIE Diagrams," *J. Opt. Soc. Amer.*, Vol. 53, p. 999, August, 1963.
19. Judd, D. B., MacAdam, D. L., and Wyszecki, G.: "Spectral Distribution of Typical Daylight as a Function of Correlated Color Temperature," *J. Opt. Soc. Amer.*, Vol. 54, p. 1031, August, 1964: Summary in *Illum. Eng.*, Vol. LX, p. 271, April, 1965.
20. Evans, R. M.: *An Introduction to Color*, John Wiley & Sons, Inc. New York, 1948.
21. Burnham, R. W., Hanes, R. M., and Bartleson, C. J.: *Color: a Guide to Basic Facts and Concepts*, John Wiley & Sons, Inc. New York, 1963.
22. Stiks, W. S. and Wyszecki, G.: "Intersections of the Spectral Reflectance Curves of Metameric Object Colors," *J. Opt. Soc. Amer.*, Vol. 58, p. 32, January, 1968.
23. Barnes, B. T.: "A Four Filter Colorimeter," *J. Opt. Soc. Amer.*, Vol. 29, p. 448, October, 1939.
24. Thorington, L., Parascandola, J., and Schiazzano, G.: "Chromaticity and Color Rendition of Light Sources from Fundamental Spectroradiometry," *Illum, Eng.*, Vol. LX. p. 227, April, 1965.
25. Sanders, C. L. and Gaw, W.: "A Versatile Spectroradiometer and Its Application," *Appl. Opt.*, Vol. 6, p. 1639, October, 1967.
26. Hammond, H. K., Holford, W. L. and Kuder, M. L.: "Ratio Recording Spectroradiometer," *J. Res. NBS*, Vol. 64C, p. 151, 1960.
27. Brown, R. L.: "Direct Recording Spectroradiometer for Light Sources," *Illum. Eng.*, Vol. LXI, p. 230, April, 1966.
28. Robertson, A. R.: "Composition of Correlated Color Temperature," *J. Opt. Soc. Amer.*, Vol. 58, p. 1528, 1968.
29. Brainerd, A. A. and Massey, R. A.: "Salvaging Waste Light for Victory," *Illum. Eng.*, Vol. XXXVII, p. 738, December, 1942.
30. Moon, P.: "Wall Materials and Lighting," *J. Opt. Soc. Amer.*, Vol. 31, p. 723, December, 1941.
31. Spencer, D. E. and Sanborn, S. E.: "Interflections and Color," *J. Franklin Inst.*, Vol. 252, p. 413, November, 1961.
32. O'Brien, P. F.: "Lighting Calculations for Thirty-Five Thousand Rooms," *Illum. Eng.*, Vol. LV, p. 215, April, 1960.
33. Jones, B. F. and Jones, J. R.: "A Versatile Method of Calculating Illumination and Brightness," *Illum. Eng.*, Vol. LIV, p. 113, February, 1959.
34. Barthès, M. E.: "Etudes Expérimental de Mm. Tarnay et Barthès pour la Mise au Point d'une Méthode de Calcul de Point de Couleur de la Lumière Reçue par le Plan Utile dans un Local a Parois Colorées," *Bull. Soc. Franc. Elec.*, Ser. 7, Vol. VII, p. 546, September, 1957.
35. Krossawa, R.: "Color Shift of Room Interior Surfaces Due to Interreflection," *Die Farbe*, Vol. 12, p. 117, 1963.
36. Yamanaka, T. and Nayatani, Y.: "A Note on Predetermination of Color Shift Due to Interreflection in a Colored Room," *Acta Chromatica*, Vol. 1, p. 111, October, 1964.
37. Nayatani, Y. and Wyszecki, G.: "Color of Daylight from North Sky," *J. Opt. Soc. Amer.*, Vol. 53, p. 626, May, 1963.
38. Macbeth, N. and Reese, W. B.: "Color Matching," *Illum. Eng.*, Vol. LIX, p. 461, June, 1964.
39. Nickerson, D.: "The Illuminant in Textile Color Matching," *Illum. Eng.*, Vol. XLIII, p. 416, April, 1948.
40. Evans, R. M., Hanson, W. T., Jr. and Brewer, W. L.: *Principles of Color Photography*, John Wiley & Sons, Inc., New York, 1953.

41. Evans, R. M.: *Eye, Film, and Camera in Color Photography,* John Wiley & Sons, Inc., New York, 1959.
42. American Society for Testing and Materials: *Recommended Practice for Lighting Cotton Classing Rooms for Color Grading D1684-61,* American Society for Testing and Materials, Philadelphia, Pa., 1961.
43. "Viewing Conditions for the Appraisal of Color Quality and Color Uniformity in the Graphic Arts Industry," ANSI Standard PH2.33-1969, American National Standards Institute, New York.
44. Helson, H.: "Color and Vision," *Illum. Eng.,* Vol. XLIX, p. 92, February, 1954.
45. Helson, H.: "Color and Seeing," *Illum. Eng.,* Vol. L, p. 271, June, 1955.
46. Helson, H., Judd, D. B., and Wilson, M.: "Color Rendition with Fluorescent Sources of Illumination," *Illum. Eng.,* Vol. LI, p. 329, April, 1956.
47. Helson, H.: "Role of Sources and Backgrounds on Pleasantness of Object Colors," a paper presented at the IES National Technical Conference, New York, September, 1965.
48. Helson, H. and Lansford, T.: "The Role of Spectral Energy of Source and Background Color in the Pleasantness of Object Colors," *Appl. Opt.,* Vol. 9, p. 1513, July, 1970.
49. Judd, D. B.: "Choosing Pleasant Color Combinations," *Lighting Design & Appl.,* Vol. 1, p. 31, August, 1971.

SECTION

6

LIGHT CONTROL AND LUMINAIRE DESIGN

Physical Principles 6–1 *Shielding 6–9* *Reflector Design 6–10* *Refractor Design 6–13*
Luminaire Materials 6–14 *Luminaire Finishes 6–19* *Equipment Design 6–21*

The control of light is of primary importance once the light has been produced. Light sources, such as flames or arcs, or incandescent, electric discharge, or fluorescent lamps, rarely are found to have the inherent characteristics of candlepower distribution, luminance (photometric brightness), and color suited to direct application without control or modification. Also, certain uncontrollable application conditions such as condensation of moisture, collection of dust, and so forth may alter the characteristics of either lamp or luminaire in service.

PHYSICAL PRINCIPLES OF LIGHT CONTROL[1]

Light control may be provided in a number of ways, all of which are applications of one or more of the following phenomena: reflection, refraction, polarization, interference, diffraction, diffusion, and absorption.

Reflection

Reflection is the process by which a part of the light falling on a medium leaves that medium from the incident side. Reflection may be specular, spread, diffuse, or compound, and selective or non-selective. Reflection from the front of a glass plate is called first-surface reflection and that from the back second-surface reflection. Refraction, diffusion, and absorption by supporting media are avoided in first-surface reflectors.

NOTE: References are listed at the end of each section.

Specular Reflection. If a surface is polished, it reflects specularly; that is, the angle between the reflected ray and the normal to the surface will equal the angle between the incident ray and the normal as shown in Fig. 6–1. If two or more rays are reflected, these may form a virtual, erect, or inverted image of the source. A lateral reversal of the image occurs when an object is reflected in an odd number of plane mirrors.

Specular Reflectors. Examples of specular reflectors are:

1. Polished and electroplated metals, such as gold or copper, and first-surface silvered glass or plastic mirrors. Inside-aluminized, sealed-beam lamps utilize first-surface reflectors in which the incident light strikes the thin metal reflecting surface without passing through the glass, as shown in Fig. 6–2b.

Light reflected from the upper surface of a glass plate, as in Figs. 6–2a and 6–2b, also is an example of first-surface reflection. As shown in Fig. 6–3, less than 10 per cent of the incident light is reflected at the first surface unless it strikes the surface at wide angles from the normal. The sheen of silk and the shine from smooth or coated paper are images of light sources reflected in the first surface.

2. Rear-surface mirrors. Some light, the quantity

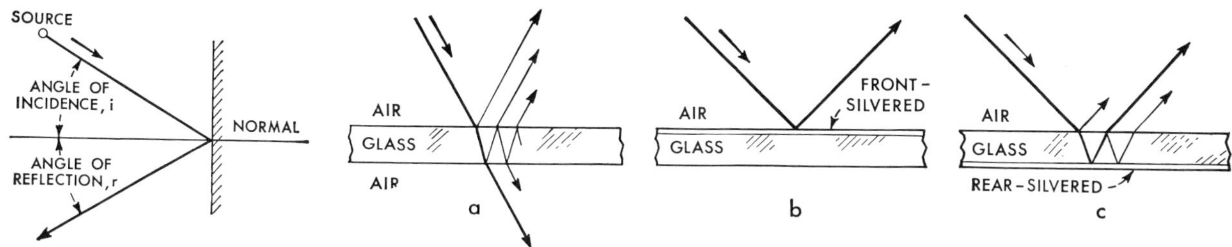

Fig. 6–1. The law of reflection states that the angle of incidence i = angle of reflection r.

Fig. 6–2. Reflections from (a) clear plate glass and (b) from front and (c) rear silvered mirrors.

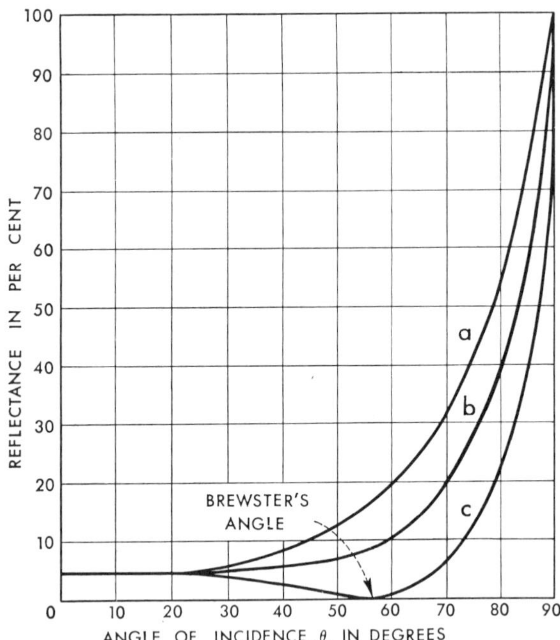

Fig. 6–3. Effect of angle of incidence and state of polarization on per cent of light reflected at an air-glass* surface: a. Light that is polarized in the plane of incidence. b. Nonpolarized light. c. Light that is polarized in plane perpendicular to plane of incidence.

* For spectacle crown glass, n = 1.523.

depending on the incident angle, is reflected by the first surface. The rest goes through to the silvered backing and is reflected back through the glass, as shown in Fig. 6–2c, parallel to the ray reflected by the first surface.

Reflection from Regular Curved Surfaces. Fig. 6–4 shows the reflection of a beam of light by a concave surface and by a convex surface. A ray of light striking the surface at T obeys the law of reflection, and by taking each ray separately, the paths of the reflected rays may be constructed.

In the case of parallel rays reflected from a concave surface, all the rays can be directed through a common point F by properly designing the curvature of the surface. This is called the *focal point.* The *focal length* is f (FA).

Spread Reflection. If a surface is figured in any way (corrugated, deeply etched, or hammered) it spreads any rays it reflects; that is, a pencil of incident rays is spread out into a cone of reflected rays, as shown in Fig. 6–5b.

Spread Reflectors. Depolished metals and similar surfaces reflect individual rays at slightly different angles but all in the same general direction. These are used where smooth beam and moderate control are required.

Corrugated, brushed, dimpled, etched, or pebbled surfaces consist of small specular surfaces in irregular planes. Brushing the surface spreads the image at right angles to the brushing. Pebbled, lightly hammered, or etched surfaces produce a random patch of highlights. These are used where beams free from striations and filament images are required; widely used for sparkling displays.

Diffuse Reflection. If a material has a rough surface or is composed of minute crystals or pigment particles, the reflection is diffuse. Each single ray falling on an infinitesimal particle obeys the law of reflection, but as the surfaces of the particles are in different planes, they reflect the light at many angles, as shown in Fig. 6–5c.

Diffuse Reflectors. Flat paints and other matte finishes and materials reflect at all angles and exhibit little directional control. These are used where wide distribution of light is desired.

Compound Reflection. Most common materials are compound reflectors and exhibit all three reflection components (specular, spread, and diffuse). In some, one or two components predominate, as shown in Fig. 6–6. Specular and narrowly spread reflection (usually surface reflection) cause the "sheen" on etched or embossed aluminum and semi-gloss paint.

Diffuse-Specular Reflectors. Porcelain enamel, glossy synthetic finishes, and other surfaces with a shiny transparent finish over a matte base exhibit no directional control except for the specularly reflected ray that is shown in Fig. 6–6a, which usually amounts to from 5 to 15 per cent of the incident light.

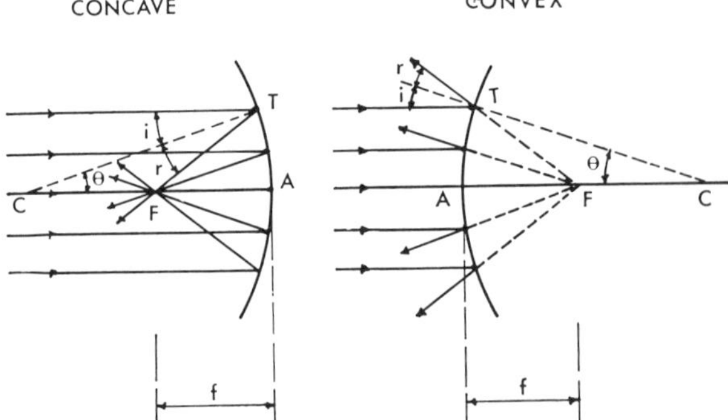

Fig. 6–4. Focal point and focal length of curved surfaces.

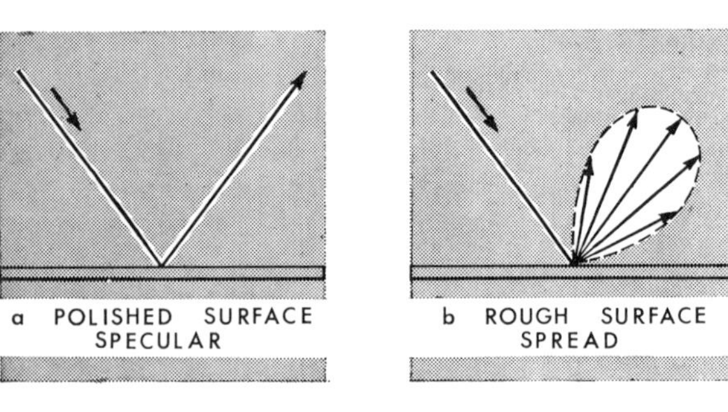

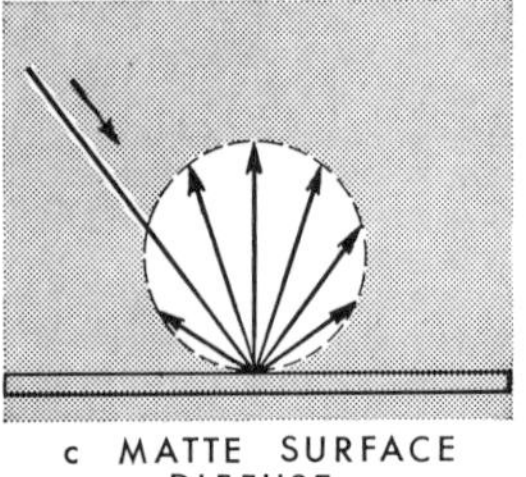

Fig. 6–5. The type of reflection varies with different surfaces: (a) polished surface (specular); (b) rough surface (spread); (c) matte surface (diffuse).

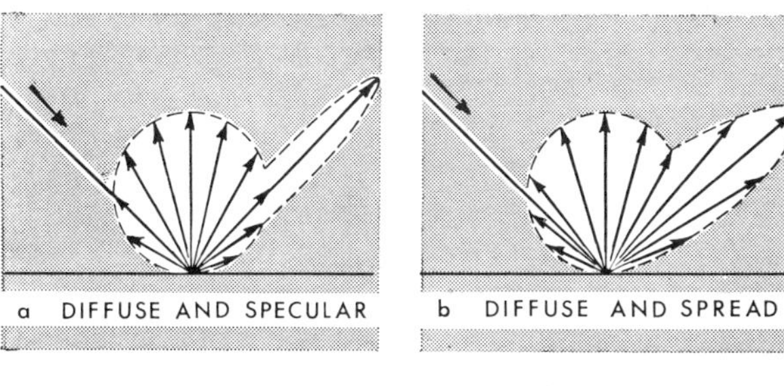

Fig. 6-6. Examples of compound reflection: (a) diffuse and specular; (b) diffuse and spread; (c) specular and spread.

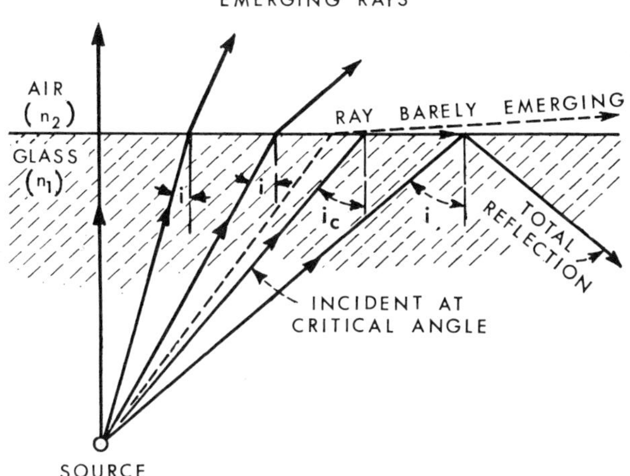

Fig. 6-7. Total reflection occurs when sin r = 1. The critical angle i_c, varies with the media.

Total Reflection. Total reflection of a light ray at a surface of a transmitting medium (see Fig. 6-7) occurs when the angle of incidence (i) exceeds a certain value such that its sine equals or exceeds n_2/n_1. If the index of refraction of the first medium (n_1) is greater than that of the second medium (n_2), sin r will become unity when sin i is equal to n_2/n_1. At angles of incidence greater than this critical angle (i_c) the incident rays are reflected totally, as in Fig. 6-7. In glass total reflection occurs whenever sin i is greater than 0.66, that is, for all angles of incidence greater than 41.8 degrees (glass to air). Both edge lighting and efficient light transmission through rods and tubes are examples of total reflection.

When light, passing through air, strikes a piece of ordinary glass ($n_2/n_1 = 1.5$) normal to its surface, about 4 per cent is reflected from the upper surface and about 3 or 4 per cent from the lower surface. Approximately 85 to 90 per cent of the light is transmitted and 2 to 8 per cent absorbed. The proportion of reflected light increases as the angle of incidence is increased. See Fig. 6-3.

Fiber Optics. Fiber optics[4] is a term given to that phase of optical science concerned with thin, cylindrical glass or plastic fibers of optical quality. Light entering one end of the fiber is transmitted to the other end by a process of total internal reflection. See Fig. 6-8. Applications for a single fiber are more theoretical than practical. Therefore, a large number of fibers (from 100 to 1,000,000) are clustered together to form a bundle. The ends of the bundle are bonded together, ground, and polished. In order to prevent light leaking from one fiber to another, each is insulated with a glass coating of lower refractive index than that of the fiber. The entire bundle is encased in a flexible tubing to protect the fibers. Fiber bundles are of two major types: coherent and non-coherent. The first is used for transmitting images and the second for transmitting light from a source to some area of investigation.

Refraction

A change in the velocity of light (speed of propagation, not frequency) occurs when a ray leaves one material and enters another of greater or less optical density. The speed will be reduced if the medium entered is more dense and increased if it is less dense.

Except when light enters at an angle normal to the surface of a medium of different density, the change in speed always is accompanied by a bending of the light from its original path at the point of entrance, as shown in Fig. 6-9. This is known as refraction. The degree of bending depends on the relative densities of the two substances, on the wavelength of the light, and on the angle of incidence, being greater for large differences in density than for small. The light is bent toward the normal to the surface when it enters a more dense medium and away from the normal when it enters a less dense material.

When light is transmitted from one medium to another, each single ray follows the law of refraction. When a pencil of rays strikes or enters a new medium, the pencil may be broken up and scattered in many directions because of irregularities of the surface, such as fine cracks, mold marks, scratches,

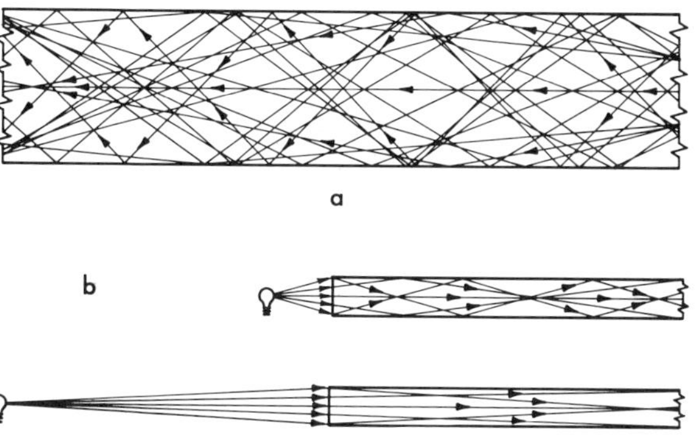

Fig. 6-8. Representation of light transmission through a single fiber of a fiber optics system showing internal reflections (a), and effect of light source location on collimation of light (b).

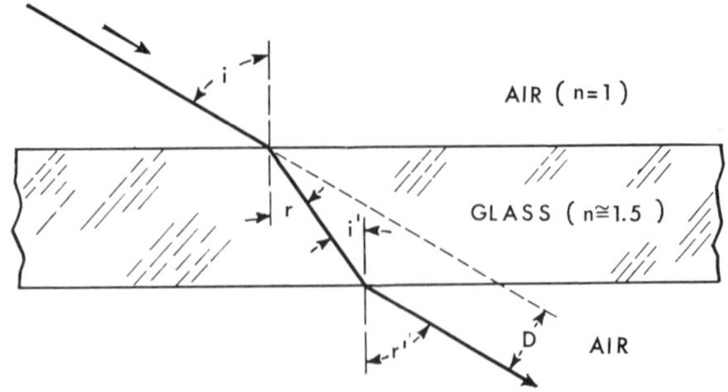

Fig. 6-9. Refraction of light rays at a plane surface causes blending of the incident rays and displacement of the emergent rays. A ray passing from a rare to a denser medium is bent toward the normal to the interface, while a ray passing from a dense to a rarer medium is bent away from the normal.

or changes in contour, or because of foreign deposits of dirt, grease, or moisture.

The law of refraction (Snell's Law) is expressed:

$$n_1 \sin i = n_2 \sin r$$

where n_1 = the index of refraction of the first medium

i = the angle the incident light ray forms with the normal to the surface

n_2 = the index of refraction of the second medium

r = the angle the refracted light ray forms with the normal to the surface.

When the first medium is air, of which the index of refraction usually is assumed to be 1 (correct to three decimal places but actually the index for a vacuum) the formula becomes:

$$\sin i = n_2 \sin r$$

The two interfaces of the glass plate shown in Fig. 6-9 are parallel and therefore the entering and emerging rays also are parallel. The rays are displaced from each other (distance D) because of refraction.

Examples of Refraction. A common example of refraction is the apparent bending of a straw at the point where it enters the water in a drinking glass. Although the straw is straight, light rays coming from that part of the straw under water are refracted when they pass from the water into the air and appear to come from higher points. These irregularities cause irregular refraction of transmitted rays and distortion of the images of objects at which the rays originate.

Prismatic light directors, such as shown in Fig. 6-10, a and b, may be designed to provide a variety of light distributions using the principles of refraction. Lens systems controlling light by refraction are used in automobile headlights, and in beacon, floodlight, and spotlight Fresnel lenses as shown in Fig. 6-10.

Prisms. Many devices use total internal reflection by use of prisms (see Fig. 6-10d) for the redirection of light beams. Performance quality depends on flatness of reflecting surfaces, accuracy of prism angles, elimination of back surface dirt in optical contact with the surface, and elimination (in manufacture) of prismatic error.

Dispersion of Light by a Prism. Consideration of Snell's Law:

$$n_2 = \frac{\sin i}{\sin r} = \frac{\text{velocity of light in air}}{\text{velocity in prism}}$$

suggests, since the velocity of light is a function of the index of refraction of the media involved and

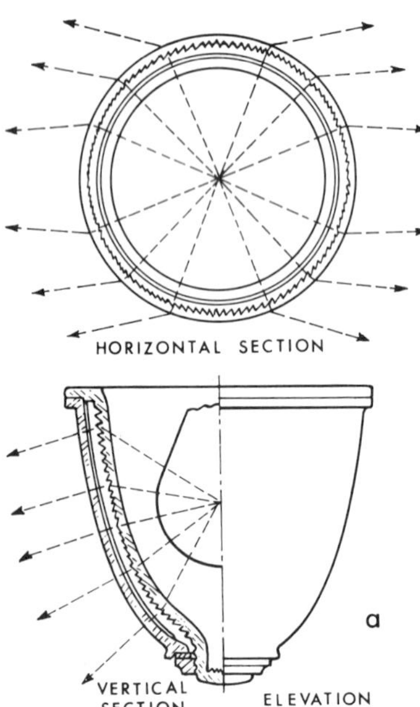

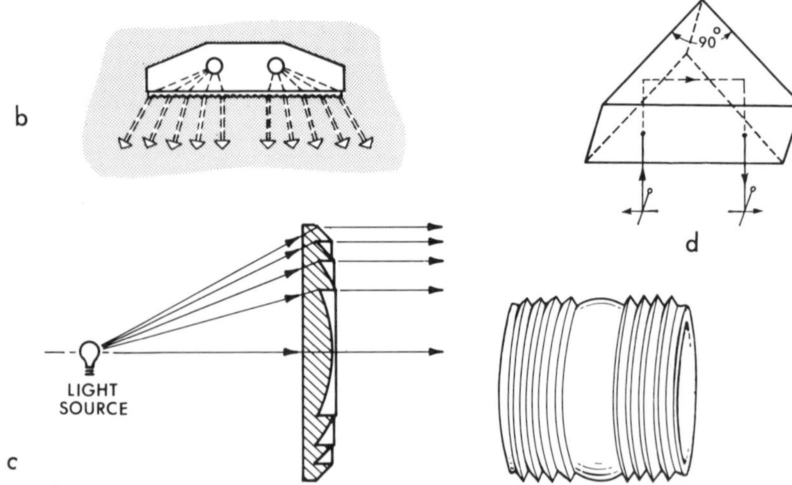

Fig. 6-10. Optical systems utilizing the refractive properties of prisms and lenses: a. Street lighting unit in which the inner piece controls the light in vertical directions (concentrating the rays into a narrow beam at about 75 degrees from the vertical) and the outer piece redirects the light in the horizontal plane. The result is a "two-way" type of candlepower distribution. b. Prismatic lens for fluorescent lamp luminaire intercepts as much light as possible, redirecting part from the glare zone to more useful directions. c. Cylindrical and flat Fresnel lenses. d. Reflecting prism.

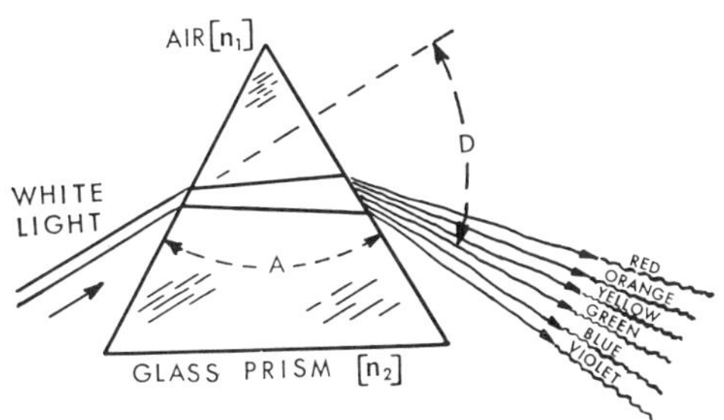

Fig. 6–11. White light is dispersed into its component colors by refraction when passed through a prism. The angle of deviation, *D*, (illustrated for green light) varies with wavelength.

also of wavelength, that the exit path from a prism will be different for each wavelength of incident light and for each angle of incidence. See Fig. 6–11. This orderly separation of incident light into its spectrum of component wavelengths is called dispersion.

Refractors and Refractor Materials

Glass, transparent plastics, and quartz are used in the manufacture of refractive devices.

Refracting Prisms. The degree of bending of light at each prism surface is a function of the refractive indexes of the media and the prism angle (*A* in Fig. 6–11). Light can be directed accurately within certain angles by having the proper angle between the prism faces.

In the design of refracting equipment, the same general considerations of proper flux distribution hold good as for the design of reflectors. Following Snell's Law of refraction, the prism angles can be computed to provide the proper deviation of the light rays from the source. For most commercially available transparent material like glass or plastics, the index of refraction used lies between 1.4 and 1.6.

Often by proper placement of the prisms, it is possible to limit the prismatic structure to only one surface of the refractor, leaving the other surface entirely smooth for easy maintenance. The number and the sizes of prisms used are governed by several considerations. Among them are ease of manufacture and convenient maintenance of lighting equipment in service. A large number of small prisms may suffer from prisms rounding in actual manufacture; on the other hand, small prisms produce greater accuracy of light control. Refracting prisms are used in headlight lenses, refracting luminaires, etc. See Fig. 6–10.

Ribbed and Prismed Surfaces. These can be designed to spread rays in one plane or scatter them in all directions. These surfaces are used in luminaires, footlight lenses, luminous elements, glass blocks, windows, and skylights.

Reflecting Prisms. These reflect light internally, as shown in Fig. 6–10d, and are used in luminaires and retrodirective markers.

Fresnel Lenses. Excessive weight and cost of glass in large lenses used in illumination equipment can be reduced by a method developed by Fresnel, as shown in Fig. 6–10c. The use of lens surfaces parallel to those replaced brings about a reduction in thickness and the optical action is approximately the same. Although outside prisms are slightly more efficient, they are likely to collect more dust. Therefore, prismatic faces often are formed on the inside.

Lenses. Positive lenses form convergent beams and real inverted images as in Fig. 6–12a. Negative lenses form divergent beams and virtual, inverted images as in Fig. 6–12b.

Lens Aberrations. There are, in all, seven principal lens aberrations: spherical, coma, axial and lateral chromatism, astigmatism, curvature, and dis-

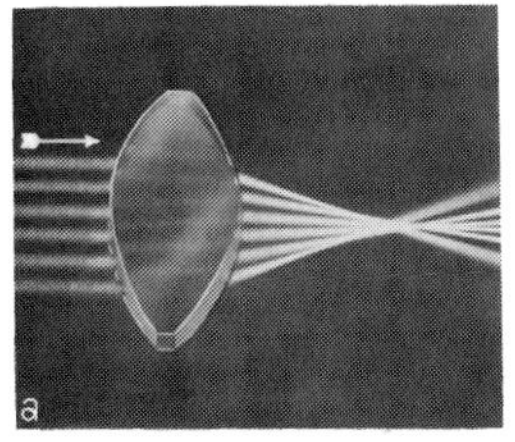

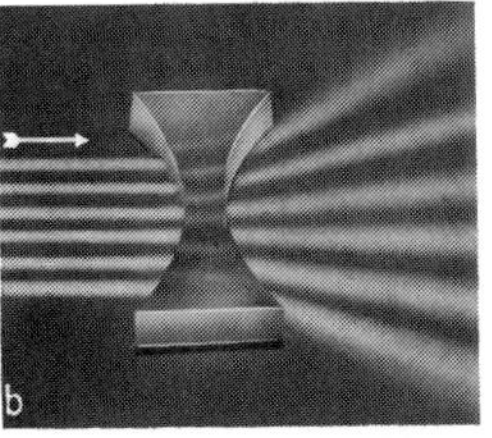

Fig. 6–12. Ray path traces through lenses: a. positive, b. negative.

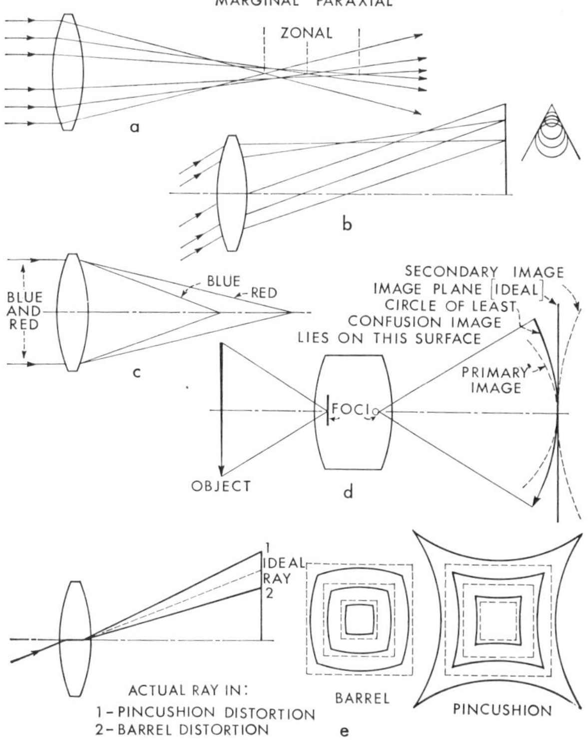

Fig. 6–13. Lens aberrations. a. Spherical aberration: conversion at different focal points of parallel rays at varying distances from the axis of a lens. b. Coma: difference in the lateral magnification of rays passing through different zones of a lens. c. Chromatism: a difference in focal length for rays of different wavelengths. d. Astigmatism and curvature: existence in two parallel planes of two mutually perpendicular line foci and a curved image plane. e. Distortion: a difference in the magnification of rays passing through a lens at different angles.

Fig. 6-14. Reflecting and Transmitting Materials

Material	Reflectance or Transmittance (per cent)	Characteristics
	Reflecting	
Specular		
Mirrored glass	80 to 90	Provide directional control of light and brightness at specific viewing angles. Effective as efficient reflectors and for special decorative lighting effects.
Metalized plastic	75 to 85	
Processed aluminum*	75 to 85	
Polished aluminum	60 to 70	
Chromium	60 to 65	
Stainless steel	55 to 65	
Black structural glass	5	
Spread		
Processed aluminum (diffuse)*	70 to 80	General diffuse reflection with a high specular surface reflection of from 5 to 10 per cent of the light.
Etched aluminum	70 to 85	
Satin chromium	50 to 55	
Brushed aluminum	55 to 58	
Porcelain enamel	60 to 90	
Aluminum paint	60 to 70	
Diffuse		
White plaster	90 to 92	Diffuse reflection results in uniform surface brightness at all viewing angles. Materials of this type are good reflecting backgrounds for coves and luminous forms.
White paint (mat)	75 to 90	
White terra-cotta	65 to 80	
White structural glass	75 to 80	
Limestone	35 to 65	
	Transmitting†	
Glass‡		
Clear	80 to 90	Low absorption; no diffusion; high concentrated transmission. Used as protective cover plates for concealed light sources.
Configurated, obscure, etched, ground, sandblasted, and frosted	70 to 85	Low absorption; high transmission; poor diffusion. Used only when backed by good diffusing glass or when light sources are placed at edges of panel to light the background.
Opalescent and alabaster	55 to 80	Lower transmission than above glasses; fair diffusion. Used for favorable appearance when indirectly lighted.
Flashed (cased) opal	30 to 65	Low absorption; excellent diffusion. Used for panels of uniform brightness with good efficiency.
Solid opal glass	15 to 40	Higher absorption than flashed opal glass; excellent diffusion. Used in place of flashed opal where a whiter appearance is required.
Plastics		
Clear prismatic lens	70 to 92	Low absorption; no diffusion; high concentrated transmission. Used as shielding for fluorescent luminaires, outdoor signs and luminaires.
White	30 to 70	High absorption; excellent diffusion. Used to diffuse lamp images and provide even appearance in fluorescent luminaires.
Colors	0 to 90	Available in any color for special color rendering lighting requirements or esthetic reasons.
Marble (impregnated)	5 to 30	High absorption; excellent diffusion; used for panels of low brightness. Seldom used in producing general illumination because of the low efficiency.
Alabaster	20 to 50	High absorption; good diffusion. Used for favorable appearance when directly lighted.

* See section on aluminum.

† Inasmuch as the amount of light transmitted depends upon the thickness of the material, the figures given are based on thicknesses generally used in lighting applications.

‡ See also Fig. 6-33.

tortion. See Fig. 6–13. Usually they are of little importance in lenses used in common types of lighting equipment. The simpler the lens system, the more difficult is the correction of the aberrations.

Transmission and Transmitting Materials

Transmission is a characteristic of many materials: glass, plastics, textiles, crystals, and so forth. The luminous transmittance τ of a material is the ratio of the total emitted light to the total incident light; it is affected by reflections at each surface of the material, as explained above, and by absorption within the material. Fig. 6–14 lists characteristics of several materials.

Bouguer's or Lambert's Law. Absorption in a clear transmitting medium is an exponential function of the thickness of the medium traversed:

$$I = I_0 \, \tau^x$$

where I = intensity of transmitted light
I_0 = intensity of light entering the medium after surface reflection
τ = transmittance of unit thickness
x = thickness of sample traversed.

Optical density (D) is the common logarithm of the reciprocal of transmittance:

$$D = \log_{10} (1/\tau).$$

Spread Transmission. Spread transmission materials offer a wide range of textures. They are used for brightness control as in frosted lamp bulbs, in luminous elements where accents of brilliance and sparkle are desired, and in moderately uniform brightness luminaire enclosing globes. Care should be used in placing lamps to avoid glare and spotty appearance.

Fig. 6–15a shows a beam of light striking the smooth side of a piece of etched glass. In Fig. 6–15b the frosted side is toward the source, a condition that with many ground or otherwise roughened glasses results in appreciably higher transmittance. For outdoor use the rough surface usually must be enclosed to avoid excessive dirt collection.

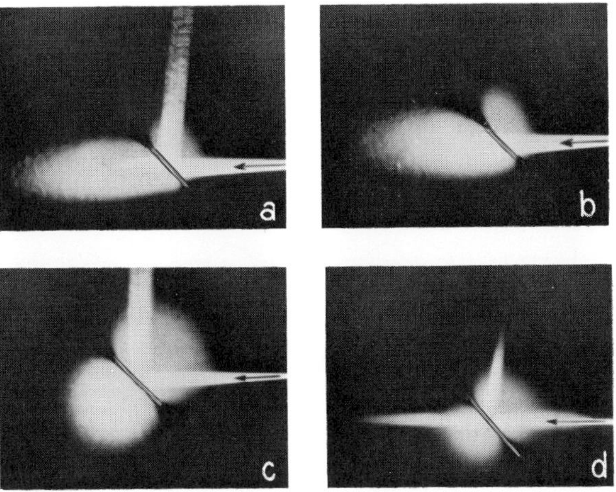

Fig. 6–15. a. Spread transmission of light incident on *smooth* surface of figured, etched, ground, or hammered glass samples. b. Spread transmission of light incident on *rough* surface of the same samples. c. Diffuse transmission of light incident on solid opal and flashed opal glass, white plastic or marble sheet. d. Mixed transmission through opalescent glass.

Diffuse Transmission. Diffusing materials scatter light in all directions, as shown in Fig. 6–15c. White, opal and prismatic plastics and glass are widely used where uniform brightness is desired.

Mixed Transmission. Mixed transmission is a result of a spectrally selective diffusion characteristic exhibited by certain materials such as fine opal glass, which permits the regular transmission of certain colors (wavelengths) while diffusing other wavelengths. This characteristic in glass varies greatly, depending on such factors as its heat treatment, composition, and thickness and the wavelengths of the incident light.

Polarization

Unpolarized light consists of visible electromagnetic waves having transverse vibrations of equal magnitude in an infinite number of planes, all of which contain the line representing the direction of propagation. See Fig. 6–16. In explaining the properties of polarized light, it is common to resolve the amplitude of the vibrations of any light ray into components vibrating in two orthogonal planes each containing the light ray. These two principal directions are usually referred to as the horizontal and vertical vibrations. The horizontal component of light is the summation of the horizontal components of the infinite number of vibrations making up the light ray. When the horizontal and vertical components are equal, the light is unpolarized. When these two components are not equal, the light is partially or totally polarized as shown in Fig. 6–16.

The percentage polarization of light from a source or luminaire at a given angle is defined by the following relation:[5]

$$\text{Per cent vertical polarization} = \frac{I_v - I_h}{I_v + I_h} \times 100$$

where I_v and I_h are the intensities of the vertical and horizontal components of light, respectively, at the given angle.

Reference to vertically polarized light or horizontally polarized light can be misleading in that it suggests that all light waves vibrate either horizontally or vertically. A better notation would be to refer to light at a given instant as consisting of one component vibrating in a horizontal plane and another component vibrating in a vertical plane. A general notation would identify the light components in terms of two reference planes as shown in Fig. 6–17. One plane is the plane of the task at the

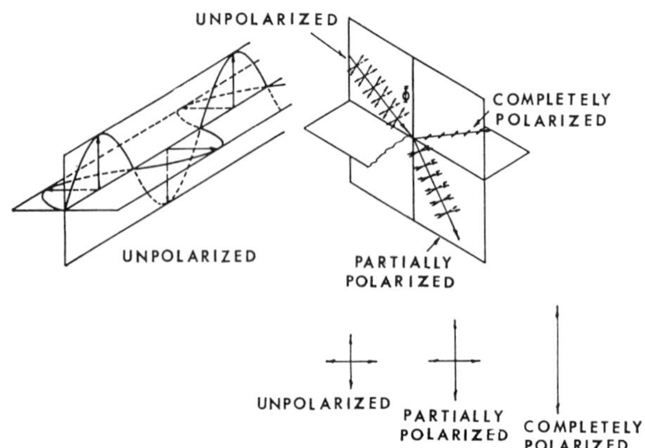

Fig. 6–16. Graphic representations of polarized and unpolarized light.

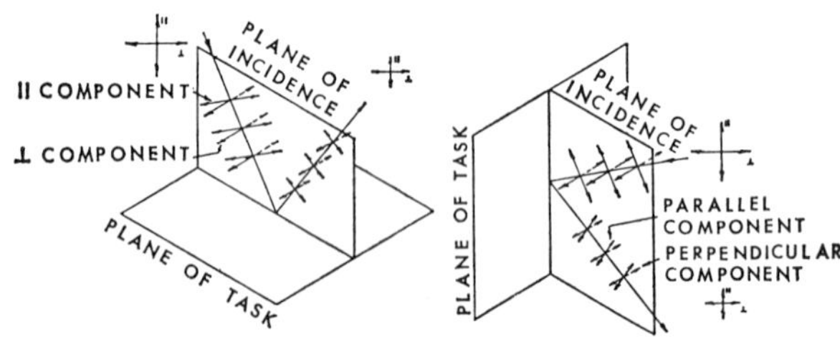

Fig. 6–17. Reference planes of task.

point of the incident light ray; and the second plane is the plane of incidence or that plane perpendicular to the plane of the task and containing the incident light ray. Then the two components of light would be referred to as the parallel component or component in the plane of incidence and the perpendicular component. This notation would apply to any task position and would not have ambiguity with space references.

Polarized light can be produced in four ways: (1) scattering, (2) birefringence, (3) absorption, and (4) reflection and refraction.

Scattering is applicable to daylighting; that is, light from a clear blue sky is partially polarized due to the scattering of light by dust particles in the air.

The *birefringence* or double refraction property of certain crystals can be utilized to achieve polarization. However, the size of these crystals limits this technique to scientific applications and is not suitable to general lighting.

Polarization by *absorption* can be achieved by using dichroic polarizers. These polarizers absorb all of the light in one particular plane and transmit a high percentage of the light in a perpendicular plane. High percentage of polarization can be obtained by this method but at a loss of total luminous transmittance. This type of polarizer is commonly used in sunglasses where it is oriented to transmit the vertical component of light while suppressing the horizontal component.

Light may be polarized by utilizing the *reflection* characteristics of dielectric materials. When light is reflected from a glass surface, it is partially polarized; *i.e.*, a larger percentage of the horizontal component is reflected than of the vertical component. See Fig. 6–3. At approximately 57 degrees, or Brewster's angle, the reflected light is composed of only the horizontal component. See Fig. 6–18. However, for one surface only, 15 per cent of the incident horizontal component is reflected. The light transmitted through a plate at this angle is made up of the remaining portion of the horizontal component and all the vertical component of the original beam. The resulting light is partially polarized. As additional glass plates are added to the system, more and more of the horizontal component is reflected and the transmitted light is more completely vertically polarized. A pile of glass plates, as shown in Fig. 6–19, then becomes a method of producing polarization, and the polarizing effect is greatest at Brewster's angle. The percentage polarization is less at all other angles and is zero for a light ray at normal incidence. Polarization by this method can be obtained by arranging glass or plastic flakes in a suitable material.

Interference

When two light waves come together at different phases of their vibration, they combine to make up a single wave whose amplitude equals the sum of the amplitudes of the two. Fig. 6–20 shows interference. Part of the incident light *ab* is first reflected as *bc*. Part is refracted as *bd*, which again reflects as *de*, and finally emerges as *ef*. If waves *bc* and *ef* have appreciable width of wave fronts, they will overlap and interfere. This interference phenomenon is utilized to increase luminous transmittance, and for extremely accurate thickness measurements.[3] Interference is the cause of the diffraction pattern which is sometimes seen around a pin hole or at the edge of a shadow cast by the sharp edge of an opaque screen. It produces iridescence in bubbles, oil slicks, and other thin films.

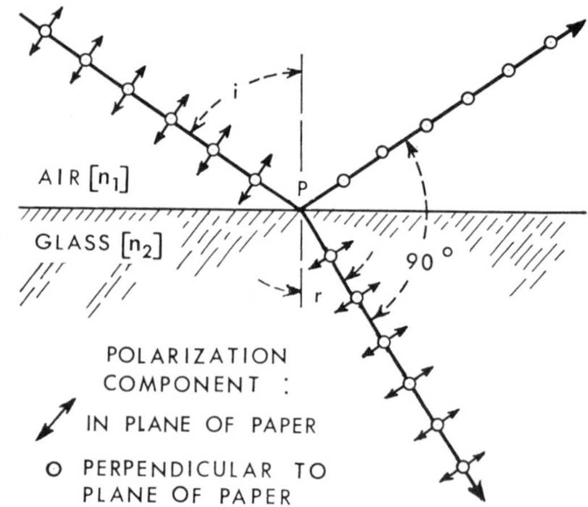

Fig. 6–18. Polarization by reflection at a glass-air surface is at a maximum when the sum of the angle of incidence *i* plus the angle of refraction *r* equals 90 degrees.

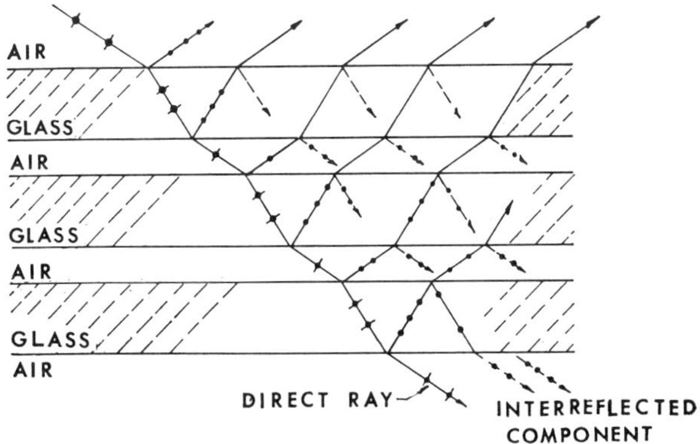

Fig. 6–19. Principle of multilayer polarizer.

Low Reflectance Films. These films are applied to surfaces to reduce reflectance, increase transmittance, and consequently improve contrast relationships. Films a quarter wavelength thick with an index of refraction between that of the medium surrounding the glass and that of the glass are used. The hardest and most permanent films are those of magnesium fluoride condensed on the transmitting surface after thermal evaporation in vacuum, and protected by a thin layer of zircon or quartz applied in the same manner.

The normal, uncoated, 4 per cent reflection at air-glass surfaces may be reduced to less than 1 per cent at each filmed surface as a result of the canceling interference between the waves reflected at the air-to-film and film-to-glass surfaces.

Diffraction

When a wave front is obstructed partially, as by the edge of a reflector or a louver, the shadow cast by the reflector or louver may be sharp or "soft," depending on the geometrical relationship and size of the source, reflector, and illuminated surface. When a series of fuzzy and ill-defined shadows is produced, this phenomenon is known as diffraction.[3]

Diffusion

Diffusion is the breaking up of a beam of light and the spreading of its rays in many directions by irregular reflection and refraction from microscopic crystalline particles, droplets, or bubbles within a transmitting medium, or from microscopic irregularities of a reflecting surface. Perfect diffusion seldom is attained in practice but sometimes is assumed in calculations in order to simplify the mathematics.

Absorption

Absorption occurs when a light beam passes through a transparent or translucent medium or meets a dense body such as an opaque reflector surface. If the intensity of all wavelengths of the light passing through a transparent body is reduced by nearly the same amount, the substance is said to show general absorption. The absorption of certain wavelengths of light in preference to others is called selective absorption. Practically all colored objects owe their color to selective absorption in some part of the visible spectrum.

LAMP CONCEALMENT BY SHIELDING[1]

In general, the term *shielding angle* is assumed to denote the angle between the horizontal and a plane which intersects the lower edge of one row of shielding element and the upper edge of the next. See Fig. 6–21.

Louver Grid or Baffles

A baffle is a single shielding element. A louver, or louver grid, is a series of baffles which may be arranged in a geometric pattern. The more common patterns are square, hexagonal, circular, and sine-wave.

Louver Shielding Angle (θ). The angle between the horizontal plane of the baffles or louver grid and the plane at which the louver conceals all objects above is called the louver shielding angle. The planes are generally so chosen that their intersection is parallel with the louver blades. See Fig. 6–21.

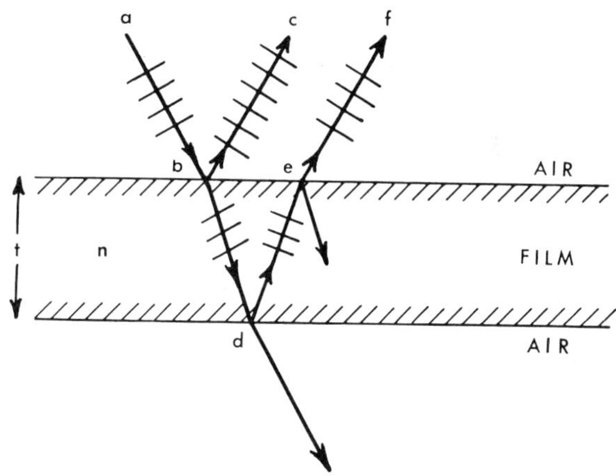

Fig. 6–20. Interference.

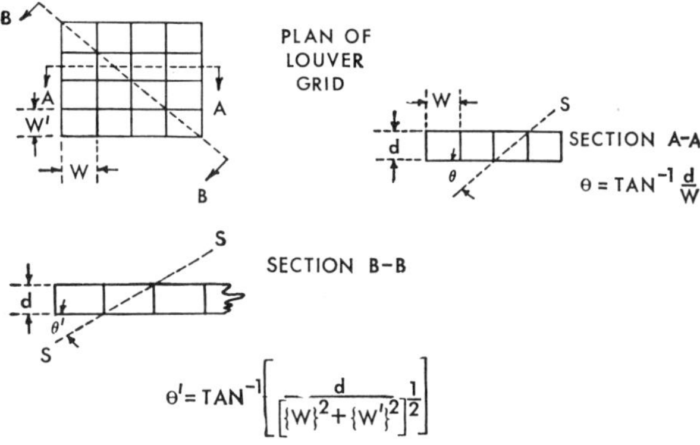

Fig. 6–21. Louver shielding angles θ and θ'.

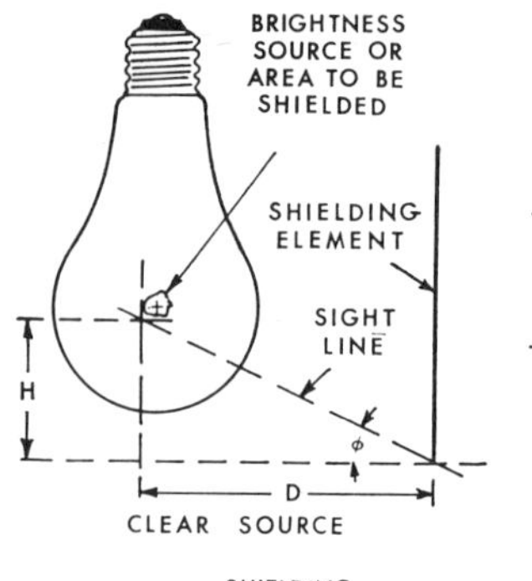

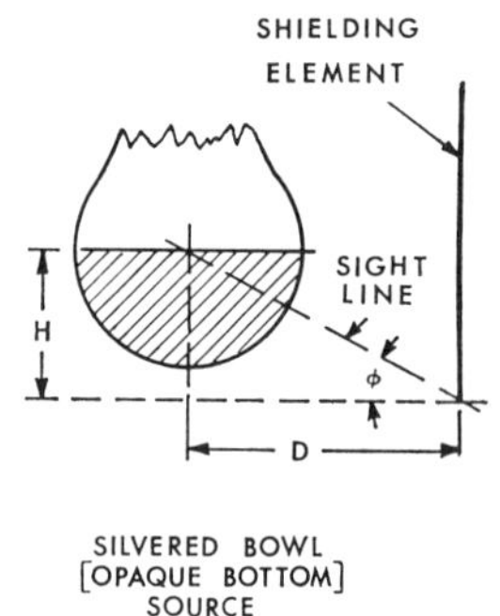

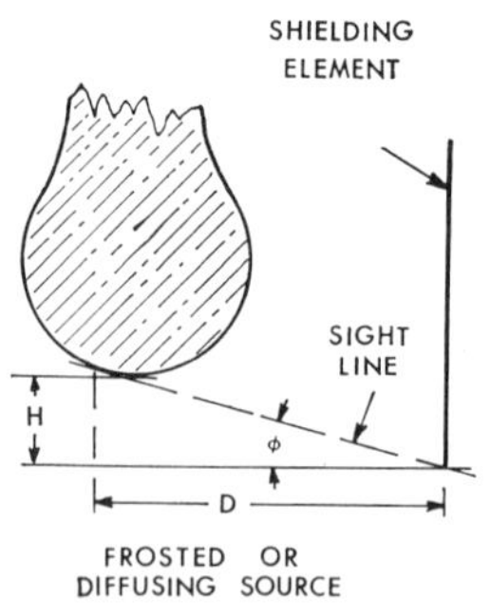

Fig. 6–22. The lamp shielding angle is formed by a sight line tangent to the lowest part of the brightness area to be shielded. *H* is the vertical distance from the brightness source to the bottom of the shielding element. *D* is the horizontal distance from the brightness source to the shielding element. Lamp shielding angle $\phi = \tan^{-1} H/D$.

Lamp Shielding Angle (ϕ). The angle between the plane of the baffles or louver grid and the plane most nearly horizontal which is tangent to both the lamps and the louver blades as illustrated in Fig. 6–22 is called the lamp shielding angle. It is frequently larger than the louver shielding angle, but never smaller.

DESIGN OF REFLECTORS[1]

The design of reflector contour is an extensive subject because the possible shapes for a particular application are almost limitless. However, the end use usually limits the choice.

For design purposes, reflector contours can be divided into two classes: *basic contours* and *general contours*. Basic contours may be defined as those which are mathematically predictable as to action and can be designed mathematically. General contours are those required to satisfy many candlepower distribution curves, but which do not conform to any of the basic contours.

Basic Reflector Contours

Basic contours which are used very frequently are the conic sections and the spherical reflector.

A conic section is by definition the locus of a point whose distance from a fixed point is in a constant ratio to its distance from a fixed straight line. The fixed point is called the *focus* of the conic section; the fixed line is its *directrix*. The constant ratio is the *eccentricity* of the conic section. If the eccentricity, *e*, of the conic section is equal to one, the section is called a parabola; if *e* is less than one, an ellipse. A third conic section, the hyperbola, occurs when *e* is greater than one.

Parabolic Reflectors. An inherent property of the parabola is its ability to redirect a ray of light originating at the focal point in a direction parallel to the axis. The proof of the property is shown in Fig. 6–23 where $A'A''$ is tangent to the curve at A, BA is perpendicular to DD', and $BA = FA$. Assuming a point source at the focus of a perfect parabolic mirror, all light from the source striking the mirror would be redirected as a beam of light parallel to the axis. The ideal conditions of the perfect mirror and the point source cannot be reached in practice nor, in most cases, would this be desirable. The further conditions deviate from the ideal, the greater will be the deviation of the light from a parallel beam. Formulas have been derived expressing the light divergence from shallow mirrors when sources of various shapes are used. Fig. 6–24 illustrates the action of a point source lying on the axis of the parabola but ahead of or behind the focal point.

Ellipsoidal Reflectors. Ellipsoidal reflectors are an efficient means of producing beams of controlled divergence and for collecting light to be controlled by a lens or lens system. The ellipse, Fig. 6–25, can be described by the following equation:

$$\frac{x^2}{a^2} + \frac{y^2}{b^2} = 1$$

Useful working equations are:

$$y = \pm\frac{b}{a}\sqrt{a^2 - x^2}$$

$$x = \pm\frac{a}{b}\sqrt{b^2 - y^2}$$

$$b^2 = a^2 - \left(\frac{F_1F_2}{2}\right)^2$$

Fig. 6–26 illustrates the action of a perfect complete ellipsoidal mirror with a light source at F_1 and a ray of light striking the mirror at P_1, passing through F_2 to P_2, and being reflected again through F_1 to P_3 and on to P_4. If the mirror is "chopped off" at F_2 (the conjugate focus) or at a point closer to F_1, all of the light from the theoretical source at F_1 will leave the mirror either directly or after one reflection. By moving the opening back along the major axis, the beam limits of the reflected light are narrowed. When the plane of the opening reaches the center of the ellipse, the maximum angle of the reflected light coincides with the angle of the direct light from the source.

Hyperbolic Reflectors. The diverging beam typical of the ellipsoidal reflector can also be produced by a reflector having a hyperbolic contour. The main difference, as shown in Fig. 6–27, is that the hyperbolic reflector will form a virtual image F_2 behind the focus, whereas the ellipsoidal reflector produces a real image in front of the focus. The equation of a hyperbola is:

$$\frac{x^2}{a^2} - \frac{y^2}{b^2} = 1.$$

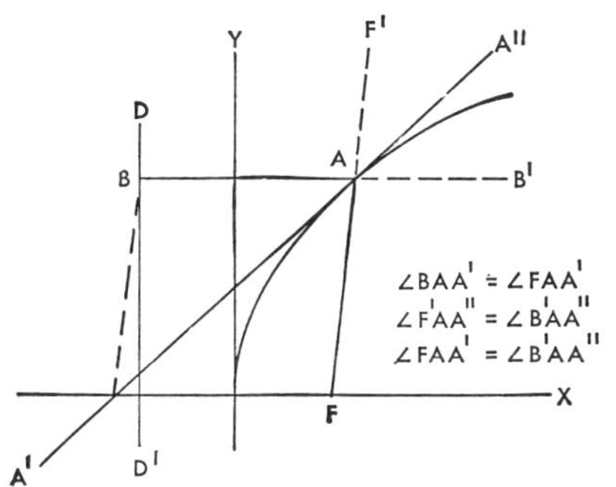

Fig. 6–23. Parabolic conic section. *DD* is the directrix. *F* is the focus.

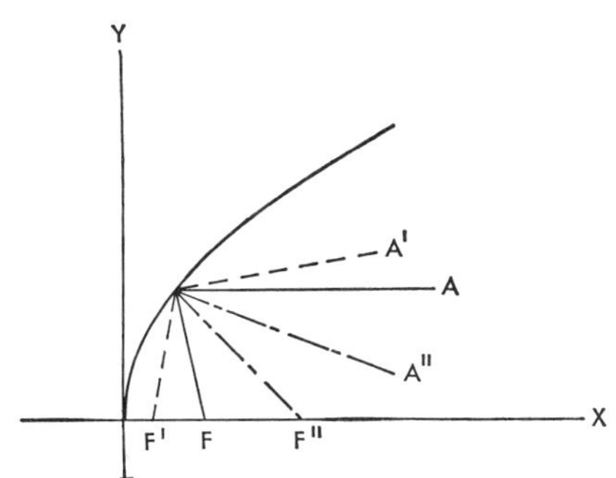

Fig. 6–24. Diverging or converging rays from point source on axis of parabolic reflector, but behind or ahead of focal point *F*.

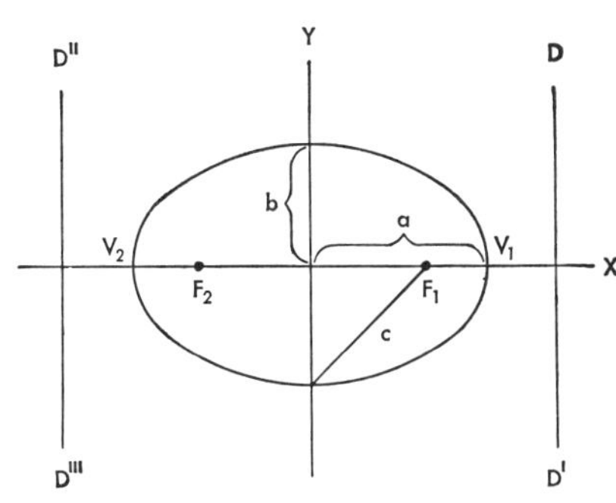

Fig. 6–25. The two foci, directrices, and axes of an ellipse.

Spherical Reflectors. A spherical reflector can be considered as a special form of the ellipsoidal reflector where the two foci are coincident. Any light leaving the source located at the focus would return and pass through the same point. This has obvious disadvantages when dealing with practical sources since the concentration of energy can often have a damaging effect on the source or the bulb wall surrounding it. The principle is, however, often used in projecting devices to increase the amount of light collected by a lens as shown in Fig. 6–28.

General Reflector Contours

For many applications, reflectors can be designed which are mathematically predictable as to contour and action. A more general problem is that of determining the contour of a reflector which does not meet these conditions. Usually there are two known factors; an approximate candlepower distribution curve for the finished reflector, and the material from which it is to be manufactured.

Diffuse Reflecting Surfaces. In the rare case of surfaces with perfectly diffuse reflection, the contour of the reflector will have very little effect on the distribution of light. The candlepower distribution curve of such a reflector (after that portion due to the bare lamp has been deducted) would be very nearly a circle, with the maximum ordinate normal to the plane of the opening. The distribution curve would be even closer to a circle except that all portions of the reflector would not be uniformly illuminated. A unit strip of area near the lower edge of the reflector, being farther from the source, would receive less light than a similar strip higher in the reflector. Since these lower portions are the parts directing light out at the upper angles, the candlepower falls off faster than the cosine of the angle. In this case, there is little that can be done by the designer in the way of light control except by aiming the entire reflector.

Reflectors with Specular and Semi-Specular Surfaces. General contours for specular reflectors are usually obtained through graphical rather than strictly mathematical methods. The problem consists of determining what reflector shape is necessary to redirect luminous flux from the lamp into the proper directions to achieve a predetermined candlepower distribution curve.

The basic steps in one method of determining reflector contour are as follows:[6]

1. Select the width, in degrees, of zones to be considered. From the required candlepower distribution curve (polar curve), calculate the luminous flux required in each zone (candlepower × zonal constant for each zone). See Fig. 4–13, Section 4. Tabulate.

2. Calculate the luminous flux emitted by the bare lamp in each zone (using the polar curve of the bare lamp and zonal constants), and tabulate.

3. Find the reflected lumens needed in each zone by subtracting the bare lamp lumens (data from

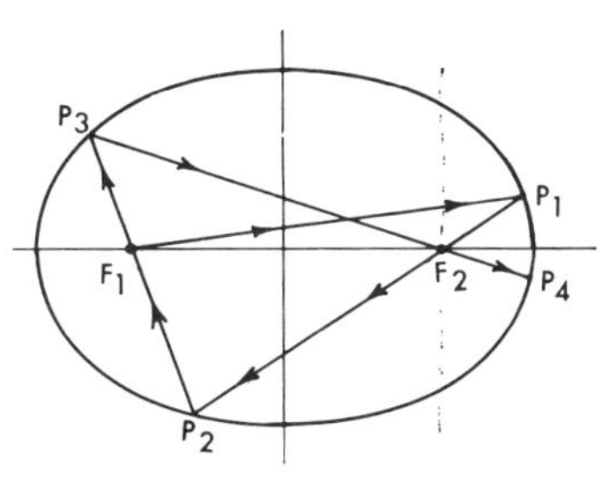

Fig. 6–26. The action of a perfect ellipsoidal mirror.

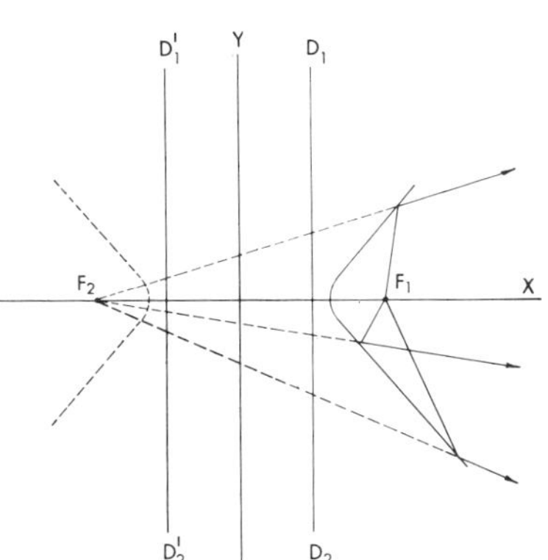

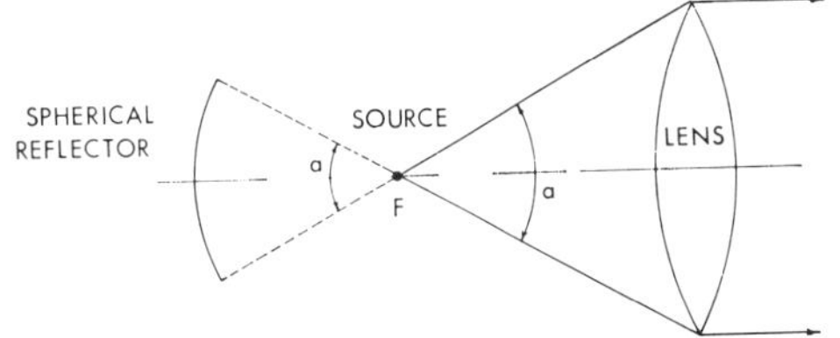

Fig. 6–28. Projecting device with spherical reflector.

Fig. 6–27 (Left). Hyperbolic reflector action.

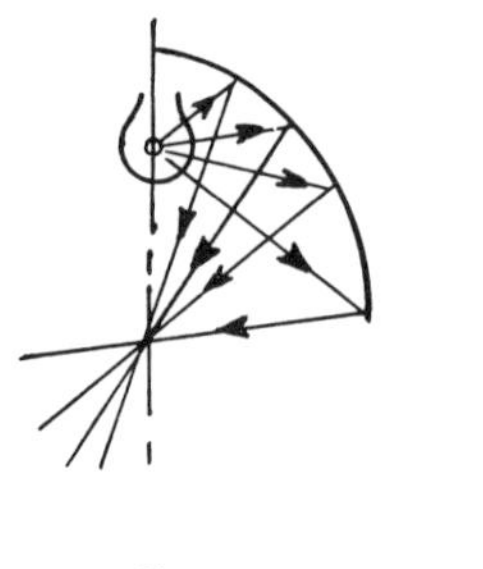

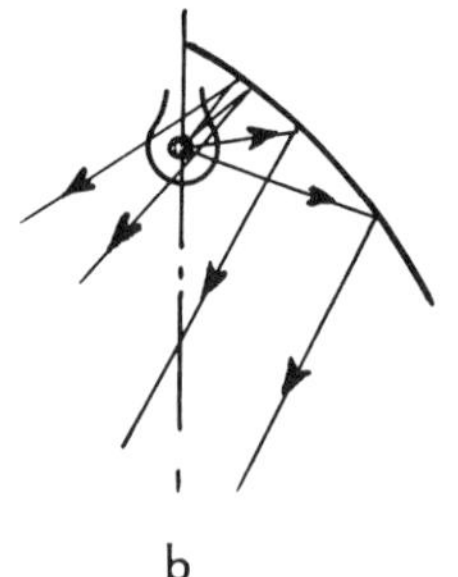

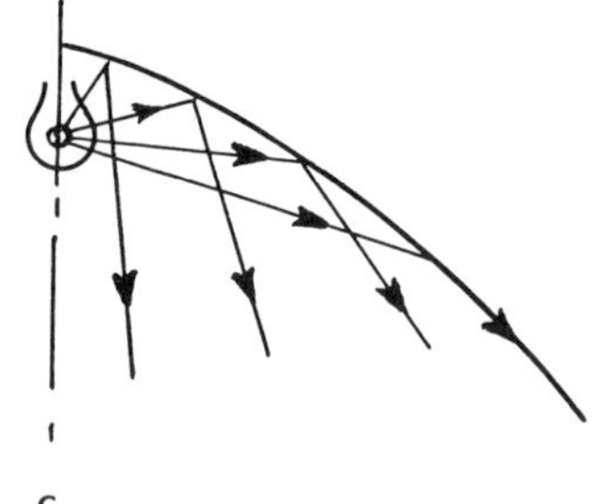

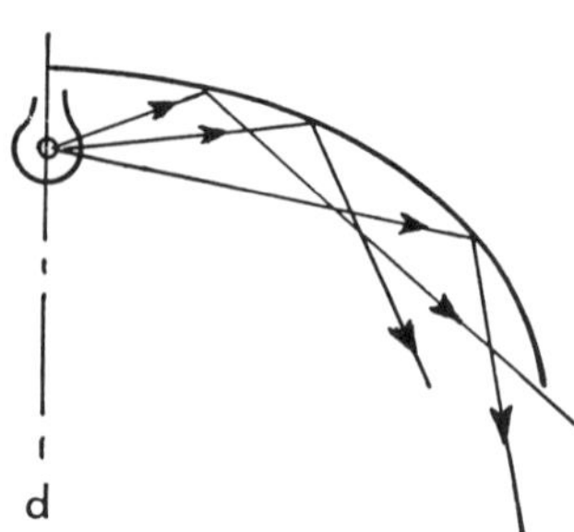

Fig. 6–29. Four basic reflector actions.

Step 2) from the required luminous flux (data from Step 1) in those zones where no reflection will take place, *i.e.*, from nadir up to cut-off. Tabulate.

4. Decide upon the general action of the reflector. There are, in general, four basic actions of the reflectors as shown in Fig. 6–29. For a given cut-off, the forms shown in Fig. 6–29 (c and d) usually require a very large reflector. The form Fig. 6–29b has the disadvantage that much of the light is passed through the lamp bulb. For most cases, the form Fig. 6–29a results in the smallest reflector and redirects the least light back through the lamp.

5. Plot a curve of reflected flux obtained in Step 3. Starting at 0 degrees, show cumulative sums of lumens required to be reflected into each zone from nadir up to cut-off, see Fig. 6–30. Similarly, plot a curve of available lamp flux (from Step 2 data). Starting at cut-off, show how many lumens are incident on the reflector progressively from cut-off to 180 degrees (Fig. 6–30b). Since all flux considered here must be reflected from the reflector surface, the available lamp lumens must first be multiplied by the reflectance of the surface and by other loss factors. It is convenient to work with rectangular coordinates, plotting the lumens along the horizontal axis and degrees along the vertical axis. Plot the reflected flux curve (Fig. 6–30a) to the same scale and directly below the available lamp flux (bare lamp flux corrected for losses) curve. If the reflector action illustrated in Fig. 6–29 (b or d) was selected, the horizontal (lumen) scales of the two curves will be in the same direction. If Fig. 6–29 (a and c) was selected, the scales will be reversed as shown in Fig. 6–30. Take intercepts at intervals (one for each zone) along the reflected flux curve (Fig. 6–30a) and project upward to the available lamp flux curve as shown in Fig. 6–30. At the point where each intercept cuts the available lamp flux curve (Fig. 6–30b), project horizontally to degrees on the vertical axis. The spacings between intercepts on the vertical axis indicate how large an angular, zonal segment of the reflector is required to direct enough light at a particular angle to satisfy the requirements of the reflected flux curve.

6. Lay out the reflector contour by starting at the point nearest the lamp and progressively drawing small, straight segments (one segment for each zone). Each segment must be of the required length, and placed in the correct position for the flux striking it to be reflected in the amount and direction determined in Step 5. Obtain the final contour by drawing a smooth curve tangent to the segments.

The procedure outlined above assumes a point source and true specular reflection. The effect of a normal filament would be to smooth out the final distribution, rounding off sharp points. The effect will be more noticeable as the reflector gets smaller in comparison with the source.

Departures from true specularity will have a similar effect which will increase as the proportion of diffuse to total reflection increases. With semi-specular surfaces, the effect of specularity of the surface increases as the angle of incident light increases. Hence, for accurate control of the light, it is well to keep the angle of incidence as large as possible.

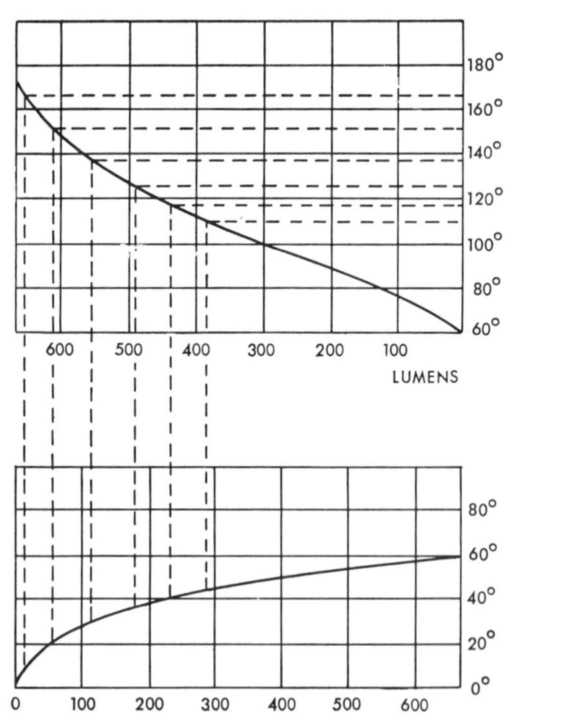

Fig. 6–30. Reflection plan of available lamp lumens (shown for 60-degree cut-off). a. Reflected flux curve. b. Incident or available flux curve (corrected for losses).

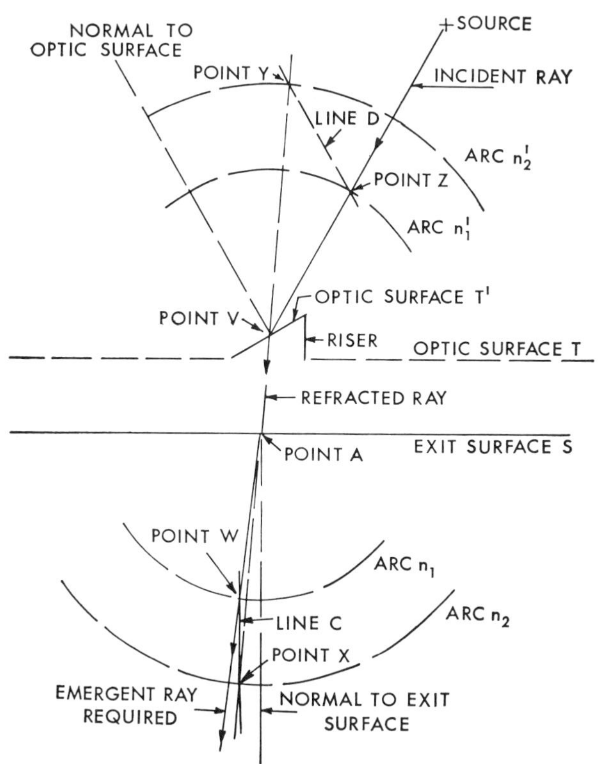

Fig. 6-31. Design of optic to redirect ray.

DESIGN OF REFRACTORS[1]

Optic Surface to Redirect Light Rays

A common problem, illustrated in Fig. 6-30, is the determination of an optic surface to cause the eventual refraction of a light ray from a source to a required direction. Initially, five elements of this problem are known:

1. Exit surface S, which is the side of the lens away from the source. This surface could be flat, spherical, cylindrical, or irregularly shaped; the basic fundamentals being the same.
2. Optic surface T to which an optical configuration must be designed in order to create the required distribution of light rays. Surface T is normally parallel to exit surface S.
3. The location of the light source.
4. The incident ray from the source to the lens.
5. The required emergent ray. The angle θ of the optic surface T' which will change the incident ray to the direction of the emergent ray is the unknown quantity in this problem.

The first step is to draw all the known components of the problem: exit surface S, optic surface T, incident ray, emergent ray required, and the source. Note that the surface T is a dashed line indicating that at the end of the complete design this surface will no longer be smooth, but will be configurated with optic elements.

To point A, the intersection of the emergent ray and exit surface S, draw the normal to exit surface S; with point A as center strike arc n_1 with radius proportional to n_1, the refractive index of the outside medium (probably air where $n_1 = 1$). See Refraction, page 6-3. Again with a point A as center, strike arc n_2 with radius proportional to the refractive index of the optical medium. Though point W, the intersection of arc n_1 and the emergent ray required, draw line C parallel to the normal to exit surface S. Through the intersection of arc n_2 and line C, which is indicated as point X, draw the construction line to point A on exit surface S. Extended, this becomes a refracted ray through the optic surface T. Further extended, it intersects arc n_2' at point Y. The radius of arc n_2' is proportional to the refractive index of the optical medium. Note that the center of this arc, point V, lies on the optic surface which has not yet been established. This fact indicates that a few trials are necessary for this solution. The initial steps are repeated until point V coincides with the intersection of the incident ray and optic surface T'. With point V as center, strike arc n_1', whose radius is proportional to the refractive index of the outside medium. Draw line D between point Y and point Z, the intersection of arc n_1' and the incident ray. The normal to the optic surface T' is then drawn parallel to line D and intersecting point V. The optic surface T' is then a line perpendicular to this normal.

All rays at an internal angle to the normal greater than critical angle ($\sin \theta = n_1/n_2$) are internally reflected from the inside of the surface.

Refraction-Reflection System

Total internal reflection is used extensively in lens design when rays must be bent at an angle larger than is possible to accomplish with refraction alone. Total internal reflection problems can be solved graphically by methods similar to those used in solving refraction problems.

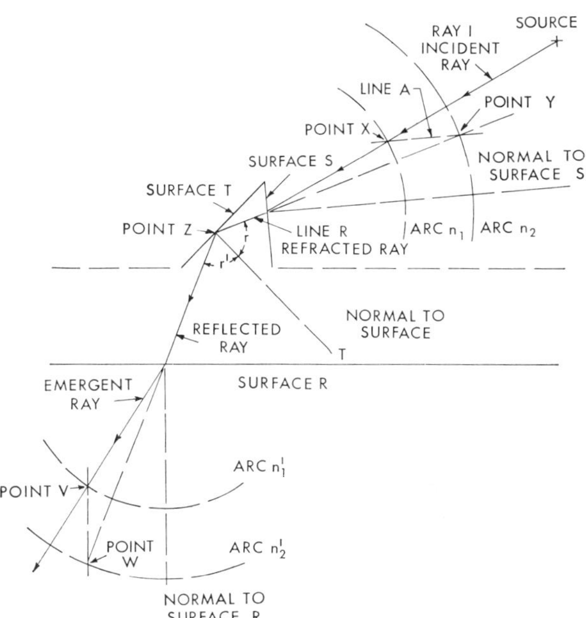

Fig. 6-32. Path of a light ray through a refraction-reflection system.

Fig. 6–32 illustrates the solution of a typical problem involving both refraction and total internal reflection. A ray of light from the source passes through the system in the following manner:

1. Ray 1 is refracted at surface S.
2. Refracted ray becomes incident to surface T at an angle greater than the critical angle of the medium.
3. Total internal reflection occurs at surface T so that $r' = r$.
4. Reflected ray becomes incident on surface R.
5. Ray is refracted at surface R and emerges from system.

The graphical method of determining the refracted rays has been described in the previous section. The graphical method of determining the reflected ray is accomplished by constructing the angle of reflected ray to normal equal to the angle of the incident ray to normal.

MATERIALS USED IN LIGHT CONTROL[8]

The materials most commonly used in light control are glass, plastics, metals, and applied finishes and coatings. Each of these materials is briefly described here along with comments on manufacturing processes and techniques. The use of these materials in lighting as well as some general information to aid in the selection and evaluation of the most suitable light control materials is also discussed in this section. For more specific applications the material manufacturer's data should always be consulted.

Glass

Glass is an inorganic product of fusion which has been cooled to a rigid condition without crystallizing. Chemically, glasses are mixtures of oxides. The oxides most commonly used in glass composition are silicon, boron, aluminum, lead, sodium, magnesium, calcium, and potassium.

The chemical and physical properties of glasses, such as color, refractive index, thermal expansion, hardness, corrosion resistance, dielectric strength, and elasticity, are obtained by varying composition, heat treatment, and surface finish.

Glasses which are important to illumination can be classified into several groups with characteristic properties.

Soda-Lime. Soda-lime glasses (or lime glasses), used for window glass, lamp envelopes, lens covers and cover glasses (tempered), etc., are easily hot-worked, and are usually specified for service where high heat resistance and chemical stability are not required.

Lead-Alkali. Lead-alkali glasses (or lead glasses) are used for electric light bulb stems, neon sign tubing, and certain optical components. They are useful because of their good hot-workability, high electrical resistivity, and high refractive indices. They will not withstand high temperatures or sudden temperature changes.

Borosilicate. Borosilicate glasses are used for refractors, reflectors, lenses, sealed-beam lamp parts, etc. because of their high chemical stability, high heat shock resistance, and excellent electrical resistivity. They may be used at higher temperatures (about 230° C) than soda-lime or lead glasses. They are not as convenient to fabricate as either soda-lime or lead-alkali glasses.

Aluminosilicate. Aluminosilicate glasses are used where high thermal shock resistance is required. They have good chemical stability, high electrical resistivity, and a high softening temperature enabling use at moderately high temperatures (about 400°C).

Ninety-Six Per Cent Silica. Ninety-six per cent silica glasses are used where high operating temperatures are required. They may be regularly used at 800°C. They are also useful because of their extremely high chemical stability, good transmittance to ultraviolet and infrared radiation, and resistance to severe thermal shock. They are considerably more expensive than soda-lime and borosilicates, and are more difficult to fabricate. Due to their low coefficient of thermal expansion, these glasses cannot be tempered to increase mechanical strength.

Vitreous Silica. Vitreous silica is a glass composed essentially of SiO_2. It is used for lamp envelopes where high temperature operation and excellent chemical stability are required. It has high resistance to severe thermal shock, high transmission to ultraviolet, visible, and infrared radiation, and excellent electrical properties. However, due to its low coefficient of thermal expansion, it cannot be tempered to increase mechanical strength. Depending on the method of manufacture, this glass may be known as fused silica, synthetic fused silica, or fused quartz.

Manufacturing Techniques. Glass may be formed by several techniques: pressing, blowing, rolling, drawing, centrifugal casting, and sagging. These operations may be performed by hand methods or by automatic machines, depending on volumes.

After being formed, many glass products must be annealed to relieve excessive stress and strain. Also additional finishing operations are often required. Finishing methods include cutting, grinding, polishing, and drilling. Further treatment, such as tempering or chemical strengthening, may be required to obtain the desired physical properties.

The finishes for glass surfaces are varied in nature, depending on the forming technique. The surfaces may be subsequently altered by chemical etching, sandblasting or shot blasting, polishing, staining, and coating. These operations are used to obtain reflection, control radiation, or to make a surface electrically conductive.

The function of glass in lighting may be divided into the following general categories:

1. Control of light and other radiant energy.
2. Protection of the light source.
3. Safety.
4. Decoration.

These functions may be combined for a particular application. Typical properties of glasses are shown in Fig. 6–33.

Plastics

Plastics generally are high molecular weight organic compounds that can be, or have been, changed by application of heat and pressure, or by pressure alone, and once formed retain their shape under normal conditions.

They can be broadly classified as thermoplastic or thermosetting. Thermoplastic resins may be repeatedly softened and hardened by heating and cooling. No chemical change takes place during such actions. Thermosetting resins cannot be softened and reshaped once they have been heated and set since their chemical structure has changed. Some of the commercially important thermoplastics are: acrylonitrile butadiene styrene (ABS), acrylic, cellulosics, acetal, fluorocarbon, nylon, polyethylene, polycarbonate, polypropylene, polystyrene, and vinyl. In the thermosetting group resins of importance are epoxy, melamine, phenolic, polyester, urea, and silicone.

Most resins, whether thermoplastic or thermosetting, can be processed into structural or low ratio expanded foams. Among the important properties of most foams are: stress free parts, improved insulation characteristics, lighter weight, and greater strength and toughness than the unfoamed form of the material.

Fillers and reinforcing agents are frequently added to plastics to obtain improved heat resistance, strength, toughness, electrical properties, chemical resistance, and to alter formability characteristics. Some of the fillers and reinforcements in general use are: aluminum powder, asbestos, calcium carbonate, clay, cotton fibers, fibrous glass, graphite, nylon, powdered metals, and wood flour.

The basic compounds from which today's plastics are produced are obtained from such sources as air, water, natural gas, petroleum, coal, and salt. The involved chemicals are reacted in large closed vessels under controlled heat and pressure with the aid of catalysts. The resultant solid product is then subjected to such further operations as: reduction of particle size; addition of fillers, softeners, and modifiers; and conversion of form to granules, pellets, film, etc.

To protect plastic surfaces from ultraviolet energy, a film compounded from a material resistant to this energy is sometimes laminated to the surface facing the energy source.

The forming and converting of plastics into end products is a highly specialized field. Some of the more important processes are: injection molding (fluid plastic forced under pressure into a controlled temperature mold); compression molding (resin placed in mold and the cavity filled by application of heat and pressure); blow molding (a thin cylinder of plastic is placed in a mold and inflated by air pressure to conformity with the mold cavity); extrusion (fluid plastic is screw driven through a die); thermoforming (a limp sheet is draped over a mold and forced into close conformity with the mold by pressure or vacuum); spray coating (a solution or emulsion is sprayed on a prepared surface); and machining (solid plastic is shaped by basic wood or metal working operations); rotational molding (resin is charged into a mold rotated in an oven, centrifugal forces distribute the resin, and the heat melts and fuses the charge to the shape of the cavity); cold stamping (stamping and/or forming of cold plastic sheet); ultrasonic welding (bonding of plastics by conversion of sonic vibrations to heat).

Steel

In the fabrication of lighting equipment, steel serves primarily in a structural capacity. Sheet steel, while having the greater strength and lower cost needed for a large volume material, must be processed additionally with platings or applied coatings before it can serve as a light controlling media. Many grades and types of sheet steel are available and should be selected for the proper use. In certain instances, other forms of steel are used in lighting equipment and many of these sizes of bars, rods, etc., are available readily from steel warehouses.

Sheet steel used in lighting equipment is of three basic types: *hot rolled steel*, *cold rolled steel*, and *porcelain enameling sheets*.

Hot rolled steel, because of the rolling process at elevated temperatures, carries an oxide coating or mill scale. It is not normally used where a smooth appearance is desired. The scale can be removed in an acid bath (pickling) but the surface still is somewhat rough.

Cold rolled steel is used primarily because of its smooth surface appearance. It is also available in thinner gauges than hot rolled steel. It is usually obtained by pickling hot rolled steel coil and further reducing the thickness without elevating the temperatures.

Porcelain enameling sheets are similar to hot rolled steel or cold rolled steel, but have very low carbon content. The low carbon content is required, because porcelain is normally fused at temperatures which would cause cold rolled steel to mill scale beneath the porcelain coating, creating an unsatisfactory finish.

Each of these steels is available from the mill in several grades. The requirements of a material for spinning are different than that of a material which will be deep drawn or simply formed. Therefore, grades are designed for each of these requirements.

Steel sheet is also available with any of a number of different finish treatments already applied.

Fig. 6–33. Properties of Lighting Glasses

Type of Glass	Color[a]	Coefficient of Thermal Expansion per °C[b] (× 10^{-7})	Upper Working Temperatures °C(°F) (Mechanical Considerations Only)				Thermal Shock[e, f] Resistance Plates 6 in x 6 in ¼ in thick	Impact Abrasion Resistance[g]	Density (grams per cc)	Young's Modulus (10^6 lb/in²)	Poisson's Ratio[h]	Refractive Index (589.3 nm)
			Annealed		Tempered							
			Normal Service [c, e]	Extreme imit [d, e]	Normal Service[c, e]	Extreme Limit[d, e]						
Soda-Lime	Clear	85–97	110 (230)	430–460 (806–860)	200–240 (392–464)	250 (482)	50	1.0–1.2	2.47–2.49	10–10.2	.24	1.512–1.514
Lead-alkali	Clear	85–91	110 (230)	370–400 (698–752)	220 (428)	240 (464)	45	.56	2.85–3.05	8.7–8.9	.22	1.534–1.56
Borosilicate	Clear	34–52	230 (446)	460–490 (860–914)	250–260 (482–500)	250–290 (482–554)	100–150	3.1–3.2	2.13–2.43	9.3–9.5	.20	1.474–1.488
Aluminosilicate	Clear	32–46	200 (392)	650 (1202)	400 (752)	450 (842)	115	2.0	2.43–2.64	12.5–12.7	.25	1.524–1.547
96% Silica	Clear	8	800–900 (1472–1652)	1090–1200 (1994–2192)			1000	3.5–3.53	2.18	9.6	.18	1.458
Vitreous Silica	Clear	5.5	1000 (1832)	1200 (2192)			1000	3.6	2.2	10.4	.16	1.458

[a] All glasses can be colored by the addition of metallic oxides that become suspended or dissolved in the parent glass usually without substantially changing its chemical composition or physical properties.
[b] From 0° to 300°C, in/in/°C, or cm/cm/°C.
[c] Normal Service: no breakage from excessive thermal shock is assumed.
[d] Extreme Limits: depends on the atmosphere in which the material operates. Glass will be very vulnerable to thermal shock; tests should be made before adopting final designs.
[e] These data approximate only.
[f] Based on plunging sample into cold water after oven heating. Resistance of 100°C means no breakage if heated to 110°C and plunged into water at 10°C. Tempered samples have over twice the resistance of annealed glass.
[g] Data show relative resistance to sandblasting.
[h] Value applies to only one glass of group.

Fig. 6–34. Lighting Uses of Important Resins

Resin	Uses
Thermoplastics	
Acrylic	louvers, formed light diffusers, prismatic lenses, diffusers, film
Cellulosics	sign faces, vacuum formed diffusers, globes, light shades, light supports
Flexible vinyl	gaskets, wire coating
Nylon	electro-mechanical parts, wire insulation, coil forms
Polycarbonate	insulators, globes, diffusers, anti-vandalism street lighting globes
Polyethylene	(high density)—wire coatings, housings
Polyethylene	(low density)—formed light diffusers, blow molded globes
Polystyrene	same as acrylic
Rigid vinyl	formed lighting diffusers, corrugated sheet for luminous ceilings
Thermosetting	
Melamine	switches, insulators
Phenolic	wire connectors, switches, sockets, shades
Polyester	(glass reinforcing)—shades, reflector housings, diffusers
Urea	louvers, lamp holders, shades
Filled reinforced plastics	insulators, reflectors, housings, globes, light shades, switch bases

These types of finish include: galvanized sheet, pre-painted sheet, aluminum clad sheet, and plastic coated sheet. Pre-plated sheets are also available including chrome, brass, and copper.

Steel sheet used in lighting is often referred to only in terms of gauge thickness. The metal gauge most commonly used is the Manufacturer's Standard Gauge for Steel Sheets.

Aluminum

Aluminum is a non-ferrous, corrosion-resistant, lightweight, non-magnetic metal having good thermal and electrical conductivity. It is high in the electro-chemical series and resists attacks by either air or water because of the formation of an invisible protective covering of aluminum oxide.

Uses in Lighting. Aluminum is used in lighting for: structural parts such as tubes or poles, fitters, holders, housings, channels, hardware, mechanical parts, and trim; and light controlling surfaces such as reflectors, louvers, baffles, and decorative surfaces.

Aluminum is also used as a reflecting surface when vaporized on glass and plastic, and as a paint when in fine powder form and suspended in a suitable liquid vehicle.

An aluminum reflector can have a high-permanence, high reflectance, diffuse or semi-diffuse surface of graduated brightness; the value of the brightness depending on the intensity and angle at which the light reaches the surface and the angle at which the surface is viewed.

Types of Aluminum. Aluminum is used in its near-pure state or may be alloyed by the addition of other elements to improve its mechanical, physical, and chemical properties. Silicon, iron, copper, manganese, magnesium, chromium, nickel, and zinc are the most common elements used. Aluminum alloys may be cast, extruded, and rolled as shapes or sheets. In sheet form, aluminum is available as reflector, homogeneous, clad, and lighting.

Finishes. The final finish on aluminum parts will depend on service requirements but structural members often require only cleaning. Aluminum may be etched, polished, brushed, plated, anodized, color anodized, brightened, plastic coated with or without vaporization, coated with clear or dye lacquers, finished with baked or porcelain enamel, or some combination of these finishes. Reflector finishes may range from diffuse, such as baked enamel or etched surfaces, to highly specular such as of polished and anodic coated. Aluminum paint, made of aluminum as a fine powder suspended in a suitable liquid vehicle, has found wide use as an attractive and practical finish for many surfaces.

Processing. Anodizing, an electrochemical process, is used to form a protective surface of aluminum oxide of thickness greater than 100 times that formed naturally in air. The aluminum oxide surface is smooth, continuous, inseparable, and has a particle hardness with a Mohs' value of 9. Anodizing combined with chemical or electrochemical brightening provides surface finishes of uniformly high reflectance and permanence.

A high purity aluminum must be used if a clear, colorless transparent, high reflectance oxide surface is required. Impurities and alloying materials will result in lower reflectance, cloudiness, dullness, or streaking of the oxide surface.

In the anodizing process, colored surfaces may be obtained by depositing dyes or pigments within the open pores of the aluminum oxide just before the final sealing of the surface. A wide range of colors and tints are available.

Physical Characteristics. Properties for several alloys of aluminum that may be of interest to the designer are shown in Fig. 6–36. The types and values shown are intended as typical illustrations. New alloys are being developed frequently, and any contemporary listing becomes rapidly out-of-date.

Other Metals

Stainless Steel. Stainless steel includes those iron-base alloys which contain sufficient chromium to render them corrosion resistant. The classification "stainless" is usually reserved for those steels having 12 to 30 per cent chromium; those with

Fig. 6-35. Properties of Plastics Used in Lighting

Materials	Common Forms										Reasons for Use and General Data										
	Castings	Compression Moldings	Extrusions	Fiber	Film	Foam	Injection Moldings	Sheet	Reinforced Plastic Moldings	Industrial Laminates	Chemical Resistance	Colorability	Flammability*	Flexibility	High Dielectrics	Low Moisture Absorption	Clarity	Strength and Rigidity	Toughness	Effect of Ultraviolet†	Resistance to Heat °F Continuous
Thermoplastics																					
ABS‡			X				X	X			X	X	S		X			X	X	NP	140–230
Acetal‡			X				X				X	X	S		X			X	X	C	180
Acrylics	X		X	X			X	X			X	X	S				X	X		N	140–190
Acrylic-styrene copolymer			X				X	X			X	X	S		X	X	X	X		NP-SL	180–200
Cellulose acetate			X	X	X		X	X				X	S-SX		X		X	X	X	SL	150–220
Cellulose acetate butyrate			X		X		X	X				X	S		X	X	X	X	X	NL	140–220
Cellulose propionate			X		X		X	X				X	S		X	X	X	X	X	SL	155–220
TFE-fluorocarbon			X		X			X			X		N		X	X				N	500
Nylon‡			X	X	X		X	X			X		SX		X			X	X	CO	180–300
Polycarbonate‡			X				X				X	X	SX			X		X	X	SL	250
Polyethylene			X	X	X	X	X	X			X	X	VS	X	X	X		X	X	NP	180–250
Polypropylene‡			X	X	X		X	X			X	X	S		X	X		X	X	NP	250–320
Polystyrene‡	X		X	X	X		X	X			X	X	S		X		X	X	X§	NP	170–200
Styrene acrylonitrile copolymer‡			X	X			X	X			X	X	S		X		X	X		NP	180–200
Vinyl	X		X	X	X	X	X	X			X	X	SX	X	X	X		X	X	SL	150–200
Polysulfone			X			X	X				X	X	SX		X			X	X		300–345
Thermosetting Plastics																					
Epoxy	X					X			X	X	X	X	S-N		X	X		X		SL	250–550
Melamine		X							X	X	X	X	SX		X	X		X	X	SL	210
Phenolic	X	X				X			X	X	X		VS		X			X	X	D	250
Polyester (other than molding compounds)	X			X	X				X	X	X	X	S-N		X		X		X	SL	300–410
Polyester		X							X	X			S-N		X	X				SL	300–350
Silicone	X	X				X			X	X			S-N		X	X				SL-N	600
Urea		X									X	X	SL							G	170

* S—slow, SX—self extinguishing, N—none, VS—very slow.
† NP—needs protection, C—chalks, N—none, SL—slight, NL—nil, CO—colors, D—darkens, G—grays.
‡ Available in glass filled forms which, in general, yield parts with improved toughness, higher rigidity, significantly higher heat resistance and slower burning rates.
§ Rubber modified form.

Fig. 6–36. Typical Physical Properties of Aluminum

Type	Alloy	Federal Specification Number	Average Coefficient of Thermal Expansion*	Specific Gravity	Weight (lb/in³)	Thermal Conductivity at 25° C (CGS units)	Reflectance (per cent)
Specular, processed sheet	#12 Reflector sheet		13.1	2.71	0.098	0.53	80–85
Diffuse, processed sheet	#31 Reflector sheet		13.1	2.71	0.098	0.53	75–80
Mill finish sheet	#1100-H14	QQ-A-561c	13.1	2.71	0.098	0.53	70
Extruded	#6061-T4	QQ-A-270a	13.0	2.7	0.098	0.37	
Extruded	#6063-T4	QQ-A-274	13.0	2.7	0.098	0.46	
Extruded	#6463-T4		13.0	2.7	0.098	0.52	
Cast, sand, or permanent	#43-F	QQ-A-371c	12.3	2.69	0.097	0.34	
Cast, sand, or permanent	#214-F	QQ-A-371c	12.4	2.89	0.104	0.29	
Cast, sand (heat treat)	#220-T4	QQ-A-371c	13.7	2.57	0.093	0.29	
Cast, die	#360	QQ-A-591a-2	11.6	2.64	0.095	0.27	
Cast, die	#380	QQ-A-591a-2	11.6	2.72	0.098	0.23	

* in/°F × 10^{-6}, 68° to 212°.

greater than 30 per cent are classed as heat-resisting alloys and not as stainless.

The family of stainless steels may be divided into three main groups:

Straight Chrome Group. The steels in this group are all magnetic. They may exhibit characteristics of rusting on exposure to corrosive atmospheres; however, the rusting is only a superficial film and acts as a barrier to further corrosive action. These are identified as AISI Type 400 Series.

Chrome-Nickel Group. These steels are non-magnetic. They do not exhibit the characteristics of rusting, but some alloys may not be satisfactory in certain corrosive atmospheres. These steels are designated as AISI Type 300 Series.

Chrome-Nickel-Manganese Group. These steels are non-magnetic. In this group, manganese substitutes for part of the nickel and this type of stainless was developed during the nickel shortage as a substitute for the Chrome-Nickel Group. These steels do not exhibit the characteristics of rusting, but some alloys may not be satisfactory in certain corrosive atmospheres. These steels are designated as AISI Type 200 Series.

Stainless steels are widely used in luminaires intended for installation outdoors or in other corrosive atmospheres. Some applications of stainless steel are for reflectors, housings, springs, latches, mounting straps, hinges, fittings, fasteners, and lampholder screw shells.

Copper. Copper is used extensively for the conductors, bus bars, and associated switchgear necessary for the distribution and control of electrical energy used for lighting. Copper is ductile, malleable, flexible, fairly strong, and may be formed by a variety of standard machines and processes.

Non-Ferrous Alloys. Bronze, an alloy of copper and tin, and brass, an alloy of copper and zinc, are often used in specialty luminaires where appearance of the attractive color is a prime consideration. A more utilitarian use of bronze and brass is in luminaries for marine use where strength and resistance to salt water corrosion are highly important.

Chromium copper or beryllium copper are often used for conducting springs, contacts, and similar highly stressed members that have to be formed in manufacture. Parts are shaped soft and then strengthened by heat treatment.

FINISHES USED IN LUMINAIRES[15]

A finish is the final treatment given to the surface of a material in the course of manufacture to render it ready for use. Three major purposes for finishes on lighting equipment are: the control of light, the protection of material, and the enhancement of appearance. In addition, there are several special applications such as flame retardant and color stabilizing treatments.

Types of Finishes

Finishes are classified both by the method of application and by the kind of material applied. Three basic types are coatings, laminates, and chemical conversion finishes.

Coatings can be separated into three general classes as organic, ceramic, and metallic.

Organic Coatings.

Lacquers may be clear, transparent, or opaque and will cure rapidly at room temperatures. They may be used for decoration or protection.

Enamels are pigmented coatings and are applied for protection, decoration, or reflectance. They cure by oxidation (air or forced drying) or polymerization (baking or catalytic action) and result in tougher finishes than lacquers.

Baked clear coatings, sometimes called baking laquers, are used for decoration and protection.

Fig. 6-37. Properties of Finishes

Type of Finish	Method of Application[a]	Principal Uses[b]	Colors Possible	Character of Reflected Light	Per Cent Reflectance[d]	Resistance[c]				Stability[c]	Flammability
						Heat	Corrosion	Abrasion	Impact		
Organic coatings											
Lacquers	D, B, S	A, P	Colorless or any color	Mixed to diffuse	10–90	F	F	P	F	F	Slow burn
Emulsions	D, B, S	A, P	All colors	Mixed to diffuse	10–90	G	G	G	G	G	Slow burn
Enamels	D, B, S	A, P, R	All colors	Mixed to diffuse	10–90	G	G	G	G	G	Slow burn
Baked clear coatings	D, B, S	A, P	Colorless, clear color	Diffuse to specular	0	G	G	G	G	G	Slow burn
Organisols	D, S	A, P	All colors	Mixed to diffuse	10–90	F	E	G	G	F	None
Ceramic coatings											
Vitreous enamels	D, S	A, P, R	All colors	Diffuse to specular	10–90	E	E	E	P	E	None
Ceramic enamels	D, S, B	A, R	All colors	Mixed to specular	10–90	E	E	E	P	E	None
Metallic coatings											
Chrome plate	Electrochemical	A, P	Fixed; depending on color of plated metal	Specular to diffuse	60–88	E	E	E	E	E	None
Nickel plate	Electrochemical	A, P		Specular to diffuse	55	E	G	E	E	E	None
Cadmium plate	Electrochemical	P		Specular to diffuse	85	G	G	F	P	E	None
Brass plate	Electrochemical	A		Specular to diffuse	55–80	P	P	F	P	F	None
Silver plate	Electrochemical	A, R		Specular	85–95	P	P	F	P	E	None
Laminates	Laminate	A, P, R	All colors of metallic effects	Mixed	10–90	Depends on nature of laminate					Slow burn
Conversion coatings											
Anodized aluminum	Electrochemical	A, P, R	Natural aluminum (or a wide variety of colors)	Diffuse to specular	60–90	E	E	E	E	E	None
Vacuum deposition	Vacuum chamber	A, R	Natural aluminum (or a wide variety of colors)	Specular	10–70	Depends on nature of protective coating					None

[a] D—dip, S—spray, B—brush.
[b] A—appearance, P—protection, R—reflectance.
[c] P—poor, F—fair, G—good, E—excellent.
[d] Depends upon color.

They cure by polymerization (baking or catalytic action).

Organisols and plastisols are usually applied by dipping and spraying. These plastic dispersion coatings offer good exterior corrosion as well as scratch and abrasion resistance and are also able to conceal many surface defects.

Ceramic Coatings. Ceramic coatings, including porcelain enamels, are fired on glass and metals at temperatures in excess of 1000° F. Primary features include high resistance to corrosion, good reflectance, and easy maintenance on metals; and reduction of brightness and increase of diffusion on glass.

Metallic Coatings.

Electrochemical deposition, commonly called electroplating, causes a second metal to be deposited over the first by means of an electrolytic action. Zinc, cadmium, or nickel is used to provide protection. Brass and silver plated finishes are primarily decorative.

Vacuum deposition, also called vacuum metallizing, consists of vaporizing a metal, usually aluminum, in a vacuum chamber and depositing it on surfaces of plastic, glass, or metal. Finishes of high specular reflectance are obtained and can be used for either light control or decorative purposes.

Dip and spray coatings, such as galvanizing and sherardizing, deposit a second metal to protect the base metal against corrosion.

Laminates. This type of finish is created by bonding a thin layer to a base material, such as a plastic film to sheet metal. The laminate can be a decorative material or it can be a light-controlling material.

Chemical Conversion Finishes. Anodizing converts an aluminum surface by an anodic process to aluminum oxide which has outstanding protective qualities against corrosion and abrasion. The resultant finish may be clear or can be dyed in a variety of colors.

Typical characteristics of finishes are indicated in Fig. 6–37. Because of the great number of possible variations in composition and application of all types of finishes, numerical values are not shown and relative gradings only are used. For more details on finishes, technical assistance should be obtained from suppliers and available literature.

PRACTICAL CONCEPTS OF EQUIPMENT DESIGN[18]

Some of the factors that must be considered in luminaire design include: (1) codes and standard practices relative to the construction and installation of luminaires, (2) physical and environmental characteristics of luminaires, (3) electrical and mechanical considerations that affect luminaire design, (4) thermal considerations relative to the light source, the environment, and the luminaire, (5) safety considerations as they apply to all areas of luminaire design, and (6) economic factors which may affect the design of luminaires.

Codes and Standard Practices

Local codes, national codes, international codes, federal standards, professional standards, and manufacturers' standards relate to specific requirements which must be met in the construction and installation of a luminaire. Standards usually relate to minimum requirements (safety, construction or performance) which may be exceeded to provide a better product.

Some codes and standards deal with fire and safety (electrical, mechanical and thermal); others relate to performance (photometric) and construction (materials and finishes). They will vary to some extent depending on geographic location and end use of the equipment. Conformance to the appropriate set of specifications is often determined by certified laboratory tests. Certification is often denoted by an identifying label. Local inspection agencies may or may not rely on conformance to national, federal or industrial codes and standards.

Local Codes. Local codes are normally, but not always, patterned after national codes. Information regarding local codes may be obtained from electrical inspection departments. Several other code jurisdictions may apply to regional and state codes. Building Officials Conference of America (BOCA), Southern Building Officials Conference (SBOC), International Committee of Building Officials (ICBO), all promulgate codes applicable to lighting installations. ICBO issues the Uniform Building Code (UBC), which is widely used on the west coast and elsewhere, to control the use of *materials* in luminaires and luminous ceilings. These requirements are usually legally binding on the contractor.

National Codes. The NEC (National Electrical Code), the CEC (Canadian Electrical Code), and similar ones in most major countries throughout the world, state specific electrical requirements which must be met by all electrical equipment, including luminaires. They have been developed by safety protection and inspection agencies in conjunction with fire protection agencies. Although based on good manufacturing practices, these codes may vary considerably in many respects. The National Electrical Code may or may not be accepted in total by local agencies throughout the United States.

National and International Standards. The UL (Underwriters' Laboratories, Inc.), the CSA (Canadian Standards Association), and similar ones in other countries publish minimum safety standards for electrical and associated products which are in conformance with the respective electrical codes of their country. They have testing laboratories to which equipment must be submitted for listing. Most manufacturers design luminaires to meet these standards.

Industry Standards. Industry standards are published by professional societies, associations and institutes, in most cases, utilizing national technical committees. Represented on these committees are inspection and safety protection agencies, manufacturers, professionals, and consumers. Conformance to such standards may be desirable or specified, but are otherwise not binding on the designer. Such organizations include:

ASTM	(American Society for Testing Materials)
CBM	(Certified Ballast Manufacturers)
IEEE	(Institute of Electrical and Electronic Engineers)
IES	(Illuminating Engineering Society)
LEMA	(Lighting Equipment Manufacturers' Association)
NEMA	(National Electrical Manufacturers' Association)
ANSI	(American National Standards Institute)

Federal Standards (U. S. Government). Standards written by the following and similar federal agencies are almost exclusively for their own requirements. They are rarely specified for non-federal commercial applications.

BU Ships	(Bureau of Ships)
CAB	(Civil Aeronautics Board)
FAA	(Federal Aviation Administration)
GSA	(General Services Administration)
USCG	(U.S. Coast Guard)
CE	(Corps of Engineers)

Manufacturers' Standards. Since codes and standard practices deal primarily with safety and performance but not with quality, the specifier should be aware of the quality standards practiced by the manufacturer.

Luminaire Characteristics

The design of a luminaire and the selection of the light source to be employed for a given application depend upon many factors. In addition to the illumination aspects (luminance, glare, uniformity, illumination level, etc.), consideration should be given to appearance, color of light, heating effect, noise level, efficiency, life, and economics.

Reflector Design Considerations. Most lamps do not act as point sources in the size of reflectors into which they are placed. Physically large lamps such as phosphored mercury require larger reflectors to control light to a given degree than do small filament sources. Generally, the larger the source, the larger the reflector required for equivalent control.

Secondary effects of reflector or housing design can often be detrimental to the performance of a luminaire. As an example, if reflected energy is concentrated on the lamp, lamp parts may fail; if the beam is concentrated on a lens front, the glass may fail from thermal stress.

When fluorescent lamps are used in confining units, two effects take place. The build-up of heat in the lamp compartment raises bulb wall temperature, reducing light output. As lamps are moved closer, a mutual heating occurs and, beyond this, light which might be redirected out of the unit is trapped between lamps, or between lamps and reflector. Both conditions lower luminaire efficiency.

Lamping and Lamp Position. Lamp operating position is important. Many lamps are designed only for base up, base down, or some other specified operating position. Ignoring such limitations will normally result in unsatisfactory lamp performance, lamp life, or both.

A basic consideration is that of ready lamp insertion and removal. Recognition should be made of possible lamp changing devices that might be used and space allowed for clearance.

Effects of Radiant Energy. Consideration must be given to the effects of lamp energy in the nonvisible regions on luminaire materials and performance. Some plastics, for example, can be altered by ultraviolet energy. Some reflector materials are excellent for visible light but absorb infrared, thus creating thermal problems.

If the intended lamp can present a hazard to people or objects under some conditions of operation (for example, some tungsten halogen lamps operated at high color temperature), protective devices should be designed into the luminaire.

Lamp Wattages. Lamp dimensions are rarely the sole criteria in determining luminaire size. Careful consideration must be given to adequate heat dissipation to insure normal lamp life and luminaire performance.

Luminaire Efficiency. This is normally a function of physical configuration and the selection of materials used. It should be recognized that many materials will change to some extent with use.

Appearance. The luminaire designer must coordinate technical, safety, and economic considerations with the final luminaire appearance. Where the lighting is primarily functional, performance has maximum importance. Design efforts will probably be concentrated on reflectors, refractors, and shielding elements.

Decorative luminaires are often selected because of their appearance. In this case, not unlike a piece of jewelry, they may serve to compliment or accent a decorative scheme. It may be desirable to sacrifice optimum performance in order to attain pleasing proportions and shapes. Sometimes both can be coordinated into a single luminaire.

Color, texture and form all play an important role in the attempt to achieve either of the above goals.

Glare. Light sources and luminaires are potential glare sources. The degree of brightness control to be designed into a luminaire depends upon the intended use of the luminaire and luminous environment within which it will be used. Frequently, design compromises are required between visual comfort, utilization and esthetics.

Systems exist to evaluate the potential glare from luminaires. See Section 3. These establish criteria commonly used to guide the luminaire designer in determining acceptable limits of maximum and average luminances. A thorough understanding of the principles involved is essential.

The problem of glare should be recognized for both interior and exterior luminaires.

Thermal Distribution. The integration of lighting units and the air handling and/or architectural aspects of a building greatly influence the basic construction of a luminaire. Some materials used in luminaires can be good reflectors of light and good absorbers of far infrared. See page 16-4 for spectral reflectance data.

Ventilation and Circulation. Air movement through the lamp compartment of a luminaire may result in the lowering of light output due to accumulation of dust and dirt, or it may maintain the lumen output by the cleaning action of air moving past the lamps.

In fluorescent heat transfer luminaires a reverse air flow is sometimes used to trap dirt and dust before it enters the lamp compartment. Consideration must also be given to the effect air currents have on bulb wall temperature since this effects light output. See Section 8. Ambient temperature affects the striking voltage of all discharge lamps.[19]

Acoustical. There are two acoustical problems associated with lighting equipment (usually more critical in spaces that are intended to maintain exceptionally low ambient sound levels). The first is the introduction of sound, generated at the unit. The second problem involves sound transmission from room to room via the luminaire and plenum.

Undesirable sound generation is normally a problem only with fluorescent or other discharge devices. As long as these lamps are ballasted with electromagnetic devices, it is likely that noise can occasionally be a factor. Luminaires can transmit this hum to the rest of the space and, in some cases, add luminaire vibration to it. Large, flat surfaces and loose parts tend to amplify the condition. Steps taken to minimize transmission of sound from the ballast to the luminaire may affect heat transfer characteristics.

Where luminaires are used as air supply or air return devices, the air controlling surfaces should be designed with full consideration for air noise. In this case, there are well accepted criteria for permissible sound levels.[20, 21]

Where luminaires are designed for recessing, thought should be given to the possibility of sound transfer from one room to another through the luminaire and then through the plenum. This can be minimized by making the luminaire as "leak-proof" as possible. Here, too, there is an established testing and rating procedure.[22, 23]

Vibration. Incandescent lamps are normally made in the smaller sizes up to 150 watts for vibration service and up to 500 watts for rough service. Where more severe vibration is present or larger wattages are required, vibration resisting sockets or shock absorbing mountings may be used. For high bay mounting on a building's steel structure, a spring steel loop of proper size and tension for the specific luminaire's weight usually solves the most difficult vibration problems.

For fluorescent luminaires that may be installed in vibration areas, spring loaded lampholders should be used—rather than those of the twist type. There are conditions of very high frequencies such as turbine deck areas where special lampholders may be required.

Radiation Interference. Electromagnetic radiation from gaseous discharge lamps, especially of the fluorescent type and auxiliary components such as starters and phase control devices may be sufficient to cause interference with nearby radios, television receivers, sound amplifying equipment, electro-cardiograph devices, sensitive electronic equipment, and certain military radar and tracking equipment. This interference is transmitted in two ways: (1) by direct radiation through the face of the luminaire; (2) by conduction through ballast and supply line.

To eliminate direct radiation the luminaire should be entirely metal enclosed, except for the light transmitting opening. Normal tolerances on the fit of parts are acceptable but there should be no appreciably large open holes in any of the metal parts and the electrical service should be brought in through grounded conduit or shielded cable. See page 8-28 on Radio Interference.

Life and Maintenance. The life of any luminaire is dependent primarily upon the luminaire's ability to withstand the environmental conditions in which it is installed. To ensure reasonable life, appropriate treatment of all materials should be considered. Where conditions are such that electrolytic action may persist, the use of dissimilar metals and high copper content alloys of aluminum should be restricted.

Maintenance may be a problem unless consideration is given to the luminaire design. Ventilation can sometimes be utilized to reduce dirt accumulation. High bay luminaires and floodlights, especially, should not have parts that can fall during relamping. The way doors of luminaires open should be predetermined to provide ease of accessibility to lamps or ballasts after installation.

Environmental Conditions

Ambient Temperature. The most obvious influence of ambient temperatures is its effect on the starting and operating of discharge lamps. Lighting equipment should be designed to keep the operating temperatures of all components within the authorized limits. In practice, this means that excess heat generated by internal heat sources such as lamps, and in some cases ballasts, must be transferred to the outer surfaces in the environment.

Surface mounted luminaires and recessed equipment, frequently operating in non-ventilated plenums, are often exposed to high ambients. Some industrial areas have shown 140°F (60°C), or even higher, to be realistic. Refrigerated storage areas, and in some climates unheated storage areas, attain temperatures of −20°F (−29°C) and frequently lower. A given luminaire should be expected to operate efficiently through these extremes.

Incandescent Filament Lamps. An incandescent filament operates at a very high temperature, near 4750°F for a 200-watt lamp. Normal changes in ambient temperature will not appreciably affect its light output or life.

Fluorescent Lamps. The performance of fluorescent lamps is affected by ambient temperature. See page 8–25. In totally enclosed luminaires, at normal room ambients, fluorescent lamps may operate at temperatures above their optimum. Designs which "bottle-up" lamp and ballast energy will reduce performance. Ventilation or other means of heat dissipation are helpful and should be considered.

Where it is not possible to reduce temperature by luminaire design, the use of components with higher temperature ratings is indicated.

High Intensity Discharge Lamps. Ambient temperature is not normally a significant factor in the light output of conventional high intensity discharge lamps. Arc-tube temperatures are such that modest variations due to varying ambients will not make much difference. Starting requirements are sometimes affected at low temperatures and ballasts meeting such requirements must be used. Special auxiliary equipment also may be required at elevated temperatures. The relationships between temperatures and service life of wiring and component parts must be studied. There are UL and National Electrical Code requirements which apply in some cases.

Wet Locations. The National Electrical Code requires that luminaires for wet locations be so labeled by the manufacturer under authority from the Underwriters' Laboratories. Booklet UL57 covers the minimum design requirements of such luminaires in detail. Article 680 of the National Electrical Code[24] specifically covers swimming pools. See also National Electrical Code, Article 410-4.

Corrosive Locations. In some areas, lighting equipment is subject to unusual* corrosive atmospheres. Many outdoor locations subject to a windborne corrosive agent, abrasives, or salt water, have corrosive action that can rapidly depreciate exposed or unprotected surfaces.

* In some cases, these might also be classified as "Hazardous Locations" and require National Electrical Code Class I equipment.

The characteristics of all materials should be carefully considered where possible corrosive or severe local atmospheric conditions (as near salt water) can be anticipated.

Hazardous Locations. All equipment for hazardous locations must be labeled by the Underwriters' Laboratories for the environment in which it is to be used or installed. To obtain approval for such labeling, the construction must comply with *Standards for Safety*, "Electric Lighting Fixtures,"[19] and "Electric Lighting Fixtures for Use in Hazardous Locations."[25] General requirements for hazardous locations are covered in articles 500 through 517 of the National Electrical Code.[24]

Some industries have additional requirements. For example, flour mills and elevators are covered by the *Electrical Code of the Mill Mutual Fire Prevention Bureau*, 400 W. Madison Street, Chicago, Illinois.

There are also more restrictive local codes, such as in New York City, Chicago and Los Angeles. For Marine use, the Underwriters' Laboratories Booklet UL595 should be consulted.[26]

Classification. The design of lighting equipment for hazardous locations is a highly specialized topic. It is imperative that the designer be completely familiar with the latest versions of the applicable documents and related writings.

The National Electric Code classifies the various hazardous locations by the type of hazard (Class I, Class II, etc.) and by the degree of probability of hazard (Division 1 and Division 2).[24] There are also Canadian equivalents,[27] requiring labeling by the Canadian Standards Association (CSA).

"Enclosed and Gasketed" luminaires are not recognized by UL as being suitable for hazardous locations.

Acoustics.* Room acoustics and sound level sometimes determine the type of luminaire to be utilized. In some areas which require very low ambient noise levels, fluorescent luminaires require the use of very quiet ballasts and assembly methods which prevent sound build-up.[20, 21] These applications may warrant the incorporation of acoustic properties and material into the luminaire proper, or the design of recessed equipment.

Vibration.* There are spaces within a building where the operation of equipment creates a condition of high vibration. The solution to this problem is discussed under Luminaire Characteristics.

Radiation.* Certain areas of buildings, especially laboratories and specialized equipment, may emit radiation which will have a detrimental effect on the luminaire. These are special cases, and the possible effects of fading or color change of paints, and degradation of plastics should be studied. In these cases, special thought should be given to selection of luminaire materials.

* See also Luminaire Characteristics, page 6-22.

Air Movement. Outdoor luminaires should be designed to withstand wind-loading to which they might be subjected. In indoor locations, the air pressure on a fluorescent troffer lens or a luminous ceiling component can dislodge a diffuser when a door to a small office is suddenly opened.

Dirt. The effect that dirt will have on a luminaire configuration will depend on the intended end-use of the luminaire. (For further details on Lighting Maintenance, see Section 16.)

Miscellaneous Environmental Effects. A complete listing of the possible effects of environment on luminaire design is impossible. The designer, however, should be alert to the possibilities of many others not specifically covered, such as, the effect of ultraviolet energy on outdoor components, thermal shock, extremes in humidity, and foreign substances in the atmosphere.

Electrical Considerations

Sockets and lampholders, wire, ballasts, and other controls are parts in electrical circuits of luminaires. All gaseous discharge lamps require some form of ballast or control equipment to provide adequate starting voltage, and to limit the current after the lamp has started.

Fluorescent Lamp Circuits. Each type of lamp requires a specific ballast circuit. See Section 8.

Ballast Quality. Rated lamp life and light output for fluorescent lamps are directly dependent on the ballast's ability to meet specified limits set by the American National Standards Institute (ANSI) Standards.* Ballasts not meeting ANSI Standards, can reduce lamp life as much as 50 per cent and light output by more than 30 per cent.

The life of ballasts made with 105°C (Class A) insulation is approximately 45,000 hours (continuous operation not exceeding 105°C). Ballast coil temperatures will not exceed the 105°C rating in properly designed ballasts when luminaires maintain a maximum ballast case temperature of 90°C. Other factors affecting ballast life are input voltage, luminaire heat dissipation characteristics, luminaire mounting and environment. Field data indicates a 12 year median ballast life for a duty cycle of 16 hours per day, 6 days per week, or 5000 hours per year. This life rating considers the fact that ballast warm-up time allows the ballast to operate at peak temperatures 12 out of the 16 hours per day or 3,750 hours per year.

Ballast safety standards are set by Underwriters' Laboratories. Ballast case temperature is limited to 90°C with the ballast mounted in a luminaire and the luminaire in a 25°C ambient.

Excessive ballast operating temperatures cause ballast insulation deterioration resulting in short life and possible actuation of the ballast protective device. A convenient rule of thumb is that every 10°C increment above 90°C results in a 50 per cent reduction in life. Luminaire design should prevent undesirable heat rise within the luminaire.

Ballast hum is a natural result of an electromagnetic device. Various ballast circuits produce varying noise levels and the ballast manufacturer should be consulted for specific sound ratings. The lighting equipment designer must consider the most common application of his product when selecting specific ballast types.

Remote locations of ballasts may involve complications of wiring, voltage, thermal considerations, and code restrictions. (See the National Electrical Code[24] for details.)

Ballast Protection—Class P Ballasts. The National Electrical Code requires that protected ballasts be used in fluorescent luminaires (except reactor types). Protected ballasts are called "Class P" ballasts and are listed by UL as such. All new fluorescent luminaires must use UL listed Class P ballasts in order to carry the UL label. The intent is to limit the maximum temperature a ballast case can reach in a luminaire under both normal and abnormal conditions.

UL requires that the protector open within two hours after the ballast case temperature has reached 110°C. The ballast-luminaire combination must be such, thermally, that the luminaire can be installed as intended without the ballast protector being actuated.

The maximum ballast case hot spot temperature should not exceed 90°C when the luminaire is in service.

High Frequency Operation. Most fluorescent lamps can be operated on higher frequencies and with increased luminous efficacy. The 40-watt rapid start lamp reacts most favorably with a gain in efficacy of approximately five per cent at 420 Hz, six per cent at 840 Hz, and eight per cent at 3000 Hz. As system frequency is increased, ballasting devices, can be smaller, lighter, and more efficient.

Low Temperature Operation. Fluorescent lamps require higher open circuit voltages for low temperature starting. Low temperature ballasts should be used where ambient temperatures are below 10°C.

Gaseous Discharge Lamp Circuits. Mercury, metal halide, or high pressure sodium ballasts (see Section 8) are usually classified as:

1. *Reactor*—consists of a series reactor to limit lamp current. Line voltage must be high enough to start lamp, and line regulation must be within ±5 per cent.

2. *Autotransformer*—used when line voltage is higher or lower than required starting voltage for the lamp; it is combined with a reactor to limit lamp current and a capacitor for power factor correction. Primary taps can be provided to extend line voltage range. Line voltage regulation limited to ±5 per cent of tap voltage range.

* Such ballasts bear the CBM/ETL label which means that they have met the Certified Ballast Manufacturers' requirements which includes the ANSI C82 standard and have been tested and certified by Electrical Testing Laboratory.

3. *Constant Wattage or Regulator Type*—consists of a special high power factor circuit to provide improved lamp current regulation over an extended line voltage range.

Sockets and Lampholders. For *incandescent lamp sockets,* maximum wattage ratings should not be exceeded and careful consideration should be given to the following:

1. Non-metallic housings and paper insulating liners should be used for low wattage lamps in well ventilated luminaires.
2. Porcelain sockets should be used in recessed or totally enclosed luminaires and in spotlights.
3. Screwshells of aluminum or copper should be used only for low wattage applications.
4. Nickel plated brass screwshells should be used for high heat conditions.
5. High heat sockets should be used for tungsten-halogen and other lamps of high wattage in small envelopes.
6. See Underwriters' Laboratories and Canadian Standards Association standards for sockets to be used for outdoor luminaires.

For *fluorescent lampholders,* a good connection, mechanical and electrical, between lamp pins and contacts is imperative, particularly for rapid start and trigger start circuits. Silver plated contacts reduce electrical contact resistance. Positive lamp seating is imperative to prevent destructive arcing at contact points.

Wire.

Total current. Wire smaller than No. 18 AWG may not be used for line voltages of 120 V or higher.

Voltage. Thermoplastic insulation is limited to 1000V, Type AF wire limited to 250V.

Temperature on insulation. 60°C to 200°C ratings are generally available.

Application will determine use of stranded or solid wire. Braided insulation usually is required where flexing is expected. Third (grounding) conductor and three-prong plug may be required for cord sets. Chain suspended luminaires may require grounding wires.

Codes, both national and local, should be carefully consulted before specifying a given wire size and insulation. There are many special requirements, for example, supply leads for a recessed luminaire must be No. 14 AWG minimum regardless of how little current they will be required to carry.

Starters. Protective starters for fluorescent lamp operation provide additional safety by disconnecting lamps which have failed. Other starter types include thermal, glow discharge, manual reset and automatic reset. See Section 8. Choice is usually dictated by specifier.

Mechanical Considerations

Tolerances. It is necessary to establish suitable dimensional tolerances for both the component parts and the final assembly. An industry standard[28] defines tolerances of recessed luminaires. Surface and pendant mounted luminaires may have less critical tolerances. Particular attention should be given to such components as spinnings, plastics, glass, and lamp tolerances.* Plenum space available should be considered in case of recessed luminaires.

Strength. Strength considerations will depend on the luminaire type. All luminaires should have housings of sufficient rigidity to withstand normal handling and installation. Luminaires intended for outdoor use should incorporate mounting and design features suitable to withstand high winds, rain, and snow accumulation.

Recessed luminaires in poured concrete should have an enclosure of suitable strength, tightness, and rigidity for the application.

Surface mounted luminaires should be strong enough so that they will not excessively distort when mounted on uneven ceilings.

Suspended luminaires should have adequate strength to minimize vertical sag between supports as well as lateral distortion and twist. Provision must be made for attachment of supports at suitable locations. Vibration should also be considered.

Local conditions sometimes require conformity with special local codes. Local codes should be investigated for these rulings.

Component Support. Lampholders and sockets should be held in place to prevent movement and to insure maintenance of proper spacing lengthwise for good lamp contact in fluorescent luminaires. To hold large, high wattage incandescent and mercury lamps, gripping sockets are often helpful. Ballasts should be securely fastened to the luminaire housing for good thermal contact.

Enclosure Fastening. Luminaire enclosures should be designed to incorporate a secure method for retaining them in the luminaire. Attention must be given to wind, rain, vibration and shock. (See also Environmental Considerations.) *Tamper-proof fastenings* may be required in correctional institutions and mental hospitals. *Wire guards* may be required around exposed bulbs in accessible locations.

Assembly. Factory assembly should be as complete as possible for minimum field assembly. Refer also to Canadian Standard Association (CSA) and Underwriters' Laboratories (UL) requirements. Instruction sheets should describe how components are to be assembled and installed on the job site.

Ultimate Mounting. Mounting components should satisfy the requirements of UL, CSA, local codes and end-use environment. Local codes should be considered with regard to power supply, continuous row wiring and grounding of the luminaire. Recessed luminaires intended for use in forced air return plenums may be subject to special requirements. Instructions and warnings should be marked

* See the applicable American National Standards Institute (ANSI) standards.

in a permanent manner. The text and physical dimensions may be governed by code requirements.

Maintenance. Ease of maintenance is an important consideration. Instructions for maintenance and correct lamping should be provided to prevent damage to and insure continuous maximum utilization of the luminaire. Electrical components should be replaceable without removing the luminaire from its mounting. Where ceiling plenums are not otherwise accessible, provision must be made for access through the luminaire to splice boxes connecting the recessed luminaire to the branch circuit.

Size. The size of a luminaire is controlled by components such as lamps, ballasts and reflectors. Mounting limitations and building modules can also determine size.

Thermal Considerations

Heat is a natural by-product of most light sources and the amount of heat generated by a particular source depends on its luminous efficacy, total wattage consumed, and type of source. Ballasts and transformers may also contribute heat. Environmental temperature is an added consideration.

Code Requirements. The National Electrical Code,[24] Canadian Electrical Code,[27] and their respective testing laboratories set specific temperature test limits for electrical components and critical areas immediately surrounding a luminaire. Most of their testing is performed at an ambient temperature of 77°F (25°C). Code requirements usually do not relate to performance characteristics and thermal considerations are regarded from a safety standpoint.

Thermal Environment. A theoretical thermal environment of 77°F (25°C) is usually not adequate. Temperatures at the luminaire are frequently higher and ambient temperatures in ceiling plenums of 140°F (60°C) are not uncommon. In cold storage areas, temperatures may be −20°F or lower. High or low temperature differentials will directly affect the thermal dissipating capabilities of a luminaire. A luminaire may be expected to operate without breakdown or reduced component life through both extremes.

Very high or low ambient temperatures may cause electrical components to fail. For example, contact of hot glass with cold air or water may result in breakage and excess heat may cause thermoplastics to distort. Satisfactory performance at the luminaire interior temperature is a major consideration in selecting components and finishes.

Supply Voltage Effect. Higher than nominal line voltages usually cause higher temperatures within electrical components. Lower voltages do not necessarily produce lower temperatures.

Luminaire Components. Ballasts and other auxiliary equipment, sockets and wires have definite safe operating temperature limits defined by UL, CSA, ANSI, and the manufacturer. Special electrical components may have to be specified for extreme temperatures.

Metal components and their finishes may be affected by temperature within a luminaire. Thought must also be given to metals as conductors of heat either to or from component parts.

Glass and plastic components should be very carefully chosen to prevent cracking, shattering, deformation or other deterioration. This may be either or both long and short term.

Expansion or contraction of components due to thermal changes should be considered. Different coefficients of expansion between materials in intimate contact can lead to serious problems.

Safety Considerations

The need for attention to safety considerations, particularly in the design stage, cannot be overemphasized. Usually, if the design meets applicable code requirements, it will be in conformance with the following criteria.

Electrical. The safe transfer of electrical energy from the source, to the light control equipment is of utmost importance. In this context important wiring considerations are:

1. Current carrying capacity of the conductors.
2. Insulation rating of the conductors.
3. Grounding. A ground conductor is often required for portable lighting equipment. Permanently installed lighting equipment should be electrically grounded to eliminate shock hazards.
4. Temperature rating of the conductors.
5. Connections to the junction boxes. Where supplied, connections must be in conformance with local codes.
6. Wire termination color coding in conformance with code requirements for safe field installation.
7. Mechanical strength and flexing requirements.

Safety interlocks are advisable in equipment where high open circuit voltages are present for protection during servicing and are often advisable in damp areas and basements.

In those applications, where required, fuses and thermal protectors must be included in the design of the equipment.

Splices, clearances and sockets should conform to applicable UL, CSA, and ANSI requirements.

Low voltage units should be considered for outdoor applications.

Thermal. Luminaires should be designed to safely meet the requirements of the thermal environment and use. If the unit is to be firmly mounted, and will not be handled during operation, it can operate at a higher temperature than a unit which must be held or moved by hand. (See also Thermal Considerations.)

Mechanical. Mechanical design of light control equipment for safety involves consideration of the components of the equipment, both individually and collectively, and may include the effect of vibration, weight, wind-load, shock, impact, snow, and ice. Materials, castings, sheets, and extrusions should meet the unit requirements for strength as follows:

1. Lenses, diffusers, or louvers should be chosen to withstand the expected mechanical loading (wind or vibration) and secured to the lighting equipment in such a way that they are not a hazard by falling.
2. Reflector assemblies should be securely fastened to the mounting members of the equipment.
3. The method of attaching the ballast should permit field replacement while still retaining the same positive holding means.
4. Trunnions, ceilings, or other mounting assemblies should support the weight of the equipment plus any normal loading such as wind, vibration, and shock.
5. Pipe stems should have adequate wall thickness for full threads. Thin wall tubing should not be used for threaded connections. Breakable couplings are required in some applications such as airport stem mounted runway lights or standards for street lighting luminaires.

Environmental. Hazardous area requirements for safety are covered by Underwriters' Laboratories Booklet UL844[25] and also under Environmental Considerations in this Section.

There are, in addition, some general safety considerations to be given to the environment when designing a luminaire and its equipment. The unit must withstand the environmental conditions without its mounting breaking loose. In outdoor applications, exposure to rain, snow, water, ice, and condensation may necessitate the use of waterproof seals and weep holes.

The nameplate or other permanent marking should include data on the lamp to be used, the voltage to be supplied, safe operating conditions, catalog number, and other pertinent information. A complete instruction sheet should be included.

Economic Considerations

The choice of components and performance level used in designing a luminaire is governed by its intended end use. The design may have to be altered to provide the most economical product to meet a given set of design requirements.

Ready-Made Components. Standard components are generally less expensive than those made only on special order or in low volume. Luminaires which use all commercially available parts tend to be less distinctive than those made of specially built parts.

Fabrication. Frequently decisions will be dependent on the equipment available to manufacture the product. Many lighting companies are specialists in sheet metal fabrication and work in that area, others specialize in castings and use those extensively. If a part can be completely fabricated in one piece, a casting may prove to be cheaper; however, if additional pieces are to be attached, sheet metal materials may be spot welded whereas castings must be jointed by screws, rivets, adhesives, or similar methods.

Inventory and Packaging. In the economics of a company and the services offered on customer orders, inventory becomes a major consideration. Many types of lighting equipment offer multiple lenses, or choice of finish, or variation in trim design on a common housing.

In some instances large inventories of finished, but not assembled, parts are desired by a manufacturer. This provides the flexibility of assembling these items in the proper combination to fill any requirement. The difficulty with this plan is the time delay necessary to assemble saleable items from stocked parts. To overcome this weakness, the method of packaging can be varied. It is best to check the Underwriters' Laboratories or CSA requirements when considering packaging of subassemblies instead of complete units.

The most desirable, but not necessarily the most economical stocking for ease in filling an order, would be to have a complete unit in one carton. There is less chance of error in shipment or breakage from handling in this type of stocking, but considerable inventory is required.

Tooling. Generally, the amount of money available for the purchase of tooling for a given product is dependent on the anticipated volume of sales for that item. Proper selection of tooling may also reduce unit labor costs and/or improve product acceptance.

Standards vs. Specials. The term "standard fixture" usually indicates the cataloged and warehoused luminaires manufactured by a company which are available from stock. All expenses of producing the unit such as engineering, tooling, photometric testing and UL listing are included in the manufacturing costs for a relatively large volume over an extended period of time. The fixed costs are spread over the maximum number of luminaires to obtain the most economic cost per unit. A special unit is not usually cataloged or warehoused.

If the same type of unit is made on a "special basis," the expenses of producing the unit may be chargeable to the items on the specific order, therefore, a special may be more costly to manufacture.

Finishing. Decisions on finishes are frequently based on the type and amount of equipment available in the manufacturing facility; other factors are: volume, quality of finish, appearance, durability, and amount of light control desired.

Material is available in a variety of preapplied finishes; such as, porcelain enamel, baked enamel, and plated sheet steel. Textured surfaces of adhesive vinyl or embossed patterns may be desirable. In the use of these materials, special attention should be given to the methods of fabricating and joining which are necessary due to the nature of the finish. Equipment required to apply porcelain coatings is much more expensive than required for baked enamel.

Protective finishes require investigation into the end-use of the product and its location. A protective finish must be compatible with the base material and strongly adhere to it. It should act as a barrier to attack or degradation from elements surrounding it.

Fasteners. If it is not essential that parts be disassembled, use of a permanent fastener such as a rivet is desirable. If components are to be removed for installation or maintenance, the use of screws, or nuts and bolts should be considered. The use of snap fasteners and clips formed of sheet metal or wire has become common. These devices can be designed to perform multiple functions.

REFERENCES

Light Control

1. Committee on Light Control and Equipment Design of the IES: "IES Guide to Design of Light Control," Parts I and II, *Illum. Eng.*, Vol. LIV, p. 722, November and p. 778, December, 1959.
2. Halliday, D. and Resnick, R.: *Physics for Students of Science and Engineering*, John Wiley & Sons, Inc., New York, 1962.
3. Hardy, A. C. and Perrin, F. H.: *The Principles of Optics*, McGraw-Hill Book Co., Inc., New York, 1932.
4. *Light Wires*, Bausch & Lomb Publication No. D-2045.
5. Committee on Testing Procedures for Illumination Characteristics of the IES: "Resolution on Reporting Polarization," *Illum. Eng.*, Vol. LVIII, p. 386, May, 1963.
6. Jolley, L. B. W., Waldram, J. M., and Wilson, G. H.: *The Theory and Design of Illuminating Engineering Equipment*, Chapman & Hall, Ltd., London, 1930.
7. Potter, W. M. and Oetting, R. L.: "Design and Illumination Characteristics of Louver and Louverall Systems," *Illum. Eng.*, Vol. XLIV, p. 16, January and p. 341, June, 1949.

Materials Used in Light Control

8. Committee on Light Control and Equipment Design of the IES: "IES Guide to Design of Light Control," Part III, *Illum. Eng.*, Vol. LXII, p. 483, August, 1967.
9. Shand, E. B.: *Glass Engineering Handbook*, McGraw-Hill Book Co., Inc., New York, 1958.
10. Phillip, C. J.: *Glass the Miracle Maker*, Pitman Publishing Co., New York, 1948.
11. Morey, G. W.: *Properties of Glass*, Reinhold Publishing Co., New York, 1938.
12. Condon, E. U. and Odishaw, H.: *Handbook of Physics*, McGraw-Hill Book Co., New York, Chapter 8, Part 8, "Glass," by Lillie, H. R., 1958.
13. *Metals Handbook*, Eighth Edition, Volume 1, "Properties and Selection of Metals," American Society for Metals, Metals Park, Ohio, 8th Edition, 1961.
14. Baer, E.: *Engineering Design for Plastics*, Reinhold Publishing Co., New York, 1964.

Finishes for Luminaires

15. Committee on Light Control and Equipment Design of the IES: "IES Guide to Design of Light Control," Part III, *Illum. Eng.*, Vol. LXII, p. 483, August, 1967.
16. Koppes, W. F.: *Metal Finishes Manual*, National Association of Architectural Metal Manufacturers, Chicago, 1964.
17. Wernick, S. and Pinner, R.: *Surface Treatment and Finishing of Aluminum and Its Alloys*, Second Edition, Robert Draper, Ltd., Teddington, England, 1959.

Equipment Design

18. Committee on Light Control and Equipment Design of the IES: "IES Guide to Design of Light Control," Part IV, *Illum. Eng.*, Vol. 65, p. 479, August, 1970.
19. "Electrical Lighting Fixtures," *Stand. for Safety*, UL.57, Underwriters' Laboratories, Inc., Chicago, Illinois, May, 1969.
20. "ASHRAE Standard Method of Testing for Rating the Acoustic Performance of Air Control and Terminal Devices and Similar Equipment," No. 36B-63, American Society of Heating, Refrigerating and Air-Conditioning Engineers, New York, N. Y., 1963.
21. *ASHRAE Guide and Data Book*, American Society of Heating, Refrigerating and Air-Conditioning Engineers, New York, N. Y., Chapter 31, 1967.
22. "Measurement of Room-to-Room Sound Transmission through Plenum Air Systems," No. ADC AD-63, Air Diffusion Council, Chicago, Illinois, 1963.
23. "Tentative Method of Test by the Two-Room Method," Acoustical Materials Association, New York, N. Y., March, 1969.
24. *National Electrical Code* (NFPA 70), National Fire Protection Association, Boston, Massachusetts, 1968.
25. "Electric Lighting Fixtures for Use in Hazardous Locations," *Stand. for Safety*, UL844, Underwriters' Laboratories, Inc., Chicago, Illinois, January, 1966.
26. "Marine-Type Electric Lighting Fixtures," *Stand. for Safety*, UL595, Underwriters' Laboratories, Inc., Chicago, Illinois, April, 1966.
27. *Canadian Electrical Code*, Canadian Standards Association, Ottawa, Canada, 9th Edition, February, 1966.
28. "Dimensions of Recessed Fluorescent Luminaires," NEMA Standard LE 1-1.02, National Electrical Manufacturers Association, New York, N. Y., Part I, p. 2, 1969.

SECTION

7

DAYLIGHTING[1]

Daylight 7–1 Architectural Design 7–5 Building Types 7–9
Maintenance 7–12 Measurements 7–12

Daylight, skillfully employed, provides the architect with one of his most effective modes of esthetic expression. Considerations of site, orientation, proportion, and fenestration are all influenced or determined by the degree of importance attached to the utilitarian and esthetic aspects of daylighting in the structure.

DAYLIGHT

To use daylight to advantage, the following design factors should be taken into account:

1. Variations in the amount and direction of the incident daylight.
2. Luminance (photometric brightness) and luminance distribution of clear, partly cloudy, and overcast skies.
3. Variations in sunlight intensity and direction.
4. Effect of local terrain, landscaping, and nearby buildings on the available light.

Daylight Source and Distribution

The daily and seasonal motions of the sun with respect to a particular building location produce a regular and predictable pattern of gradual variation in the amount and direction of available light. Superimposed, also, is a variable pattern caused by less regular changes in the weather, particularly to the degree of cloudiness.

The availability and variability of daylight throughout the U.S. are given in Figs. 7–1 through 7–3. These figures show the number of clear and cloudy days and the annual number of sunshine hours. Sunshine hours vary with the season as shown in Figs. 7–4 and 7–5.

The Sun as a Light Source

The sun is an abundant source of radiant energy; however, about half of this energy reaches the earth's surface as sunlight or visible radiation. The other half of the radiant solar energy contains invisible shorter wavelength components (ultraviolet) and invisible longer wavelength components (infrared). When absorbed, virtually all the radiant energy from the sun is converted to heat, whether this energy happens to be visible or invisible. Therefore, solar energy, sunlight, and solar heat are merely different names for radiant solar energy. The amount of usable visible energy in the solar spectrum, varies depending on the depth of the atmosphere the light traverses. It depends on the elevation of the sun above the horizon and on variable atmospheric conditions such as moisture and dust.

The rotation of the earth about its own axis, as well as its revolution about the sun, produces a continual apparent motion of the sun with respect to any reference point on the earth's surface. The position of the sun with respect to such a reference point at any instant, is usually expressed in terms of two angles—the *solar altitude,* which is the vertical angle of the sun above the horizon, and the *solar azimuth,* which is usually taken as the horizontal angle of the sun from the due South Line. See Fig. 7–6.

The illumination produced on an exterior surface by the sun is influenced by the altitude angle of the sun, the amount of haze and dust in the atmosphere and the angle between the incident sunlight and the surface on which the sunlight falls. Data on solar azimuth and altitude are shown in Fig. 7–7. Data on solar illumination produced on exterior surfaces are shown in Fig. 7–8.

The Sky as a Light Source

In Daylighting design, three conditions should be considered:

1. Incident light from overcast sky.
2. Incident light from clear sky.
3. Incident light from clear sky plus direct sunlight.

The amount of light received from an overcast sky reaching the windows of a building depends on the cloud pattern of the sky; the cloud pattern de-

---

NOTE: References are listed at the end of each section.

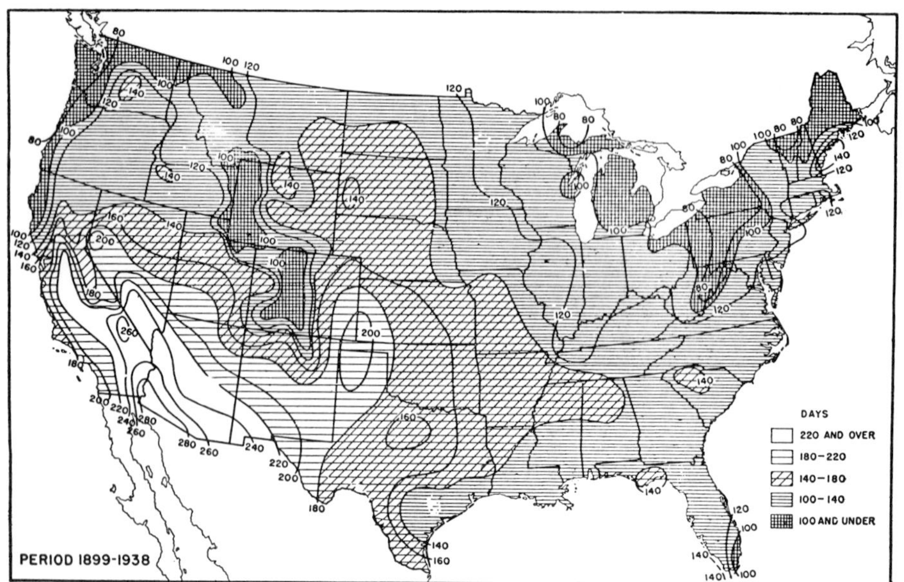

Fig. 7–1. Average number of clear days in the United States.

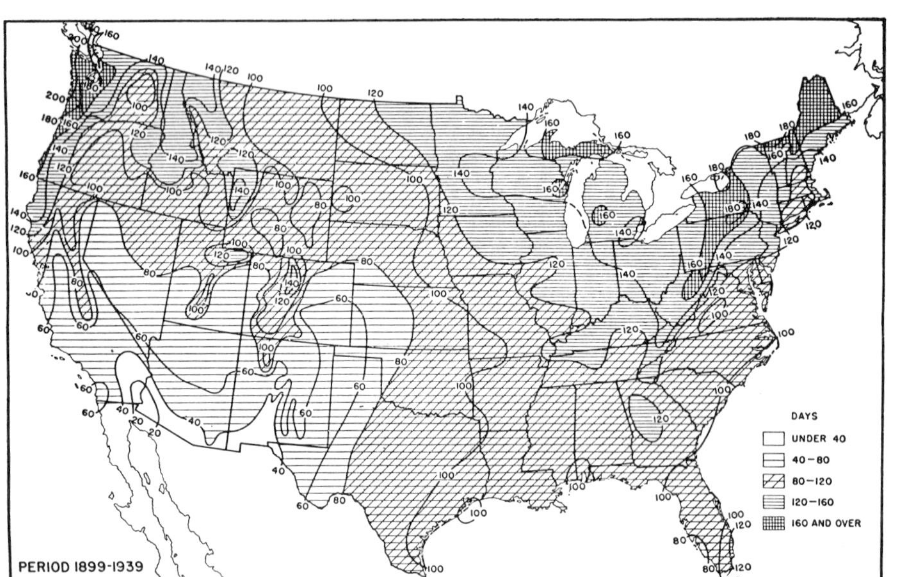

Fig. 7–2. Average number of cloudy days in the United States.

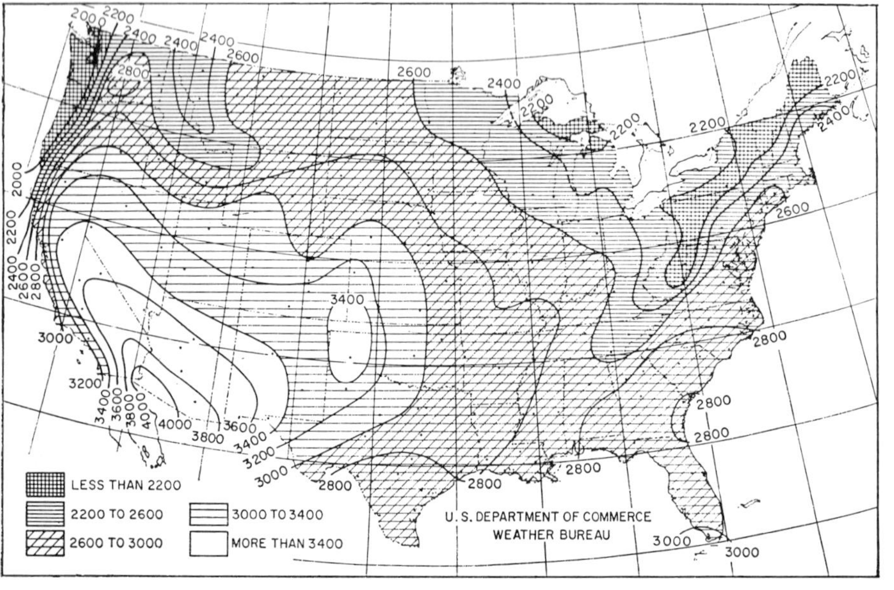

Fig. 7–3. Average annual amount of sunshine, in hours.

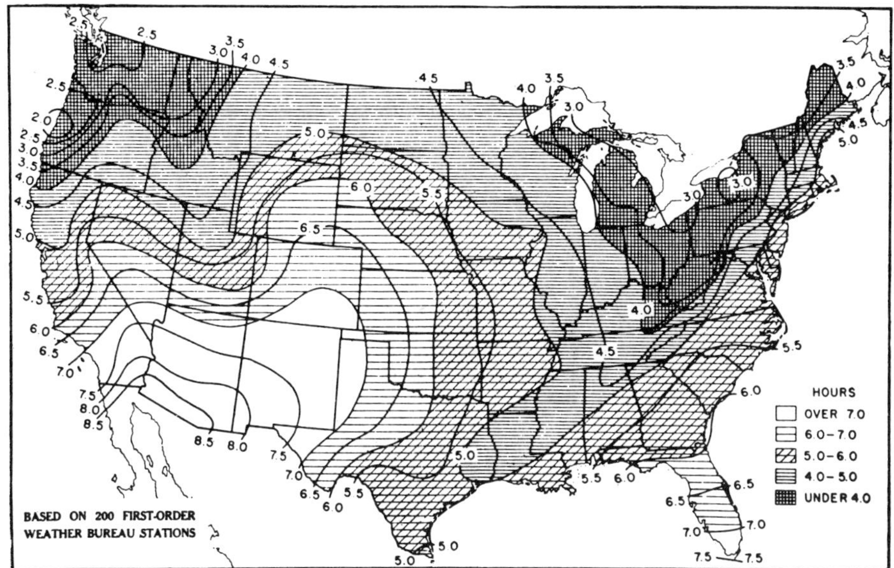

Fig. 7–4. Average number of hours of sunshine, daily, winter (December–February).

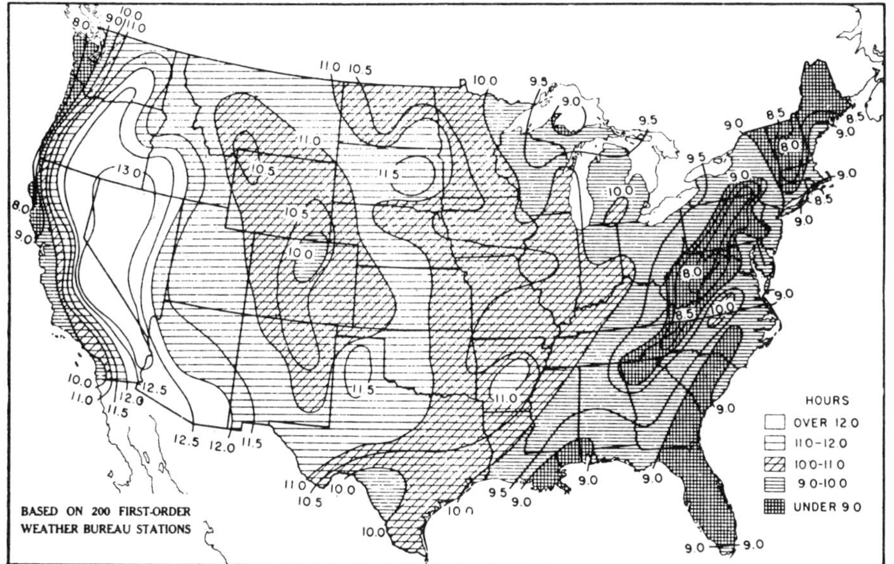

Fig. 7–5. Average number of hours of sunshine, daily, summer (June–August).

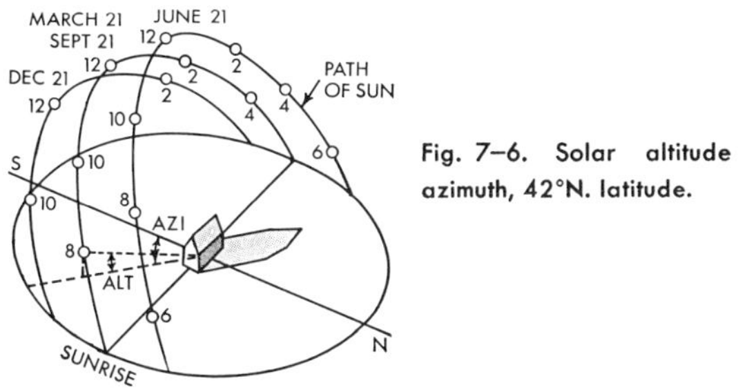

Fig. 7–6. Solar altitude and azimuth, 42°N. latitude.

fining the luminous distribution. The luminance and luminous distribution vary with the geographical location, time, density, and uniformity of the overcast.[2] "Uniformily" overcast skies are 2½ to 3 times as bright overhead as at the horizon. However, for design purposes, a single value representing equivalent uniform sky luminance may be used. See Fig. 7–9.

On clear days, the sky luminance varies with the position of the sun.[3] Except in the immediate vicinity of the sun, a clear sky is normally brighter near the horizon than overhead. For design purposes, the concept of equivalent sky luminance may also be used for clear skies. See Fig. 7–10. In daylight calculations for clear days, sky light only is included on non-sun exposures, while light from both sun and sky is included in calculations for sun exposures. The illumination received on a horizontal surface from the sky can be correlated with solar altitude. See Figs. 7–11 and 7–12. On vertical or tilted surfaces, this illumination depends on orientation of the particular surface. Data showing illumination received on vertical surfaces is given in Figs. 7–13 and 7–14. Average illumination with cloudy sky is given in Fig. 7–15 for Plains States and Atlantic Coast States.

The Ground as a Light Source

Light reflected from the ground, or from other exterior surfaces, is important in daylighting de-

Fig. 7-7. Solar Altitude and Azimuth for Different Latitudes

		Date	AM: 6 PM: 6	7 5	8 4	9 3	10 2	11 1	Noon
30°N	ALTITUDE	June 21	12	24	37	50	63	75	83
		Mar.-Sept. 21	—	13	26	38	49	57	60
		Dec. 21	—	—	12	21	29	35	37
	AZIMUTH	June 21	111	104	99	92	84	67	0
		Mar.-Sept. 21	90	83	74	64	49	28	0
		Dec. 21	—	60	54	44	32	17	0
34°N	ALTITUDE	June 21	13	25	37	50	62	74	79
		Mar.-Sept. 21	—	12	25	36	46	53	56
		Dec. 21	—	—	9	18	26	31	33
	AZIMUTH	June 21	110	103	95	90	78	58	0
		Mar.-Sept. 21	90	82	72	61	46	26	0
		Dec. 21	—	—	54	43	30	16	0
38°N	ALTITUDE	June 21	14	26	37	49	61	71	75
		Mar.-Sept. 21	—	12	23	34	43	50	52
		Dec. 21	—	—	7	16	23	27	28
	AZIMUTH	June 21	109	101	90	83	70	46	0
		Mar.-Sept. 21	90	81	71	58	43	24	0
		Dec. 21	—	—	54	43	30	16	0
42°N	ALTITUDE	June 21	16	26	38	49	60	68	71
		Mar.-Sept. 21	—	11	22	32	40	46	48
		Dec. 21	—	—	4	13	19	23	25
	AZIMUTH	June 21	108	99	89	78	63	39	0
		Mar.-Sept. 21	90	80	69	56	41	22	0
		Dec. 21	—	—	53	42	29	15	0
46°N	ALTITUDE	June 21	17	27	37	48	57	65	67
		Mar.-Sept. 21	—	10	20	30	37	42	44
		Dec. 21	—	—	2	10	15	20	21
	AZIMUTH	June 21	107	97	88	74	58	34	0
		Mar.-Sept. 21	90	79	67	54	39	21	0
		Dec. 21	—	—	52	41	28	14	0
48°N	ALTITUDE	June 21	17	27	37	47	56	63	65
		Mar.-Sept. 21	—	10	20	29	36	40	42
		Dec. 21	—	—	1	8	14	17	19
	AZIMUTH	June 21	106	95	85	72	55	31	0
		Mar.-Sept. 21	90	79	67	53	38	20	0
		Dec. 21	—	—	52	41	28	14	0

Solar Time* (column headings AM/PM hours)

* Time measured by the daily motion of the sun. Noon is taken as the instant in which the center of the sun passes the observer's meridian.

Fig. 7-8. Average Solar Illumination as a Function of Altitude[4, 5, 10]

Latitude	Plane	Illumination (footcandles)								
		December 21			March, September 21			June 21		
		8 AM 4 PM	10 AM 2 PM	Noon	8 AM 4 PM	10 AM 2 PM	Noon	8 AM 4 PM	10 AM 2 PM	Noon
30° N	Perp.*	4200	7000	7700	6400	8300	8600	7700	8600	8900
	Horiz.	700	3400	4400	2600	5900	7000	4400	7200	8500
34° N	Perp.*	3100	6500	7100	6300	8100	8400	7600	8600	8900
	Horiz.	400	2700	3700	2400	5600	6700	4700	7100	8400
38° N	Perp.*	2500	6000	6900	6100	8000	8300	7600	8500	8900
	Horiz.	100	2000	3000	2100	5400	6200	4400	7000	8300
42° N	Perp.*	2000	5500	6400	6000	7800	8200	7600	8400	8800
	Horiz.	100	1600	2700	2000	4800	5800	4700	6800	7900
46° N	Perp.*	500	4500	5800	5800	7600	8100	7600	8100	8800
	Horiz.	—	1000	1800	1800	4400	5500	4400	6700	7400

* Perpendicular to sun's rays.

sign. As with other light sources, it may require brightness control. On sunny elevations, the light reflected from the ground may represent ten to fifteen per cent of the total daylight reaching a window area. It frequently exceeds this proportion with light-colored, sandy soils, light-colored vegetation, or snow cover. On non-sun exposures, the light reflected from the ground accounts for more than half the total light reaching the window.

The direction from which the ground light is received at the fenestration is such that it can be utilized most effectively in the interior of the room, particularly at points well removed from the window area. Furthermore, ground light may be under the partial control of the architect or engineer to a considerable extent. By use of light-colored ground surfacing materials near the building, the daylight incident on the window areas and penetrating the inner portions of the rooms inside the building can be increased significantly.

ARCHITECTURAL DESIGN

Building Sections

Unilateral. This design (see Fig. 7-16) lends itself to continuous fenestration and curtain wall construction. Window heads are usually placed close to the ceiling line. For good daylight distribution, the distance of the inner wall to the outdoor wall should be limited to 2 to 2½ times the room height measured from the floor to the head of the window. Extreme luminance at the window, due to sun or sky, should be reduced using shades or tinted glass.

Bilateral. Bilateral daylighting design (see Fig. 7-17) permits doubling the room width. The second set of windows often occupies only the upper part of the wall. A reflecting roof under the secondary windows, acts like ground light and contributes materially to the light entering the room. At least one set of windows faces a sun exposure, necessitating brightness control. Sloping ceilings sometimes employed with this design generally have little effect on quantity or distribution of illumination. High reflectance materials used in the ceiling, however, do contribute to the utilization of light entering the space.

Roof Monitor. This fenestration system is most frequently used in industrial design where a central high bay area is set between two low bay areas. See Fig. 7-18. Brightness controls may be necessary. The roof surfaces below the monitors should be reflecting surfaces for improved light distribution.

Clerestory. The additional fenestration on the roof facing in the same direction as the main window (see Fig. 7-19) aids in overcoming the room width limitations of the unilateral section. Brightness control must be used on some exposures; brightness control is not as prominent a problem as with bilateral designs. The roof adjacent to the clerestory window should be a reflecting-type roof.

Sawtooth. This fenestration is used principally in low roof, large area, industrial buildings. See Fig. 7-20. The windows usually face north in northern latitudes; brightness controls are not then required. Slanting the windows increases the admission of daylight, but may increase dirt collection on the glazing material, in addition to increased thermal stresses in the tinted glazing material.

Skylight. Modern skylights assume many forms and are widely used in contemporary architecture. See Fig. 7-21. These are domes, panels with integral sun and brightness control, panels of glass-fiber reinforced plastic, and louvered-louvers for heat and brightness control. In addition to effective light distribution, skylight design should be carefully detailed to provide for effective seals against moisture penetration and possible dripping from condensa-

Fig. 7-9. Equivalent Sky Luminance in Footlamberts—Average Overcast day[5, 9]

Latitude	8 AM 4 PM	9 AM 3 PM	10 AM 2 PM	11 AM 1 PM	Noon
December 21					
30° N	420	740	1020	1210	1270
32	350	700	960	1150	1200
34	320	650	910	1100	1140
36	260	600	840	1020	1070
38	230	550	790	940	1000
40	190	500	740	900	930
42	150	450	660	820	860
44	100	380	600	760	790
46	60	340	550	680	730
48	40	290	470	630	650
50	0	240	420	560	580
March 21 or September 21					
30° N	910	1320	1710	2010	2140
32	880	1290	1650	1940	2070
34	860	1250	1600	1870	1980
36	840	1220	1560	1800	1900
38	800	1200	1500	1740	1840
40	790	1140	1460	1670	1760
42	760	1120	1410	1600	1690
44	740	1080	1340	1540	1620
46	710	1030	1229	1470	1550
48	690	990	1240	1410	1480
50	650	940	1180	1330	1400
June 21					
30° N	1270	1730	2250		
32	1280	1730	2240		
34	1290	1730	2220		
36	1290	1730	2200	2960	
38	1290	1720	2160	2840	
40	1290	1700	2120	2650	3060
42	1300	1690	2080	2540	2860
44	1290	1670	2050	2430	2660
46	1290	1640	2010	2330	2520
48	1290	1620	1960	2250	2400
50	1260	1590	1900	2160	2280

tion. They also may be used to provide heat control and ventilation.

Window Orientation

In northern latitudes in the United States and Canada, windows oriented to the South are usually preferred. This orientation not only offers higher levels from daylight, particularly in the winter months, but permits the use of solar heat as an aid to room heating. As a general rule, east-west orientations present the most difficult problems of daylight control and complicate heating and cooling design.

Materials and Control Elements

The various materials and control elements used in daylighting are selected for their ability to transmit, absorb, reflect, diffuse, or refract light. See Figs. 7-22 and 7-23.

Transparent (High Transmittance Materials). These include sheet, polished plate, float and molded glass, also rigid plastic materials and formed panels which transmit light without appreciably changing its direction or color and allow vision in either direction. Some of these materials are available as sealed double glazed units which reduce conductive heat flow.

Transparent (Low Transmittance Materials). Low transmittance glasses and plastics offer a measure of brightness control which increases as their transmittance is decreased. Reduction in radiant solar heat accompanies the reduction in visible light. During daylight hours with such materials, the view into the room is reduced while the view to the outdoors from the room is not noticeably affected. At night, the view into the room is apparent while the view from the room to the outdoors is reduced.

Reflective (High Reflectance, Low Transmittance Materials). Reflective glasses and plastics also offer a measure of brightness control which increases as their reflectance is increased. These materials act as one-way mirrors, depending on the ratios of the levels, indoors to outdoors. Also, they may reflect more heat while transmitting more light than non-reflective, low transmittance material.

Diffusing Materials. These include translucent and surface coated or patterned glass, plastics, and diffusing glass block. Transmittance is directionally non-selective. The amount of diffusion varies over a wide range depending on the material and surface. As a rule, transmittance and brightness decrease as diffusion increases. Some types may become excessively bright under sun exposure, requiring brightness control. The brightness of highly diffusing materials is nearly constant from all viewing angles.

Directional Transmitting Materials. These include prismatic surfaced glass and plastics to obtain the desired directional control of light and brightness. They are used in either horizontal or vertical panels.

Specularly Selective Transmitting Materials. These include the various heat absorbing and reflecting materials which are designed to pass most of the visible light but absorb or reflect a portion of the infrared radiation. The absorbed heat is then reradiated in approximately equal proportions inside and outside the building. Stained glass comes under this classification as it is selective in the visible portion of the spectrum. However, the primary use of stained glass is esthetic rather than as illumination control on visual tasks.

Fig. 7-10. Equivalent Sky Luminance in Footlamberts—Clear Days*

Latitude	December 21					March and September 21					June 21				
	8 AM	10 AM	Noon	2 PM	4 PM	8 AM	10 AM	Noon	2 PM	4 PM	8 AM	10 AM	Noon	2 PM	4 PM
North															
30° N	450	600	600	600	450	700	1000	1050	1000	700	1550	1400	1000	1400	1550
34° N	350	550	550	550	350	800	800	900	800	800	1350	1400	950	1400	1350
38° N	300	550	550	550	300	750	800	900	800	750	1350	1300	950	1300	1350
42° N	250	500	500	500	250	700	750	800	750	700	1300	1300	950	1300	1300
46° N	150	450	500	450	150	700	750	750	750	700	1300	1250	950	1250	1300
South															
30° N	1100	1950	2250	1950	1100	1700	2300	2800	2300	1700	1200	1600	2400	1600	1200
34° N	1100	1900	2200	1900	1100	1700	2650	2900	2650	1700	1350	1650	2300	1650	1350
38° N	900	2300	2200	2300	900	1700	2700	2950	2700	1700	1350	1650	2300	1650	1350
42° N	600	2100	2150	2100	600	1700	2700	2450	2700	1700	1350	2000	2500	2000	1350
46° N	400	1900	2100	1900	400	1700	2700	2900	2710	1700	1350	2100	2700	2100	1350
East															
30° N	1550	1500	1000	700	400	2000	2500	1500	900	700	2800	2650	1400	1000	700
34° N	1350	1400	950	700	400	2400	2600	1600	950	650	2800	2700	1450	1000	700
38° N	1200	1300	900	650	350	2500	2600	1500	900	600	2800	2700	1400	1050	700
42° N	750	1200	850	600	250	2400	2400	1450	800	600	2900	2600	1400	1000	700
46° N	500	1100	800	500	150	2300	2100	1400	700	600	2850	2600	1400	1000	700
West															
30° N	400	700	1000	1500	1550	700	900	1500	2500	2000	700	1000	1440	2650	2800
34° N	400	700	950	1400	1350	650	900	1600	2600	2400	700	1000	1400	2700	2800
38° N	350	650	900	1300	1200	600	900	1500	2600	2500	700	1050	1400	2700	2800
42° N	250	600	850	1200	750	600	800	1450	2400	2400	700	1000	1400	2600	2900
46° N	150	500	800	1100	500	600	700	1400	2100	2300	700	1000	1400	2600	2850

* Average values, direct sunlight excluded.

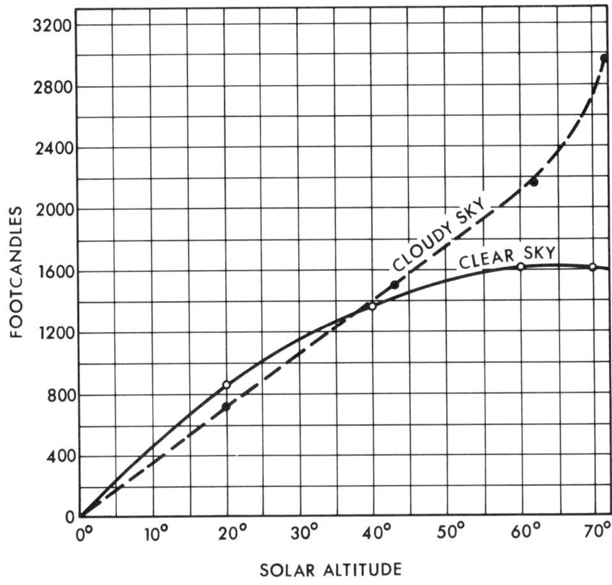

Fig. 7-11. Curves of clear and cloudy summer sky light illumination on horizontal surface.

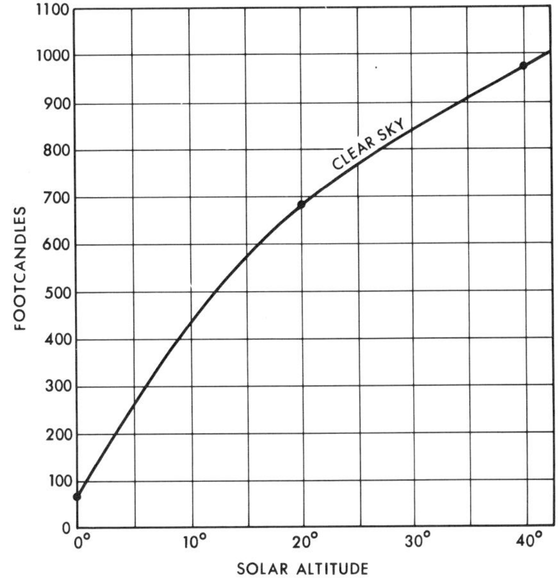

Fig. 7-12. Curve of clear winter sky light illumination on horizontal surface.

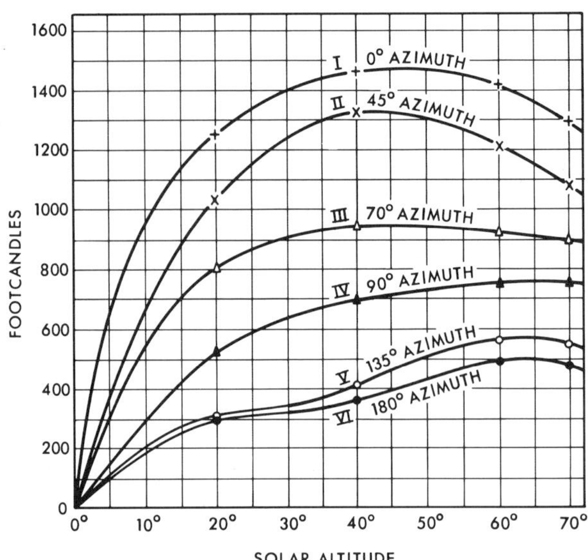

Fig. 7–13. Curves of clear summer sky light illumination on vertical surfaces:

Curve I. Vertical surface facing 0° in azimuth from sun.

Curve II. Vertical surface facing 45° in azimuth from sun.

Curve III. Vertical surface facing 70° in azimuth from sun.

Curve IV. Vertical surface facing 90° in azimuth from sun.

Curve V. Vertical surface facing 135° in azimuth from sun.

Curve VI. Vertical surface facing 180° in azimuth from sun.

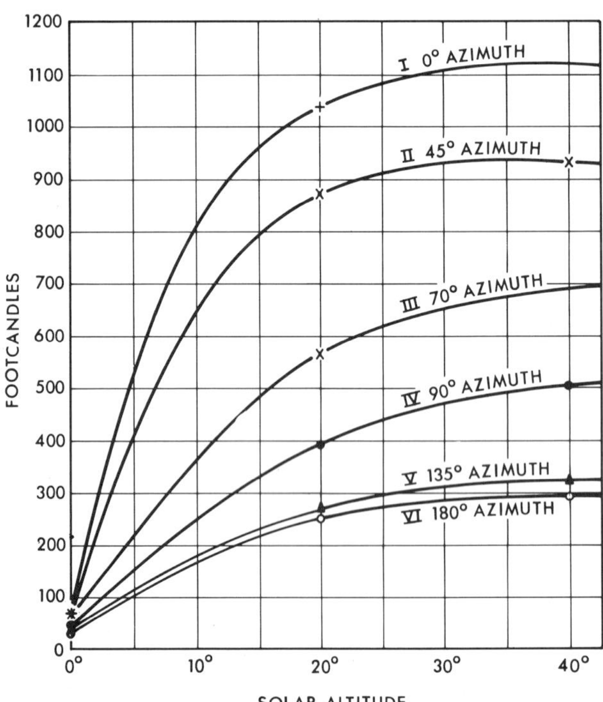

Fig. 7–14. Curves of clear winter sky light illumination on vertical surfaces:

Curve I. Vertical surface facing 0° in azimuth from sun.

Curve II. Vertical surface facing 45° in azimuth from sun.

Curve III. Vertical surface facing 70° in azimuth from sun.

Curve IV. Vertical surface facing 90° in azimuth from sun.

Curve V. Vertical surface facing 135° in azimuth from sun.

Curve VI. Vertical surface facing 180° in azimuth from sun.

Overhangs. These shade the window from direct sunlight and reduce the luminance of the upper part of the window at a sacrifice in the amount of light reaching the room side. Overhangs of practical width do not provide complete shielding at all times. In multi-story buildings, projecting balconies serve as overhangs.

Vertical Elements. Vertical opaque elements are effective on east and west walls as sun controls. Matte textures and suitable reflectances should be used. Combinations of vertical and horizontal elements as sun controls are most common in the tropics.

Shades and Draperies. These include opaque or diffusing shades and draperies for excluding or moderating daylight. For darkening a room the material must be opaque and should tightly cover the entire window.

Louvers. These are widely used as shielding elements in daylighting design. The slats may be fixed or adjustable, horizontal or vertical. They may prevent entrance of direct sunlight and reduce radiant heat, while reflecting a high proportion of sun, sky, and ground light into the interior. With fixed louvers, spacing and height of the slats should be determined to shield the light source at normal viewing angles. Overhangs for sun control are often made with louver elements so that more of the sky light can reach the windows. Louvers are also employed in top lighting arrangements, sometimes with two sets of slats set at right angles to form an egg crate. Matte textures and suitable reflectances should be used where possible.

Landscaping. Trees are effective shading devices when properly located with respect to the building and its fenestration. Deciduous trees provide shade and protect against sun glare during the warm months but allow the sun to reach the building during the winter. Deciduous vines on louvered overhangs or arbors provide a similar seasonal shade.

Exterior Reflecting Elements. Reflective pavements and similar treatment of the immediate terrain increase the amount of ground light entering the building. Reflecting materials or finishes on roofs below windows have the same effect.

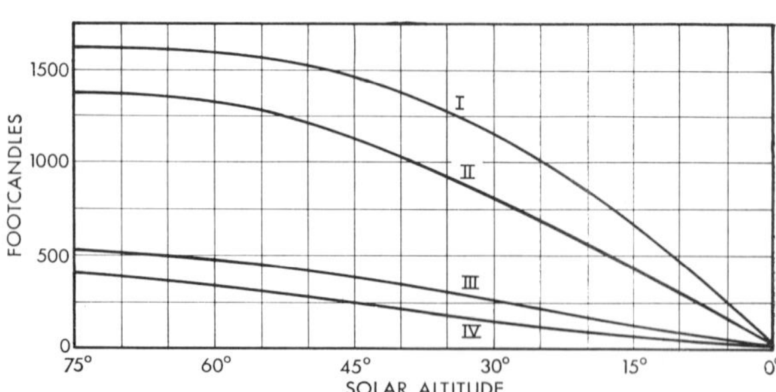

Fig. 7–15. Illumination with a cloudy sky. Curve I, average for Plains states. Curve II, average for Atlantic coast states. Curve III, minimum for Plains states. Curve IV, minimum for Atlantic coast states.

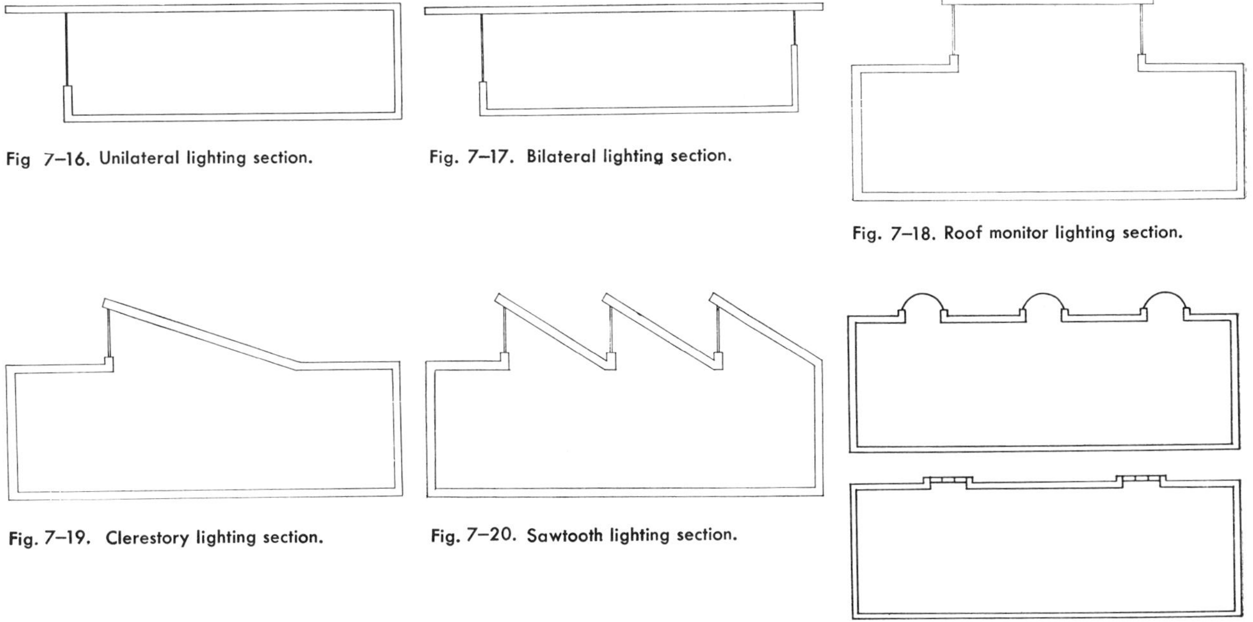

Fig 7–16. Unilateral lighting section.

Fig. 7–17. Bilateral lighting section.

Fig. 7–18. Roof monitor lighting section.

Fig. 7–19. Clerestory lighting section.

Fig. 7–20. Sawtooth lighting section.

Fig. 7–21. Skylight sections.

Interior Reflecting Elements. Ceilings of higher reflectance contribute to the utilization of ground and sky light entering the space. Suitable reflectances of walls, floors, and furnishings contribute to interreflections of light and to the luminance ratios required for visual tasks. It is important that the texture of all these surfaces be matte. See recommended reflectance values for various occupancies in appropriate Sections of this Handbook, *e.g.*, Offices and Schools, Industries. The importance of interior reflectances to daylighting design is indicated in Fig. **7–24**, showing increases in quantity and uniformity of daylight illumination produced by higher reflectances of the principal room surfaces.

Fig. 7–22. Transmittance Data of Glass and Plastic Materials

Material	Approximate Transmittance (in per cent)
Polished Plate/Float Glass	80–90
Sheet Glass	85–91
Heat Absorbing Plate Glass	70–80
Heat Absorbing Sheet Glass	70–85
Tinted Polished Plate	40–50
Reflective Glass	23–30
Figure Glass	70–90
Corrugated Glass	80–85
Glass Block	60–80
Clear Plastic Sheet	80–92
Tinted Plastic Sheet	90–42
Colorless Patterned Plastic	80–90
White Translucent Plastic	10–80
Glass Fiber Reinforced Plastic	5–80
Double Glazed—2 Lights Clear Glass	77
Tinted Plus Clear	37–45
Reflective Glass*	5–25

* Includes single glass, double glazed units and laminated assemblies. Consult manufacturer's material for specific values.

BUILDING TYPES

Fenestration, the location of windows and openings in buildings provides: (1) for the admission, control, and distribution of daylight;* (2) a distant focus for the eyes which relaxes the eye muscles; (3) eliminates the dissatisfaction many people experience in completely closed-in areas; and (4) may provide for emergency exits or entrances.

An adequate electric lighting system should always be provided because of the wide variation in daylight, from several thousand footcandles down to zero.

Effective daylighting compliments a well designed electric lighting system. Such a combination can provide the recommended quantities and quality of lighting and also add desirable highlighting and variations.

Daylight illumination distribution is improved when light is introduced through more than one room surface. Unilateral fenestration limits the width of the space in which daylighting can be effective to about 2½ times the height from floor to window head. For wall fenestration, it is recommended that the light transmitting areas be continuous and preferably extend the full length of the room and to the ceiling. Piers, where required,

* Because light is only one portion of the solar spectrum, daylighting design inevitably affects the thermal environment of the interior space. For information on this important subject, refer to publications of the American Society of Heating, Refrigerating, and Air-Conditioning Engineers.

Fig. 7-23. Reflectances of Building Materials and Outside Surfaces

Material	Reflectance (in per cent)	Material	Reflectance (in per cent)
Bluestone, sandstone	18	Asphalt (free from dirt)	7
Brick		Earth (moist cultivated)	7
light buff	48	Granolite pavement	17
dark buff	40	Grass (dark green)	6
dark red glazed	30	Gravel	13
Cement	27	Macadam	18
Concrete	40	Slate (dark clay)	8
Marble (white)	45	Snow	
Paint (white)		new	74
new	75	old	64
old	55	Vegetation (mean)	25
Glass			
clear	7		
reflective	20-30		
tinted	7		

Fig. 7-24. Effect of Surface Reflectance—Room 30 x 32 x 12 Feet—Unilateral Fenestration—6 x 32 Feet—Directional Glass Block

Reflectance (per cent)			Relative Illumination at Various Distances from Fenestration		
Walls	Floor	Ceiling	3 Feet	15 Feet	17 Feet
6	6	6	6.57	2.66	1.00
28	28	28	8.55	3.62	1.60
62	28	62	11.98	5.75	3.16

should be of minimum width. When deep window reveals are necessary, they should be splayed.

Offices and Schools. Detailed discussion of quantity and quality of illumination for the wide variety of visual tasks encountered in these occupancies will be found in Sections 3 and 11 of this Handbook. While the treatment for the various occupancies is from the viewpoint of electric lighting, the basic requirements of quantity, luminance ratios, and the reflectances of the principal architectural and work surfaces are the same whether the lighting be daylight, electric, or a combination of the two. Consideration of direct, reflected glare, and veiling reflections are of equal importance whether the source be a window or a luminaire.

In offices and schools, critical seeing is done over protracted periods of time from fixed positions involving frequent viewing of the fenestration. Direct and reflected glare from these sources should be minimized by brightness controls for all fenestration areas. In school rooms, to meet current educational practices, it is necessary to provide comfortable fenestration luminance for any orientation of the occupants throughout 360 degrees. Unfortunately, an accepted criterion for luminance limitation for vertical windows in the direct glare zone does not exist. However, the limitations of the luminance of luminaires established for schools and offices (see page 11-4), might well be used as a guide in the design of top-lighting elements for buildings with these occupancies.

When light-directing panels are used, they should begin approximately six feet above the floor. In conjunction with the light-directing panel, a vision strip should be incorporated below it. Provision should be made for the brightness control of this vision area, using draperies or blinds or other shading devices. Another design provides an eye-level window area of low transmission glass for outward vision and above an area of clear, high transmission glass to admit ground light. The sky brightness through the upper area must be shielded by louvers or overhangs.

Industrial. Detailed discussion of quantity and quality of illumination for the wide variety of visual tasks encountered in industry will be found in Sections 3 and 14 of this Handbook. A variety of fenestration treatments suited to industrial buildings and brightness control devices adapted to the problems encountered are described above. Figs. 7-25 through 7-30 show examples of illumination distribution from typical industrial building sections.

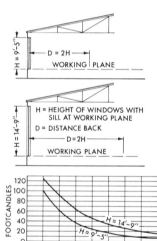

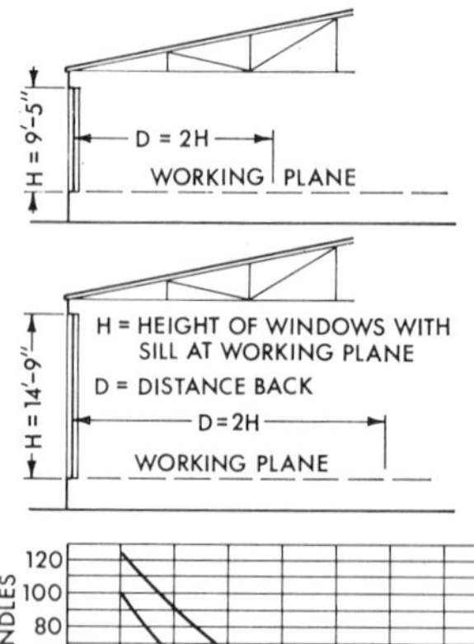

FOOTCANDLES
120
100
80
60
40
20
0
0 5 10 15 20 25 30 35 40
DISTANCE BACK FROM WINDOW
H = 14′-9″
H = 9′-5″

Fig. 7-25. Illumination distribution from sidewall windows assuming no reflection, clean glass, 1000 footlamberts uniform sky luminance, and window length equal to or greater than 5 times the window height.

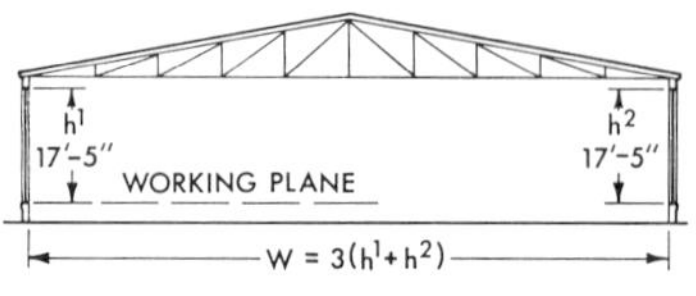

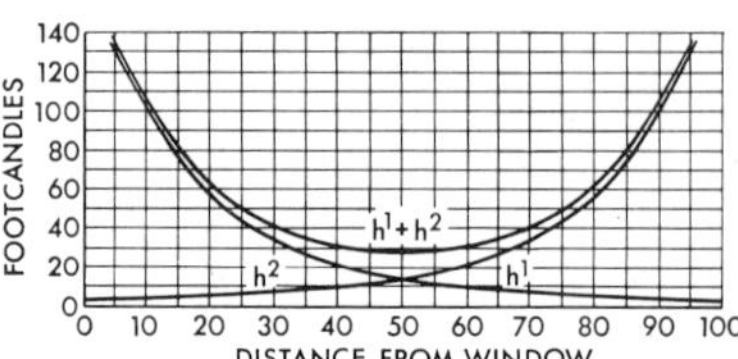

Fig. 7-26. Illumination distribution from sidewall windows assuming no reflection, clean glass, 1000 footlamberts uniform sky luminance, and window length equal to or greater than 5 times the window height.

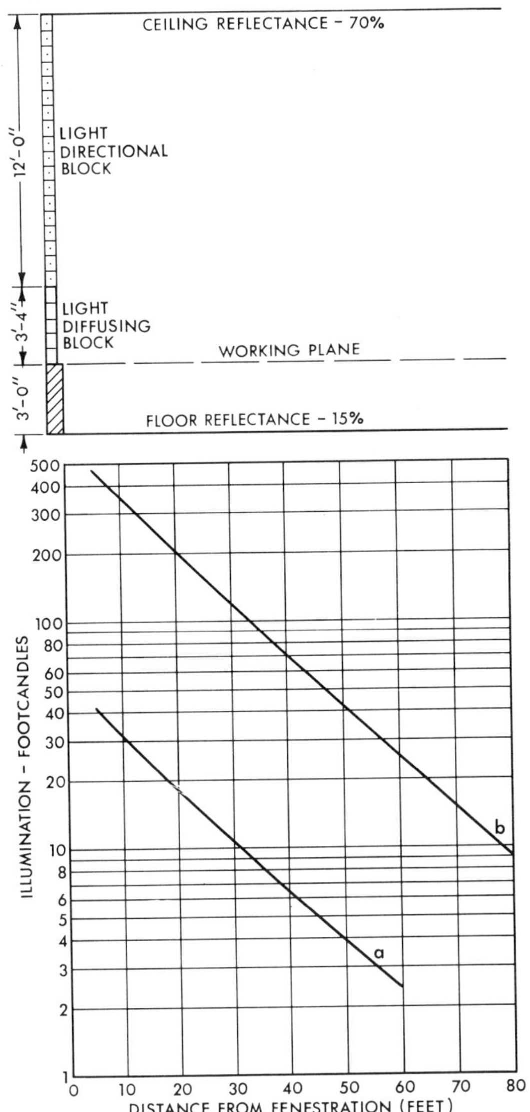

Fig. 7–27. Unilateral daylighting using glass block: a. overcast sky, equivalent sky luminance 1000 footlamberts; b. sun and clear sky, south exposure, 40 degrees N. latitude, equinox, 10 A. M. or 2 P. M.

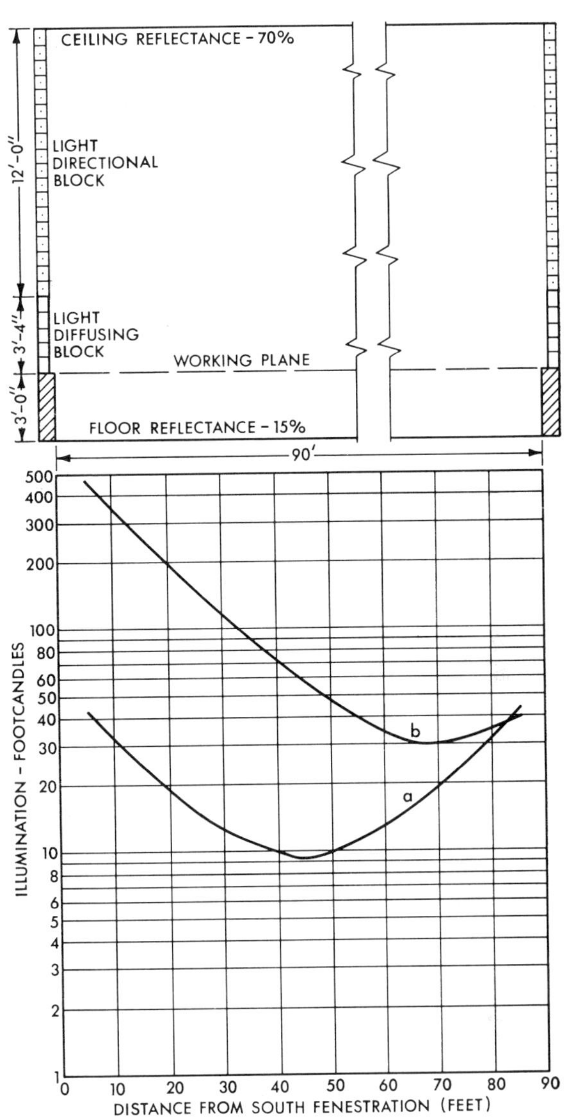

Fig. 7–28. Bilateral daylighting using glass block; a. overcast sky, equivalent sky luminance 1000 footlamberts; b. sun and clear sky, south and north exposures, 40 degrees N. latitude, equinox 10 A. M. or 2 P. M.

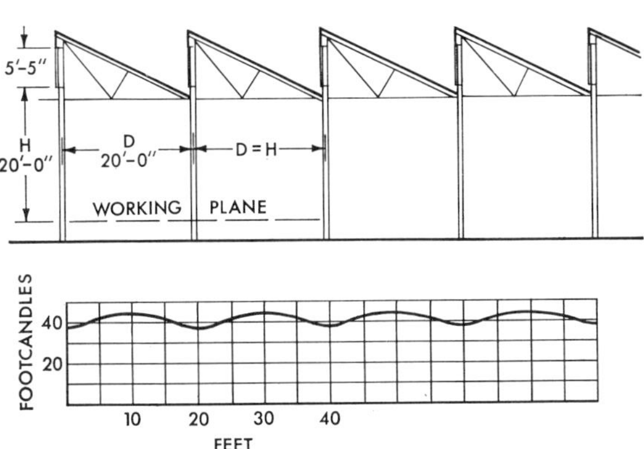

Fig. 7–29. Illumination distribution from sawtooth windows assuming no reflection, clean glass, and 1000 footlamberts uniform sky luminance; spacing equal to mounting height.

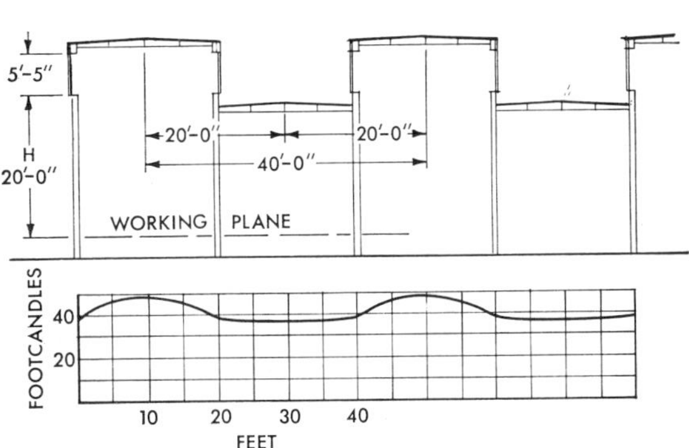

Fig. 7–30. Illumination distribution from monitor windows assuming no reflection, clean glass, and 1000 footlamberts uniform sky luminance; spacing equal to twice mounting height.

MAINTENANCE

If the planned results of the daylighting system are to be permanently achieved, maintenance is required. This will involve periodic scheduled cleaning to meet the demands and criteria of the area enclosed. See Fig. 7–31 for typical light loss factor.

DAYLIGHTING MEASUREMENTS

Daylighting surveys of illumination and luminance may be made using established procedures with scale models prior to construction or surveys may be made in actual buildings after construction.[11]

Survey Procedure. It is recommended that surveys be made by experienced persons during relatively representative daylight conditions. For surveys in a room, all electric lighting should be turned off, adjustments of all daylighting control devices should be noted, and exterior illumination and luminance readings should be taken at the same time. Illumination values excluding direct sunlight or reflected light from the building wall may be obtained by placing black velvet or similar material in a position to shield the cell or test plate from the unwanted light flux. Vertical illumination from the sun only may be obtained by subtracting the illumination from the sky and ground from the total illumination.[12]

Fig. 7–31. Typical Light Loss Factors for Daylighting Design

Locations	Light Loss Factor Glazing Position		
	Vertical	Sloped	Horizontal
Clean Areas	0.9	0.8	0.7
Industrial Areas	0.8	0.7	0.6
Very Dirty Areas	0.7	0.6	0.5

Instruments. Luminance measurements may be taken with a visual comparison or color corrected photoelectric meter. Illumination readings may be taken with a color and cosine corrected meter or with a luminance meter and test plate of known reflectance. Reflectances may be obtained with the luminance meter and test plate or by comparison with standard color chips.

REFERENCES

1. Daylighting Committee of the IES: "Recommended Practice of Daylighting," *Ilum. Eng.*, Vol. LVII, p. 517, August, 1962.*
2. Proposed standardized "C.I.E. Overcast Sky," Proceedings 13th Session CIE, Vol. 11, Part 3-2, Zurich, 1955.
3. CIE Committee E-3.2: "CIE Clear Sky," Published in Proceedings, Bouwcentrum, Hague, 1967.
4. Johnson, H. L.: "Daylight Variations," *Illum. Eng.*, Vol. XXXIV, p. 783, July, 1939.
5. Kimball, H. H.: "The Determination of Daylight Intensity at a Window Opening," *Trans. Illum. Eng. Soc.*, Vol. XIX, p. 217, March, 1924.
6. Kimball, H. H.: "Records of Total Solar Radiation Intensity and Their Relations to Daylight Intensity," *Trans. Illum. Eng. Soc.*, Vol. XX, p. 477, May, 1925.
7. Kimball, H. H.: "Variations in the Total and Luminous Solar Radiation with Geographical Position in the United States," *U.S. Weather Rev.*, Vol. 47, p. 769, 1919.
8. Kimball, H. H. and Associates: "Daylight Illumination on Horizontal, Vertical, and Sloping Surfaces," *U.S. Weather Rev.*, Vol. 50, p. 615, 1923; also *Trans. Illum. Eng. Soc.*, Vol. XVIII, p. 434, May, 1923.
9. Kimball, H. H. and Hand, I. F.: "Sky Brightness and Daylight Illumination Measurement," *Monthly Weather Rev.*, Vol. 48, p. 481, 1921; also *Trans. Illum. Eng. Soc.*, Vol. XVI, p. 255, July, 1921.
10. Kunerth, W. and Miller, R. D.: "Variations of Intensities of the Visible and of the Ultraviolet in Sunlight and Skylight," *Trans. Illum. Eng. Soc.*, Vol. XXVII, p. 82, January, 1932.
11. Subcommittee on Measurement and Reporting on Daylighting of Interiors of the Daylighting Committee of the IES: "IES Guide for Measuring and Reporting Daylight Illumination," *Illum. Eng.*, Vol. LIII, p. 213, April, 1958.
12. CIE Committee E-3.2: "Daylight—International Recommendations for the Calculation of Natural Daylight," CIE Publication No. 16, 1970.
13. Walsh, J. W. T.: *The Science of Daylight*, Pitman Publishing Co., New York, N. Y., 1961.
14. Turner, D. P., (Editor): *Windows and Environment*, Pilkington Bros. Ltd., McCorquodale and Co., Newton-le-Willows, 1969.
15. Henderson, S. T.: *Daylight and Its Spectrum*, American Elsevier Publishing Co., Inc., New York, N. Y., 1970.

* Contains a very comprehensive daylighting bibliography.

SECTION

8

LIGHT SOURCES

Incandescent Filament 8–1 *Fluorescent 8–17* *High-Intensity Discharge (Mercury, Metal Halide, High Pressure Sodium) 8–33* *Short Arc 8–40* *Miscellaneous Discharge 8–43* *Electroluminescent 8–47* *Light Emitting Diodes 8–48* *Nuclear 8–49* *Carbon Arc 8–50* *Gaslights 8–52*

This section is devoted almost exclusively to a description of the various light sources now available.* Fundamental information concerning the Physics of Light is discussed in Section 2. Daylighting is covered in Section 7.

INCANDESCENT FILAMENT LAMPS

The primary consideration of lamp design is that the lamp will produce the spectral radiation desired (light, infrared, ultraviolet) most economically for the application intended, or, in other words, obtain the best balance of over-all lighting cost in terms of the lighting results. Realization of this objective in an incandescent filament lamp, requires the specification of the following: filament material, length, diameter, form, coil spacing, mandrel size (the mandrel is the form on which the filament is wound); lead-in wires; number of filament supports; method of mounting filament; the vacuum or filling gas; gas pressure; and bulb size, shape, and finish.

The manufacture of good quality lamps requires adherence to these specifications and necessitates careful production control.[1]

The Filament

The efficacy of light production by incandescent filament lamps depends on the temperature of the filament. An iron rod heated in a furnace will first glow a dull red, becoming brighter and whiter as its temperature increases. Iron, however, melts at about 1800 K and, with such a low melting point, is not an efficient source of light. The higher the temperature of the filament, the greater will be the portion of the radiated energy that falls in the visible region; for this reason it is important in the design of a lamp that the filament temperature be kept as high as is consistent with satisfactory life. Considerable research and investigation has been carried out on a variety of materials and metals in search of a suitable filament. The desirable properties of filament materials are high melting point, low vapor pressure, high strength, high ductility, and suitable radiation and electrical resistance characteristics.

Tungsten for Filaments. Tungsten has many desirable properties for use as an incandescent light source. (Early incandescent lamps utilized carbon osmium, and tantalum filaments.) Its low vapor pressure and high melting point (3655 K) permit high operating temperatures and consequently high efficacies. Drawn tungsten wire has high strength and ductility, allowing the drawing of wire of the high degree of uniformity necessary for the exacting specifications of present-day lamps. Tungsten has been alloyed with metals such as rhenium and used in lamp designs which can utilize the characteristics of the alloy. Thoriated tungsten wire is used in filaments for rough service applications.

Radiating Characteristics of Tungsten.[2] The emissivity of a blackbody is 1.0 for all wavelengths (see Section 2). Since the emissivity of tungsten is a function of the wavelength, tungsten is a selective radiator. Fig. 8–1 illustrates the radiation characteristics of tungsten and a blackbody. Curves A and B show the spectral radiant intensity for an area of one square centimeter of a blackbody and tungsten respectively at 3000 K. Curve B′ is for 2.27 square centimeters of tungsten at 3000 K and has the same amount of radiation in the visible region as the blackbody curve A. This illustrates that for the same amount of visible radiation, tungsten radiates only 76 per cent as much as the total radiation from a blackbody at the same temperature.

* For history and references, see end of this section.

Only a small percentage of the total radiation from an incandescent source is in the visible region of the spectrum. As the temperature of a tungsten filament is raised, the radiation at shorter wavelengths increases more rapidly than that at longer wavelengths. Thus the radiation in the visible region (see Fig. 8–2) increases more rapidly than radiation in the infrared region. Therefore, the luminous efficacy increases. The luminous efficacy of an uncoiled tungsten wire at its melting point (3655 K) is approximately 53 lumens per watt. In order to obtain life, it is necessary to operate a filament at a temperature well below the melting point.

Resistance Characteristics of Tungsten. Tungsten has a positive resistance characteristic so that the resistance at operating temperature is much greater than its cold resistance. In general service lamps, the hot resistance is 12 to 16 times the cold resistance. Fig. 8–3 illustrates the change in resistance of the tungsten filament with temperature for various lamps. The low cold resistance of tungsten filaments results in an intial inrush of current which, because of the reactive impedance characteristic of the circuit, does not reach the theoretical value indicated by the ratio of the hot to cold resistance. Fig. 8–77 (page 8–56) gives the effect of the change in resistance on the current in incandescent filament lamps. The inrush current due to incandescent-filament-lamp loads is important in the design and adjustment of circuit breakers, in circuit fusing, and in the design of lighting-circuit switch contacts.

Color Temperature. In many applications, it is useful to know the apparent color temperature of an incandescent lamp. Fig. 8–4 expresses the approximate relationship between color temperature and luminous efficacy for a fairly wide range of wattages of gas filled lamps. Efficacy is frequently published, or it may be calculated from published lumen and wattage data. From this value it is possible to estimate the average color temperature of the filament.

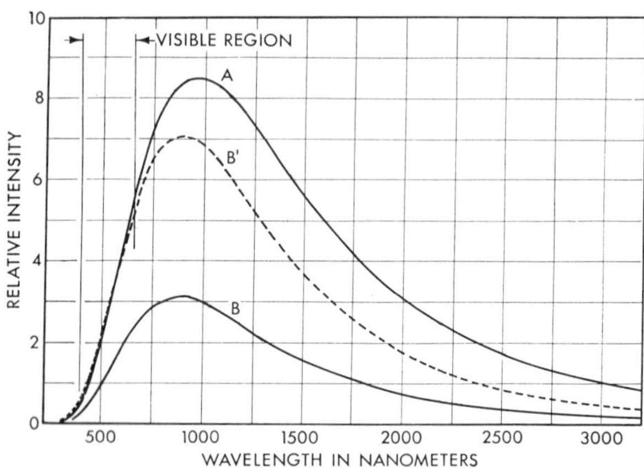

Fig. 8–1. Radiating characteristics of tungsten. Curve A: radiant flux from one square centimeter of a blackbody at 3000 K. Curve B: radiant flux from one centimeter of tungsten at 3000 K. Curve B′: radiant flux from 2.27 square centimeters of tungsten at 3000 K (equal to curve A in visible region). (The 500-watt 120-volt general service lamp operates at about 3000 K.)

Construction and Assembly. Fig. 8–5 shows the basic parts and steps in the assembly of a typical incandescent filament lamp. In miniature lamps three methods of construction are generally used: flange seal, butt seal, and pinch seal (see Fig. 8–6).

The *flange seal* is used generally with lamps ¾ inch and larger in size. This construction features a glass stem with a flange at the bottom which is sealed to the neck of the bulb. When used with bayonet bases the plane of the filament and lead wires are normally at right angles to the plane of the base pins. However, a tolerance of ±15 degrees is generally permitted. The advantages of this construction are: (1) sufficiently heavy lead-in wires can be used for lamp currents up to 12 amperes; (2) filament can be accurately positioned; and (3) sturdy stem construction resists filament displacement and damage from shock and vibration.

Butt seal refers to the method of construction. A mount consisting of lead-in wires, bead, and filament

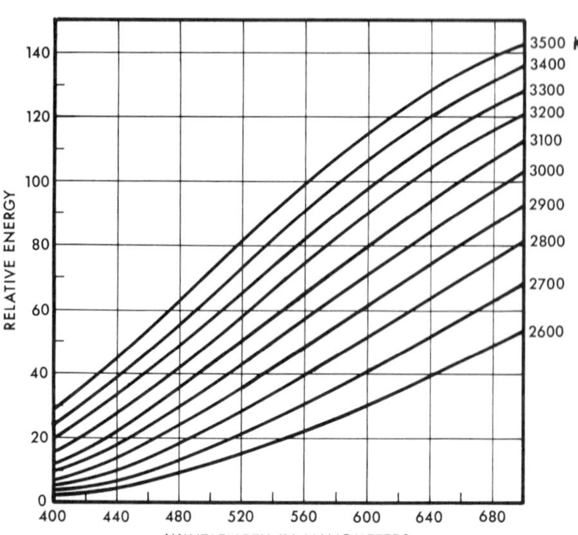

Fig. 8–2. Spectral energy distribution in the visible region from tungsten filaments of equal wattage but different temperatures.

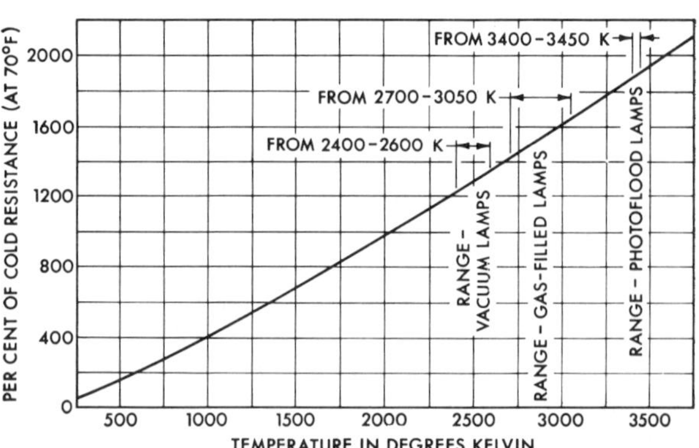

Fig. 8–3. Variation of tungsten filament hot resistance with temperature.

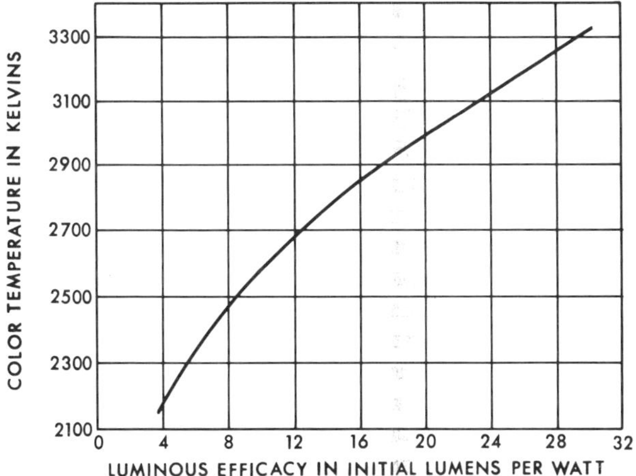

Fig. 8–4. Variation of color temperature with lamp efficacy.

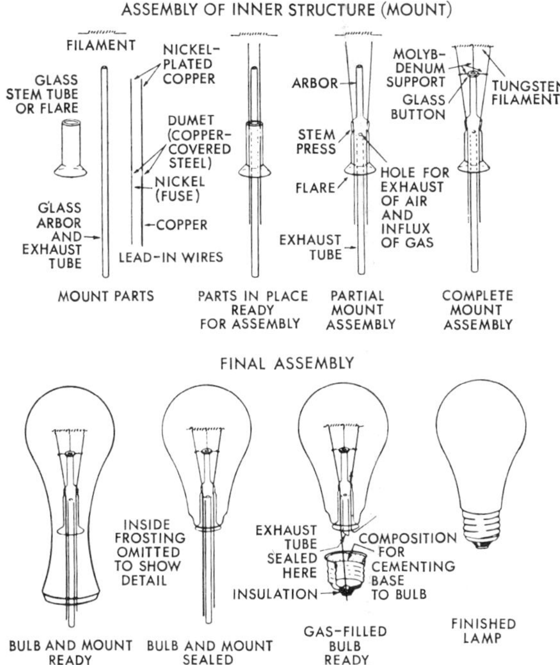

Fig. 8–5. Steps in the manufacture of a typical incandescent filament lamp.

is dropped into the open end of the bulb. The lead-in wires are bent to locate the filament at approximately the desired distance from the bulb end. An exhaust tube is then dropped down and butted against the lead-in wire and glass bulb just prior to sealing and exhausting. The base, applied later, together with the basing cement, must not only provide the lamp contacts but also protect the delicate seal. Because of seal limitations, butt seal lamps are restricted to relatively small wire sizes with a current limit of about 1.0 ampere. The filament position varies considerably more than in flange seal lamps since there is no definite relationship between the plane of the filament and base pins. Occasionally butt seal lamps are used without bases; when done so these lamps should be carefully handled. When

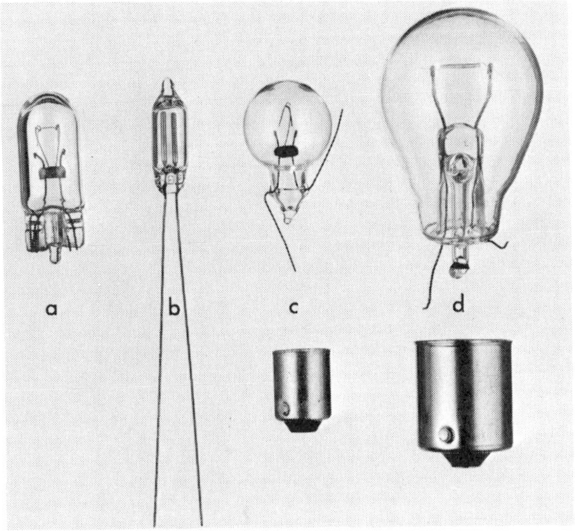

Fig. 8–6. Primary type of bulb construction: a. pinch seal with wedge base, b. pinch seal with lead-in wire terminals, c. butt seal, and d. flange seal.

used with a base the advantages of butt seal construction are: (1) low cost and (2) small size (usually ¾ inch and below).

The *pinch seal* is so named because glass is pinched or formed around the lead-in wires. Two forms are used: wire terminals or wedge base construction. For the smaller types of glow lamps the bulb is exhausted and tipped off at the end opposite the lead-in wires. With the newer wedge base lamps the exhaust tip is at the bottom rather than the top. Pinch seal construction eliminates the need for a conventional base. The advantages are: (1) low cost; (2) small size; (3) with filament lamps, the elimination of solder and cement which allows operation up to 600° F; and (4) small space required for the wedge base lamps.

Filament Forms and Designations. Filament design is a careful balance of light output and life. Filament forms, sizes, and support constructions vary widely with different types of lamps. See Fig. 8–7. Their designs are determined largely by service requirements. Filament forms are designated by a letter or letters followed by an arbitrary number. Most commonly used letters are: S—straight, meaning the wire is uncoiled; C—coiled, the wire is wound into a helical coil; CC—coiled coil, the wire is wound into a double helical coil. Coiling the filament increases its luminous efficacy; forming a coiled-coil further increases efficacy. More filament supports are required in lamps designed for rough service and vibration service than for general service lamps.

The Bulb

Shapes and Sizes. Common lamp-bulb shapes are shown in Fig. 8–8. Bulbs are designated by a letter referring to the shape classification and a number which is the maximum diameter in eights of

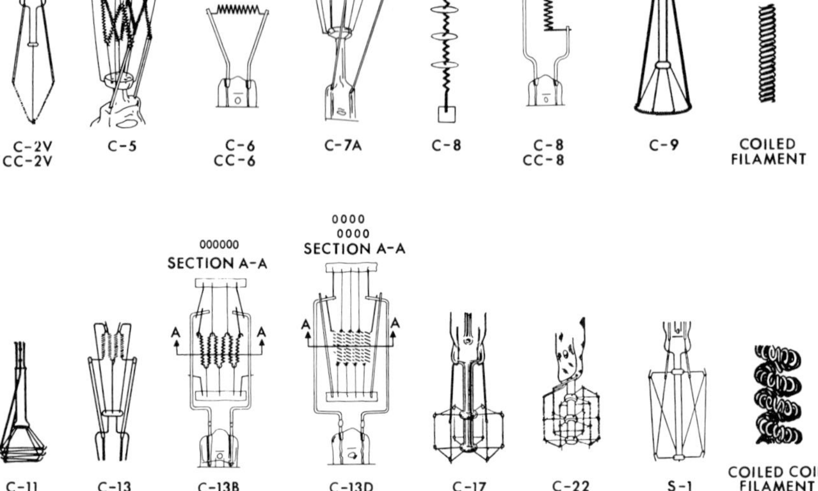

Fig. 8–7. Typical incandescent lamp filament construction (not to scale).

an inch. Thus, "R-40" designates a bulb of the "R" shape which is $^{40}/_{8}$ inches or 5 inches in diameter.

Types of Glass. Most bulbs are made of regular lead or lime "soft" glass; other types are of "heat-resisting" "hard" glass. The latter withstand higher temperatures and are used for highly loaded lamps. Under most circumstances they will better withstand exposure to moisture or luminaire parts touching the bulb. (See Bulb and Socket Temperatures, page 8-6.) Two specialized forms of glass are used as lamp envelopes: fused quartz and high silica glass. These materials both have high temperature resistance.

Bulb Finishes and Colors. Inside frosting is widely applied to many types and sizes of bulbs. It produces moderate diffusion of the light with scarcely measurable reduction in output and with little reduction in bulb strength. The extremely high filament luminance (photometric brightness) of clear lamps is reduced and striations and shadows practically eliminated with the use of frosting. White lamps having an inside coating of finely powdered white silica provide a substantially greater degree of diffusion with very little absorption of light.

Daylight lamps have bluish glass bulbs which absorb some of the red and yellow light produced by the filament. The resulting light output is of a higher color temperature. This color, achieved at the expense of about 35 per cent reduction in light output through absorption, varies between 3500 and

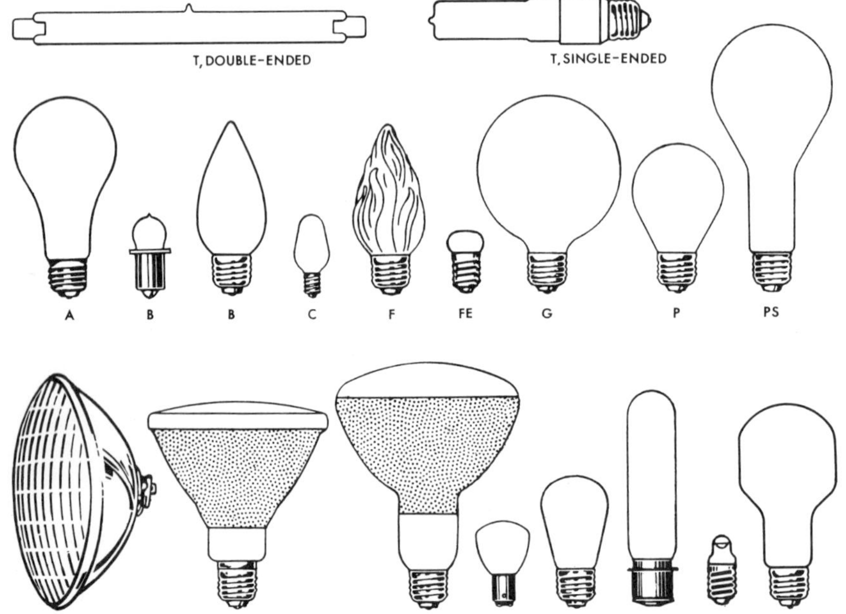

Fig. 8–8. Typical bulb shapes and designations (not to scale).

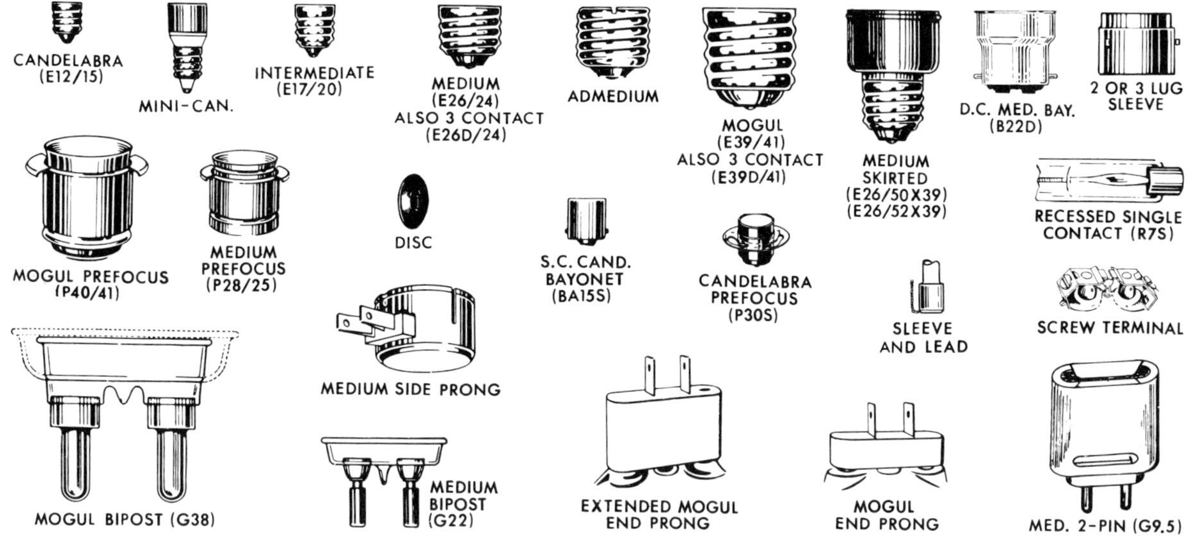

Fig. 8–9. Common lamp bases (not to scale). IEC Designations are shown, where available.

4000 K. This is about midway between tungsten-filament light and natural daylight.

Colored lamps are available with inside and outside-spray-coated, outside-ceramic, transparent plastic coated, and natural-colored bulbs. Outside-spray-coated lamps are generally used for indoor use where not exposed to the weather. Their surfaces collect dirt readily and are not easily cleaned. Inside-coated bulbs have smooth outside surfaces which are easily cleaned; the pigments are not exposed to weather and, therefore, have the advantage in permanence of color. Ceramic-coated bulbs have the colored pigments fused into the glass, providing a permanent finish to the bulbs. They are suitable for indoor or outdoor use. Most transparent plastic coated bulbs also can be used both indoors and outside. The coating permits the filament to be observed directly.

Natural colored bulbs are made of colored glass. Reflectorized colored lamps utilize ceramic coated bulbs, stained bulbs, plastic coated bulbs, and dichroic interference filters to obtain the desired color characteristics.

The Base

Fig. 8–9 shows the most common types of lamp bases.

Most lamps for general-lighting purposes employ one of the various types of screw bases. Where a high degree of accuracy in positioning of light sources with relation to optical elements is important, as is the case in projection systems, bipost and prefocus bases insure proper filament location. The bipost construction eliminates the conventional stem seal and lead assembly and substitutes a supporting structure that is made of one piece of channel nickel. Such construction gives maximum strength to the long leads supporting the heavy filament and provides a greater exposed surface to dissipate heat. The metal prongs which hold the lamp in the socket carry the weight of the entire metal structure instead of putting the burden on the glass. Lamp wattage is often a factor in determining the base type.

Most bases are secured to the bulbs by cement, cured by heat when the lamp is manufactured. Since this cement becomes weaker with age, particularly if exposed to excessive heat, some lamps intended for service which causes the base to run hot use a special heat-resisting basing cement or "mechanical" bases which employ no cement.

The Fill Gas

Up to about 1913, all lamps were of the vacuum type with uncoiled wire filaments. Attempts were made to reduce the rate of evaporation of the filament by the use of gas-filled bulbs which would allow higher filament temperatures and hence higher efficacies without a sacrifice in life. It was found that with uncoiled wire filaments the heat taken from the filament by the gas more than offset the advantage of the higher filament temperature. An incandescent filament operating in an inert gas is surrounded by a more or less stationary thin sheath of the gas and the percentage of the input energy that is lost becomes less as the filament diameter is increased. When the filament was coiled in a small helix it was found that the gas sheath surrounded the entire coil and greatly reduced the loss. The use of coiled filaments and gas-filled bulbs was a major improvement in incandescent lamp efficacies. A further improvement in efficacy has been obtained by the use of coiled-coil filaments in an increasing number of lamp types.

Inert gases are necessary for filling incandescent filament lamps in order to prevent chemical action between the gas and the internal parts of the lamp. Nitrogen was first used for the purpose. It was known that argon, due to its lower heat conductivity, was superior to nitrogen, but it was some years after the development of the gas-filled lamps before argon became available in sufficient quantity and purity, and at reasonable cost. Most present-day lamps are filled with argon with a small percentage of nitrogen, the nitrogen being necessary to suppress the tendency of arcing to take place between the lead-in wires.

The proportion of argon and nitrogen used depends on the voltage rating, the filament construction and temperature, and the lead-tip spacing. Typical amounts of argon that have been used are 99.6 per cent for 6-volt lamps, 95 per cent for 120-volt general service coiled-coil lamps, 90 per cent for 230-volt lamps having fused lead wires, and 50 per cent or less for 230-volt lamps when no fuses are used in leads. Projection lamps are 100 per cent nitrogen filled.

Krypton gas has lower heat conductivity than argon; also, the krypton molecule is larger than that of argon and therefore further retards the evaporation of the filament. Krypton is used in some lamps where the increase in cost due to the use of krypton is justified by the increased efficacy or increased life. Depending on the filament form, bulb size, and mixture of nitrogen and argon, conversion to krypton fill may increase efficacy by 7 to 20 per cent. Krypton is used in some special lamps such as marine signal and miner's cap lamps where high efficacy means less drain on the power source for a given amount of light produced.

Hydrogen gas has a high heat conductivity and is therefore useful for filling signaling lamps where quick flashing is desired.[3]

Tungsten-Halogen Cycle. Tungsten-halogen lamps utilize a halogen regenerative cycle to provide excellent lumen maintenance, together with lamp compactness. Tungsten-halogen lamps are a variation of incandescent filament lamps. The term halogen is the name given to a family of negative elements, *i.e.*, bromine, chlorine, fluorine, and iodine.

Although the tungsten-halogen regenerative cycle has been known for many years, no practical method of using it was established until the use of small diameter quartz envelopes for filament lamps provided the proper temperature parameters. Iodine was the halogen used in the first tungsten halogen lamp; today, other halogens are also being used.

It is generally accepted that the regenerative cycle starts with the tungsten filament operating at incandescence evaporating tungsten off the filament. Normally the tungsten particles would collect on the bulb wall, resulting in bulb blackening, common with incandescent lamps, and most evident near end of life. However, in halogen lamps, the temperature of the bulb is high enough (controlled minimum 500° F) so that the tungsten combines with the halogen within the lamp. This forms a tungsten-halogen compound which is gas. The gas continues to circulate inside the lamp until it comes in contact with the incandescent filament. Here, the heat is sufficient to break down the compound into tungsten, which is redeposited on the filament, and into the halogen, which continues its role in the regenerative cycle. It should be noted, however, that since the tungsten does not necessarily redeposit exactly at the spot from where it came, the tungsten-halogen lamp has a definitive life.

Energy Characteristics

The manner in which the energy input to a lamp is dissipated can be seen by reference to Fig. 8–78 (page 8–56) for typical general service lamps. The radiation in the visible spectrum (column 2) is the percentage of the input watts that is actually converted to visible radiation. The gas loss (column 4) indicates the amount of heat that is lost by the filament due to the conduction through and convection by the surrounding gas in gas-filled lamps. The end loss (column 6) is the heat lost from the filament by the lead-in wires and support hooks which conduct heat from the filament. Column 3 shows the total radiation beyond the bulb, which is less than the actual filament radiation due to absorption by the glass bulb and the lamp base.

Bulb and Socket Temperatures

Incandescent filament lamp operating temperatures are important for various reasons. Excessive lamp temperatures may affect the performance of the lamp itself. They may shorten the life of the luminaire or of the electrical supply circuit. Excessive lamp temperatures may also result in unsafe temperatures on combustible materials that form a part of the luminaire or of the material adjacent to the luminaire. Under certain atmosphere or dust conditions, high bulb temperatures (above 160° C) may be hazardous. Bulb and socket temperatures for a 100-watt A-19 lamp and a 500-watt PS-35 lamp for different operating positions and input voltages are shown in Fig. 8–10.

General service incandescent filament lamps are made of regular lead or lime "soft" glass, the maximum safe operating temperature of which is 700° F. Some lamps for special applications, such as outdoor floodlighting lamps, have "hard" glass bulbs which have a safe temperature limit of 885° F. Lamps with "still harder" glass bulbs can be operated up to 975° F. However, there are many low wattage lamps which have inner parts that will not operate satisfactorily with such high bulb temperatures.

From a lamp performance standpoint, the maximum safe base temperature for general service lamps is 345° F measured at the junction of the base and the bulb. Excessive temperature may cause failure of the basing cement as well as softening of the solder used to connect the lead wires to

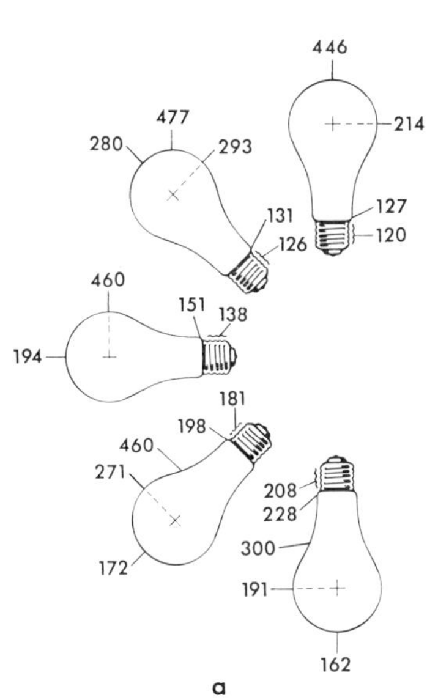

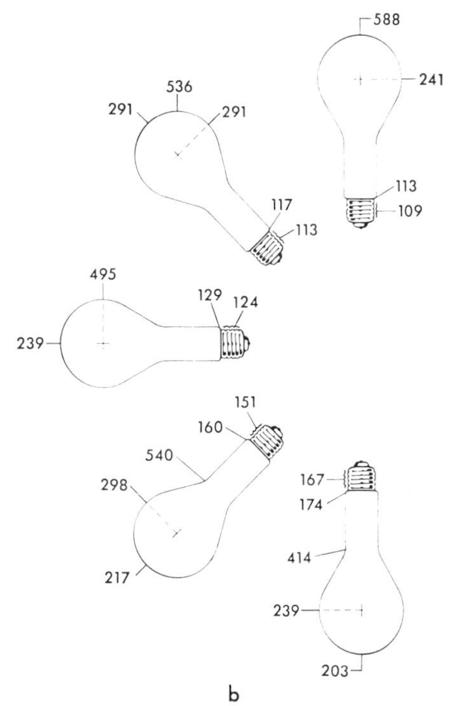

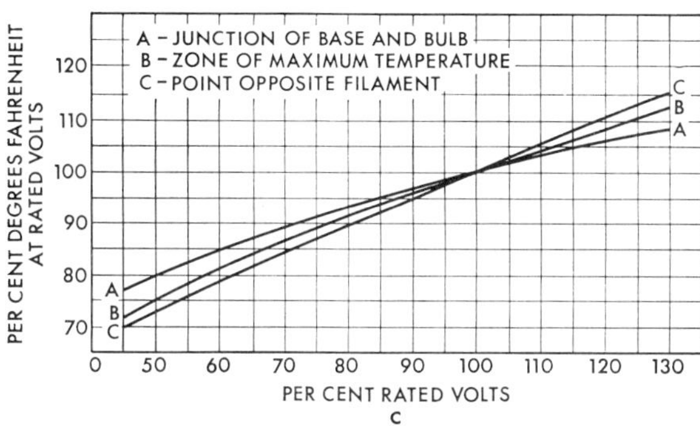

Fig. 8–10. Incandescent filament lamp operating temperatures in still air at 77°F ambient: (a) 100-watt cc-8, A-19 lamp; (b) 500-watt cc-8, PS-35 lamp; (c) approximate effect of voltage on the bulb temperature of a gas-filled lamp burning base up. All temperatures shown are Fahrenheit.

the base. Silicone-type high-temperature basing cement, combined with a higher-melting-point solder, permits a base temperature up to 500° F. Mechanical bases which are used in some lamps without basing cement generally should not be operated above 390° F. There are some incandescent filament lamps which have internal parts which will withstand operation of the mechanical base up to 410° F. Bipost bases carry considerable heat to the socket through the base pins, and the parts of the socket in contact with the base pins should be capable of withstanding a temperature of 550° F. Tubular quartz infrared and tungsten-halogen cycle lamps have a maximum seal temperature limit of 650° F. In addition, the tungsten-halogen cycle lamps have a *minimum* bulb temperature limit of 500° F to insure proper functioning of the regenerative cycle.

Some of the factors affecting base temperature are filament type, light center length, use of or lack of a heat shield, bulb shape and size, and gas fill. Lamps of lower wattage do not necessarily have lower base temperatures. When lamps of similar bases are placed in luminaires for the illumination effect desired, rather than by choice of lamp type or lamp wattage, luminaires should be capable of accepting lamp base temperatures in the order of 155° C. (This is a CSA* and IEC† limit for base temperature for A-type lamps.) If the luminaire will accept R- and PAR-type lamps then the temperature capability of the luminaire should be for 185° C if reasonable electrical insulation life is desired. Dichroic-reflector lamps should be placed only in luminaires specifically designed for them; this is normally indicated on the luminaire.

For base up burning where only slight enclosing of the lamp is provided by the luminaire, base temperature is the major factor affecting luminaire temperatures and lamp wattage is a minor consideration. As the luminaire provides more and more enclosing of the lamp, base temperatures become a less dominant factor and wattage assumes more importance.

Measurements of the thermal effect of radiation from incandescent lamps indicates a variation with direction at right angles to the bulb axis to a degree which cannot be ignored if the life of the electrical insulation in luminaire components is not to be impaired. Fig. 8–11 illustrates the type and extent of variation that may occur.

---

* Canadian Standards Association.
† International Electrotechnical Commission.

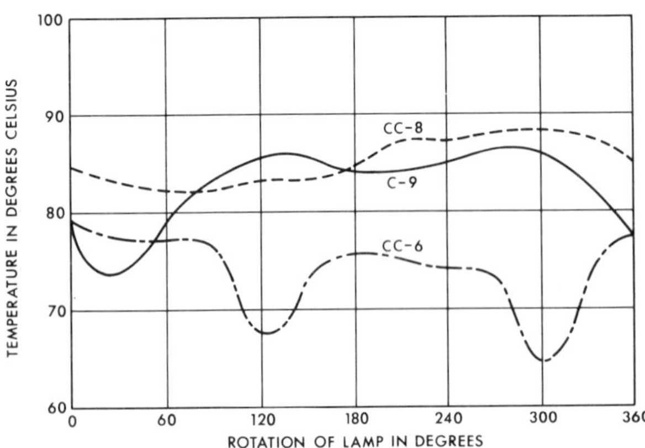

Fig. 8-11. Graphic illustration of how profile temperatures vary with the direction of radiation due to filament type.

Lamp Characteristics

Life, Efficacy, and Voltage Relationships. If the voltage applied to an incandescent filament lamp is varied, there is a resulting change in the filament resistance and temperature, current, watts, light output, efficacy, and life. These characteristics are interrelated and one cannot be changed without affecting the others. The following characteristic equations can be used to calculate the effect of a change from the design conditions on lamp performance (capital letters represent normal rated values):

$$\frac{\text{life}}{\text{LIFE}} = \left(\frac{\text{VOLTS}}{\text{volts}}\right)^d$$

$$\frac{\text{lumens}}{\text{LUMENS}} = \left(\frac{\text{volts}}{\text{VOLTS}}\right)^k$$

$$\frac{\text{LPW}}{\text{lpw}} = \left(\frac{\text{VOLTS}}{\text{volts}}\right)^g$$

$$\frac{\text{watts}}{\text{WATTS}} = \left(\frac{\text{volts}}{\text{VOLTS}}\right)^n$$

For *approximate* calculations, the following exponents may be used in the above equations: $d = 13$, $g = 1.9$, $k = 3.4$, and $n = 1.6$. For more accuracy, the exponents must be determined by each lamp manufacturer from a comparison of normal and over or under voltage tests of many lamp groups. Exponents will vary for different lamp types, for different lamp wattages, and for various ranges of per cent voltage variation. The values given above are roughly applicable to vacuum lamps of about 10 lumens per watt and gas-filled lamps of about 16 lumens per watt in a voltage range of 90 to 110 per cent of rated volts. For information outside this range, refer to Fig. 8-12.

The curves of Fig. 8-12a show the effect of voltage variations[4] on the characteristics of multiple lamps while the curves of Fig. 8-12b show similar characteristics for series lamps for variations in current.

In some instances, such as with flashlight lamps, life is modified by the lamp manufacturer because the most economical life would be so short that the frequency of lamp replacements would become a nuisance. Lamps designed for between 6- and 12-volt operation are usually the most efficient. Below 6 volts the lead-in wires conduct heat from the filament at a relatively greater rate. For example, a 5-volt flashlight lamp has an efficacy of 0.87; a 2.5-volt lamp has an efficacy of 0.67; and a 1.25-volt lamp has an efficacy of 0.25 candlepower per watt.

Filament Notching. Ordinarily, for laboratory test operation, normal tungsten filament evaporation controls incandescent lamp life. Where normal filament evaporation is the dominant failure mechanism, lamps should reach their design-predicted life. In recent years another factor influencing filament life has become prominant. This is commonly referred to as "filament notching". Filament notching is the appearance of step-like or saw-tooth irregularities on all or part of the tungsten filament surface after substantial burning. These notches reduce the filament wire diameter at random points. In some cases, especially in fine-wire diameter filament lamps, the notching is so severe as to almost penetrate the entire wire diameter. Thus, faster spot evaporation due to this notching and reduced filament strength become the dominant factors influencing lamp life. Lamp life due to filament notching may be as much as one half of so-called ordinary or normal, predicted lamp life.

The prominence of filament notching is due to at least three factors (primarily associated with fine-

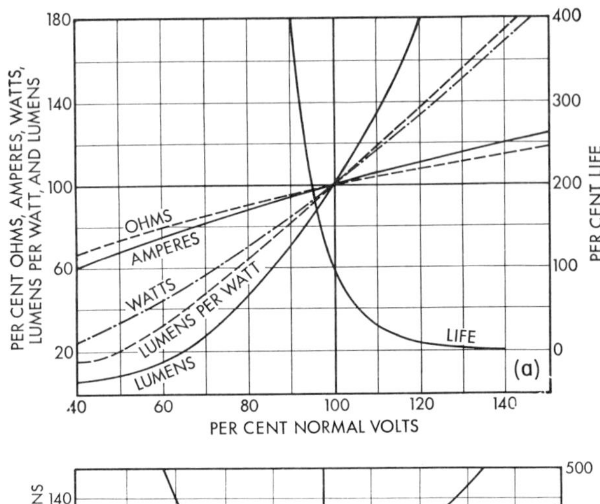

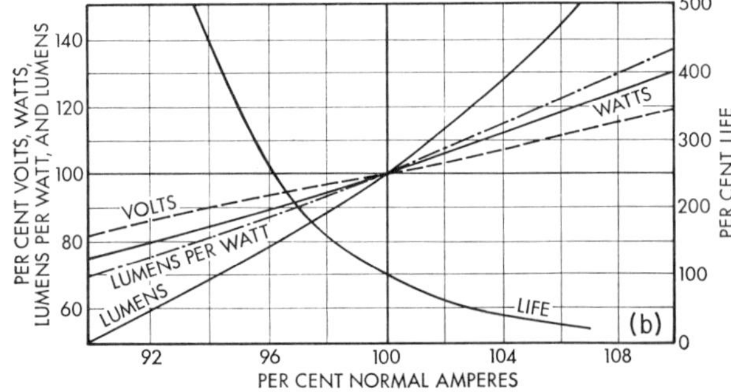

Fig. 8-12. Effect of current and voltage variation on the operating characteristics of street lighting lamps: (a) multiple lamps; (b) series lamps.

wire filament lamps): (1) low temperature filament operation, less than that for significant normal tungsten evaporation—these are long life lamp designs such as 10,000- to 100,000-hour designs; (2) small filament wire sizes—less than 0.001-inch diameter in many cases; and (3) increased use of dc voltage operation—generally as a result of advances in solid state technology.

Depreciation During Life

Over a period of time incandescent filaments evaporate and become smaller, which increases their resistance. In multiple circuits, the increase in filament resistance causes a reduction in amperes, watts, and lumens. A further reduction in lumen output is caused by the absorption of light by the deposit of the evaporated tungsten particles on the bulb. Fig. 8–13a shows the change in watts, lumens, and lumens per watt for a 200-watt lamp.

In series circuits having constant current regulators the increase in filament resistance during life causes an increase in the voltage across the lamp and a consequent increase in wattage and generated lumens. This increase in lumens is offset to varying degrees by the absorption of light by the tungsten deposit on the bulb. In low-current lamps the net depreciation in light output during life is very small, or in the smaller sizes there may be an actual increase in light output during life. In lamps of 15- and 20-ampere ratings, the bulb blackening is much greater and more than offsets throughout life the increase in lumens due to the increased wattage. Fig. 8–13b shows the change in watts, lumens, and lumens per watt for high-current series lamps.

The blackening in vacuum lamps is uniform over the bulb. In gas-filled lamps the evaporated tungsten particles are carried by convection currents to the upper part of the bulb. When gas-filled lamps are burned base-up most of the blackening occurs on the neck area where some of the light is normally intercepted by the base. Consequently, the lumen maintenance for base-up burning is better than for base-down or horizontal burning.

An appreciable gain in lumen maintenance can be obtained through the use of a coiled-coil filament located on or parallel to the axis in a base-up burning lamp.

To reduce blackening from impurities in the gas filling, an active chemical, known as a "getter," is used inside the bulb. This "getter," during the first burning hours of the lamp, tends to combine with and absorb any traces of oxygen or water vapor remaining in the bulb. In certain other lamps where the blackening would not be reduced sufficiently by "getters" alone, various other means are employed. Some high-wattage lamps have a small amount of tungsten powder loose in the bulb, which, when shaken about, wipes off much of the blackening. Other lamps have a wire mesh "collector grid," mounted above the filament, to attract and con-

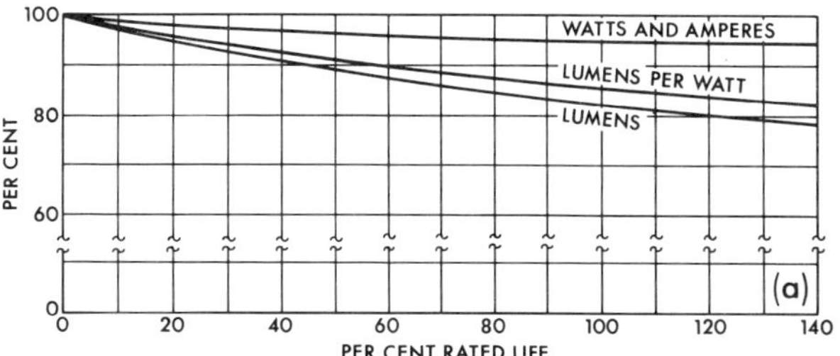

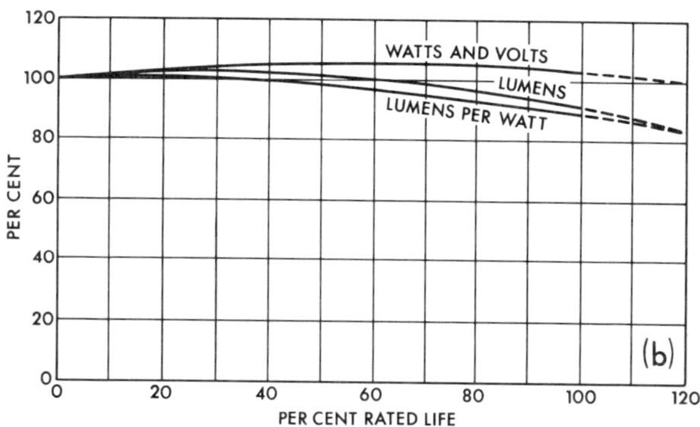

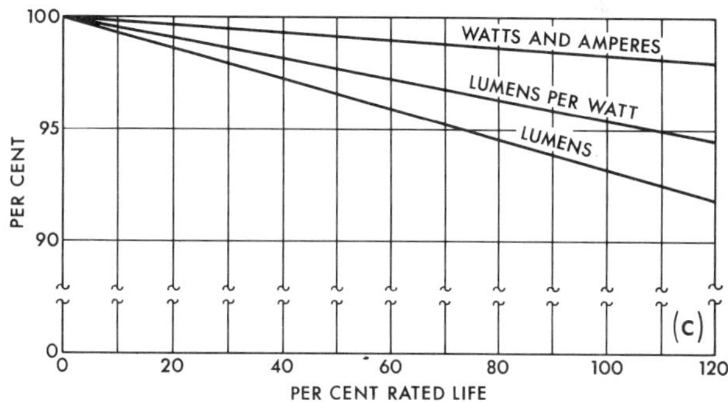

Fig. 8–13. Typical operating characteristics of lamps as a function of burning time: (a) multiple lamps and (b) series lamps. Operating characteristics of a tungsten-halogen lamp are shown in (c).

dense the tungsten vapor and hold the tungsten particles, thereby reducing bulb blackening.

Tungsten-halogen cycle lamps have significantly less depreciation during life due to the regenerative cycle which removes the evaporated tungsten from the bulb and redeposits it on the filament. Fig. 8–13c shows the change in lumens, watts, and lumens per watt for the 500-watt tungsten-halogen lamp.

Lamp Mortality and Renewal Rate

Many factors inherent in lamp manufacture and lamp materials make it impossible to have each individual lamp operate for exactly the life for which it was designed. For this reason lamp life is rated as the average of a large group. A typical mortality curve of a large group of lamps is illustrated in Fig. 8–14 and is representative of the performance of good quality lamps.

The mortality curve influences the rate of lamp replacements for installations involving large numbers of sockets. If individual lamps are replaced as they burnout the replacement rate is shown in Fig.

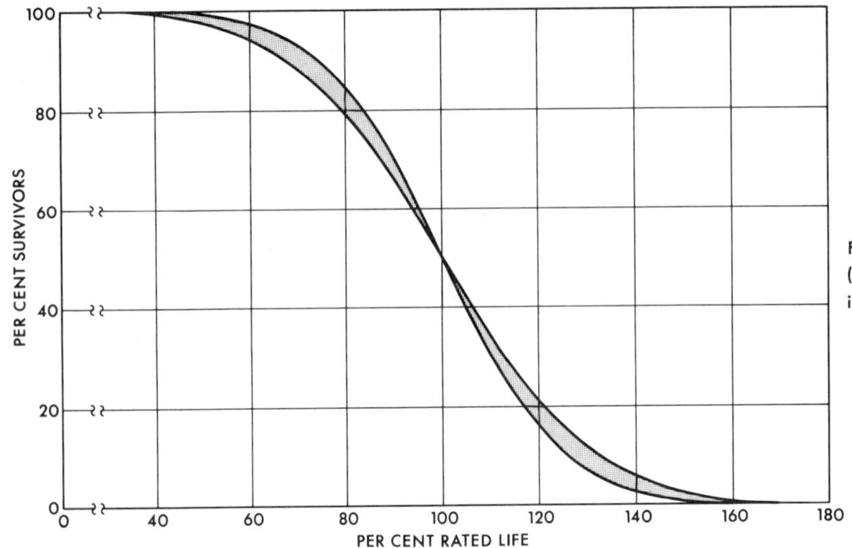

Fig. 8–14. Range of typical mortality curves (average for a large group of good quality incandescent filament lamps).

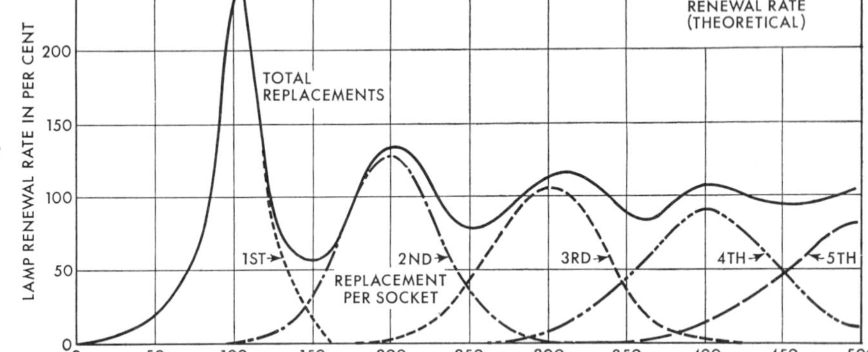

Fig. 8–15. Renewal rate curves applicable to all types of lamps.

8–15. Starting with a new installation, relatively few burnouts would be expected during the first several hundred hours of operation. As design life is approached the rate of burnout increases rapidly. After elapse of a burning period equivalent to some four or five times the average life of the lamps involved, the renewal rate fluctuation finally reaches a fairly steady or normal rate. In an "old" installation, the number of lamp failures to be expected for a given period may be determined by multiplying number-of-lamps-in-installation by number of hours actually burned during the period and dividing the result by average lamp life in hours.

Classes of Lamps

Incandescent filament lamps are divided into three major groups: large lamps, miniature lamps, and photographic lamps. These are cataloged separately by lamp manufacturers. There is no sharp dividing line between the groups. The large lamp classification generally refers to those with larger bulbs, with medium or mogul bases, and for operation in circuits of 30 volts or higher. The miniature classification generally includes lamps referred to as automotive, aircraft, flashlight, Christmas tree, radio panel, telephone switchboard, bicycle, toy train, and many other small lamps generally operated from circuits of less than 50 volts. The photographic classification includes lamps designed for photographic or projection service. The following gives a brief description of a few of the many types of lamps that are regularly manufactured. More complete details are available in manufacturers' catalogs.

General Service. These are large lamps made for general lighting use on 120-volt circuits. General service lamps are made in sizes from 15 watts to 1500 watts and satisfy the majority of lighting applications. The larger sizes are made in both clear and frosted bulbs. Below 150 watts, inside frosted bulbs are standard. Performance data are shown in Fig. 8–79 on page 8–57.

High Voltage (220–300 Volts).* This voltage class refers to lamps designed to operate directly on

* It should be noted that lampholders should be Underwriter's Laboratories or Canadian Standards Association "approved" for voltage level appropriate to the voltage rating of the lamps being used, i.e., 250 volts for lamps up to 250 volts, 600 volts for lamps above 250 volts.

circuits of 220 to 300 volts and represents a very small portion of the lamp demand.

High-voltage lamps have filaments of small diameter and longer length and the filaments require more supports than corresponding 120-volt lamps. They are, therefore, less rugged and produce 25 to 30 per cent fewer lumens per watt because of greater heat losses. The higher operating voltage causes these lamps to take less current for the same watts, permitting some wiring economy.[5] Operating characteristics and physical data are shown in Fig. 8-80.

Extended Service. Extended service lamps are intended for use in applications where a lamp failure causes great inconvenience, a nuisance, or a hazard, or where replacement labor cost is high or power cost unusually low. Therefore, for such applications where longer life is most important and a reduction of approximately 15 per cent in light output is acceptable, available extended service lamps with 2500 hours rated life are recommended. Where replacement of burned-out lamps is an easy, convenient operation, as in residential use, long life lamps are not usually recommended.[6] For most general use, incandescent lamps with the usual 750 or 1000 hours design life give a lower cost of light than extended service lamps.† See Fig. 8-82.

General Lighting Tungsten-Halogen. These lamps compete with the regular incandescent sources. The advantages over regular incandescent lamps include excellent lumen maintenance and compactness. The source can also be described as providing a whiter light (higher color temperature) as well as longer life at a given light output. The lamps shown in Fig. 8-85 are tungsten-halogen lamps without an external bulb and are intended for use in general lighting applications. Tungsten-halogen lamps intended for specialty use and lamps with an external bulb are shown in the appropriate specialty tabulations.

There is relatively more ultraviolet radiation from tungsten-halogen lamps than from regular incandescent lamps due to higher filament temperature combined with a transparent quartz envelope. Precaution against hazard to both people and objects are recommended in applications where tungsten-halogen lamps are operated at color temperatures above 3100 K because both the visible light output and the relative ultraviolet output are sharply increased.[7]

Train and Locomotive. Lamps designated as train and locomotive service are designed for several classes of low-voltage (30, 60, 75, etc.) service usually provided by generators, with a battery floated across the line. Data for these lamps are shown in Fig. 8-88. Low-voltage lamps have shorter and heavier filaments than 120-volt lamps of the same wattage and, consequently, are more rugged and generally produce more lumens per watt.

† Longer life is obtained by operating the lamp's filament at lower temperature than normal. This, however, lowers the lamp's luminous efficacy. In most general service use the cost of power used during the lamp life runs many times the lamp cost and, therefore, efficacy is important.

DC Series. Transit system voltages and some railway shop and yard voltages range from 525 to 625 volts. Lamps for this service (see Fig. 8-81) are operated five to twenty in series on these voltages. The design of voltages of individual lamps operated five in series are nominally 115, 120, and 125 volts. To identify dc series lamps, they are rated in odd wattages (36, 56, 94, 101, etc.).

Thirty-volt gas-filled lamps also are used for car lighting. The trolley voltage divided by 30 determines the number of lamps connected in series across the line. These lamps are equipped with short-circuiting cut-outs which short-circuit the lamps on burnout, thus preventing arcing and leaving the remainder of the lamps in a given circuit operating. These 30-volt lamps are rated in amperes, instead of the usual watt rating.

Spotlight and Floodlight. Lamps used in spotlights, floodlights, and other specialized luminaires for lighting theater stages, motion picture studios, and television studios have concentrated filaments accurately positioned with respect to the base. When the filament is placed at the focal point of a reflector or lens system, a precisely controlled beam is obtained. These lamps, listed in Fig. 8-90, are intended for use with external reflector systems. Because of their construction, these lamps must be burned in positions for which they are designed, to avoid premature failures.

Reflectorized. These lamps include those made in standard and special bulb shapes and which have a reflecting coating directly applied to part of the bulb surface. Both silver and aluminum coatings are used.[8] Silver coatings may be applied internally or externally, and in the latter case the silver coating is protected by an electrolytically-applied copper coating and sprayed aluminum finish. Aluminum coatings are applied internally by condensation of vaporized aluminum on the bulb surface. The following reflectorized lamps are regularly available: *bowl reflector lamps*, data for which are given in Fig. 8-84; *neck reflector lamps; reflector spot and reflector flood lamps* in R-type bulbs (certain sizes of reflectorized lamps are available with heat resisting glass bulbs; performance data are given in Fig. 8-91); *PAR spot and PAR flood lamps* (PAR bulbs typically are constructed from two molded glass parts, the reflector and the lens, which are fused together; see Fig. 8-91 for performance data and Fig. 8-16 for typical candlepower distribution curves); *reflector show case lamps* in T-type bulbs (see Fig. 8-83 for performance data). Colored R and PAR lamps are available. PAR lamps with interference-filter reflectors are available for applications where it is desirable to reduce the infrared energy in the beam. General lighting PAR lamps utilizing tungsten-halogen cycle filament tubes are included in Fig. 8-91.

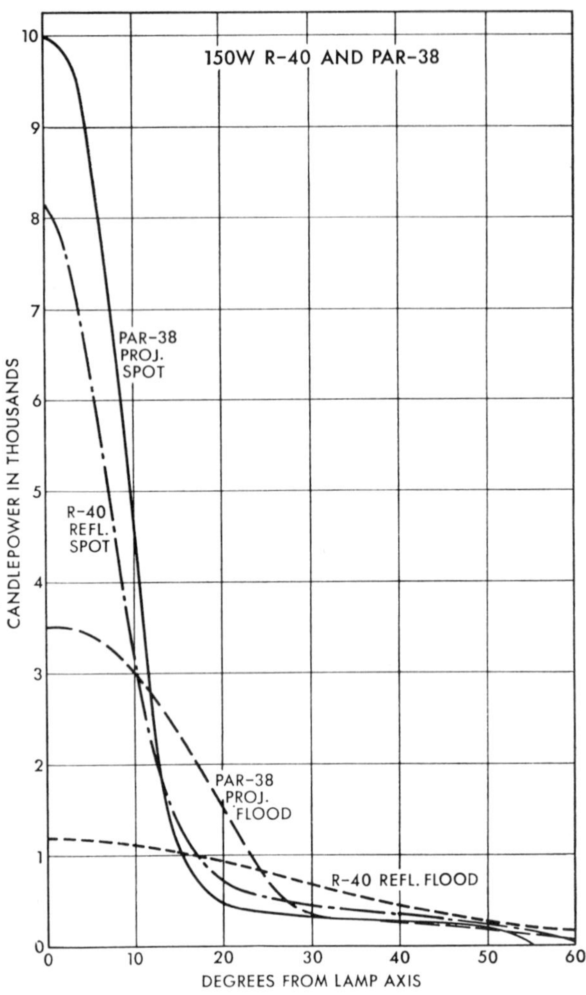

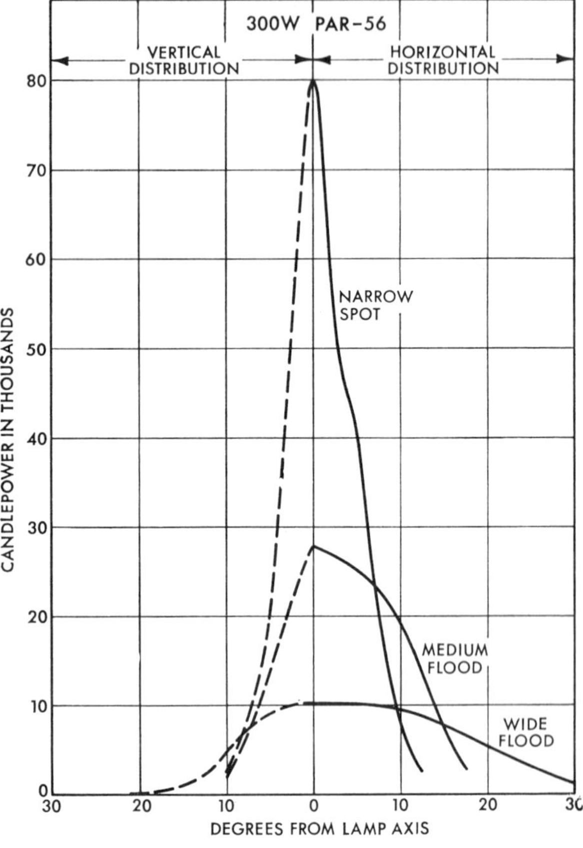

Fig. 8–16. Typical candlepower distribution characteristics of representative types of reflector and PAR lamps.

High Temperature Appliance. These lamps are specially constructed for high temperature service. The most common types are clear, medium base, 40-watt A-15 appliance and oven; 50-watt, A-19; and 100-watt, A-23 bake oven lamps. Range oven lamps are designed to withstand oven temperatures up to 475° F and bake oven up to 600° F.

Rough Service. To provide the resistance to filament breakage as required for portable extension cords, etc., rough service lamps employ special, multiple-filament-support construction. See Fig. 8–7, C-22. Because of the number of supports, the heat loss is higher and the efficacy lower than for general service lamps. Performance data are given in Fig. 8–89.

In using miniature lamps where rough service conditions are encountered, bayonet and wedge base lamps should be chosen instead of screw base lamps. Bayonet and wedge base lamps lock in the socket whereas screw base lamps tend to work loose.

Vibration Service. Most lamps have coiled filaments made of tungsten having high sag resistance. Vibration lamps, designed for use where high-frequency vibrations would cause early failure of general service lamps, are made with a more flexible tungsten filament. The sagging characteristics of the wire used allow the coils to open up under vibration, thus preventing short circuits between coils. See Fig. 8–100 for data.

Miniature radio panel lamps of 6.3 volts and under, incorporate mounts whose resonant frequency has been synchronized with that of the coiled filament to withstand shock and vibration.

Vibration and shock frequently accompany each other and sometimes only experiment will determine the best lamp for the purpose. Vibration-resisting sockets or equipment, utilizing a coiled spring or other flexible material to deaden vibration, have been employed where general service lamps are used under conditions of severe vibration.

Lumiline. The lumiline lamp has a tubular bulb and two disk bases: one at each end of the lamp, with the filament connected between them. The filament, in the form of a stretched coil, is supported on glass insulating beads along a small metal channel within the bulb. The 30- and 60-watt sizes are available in the 18-inch length. The 40-watt lamp is made in a 12-inch length. All sizes are available in either clear or inside-frosted tubes as well as white and various color coatings. See Fig. 8–97 for data.

Showcase. These use tubular bulbs and conventional screw bases. The longer lamps have elongated filaments with filament supports similar to lumiline lamps. The common sizes are 25- and 40-watt, but sizes up to 75 watts are available. See Fig. 8-83.

Three-Lite. These lamps employ two filaments, operated separately or in combination to provide three levels of illumination. The common lead-in wire is connected to the shell of the base; the other end of one filament is connected to a ring contact, and the end of the other filament to a center contact. Three-lite lamps are available in several different wattage ratings. See Fig. 8-86.

Sign. While large numbers of gas-filled lamps are used in enclosed and other types of electric signs, those designated particularly as "sign" lamps are mostly of the vacuum type. Lamps of this type are best adapted for exposed sign and festoon service because the lower bulb temperature of vacuum lamps minimizes the occurrence of thermal cracks resulting from rain and snow. Some low-wattage lamps, however, are gas-filled for use in flashing signs. Bulb temperatures of these low-wattage, gas-filled lamps are sufficiently low to permit exposed outdoor use.

Decorative. A wide variety of lamps for decorative applications is available from lamp manufacturers. Different bulb shapes, together with numerous colors and finishes, are used to achieve the desired appearance. Lamp manufacturers' catalogs should be consulted for information on the various decorative types.

Series Street Lighting. Street series lamps are designed to operate in series on constant current circuits. The most common circuit uses 6.6 amperes, automatically regulated to maintain this value regardless of the number and size of the lamps in the circuit. Lamps for 15- and 20-ampere ratings usually are connected in the secondary of a transformer, the primary being in the 6.6-ampere circuit.

These lamps are designated by their initial lumen output and ampere rating; for example, the 2500-lumen, 6.6-ampere lamps, or the 6000-lumen, 20-ampere lamps. Watt and volt ratings, as used to designate multiple lamps, are not commonly employed. Because the lumen output of series lamps is generally specified in street lighting contracts, improvements in light output are reflected by reductions in watts and volts. This usually results in odd numbers and fractions for volts and watts; for example, the present 2500-lumen, 6.6-ampere lamp has an average rating of 21.5 volts and 142 watts. See Fig. 8-92.

Multiple Street Lighting. Multiple street lighting lamps are designed so that their mean lumens correspond approximately to the mean lumens of series lamps of the same initial lumen rating. Since the lamp's voltage is established by a multiple circuit, the watts come out in odd values in order to obtain the desired nominal lumen ratings. See Fig. 8-93 for performance data.

Traffic Signal Lamps. Lamps used in traffic signals are subjected to more severe service requirements than most applications of incandescent lamps. In order to provide uniformity of application, lamps must be compatible with the design requirements of optical systems of standard traffic signals. Data on typical traffic signal lamps are shown in Fig. 8-94.

Aviation. Lighting for aviation is divided into two classes: lighting on and around airports, and lighting on aircraft.

In airport lighting, both multiple and series type lamps are used. Most new systems being installed use series type lamps of 6.6-ampere and 20-ampere designs for airport approach, runway, and taxiway lighting, whereas multiple lamps are used for obstruction, hazard beacon, and airport identification beacon lighting. On aircraft, small and miniature lamps are used exclusively for both interior and exterior lighting.

Most of the lamps used in airport lighting are designed to be used in precise projection type equipment to produce a controlled beam of light complying with required standards.

Hazard beacons and airport identification beacons signaling the presence of high obstructions or the whereabouts of the airport, use lamps of wattages ranging from 500 to 1200 watts. Lamps used on the airport proper range in size from 10 to 500 watts. Lamps used for aircraft lighting are in the miniature classification, although landing lamps as large as 1000 watts are used.

Lamps widely used for aviation service are listed in Figs 8-95 and 8-96. These tabulations include those lamps currently recommended for use in aircraft and airport lighting. In addition, tungsten-halogen lamps can be provided in place of many regular incandescent types. Tungsten-halogen lamps have the advantages of better lumen maintenance and longer life. Many of the types of lamps available but seldom used are not shown. Refer to Section 21 for information on the application of lamps in the airport and aircraft lighting fields.

Ribbon Filament Lamps. Incandescent lamps made with ribbons or strips of tungsten for the filaments have been used in special applications where it is desirable to have a substantial area of fairly uniform luminance.[9] Ribbon dimensions vary from 0.7 to 4 millimeters in width and up to 50 millimeters in length. The 5- to 20-ampere ribbon filament lamps are usually employed in recorders, instruments, oscillographs, and microscope illuminators. The 30- to 75-ampere lamps are used for pyrometer calibration standards and for spectrographic work. Fig. 8-17 shows typical lamps.

Miniature Lamps

The term "miniature" applied to light sources is really a lamp manufacturer's designation deter-

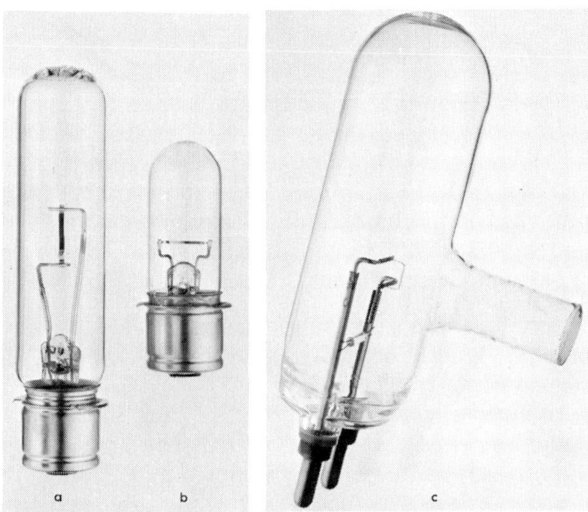

Fig. 8–17. Typical ribbon filament lamps: (a) 6-volt, 18-ampere, T-10, 2 mm., 3000 K microscope illuminator; (b) 6-volt, 9-ampere, T-8½, 1 mm., 3000 K optical source; and (c) 3.5-volt, 30-ampere, T-24 with quartz window, 3 mm., U-shaped filament 2300 K pyrometer and spectroscope source.

mined by the trade channels through which the lamps so identified are distributed, rather than by the size or characteristics of the lamps. In general, however, it *is* true that most miniature lamps *are* small, and consume relatively little power. The most notable exceptions to this generalization are the sealed-beam type lamps, some of which are classed as miniature lamps, even though they may be as large as eight inches in diameter and consume up to 1000 watts.

The great majority of all miniature lamps are either incandescent filament lamps or glow lamps. (Glow lamps are covered under Miscellaneous Discharge Lamps, see page 8–44.) Some low wattage and relatively low voltage fluorescent lamps are listed as miniature types by at least one lamp manufacturer. Also, electroluminescent lamps and light emitting diodes (see page 8–48) are included in the miniature lamp family. Incandescent miniature lamps[10] and glow lamps[11] are designated completely by numbers standardized and issued by the American National Standards Institute.

With the notable exception of multiple type Christmas lamps, miniature incandescent lamps are designed to operate under 50 volts. These voltages may be obtained from batteries, generators, or circuits with low voltage transformers.

Miniature lamps are used chiefly when conditions require a light source to be of a small size or consume little power. They have many uses, such as in the following principal fields: automotive, aircraft, and decorative. Glow lamps are also used as components in electronic circuitry.

Sub-miniature lamps have increased in popularity. They range in size from T-2 down to T-⅛. Since early in World War II the T-1¾ has been used extensively for instruments and indicators. The T-1 size has become popular for aircraft instruments and indicators. The T-⅝ and smaller sizes down to T-⅛ are used chiefly in novelty applications such as tiny flashlights, jewelry, etc. Examples of variations in range and specifications are given in Fig. 8–98.

Power Sources. Most miniature incandescent filament lamps operate at voltages under 50 volts. However, miniature lamps may be used on 120-volt circuits when transformers, rectifiers, or resistors are used to reduce the voltage.

The mean effective voltage delivered by the battery or circuit is generally higher than average voltage and should be the design voltage of the lamp. Design voltages for flashlight lamps have been determined by extensive tests. Proper lamp and battery combinations are shown in manufacturers' catalogs.

With transformers or resistors, delivered rather than rated voltage must be precisely known in order to obtain proper lamp life and output. On resistor operation, because the voltage increases as the filament evaporates, lamp life will generally be one-half of transformer operation.

Where space permits, larger rather than smaller dry cells should be chosen to reduce power cost. Fig. 8–18 gives sizes of batteries used with miniature lamps.

Flashlight, Handlantern, and Bicycle. These lamps are commonly operated from dry cell batteries having an open circuit voltage of 1.5 volts per cell for new batteries and dropping to approximately 0.9 volts per cell at the end of battery life. This results in a big difference in light output, depending upon whether the batteries are new or old. Typical lamps of this group are listed in Fig. 8–87.

Automotive. Lamps for most passenger vehicles, trucks, and coaches presently operate at 12 to 16 volts. The power source is a storage battery-rectified alternator system. Performance data of typical 6- and 12-volt lamps are given in Figs. 8–101 and 8–102 respectively.

Sealed Beam. These lamps contain filaments, lens, and reflectors in a precise rugged optical package available in a wide variety of sizes in voltages ranging from 6 to 120 volts. Sealed beam or PAR lamp lenses are made of an oven-ware type glass.

Fig. 8–18. American Standard Sizes of Dry Cells*

Designation	Nominal Cell Dimensions	
	Diameter	Height
AAA	25⁄64 in. (9.9 mm)	1 11⁄16 in. (42.8 mm)
AA	17⁄32 in. (13.5 mm)	1⅞ in. (47.6 mm)
C	15⁄16 in. (23.8 mm)	1 13⁄16 in. (46.0 mm)
D	1¼ in. (31.8 mm)	2¼ in. (57.2 mm)
F	1¼ in. (31.8 mm)	3 7⁄16 in. (87.3 mm)
No. 6	2½ in. (63.5 mm)	6 in. (152.4 mm)

* *American National Standard Specifications for Dry Cells and Batteries,* C18.1-1969 (IEC 86-1, 86-2, and 86-3) American National Standards Institute, 1430 Broadway, New York, N.Y. 10018.

The reflector is vaporized aluminum on glass, hermetically sealed to the lens cover. The advantages are: (1) accurate reflector contour for accurate beam control; (2) precise filament positioning on rugged filament supports; and (3) high efficacy and excellent candlepower maintenance. Vaporized aluminum on glass is an excellent reflector, does not deteriorate, and the normal bulb blackening has little effect on the candlepower output throughout lamp life. The sealed beam lamp is particularly suitable where a large amount of concentrated light at low voltage is required. Fig. 8–103 gives data on typical sealed beam lamps for automotive use.

Indicator, Radio, and Television. Lamps for indicator, radio, and television service are usually operated from low voltage transformers. Performance data on typical lamps are given in Fig. 8–104.

Flasher Lamps. Incandescent lamps which flash automatically (see Fig. 8–19), because of a built-in bi-metal strip similar to those used in thermostats, are available in several sizes. When the lamp lights, heat from the filament causes the bi-metal strip to bend away from the lead-in wire. This breaks the circuit and the lamp goes out. As the bi-metal strip cools it bends back to its original position against the lead-in wire and lights the lamp. This alternating cooling and heating keeps the lamp flashing. *Caution:* An exception to this is found in certain miniature screw base foreign made lamps. Some operate as described above, however, a few are of the "shorting" type. The bi-metal in the shorting type is so mounted that when it heats up it shorts across the lead-in wire. If these lamps, which are difficult to distinguish from the "opening" type, are inserted in sockets intended for the normal flasher lamps, they may cause run-down batteries, blown fuses, over-heated wires, or burned out transformers.

Some of the lamps listed in Fig. 8–99 (120-volt D26 and D27) are available also in transparent colors. The latter are designed particularly for Christmas and decorative effects.

Typical Uses. Indicator lamps provide a visual indication of existing circuit conditions. They are widely used in fire and police signaling systems, power plant switchboards, production machinery, motor switches, furnaces, and innumerable other devices requiring warning or pilot lights.

Miniature lamps may be wired in various ways with motor or heating elements and are used in many appliances to indicate that the current is flowing to the appliance or that it is functioning properly. They are used in many ways in instrumentation and in connection with photocells and relays, and play a vital part in computers and automation in general. Flashlights, radios, clocks, bicycles, and toys account for many more uses.

Other applications include the use of miniature lamps for Christmas and other festive occasions, and for colorful patio and garden lighting. For garden lighting low voltage miniature equipment is available. It should be noted that not all string sets and devices have Underwriters' Laboratories' approval. See also Section 17.

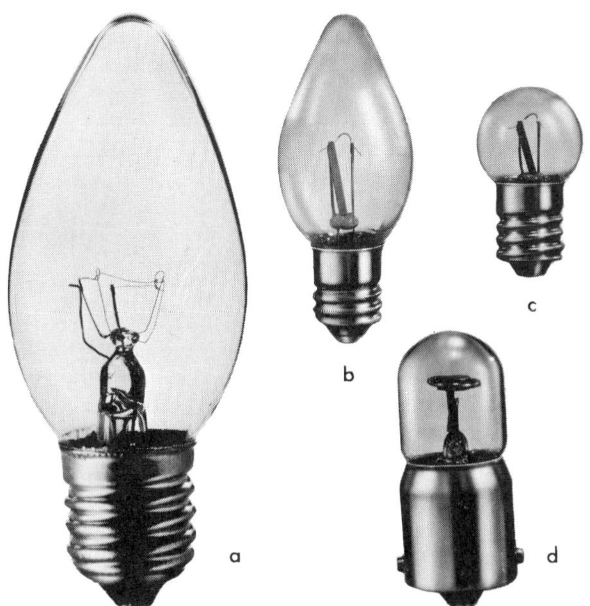

Fig. 8–19. Typical lamps for flashing: a. D27, b. 405, c. 407 (incandescent lamps with integral flasher); d. B6A (NE-21) (a glow lamp). (Shown here at approximately full size.)

Incandescent Filament Lamps for Photography

The application of lamps to photography is covered in Section 24. The design of lamps specifically for photographic service is concerned with actinic quality—that is, providing sources which are best adapted to the response or sensitivity of several classes of film emulsions. Some lamps are specified in terms of color temperature, which serves as a basic rating for film exposure data. Thus, several lines of lamps are made available for the requirements of commercial studios as well as for home movies and still hobby photography. Photographic efficiency and unvarying spectral quality are of major concern to the photographer. Comparatively, life is of less importance. Lamps of various sizes are often matched for color temperature, and rated life varies as necessary with wattage to achieve the specified color temperature.

Lamps with Color Temperature Ratings. Lamps rated at 3200 K color temperature are used primarily in professional photography, with both black-and-white films and many types of color films. Another group of lamps, rated at 3350 K, is used primarily in professional color motion picture work, sometimes in conjunction with blue filters to simulate daylight color quality. Lamps rated at 3400 K are widely used with color film by hobbyists and some professional photographers. So-called "daylight blue" photoflood lamps provide light of approximately 5000 K. Their blue bulbs act as filters, absorbing red and yellow rays, producing a whiter light. The "photographic blue" lamps produce a spectral quality that approximates sunlight at approximately 5500 K color temperature.

Maintenance of color temperature throughout lamp life is important in color photography. Typical color-temperature rated lamps of conventional construction may drop about 100 K through life. There is negligible change in the color temperature of tungsten-halogen lamps during their life.

Photoflood. These are high efficacy sources of the same character as other incandescent filament lamps for picture taking, with color temperatures ranging from 3200 to 3400 K. Because of their high filament temperature, these lamps generally produce about twice the lumens and three times the photographic effectiveness of similar wattages of general service lamps. Relatively small bulb sizes are employed (*e.g.*, the 250-watt No. 1 photoflood has the size bulb formerly used for the 60-watt general service lamp) so that these lamps may be conveniently used in less bulky reflecting equipment or for certain effects in ordinary residential or commercial luminaires.

The photoflood family includes reflector (R) and projector (PAR) lamps with various beam spreads. Some of these have tungsten-halogen light sources; some have built-in 5000 K daylight filters. In addition, tungsten-halogen lamps in several sizes and color temperatures are classed as photofloods and used in especially-designed reflectors. See Fig. 8–105.

Photoflash. These are physically patterned after standard incandescent filament lamps but are actually "combustion" sources. The lamp bulb is simply a container for the flammable material (usually metallic aluminum or zirconium or a metallic compound), a tungsten filament with a small amount of chemical applied or a pressure sensitive chemical igniter, and oxygen. The bulbs are coated with lacquer to safeguard against shattering the glass. These lamps are designed to function over specified voltage ranges or within fixed force and energy requirements (pressure sensitive igniters).

Photoflash lamps have a burning life of only a few hundredths of a second and the design in predicated on the service, type of camera, and the necessary synchronism of flash and shutter opening.

The use of more efficient flammable material such as zirconium has made possible the popular reflector lamp cube. Each small cube contains four miniature flash lamps complete with a tiny reflector for each lamp. These lamps may be ignited either electrically or mechanically.

Photoflash lamps are rated in lumen-seconds while reflectored cube units are rated in beam-candlepower-seconds, which is a measure of their photographic effectiveness. Color-corrected (blue-bulb) photoflash lamps and cubes are used to simulate daylight at approximately 5500 K by filtering out red and yellow. This correction reduces the light output by about one-fourth. Deep red purple bulbs are used to filter out all visible light for taking pictures with infrared film. See Fig. 8–106 for lamp data.

Photo-Enlarger. Most equipment used in the process of enlarging requires a highly diffused light source. For this service incandescent filament enlarger lamps in white glass bulbs are available. These lamps are designed for high efficacy at a short life. See Fig. 8–107. Fluorescent lamps, especially the "Circline" types, are used in some enlargers.

Projection. Lamps for projectors have carefully-positioned filaments and prefocus-type bases so the filament will be properly aligned with the optical system. The filaments are very compact and operate at relatively high temperatures. The efficacy, therefore, is high with consequent short life. Forced-air cooling is frequently required because of the high lamp temperatures.

The "4-pin" base is widely used, because it provides accurate light source positioning, allows minimum bulb height in "low profile" projectors, and facilitates the use of internal reflectors. Internal "proximity" reflectors, usually of metal, eliminate the need for the spherical mirror formerly used behind the lamp in conventional projectors. Internal ellipsoidal reflectors—metal mirrors or glass with dichroic heat-transmitting mirror coating—focus light through the film aperture into the projection lens, replacing both the conventional external spherical mirror and condenser lenses. For projection lamp data, see Fig. 8–108.

The use of tungsten-halogen lamps is increasing greatly in the projection area. They are available in double ended and single ended lamps in almost every type of base used with regular incandescent lamps. There is an additional tungsten-halogen type which is composed of a tiny lamp within a specially designed integral condenser-reflector which eliminates the need for condensers and reflectors within the projector itself.

Infrared Lamps

Any incandescent filament lamp is a very effective and efficient generator of infrared radiation. From 75 to 80 per cent of the wattage input to an incandescent filament lamp is radiated as infrared energy, and the greater portion of this energy is emitted in the invisible wavelength range from 780 to 5000 nanometers. Wavelengths longer than 5000 nanometers are absorbed to a large extent by the glass or quartz envelope. Lamps for heating applications are specially designed for low light output and long life.

Tungsten filament infrared lamps are available with ratings up to 5000 watts. Generally speaking, tungsten filament lamps for industrial, commercial, and residential service operate at a filament color temperature of 2500 K. At this relatively low operating temperature compared to lighting lamp

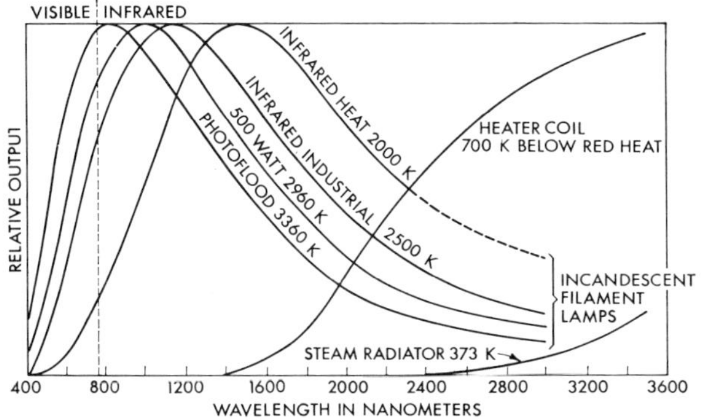

Fig. 8-20. Spectral distribution of energy from various infrared sources.

filament temperatures, the service life is well in excess of 5000 hours. Frequently, lamps using tungsten filaments have provided many years of operation because the service life is generally determined by mechanical breakage or rupture of the filament due to vibration or handling, rather than due to the rate of evaporation of tungsten as is the case with lighting lamps. Lamps having heat-resisting glass bulbs or tubular quartz envelopes are recommended where liquids may drop or be splashed on the bulb surface.

The distribution of energy radiated by various infrared sources is shown in Fig. 8-20, while Fig. 8-109 tabulates the most popular types of infrared lamps and metal sheathed heaters. For application information, see Section 25.

RF (Radio Frequency) Lamp

The RF (radio frequency) lamp is an incandescent source, in which a disk of highly refractory material is inductively heated to incandescence by radio-frequency power. This disk or target is a small, bright, and uniform light emitting surface 5/16-inch in diameter. The lamp can be operated at color temperatures up to 3700 K with a luminance of 40,000 candelas per square inch. Circumferential non-uniformity is held to within four per cent. Lamp

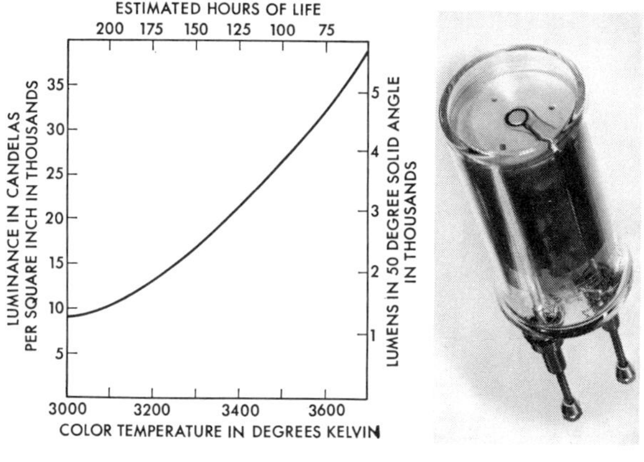

Fig. 8-21. Radio-frequency lamp (photograph and curve of performance characteristics as a function of color temperature).

performance characteristics as a function of color temperature operation and a photograph of a lamp are shown in Fig. 8-21. Concentration of the radio-frequency field on the light emitting target is accomplished by means of a split copper cylinder inside the glass envelope.

The RF lamp is used principally in motion picture printing but has other areas of use where a narrow pencil of light rays from a small but intense source is required.

FLUORESCENT LAMPS

The fluorescent lamp is an electric discharge source, in which light is produced predominantly by fluorescent powders activated by ultraviolet energy generated by a mercury arc. The lamp, usually in the form of a long tubular bulb with an electrode sealed into each end, contains mercury vapor at low pressure with a small amount of inert gas, principally argon, for starting. The inner walls of the bulb are coated with fluorescent powders commonly called phosphors (see Fig. 8-22). When the proper voltage is applied an "arc" is produced by current flowing between the electrodes through the mercury vapor. This discharge generates some visible radiation, or light, but mostly invisible ultraviolet radiation. The ultraviolet in turn excites the phosphors to emit light. See Fluorescence, page 2-8.

The phosphors in general use have been selected and blended to respond most efficiently to ultraviolet energy at a wavelength of 253.7 nanometers,[12] the primary wavelength generated in a mercury discharge.

Like most electric discharge lamps, fluorescent lamps must be operated in series with a current limiting device. This auxiliary, commonly called a *ballast*, limits the current to the value for which each lamp is designed. It also provides the required starting and operating lamp voltages.

Lamp Construction

Bulbs. Fluorescent lamps are most commonly made with tubular bulbs varying in diameter from 5/8 inches (T-5) to 2 1/8 inches (T-17) and in over-all length from a nominal 6 to 96 inches. The bulb is designated by a letter indicating the shape followed by a number indicating the maximum diameter in eighths of an inch. Hence "T-17" indicates a Tubular bulb 17/8 inches, or 2 1/8 inches, in diameter. The nominal length of the lamp includes the thickness of the standard lampholders. It is the back-to-back dimension of the lampholders with a seated lamp.

All fluorescent lamps are basically tubular bulbs of small cross sectional diameter. Shapes other than straight tubes are available. These include: a circle (circline), where the tube has been uniformly bent with the two ends adjacent to each other; a helicoid, in which the tube has a spiral groove running the length of the bulb; an intermittently grooved tube;

and a letter U, where the tube has been bent back upon itself.

Electrodes. Two electrodes are hermetically sealed into the bulb, one at each end. These electrodes are designed for operation as either "cold" or "hot" electrodes, more correctly called glow or arc modes of discharge operation.

Electrodes for glow or *cold cathode* operation may consist of closed-end metal cylinders, generally coated on the inside with an emissive material. Cold cathode lamps operate at a current of the order of a few hundred milliamperes, with a high cathode fall, something in excess of 50 volts.

The arc mode or *hot cathode* electrode is generally constructed from a tungsten wire or a tungsten wire around which another very fine tungsten wire has been uniformly wound. The larger tungsten wire is coiled producing a triple coil electrode. When the fine wire is absent, the electrode is referred to as a coiled-coil electrode. This coiled-coil or triple-coiled tungsten wire is coated with a mixture of alkaline earth oxides. During lamp operation the coil and coating reach temperatures of about 1100° C where the coil/coating combination thermally emits large quantities of electrons at a low cathode fall of the order of 10 to 12 volts. The normal operating current of hot cathode lamps presently range upwards to 1.5 amperes. As a consequence of the lower cathode fall associated with the "hot" cathode, more efficient lamp operation is obtained and, therefore, most fluorescent lamps are designed for "hot" cathode operation.

Gas Fill. The operation of the fluorescent lamp depends upon the development of a discharge between the two electrodes sealed at the extremities of the lamp bulb. This discharge is developed by ionization of mercury gas contained in the bulb. The mercury gas is maintained at a pressure of about .008 torr which corresponds to a condensed or liquid mercury temperature of 40° C (104° F), the optimum temperature of operation for which most lamps are designed. In addition to the mercury, a rare gas at low pressure (of the order of 1 to 3 torr) is added to the lamp to facilitate ignition of the discharge. As a consequence of ionization and mercury atom excitation, ultraviolet radiation is generated, particularly at a wavelength of 253.7 nanometers.

Phosphors. The color produced by a fluorescent lamp depends upon the blend of fluorescent chemicals (phosphors) used to coat the wall of the tube. (See Fig. 2-12 for a list of important phosphors.) There are different "white" and colored fluorescent lamps available, each having its own characteristic spectral power distribution (see Fig. 8-23). When not lighted, most fluorescent powders appear as a matte white, translucent coating; however, suitable fluorescent materials are not available to produce the red and gold lamps and an inner coating of that pigment is applied before the phosphor coating.

Bases. For satisfactory performance, each fluorescent lamp must be connected to a ballasted electrical circuit with proper voltage and current characteristics for its type. A number of different fluorescent lamp base designs are used. The bases physically support the lamp in most cases and provide a means of electrical connection (see Fig. 8-22).

The design of the base is dependent upon the lamp type. Lamps designed for instant start operation (see page 8-29) generally have a single connection at each end. As a consequence a single pin base is satisfactory. Such bases are shown in Fig. 8-24, numbers 8 through 10.

Preheat and rapid start lamps (see page 8-28) have four electrical connections; two at each end of the tube and, therefore, require dual contact bases. In the case of the circline lamp, a single four-pin connector is required. Examples of such bases are shown in Fig. 8-24. The base shown as number 1 is designed for use in a circline lamp, the bipin bases from 4 through 7 are used in low current applications (less than about 0.5 ampere), 11 and 12 in higher current applications, and numbers 13 and 14 for high current (0.800 to 1.50 amperes) rapid start lamp applications.

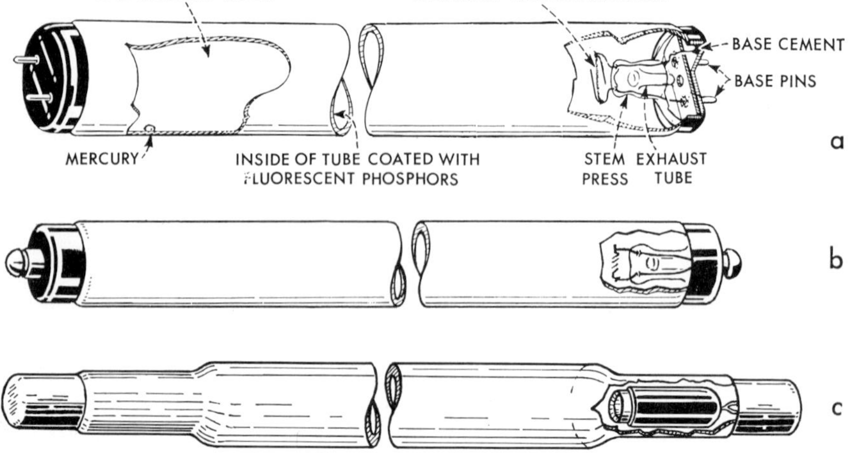

Fig. 8-22. Cutaway view of fluorescent lamps showing typical electrodes: (a) hot-cathode (filamentary) preheat-starting; (b) hot-cathode (filamentary) instant-starting; (c) cold-cathode (cylindrical) instant-starting.

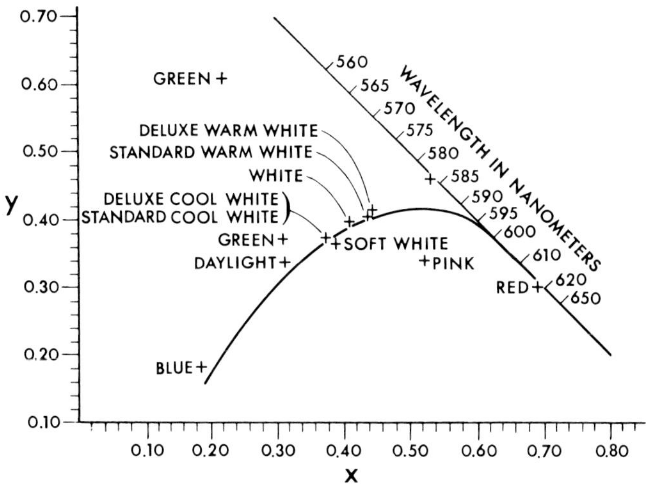

Fig. 8–23. (Left) CIE chromaticity diagram showing some white and colored fluorescent lamps in relation to the blackbody curve. (Below) Spectral power curves of light from typical fluorescent lamps. (Next page) Spectral power curves of high-intensity discharge and incandescent lamps (colors shown are approximate representations.) Consult Section 5 for details on the significance of these diagrams.

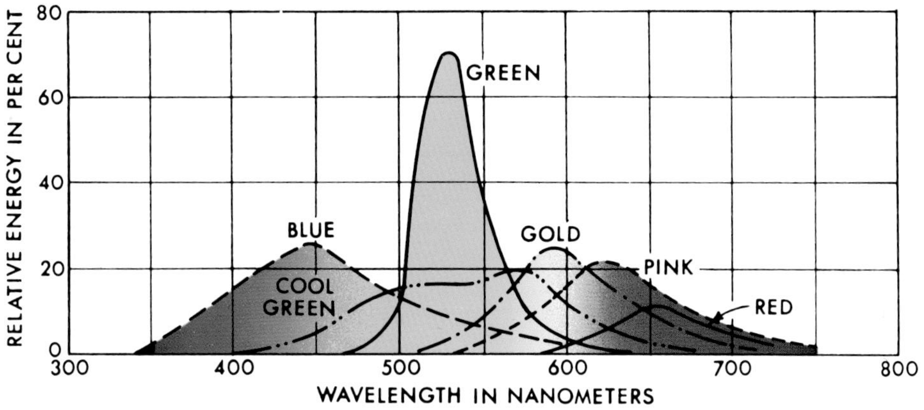

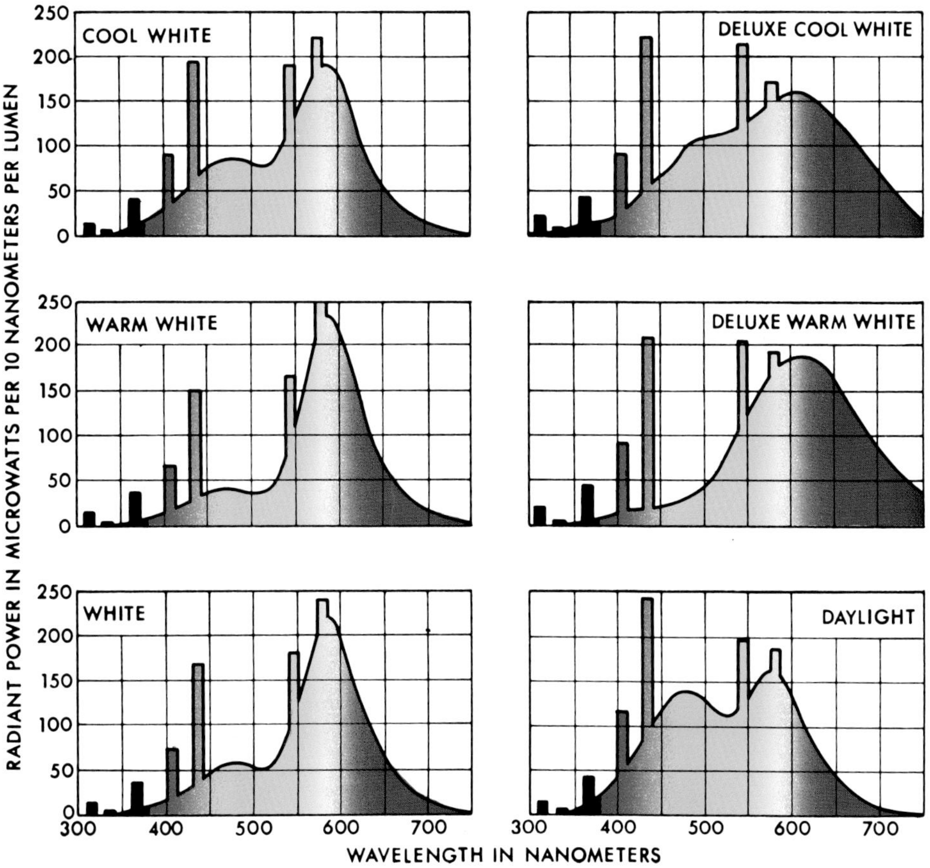

Fig. 8-23. Continued

MERCURY LAMPS

Clear mercury

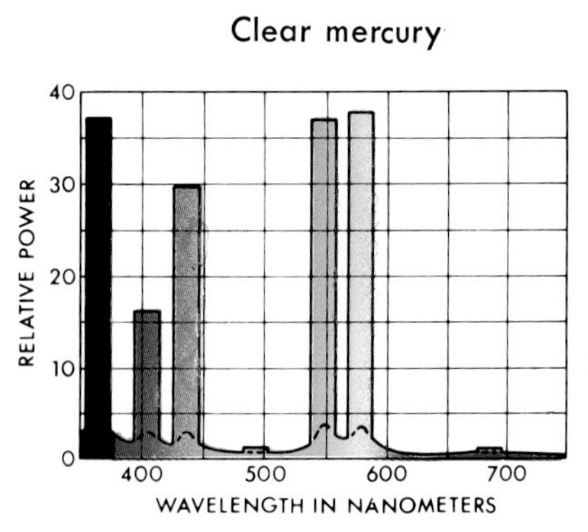

Increased-efficacy phosphor-coated

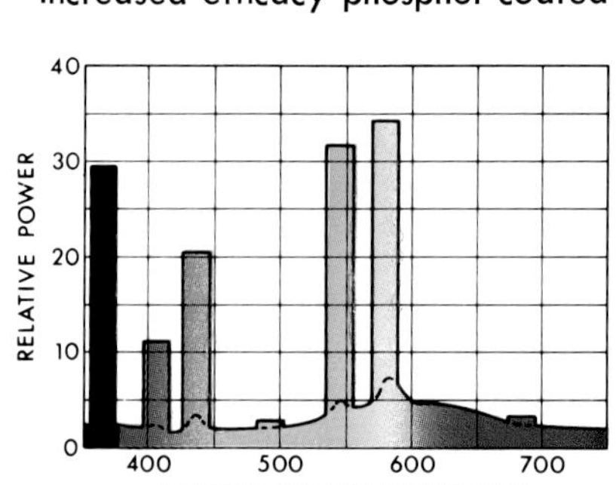

Improved-color phosphor-coated

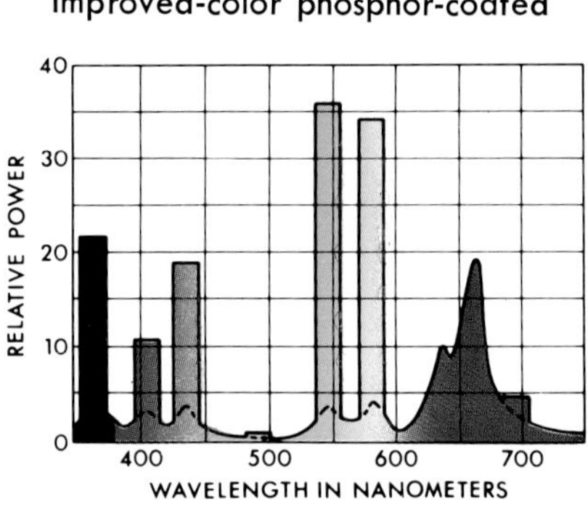

METAL HALIDE LAMPS

Na-SCANDIUM IODIDE

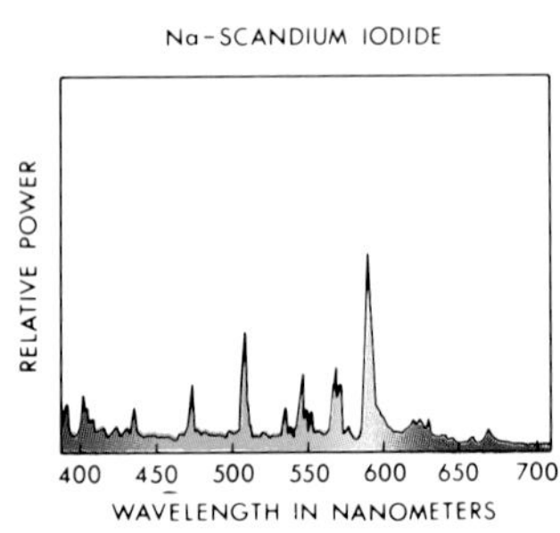

Na-Tl-In IODIDE

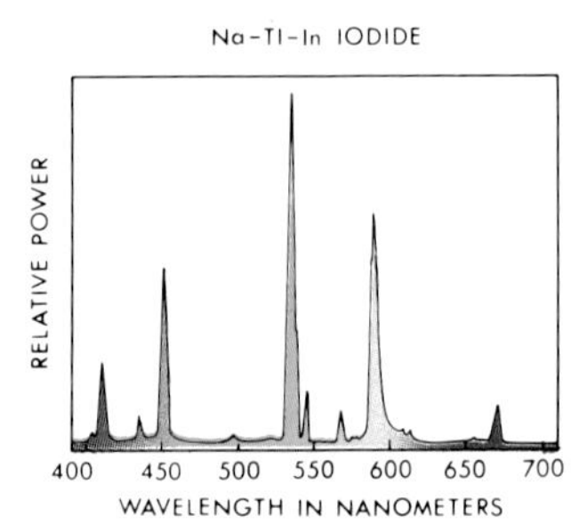

Tl-DYSPROSIUM IODIDE

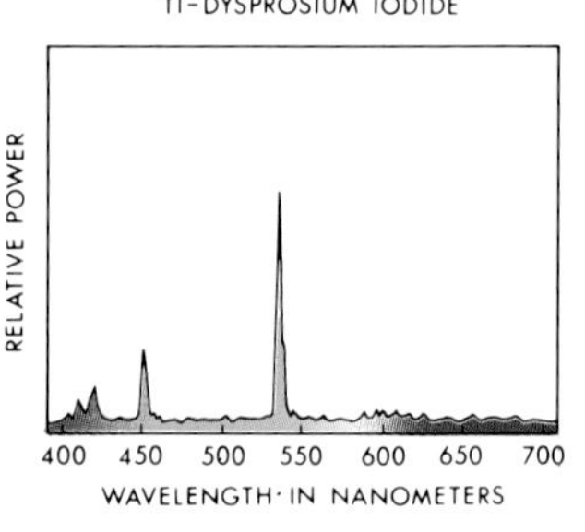

HIGH PRESSURE SODIUM LAMPS

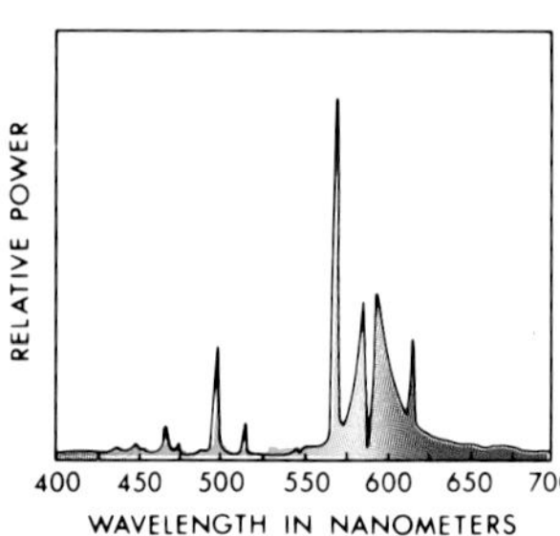

LOW PRESSURE SODIUM LAMPS

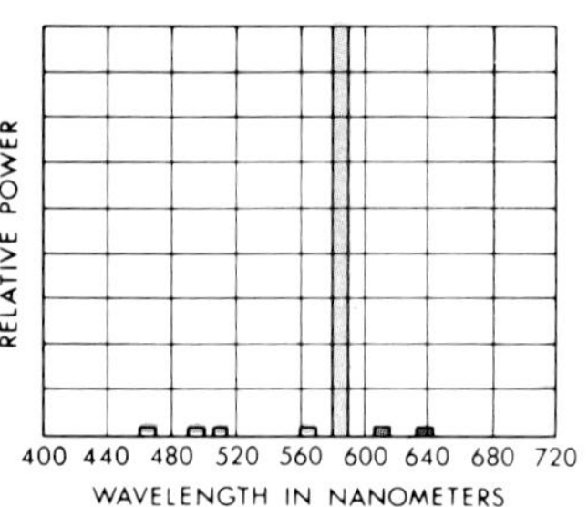

INCANDESCENT LAMPS (INCLUDING TUNGSTEN-HALOGEN)

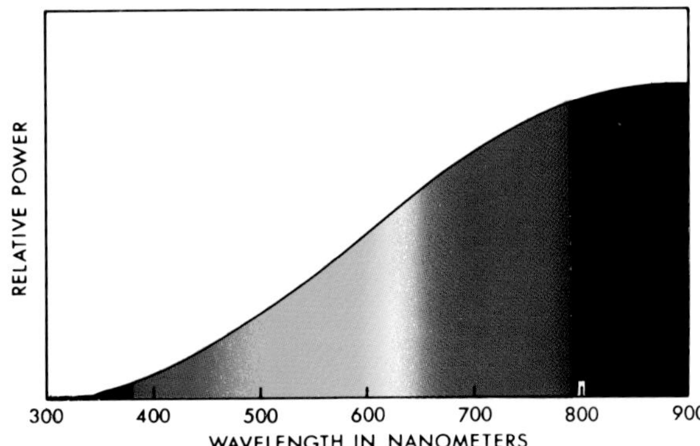

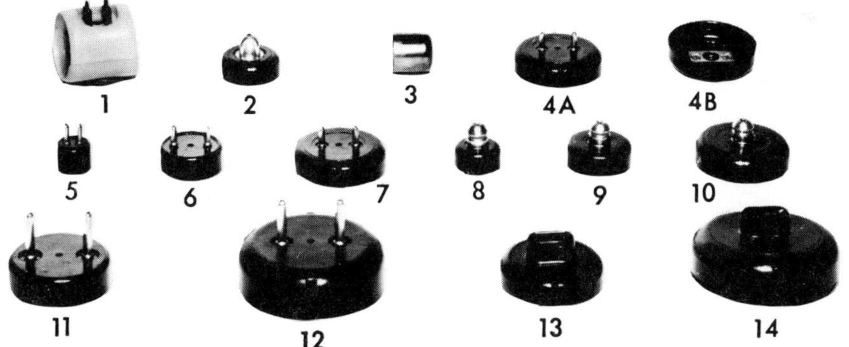

Fig. 8–24. Typical bases for fluorescent lamps: (1) Circular lamp; (2) Fluted single pin (K or clover leaf) cold cathode; (3) Ferrule type, cold cathode; (4) A. Medium bipin for T-12 lamps, showing B. internal connection between pins for instant-start lamp; (5) Miniature bipin for T-5 lamps; (6) Medium bipin for T-8 lamp; (7) Medium bipin for T-12 lamp; (8) Single pin for T-6 lamp; (9) Single pin for T-8 lamp; (10) Single pin for T-12 lamp; (11) Mogul bipin for T-12 lamp; (12) Mogul bipin for T-17 lamp; (13) Recessed double contact for T-12 lamp; (14) Recessed double contact for T-17 lamp.

The 40-watt instant start lamp is manufactured with the medium bipin base; however, there is an internal connection between the pins, in effect producing a single contact for each cathode. Because of this construction these lamps should not be operated in preheat or rapid start circuits as the auxiliary equipment will be damaged.

Special Fluorescent Lamps

In addition to the lamps described above, there are special lamps that utilize internal light control devices and those designed to produce radiation in the ultraviolet region of the spectrum. Lamps in the latter category use special phosphors or special glass without phosphor coatings and are described more fully in Section 25. A third group of special lamps is designed for optimum light generation at temperatures other than 40° C.

There are two types of lamps with light control from within the lamp: the reflector fluorescent lamp and the aperture fluorescent lamp. The reflector fluorescent lamp has an internal white powder layer between the phosphor and the envelope glass. This coating, which covers a major angular portion of the envelope wall, reflects a high percentage of the visible radiation striking it. The major portion of the light is emitted through the strip coated with just the fluorescent phosphor. A cross-sectional diagram and relative candlepower distribution for a 235-degree reflector lamp are shown in Fig. 8–25a. Reflector lamps with other angular width reflectors are available.

The aperture lamp is constructed similar to the reflector lamp except a clear window (no phosphor) exists for an angular aperture along the length of the tube. Elsewhere the phosphor coat is underlayed with a reflective layer. This results in a linear window of high luminance, up to ten times that of standard fluorescent lamps. The cross sectional diagram and candlepower distribution curve for a 60-degree aperture lamp are shown in Fig. 8–25b.

As a consequence of the reflector layer, absorption of generated light is somewhat higher in both types of lamps than in standard fluorescent lamps producing a somewhat reduced total lumen output. However, they are designed for applications which can effectively utilize the resulting light output distribution pattern.

Other special lamps are available for unusual ambient temperature applications. One family designed for low temperatures incorporates a jacket to conserve heat. Another group incorporates a mercury amalgam to optimize mercury vapor pressure at elevated temperatures. In both cases, these lamps are designed for optimization of the mercury vapor pressures at unusual temperatures.

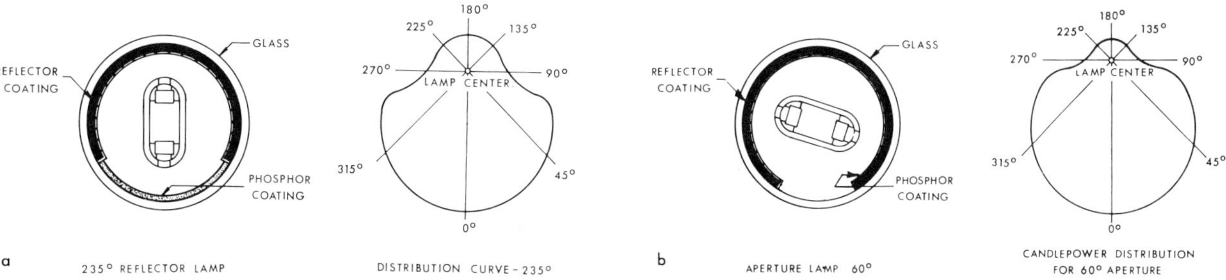

Fig. 8–25. Cross-section diagrams and relative candlepower distribution curves for (a) 235-degree reflector lamp and (b) 60-degree aperture lamp.

Performance Parameters

To fully describe a light source, the performance parameters of the device need to be defined. Some of the more important parameters are:

1. Luminous efficacy.
2. Lamp life.
3. Light output maintenance with life.
4. Color (spectral power distribution and chromaticity).
5. Energy distribution.
6. Temperature effect on operation.
7. Flicker and stroboscopic effect.
8. Radio interference.

Luminous Efficacy: Light Output. During the process of generating light by a fluorescent lamp three main energy conversions occur. Initially, electrical energy is converted into kinetic energy by accelerating elemental particles. These in turn yield their energy during particle collision to electromagnetic radiation, particularly ultraviolet. This ultraviolet energy in turn is converted to visible energy or light by the lamp phosphor. During each of these conversions some energy is lost so that only a small percentage of the input is converted into visible radiation light as shown in Fig. 8–26.

The geometric design and operating conditions of a lamp influence the efficacy with which energy conversions take place. Figs. 8–27, 8–28, 8–29, 8–30, and 8–31 present data depicting the effect of bulb design and operating current.

Figs. 8–27 and 8–28 show the effect of the bulb design on lamp operation. As seen in Fig. 8–27, as lamp diameter increases efficacy increases, passes through a maximum, and decreases. There are at least two important reasons for this phenomenon. In lamps of small diameter, an excessive amount of energy is lost by recombinations of electrons with ions at the bulb wall. As the bulb diameter is increased, this loss becomes progressively smaller, but losses due to "imprisonment of radiation" become correspondingly larger.

As shown in Figs. 8–28 and 8–29, the length of a lamp influences its efficacy, *i.e.*, the longer the length the higher the efficacy. This is based on two separate energy losses within the lamp; the energy absorbed by the electrodes, which do not generate any appreciable light, and the energy loss directly associated with the generation of light. The electrode loss is essentially a constant, whereas the loss associated with light generation is a function of lamp length. As the lamp length increases, the effect of the electrode loss becomes less in comparison to the total losses.

Lamp operating voltage, similar to efficacy, is a function of lamp length. This effect is shown in Figs. 8–30 and 8–31. Fig. 8–30 presents data for hot cathode T-12 operation at four different current

LUMENS PER WATT
If the energy in any light source could be converted without loss into yellow-green light (555 nanometers) the efficacy of the source would be 673 lumens per watt (100 per cent of the theoretical maximum).
673
But phosphors produce light over a range of wavelengths. When combined to produce the standard cool white color the average relative luminous efficiency is 51 per cent of the maximum.
343
Of the input watts delivered to the lamp, 53 per cent is converted to exciting ultraviolet. Most of the balance goes into electrode heating and bulb warmth.
102
The conversion from the ultraviolet wavelength (253.7 nanometers) to the visible wavelengths which make up the cool white color, is accomplished by the phosphor at the theoretical maximum efficiency (47 per cent) known as the quantum ratio.
86
Losses from phosphor imperfections, non-utilization of 253.7 nanometer radiation, bulb absorption, and coating absorption of light total 7 per cent (93 per cent efficiency).
80
74
74 Phosphor
5 Visible mercury lines
79 Rated Efficacy
The light from the visible mercury lines adds 5 lumens per watt to the light from the phosphor. The 7 per cent loss from 80 to 74 lumens per watt in phosphor output results from losses in manufacturing operations, such as milling and baking, and losses due to deprecation in the first 100 hours of burning.

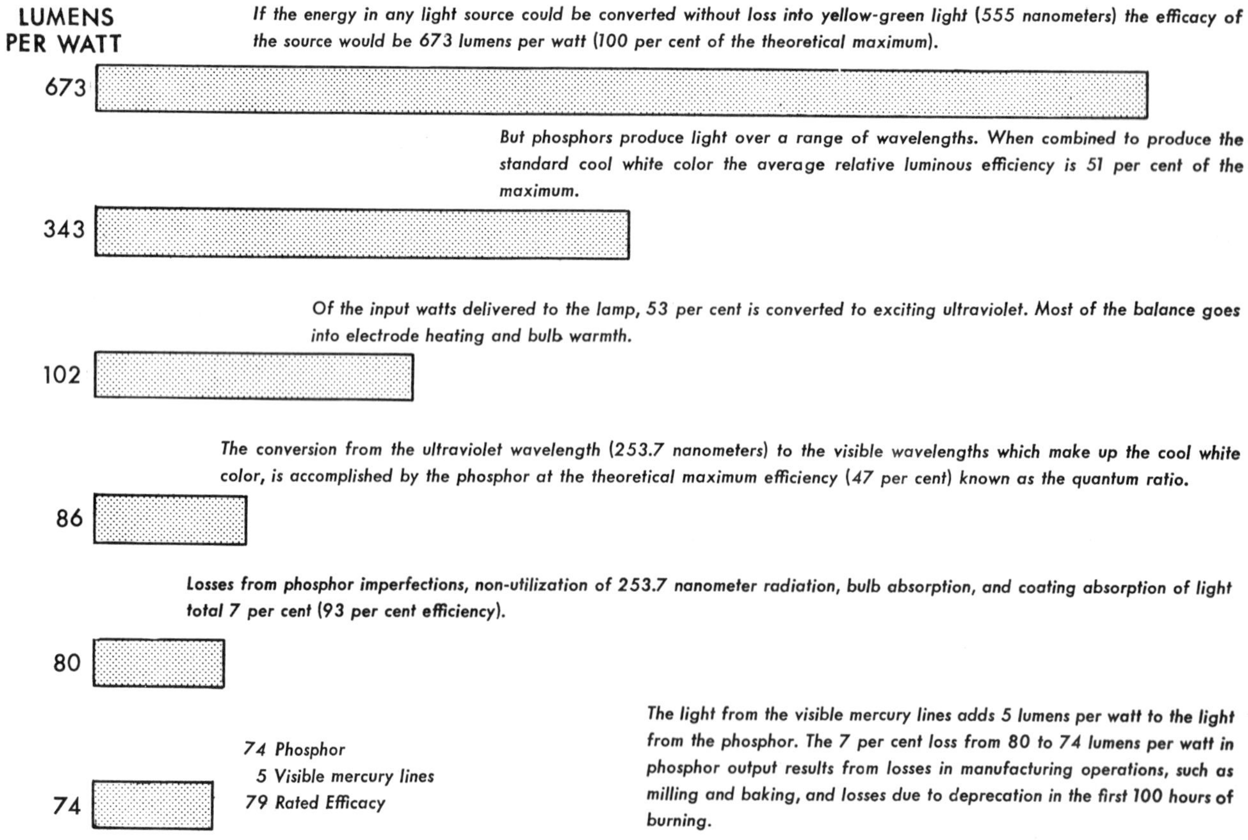

Fig. 8–26. Energy conversion efficacy in a typical cool white fluorescent lamp.

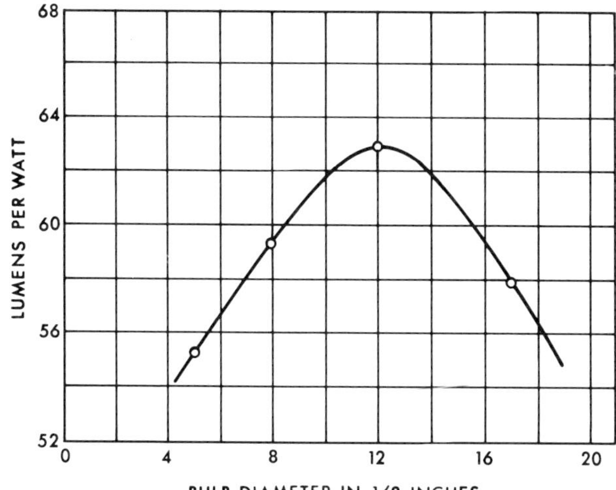

Fig. 8-27. Efficacy of typical fluorescent lamps as a function of bulb diameter.[13]

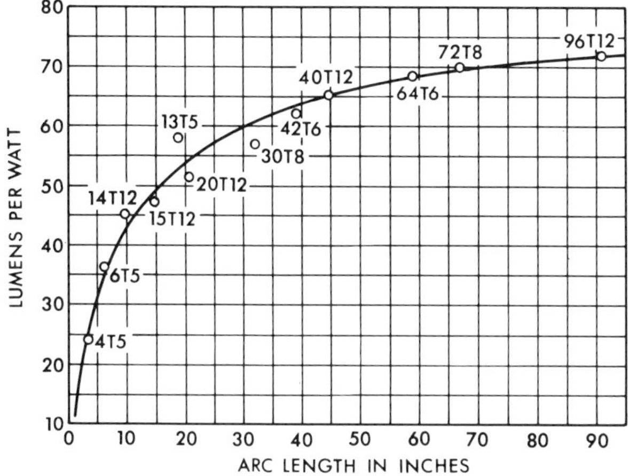

Fig. 8-28. Efficacy of typical fluorescent lamp as function of lamp length.[14]

levels. Fig. 8-31 presents similar data for both hot and cold cathode T-8 lamp operation. The characteristic electrode drop for the two types of cathodes is indicated by the intersection of the curves with the ordinate corresponding to zero length.[14]

Figs. 8-110, 8-111, and 8-112 show the light output data on typical fluorescent lamps. So far as possible, data presented on types produced by several manufacturers represent industry averages. Since these data are likely to differ slightly with specific figures for one manufacturer's product, it is advisable to check the manufacturers' sheets for detailed information on current production.

Lamp Life. Lamp life of hot cathode lamps is determined by the rate of loss of the electron emissive coating on the electrodes. Some of the coating is eroded from the filaments each time the lamp is started. Also, during lamp operation there is evaporation of emissive material. Although electrodes are designed to minimize both of these effects, the end of lamp life is reached when either the coating is completely removed from one or both electrodes or the remaining coating becomes non-emissive.

Because some of the emissive coating is lost from the electrodes during each start, the frequency of starting hot cathode lamps influences lamp life. The rated average life of fluorescent lamps is usually based on three hours of operation per start. The effect of starting frequency on life is presented in Fig. 8-32. These data have been normalized to 100

Fig. 8-29. Relationship of Arc Length and Lumens per Watt for Typical Cool White Fluorescent Lamps

Approximate Arc Length (inches)	Hot Cathode				Low Pressure T-8 Cold Cathode
	Approximate Efficacy (lumens per watt)	Approximate Lamp and Auxiliary Efficacy (lumens per watt)			Approximate Efficacy (lumens per watt)
		Preheat Start	Instant Start	Rapid Start	
5	33.0	22			
10	45.0	34			
15	53.0	42			
20	59.0	48			
25	63.0	53			
30	66.0	57			
35	68.0	59			
40	69.0	61	52	60	38
45	70.0	62			
50	71.0	63	54	63	41
55	72.0				
60	72.5		56	65	44
65	73.0				
70	74.0		59	67	47
75	75.0				
80	76.0		61	69	50
85	76.5				
90	77.0		63	71	52

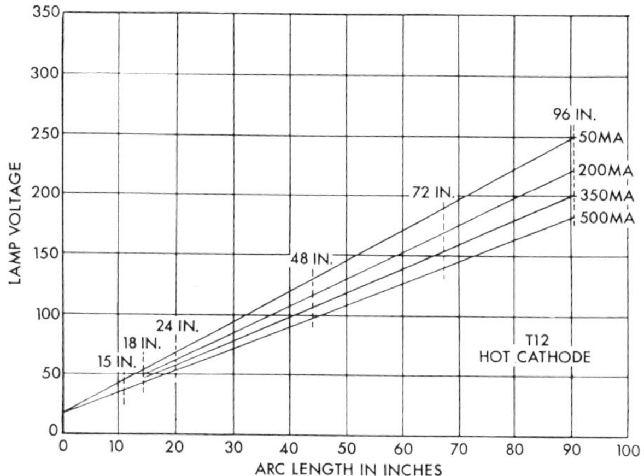

Fig. 8-30. Operating voltage of typical T-12 hot cathode fluorescent lamps as a function of arc length.

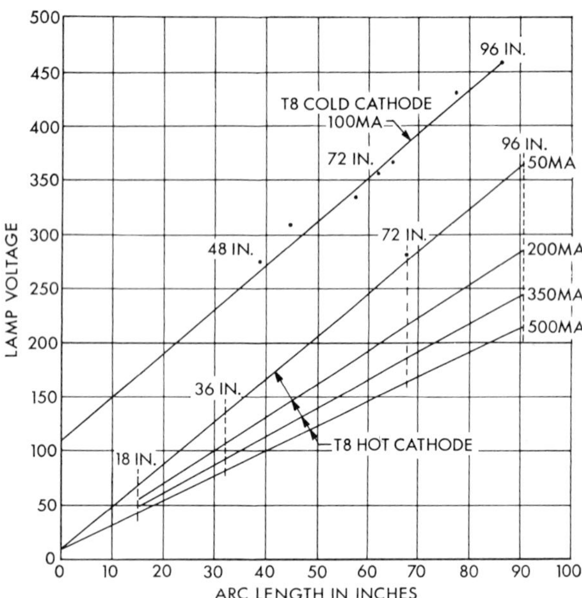

Fig. 8–31. Operating voltage of typical T-8 (hot and cold cathode) fluorescent lamps as a function of arc length.

per cent for life at three hours per start. Cold cathode lamps are not appreciably affected by starting frequency because of the type of electrode used.

There are many other variable conditions that affect lamp life in actual use. Ballast characteristics and starter design are key factors in the case of preheat circuits. Ballasts which do not provide specified starting requirements, or which do not operate lamps at proper voltage levels, can greatly affect lamp life. For preheat circuits, starters must also be designed to meet specified characteristics.

Proper electrode (cathode) heating current in rapid start lamps is critical and is affected not only by ballasts, but also by poor lamp to lampholder contact or improper circuit wiring. Improper seating of a lamp in a lampholder may result in no electrode (cathode) heating. Lamps operating in this mode will fail within a few hours—50 to 500 hours, depending on lamp type. Another factor in lamp life is line voltage. If line voltage is too high, it may cause instant starting of lamps in preheat and rapid start circuits. If it is low, slow starting of rapid start or instant start type lamps, or recycling of starters in preheat circuits may result. All of these conditions adversely affect lamp life.

Ballasts are available for low temperature starting of rapid start lamps. With higher temperatures encountered during the summer months, lamps operating on these ballasts will start before the electrodes are properly heated. This has an adverse effect on lamp life. Time delay relays are available to insure proper electrode heating prior to application of high ignition voltage to the lamp.

Light Output Maintenance with Life. The lumen output of fluorescent lamps decreases with accumulated burning time. Although the exact nature of the change in the phosphor which causes the phenomenon is not fully understood, it is known that at least during the first 4000 hours of operation the reduction in efficacy is related to arc-power to phosphor-area ratios. This relationship in several typical lamps is shown in Fig. 8–33. As depicted in Fig. 8–33a, lamps with different arc power to phosphor-area ratios have different lumen maintenance curves. This effect of lamp loading on lumen maintenance is presented in Fig. 8–33.[15, 16]

Another important effect that reduces light output is end darkening, resulting from emission coating deposition on the phosphor.

Color, Spectral Power Distribution, and Chromaticity. Spectral power distribution curves for typical colors of fluorescent lamps are shown in Fig. 8–23. The CIE chromaticity diagram on which the position of various lamp colors are noted in conjunction with the blackbody curve is also shown in this figure. For a discussion of color and Color Rendering Index, see Section 5.

Energy Distribution. Fig. 8–34 shows the approximate distribution of energy in a typical cool white fluorescent lamp.

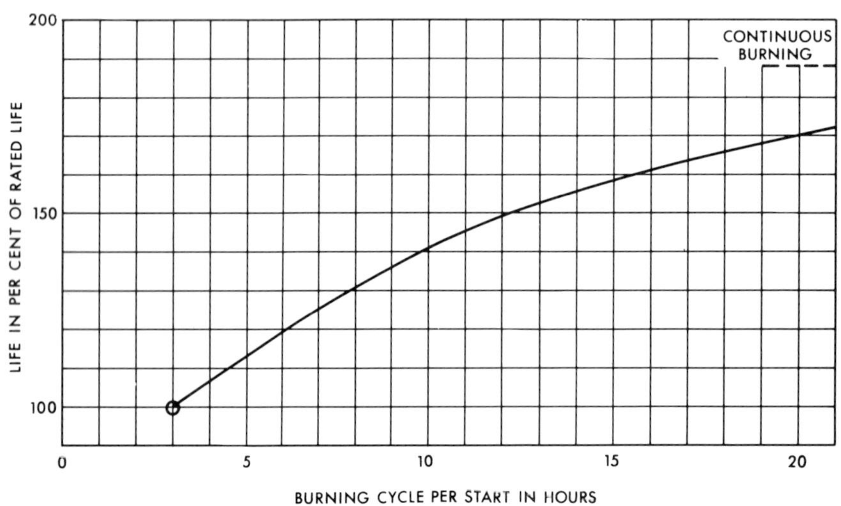

Fig. 8–32. Life of typical fluorescent lamps as a function of burning cycle. Variations from this curve can be expected with lamp loading.

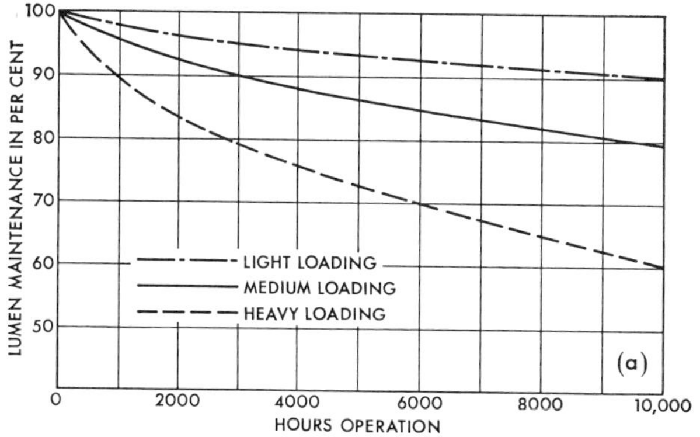

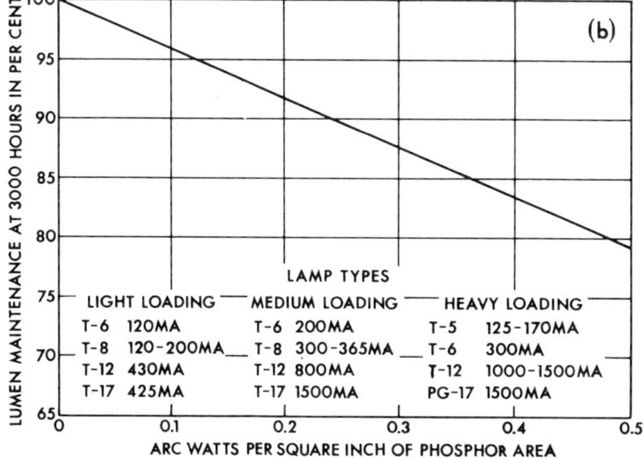

Fig. 8-33. Lumen maintenance curves for typical fluorescent lamps as a function of: (a) hours operation and (b) incident radiant power density on the phosphor surface.

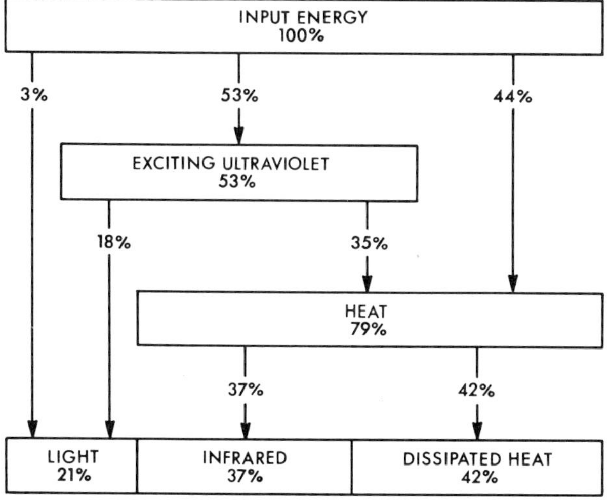

Fig. 8-34. Energy distribution in a typical cool white fluorescent lamp.

Temperature Effect On Operation[17]

The fluorescent lamp is dependent upon mercury vapor for its operation as has been discussed. Thus the characteristics of a fluorescent lamp are very much dependent upon the concentration of mercury vapor, or more specifically, the mercury vapor pressure within the lamp. This mercury vapor pressure depends upon temperature.

A fluorescent lamp contains a larger quantity of liquid mercury than will become vaporized at any one time. This excess liquid mercury tends to condense at the coolest point or points on the lamp. If any particular location is significantly cooler than others in the lamp all the liquid mercury will collect at this point. The mercury pressure within the lamp will depend upon the temperature of this coolest point or points, the temperature depending upon lamp construction, lamp wattage, ambient temperature, luminaire design and wind or draft conditions.

The effects of temperature on mercury vapor pressure manifests itself as variations in light output and color.

Lamps using mercury amalgams are available for extending the usable ambient temperatures to higher values. The amalgam functions to stabilize and control the mercury pressure.

Effect of Temperature on Light Output. Light output depends upon mercury vapor pressure which in turn depends upon the temperature of the coolest point on the fluorescent lamp. Electrical characteristics are also affected. Typical characteristics of a lamp operated at constant current are shown in Fig. 8-35. The exact shape of the curves depend upon the lamp cross section, loading, and type of ballasting. However, all fluorescent lamps have essentially the same relation between light output and minimum bulb wall temperature, since this relation depends primarily upon mercury vapor pressure, which is a common function of temperature.

Light output and luminous efficacy reach optimum values at about 100°F. Fluorescent lamps intended primarily for indoor use are designed so that their coolest bulb-wall temperature will be near 100°F when the lamp is operated in typical well-designed luminaires under usual indoor temperatures.

Curves for an 800-mA high output fluorescent lamp are shown in Fig. 8-36 (left). As these curves indicate, light output falls to very low values at temperatures below freezing. Lamps intended for indoor operation will produce poor low-temperature performance unless protected by suitable enclosures.

Fig. 8-36 (right) shows the relationship between ambient temperature and light output for a typical outdoor floodlight using 800-mA high output lamps. While considerable variation occurs with temperature change, satisfactory illumination is obtained for temperatures commonly encountered in most areas of the United States and Canada.

Each particular lamp-luminaire combination has its own distinctive characteristic of light output vs. ambient temperature. In general, the shape of the curve will be quite similar for all luminaires, but the temperature at which the highest light output will be reached may be quite different.

In areas with fluorescent lighting systems, high ambient temperatures will cause losses in the footcandle level. The lighting designer should take this into account in making his calculations.

Fig. 8–36 shows that the loss in light as the lamp is heated beyond the optimum temperature is nearly linear. From this fact an estimating rule can be derived which enables the designer to compensate for high ambient temperature condition: there will be a 1 per cent loss in light for every 2° F that the ambient temperature around the lamp exceeds 77° F.

Effects of Temperature on Color. Just as temperature affects the light output of a lamp, temperature also affects the color of the light produced. The color of light from a fluorescent lamp has two components: that from the phosphor coating on the lamp and that from the mercury arc discharge within the lamp. Each of these components reacts independently to temperature changes. In general, the lamps shift toward the blue-green with increasing temperature due to the increasing contribution from the mercury arc visible spectrum.

Fig. 8–37 shows a typical color shift characteristic of a fluorescent lamp. In addition, the MacAdam four step ovals are noted for each color. These ovals depict the color tolerance limits for lamps at 77° F. These variations are due to manufacturing, color changes through life, etc. In a new interior lighting installation, color is much more uniform than this since all lamps are of the same manufacture, age, etc. Color shift becomes a concern to the lighting designer chiefly in cases where substantial differences in internal temperature may exist between adjacent luminaires. This may arise from the proximity of certain luminaires to air diffusers, open windows, etc.; differences in ceiling cavity conditions or ceiling material with surface and recessed equipment; differences in the tightness of enclosures with enclosed equipment; differences in lamp loading or number of lamps in identical luminaires; and use of some of the luminaires as air diffusers in conjunction with the air-conditioning system.

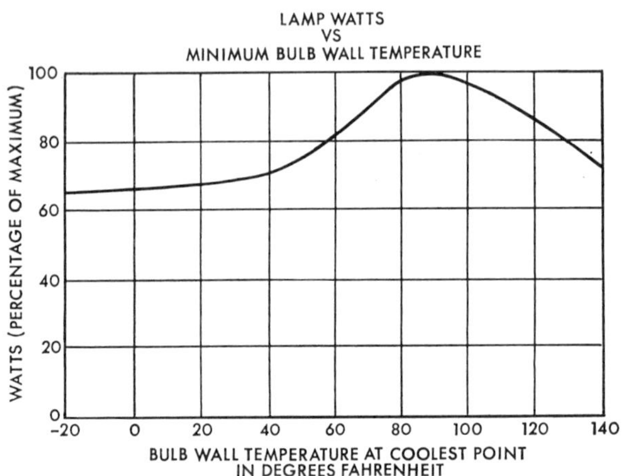

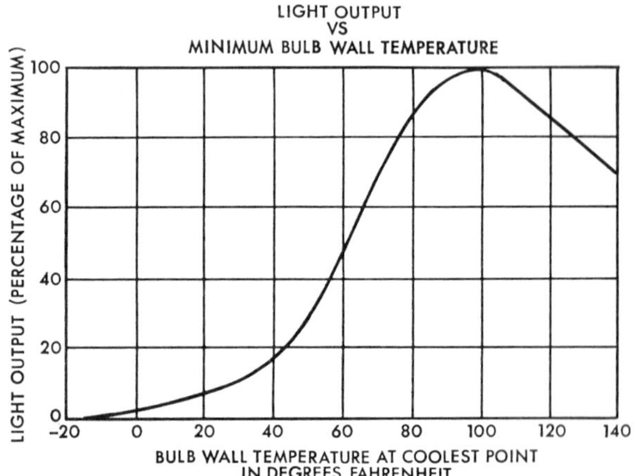

Fig. 8–35. Typical fluorescent lamp temperature characteristics. Exact shape of curves will depend upon lamp and ballast type; however, all fluorescent lamps have curves of the same general shape, since this depends upon mercury vapor pressure.

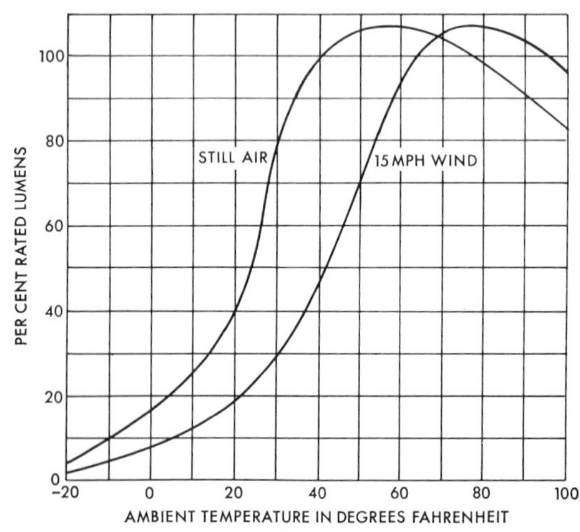

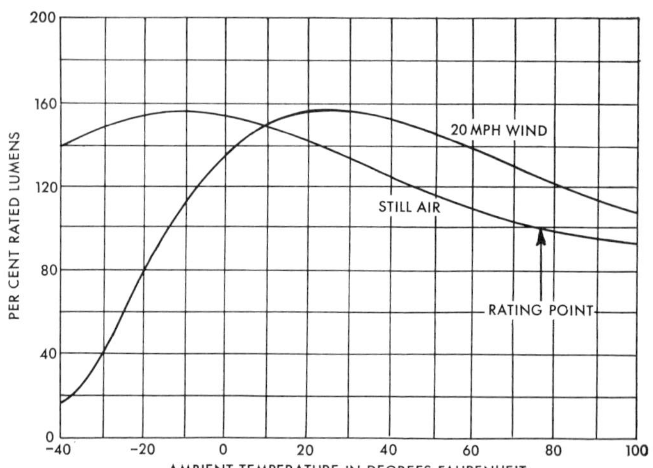

Fig. 8–36. Light output vs ambient temperature. (left) F96T12/HO fluorescent lamp. Light output falls to low values at temperatures below freezing. Loss in light at high ambient temperatures is much less. (right) Typical outdoor floodlight - two F72T12/HO. By a suitable enclosure, the lamp is warmed to a high enough temperature to produce good light output under cold windy conditions.

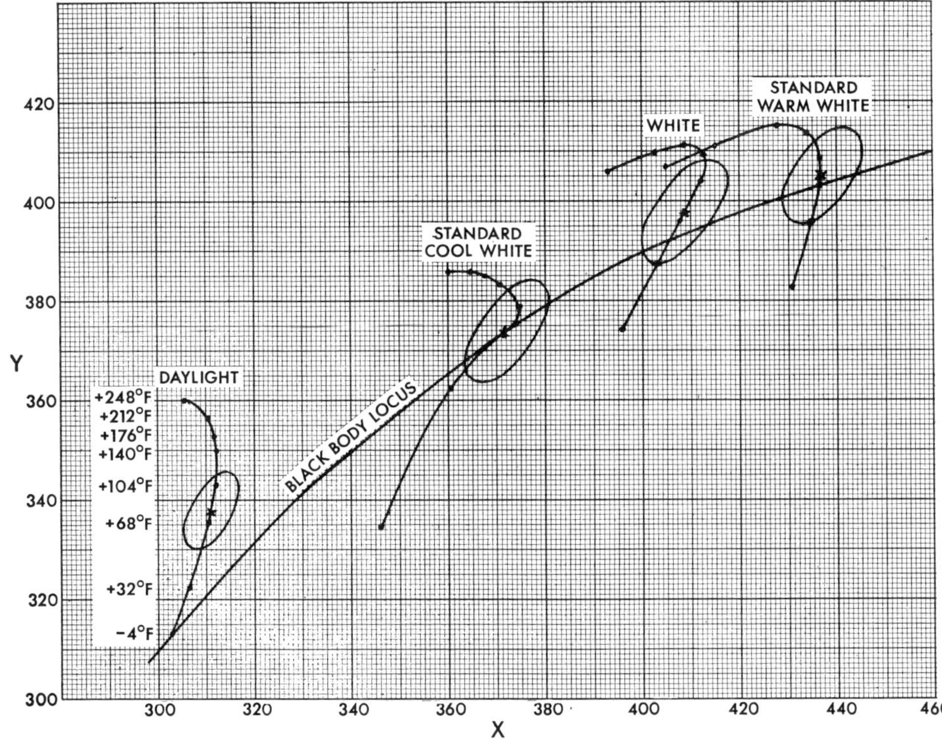

Fig. 8-37. Typical color shift characteristic of a fluorescent lamp with changes in ambient temperature. The MacAdam 4-step ovals illustrate the tolerance limits for the particuar white. Color shift over a normal temperature range is in the same order of magnitude as that which may be experienced between lamps of the same nominal color due to manufacturing variations, depreciation through life, etc. (The lowest point on each curve is at −4°F, the top point at +248°F, intermediate points are 36°F apart; see Daylight Lamp.)

Flicker and Stroboscopic Effect

The ultraviolet energy generated by an arc discharge is a function of the instantaneous power input. As a consequence, the generated ultraviolet energy shows cyclic changes similar to the power input. The frequency of this variation is twice the input frequency. The cyclic variation of the ultraviolet energy is transferred to the emitted light output where the phosphors show both fluorescence and phosphoresence. As a consequence, some smoothing of the instantaneous light output occurs. Even so, cyclic variation in instantaneous light output results. This variation in light output is called *flicker.*

With a 60 hertz input frequency, the resulting 120-cycle variation is too fast to be noticed by the eye. However, when rapidly moving objects are viewed under these fluorescent systems, blurred "ghost" images may be observed. This effect is known as *stroboscopic effect.* The greater the flicker, the more noticeable is the stroboscopic effect.

Flicker Index[18] has been established as a reliable, relative measure of the cyclic variation in output of various sources at a given power frequency. Previously this cyclic variation has been defined by *Per Cent Flicker.* However, Flicker Index is now considered to be a more reliable measure since it takes into account the wave form of the light output. It is calculated by dividing the area above the line of average light output by the total area under the light output curve for a single cycle (see Fig. 8-38). In Fig. 8-38, Per Cent Flicker is equal to $100(A - B)/(A + B)$, (per cent modulation).

The Flicker Index gives ratings from 0 to 1.0 with 0 for steady light output. The higher values of Flicker Index indicate an increased possibility of noticeable stroboscopic effect, as well as lamp flicker. In Fig. 8-39 the Flicker Index and Per Cent Flicker are listed for seven "white" lamps when operated from typical circuits.

Some flicker can be seen from the ends of a lamp when viewed in the periphery of the retina. This is due to a slow movement of the discharge attachment position on the electrode. It occurs at random and is generally not objectionable.

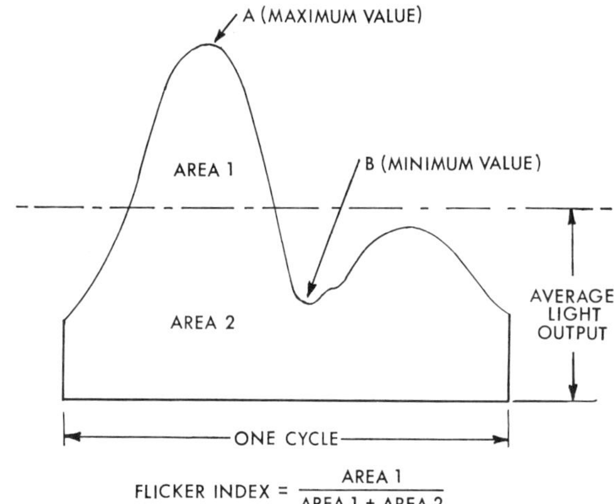

Fig. 8-38. Curve of the light output variation from a fluorescent lamp during each cycle, showing the method of calculating the Flicker Index.

Fig. 8-39. Flicker Index and Per Cent Flicker Values for "White" Fluorescent Lamps

	Single Lamp		2-Lamp Lead-Lag Instant Start		2-Lamp Lead-Lag Preheat Switch Start	
	Flicker Index	Per Cent Flicker	Flicker Index	Per Cent Flicker	Flicker Index	Per Cent Flicker
Cool white	.079	34	.071	26	.056	16
Cool white, improved-color	.078	34	.075	27	.046	14
Warm white	.048	20	.044	16	.029	10
Warm white, improved-color	.049	20	.043	16	.030	10
Daylight	.119	50	.107	41	.075	24
White	.058	25	.054	20	.042	12
Soft white	.079	30	.064	25	.044	13

Radio Interference

The mercury arc in a fluorescent lamp emits electromagnetic radiation. This radiation may be picked up by nearby radios causing an audible sound. Because of the frequencies generated by the fluorescent lamp, interference is ordinarily limited to the AM broadcast band and nearby amateur and communications bands. FM, television, and higher frequencies are very rarely affected. Most instant start ballasts, rapid start ballasts, and starters for preheat circuits have capacitors for reduction of radio interference.

Radio noise reaches the radio either by radiation to the antenna or by conduction over the power lines. Radiated interference may be eliminated by moving the antenna farther from the lamp. Ten feet is usually sufficient. Where this is not practical, shielding media, such as electrically conducting glass or certain louver materials, will suppress the noise below the interference level. Conducted interference may be suppressed by an electric filter in the line at the luminaire. Fig. 8-40 shows a typical design. Luminaires with this type of filtering and appropriate shielding material have been qualified under pertinent military specifications for sensitive areas.

Lamp Circuits and Auxiliary Equipment

Like most arc discharge lamps fluorescent lamps have a negative volt-ampere characteristic and therefore, they require an auxiliary device to limit current flow. This device, normally called a ballast, may incorporate an added function which provides a voltage sufficient to insure ignition of the arc discharge. This voltage may vary between 1.5 to 4 times the normal lamp operating voltage.

The life and light output ratings of fluorescent lamps are based on their use with ballasts providing proper operating characteristics. The required operating characteristics have been established in the American National Standards for Dimensional and Electrical Characteristics of Fluorescent Lamps (C78 group). Ballasts that do not provide proper electrical values may reduce either lamp life or light output, or both. This auxiliary equipment consumes power and therefore reduces the over-all lumens-per-watt rating below that based on the power consumed by the lamp. Typical data are presented in Fig. 8-115.

Lamp Starting

The starting of a fluorescent lamp occurs in two stages. First, a sufficient voltage must exist between one lamp electrode and a nearby grounded conductor such as the metal surface of the luminaire. This will initiate ionization of the gas in the lamp. Secondly a sufficient voltage must exist across the lamp to extend the ionization throughout the lamp and develop an arc.

As ambient temperature is reduced, starting of all fluorescent lamps becomes more difficult. For reliable starting at low temperatures higher available output voltages are required. For more efficient ballast/lamp operation ballasts are available for each of the following temperature ranges:

Above 50° F for indoor applications.

Above 0° F for outdoor temperature applications.

Above −20° F for outdoor temperature applications.

A number of different means of lamp starting have been developed since the advent of the fluorescent lamp. The first was preheat starting which required an automatic or manual starting switch. Then came instant starting which required higher voltage. The most recent and probably the most important development was rapid starting where the use of continuously heated electrodes resulted in lamp starting without high voltage or starting switches. Several circuits for operating fluorescent lamps are shown in Fig. 8-41.

Preheat Lamp/Ballast Operation. Early fluorescent lamp systems were all of the preheat type. As the name implies, the lamp electrodes are heated before application of the high voltage across the lamp. Lamps designed for such operation have bipin bases to facilitate electrode heating.

The preheating requires a few seconds and this is usually accomplished by an automatic switch which places the lamp electrodes in series across the output of the ballast. Current flows through both electrode

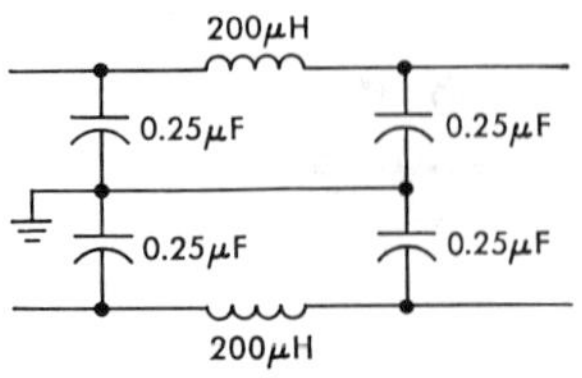

Fig. 8-40. Typical radio interference filter used in critical applications.

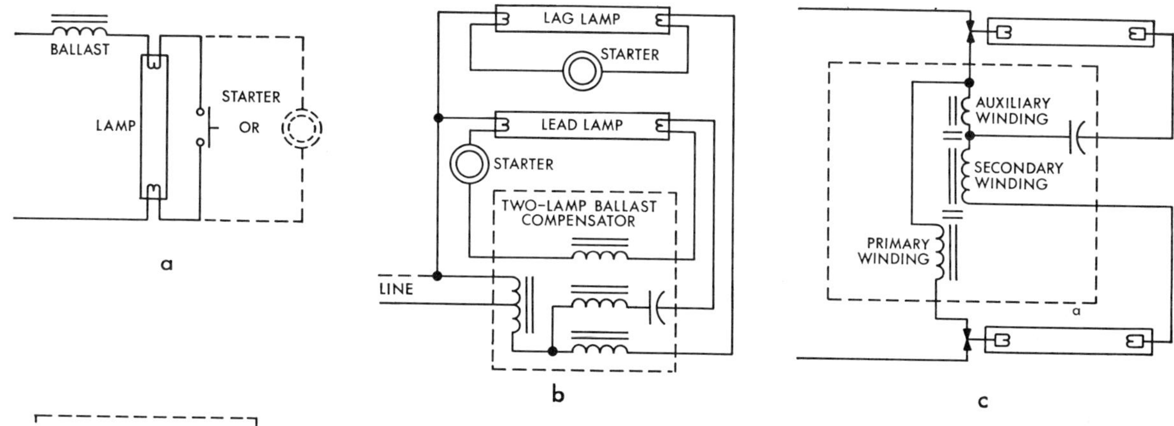

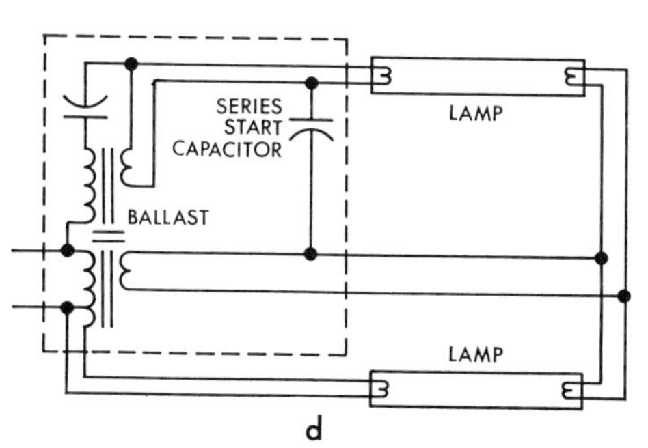

Fig. 8-41. Typical fluorescent lamp circuit: (a) Simple preheat circuit. Low power factor. Used for appliances, desk lamps, etc. (b) Two-lamp lead-lag preheat. High power factor. Used in 40-watt and 90-watt preheat type general luminaires. Note compensator winding which is needed to produce sufficient preheat current in the lead circuit. (c) Two lamp series-sequence instant start. High power factor. Used in slimline indoor units. Lamps start in sequence with auxiliary winding helping to start first lamp. Note disconnect lampholder connection which removes power from the ballast primary when lamps are being changed, thus preventing electric shock. (d) Two lamp series rapid start. High power factor. Used for rapid start, high output, and extra high output both indoors and for low temperature application. The small capacitor shunted across one lamp momentarily applies nearly all of the ballast secondary voltage across the other lamp.

filaments causing a temperature rise in the filaments. Subsequently, the switch opens applying the voltage across the lamp. Due to the opening of the switch under load, a transient voltage (an inductive kick) is developed in the circuit which aids in ignition of the lamp. If the lamp does not ignite, the switch will reclose and reheat the filaments. In some applications this preheating is accomplished by a manual switch.

The automatic switch is commonly called a *starter*. It may incorporate a small capacitor (0.006 microfarads) across the switch contacts to shunt high frequency oscillations which may cause radio interference.

Ballasts are available to operate some preheat lamps without the use of starters. These ballasts use the rapid start principle of lamp starting and operation and are popularly called *trigger start* ballasts.

Instant Start Lamp/Ballast Operation. Arc initiation in instant start lamps depends solely on the application of a high voltage across the lamp. This high voltage (400 to 1000 volts) ejects electrons from the electrodes by field emission. These electrons flow through the tube, ionizing the gas and initiating an arc discharge. Thereafter the arc current provides electrode heating. Because no preheating of electrodes is required, instant start lamps need only a single contact at each end of the lamp. Thus, the single pin lamp is used on most instant start lamps. These are commonly called "slimline" lamps. A few instant start lamps use bipin bases with the pins connected internally. In the case of lamps designed for instant starting at 400 to 1000 volts open circuit, it is necessary to provide some means of counteracting the effect of humidity on the capacitive lamp-ground current which initiates the necessary glow discharge. Most manufacturers coat the outside of the bulb of this type of lamp with a transparent, non-wetting material; others apply a narrow conducting strip along the bulb. A grounded conducting plate, such as a metal reflector near the lamp, commonly known as a "starting aid," is necessary to obtain the lowest lamp starting.[19]

Rapid Start Lamp/Ballast Operation. The rapid start principle has been used in most recent fluorescent lamp developments. It can use low or high resistance electrodes which are heated continuously. Heating is accomplished through low voltage windings built into the ballast or through separate low voltage transformers designed for this purpose. This results in a starting voltage requirement similar to that of preheat lamps. Lamps usually start in 1 to 2 seconds, the time required to bring the filaments up to proper temperature.

A starting aid, consisting of a grounded conducting plate extending the length and adjacent to the lamp is a prerequisite to reliable starting. For lamps operating at 500 mA or less the nominal distance between the lamp and a 1 inch wide conducting plate is ½ inch; for lamps operating at currents greater than 500 mA the nominal distance to the conducting strip is 1 inch. Peak voltage recommendations for lamp ignition using starting aids other than the nominal are listed in Fig. 8-42.

Rapid start lamps are coated with a transparent non-wetting material to counteract the adverse effect of humidity in lamp starting. All 800 mA and most 1500 mA lamp types operate on the rapid start

Fig. 8-42. Effect of Starting Aid Dimensions on Peak Voltage Requirements for Reliable Starting of Rapid Start Lamps

Starting Aid Dimensions (Inches)		Per Cent Change in Peak Voltage	
Distance to Lamp	Width of Aid	Lamp Operating Current Equal to or Less Than 500 mA	Lamp Operating Current Greater Than 500 mA
Nominal*	8 or greater	−8	−8
"	2	−4	−4
"	1	−0	0
"	½	+4	+4
"	¼	+8	+8
"	⅛	+12	+12
"	1/16	+20	+20
3	1	+32	+25
2	1	+22	+15
1½	1	+15	+8
1	1	+7	0
½	1	0	−7
¼	1	−8	−15
⅛ or less	1	−12	−20

* Nominal distance from starting aid to lamp is ½ inch for operating lamp current of 500 mA or less and 1 inch for operating lamp current of greater than 500 mA.

principle. Forty-watt and circline lamps designed for rapid start service can also be used in comparable preheat circuits.

Ballasts

Ballast Construction. The construction of a typical thermally protected rapid start ballast is shown in Fig. 8-43. The components consist of a transformer type core and coil. Depending upon the circuit, a capacitor may be part of the ballast. These components are the heart of the ballast, providing sufficient voltage for lamp ignition and lamp current regulation through their reactance.

The core and core assembly is made of laminated transformer steel wound with copper or aluminum magnet wire. The assembly is impregnated with a non-electrical conducting material that provides electrical insulation while aiding in heat dissipation, and with leads attached, is placed into a case. The case is filled with a potting material (hot asphalt for example) containing a filler such as silica. This compound completely fills the case encapsulating the core and coil and capacitor. The base is then attached.

Average ballast life at a 50 per cent duty cycle and at proper ballast operating temperature is normally estimated at about twelve years. Shorter ballast life will result at higher ballast temperature or longer duty cycle.

In the United States and Canada, it is now mandatory that all fluorescent lamp ballasts used indoors be internally thermally protected. This was done to prevent misapplication of the ballast as well as to protect against undesirable failure and conditions which can occur at end of ballast life. In the United States the thermally protected Underwriters' Laboratories approval ballast is known and marked or labeled as "Class P."

Because of the magnetic elements in a ballast, vibrations are set up in the luminaire based on the input power frequency. This may produce an audible hum which is undesirable. The sound level produced will depend upon the ballast and luminaire construction and mounting. The acoustical characteristics of the space and the number of luminaires will have significant effect on the degree of audibility. Ballast manufacturers publish "sound ratings" which indicate the relative sound producing potential of their different models. However, no industry standards have as yet evolved that make possible the comparison of different brands. Some luminaire manufacturers also publish sound ratings for their units. (See Section 6.)

Ballasts Power Factor. Characteristics of ballasts may result in low power factor. The measured watts of a low-power-factor ballast are approximately the same as the measured watts of the high-power-factor type when connected to the same load. The low-power-factor type draws more current from the power supply, and therefore, larger supply conductors may be necessary. The use of high-power-factor ballasts permits greater loads to be carried by existing wiring systems. Some public utilities have established penalty clauses in their rate schedules for low-power-factor installations. In some localities utilities require the use of equipment providing a high power factor.

Starters for Preheat Circuits

The operation of a preheat circuit requires heating of the electrodes prior to application of voltage across the lamp. Preheating can be effected by use of a manual switch or one which is activated by ap-

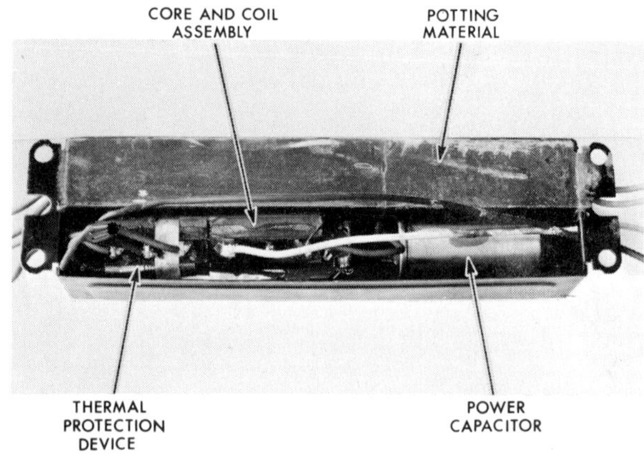

Fig. 8-43. Construction of typical rapid start ballast.

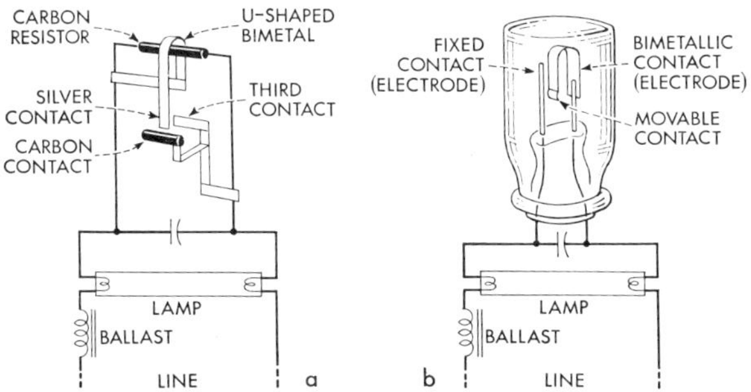

Fig. 8-44. Starter switches for preheat cathode circuits: (a) thermal type; (b) glow switch type.

plication of voltage to the ballast circuit. A number of designs of automatic switches are commercially available. Circuit diagrams of designs are presented in Fig. 8-44.

Thermal Switch Starter. The circuit diagramming this type starter is presented in Fig. 8-44. Initially the silver-carbon contact of the thermal starter is closed, placing the electrodes in series with the parallel combination of the bimetal and the carbon resistor. Upon closing the ballast supply circuit, output voltage of the ballast is applied to this series/parallel combination. The current heats the bimetallic strip in the starter, causing it to expand differentially, resulting in opening the silver/carbon contact. The time of opening is sufficient to raise the temperature of the electrodes to approximately their normal operating value. Upon opening the circuit, the ballast output voltage in series with an inductive spike (kick) voltage is applied to the lamp. If the lamp ignites, its normal operating voltage maintains a low current through the carbon resistor, developing and transferring sufficient heat to the bimetal to hold its contact open thereafter.

Should the lamp fail to start on the first attempt, the ballast open circuit voltage applied to the carbon resistor heats the bimetal sufficiently to cause the silver contact to move over against the third contact. This then short-circuits the carbon resistor, permitting preheating current to flow through the electrodes. As the bimetal cools, the circuit through the third contact is opened resulting in the application of the circuit voltage to the lamp again. This making and breaking of the circuit through the third contact continues until the lamp ignites. The bimetal circuit is held open thereafter as noted above. The carbon contact circuit functions only when the line voltage is initially applied to the ballast.

Thermal-switch starters consume some power (½ to 1½ watts) during lamp operation, but their design insures positive starting by providing: an adequate preheating period; a high induced starting voltage; and characteristics inherently less susceptible to line voltage variations. For these reasons they give good all-around performance of lamps under adverse conditions, such as direct-current operation, low ambient temperature, and varying voltage.

Glow-Switch Starter. The circuit for this starter is presented in Fig. 8-44. The glass bulb shown is filled with an inert gas chosen for the voltage characteristics desired. On starting, the line switch is closed. There is practically no voltage drop in the ballast and the voltage at the starter is sufficient to produce a glow discharge between the contacts. The heat from the glow distorts the bimetallic strip, the contacts close, and electrode preheating begins. This short-circuits the glow discharge so that the bi-metal cools and in a short time the contacts open. The open circuit voltage in series with an inductive spike voltage is applied to the lamp. If the lamp fails to ignite, the ballast open circuit voltage again develops a glow in the bulb and the sequence of events are repeated. This continues until the lamp ignites. During normal operation, there is not enough voltage across the lamp to produce further starter glow, so the contacts remain open and the starter consumes no power.

Cutout Starter. These starters may be made to reset either manually or automatically. They are designed to prevent the repeated blinking or attempts to start a deactivated lamp. This type of starter should be good for at least ten or more renewals.

Lamp Failure in Preheat Circuit. Starters which provide no means for deactivation when a lamp fails, will continue to function and attempt to start the lamp. A blinking of the lamp may result. This may lead to either ballast failure or starter failure. Thus it is important to remove a failed preheat lamp without significant delay.

Fluorescent Lampholders

Lampholders are designed for each base style. Several types are available for various spacings and mounting methods in luminaires. See Fig. 8-45. It is important that proper spacing be maintained between lampholders in luminaires to insure satisfactory electrical contact. Manufacturers' catalogs

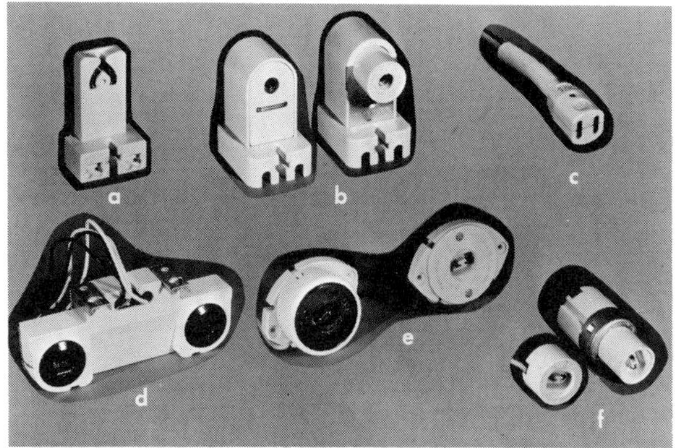

Fig. 8-45. Typical lampholder designs: (a) standard medium bipin lampholder; (b) standard single pin lampholder; (c) circline connector; (d) turret lampholder; (e) flange-mounted lampholder; (f) end-mounted lampholder.

should be consulted for dimension and spacing information on any particular lampholder type.

When fluorescent lamps are used in circuits which may provide an open circuit voltage in excess of 300 volts, or in circuits which may permit a lamp to ionize and conduct current with only one end inserted in the lampholder, electrical codes usually require some automatic means for de-energizing the circuit when the lamp is removed. This is usually accomplished by the lampholder so that when the lamp is removed, the ballast primary circuit is opened. The use of recessed contact bases on 800 and 1500 mA fluorescent lamps has eliminated the need for this disconnect feature in lampholders for these lamps.

Dimming of Fluorescent Lamps[20]

Certain types of fluorescent lamps can be controlled in luminous intensity with dimming equipment of standard design. See also Section 16 Dimming is achieved by reducing the effective current through the lamp. This is accomplished by either: lowering the primary voltage to the transformer or the ballast; adding impedance in the arc circuit; or shortening the time that the arc current flows each half cycle by gating action. Best dimming results when the open circuit voltage is high with respect to the lamp running voltage. This permits adequate firing voltage at dimmer settings for low luminous output.

Firing voltage can also be maintained by use of a peaking transformer incorporated into the dimming ballast. This permits dimming to low intensities with arc voltage maintained and starting qualities unimpaired, while at the same time without requiring excessively high open circuit voltages.

The arrangement most readily adaptable for dimming the cold-cathode fluorescent lamp is the series-circuit. A diagram showing typical arrangements for this circuit is shown in Fig. 8–46a. When using this type of circuit, it is recommended that the lamp loading be about 20 per cent less than that recommended for general lighting requirements. With this arrangement, it is possible to reduce luminous intensity to a value of 10 per cent of the rated light output. It must be remembered that the constant wattage type transformer cannot be employed in dimming. High-power-factor cold-cathode lamp transformers and ballasts cannot be used with saturable reactor or magnetic amplifier dimmers.

Suitable stops may be placed on auto-transformer dimmers to prevent flicker at low intensities.

In the arrangement for dimming hot-cathode rapid start lamps, the ballast provides for the operation of the cathode-heater circuits at full voltage at all times. The dimmer control, either in the form of a solid state circuit, an adjustable voltage transformer circuit, or adjustable reactor circuit, influences the current of the arc-circuit only, while maintaining sufficient arc voltage to insure positive starting and firing. This circuiting permits full control of the luminous intensity of these lamps from full brightness to nearly blackout. Various numbers of lamps can be dimmed from one control unit. Typical ballast connections are shown in Fig. 8–46.

Flashing of Fluorescent Lamps[21]

Cold cathode and rapid start or preheat start hot cathode fluorescent lamps can be flashed with good performance. Cold cathode lamps are flashed through control of either transformer primary or secondary voltage. Hot cathode lamps can be flashed by means of special single-lamp or two-lamp ballasts designed to control lamp arc current while retaining cathode circuit voltage. An external flashing device is required with either system. This unit must be rated for the voltage and current involved. The single-lamp circuit is similar to the single-lamp dimming circuit, shown in Fig. 8–46b except that a flashing device is required in place of the dimming element. Flashing of fluorescent lamps has wide application in electrical advertising.

High Frequency Operation of Fluorescent Lamps

Operation of fluorescent lamps at frequencies above 60 hertz yields increased lamp efficacy. This increase is dependent upon lamp types as shown in Fig. 8–47, where data are presented for the 4-foot 40-watt, 8-foot slimline, and 8-foot 800 mA lamps. In addition to the increased lamp efficacy, the ballast can often be reduced in size and weight.

At low frequencies, a reactor or a combination of reactor and capacitor is required for satisfactory lamp operation. At higher frequencies a low loss capacitor can be used without introducing significant distortion of the wave shape of lamp electrical parameters. In addition, a reactor can be constructed with a low loss powdered iron (ferrite) core.

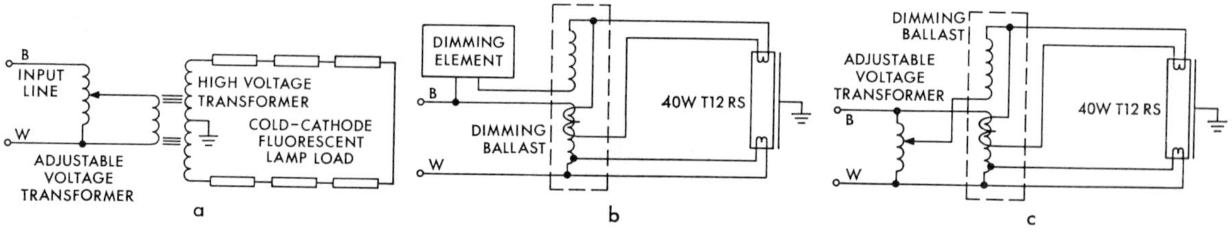

Fig. 8–46. Typical dimming circuits for: (a) series-connected cold cathode lamps; (b) hot cathode rapid start lamps; (c) alternate for (b).

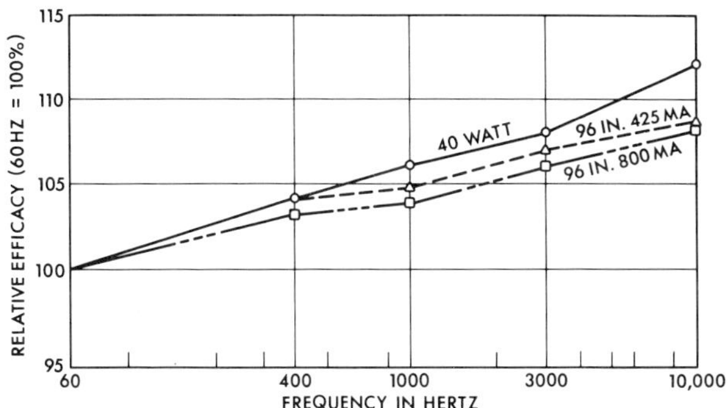

Fig. 8–47. Fluorescent lamp efficacy versus frequency for three different lamp types.

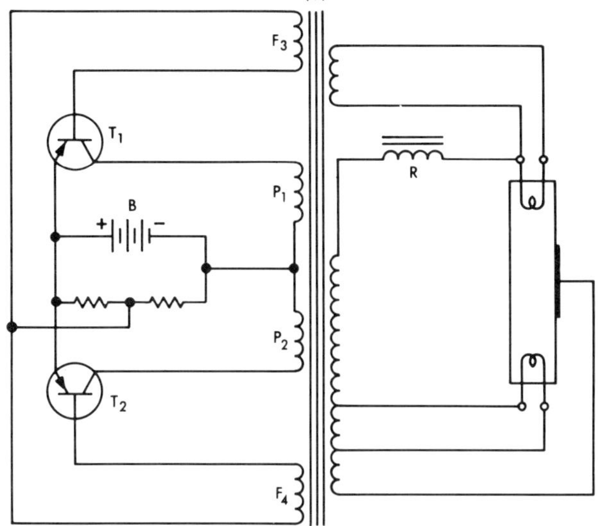

Fig. 8–48. Transistor inverter for high frequency operation of a fluorescent lamp.

The high frequency voltage for operation of the lamps can be generated by means of rotary converters, thyratron inverters, or more recently by means of solid state inverters incorporating transistors or silicon controlled rectifiers. A transistor inverter changes low voltage dc to high frequency ac. A basic circuit is shown in Fig. 8–48. Transistors *T1* and *T2* of Fig. 8–48 act as switches alternately connecting the dc supply (B) across the primary winding *P1* and *P2*. Feedback windings *F3* and *F4* are arranged in such a way that the base of the conducting transistor is negative whereas the base of the non-conducting transistor is positive. The collector current through the primary winding initially increases with time, thereby inducing voltage in the secondary windings. As soon as this current becomes constant (mainly determined by the inductance and resistance of the transformer) the induced voltage will be reduced to zero and the polarity of the feedback windings reversed. The first transistor will now be blocked and the second one will conduct. The ac voltage generated is transformed into the lamp circuit by transformer (*T*). This voltage supplies power for heating the electrodes and operating a discharge within the lamp. A small reactance (R) is used to limit the current flow in the lamp circuit. The efficiency of such a simple inverter is rather low but can be improved to over 80 per cent by reducing the commutation loss in the transistors (pulsed base currents) and by capacitive tuning of the frequency.

The circuit of Fig. 8–48 is used when the inverter is contained within the luminaire. Other possibilities are a common inverter operating several luminaires with individual reactive or capacitive components to limit current in each lamp circuit.

The operating frequency should be high in order to increase the efficacy and to shift ballast noise to the inaudible range of the noise spectrum. A high frequency presents problems in the distribution of power in the case of central inverters. Dc to ac inverters are used extensively for illumination in buses, trains, boats, trailers, and planes, and for advertising signs on cars and for battery operated emergency lights.

The conversion efficiency of low frequency ac to high frequency ac is at present low; therefore, practical applications are limited.

HIGH-INTENSITY DISCHARGE LAMPS

High-Intensity Discharge Lamps (HID) include the groups of lamps commonly known as mercury, metal halide, and high pressure sodium lamps. These lamp types are characterized as discharge devices which are wall stabilized and whose light producing envelopes have a bulb wall loading in excess of 3 watts per square centimeter.

Lamp Construction and Operation

Mercury Lamps. In mercury lamps, light is produced by the passage of an electric current through mercury vapor. Since mercury has a low vapor pressure at room temperature, and even lower at cold temperatures, a small amount of more readily ionized argon gas is introduced to facilitate starting. The original arc is struck through the ionization of this argon gas. Once the arc strikes, its heat begins to vaporize the mercury, and this process continues until all of the mercury is evaporated. The amount of mercury in the lamp essentially determines the final operating pressure, which is usually 2 to 4 atmospheres in the majority of lamps.

The operating electrodes used in mercury lamps are usually of the metal-oxide type, in which the emission material, composed of several metallic oxides, is embedded within the turns of a tungsten coil protected by an outer tungsten coil. The electrodes are heated to the proper electron-emissive temperature by bombardment energy received from the arc.

Most mercury lamps are constructed with two envelopes, an inner envelope (arc tube) which contains the arc, and an outer envelope which (a) shields the arc tube from outside drafts and changes in temperature, (b) usually contains an inert gas (generally nitrogen) which prevents oxidation of internal parts, and also maintains a relatively high breakdown voltage across the outer bulb parts, (c) provides an inner surface for coating of phosphors, and (d) in some cases acts as a filter to remove certain wavelengths of arc radiation. In most cases the arc tube is made of fused silica with thin molybdenum ribbons sealed into the ends as current conductors. The outer bulb is made of "hard" (borosilicate) glass in most cases, but may also be of other glasses for special transmission or where thermal shock is not a problem.

The essential construction details shown in Fig. 8–49 are typical of lamps with fused silica inner arc tubes within an outer envelope. (Other lamps such as 1000-watt capillary lamps, 55-inch 3000-watt lamps, those for special photochemical application, and self-ballasted types have different constructions.)

The pressure at which a mercury lamp operates accounts in a large measure for its characteristic spectral power distribution. In general, higher operating pressure tends to shift a larger proportion of emitted radiation into longer wavelengths. At extremely high pressure there is also a tendency to spread the line spectrum into wider bands. Within the visible region the mercury spectrum consists of five principal lines (404.7, 435.8, 546.1, 557, and 559 nanometers) which result in the greenish-blue light at efficacies of 30 to 65 lumens per watt. While the light source itself appears to be bluish-white, there is an absence of red radiation, especially in the low and medium pressure lamps, and most colored objects appear distorted in color rendition. Blue, green, and yellow colors in objects are emphasized; orange and red appear brownish. Where color rendering is of great importance, mercury lamps are sometimes combined with incandescent lamps in an installation.

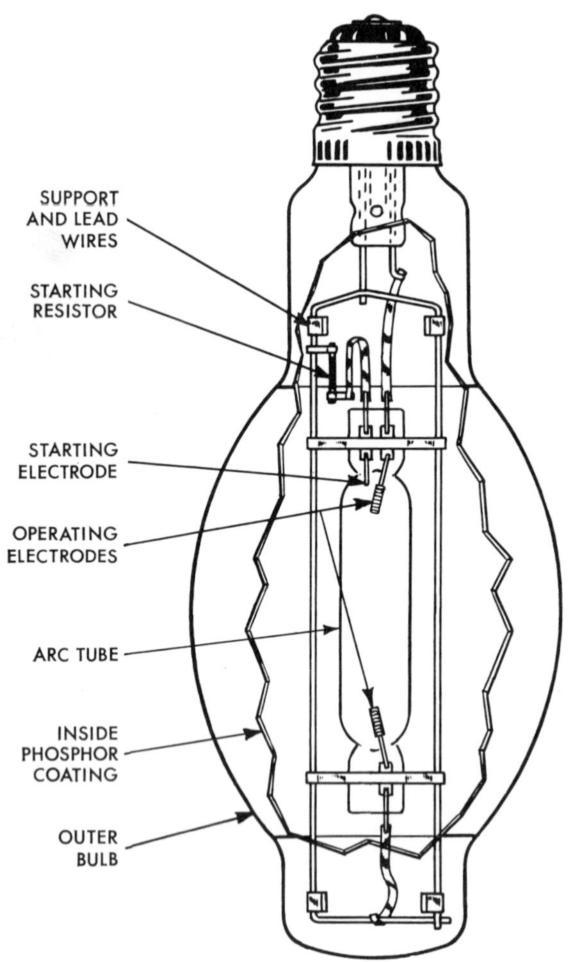

Fig. 8–49. A 400-watt phosphor-coated mercury lamp. Lamps of other sizes are constructed similarly.

A significant portion of the energy radiated by the mercury arc is in the ultraviolet region. Through the use of phosphor coatings on the inside of the outer envelope some of this ultraviolet energy is converted to visible light by the same mechanism employed in fluorescent lamps. See Section 2. One group of mercury lamps employs a fluorogermanate phosphor (/C*) which emits most of its energy in the red region. As a result, color rendering is greatly improved over clear mercury types. Through the use of an orthophosphate phosphor (W*), lamp efficacy can be increased above clear lamp values. Although color rendering is improved with this type of lamp, as compared with clear mercury lamps, its color rendering properties are not as good as that of the previously mentioned lamp. A third phosphor emitting orange-red radiation, vanadate (/DX*), improves efficacy and color rendition, rendering skin tones reasonable well. Fig. 8–23 shows the spectral distribution of a clear lamp and lamps using these phosphors.

Metal Halide Lamps. Metal halide lamps are very similar in construction to the mercury lamp, the major difference being that the metal halide arc tube contains various metal halides in addition to mercury. When the lamp has attained full operating temperature, the metal halides in the arc tube are partially vaporized. When the halide vapor approaches the high temperature central core of the discharge it is dissociated into the halogen and the metal, with the metal radiating its appropriate spectrum. As the halogen and metal move near the cooler arc tube wall by diffusion and convection, they recombine, and the cycle starts over again. This cycle provides two desirable advantages. First, some metals, which in their metallic form cannot be vaporized at the temperatures which a fused silica arc tube can withstand, can be introduced into the discharge by dissociation of halides which vaporize at much lower temperatures. Typical metals intro-

* Commonly used color designations.

duced in this manner into metal halide lamps are thallium, indium, scandium and dysprosium. Secondly, some other metals, which react chemically with the arc tube, can be used in the form of a halogen which does not readily react with fused silica, since the metal is in the halogen form at the wall temperature, and the metal plus halogen form at the center of the discharge. A common example of a fused silica reactive metal which can be used in the halide form is sodium.

Compared with mercury lamps the efficacy of metal halide lamps is greatly improved. Commercially available metal halide lamps have efficacies 1.5 to almost 2 times that of mercury lamps. Almost all varieties of available "white"-light metal halide lamps produce color rendering which is equal to or superior to the presently available phosphor coated mercury lamps.

Three typical combinations of halide used in metal halide lamps are: (1) sodium, thallium, and indium iodides, (2) sodium and scandium iodides, and (3) dysprosium and thallium iodides. Their spectral power distributions are shown in Fig. 8–23. Some halides, such as, sodium (589nm), thallium (535nm), and Indium (435nm) principally produce line spectra, while others, such as scandium, thorium, dysprosium and other rare earths produce multiline spectra across the full visible region. Halides, such as tin, may produce a continuous spectra across the visible region.

Improved color balance can be produced by combining the spectra of elements which radiate in various regions of the spectrum as in the sodium-thallium-indium lamp; while it can also be achieved by use of the multiline spectra of scandium or dysprosium. High efficacy is produced by the use of iodides, such as thallium and sodium.

High-intensity discharge lamps of selected colors or for producing near ultraviolet can also be made by using the metal halide technique; for example, sodium for orange, thallium for green, indium for blue, and lead for ultraviolet.

Metal halide lamps are also available with phosphors applied to the outer envelopes to further modify the color and generally to lower the color temperature of the lamp.

A close look at a metal halide lamp will reveal several construction features which differ from mercury lamps. (1) The arc tubes are usually smaller than in mercury lamps for equivalent wattage, with a coating or reflector at the ends of the arc tube. This end reflector and small arc tube size serves to increase the temperature at the end of the arc tube to assure vaporization of the halides. (2) Some lamps include a system for either shorting the starting electrode to the operating electrode or opening the starting electrode circuit. This is required to prevent electrolysis in the fused silica between the starting and operating electrodes especially when a halide such as sodium iodide is used in the lamp. (3) In some lamps the electrical connection to the electrode at the dome of the lamp is made by means of a small wire remote from the arc tube. This is to prevent diffusion of sodium through the arc tube by electrolysis caused by a photoelectric effect when the current lead is close to the arc tube.

High Pressure Sodium Lamps. In the high pressure sodium lamp light is produced by electricity passing through sodium vapor. This lamp is constructed with two envelopes, the inner being polycrystalline alumina which has the properties of resistance to sodium attack at high temperatures as well as a high melting point, and good light transmission (more than 90 per cent) even though this material is translucent. The construction of a typical high pressure sodium lamp is shown in Fig. 8–50. Polycrystalline alumina cannot be fused to metal by melting the alumina without causing the material to crack. Therefore, the seal to metal is made by means of an intermediate solder glass or metal between the niobium end caps and the alumina. The arc tube contains xenon as a starting gas, and a small quantity of sodium-mercury amalgam which is partially vaporized when the lamp attains operating temperature. The mercury acts as a buffer gas to raise the gas pressure and operating voltage of the lamp to a practical level.

The outer borosilicate glass envelope is evacuated and serves to prevent chemical attack of the arc tube metal parts as well as maintaining the arc tube temperature by isolating it from ambient temperature effects and drafts.

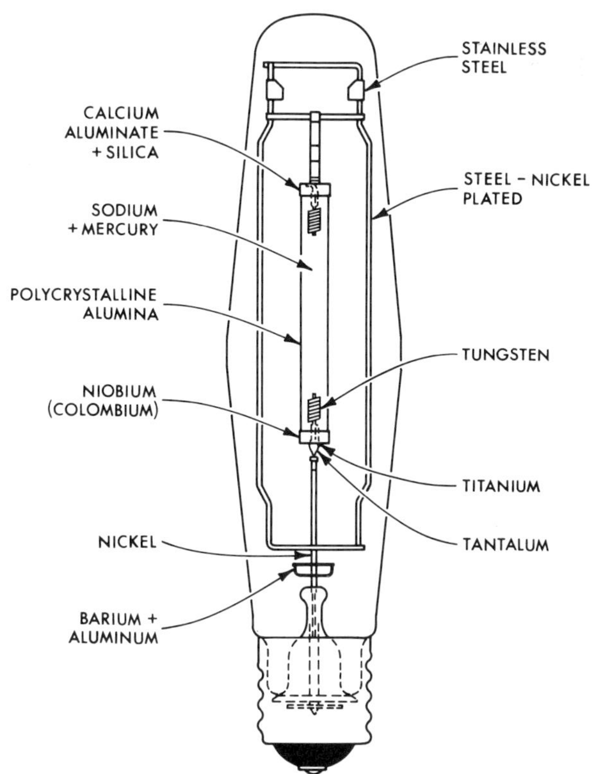

Fig. 8–50. Construction of a typical high pressure sodium lamp.

High pressure sodium lamps radiate energy across the visible spectrum. This is in contrast to the low pressure sodium lamp (see page 8-43) which radiates principally the doublet "D" lines of sodium at 589 nanometers. At higher sodium pressures, about 200 torr, in the high pressure sodium lamp, sodium radiation of the "D" line is self absorbed by the gas and is radiated as continuum on both sides of the "D" line frequency. This results in the "dark" region at 589 nm as shown in the typical spectrum in Fig. 8-23. The light produced by this lamp is consequently golden-white in color with all visible frequencies present. These lamps have efficacies of about 110 lumens per watt.

Because of the small diameter of a high pressure sodium lamp arc tube, no starting electrode is built into the arc tube as in the mercury lamp. A high voltage high frequency pulse is required to start these lamps.

Lamp Designations

The current identifying designations of mercury lamps generally follow a system which is authorized and administered by the American National Standards Institute (ANSI). All designations start with the letter "H" (to indicate mercury). This is followed by an ANSI assigned number which identifies the electrical characteristics of the lamp and consequently the ballast. After the number there are two arbitrary letters which identify the bulb size, shape, finish, etc., excluding color. After this section, the individual manufacturers may add, at their discretion any additional letters or numbers which they desire to indicate information not covered by the standard section of the designation, such as lamp wattage or color. Another system being used starts with the letter "H" followed by lamp wattage, then an arbitrary letter identifying the bulb, and finish, and finally the numbers defining electrical characteristics.

An extension of the present system is being considered in which the letter "H" will be replaced by "M" for metal halide lamps and the letter "S" for high pressure sodium lamps.

Lamp Starting

Mercury Lamps. The two-electrode types of mercury lamps, such as the H9FJ, and also many photochemical types, require a high open circuit voltage to ionize the argon gas and permit the arc to strike. In the more common three-electrode type of lamp an auxiliary starting electrode placed close to one of the main electrodes makes it possible to start the lamp at a lower voltage. Here an electric field is first set up between the starting electrode (which is connected to the opposite main electrode through a current limiting resistor) and the adjacent main electrode. This causes an emission of electrons which develops a local glow discharge and ionizes the starting gas. The arc then starts between the main electrodes. The mercury gradually becomes vaporized from the heat of the arc and carries an increasing portion of the current. During this process the arc stream changes from the diffuse bluish glow characteristic of the argon arc, to the blue-green of mercury, increasing greatly in luminance and becoming concentrated in the center of the tube. At the instant the arc strikes, the current is high and the voltage is low. Normal operating values are reached after a warm-up period of several minutes during which the current drops and the voltage rises until the arc attains a point of stabilization in vapor pressure, as shown in Fig. 8-51 for one type of 400-watt lamp operated from a high reactance or inductive ballast; the mercury is then entirely evaporated.

If the arc is extinguished, the lamp will not relight until it is cooled sufficiently to lower the vapor pressure to a point where the arc will restrike with the voltage available. The time from initial starting to full light output at ordinary room temperatures, with no enclosing lighting unit, and also the restriking time or cooling time until the lamp will restart, vary between 3 and 7 minutes for the various types of lamps.

Metal Halide Lamps. The method of starting metal halide lamps is the same as mercury lamps. However, because of the presence of the halide in the lamps, the starting voltage required is sometimes higher than that of mercury lamps. As the lamp warms up it may go through several color changes as the various halides begin to vaporize until it reaches its equilibrium color and electrical characteristics after several minutes.

Since a metal halide arc tube operates at higher temperatures than a mercury lamp arc tube the time to cool and lower the vapor pressure is generally longer than that of the mercury lamp, consequently the restrike time may be up to 15 minutes.

High Pressure Sodium Lamps. Since the high pressure sodium lamp does not contain a starting electrode, a high voltage, high frequency pulse is used to ionize the xenon starting gas to facilitate

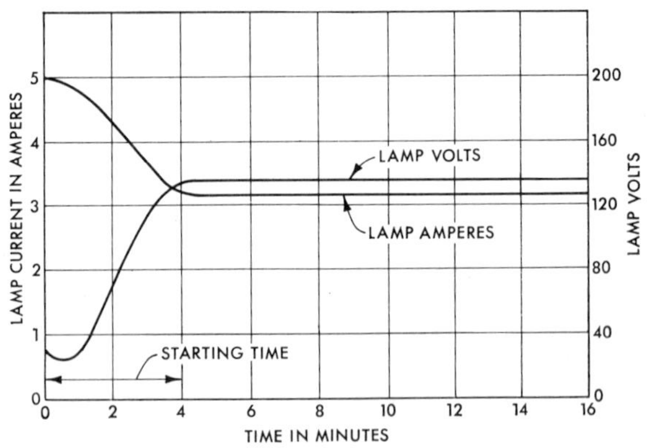

Fig. 8-51. Warm-up characteristics for a 400-watt quartz arc tube lamp on an inductive ballast. This is typical of most mercury lamps.

starting. Once started, the lamp warms up in approximately 15 minutes during which time the color rendition will change from poor when the lamp first starts to its normal broad spectrum when stable operating conditions are achieved.

Because the operating pressure of a high pressure sodium lamp is lower than that of a mercury lamp the restrike time is shorter. It will restrike in 1½ to 2 minutes.

Lamp Operating Position

Mercury Lamps. When a mercury lamp is operated horizontally, the arc tends to bow upward because of the convection currents in the gas. This bowing of the arc will generally cause a small change in the electrical characteristics of the lamp as well as a small reduction of lamp output caused by a reduction in lamp wattage and efficacy. Many ballasts designed for horizontally operating lamps compensate for this wattage decrease by increasing the current through the lamp, but do not compensate for the loss in efficacy.

Metal Halide Lamps. When a metal halide lamp is operated horizontally the arc also bows upward, but the effect of change of operating position can be much greater than in the case of the mercury lamp especially with regard to color. Because a portion of the halides in metal halide lamps is not vaporized during lamp operation, any change in the cold spot temperature of the lamp will change the pressure of the halide and therefore the lamp color. Generally, the color temperature of these lamps is higher when they are operated in the horizontal position than when operated in the vertical position. When the burning position of a metal halide lamp is changed, as much as six hours may be required before the lamp characteristics, color, electrical characteristics, and efficacy are stabilized. Some lamps have restricted burning positions. These should be observed if optimum performance is to be obtained.

High Pressure Sodium Lamps. High pressure sodium lamps having small arc tube diameters permitting a very small arc bow, exhibit very little change in efficacy or electrical characteristics when operated in a horizontal vs a vertical position. However, since the position of the cold spot at the end of the arc tube controls the sodium pressure in the lamp, some lamps are restricted as to burning position.

Lamp Life and Lumen Maintenance

Mercury Lamps. General service mercury lamps have a long average life. These lamps usually employ an electrode with a mixture of metal oxides embedded in the turns of tungsten coils from which the electrode is assembled. During the life of the lamp, this emission material is very slowly evaporated or sputtered from the electrode and is deposited on the inner surface of the arc tube. This process results first in a white deposit on the inner surface of the arc tube, eventually in a blackening of the arc tube, and ultimately in the exhaustion of the emission material in the electrodes and the end of lamp life, when the starting voltage exceeds the open circuit voltage.

Mercury lamps are usually rated for "initial" lumens after 100 hours burning. Initial lumens for the various lamp types are given in Fig. 8-116. Because of the very slow evaporation of emission material from the electrodes, the maintenance of mercury lamps is good. Fig. 8-52 illustrates a typical lumen maintenance curve. Neither life nor maintenance is materially affected by reasonable changes in burning cycle. (Two hours or more per start).

Metal Halide Lamps. Chemical reaction between the iodine in a metal halide lamp and the emission materials included in mercury lamp electrodes prevents the use of those electrodes. Because the alternative electrodes used evaporate more rapidly than corresponding mercury lamp electrodes the maintenance of metal halide lamps is not as good as that of the mercury lamp.

As in the case of the mercury lamp rated "initial" lumens are based on 100 hours operation; however because of the poorer maintenance of metal halide lamps, the life ratings are shorter.

High Pressure Sodium. High pressure sodium lamps employ electrodes very similar to those used in mercury lamps. This fact plus the smaller diameter arc tube combine to give high pressure sodium lamps excellent maintenance. Initial ratings also apply at 100 hours of operation.

The life of high pressure sodium lamps is limited by a slow rise in operating voltage which occurs as small amounts of sodium in the arc tube clean up, and as seals at the ends of the arc tube deteriorate during life. This limits the life to the ratings shown in Fig. 8-116.

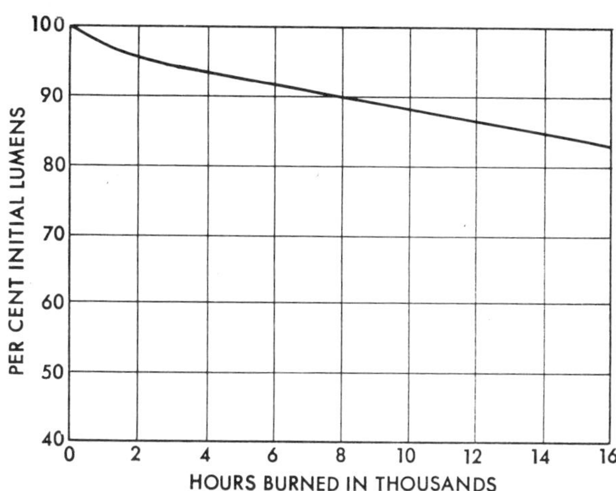

Fig. 8-52. Lumen maintenance curve for a 400-watt clear mercury lamp with a quartz arc tube. Per cent of rated initial lumens determined at 100 hours. Typical of all mercury lamps.

Effect of Ambient Temperature

The light output of the double-envelope type of high-intensity discharge lamp is little affected by the ambient temperature. Experience has shown that these lamps are satisfactory for temperatures down to −20° F or lower. On the other hand, the single-envelope lamps, intended primarily for use as ultraviolet sources, are critically affected by low temperatures, particularly if the surrounding air is moving, and are not considered suitable for use below 32° F without special protection. Ambient temperature affects the striking voltage of all discharge lamps and in some cases higher starting voltages than those listed in Fig. 8–116 for indoor use are recommended for roadway and floodlighting installations in cold climates. Ballasts for roadway lighting service and other low temperature applications are designed to provide the necessary voltage to start and operate each particular lamp at temperatures as low as −20° F. Recommendations for starting voltages have been developed by the American National Standards Institute.

Lamp Operating Temperature

Because most high-intensity discharge lamps are long-lived, operating temperatures are particularly important. The effect of heat is partly a function of time, and the longer the rated life of the lamp the greater the possibility of damage from high temperatures. Excessive envelope and base temperatures may cause failures or unsatisfactory performance due to softening of the glass, damage to the arc tube by moisture driven out of the outer envelope, softening of the basing cement or solder, or corrosion of the base, socket, or lead-in wires. The use of reflecting equipment which concentrates heat and light rays on either the inner arc tube or the outer envelope should be avoided.

In the case of metal halide and high pressure sodium lamps in which all the material is not vaporized, concentrated heat on the arc tube can materially effect their color as well as their electrical characteristics.

Auxiliary Equipment

High intensity discharge lamps have negative volt-ampere characteristics and, therefore, some current limiting device, usually in the form of a transformer and reactor ballast, must be provided to prevent excessive lamp and line currents. Lamps are operated on either multiple or series circuits. Fig. 8–53 illustrates schematic diagrams of several typical ballast types.

Mercury Lamps. There are a number of lamp ballasts in use for operating mercury lamps. Wattage losses in ballasts are usually in the order of 5 to 15 per cent of lamp wattage depending upon ballast and lamp type.

Normal Power Factor Reactor. The simplest ballast is the reactor consisting of a single coil wound on an iron core placed in series with the lamp. The only function of the reactor is to limit the current delivered to the lamp. The reactor can only be used when the line voltage is within the specified lamp starting voltage requirement. Inherently the power factor of this circuit is about 50 per cent lagging and is commonly referred to as being normal or low power factor.

Since it only performs the function of current control, This reactor is the most economical, smallest, and most efficient ballast. However, there are shortcomings which should be considered in application. The reactor provides little regulation for fluctuations in line voltage; for example, a 3 per cent change in line voltage would cause a 6 per cent change in lamp wattage. Therefore, the reactor is not recommended where line fluctuations exceed 5 per cent. Also, the line current under starting conditions is approximately 50 per cent higher than normal operating current and, therefore, it is recommended that supply wiring be sized for approximately twice the normal operating current.

High Power Factor Reactor. This circuit consists of a reactor in series with the lamp and a capacitor connected across the line. Both lamp current and regulation are essentially the same as described for the normal power factor reactor. The capacitor connected across the line does not affect the lamp circuit, but increases the power factor of the system to better than 90 per cent. It also reduces the input current under starting and operating conditions almost 50 per cent over the normal power factor system, allowing the use of a larger number of ballasts on a given line.

Normal Power Factor Autotransformer. Where the line voltage is below or above the specified lamp starting voltage range, a transformer is used in conjunction with the reactor to provide proper starting voltage. This is normally accomplished with the combination of primary and secondary coils forming a one-piece single autotransformer, also known as a "high reactance" ballast. The power factor of this circuit is about 50 per cent lagging and has the same features and shortcomings as the normal power factor reactor circuit.

High Power Factor Autotransformer. The high power factor autotransformer circuit is the same as the normal power factor autotransformer, except a capacitor is added to the primary circuit. In order to provide a more economical system, an extra capacitor winding is normally added to the autotransformer primary. This economical combination of extended windings with the capacitor increases the system power factor to above 90 per cent. The effect on input current is the same as in the high power factor reactor. Regulation and lamp performance are unchanged.

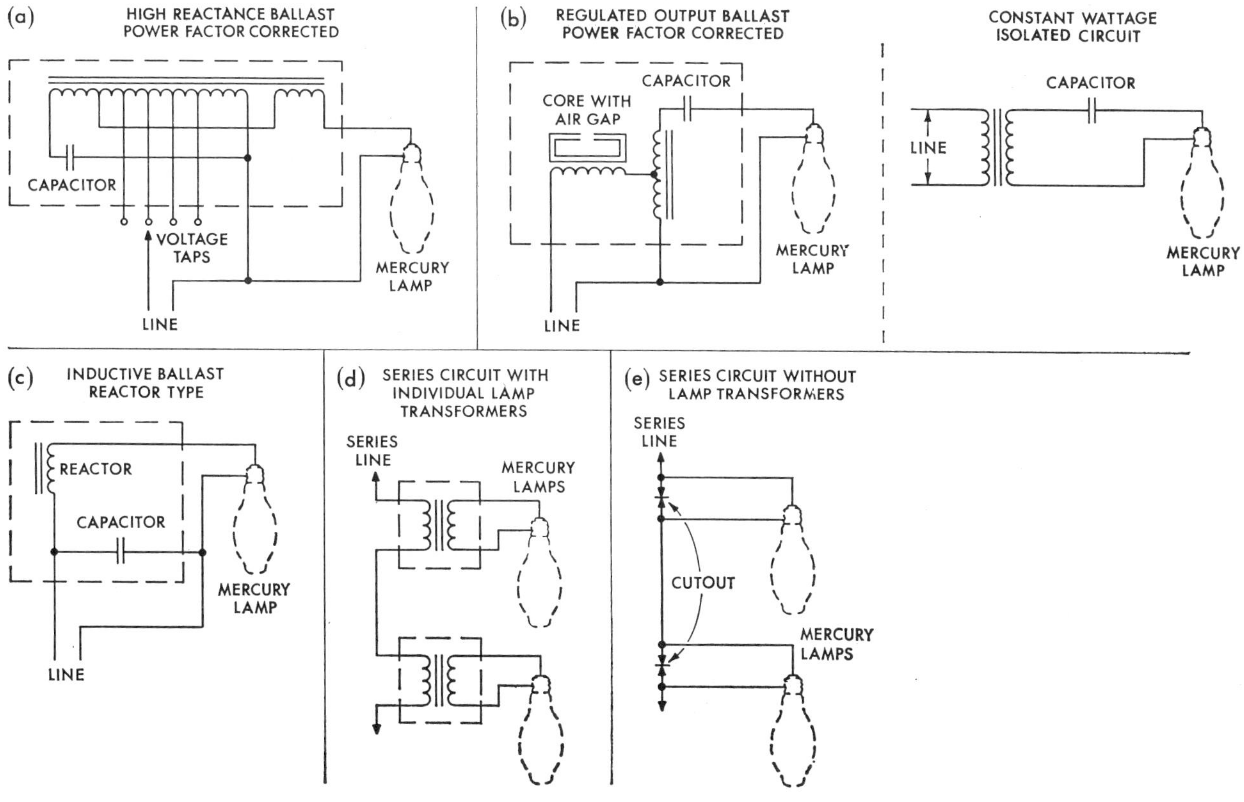

Fig. 8-53. Typical circuits for operating mercury lamps.

Constant Wattage Autotransformer (CWA). This type of ballast circuit, commonly referred to as "CWA" type, is the most widely used in modern mercury lighting systems. It consists of a high reactance autotransformer with a capacitor in series with the lamp. Use of the capacitor allows the lamp to operate with better wattage stability when the voltage on the branch circuit fluctates. This ballast is used when line voltage is expected to vary more than 5 per cent. For example, a 10 per cent change in line voltage would result in only a 5 per cent change in lamp watts. Other advantages with the CWA ballast are high power factor, low line extinguishing voltage, and line starting currents that are lower than normal line currents. The CWA features allow maximum loading on branch circuits and provide an economical and efficient mercury lighting system.

The capacitor used with the CWA ballast performs an important ballasting function as in all lead-type circuits. The capacitor used in lag-type circuits such as reactor and high reactance autotransformer ballasts is purely a power factor correction component and has no ballasting property.

Premium Constant Wattage (PCW). This type of ballast, also referred to as regulated or stabilized type, has operating characteristics similar to the CWA, except that there is more control over lamp watts. For applications where a constant light output is required within ±2 per cent from a varying line voltage with a range of ±13 per cent, the premium constant wattage ballast should be used. The PCW ballast, like the CWA ballast, uses a lead circuit; but unlike the CWA, the lamp circuit is completely isolated from the primary winding. It also has the same advantages of the CWA ballast, such as high power factor, low line extinguishing voltage, and low line starting currents.

Two Lamp Lead-Lag Reactor. The lead-lag design approach is commonly used with encased type two-lamp, 400-watt and 1000-watt mercury ballasts and consists of two independent circuits. A current limiting reactor operates one lamp, and a combination reactor and a capacitor connected in series operates the second. The lamps operate independently so that a failure of one has no effect on the other. The input current of the combination of capacitors and reactors is lower than the operating current. These elements provide a high power factor and reduce stroboscopic effect. This circuit can only be used when the line voltage is within the specified lamp starting voltage. It is the most economical two lamp system with regulation similar to the reactor and high reactance ballasts.

Two Lamp Series (Isolated) Constant Wattage. This circuit is essentially the same as single lamp premium constant wattage, except it operates two lamps in series. The most effective use of this circuit is with applications where the ambient temperature is 0° F or above. It is most popular with indoor 400 watt applications. Electrical characteristics are similar to the single lamp, premium constant wattage circuit, except for regulation which is equal to CWA. A two-lamp in series autotransformer—connected circuits—is available. However, when two lamps are connected in series with the mid-point grounded in compliance with the National Electrical Code, an isolated transformer is required.

Mercury lamps are also operated in series on constant current series regulators. The most commonly used method employs a current transformer for each lamp (Fig. 8–53d). It differs in design from the more common multiple type of ballast. The usual design is a two-winding transformer as illustrated in Fig. 8–53d. Since the inherent regulator reactance limits current in the circuit, the individual lamp current transformer is not designed to limit current but rather to transform it from the regulator secondary current to the rated lamp current. In addition, the transformer is made to limit secondary open circuit voltage so that no cutout is necessary in case of a lamp failure. Series transformers are available for the more popular lamps to operate from either 6.6, 7.5, or 20-ampere series circuits and can be operated on all types of constant current transformers. These circuits will normally be satisfactory for metal halide lamps; but not high pressure sodium lamps.

Metal Halide Lamps. The metal halide lamps will not operate reliably on most standard mercury lamp ballasts. Certain metal halide lamps may be used with certain lag-type ballasts designed for mercury lamps. To insure proper starting and operating, a lead-peaked autotransformer ballast circuit specifically designed for all commercially available metal halide lamps is normally used. This ballast will provide improved regulation in comparison to the conventional nonregulated lag type. For example a 10 per cent change in line voltage will result in a 10 per cent change in lamp watts. Aside from regulation, the lead-peaked auto ballast performs much like the CWA with similar operating features. Standard mercury lamps may be operated on this ballast.

High Pressure Sodium Lamps. A specially designed ballast is required to meet the electrical characteristics of high pressure sodium lamps. Several different ballast design approaches are available. Two of these are a magnetic design and a phase control design.

The magnetic design has three basic parts: the magnetic component, ac capacitors, and an electronic starting circuit. In addition to providing suitable voltage for striking the lamp, the magnetic component limits the lamp current to the desired value, similar to any magnetic ballast for mercury lamps. Capacitors used with magnetic components provide high power factor in addition to aiding regulation. The electronic starting circuit provides a pulse voltage which, when stepped up by the magnetic component, provides approximately 2500 peak volts or more for initially starting the arc in the lamp.

Fig. 8–54. Flicker Index for Various Lamps Under Different Operating Conditions

Type of Circuit or Ballast	Number of Lamps	Typical Flicker Index	Per Cent Flicker
60 cycle standard reactor 1 phase	1	0.235	78
60 cycle regulated output 1 phase	1	0.196	83
60 cycle lead-lag 1 phase	2	0.136	50
60 cycle reactor 3 phase	3	0.272	5

The phase control ballast uses a solid state component in series with the magnetic choke to provide wattage regulation. A solid state switch is controlled by an electronic circuit which senses line voltage and arc voltage changes, and varies the conduction time of the switch during each half cycle of applied voltage. The starting voltage is also generated each half cycle by this electronic control.

Self-Ballasted Lamps

Self-ballasted mercury lamps are available in various wattages. These lamps have a mercury vapor arc tube in series with a current limiting tungsten filament. In some types, phosphors coated on the outer envelope are used to provide additional color improvement. The over-all efficacy is lower than that of other mercury lamps because of the lower efficacy tungsten filament. As the title denotes, these lamps do not require an auxiliary ballast as standard mercury lamps. Typical self-ballasted mercury lamps are shown in Fig. 8–117.

Stroboscopic Effect and Flicker

Since the arc current varies with the cyclic changes of the line voltage it could be expected that the light output will vary similarly. This is true to a degree. The Flicker Index has been established as a method of measuring this effect. The type ballast from which the lamp is operated has considerable effect on the Flicker Index. Operating from a lead-lag type ballast or from a three-phase circuit reduces the flicker considerably. Typical Flicker Indices of lamps operated on various circuits are shown in Fig. 8–54. On 60-cycle power no visible flicker of the lamp is evident, as the light varies at a 120-cycle frequency. This is well beyond the flicker fusion frequency of the eye. High-intensity discharge lamps have been found suitable for use where very fast motion occurs, such as in tennis courts, other sports areas, or machine shops.

SHORT-ARC LAMPS

High pressure gas discharge lamps having an arc length which is small compared with the size of the electrodes are called short-arc or compact arc lamps. Depending on rated wattage and intended application, the arc length of short-arc lamps may vary from about a third of a millimeter to about a centimeter. These arcs have the highest luminance and radiance of any continuously operating light source and are the closest approach to a true "point" source.

Some typical short-arc lamps are shown in Fig. 8-55. Compact arc lamps have a clear fused quartz bulb of spherical or ellipsoidal shape with two diametrically opposite current carrying seals.

Today most short-arc lamp types are designed for dc operation. Better arc stability and substantially longer life of the dc lamps have limited the use of ac compact arc lamps to a few special applications.

Mercury and Mercury-Xenon Lamps

To facilitate lamp starting, short-arc mercury lamps contain argon or another rare gas at a pressure of some torrs,* in the same way as the standard

* A standard atmosphere has a pressure of 760 torrs.

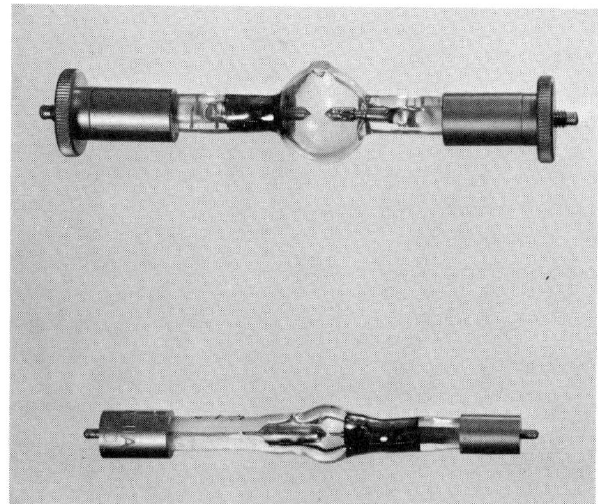

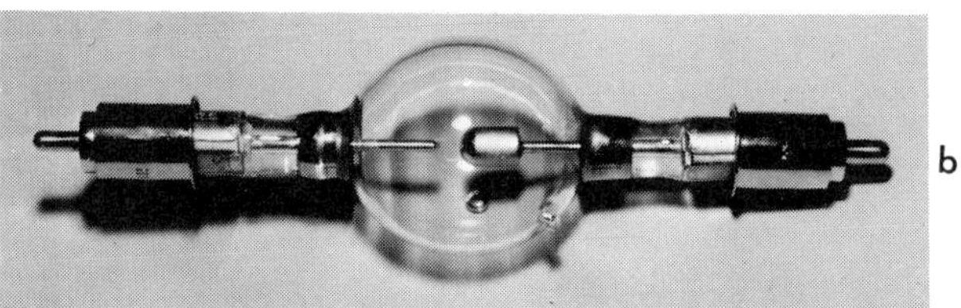

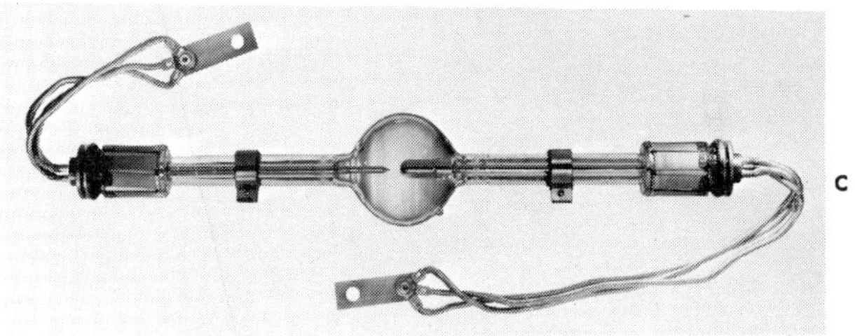

Fig. 8-55. Typical short-arc lamps. (a) Two types of mercury-argon lamps—a 100-watt (lower) and a 200-watt (upper). (b) A 5000-watt mercury xenon lamp. (c) A 6000-watt xenon lamp.

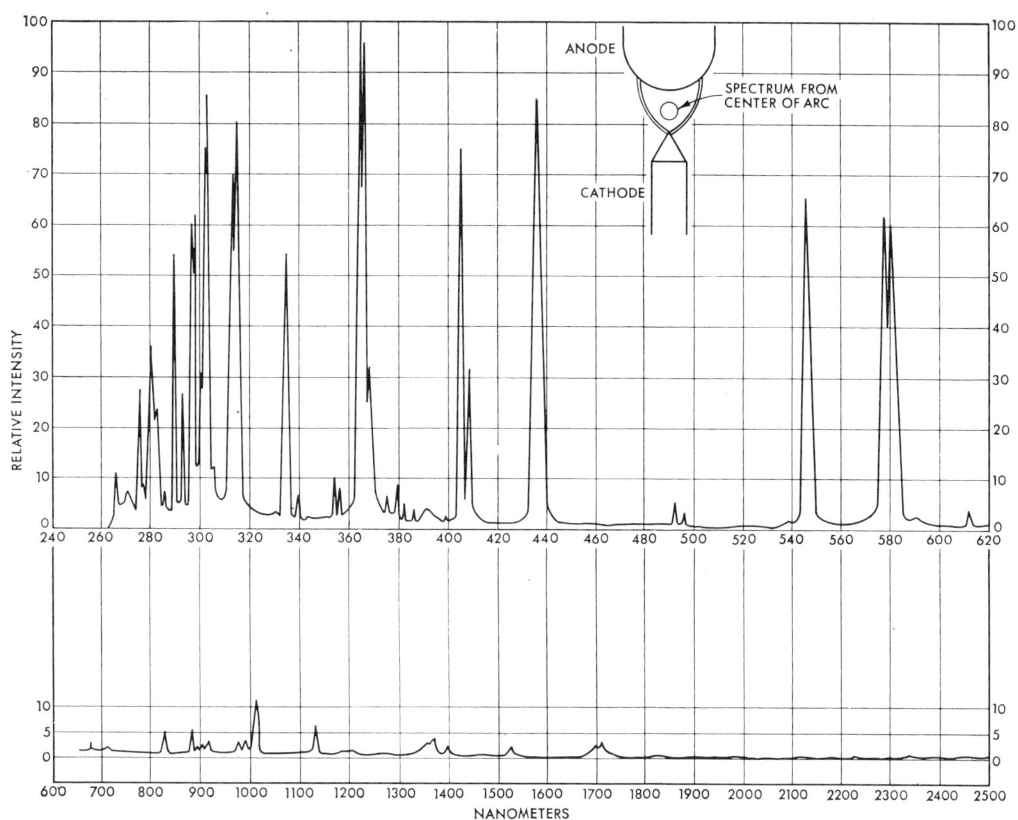

Fig. 8-56. Spectral distribution of a 2.5 kW mercury xenon lamp.

mercury lamps. After the initial arc is struck, the lamp gradually warms up, the voltage increases and stabilizes as the mercury is completely vaporized. A mercury lamp requires several minutes to warm up to full operating pressure and light output. This warm-up time is reduced by about 50 per cent if xenon at a pressure exceeding one atmosphere is added to the mercury. Lamps with this type of filling are known as mercury-xenon short-arc or compact arc lamps. The output in the visible is the same for both types, consisting mainly of the four mercury lines and some continuum, due to the high operating pressure. The luminous efficacy for these lamps is approximately 50 lumens per watt at a rated wattage of 1000 watts and about 55 lumens per watt at 5000 watts.

Mercury and mercury-xenon lamps are available for wattages from 30 to 5000 watts. For technical data see Fig. 8–121. A spectrum of a typical mercury-xenon lamp is shown in Fig. 8–56.

Xenon Lamps

Xenon short-arc lamps are filled with several atmospheres of xenon gas. They reach some 80 per cent of the final output immediately after the start. The arc color approximates daylight very closely (color temperature approximately 6000 K). The spectrum is continuous in the visible. It exhibits strong lines in the near infrared between 800 and 1000 nanometers, some weak lines in the blue, and extends into the far ultraviolet. A spectral energy distribution curve is shown in Fig. 8–57.

Xenon short-arc lamps are made with rated wattages from 30 watts to 30,000 watts. Lamps designed to be operated at wattages above 10 kW require liquid cooling of the electrodes.

The luminous efficacy of the xenon short-arc is approximately 30 lumens per watt at 1000 watts, 45 lumens per watt at 5000 watts, and over 50 lumens per watt at 20 kW input.

Lamp Operating Enclosure

Short-arc mercury, mercury-xenon, and xenon lamps are, during operation, under considerable pressure (up to 50 atmospheres for small lamps and about 10 atmospheres for large units) and therefore must be operated in an enclosure at all times. Further, precaution must be taken to insure protection from the powerful ultraviolet radiation emitted from the lamps.

In general, short-arc lamps up to about 2.5 kW are designed to operate with natural air draft cooling. No special ventilation is required unless the lamps are operated in a closely confined enclosure or at excessive ambient temperature.

In order to eliminate a possible hazard during shipment, storage, or handling of xenon or mercury-xenon lamps, special protection cases are provided. These cases are made of metal or plastic and are so arranged around the bulb that the lamp can

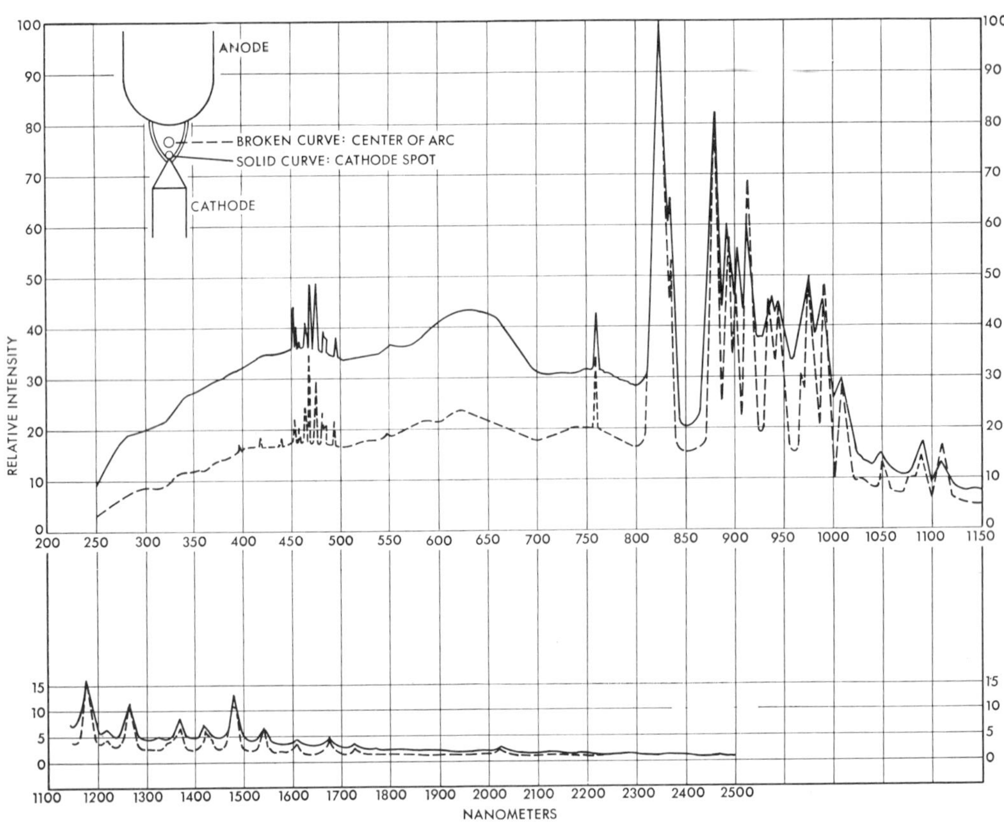

Fig. 8–57. Spectral distribution of a 2.2 kW xenon lamp.

be electrically connected without removing the case. The case should not be removed until immediately before the lamp is put into operation.

Auxiliary Equipment

Like most vapor discharge lamps, short-arc lamps require auxiliary devices to start the arc and limit the current. For ac lamps either resistive or induction ballasts may be used. Direct current lamps are best operated from specifically designed power systems which provide, with high efficiency, the high voltage pulses (up to 50,000 volts) to break down the gap between the electrodes, to ionize it and heat the cathode tip to thermionic emitting temperatures (starter), to provide enough open dc voltage to assure the transition from the low-current high-voltage spark discharge initiated by the starter to the high-current low-voltage arc, and finally to deliver the required current at the relatively low lamp voltage. With a properly designed system, a short-arc dc lamp will start within a fraction of a second. Many power supplies are regulated, so that the lamp current is independent of line voltage fluctuations.

Applications

Short-arc lamps combine the high luminance of carbon arcs with the maintenance free and clean operation of regular gas discharge or incandescent lamps. They are primarily used for searchlights, projectors, display systems, optical instruments as spectrophotometers, recording instruments, and for simulation of solar radiation in space. Compact arc lamps are specially suited as sources of modulated light, generated through current modulation.

MISCELLANEOUS DISCHARGE LAMPS

Low Pressure Sodium Lamps

In low pressure sodium discharge lamps, the arc is carried through vaporized sodium. The starting gas is neon with small additions of argon, xenon, or helium. In order to obtain the maximum efficacy of the conversion of the electrical input to the arc discharge into light, the vapor pressure of the sodium must be in the order of 5×10^{-3} millimeters of mercury which corresponds to an arc tube bulb wall temperature of approximately 500° F. Any appreciable deviation from this pressure results in a very undesirable loss in the lamp efficacy. To maintain this proper operating temperature, the arc tube must be enclosed in a vacuum flask or in an outer bulb at high vacuum.

The light produced by the low pressure sodium arc is almost monochromatic, consisting of a double line in the yellow region of the spectrum at 589 and 589.6 nanometers as shown in Fig. 8–23.

The starting time to full light output is 7 to 15 minutes. When first started, the light output is the characteristic red of the neon discharge and this gradually gives way to the characteristic yellow as the sodium is vaporized. The sodium lamp will restart immediately after interruption of the power supply. Lamp performance during starting is shown in Fig. 8–58.

Two types of arc tube construction are used in modern low pressure sodium lamps—the hairpin, or "U" tube, and the linear. In the hairpin type of construction, the arc tube is doubled back on itself with its limbs being very close together. The electrodes are sealed in at the pinches of the arc tube mounted inside an outer vacuum jacket. The lamp is completed by a two pin bayonet base. The construction of this lamp is shown in Fig. 8–59. The electrodes in this lamp are of a metal oxide type and are heated to electron-emissive temperature by ion bombardment. Electrical and light output values are given in Fig. 8–118.

In the linear lamp, the arc tube is double ended and is dimpled at regular intervals. Preheat type electrodes are sealed into each end of the arc tube. This arc tube is sealed in turn into an outer vacuum jacket and completed by a medium bipin base at each end. See Fig. 8–59. After the arc is struck, the electrodes are maintained at an electron-emissive temperature by ion bombardment. Electrical and light output values are given in Fig. 8–118.

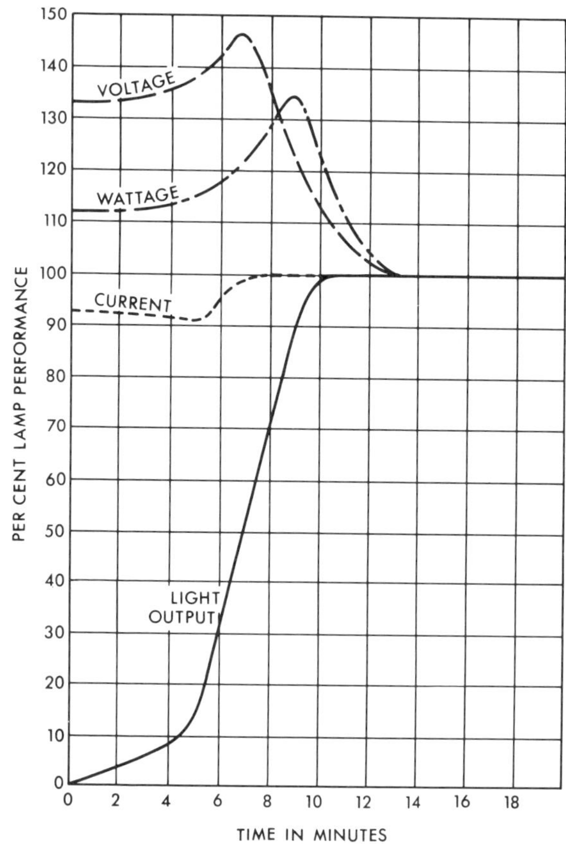

Fig. 8–58. Low pressure sodium lamp performance during starting.

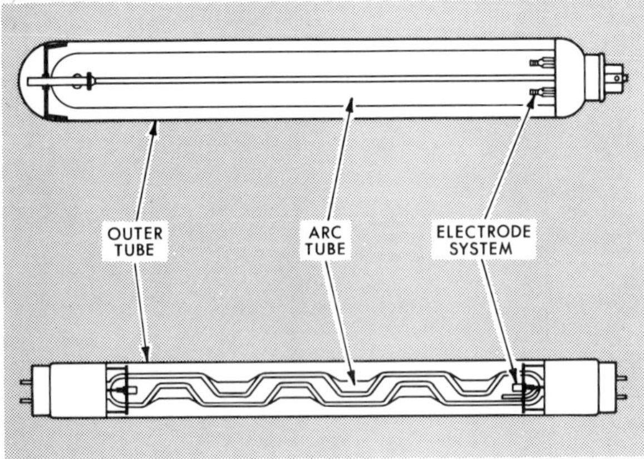

Fig. 8-59. Basic construction of two types of low pressure sodium lamps: hairpin or "U" tube (upper) and linear (lower).

Efficacy. Low current density is vital to efficient generation of resonance radiation, as high densities result in higher excitation phenomena and loss of resonance radiation. It is in the field of thermal insulation that greatest strides have been made in recent years, resulting in present day efficacies in excess of 170 lumens per watt for the 180-watt "U" type low pressure sodium lamp. This thermal insulation consists of a light transparent infrared reflecting layer on the inside of the outer envelope. In current designs, this is an indium oxide layer, replacing the formally used tin oxide layer.

Glow Lamps

These are low wattage, long life lamps designed primarily for use as indicator or pilot lamps, but are also used for night-lights and location markers. They range from $1/25$ to 3 watts and have an efficacy of approximately 0.3 lumens per watt. A group of typical glow lamps is shown in Fig. 8-60. These emit light having the spectral character of the gas with which they are filled. The most commonly used gas is neon, having a characteristic orange light output. The glow is confined to the negative electrode.

Glow lamps have a critical starting voltage, below which they are, in effect, an open circuit. The starting voltages for several glow lamps are shown in Fig. 8-122.

Like other discharge lamps, glow lamps require a current limiting resistance in series. Glow lamps with screw bases have this resistor built into the base, while for unbased lamps or lamps with bayonet bases a resistor of the proper value must be employed external to the lamps.

Glow lamps filled with an argon mixture rather than neon radiate chiefly in the near ultraviolet region around 360 nanometers and are therefore used mainly to excite fluorescence in minerals and other materials as well as for some photographic applications.

Zirconium Concentrated Arc Lamps, Enclosed Type

These lamps are direct current arcs typified by a small concentrated point source of light of high intrinsic luminance. They are made with permanently fixed electrodes sealed into an argon-filled glass bulb. The light source is a small spot, 0.005 to 0.11 inches in diameter (depending on lamp wattage) which forms on the end of a zirconium oxide-filled tantalum tube which serves as the negative electrode. The spectral power distribution is similar to that of a blackbody with a color temperature of 3200 K. The spatial distribution is a cosine type.

These lamps require special circuits which generate a high voltage pulse for starting and a well filtered and ballasted operating current. Suitable power supplies are recommended by the manufacturer.

Fig. 8-61 illustrates various examples of side and end emission lamps. Fig. 8-123 gives essential characteristics of representative lamps.

Fig. 8-60. Typical glow lamps with American National Standards Institute numbers (old trade numbers).

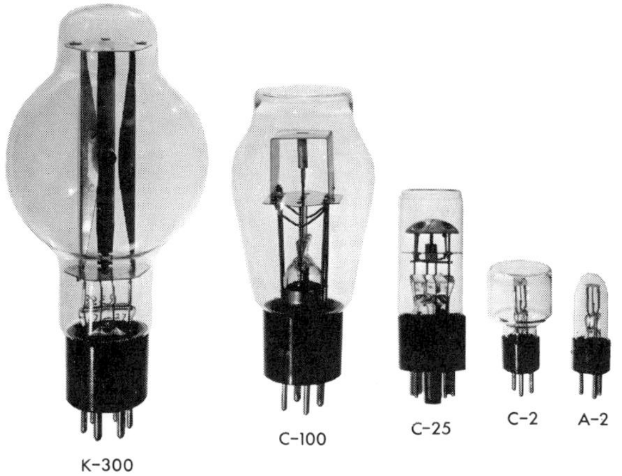

Fig. 8-61. Side and end-emission concentrated arc lamps.

Pulsed Xenon Arc (PXA) Lamps

These are ac non-polarized xenon lamps with two active cathodes. A switching reactor in series with the low pressure lamp forces 50 to 100 peak amperes (pulsed 120 per second) through the lamp to achieve high efficacy. The reactor also supplies a continuous current of 7 to 8 amperes to keep the lamp operating between pulses. The daylight spectrum produced is characteristic of xenon, *i.e.*, typically 6000 K. PXA lamps are available in linear or helical types.

The efficacy of these sources is about 35 to 40 lumens per watt. Available lamp wattages range from 750 to 8000 watts with forced air cooling a requirement during operation.

PXA lamps are used in the graphic arts industry where instant start, high stable light output, and daylight quality color temperature are required.

Flashtubes

These sources are designed to produce high luminance flashes of light of extremely short duration. They are used for photographic applications, timing of reciprocating and rotating machinery, airport approach lighting systems, and other applications where flashing lights are required. Most recently, xenon flash tubes have generated additional interest for their ability to pump solid laser materials.

A flashtube is a tube of glass or quartz which has an electrode at each end and is filled with gas. The spectral distribution of light from a discharge in a tube filled with xenon is similar to that of average daylight. See Fig. 8-62. Other gases employed include argon, hydrogen, and krypton. These are used to obtain other spectral power distributions or different electrical characteristics.

In addition, there are flashtubes designed for repetitive flashing at short intervals. There are two types available. One is basically a xenon flashtube filled at a lower pressure, allowing it to be fired at very high repetition rates (100 flashes per second). The other is a neon-filled cold cathode thyratron which has a cathode, two grids, and an anode. The advantage of the latter type over most stroboscopic lamps is that electrically it is a trigger tube and can be fired directly from a multivibrator or blocking oscillator, requiring only 125 volts to fire.

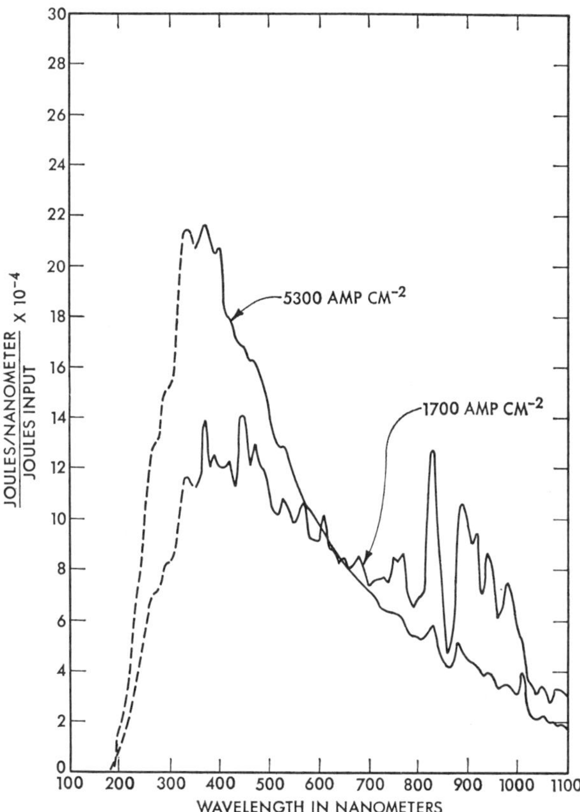

Fig. 8-62. Spectral energy distribution curve of a typical xenon-filled flashtube (radiation in direction perpendicular to helix) for two current densities.

Straight flashtubes are available. However, the arc tube is frequently coiled into a helix for higher concentration of light and may be provided with an outer glass envelope. Fig. 8-63 shows several sizes

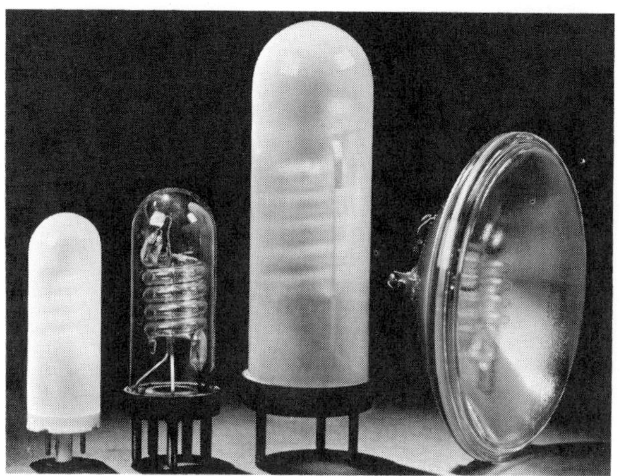

Fig. 8-63. Typical helical-type flashtubes.

of helices, and bulb and base combinations of typical lamps. A small U-shaped flashtube, Fig. 8–64, is also available for use in low-powered portable equipment.

The amount of light furnished by the tube depends upon its loading which in turn depends upon the discharge capacitance and the voltage across the tube, in accordance with the formula:

$$\text{Loading in watt-seconds (joules)} = \frac{CV^2}{2}$$

where C is the capacity of the capacitor in microfarads and V is the voltage across the tube (and capacitor) in kilovolts.

The flash duration depends upon the resistance of the tube and the discharge capacitance. The effective discharge time, in seconds, is given by the product RC, where R is the tube resistance in ohms and C is the capacity of the capacitor in microfarads. The resistance of the flashtube is calculated from the formula:

$$\text{Flashtube resistance} = \frac{\gamma l}{A}$$

where γ is the plasma resistivity in ohm-cm, l is the arc length in cm, and A is the cross section area in cm^2. The resistivity, γ, is not strictly a constant. It varies inversely as the square root of the current density. However, at current densities encountered in usual practice, γ has a value between 0.012 and 0.020 ohm-cm.

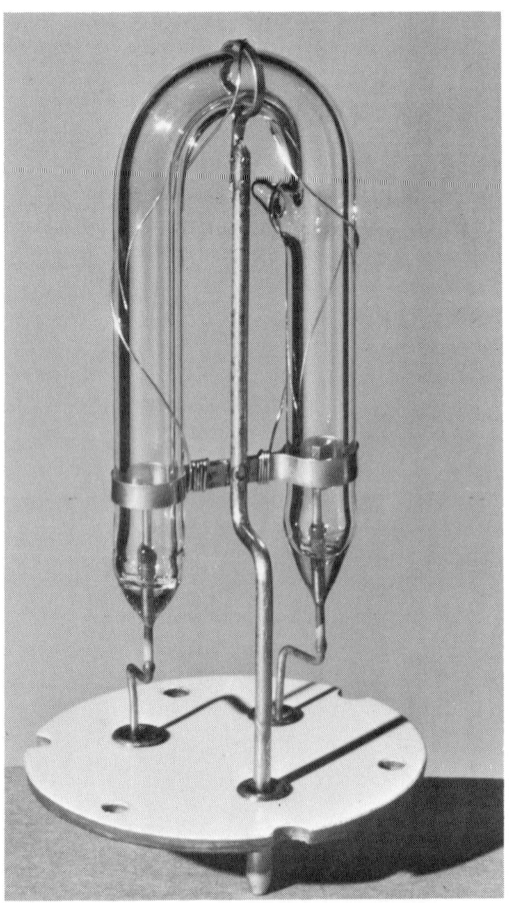

Fig. 8–64. A highly efficient form of flashtube requiring careful selection of reflector contour and surface to obtain maximum utilization and beam uniformity.

Power Supply. All flashtube power supplies include a source of high voltage direct current to charge the energy storage capacitors. If the voltage at which energy is stored in the capacitor is high enough, any flashtube will flash over when that voltage is applied to the tube terminals. Operation under these conditions requires the use of some form of mechanical switch or switch tube in the discharge circuit to alternately connect the capacitors to the charging source and the flashtube. Both types of switches have been used, the second particularly for those applications where repetitive operation at high flashing rates or extremely short flashing duration is required.

The most generally used method of initiating the flash is to employ a combination of flashtube characteristics and capacitor voltage such that the tube will flash only when ionized. In such cases the tube may be continuously connected to the energy storage capacitors. Ionization may be accomplished by applying a momentary high voltage (of the order of 10 kilovolts or more) to the external wall of the flashtube and to one terminal. This high voltage pulse causes the gas in the tube to become conductive. When the capacitor charge has been almost entirely expended, the voltage across the terminals is at a low value and the tube ceases to conduct. The rate at which the energy storage capacitors are then automatically recharged may be any value required by conditions of use provided; however, it is not so rapid as to cause the tube to self-flash before deionization is complete. The circuit for a typical power supply in which ionization is employed is shown in Fig. 8–65. Since no switch is re-

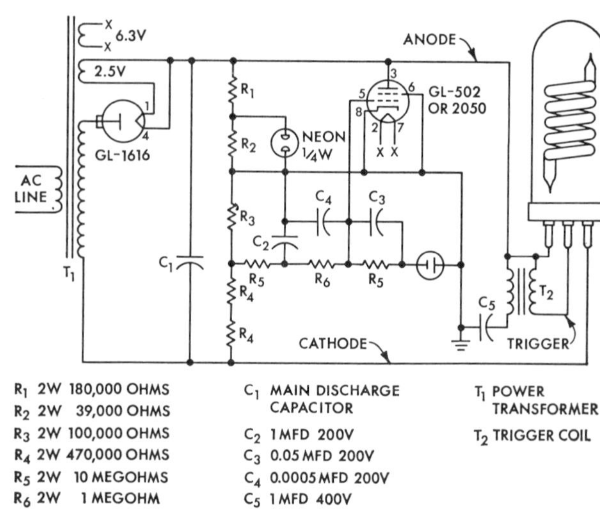

Fig. 8–65. Basic elements of a typical flashtube power supply.

quired in the discharge circuit this method has obvious advantages and is limited only by the deionization of the flashtube which precludes operation at high repetition rates.

Limits of Energy Input. For single flash operation, the limit to the amount of energy which can be consumed depends upon the desired tube life, that is, the total number of useful flashes. This is affected by the rate of tube blackening and destruction of the tube or its parts.

Maximum loadings (in watt-seconds) for flashtubes presuppose operation at design voltage. For operation above design voltage, maximum loadings for satisfactory operation are less. Flashtubes designed for very high loading have quartz tubes and some of these tubes have closed end outer bulbs. The peak current encountered during the discharge produces a shock wave which may be of sufficient magnitude to shatter a closed end bulb. To limit the peak current and shock wave, it is necessary to include ½ millihenry of inductance in series with each 100 microfarads of capacitance in the operating circuit. This also aids in reducing the noise accompanying the discharge at very high loadings.

The recommended operating voltages and maximum loadings for typical flashtubes are shown in Fig. 8–124.

The maximum amount of energy can also be increased by lengthening the flash duration. Techniques for producing longer discharges include operation at lower voltages, greater capacitance, use of an air core choke in series with the capacitor, and excitation of the flashtube from a pulse forming network.

If a flashtube is operated repetitively and rapidly at energy input levels such that its temperature rises excessively, it will either miss (fail to flash) or become continuously conductive. In the latter case the tube may be damaged. The total watts consumed are the product of the watt-seconds per flash and the number of flashes per second. The figures for a tube operating at 2000 volts and 112 microfarads (224 watt-seconds per flash) are tabulated below for different rates of flash:

Flashing Rate	Input to Lamp (watts)
One flash per minute	3.7
One flash per 10 seconds	22.4
One flash per second	224.0
Ten flashes per second	2240.0
One hundred flashes per second	22,400.0

The highest flashing rate tabulated above obviously presents a difficult, if not impossible, problem of heat dissipation. When flashing rates of this order of magnitude are required, lower values of watt-seconds per flash must be employed to avoid excessive tube temperatures or steps taken to cool the flashtube.

Flashtubes can be effectively cooled by forced air or forced liquid coolants. Use of reasonably pure water permits the design of flashtubes with simple transparent water jackets that can be externally triggered. If impure water is used its resistivity may be too low and preclude the use of an external trigger wire. Water cooled flashtubes have been run at average power as high as 30 kW (typically 5000 watt-seconds per flash and 6 flashes per second).

Glow Modulator Lamp

This is a cold cathode type lamp with a hollow or crater cathode. Its particular value lies in its linear relationship between lamp current and light output, permitting modulation of the light output at frequencies up to a megacycle per second. The light is emitted from the crater giving a source that approaches a point source.

The glow modulator has been used principally for facsimile picture transmission applications where the tube serves as a modulated light source in the receiver unit.

Metallic Iodide Vapor Arc Lamp

These are very small (3-inch) high luminance dc arc lamps particularly suitable for very high performance in compact portable motion picture or slide projection equipment. They typically operate from special light weight power supplies at 7½ to 8 amperes and 250 to 300 watts. A basic arc tube is available and special models incorporating integral dichroic coated prefocused glass reflectors have been designed for easy replacement in many 16 millimeter motion picture projectors.

Although total luminous efficacy is about 50 lumens per watt, the chamber filling nature of the water drop sized arc permits simplified optical reflector design resulting in projector light output as much as 400 per cent more than obtained with 1000-watt incandescent projection lamps. Color temperature is approximately 5000 K.

ELECTROLUMINESCENT LAMPS

An electroluminescent lamp is a thin area source in which light is produced by a phosphor excited by a pulsating electrical field. In essence, the lamp is a plate capacitor with a phosphor embedded in its dielectric and with one or both of its plates translucent. Green, blue, yellow, or white light may be produced by choice of phosphor. The green phosphor has the highest luminance. These lamps are available in ceramic and plastic form, flexible or with stiff backing, and are easily fabricated into simple or complex shapes. They have been used in decorative lighting, night lights, switchplates, instrument panels, clock faces, telephone dials, thermometers, and signs. Their application is limited to locations where the general illumination is low.

Luminance varies with applied voltage, frequency, and temperature, as well as with the type of phosphor. See Fig. 8-66. At 120 volts, 60 cycles, the luminance of the ceramic form with the green phosphor is approximately one footlambert; the luminance of the plastic form may be as high as eight footlamberts under these conditions or up to 30 footlamberts at 120 volts, 400 cycles. With the ceramic form at 600 volts, 400 cycles, a luminance of 20 footlamberts has been achieved. These higher luminances are at the expense of useful lamp life.

Life is long and power consumption low. There is no abrupt point at which the lamp fails; the time at which the luminance has fallen to 50 per cent of initial is sometimes used as a measure of useful life. For the ceramic form, this is approximately 20,000 hours at 120 volts, 60 cycles. See Fig. 8-67. Approximate initial current and wattage values per square foot of lamp under these operating conditions is 60 milliamperes and 3.5 watts.

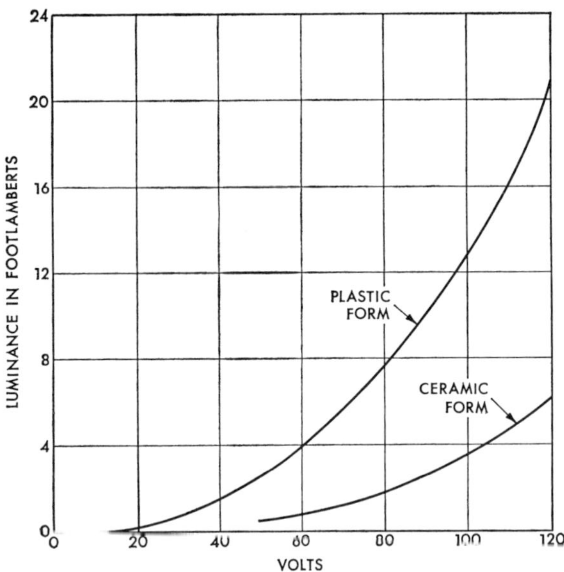

Fig. 8-66. Luminance of green ceramic and plastic forms of electroluminescent lamps operated at 400 cycles as effected by voltage.

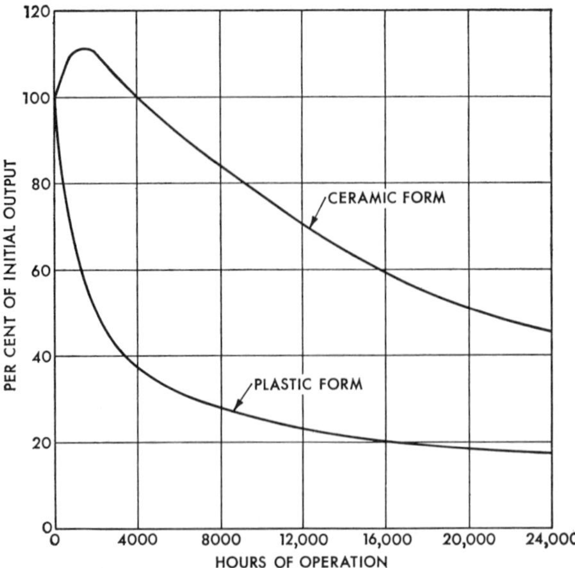

Fig. 8-67. Light output versus burning hours for green ceramic and plastic forms of electroluminescent lamps.

LIGHT EMITTING DIODES[38]

The light emitting diode (LED) is a p-n junction semiconductor lamp which emits radiation when biased in a forward direction. The emitted radiation can be either invisible (infrared) or in the visible spectrum. Semiconductor light sources are available in a wide range of wavelengths, extending from the green region of the visible spectrum to the far-infrared region.

Visible solid state lamps are used for long life indicator service. Infrared lamps have spectral outputs closely matched to the response of silicon photoreceivers. They are used in conjunction with these receivers for counting, sorting, sensing and positioning in applications as diverse as computer equipment, optical radar, and burglar alarms.

The light producing material in a LED is a specially prepared semiconductor material of high purity, having small amounts of other elements added as controlled "impurities." Two classes of impurities are added: one to produce material having an excess of electrons and is called n-type material, the other to produce material having a shortage of electrons or "holes" which act as positive charges and is called p-type material. The impurities of the two classes are diffused into the same piece of semiconductor material so that an interface or "junction" is created between the p-type and the n-type materials. See Fig. 8-68a.

When a dc voltage is applied to the semiconductor material with polarity so that the n-type is negative and the p-type is positive the "holes" and the electrons are forced to meet at the junction where they combine to produce photons of light. The special characteristic of the energy or light is dependent upon the semiconductor material and the controlled impurities.

A typical LED construction using a metal header similar to that used for a transistor is shown in Fig. 8-68b. The lens is used to distribute the radiated energy.

The spectral output characteristics of four solid state lamp semiconductor materials are shown in Fig. 8-68c: gallium arsenide and gallium arsendide with silicon producing infrared radiation, gallium phosphide producing red light, and silicon carbide producing yellow light. A phosphor, which absorbs infrared radiation at about 900 to 1000 nanometers and re-radiates at about 540 nanometers, can be directly applied to gallium arsenide (with silicon) semiconductor material to produce green light.

The size of the "chip" (the piece of semiconductor material used in a solid state lamp) is generally 0.010 to .040 inch square so lamps can be made very small in size. Present lamp sizes are of the order of

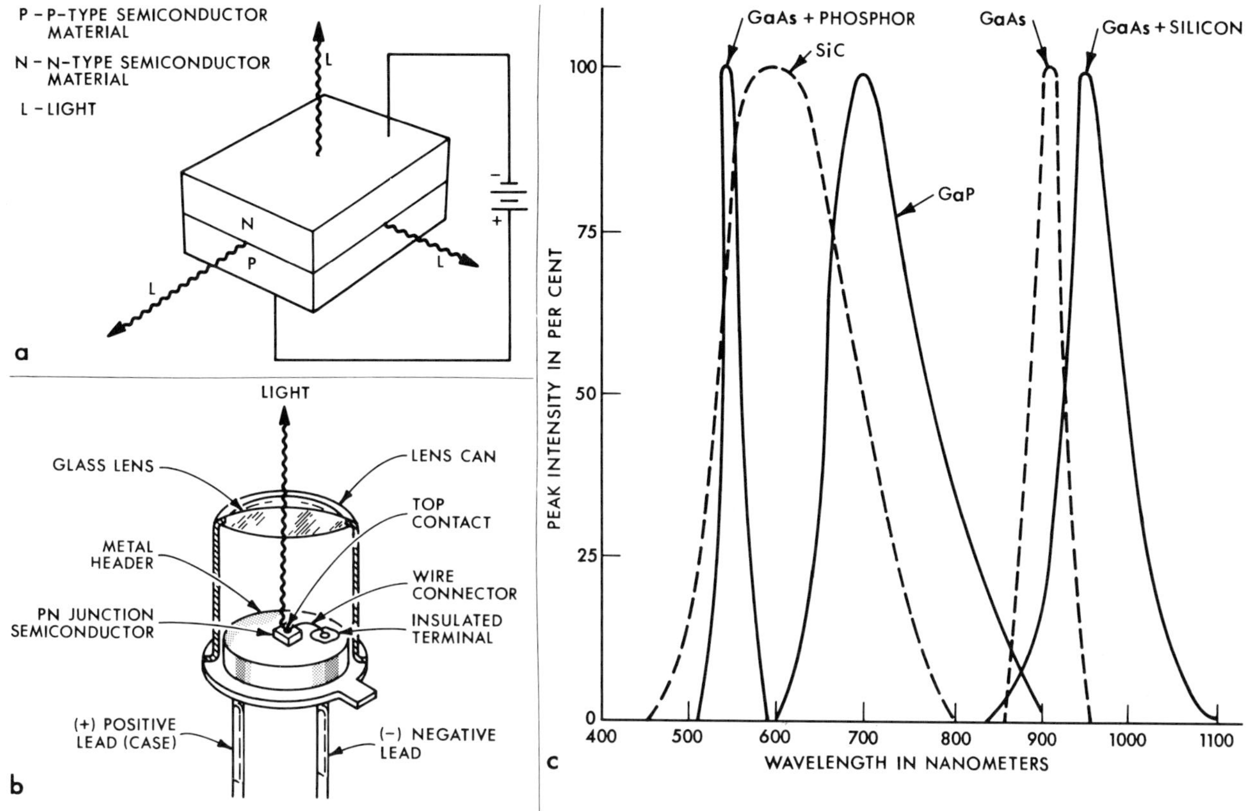

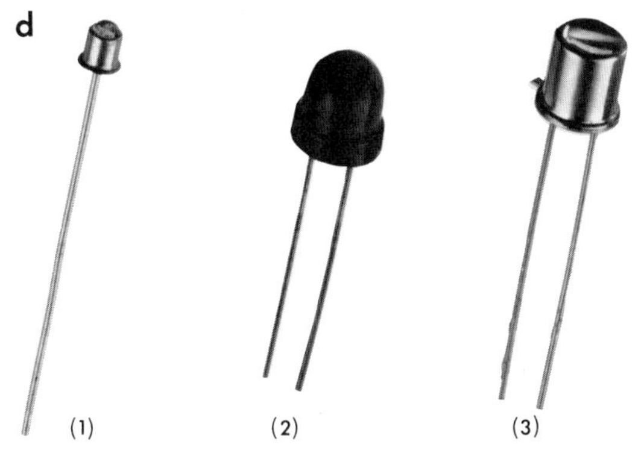

Fig. 8–68. a. A semiconductor junction. b. A cross-section view of a typical light emitting diode. c. Spectral output of several light emitting diode semiconductor materials (peak intensities are not equal for the different materials shown). d. Typical light emitting diodes: (1) co-axial, (2) plastic encapsulated, (3) header and lens cap.

1⁄10 to 1⁄4 inch in diameter. Some lamps have the semiconductor chip and connections imbedded in plastic, such as epoxy, to form a single molded piece rather than having separate header, cap, and lens construction. See Fig. 8–68d.

LED generally operate in the range of 1 to 3 volts at currents below 10 mA to above 100 mA continuous. Output of visible solid state lamps is of the order of 0.015 candela. Infrared lamps, operating continuously, emit up to 5.5 milliwatts at 100 mA input current. Some lamps may be pulsed at currents exceeding 70 amperes to produce extremely short bursts of high infrared output with peak energies up to 0.5 watts.

NUCLEAR LIGHT SOURCES

Nuclear light sources are self-powered, self-contained light sources requiring no power supply. The light output is such that they can be easily seen by a person whose eyes are not dark adapted; some forms are visible at considerable distances and, therefore, provide a ready means for illuminating instrument panels, controls, locks, and other devices.

These sources consist of a sealed glass tube or bulb internally coated with a phosphor and filled with tritium gas. Low energy beta particles (electrons) from the tritium, an isotope of hydrogen, strike the phosphor, which in turn emits light of a color characteristic of the type of phosphor used. Thus, the mechanism of light production is very similar to that in a conventional television tube where electrons strike a phosphor and cause it to emit visible light. With tritium having a half life of 12.3 years, the resulting light intensity should follow this decay. In reality, the time when half-intensity is reached is about six to seven years, and the useful life of the present lamp is approximately fifteen years.

The glass wall is impervious to tritium and completely absorbs any beta radiation not already absorbed by the phosphor. The unit is thus a completely sealed source and does not present any radiation hazard. Glass capsules can be produced in a wide variety of shapes and sizes and are usually made to normal glassworking tolerances.

The apparent luminance of a nuclear light source is determined by the beta-flux incident on the phosphor surface and by the color. The higher the tritium-phosphor area ratio, the greater the luminance. Luminance can range up to 2 footlamberts, with a typical average of 0.5 footlamberts (this level is approximately that of an illuminated car instrument panel) and the sources can be supplied in a variety of colors. Highest apparent luminance is obtained in the green/yellow range and green is supplied as a standard color. Light sources can be supplied in very small sizes, down to 5 millimeters in diameter by 2 millimeters in length.

Use of these lamps in all applications is monitored by the U.S. Atomic Energy Commission.

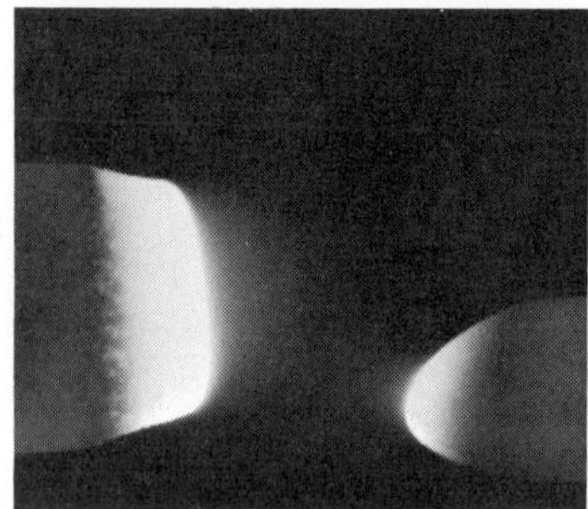

LOW-INTENSITY ARC

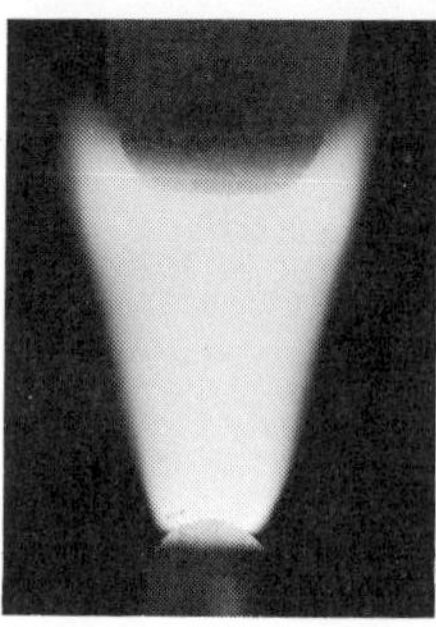

FLAME TYPE CARBON ARC

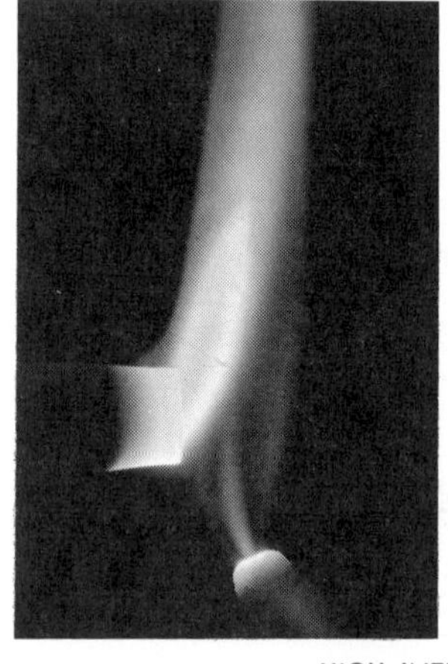

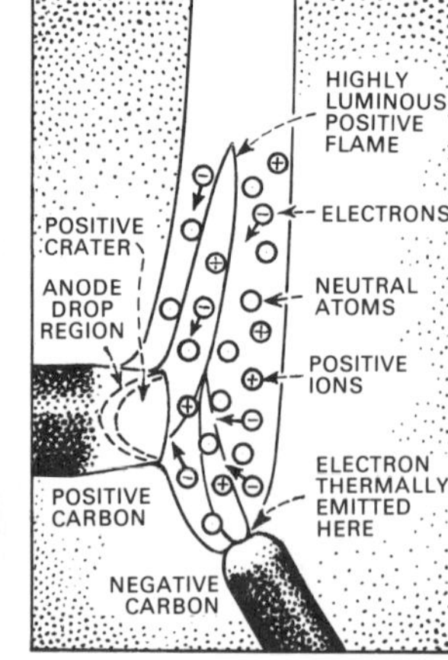

HIGH-INTENSITY ARC

Fig. 8–69. Low-intensity arc, 30 amperes, 55 volts, direct current. Flame arc, 60 amperes, 50 volts, alternating current. (Direct current flame arcs are very similar.) High-intensity arc, 125 amperes, 70 volts, direct current (rotating positive carbon).

CARBON ARC SOURCES

Carbon arc sources (which were the first commercially practical electric light sources) now are used where extremely high luminance is necessary, where large amounts of radiant energy are required, and where their radiant energy spectrum is advantageous. The three distinct differences between the three basic types of carbon arcs (the low intensity, the high intensity, and the flame arc), are discussed below.

Types of Carbon Arcs

Low Intensity Arcs. Of the three principal types of carbon arcs in commercial use, the low intensity arc is the simplest. In this arc, the light source is the white-hot tip of the positive carbon. This tip is heated to a temperature near its sublimation point (3700°C) by the concentration of a large part of the electrical energy of the discharge in a narrow region close to the anode surface. See Fig. 8–69.

The gas in the main part of the arc stream is extremely hot (in the neighborhood of 6000°C) and so has a relatively high ion density and good electrical conductivity. The current is carried through this region largely by the electrons since they move much more readily, because of their small mass, than the positive ions. However, equal numbers of positive ions and negative electrons are interspersed throughout the arc stream, so no net space charge exists, and the only resistance to the motion of the electrons is that supplied by collisions with inert atoms and molecules. Near the anode surface, the conditions are not as favorable for the conduction of current. The electrode tip is about 2000 degrees cooler than the arc stream, and the gas immediately adjacent consists largely of carbon vapor in temperature equilibrium with the surface. At 3700° C, this carbon vapor is a very poor conductor of electricity. It therefore requires a high voltage to force the current-carrying electrons through this vapor layer and into the anode. In a pure carbon arc, this anode drop is about 35 volts. Most of the heat so developed is transferred to the surface of the positive carbon, part by the impact of the highly accelerated electrons and part by thermal conduction. Finally, as the electrons reach the anode surface, they release their heat of condensation, contributing further to the high temperature of the electrode tip.

The positive electrode of the low intensity arc may contain a core consisting of a mixture of soft carbon and a potassium salt. The potassium does not contribute to the light, but does increase the steadiness of the arc by lowering the effective ionization potential of the arc gas.

Flames Arcs. A flame arc is obtained by enlarging the core in the electrodes of a low intensity arc and replacing part of the carbon with chemical compounds known as flame materials, capable of radiating efficiently in a highly heated gaseous form. These compounds are vaporized along with the carbon and diffuse throughout the arc stream, producing a flame of a color determined by the compounds used. Typical flame materials are iron for the ultraviolet, rare earths of the cerium group for white light, calcium compounds for yellow, and strontium for red. See Fig. 8–69.

Such flame materials have a considerably lower ionization potential than carbon. This greater ease of ionization reduces the temperature of the anode layer necessary for the conduction of current into the anode and results in a lower anode voltage drop (about 15 volts). The lower anode power input reduces the area and luminance of the anode spot so that its contribution to the total light becomes unimportant. The radiation from the flame arc consists chiefly of the characteristic line spectra of the elements in the flame material and the band spectra of the compounds formed. The excitation of the line and band spectra is thermal in nature, caused by the high temperature of the arc stream gas. The concentration of flame materials in the arc stream is not very high, so that the flame arc is considerably less bright than either the low or the high intensity arc. Since the whole arc flame is made luminous, the light source is one of large area and has high radiating efficacies (up to 80 lumens per watt).

High Intensity Arcs. The high intensity arc is obtained from the flame arc by increasing the size and the flame material content of the core of the anode, and at the same time greatly increasing the current density, to a point where the anode spot spreads over the entire tip of the carbon. This results in a rapid evaporation of flame material and carbon from the core so that a crater is formed. The principal source of light is the crater surface and the gaseous region immediately in front of it. See Fig. 8–69. Since the flame material is more easily ionized than the carbon, a lower anode drop exists at the core area than at the shell of the carbon. This tends to concentrate the current at the core surface, and so encourages the formation of the crater.

The increased luminance of the high intensity arc is produced by radiation resulting from the combination of the heavy concentration of flame materials and the high current density within the confines of the crater. The spectrum of the radiation from the crater consists of a continuous portion plus line and band spectra characteristic of the gaseous components in the crater. This line and band radiation is greatly broadened, indicative of the many collisions, high temperature, and other processes and conditions existing in the crater. When compounds of the cerium family of rare earths are used as core materials, the combined effects produce radiation that is effectively continuous through the visual region of the spectrum and yield a luminance over ten times that of the low intensity arc.

Carbon Arc Lamps and Power Sources

Carbon arcs are operated in lamphouses which shield the outside from stray radiation; which incorporate optical components, such as lenses and reflectors; which provide means to conduct the electrical current to the carbon electrodes and to feed the electrodes together to compensate for the portions consumed; and which provide for removal of the products of combustion. Some lamps employ directed streams of air to control the position of the flame from the positive electrode and to remove the combustion products.

Arcs for projection of motion pictures generally operate on direct current to prevent disturbing stroboscopic effects of the projector shutter. Both motor generators and rectifiers are employed. The flame arcs are widely used both on direct and alternating current. In some cases, alternating current arcs are operated directly from power lines and in others, from special transformers. Low current arcs have a negative volt-ampere characteristic and therefore must be operated from circuits which include ballast resistances or reactances (either in the generating or rectifying equipment or as separate units in the arc circuits). High intensity carbon arcs have positive volt-ampere characteristics and may be operated without ballasts. Suitable power supply equipment is available for all principal types of carbon arc lamps.

Carbon arcs may be started with a third (starting) electrode or the electrodes may be brought together momentarily after which they are separated to the proper distance to maintain the correct arc voltage and current. These conditions can be maintained and the carbons fed manually but in modern carbon arc lamps, automatic mechanisms feed the carbons as they are consumed and regulate the arc length and position of the light source. Some carbon arc lamps, such as those used to illuminate space environmental simulation chambers, are designed for continuous operation, without the necessity of shut-down for re-carboning. Carbons are automatically dispensed from a magazine and joined to form a continuously burning electrode.

Low Intensity Arcs. Fig. 8–125 shows the characteristics of a typical low intensity carbon arc used for microscope illumination and projection. The light from a low intensity carbon arc has a luminance (150 to 180 candelas per square centimeter) and a color temperature (3600 to 3800 K) which exceed those from incandescent tungsten. When operated under prescribed conditions,[33] the low intensity carbon arc produces radiation which closely approximates that from a blackbody at 3800 K and is widely used as a radiation standard.

High Intensity Arcs. Fig. 8–125 shows the characteristics of a variety of typical high intensity carbon arcs ranging in power input from 2 to 30 kilowatts, in crater candlepower from 10,500 to 185,000 candelas, and in crater luminance from 55,000 to 145,000 candelas per square centimeter. The color temperature of the light ranges from as low as 2900 K to as high as 6500 K. These values will be further modified by the characteristics of the optical system used (see Section 24).

Fig. 8–70 shows the luminance distribution across the crater for typical low and high intensity carbons and illustrates two basic differences in the luminance characteristics: first, that the luminance is lower for the low intensity than the high intensity arcs; and secondly, that the luminance depends on the current, very markedly with the high intensity, but very little with the low intensity arc.

The high intensity arc is in practically universal use for theatre motion picture projection. It has also been adapted to projection of 16mm film before large groups, where large screens are necessary (see Section 24). The high intensity arc is the traditional light source for military searchlights (see Section 18) because of its high luminance. Because of a close duplication of the spectral energy distribution of solar radiation outside the earth's atmosphere, the high intensity carbon arc is used for solar simulation in space environmental test chambers.

Flame Arcs. Fig. 8–126 gives characteristics of typical flame arcs. This compilation covers arcs with power input ranging from 1 to 4 kilowatts and shows the effect of different currents, voltages, and flame materials. All arcs shown operate on alternating current with one exception. The wavelength intervals shown have been chosen to coincide with those important to various applications. The radiation in the indicated wavelength intervals is shown in two ways: first, as radiant intensity, expressed as microwatts incident on one square centimeter of area one meter distance from the arc; and secondly, as per cent of power input to the arc radiated over the entire sphere.

Rare earth flame materials exhibit the highest luminous efficacy, readily exceeding 50 lumens per watt, and result in 15 to 20 per cent of the arc input energy radiated in the band of visible wavelengths (400 to 700 nanometers). The polymetallic flame materials with 3.6 to 7.2 per cent of the input power radiated below 320 nanometers are important to photochemical and therapeutic applications. Both rare earth flame arcs and the enclosed arc radiate efficiently in the 320 to 450 nanometer band so useful to various photo-reproduction processes. Strontium concentrates the radiated energy in the red and near infrared regions of the spectrum.

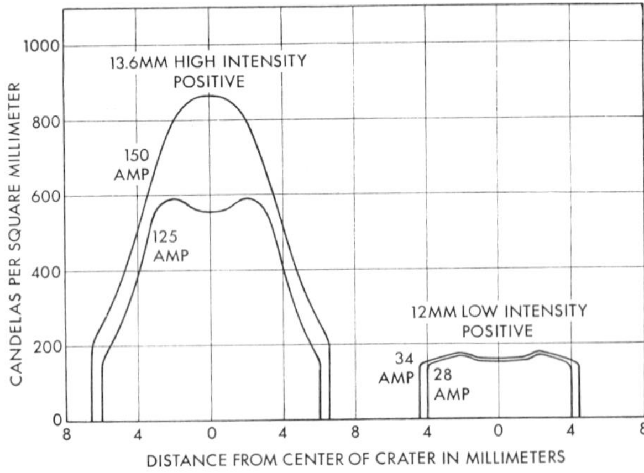

Fig. 8–70. Crater luminance distribution in forward direction for typical low and high intensity carbons.

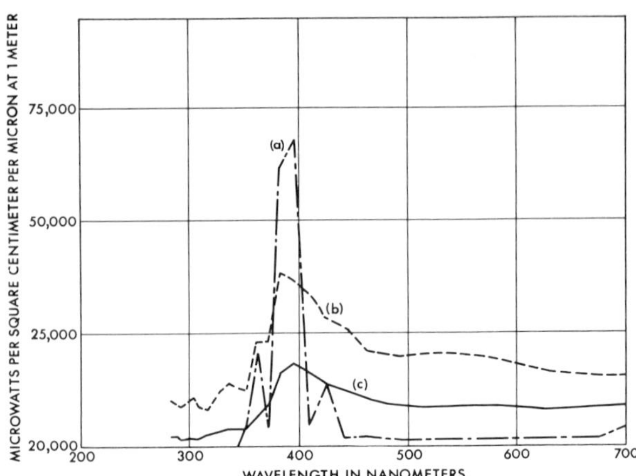

Fig. 8–71. Spectral energy distribution of arcs used for graphic arts. (a) Half-inch enclosed arc carbons, 16 amperes, 138 volts. (b) Nine millimeter high-intensity photo carbons, 95 amperes, 30 volts. (c) Half-inch photographic white flame carbons, 38 amperes, 50 volts.

Fig. 8–71 shows the spectral energy distribution for flame arcs used for graphic arts applications. All of these show a concentration of radiation in the region 320 to 450 nanometers.

GASLIGHTS

Gaslights are devices which use gaseous fuels for lighting and decorative purposes. They use incandescent mantles of the upright and inverted types, although the latter is used to a lesser degree. See Fig. 8–72.

Mantle Construction

A fabric such as rayon, silk, cotton ramie, or viscose is woven into fabric tubing of the desired stitch and impregnated with a mixed solution of the nitrates of cerium and one or more of the following: thorium, beryllium, aluminum, and magnesium. After impregnation, the knitted tubing is cut into short lengths and attached to individual rings and mountings.

Mantles are available in two forms: soft and hard. Soft mantles are sold "unburned" and must be shaped on the burner, a familiar task for those who own gasoline lanterns.

Hard mantles are pre-shaped and pre-shrunk by burning out the fabric during manufacture. The remaining ash consists of oxides of the impregnating metals. A collodion coating is then applied to strengthen the burned mantle for handling and shipping; the coating, in turn, burns out when the mantle is first put to use.

Mantle Performance

Incandescence of a gas mantle is mainly dependent on an *exacting* cerium content, and on flame temperature and flame velocity. The type of gas, injection pressure, and burner design may contribute to the candlepower obtained. The weave of the mantle is well established for each type of gas and is not a factor in mantle life. The single stitch weave gives the same light output as the double stitch; however, light from a double stitch mantle may appear more yellow to the eye.

Light Output. The light output, within design limits, is a direct function of the Btus consumed in the mantle, and the corresponding mantle temperature. Gaslights are built so that a change in orifice size or the use of a properly adjusted pressure regulator is all that is required to accommodate the available gas pressures.

According to an ANSI standard which covers requirements for gas-fired illuminating appliances,[39] units incorporating a mantle should produce an average light output of at least one candela for each 65 Btu per hour of input rating. Fig. 8–73 gives performance data for three mantle arrangements. As it can be seen, the light output of a gaslight with a type A upright mantle (434 lumens) is about the same as that of a 40-watt incandescent lamp; the arrangement of two No. 222 inverted mantles has about the same light output (850 lumens) as a 60-watt lamp. Fig. 8–74 gives the candlepower distribution for a single upright mantle and a two-cluster inverted mantle. In the span of normal operating temperatures, light output varies as the tenth power of mantle temperature in degrees F. The luminous efficacy of a gas mantle is the equivalent of 0.67 to 1.32 lumens per watt.

Fig. 8–72. Gas yard light with upright incandescent mantle.

Fig. 8–73. Light Output and Efficacies of Various Mantle Arrangements

Arrangement and Type of Mantle(s)	Total Input (Btu per hour)	Total Light Output (lumens)	Luminous Efficacy		Can-delas Per 65 Btu Per Hour*
			(lumens per Mbtu per hour)	(lumens per watt)	
One A† upright	2200	434	197	0.637	1.0
Two No. 222‡ inverted	2200	851	387	1.320	2.0
Three No. 222‡ inverted	3300	1090	330	1.130	1.7

* 1.0 is minimum, per ANSI Z21.42-1963 (R1969).
† Has a coated single wire top support and a single stitch weave.
‡ Representative of most inverted mantles.

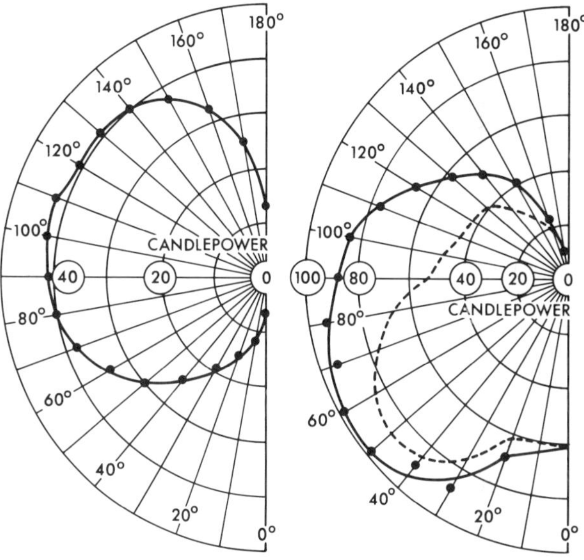

Fig 8–74. Vertical candlepower distribution,* in candelas, for mantles operated with optimum adjustment at total gas inputs of 2200 Btu per hour: (left) a single type A upright mantle—view unobstructed by mantle support bracket; (right) a two-cluster No. 222 inverted mantle (solid line—both mantles foremost, dotted line—one mantle foremost.)

* Note difference in scale; 180 degrees shown—the other 180 degrees are symmetrical.

Inverted mantles are generally more efficient than upright mantles because of superheating of the gas-air mixture in the mantle. The flame burns back on itself, concentrating more heat on the mantle fabric, and thus more light is generated.

A properly designed clear glass chimney (about 3-inch diameter) surrounding an upright mantle (not used with inverted mantles) will aid in directing the hot gases through the chimney as much as 35 per cent above that of a bare mantle with no chimney. Opal domes may be used to produce a low-intensity upward light.

Life. Unless mechanically abused, mantles have an indefinite life.

High Intensity Gaslights

Recent advances in gaslight technology make it possible to get brighter light from inverted gas mantles for a small increase in natural gas consumed in the mantle. The innovation is higher natural gas pressure, 2 to 5 pounds per square inch gauge (psig), rather than the usual pressure of a few inches of water column (w.c.). Proper design as well as increased pressure provide significant increases in light output. Typical photometric data for a 2 psig gaslight are shown in Fig. 8–75.

Gas, at increased pressures, aspirates more primary air and combines more thoroughly with that primary air. The resultant higher intensity flame, if matched to a suitable mantle, results in an appreciable increase in light output. Liquefied petroleum gas (propane and butane) at medium pressure (up to 15 psig) has been used for many years in floodlights which use incandescent mantles.

Automatic Controls

When desired, gaslights may be turned down or off by clock or photocell actuated devices. However, most gaslights in use burn continuously at full light output.

Open Flame Gaslights

Open flame gaslights have no mantles and use non-primary aerated burners. They are used mainly for ornamental effects rather than for their lighting ability.

Fishtail Burners. These non-primary aerated burners yield a fan-shaped or fishtail-shaped flame. The head can either be slotted or have two ports angled so that their respective gas streams impinge. Such burners may be installed in gaslight assemblies if the maximum height of the flame is controlled and flame impingement is avoided (to minimize carbon deposits).

Gas Torches, Torch Lights. Also known as Luau torches, these units are often shaped like frustrums of right circular cones. Targets or flame spreaders, vanes to redirect drafts, and wire screens or porous plugs to coalesce and harden flames are used to improve the flame stability and wind resistance of these torches.

Portable Candle Lights. Another type of contemporary yellow flame burner is designed to operate with small portable cans of liquefied petroleum fuel and is meant to simulate a candle flame. A yellow flame about 1 inch high usually is adequate to create the desired effect. The light is normally operated in small lamp assemblies, which are basically miniature models of outdoor gaslight assemblies, and is therefore afforded some protection from drafts.

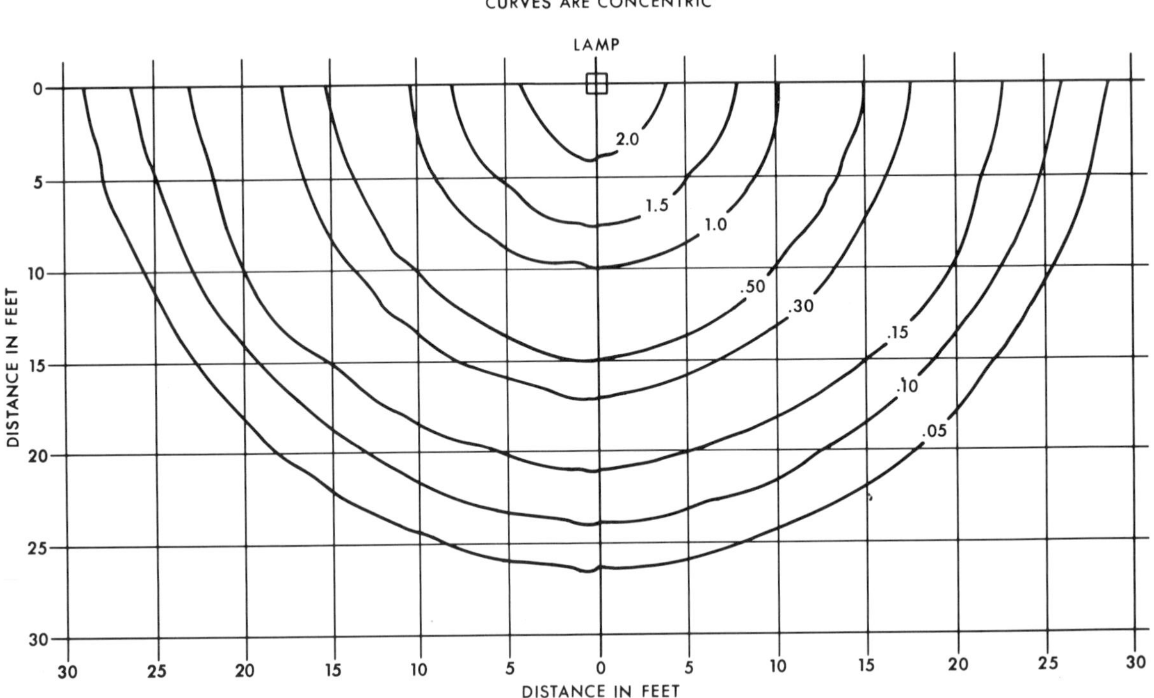

Fig. 8–75. Isofootcandle diagram (calculated) for 2 psig gaslight. Orifice size—74, 4 inverted mantles (ceramic tips on burners), 6-foot 10-inch post (mantle level 8 feet 4 inches above ground).

Fig. 8-76. Appoximate Luminance (Photometric Brightness) of Various Light Sources

Light Source		Approximate Average Luminance (cd cm^{-2})
Natural Light Sources:		
Sun (at its surface)		225,000
Sun (as observed from earth's surface)	At meridian	160,000
Sun (as observed from earth's surface)	Near horizon	600
Moon (as observed from earth's surface)	Bright spot	0.25
Clear sky	Average brightness	0.8
Overcast sky		0.2
Lightning flash		8,000,000
Combustion Sources:		
Candle flame (sperm)	Bright spot	1.0
Kerosene flame (flat wick)	Bright spot	1.2
Illuminating gas flame	Fish-tail burner	0.4
Welsbach mantle	Bright spot	6.2
Acetylene flame	Mees burner	10.5
Photoflash lamps		16,000–40,000 peak
Nuclear Sources:		
Atomic fission bomb	0.1 millisecond after firing—90 foot diameter ball	200,000,000
Self-luminous paints		0.00003–0.000017
Incandescent Electric Lamps:		
Carbon filament	3.15 lumens per watt	52
Tantalum filament	6.3 lumens per watt	70
Tungsten filament	Vacuum lamp 10 lumens per watt	200
Tungsten filament	Gas filled lamp 20 lumens per watt	1,200
Tungsten filament	750-watt projection lamp 26 lumens per watt	2,400
Tungsten filament	1200-watt projection lamp 31.5 lumens per watt	3,300
RF (radio frequency) lamp	5/16-inch diameter disk	6,200
Blackbody at 6500 K		300,000
Blackbody at 4000 K		25,000
Blackbody at 2042 K		60
60-watt inside frosted bulb		12
25-watt inside frosted bulb		5
15-watt inside frosted bulb		3
10-watt inside frosted bulb		2
60-watt "white" bulb		3
Fluorescent Lamps:		
T-17 bulb cool white	420 mA low loading	0.43
T-12 bulb cool white	430 mA medium loading	0.82
T-12 bulb cool white	800 mA high loading	1.13
T-17 grooved bulb cool white	1500 mA extra high loading	1.50
T-12 bulb cool white	1500 mA extra high loading	1.70
Electroluminescent Lamps:		
Green color at 120 volts 60 cycles		0.0027
Green color at 600 volts 400 cycles		0.0068
Electric Arcs:		
Plain carbon arc	Positive crater	13,000–16,000
High intensity carbon arc	13.6 mm rotating positive carbon	70,000–150,000
Flaming carbon arc	Arc stream	750
Air-blown carbon arc	At positive crater	195,000
Electric Arc Lamps:		
Low pressure mercury arc	50-inch ac rectifier tube	2.0
High intensity mercury arc	Type H33, 35 atmospheres	150
High pressure mercury arc	Type H38, 10 atmospheres	180
High intensity mercury short arc	Type SAH1000A, 30 atmospheres	24,000 (425,000 peak)
Water cooled high pressure mercury arc	Type H6, 75 atmospheres	13,000 (30,000 peak)
Xenon short arc	900W dc	18,000
Clear glass neon tube	15 mm 60 mA	0.16
Clear glass blue tube	15 mm 60 mA	0.08
Clear colored glass tube		
red and green	15 mm 60 mA	0.04
blue	15 mm 60 mA	0.01
Clear glass fluorescent		
daylight and white	15 mm 60 mA	0.50
green	15 mm 60 mA	0.95
blue and gold	15 mm 60 mA	0.30
pink and coral	15 mm 60 mA	0.20
Colored glass fluorescent		
blue and red	15 mm 60 mA	0.05
green and yellow	15 mm 60 mA	0.35
orange	15 mm 60 mA	0.20
Sodium Arc lamp	10,000 lumen size	5.50
Zirconium concentrated arc	300-watt size	4500
Tungsten concentrated arc		2000–2600
Electronic flash tubes		100,000–300,000

Fig. 8–77. Effect of Hot-Cold Resistance on Current in an Incandescent Filament (Laboratory Conditions)

Lamp Wattage	Voltage	Normal Current (amperes)	Theoretical Inrush: Basis Hot-to-Cold Resistance (amperes)*	Time for Current to Fall to Normal Value (seconds)
General Service Incandescent				
15	120	0.125	2.30	0.05
25	120	0.208	3.98	0.06
40	120	0.333	7.00	0.07
50	120	0.417	8.34	0.07
60	120	0.500	10.2	0.08
75	120	0.625	13.1	0.09
100	120	0.835	17.9	0.10
150	120	1.25	26.1	0.12
200	120	1.67	39.5	0.13
300	120	2.50	53.0	0.13
500	120	4.17	89.5	0.15
750	120	6.25	113	0.17
1000	120	8.3	195	0.18
1500	120	12.5	290	0.20
2000	120	16.7	378	0.23
Tungsten-halogen Lamps (C-8 Filament)				
300	120	2.50	62	**
500	120	4.17	102	**
1000	240	4.17	100	**
1500	240	6.24	147	**
1500	277	5.42	129	**

* The current will reach the peak value within the first peak of the applied voltage. Thus the time approaches zero if the instantaneous supplied voltage is at peak, or it could be as much as 0.006 second.
** Not established. Estimated time is 5 to 20 cycles.

Fig. 8–78. Luminous and Thermal Characteristics of Typical Vacuum and Gas-Filled Incandescent Filament Lamps

1	2	3	4	5	6	7	8	9	10	
Watts	Radiated in Visible Spectrum (per cent of input wattage)	Total Filament Radiation Beyond Bulb (per cent of input wattage)	Gas Loss (per cent of input wattage)	Base and Bulb Loss (per cent of input wattage)	End Loss (Loss by conduction at filament ends) (per cent of input wattage)	Filament Heat Content (joules)	Heating Time to 90 Per Cent Lumens (seconds)	Cooling Time to 10 Per Cent Lumens (seconds)	Per Cent Flicker (see page 8–27)	
									60 hertz	25 hertz
6*	6.0	93.0	—	5.5	1.5	0.25	0.04	0.01	29	69
10*	7.1	93.5	—	5.0	1.5	0.62	.06	.02	17	40
25*	8.7	94.0	—	4.5	1.5	2.8	.10	.03	10	28
40†	7.4	71.3	20.0	7.1	1.6	2.5	.07	.03	13	29
60†‡	7.5	80.8	13.5	4.5	1.2	5.5	.10	.04	8	19
100†‡	10.0	82.0	11.5	5.2	1.3	14.1	.13	.06	5	14
200†	10.2	77.4	13.7	7.2	1.7	39.5	.22	.09	4	11
300†	11.1	79.8	11.6	6.8	1.8	80.0	.27	.13	3	8
500†	12.0	82.3	8.8	7.1	1.8	182.0	.38	.19	2	6
10000†	12.1	87.4	6.0	4.7	1.9	568.0	.67	.30	1	4

* Vacuum.
† Gas filled.
‡ Coiled-coil filament.

Fig. 8–79. General Service Lamps for 115, 120, and 125 Volt Circuits (Will Operate in Any Position but Lumen Maintenance is Best for 40 to 1500 Watts When Burned Vertically Base-Up)

Watts	Bulb and Other Description	Base (see Fig. 8–9)	Filament (see Fig. 8–7)	Rated Average Life (hours)	Maximum Over-All Length (inches)	Average Light Center Length (inches)	Approximate Initial Filament Temperature (K)	Maximum Bare Bulb Temperature* (°F)	Base Temperature† (°F)	Approximate Initial Lumens	Rated Initial Lumens Per Watt‡	Lamp Lumen Depreciation** (per cent)
10	S-14 inside frosted or clear	Med.	C-9	1500	3½	2½	2420	106	106	80	8.0	89
15	A-15 inside frosted	Med.	C-9	2500	3½	2⅜	—	—	—	126	8.4	83
25	A-19 inside frosted	Med.	C-9	2500	3⅞	2½	2550	110	108	230	9.2	79
40	A-19 inside frosted and white¶	Med.	C-9	1500	4¼	2 15/16	2650	260	221	455	11.4	87.5
40	S-11 clear	Inter-med.	CC-2V or C-7A	350 500	2 5/16	1⅝	2800 —	570 —	390 —	477 —	11.9 —	— —
50	A-19 inside frosted	Med.	CC-6	1000	4 7/16	3⅛	—	—	—	680	13.6	—
60	A-19 inside frosted and white¶	Med.	CC-6	1000	4 7/16	3⅛	2790	255	200	860	14.3	93
75	A-19 inside frosted and white¶	Med.	CC-6	750	4 7/16	3⅛	2840	275	205	1180	15.7	92
100	A-19 inside frosted and white¶	Med.	CC-8	750	4 7/16	3⅛	2905	300	208	1740	17.4	90.5
100§	A-19 inside frosted and white	Med.	CC-8	1000	4 7/16	3⅛	—	—	—	1680	16.8	—
100	A-21 inside frosted	Med.	CC-6	750	5¼	3⅞	2880	260	194	1690	16.9	90
100§	A-23 inside frosted or clear	Med.	C-9	1000	5 15/16	4 7/16	—	—	—	1480	14.8	—
150	A-21 inside frosted	Med.	CC-8	750	5½	4	2960	—	—	2880	19.2	89
150	A-21 white	Med.	CC-8	750	5½	4	2930	—	—	2790	18.6	89
150	A-23 inside frosted or clear or white	Med.	CC-6	750	6 3/16	4⅝	2925	280	210	2780	18.5	89
150	PS-25 clear or inside frosted	Med.	C-9	750	6 15/16	5¼	2910	290	210	2660	17.7	87.5
200	A-23 inside frosted white or clear	Med.	CC-8	750	6 5/16	4⅝	2980	345	225	4000	20.0	89.5
200	PS-25 clear or inside frosted	Med.	CC-6	750	6 15/16	5¼	—	—	—	3800	19.0	—
200	PS-30 clear or inside frosted	Med.	C-9	750	8 1/16	6	2925	305	210	3700	18.5	85
300	PS-25 clear or inside frosted	Med.	CC-8	750	6 15/16	5 3/16	3015	401	234	6360	21.2	87.5
300	PS-30 clear or inside frosted	Med.	C-9	750	8 1/16	6	3000	275	175	6100	20.3	82.5
300	PS-30 clear or inside frosted	Mog.	CC-8	1000	8⅝	7	—	—	—	5960	19.8	—
300	PS-35 clear or inside frosted	Mog.	C-9	1000	9⅜	7	2980	330	215	5860	19.6	86
500	PS-35 clear or inside frosted	Mog.	CC-8	1000	9⅜	7	3050	415	175	10600	21.2	89
500	PS-40 clear or inside frosted	Mog.	C-9	1000	9¾	7	2945	390	215	10140	20.3	—
750	PS-52 clear or inside frosted	Mog.	C-7A	1000	13 1/16	9½	2990	—	—	15660	20.9	—
750	PS-52 clear or inside frosted	Mog.	CC-8 or 2CC-8	1000	13 1/16	9½	3090	—	—	17000	22.6	89
1000	PS-52 clear or inside frosted	Mog.	C-7A	1000	13 1/16	9½	2995	480	235	21800	21.8	—
1000	PS-52 clear or inside frosted	Mog.	CC-8 or 2CC-8	1000	13 1/16	9½	3110	—	—	23600	23.6	89
1500	PS-52 clear or inside frosted	Mog.	C-7A	1000	13 1/16	9½	3095	510	265	34000	22.6	78

* Lamp burning base up in ambient temperature of 77°F.
† At junction of base and bulb.
‡ For 120-volt lamps.
§ Used mainly in Canada.
¶ Lumen and lumen per watt values of white lamps are generally lower than for inside frosted.
** Per cent initial light output at 70 per cent of rated life.

Fig. 8–80. Lamps for High Voltage Service (May be Burned in Any Position)

Watts	Bulb and Other Description	Base (see Fig. 8–9)	Filament (see Fig. 8–7)	Rated Average Life (hours)	Maximum Over-All Length (inches)	Average Light Center Length (inches)	230–250 Volts: Approximate Initial Lumens**	230–250 Volts: Rated Initial Lumens Per Watt**	230–250 Volts: Lamp Lumen Depreciation*** (per cent)	277 Volts: Approximate Initial Lumens	277 Volts: Rated Initial Lumens Per Watt	277 Volts: Lamp Lumen Depreciation*** (per cent)
25	A-19 inside frosted	Med.	C-17 or C-17A	1000	3⅞	2 9/16	220	8.8	86	210	8.4	—
50	A-19 inside frosted	Med.	C-17 or C-17A	1000	3⅞	2 9/16	485	9.7	79	460	9.2	—
100	A-21 inside frosted	Med.	C-7A or C-9	1000	5 5/16	3⅞	1280	12.8	90	1200	12.0	—
150	PS-25 clear or inside frosted	Med.	C-7A or C-9	1000	6 15/16	5 3/16	2080	13.9	—	1950	13.0	—
200	PS-30 clear or inside frosted*	Med.	C-9	1000	8 1/16	6	3040	15.2	90	2700	13.5	—
300	PS-30 clear	Med.	C-7A	1000	8 1/16	6	4735	15.8	—	4450	14.8	—
300	PS-35 clear or inside frosted*	Mog.	C-7A	1000	9⅜	7	4860	16.2	89	4480	14.9	—
500	PS-40 clear or inside frosted*	Mog.	C-7A	1000	9¾	7	9270	18.5	87	8900	17.8	—
750	PS-52 clear or inside frosted*	Mog.	C-7A	2000†	13 1/16	9½	13600	18.1	—	13500	18.0	—
1000	PS-52 clear or inside frosted*	Mog.	C-7A	2000†	13 1/16	9½	18800	18.8	82	18000	18.0	—
1500	PS-52 clear or inside frosted*	Mog.	C-7A	2000†	13 1/16	9½	27000	18.0	—	26000	17.3	—

* Clear only for 277-volt lamps. ** For 230-volt lamps. *** Per cent initial light output at 70 per cent of rated life.
† 1000 hour life for 277-volt lamps.

Fig. 8–81. Incandescent Filament Lamps for Series Operation on 600-Volt DC Circuits

Amperes or Watts	Volts	Bulb and Other Description	Base (see Fig. 8–9)	Filament (see Fig. 8–7)	Rated Average Life (hours)	Maximum Over-All Length (inches)	Average Light Center Length (inches)	Approximate Initial Lumens	Rated Initial Lumens Per Watt	Lamp Lumen Depreciation§ (per cent)
These Lamps Are Used Chiefly in Trolley Car, Trolley Bus, and Subway Car Lighting Operated 20 in Series										
Amperes										
1.0	30	A-19 inside frosted	Med.	C-2R	2000	3⅞	2 9/16	390	13.0	91.5
1.6	30	A-21 inside frosted*	Med.	C-2R	2000	4⅜	2 15/16	720	15.0	91.5
2.5	30	A-21 inside frosted and white*	Med.	C-2R	2000	4 7/16	2⅞	1120	14.9	—
These Lamps Are Used Chiefly in Trolley Car, Trolley Bus, and Subway Car Lighting Operated 5 in Series										
Watts***										
36‡	120**	A-21 inside frosted†	Med.	C-9	2000	4⅜	2 15/16	390	9.5	—
56 #	120**	A-21 inside frosted†	Med.	C-9	2000	4⅜	2 15/16	630	10.1	—
94 # #	120**	P-25 St. Rwy. Headlight†	Med.	C-5	1000	4¾	2 1/16	985	9.5	—
101	120**	A-23 inside frosted	Med.	C-9	1500	5 15/16	4 7/16	1150	11.9	—

* Cut-out in lamp.
** Also available in 115- and 125-volt designs.
*** Nominal watts.
† Vacuum.
‡ Design current 0.342 amps.
§ Per cent initial light output at 70 per cent of rated life.
\# Design current 0.519 amps.
\# # Design current 0.863 amps.

Fig. 8–82. Extended Service (2500 Hours Rated Life) Incandescent Filament Lamps†

Watts	Bulb and Finish	Base (see Fig. 8–9)	Filament (see Fig. 8–7)	Maximum Over-All Length (inches)	Average Light Center Length (inches)	Approximate Initial Lumens	Initial Lumens Per Watt	Lamp Lumen Depreciation (per cent)
For 120, 125, and 130 Volts								
15	A-15 inside frosted	Med.	C-9	3½	2⅜	125	8.3	83
25	A-19 inside frosted	Med.	C-9	3$\frac{15}{16}$	2½	235	9.4	79
40	A-19 inside frosted	Med.	C-9	4¼	2⅞	415	10.4	87.5
60	A-19 inside frosted	Med.	CC-6	4$\frac{7}{16}$	3⅛	760	12.7	91.5
100	A-19 inside frosted	Med.	CC-8	4$\frac{7}{16}$	3⅛	1480	14.8	92.5
100	A-21 inside frosted	Med.	CC-6	5$\frac{5}{16}$	3⅞	1450	14.5	—
150	A-23 inside frosted and clear	Med.	CC-6	6$\frac{5}{16}$	4⅝	2350	15.7	89
150	PS-25 inside frosted and clear	Med.	C-9	6$\frac{15}{16}$	5¼	2300	15.3	85.5
200	A-23 inside frosted and clear	Med.	CC-8	6$\frac{3}{16}$	4⅝	3350	16.8	87.5
200	A-25 inside frosted and clear	Med.	CC-6	6$\frac{15}{16}$	5¼	3250	16.2	—
200	PS-25 inside frosted and clear	Med.	CC-6	6$\frac{15}{16}$	5¼	3220	16.1	—
200	PS-30 inside frosted and clear	Med.	C-9	8$\frac{1}{16}$	6	3200	16.0	81
300	PS-30 inside frosted and clear	Med.	C-9	8$\frac{1}{16}$	6	5100	17.0	79
300	PS-35 inside frosted and clear	Mog.	C-9	9⅜	7	5100	17.0	84
500	PS-40 inside frosted and clear	Mog.	C-9	9¾	7	9000	18.0	80
750	PS-52 clear	Mog.	C-7A	13$\frac{1}{16}$	9½	14000	18.7	83
1000	PS-52 clear	Mog.	C-7A	13$\frac{1}{16}$	9½	19500	19.5	78
1500	PS-52 clear	Mog.	C-7A	13$\frac{1}{16}$	9½	30000	20.0	—
For 230, 240, and 250 Volts								
100	A-21 inside frosted	Med.	C-7A or C-9	5¼	3⅞	1070	10.7	—
150	PS-25 inside frosted and clear	Med.	C-7A or C-9	6$\frac{15}{16}$	5¼	1840	12.3	—
200	PS-30 inside frosted and clear	Med.	C-7A or C-9	8$\frac{1}{16}$	6	2590	13.0	—
300	PS-35 inside frosted and clear	Mog.	C-7A	9⅜	7	4290	14.3	—
500	PS-40 inside frosted and clear	Mog.	C-7A	9¾	7	7930	15.9	—
750*	PS-52 clear	Mog.	C-7A	13$\frac{1}{16}$	9½	13600	18.2	—
1000*	PS-52 clear	Mog.	C-7A	13$\frac{1}{16}$	9½	18900	18.9	78
1500*	PS-52 clear	Mog.	C-7A	13$\frac{1}{16}$	9½	27000	18.0	—

* Life is 2000 hours.

† Lamps of 3000–3500 hours rated life for 120, 125 and 130 volts also are available in some wattages.

Fig. 8–83. Showcase Lamps for 120 and 130-Volt Circuits

Watts	Bulb and Other Description	Base (see Fig. 8–9)	Filament (see Fig. 8–7)	Rated Average Life (hours)	Maximum Over-All Length (inches)	Approximate Initial Lumens	Lamp Lumen Depreciation‡ (per cent)
25*	T-6½ clear or inside frosted	Intermed.	C-8	1000	5½	240	76
25*	T-10 clear or inside frosted	Med.	C-8	1000	5⅝	255	79
25*	T-10 light inside frosted and side reflectorized	Med.	CC-8	1000	5⅝**	230	80
40*	T-8 clear or inside frosted	Med.	C-23 or C-8	1000	11⅞	420	77
40*	T-10 clear or inside frosted	Med.	C-8	1000	5⅝	445	77
40*	T-10 light inside frosted and side reflectorized	Med.	CC-8	1000	5⅝**	430	80
60†	T-10 clear	Med.	C-8	1000	5⅝	745	—
75*	T-10 clear	Med.	C-23 or C-8	1000	11⅞	800	—

* May be burned in any position.

** Exclusive of spring contact.

† Must be burned from base down to horizontal.

‡ Per cent initial light output at 70 per cent of rated life.

Fig. 8–84. Silvered-Bowl Lamps for 115, 120, and 125-Volt Circuits

Watts	Bulb	Base (see Fig. 8–9)	Approximate Initial Lumens**	Rated Average Life (hours)	Maximum Over-All Length (inches)
60	A-19	Med. Sc.	750	1000	4 7/16
100*	A-21	Med. Sc.	1430	1000	5 1/4
150*	PS-25	Med. Sc.	2320	1000	6 15/16
200*	PS-30	Med. Sc.	3320	1000	8 1/16
300*	PS-35	Mog. Sc.	5325	1000	9 3/8
500*	PS-40	Mog. Sc.	9420	1000	9 3/4
750*	PS-52	Mog. Sc.	14400	1000	13
1000*	PS-52	Mog. Sc.	20200	1000	13

* Base-up burning only. Use only porcelain sockets. For light center length see Fig. 8–79.
** Lumens based on 120 volts.

Fig. 8–85. Tungsten-Halogen Lamps for General Lighting (Burning Position—Any, Except as Noted.)

Watts	Volts	Bulb and Finish	Base*	Filament	Maximum Over-All Length (inches)	Average Light Center Length (inches)	Rated Life (hours)	Approximate Initial Lumens	Approximate Initial Lumens Per Watt	Lamp Lumen Depreciation† (per cent)	Approximate Color Temperature (K)
Double-Ended Types											
200	120	T-2½ clear	RSC	C-8	4 11/16	2	1000	3650	18.3	—	—
200	120	T-3 clear	RSC‡	CC-8	3 1/8	3/4	1500	3460	17.3	96	2900
300	120	T-3 clear	RSC	C-8	4 11/16	2 5/16	2000	5950	19.9	96	2950
300	120	T-3 frosted	RSC‡	C-8	4 11/16	2 5/16	2000	5800	19.3	96	2950
300	120	T-3 clear	RSC	CC-8	3 1/8	9/16	2000	5550	18.5	—	2900
300	120	T-4 clear	RSC	CC-8	3 1/8	5/8	2000	5650	18.9	96	3000
300	130	T-4 clear	RSC	CC-8	3 1/8	5/8	2000	5650	18.9	—	3000
300	120	T-4 frosted	RSC	CC-8	3 1/8	5/8	2000	5300	17.7	—	3000
400	120	T-4 clear	RSC	CC-8	3 1/8	5/8	2000	7750	19.4	96	2950
400	120	T-4 frosted	RSC	CC-8	3 1/8	5/8	2000	—	—	—	2900
400	125	T-4 clear	RSC	CC-8	3 1/8	5/8	2000	7500	18.7	—	3000
400	130	T-4 clear	RSC	CC-8	3 1/8	5/8	2000	7800	19.5	—	3000
500	120	T-3 clear	RSC‡	C-8§	4 11/16	2 3/8	1500	10400	20.8	96	—
500	120	T-3 clear	RSC‡	C-8	4 11/16	2 1/4	2000	10950	21.9	96	3000
500	120	T-3 frosted	RSC‡	C-8	4 11/16	2 1/4	2000	10700	21.4	96	3000
500	125	T-3 clear	RSC	C-8	4 11/16	2	2000	10500	21.0	—	3000
500	130	T-3 clear	RSC‡	C-8	4 11/16	2 7/16	2000	10750	21.5	96	3000
1000	120	T-6 clear	RSC or RSC (Rect)	CC-8	5 5/8	1	2000	23400	23.4	96	3000
1000	120	T-6 clear	RSC or RSC (Rect)	CC-8	5 5/8	1 3/32	4000	19800	19.8	93	2950
1000	120	T-6 frosted	RSC or RSC (Rect)	CC-8	5 5/8	1	2000	22700	22.7	96	3000
1000	120	T-6 clear	RSC or RSC (Rect)	CC-8	5 5/8	1 1/4	500	24500	24.5	—	3100
1000	220	T-3 clear	RSC‡	C-8	10 1/16	5 15/16	2000	21400	21.4	96	3000
1000	240	T-3 clear	RSC‡	C-8	10 1/16	6 3/16	2000	21400	21.4	96	3000
1250	208	T-3 clear	RSC‡	C-8	10 1/16	6	2000	28000	22.4	96	3050
1250	208	T-3 frosted	RSC	C-8	10 1/16	6	2000	26400	21.1	—	3050
1500	208	T-3 clear	RSC‡	C-8	10 1/16	6 3/4	2000	34500	23.0	96	3050
1500	208	T-3 frosted	RSC	C-8	10 1/16	6 3/4	2000	31000	20.6	—	3050
1500	220	T-3 clear	RSC‡	C-8	10 1/16	6 3/4	2000	34800	23.2	96	3050
1500	240	T-3 clear	RSC	C-8§	10 1/16	6 3/4	1000	35800	23.9	96	—
1500	240	T-3 clear	RSC‡	C-8	10 1/16	6 3/4	2000	34800	23.2	96	3050
1500	240	T-3 frosted	RSC	C-8	10 1/16	6 3/4	2000	32000	21.3	—	3050
1500	277	T-3 clear	RSC	C-8§	10 1/16	6 3/4	1000	34400	22.9	96	—
1500	277	T-3 clear	RSC‡	C-8	10 1/16	6 3/4	2000	33700	22.5	96	3050
1500	277	T-3 frosted	RSC	C-8	10 1/16	6 3/4	2000	31600	21.0	—	3050

See footnotes on page 8-61.

Continued on next page.

Fig. 8–85. *Continued*

Watts	Volts	Bulb and Finish	Base*	Filament	Maximum Over-All Length (inches)	Average Light Center Length (inches)	Rated Life (hours)	Approximate Initial Lumens	Approximate Initial Lumens Per Watt	Lamp Lumen Depreciation† (per cent)	Approximate Color Temperature (K)
Single-Ended Types											
75	28	T-3 clear	Min.	CC-6	2⅜	1$^{3}/_{16}$	2000	1600	21.4	—	3000
150	120	T-4 clear	Minican.	CC-2V	2¾	—	1500	2900	19.3	—	3000
150	120	T-4 clear	D.C. Bay.	CC-2V	2$^{7}/_{16}$	—	1500	2900	19.3	—	3000
250	120	T-4 clear	Minican.	CC-8	3⅛	1⅝	2000	4850	19.4	96	2950
250	120	T-4 frosted	Minican.	CC-8	3⅛	1⅝	2000	4700	18.8	96	2950
250	120	T-4 clear	D.C. Bay.	CC-8	2$^{13}/_{16}$	1½	2000	4850	19.4	—	2950
250	120	T-4 clear	D.C. Bay.	CC-8	2$^{13}/_{16}$	1⅝	2000	4850	19.4	96	2950
250	120	T-4 frosted	D.C. Bay.	CC-8	2$^{13}/_{16}$	1⅝	2000	4700	18.8	96	2950
250	130	T-4 clear	Minican.	CC-8	3⅛	1⅝	2000	4850	19.4	—	2950
400	120	T-4 frosted	Minican.	CC-8	3⅝	2	2000	8550	21.4	96	3050
400	120	T-4 clear	Minican.	CC-8	3⅝	2	2000	8800	22.0	96	3050
500	120	T-4 clear	Minican.	CC-8	3¾	2	2000	11500	23.0	—	3000
500	120	T-4 frosted	Minican.	CC-8	3¾	2	2000	10000	20.0	—	3000
500	120	T-4 frosted	D.C. Bay.	CC-8	3$^{7}/_{16}$	2⅛	2000	10100	20.2	96	3000
500	120	T-4 clear	D.C. Bay.	CC-8	3$^{7}/_{16}$	2⅛	2000	10450	20.9	96	3000
500	130	T-4 clear	Minican.	CC-8	3¾	2	2000	11500	23.0	—	3000
750	120	T-24 frosted	Med. Bipost	CC-8	9$^{3}/_{16}$	5½	3000	16100	21.4	93	3000
1000	120	E-37 clear	Mog.	CC-8	11$^{5}/_{16}$	7	3000	20900	20.9	93	3000
1000	120	T-24 frosted	Med. Bipost	CC-8	9$^{3}/_{16}$	5½	3000	22400	22.4	93	3050

* RSC = recessed single contact, RSC (Rect) = rectangular recessed single contact.
† Per cent initial light output at 70 per cent of rated life.
‡ Burning position—horizontal.
§ Lamp provides maximum filament straightness under severe operating conditions.

Fig. 8–86. Three-Lite Lamps for 115, 120, and 125 Volt Circuits (For Base-Down Burning Only)

Watts	Bulb and Other Description	Base	Filament (see Fig. 8–7)	Rated Average Life (hours)	Maximum Over-All Length (inches)	Light Center Length (inches)	Approximate Initial Lumens	Approximate Initial Lumens Per Watt	Lamp Lumen Depreciation* (per cent)
30 70 100	A-21 or T-19 white	3 contact Med.	C-8 or 2CC-6 or 2CC-8 or C-2R/CC-8	1200	5$^{5}/_{16}$	3¾	290 980 1270	9.7 14.0 12.7	86
50 100 150	A-23 or A-21 white	3 contact Med.	2CC-6 or 2CC-8 or C-2R/CC-8	1200	5$^{15}/_{16}$	3⅞	600 1600 2200	12.0 16.0 14.7	85
50 100 150	PS-25 or T-19 white	3 contact Med.	2C-2R or 2CC-6 or 2CC-8	1200	5$^{15}/_{16}$	3⅞	575 1450 2025	11.5 14.5 13.5	—
50 100 150	PS-25 inside frosted	3 contact Mog.	2C-2R or 2CC-8	1200	6$^{13}/_{16}$	5	640 1630 2270	12.8 16.3 15.1	—
100 200 300	PS-25 white	3 contact Mog.	2CC-6 or 2CC-8 or C-2R/CC-8	1200	6$^{13}/_{16}$	4⅛	1360 3400 4760	13.6 17.0 15.9	72
50 200 250	PS-25, A-23 or T-21 white	3 contact Med.	2C-2R or 2CC-6 or 2CC-8 or C-2R/CC-8	1200	5$^{15}/_{16}$	3⅞	600 3550 4150	12.0 17.8 16.6	72

* Per cent initial light output at 70 per cent of rated life.

Fig. 8-87. Lamps for Flashlight, Handlantern, Bicycle, and Other Services

Lamp Number	Bulb (see Fig. 8-8)	Base (see Fig. 8-9)	Design Volts	Amperes	Filament (see Fig. 8-7)	Light Center Length (inches)	Maximum Over-All Length (inches)	Rated† Average Life (hours)	Service—Use with Cell
112	TL-3	Min. Sc.	1.20	0.22	S-2	—	59/64	5	1-AA
401**	G-4½	Min. Sc.	1.25	0.22	S-2	⅞	1 1/16	15	1-D
131	G-3½	Min. Sc.	1.30	0.10	S-2	23/32	15/16	50	1-D
222	TL-3	Min. Sc.	2.25	0.25	C-6	—	59/64	5	2-AA
223	FE-3¾	Min. Sc.	2.25	0.25	C-6	21/32	13/16	5	2-AA
233	G-3½	Min. Sc.	2.33	0.27	C-2R	23/32	15/16	10	2-C
PR4	B-3½	S.C. Min. Fl.	2.33	0.27	C-2R	¼	1¼	10	2-C
PR2	B-3½	S.C. Min. Fl.	2.38	0.50	C-2R	¼	1¼	15	2-D
PR6	B-3½	S.C. Min. Fl.	2.47	0.30	C-2R	¼	1¼	30	2-D
14	G-3½	Min. Sc.	2.47	0.30	C-2R	23/32	15/16	15	2-D
406**	G-4½	Min. Sc.	2.60	0.30	C-2R	⅞	1 1/16	50	2-D
PR3	B-3½	S.C. Min. Fl.	3.57	0.50	C-2R	¼	1¼	15	3-D
PR7	B-3½	S.C. Min. Fl.	3.70	0.30	C-2R	¼	1¼	30	3-D
13	G-3½	Min. Sc.	3.70	0.30	C-2R	23/32	15/16	15	3-D
4546	PAR-36	Sc. Term.	4.70	0.50	C-2R	—	2¾	100	4-F
PR13	B-3½	S.C. Min. Fl.	4.75	0.50	C-2R	¼	1¼	15	4-F
PR15	B-3½	S.C. Min. Fl.	4.82	0.50	C-2R	¼	1¼	30	8-D*
27	G-4½	Min. Sc.	4.90	0.30	C-2R	23/32	1 1/16	30	4-F
407**	G-4½	Min. Sc.	4.90	0.30	C-2R	13/16	1 1/16	50	4-F
502	G-4½	Min. Sc.	5.10	0.15	C-2R	23/32	1 1/16	100	4-F
PR12	B-3½	S.C. Min. Fl.	5.95	0.50	C-2R	¼	1¼	15	5-D
605	G-4½	Min. Sc.	6.15	0.50	C-2R	23/32	1 1/16	15	5-D
PR18	B-3½	S.C. Min. Fl.	7.2	0.50	C-2R	¼	1¼	3	6-D
PR20	B-3½	S.C. Min. Fl.	8.63	0.50	C-2R	¼	1¼	15	7-D

* Two—4 cell groups in parallel. ** Flasher lamp. † Laboratory.

Fig. 8-88. Lamps for Train and Locomotive Service*

Watts	Volts	Bulb and Other Description	Base (see Fig. 8-9)	Filament (see Fig. 8-7)	Rated Average Life (hours)	Maximum Over-All Length (inches)	Average Light Center Length (inches)	Approximate Initial Lumens	Rated Initial Lumens Per Watt	Lamp Lumen Depreciation† (per cent)
					Train					
6	30	S-6 Indicator (A)	Cand.	C-2V	1500	1⅞	—	50	8.3	—
	60		Sc.	C-7A			—	48	8.0	—
6	30	S-6 Blue-Night	Cand.	C-2V	1500	1⅞	—	—	—	—
	60	Light (A)	Sc.	C-7A			—	—	—	—
15	30	A-17 inside frosted	Med.	C-9	1000	3⅝	2⅜	179	11.9	89
	60	(A)			—	—	—	147	9.8	—
25	30	A-19 inside frosted	Med.	C-9	1000	3 15/16	2½	340	13.6	89
	60				—	—	—	282	11.3	90
25	30	T-8½ inside	Med.	C-8	1000	5½	—	325	13.0	—
	60	frosted					—	280	11.2	—
40	30	A-19 inside frosted	Med.	C-9	1000	4¼	2⅞	590	14.8	—
	60							525	13.1	—
50	30	A-21 inside frosted	Med.	C-9	1000	4 15/16	3⅜	810	16.2	86
	60							700	14.0	—
75	30	A-23	Med.	C-9	1000	6 1/16	4⅜	1230	16.4	—
	60							1140	15.2	—
100	30	A-23	Med.	C-9	1000	6 1/16	4⅜	1800	18.0	—
	60							1650	16.5	—

See footnotes om page 8-63.

Continued on next page.

Fig. 8–88. *Continued*

Watts	Volts	Bulb and Other Description	Base (see Fig. 8–9)	Filament (see Fig. 8–7)	Rated Average Life (hours)	Maximum Over-All Length (inches)	Average Light Center Length (inches)	Approximate Initial Lumens	Rated Initial Lumens Per Watt	Lamp Lumen Depreciation† (per cent)
				Locomotive						
6	60	S-6 Indicator (A)	S.C. Bay.	C-7A	1500	1 13/16	1 1/16	48	8.0	—
15	75	S-11 Marker (A)	D.C. Bay.	C-1	1000	2 3/8	1 1/4	142	9.5	—
15	34	S-14 inside frosted Cab (A)	Med.	C-9	1000	3 1/2	2 1/2	142	9.5	72
25	60	A-19 inside frosted	Med.	C-9	1000	3 7/8	2 9/16	265	10.6	—
	75	A-17 RS Cab (A)	Med.	C-9	1000	3 5/8	2 1/2	245	9.8	75
30	64 75	S-11 Marker	Bay. S.C.-D.C.	C-7A	500	2 3/8	1 1/4	350 336	11.7 11.2	— —
50	60	A-21 inside frosted	Med.	C-9	1000	4 7/8	3 7/16	535	10.7	—
	75	A-19 RS Cab-Engine (A)	Med.	C-9	1000	3 7/8	2 1/2	545	10.9	57
75	75	PAR-36 RS Train Warning	Sc. Term.	CC-6	500	2 3/4	—	—	—	—
100	32	A-21 Headlight	Med.	C-5	500	4 7/16	3	1600	16.0	87
200	12 30	PAR-56 Headlight	Sc. Term.	CC-8 CC-8	500 500	4 1/2 4 1/2		(B) (C)	— —	— —
250	32	P-25 Headlight	Med. Pref. Med.	C-5A	500 500	5 4 3/4	2 3/16 3	4550 4550	18.2 18.2	89 89

* **Lamps may be burned in any position except the Headlight Lamps. They can be burned in any position except within 45 degrees of vertically base-up.**
† Per cent initial light output at 70 per cent of rated life.
(A) Vacuum.
(B) 180,000 max. candlepower.
(C) 260,000 max. candlepower.

Fig. 8–89. Rough Service Lamps for 115, 120, and 130-Volt Circuits (May Be Burned in Any Position)

Watts	Bulb and Other Description	Base (see Fig. 8–9)	Filament (see Fig. 8–7)	Rated Average Life (hours)	Maximum Over-All Length (inches)	Light Center Length (inches)	Approximate Initial Lumens	Approximate Initial Lumens per Watt	Lamp Lumen Depreciation† (per cent)
25*	A-19 inside frosted	Med.	C-17	1000	3 15/16	2 1/2	230	9.2	—
50	A-19 inside frosted	Med.	C-22	1000	3 15/16	2 1/2	480	9.6	76
75*	A-21 inside frosted	Med.	C-22 or C-17	1000	4 7/16	2 7/8	750	10.0	
100	A-21 inside frosted	Med.	C-17	1000	5 5/16	3 7/8	1250	12.5	79
150*	A-23 inside frosted	Med.	C-17	1000	6 1/16	4 3/8	2130	14.2	—
150	PS-25 inside frosted	Med.	C-17	1000	6 15/16	5 1/4	2130	14.2	80
200*	PS-30 inside frosted or clear	Med.	C-9	1000	8 1/16	6	3380	16.9	82.5
200*	A-23 inside frosted or clear	Med.	C-17	1000	6 1/16	4 3/8	3380	16.9	—

* 115-volt not available.
† Per cent initial light output at 70 per cent of rated life.

Fig. 8–90. Incandescent Filament and Tungsten-Halogen Lamps for Theatre Stages and Motion Picture and Television Studios (All Data for 120 Volt Lamps Except as Noted)

Watts	Manufacturer's Ordering Code	ANSI Code	Color Temperature (K)	Average Life (hours)	Approximate Initial Lumens	Lighted Length (inches)	Bulb Finish
I. Tungsten-Halogen Double-Ended Lamps							
(A) T-3 or T-4 Bulb; Compact CC-8 Filament[1]; Recessed Single Contact Bases; Maximum Over-All Length: 3 1/8 Inches; Burning Position: Any							
200	Q200T3/CL	—	2850	1500	3460	3/4	Clear
300	Q300T4/CL	EHP	2900	2000	5650	9/16	Clear
400	Q400T4/CL	EHR	2900	2000	7750	5/8	Clear
650	FAD	FAD	See Fig. 8–105.				
650	FBX	FBX					
650	DWY	DWY					
650	FCA	FCA					
800	DXX (230, 240V)	DXX	3200	75	20500	1	*Clear
800	DXV (230V)	DXV	3400	16	24000	7/8	*Clear
(B) T-4 or T-5 Bulb; Compact CC-8 Filament[1]; Recessed Single Contact Bases; Maximum Over-All Length: 3 3/4 Inches; Burning Position: Any							
500	FGD	FGD	3200	100	13000	1/2	*Clear
600	FEA (240V)	FEA	3200	60	15000	7/8	Clear
600	FEB (220V)	FEB	3200	60	15000	7/8	Clear
600	FCB	FCB	3250	75	17000	11/16	Clear
750	FGB	FGB	3200	125	19500	5/8	*Clear
1000	DXW	DXW	See Fig. 8–105.				
1000	FBY	FBY					
1000	DXN	DXN					
(C) T-5 Bulb; Compact CC-8 Filament[1]; Recessed Single Contact Bases; Maximum Over-All Length: 4 3/8 Inches; Burning Position: Any							
1000	DYA	DYA	3200	150	28000	7/8	Clear
1000	DYN	DYN	3200	150	28000	—[15]	Frosted
(D) T-3 or T-5 Bulb; Linear C-8 Filament[1]; Recessed Single Contact Bases; Maximum Over-All Length: 4 11/16 Inches; Burning Position: ±4°							
300	Q300T3/CL	EHM	2950	2000	5950	2 5/16	Clear
300	Q300T3	EHZ	2950	2000	5950	—[15]	Frosted
500	Q500T3/CL	FCL	3000	2000	10950	2 1/4	Clear
500	Q500T3/CL (125–130V)	—	3000	2000	10550	2 7/16	Clear
500	Q500T3	FCZ	3000	2000	10700	—[15]	Frosted
500	Q500T3 (125–130V)	—	3000	2000	10300	—[15]	Frosted
500	Q500T3/4CL	FDF	3200	400	13250	2 3/8	Clear
500	Q500T3/4	FDN	3200	400	12800	—[15]	Frosted
500	Q500T3/CL/6	—	3000	—[9]	10400	2 3/8	Clear
750	Q750T3/4CL	EJG	3200	400	20200	2 7/16	Clear
750	Q750T3/4	—	3200	400	19500	—[15]	Frosted
800	Q800T3/4CL (240V)	—	3200	250	22000	2 3/4	Clear
800	Q800T3/4 (240V)	—	3200	250	21400	—[15]	Frosted

See footnotes on page 8–72.

Continued on next page.

Fig. 8–90. *Continued*

Watts	Manufacturer's Ordering Code	ANSI Code	Color Temperature (K)	Average Life (hours)	Approximate Initial Lumens	Lighted Length (inches)	Bulb Finish
I. Tungsten-Halogen Double-Ended Lamps (*Continued*)							
1000	Q1000T3/3CL(185V)	EJD	3350	100	33600	2 11/16	Clear
1000	Q1000T3/3(185V)	EJE	3350	100	32800	—[15]	Frosted
1000	Q1000T3/4CL	FCM	3200	500	28000	2 9/16	Clear
1000	Q1000T3/4	FHM	3200	500	28000	—[15]	Frosted
1100	EKV	EKV	See Fig. 8–105.				
(E) T-6 or T-8 Bulb; Compact CC-8 Filament[1]; Recessed Single Contact Bases; Maximum Over-All Length: 5 5/8 Inches; Burning Position: Any							
1000	Q1000T6/CL/2	—	2950	4000	19800	1 3/32	Clear
1000	Q1000T6/3CL	DWS[11]	3350	75	32000	7/8	Clear
1000	Q1000T6/4CL	EHS or FER[11]	3200	500	27500	3/4	Clear
1000	Q1000T6/CL	DWT	3000	2000	23400	1	Clear
1000	Q1000T6	—	3000	2000	22700	—[15]	Frosted
1500	Q1500T8/3CL	—	3350	100	47500	1 1/8	Clear
1500	Q1500T8/4CL	DVV[11]	3200	500	42700	1 1/4	Clear
2000	Q2000T8/3CL	—	3350	100	65000	1 1/8	Clear
2000	Q2000T8/4CL	FEY[12]	3200	500	56600	1 7/16	Clear
(F) T-3 or T-4 Bulb; Linear C-8 Filament[1]; Recessed Single Contact Bases; Maximum Over-All Length: 6 9/16 Inches; Burning Position: ±4°							
1000	Q1000T3/1CL	FFT	3200	500	26400	2 9/16	Clear
1500	Q1500T4/4CL	—	3200	400	41200	2 1/4	Clear

Watts	Bulb	Manufacturer's Ordering Code	ANSI Code	Color Temperature (K)	Average Life (hours)	Approx. Initial Lumens	Filament Form	Maximum Over-All Length (inches)	Bulb Finish
II. Tungsten-Halogen Single-Ended Lamps									
(A) Medium 2-Pin Base; Light Center Length: 2 3/8 Inches; Burning Position: Any									
500	T-4	Q500/5CL	EHB	3150	500	12700	CC-8[1]	4	Clear
500	T-4	Q500CL/TP	EHD	3000	2000	10450	CC-8[1]	4	Clear
750	T-6	Q750/4CL	EHF	3200	500	20400	CC-8[1]	4 1/8	Clear
1000	T-6	Q1000/4CL	EHK or FEL[11]	3200	500	27500	CC-8[1]	4 1/8	Clear
1000	T-6	Q1000/4	FCV[11]	3200	500	26500	CC-8[1]	4 1/8	Frosted
(B) Miniature 2-Pin Base; Light Center Length: 1 11/16 Inches; Burning Position: Base Down to Horizontal									
420	G-7	DYM	DYM	3200	75	11000	CC-6[1]	2 3/8	*Clear
600	G-7	DYH	DYH	3200	75	17000	CC-6[1]	2 3/8	*Clear
650	G-6	DVY	DVY	3400	25	20000	CC-6[1]	2 3/8	*Clear
650	G-7	DYJ (230V)	DYJ	3400	20	20000	CC-2P[1]	2 1/2	*Clear

See footnotes on page 8–72.

Continued on next page.

Fig. 8–90. *Continued*

Watts	Bulb	Manufacturer's Ordering Code	ANSI Code	Color Temper-ature (K)	Average Life (hours)	Approx. Initial Lumens	Filament Form	Maximum Over-All Length (inches)	Bulb Finish
II. Tungsten-Halogen Single-Ended Lamps (*Continued*)									
(C) 2-Pin Prefocus Base; Light Center Length: 1 7/16 Inches; Burning Position: Burn with Coil Horizontal									
420	G-7	EKB	EKB	3200	75	11000	CC-6[1]	2½	*Clear
600	G-7	DYS	DYS	3200	75	17000	CC-6[1]	2½	*Clear
600	G-7	DYV[18]	DYV	3200	75	17000	CC-6[1]	2½	*Clear
650	G-6	EKD	EKD	3400	25	20000	CC-6[1]	2½	*Clear
650	G-7	DYR (220, 240V)	DYR	3200	50	16500	2CC-8[1]	2½	*Clear

Watts	Manufacturer's Ordering Code	ANSI Code	Color Temper-ature (K)	Average Life (hours)	Beam Shape	Field Angle (degrees)[5]	Beam Angle (degrees)[3]	Approximate Candle-Power[6]	Approximate "Beam Lumens"[7]
III. Tungsten-Halogen and Incandescent Lamps									
(A) PAR-36 Bulb; Bases: Ferrule Contacts or Screw Terminals[13] or 2-Prong[16]; "Q" After "WATTS" Indicates Tungsten-Halogen Types, Burning Position: Any									
650	DWA	—	See Fig. 8–105.						
650Q	DWE[13]	—	3200	100	MFL	—	—	24000	—
650Q	DXK	—	See Fig. 8–105.						
650	DXY[13]	—							
650	FAF (220, 240V)	—	3400	8	MFL	—	—	23000	—
650	FAP	—	See Fig. 8–105.						
650Q	FAY	—							
650Q	FBE[13]	—	5000	20	SP	—	15 x 22	35000	—
650Q	FBJ	—	See Fig. 8–105.						
650	FBM[16]	—	3400	8	MFL	—	30 x 40	25500	—
650Q	FBO[13]	—	3400	20	SP	—	15 x 22	75000	—
650Q	FCW	—	See Fig. 8–105.						
650Q	FCX	—							
650Q	FGJ	—	5000	20	MFL	—	30 x 40	24000	—
650Q	FGK[13]	—	5000	20	MFL	—	30 x 40	24000	—
(B) PAR-64 Bulb; Bases: Extended Mogul End Prong or Screw Terminals[13]; Burning Position: Any Except Where Noted									
1000	Q1000PAR64/1	FFN	3200	400[14]	VNSP	10 x 24	6 x 12	400000	11000
1000	Q1000PAR64/2	FFP	3200	400	NSP	14 x 26	7 x 14	330000	12000
1000	Q1000PAR64/5	FFR	3200	400	MFL	21 x 44	12 x 28	125000	14000
1000	Q1000PAR64/6	FFS	3200	400	WFL	45 x 71	24 x 48	40000	19000
1000	Q1000PAR64/7	FFD[11]	3350	60	MFL	20 x 43	11 x 27	140000	16500
1000	Q1000PAR64/3D	—	5200	60	NSP	12 x 24	6 x 13	200000	7000
1000	Q1000PAR64/7D	—	5200	60	MFL	20 x 43	11 x 27	70000	8200
1000	Q1000PAR64/8D	—	5200	60	WFL	44 x 70	23 x 47	25000	11200
(C) R-30 Bulb; Base: Medium Screw; Burning Position: Any									
300	BEP	BEP	See Fig. 8–105.						
375	EBR	EBR							

See footnotes on page 8–72.

Continued on next page.

Fig. 8–90. *Continued*

Watts	Manufacturer's Ordering Code	ANSI Code	Color Temper-ature (K)	Average Life (hours)	Beam Shape	Field Angle (degrees)[5]	Beam Angle (degrees)[3]	Approximate Candle-Power[6]	Approximate "Beam Lumens"[7]
III. Tungsten-Halogen and Incandescent Lamps (*Continued*)									
(D) R-40 Bulb; Medium Screw Base; Burning Position: Any									
250	BFC (30V)	—	3400	3/4	MFL	—	—	—	—
375	BFA	—	See Fig. 8–105.						
500	EAL	—	3200	15	MFL	—	60	8200	—
500	DXB	—	See Fig. 8–105.						
500	DXC	—							
500	EAH (220–240V)	—	3400	6	FL	—	90	5000	—

Watts	Bulb	Manufacturer's Ordering Code	ANSI Code	Color Temper-ature (K)	Average Life (hours)	Approxi-mate Initial Lumens	Filament Form	Maximum Over-All Length (inches)	Recom-mended Operating Position
IV. Bipost-Based Lamps—Tungsten-Halogen and Incandescent									
(A) Medium Bipost Base; Light Center Length: 2 1/2 Inches									
500	T-20	500T20/70	—	3000	200	11000	C-13[1]	6 1/2	BD45
500	T-20	500T20/63	DVG, EDC	3200	50	14500	C-13[1]	6 1/2	BD45
750	T-24	750T24/16	DVH, EDG	3200	50	22000	C-13[1]	6 1/2	BD45
750	T-24	750T24/4 (230V)	—	3200	20	20000	CC-13	6 1/2	BD45
750	T-24	750T24/13	EDH	3350	12	25000	C-13[1]	6 1/2	BD45
1000	T-24	1M/T24/16	DXP	3200	50	29000	C-13[1]	6 1/2	BD45
1000	T-24	1M/T24/13	EBB	3350	12	33500	C-13[1]	6 1/2	BD45
750	T-20	Q750T20/4CL	—	3200	200	20900	C-13[1]	6 1/2	BD45
1000	T-20	Q1000T20/4CL	—	3200	250	28500	C-13[1]	6 1/2	BD45
(B) Mogul Bipost Base; Light Center Length: 5 Inches									
1000	G-48	1M/G48/11	—	3050	200	24000	C-13[1]	9 3/8	BDTH
1000	G-48	1M/G48/8	DVB	3100	100	27000	C-13[1]	9 3/8	BDTH
2000	G-48	2M/G48/17	—	3150	200	58000	C-13[1]	9 3/8	BD45
2000	G-48	2M/G48/10	—	3200	100	60000	CC-8[2]	9 3/8	BD45
2000	G-48	2M/G48/18	DVF, ECG	3200	100	61000	C-13[1]	9 3/8	BD45
2000	G-48	2M/G48/14	ECK	3350	25	65000	C-13[1]	9 3/8	BD45
2000	T-48	2M/T48/1	DVJ, DVK	3200	100[10]	61000	C-13[1]	10 3/8	BD45
2000	T-48	2M/T48/4	EDL, EDM	3350	25[10]	65000	C-13[1]	10 3/8	BD45
1500	T-11	Q1500T11/4CL	CXZ	3200	—	44500	C-13[1]	8 1/2	BDTH
2000	T-11	Q2000T11/4CL	CYX[11]	3200	250	59000	C-13[1]	8 1/2	BDTH
2000	T-11	Q2000T11/3CL	CYW[11]	3350	75	65000	C-13[1]	8 1/2	BDTH
2000	T-30	Q2000T30/4CL	—	3200	750	58500	C-13[1]	10 1/4	BDTH
2000	T-30	Q2000T30/3CL	—	3350	250	64000	C-13[1]	10 1/4	BDTH

See footnotes on page 8–72.

Continued on next page.

Fig. 8–90. *Continued*

Watts	Bulb	Manufacturer's Ordering Code	ANSI Code	Color Temperature (K)	Average Life (hours)	Approximate Initial Lumens	Filament Form	Maximum Over-All Length (inches)	Recommended Operating Position
IV. Bipost-Based Lamps—Tungsten-Halogen and Incandescent (*Continued*)									
(C) Mogul Bipost Base; Light Center Length: 6 1/2 Inches									
5000	G-64	5M/G64/7	ECN	3200	150[10]	145000	C-13[1]	11 7/8	BD45
5000	G-64	5M/G64/3	ECM	3350	75[10]	161500	C-13[1]	11 7/8	BD45
5000	T-64	5M/T64	—	3050	500[10]	120000	C-13D[1]	13 3/8	BD30
5000	T-64	5M/T64/3	DPK	3200	150[10]	145000	C-13[1]	13 3/8	BD45
5000	T-64	5M/T64/1	EDN	3350	75[10]	161500	C-13[1]	13 3/8	BD45
5000	T-20	Q5000T20/4CL	DPY[11]	3200	500	143000	C-13[1]	11	BDTH
5000	T-20	Q5000T20/3CL	DPZ[11]	3350	250	168000	C-13[1]	11	BDTH
(D) Mogul Bipost Base; Light Center Length: 10 Inches									
10000	G-96	10K/G96/1	EBA	3200	150[10]	295000	C-13	17 7/16	BD45
10000	G-96	10K/G96	ECP	3350	75[10]	335000	C-13	17 7/16	BD45
10000	T-24	Q10M/T24/4CL	DTY	3200	300[8]	295000[8]	C-13	16[8]	BD45
10000	T-24	Q10M/T24/3CL	DTZ	3350	150[8]	335000[8]	C-13	16[8]	BD45
V. Incandescent Prefocus-Base Lamps									
(A) Medium Prefocus Base; Light Center Length: 2 3/16 Inches									
250	T-20	250T20/47	—	2900	200	4600	C-13	5 3/4	BDTH
250	T-14	DLC	DLC	3050	50	5500	C-13	5 3/4	BD30
300	T-20	DRL	DRL	2700	500	4800	C-13	5 3/4	BD30
500	T-20	500T20/64	—	2900	500	10000	C-13	5 3/4	BDTH
500	T-20	500T20/48	—	3000	200	11000	C-13	5 3/4	BDTH
500	T-20	DMX	DMX	See Fig. 8–105.					
500	T-20	DWB (230V)	DWB	3000	50	11500	CC-13	5 3/4	BD30
750	T-20	750T20P/SP	BFE	3000	200	17000	C-13	5 3/4	BD45
750	T-20	BFL	BFL	3050	200	13500[19]	C-13	5 3/4	BD30
750	T-20	DPJ	DPJ	3250	25	19000	C-13D	5 3/4	BD30
1000	T-20	1M/T20P/SP	—	3050	200	23400	C-13D	5 3/4	BD45
1000	T-20	DRC	DRC	3250	50	30000	C-13	5 3/4	BD30
1000	T-20	DRS	DRS	3325	25	28500	C-13D	5 3/4	BD30
1000	T-20	DRB	DRB	3350	25	32000	C-13	5 3/4	BD30
1000	T-20	DWK (230V)	DWK	3100	50	25000	CC-13	5 3/4	BD30
VI. Prefocus and Screw-Base Lamps—Tungsten-Halogen and Incandescent									
(A) Mogul Prefocus Base; Light Center Length: 3 15/16 Inches									
1000	G-40	1M/G40/20	—	3000	200	23400	C-13D	8 7/16	BD30
1000	G-40	1M/G40/23	—	3050	200	24000	C-13[1]	8 7/16	BDTH
1000	G-40	1M/G40PSP	—	3050	200	25000	C-5	8 7/16	BDTH
1000	G-40	1M/G40/25	ECC	3200	50	29000	C-13	8 7/16	BDTH
1500	G-40	1500G40/21	—	3000	200	35500	C-13D	8 7/16	BD30
1500	G-40	1500G40/15	—	3050	200	36000	C-5	8 7/16	BDTH
2000	G-48	2M/G48/5	—	3100	200	58000	C-13	9	BD45
2000	G-48	2M/G48/21	ECH	3200	60	61000	C-13	9	BD45

See footnotes on page 8–72.

Continued on next page.

Fig. 8–90. *Continued*

Watts	Bulb	Manufacturer's Ordering Code	ANSI Code	Color Temperature (K)	Average Life (hours)	Approximate Initial Lumens	Filament Form	Maximum Over-All Length (inches)	Recommended Operating Position
VI. Prefocus and Screw-Base Lamps—Tungsten-Halogen and Incandescent (*Continued*)									
(B) Medium Screw Base; Light Center Length: 3 Inches									
100	A-21	100A21/SP	—	2750	200	1340	C-5	4⅜	Any
100	A-21	100A21/SP (230V)	—	2650	200	1050	C-5	4⅜	Any
150	P-25	150P25/2	—	2750	200	2100	C-5	4¾	Any
250	P-25	250P25	—	2800	500	3700	C-5	4¾	Any
250	G-30	250G/FL	—	2800	800	3650	C-5	5⅛	BDTH
250	G-30	250G/SP	—	2900	200	4500	C-5	5⅛	BDTH
400	G-30	400G/FL	—	2850	800	6800	C-5	5⅛	BDTH
400	G-30	400G/SP	—	2950	200	8400	C-5	5⅛	BDTH
400	G-30	400G/FL (230 V)	—	2700	800	5000	C-5	5⅛	BDTH
500	T-20	DMS	—	See Fig. 8–105.					
(C) Mogul Screw Base; Light Center Length: 4 1/4 Inches									
500	G-40	500G/FL	—	2850	800	9300	C-5	7 1/16	BDTH
1000	G-40	1M/G40SP4-1/4	—	3050	200	24500	C-5	7 1/16	BDTH
(D) Mogul Screw Base; Light Center Length: 5 1/4 Inches									
1000	G-40	1M/G40FL	—	2950	800	20000	C-5	7⅞	BDTH
1500	G-48	1500G48/6	—	2950	800	32200	C-5	8 9/16	BDTH
1500	G-48	1500G48/FL (230, 250V)	—	—	800	—	C-5	8 9/16	BDTH
2000	T-8	Q2000/4CL	—	3200	750	59000	CC-8[1]	7½	Any
2000	T-8	Q2000/4 (frosted)	—	3200	750	57200	CC-8[1]	7½	Any
2000	T-8	Q2000/3CL	—	3350	200	65000	CC-8[1]	7½	Any
VII. Lamps for Ellipsoidal Reflector Spotlights—Tungsten-Halogen and Incandescent									
(A) Medium Prefocus Base; Light Center Length: 3 1/2 Inches									
250	T-12	250T12/8	—	2800	800	3750	C-13	6⅛	BU30
500	T-12	500T12/8	—	2850	800	9000	C-13D	6⅛	BU30
500	T-12	500T12/9	—	2950	200	11000	C-13D	6⅛	BU30
500	T-12	Q500T12CL	—	2950	2000	8000	CC-8	6 13/16	Any
500	T-12	Q500T12/4CL	—	3200	300	12200	CC-8	6 13/16	Any
750	T-12	750T12/9	—	3000	200	17000	C-13D	6⅛	BU30
750	T-12	Q750T12CL	—	2950	2000	14900	CC-8	6 13/16	Any
750	T-12	Q750T12/4CL	—	3200	300	19900	CC-8	6 13/16	Any
1000	T-12	1M/T12/2	—	3050	200	25000	C-13D	6⅛	BU30
1000	T-6	Q1000/4CL/P	EGJ[11]	3200	500	27500	CC-8[1]	6	Any
1000	T-6	Q1000/4/P (frosted)	EGK[11]	3200	500	26500	CC-8[1]	6	Any

See footnotes on page 8–72.

Continued on next page.

Fig. 8–90. *Continued*

Watts	Bulb	Manufacturer's Ordering Code	ANSI Code	Color Temperature (K)	Average Life (hours)	Approximate Initial Lumens	Filament Form	Maximum Over-All Length (inches)	Recommended Operating Position
VII. Lamps for Ellipsoidal Reflector Spotlights—Tungsten-Halogen and Incandescent (*Continued*)									
(B) Medium Bipost Base; Light Center Length: 4 Inches									
250	T-14	250T14/9	—	2800	800	3750	C-13[1]	6 7/16	BU30
500	T-14	500T14/7	—	2850	800	9000	C-13D[1]	6 7/16	BU30
500	T-14	500T14/8	—	2950	200	11000	C-13D[1]	6 7/16	BU30
750	T-14	750T14	—	3000	200	17000	C-13D[1]	6 7/16	BU30
(C) Mogul Bipost Base; Light Center Length: 6 1/2 Inches									
1000	T-24	1M/T24/5	—	3000	200	21500	C-13D	10	BU30
1500	T-24	1500T24/6	—	3000	200	33700	C-13D	10	BU30
2000	T-30	2M/T30/1	—	3050	200	48000	C-13D	10	BU30
2100	T-24	2100T24/9 (60V)	—	3100	50	54100	C-13D	10 1/2	BU30
(D) Mogul Bipost Base; Light Center Length: 7 1/2 Inches									
3000	T-32	3M/T32/2	—	3150	100	81000	C-13D[1]	11 1/2	BU30
(E) Mogul Bipost Base; Light Center Length: 8 1/2 Inches									
5000	T-32	5M/T32/1	—	3250	100	150000	C-13D[1]	13	BU30

Watts	Bulb	Manufacturer's Ordering Code	Color Temperature (K)	Average Life (hours)	Approximate Initial Lumens	Filament Form	Maximum Over-All Length (inches)	Recommended Operating Position
VIII. Low-Wattage Lamps for Spotlights and Projectors								
(A) Single-Contact Bayonet Candelabra Base; Light Center Length: 1 3/8 Inches								
50	S-11	BLR	See Fig. 8–108.					
50	T-8	BZW	2800	50	900	CC-2V	3 1/8	BD30
50	T-8	CAS/CAV	See Fig. 8–108.					
50	T-8	CHW/CHS (230V)	2550	50	650[19]	CC-13	3 1/8	BD30
75	T-8	CBJ/CBC	See Fig. 8–108.					
100	G16 1/2	100G16 1/2/29SC	2850	200	1660	CC-13	3	BDTH
100	S-11	BMD	See Fig. 8–108.					
100	S-11	BMY						
100	T-8	CDD						
100	T-8	CDS/CDX						
100	T-8	CJW/CJT (230V)	2850	50	1550[19]	CC-13	3 1/8	BD30
125	G-16 1/2	125G16 1/2/1SC	2700	600	1600	CC-13B	3	BDTH
150	G16 1/2	150G16 1/2SC**	2900	200	2550	2CC-8	3	BDTH

See footnote on page 8–72.

Continued on next page.

Fig. 8–90. *Continued*

Watts	Bulb	Manufacturer's Ordering Code	Color Temperature (K)	Average Life (hours)	Approximate Initial Lumens	Filament Form	Maximum Over-All Length (inches)	Recommended Operating Position
VIII. Low-Wattage Lamps for Spotlights and Projectors (*Continued*)								
150	T-8	CEW/CFC	See Fig. 8–108.					
150	T-8	CGE	3000	200	2500[19]	2CC-8	3⅝	BD30
150	T-8	CKK (230V)	2850	50	2700	CC-13	3⅝	BD30
200	T-8	CGW/CGT	3050	25	4200[19]	2CC-8	3⅝	BD30
200	T-10	CTM	3050	25	4200[19]	2CC-8	3 3/16	BD30
(B) Double-Contact Bayonet Candelabra Base; Light Center Length: 1 3/8 Inches								
50	S-11	BLX	See Fig. 8–108.					
50	T-8	CAC	2800	50	900	CC-2V	3⅛	BD30
50	T-8	CAJ	2800	50	850[19]	CC-2V	3⅛	BD30
50	T-8	CAX	See Fig. 8–108.					
50	T-8	CHY (230V)	2550	50	650	CC-13	3⅛	BD30
75	G-16½	75G16½DC	2700	200	800	C-5	3	BDTH
75	S-11	BNF	2900	25	1120	CC-2V	2⅜	BDTH
75	T-8	CBX/CBS	See Fig. 8–108.					
100	G-16½	100G16½/1DC[4]	2850	200	1660	CC-13	3	BDTH
100	G-16½	100G16½/29DC	2850	200	1660	CC-13	3	BDTH
100	S-11	BMG/BMH	3000	25	2000	CC-2V	2⅜	BDTH
100	T-8	CDJ	See Fig. 8–108.					
100	T-8	CDK/CEA	2950	50	2000[19]	CC-2V	3⅛	BD30
100	T-8	CEB	See Fig. 8–108.					
100	T-8	CJX (230V)	2850	50	1650	CC-13	3⅛	BD30
150	G-16½	150G16½DC**	2900	200	2550	2CC-8	3	BDTH
150	G-16½	150G16½/3DC**,[4]	2900	200	2550	2CC-8	3	BDTH
150	B-12	BEC	See Fig. 8–108.					
150	T-8	CGP/CGF	3050	25	3500[19]	2CC-8	3⅝	BD30
200	B-12	BDJ	3200	20	5000	2CC-8	2⅝	BD30
200	B-12	BEJ	See Fig. 8–108.					
200	T-8	CCM/CHD						
200	T-10	CWG**	3050	25	4600	2CC-8	3⅛	BD30

Watts	Bulb	Manufacturer's Ordering Code	ANSI Code	Color Temperature (K)	Average Life (hours)	Approximate Initial Lumens	Bulb Finish	Filament Form	Maximum Over-All Length (inches)	Recommended Operating Position
IX. Screw-Base Lamps for Floodlights and Miscellaneous Special Effects—Tungsten-Halogen and Incandescent										
(A) Mogul Screw Base; Light Center Length: 9 1/2 Inches										
750	PS-52	750/IF	—	See Fig. 8–79.						
1000	PS-52	1000/1/IF	—							
1000	PS-52	1000/IF	—							

See footnote on page 8–72.

Continued on next page.

Fig. 8-90. *Continued*

Watts	Bulb	Manufacturer's Ordering Code	ANSI Code	Color Temperature (K)	Average Life (hours)	Approximate Initial Lumens	Bulb Finish	Filament Form	Maximum Over-All Length (inches)	Recommended Operating Position
IX. Screw-Base Lamps for Floodlights and Miscellaneous Special Effects—Tungsten-Halogen and Incandescent (*Continued*)										
1000	PS-52	1M/PS52/77	ECX	3200	75	26500	Clear	C-7A[1]	13	Any
1000	PS-52	DKE	DKE	5000[17]	300	9400	I.F.	C-7A	13	Any
1500	PS-52	1500/IF	—	See Fig. 8-79.						
1500	PS-52	1500PS52/78	ECY	3200	100	41000	Clear	C-7A[1]	13	BDBU
1500	PS-52	1500PS52/79	EFY	3200	100	41000	I.F.	C-7A[1]	13	BDBU
2000	PS-52	2M/PS52/34**	—	3000	1000	44000	I.F.	C-7A	13	BDBU
2000	PS-52	2M/PS52/76	ECZ	3350	15[10]	66000	Clear	C-7A	13	BDBU
(B) Mogul Screw Base; Light Center Length: 7 Inches										
300	PS-35	300/IF	—	See Fig. 8-79.						
500	PS-40	500PS40/2IF	—	2950	1000	9900	I.F.	C-9[1]	9¾	Any
500	PS-40	500PS40/IF	—	See Fig. 8-79.						
1000	E-37	Q1000E37/CL	—	3000	3000	20900	Clear	CC-8	11 5/16	Any
1000	PS-40	ECV	ECV	3200	60	26500	I.F.	C-7A	9¾	Any
1000	PS-35	DXT	DXT	4800[17]	10	19200	I.F.	C-9	9⅜	Any
(C) Mogul Screw Base; Light Center Length: 3 7/16 Inches										
1000	T-20	DPW	DPW	3200	50	28000	Clear	C-13	9½	BD30
1000	T-20	DSB	DSB	3225	25	28500	Clear	C-13D	9½	BD30
1500	T-20	DTJ	DTJ	3225	25	42500	Clear	C-13D	9½	BD30
(D) Mogul Bipost Base; Light Center Length: 4 Inches										
2100	T-24	2100T24/8 (60V)	—	3100	50[10]	54100	Clear	C-13D	10½	BD30
4000	T-32	4M/T32 (110V)	—	3100	50[10]	114000	—	C-13D	14	BD30

* Ultraviolet-absorbing bulb.
** Hard glass bulb.
[1] Low noise construction.
[2] Very low noise construction.
[3] Beam spread to 50 per cent peak candlepower.
[4] Brass contacts soldered to base eyelets, to facilitate lamp removal from socket.
[5] Beam spread to 10 per cent peak candlepower.
[6] Candlepower average in central 5° cone for "spot" beam lamps; average in central 10° cone for "flood" beam lamps.
[7] Measured within field angle.
[8] Data are tentative.
[9] Lamp has six filament supports for increased resistance to shock and vibration. Life is dependent on service conditions.
[10] Tungsten powder cleaner in bulb. Useful lamp life and maintenance of output depend upon periodic removal from socket and scouring of bulb wall with tungsten powder to remove dark film that normally accumulates.
[11] Similar to ANSI-coded lamp indicated.
[12] Can replace lamp with ANSI code FEY, provided lamp seal temperature does not exceed 400°C limit under service conditions.
[13] Screw terminals.
[14] Burn between 45° base down and 45° base up for best performance.
[15] Apparent lighted length slightly longer than corresponding clear lamp.
[16] 2-prong base.
[17] Blue glass bulb. Apparent color temperature may vary among lamps.
[18] Precision coil location.
[19] Top end of bulb is opaque-coated.

Fig. 8–91. Reflectorized R and PAR Lamps for 120-Volt Circuits

Watts	Bulb	Description	Base (see Fig. 8–9)	Maximum Over-All Length (inches)	Approximate Beam Spread (degrees)[b]	Approximate Beam Lumens	Approximate Total Lumens	Approximate Average Candlepower in Central 10 Degree Cone[e]	Rated Average Life (hours)
R Lamps for Spotlighting and Floodlighting									
30	R-20	Flood	Med.	3 15/16	85	150	205	290	2000
50	R-20	Flood	Med.	3 15/16	90	430	435	550	2000
75	R-30	Spot	Med.	5 3/8	50	400	850	1730	2000
75	R-30	Flood	Med.	5 3/8	130	610	850	430	2000
150	R-40	Spot	Med.	6 1/2	37	835	1825	7000	2000
150	R-40	Flood	Med.	6 1/2	110	1550	1825	1200	2000
300	R-40	Spot	Med.	6 1/2	35	1800	3600	13500	2000
300	R-40	Flood	Med.	6 1/2	115	3000	3600	2500	2000
300	R-40[a]	Spot	Med.	6 7/8	35	1800	3600	13500	2000
300	R-40[a]	Flood	Med.	6 7/8	115	3000	3600	2500	2000
300	R-40[a]	Spot	Mog.	7 1/4	35	1800	3600	14000	2000
300	R-40[a]	Flood	Mog.	7 1/4	115	3000	3600	2500	2000
500	R-40[a]	Spot	Mog.	7 1/4	60	3300	6500	22000	2000
500	R-40[a]	Flood	Mog.	7 1/4	120	5700	6500	4750	2000
1000	R-60[a, f]	Spot	Mog.	10 1/8	32	11500	18300	135000	3000
1000	R-60[a, f]	Flood	Mog.	10 1/8	110	15500	18300	15500	3000
R Lamps for General Lighting[c]									
500	R-52	Wide Beam	Mog.	11 3/4	90	—	7750	—	2000
750	R-52	Wide Beam	Mog.	11 3/4	110	—	13000	—	2000
1000	R-52[a]	Wide Beam	Mog.	11 3/4	110	—	16300	—	2000
1000	RB-52	Wide Beam	Mog.	12 11/16	130	—	18900	—	2000
PAR Lamps for Spotlighting and Floodlighting[d]									
75	PAR-38	Spot	Med. Skt.	5 5/16	30 x 30	465	750	3800	2000
75	PAR-38	Flood	Med. Skt.	5 5/16	60 x 60	570	750	1500	2000
150	PAR-38[h]	Spot	Med. Skt.	5 5/16	30 x 30	1100	1735	11000	2000
150	PAR-38[h]	Flood	Med. Skt.	5 5/16	60 x 60	1350	1735	3700	2000
150	PAR-38	Spot	Med. Side Prong	4 5/16	30 x 30	1100	1735	11000	2000
150	PAR-38	Flood	Med. Side Prong	4 5/16	60 x 60	1350	1735	3700	2000
200	PAR-46	Narrow Spot	Med. Side Prong	4	17 x 23	1200	2325	32500[g]	2000
200	PAR-46	Med. Flood	Med. Side Prong	4	20 x 40	1300	2325	11200[g]	2000
250	PAR-38[f]	Spot	Med. Skt.	5 5/16	26 x 26	1600	3180	25000	4000
250	PAR-38[f]	Flood	Med. Skt.	5 5/16	60 x 60	2400	3180	6500	4000
300	PAR-56[h]	Narrow Spot	Mog. End Prong	5	15 x 20	1800	3750	70000[g]	2000
300	PAR-56[h]	Med. Flood	Mog. End Prong	5	20 x 35	2000	3750	24000[g]	2000
300	PAR-56[h]	Wide Flood	Mog. End Prong	5	30 x 60	2100	3750	10000[g]	2000
500	PAR-64	Narrow Spot	Extended Mog. End Prong	6	13 x 20	3000	6000	110000[g]	2000
500	PAR-64	Med. Flood	Extended Mog. End Prong	6	20 x 35	3400	6000	35000[g]	2000
500	PAR-64	Wide Flood	Extended Mog. End Prong	6	35 x 65	3500	6000	12000[g]	2000
500	PAR-56[f]	Narrow Spot	Mog. End. Prong	5	15 x 32	4900	7650	96000	4000
500	PAR-56[f]	Med. Flood	Mog. End. Prong	5	20 x 42	5700	7650	43000	4000
500	PAR-56[f]	Wide Flood	Mog. End. Prong	5	34 x 66	5725	7650	19000	4000
1000	PAR-64[f]	Narrow Spot	Extended Mog. End Prong	6	14 x 31	8500	19400	180000	4000
1000	PAR-64[f]	Med. Flood	Extended Mog. End Prong	6	22 x 45	10000	19400	80000	4000
1000	PAR-64[f]	Wide Flood	Extended Mog. End Prong	6	45 x 72	13500	19400	33000	4000

[a] Heat-Resistant glass bulb.
[b] To 10 per cent of maximum candlepower.
[c] Some of these types are also available for 230 to 260-volt circuits.
[d] All PAR lamps have bulbs of molded heat-resistant glass.
[e] Central cone defined as 5-degree cone for all spots and 10-degree cone for all floods.
[f] Halogen cycle lamps.
[g] Horizontal operation. May be slightly lower for vertical operation.
[h] Also available with an interference filter reflector.

Fig. 8–92. Incandescent Series Street Lighting Lamps (Mogul Screw Base)

Rated Initial Lumens	Amperes	Bulb	Burning Position*	Filament (see Fig. 8–7)	Maximum Over-All Length (inches)	Average Light Center Length (inches)	Approximate Initial Volts	Approximate Initial Watts	Approximate Mean Lumens Throughout Rated Life
A. 2000 Hours Life									
600	6.6	PS-25	Any	C-8	$7\frac{1}{8}$	$5\frac{3}{8}$	6.4	42	600
1000	6.6	PS-25	Any	C-8	$7\frac{1}{8}$	$5\frac{3}{8}$	9.6	63	1000
1000	7.5	PS-25	Any	C-8	$7\frac{1}{8}$	$5\frac{3}{8}$	8.3	62	1000
2500	6.6	PS-25	B.U.	C-2V	$7\frac{1}{8}$	$5\frac{3}{8}$	21.2	140	2500
2500	6.6	PS-35	Any	C-2V	$9\frac{3}{8}$	7	21.6	142	2500
2500	7.5	PS-35	Any	C-2V	$9\frac{3}{8}$	7	19.0	142	2500
4000	6.6	PS-35	Any	C-2V	$9\frac{3}{8}$	7	32.8	215	4000
4000	7.5	PS-35	Any	C-2V	$9\frac{3}{8}$	7	28.6	214	4000
4000	15	PS-35	B.U.	C-2V	$9\frac{3}{8}$	7	13.8	205	3900
4000	15	PS-35	B.D.	C-2V	$9\frac{3}{8}$	$6\frac{1}{4}$	13.8	205	3600
6000	6.6	PS-40	Any	C-2V	$9\frac{3}{4}$	7	47.8	316	6000
6000	20	PS-40	B.U.	C-2V	$9\frac{3}{4}$	7	15	300	5900
6000	20	PS-40	B.D.	C-2V	$9\frac{3}{4}$	$6\frac{1}{4}$	15.2	300	5400
10000	6.6	PS-40	Any	C-7A	$9\frac{3}{4}$	7	79.7	525	9200
10000	20	PS-40	B.U.	C-7	$9\frac{3}{4}$	7	24.8	496	9400
10000	20	PS-40	B.D.	C-7	$9\frac{3}{4}$	$6\frac{1}{4}$	24.8	496	8600
15000	20	PS-40	B.U.	C-7	$9\frac{3}{4}$	7	35.9	712	13800
15000	20	PS-40	B.D.	C-7	$9\frac{3}{4}$	$6\frac{1}{4}$	36.4	729	12900
B. 3000 Hours Life									
600	6.6	PS-25	Any	C-8	$7\frac{1}{8}$	$5\frac{3}{8}$	6.7	44	600
800	6.6	PS-25	Any	C-8	$7\frac{1}{8}$	$5\frac{3}{8}$	8.2	54	
1000	6.6	PS-25	Any	C-8	$7\frac{1}{8}$	$5\frac{3}{8}$	10.0	66	1000
1500	6.6	PS-25	Any	C-8	$7\frac{1}{8}$	$5\frac{3}{8}$	14.2	94	
2500	6.6	PS-25	B.U.	C-2V	$7\frac{1}{8}$	$5\frac{3}{8}$	22.2	146	2500
2500	6.6	PS-35	Any	C-2V	$9\frac{3}{8}$	7	22.4	147	2500
2500	7.5	PS-25	Any	C-2V	$9\frac{3}{8}$	7	19.6	147	2500
4000	6.6	PS-35	Any	C-2V	$9\frac{3}{8}$	7	34.2	225	4000
4000	15	PS-35	B.U.	C-2V	$9\frac{3}{8}$	7	14.6	219	3900
4000	15	PS-35	B.D.	C-2V	$9\frac{3}{8}$	$6\frac{1}{4}$	14.6	220	3600
6000	6.6	PS-40	Any	C-2V	$9\frac{3}{4}$	7	50.1	330	6000
6000	20	PS-40	B.U.	C-2V	$9\frac{3}{4}$	7	15.7	315	5900
6000	20	PS-40	B.D.	C-2V	$9\frac{3}{4}$	$6\frac{1}{4}$	16.0	319	5400
10000	6.6	PS-40	Any	C-7A	$9\frac{3}{4}$	7	86.2	570	9200
10000	20	PS-40	B.U.	C-7	$9\frac{3}{4}$	7	26.1	522	9400
10000	20	PS-40	B.D.	C-7	$9\frac{3}{4}$	$6\frac{1}{4}$	26.5	530	8600
15000	20	PS-40	B.U.	C-7	$9\frac{3}{4}$	7	37.5	750	13500
C. 6000 Hours Life									
1000	6.6	PS-25	Any	C-8	$7\frac{1}{8}$	$5\frac{3}{8}$	11.1	73	
1500	6.6	PS-25	Any	C-8	$7\frac{1}{8}$	$5\frac{3}{8}$	15.7	104	
2500	6.6	PS-25	B.U.	C-2V	$7\frac{1}{8}$	$5\frac{3}{8}$	24.2	159	2500
2500	6.6	PS-35	Any	C-2V	$9\frac{3}{8}$	7	24.6	162	2500
4000	6.6	PS-35	Any	C-2V	$9\frac{3}{8}$	7	37.0	244	4000
4000	15	PS-35	B.U.	C-2V	$9\frac{3}{8}$	7	15.7	235	4000
6000	6.6	PS-40	Any	C-2V	$9\frac{3}{4}$	7	54.3	358	6000
6000	20	PS-40	B.U.	C-2V	$9\frac{3}{4}$	7	17.3	346	6000
10000	6.6	PS-40	Any	C-7A	$9\frac{3}{4}$	7	92.0	608	10000

* B.U.—Base-up. B.D.—Base-down.

Fig. 8-93. Incandescent Multiple Street Lighting Lamps* (Any Burning Position)

Watts	Nominal Lumens	Clear Bulb	Screw Base	Maximum Over-All Length (inches)	Average Light Center Length (inches)	Filament Form (see Fig. 8-7)	Approximate Initial Lumens**	Lamp Lumen Depreciation† (per cent)
A. 1500 Hour Life								
85	1000	A-23	Med.	6	4 7/16	C-9	1140	—
175	2500	PS-25	Med.	6 15/16	5 1/4	C-9	2800	—
268	4000	PS-35	Mog.	9 3/8	7	C-9	4700	—
370	6000	PS-40	Mog.	9 3/4	7	C-9	6700	—
575	10000	PS-40	Mog.	9 3/4	7	C-7A	11000	—
B. 3000 Hour Life								
58	600	A-19	Med.	4 1/4	2 7/8	C-9	665	—
92	1000	A-23	Med.	6	4 7/16	C-9	1185	89
189	2500	PS-25	Mog.	7 1/8	5 3/8	C-9	2900	86.5
189	2500	PS-25	Med.	6 15/16	5 1/4	C-9	2900	82
295	4000	PS-35	Mog.	9 3/8	7	C-9	4840	82
405	6000	PS-40	Mog.	9 3/4	7	C-9	6850	80.5
620	10000	PS-40	Mog.	9 3/4	7	C-7A	11000	79.5
860	15000	PS-52	Mog.	13 1/16	9 1/2	C-7A	15700	79.5
C. 6000 Hour Life								
103	1000	A-23	Med.	6 1/16	4 3/8	C-9	1150	—
202	2500	PS-25	Med.	6 15/16	5 1/4	C-9	2800	—
202	2500	PS-25	Mog.	7 1/8	5 3/8	C-9	2800	—
327	4000	PS-35	Mog.	9 3/8	7	C-9	4850	—
448	6000	PS-40	Mog.	9 3/4	7	C-9	6820	—
690	10000	PS-40	Mog.	9 3/4	7	C-7A	11000	—

* 120-volt operation. For mercury lamps see Fig. 8-116.

** Values apply to vertical base-up burning only.

† Per cent initial light output at 70 per cent of rated life.

Fig. 8-94. Traffic Signal Lamps
(Medium screw base. Voltage specified by user.)

Nominal Traffic Signal Diameter (inches)	Wattage Group	Approximate Watts	Bulb	Light Center Length (inches)	Rated Average Life* (hours)	Average Initial Lumens
8	60	60	A-21	2 7/16	2000	675
8	60	63-64	A-21	2 7/16	3000	675
8	60	69-73	A-21	2 7/16	8000	675
8†	60	60	A-21	2 7/16	4000	585
8†	60	67	A-21	2 7/16	3000-4000	690
8†	60	67	A-21	2 7/16	6000-8000	600
8	100	100	A-21	2 7/16	2000	1275
8	100	100-107	A-21	2 7/16	3000	1275
8	100	114-116	A-21	2 7/16	8000	1275
12‡		150-165	P-25	3	2000-3000	2250
12‡		150-165	P-25	3	6000-8000	1950

* Under laboratory test conditions.

† Used mainly in Canada.

‡ Listed by lumen rating rather than watts.

Fig. 8–95. Lamps for Airport and Airway Lighting

A. Incandescent Filament Types

Typical Service	Watts, Lumens, Amps, or CP	Volts or Amps	Bulb[a]	Base	Filament	Light Center Length (inches)	Maximum Over-All Length (inches)	Rated[b] Average Lab Life (hours)	Approximate Initial Lumens	Approximate Initial Maximum CP	Beamspread to 10 Per Cent Maximum CP (degrees)		Lamp Specifications		Application Specifications		Users
											Hor.	Vert.	FIIN 6240-	MIL Standards	FAA	MIL Specifications	
Flush Appr,* Thres, Rnwy	100W	20A	PAR-36	Sc. Term.	C-6		2¾	100		90000	9	7		21999		L-26202	A, C
Appr	300W	20A	PAR-56[e]	Mog. End Prong	C-6[k]		5	500		28000	50	20	929–8003		C-5407-1	L-26764	A, C
Flush Appr, Thres	300W	20A	PAR-56[i]	Sc. Term.	C-6[k]		4½	500		250000	20	8	869–5077		L-1199, L-838	L-26202	A, C
Flush Appr, Thres, Rnwy	499W	20A	PAR-56	Sc. Term.	CC-6[k]		4½	500		355000	15	11		24488–5		L-26202	A, N, C
Appr	500W	20A	PAR-56[e]	Mog. End Prong	CC-6[k]		5	500		47000	60	20	869–5079	24348–3		L-26764	A, N, C
Thres, Rnwy	503W	20A	T-20[j]	Med. Bipost	C-13	2 21/32	6½	500	11300				914–2549	24321–4		L-22252 L-26990	A, N
Med Int Rnwy, Txwy	30W	6.6A	T-10	Med. Pref.	C-2V	1½	3 15/16	1000	450				196–4470	25012–1	L-802, L-820 L-822, L-829	L-7082	A, N, C
Rnwy Dist Mkr	45W	6.6A	PAR-38[e]	Med. Sc.	C-6		5 5/16	800		1340	60	60	889–1777	24479–1			A
Lo Int Rnwy, Txwy	45W	6.6A	PAR-56	Sc. Term.	C-8		4½	1000		65000	8	8	914–2546	24488–4		L-26202	A, N
Ctrln Rnwy, Txwy	45W	6.6A	T-2½	RSC	C-8		2 1/16	1500	630				889–1801		L-842		A, N, C
Ctrln Rnwy, Txwy	45W	6.6A	T-2½	Leads	C-8		1¾	1500	630						L-842		A, N, C
Med Int Rnwy, Txwy	45W	6.6A	T-10	Med. Pref.	C-2V	1½	3 15/16	1000	740				196–4472	25012–2	L-802 L-820, L-822	L-7082	A, N, C
Lo Int Rnwy,* Txwy, Obstr	1020lm	6.6A	A-21	Med. Pref.	C-8	2¾	5 5/16	2000	1000				156–1563		L-810	L-7830	A, N, C
Rnwy*	95W	6.6A	PAR-46	Mog. End Prong	CC-8		4⅜	500		90000	9	11			L-820		C
Rnwy*	100W	6.6A	PAR-56[e]	Mog. End Prong	C-6		5	75		100000	15	10			L-820		C
Ctrln Rnwy	100W	6.6A	T-3	RSC	C-8		2⅝	1000	1950								N
Ctrln Rnwy	65W	6.6A	T-2½	Leads	C-8		1¾	1000	1000						L-852		C
Ctrln Rnwy	200W	6.6A	T-4	D. C. Ring	CC-6	1 1/16	2¼	600	4400						L-850		C
Appr Gl Path	100W	6.6A	T-10	Med. Pref.	C-6	1¾	3 15/16	250	2140				688–6351				A, N
Ship Fldltg	200W	6.6A	PAR-46	Sc. Term.	CC-6		3¾	500	(Filament behind reflector focus)				752–2423	17994–1		F-23528	N
Flush Rnwy	200W	6.6A	PAR-56[i]	Sc. Term.	CC-6		4½	300		150000	11	9	889–1783	24488–1	L-838, L-845	L-26202	A, N, C
Appr	200W	6.6A	PAR-56[e]	Mog. End Prong	CC-6		5	300		13500	45	20	538–8853	24348–1		L-26764	A, N
Flush Rnwy	200W	6.6A	T-4	RSC	CC-8		2 9/16	500	4400				892–1580		L-843, L-850		A, N, C
Flush Rnwy	200W	6.6A	T-4	Leads	CC-8		2⅛	500	4400						L843, L-850		A, N, C
Rnwy	210W	6.6A	T-14	Med. Pref.	C-13	2 3/16	5¾	300	4500					25013–3	L-819	L-5904	C
Rnwy	204W	6.6A	T-14	Med. Pref.	C-13	2 3/16	5¾	500	4200				299–6753	25013–2		L-5904	A, N
Fldltg	500W	20A	PAR-56	Sc. Term.	CC-6[k]		4½	500	(Filament behind reflector focus)								N
VASI	200W	6.6A	PAR-64	Mog. End Prong	CC-6[k]		4½	2000	(Filament ahead of reflector focus)						E-2351		C
VASI	300W	6.6A	PAR-64	Mog. End Prong	CC-6[k]		4½	2000	(Filament ahead of reflector focus)				901–8612		E-2351	L-27504	A, N, C
Angle Appr	0.50A	5.95V	B-3½	Min. Sc. Flange	C-2R	¼	1¼	15					299–6767			L-19661	N
Air Traf Sig*	50CP	6V	RP-11	S. C. Pref.	C-2V	⅞	2¼	200					155–7901			L-6617	N
Air Traf Sig*	50CP	6–8V	RP-11	S. C. Pref.	C-6	⅞	2¼	200					018–6836			L-6617	N

Deck Edge	5.10A	11.75V	G-16½	D. C. Bay.	C-2V	1¼	3	100	100				019–3066	15585–3			N
Deck Guide	3.04A	12.5V	RP-11	D. C. Bay.	C-2V	1¼	2¼	300	50				155–7940	15564–6			N
Wheels Watch Sig	5.3A	26V	PAR-46	Sc. Term.	4CC-8[d]		3¾	50		130000	10	10	155–7780	25240			N
Rnwy, Txwy	15W	115–125V	T-7	Inter.	C-7A		2¼	1000	115				617–1720		L-840		C
Thres, Rnwy, Txwy	40W	120V	T-10	Med. Pref.	CC-2V	1½	3 15/16	1000	425				295–2862		L-802, 822		C
Rnwy Dist Mkr	75W	75V	PAR-38[c]	Med. Sc.	CC-6		5 5/16	1000		1700	60	60					N
Nt Vis Fldlt	100W	6.5V	PAR-36	Mog. End Prong	C-6		3⅜	100		95000	8	6	174–1991				N
Mirr Ldg Sys	100W	12V	PAR-56	Sc. Term.	C-6		4½	100		225000	9	5	635–9953				N
Rnwy, Thres	100W	90, 120V	T-10	Med. Pref.	CC-2V	1½	3 15/16	1000	1170							L-7802	N, C
Obstr*	100W	120V	A-21	Med. Sc.	C-9	2 7/16	4 7/16	2000	1260				143–7437		L-810		C
Obstr*	107W	120V	A-21	Med. Sc.	C-9	2 7/16	4⅜	3000	1260				617–1824		L-810		C
Obstr	116W	120V	A-21	Med. Sc.	C-9	2 7/16	4⅜	6000	1260				842–2887		L-810	L-7830	N, C
T'missomtr	120W	6V	PAR-64	Sc. Term.	C-6[d]		4	3000		180000	9	5	299–6769			T-4663	A, N, C
Airpt Ltg	399W	115V	PAR-56[c]	Mog. End Prong	CC-13		5	100		30000	60	20	299–4740	24348–4			N
Rotating Bm* Ceilometer	420W	12V	T-20[j]	Med. Bipost	C-8	2½	6½	1000	7560					28933–1			A, N, C
Flshg Obstr*	500W	120V	PS-40	Mog. Pref.	C-9	5 11/16	10 1/16	1000	9900				269–0948			L-6273	A, N, C
Airpt Bcn	500W	120V	T-20[j]	Med. Bipost	C-13B	3	7½	500	9250				244–5364				C
Flshg Obstr*	600W	240V	PS-40	Mog. Pref.	C-7A	5 11/16	10 1/16	1000	10600								C
Flshg Obstr*	620W	120V	PS-40	Mog. Pref.	C-7A	5 11/16	10 1/16	3000	11200				295–0901	27269	L-606	L-6273	A, N, C
Flshg Obstr	700W	120V	PS-40[j]	Mog. Pref.	C-7A	5 11/16	10 1/16	6000	11200					27269–1		L-6273	A, N, C
Airpt Bcn*	1000W	120V	T-20[j]	Mog. Bipost	C-13	4	9½	500	20500				250–6435	25015–1	L-291		C
Airpt Bcn	1200W	115V	T-20[j]	Mog. Bipost	CC-8	4	9½	750	27500				556–8012	25015–2		L-7158	A, N

B. Gaseous Discharge Types

Typical Service	Watts or Lumens	Volts or Amps	Bulb[a]	Base	Lamp Type	Light Center Length (inches)	Maximum Over-All Length (inches)	Approximate Initial Lumens	Rated[b] Average Lab Life (hours)	Lamp Specifications		Application Specifications		Users
										FIIN 6240-	MIL Standards	FAA	MIL Specifications	
Chnl Thres	6W	[e]	T-5 (gold)	Min. Bipin	Fluorescent		9	103	6000[f]	151–7631				N
Chnl Thres	6W	[e]	T-5 (green)	Min. Bipin	Fluorescent		9	368	6000[f]				L-7835	N
Clg Proj*	900W	[e]	T-2	Sleeve	Mercury		3¼[g]	65000	100					A, N, C
Seq Flshr Appr	60W-sec/fl	2000[e]	Helix	Giant 5-Pin	Xenon	4	6		500[h]	752–1938		L-1106, L-1107, L-847	L-26311	A, N, C
Seq. Flshr Appr	60W-sec/fl	2000[e]	PAR-56	Sc. Term.	Xenon		4½	13,700[l]	500[h]			L-847		

[a] Clear bulb unless otherwise specified.
[b] Under specified test conditions.
[c] Lens cover, prismatic.
[d] Shielded filament.
[e] Lamp to be used with auxiliary of proper design.
[f] Life based on three hours per start.
[g] Nominal length.
[h] Life based on two flashes per second.
[i] Stippled cover.
[j] Heat resistant glass.
[k] Tungsten-halogen lamp within an outer bulb.
[l] Effective candlepower.
* Lamps now in use but not recommended for new designs.

Users:
A—United States Air Force or Army Air Bases
N—United States Navy Air Bases
C—Civil Airports

Fig. 8–96. Lamps for Aircraft

Service	Design Volts	Trade Number	Amperes or Watts	Mean Spherical Candle-power	Bulb	Base	Filament	Light Center Length (inches)	Maximum Over-All Length (inches	Amperes[a] at Design Volts	Rated[b] Average Lab Life (hours)	Approx-imate Weight (pounds)	FIIN 6240	Military Standard
Landing	13.0	4509[c]	100 W		PAR-36	Sc. Term.	C-6		2¾	7.69	25	0.52	237–7867	MS25243-4509
	13.0	4537[c]	100 W		PAR-46	Sc. Term.	C-6		3⅛	7.69	25	0.92		
	13.0	4313[c, d]	250 W		PAR-36	Sc. Term.	C-6		2¾	19.23	25	0.52		
	13.0	4522[c]	250 W		PAR-46	Sc. Term.	C-2		3⅛	19.23	25	0.92	155–7920	MS25241-4522
	28.0	4553[c]	250 W		PAR-46	Sc. Term.	CC-8		3⅛	8.92	25	0.92	725–4683	MS25241-4553
	28.0	4559[c]	600 W		PAR-64	Sc. Term.	CC-8		3¾	21.41	25	2.00	283–9598	MS25242-4559
	28.0	4556[c, d]	1000 W		PAR-64	Sc. Term.	CC-8		3¾	35.70	25	2.00	725–2274	MS25242-4556
	28.0	4581[c, d]	450 W		PAR-46	Sc. Term.	CC-8		3⅛	16.08	10	0.92	557–3065	MS25241-4581
Taxiing	28.0	4551[c]	250 W		PAR-46	Sc. Term.	CC-6		3¾	8.92	25	0.92	583–3334	
	28.0	4570[c]	150 W		PAR-46	Sc. Term.	CC-6		3¾	5.36	300	0.92	132–5328	
Navigation	6.0	1680	4.1A	32	S-8	S.C. Bay.	C-6	1¼	2	4.1	300	0.021	870–7778	MS35478-1680
	6.0	4614[e]	100 W		PAR-36	Sc. Term.	C-6		2¾	16.67	300	0.52		
	6.2	1105	5.72A	50	R.P.-11	S.C. Bay.	C-2V	1¼	2¼	5.72	125	0.022		
	6.2	1163[e, f]	40 W		GG-12	S.C. Index	C-6		2½	6.45	300	0.025		
	6.2	600[e, f]	26 W		GG-10	S.C. Index	C-2R		2 3/16	4.19	300	0.019		MS25309-600
	12.8	93	1.04A	15	S-8	S.C. Bay.	C-2R	1⅛	2	1.04	500	0.021	014–2455	
	12.8	1777	1.52A	26[l]	S-8	S.C. Bay.	C-2R	1⅛	2	1.52	400	0.021		
	14.0	7512-12[e, f]	26 W		GG-10	S.C. Index	C-2R		2 3/16	2.0	300	0.019		
	14.0	4174-12[e, f]	40 W		GG-12	S.C. Index	C-6		2½	3.08	300	0.025		
	28	7512[e, f]	26 W		GG-10	S.C. Index	CC-6		2 3/16	.93	300	0.019	504–2090	MS25309-7512
	28	4174[e, f]	40 W		GG-12	S.C. Index	CC-6		2½	1.43	300	0.025	519–0854	MS24513-4174
	28	1683	1.02A	32	S-8	S.C. Bay.	CC-6	1¼	2	1.02	500	0.021	044–6914	MS35478-1683
	6.2	1687[e, f]	40 W		GG-12	S.C. Index	C-6		2½	6.45	150	0.025	870–0799	MS25338-1687
	14	7079B-12[e, f]	40 W		GG-12	S.C. Index	C-6		2½	3.08	150	0.025		
	28	7079[e, f]	40 W		GG-12	Sc. Term.	CC-6		2½	1.43	150	0.025	789–2260	MS25338-7079
	28	4594[c]	100 W		PAR-36	S.C. Term.	CC-6		2¾	3.57	300	0.52	295–2118	MS25243-4594
	28	4613[c, m]	100 W		PAR-36	S.C. Term.	CC-6		2¾	3.57	300	0.52		
Wing Inspection	28	4502[c]	50 W		PAR-36	S.C. Term.	CC-6		2¾	1.78	400	0.52	196–4518	MS25243-4502
Interior	12.8	1141	1.44A	21	S-8	S.C. Bay.	C-6	1¼	2	1.44	500	0.021	155–7799	MS25478-1141
	13.0	77	.37A	3	G-5	S.C. Bay.	C-2V	11/16	1⅜	.37	1000	0.012	155–7954	
	13.0	89	.58A	6	G-6	S.C. Bay.	C-2R	¾	1 7/16	.58	750	0.012	143–3159	MS15570-5
	13.0	1383[e]	20 W		R-12	S.C. Bay.	C-8		2⅝	1.54	300	0.026	539–9659	MS35480-1
	28	301	.17A	3	G-5	S.C. Bay.	C-2F	11/16	1⅜	.17	500	0.012	155–7947	MS25238-301
	28	303	.30A	6	G-6	S.C. Bay.	C-2F	¾	1 7/16	.30	500	0.012	155–7848	MS15570-6
	28	1309	.52A	15	B-6	S.C. Bay.	2C-2R	1 1/16	1¾	.52	300	0.016		
	28	305	.51A	15	S-8	S.C. Bay.	C-2V	1⅛	2	.51	300	0.021	155–7791	MS35478-305
	28	1691	.61A	15	S-8	S.C. Bay.	2C-2R	1⅛	2	.61	1000	0.021	295–2668	MS35478-1691
	28	305IF	.51A	15	S-8IF	S.C. Bay.	C-2V		2	.51	300	0.021	295–1680	
	28	1691IF	.61A	15	S-8IF	S.C. Bay.	2C-2R		2	.61	1000	0.021		
	28	307	.67A	21	S-8	S.C. Bay.	C-2V	1⅛	2	.67	300	0.021	155–7784	MS35478-307
	28	307IF[h]	.67A	21	S-8IF	S.C. Bay.	C-2V		2	.67	300	0.021	122–0264	
	28	307SB[i]	.67A		S-8	S.C. Bay.	C-2V	1⅛	2	.67	300	0.021	155–7787	MS35478-307SB
	28	1665	.80A	21	S-8	S.C. Bay.	C-2V	1⅛	2	.80	1000	0.021		
	28	1665IF	.80A	21	S-8IF	S.C. Bay.	C-2V		2	.80	1000	0.021		
	28	311	1.29A	50	S-11	S.C. Bay.	C-2V	1¼	2⅜	1.29	300	0.022	155–7924	MS25235-311
	28	311SB[i]	1.29A		S-11	S.C. Bay.	C-2V	1¼	2⅜	1.29	300	0.022	155–7966	MS25235-SB311

	28	1385[e]	20W		R-12	S.C. Bay.	CC-8		2⅝	.71	300	0.026	186–6276	MS35480-2
	28	8115-24[e, f]	20W		GG-12	S.C. Index	CC-6		2½	.71	300	0.025		
	[j]	5004[j]	4W		T-5	Min. Pinless			6			0.045		
	[j]	5008[j]	8W		T-5	Min. Pinless			12			0.075		
	[j]	5013[j]	13W		T-5	Min. Pinless			21			0.125		
	[j]	5108[j]	8W		T-5	Min. Bipin			12			0.075		
	[j]	5113[j]	13W		T-5	Min. Bipin			21			0.125		
Indicator	5.0	680	.06A	.032	T-1	Wire Term.	C-2R		¼	.06	Indef. Long	0.0004	878–1965	MS24367-680
	5.0	682	.06A	.029	T-1	Sub-mid. Flange	C-2R	3/16	⅜	.06	Indef. Long	0.0006	879–4980	MS24515-682
	5.0	683	.06A	.053	T-1	Wire Term.	C-2R		¼	.06	Indef. Long	0.0004		MS24367-683
	5.0	685	.06A	.048	T-1	Sub-mid. Flange	C-2R	3/16	⅜	.06	Indef. Long	0.0006	752–2581	MS24515-685
	5.0	715	.115A	.147	T-1	Wire Term.	C-2R		¼	.115	Indef. Long	0.0004		MS24367-715
	5.0	718	.115A	.132	T-1	Sub-mid. Flange	C-2R	3/16	⅜	.115	Indef. Long	0.0006		MS24515-718
OR	6	316	.7A	3.4[l]	T-3¼	Min. Bay.	C-2R	⅝	1 3/16	.7	500	0.006		MS25231-316
	6	328	.2A	.34[k]	T-1¾	Mid. Flange	C-6	⅜	⅝	.2	1000	0.0012	155–7857	MS25237-328
	13	77	.37A	3	G-5	S.C. Bay.	C-2V	11/16	1⅜	.37	1000	0.012	155–7954	
Instrument	13	89	.58A	6	G-6	S.C. Bay.	C-2R	¾	1 7/16	.58	750	0.012	143–3159	MS15570-5
	13	1816	.33A	3[l]	T-3¼	Min. Bay.	C-2V	⅝	1 3/16	.33	1000	0.006	155–7949	
	14	330	.08A	0.5[l]	T-1¾	Mid. Flange	C-2F	⅜	⅝	.08	750	0.0012	196–4491	
	28	301	.17A	3	G-5	S.C. Bay.	C-2F	11/16	1⅜	.17	500	0.012	155–7947	MS25238-301
	28	303	.30A	6	G-6	S.C. Bay.	C-2F	¾	1 7/16	.30	500	0.012	155–7848	MS15570-6
	28	313	.17A	3.5[l]	T-3¼	Min. Bay.	C-2F	⅝	1 3/16	.17	500	0.006	155–8714	MS25231-313
	28	1864	.17A	3.0[l]	T-3¼	Min. Bay.	C-2F	⅝	1 3/16	.17	1500	0.006	765–8443	
	28	327	.04A	.34	T-1¾	Mid. Flange	C-2F	⅜	⅝	.04	1000	0.0012	155–7836	MS25237-327
	28	1495	.30A	6	T-4½	Min. Bay.	C-2F	⅝	1⅜	.30	500	0.006	299–4742	MS25069-1495

[a] For purposes of wiring design maximum amperes at design volts may be approximately 15 per cent greater than design amperes.

[b] Under specified test conditions.

[c] Beam data for PAR lamps:

Lamp Number	Type Cover Glass	Approximate Initial Maximum Candlepower	Approximate Spread to 10 Per Cent Maximum (degrees)	
			Horizontal	Vertical
4313	Clear	150,000	15	7
4502	Lens	10,000	40	7
4509	Clear	110,000	11	6
4522	Clear	290,000	12	10
4537	Clear	200,000	11	6
4551	Lens	75,000	50	10
4553	Clear	300,000	11	12
4559	Clear	600,000	11	12
4556	Clear	800,000	12	13
4570	Lens	32,000	50	9
4581	Clear	400,000	13	14
4594	Clear	75,000	13	7
4613	Clear	75,000	13	7
4614	Clear	85,000	11	6

[d] Consult lamp manufacturer before using.

[e] Reflectorized bulb.

[f] Available from Grimes Manufacturing Co.

[h] Inside frosted.

[i] Silvered bowl.

[j] Fluorescent lamp to be used with auxiliary of proper design. Available in warm white (WW), cool white (CW), and red (R).

[k] At 5.0 volts.

[l] Approximated candlepower.

[m] Ruggedized version of #4594.

Fig. 8–97. Lumiline Lamps for 115–125 Volt Circuits (May Be Burned in Any Position)

Watts	Bulb and Other Description	Base (see Fig. 8–9)	Filament (see Fig. 8–7)	Rated Average Life (hours)	Average Over-All Length (inches)	Approximate Initial Lumens	Approximate Initial Lumens per Watt
30	T-8 clear	Disk	C-8	1500	17¾	255	8.5
30	T-8 inside frosted	Disk	C-8	1500	17¾	255	8.5
40	T-8 clear	Disk	C-8	1500	11¾	350	8.8
40	T-8 inside frosted	Disk	C-8	1500	11¾	350	8.8
60	T-8 clear	Disk	C-8	1500	17¾	540	9.0
60	T-8 inside frosted	Disk	C-8	1500	17¾	540	9.0

8–98. Examples of Variations in Range and Specifications of Subminiature Lamps

Lamp No.	Candlepower	Volts Rated/Design	Amperes Design	Watts	Battery	Bulb Shape	Base
A9A (NE-2E)	.003	105–125	.0007	1⁄12	—	T-2	Wire Term
B7A (NE-45)	.01	105–125	.002	¼	—	T-4½	Cand. S.C.
327	.34	28	.04	—	—	T-1¾	S.C. Mid. Fls.
14	.5	2.47	.3	—	2-D Cells	G-3½	Min. S.C.
PR2	.8	2.38	.5	—	2-D Cells	B-3½	S.C. Min. Fl.
44	.9	6.3	.25	—	—	T-3¼	Min. Bay.
1157	32/3	12.8/14.0	2.10/.59	—	—	S-8	D.C. Indexing
51	1	6/7.5	.22	—	—	G-3½	Min. Bay.
4535	95000 B.C.P.	6/6.4	—	30	—	PAR-46	S.C. Term.
222	5 B.C.P.	2.25	.25	—	2-AA Cells	TL-3	Min. S.C.
194	2.	14.	.27	—	—	T-3¼	Wedge
259	.65	6.3	.25	—	—	T-3¼	Wedge

Lamp No.	Length (inches)		Filament or Electrode	Approximate Life (hours)	Maximum Safe Temperature (°F)	Pre-focused
	Light Center	Over-All				
A9A (NE-2E)	—	¾	W-11	25,000+	300	No
B7A (NE-45)	—	1⅝	P-3	7,500+	300	No
327	⅜	⅝	C-2F	4,000	300	No
14	23⁄32	15⁄16	C-2R	15	350	No
PR2	¼	1¼	C-2R	15	—	Yes
44	25⁄32	1 3⁄16	C-2R	3,000	350	No
1157	1¼	2	C-6/C-6	1200/5000	350	No
51	½	15⁄16	C-2R	1,000	350	No
4535		3¾	C-6	100	350	Yes
222		59⁄64	C-6	5	350	No
194	.562	1 1⁄16	C-2F	2,500	600	No
259	11⁄16	1 1⁄16	C-2R	indf. long	600	No

Fig. 8–99. Specifications of Flasher-Filament Lamps

Lamp No.	Design Volts	Design Amps or Watts	Approximate Flashes Per Minute	Average Useful Life (hours)	Bulb	Bulb Diameter (inches)	Base	Maximum Over-All Length (inches)
405	6.5	0.50	40–160	500	G-4½	9⁄16	Min. S.C.	1 1⁄16
406	2.6	0.30	40–160	50	G-4½	9⁄16	Min. S.C.	1 1⁄16
407	4.9	0.30	40–160	50	G-4½	9⁄16	Min. S.C.	1 1⁄16
D26	120	7W	45	750	C-7½	31⁄32	Cand. S.C.	2⅛
D27	120	7W	45	750	C-9½	1 3⁄16	Inter. S.C.	3⅛

Fig. 8–100. Vibration Service Lamps for 120 and 130-Volt Circuits

Watts	Bulb and Other Description	Burning Position	Base (see Fig. 8–9)	Filament (see Fig. 8–7)	Rated Average Life (hours)	Maximum Over-All Length (inches)	Light Center Length (inches)	Approximate Initial Lumens	Approximate Initial Lumens Per Watt	Lamp Lumen Depreciation* (per cent)
25	A-19 inside frosted or clear	Any	Med.	C-9	1000	3 15/16	2½	260	10.4	—
50	A-19 inside frosted or clear	Not horizontal	Med.	C-9	1000	3 15/16	2½	550	11.0	72
75	A-21 inside frosted	Not horizontal	Med.	C-9	1000	5¼	3⅞	935	12.5	—
100	A-23 inside frosted or clear	Any	Med.	C-9	1000	5 15/16	4 7/16	1340	13.4	83
100	A-21 inside frosted or clear	Not horizontal	Med.	C-9	1000	5 5/16	3⅞	1400	14.0	—
150	PS-25 inside frosted	Any	Med.	C-9	1000	6 15/16	5¼	2390	15.9	—

* Per cent initial light output at 70 per cent of rated life.

Fig. 8–101. Lamps for 6-Volt Automotive Service

Lamp Number	Candlepower	Bulb (see Fig. 8–8)	Base (see Fig. 8–9)	Design Volts	Design Amperes	Filament (see Fig. 8–7)	Light Center Length (inches)	Maximum Over-All Length (inches)	Service*
51	1	G-3½	Min. Bay.	7.5	.22	C-2R	½	15/16	A, C
55	2	G-4½	Min. Bay.	7.0	.41	C-2R	9/16	1 1/16	A, C
147	2	T-3¼	Wedge	7.0	.43	C-2R	0.562	1 1/16	A, C, E, M
63	3	G-6	S.C. Bay.	7.0	.63	C-2R	¾	1 7/16	L, M, P, T
64	3	G-6	D.C. Bay.	7.0	.63	C-2R	¾	1 7/16	A, C, E
81	6	G-6	S.C. Bay.	6.5	1.02	C-2R	¾	1 7/16	E
82	6	G-6	D.C. Bay.	6.5	1.02	C-2R	¾	1 7/16	E
87	15	S-8	S.C. Bay.	6.8	1.91	C-2R	1⅛	2	E
88	15	S-8	D.C. Bay.	6.8	1.91	C-2R	1⅛	2	E
209	15	B-6	S.C. Bay.	6.5	1.78	C-6	1 1/16	1¾	E
210	15	B-6	D.C. Bay.	6.5	1.78	C-6	1 1/16	1¾	E
1129	21	S-8	S.C. Bay.	6.4	2.63	C-6	1¼	2	E, D, S
1133	32	RP-11	S.C. Bay.	6.2	3.91	C-2R	1¼	2¼	B
1154	21/3	S-8	D.C. Index	6.4/7.0	2.63/.75	C-6/C-6	1¼	2	D, L, S, T, P
1158	21/3	S-8	D.C. Bay.	6.4/7.0	2.63/.75	C-6/C-6	1¼	2	D, L, S, T, P

* Letter designations are defined as follows:

A—Instrument L—License
B—Back-up M—Marker, clearance, and identification
C—Indicator P—Parking
D—Turn signal S—Stop
E—Interior T—Tail

Fig. 8–102. Lamps for 12-Volt Automotive and Heavy Duty Truck Service

Lamp Number	Candle-Power	Bulb (see Fig. 8–8)	Base (see Fig. 8–9)	Design Volts	Design Amperes	Filament (see Fig. 8–7)	Light Center Length (inches)	Maximum Over-All Length (inches)	Service*
1445	0.7	G-3½	Min. Bay.	14.4	.135	C-2V	½	15/16	A, C, H
53	1	G-3½	Min. Bay.	14.4	.12	C-2V	½	15/16	A, C
182	1	G-3½	Min. Bay.	14.4	.18	C-2F	½	15/16	A, C, H
53X	1	G-3½	Min. Bay.	14.4	.12	C-2V	½	15/16	A, C
161	1	T-3¼	Wedge	14.0	.19	C-2F	½	1 1/16	A, C, H
184	1	T-3¼	Wedge	14.4	.24	C-2F	9/16	1 1/16	A, C, H
257	1.6	G-4½	Min. Bay.	14	.27	C-2R	9/16	1 1/16	C (Flasher)
57	2	G-4½	Min. Bay.	14.0	.24	C-2V	9/16	1 1/16	A, C
57X	2	G-4½	Min. Bay.	14.0	.24	C-2F	9/16	1 1/16	A, C, H
158	2	T-3¼	Wedge	14.0	.24	C-2V	½	1 1/16	A, C
194	2	T-3¼	Wedge	14.0	.27	C-2F	9/16	1 1/16	A, C, H
193	2	T-3¼	Wedge	14.0	.33	C-2F	9/16	1 1/12	A, C, H
1895	2	G-4½	Min. Bay.	14.0	.27	C-2F	9/16	1 1/16	A, C, H
293	2	G-4½	Min. Bay.	14.0	.33	C-2F	9/16	1 1/16	A, C, H
557	2.5	T-3¼	Wedge	14	.42	C-2R	0.562	1 1/16	C (Flasher)
168	3	T-3¼	Wedge	14	.36	C-2F	0.562	1 1/16	A, C, H, M
1247	3	G-6	S.C. Bay.	14.0	.43	2C-2R	13/16	1 7/16	A, L, M, T, H
97	4	G-6	S.C. Bay.	13.5	.69	C-2V	13/16	1 7/16	A, C, H, L, M, T, P
67	4	G-6	S.C. Bay.	13.5	.59	C-2R	13/16	1 7/16	A, M, T, L, P
97	4	G-6	S.C. Bay.	13.5	.69	C-2V	13/16	1 7/16	A, M, T, L, H
68	4	G-6	D.C. Bay.	13.5	.59	C-2R	13/16	1 7/16	A, C, M, E
96	4	G-6	D.C. Bay	13.5	.69	C-2V	13/16	1 7/16	A, C, M, E, H
1155	4	G-6	S.C. Bay.	13.5	.59	2C-2R	13/16	1 7/16	A, H, M, T, L
89	6	G-6	S.C. Bay.	13.0	.58	C-2R	¾	1 7/16	E, M
631	6	G-6	S.C. Bay.	14.0	.63	2C-2R	¾	1 7/16	A, E, M, T, H
98	6	G-6	S.C. Bay.	13.0	.62	C-2V	¾	1 7/16	E, M, H
90	6	G-6	D.C. Bay.	13.0	.58	C-2R	¾	1 7/16	A, E, M
99	6	G-6	D.C. Bay.	13.0	.62	C-2V	¾	1 7/16	A, E, M, H
105	12	B-6	S.C. Bay.	12.8	1.0	C-6	1 1/16	1¾	E, F, H
104	12	B-6	D.C. Bay.	12.8	1.0	C-6	1 1/16	1¾	E, F, H
93	15	S-8	S.C. Bay.	12.8	1.04	C-2R	1⅛	2	D, E
1093	15	S-8	S.C. Bay.	12.8	1.19	C-2R	1⅛	2	D, E, H
94	15	S-8	D.C. Bay.	12.8	1.04	C-2R	1⅛	2	E
1003	15	B-6	S.C. Bay.	12.8	.94	C-6	1 1/16	1¾	E
1004	15	B-6	D.C. Bay.	12.8	.94	C-6	1 1/16	1¾	E
1034	32/3	S-8	D.C. Index	12.8/14.0	1.80/.59	C-6/C-6	1¼	2	D, L, S, T, P
1073	32	S-8	S.C. Bay.	12.8	1.80	C-6	1¼	2	D, S, B
1141	21	S-8	S.C. Bay.	12.8	1.44	C-6	1¼	2	B, D, S
1159	21	S-8	S.C. Bay.	12.8	1.6	C-6	1¼	2	D, S, H
1142	21	S-8	D.C. Bay.	12.8	1.44	C-6	1¼	2	B, D, S
1176	21/6	S-8	D.C. Bay.	12.8/14.0	1.26/.57	C-6	1¼	2	E
1376	21/6	S-8	D.C. Bay.	12.8/14.0	1.6/.64	C-6/C-6	1¼	2	E, H
1156	32	S-8	S.C. Bay.	12.8	2.10	C-6	1¼	2	B, D, H, S
199	32	S-8	S.C. Bay.	12.8	2.25	C-6	1¼	2	B, D, S, H
1157	32/3	S-8	D.C. Index	12.8/14.0	2.10/.59	C-6/C-6	1¼	2	D, H, L, P, S, T
198	32/3	S-8	D.C. Index	12.8/14.0	2.25/.68	C-6/C-6	1¼	2	D, L, P, S, T, H
1293	50	RP-11	S.C. Bay.	12.5	3.00	C-2R	1¼	2¼	B, G
1195	50	RP-11	S.C. Bay.	12.5	3.00	C-2V	1¼	2¼	G
1295	50	S-8	S.C. Bay.	12.5	3.00	C-2R	1¼	2	B, G

* Letter designations are defined as follows:

A—Instrument
B—Back-up
C—Indicator or pilot
D—Turn-signal
E—Interior
G—Auxiliary service
H—Heavy duty
L—License
M—Marker, clearance, and identification
P—Parking
S—Stop
T—Tail

Fig. 8–103. Sealed Beam Lamps for Land Vehicles

Type of Service	Trade Number	Design Watts	Design Volts	Rated Average Laboratory Life (hours)	Filament	Bulb		Maximum Over-All Length (inches)	Base (see Fig. 8–9)		
						Type	Maximum Diameter		S.A.E. Type	Terminal	Contacts
Headlight	6006	{50 / 40	6.1 / 6.2	60‡ / 100‡	C-6 / C-6	PAR-56	7.031	5	H-3	Lugs	3
Fog*	4012	35	6.2	80‡	C-6	PAR-46	5.70	3¾	G-2	Screw	2
Spotlight	4535	30	6.4	35‡	C-6	PAR-46	5.70	3¾	G-2	Screw	2
Spotlight	4515	30	6.4	35‡	C-6	PAR-36	4.46	2¾	G-2	Screw	2
Farm Tractor	4013	25	6.4	100‡	C-6	PAR-46	5.70	3¾	G-2	Screw	2
Farm Tractor	4019	30	6.2	80‡	C-6	PAR-46	5.70	3¾	G-2	Screw	2
Fog*	4015	35	6.2	80‡	C-6	PAR-36	4.46	2¾	G-2	Screw	2
Cycle Headlight	4568	25	12.8	300	C-6	PAR-36	4.46	2¾	—	Lugs	Slip-on 2
Farm Tractor	4419	35	12.8	300†	C-6	PAR-46	5.70	3¾	G-2	Screw	2
Farm Tractor	4413	35	12.8	300†	C-6	PAR-46	5.70	3¾	G-2	Screw	2
Headlight**	4001	37.5	12.8	300†	C-6	PAR-46	5.70	4	H-2	Lugs	2
Headlight	4101	55	12.8	200	C-6	PAR-46	5.7	4	—	Lugs	2
Headlight**	4002	{50 / 37.5	12.8 / 12.8	500† / 300†	C-6 / C-6	PAR-46	5.70	4	H-3	Lugs	3
Headlight**	4000	{60 / 37.5	12.8 / 12.8	320† / 200†	C-6 / C-6	PAR-46	5.70	4	H-3	Lugs	3
Headlight**	4005	{50 / 37.5	12.8 / 12.8	500† / 300†	C-6 / C-6	PAR-46	5.70	4	H-3	Lugs	3
Headlight**	4040	{60 / 37.5	12.8 / 12.8	320† / 200†	C-6 / C-6	PAR-46	5.70	4	H-3	Lugs	3
Headlight**	4006	37.5	12.8	300†	C-6	PAR-46	5.70	4	H-2	Lugs	2
Headlight	6012	{50 / 40	12.8 / 12.8	200† / 500†	C-6 / C-6	PAR-56	7.031	5	H-3	Lugs	3
Headlight	6013	{50 / 40	12.8 / 12.8	200† / 500†	C-6 / C-6	PAR-56	7.031	5	H-3	Lugs	3
Headlight	6014	{60 / 50	12.8 / 12.8	200† / 320†	C-6 / C-6	PAR-56	7.031	5	H-3	Lugs	3
Headlight	6015	{60 / 50	12.8 / 12.8	200† / 320†	C-6 / C-6	PAR-56	7.031	5	H-3	Lugs	3
Headlight	6016	{60 / 50	12.8 / 12.8	300 / 500	C-6 / C-6	PAR-56	7.031	5	—	Lugs	3
Spotlight	4404	30	12.8	300	C-6	PAR-36	4.46	2¾	G-2	Screw	2
Spotlight	4405	30	12.8	100†	C-6	PAR-36	4.46	2¾	G-2	Screw	2
Cycle Headlight	4420	{30 / 30	12.8 / 12.8	300 / 300	C-6 / C-6	PAR-46	5.70	4	H-3	Lugs	3
Fog*	4412	35	12.8	100†	C-6	PAR-46	5.70	3⅞	G-2	Screw	2
Fog*	4415	35	12.8	100†	C-6	PAR-36	4.46	2¾	G-2	Screw	2
Aux. Headlight	Q4051	50	12.8	200†	C-6	PAR-46	5.70	3	—	Lugs	Slip-on 2

* Available with clear or amber lenses.

** Dual headlamp system (one unit of each type installed and used as a pair at each side on the front of vehicle).

† Rated life at 14.0 volts in these types. ‡ Rated life at 7.0 volts in these types.

Fig. 8–104. Lamps for Indicator, Radio, TV, Toy Train, and Other Services

Lamp Number	Bulb	Base	Design Volts	Amperes	Filament (see Fig. 8–7)	Light Center Length (inches)	Maximum Over-All Length (inches)	Rated Average Life (hours)
49	T-3¼	Min. Bay.	2.0	0.06	S-2	—	1 3/16	1000
253	TL-1¾	Midget Grooved	2.5	0.35	C-2R	—	11/16	10000
41	T-3¼	Min. Sc.	2.5	0.50	C-2R	31/32	1 3/16	3000
2158-2158D	T-1¾	Wire Term.	3.0	0.013–0.017	C-6	—	0.520	20000
1490	T-3¼	Min. Bay.	3.2	0.16	C-2R	25/32	1 3/16	3000
680	T-1	Wire Term.	5.0	0.06	C-2R	—	¼	60000
682	T-1	Sub-Midget Flanged	5.0	0.06	C-2R	3/16	⅜	60000
683	T-1	Wire Term.	5.0	0.06	C-2R	—	¼	50000
685	T-1	Sub-Midget Flanged	5.0	0.06	C-2R	3/16	⅜	50000
1850	T-3¼	Min. Bay.	5.0	0.09	C-2R	⅝	1 3/16	1500
715	T-1	Wire Term.	5.0	0.115	C-2R	—	¼	40000
718	T-1	Sub-Midget Flanged	5.0	0.115	C-2R	3/16	⅜	40000
328	T-1¾	S.C. Midget Flanged	6.0	0.20	C-2R	⅜	⅝	1000
1784-1784D	T-1¾	Wire Term.	6.0	0.20	C-2R	—	0.520	1000
1847	T-3¼	Min. Bay.	6.3	0.15	C-2R	25/32	1 3/16	5000†
1866	T-3¼	Min. Bay.	6.3	0.25	C-2R	25/32	1 3/16	5000†
159	T-3¼	Wedge	6.3	0.15	C-2R	½	1 1/16	5000†
259	T-3¼	Wedge	6.3	0.25	C-2R	11/16	1 1/16	5000†
755	T-3¼	Min. Bay.	6.3	0.15	C-2R	25/32	1 3/16	20000
40	T-3¼	Min. Sc.	6.3	0.15	C-2R	31/32	1 3/16	3000
47	T-3¼	Min. Bay.	6.3	0.15	C-2R	25/32	1 3/16	3000
44	T-3¼	Min. Bay.	6.3	0.25	C-2R	25/32	1 3/16	3000
46	T-3¼	Min. Sc.	6.3	0.25	C-2R	31/32	1 3/16	3000
380	T-1¾	S.C. Midget Flanged	6.3	0.04	C-2V	⅜	⅝	20000
2180-2180D	T-1¾	Wire Term.	6.3	0.04	C-2V	—	0.520	20000
381	T-1¾	S.C. Midget Flanged	6.3	0.20	C-2F	⅜	⅝	20000
2181-2181D	T-1¾	Wire Term.	6.3	0.20	C-2F	—	0.520	20000
455*	G-4½	Min. Bay.	6.5	0.50	C-2R	9/16	1 1/16	500
50	G-3½	Min. Sc.	7.5	0.22	C-2R	23/32	15/16	1000
344	T-1¾	S.C. Midget Flanged	10.0	0.0126–0.0154	C-2F	⅜	⅝	50000†
1869-1869D	T-1¾	Wire Term.	10.0	0.0126–0.0154	C-2F	—	0.520	50000†
428	G-4½	Min. Sc.	12.5	0.25	C-2V	23/32	1 1/16	250
382	T-1¾	S.C. Midget Flanged	14.0	0.08	C-2F	⅜	⅝	15000
2182–2182D	T-1¾	Wire Term.	14.0	0.08	C-2F	—	0.520	15000
430	G-4½	Min. Sc.	14.0	0.25	C-2V	23/32	1 1/16	250
1891	T-3¼	Min. Bay.	14.0	0.24	C-2F	⅝	1 3/16	500
1893	T-3¼	Min. Bay.	14.0	0.33	C-2F	⅝	1 3/16	7500
257*	G-4½	Min. Bay.	14.0	0.27	C-2R	9/16	1 1/16	500
256*	T-3¼	Min. Bay.	14.0	0.27	C-2R	⅝	1 3/16	500
1892	T-3¼	Min. Bay.	14.4	0.12	C-2F	⅝	1 3/16	1000
1458	G-5	Min. Bay.	20.0	0.25	C-2V	⅝	1 3/16	250
327	T-1¾	S.C. Midget Flanged	28.0	0.04	C-2F	⅜	⅝	4000
1762-1762D-1762U	T-1¾	Wire Term.	28.0	0.04	C-2F	—	0.520	4000
387	T-1¾	S.C. Midget Flanged	28.0	0.04	C-2F	⅜	⅝	7000
1829	T-3¼	Min. Bay.	28.0	0.07	C-2F	⅝	1 3/16	1000
757	T-3¼	Min. Bay.	28.0	0.08	C-2F	⅝	1 3/16	7500

* Flasher Lamp.
† At 6.6 volts.

Fig. 8–105. Incandescent Filament Photoflood or Superflood Lamps (115–120 Volts)

Lamp Code	Approximate Watts	Approximate Life At Rated Voltage (hours)	Approximate Initial Lumens	Center Beam Candlepower[a]	Beam Spread	Bulb	Maximum Over-All Length (inches)	Base[b]	Color Temperature (K)
DAN	200	4	—	10500	Medium	R-20	3 15/16	Med.	3400
BBA[h]	250	4	8800	—	—	A-21	4 15/16	Med.	3400
BCA[h]	250	4	5600	—	—	A-21	4 15/16	Med.	4800
ECA	250	20	6500	—	—	A-23	6 1/16	Med.	3200
BEP[h]	300	4	—	14000	Medium	R-30	5 1/4	Med.	3400
BFA	375	4	—	15000	Medium	R-40	6 5/8	Med.	3400
EBR	375	4	—	16000	Medium	R-30	5 1/4	Med.	3400
DMS	500	50	13200	—	—	T-20	5 1/2	Med.	3200
DMX	500	50	13200	—	—	T-20	5 3/4	Med. Pf.	3200
DXB	500	6	—	50000	Spot	R-40	6 5/8	Med.	3400
DXC	500	6	—	6500	Flood	R-40	6 5/8	Med.	3400
EBV[h]	500	8	17700	—	—	PS-25	6 15/16	Med.	3400
EBW	500	8	11000	—	—	PS-25	6 15/16	Med.	4800
ECT	500	60	13650	—	—	PS-25	6 15/16	Med.	3200
EAL	500	15	8200	—	Medium	R-40	6 5/8	Med.	3200
FAP	650	8	30000	—	Medium	PAR-36	2 7/16	FC	Daylight[c]
DXY	650	8	[f]	—	Medium	PAR-36	2 3/4	Med.	—
FDL	650	8	[f]	—	Medium	PAR-36	2 3/4	2-Lug.	—
FBN	650	8	[g]	—	Medium	PAR-36	2 3/4	Med.	—
DWA	650	8	[f]	25500	Medium	PAR-36	2 7/16	FC	3400
FBC dual beam	650	8	—	40000 20000	Spot Flood	PAR-36	2 1/2	3-FC	3400
FBF dual beam	650	8	—	40000 20000	Spot Flood	PAR-36	2 1/2	3-P	3400
FBK dual beam	650	8	—	17000 7000	Spot Flood	PAR-36	2 1/2	3-FC	Daylight[c]
FBL dual beam	650	8	—	17000 7000	Spot Flood	PAR-36	2 1/2	3-P	Daylight[c]
DXK	650[e]	16	—	30000	Medium	PAR-36	2 7/16	FC	3400
FAY	650[e]	16	—	35000	Medium	PAR-36	2 7/16	FC	Daylight[c]
FAZ	650[e]	16	—	9500	Wide	PAR-36	2 7/16	FC	3400
FBJ	650[e]	16	—	60000	Spot	PAR-36	2 7/16	FC	3400
FCW	650[e]	100	—	7000	Wide	PAR-36	2 7/16	FC	3200
FCX	650[e]	100	—	20000	Medium	PAR-36	2 7/16	FC	3200
FCY	650[e]	100	—	50000	Spot	PAR-36	2 7/16	FC	3200
FAD, FBX[d]	650[e]	100	16500	—	—	T-4	3 1/16	RSC	3200
DWY, FCA[d]	650[e]	25	20000	—	—	T-4	3 1/16	RSC	3400
DXN, FBZ[d]	1000[e]	50	33000	—	—	T-5	3 3/4	RSC	3400
DXW, FBY[d]	1000[e]	150	28000	—	—	T-5	3 3/4	RSC	3200
ECV	1000	60	26500	—	—	PS-40	9 3/4	Mog.	3200
DKE	1000	300	9400	—	—	PS-52	13 1/16	Mog.	5000
DXR	1000	10	32200	—	—	PS-35	9 3/8	Mog.	3400
DXT	1000	10	20500	—	—	PS-35	9 3/8	Mog.	4800
EKV	1100	25	39000	—	—	T-5	4 11/16	RSC	3400

a Mean candlepower in 10-degree zone.
b Bases, Ferrule Contact, 3 Ferrule Contact, 3 Prong, Recessed Single Contact, Medium Screw, and Mogul Screw.
c Daylight—balanced for daylight type color film, new Kodacolor-X, Polacolor, black and white.
d Frosted bulb version.
e Iodine cycle source.
f Guide-number 44 with ASA film speed 40.
g Guide-number 33 with ASA film speed 40.
h Watts, life and lumens at 118 volts.

Fig. 8–106. Typical Photoflash Lamps*

Class	Designations	Approximate Time to Peak†	Approximate Lumen Seconds (or Beam Candle-power Seconds)	Approximate Peak Lumens‡ (or Peak Candle-power)	Bulb (or Shape)	Maximum Over-All Length (inches)	Base
Clear Lacquer-Coated for Black and White Film (Approximate Color Temperature—3800 K)							
F	SM	5	4900	830	B-11	2⅝	S.C. Bay.
MF	AG-1	12	7200	450	T-3¾	1 5/16	Glass Groove
Spec.	M2	13–15	7500	800	T-6½	1 13/16	Min. Bay. Pinless
M	M3	17	16000	1000	T-6½	1 13/16	Min. Bay. Pinless
M	5, Press 25	20	21000	1600	B-11, B-12	2⅝	S.C. Bay.
M	11, Press 40	20	33000	1900	A-15	4	Med.
M	2, 22	20	70000	4200	A-19	4¾	Med.
S	3, 50	30	100000	6000	A-23, A-21	5 15/16	Med.
FP	6, FP26	—	18000	630	B-11, B-12	2⅝	S.C. Bay.
FP	31	—	81000	1500	A-21	5⅜	Med.
Blue Lacquer-Coated for Daylight Color Film (Approximate Color Temperature—5500 to 6000 K)							
MF	AG-1B	12	5300	250	T-3¾	1 5/16	Glass Groove
MF	AG-3B	12	7200	430	T-3¾	1 5/16	Glass Groove
Spec.	M2B	13–15	4400	420	T-6½	1 13/16	Min. Bay. Pinless
M	M3B	17	10000	550	T-6½	1¾	Min. Bay. Pinless
M	5B, 25B	20	10000	760	B-11, B-12	2⅝	S.C. Bay.
M	11B	20	16200	1100	A-15	4	Med.
M	2B, 22B	20	33500	2000	A-19	4¾	Med.
S	3B, 50B	30	53000	2900	A-23, A-21	5 15/16	Med.
FP	6B, FP26B	—	9800	270	B-11, B-12	2⅝	S.C. Bay.
Multiple Flash (5500 K)							
MF	Flashcube	13	(2000)	(130,000)	(cube)	1 13/32	Plastic
	Supercube						
—	Hi-Power	13	(4800)	—	(cube)	1 13/32	Plastic
MF	Magicube	7	(2000)	(150,000)	(cube)	1 19/32	Plastic
						1⅝	W/Index Post

* Voltage ranges usually 3, 3–45, or 3–125 (see manufacturers' catologs). † Milliseconds. ‡ Thousands of lumens.

Fig. 8–107. Typical White Diffusing Bulb Incandescent Filament and Fluorescent Enlarger Lamps (115 to 120 Volts)

Manufacturers' Designations	Rated Watts	Approximate Lumens (at 115V)	Rated Life (hours at 115V)	Bulb	Maximum Over-All Length (inches)	Light Center Length (inches)	Base
PH/111	75	1120	25	S-11	2⅜	1⅜	S.C. Bay
PH/111A	75	1120**	25**	S-11	2⅜	1⅜	S.C. Bay
PH/211	75	1200	100	A-21	4 15/16	3⅜	Med.
PH/212	150	2850	100	A-21	4 15/16	3⅜	Med.
PH/213	250	7500	3	A-21	4 15/16	3⅜	Med.
PH/300	150	3300	100	PS-30	8 3/16	6	Med.
PH/301	300	6650	100	PS-30	8 3/16	6	Med.
PH/302	500	11000	100	PS-30	8 3/16	6	Med.
FC8T9/CW*	22	1050	7500	T-9			4 Pin Circline

* Fluorescent, use with auxiliary ballast—life rated at 3 hours per start. ** At 125 volts.

Fig. 8–108. Projection Lamps, 21.5-, 30-, 115-, 120-, and 125-Volt
(Bulb Glass Heat-Resistant and Burning Position Base Down, Except as Noted)

ANSI Code		Watts	Bulb	Base (see Fig. 8–9)	Light Center Length (inches)	Maximum Over-All Length (inches)	Approximate Initial Lumens[a]	Average Life (hours)	Filament (see Fig. 8–7)	Footnotes
Clear	Opaque									
BKR		30	S-11	Cand.	1 5/8	2 1/4	400	50	CC-2V	b
BKV		30	S-11	S.C. Bay.	1 3/8	2 3/8	400	50	CC-2V	b
BLC		30	S-11	D.C. Bay.	1 3/8	2 3/8	400	50	CC-2V	b
BVR		30	T-7	D.C. Bay.	1 3/8	2 5/8	465	25	CC-2V	d
	BVB	30	T-7	S.C. Bay.	1 3/8	2 5/8	435	25	CC-2V	b
DKN		35	T-12	4-Pin	1 9/16	3 1/2	—	500	C-6	c
BLR		50	S-11	S.C. Bay.	1 3/8	2 3/8	780	50	CC-2V	b, c
BVJ		50	T-7	D.C. Bay.	1 3/8	2 3/8	900	25	CC-2V	b, c
BLX		50	S-11	D.C. Bay.	1 3/8	2 3/8	780	50	CC-2V	b, c
CAS	CAV	50	T-8	S.C. Bay.	1 3/8	3 1/8	800	50	CC-13	b
CXL		50	T-9 1/2	S.C. Pref.	1 13/16	4	—	25	C-6	d, g
CAX	CAW	50	T-8	D.C. Bay.	1 3/8	3 1/8	790	50	CC-13	b
CBC	CBJ	75	T-8	S.C. Bay.	1 3/8	3 1/8	1300	50	CC-13	b
CBS	CBX	75	T-8	D.C. Bay.	1 3/8	3 1/8	1300	50	CC-13	b
CAD		75	T-10	4-Pin	1 5/16	3 1/8	—	50	2CC-8	d, f
DFE		80	T-12	4-Pin	1 9/16	3 3/16	—	15	CC-8	h
DFZ		80	T-12	4-Pin	1 9/16	3 1/4	—	10	CC-8	h
DGB		80	T-12	4-Pin	1 9/16	3 9/16	—	15	CC-6	d
DYT		80	T-3	2-Pin	1 1/16	1 27/32	—	25	CC-6	l
ELE		80	T-3	2-Pin	5/8	1 1/2	—	20	CC-6	l
DMD		80	TB-12	4-Pin	1 9/16	3 9/16	—	15	CC-6	b, g
EKP		80	MA-16	2-Pin	—	—	—	25	CC-6	b, k
ENA		80	MA-16	2-Pin	—	—	—	25	CC-6	b, k
BMD		100	S-11	S.C. Bay.	1 3/8	2 3/8	2000	25	CC-2V	b, c
BMY		100	S-11	S.C. Bay.	1 3/8	2 3/8	1860	50	CC-13	b, c
CDD		100	T-8	S.C. Bay.	1 3/8	3 1/8	2000	50	CC-2V	b
CDX	CDS	100	T-8	S.C. Bay.	1 3/8	3 1/8	1950	50	CC-13	b
CEB	CDK	100	T-8	D.C. Bay.	1 3/8	3 1/8	1950	50	CC-13	b
	CEM	100	T-8	S.C. Bay.	1 3/8	3 1/8	1950	100	CC-2V	b
	CEL	100	T-8	D.C. Bay.	1 3/8	3 1/8	1950	100	CC-2V	b
DJR		100	T-12	4-Pin	1 5/16	3 3/16	—	50	C-6	c, f
FCR		100	T-4	2-Pin	1 1/4	1 3/4	2900	50	C-6	l
FDT		100	T-4	2-Pin PF	1 5/6	2 1/4	2900	50	C-6	l
FDX		100	T-4	2-Pin Ceramic	1 3/16	1 3/4	2900	50	C-6	l
CED		100	T-8S	S.C. Bay.	1 3/8	2 3/8	1950	50	CC-13	b
CDJ		100	T-8	D.C. Bay.	1 3/8	3 1/8	2000	50	CC-2V	b
EDR		100	T-8 1/2	Med.	3	5 1/2	1950	50	CC-13	b
CLD		100	T-8 1/2	Med. Pref.	2 3/16	5 3/4	1950	50	CC-13	b
CEL		120	T-8	D.C. Bay.	1 3/8	3 1/8	1950	200	2CC-8	d
CEM		120	T-8	S.C. Bay.	1 3/8	3 1/8	1950	200	2CC-8	d
EKN		120	T-3	Special 2 Pin	—	—	—	25	—	l
CFC	CEW	150	T-8	S.C. Bay.	1 3/8	3 5/8	3500	25	2CC-8	b
CFK	CGP	150	T-8	D.C. Bay.	1 3/8	3 5/8	3500	25	2CC-8	b
	CHK	150	T-8	S.C. Bay.	1 3/8	3 5/8	2295	500	2CC-8	b
	CSH	150	T-10	4-Pin	1 9/16	3 9/16	2295	500	2CC-8	b
	CTD	150	T-10	4-Pin	1 5/16	3 1/8	1950	25	CC-13	b
	CXG	150	T-10	4-Pin	1 5/16	3 1/8	3200	15	CC-6	b, g
	BEH	150	T-10	4-Pin	1 5/16	3 1/8	3500	15	2CC-8	b
	CAR	150	T-10	4-Pin	1 5/16	3 1/8	3300	15	2CC-8	b, f
	DAC	150	T-10	4-Pin	1 9/16	4	2600	500	C-13	a, b
DHX		150	TB-14	4-Pin	1 9/16	3 7/16	—	15	CC-8	b, h

See footnotes on page 8–91.

Continued on next page.

Fig. 8–108. *Continued*

ANSI Code		Watts	Bulb	Base (see Fig. 8–9)	Light Center Length (inches)	Maximum Over-All Length (inches)	Approximate Initial Lumens[a]	Average Life (hours)	Average (see Fig. 8–7)	Footnotes
Clear	Opaque									
EJA		150	MA-16	2-Pin	—	—	—	40	CC-6	b, k
EJM		150	MA-16	2-Pin	—	—	—	40	CC-6	b, k
EJV		150	MA-16	2-Pin	—	—	—	40	CC-6	b, k
BEC		150	B-12	D.C. Bay.	1 3/8	2 5/8	3500	25	2CC-8	
	DCF	150	T-12	4-Pin	1 9/16	3 9/16	—	10	CC-6	b, g
DCH	DFA	150	T-12	4-Pin	1 9/16	3 9/16	—	15	CC-6	b, g
DFC		150	T-12	4-Pin	1 9/16	3 3/16	—	15	CC-8	b, g
DJL		150	T-12	4-Pin	1 9/16	3 9/16	—	15	CC-8B	b, g
DJA		150	T-12	4-Pin	1 9/16	3 9/16	—	15	CC-6	b, g
BEH		150	T-10	4-Pin	1 5/16	3 1/8	3500	15	2CC-8	
DCW		150	T-12	4-Pin	1 5/16	3 3/16	—	50	C-8	c
DNE		150	T-3 1/4	2-Pin	5/8	2	—	12	—	l
DNF		150	T-3	2-Pin	5/8	2	—	25	—	l
DZE		150	T-4	2-Pin PF	1 5/16	2 15/16	—	100	—	l
EJN		150	T-3	Special 2-Pin	—	—	—	40	CC-6	l
ENB/EKE		150	T-3	Special 2-Pin	—	—	—	100	CC-6	l
FCS		150	T-4	2-Pin	1 1/4	2	4500	50	C-6	c, l
FDS		150	T-4	2-Pin PF	1 5/16	2 1/4	4500	50	C-6	l
FDV		150	T-4	2-Pin	1 1/4	2	—	100	C-6	l
FDW		150	T-5	2-Pin Ceramic	1 1/4	2	4500	50	C-6	l
	DFG	150	T-12	4-Pin	1 9/16	3 9/16	—	15	CC-6B	b, g
DCL		150	T-12	4-Pin	1 9/16	3 9/16	—	15	CC-6	b, h
DCR		150	T-12	4-Pin	1 9/16	3 9/16	—	15	CC-6	b, g
	DFF	150	T-12	4-Pin	1 5/16	3 1/32	—	500	2CC-8	b, f
DKR		150	T-14	4-Pin	1 9/16	3 3/8	—	15	CC-6	b, h
DLG		150	T-14	4-Pin	1 9/16	3 3/8	—	15	CC-8	b, c, g
DLS		150	T-14	4-Pin	1 9/16	3 3/8	—	15	CC-8	b, c, h
DCA		150	T-12	4-Pin	1 9/16	3 9/16	—	15	CC-6	b, c, g
DEF		150	T-12	4-Pin	1 9/16	3 9/16	—	15	CC-6	b, c, h
DFN		150	TB-12	4-Pin	1 9/16	3 3/16	—	15	CC-8	b, c, g
DFP		150	TB-12	4-Pin	1 9/16	3 9/16	—	15	CC-6	b, c, g
BFB		170	T-10	S.C. Bay.	1 3/8	3 3/16	—	25	2CC-6	
CGT	CGW	200	T-8	S.C. Bay.	1 3/8	3 5/8	4700	25	2CC-8	b
CHD	CCM	200	T-8	D.C. Bay.	1 3/8	3 5/8	4700	25	2CC-8	b
CHG	CET	200	T-8	D.C. Pref.	1 1/4	3 5/8	4700	25	2CC-8	b
DSW		200	T-14	2-Pin	1 9/16	3 7/16	—	25	CC-8	c
EJL		200	T-3	Special 2-Pin	—	—	—	25	CC-6	l
CVS	CVX	200	T-10	Med. Pref.	2 3/16	5 3/4	4250	50	CC-13	b
BEJ		200	B-12	D.C. Bay.	1 3/8	2 5/8	4700	25	2CC-8	b
DMH		250	T-14	4-Pin	1 9/16	3 7/16	—	15	CC-6	b, h
	DLH	250	T-14	4-Pin	1 9/16	3 3/8	—	15	CC-6	b, g, h
	DLR	250	T-14	4-Pin	1 9/16	3 1/2	—	10	CC-6	b, g, h
DKP		250	T-14	4-Pin	1 9/16	3 15/16	—	25	CC-6	d
EKK		250	T-4	2-Pin Min	1 3/4	3	5500	1000	2CC-8	l
EKS		250	T-3 1/2	2-Pin	—	—	—	25	CC-6	l
	DKM	250	T-14	4-Pin	1 9/16	3 1/2	--	25	CC-6	b, g, h
	CNP	300	T-8 1/2	S.C. Bay.	1 3/8	4 1/16	6600	25	CC-13	
CLG	CLS	300	T-8 1/2	S.C. Bay.	1 3/8	4 1/8	7850	25	C-13	b
CMB	CLX	300	T-8 1/2	D.C. Bay.	1 3/8	4 1/8	7850	25	C-13	b
CMT	CMV	300	T-8 1/2	S.C. Bay.	1 3/8	4 1/8	7250	25	C-13D	b
CNC	CMS	300	T-8 1/2	D.C. Bay.	1 3/8	4 1/8	7250	25	C-13D	b
	CWD	300	T-10	4-Pin	1 9/16	4	7750	25	C-13	b
	CRA	300	T-10	4-Pin	1 3/4	4	7750	25	C-13	b, c, e, f
	CXK	300	T-10	Med. Pref.	2 3/16	5 3/4	7200	25	C-13	b
	CXY	300	T-10	Med. Pref.	2 3/16	5 3/4	7300	25	C-13D	b
	DGA	300	T-10	4-Pin	1 3/4	4	7750	25	C-13	b, c

Continued on next page.

See footnotes on page 8–91.

Fig. 8–108. *Continued*

ANSI Code		Watts	Bulb	Base (see Fig. 8–9)	Light Center Length (inches)	Maximum Over-all Length (inches)	Approximate Initial Lumens[a]	Average Life (hours)	Filament (see Fig. 8–7)	Footnotes
Clear	Opaque									
	CYC	300	T-10	S.C. Bay.	1 3/8	3 3/16	7750	25	C-13	b
	BEN	300	T-10	4-Pin	1 3/4	4	6600	25	CC-13	b
	BEL	300	T-10	S.C. Bay.	1 3/8	3 3/16	6600	25	CC-13	b
BCL		300	T-6	4-Pin	1 9/16	3 3/16	—	1000	C-13D	f
CNP		300	T-8 1/2	S.C. Bay.	1 3/8	4 1/8	6600	25	CC-13	d
CXH		300	T-10	4-Pin	1 9/16	4	—	25	CC-13	d, f
CYM/CYF		300	T-10	S.C. Bay.	1 3/8	3 3/16	—	300	C-13	d
		300	3 RFL	Special 2-Pin	—	—	—	25	—	
		300	2 1/2 RFL	Special 2-Pin	—	—	—	25	—	
	CSF	300	T-10	4-Pin	1 9/16	4	6600	25	CC-13	b
	CAL	300	T-10	4-Pin	1 9/16	4	6500	25	C-13	b, f
	CXP	300	T-10	4-Pin	1 9/16	4	7200	25	C-13D	
	CYE	300	T-10	S.C. Bay.	1 3/8	4 1/8	4600	300	C-13D	b
	DAF	300	T-10	4-Pin	1 5/16	4	5850	300	C-13	
	BEM	400	T-8 1/2	S.C. Bay.	1 3/8	4 1/8	9800	25	C-13D	
	—	400	T-10	Med. Pref.	2 3/16	5 3/4	—	200	C-13D	
CYK		400	T-10HG	4-Pin	1 9/16	4 3/32	—	200	C-13D	d
FEM		400	T-6	Special 2-Pin	1 7/16	3	—	50	C-13D	l
FAL		420	T-4	RSC	—	2 5/8	11000	75	CC-8	l
DWW		460	T-5	RSC	—	2 5/8	15000	18	CC-8	l
	CBA	500	T-6	4-Pin	1 3/4	3 5/8	—	50	C-13D	b, f
DMK		500	T-14	4-Pin	1 5/8	3 5/8	—	75	C-13D	f
EGH		500	T-6	4-Pin	1 9/16	3 1/4	14000	50	C-13D	c
EHA		500	T-5 1/2	2-Pin Pref.	1 7/16	3 1/4	—	50	C-13D	c, f
	CZS	500	T-10	Med.	3	5 3/4	13100	25	C-13D	
	CZX	500	T-10	Med. Pref.	2 3/16	5 3/4	13100	25	C-13D	
	DAR	500	T-10	S1-Ring	2 5/16	5 1/2	13100	25	C-13D	
	DAG	500	T-10	L1-Ring	2 5/16	5 1/2	13300	25	C-13D	
	DAK	500	T-10	4-Pin	1 9/16	4	13400	25	C-13D	b
	CZB	500	T-10	4-Pin	1 9/16	4	13800	10	C-13D	f
	CZA	500	T-10	4-Pin	1 9/16	4	11900	25	C-13D	b, f
	DBJ	500	T-12	4-Pin	1 3/4	3 7/8	12600	25	C-13D	b, j
	DAY	500	T-10	4-Pin	1 9/16	4 3/32	13200	30	C-13D	
	EDK	500	T-10	S.C. Bay.	1 5/8	4 1/8	13200	25	C-13D	b
BCK		500	T-6	4-Pin	1 9/16	3 1/4	—	50	C-13D	f
CBF/DEL		500	T-6	4-Pin	1 3/4	3 5/8	—	10	C-13D	f
DBT		500	T-10HG	Med. Pref.	2 3/16	5 3/4	—	25	C-13D	d
DFX		500	T-12	4-Pin	1 3/4	3 3/4	—	25	C-13D	c, f
DHN		500	T-12	4-Pin	1 3/4	3 3/4	—	25	C-13D	c, f
DJH		500	T-12	4-Pin	1 3/4	3 3/4	10000	200	C-13D	c
DZG		500	T-6	D.C.Med. Ring	3 1/2	5 1/2	—	100	C-13D	
FBD		500	T-4	2-Pin	1 3/4	3	—	50	2CC-8	l
	DET	500	T-10	Med. Pref.	2 3/16	5 2/3	12300	25	C-13D	b
	DBS	500	T-10	4-Pin	1 9/16	4	11700	25	C-13D	
	CZG	500	T-10	4-Pin	1 9/16	4	14500	25	C-13D	b, f
	DCC	500	T-12	Med. Pref.	2 3/16	5 3/4	13800	25	C-13D	
	DFR	500	T-12	4-Pin	1 3/4	3 7/8	—	25	C-13D	c, e, f, j
	DAH	500	T-12	4-Pin	1 3/4	3 3/4	10000	200	C-13D	e, f
	DBN	500	T-12	4-Pin	1 3/4	4	11700	25	C-13D	b
	DFW	500	T-12	4-Pin	1 3/4	4	14500	25	C-13D	b, c, e, f
	DEK	500	T-12	4-Pin	1 3/4	3 1/2	—	25	C-13D	e, f
	DER	500	T-12	4-Pin	1 3/4	3 3/4	—	25	C-13D	b, e, f
	DGF	500	T-12	4-Pin	1 3/4	4	12600	25	C-13D	b, e, f
	DHJ	500	T-12	4-Pin	1 9/16	4 5/8	—	10	C-13D	c, g

See footnotes on page 8–91.

Continued on next page.

Fig. 8–108. *Continued*

ANSI Code		Watts	Bulb	Base (see Fig. 8–9)	Light Center Length (inches)	Maximum Over-All Length (inches)	Approximate Initial Lumens[a]	Average Life (hours)	Filament (see Fig. 8–7)	Footnotes
Clear	Opaque									
	DFH	500	T-12	4-Pin	1¾	3⅞	—	25	C-13D	f, j
DCS		500	T-12	L1-Ring	3½	5⅞	13200	25	C-13D	d
	EEB	500	T-14	Med. Pref.	2 3/16	5¾	10900	200	C-13	
DMS		500	T-20	Med. Sc.	3	5½	13200	50	C-13	
DMX		500	T-20	Med. Pref.	2 3/16	5¾	13200	50	C-13	
DYH		600	G-7	2-Pin	1 11/16	2⅜	17000	75	CC-6	l
DYP		600	G-7	2 Button	1 9/16	2¼	17000	75	CC-6	l
DYS		600	G-7	2-Pin Pref.	1 7/16	2½	17000	75	CC-6	l
DYV		600	G-6	2-Pin Pref.	1 7/16	2½	17000	75	CC-6	l
FCB		600	T-4	RSC	—	3¾	16500	75	CC-8	l
FFJ		600	T-4	RSC	—	2⅝	17000	75	CC-8	l
	DDB	750	T-12	Med. Pref.	2 3/16	5¾	21000	25	C-13D	
	DDY	750	T-12	Med. Pref.	2 3/16	5¾	18000	200	C-13D	
	DEP	750	T-12	4-Pin	1 9/16	4⅝	20500	25	C-13D	b
BFK		750	T-20HG	Med. Pref.	2 3/16	5¾	—	200	C-13	
BRP		750	T-7	4-Pin	1 9/16	3¾	18000	50	C-13D	f
BVN		750	T-7	4-Pin	1¾	4	18000	50	C-13D	f
DKK		750	T-12HG	Med. Pref.	2 5/16	5¾	18300	25	C-13D	d
	DEJ	750	T-12	L1-Ring	2 5/16	5½	20500	25	C-13D	
	DET	750	T-12	L1-Ring	2 5/16	5½	18000	200	C-13D	
DEC		750	T-12	D.C. Med. Ring	3½	5⅞	20500	25	C-13D	d
	DGH	750	T-12	Med. Pref.	2 3/16	5¾	16000	500	C-13D	
	DKK	750	T-12	Med. Pref.	2 3/16	5¾	18000	25	C-13D	
	DCX	750	T-12	Med.	3	5½	21000	25	C-13D	
	CWA	750	T-12	4-Pin	1 9/16	4⅝	20100	25	C-13D	f
	DGR	750	T-12	4-Pin	1 9/16	4⅝	18300	25	C-13D	
	DDW	750	T-12	Med. Pref.	2 3/16	5¾	22500	10	C-13D	
DES		750	T-12	L1-Ring	3½	5⅞	20100	25	C-13D	d
—	—	750	T-20	Med. Pref.	2 3/16	5¾	17000	500	C-13	b
DPJ		750	T-20	Med. Pref.	2 3/16	5¾	20500	25	C-13D	
DPB		750	T-20	Med.	3	5½	20500	25	C-13D	
—	—	750	T-20	Med. Pref.	2 3/16	5¾	18000	200	C-13D	b
BVA		900	T-7	2-Pin Pref.	1¾	3½	—	75	C-13D	f
BRH		1000	T-5	—	—	3¾	30000	75	CC-8	i
	DFD	1000	T-12	Med. Pref.	2 3/16	5¾	30500	10	C-13D	a
	DFT	1000	T-12	Med. Pref.	2 3/16	5¾	28500	25	C-13D	a, e
	DFY	1000	T-12	L1-Ring	2 5/16	5½	28500	25	C-13D	a, e
DGS		1000	T-12	D.C. Med.	3½	5⅞	33000	10	C-13D	d
	DFT	1000	T-12	Med. Pref.	2 3/16	5¾	28700	25	C-13D	
	DFD	1000	T-12	Med. Pref.	2 3/16	5¾	31500	10	C-13D	
	DFY	1000	T-12	L1-Ring	2 5/16	5½	28700	25	C-13D	
	DFK	1000	T-12	L1-Ring	2 5/16	5½	31500	10	C-13D	
	EBP	1000	T-12	Med.	2 5/16	5 7/16	—	25	C-13D	
DGS		1000	T-12	D.C. Med. Ring	3½	5⅞	31500	10	C-13D	d
	CTT	1000	T-12	4-Pin	1 9/16	4⅝	25200	25	C-13D	e, f
	DAX	1000	T-12	4-Pin	1 9/16	4⅝	27900	10	C-13D	e, f
	DEM	1000	T-12	4-Pin	1 9/16	4⅝	31000	10	C-13D	
DHB		1000	T-12	L1-Ring	3½	5⅞	28000	25	C-13D	
DPT		1000	T-20	Mog.	4¾	9 1/16	28000	50	C-13	
DPW		1000	T-20	Mog. Pref.	3 7/16	9½	28000	50	C-13	
DRG		1000	T-20	Med. Pref.	3	5½	28500	25	C-13D	
DRB		1000	T-20	Med. Pref.	2½	5¾	34000	25	C-13	e
BRR		1000	T-7	4-Pin	1 9/16	3¾	25200	50	C-13D	f

See footnotes on page 8–91.

Continued on next page.

Fig. 8–108. *Continued*

ANSI Code		Watts	Bulb	Base (see Fig. 8–9)	Light Center Length (inches)	Maximum Over-All Length (inches)	Approximate Initial Lumens[a]	Average Life (hours)	Filament (see Fig. 8–7)	Footnotes
Clear	Opaque									
BTA		1000	T-7	2-Pin Pref.	1¾	3½	25200	50	C-13D	
DGL		1000	T-12HG	D.C. Med. Ring	3½	5⅞	28000	25	C-13D	d
DRS		1000	T-20HG	Med. Pref.	2³⁄₁₆	5¾	28000	25	C-13D	d
DRS		1000	T-20	Med. Pref.	2³⁄₁₆	5¾	28500	25	C-13D	
DRC		1000	T-20	Med. Pref.	2³⁄₁₆	5¾	28000	50	C-13	e
DRW		1000	T-20	Mog.	4¾	9¹⁄₁₆	28500	25	C-13D	
DSB		1000	T-20	Mog. Pref.	3⁷⁄₁₆	9½	28500	25	C-13D	
DRE		1000	T-20	Mog. Bipost	4	9½	28500	25	C-13D	
	DHT	1200	T-12	Med. Pref.	2³⁄₁₆	5¾	38000	10	C-13D	e
	DHV	1200	T-12	L1-Ring	2⁵⁄₁₆	5½	38200	10	C-13D	
BRN		1200	T-7	4-Pin	1⁹⁄₁₆	3¾	38200	20	C-13D	f
	CYS	1200	T-12	4-Pin	1⁹⁄₁₆	4⅝	38200	10	C-13D	e, f
	DBH	1200	T-12	4-Pin	1⁹⁄₁₆	4⅝	38200	10	C-13D	
DTJ		1500	T-20	Mog. Pref.	3⁷⁄₁₆	9½	42500	25	C-13D	
DTS		1500	T-20	Mog. Bipin	4	9½	42500	25	C-13D	
EAN		5000	T-32	Mog. Bipin	6½	14	143000	150	C-13	
	FCB	600	T-4	S.C. Cont. Recess	—	3¾	17000	75	CC-6	i

[a] Readings based on photometry of clear bulb lamps.
[b] Non-heat-resistant bulb glass.
[c] Base-down to horizontal.
[d] Base-up.
[e] Collector grid.
[f] Proximity reflector.
[g] Internal reflector.
[h] Internal dichroic reflector.
[i] Iodine cycle lamp, double ended.
[j] 4-Pin Lock.
[k] External reflector.
[l] Quartz or silica.

Fig. 8–109. Infrared Energy Sources
A. Sheathed Resistance Radiant Heaters* (1500–14,000 nm)
(For Use with External Reflectors)

Watts	Watts Per Square Inch	Volts	Terminals	Description	Intended Sheath Temperature (° F)	Heated Length (inches)
1000**	14	230	Bayonet plug-in	Hairpin	850	70
2000**	44	230	Bayonet plug-in	Hairpin	1500	46
600†	30	115 or 230	Screw	Straight Rod	1300	17
800†	40	115 or 230	Screw	Straight Rod	1450	17
1000†	37	230	Screw	Straight Rod	1400	23
1500†	33	230	Screw	Straight Rod	1340	39
2500†	40	230 or 460	Screw	Straight Rod	1450	53
3000†	40	230 or 460	Screw	Straight Rod	1450	65
3600†	40	230 or 460	Screw	Straight Rod	1450	77

* Rated life in excess of 7000 hours; may be burned in any position.
** General Electric, Westinghouse, Tuttle & Kift.
† Edwin L. Wiegand Company.

B. Tungsten Filament Lamps* (500–4000 nm)

Watts	Bulb and Other Description	Base (see Fig. 8–9)	Service	Filament (see Fig. 8–7)	Maximum Over-All Length (inches)	Light Center Length (inches)	Voltage Range
125	G-30 clear	Med. Skt.	Industrial	C-7A C-11	$7\frac{1}{8}$ $7\frac{5}{16}$	5 $4\frac{7}{8}$	115–125
125	R-40 light inside frosted reflector	Med. Skt.	Industrial	C-9 C-11	$7\frac{1}{4}$ $7\frac{3}{8}$	—	115–125
250	G-30 clear	Med. Skt.	Industrial	C-7A C-11	$7\frac{3}{16}$ $7\frac{5}{16}$	5 $4\frac{7}{8}$	115–125
250	R-40 light inside frosted reflector	Med. Skt.	Industrial	C-9 C-11	$7\frac{1}{4}$ $7\frac{3}{8}$	—	115–125
250	R-40 light inside frosted reflector, heat resistant bulb	Med. Skt.	Industrial	C-9 C-11	$7\frac{1}{2}$ $7\frac{1}{4}$	—	115–125
250	R-40 light inside frosted reflector	Med.	Home and farm	CC-6 C-9	$6\frac{1}{2}$	—	115–125
250	R-40 reflector, red bowl, heat resistant bulb	Med.	Home and farm	C-9 C-11	$6\frac{7}{8}$	—	115–125
250	PS-30 reflector	Med.	Chicken brooding	C-9	$8\frac{1}{16}$	—	115–125
375	G-30 clear	Med. Skt.	Industrial	C-7A C-11	$7\frac{1}{8}$ $7\frac{5}{16}$	5 $4\frac{7}{8}$	115–125
375	R-40 light inside frosted reflector	Med. Skt.	Industrial	C-9 C-7A C-11	$7\frac{1}{4}$ $7\frac{3}{8}$	—	115–125
375	R-40 light inside frosted reflector heat resistant bulb	Med. Skt.	Industrial	C-9 C-7A C-11	$7\frac{5}{8}$ $7\frac{1}{2}$	—	115–125
375	R-40 reflector, red bowl, heat resistant bulb	Med. Skt.	Industrial	C-9	$7\frac{1}{2}$ $7\frac{5}{8}$	—	115–125
500	G-30 clear	Med. Skt.	Industrial	C-7A C-11	$7\frac{5}{16}$ $7\frac{3}{16}$	5	115–125
1000	T-40 clear	Med. Bipin**	Industrial	Triang.	$7\frac{5}{16}$	$3\frac{1}{4}$	115–125

* Rated average life is excess of 5000 hours.
** With 6-inch flexible connectors.

Fig. 8–109. *Continued*

C. Tubular Quartz Heat Lamps

(Burning Position—Horizontal, except as noted. All lamps listed are translucent quartz except as noted.)

Watts	Bulb and Other Description	Base (see Fig. 8–9)	Service	Filament (see Fig. 8–7)	Maximum Over-All Length (inches)	Lighted Length	Voltage Range
375	T-3(1)	Sleeve**	Industrial	C-8	8 13/16	5	115–125
500	T-3(1)	Sleeve**	Industrial	C-8	8 13/16	5	115–125
500	T-3 clear(1)	Sleeve**	Industrial	C-8	8 13/16	5	115–125
500	T-3 clear(1, 2)	Sleeve**	Special equip.	C-8	8 13/16	5	115–125
500	T-3(1)	R.S.C.***	Industrial	C-8	8 5/8	5	115–125
800	T-3	Sleeve**	Industrial	C-8	11 15/16	8	115–125
1000	T-3	Sleeve**	Industrial	C-8	13 13/16	10	200–220, 230–250
1000	T-3 clear	Sleeve**	Industrial	C-8	11 15/16	10	230–250
1000	T-3 clear(2)	Sleeve**	Special equip.	C-8	11 15/16	10	230–250
1000	T-3 clear, bent end construction	Sleeve & tab	Special equip.	C-8	11 9/16	10	230–250
1000	T-3 clear(2)	Sleeve**	Special equip.	C-8	13 13/16	10	230–250
1200	T-3 clear	Sleeve**	Special equip.	C-8	8 13/16	6	144
1600	T-3	Sleeve**	Industrial	C-8	19 13/16	16	200–220, 230–250, 277
1600	T-3	R.S.C.***	Industrial	C-8	19 5/8	16	200–220, 230–250, 277
1600	T-3 clear	Sleeve**	Industrial	C-8	19 13/16	16	230–250
1600	T-3 tubular red Vycor	Sleeve**	Comfort heating	C-8	19 13/16	16	200–220, 230–250, 277
1600	T-3 vertical burning	Sleeve**	Industrial	C-8	19 13/16	16	230–250
1600	T-3 clear	Sleeve**	Special equip.	C-8	17 15/16	16	230–250
1600	T-3 clear(2)	Sleeve**	Special equip.	C-8	17 15/16	16	230–250
2000	T-3 clear	Sleeve**	Special equip.	C-8	11 15/16	9 3/4	230–250
2000	T-3 clear(2)	Sleeve**	Special equip.	C-8	11 15/16	9 3/4	230–250
2000	T-3 clear, bent end construction	Sleeve & tab	Special equip.	C-8	11 9/16	10	230–250
2000	T-3 clear(2)	Sleeve**	Special equip.	C-8	13 13/16	10	230–250
2500	T-3	Sleeve**	Industrial	C-8	28 13/16	25	460–500, 575–625
2500	T-3	R.S.C.***	Industrial	C-8	28 5/8	25	460–500
2500	T-3 vertical burning	Sleeve**	Industrial	C-8	28 13/16	25	460–500
2500	T-3 tubular red Vycor	Sleeve**	Comfort heating	C-8	28 13/16	25	460–500
2500	T-3 clear	Sleeve**	Industrial	C-8	28 13/16	25	460–500
2500	T-3 clear(2)	Sleeve**	Special equip.	C-8	28 13/16	25	460–500
3800	T-3	Sleeve**	Industrial	C-8	41 13/16	38	550–600
3800	T-3 clear	Sleeve**	Industrial	C-8	41 13/16	38	550–600
3800	T-3 vertical burning	Sleeve**	Industrial	C-8	41 13/16	38	550–600
5000	T-3 clear(2)	Sleeve**	Special equip.	C-8	28 13/16	25	575–625
5000	T-3 clear	Sleeve**	Industrial	C-8	53 13/16	50	920–1000

* Rated average life in excess of 5000 hours. ** With flexible connectors. *** Recessed single contact.

(1) Burning position: any.

(2) High temperature construction.

Fig. 8-110. Typical Hot-Cathode Fluorescent Lamps (Rapid Starting)*

	Circline			U-Shaped	Lightly Loaded Lamps					Medium Loaded Lamps											
Nominal length (inches) Bulb Base	8¼ dia. T-9 4-Pin	12 dia. T-10 4-Pin	16 dia. T-10 4-Pin	24 T-12 Med. Bipin	24 T-12 Med. Bipin	36 T-12 Med. Bipin	48 T-12 Med. Bipin	48 T-12 Med. Bipin	48 T-10 Med. Bipin	24 T-12 Recess D.C.		30 T-12 Recess D.C,		36 T-12 Recess D.C.		42 T-12 Recess D.C.		48 T-12 Recess D.C.		60 T-12 Recess D.C.	
Approx. lamp amperes	0.39	0.435	0.42	0.42		0.43	0.43		0.43	0.8	1.0	0.8	1.0	0.8	1.0	0.8	1.0	0.8	1.0	0.8	1.0
Approx. lamp volts	62	80	109	103		81	100		101	39.5	38			58	55	67	63	76	71	94	88
Approx. lamp watts[a]	22.5	33	41.5	40		32.4	40.7		41	35.5	41			47.5	54.5	56	64	61	70	74	85
Rated life (hours)[b]	7500	7500	7500	12000	7500	15000	18000		18000	9000	7500	9000		9000	7500	9000	7500	12000	7500	12000	7500
Lamp lumen depreciation (LLD)[c]	72	82	77	84		81	84			77	72			77	72	77	72	82	72	82	72
Initial lumens[d]																					
Cool White	980	1750	2450	2900	1230	2300	3200	3350	3200	1650	1800	2290	2500	2810	3060	3520	3840	4150	4520	5300	5780
Deluxe Cool White	800	1250	1780			1560	2220		2270									2880			
Warm White[e]	980	1800	2500	2950		2260	3250		3250	1690				2850				4200		5400	
Deluxe Warm White	745	1240	1760			1530	2190											2920			
White		1810	2300	2950		2260	3250		3250									4200			
Daylight	875	1525	2100			1850	2660		2700	1385	1510			2330	2850	2870	3150	3500	3850	4300	4920
Blue							1160														
Green							4400											5500			
Pink							1160														
Gold		1220	1600				2200														
Red							200														
5000 K, high CRI[f]						1630	2100	2450													
7500 K, high CRI[g]							2050														
Soft White/Natural						1550	2080											2800			
Cool Green							2650														
Sign White							2440			1200				2190		2600		3300		4150	
Design White							2300			1275				2200		2800		3200		4100	
Living White							2410														
Supermarket White							2370														
High Efficiency							3680														
Incandescent/Fluorescent							1600											2000			
Glacial White							2800														
Super Soft White, Soft-Glo							2320														
Vita-Lite, Vita-Lux, Vitama						1550	2180	2400										3140			
Turquoise, Aqualux, Cool-Tone							2900	3130													
Super White, North-Lux, Day-Tone							2860	3090													
Super Delux 45							2450														
Snowlite							3000	3240													
Ultra-White							2420	2680													
Lamps for plant growth[h]	340	525					925														
Lamps for plant growth[i]	600						1700														
Footlamberts[d]																					
Cool White	2200	2290	2320	2200	1970	2330	2435			2860	3120			2860	3115	3380	3690	3320	3620	3540	3850
Deluxe Cool White	1800	1630	1690			1580	1690											2305			
Warm White	2200	2355	2370	2240		2285	2470			2925				2900				3360		3600	
Deluxe Warm White	1675	1620	1670			1550	1675											2335			
White		2365	2180	2240		2285	2470											3360			
Daylight	1970	1995	1990			1870	2020			2400	2620			2370	2900	2755	3025	2800	3080	2870	3280
Blue							880														
Green							3340											4400			
Pink							880														
Gold							1675														
Red							150														
5000 K, high CRI						1650	1600														
7500 K, high CRI							1560														
Soft White/Natural						1570	1580											2240			
Cool Green							2020														
Sign White							1855			2080				2230		2500		2640		2770	
Design White							1750			2210				2240		2690		2560		2740	
Living White							1830														
Supermarket White							1800														
High Efficiency							2800														
Incandescent/Fluorescent							1220											1600			
Glacial White							2115														
Super Soft White, Soft-Glo							1765														
Vita-Lite, Vita-Lux, Vitama						1570	1660											2515			
Turquoise, Aqualux, Cool-Tone							2205														
Super White, North-Lux, Day-Tone							2175														
Super Delux 45							1860														
Snowlite							2280														
Ultra-White							1840														

Medium Loaded Lamps

Nominal length (inches) / Bulb / Base	64 T-12 Recess D.C.		72 T-12 Mog. Bipin[j]	72 T-12 Recess D.C.		84 T-12 Recess D.C.		96 T-12 Recess D.C.		48[k] T-10 Rec. D.C.	72[k] T-10 Rec. D.C.	96[k] T-10 Rec. D.C.	48[m] T-12 Rec. D.C.	60 T-12 Rec. D.C.	72[m] T-12 Rec. D.C.	96[m] T-12 Rec. D.C.	48 PG-17 Rec. D.C.	72 PG-17 Rec. D.C.	96 PG-17 Rec. D.C.	96 TH-17 Rec. D.C.
Approx. lamp amperes	0.8	1.0	1.0	0.8	1.0	0.8	1.0	0.8	1.0	1.5	1.5	1.5	1.5	1.5	1.5	1.5	1.5	1.5	1.5	
Approx. lamp volts	100	93	104	112	104	132	121	149	142	80	120	160	80		120	155	85	120	175	
Approx. lamp watts[a]	78	89.5	98.5	86	98	98	111	111	128	105[i]	150[i]	205[i]	110		165	215	110	165	215	
Rated life (hours)[d]	12000	7500	7500	12000	7500	12000	7500	12000	7500	9000	9000	9000	9000	9000	9000	9000	9000	9000	9000	15000
Lamp lumen depreciation (LLD)[c]	82	72	72	82	72	82	72	82	79	66	66	66	69		72	72	69	69	69	
Initial lumens[d]																				
Cool White	5650	6150	6900	6450	7040	7700	8400	9150	9980	6000[i]	9500[i]	13000[i]	6800	9450	11000	15300	7000	11500	16000	16500
Deluxe Cool White				4500	5050			6350		4400[i]	7100[i]	9700[i]	4300		7200	10800	4800	8000	11000	
Warm White[e]				6500	7300			9150	9800				6500		11000	15300	6400	10600	15000	
Deluxe Warm White				4500	5000			6350												
White				6500	7300			9150					7200		10900	15750	6600	10900	15500	
Daylight	4650			5450	6100	6500	7100	7650	8500				5700		9300	13000	5600	9300	13300	
Blue																				
Green				9000																
Pink																				
Gold				4400				6200												
Red																				
5000 K, high CRI[f]								5900												
7500 K, high CRI[g]								5760												
Soft White/Natural				4300				6000	6250				5000		8200	11000				
Cool Green																				
Sign White	4450			5100		6000		6600												
Design White	4400			4900		6000		6900												
Living White																				
Supermarket White				4770				6660												
High Efficiency																				
Incandescent/Fluorescent				3350				4600												
Glacial White								8500												
Super Soft White, Soft-Glo																				
Vita-Lite, Vita-Lux, Vitama				4560				6800												12500
Turquoise, Aqualux, Cool-Tone																				
Super White, North-Lux, Day-Tone																				
Super Delux 45																				
Snowlite																				
Ultra-White																				
Lamps for plant growth[h]								2900					2200		3800	5000				
Lamps for plant growth[i]								5500							5500	8500				
Footlamberts[d]																				
Cool White	3490	3800	3450	3270	3565	3430	3730	3380	3680	5660	5850	5940	5810		6050	5800	4575[n]	4850[n]	5065[n]	
Deluxe Cool White				2280	2560			2340		4155	4370	4430	3680		3950	4100	3140[n]	3375[n]	3480[n]	
Warm White				3300	3700			3380	3620				5555		6050	5800	4190[n]	4475[n]	4750[n]	
Deluxe Warm White				2280	2535			2340												
White				3300	3700			3380					6155		6000	5965	4320[n]	4600[n]	4805[n]	
Daylight	2875			2760	3090	2880	3150	2820	3140				4880		5110	4930	3660[n]	3925[n]	4210[n]	
Blue																				
Green				4560																
Pink																				
Gold				2230				2290												
Red																				
5000 K, high CRI								2180												
7500 K, high CRI								2125												
Soft White/Natural				2180				2215	2310				4280		4500	4170				
Cool Green																				
Sign White	2750			2585		2660		2440												
Design White	2720			2485		2660		2550												
Living White																				
Supermarket White				2420				2460												
High Efficiency																				
Incandescent/Fluorescent				1700				1700												
Glacial White								3140												
Super Soft White, Soft-Glo																				
Vita-Lite, Vita-Lux, Vitama				2310				2510												

* The life and light output ratings of fluorescent lamps are based on their use with ballasts that provide proper operating characteristics. Ballasts that do not provide proper electrical values may substantially reduce either lamp life or light output, or both.

[a] Includes watts for cathode heat.

[b] Rated life under specified test conditions with three hours per start. At longer burning intervals per start, longer life can be expected.

[c] Per cent of initial light output at 70 per cent rated life at 3 hours per start. Average for cool white lamps. Approximate values.

[d] At 100 hours. Where color is made by more than one manufacturer, lumens and footlamberts represent average of manufacturers.

[e] Also called Candelite, Charm-Tone, Ember Glo.

[f] Chroma 50, Color Classer 50, Color-Matcher 50, Color Rater 50, Optima, Magnalux, Ultima.

[g] Chroma 75, Color Classer 75, Color-Matcher 75, Color Rater 75.

[h] Gro-Lux, Plant Light, Plant-Gro.

[i] Gro-Lux Wide-Spectrum.

[j] This lamp is primarily used in street lighting applications.

[k] A jacketed T-10 design is also available for use in applications where lamps are directly exposed to cold temperatures.

[l] Peak value. At 77°F, lumen and wattage values are lower.

[m] These lamps available in several variations (outdoor, low temperature, jacketed) with the same, or slightly different, ratings.

[n] At 0-degree angle.

Note: All electrical and lumen values apply only under standard photometric conditions.

Fig. 8–111. Typical Hot Cathode Fluorescent Lamps (Preheat Starting)*

Nominal lamp watts	4	6	8	13	14	15	15	20	25	25	30	40[a]	90[b]
Nominal length (inches)	6	9	12	21	15	18	18	24	28	33	36	48	60
Bulb	T-5	T-5	T-5	T-5	T-12	T-8	T-12	T-12	T-12	T-12	T-8	T-12	T-17
Base (bipin)	Min.	Min.	Min.	Min.	Med.	Med.	Med.	Med.	Med.	Med.	Med.	Med.	Mog.
Approx. lamp amperes	.17	.16	.145	.165	.38	.3	.33	.38	.48	.48	.36	.43	1.52
Approx. lamp volts	29	42	57	95	40	55	47	57	61	60	98	103	63
Approx. lamp watts	4.5	6.0	7.2	13.0	14.5	13.0	13.0	20.5	27.4	26.5	30.0	39.8	87.5
Preheat amperes max.	.25	.25	.25	.27	.65	.65	.65	.65	.85	.85	.65	.75	2.2
Preheat amperes min.	.16	.16	.16	.18	.44	.44	.44	.44	.65	.65	.40	.55	1.45
Rated life[c] (hours)	4000–7500	6000–7500	6000–7500	6000–7500	6000–7500	7500	7500	7500–9000	7500	7500	7500	18000	9000
Lamp lumen depreciation (LLD)[d, e]	67	67	75	72	82	79	81	85	79	79	79	82	85
Initial lumens[f]													
Cool White	125	265	400	850	675	860	775	1250	1600	1815	2190	3200	6350
Deluxe Cool White	85	180	280	590	470	600	865	865	1110	1260	1525	2220	4410
Warm White[i]	127	269	405	860	685	870	785	1270	1620	1840	2220	3250	6450
Deluxe Warm White	85	180	275	580	465	590	530	855	1100	1245	1500	2190	4350
White	127	269	405	860	685	870	785	1270	1620	1840	2220	3250	6450
Daylight	105	220	330	700	560	715	645	1040	1330	1505	1820	2660	5270
Blue	50	110	140		320	330	340	455		725	840	1160	
Green	140	300	550		890	1230	1000	1820			3150	4400	
Pink	50	110	140		300	330	340	410			840	1160	
Gold	70	175	290		410	650	470	800			1470	2200	
Red					40	40	45	80			105	200	
5000 K, high CRI[g]					490	600	555	870			1550	2100	
7500 K, high CRI[h]						590		850				2050	
Soft White/Natural						575	500	825			1470	2080	
Cool Green												2650	
Sign White												2440	
Design White					520	660		940			1650	2300	
Living White						620		925				2410	
Supermarket White						610	750	900			1560	2370	
High Efficiency												3680	6400
Incandescent/Fluorescent					350	450		650			1170	1600	
Glacial White						785		1070			1820	2800	
Super Soft White, Soft Glo						660		900			1650	2320	
Vita-Lite, Vita-Lux, Vitama						640		870				2180	
Turquoise, Aqualux, Cool-Tone						840		1100			1950	2900	
Super White, North-Lux, Day-Tone						800		1100			2000	2860	
Super Delux 45												2450	
Snowlite							800	1060				3000	
Ultra-White						740		975				2420	
Enhance					190	240		340					
Lamps for plant growth[j]			130	260	230	265		350			665	925	
Lamps for plant growth[k]												1700	

Footlamberts[f]													
Cool White	2620	3290	3510	3860	1925	2920	1750	2000	2110	2020	3365	2435	2810
Delux Cool White	1780	2235	2455	2680	1340	2040	1210	1385	1460	1400	2345	1690	1950
Warm White	2660	3340	3550	3910	1955	2960	1775	2030	2140	2045	3420	2470	2860
Deluxe Warm White	1780	2235	2410	2640	1330	2005	1200	1370	1450	1385	2310	1674	1920
White	2660	3340	3550	3910	1955	2960	1775	2030	2140	2045	3420	2470	2860
Daylight	2200	2730	2890	3180	1600	2430	1455	1665	1750	1670	2800	2020	2330
Blue	1050	1365	1230		915	1120	770	730		805	1290	880	
Green	2935	3725	4820		2540	4180	2260	2920			4850	3340	
Pink	1050	1365	1230		860	1120	770	655			1290	880	
Gold	1470	2175	2540		1170	2210	1060	1280			2260	1675	
Red					115	135	100	128			160	150	
5000 K, high CRI					1400	2040	1250	1395				1600	
7500 K, high CRI						2005		1360				1560	
Soft White/Natural						1955	1130	1320			2260	1580	
Cool Green												2020	
Sign White												1860	
Design White					1485	2250		1500			2540	1750	
Living White						2110		1480				1835	
Supermarket White						2080	1290	1440			2400	1810	
High Efficiency												2800	2835
Incandescent/Fluorescent					1000	1530		1040			1800	1215	
Glacial White						2670		1710			2800	2130	
Super Soft White, Soft Glo						2245		1440			2540	1765	
Vita-Lite, Vita-Lux, Vitama						2180		1390				1660	
Turquoise, Aqualux, Cool-Tone						2860		1760			3000	2210	
Super White, North-Lux, Day-Tone						2720		1760			3080	2180	
Super Delux 45												1865	
Snowlite							1805	1695				2280	
Ultra-White						2520		1560				1840	
Enhance					545	815		545					
Lamps for plant growth[j]			1140	1180	655	900		560			1025	705	
Lamps for plant growth[k]												1295	

* The life and light output ratings of fluorescent lamps are based on their use with ballasts that provide proper operating characteristics. Ballasts that do not provide proper electrical values may substantially reduce either lamp life or light output or both.

[a] For data on 40-watt U-shaped lamps, see table on rapid start lamps, Fig. 8-110.

[b] 90 and 100 watt T-17 lamps are interchangeable on same ballast.

[c] Rated life under specified test conditions with three hours per start. At longer burning intervals per start longer life can be expected.

[d] Per cent of initial light output at 70 per cent rated life at 3 hours per start.

[e] Average for cool white lamps. Approximate values.

[f] At 100 hours. Where color is made by more than one manufacturer, lumens and footlamberts represent average of manufacturers.

[g] Chroma 50, Color Classer 50, Color-Matcher 50, Color Rater 50, Optima, Magnalux, Ultima.

[h] Chroma 75, Color Classer 75, Color-Matcher 75, Color Rater 75.

[i] Also called Candelite, Charm-Tone, Ember Glo.

[j] Gro-Lux, Plant Light, Plant-Gro.

[k] Gro-Lux Wide-Spectrum.

Note: All electrical and lumen values apply only under standard photometric conditions.

Fig. 8–112. Typical Hot-Cathode Fluorescent Lamps (Instant Starting)*

Nominal length (inches)	42	64	72	96	48	60	24	36	42	48	60	64	72	84	96
Bulb	T-6	T-6	T-8	T-8	T-12	T-17	T-12	T-12	T-12	T-12	T-12	T-12	T-12	T-12	T-12
Base	Single Pin	Single Pin	Single Pin	Single Pin	Medium Bipin	Mogul Bipin	Single Pin	Single Pin	Single Pin	Single Pin	Single Pin	Single Pin	Single Pin	Single Pin	Single Pin
Lamp amperes	0.200	0.200	0.200	0.200	0.425	0.425	0.425	0.425	0.425	0.425	0.425	0.425	0.425	0.425	0.425
Approx. lamp volts	147	227	215	290	104	107.5	53	77	88	100	123	131	147	172	200
Approx. lamp watts	25	38	37.5	50	41	41.5	21.5	30	34.5	39	48	50.5	57.5	66.5	75
Rated life[a] (hours)	7500	7500	7500	7500	7500–12000	7500–9000	7500–9000	7500–9000	7500–9000	9000–12000	7500–12000	7500–12000	9000–12000	7500–12000	12000
Lamp lumen depreciation (LLD)[b]	76	77	83	89	83	89	81	82	80	82	78	78	89	91	89
Initial lumens[c]															
Cool White	1800	2900	3000	4125	2850	2900	1150	1975	2325	2950	3450	3700	4500	5200	6200
Deluxe Cool White	1250	2020	2085	2870	1980	2020	800	1370	1615	2050	2400	2570	3130	3615	4310
Warm White[f]	1830	2945	3045	4190	2890	2940	1170	2000	2360	2995	3505	3760	4570	5280	6300
Deluxe Warm White	1230	1990	2055	2825	1950	1990	790	1350	1590	2020	2365	2540	3085	3560	4250
White	1830	2945	3045	4190	2890	2940	1170	2000	2360	2995	3505	3760	4570	5280	6300
Daylight	1495	2410	2495	3430	2370	2410	955	1640	1930	2450	2865	3075	3740	4320	5150
Blue													1750		2400
Green	2600	4200								4100			6100		9000
Pink													1750		2400
Gold										2030			3040		4250
Red													220		300
5000 K, high CRI[d]															4200
7500 K, high CRI[e]										2150			3250		4100
Soft White/Natural	1180	1880		2620						1920			2900		4175
Cool Green		2900		3500	2720					2400			4000		5400
Sign White										2100			3200		4550
Design White		2900								2180					4800
Living White															4750
Supermarket White										2150			3250		4660
High Efficiency															7250
Incandescent/Fluorescent										1560			2340		3280
Glacial White										2440			4000		5500
Super Soft White, Soft-Glo															4670
Vita-Lite, Vita-Lux, Vitama				3220						2120			3470		4600
Turquoise, Aqualux, Cool-Tone										2680					6040
Super White, North-Lux, Day-Tone				4050						2570			4400		5870
Super Delux 45															4960
Snowlite															6000
Ultra-White				3390	2300								3460		5000
Gro-Lux															1870
Gro-Lux Wide-Spectrum															3270

Footlamberts[c]															
Cool White	3280	3375	2220	2295	2190	1270	2000	2010	2070	2305	2155	2180	2250	2200	2275
Deluxe Cool White	2280	2350	1540	1595	1520	885	1390	1390	1435	1600	1500	1515	1565	1525	1580
Warm White	3335	3425	2255	2330	2225	1290	2040	2030	2100	2340	2195	2215	2285	2230	2310
Deluxe Warm White	2240	2315	1520	1570	1500	875	1375	1370	1410	1580	1480	1495	1545	1500	1560
White	3335	3425	2255	2330	2225	1290	2040	2030	2100	2340	2195	2215	2285	2230	2310
Daylight	2720	2800	1845	1915	1820	1055	1660	1665	1715	1915	1790	1810	1870	1825	1890
Blue													875		880
Green	4735	4890								3205			3050		3300
Pink													875		880
Gold										1590			1520		1560
Red													110		110
5000 K, high CRI															1540
7500 K, high CRI													1625		1500
Soft White/Natural	2150	2185		1455						1500			1450		1530
Cool Green		3375		1945	2090					1875			2000		1980
Sign White										1640			1600		1670
Design White		3375								1700					1760
Living White															1740
Supermarket White										1680			1625		1710
High Efficiency															2660
Incandescent/Fluorescent										1220			1170		1200
Glacial White										1910			2000		2015
Super Soft White, Soft-Glo															1710
Vita-Lite, Vita-Lux, Vitama				1790						1655			1735		1685
Turquoise, Aqualux, Cool-Tone										2095					2215
Super White, North-Lux, Day-Tone				2250						2010			2200		2150
Super Delux 45															1820
Snowlite															2200
Ultra-White				1885	1770								1730		1830
Gro-Lux															685
Gro-Lux Wide-Spectrum															1200

* The life and light output ratings of fluorescent lamps are based on their use with ballasts that provide proper operating characteristics. Ballasts that do not provide proper electrical values may substantially reduce either lamp life, or light output, or both.

[a] Rated life under specified test conditions with three hours per start. At longer burning intervals per start, longer life can be expected.

[b] Per cent of initial light output at 70 per cent rated life at 3 hours per start. Average for cool white lamps. Approximate values.

[c] At 100 hours. Where color is made by more than one manufacturer, lumens and footlamberts represent average of manufacturers.

[d] Chroma 50, Optima, Magnalux, Ultima, Color Classer 50, Color-Matcher 50, Color Rater 50.

[e] Chroma 75, Color Classer 75, Color-Matcher 75, Color Rater 75.

[f] Also called Candelite, Charm-Tone, Ember Glo.

Note: All electrical and lumen values apply only under standard photometric conditions.

Fig. 8-113. Typical Cold-Cathode Instant-Starting Fluorescent Lamps

Lamp Designation	48T8			72T8			96T8			Hairpin (6-inch) U48T8		
Nominal length (inches)[a]		48			72			96			45	
Actual lamp length (inches)		45			69			93			[e]	
Lamp base		Cap			Cap			Cap			Cap	
Bulb		T-8			T-8			T-8			T-8	
Starting volts—lamp type L. P.[b]		450			600			750			750	
Starting volts—lamp type H. P.[b]		600			750			835			835	
Lamp current (milliamperes)[c]	120	150	200	120	150	200	120	150	200	120	150	200
Lamp voltage—lamp type L. P.	250	240	220	330	310	300	420	400		420	400	
Lamp voltage—lamp type H. P.	270	250	240	350	330	310	450	425	400	450	425	400
Lamp wattage—lamp type L. P.[c]	26	30		34	40		42	49	59	42	49	60
Lamp wattage—lamp type H. P.[c]	28	33	40	37	43	52	46	54	65	46	54	65
Initial lumens—warm white[d]	1100	1300	1600	1700	2000	2350	2300	2700	3400	2300	2700	3400
Initial lumens—3500° white[d]	1050	1250	1550	1650	1900	2300	2250	2650	3300	2225	2650	3300
Initial lumens—4500° white[d]	1000	1200	1500	1600	1850	2200	2200	2600	3200	2200	2600	3200
Initial lumens—daylight[d]	950	1150	1450	1550	1800	2150	2150	2550	3100	2150	2550	3100
Luminance (footlamberts) warm white	1180	1400	1710	1270	1500	1780	1310	1530	1930	1310	1530	1930
Luminance (footlamberts) 3500° white	1150	1340	1660	1240	1470	1730	1280	1500	1880	1280	1500	1880
Luminance (footlamberts) 4500° white	1100	1280	1600	1210	1440	1650	1250	1480	1700	1250	1480	1700
Luminance (footlamberts) daylight	1090	1230	1550	1190	1410	1600	1230	1450	1650	1230	1450	1650
Rated lamp life (thousands of hours)[f]												
type L. P.	15	12.5		15	12.5		18	16	13	18	16	13
type H. P.	25	20	15	25	20	15	30	25	20	30	25	20

[a] Nominal length of lamps and sockets.
[b] Minimum ballast or transformer volts.
[c] Lamps can be operated up to 200 mA.
[d] Initial rating after 100 hours for types LP and HP. Other color lamps are available.
[e] Extended lamp length 93 inches formed to U shape with 180-degree, 6-inch arc.
[f] Life not affected by number of starts.

Fig. 8-114. Sunlamps

Designation	FS-20 T-12[a]	FS-40 T-12[b]	RS
Rated power input (watts)	20	40	275
Rated life[c] (hours)			
at 3 hours per start	2500	2500	—
at 6 hours per start	4000	4000	—
at 10 hours per start	—	—	2000
Over-all length (inches)	24 (nominal)	48 (nominal)	7.0 (maximum)
Useful arc length (inches)	20	44	—
Maximum diameter (inches)	1.5	1.5	5
Base	Med. Bipin	Med. Bipin	Med. Screw
Transformer secondary voltage (no load)	118	205 (minimum)	[d]
Potential drop between arc electrode (volts)	57	103	[d]
Arc current (amperes)	0.38	0.43	[d]
Total E-viton output (bare lamp)	40000	125000	35000[e]

[a] Preheat starting fluorescent sunlamp.
[b] Rapid starting fluorescent sunlamp.
[c] Fluorescent sunlamps are also rated at 5000 15-minute applications. RS lamps at 1200 15-minute applications.
[d] This lamp has a tungsten-filament resistance and a thermal switch enclosed in its reflector-type outer envelope. It is operated without external ballast on 110 to 125 volts, 50 to 60 hertz alternating current only.
[e] These lamps contain their own internal reflector; output figures apply to the energy directed out the face of the lamp.

Fig. 8–115. Fluorescent Lamp Power Requirements

Lamp Designation	Nominal Length (inches)	Minimum Required RMS Voltage across lamp for reliable starting[a]	Operating Current (milliamperes)	Approximate Watts Consumed				
				Lamp	Single Lamp Circuit		Two Lamp Circuit	
					Ballast	Total	Ballast	Total
Switch Start Types								
4T5	6	—[b]	170	4.5	2	6.5		
6T5	9	—[b]	160	6	2	8		
8T5	12	—[b]	145	7.2	2	9.2		
13T5	21	180	165	13	5	18		
15T8	18	106	305	15	4.5	19.5	10	40
30T8	36	176	355	30	10.5	40.5	17	77
14T12	15	106	380	14.3	5.5	19.8		
15T12	18	106	325	15	4.5	19.5	10	39
20T12	24	106	380	20	5	25	10	50
25T12	33	—	490	26	6			
40T12	48	176	430	40	12	52	16	96
90T17	60	132	1520	90	20	110	24	204
Instant Start Types								
40T12	48	385	425	40	24	64	26	106
40T17	60	385	425	40	24	64	26	106
Slimline Types								
42T6*	42	405	200	25.5	13	38.5	21	72
64T6*	64	540	200	38.5	16	54.5	20	99
72T8*	72	540	200	37.5	16	53.5	31	106
96T8*	96	675	200	50	30	80	23	123
48T12 (lead-lag)	48	385	425	39			26	104
(series)	48	385	425	39			15	93
72T12 (lead-lag)	72	475	425	57			47	161
(series)	72	475	425	57			25	139
96T12 (lead-lag)	96	565	425	75			40	190
(series)	96	565	425	75			22	172
Rapid Start Lightly Loaded Types[c]								
22T9	8¼**	185	380	22.5		34†		
32T10	12**	205	430	33		45†		90†
40T10	16**	205	415	41.5		54†		108†
30T12	36	150	430	32		44†		72†
40T12	48	200	430	41		54†		92†
Rapid Start Medium Loaded Types[c]								
100T12	72	—	1000	100				258†
24T12	24	85	800	35		70†		111†
			1000	40				125†
48T12	48	155	800	60		85†		140†
		155	1000	69				175†
72T12	72	260	800	87		106†		200†
		260	1000	100				258†
96T12	96	280	800	112		130†		250†
		280	1000	129				260†
Rapid Start Highly Loaded Types[c]								
48T12, 48PG17	48	160	1500	110		140†		240†
72T12, 72PG17	72	225	1500	165		210†		320†
96T12, 96PG17	96	300	1500	215		260†		450†
Cold Cathode Types								
96T8LP	96	750	120	42	9	51	17	101
			150	49				
			200	59	19	78	30	148
96T8HP	96	835	120	46	12	58	24	115
			150	54			28	136
			200	65			30	160

* T-6 and T-8 slimline also operate at 120 and 300 mA.
** Circular lamps, dimension given is nominal outside diameter.
† Total watts consumed includes power required to heat lamp cathodes.
[a] Between 50°F (10°C) and 110°F (43.3°C).
[b] Suitable for operation on 110–125V ac lines with series-type ballast.
[c] Requires starting aid.

Fig. 8–116. Typical High-Intensity Discharge Lamps*

A. Mercury Lamps

ANSI Lamp Designation[k]	Manufacturer's Lamp Designation	ANSI Ballast Number	Lamp Watts	Approximate Ballast Watts	Approximate Lumens[f]: Initial (100 hours)		Lamp Lumen Depreciation** (in per cent)		Bulb[h]	Outer Bulb Finish	Maximum Over-All Length (inches)	Light Center Length (inches)	Arc Length (inches)	Base	Rated Life (hours)	Lamp Voltage (volts): Minimum Starting for −20° F		Operating	Lamp Current (amperes): Starting	Operating
					Vert.	Horiz.[g]	Vert.	Horiz.[g]								RMS	Peak			
H45AZ	H45AZ[c]	45	40	7–12	1200	1145	—	—	B-21	Clear	6½	3¾	1 1/16	Med.	16000	180	255	90	.80	.53
H45AY/C	H45AY/C[c]	45	40	7–12	1165	1110	—	—	B-21	Phos. Coat	6½	3¾	1 1/16	Med.	16000	180	255	90	.80	.53
H45AY/DX	H45AY/DX[c]	45	40	7–12	1350	—	—	—	B-21	Phos. Coat	6½	3¾	1 1/16	Med.	16000	180	255	90	.80	.53
H46DL/DX[l]	H50DX46[a]	46	50	—	1575	—	61	—	E-17	Phos. Coat	5⅛	3⅛	—	Med.	10000	—	—	90	—	.62
H43AZ	H43AZ[b,c]	43	75	8.0	2800	2650	—	—	B-21 or E-21	Clear	6½	3¾	1 1/16	Med.	16000	225	320	130	.92	.64
H43AN[l]	H75A43[a]	43	75	8.0	2700	—	83	—	E-17	Clear	5 7/16	3½	1 1/16	Med.	10000	225	320	130	.92	.64
H43AY/C	H43AY/C[b,c]	43	75	8.0	2650	2505	—	—	B-21 or E-21	Phos. Coat	6½	3¾	1 1/16	Med.	16000	225	—	130	.92	.64
H43AY/DX	H43AY/DX[c]	43	75	8.0	3150	3000	—	—	B-21	Phos. Coat	6½	3¾	1 1/16	Med.	16000	225	320	130	.92	.64
H43AV/C[l]	H75C43[a]	43	75	8.0	2600	—	72	—	E-17	Phos. Coat	5 7/16	3½	1 1/16	Med.	10000	225	—	130	.92	.64
H43AV/DX[l]	H75DX43[a]	43	75	8.0	2800	—	63	—	E-17	Phos. Coat	5 7/16	3½	—	Med.	10000	—	—	—	—	—
H38-4LL[l]	H100A38-4/A23[a] H38-4LL[b]	38-4	100	10–35	3700	—	83	—	A-23	Clear	5 7/16	3½	1⅛	Med.	10000[m]	225	320	130	1.3	.85
H38-4MP/C[l]	H100C38-4A23[a] H38-4MP/C[b]	38-4	100	10–35	3400	—	85	—	A-23	Phos. Coat	5 7/16	3½	—	Med.	10000	225	320	130	1.3	.85
H38-4MP/DX[l]	H100DX38-4/A23[a]	38-4	100	10–35	4000	—	78	—	A-23	Phos. Coat	5 7/16	3½	—	Med.	10000[m]	225	320	130	1.3	.85
H38-4HT[n]	H100A38-4[a] H38-4HT[b,c,d,e]	38-4	100	10–35	3975	3775	62	55	BT-25 or E-23½	Clear	7½	5	1⅛	Mog.	24000+	225	320	130	1.3	.85
H38-4JA/C[n]	H100C38-4[a] H38-4JA/C[b,c,d,e]	38-4	100	10–35	3925	3725	62	54	E-23½	Phos. Coat	7½	5	—	Mog.	24000+	225	320	130	1.3	.85
H38-4JA/W[n]	H100W38-4[a] H38-4JA/W[b,c,d,e]	38-4	100	10–35	4225	4025	47	37	E-23½	Phos. Coat	7½	5	—	Mog.	24000+	225	320	130	1.3	.85
H38-4JA/DX[n]	H100DX38-4[a] H38-4JA/DX[b,c,d]	38-4	100	10–35	4350	4100	67	61	E-23½ BT-25	Phos. Coat	7½	5	—	Mog.	24000+	225	320	130	1.3	.85
H38JA/R	H38JA/R[b]	38-4	100	10–35	4400	4270	—	—	BT-25	Phos. Coat	7½	5	—	Mog.	24000	225	320	130	1.3	.90
H38-4FK	H38-4FB[e]	38-4	100	10–35	3650	—	—	—	BT-25	Clear	7	4⅜	—	Admed.	16000	225	320	130	1.3	.90
H38-4FL/C	H38-4FL/C[e]	38-4	100	10–35	3350	—	—	—	BT-25	Phos. Coat	7	4⅜	—	Sktd.	16000	225	320	130	1.3	.90
H38-4FL/W	H38-4FL/W[e]	38-4	100	10–35	4000	—	—	—	BT-25	Phos. Coat	7	4⅜	—	Sktd.	16000	225	320	130	1.3	.90
H38-4BT	R40-H38-4/SP[e]	38-4	100	10–35	3200	—	—	—	R-40 Spot	Frost	7	—	—	Mog.	16000	225	320	130	1.3	.90
H38-4BS	R40-H38-4/FL[e]	38-4	100	10–35	3200	—	—	—	R-40	Frost	7	—	—	Mog.	16000	225	320	130	1.3	.90
H38-4BM[l]	H100RFL38-4[a]	38-4	100	10–35	2850	—	58	—	R-40 Flood	Frost	7	—	—	Med.	24000+	225	320	130	1.3	.85

H38-4BP/DX[l]	H100RDXFL38-4[a]	38-4	100	10–35	2850	—	61	—	R-40 Flood	Phos. Coat & Frost	7	—	—	Med.	24000+	225	320	130	1.3	.85
H38-4BR	R40-C-H38-4[e]	38-4	100	10–35	2850	—	—	—	R-40	Phos. Refl.	7	—	—	Mog.	16000	225	320	130	1.3	.90
H38-4HX	R40-H38-4/SP/A[e]	38-4	100	10–35	3200	—	—	—	R-40	Frost	6⅜	—	—	Admed.	16000	225	320	130	1.3	.90
H38-4HW	R40-H38-4/FL/A[e]	38-4	100	10–35	3200	—	—	—	R-40	Frost	6⅜	—	—	Admed.	16000	225	320	130	1.3	.90
H38-4HV	R40-H38-4/A[e]	38-4	100	10–35	2850	—	—	—	R-40	Phos. Refl.	6⅜	—	—	Admed.	16000	225	320	130	1.3	.90
H44-4GS[l]	H44-4GS[b,c] H100PSP44-4[a]	44-4	100	10–35	2475	—	—	—	Spot PAR-38[h]	Refl.	5 7/16	—	—	Admed.	12000	225	320	130	1.3	.85
H44-4JM[l]	H100PFL44-4[a] H44-4dM[b,c]	44-4	100	10–35	2475	—	—	—	Flood PAR-38[h]	Refl.	5 7/16	—	—	Admed.	12000	225	320	130	1.3	.85
H38-AW	H38-AW[e]	44-4	100	10–35	3650	—	—	—	T-10	Clear	5⅝	3 7/16	—	Admed.	12000	225	320	130	1.3	.90
H38-AY/DX	H38-AY/DX[c]	38-4	100	10–35	4400	4200	—	—	B-21	Phos. Coat	6½	3¾	1 1/16	Med.	16000	225	320	130	1.3	.90
H4AB[l]	H100A4/T[a]	4	100	—	3700	—	—	—	T-10	Clear	5⅝	3 7/16	—	Admed.	6000	—	—	—	—	.85
H42-HT	H42-HT[e]	—	125	—	5600	—	—	—	BT-28	Clear	7⅜	5	—	Mog.	16000	—	—	—	—	—
H42-JA/C	H42-JA/C[e]	—	125	—	5300	—	—	—	BT-28	Phos. Coat	7⅜	5	—	Mog.	16000	—	—	—	—	—
H42-JA/W	H42-JA/W[e]	—	125	—	6000	—	—	—	BT-28	Phos. Coat	7⅜	5	—	Mog.	16000	—	—	—	—	—
H39-22KB[n]	H175A39-22[a] H39-22KB[b,c,d,e]	39-22	175	15–35	7470	7100	80	75	E-28 or BT-28	Clear	8¼	5	2	Mog.	24000+	225	320	130	2.5	1.5
H39-22KC/C[n]	H175-C39-22[a] H39-22KC/C[c,d,e] H39-22KC/C/RS	39-22	175	15–35	7535	7033	79	73	E-28 or BT-28	Phos. Coat	8¼	5	2	Mog.	24000+	225	320	130	2.5	1.5
H39KC/R	H39-KC/R[b]	39-22	175	15–35	8500	8290	—	—	BT-28	Phos. Coat	8 5/16	5	2	Mog.	24000	225	320	130	2.5	1.45
H39-22KC/W[n]	H175W39-22[a] H39-22KC/W[b,c,d,e]	39-22	175	15–35	8270	8165	62	55	E-28 or BT-28	Phos. Coat	8¼	5	—	Mog.	24000+	225	320	130	2.5	1.5
H39-22KC/DX[n]	H175DX39-22[a] H39-22KC/DX[b]	39-22	175	15–35	8380	8000	75	66	E-28 or BT-28	Phos. Coat	8¼ 8 5/16	5	—	Mog.	24000+	225	320	130	2.5	1.5
H39-22BN/C[l]	H175RCFL39-22[a]	39-22	175	15–35	5250	—	78	—	R-40	Phos. Refl.	7	—	—	Med.	24000+	225	320	130	2.5	1.5
H39-22BN/DX	H39-22BN/DX[c]	39-22	175	15–35	5800	—	—	—	R-40	Phos. Coat	7½	—	—	Med.	24000	225	320	130	2.5	1.5
H39-22BP/DX[l]	H175RDXFL39-22[a]	39-22	175	15–35	5700	—	61	—	R-40	Phos. Refl.	7	—	—	Med.	24000+	225	320	130	2.5	1.5
H39-22BV/DX[l]	H175RDXFL39-22/M[a]	39-22	175	15–35	5700	—	61	—	R-40	Phos. Refl.	7½	—	—	Mog.	24000+	225	320	130	2.5	1.5
H39-22BM[l]	H175RFL39-22[a]	39-22	175	15–35	5700	—	76	—	R-40	I. F. Refl.	7	—	—	Med.	24000+	225	320	130	2.5	1.5
	H175RFL39-22/M[a]	39-22	175	15–35	5700	—	76	—	R-40	I. F. Refl.	7½	—	—	Mog.	24000+	225	320	130	2.5	1.5
H39-22BT	R40-H39-22/SP[e]	39-22	175	15–35	6500	—	—	—	R-40	Frost	7½	—	—	Mog.	16000	225	320	130	2.5	1.45
H39-22BS	R40-H39-22/FL[e]	39-22	175	15–35	6500	—	—	—	R-40	Frost	7½	—	—	Mog.	16000	225	320	130	2.5	1.45
H39-22BR	R40C-H39-22[e]	39-22	175	15–35	5900	—	—	—	R-40	Phos. Refl.	7½	—	—	Mog.	16000	225	320	130	2.5	1.45
H37-5KB[n]	H250A37-5[a] H37-5KB[b,c,d,e]	37-5	250	25–35	11570	10700	81	74	E-28 or BT-28	Clear	8¼	5	2¼	Mog.	24000+	225	320	130	3.1	2.1
H37-5KC/C[n]	H250C37-5[a] H37-5KC/C[b,c,d,e]	37-5	250	25–35	11270	10650	80	74	E-28 or BT-28	Phos. Coat	8¼	5	—	Mog.	24000+	225	320	130	3.1	2.1
H37-5KC/DX[n]	H250DX37-5[a] H37-5KC/DX[b,c]	37-5	250	25–35	12700	12100	74	66	E-28	Phos. Coat	8¼	5	—	Mog.	24000+	225	320	130	3.1	2.1

Continued on next page.

Fig. 8–116. *Continued*

ANSI Lamp Designation[k]	Manufacturer's Lamp Designation	ANSI Ballast Number	Lamp Watts	Approximate Ballast Watts	Approximate Lumens[f] Initial (100 hours) Vert.	Initial (100 hours) Horiz.[g]	Lamp Lumen Depreciation** (in per cent) Vert.	Lamp Lumen Depreciation** Horiz.[g]	Bulb[h]	Outer Bulb Finish	Maximum Over-All Length (inches)	Light Center Length (inches)	Arc Length (inches)	Base	Rated Life (hours)	Lamp Voltage (volts) Minimum Starting for −20° F RMS	Minimum Starting for −20° F Peak	Operating	Lamp Current (amperes) Starting	Operating
H37-5KC/W[n]	H250-W37-5[a] H37-5KC/W[b,c,d,e]	37-5	250	25–35	12920	11870	62	47	E-28 or BT-28	Phos. Coat	8¼	5	—	Mog.	24000+	225	320	130	3.1	2.1
H37-5KC/R	H375KC/R[b]	37-5	250	25–35	13000	12500	—	—	BT-28	Phos. Coat	8 5/16	5	2¼	Mog.	24000	225	320	130	3.1	2.1
H37-5FP	R60-H37-5/SP[e]	37-5	250	25–35	10200	—	—	—	R-60	Frost	10¾	—	—	Mog.	16000	225	320	130	3.1	2.1
H37-5LN	R60-H37-5/FL[e]	37-5	250	25–35	10200	—	—	—	R-60	Frost	10¾	—	—	Mog.	16000	225	320	130	3.1	2.1
H37-5FS	R60C-H37-5[e]	37-5	250	25–35	8650	—	—	—	R-60	Phos. Refl.	10¾	—	—	Mog.	16000	225	320	130	3.1	2.1
H33-1CD[n]	H400A33-1[a] H33-1CD[b,c,d,e]	33-1	400	20–55	20770	19667	79	73	BT-37 or E-37	Clear	11½	7	2¾	Mog.	24000+	225	320	135	5.0	3.2
H33-1GL/C[n]	H400C33-1[a] H33-1GL/C[b,c,d,e]	33-1	400	20–55	20750	19700	76	70	BT-37 or E-37	Phos. Coat	11½	7	—	Mog.	24000+	225	225	135	5.0	3.2
H33-1GL/DX[n]	H33-1GL/DX[b,c] H400DX33-1[a]	33-1	400	20–55	22830	21780	70	64	BT-37 or E-37	Phos. Coat	11½	7	—	Mog.	24000+	225	225	135	5.0	3.2
H33-1GL/W[n]	H400W33-1[a] H33-1GL/W[b,c,d,e]	33-1	400	20–55	22630	21200	62	55	BT-37 or E-37	Phos. Coat	11½	7	—	Mog.	24000+	225	225	135	5.0	3.2
H33GL/R	H33GL/R[b]	33-1	400	20–55	23000	22100	—	—	BT-27 or E-37	Phos. Coat	11½	7	—	Mog.	24000	225	320	135	5.0	3.2
H33-1GL/X	H33-1GL/X[d]	33-1	400	20–55	15000	14200	—	—	BT-37 or E-37	Filt.[i] Phos.	11½	7	—	Mog.	16000	225	320	135	5.0	3.2
H33-1GL/Y[j]	H400Y33-1[a]	33-1	400	20–55	12000	11500	—	—	BT-37[h]	Phos. Coat + Filter[i]	11½	7	—	Mog.	24000+	225	320	135	5.0	3.2
H33-1BA[n]	H400A33-1/3.2A[a]	33-1	400	20–55	20750	19750	79	73	BT or E-37	Clear	11 5/16	7	—	Mog.	24000+	225	320	135	5.0	3.2
H33-1BB/C[n]	H33-1BA H400C33-1/3.2A[a]	33-1	400	20–55	20250	19250	76	70	BT or E-37	Phos. Coat	11 5/16	7	—	Mog.	24000+	225	320	135	5.0	3.2
H33-1FY[n]	H33-1BB/C H400R33-1[a] H33-1FY[b,c]	33-1	400	20–55	18650	17100	—	—	R-52 or R-57	Clear Refl.	11¾	—	—	Mog.	24000+	225	320	135	5.0	3.2
H33-1DN/C[l]	H400RC33-1[a] H33-1DN/C[b,c,e]	33-1	400	20–55	20885	19500	—	—	R-52 or R-57	Phos. Refl.	11¾	7	—	Mog.	24000+	225	320	135	5.0	3.2
H33-1DN/W	H33-1DN/W[b,e]	33-1	400	20–55	22500	20500	—	—	R-52 or R-57	Phos. Refl.	11¾	—	—	Mog.	24000	225	320	135	5.0	3.2
H33-1DN/DX[l]	H33-1DN/DX[b,c] H400RDX33-1[a]	33-1	400	20–55	22665	22100	—	—	R-52 or R-57	Phos. Refl.	11¾	—	—	Mog.	24000+	225	320	135	5.0	3.2
H33-1HS	H33-1HS[b]	33-1	400	20–55	17500	—	—	—	R-57	Clear Refl.	11¾	—	—	Mog.	24000	225	320	135	5.0	3.2
H33-1FP[l]	R60-H33-1SP[e]	33-1	400	20–55	15300	—	—	—		—	—	—	—	Mog.	24000	225	320	135	5.0	3.2
	H400RSP33-1[a]								R60 Spot	Clear Refl.	10⅛	—	—							
H33-1LN[l]	H400RFL33-1[a] H33-1LN[b,c]	33-1	400	20–55	15300	—	—	—	R-60 Flood	Clear Refl.	10⅛	—	—	Mog.	24000+	225	320	135	5.0	3.2

H33-1FS	R60-H33-1[e]	33-1	400	20–55	16500	15700	—	—	R60 Wide Flood	Phos. Refl.	10¾	—	—	Mog.	16000	225	320	135	5.0	3.2
H33-1FS/C[1]	H33-1FS/C[b] H400RCFL-33-1[a]	33-1	400	20–55	15650	—	—	—	R60 Wide Flood	Phos. Refl.	10¾		—	Mog.	24000+	225	320	135	5.0	3.2
H33-1FS/DX[1]	H33-1FS/DX[b,c] H400RDXFL33-1[a]	33-1	400	20–55	15800	14900	—	—	R60 Wide Flood	Phos. Refl.	10¾	—	—	Mog.	24000+	225	320	135	5.0	3.2
H33-1HL	R60C-H33-1/HB[e]	33-1	400	20–55	16500	15700	—	—	R-60 Flood	Clear Phos. Refl.	10¾	—	—	Mog.	16000	225	320	135	5.0	3.2
H33-1HL/DX	H33-1HL/DX[c]	33-1	400	20–55	16400	—	—	—	R-60	Phos. Coat	10¾	—	—	Mog.	24000	225	320	135	5.0	3.2
H33-1AR[n]	H400A33-1/T-16[a] H33-1AR[c]	33-1	400	20–55	19750	18750	—	—	T-16	Clear	11	7	2¾	Mog.	12000	225	320	135	5.0	3.2
H40-17MA	H40-17MA[b]	—	425	25–55	21000	19500	—	—	BT-37	Clear	11½	7	3¾	Mog.	24000	440	620	265	2.5	1.7
H40-17GL/C	H40-17GL/C[b]	—	425	25–55	20750	19500	—	—	BT-37	Phos.	11½	7	3¾	Mog.	24000	440	620	265	2.5	1.7
H40-17GL/W	H40-17GL/W[b]	—	425	25–55	24000	—	—	—	BT-37	Phos.	11½	7	3¾	Mog.	24000	440	620	265	2.5	1.7
H40-17DN/C	H40-17DN/C[b]	—	425	25–55	21000	—	—	—	R-57	Phos.	12¾	—	3¾	Mog.	24000	440	620	265	2.5	1.7
H40-17DN/W	H40-17DN/W[b]	—	425	25–55	24000	—	—	—	R-57	Phos.	12¾	—	3¾	Mog.	24000	440	620	265	2.5	1.7
H35-18NA[n]	H35-18NA[b,c,d,e] H700A35-18[a]	35-18	700	35–65	39000	36400	79	71	BT-46	Clear	14½	9½	5	Mog.	24000+	375	530	265	5.0	2.8
H35-18ND/C[n]	H35-18ND/C[b,c,e] H700C35-18[a]	35-18	700	35–65	37835	35830	73	64	BT-46	Phos.	14½	9½	5	Mog.	24000+	375	530	265	5.0	2.8
H35-18ND/DX[n]	H35-18ND/DX[b,c] H700DX35-18[a]	35-18	700	35–65	43170	40500	64	55	BT-46	Phos.	14½	9½	5	Mog.	24000+	375	530	265	5.0	2.8
H35-18ND/W[n]	H700W35-18[a] H35-18ND/W[c,e]	35-18	700	35–65	43500	40500	62	55	BT-37	Phos.	14½	9½	5	Mog.	24000+	375	530	265	5.0	2.8
H35-18ND/X	H35-18ND/X[d]	35-18	700	35–65	28000	—	—	—	BT-46	Filt. Phos.[i]	14½	9½	5	Mog.	24000	375	530	265	5.0	2.8
H35-18FC	R80-H35-18/SP[e]	35-18	700	35–65	34500	—	—	—	R-80	Refl.	13⅞	—	—	Mog.	16000	375	530	265	5.0	2.8
H35-18FB	R80-H35-18/FL[e]	35-18	700	35–65	34500	—	—	—	R-80	Refl.	13⅞	—	—	Mog.	16000	375	530	265	5.0	2.8
H35-18FA	R80-H35-18[e]	35-18	700	35–65	32500	—	—	—	R-80	Refl.	13⅞	—	—	Mog.	16000	375	530	265	5.0	2.8
H36-15GV[1]	H1000A36-15[a] H36-15GV[b,c,d,e]	36-15	1000	40–90	57000	54000	68	69	BT-56	Clear	15⅜	9½	6	Mog.	24000+	375	530	265	6.0	4.0
H36-15GW/C[1]	H1000C36-15[a] H36-15GW/C[b,c,d,e]	36-15	1000	40–90	55750	53000	59	63	BT-56	Phos.	15⅜	9½	6	Mog.	24000+	375	330	265	6.5	4.0
H36-GW/R	H36-GW/R[b]	36-15	1000	40–90	63000	60800	—	—	BT-56	Phos.	15⅜	9½	6	Mog.	24000	375	375	265	6.5	4.0
H36-15GW/DX[1]	H36-15GW/DX[b,c] H1000DX36-15[a]	36-15	1000	40–90	63000	60000	49	61	BT-56	Phos.	15⅜	9½	6	Mog.	24000+	375	375	265	6.5	4.0
H36-15GW/W[1]	H1000W36-15[a] H36-15GW/W[b,c,d,e] H36-15GW/W/RS	36-15	1000	40–90	62800	59735	37	61	BT-56	Phos.	15⅜	9½	6	Mog.	24000+	375	375	265	6.5	4.0
H36-15GW	H36-15GW/X[b,d]	36-15	1000	40–90	42000	—	—	—	BT-56	Filt. Phos.	15⅜	9½	6	Mog.	16000	375	375	265	6.5	4.0
H36-15KY/C[1]	H1000RC36-15[a] H36-15KY/C[b,c,e]	36-15	1000	40–90	56000	53250	—	—	BT-56	Semi. Phos.	15⅜	9½	6	Mog.	24000+	375	375	265	6.5	4.0
H36-15KY/DX[1]	H36-15KY/DX[c] H1000RDX36-15[a]	36-15	1000	40–90	60250	—	—	—	BT-56	Semi. Phos.	15⅜	9½	6	Mog.	24000+	375	375	265	6.5	4.0
H36-15KY/W	H36-15KY/W[b]	36-15	1000	40–90	58700	56000	—	—	BT-56	Semi. Phos.	15⅜	9½	6	Mog.	24000	375	375	265	6.5	4.0
H34-12GV[1]	H34-12GV[b,c,d,e] H1000A34-12[a]	34-12	1000	40–100	55500	52500	81	80	BT-56	Clear	—	9½	5½	Mog.	16000+	300	425	135	12.0	8.0

Continued on next page.

Fig. 8–116. *Continued*

ANSI Lamp Designation[k]	Manufacturer's Lamp Designation	ANSI Ballast Number	Lamp Watts	Approximate Ballast Watts	Approximate Lumens: Initial (100 hours) Vert.	Initial Horiz.[g]	Lamp Lumen Depreciation** (in per cent) Vert.	Depreciation Horiz.[g]	Bulb[h]	Outer Bulb Finish	Maximum Over-All Length (inches)	Light Center Length (inches)	Arc Length (inches)	Base	Rated Life (hours)	Lamp Voltage (volts): Minimum starting for −20° F RMS	Minimum starting Peak	Operating	Lamp Current (amperes): Starting	Operating
H34-12GW/C[l]	H1000C34-12[a]	34-12	1000	40–100	—	—	—	—	—	—	—	—	—	Mog.						
	H34-12GW/C[b,c,d,e]																			
H34-12GW/DX[l]	H34-12GW/DX[b,c]	34-12	1000	40–100	59670	59000	68	74	BT-56	Phos. Coat	15⅝	9½	5½	Mog.	16000+	300	475	135	12.0	8.0
	H1000DX34-12[a]																			
H34-12GW/W	H34-12GW/W[b,c,d]	34-12	1000	40–100	60500	57250	—	—	BT-56	Phos.	15⅝	9½	5½	Mog.	16000	300	475	135	12.0	8.0
	H34-12GW/W[e]																			
H34-12GW/X	H34-12GW/X[d]	34-12	1000	40–100	40000	—	—	—	BT-56	Filt. Phos.[i]	15⅝	9½	5½	Mog.	16000	300	425	135	12.0	8.0
H34-12KY/C[l]	H1000RC34-12[a]	34-12	1000	40–100	55000	51500	—	—	BT-56	Semi. Phos.	15⅝	9½	5½	Mog.	16000+	300	425	135	12.0	8.0
H34-12KY/W	H34-12KY/W[b,c]	34-12	1000	40–100	57500	55000	—	—	BT-56	Semi. Phos.	15⅝	9½	5½	Mog.	16000	300	425	135	12.0	8.0
H34-12KY/DX	H34-12KY/DX	—	1000	—	57500	55000	—	—	BT-56	Semi. Phos.	15⅝	9½	5⅝	Mog.	16000	300	—	—	—	—
H36-15FC[l]	H1000RSP36-15[a]	36-15	1000	40–90	41000	—	—	—	R-80	Clear Refl.	16 9/16	—	6	Mog.	24000+	375	530	265	6.0	4.0
	R80-H36-15SP[e]				53750	51000					13⅞	—	6	Mog.	16000	375	530	265	6.0	4.0
H36-15FB[l]	H1000RFL36-15[a]	36-15	1000	40–90	42000	—	—	—	—	Frost. Refl.	16 9/16	—	6	Mog.	24000+	375	530	265	6.0	4.0
	R80-H36-15/FL[e]																			
	H36-15FB[b]				53750	51000					13⅞	—	6	Mog.	16000	375	530	265	6.0	4.0
H36-15FA[l]	H1000RSC36-15[a]	36-15	1000	40–90	41500	—	—	—	R-80	Phos. Refl.	16 9/16	—	6	Mog.	24000+	375	530	265	6.0	4.0
	R80C-H36-15[e]	—	—	—	—	—	—	—	—	—	—	—	—	—	24000	375				
H36-15FA/C[l]	H1000RCFL36-15[a]	36-15	1000	40–90	46000	—	—	—	R-80	Phos. Refl.	16 9/16	—	6	Mog.	24000+	375	530	265	6.0	4.0
	H36-15FA/C[b]				40000	—	—	—			13⅞				16000					
	H1000RDXFL36-15[a]				48000	—	—	—			16 9/16				24000+					
H36-15HR	R80C-H36-15/HB[e]	36-15	1000	40–90	51500	49000	—	—	R-80	Phos. Refl.	13⅞	—	6	Mog.	16000	375	530	265	6.0	4.0
H34-12FC	R80-H34-12/SP[e]	34-12	1000	40–100	53500	50500	—	—	R-80	Clear, Refl.	13⅞	—	5½	Mog.	16000	300	425	135	12.0	8.0
H34-12FB	R80-H34-12/FL[e]	34-12	1000	40–100	53500	50500	—	—	R-80	Frost, Refl.	13⅞	—	5½	Mog.	16000	300	425	135	12.0	8.0
H34-12FA	R80-H34-12[e]	34-12	1000	40–100	51500	48500	—	—	R-80	Phos. Refl.	13⅞	—	5½	—	16000	300	425	135	12.0	8.0
H34-12HR	R80-H34-12/HB[e]	34-12	1000	40–100	57500	48500	—	—	R-80	Phos. Refl.	13⅞	—	5½	—	16000	300	425	135	12.0	8.6
H9FJ[l]	H3000A9[a]	9	3000	90	—	131500	—	—	T-9½	Clear	55	—	48	S. C. Term.	6000	850	—	535	9.3	6.1
	H9-FJ[b]																			
	H1000RSDX36-15[a]	36-15	1000	40–90	43000	—	—	—	R-80	Phos. Refl.	16 9/16	—	6	Mog.	24000+	375	530	265	6.0	4.0

* The life and light output ratings of mercury lamps are based on their use with ballasts that provide proper operating characteristics. Ballasts that do not provide proper electrical values may substantially reduce either lamp life or light output, or both. Unless otherwise noted, ratings apply to operation in ac circuits.

** Per cent initial light output at 70 per cent of rated life.

[a] General Electric. [b] Westinghouse. [c] Sylvania. [d] ITT. [e] North American Philips Lighting Corporation.

[f] Average of manufacturers' ratings. For approximate lumen maintenance data, refer to Fig. 8–52.

[g] Horizontal lumen data are based on operation on uncompensated ballasts which operate a lamp at rated watts. If lumen compensated ballasts are used, lumen data will be the same as for vertical operation.

[h] All bulbs are of heat resistant glass except H33-1GL/Y and PAR types.

[i] Pigment on bulb surface to improve color. [j] Recommended for street lighting applications.

[k] The basic ANSI designation system does not distinguish between different bulb coatings for the same lamp type. Each manufacturer adds a suffix to identify the coating.

Fig. 8–116. *Continued*

B. Metal Halide Lamps*

Manufacturer's Lamp Designation	Watts	Bulb[a]	Approximate Initial Lumens (100 hours) (10 hours per start)	Lamp Lumen Depreciation (in per cent)	Lamp Current (amperes)		Rated Life (hours)	Burning Position	Base	Maximum Over-All Length (inches)	Light Center Length (inches)	Minimum Starting Voltage		Operating Lamp Voltage	Arc Length (inches)
					Starting	Operating						RMS	Peak		
M-175[l]	175	BT-28	12000[v]	—	—	1.5	7500		Mog.	8 15/16	5	382	540	130	1 1/8
M-175/C[l]	175	BT-28[b]	12000[v]	—	—	1.5	7500		Mog.	8 15/16	5	382	540	130	1 1/8
M400[l]	400	BT-37	32000[v] 30000[h]	—	—	3.3	10500		Mog.	11 1/2	7	382	540	130	1 7/8
M400/C[l]	400	BT-37[b]	32000[v] 30000[h]	—	—	3.3	10500		Mog.	11 1/2	7	382	540	130	1 7/8
MV400/VBU/O[m]	400	E-37	34000[v]	72	5.0	3.2	8000	—[d]	Mog.	11 5/16	7	—	—	—	—
MV400/VBD/O[m]	400	E-37	34000[v]	72	5.0	3.2	8000	—[e]	Mog.	11 5/16	7	—	—	—	—
MV400/HV/E[m, c]	400	E-37	32000	72	5.0	3.2	8000	—[f]	Mog.	11 5/16	7	—	—	—	—
MH400BU[n]	400	BT-37	34000[v]	70[h] 58[v]	—	—	10500	BU to Horiz.	Mog.	11 1/2	7	—	—	—	—
MH400BD[n]	400	BT-37	34000[v]	70[h] 58[v]	—	—	10500	BD to Horiz.	Mog.	11 1/2	7	—	—	—	—
M1000[l]	1000	BT-56	100000[v] 95000[h]	—	—	4.3	7500		Mog.	15 3/8	9 1/2	530	750	252	3 13/16
MV1000/VBU/O[m]	1000	BT-56	91500[v]	64	6.0	4.0	7500	[d]	Mog.	15 1/16	9 3/8	—	—	—	—
MV1000/HBU/E[m, c]	1000	BT-56	85500[h]	77	6.0	4.0	6000	[i]	Mog.	15 1/16	9 3/8	—	—	—	—
MV1000/HBD/E[m, c]	1000	BT-56	85500[h]	77	6.0	4.0	6000	[j]	Mog.	15 1/16	9 3/8	—	—	—	—
MH1000[n]	1000	BT-46	90000[v]	—	—	—	8000	Horiz. ±30°	Mog.	14 1/2	9 1/2	—	—	—	—
MH1000BU[n]	1000	BT-46	95000[v, h]	72[h] 62[v]	—	—	8000	BU to Horiz.	Mog.	14 1/2	9 1/2	—	—	—	—
MH1000BD[n]	1000	BT-46	95000[v, h]	—	—	—	8000	BD to Horiz.	Mog.	14 1/2	9 1/2	—	—	—	—
M1500[l, c]	1500	BT-56	155000[v] 150000[h]	—	—	6.3	1500		Mog.	15 3/8	9 1/2	530	750	275	3 13/16
MV1500/HBU/E[m, c]	1500	BT-56	153000	91	9.0	5.9	1500[k]	[i]	Mog.	15 1/16	9 3/8	—	—	—	—
MV1500/HBD/E[m, c]	1500	BT-56	153000	91	9.0	5.9	1500[k]	[j]	Mog.	15 1/16	9 3/8	—	—	—	—

C. High Pressure Sodium

Manufacturer's Lamp Designation	Watts	Bulb (Clear) (A)	Approximate Initial Lumens (100 hours) (E)	Lamp** Lumen Depreciation (in per cent)	Lamp Current (amperes)		Lamp Voltage (Nom.) (B)	Rated Life (hours) (F)	Burning Position	Base	Maximum Over-All Length (inches)	Light Center Length (inches)
					Starting	Operating						
LU250/BU[m]	250	E-18	25500	88	4.5	3.0	100	10000	(C)	Mog.	9 3/4	5 3/4
LU250/BD[m]	250	E-18	25500	88	4.5	3.0	100	10000	(D)	Mog.	9 3/4	5 3/4
C250[n]	250	E-18	25500	88	4.5	3.0	100	10000	Any	Mog.	9 3/4	5 3/4
LU400/BU[l, m]	400	E-18	47000	86	7.0	4.7	100	15000	(C)	Mog.	9 3/4	5 3/4
LU400/BD[l, m]	400	E-18	47000	86	7.0	4.7	100	15000	(D)	Mog.	9 3/4	5 3/4
C400[n]	400	E-18	47000	86	7.0	4.7	100	15000	Any	Mog.	9 3/4	5 3/4

* The life and light output ratings are based on lamp use with ballasts that provide proper operating characteristics. Ballasts that do not provide proper electrical values may substantially reduce either lamp life or light output or both. Ratings apply to operation in ac circuits.

[a] Clear bulb unless footnoted. All bulbs of heat resistant glass. [b] Phosphor coated outer bulb.

[c] For use in enclosed luminaires only.

[d] Burning position: vertical base up to 15° of vertical base up. Use in open or enclosed luminaires.

[e] Burning position: vertical base down to 15° of vertical base down. Use in open or enclosed luminaires.

[f] Burning position: horizontal to vertical.

[g] Per cent initial light output at 70 per cent of rated life.

[h] Horizontal lumens. [v] Vertical lumens.

[i] Burning position: base up to base 15° below horizontal. [j] Burning position: base down to base 15° above horizontal.

[k] At 5 hours per start.

[l] Sylvania. [m] General Electric. [n] Westinghouse.

** Per cent Initial Light Output at 70 per cent of Rated Life.

(A) All bulbs are heat and shock resistant glass.

(B) Ballast must provide a high-voltage starting pulse.

(C) Burning position: base up to horizontal.

(D) Burning position: base down to horizontal.

(E) Lumen values are for any burning position.

(F) 10 hours per start.

Fig. 8–117. Self-Ballasted Mercury Lamps Used for General Lighting

Watts	Bulb	Volts[a]	Bulb Finish[c]	Approximate Initial Lumens (100 hours)[b]	Lamp Current (amperes)[b]		Rated Life (hours)[d]	Base	Burning Position	Max. Over-All Length (inches)	Light Center Length (inches)
					Starting	Operating					
160	PS-30	120	White	2400	1.65	1.45	5,000–8,000	Med.	Any	7¼	5³⁄₁₆
	PS-30	208, 230	White	3250	.96	.77	12,000	Med.	Any	7⁵⁄₁₆	6⁵⁄₁₆
	R-40	120	Refl. Fl.	—	1.65	1.45	5,000–8,000	Med.	Any	6⁹⁄₁₆	—
250	PS-35	120	White	5200	3.0	2.4	8,000–10,000	Med.	Any	9⅜	6⅞
	PS-35	230	White	6250	1.3	1.1	12,000	Med.	Any	9	8
	R-40	120	Refl. Fl.	—	3.0	2.4	8,000–10,000	Med.	Any	6⁹⁄₁₆	—
300	PS-35	208, 230	Clear	6700	2.5	1.65	8,000–10,000	Mog.	Any	9⅜	7
	PS-35	208, 230	Del. White	6700	2.5	1.65	8,000–10,000	Mog.	Any	9⅜	7
450	PS-40	120	Clear	10,000	5.9	4.1	8,000–10,000	Mog.	Base Up ±70°	9¾	7
	PS-40	208, 230	Delux White	12,200	3.5	2.3	8,000–10,000	Mog.	Any	9¾	7
	E-37	230	White	14,500	2.4	1.9	16,000	Mog.	Any	11½	10½
	BT-37	208, 230	½ Coated White	12,000	3.5	2.3	14,000–16,000	Mog.	Any	11½	7
	R-60	230	Refl. Fl.	11,600	2.4	1.9	16,000	Mog.	Any	10¾	—
	R-60	208	White Refl.	12,200	3.5	1.3	14,000–16,000	Mog.	Any	10¾	—
500	PS-40	208, 230	White	14,700	3.0	2.2	14,000	Mog.	Any	10¾	9¾
750	R-57	120	White	17,700	8.8	6.6	14,000–16,000	Mog.	Any	12¾	8¾
	R-57	208, 230	Clear	22,600	5.1	3.6	18,000–20,000	Mog.	Any	12¾	8¾
	R-57	208, 230	White Refl.	22,100	5.1	3.6	18,000–20,000	Mog.	Any	12¾	8¾
	R-60	208, 230	Refl. Fl.	18,500	5.1	3.6	16,000	Mog.	Any	10¾	—
	R-60	208, 230	White Refl. Fl.	17,000	5.1	3.6	16,000	Mog.	Any	10¾	—
1250	BT-56	208, 230	Clear	37,500	10.0	6.4	14,000–16,000	Mog.	Any	15⅜	9½
	BT-56	208, 230	White	35,500	10.0	6.4	14,000–16,000	Mog.	Any	15⅜	9½
	E-56	230, 250	White	41,000	6.8	5.0	12,000	Mog.	Any	15⅜	14

[a] 120-volt lamps for 110–130V circuits; 208-volt lamps for 200–224V circuits; 230-volt lamps for 220–240V circuits; 250 volt lamps for 240–260V circuits. All 50 to 60 hertz.
[b] Values apply to lower voltage where two different voltage lamps are indicated.
[c] White—Phosphor coated; White Refl. Fl.—Phosphor coated on metallic reflector.
[d] Life expectancy dependent on line voltage.

Fig. 8–118. Typical Low Pressure Sodium Lamps for Street and Floodlighting Applications*

Lamp Designation	Type	Rated Watts	Nominal Volts	Amperes	Bulb Diameter (inches)	Nominal Length (inches)	Base	Initial Lumens
SLI/H 60	Linear	60	80	.83	1½	16¼	G13/10 × 35 (Med. Bipin)	6000
SLI/H 200	Linear	200	135	1.6	1½	35¼	G13/10 × 35 (Med. Bipin)	25000
SOX 35W	"U" Tube	35	70	.60	2⅛	12 3/16	D.C. Bay.	4650
SOX 55W	"U" Tube	55	105	.59	2⅛	16¾	D.C. Bay.	7700
SOX 90W	"U" Tube	90	115	.94	2⅝	20 13/16	D.C. Bay.	12500
SOX 135W	"U" Tube	135	160	.95	2⅝	30½	D.C. Bay.	21500
SOX 180W	"U" Tube	180	245	.91	2⅝	44⅛	D.C. Bay.	32000

* Used mainly in Canada. 15000 hours life.

Fig. 8–119. Characteristics of Typical Low-Pressure Mercury-Vapor Sources of Ultraviolet Energy

A. Near Ultraviolet Output (320–420 Nanometers)

Designation	Nominal Lamp Watts	Rated Life[a] (hours)	Relative Black Light Energy[d] (100 hours)	Nominal Over-All Length[e] (inches)	Useful Arc Length (inches)	Diameter (inches)	Base	Minimum RMS Starting Voltage[t]	Lamp Volts	Lamp Arc Current (amperes)
Black Light-Preheat Start Lamps										
MF-3RP12/BL[u]	3	—	3	2 9/16[f]	—	1½	D.C. Bay.	10–16 DC	8	.500 max.
MF-5000[u]	4	—	3	2 9/16[f]	—	1½	D.C. Bay.	24–28 DC	10	.500 max.
F4T4/BL[x]	4–7	4000	4	5¼	6	½[h]	Oval Small 4 Pin	120	51–55	.09–.16
F4T5/BL F4T5/BLB[b] [x]	4	4000	4 3	6	2½	⅝	Min. Bipin	120	29	.17
F6T5/BL F6T5/BLB[b] [x]	6	6000	7 6	9	5½	⅝	Min. Bipin	120	42	.16
F8T5/BL F8T5/BLB[b] [x]	8	6000	13 8	12	8½	⅝	Min. Bipin	120	57	.145
F14/T8/BL	14	7500	21	15	10	1	Med. Bipin	120	45	.365
F15T8/BL F15T8/BLB[b]	15	7500	25 20	18	13	1	Med. Bipin	120	55	.305
F20T12/BL F20T12/BLB[b]	20	7500	42 31	24	19	1½	Med. Bipin	120	56	.38
F25T12/BL/28	25	7500	53	28	23	1½	Med. Bipin	120	63	.46
F25T12/BL/33	25	7500	57	33	28	1½	Med. Bipin	120	61	.48
F30T8/BL F30T8/BLB[b]	30	7500	65 54	36	31	1	Med. Bipin	176	98	.355
F40BL F40BLB[b]	40	12000	100 81	48	43	1½	Med. Bipin	176	102	.43
F40BL/5	65	7500	130	48	43	1½	Med. Bipin	195	—	—
Instant Start Lamps										
F64T6/BL	50	7500	110	64	57	¾	Single Pin	540	201	.30
F96T8/BL	65	7500	150	96	89	1	Single Pin	675	263	.30
Rapid Start Lamps										
FCT9/BL[c]	22.5	7500	38	8¼[y]	19	1⅛	4-Pin	185	62	.39
FC12T10/BL[c]	33	7500	63	12[y]	31	1¼	4-Pin	205	80	.43
F40BL F40BLB[b]	40	12000	100 81	48	43	1½	Med. Bipin	200	100	.43
F72T12/BL/HO	85	12000	190	72	65	1½	Recessed D.C.	260	113	.80

See footnotes on page 8–110.

Continued on next page.

Fig. 8–119. *Continued*

B. Bactericidal Ultraviolet Output (253.7 Nanometers)

Designation	Rated Power Input (watts)	Rated Life (hours)	253.7 nm Ultra-violet Power Output after 100-Hours Oper-ation (watts)	253.7 nm Ultra-violet Power Density at Point 1 Meter from Arc Axis after 100 Hours Oper-ation (mi-crowatts/sq cm)	Nominal Over-All Length (inches)	Useful Arc Length (inches)	Nominal Diameter (inches)	Outside Wall Tem-perature at Center in Still Air (80°F)	Base	Minimum RMS Starting Voltage[t]	Potential Drop between Arc Electrodes for Rated Output Opera-tion (volts)	Arc current (amperes)
Preheat Starting												
G4T4/1[p]	7	7500[j]	0.9	15	5¼	6	0.5[h]	115	Oval Small 4 Pin	120	52	0.16
G8T5[p, r]	8	7500[g]	1.3–1.5	15	12[l]	8.5	0.625	115	Min. Bipin	120	60	0.155
G15T8[p, r]	15	7500[g]	3.3–3.6	35–41	18[l]	14	1	110	Med. Bipin	120	55	0.30
G25T8[p, r]	25	7500[g]	5–5.2[k]	54[k]	18[l]	14	1	—	Med. Bipin	120	48.5	0.6
G30T8[p, r]	30	7500[g]	8.3–9.0	80–90	36[l]	32	1	110	Med. Bipin	200	98	0.355
Instant Starting												
794[s]	3.5	4000[g]	0.1	1.2	2.25	0.375	1.375		Int. Screw	24	10–12	0.35
G4-S11[p, r]	3.8	6000[w]	0.1	2.0	2.25	0.375	1.375		Int. Screw	24	10–12	0.35
782-H-10[s, v]	12	17500[g]	2.0	20	16.875[l]	10	0.625	120	Single Pin	950	240	0.060
782-H-10[s, v]	20	12000[g]	2.8	28	16.875[l]	10	0.625	120	Single Pin	900	235	0.090
782-L-10[s, v]	12	17500[g]	2.0	20	16.875[l]	10	0.625	120	Single Pin	950	240	0.060
782-L-10[s, v]	20	12000[g]	2.8	28	16.875[l]	10	0.625	120	Single Pin	900	235	0.090
782-L-20[s, v]	14	17500[g]	3.9	35	26.875[l]	20	0.625	120	Single Pin	900	325	0.050
782-L-30[s, v]	17	17500[g]	5.2	46	36.875[l]	30	0.625	120	Single Pin	950	410	0.050
782-L-30[s, v]	29	12000[g]	8.3	73	36.875[l]	30	0.625	120	Single Pin	900	390	0.090
782-H-30[s, v]	17	17500[g]	5.2	46	36.875[l]	30	0.625	120	Single Pin	950	410	0.050
782-H-30[s, v]	29	12000[g]	8.3	73	36.875[l]	30	0.625	120	Single Pin	900	390	0.090
782-VH-29[s, v]	29	12000[g]	8.3	73	36.875[l]	30	0.625	120	Single Pin	900	390	0.090
93A-1 (2851)[q]	8	17500	1.9	21	16.250	11.500	0.520	120	Single Pin	3000	300	0.030
84A-1 (2852)[q]	14	17500	4.1	46	29.375	24.750	0.520	120	Single Pin	3000	450	0.030
83A-1[q]	22	17500	3.1	35	16.250	10.750	0.6875	115	Single Pin	600	200	0.120
94A-1[q]	32	17500	7.2	80	30.250	24.750	0.6875	115	Single Pin	600	300	0.120
86A-45[q]	8	17500	1.4	16	5.250	4.500	0.6875	115	Bipin	600	200	0.120
87A-45[q]	22	17500	4.3	47	8.125	10.500	0.6875	115	Bipin	600	200	0.120
88A-45[q]	32	17500	10.4	113	15.125	24.500	0.6875	115	Bipin	600	300	0.120
G10T5-½H[s, v]	16	7500[g]	5.3	55	16.875[l]	10.0	0.625		Single Pin	450	55	0.400
G36T6[p, s, v] & G36T6H[s, v]	16–17	7500[g]	6.3–8.0	65–70	36[l]	30	0.625–0.75	100	Single Pin	450–660	150–180	0.100–0.120
G36T6[p, s, v] & G36T6H[s, v]	22–25	7500[g]	9.7–11.4	100	36[l]	30	0.625–0.75	105	Single Pin	450–660	135–150	0.200
G36T6[p, s, v] & G36T6H[s, v]	29–32	7500[g]	11.6–13.1[k]	115–120[k]	36[l]	30	0.625–0.75	115	Single Pin	450–660	115–130	0.300
G36T6[p, s, v] & G36T6H[s, v]	36–39	7500[g]	13.1–13.8[k]	120–135[k]	36[l]	30	0.625–0.75	125[i]	Single Pin	450–660	105–120	0.425
G37T6VH[s, v]	36–39	7500[g]	13.1–13.8[k]	120–135[k]	36[l]	30	0.625–0.75	125[i]	Single Pin	440–460	105–120	0.420
G64T6[p]	65	7500[g]	48.0	150	64[l]	57	0.75	—	Single Pin		180	0.425

[a] At 3 hours per start. Useful life will depend on the requirements of the application.
[b] Blue glass bulb—requires no external filter.
[c] Circline lamp.
[d] Relative value of 100 equals 8100 fluorens.
[e] Including one lamp plus two standard lampholders.
[f] Maximum over-all length.
[g] At 8 hours per start.
[h] Bent tube construction makes tube approximately one inch wide.
[i] This must be in moving air.
[j] At 3 hours per start.
[k] Approximate output in still air at 77° F; output increased in cool or moving air.
[l] Length includes two standard lampholders.
[p] General Electric
[q] Hanovia
[r] Sylvania
[s] Westinghouse
[t] Minimum required RMS voltage between lamp terminals for reliable starting from 50°F (10°C) to 110°F (43.3°C).
[u] RP-12 bulb.
[v] Voltarc Tubes.
[w] Useful lamp life. Actual burning hours longer.
[x] Operation on same "universal" ballast.
[y] Nominal O.D. of lamp.

Lamp Designation	Lamp Watts (rated)	Outer Bulb	Base	Rated Initial Lumens	Rated Average Life (hours at 5 hours per start)	Operating Volts	Starting Amperes	Operating Amperes	Warm-up Time[d] (minutes)	Maximum Over-All Length (inches)	Light Center Length (inches)	Lighted Length (inches)	Recommended Applications: Blacklight	Blue & White Printing	Copyboard Lighting	Vacuum Frame Printing	General Ultraviolet (300–400 nm)	General Ultraviolet (220–400 nm)	Therapeutic	Photochemical
Short Lamps (Less Than 25 Inches Over-All Length)																				
H75A43[f] H43-AZ[i,g]	75	B-21, E-17 or E-21	Med.	2700	10000	130	0.95	0.66	3	6	3 5/8	1 1/16	X				X			X
H85A3/UV[f]	85	T-10	Med.	3050	500	250	0.6	0.4	6	5 3/8	3	1 1/16					X	X	X	X
H100A4/T[f] A-H4[g] H38-AW[j]	100	T-10	Admed.	3700	6000	130	1.3	0.9	3–6	5 5/8	3 7/16	1 1/8	X				X			X
H100BL38-4[f] B-H4[g]	100	T-16	Admed.	—	6000	130	1.3	0.9	3–6	5 1/2	3 7/16	1 1/8	X							
H38-4HT[i,j]	100	BT-25	Mog.	3830	16000	130	1.3	0.9	3	7 1/2	5	1 1/16	X				X			X
H100PSP44-4[f] H100PFL44-4[f] H44-4GS[g,i] H44-4JM[g,i]	100	PAR-38	Ad. Sk.	2450	12000	130	1.3	0.9	3–8	5 7/16	—	1	X				X			X
16A[h]	100	T-4[b]	Pin. Term.	—	1000	115	1.5	1.15	5	2 1/2	—	1 3/4					X	X	X	
H175A39-22[f] H39-22KB[g,i,j]	175	BT-28 or E-28	Mog.	7500	24000+	130	2.15	1.5	4–5	8 5/16	5	2	X				X			X
54A[h]	200	T-5 1/2[b]	Wire Term.	—	1000	125	2.5	1.9	5	11	—	4 1/2					X	X	X	X
H250A37-5[f] H37-5KB[g,i,j]	250	BT-28 or E-28	Mog.	11500	24000+	130	2.9	2.1	4	8 5/16	5	2 3/8	X				X			X
79A[h]	325	T-6 1/2[b]	Wire Term.	—	1000	135	3.5	2.8	5	11	—	4 1/2					X	X	X	X
H100A33[f] H33-1CD[g,i,j]	400	E-37 or BT-37	Mog.	21000	24000+	135	5.0	3.2	4–5	11 1/2	7	2 3/4	X				X			X
H33-1AR[i]	400	T-16	Mog.	19500	6000	135	5.0	3.2	4	11	7	2 3/4	X				X			X
H2T5½[f]	400	T-5 1/2[b]	RSC[l]	—	1000	135	5.0	3.3	2–3	4 1/2	—	2 3/8	X		X	X	X	X		X
64A[h]	400	—	—	—	1000	195	3.3	2.3	—	15 1/4	—	11 3/4		X	X	X	X	X		X
H33-1FP[i,j]	400	R-60	Mog.	18500	16000	135	5.0	3.2	6	10 3/4	—	2 13/16	X				X			X
VMH-400A[e]	400	T-4 1/2[c]	Wire Term.	—	1000	195	3.3	2.3	—	15 5/8	—	11 3/8	X	X	X	X	X	X		X
VMH-400B/35[e]	400	T-5 1/2[b]	3/8 Ferrule	—	200	95	—	4.8	—	4 13/16	—	1 3/8	X	X	X	X	X	X		X
H40-17MA[i]	425	BT-37	Mog.	25000	16000	265	2.5	1.7	4	11 1/2	7	3 13/16	X				X			X
73A[h]	560	T-7 1/2[b]	Wire Term.	—	1000	145	6.0	4.4	5	11	—	4 1/2					X	X	X	
51B[h]	600	T-8	3/8 Sleeve	—	1000	500	1.9	1.34	5	17 5/8	—	12	X	X	X	X	X	X		X
VMH-600A[e]	600	T-4 1/2[c]	Wire Term.	—	1000	500	1.9	1.34	—	16	—	11 3/4	X	X	X	X	X	X		X

See footnotes on page 8-113.

Continued on next page.

Fig. 8–120. *Continued*

Lamp Designation	Lamp Watts (rated)	Outer Bulb	Base	Rated Initial Lumens	Rated Average Life (hours at 5 hours per start)	Operating Volts	Starting Amperes	Operating Amperes	Warm-up Time[d] (minutes)	Maximum Over-All Length (inches)	Light Center Length (inches)	Lighted Length (inches)	Recommended Applications: Blacklight	Blue & White Printing	Copy board Lighting	Vacuum Frame Printing	General Ultraviolet (300–400 nm)	General Ultraviole (220–400 nm)	Therapeutic	Photochemical
74A[h]	700	T-7½[b]	Wire Term.	—	1000	155	7.5	5.3	5	13½	—	7¼					X	X	X	X
H700A35-18[f] / H35-18NA[i,j]	700	BT-46	Mog.	39000	24000+	265	4.5	2.8	4–5	14½	9½	5⅞	X		X	X	X			X X
VMH-700A[e]	700	T-4[e]	Wire Term.	—	1000	500	2.6	1.6	—	17⅜	—	13⅛	X	X	X	X	X	X		X
H12T3[f]	750	T-3[l,n]	RSC	30000	750	500	2.65	1.65	2–3	14 3/16	—	12	X	X	X	X	X	X		X
4A[h]	750	—	—	—	1000	500	2.7	1.7	5	18	—	13½	X	X	X	X	X	X		X
H750T3/12A,B,C[i]	750	T-3[b]	[m]	30000	1000	500	2.45	1.65	4	14 3/16	—	12	X		X	X	X	X		X
H3T7[f]	750	T-7[b]	RSC	—	1000	135	10.0	6.0	2–3	5¼	—	2 11/16	X		X	X	X	X		X
VMH-800A/1[e]	800	T-10	15 mm Ferrule	—	1000	660	1.95	1.5	—	17½	—	12¼	X	X	X	X	X	X		X
H14T3[f]	850	T-3[l,n]	RSC	34000	750	500	3.10	1.9	2–3	16 3/16	—	14	X	X	X	X	X	X		X
H850T3/14B,C[i]	850	T-3[b]	[m]	34000	1000	500	2.8	1.9	4	16 3/16	—	14						X		X
H1000A36-15[f] / H36-15GV[g,i,j]	1000	BT-56	Mog.	57000	24000+	265	6.0	4.0	4	15 1/16	9⅜	5⅞	X		X	X	X			X
H1000A34-12[f,j] / H34-12GV[g,i]	1000	BT-56	Mog.	56000	16000+	135	12.0	8.0	4	15 1/16	9⅜	5⅝	X		X	X	X			X
H34-12FC[i,j]	1000	R-80	Mog.	53500	16000	135	12.0	8.0	6	13⅞	—	5⅝	X		X	X	X			X
H36-15FC[i,j]	1000	R-80	Mog.	54000	16000	265	6.0	4.0	6	13⅞	—	5⅞	X		X	X	X			X
H18T3[f]	1200	T-3[l,n]	RSC	48000	1000	1200	1.75	1.1	2–3	20 3/16	—	18	X	X	X	X	X	X		X
H1200T3/18B,C[i]	1200	T-3[b]	[m]	48000	1000	1200	1.6	1.1	4	20 3/16	—	18						X		X
67A[h]	1200	—	—	—	1000	450	5.0	3.1	—	22	—	17½		X	X	X	X	X		X
89A[h]	1200	T-6½	Wire Term.	—	1000	270	6.3	5.1	—	17	—	12		X	X	X	X	X		X
VMH-1200A[e]	1200	T-6¼[e]	Wire Term.	—	1000	450	5.0	3.1	—	21¼	—	17	X	X	X	X	X	X		X
Long Lamps (Over 25 Inches Over-All Length)																				
71A[h]	1200	—	—	—	1000	1200	1.8	1.2	—	38½	—	35		X	X	X	X	X		X
VMH-1200B[e]	1200	T-3¾[e]	Wire Term.	—	1000	1200	1.8	1.2	—	37½	—	33¼	X	X	X	X	X	X		X
VMH-1400/A[e]	1400	T-3¾[e,n]	Wire Term.	—	1000	1700	1.7	1.0	—	51¾	—	47½	X	X	X	X	X	X		X
H24T3[f]	1440	T-3[b,n]	RSC	58000	1000	1200	2.2	1.35	2–3	26 3/16	—	24		X	X	X	X	X		X
H1440T3/24B,C[i]	1440	T-3[b]	[m]	58000	1000	1200	2.0	1.35	4	26 3/16	—	24						X		X
33D[h]	1500	—	—	—	1000	1500	1.8	1.2	—	54	—	48		X	X	X	X	X		X
VMH-1500/A[e]	1500	T-3¾[e]	Wire Term.	—	1000	1500	1.8	1.1	—	51¾	—	47½	X	X	X	X	X	X		X
70A[h]	1800	—	—	—	1000	1000	3.1	2.2	—	27½	—	22		X	X	X	X	X		X
78A[h]	2000	T-7½[b]	Wire Term.	—	1000	550	6.0	4.0	5	25⅝	—	21½		X	X	X	X	X		X
30A[h]	2000	—	—	—	1000	550	6.0	4.2	—	54	—	48			X	X		X		

34A[h]	2200	—	—	—	1000	1000	3.9	2.4	5	29½	—	25	X	X	X	X	X	X		X
462A[h]	2200	—	—	—	1000	1000	3.9	2.4	5	27	—	22	X	X	X	X	X	X		X
VMH-2200/A[e]	2200	T-6¼[c]	Wire Term.	—	1000	1000	3.9	2.4	—	26¼	—	22	X	X	X	X	X	X		X
VMH-2250/A[e]	2250	T-7[c]	Wire Term.	—	1000	1000	4.0	2.5	—	29⅛	—	24⅞	X	X	X	X	X	X		X
158C[h]	2750	—	—	—	1000	1700	2.8	1.9	—	54	—	48		X	X	X	X	X		X
24A[h]	3000	—	—	—	1000	550	9.0	6.3	—	54	—	48		X	X	X	X	X		X
B-H9[g]	3000	T-9½[b]	Spade Term.	120000	3000–4500	535	9.3	6.1	7	57	—	50		X	X	X	X			X
153A[h]	3000	T-4½[b]	19/32 Sleeve	—	1000	1500	3.0	2.15	—	52 7/16	—	48		X	X	X	X	X		X
166B[h]	3500	—	—	—	1000	1500	3.9	2.6	—	54	—	48		X	X	X	X	X		X
47A[h]	3500	—	—	—	1000	925	6.0	4.3	—	54	—	48		X	X	X	X	X		X
61A[h]	4000	—	—	—	1000	1500	4.5	3.0	—	54	—	48		X	X	X	X	X		X
77A[h]	4500	—	—	—	1000	955	6.6	5.0	—	48½	—	42		X	X	X	X	X		X
57A[h]	4800	T-7½[b]	Wire Term.	—	1000	1250	6.0	4.5	—	54	—	48		X	X	X	X	X		X
175B[h]	5000	—	—	—	1000	1500	5.8	3.8	—	54	—	48		X	X	X	X	X		X
59B[h]	5500	—	—	—	1000	1250	6.6	5.0	—	64	—	59		X	X	X	X	X		X
56B[h]	7500	—	—	—	1000	1650	7.3	4.7	—	54	—	48		X	X	X	X	X		X
40B[h]	9000	—	—	—	1000	1650	8.8	5.7	—	64	—	59		X	X	X	X	X		X

[a] May be burned in any position. All outer bulbs have clear finish except B-H4 and H100BL38-4 which have natural purple and H44-4GS, H100PSP44-4, H44-4JM and H100PFL44-4 which have reflector bulbs.
[b] Only single bulb used.
[c] Single bulb. Can be furnished with outer bulb and fittings.
[d] Approximate—depends upon ballast, luminaire, and ambient temperatures.
[e] Voltarc Tubes.
[f] General Electric.
[g] Westinghouse.
[h] Hanovia.
[i] Sylvania.
[j] Radiant.
[k] Seconds.
[l] Recessed Single Contact—RSC.
[m] A—ceramic tubular, B—RSC, C—axial lead with ceramic support.
[n] Ozone free.

Fig. 8–121. Footnotes (see next page.)

[a] Designed for ac but may be used on dc.
[b] Anode up ±10 degrees.
[c] These lamps are double bulb lamps. Dimensions are for outer bulb medium prefocus base. Light center length is 2 13/16 inches.
[d] Not including bulb tip—¼-inch additional clearance required at tip location.
[e] These lamps may be operated in any position but best over-all performance will be obtained with vertical operation.
[f] Available with third starting electrode.
[g] Hanovia.
[h] Westinghouse.
[i] Sylvania.
[j] PEK Labs.
[k] Duro-Test.
[l] General Electric.
[m] Arc tube enclosed by outer bulb. Dimensions are for outer bulb. Special medium bipost base. Light center length 2 11/16 inches.
[n] Ellipsoidal cold-mirror reflector outer bulb. Anode terminal on lamp face, cathode terminal at base. Beam converges approximately 10 inches in front of bulb rim.
[o] Ushio Electric.
[p] ac or dc.

Fig. 8-121. Approximate Operating Characteristics of Mercury-Argon, Mercury-Xenon, and Xenon Short-Arc Lamps

Designations	Arc Tube Maximum Diameter[d] (inches)	Gas	Watts	Arc Length (mm)	Average Luminance Along Arc Axis (cd/sq mm)	Rated Average Life (hours)	Burning Position	Maximum Over-All Length (inches)	Operating Volts	Current (approximate amperes)		Operating Vapor Pressure (approximate atmospheres)	Initial Lumens	Time To Full Output (minutes)
										Start	Operating			
HGK100[i, f]	1/2	Hg-Ar	100 dc	0.5	1400	100	Vert.	3 1/2	16–24	10.0	5.0	—	2200	—
PEK110[j, f]	1/2	Hg	100 dc	0.3	1400	100	Horiz. or Vert.	3 1/2	16–24	—	5.2	—	2200	—
USH102[o]	1/2	Hg	100 dc	0.25	1700	200	Vert.	3 1/2	16–24	—	3.5	—	2200	—
PEK107[j]			100[p]	0.3	1400	250							2200	
PEK112[j]		Hg	100[p]	0.3	1400	250							2200	
HGK200[a, i, f]	3/4	Hg-Ar	200 ac	2.5	250	200	Vert.	4 1/4	49–65	12.0	3.5	—	9500	—
PEK202[j, f]	11/16	Hg	200 dc	1.8	330	400	Vert.	4 3/4	50–65	—	3.5	—	9500	—
USH200[o]	23/32	Hg	200 dc	2.2	330	400	Vert.	5	49–65	—	3.5	—	10000	—
PEK203[j]		Hg	200[p]	1.8	330	400							9500	
PEK212[j]		Hg	200[p]	1.8	330	700							9500	
SAH-250A[h]	1/2[e]	Hg-Ar	250 ac	2.5	175	300	Vert.[e]	6 1/2	37	10.0	8.0	30	10000	6.5
SAH-250B[h]	1 1/2[e]	Hg-Ar	250 dc	2.5	175	600	Vert.	6 1/4	42	10.0	6.0	30	10000	6.5
USH250[o]	7/8	Hg	250 dc	2.0	310	600	Vert.	6	34–46	—	6.5	—	12500	—
PEK500-2[j, f]	1 1/8	Hg	500 dc	3.9	220	500	Vert.	6 3/4	70–85	—	6.5	—	22500	—
PEK500-3[j]		Hg	500[p]	3.9	220	500							22500	
PEK500-4[j]		Hg	500[p]	3.9	220	500							22500	
USH500[o]	1 1/8	Hg	500 dc	4.1	300	600	Vert.	7 1/2	53–67	—	8.5	—	25000	
517A[g]	1 17/32	Hg-Ar	1000 ac	6.5	240	300	Vert.	7	65	32.0	18.0	25	52000	6.5
USH1000[o]	1 9/16	Hg	1000 dc	6.5	180	600	Vert.	9 1/2	59–67	—	18.0	—	58760	—
919B1[g]	1/2	Hg-Xe	30 dc	0.3–0.4	36	100	—	2	10–16	—	2.7	—	500	—
901B1[g]	25/32	Hg-Xe	200 dc	1.4	190	1000	—	4 1/2	20–25	—	9.0	—	5000	—
510B1[g]	25/32	Hg-Xe	250 dc	3.0	84	300	—	4 1/2	18–24	—	9.5	—	4600	—
537B9[g]	1 1/2	Hg-Xe	1000 ac	5.0	167	300	—	7 1/2	60–70	—	18.0	—	50000	—
528B9[g]	1 1/2	Hg-Xe	1000 dc	5.0	150	1000	—	7 1/2	58–72	—	16.0	—	40000	—
USH1005[o]	1 9/16	Hg	1000 dc	3.0	590	600	Vert.	9 1/2	34–42	—	28.0	—	45000	—
929B9U[g]	2 7/16	Hg-Xe	2500 dc	5.0	540	1000	Anode down	10 3/4	45–55	—	50.0	—	120000	—
973BU[g]	2 7/8	Hg-Xe	3500 dc	5.5	650	1000	Anode down	13	51–61	75–100	62.5	18	190000	6.5
932B39[g]	3 3/8	Hg-Xe	5000 dc	5.5	780	500	—	13.0	54–66	—	83.0	—	230000	—
X-15[j]		Xe	15	0.15	400	50							250	
X-16[j]		Xe	15	0.15	400	30							250	
X-35[j, f]	3/8	Xe	35 dc	0.3	400	200	Horiz. or Vert.	3 3/8	10–14	—	3.0	—	700	—
X-36[j]		Xe	35	0.3	400	200							700	
X-75[j, f]	1/2	Xe	75 dc	0.4	800	400	Horiz. or Vert.	3 1/2	12–16	—	5.4	—	1388	—
X-76[j]		Xe	75	0.4	800	400							1388	
UXL75HK[o]	1/2	Xe	75 dc	0.85	80	300	Horiz.	3 1/2	10–12	—	6.8	—	1600	—
UXL75[o]	15/16	Xe	75 dc	0.85	80	300	Vert.	4 1/4	15.2	—	5	—	1600	—
971C[g]	3/8	Xe	80 dc	0.3–0.4	250	100	—	3 1/8	10–14	—	6.6	—	900	—
XE100[k]	7/16	Xe	100 dc	2.0	170	100	Vert.	4 1/4	24	—	4.2	40	1500	—
XE100A[k]	7/16	Xe	100 dc	0.2	1700	100	Vert.	4 1/4	13.5	—	7.4	40	750	—
X-150[j, f]	3/4	Xe	150 dc	1.3	300	1200	Vert.	4 3/4	17–23	—	7.7	—	3150	—

X-151[j]		Xe	150	1.3	300	1200							3150	
901C1[g]	25/32	Xe	150 dc	1.4	96	1000	—	4½	17–23	—	7.5	—	3200	
510C1[g]	25/32	Xe	150 ac	3.0	20	300	—	4½	18–24	—	8.5	—	3200	—
UXL155HK[o]	¾	Xe	150 dc	2.1	135	1000	Horiz.	5 15/16	16–20	—	7.5	—	3200	—
UXL155[o]	¾	Xe	150 dc	1.8	135	1000	Vert.	5 15/16	20	—	7.5	—	3200	—
914C1[g]	25/32	Xe	300 dc	1.5	540	1000		5¼	15–20	—	17.5	—	9000	—
X-300[j]	¾	Xe	300 dc	1.3	350	1200	Vert.	4¾	14–24	—	17.0	—	6300	—
X-301[j]		Xe	300	1.3	350	1200							6300	
UXL300HK[o]	1	Xe	300 dc	2.6	—	1000	Horiz.	6 15/16	15–19	—	15.0	—	—	
UXL305[o]	1	Xe	300 dc	2.3	220	1000	Vert.	6 15/16	20	—	15.0		7600	
UXL451[o]	1⅛	Xe	450 dc	2.7	290	2000	Vert.	10⅜	18.2	—	25.0	—	13000	
959C[g]	1	Xe	500 dc	0.6–0.9	4000	200	—	4⅞	14–20	—	30.0	—	18000	—
XE500T14[l]	1¾[m]	Xe	500 dc	2.5	—	1500	Vert. ±30°	6½[m]	20	—	25.0	—	15000	—
XE500EAR46[l]	5¾[n]	Xe	500 dc	2.5	—	1500	Base down to horiz.	5½[n]	20	—	25.0	—	—	—
UXL500HK[o]	1 3/16	Xe	500 dc	4.0	—	1000	Horiz.	9¼	16–20	—	25	—	—	
X-500A[j]		Xe	500	2.7	350								15000	
418C9[g]	1 13/32	Xe	800 ac	5.0	57	300	—	7½	27–31	—	30.0	—	26000	—
538C9[g]	1½	Xe	900 dc	5.0	110	1000	—	7½	29–35	—	28.0	—	35000	—
UXL900[o]	1¾	Xe	900 dc	3.3	460	1000	Vert.	12¾	22	—	42	—	30500	—
XE1000[k]	1 13/16	Xe	1000 dc	4.0	380	1000	Vert.[b]	9¾	24	—	41.6	17	35000	—
X-1000A[j]		Xe	1000	3.6	520								37000	
SSX-1000A[j]		Xe	1000	3.6	450								33500	
UXL 1001HK[o]	1 9/16	Xe	1000 dc	5.0	260	800	Horiz.	12½	19–23	—	40	—	26000	
UXL 1600[o]	2¼	Xe	1600 dc	4.2	430	2000	Vert.	14½	26	—	63	—	60000	
X-1600A[j]		Xe	1600	4.2	650								64000	
UXL 2000HK[o]	1¾	Xe	2000 dc	6.0	350	800	Horiz.	14 11/32	23–27	—	70	—	61000	
XE2200[k]	2⅛	Xe	2200 dc	4.5	440	1000	Vert.[b]	10½	23.2	—	95.0	17	80000	—
491C39[g]	2¼	Xe	2200 dc	4.5	500	1000	—	13 3/16	20–22	—	100.0	—	80000	—
X-2500A[j]		Xe	2500	5.8	620								100000	
UXL3000[o]	2 3/16	Xe	3000 dc	6.4	570	1500	Vert.	16	30	—	100	—	120000	
UXL4000HK[o]	2 3/16	Xe	4000 dc	6.5	550	1000	Horiz.	16	26–33	—	120	—	130000	
UXL4200[o]	2⅜	Xe	4200 dc	6.3	660	1000	Vert.	17⅛	30	—	140	—	180000	
XE5000[k]	3⅝	Xe	5000 dc	7.0	870	500	Vert.	19½	34.5	—	145.0	10	240000	—
XE5000/2[l] XE5000/4[l]	3½	Xe	5000 dc	8.0	1500		Vert.	19½	34.5	145.0	145.0	—	275000	0.05
966C39[g]	2 13/16	Xe	5000 dc	7.5	700	1000	—	17⅛	32–36	—	150.0	—	220000	—
X-5000A[j]		Xe	5000	7.0	720								250000	
UXL6000HK[o]	2¾	Xe	6000 dc	9.0	860	800	Horiz.	18	34	—	160	—	250000	
UXL6500[o]	2¾	Xe	6500 dc	8.0	920	1000	Vert.	22¼	40	—	165	—	325000	
L5029[g]	2⅞	Xe	10000 dc	9.0	1000	400	Anode up	22	40	—	250.0	12	500000	—
XE10000[k]	4	Xe	10000 dc	9.0	1250	500	Vert.[b]	24	40	—	250.0	8	550000	—
XE20000[k]	5	Xe	20000 dc	12.5	1700	500	Vert. or Horiz.	19½	45.0	—	450.0	6	1100000	—
X-20KW[j]		Xe	20000	12.5	3000								1150000	
L5020[g]	4¾	Xe	20000 dc	11.0	2000	500	Anode up or Horiz.	19	45.0	—	450.0	8	1000000	—
L5020[g]	4¾	Xe	30000 dc	11.0	2500	300	Anode up	—	49.0	—	62.00	9	1600000	—
X-30KW[j]			30000	12.5	7300								1750000	

See footnotes on page 8–113.

Fig. 8–122. Typical Glow Lamps

Lamp Designation[a]	A9A (NE-2E)	C2A (NE-2H)	C9A (NE-2J)	A1B	A1C	B1A (NE-51)	B2A (NE-51H)
Gas	Neon	Neon	Neon	Neon	Neon	Neon	Neon
Brightness class	Standard	High	High	Standard	High	Standard	High
Nominal watts	1/12	¼	¼	1/25	1/7	1/25	1/7
Circuit volts	105–125 ac or dc	105–125 ac 150 dc	105–125 ac 150 dc	105–125 ac or dc	105–125 ac 150 dc	105–125 ac or dc	105–125 ac 150 dc
Bulb (clear)	T-2	T-2	T-2	T-2	T-2	T-3¼	T-3¼
Base[b] (see Fig. 8–9)	Wire Term.	Wire Term.	S.C. Mid. Fl.	Wire Term.	Wire Term.	Min. Bay.	Min. Bay.
Max. over-all length (inches)	3/4	¾[d]	15/16	½[d]	½[d]	13/16	13/16
Max. breakdown voltage (volts) initial							
ac	65	95	95	65	95	65	95
dc	90	135	135	90	135	90	135
Series resistance (ohms)[e]	100,000	30,000	30,000	220,000	47,000	220,000	47,000
Average life (hours)[c]	25,000	25,000	25,000	25,000	25,000	15,000	25,000

Lamp Designation[a]	B7A (NE-45)	B9A (NE-48)	J2A (AR-3)	J3A (AR-4)	F4A (NE-58)	J5A (NE-30)	L5A (NE-32)	R6A (NE-40)
Gas	Neon	Neon	Argon	Argon	Neon	Neon	Neon	Neon
Brightness class	Standard	Standard	Standard	Standard	Standard	Standard	Standard	Standard
Nominal watts	¼	¼	¼	¼	½	1	1	3
Circuit volts	105–125 ac or dc	105–125 ac or dc	105–125 ac 135 dc	105–125 ac 135 dc	210–250 ac or dc	105–125 ac or dc	105–125 ac or dc	105–125 ac or dc
Bulb (clear)	T-4½	T-4½	T-4½	T-4½	T-4½	S-11	G-10	S-14
Base[b] (see Fig. 8–9)	Cand. Sc.	D.C. Bay.	Cand. Sc.	D.C. Bay.	Cand. Sc.	Med. Sc.	D.C. Bay.	Med. Sc.
Max. over-all length (inches)	1 7/32	1½	1 7/32	1½	1 7/32	2¼	2⅛	3½
Max. breakdown voltage (volts) initial								
ac	65	65	85	85	75	65	75	60
dc	90	90	120	120	100	90	100	85
Series resistance (ohms)[e]	None[e]	30,000	None[e]	15,000	None[e]	None[e]	7,500	None[e]
Average life (hours)[c]	7,500	7,500	150[f]	150[f]	7,500	10,000	10,000	10,000

[a] ANSI. Existing designation in parenthesis.

[b] Wire Term—Wire terminal (generally available with attached resistor, if desired); Min. Bay—Single contact, miniature bayonet; Cand. Sc.—Candelabra screw; D.C. Bay.—Double contact, bayonet candelabra; S.C. Mid. Fl.—Single contact midget flanged; Med. Sc.—Medium screw.

[c] Ac life ratings shown. For standard brightness lamps, life on dc is approx. 60 per cent of ac values. For high brightness lamps, where a minimum of 150 volts is recommended for dc operation, life will be somewhat lower than the 60 per cent figure and current and wattage will increase.

[d] Glass parts only.

[e] Screw base lamps have resistors built into base; all others must be added externally.

[f] Ultraviolet output drops to 50 per cent of initial during life shown in table. Visible light decreases to 50 percent of initial in 1000 hours.

Fig. 8-123. Enclosed Concentrated-Arc Lamps*

Watts	Volts	Amperes	Average Light Source Diameter (inches)	Average Luminance (candelas per sq cm)	Average Candlepower	Rated Life (hours)	Maximum Temperature (°F)	
							Bulb	Base
2**	27	.099	.007	1800	.30	150	140	100
2	38	.055	.005	2400	.30	150	140	100
10	20	.500	.015	4500	4.7	450	225	130
25	20	1.25	.030	3500	16.0	325	355	145
100	16	6.25	.072	3800	100	375	470	160
300	20	15.00	.110	4500	275	250	520	180

* Nominal color temperature 3200 K on all lamps except 2**.
Various types of bulbs are available for all wattages of lamps.
** Tungsten arc 3085 K color temperature; all others—zirconium arcs.

Fig. 8-124. Typical Flashtubes

Number	Watt-Seconds Per Flash	Outer Bulb	Base	Average Anode Voltage	Lumen-Seconds Per Flash	Peak Lumens[a]	Maximum Watts Dissipation	Configuration	Application
4350	7.5	T-10	Octal 3-pin	850	300	.03	15		
4306	40	None[b]	None	450	1100	4	2		
4307	50	None[b]	3-pin wafer	400	1400	5	2.5		
4336	60	None[b]	Giant 5-pin	2000	2100	15	120		
4330	100	T-11	Octal 3-pin	2250	3500	12	10		
4340	400	T-16	Giant 5-pin	2250	18000	12	26.6		
FT-30	—	None[b]	Wire term.	150	—	—	3	Straight tube	
FT-106	40	None[b]	3-pin wafer	300	1100	—	5	U-Shaped	
FT-110	100	None[b]	3-pin wafer	1000	4400	—	5	U-Shaped	
FT-118	125	None[b]	3-pin wafer	500	4500	4	15	Helix	
FT-120	125	T-10	Radio 4-pin	500	4500	4	5	Helix	
FT-214	200	T-12½	Giant 5-pin	2000	7000	25	7	Helix	
FT-217	200	T-10	Radio 4-pin	1000	10000	25	5	Helix	
FT-218	200	None[b]	3-pin wafer	1000	9000	25	15	Hexlix	
FT-220	200	PAR-46	3 screw term.	2000	—	—	7	Helix	
FT-221	200	T-10	Radio 4-pin	1000	10000	25	5	Hexlix	
FT-306B	500	T-18	Large 3-pin	900	32000	—	20	Helix	
FT-403	480	T-18	Large 3-pin	2000	19000	50	15	Helix	
FT-503	2000	T-18	Large 3-pin	4000	100000	250	35	Helix	
FT-506	1000	T-11	Octal 5-pin	900	40000	15	35	Helix	
FX-1	200	T-18	End cap	2000	—	—	40	Straight tube	
VFT-U106	0.25	None[b]	Wire term.	450	—	—	5	U-Shaped	Auto Timing
VFT-U106A	0.25	None[b]	Wire term.	150	—	—	5	U-Shaped	Auto Timing
VFT-C1000	1000	None[b]	Wire term.	400	—	—	100	2½″ Circular	Photographic
VFT-C4502	450	None[b]	Wire term.	500	—	—	30	3″ Circular	Photographic
VFT-C6501	650	None[b]	Wire term.	275	—	—	45	3″ Circular	Photographic
VFT-P4004K	4000	None[b]	Wire term.	2500	—	—	130	Grid	Reproduction
VFT-H5007K	5000	None[b]	Wire term.	4500	—	—	250	Helix	Laser Stimulation
VFT-H10000K	10000	None[b]	Wire term.	4500	—	—	160	Helix	Laser Stimulation
VFT-H160	16	None[b]	Wire term.	500	—	—	12	Helix	Beacon
VFT-H351	35	None[b]	Wire term.	575	—	—	35	Helix	Beacon
VFT-S380	38	None[b]	Wire term.	280	—	—	4	2″ straight	Flash Guns
VFT-S3001B	300	None[b]	Wire term.	850	—	—	300	12″ straight	Photocopiers

[a] In millions of lumens.
[b] Intended for use only in covered reflectors or housings designed to protect user from high voltage.

Fig. 8–125. Direct-Current Carbon Arc

	Low Intensity	Non-Rotating High Intensity		Rotating High Intensity						
	Application Number*									
	1	2	3	4	5	6	7	8	9	10
Type of Carbon	Microscope	Projector	Projector	Projector	Projector	Projector	Projector	Searchlight	Studio "Yellow Flame"	
Positive Carbon										
Diameter, mm	5	7	8	10	11	13.6	13.6	16	16	16
Length, inches	8	12–14	12–14	20	20	22	22	22	22	22–30
Negative Carbon										
Diameter	6 mm	6 mm	7 mm	11/32 in.	3/8 in.	1/2 in.	1/2 in.	11 mm	17/32 in.	7/16 in.
Length, inches	4½	9	9	9	9	9	9	12	9	12–48
Arc Current, amperes	5	50	70	105	120	160	180	150	225	400
Arc Volts, dc	59	40	42	59	57	66	74	78	70	80
Arc Power, watts	295	2000	2940	6200	6840	10600	13300	11700	15800	32000
Burning Rate, inches per hour										
Positive Carbon	4.5	11.6	13.6	21.5	16.5	17.0	21.5	8.9	20.2	55.0
Negative Carbon	2.1	4.3	4.3	2.9	2.4	2.2	2.5	3.9	2.2	3.5
Approximate Crater Diameter, inches	.12	.23	.28	.36	.39	.50	.50	.55	.59	.59
Maximum Luminance of Crater, candelas per sq cm	15000	55000	83000	90000	85000	96000	95000	65000	68000	145000
Forward Crater Candlepower	975	10500	22000	36000	44000	63000	78000	68000	99000	185000
Crater Lumens**	3100	36800	77000	126000	154000	221000	273000	250000	347000	660000
Total Lumens†	3100	55000	115000	189000	231000	368000	410000	374000	521000	990000
Total Lumens per Arc Watt	10.4	29.7	39.1	30.5	33.8	34.7	30.8	32.0	33.0	30.9
Color Temperature, degrees Kelvin‡	3600	5950	5500–6500	5500–6500	5500–6500	5500–6500	5500–6500	5400	4100	5800–6100

* **Typical applications:** 1, microscope illumination and projection; 2, 3, 4, 5, 6 and 7, motion picture projection; 8, searchlight projection; 9, motion picture set lighting and motion picture and television background projection.

** Includes light radiated in forward hemisphere.

† Includes light from crater and arc flame in forward hemisphere.

‡ Crater radiation only.

Fig. 8–126. Flame-Type Carbon Arcs

	Application Number[a]									
	1	2	3	4	5	6	7[b]	8[c,g]	9[d,g]	10[e]
Type of Carbons	"C"	"E"	"Sunshine"	"Sunshine"	"W"	Enclosed arc	Photo[b]	"Sunshine"[c]	Photo[d]	Studio[e]
Flame Materials	Poly-metallic	Strontium	Rare earth	Rare earth	Poly-metallic	None	Rare earth	Rare earth	Rare earth	Rare earth
Burning Position[f]	Vertical	Vertical	Vertical	Vertical	Vertical	Vertical	Vertical	Horizontal	Horizontal	Vertical
Upper Carbon[g]										
Diameter	22 mm	22 mm	22 mm	22 mm	22 mm	½ in.	½ in.	6 mm	9 mm	8 mm
Length	12 in.	12 in.	12 in.	12 in.	12 in.	3 to 16 in.	12 in.	6⅝ in.	8 in.	12 in.
Lower Carbon[g]										
Diameter	13 mm	13 mm	13 mm	13 mm	13 mm	½ in.	½ in.	6 mm	9 mm	7 mm
Length	12 in.	12 in.	12 in.	12 in.	12 in.	3 to 16 in.	12 in.	6⅝ in.	8 in.	9 in.
Arc Current, amperes	60	60	60	80	80	16	38	40	95	40
Arc Voltage, ac[h]	50	50	50	50	50	138	50	24	30	37dc[h]
Arc Power, kW	3.0	3.0	3.0	4.0	4.0	2.2	1.9	1.0	2.85	1.5
Candlepower[i]	2100	6300	9100	10000	8400	1170	6700	4830	14200	11000
Lumens	23000	69000	100000	110000	92000	13000	74000	53000	156000	110000
Lumens per Arc Watt	7.6	23.0	33.3	27.5	23.0	5.9	39.8	53.0	54.8	73.5
Color Temperature, K	j	j	12800[j]	24000[j]	j	j	7420[j]	6590	8150	4700
Spectral Intensity (microwatts/cm² one meter from arc axis[k])										
Below 270 nm	540	180	102	140	1020	—	95	11	—	12
270–320 nm	540	150	186	244	1860	—	76	49	100	48
320–400 nm	1800	1200	2046	2816	3120	1700	684	415	1590	464
400–450 nm	390	1100	1704	2306	1480	177	722	405	844	726
450–700 nm	600	4050	3210	3520	2600	442	2223	1602	3671	3965
700–1125 nm	1580	2480	3032	3500	3220	1681	1264	1368	5632	2123
Above 1125 nm	9480	10290	9820	11420	14500	6600	5189	3290	8763	4593
Total	14930	19460	20100	24000	27800	10600	10253	7140	20600	11930
Spectral Radiation (per cent of input power)										
Below 270 nm	1.8	.6	.34	.35	2.55	—	.5	.11	—	.08
270–320 nm	1.8	.5	.62	.61	4.65	—	.4	.49	.35	.32
320–400 nm	6.0	4.0	6.82	7.04	7.80	7.7	3.6	4.15	5.59	3.09
400–450 nm	1.3	3.7	5.68	5.90	3.70	.8	3.8	4.05	2.96	4.84
450–700 nm	2.0	13.5	10.70	8.80	6.50	2.0	11.7	16.02	12.86	26.43
700–1125 nm	5.27	8.27	10.10	8.75	8.05	7.6	6.7	13.68	19.75	14.15
Above 1125 nm	31.6	34.3	32.70	28.55	36.25	29.9	27.3	32.90	30.69	30.62
Total	49.77	64.87	67.00	60.00	69.50	48.0	54.0	71.40	72.20	79.53

[a] Typical applications: 1, 2, 3, 4, 5, and 8, photochemical, therapeutic, accelerated exposure testing, or accelerated plant growth; 6, 7, and 9 blueprinting diazo printing, photo copying, and graphic arts; 10, motion picture and television studio lighting.
[b] Photographic white flame carbons.
[c] High intensity copper coated sunshine carbons.
[d] High intensity photo 98 carbons.
[e] Motion picture studio carbons.
[f] All combinations shown are operated coaxially.
[g] Both carbons are same in horizontal, coaxial ac arcs.
[h] All operated on alternating current except Item 10.
[i] Horizontal candlepower, transverse to arc axis.
[j] Deviate enough from blackbody colors to make color temperature of doubtful meaning.
[k] See Section 24 for spectral energy distribution curves.

REFERENCES

1. Hall, J. D.: "The Manufacture of Incandescent Lamps," *Elect. Eng.*, Vol. 60, p. 575, December, 1941. Millar, P. S.: "Safeguarding the Quality of Incandescent Lamps," *Trans. Illum. Eng. Soc.*, Vol. XXVI, p. 948, November, 1931.
2. Coolidge, W. D.: "Ductile Tungsten," *Trans. Am. Inst. Elect. Engrs.*, Vol. XXIX, p. 961, May, 1910. Forsythe, W. E., and Adams, E. Q.: "The Tungsten Filament Incandescent Lamp," *J. Sci. Lab.*, Denison University, April, 1937.
3. Morris, R. W.: "Considerations Affecting the Design of Flashing Signal Filament Lamps," *Illum. Eng.*, Vol. XLII, p. 625, June, 1947.
4. Forsythe, W. E., Adams, E. Q., and Cargill, P. D.: "Some Factors Affecting the Operation of Incandescent Lamps," *J. Sci. Lab.*, Denison University, April, 1939. Merrill, G. S.: "Voltage and Incandescent Electric Lighting," *Proc. Intern. Illum. Congr.*, Vol. II, 1931.
5. The Industry Committee on Interior Wiring Design, *Handbook of Interior Wiring Design*, 420 Lexington Ave., New York, 1941.
6. Potter, W. M. and Reid, K. M.: "Incandescent Lamp Design Life for Residential Lighting," *Illum. Eng.* Vol. LIV, p. 751, December, 1959.
7. "Questions and Answers on Light Sources," *Illum. Eng.*, Vol. LXII, p. 139, March, 1967.
8. Whittaker, J. D.: "Applications of Silver Processed Incandescent Lamps with Technical Data," *Trans. Illum. Eng. Soc.*, Vol. XXVIII, p. 418, May, 1933.
9. Leighton, L. G.: "Characteristics of Ribbon Filament Lamps," *Illum. Eng.*, Vol. LVII, p. 121, March, 1962.
10. *Method for the Designation of Miniature Incandescent Lamps*, C78.390-1964 (R1969), American National Standards Institute, New York.
11. *Method for the Designation of Glow Lamps*, C78.381-1961 (R1969), American National Standards Institute, New York.
12. Townsend, M. A.: "Electronics of the Fluorescent Lamp," *Trans. Am. Inst. Elect. Engrs.*, Vol. 61, p. 607, August, 1942.
13. Lowry, E. F., Gungle, W. C., and Jerome, C. W.: "Some Problems Involved In The Design of Fluorescent Lamps," *Illum. Eng.*, Vol. XLIX, p. 545, November, 1954.
14. Lowry, E. F., Frohock, W. S., and Meyers, G. A.: "Some Fluorescent Lamp Parameters and Their Effect on Lamp Performances," *Illum. Eng.*, Vol. XLI, p. 859, December, 1946. Lowry, E. F.: "The Physical Basis for Some Aspects of Fluorescent Lamp Behavior," *Illum. Eng.*, Vol. XLVII, p. 639, December, 1952.
15. Lowry, E. F.: "A Study of Fluorescent Lamp Maintenance," *Illum. Eng.*, Vol. XLIII, p. 141, February, 1948.
16. Lowry, E. F. and Mager, E. L.: "Some Factors Affecting the Life and Lumen Maintenance of Fluorescent Lamps," *Illum. Eng.*, Vol. XLIV, p. 98, February, 1949.
17. Committee on Light Sources of the IES: "Effect of Temperature on Fluorescent Lamps," *Illum. Eng.*, Vol. LVIII, p. 101, February, 1963.
18. Eastman, A. A., and Campbell, J. H.: "Stroboscopic and Flicker Effects from Fluorescent Lamps," *Illum. Eng.*, Vol. XLVII, p. 27, January, 1952.
19. McFarland, R. H., and Sargent, T. C.: "Humidity Effect on Instant Starting of Fluorescent Lamps," *Illum. Eng.*, Vol. XLV, p. 415, July, 1950.
20. Carpenter, W. P.: "Application Data for Proper Dimming of Cold Cathode Fluorescent Lamp," *Illum. Eng.*, Vol. XLVI, p. 306, June, 1951. Campbell, J. H., Schultz, H. E., and Abbott, W. H.: "Dimming Hot Cathode Fluorescent Lamps," *Illum. Eng.*, Vol. XLIX, p. 7, January, 1954. Von Zastrow, E. E.: "Fluorescent Lamp Dimming with Semiconductors," *Illum. Eng.*, Vol. LVIII, p. 312, April, 1963.
21. Campbell, J. H., and Kershaw, D. C.: "Flashing Characteristics of Fluorescent Lamps," *Illum. Eng.*, Vol. LI, p. 755, November, 1956. Bunner, R. W., and Dorsey, R. T.: "Flashing Applications of Fluorescent Lamps," *Illum. Eng.*, Vol. LI, p. 761, November, 1956.
22. Meyers, G. A., and Strojny, F. M.: "Design of Fluorescent Lamps for High Frequency Service," *Illum. Eng.*, Vol. LIV, p. 65, January, 1959. Burnham, R. D.: "Economics of High Frequency Lighting," *Illum. Eng.*, Vol. LIV, p. 419, July, 1959. Campbell, J. H.: "New Parameters for High Frequency Lighting Systems," *Illum. Eng.*, Vol. LV, p. 247, May, 1960. Gilleard, G.: "High-Frequency Operation of Fluorescent Lighting Systems from Gas-Driven Turbines," *Illum. Eng.*, Vol. LIX, p. 163, March, 1964. Tugman, J. L.: "Progress Report: High-Frequency Fluorescent Lighting," *Progr. Architect.*, Vol. XLI, p. 174, March, 1960.
23. Till, W. S., and Pisciotta, M.: "New Designations for Mercury Lamps," *Illum. Eng.*, Vol. LIV, p. 594, September, 1959.
24. Till, W. S., and Unglert, M. C.: "New Designs for Mercury Lamps Increase Their Usefulness," *Illum. Eng.*, Vol. LV, p. 269, May, 1960.
25. Jerome, C. W.: "Color of High Pressure Mercury Lamps," *Illum. Eng.*, Vol. LVI, p. 209, March, 1961.
26. Larson, D. A., Fraser, H. D., Cushing, W. V., and Unglert, M. C.: "Higher Efficiency Light Source Through Use of Additives to Mercury Discharge," *Illum. Eng.*, Vol. LVIII, p. 434, June, 1963.
27. Martt, E. C., Smialek, L. J., and Green, A. C.: "Iodides in Mercury Arcs—For Improved Color and Efficacy," *Illum. Eng.*, Vol. LIX, p. 34, January, 1964.
28. Waymouth, J. F., Gungle, W. C., Harris, J. M., and Koury, F.: "A New Metal Halide Arc Lamp," *Illum. Eng.*, Vol. LX, p. 85, February, 1965.
29. Retzer, T. C.: "Circuits for Short-Arc Lamps," *Illum. Eng.*, *Vol.* LIII, p. 606, November, 1958.
30. Thouret, W. E., and Strauss, H. S.: "New Designs Demonstrate Versatility of Xenon High-Pressure Lamps," *Illum. Eng.*, Vol. LVII, p. 150, March, 1962.
31. Lienhard, O. E., and McInally, J. A.: "New Compact-Arc Lamps of High Power and High Brightness," *Illum. Eng.*, Vol. LVII, p. 173, March, 1962.
32. Lienhard, O. E.: "Xenon Compact-Arc Lamps with Liquid-Cooled Electrodes," *Illum. Eng.*, Vol. LX, p. 348, May, 1965.
33. Null, M. R. and Lozier, W. W.: "Carbon Arc as a Radiation Standard," *J. Opt. Soc. Amer.*, Vol. 52, p. 1156, October, 1962.
34. Payne, E. C., Mager, E. L., and Jerome, C. W.: "Electroluminescence—A New Method of Producing Lighting," *Illum. Eng.*, Vol. XLV, p. 688, November, 1950.
35. Ivey, H. F.: "Problems and Progress In Electroluminescent Lamps," *Illum. Eng.*, Vol. LV, p. 13, January, 1960.
36. Blazek, R. J.: "High Brightness Electroluminescent Lamps of Improved Maintenance," *Illum. Eng.*, Vol. LVII, p. 726, November, 1962.
37. Weber, K. H.: "Electroluminescence—An Appraisal of Its Short-Term Potential," *Illum. Eng.*, Vol. LIX, p. 329, May, 1964.
38. Hall II, J. W.: "Solid State Lamps—How They Work and Some of Their Applications," *Illum. Eng.*, Vol. LXIV, p. 88, February, 1969.
39. *Approval Requirements for Gas-Fired Illuminating Appliances*, Z21.42-1963 (R1969), American National Standards Institute, New York.

History of Light Sources

40. Hammer, W. J.: "The William J. Hammer Historical Collection of Incandescent Electric Lamps," *Trans. New York Elec. Soc.*, New Series, No. 4, 1913.
41. Schroeder, Henry: "History of Electric Light," Smithsonian Miscellaneous Collections, Vol. 76, No. 2, August 15, 1923.
42. Howell, J. W., and Schroeder, Henry: "History of the Incandescent Lamp," The Maqua Company, 1927.
43. "The Development of the Incandescent Electric Lamp Up to 1879," *Trans. Illum. Eng.*, Vol. XXIV, p. 717, October, 1929.
44. "The Lamp Makers' Story," *Illum. Eng.*, Vol. LI, p. 1, January, 1956.

SECTION

9

LIGHTING CALCULATIONS

Calculation Procedure 9–1 Average Illumination 9–6 Average Luminance 9–33 Luminaire Spacing 9–33 Flux Transfer 9–36 Point Illumination 9–44 Specific Application Methods 9–63 Floodlighting 9–68 Searchlighting 9–69 Roadway Illumination 9–71 Daylighting 9–74 Recommended Illumination Levels 9–81

The design of lighting systems, as in other fields of engineering application, requires the use of mathematical or graphical techniques to obtain a solution. This section presents the most up-to-date and accurate methods of solving lighting application problems available at this time and these methods are recommended to the designer. A number of short-cut methods are available which, in general, save time in calculation but sacrifice accuracy. For this reason, it is important to choose a method commensurate with the requirements of time and accuracy.

INTERIOR CALCULATIONS

Recommended levels of illumination listed in Fig. 9–80, page 9–81, are the minimum values required on the tasks found in various types of interiors. It is usually more practical, however, to design lighting systems to provide an average illumination level with a reasonable degree of uniformity throughout the room. Such calculations may be made by *The Lumen Method.* If illumination at specific task locations is required, a *Point Method* should be used.

General Procedure for Maintained Illumination and Luminance

The procedure[1] which follows is intended to be as comprehensive as possible. It is planned to present needed data and to bring to mind items affecting the design calculations, and ultimately the completed project. Not all design specifications may be known initially, such as room surface reflectances; but all unknowns must be assumed for a complete design, the resulting design being only as accurate as the assumptions.

The general procedure consists of 18 steps, divided into four major groups as shown in Fig. 9–1: *A. Objectives and Specifications; B. Light Loss Factors Not to be Recovered; C. Light Loss Factors to be Recovered;* and *D. Calculations.*

A. Objectives and Specifications.

1. Seeing Task. Providing proper lighting for the seeing task is the basic reason for the design calculation and the procedure presented here. (In some instances, for example, where there are no significant seeing tasks performed in an area, the design may begin with the luminance distribution desired.) A complete knowledge and understanding of the visual requirements is important. If, however, the seeing task to be performed is unknown to the designer, it should be stated what the design does cover. Seeing tasks relating to different areas, such as offices, schools, industries, and institutions, can be found in later sections of this Handbook, American National Standards Institute (ANSI) and IES Practices or Guides, and IES Committee Reports, in which analyses of seeing tasks are included.*

2. Quality Required. After learning as much as possible about the seeing task, a knowledge and understanding of the quality of illumination required for the particular task becomes important. It is necessary to comply with visual requirements such as recommended luminance ratios, luminaire luminance, luminaire comfort criteria, veiling reflections, allowable variations between maximum and minimum levels of illumination and physical requirements such as the allowable percentage of burned-out lamps, and standards of cleanliness that will influence cleaning cycles. Since lighting quality requirements vary with the application, specific recommendations and suggestions are given in the same publications as listed in Step 1 above.

---

* Information on most recent recommendations can be obtained from the Illuminating Engineering Society, 345 E. 47th Street, New York, N. Y. 10017.

NOTE: References are listed at the end of each section.

Fig. 9-1. Steps in Calculating Maintained Illumination and Luminance

A. Objectives and Specifications	B. Light Loss Factors Not to be Recovered*	C. Light Loss Factors to be Recovered*	D. Calculations
1. Seeing Task 2. Quality Required 3. Quantity Required 4. Area Atmosphere 5. Area Description and Use 6. Selection of Luminaire	7. Luminaire Ambient Temperature 8. Voltage to Luminaire 9. Ballast Factor 10. Luminaire Surface Depreciation	11. Room Surface Dirt Depreciation 12. Burn-Outs 13. Lamp Lumen Depreciation 14. Luminaire Dirt Depreciation	15. Total Light Loss Factor 16. Calculations 17. Layout 18. Review Compliance with Objectives†

* If losses are too great, reselect luminaire for smaller losses.
† If neither the design objectives nor the budget requirements have been met, the design process should begin again.

3. Quantity Required. The quantity of illumination for the seeing task can be found in Fig. 9-80 and the same publications as listed in Step 1 above. Also, consideration should be given to allowable variations between maximum and minimum levels of illumination and the tolerable percentage of burned-out lamps. For example, to help assure that the recommended level is provided on the task, the design average level of illumination may be higher than the recommended level by the factor of the ratio of average to minimum allowable.

4. Area Atmosphere. Next to be considered is an analysis of the environment in which the lighting system will operate. For example, are dirt, water vapor, explosive gases or corrosive vapor present?

Dirt in the atmosphere will have come from two sources: that passed from adjacent atmosphere(s) and that generated by work done in the surrounding atmosphere. Dirt may be classified as adhesive, attracted or inert and it may come from intermittent or constant sources. Adhesive dirt will cling to luminaire surfaces by its stickiness, while attracted dirt is held by electrostatic force. Inert dirt will vary in accumulation from practically nothing on vertical surfaces, to as much as a horizontal surface will hold before the dirt is dislodged by gravity or air circulation. Examples of adhesive dirt are: grease from cooking, particles from machine operation borne by oil vapor, particles borne by water vapor as in a laundry, fumes from metal-pouring operations or plating tanks. Examples of attracted dirt are: hair, lint, fibers or dry particles which are electrostatically charged from machine operations. Inert dirt is represented by non-sticky, uncharged particles such as dry flour, sawdust, fine cinders and the like. Figs. 9-2 and 9-3 are useful for evaluating the atmosphere.

5. Area Description and Use. A complete description is required for each area to be lighted. This should include: the physical characteristics such as room dimensions, room reflectances, work locations or location of work-plane, and the operating characteristics of the lighting system such as the hours of operations per day (hours per start for fluorescent lamps) and annual hours of use of the system.

6. Selection of Luminaire. Selecting the specific luminaire requires the almost simultaneous consideration of many factors. Selection of the type of luminaire for a given application depends upon the requirements and conditions found in Steps 2 through 5 above, such as visual comfort criteria, veiling reflections and atmospheric conditions and such factors (whose relative importance will vary from project to project) as: mounting height; luminaire dirt depreciation classification (see Step 14); lamp choice; maintenance considerations, including cleaning and lamp replacement; luminaire and installation appearance; color of light; heating effect; lighting and relighting time; noise level; radio interference; vibration; cost of equipment; etc. All factors should be examined in detail first, then reviewed so that proper weight will be given to everything that might affect luminaire selections. See Section 10 for luminaire classification.

B. Light Loss Factors Not to Be Recovered.* Once the basic facts about the task, the area, and the chosen luminaire are known, light loss factors can be studied. The factors immediately following are those usually subject to very little correction. Some will exist initially and continue through the life of the installation—either being of such little effect as to make correction impractical, or being too costly to correct. However, all should be studied, because they can diminish the planned output of the lighting system.

7. Luminaire Ambient Temperature. The effect of ambient temperature on the output of some luminaires is considerable. Variations in temperature, above or below those normally encountered in interiors, have little effect on the light output of incandescent and high-intensity discharge lamp luminaires but have an effect on light output of fluorescent luminaires. Each particular lamp-luminaire combination has its own distinctive characteristic of light output *vs* ambient temperature. To apply a factor for light loss due to ambient temperature, the designer needs to know the highest or lowest temperature expected and to have data showing the variation in light output *vs* ambient temperature for the specific luminaire to be used, and the application (mounting) conditions and effect of heat transfer (if applicable).

* Usually not controlled by lighting maintenance procedures.

Fig. 9-2. Five Degrees of Dirt Conditions

	Very Clean	Clean	Medium	Dirty	Very Dirty
Generated Dirt	None	Very little	Noticeable but not heavy	Accumulates rapidly	Constant accumulation
Ambient Dirt	None (or none enters area)	Some (almost none enters)	Some enters area	Large amount enters area	Almost none excluded
Removal or Filtration	Excellent	Better than average	Poorer than average	Only fans or blowers if any	None
Adhesion	None	Slight	Enough to be visible after some months	High—probably due to oil, humidity, or static	High
Examples	High grade offices, not near production; laboratories; clean rooms	Offices in older buildings or near production; light assembly; inspection	Mill offices; paper processing; light machining	Heat treating; high speed printing; rubber processing	Similar to **Dirty** but luminaires within immediate area of contamination

Fig. 9-3. Evaluation of Operating Atmosphere[2]

Factors for Use in Table Below

1 = Cleanest conditions imaginable
2 = Clean, but not the cleanest
3 = Average
4 = Dirty, but not the dirtiest
5 = Dirtiest conditions imaginable

Type of Dirt	Area Adjacent to Task Area						Filter Factor (per cent of dirt passed)		Area Surrounding Task						Sub Total
	Intermittent Dirt		Constant Dirt		Total				From Adjacent		Intermittent Dirt		Constant Dirt		
Adhesive Dirt		+		=		×		=		+		+		=	
Attracted Dirt		+		=		×		=		+		+		=	
Inert Dirt		+		=		×		=		+		+		=	

Total of Dirt Factors

0–12 = Very Clean	13–24 = Clean	25–36 = Medium	37–48 = Dirty	49–60 = Very Dirty

8. Voltage to Luminaire. In-service voltage is difficult to predict, but high or low voltage at the luminaire will affect the output of most luminaires. For incandescent units, small deviations from rated lamp voltage cause approximately three per cent change in lumens for each one per cent of voltage deviation. For mercury luminaires with high reactance ballasts there is a change of approximately three per cent in lamp lumens for each one per cent change in primary voltage deviation from rated ballast voltage. When regulated output ballasts are used, the lamp lumen output is relatively independent of primary voltage within the design range. Fluorescent luminaire output changes approximately one per cent for each two and a half per cent change in primary voltage. Fig. 9-4 shows these variations in graph form. See also Section 8.

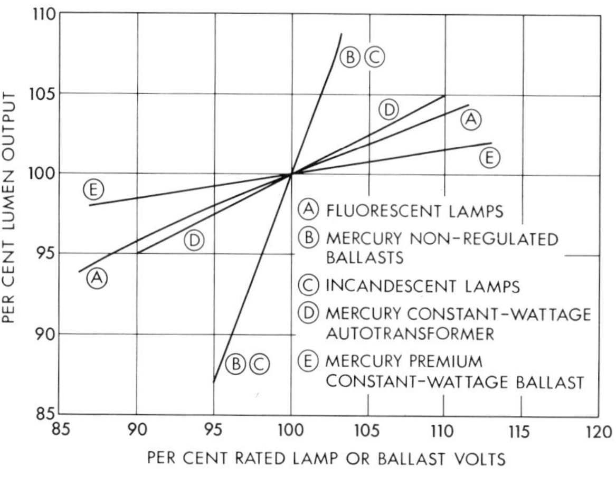

Fig. 9-4. Light output change due to voltage change.

9. Ballast Factor. If the ballast factor of the ballast used in a luminaire (fluorescent or high-intensity discharge) differs from that of the ballast used in the actual photometry of the luminaire, the light output will differ by the same amount.[4] The manufacturer should be consulted for necessary factors. Note that when uncertified ballasts are used, there may be no reliable data available.

10. Luminaire Surface Depreciation. Luminaire surface depreciation results from adverse changes in metal, paint, and plastic components which result in reduced light output. Surfaces of glass, porcelain or processed aluminum have negligible depreciation and can be restored to original reflectance. Baked enamel and other painted surfaces have a permanent depreciation due to all paints being porous to some degree. For plastics, acrylic is least susceptible to change, but its transmittance may be reduced by usage over a period of 15 to 20 years in certain atmospheres. For the same usage, polystyrene will have lower transmittance than acrylic and will depreciate faster.

Because of the complex relationship between the light-controlling elements of luminaires using more than one type of material (such as a lensed troffer) it is difficult to predict losses due to deterioration of materials. Also, for luminaires with one type of surface the losses will be affected by the type of atmosphere in the installation. No factors are available at present.

C. Light Loss Factors to be Recovered.*

11. Room Surface Dirt Depreciation. The accumulation of dirt on room surfaces reduces the amount of luminous flux reflected and interreflected to the work plane. To take this into account, Fig. 9-5 has been developed to provide Room Surface Dirt Depreciation (RSDD) factors for use in calculating maintained average illumination levels. These factors are determined as follows:

(*1*) Find the expected dirt depreciation using the Area Atmosphere from Step 4, the time between cleaning and the curves in Fig. 9-5. For example, if the atmosphere is dirty and room surfaces are cleaned every 24 months, the expected dirt depreciation would be approximately 30 per cent.

(*2*) Knowing the expected dirt depreciation, the type of luminaire distribution (see Step 6) and the room cavity ratio (see page 9-9), determine the RSDD factor from Fig. 9-5. For example, for a dirt depreciation of 30 per cent, a direct luminaire and a Room Cavity Ratio (RCR) of 4, the RSDD would be .92.

12. Burnouts. Unreplaced burned-out lamps will vary in quantity, depending on the kinds of lamps and the relamping program used. Manufacturers' mortality statistics should be consulted for the performance of each lamp type to determine the number expected to burn out before the time of planned replacement is reached. Practically, quantity of lamp burnouts is determined by the quality of the lighting services program incorporated in the initial design procedure and by the quality of the physical performance of the program.

Lamp burnouts contribute to loss of light. If lamps are not replaced promptly after burnout, the average illumination level will be decreased proportionately. In some instances, more than just the faulty lamp may be lost. For example, when series sequence fluorescent ballasts are used and one lamp fails, both lamps go out. The Lamp Burnouts (LBO) factor is the ratio of the lamps remaining lighted to the total, for the maximum number of burnouts permitted.

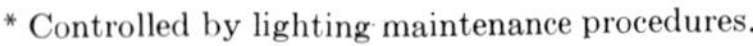

* Controlled by lighting maintenance procedures.

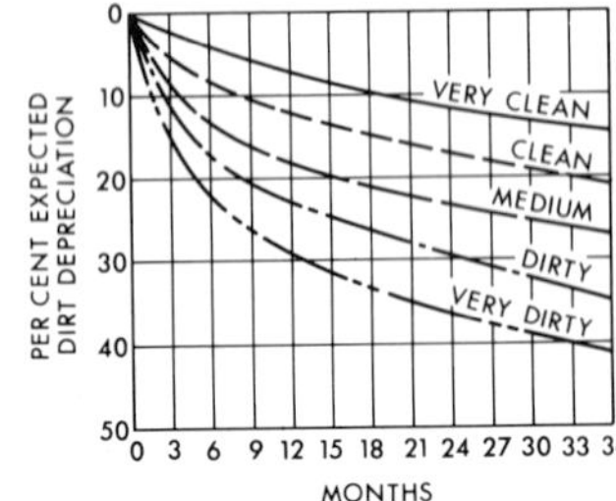

Fig. 9-5. Room Surface Dirt Depreciation Factors

	Luminaire Distribution Type																			
	Direct				Semi-Direct				Direct-Indirect				Semi-Indirect				Indirect			
Per Cent Expected Dirt Depreciation	10	20	30	40	10	20	30	40	10	20	30	40	10	20	30	40	10	20	30	40
Room Cavity Ratio																				
1	.98	.96	.94	.92	.97	.92	.89	.84	.94	.87	.80	.76	.94	.87	.80	.73	.90	.80	.70	.60
2	.98	.96	.94	.92	.96	.92	.88	.83	.94	.87	.80	.75	.94	.87	.79	.72	.90	.80	.69	.59
3	.98	.95	.93	.90	.96	.91	.87	.82	.94	.86	.79	.74	.94	.86	.78	.71	.90	.79	.68	.58
4	.97	.95	.92	.90	.95	.90	.85	.80	.94	.86	.79	.73	.94	.86	.78	.70	.89	.78	.67	.56
5	.97	.94	.91	.89	.94	.90	.84	.79	.93	.86	.78	.72	.93	.86	.77	.69	.89	.78	.66	.55
6	.97	.94	.91	.88	.94	.89	.83	.78	.93	.85	.78	.71	.93	.85	.76	.68	.89	.77	.66	.54
7	.97	.94	.90	.87	.93	.88	.82	.77	.93	.84	.77	.70	.93	.84	.76	.68	.89	.76	.65	.53
8	.96	.93	.89	.86	.93	.87	.81	.75	.93	.84	.76	.69	.93	.84	.76	.68	.88	.76	.64	.52
9	.96	.92	.88	.85	.93	.87	.80	.74	.93	.84	.76	.68	.93	.84	.75	.67	.88	.75	.63	.51
10	.96	.92	.87	.83	.93	.86	.79	.72	.93	.84	.75	.67	.92	.83	.75	.67	.88	.75	.62	.50

13. Lamp Lumen Depreciation. Information about lamp lumen depreciation is available from manufacturers' tables and graphs for lumen depreciation and mortality of the chosen lamp. Rated average life should be determined for the specific hours per start; it should be known when burnouts will begin in the lamp life cycle. From these facts, a practical group relamping cycle will be established and then, based on the hours elapsed to lamp removal, the specific Lamp Lumen Depreciation (LLD) factor can be determined. Consult the tables at the end of Section 8 or manufacturers' data for LDD factors. Seventy per cent of average rated life is the minimum reached in an installation where burnouts are promptly replaced, whether the planned relamping is to be a group or a random replacement program.

14. Luminaire Dirt Depreciation. The accumulation of dirt on luminaires results in a loss in light output, and therefore a loss on the work plane. This loss is known as the Luminaire Dirt Depreciation (LDD) factor and is determined as follows:

1. The luminaire maintenance category is selected from manufacturers' data or by using Fig. 9-6.

2. The atmosphere (one of five degrees of dirt conditions) in which the luminaire will operate is found from Step 4 and Figs. 9-2 or 9-3.

3. From the appropriate luminaire maintenance category curve of Fig. 9-7, the applicable dirt condition curve and the proper elapsed time in months of the planned cleaning cycle, the LDD factor is found. For example, if the Category is I, the atmosphere dirty and cleaning every 20 months, LDD is .80.

D. Calculations.

15. Total Light Loss Factor. The total Light Loss Factor is simply the product of multiplying all the contributing factors described above such as: LLF = Nos. 7 x 8 x 9 x 10 x 11 x 12 x 13 x 14. Where factors are not known, or applicable, they are assumed to be unity. At this point, if it is found that the total light loss factor is excessive it may be desirable to reselect the luminaire.

16. Average and Point Illumination and Luminance. During the preceding steps, the seeing task has been established along with its requirements of quality and quantity. A value has been calculated to represent the light losses which must be overcome. To calculate:

Average Illumination, see page 9-6.
Point Illumination, see page 9-44.
Average Luminance, see page 9-33.

17. Layout. In making a luminaire layout, the task, quality requirements, the area, the type and distribution of the luminaire, the requirements for uniformity of illumination, and appearance have to be considered.

When uniformity of illumination is desirable, proper spacing-to-mounting height ratio should be followed. See Fig. 9-14.

Fig. 9-6. Procedure for Determining Luminaire Maintenance Categories

To assist in determining Luminaire Dirt Depreciation (LDD) factors, luminaires are separated into six maintenance categories (I through VI). To arrive at categories, luminaires are arbitrarily divided into sections, a *Top Enclosure* and a *Bottom Enclosure*, by drawing a horizontal line through the light center of the lamp or lamps. The characteristics listed for the enclosures are then selected as best describing the luminaire. Only one characteristic for the top enclosure and one for the bottom enclosure should be used in determining the category of a luminaire. Percentage of uplight is based on 100 per cent for the luminaire.

The maintenance category is determined when there are characteristics in both enclosure columns. If a luminaire falls into more than one category, the lower numbered category is used.

Maintenance Category	Top Enclosure	Bottom Enclosure
I	1. None.	1. None
II	1. None 2. Transparent with 15 per cent or more uplight through apertures. 3. Translucent with 15 per cent or more uplight through apertures. 4. Opaque with 15 per cent or more uplight through apertures.	1. None 2. Louvers or baffles
III	1. Transparent with less than 15 per cent upward light through apertures. 2. Translucent with less than 15 per cent upward light through apertures. 3. Opaque with less than 15 per cent uplight through apertures.	1. None 2. Louvers or baffles
IV	1. Transparent unapertured. 2. Translucent unapertured. 3. Opaque unapertured.	1. None 2. Louvers
V	1. Transparent unapertured. 2. Translucent unapertured. 3. Opaque unapertured.	1. Transparent unapertured 2. Translucent unapertured
VI	1. None. 2. Transparent unapertured. 3. Translucent unapertured. 4. Opaque unapertured.	1. Transparent unapertured 2. Translucent unapertured 3. Opaque unapertured

18. Review Compliance with Objectives. At this point in the design procedure, it is incumbent upon the designer to review the solution to the problem given to determine that the derived objectives

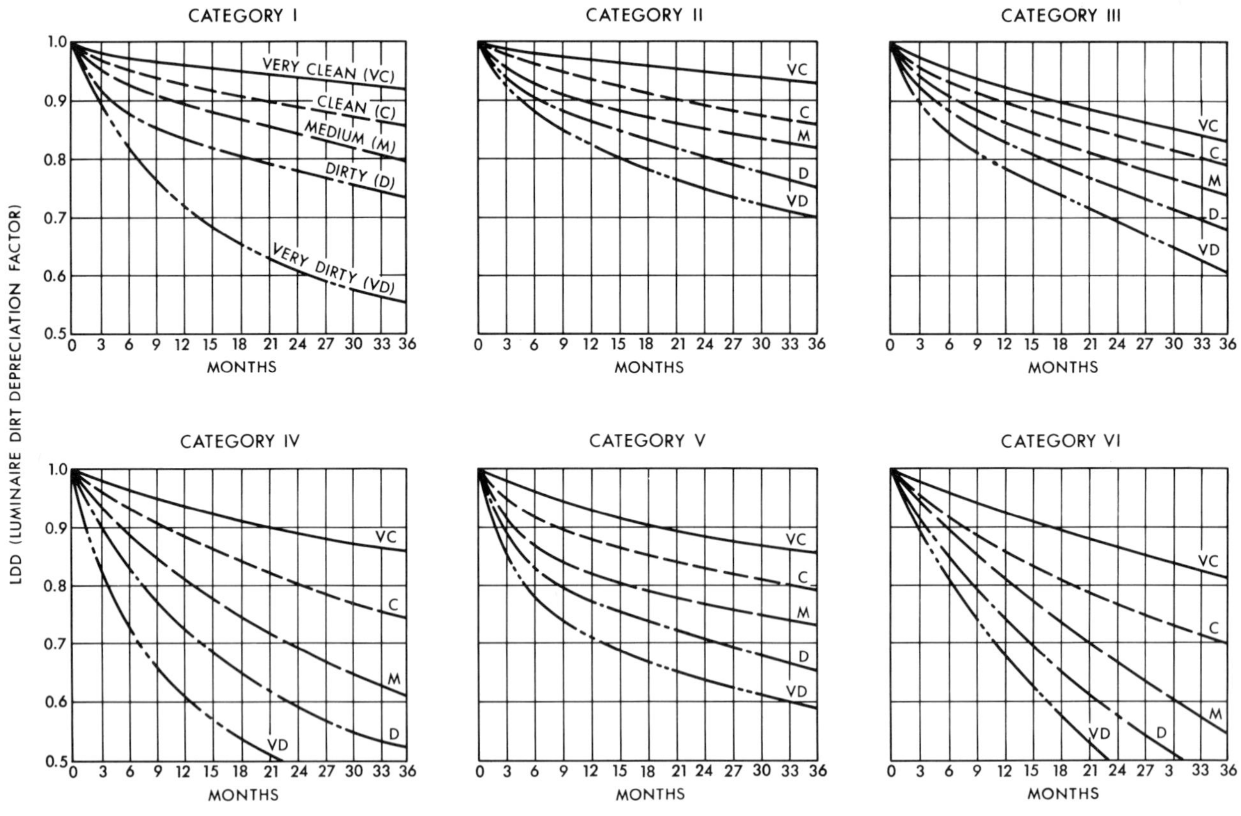

Fig. 9–7. Luminaire Dirt Depreciation factors (LDD) for six luminaire categories (I to VI) and for five degrees of dirtiness as determined from either Figs. 9–2 or 9–3.

for quantity and quality of illumination have been met, and that the solution meets the job specifications in all respects. Also, it is important to review the solution to determine if it is within budget considerations and is an economical system. (See page 16–20.) If neither the objectives nor budget requirements have been met, the design process should begin again.

It should be recognized that if the completed lighting installation is not maintained (cleaned, relamped, etc.) in accordance with the facts or assumptions used in arriving at the light loss factors, the level of illumination will not be as calculated. It is therefore important that the designer inform the user of the facts used in the design.

Average Illumination Calculations

The lumen method is used in calculating the illumination that represents the average of all points on the work plane in an interior. It is based on the definition of illumination as luminous flux per unit area, or:

$$\text{Illumination} = \frac{\text{Luminous Flux}}{\text{Area}}$$

where luminous flux is expressed in lumens. If the area is in square feet, the illumination is in footcandles (lumens per square foot): if the area is in square meters, the illumination is in lux (lumens per square meter).

Because not all the lamp lumens will reach the work plane due to losses in the luminaire and at the room surfaces, they must be multiplied by a coefficient of utilization which represents the portion that reaches the work plane. Thus:

$$\text{Initial Illumination} = \frac{\text{Lamp Lumens} \times \text{Coefficient of Utilization}}{\text{Area}}$$

Since the design objective is usually the minimum maintained illumination, factors must be applied to account for the estimated depreciation in lamp lumens, the estimated losses from dirt collection on the luminaire surfaces (including lamps), etc. The formula thus becomes:

$$\text{Maintained Illumination} = \frac{\text{Lamp Lumens} \times \text{CU} \times \text{LLF}}{\text{Area}}$$

where
CU = the Coefficient of Utilization
LLF = the Light Loss Factor (see page 9–5)

The lamp lumens in the formula are most conveniently taken as the total rated lamp lumens in

Fig. 9-8. Average Illumination Calculation Sheet

GENERAL INFORMATION

Project identification: ______________________________
(*Give name of area and/or building and room number*)

Average maintained illumination for design: ____ footcandles

Luminaire data:

Manufacturer: ______________________

Catalog number: ______________________

Lamp data:

Type and color: ________

Number per luminaire: ________

Total lumens per luminaire: ________

SELECTION OF COEFFICIENT OF UTILIZATION

Step 1: Fill in sketch at right.

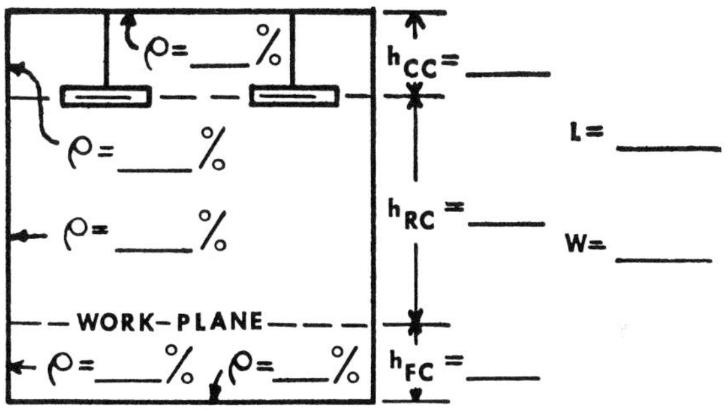

Step 2: Determine Cavity Ratios from Fig. 9-10, or by formulas.

Room Cavity Ratio, RCR = ____________

Ceiling Cavity Ratio, CCR = ____________

Floor Cavity Ratio, FCR = ____________

ρ= ___%
ρ= ___%
ρ= ___%
WORK-PLANE
ρ= ___% ρ= ___%
h_{CC}= ____
h_{RC}= ____
h_{FC}= ____
L= ____
W= ____

Step 3: Obtain Effective Ceiling Cavity Reflectance (ρ_{CC}) from Fig. 9-11. ρ_{CC} = ________

Step 4: Obtain Effective Floor Cavity Reflectance (ρ_{FC}) from Fig. 9-11. ρ_{FC} = ________

Step 5: Obtain Coefficient of Utilization (CU) from Manufacturer's Data. CU = ________

SELECTION OF LIGHT LOSS FACTORS

Unrecoverable		Recoverable	
Luminaire ambient temperature (See page 9-2.)	______	Room surface dirt depreciation RSDD (See page 9-4.)	______
Voltage to luminaire (See page 9-3.)	______	Lamp lumen depreciation LLD (See page 9-5.)	______
Ballast factor (See page 9-4.)	______	Lamp burnouts factor LBO (See page 9-4.)	______
Luminaire surface depreciation (See page 9-4.)	______	Luminaire dirt depreciation LDD (See page 9-5.)	______

Total light loss factor, LLF (product of individual factors above): ___

CALCULATIONS

(Average Maintained Illumination Level)

$$\text{Number of Luminaires} = \frac{(\text{Footcandles}) \times (\text{Area in Square Feet})}{(\text{Lumens per Luminaire}) \times (\text{CU}) \times (\text{LLF})}$$

$$= \frac{\qquad\qquad}{\qquad\qquad} =$$

$$\text{Footcandles} = \frac{(\text{Number of Luminaires}) \times (\text{Lumens per Luminaire}) \times (\text{CU}) \times (\text{LLF})}{(\text{Area in Square Feet})}$$

$$= \frac{\qquad\qquad}{\qquad\qquad} =$$

Calculated by: ______________________ Date: ____________

the luminaire, and the area then becomes the area per luminaire. Thus:

$$\text{Maintained Illumination} = \frac{\text{Lamp Lumens per Luminaire} \times \text{CU} \times \text{LLF}}{\text{Area per Luminaire}}$$

or, if the desired illumination is known, the area per luminaire (and hence the spacing between luminaires) to produce this illumination may be obtained by:

$$\text{Area per Luminaire} = \frac{\text{Lamp Lumens per Luminaire} \times \text{CU} \times \text{LLF}}{\text{Maintained Illumination}}$$

A lighting system can be designed with spacings between units to approximate this area, but if the total number of luminaires is also desired, then:

$$\text{Total Number of Luminaires} = \frac{\text{Total Room Area}}{\text{Area per Luminaire}}$$

For a typical form for calculating illumination, see Fig. 9-8.

Cavity Ratios. In the Zonal-Cavity Method,[5, 6, 7] the effects of room proportions, luminaire suspension length, and work-plane height upon the coefficient of utilization are respectively accounted for by the *Room Cavity Ratio, Ceiling Cavity Ratio,* and *Floor Cavity Ratio.* These ratios are determined by dividing the room into three cavities as shown by Fig. 9-9 and substituting dimensions (in feet or meters) in the following formula:

$$\text{Cavity Ratio} = \frac{5h(\text{Room Length} + \text{Room Width})}{(\text{Room Length}) \times (\text{Room Width})}$$

where

$h = h_{RC}$ for the Room Cavity Ratio, RCR
$= h_{CC}$ for the Ceiling Cavity Ratio, CCR
$= h_{FC}$ for the Floor Cavity Ratio, FCR

Note that

$$\text{CCR} = \text{RCR}\,\frac{h_{CC}}{h_{RC}}$$

and

$$\text{FCR} = \text{RCR}\,\frac{h_{FC}}{h_{RC}}$$

Cavity Ratios may also be obtained from Fig. 9-10 (page 9-9).

The illumination in rooms of irregular shape can be determined by calculating the Room Cavity Ratio using the following formula and solving the problem in the usual manner:

$$\text{Cavity Ratio} = \frac{2.5 \times (\text{Cavity Height}) \times (\text{Cavity Perimeter})}{(\text{Area of Cavity Base})}$$

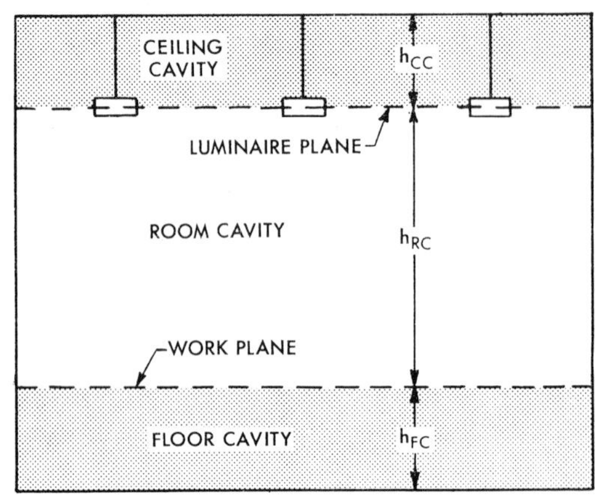

Fig. 9-9. The three cavities used in the Zonal-Cavity Method.

Effective Cavity Reflectances. Fig. 9-11 provides a means of converting the combination of wall and ceiling or wall and floor reflectances into a single *Effective Ceiling Cavity Reflectance,* ρ_{CC}, and a single *Effective Floor Cavity Reflectance,* ρ_{FC}. In calculations, ceiling, wall, and floor reflectances should be initial values. The RSDD factor (see page 9-4) compensates for the decrease of reflectance with time. Note that for surface-mounted and recessed luminaires, CCR = 0 and the ceiling reflectance may be used as ρ_{CC}.

The effective ceiling cavity ratio of non-horizontal ceilings can be determined by the following formula and solving the problem in the usual manner:

$$\rho_{CC} = \frac{\rho A_0}{A_s - \rho A_s + \rho A_0}$$

where

A_0 = area of ceiling opening
A_s = area of ceiling surface
ρ = reflectance of ceiling surface

If the ceiling surface reflectance is not the same for all parts of the ceiling, use an area weighted average. Thus, if the ceiling has several sections 1, 2, 3, $\cdots$

$$\rho = \frac{\rho_1 A_1 + \rho_2 A_2 + \rho_3 A_3 + \cdots}{A_1 + A_2 + A_3 + \cdots}$$

This formula for ρ_{CC} applies to concave ceilings such as a hemispherical dome where all parts of the ceiling are exposed to all other parts of the ceiling.

Luminaire Coefficients of Utilization. Absorption of light in a luminaire is accounted for in the computation of coefficients of utilization (CU) for that particular luminaire. Fig. 9-12 is a tabulation of coefficients of utilization calculated by the Zonal-Cavity Method for representative luminaire types. These coefficients are for an Effective Floor Cavity Reflectance of 20 per cent, but any CU obtained from the table may be corrected for a different value of ρ_{FC} by applying the appropriate multiplier from Fig. 9-13.

Since the Light Loss Factor includes the effect of dirt depositing on wall surfaces, the selection of the

(Continued on page 9-32.)

Fig. 9-10. Cavity Ratios

(For cavity dimensions other than those shown below the cavity ratio can be calculated by the formulas on page 9-8.)

Room Dimensions		Cavity Depth																			
Width	Length	1.0	1.5	2.0	2.5	3.0	3.5	4.0	5.0	6.0	7.0	8	9	10	11	12	14	16	20	25	30
8	8	1.2	1.9	2.5	3.1	3.7	4.4	5.0	6.2	7.5	8.8	10.0	11.2	12.5	—	—	—	—	—	—	—
	10	1.1	1.7	2.2	2.8	3.4	3.9	4.5	5.6	6.7	7.9	9.0	10.1	11.3	12.4	—	—	—	—	—	—
	14	1.0	1.5	2.0	2.5	3.0	3.4	3.9	4.9	5.9	6.9	7.8	8.8	9.7	10.7	11.7	—	—	—	—	—
	20	0.9	1.3	1.7	2.2	2.6	3.1	3.5	4.4	5.2	6.1	7.0	7.9	8.8	9.6	10.5	12.2	—	—	—	—
	30	0.8	1.2	1.6	2.0	2.4	2.8	3.2	4.0	4.7	5.5	6.3	7.1	7.9	8.7	9.5	11.0	—	—	—	—
	40	0.7	1.1	1.5	1.9	2.3	2.6	3.0	3.7	4.5	5.3	5.9	6.5	7.4	8.1	8.8	10.3	11.8	—	—	—
10	10	1.0	1.5	2.0	2.5	3.0	3.5	4.0	5.0	6.0	7.0	8.0	9.0	10.0	11.0	12.0	—	—	—	—	—
	14	0.9	1.3	1.7	2.1	2.6	3.0	3.4	4.3	5.1	6.0	6.9	7.8	8.6	9.5	10.4	12.0	—	—	—	—
	20	0.7	1.1	1.5	1.9	2.3	2.6	3.0	3.7	4.5	5.3	6.0	6.8	7.5	8.3	9.0	10.5	12.0	—	—	—
	30	0.7	1.0	1.3	1.7	2.0	2.3	2.7	3.3	4.0	4.7	5.3	6.0	6.6	7.3	8.0	9.4	10.6	—	—	—
	40	0.6	0.9	1.2	1.6	1.9	2.2	2.5	3.1	3.7	4.4	5.0	5.6	6.2	6.9	7.5	8.7	10.0	12.5	—	—
	60	0.6	0.9	1.2	1.5	1.7	2.0	2.3	2.9	3.5	4.1	4.7	5.3	5.9	6.5	7.1	8.2	9.4	11.7	—	—
12	12	0.8	1.2	1.7	2.1	2.5	2.9	3.3	4.2	5.0	5.8	6.7	7.5	8.4	9.2	10.0	11.7	—	—	—	—
	16	0.7	1.1	1.5	1.8	2.2	2.5	2.9	3.6	4.4	5.1	5.8	6.5	7.2	8.0	8.7	10.2	11.6	—	—	—
	24	0.6	0.9	1.2	1.6	1.9	2.2	2.5	3.1	3.7	4.4	5.0	5.6	6.2	6.9	7.5	8.7	10.0	12.5	—	—
	36	0.6	0.8	1.1	1.4	1.7	1.9	2.2	2.8	3.3	3.9	4.4	5.0	5.5	6.0	6.6	7.8	8.8	11.0	—	—
	50	0.5	0.8	1.0	1.3	1.5	1.8	2.1	2.6	3.1	3.6	4.1	4.6	5.1	5.6	6.2	7.2	8.2	10.2	—	—
	70	0.5	0.7	1.0	1.2	1.5	1.7	2.0	2.4	2.9	3.4	3.9	4.4	4.9	5.4	5.8	6.8	7.8	9.7	12.2	—
14	14	0.7	1.1	1.4	1.8	2.1	2.5	2.9	3.6	4.3	5.0	5.7	6.4	7.1	7.8	8.5	10.0	11.4	—	—	—
	20	0.6	0.9	1.2	1.5	1.8	2.1	2.4	3.0	3.6	4.2	4.9	5.5	6.1	6.7	7.3	8.6	9.8	12.3	—	—
	30	0.5	0.8	1.0	1.3	1.6	1.8	2.1	2.6	3.1	3.7	4.2	4.7	5.2	5.8	6.3	7.3	8.4	10.5	—	—
	42	0.5	0.7	1.0	1.2	1.4	1.7	1.9	2.4	2.9	3.3	3.8	4.3	4.7	5.2	5.7	6.7	7.6	9.5	11.9	—
	60	0.4	0.7	0.9	1.1	1.3	1.5	1.8	2.2	2.6	3.1	3.5	3.9	4.4	4.8	5.2	6.1	7.0	8.8	10.9	—
	90	0.4	0.6	0.8	1.0	1.2	1.4	1.6	2.0	2.5	2.9	3.3	3.7	4.1	4.5	5.0	5.8	6.6	8.3	10.3	12.4
17	17	0.6	0.9	1.2	1.5	1.8	2.1	2.3	2.9	3.5	4.1	4.7	5.3	5.9	6.5	7.0	8.2	9.4	11.7	—	—
	25	0.5	0.7	1.0	1.2	1.5	1.7	2.0	2.5	3.0	3.5	4.0	4.5	5.0	5.5	6.0	7.0	8.0	10.0	12.5	—
	35	0.4	0.7	0.9	1.1	1.3	1.5	1.7	2.2	2.6	3.1	3.5	3.9	4.4	4.8	5.2	6.1	7.0	8.7	10.9	—
	50	0.4	0.6	0.8	1.0	1.2	1.4	1.6	2.0	2.4	2.8	3.1	3.5	3.9	4.3	4.5	5.4	6.2	7.7	9.7	11.6
	80	0.4	0.5	0.7	0.9	1.1	1.2	1.4	1.8	2.1	2.5	2.9	3.3	3.6	4.0	4.3	5.1	5.8	7.2	9.0	10.9
	120	0.3	0.5	0.7	0.8	1.0	1.2	1.3	1.7	2.0	2.3	2.7	3.0	3.4	3.7	4.0	4.7	5.4	6.7	8.4	10.1
20	20	0.5	0.7	1.0	1.2	1.5	1.7	2.0	2.5	3.0	3.5	4.0	4.5	5.0	5.5	6.0	7.0	8.0	10.0	12.5	—
	30	0.4	0.6	0.8	1.0	1.2	1.5	1.7	2.1	2.5	2.9	3.3	3.7	4.1	4.5	4.9	5.8	6.6	8.2	10.3	12.4
	45	0.4	0.5	0.7	0.9	1.1	1.3	1.4	1.8	2.2	2.5	2.9	3.3	3.6	4.0	4.3	5.1	5.8	7.2	9.1	10.9
	60	0.3	0.5	0.7	0.8	1.0	1.2	1.3	1.7	2.0	2.3	2.7	3.0	3.4	3.7	4.0	4.7	5.4	6.7	8.4	10.1
	90	0.3	0.5	0.6	0.8	0.9	1.1	1.2	1.5	1.8	2.1	2.4	2.7	3.0	3.3	3.6	4.2	4.8	6.0	7.5	9.0
	150	0.3	0.4	0.6	0.7	0.8	1.0	1.1	1.4	1.7	2.0	2.3	2.6	2.9	3.2	3.4	4.0	4.6	5.7	7.2	8.6
24	24	0.4	0.6	0.8	1.0	1.2	1.5	1.7	2.1	2.5	2.9	3.3	3.7	4.1	4.5	5.0	5.8	6.7	8.2	10.3	12.4
	32	0.4	0.5	0.7	0.9	1.1	1.3	1.5	1.8	2.2	2.6	2.9	3.3	3.6	4.0	4.3	5.1	5.8	7.2	9.0	11.0
	50	0.3	0.5	0.6	0.8	0.9	1.1	1.2	1.5	1.8	2.2	2.5	2.8	3.1	3.4	3.7	4.4	5.0	6.2	7.8	9.4
	70	0.3	0.4	0.6	0.7	0.8	1.0	1.1	1.4	1.7	2.0	2.2	2.5	2.8	3.0	3.3	3.8	4.4	5.5	6.9	8.2
	100	0.3	0.4	0.5	0.6	0.8	0.9	1.0	1.3	1.6	1.8	2.1	2.4	2.6	2.9	3.1	3.7	4.2	5.2	6.5	7.9
	160	0.2	0.4	0.5	0.6	0.7	0.8	1.0	1.2	1.4	1.7	1.9	2.1	2.4	2.6	2.8	3.3	3.8	4.7	5.9	7.1
30	30	0.3	0.5	0.7	0.8	1.0	1.2	1.3	1.7	2.0	2.3	2.7	3.0	3.3	3.7	4.0	4.7	5.4	6.7	8.4	10.0
	45	0.3	0.4	0.6	0.7	0.8	1.0	1.1	1.4	1.7	1.9	2.2	2.5	2.7	3.0	3.3	3.8	4.4	5.5	6.9	8.2
	60	0.3	0.4	0.5	0.6	0.7	0.9	1.0	1.2	1.5	1.7	2.0	2.2	2.5	2.7	3.0	3.5	4.0	5.0	6.2	7.4
	90	0.2	0.3	0.4	0.6	0.7	0.8	0.9	1.1	1.3	1.6	1.8	2.0	2.2	2.5	2.7	3.1	3.6	4.5	5.6	6.7
	150	0.2	0.3	0.4	0.5	0.6	0.7	0.8	1.0	1.2	1.4	1.6	1.8	2.0	2.2	2.4	2.8	3.2	4.0	5.0	5.9
	200	0.2	0.3	0.4	0.5	0.6	0.7	0.8	1.0	1.1	1.3	1.5	1.7	1.9	2.0	2.2	2.6	3.0	3.7	4.7	5.6
36	36	0.3	0.4	0.6	0.7	0.8	1.0	1.1	1.4	1.7	1.9	2.2	2.5	2.8	3.0	3.3	3.9	4.4	5.5	6.9	8.3
	50	0.2	0.4	0.5	0.6	0.7	0.8	1.0	1.2	1.4	1.7	1.9	2.1	2.5	2.6	2.9	3.3	3.8	4.8	5.9	7.2
	75	0.2	0.3	0.4	0.5	0.6	0.7	0.8	1.0	1.2	1.4	1.6	1.8	2.0	2.3	2.5	2.9	3.3	4.1	5.1	6.1
	100	0.2	0.3	0.4	0.5	0.6	0.7	0.8	0.9	1.1	1.3	1.5	1.7	1.9	2.1	2.3	2.6	3.0	3.8	4.7	5.7
	150	0.2	0.3	0.3	0.4	0.5	0.6	0.7	0.9	1.0	1.2	1.4	1.6	1.7	1.9	2.1	2.4	2.8	3.5	4.3	5.2
	200	0.2	0.2	0.3	0.4	0.5	0.6	0.7	0.8	1.0	1.1	1.3	1.5	1.6	1.8	2.0	2.3	2.6	3.3	4.1	4.9
42	42	0.2	0.4	0.5	0.6	0.7	0.8	1.0	1.2	1.4	1.6	1.9	2.1	2.4	2.6	2.8	3.3	3.8	4.7	5.9	7.1
	60	0.2	0.3	0.4	0.5	0.6	0.7	0.8	1.0	1.2	1.4	1.6	1.8	2.0	2.2	2.4	2.8	3.2	4.0	5.0	6.0
	90	0.2	0.3	0.3	0.4	0.5	0.6	0.7	0.9	1.0	1.2	1.4	1.6	1.7	1.9	2.1	2.4	2.8	3.5	4.4	5.2
	140	0.2	0.2	0.3	0.4	0.5	0.5	0.6	0.8	0.9	1.1	1.2	1.4	1.5	1.7	1.9	2.2	2.5	3.1	3.9	4.6
	200	0.1	0.2	0.3	0.4	0.4	0.5	0.6	0.7	0.9	1.0	1.1	1.3	1.4	1.6	1.7	2.0	2.3	2.9	3.6	4.3
	300	0.1	0.2	0.3	0.3	0.4	0.5	0.5	0.7	0.8	0.9	1.1	1.3	1.4	1.5	1.7	1.9	2.2	2.8	3.5	4.2
50	50	0.2	0.3	0.4	0.5	0.6	0.7	0.8	1.0	1.2	1.4	1.6	1.8	2.0	2.2	2.4	2.8	3.2	4.0	5.0	6.0
	70	0.2	0.3	0.3	0.4	0.5	0.6	0.7	0.9	1.0	1.2	1.4	1.5	1.7	1.9	2.0	2.4	2.7	3.4	4.3	5.1
	100	0.1	0.2	0.3	0.4	0.4	0.5	0.6	0.7	0.9	1.0	1.2	1.3	1.5	1.6	1.8	2.1	2.4	3.0	3.7	4.5
	150	0.1	0.2	0.3	0.3	0.4	0.5	0.5	0.7	0.8	0.9	1.1	1.2	1.3	1.5	1.6	1.9	2.1	2.7	3.3	4.0
	300	0.1	0.2	0.2	0.3	0.3	0.4	0.5	0.6	0.7	0.8	0.9	1.0	1.1	1.3	1.4	1.6	1.9	2.3	2.9	3.5
60	60	0.2	0.2	0.3	0.4	0.5	0.6	0.7	0.8	1.0	1.2	1.3	1.5	1.7	1.8	2.0	2.3	2.7	3.3	4.2	5.0
	100	0.1	0.2	0.3	0.3	0.4	0.5	0.5	0.7	0.8	0.9	1.1	1.2	1.3	1.5	1.6	1.9	2.1	2.7	3.3	4.0
	150	0.1	0.2	0.2	0.3	0.3	0.4	0.5	0.6	0.7	0.8	0.9	1.0	1.2	1.3	1.4	1.6	1.9	2.3	2.9	3.5
	300	0.1	0.1	0.2	0.2	0.3	0.3	0.4	0.5	0.6	0.7	0.8	0.9	1.0	1.1	1.2	1.4	1.6	2.0	2.5	3.0
75	75	0.1	0.2	0.3	0.3	0.4	0.5	0.5	0.7	0.8	0.9	1.1	1.2	1.3	1.5	1.6	1.9	2.1	2.7	3.3	4.0
	120	0.1	0.2	0.2	0.3	0.3	0.4	0.4	0.5	0.6	0.8	0.9	1.0	1.1	1.2	1.3	1.5	1.7	2.2	2.7	3.3
	200	0.1	0.1	0.2	0.2	0.3	0.3	0.4	0.5	0.5	0.6	0.7	0.8	0.9	1.0	1.1	1.3	1.5	1.8	2.3	2.7
	300	0.1	0.1	0.2	0.2	0.2	0.3	0.3	0.4	0.5	0.6	0.7	0.7	0.8	0.9	1.0	1.2	1.3	1.7	2.1	2.5
100	100	0.1	0.1	0.2	0.2	0.3	0.3	0.4	0.5	0.6	0.7	0.8	0.9	1.0	1.1	1.2	1.4	1.6	2.0	2.5	3.0
	200	0.1	0.1	0.1	0.2	0.2	0.3	0.3	0.4	0.4	0.5	0.6	0.7	0.7	0.8	0.9	1.0	1.2	1.5	1.9	2.2
	300	0.1	0.1	0.1	0.2	0.2	0.2	0.3	0.3	0.4	0.5	0.5	0.6	0.7	0.7	0.8	0.9	1.1	1.3	1.7	2.0
150	150	0.1	0.1	0.1	0.2	0.2	0.2	0.3	0.3	0.4	0.5	0.5	0.6	0.7	0.7	0.8	0.9	1.1	1.3	1.7	2.0
	300	—	0.1	0.1	0.1	0.1	0.2	0.2	0.2	0.3	0.3	0.4	0.5	0.5	0.6	0.6	0.7	0.8	1.0	1.2	1.5
200	200	—	0.1	0.1	0.1	0.1	0.2	0.2	0.2	0.3	0.3	0.4	0.5	0.5	0.6	0.6	0.7	0.8	1.0	1.2	1.5
	300	—	0.1	0.1	0.1	0.1	0.1	0.2	0.2	0.2	0.3	0.3	0.4	0.4	0.5	0.5	0.6	0.7	0.8	1.0	1.2
300	300	—	—	0.1	0.1	0.1	0.1	0.1	0.2	0.2	0.2	0.3	0.3	0.3	0.4	0.4	0.5	0.5	0.6	0.7	0.8
500	500	—	—	—	—	0.1	0.1	0.1	0.1	0.1	0.1	0.2	0.2	0.2	0.2	0.2	0.3	0.3	0.4	0.5	0.6

Fig. 9–11. Per Cent Effective Ceiling or Floor Cavity Reflectances for Various Reflectance Combinations

Per Cent Base* Reflectance	90									
Per Cent Wall Reflectance	90	80	70	60	50	40	30	20	10	0
Cavity Ratio										
0.2	89	88	88	87	86	85	85	84	84	82
0.4	88	87	86	85	84	83	81	80	79	76
0.6	87	86	84	82	80	79	77	76	74	73
0.8	87	85	82	80	77	75	73	71	69	67
1.0	86	83	80	77	75	72	69	66	64	62
1.2	85	82	78	75	72	69	66	63	60	57
1.4	85	80	77	73	69	65	62	59	57	52
1.6	84	79	75	71	67	63	59	56	53	50
1.8	83	78	73	69	64	60	56	53	50	48
2.0	83	77	72	67	62	56	53	50	47	43
2.2	82	76	70	65	59	54	50	47	44	40
2.4	82	75	69	64	58	53	48	45	41	37
2.6	81	74	67	62	56	51	46	42	38	35
2.8	81	73	66	60	54	49	44	40	36	34
3.0	80	72	64	58	52	47	42	38	34	30
3.2	79	71	63	56	50	45	40	36	32	28
3.4	79	70	62	54	48	43	38	34	30	27
3.6	78	69	61	53	47	42	36	32	28	25
3.8	78	69	60	51	45	40	35	31	27	23
4.0	77	69	58	51	44	39	33	29	25	22
4.2	77	62	57	50	43	37	32	28	24	21
4.4	76	61	56	49	42	36	31	27	23	20
4.6	76	60	55	47	40	35	30	26	22	19
4.8	75	59	54	46	39	34	28	25	21	18
5.0	75	59	53	45	38	33	28	24	20	16
6.0	73	61	49	41	34	29	24	20	16	11
7.0	70	58	45	38	30	27	21	18	14	08
8.0	68	55	42	35	27	23	18	15	12	06
9.0	66	52	38	31	25	21	16	14	11	05
10.0	65	51	36	29	22	19	15	11	09	04

Per Cent Base* Reflectance	80									
Per Cent Wall Reflectance	90	80	70	60	50	40	30	20	10	0
Cavity Ratio										
0.2	79	78	78	77	77	76	76	75	74	72
0.4	79	77	76	75	74	73	72	71	70	68
0.6	78	76	75	73	71	70	68	66	65	63
0.8	78	75	73	71	69	67	65	63	61	57
1.0	77	74	72	69	67	65	62	60	57	55
1.2	76	73	70	67	64	61	58	55	53	51
1.4	76	72	68	65	62	59	55	53	50	48
1.6	75	71	67	63	60	57	53	50	47	44
1.8	75	70	66	62	58	54	50	47	44	41
2.0	74	69	64	60	56	52	48	45	41	38
2.2	74	68	63	58	54	49	45	42	38	35
2.4	73	67	61	56	52	47	43	40	36	33
2.6	73	66	60	55	50	45	41	38	34	31
2.8	73	65	59	53	48	43	39	36	32	29
3.0	72	65	58	52	47	42	37	34	30	27
3.2	72	65	57	51	45	40	35	33	28	25
3.4	71	64	56	49	44	39	34	32	27	24
3.6	71	63	54	48	43	38	32	30	25	23
3.8	70	62	53	47	41	36	31	28	24	22
4.0	70	61	53	46	40	35	30	26	22	20
4.2	69	60	52	45	39	34	29	25	21	18
4.4	69	60	51	44	38	33	28	24	20	17
4.6	69	59	50	43	37	32	27	23	19	15
4.8	68	58	49	42	36	31	26	22	18	14
5.0	68	58	48	41	35	30	25	21	18	14
6.0	66	55	44	38	31	27	22	19	15	10
7.0	64	53	41	35	28	24	19	16	12	07
8.0	62	50	38	32	25	21	17	14	11	05
9.0	61	49	36	30	23	19	15	13	10	04
10.0	59	46	33	27	21	18	14	11	08	03

Per Cent Base* Reflectance	70									
Per Cent Wall Reflectance	90	80	70	60	50	40	30	20	10	0
Cavity Ratio										
0.2	70	69	68	68	67	67	66	66	65	64
0.4	69	68	67	66	65	64	63	62	61	58
0.6	69	67	65	64	63	61	59	58	57	54
0.8	68	66	64	62	60	58	56	55	53	50
1.0	68	65	62	60	58	55	53	52	50	47
1.2	67	64	61	59	57	54	50	48	46	44
1.4	67	63	60	58	55	51	47	45	44	41
1.6	67	62	59	56	53	47	45	43	41	38
1.8	66	61	58	54	51	46	42	40	38	35
2.0	66	60	56	52	49	45	40	38	36	33
2.2	66	60	55	51	48	43	38	36	34	32
2.4	65	60	54	50	46	41	37	35	32	30
2.6	65	59	54	49	45	40	35	33	30	28
2.8	65	59	53	48	43	38	33	30	28	26
3.0	64	58	52	47	42	37	32	29	27	24
3.2	64	58	51	46	40	36	31	28	25	23
3.4	64	57	50	45	39	35	29	27	24	22
3.6	63	56	49	44	38	33	28	25	22	20
3.8	63	56	49	43	37	32	27	24	21	19
4.0	63	55	48	42	36	31	26	23	20	17
4.2	62	55	47	41	35	30	25	22	19	16
4.4	62	54	46	40	34	29	24	21	18	15
4.6	62	53	45	39	33	28	24	21	17	14
4.8	62	53	45	38	32	27	23	20	16	13
5.0	61	52	44	36	31	26	22	19	16	12
6.0	60	51	41	35	28	24	19	16	13	09
7.0	58	48	38	32	26	22	17	14	11	06
8.0	57	46	35	29	23	19	15	13	10	05
9.0	56	45	33	27	21	18	14	12	09	04
10.0	55	43	31	25	19	16	12	10	08	03

Per Cent Base* Reflectance	60									
Per Cent Wall Reflectance	90	80	70	60	50	40	30	20	10	0
Cavity Ratio										
0.2	60	59	59	59	58	57	56	56	55	53
0.4	60	59	59	58	57	55	54	53	52	50
0.6	60	58	57	56	55	53	51	51	50	46
0.8	59	57	56	55	54	51	48	47	46	43
1.0	59	57	55	53	51	48	45	44	43	41
1.2	59	56	54	51	49	46	44	42	40	38
1.4	59	56	53	49	47	44	41	39	38	36
1.6	59	55	52	48	45	42	39	37	35	33
1.8	58	55	51	47	44	40	37	35	33	31
2.0	58	54	50	46	43	39	35	33	31	29
2.2	58	53	49	45	42	37	34	31	29	28
2.4	58	53	48	44	41	36	32	30	27	26
2.6	58	53	48	43	39	35	31	28	26	24
2.8	58	53	47	43	38	34	29	27	24	22
3.0	57	52	46	42	37	32	28	25	23	20
3.2	57	51	45	41	36	31	27	23	22	18
3.4	57	51	45	40	35	30	26	23	20	17
3.6	57	50	44	39	34	29	25	22	19	16
3.8	57	50	43	38	33	29	24	21	19	15
4.0	57	49	42	37	32	28	23	20	18	14
4.2	56	49	42	37	32	27	22	19	17	14
4.4	56	49	42	36	31	27	22	19	16	13
4.6	56	49	41	35	30	26	21	18	16	13
4.8	56	48	41	34	29	25	21	18	15	12
5.0	56	48	40	34	28	24	20	17	14	11
6.0	55	45	37	31	25	21	17	14	11	07
7.0	54	43	35	30	24	20	15	12	09	05
8.0	53	42	33	28	22	18	14	11	08	04
9.0	52	40	31	26	20	16	12	10	07	03
10.0	51	39	29	24	18	15	11	09	07	02

Per Cent Base* Reflectance	50									
Per Cent Wall Reflectance	90	80	70	60	50	40	30	20	10	0
Cavity Ratio										
0.2	50	50	49	49	48	48	47	46	46	44
0.4	50	49	48	48	47	46	45	45	44	42
0.6	50	48	47	46	45	44	43	42	41	38
0.8	50	48	47	45	44	42	40	39	38	36
1.0	50	48	46	44	43	41	38	37	36	34
1.2	50	47	45	43	41	39	36	35	34	29
1.4	50	47	45	42	40	38	35	34	32	27
1.6	50	47	44	41	39	36	33	32	30	26
1.8	50	46	43	40	38	35	31	30	28	25
2.0	50	46	43	40	37	34	30	28	26	24
2.2	50	46	42	38	36	33	29	27	24	22
2.4	50	46	42	37	35	31	27	25	23	21
2.6	50	46	41	37	34	30	26	23	21	20
2.8	50	46	41	36	33	29	25	22	20	19
3.0	50	45	40	36	32	28	24	21	19	17
3.2	50	44	39	35	31	27	23	20	18	16
3.4	50	44	39	35	30	26	22	19	17	15
3.6	50	44	39	34	29	25	21	18	16	14
3.8	50	44	38	34	29	25	21	17	15	13
4.0	50	44	38	33	28	24	20	17	15	12
4.2	50	43	37	32	28	24	20	17	14	12
4.4	50	43	37	32	27	23	19	16	13	11
4.6	50	43	36	31	26	22	18	15	13	10
4.8	50	43	36	31	26	22	18	15	12	09
5.0	50	42	35	30	25	21	17	14	12	09
6.0	50	42	34	29	23	19	15	13	10	06
7.0	49	41	32	27	21	18	14	11	08	05
8.0	49	40	30	25	19	16	12	10	07	03
9.0	48	39	29	24	18	15	11	09	07	03
10.0	47	37	27	22	17	14	10	08	06	02

* Ceiling, floor, or floor of cavity.

Fig. 9–11. *Continued*

Per Cent Base* Reflectance	40									
Per Cent Wall Reflectance	90	80	70	60	50	40	30	20	10	0
Cavity Ratio										
0.2	40	40	39	39	39	38	38	37	36	36
0.4	41	40	39	39	38	37	36	35	34	34
0.6	41	40	39	38	37	36	34	33	32	31
0.8	41	40	38	37	36	35	33	32	31	29
1.0	42	40	38	37	35	33	32	31	29	27
1.2	42	40	38	36	34	32	30	29	27	25
1.4	42	39	37	35	33	31	29	27	25	23
1.6	42	39	37	35	32	30	27	25	23	22
1.8	42	39	36	34	31	29	26	24	22	21
2.0	42	39	36	34	31	28	25	23	21	19
2.2	42	39	36	33	30	27	24	22	19	18
2.4	43	39	35	33	29	27	24	21	18	17
2.6	43	39	35	32	29	26	23	20	17	15
2.8	43	39	35	32	28	25	22	19	16	14
3.0	43	39	35	31	27	24	21	18	16	13
3.2	43	39	35	31	27	23	20	17	15	13
3.4	43	39	34	30	26	23	20	17	14	12
3.6	44	39	34	30	26	22	19	16	14	11
3.8	44	38	33	29	25	22	18	16	13	10
4.0	44	38	33	29	25	21	18	15	12	10
4.2	44	38	33	29	24	21	17	15	12	10
4.4	44	38	33	28	24	20	17	14	11	09
4.6	44	38	32	28	23	19	16	14	11	08
4.8	44	38	32	27	22	19	16	13	10	08
5.0	45	38	31	27	22	19	15	13	10	07
6.0	44	37	30	25	20	17	13	11	08	05
7.0	44	36	29	24	19	16	12	10	07	04
8.0	44	35	28	23	18	15	11	09	06	03
9.0	44	35	26	21	16	13	10	08	05	02
10.0	43	34	25	20	15	12	08	07	05	02

Per Cent Base* Reflectance	30									
Per Cent Wall Reflectance	90	80	70	60	50	40	30	20	10	0
Cavity Ratio										
0.2	31	31	30	30	29	29	29	28	28	27
0.4	31	31	30	30	29	28	28	27	26	25
0.6	32	31	30	29	28	27	26	26	25	23
0.8	32	31	30	29	28	26	25	25	23	22
1.0	33	32	30	29	27	25	24	23	22	20
1.2	33	32	30	28	27	25	23	22	21	19
1.4	34	32	30	28	26	24	22	21	19	18
1.6	34	33	29	27	25	23	22	20	18	17
1.8	35	33	29	27	25	23	21	19	17	16
2.0	35	33	29	26	24	22	20	18	16	14
2.2	36	32	29	26	24	22	19	17	15	13
2.4	36	32	29	26	24	22	19	16	14	12
2.6	36	32	29	25	23	21	18	16	14	12
2.8	37	33	29	25	23	21	17	15	13	11
3.0	37	33	29	25	22	20	17	15	12	10
3.2	37	33	29	25	22	19	16	14	12	10
3.4	37	33	29	25	22	19	16	14	11	09
3.6	38	33	29	24	21	18	15	13	10	09
3.8	38	33	28	24	21	18	15	13	10	08
4.0	38	33	28	24	21	18	14	12	09	07
4.2	38	33	28	24	20	17	14	12	09	07
4.4	39	33	28	24	20	17	14	11	09	06
4.6	39	33	28	24	20	17	13	10	08	06
4.8	39	33	28	24	20	17	13	10	08	05
5.0	39	33	28	24	19	16	13	10	08	05
6.0	39	33	27	23	18	15	11	09	06	04
7.0	40	33	26	22	17	14	10	08	05	03
8.0	40	33	26	21	16	13	09	07	04	02
9.0	40	33	25	20	15	12	09	07	04	02
10.0	40	32	24	19	14	11	08	06	03	01

Per Cent Base* Reflectance	20									
Per Cent Wall Reflectance	90	80	70	60	50	40	30	20	10	0
Cavity Ratio										
0.2	21	20	20	20	20	20	19	19	19	17
0.4	22	21	20	20	20	19	19	18	18	16
0.6	23	21	21	20	19	19	18	18	17	15
0.8	24	22	21	20	19	19	18	17	16	14
1.0	25	23	22	20	19	18	17	16	15	13
1.2	25	23	22	20	19	17	17	16	14	12
1.4	26	24	22	20	18	17	16	15	13	12
1.6	26	24	22	20	18	17	16	15	13	11
1.8	27	25	23	20	18	17	15	14	12	10
2.0	28	25	23	20	18	16	15	13	11	09
2.2	28	25	23	20	18	16	14	12	10	09
2.4	29	26	23	20	18	16	14	12	10	08
2.6	29	26	23	20	18	16	14	11	09	08
2.8	30	27	23	20	18	15	13	11	09	07
3.0	30	27	23	20	17	15	13	11	09	07
3.2	31	27	23	20	17	15	12	11	09	06
3.4	31	27	23	20	17	15	12	10	08	06
3.6	32	27	23	20	17	15	12	10	08	05
3.8	32	28	23	20	17	15	12	10	07	05
4.0	33	28	23	20	17	14	11	09	07	05
4.2	33	28	23	20	17	14	11	09	07	04
4.4	34	28	24	20	17	14	11	09	07	04
4.6	34	29	24	20	17	14	11	09	07	04
4.8	35	29	24	20	17	13	10	08	06	04
5.0	35	29	24	20	16	13	10	08	06	04
6.0	36	30	24	20	16	13	10	08	05	02
7.0	36	30	24	20	15	12	09	07	04	02
8.0	37	30	23	19	15	12	08	06	03	01
9.0	37	29	23	19	14	11	08	06	03	01
10.0	37	29	22	18	13	10	07	05	03	01

Per Cent Base* Reflectance	10									
Per Cent Wall Reflectance	90	80	70	60	50	40	30	20	10	0
Cavity Ratio										
0.2	11	11	11	10	10	10	10	09	09	09
0.4	12	11	11	11	11	10	10	09	09	08
0.6	13	13	12	11	11	10	10	09	08	08
0.8	15	14	13	12	11	10	10	09	08	07
1.0	16	14	13	12	12	11	10	09	08	07
1.2	17	15	14	13	12	11	10	09	07	06
1.4	18	16	14	13	12	11	10	09	07	06
1.6	19	17	15	14	12	11	09	08	07	06
1.8	19	17	15	14	13	11	09	08	06	05
2.0	20	18	16	14	13	11	09	08	06	05
2.2	21	19	16	14	13	11	09	07	06	05
2.4	22	19	17	15	13	11	09	07	06	05
2.6	23	20	17	15	13	11	09	07	06	04
2.8	23	20	18	16	13	11	09	07	05	03
3.0	24	21	18	16	13	11	09	07	05	03
3.2	25	21	18	16	13	11	09	07	05	03
3.4	26	22	18	16	13	11	09	07	05	03
3.6	26	22	19	16	13	11	09	06	04	03
3.8	27	23	19	17	14	11	09	06	04	02
4.0	27	23	20	17	14	11	09	06	04	02
4.2	28	24	20	17	14	11	09	06	04	02
4.4	28	24	20	17	14	11	08	06	04	02
4.6	29	25	20	17	14	11	08	06	04	02
4.8	29	25	20	17	14	11	08	06	04	02
5.0	30	25	20	17	14	11	08	06	04	02
6.0	31	26	21	18	14	11	08	06	03	01
7.0	32	27	21	17	13	11	08	06	03	01
8.0	33	27	21	17	13	10	07	05	03	01
9.0	34	28	21	17	13	10	07	05	02	01
10.0	34	28	21	17	12	10	07	05	02	01

Per Cent Base* Reflectance	0									
Per Cent Wall Reflectance	90	80	70	60	50	40	30	20	10	0
Cavity Ratio										
0.2	02	02	02	01	01	01	01	00	00	0
0.4	04	03	03	02	02	02	01	01	00	0
0.6	05	05	04	03	03	02	02	01	01	0
0.8	07	06	05	04	04	03	02	02	01	0
1.0	08	07	06	05	04	03	02	02	01	0
1.2	10	08	07	06	05	04	03	02	01	0
1.4	11	09	08	07	06	04	03	02	01	0
1.6	12	10	09	07	06	05	03	02	01	0
1.8	13	11	09	08	07	05	04	03	01	0
2.0	14	12	10	09	07	05	04	03	01	0
2.2	15	13	11	09	07	06	04	03	01	0
2.4	16	13	11	09	08	06	04	03	01	0
2.6	17	14	12	10	08	06	05	03	02	0
2.8	17	15	13	10	08	07	05	03	02	0
3.0	18	16	13	11	09	07	05	03	02	0
3.2	19	16	14	11	09	07	05	03	02	0
3.4	20	17	14	12	09	07	05	03	02	0
3.6	20	17	15	12	10	08	05	04	02	0
3.8	21	18	15	12	10	08	05	04	02	0
4.0	22	18	15	13	10	08	05	04	02	0
4.2	22	19	16	13	10	08	06	04	02	0
4.4	23	19	16	13	10	08	06	04	02	0
4.6	23	20	17	13	11	08	06	04	02	0
4.8	24	20	17	14	11	08	06	04	02	0
5.0	25	21	17	14	11	08	06	04	02	0
6.0	27	23	18	15	12	09	06	04	02	0
7.0	28	24	19	15	12	09	06	04	02	0
8.0	30	25	20	15	12	09	06	04	02	0
9.0	31	25	20	15	12	09	06	04	02	0
10.0	31	25	20	15	12	09	06	04	02	0

* Ceiling, floor, or floor of cavity.

Fig. 9–12. Coefficients of Utilization, Wall Luminance Coefficients, Ceiling Cavity Luminance Coefficients, Guide

In some cases, luminaire data in this table are based on an actual typical luminaire; in other cases, the data represent a composite of generic luminaire types. Therefore, whenever possible, specific luminaire data should be used in preference to this table of typical luminaires.

The polar intensity sketch (candlepower distribution curve) and the corresponding spacing-to-mounting height guide are representative of many luminaires of each type shown. A specific luminaire may differ in perpendicular plane (crosswise) and parallel plane (lengthwise) intensity distributions and in S/MH guide from the values shown. However, the various coefficients depend only on the average intensity at each polar angle from nadir. The tabulated coefficients can be applied to any luminaire whose average intensity distribution matches the values used to generate the coefficients. The average intensity values used to generate the coefficients are given at the end of the table, normalized to a per thousand lamp lumen basis.

The various coefficients below depend on the shape of the average intensity distribution curve and are linearly related

Typical Luminaire	Typical Distribution and Per Cent Lamp Lumens	Maint. Cat.	Maximum S/MH Guide[d]	ρ_{CC}[a] →	80			70			50			30			10			0
				ρ_W[b] →	50	30	10	50	30	10	50	30	10	50	30	10	50	30	10	0
				RCR[c] ↓	Coefficients of Utilization for 20 Per Cent Effective Floor Cavity Reflectance (ρ_{FC} = 20)															
1		V	1.5	0	.87	.87	.87	.81	.81	.81	.69	.69	.69	.59	.59	.59	.49	.49	.49	.44
				1	.71	.67	.63	.66	.62	.59	.56	.53	.50	.47	.45	.43	.39	.37	.35	.31
				2	.61	.54	.49	.56	.50	.46	.47	.43	.39	.39	.36	.33	.32	.29	.27	.23
				3	.52	.45	.39	.48	.42	.37	.41	.36	.31	.34	.30	.26	.27	.24	.22	.18
	35½% ↑			4	.46	.38	.33	.42	.36	.30	.36	.30	.26	.30	.26	.22	.24	.21	.18	.15
				5	.40	.33	.27	.37	.30	.25	.32	.26	.22	.26	.22	.19	.21	.18	.15	.12
	45% ↓			6	.36	.28	.23	.33	.26	.21	.28	.23	.19	.23	.19	.16	.19	.15	.13	.10
				7	.32	.25	.20	.29	.23	.18	.25	.20	.16	.21	.16	.13	.17	.13	.11	.09
				8	.29	.22	.17	.27	.20	.16	.23	.17	.14	.19	.15	.12	.15	.12	.09	.07
Pendant diffusing sphere with incandescent lamp				9	.26	.19	.15	.24	.18	.14	.20	.15	.12	.17	.13	.10	.14	.11	.08	.06
				10	.23	.17	.13	.22	.16	.12	.19	.14	.10	.16	.12	.09	.13	.09	.07	.05
2		II	1.5	0	.83	.83	.83	.71	.71	.71	.49	.49	.49	.30	.30	.30	.12	.12	.12	.03
				1	.72	.69	.66	.62	.60	.57	.43	.42	.40	.26	.25	.25	.10	.10	.10	.03
				2	.63	.58	.54	.54	.50	.47	.38	.36	.33	.23	.22	.21	.09	.09	.08	.02
				3	.55	.49	.45	.48	.43	.39	.33	.30	.28	.20	.19	.17	.08	.08	.07	.02
				4	.48	.42	.37	.42	.37	.33	.29	.26	.24	.18	.16	.15	.07	.07	.06	.02
	83% ↑			5	.43	.36	.32	.37	.32	.28	.26	.23	.20	.16	.14	.13	.06	.06	.05	.01
				6	.38	.32	.27	.33	.28	.24	.23	.20	.17	.14	.12	.11	.06	.05	.04	.01
	3½% ↓			7	.34	.28	.23	.30	.24	.21	.21	.17	.15	.13	.11	.09	.05	.04	.04	.01
Concentric ring unit with incandescent silvered-bowl lamp				8	.31	.25	.20	.27	.21	.18	.19	.15	.13	.12	.10	.08	.05	.04	.03	.01
				9	.28	.22	.18	.24	.19	.16	.17	.14	.11	.10	.09	.07	.04	.03	.03	.01
				10	.25	.20	.16	.22	.17	.14	.16	.12	.10	.10	.08	.06	.04	.03	.03	.01
3		IV	1.3	0	.99	.99	.99	.97	.97	.97	.92	.92	.92	.88	.88	.88	.85	.85	.85	.83
				1	.88	.85	.82	.86	.83	.81	.83	.80	.78	.79	.78	.76	.77	.75	.73	.72
	0% ↑			2	.78	.73	.68	.76	.72	.67	.73	.69	.66	.71	.67	.64	.68	.65	.63	.61
				3	.69	.62	.57	.67	.61	.57	.65	.60	.56	.63	.58	.55	.61	.57	.54	.52
				4	.61	.54	.49	.60	.53	.48	.58	.52	.48	.56	.51	.47	.54	.50	.46	.45
				5	.54	.47	.41	.53	.46	.41	.51	.45	.41	.50	.44	.40	.48	.43	.40	.38
	83½% ↓			6	.48	.41	.35	.47	.40	.35	.46	.39	.35	.44	.39	.34	.43	.38	.34	.32
				7	.43	.35	.30	.42	.35	.30	.41	.34	.30	.39	.34	.30	.38	.33	.29	.28
Porcelain-enameled ventilated standard dome with incandescent lamp				8	.38	.31	.26	.38	.31	.26	.37	.30	.26	.36	.30	.26	.35	.30	.26	.24
				9	.35	.28	.23	.34	.27	.23	.33	.27	.23	.32	.27	.23	.31	.26	.22	.21
				10	.31	.25	.20	.31	.24	.20	.30	.24	.20	.29	.24	.20	.29	.23	.20	.18
4		V	1.3	0	.89	.89	.89	.85	.85	.85	.77	.77	.77	.70	.70	.70	.63	.63	.63	.60
				1	.78	.75	.72	.74	.72	.69	.68	.66	.64	.62	.60	.58	.56	.55	.54	.51
				2	.69	.65	.61	.66	.62	.58	.61	.57	.54	.56	.53	.50	.51	.49	.47	.44
	18½% ↑			3	.62	.57	.52	.60	.55	.50	.55	.51	.47	.50	.47	.44	.46	.44	.41	.39
				4	.56	.50	.46	.54	.49	.44	.50	.45	.42	.46	.42	.39	.42	.39	.37	.35
				5	.51	.45	.40	.49	.43	.39	.45	.41	.37	.42	.38	.35	.39	.36	.33	.31
	60½% ↓			6	.46	.40	.36	.45	.39	.35	.42	.37	.33	.39	.35	.31	.36	.32	.30	.28
				7	.42	.36	.32	.41	.35	.31	.38	.33	.29	.35	.31	.28	.33	.29	.27	.25
				8	.39	.32	.28	.37	.32	.28	.35	.30	.26	.32	.28	.25	.30	.27	.24	.22
Prismatic square surface drum				9	.35	.29	.25	.34	.29	.25	.32	.27	.24	.30	.26	.23	.28	.24	.22	.20
				10	.32	.27	.23	.31	.26	.22	.29	.25	.21	.27	.23	.20	.26	.22	.20	.18

[a] ρ_{CC} = per cent effective ceiling cavity reflectance.
[b] ρ_W = per cent wall reflectance.
[c] RCR = Room Cavity Ratio.
[d] Maximum S/MH guide—ratio of maximum luminaire spacing to mounting or ceiling height above work-plane.

for Maximum Spacing-to-Mounting Height Ratios, and Maintenance Categories of Typical Luminaires

to the total luminaire efficiency. Consequently, the tabulated coefficients can be applied to luminaires with similarly shaped average intensity distributions using a correcting multiplier equal to the new luminaire total efficiency divided by the tabulated luminaire total efficiency.

Satisfactory installations depend on the environment and space utilization as well as upon the luminaire. Consequently, a definitive S/MH limit cannot be assigned independently to the luminaire. The values for S/MH limits given below are only general guides (see page 9-33 for a discussion of spacing limitations). The S/MH guide for incandescent and high-intensity discharge lamp luminaires below are based on Fig. 9-14. By convention, scale A of Fig. 9-14 is used for fluorescent units as being somewhat representative for rows of luminaires. Values for semi-indirect and indirect luminaires are based on historical precedence. A maximum of 1.5 is shown since the use of larger values usually does not produce acceptable installations when all criteria are examined.

Wall Luminance Coefficients[e] for 20 Per Cent Effective Floor Cavity Reflectance (ρ_{FC} = 20)

ρ_{CC} →	80			70			50			30			10		
ρ_W →	50	30	10	50	30	10	50	30	10	50	30	10	50	30	10
RCR ↓															
1	.32	.18	.06	.30	.17	.05	.27	.15	.05	.24	.14	.04	.21	.12	.04
2	.27	.15	.05	.25	.14	.04	.23	.13	.04	.20	.11.	04	.18	.10	.03
3	.24	.13	.04	.22	.12	.04	.20	.11	.03	.17	.09	.03	.15	.08	.03
4	.21	.11	.03	.20	.10	.03	.17	.09	.03	.15	.08	.02	.13	.07	.02
5	.19	.10	.03	.18	.09	.03	.16	.08	.02	.14	.07	.02	.12	.06	.02
6	.18	.09	.03	.16	.08	.02	.14	.07	.02	.13	.07	.02	.11	.06	.02
7	.16	.08	.02	.15	.08	.02	.13	.07	.02	.12	.06	.02	.10	.05	.02
8	.15	.07	.02	.14	.07	.02	.12	.06	.02	.11	.05	.02	.09	.05	.01
9	.14	.07	.02	.13	.06	.02	.12	.06	.02	.10	.05	.01	.09	.04	.01
10	.13	.06	.02	.12	.06	.02	.11	.05	.01	.09	.05	.01	.08	.04	.01
1	.23	.13	.04	.19	.11	.04	.14	.08	.03	.08	.05	.02	.03	.02	.01
2	.21	.11	.03	.18	.10	.03	.13	.07	.02	.08	.04	.01	.03	.02	.01
3	.19	.10	.03	.16	.09	.03	.12	.06	.02	.07	.04	.01	.03	.02	.00
4	.18	.09	.03	.15	.08	.02	.11	.06	.02	.07	.04	.01	.03	.01	.00
5	.16	.08	.02	.14	.07	.02	.10	.05	.02	.06	.03	.01	.02	.01	.00
6	.15	.08	.02	.13	.07	.02	.09	.05	.01	.06	.03	.01	.02	.01	.00
7	.14	.07	.02	.12	.06	.02	.09	.04	.01	.05	.03	.01	.02	.01	.00
8	.13	.07	.02	.12	.06	.02	.08	.04	.01	.05	.03	.01	.02	.01	.00
9	.13	.06	.02	.11	.05	.01	.08	.04	.01	.05	.02	.01	.02	.01	.00
10	.12	.06	.02	.10	.05	.01	.07	.04	.01	.04	.02	.01	.02	.01	.00
1	.23	.13	.04	.23	.13	.04	.21	.12	.04	.20	.12	.04	.20	.11	.04
2	.22	.12	.04	.22	.12	.04	.21	.11	.04	.20	.11	.03	.19	.11	.03
3	.21	.11	.03	.21	.11	.03	.20	.11	.03	.19	.10	.03	.18	.10	.03
4	.20	.10	.03	.19	.10	.03	.19	.10	.03	.18	.10	.03	.17	.09	.03
5	.19	.09	.03	.18	.09	.03	.18	.09	.03	.17	.09	.03	.16	.09	.03
6	.18	.09	.03	.17	.09.	03	.17	.09	.02	.16	.08	.02	.16	.08	.02
7	.17	.08	.02	.16	.08	.02	.16	.08	.02	.15	.08	.02	.15	.08	.02
8	.16	.08	.02	.15	.08	.02	.15	.07	.02	.14	.07	.02	.14	.07	.02
9	.15	.07	.02	.15	.07	.02	.14	.07	.02	.14	.07	.02	.13	.07	.02
10	.14	.07	.02	.14	.07	.02	.13	.07	.02	.13	.06	.02	.13	.06	.02
1	.23	.13	.04	.22	.13	.04	.20	.11	.04	.18	.10	.03	.16	.09	.03
2	.20	.11	.03	.19	.11	.03	.18	.10	.03	.16	.09	.03	.14	.08	.03
3	.19	.10	.03	.18	.09	.03	.16	.09	.03	.14	.08	.02	.13	.07	.02
4	.17	.09	.03	.16	.08	.02	.15	.08	.02	.13	.07	.02	.12	.06	.02
5	.16	.08	.02	.15	.08	.02	.14	.07	.02	.12	.07	.02	.11	.06	.02
6	.15	.07	.02	.14	.07	.02	.13	.07	.02	.12	.06	.02	.10	.06	.02
7	.14	.07	.02	.13	.07	.02	.12	.06	.02	.11	.06	.02	.10	.05	.02
8	.13	.06	.02	.13	.06	.02	.11	.06	.02	.10	.05	.02	.09	.05	.01
9	.12	.06	.02	.12	.06	.02	.11	.05	.02	.10	.05	.01	.09	.05	.01
10	.12	.06	.02	.11	.05	.02	.10	.05	.01	.09	.05	.01	.09	.04	.01

Ceiling Cavity Luminance Coefficients[e] for 20 Per Cent Effective Floor Cavity Reflectance (ρ_{FC} = 20)

ρ_{CC} →	80			70			50			30			10		
ρ_W →	50	30	10	50	30	10	50	30	10	50	30	10	50	30	10
RCR ↓															
0	.42	.42	.42	.36	.36	.36	.25	.25	.25	.14	.14	.14	.05	.05	.05
1	.42	.40	.37	.36	.34	.32	.25	.23	.22	.14	.14	.13	.05	.04	.04
2	.42	.38	.35	.36	.33	.30	.24	.23	.21	.14	.13	.12	.05	.04	.04
3	.41	.37	.33	.35	.32	.29	.24	.22	.20	.14	.13	.12	.04	.04	.04
4	.41	.36	.32	.35	.31	.28	.24	.22	.20	.14	.13	.12	.04	.04	.04
5	.40	.35	.31	.34	.30	.27	.24	.21	.19	.14	.12	.11	.04	.04	.04
6	.39	.34	.31	.34	.30	.27	.23	.21	.19	.14	.12	.11	.04	.04	.04
7	.39	.34	.31	.33	.29	.27	.23	.21	.19	.13	.12	.11	.04	.04	.04
8	.38	.34	.30	.33	.29	.26	.23	.20	.19	.13	.12	.11	.04	.04	.04
9	.38	.33	.30	.33	.29	.26	.23	.20	.19	.13	.12	.11	.04	.04	.04
10	.37	.33	.30	.32	.29	.26	.22	.20	.19	.13	.12	.11	.04	.04	.04
0	.79	.79	.79	.68	.68	.68	.46	.46	.46	.27	.27	.27	.09	.09	.09
1	.79	.77	.76	.68	.66	.65	.46	.46	.45	.27	.26	.26	.09	.08	.08
2	.78	.76	.73	.67	.65	.63	.46	.45	.44	.27	.26	.26	.09	.08	.08
3	.78	.74	.71	.67	.64	.62	.46	.45	.43	.26	.26	.26	.09	.08	.08
4	.77	.73	.70	.66	.63	.61	.46	.44	.43	.26	.26	.25	.08	.08	.08
5	.77	.73	.70	.66	.63	.60	.45	.44	.43	.26	.26	.25	.08	.08	.08
6	.76	.72	.69	.66	.62	.60	.45	.44	.43	.26	.26	.25	.08	.08	.08
7	.76	.72	.69	.65	.62	.60	.45	.44	.42	.26	.26	.25	.08	.08	.08
8	.75	.71	.68	.65	.62	.60	.45	.43	.42	.26	.26	.25	.08	.08	.08
9	.75	.71	.68	.65	.61	.59	.45	.43	.42	.26	.26	.25	.08	.08	.08
10	.75	.71	.68	.64	.61	.59	.45	.43	.42	.26	.25	.25	.08	.08	.08
0	.15	.15	.15	.13	.13	.13	.09	.09	.09	.05	.05	.05	.02	.02	.02
1	.15	.13	.11	.13	.11	.10	.09	.08	.07	.05	.04	.04	.02	.01	.01
2	.14	.11	.08	.12	.09	.07	.08	.07	.05	.05	.04	.03	.02	.01	.01
3	.13	.10	.06	.12	.08	.06	.08	.06	.04	.05	.03	.02	.01	.01	.01
4	.13	.08	.05	.11	.07	.04	.08	.05	.03	.04	.03	.02	.01	.01	.01
5	.12	.08	.04	.11	.07	.04	.07	.05	.02	.04	.03	.01	.01	.01	.00
6	.12	.07	.03	.10	.06	.03	.07	.04	.02	.04	.02	.01	.01	.01	.00
7	.11	.06	.03	.10	.05	.02	.07	.04	.02	.04	.02	.01	.01	.01	.00
8	.11	.06	.02	.09.	05	.02	.06	.04	.01	.04	.02	.01	.01	.01	.00
9	.10	.05	.02	.09	.05	.02	.06	.03	.01	.04	.02	.01	.01	.01	.00
10	.10	.05	.02	.09	.04	.02	.06	.03	.01	.03	.02	.01	.01	.01	.00
0	.29	.29	.29	.24	.24	.24	.17	.17	.17	.10	.10	.10	.03	.03	.03
1	.28	.26	.25	.24	.23	.21	.17	.16	.15	.10	.09	.09	.03	.03	.03
2	.27	.25	.22	.24	.21	.19	.16	.15	.13	.09	.09	.08	.03	.03	.03
3	.27	.23	.21	.23	.20	.18	.16	.14	.12	.09	.08	.07	.03	.03	.02
4	.26	.22	.19	.22	.19	.17	.15	.13	.12	.09	.08	.07	.03	.03	.02
5	.25	.21	.18	.22	.19	.16	.15	.13	.11	.09	.08	.07	.03	.02	.02
6	.25	.21	.18	.21	.18	.16	.15	.13	.11	.09	.07	.07	.03	.02	.02
7	.24	.20	.17	.21	.18	.15	.15	.12	.11	.08	.07	.06	.03	.02	.02
8	.24	.20	.17	.21	.17	.15	.14	.12	.10	.08	.07	.06	.03	.02	.02
9	.24	.19	.17	.20	.17	.15	.14	.12	.10	.08	.07	.06	.03	.02	.02
10	.23	.19	.16	.20	.17	.14	.14	.12	.10	.08	.07	.06	.03	.02	.02

[e] Although it is recommended that luminance coefficients and wall direct radiation coefficients be published to three decimal places, only two are shown here for these typical luminaires. Three decimal place data should be obtained from manufacturers of actual luminaires used.

Fig. 9-12. *Continued*

Typical Luminaire	Typical Distribution and Per Cent Lamp Lumens: Maint. Cat.	Maximum S/MH Guide[d]	ρ_{CC}[a] →	80			70			50			30			10			0
			ρ_W[b] →	50	30	10	50	30	10	50	30	10	50	30	10	50	30	10	0
			RCR[c] ↓	Coefficients of Utilization for 20 Per Cent Effective Floor Cavity Reflectance (ρ_{FC} = 20)															
5	IV	0.8	0	1.18	1.18	1.18	1.16	1.16	1.16	1.11	1.11	1.11	1.06	1.06	1.06	1.01	1.01	1.01	.99
			1	1.09	1.07	1.04	1.07	1.05	1.02	1.03	1.01	.99	.99	.98	.96	.96	.95	.94	.92
	0% ↑		2	1.01	.97	.93	.99	.95	.92	.96	.93	.90	.93	.90	.88	.90	.88	.86	.84
			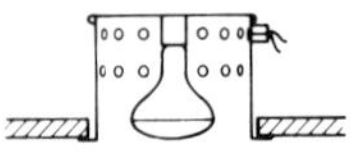  3	.93	.88	.84	.92	.87	.83	.89	.85	.81	.87	.83	.80	.84	.82	.79	.77
			4	.87	.81	.76	.85	.80	.75	.83	.78	.75	.81	.77	.74	.79	.76	.73	.71
	100% ↓		5	.80	.74	.69	.79	.73	.69	.77	.72	.68	.76	.71	.67	.74	.70	.67	.65
			6	.74	.68	.63	.73	.67	.63	.72	.66	.62	.70	.66	.62	.69	.65	.61	.60
			7	.69	.62	.57	.68	.62	.57	.67	.61	.57	.65	.60	.56	.64	.60	.56	.55
			8	.64	.57	.53	.63	.57	.52	.62	.56	.52	.61	.56	.52	.60	.55	.52	.50
			9	.59	.52	.48	.59	.52	.48	.58	.52	.48	.57	.51	.48	.56	.51	.47	.46
R-40 flood without shielding			10	.55	.49	.44	.55	.48	.44	.54	.48	.44	.53	.48	.44	.52	.47	.44	.42
6	IV	0.7	0	1.00	1.00	1.00	.98	.98	.98	.94	.94	.94	.90	.90	.90	.86	.86	.86	.84
			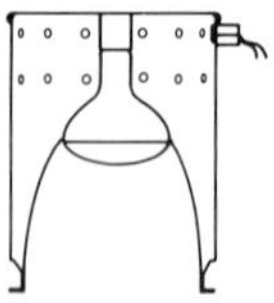 1	.96	.94	.92	.94	.92	.91	.90	.89	.88	.87	.86	.85	.84	.84	.83	.82
	0% ↑		2	.91	.88	.86	.90	.87	.85	.87	.85	.83	.84	.83	.82	.82	.81	.80	.79
			 3	.87	.84	.81	.86	.83	.81	.84	.81	.79	.82	.80	.78	.80	.78	.77	.76
			4	.83	.80	.77	.82	.79	.77	.81	.78	.76	.79	.77	.75	.78	.76	.74	.73
	85% ↓		5	.79	.76	.73	.79	.75	.73	.77	.74	.72	.76	.73	.71	.75	.73	.71	.70
			6	.76	.73	.70	.76	.72	.70	.75	.72	.69	.74	.71	.69	.73	.70	.68	.67
			7	.73	.69	.66	.73	.69	.66	.72	.68	.66	.71	.68	.66	.70	.67	.65	.64
R-40 flood with specular anodized reflector skirt; 45° cutoff			8	.70	.66	.63	.70	.66	.63	.69	.65	.63	.68	.65	.63	.67	.65	.63	.62
			9	.67	.63	.60	.67	.63	.60	.66	.62	.60	.65	.62	.60	.65	.62	.60	.59
			10	.64	.60	.58	.64	.60	.58	.63	.60	.58	.63	.60	.57	.62	.59	.57	.56
7	IV	0.7	0	.51	.51	.51	.50	.50	.50	.48	.48	.48	.46	.46	.46	.44	.44	.44	.43
			1	.49	.48	.48	.48	.48	.47	.47	.46	.46	.45	.45	.44	.44	.43	.43	.42
	0% ↑		2	.47	.46	.45	.46	.45	.44	.45	.44	.43	.44	.43	.42	.43	.42	.42	.41
			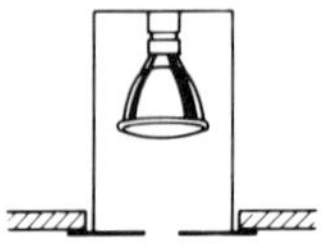  3	.45	44	.43	.45	.43	.42	.44	.42	.42	.43	.42	.41	.42	.41	.40	.40
			4	.43	.42	.41	.43	.41	.40	.42	.41	.40	.41	.40	.39	.41	.40	.39	.38
			5	.42	.40	.39	.41	.40	.38	.41	.39	.38	.40	.39	.38	.39	.38	.38	.37
	43½% ↓		6	.40	.39	.37	.40	.38	.37	.39	.38	.37	.39	.38	.37	.38	.37	.36	.36
			7	.39	.37	.36	.39	.37	.36	.38	.37	.35	.38	.36	.35	.37	.36	.35	.35
			8	.37	.36	.34	.37	.35	.34	.37	.35	.34	.36	.35	.34	.36	.35	.34	.33
EAR-38 lamp above 2″ diameter aperture			9	.36	.34	.33	.36	.34	.33	.35	.34	.33	.35	.34	.33	.35	.34	.33	.32
			10	.35	.33	.32	.35	.33	.32	.34	.33	.32	.34	.33	.32	.34	.32	.31	.31
8	IV	0.7	0	.53	.53	.53	.52	.52	.52	.49	.49	.49	.47	.47	.47	.45	.45	.45	.44
			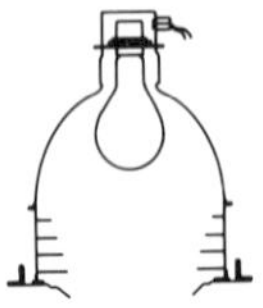 1	.51	.50	.49	.50	.49	.48	.48	.47	.47	.46	.46	.45	.45	.44	.44	.43
			2	.48	.47	.46	.48	.46	.45	.46	.45	.44	.45	.44	.44	.44	.43	.43	.42
	0% ↑		3	.47	.45	.44	.46	.45	.43	.45	.44	.43	.44	.43	.42	.43	.42	.41	.41
			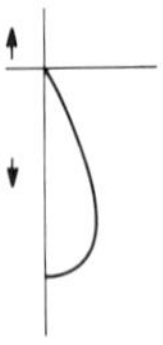 4	.45	.43	.42	.44	.43	.42	.43	.42	.41	.43	.41	.41	.42	.41	.40	.40
	44½% ↓		5	.43	.41	.40	.43	.41	.40	.42	.40	.39	.41	.40	.39	.41	.40	.39	.38
			6	.42	.40	.39	.41	.40	.38	.41	.39	.38	.40	.39	.38	.40	.39	.38	.37
			7	.40	.38	.37	.40	.38	.37	.39	.38	.37	.39	.38	.37	.38	.37	.36	.36
Reflector downlight with baffles and inside frosted lamp			8	.39	.37	.36	.38	.37	.36	.38	.37	.35	.38	.36	.35	.37	.36	.35	.35
			9	.37	.36	.34	.37	.35	.34	.37	.35	.34	.36	.35	.34	.36	.35	.34	.33
			10	.36	.34	.33	.36	.34	.33	.36	.34	.33	.35	.34	.33	.35	.34	.33	.32
9	V	1.0	0	.64	.64	.64	.63	.63	.63	.60	.60	.60	.57	.57	.57	.55	.55	.55	.54
			1	.60	.58	.57	.58	.57	.56	.56	.55	.54	.54	.53	.52	.52	.52	.51	.50
	0% ↑		2	.55	.53	.51	.54	.52	.50	.52	.50	.49	.51	.49	.48	.49	.48	.47	.46
			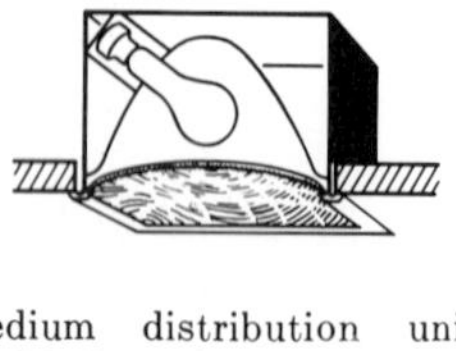  3	.51	.48	.46	.50	.47	.45	.49	.46	.44	.47	.45	.44	.46	.44	.43	.42
			4	.47	.44	.41	.47	.44	.41	.45	.43	.41	.44	.42	.40	.43	.41	.40	.39
			5	.44	.40	.38	.43	.40	.38	.42	.39	.37	.41	.39	.37	.40	.38	.37	.36
	54½% ↓		6	.41	.37	.35	.40	.37	.35	.39	.36	.34	.39	.36	.34	.38	.36	.34	.33
			7	.38	.34	.32	.37	.34	.32	.37	.34	.31	.36	.33	.31	.35	.33	.31	.30
Medium distribution unit with lens plate and inside frost lamp			8	.35	.32	.29	.35	.31	.29	.34	.31	.29	.34	.31	.29	.33	.30	.29	.28
			9	.33	.29	.27	.32	.29	.27	.32	.29	.26	.31	.28	.26	.31	.28	.26	.25
			10	.30	.27	.25	.30	.27	.24	.30	.27	.24	.29	.26	.24	.29	.26	.24	.23

[a] ρ_{CC} = per cent effective ceiling cavity reflectance.
[b] ρ_W = per cent wall reflectance.
[c] RCR = Room Cavity Ratio.
[d] Maximum *S/MH* guide—ratio of maximum luminaire spacing to mounting or ceiling height above work-plane.

Fig. 9-12. *Continued*

ρ_{CC} →	80			70			50			30			10		
ρ_W →	50	30	10	50	30	10	50	30	10	50	30	10	50	30	10
RCR ↓	Wall Luminance Coefficients[e] for 20 Per Cent Effective Floor Cavity Reflectance (ρ_{FC} = 20)														
1	.20	.11	.04	.19	.11	.04	.18	.10	.03	.17	.10	.03	.16	.09	.03
2	.19	.10	.03	.18	.10	.03	.17	.10	.03	.16	.09	.03	.15	.09	.03
3	.18	.10	.03	.17	.09	.03	.17	.09	.03	.16	.09	.03	.15	.08	.03
4	.17	.09	.03	.17	.09	.03	.16	.08	.03	.15	.08	.02	.14	.08	.02
5	.16	.08	.02	.16	.08	.02	.15	.08	.02	.15	.08	.02	.14	.07	.02
6	.16	.08	.02	.15	.08	.02	.15	.07	.02	.14	.07	.02	.13	.07	.02
7	.15	.07	.02	.15	.07	.02	.14	.07	.02	.14	.07	.02	.13	.07	.02
8	.14	.07	.02	.14	.07	.02	.14	.07	.02	.13	.07	.02	.13	.07	.02
9	.14	.07	.02	.14	.07	.02	.13	.07	.02	.13	.06	.02	.12	.06	.02
10	.13	.06	.02	.13	.06	.02	.13	.06	.02	.12	.06	.02	.12	.06	.02
1	.12	.07	.02	.11	.07	.02	.10	.06	.02	.09	.05	.02	.09	.05	.02
2	.11	.06	.02	.11	.06	.02	.10	.05	.02	.09	.05	.02	.08	.05	.01
3	.10	.05	.02	.10	.05	.02	.09	.05	.02	.09	.05	.01	.08	.04	.01
4	.10	.05	.01	.09	.05	.01	.09	.05	.01	.08	.04	.01	.08	.04	.01
5	.09	.05	.01	.09	.05	.01	.09	.04	.01	.08	.04	.01	.08	.04	.01
6	.09	.04	.01	.09	.04	.01	.08	.04	.01	.08	.04	.01	.07	.04	.01
7	.09	.04	.01	.08	.04	.01	.08	.04	.01	.08	.04	.01	.07	.04	.01
8	.08	.04	.01	.08	.04	.01	.08	.04	.01	.08	.04	.01	.07	.04	.01
9	.08	.04	.01	.08	.04	.01	.08	.04	.01	.07	.04	.01	.07	.04	.01
10	.08	.04	.01	.08	.04	.01	.07	.04	.01	.07	.04	.01	.07	.03	.01
1	.06	.03	.01	.06	.03	.01	.05	.03	.01	.05	.03	.01	.04	.02	.01
2	.05	.03	.01	.05	.03	.01	.05	.03	.01	.04	.02	.01	.04	.02	.01
3	.05	.03	.01	.05	.03	.01	.04	.02	.01	.04	.02	.01	.04	.02	.01
4	.05	.02	.01	.05	.02	.01	.04	.02	.01	.04	.02	.01	.04	.02	.01
5	.05	.02	.01	.04	.02	.01	.04	.02	.01	.04	.02	.01	.04	.02	.01
6	.04	.02	.01	.04	.02	.01	.04	.02	.01	.04	.02	.01	.04	.02	.01
7	.04	.02	.01	.04	.02	.01	.04	.02	.01	.04	.02	.01	.03	.02	.01
8	.04	.02	.01	.04	.02	.01	.04	.02	.01	.04	.02	.01	.03	.02	.01
9	.04	.02	.01	.04	.02	.01	.04	.02	.01	.03	.02	.01	.03	.02	.00
10	.04	.02	.00	.04	.02	.00	.04	.02	.00	.03	.02	.00	.03	.02	.00
1	.06	.03	.01	.06	.03	.01	.05	.03	.01	.05	.03	.01	.04	.02	.01
2	.05	.03	.01	.05	.03	.01	.05	.03	.01	.04	.02	.01	.04	.02	.01
3	.05	.03	.01	.05	.03	.01	.04	.02	.01	.04	.02	.01	.04	.02	.01
4	.05	.02	.01	.05	.02	.01	.04	.02	.01	.04	.02	.01	.04	.02	.01
5	.05	.02	.01	.04	.02	.01	.04	.02	.01	.04	.02	.01	.04	.02	.01
6	.04	.02	.01	.04	.02	.01	.04	.02	.01	.04	.02	.01	.04	.02	.01
7	.04	.02	.01	.04	.02	.01	.04	.02	.01	.04	.02	.01	.03	.02	.01
8	.04	.02	.01	.04	.02	.01	.04	.02	.01	.04	.02	.01	.03	.02	.01
9	.04	.02	.01	.04	.02	.01	.04	.02	.01	.03	.02	.00	.03	.02	.00
10	.04	.02	.00	.04	.02	.00	.04	.02	.00	.03	.02	.00	.03	.02	.00
1	.11	.06	.02	.11	.06	.02	.10	.06	.02	.09	.05	.02	.09	.05	.02
2	.10	.06	.02	.10	.05	.02	.09	.05	.02	.09	.05	.02	.08	.05	.01
3	.10	.05	.02	.09	.05	.02	.09	.05	.01	.08	.05	.01	.08	.04	.01
4	.09	.05	.01	.09	.05	.01	.09	.05	.01	.08	.04	.01	.08	.04	.01
5	.09	.04	.01	.09	.04	.01	.08	.04	.01	.08	.04	.01	.08	.04	.01
6	.08	.04	.01	.08	.04	.01	.08	.04	.01	.08	.04	.01	.07	.04	.01
7	.08	.04	.01	.08	.04	.01	.08	.04	.01	.07	.04	.01	.07	.04	.01
8	.08	.04	.01	.08	.04	.01	.07	.04	.01	.07	.04	.01	.07	.04	.01
9	.07	.04	.01	.07	.04	.01	.07	.04	.01	.07	.03	.01	.07	.03	.01
10	.07	.03	.01	.07	.03	.01	.07	.03	.01	.07	.03	.01	.06	.03	.01

ρ_{CC} →	80			70			50			30			10		
ρ_W →	50	30	10	50	30	10	50	30	10	50	30	10	50	30	10
RCR ↓	Ceiling Cavity Luminance Coefficients[e] for 20 Per Cent Effective Floor Cavity Reflectance (ρ_{FC} = 20)														
0	.19	.19	.19	.16	.16	.16	.11	.11	.11	.06	.06	.06	.02	.02	.02
1	.17	.16	.14	.15	.13	.12	.10	.09	.08	.06	.05	.05	.02	.02	.02
2	.16	.13	.11	.14	.11	.10	.09	.08	.07	.05	.05	.04	.02	.01	.01
3	.15	.11	.09	.13	.10	.08	.09	.07	.05	.05	.04	.03	.02	.01	.01
4	.14	.10	.07	.12	.09	.06	.08	.06	.04	.05	.03	.03	.02	.01	.01
5	.13	.09	.06	.11	.08	.05	.08	.05	.04	.04	.03	.02	.01	.01	.01
6	.12	.08	.05	.11	.07	.04	.07	.05	.03	.04	.03	.02	.01	.01	.01
7	.12	.07	.04	.10	.06	.03	.07	.04	.02	.04	.03	.01	.01	.01	.00
8	.11	.07	.03	.10	.06	.03	.07	.04	.02	.04	.02	.01	.01	.01	.00
9	.11	.06	.03	.09	.05	.03	.06	.04	.02	.04	.02	.01	.01	.01	.00
10	.10	.06	.03	.09	.05	.02	.06	.03	.02	.04	.02	.01	.01	.01	.00
0	.16	.16	.16	.13	.13	.13	.09	.09	.09	.05	.05	.05	.02	.02	.02
1	.14	.13	.13	.12	.11	.11	.08	.08	.07	.05	.05	.04	.02	.01	.01
2	.13	.11	.10	.11	.10	.09	.07	.07	.06	.04	.04	.04	.01	.01	.01
3	.12	.10	.08	.10	.08	.07	.07	.06	.05	.04	.03	.03	.01	.01	.01
4	.10	.08	.07	.09	.07	.06	.06	.05	.04	.04	.03	.02	.01	.01	.01
5	.10	.07	.05	.08	.06	.05	.06	.04	.03	.03	.03	.02	.01	.01	.01
6	.09	.06	.05	.08	.06	.04	.05	.04	.03	.03	.02	.02	.01	.01	.01
7	.08	.06	.04	.07	.05	.03	.05	.04	.02	.03	.02	.01	.01	.01	.00
8	.08	.05	.03	.07	.05	.03	.05	.03	.02	.03	.02	.01	.01	.01	.00
9	.07	.05	.03	.06	.04	.03	.04	.03	.02	.03	.02	.01	.01	.01	.00
10	.07	.04	.03	.06	.04	.02	.04	.03	.02	.02	.02	.01	.01	.01	.00
0	.08	.08	.08	.07	.07	.07	.04	.04	.04	.02	.02	.02	.01	.01	.01
1	.07	.07	.06	.06	.06	.06	.04	.04	.04	.02	.02	.02	.01	.01	.01
2	.07	.06	.05	.06	.05	.04	.04	.03	.03	.02	.02	.02	.01	.01	.01
3	.06	.05	.04	.05	.04	.04	.03	.03	.03	.02	.02	.02	.01	.01	.00
4	.05	.04	.03	.05	.04	.03	.03	.03	.02	.02	.02	.01	.01	.00	.00
5	.05	.04	.03	.04	.03	.03	.03	.02	.02	.02	.01	.01	.01	.00	.00
6	.04	.03	.02	.04	.03	.02	.03	.02	.01	.02	.01	.01	.00	.00	.00
7	.04	.03	.02	.04	.03	.02	.02	.02	.01	.01	.01	.01	.00	.00	.00
8	.04	.03	.02	.03	.02	.02	.02	.02	.01	.01	.01	.01	.00	.00	.00
9	.04	.02	.02	.03	.02	.01	.02	.01	.01	.01	.01	.01	.00	.00	.00
10	.04	.02	.01	.03	.02	.01	.02	.01	.01	.01	.01	.00	.00	.00	.00
0	.08	.08	.08	.07	.07	.07	.04	.04	.04	.03	.03	.03	.01	.01	.01
1	.08	.07	.07	.06	.06	.06	.04	.04	.04	.03	.02	.02	.01	.01	.01
2	.07	.06	.05	.06	.05	.05	.04	.04	.03	.02	.02	.02	.01	.01	.01
3	.06	.05	.04	.05	.04	.04	.04	.03	.03	.02	.02	.02	.01	.01	.01
4	.05	.04	.04	.05	.04	.03	.03	.03	.02	.02	.02	.01	.01	.01	.00
5	.05	.04	.03	.04	.03	.03	.03	.02	.02	.02	.01	.01	.01	.00	.00
6	.05	.03	.03	.04	.03	.02	.03	.02	.02	.02	.01	.01	.01	.00	.00
7	.04	.03	.02	.04	.03	.02	.03	.02	.01	.01	.01	.01	.00	.00	.00
8	.04	.03	.02	.03	.02	.02	.02	.02	.01	.01	.01	.01	.00	.00	.00
9	.04	.02	.02	.03	.02	.01	.02	.02	.01	.01	.01	.01	.00	.00	.00
10	.04	.02	.01	.03	.02	.01	.02	.01	.01	.01	.01	.01	.00	.00	.00
0	.10	.10	.10	.08	.08	.08	.06	.06	.06	.03	.03	.03	.01	.01	.01
1	.09	.09	.08	.08	.07	.07	.05	.05	.05	.03	.03	.03	.01	.01	.01
2	.09	.07	.06	.07	.06	.05	.05	.04	.04	.03	.02	.02	.01	.01	.01
3	.08	.06	.05	.07	.05	.04	.05	.04	.03	.03	.02	.02	.01	.01	.01
4	.07	.05	.04	.06	.05	.03	.04	.03	.02	.03	.02	.01	.01	.01	.00
5	.07	.05	.03	.06	.04	.03	.04	.03	.02	.02	.02	.01	.01	.01	.00
6	.07	.04	.03	.06	.04	.02	.04	.03	.02	.02	.02	.01	.01	.00	.00
7	.06	.04	.02	.05	.03	.02	.04	.02	.01	.02	.01	.01	.01	.00	.00
8	.06	.04	.02	.05	.03	.02	.04	.02	.01	.02	.01	.01	.01	.00	.00
9	.06	.03	.02	.05	.03	.01	.03	.02	.01	.02	.01	.01	.01	.00	.00
10	.06	.03	.01	.05	.03	.01	.03	.02	.01	.02	.01	.01	.01	.00	.00

[e] Although it is recommended that luminance coefficients and wall direct radiation coefficients be published to three decimal places, only two are shown here for these typical luminaires. Three decimal place data should be obtained from manufacturers of actual luminaires used.

Fig. 9-12. *Continued*

Coefficients of Utilization for 20 Per Cent Effective Floor Cavity Reflectance (ρ_{FC} = 20)

Typical Luminaire	Maint. Cat.	Maximum S/MH Guide[d]	Typical Distribution and Per Cent Lamp Lumens	RCR[c] ↓	ρ_{CC}[a] 80, ρ_W[b] 50	80, 30	80, 10	70, 50	70, 30	70, 10	50, 50	50, 30	50, 10	30, 50	30, 30	30, 10	10, 50	10, 30	10, 10	0, 0
10	V	1.4		0	.63	.63	.63	.62	.62	.62	.59	.59	.59	.56	.56	.56	.54	.54	.54	.53
				1	.58	.56	.54	.57	.55	.54	.54	.53	.52	.52	.51	.50	.50	.50	.49	.48
			0% ↑	2	.53	.50	.48	.52	.49	.47	.50	.48	.46	.48	.47	.45	.47	.45	.44	.43
				3	.48	.45	.42	.47	.44	.42	.46	.43	.41	.44	.42	.40	.43	.41	.40	.39
				4	.44	.40	.37	.43	.40	.37	.42	.39	.37	.41	.38	.36	.40	.38	.36	.35
			53½% ↓	5	.40	.36	.33	.39	.36	.33	.38	.35	.33	.37	.35	.32	.36	.34	.32	.31
				6	.36	.32	.30	.36	.32	.29	.35	.32	.29	.34	.31	.29	.33	.31	.29	.28
				7	.33	.29	.26	.33	.29	.26	.32	.28	.26	.31	.28	.26	.30	.28	.26	.25
Wide distribution unit with				8	.30	.26	.23	.30	.26	.23	.29	.26	.23	.28	.25	.23	.28	.25	.23	.22
lens plate and inside frost				9	.27	.23	.21	.27	.23	.21	.26	.23	.21	.26	.23	.20	.25	.22	.20	.19
lamp				10	.25	.21	.18	.25	.21	.18	.24	.21	.18	.24	.20	.18	.23	.20	.18	.17
11	V	1.3		0	.61	.61	.61	.60	.60	.60	.57	.57	.57	.54	.54	.54	.51	.51	.51	.50
				1	.53	.51	.48	.52	.50	.47	.49	.47	.46	.47	.45	.44	.45	.44	.42	.41
			1½% ↑	2	.46	.42	.39	.45	.42	.39	.43	.40	.38	.41	.39	.37	.39	.37	.35	.34
				3	.40	.36	.33	.40	.35	.32	.38	.34	.31	.36	.33	.31	.35	.32	.30	.29
				4	.36	.31	.28	.35	.31	.28	.34	.30	.27	.32	.29	.26	.31	.28	.26	.25
			50½% ↓	5	.32	.27	.24	.31	.27	.24	.30	.26	.23	.29	.25	.23	.28	.25	.22	.21
				6	.29	.24	.20	.28	.24	.20	.27	.23	.20	.26	.22	.20	.25	.22	.19	.18
				7	.26	.21	.18	.25	.21	.18	.24	.20	.17	.23	.20	.17	.22	.19	.17	.16
				8	.23	.19	.16	.23	.18	.15	.22	.18	.15	.21	.18	.15	.20	.17	.15	.14
Recessed unit with dropped				9	.21	.17	.14	.21	.16	.14	.20	.16	.13	.19	.16	.13	.19	.15	.13	.12
diffusing glass				10	.19	.15	.12	.19	.15	.12	.18	.14	.12	.18	.14	.12	.17	.14	.12	.11
12	V	1.3		0	.78	.78	.78	.76	.76	.76	.73	.73	.73	.70	.70	.70	.67	.67	.67	.65
				1	.71	.69	.68	.70	.68	.66	.67	.66	.64	.65	.64	.62	.62	.62	.61	.59
				2	.65	.62	.59	.64	.61	.58	.62	.59	.57	.60	.58	.56	.58	.56	.55	.54
			0% ↑	3	.59	.55	.52	.58	.55	.52	.57	.53	.51	.55	.52	.50	.53	.51	.49	.48
				4	.54	.50	.47	.54	.50	.46	.52	.49	.46	.51	.48	.45	.49	.47	.45	.43
				5	.50	.45	.42	.49	.45	.41	.48	.44	.41	.47	.43	.41	.46	.43	.40	.39
			66% ↓	6	.46	.41	.37	.45	.40	.37	.44	.40	.37	.43	.39	.37	.42	.39	.36	.35
				7	.41	.37	.33	.41	.36	.33	.40	.36	.33	.39	.35	.33	.38	.35	.32	.31
Clear HID lamp and glass re-				8	.38	.33	.30	.38	.33	.30	.37	.33	.30	.36	.32	.29	.35	.32	.29	.28
fractor above plastic lens				9	.35	.30	.27	.34	.30	.27	.34	.29	.26	.33	.29	.26	.32	.29	.26	.25
panel				10	.32	.27	.24	.31	.27	.24	.31	.27	.24	.30	.26	.24	.30	.26	.23	.22
13	V	1.4		0	.85	.85	.85	.83	.83	.83	.79	.79	.79	.76	.76	.76	.73	.73	.73	.71
				1	.78	.76	.74	.76	.74	.73	.73	.72	.70	.71	.69	.68	.68	.67	.66	.65
			0% ↑	2	.71	.68	.65	.70	.67	.64	.68	.65	.63	.65	.63	.61	.63	.62	.60	.59
				3	.65	.61	.57	.64	.60	.57	.62	.59	.56	.60	.57	.55	.59	.56	.54	.53
				4	.60	.55	.51	.59	.54	.51	.57	.53	.50	.55	.52	.50	.54	.51	.49	.48
			71½% ↓	5	.54	.49	.45	.54	.49	.45	.52	.48	.45	.51	.47	.44	.50	.46	.44	.43
				6	.49	.44	.40	.49	.44	.40	.47	.43	.40	.46	.42	.40	.45	.42	.39	.38
				7	.44	.39	.35	.44	.39	.35	.43	.38	.35	.42	.38	.35	.41	.37	.35	.33
				8	.40	.35	.31	.40	.35	.31	.39	.35	.31	.38	.34	.31	.38	.34	.31	.30
Enclosed reflector with an in-				9	.37	.31	.28	.36	.31	.28	.36	.31	.28	.35	.31	.28	.34	.30	.27	.26
candescent lamp				10	.33	.28	.25	.33	.28	.25	.32	.28	.25	.32	.28	.25	.31	.27	.24	.23
14	III	0.7		0	.92	.92	.92	.90	.90	.90	.86	.86	.86	.82	.82	.82	.78	.78	.78	.76
				1	.87	.85	.83	.85	.83	.82	.81	.80	.79	.78	.77	.76	.75	.75	.74	.72
			1½% ↑	2	.81	.79	.76	.80	.77	.75	.77	.75	.73	.75	.73	.72	.72	.71	.70	.69
				3	.77	.73	.71	.76	.72	.70	.73	.71	.69	.71	.69	.67	.70	.68	.66	.65
				4	.73	.69	.66	.72	.68	.65	.70	.67	.64	.68	.66	.64	.67	.65	.63	.62
				5	.69	.65	.62	.68	.64	.61	.66	.63	.61	.65	.62	.60	.64	.61	.59	.58
			77% ↓	6	.65	.61	.58	.64	.61	.58	.63	.60	.57	.62	.59	.57	.61	.58	.56	.55
				7	.62	.57	.54	.61	.57	.54	.60	.56	.54	.59	.56	.53	.58	.55	.53	.52
Narrow distribution venti-				8	.58	.54	.51	.58	.54	.51	.57	.53	.51	.56	.53	.51	.55	.52	.50	.49
lated reflector with clear				9	.55	.51	.48	.55	.51	.48	.54	.50	.48	.53	.50	.48	.53	.50	.48	.47
HID lamp				10	.53	.49	.46	.52	.48	.46	.52	.48	.46	.51	.48	.45	.50	.47	.45	.44

[a] ρ_{CC} = per cent effective ceiling cavity reflectance.
[b] ρ_W = per cent wall reflectance.
[c] RCR = Room Cavity Ratio.
[d] Maximum S/MH guide—ratio of maximum luminaire spacing to mounting or ceiling height above work-plane.

Fig. 9–12. *Continued*

ρ_{CC} →	80			70			50			30			10			ρ_{CC} →	80			70			50			30			10		
ρ_W →	50	30	10	50	30	10	50	30	10	50	30	10	50	30	10	ρ_W →	50	30	10	50	30	10	50	30	10	50	30	10	50	30	10
RCR ↓	Wall Luminance Coefficients[e] for 20 Per Cent Effective Floor Cavity Reflectance (ρ_{FC} = 20)															RCR ↓	Ceiling Cavity Luminance Coefficients[e] for 20 Per Cent Effective Floor Cavity Reflectance (ρ_{FC} = 20)														
																0	.10	.10	.10	.08	.08	.08	.05	.05	.05	.03	.03	.03	.01	.01	.01
1	.12	.07	.02	.12	.07	.02	.11	.06	.02	.10	.06	.02	.10	.06	.02	1	.09	.08	.08	.08	.07	.06	.05	.05	.04	.03	.03	.03	.01	.01	.01
2	.11	.06	.02	.11	.06	.02	.10	.06	.02	.10	.06	.02	.09	.05	.02	2	.09	.07	.06	.07	.06	.05	.05	.04	.03	.03	.02	.02	.01	.01	.01
3	.11	.06	.02	.11	.06	.02	.10	.05	.02	.10	.05	.02	.09	.05	.02	3	.08	.06	.04	.07	.05	.04	.05	.04	.03	.03	.02	.02	.01	.01	.01
4	.10	.05	.02	.10	.05	.02	.10	.05	.02	.09	.05	.01	.09	.05	.01	4	.08	.05	.04	.07	.05	.03	.04	.03	.02	.03	.02	.01	.01	.01	.00
5	.10	.05	.01	.10	.05	.01	.09	.05	.01	.09	.05	.01	.09	.05	.01	5	.07	.05	.03	.06	.04	.03	.04	.03	.02	.02	.02	.01	.01	.01	.00
6	.09	.05	.01	.09	.05	.01	.09	.05	.01	.09	.04	.01	.08	.04	.01	6	.07	.04	.02	.06	.04	.02	.04	.03	.01	.02	.02	.01	.01	.00	.00
7	.09	.04	.01	.09	.04	.01	.09	.04	.01	.08	.04	.01	.08	.04	.01	7	.07	.04	.02	.06	.03	.02	.04	.02	.01	.02	.01	.01	.01	.00	.00
8	.09	.04	.01	.09	.04	.01	.08	.04	.01	.08	.04	.01	.08	.04	.01	8	.06	.04	.02	.05	.03	.01	.04	.02	.01	.02	.01	.01	.01	.00	.00
9	.08	.04	.01	.08	.04	.01	.08	.04	.01	.08	.04	.01	.07	.04	.01	9	.06	.03	.01	.05	.03	.01	.04	.02	.01	.02	.01	.01	.01	.00	.00
10	.08	.04	.01	.08	.04	.01	.08	.04	.01	.07	.04	.01	.07	.04	.01	10	.06	.03	.01	.05	.03	.01	.04	.02	.01	.02	.01	.00	.01	.00	.00
																0	.11	.11	.11	.09	.09	.09	.06	.06	.06	.04	.04	.04	.01	.01	.01
1	.18	.10	.03	.17	.10	.03	.16	.09	.03	.16	.09	.03	.15	.09	.03	1	.11	.09	.08	.09	.08	.07	.06	.06	.05	.04	.03	.03	.01	.01	.01
2	.16	.09	.03	.15	.08	.03	.15	.08	.03	.14	.08	.02	.13	.08	.02	2	.10	.08	.06	.09	.07	.05	.06	.05	.04	.03	.03	.02	.01	.01	.01
3	.14	.08	.02	.14	.08	.02	.13	.07	.02	.13	.07	.02	.12	.07	.02	3	.10	.07	.05	.09	.06	.04	.06	.04	.03	.03	.03	.02	.01	.01	.01
4	.13	.07	.02	.13	.07	.02	.12	.06	.02	.12	.06	.02	.11	.06	.02	4	.09	.07	.04	.08	.06	.04	.06	.04	.03	.03	.02	.02	.01	.01	.01
5	.12	.06	.02	.12	.06	.02	.11	.06	.02	.11	.06	.02	.10	.06	.02	5	.09	.06	.04	.08	.05	.03	.05	.04	.02	.03	.02	.01	.01	.01	.00
6	.11	.06	.02	.11	.06	.02	.11	.05	.02	.10	.05	.02	.10	.05	.02	6	.09	.06	.03	.08	.05	.03	.05	.03	.02	.03	.02	.01	.01	.01	.00
7	.11	.05	.01	.10	.05	.01	.10	.05	.01	.10	.05	.01	.09	.05	.01	7	.08	.05	.03	.07	.05	.03	.05	.03	.02	.03	.02	.01	.01	.01	.00
8	.10	.05	.01	.10	.05	.01	.09	.05	.01	.09	.05	.01	.09	.04	.01	8	.08	.05	.03	.07	.04	.02	.05	.03	.02	.03	.02	.01	.01	.01	.00
9	.09	.04	.01	.09	.04	.01	.09	.04	.01	.08	.04	.01	.08	.04	.01	9	.08	.05	.03	.07	.04	.02	.05	.03	.02	.03	.02	.01	.01	.01	.00
10	.09	.04	.01	.09	.04	.01	.08	.04	.01	.08	.04	.01	.08	.04	.01	10	.08	.04	.02	.06	.04	.02	.04	.03	.02	.03	.02	.01	.01	.01	.00
																0	.12	.12	.12	.10	.10	.10	.07	.07	.07	.04	.04	.04	.01	.01	.01
1	.15	.08	.03	.14	.08	.03	.14	.08	.02	.13	.07	.02	.12	.07	.02	1	.12	.10	.09	.10	.09	.08	.07	.06	.06	.04	.04	.03	.01	.01	.01
2	.14	.08	.02	.14	.07	.02	.13	.07	.02	.12	.07	.02	.12	.07	.02	2	.11	.09	.07	.09	.08	.06	.06	.05	.04	.04	.03	.03	.01	.01	.01
3	.13	.07	.02	.13	.07	.02	.12	.07	.02	.12	.06	.02	.11	.06	.02	3	.10	.08	.06	.09	.06	.05	.06	.05	.03	.03	.03	.02	.01	.01	.01
4	.13	.07	.02	.12	.06	.02	.12	.06	.02	.11	.06	.02	.11	.06	.02	4	.09	.07	.04	.08	.06	.04	.06	.04	.03	.03	.02	.02	.01	.01	.01
5	.12	.06	.02	.12	.06	.02	.11	.06	.02	.11	.06	.02	.10	.06	.02	5	.09	.06	.04	.08	.05	.03	.05	.04	.02	.03	.02	.01	.01	.01	.00
6	.11	.06	.02	.11	.06	.02	.11	.06	.02	.10	.05	.02	.10	.05	.02	6	.08	.05	.03	.07	.05	.03	.05	.03	.02	.03	.02	.01	.01	.01	.00
7	.11	.05	.02	.11	.05	.02	.10	.05	.02	.10	.05	.01	.10	.05	.01	7	.08	.05	.02	.07	.04	.02	.05	.03	.02	.03	.02	.01	.01	.01	.00
8	.11	.05	.01	.10	.05	.01	.10	.05	.01	.10	.05	.01	.09	.05	.01	8	.08	.04	.02	.07	.04	.02	.05	.03	.01	.03	.02	.01	.01	.01	.00
9	.10	.05	.01	.10	.05	.01	.10	.05	.01	.09	.05	.01	.09	.05	.01	9	.07	.04	.02	.06	.04	.02	.04	.03	.01	.03	.01	.01	.01	.00	.00
10	.10	.05	.01	.10	.05	.01	.09	.05	.01	.09	.04	.01	.09	.04	.01	10	.07	.04	.02	.06	.03	.01	.04	.02	.01	.03	.01	.01	.01	.00	.00
																0	.13	.13	.13	.11	.11	.11	.08	.08	.08	.04	.04	.04	.01	.01	.01
1	.15	.09	.03	.15	.09	.03	.14	.08	.03	.13	.08	.02	.12	.07	.02	1	.13	.11	.10	.11	.10	.09	.07	.07	.06	.04	.04	.04	.01	.01	.01
2	.15	.08	.02	.14	.08	.02	.13	.07	.02	.13	.07	.02	.12	.07	.02	2	.12	.10	.08	.10	.08	.07	.07	.06	.05	.04	.03	.03	.01	.01	.01
3	.14	.07	.02	.14	.07	.02	.13	.07	.02	.12	.07	.02	.12	.07	.02	3	.11	.08	.06	.09	.07	.05	.06	.05	.04	.04	.03	.02	.01	.01	.01
4	.13	.07	.02	.13	.07	.02	.13	.07	.02	.12	.06	.02	.11	.06	.02	4	.10	.07	.05	.09	.06	.04	.06	.04	.03	.03	.03	.02	.01	.01	.01
5	.13	.07	.02	.13	.06	.02	.12	.06	.02	.12	.06	.02	.11	.06	.02	5	.10	.06	.04	.08	.06	.03	.06	.04	.02	.03	.02	.01	.01	.01	.00
6	.12	.06	.02	.12	.06	.02	.12	.06	.02	.11	.06	.02	.11	.06	.02	6	.09	.06	.03	.08	.05	.03	.06	.04	.02	.03	.02	.01	.01	.01	.00
7	.12	.06	.02	.12	.06	.02	.11	.06	.02	.11	.06	.02	.11	.06	.02	7	.09	.05	.03	.08	.05	.02	.05	.03	.02	.03	.02	.01	.01	.01	.00
8	.12	.06	.02	.11	.06	.02	.11	.06	.02	.11	.05	.02	.10	.05	.02	8	.09	.05	.02	.07	.04	.02	.05	.03	.01	.03	.02	.01	.01	.01	.00
9	.11	.05	.02	.11	.05	.02	.11	.05	.01	.10	.05	.01	.10	.05	.01	9	.08	.05	.02	.07	.04	.02	.05	.03	.01	.03	.02	.01	.01	.01	.00
10	.11	.05	.01	.11	.05	.01	.10	.05	.01	.10	.05	.01	.10	.05	.01	10	.08	.04	.02	.07	.04	.02	.05	.03	.01	.03	.02	.01	.01	.01	.00
																0	.15	.15	.15	.13	.13	.13	.09	.09	.09	.05	.05	.05	.02	.02	.02
1	.13	.07	.02	.13	.07	.02	.11	.07	.02	.10	.06	.02	.10	.06	.02	1	.14	.13	.12	.12	.11	.11	.08	.08	.07	.05	.05	.04	.02	.01	.01
2	.12	.07	.02	.12	.06	.02	.11	.06	.02	.10	.06	.02	.09	.05	.02	2	.13	.11	.10	.11	.10	.09	.08	.07	.06	.04	.04	.04	.01	.01	.01
3	.11	.06	.02	.11	.06	.02	.10	.06	.02	.10	.05	.02	.09	.05	.02	3	.12	.10	.08	.10	.08	.07	.07	.06	.05	.04	.03	.03	.01	.01	.01
4	.11	.06	.02	.10	.05	.02	.10	.05	.02	.09	.05	.01	.09	.05	.01	4	.11	.09	.07	.10	.08	.06	.07	.05	.04	.04	.03	.02	.01	.01	.01
5	.10	.05	.02	.10	.05	.01	.09	.05	.01	.09	.05	.01	.08	.04	.01	5	.10	.08	.06	.09	.07	.05	.06	.05	.04	.04	.03	.02	.01	.01	.01
6	.10	.05	.01	.10	.05	.01	.09	.05	.01	.09	.04	.01	.08	.04	.01	6	.10	.07	.05	.08	.06	.04	.06	.04	.03	.03	.03	.02	.01	.01	.01
7	.09	.05	.01	.09	.05	.01	.09	.04	.01	.08	.04	.01	.08	.04	.01	7	.09	.06	.04	.08	.06	.04	.06	.04	.03	.03	.02	.02	.01	.01	.01
8	.09	.04	.01	.09	.04	.01	.08	.04	.01	.08	.04	.01	.08	.04	.01	8	.09	.06	.04	.08	.05	.03	.05	.04	.02	.03	02	.01	.01	.01	.00
9	.09	.04	.01	.09	.04	.01	.08	.04	.01	.08	.04	.01	.08	.04	.01	9	.08	.06	.04	.07	.05	.03	.05	.03	.02	.03	.02	.01	.01	.01	.00
10	.08	.04	.01	.08	.04	.01	.08	.04	.01	.08	.04	.01	.07	.04	.01	10	.08	.05	.03	.07	.05	.03	.05	.03	.02	.03	.02	.01	.01	.01	.00

[e] Although it is recommended that luminance coefficients and wall direct radiation coefficients be published to three decimal places, only two are shown here for these typical luminaires. Three decimal place data should be obtained from manufacturers of actual luminaires used.

Fig. 9–12. *Continued*

Typical Luminaire	Typical Distribution and Per Cent Lamp Lumens: Maint. Cat.	Maximum S/MH Guide[d]	ρ_{CC}[a] → / ρ_W[b] → / RCR[c] ↓	80			70			50			30			10			0
			ρ_W →	50	30	10	50	30	10	50	30	10	50	30	10	50	30	10	0
				Coefficients of Utilization for 20 Per Cent Effective Floor Cavity Reflectance (ρ_{FC} = 20)															
15	III	1.0	0	.91	.91	.91	.89	.89	.89	.84	.84	.84	.81	.81	.81	.77	.77	.77	.75
			1	.84	.81	.79	.82	.80	.78	.79	.77	.76	.76	.74	.73	.73	.72	.71	.69
			2	.77	.73	.70	.76	.72	.70	.73	.70	.68	.70	.68	.66	.68	.66	.65	.63
	1% ↑		3	.71	.66	.63	.69	.65	.62	.67	.64	.61	.65	.62	.60	.63	.61	.59	.57
			4	.65	.60	.56	.64	.59	.56	.62	.58	.55	.60	.57	.54	.59	.56	.54	.52
	76% ↓		5	.59	.54	.50	.59	.54	.50	.57	.53	.50	.56	.52	.49	.54	.51	.48	.47
			6	.54	.49	.45	.54	.49	.45	.52	.48	.45	.51	.47	.44	.50	.47	.44	.42
			7	.50	.44	.40	.49	.44	.40	.48	.43	.40	.47	.43	.39	.46	.42	.39	.38
Intermediate distribution ventilated reflector with clear HID lamp			8	.45	.40	.36	.45	.40	.36	.44	.39	.36	.43	.39	.35	.42	.38	.35	.34
			9	.41	.36	.32	.41	.36	.32	.40	.35	.32	.39	.35	.32	.38	.35	.32	.30
			10	.38	.33	.29	.37	.32	.29	.37	.32	.29	.36	.32	.29	.35	.31	.28	.27
16	III	1.5	0	.92	.92	.92	.90	.90	.90	.86	.86	.86	.82	.82	.82	.79	.79	.79	.77
			1	.85	.82	.80	.83	.81	.79	.79	.78	.76	.76	.75	.74	.74	.72	.71	.70
			2	.77	.73	.70	.75	.72	.69	.73	.70	.67	.70	.68	.66	.68	.66	.64	.63
	½% ↑		3	.70	.65	.61	.68	.64	.60	.66	.62	.59	.64	.61	.58	.62	.59	.57	.56
			4	.63	.58	.53	.62	.57	.53	.60	.56	.52	.58	.55	.52	.57	.54	.51	.49
			5	.57	.51	.47	.56	.51	.47	.55	.50	.46	.53	.49	.46	.52	.48	.45	.44
	77½% ↓		6	.51	.45	.41	.51	.45	.41	.49	.44	.40	.48	.43	.40	.47	.43	.40	.38
			7	.46	.40	.35	.45	.39	.35	.44	.39	.35	.43	.38	.35	.42	.38	.34	.33
Wide distribution ventilated reflector with clear HID lamp			8	.41	.35	.31	.41	.35	.31	.40	.34	.31	.39	.34	.30	.38	.33	.30	.29
			9	.37	.31	.27	.37	.31	.27	.36	.30	.27	.35	.30	.27	.34	.30	.26	.25
			10	.33	.27	.24	.33	.27	.23	.32	.27	.23	.31	.27	.23	.31	.26	.23	.22
17	III	1.0	0	.96	.96	.96	.93	.93	.93	.87	.87	.87	.82	.82	.82	.77	.77	.77	.75
			1	.89	.87	.84	.86	.84	.83	.82	.80	.79	.78	.76	.75	.74	.73	.72	.70
	6½% ↑		2	.82	.79	.76	.80	.77	.74	.76	.74	.72	.73	.71	.69	.70	.68	.67	.65
			3	.76	.72	.68	.74	.70	.67	.71	.68	.65	.68	.66	.63	.66	.63	.61	.60
			4	.70	.66	.62	.69	.65	.61	.66	.63	.60	.64	.61	.58	.62	.59	.57	.55
			5	.65	.60	.56	.64	.59	.56	.62	.58	.54	.60	.56	.53	.58	.55	.52	.51
	75½% ↓		6	.60	.55	.51	.59	.55	.51	.57	.53	.50	.56	.52	.49	.54	.51	.48	.47
			7	.56	.51	.47	.55	.50	.46	.53	.49	.46	.52	.48	.45	.50	.47	.44	.43
Intermediate distribution ventilated reflector with phosphor coated HID lamp			8	.52	.47	.43	.51	.46	.43	.50	.45	.42	.48	.44	.41	.47	.43	.41	.40
			9	.48	.43	.39	.47	.42	.39	.46	.42	.39	.45	.41	.38	.44	.40	.38	.36
			10	.45	.40	.36	.44	.39	.36	.43	.39	.36	.42	.38	.35	.41	.37	.35	.34
18	III	1.5	0	.93	.93	.93	.89	.89	.89	.83	.83	.83	.77	.77	.77	.71	.71	.71	.68
			1	.85	.83	.81	.82	.80	.78	.77	.75	.74	.72	.71	.69	.67	.66	.65	.63
	12% ↑		2	.78	.74	.71	.76	.72	.69	.71	.68	.66	.67	.65	.63	.63	.61	.60	.58
			3	71	.67	.63	.69	.65	.62	.65	.62	.59	.62	.59	.57	.58	.56	.54	.53
			4	.65	.60	.56	.64	.59	.55	.60	.56	.53	.57	.54	.51	.54	.52	.50	.48
	69% ↓		5	.60	.54	.50	.58	.53	.49	.55	.51	.48	.53	.49	.46	.50	.47	.45	.43
			6	.54	.49	.45	.53	.48	.44	.51	.46	.43	.48	.45	.42	.46	.43	.40	.39
			7	.49	.44	.40	.48	.43	.39	.46	.41	.38	.44	.40	.37	.42	.39	.36	.34
Wide distribution ventilated reflector with phosphor coated HID lamp			8	.45	.39	.35	.44	.38	.35	.42	.37	.34	.40	.36	.33	.38	.35	.32	.31
			9	.41	.35	.31	.40	.34	.31	.38	.33	.30	.36	.32	.29	.35	.31	.28	.27
			10	.37	.31	.27	.36	.31	.27	.34	.30	.26	.33	.29	.26	.32	.28	.25	.24
19	III	1.3	0	1.00	1.00	1.00	.96	.96	.96	.89	.89	.89	.82	.82	.82	.76	.76	.76	.73
			1	.88	.85	.82	.85	.82	.79	.79	.77	.74	.73	.72	.70	.68	.67	.66	.63
			2	.78	.72	.67	.75	.70	.66	.70	.66	.62	.65	.62	.59	.61	.58	.56	.53
	13% ↑		3	.69	.62	.57	.66	.60	.56	.62	.57	.53	.58	.54	.51	.54	.51	.48	.46
			4	.61	.54	.48	.59	.52	.47	.55	.50	.45	.52	.47	.43	.49	.45	.42	.39
			5	.54	.46	.41	.52	.45	.40	.49	.43	.39	.46	.41	.37	.43	.39	.36	.33
	74% ↓		6	.48	.41	.35	.47	.40	.35	.44	.38	.34	.41	.36	.32	.39	.34	.31	.29
			7	.43	.36	.31	.42	.35	.30	.40	.34	.29	.37	.32	.28	.35	.31	.27	.25
			8	.39	.32	.27	.38	.31	.26	.36	.30	.25	.34	.28	.24	.32	.27	.24	.22
Porcelain-enameled reflector with 14°CW shielding			9	.35	.28	.23	.34	.27	.23	.32	.26	.22	.30	.25	.21	.28	.24	.20	.19
			10	.32	.25	.20	.31	.24	.20	.29	.23	.19	.28	.22	.19	.26	.21	.18	.17

[a] ρ_{CC} = per cent effective ceiling cavity reflectance.
[b] ρ_W = per cent wall reflectance.
[c] RCR = Room Cavity Ratio.
[d] Maximum *S/MH* guide—ratio of maximum luminaire spacing to mounting or ceiling height above work-plane.

Fig. 9–12. *Continued*

ρ_{CC} →	80			70			50			30			10			ρ_{CC} →	80			70			50			30			10		
ρ_W →	50	30	10	50	30	10	50	30	10	50	30	10	50	30	10	ρ_W →	50	30	10	50	30	10	50	30	10	50	30	10	50	30	10
RCR ↓	Wall Luminance Coefficients[e] for 20 Per Cent Effective Floor Cavity Reflectance (ρ_{FC} = 20)															RCR ↓	Ceiling Cavity Luminance Coefficients[e] for 20 Per Cent Effective Floor Cavity Reflectance (ρ_{FC} = 20)														
																0	.15	.15	.15	.13	.13	.13	.08	.08	.08	.05	.05	.05	.02	.02	.02
1	.16	.09	.03	.15	.09	.03	.14.	08	.03	.13	.08	.02	.12	.07	.02	1	.14	.13	.12	.12	.11	.10	.08	.07	.07	.05	.04	.04	.02	.01	.01
2	.15	.08	.03	.15	.08	.02	.14	.08	.02	.13	.07	.02	.12	.07	.02	2	.13	.11	.09	.11	.09	.08	.08	.06	.05	.04	.04	.03	.01	.01	.01
3	.14	.08	.02	.14	.08	.02	.13	.07	.02	.13	.07	.02	.12	.07	.02	3	.12	.09	.07	.10	.08	.06	.07	.06	.04	.04	.03	.03	.01	.01	.01
4	.14	.07	.02	.13	.07	.02	.13	.07	.02	.12	.07	.02	.12	.06	.02	4	.11	.08	.06	.10	.07	.05	.07	.05	.04	.04	.03	.02	.01	.01	.01
5	.13	.07	.02	.13	.07	.02	.12	.06	.02	.12	.06	.02	.11	.06	.02	5	.11	.07	.05	.09	.06	.04	.06	.04	.03	.04	.03	.02	.01	.01	.01
6	.13	.06	.02	.13	.06	.02	.12	.06	.02	.11	.06	.02	.11	.06	.02	6	.10	.07	.04	.09	.06	.04	.06	.04	.03	.04	.02	.02	.01	.01	.01
7	.12	.06	.02	.12	.06	.02	.12	.06	.02	.11	.06	.02	.11	.06	.02	7	.10	.06	.04	.09	.05	.03	.06	.04	.02	.03	.02	.01	.01	.01	.00
8	.12	.06	.02	.12	.06	.02	.11	.06	.02	.11	.06	.02	.10	.05	.02	8	.10	.06	.03	.08	.05	.03	.06	.04	.02	.03	.02	.01	.01	.01	.00
9	.11	.06	.02	.11	.06	.02	.11	.05	.02	.11	.05	.02	.10	.05	.01	9	.09	.05	.03	.08	.05	.02	.06	.03	.02	.03	.02	.01	.01	.01	.00
10	.11	.05	.01	.11	.05	.01	.11	.05	.01	.10	.05	.01	.10	.05	.01	10	.09	.05	.03	.08	.04	.02	.05	.03	.02	.03	.02	.01	.01	.01	.00
																0	.15	.15	.15	.13	.13	.13	.08	.08	.08	.05	.05	.05	.02	.02	.02
1	.18	.10	.03	.17	.10	.03	.16	.09	.03	.15	.09	.03	.14	.08	.03	1	.14	.13	.11	.12	.11	.10	.08	.08	.07	.05	.04	.04	.02	.01	.01
2	.17	.09	.03	.16	.09	.03	.15	.09	.03	.15	.08	.03	.14	.08	.02	2	.13	.11	.09	.11	.09	.08	.08	.06	.05	.04	.03	.03	.01	.01	.01
3	.16	.09	.03	.16	.08	.03	.15	.08	.02	.14	.08	.02	.14	.08	.02	3	.12	.09	.07	.11	.08	.06	.07	.06	.04	.04	.03	.03	.01	.01	.01
4	.16	.08	.02	.15	.08	.02	.14	.08	.02	.14	.07	.02	.13	.07	.02	4	.12	.08	.06	.10	.07	.05	.07	.05	.03	.04	.04	.02	.01	.01	.01
5	.15	.08	.02	.15	.07	.02	.14	.07	.02	.13	.07	.02	.13	.07	.02	5	.11	.07	.05	.10	.06	.04	.07	.04	.03	.04	.03	.02	.01	.01	.01
6	.14	.07	.02	.14	.07	.02	.14	.07	.02	.13	.07	.02	.13	.07	.02	6	.11	.07	.04	.09	.06	.03	.06	.04	.02	.04	.02	.01	.01	.01	.00
7	.14	.07	.02	.14	.07	.02	.13	.07	.02	.13	.07	.02	.12	.06	.02	7	.10	.06	.03	.09	.05	.03	.06	.04	.02	.04	.02	.01	.01	.01	.00
8	.13	.07	.02	.13	.06	.02	.13	.06	.02	.12	.06	.02	.12	.06	.02	8	.10	.06	.03	.09	.05	.03	.06	.04	.02	.03	.02	.01	.01	.01	.00
9	.13	.06	.02	.13	.06	.02	.12	.06	.02	.12	.06	.02	.11	.06	.02	9	.10	.05	.03	.08	.05	.02	.06	.03	.02	.03	.02	.01	.01	.01	.00
10	.12	.06	.02	.12	.06	.02	.12	.06	.02	.11	.06	.02	.11	.06	.02	10	.09	.05	.02	.08	.04	.02	.06	.03	.01	.03	.02	.01	.01	.01	.00
																0	.20	.20	.20	.17	.17	.17	.12	.12	.12	.07	.07	.07	.02	.02	.02
1	.16	.09	.03	.15	.09	.03	.14	.08	.03	.13	.07	.02	.11	.07	.02	1	.19	.18	.17	.17	.16	.15	.11	.11	.10	.07	.06	.06	.02	.02	.02
2	.15	.08	.03	.14	.08	.02	.13	.07	.02	.12	.07	.02	.11	.06	.02	2	.18	.16	.14	.16	.14	.12	.11	.10	.09	.06	.06	.05	.02	.02	.02
3	.14	.08	.02	.14	.07	.02	.13	.07	.02	.12	.06	.02	.11	.06	.02	3	.17	.15	.12	.15	.13	.11	.10	.09	.08	.06	.05	.04	.02	.02	.01
4	.14	.07	.02	.13	.07	.02	.12	.06	.02	.11	.06	.02	.10	.06	.02	4	.16	.13	.11	.14	.12	.10	.10	.08	.07	.06	.05	.04	.02	.02	.01
5	.13	.07	.02	.13	.06	.02	.12	.06	.02	.11	.06	.02	.10	.05	.02	5	.16	.12	.10	.14	.11	.09	.09	.08	.06	.05	.04	.04	.02	.01	.01
6	.13	.06	.02	.12	.06	.02	.11	.06	.02	.11	.06	.02	.10	.05	.02	6	.15	.12	.09	.13	.10	.08	.09	.07	.06	.05	.04	.03	.02	.01	.01
7	.12	.06	.02	.12	.06	.02	.11	.06	.02	.10	.05	.02	.10	.05	.01	7	.15	.11	.09	.13	.10	.07	.09	.07	.05	.05	.04	.03	.02	.01	.01
8	.12	.06	.02	.11	.06	.02	.11	.05	.02	.10	.05	.01	.09	.05	.01	8	.14	.11	.08	.12	.09	.07	.09	.07	.05	.05	.04	.03	.02	.01	.01
9	.11	.05	.02	.11	.05	.01	.10	.05	.01	.10	.05	.01	.09	.05	.01	9	.14	.10	.08	.12	.09	.07	.08	.06	.05	.05	.04	.03	.02	.01	.01
10	.11	.05	.01	.10	.05	.01	.10	.05	.01	.09	.05	.01	.09	.04	.01	10	.14	.10	.07	.12	.09	.07	.08	.06	.05	.05	.04	.03	.02	.01	.01
																0	.24	.24	.24	.20	.20	.20	.14	.14	.14	.08	.08	.08	.02	.02	.02
1	.17	.10	.03	.16	.09	.03	.14	.08	.03	.13	.07	.02	.11	.07	.02	1	.23	.22	.21	.20	.19	.18	.14	.13	.12	.08	.07	.07	.02	.02	.02
2	.16	.09	.03	.15	.08	.03	.14	.08	.02	.12	.07	.02	.11	.06	.02	2	.22	.20	.18	.19	.17	.16	.13	.12	.11	.07	.07	.06	.02	.02	.02
3	.15	.08	.02	.15	.08	.02	.13	.07	.02	.12	.07	.02	.11	.06	.02	3	.21	.18	.16	.18	.16	.14	.13	.11	.10	.07	.06	.06	.02	.02	.02
4	.15	.08	.02	.14	.07	.02	.13	.07	.02	.12	.06	.02	.11	.06	.02	4	.21	.17	.15	.18	.15	.13	.12	.10	.09	.07	.06	.05	.02	.02	.02
5	.14	.07	.02	.13	.07	.02	.12	.06	.02	.11	.06	.02	.10	.06	.02	5	.20	.16	.14	.17	.14	.12	.12	.10	.08	.07	.06	.05	.02	.02	.02
6	.13	.07	.02	.13	.07	.02	.12	.06	.02	.11	.06	.02	.10	.05	.02	6	.20	.16	.13	.17	.14	.11	.12	.10	.08	.07	.06	.05	.02	.02	.02
7	.13	.06	.02	.13	.06	.02	.12	.06	.02	.11	.06	.02	.10	.05	.02	7	.19	.15	.12	.16	.13	.11	.11	.09	.08	.07	.05	.05	.02	.02	.02
8	.13	.06	.02	.12	.06	.02	.11	.06	.02	.10	.05	.02	.10	.05	.01	8	.19	.15	.12	.16	.13	.10	.11	.09	.07	.07	.05	.04	.02	.02	.01
9	.12	.06	.02	.12	.06	.02	.11	.05	.02	.10	.05	.01	.10	.05	.01	9	.18	.14	.12	.16	.12	.10	.11	.09	.07	.06	.05	.04	.02	.02	.01
10	.12	.06	.02	.11	.05	.02	.11	.05	.01	.10	.05	.01	.09	.05	.01	10	.18	.14	.11	.16	.12	.10	.11	.09	.07	.06	.05	.04	.02	.02	.01
																0	.26	.26	.26	.22	.22	.22	.15	.15	.15	.09	.09	.09	.03	.03	.03
1	.24	.14	.04	.23	.13	.04	.22	.12	.04	.20	.11	.04	.18	.11	.03	1	.25	.23	.22	.22	.20	.19	.15	.14	.13	.09	.08	.07	.03	.03	.02
2	.23	.13	.04	.22	.12	.04	.21	.11	.04	.19	.11	.03	.18	.10	.03	2	.25	.21	.19	.21	.18	.16	.14	.13	.11	.08	.07	.07	.03	.02	.02
3	.21	.11	.03	.21	.11	.03	.19	.10	.03	.18	.10	.03	.16	.09	.03	3	.24	.20	.17	.20	.17	.14	.14	.12	.10	.08	.07	.06	.03	.02	.02
4	.20	.10	.03	.19	.10	.03	.18	.10	.03	.17	.09	.03	.16	.08	.03	4	.23	.19	.15	.20	.16	.13	.14	.11	.09	.08	.07	.06	.03	.02	.02
5	.19	.10	.03	.18	.09	.03	.17	.09	.03	.16	.08	.02	.15	.08	.02	5	.23	.18	.14	.19	.15	.12	.13	.11	.09	.08	.06	.05	.03	.02	.02
6	.18	.09	.03	.17	.09	.02	.16	.08	.02	.15	.08	.02	.14	.07	.02	6	.22	.17	.14	.19	.15	.12	.13	.10	.08	.08	.06	.05	.02	.02	.02
7	.17	.08	.02	.16	.08	.02	.15	.08	.02	.14	.07	.02	.13	.07	.02	7	.22	.17	.13	.19	.14	.11	.13	.10	.08	.07	.06	.05	.02	.02	.02
8	.16	.08	.02	.15	.07	.02	.14	.07	.02	.13	.07	.02	.13	.06	.02	8	.21	.16	.13	.18	.14	.11	.13	.10	.08	.07	.06	.05	.02	.02	.02
9	.15	.07	.02	.14	.07	.02	.14	.07	.02	.13	.06	.02	.12	.06	.02	9	.21	.16	.12	.18	.14	.11	.12	.10	.08	.07	.06	.05	.02	.02	.02
10	.14	.07	.02	.14	.07	.02	.13	.06	.02	.12	.06	.02	.11	.06	.02	10	.20	.15	.12	.17	.13	.11	.12	.09	.08	.07	.06	.04	.02	.02	.01

[e] Although it is recommended that luminance coefficients and wall direct radiation coefficients be published to three decimal places, only two are shown here for these typical luminaires. Three decimal place data should be obtained from manufacturers of actual luminaires used.

Fig. 9–12. *Continued*

Typical Luminaire	Typical Distribution and Per Cent Lamp Lumens		ρ_{CC}[a] →	80			70			50			30			10			0
			ρ_W[b] →	50	30	10	50	30	10	50	30	10	50	30	10	50	30	10	0
	Maint. Cat.	Maximum S/MH Guide[d]	RCR[c] ↓	Coefficients of Utilization for 20 Per Cent Effective Floor Cavity Reflectance (ρ_{FC} = 20)															
20	II	1.3	0	.99	.99	.99	.94	.94	.94	.84	.84	.84	.76	.76	.76	.68	.68	.68	.65
			1	.88	.85	.82	.84	.81	.78	.76	.74	.72	.69	.67	.66	.62	.61	.60	.57
			2	.78	.73	.68	.74	.70	.66	.68	.64	.61	.62	.59	.56	.56	.54	.52	.49
	22½% ↑		3	.69	.63	.58	.66	.61	.56	.61	.56	.53	.56	.52	.49	.51	.48	.46	.43
			4	.62	.55	.50	.60	.53	.49	.55	.50	.46	.50	.46	.43	.46	.43	.40	.37
			5	.55	.48	.43	.53	.47	.42	.49	.44	.39	.45	.41	.37	.41	.38	.35	.32
	65% ↓		6	.50	.43	.38	.48	.41	.37	.44	.39	.35	.41	.36	.33	.37	.34	.31	.29
			7	.45	.38	.33	.43	.37	.32	.40	.34	.30	.37	.32	.29	.34	.30	.27	.25
			8	.40	.34	.29	.39	.32	.28	.36	.30	.27	.33	.28	.25	.31	.27	.24	.22
Porcelain-enameled reflector			9	.36	.30	.25	.35	.29	.24	.32	.27	.23	.30	.25	.22	.28	.24	.21	.19
with 35°CW shielding			10	.33	.27	.22	.32	.26	.22	.29	.24	.20	.27	.23	.19	.25	.21	.18	.17
21	II	1.5/1.3	0	.94	.94	.94	.90	.90	.90	.82	.82	.82	.75	.75	.75	.69	.69	.69	.66
			1	.85	.82	.80	.82	.79	.77	.75	.73	.72	.69	.68	.66	.64	.63	.62	.59
			2	.76	.72	.68	.74	.70	.66	.68	.65	.62	.63	.61	.58	.58	.56	.55	.52
	17% ↑		3	.69	.63	.59	.66	.61	.57	.62	.58	.54	.57	.54	.51	.53	.51	.48	.46
			4	.62	.56	.51	.60	.54	.50	.56	.51	.47	.52	.48	.45	.48	.45	.43	.41
			5	.55	.49	.44	.53	.48	.43	.50	.45	.41	.47	.43	.39	.44	.40	.38	.36
	66% ↓		6	.50	.43	.39	.48	.42	.38	.45	.40	.36	.42	.38	.35	.40	.36	.33	.31
			7	.45	.38	.34	.43	.37	.33	.41	.36	.32	.38	.34	.30	.36	.32	.29	.27
			8	.40	.34	.29	.39	.33	.29	.37	.31	.28	.34	.30	.26	.32	.28	.25	.24
Diffuse aluminum reflector			9	.36	.30	.25	.35	.29	.25	.33	.28	.24	.31	.26	.23	.29	.25	.22	.20
with 35°CW shielding			10	.33	.26	.22	.32	.26	.22	.30	.25	.21	.28	.23	.20	.26	.22	.19	.18
22	II	1.0	0	.90	.90	.90	.85	.85	.85	.76	.76	.76	.68	.68	.68	.60	.60	.60	.57
			1	.81	.78	.76	.77	.74	.72	.69	.67	.66	.62	.61	.60	.56	.55	.54	.57
			2	.72	.68	.64	.69	.65	.62	.62	.59	.57	.56	.54	.52	.51	.49	.47	.45
	23½% ↑		3	.65	.59	.55	.62	.57	.53	.56	.52	.49	.51	.48	.46	.46	.44	.42	.39
			4	.58	.52	.48	.56	.50	.46	.51	.46	.43	.46	.43	.40	.42	.39	.37	.35
			5	.52	.46	.41	.50	.44	.40	.46	.41	.38	.42	.38	.35	.38	.35	.33	.30
	57% ↓		6	.47	.41	.36	.45	.39	.35	.41	.37	.33	.38	.34	.31	.35	.31	.29	.27
			7	.43	.36	.32	.41	.35	.31	.38	.33	.29	.34	.30	.27	.32	.28	.26	.24
Porcelain-enameled reflector			8	.38	.32	.28	.37	.31	.27	.34	.29	.26	.31	.27	.24	.29	.25	.23	.21
with 30°CW x 30°LW shield-			9	.35	.29	.24	.33	.28	.24	.31	.26	.22	.28	.24	.21	.26	.22	.20	.18
ing			10	.32	.26	.22	.30	.25	.21	.28	.23	.20	.26	.22	.19	.24	.20	.18	.16
23	II	1.5/1.1	0	.83	.83	.83	.79	.79	.79	.71	.71	.71	.65	.65	.65	.59	.59	.59	.56
			1	.75	.72	.70	.72	.69	.68	.65	.64	.62	.60	.59	.58	.55	.54	.53	.50
			2	.67	.63	.60	.65	.61	.58	.59	.57	.54	.55	.53	.51	.50	.49	.47	.45
	17% ↑		3	.61	.56	.52	.58	.54	.51	.54	.50	.48	.50	.47	.45	.46	.44	.42	.40
			4	.55	.49	.45	.53	.48	.44	.49	.45	.42	.45	.42	.40	.42	.39	.37	.36
			5	.49	.44	.40	.47	.42	.39	.44	.40	.37	.41	.38	.35	.38	.35	.33	.31
			6	.45	.39	.35	.43	.38	.34	.40	.36	.33	.37	.34	.31	.35	.32	.30	.28
	56½% ↓		7	.40	.35	.31	.39	.34	.30	.36	.32	.29	.34	.30	.27	.32	.29	.26	.25
Diffuse aluminum reflector			8	.36	.31	.27	.35	.30	.26	.33	.28	.25	.31	.27	.24	.29	.25	.23	.21
with 35°CW x 35°LW shield-			9	.33	.27	.23	.32	.26	.23	.29	.25	.22	.28	.24	.21	.26	.22	.20	.19
ing			10	.30	.24	.21	.29	.24	.20	.27	.22	.19	.25	.21	.19	.23	.20	.18	.16
24	II	1.1	0	.75	.75	.75	.68	.68	.68	.57	.57	.57	.46	.46	.46	.36	.36	.36	.31
			1	.67	.64	.62	.61	.59	.57	.51	.50	.49	.42	.41	.40	.34	.33	.32	.29
			2	.59	.55	.52	.55	.51	.49	.46	.44	.42	.38	.36	.35	.31	.30	.29	.25
			3	.53	.48	.45	.49	.45	.42	.41	.39	.36	.35	.32	.31	.28	.27	.26	.23
	39% ↑		4	.47	.42	.39	.44	.40	.36	.37	.34	.32	.31	.29	.27	.26	.24	.23	.20
			5	.43	.37	.33	.40	.35	.31	.34	.30	.28	.28	.26	.24	.23	.22	.20	.18
			6	.39	.33	.29	.36	.31	.28	.31	.27	.25	.26	.23	.21	.22	.20	.18	.16
	32% ↓		7	.35	.30	.26	.33	.28	.25	.28	.24	.22	.24	.21	.19	.20	.18	.16	.15
Metal or dense diffusing sides			8	.32	.27	.23	.30	.25	.22	.25	.22	.19	.22	.19	.17	.18	.16	.15	.13
with 45°CW x 45°LW shield-			9	.29	.24	.20	.27	.22	.19	.23	.20	.17	.20	.17	.15	.16	.15	.13	.12
ing			10	.26	.21	.18	.25	.20	.17	.21	.18	.15	.18	.15	.14	.15	.13	.12	.10

[a] ρ_{CC} = per cent effective ceiling cavity reflectance.
[b] ρ_W = per cent wall reflectance.
[c] RCR = Room Cavity Ratio.
[d] Maximum S/MH guide—ratio of maximum luminaire spacing to mounting or ceiling height above work-plane.

Fig. 9–12. *Continued*

Wall Luminance Coefficients[e] for 20 Per Cent Effective Floor Cavity Reflectance (ρ_{FC} = 20)

ρ_{CC} →	80			70			50			30			10		
ρ_W →	50	30	10	50	30	10	50	30	10	50	50	10	50	30	10
RCR ↓															
1	.23	.13	.04	.22	.12	.04	.19	.11	.04	.17	.10	.03	.15	.09	.03
2	.22	.12	.04	.21	.11	.03	.18	.10	.03	.16	.09	.03	.14	.08	.03
3	.20	.11	.03	.19	.10	.03	.17	.09	.03	.15	.08	.03	.14	.08	.02
4	.19	.10	.03	.18	.09	.03	.16	.09	.03	.15	.08	.02	.13	.07	.02
5	.18	.09	.03	.17	.09	.03	.15	.08	.02	.14	.07	.02	.12	.07	.02
6	.17	.08	.02	.16	.08	.02	.15	.07	.02	.13	.07	.02	.12	.06	.02
7	.16	.08	.02	.15	.08	.02	.14	.07	.02	.12	.06	.02	.11	.06	.02
8	.15	.07	.02	.14	.07	.02	.13	.07	.02	.12	.06	.02	.11	.06	.02
9	.14	.07	.02	.14	.07	.02	.12	.06	.02	.11	.06	.02	.10	.05	.02
10	.14	.07	.02	.13	.06	.02	.12	.06	.02	.11	.05	.02	.10	.05	.01
1	.20	.11	.04	.19	.11	.03	.17	.10	.03	.15	.09	.03	.13	.08	.02
2	.19	.10	.03	.18	.10	.03	.16	.09	.03	.15	.08	.03	.13	.07	.02
3	.18	.10	.03	.17	.09	.03	.15	.08	.03	.14	.08	.02	.13	.07	.02
4	.17	.09	.03	.16	.09	.03	.15	.08	.02	.13	.07	.02	.12	.07	.02
5	.16	.08	.02	.16	.08	.02	.14	.07	.02	.13	.07	.02	.12	.06	.02
6	.15	.08	.02	.15	.07	.02	.14	.07	.02	.12	.06	.02	.11	.06	.02
7	.15	.07	.02	.14	.07	.02	.13	.07	.02	.12	.06	.02	.11	.06	.02
8	.14	.07	.02	.13	.07	.02	.12	.06	.02	.11	.06	.02	.11	.05	.02
9	.13	.06	.02	.13	.06	.02	.12	.06	.02	.11	.06	.02	.10	.05	.01
10	.13	.06	.02	.12	.06	.02	.11	.06	.02	.11	.05	.01	.10	.05	.01
1	.20	.11	.04	.19	.11	.03	.16	.09	.03	.14	.08	.03	.12	.07	.02
2	.19	.10	.03	.18	.10	.03	.16	.09	.03	.14	.08	.02	.12	.07	.02
3	.18	.09	.03	.17	.09	.03	.15	.08	.02	.13	.07	.02	.11	.06	.02
4	.17	.09	.03	.16	.08	.02	.14	.07	.02	.12	.07	.02	.11	.06	.02
5	.16	.08	.02	.15	.08	.02	.13	.07	.02	.12	.06	.02	.10	.06	.02
6	.15	.07	.02	.14	.07	.02	.13	.06	.02	.11	.06	.02	.10	.05	.02
7	.14	.07	.02	.13	.07	.02	.12	.06	.02	.11	.06	.02	.10	.05	.01
8	.13	.07	.02	.13	.06	.02	.12	.06	.02	.10	.05	.02	.09	.05	.01
9	.13	.06	.02	.12	.06	.02	.11	.05	.02	.10	.05	.01	.09	.04	.01
10	.12	.06	.02	.12	.06	.02	.10	.05	.01	.09	.05	.01	.08	.04	.01
1	.17	.10	.03	.16	.09	.03	.14	.08	.03	.13	.07	.02	.11	.06	.02
2	.16	.09	.03	.15	.09	.03	.14	.08	.02	.12	.07	.02	.11	.06	.02
3	.15	.08	.02	.15	.08	.02	.13	.07	.02	.12	.06	.02	.10	.06	.02
4	.15	.08	.02	.14	.07	.02	.13	.07	.02	.11	.06	.02	.10	.05	.02
5	.14	.07	.02	.13	.07	.02	.12	.06	.02	.11	.06	.02	.10	.05	.02
6	.13	.07	.02	.13	.06	.02	.11	.06	.02	.10	.05	.02	.09	.05	.01
7	.13	.06	.02	.12	.06	.02	.11	.06	.02	.10	.05	.01	.09	.05	.01
8	.12	.06	.02	.12	.06	.02	.11	.05	.02	.10	.05	.01	.09	.04	.01
9	.12	.06	.02	.11	.05	.02	.10	.05	.01	.09	.05	.01	.08	.04	.01
10	.11	.05	.01	.11	.05	.01	.10	.05	.01	.09	.04	.01	.08	.04	.01
1	.18	.10	.03	.16	.09	.03	.13	.07	.02	.10	.06	.02	.07	.04	.01
2	.16	.09	.03	.15	.08	.03	.12	.07	.02	.09	.05	.02	.07	.04	.01
3	.15	.08	.02	.14	.07	.02	.11	.06	.02	.09	.05	.01	.07	.04	.01
4	.14	.07	.02	.13	.07	.02	.11	.06	.02	.08	.05	.01	.06	.03	.01
5	.13	.07	.02	.12	.06	.02	.10	.05	.02	.08	.04	.01	.06	.03	.01
6	.13	.06	.02	.12	.06	.02	.09	.05	.01	.08	.04	.01	.06	.03	.01
7	.12	.06	.02	.11	.05	.02	.09	.05	.01	.07	.04	.01	.06	.03	.01
8	.11	.05	.02	.10	.05	.01	.09	.04	.01	.07	.03	.01	.05	.03	.01
9	.11	.05	.01	.10	.05	.01	.08	.04	.01	.07	.03	.01	.05	.03	.01
10	.10	.05	.01	.09	.04	.01	.08	.04	.01	.06	.03	.01	.05	.02	.01

Ceiling Cavity Luminance Coefficients[e] for 20 Per Cent Effective Floor Cavity Reflectance (ρ_{FC} = 20)

ρ_{CC} →	80			70			50			30			10		
ρ_W →	50	30	10	50	30	10	50	30	10	50	30	10	50	30	10
RCR ↓															
0	.33	.33	.33	.28	.28	.28	.19	.19	.19	.11	.11	.11	.03	.03	.03
1	.33	.31	.29	.28	.27	.25	.19	.18	.17	.11	.11	.10	.04	.03	.03
2	.32	.29	.26	.27	.25	.23	.19	.17	.16	.11	.10	.09	.03	.03	.03
3	.31	.28	.25	.27	.24	.21	.18	.16	.15	.11	.10	.09	.03	.03	.03
4	.31	.26	.23	.26	.23	.20	.18	.16	.14	.10	.09	.08	.03	.03	.03
5	.30	.25	.22	.26	.22	.19	.18	.15	.14	.10	.09	.08	.03	.03	.03
6	.30	.25	.21	.25	.21	.19	.18	.15	.13	.10	.09	.08	.03	.03	.03
7	.29	.24	.21	.25	.21	.18	.17	.15	.13	.10	.09	.08	.03	.03	.03
8	.29	.24	.20	.25	.21	.18	.17	.14	.13	.10	.09	.08	.03	.03	.02
9	.28	.23	.20	.24	.20	.18	.17	.14	.12	.10	.08	.07	.03	.03	.02
10	.28	.23	.20	.24.	.20	.17	.17	.14	.12	.10	.08	.07	.03	.03	.02
0	.28	.28	.28	.24	.24	.24	.16	.16	.16	.09	.09	.09	.03	.03	.03
1	.27	.26	.24	.23	.22	.21	.16	.15	.15	.09	.09	.08	.03	.03	.03
2	.27	.24	.22	.23	.21	.19	.16	.14	.13	.09	.08	.08	.03	.03	.03
3	.26	.22	.20	.22	.19	.17	.15	.13	.12	.09	.08	.07	.03	.03	.02
4	.25	.21	.18	.22	.18	.16	.15	.13	.11	.09	.08	.07	.03	.02	.02
5	.25	.21	.17	.21	.18	.15	.15	.12	.11	.08	.07	.06	.03	.02	.02
6	.24	.20	.17	.21	.17	.15	.14	.12	.10	.08	.07	.06	.03	.02	.02
7	.24	.19	.16	.20	.17	.14	.14	.12	.10	.08	.07	.06	.03	.02	.02
8	.23	.19	.16	.20	.16	.14	.14	.11	.10	.08	.07	.06	.03	.02	.02
9	.23	.19	.15	.20	.16	.13	.14	.11	.10	.08	.07	.06	.03	.02	.02
10	.23	.18	.15	.20	.16	.13	.14	.11	.09	.08	.07	.06	.03	.02	.02
0	.33	.33	.33	.28	.28	.28	.19	.19	.19	.11	.11	.11	.03	.03	.03
1	.32	.31	.29	.28	.26	.25	.19	.18	.18	.11	.11	.10	.03	.03	.03
2	.32	.29	.27	.27	.25	.23	.19	.17	.16	.11	.10	.09	.03	.03	.03
3	.31	.28	.25	.27	.24	.22	.18	.17	.15	.11	.10	.09	.03	.03	.03
4	.30	.27	.24	.26	.23	.21	.18	.16	.15	.10	.09	.09	.03	.03	.03
5	.30	.26	.23	.26	.22	.20	.18	.16	.14	.10	.09	.08	.03	.03	.03
6	.29	.25	.22	.25	.22	.19	.17	.15	.14	.10	.09	.08	.03	.03	.03
7	.29	.25	.22	.25	.21	.19	.17	.15	.13	.10	.09	.08	.03	.03	.03
8	.28	.24	.21	.25	.21	.18	.17	.15	.13	.10	.09	.08	.03	.03	.03
9	.28	.24	.21	.24	.21	.18	.17	.15	.13	.10	.09	.08	.03	.03	.03
10	.28	.23	.21	.24	.20	.18	.17	.14	.13	.10	.08	.08	.03	.03	.03
0	.26	.26	.26	.22	.22	.22	.15	.15	.15	.09	.09	.09	.03	.03	.03
1	.26	.24	.23	.22	.21	.20	.15	.14	.14	.09	.08	.08	.03	.03	.03
2	.25	.23	.21	.21	.20	.18	.15	.14	.13	.08	.08	.07	.03	.03	.02
3	.24	.21	.19	.21	.19	.17	.14	.13	.12	.08	.08	.07	.03	.02	.02
4	.24	.21	.18	.20	.18	.16	.14	.12	.11	.08	.07	.07	.03	.02	.02
5	.23	.20	.17	.20	.17	.15	.14	.12	.11	.08	.07	.06	.03	.02	.02
6	.23	.19	.17	.20	.17	.14	.14	.12	.10	.08	.07	.06	.03	.02	.02
7	.23	.19	.16	.19	.16	.14	.13	.11	.10	.08	.07	.06	.03	.02	.02
8	.22	.18	.16	.19	.16	.14	.13	.11	.10	.08	.07	.06	.02	.02	.02
9	.22	.18	.15	.19	.16	.13	.13	.11	.10	.08	.06	.06	.02	.02	.02
10	.22	.18	.15	.19	.15	.13	.13	.11	.09	.08	.06	.06	.02	.02	.02
0	.43	.43	.43	.37	.37	.37	.25	.25	.25	.14	.14	.14	.05	.05	.05
1	.43	.41	.40	.36	.35	.34	.25	.24	.24	.14	.14	.14	.05	.05	.04
2	.42	.40	.38	.36	.34	.33	.25	.24	.23	.14	.14	.13	.05	.04	.04
3	.41	.39	.36	.35	.33	.31	.24	.23	.22	.14	.13	.13	.05	.04	.04
4	.41	.38	.35	.35	.33	.31	.24	.23	.22	.14	.13	.13	.04	.04	.04
5	.40	.37	.34	.35	.32	.30	.24	.22	.21	.14	.13	.13	.04	.04	.04
6	.40	.36	.34	.34	.32	.30	.24	.22	.21	.14	.13	.12	.04	.04	.04
7	.40	.36	.33	.34	.31	.29	.24	.22	.21	.14	.13	.12	.04	.04	.04
8	.39	.36	.33	.34	.31	.29	.23	.22	.21	.14	.13	.12	.04	.04	.04
9	.39	.35	.33	.34	.31	.29	.23	.22	.20	.14	.13	.12	.04	.04	.04
10	.39	.35	.33	.33	.30	.29	.23	.21	.20	.13	.13	.12	.04	.04	.04

[e] Although it is recommended that luminance coefficients and wall direct radiation coefficients be published to three decimal places, only two are shown here for these typical luminaires. Three decimal place data should be obtained from manufacturers of actual luminaires used.

Fig. 9–12. *Continued*

Typical Luminaire	Typical Distribution and Per Cent Lamp Lumens	Maint. Cat.	Maximum S/MH Guide[d]	ρ_{CC}[a] →	80			70			50			30			10			0
				ρ_W[b] →	50	30	10	50	30	10	50	30	10	50	30	10	50	30	10	0
				RCR[c] ↓	Coefficients of Utilization for 20 Per Cent Effective Floor Cavity Reflectance (ρ_{FC} = 20)															
25		IV	1.0	0	.60	.60	.60	.58	.58	.58	.54	.54	.54	.50	.50	.50	.47	.47	.47	.46
				1	.54	.52	.51	.52	.51	.49	.49	.48	.47	.46	.45	.44	.43	.43	.42	.40
				2	.49	.46	.43	.47	.44	.42	.44	.42	.40	.42	.40	.38	.39	.38	.37	.35
	6% ↑			3	.44	.40	.37	.43	.39	.37	.40	.38	.35	.38	.36	.34	.36	.34	.33	.31
				4	.40	.36	.33	.39	.35	.32	.37	.34	.31	.35	.32	.30	.33	.31	.29	.28
				5	.36	.32	.29	.35	.31	.28	.33	.30	.28	.32	.29	.27	.30	.28	.26	.25
	46% ↓			6	.33	.29	.26	.32	.28	.25	.30	.27	.25	.29	.26	.24	.28	.25	.23	.22
				7	.30	.26	.23	.29	.25	.23	.28	.25	.22	.27	.24	.22	.25	.23	.21	.20
				8	.27	.23	.20	.27	.23	.20	.25	.22	.20	.24	.21	.19	.23	.21	.19	.18
Same as unit #24 except with top reflectors				9	.25	.21	.18	.24	.21	.18	.23	.20	.18	.22	.19	.17	.21	.19	.17	.16
				10	.23	.19	.16	.22	.19	.16	.21	.18	.16	.20	.18	.16	.20	.17	.15	.14
26		IV	1.1/0.8	0	.50	.50	.50	.49	.49	.49	.47	.47	.47	.45	.45	.45	.43	.43	.43	.42
				1	.46	.45	.44	.45	.44	.43	.44	.43	.42	.42	.41	.41	.41	.40	.40	.39
				2	.43	.41	.39	.42	.40	.38	.40	.39	.38	.39	.38	.37	.38	.37	.36	.35
	0% ↑			3	.39	.37	.35	.39	.36	.34	.37	.35	.34	.36	.35	.33	.35	.34	.33	.32
				4	.36	.33	.31	.35	.33	.31	.35	.32	.31	.34	.32	.30	.33	.31	.30	.29
				5	.33	.30	.28	.33	.30	.28	.32	.29	.28	.31	.29	.27	.30	.29	.27	.26
	42½% ↓			6	.31	.28	.26	.30	.28	.26	.30	.27	.25	.29	.27	.25	.28	.26	.25	.24
1′ wide aluminum troffer with 40°CW x 45°LW shielding and single extra-high-output lamp				7	.28	.25	.23	.28	.25	.23	.27	.25	.23	.27	.25	.23	.26	.24	.23	.22
				8	.26	.23	.21	.26	.23	.21	.25	.23	.21	.25	.23	.21	.24	.22	.21	.20
				9	.24	.21	.19	.24	.21	.19	.23	.21	.19	.23	.20	.19	.22	.20	.19	.18
				10	.22	.19	.17	.22	.19	.17	.21	.19	.17	.21	.19	.17	.21	.19	.17	.16
27		I	1.3	0	1.02	1.02	1.02	.98	.98	.98	.92	.92	.92	.86	.86	.86	.80	.80	.80	.78
				1	.86	.82	.78	.83	.79	.75	.78	.74	.71	.73	.70	.67	.68	.66	.64	.61
				2	.74	.67	.61	.71	.65	.60	.67	.61	.57	.62	.58	.54	.58	.55	.52	.49
	9½% ↑			3	.64	.56	.50	.62	.55	.49	.58	.52	.47	.54	.49	.45	.51	.47	.43	.41
				4	.56	.48	.42	.55	.47	.41	.51	.45	.39	.48	.42	.38	.45	.40	.36	.34
				5	.49	.41	.35	.48	.40	.34	.45	.38	.33	.42	.36	.32	.39	.34	.30	.28
2 lamp, surface mounted, bare lamp unit—Photometry with 18″ wide panel above luminaire (lamps on 6″ centers)	78% ↓			6	.44	.36	.30	.43	.35	.29	.40	.33	.28	.38	.32	.27	.35	.30	.26	.24
				7	.39	.31	.25	.38	.30	.25	.36	.29	.24	.34	.28	.23	.32	.27	.23	.21
				8	.35	.27	.22	.34	.27	.22	.32	.26	.21	.30	.24	.20	.29	.23	.19	.18
				9	.32	.24	.19	.31	.23	.18	.29	.22	.18	.27	.21	.17	.26	.20	.17	.15
				10	.29	.21	.17	.28	.21	.16	.26	.20	.16	.25	.19	.15	.23	.18	.15	.13
28		VI	1.5	0	.77	.77	.77	.67	.67	.67	.49	.49	.49	.33	.33	.33	.18	.18	.18	.11
				1	.67	.64	.62	.59	.57	.54	.44	.42	.41	.30	.29	.28	.17	.16	.16	.10
				2	.59	.54	.50	.51	.48	.45	.38	.36	.34	.26	.25	.23	.15	.14	.13	.09
				3	.51	.46	.42	.45	.41	.37	.34	.31	.28	.23	.21	.20	.13	.12	.12	.07
				4	.45	.40	.35	.40	.35	.31	.30	.27	.24	.20	.18	.17	.12	.11	.10	.06
	66% ↑			5	.40	.34	.30	.35	.30	.27	.26	.23	.20	.18	.16	.14	.10	.09	.08	.05
				6	.36	.30	.26	.32	.27	.23	.24	.20	.18	.16	.14	.12	.09	.08	.07	.05
				7	.32	.26	.22	.28	.23	.20	.21	.18	.15	.15	.12	.11	.08	.07	.06	.04
Luminous bottom suspended unit with extra-high-output lamp	12% ↓			8	.29	.23	.19	.25	.21	.17	.19	.16	.13	.13	.11	.09	.08	.06	.06	.03
				9	.26	.20	.17	.23	.18	.15	.17	.14	.12	.12	.10	.08	.07	.06	.05	.03
				10	.24	.18	.15	.21	.16	.13	.16	.12	.10	.11	.09	.07	.06	.05	.04	.03
29		VI	1.4/1.2	0	.90	.90	.90	.84	.84	.84	.73	.73	.73	.63	.63	.63	.54	.54	.54	.49
				1	.80	.77	.74	.75	.73	.70	.66	.64	.62	.57	.56	.54	.49	.48	.47	.43
				2	.71	.66	.62	.67	.63	.59	.59	.56	.53	.51	.49	.47	.44	.43	.41	.38
	33% ↑			3	.63	.58	.53	.60	.55	.50	.53	.49	.45	.46	.43	.41	.40	.38	.36	.33
				4	.57	.50	.46	.53	.48	.43	.47	.43	.39	.41	.38	.35	.36	.34	.32	.29
				5	.50	.44	.39	.48	.42	.37	.42	.38	.34	.37	.34	.31	.33	.30	.28	.25
	50% ↓			6	.45	.39	.34	.43	.37	.33	.38	.33	.30	.34	.30	.27	.30	.27	.24	.22
Prismatic bottom and sides, open top, 4 lamp suspended unit—multiply by 1.05 for 2 lamps				7	.41	.34	.30	.39	.33	.28	.34	.30	.26	.30	.27	.24	.27	.24	.21	.19
				8	.37	.30	.26	.35	.29	.25	.31	.26	.23	.27	.24	.21	.24	.21	.19	.17
				9	.33	.27	.22	.31	.26	.22	.28	.23	.20	.25	.21	.18	.22	.19	.16	.15
				10	.30	.24	.20	.28	.23	.19	.25	.21	.18	.23	.19	.16	.20	.17	.14	.13

[a] ρ_{CC} = per cent effective ceiling cavity reflectance.
[b] ρ_W = per cent wall reflectance.
[c] RCR = Room Cavity Ratio.
[d] Maximum *S/MH* guide—ratio of maximum luminaire spacing to mounting or ceiling height above work-plane.

Fig. 9–12. *Continued*

ρ_{CC} →	80			70			50			30			10		
ρ_W →	50	30	10	50	30	10	50	30	10	50	30	10	50	30	10
RCR ↓	Wall Luminance Coefficients[e] for 20 Per Cent Effective Floor Cavity Reflectance (ρ_{FC} = 20)														
1	.13	.08	.02	.13	.07	.02	.12	.07	.02	.11	.06	.02	.10	.06	.02
2	.13	.07	.02	.12	.07	.02	.11	.06	.02	.10	.06	.02	.10	.05	.02
3	.12	.06	.02	.11	.06	.02	.10	.06	.02	.10	.05	.02	.09	.05	.02
4	.11	.06	.02	.10	.05	.02	.10	.05	.02	.09	.05	.01	.08	.05	.01
5	.10	.05	.02	.10	.05	.01	.09	.05	.01	.09	.05	.01	.08	.04	.01
6	.10	.05	.01	.09	.05	.01	.09	.04	.01	.08	.04	.01	.08	.04	.01
7	.09	.04	.01	.09	.04	.01	.08	.04	.01	.08	.04	.01	.07	.04	.01
8	.09	.04	.01	.08	.04	.01	.08	.04	.01	.07	.04	.01	.07	.04	.01
9	.08	.04	.01	.08	.04	.01	.08	.04	.01	.07	.04	.01	.07	.03	.01
10	.08	.04	.01	.08	.04	.01	.07	.04	.01	.07	.03	.01	.06	.03	.01
1	.09	.05	.02	.09	.05	.02	.08	.05	.01	.07	.04	.01	.07	.04	.01
2	.08	.05	.01	.08	.05	.01	.08	.04	.01	.07	.04	.01	.07	.04	.01
3	.08	.04	.01	.08	.04	.01	.07	.04	.01	.07	.04	.01	.07	.04	.01
4	.08	.04	.01	.07	.04	.01	.07	.04	.01	.07	.04	.01	.06	.04	.01
5	.07	.04	.01	.07	.04	.01	.07	.04	.01	.07	.03	.01	.06	.03	.01
6	.07	.03	.01	.07	.03	.01	.07	.03	.01	.06	.03	.01	.06	.03	.01
7	.07	.03	.01	.07	.03	.01	.06	.03	.01	.06	.03	.01	.06	.03	.01
8	.06	.03	.01	.06	.03	.01	.06	.03	.01	.06	.03	.01	.06	.03	.01
9	.06	.03	.01	.06	.03	.01	.06	.03	.01	.06	.03	.01	.06	.03	.01
10	.06	.03	.01	.06	.03	.01	.06	.03	.01	.05	.03	.01	.05	.03	.01
1	.32	.18	.06	.31	.18	.06	.30	.17	.05	.28	.16	.05	.27	.15	.05
2	.29	.16	.05	.28	.15	.05	.26	.15	.05	.25	.14	.04	.24	.13	.04
3	.26	.14	.04	.25	.13	.04	.23	.13	.04	.22	.12	.04	.21	.12	.04
4	.23	.12	.04	.23	.12	.03	.21	.11	.03	.20	.11	.03	.19	.10	.03
5	.21	.11	.03	.21	.11	.03	.20	.10	.03	.19	.10	.03	.18	.09	.03
6	.20	.10	.03	.19	.10	.03	.18	.09	.03	.17	.09	.03	.16	.09	.03
7	.18	.09	.03	.18	.09	.03	.17	.09	.02	.16	.08	.02	.15	.08	.02
8	.17	.08	.02	.17	.08	.02	.16	.08	.02	.15	.08	.02	.14	.07	.02
9	.16	.08	.02	.16	.08	.02	.15	.07	.02	.14	.07	.02	.13	.07	.02
10	.15	.07	.02	.15	.07	.02	.14	.07	.02	.13	.07	.02	.13	.06	.02
1	.20	.12	.04	.18	.10	.03	.13	.08	.02	.09	.05.	02	.05	.03	.01
2	.19	.10	.03	.17	.09	.03	.12	.07	.02	.08	.05	.01	.04	.03	.01
3	.17	.09	.03	.15	.08	.02	.11	.06	.02	.08	.04	.01	.04	.02	.01
4	.16	.08	.02	.14	.07	.02	.11	.06	.02	.07	.04	.01	.04	.02	.01
5	.15	.08	.02	.13	.07	.02	.10	.05	.02	.07	.04	.01	.04	.02	.01
6	.14	.07	.02	.12	.06	.02	.09	.05	.01	.06	.03	.01	.03	.02	.01
7	.13	.07	.02	.12	.06	.02	.09	.04	.01	.06	.03	.01	.03	.02	.01
8	.12	.06	.02	.11	.05	.02	.08	.04	.01	.06	.03	.01	.03	.02	.00
9	.12	.06	.02	.10	.05	.01	.08	.04	.01	.05	.03	.01	.03	.02	.00
10	.11	.05	.01	.10	.05	.01	.07	.04	.01	.05	.02	.01	.03	.01	.00
1	.22	.12	.04	.20	.12	.04	.17	.10	.03	.15	.08	.03	.12	.07	.02
2	.20	.11	.03	.19	.10	.03	.16	.09	.03	.14	.08	.02	.11	.06	.02
3	.19	.10	.03	.18	.09	.03	.15	.08	.02	.13	.07	.02	.11	.06	.02
4	.18	.09	.03	.16	.09	.03	.14	.08	.02	.12	.07	.02	.10	.06	.02
5	.17	.08	.02	.16	.08	.02	.13	.07	.02	.12	.06	.02	.10	.05	.02
6	.16	.08	.02	.15	.07	.02	.13	.07	.02	.11	.06	.02	.09	.05	.01
7	.15	.07	.02	.14	.07	.02	.12	.06	.02	.10	.05	.02	.09	.05	.01
8	.14	.07	.02	.13	.06	.02	.11	.06	.02	.10	.05	.01	.08	.04	.01
9	.13	.06	.02	.12	.06	.02	.11	.05	.02	.09	.05	.01	.08	.04	.01
10	.13	.06	.02	.12	.06	.02	.10	.05	.01	.09	.04	.01	.08	.04	.01

ρ_{CC} →	80			70			50			30			10		
ρ_W →	50	30	10	50	30	10	50	30	10	50	30	10	50	30	10
RCR ↓	Ceiling Cavity Luminance Coefficients[e] for 20 Per Cent Effective Floor Cavity Reflectance (ρ_{FC} = 20)														
0	.14	.14	.14	.12	.12	.12	.08	.08	.08	.05	.05	.05	.01	.01	.01
1	.14	.13	.12	.12	.11	.10	.08	.08	.07	.05	.04	.04	.01	.01	.01
2	.13	.12	.10	.11	.10	.09	.08	.07	.06	.05	.04	.04	.01	.01	.01
3	.13	.11	.09	.11	.09	.08	.08	.06	.05	.04	.04	.03	.01	.01	.01
4	.12	.10	.08	.11	.09	.07	.07	.06	.05	.04	.04	.03	.01	.01	.01
5	.12	.09	.07	.10	.08	.06	.07	.06	.05	.04	.03	.03	.01	.01	.01
6	.12	.09	.07	.10	.08	.06	.07	.05	.04	.04	.03	.03	.01	.01	.01
7	.11	.09	.07	.10	.07	.06	.07	.05	.04	.04	.03	.02	.01	.01	.01
8	.11	.08	.06	.10	.07	.06	.07	.05	.04	.04	.03	.02	.01	.01	.01
9	.11	.08	.06	.09	.07	.05	.06	.05	.04	.04	.03	.02	.01	.01	.01
10	.11	.08	.06	.09	.07	.05	.06	.05	.04	.04	.03	.02	.01	.01	.01
0	.08	.08	.08	.06	.06	.06	.04	.04	.04	.02	.02	.02	.01	.01	.01
1	.07	.07	.06	.06	.06	.05	.04	.04	.04	.02	.02	.02	.01	.01	.01
2	.07	.06	.05	.06	.05	.04	.04	.03	.03	.02	.02	.02	.01	.01	.01
3	.06	.05	.04	.05	.04	.03	.04	.03	.02	.02	.02	.01	.01	.01	.00
4	.06	.04	.03	.05	.04	.03	.04	.03	.02	.02	.01	.01	.01	.00	.00
5	.06	.04	.02	.05	.03	.02	.03	.02	.01	.02	.01	.01	.01	.00	.00
6	.05	.03	.02	.05	.03	.02	.03	.02	.01	.02	.01	.01	.01	.00	.00
7	.05	.03	.02	.04	.03	.01	.03	.02	.01	.02	.01	.01	.01	.00	.00
8	.05	.03	.01	.04	.02	.01	.03	.02	.01	.02	.01	.01	.01	.00	.00
9	.05	.03	.01	.04	.02	.01	.03	.02	.01	.02	.01	.00	.01	.00	.00
10	.04	.02	.01	.04	.02	.01	.03	.01	.01	.02	.01	.00	.01	.00	.00
0	.23	.23	.23	.20	.20	.20	.14	.14	.14	.08	.08	.08	.03	.03	.03
1	.23	.21	.19	.20	.18	.16	.14	.12	.11	.08	.07	.06	.03	.02	.02
2	.23	.19	.16	.20	.16	.13	.13	.11	.09	.08	.07	.05	.02	.02	.02
3	.22	.17	.14	.19	.15	.12	.13	.10	.08	.08	.06	.05	.02	.02	.02
4	.22	.16	.12	.18	.14	.11	.13	.10	.08	.07	.06	.04	.02	.02	.01
5	.21	.15	.11	.18	.13	.10	.12	.09	.07	.07	.05	.04	.02	.02	.01
6	.20	.15	.11	.17	.13	.09	.12	.09	.07	.07	.05	.04	.02	.02	.01
7	.20	.14	.10	.17	.12	.09	.12	.09	.06	.07	.05	.04	.02	.02	.01
8	.19	.14	.10	.16	.12	.09	.11	.08	.06	.07	.05	.04	.02	.02	.01
9	.19	.13	.10	.16	.12	.08	.11	.08	.06	.06	.05	.04	.02	.02	.01
10	.18	.13	.09	.16	.11	.08	.11	.08	.06	.06	.05	.03	.02	.02	.01
0	.65	.65	.65	.55	.55	.55	.38	.38	.38	.22	.22	.22	.07	.07	.07
1	.65	.63	.62	.55	.54	.53	.38	.37	.37	.22	.22	.21	.07	.07	.07
2	.64	.61	.59	.55	.53	.51	.38	.37	.36	.22	.21	.21	.07	.07	.07
3	.64	.60	.58	.55	.52	.50	.37	.36	.35	.22	.21	.21	.07	.07	.07
4	.63	.59	.57	.54	.51	.49	.37	.36	.35	.22	.21	.21	.07	.07	.07
5	.63	.59	.56	.54	.51	.49	.37	.36	.34	.21	.21	.20	.07	.07	.07
6	.62	.58	.55	.54	.50	.48	.37	.35	.34	.21	.21	.20	.07	.07	.07
7	.62	.58	.55	.53	.50	.48	.37	.35	.34	.21	.21	.20	.07	.07	.07
8	.61	.57	.55	.53	.50	.48	.37	.35	.34	.21	.21	.20	.07	.07	.07
9	.61	.57	.55	.53	.50	.48	.36	.35	.34	.21	.21	.20	.07	.07	.07
10	.61	.57	.54	.52	.49	.47	.36	.35	.34	.21	.21	.20	.07	.07	.07
0	.40	.40	.40	.35	.35	.35	.23	.23	.23	.13	.13	.13	.04	.04	.04
1	.40	.38	.37	.34	.33	.32	.23	.22	.22	.13	.13	.13	.04	.04	.04
2	.39	.37	.34	.34	.31	.30	.23	.22	.21	.13	.13	.12	.04	.04	.04
3	.39	.35	.32	.33	.30	.28	.23	.21	.20	.13	.12	.12	.04	.04	.04
4	.38	.34	.31	.33	.29	.27	.22	.21	.19	.13	.12	.11	.04	.04	.04
5	.38	.33	.30	.32	.29	.26	.22	.20	.19	.13	.12	.11	.04	.04	.04
6	.37	.33	.30	.32	.28	.26	.22	.20	.18	.13	.12	.11	.04	.04	.04
7	.37	.32	.29	.31	.28	.25	.22	.20	.18	.13	.12	.11	.04	.04	.04
8	.36	.32	.29	.31	.28	.25	.22	.19	.18	.13	.11	.11	.04	.04	.03
9	.36	.31	.28	.31	.27	.25	.21	.19	.18	.12	.11	.10	.04	.04	.03
10	.35	.31	.28	.31	.27	.25	.21	.19	.17	.12	.11	.10	.04	.04	.03

[e] Although it is recommended that luminance coefficients and wall direct radiation coefficients be published to three decimal places, only two are shown here for these typical luminaires. Three decimal place data should be obtained from manufacturers of actual luminaires used.

Fig. 9–12. *Continued*

Typical Luminaire	Typical Distribution and Per Cent Lamp Lumens	Maint. Cat.	Maximum S/MH Guide[d]	RCR[c] ↓	ρ_{CC}[a] → 80 / ρ_W[b] → 50	80 / 30	80 / 10	70 / 50	70 / 30	70 / 10	50 / 50	50 / 30	50 / 10	30 / 50	30 / 30	30 / 10	10 / 50	10 / 30	10 / 10	0 / 0
					Coefficients of Utilization for 20 Per Cent Effective Floor Cavity Reflectance (ρ_{FC} = 20)															
30		V	1.5/1.2	0	.80	.80	.80	.77	.77	.77	.71	.71	.71	.66	.66	.66	.60	.60	.60	.58
				1	.71	.69	.66	.69	.66	.64	.64	.62	.60	.59	.58	.56	.55	.54	.53	.50
	11½% ↑			2	.64	.59	.56	.61	.58	.54	.57	.54	.51	.53	.51	.49	.49	.48	.46	.44
				3	.57	.52	.48	.55	.50	.47	.51	.48	.45	.48	.45	.42	.45	.42	.40	.38
				4	.51	.46	.41	.49	.44	.40	.46	.42	.39	.43	.40	.37	.41	.38	.35	.34
	58½% ↓			5	.46	.40	.36	.44	.39	.35	.41	.37	.34	.39	.35	.32	.37	.33	.31	.29
				6	.41	.35	.31	.40	.35	.31	.38	.33	.30	.35	.31	.28	.33	.30	.27	.26
				7	.37	.31	.27	.36	.31	.27	.34	.29	.26	.32	.28	.25	.30	.27	.24	.23
2 lamp prismatic wraparound —multiply by 0.95 for 4 lamps				8	.33	.28	.24	.32	.27	.23	.30	.26	.22	.29	.25	.22	.27	.24	.21	.19
				9	.30	.24	.20	.29	.24	.20	.27	.23	.19	.26	.22	.19	.24	.21	.18	.17
				10	.27	.22	.18	.26	.21	.18	.25	.20	.17	.23	.19	.16	.22	.18	.16	.15
31		V	1.2	0	.82	.82	.82	.77	.77	.77	.68	.68	.68	.60	.60	.60	.53	.53	.53	.49
				1	.71	.68	.65	.67	.65	.62	.60	.58	.56	.53	.51	.50	.47	.45	.44	.41
				2	.63	.58	.54	.59	.55	.52	.53	.50	.47	.47	.45	.42	.42	.40	.38	.35
	24% ↑			3	.56	.50	.46	.53	.48	.44	.47	.44	.40	.42	.39	.37	.38	.35	.33	.31
				4	.50	.44	.40	.48	.42	.38	.43	.39	.35	.38	.35	.32	.34	.32	.29	.27
				5	.45	.39	.34	.43	.37	.33	.38	.34	.31	.35	.31	.28	.31	.28	.26	.24
	50% ↓			6	.41	.35	.30	.39	.33	.29	.35	.30	.27	.32	.28	.25	.28	.25	.23	.21
				7	.37	.31	.27	.35	.30	.26	.32	.27	.24	.29	.25	.22	.26	.23	.20	.19
2 lamp prismatic wraparound —multiply by 0.95 for 4 lamps				8	.33	.27	.23	.32	.26	.23	.29	.24	.21	.26	.22	.20	.23	.20	.18	.16
				9	.30	.24	.20	.29	.23	.20	.26	.22	.18	.24	.20	.17	.21	.18	.16	.14
				10	.27	.22	.18	.26	.21	.18	.24	.19	.16	.22	.18	.15	.19	.16	.14	.13
32		V	1.3	0	.52	.52	.52	.50	.50	.50	.46	.46	.46	.42	.42	.42	.39	.39	.39	.37
				1	.45	.43	.41	.43	.41	.39	.40	.38	.37	.36	.35	.34	.34	.33	.32	.30
				2	.39	.35	.33	.37	.34	.32	.34	.32	.30	.32	.30	.28	.29	.28	.26	.25
	8% ↑			3	.34	.30	.27	.33	.29	.26	.30	.27	.25	.28	.25	.23	.26	.24	.22	.21
				4	.30	.26	.23	.29	.25	.22	.27	.24	.21	.25	.22	.20	.23	.21	.19	.18
	37½% ↓			5	.26	.22	.19	.25	.21	.19	.23	.20	.18	.22	.19	.17	.20	.18	.16	.15
				6	.23	.19	.16	.23	.19	.16	.21	.18	.15	.19	.17	.14	.18	.16	.14	.13
				7	.21	.17	.14	.20	.16	.14	.19	.16	.13	.17	.15	.13	.16	.14	.12	.11
2 lamp white diffuse wraparound—multiply by 0.90 for 4 lamps				8	.19	.15	.12	.18	.14	.12	.17	.14	.11	.16	.13	.11	.15	.12	.10	.09
				9	.17	.13	.10	.16	.13	.10	.15	.12	.10	.14	.11	.09	.13	.11	.09	.08
				10	.15	.12	.09	.15	.11	.09	.14	.11	.09	.13	.10	.08	.12	.10	.08	.07
33		IV	1.0	0	.54	.54	.54	.53	.53	.53	.51	.51	.51	.48	.48	.48	.46	.46	.46	.45
				1	.49	.48	.46	.48	.47	.46	.46	.45	.44	.45	.44	.43	.43	.42	.42	.41
				2	.44	.42	.40	.43	.41	.39	.42	.40	.38	.40	.39	.37	.39	.38	.37	.36
	0% ↑			3	.40	.37	.34	.39	.36	.34	.38	.36	.34	.37	.35	.33	.36	.34	.33	.32
				4	.36	.33	.30	.36	.32	.30	.35	.32	.30	.34	.31	.29	.33	.31	.29	.28
	46% ↓			5	.33	.29	.26	.32	.29	.26	.31	.28	.26	.30	.28	.26	.30	.27	.26	.25
				6	.30	.26	.24	.29	.26	.24	.29	.26	.23	.28	.25	.23	.27	.25	.23	.22
				7	.27	.24	.21	.27	.23	.21	.26	.23	.21	.26	.23	.21	.25	.22	.21	.20
2 lamp, 1′ wide troffer with 45° plastic louver—multiply by 0.90 for 3 lamps				8	.25	.21	.19	.24	.21	.19	.24	.21	.19	.23	.21	.18	.23	.20	.18	.18
				9	.22	.19	.17	.22	.19	.17	.22	.19	.17	.21	.18	.16	.21	.18	.16	.16
				10	.21	.17	.15	.20	.17	.15	.20	.17	.15	.20	.17	.15	.19	.17	.15	.14
34		IV	0.9	0	.50	.50	.50	.49	.49	.49	.47	.47	.47	.45	.45	.45	.43	.43	.43	.42
				1	.46	.45	.44	.45	.44	.43	.43	.42	.42	.42	.41	.40	.40	.40	.39	.38
				2	.42	.40	.38	.41	.39	.37	.40	.38	.36	.38	.37	.36	.37	.36	.35	.34
	0% ↑			3	.38	.35	.33	.37	.35	.33	.36	.34	.32	.35	.33	.32	.34	.33	.31	.31
				4	.35	.32	.29	.34	.31	.29	.33	.31	.29	.32	.30	.28	.31	.30	.28	.27
	42½% ↓			5	.31	.28	.26	.31	.28	.26	.30	.28	.26	.29	.27	.25	.29	.27	.25	.24
				6	.29	.26	.23	.29	.26	.23	.28	.25	.23	.27	.25	.23	.27	.24	.23	.22
				7	.27	.23	.21	.26	.23	.21	.26	.23	.21	.25	.23	.21	.24	.22	.21	.20
2 lamp, 1′ wide troffer with 45° white metal louver—multiply by 0.90 for 3 lamps				8	.24	.21	.19	.24	.21	.19	.23	.21	.19	.23	.20	.19	.22	.20	.19	.18
				9	.22	.19	.17	.22	.19	.17	.21	.19	.17	.21	.19	.17	.21	.18	.17	.16
				10	.20	.17	.15	.20	.17	.15	.20	.17	.15	.19	.17	.15	.19	.17	.15	.14

[a] ρ_{CC} = per cent effective ceiling cavity reflectance.
[b] ρ_W = per cent wall reflectance.
[c] RCR = Room Cavity Ratio.
[d] Maximum *S/MH* guide—ratio of maximum luminaire spacing to mounting or ceiling height above work-plane.

Fig. 9-12. *Continued*

Wall Luminance Coefficients[e] for 20 Per Cent Effective Floor Cavity Reflectance (ρ_{FC} = 20)

ρ_{CC} →	80			70			50			30			10		
ρ_W →	50	30	10	50	30	10	50	30	10	50	30	10	50	30	10
RCR ↓															
1	.19	.11	.03	.18	.10	.03	.17	.10	.03	.15	.09	.03	.14	.08	.03
2	.18	.10	.03	.17	.09	.03	.15	.09	.03	.14	.08	.02	.13	.07	.02
3	.16	.09	.03	.16	.08	.03	.14	.08	.02	.13	.07	.02	.12	.07	.02
4	.15	.08	.02	.15	.08	.02	.14	.07	.02	.12	.07	.02	.12	.06	.02
5	.14	.07	.02	.14	.07	.02	.13	.07	.02	.12	.06	.02	.11	.06	.02
6	.14	.07	.02	.13	.07	.02	.12	.06	.02	.11	.06	.02	.10	.05	.02
7	.13	.06	.02	.12	.06	.02	.12	.06	.02	.11	.05	.02	.10	.05	.02
8	.12	.06	.02	.12	.06	.02	.11	.05	.02	.10	.05	.01	.10	.05	.01
9	.12	.06	.02	.11	.05	.02	.11	.05	.01	.10	.05	.01	.09	.05	.01
10	.11	.05	.01	.11	.05	.01	.10	.05	.01	.09	.05	.01	.09	.04	.01
1	.22	.13	.04	.21	.12	.04	.19	.11	.03	.16	.09	.03	.14	.08	.03
2	.20	.11	.03	.19	.10	.03	.17	.09	.03	.15	.08	.03	.13	.07	.02
3	.18	.10	.03	.17	.09	.03	.15	.08	.02	.13	.07	.02	.12	.06	.02
4	.16	.09	.03	.15	.08	.02	.14	.07	.02	.12	.07	.02	.11	.06	.02
5	.15	.08	.02	.14	.07	.02	.13	.07	.02	.11	.06	.02	.10	.05	.02
6	.14	.07	.02	.13	.07	.02	.12	.06	.02	.11	.06	.02	.09	.05	.01
7	.13	.07	.02	.13	.06	.02	.11	.06	.02	.10	.05	.01	.09	.05	.01
8	.13	.06	.02	.12	.06	.02	.11	.05	.02	.09	.05	.01	.08	.04	.01
9	.12	.06	.02	.11	.06	.02	.10	.05	.01	.09	.05	.01	.08	.04	.01
10	.11	.05	.02	.11	.05	.01	.10	.05	.01	.09	.04	.01	.08	.04	.01
1	.15	.09	.03	.15	.08	.03	.14	.08	.02	.13	.07	.02	.12	.07	.02
2	.14	.08	.02	.13	.07	.02	.12	.07	.02	.11	.06	.02	.11	.06	.02
3	.12	.07	.02	.12	.06	.02	.11	.06	.02	.10	.06	.02	.10	.05	.02
4	.11	.06	.02	.11	.06	.02	.10	.05	.02	.09	.05	.02	.09	.05	.01
5	.11	.05	.02	.10	.05	.02	.09	.05	.01	.09	.05	.01	.08	.04	.01
6	.10	.05	.01	.09	.05	.01	.09	.05	.01	.08	.04	.01	.08	.04	.01
7	.09	.05	.01	.09	.04	.01	.08	.04	.01	.08	.04	.01	.07	.04	.01
8	.09	.04	.01	.08	.04	.01	.08	.04	.01	.07	.04	.01	.07	.03	.01
9	.08	.04	.01	.08	.04	.01	.07	.04	.01	.07	.03	.01	.06	.03	.01
10	.08	.04	.01	.07	.04	.01	.07	.03	.01	.06	.03	.01	.06	.03	.01
1	.11	.06	.02	.11	.06	.02	.10	.06	.02	.10	.06	.02	.09	.05	.02
2	.11	.06	.02	.10	.06	.02	.10	.06	.02	.09	.05	.02	.09	.05	.02
3	.10	.05	.02	.10	.05	.02	.09	.05	.02	.09	.05	.02	.09	.05	.01
4	.10	.05	.01	.09	.05	.01	.09	.05	.01	.09	.05	.01	.08	.04	.01
5	.09	.05	.01	.09	.05	.01	.09	.04	.01	.08	.04	.01	.08	.04	.01
6	.09	.04	.01	.08	.04	.01	.08	.04	.01	.08	.04	.01	.08	.04	.01
7	.08	.04	.01	.08	.04	.01	.08	.04	.01	.07	.04	.01	.07	.04	.01
8	.08	.04	.01	.08	.04	.01	.07	.04	.01	.07	.04	.01	.07	.04	.01
9	.07	.04	.01	.07	.04	.01	.07	.03	.01	.07	.03	.01	.07	.03	.01
10	.07	.03	.01	.07	.03	.01	.07	.03	.01	.07	.03	.01	.06	.03	.01
1	.10	.06	.02	.10	.05	.02	.09	.05	.02	.08	.05	.02	.08	.05	.01
2	.09	.05	.02	.09	.05	.02	.09	.05	.02	.08	.05	.01	.08	.04	.01
3	.09	.05	.01	.09	.05	.01	.08	.04	.01	.08	.04	.01	.08	.04	.01
4	.08	.04	.01	.08	.04	.01	.08	.04	.01	.08	.04	.01	.07	.04	.01
5	.08	.04	.01	.08	.04	.01	.08	.04	.01	.07	.04	.01	.07	.04	.01
6	.08	.04	.01	.07	.04	.01	.07	.04	.01	.07	.04	.01	.07	.04	.01
7	.07	.04	.01	.07	.04	.01	.07	.03	.01	.07	.03	.01	.06	.03	.01
8	.07	.03	.01	.07	.03	.01	.07	.03	.01	.06	.03	.01	.06	.03	.01
9	.07	.03	.01	.07	.03	.01	.06	.03	.01	.06	.03	.01	.06	.03	.01
10	.06	.03	.01	.06	.03	.01	.06	.03	.01	.06	.03	.01	.06	.03	.01

Ceiling Cavity Luminance Coefficients[e] for 20 Per Cent Effective Floor Cavity Reflectance (ρ_{FC} = 20)

ρ_{CC} →	80			70			50			30			10		
ρ_W →	50	30	10	50	30	10	50	30	10	50	30	10	50	30	10
RCR ↓															
0	.22	.22	.22	.18	.18	.18	.12	.12	.12	.07	.07	.07	.02	.02	.02
1	.21	.20	.18	.18	.17	.16	.12	.12	.11	.07	.07	.06	.02	.02	.02
2	.21	.18	.16	.18	.16	.14	.12	.11	.10	.07	.06	.06	.02	.02	.02
3	.20	.17	.14	.17	.15	.13	.12	.10	.09	.07	.06	.05	.02	.02	.02
4	.19	.16	.13	.17	.14	.12	.11	.10	.08	.07	.06	.05	.02	.02	.02
5	.19	.15	.13	.16	.13	.11	.11	.09	.08	.06	.05	.05	.02	.02	.02
6	.18	.15	.12	.16	.13	.10	.11	.09	.07	.06	.05	.04	.02	.02	.01
7	.18	.14	.11	.16	.12	.10	.11	.09	.07	.06	.05	.04	.02	.02	.01
8	.18	.14	.11	.15	.12	.10	.11	.08	.07	.06	.05	.04	.02	.02	.01
9	.17	.13	.11	.15	.12	.09	.10	.08	.07	.06	.05	.04	.02	.02	.01
10	.17	.13	.11	.15	.11	.09	.10	.08	.07	.06	.05	.04	.02	.02	.01
0	.32	.32	.32	.27	.27	.27	.19	.19	.19	.11	.11	.11	.03	.03	.03
1	.32	.30	.28	.27	.26	.24	.19	.18	.17	.11	.10	.10	.03	.03	.03
2	.31	.28	.26	.27	.24	.22	.18	.17	.16	.11	.10	.09	.03	.03	.03
3	.30	.27	.24	.26	.23	.21	.18	.16	.15	.10	.09	.09	.03	.03	.03
4	.30	.26	.23	.26	.23	.20	.18	.16	.14	.10	.09	.08	.03	.03	.03
5	.29	.25	.23	.25	.22	.20	.17	.15	.14	.10	.09	.08	.03	.03	.03
6	.29	.25	.22	.25	.22	.19	.17	.15	.14	.10	.09	.08	.03	.03	.03
7	.28	.24	.22	.24	.21	.19	.17	.15	.13	.10	.09	.08	.03	.03	.03
8	.28	.24	.21	.24	.21	.19	.17	.15	.13	.10	.09	.08	.03	.03	.03
9	.28	.24	.21	.24	.21	.18	.17	.14	.13	.10	.09	.08	.03	.03	.03
10	.27	.23	.21	.24	.20	.18	.16	.14	.13	.10	.08	.08	.03	.03	.03
0	.14	.14	.14	.12	.12	.12	.08	.08	.08	.05	.05	.05	.02	.02	.02
1	.14	.13	.12	.12	.11	.10	.08	.08	.07	.05	.04	.04	.02	.01	.01
2	.14	.12	.10	.12	.10	.09	.08	.07	.06	.05	.04	.04	.02	.01	.01
3	.14	.11	.09	.12	.10	.08	.08	.07	.06	.05	.04	.03	.01	.01	.01
4	.13	.11	.09	.11	.09	.08	.08	.06	.05	.05	.04	.03	.01	.01	.01
5	.13	.10	.08	.11	.09	.07	.08	.06	.05	.04	.04	.03	.01	.01	.01
6	.13	.10	.08	.11	.09	.07	.07	.06	.05	.04	.04	.03	.01	.01	.01
7	.12	.10	.08	.11	.08	.07	.07	.06	.05	.04	.03	.03	.01	.01	.01
8	.12	.09	.07	.10	.08	.07	.07	.06	.05	.04	.03	.03	.01	.01	.01
9	.12	.09	.07	.10	.08	.06	.07	.06	.05	.04	.03	.03	.01	.01	.01
10	.12	.09	.07	.10	.08	.06	.07	.05	.04	.04	.03	.03	.01	.01	.01
0	.08	.08	.08	.07	.07	.07	.05	.05	.05	.03	.03	.03	.01	.01	.01
1	.08	.07	.06	.07	.06	.06	.05	.04	.04	.03	.02	.02	.01	.01	.01
2	.08	.06	.05	.06	.05	.04	.04	.04	.03	.03	.02	.02	.01	.01	.01
3	.07	.05	.04	.06	.05	.03	.04	.03	.02	.02	.02	.01	.01	.01	.00
4	.07	.05	.03	.06	.04	.03	.04	.03	.02	.02	.02	.01	.01	.01	.00
5	.06	.04	.02	.06	.04	.02	.04	.02	.01	.02	.01	.01	.01	.00	.00
6	.06	.04	.02	.05	.03	.02	.04	.02	.01	.02	.01	.01	.01	.00	.00
7	.06	.03	.02	.05	.03	.01	.03	.02	.01	.02	.01	.01	.01	.00	.00
8	.06	.03	.01	.05	.03	.01	.03	.02	.01	.02	.01	.01	.01	.00	.00
9	.05	.03	.01	.05	.03	.01	.03	.02	.01	.02	.01	.00	.01	.00	.00
10	.05	.03	.01	.04	.02	.01	.03	.02	.01	.02	.01	.00	.01	.00	.00
0	.08	.08	.08	.06	.06	.06	.04	.04	.04	.03	.03	.03	.01	.01	.01
1	.07	.07	.06	.06	.06	.05	.04	.04	.04	.03	.02	.02	.01	.01	.01
2	.07	.06	.05	.06	.05	.04	.04	.03	.03	.02	.02	.02	.01	.01	.01
3	.07	.05	.04	.06	.04	.03	.04	.03	.02	.02	.02	.01	.01	.01	.00
4	.06	.04	.03	.05	.04	.02	.04	.03	.02	.02	.02	.01	.01	.00	.00
5	.06	.04	.02	.05	.03	.02	.03	.02	.01	.02	.01	.01	.01	.00	.00
6	.06	.03	.02	.05	.03	.02	.03	.02	.01	.02	.01	.01	.01	.00	.00
7	.05	.03	.02	.05	.03	.01	.03	.02	.01	.02	.01	.01	.01	.00	.00
8	.05	.03	.01	.04	.02	.01	.03	.02	.01	.02	.01	.01	.01	.00	.00
9	.05	.03	.01	.04	.02	.01	.03	.02	.01	.02	.01	.00	.01	.00	.00
10	.05	.02	.01	.04	.02	.01	.03	.02	.01	.02	.01	.00	.01	.00	.00

[e] Although it is recommended that luminance coefficients and wall direct radiation coefficients be published to three decimal places, only two are shown here for these typical luminaires. Three decimal place data should be obtained from manufacturers of actual luminaires used.

Fig. 9–12. *Continued*

Typical Luminaire	Maint. Cat.	Maximum S/MH Guide[d]	Typical Distribution and Per Cent Lamp Lumens	ρ_{CC}[a] → / ρ_W[b] → / RCR[c] ↓	80			70			50			30			10			0
					50	30	10	50	30	10	50	30	10	50	30	10	50	30	10	0
					Coefficients of Utilization for 20 Per Cent Effective Floor Cavity Reflectance (ρ_{FC} = 20)															
35	IV	1.0		0	.59	.59	.59	.58	.58	.58	.55	.55	.55	.53	.53	.53	.51	.51	.51	.50
				1	.54	.52	.50	.52	.51	.49	.50	.49	.48	.48	.47	.46	.47	.46	.45	.44
			0%↑	2	.48	.45	.43	.47	.44	.42	.45	.43	.41	.44	.42	.40	.42	.41	.39	.39
				3	.43	.40	.37	.42	.39	.37	.41	.38	.36	.40	.37	.36	.39	.37	.35	.34
				4	.39	.35	.32	.38	.35	.32	.37	.34	.32	.36	.33	.31	.35	.33	.31	.30
			50%↓	5	.35	.31	.28	.35	.31	.28	.34	.30	.28	.33	.30	.28	.32	.29	.27	.26
				6	.32	.28	.25	.32	.28	.25	.31	.27	.25	.30	.27	.25	.29	.26	.24	.23
4 lamp, 2′ wide troffer with 45° plastic louver—multiply by 1.05 for 2 lamps and 0.95 for 6 lamps				7	.29	.25	.22	.29	.25	.22	.28	.25	.22	.27	.24	.22	.27	.24	.22	.21
				8	.26	.22	.20	.26	.22	.20	.25	.22	.20	.25	.22	.19	.24	.21	.19	.18
				9	.24	.20	.17	.24	.20	.17	.23	.20	.17	.23	.19	.17	.22	.19	.17	.16
				10	.22	.18	.16	.22	.18	.16	.21	.18	.16	.21	.18	.15	.20	.17	.15	.15
36	IV	0.9		0	.49	.49	.49	.48	.48	.48	.46	.46	.46	.44	.44	.44	.42	.42	.42	.41
				1	.45	.44	.43	.44	.43	.42	.43	.42	.41	.41	.40	.39	.40	.39	.38	.38
			0%↑	2	.41	.39	.37	.40	.38	.36	.39	.37	.36	.37	.36	.35	.36	.35	.34	.33
				3	.37	.35	.32	.37	.34	.32	.35	.33	.32	.34	.33	.31	.33	.32	.31	.30
				4	.34	.31	.29	.33	.31	.29	.32	.30	.28	.32	.29	.28	.31	.29	.27	.27
				5	.31	.28	.25	.30	.27	.25	.30	.27	.25	.29	.27	.25	.28	.26	.25	.24
			42%↓	6	.28	.25	.23	.28	.25	.23	.27	.25	.23	.27	.24	.22	.26	.24	.22	.22
4 lamp, 2′ wide troffer with 45° white metal louver—multiply by 1.05 for 2 lamps and 0.95 for 6 lamps				7	.26	.23	.21	.26	.23	.21	.25	.22	.20	.24	.22	.20	.24	.22	.20	.19
				8	.24	.21	.18	.23	.21	.18	.23	.20	.18	.22	.20	.18	.22	.20	.18	.17
				9	.22	.19	.17	.21	.18	.16	.21	.18	.16	.21	.18	.16	.20	.18	.16	.16
				10	.20	.17	.15	.20	.17	.15	.19	.17	.15	.19	.17	.15	.19	.16	.15	.14
37	V	1.2		0	.66	.66	.66	.64	.64	.64	.61	.61	.61	.58	.58	.58	.56	.56	.56	.54
				1	.58	.56	.53	.57	.54	.52	.54	.52	.51	.52	.50	.49	.49	.48	.47	.46
				2	.51	.47	.44	.50	.46	.43	.48	.45	.42	.46	.43	.41	.44	.42	.40	.39
			1%↑	3	.45	.41	.37	.44	.40	.37	.42	.39	.36	.41	.38	.35	.39	.36	.34	.33
				4	.40	.35	.32	.39	.35	.31	.38	.34	.31	.36	.33	.30	.35	.32	.30	.28
			55%↓	5	.35	.30	.27	.35	.30	.27	.33	.29	.26	.32	.29	.26	.31	.28	.25	.24
				6	.32	.27	.23	.31	.26	.23	.30	.26	.23	.29	.25	.22	.28	.25	.22	.21
Fluorescent unit with dropped white diffuser, 2 lamp 1′ wide—multiply by 0.90 for 3 lamps				7	.29	.24	.20	.28	.23	.20	.27	.23	.20	.26	.22	.20	.25	.22	.19	.18
				8	.26	.21	.18	.25	.21	.17	.24	.20	.17	.24	.20	.17	.23	.19	.17	.16
				9	.23	.18	.15	.23	.18	.15	.22	.18	.15	.21	.17	.15	.21	.17	.15	.14
				10	.21	.16	.13	.21	.16	.13	.20	.16	.13	.19	.16	.13	.19	.15	.13	.12
38	V	1.2		0	.72	.72	.72	.70	.70	.70	.67	.67	.67	.64	.64	.64	.61	.61	.61	.60
				1	.64	.61	.59	.62	.60	.58	.60	.58	.56	.57	.56	.54	.55	.54	.52	.51
				2	.56	.52	.49	.55	.51	.48	.52	.49	.47	.50	.48	.46	.48	.46	.44	.43
			1%↑	3	.50	.45	.41	.49	.44	.41	.47	.43	.40	.45	.42	.39	.43	.41	.38	.37
				4	.44	.39	.35	.43	.38	.35	.42	.37	.34	.40	.36	.33	.39	.36	.33	.32
				5	.39	.34	.30	.38	.33	.29	.37	.32	.29	.36	.32	.29	.34	.31	.28	.27
			60½%↓	6	.35	.30	.26	.34	.29	.25	.33	.29	.25	.32	.28	.25	.31	.27	.25	.23
Fluorescent unit with dropped white diffuser, 4 lamp 2′ wide—multiply by 1.10 for 2 lamps and 0.90 for 6 lamps				7	.31	.26	.22	.31	.26	.22	.30	.25	.22	.29	.25	.22	.28	.24	.22	.20
				8	.28	.23	.19	.28	.23	.19	.27	.22	.19	.26	.22	.19	.25	.22	.19	.18
				9	.25	.20	.17	.25	.20	.17	.24	.20	.17	.23	.19	.16	.23	.19	.16	.15
				10	.23	.18	.15	.23	.18	.15	.22	.18	.15	.21	.17	.15	.21	.17	.14	.13
39	V	1.2		0	.68	.68	.68	.66	.66	.66	.63	.63	.63	.61	.61	.61	.58	.58	.58	.57
				1	.61	.58	.56	.59	.57	.56	.57	.55	.54	.55	.53	.52	.53	.52	.51	.49
				2	.53	.50	.47	.52	.49	.46	.50	.48	.45	.49	.46	.44	.47	.45	.43	.42
			0%↑	3	.47	.43	.40	.47	.42	.39	.45	.41	.38	.43	.40	.38	.42	.39	.37	.36
				4	.42	.37	.34	.41	.37	.33	.40	.36	.33	.39	.35	.33	.37	.35	.32	.31
			57½%↓	5	.37	.32	.29	.37	.32	.28	.35	.31	.28	.34	.31	.28	.33	.30	.27	.26
Fluorescent unit with flat bottom white diffuser, 4 lamp 2′ and 2 lamp 1′ wide—multiply by 1.08 for 2 lamp 2′ and by 0.92 for 3 lamp 1′				6	.33	.28	.25	.33	.28	.25	.32	.28	.24	.31	.27	.24	.30	.27	.24	.23
				7	.30	.25	.22	.30	.25	.21	.29	.24	.21	.28	.24	.21	.27	.24	.21	.20
				8	.27	.22	.19	.27	.22	.19	.26	.22	.18	.25	.21	.18	.24	.21	.18	.17
				9	.24	.19	.16	.24	.19	.16	.23	.19	.16	.23	.19	.16	.22	.18	.16	.15
				10	.22	.17	.14	.22	.17	.14	.21	.17	.14	.21	.17	.14	.20	.17	.14	.13

[a] ρ_{CC} = per cent effective ceiling cavity reflectance.

[b] ρ_W = per cent wall reflectance.

[c] RCR = Room Cavity Ratio.

[d] Maximum S/MH guide—ratio of maximum luminaire spacing to mounting or ceiling height above work-plane.

Fig. 9–12. *Continued*

Wall Luminance Coefficients[e] for 20 Per Cent Effective Floor Cavity Reflectance (ρ_{FC} = 20)

ρ_{CC} →	80			70			50			30			10		
ρ_W → / RCR ↓	50	30	10	50	30	10	50	30	10	50	30	10	50	30	10
1	.13	.07	.02	.12	.07	.02	.12	.07	.02	.11	.06	.02	.10	.06	.02
2	.12	.07	.02	.12	.06	.02	.11	.06	.02	.11	.06	.02	.10	.06	.02
3	.11	.06	.02	.11	.06	.02	.11	.06	.02	.10	.06	.02	.10	.05	.02
4	.11	.06	.02	.10	.05	.02	.10	.05	.02	.10	.05	.02	.09	.05	.02
5	.10	.05	.01	.10	.05	.01	.10	.05	.01	.09	.05	.01	.09	.05	.01
6	.10	.05	.01	.09	.05	.01	.09	.05	.01	.09	.04	.01	.08	.04	.01
7	.09	.04	.01	.09	.04	.01	.09	.04	.01	.08	.04	.01	.08	.04	.01
8	.09	.04	.01	.08	.04	.01	.08	.04	.01	.08	.04	.01	.08	.04	.01
9	.08	.04	.01	.08	.04	.01	.08	.04	.01	.08	.04	.01	.07	.04	.01
10	.08	.04	.01	.08	.04	.01	.07	.04	.01	.07	.04	.01	.07	.04	.01
1	.10	.06	.02	.09	.05	.02	.09	.05	.02	.08	.05	.02	.08	.05	.01
2	.09	.05	.02	.09	.05	.02	.09	.05	.01	.08	.05	.01	.08	.04	.01
3	.09	.05	.01	.09	.05	.01	.08	.04	.01	.08	.04	.01	.07	.04	.01
4	.08	.04	.01	.08	.04	.01	.08	.04	.01	.07	.04	.01	.07	.04	.01
5	.08	.04	.01	.08	.04	.01	.07	.04	.01	.07	.04	.01	.07	.04	.01
6	.07	.04	.01	.07	.04	.01	.07	.04	.01	.07	.04	.01	.07	.03	.01
7	.07	.04	.01	.07	.03	.01	.07	.03	.01	.06	.03	.01	.06	.03	.01
8	.07	.03	.01	.07	.03	.01	.06	.03	.01	.06	.03	.01	.06	.03	.01
9	.07	.03	.01	.06	.03	.01	.06	.03	.01	.06	.03	.01	.06	.03	.01
10	.06	.03	.01	.06	.03	.01	.06	.03	.01	.06	.03	.01	.06	.03	.01
1	.17	.10	.03	.17	.10	.03	.16	.09	.03	.15	.09	.03	.15	.08	.03
2	.16	.09	.03	.16	.09	.03	.15	.08	.03	.14	.08	.02	.14	.08	.02
3	.15	.08	.02	.14	.08	.02	.14	.07	.02	.13	.07	.02	.12	.07	.02
4	.14	.07	.02	.13	.07	.02	.13	.07	.02	.12	.07	.02	.12	.06	.02
5	.13	.06	.02	.12	.06	.02	.12	.06	.02	.11	.06	.02	.11	.06	.02
6	.12	.06	.02	.12	.06	.02	.11	.06	.02	.11	.06	.02	.10	.05	.02
7	.11	.05	.02	.11	.05	.02	.10	.05	.02	.10	.05	.02	.10	.05	.01
8	.10	.05	.01	.10	.05	.01	.10	.05	.01	.10	.05	.01	.09	.05	.01
9	.10	.05	.01	.10	.05	.01	.09	.05	.01	.09	.05	.01	.09	.04	.01
10	.09	.04	.01	.09	.04	.01	.09	.04	.01	.09	.04	.01	.08	.04	.01
1	.18	.10	.03	.18	.10	.03	.17	.10	.03	.16	.09	.03	.15	.09	.03
2	.17	.09	.03	.17	.09	.03	.16	.09	.03	.15	.09	.03	.15	.08	.03
3	.16	.08	.03	.15	.08	.02	.15	.08	.02	.14	.08	.02	.14	.08	.02
4	.15	.08	.02	.14	.08	.02	.14	.07	.02	.13	.07	.02	.13	.07	.02
5	.14	.07	.02	.14	.07	.02	.13	.07	.02	.13	.07	.02	.12	.06	.02
6	.13	.06	.02	.13	.06	.02	.12	.06	.02	.12	.06	.02	.11	.06	.02
7	.12	.06	.02	.12	.06	.02	.11	.06	.02	.11	.06	.02	.11	.06	.02
8	.11	.06	.02	.11	.06	.02	.11	.05	.02	.10	.05	.02	.10	.05	.02
9	.11	.05	.01	.11	.05	.01	.10	.05	.01	.10	.05	.01	.10	.05	.01
10	.10	.05	.01	.10	.05	.01	.10	.05	.01	.09	.05	.01	.09	.05	.01
1	.16	.09	.03	.16	.09	.03	.15	.09	.03	.14	.08	.03	.14	.08	.03
2	.15	.08	.03	.15	.08	.03	.14	.08	.02	.14	.08	.02	.13	.08	.02
3	.14	.08	.02	.14	.08	.02	.14	.07	.02	.13	.07	.02	.13	.07	.02
4	.14	.07	.02	.13	.07	.02	.13	.07	.02	.12	.07	.02	.12	.06	.02
5	.13	.07	.02	.13	.06	.02	.12	.06	.02	.12	.06	.02	.11	.06	.02
6	.12	.06	.02	.12	.06	.02	.11	.06	.02	.11	.06	.02	.11	.06	.02
7	.11	.06	.02	.11	.06	.02	.11	.05	.02	.10	.05	.02	.10	.05	.02
8	.11	.05	.01	.11	.05	.01	.10	.05	.01	.10	.05	.01	.10	.05	.01
9	.10	.05	.01	.10	.05	.01	.10	.05	.01	.09	.05	.01	.09	.05	.01
10	.10	.05	.01	.09	.05	.01	.09	.04	.01	.09	.04	.01	.09	.04	.01

Ceiling Cavity Luminance Coefficients[e] for 20 Per Cent Effective Floor Cavity Reflectance (ρ_{FC} = 20)

ρ_{CC} →	80			70			50			30			10		
ρ_W → / RCR ↓	50	30	10	50	30	10	50	30	10	50	30	10	50	30	10
0	.09	.09	.09	.08	.08	.08	.05	.05	.05	.03	.03	.03	.01	.01	.01
1	.09	.08	.07	.08	.07	.06	.05	.05	.04	.03	.03	.02	.01	.01	.01
2	.08	.07	.05	.07	.06	.05	.05	.04	.03	.03	.02	.02	.01	.01	.01
3	.08	.06	.04	.07	.05	.03	.05	.03	.02	.03	.02	.01	.01	.01	.00
4	.07	.05	.03	.06	.04	.03	.04	.03	.02	.03	.02	.01	.01	.01	.00
5	.07	.04	.03	.06	.04	.02	.04	.03	.02	.02	.02	.01	.01	.01	.00
6	.07	.04	.02	.06	.04	.02	.04	.02	.01	.02	.01	.01	.01	.00	.00
7	.06	.04	.02	.06	.03	.02	.04	.02	.01	.02	.01	.01	.01	.00	.00
8	.06	.03	.02	.05	.03	.01	.04	.02	.01	.02	.01	.01	.01	.00	.00
9	.06	.03	.01	.05	.03	.01	.04	.02	.01	.02	.01	.00	.01	.00	.00
10	.06	.03	.01	.05	.03	.01	.03	.02	.01	.02	.01	.00	.01	.00	.00
0	.07	.07	.07	.06	.06	.06	.04	.04	.04	.02	.02	.02	.01	.01	.01
1	.07	.07	.06	.06	.06	.05	.04	.04	.04	.02	.02	.02	.01	.01	.01
2	.07	.06	.04	.06	.05	.04	.04	.03	.03	.02	.02	.02	.01	.01	.01
3	.06	.05	.03	.05	.04	.03	.04	.03	.02	.02	.02	.01	.01	.01	.00
4	.06	.04	.03	.05	.04	.02	.04	.03	.02	.02	.01	.01	.01	.00	.00
5	.06	.04	.02	.05	.03	.02	.03	.02	.01	.02	.01	.01	.01	.00	.00
6	.05	.03	.02	.05	.03	.02	.03	.02	.01	.02	.01	.01	.01	.00	.00
7	.05	.03	.02	.04	.03	.01	.03	.02	.01	.02	.01	.01	.01	.00	.00
8	.05	.03	.01	.04	.02	.01	.03	.02	.01	.02	.01	.00	.01	.00	.00
9	.05	.03	.01	.04	.02	.01	.03	.02	.01	.02	.01	.00	.01	.00	.00
10	.05	.02	.01	.04	.02	.01	.03	.01	.01	.02	.01	.00	.01	.00	.00
0	.11	.11	.11	.09	.09	.09	.06	.06	.06	.04	.04	.04	.01	.01	.01
1	.11	.10	.08	.09	.08	.07	.06	.06	.05	.04	.03	.03	.01	.01	.01
2	.11	.08	.07	.09	.07	.06	.06	.05	.04	.04	.03	.02	.01	.01	.01
3	.10	.07	.05	.09	.06	.05	.06	.04	.03	.03	.03	.02	.01	.01	.01
4	.10	.07	.04	.08	.06	.04	.06	.04	.03	.03	.02	.02	.01	.01	.01
5	.09	.06	.04	.08	.05	.03	.05	.04	.02	.03	.02	.01	.01	.01	.00
6	.09	.06	.03	.08	.05	.03	.05	.03	.02	.03	.02	.01	.01	.01	.00
7	.09	.05	.03	.07	.05	.02	.05	.03	.02	.03	.02	.01	.01	.01	.00
8	.08	.05	.03	.07	.04	.02	.05	.03	.02	.03	.02	.01	.01	.01	.00
9	.08	.05	.02	.07	.04	.02	.05	.03	.01	.03	.02	.01	.01	.01	.00
10	.08	.04	.02	.07	.04	.02	.05	.03	.01	.03	.02	.01	.01	.01	.00
0	.12	.12	.12	.10	.10	.10	.07	.07	.07	.04	.04	.04	.01	.01	.01
1	.12	.10	.09	.10	.09	.08	.07	.06	.05	.04	.03	.03	.01	.01	.01
2	.11	.09	.07	.09	.08	.06	.07	.05	.04	.04	.03	.02	.01	.01	.01
3	.11	.08	.05	.09	.07	.05	.06	.05	.03	.04	.03	.02	.01	.01	.01
4	.10	.07	.04	.09	.06	.04	.06	.04	.03	.03	.02	.01	.01	.01	.01
5	.10	.06	.04	.08	.05	.03	.06	.04	.02	.03	.02	.01	.01	.01	.00
6	.09	.06	.03	.08	.05	.03	.06	.03	.02	.03	.02	.01	.01	.01	.00
7	.09	.05	.03	.08	.05	.02	.05	.03	.02	.03	.02	.01	.01	.01	.00
8	.09	.05	.02	.07	.04	.02	.05	.03	.01	.03	.02	.01	.01	.01	.00
9	.08	.05	.02	.07	.04	.02	.05	.03	.01	.03	.02	.01	.01	.01	.00
10	.08	.04	.02	.07	.04	.02	.05	.03	.01	.03	.02	.01	.01	.01	.00
0	.10	.10	.10	.09	.09	.09	.06	.06	.06	.03	.03	.03	.01	.01	.01
1	.10	.09	.08	.09	.08	.07	.06	.05	.05	.03	.03	.03	.01	.01	.01
2	.10	.08	.06	.08	.07	.05	.06	.05	.03	.03	.03	.02	.01	.01	.01
3	.09	.07	.04	.08	.06	.04	.05	.04	.03	.03	.02	.02	.01	.01	.01
4	.09	.06	.03	.08	.05	.03	.05	.03	.02	.03	.02	.01	.01	.01	.00
5	.08	.05	.03	.07	.04	.02	.05	.03	.02	.03	.02	.01	.01	.01	.00
6	.08	.05	.02	.07	.04	.02	.05	.03	.01	.03	.02	.01	.01	.01	.00
7	.08	.04	.02	.07	.04	.02	.05	.03	.01	.03	.02	.01	.01	.01	.00
8	.07	.04	.02	.06	.03	.01	.04	.02	.01	.03	.01	.01	.01	.00	.00
9	.07	.04	.01	.06	.03	.01	.04	.02	.01	.02	.01	.01	.01	.00	.00
10	.07	.03	.01	.06	.03	.01	.04	.02	.01	.02	.01	.00	.01	.00	.00

[e] Although it is recommended that luminance coefficients and wall direct radiation coefficients be published to three decimal places, only two are shown here for these typical luminaires. Three decimal place data should be obtained from manufacturers of actual luminaires used.

Fig. 9–12. *Continued*

Typical Luminaire	Typical Distribution and Per Cent Lamp Lumens	Maint. Cat.	Maximum S/MH Guide[d]	ρ_{CC}[a] →	80			70			50			30			10			0
				ρ_W[b] →	50	30	10	50	30	10	50	30	10	50	30	10	50	30	10	0
				RCR[c] ↓	Coefficients of Utilization for 20 Per Cent Effective Floor Cavity Reflectance (ρ_{FC} = 20)															
40 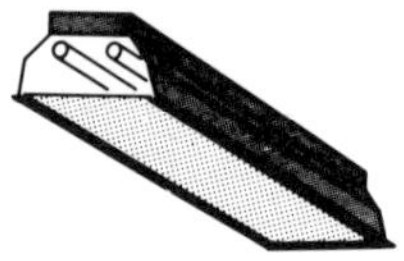		V	1.4/1.2	0	.69	.69	.69	.67	.67	.67	.64	.64	.64	.61	.61	.61	.59	.59	.59	.58
				1	.62	.60	.58	.61	.59	.57	.58	.57	.55	.56	.55	.54	.54	.53	.52	.51
				2	.55	.52	.49	.54	.51	.49	.52	.50	.48	.51	.48	.47	.49	.47	.46	.44
	0% ↑			3	.50	.46	.43	.49	.45	.42	.47	.44	.41	.46	.43	.41	.44	.42	.40	.39
				4	.45	.40	.37	.44	.40	.37	.43	.39	.36	.41	.38	.36	.40	.37	.35	.34
	58% ↓ 60°			5	.40	.35	.32	.39	.35	.32	.38	.34	.31	.37	.34	.31	.36	.33	.31	.30
				6	.36	.31	.28	.36	.31	.28	.35	.31	.28	.34	.30	.27	.33	.30	.27	.26
				7	.33	.28	.25	.32	.28	.24	.31	.27	.24	.30	.27	.24	.30	.26	.24	.23
				8	.29	.25	.21	.29	.24	.21	.28	.24	.21	.27	.24	.21	.27	.23	.21	.20
Fluorescent unit with flat prismatic lens, 2 lamp 1′ wide				9	.26	.22	.18	.26	.21	.18	.25	.21	.18	.25	.21	.18	.24	.21	.18	.17
				10	.24	.19	.16	.24	.19	.16	.23	.19	.16	.22	.19	.16	.22	.19	.16	.15
41 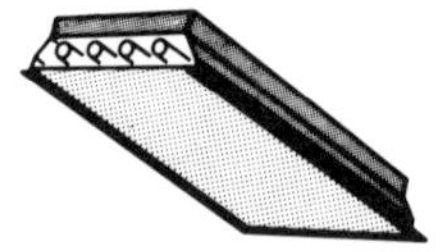		V	1.4/1.2	0	.73	.73	.73	.72	.72	.72	.68	.68	.68	.66	.66	.66	.63	.63	.63	.62
				1	.66	.64	.62	.65	.63	.61	.62	.60	.59	.60	.58	.57	.57	.56	.55	.54
				2	.59	.55	.52	.58	.54	.52	.56	.53	.50	.54	.51	.49	.52	.50	.48	.47
	0% ↑			3	.53	.48	.45	.52	.48	.44	.50	.46	.44	.48	.45	.43	.47	.44	.42	.41
				4	.47	.42	.39	.46	.42	.38	.45	.41	.38	.43	.40	.37	.42	.39	.37	.36
	62% ↓ 60°			5	.42	.37	.33	.41	.37	.33	.40	.36	.33	.39	.35	.32	.38	.35	.32	.31
				6	.38	.33	.29	.37	.32	.29	.36	.32	.29	.35	.31	.28	.34	.31	.28	.27
				7	.34	.29	.25	.33	.29	.25	.33	.28	.25	.32	.28	.25	.31	.27	.25	.23
Fluorescent unit with flat prismatic lens, 4 lamp 2′ wide—multiply by 1.10 for 2 lamp				8	.30	.25	.22	.30	.25	.22	.29	.25	.22	.28	.24	.21	.28	.24	.21	.20
				9	.27	.22	.19	.27	.22	.19	.26	.22	.19	.25	.21	.19	.25	.21	.18	.17
				10	.25	.20	.17	.24	.20	.16	.24	.19	.16	.23	.19	.16	.23	.19	.16	.15
42 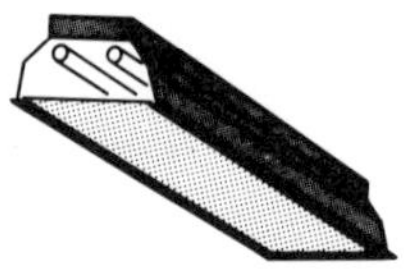		V	1.4/1.2	0	.66	.66	.66	.65	.65	.65	.62	.62	.62	.59	.59	.59	.57	.57	.57	.56
				1	.61	.59	.57	.59	.58	.56	.57	.56	.54	.55	.54	.53	.53	.52	.51	.50
				2	.55	.52	.50	.54	.51	.49	.52	.50	.48	.50	.48	.47	.49	.47	.46	.45
	0% ↑			3	.50	.46	.43	.49	.46	.43	.47	.45	.42	.46	.44	.42	.45	.43	.41	.40
				4	.45	.41	.38	.45	.41	.38	.43	.40	.38	.42	.39	.37	.41	.39	.37	.36
	56% ↓ 60°			5	.41	.37	.34	.40	.36	.34	.39	.36	.33	.38	.35	.33	.37	.35	.33	.32
				6	.37	.33	.30	.37	.33	.30	.36	.32	.30	.35	.32	.29	.34	.31	.29	.28
				7	.34	.30	.27	.34	.29	.27	.33	.29	.26	.32	.29	.26	.31	.28	.26	.25
				8	.31	.26	.24	.30	.26	.23	.30	.26	.23	.29	.26	.23	.28	.25	.23	.22
Fluorescent unit with flat prismatic lens, 2 lamp 1′ wide				9	.28	.23	.21	.27	.23	.21	.27	.23	.20	.26	.23	.20	.26	.23	.20	.19
				10	.25	.21	.18	.25	.21	.18	.24	.21	.18	.24	.21	.18	.23	.20	.18	.17
43 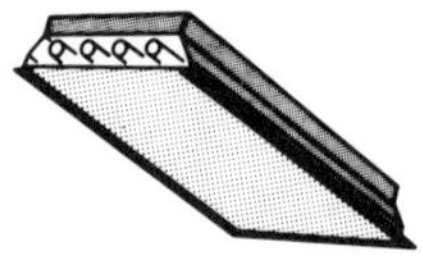		V	1.4/1.2	0	.71	.71	.71	.69	.69	.69	.66	.66	.66	.63	.63	.63	.61	.61	.61	.60
				1	.65	.63	.61	.63	.62	.60	.61	.59	.58	.59	.57	.56	.57	.56	.55	.54
				2	.59	.55	.53	.57	.55	.52	.55	.53	.51	.54	.52	.50	.52	.50	.49	.48
	0% ↑			3	.53	.49	.46	.52	.49	.46	.50	.47	.45	.49	.46	.44	.47	.45	.43	.42
				4	.48	.44	.40	.47	.43	.40	.46	.42	.40	.45	.42	.39	.43	.41	.39	.38
	60% ↓ 60°			5	.43	.39	.35	.43	.38	.35	.42	.38	.35	.40	.37	.34	.39	.36	.34	.33
				6	.39	.35	.31	.39	.34	.31	.38	.34	.31	.37	.33	.31	.36	.33	.31	.29
				7	.36	.31	.28	.35	.31	.28	.34	.30	.27	.33	.30	.27	.33	.30	.27	.26
Fluorescent unit with flat prismatic lens, 4 lamp 2′ wide—multiply by 1.10 for 2 lamp				8	.32	.27	.24	.32	.27	.24	.31	.27	.24	.30	.27	.24	.30	.26	.24	.23
				9	.29	.24	.21	.29	.24	.21	.28	.24	.21	.27	.24	.21	.27	.23	.21	.20
				10	.26	.22	.19	.26	.22	.19	.25	.21	.19	.25	.21	.19	.24	.21	.18	.17
44				0	.65	.65	.65	.63	.63	.63	.60	.60	.60	.58	.58	.58	.55	.55	.55	.54
				1	.58	.56	.54	.56	.54	.53	.54	.53	.51	.52	.51	.49	.50	.49	.48	.47
	0% ↑			2	.51	.47	.44	.50	.47	.44	.48	.45	.43	.46	.44	.42	.45	.43	.41	.40
				3	.45	.41	.37	.44	.40	.37	.43	.39	.36	.41	.38	.36	.40	.37	.35	.34
	55% ↓			4	.40	.35	.32	.39	.35	.32	.38	.34	.31	.37	.33	.31	.35	.33	.30	.29
At press time, "Batwing" fluorescent units are new and evolving rapidly. These coefficients are included to indicate form; more recent data is preferable	45°			5	.35	.30	.27	.35	.30	.27	.33	.29	.26	.32	.29	.26	.31	.28	.26	.25
				6	.31	.26	.23	.31	.26	.23	.30	.26	.23	.29	.25	.22	.28	.25	.22	.21
				7	.28	.23	.19	.27	.23	.19	.26	.22	.19	.26	.22	.19	.25	.22	.19	.18
				8	.25	.20	.16	.24	.20	.16	.23	.19	.16	.23	.19	.16	.22	.19	.16	.15
				9	.22	.17	.14	.21	.17	.14	.21	.17	.14	.20	.16	.14	.19	.16	.13	.12
				10	.19	.15	.12	.19	.15	.12	.19	.15	.12	.18	.14	.12	.18	.14	.12	.11

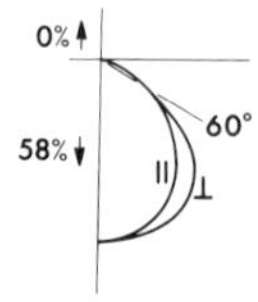

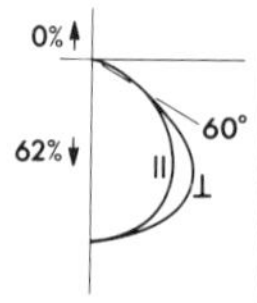

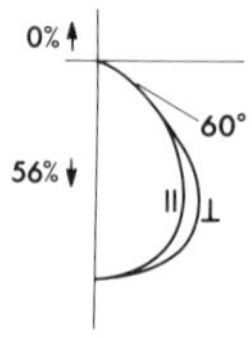

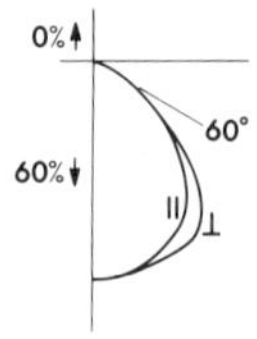

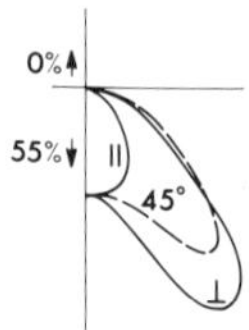

[a] ρ_{CC} = per cent effective ceiling cavity reflectance.
[b] ρ_W = per cent wall reflectance.
[c] RCR = Room Cavity Ratio.
[d] Maximum S/MH guide—ratio of maximum luminaire spacing to mounting or ceiling height above work-plane.

Fig. 9–12. *Continued*

Wall Luminance Coefficients[e] for 20 Per Cent Effective Floor Cavity Reflectance (ρ_{FC} = 20)

ρ_{CC} →	80			70			50			30			10		
ρ_W →	50	30	10	50	30	10	50	30	10	50	30	10	50	30	10
RCR ↓															
1	.15	.08	.03	.14	.08	.03	.14	.08	.03	.13	.08	.02	.12	.07	.02
2	.14	.08	.02	.14	.08	.02	.13	.07	.02	.12	.07	.02	.12	.07	.02
3	.13	.07	.02	.13	.07	.02	.12	.07	.02	.12	.07	.02	.11	.06	.02
4	.13	.07	.02	.12	.06	.02	.12	.06	.02	.11	.06	.02	.11	.06	.02
5	.12	.06	.02	.12	.06	.02	.11	.06	.02	.11	.06	.02	.10	.06	.02
6	.11	.06	.02	.11	.06	.02	.11	.05	.02	.10	.05	.02	.10	.05	.02
7	.11	.05	.02	.11	.05	.01	.10	.05	.01	.10	.05	.01	.10	.05	.01
8	.10	.05	.01	.10	.05	.01	.10	.05	.01	.09	.05	.01	.09	.05	.01
9	.10	.05	.01	.10	.05	.01	.09	.05	.01	.09	.05	.01	.09	.04	.01
10	.09	.04	.01	.09	.04	.01	.09	.04	.01	.09	.04	.01	.08	.04	.01
1	.16	.09	.03	.16	.09	.03	.15	.09	.03	.14	.08	.03	.14	.08	.03
2	.15	.08	.03	.15	.08	.03	.14	.08	.02	.14	.08	.02	.13	.07	.02
3	.15	.08	.02	.14	.08	.02	.14	.07	.02	.13	.07	.02	.13	.07	.02
4	.14	.07	.02	.13	.07	.02	.13	.07	.02	.12	.07	.02	.12	.06	.02
5	.13	.07	.02	.13	.07	.02	.12	.06	.02	.12	.06	.02	.11	.06	.02
6	.12	.06	.02	.12	.06	.02	.12	.06	.02	.11	.06	.02	.11	.06	.02
7	.12	.06	.02	.12	.06	.02	.11	.06	.02	.11	.06	.02	.10	.05	.02
8	.11	.05	.02	.11	.05	.02	.11	.05	.02	.10	.05	.02	.10	.05	.01
9	.11	.05	.01	.10	.05	.01	.10	.05	.01	.10	.05	.01	.10	.05	.01
10	.10	.05	.01	.10	.05	.01	.10	.05	.01	.09	.05	.01	.09	.05	.01
1	.13	.07	.02	.12	.07	.02	.12	.07	.02	.11	.06	.02	.10	.06	.02
2	.12	.07	.02	.12	.07	.02	.11	.06	.02	.11	.06	.02	.10	.06	.02
3	.12	.06	.02	.11	.06	.02	.11	.06	.02	.10	.06	.02	.10	.05	.02
4	.11	.06	.02	.11	.06	.02	.10	.06	.02	.10	.05	.02	.10	.05	.02
5	.11	.05	.02	.10	.05	.02	.10	.05	.02	.10	.05	.02	.09	.05	.01
6	.10	.05	.01	.10	.05	.01	.10	.05	.01	.09	.05	.01	.09	.05	.01
7	.10	.05	.01	.09	.05	.01	.09	.05	.01	.09	.05	.01	.09	.04	.01
8	.09	.05	.01	.09	.04	.01	.09	.04	.01	.09	.04	.01	.08	.04	.01
9	.09	.04	.01	.09	.04	.01	.09	.04	.01	.08	.04	.01	.08	.04	.01
10	.09	.04	.01	.08	.04	.01	.08	.04	.01	.08	.04	.01	.08	.04	.01
1	.14	.08	.03	.14	.08	.02	.13	.07	.02	.12	.07	.02	.12	.07	.02
2	.13	.07	.02	.13	.07	.02	.12	.07	.02	.12	.07	.02	.11	.06	.02
3	.13	.07	.02	.12	.07	.02	.12	.06	.02	.11	.06	.02	.11	.06	.02
4	.12	.06	.02	.12	.06	.02	.11	.06	.02	.11	.06	.02	.10	.06	.02
5	.12	.06	.02	.11	.06	.02	.11	.06	.02	.11	.06	.02	.10	.05	.02
6	.11	.06	.02	.11	.05	.02	.10	.05	.02	.10	.05	.02	.10	.05	.02
7	.11	.05	.01	.10	.05	.01	.10	.05	.01	.10	.05	.01	.09	.05	.01
8	.10	.05	.01	.10	.05	.01	.10	.05	.01	.09	.05	.01	.09	.05	.01
9	.10	.05	.01	.10	.05	.01	.09	.05	.01	.09	.05	.01	.09	.04	.01
10	.09	.04	.01	.09	.04	.01	.09	.04	.01	.09	.04	.01	.08	.04	.01
1	.16	.09	.03	.15	.09	.03	.15	.08	.03	.14	.08	.03	.13	.08	.02
2	.15	.08	.02	.14	.08	.02	.14	.08	.02	.13	.07	.02	.13	.07	.02
3	.14	.07	.02	.14	.07	.02	.13	.07	.02	.12	.07	.02	.12	.07	.02
4	.13	.07	.02	.13	.07	.02	.12	.07	.02	.12	.06	.02	.11	.06	.02
5	.12	.06	.02	.12	.06	.02	.12	.06	.02	.11	.06	.02	.11	.06	.02
6	.12	.06	.02	.11	.06	.02	.11	.06	.02	.11	.06	.02	.10	.05	.02
7	.11	.05	.02	.11	.05	.02	.10	.05	.02	.10	.05	.02	.10	.05	.01
8	.10	.05	.01	.10	.05	.01	.10	.05	.01	.10	.05	.01	.09	.05	.01
9	.10	.05	.01	.10	.05	.01	.10	.05	.01	.09	.05	.01	.09	.05	.01
10	.09	.05	.01	.09	.04	.01	.09	.04	.01	.09	.04	.01	.09	.04	.01

Ceiling Cavity Luminance Coefficients[e] for 20 Per Cent Effective Floor Cavity Reflectance (ρ_{FC} = 20)

ρ_{CC} →	80			70			50			30			10		
ρ_W →	50	30	10	50	30	10	50	30	10	50	30	10	50	30	10
RCR ↓															
0	.11	.11	.11	.09	.09	.09	.06	.06	.06	.03	.03	.03	.01	.01	.01
1	.10	.09	.08	.09	.08	.07	.06	.05	.05	.03	.03	.03	.01	.01	.01
2	.10	.08	.06	.08	.07	.05	.06	.05	.04	.03	.03	.02	.01	.01	.01
3	.09	.07	.05	.08	.06	.04	.05	.04	.03	.03	.02	.02	.01	.01	.01
4	.09	.06	.04	.07	.05	.03	.05	.04	.02	.03	.02	.01	.01	.01	.00
5	.08	.05	.03	.07	.05	.03	.05	.03	.02	.03	.02	.01	.01	.01	.00
6	.08	.05	.02	.07	.04	.02	.05	.03	.01	.03	.02	.01	.01	.01	.00
7	.08	.04	.02	.06	.04	.02	.04	.03	.01	.03	.02	.01	.01	.01	.00
8	.07	.04	.02	.06	.03	.02	.04	.02	.01	.03	.01	.01	.01	.00	.00
9	.07	.04	.01	.06	.03	.01	.04	.02	.01	.02	.01	.01	.01	.00	.00
10	.07	.03	.01	.06	.03	.01	.04	.02	.01	.02	.01	.00	.01	.00	.00
0	.11	.11	.11	.10	.10	.10	.06	.06	.06	.04	.04	.04	.01	.01	.01
1	.11	.10	.09	.09	.08	.07	.06	.06	.05	.04	.03	.03	.01	.01	.01
2	.10	.08	.06	.09	.07	.06	.06	.05	.04	.03	.03	.02	.01	.01	.01
3	.10	.07	.05	.08	.06	.04	.06	.04	.03	.03	.02	.02	.01	.01	.01
4	.09	.06	.04	.08	.05	.03	.05	.04	.02	.03	.02	.01	.01	.01	.00
5	.09	.06	.03	.08	.05	.03	.05	.03	.02	.03	.02	.01	.01	.01	.00
6	.09	.05	.03	.07	.04	.02	.05	.03	.02	.03	.02	.01	.01	.01	.00
7	.08	.05	.02	.07	.04	.02	.05	.03	.01	.03	.02	.01	.01	.01	.00
8	.08	.04	.02	.07	.04	.02	.05	.03	.01	.03	.02	.01	.01	.01	.00
9	.08	.04	.02	.07	.03	.01	.05	.02	.01	.03	.01	.01	.01	.00	.00
10	.07	.04	.01	.06	.03	.01	.04	.02	.01	.03	.01	.01	.01	.00	.00
0	.10	.10	.10	.09	.09	.09	.06	.06	.06	.03	.03	.03	.01	.01	.01
1	.10	.09	.08	.08	.08	.07	.06	.05	.05	.03	.03	.03	.01	.01	.01
2	.09	.07	.06	.08	.06	.05	.05	.04	.04	.03	.03	.02	.01	.01	.01
3	.09	.06	.05	.07	.06	.04	.05	.04	.03	.03	.02	.02	.01	.01	.01
4	.08	.06	.04	.07	.05	.03	.05	.03	.02	.03	.02	.01	.01	.01	.00
5	.08	.05	.03	.07	.04	.03	.05	.03	.02	.03	.02	.01	.01	.01	.00
6	.07	.04	.02	.06	.04	.02	.04	.03	.02	.03	.02	.01	.01	.01	.00
7	.07	.04	.02	.06	.04	.02	.04	.02	.01	.02	.01	.01	.01	.00	.00
8	.07	.04	.02	.06	.03	.02	.04	.02	.01	.02	.01	.01	.01	.00	.00
9	.07	.04	.02	.06	.03	.01	.04	.02	.01	.02	.01	.01	.01	.00	.00
10	.06	.03	.01	.05	.03	.01	.04	.02	.01	.02	.01	.00	.01	.00	.00
0	.11	.11	.11	.09	.09	.09	.06	.06	.06	.04	.04	.04	.01	.01	.01
1	.11	.09	.08	.09	.08	.07	.06	.06	.05	.04	.03	.03	.01	.01	.01
2	.10	.08	.06	.08	.07	.06	.06	.05	.04	.03	.03	.02	.01	.01	.01
3	.09	.07	.05	.08	.06	.04	.05	.04	.03	.03	.02	.02	.01	.01	.01
4	.09	.06	.04	.07	.05	.03	.05	.04	.02	.03	.02	.01	.01	.01	.00
5	.08	.05	.03	.07	.05	.03	.05	.03	.02	.03	.02	.01	.01	.01	.00
6	.08	.05	.03	.07	.04	.02	.05	.03	.02	.03	.02	.01	.01	.01	.00
7	.08	.04	.02	.07	.04	.02	.05	.03	.01	.03	.02	.01	.01	.01	.00
8	.07	.04	.02	.06	.04	.02	.04	.02	.01	.03	.01	.01	.01	.00	.00
9	.07	.04	.02	.06	.03	.01	.04	.02	.01	.02	.01	.01	.01	.00	.00
10	.07	.04	.01	.06	.03	.01	.04	.02	.01	.02	.01	.01	.01	.00	.00
0	.10	.10	.10	.08	.08	.08	.06	.06	.06	.03	.03	.03	.01	.01	.01
1	.10	.09	.07	.08	.07	.06	.06	.05	.04	.03	.03	.03	.01	.01	.01
2	.09	.07	.06	.08	.06	.05	.05	.04	.03	.03	.03	.02	.01	.01	.01
3	.09	.06	.04	.08	.05	.04	.05	.04	.03	.03	.02	.02	.01	.01	.00
4	.08	.06	.03	.07	.05	.03	.05	.03	.02	.03	.02	.01	.01	.01	.00
5	.08	.05	.03	.07	.04	.02	.05	.03	.02	.03	.02	.01	.01	.01	.00
6	.08	.05	.02	.07	.04	.02	.05	.03	.01	.03	.02	.01	.01	.01	.00
7	.07	.04	.02	.06	.04	.02	.04	.03	.01	.03	.01	.01	.01	.00	.00
8	.07	.04	.02	.06	.03	.01	.04	.02	.01	.02	.01	.01	.01	.00	.00
9	.07	.04	.01	.06	.03	.01	.04	.02	.01	.02	.01	.00	.01	.00	.00
10	.07	.03	.01	.06	.03	.01	.04	.02	.01	.02	.01	.00	.01	.00	.00

[e] Although it is recommended that luminance coefficients and wall direct radiation coefficients be published to three decimal places, only two are shown here for these typical luminaires. Three decimal place data should be obtained from manufacturers of actual luminaires used.

Fig. 9–12. *Continued*

Typical Luminaires	ρ_{CC}[c] →	80			70			50			30			10			0
	ρ_W[d] →	50	30	10	50	30	10	50	30	10	50	30	10	50	30	10	0
	RCR[b] ↓	Coefficients of Utilization for 20 Per Cent Effective Floor Cavity Reflectance, ρ_{FC}															
45 Single row fluorescent lamp cove without reflector, mult. by 0.93 for 2 rows and by 0.85 for 3 rows.	1	.42	.40	.39	.36	.35	.33	.25	.24	.23	Coves are not recommended for lighting areas having low reflectances.						
	2	.37	.34	.32	.32	.29	.27	.22	.20	.19							
	3	.32	.29	.26	.28	.25	.23	.19	.17	.16							
	4	.29	.25	.22	.25	.22	.19	.17	.15	.13							
	5	.25	.21	.18	.22	.19	.16	.15	.13	.11							
	6	.23	.19	.16	.20	.16	.14	.14	.12	.10							
	7	.20	.17	.14	.17	.14	.12	.12	.10	.09							
	8	.18	.15	.12	.16	.13	.10	.11	.09	.08							
	9	.17	.13	.10	.15	.11	.09	.10	.08	.07							
	10	.15	.12	.09	.13	.10	.08	.09	.07	.06							
46 ρ_{CC} from below ~65%. Diffusing plastic or glass 1) Ceiling efficiency ~60%; diffuser transmittance ~50%; diffuser reflectance ~40%. Cavity with minimum obstructions and painted with 80% reflectance paint—use $\rho_c = 70$. 2) For lower reflectance paint or obstructions—use $\rho_c = 50$.	1				.60	.58	.56	.58	.56	.54							
	2				.53	.49	.45	.51	.47	.43							
	3				.47	.42	.37	.45	.41	.36							
	4				.41	.36	.32	.39	.35	.31							
	5				.37	.31	.27	.35	.30	.26							
	6				.33	.27	.23	.31	.26	.23							
	7				.29	.24	.20	.28	.23	.20							
	8				.26	.21	.18	.25	.20	.17							
	9				.23	.19	.15	.23	.18	.15							
	10				.21	.17	.13	.21	.16	.13							
47 ρ_{CC} from below ~60%. Prismatic plastic or glass. 1) Ceiling efficiency ~67%; prismatic transmittance ~72%; prismatic reflectance ~18%. Cavity with minimum obstructions and painted with 80% reflectance paint—use $\rho_c = 70$. 2) For lower reflectance paint or obstructions—use $\rho_c = 50$.	1				.71	.68	.66	.67	.66	.65	.65	.64	.62				
	2				.63	.60	.57	.61	.58	.55	.59	.56	.54				
	3				.57	.53	.49	.55	.52	.48	.54	.50	.47				
	4				.52	.47	.43	.50	.45	.42	.48	.44	.42				
	5				.46	.41	.37	.44	.40	.37	.43	.40	.36				
	6				.42	.37	.33	.41	.36	.32	.40	.35	.32				
	7				.38	.32	.29	.37	.31	.28	.36	.31	.28				
	8				.34	.28	.25	.33	.28	.25	.32	.28	.25				
	9				.30	.25	.22	.30	.25	.21	.29	.25	.21				
	10				.27	.23	.19	.27	.22	.19	.26	.22	.19				
48 ρ_{CC} from below ~45%. Louvered ceiling. 1) Ceiling efficiency ~50%; 45° shielding opaque louvers of 80% reflectance. Cavity with minimum obstructions and painted with 80% reflectance paint—use $\rho_c = 50$. 2) For other conditions refer to Fig. 9–55.	1							.51	.49	.48				.47	.46	.45	
	2							.46	.44	.42				.43	.42	.40	
	3							.42	.39	.37				.39	.38	.36	
	4							.38	.35	.33				.36	.34	.32	
	5							.35	.32	.29				.33	.31	.29	
	6							.32	.29	.26				.30	.28	.26	
	7							.29	.26	.23				.28	.25	.23	
	8							.27	.23	.21				.26	.23	.21	
	9							.24	.21	.19				.24	.21	.19	
	10							.22	.19	.17				.22	.19	.17	

49

3′ x 3′ fluorescent troffer with 48″ lamps mounted along diagonals—use units 38, 41, or 43 as appropriate

50

2′ x 2′ fluorescent troffer with two "U" lamps—use units 38, 41, or 43 as appropriate

[b] RCR = Room Cavity Ratio.
[c] ρ_{CC} = Per cent effective ceiling cavity reflectance.
[d] ρ_W = Per cent wall reflectance.

Luminaire No. →	1	2	3	4	5	6	7	8	9	10	11	12	13	14	15	16	17	18	19	20	21	22
Angle from Nadir ↓	Normalized Average Intensity (Candelas per 1000 Lamp Lumens)																					
5°	72.5	6.4	256.0	238.0	808.0	1320.0	695.0	759.0	374.0	208.0	152.0	316.0	288.0	999.0	470.0	294.0	576.0	274.0	261.0	263.0	246.0	284.0
15°	72.5	8.1	246.0	264.0	671.0	1010.0	630.0	650.0	357.0	220.0	148.0	311.0	321.0	775.0	384.0	282.0	519.0	302.0	256.0	258.0	260.0	262.0
25°	72.5	9.4	238.0	248.0	494.0	584.0	286.0	308.0	305.0	254.0	141.0	301.0	331.0	475.0	344.0	294.0	426.0	344.0	234.0	236.0	264.0	226.0
35°	72.5	9.8	238.0	191.0	340.0	236.0	88.0	55.0	212.0	220.0	125.0	271.0	260.0	188.0	290.0	294.0	274.0	321.0	206.0	210.0	248.0	187.0
45°	72.5	8.1	203.0	122.0	203.0	22.0	5.2	13.0	81.0	130.0	106.0	156.0	202.0	90.7	210.0	246.0	127.0	209.0	172.0	163.0	192.0	145.0
55°	72.1	6.6	168.0	62.5	91.1	0	0	4.0	40.5	59.2	87.5	63.0	114.0	32.1	86.5	137.0	69.3	45.7	126.0	97.9	98.0	83.2
65°	71.7	4.6	130.0	45.7	33.2	0	0	0	20.6	25.9	69.5	31.7	13.6	8.4	18.0	26.0	19.8	8.1	94.8	55.5	32.7	36.7
75°	70.6	2.6	34.2	38.0	12.4	0	0	0	9.5	11.1	47.0	17.5	5.9	5.8	5.0	6.5	2.6	2.9	45.1	29.6	12.7	18.4
85°	70.0	1.8	6.8	32.0	4.1	0	0	0	2.4	3.7	23.5	4.2	2.0	1.1	1.1	1.2	1.4	2.4	13.7	10.8	3.9	5.6
95°	67.0	14.9	0	28.0	0	0	0	0	0	0	9.4	0	0.9	0.3	0.3	0.3	0.7	3.4	6.4	7.8	3.5	3.6
105°	62.5	147.0	0	28.0	0	0	0	0	0	0	4.7	0	0	0.4	0.4	0.3	0.7	8.2	8.5	14.4	6.6	11.2
115°	58.0	170.0	0	41.1	0	0	0	0	0	0	1.2	0	0	0.4	0.4	0.3	4.5	15.3	11.3	21.4	11.9	21.0
125°	54.7	168.0	0	42.5	0	0	0	0	0	0	0	0	0	1.1	0.4	0.4	10.5	22.4	15.3	31.2	21.6	34.4
135°	51.2	183.0	0	32.8	0	0	0	0	0	0	0	0	0	1.7	0.8	0.4	16.5	28.8	21.6	46.9	33.5	51.4
145°	48.2	159.0	0	22.6	0	0	0	0	0	0	0	0	0	8.1	3.2	1.4	20.7	33.5	32.8	59.4	49.8	71.5
155°	46.4	139.0	0	9.1	0	0	0	0	0	0	0	0	0	8.7	8.2	7.4	32.2	42.0	51.5	82.5	70.4	92.1
165°	45.0	94.5	0	3.1	0	0	0	0	0	0	0	0	0	0.7	0.7	0.6	33.1	27.5	67.3	105.0	92.0	109.0
175°	43.9	50.5	0	1.1	0	0	0	0	0	0	0	0	0	0.7	0.7	0.6	16.5	2.6	73.0	111.0	102.0	115.0

Luminaire No. →	23	24	25	26	27	28	29	30	31	32	33	34	35	36	37	38	39	40	41	42	43	44
Angle From Nadir ↓	Normalized Average Intensity (Candelas per 1000 Lamp Lumens)																					
5°	244.0	189.0	270.0	326.0	199.0	41.4	194.0	210.0	206.0	107.0	272.0	292.0	272.0	284.0	193.0	218.0	206.0	232.0	231.0	272.0	271.0	114.0
15°	248.0	176.0	249.0	292.0	194.0	38.5	192.0	211.0	199.0	104.0	238.0	252.0	244.0	244.0	188.0	207.0	202.0	227.0	226.0	266.0	264.0	123.0
25°	242.0	147.0	200.0	231.0	184.0	35.3	187.0	212.0	185.0	98.5	192.0	197.0	202.0	194.0	170.0	187.0	183.0	217.0	216.0	252.0	251.0	154.0
35°	218.0	110.0	144.0	151.0	170.0	32.3	169.0	204.0	158.0	90.0	143.0	137.0	156.0	135.0	146.0	164.0	162.0	195.0	199.0	214.0	228.0	192.0
45°	152.0	64.0	86.5	73.4	154.0	28.8	123.0	164.0	108.0	79.4	93.4	81.0	106.0	78.9	121.0	135.0	133.0	155.0	170.0	141.0	161.0	166.0
55°	70.1	34.4	53.5	30.1	137.0	22.2	77.5	78.3	51.6	66.4	60.1	47.5	68.1	46.5	94.0	106.0	104.0	88.0	101.0	60.9	75.5	93.2
65°	26.2	20.4	34.2	15.3	117.0	14.5	37.4	36.4	35.6	51.8	37.2	28.1	41.9	27.5	66.8	74.2	70.5	40.8	51.1	26.6	28.4	58.5
75°	9.8	9.8	20.4	6.7	88.4	7.2	18.3	26.0	34.5	35.8	18.1	13.6	21.4	13.9	38.0	42.5	36.3	22.4	27.0	12.0	13.4	28.6
85°	2.9	2.7	10.2	2.7	59.2	1.7	10.6	17.6	32.2	21.4	4.2	2.9	6.0	3.6	19.9	15.3	7.2	7.2	7.2	4.4	5.4	9.7
95°	2.7	4.1	6.9	0	49.5	11.1	14.5	15.5	32.7	14.4	0	0	0	0	9.0	5.3	0	0	0	0	0	0
105°	5.9	18.9	8.5	0	32.7	49.5	40.0	21.8	48.8	14.4	0	0	0	0	2.5	2.4	0	0	0	0	0	0
115°	12.8	40.7	9.6	0	6.4	96.0	57.1	26.8	48.8	13.8	0	0	0	0	0	0	0	0	0	0	0	0
125°	24.0	67.0	10.2	0	0	130.0	68.6	22.8	44.6	13.1	0	0	0	0	0	0	0	0	0	0	0	0
135°	36.2	93.1	10.8	0	0	155.0	71.6	18.7	35.8	11.8	0	0	0	0	0	0	0	0	0	0	0	0
145°	49.6	117.0	11.1	0	0	172.0	67.5	12.1	28.6	10.3	0	0	0	0	0	0	0	0	0	0	0	0
155°	69.8	136.0	11.4	0	0	183.0	65.1	7.3	24.0	8.4	0	0	0	0	0	0	0	0	0	0	0	0
165°	88.6	151.0	12.0	0	0	189.0	67.5	4.7	20.8	6.6	0	0	0	0	0	0	0	0	0	0	0	0
175°	95.6	155.0	13.1	0	0	201.0	73.4	3.9	17.8	5.7	0	0	0	0	0	0	0	0	0	0	0	0

Fig. 9-13. Multiplying Factors for Other than 20 Per Cent Effective Floor Cavity Reflectance

% Effective Ceiling Cavity Reflectance, ρ_{CC}	80				70				50			30			10		
% Wall Reflectance, ρ_W	70	50	30	10	70	50	30	10	50	30	10	50	30	10	50	30	10
For 30 Per Cent Effective Floor Cavity Reflectance (20 Per Cent = 1.00)																	
Room Cavity Ratio																	
1	1.092	1.082	1.075	1.068	1.077	1.070	1.064	1.059	1.049	1.044	1.040	1.028	1.026	1.023	1.012	1.010	1.008
2	1.079	1.066	1.055	1.047	1.068	1.057	1.048	1.039	1.041	1.033	1.027	1.026	1.021	1.017	1.013	1.010	1.006
3	1.070	1.054	1.042	1.033	1.061	1.048	1.037	1.028	1.034	1.027	1.020	1.024	1.017	1.012	1.014	1.009	1.005
4	1.062	1.045	1.033	1.024	1.055	1.040	1.029	1.021	1.030	1.022	1.015	1.022	1.015	1.010	1.014	1.009	1.004
5	1.056	1.038	1.026	1.018	1.050	1.034	1.024	1.015	1.027	1.018	1.012	1.020	1.013	1.008	1.014	1.009	1.004
6	1.052	1.033	1.021	1.014	1.047	1.030	1.020	1.012	1.024	1.015	1.009	1.019	1.012	1.006	1.014	1.008	1.003
7	1.047	1.029	1.018	1.011	1.043	1.026	1.017	1.009	1.022	1.013	1.007	1.018	1.010	1.005	1.014	1.008	1.003
8	1.044	1.026	1.015	1.009	1.040	1.024	1.015	1.007	1.020	1.012	1.006	1.017	1.009	1.004	1.013	1.007	1.003
9	1.040	1.024	1.014	1.007	1.037	1.022	1.014	1.006	1.019	1.011	1.005	1.016	1.009	1.004	1.013	1.007	1.002
10	1.037	1.022	1.012	1.006	1.034	1.020	1.012	1.005	1.017	1.010	1.004	1.015	1.009	1.003	1.013	1.007	1.002
For 10 Per Cent Effective Floor Cavity Reflectance (20 Per Cent = 1.00)																	
Room Cavity Ratio																	
1	.923	.929	.935	.940	.933	.939	.943	.948	.956	.960	.963	.973	.976	.979	.989	.991	.993
2	.931	.942	.950	.958	.940	.949	.957	.963	.962	.968	.974	.976	.980	.985	.988	.991	.995
3	.939	.951	.961	.969	.945	.057	.966	.973	.967	.975	.981	.978	.983	.988	.988	.992	.996
4	.944	.958	.969	.978	.950	.963	.973	.980	.972	.980	.986	.980	.986	.991	.987	.992	.996
5	.949	.964	.976	.983	.954	.968	.978	.985	.975	.983	.989	.981	.988	.993	.987	.992	.997
6	.953	.969	.980	.986	.958	.972	.982	.989	.977	.985	.992	.982	.989	.995	.987	.993	.997
7	.957	.973	.983	.991	.961	.975	.985	.991	.979	.987	.994	.983	.990	.996	.987	.993	.998
8	.960	.976	.986	.993	.963	.977	.987	.993	.981	.988	.995	.984	.991	.997	.987	.994	.998
9	.963	.978	.987	.994	.965	.979	.989	.994	.983	.990	.996	.985	.992	.998	.988	.994	.999
10	.965	.980	.989	.995	.967	.981	.990	.995	.984	.991	.997	.986	.993	.998	.988	.994	.999
For 0 Per Cent Effective Floor Cavity Reflectance (20 Per Cent = 1.00)																	
Room Cavity Ratio																	
1	.859	.870	.879	.886	.873	.884	.893	.901	.916	.923	.929	.948	.954	.960	.979	.983	.987
2	.871	.887	.903	.919	.886	.902	.916	.928	.926	.938	.949	.954	.963	.971	.978	.983	.991
3	.882	.904	.915	.942	.898	.918	.934	.947	.936	.950	.964	.958	.969	.979	.976	.984	.993
4	.893	.919	.941	.958	.908	.930	.948	.961	.945	.961	.974	.961	.974	.984	.975	.985	.994
5	.903	.931	.953	.969	.914	.939	.958	.970	.951	.967	.980	.964	.977	.988	.975	.985	.995
6	.911	.940	.961	.976	.920	.945	.965	.977	.955	.972	.985	.966	.979	.991	.975	.986	.996
7	.917	.947	.967	.981	.924	.950	.970	.982	.959	.975	.988	.968	.981	.993	.975	.987	.997
8	.922	.953	.971	.985	.929	.955	.975	.986	.963	.978	.991	.970	.983	.995	.976	.988	.998
9	.928	.958	.975	.988	.933	.959	.980	.989	.966	.980	.993	.971	.985	.996	.976	.988	.998
10	.933	.962	.979	.991	.937	.963	.983	.992	.969	.982	.995	.973	.987	.997	.977	.989	.999

proper column of wall reflectances, ρ_W, should be based upon the initial values expected. The wall reflectance should also represent the weighted average of the painted areas, glass fenestration or daylight controls, chalkboards, shelves, etc., in the area to be lighted. The weighting should be based on the relative areas of each type of surface.

In using Fig. 9-12, it will often be necessary to interpolate between Room Cavity Ratios and/or Effective Ceiling Cavity Reflectances. This is most easily accomplished by interpolating first between RCRs to obtain CUs for Effective Ceiling Cavity

Reflectances that straddle the actual ρ_{CC} and then interpolating between these CUs.

Limitations. The illumination computed by the lumen method is an average value that will be representative only if the luminaires are spaced to obtain reasonably uniform illumination. The coefficients of utilization are based on empty interiors, but it is possible to comprehend the effects of obstructions in the ceiling or floor cavities by calculating ratios for the cavities bounded by the obstructions to obtain effective cavity reflectances that can be weighted by relative area to obtain an average effective cavity reflectance.[6] All calculations also assume that installation conditions (voltage, temperature, etc.) are such that luminaires will provide their rated output.

Average Luminance Calculations

The ability to predict the luminance[8] of the visual environment is needed in order to (1) design lighting that promotes both visual comfort and good visual performance, (2) predict illumination as specific points within the environment, and (3) evaluate various criteria of the lighting system such as Visual Comfort Probability and Veiling Reflections. Luminance calculations are greatly simplified through the use of luminance coefficients (LC). These coefficients, like coefficients of utilization, may be computed for any specific luminaire. The wall and ceiling cavity luminance coefficients for certain generic luminaires are found in Fig. 9–12.

Luminance coefficients are similar to coefficients of utilization. They may be substituted into a variation of the lumen method formula in place of the coefficient of utilization. The result obtained is either the average wall luminance or the average ceiling cavity luminance, rather than illumination on the work plane. Thus:

$$\text{Average Initial Wall Luminance} = \frac{\text{Total Bare Lamp Lumens} \times \text{Wall Luminance Coefficient}}{\pi^{*} \times \text{Floor Area}}$$

and

$$\text{Average Initial Ceiling Cavity Luminance} = \frac{\text{Total Bare Lamp Lumens} \times \text{Ceiling Cavity Luminance Coefficient}}{\pi^{*} \times \text{Floor Area}}$$

If the area is expressed in square feet, the luminance is in candela per square foot; if the area is in square meters, the luminance is in candela per square meter. Luminance can be obtained in footlamberts by expressing area in square feet and omitting the factor of π in the denominators.

If the maintained average wall luminance or the maintained average ceiling cavity luminance is required, a Light Loss Factor is introduced into these equations in the same manner as it is used for maintained average illumination.

For suspended luminaires the average ceiling cavity luminance obtained is the average luminance of the imaginary plane at the level of the luminaires. This luminance does not include the weighted average luminance of the luminaires as seen from below. It is, rather, the average luminance of the background against which the luminaires are seen. In the case of recessed or ceiling mounted luminaires the average ceiling cavity luminance obtained is the average luminance of that part of the ceiling between luminaires.

Limitations. The limitations for luminance calculations are similar to those for average illumination. In addition, the wall reflectance used to enter the tables is a weighted average of the reflectances for the various parts of the walls: the wall luminance found by the wall luminance coefficient is the value that would occur if the walls were of a uniform reflectance equal to the average reflectance used. Thus, many parts of the wall may have luminance values that differ from the calculated average value. A correction can be applied to determine the approximate luminance of any part of the wall. For any area on the wall,

$$\text{Luminance} = \frac{\begin{pmatrix}\text{Average Wall} \\ \text{Luminance}\end{pmatrix} \times \begin{pmatrix}\text{Reflectance} \\ \text{of Area}\end{pmatrix}}{(\text{Average Wall Reflectance})}$$

Luminaire Spacing

The illumination levels recommended in Fig. 9–80 are minimum footcandles on the task. In most instances, the visual task is likely to be located (or relocated) at any point in the room being lighted. It follows, therefore, that the illumination at any point should not differ materially from the calculated average. Illumination uniformity is generally not feasible, but uniformity is considered acceptable if the maximum and minimum values in the room are not more than 1/6 above and below the average.

To achieve acceptable uniformity, luminaires should not be spaced too far apart or too far from the walls. Spacing limitations between luminaires are related to the intensity distribution of the luminaires, placement of the luminaires within the room, and the reflectances of the room surfaces. The principal factor for direct, semi-direct, and general diffuse luminaires is the mounting height above the work-plane; for semi-indirect and indirect luminaires, it is the ceiling height above the work-plane.

A guide for the maximum spacing-to-mounting height (S/MH) ratio of direct, semi-direct, and general diffuse luminaires can be based on the luminaire's intensity distribution. This guide is determined by the spacing for equality of the direct component of illumination at two points assuming that the inverse-square law is valid. The maximum S/MH ratio often quoted for luminaires is based on the comparison of a point directly under one luminaire

* π omitted if luminance is to be obtained in footlamberts.

Fig. 9-14. Guide for Maximum Spacing-to-Mounting Height Ratio for Direct, Semi-Direct, and General Diffuse Luminaires

This procedure is directly applicable to direct and semi-direct luminaires since it is based on the direct component of illumination. It may be used, however, as a basis for determining the ratio for general diffuse luminaires by increasing the ratio obtained for the direct component 10 to 20 percent to recognize the effect of the reflected component.

Maximum permissible spacings between semi-indirect and indirect luminaires should be relative to the ceiling height above the work-plane. They are best determined from experience with measured data in actual installations or by illumination measurements from a single luminaire or row operating in simulated installation conditions.

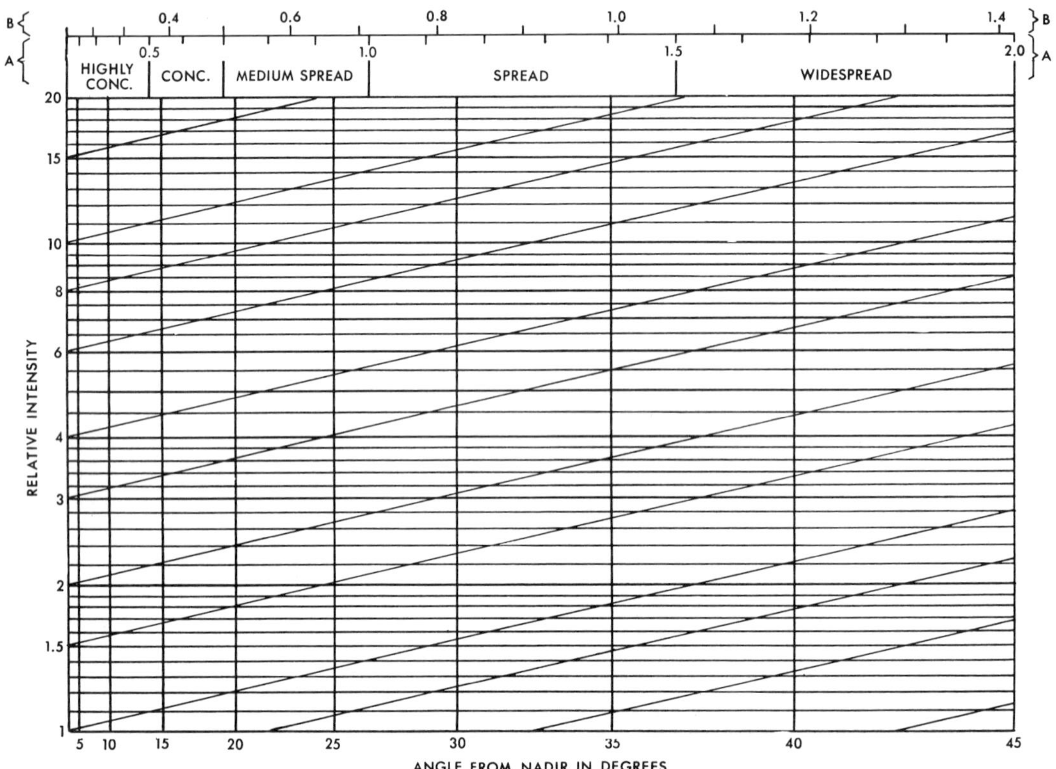

I. Use of the chart:

(a) Plot the relative intensity (candlepower) of the luminaire in rectangular coordinates on this chart.

(b) Smooth out any pronounced irregularities in the curve.

(c) On the left vertical edge of the chart, take a point at ½ the intensity at 0 degrees, and draw a line from this point parallel to the diagonal lines.

(d) From the point at which this diagonal line crosses the intensity curve, follow a vertical line up to the top of the chart and read the S/MH ratio and distribution classification on scale A.

Scale A is established on the basis of two identical luminaires, M and N, located at height H above the work-plane. The maximum spacing-to-mounting height ratio is determined for the following condition: the direct illumination at point P due only to luminaire M equals the direct illumination at point Q due to luminaires M and N (as calculated by the inverse-square law).

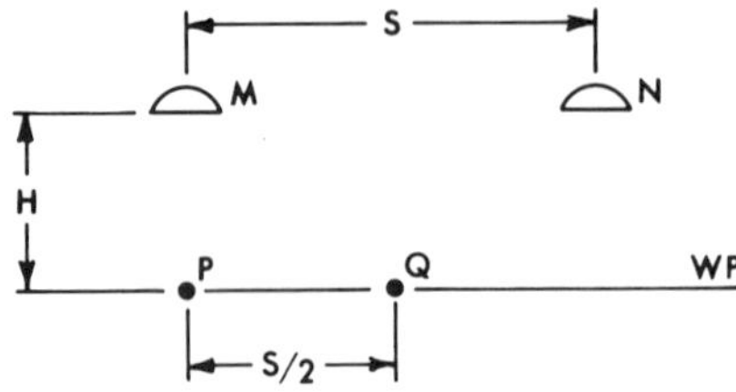

II. A point R on the work-plane centered between four adjacent identical luminaires will provide an additional check on maximum permissible spacing.

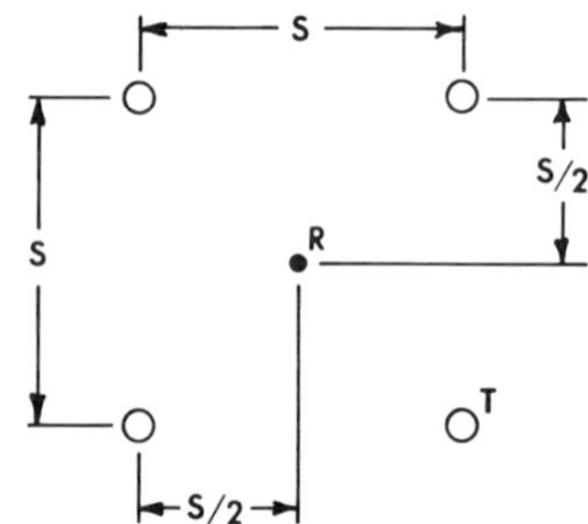

Here, the direct illumination at point R due to all four luminaires equals the direct illumination on the work-plane due to luminaire T directly below that luminaire. Procedure:

(e) Use the curve drawn in (a) and (b).

(f) On the left vertical edge of the chart, take a point ¼ the intensity at 0 degrees, and draw a line from this point parallel to the diagonal lines.

(g) From the point at which this diagonal line crosses the intensity curve, follow a vertical line up to the top of the chart and read the S/MH ratio on scale B.

In the event that this value is less than that determined in part I (d), it should be used to describe the luminaire.

with a point half way between adjacent luminaires considering only the effect of the two adjacent luminaires. The chart of Fig. 9–14 permits rapid determination of this S/MH guide. Also, it simultaneously gives a similar S/MH guide using a point at the center of a square arrangement of four luminaires.

Guides of this type for maximum S/MH ratio tend to be conservative when there is a large reflected component of illumination and when the intensity distribution decreases gradually as the angle from the nadir increases. They may be invalid for luminaires with a rapid decrease of intensity within a small angular range, such as may be used for luminance control. The interpretation of these S/MH ratios as currently used with photometric descriptions of luminaires is illustrated in Fig. 9–15. In general, uniformity of illumination should be considered one of the design factors of a complete lighting system to be explored by point illumination calculations rather than to be a factor uniquely specified by a single number based only on the luminaire characteristics.

The commonly used practice of letting the distance from the luminaires to the wall equal one-half the distance between rows (see Fig. 9–16a) results in inadequate illumination near the walls. Since desks and benches are frequently located along the walls, a distance of 2½ feet from the wall to the center of the luminaire should be employed to avoid excessive drop-off in illumination. This will locate the luminaires over the edge of desks facing the wall or over the center of desks that are perpendicular to the wall (see Fig. 9–16b). To further improve illumination uniformity across the room, it is often desirable to use somewhat closer spacings between outer rows of luminaires than between central rows, taking care to be sure that no spacing exceeds the maximum permissible spacing.

To prevent excessive reduction in illumination at the ends of the room, the ends of fluorescent luminaire rows should preferably be 6 to 12 inches from the walls, or in no case more than 2 feet from the walls. Even the 6- to 12-inch spacing leaves much to be desired from the standpoint of uniformity and, where practicable, the arrangement shown in Fig. 9–16c is much more satisfactory. With this arrangement, the units at each end of the row are replaced by a continuous row parallel to, and 2½ feet from, the end wall. In the example shown, 5 units were replaced by 9 units providing a potential increase in the illumination at the end of the room of 80 per cent over what it would be with the layout shown by Fig. 9–16b. This technique not only improves uniformity but also eliminates scallops of light on the end walls and provides a uniform wash of light on all four walls.

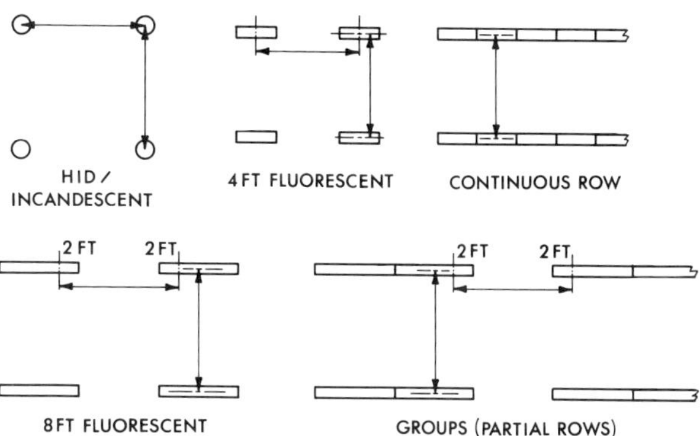

Fig. 9–15. Spacing dimensions to be used in relation to spacing-to-mounting height ratios. Mounting height is from luminaire to work plane for direct, semi-direct and general-diffuse luminaires and from ceiling to work plane for semi-indirect and indirect luminaires.

Another excellent method of compensating for the normal reduction in illumination that may be expected at the ends of rows is to use a greater number of lamps in the end units. Still another technique is to provide additional units between the rows at each end. The units could be either parallel or at right angles to the rows.

Spacings closer than the maximum permissible are often highly desirable to reduce harsh shadows and veiling reflections in the task as well as further improve uniformity. This is particularly true for direct and semi-direct equipment, and spacings that are substantially less than the maximum permissible spacing should be seriously considered.

The following formulas can be used to calculate the values:

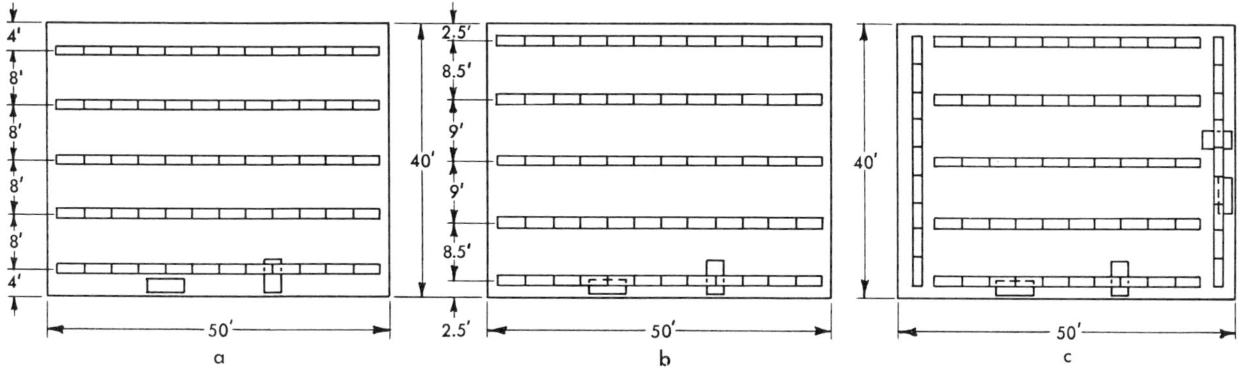

Fig. 9–16. (a) Lighting layout using equal spacing between continuous rows of luminaires. (b) Layout is changed to provide more illumination near side walls. (c) By adding four more four-foot units on each end, layout (b) can be modified to provide 80 per cent more light near the end walls and prevent possible scallop effects.

(1) For individually mounted luminaires, the wall-to-luminaire spacing should be:

$$\text{Wall-to-Luminaire Spacing} = \frac{\text{Luminaire-to-Luminaire Spacing}}{3}$$

(2) For individual units or crosswise spacing of continuous rows:

$$\text{Minimum number of rows} = \frac{\text{Room Width}}{\text{Maximum spacing allowed}}$$

(3) For lengthwise spacing in continuous rows:

$$\text{Maximum number of units per row} = \frac{\text{Room Length} - 1}{\text{Luminaire Length}}$$

(allows one-half-foot end spacing)

$$\text{Minimum number of units per row} = \frac{\text{Room Length} - 4}{\text{Luminaire Length}}$$

(allows two-foot end spacing)

Elements of Flux Transfer Theory

Assume a surface A (Fig. 9–17a) that emits or reflects flux in a completely diffuse manner. Some of this flux falls on surface B. The percentage of the total flux emitted from A that falls on B is called the *form factor* of B with respect to A, of $f_{A\text{-}B}$. This is the only new concept involved in the theory.

Now consider a normal room cavity as shown in Fig. 9–17b. Assume all walls have the same reflectance and that the lighting system is symmetrical. This allows all four walls to be lumped together and treated as one surface. Assume that the three inside surfaces are completely diffuse and have initial luminances L_{01}, L_{02}, and L_{03} respectively. These initial luminances might, for example, be due to the flux falling directly on them from a luminaire within the enclosure. Express luminance in candela per unit area and illumination in lumens per unit area.

Since there will be reflections, the final luminances L_1, L_2, and L_3 will be higher than the initial ones. The increase will be due to light falling on the surfaces from other surfaces in the room. The final luminance of surface 3, then, can be expressed as follows:

$$L_3 = L_{03} + \frac{1}{\pi}\rho_3 E_3$$

where ρ_3 is the reflectance of surface 3, and E_3 is the illumination on that surface due to the other surfaces in the room. What is now needed is to find E_3.

This is where the form factor comes in. From the first paragraph above, the flux falling on surface 3 from surface 2 will equal the total emission of surface 2 times the form factor $f_{2\text{-}3}$. The total emission of surface 2 will equal $\pi L_2 A_2$, where L_2 is the final

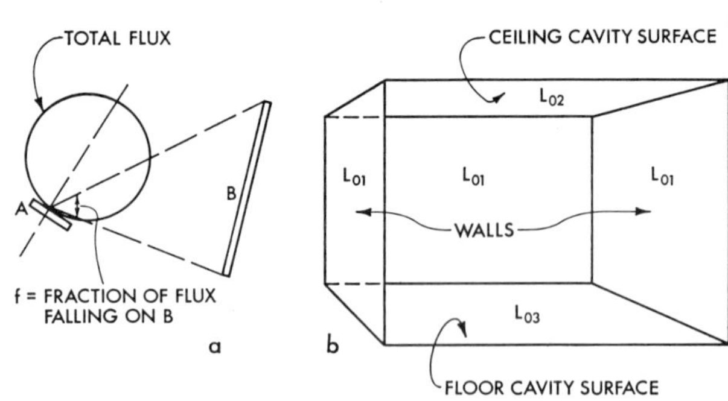

Fig. 9–17. Diagrams for explanation of Flux Transfer Theory, including form factor and surface designations.

Fig. 9–18. Form Factors for Zonal-Cavity System

RCR	$f_{1\text{-}1}$	$f_{1\text{-}2}$ $f_{1\text{-}3}$	$f_{2\text{-}1}$ $f_{3\text{-}1}$	$f_{2\text{-}3}$ $f_{3\text{-}2}$
0	0.000	0.500	0.000	1.000
1	.135	.432	.173	.827
2	.225	.387	.310	.690
3	.300	.350	.420	.580
4	.360	.320	.511	.489
5	.416	.292	.585	.415
6	.460	.270	.646	.354
7	.502	.249	.696	.304
8	.540	.230	.737	.263
9	.572	.214	.771	.229
10	.600	.200	.800	.200

luminance of surface 2 and A_2 is its area. This can be said because the surface emits diffusely, and the luminance of a diffuse emitter is the exitance (flux per unit area emitted) divided by π.

The illumination on surface 3 due to surface 2, therefore, is

$$E_{3(2)} = \text{Flux/Area} = \frac{\pi L_2 A_2 f_{2\text{-}3}}{A_3}$$

This equation, and those to follow, can be simplified by making use of what is called the "Reciprocity Relationship." * When this is done, the equation reduces to:

$$E_{3(2)} = \pi L_2 f_{3\text{-}2}$$

Applying the same reasoning to the flux from surface 1, an equation results for L_3 in these terms:

$$L_3 = L_{03} + \frac{1}{\pi}\rho_3(L_2 f_{3\text{-}2} + L_1 f_{3\text{-}1})$$

All of the above applies equally well to surface 2, giving a similar equation for it, which is:

$$L_2 = L_{02} + \frac{1}{\pi}\rho_2(L_3 f_{2\text{-}3} + L_1 f_{2\text{-}1})$$

Surface 1, the walls, has one difference from those mentioned above, however. Not all of the flux from it goes to the other two surfaces. Some of it goes to other parts of itself. This flux interchange between

* $A_m f_{m\text{-}n} = A_n f_{n\text{-}m}$.

the walls must be taken into account. There is an illumination component on the walls due to the walls themselves, or

$$E_{1(1)} = \pi L_1 f_{1\text{-}1}$$

When this is added, the result is:

$$L_1 = L_{01} + \frac{1}{\pi}\rho_1(L_1 f_{1\text{-}1} + L_2 f_{1\text{-}2} + L_3 f_{1\text{-}3})$$

With these three equations and the form factors, any interior illumination or luminance can be determined since L_1, L_2, and L_3 are the unknowns in a set of three simultaneous equations. The illumination on surface 3 (the work-plane) is obtained by dividing πL_3 by ρ_3.

The above set of simultaneous equations is the basis of generating the tables of coefficients of utilization, the wall luminance coefficients, and the ceiling cavity luminance coefficients. Form factors are used for a square room of varying height, and the room cavity ratio reduces rectangular rooms to equivalent square rooms. Fig. 9–18 gives numerical values for the various form factors. Some useful relations are:

$$f_{2\text{-}1} = f_{3\text{-}1} = 1 - f_{2\text{-}3}$$

$$f_{1\text{-}2} = f_{1\text{-}3} = \frac{A_2}{A_1}(1 - f_{2\text{-}3}) = \frac{1 - f_{2\text{-}3}}{0.4(\text{RCR})}$$

$$f_{1\text{-}1} = 1 - 2\frac{A_2}{A_1}(1 - f_{2\text{-}3}) = 1 - \frac{1 - f_{2\text{-}3}}{0.2(\text{RCR})}$$

Fig. 9–19. Direct Ratio Work Sheets

WORK SHEET a—For Computing Direct Ratios (Spacing-to-Mounting Height Ratio = 0.4)

ZONE (DEG.)	ROOM CAVITY RATIO / LUMINAIRE LUMENS	1	2	3	4	5	6	7	8	9	10
		ZONAL MULTIPLIERS									
0-10		1.0	1.0	1.0	1.0	1.0	1.0	1.0	1.0	1.0	1.0
10-20		.98	.92	.89	.85	.81	.79	.76	.73	.70	.66
20-30		.94	.89	.84	.78	.73	.68	.64	.59	.53	.48
30-40		.91	.82	.72	.64	.56	.50	.43	.36	.30	.25
40-50		.88	.78	.67	.57	.47	.38	.28	.19	.11	.06
50-60		.83	.67	.53	.39	.28	.16	.09	.04	.01	0
60-70		.77	.53	.35	.21	.05	.01	0	0	0	0
70-80		.60	.22	.05	0	0	0	0	0	0	0
80-90		.14	0	0	0	0	0	0	0	0	0
TOTAL											
DIRECT RATIO											

WORK SHEET b—For Computing Direct Ratios (Spacing-to-Mounting Height Ratio = 0.7)

ZONE (DEG.)	ROOM CAVITY RATIO / LUMINAIRE LUMENS	1	2	3	4	5	6	7	8	9	10
		ZONAL MULTIPLIERS									
0-10		1.0	1.0	1.0	1.0	1.0	1.0	1.0	1.0	1.0	1.0
10-20		.98	.96	.95	.93	.91	.90	.88	.86	.85	.83
20-30		.96	.93	.90	.87	.83	.80	.76	.73	.68	.66
30-40		.91	.82	.72	.63	.54	.46	.38	.30	.23	.15
40-50		.88	.77	.65	.55	.46	.36	.27	.18	.10	.03
50-60		.83	.67	.52	.39	.28	.16	.05	.02	0	0
60-70		.76	.52	.32	.17	.04	0	0	0	0	0
70-80		.58	.21	.02	0	0	0	0	0	0	0
80-90		.12	0	0	0	0	0	0	0	0	0
TOTAL											
DIRECT RATIO											

WORK SHEET c—For Computing Direct Ratios (Spacing-to-Mounting Height Ratio = 1.0)

ZONE (DEG.)	ROOM CAVITY RATIO / LUMINAIRE LUMENS	1	2	3	4	5	6	7	8	9	10
		ZONAL MULTIPLIERS									
0-10		1.0	1.0	1.0	1.0	1.0	1.0	1.0	1.0	1.0	1.0
10-20		1.0	1.0	1.0	1.0	1.0	1.0	1.0	1.0	1.0	1.0
20-30		.98	.97	.96	.94	.92	.91	.89	.86	.82	.68
30-40		.90	.81	.72	.61	.52	.42	.33	.24	.16	.07
40-50		.87	.75	.64	.53	.43	.34	.25	.17	.08	0
50-60		.83	.67	.53	.39	.27	.16	.06	0	0	0
60-70		.74	.51	.30	.13	.02	0	0	0	0	0
70-80		.55	.20	.04	0	0	0	0	0	0	0
80-90		.10	0	0	0	0	0	0	0	0	0
TOTAL											
DIRECT RATIO											

Thus, all of the form factors can be expressed in terms of $f_{2\text{-}3}$. The analytical expression for $f_{2\text{-}3}$ is complex; however, it can be approximated by the expression (error $\gtrsim \frac{1}{2}$ per cent for $1 \leq \text{RCR} \leq 10$):

$$f_{2\text{-}3} \cong e^{-0.184(\text{RCR})} + 0.00535\ (\text{RCR}) - 0.011$$

Coefficient Tables

Tables of coefficients of utilization, wall luminance coefficients, and ceiling cavity luminance coefficients can be prepared by systemized procedures.[9] It is desirable to standardize published tables of these values to prevent misunderstandings and to facilitate direct comparisons of the data for different luminaires.

Fig. 9–12 shows the recommended form for unabridged tables. It is recognized that space limitations often necessitate abridgements, and when necessary, only the columns for 80, 50, and 10 per cent ρ_{CC} are recommended for luminaires having 0 to 35 per cent of their output in the 90- to 180-degree zone; and 80, 70, and 50 per cent for luminaires having over 35 per cent of their output in the 90- to 180-degree zone. Also, 10 per cent ρ_W columns are not required for luminance coefficient abridged tables. It is recommended that coefficients of utilization be published to two decimal places, wall luminance coefficients to three decimal places, and ceiling cavity luminance coefficients to three decimal places when their value is less than 0.100 and to two decimal places when their value is greater than 0.100. A wall direct radiation coefficient (WDRC) should be published to three decimal places for each room cavity ratio adjacent to the wall luminance coefficient table.

Computation.

Step 1. The direct ratio (DR), the fraction of luminaire flux initially reaching the work-plane, is required for each room cavity ratio. Since these values depend on luminaire spacing, arbitrary typical values are used to standardize coefficient tables; $S/MH = 0.4$ is used for fluorescent luminaires and $S/MH = 0.7$ is used for incandescent and high-intensity discharge luminaires. If another value is used on the basis of specific application information, it should be so stated on the tables. (a) Select the appropriate work sheet in Fig. 9–19 and enter the luminaire lumens for each ten-degree zone in the appropriate column. (b) Multiply these zonal lumens by their respective zonal multipliers for each room cavity ratio. (c) Total the columns under each value of room cavity ratio. (d) Divide the totals obtained in Step 1(c) by the total luminaire lumens in the 0- to 90-degree zone. These quotients are the direct ratios (DR).

Step 2. (a) Enter the direct ratios in the appropriate spaces on a work sheet of the type shown in Fig. 9–20. (b) For each room surface reflectance combination determine the downward reflected component (DRC) from Fig. 9–21 and insert opposite

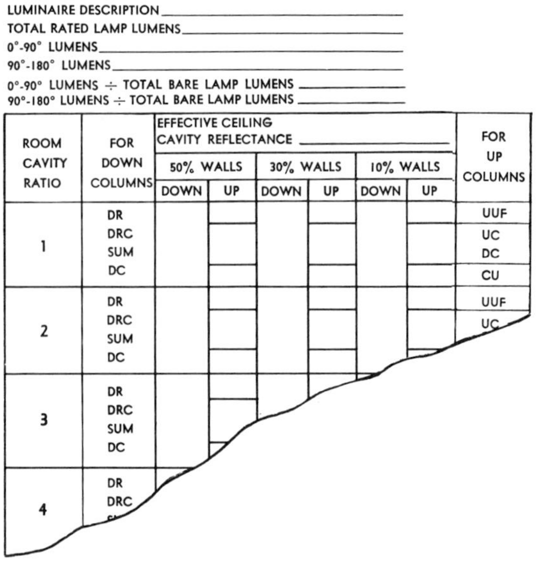

LUMINAIRE DESCRIPTION________
TOTAL RATED LAMP LUMENS________
0°-90° LUMENS________
90°-180° LUMENS________
0°-90° LUMENS ÷ TOTAL BARE LAMP LUMENS________
90°-180° LUMENS ÷ TOTAL BARE LAMP LUMENS________

ROOM CAVITY RATIO	FOR DOWN COLUMNS	EFFECTIVE CEILING CAVITY REFLECTANCE ____ 50% WALLS		30% WALLS		10% WALLS		FOR UP COLUMNS
		DOWN	UP	DOWN	UP	DOWN	UP	
1	DR							UUF
	DRC							UC
	SUM							DC
	DC							CU
2	DR							UUF
	DRC							UC
	SUM							
	DC							
3	DR							
	DRC							
	SUM							
	DC							
4	DR							
	DRC							

Fig. 9–20. Portion of a typical Work Sheet for computing coefficients of utilization (for effective floor cavity reflectance of 20 per cent).

DRC on the work sheet of Fig. 9–20. (c) Add the direct ratio and the downward reflected component. This sum is the downward utilization factor (DUF), the total per cent of downward luminaire lumens that reach the work plane. (d) Multiply the sums (DUF's) by the per cent of lamp lumens emitted downward by the luminaire. *These results are the downward coefficients (DC) and, if the luminaire is totally direct, are also the coefficients of utilization.*

Step 3. (a) Multiply the upward utilization factor (UUF) for each reflectance combination for each room cavity ratio by the per cent of bare lamp lumens emitted upward by the luminaire. Upward utilization factors are found in Fig. 9–22. The products obtained are the upward coefficients (UC). For totally indirect luminaires, they are the coefficients of utilization. (b) Enter the upward coefficients in the appropriate positions in Fig. 9–20. (c) Add the upward coefficients and the downward coefficients for each wall and ceiling reflectance combination for each room cavity ratio. *The sums are the coefficients of utilization for luminaires that emit both upward and downward light.*

Step 4. (a) Fill out the top of the work sheet in Fig. 9–23. (b) Enter the direct ratios from Fig. 9–19 for each room cavity ratio in the space for DR in data column 1. (c) Multiply (1 − DR) by ϕ_D to obtain ϕ_1, and enter these values in data column 1. (d) Multiply ϕ_1 by the number proceding ϕ_1 that is shown in the top line for each room cavity ratio in data column 2. Enter the results immediately to the right. This is also the wall direct radiation coefficient. (e) Enter ϕ_2 from the top of the work sheet at the indicated position in data column 2. (f) Multiply ϕ_D by the direct ratio. This is ϕ_3; enter in data column 2. (g) Multiply the ϕ values (2.5 ϕ_1, ϕ_2, ϕ_3) by the number shown on the same line in the columns at

Fig. 9-21. Downward Reflected Component vs Direct Ratio

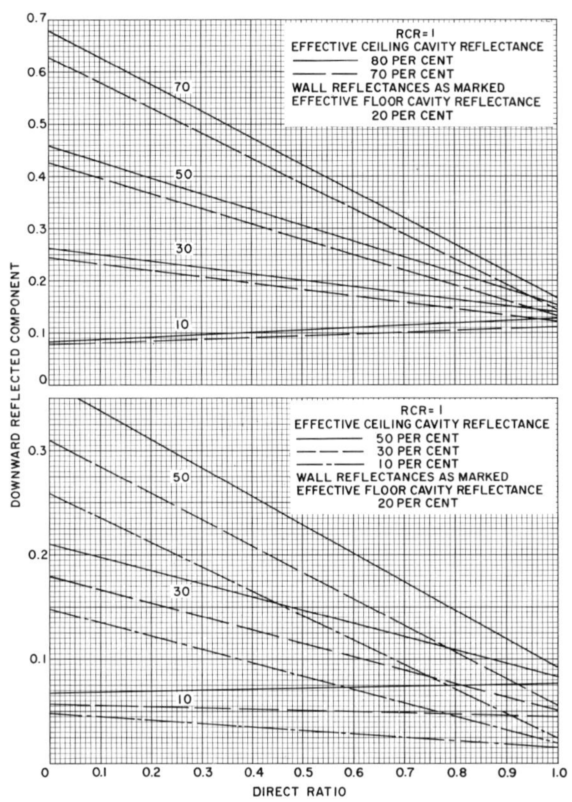

a. Downward Reflected Component vs Direct Ratio for Room Cavity Ratio = 1.0.

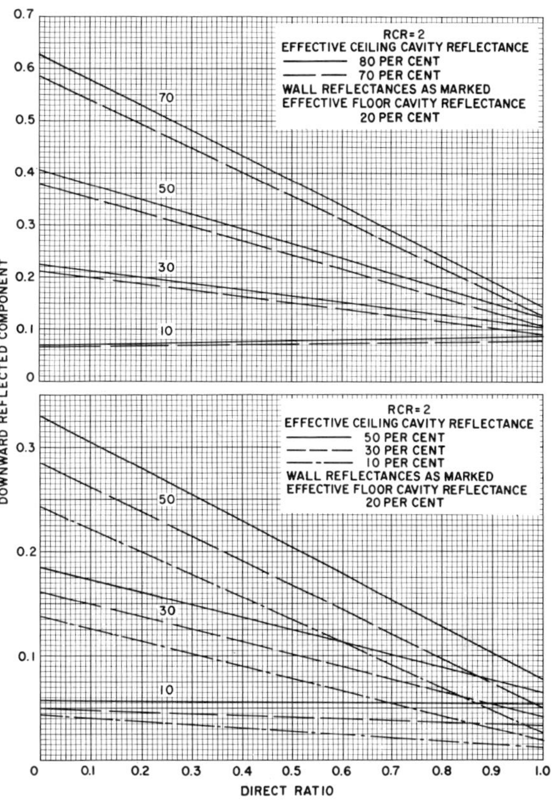

b. Downward Reflected Component vs Direct Ratio for Room Cavity Ratio = 2.0.

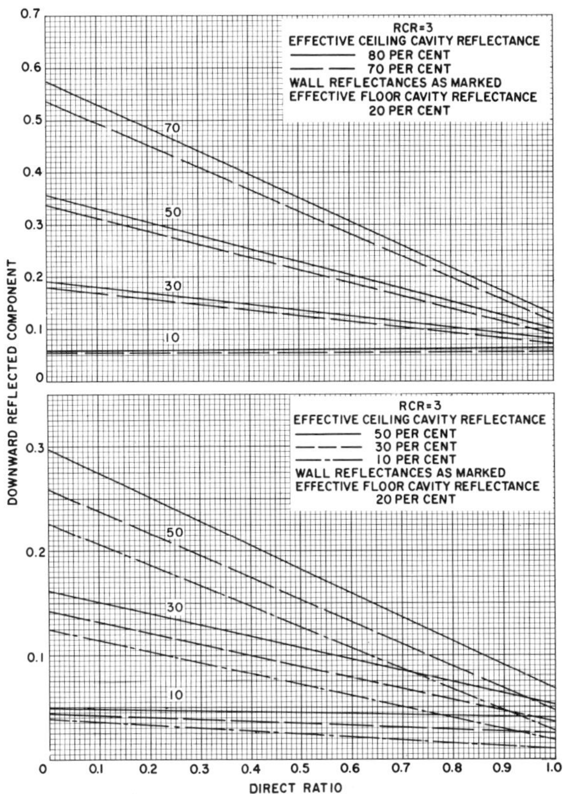

c. Downward Reflected Component vs Direct Ratio for Room Cavity Ratio = 3.0.

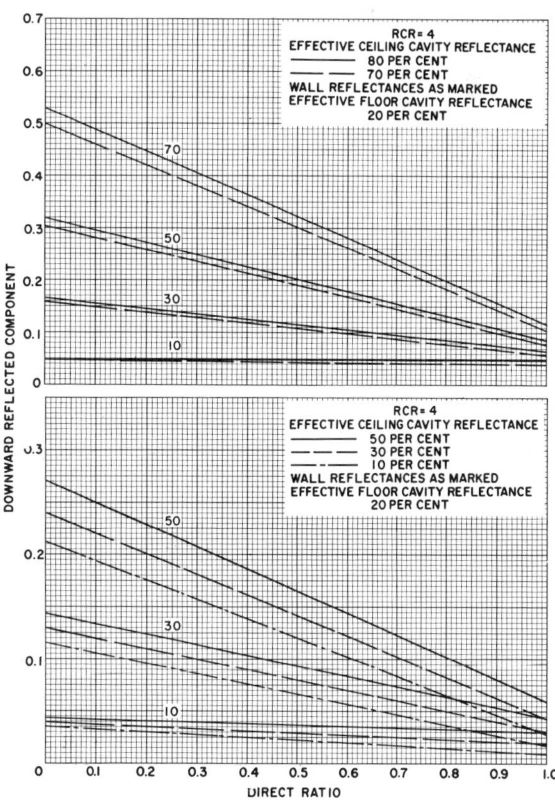

d. Downward Reflected Component vs Direct Ratio for Room Cavity Ratio = 4.0.

Fig. 9–21. *Continued*

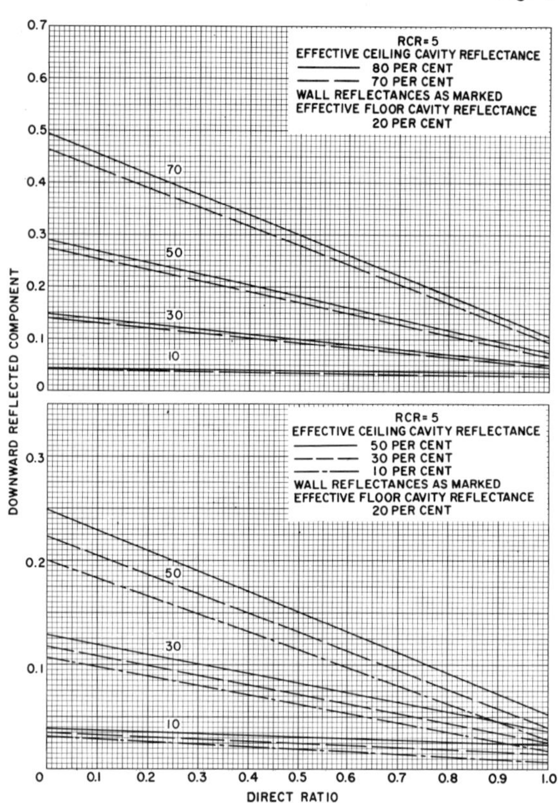

e. Downward Reflected Component vs Direct Ratio for Room Cavity Ratio = 5.0.

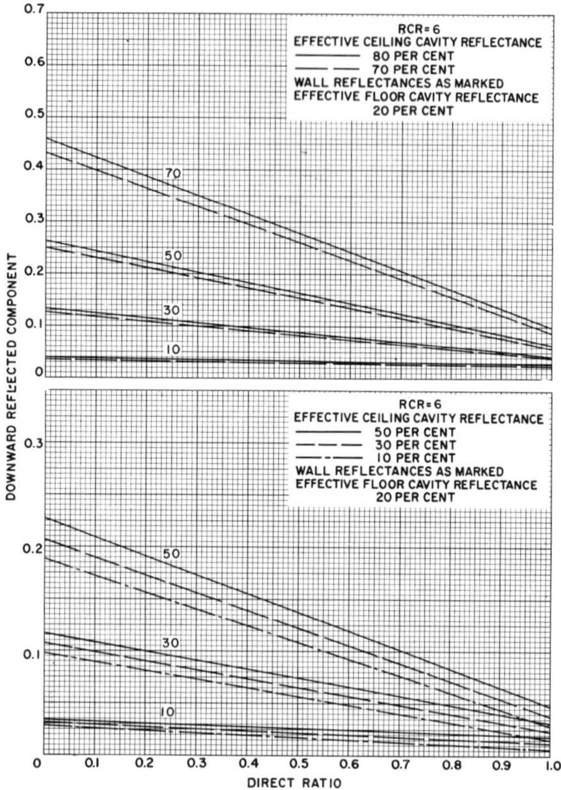

f. Downward Reflected Component vs Direct Ratio for Room Cavity Ratio = 6.0.

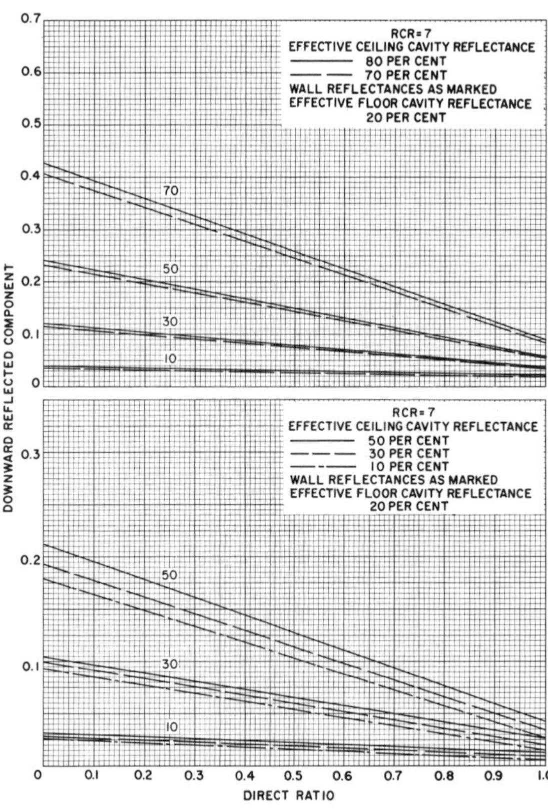

g. Downward Reflected Component vs Direct Ratio for Room Cavity Ratio = 7.0.

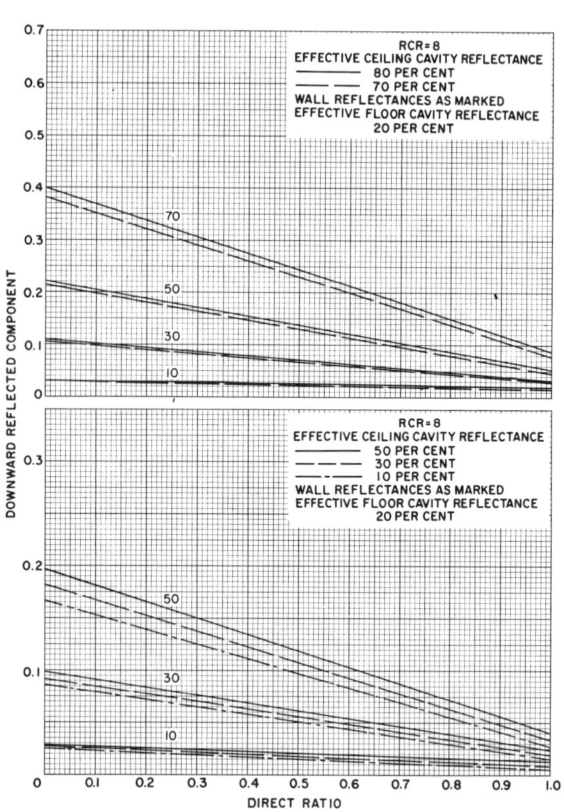

h. Downward Reflected Component vs Direct Ratio for Room Cavity Ratio = 8.0.

Fig. 9-21. *Continued*

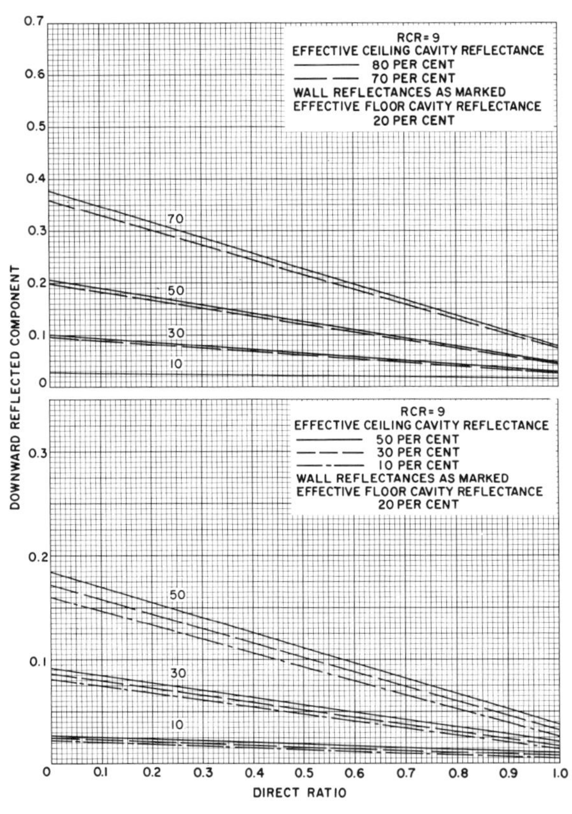

i. Downward Reflected Component vs Direct Ratio for Room Cavity Ratio = 9.0.

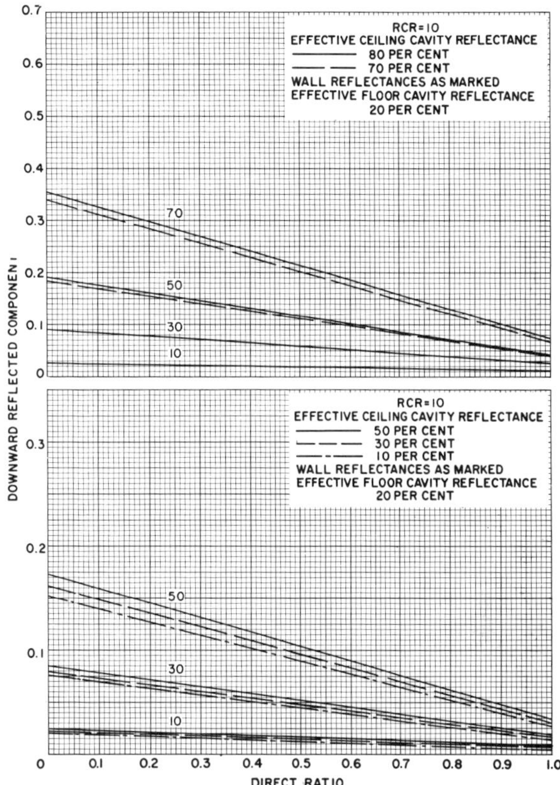

j. Downward Reflected Component vs Direct Ratio for Room Cavity Ratio = 10.0.

Fig. 9-22. Upward Utilization Factors (20 Per Cent Effective Floor Cavity Reflectance)

Per Cent Effective Ceiling Cavity Reflectance	80				70				50			30			10		
Per Cent Wall Reflectance	70	50	30	10	70	50	30	10	50	30	10	50	30	10	50	30	10
Room Cavity Ratio																	
0	.952	.952	.952	.952	.814	.814	.814	.814	.555	.555	.555	.319	.319	.319	.102	.102	.102
1	.866	.826	.790	.758	.739	.706	.678	.643	.484	.467	.451	.278	.270	.262	.089	.087	.085
2	.786	.718	.662	.614	.670	.615	.569	.530	.422	.394	.370	.243	.229	.217	.078	.074	.071
3	.718	.632	.566	.513	.611	.541	.488	.445	.372	.339	.312	.215	.200	.184	.069	.064	.060
4	.654	.556	.484	.430	.556	.478	.418	.373	.328	.291	.263	.190	.171	.155	.061	.056	.051
5	.598	.482	.418	.364	.509	.423	.362	.316	.291	.253	.223	.169	.148	.133	.054	.048	.044
6	.548	.438	.364	.311	.466	.376	.315	.271	.260	.221	.192	.151	.130	.114	.049	.042	.038
7	.504	.391	.318	.267	.428	.336	.276	.233	.232	.193	.165	.135	.114	.098	.044	.037	.033
8	.464	.352	.280	.232	.395	.302	.243	.202	.209	.171	.143	.124	.101	.085	.039	.033	.028
9	.429	.317	.248	.202	.365	.273	.216	.176	.189	.152	.125	.110	.090	.075	.036	.029	.025
10	.399	.288	.222	.178	.340	.248	.193	.155	.173	.136	.110	.101	.080	.066	.033	.026	.022

the right for the various room surface reflectance combinations. (h) Add the results of the three multiplications performed in Step (g) and enter immediately below. *These values are the wall luminance coefficients.*

Step 5. (a) Copy the values from data column 2 of Fig. 9-23 in data column 2 of Fig. 9-24. (b) Multiply the ϕ values ($2.5\,\phi_1$, ϕ_2, ϕ_3) by the number shown in the same line in the columns at the right for the various room surface reflectance combinations. (c) Add the results of the three multiplications performed in Step (b) and enter immediately below.

These values are the ceiling cavity luminance coefficients.

Fig. 9–23. Work Sheet for Computing Wall Luminance Coefficients

Luminaire Description________

Total Rated Lamp Lumens________

ϕ_D = 0° to 90° luminaire lumens ÷ rated lamp lumens = ________

$\phi_1 = \phi_D$ (1-DR)

ϕ_2 = 90° to 180° luminaire lumens ÷ rated lamp lumens = ________

$\phi_3 = \phi_D \times$ DR

DR = Direct Ratio

WDRC = Wall Direct Radiation Coefficient. This is found for each room cavity ratio by multiplying ϕ_1 for that room cavity ratio by the number preceding it in Data Column 2.

			Effective Ceiling Cavity Reflectance 80%			Effective Ceiling Cavity Reflectance 70%			Effective Ceiling Cavity Reflectance 50%			Effective Ceiling Cavity Reflectance 30%			Effective Ceiling Cavity Reflectance 10%		
Room Cavity Ratio	Data Column 1	Data Column 2	50% Walls	30% Walls	10% Walls	50% Walls	30% Walls	10% Walls	50% Walls	30% Walls	10% Walls	50% Walls	30% Walls	10% Walls	50% Walls	30% Walls	10% Walls
	1.000	2.5 ϕ_1	.568	.323	.102	.564	.322	.102	.557	.319	.102	.550	.317	.102	.543	.315	.102
	DR	ϕ_2	.256	.146	.046	.221	.126	.040	.150	.086	.027	.087	.050	.016	.028	.016	.005
1	1-DR	ϕ_3	.092	.052	.017	.085	.049	.015	.073	.042	.013	.062	.036	.011	.052	.030	.010
	ϕ_1		LC_W	LC_W	LC_W	LC_W	LC_W	LC_W	LC_W	LC_W	LC_W	LC_W	LC_W	LC_W	LC_W	LC_W	LC_W
	1.000	1.25 ϕ_1	.616	.338	.104	.609	.336	.104	.596	.332	.103	.584	.328	.103	.572	.325	.103
	DR	ϕ_2	.238	.131	.040	.204	.113	.035	.140	.078	.024	.081	.045	.014	.026	.015	.005
2	1-DR	ϕ_3	.081	.045	.014	.076	.042	.013	.066	.037	.011	.057	.032	.010	.049	.028	.009
	ϕ_1		LC_W	LC_W	LC_W	LC_W	LC_W	LC_W	LC_W	LC_W	LC_W	LC_W	LC_W	LC_W	LC_W	LC_W	LC_W
	1.000	.8333 ϕ_1	.660	.351	.105	.651	.349	.105	.634	.344	.104	.619	.339	.104	.605	.335	.104
	DR	ϕ_2	.218	.116	.035	.187	.100	.030	.128	.069	.021	.074	.041	.012	.024	.013	.004
3	1-DR	ϕ_3	.071	.038	.011	.067	.036	.011	.059	.032	.010	.052	.028	.009	.045	.025	.008
	ϕ_1		LC_W	LC_W	LC_W	LC_W	LC_W	LC_W	LC_W	LC_W	LC_W	LC_W	LC_W	LC_W	LC_W	LC_W	LC_W
	1.000	.625 ϕ_1	.693	.360	.106	.683	.358	.106	.664	.352	.105	.647	.347	.105	.630	.343	.104
	DR	ϕ_2	.202	.105	.031	.174	.091	.027	.120	.064	.019	.069	.037	.011	.022	.012	.004
4	1-DR	ϕ_3	.064	.033	.010	.061	.032	.009	.054	.029	.009	.048	.026	.008	.042	.023	.007
	ϕ_1		LC_W	LC_W	LC_W	LC_W	LC_W	LC_W	LC_W	LC_W	LC_W	LC_W	LC_W	LC_W	LC_W	LC_W	LC_W
	1.000	.500 ϕ_1	.721	.368	.107	.711	.365	.106	.690	.360	.106	.671	.354	.105	.653	.350	.105
	DR	ϕ_2	.188	.096	.028	.162	.083	.025	.111	.058	.017	.065	.034	.010	.021	.011	.003
5	1-DR	ϕ_3	.058	.029	.009	.055	.028	.008	.050	.026	.008	.045	.023	.007	.040	.021	.006
	ϕ_1		LC_W	LC_W	LC_W	LC_W	LC_W	LC_W	LC_W	LC_W	LC_W	LC_W	LC_W	LC_W	LC_W	LC_W	LC_W
	1.000	.4167 ϕ_1	.746	.374	.107	.734	.371	.107	.713	.365	.106	.693	.360	.106	.674	.355	.105
	DR	ϕ_2	.175	.088	.026	.151	.076	.022	.104	.053	.015	.060	.032	.009	.019	.010	.003
6	1-DR	ϕ_3	.053	.026	.008	.050	.027	.007	.046	.023	.007	.042	.022	.006	.038	.020	.006
	ϕ_1		LC_W	LC_W	LC_W	LC_W	LC_W	LC_W	LC_W	LC_W	LC_W	LC_W	LC_W	LC_W	LC_W	LC_W	LC_W
	1.000	.357 ϕ_1	.766	.380	.107	.755	.376	.107	.733	.370	.107	.712	.365	.106	.693	.360	.106
	DR	ϕ_2	.164	.082	.023	.141	.071	.020	.097	.049	.014	.057	.029	.008	.018	.009	.003
7	1-DR	ϕ_3	.048	.024	.007	.046	.023	.007	.042	.021	.006	.039	.020	.006	.036	.018	.005
	ϕ_1		LC_W	LC_W	LC_W	LC_W	LC_W	LC_W	LC_W	LC_W	LC_W	LC_W	LC_W	LC_W	LC_W	LC_W	LC_W
	1.000	.3125 ϕ_1	.784	.383	.108	.773	.380	.108	.750	.375	.107	.729	.370	.107	.710	.365	.106
	DR	ϕ_2	.154	.075	.021	.132	.065	.018	.091	.046	.013	.053	.026	.008	.017	.009	.002
8	1-DR	ϕ_3	.044	.021	.006	.043	.021	.006	.039	.020	.006	.036	.018	.005	.034	.017	.005
	ϕ_1		LC_W	LC_W	LC_W	LC_W	LC_W	LC_W	LC_W	LC_W	LC_W	LC_W	LC_W	LC_W	LC_W	LC_W	LC_W
	1.000	.2778 ϕ_1	.800	.387	.108	.788	.384	.108	.766	.379	.107	.745	.374	.107	.726	.369	.107
	DR	ϕ_2	.145	.070	.019	.125	.061	.017	.086	.042	.012	.050	.025	.007	.016	.008	.002
9	1-DR	ϕ_3	.041	.020	.006	.040	.019	.005	.037	.018	.005	.034	.017	.005	.032	.016	.005
	ϕ_1		LC_W	LC_W	LC_W	LC_W	LC_W	LC_W	LC_W	LC_W	LC_W	LC_W	LC_W	LC_W	LC_W	LC_W	LC_W
	1.000	.250 ϕ_1	.815	.390	.108	.802	.388	.108	.780	.382	.108	.760	.377	.107	.740	.372	.107
	DR	ϕ_2	.136	.066	.018	.118	.057	.016	.081	.040	.011	.047	.024	.007	.015	.008	.002
10	1-DR	ϕ_1	.038	.018	.005	.037	.018	.005	.036	.017	.005	.032	.016	.005	.030	.015	.004
	ϕ_1		LC_W	LC_W	LC_W	LC_W	LC_W	LC_W	LC_W	LC_W	LC_W	LC_W	LC_W	LC_W	LC_W	LC_W	LC_W

Fig. 9–24. Work Sheet for Computing Ceiling Cavity Luminance Coefficients

Room Cavity Ratio	Data Column 1	Data Column 2	Effective Ceiling Cavity Reflectance 80%			Effective Ceiling Cavity Reflectance 70%			Effective Ceiling Cavity Reflectance 50%			Effective Ceiling Cavity Reflectance 30%			Effective Ceiling Cavity Reflectance 10%		
			50% Walls	30% Walls	10% Walls	50% Walls	30% Walls	10% Walls	50% Walls	30% Walls	10% Walls	50% Walls	30% Walls	10% Walls	50% Walls	30% Walls	10% Walls
	1.000	2.5 ϕ_1	.103	.059	.019	.088	.050	.016	.061	.035	.011	.035	.020	.006	.011	.006	.002
	DR	ϕ_2	.945	.924	.906	.807	.793	.780	.552	.546	.539	.318	.315	.314	.102	.102	.102
1	1-DR	ϕ_3	.165	.158	.152	.141	.136	.130	.097	.093	.090	.056	.054	.052	.018	.017	.017
	ϕ_1		LC_{CC}	LC_{CC}	LC_{CC}	LC_{CC}	LC_{CC}	LC_{CC}	LC_{CC}	LC_{CC}	LC_{CC}	LC_{CC}	LC_{CC}	LC_{CC}	LC_{CC}	LC_{CC}	LC_{CC}
	1.000	1.25 ϕ_1	.191	.105	.032	.164	.090	.028	.112	.062	.019	.065	.036	.011	.021	.012	.004
	DR	ϕ_2	.940	.905	.877	.803	.780	.759	.551	.539	.528	.318	.314	.310	.102	.102	.101
2	1-DR	ϕ_3	.144	.132	.122	.123	.114	.106	.084	.079	.074	.049	.046	.043	.016	.015	.014
	ϕ_1		LC_{CC}	LC_{CC}	LC_{CC}	LC_{CC}	LC_{CC}	LC_{CC}	LC_{CC}	LC_{CC}	LC_{CC}	LC_{CC}	LC_{CC}	LC_{CC}	LC_{CC}	LC_{CC}	LC_{CC}
	1.000	.8333 ϕ_1	.262	.139	.042	.224	.120	.036	.154	.083	.025	.089	.049	.015	.029	.016	.005
	DR	ϕ_2	.932	.890	.859	.800	.768	.745	.548	.534	.522	.317	.312	.308	.102	.101	.101
3	1-DR	ϕ_3	.126	.113	.103	.108	.098	.089	.074	.068	.062	.043	.040	.037	.014	.013	.012
	ϕ_1		LC_{CC}	LC_{CC}	LC_{CC}	LC_{CC}	LC_{CC}	LC_{CC}	LC_{CC}	LC_{CC}	LC_{CC}	LC_{CC}	LC_{CC}	LC_{CC}	LC_{CC}	LC_{CC}	LC_{CC}
	1.000	.625 ϕ_1	.325	.168	.049	.278	.145	.043	.191	.101	.030	.111	.059	.018	.036	.019	.006
	DR	ϕ_2	.925	.880	.845	.794	.760	.734	.546	.530	.517	.316	.311	.307	.102	.101	.101
4	1-DR	ϕ_2	.113	.097	.086	.095	.084	.075	.006	.058	.053	.038	.034	.031	.012	.011	.010
	ϕ_1		LC_{CC}	LC_{CC}	LC_{CC}	LC_{CC}	LC_{CC}	LC_{CC}	LC_{CC}	LC_{CC}	LC_{CC}	LC_{CC}	LC_{CC}	LC_{CC}	LC_{CC}	LC_{CC}	LC_{CC}
	1.000	.500 ϕ_1	.376	.191	.056	.323	.166	.048	.223	.116	.034	.129	.068	.020	.042	.022	.007
	DR	ϕ_2	.921	.871	.836	.790	.754	.727	.544	.527	.514	.316	.310	.305	.102	.101	.101
5	1-DR	ϕ_3	.098	.083	.073	.085	.072	.063	.058	.051	.045	.034	.030	.027	.011	.010	.009
	ϕ_1		LC_{CC}	LC_{CC}	LC_{CC}	LC_{CC}	LC_{CC}	LC_{CC}	LC_{CC}	LC_{CC}	LC_{CC}	LC_{CC}	LC_{CC}	LC_{CC}	LC_{CC}	LC_{CC}	LC_{CC}
	1.000	.4167 ϕ_1	.421	.211	.060	.367	.183	.053	.250	.128	.037	.145	.075	.022	.047	.025	.007
	DR	ϕ_2	.915	.867	.829	.786	.750	.722	.542	.526	.512	.315	.309	.304	.102	.101	.101
6	1-DR	ϕ_3	.087	.073	.062	.075	.063	.054	.052	.044	.038	.030	.026	.023	.010	.009	.008
	ϕ_1		LC_{CC}	LC_{CC}	LC_{CC}	LC_{CC}	LC_{CC}	LC_{CC}	LC_{CC}	LC_{CC}	LC_{CC}	LC_{CC}	LC_{CC}	LC_{CC}	LC_{CC}	LC_{CC}	LC_{CC}
	1.000	.357 ϕ_1	.459	.227	.964	.396	.197	.056	.273	.138	.040	.159	.081	.024	.052	.027	.008
	DR	ϕ_2	.910	.860	.825	.781	.745	.718	.541	.523	.510	.314	.308	.304	.102	.101	.100
7	1-DR	ϕ_3	.078	.064	.053	.067	.055	.047	.046	.039	.033	.027	.023	.020	.009	.008	.007
	ϕ_1		LC_{CC}	LC_{CC}	LC_{CC}	LC_{CC}	LC_{CC}	LC_{CC}	LC_{CC}	LC_{CC}	LC_{CC}	LC_{CC}	LC_{CC}	LC_{CC}	LC_{CC}	LC_{CC}	LC_{CC}
	1.000	.3125 ϕ_1	.493	.240	.068	.424	.209	.059	.299	.147	.041	.171	.087	.025	.055	.028	.008
	DR	ϕ_2	.906	.855	.822	.780	.742	.716	.539	.522	.508	.314	.307	.303	.101	.101	.100
8	1-DR	ϕ_3	.070	.056	.046	.061	.049	.040	.042	.034	.029	.024	.020	.017	.008	.007	.006
	ϕ_1		LC_{CC}	LC_{CC}	LC_{CC}	LC_{CC}	LC_{CC}	LC_{CC}	LC_{CC}	LC_{CC}	LC_{CC}	LC_{CC}	LC_{CC}	LC_{CC}	LC_{CC}	LC_{CC}	LC_{CC}
	1.000	.2778 ϕ_1	.521	.252	.070	.449	.218	.061	.311	.154	.044	.181	.090	.026	.059	.030	.009
	DR	ϕ_2	.900	.851	.820	.775	.740	.714	.537	.520	.507	.313	.306	.303	.101	.101	.100
9	1-DR	ϕ_3	.063	.050	.040	.055	.043	.035	.038	.030	.025	.022	.018	.015	.007	.006	.005
	ϕ_1		LC_{CC}	LC_{CC}	LC_{CC}	LC_{CC}	LC_{CC}	LC_{CC}	LC_{CC}	LC_{CC}	LC_{CC}	LC_{CC}	LC_{CC}	LC_{CC}	LC_{CC}	LC_{CC}	LC_{CC}
	1.000	.250 ϕ_1	.545	.262	.073	.470	.227	.063	.326	.160	.045	[illegible]	.095	.027	.062	.031	.009
	DR	ϕ_2	.896	.847	.818	.773	.737	.713	.536	.518	.506	.313	.306	.302	.101	.101	.100
10	1-DR	ϕ_2	.058	.045	.036	.050	.039	.031	.034	.027	.022	.020	.016	.013	.007	.005	.004
	ϕ_1		LC_{CC}	LC_{CC}	LC_{CC}	LC_{CC}	LC_{CC}	LC_{CC}	LC_{CC}	LC_{CC}	LC_{CC}	LC_{CC}	LC_{CC}	LC_{CC}	LC_{CC}	LC_{CC}	LC_{CC}

Fig. 9-25. Methods of Determining the Direct Illumination Component of Illumination at a Point

Source	Point			Linear			Area		
Plane	Horiz.	Vert.	Inclined	Horiz.	Vert.	Inclined	Horiz.	Vert.	Inclined
1. Inverse Square	×	×	×						
2. Plan-Scale Method	×	×							
3. Angular Coord-DIC Method				×	×				
4. IES-London-Aspect Factor Method				×	×	×			
5. Illumination Charts and Tables	×	×	×	×	×	×	×	×	×
6. Idealized Source Chart							×	×	
7. Configuration Factor							×	×	

Point Calculation Methods[10]

The calculation of the illumination at a point, whether on a horizontal, a vertical, or an inclined plane consists of two parts: the Direct Component and the Reflected Component. The total of these two components is the illumination at the point in question.

Fig. 9-25 is a matrix of the most commonly used systems of computing the direct component. Although each of the methods can be used to make calculations under all the conditions if sufficient data is available, the × in the matrix indicated the conditions under which the method is easier to use. The following is a brief description of each method.

Inverse-Square Method. Variations in the formula involving the inverse-square law are used to determine the illumination at definite points where the distance from the source is at least five times the maximum dimension of the source. In such situations the illumination is proportional to the candlepower of the source in the given direction, and inversely proportional to the square of the distance from the source:

$$\text{Illumination on Plane Normal to Light Ray} = \frac{\text{Candlepower of Source in Direction of Ray}}{\left(\text{Distance from Source to Plane}\right)^2}$$

$$= \frac{I}{D^2} \text{ (see Fig. 9-26a).}$$

If the surface on which the illumination to be determined is tilted, instead of normal to the rays, the light will be spread over a greater area, reducing the illumination in the ratio of the area of plane A to the area of plane B, as shown in Fig. 9-26b. This ratio is equal to the cosine of the angle of incidence or tilt, thus:

$$\text{Illumination on Plane } B = \frac{\text{Candlepower of Source in Direction of Ray}}{\left(\text{Distance from Source to Point in Plane}\right)^2} \times \cos\beta$$

$$= \frac{I}{D^2} \times \cos\beta$$

where β equals the angle between the light ray and a perpendicular to the plane at that point. Fig. 9-27 is a convenient table of trigonometric functions.

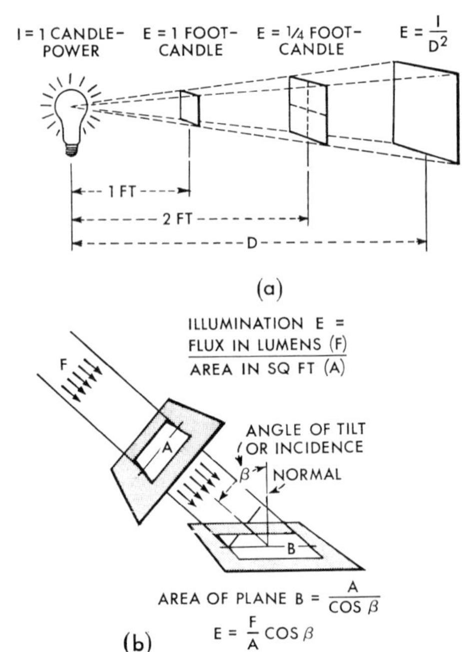

Fig. 9-26. Point calculations assume a point source and involve applications of the inverse-square and cosine laws.

Fig. 9-28 illustrates the particular cases where the plane on which the illumination is to be determined is either horizontal or vertical. In Fig. 9-28 H is the vertical mounting height of the light source above the plane of measurement; R is the horizontal distance from the light source to the point whose illumination is being computed; D is the actual distance from the light source to the point; and I is the candlepower of the source in the direction of the point (from the distribution curve). Where H and R are known, the angle θ may be obtained from Fig. 9-29. For the horizontal plane, angle β equals angle θ. Furthermore, since $\cos\theta = \cos\beta = H/D$, the formula in Fig. 9-28a may be expressed as follows:

$$\text{Footcandles on Horizontal Plane } (E_h) = \frac{I \times \cos\theta}{D^2} = \frac{I \times \cos\beta}{D^2}$$

$$= \frac{I \times H}{D^3} = \frac{I \times \cos^3\theta}{H^2}$$

Fig. 9-27. Trigonometric Functions

θ°	sin θ	cos θ	cos² θ	cos³ θ	tan θ	θ°	sin θ	cos θ	cos² θ	cos³ θ	tan θ
0	0.0	1.000	1.000	1.000	0.0	46	0.719	0.695	0.483	0.335	1.035
1	.0175	1.000	1.000	1.000	.0174	47	.731	.682	.465	.317	1.072
2	.0349	0.999	0.999	0.998	.0349	48	.743	.669	.448	.300	1.110
3	.0523	.999	.997	.996	.0524	49	.755	.656	.430	.282	1.150
4	.0698	.998	.995	.993	.0699	50	.766	.643	.413	.266	1.191
5	.0872	.996	.992	.989	.0874	51	.777	.629	.396	.249	1.234
6	.105	.995	.989	.984	.1051	52	.788	.616	.379	.233	1.279
7	.122	.993	.985	.978	.1227	53	.799	.602	.362	.218	1.327
8	.139	.990	.981	.971	.1405	54	.809	.588	.345	.203	1.376
9	.156	.988	.976	.964	.1583	55	.819	.574	.329	.189	1.428
10	.174	.985	.970	.955	.1763	56	.829	.559	.313	.175	1.482
11	.191	.982	.964	.946	.1943	57	.839	.545	.297	.162	1.539
12	.208	.978	.957	.936	.2125	58	.848	.530	.281	.149	1.600
13	.225	.974	.949	.925	.2308	59	.857	.515	.265	.137	1.664
14	.242	.970	.941	.913	.2493	60	.866	.500	.250	.125	1.732
15	.259	.966	.933	.901	.2679	61	.875	.485	.235	.114	1.804
16	.276	.961	.924	.888	.2867	62	.883	.470	.220	.103	1.880
17	.292	.956	.915	.875	.3057	63	.891	.454	.206	.0936	1.962
18	.309	.951	.905	.860	.3249	64	.899	.438	.192	.0842	2.050
19	.326	.946	.894	.845	.3443	65	.906	.423	.179	.0755	2.144
20	.342	.940	.883	.830	.3639	66	.914	.407	.165	.0673	2.246
21	.358	.934	.872	.814	.3838	67	.921	.391	.153	.0597	2.355
22	.375	.927	.860	.797	.4040	68	.927	.375	.140	.0526	2.475
23	.391	.921	.847	.780	.4244	69	.934	.358	.128	.0460	2.605
24	.407	.914	.835	.762	.4452	70	.940	.342	.117	.0400	2.747
25	.423	.906	.821	.744	.4663	71	.946	.326	.106	.0347	2.904
26	.438	.899	.808	.726	.4877	72	.951	.309	.0955	.0295	3.077
27	.454	.891	.794	.707	.5095	73	.956	.292	.0855	.0250	3.270
28	.470	.883	.780	.688	.5317	74	.961	.276	.0762	.0211	3.487
29	.485	.875	.765	.669	.5543	75	.966	.259	.0670	.0173	3.732
30	.500	.866	.750	.650	.5773	76	.970	.242	.0585	.0142	4.010
31	.515	.857	.735	.630	.6008	77	.974	.225	.0506	.0114	4.331
32	.530	.848	.719	.610	.6248	78	.978	.208	.0432	.0090	4.704
33	.545	.839	.703	.590	.6494	79	.982	.191	.0364	.0070	5.144
34	.559	.829	.687	.570	.6745	80	.985	.174	.0302	.0052	5.671
35	.574	.819	.671	.550	.7002	81	.988	.156	.0245	.0038	6.313
36	.588	.809	.655	.530	.7265	82	.990	.139	.0194	.0027	7.115
37	.602	.799	.638	.509	.7535	83	.993	.122	.0149	.0018	8.144
38	.616	.788	.621	.489	.7812	84	.995	.105	.0109	.0011	9.514
39	.629	.777	.604	.469	.8097	85	.996	.0872	.0076	.0007	11.430
40	.643	.766	.587	.450	.8391	86	.9976	.0698	.0048	.0003	14.300
41	.656	.755	.570	.430	.8692	87	.9986	.0523	.0027	.0001	19.080
42	.669	.743	.552	.410	.9004	88	.9993	.0349	.0012	.0000	28.630
43	.682	.731	.535	.391	.9325	89	.9998	.0175	.0003	.0000	57.280
44	.695	.719	.517	.372	.9656	90	1.0000	0.0000	.0000	.0000	Infinite
45	.707	.707	.500	.354	1.0000						

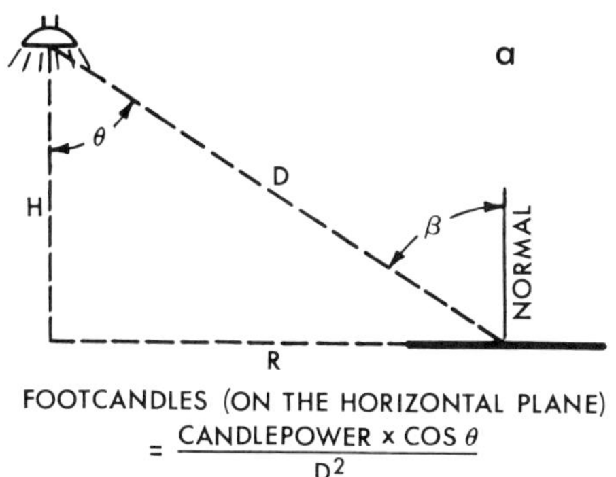

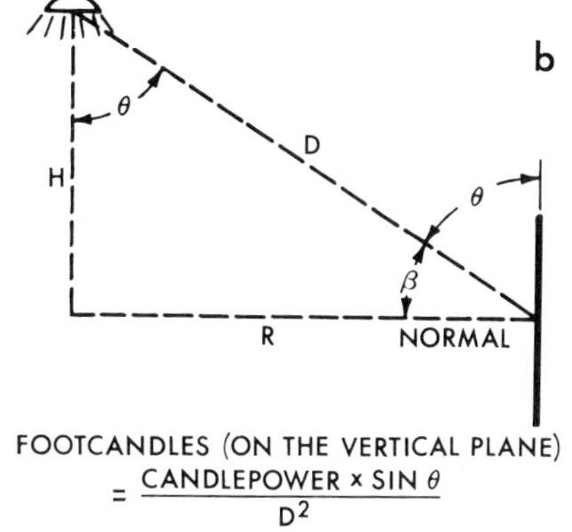

Fig. 9-28. Fundamental relationships for point calculations where the inverse-square law applies.

For the vertical plane $\sin\theta = \cos\beta = R/D$, the formula in Fig. 9–28b may be expressed as follows:

$$\text{Footcandles on Vertical Plane } (E_v) = \frac{I \times \sin\theta}{D^2} = \frac{I \times \cos\beta}{D^2} = \frac{I \times R}{D^3} = \frac{I \times \cos^2\theta \sin\theta}{H^2}$$

It is often more convenient to obtain the illumination from tables than by computation from formulas. Fig. 9–30 gives footcandles on horizontal and vertical planes calculated for various mounting heights and distances from a light source of 100 and 100,000 candelas. Illumination for other values of candlepower is simply a matter of percentages.

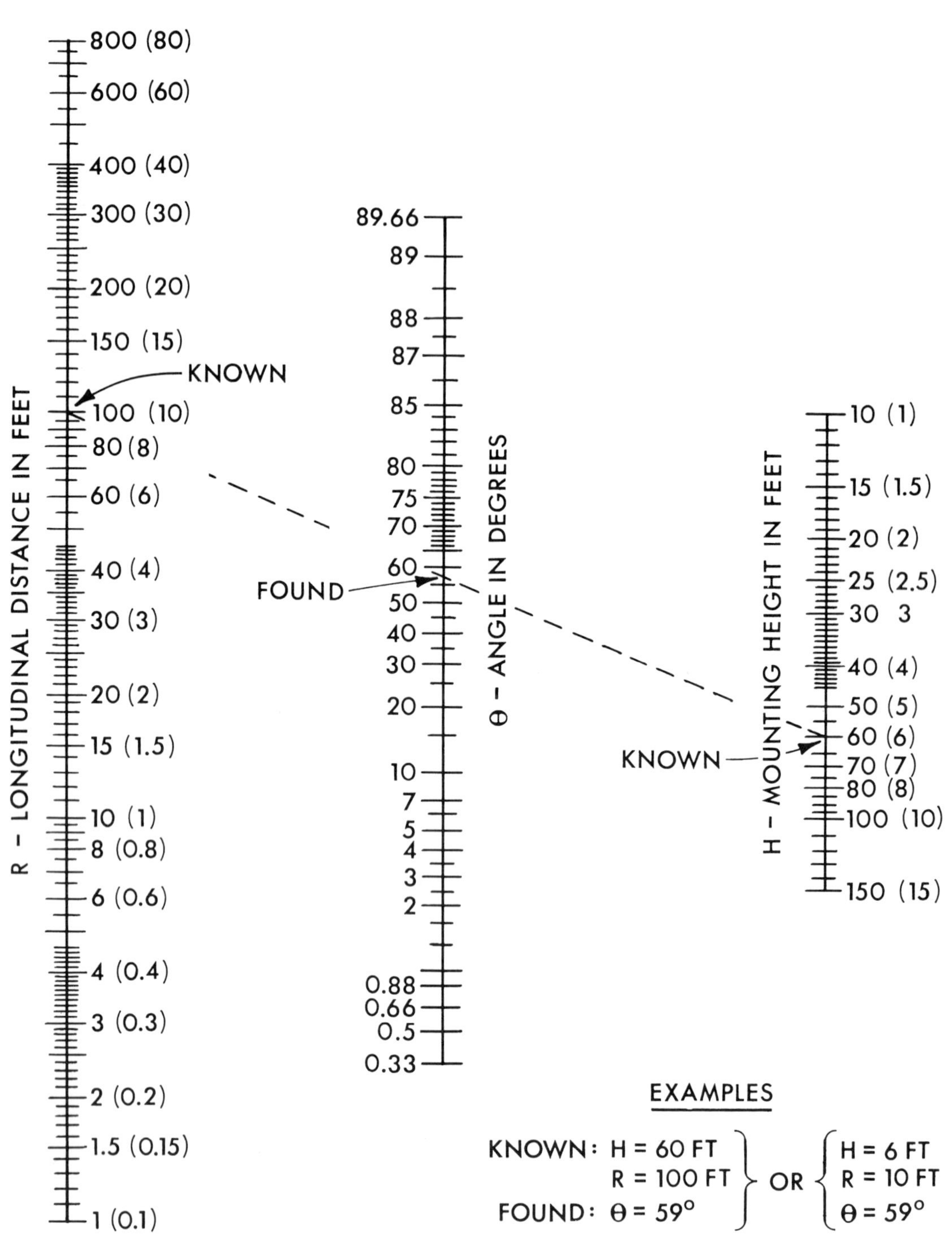

Fig. 9–29. Nomogram for determining vertical angle θ in terms of longitudinal distance R and mounting height H.

Fig. 9-30. Initial Illumination Computed for Points at Various Locations on a Horizontal Plane in Terms of 100 and 100,000 Candlepower Sources*

Horizontal Distance From Axis of Light Source—Feet

Footcandles for Each 100 Candlepower

Height of Light Source Above Surface—Feet	0	1	2	3	4	5	6	7	8	9	10	11	12	13	14	15	16	18	20	22	24	26	28	30	35	40	50
2	0°0′ 25.00	27° 17.85	45° 8.850	56° 4.275	63° 2.245	68° 1.298	71° .802	74° .528	76° .355	78° .255	79° .190	80° .142	81° .113	81° .090	82° .070	82° .058	83° .048	84° .038	84° .025	85° .020	85° .015	86° .013	86° .008	86°. .007	87° .004	87° .000	87° .000
3	0°0′ 11.11	18° 9.500	34° 6.400	45° 3.933	53° 2.400	59° 1.522	63° 1.000	67° .680	69° .477	72° .356	73° .264	75° .205	76° .161	77° .126	78° .100	79° .084	80° .070	81° .050	81° .036	82° .027	83° .021	83° .016	84° .012	84° .011	85° .007	86° .004	87° .002
4	0°0′ 6.250	14° 5.707	27° 4.472	37° 3.200	45° 2.210	51° 1.524	56° 1.066	60° .764	63° .559	66° .419	68° .320	70° .249	72° .198	73° .159	74° .130	75°. .107	76° .090	78° .064	79° .047	80° .037	81° .028	81° .022	82° .018	82°. .015	84° .009	84° .006	86° .003
5	0°0′ 4.000	11° 3.771	22° 3.202	31° 2.522	39° 1.904	45° 1.414	50° 1.050	54° .785	58° .595	61° .458	63° .358	66° .283	67° .228	69° .185	70° .152	72° .126	73° .106	74° .077	76° .057	77° .044	78° .034	79° .027	80° .022	81° .017	82° .010	83° .008	84° .004
6	0°0′ 2.778	9° 2.673	18° 2.372	27° 1.987	34° 1.600	40° 1.260	45° .982	49° .766	53° .600	56° .474	59° .378	61° .305	63° .249	66° .205	67° .170	68° .142	69° .120	71° .088	73° .066	75° .051	76° .040	77° .032	78° .026	79° .021	80° .013	81° .009	83° .005
7	0°0′ 2.041	8 1.980	16° 1.814	23° 1.585	30° 1.336	36° 1.100	41° .893	45° .722	49° .583	52° .473	55° .385	58° .316	60° .261	62° .218	63° .183	65° .154	66° .131	69° .097	71° .074	72° .057	74° .045	75° .036	76° .029	77° .024	79° .016	80° .010	82° .006
8	0°0′ 1.563	7° 1.527	14° 1.427	21° 1.283	27° 1.118	32° .958	37° .800	41° .672	45° .552	48° .458	51° .381	54° .318	56° .267	58° .225	60° .191	62° .163	63° .140	66° .105	68° .080	70° .063	72° .050	73° .040	74° .032	75° .026	77° .018	79° .012	81° .007
9	0°0′ 1.235	6° 1.212	13° 1.148	18° 1.054	24° .943	29° .825	34° .711	38° .697	42° .515	45° .437	48° .370	51° .314	53° .267	55° .228	57° .196	59° .168	61° .146	63° .110	66° .085	68° .067	69° .053	71° .043	72° .035	73° .029	76° .019	77° .013	80° .008
10	0°0′ 1.000	5°43′ .985	11° .943	17° .879	22° .801	27° .716	31° .631	35° .550	39° .476	42° .411	45° .354	48° .305	50° .263	52° .227	54° .196	56° .171	58° .149	61° .115	63° .039	66° .071	67° .057	69° .046	70° .038	72° .032	74° .021	76° .014	79° .008
11	0°0′ .826	5°12′ .816	10° .787	15° .742	20° .686	24° .623	29° .559	32° .496	36° .437	39° .383	42° .335	45° .292	48° .255	50° .223	52° .195	54° .171	56° .150	59° .117	61° .092	63° .074	65° .060	67° .049	69° .040	70° .034	73° .023	75° .015	78° .009
12	0°0′ .694	4°46′ .687	9° .668	14° .634	18° .593	23° .546	27° .497	30° .448	34° .400	37° .356	40° .315	43° .278	45° .246	47° .217	49° .191	51° .169	53° .150	56° .119	59° .094	61° .076	63° .065	65° .051	67° .043	68° .036	71° .024	73° .017	77° .009
13	0°0′ .592	4°24′ .587	9° .571	13° .547	17° .517	21° .481	25° .447	28° .404	32° .366	35° .329	38° .295	40° .263	43° .235	45° .209	47° .187	49° .166	51° .148	54° .119	57° .096	59° .078	62° .064	63° .053	65° .044	67° .037	70° .025	72° .017	76° .010
14	0°0′ .510	4°5′ .506	8° .495	12° .477	16° .454	20° .426	23° .396	27° .365	30° .334	33° .304	36° .275	38° .248	41° .223	43° .201	45° .180	47° .162	49° .146	52° .118	55° .096	58° .079	60° .065	62° .054	63° .046	65° .039	68° .026	71° .018	75° .011
15	0°0′ .444	3°49′ .442	8° .433	11° .419	15° .401	18° .380	22° .356	25° .331	28° .305	31° .280	34° .256	36° .233	39° .212	41° .192	43° .174	45° .157	47° .142	50° .117	53° .096	56° .079	58° .066	60° .055	62° .047	63° .040	67° .027	69° .019	73° .011
16	0°0′ .391	3°35′ .388	7° .382	11° .371	14° .357	17° .339	21° .321	24° .300	27° .280	29° .259	32° .238	35° .219	37° .200	39° .183	41° .167	43° .152	45° .138	48° .115	51° .095	54° .080	56° .067	58° .056	60° .048	62° .041	66° .028	68° .020	72° .012
17	0°0′ .346	3°22′ .344	7° .339	10° .331	13° .319	16° .306	19° .290	22° .274	25° .256	28° .239	30° .222	33° .205	35° .189	37° .174	39° .159	41° .146	43° .134	47° .112	50° .094	52° .079	55° .069	57° .057	59° .048	60° .042	64° .029	67° .021	71° .012
18	0°0′ .309	3°11′ .307	6° .303	9° .297	13° .287	16° .276	18° .264	21° .250	24° .236	27° .221	29° .206	31° .192	34° .178	36° .165	38° .152	40° .140	42° .129	45° .109	48° .092	51° .079	53° .067	55° .057	57° .049	59° .042	63° .030	66° .021	70° .012
19	0°0′ .277	3°1′ .276	6° .273	9° .267	12° .260	15° .251	18° .240	20° .229	23° .217	25° .205	28° .192	30° .180	32° .167	34° .156	36° .145	38° .134	40° .124	43° .106	46° .090	49° .077	52° .066	54° .057	56° .049	58° .042	62° .030	65° .022	69° .013
20	0°0′ .250	2°51′ .249	5°43 .246	9° .242	11° .236	14° .228	17° .219	19° .210	22° .200	24° .190	27° .179	29° .168	31° .158	33° .147	35° .137	37° .128	39° .119	42° .103	45° .088	48° .076	50° .066	52° .057	54° .049	56° .043	60° .030	63° .022	68° .013
21	0°0′ .227	2°44′ .226	5°26′ .224	8° .220	11° .215	13° .210	16° .201	18° .194	21° .185	23° .176	25° .167	28° .158	30° .144	32° .139	34° .131	36° .122	37° .114	41° .099	44° .086	46° .075	49° .065	51° .056	53° .049	55°. .043	59° .031	62° .023	67° .014
22	0°0′ .207	2°36′ .206	5°10′ .205	8° .201	10° .196	13° .192	15° .185	18° .179	20° .171	22° .164	25° .155	27° .148	29° .140	31° .132	33° .124	34° .114	36° .109	39° .096	42° .084	45° .073	47° .064	50° .056	52° .049	54° .043	58° .031	61° .023	66° .014

* Upper figures—angle between light ray and vertical. Lower figures—footcandles on a horizontal plane produced by source. Footcandles on the vertical surface—at a point that lies in a vertical plane which also includes the light source—may be determined by using the multiplying factor found when the table headings are reversed, i.e., the height of the light source is read on the horizontal distance scale, etc.

Fig. 9–30. *Continued*

Height of Light Source Above Surface—Feet	0	1	2	3	4	5	6	7	8	9	10	11	12	13	14	15	16	18	20	22	24	26	28	30	35	40	50
Footcandles for Each 100 Candlepower																											
23	0°0′ .189	2°29′ .189	4°58′ .187	7° .184	10° .181	12° .176	15° .171	17° .165	19° .159	21° .153	24° .146	26° .139	28° .132	29° .125	31° .118	33° .111	35° .105	38° .092	41° .081	44° .071	46° .063	49° .055	51° .049	53° .043	57° .031	60° .023	65° .014
24	0°0′ .174	2°23′ .173	4°45′ .172	7° .170	10° .166	12° .163	14° .158	16° .154	18° .148	21° .143	23° .137	25° .130	27° .124	28° .118	30° .112	32° .106	34° .100	37° .089	40° .079	43° .070	45° .061	47° .054	49° .048	51° .042	56° .031	59° .024	64° .014
25	0°0′ .160	2°17′ .160	4°34′ .158	7° .157	9° .154	11° .151	14° .147	16° .143	18° .138	20° .133	22° .128	24° .123	26° .117	27° .112	29° .106	31° .101	33° .096	36° .086	39° .076	41° .068	44° .060	46° .053	48° .047	50° .042	55° .031	58° .024	63° .015
27	0°0′ .137	2°7′ .137	4°14′ .136	6° .135	8° .133	10° .130	12° .128	15° .124	17° .121	18° .117	20° .113	22° .109	24° .105	26° .100	27° .096	29° .092	31° .087	34° .079	37° .071	39° .064	42° .057	44° .051	46° .046	48° .041	52° .031	56° .024	62° .015
30	0°0′ .111	1°54′ .111	3°50′ .111	5°43′ .109	8° .108	9° .107	11° .105	13° .103	15° .100	17° .098	18° .095	20° .092	22° .089	23° .086	25° .083	27° .080	28° .077	31° .070	34° .064	36° .058	39° .053	41° .048	43° .043	45° .039	49° .031	53° .024	59° .015
33	0°0′ .092	1°44′ .092	3°28′ .091	5°12′ .091	7° .090	9° .089	10° .087	12° .086	14° .084	15° .082	17° .080	18° .078	20° .076	22° .074	23° .072	24° .069	26° .067	29° .062	31° .058	34° .053	36° .049	38° .045	40° .041	42° 037	47° 030	50° .024	57° .015
36	0°0′ .077	1°36′ .077	3°11′ .077	4°46′ .076	6° .076	8° .075	9° .074	11° .073	13° .072	14° .070	16° .069	17° .067	18° .066	20° .064	21° .062	23° .061	24° .059	27° .055	29° .052	31° .048	34° .044	36° .041	38° .038	40° .035	44° .029	48° .023	54° .015
40	0°0′ .063	1°26′ .062	2°52′ .062	4°17′ .062	5°43′ .062	7° .061	9° .060	10° .060	11° .059	13° .058	14° .057	15° .056	17° .055	18° 054	19° .053	21° .051	22° .050	24° .047	27° .045	29° .042	31° .039	33° .037	35° .034	37° .032	41° .027	45° .022	51° .015
45	0°0′ .049	1°16′ .049	2°33′ .049	3°49′ .049	5°5′ .049	6° .049	8° .048	9° .048	10° .047	11° .047	13° .046	14° .045	15° .045	16° .044	17° .043	18° .042	20° .041	22° .040	24° .038	26° .036	28° .034	30° .032	32° .030	34° .028	38° .025	42° .021	48° .013
50	0°0′ .040	1°9′ .040	2°17′ .040	3°26′ .040	4°34′ .040	5°43′ .039	7° .039	8° .039	9° .039	10° .038	11° .038	12° .037	14° .037	15° .036	16° .036	16° .035	18° .035	20° .033	22° .032	24° .031	26° .029	27° .028	29° .027	31° .025	35° .022	39° .019	45° .013
55	0°0′ .033	1°2′ .033	2°5′ .033	3°7′ .033	4°10′ .033	5°9′ .033	6° .032	7° .032	8° .032	9° .032	10° .032	11° .031	12° .031	13° .031	14° .030	15°. .030	16° .029	18° .028	20° .027	22° .026	24° .025	25° .024	27° .023	29° .022	33° .020	36° .018	42° .013
60	0°0′ .028	0°57′ .028	1°55′ .028	2°52′ .028	3°50′ .028	4°46′ .027	5°43′ .027	7° .027	8° .027	9° .027	9° .027	10° .026	11° .026	12° .026	13° .026	14° .025	15° .025	17° .024	18° .024	20° .023	22° .022	23° .021	25° .021	27° .020	30° .018	34° .016	40° .013
70	0°0′ .020	0°49′ .020	1°38′ .020	2°34′ .020	3°16′ .020	4°5′ .020	4°54′ .020	5°43′ .020	7° .020	7° .020	8° .020	9° .020	10° .020	11° .019	11° .019	12° .019	13° .019	14° .019	16° .018	17° .018	19° .017	20° .017	22° .016	23° .016	27° .015	30° .013	36° .011
Footcandles for Each 100,000 Candlepower																											
80	0°0′ 15.63	0°43′ 15.62	1°26′ 15.61	2°9′ 15.59	2°52′ 15.57	3°35′ 15.53	4°17′ 15.49	5°0′ 15.45	5°43′ 15.39	6° 15.35	7° 15.27	8° 15.19	9° 15.09	9° 15.04	10° 14.93	11° 14.82	11° 14.75	13° 14.49	14° 14.27	15° 14.03	17° 13.71	18° 13.44	19° 13.16	21° 12.79	24° 12.02	27° 11.14	32° 9.531
100	0°0′ 10.00	0°34′ 9.999	1°9′ 9.994	1°43′ 9.987	2°17′ 9.976	2°52′ 9.963	3°26′ 9.946	4°0′ 9.927	4°34′ 9.905	5°9′ 9.880	5°43′ 9.852	6° 9.826	7° 9.785	7° 9.761	8° 9.712	9° 9.660	9° 9.630	10° 9.539	11° 9.439	12° 9.330	14° 9.175	15° 9.048	16° 8.914	16° 8.819	19° 8.440	22° 7.993	27° 7.140
125	0°0′ 6.400	0°28′ 6.399	0°55′ 6.398	1°22′ 6.395	1°50′ 6.390	2°17′ 6.385	2°45′ 6.378	3°12′ 6.370	3°40′ 6.361	4°7′ 6.351	4°34′ 6.339	5°2′ 6.326	5°29′ 6.313	6° 6.297	6° 6.286	7° 6.262	7° 6.250	8° 6.209	9° 6.163	10° 6.113	11° 6.059	12° 6.001	13° 5.938	14° 5.872	16° 5.708	18° 5.521	22° 5.120
150	0°0′ 4.444	0°23′ 4.444	0°46′ 4.443	1°9′ 4.442	1°32′ 4.440	1°55′ 4.437	2°17′ 4.434	2°40′ 4.430	3°2′ 4.421	3°26′ 4.416	3°49′ 4.415	4°11′ 4.409	4°34′ 4.402	4°57′ 4.395	5°20′ 4.387	5°43′ 4.379	6° 4.370	7° 4.349	8° 4.324	8° 4.309	9° 4.280	10° 4.249	11° 4.216	11° 4.195	13° 4.102	15° 4.008	19° 3.813
175	0°0′ 3.265	0°20′ 3.265	3°39′ 3.265	0°59′ 3.264	1°19′ 3.263	1°38′ 3.261	1°58′ 3.260	2°17′ 3.258	2°37′ 3.255	2°57′ 3.252	3°16′ 3.249	3°36′ 3.246	3°55′ 3.242	4°15′ 3.238	4°34′ 3.234	4°54′ 3.230	5°13′ 3.225	6° 3.213	7° 3.199	7° 3.191	8° 3.174	8° 3.164	9° 3.145	10° 3.124	11° 3.076	13° 3.024	16° 2.899
200	0°0′ 2.500	0°17′ 2.500	0°34′ 2.500	0°52′ 2.499	1°9′ 2.499	1°26′ 2.498	1°43′ 2.497	2°0′ 2.495	2°17′ 2.494	2°35′ 2.492	2°52′ 2.490	3°9′ 2.489	3°26′ 2.487	3°43′ 2.484	4°0′ 2.482	4°17′ 2.479	4°34′ 2.476	5°9′ 2.470	5°43′ 2.463	6° 2.457	7° 2.446	7° 2.440	8° 2.428	9° 2.415	10° 2.390	11° 2.360	14° 2.282

* Upper figures—angle between light ray and vertical. Lower figures—footcandles on a horizontal plane produced by source. Footcandles on the vertical surface—at a point that lies in a vertical plane which also includes the light source—may be determined by using the multiplying factor found when the table headings are reversed, i.e., the height of the light source is read on the horizontal distance scale, etc.

Plan-Scale Method.[11] The Plan-Scale Method provides a rapid means for finding the illumination at a point due to several luminaires. A single candlepower-to-illumination conversion chart permits the quick construction of a grid of equal illumination contour lines to the scale of the plan of the lighting system or, when the luminaire is symmetrical, an illumination-distance scale. See Fig. 9-31.

Laying these grids or scales directly over the plan drawing of the lighting system, the contributions of all the luminaires to the illumination at any desired point are quickly determined. A correction factor to take into account the scale of the candlepower distribution curve and the height is introduced at the end. See Fig. 9-32 for an example of the use of the method.

Illumination on horizontal planes, or on vertical planes parallel to rows of luminaires, is readily determined.

Angular Coordinate—DIC* Method.[12] The Angular Coordinate Method is most applicable to continuous rows of fluorescent luminaires. Two angles are involved in this calculation, a longitudinal angle α and a lateral angle β. See Fig. 9-33. Angle α is the angle between a vertical line (perpendicular to the ceiling) passing through the seeing task (Point P), and a line from the seeing task to the end of the rows of the luminaires. If the seeing task is not in the vertical plane of a row of luminaires, a parallel reference plane is created for the specification of angle α.

Angle α is easily determined graphically from a chart showing angles α and β for various combinations of V, the vertical distance from the seeing task to the plane of the luminaire, and H, the horizontal distance parallel to the luminaires from the seeing task to the end of the row of luminaires. See Fig.

* Direct Illumination Component.

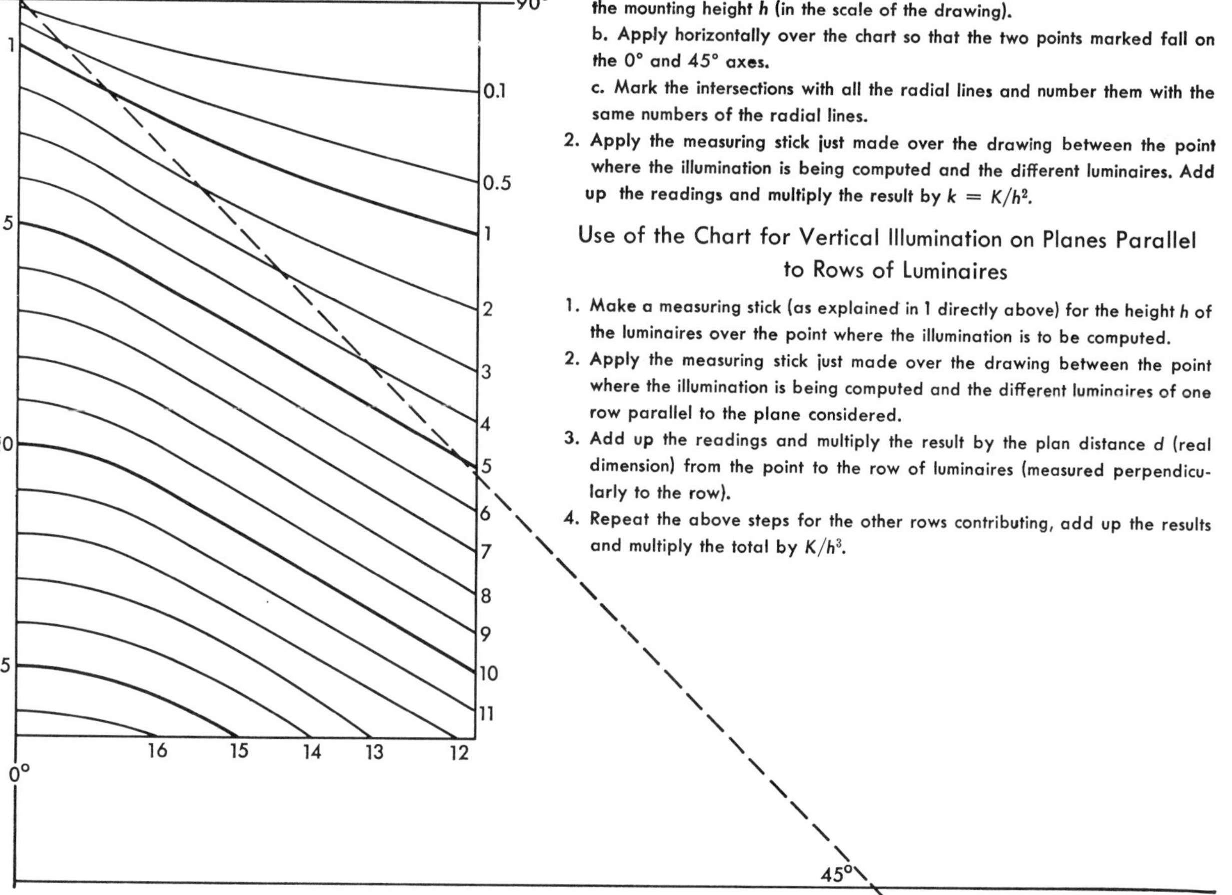

Construction of Chart for any Luminaire
1. Lay a tracing of the candlepower curve over this chart so that the 0° and 90° axes of both coincide.
2. Draw radial lines from point O (origin) passing through all the points where the candlepower curve intersects the chart curves. Number these radial lines with the numbers of the chart curves. Draw the 0° and 45° axes.
3. Read the candlepower on the 0° axis over point 10. Divide by 10 to obtain the value K.
Use of the Chart for Horizontal Illumination on any Plan Drawing
1. Make a measuring stick as follows:
a. Mark two points on the edge of a piece of paper at a distance equal to the mounting height h (in the scale of the drawing).
b. Apply horizontally over the chart so that the two points marked fall on the 0° and 45° axes.
c. Mark the intersections with all the radial lines and number them with the same numbers of the radial lines.
2. Apply the measuring stick just made over the drawing between the point where the illumination is being computed and the different luminaires. Add up the readings and multiply the result by k = K/h².
Use of the Chart for Vertical Illumination on Planes Parallel to Rows of Luminaires
1. Make a measuring stick (as explained in 1 directly above) for the height h of the luminaires over the point where the illumination is to be computed.
2. Apply the measuring stick just made over the drawing between the point where the illumination is being computed and the different luminaires of one row parallel to the plane considered.
3. Add up the readings and multiply the result by the plan distance d (real dimension) from the point to the row of luminaires (measured perpendicularly to the row).
4. Repeat the above steps for the other rows contributing, add up the results and multiply the total by K/h³.
0
1
5
10
15
90°
0.1
0.5
1
2
3
4
5
6
7
8
9
10
11
16
15
14
13
12
0°
45°

Fig. 9-31. Chart for use in a simplified procedure for point calculations.

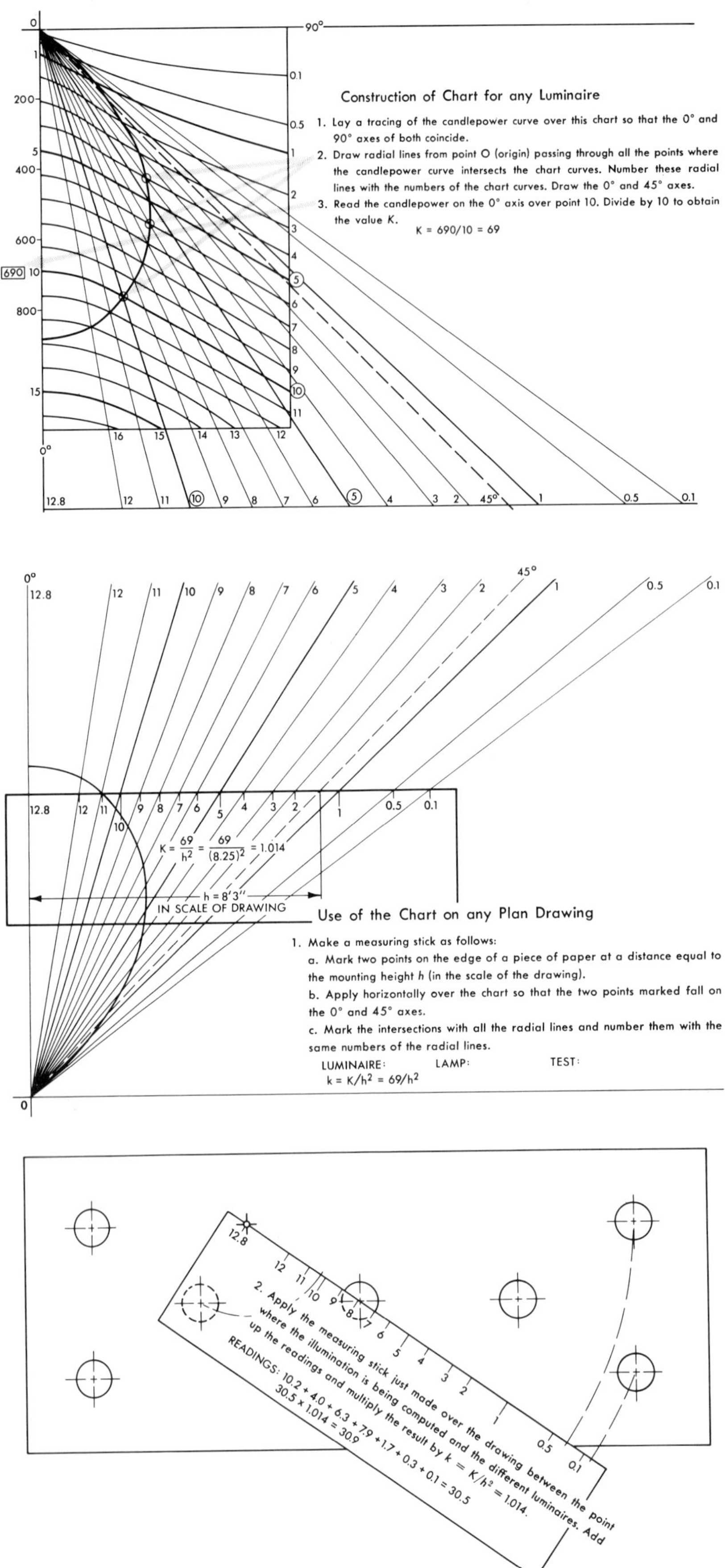
Construction of Chart for any Luminaire
1. Lay a tracing of the candlepower curve over this chart so that the 0° and 90° axes of both coincide.
2. Draw radial lines from point O (origin) passing through all the points where the candlepower curve intersects the chart curves. Number these radial lines with the numbers of the chart curves. Draw the 0° and 45° axes.
3. Read the candlepower on the 0° axis over point 10. Divide by 10 to obtain the value K.
K = 690/10 = 69
Use of the Chart on any Plan Drawing
1. Make a measuring stick as follows:
a. Mark two points on the edge of a piece of paper at a distance equal to the mounting height h (in the scale of the drawing).
b. Apply horizontally over the chart so that the two points marked fall on the 0° and 45° axes.
c. Mark the intersections with all the radial lines and number them with the same numbers of the radial lines.
LUMINAIRE: LAMP: TEST:
k = K/h² = 69/h²
K = 69/h² = 69/(8.25)² = 1.014
h = 8′3″
IN SCALE OF DRAWING
2. Apply the measuring stick just made over the drawing between the point where the illumination is being computed and the different luminaires. Add up the readings and multiply the result by k = K/h² = 1.014.
READINGS: 10.2 + 4.0 + 6.3 + 7.9 + 1.7 + 0.3 + 0.1 = 30.5
30.5 × 1.014 = 30.9

Fig. 9–32. Example of use of chart of Fig. 9–31.

9–33. Usually, all rows of luminaires have the same α coordinates, one coordinate for each end of the row.

Angle β is the angle between the vertical plane of the row of luminaires and a tilted plane containing both the seeing task and the luminaire or row of luminaires. This angle is determined from the same chart as angle α, again using V, the mounting height of the luminaires above the seeing task, and H, which in determining angle β, is the horizontal distance from the seeing task to the row of luminaires, measured perpendicular to the luminaires. Angle β is different for each row of luminaires. Each row has only one β coordinate.

The direct illumination component for each luminaire or row of luminaires is determined by referring to a table of direct illumination components for the specific luminaire. Tables such as shown in Fig. 9–34 list direct illumination components for specific luminaires for determining the illumination on the horizontal and on vertical planes either parallel to, or perpendicular to the luminaires. The direct illumination components are based on the assumption that the luminaire is mounted six feet above the seeing task. If this mounting height above the task is other than six feet, the direct illumination component shown in Fig. 9–34 must be multiplied by 6/V where V is the mounting height above the task. Thus, the total direct illumination component would be the product of 6/V and the sum of the individual direct illumination components for each row.

As an example of the method, assume four rows of six four-foot luminaires for which data are shown in Fig. 9–34. They are surface mounted on 8-foot cen-

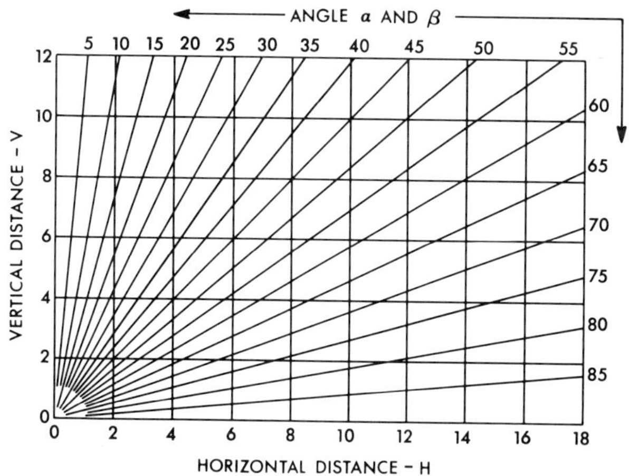

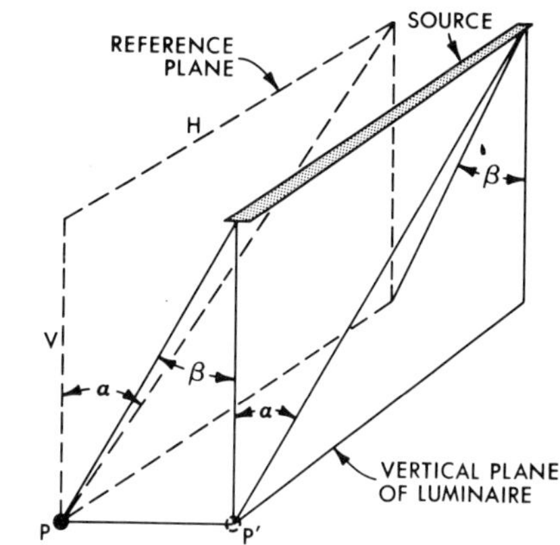

Fig. 9–33. Relationship of angles and distances used in the Angular Coordinate Method.

Fig. 9–34. Typical Presentation of Angular Coordinate Method Data

β	Direct Illumination Components															
	5	15	25	35	45	55	65	75	5	15	25	35	45	55	65	75
α	Vertical Surface Illumination Footcandles at a Point On a Plane Parallel to Luminaires								Vertical Surface Illumination Footcandles at a Point On a Plane Perpendicular to Luminaires							
0–10	1.0	2.8	3.8	3.4	2.0	1.1	.4	...	1.0	.9	.7	.4	.2	.1	...	...
0–20	2.0	5.4	7.4	6.7	4.0	2.2	.9	...	4.0	3.5	2.7	1.6	.7	.3	.1	...
0–30	2.8	7.6	10.6	9.6	5.9	3.3	1.4	...	8.3	7.4	5.9	3.5	1.6	.7	.2	...
0–40	3.4	9.3	12.9	12.0	7.5	4.4	1.9	...	13.3	11.9	9.4	5.9	2.7	1.2	.4	...
0–50	3.8	10.4	14.4	13.6	8.7	5.4	2.5	...	17.7	15.9	12.7	8.2	3.9	1.9	.7	...
0–60	4.0	10.9	15.2	14.6	9.6	6.2	3.1	...	20.7	18.7	15.2	10.2	5.1	2.7	1.1	...
0–70	4.1	11.1	15.5	15.1	10.1	6.8	3.7	...	22.3	20.2	16.7	11.6	6.2	3.5	1.7	...
0–80	4.1	11.2	15.6	15.2	10.3	7.1	4.1	...	23.0	20.9	17.4	12.3	6.9	4.2	2.3	.1
0–90	4.1	11.2	15.6	15.2	10.3	7.1	4.2	.2	23.2	21.1	17.6	12.5	7.1	4.4	2.5	.3
	Footcandles at a Point on Work-Plane															
0–10	11.5	10.6	8.1	4.9	2.0	.8	.2	...	Category V							
0–20	22.7	20.4	15.7	9.6	4.0	1.6	.4	...								
0–30	31.9	28.7	22.5	13.7	5.9	2.4	.6	...								
0–40	39.1	35.1	27.5	17.1	7.5	3.2	.8	...								
0–50	43.5	39.1	30.8	19.4	8.7	3.9	1.1	...								
0–60	45.6	41.1	32.5	20.8	9.6	4.4	1.4	...								
0–70	46.3	41.8	33.2	21.5	10.1	4.8	1.7	...	2 T-12 Lamps—430 MA							
0–80	46.5	42.0	33.4	21.7	10.3	5.0	1.9	...	1' Wide Prismatic Wrap-Around							
0.90	46.5	42.0	33.4	21.7	10.3	5.0	1.9	...								

Row	α_1	α_2	β	Direct Illum. Component from Left End	Direct Illum. Component from Right End	Total
A	50	60*	55	3.9	4.4	8.3
B	50	60*	25	30.8	32.5	63.3
C	50	60*	25	30.8	32.5	63.3
D	50	60*	55	3.9	4.4	8.3
					Total	143.2

* Actually, α_2 is 59° but is rounded off to 60°.

Fig. 9–35. Layout and data obtained using the Angular Coordinate Method.

ters in a room 28 by 30 feet. Assume the ceiling reflectance is 80 per cent and that of the walls is 50 per cent. Floor cavity reflectance is 20 per cent. The mounting height of the luminaires is 8½ feet above the work-plane. The initial illumination on the horizontal work-plane at Point *P* in Fig. 9–35 is desired.

It is first necessary to determine angle α for both ends of the rows of luminaires and angle β. For angle α, H is 10 feet for α_1; and 14 feet for α_2. The vertical distance, V, is 8½ feet. For angle β, H is 12 feet for rows A and D and is four feet for rows B and C. The vertical distance, V, still is 8½ feet. Fig. 9–33 is used to determine the angles and values shown in Fig. 9–35. Since the direct illumination component tables are all based on an assumed mounting height of 6 feet above the point, and in this case the luminaires are actually 8½ feet above Point *P*, it is necessary to multiply the total of 143.2 by 6/8½. The resultant direct component is 101 footcandles.

Aspect-Factor Method—IES (London).[13] In this method there are published tables of multipliers known as aspect factors, based on angular coordinates, which permit rapid and simple calculation of illumination on horizontal, vertical, and inclined planes from linear rows of various generic luminaires. These generic luminaires represent many common luminaire types within practical accuracy.

The method, as presently developed, involves assumptions that are adequate for many applications. The principal assumption is that the luminaire can be categorized by the shape of the parallel plane intensity distribution. A luminaire is identified with one of the generic types by plotting its parallel plane intensity distribution on a special graph. In the event that the luminaire is considerably different from the generic types, a method is available to generate a table of factors for that particular luminaire. This method differs from other angular coordinate techniques, such as Jones et al.[12] and McPhail,[14] in that (1) generic luminaire data are available and (2) a method exists for classifying a particular luminaire with respect to generic types.

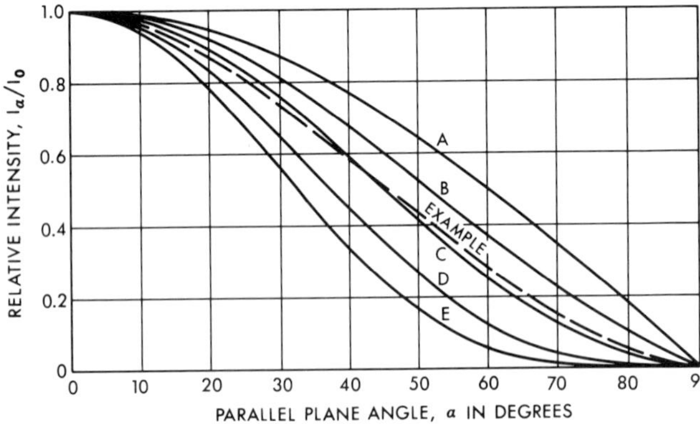

Fig. 9–36. Plot of five curves to classify luminaire types for the Aspect Factor Method.

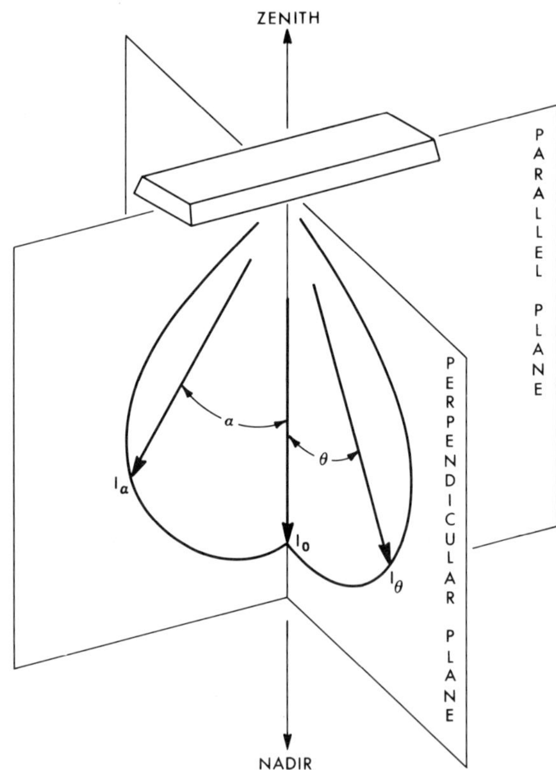

Fig. 9–37. Planes of intensity distributions used in the Aspect Factor Method.

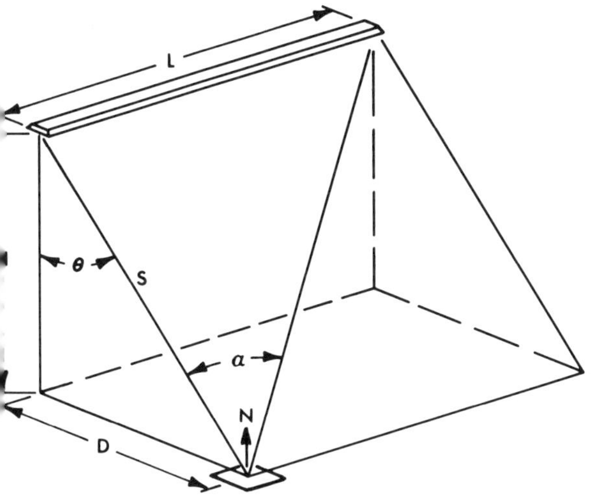

HORIZONTAL ILLUMINATION E_a

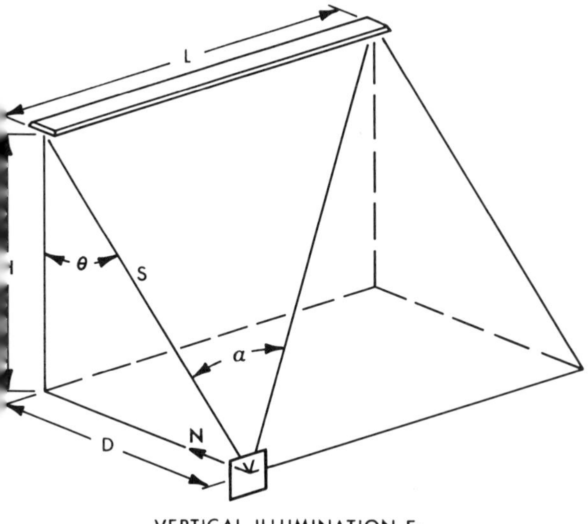

VERTICAL ILLUMINATION E_b

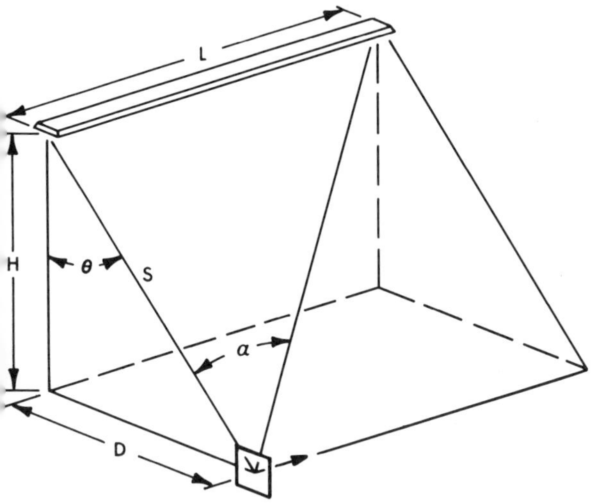

L
H
θ
S
α
D

VERTICAL ILLUMINATION E_c

Fig. 9-38. Diagrams showing symbols used in the Aspect Factor Method. N = normal to illuminated surface.

Five hypothetical luminaires that reasonably well represent the majority of linear luminaires are used. Luminaires with open bottoms, diffusers, lenses, and diffuse louvers normally can be fit to these generic types; luminaires with specular reflectors or louvers usually do not fit. The five generic types are identified by the parallel plane intensity distributions in Fig. 9-36. In the method, the procedure followed is to divide the parallel plane intensity of the luminaire being used by the nadir intensity (I_0), and to plot this on Fig. 9-36 to identify the luminaire type (A, B, C, D, or E). Occasionally, the luminaire intensity curve will match one generic type in one region and another elsewhere.

The method utilizes the following equations for calculating the direct components of vertical and horizontal illumination from linear sources:

$$E_a = \frac{I_\theta}{S} \cos \theta \, (AF),$$

$$E_b = \frac{I_\theta}{S} \sin \theta \, (AF), \quad \text{and}$$

$$E_c = \frac{I_\theta}{S} (af)$$

where I_θ is the perpendicular plane intensity of the luminaire at angle θ (see Fig. 9-37), the distance S (see Fig. 9-38) is calculated by

$$S = \sqrt{H^2 + D^2},$$

the angles are given by

$$\theta = \tan^{-1} \frac{D}{H} \quad \text{and} \quad \alpha = \tan^{-1} \frac{L}{S}.$$

The aspect (AF) and (af) are constants derived from the distributions of generic intensities and from the geometric relations determined by angle α. Fig. 9-39 presents a tabulation of these factors, and the derivations can be found in Reference 13. Note that it is convenient to determine the angles and the distance S by graphical means.

A row of luminaires with small equal spacings between the individual units can be considered as a single linear source. To do this the single luminaire intensity is multiplied by the ratio of (luminaire length times number of luminaires) divided by (total length of the row). The separation of luminaires should not exceed $H/(4 \cos \theta)$ if the error is to be less than 10 per cent. Equations for determining the illumination on inclined planes, a method for developing aspect factors of luminaires that do not match the generic models, and graphic methods for simply determining various angles are given in Reference 13.

Example. Fig. 9-40 illustrates a 12-foot long row of luminaires mounted 8 feet above point P. Point P is 2 feet beyond the end of the row and displaced 5 feet horizontally. The aspect factors apply for a point that is at the end of the row. Consequently, the illumination due to a 2-foot row is subtracted from the illumination due to a 14-foot row. Fig. 9-36

Fig. 9–39. Aspect Factors for Parallel (AF) and Perpendicular (af) Planes

a. Aspect Factors (AF)

	Luminaire Classification						Luminaire Classification				
a (degrees)	A	B	C	D	E	a (degrees)	A	B	C	D	E
0	0.000	0.000	0.000	0.000	0.000	46	0.652	0.623	0.595	0.549	0.510
1	0.017	0.017	0.017	0.018	0.018	47	0.660	0.630	0.601	0.553	0.512
2	0.035	0.035	0.035	0.035	0.035	48	0.668	0.637	0.606	0.556	0.515
3	0.052	0.052	0.052	0.052	0.052	49	0.675	0.643	0.612	0.560	0.517
4	0.070	0.070	0.070	0.070	0.070	50	0.683	0.649	0.616	0.563	0.519
5	0.087	0.087	0.087	0.087	0.087						
6	0.105	0.104	0.104	0.104	0.104	51	0.690	0.655	0.621	0.566	0.521
7	0.122	0.121	0.121	0.121	0.121	52	0.697	0.661	0.625	0.566	0.523
8	0.139	0.138	0.138	0.138	0.137	53	0.703	0.666	0.629	0.571	0.524
9	0.156	0.155	0.155	0.155	0.154	54	0.709	0.671	0.633	0.573	0.525
10	0.173	0.172	0.172	0.171	0.170	55	0.715	0.675	0.636	0.575	0.527
11	0.190	0.189	0.189	0.187	0.186	56	0.720	0.679	0.639	0.577	0.528
12	0.206	0.205	0.205	0.204	0.202	57	0.726	0.684	0.642	0.578	0.528
13	0.223	0.222	0.221	0.219	0.218	58	0.731	0.688	0.645	0.580	0.529
14	0.239	0.238	0.237	0.234	0.233	59	0.736	0.691	0.647	0.581	0.530
15	0.256	0.254	0.253	0.250	0.248	60	0.740	0.695	0.650	0.582	0.530
16	0.272	0.270	0.269	0.265	0.262	61	0.744	0.698	0.652	0.583	0.531
17	0.288	0.286	0.284	0.280	0.276	62	0.748	0.701	0.654	0.584	0.531
18	0.304	0.301	0.299	0.295	0.290	63	0.752	0.703	0.655	0.585	0.532
19	0.320	0.316	0.314	0.309	0.303	64	0.756	0.706	0.657	0.586	0.532
20	0.335	0.332	0.329	0.322	0.316	65	0.759	0.708	0.658	0.586	0.532
21	0.351	0.347	0.343	0.336	0.329	66	0.762	0.710	0.659	0.587	0.533
22	0.366	0.361	0.357	0.349	0.341	67	0.764	0.712	0.660	0.587	0.533
23	0.380	0.375	0.371	0.362	0.353	68	0.767	0.714	0.661	0.588	0.533
24	0.396	0.390	0.385	0.374	0.364	69	0.769	0.716	0.662	0.588	0.533
25	0.410	0.404	0.398	0.386	0.375	70	0.772	0.718	0.663	0.588	0.533
26	0.424	0.417	0.410	0.398	0.386	71	0.774	0.719	0.664	0.588	0.533
27	0.438	0.430	0.423	0.409	0.396	72	0.776	0.720	0.664	0.589	0.533
28	0.452	0.443	0.435	0.420	0.405	73	0.778	0.721	0.665	0.589	0.533
29	0.465	0.456	0.447	0.430	0.414	74	0.779	0.722	0.665	0.589	0.533
30	0.478	0.473	0.458	0.440	0.423	75	0.780	0.723	0.666	0.589	0.533
31	0.491	0.480	0.469	0.450	0.431	76	0.781	0.723	0.666	0.589	0.533
32	0.504	0.492	0.480	0.459	0.439	77	0.782	0.724	0.666	0.589	0.533
33	0.517	0.504	0.491	0.468	0.447	78	0.782	0.724	0.666	0.589	0.533
34	0.529	0.515	0.501	0.476	0.454	79	0.783	0.724	0.666	0.589	0.533
35	0.541	0.526	0.511	0.484	0.460	80	0.784	0.725	0.666	0.589	0.533
36	0.552	0.537	0.520	0.492	0.466	81	0.784	0.725	0.667	0.589	0.533
37	0.564	0.546	0.528	0.499	0.472	82	0.785	0.725	0.667	0.589	0.533
38	0.574	0.556	0.538	0.506	0.478	83	0.785	0.725	0.667	0.589	0.533
39	0.585	0.565	0.546	0.513	0.483	84	0.785	0.725	0.667	0.589	0.533
40	0.596	0.575	0.554	0.519	0.488	85	0.786	0.725	0.667	0.589	0.533
41	0.606	0.584	0.562	0.525	0.492	86					
42	0.615	0.591	0.569	0.530	0.496	87					
43	0.625	0.598	0.576	0.535	0.500	88	Same values as for 85 degrees				
44	0.634	0.608	0.583	0.540	0.504	89					
45	0.643	0.616	0.589	0.545	0.507	90					

Fig. 9-39. *Continued*

b. Aspect Factors (af)

	Luminaire Classification						Luminaire Classification				
α (degrees)	A	B	C	D	E	α (degrees)	A	B	C	D	E
0	0.000	0.000	0.000	0.000	0.000	46	0.259	0.240	0.221	0.192	0.168
1	0.000	0.000	0.000	0.000	0.000	47	0.267	0.247	0.227	0.196	0.171
2	0.001	0.001	0.001	0.001	0.001	48	0.276	0.254	0.233	0.200	0.173
3	0.001	0.001	0.001	0.001	0.001	49	0.285	0.262	0.239	0.204	0.176
4	0.002	0.002	0.002	0.002	0.002	50	0.293	0.268	0.244	0.207	0.178
5	0.004	0.003	0.003	0.004	0.004						
6	0.005	0.005	0.005	0.005	0.005	51	0.302	0.276	0.250	0.211	0.180
7	0.007	0.007	0.007	0.007	0.007	52	0.310	0.282	0.255	0.214	0.182
8	0.010	0.009	0.009	0.010	0.010	53	0.319	0.289	0.260	0.217	0.184
9	0.012	0.012	0.012	0.012	0.012	54	0.327	0.296	0.265	0.220	0.186
10	0.015	0.015	0.015	0.015	0.015	55	0.335	0.302	0.270	0.223	0.188
11	0.018	0.018	0.018	0.018	0.018	56	0.344	0.309	0.275	0.226	0.189
12	0.022	0.021	0.021	0.021	0.021	57	0.352	0.315	0.279	0.228	0.190
18	0.025	0.025	0.025	0.025	0.024	58	0.360	0.321	0.283	0.230	0.192
14	0.029	0.029	0.029	0.028	0.028	59	0.367	0.327	0.287	0.232	0.193
15	0.033	0.033	0.033	0.032	0.032	60	0.375	0.333	0.291	0.234	0.194
16	0.038	0.037	0.037	0.037	0.036	61	0.383	0.339	0.295	0.236	0.195
17	0.043	0.042	0.041	0.041	0.040	62	0.390	0.344	0.299	0.238	0.195
18	0.048	0.047	0.046	0.046	0.044	63	0.397	0.349	0.302	0.239	0.196
19	0.053	0.052	0.051	0.040	0.049	64	0.404	0.354	0.305	0.241	0.197
20	0.059	0.057	0.056	0.055	0.054	65	0.410	0.359	0.308	0.242	0.197
21	0.064	0.063	0.062	0.060	0.058	66	0.417	0.364	0.311	0.243	0.198
22	0.070	0.068	0.067	0.065	0.063	67	0.424	0.368	0.313	0.244	0.198
23	0.076	0.074	0.073	0.071	0.068	68	0.430	0.372	0.315	0.245	0.199
24	0.083	0.081	0.079	0.076	0.073	69	0.436	0.377	0.318	0.246	0.199
25	0.089	0.087	0.085	0.081	0.078	70	0.442	0.381	0.320	0.247	0.199
26	0.096	0.093	0.091	0.087	0.083	71	0.447	0.384	0.322	0.247	0.199
27	0.103	0.100	0.097	0.092	0.088	72	0.452	0.387	0.323	0.248	0.199
28	0.110	0.107	0.104	0.098	0.093	73	0.457	0.391	0.323	0.248	0.200
29	0.118	0.113	0.110	0.104	0.098	74	0.462	0.394	0.326	0.249	0.200
30	0.125	0.120	0.116	0.109	0.103	75	0.466	0.396	0.327	0.249	0.200
31	0.132	0.127	0.123	0.115	0.108	76	0.470	0.399	0.328	0.249	0.200
32	0.140	0.135	0.130	0.121	0.112	77	0.474	0.401	0.329	0.249	0.200
33	0.148	0.142	0.136	0.126	0.117	78	0.478	0.404	0.330	0.250	0.200
34	0.156	0.149	0.143	0.132	0.122	79	0.482	0.406	0.331	0.250	0.200
35	0.165	0.157	0.150	0.137	0.126	80	0.485	0.408	0.331	0.250	0.200
36	0.173	0.164	0.156	0.143	0.131	81	0.488	0.410	0.332	0.250	0.200
37	0.181	0.172	0.163	0.148	0.135	82	0.490	0.411	0.332	0.250	0.200
38	0.190	0.180	0.170	0.154	0.139	83	0.492	0.412	0.332	0.250	0.200
39	0.198	0.187	0.177	0.159	0.143	84	0.494	0.413	0.333	0.250	0.200
40	0.207	0.195	0.183	0.164	0.147	85	0.496	0.414	0.333	0.250	0.200
41	0.216	0.203	0.190	0.169	0.151	86	0.498	0.415	0.333	0.250	0.200
42	0.224	0.210	0.196	0.174	0.155	87	0.499	0.416	0.333	0.250	0.200
43	0.233	0.218	0.203	0.179	0.158	88	0.499	0.416	0.333	0.250	0.200
44	0.242	0.224	0.209	0.183	0.162	89	0.500	0.416	0.333	0.250	0.200
45	0.250	0.232	0.215	0.188	0.165	90	0.500	0.416	0.333	0.250	0.200

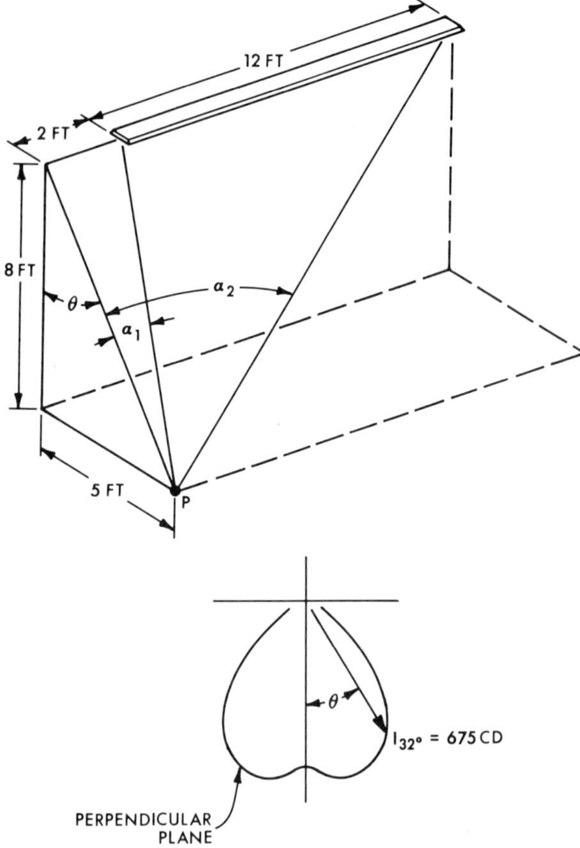

Fig. 9–40. Layout and symbols for example of use of Aspect Factor Method.

shows that the parallel plane intensity distribution approximately matches luminaire class C.

$$S = \sqrt{8^2 + 5^2} = 9.44$$

$$\tan\theta = \frac{5}{8} \qquad \theta = 32.0°$$

$$\tan\alpha_1 = \frac{2}{9.44} \qquad \alpha_1 = 12.0°$$

$$\tan\alpha_2 = \frac{14}{9.44} \qquad \alpha_2 = 56.0°$$

The perpendicular plane intensity for $\theta = 32.0°$ is 675 candelas. Thus,

$$E_a = \frac{675}{9.44}\cos 32.0°\,(.639) - \frac{675}{9.44}\cos 32.0°\,(.205)$$

$$= 26.3 \text{ footcandles}$$

$$E_c = \frac{675}{9.44}(.275) - \frac{675}{9.44}(.021) = 18.2 \text{ footcandles}$$

Illumination Charts and Tables. The isofootcandle method of calculating illumination at a point makes use of predetermined plots of lines of equal footcandles on horizontal or vertical surfaces. These charts are also known as isolux diagrams and are sometimes referred to as "Isoillumination Charts." By referring to a set of coordinates on a graph whic[h] coincide with the location of the point at which th[e] illumination is to be determined, the contribution o[f] footcandles from one luminaire to that point may b[e] determined.

The use of these charts is illustrated by referenc[e] to Fig. 9–41a, where the locus of equal footcandle[s] from a single fluorescent luminaire mounted 10 fee[t] above the work-plane is shown. By locating [a] point at certain distances with reference to the lumi[-]naire, the direct illumination at the point can be rea[d] directly. If many luminaires are involved, all of th[e] various contributions can be read and added to[-]gether to obtain total direct illumination at a give[n] point. Charts such as these are drawn on the basis o[f] ratios of distances to mounting heights and may b[e] used at other mounting heights than those for whic[h] they are measured by applying appropriate conver[-]sion factors given on the chart. An example of the us[e] of an isofootcandle chart is shown on page 9–73.

Isofootcandle charts are also prepared to sho[w] vertical illumination at a point (see Fig. 9–41b)[.] These may show lines of equal footcandles on a wal[l] produced by a luminaire mounted from the ceiling[.]

This method is also advantageous in determinin[g] the spacing of luminaires for uniform illuminatio[n] or for a given degree of departure from uniformity[.] The data also can be displayed in tabular form.

Idealized Source Charts. Exact calculations o[f] the direct illumination component due to linear an[d] area sources are possible when the luminance distri[-]bution of the luminaire is known. In general, th[e] variation of luminance is too complicated to mak[e] such calculations practical. However, certain specia[l] cases are useful; for example, rectangular are[a] sources with constant luminance (lambertian sur[-]face) and rectangular area sources with a typica[l] luminance distribution of the overcast sky (see Day[-]light Design, page 9–74). The lambertian surface ca[n] be used to approximate luminaires with diffusin[g] panels or to approximate diffuse reflecting areas suc[h] as walls. The overcast sky distribution is used t[o] estimate the effect of daylighting through windows[.]

Calculations with Linear Sources.[15] For lin[e] sources whose intensity characteristics are either [a] lambertian or a toroidal distribution* the followin[g] simplified expression may be used for determinin[g] the illumination at point P lying on line XY or it[s] extension:

$$E_p = \frac{L \times W}{2D} \quad \text{(see Fig. 9–42)}$$

where

E_p = illumination at P in footcandles
L = luminance of source in footlamberts
W = width of source in feet
D = distance from source to point P in feet

* Either of these distributions is approximated by the direct component of most linear source installations, with the exception of clear glass tube sources such as neon tubes, etc.

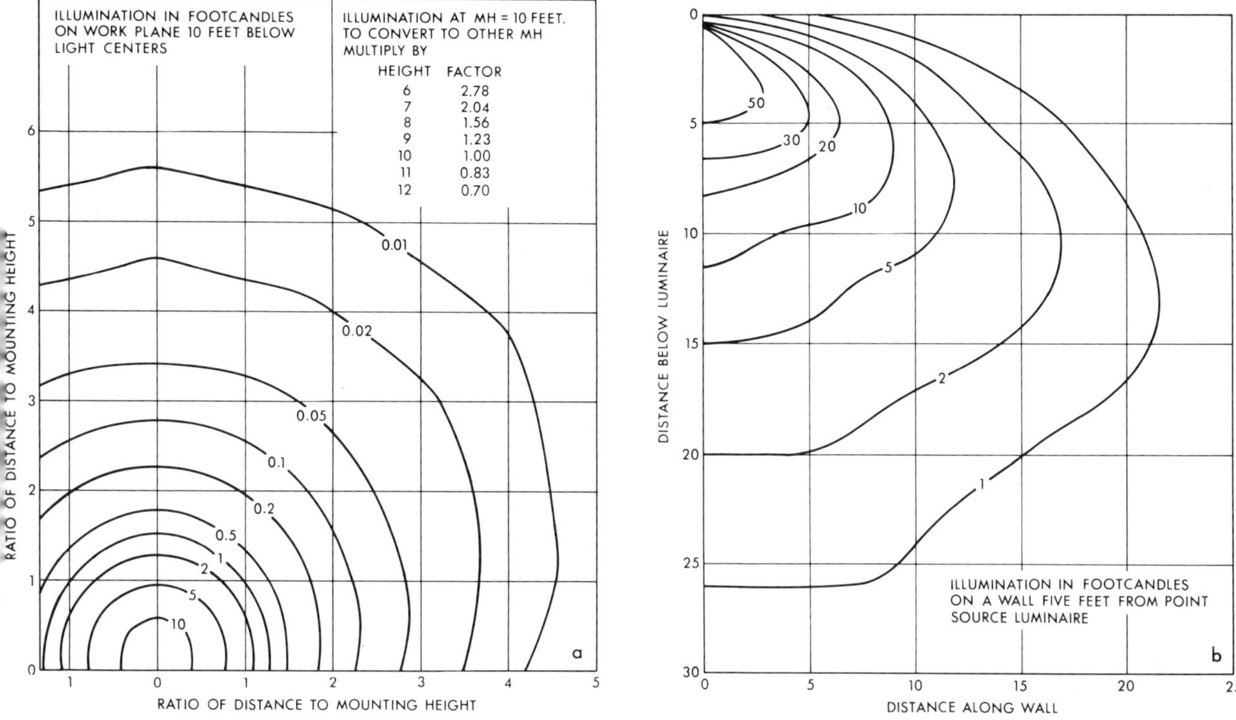

Fig. 9–41. Examples of isofootcandle charts useful in determining the direct component of illumination. a. For horizontal surfaces (or plane perpendicular to nadir or axis of luminaire). b. For vertical surfaces.

This expression is exact only in the case of an infinitely long source, but will be accurate to within 10 per cent if both d_1 and d_2 are greater than $1.5D$, and accurate to within 5 per cent if both d_1 and d_2 are greater than $2D$.

Note that the illumination from a line source of infinite length varies inversely as the distance to the source, *not* inversely as the square of the distance as is the case with point sources.

The following equations provide a means of determining the illumination, at point P not directly below the luminaire (see Fig. 9–43), produced by a source with either a toroidal (as approximated by a fluorescent lamp) or lambertian distribution. The formulas are applicable when P is in any of the three major planes provided line M is perpendicular to the axis of the source and passes through one end of the source. If the source extends beyond M the illumination may be obtained by considering the source to consist of two parts, separated by line M, and adding the results of each computation. Conversely, if the source does not extend far enough to reach line M, the problem is solved by assuming a longer source and subtracting the illumination that would be provided by the imaginary source that was added, to reach line M.

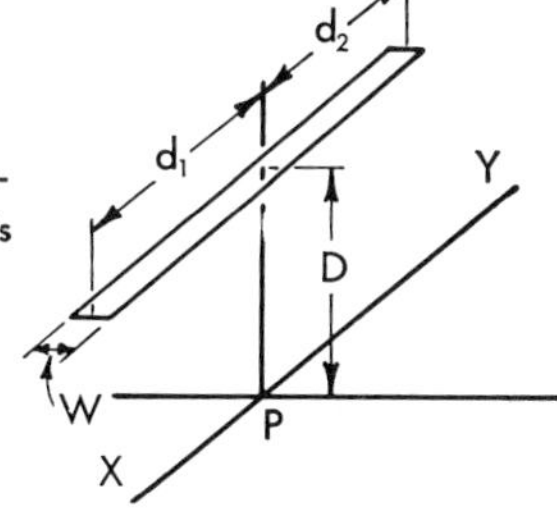

Fig. 9–42. Symbols used in computing illumination at specific points with line sources.

For source and P locations shown in Fig. 9–43a:

Lambertian distribution

$$E = \frac{LWV^2}{2\pi M^2}\left(\frac{1}{M}\,\text{Tan}^{-1}\frac{d}{M} + \frac{d}{R^2}\right)$$

Toroidal distribution

$$E = \frac{LWV}{2\pi M}\left(\frac{1}{M}\,\text{Tan}^{-1}\frac{d}{M} + \frac{d}{R^2}\right)$$

In the above formulas, and others which follow,
R, W, b, V, d and M are in feet
Angles are in radians
L = luminance of source in footlamberts
E = illumination in footcandles

For source and P locations shown in Fig. 9–43b:

Lambertian sphere distribution

$$E = \frac{LWbV}{2\pi M^2}\left(\frac{1}{M}\,\text{Tan}^{-1}\frac{d}{M} + \frac{d}{R^2}\right)$$

Toroidal distribution

$$E = \frac{LWb}{2\pi M}\left(\frac{1}{M}\,\text{Tan}^{-1}\frac{d}{M} + \frac{d}{R^2}\right)$$

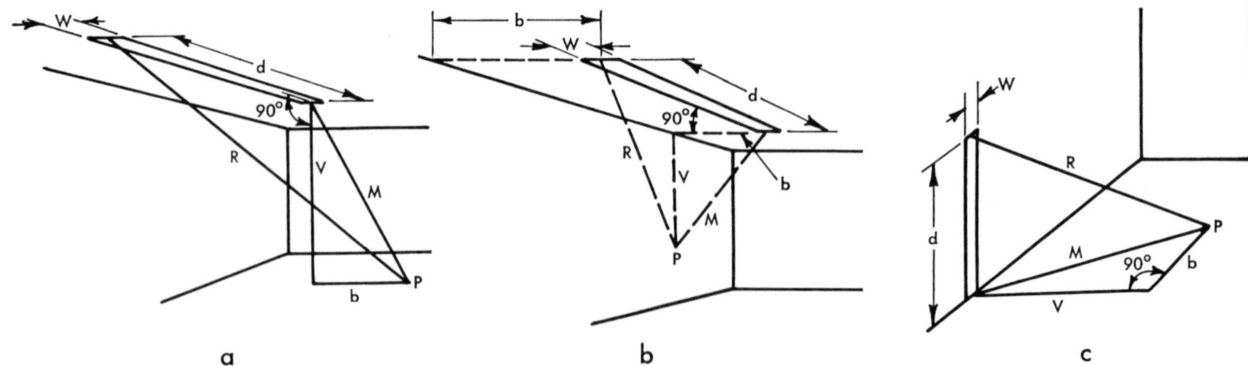

Fig. 9–43. Geometric relationships for calculating illumination at point P from a linear source $d \times W$. ($M = \sqrt{V^2 + b^2}$, $R = \sqrt{V^2 + b^2 + d^2}$)

For source and P locations shown in Fig. 9–43c:

Lambertian sphere distribution

$$E = \frac{LWV}{2\pi M^2}\left(\frac{d^2}{R^2}\right)$$

Toroidal distribution

$$E = \frac{LW}{2\pi M}\left(\frac{d^2}{R^2}\right)$$

Fig. 9–44 gives illumination values for Lambertian distribution.

Calculations with Area Sources. An infinitely large plane source radiating light to a parallel work-plane produces illumination as follows:

$$\text{Footcandles on Working Plane} = \text{Footlamberts Luminance of Infinite Source}$$

With such a source (large skylight or uniformly bright ceiling in large room), the illumination is theoretically independent of the distance. For surface sources whose greatest dimension is less than one-fifth the distance from the source to the working plane, point source formulas may be used with reasonable accuracy.

The illumination from a rectangular source is usually computed at a point perpendicular to the corner of the source. The variables involved are the distance from the source d, the height of the source H, the width W, and the luminance L. (See Fig. 9–45.) The illumination usually calculated for horizontal planes is E_h, and for a vertical plane, E_v.

The illumination can be determined for any point on a plane, in relation to the source, by the principle of superposition. For instance, if it is desired to calculate the illumination at point P, from source D, Fig. 9–46, the illumination is first calculated from source $ABCD$ and then from sources A, AC, and AB. The illumination from source D is then

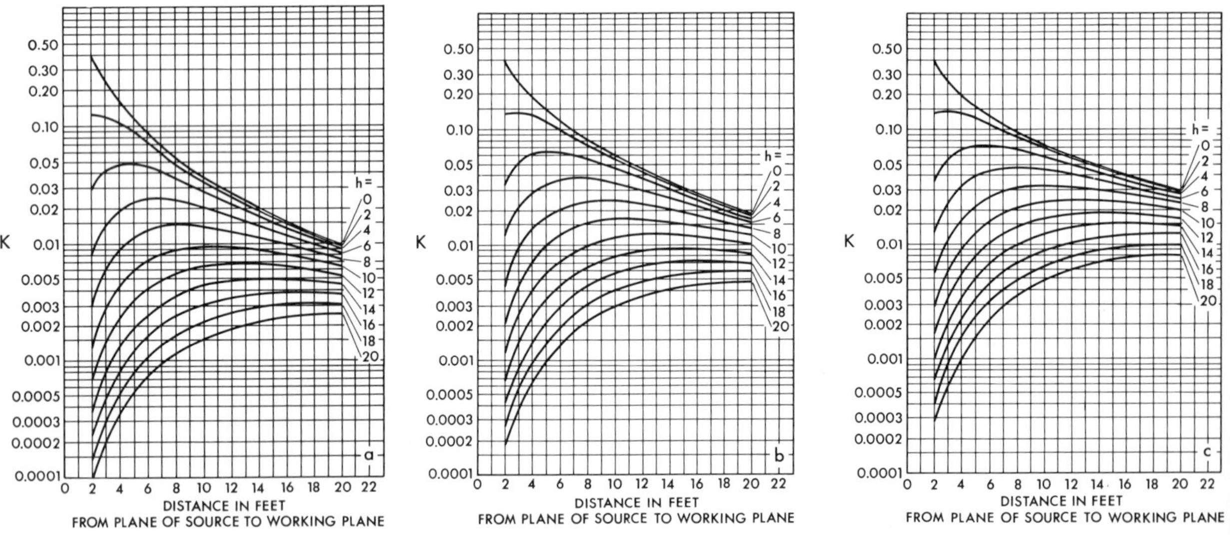

Fig. 9–44. Maximum candlepower per foot times K equals the illumination in footcandles at a point h (=b in Fig. 9–43a) feet distant from the projection of the linear source on the working plane (which is parallel to the source plane)[16]: a. 4-foot source, b. 8-foot source, c. 16-foot source. Note: These data hold only for sources terminating at the plane VbPM in Fig. 9–43a. For sources extending on both sides of the plane determine the illumination components from each side of the plane and add the two. For sources which do not extend to the plane calculate the contribution for the over-all length to the plane and subtract the hypothetical contribution of the portion missing.

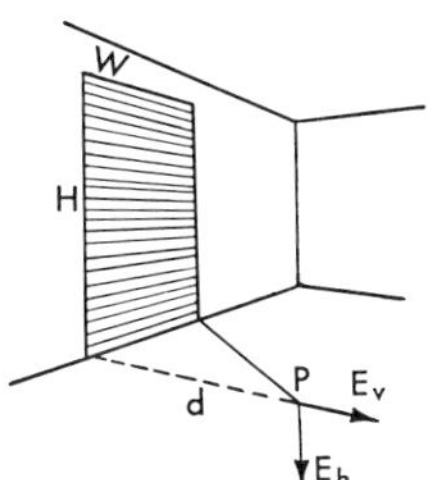

Fig. 9–45. Rectangular surface source perpendicular to illuminated surface.

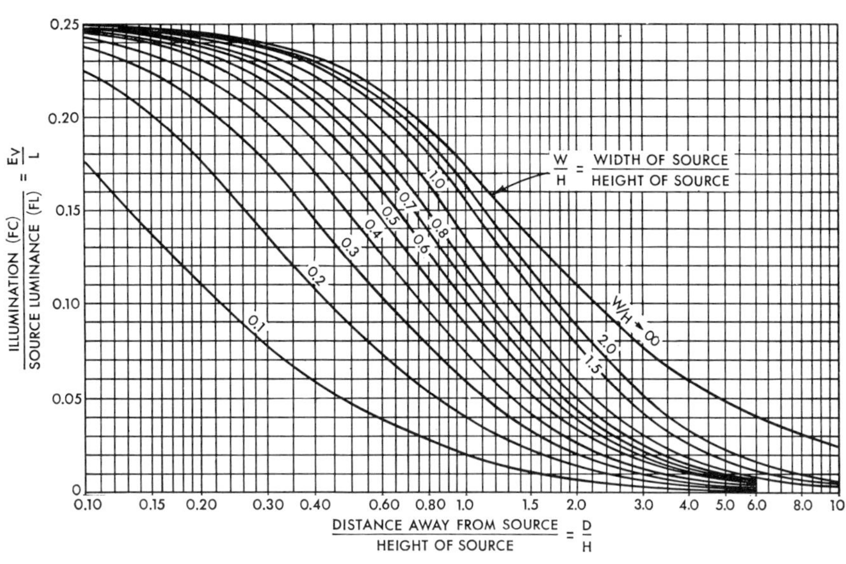

Fig. 9–47. Illumination from a rectangular uniform source parallel to illuminated plane.

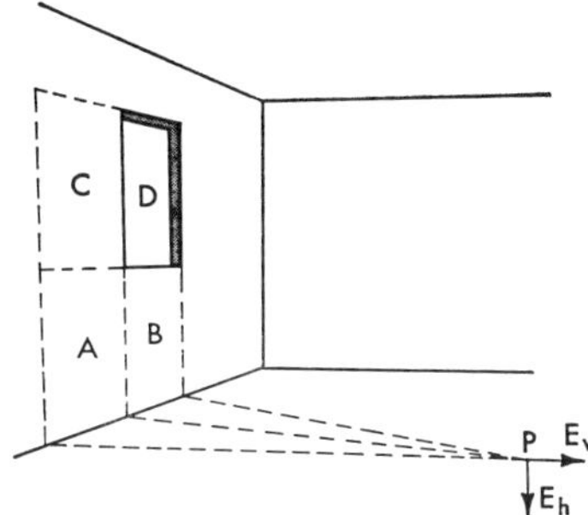

Fig. 9–46. Example of illumination calculation from source D.

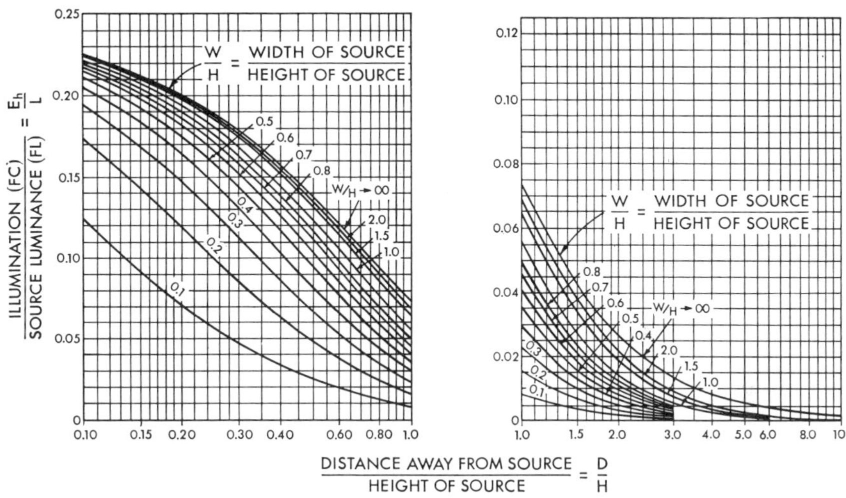

Fig. 9–48. Illumination from a rectangular uniform source perpendicular to illuminated plane.

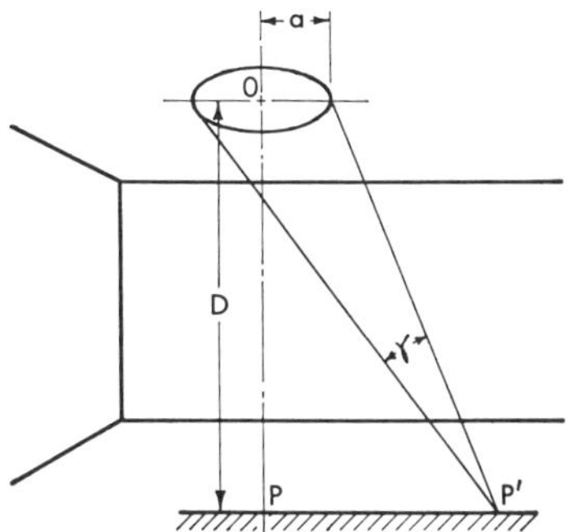

Fig. 9–49. Geometric relationships for calculating illumination at point P from a circular surface source.

equal to the illumination from source *ABCD* minus the sum of the illumination from source *AB* and *AC*, plus the illumination from source *A*.

The illumination from a *rectangular source of uniform luminance*, assuming no contributions from walls and ceilings, can be obtained from Figs. 9–47 and 9–48. For example, to obtain the illumination at distance 10 feet from a rectangular uniform source 20 feet wide and 10 feet high, which is perpendicular to the illuminated plane, obtain the values of source width to height (20/10) and the distance from the source to source height (10/10). From Fig. 9–48 follow the curve $W/H = 2$ until it strikes the ordinate $D/H = 1$; the ratio of illumination to source luminance for this condition is .07. Assuming a uniform source luminance of 1000 footlamberts then the illumination is 70 footcandles.

For uniformly bright circular surface sources, the illumination at points along the axis of the source can be determined from the following formulas (see Fig. 9–49):

Footcandles at Point on Axis

$$= (\text{Source Luminance in Footlamberts}) \times \left(\frac{a^2}{a^2 + D^2}\right)$$

$$\text{or} = \frac{\text{Lumen Output of Circular Surface Sources}}{\pi\ (\text{Distance From Point } P \text{ to Edge of Source})^2}$$

or = (Source Luminance in Footlamberts) × (Sine of Angle Between *P* and Edge of Source)^2

If the point lies off the axis, as *P′* of Fig. 9–49, the general formula may be employed:

$$\text{Footcandles} = \tfrac{1}{2}\ (\text{Source Luminance in Footlamberts}) \times (1 - \cos\gamma)$$

Configuration-Factor Method.* Use of configuration factors permits the calculation of illumination at a point from a diffuse surface. It is necessary to know only the luminance of the surface and the spacial relationship of the surface to the point. Fig. 9–50 is a precalculated table of Room Position Multipliers (RPM) for this method.

Example. Fig. 9–52 shows a room with one row of 2-foot by 4-foot lambertian distribution luminaires (I_0 = 3000 candelas) mounted 7 feet above the work-plane. The illumination at P is desired. In the solution the first step is to determine the room cavity ratios for areas $ACDF$ and $BCDE$:

$$\text{RCR}_{ACDF} = \frac{5(7)(12 + 20)}{12 \times 20} = 4.67$$

$$\text{RCR}_{BCDE} = \frac{5(7)(10 + 20)}{10 \times 20} = 5.25$$

The next step is to determine the Room Position Multipliers for these areas using Fig. 9–50. This means that the location of P in terms of letter and number in Fig. 9–51 must be determined first. From Fig. 9–51 it is found that P is at location $A, 3$ or $D, 0$. Then from Fig. 9–50, RPM values are found:

$ACDF$ – .33
$BCDE$ – .30

Using the RPM values and the luminance of the luminaires the illumination at P is found.

$$L = \frac{I_0 \times \pi}{A} = \frac{3000 \times 3.14}{2 \times 4} = 1178\,fL$$

Illumination at point P

$$= L \times (\text{RPM}_{\text{ACDF}} - \text{RPM}_{\text{BCDE}})$$
$$= 1178\,(.33 - .30) = 35 \text{ fc}$$

Reflected Components.

For Horizontal Surfaces. The reflected illumination component on horizontal surfaces is calculated in exactly the same manner as the average illumination is computed using the lumen method except that the RRC, the reflected radiation coefficient, is substituted for the CU, coefficient of utilization.

$$\begin{array}{c}\text{Reflected Illumination} \\ \text{(Horizontal)}\end{array} = \frac{\text{Lamp Lumens per Luminaire} \times \text{RRC}}{\text{Area per Luminaire (on Work-Plane)}}$$

where

$$\text{RRC} = \text{LC}_{\text{W}} + \text{RPM}(\text{LC}_{\text{CC}} - \text{LC}_{\text{W}})$$

LC_{W} = wall luminance coefficient (see Fig. 9–12)
LC_{CC} = ceiling cavity luminance coefficient (see Fig. 9–12)
RPM = room position multiplier (see Fig. 9–50)

For Vertical Surfaces. To determine the illumination reflected to vertical surfaces the approximate

* For a more precise method see Elements of Flux Transfer Theory, page 9-36.

Fig. 9–50. Room Position Multipliers
(For All Room Cavity Ratios and for All Points Designated by a Number and a Letter as Illustrated in Fig. 9–51.)

Points Along Width of Room	RCR	Points Along Length of Room					
		A	B	C	D	E	F
0	1	.24	.42	.47	.48	.48	.48
	2	.24	.36	.42	.44	.46	.46
	3	.23	.32	.37	.40	.42	.42
	4	.22	.28	.32	.35	.37	.37
	5	.21	.25	.28	.31	.33	.33
	6	.20	.23	.26	.28	.29	.30
	7	.18	.21	.23	.25	.26	.27
	8	.17	.18	.21	.22	.22	.23
	9	.15	.17	.18	.19	.20	.20
	10	.14	.16	.16	.17	.18	.18
1	1	.42	.74	.81	.83	.84	.84
	2	.36	.51	.60	.63	.66	.68
	3	.32	.40	.48	.51	.53	.57
	4	.28	.33	.40	.42	.44	.48
	5	.25	.29	.33	.36	.38	.42
	6	.23	.26	.29	.31	.33	.36
	7	.21	.23	.26	.28	.29	.30
	8	.18	.20	.23	.25	.26	.26
	9	.17	.18	.20	.21	.22	.23
	10	.16	.17	.18	.19	.19	.20
2	1	.47	.81	.90	.92	.93	.93
	2	.42	.60	.68	.72	.78	.83
	3	.37	.48	.58	.61	.64	.67
	4	.32	.40	.48	.50	.52	.57
	5	.28	.33	.40	.42	.44	.48
	6	.26	.29	.35	.37	.38	.40
	7	.23	.26	.30	.32	.33	.34
	8	.21	.23	.26	.27	.28	.29
	9	.18	.20	.23	.24	.25	.25
	10	.16	.18	.19	.21	.22	.22
3	1	.48	.83	.92	.94	.95	.95
	2	.44	.63	.72	.77	.82	.85
	3	.40	.51	.61	.65	.69	.71
	4	.35	.42	.50	.54	.58	.61
	5	.31	.36	.42	.46	.49	.52
	6	.28	.31	.37	.39	.41	.43
	7	.25	.28	.32	.34	.35	.36
	8	.22	.25	.27	.29	.30	.30
	9	.19	.21	.24	.25	.26	.26
	10	.17	.19	.21	.22	.23	.23
4	1	.48	.84	.93	.95	.96	.97
	2	.46	.66	.78	.82	.85	.86
	3	.42	.53	.64	.69	.73	.75
	4	.37	.44	.52	.58	.62	.64
	5	.33	.38	.44	.49	.52	.54
	6	.29	.33	.38	.41	.43	.45
	7	.26	.29	.33	.35	.37	.37
	8	.22	.26	.28	.30	.31	.32
	9	.20	.22	.25	.26	.26	.27
	10	.18	.19	.22	.23	.23	.24
5	1	.48	.84	.93	.95	.97	.97
	2	.46	.68	.83	.85	.86	.87
	3	.42	.57	.67	.71	.75	.77
	4	.37	.48	.57	.61	.64	.66
	5	.33	.42	.48	.52	.54	.56
	6	.30	.36	.40	.43	.45	.47
	7	.27	.30	.34	.36	.37	.38
	8	.23	.26	.29	.30	.31	.32
	9	.20	.23	.25	.26	.27	.27
	10	.18	.20	.22	.23	.24	.25

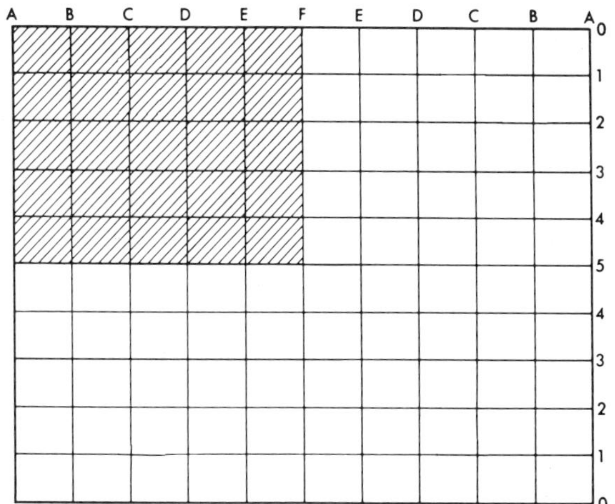

Fig. 9-51. Chart for locating points on the work plane in a room. Letters are placed along length of room and numbers along width. Each block represents ten per cent of length or width. In a square room each position could have two designations. Position A0 is corner of a room and position F5 is the center.

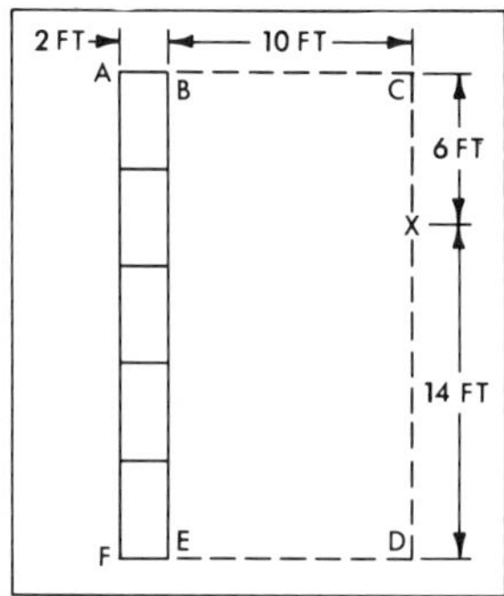

Fig. 9-52. Layout for example of use of the Configuration Factor Method. Luminaires are 2- by 4-foot units mounted 7 feet above the work-plane.

average value is determined using the same general formula, but substituting the WRRC, wall reflected radiation coefficient, for the CU, coefficient of utilization.

$$\text{Reflected Illumination (Vertical)} = \frac{\text{Lamp Lumens per Luminaire} \times \text{WRRC}}{\text{Area per Luminaire (on Work-Plane)}}$$

where

$$\text{WRRC} = \frac{LC_W}{\rho_W} - \text{WDRC}$$

ρ_W = average wall reflectance
WDRC = the wall direct radiation coefficient (see Fig. 9-12)

Luminous and Louvered Ceiling Systems

Luminous ceiling systems employ translucent media with lamps installed in the cavity above that media. Luminous ceiling systems employing highly diffusing media produce about the same results as well designed indirect lighting from conventional suspended luminaires (or coves), but with generally greater uniformity of ceiling luminance than is possible with the latter method. When luminous ceilings employ prismatic materials they give generally higher utilization than those utilizing diffusing materials; they also have better direct glare control.

Either glass or plastic of proper optical characteristics may be used as the translucent medium. This material is usually supported in a structural framework and is formed in panels or rolls of a size which is convenient for handling. The diffusing material should have the highest possible transmittance consistent with adequate concealment of the lamps. Experience indicates that a material should be used having a transmittance of about 50 per cent and as low an absorptance as possible. Prismatic materials should have as high a transmittance as possible with the degree of light control desired. In general, such materials will have an effective transmittance of 70 to 75 per cent.

Ideally, the cavity above the diffuser should be of just sufficient depth to obtain the proper relation between spacing of lamps and vertical distance from the diffuser to produce acceptable uniform brightness. The luminous efficiency of the system decreases as the depth of the cavity is increased in respect to its length and width. The cavity should be relatively unobstructed and all surfaces painted in a white paint of at least 80 per cent reflectance. All dust leaks in cavity walls and ceilings should be sealed. Where deep heavily obstructed cavities occur, it is wise to consider the possibility of furring down the cavity ceiling to a reasonable level. The cost of such construction may be less than the additional cost of equipment, power, and maintenance required to force an otherwise inefficient system to deliver a given level of illumination.

Successful luminous ceilings have been designed using practically all types of lamps, however fluorescent are generally used. A typical arrangement of lamps within the cavity is shown in Fig. 9-53. For uniform brightness appearance over a diffusing ceiling, the spacing of the lamps should not exceed 1.5 to 2 times their height above the diffuser. Where the plenum is shallow, the plenum ceiling painted white, and the lamps mounted against the plenum ceiling, S/L can be greater than is the case when the plenum above the lamps is deep, or dark, or obstructed. Also, when it is necessary to use units with reflectors such as those in Fig. 9-54, the ratio S/L should not exceed 1.5. With ceilings of prismatic materials, S/L should rarely exceed 1.0 if streaks of light and dark are to be avoided.

For greater uniformity of brightness over the entire ceiling, it is recommended that the lamp rows

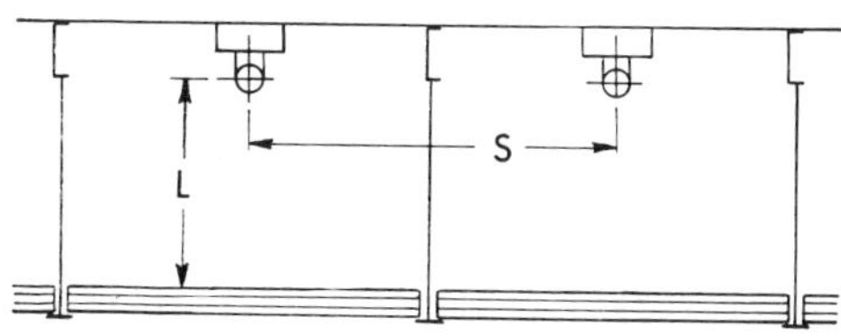

Fig. 9-53. Section of a portion of a luminous ceiling. The distance S should not exceed 1.5 to 2*L*.

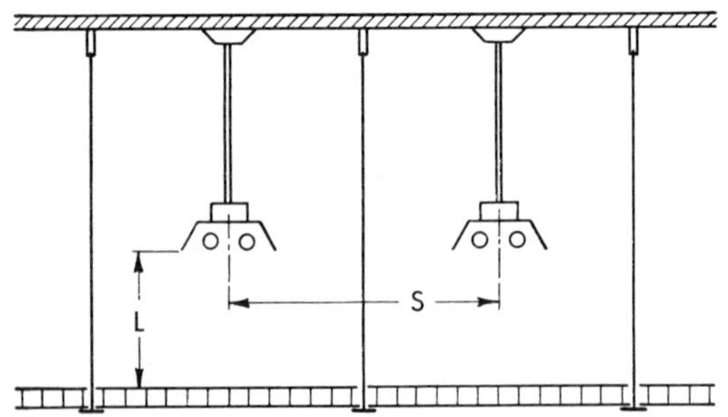

Fig. 9-54. Section of a portion of a louvered ceiling. For translucent louver vanes the ratio of S to L should not be over 1½ to 1 and for opaque louvers, not over 1½ or 1 to 1, depending upon the louver finish.

be spaced closer together at the sides of the cavity. The ends of the rows should be carried all the way to the ends of the cavity. In some instances it is desirable to install extra lamps between the rows at the end of the cavity, or across the ends of the rows, to increase the luminance at this point.

Luminous ceiling systems are often used for the dual purpose of providing light and concealing the visual clutter of beams and building services. Aside from the reduction in system efficiency caused by these devices, there is the possibility that they will cast unsightly shadows on the translucent medium. It is well to place the prime sources of light below, or at right angles to, such obstructions wherever possible.

Obtaining proper acoustical treatment of a room with a luminous ceiling can be difficult. In many cases, it is better to obtain the required results by treating other room surfaces. Generally this means wall surfaces in small rooms, and floors in large ones.

Louvered ceiling systems consist of an open network of translucent or opaque vanes through which light passes from the lighting recess above to the area below. The louver cells can be made in a number of shapes and sizes, depending upon mechanical design relationships and desired architectural effects, with various degrees of shielding. Fluorescent light sources are generally used with this type of ceiling. Here, the lighted appearance of the louvers is dependent on the conditions within the cavity, the spacing of the lamps, the height of the lamps above the louvers, the size of the louver cells, and the characteristics of the louver blades themselves. The transmittance (if any) of the blades, and both the type of reflectance (specular, spread, or diffuse) and the amount (per cent) are important. Louvers of the parabolic wedge type have yet different requirements.

Since characteristics and requirements differ so greatly, it is difficult to give definite recommendations. In general, it should be remembered that wide spacings of luminaires above the louvers will result in streaks of varied brightness, and shadow patterns on the blades themselves which may or may not be desirable. Close spacings may result in a "flat" appearance of the louver.

As a guide, in unobstructed, shallow plenums with white-painted surfaces and ceiling-mounted strip luminaires, the ratio of spacing to height above the louver should not exceed 1½ to 1 for translucent louvers, and 1 to 1 for opaque louvers. See Fig. 9-54. For other conditions, closer spacings are the rule.

Pipes, ducts, and beams should be considered in their relationship to the louvered ceiling system in order to avoid the forming of unsightly shadows.

The lighting equipment should be mounted high enough above the louvers to permit ease of access and relamping. Means for removing the louvers in sections small enough for ease of handling facilitates the relatively infrequent cleaning necessary.

The acoustical effect of louvers depends on the cell size and the louver material. In most cases, however, it is best to ignore the louvers, and apply acoustical material to the ceiling of the plenum, or to other room surfaces.

The efficiency and effective reflectance of luminous and louvered ceilings vary with the cavity proportions and reflectances, the type of lighting equipment used, the reflectance and transmittance of the ceiling material, the shielding angle of louvers, and the amount and kind of obstruction within the cavity. Fig. 9-12 gives coefficients of utilization for typical ceilings of these types. Performance data for other ceilings can be determined by photometric tests or by zonal-cavity calculations for coupled enclosures. The ceiling cavity reflectance for a room with a luminous or louvered ceiling can be found by

$$\rho_{eff} = \rho_d + \tau^2 \left(\frac{\rho_{CC}}{1 - \rho_d \rho_{CC}} \right)$$

where

τ = ceiling material transmittance
ρ_d = ceiling material reflectance
ρ_{CC} = effective reflectance of ceiling cavity
ρ_{eff} = effective reflectance of luminous ceiling seen from below.

Effective transmittance and reflectance for opaque louvers can be estimated using the curves of Fig. 9-55.

Louvers made from translucent materials vary tremendously in performance, and cannot be accurately evaluated by Fig. 9-55. An approximation can be made, however, by measuring the effective transmittance of the louver in diffuse light, and using it along with the shielding angle to find the equivalent louver blade reflectance from Fig. 9-55. Using this value, the effective reflectance can be estimated from Fig. 9-55.

Coves[17]

Coves may be for the production of illumination or the creation of a decorative effect. In most instances, an efficient lighting system and a more pleasing brightness pattern are obtained if the brightness of the ceiling and the brightness of the wall above the cove are relatively uniform in ap-

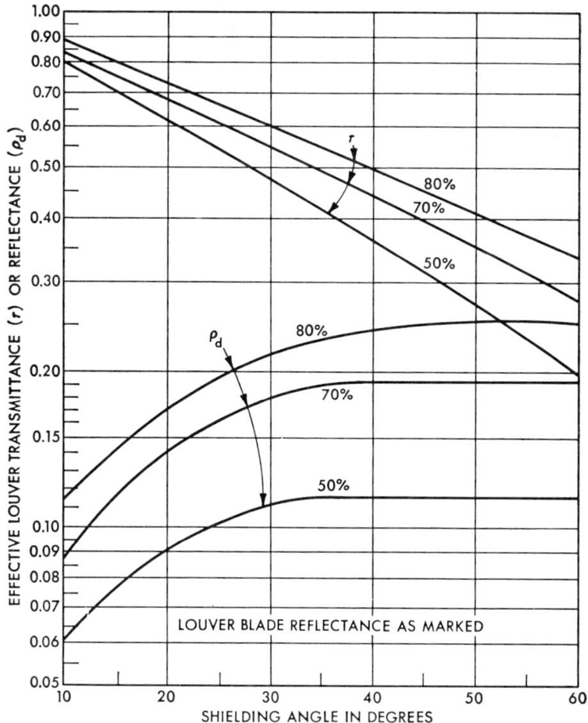

Fig. 9–55 Graph for obtaining effective transmittance and reflectance for square-cell opaque louvers.

pearance. Usually these objectives are more easily attained if the cove is not located too close to the ceiling. The brightness of the wall above the cove and that of the ceiling near the cove are produced largely by direct illumination from the light source and from reflections between the cove, wall, and ceiling. The brightness of the ceiling at the far end of the room is largely dependent upon the general room surface reflectances. To attain more uniform wall and ceiling brightness, it is recommended that the lamps be mounted several inches away from the wall and that the wall adjacent to the source be shielded in part from the direct radiation from the lamp. The channel serves this purpose very well. In general, the further the cove is located from the ceiling, the further the lamp and channel should be located from the wall. See Fig. 9–56. If the cove is 12 inches from the ceiling (H), tests have shown that the lamp center should be 2½ inches or more from the wall (S); if the cove is 15 to 20 inches from the ceiling, the lamp center should be 3½ inches or more from the wall; and for coves that are located 21 or 30 inches from the ceiling, it is recommended that the lamp center be 4½ inches or more from the wall.

Reflectors can be used in fluorescent coves to obtain very pleasing results when the reflector is aimed about 20 degrees above the horizontal.

Specific Application Methods

Empirical data are available to compute illumination levels for a great many specialized applications. The data shown are for typical lighting equipment and will vary according to actual equipment design.

Shelf and Garment Case Illumination. To estimate the illumination normal to a vertical surface from a continuous row of fluorescent lamps located in close proximity, the charts shown in Fig. 9–57 may be used. Such data, while particularly adapted to shelf and garment case illumination, can be used for any similar type of vertical surface lighting installations.

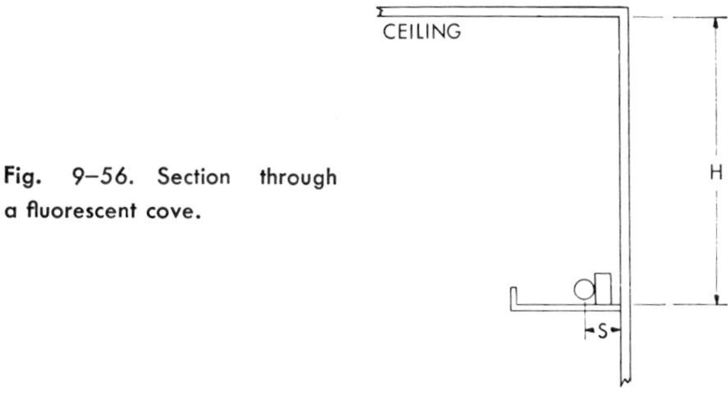

Fig. 9–56. Section through a fluorescent cove.

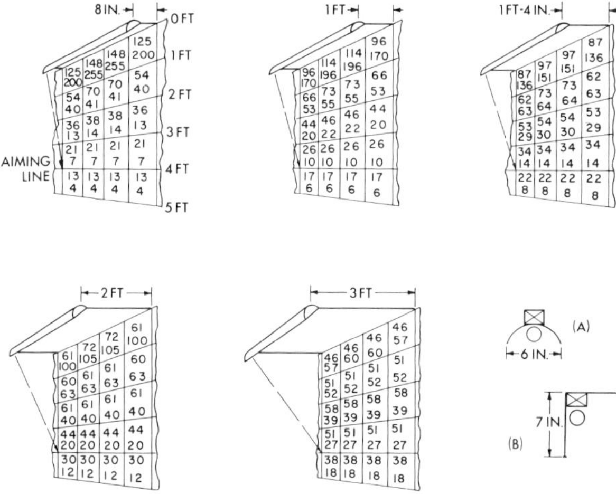

Fig. 9–57. Chart showing a four-foot wide section of a vertical area such as a wall, display background, garment rack, shelves, etc., with a height of five feet, lighted by a single continuous row of T-12 fluorescent lamps of 2500 lumens each. Figures shown are initial footcandles in the center of one-foot squares. Upper figures: typical polished parabolic trough reflector (A), aimed as shown in the diagram. Lower figures: typical white cornice (B), with a 75 per cent reflectance.

1. Illumination can be approximately doubled using two rows of lamps.

2. Charts may be inverted for lamps mounted below the vertical surface.

3. Improved uniformity as well as increased illumination can be obtained using lamps top and bottom. For areas up to nine feet high overlap the upright and inverted charts as necessary to obtain required height. Add together the overlapping footcandles.

4. Illumination from other types of T-12 lamps may be determined by multiplying the footcandles shown by a factor of

$$\frac{\text{lamp lumens per foot}}{625}$$

This factor also applies for other diameter lamps in a white cornice or in a concentrating parabolic reflector having a width approximately four times the lamp diameter.

Show Window Illumination. For estimating average illumination in show windows, two planes are generally used to represent the average display surfaces as shown in Fig. 9-58. These vary in size and position in different windows. They are divided into zones A, B and C (B being of equal width to C) to permit designing for either a variation in illumination level between parts of the display or for a uniform level equally effective throughout. In selecting a zone for use in estimating average illumination consider the nature of the trim and whether the back will be open or closed and proceed as follows:

A. For lamps in external reflectors (Fig. 9-59).

1. Determine how many lamp lumens per linear foot of window must be provided.

Lamp Lumens Needed per Foot = Footcandles Desired × Height in Feet × $K \times P \times Q$

where

K = a constant that takes into account the height-to-depth (H/D) ratio of the window, and the light distribution

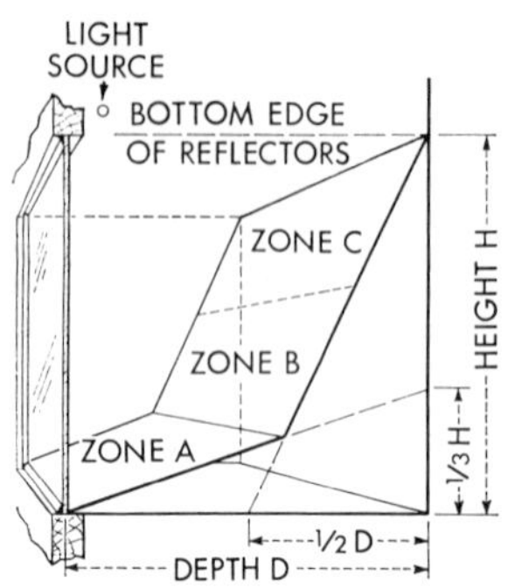

Fig. 9-58. Show windows are divided into three zones (A, B, and C) when it is desired to estimate the number of lamps necessary for a given illumination level.

P = length factor [introduces the effects of the length-to-height (L/H) ratio and of the kinds of ends the window has]

Q = shielding factor (adjusts for the effects of type and orientation of louvering used).

The K, P, and Q factors for incandescent filament and fluorescent lamps can be found in Figs. 9-59A and 9-59B respectively. For each H/D ratio, the

Fig. 9-59. Multiplying Factors for Lamps in External Reflectors

A. For Incandescent Filament Lamps

K Factors*

Cross Section	Zone A			Zone B			Zone C		
H/D	Wide	Semi-Conc.	Conc.	Wide	Semi-Conc.	Conc.	Wide	Semi-Conc.	Conc.
4.0	4.2	3.4	2.0	6.8	5.5	3.9	1.6	2.0	3.0
3.5	3.6	3.0	1.8	6.0	5.0	4.1	1.8	2.3	3.6
3.0	3.2	2.6	1.7	5.5	4.6	4.5	2.0	2.8	4.1
2.5	3.0	2.4	1.7	4.6	4.0	5.0	2.4	3.7	5.0
2.0	2.9	2.3	1.7	4.3	3.7	5.5	3.1	5.1	6.1
1.5	3.0	2.4	1.9	4.1	3.6	6.3	4.1	8.3	9.4
1.0	3.3	2.9	2.3	4.6	4.1	7.5	7.5	20.3	20.0

P (Length) Factors

Type of Equipment	Window Length Divided by Height (L/H)							
	0.5		1.0		1.5		2.0	
	1 Glass End	Solid Ends	1 Glass End	Solid Ends	1 Glass End	Solid Ends	1 Glass End	Solid Ends
Wide	1.40	1.25	1.10	1.05	1.00	1.00	1.00	0.95
Semi-conc.	1.30	1.20	1.05	1.00	1.00	1.00	1.00	0.95
Conc.	1.20	1.10	1.00	1.00	1.00	1.00	1.00	1.00

Q (Shielding) Factors (Factor for Unlouvered Lamps = 1.00)

	Louvers at Right Angles to Plate Glass			Louvers Parallel to Plate Glass			Eccentric Ring Louvers		
H/D	Zone A	Zone B	Zone C	Zone A	Zone B	Zone C	Zone A	Zone B	Zone C
4.0	1.3	1.4	1.4	1.2	1.4	2.2	Not Usually Employed		
3.5	1.3	1.4	1.4	1.2	1.4	2.3	Not Usually Employed		
3.0	1.3	1.4	1.5	1.2	1.5	2.6	Not Usually Employed		
2.5	1.3	1.4	1.6	1.2	1.6	3.0	Not Usually Employed		
2.0	1.4	1.4	1.6	1.2	1.8	3.7	1.4	1.6	2.9
1.5	1.4	1.5	1.7	1.2	1.9	4.6	1.4	1.8	3.3
1.0	1.4	1.5	1.8	1.3	2.1	5.3	1.4	2.2	4.0

Fig. 9-59. *Continued*

B. For Fluorescent Lamps

K Factors**

Cross Section	Zone A			Zone B			Zone C		
H/D	Wide	Semi-Conc.	Conc.	Wide	Semi-Conc.	Conc.	Wide	Semi-Conc.	Conc.
4.0	4.9	5.7	4.2	11.3	10.3	7.6	2.7	1.5	1.7
3.5	4.6	5.2	3.9	9.2	9.2	6.8	2.3	1.6	1.9
3.0	4.2	4.2	3.4	7.9	6.9	5.6	2.5	1.8	2.2
2.5	4.1	3.9	3.0	6.2	5.0	4.7	2.8	2.2	2.7
2.0	4.1	3.8	2.7	5.7	4.3	3.7	3.3	2.8	3.7
1.5	4.0	3.6	2.4	5.2	4.4	4.1	4.3	4.1	5.7
1.0	4.3	3.6	2.3	5.0	5.0	6.8	7.6	7.6	15.1

P (Length) Factors

Window Length Divided by Height (L/H)

0.5		1.0		1.5		2.0	
1 Glass End	Solid Ends	1 Glass End	Solid Ends	1 Glass End	Solid Ends	1 Glass End	Solid Ends
1.55	1.45	1.20	1.10	1.00	0.95	0.95	0.90

Q (Shielding) Factors (Factor for Unlouvered Lamps = 1.00)

H/D Ratio	Shielding Factors (S) — Cellular Louvers (Matte White) Shielding Lamps to 45 degree Crosswise and 25 degree Lengthwise		
	Zone A	Zone B	Zone C
4.0	1.11	1.20	1.14
3.5	1.11	1.20	1.14
3.0	1.11	1.18	1.14
2.5	1.11	1.18	1.14
2.0	1.11	1.18	1.16
1.5	1.09	1.15	1.19
1.0	1.09	1.10	1.27

* (Based on typical commercial equipments. A light loss factor of 0.75 has been assumed.)

** (Based on equipments of typical distribution. Efficiencies: Wide 65%, Semi-Concentrating 85%, Concentrating 80%. . A light loss factor of 0.75 has been assumed.)

Note: This simplified table is based on four rows of fluorescent lamps in reflectors of typical widths, at typical angles of tilt. Some variation in results is to be expected depending on type of equipment selected, number of rows, and mounting angle. As the number of rows is increased, rear rows contribute progressively less light to the A and B zones, and more to the upper background.

lowest K factor in the selected zone identifies the reflector distribution that will provide the desired illumination most efficiently.

2. (*For incandescent filament lamps.*) Make tentative selection of lamp size. One way to do it is to choose the smallest lamp size that has a lumen output rating higher than the lamp lumens needed per foot, found in step 1. Then:

$$\text{Lamp Spacing in Inches} = 12 \times \frac{\text{Rated Lumens per Lamp}}{\text{Lamp Lumens Needed per Foot}}$$

If spotlighting equipment is to be mounted between the general lighting units, the spacing found from this formula may not always be great enough to accommodate them. In this case, select the next-higher-wattage lamp, and recalculate the spacing.

(*For fluorescent lamps.*) Determine the lumens per foot of window produced by a single row of fluorescent lamps of the size, type, and color selected:

$$\text{Lumens per Foot for One Row} = \frac{\text{Total Lamp Lumens for One Row}}{\text{Length of Window}}$$

$$\text{Number of Rows Needed} = \frac{\text{Lamp Lumens Needed per Foot (from Step 1)}}{\text{Lumens per Foot for One Row}}$$

B. For lamps with internal reflectors (Fig. 9-60). Design data for show window lighting with reflector lamps must take into account the various possible lamp aiming angles. The following procedure includes this element:

1. Determine window height in feet from floor to lamps (H), and window depth in feet from glass to background (D). [The ratio of these items

Fig. 9–60. Illumination Factors† for Lamps with Internal Reflectors

H/D	Aiming Angle (degrees from vertical)	PAR-38 Spot			150-Watt R-40 Spot*			PAR-46 Medium Flood**			PAR-38 Flood			150-Watt R-40 Flood*		
		F_A	F_B	F_C	F_A	F_B	F_C	F_A	F_B	F_C	F_A	F_B	F_C	F_A	F_B	F_C
1.0	0	55	10	5	55	15	5	80	10	5	75	10	5	50	20	5
	10	100	15	5	75	20	10	110	10	5	85	15	10	55	25	5
	20	125	20	5	85	25	10	115	15	5	90	30	10	60	35	10
	30	90	40	10	65	45	15	90	65	10	75	55	10	50	45	20
	35	70	70	15	55	60	20	70	85	10	65	70	15	**45**	**50**	**25**
	40	50	105	15	45	80	20	45	100	15	50	85	20	**40**	**55**	**30**
	45	30	125	20	30	90	25	30	115	20	35	90	30	**35**	**60**	**35**
	50	20	135	25	20	90	30	20	130	25	25	95	40	30	55	40
	55	15	130	35	10	90	40	15	125	45	20	90	50	30	50	45
1.5	0	75	15	5	80	20	10	120	15	15	115	20	10	70	35	10
	10	135	25	5	100	30	15	140	20	15	115	35	10	65	40	15
	20	130	50	10	90	45	20	105	55	20	100	60	15	60	50	25
	25	85	90	15	65	70	25	75	90	20	80	80	20	**55**	**55**	**30**
	30	50	130	20	50	90	30	50	115	25	60	90	30	**50**	**55**	**40**
	35	40	140	30	35	95	40	35	125	35	40	100	40	**40**	**55**	**50**
	40	30	125	45	30	90	50	25	115	50	30	95	55	35	55	60
	45	25	100	60	25	80	65	20	95	70	25	85	75	30	55	70
2.0	0	125	20	5	100	25	10	140	15	20	125	25	10	70	35	15
	10	150	40	10	110	40	15	140	40	20	115	45	15	65	40	25
	15	145	60	15	95	50	20	120	60	25	100	60	20	**60**	**45**	**30**
	20	100	90	25	80	70	30	85	75	30	85	75	30	**60**	**45**	**40**
	25	60	120	35	50	80	40	45	90	40	**65**	**80**	**40**	**50**	**50**	**50**
	30	40	125	45	40	85	55	25	100	55	45	80	55	45	50	65
	35	30	100	60	30	70	75	20	105	70	30	75	85	40	45	80
	40	25	60	85				20	90	95						
2.5	0	185	20	10	120	25	20	150	20	25	125	30	15	70	35	25
	5	175	30	10	115	35	20	145	30	25	120	40	15	70	40	30
	10	155	50	15	100	45	25	125	60	30	110	55	20	**65**	**40**	**35**
	15	120	80	25	**80**	**60**	**35**	**90**	**75**	**40**	95	65	30	**60**	**45**	**45**
	20	70	105	35	**55**	**80**	**45**	**50**	**80**	**50**	**70**	**75**	**45**	55	45	55
	25	40	110	50	40	75	65	30	80	65	50	70	65	45	40	70
	30	30	90	75	35	60	85	20	70	85	35	65	90	40	40	85
3.0	0	185	20	15	130	25	25	155	25	30	125	35	20	65	35	30
	5	170	30	20	115	35	30	140	40	30	115	45	25	**65**	**40**	**35**
	10	145	60	25	95	50	35	115	60	40	100	55	35	60	40	45
	15	100	85	35	**70**	**65**	**45**	**80**	**75**	**55**	**85**	**65**	**50**	55	40	55
	20	50	105	50	45	70	60	40	80	70	**60**	**70**	**65**	50	40	70
	25	30	95	75	35	60	85	25	65	90	40	65	90	45	35	85
3.5	0	180	25	20	130	25	25	160	40	30	125	40	20	65	30	35
	5	170	40	25	110	35	30	140	50	40	110	50	30	60	35	45
	10	**135**	**65**	**30**	**85**	**50**	**40**	**105**	**65**	**50**	95	55	40	60	35	55
	15	**75**	**85**	**40**	**60**	**65**	**55**	**65**	**75**	**70**	**75**	**60**	**60**	55	35	70
	20	45	85	65	40	60	75	35	70	95	55	60	85	45	35	85
4.0	0	175	25	20	125	25	30	155	35	30	120	40	25	60	30	40
	5	160	35	25	105	35	35	130	55	45	105	50	35	60	35	50
	10	**120**	**65**	**35**	**70**	**50**	**45**	**100**	**65**	**65**	**90**	**55**	**50**	55	35	65
	15	**65**	**80**	**50**	50	55	65	60	65	90	70	50	75	50	30	80
	20	40	70	75												

† These numbers are not footcandles, but must be divided by window height to find lamp spacing in inches that will produce 200 footcandles in a selected zone. Based on assumed light loss factor of 0.80.

* Multiply illumination factors by 2.0 for 300-watt R-40, 3.1 for 500-watt R-40, 0.4 for 75-watt R-30.

** Wide section of beam perpendicular to glass. Multiply illumination factors by 1.7 for 300-watt PAR-56 medium flood.

(H/D) provides an index of window proportions for use in the first column of Fig. 9–60.]

2. Find the most suitable lamp type and aiming angle by referring to Fig. 9–60 and finding the illumination factors in bold type appearing opposite the proper value of H/D. (Suitable lamp types are indicated by the column in which the bold numbers appear, and suitable aiming angles are indicated by the angle on the same line in which the bold numbers appear.) If more than one suitable lamp type or aiming angle is found in this manner, then select either the one that produces the desired de-

gree of emphasis in the most important zone, or provides greatest uniformity among zones. Relative illumination in the three zones will be in proportion to the illumination factors (F_A, F_B, and F_C) for the zones.

3. Find the center-to-center lamp spacing in inches that will produce an average of 200 footcandles maintained in the zone considered most important (usually zone A or zone B). The formula is:

$$\text{Spacing} = \frac{F}{H}$$

(Where F is either F_A or F_B from Fig. 9–60 for the selected lamp type and aiming angle, and H is window height in feet.)

4. Check to be sure that the lamp spacing just found does not exceed the maximum spacing in inches for generally satisfactory lateral uniformity (absence of streaks and scallops) in all zones.

$$\text{Maximum Permissible Spacing (Inches)} = U \times D$$

(Where U is the uniformity factor from Fig. 9–61 for the lamp type used, and D is window depth in feet. Unless this maximum spacing is radically exceeded, the effect may not be objectionable, because the streaks or scallops would be at the top of the background in a closed-back window, and would probably be unnoticed in an open-back one.)

5. Illumination in the other zones will be in accordance with these relationships:

Footcandles in Zone A =

$$\frac{F_A \times (\text{Footcandles in Zone } B)}{F_B} \quad \text{or} \quad \frac{F_A \times (\text{Footcandles in Zone } C)}{F_C}$$

Footcandles in Zone B =

$$\frac{F_B \times (\text{Footcandles in Zone } A)}{F_A} \quad \text{or} \quad \frac{F_B \times (\text{Footcandles in Zone } C)}{F_C}$$

Footcandles in Zone C =

$$\frac{F_C \times (\text{Footcandles in Zone } B)}{F_B} \quad \text{or} \quad \frac{F_C \times (\text{Footcandles in Zone } A)}{F_A}$$

6. If a value of illumination other than 200 footcandles is desired in a particular zone, proceed as in steps 1 through 3 above. Then multiply the lamp spacing found in step 3 by 200/desired footcandles. Illumination in the other zones will be greater or less in the ratio of desired footcandles/200. Uniformity should be checked as in step 4.

Showcase Illumination. When lighting is supplied from within the showcase by lamps at the top front edge, footcandles in each of three typical trim-planes can be estimated. (Fig. 9–58.) Plane A extends from the lower front edge to a line at ⅓ the height of the case. Plane B-C runs from the top rear edge to a line at ½ the case depth on the bottom. Zones B and C are equal. Any or all of these zones may be of prime importance, depending on the method of displaying merchandise.

In Fig. 9–62 typical glass showcases 20 inches deep are assumed. Constants K are given for each zone, for three typical case heights, and for four lighting methods. The values of K are based on an assumed light loss factor of 0.75.

Fig. 9–61. Uniformity Factors for Lamps with Internal Reflectors

Lamp Type	Uniformity Factor
PAR-38 spot	2
R-30 spot	2
R-40 spot	2
PAR-46 med. flood	2
PAR-56 med. flood	2
PAR-38 flood	6
R-40 flood	12

Fig. 9–62. Constants K for Showcases*
(D = 20 inches)

Lamp	Lighting Method	H = 10 (inches)			H = 20 (inches)			H = 27 (inches)		
		Zone			Zone			Zone		
		A	B	C	A	B	C	A	B	C
Fluorescent	White Diffusing Reflector	.199	.091	.038	.128	.103	.061	.097	.084	.071
	Concentrating Reflector	.241	.148	.070	.124	.144	.104	.092	.109	.113
Incandescent filament	Clear T-10 in Semidiffusing Reflector	.164	.103	.071	.107	.089	.091	.081	.070	.091
	T-10 Reflector Showcase	.267	.269	.096	.170	.250	.138	.134	.178	.155

* Assumed 0.75 light loss factor.

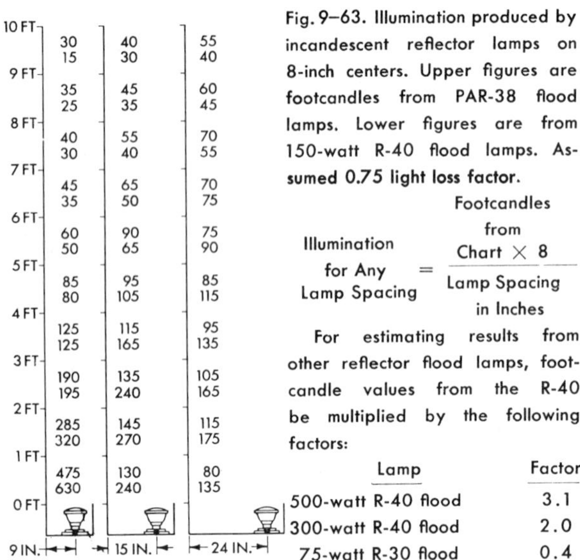

Fig. 9–63. Illumination produced by incandescent reflector lamps on 8-inch centers. Upper figures are footcandles from PAR-38 flood lamps. Lower figures are from 150-watt R-40 flood lamps. Assumed 0.75 light loss factor.

$$\text{Illumination for Any Lamp Spacing} = \frac{\text{Footcandles from Chart} \times 8}{\text{Lamp Spacing in Inches}}$$

For estimating results from other reflector flood lamps, footcandle values from the R-40 be multiplied by the following factors:

Lamp	Factor
500-watt R-40 flood	3.1
300-watt R-40 flood	2.0
75-watt R-30 flood	0.4

When lamp and reflector have been selected, the footcandles in any zone can be estimated:

$$\text{Footcandles} = K \times (\text{Lumens-per-Foot of Case Length})$$

The lumens-per-foot value for a continuous row of fluorescent lamps running the entire case length will be the lumens-per-foot of the lamps used. Otherwise, lumens-per-foot of case length for either fluorescent or incandescent filament lamps will be:

$$\text{Lumens-per-Foot of Case Length} = \frac{\text{Lumens per Lamp} \times \text{Number of Lamps}}{\text{Length of Case in Feet}}$$

Vertical Surface Illumination. Fig. 9–63 may be used to calculate the illumination normal to a large vertical surface from a row of reflector lamps mounted close by. Care should be exercised to avoid a scalloping effect caused by having the lamps too far apart in relation to their distance from the lighted surface.

FLOODLIGHTING CALCULATIONS

The two most commonly used systems for floodlight calculations are the point method and the beam-lumen method. The point method (page 9–44) permits the determination of footcandles at any point and orientation on a surface. This method is valuable since it permits a visualization of the degree of lighting uniformity realized for any given set of conditions. The beam-lumen method is quite similar to the method for interior lighting except that it must take into consideration the fact that floodlights are not usually directly above the surface, but instead are aimed at various angles to the surface.

The Beam-Lumen Method for Floodlights

The basic formula is:

$$\text{Footcandles (Average Maintained)} = \frac{\text{Quantity of Floodlights} \times \text{Beam Lumens} \times \text{Coefficient of Beam Utilization} \times \text{Light Loss Factor}}{\text{Area in Square Feet}}$$

or

$$\text{Quantity of Floodlights} = \frac{\text{Area in Square Feet} \times \text{Average Maintained Footcandles}}{\text{Beam Lumens} \times \text{Coefficient of Beam Utilization} \times \text{Light Loss Factor}}$$

Beam Lumens. Beam lumens are defined as the quantity of light that is contained within the beam limits as described on page 1–4 under "beam spread." Beam lumens also equal the lamp lumens multiplied by the *beam efficiency* of the floodlight. Outdoor floodlight luminaire designations and minimum beam efficiencies are given in Fig. 9–64.

It may be more economical to operate incandescent filament lamps above rated voltage and mercury lamps above rated wattage. Multipliers resulting from operating two typical lamps above rated values are tabulated below:

Parameter	*Multiplier*	
	Incandescent Filament	*Mercury*
Voltage	1.10	—
Wattage	1.16	1.50
Life	0.30	0.12
Lumens	1.35	1.50

When using the equations above, the "beam lumens" should be increased by the appropriate lamp-lumens multiplier.

Coefficient of Beam Utilization. This factor, CBU, written as a decimal fraction, expresses the following ratio:

$$\text{Coefficient of Beam Utilization} = \frac{\text{Lumens Initially Reaching the Specified Area Directly from the Floodlight}}{\text{Total Beam Lumens}}$$

To determine the number of lumens from each floodlight beam that initially strikes the area, it is necessary to first locate and determine the aiming of each floodlight with respect to the area. For each floodlight location and direction the area is replotted in angular coordinates upon the isocandela curve of the floodlight, as in Fig. 9–65, and the lumens within the area boundaries summed. In this example the floodlight at F is directed at point O on building $ABCD$, and angle LFO is the angle

Fig. 9-64. Outdoor Floodlight Luminaire Designations*

Beam Spread Degrees	NEMA Type	Minimum Efficiencies (per cent) — Effective Reflector Area (square inches)				
		Incandescent Lamps		Mercury Lamps		Fluorescent Lamps
		Under 227	Over 227	Under 227	Over 227	Any
10 up to 18	1	34	35	—	—	20
18 up to 29	2	36	36	22	30	25
29 up to 46	3	39	45	24	34	35
46 up to 70	4	42	50	35	38	42
70 up to 100	5	46	50	38	42	50
100 up to 130	6	—	—	42	46	55
130 and up	7	—	—	46	50	55

* Taken from National Electrical Manufacturers' Association, 155 East 44th Street, New York, New York 10017, Publication #FL 1-1964.

Asymmetrical beam floodlights may be designated by a combination type designation which indicates the horizontal and vertical beam spreads in that order; *e.g.*, a floodlight with a horizontal beam spread of 75 degrees (Type 5) and a vertical beam of 35 degrees (Type 3) would be designated as a Type 5 X 3 floodlight.

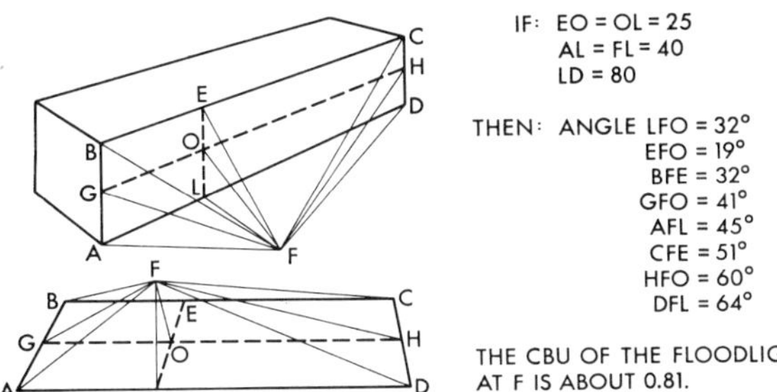

NOMINAL LAMP DATA

WATTS	1500
VOLTS	115
BULB	PS-52 Clear
SERVICE	General
LUMENS	33,000

TEST RESULTS

AVG. MAX. CANDLEPOWER	23,500
BEAM LUMENS	18,790
BEAM EFFICIENCY	57%
HOR. BEAM SPREAD	98 degrees
VERT. BEAM SPREAD	91 degrees

AVERAGE OF RIGHT AND LEFT SIDES

2350 C.P. 10% MAX. 10
5000 C.P.
10,000 C.P.
15,000 C.P.
20,000 C.P.
25,000 C.P.
OUTLINE OF AREA

LUMENS — AVERAGE OF RIGHT AND LEFT SIDES

	0°–10°	10°–20°	20°–30°	30°–40°	40°–50°		TOTALS - ½ OF HORIZONTAL ZONES
50°–40°	66	58	44	27	14		209
40°–30°	146	125	88	53	26		438
30°–20°	280	228	158	86	40		792
20°–10°	502	388	232	122	53		1297
10°–0°	784	581	310	145	62		1882
0°–10°	884	630	332	151	65		2062
10°–20°	654	481	265	128	56		1584
20°–30°	351	274	167	90	41		923
30°–40°	167	140	98	57	27		489
40°–50°	80	70	50	32	14		246
TOTALS OF VERTICAL ZONES	3914	2975	1744	891	398		9922

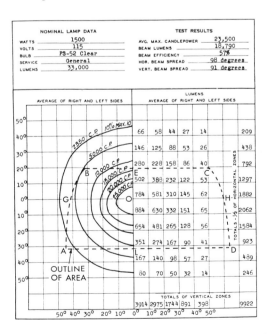

Fig. 9-65. Procedure for determining the coefficient of beam utilization for floodlighting equipment.

whose tangent is 25/40 or 32 degrees. Angle *EFL* is the angle whose tangent is 50/40 or 51 degrees. Angle *EFO* must then be 19 degrees. By similar calculations all the angles in Fig. 9-65 can be determined and the building can then be plotted on the grid of the isocandela curve of the floodlight. See Figs. 9-29 and 9-66 for nomograms of vertical and lateral angles. Because of the manner in which floodlights are photometered, all horizontal lines parallel to a line perpendicular to the beam axis appear as straight horizontal lines on the grid. All vertical lines except the one through the beam axis appear slightly curved.

By calculating the coefficients of beam utilization for a few representative floodlight locations and aiming points, a fairly accurate average CBU can be established for that particular installation.[18]

Coverage.[19] It is recommended that sufficient point calculations be made for each job to check uniformity and coverage. A check for coverage alone can be made either by means of a scale drawing and protractor or by a coverage chart provided by the floodlight manufacturer.

Light Loss Factor. The light loss factor is an allowance for depreciation of lamp output with life and depreciation of floodlight efficiency due to the collection of dirt on lamp, reflector, and cover glass. Depreciation of lamp output with life is dependent upon the type of lamp used. Dirt factors will differ between open and enclosed floodlights. The total factor then may vary from .65 to .85.

SEARCHLIGHT CALCULATIONS

Useful Range of Searchlights

Searchlight applications fall into two general categories: signal lights which are to be seen, such as lighthouse or airway beacons; and illumination sources by which targets are to be seen, such as anti-aircraft searchlights. The range for lights which are themselves to be seen is discussed starting on page **3-33**. For the second category, range is dependent not only on physical properties of the searchlight but on many other physical, physiological, and psychological parameters of the seeing situation. The following factors must all be specified: beam intensity; intensity distribution; atmospheric transmittance; target size, shape, and reflectance; background luminance and offset distance of the observer from the searchlight.

The useful range of searchlights used for illuminating targets may be estimated approximately from Fig. **9-67**, which has been calculated from formulas and data by researchers.[20] These data apply to naked-eye observation with the observer located in the vicinity of the searchlight. It will be seen that the range varies considerably with the atmos-

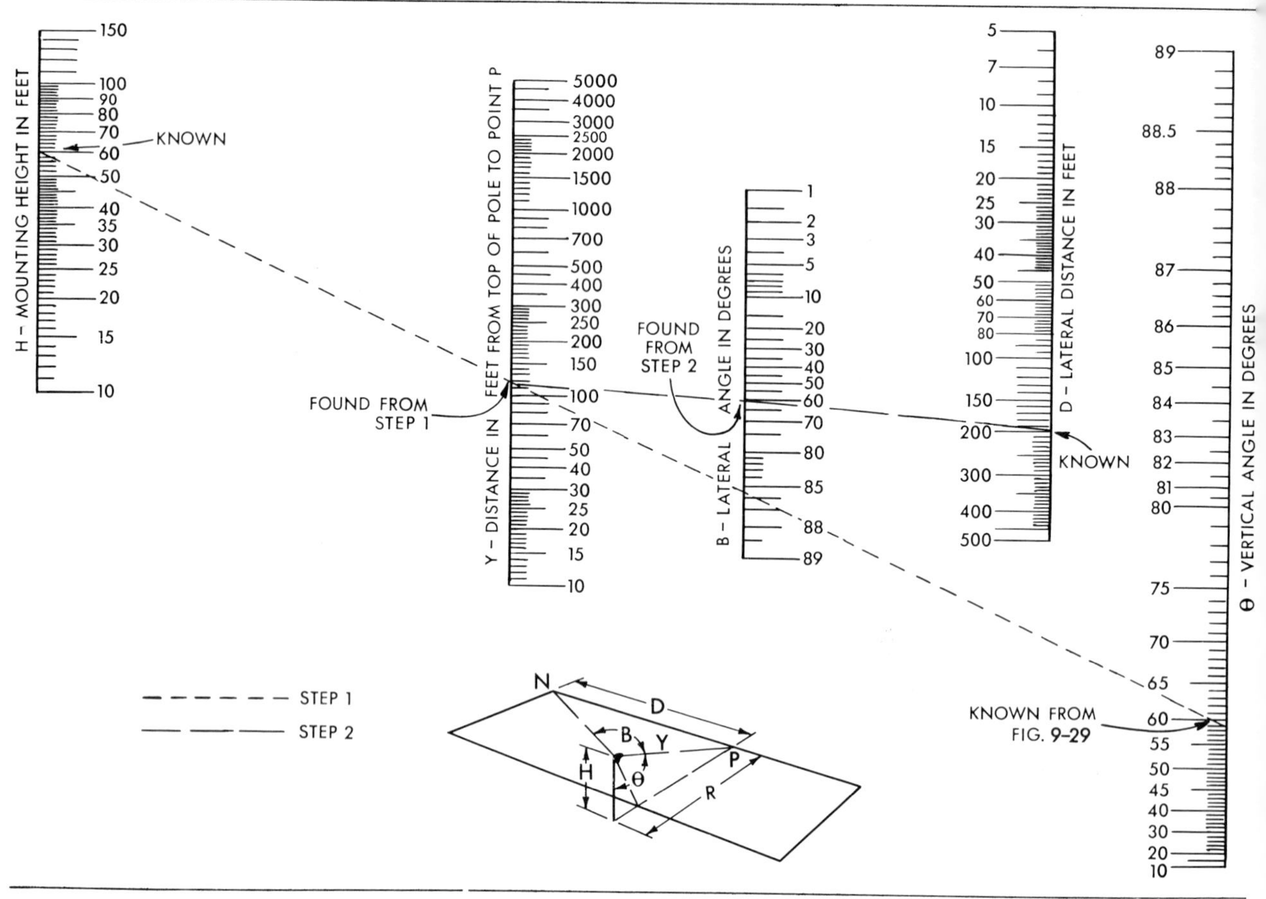

Fig. 9–66. Nomogram for determining lateral angle B in terms of vertical angle θ, mounting height H, and distances Y and D.

pheric transmission and target size and contrast. The observer should be separated laterally from the searchlight as far as possible to minimize the amount of beam through which the target must be seen. The visibility of the target will increase quite rapidly for the first few feet of separation and will continue to improve as the separation is increased. The effect of separation on visibility and the rate of improvement with increasing separation is greater for lower atmospheric transmittance than for higher, although adequate separation is always important. As an example, an observer looking at a target 10,000 yards from a 60-inch searchlight should be at least 100 feet to one side of the beam axis. Still better visibility[21] will result if the observer is 300 feet from the light. At a range of one mile, the observer should have a lateral separation from the searchlight of at least 20 feet. The use of binoculars will greatly increase the range at which small targets can be seen.

Peak Candlepower of Searchlights

A useful approximation of the peak candlepower of a searchlight may be obtained from the formula:

$$I = L\rho A$$

where

I = the peak candlepower

L = the average peak luminance of the source within the collecting angle of the optic

ρ = the reflectance of the reflector or transmittance of the refractor

A = the effective aperture area of the searchlight, or the total flashed area of the aperture.

Beam Spread (Divergence) of Searchlights

To a first approximation, the beam divergences, θ, of a searchlight can be obtained from the formula:

$$\theta = 2 \tan^{-1} \frac{r}{f}$$

where

r = the radius of the source

f = the focal length of the optic

Aberrations, imperfections, and incorrect adjustments tend to increase θ.

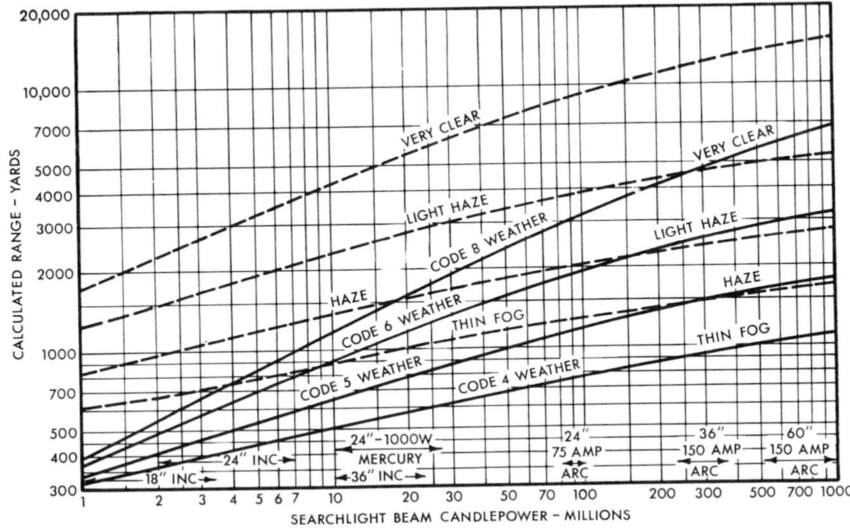

Fig. 9-67. The effect of atmospheric absorption on the useful range of searchlight beams.

Solid lines—Medium target such as group of men or dark colored car against slightly lighter background such as field or buildings. Broken lines—Large target such as large aircraft, a battleship, or a church steeple as viewed against the sky.

Code Number	Weather	Maximum Daylight Visual Range	Maximum Transmittance per Kilometer (per cent)	Maximum Transmittance per Mile (per cent)	Maximum Transmittance per Nautical Mile (per cent)
0	dense fog	50 meters			
1	thick fog	200 meters			
2	mod. fog	500 meters	0.004	0.0003	0.00005
3	light fog	1000 meters	2.0	0.19	0.07
4	thin fog	2 km	14	4.4	2.7
5	haze	4 km	37	21	16
6	light haze	10 km	68	53	48
7	clear	20 km	82	73	70
8	very clear	50 km	93	88	87
9	except. clear	over 50 km	over 93	over 88	over 87

ROADWAY ILLUMINATION DATA AND CALCULATIONS

Roadway illumination calculations of horizontal footcandles fall into two general types:

1. Determination of the average illumination on the roadway pavement.
2. Determination of the illumination at a specific point on the roadway.

Determination of Average Illumination

The average illumination over a large pavement area in terms of horizontal footcandles may be calculated by means of a "utilization curve" of the type shown in Fig. 9-68.

Utilization Curves. Utilization curves, available for various types of luminaires, afford a practical method for the determination of lumens per square foot (average footcandles) over the roadway surface where lamp size, mounting heights, width of roadway, overhang, and spacing between luminaires are known or assumed. Conversely, the desired spacing or any other unknown factor may readily be determined if the other factors are given.

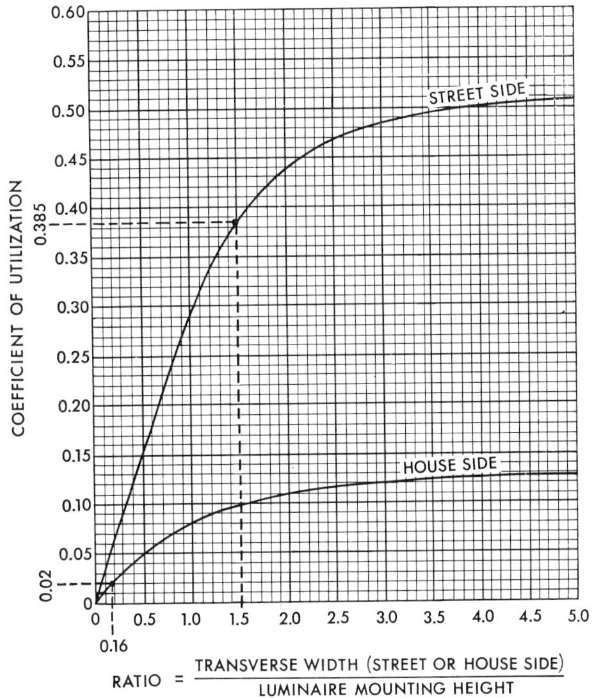

Fig. 9-68. Example of coefficient of utilization curves for luminaire providing Type III-M light distribution.

The Coefficient of Utilization, as shown in Fig. 9-68, is the percentage of rated lamp lumens which will fall on either of two strip-like areas of infinite length, one extending in front of the luminaire (street side), and the other behind the luminaire (house side), when the luminaire is level and oriented over the roadway in a manner equivalent to that in which it was tested. Since roadway width is expressed in terms of a ratio of luminaire mounting height to roadway width, the term has no dimensions.

Depreciation Factors. The several causes of light loss from street lighting luminaires are illustrated in Fig. 9-69. For further clarification, see page 16-11. These deteriorating conditions always exist, in varying degrees. Thus each circumstance should be considered separately to apply reasonable depreciation values for it.

Formulas for Computation. The basic formula for determination of average horizontal footcandles is as follows:

$$\text{Average Footcandles} = \frac{\text{Lamp Lumens} \times \text{Coefficient of Utilization}}{\text{Pavement Area per Luminaire in Square Feet}}$$

This formula is usually expanded as follows:

Average Footcandles (Lumens per Square Foot) =

$$\frac{(\text{Lamp Lumens}) \times (\text{Coefficient of Utilization})}{(\text{Spacing Between Luminaires in Feet})^{*} \times (\text{Width of Roadway in Feet})}$$

It can be seen that with this expression of the formula, it is possible to solve horizontal average footcandles, or spacing, or lamp lumens as desired. A further modification of this formula is necessary to determine the average illumination on the roadway when the illuminating source is at its dirtiest condition (see Depreciation Factors above). For such a calculation, the formula is expressed as follows:

Average Footcandles (Lumens per Square Foot) =

$$\frac{\left(\text{Lamp Lumens}\right) \times \left(\begin{matrix}\text{Coefficient of}\\ \text{Utilization}\end{matrix}\right) \times (\text{Light Loss Factor})^{**}}{(\text{Spacing Between Luminaires in Feet})^{*} \times (\text{Width of Roadway in Feet})}$$

* This is the longitudinal distance between luminaires if spaced in staggered or one-side arrangement. This distance is one-half the longitudinal distance between luminaires if luminaires are arranged in opposite spacing.

** This value may be experimentally determined or estimated if not known.

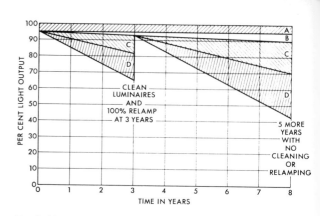

Fig. 9-69. Causes of Light Loss shown* for a typical (400-watt mercury) roadway lighting system.
A. Variation in Temperature and/or Voltage
B. Deterioration of Luminaire Surfaces or Refractors
C. Lamp Lumen Depreciation
D. Luminaire Dirt Depreciation

* Values shown are illustrative of losses. Relative amounts will differ for specific installation. If lamp outages are not replaced the end values shown will be further reduced.

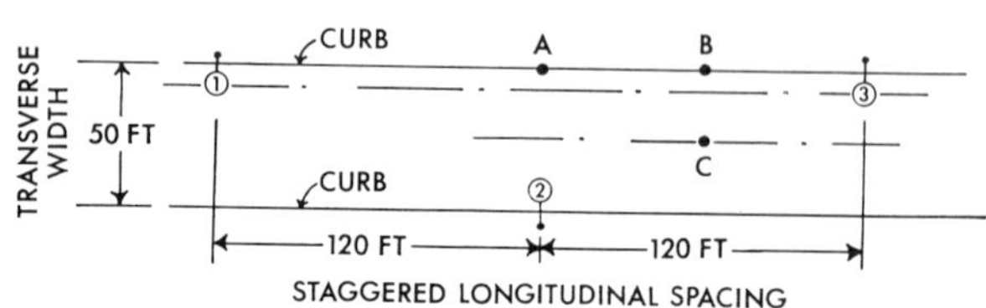

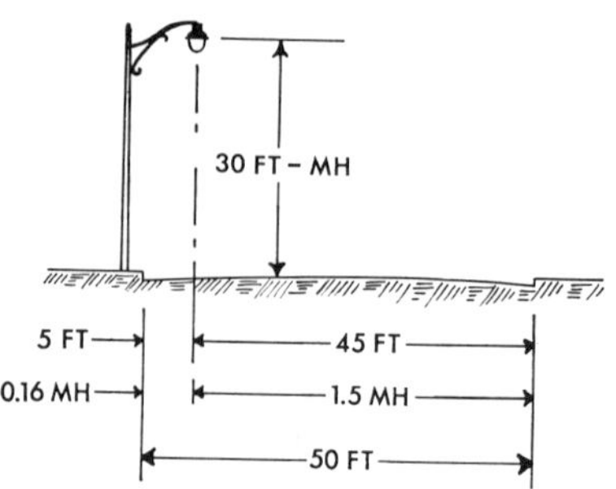

Fig. 9-70. Layout of luminaire and roadway assumed for typical computation.

Typical Computations. To illustrate the use of a utilization curve, Fig. 9-68, a typical calculation is provided as follows:

Given: Roadway with layout as in Fig. 9-70.
Staggered luminaire spacing, 120 ft
Roadway width curb-to-curb (pavement), 50 ft
Luminaire mounting height, 30 ft
Luminaire overhang, 5 ft
Light loss factor, (0.6)
Mercury lamp with 20,000 lumens initial.

Required: To calculate the minimum average lumens per square foot (average footcandles) for the above roadway.

Solution: For average illumination:

1. Determine the coefficient of utilization for the "street side" of the luminaire: Ratio (street side) (from Fig. 9–70)

$$\frac{50 \text{ feet} - 5 \text{ feet*}}{30 \text{ feet}} = \frac{45 \text{ feet}}{30 \text{ feet}} = 1.50$$

Coefficient of utilization from Fig. 9–68 for ratio 1.50 is 0.385.

2. To determine "house side" coefficient of utilization: Ratio (house side) (from Fig. 9–70)

$$\frac{5 \text{ feet*}}{30 \text{ feet}} = 0.16$$

Coefficient of utilization from Fig. 9–68 for ratio 0.16 is 0.02.

3. Total coefficient for "street side" plus "house side" is 0.405.

4. To determine average illumination on roadway use the formula from above, giving:

$$\text{Average Footcandles} = \frac{20{,}000 \times 0.405 \times 0.6}{120 \times 50}$$

$$= 0.8 \text{ footcandle}$$

Determination of the Illumination at a Specific Point

The determination of the horizontal illumination in footcandles at a specific point may be determined from an "isofootcandle" curve, Fig. 9–71, or by means of the classical point method of calculation.

Isofootcandle Diagram. An isofootcandle diagram is a graphical representation of points of equal illumination connected by a continuous line. These lines may show footcandle values on a horizontal plane from a single unit having a definite mounting height, or they may show a composite picture of the illumination from a number of sources arranged in any manner or at any mounting height. They are useful in the study of uniformity of the illumination and in the determination of the level of illumination at any specific point. In order to make these curves applicable to all conditions, they are computed for a given mounting height, but horizontal distances are expressed in ratios of the actual distance to the mounting height. Correction factors for other mounting heights are usually given in the tabulation alongside the isofootcandle curves.

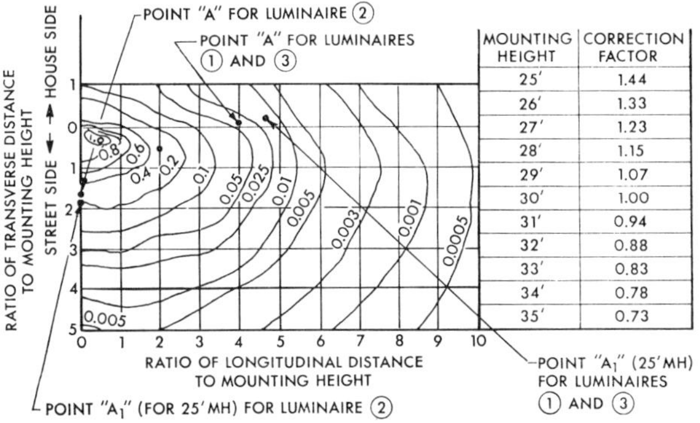

MOUNTING HEIGHT	CORRECTION FACTOR
25′	1.44
26′	1.33
27′	1.23
28′	1.15
29′	1.07
30′	1.00
31′	0.94
32′	0.88
33′	0.83
34′	0.78
35′	0.73

Fig. 9–71. Example of an isofootcandle diagram of horizontal footcandles on pavement surface for a luminaire providing a Type III-M light distribution; per 1000 lumens of lamp output times 10.

Typical Computations. To illustrate the use of the isofootcandle diagram, a typical calculation is as follows:

Given: Roadway with layout as in Fig. 9–70
Staggered luminaire spacing, 120 ft
Roadway width curb-to-curb, 50 ft
Luminaire mounting height, 30 ft
Luminaire overhang, 5 ft
Light loss factor, (0.6)
Mercury lamp with 20,000 lumens initial.

Required: To determine the footcandle level at point "A" on Fig. 9–70 which is the total of contributions from luminaires 1, 2, and 3.

Solution:

1. The location of point "A" with respect to a point on the pavement directly under the luminaire is dimensioned in transverse and longitudinal multiples of the mounting height. Assume that the luminaire distribution provides isofootcandle lines (horizontal footcandles) as shown in Fig. 9–71. Point "A" is then located on this isofootcandle diagram for its position with respect to each luminaire.

2. To determine the contribution of each luminaire to point "A":

a. Luminaires Numbers 1 and 3:

Locate point "A"—Transverse 5 feet to "house side":

$$\frac{5}{30} = 0.16 \text{ times mounting height}$$

Longitudinal 120 feet along pavement

$$\frac{120}{30} = 4.0 \text{ times mounting height}$$

At point "A" for these luminaires, the estimated footcandle value from Fig. 9–71 isofootcandle diagram is 0.04 footcandle. This is from each luminaire, Numbers 1 and 3. Both luminaires together provide 0.08 footcandle.

b. Luminaire Number 2: Locate point "A"—Transverse 45 feet to "street side":

$$\frac{45}{30} = 1.5 \text{ times mounting height}$$

Longitudinal location is 0, directly across from the luminaire. At point "A" for this luminaire, the es-

* Use actual overhang distance from curb to point below luminaire.

timated footcandle value from Fig. 9–71 is 0.3 footcandle.

3. The total at point "A" from the three luminaires is 0.08 plus 0.3 = 0.38 footcandle. The value of 0.38 footcandle is based on 1000 lamp lumens times 10 and clean luminaires with a lamp producing rated output. The initial footcandle level, therefore, is 0.38 × 2 = 0.76 footcandles. If it is desired to express the footcandle level in terms of the footcandles when the illuminating source is at its lowest output and when the luminaire is in its dirtiest condition, this can be done by utilizing the procedure covered above (0.76 × 0.6 = 0.46 fc.)

4. To use these data for a mounting height other than the one for which the isofootcandle curves are made, it is necessary to find the correct new location on the diagram as well as apply a correction factor to the footcandle value at this new location. The following procedure may be used.

a. Compute new transverse and longitudinal distance-to-mounting height, and locate points on the diagram as outlined in the following calculation:

Example for 25-foot mounting height;
Luminaire Numbers 1 and 3—Point "A_1";
Transverse 5 feet to "house side":

$$\frac{5}{25} = 0.2 \text{ MH}$$

Longitudinal 120 feet along pavement:

$$\frac{120}{25} = 4.8 \text{ MH}$$

Point "A_1" is located on the isofootcandle diagram, Fig. 9–71, with these new dimensions.

b. Obtain estimated footcandle values at the new locations and multiply these values by the correction factor for the new mounting height.

Footcandle value estimated at point "A_1," Fig. 9–71, is 0.015 footcandle. This is multiplied by the correction factor for 25 feet which is 1.44.

0.015 × 1.44 = 0.0216 footcandle from each luminaire, Numbers 1 and 3. Both luminaires provide 0.043 footcandle.

Luminaire Number 2—Point "A_1."
Transverse 45 feet to "street side":

$$\frac{45}{25} = 1.8 \text{ MH}$$

Longitudinal location is still 0, directly across from luminaire. The estimated footcandles from Fig. 9–71 is 0.2 footcandle. This is multiplied by the correction factor, 1.44.

$$0.2 \times 1.44 = 0.288 \text{ footcandle.}$$

The total at point "A_1" is

$$0.043 + 0.288 = 0.331 \text{ footcandle.}$$

As before this must then be multiplied by the ratio of the actual lamp lumens to the lamp lumens of the isofootcandle diagram (20,000/10,000 = 2) for the initial footcandle level.

Uniformity Ratios

The uniformity of illumination requirements (see page 20–7 should be determined by computing the ratio:

$$\frac{\text{Minimum Horizontal Footcandles}}{\text{Average Horizontal Footcandles}}$$

It can also be expressed as the ratio:

$$\frac{\text{Average Horizontal Footcandles}}{\text{Minimum Horizontal Footcandles}}$$

Sufficient number of specification points over the roadway should be checked, as outlined above, to ascertain accurately the location and value of the minimum point. If the values at points "A," "B," and "C" as shown in Fig. 9–70, are first determined, the approximate location of the minimum point may be located or its location will become more apparent.

The average illumination on the roadway pavement should be computed as shown above, taking care to use the same lamp lumen output and other conditions as used in determining the minimum illumination value.

DAYLIGHTING DESIGN

Computation of Illumination

The procedure involves: (1) determination of the daylight (sun, sky, and ground light) incident on the windows, (2) the light flux actually entering the lighted space, (3) the distribution of the light flux within the room, and (4) the illumination produced on the surfaces of interest.

Sky Light. The daylight received from the sky in the plane of vertical or horizontal fenestration on various days of the year and times of day for a number of orientations is obtained as follows: (1) from Fig. 7–7 (page 7–4) determine the solar altitude for the date and time, (2) from the appropriate curves of Figs. 7–13 or 7–14 read the illumination from the sky incident on the fenestration.

Sunlight, Horizontal Fenestration. For various days of the year and times of day, find the solar illumination on the horizontal plane from Fig. 7–8.

Sunlight, Vertical Fenestration. (1) For various days of the year and times of day, find the solar illumination on a plane perpendicular to the sun's rays in Fig. 7–8, (2) find the solar altitude θ_1 and azimuth θ_2 in Fig. 7–7, (3) determine the azimuth of the plane normal to the window θ_3, (4) sunlight on vertical window = footcandles (from Fig. 7–8) $\times \sin \theta_1 \cos (\theta_2 - \theta_3)$.

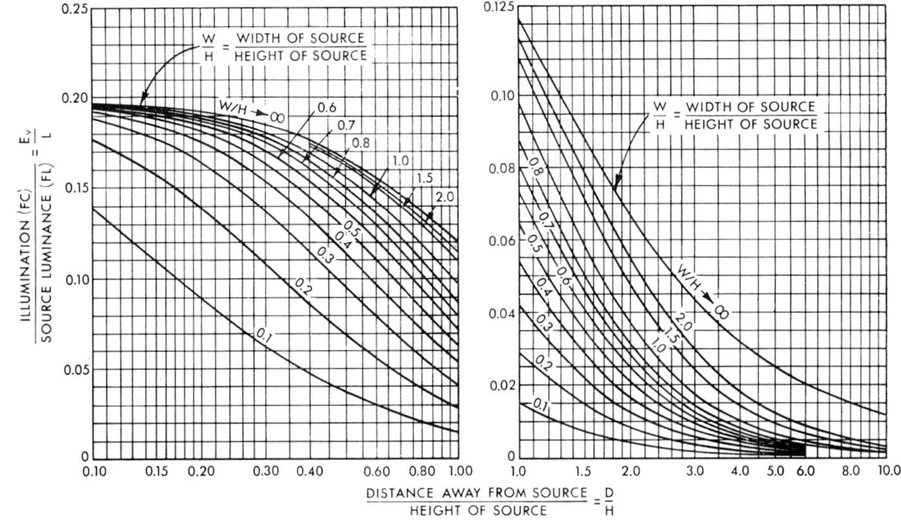

Fig. 9-72. Illumination on a plane parallel to a rectangular non-uniform source (equivalent to an overcast sky).[22]

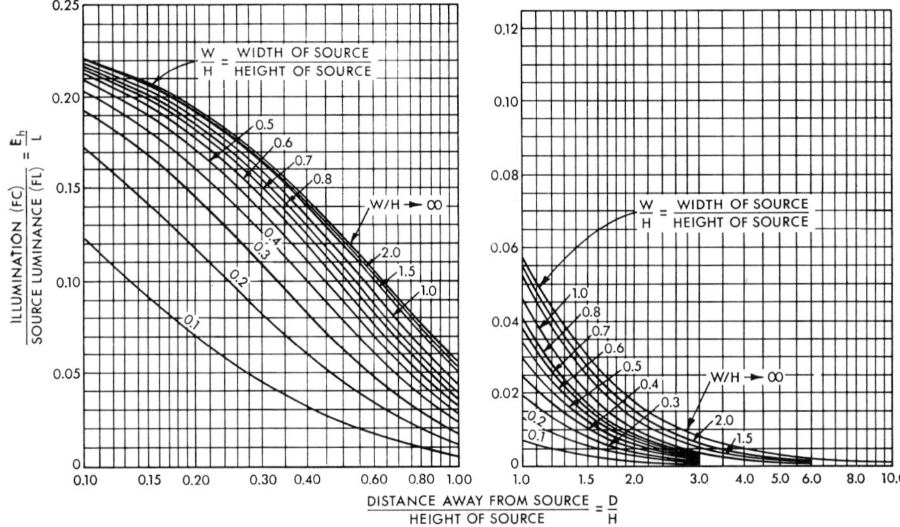

Fig. 9-73. Illumination on a plane perpendicular to a rectangular non-uniform source (equivalent to an overcast sky).[22]

Ground Light. The illumination on a vertical window produced by reflection from the ground is dependent on the distance from the ground, its area and luminance, the latter being the product of the horizontal illumination times the reflectance of the ground. A window receives illumination from a uniformly bright ground of infinite extent equal in footcandles to one-half the ground luminance regardless of the height of the window above the ground. For ground of limited area, the illumination may be calculated by the methods outlined on page 9-59 for area sources.

Determination of Light Flux Entering a Room. Only a portion of the light flux incident on the fenestration area actually enters the lighted space. The gross fenestration area is reduced by the mechanical supports of the glazing material, such as mullions, window members, or mortar joints. The reduction in transmission area can be readily determined.

There is also a transmission loss in the glazing material itself. Data on transmittance of glazing materials are included in Fig. 7-22. There is a further loss attributable to dirt collection on the glazing material (see Fig. 7-31), which should be included in computations as a light loss factor. Finally, there is absorption or reflection by daylight controls, such as shades, louvers, prismatic devices, overhangs, or by the light wells of toplighting arrangements.

Daylight is generally abundant, and efficiency of transmission into the room can and must often be compromised with other factors, such as control of sunlight, heat gain, or glare, in the over-all daylighting design.

Rectangular Source Computation. The illumination from windows or skylights may be calculated by the methods described for area sources on page 9-59, assuming a uniform sky luminance and applying suitable reduction factors as described directly above. No allowance is made in the calculations for inter-reflections.

Non-Uniform Overcast Sky

The luminance distribution of overcast skies can be approximated by the simple equation:

$$L_\theta = 3/7\ E_h\ (1 + 2 \sin \theta)$$

where

L_θ = sky luminance at angle θ above the horizon

E_h = illumination on horizontal plane from unobstructed sky.

This equation can then be applied to the general method for the calculation of the illumination from a rectangular source of any luminance distribution and new equations derived. Graphs of the results of this procedure are shown in Figs. 9–72 and 9–73 for the luminance distribution approximating an overcast sky. The results obtained from these curves when multiplied by the glass transmittance (Fig. 7–22), the equivalent sky luminance (Fig. 7–9), and the ratio of glass area to the area of the window give values which closely approximate test results for the conditions of dark walls and ceilings on an overcast day.

Lumen Method, Toplighting

For systems employing lighting elements, horizontal or in the plane of a nearly level roof, the average illumination in the lighted space can be found by much the same principles as are used for electric lighting practice. The basic formula is:

$$E_{tl} = \frac{E_h \times A_{tl} \times K_u \times K_m}{A_r}$$

where

E_{tl} = Average illumination produced on the work plane by the toplighting system, in lumens per square foot (footcandles).

E_h = Incident illumination on exterior of toplighting element, in lumens per square foot.

A_{tl} = Gross area of toplighting element, square feet.

K_u = Utilization coefficient, dependent on room proportions, reflectances, and design of toplighting element. From Figs. 9–74 and 9–48.

K_m = Light loss, from Fig. 7–31.

A_r = Room area, square feet.

Utilization Coefficient. The utilization coefficient to be used in the basic relationship above can be obtained as follows. From Fig. 9–74 determine the room ratio, which is a factor dependent on the room proportions. Using this factor, enter Fig. 9–75, and determine the utilization factor, K_w for the net transmission of the toplight, and the reflectances of the room.

Fig. 9–74. Room Ratios—Toplighting

Room Length (feet)	Room Width (feet)	Ceiling Height Above Floor (feet)					
		8	10	12	15	19	23
12	12	1.1	0.8	0.6	0.5		
	16	1.3	0.9	0.7	0.6		
	20	1.4	1.0	0.8	0.6	0.5	
	24	1.5	1.1	0.8	0.7	0.5	
	30	1.6	1.1	0.9	0.7	0.5	
	40	1.7	1.2	1.0	0.7	0.6	0.5
20	20	1.8	1.3	1.1	0.8	0.6	0.5
	24	2.0	1.5	1.2	0.9	0.7	0.5
	30	2.2	1.6	1.3	1.0	0.7	0.6
	40	2.4	1.8	1.4	1.1	0.8	0.7
	60	2.7	2.0	1.6	1.2	0.9	0.7
	80	2.9	2.1	1.7	1.3	1.0	0.8
30	30	2.7	2.0	1.6	1.2	0.9	0.7
	40	3.1	2.3	1.8	1.4	1.0	0.8
	60	3.6	2.7	2.1	1.6	1.2	1.0
	80	4.0	2.9	2.3	1.7	1.3	1.1
	100	4.2	3.1	2.4	1.9	1.4	1.1
	120	4.4	3.2	2.5	1.9	1.5	1.2
40	40	3.6	2.7	2.1	1.6	1.2	1.0
	60	4.4	3.2	2.5	1.9	1.5	1.2
	80	4.9	3.6	2.8	2.1	1.6	1.3
	100	5+	3.8	3.0	2.3	1.7	1.4
	120	5+	4.0	3.2	2.4	1.8	1.5
	140	5+	4.1	3.3	2.5	1.9	1.5

Toplight Transmission. The net transmittance of a toplight can be obtained from the manufacturer of one of the prefabricated units, or can be determined from the following formula:

$$T_{tl} = (A'_{tl}/A_{tl}) \times T_g \times T_{lw} \times T_c$$

where

T_{tl} = Net transmittance of the tonlight.

A'_{tl}/A_{tl} = Ratio of net to gross transmission area of the toplight (normally 0.6 to 0.9).

T_g = Transmittance of the glazing medium employed (from Fig. 7–22).

T_{lw} = The light well effect (discussed below).

T_c = Transmission effectiveness of louvers or other controls (normally 0.6 to 0.8).

Light Well Effect. A toplight panel is usually located at the roof level, with a light well below it extending down to the ceiling level of the room. This depth may be in a range from a few inches to several feet, depending upon the depth of the roof trusses and other factors. The interflections of the light in this well result in a decrease in the net transmission of the total assembly. The magnitude of the light well effect is shown in Fig. 9–76.

Fig. 9-75. Coefficients of Utilization for Toplighting

Net Toplight Transmittance (Including Light Well Effect, Controls, Etc.)	Room Ratio	Ceiling Reflectance 75 Per Cent		Ceiling Reflectance 50 Per Cent	
		Wall Reflectance			
		50 Per Cent	30 Per Cent	50 Per Cent	30 Per Cent
70 Per Cent	0.6	.37	.34	.36	.34
	0.8	.45	.42	.44	.41
	1.0	.49	.46	.48	.45
	1.25	.52	.50	.51	.49
	1.5	.55	.53	.53	.51
	2.0	.58	.56	.57	.55
	2.5	.61	.59	.60	.58
	3.0	.63	.61	.62	.60
	4.0	.65	.62	.63	.61
50 Per Cent	0.6	.26	.24	.26	.24
	0.8	.32	.30	.31	.29
	1.0	.35	.33	.34	.32
	1.25	.37	.36	.36	.35
	1.5	.39	.38	.38	.36
	2.0	.41	.40	.41	.39
	2.5	.44	.42	.43	.41
	3.0	.45	.44	.44	.43
	4.0	.46	.44	.45	.44
30 Per Cent	0.6	.16	.15	.16	.15
	0.8	.19	.18	.19	.18
	1.0	.21	.20	.21	.19
	1.25	.22	.21	.22	.21
	1.5	.24	.23	.23	.22
	2.0	.25	.24	.24	.24
	2.5	.26	.25	.26	.25
	3.0	.27	.26	.27	.26
	4.0	.28	.27	.27	.26
10 Per Cent	0.6	.05	.05	.05	.05
	0.8	.06	.06	.06	.06
	1.0	.07	.07	.07	.06
	1.25	.07	.07	.07	.07
	1.5	.08	.08	.08	.07
	2.0	.08	.08	.08	.08
	2.5	.09	.08	.09	.08
	3.0	.09	.09	.09	.09
	4.0	.09	.09	.09	.09

Uniformity. To achieve acceptable uniformity of illumination on the horizontal work-plane, the ratio of center-to-center spacing of the toplighting elements to the ceiling height above the work-plane should not exceed 2 for large elements or 1 for those of smaller area.

Lumen Method, Sidelighting

The calculation of illumination produced by sidelighting is somewhat more complex than for toplighting. First, the light source is located asymmetrically with respect to the work plane. Second, light reaching the fenestration from above the horizontal is affected differently than light from below the horizontal in most daylighting systems. Because of this complexity, the lumen methods for side wall fenestration are developing slowly, and tables are still somewhat restricted as to types of daylight control systems included. The tables as presented here are limited to windows extending from a nominal three-foot sill to the ceiling.

The basic approach of the lumen method for side wall lighting is the same as for toplighting. A coefficient is applied to the light flux incident on the fenestration, to determine the illumination on the work plane in the room. Since the work-plane illumination varies with the distance from the window, coefficients are determined for three points on the work plane in a line at right angles to the middle of the window. Five feet from the window gives a value E_{max}. The midpoint of the room gives a value E_{mid}. Five feet from the inner wall gives a value E_{min}. For each point, separate computations must be made for sky light and ground light, and the results added to obtain the total work-plane illumination at the point.

Side Wall Illumination Computations. The basic equation is:

$$E_p = E_i \times A_w \times K_u \times K_m$$

where

E_p = Work plane illumination at point P, in lumens per square foot.

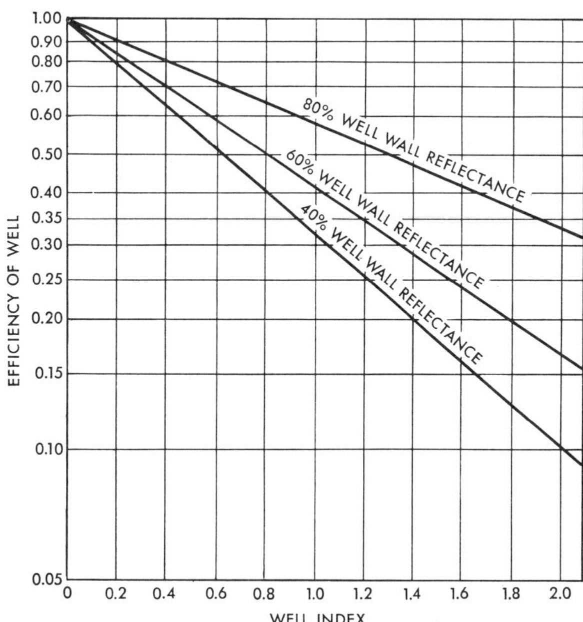

Fig. 9-76. Efficiency factors for various depths of light wells, based on well interflectance values where:

$$\text{Well Index} = \frac{\text{Well Height} \times (\text{Well Width} + \text{Well Length})}{2 \times \text{Well Length} \times \text{Well Width}}$$

Fig. 9–77. Coefficients of Utilization for Sidelighting for Rooms with a Ceiling Reflectance of 75 Per Cent and a Floor Reflectance of 30 Per Cent

A. 0.8 Ratio of Net Transmission Area to Gross Window Area—80 Per Cent Transmittance of Clear Glazing Medium—No Other Daylight Control

K_u	Room Length in Feet	Room Width in Feet	Light from Clear Sy				Overcast Sky				Uniform Ground			
			Ceiling Height in Feet				Ceiling Height in Feet				Ceiling Height in Feet			
			10		14		10		14		10		14	
			Wall Reflectance in Per Cent				Wall Reflectance in Per Cent				Wall Reflectance in Per Cent			
			70	30	70	30	70	30	70	30	70	30	70	30
Max.	20	20	.00191	.00165	.00149	.00123	.00240	.00223	.00197	.00176	.00147	.00109	.00096	.00070
		30	183	163	139	126	239	216	185	173	142	116	94	71
		40	194	162	139	120	254	223	189	172	138	115	92	71
	30	20	.00133	.00118	.00104	.00087	.00167	.00154	.00137	.00122	.00103	.00086	.00067	.00054
		30	123	113	93	87	165	150	128	121	98	90	65	55
		40	127	115	91	85	171	155	127	121	94	78	63	59
	40	20	.00102	.00094	.00080	.00070	.00125	.00121	.00102	.00095	.00082	.00062	.00053	.00044
		30	87	86	67	67	120	114	93	91	78	72	50	44
		40	93	88	66	65	126	118	94	92	73	70	49	43
Mid.	20	20	.00128	.00094	.00113	.00081	.00122	.00097	.00120	.00102	.00112	.00084	.00092	.00062
		30	65	50	66	47	39	41	51	48	64	46	66	43
		40	46	27	47	28	34	20	38	24	40	28	47	31
	30	20	.00084	.00072	.00074	.00062	.00078	.00071	.00077	.00076	.00083	.00066	.00068	.00048
		30	50	40	50	37	37	33	47	39	48	38	50	36
		40	35	24	36	24	27	18	37	22	32	26	38	28
	40	20	.00070	.00060	.00062	.00052	.00062	.00059	.00062	.00062	.00065	.00056	.00053	.00041
		30	36	34	37	31	23	27	30	32	38	33	39	31
		40	26	20	26	21	20	15	22	18	23	20	27	23
Min.	20	20	.00085	.00056	.00084	.00050	.00059	.00043	.00073	.00051	.00083	.00058	.00082	.00054
		30	42	25	42	25	23	16	29	18	36	19	45	22
		40	29	12	26	12	18	07	19	08	20	13	26	17
	30	20	.00064	.00045	.00063	.00041	.00044	.00035	.00054	.00042	.00065	.00045	.00063	.00041
		30	36	21	36	21	21	14	26	15	29	18	36	19
		40	25	11	23	11	14	07	15	08	18	09	23	12
	40	20	.00054	.00040	.00053	.00036	.00034	.00030	.00042	.00035	.00053	.00037	.00051	.00034
		30	25	19	25	19	15	12	18	13	24	16	29	18
		40	19	10	18	10	11	06	12	07	15	08	20	10

B. With Daylight Controls

K_u	Room Length in Feet	Room Width in Feet	Light from Sun and Sky—Sun Altitude 30°—Matte White Horizontal Louvers at 45° Tilt		Light from Uniform Ground—Matte White Horizontal Louvers at 45° Tilt		Light from Sun, Sky, and Ground—Translucent Diffusing Drapes or Shades—30 Per Cent Transmittance	
			Ceiling Height in Feet		Ceiling Height in Feet		Ceiling Height in Feet	
			10	14	10	14	10	14
			Wall Reflectance in Per Cent		Wall Reflectance in Per Cent		Wall Reflectance in Per Cent	

Max.	20	20 30 40	.000205 192 186	.000162 157 155	.000142 136 129	.000113 110 108	.00080 76 73	.00064 63 63	.00051 49 46	.00037 37 37	.00055 54 56	.00046 47 46	.00041 39 39	.00033 33 33
	30	20 30 40	.000145 136 132	.000118 115 112	.000101 95 92	.000080 79 77	.00057 53 52	.00049 49 49	.00036 34 33	.00029 29 29	.00039 37 38	.00034 34 34	.00029 27 26	.00024 24 25
	40	20 30 40	.000110 101 98	.000093 92 90	.000076 73 69	.000065 63 62	.00044 42 39	.00040 40 39	.00028 27 26	.00023 23 23	.00029 28 28	.00029 27 26	.00022 20 20	.00019 19 18
Mid.	20	20 30 40	.000135 78 40	.000092 47 28	.000122 84 49	.000078 47 29	.00052 28 16	.00037 20 11	.00042 28 19	.00028 18 12	.00038 18 13	.00029 15 08	.00033 19 14	.00024 14 09
	30	20 30 40	.000103 59 38	.000067 40 24	.000092 62 43	.000059 40 26	.00039 22 13	.00031 17 11	.00032 21 15	.00023 15 11	.00026 15 10	.00023 12 07	.00022 15 11	.00018 11 08
	40	20 30 40	.000078 40 27	.000060 36 21	.000070 49 30	.000052 36 22	.00030 17 10	.00025 15 09	.00025 17 12	.00019 14 09	.00021 11 07	.00019 10 06	.00018 11 08	.00015 10 07
Min.	20	20 30 40	.000090 42 20	.000051 19 09	.000070 48 22	.000051 22 10	.000353 131 61	.000220 72 31	.000345 161 85	.000202 85 40	.000250 108 66	.000173 62 35	.000253 124 78	.000165 65 43
	30	40 30 40	.000070 32 18	.000042 17 08	.000070 36 22	.000042 19 10	.000276 108 54	.000188 63 31	.000269 138 77	.000180 76 38	.000191 90 56	.000136 56 28	.000192 107 67	.000128 60 35
	40	20 30 40	.000052 26 16	.000038 16 08	.000052 30 20	.000038 18 09	.000207 85 46	.000157 58 27	.000208 108 61	.000148 72 36	.000152 68 47	.000116 49 25	.000154 79 55	.000110 54 30

E_i = Illumination from sky or ground incident on vertical windows, lumens per square foot.

A_w = Gross area of fenestration, square feet.

K_m = Light loss coefficient.

K_u = Utilization coefficient, which includes the effect of fenestration design, daylight controls, interior reflectances, and room proportions.

Utilization Coefficient Tables. Values of the utilization coefficients $K_{u\,max}$, $K_{u\,mid}$, and $K_{u\,min}$ are given for several typical fenestration and control systems in Fig. 9-77. The values for the utilization coefficient given in these tables have been derived for particular ratios of the net transmitting area to the gross fenestration area, as noted, and for a particular transmittance of the glazing medium. For other ratios or transmittances, the utilization coefficient must be adjusted proportionately. For values not included in the tables, straight line interpolation will be sufficiently accurate for most evaluation computations.

Fig. 9-77 is limited to rooms of conventional size and shape, with fenestration extending from a normal sill height to near the ceiling and covering most of the length of the window wall. These values were determined for a ceiling reflectance of 75 per cent and a floor reflectance of 30 per cent. Variation in floor reflectance has been found to have only a minor effect on the illumination produced on the work-plane in daylighting design.

Combination Designs. Daylighting designs which combine side wall fenestration in opposite walls or side and toplighting can be treated by superposition and the direct addition of the component illuminations at corresponding points in the working plane.

Sidelighting Computations, Glass Block. Consult the respective glass block manufacturers for data on illumination from glass block walls of various designs.

Computation of Luminance

The methods for computing luminance in a lighted space are not too well advanced, particularly for daylighting designs. However, certain luminance values can be determined for almost any design.

Task Luminance. Task luminance in footlamberts can be computed by multiplying the work plane illumination in footcandles by the diffuse reflectance of the task.

Surround Luminance. The surround luminance is usually somewhat more difficult to determine. Where the immediate surround is in the work-plane, such as a desk or table top, its luminance will be

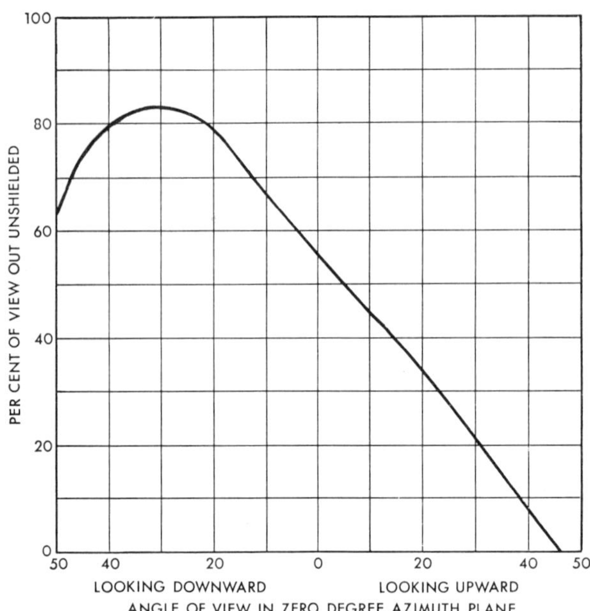

Fig. 9–78. Shielding of exterior view by using darkened bronze minute-louvered material.

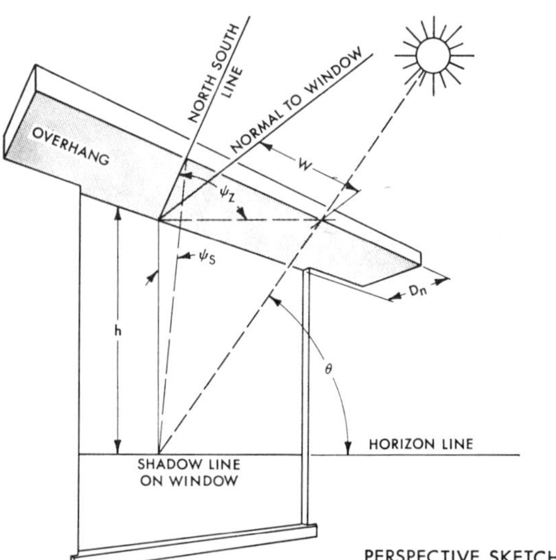

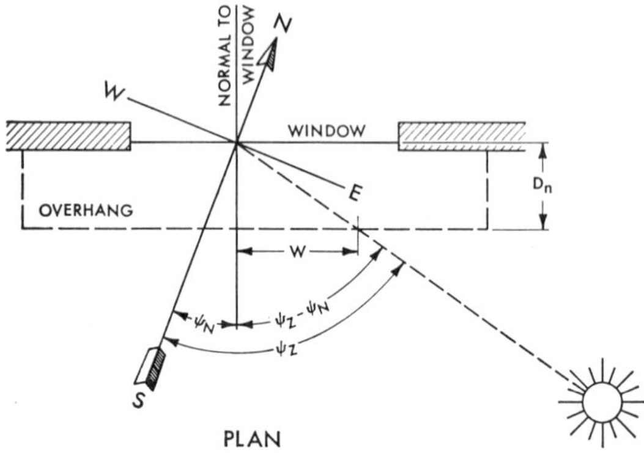

Fig. 9–79. Sunlight shading from a horizontal overhang.

the work-plane illumination multiplied by the diffuse reflectance of the desk surface.

For computing the luminance of floors, walls, or ceiling due to daylighting, there is no established method.

Source Luminance. Luminance of fenestration areas can be approximated in one of several ways. For clear glazing, luminance of the sky visible through the windows can be taken as the sky luminance assumed for the illumination computation, multiplied by the transmittance of the glazing material.

A light control medium with perfectly diffuse transmission will have a uniform luminance at any angle when viewed from the room interior, equal in footlamberts to the lumens per square foot leaving its visible surface; this in turn is equal to the illumination in footcandles on the outer surface multiplied by the transmittance of the material. As most practical diffusing media are not perfectly diffusing, the luminance will vary somewhat at different angles of view.

For the luminance of directional transmitting materials, consult the manufacturers for data.

The luminance of dark-colored, minute-louvered materials at any vertical angle can be closely approximated by multiplying the luminance of the unshielded view at that angle by the proportion of the view unimpeded by the louver slats. See Fig. 9–78 for values.

Overhangs

The amount of window shaded from a long horizontal overhang mounted above the window (Fig. 9–79) can be calculated from the formula:

$$h = D_n \sec (\psi_z - \psi_n) \tan \theta$$

where

h = height of shadow below overhang
D_n = depth of overhang from the window surface
ψ_z = solar azimuth
ψ_n = angle between a line drawn (outward) normal to the window and true south
θ = solar altitude above the horizon.

The amount of shading from a vertical projection mounted at the side of a window can be obtained from the formula:

$$W = D_v \tan (\psi_z - \psi_n)$$

where

W = width of the shadow from the side projection
D_v = depth of vertical projection.

In planning fixed horizontal and vertical projections for sunlight control, it is advisable to calculate the window shading for several different months and times.

Fig. 9–80. Levels of Illumination Currently Recommended (in Footcandles and Dekalux#)

ile for convenience of use this table sometimes lists loca- rather than tasks, the recommended footcandle values been arrived at for specific visual tasks. The tasks se- l for this purpose have been the more difficult ones which only occur in the various areas.

order to assure these values at all times, higher initial should be provided as required by the maintenance con- s (see page 9–2).

ere tasks are located near the perimeter of a room special deration should be given to the arrangement of the lumi- s in order to provide the recommended level of illumina- n the task (see page 9–33).

e illumination levels shown in the table are intended to nimum on the task irrespective of the plane in which it ated. In some instances, denoted by a (†), the values will be for equivalent sphere illumination, E_s (see page The commonly used lumen method of illumination calculation (see page 9–6) which gives results only for a horizontal work plane, cannot be used to calculate or predetermine E_s values. For use of E_s values, see page 3–33. The ratio of vertical to horizontal illumination will generally vary from 1/3 for luminaires having narrow distribution to 1/2 for luminaires of wide distribution. For a more specific determination one of the point calculation methods should be used (see page 9–44). Where the levels thus achieved are inadequate, special luminaire arrangements should be used or supplemental lighting equipment employed.

Supplementary luminaires may be used in combination with general lighting to achieve these levels. The general lighting should be not less than 20 footcandles and should contribute at least one-tenth the total illumination level.

Many of the following values have appeared, or in the future will appear, in other reports of the Society, some of which are jointly sponsored with other agencies and organizations.

Interior Lighting

	Footcandles on Tasks*	Dekalux# on Tasks*
ft manufacturing		
Stock parts		
Production	100	110
Inspection	200	220
Parts manufacturing		
Drilling, riveting, screw fastening	70	75
Spray booths	100	110
Sheet aluminum layout and template work, shaping, and smoothing of small parts for fuselage, wing sections, cowling, etc.	100[j]	110[j]
Welding		
General illumination	50	54
Precision manual arc welding	1000[a]	1080[a]
Subassembly		
Landing gear, fuselage, wing sections, cowling, and other large units	100	110
Final assembly		
Placing of motors, propellers, wing sections, landing gear	100	110
Inspection of assembled ship and its equipment	100	110
Machine tool repairs	100	110
ft hangars		
Repair service only	100	110
ries		
Drill	20	22
Exhibitions	30	32
alleries		
General	30	32
On paintings (supplementary)	30[b]	32[b]
On statuary and other displays	100[c]	110[c]
bly		
Rough easy seeing	30	32
Rough difficult seeing	50	54
Medium	100	110
Fine	500[a]	540[a]
Extra fine	1000[a]	1080[a]

Area	Footcandles on Tasks*	Dekalux# on Tasks*
Auditoriums		
Assembly only	15	16
Exhibitions	30	32
Social activities	5	5.4
Automobile showrooms (see **Stores**)		
Automobile manufacturing		
Frame assembly	50	54
Chassis assembly line	100	110
Final assembly, inspection line	200	220
Body manufacturing		
Parts	70	75
Assembly	100	110
Finishing and inspecting	200	220
Bakeries		
Mixing room	50	54
Face of shelves (vertical illumination)	30	32
Inside of mixing bowl (vertical mixers)	50	54
Fermentation room	30	32
Make-up room		
Bread	30	32
Sweet yeast-raised products	50	54
Proofing room	30	32
Oven room	30	32
Fillings and other ingredients	50	54
Decorating and icing		
Mechanical	50	54
Hand	100	110
Scales and thermometers	50	54
Wrapping room	30	32
Banks		
Lobby		
General	50	54
Writing areas	70†	75†
Tellers' stations	150†	160†
Posting and keypunch	150†	160†
Barber shops and beauty parlors	100	110

inimum on the task at any time for young adults with normal and better than 20/30 corrected vision. For general notes see beginning of tabulation. For other ee end of tabulation.

ekalux is an SI unit equal to 1.076 footcandles. 1 dekalux = 10 lux.

uivalent sphere illumination. See general notes at beginning of tabulation.

Fig. 9-80. *Continued*

Area	Footcandles on Tasks*	Dekalux# on Tasks*
Book binding		
Folding, assembling, pasting, etc.	70	75
Cutting, punching, stitching	70	75
Embossing and inspection	200	220
Breweries		
Brew house	30	32
Boiling and keg washing	30	32
Filling (bottles, cans, kegs)	50	54
Candy making		
Box department	50	54
Chocolate department		
Husking, winnowing, fat extraction, crushing and refining, feeding	50	54
Bean cleaning, sorting, dipping, packing, wrapping	50	54
Milling	100	110
Cream making		
Mixing, cooking, molding	50	54
Gum drops and jellied forms	50	54
Hand decorating	100	110
Hard candy		
Mixing, cooking, molding	50	54
Die cutting and sorting	100	110
Kiss making and wrapping	100	110
Canning and preserving		
Initial grading raw material samples	50	54
Tomatoes	100	110
Color grading (cutting rooms)	200[a]	220[a]
Preparation		
Preliminary sorting		
Apricots and peaches	50	54
Tomatoes	100	110
Olives	150	160
Cutting and pitting	100	110
Final sorting	100	110
Canning		
Continuous-belt canning	100	110
Sink canning	100	110
Hand packing	50	54
Olives	100	110
Examination of canned samples	200[f]	220[f]
Container handling		
Inspection	200[a]	220[a]
Can unscramblers	70	75
Labeling and cartoning	30	32
Central station		
Air-conditioning equipment, air preheater and fan floor, ash sluicing	10	11
Auxiliaries, battery rooms, boiler feed pumps, tanks, compressors, gauge area	20	22
Boiler platforms	10	11
Burner platforms	20	22
Cable room, circulator, or pump bay	10	11
Chemical laboratory	50	54
Coal conveyor, crusher, feeder, scale areas, pulverizer, fan area, transfer tower	10	11
Condensers, deaerator floor, evaporator floor, heater floors	10	11
Control rooms (see **Control rooms**)		
Hydrogen and carbon dioxide manifold area	20	22
Precipitators	10	11
Screen house	20	22
Soot or slag blower platform	10	11
Steam headers and throttles	10	11
Switchgear, power	20	22
Telephone equipment room	20	22
Tunnels or galleries, piping	10	11
Turbine bay sub-basement	20	22
Turbine room	30	32
Visitor's gallery	20	22
Water treating area	20	22
Chemical works		
Hand furnaces, boiling tanks, stationary driers, stationary and gravity crystallizers	30	32
Mechanical furnaces, generators and stills, mechanical driers, evaporators, filtration, mechanical crystallizers, bleaching	30	32
Tanks for cooking, extractors, percolators, nitrators, electrolytic cells	30	32
Churches and synagogues		
Altar, ark, reredos	100[e]	110[e]
Choir[d] and chancel	30[e]	32[e]
Classrooms	30†	32†
Pulpit, rostrum (supplementary illumination)	50[e]	54[e]
Main worship area[d]		
Light and medium interior finishes	15[e]	16[e]
For churches with special zeal	30[d]	32[d]
Art glass windows (test recommended)		
Light color	50	54
Medium color	100	110
Dark color	500	540
Especially dense windows	1000	1080
Clay products and cements		
Grinding, filter presses, kiln rooms	30	32
Molding, pressing, cleaning, trimming	30	32
Enameling	100	110
Color and glazing—rough work	100	110
Color and glazing—fine work	300[a]	320[a]
Cleaning and pressing industry		
Checking and sorting	50	54
Dry and wet cleaning and steaming	50	54
Inspection and spotting	500[a]	540[a]
Pressing	150	160
Repair and alteration	200[a]	220[a]
Cloth products		
Cloth inspection	2000[a]	2150[a]
Cutting	300[a]	320[a]
Sewing	500[a]	540[a]
Pressing	300[a]	320[a]
Clothing manufacture (men's)		
Receiving, opening, storing, shipping	30	32
Examining (perching)	2000[a]	2150[a]
Sponging, decating, winding, measuring	30	32
Piling up and marking	100	110
Cutting	300[a]	320[a]
Pattern making, preparation of trimming, piping, canvas and shoulder pads	50	54
Fitting, bundling, shading, stitching	30	32
Shops	100	110
Inspection	500[a]	540[a]
Pressing	300[a]	320[a]
Sewing	500[a]	540[a]
Club and lodge rooms		
Lounge and reading rooms	30	32
Auditoriums (see **Auditoriums**)		
Coal tipples and cleaning plants		
Breaking, screening, and cleaning	10	11
Picking	300[a]	320[a]

* Minimum on the task at any time for young adults with normal and better than 20/30 corrected vision. For general notes see beginning of tabulation. For other notes see end of tabulation.

Dekalux is an SI unit equal to 1.076 footcandles. 1 dekalux = 10 lux.

† Equivalent sphere illumination. See general notes at beginning of tabulation.

Fig. 9-80. *Continued*

Area	Footcandles on Tasks*	Dekalux# on Tasks*
ontrol rooms and dispatch rooms		
Control rooms		
Vertical face of switchboards		
Simplex or section of duplex facing operator:		
Type A—Large centralized control room 66 inches above floor	50	54
Type B—Ordinary control room 66 inches above floor	30	32
Section of duplex facing away from operator	30	32
Bench boards (horizontal level)	50	54
Area inside duplex switchboards	10	11
Rear of all switchboard panels (vertical)	10	11
Emergency lighting, all areas	3	3.2
Dispatch boards		
Horizontal plane (desk level)	50	54
Vertical face of board (48 inches above floor, facing operator):		
System load dispatch room	50	54
Secondary dispatch room	30	32
otton gin industry		
Overhead equipment—separators, driers, grid cleaners, stick machines, conveyers, feeders and catwalks	30	32
Gin stand	50	54
Control console	50	54
Lint cleaner	50	54
Bale press	30	32
ourt rooms		
Seating area	30	32
Court activity area	70†	75†
airy farms (see **Farms**)		
airy products		
Fluid milk industry		
Boiler room	30	32
Bottle storage	30	32
Bottle sorting	50	54
Bottle washers	f	f
Can washers	30	30
Cooling equipment	30	30
Filling: inspection	100	110
Gauges (on face)	50	54
Laboratories	100	110
Meter panels (on face)	50	54
Pasteurizers	30	32
Separators	30	32
Storage refrigerator	30	32
Tanks, vats		
Light interiors	20	22
Dark interiors	100	110
Thermometer (on face)	50	54
Weighing room	30	32
Scales	70	75
ance halls	5	5.4
epots, terminals, and stations		
Waiting room	30	32
Ticket offices		
General	100†	110†
Ticket rack and counters	100	110†
Rest rooms and smoking room	30	32
Baggage checking	50	54
Concourse	10	11
Platforms	20	22
Toilets and washrooms	30	32
Dispatch boards (see **Control rooms**)		
Drafting rooms (see **Offices**)		
Electrical equipment manufacturing		
Impregnating	50	54
Insulating: coil winding	100	110
Testing	100	110
Electrical Generating Station (see **Central Station**)		
Elevators, freight and passenger	20	22
Engraving (wax)	200ᵃ	220ᵃ
Explosives		
Hand furnaces, boiling tanks, stationary driers, stationary and gravity crystallizers	30	32
Mechanical furnace, generators and stills, mechanical driers, evaporators, filtration, mechanical crystallizers	30	32
Tanks for cooking, extractors, percolators, nitrators	30	32
Farms—dairy		
Milking operation area (milking parlor and stall barn)		
General	20	22
Cow's udder	50	54
Milk handling equipment and storage area (milk house or milk room)		
General	20	22
Washing area	100	110
Bulk tank interior	100	110
Loading platform	20	22
Feeding area (stall barn feed alley, pens, loose housing feed area)	20	22
Feed storage area—forage		
Haymow	3	3.2
Hay inspection area	20	22
Ladders and stairs	20	22
Silo	3	3.2
Silo room	20	22
Feed storage area—grain and concentrate		
Grain bin	3	3.2
Concentrate storage area	10	11
Feed processing area	10	11
Livestock housing area (community, maternity, individual calf pens, and loose housing holding and resting areas)	7	7.5
Machine storage area (garage and machine shed)	5	5.4
Farm shop area		
Active storage area	10	11
General shop area (machinery repair, rough sawing)	30	32
Rough bench and machine work (painting, fine storage, ordinary sheet metal work, welding, medium benchwork)	50	54
Medium bench and machine work (fine woodworking, drill press, metal lathe, grinder)	100	110
Miscellaneous areas		
Farm office	70†	75†
Restrooms	30	32
Pumphouse	20	22

* Minimum on the task at any time for young adults with normal and better than 20/30 corrected vision. For general notes see beginning of tabulation. For other notes see end of tabulation.

\# Dekalux is an SI unit equal to 1.076 footcandles. 1 dekalux = 10 lux.

† Equivalent sphere illumination. See general notes at beginning of tabulation.

Fig. 9–80. *Continued*

Area	Footcandles on Tasks*	Dekalux# on Tasks*
Farms—poultry (see **Poultry industry**)		
Fire hall (see **Municipal buildings**)		
Flour mills		
Rolling, sifting, purifying	50	54
Packing	30	32
Product control	100	110
Cleaning, screens, man lifts, aisleways and walkways, bin checking	30	32
Forge shops	50	54
Foodservice facilities		
Dining areas		
Cashier	50	54
Intimate type		
Light environment	10	11
Subdued environment	3	3.2
For cleaning	20	22
Leisure type		
Light environment	30	32
Subdued environment	15	16
Quick service type		
Bright surroundings[n]	100	110
Normal surroundings[n]	50	54
Food displays—twice the general levels but not under	50	54
Kitchen, commercial		
Inspection, checking, preparation, and pricing	70	75
Entrance foyer	30	32
Marquee		
Dark surroundings	30	32
Bright surroundings	50	54
Foundries		
Annealing (furnaces)	30	32
Cleaning	30	32
Core making		
Fine	100	110
Medium	50	54
Grinding and chipping	100	110
Inspection		
Fine	500[a]	540[a]
Medium	100	110
Molding		
Medium	100	110
Large	50	54
Pouring	50	54
Sorting	50	54
Cupola	20	22
Shakeout	30	32
Garages—automobile and truck		
Service garages		
Repairs	100	110
Active traffic areas	20	22
Parking garages		
Entrance	50	54
Traffic lanes	10	11
Storage	5	5.4
Gasoline station (see **Service station**)		
Glass works		
Mix and furnace rooms, pressing and lehr, glassblowing machines	30	32
Grinding, cutting glass to size, silvering	50	54
Fine grinding, beveling, polishing	100	110
Inspection, etching and decorating	200[f]	220[f]
Glove manufacturing		
Pressing	300[a]	320[a]
Knitting	100	110
Sorting	100	110
Cutting	300[a]	320[a]
Sewing and inspection	500[a]	540[a]
Hangars (see **Aircraft hangars**)		
Hat manufacturing		
Dyeing, stiffening, braiding, cleaning, refining	100	110
Forming, sizing, pouncing, flanging, finishing, ironing	200	220
Sewing	500[a]	540[a]
Homes (see **Residences**)		
Hospitals		
Anesthetizing and preparation room	30	32
Autopsy and morgue		
Autopsy room	100	110
Autopsy table	1000	1080
Museum	50	54
Morgue, general	20	22
Central sterile supply		
General, work room	30	32
Work tables	50	54
Glove room	50	54
Syringe room	150	160
Needle sharpening	150	160
Storage areas	30	32
Issuing sterile supplies	50	54
Corridor		
General in nursing areas—daytime	20	22
General in nursing areas—night (rest period)	3	3.2
Operating, delivery, recovery, and laboratory suites and service areas	30	32
Cystoscopic room		
General	100	110
Cystoscopic table	2500	2690
Dental suite		
Operatory, general	70	75
Instrument cabinet	150	160
Dental entrance to oral cavity	1000	1080
Prosthetic laboratory bench	100	110
Recovery room, general	5	5.4
Recovery room, local for observation	70	75
(EEG) encephalographic suite		
Office (see **Offices**)		
Work room, general	30	32
Work room, desk or table	100	110
Examining room	30	32
Preparation rooms, general	30	32
Preparation rooms, local	50	54
Storage, records, charts	30	32
Electromyographic suite		
Same as EEG but provisions for reducing level in preparation area to 1		
Emergency operating room		
General	100	110
Local	2000	2150
EKG, BMR, and specimen room		
General	30	32
Specimen table	50	54
EKG machine	50	54
Examination and treatment room		
General	50	54
Examining table	100	110

* Minimum on the task at any time for young adults with normal and better than 20/30 corrected vision. For general notes see beginning of tabulation. For other notes see end of tabulation.

Dekalux is an SI unit equal to 1.076 footcandles. 1 dekalux = 10 lux.

† Equivalent sphere illumination. See general notes at beginning of tabulation.

Fig. 9-80. *Continued*

Area	Footcandles on Tasks*	Dekalux# on Tasks*
Exits, at floor	5	5.4
Eye, ear, nose, and throat suite		
Darkroom (variable)	0-10	0-11
Eye examination and treatment	50	54
Ear, nose, throat room	50	54
Flower room	10	11
Formula room		
Bottle washing	30	32
Preparation and filling	50	54
Fracture room		
General	50	54
Fracture table	200	220
Splint closet	50	54
Plaster sink	50	54
Intensive care nursing areas		
General	30	32
Local	100	110
Laboratories		
General	50	54
Close work areas	100	110
Linens (see **Laundries**)		
Sorting soiled linen	30	32
Central (clean) linen room	30	32
Sewing room, general	30	32
Sewing room, work area	100	110
Linen closet	10	11
Lobby (or entrance foyer)		
During day	50	54
During night	20	22
Locker rooms	20	22
Medical records room	100†	110†
Nurses station		
General—day	70†	75†
General—night	30	32
Desk for records and charting	70†	75†
Table for doctor's making or viewing reports	70†	75†
Medicine counter	100†	110†
Nurses gown room		
General	30	32
Mirror for grooming	50	54
Nurseries, infant		
General	30	32
Examining, local at bassinet	100	110
Examining and treatment table	100	110
Nurses station and work space (see Nurses Station)		
Obstetrical suite		
Labor room, general	20	22
Labor room, local	100	110
Scrub-up area	30	32
Delivery room, general	100	110
Substerilizing room	30	32
Delivery table	2500	2690
Clean-up room	30	32
Recovery room, general	30	32
Recovery room, local	100	110
Patients' rooms (private and wards)		
General	20	22
Reading	30	32
Observation (by nurse)	2	2.2
Night light, maximum at floor (variable)	0.5	0.5
Examining light	100	110
Toilets	30	32
Pediatric nursing unit		
General, crib room	20	22
General, bedroom	10	11
Reading	30	32
Playroom	30	32
Treatment room, general	50	54
Treatment room, local	100	110
Pharmacy		
Compounding and dispensing	100	110
Manufacturing	50	54
Parenteral solution room	50	54
Active storage	30	32
Alcohol vault	10	11
Radioisotope facilities		
Radiochemical laboratory, general	30	32
Uptake or scanning room	20	22
Examining table	50	54
Retiring room		
General	10	11
Local for reading	30	32
Solarium		
General	20	22
Local for reading	30	32
Stairways	20	22
Surgical suite		
Instrument and sterile supply room	30	32
Clean-up room, instrument	100	110
Scrub-up area (variable)	200	220
Operating room, general (variable)	200	220
Operating table	2500	2690
Recovery room, general	30	32
Recovery room, local	100	110
Anesthesia storage	20	22
Substerilizing room	30	32
Therapy, physical		
General	20	22
Exercise room	30	32
Treatment cubicles, local	30	32
Whirlpool	20	22
Lip reading	150	160
Office (see **Offices**)		
Therapy, occupational		
Work area, general	30	32
Work tables or benches, ordinary	50	54
Work tables or benches, fine work	100	110
Toilets	30	32
Utility room		
General	20	22
Work counter	50	54
Waiting rooms, or areas		
General	20	22
Local for reading	30	32
X-ray suite		
Radiographic, general	10	11
Fluoroscopic, general (variable)	0-50	0-54
Deep and superficial therapy	10	11
Control room	10	11
Film viewing room	30	32
Darkroom	10	11
Light room	30	32
Filing room, developed films	30	32
Storage, undeveloped films	10	11
Dressing rooms	10	11
Hotels		
Bathrooms		
Mirror	30[g]	32[g]
General	10	11
Bedrooms		
Reading (books, magazines, newspapers)	30	32
Inkwriting	30[h]	32[h]
Make-up	30[i]	32[i]
General	10	11

* Minimum on the task at any time for young adults with normal and better than 20/30 corrected vision. For general notes see beginning of tabulation. For other notes see end of tabulation.

\# Dekalux is an SI unit equal to 1.076 footcandles. 1 dekalux = 10 lux.

† Equivalent sphere illumination. See general notes at beginning of tabulation.

Fig. 9-80. *Continued*

Area	Footcandles on Tasks*	Dekalux# on Tasks*
Corridors, elevators, and stairs	20	22
Entrance foyer	30	32
Front office	50	54
Linen room		
Sewing	100	110
General	20	22
Lobby		
General lighting	10	11
Reading and working areas	30	32
Marquee		
Dark surroundings	30	32
Bright surroundings	50	54
Ice making—engine and compressor room	20	22
Inspection		
Ordinary	50	54
Difficult	100	110
Highly difficult	200	220
Very difficult	500[a]	540[a]
Most difficult	1000[a]	1080[a]
Iron and steel manufacturing		
Open hearth		
Stock yard	10	11
Charging floor	20	22
Pouring slide		
Slag pits	20	22
Control platforms	30	32
Mold yard	5	5.4
Hot top	30	32
Hot top storage	10	11
Checker cellar	10	11
Buggy and door repair	30	32
Stripping yard	20	22
Scrap stockyard	10	11
Mixer building	30	32
Calcining building	10	11
Skull cracker	10	11
Rolling mills		
Blooming, slabbing, hot strip, hot sheet	30	32
Cold strip, plate	30	32
Pipe, rod, tube, wire drawing	50	54
Merchant and sheared plate	30	32
Tin plate mills		
Tinning and galvanizing	50	54
Cold strip rolling	50	54
Motor room, machine room	30	54
Inspection		
Black plate, bloom and billet chipping	100	110
Tin plate and other bright surfaces	200[j]	220[j]
Jewelry and watch manufacturing	500[a]	540[a]
Kitchens (see **Foodservice facilities** or **Residences**)		
Laundries		
Washing	30	32
Flat work ironing, weighing, listing, marking	50	54
Machine and press finishing, sorting	70	75
Fine hand ironing	100	110
Leather manufacturing		
Cleaning, tanning and stretching, vats	30	32
Cutting, fleshing and stuffing	50	54
Finishing and scarfing	100	110
Leather working		
Pressing, winding, glazing	200	220
Grading, matching, cutting, scarfing, sewing	300[a]	320[a]

Area	Footcandles on Tasks*	Dekalux# on Tasks*
Library		
Reading areas		
Reading printed material	30†	32†
Study and note taking	70†	75†
Conference areas	30†	32†
Seminar rooms	70†	75†
Book stacks (30 inches above floor)		
Active stacks	30[r]	32[r]
Inactive stacks	5[r]	5.
Book repair and binding	70	75
Cataloging	70†	75†
Card files	100†	110†
Carrels, individual study areas	70†	75†
Circulation desks	70†	75†
Rare book rooms—archives		
Storage areas	30	32
Reading areas	100†	110†
Map, picture, and print rooms		
Storage areas	30	32
Use areas	100†	110†
Audiovisual areas		
Preparation rooms	70	75
Viewing rooms (variable)	70	75
Television receiving room (shield viewing screen)	70	75
Audio listening areas		
General	30	32
For note taking	70†	75†
Record inspection table	100[a]	110[a]
Microform areas		
Files	70†	75†
Viewing areas	30	32
Locker rooms	20	22
Machine shops		
Rough bench and machine work	50	54
Medium bench and machine work, ordinary automatic machines, rough grinding, medium buffing and polishing	100	110
Fine bench and machine work, fine automatic machines, medium grinding, fine buffing and polishing	500[a]	540[a]
Extra-fine bench and machine work, grinding, fine work	1000[a]	1080[a]
Materials handling		
Wrapping, packing, labeling	50	54
Picking stock, classifying	30	32
Loading, trucking	20	22
Inside truck bodies and freight cars	10	11
Meat packing		
Slaughtering	30	32
Cleaning, cutting, cooking, grinding, canning, packing	100	110
Municipal buildings—fire and police		
Police		
Identification records	150†	160†
Jail cells and interrogation rooms	30	32
Fire hall		
Dormitory	20	22
Recreation room	30	32
Wagon room	30	32
Museums (see **Art galleries**)		
Nursing homes		
Corridors and interior ramps	20	22
Stairways other than exits	30	32

* Minimum on the task at any time for young adults with normal and better than 20/30 corrected vision. For general notes see beginning of tabulation. For other notes see end of tabulation.

\# Dekalux is an SI unit equal to 1.076 footcandles. 1 dekalux = 10 lux.

† Equivalent sphere illumination. See general notes at beginning of tabulation.

Fig. 9-80. *Continued*

Area	Footcandles on Tasks*	Dekalux# on Tasks*
Exit stairways and landings, on floor	5	5.4
Doorways	10	11
Administrative and lobby areas, day	50	54
Administrative and lobby areas, night	20	22
Chapel or quiet area, general	5	5.4
Chapel or quiet area, local for reading	30	32
Physical therapy	20	22
Occupational therapy	30	32
Work table, course work	100	110
Work table, fine work	200	220
Recreation area	50	54
Dining area	30	32
Patient care unit (or room), general	20	22
Patient care room, reading	30	32
Nurse's station, general		
Day	50†	54†
Night	20	22
Nurse's desk, for charts and records	70†	75†
Nurse's medicine cabinet	100†	110†
Utility room, general	20	22
Utility room, work counter	50	54
Pharmacy area, general	30	32
Pharmacy, compounding, and dispensing area	100	110
Janitor's closet	15	15
Toilet and bathing facilities	30	30
Barber and beautician areas	50	50
Offices		
Drafting rooms		
Detailed drafting and designing, cartography	200†	220†
Rough layout drafting	150†	160†
Accounting offices		
Auditing, tabulating, bookkeeping, business machine operation, computer operation	150†	160†
General offices		
Reading poor reproductions, business machine operation, computer operation	150†	160†
Reading handwriting in hard pencil or on poor paper, reading fair reproductions, active filing, mail sorting	100†	110†
Reading handwriting in ink or medium pencil on good quality paper, intermittent filing	70†	75†
Private offices		
Reading poor reproductions, business machine operation	150†	160†
Reading handwriting in hard pencil or on poor paper, reading fair reproductions	100†	110†
Reading handwriting in ink or medium pencil on good quality paper	70†	75†
Reading high contrast or well-printed materials	30†	33†
Conferring and interviewing	30	33
Conference rooms		
Critical seeing tasks	100†	110†
Conferring	30	33
Note-taking during projection (variable)	30†	33†
Corridors	20[k]	22[k]
Packing and boxing (see Materials handling)		
Paint manufacturing		
General	30	32
Comparing mix with standard	200[a]	220[a]
Paint shops		
Dipping, simple spraying, firing	50	54
Rubbing, ordinary hand painting and finishing art, stencil and special spraying	50	54
Fine hand painting and finishing	100	110
Extra-fine hand painting and finishing	300[a]	320[a]
Paper-box manufacturing		
General manufacturing area	50	54
Paper manufacturing		
Beaters, grinding, calendering	30	32
Finishing, cutting, trimming, papermaking machines	50	54
Hand counting, wet end of paper machine	70	75
Paper machine reel, paper inspection, and laboratories	100	110
Rewinder	150	160
Plating	30	32
Polishing and burnishing	100	110
Power plants (see Central station)		
Post Offices		
Lobby, on tables	30	32
Sorting, mailing, etc.	100	110
Poultry industry (see also Farm—dairy)		
Brooding, production, and laying houses		
Feeding, inspection, cleaning	20	22
Charts and records	30	32
Thermometers, thermostats, time clocks	50	54
Hatcheries		
General area and loading platform	20	22
Inside incubators	30	32
Dubbing station	150	160
Sexing	1000	1080
Egg handling, packing, and shipping		
General cleanliness	50	54
Egg quality inspection	50	54
Loading platform, egg storage area, etc.	20	22
Egg processing		
General lighting	70	75
Fowl processing plant		
General (excluding killing and unloading area)	70	75
Government inspection station and grading stations	100	110
Unloading and killing area	20	22
Feed storage		
Grain, feed rations	10	11
Processing	10	11
Charts and records	30	32
Machine storage area (garage and machine shed)	5	5.4
Printing industries		
Type foundries		
Matrix making, dressing type	100	110
Font assembly—sorting	50	54
Casting	100	110
Printing plants		
Color inspection and appraisal	200[a]	220[a]
Machine composition	100	110
Composing room	100	110
Presses	70	75
Imposing stones	150	160
Proofreading	150	160
Electrotyping		
Molding, routing, finishing, leveling molds, trimming	100	110
Blocking, tinning	50	54
Electroplating, washing, backing	50	54

* Minimum on the task at any time for young adults with normal and better than 20/30 corrected vision. For general notes see beginning of tabulation. For other notes see end of tabulation.

\# Dekalux is an SI unit equal to 1.076 footcandles. 1 dekalux = 10 lux.

† Equivalent sphere illumination. See general notes at beginning of tabulation.

Fig. 9–80. *Continued*

Area	Footcandles on Tasks*	Dekalux# on Tasks*
Photoengraving		
Etching, staging, blocking	50	54
Routing, finishing, proofing	100	110
Tint laying, masking	100	110
Professional offices (see **Hospitals**)		
Receiving and shipping (see **Materials handling**)		
Residences		
Specific visual tasks[l]		
Dining	15	16
Grooming, shaving, make-up	50	54
Handcraft		
Ordinary seeing tasks	70	75
Difficult seeing tasks	100	110
Very difficult seeing tasks	150	160
Critical seeing tasks	200	220
Ironing (hand and machine)	50	50
Kitchen duties		
Food preparation and cleaning	150	160
Serving and other non-critical tasks	50	54
Laundry		
Preparation, sorting, inspection	50	54
Tub area-soaking, tinting	50	54
Washer and dryer areas	30	32
Reading and writing		
Handwriting, reproductions, and poor copies	70†	75†
Books, magazines, newspapers	30†	32†
Reading piano or organ scores		
Advanced (substandard size)	150	160
Advanced	70	75
Simple	30	32
Sewing (hand and machine)		
Dark fabrics	200	220
Medium fabrics	100	110
Light fabrics	50	54
Occasional-high contrast	30	32
Study	70†	75†
Table games	30	32
General lighting		
Conversation, relaxation, entertainment	10[m]	11[m]
Passage areas, for safety	10[m]	11[m]
Areas involving visual tasks, other than kitchen	30	32
Kitchen	50	54
Restaurants (see **Foodservice facilities**)		
Rubber goods—mechanical		
Stock preparation		
Plasticating, milling, Banbury	30	32
Calendering	50	54
Fabric preparation, stock cutting, hose looms	50	54
Extruded products	50	54
Molded products and curing	50	54
Inspection	200[a]	220[a]
Rubber tire manufacturing		
Banbury	30	32
Tread stock		
General	50	54
Booking and inspection, extruder, check weighing, width measuring	100[q]	110[q]

Area	Footcandles on Tasks*	Deka on T
Calendering		
General	30	3
Letoff and windup	50	5
Stock cutting		
General	30	3
Cutters and splicers	100[q]	11
Bead Building	50	5
Tire Building		
General	50	5
At machines	150[a]	16
In-process stock	30	3
Curing		
General	30	3
At molds	70[a]	7
Inspection		
General	100	11
At tires	300[q]	32
Storage	20	2
Sawmills		
Grading redwood lumber	300[a]	32
Schools		
Tasks		
Reading printed material	30†	3
Reading pencil writing	70†	7
Spirit duplicated material		
Good	30†	3
Poor	100†	11
Drafting†, benchwork	100[a]	11
Lip reading, chalkboards†, sewing	150[a]	16
Classrooms		
Art rooms	70	7
Drafting rooms	100[a]†	11
Home economics rooms		
Sewing	150[a]	16
Cooking	50	54
Ironing	50	54
Sink activities	70	75
Note-taking areas	70†	75
Laboratories	100	110
Lecture rooms		
Audience area	70†	75
Demonstration area	150[a]	160
Music rooms		
Simple scores	30†	32
Advanced scores	70[q]†	75
Shops	100[a]	110
Sight-saving rooms	150[a]†	160
Study halls	70†	75
Typing	70†	75
Corridors and stairways	20	22
Dormitories		
General	10	11
Reading books, magazines, newspapers	30†	32
Study desk	70†	75
Service space (see also **Storage rooms**)		
Stairways, corridors	20	22
Elevators, freight and passenger	20	22
Toilets and wash rooms	30	32
Service stations		
Service bays	30	32
Sales room	50	54
Shelving and displays	100	110
Rest rooms	15	16
Storage	5	5

* Minimum on the task at any time for young adults with normal and better than 20/30 corrected vision. For general notes see beginning of tabulation. For notes see end of tabulation.

\# Dekalux is an SI unit equal to 1.076 footcandles. 1 dekalux = 10 lux.

† Equivalent sphere illumination. See general notes at beginning of tabulation.

Fig. 9-80. *Continued*

	Footcandles on Tasks*	Dekalux# on Tasks*
etal works		
iscellaneous machines, ordinary bench work	50	54
resses, shears, stamps, spinning, medium bench work	50	54
unches	50	54
in plate inspection, galvanized	200[j]	220[j]
ribing	200[j]	220[j]
anufacturing—leather		
utting and stitching		
Cutting tables	300[a]	320[a]
Marking, buttonholing, skiving, sorting, vamping, counting	300[a]	320[a]
Stitching, dark materials	300[a]	320[a]
aking and finishing, nailers, sole layers, welt beaters and scarfers, trimmers, welters, lasters, edge setters, sluggers, randers, wheelers, treers, cleaning, spraying, buffing, polishing, embossing	200	220
anufacturing—rubber		
ashing, coating, mill run compounding	30	32
arnishing, vulcanizing, calendering, upper and sole cutting	50	54
le rolling, lining, making and finishing processes	100	110
indows°		
aytime lighting		
General	200	220
Feature	1000	1080
ighttime lighting		
Main business districts—highly competitive		
General	200	220
Feature	1000	1080
Secondary business districts or small towns		
General	100	110
Feature	500	540
Open-front stores (see display lighting under **Stores**)		
anufacturing		
ettle houses, cutting, soap chip and powder	30	32
tamping, wrapping and packing, filling and packing soap powder	50	54
ays (see Service space)		
ee Iron and steel)		
rushing and screening		
elt conveyor tubes, main line shafting spaces, chute rooms, inside of bins	10	11
rimary breaker room, auxiliary breakers under bins	10	11
creens	20	22
e battery manufacturing		
olding of grids	50	54
e rooms or warehouses		
nactive	5	5.4
ctive		
Rough bulky	10	11
Medium	20	22
Fine	50	54

Area	Footcandles on Tasks*	Dekalux# on Tasks*
Stores°		
Circulation areas	30	32
Merchandising areas		
Service	100	110
Self-service	200	220
Showcases and wall cases		
Service	200	220
Self-service	500	540
Feature displays		
Service	500	540
Self-service	1000	1080
Alteration room		
General	50	54
Pressing	150	160
Sewing	200	220
Fitting room		
Dressing areas	50	54
Fitting areas	200	220
Stockrooms	30	32
Structural steel fabrication	50	54
Sugar refining		
Grading	50	54
Color inspection	200[a]	220[a]
Television (see Section 24)		
Testing		
General	50	54
Extra-fine instruments, scales, etc.	200[a]	220[a]
Textile mills—cotton		
Opening, mixing, picking	30	32
Carding and drawing	50	54
Slubbing, roving, spinning, spooling	50	54
Beaming and splashing on comb		
Gray goods	50	54
Denims	150	160
Inspection		
Gray goods (hand turning)	100	160
Denims (rapidly moving)	500[a]	540[a]
Automatic tying-in	150	160
Weaving	100	110
Drawing-in by hand	200	220
Textile mills—silk and synthetics		
Manufacturing		
Soaking, fugitive tinting, and conditioning or setting of twist	30	32
Winding, twisting, rewinding and coning, quilling, slashing		
Light thread	50	54
Dark thread	200	220
Warping (silk or cotton system)		
On creel, on running ends, on reel, on beam, on warp at beaming	100	110
Drawing-in on heddles and reed	200	220
Weaving	100	110
Textile mills—woolen and worsted		
Opening, blending, picking	30	32
Grading	100[a]	110[a]
Carding, combing, recombing and gilling	50	54
Drawing		
White	50	54
Colored	100	110
Spinning (frame)		
White	50	54

nimum on the task at any time for young adults with normal and better than 20/30 corrected vision. For general notes see beginning of tabulation. For other e end of tabulation.

kalux is an SI unit equal to 1.076 footcandles. 1 dekalux = 10 lux.

uivalent sphere illumination. See general notes at beginning of tabulation.

Fig. 9-80. *Continued*

Area	Footcandles on Tasks*	Dekalux# on Tasks*
Colored	100	110
Spinning (mule)		
White	50	54
Colored	100	110
Twisting		
White	50	54
Winding		
White	30	32
Colored	50	54
Warping		
White	100	110
White (at reed)	100	110
Colored	100	110
Colored (at reed)	300[a]	320[a]
Weaving		
White	100	110
Colored	200	220
Gray-goods room		
Burling	150	160
Sewing	300[a]	320[a]
Folding	70	75
Wet finishing, fulling, scouring, crabbing, drying	50	54
Dyeing	100[a]	110[a]
Dry finishing, napping, conditioning, pressing	70	75
Dry finishing, shearing	100	110
Inspecting (perching)	2000[a]	2150[b]
Folding	70	75
Theatres and motion picture houses		
Auditoriums		
During intermission	5	5.4
During picture	0.1	0.1
Foyer	5	5.4
Lobby	20	22
Tobacco products		
Drying, stripping, general	30	32
Grading and sorting	200[a]	220[a]
Toilets and wash rooms	30	32
Upholstering—automobile, coach, furniture	100	110
Warehouse (see **Storage rooms**)		
Welding		
General illumination	50	54
Precision manual arc welding	1000[a]	1080[a]
Woodworking		
Rough sawing and bench work	30	32
Sizing, planing, rough sanding, medium quality machine and bench work, gluing, veneering, cooperage	50	54
Fine bench and machine work, fine sanding and finishing	100	110

Exterior Lighting

Area	Footcandles on Tasks*	Dekalux# on Tasks*
Building (construction)		
General construction	10	11
Excavation work	2	2.2
Building exteriors		
Entrances		
Active (pedestrian and/or conveyance)	5	5.4
Inactive (normally locked, infrequently used)	1	1.1
Vital locations or structures	5	5.4
Building surrounds	1	1.1
Buildings and monuments, floodlighted		
Bright surroundings		
Light surfaces	15	16
Medium light surfaces	20	22
Medium dark surfaces	30	32
Dark surfaces	50	54
Dark surroundings		
Light surfaces	5	5.4
Medium light surfaces	10	11
Medium dark surfaces	15	16
Dark surfaces	20	22
Bulletin and poster boards		
Bright surroundings		
Light surfaces	50	54
Dark surfaces	100	110
Dark surroundings		
Light surfaces	20	22
Dark surfaces	50	54
Central station		
Catwalks	2	2.2
Cinder dumps	0.1	0.1
Coal storage area	0.1	0.1
Coal unloading		
Dock (loading or unloading zone)	5	5.4
Barge storage area	0.5	0.5
Car dumper	0.5[a]	0.5[a]
Tipple	5	5.4
Conveyors	2	2.2
Entrances		
Generating or service building		
Main	10	11
Secondary	2	2.2
Gate house		
Pedestrian entrance	10	11
Conveyor entrance	5	5.4
Fence	0.2	0.2
Fuel-oil delivery headers	5	5.4
Oil storage tanks	1	1.1
Open yard	0.2	0.2
Platforms—boiler, turbine deck	5	5.4
Roadway		
Between or along buildings	1	1.1
Not bordered by buildings	0.5	0.5
Substation		
General horizontal	2	2.2
Specific vertical (on disconnects)	2	2.2
Coal yards (protective)	0.2	0.2
Dredging	2	2.2
Farms—dairy and poultry		
General inactive areas (protective lighting)	0.2	0.2
General active areas (paths, steps, rough storage, barn lots)	1	1.1
Service areas (fuel storage, shop, feed lots, building entrances)	3	3.2

* Minimum on the task at any time for young adults with normal and better than 20/30 corrected vision. For general notes see beginning of tabulation. For other notes see end of tabulation.

Dekalux is an SI unit equal to 1.076 footcandles. 1 dekalux = 10 lux.

† Equivalent sphere illumination. See general notes at beginning of tabulation.

Fig. 9-80. *Continued*

ea	Footcandles on Tasks*	Dekalux# on Tasks*
lags, floodlighted (see **Bulletin and poster boards)**		
ardens[p]		
General lighting	0.5	0.5
Path, steps, away from house	1	1.1
Backgrounds—fences, walls, trees, shrubbery	2	2.2
Flower beds, rock gardens	5	5.4
Trees, shrubbery, when emphasized	5	5.4
Focal points, large	10	11
Focal points, small	20	22
asoline station (see Service stations)		
ighways (see Fig. 20-4)		
oading and unloading platforms	20	22
Freight car interiors	10	11
umber yards	1	1.1
arking areas		
Self-parking area	1	1.1
Attendant-parking area	2	2.2
iers		
Freight	20	22
Passenger	20	22
Active shipping area surrounds	5	5.4
rison yards	5	5.4
rotective lighting (see Fig. 18-11)		
uarries	5	5.4
ailroad yards		
Retarder classification yards		
Receiving yard		
Switch points	2	2.1
Body of yard	1	1.1
Hump area (vertical)	20	22
Control tower and retarder area (vertical)	10	11
Head end	5	5.4
Body	1	1.1
Pull-out end	2	2.2
Dispatch or forwarding yard	1	1.1

Area	Footcandles on Tasks*	Dekalux# on Tasks*
Hump and car rider classification yard		
Receiving yard		
Switch points	2	2.2
Body of yard	1	1.1
Hump area	5	5.4
Flat switching yards		
Side of cars (vertical)	5	5.4
Switch points	2	2.2
Trailer-on-flatcars		
Horizontal surface of flatcar	5	5.4
Hold-down points (vertical)	5	5.4
Container-on-flatcars	3	3.2
Roadways (see Fig. 20-4)		
Service station (at grade)		
Dark surrounding		
Approach	1.5	1.6
Driveway	1.5	1.6
Pump island area	20	22
Building faces (exclusive of glass)	10[r]	11[r]
Service areas	3	3.2
Landscape highlights	2	2.2
Light surrounding		
Approach	3	3.2
Driveway	5	5.4
Pump island area	30	32
Building faces (exclusive of glass)	30[r]	32[r]
Service areas	7	7.5
Landscape highlights	5	5.4
Ship yards		
General	5	5.4
Ways	10	11
Fabrication areas	30	32
Smokestacks with advertising messages (see **Bulletin and poster boards)**		
Storage yards		
Active	20	22
Inactive	1	1.1
Streets (see Fig. 20-4)		
Water tanks with advertising messages (see **Bulletin and poster boards)**		

Sports Lighting

rchery (indoor)		
Target, tournament	50[r]	54[r]
Target, recreational	30[r]	32[r]
Shooting line, tournament	20	22
Shooting line, recreational	10	11
rchery (outdoor)		
Target, tournament	10[r]	11[r]
Target, recreational	5[r]	5.4[r]
Shooting line, tournament	10	11
Shooting line, recreational	5	5.4
adminton		
Tournament	30	32
Club	20	22
Recreational	10	11

Baseball		
Major league		
Infield	150	160
Outfield	100	110
AA and AAA league		
Infield	70	75
Outfield	50	54
A and B league		
Infield	50	54
Outfield	30	32
C and D league		
Infield	30	32
Outfield	20	22
Semi-pro and municipal league		
Infield	20	22
Outfield	15	16

* Minimum on the task at any time for young adults with normal and better than 20/30 corrected vision. For general notes see beginning of tabulation. For her notes see end of tabulation.

Dekalux is an SI unit equal to 1.076 footcandles. 1 dekalux = 10 lux.

† Equivalent sphere illumination. See general notes at beginning of tabulation.

Fig. 9-80. *Continued*

Area	Footcandles on Tasks*	Dekalux# on Tasks*
Recreational		
Infield	15	16
Outfield	10	11
Junior league (Class I and Class II)		
Infield	30	32
Outfield	20	22
On seats during game	2	2.2
On seats before and after game	5	5.4
Basketball		
College and professional	50	54
College intramural and high school	30	32
Recreational (outdoor)	10	11
Bathing beaches		
On land	1	1.1
150 feet from shore	3[r]	3.2[r]
Billiards (on table)		
Tournament	50	54
Recreational	30	32
Bowling[a]		
Tournament		
Approaches	10	11
Lanes	20	22
Pins	50[r]	54[r]
Recreational		
Approaches	10	11
Lanes	10	11
Pins	30[r]	32[r]
Bowling on the green		
Tournament	10	11
Recreational	5	5.4
Boxing or wrestling (ring)		
Championship	500	540
Professional	200	220
Amateur	100	110
Seats during bout	2	2.2
Seats before and after bout	5	5.4
Casting—bait, dry-fly, wet-fly		
Pier or dock	10	11
Target (at 80 feet for bait casting and 50 feet for wet or dry-fly casting)	5[r]	5.4[r]
Combination (outdoor)		
Baseball/football		
Infield	20	22
Outfield and football	15	16
Industrial softball/football		
Infield	20	22
Outfield and football	15	16
Industrial softball/6-man football		
Infield	20	22
Outfield and football	15	16
Croquet or Roque		
Tournament	10	11
Recreational	5	5.4
Curling		
Tournament		
Tees	50	54
Rink	30	32

Area	Footcandles on Tasks*	Dek on
Recreational		
Tees	20	2
Rink	10	1
Fencing		
Exhibitions	50	5
Recreational	30	3
Football		
Distance from nearest sideline to the farthest row of spectators		
Class I Over 100 feet	100	11
Class II 50 feet to 100 feet	50	5
Class III 30 feet to 50 feet	30	3
Class IV Under 30 feet	20	2
Class V No fixed seating facilities	10	1

It is generally conceded that the distance between the spectactors and the play is the first consideration in determining the class and lighting requirements. However, the potential seating capacity of the stands should also be considered and the following ratio is suggested: Class I for over 30,000 spectators; Class II for 10,000 to 30,000; Class III for 5000 to 10,000; and Class IV for under 5000 spectators.

Area	Footcandles on Tasks*	Dek on
Football, Canadian—rugby (see **Football**)		
Football, six-man		
High school or college	20	2
Jr. high and recreational	10	1
Golf		
Tee	5	[illegible]
Fairway	1, 3[r]	1.1, [illegible]
Green	5	[illegible]
Driving range		
At 200 yards	5[r]	[illegible]
Over tee area	10	1
Miniature	10	1
Practice putting green	10	1
Gymnasiums (refer to individual sports listed)		
Exhibitions, matches	50	5
General exercising and recreation	30	3
Assemblies	10	1
Dances	5	[illegible]
Lockers and shower rooms	20	2
Handball		
Tournament	50	5
Club		
Indoor—four-wall or squash	30	3
Outdoor—two-court	20	2
Recreational		
Indoor—four-wall or squash	20	22
Outdoor—two-court	10	11
Hockey, field	20	22
Hockey, ice (indoor)		
College or professional	100	110
Amateur	50	54
Recreational	20	22
Hockey, ice (outdoor)		
College or professional	50	54
Amateur	20	22
Recreational	10	11

* Minimum on the task at any time for young adults with normal and better than 20/30 corrected vision. For general notes see beginning of tabulation. For notes see end of tabulation.

\# Dekalux is an SI unit equal to 1.076 footcandles. 1 dekalux = 10 lux.

† Equivalent sphere illumination. See general notes at beginning of tabulation.

Fig. 9-80. *Continued*

	Footcandles on Tasks*	Dekalux# on Tasks*
shoes		
ournament	10	11
ecreational	5	5.4
shows	20	22
i		
rofessional	100	110
mateur	70	75
se	20	22
ounds	5	5.4
	5	5.4
(outdoor)		
uto	20	22
icycle		
Tournament	30	32
Competitive	20	22
Recreational	10	11
og	30	32
ragstrip		
Staging area	10	11
Acceleration, 1320 feet	20	22
Deceleration, first 660 feet	15	16
Deceleration, second 660 feet	10	11
Shutdown, 820 feet	5	5.4
orse	20	22
otor (midget of motorcycle)	20	22
50 yards—outdoor)		
n targets	50ʳ	54ʳ
iring point	10	11
ange	5	5.4
nd pistol range (indoor)		
n targets	100ʳ	110ʳ
iring point	20	22
ange	10	11
rena		
Professional	50	54
Amateur	30	32
Recreational	10	11
ens and chutes	5	5.4
(see Croquet)		
eboard (indoor)		
Tournament	30	32
Recreational	20	22
eboard (outdoor)		
Tournament	10	11
Recreational	5	5.4
ng		
Roller rink	10	11
Ice rink, indoor	10	11
Ice rink, outdoor	5	5.4
Lagoon, pond, or flooded area	1	1.1

Area	Footcandles on Tasks*	Dekalux# on Tasks*
Skeet		
Targets (at 60 feet)	30ʳ	32ʳ
Firing points	5	5.4
Skeet and trap (combination)		
Targets (at 100 feet for trap, 60 feet for skeet)	30ʳ	32ʳ
Firing points	5	5.4
Ski slope	1	1.1
Soccer (see **Football**)		
Softball		
Professional and championship		
Infield	50	54
Outfield	30	32
Semi-professional		
Infield	30	32
Outfield	20	22
Industrial league		
Infield	20	22
Outfield	15	16
Recreational (6-pole)		
Infield	10	11
Outfield	7	7.5
Slow pitch, tournament—see industrial league		
Slow pitch, recreational (6-pole)—see recreational (6-pole)		
Squash (see **Handball**)		
Swimming (indoor)		
Exhibitions	50	54
Recreational	30	32
Underwater—100 lamp lumens per square foot of surface area		
Swimming (outdoor)		
Exhibitions	20	22
Recreational	10	11
Underwater—60 lamp lumens per square foot of surface area		
Tennis, lawn (indoor)		
Tournament	50	54
Club	30	32
Recreational	20	22
Tennis, lawn (outdoor)		
Tournament	30	32
Club	20	22
Recreational	10	11
Tennis, table		
Tournament	50	54
Club	30	32
Recreational	20	22
Trap		
Targets (at 100 feet)	30ʳ	32ʳ
Firing points	5	5.4
Volley ball		
Tournament	20	22
Recreational	10	11

inimum on the task at any time for young adults with normal and better than 20/30 corrected vision. For general notes see beginning of tabulation. For other see end of tabulation.

ekalux is an SI unit equal to 1.076 footcandles. 1 dekalux = 10 lux.

quivalent sphere illumination. See general notes at beginning of tabulation.

Fig. 9–80. *Continued*

Transportation Lighting

Area	Footcandles on Tasks*	Dekalux# on Tasks*
Aircraft		
Passenger compartment		
General	5	5.4
Reading (at seat)	20	22
Airports		
Hangar apron	1	1.1
Terminal building apron		
Parking area	0.5	0.5
Loading area	2[r]	2.2[r]
Automobiles		
License plates	0.5	0.5
Rail conveyances		
Boarding or exiting	10	11
Fare box (rapid transit train)	15	16
Vestibule (commuter and inter-city trains)	10	11
Aisles	10	11
Advertising cards (rapid transit and commuter trains)	30	32
Back-lighted advertising cards (rapid transit and commuter trains)—250 fL (857 cd/m²) maximum		
Reading	30†	32†
Rest room (inter-city train)	20	22
Dining area (inter-city train)	50	54
Food preparation (inter-city train)	70	75
Lounge (inter-city train)		
General lighting	20	22
Table games	30	32
Sleeping car		
General lighting	10	11
Normal reading	30†	32†
Prolonged seeing	70†	75†
Railway mail cars		
Mail bag racks and letter cases	30	32
Mail storage	15	16
Road Conveyances		
Step well and adjacent ground area	10	11
Fare box	15	16
General lighting (for seat selection and movement)		
City and inter-city buses at city stop	10	11
Inter-city bus at country stop	2	2.2
School bus while moving	15	16
School bus at stops	30	32
Advertising cards	30	32
Back-lighted advertising cards (see Rail conveyances)		
Reading	30†	32†
Emergency exit (school bus)	5	5.4
Ships		
Living Areas		
Staterooms and Cabins		
General lighting	10	11
Reading and writing	30[w]†	32[w]†
Prolonged seeing	70[u]†	75[u]†
Baths (general lighting)	10	11
Mirrors (personal grooming)	50	54
Barber shop and beauty parlor	50	54
On subject	100	110
Day rooms		
General lighting	20[w]	22[w]
Desks	50[u]†	54[u]†
Dining rooms and messrooms	20	22
Enclosed promenades		
General lighting	10	11
Entrances and passageways		
General	10	11
Daytime embarkation	30	32
Gymnasiums		
General lighting	30	32
Hospital		
Dispensary (general lighting)	30[u]	32[u]
Operating room		
General lighting	50[u]	54[u]
Doctor's office	30[u]	32[u]
Operating table	2000	2200
Wards		
General lighting	10	11
Reading	30	32
Toilets	20	22
Libraries and lounges		
General lighting	20	22
Reading	30[u]†	32[u]†
Prolonged seeing	70[u]†	72[u]†
Purser's office	20[u]	22[u]
Shopping areas	20	22
Smoking rooms	15	16
Stairs and foyers	20	22
Recreation areas		
Ball rooms	15[w]	16[w]
Cocktail lounges	15[w]	16[w]
Swimming pools		
General	15[w]	16[w]
Underwater		
Outdoors—60 lamp lumens/square of foot surface area		
Indoors—100 lamp lumens/square of surface area		
Theatre		
Auditorium		
General	10[w]	11[w]
During picture	0.1	1.1
Navigating Areas		
Chart room		
General	10	11
On chart table	50[u]†	54[u]†
Gyro room	20	22
Radar room	20	22
Radio room	10[u]	11[u]
Radio room, passenger foyer	10	11
Ship's offices		
General	20[u]	22[u]
On desks and work tables	50[u]†	54[u]†
Wheelhouse	10[u]	11[u]
Service Areas		
Food preparation		
General	20[u]	22[u]
Butcher shop	20[u]	22[u]
Galley	30[u]	32[u]
Pantry	20[u]	22[u]
Thaw room	20[u]	22[u]
Sculleries	20[u]	22[u]
Food storage (non-refrigerated)	10	11
Refrigerated spaces (ship's stores)	5	5.4
Laundries		
General	20[u]	22[u]
Machine and press finishing, sorting	50	54
Lockers	5	5.4

* Minimum on the task at any time for young adults with normal and better than 20/30 corrected vision. For general notes see beginning of tabulation. For other notes see end of tabulation.

\# Dekalux is an SI unit equal to 1.076 footcandles. 1 dekalux = 10 lux.

† Equivalent sphere illumination. See general notes at beginning of tabulation.

Fig. 9-80. Continued

Area	Footcandles on Tasks*	Dekalux# on Tasks*
Offices		
General	20	22
Reading	50[u,†]	54[u,†]
Passenger counter	50[u,†]	54[u,†]
Storerooms	5	5.4
Telephone exchange	20	22
Operating Areas		
Access and casing	10	11
Battery room	10	11
Boiler rooms	20[u]	22[u]
Cargo handling (weather deck)	5[u]	5.4[u]
Control stations (except navigating areas)		
General	20	22
Control consoles	30	32
Gage and control boards	30	32
Switchboards	30	32
Engine rooms	20[u]	22[u]
Generator and switchboard rooms	20[u]	22[u]
Fan rooms (ventilation & air conditioning)	10	11
Motor rooms	20	22
Motor generator rooms (cargo handling)	10	11
Pump room	10	11
Shaft alley	10	11
Shaft alley escape	3	3.2
Steering gear room	20	22
Windlass rooms	10	11
Workshops		
General	30[u]	32[u]
On top of work bench	50[u]	54[u]
Tailor shop	50[u]	54[u]
Cargo holds		
Permanent luminaires	3[u]	3.2[u]
Passageways and trunks	10	11

inimum on the task at any time. For general notes see beginning of tabulation.
btained with a combination of general lighting plus specialized supplementary lighting. Care should be taken to keep within the recommended luminance ratios. seeing tasks generally involve the discrimination of fine detail for long periods of time and under conditions of poor contrast. The design and installation of the nation system must not only provide a sufficient amount of light, but also the proper direction of light, diffusion, color, and eye protection. As far as possible it eliminate direct and reflected glare as well as objectionable shadows.
)ark paintings with fine detail should have 2 to 3 times higher illumination.
ı some cases, much more than 100 footcandles is necessary to bring out the beauty of the statuary.
educed or dimmed during sermon, prelude or meditation.
wo-thirds this value if interior finishes are dark (less than 10 per cent reflectance) to avoid high luminance ratios, such as between hymnbook pages and the sur- Careful planning is essential for good design.
pecial lighting such that (1) the luminous area shall be large enough to cover the surface which is being inspected and (2) the luminance be within the limits neces- obtain comfortable contrast conditions. This involves the use of sources of large area and relatively low luminance in which the source luminance is the principal rather than the footcandles produced at a given point.
or close inspection, 50 footcandles (54 dekalux).
encil handwriting, reading of reproductions and poor copies require 70 footcandles (75 dekalux).
or close inspection, 50 footcandles (54 dekalux). This may be done in the bathroom, but if a dressing table is provided, local lighting should provide the level recom- d.
he specular surface of the material may necessitate special consideration in selection and placement of lighting equipment, or orientation of the work.
r not less than ⅕ the level in adjacent areas.
or size of task plane see pages 15-15 through 15-25.
General lighting for these areas need not be uniform in character.
ncluding street and nearby establishments.
) Values are illumination on the merchandise on display or being appraised. The plane in which lighting is important may vary from horizontal to vertical. (b) c appraisal areas involving difficult seeing may be lighted to substantially higher levels. (c) Color rendering of fluorescent lamps is important. Incandescent and cent usually are combined for best appearance of merchandise. (d) Illumination may often be made non-uniform to tie in with merchandising layout.
alues based on a 25 per cent reflectance, which is average for vegetation and typical outdoor surfaces. These figures must be adjusted to specific reflectances of ma- lighted for equivalent brightnesses. Levels give satisfactory brightness patterns when viewed from dimly lighted terraces or interiors. When viewed from dark areas nay be reduced by at least ½; or they may be doubled when a high key is desired.
ocalized general lighting.
ertical.
evels shown are based on visual considerations. Otherwise for public attraction and increased business considerations, practice is as follows:

Class	*Approaches*	*Lanes*	*Pins*
Tournament	70fc (75 dalx)	100fc (110 dalx)	200fc (220 dalx) Vertical
Recreational	50fc (54 dalx)	70fc (75 dalx)	150fc (160 dalx) Vertical

upplementary lighting should be provided in this space to produce the higher levels of lighting required for specific seeing tasks involved.
The footcandle values may vary widely, depending upon the effect desired, the decorative scheme, and the use made of the room; however, the lighting system provide at least the recommended minimum illumination levels.

(See next page for References)

REFERENCES

1. Lighting Design Practice Committee of the IES: "General Procedure for Calculating Maintained Illumination," *Illum. Eng.*, Vol. 65, p. 602, October, 1970.
2. Clark, F.: "Accurate Maintenance Factors," *Illum. Eng.*, Vol. LVIII, p. 124, March, 1963. Part Two, Vol. LXI, p. 37, January, 1966.
3. Clark, F.: "Light Loss Factors in the Design Process," *Illum. Eng.*, Vol. 63, p. 515, November, 1968.
4. Testing Procedures Committee of the IES: "IES Approved Method for Determining Luminaire—Lamp—Ballast Combination Operating Factors for High Intensity Discharge Luminaires," *Illum. Eng.*, Vol. 65, p. 718, December, 1970.
5. Lighting Design Practice Committee of the IES: "Zonal-Cavity Method of Calculating and Using Coefficients of Utilization," *Illum. Eng.*, Vol. LIX, p. 309, May, 1964.
6. Jones, J. R. and Jones, B. F.: "Using the Zonal-Cavity System in Lighting Calculations," *Illum. Eng.*, Vol. LIX; Part I, p. 413, May, 1964; Part II, p. 448, June, 1964; Part III, p. 501, July, 1964; Part IV, p. 556, August, 1964.
7. Jones, B. F.: "Zonal-Cavity—A Three-Level Approach," *Illum. Eng.*, Vol. 64, p. 149, March, 1969.
8. Levin, R. E.: "Luminance—A Tutorial Paper," *J. of the SMPTE*, Vol. 77, p. 1005, October, 1968.
9. Lighting Design Practice Committee of the IES: "Zonal-Cavity Method of Calculating and Using Coefficients of Utilization," *Illum. Eng.*, Vol. LIX, Section I, p. 309, May 1964. Lighting Design Practice Committee of the IES: "Calculation of Luminance Coefficients Based Upon the Zonal-Cavity Method," *Illum. Eng.*, Vol. 63, p. 423, August, 1968. O'Brien, P. F.: "Numerical Analysis for Lighting Design," *Illum. Eng.*, Vol. LX, Section I, p. 169, April, 1965.
10. Lighting Design Practice Committee of the IES: "Methods of Calculating Illumination at a Point," to be published.
11. Goodbar, I.: "New Methods for Point by Point Calculations," *Illum. Eng.*, Vol. XLI, p. 39, January, 1946.
12. Jones, J. R., LeVere, R. C., Ivanicki, N., and Chesebrough, P.: "Angular Coordinate System for Computing Illumination at a Point," *Illum. Eng.*, Vol. 64, Section I, p. 296, April, 1969.
13. "The Calculation of Direct Illumination from Linear Sources," IES (London) Technical Report No. 11, The Illuminating Engineering Society, York House, Westminster Bridge Road, London S.E. 1, England.
14. McPhail, R. G.: "A Method of Calculating Direct Illumination from Linear Sources," *Illum. Eng.*, Vol. XLVI, p. 511, October, 1951.
15. Spencer, D. E.: "Exact and Approximate Formulae for Illumination from Troffers," *Illum. Eng.*, Vol. XXXVII, p. 596, November, 1942. Wakefield, E. H., and McCord, C.: "Discussion of Illumination Distribution from Linear Strip and Surface Sources," *Illum. Eng.*, Vol. XXXVI, p. 1330, December, 1941. Wakefield, E. H.: "A Simple Graphical Method of Finding Illumination Values from Tubular, Ribbon, and Surface Sources," *Illum. Eng.*, Vol. XXXV, p. 142, February, 1940. Woblauer, A. A.: "The Flux from Lines of Light," *Trans. Illum. Eng. Soc.*, Vol. XXXI, p. 694, July, 1936. Whipple, R. R.: "Rapid Computation of Illumination from Certain Line Sources," *Trans. Illum. Eng. Soc.*, Vol. XXX, p. 492, June, 1935.
16. Burnham, R. D.: "The Illumination at a Point from an Industrial Fluorescent Luminaire," *Illum. Eng.*, Vol. XLV, p. 753, December, 1950.
17. Brown, L. C., and Jones, J. R.: "Engineering Aspects of Cove Lighting Design," *Illum. Eng.*, Vol. XLIV, p. 233, April, 1949.
18. Keck, M. E.: "Utilization Curves for Floodlights," *Illum. Eng.*, Vol. XLV, p. 155, March, 1950.
19. Hallman, E. D.: "Floodlighting Design Procedure as Applied to Modern Setback Construction," *Trans. Illum. Eng. Soc.*, Vol. XXIX, p. 287, April, 1934. Dearborn, R. L.: "Floodlighting Design by Graphical Method," *Illum. Eng.*, Vol. XL, p. 514, September, 1945.
20. Blondell, A. and Rey, J.: "The Perception of Lights of Short Duration at Their Range Limits," *Trans. Illum. Eng. Soc.*, Vol. VII, p. 625, November, 1912.
21. Symposium on Searchlights, Illuminating Engineering Society (London), 1948.
22. Higbie, H. H. and Randall, W. C.: "A Method for Predicting Daylight from Windows," *Eng. Res. Bull.*, No. 6, Dept. Eng. Res., U. of Mich., January, 1927.
23. *International Recommendations for the Calculation of Natural Daylight*, Publication CIE No. 16 (E-3.2) 1970, CIE Bureau Central 4, Ave. du Recteur Poincaré 75, Paris 16.

SECTION

10

INTERIOR LIGHTING DESIGN APPROACHES

Luminous Environment Design Approach 10–3 Visual Task Oriented Design Approach 10–10 Types of Lighting Systems 10–11

The basic purpose of lighting is nominally considered to be that of enabling people to see; however, its function is actually much broader in scope. Lighting is also a dominant factor in environmental design. It affects the usefulness and enjoyment of a building interior, and is often an inseparable part of the architectural concept.

Thus, the art and science of modern interior lighting design is broad in scope and involves many factors. Some of these relate to light and vision and the visual response of the eye under varying light and environmental conditions. Others relate to the production, control, and distribution of electric light. Still others relate to the enclosure itself, or building structure, including size, shape, color, and related decorative considerations.

This section covers two approaches to the design process—the luminous environment viewpoint and the visual task viewpoint (both entirely compatible) and the types of lighting systems used. For design factors such as economic analysis, thermal considerations, maintenance, and wiring, see Section 16.

Continuing light and vision research has added much new knowledge to the art and science of lighting. Many new technological developments have been made in the field of design and production of light sources, lighting equipment, and lighting accessories. Hence, the lighting designer and the total environmental design team have added new tools and further guidance for the complex and variable problems of interior lighting design. Most of these factors are fundamental, and have been well documented by lighting scientists and practitioners through the years. These, and some of the more recent concepts, are discussed in this and other sections; for the sake of convenience and clarity, some of the major ones are considered and discussed briefly here.

NOTE: References are listed at the end of each section.

The lighting designer needs to know and understand fully the visual sense, or *how we see* (Section 3). Such knowledge is basic in selecting the actual luminance (photometric brightness) of the visual task and the relative luminance of the task, its immediate surroundings and anything else in the peripheral field of view. These luminances affect visual comfort and the performance of the task. Research indicates that desirable seeing conditions exist when the luminances of the surroundings and the visual task are relatively uniform, and veiling reflections are eliminated or effectively reduced and diminished. Since this condition is not always practical or easy to attain, especially in view of decorative and other considerations which are usually involved, luminance limitation recommendations have been developed. When followed, these recommendations will provide a generally satisfactory visual environment. See later application Sections.

The general approach to low luminance ratios over an entire visual field is to limit the luminance of the luminaires and fenestration, and to build up the luminance of all other interior surfaces (ceiling, walls. trim, floor, work surfaces, and equipment) by suitable reflectances, textures, and distribution of light.

The use of architectural and structural design and color in the visual field is appreciated for the stimulating benefits which they produce. These factors are therefore discussed in detail, as they relate to interior lighting design.

The lighting designer has considerable latitude in the selection of a lighting solution for a specific area. Modern standards call for provision of both quantity and quality of lighting, which are commensurate with the degree of severity of the seeing tasks which will exist in the area, and which will aid performance and minimize the fatigue resulting from from visual effort. These standards are in the form of minimum recommended values (see Section 9 and later application sections).

On the other hand, the lighting designer has at his disposal a wide range of types and sizes of light sources, of luminaires, and of lighting equipment and lighting components. Here, he has an opportunity to exercise his professional judgment, personal taste, or choice based on economic analyses, and come up with any of several solutions, each of which may be justified on its own merit.

Further, the lighting designer, working in cooperation with the architect and interior designer, has an opportunity to coordinate the lighting design with the decorative design, to enhance the over-all appearance and decorative mood desired. The lighting can and should be a dominant factor in the decorative treatment of any interior, for without light, or with an inadequate and improperly designed lighting system, the desired decorative effects will not be realized.

The lighting designer may also have a choice in the method of installation of the lighting system; or, architectural design and structural conditions may dictate a particular method of installation. In either case, a competent knowledge of the principles of light control and of the lighting tools and devices which are available for such control, will be most helpful in the design of the most satisfactory and efficient lighting system possible for the problem involved.

The following considerations are deemed pertinent to the analysis and solution of most interior lighting design problems:

Light and Architecture. Since architecture has to do with the enclosure of space, it involves structure, form, color, brightness patterns, etc. These, in turn, require light in order to be seen. Thus light automatically becomes an element of architectural design. Traditionally, architects have used light, especially daylight, to give form to planes, curved surfaces, and structural design. More recently, they have used electric light for the same purposes, but on a broadly expanded basis. Now, with the great variety of light sources, luminaires, lighting equipment, and light control devices that is available, the architect has at his disposal an almost unlimited range of lighting treatments. Using lighting as an element of architectural design, it is possible and practical to develop almost any design appearance, or visual mood, and to satisfy the most critical of spatial needs for any enclosure, either by day or at night.

Architecture comprises the esthetic as well as the physical and economic aspects of structures, and this is equally true of lighting. The lighting art is influenced in all its aspects by individual interpretation. Lighting can be used by the architect to dramatize the many features of his architectural plan.

Building Function. The function of a building or other structure greatly influences the way in which lighting is applied. In an office building, for example, the typical visual task may be considered as reading at desk top level. However, the same type of visual task (reading) may be encountered, regardless of location, in a factory, in a store, or in a home. But such factors as economics, appearance, continuity of effort, and quality of lighting results desired influence the lighting design developed for the task. Thus application techniques generally designated as industrial lighting, store lighting, office lighting, and so on have developed based on lighting solutions for the types of visual tasks encountered generally in each type of occupancy. Each of these is a synthesis of engineering theory, application experience, and consumer acceptance and desire in a particular field. Because these include more than an objective assessment of engineering considerations, it is necessary to relate the design of a lighting installation to the particular occupancy of the space it is to serve.

Lighting Systems. Lighting systems have been simply classified by illuminating engineers to reflect the general type of lighting produced, and the general layout of luminaires. They are generically described as general, local, localized general, or supplementary. Also, luminaires have been divided into five standard classifications, based on light distribution characteristics, as follows: direct, semi-direct, general diffuse, semi-indirect, and indirect (see page 10–11). These classifications are useful primarily in that they simplify professional discussion relating to lighting techniques as employed for lighting any specific area.

Light Sources. The choice of light source, of luminaire characteristics, and of the system layout are closely interrelated in an application technique. A method easily applied with one type of light source may be equally applicable, or most impractical, with another. Frequently local conditions of vibration, ambient temperature, dust and dirt, continuity of service, and color influence light source selection, application, operation, output, and maintenance and, indirectly therefore, the lighting application technique. See also Section 8. Where various light sources are equally applicable, economics may be the deciding factor.

Economics. The design of a lighting system is affected by both initial and operating costs. There is no sharp line of demarcation between excellent and good lighting, between good and average, or between average and poor. Also, there is no easy way to predict the exact value of commercial or industrial lighting in terms of production, safety, quality control, employee morale, or employee health; or to weigh the importance of home lighting in dollars and cents. Nevertheless, illuminating engineers must balance costs against the attainable results in developing any lighting design, relying to a great extent upon experience gained in the solution of comparable problems if such is available. See Section 16.

Coordination with Mechanical and Acoustical Systems. The enjoyment of a space or activity involves all of the senses, and though lighting systems and the visual implications of light can be discussed separately as an abstraction, the final design should reflect the needs of the total environment. The development of air-conditioning systems and the evolution of acoustical control techniques have combined with electric light to provide an extensive and unprecedented capacity for environmental control. As a part of this more comprehensive mechanism, the lighting system should be analyzed in terms of its secondary characteristics and for its total influence on system operation and room comfort, see page 16-1. At the same time the detailing and coordination of various mechanical and electrical elements should reflect careful design judgment.

Coordinated Components and Modules. It has become increasingly critical that electrical and mechanical components be compatible with other elements in the same building system so that the resulting assembly is an efficient and economical operating unit *and* an architecturally coordinated design. In this regard, an increasing number of luminaire types has been developed to supply both light and air to a space. See page 16-1. This equipment offers the initial advantage of a simplified ceiling appearance, due to the elimination of overlapping (and sometimes conflicting) mechanical and electrical ceiling patterns. These units may also offer considerable flexibility in space planning, where the "mechanically coordinated module" is employed. Within each standard dimensional module, there should be: (a) lighting, (b) air supply, (c) air return, and (d) sound absorption. Where this can be achieved, it means that modular floor and partition systems can be used (initially or in future modifications) to set off any module or any combination of modules as a separate room. Each room, then, would automatically include all of the basic elements of the mechanical-electrical environment. To achieve this degree of extreme flexibility with conventional equipment would require a luminaire, an air diffuser, and an air return in each module.

Interior-Exterior Relationships. With increasingly widespread circulation and activity at night, the problem of building esthetics extends beyond daytime architecture. Particularly where there is an extensive use of glass, the organization and design of the interior lighting system will affect the exterior impression of the building at night and on days that are relatively dull and overcast. The brightness of visible interior surfaces, the pattern of luminaires, and the color of the light source may exert an important influence on the exterior appearance of the building during these periods. See Fig. 17-9.

The interaction of light and form is a major factor that affects both interior or exterior architectural relationships. The level, color, and distribution of light become fundamental considerations in providing for visual identification, orientation, and enjoyment of an activity and space. In this sense, the lighting system is a basic factor in architectural analysis and design.

LUMINOUS ENVIRONMENT DESIGN APPROACH

The illuminating engineer, architect, contractor, interior designer, and consulting engineer are all concerned with the physical organization of human activities and with the problem of providing pleasant, protective shelter to house these activities. This problem includes the search for an understanding of optimum environmental conditions. And in this search, it has been recognized that a major criterion in building design for optimum environments is to properly solve the problem of bringing light into an enclosed, protected space.

In this respect, the contemporary design team has a greater degree of freedom and flexibility than its predecessors. The contemporary builder is free to use available daylight, as architects have done throughout history; or he can solve the lighting problem somewhat independent of structure by using electric light to achieve his objectives.

But this increased freedom means that the designer must make decisions and selections in areas where until now little guidance has been offered. If he can now control and specify light within narrow limits, how should he use it? How much does he need? Where does he need it? What color should it be? What fundamental combination of components and systems will give the results he needs for a particular job?

For the design team, the answers are complex and variable, because the criteria for lighting design is variable. A decision or design judgment will vary, depending on the visual priorities inherent in each activity.

In this regard, it is suggested that the designer adapt his analysis to include the following steps: (1) determine the desired visual composition in the space; (2) determine the desired appearance of visual tasks and objects in the space; (3) select luminaires that fit the concept of visual composition and implement the desired appearance of objects.

Determining the Visual Composition of the Space

Light can influence an observer's unconscious interpretation of a space—for his judgment is based not only on form, but on form *as modified by light*. There are both esthetic and psychological implications in this. Through the design and placement of lighting elements, the designer can specify the combination of surfaces to be lighted, or left in darkness. In this sense, the designer can specify how the pattern of brightness is to merge with the structural pattern. Looking at it in another way, he can specify how the visual perception of space is to merge with the activity involved—the use of light to identify centers of interest and attention, and otherwise to complement the basic mood of the activity as this is understood and interpreted by the individual designer.

In practice this responsibility seldom lies entirely in the hands of a single individual. Usually, the design is the result of some combination of skills in architecture, interior design, illuminating engineering, and electrical engineering. The specific responsibility of the illuminating engineer, then, is to recognize the importance of both spatial and task-oriented attributes so that his technical efforts will contribute to the total result. To do this, the following should be considered:

1. *Focal Centers.* Does the lighting system properly identify centers of primary and secondary attention? A display, a picture, planting, or a featured wall are examples. (See Fig. 10–1.) Depending on the relative dominance desired, the luminance of these focal areas should range from 5 to 10 times that of other nearby surfaces.

2. *The Overhead Zone.* Since the ceiling is usually of secondary or subordinant interest relative to the activity, the appropriate influence of form, pattern, and brightness in this area should be considered. (See Figs. 10–2 and 10–3.)

3. *The Perimeter Zone.* In many cases (particularly where a sense of relaxation is desired), perimeter brightness should be greater than the brightness of the overhead zone. (See Fig. 10–4.) In general, simplicity is desirable in the perimeter zone.

Fig. 10–1. An example of a very strong focal area. The ratio of picture area luminance to that of the walls is in the order of 20 to 1. If both the table tops and the perimeter are very dark, the space tends to appear gloomy rather than dramatic.

Fig. 10–2. The spatial influence of the lighting system affects orientation and comprehension of the room. The system on the left is spatially confusing despite the fact that work surface illumination and visual comfort are both relatively good. By way of contrast, the lighting system on the right functions in a sense of unity with the basic rhythm of the space. Furthermore, because of the placement of the luminaires, note that the ceiling gradients produced by the electric lighting tend to complement and reinforce the natural gradients of the curved form.

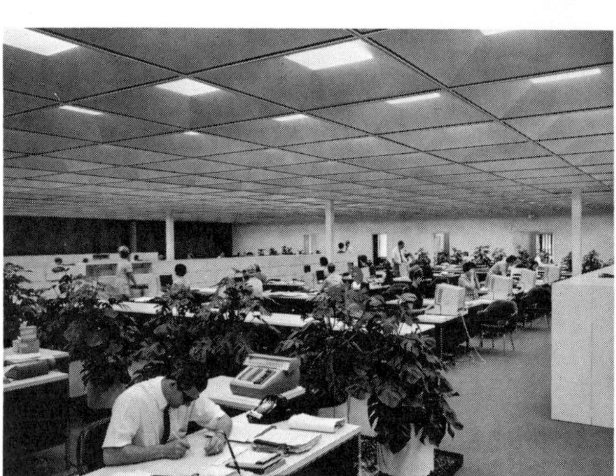

Fig. 10-3. The upper left shows a space in which very simple surfaces are involved. Note how the simplicity of the perimeter emphasizes the people. The upper right shows a somewhat similar space, but the introduction of pattern on the ceiling tends to create attention in the overhead zone. Reduction of brightness in the overhead zone in the room shown in the lower left is accomplished with louvers providing a high degree of brightness control. The slight configuration of the ceiling provides relief from a flat plane. Note how the coordination of the ceiling pattern and the wall pattern helps create unity. In the office shown in the lower right, again, attention is removed from the overhead zone by reduction of brightness—here by ceiling configuration which causes the lighting to disappear from the field of view as one looks across the room. Note how the simplicity of the perimeter helps emphasize the people. Note, too, the depressing effect of the large, dark wall area as contrasted to the lighter wall on the right. Used in combination the dark and the light walls provide relief. If all the walls had been dark or all had been light, the effect would have been less pleasing in this large space.

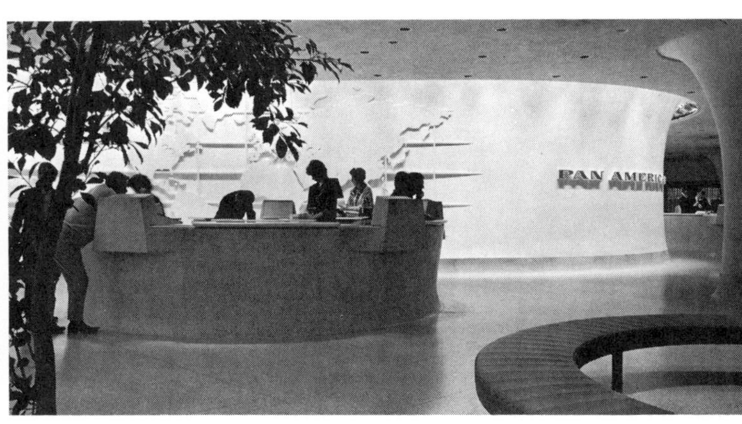

Fig. 10-4. A strong, very large, focal area which provides a powerful sense of enclosure. The perimeter wall luminance of 100 footlamberts and more is sufficient to overcome reflections in the window of this ticket lobby as viewed from outdoors in the daytime. When people tend to appear in silhouette, they are generally more comfortable than as if they appear bright against the background.

"Visual clutter" here may cause confusion in spatial comprehension and orientation. It may also complicate identification of meaningful focal centers in merchandise and display areas. (See Figs. 10–5 and 10–6.)

4. *The Occupied Zone.* The functional or activity emphasis is usually in this zone. For this reason, illumination objectives here are generally those listed in Fig. 9–80. The balance among this zone and the two previously mentioned determines overall effectiveness. (See Figs. 10–7 and 10–8.)

5. *Transitional Considerations.* When the luminance ratio between adjoining spaces and activities approximates 2 or 3 to 1, or less, the sense of visual change tends to be subliminal. When a sense of "change" is desired, rather than "continuity," luminance differences should substantially exceed this 3 to 1 ratio. In general, a clearly noticeable transition is approximately a 10 to 1 difference. A dominant or abrupt change is approximately 100 to 1 or greater.

Color continuity or change ("whiteness") should also be considered in the lighting of adjacent spaces.

6. *Level of Stimulation.* There are few absolutes here; but some significant variables can be outlined—along with the direction of their influence.

General Luminance Level. Levels below 10 footlamberts are associated with twilight conditions. These levels may appear "dingy" unless higher brightness accents are provided.

Color and Shade. Warm colors tend to stimulate; cool tones tend to soothe. Saturated colors are more stimulating than tints.

Areas of darkness tend to be subduing or relaxing, and may be dramatic if focal accents are provided.

Information Content. When a person is understimulated, he may become bored with the task or activity. In these situations, techniques involving spatial pattern (brightness, color, sparkle) often tend to stimulate interest and vitality.

Fig. 10–5. A lighted wall surface should generally be perceived as an integrated form, not as a form or surface intersected by irrelevant patterns of light.

Fig. 10–6. Where scallops exist in perimeter lighting, they should relate to the basic rhythm of the design. In this way, the unity of the design and the unity of the surface is preserved.

Determining the Desired Appearance of Objects in the Space

In addition to the above spatial factors, parallel lighting criteria relate to the desired appearance of objects in the space.

1. *Diffusion.* Diffuse light tends to reduce the variations that relate to form (contour gradients), pattern, and texture. (Consider that a tennis ball on an overcast day may actually seem to disappear.) However, diffusion is desired in most work areas to prevent distracting shadows at the task itself.

Sculpture should generally be developed with some directionality in lighting (see page 12–19)—but with significant diffusion to relieve the harshness. The appearance of facial forms and expressions are best treated with a similar combination of directional and diffuse lighting.

2. *Sparkle.* Sparkle and highlight can enhance the sense of vitality in a space. For this reason, consideration should be given to the value of concentrated or "point" light sources interacting with polished and refractive surfaces.

3. *Color Rendition.* Color schemes (paint, fabrics, etc.) should be chosen under the lighting to be used in the space. In choosing light sources, consider the importance of appearance of people, merchandise, etc. Visual contact is nearly always important. Light sources should not be compared side-by-side because in a given space the eye tends to adapt over a wide range. The eye is extremely sensitive to color differences in transition from one space to another or from one display to another.

Fig. 10–7. For an effective appearance there should be a balance among the brightnesses in the overhead, perimeter and occupied zones. In the store at the upper left, the ceiling dominates tending to compete with merchandise. In the upper right, attention has been removed from the ceiling by low brightness luminaires but space tends to be dingy because of low perimeter brightness. An example of proper combination of minimum attention on ceiling and accent on perimeter in a general merchandise operation is shown at the left.

Fig. 10–8. Two examples of lighting in a private office. In the one at the left, the ceiling dominates and there is no focal emphasis for the occupant. A carefully calculated three dimensional composition in light, texture, form, and color is shown in the right. Note the placement of the ceiling element and its low brightness, the drapery and column lighting, the wall lighting (incandescent to bring out the wood paneling), and the portable lamp which provides an accent, all producing a balanced, interesting interior.

Selection of Luminaires to Fit the Concept of Visual Composition and to Implement the Desired Appearance of Objects

This section has, so far, stressed the visual implications of light in a space. But lighting is a design objective—an end result to be achieved through the careful selection and placement of lighting components.

The terms "luminaire" and "lighting element" imply the ingredient of light control. Such factors as brightness control (for glare) and appropriate beam distribution (for direction or diffusion of light) should be considered. It is in the design and placement of these devices and elements that the designer specifies the luminous environment.

The Engineering Study. An initial engineering study is needed to determine the alternative techniques available to achieve the design objectives. The background for this study may come from the experience of the designer or from a consultant in this field. From these alternative techniques, a preliminary selection can be made that reflects such considerations as initial and operating costs, maintenance, ruggedness, candlepower distribution, and brightness control.

Once this preliminary selection of techniques is made, then, the problem becomes primarily one of testing and design assimilation. In this regard, the following engineering-oriented criteria are important in guiding the final selection of lighting equipment from among the alternatives available:

1. *Distribution Characteristics and Color of Light.* Is the selection suitable to achieve the desired visual result (spatial illumination and illumination at the various task centers, as discussed under Steps 1 and 2)?

2. *Dimensional Characteristics and Form.* Are the physical characteristics regarding shape and size generally appropriate to meet the needs of the illumination concept? Reflector size and finish, lamp-to-diffuser distance, shielding and cut-off angles, lamp ventilation, etc. should be considered as well as the dimensional intrusion into the space when pendant or bracket equipment is involved.

3. *Lighted and Unlighted Appearance of Lighting Materials.* Is the appearance and quality of detail compatible with the general quality of other materials in the building design?

4. *Initial and Operating Costs.* Is the cost structure compatible with the general quality of other materials and systems in the building design?

5. *Maintenance.* Is the system accessible for lamp replacement and cleaning? Are the characteristics regarding dirt collection and deterioration compatible with the use of the space? What is the recommended maintenance interval? See Section 16.

The Architectural Study. In addition to the engineering-oriented considerations, it should be recognized that luminaires and other lighting components become potentially prominent factors in the architectural composition. In this sense, modern lighting techniques should be carefully assimilated into the basic architectural design concept, because this system assumes an inherent esthetic significance far beyond the normal connotations of electrical-mechanical design. For this reason, lighting devices and elements should be analyzed by architectural standards, in addition to their engineering function and performance.

1. *Brightness, Color, Scale, and Form.* Does the equipment assume an appropriate textural or pattern role in the architectural composition? Does it contribute to a sense of unity?

2. *Compatability with "Period" Designs.* Is the detailing in appropriate harmony with the architectural period of the building? See Fig. 10-9. The use of wall urns, chandeliers, etc., are generally decorative, and require another unobtrusive system to produce comfortable illumination without glare.

3. *Space Requirements and Architectural Detailing.* Are the physical space allowances sufficient? Is the building cubage and design sufficient to provide for necessary lighting cavities and recesses? Is the detailing and use of materials compatible with the detailing of other elements and systems in the building?

4. *Coordination with Other Environmental Systems.* Is the lighting system functionally and physically compatible with other environmental systems in the building design? Consider coordination with other ceiling elements and wall systems.

The Architectural Context of Luminaires. A study of architectural history reveals two basic alternatives in the approach to lighting systems and lighting design: (1) the visually-subordinant system, and (2) the visually-prominent system.

1. *Visually-Subordinant Lighting Systems.* Throughout the history of building, some designers have attempted to introduce light in a way that the observer will be conscious of the effect of the light, while the light source itself is played down in the architectural composition. For example, in some Byzantine churches, small unobtrusive windows were placed at the base of a dome to light this large structural element. The brilliant dome then became a major focal center; and serving as a huge reflector, the dome (not the windows) became the apparent primary light source for the interior space. Similarly, the windows of some Baroque interiors were placed so that they were somewhat concealed from the normal view of the observer, and the observer's attention was focused on a brightly lighted adjacent decorative wall. In both cases, the objective was to place emphasis on the surfaces to be lighted while minimizing any distracting influence from the lighting system itself.

Fig. 10-9. Styles and Lighting Effects of Architectural Periods

Period	Architectural Style	Natural Lighting Effects†
Greek 700–146 B.C. Orders: Doric, Ionic, Corinthian Important buildings: Temples	Column and lintel, with entablature. Harmony of design so as to obtain perfect balance between horizontal and vertical elements. Perfect proportion, simple decoration	Emphasis on the statue of the god or goddess to whom the temple was dedicated. Light was obtained from roof openings usually over the statue, or from clerestory openings, or from doorways. Temples were usually oriented so that the rising sun might light up the statue. Direction of incident light mainly from above, at oblique angles
Roman 146 B.C.–365 A.D. Orders: Tuscan, Doric, Ionic, Corinthian, Composite Important buildings: Temples, basilicas, thermae (baths), palaces	Column and lintel, with entablature. Arch developed. Vault and dome evolved. Elaborate decoration	The Romans used windows extensively. They obtained light by means of clerestories, openings in the center of domes, or windows at the base of domes. Direction of incident light mainly from above, at oblique angles. Light used to enhance the elaborate decoration and majestic proportions of interiors
Early Christian 300–900 A.D. Important buildings: Basilican churches	Column and lintel, with a long interior perspective. Occasional domes and rotundas supported on arched colonnades	Oblique lighting from upper angles obtained through clerestories and window openings, usually small. Emphasis on altar obtained by columnar perspective as well as the convergent perspective of windows in clerestories. Glass mosaics reflecting light often used for the high altar
Byzantine 324 A.D. Important buildings: Churches	The dome on pendentives is the main feature of Byzantine architecture. In Roman architecture domes were used only over circular or polygonal buildings, but in Byzantine architecture domes were placed also over square structures. Here the earlier horizontal motif changes almost imperceptibly to a vertical motif	Lighting from upper angles obtained through windows at the base of domes. The dome being highly illuminated acted as a huge reflector. Small glass and translucent marble windows prevented glare and added color to the interior. Brilliant mosaics glowed with numerous subdued reflections. To relieve their flat wall decoration, the Byzantine builders obtained "depth" by means of arcades
Romanesque 800–1200 A.D. Important buildings: Churches, castles	Massive Roman walls coupled with the round arch	The effect of solemnity and vastness was produced by the contrast between great wall spaces and small windows. Such windows, single or grouped together, admitted rays of light through clerestories
Gothic 1200–1500 A.D. Important buildings: Churches, monasteries, castles, mansions, town halls	This aspiring style with its pointed arches definitely introduced the vertical motif. Solids prevailed in Roman architecture, but in Gothic architecture, voids prevailed instead, since slender buttresses were used instead of massive walls	In churches the mood of solemnity was produced by the lofty, dimly illuminated ceiling, while long rays of light penetrated stained glass windows. In castles and manor houses larger windows than ever had been used before in domestic architecture became the vogue
Renaissance 1400–present day Important buildings: Churches, castles, town halls, palaces, villas, chateaux, civic buildings	The rebirth of classical ideals brought the ideal of architectural harmony again into vogue. Buildings were so designed that the vertical and horizontal members obeyed the classical laws of proportion. For decoration Greek and Roman details were copied	Lighting effects became more numerous to suit different types of buildings. Domes were supported on "drums" which were pierced with large windows. The dome lighting of the Byzantine period was revived and improved. The direction of incident light was still mainly from above, though lower windows also were enlarged. Windows became more numerous, and more light was sought than before

† Some use was made of flame sources (wooden torches, tapers, candles, and oil and gas lamps) even in very early periods. The design of luminaire in period interiors frequently follows the pattern established by the characteristics of these early lamps.

Continued on next page

Fig. 10-9. *Continued*

Period	Architectural Style	Natural Lighting Effects†
Modern (twentieth century) All types of buildings	The twentieth-century style strives for structural logic. For skyscraper design the vertical motif is emphasized. For smaller buildings the supporting steel structure is not camouflaged but rather is indicated by simple "wall lines" and other decorative devices. Stone, glass, chromium, and other metals are used without elaborate ornamentation	Electric illumination now is recognized as an architectural medium. Modern lighting systems vary from the layout with outlets located with mathematical symmetry to the decorative system with light sources in arcades, columns, recesses, panels, cornices, coves, wall pockets, urns, etc. Luminaires differ widely in design and in material

† See footnote on previous page.

This design attitude can be seen in the development of some electric lighting systems (see Figs. 10–3 (lower left and right), 10–4 and 10–8 (right). In this regard, indirect systems and concentrating direct lighting systems are useful design tools. With appropriate shielding and careful control of luminaire brightness, these devices direct light toward a specific surface, plane, or object—emphasizing these objects or areas, with little distracting influence from the lighting device itself. Inherently, then, the space is visually organized as a composition of *reflected* brightness patterns (horizontal and vertical).

2. *Visually-Prominent Lighting Systems.* A light source or luminaire may itself compel attention, even to the extent that such elements become dominant factors in the visual environment. In architectural history, the large stained glass windows of the Gothic period are probably the most obvious examples of this approach. In contemporary building, transilluminated ceilings and walls are a similar dominant influence as seen in Figs. 10–3 and 10–7 (upper left).

Where light-transmitting (rather than opaque) materials are prominently involved in the lighting unit itself, the units become architectural forms and surfaces, as well as lighting elements. Such self-luminous elements help to visually define a space and are important to the general architectural organization of the room.

VISUAL TASK ORIENTED DESIGN APPROACH

In areas where the primary function of the lighting installation is to provide illumination for the quick, accurate performance of visual tasks, the task itself is the starting point in the lighting design. Some of the factors to be considered in the design approach are briefly listed below; however, more detailed information can be found in Section 3 under the IES Method for Prescribing Illumination and in Section 9 and in the later application sections.

1. *Visual Task.*

a. What are the commonly found visual tasks?
b. How should the task be portrayed by the lighting? Should the lighting be diffuse or directional? Are shadows important for a three dimensional effect? Will the task be susceptible to veiling reflections? Is color important?
c. What level of illumination should be provided?

2. *Area in Which Task is Performed.*

a. What are the dimensions of the area and reflectances of surfaces?
b. What should the surface luminances be to minimize transient adaptation effects without creating a bland environment?
c. Might the surfaces produce reflected glare?
d. Is illumination uniformity desirable?

3. *Luminaire Selection.*

a. What type of distribution and spectral quality is needed to properly portray the task (for diffusion, shadows, or avoiding veiling reflections) and provide a comfortable environment (visually, thermally, and sonically)?
b. What type is needed to illuminate the area surfaces (for transient adaptation, for avoiding reflected glare)?
c. What should the luminaire look like and how should it be mounted (see previous design approach)?
d. What is the area atmosphere and therefore the type of maintenance characteristics needed?
e. What are the economics of the lighting system?

4. *Calculation, Layout, and Evaluation.*

a. What layout of luminaires will portray the task best (illumination level, direction of illumination, veiling reflections, disability glare)?
b. What layout will be most comfortable (visually—direct and reflected glare, thermally)?
c. What layout will be most pleasing esthetically?

TYPES OF LIGHTING SYSTEMS

Lighting systems are often classified in accordance with their layout or location with respect to the visual task or object lighted—general lighting, localized general lighting and local (supplementary) lighting. They are also classified in accordance with the CIE* type of luminaire used—direct, semi-direct, general diffuse (direct-indirect), semi-indirect, and indirect. See Fig. 10-10.

Classification by Layout and Location

General Lighting. Lighting systems which provide an approximately uniform level of illumination on the work-plane over the entire area are called general lighting systems. The luminaires are usually arranged in a symmetrical plan fitted into the physical characteristics of the area. See Fig. 10-8 (left). Such installations thus blend well with the room architecture. They are relatively simple to install and require no coordination with furniture or machinery that may not be in place at the time of the installation. Perhaps the greatest advantage of general lighting systems is that they permit complete flexibility in task location.

Localized General Lighting. A localized general lighting system consists of a functional arrangement of luminaires with respect to the visual task or work areas. See Fig. 10-8 (right) where the recessed low brightness ceiling unit illuminates the desk. It also provides illumination for the entire room area. Such a lighting system requires special coordination in installation and careful consideration to ensure adequate general lighting for the room. This system has the advantages of better utilization of the light on the work area and the opportunity to locate the luminaires so that annoying shadows and direct and reflected glare are prevented.

Local Lighting. A local lighting system provides lighting only over a relatively small area occupied by the task and its immediate surround. The illumination may be from luminaires mounted near the task or from remote spotlights (portable table lamp and wall washing units in Fig. 10-8 (right). It is an economical means of providing higher illumination levels over a small area, and it usually permits some adjustment of the lighting to suit the requirements of the individual. Improper adjustments may, however, cause annoying glare for nearby workers. Local lighting, by itself, is seldom desirable. To prevent excessive changes in adaptation, it should be used in conjunction with general lighting that is at least 20 per cent of the local lighting level; it then becomes *supplementary lighting.*

Classification by CIE Luminaire Type

Direct Lighting. When luminaires direct 90 to 100 per cent of their output downward, they form a direct lighting system. The distribution may vary from wide spread to highly concentrating depending on the reflector material, finish, and contour and on the shielding or control media employed.

Direct lighting units can have the highest utilization of all types, but this utilization may be reduced in varying degrees by brightness control media required to minimize direct glare. Direct glare may also be reduced by using large area units with minimum lampings, *e.g.*, a two-foot-wide fluorescent unit with just two lamps.

Veiling reflections may be excessive unless distribution of light is designed to reduce the effect.

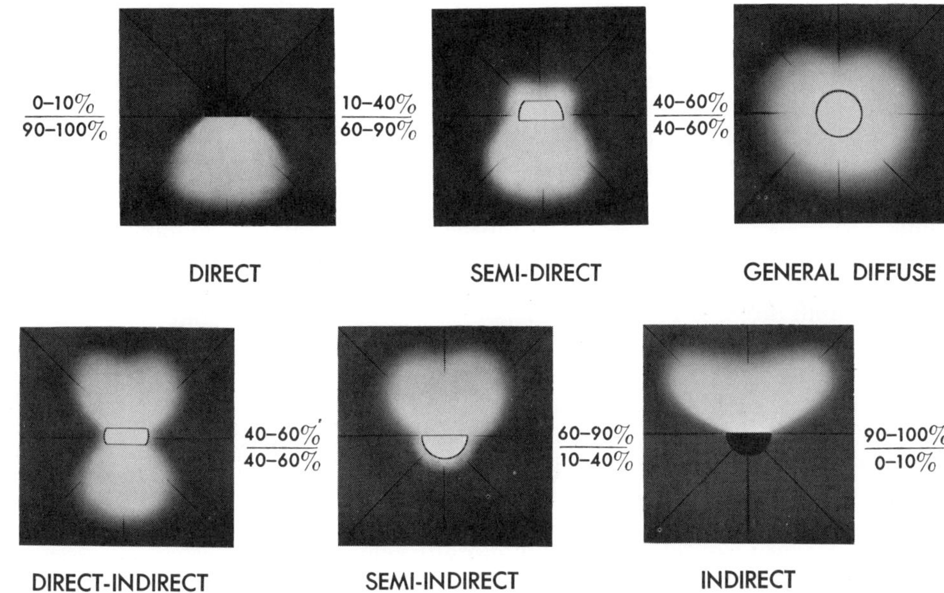

Fig. 10-10. Luminaires for general lighting are classified by the CIE* in accordance with the percentages of total luminaire output emitted above and below horizontal. The light distribution curves may take many forms within the limits of upward and downward distribution, depending on the type of light source and the design of the luminaire.

* International Commission on Illumination.

Reflected glare and shadows may be problems with direct lighting unless close spacings are employed. Large area units are also advantageous in this respect.

High reflectance room surfaces are particularly important with direct lighting to improve brightness relationships, and higher illumination levels provided by controlled brightness equipment will also tend to improve the brightness relationships throughout the room. With very concentrating distributions, care should be taken to ensure adequate wall luminances and illumination on vertical surfaces.

Luminous ceilings, louverall ceilings, and large-area modular lighting elements are forms of direct lighting having characteristics similar to those of indirect lighting discussed in later paragraphs below. These forms of lighting are frequently used to obtain the higher illumination levels, but care should be taken to limit the luminance of the shielding medium to 250 footlamberts or less to prevent direct glare if critical, prolonged seeing is involved. Reflected glare may be a problem with systems employing cellular louvers as the shielding medium since the images of the light sources above the louvers may be reflected by shiny surfaces at the work-plane.

Semi-Direct Lighting. The distribution from semi-direct units is predominantly downward (60 to 90 per cent) but with a small upward component to illuminate the ceiling and upper walls. The characteristics are essentially the same as for direct lighting except that the upward component will tend to soften shadows and improve room brightness relationships. Care should be exercised with close-to-ceiling mounting of some types to prevent overly bright ceilings directly above the luminaire. Utilization can approach, or even sometimes exceed, that of well-shielded direct units.

General Diffuse Lighting. When downward and upward components of light from luminaires are about equal (each 40 to 60 per cent of total luminaire output), the system is classified as general diffuse. **Direct-Indirect** is a special (non-CIE) category within the classification for luminaires which emit very little light at angles near the horizontal. Since this characteristic results in lower luminances in the direct-glare zone, direct-indirect luminaires are usually more suitable than general diffuse luminaires which distribute the light about equally in all directions.

General diffuse units combine the characteristics of direct lighting described above and those of indirect lighting described below. Utilization is somewhat lower than for direct or semi-direct units, but it is still quite good in rooms with high reflectance surfaces. Brightness relationships throughout the room are generally good, and shadows from the direct component are softened by the upward light reflected from the ceiling. Excellent direct glare control can be provided by well-shielded units, but short suspensions can result in ceiling luminances that exceed the luminaire luminances. Reflected glare from the downward component can be a problem, but it is mitigated by the reflected upward light; close spacings or layouts that locate units so that they are not reflected in the task result in further reductions.

Luminaires designed to provide a general-diffuse or direct-indirect distribution when pendant mounted are frequently installed on or very close to the ceiling. It should be recognized that such mountings change the distribution to direct or semi-direct since the ceiling acts as a top reflector redirecting the upward light back through the luminaire. Photometric data obtained with the luminaire equipped with top reflectors or installed on a simulated ceiling board should be employed to determine the luminaire characteristics for such application conditions.

Semi-Indirect Lighting. Lighting systems which emit 60 to 90 per cent of their output upward are defined as semi-indirect. The characteristics of semi-indirect lighting are similar to those of indirect systems discussed below except that the downward component usually produces a luminaire luminance that closely matches that of the ceiling. However, if the downward component becomes too great and is not properly controlled, direct or reflected glare may result. An increased downward component improves utilization of light somewhat over that of indirect lighting. This factor makes somewhat higher illumination levels possible with fewer semi-indirect luminaires and without excessive ceiling luminance.

Indirect Lighting. Lighting systems classified as indirect are those which direct 90 to 100 per cent of the light upward to the ceiling and upper side walls. In a well-designed installation the entire ceiling becomes the primary source of illumination, and shadows will be virtually eliminated. Also, since the luminaires direct very little light downward, both direct and reflected glare will be minimized if the installation is well planned. Luminaires whose luminance approximates that of the ceiling have some advantages in this respect. It is also important to suspend the luminaires a sufficient distance below the ceiling to obtain reasonable uniformity of ceiling luminance without excessive luminance immediately above the luminaires.

Since with indirect lighting the ceiling and upper walls must reflect light to the work-plane, it is essential that these surfaces have high reflectances. Even then, utilization is relatively low when compared to other systems. Care should also be exercised to prevent over-all ceiling luminance from becoming too high and thus glaring.

REFERENCES

1. Flynn, J. E. and Mills, S. M.: *Architectural Lighting Graphics*, Reinhold Publishing Corp., New York, 1962.
2. Kohler, W. and Luckhardt, W.: *Lighting in Architecture*, Reinhold Publishing Corp., New York, 1959.

SECTION
11

OFFICES AND SCHOOLS

Office Lighting 11–1 *School Lighting 11–10*

In offices and classrooms the room surfaces and lighting can be controlled to provide the proper visual environment for the performance of the critical and non-critical seeing tasks present. This Section deals specifically with the environment and lighting of offices and schools, but for general information on lighting systems, see Section 10, and for maintenance and air conditioning, see Section 16. For more on Daylighting, Libraries and Stages, and Gymnasiums, see Sections 7, 12, and 19 respectively.

OFFICE LIGHTING

Lighting in offices should provide people of varying visual ability with efficient, accurate, comfortable seeing in a pleasant environment.

Office Tasks

Modern office routines require seeing at close range. Severe visual tasks are almost universally encountered. Poor contrast in duplicated material, pencilled stenographic notes, multiple carbon copies, fine print, and handwriting, represent some of the typical visual problems in offices. All of these tasks should be made quickly and easily decipherable. The visibility of the details is determined by the *size* of the characters, *contrast* with the background, and the *luminance* (photometric brightness). These factors are interdependent. A deficiency in one may, within limits, be compensated for by augmenting one or more of the remaining factors. Another factor, *readability,* also influences the difficulty of the work.

Size. Both printing and writing vary considerably in size. As size increases, visibility increases. Up to a certain point, readability improves, seeing becomes easier, and worker output increases.

Type size ranges from this 6 point type, which is not desirable for continuous reading—even by persons with normal vision...

Up through this 8 point type which may be regarded as the practical minimum size, if visibility and ease of seeing are of any consequence...

To this 10 point type which is, fortunately, finding ever-increasing use for work where prolonged or continuous reading is expected.

Contrast. To be visible, each critical detail of a seeing task must differ in luminance, texture, or color with the background. Maximum visibility is obtained when the luminance contrast (and color contrast if present) of the detail with its background is greatest (see Section 3).

Visibility may be reduced with a lower reflectance background as shown in this paragraph where the background has been purposely decreased in reflectance while the reflectance of the printed text has been kept constant. Note that the lower reflectance background reduces the brightness.

Most office tasks generally considered as black ink on white paper, actually consist of gray print superimposed on a lighter gray background. Thin papers lacking opacity have a fairly low reflectance, will appear less bright, and will have a low luminance contrast with the characters indicated on them. Highly color-saturated or dark papers should not be used. Light tints of color are usually sufficient for identification of forms. Color bands along the edge or across the corner of a paper also can serve as identification of forms.

Fig. 11–1 illustrates some of the variations in contrast and legibility found in typical office seeing tasks. The poor quality of characters resulting from hard-pencil shorthand notes, worn typewriter ribbons, and poor carbon paper is illustrated here. Care should be taken to use highly contrasting ink, ribbons, carbons, and pencils, to produce highest contrast. Low gloss inks should be used instead of pencils when practicable to minimize losses in contrast due to veiling reflections.

Time. The time taken to decipher a given message and to take action on it, is an important measure of office efficiency. If, because of inadequate task luminance, poor size of detail to be assimilated, or poor contrast of the text with its background, it takes longer to perform the task, the efficiency of the office routine suffers. Consequently, operating costs rise and the attendant possibility of increased errors multiplies.

NOTE: References are listed at the end of each section.

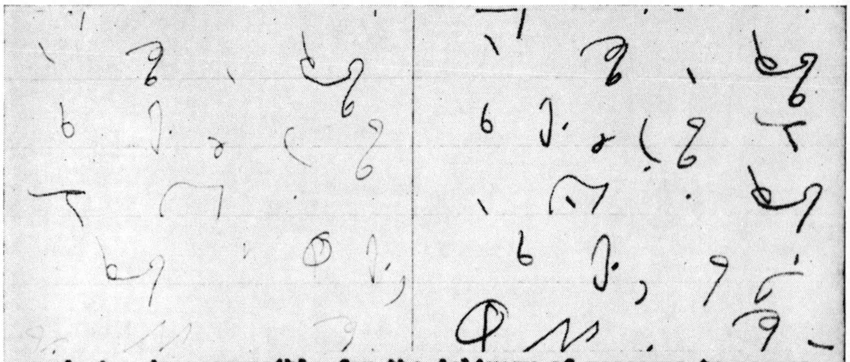

conductor is responsible for the delivery of messages to passengers on his train. If he is unable to locate a passenger on the train, he will return the message to the telegraph company at the next station. The telegraph company will then notify the sender

conductor is responsible for the delivery of messages to passengers on his train. If he is unable to locate a passenger on the train, he will return the message to the telegraph company at the next station. The telegraph company will then notify the sender

Fig. 11-1. Typical office seeing tasks. Shorthand: left, with No. 3 pencil; right, with No. 2 pencil. Typed copy (electric): upper, fifth copy from used carbon; lower, original from slightly worn ribbon.

Luminance. Research in vision has shown the relationship between contrast and luminance (see Fig. 3–24), size and luminance (see Fig. 3–13), and speed of vision and luminance (see Fig. 3–14). The common controllable factor in all of these relationships is the luminance of the task. The illumination on the task multiplied by the task reflectance determines this luminance.

Readability. Most office work involves visual interpretation of printed words, numbers, or characters. Readability depends on size, contrast, texture, color, type face of printing, legibility of writing, physical layout of the task—such as length of lines, spacing between lines, margin width, and illustration legibility. Individual rules, lines, guides, illustrations, and text having varying degrees of legibility combine to make a form or an office task meaningful. The coordination of all elements contributes immeasurably to its readability.

Influence of Lighting

Most office work requires the use of eyes focused at short distances (often about 14 inches). Such use may be intermittent or prolonged. Youthful or older eyes, and those with perfect or impaired vision may be called on to perform the seeing and interpretive tasks. In order to see the tasks quickly, accurately, and efficiently, there must be a sufficient quantity of illumination. The illumination must also be of a quality which will aid and not distract or interfere with seeing. It should help create a comfortable environment that is esthetically pleasing.

Quantity of Illumination

The illumination levels shown in Fig. 9–80 are based on conservative interpretations of presently available laboratory data. They are graded according to the difficulty of the visual tasks encountered in offices and drafting rooms and represent levels that are practical from the standpoint of lighting techniques and economics. It should be noted that some of the values shown in Fig. 9–80 are indicated as equivalent sphere illumination. See the general notes for Fig. 9–80 and Section 3 for the principles involved.

The illumination values specified should be provided *on the work*. The lighting system should be designed so that illumination will not fall below the specified level. Regular cleaning of luminaires and all room surfaces, and group relamping or prompt replacement of burned-out lamps, will help to maintain the necessary level with the minimum total investment.

Usual methods of illumination calculations are for horizontal surfaces; however, much office work is done on slanting or vertical surfaces, and illumination on these surfaces may be only one-half to one-third that on horizontal surfaces. To obtain the specified recommended illumination levels on slanting and vertical work surfaces, use should be made of point illumination calculation methods (see Section 9). The lighting system also should be designed to provide the required illumination on tasks located at the sides of the room where the illumination is otherwise apt to be considerably less than the average calculated level. For equivalent sphere illumination determinations, see Section 3.

Quality of Illumination

Through research and field experience principles of lighting quality have been established which can provide an efficient, comfortable, and pleasing visual environment. The concept envisions the tasks as the focal points of design, with the appropriate surroundings contributing to the efficiency of seeing the task, the elimination of hindering glare, and the positive feeling of pleasantness in the interior.

Of first consideration is the luminance relationship of the various surfaces in the visual field. In order to see immediately anywhere in the visual environment the luminance ratios of *appreciable* areas should be kept low, preferably not more than a change of 3 to 1. Luminaires, unless they are large, occupy less than significant areas for adaptation, and therefore discomfort glare criteria apply.

Of second consideration, but of equal importance, is the downward component of light in the desk viewing zone, which causes serious visibility loss of the details in handwritten, printed, and duplicated material. This light in the zone between the vertical and 40 degrees out from the vertical should be reduced to a minimum, and most of the light for the task should come from beyond the offending zone.

Overly bright luminaires should be avoided from the viewpoint of discomfort glare, which involuntarily inhibits motivation to perform the visual tasks. Discomfort has been found by recent research to be caused by the strain in the muscle that closes down the pupil of the eye to protect it from the overly bright light sources in the field of view, electric or daylighting.

A positive contribution to the pleasantness of any room interior is the proper selection of color combinations. These should be selected in consultation with the architect or interior designer, but should have the reflectances necessary to provide the proper luminance ratios for efficient seeing. (See Section 5, Color.) Research in the field of preferred color combinations has provided psychological ratings of various color combinations under five different light sources.

In applying the principles to field application the following is recommended:

1. Any office lighting system should be designed in such a manner that it will contribute to a cheerful, friendly, and esthetically pleasing environment. The elimination of glare and the use of tasteful color combinations should be an important component of it.

2. Under optimum conditions for visual efficiency and comfort the task luminance should be equal to, or not greater than, 3 times the luminances of the entire visual environment.

3. In an office the luminance of any appreciable area viewed from the horizontal to 40 degrees above the horizontal, from any normal standing or sitting position, should not differ excessively from the luminance of the visual task. As higher or lower luminances of surfaces in the visual field approach the luminance of the task, visual comfort and performance increase. Present research indicates that for the best results the highest acceptable luminance of any significant surface in the visual field should not be greater than 5 times the luminance of the task. The lowest acceptable luminance of any significant surface in the visual field should not be less than one-third the luminance of the task. See Fig. 11-2.

4. The luminance of surfaces immediately adjacent to the visual task is more critical in terms of visual comfort and performance than that of more remote surfaces in the visual surround. These adjacent surfaces have lower acceptable luminance limits than surfaces farther removed from the task. Present research indicates that surfaces immediately adjacent to the visual task should not exceed the luminance of the task and should be at least one-third the luminance of the task.

Fig. 11-2. Recommended Luminance Ratios

To achieve a comfortable balance in the office, it is desirable and practical to limit luminance ratios between areas of appreciable size from normal viewpoints as follows:

1 to 1/3 Between task and adjacent surroundings
1 to 1/5 Between task and more remote darker surfaces
1 to 5 Between task and more remote lighter surfaces

These ratios are recommended as maximums; reductions are generally beneficial.

5. The *luminance*-difference between adjacent surfaces in the visual surround should be reduced to an acceptable minimum. See (3) above.

6. The distribution of any lighting system, together with wall reflectances, should minimize the light in the downward zone from 0 to 40 degrees to reduce serious loss of visibility due to veiling reflections, and in the direct discomfort glare zone there should be a suitable limitation for eliminating glare.

7. Daylight and electric lighting systems should conform to the same *luminance* and *luminance*-difference principles, and both systems should be coordinated in design to assure the effective contribution of both.

Visual Comfort. Visual comfort may occur when no excessively high luminances are within the office worker's visual field. High luminances can also distract and can even reduce visibility.

Windows without luminance control are a frequent cause of visual discomfort and disability glare. They usually permit a direct view of the sky and the brightness of adjacent buildings. This can result in large areas of very high luminance in the usual fields of view. Direct sunlight may produce areas of excessively high luminance within the room itself. These conditions should be controlled by a proper combination of reduced-transmission glass, shades, louver or baffle systems on windows, or the interior luminance should be increased to bring the luminance ratios into a comfortable range.

Luminaires which have luminances that are too high for the environment in which they are located will produce discomfort.

Evaluation of Visual Discomfort. Both maximum and average luminances of a potential glare source are significant factors in the control of glare, but average luminances are recognized as being the more pertinent. A rating system based on the degree of freedom from discomfort glare in a proposed lighting installation, called Visual Comfort Probability (VCP), uses these luminances. See page 3-24.

This evaluation of comfort is based on the following factors which influence subjective judgments of visual comfort: room size and shape; room surface reflectances; illumination levels; luminaire type, size, luminance, maximum luminance, and light distribution; number of luminaires; luminance of the field of view; observer location and line of sight; and differences in individual glare sensitivity. Since each of these factors can vary considerably, a standard set of conditions has been established and used as a basis for Visual Comfort Probability (VCP) tables.

Direct discomfort glare will not be considered a problem in office lighting installations if all three of the conditions shown on page 3–26 are satisfied.

An alternate simplified method of providing an acceptable degree of comfort (see page 3–26) has been derived from the formulas for discomfort glare. This simplified method is based on the premise that luminaire designers do not design different units for rooms of different sizes, but consider the probable range of room sizes and design for the "commonly found more difficult" potential glare situation. (In rooms less than 20 feet in length and width, the luminaires are largely out of the field of view.) Accordingly a room size of 60 by 60 by 10 feet with 100 footcandles on the task has been selected as representative of general offices. Luminaires designed for acceptable visual comfort in these large spaces will be more comfortable in smaller spaces, and if those smaller spaces again become larger spaces the protection is still there.

In the method a factor $\bar{L}$ of 320 footlamberts represents a VCP of 70 when the room surface reflectances are in accordance with those in Fig. 11–5. Units of lower luminance will have greater comfort and therefore higher VCP.

Reflected Glare. Reflected glare may occur when a bright area of a luminaire or window is reflected toward the eye by a shiny surface. Fig. 11–3 shows an example of reflected glare in the area immediately surrounding the task location (task not shown). Such reflections are emphasized by surfaces that have a low diffuse reflectance and a high specular reflectance. They are annoying and distracting since the attention of the eyes is involuntarily drawn to them.

Reflected glare can be minimized by using high reflectance matte surfaces as in Fig. 11–4. When specular surfaces cannot be avoided, large-area low-luminance luminaires are often used.

Veiling Reflections. Substantial losses in task contrast, hence in visibility and visual performance, can result when light sources are reflected in even such subtly specular (shiny) visual tasks as stenographic notes in pencil and typing on bond paper. The apparent "veil" that is cast over a task when a light source is reflected in it may be so subtle as to be undetectable by the eye. For a discussion of the factors causing contrast lost from veiling reflections, methods of evaluation, the visual significance of the losses, and guidelines for reducing the effect of veiling reflections, see page 3–28.

Illumination Uniformity. Since it is usually not possible to predict the locations or changes in locations of visual tasks in offices, it is generally desirable to provide a degree of illumination uniformity that will ensure adequate illumination on the task regardless of its location. A discussion of acceptable uniformity and recommendations for achieving it are given in Section 9, page 9–33.

In private or semi-private offices, with fixed locations of furniture and visual tasks, it is often possible and desirable to design a functional lighting system with the luminaires located for maximum illumination on the work area and minimum direct and reflected glare. Illumination need not be uniform over the entire area, but changes should be gradual. Care should also be taken to be certain that the complete visual environment is adequately

Fig. 11-3. Streaks of light are reflected images of two continuous rows of luminaires. Very severe glare condition is produced by top which is both dark and polished.

Fig. 11-4. The reflected glare from luminaires disappears when a piece of light diffuse linoleum is placed over the dark, polished desk top.

lighted to obtain luminance ratios within recommended limits.

Diffusion. Adequate diffusion is necessary to prevent shadows at or near the task that will reduce task luminance and which may be annoying or confusing. Diffuse illumination that softens or minimizes shadows can be obtained from ceiling area lighting, indirect lighting, or closely spaced equipment having predominantly direct components. High reflectance matte finishes on room surfaces and equipment also help reduce shadows by reflecting and diffusing light into shadowed areas.

Fig. 11-5. Recommended Surface Reflectances for Offices

Surface	Reflectance Equivalent Range
Ceiling finishes*	80-90%
Walls	40-60%
Furniture	25-45%
Office machines and equipment	25-45%
Floors	20-40%

* Recommended reflectances are for finish only. Over-all average reflectance of acoustic materials may be somewhat lower.

Influence of Environmental Factors

Comfortable and efficient vision is the result of the careful selection of all design elements in a room. These include: lighting equipment; surface finishes, both room and furniture; upholsteries; and decorative accents.

Room Finishes. The large masses of color, ceilings, walls, and floors, are the most important factors in determining luminance ratios between task and surroundings. Matte surfaces, which reduce specular reflections and which have reflectance values as recommended, are necessary to develop a comfortable visual environment.

The almost universal use of pre-finished acoustic materials for office ceilings has fairly well standardized ceiling reflectances at a high value. Carrying the light ceiling color down the walls to the level of the luminaires can improve utilization 10 per cent or even more while enhancing the appearance of high ceiling space. Acoustic material is sometimes used for this purpose for better sound control.

Walls should be considered as secondary light sources and finish reflectances maintained in the 40 to 60 per cent range. Small sections of walls may have reflectances higher or lower than the extremes permitted by the range given in Fig. 11-5. If these areas are thought of as accents and restricted to about 10 per cent of the total wall area in the room, little harm will be done to the efficiency of the lighting system or to the significant environmental luminance ratios, while the environment can be made to look considerably more pleasant and interesting. Entire walls having reflectances higher or lower than the extremes permitted by the range given in Fig. 11-5 should be used only with discretion and by an experienced designer who recognizes the precautions that must be taken. For more dramatic effect in the exceptional uses of color and reflectance values, the advice of an experienced designer should be sought.

Window shielding media can be treated as wall areas with reflectance comparable to wall areas, or handled as accents, as is often done.

In the consideration of lighting as one of the major decorative elements in an office, many times it is advantageous to vary illumination levels from area to area.

Furniture Finishes. No surface in the office is more important visually than the desk top, for the top occupies most of the visual field of the desk user. Because the task is usually of high reflectance, desk tops should meet recommended reflectance values to hold luminance ratios to a minimum. See Fig. 11-6. Office furniture should have a finish which harmonizes with the environment or provides an interesting color complement to it.

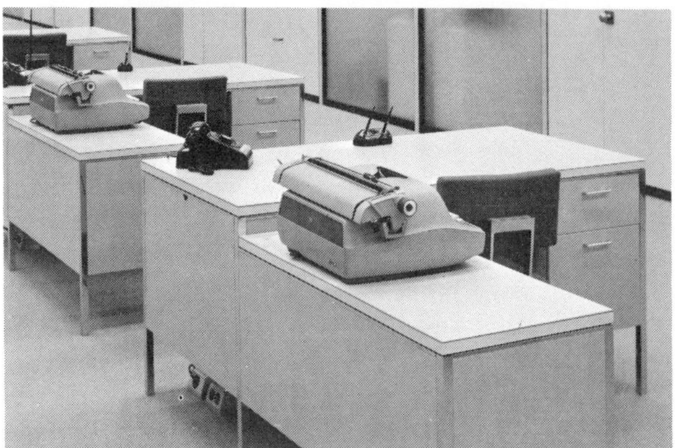

Fig. 11-6. Dark desk tops (upper) have reflectances of only four to seven per cent. The resulting excessive luminance ratios between the work and the desk top (frequently greater than 1 to $\frac{1}{10}$) create an uncomfortable seeing condition. Light desk tops (lower) provide luminance ratios between the work and these surfaces in the order of two to one. Note also light finishes on vertical surfaces.

Fig. 11-7. Desk tops and vertical surfaces of files often constitute major portions of the field of view and therefore should have recommended reflectances.

Filing cabinets and desks constitute major masses of color in an area (see Fig. 11-7) and, unless the designer has planned to use them as accent pieces, finished surfaces of these objects should have reflectances as recommended in Fig. 11-5, or their area should be limited to about 10 per cent of the visual field.

Office chair frames usually constitute unimportant small broken areas but chair upholstery and frames may offer the designer opportunity for the interesting use of accent colors and textures.

Machine Finishes. Business machines rank with desk tops in visual importance. They occupy a central and dominating part of the visual field. Their finishes therefore should meet values specified in Fig. 11-5. Shiny surfaces, trim, and dial covers should be eliminated for a small amount of specular trim can be annoying, even though it may not be viewed directly.

Color. Much has been written on the psychology of color in the office, and colors do have certain psychological and even physiological effects. However, today the major factor in the selection of office colors is style. Trends are toward light neutral colors which can be used with bright accents in smaller pieces for an over-all cohesive scheme. Somewhat deeper hues are usually, but not always, used in flooring and trim.

Newer materials, such as plastic wall coverings, can provide non-specular surfaces in interesting colors and textures. Planters, pictures, wall plaques, and bright upholstery provide color accents to neutral wall treatments.

Architects, lighting engineers, and interior designers should cooperate to develop integrated decorative schemes in which each element interrelates and complements the other. Thus lighting specifications developed in coordination with color specifications will provide a guide for the initial installation and also for subsequent redecoration. In executive offices which are frequently developed about the personality of the occupant, it is particularly important for designer, architect, and lighting engineer to work in close cooperation with each other.

Daylighting Systems for Offices

Fenestration areas in offices should be designed for admission, control, and distribution of daylight for comfortable seeing and to provide a vision strip so that important eye muscles may be relaxed by focusing on distant objects.

Information on the design of daylighting systems is given in Sections 7 and 9. Many of the fundamental principles explained pertain to office lighting. Brightness control is particularly important and should be provided in all fenestration design. The fenestration system should be developed to suit the orientation, variations in topography, and landscaping related to each exterior wall.

Electric Lighting Systems for Offices

Types of Electric Lighting Systems. All interior lighting systems are included in one of the following five classifications:

1. Indirect
2. Semi-indirect
3. General diffuse and direct-indirect
4. Semi-direct
5. Direct and ceiling area lighting

A complete description of these five classifications is given in Section 10.

Before the advent of central air conditioning in offices, building codes required high ceilings to allow sufficient ventilation by convection. Types 1, 2, and 3 above were generally employed because they were best suited to such older building practices. Today, the trend is toward maximum utilization of building cubage. As ceiling heights are often 8 to 9 feet, luminaires are surface-mounted or recessed of types 4 and 5. Moreover, in modernizing older structures, suspended ceilings are usually employed not only to conceal the old structure, ducts, and pipes but also to reduce the volume to be cooled or heated. Thus, for economic building construction and operation, either for new construction or modernization work, there is a minimum of space between the ceiling and the floor. Fortunately, advances in luminaire design make it possible to achieve current lighting recommendations for quantity and quality even under these restrictive conditions.

Two other factors affect the selection of luminaire types. One is the use of modular construction, in which the building dimensions are exact multiples of a basic modular unit. Thus, the lighting system necessarily must conform to the modular arrangement. The other factor is the inclusion of "air handling" as part of the luminaire, using slots in the sides or ends of the luminaire to supply cool or heated air or for returning room air to the plenum or exhaust ducts or for both supply and return. Luminaires are also designed to return room air through the lamp compartment. Using this principle of heat transfer the lamps operate more efficiently and a large proportion of the heat from the lighting is prevented from entering the occupied space. See also Section 16 under Lighting and Air Conditioning.

Thus, all three factors, namely, low ceiling height, modular construction, and air handling techniques favor the use of recessed lighting for modern offices. Where ceiling heights permit, types 1 through 4 above may be employed to meet decorative requirements or individual preferences.

Private Offices

Private offices are generally small to moderate in size. Any of the five lighting systems discussed above will find application, as ceiling heights and building construction permit.

If the arrangement and orientation of furniture in private offices is fixed, the luminaires may be positioned to minimize both direct and reflected glare and veiling reflections. See Figs. 10–8 (right) and 11–8. Flexibility and other requirements will frequently necessitate the use of general lighting.

The illumination required in private offices varies from that necessary for casual seeing during interviews to higher values for prolonged, difficult seeing tasks. When possible, the lighting system should be sufficiently flexible to provide the degree of illumination needed for the diversity of the visual tasks encountered. This may include dimming of general lighting plus environmental lighting elements, such as wall and accent lighting.

General Offices

Most general offices are moderate to large sized areas in which it can be assumed that many luminaires will be located directly in the field of view. In these areas it is particularly important to avoid luminaires that will produce uncomfortable direct glare either by virtue of their own characteristics or by high adjacent ceiling luminance.

Since many office operations require more than one orientation of personnel, lighting systems should be designed so that lighting and seeing are comfortable for all angles of view. General lighting systems originally designed to be comfortable from all directions need not be modified with changes in office procedure or organization. When it is known in advance that personnel will face in a certain direction and if the intended lighting system has a marked advantage in one viewing direction, the installation should be oriented to take advantage of it. Often there can be little orientation of the lighting systems in relation to the desk locations, hence special care should be taken to minimize the serious loss of visibility due to veiling reflections.

It is possible that large offices will be subdivided in the future, and effort should be made to ascertain in advance the smallest subdivided area so that an allowance can be made for an adequate level of illumination in the future small areas (the utilization of light in a small area is less than in a larger one for a given lighting system).

This planning requires close cooperation between the architect and the illuminating engineer to insure that the ceiling and lighting modules and partition placement are compatible and upon later subdivision, that the lighting system will still produce adequate levels of illumination.

Drafting Rooms

Drafting involves accurate discrimination of fine details, frequently over long periods of time and, therefore, requires higher levels of high quality illumination. The contrast between the work detail and its background may be very poor, as for example, when working with faint prints or hard-pencil drawings. Specular drawing equipment, such as polished T-squares, plastic triangles, and scales are potential sources of reflected glare.

Reflections. On nearly horizontal boards, any ceiling or luminaire brightness may be reflected by the work into the eyes of the draftsman. Such reflections may sometimes be eliminated by proper positioning of the board surface with respect to the direction of the light flux. It is usually more practical to minimize the effect of these reflections by limiting the relative proportion of luminous flux in the usual range of reflected glare angles. This may be accomplished by careful placement of the drafting tables with respect to the luminaires so that their reflection is away from the observer. Indirect, semi-indirect, or ceiling area lighting can be employed if recommended lighting levels for drafting (see Fig. 9–80) can be achieved. See Fig. 11–9.

Fig. 11–8. Peripheral lighting in a private office provides light both for working and conference purposes. Arrangement of luminaires helps minimize veiling reflections on the desk surface.

Fig. 11-9. A low-luminance ceiling area lighting system provides the illumination in this drafting room.

Fig. 11-10. Tracing by transillumination—a method suitable for areas where it is not practical or feasible to obtain adequate levels of general drafting room lighting.

Another way of combating reflections is to tilt the board toward a nearly vertical position so that specular reflections will be cast below the line of sight. In addition, a nearly vertical board may be high enough to shield the eyes of the draftsman from luminaire brightness otherwise in the field of view.

Shadows. Over-all ceiling lighting, indirect or semi-indirect lighting will minimize line shadows cast by the edges of T-squares or triangles, particularly if the boards are near horizontal. However, in modern low-ceiling installations, such systems may not be feasible. Direct lighting systems designed to meet drafting room recommended levels should not cause sharp shadows due to the close spacing required to obtain these levels. The frequency of occurrence of shadows can be minimized by orienting the drafting tables to an angle of 15 to 20 degrees between the long axes of the luminaires and any of the major straightedge positions. The use of vertical drafting boards also virtually eliminates shadows.

Transillumination. A drafting table with a diffusing surface panel illuminated from behind to a luminance of 300 footlamberts is recommended for difficult tracing. The desirable luminance of the glass depends on the nature of the work and the level of illumination from above. To avoid direct glare, the draftsman should use opaque paper to cover the portion of the glass which is not concealed by the drawing, and dimming should be provided for the lamps. See Fig. 11–10.

Filing Facilities

Filing presents a variety of tasks in planes that range from horizontal to vertical. Seeing tasks in files vary from those of good contrast, black type on white paper or cards, to pencil or ink on colored stock, poor carbon copy on low-grade paper, and finely printed cards. It is also common to use a transparent covering over active file cards or the tabs of file folders to keep them clean and legible. This mirrorlike surface reflects light sources and the veiling reflections reduce visibility.

It is significant that the illumination levels listed in Fig. 9–80 for office filing are the recommended levels on the task whether horizontal, inclined, or vertical and that these levels vary over a 3 to 1 range for active, intermittent, and inactive filing. Because active file areas require a higher level of illumination, a corresponding comfortable lighting environment should also be sought by minimizing direct glare and veiling reflections.

General illumination using indirect type luminaires, large area direct low luminance luminaires, or general diffusing type luminaires with direct component, having maximum output in angles approaching 45 degrees, will provide maximum vertical illumination. The latter, however, may increase reflected glare, and therefore, it is important to orient and position the luminaires properly to direct these reflections away from the workers' eyes.

Active file rooms having rows of files with aisles between may be most economically lighted by localized general lighting or by luminaires specially designed for the purpose. Intermittent or inactive file areas will be adequately covered by general illumination as described above. In all cases, the reflectance of the files and the floor should be high to help illuminate the lower portion of files by reflected light from the floor.

Mail Rooms

The levels of illumination recommended for office work in Fig. 9–80 apply. Where there is rapid sorting, the level of illumination given for regular office work will be required. When speed is not a factor and the mail room activity is only for a short time, a lower level of illumination may be used.

Conference and Board Rooms

The diversity of work performed in conference and board rooms suggests that the lighting should be flexible. Since meetings may extend over many hours the entire environment should be comfortable and pleasant, with general illumination provided for regular office work. Dimming or switching arrangements, installed to vary illumination levels down to the point where slides or movies may be shown, permit increased usage of the room.

Coordination with the architect and interior designer is necessary to make lighting an important part of the entire concept of the room. Richly paneled walls, murals, and wallpaper can be made more attractive by being "washed" with light. At the same time this "washing" technique will improve luminance ratios. Supplementary lighting should also be provided for charts, maps, and photographs. A change in atmosphere can be accomplished by raising or lowering the level of illumination with dimmers or circuit switching arrangements. Concealed indirect lighting, combined with direct lighting and wall lighting provides flexibility to create a changing environment as well as eye comfort for seeing tasks.

Reception Rooms

Lighting plays an important part in a reception area in creating the desired image of the company or institution. In a manner similar to the conference or board room, this area requires coordination with the architect or designer to utilize light to create an atmosphere. The receptionist at times may perform office work or serve as the switchboard operator. When the receptionist is called upon for these office duties, the recommended level of illumination for offices should be provided.

Corridors and Hallways

To attain proper adaptation levels, one-third to one-fifth of the illumination level found in adjacent offices should be provided as a minimum in corridors and hallways. These levels will also promote safety for people in motion. The illumination should be designed with consideration as to the width and length of the hallway. With well planned lighting and color treatment, tunnel effects can be reduced, narrow halls can be made to appear wider, and the perspective of long halls can be broken to make them appear shorter. A switching arrangement should be provided so that a sufficient number of lamps are left on at all times for ingress and egress unless continuous emergency lighting is provided.

Stairways, Escalators, and Elevators

The need for good lighting at these points is paramount in the interest of safety. Installations should be so designed that failure of one lamp will not leave an area inadequately lighted. Floor, wall, and overhead finishes should be of high reflectance to gain the maximum utilization of the light and to reduce shadows.

Stairways. Lighting equipment should be located so that people do not cast shadows across the steps. The location and use of shielded luminaires are important in eliminating the direct glare that might reduce the visibility of the steps. The use of small individual shielded lights at the tread level may be used to emphasize each step. Handrail lighting can also be used. A contrasting paint treatment of the stair treads and risers will emphasize the change in elevation.

Elevators and Escalators. Safety is a major consideration. Lighting is required to call attention to any difference between the elevator and the floor levels and to the moving tread of the escalator.

Personal Service

Sanitary maintenance of restrooms and washrooms is considerably improved by sufficient illumination throughout the areas. See Fig. 9–80 for recommended levels.

Mirror lighting and similar residential treatments improve the appearance and usefulness of these personal service rooms. See Section 15.

Supplementary Lighting

Supplementary lighting is used to provide an additional quantity of illumination or a quality of illumination which cannot readily or economically be obtained by a general lighting system. It is used to supplement the general lighting level for specific work requirements.

Supplementary equipment should be properly shielded to minimize glare for the user and his associates, and the design of supplementary and general lighting should be carefully coordinated to maintain recommended luminance ratios. For esthetic reasons, except for temporary or isolated tasks in a general area, the general lighting system should be designed to properly light the specific work requirements of the area.

Business Machines and Computer Rooms

Care should be taken in lighting for business machines and computers to evaluate the great variety of tasks that are performed. Whenever possible, lighting should be provided to suit the most critical task. Seeing accuracy is of paramount importance because mistakes can become costly. See Fig. 11–11.

Business Machines. Many types of equipment fall into the general category of business machines.

Fig. 11-11. Recessed 2-foot by 4-foot troffers with air return sides provide general illumination in this computer room. Located adjacent to a main reception area, the room is illuminated for both operation and display.

In general, the levels of illumination recommended in Fig. 9-80 will provide sufficient quantity. However, care should be taken to avoid veiling reflections on transparent covers over the readout areas and on the surfaces of keys on equipment such as desk top calculators. Large area, low luminance luminaires will help to minimize this problem, or locating luminaires properly with respect to the machine can eliminate the problem.

The reading task on machines with self-luminous presentations such as microfilm projectors and equipment with cathode ray tube projection screens, etc., may present some difficulty if these units are installed in areas having higher levels of illumination. Provision should be made to enable the illumination to be reduced by switching or dimming. Veiling reflections can be reduced by reorienting or shielding the projector so that the reflected images of luminaires or windows are not seen on the screens.

Computer Rooms. Computer operations can range from a time-sharing terminal up to a complex of several rooms including the machine room, key punch area and programming office. The time-sharing terminal and key-punch operators may have to read fine print or pencil writing, often on colored paper, while typing in data and reading data coming out, printed on white or colored paper. The programmer is usually working with printed forms and pencil writing.

In a large system, the key-punch room requires the levels specified in Fig. 9-80, but a lower level may be practical in the machine room. In this case, the readout data is usually analyzed in an accountant's or engineer's office. The machine room operator has a variety of short-term seeing tasks such as log book entries, reading labels on magnetic tapes, occasional reading of input cards, some typing, correcting malfunctions caused by jammed cards or tapes.

Some manufacturers of equipment require that a lower level of illumination be available when operators are servicing and making tests and adjustments where an oscilloscope is used for analysis. The lighting could be arranged so that by switching out circuits or dimming, the room lighting will not be too great for good visibility of the screen. Servicing within large computer cabinets will rquire supplementary lighting.

Care should be taken in the placing of luminaires in relation to equipment with high vertical information faces so that images of luminaires do not produce veiling reflections. Indicator lights and lighted pushbuttons will need to be sufficiently bright so that they can be easily seen with the general lighting on.

Maintenance

Proper maintenance of all elements which affect the over-all efficiency of a lighting system is essential to retention of the originally designed quality and effectiveness of the installation. The decrease in output of service accompanying use, and the accumulation of dirt and dust on lighting equipment, lamps, room surfaces, and windows can rob the effectiveness of a lighting system. A planned maintenance program will help assure that light is obtained at the minimum cost. See Section 16 for an explanation of the principles and recommended procedures for effective maintenance.

SCHOOL LIGHTING

The goal of school lighting is the creation of an effective total visual environment for learning. This should provide a level of illumination consistent with the demands of the visual tasks, and of such quality as to allow students and instructors to see comfortably and efficiently. The lighting system should provide direct aid to the educational process, supporting and enhancing the teaching aids in the environment, and indeed enhance the environment itself.

Visual Environment

Physically, the visual environment is a three-dimensional pattern of brightness and colors, visible to a person within the environment. It also includes emotional and esthetic values, less easily measured, but none the less important to the design. Much research has been carried out in the fields of light and vision. Though much more remains to be done, the data available permit recommendations to be made which school administrators, engineers, and architects can accept with confidence.

The basic considerations in the creation of the optimum visual environment for schools closely parallel, and are in many instances identical with, those which apply to offices.

Visual Tasks

Research results show that tasks of comparative size, with good contrast (for instance, 6- to 12-point printed type on good book stock) have adequate visibility at low levels of illumination; while those with poor contrast (such as handwork and/or badly reproduced material) need relativly higher levels than would be expected from the proportional decrease in contrast.

Educators, by providing better materials, may improve the visibility of tasks to be performed by students in the classroom, in the following ways:

1. Since pencil and ball-point pen are more glossy than ink writing, the use of low-gloss ink should be encouraged.

2. The minimum type size to be used in textbooks is considered to be 10-point Bodoni. Younger pupils should use a larger type size to make reading tasks easier. Also important are adequate spacing between lines of type and the use of low-gloss ink.

3. For both working paper and printed material, matte paper should be used in preference to glossy.

4. Paper should have a good degree of opaqueness and high reflectance.

5. The proper combination chalk, chalkboard, and illumination should be provided. Supplementary lighting is required for chalkboards—for light boards to compensate for reduced contrast and for dark boards to attain acceptable luminance (photometric brightness). White chalk has been found more visible than other light-colored chalks, such as yellow. Only a high grade of chalk with a minimum of clay filler should be used. Chalkboards should be kept clean and periodically restored to good condition or replaced. Teachers should be encouraged to use large letters and figures.

Of special importance is the quality of reproduced material. Since poor copies have been shown to require the higher illumination levels, it is worthwhile to consider the cost of quality duplication machinery and skilled operation against cost of increased illumination requirements.

In general, the visibility of tasks may be improved by properly designed additional illumination, and by providing tasks having maximum contrast and minimum specularity. Often this may be accomplished at a minimum of expense. All that is needed is an understanding of what constitutes a good task and a desire to provide one.

Quantity of Illumination

Research has been conducted to determine which tasks occupy more time in the classroom. In one study, 62 per cent of the total time was spent in visual tasks. Of this percentage, 64 per cent was spent on such tasks as reading, writing with pencil, and working with duplicated materials. This indicates that the greater part of a student's visual time was spent on tasks requiring 70 to 100 footcandles equivalent sphere illumination.

It is impractical to light a classroom by providing only the level needed for one given task. In a classroom, individual students may be performing different tasks in different locations and with different orientations. Also, the whole working area, as potential teaching space, should be provided with a satisfactory minimum level of illumination. The general lighting level should be designed for a commonly found, most difficult task.

Certain classrooms are used for special purposes. In these it is desirable, indeed necessary, to light for the specific task regardless of orientation and location. Where physical disabilities are encountered, special levels of illumination will be required. In school shops, sewing rooms, and art classrooms, lighting should be planned on a specific visual task basis. In some cases, supplementary lighting should be used to provide sufficient illumination for specific areas (*e.g.*, lighting chalkboards, carrels, etc.).

The most commonly occurring critical tasks in most classrooms are those that involve veiling reflections. It is therefore necessary that lighting systems for these learning areas be designed on the basis of equivalent sphere illumination. Previously used average illumination calculations do not consider the effects of veiling reflections and are therefore not entirely adequate for design purposes. Section 3 explains this principle and Fig. 9-80 in Section 9 shows illumination levels recommended for various school tasks and areas, indicating whether equivalent sphere illumination criteria applies.

Perfect general illumination uniformity is generally not feasible, but uniformity is considered acceptable if the maximum and minimum values in the work space of the room are not more than one-sixth above and below the average. (See Section 9.)

Quality of Lighting

The quality aspect of school lighting embraces those factors which render a visual task most effectively and which contribute to visual comfort. This is accomplished by a consideration of all factors affecting the seeing process such as the elimination of discomfort and transitional adaptation effects and the creation of pleasantness with proper visual emphasis.

Direct Glare. The lighting in a classroom should be free from direct glare. When the light sources, electric or daylight, become too bright they produce direct glare. Depending on the magnitude of luminance, glare sources become progressively distracting, uncomfortable, and may produce disability glare conditions where seeing is actually impaired. (See Section 3.) For visual comfort probability criteria refer to page 11-3.

Veiling Reflections. While control of direct glare is relatively simple and may be approached directly, practical methods for minimizing veiling reflections are less specific. See page 11-4. The presence of direct glare is usually obvious; a veiling reflection is much more subtle in effect and often requires instrumentation to determine its presence. The losses in visibility due to veiling reflections are real and measurable. The results of one study of veiling reflections are shown in Fig. 11–12. These indicate the losses in contrast for two illumination levels, and for tasks commonly found in the classroom. It will be easily seen that the losses for glossy background tasks are larger than those for matte background tasks, but even the latter are serious and hinder visual performance.

A recent study[29] indicates that there were similar losses in contrast in 18 schools surveyed. Of the 18 classrooms tested, 14 had a level of more than 70 footcandles measured in the position tested. Of these 14, only one had an equivalent sphere illumination level of 70. Four of the 18 rooms had a level of more than 100 footcandles and an equivalent sphere illumination level of less than 25, indicating the serious losses that can occur under common lighting systems. The study well illustrates that measured illumination levels alone cannot be used as a criterion for visual performance and that light incident on the task from wide angles outside the glare zone gives the greater visibility.

Reflected Glare. All surfaces in a classroom should be free from reflected glare whether produced by electric or daylight sources. Non-glossy furniture and room surface finishes are recommended to avoid the high brightness specular reflections causing reflected glare; however, where glossy finishes must be used, luminaire luminances should be low.

Fig. 11–12. Loss of Contrast Due to Veiling Reflections for Different Lighting Systems

Task*	Open Bottom Louver Direct System Per Cent Loss	General Diffuse System Per Cent Loss	Luminous Indirect System Per Cent Loss
55 Footcandles			
A-1	16.5	13.2	8.8
A-2	4.4	5.3	4.6
N-J	20.6	17.0	11.3
8-11	24.1	20.6	14.6
6-2	24.7	—	10.7
6-9	19.0	—	11.3
30 Footcandles			
A-1	16.6	17.6	9.0
A-2	8.2	7.7	7.2
N-J	21.6	20.2	10.8
8-11	30.1	23.6	15.7
6-2	25.3	19.4	9.9
6-9	22.9	21.3	10.8

* A-1 and A-2 are identical printed tasks except for the paper; A-1 is printed on glossy "coated book stock"; A-2 is printed on matte "offset book stock."

N-J, 8-11, 6-2 and 6-9 are pencil tasks using No. 2 pencil on ruled matte white paper having diffuse reflectance of 77 per cent.

Luminance Ratios. Of first importance is the concept that the luminance relationship of the various surfaces in the visual area must be kept within recommended limits. When the eye fixes on a task, an adaptation level is established. This adaptation level is a combination of task luminance and field luminance. As the eye changes from a field of one luminance (*i.e.* the book) to a field of another luminance (*i.e.* the chalkboard) the eye readapts to the new luminance level. If there is much difference between the two levels, there is immediately a considerable loss in ability to see and a period of time is required for the eye to adapt itself completely to the new situation. Furthermore, if the difference is too great, the reaction will be discomfort, attended by a transient change in pupillary opening. See Section 3. To avoid this, for large surface areas and wherever good visual performance is required, the difference in luminance should be kept within desirable limits. See Section 9 for calculation procedures.

Present research indicates that for good visual peformance, the luminance of any significant surface normally viewed directly should not be greater than 5 times the luminance of the task, nor less than one-third the luminance of the task. The luminance of surfaces immediately adjacent to the visual task is more critical in terms of visual comfort and performance than that of more remote surfaces in the visual surround. These adjacent surfaces have lower desirable luminance limits than surfaces farther removed from the task. Present research indicates that surfaces immediately adjacent to the visual task should not exceed the luminance of the task and should be at least one-third the luminance of the task for ideal design. The luminance-difference between adjacent surfaces in the visual surround should be reduced to an acceptable minimum.

The general approach in providing low luminance ratios over the entire visual field is to limit the luminaire and fenestration luminance (see Direct Glare above) and to build up the luminance of all other interior surfaces in the area (ceiling, walls, floor, work surfaces, equipment, etc.) by providing suitable reflectance and distribution of light. Fig. 11–13 shows recommended reflectances of surfaces in a classroom.

Lighting Systems

No one lighting system can be recommended exclusively. Each system has qualities which may meet the requirements for a given situation. Consideration should be given to whether or not the system chosen will meet the goal in school lighting—to allow students and instructors to see efficiently and without distraction—as well as further the architectural and decorative design of the school. See Section 10 for Electric Lighting and Section 7 for Daylighting.

The lamps most frequently used at present for school lighting, fluorescent sources, are employed

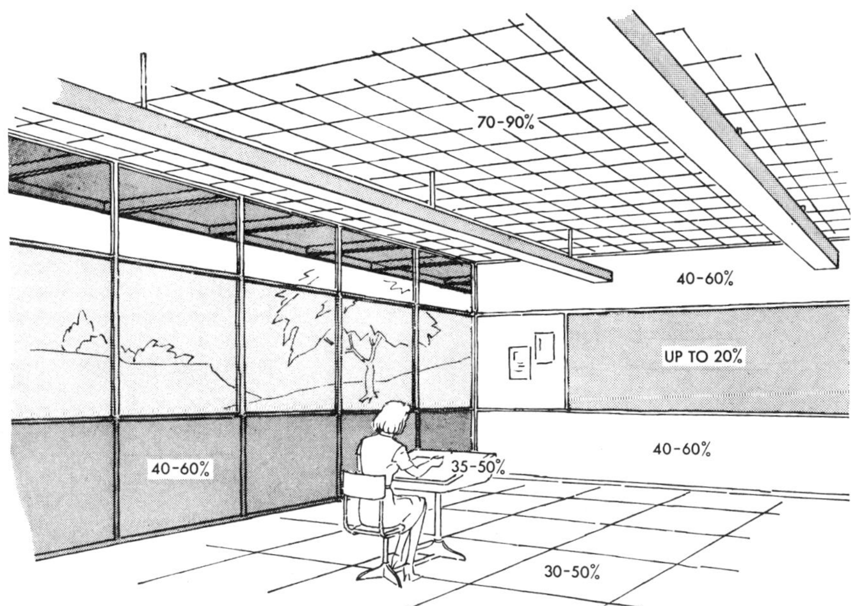

Fig. 11-13. Recommended reflectances for surfaces and furnishings in the classroom. (Note control media is used at windows to reduce exterior luminances so that they are in balance with interior luminances.)

not only for their high efficacy but also for the variety of colors available to enhance the visual environment. Cool white lamps, for instance, blend well with daylight; warm white lamps have a closer match with incandescent lamps than most other fluorescent sources. Other sources, such as high-intensity discharge and incandescent lamps may also be employed within luminaires which are properly designed for the lamp and which provide desired illumination characteristics.

Supplementary lighting should be provided where necessary for optimizing visibility and visual impact at chalkboards, demonstration areas, lecture areas, and separate study carrels. For example, in a classroom where complicated mathematical formulas are written on the chalkboard regularly, the chalkboard should be highlighted. Rooms used regularly for lecturing should have supplementary lighting for the lecturer. Demonstration areas should have special lighting to aid visibility and provide visual emphasis of the demonstration itself.

Specific Applications

Numerous areas in a school have design criteria with either more or less stringent requirements than classrooms. The criteria may put a greater emphasis on such problems as discrimination of color, three-dimensional modeling, and variation in brightness to introduce architectural interest or direct traffic. For such functions the lighting requirements dealing with luminance ratios and glare may be modified to meet these functions.

Art Rooms. Because the appearance of colors in an art room is paramount, the light source used should render the colors accurately. Lamps with high color rendering capability provide a more natural appearance of colors over a wide range even though efficacy may be lower (see Section 8).

Supplementary lighting from directional concentrating sources is useful on displays and models for improved visibility at a distance and for modeling purposes. Such lighting creates the desired highlights and shadows on art arrangements being painted or drawn. Supplementary lighting, such as from adjustable luminaires which can provide a definite directional light, is frequently desirable when students work on or study surfaces involving glaze or texture. An independent portable lighting system, separate from the general illumination system and with some degree of flexibility with respect to aiming and interchangeability of light sources, is often useful for exhibitions.

Auditoriums. The auditorium is made up of a seating area (house), and a stage area. It may serve as an assembly and lecture hall, study room, theatre, concert hall, audio-visual aid room (including television), and for many other activities. Many schools today are developing speech and drama programs, not only among students, but also among adult groups. The auditorium serves so many purposes it must be well planned and properly equipped to satisfy them all.

For general auditorium use, there is need for atmosphere lighting which can be incorporated into the architecture as ceiling or wall coves, luminous ceiling treatments or decorative suspended indirect luminaires. Facilities to provide more than one level or pattern of illumination are desirable since this area may serve several purposes.

The basic lighting for the seating area should be planned for audience assembly purposes, and at dramatic presentations for such intermission tasks as reading programs. The system should allow for the varying of lighting levels (by dimmers, step switching) to allow for transitions and for lower levels to be used in showing slides and motion pictures. If windows are used, means should be provided to darken them during projection.

Supplementary illumination is required if the seating area is used for study or testing purposes. Several circuits of this supplementary system, well distributed over the seating area and providing additional illumination for handwriting and reading, can be set up for independent switching. Special care should be given to avoiding specular reflection from any direct downlighting.

Aisle lights, with a low level of non-distracting illumination, should be provided. The aisle lights should not be visible to any of the seated audience. Because of the high luminance ratios in a semi-darkened room, they may prove irritating despite low surface luminance. If there are steps in the aisles, small and well-shielded step lights offer a safety aid. These should be located to prevent light spilling on the stage or into eyes of people going up the steps.

Particular attention should be given to local building codes regarding exit lights, panic controls, and so on. It is desirable that emergency lighting be installed which would be automatically operated and connected to an independent source of power.

For proper lighting facilities for the stage, see Theatres in Section 24.

Cafeterias and Kitchens. In eating areas the lighting should create a cheerful, comfortable area. Good lighting promotes cleanliness. Dining areas are frequently used for other activities and when so used the lighting should provide the levels of illumination recommended for the task. Where the appearance of food is of prime consideration, as in the cafeteria, pleasant colors and appropriate sources should be used. Additional incandescent lighting may be used over the serving counter to give the food eye appeal, provide heat, and to speed up the selection of food. Good general lighting is needed in the kitchen, especially at ranges, work tables and sinks, to assure cleanliness, safety and good housekeeping. For further information see page 12–4.

Carrels. Supplementary lighting should be provided for effective visibility in study carrels. The walls of carrels act as light baffles intercepting much of the general room lighting causing the carrel desk top to be considerably darker than a conventional open desk top. Without supplementary lighting, carrels are prone to having large shadows across sections of the work area. Carrels typically have electric power available for electronic teaching aides, and other electrically powered devices. Hence, adding local electric lighting is generally simple. Lighting equipment placed at the side of the user usually will be considerably better than lighting placed at the wall faced by the user (see lighting for study desks, page 15–16). Lighting equipment placed on walls of the carrel should have specific photometric control characteristics to eliminate veiling reflections (see Section 3). Another important consideration is to use relatively light finishes on the carrel surfaces to reduce absorption losses and excessive luminance differences. See page 12–16.

Chalkboards. Black chalkboards with white chalk provide the highest degree of contrast; however, a black chalkboard is poor from the standpoint of luminance ratios. To maintain the correct ratio, chalkboards require supplementary illumination.

Classrooms for the Partially Seeing and Hearing. Classrooms specifically designed for students handicapped by impaired vision and hearing should be provided with the best illumination that economic conditions will permit. Special care should be taken to keep luminous areas in the field of view, either from daylighting or electric lighting, as low in luminance as practicable. Supplementary lighting for charts and chalkboards should be provided.

Most partially seeing individuals benefit from the higher lighting levels. There are some exceptions, however, such as people with achromatopsia, cataracts having an opacity at the nodal point close to the lens, and albinos. These individuals all need lower levels of luminance which can be taken care of by suitably designed glasses.

Corridors. Corridors are the transition areas from the high luminances of the out-of-doors to the lower luminances of the classrooms. They should be well lighted to promote traffic safety and discipline and to assure good housekeeping in the lockers which frequently line the corridors. Care should be taken that lamps are not exposed to unshielded view from within classrooms, through windows common to the corridor and classroom. Where monitors are stationed in corridors, supplementary lighting should be provided at their stations. Corridors also offer an opportunity to add visual interest to school environments and visual importance to displays, bulletin boards, posters, notices, etc. Selected highlighted areas can add considerably to the pleasantness and visual vitality of corridors.

Dormitory Rooms. Dormitory rooms are commonly provided with two systems of illumination, one of relatively low level for general illumination, the other of higher value for study purposes. Such an installation is particularly desirable where a room is shared by two or more students in order that one may retire with comfort if the other wishes to continue studying. General lighting may be provided by centrally located ceiling luminaires and by decorative valance lighting along walls or over draped windows. Direct-indirect types of table or floor lamps of approved performance may be provided for effective visibility in study carrels. The type has the advantage of portability so that it may be used either at a study table or beside an easy chair. See Section 15.

Drafting Rooms. Drafting makes very serious demands upon the eyes and requires a higher level of high quality illumination. The contrast between the work detail and its background may be very poor, as for example, when tracing a faint blueprint or a worn pencil drawing. Specular drawing surfaces, polished T-Squares, plastic triangles, and scales are potential sources of veiling reflections and reflected glare. For further information on lighting for Drafting Rooms, see page 11–7.

Gymnasiums. See page 19–4.

Laboratories. These involve special laboratory tables or benches at which very detailed work is carried out in dissection, inspection of reactions, instrumentation and measurement. Appropriate levels of vertical and horizontal illumination, as well as color quality and comfort criteria should be met.

Laboratories commonly require special electrical provisions such as portable table reflectors to assist in microscope work and in reading precision instruments and meters. Numerous convenience outlets should be provided at work tables and at the sides of rooms to permit connection to electrical apparatus and to portable lighting equipment for experiments. For zoology and anatomy lecture rooms, and laboratories where dissection by the instructor is part of the class work, a large mirror at an angle of about 45 degrees, suitably located and combined with higher level concentrated illumination over the lecture table, will permit students to observe experimental work in both the top horizontal as well as the vertical plane.

Lecture Rooms. (Somewhat similar to auditoriums.) The typical lecture room should be provided with a comfortable general lighting system which is flexible enough to provide a moderately high level for general use and a subdued level for use during projection or special demonstrations. Many times downlighting is used to avoid spilling light on the screen. However, all downlighting should be carefully designed to avoid loss of visibility due to veiling reflections. If a demonstration table is used, directional downlights should be aimed down on this table and the lecturer. These lights should be located within a 45-degree to 60-degree angle above horizontal as measured from the probable location of a speaker's head when he stands behind the table. This arrangement assures minimum glare are provides good lighting on the speaker's face.

Libraries and **Reading Rooms.** See page 12–15.

Locker Rooms. The lighting of locker rooms and dressing room areas is principally a matter of arranging the lighting equipment so that the interior of the lockers is illuminated and general lighting is supplied for safe movement about the room. This usually can be accomplished by means of luminaires located between rows of lockers.

Multi-Function Spaces. This designation applies to two kinds of school spaces: those fixed spaces that house diverse activities, usually at different times, such as auditorium-gymnasium combinations, and those spaces that can be modified in size or shape to serve multiple uses simultaneously or a single common use, such as a divisible auditorium. Lighting in multi-function spaces should be designed to accommodate the several uses of the space; in the first type, the critical task demands will govern. In the flexible or divisible space, the brightness relationships and other design limitations may be critical, not any specific task.

Outdoor Areas. Since many educational institutions are used after dark, it is important to give careful consideration to the various aspects of outdoor lighting. Building facades, approaches, and outdoor activity areas should be illuminated both for the activity itself and for general safety and protection against vandalism.

Sewing Rooms. In the high school the most difficult seeing tasks are generally encountered by those involved in sewing. The process of seeing the fine stitching on cloth which matches the thread in color is inherently difficult and fatiguing. The stitching on a piece of fabric is seen in part by the reflected glint from the thread and by the variations in shades and shadows produced by the thread as it weaves in and out of the cloth. Supplementary lighting is desirable and is especially useful where the sewing is being done on dark cloth where the luminance level is inherently low. See also pages 15–24 and 15–25.

Shops. Students working with power tools are generally novices, inexperienced with the dangers of machine operation. The lighting of a school shop should follow the best industrial lighting practice for the types of industrial activities practiced in the shop (see Section 14), with special emphasis to assist accuracy of manual operations and to make all elements of danger visually obvious. The painting of special backgrounds of the work points on the machinery is desirable to make them stand out; and special colors, following the American National Standard Safety Color Code, Z53.1, should be adopted for those machine controls, and for those machine parts that represent a hazard to the worker.

Luminaires, similar to those used in the rest of the school, may be used, but consideration should be given to possible stroboscopic effects and temporary outages when high-intensity discharge sources are used.

In addition to general lighting the use of localized lighting on high speed tool points and saws is desirable to further improve seeing and safety.

Stairs. The lighting installation for the corridor frequently does not adequately light the stair landings, in which case special lighting units must be installed for each landing. Care should be taken in locating stairway luminaires so that the edges of all steps are properly illuminated. These units should be so designed and applied that their luminance in the line of vision is low.

Visual Aids. For viewing motion picture and slide projector screens or television in the classroom, the comfortable visual environment should be much the same as for any other visual work. The area immediately surrounding these screens should be slightly less bright than the screen itself, and the entire visual surroundings should be kept as light as possible without causing the image on the screen to lose contrast. Studies of lighting for television viewing in classrooms indicate that the luminance and contrast of modern receivers permit good visibility with general lighting of 70 footcandles or more. Reflections from the tube face should be avoided by the use of luminaires and fenestration areas of low luminance and by positioning the receiver to minimize such reflections. In lecture rooms where demonstrations and a variety of projection equipment are used, flexibility of the general lighting system, including dimming, together with effective means for excluding daylight are advisable. Screen luminances with motion pictures and slides are much lower than with television and require lower surround luminances for effective presentation. Some general room illumination is desirable during these types of projection for considerations of visual comfort, safety, and the ability to take notes. The optimum quantity will vary depending on the luminance of the picture and can best be determined empirically through the use of the dimming means.

For further information on Picture Projection see Section 24.

Safety

Provisions for general emergency lighting of stairways and corridors should be made to permit classrooms, auditoriums, and dormitories to be vacated in safety in case of any interruptions or failure of the regular lighting. Exit units should be located over all auditorium and building exits and along any intervening stairs or corridors. Emergency general lighting should be provided for all classrooms, offices, auditoriums, cafeterias, multi-purpose rooms, gymnasiums, and large group spaces to avoid panic and assure safe evacuation. See page 11-9.

Care should be taken to provide sufficient lighting units as part of an exit system so that a continuous lighted pathway, even though of low level, will be provided from any area of personnel concentration within the building to the exterior.

REFERENCES

1. *American Standard Practice for Office Lighting,* A132.1-1966, American Standards Association, New York. (Committee on Office Lighting of the IES: "Recommended Practice for Office Lighting," *Illum. Eng.*, Vol. LV, p. 312, June, 1960.)
2. Joint two year study made by the Public Buildings Administration and the Public Health Service, "The Influence of Lighting, Eyesight and Environment Upon Work Production," Public Buildings Administration, Washington, D. C., 1947.
3. Joint study made by the Public Buildings Service and the Public Health Service, Federal Security Agency, "Influence of Lighting, Eyesight and Environment Upon Work Production," General Service Administration, Public Buildings Service, Washington, D. C., 1950.
4. "Study of Light Finishes for Office Walls and Furnishings," Public Buildings Service, Washington, D. C., 1954.
5. "Study of Fluorescent Lighting in Corridors and Light Finishes for Corridor Walls and Floors," Public Buildings Service, Washington, D. C., 1955.
6. Fisher, W. S. and Schmies, H. M.: "Visibility of Office Tasks," *Illum. Eng.*, Vol. LI, p. 459, June, 1956.
7. Neidhart, J. J.: "An Analysis of Fluorescent Luminaire Brightness," *Illum. Eng.*, Vol. XLVI, p. 569, November, 1951.
8. Darley, W. G.: "An Analysis of Reflected Glare," *Illum Eng.*, Vol. XLIII, p. 85, January, 1948.
9. Wakefield, G. P.: "Control and Measure of Directional Flux at the Task," *Illum. Eng.*, Vol. XLVIII, p. 124, March, 1953.
10. Sharp, H. M. and Parsons, J. F.: "Loss of Visibility Due to Reflections of Bright Area," *Illus. Eng.*, Vol. XLVI, p. 581, November, 1951.
11. Neidhart, J. J.: "Selection Guide to General Office Lighting," *Architect. Rec.*, Vol. 131, p. 209, May, 1962.
12. Neidhart, J. J.: "Visual Comfort in the Office," *Ind. Med. & Surg.*, Vol. 27, p. 583, November, 1958.
13. Finn, J. F.: " 'First Class' Perimeter Lighting," *Illum. Eng.*, Vol. LV, p. 357, June, 1960.
14. Allphin, W.: "BCD Appraisals of Luminaire Brightness In a Simulated Office," *Illum. Eng.*, Vol. LVI, p. 31, January, 1961.
15. Ake, T. Jr., Neidhart, J. J.: "A Mathematical Equivalent to the 'Scissors Curve,' " *Illum. Eng.*, Vol. LVI, p. 544, September, 1961.
16. Krok, E.: "Using the Scissors Curve," *Illum. Eng.*, Vol. LVI, p. 693, December, 1961.
17. Crouch, C. L.: "New Levels and Scissors Curve Highlight Revised Office Practice," *Illum. Eng.*, Vol. LV, p. 345, June, 1960.
18. "Clarification of the Scissors Curve Luminaire Brightness Limitations for Office and School Lighting," *Illum. Eng.*, Vol. LIX, p. 390, May, 1964.
19. Meckler, G. and Meckler, M.: "Design and Evaluation of Dynamically Integrated Lighting-Air Conditioning Systems," *Illum. Eng.*, Vol. LVIII, p. 75, February, 1963.
20. Allphin, W.: "Sight Lines to Desk Tasks in Schools and Offices," *Illum. Eng.*, Vol. LVIII, p. 244, April, 1963.
21. Sampson, F. K. and Jones, B. F.: The Effects of Lighting System Geometry, Distribution and Polarization on Typical Pencil Tasks," *Illum. Eng.*, Vol. LIX, p. 171, March, 1964.
22. Griffith, J. W.: "Analysis of Reflected Glare and Visual Effect From Windows," *Illum. Eng.*, Vol. LIX, p. 184, March, 1964.
23. *American Standard Guide for School Lighting,* A23.1-1962, American Standards Association, New York, 1962. Also *Illum. Eng.*, Vol. LVII, p. 253, April, 1962.
24. Kahler, W. H.: "Visibility of Chalkboards for Classrooms," *Sch. Board J.*, p. 47, December, 1950.
25. Subcommitte on Lighting for Audio Visual Aids of the IES: "Guide for Lighting Audio-Visual Areas in Schools," *Illum. Eng.*, Vol. LXI, p. 477, July, 1966.
26. Committee on Recommendations of Quality and Quantity of Illumination of the IES: "Outline of a Standard Procedure for Computing Visual Comfort Ratings for Interior Lighting—Report No. 2," *Illum. Eng.*, Vol. LXI, p. 643, October, 1966.
27. Committee on Recommendations of Quality and Quantity of Illumination of the IES: "An Alternate Simplified Method for Determining the Acceptability of a Luminaire, from the VCP Standpoint, for Use in Large Rooms—Report No. 3," *J. Illum. Eng. Soc.*, to be published.
28. Committee on Recommendations for Quality and Quantity of Illumination of the IES: "RQQ Report No. 4—A Method of Evaluating the Visual Effectiveness of Lighting Systems," *Illum. Eng.*, Vol. 65, p. 505, August, 1970.
29. Sampson, Foster K.: "Contrast Rendition in School Lighting," EFL Technical Report No. 4, Educational Facilities Laboratories, 477 Madison Avenue, New York, January, 1970.
30. Theatre-Television Committee of the IES: "Lighting for the Vidicon Camera," *Illum. Eng.*, Vol. LVIII, p. 387, May, 1963.

SECTION

12

INSTITUTIONS AND PUBLIC BUILDINGS

Banks 12–1 *Churches and Synagogues 12–2* *Foodservice Facilities 12–4*
Health Care Facilities 12–6 *Hotels and Motels 12–13*
Libraries 12–15 *Museums and Art Galleries 12–18*

Almost all structures other than residences might be called "institutions and public buildings." However, illuminating engineers usually include only such structures as libraries, art galleries, museums, hotels, churches, hospitals, and the public portions of banks in that category. For the lighting of office areas, merchandising areas, and exteriors of institutions and public buildings, see Sections 11, 13, and 17, respectively. For general information on interior lighting design, see Sections 10 and 16.

BANKS

Perhaps of all institutions and public buildings the main floor areas of banks present the greatest number of examples of coordination between interior lighting design and architectural theme. Today's bank interiors vary from modern low-ceiling areas to traditional high ceiling areas, as shown in Fig. 12–1, and therefore a wide variety of lighting equipment is in common use.

Specific Areas

Modern banking facilities house areas where many varied tasks are performed. The following are several specific banking areas (for recommended levels of illumination, see Fig. 9–80):

Conference Rooms. Conference rooms vary from small offices for personal consultation to rooms in which meetings of bank directors are held. Although discussion is the major function, the visual task is usually typed and handwritten financial information and the designer should be careful to provide both a comfortable atmosphere and sufficient illumination for this type of task. See Section 11.

Desk (Stand-Up) Areas. In the main public area of most banks, there are the stand-up type desks at which customers prepare the paper work of their transactions before going to a teller or other station. While these are usually lighted by the general illumination system, in some cases it may be necessary to provide supplementary illumination to produce the recommended levels of illumination.

Fig. 12–1. Views of lighting installations in banks. Left is a modern low-ceiling bank with fluorescent troffers and cove lights. Right is a lighted traditional high-ceiling bank with high-intensity discharge lamp downlights.

NOTE: References are listed at the end of each section.

Executive Offices. The trend in most banking organizations appears to favor private offices for only a few key individuals. It is common for vice presidents or even presidents of banks to be seated at desks in one large room. In these large rooms, the general lighting system should provide the proper lighting requirements for reading of financial statements and similar data. In private offices, individual taste becomes important, but the visual task still remains typed and handwritten financial data. Dimming circuitry might be considered here to allow executives to change the illumination level to suit the task of the moment. See Section 11.

Public Relations. Some financial institutions offer merchandise or other promotional incentives to attract new accounts. Merchandise is generally displayed in windows or throughout the public area of the bank and should be highlighted by accent lighting to create a point of interest. See Section 13.

Records, Storage, and Retrieval. In record, storage, and retrieval areas, provision should be made for both the accurate filing and retrieval of data. A section of the storage area is usually set aside for perusing the documents. See Section 11.

Safe Depository. The safe depository area usually has a place to retrieve a box by key and a table or counter for personal examination of its contents. The general illumination should be satisfactory for safe access with local lighting provided at tables or counters.

Teller's Cages. Teller's cages are always in the public area of the bank, and it is critical that they be given careful consideration in the design of the general lighting system. The teller is required to work very quickly and accurately and make constant reference to financial data, both typed and longhand. If the general lighting does not provide adequate illumination for the teller's operations, supplementary lighting should be provided.

CHURCHES AND SYNAGOGUES[2]

Skillfully used lighting can make worship services more meaningful and enhance the architectural design of the space. The lighting can mold and give depth, and can subdue or accentuate, or perhaps change its accent, as the service proceeds. In certain interiors it can add a fourth dimension: a suggestion of the infinite.

Entrances

In the entrance vestibule or narthex, the lighting should enable the quick recognition of faces and facilitate the taking of notes of names and requests. Diffuse illumination should be used so that faces appear pleasantly lighted and are not made to appear lined or strained by highly directional harsh sources.

Main Worship Areas

There are vast differences in the service and liturgy of the many faiths and denominations and the lighting designer should be familiar with their customs in order to assure proper lighting emphasis at the proper time. For recommended levels of illumination, see Fig. 9–80.

General Lighting. There should be appropriate general lighting for reading, moving about, visual social contact, and to help the worshipper relate to the structure and its features. In many churches there is a trend away from the traditional service of listening, watching, and meditating, to a service that includes more participation. Higher general lighting in the space can encourage the feeling of being part of a body of people. Such participation also means more reading—requiring particular attention to light at the pew.

Often there are two components to the general lighting: (1) direct lighting for the pews, and (2) indirect lighting to relieve shadows and to create desired brightnesses on the structure. Sometimes, indirect lighting provides all the general illumination—particularly if lighting equipment cannot be mounted on or in the ceiling.

Lighting from one direct point source creates dark shadows and specular reflections. This may be desirable to highlight an object, but it is undesirable where people are attempting to read or follow printed material. Also it can make the leader of a service appear unpleasant through deep eye and other facial shadows. An overlap of light from direct sources or the use of indirect lighting with direct light will soften shadows and reduce specular reflections.

Accent Lighting. Certain parts of the worship area become central at different times in the service, and when so, they should be highlighted. Those areas may be where the worship leader, the choir, the Torah, the communion table, the stations of the cross, the Bible, and the Ark are located. Controlled beams of light should be used that will properly render the features in these areas and not create glare for those participating. This will mean careful choice of beam spread, intensity, and location of spotlighting.

The location and orientation of the congregation should be kept in mind so that any directional lighting will not create glare. Particular attention to the shielding of directional lighting is needed for the church-in-the-round.

Controls. Lighting can help shift attention and emphasis during a service. By switching or dimming, the appropriate changes can be made in the brightness of different parts of the worship area. When dimming is used it produces these changes much more subtly than switching. This is particularly true of general lighting or lighting of large features.

Church Architecture and Lighting

In the nave or main auditorium the quantity of light, and use of patterns of light and shade, vary widely with different architectural styles. See Figs. 12-2 through 12-5. The lighting designer should consult closely with the architect to understand the purpose behind the architectural style being used and to develop the lighting approach for it. They should cooperate through the following stages of translation: (1) the architect's concept of the space, (2) the brightness patterns desired, and (3) the lighting equipment needed. For a further discussion of the relationship of light to architecture, see Section 10.

If lighting equipment is to be concealed in or behind a structural element, space is often limited. This need for compactness will generally call for a larger number of lower output units than might otherwise be used, or much better optical control in the units. If this care is not taken, the results may be uneven illumination and excessive brightness from spill light on exposed surfaces adjacent to the equipment. For example, incandescent lighting in a small cove or cornice should use a large number of small devices rather than a few high output units unless very compact, sophisticated optics are used; otherwise the adjacent ceiling or wall could be unevenly and excessively bright.

Lanterns or other suspended decorative luminaires may be effectively used with many architectural styles. If they are to produce direct general illumination, however, care should be taken that there is sufficiently wide distribution of light for good coverage but without discomfort glare. This may be impossible if appearance dictates a very low suspension; in this case, other sources should be used to provide the illumination and the suspended equipment used as a luminous decorative element.

The reflectance of some large surfaces in the worship area—usually wood—may be very low. Such surfaces should be lighted to make them visible and to relieve an otherwise "too-dark" atmosphere, but not so much as to make them brighter than they would be expected to appear normally.

Fig. 12-2. Tent-type church. Exposed rafters conceal equipment which is aimed down and forward for lighting pews and chancel. Light from coves diffuses, balances and supplements the downlighting to create reverent mood desired during a service.

Art Windows

As in the architectural considerations for lighting, the lighting designer should work closely with the art window designer to determine the desired appearance of the lighted window. In all stained and art windows, the density, diffusion, and refractive qualities of the glass or plastic will determine the light source luminance and size to be used.

Fig. 12-3. As in Fig. 12-2, the ceiling design hides luminaires from the seated congregation. Incandescent and fluorescent sources (inset) provide varied lighting effects during a service. As the wiring is semi-exposed, lighting changes can be easily made in future years.

Fig. 12-4. Unsymmetrical design. Floodlights across court at left illuminate the ceiling of nave and part of chancel. Other exterior floodlights, small in size, aim light down at a pool in the court so that reflected rays play faintly over the nave ceiling. Chancel and pew lighting come from behind the ceiling beams.

Fig. 12-5. Colonial church. The lighting designer worked with the interior designer to arrive at the feeling and the image required. Tradition and a worshipful atmosphere are preserved. After dark three systems of nave lighting provide effectively for evening worship. On rainy days, most or all of the evening lighting is used. On sunny mornings, interior lighting may be used to retain attention indoors.

It is not necessary to achieve a perfectly flat or uniform lighting effect. In fact, it is often not desirable. It is almost always necessary to set up a trial lighting system to see how the glass responds to different lighting. In such trials, a great deal of equipment may be necessary both to get sufficient light on the glass and to have enough different lighting approaches to examine.

Generally the lighting of art windows serves two main purposes: (1) for viewing from inside during nighttime services, and (2) for viewing from outside for passing traffic.

Viewing From Inside. The window can be lighted with outside floodlighting units if the glass has sufficient diffusion and refracting qualities (from irregularities on the surface of the glass and within the glass). If the glass is not extremely diffuse, the units should be located so that they are not seen through the glass and do not produce visual "hot spots." Clear stained glass needs a luminous background such as a closed light box around the outside of the window.

Viewing From Outside. The floodlighting approach above also can be used, but with equipment located inside. Spots of brightness may be more difficult to avoid, however, since lighting equipment is usually most conveniently located on the ceiling and the viewer is usually below the window. A larger number of lower intensity floodlights can make the spots of brightness less apparent. For clear stained glass a movable screen or drape can be used on the inside, lighted (either transilluminated or lighted from the window side) to form a luminous background for clear glass; it can be moved away for times of viewing from the inside.

FOODSERVICE FACILITIES

Commercial Kitchens[3]

Kitchen Operations. Well designed illumination helps to provide a bright hygienic atmosphere in a kitchen and, by revealing the presence of debris, it can stimulate good housekeeping. Illumination can improve productivity, with fewer mistakes, and eliminate shadows and dark areas resulting in fewer accidents. All these factors are elements in the overall lighting design.

While most commercial kitchens are not departmentalized as to space (See Fig. 12-6), the total operation can be separated into specific areas or functions.

1. Receiving and storage areas.
2. Walk-in refrigerators and freezers.
3. Vegetable inspection and preparation.
4. Quantity cooking and baking.
5. Short orders.
6. Service, checking, and pricing.
7. Washing dishes, pots, and pans.
8. Refuse areas.

For recommended levels of illumination see Fig. 9-80.

Environment Luminance and Glare. Seeing in the kitchen is influenced adversely by great variations in luminance; (1) directly, if these appear in the immediate task, such as reflections of bright sources in polished surface of utensils, and (2) indirectly, if marked variations are away from the task but dominant in the general field of view. Direct and reflected glare are obstacles to employee comfort and productivity; therefore, exposed lamps are not recommended.

Direct glare may be prevented by effective shielding in the luminaire. The installation of any type

Fig. 12-6. A hotel kitchen lighted by continuous rows of glass bottom troffers.

luminaire should take into consideration the shielding of lamps from those entering and leaving the kitchen area. This is of particular importance when the adjacent dining area is illuminated to a lower level of illumination—the usual case. Even the relatively low luminance of unshielded fluorescent lamps will require the eyes to adapt themselves. This period of adaptation time is one of partially impaired vision.

Reflected glare results when surfaces within the line of vision are polished; bright images then may be seen in the surfaces. Such spots can be exceedingly objectionable, resulting in discomfort and fatigue. These reflections can be reduced by controlling the luminance and placement of luminaires.

Luminance Ratios. Current good lighting practice has established that best results are obtained when luminance variation of adjacent areas, particularly within the working field, does not exceed 3 to 1, the work itself being brighter. A maximum limit of 10 to 1 is recommended for non-adjacent areas.

Luminaire Preference. It is recommended that luminaires be totally enclosed, provide comfortable luminance control, and be convenient to maintain. In larger kitchens having quantity cooking and baking, moisture and flour dust conditions may exist; vapor-tight or dustproof luminaires are recommended in these locations. In selecting luminaires, local electrical, fire, health, and safety codes should be consulted.

Choice of Light Source. The use of incandescent filament or one of the improved-color white fluorescent lamps is recommended. Color preference is partly influenced by the experience and opinions of the users.

Serving and Dining Spaces

Lighting is a key design element in establishing the mood or atmosphere in any dining space. The mood set can vary from subdued, relaxing and cheerful, to bright and lively. These spaces are usually grouped into three types: intimate, leisure, and quick service. See Fig. 12–7. For recommended levels of illumination for each type, see Fig. 9–80.

The "intimate" type consists of those areas where people congregate as much to visit and to be entertained as to eat or drink. This includes cocktail lounges and some dining rooms and restaurants. These areas usually have subdued atmospheres obtained by a low level lighting system.

The "leisure" type refers to most restaurants and many dining rooms, places where eating is leisurely but where time is a consideration. Moderate levels of illumination can be used effectively to create a calculated atmosphere that is suitable to the type of dining facility desired.

Fig. 12–7. (Left) An intimate class dining area, where two large spheres slowly revolve to scatter reflected light on ceiling and columns. (Upper Right) Leisure type dining area illuminated by incandescent downlights with a wall wash effect in the rear. (Lower Right) Quick service class dining area lighted to a higher level from fluorescent troffers and incandescent downlights.

The "quick service" type includes lunch rooms, cafeterias, snack bars and franchised menu types, where the diner is willing to allot only enough time to eat, and the operation is based on a quick service. Higher levels of illumination are recommended here, both for seeing and to psychologically suggest a feeling of economy and efficiency.

The lighting effects depend on the level of illumination, color of light, luminaire light distribution, and source size. Well-shielded downlights, for example, can create sparkle in objects, and an intimate feeling. On the other hand, indirect lighting or large area diffuse sources, such as fluorescent lighting, create a brighter-looking space and call more attention to the whole room. Many suspended decorative luminaires, regardless of shape, size, or style, have a general diffuse distribution that can mean uniformly lighted spaces but often with uncomfortable luminaire luminances, unless low wattage lamps are used permitting these luminaires to act only as luminous ornaments.

Variation in lighting level may be needed both to change the mood for different times of the day and to permit a higher level for clean-up than is desired for dining. Dimming control is needed instead of switching when a smooth transition is desired.

The lighting of any feature in the dining area will need special attention. These can range from the highlighting of a picture or sculpture to a full luminous wall, creating effects that can range from dramatic to open and friendly.

Food displays should be so lighted that attention is attracted to them and the details are clearly seen. A good rule is that the level used here should be not less than twice that of the surrounding areas, but also not less than 50 footcandles.

The use of colored light is often overlooked as a lighting design tool. Strong colors of light can create interesting effects when lighting surfaces for decorative purposes, but should not be used to light food or people because of the resulting color distortion. Tints of white light can be used for general illumination that will create a cool or warm atmosphere, or will render colors pleasingly or poorly. The psychological aspects of colored light are more variable and require creative design considerations that are not utilized in more conventional lighting systems. For example, food and people appear more pleasing when the lighting utilizes incandescent or improved-color white fluorescent lamps.

HEALTH CARE FACILITIES[4]

The lighting of health care facilities such as hospitals, nursing homes, clinics, and extended care facilities present many problems involving a wide range of seeing conditions. Optimum seeing conditions should be provided for doctors, nurses, technicians, maintenance workers, and patients. For a better appreciation of the principles involved, a review should be made of Section 3, Light and Vision; Section 5, Color; Section 9, Lighting Calculations; Section 10, Interior Lighting Design Approaches; and Section 11, Offices and Schools.

Many activities in health care facilities are not related directly to patient care but are necessary as supportive institutional functions. Areas such as business offices and laundries are not discussed in this section (see Sections 11 and 14).

Fig. 9–80 lists minimum levels of illumination for health care facilities. Where higher levels of localized lighting are required as in surgery, obstetrics, dentistry, emergency, treatment, and autopsies, it is desirable to insure comfortable lighting conditions by limiting luminance ratios between the task and other areas in the normal field of view. Recommended luminance ratios and room surface reflectances are the same as listed in Section 11 for offices.

Color, as produced by the light source and the surroundings, is an important consideration in some areas. For example, it is important in patients' rooms and in treatment and examination rooms, where diagnosis of a patient's condition may be related to the color, or to a change in color, of the patient's skin. In nearly all parts of the hospital the colors resulting from lighting and wall finishes can be used to alleviate the institutional appearance and to suggest the impression of the home in the patient's rooms, lounge rooms, library, and entrance lobby.

Nosocomial infections endanger life, produce suffering, increase the occupancy of needed bed space, and create a substantial financial loss to the patient and his family. It is therefore most important that lighting systems and luminaires for health care facilities be designed with careful consideration of the requirements for asepsis and ventilation.

Lighting systems must not produce objectionable interference with electronic equipment used for diagnosis and treatment in operating rooms, heart catheterization procedure rooms, delivery rooms, and other specific areas. Interference-free lighting must be provided in every area where the use of electronic equipment can be anticipated on an emergency basis or as a result of rearrangement of facilities.

Specific Areas

Autopsy and Morgue. In autopsy, good lighting is needed for the benefit of the operator rather than the subject, for he must be able to do surgical "dissection" over and in all parts of a body, sometimes diseased and decomposed, with a minimum of risk to himself. He must have the same color quality of lighting as the surgeon, but does not require the illumination level needed to overcome light losses in deep cavities. He must be able to see the differences in color and texture between normal and abnormal conditions.

Large, low-luminance, nonadjustable ceiling units may be used, and there must be provision for whole-body illumination by a large area light source, a moveable table, or an adjustable ceiling unit. A unit similar to those for minor surgery, or for portable use with adjustable vertical and horizontal positions and beam spread, can be used.

Corridors. The lighting of corridors should be consistent with that of the rooms served so that there is no objectionable luminance difference in passing from one to the other. This often means provision for reducing the levels at nighttime.

Cystoscopic Room. Cystoscopic rooms should have much the same quality and flexibility in lighting and equipment as the obstetrical room. There must be provision for darkening the room for fluoroscopy and for the use of lighted exploratory instruments.

Dental Suites. In the *dental operatory* the luminance differences between the patient's mouth and face, patient's "bib", instrument tray, and the surrounding areas should be within the range recommended in Fig. 11-2 for offices.

General lighting should be provided at the level of the patient's face and the instrument tray. For an acceptable surrounding luminance in the usually rather small dental operating room, the ceiling should be a near-white color with reflectance of 70 per cent or more; walls should be of a non-glossy surface of any light color such as blue-green with a reflectance of about 60 per cent; and floor reflectance should be in the range of 20 to 40 per cent. The color quality of the general lighting must enable the dentist to match colors of filling and tooth. Daylight from the window may be very desirable for that purpose, but it may not always be available.

Lighting inside the mouth, or oral cavity, should be supplied from a luminaire easily adjustable to exclude high luminance from patients' eyes and at the same time provide such lighting as needed by the dentist to see fine details over long periods of time. This light should have color characteristics and level suitable for the dentist to judge the matching of colors of teeth and fillings and to see occlusions of dentures any place within the mouth. The dentist must be able to accurately judge the depth of drillings and the preparation for retention of fillings. See Fig. 12-8.

A luminaire for producing such a penetrating light, relatively free of shadows at the oral cavity, must produce a convergent beam, and at a distance of 3 to 4 feet should be capable of lighting a semicircular area with a cut-off to exclude the bright light from patients' eyes. To ensure such a level of interior lighting with the listed handicaps, there should be a level of 1000 footcandles on the patient's chin. Such a light source is very similar to the larger ones developed for surgical lighting,

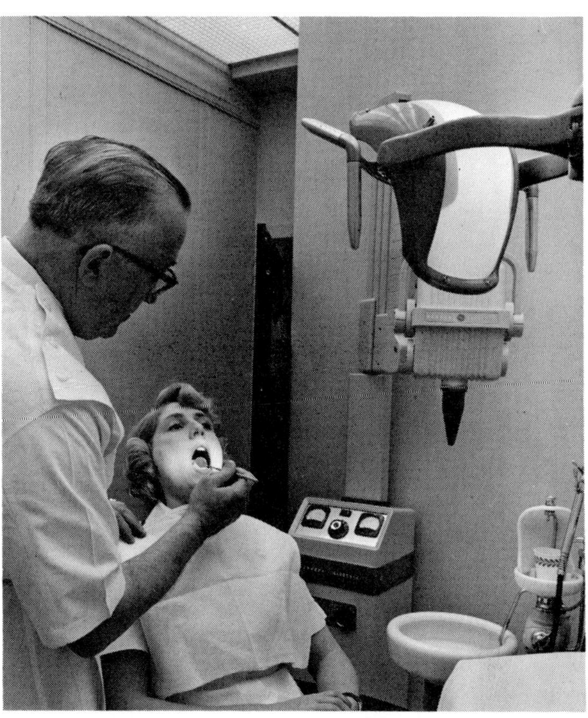

Fig. 12-8. A dental spotlight providing a beam pattern shaped to fit the patient's mouth helps to provide patient comfort. A multi-position switch is included to allow an adjustment in illumination from about 1000 to 2000 footcandles for the dentist's needs.

which may, in fact, be used for some dental surgery not requiring the other facilities of the hospital operating room.

Prosthetic work in the *laboratory* requires speed, accuracy, and close inspection. Therefore, a general level of 50 footcandles should be provided, with a supplementary level of 100 footcandles at the work bench and 200 footcandles available at one or more points depending on the number of people using the laboratory at one time.

If a dental suite is large enough to have a separate *recovery room*, a level of 5 footcandles is recommended, with provision for such higher-level lighting for emergency examination or treatment as the type of dental practice may require.

Electroencephalographic Suite (EEG). General illumination should be provided through out this suite. A system of dimming the lighting is required in the technician's workroom and in the patient's preparation and examining room. A more complete description of the EEG suite is given in a separate publication.[5]

Electromyographic Suite. Lighting layout, and electrical services for electromyographic suites are essentially the same as for electroencephalographic suites. Provision for reducing the lighting level to about 1 footcandle should be made in the electromyographic preparation area. Lighting equipment for use in the shielded room should not produce static or interference with highly sensitive electronic equipment used in electromyography.

Emergency Outpatient Suite. The emergency out patient suite should be generally self-sufficient to handle most cases without resort to the rest of the hospital. Lighting for patient examination should follow the recommendations made under Examining Rooms below. General illumination in the emergency operating room is supplemented by fixed ceiling-mounted directional luminaires or by portable lights to provide illumination at the center of the operating area. For minor surgery, a minimum of 2000 footcandles is recommended at the operative site. When major surgery is anticipated, the lighting provisions should be the same as recommended on page 12–11 under Surgical Suites.

Examining Rooms. Examining facilities may be in a hospital, clinic, or physician's office. The same facilities may be used for examination, treatment, and even for minor surgery. The lighting of examining facilities must, therefore, be planned to accommodate the possibility of a wide variety of seeing tasks.

General lighting should be coordinated with other elements of the room environment, not only to achieve a pleasant atmosphere necessary for patient assurance, but also to produce a color consistent with proper diagnosis of a patient's condition. Fluorescent lamps should be of the improved color types.

The relationship between general room lighting and local lighting for the specific investigation of areas or cavities of the body varies considerably with the types of examinations made. In many examinations, the general room lighting is complementary to localized lighting and is varied in level according to the examination. Thus, the lighting system for general illumination should provide dimming to extinction.

Lighting levels provided by supplementary luminaires will vary from the recommended 100 footcandles minimum on the examining table to levels for surgical procedures.

Waiting rooms and dressing rooms used in conjunction with examining rooms should provide a soothing, cheerful, and flattering environment. Lighting in the dressing room should facilitate grooming and cosmetic application. Design considerations and illumination levels at mirror locations are described in Section 15, page 15–20.

Eye, Ear, Nose, and Throat Suites. Lighting of variable levels, ranging down to total darkness is needed. During the use of lighted instruments, only 5 footcandles or less (sometimes complete darkness) is required. Because of the variety of lighted instruments used in these suites, duplex convenience outlets should be provided about every 5 feet around the walls of such examining rooms, with additional outlets for treatment stations.

Fluoroscopy. The lighting system should provide means to vary general illumination, from 50 footcandles minimum for pre-examination tasks, down to 1 footcandle for use during examination procedures when doctor's eyes are dark-adapted. Because of the after glow (phosphorescence) characteristics of fluorescent lamps, incandescent filament lamps are preferred for lighting in fluoroscopy rooms.

Intensive Care Units (ICU). The equipment and all physical aspects of intensive care units, including lighting, are intended to provide not only for constant observation and treatment of patients but also to facilitate emergency response to deterioration of a patient's condition. Proper illumination must be provided for a wide variety of visual tasks. Lighting systems must have both flexibility in regard to illumination levels and the capability of changing levels quickly to meet emergency conditions. See Fig. 12–9.

Different types of electronic monitoring equipment designed to measure, display, or record patient's physiological conditions frequently utilize extremely weak electrical signals which must be highly amplified. Special precautions must be taken to avoid shock hazards to patients and electrical disturbances to monitoring equipment; for example, all electrical equipment must be properly grounded and spurious electrical signals generated by electrical devices, including lighting equipment, must be controlled by means such as filtering and/or shielding. The Public Health Service Publication No. 930-D-25, "Electronics for Hospital Patient Care", offers a thorough and authoritative discussion of this subject. Local codes and NFPA No. 76BM, "Manual on the Safe Use of Electricity in Hospitals", should be consulted for safety requirements.

From the decorating and lighting standpoints, the ICU patients' rooms should be treated as nearly as possible as regular patient bedrooms to minimize the awesome appearance of the electronic monitoring

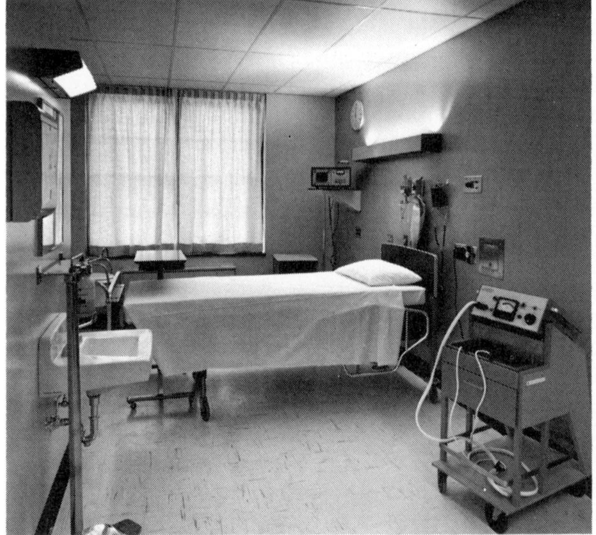

Fig. 12–9. Intensive care room. Wall brackets contain two fluorescent lamps for indirect general lighting, one fluorescent lamp as a downlight for reading, and an incandescent night-light for surveillance from the nurses station.

equipment. Lighting systems should provide flexible and positive control of illumination levels. The general illumination should be variable to permit simulation of the day/night cycle, and to allow illumination levels conducive to rest. Fluorescent luminaires should use improved-color lamps.

Supplementary luminaires should be readily available if required to provide the higher local illumination levels recommended for examination and treatment. Surgical procedures may have to be carried out in a patient's bed (for example, a tracheostomy) and for these purposes portable surgical luminaires must be quickly available. Cubicle curtains or other means should be provided to protect adjacent patients from exposure to excessive luminances.

The arrangement of lamps, receptacles, switches, and other electrical accessories should be consistent among ICU patient care areas to minimize critical time lost by the attending nurse or physician in locating the proper electrical item. Electrical circuits must be adequate in number and capacity to accommodate the supplementary luminaires. Lighting apparatus, stands, and other items should be accessible and conveniently placed, but should not conflict with other equipment used around the patients' beds.

Lighting for nurses' stations must make allowance for viewing of monitoring equipment and patients' beds without veiling reflections or uncomfortable luminances. The illumination of medication cupboards and supply areas must facilitate accurate and rapid dispensation.

Laboratories. Laboratory work areas should have both general illumination and local lighting at workbenches. Care should be taken to avoid veiling reflections and reflected glare from the local lighting equipment. See Fig. 12-10.

For critical color examinations, daylight quality of lighting has been and still is preferred. Special lamps are available and in wide use in the textile industry, graphic arts, and other areas where color matching or color rendition is important enough to warrant the use of special lamps. See page 14-25. Standard fluorescent lamps should be of the improved color type. The color and reflectance of nearby surfaces must be chosen carefully to avoid color distortion at the areas of critical color examinations.

Mental Health Facilities.[6] Lighting generally should be consistent with a residential environment to promote a tranquil atmosphere. Flickering of lights must be avoided especially. Subdued, low level lighting must be provided for those patients whose medical treatment causes a discomforting and possibly painful sensitivity to illumination levels recommended for patients' rooms.

In patient rooms where patients are not constantly supervised and where they may injure themselves or others, lighting should be provided by fixed nonadjustable wall or ceiling mounted luminaires. In detention or seclusion areas for severely disturbed patients, lighting should be provided by specially designed luminaires recessed in the ceiling. These are difficult for patients to reach, disassemble, break, injure themselves with, or fashion weapons from, and should be controlled by key switches located outside the detention area.

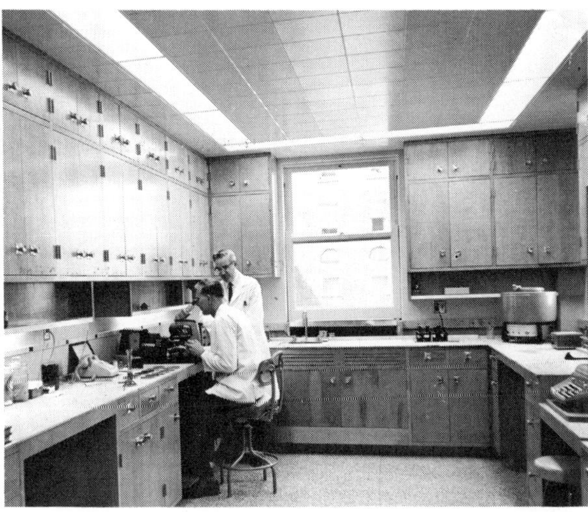

Fig. 12-10. Fluorescent luminaires for the general illumination are located to provide higher levels at the work benches. Supplementary lighting is added at the benches at the left from fluorescent lamps mounted under the cabinets.

Nuclear Medicine Facilities. While the lighting for nuclear medicine facilities is not critical, it is important that there is flexibility in the examination and treatment rooms and the offices. Dimmers in these areas will make this possible.* The lighting in laboratories and waiting areas should be the same as that used in other laboratories and waiting areas of the hospital. The patient should not be exposed to luminaire luminances exceeding 90 footlamberts for an extended period of time.

Nurseries. Nurseries for the newborn should be well lighted to permit easy detection of jaundice and cyanosis. Supplementary or local lighting is needed momentarily for examining or treating infants in either the bassinet or on the examining table. The bed areas in pediatric nurseries should have lighting comparable to that described above for nursing services for adults.

Nurses Station. At the nurses station, illumination is needed for office type work such as charting and other routine paper work. Additional lighting is needed at the medicine cabinet for reading notations, treatment instructions, drug labels, measurements, and instrument graduations. Provision for varying the general lighting is recommended. See Fig. 12-11.

* Attention should be given to the light source. Incandescent filament lamps are preferable to any of the gaseous discharge lamps such as fluorescent, because the latter may interfere with the accuracy of radiation measuring instruments. The possibility of such interference from adjacent rooms, above and below, should be anticipated.

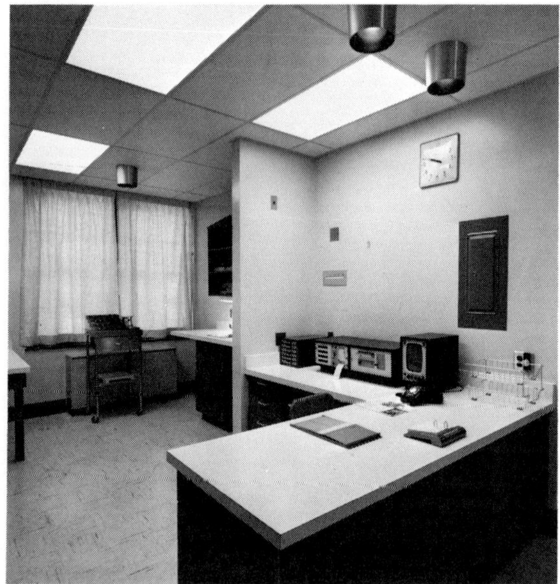

Fig. 12-11. Nurses station. Fluorescent lighting is used for general illumination. Incandescent luminaires over nurses desk can be dimmed for nighttime use, or when reading monitors.

Nursing Homes, Extended Care, and Long Term Care Facilities.[7] Exterior lighting is of great importance. Carefully planned, tasteful exterior lighting will avoid a depressing institutional appearance and enhance safety. By extending the view of patients, it promotes a feeling of well-being and security.

Safety from falls or other accidental self-injury is highly dependent on sight, and sight is dependent on adequate light for seeing. Areas in which nursing home patients reside, travel, or occupy frequently or occasionally, such as their living quarters, toilets, corridors, lounging or recreation rooms, and dining areas should have lighting adequate for the area. The lighting system should be so designed that one burned out lamp will not leave the area in darkness. To minimize eye strain and the accompanying physical and psychological tension due to poor vision, good lighting and avoidance of glare in all areas of the nursing home is mandatory for the well being and safety of the elderly.

Lighting for patients' rooms or suites in nursing homes should be similar to that recommended for Patient's Rooms. Various decorative touches can be added in the nursing home where longer periods of patient confinement make a residential atmosphere important. If a suite has a small galley for food preparation, an illumination level of at least 50 footcandles is recommended to aid depth perception when hot cooking utensils are handled. See Section 15, Residential Lighting.

Lighting systems for indoor recreation facilities should provide comfortable general lighting adequate for reading, card playing, and table games. A supplementary lighting system for color and decorative accents is desirable for creating the "party" atmosphere necessary for social events. If the general dining area also serves as a recreation area, a lighting system permitting selection and dimming of general lighting, perimeter wall washing, and overhead "punch and sparkle" lighting will provide the flexibility for most social functions.

Obstetrical Suites. The general illumination required for the delivery room is identical to that required for operating rooms. If the room is used for all deliveries, including Caesarian sections and others which require extensive surgery, lighting at the table (vertical) should be equal to that recommended for operating rooms. Where it is contemplated that the room will be used for only normal deliveries and for those which require only minor surgery, the lighting levels at the table may be somewhat less than that required for operating rooms. However, the light reaching the table area from an overhead source is not sufficiently directed for the gynecologist to see properly into the narrow horizontal cavity; hence, a floor mounted luminaire or a ceiling or wall mounted adjustable luminaire, that can be lowered to about 3 or 4 feet above the floor, may be needed. If flammable anesthetics are permitted to be used in the delivery room, floor units and ceiling or wall bracket units that can be lowered to less than 5 feet above the floor are required to be explosion proof. See applicable codes under Surgical Suite, page 12-12.

Patients' Rooms.[8] In the patient's room the patient, the nurse, the doctor, and housekeeping personnel require different illumination levels to accommodate their individual needs in the same room. The lighting should be provided in a way that it is not objectionable to other patients in the same room as well as those patients whose field of view may be only the ceiling above. See Fig. 12-12.

An illumination level of 20 footcandles is recommended for general lighting for housekeeping and between-patient aseptic procedures. The general lighting can be made variable by a control system located at the door so that lower levels can be selected for the patient's comfort or to create a more intimate atmosphere for dining or receiving visitors. When fluorescent lamps are used, they should be only the improved-color types.

For safety and gross observations, there should be a low-mounted nightlight system controlled at the door. This system should provide illumination of 0.5 footcandles maximum on the floor thirty inches from the luminaire and located so as to silhouette the objects in the room. The nightlight luminance should not exceed 20 footlamberts. Provision may also be made for an observation light, controlled near the bed, so that the nurse may observe the patient and the equipment during the night with minimum disturbance to other patients in the same room. The observation light should provide a 2-footcandle level minimum on the bed and the luminance should not exceed 60 footlamberts.

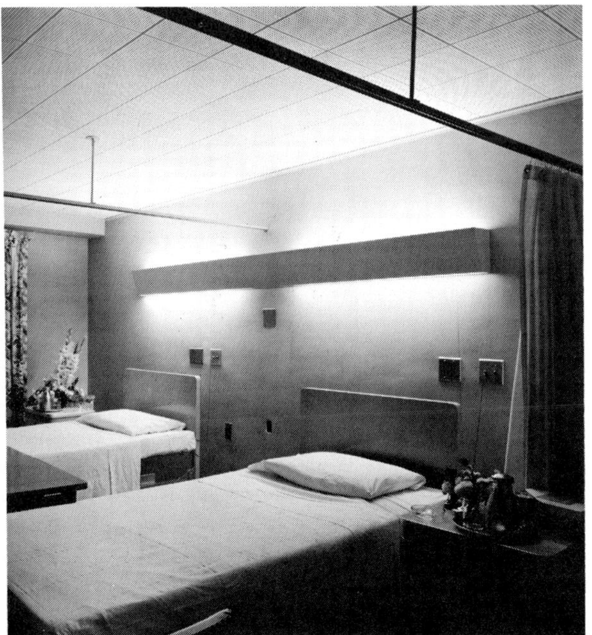

Fig. 12–12. A semi-private patient's room equipped with a wall mounted non-adjustable light for reading in bed and for general illumination.

The lighting for general examination of the patient should be at least 100 footcandles. Supplementary luminaires may be used if higher levels of illumination are required for critical examination. Color as produced by the light sources and the surroundings should be selected to avoid an incorrect diagnosis of a patient's condition.

The patient's bed reading light should have the capability of supplying 30 footcandles minimum at the reading task. The location of the task may vary greatly depending upon factors such as the type of bed, the angle to which the bed might be adjusted, condition of the patient and type of treatment, auxiliary equipment, and nature of the reading material. The luminance of the reading lamp and of any surface illuminated by it, as normally seen from any reading or bed position, should not exceed 90 footlamberts.

Pharmacy. The pharmacy should be lighted to minimize the possibility of misreading labels, and to promote accuracy in filling prescriptions. In the alcohol storage vault, lighting units and control switches may be required to be explosion-proof, depending upon the containers used and the inspection authority having jurisdiction. Consult Sections 500-2 and 500-4 of the *National Electrical Code*. In Canada, see *Canadian Electrical Code*.

Radiation, Diagnostic. All rooms in the x-ray department should have general lighting sufficient for setting up equipment, positioning patients, cleaning the room, and maintenance. There should be local lighting where needed for such procedures as locating veins and injecting dye chemicals to produce contrasts in tissue and organs, or for exact positioning for deep therapy treatment. Controls should provide for dimming lights when needed, as for fluoroscopic examinations.

Darkrooms must have safelights for use while handling undeveloped film, and ordinary lights for cleaning, maintenance, and handling developed or light protected film. Switches for the safelights and ordinary lights should be separated and so located to avoid error of turning on ordinary lights while undeveloped film is exposed.

Radiation Therapy. In the x-ray and cobalt-teletherapy suites, lower level general illumination is sufficient, except where records are processed. Wall murals are often used in these treatment rooms and they require local lighting carefully coordinated with the general illumination to provide the desired atmosphere. Controls for dimming the general lighting in the viewing room are recommended. Where either portable or installed x-ray units are used in operating rooms, the units should be approved for use in these locations.

Surgical Suite. The lighting of the *operating room* is perhaps the most important in the hospital, not in the number of people to be satisfied but in the importance of the work done there. There should be no dense shadows to prevent the surgeon from seeing past his own hands and instruments, nor to prevent him from adequately seeing the patient's tissue, organs, and blood, exactly as they are. Sometimes he must see into deep body cavities, natural or artificial. To enhance physical comfort for the surgical team, heat reaching the back of the surgeon's head and neck from the overhead surgical light must be minimized. The surgeon must be able to work for hours, if necessary, without any discomfort, and must be able to glance to and from his work without having to take time to adjust to large differences in luminances.

Color and reflectances of operating and delivery room interior surfaces and draping and gown fabrics should be somewhat as follows: ceilings, a near-white color with 80 per cent or more reflectance; walls, non-glossy surface of any light color such as blue-green with 60 per cent reflectance; conductive floor, reflectance preferably in the range of 25 to 40 per cent, but may be as low as 8 per cent depending on the limited selection of conductive flooring materials available; and fabrics for gowns and surgical drapes, should be colored, usually a dull shade of blue-green or gray with 30 per cent or less reflectance. Surgical instruments should be of a non-reflective matte finish to minimize reflected glare in the area of the operative cavity. Any plastic materials used in draping should also be of matte finish.

Equipment such as x-ray, anesthesia, and ventilation competes with the lighting system for the limited ceiling space available. Therefore, to achieve the recommended general levels, it is necessary to carefully plan the location and arrangement of the lighting system. Due to the variety of surgical procedures, it is highly desirable to control the illumination level from the general lighting system to suit visual requirements of the surgeon and staff. The general room illumination in the surgical suite should be a minimum of 200 footcandles uniformly distributed with provision for reducing the level.

The color of the general room illumination should be compatible with the illumination from the surgical light; that is, the color temperature of the two sources should match as closely as possible. For example, if the main surgical light has a color temperature of 4000 K, general room illumination should be provided by flourescent lamps with about the same color temperature. However, in all cases, if the general illumination is supplied by flourescent lamps, they should be of the color-improved type only.

As levels of general illumination become higher, luminaire shielding becomes more important. Direct glare, however, should not be a problem if the luminaire luminance limitations in Section 11, page 11–3 are followed. Asepsis requirements encountered in a surgical suite rule out the use of open louvers as a means of controlling glare.

The surgical lighting system should be capable of providing a minimum of 2500 footcandles directed to the center of a 78 square-inch (or larger) pattern on a surgical table with the top 38 inches from the floor. This pattern is defined as an area within which the level of illumination tapers from center to edge so that at the edge it is no less than 20 per cent of that at the center. For ceiling-suspended surgical lighting systems, the illumination and patterns are measured 42 inches from the face of the lamp cover glass, if a cover glass is used, or the lower edge of the outer reflectors in a multiple reflector unit used with individual covers over each unit. If the surgical lighting system is fixed in place, the light should emanate from a number of sources dispersed over a wide angular area providing illumination for all required surgical approaches.

The above is intended as minimum for general surgical procedures. In many specialized instances higher illumination levels, various pattern sizes and shapes, and control of level are desirable. All illumination measurements should be made with a color- and cosine-corrected light sensing element that will indicate the average level over a 1½-inch diameter. To prevent obscuring shadows from the surgeon's hands, head, and instruments, the light should reach the operating area from wide angles. For test purposes, the light should provide a level of 10 per cent of the unshadowed level inside and at the bottom of a tube 2 inches in diameter and 3 inches long, finished flat black inside, from a distance of 42 inches when the beam is obstructed by a disk 10 inches in diameter, 23 inches above the operating table, and normal to the axis of the tube. Where only one unit is used it should be supplied with a two-filament lamp designed for multiple operation to protect against total lamp failure.

The radiant heat produced by surgical lights must be minimized for comfort and efficiency of the surgeon and his assistants. For some operations such as in neural surgery, the radiant energy in the spectral region of 800 to 1000 nanometers reaching the patient should be kept to the absolute minimum. This is the energy of infrared absorption by flesh and water and hence results in a noticeable heat to the surgeon, or may cause a drying of exposed tissues. The surgical light assembly should limit the radiant heat energy in the beam to not more than 25,000 microwatts per square centimeter at maximum intensity in the light pattern, or at any other point along the beam of light more than 3 inches from the cover glass or lower edge of the reflector. Operations at overvoltage with a small spot may produce more than 25,000 microwatts per square centimeter.

For general surgery, the light from the luminaire should have a color within an area described by a five-sided polygon. The range of CIE coefficients is defined by the following x and y values:

x	y
.310	.310
.400	.375
.400	.415
.375	.415
.310	.365

When plotted, the above points will result in coordinated color temperatures between 3500 K and 6700 K. See Section 5 for an interpretation of these color temperature and chromaticity values.

Secondary only to its unusual optical qualities is the flexibility designed into the surgical lighting unit. This may be mechanical on units suspended from the ceiling (see Fig. 12–13) or electrical switching arrangement in stationary units in the ceiling.

Smaller "portable" or floor type units provide flexibility and supplement the larger ceiling-mounted units in lighting areas or cavities difficult to light adequately with only an overhead unit. Portable lamps if used where flammable anesthetic materials exist, are required to be explosion-proof.

In addition to the primary lights and portable surgical lights commonly used in operating rooms, provisions and/or space may be provided in the operating room for various other lighting devices of a specialty nature. These include fiber optics, grain of wheat lamps, headlamps, and various types of illuminated endoscopic equipment.

Rooms in which flammable anesthetic agents are used or stored are defined by NFPA No. 56A, "Standard for the Use of Inhalation Anesthetics," as hazardous locations. NFPA No. 70, "National Electrical Code," specifies the standards of electrical construction and equipment permitted in hazardous locations. In Canada, CSA Standard Z32-1963 "Code for Use of Flammable Anesthetics," and CSA Standard C22.1 and C22.2 "Canadian Electrical Code," Part I and Part II respectively.

Scrub Room. Scrub room areas and corridors adjacent to the operating room provide areas where personnel can accommodate themselves to the levels of illumination of the operating room. While scrubbing before an operation, the surgical team should be exposed to light of the same color and level that they will encounter in the operating room. This will not only insure a good job of scrubbing, but will allow them to enter the operating room fully adapted. These levels of illumination will also promote a cleaner scrub area and consequently more aseptic conditions. This same reasoning holds for corridors leading to the operating room. These are the areas where the surgical team adapts to the operating room environment.

Patient Holding Areas. In the patient holding areas, lower levels are recommended as being desirable for patient comfort. High luminance or spot sources should not be used for general illumination. The patient should not be exposed to luminance levels exceeding 90 footlamberts.

Surgical Recovery. Lighting in these rooms is more critical than in ordinary nursing areas of the hospital. The luminances from luminaires and background should be kept as low as practical so as not to add to the discomfort of the patient. At the same time, the lighting should be sufficient for the nurse to observe any appreciable change in the patient's color or physical condition. See Fig. 12–14. The critical problem here is the proper rendering of patient color and physical condition rather than critical detail seeing tasks as performed in the operating room. Where fluorescent lighting is used as the illumination, these lamps should be of the color-improved type.

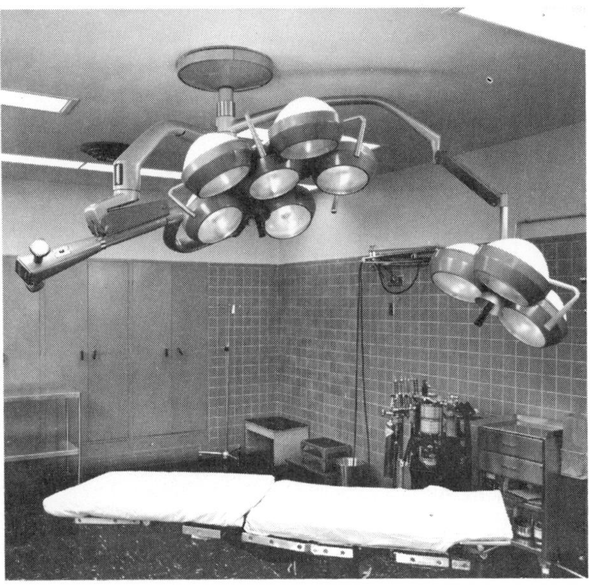

Fig. 12–13. Ceiling mounted adjustable operating lights. General room illumination is provided by recessed fluorescent luminaires.

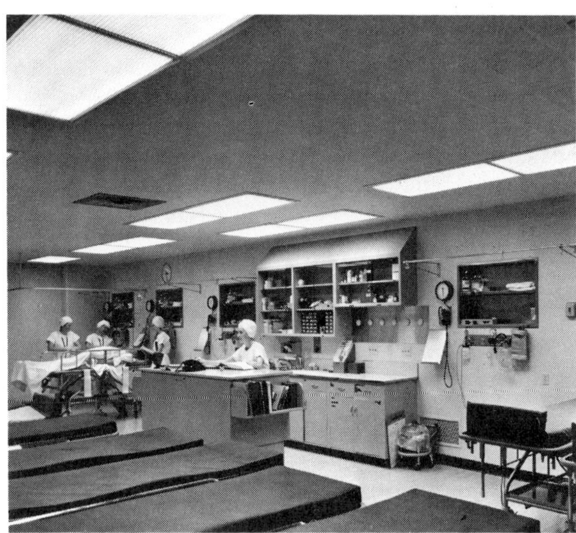

Fig. 12–14. A fluorescent troffer system provides the general lighting in a recovery room.

Emergency Lighting

Provision should be made for emergency lighting of exit ways, anesthetizing areas, and other areas where life and safety would be jeopardized in the absence of lighting. As a minimum, emergency lighting should be provided for surgical and obstetrical tables, exits and exit-direction signs, stairways, corridors between patients' rooms and exits, nurseries, recovery rooms, intensive care nursing rooms, telephone switchboard, and boiler room. Provision of this lighting is electrically inseparable from other essential emergency provisions for continued operation such as iron lungs, central suction systems, refrigeration of blood, bone, and biologicals, and many other items of mechanical equipment. For greater detail, refer to NFPA 101, "Life Safety Code," and NFPA 76A, "Essential Electrical Systems for Hospitals."

HOTELS AND MOTELS[10]

The lighting for public spaces should be designed to provide a series of pleasurable visual experiences. The following general principles should be considered in designing any lighting scheme for a hotel or motel:

1. Harmony with architectural and decorative character of the area (see Section 10).
2. Concern for visual tasks (see Section 15).
3. Control of glare (see Section 3).
4. Luminance ratios (see Section 3).
5. Recommended levels of illumination (see Fig. 9–80).
6. First cost, operation cost, and maintenance cost (see Section 16).

Specific Locations

Exterior. The total exterior lighting should identify the hotel and motel, create a favorable visual impression, and welcome guests. All building lighting, marquee, walkway, parking facilities, lighting, and signs should be interrelated for they are part of the total impression. See Sections 17 and 23.

Entrance Foyer. This is a transition area between the outdoors and the hotel or motel interior. Lighting should provide a sense of security as well as contribute to a smooth orientation to a new environment of a higher or lower illumination level.

Lobby and Lounge. The lobby should unite function and esthetics. Luggage and personal belongings must be easily identified and handled, and there are both casual and prolonged reading tasks in the lounge area. The general transient space can be made to appear different from the lounge seating area by a change in lighting application. Lighting of walls can add to the pleasantness. See Fig. 12-15.

Reception Desk. To make this area easy to locate, it is desirable to have a higher level of illumination here than in the transient area. Although the visual tasks are reading, writing, and personal communication between guest and the hotel or motel representative, the lighting should be esthetically compatible with the surrounding environment. See Fig. 12-16 and Section 11 for the lighting of desk and clerical areas.

Corridors. Light should be provided for safe passage through hallways, on stairs, and to elevators. Corridor lighting should illuminate room numbers and locks in the doors. The lighting should be designed to give the guests a feeling of security, to give special identification to elevator area and ice-making and vending machines, and to reduce the tunnel effect in long corridors. National and local codes require emergency lighting which will operate in the event of failure of the primary source of electricity.

Fig. 12-15. General illumination in this lobby is provided by a downlighting system supplemented by mural and drapery lighting systems. Local lighting is provided by table lamps.

Fig. 12-16. Lighting for attraction, identification, and visual tasks is provided at this reception desk.

Guest Room. The guest room is one of the major commodities of a hotel or motel. Since it is frequently used for small business conferences, flexibility needs to be a part of the lighting plan. General illumination from ceiling or wall-mounted luminaires provides a background for task lighting, aids in housekeeping, and gives a feeling of cheer as well as providing the needed flexibility for nonresidential use. Those visual tasks which need consideration in the guest room are: reading in chair or bed, desk work, and grooming at mirror in both the bathroom and at the dresser. The small entry or foyer to the room should have its own lighting and often this can be so designed to be used also for closet, luggage, or grooming light. For specific techniques, see Section 15.

Ballrooms and Function Rooms. These areas can be used for meetings, exhibits, dancing, and dining. The multi-purpose requirements make it necessary to consider a variety of lighting techniques. Flexibility of space can be provided by controls, such as for dimming and multiple switching, organized for easy operation by banquet and function personnel.

The lighting should be adequate for critical tasks of reading and note taking as well as to enhance the character of the space. Accent lighting should be provided in several locations for speakers, head tables, and displays. Facilities should be provided for local lighting in exhibit areas.

Dining Room and Bar. See Serving and Dining Spaces, page 12-5.

LIBRARIES[11]

In addition to reading areas, there are other locations in libraries involving many different seeing tasks. Among these are: (1) book storage areas or stacks, (2) individual study areas (carrels), (3) microform viewing areas, (4) meeting, conference or seminar rooms, and (5) technical processing areas. Each of these areas present a different or unique lighting problem.

Seeing Tasks in Libraries

Reading is by far the visual task performed most often in a library. Seeing tasks vary from children's books printed in 10 to 14 point type on good non-glossy paper, to newspapers printed in 7 point type on low contrast off-white pulp paper, to law books with long paragraphs in condensed type, to rare books with unusual type faces printed on old paper, to handwriting, both pencil or pen. For a better appreciation of the principles involved in seeing, a review should be made of Section 3, Light and Vision. Section 9, Lighting Calculations; Section 10, Interior Lighting Design Approaches; and Section 11, Offices and Schools are useful for design purposes. Fig. 9–80 lists recommended levels of illumination for the various areas in the library.

Room surface reflectances recommended for use in offices and schools will provide satisfactory luminance ratios for library room surfaces. See Section 11 for recommended reflectances and luminance ratios.

Lighting Systems

A variety of lighting systems are used in libraries. Many libraries make use of daylight through windows or skylights, but more and more are using electric lighting exclusively. In all cases the luminance comfort recommendations should be the same as for offices and schools. See page 11–3.

In areas where architectural features are dominant, design concepts may require sacrifice of efficiency for esthetics when translating the architect's concepts into practical lighting designs. In areas that do not have dominant architectural features, the lighting systems should be selected to provide comfortable seeing conditions with more emphasis placed on economics and luminaire design features. See Figs. 12–17 and 12–18.

For library lighting applications, there are three basic types of light sources in use today: incandescent, fluorescent and high-intensity discharge. See Section 8. No one lighting system can be recommended exclusively. Each system has qualities that match the requirements for a given situation. The first consideration in choosing a lighting system should be to allow the library user to see efficiently and without distraction; the second should be the appearance of the installation within the architectural and decorative design concepts of the library.

In general it is desirable to provide sufficient illumination for the most difficult seeing task throughout a large area. However, if a more difficult seeing task is being performed in a small portion of that area, additional illumination should be provided by providing additional overhead luminaires, or by

Fig. 12–17. Two lighting techniques are combined in this high-ceiling college library. General lighting is provided by a high-intensity discharge lamp downlighting system. The chandeliers, with low-wattage lamps, are used as decorative elements in keeping with the original architect's design concept.

Fig. 12–18. Low-ceiling library illuminated by large luminous areas. Same lighting system is used in both reading and stack portions.

supplementary lighting equipment located in relation to the specific seeing task. Higher levels of illumination should also be provided in areas that will be used by persons with impaired vision. When relighting existing traditional-type library reading rooms, the use of supplementary lighting equipment consistent with the decorative treatment of the room is sometimes required. It is especially important to avoid direct and reflected glare and to avoid veiling reflections when using supplementary lighting equipment.

Specific Areas

Reading Areas. Reading areas in libraries, including main reading rooms and reference rooms, occur throughout almost the entire library. See Fig. 9–80 for a list of the various tasks encountered in those areas. Reading is usually performed on either side of long tables, in lounge chairs, in study carrels, or at the circulation desk. Care should be taken to locate the luminaires to avoid veiling reflections on the seeing tasks and to use luminaires that reduce the luminance in the direct glare zones.

Individual Study Areas (Carrels). Individual study areas or carrels may be found in almost any public area of the library building, such as main reading rooms, enclosed individual rooms, and stack areas. See Fig. 12–19. One of the most serious lighting problems for carrels are the shadows produced by dividing walls. To avoid shadows it is desirable to provide lighting from as many directions as practicable. Special care should be taken to avoid veiling reflections especially from localized luminaires.

Shelving and Stack Areas. This area applies to shelving and storage units for all types of materials in addition to books. The visual tasks in book stacks are very difficult; for example, it is necessary to identify the book by number and author on the lowest shelf. As a result of studies made of typical books at actual viewing angles, it is recommended that, when practical, non-glossy plastic book jackets should be used rather than glossy; large and legible non-glossy lettering should be used for authors, titles, and index number. Dark book, shelf and floor surfaces reflect very little light, and therefore, the use of light colored surfaces should be encouraged.

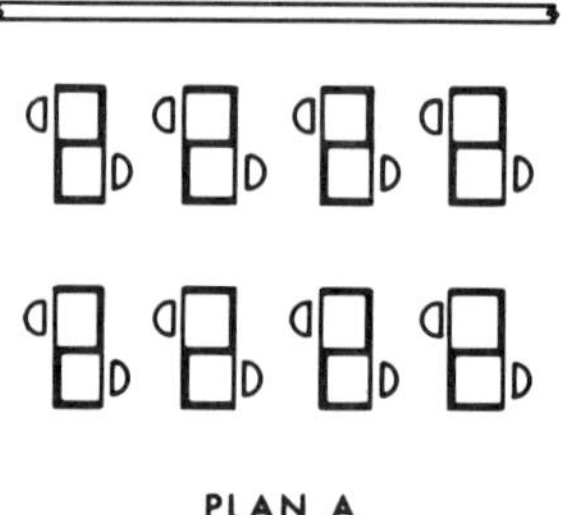

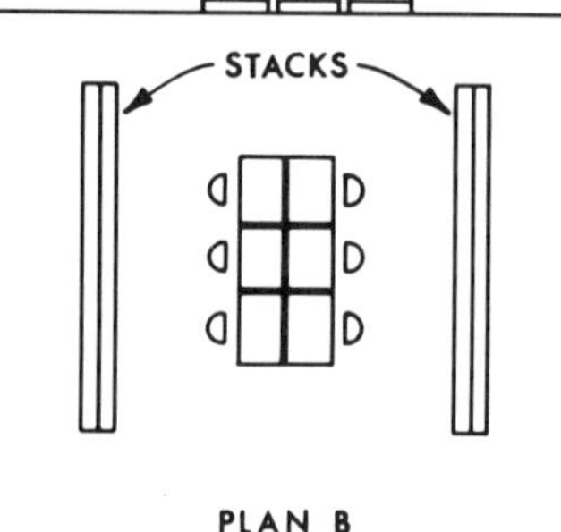

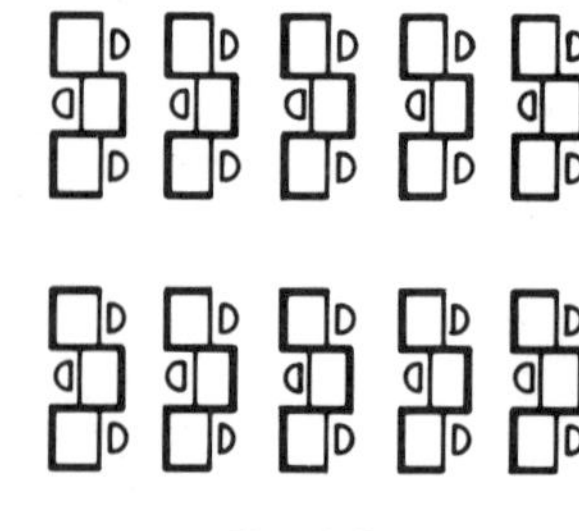

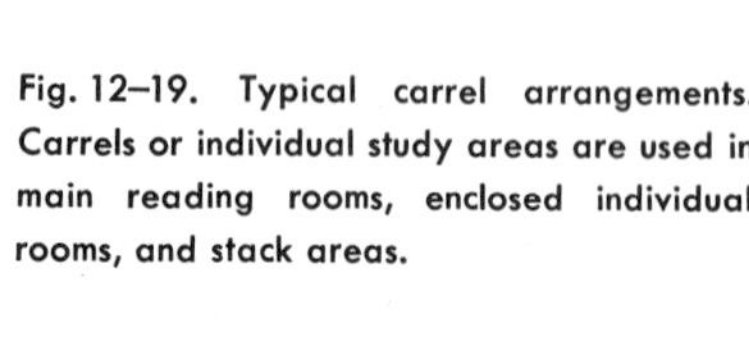
Fig. 12–19. Typical carrel arrangements. Carrels or individual study areas are used in main reading rooms, enclosed individual rooms, and stack areas.

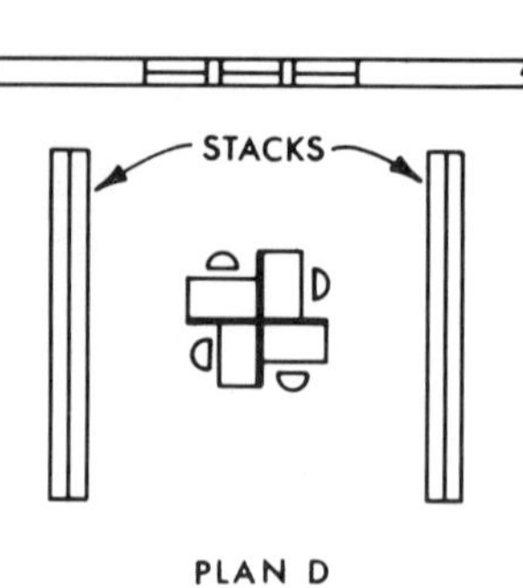

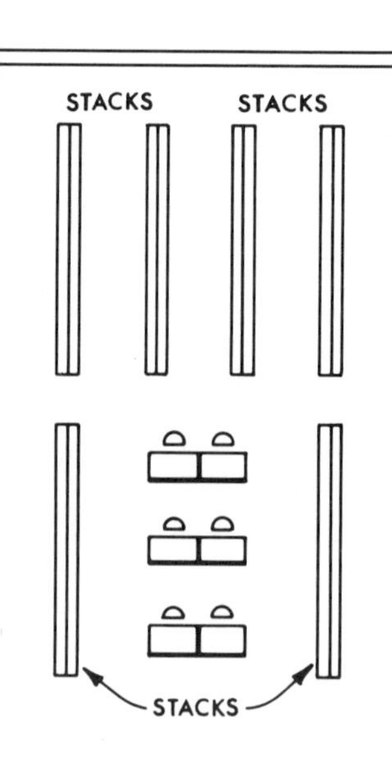

Open Access Stacks. Open access stacks are open to the public for finding their own books or for browsing. See Fig. 12–20. Book stacks are usually arranged in rows with continuous rows of fluorescent luminaires located along the center of each aisle. Illumination from these luminaires reaches the lower shelves at a glancing angle. If luminaires are located at right angles to the stacks some of the illumination reaches the vertical surface at a more satisfactory angle. Advantages to right-angle mounting are that the rows of luminaires can be spaced somewhat further apart and that the spacing of the book stacks may be changed without relocating the luminaires. Regardless of which lighting arrangement is used, the levels of illumination shown in Fig. 9–80 should be provided.

Closed Access Stacks. Closed access stacks are not open to the public and are generally used by library personnel only. The aisles are usually narrower. Luminaires controlled by delayed time switches may be considered for these stacks.

Card Catalogs. Individual files of card indexes are usually located in the main reading rooms. Location of overhead general lighting luminaires at right angles to the file cards rather than parallel to them will provide slightly better illumination on the vertical surfaces of the cards.

Circulation Desks. Circulation desks are usually located near the entrance to the main reading room. Often the general overhead lighting system will provide sufficient illumination for the desk; however, if not, sufficient supplementary illumination should be provided, and may take the form of an architectural element that will identify the circulation desk.

Technical Processing Areas. A general lighting system similar to that used for office areas is recommended for the technical processing areas. See Section 11.

Conference and Seminar Rooms. Conferences are frequentlly scheduled in libraries, and groups hold seminars on occasion. In addition to general overhead lighting, provision should be made to illuminate the speakers and their materials at the lectern and at the seminar table. Two adjustable spotlighting units should be provided for each lectern position. The location of such units is shown under Lecture Rooms for schools in Section 11, page 11–15.

Display and Exhibition Areas. Many libraries have display and exhibition areas. These may be in glass covered horizontal cases or may be mounted on vertical walls or dividers. See Museums and Art Galleries, page 12–18 for lighting such displays.

Audio Visual Rooms. There is an increasing use of listening areas for lectures, music, and other recorded material. These areas are either small rooms with individual reproducing equipment, or large rooms where head receivers may be plugged into circuits at carrels. Small rooms have poor utilization of light because more light is absorbed by the wall surfaces and therefore require a closer spacing between luminaires. Lighting similar to that required for carrels in a large room is also needed for the audio carrel system.

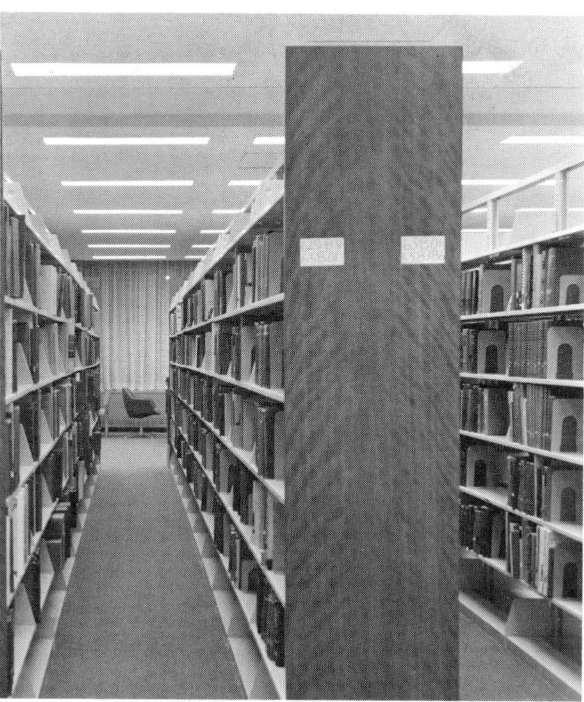

Fig. 12–20. Open access stacks with rows of fluorescent luminaires at right angles to the stacks.

Lighting for Microform Viewing Areas. Microform materials permit much larger holdings of newspaper files, rare books, special collections and technical publications. Microform materials include rolls and cartridges of microfilms on strips, aperture cards containing single frames of microfilm and microfiche cards or sheets containing a series of highly reduced micro-images. One of the most difficult seeing tasks is the reading of a microfilmed newspaper page, viewed in a machine located under a general lighting system needed for other tasks in the area. (Reflections, diffuse and specular, tend to wash out the already poor image on the screen.) When notes must be taken over long periods of time it is desirable to provide illumination on the note pad, but controlled to reduce reflections on the screen.

Higher illumination levels are needed for files of microforms (see Fig. 9–80) than are needed for viewing. Where viewers must be placed in reading areas or work areas with higher level general illumination and no controlled lighting or dimming is available, machines should be selected that are hooded and have screens which are treated to reduce reflections. A small luminaire should be provided between viewers to illuminate a fixed or sliding shelf in front of each machine for note taking. Such a luminaire should be moveable so that it may be individually located to accommodate right or left-handed operators.

Offices and Computer Machine Areas. Office areas in libraries should be illuminated in accordance with the recommendation for Office Lighting in Section 11. Overhead general lighting should be used for computer rooms; however, for those portions of the room that have console and signal lights the illumination and particularly the luminance should be lower. This may be accomplished by providing switching or dimming for the luminaires over the critical areas. This lower illumination will also provide better visibility when cathode ray tube (CRT) consoles or oscilloscopes are being used. Maximum illumination is desirable for viewing computer printouts.

Punched Card and Punched Tape Areas. The illumination in punched card or punched tape areas should be similar to that required for the operation of business machines. See Office Lighting, Section 11.

Rare Book Rooms. Higher illumination levels are recommended for rare book rooms because of the poor quality of printing often found in many rare books; however, lighting techniques such as those used in Museums and Art Galleries should be used for books displayed in glass cases. These would include means of reducing the amount of deleterious radiation.

Archives. Archives are for the storage and examination of public documents of all kinds. This would include legal documents, minutes of meeting, legislative actions, and other historical papers. Pencil writing, small letters and condensed type are used in many of these documents.

Map Rooms. Map rooms have both storage and reading areas. Storage of maps involves the use of deep cabinets which, in turn, requires aisles sufficiently wide enough to open drawers for access to the maps. Maps mounted on vertical surfaces require vertical surface lighting.

Fine Arts, Picture and Print Rooms. See Museums and Art Galleries for the proper lighting of displays, paintings and art objects.

Typing Rooms. Lighting appropriate for office work is required for typing rooms. See Section 11.

Group Study Rooms. Sometimes a group of 4 to 6 students is assigned to a project to be solved by consulting among themselves and isolated rooms may be provided for this purpose. Techniques used for classroom lighting are recommended for these rooms. See Section 11, School Lighting.

Overnight Study Halls. Sometimes students prefer to work all night when preparing for examinations and libraries may provide a portion of the building that can be isolated for this purpose. Lighting for these areas is similar to that required for Reading Areas or Individual Study Areas.

Entrance Vestibules and Lobbies. Lighting in entrance vestibules and lobbies should create an atmosphere suitable for the particular type of library. The lighting may emphasize the architectural features and provide a smooth transition to the functional areas.

MUSEUMS AND ART GALLERIES

In the past, museums and art galleries made the best use possible of daylight for the illumination of not only the interiors, but also the objects and displays being exhibited. With the abandonment of traditional masonry and ancient architectural forms, modern museum structures are free of stylistic restrictions and can concentrate on display functionality. The temple form with its columns, and the palace form with its courts, have given way to functional structures that can better meet the modern needs of displays for educational purposes and of interest to the public.

Modern architectural and structural design coupled with new lighting technology and tools have made it possible to sustain fully the intent of the artist, sculptor, or other exhibitor. Modern structures can be more responsive to the variety in form, size, and intent of displays found in today's art, science, technological and historical museums. This diversity of exhibits dictates the need for flexible lighting systems to illuminate them. The very nature of art and the artifacts of man's epic achievements are dynamic, unusual, unpredictable; precluding a single standard approach for lighting.

It is possible with modern light sources to produce a wide range of lighting effects for art galleries and museums. Individual decisions must be made on the basis of local conditions and requirements as to desired emphasis, degree of drama, textural modelling and specular effects in surfaces, level of illumination, and color.

Low-contrast surroundings create a relaxed mood; high-contrast surroundings produce "tautness" and a sense of theatrical, visual drama which should be employed only where appropriate and without visual discomfort. To choose a rather startling comparison as an example, it might be appropriate to light a room filled with starkly dramatic drawings, etchings, and paintings—such as Goya's depictions of the Inquisition—with spotlighting on the paintings only and no general illumination. (To heighten the drama even more, the walls might be of a dark value—perhaps deep red). The high contrasts from this type of lighting could cause a psychophysiologic shock to the viewer, possibly enhancing his appreciation of the subject matter. At the other extreme, a gallery of frolicsome bucolic scenes by Fragonard might be generally suffused with a soft daylight effect (from daylight or electric light) to help the viewer feel pleasantly comfortable. To reverse the effects in these two rooms would certainly be inappropriate. But such effects should be worked out with the exhibitor. Many, for example, feel that the exhibited objects should provide the only contrast in a room.

Fig. 12–21. Daylighted gallery at night. Concealed services are shown at the right. Wall lighting (from concentrated sources) is beamed through lenses and is spread evenly. Fluorescent luminaires provide the diffuse illumination.

Both general illumination and exhibit-object lighting determine the total luminous environment. The two should compliment each other—not create a handicap for the viewer.

Concentrating vs Diffuse Lighting

Flexibility of positioning all concentrating light sources is a basic requirement. The use of electrified track is now widespread for this purpose. The only exceptions to the full flexibility afforded by such means are (a) for individual, permanently installed features, and (b) for certain galleries of large fine art museums, *e.g.*, those which own collections of paintings that substantially exceed in number the available wall space (here a rotation of items, as well as the long-term exhibition of the best known of great masterpieces, require approximately full wall coverage of vertical illumination so that repeated adjustment of the lighting is not needed). See Fig. 12–21. All concentrating sources may then be in a fixed position and may be concealed.[12]

Concentrating light sources (spots or floods) work best with the softening effect of diffuse light. Concentrating sources alone, tend to obscure the esthetic quality of sculpture and to discourage examination of detail. Diffuse light alone may detract from the appreciation of sculpture, but if luminance contrasts are held within 6 to 1 limits, as is possible with both sources, subtle results may often be obtained. See Fig. 12–22. These limits conform to the special talent of many museum curators, who manage the "molding" of sculptures with electric light to equal the best which daylight affords.

The principle of three-dimensional lighting applies to forms of all kinds. A sufficient diffuse component may be frequently supplied by light reflected from adjacent or surrounding surfaces, or by general room inter-reflections. Freestanding sculpture, in central areas without some kind of backdrop to limit viewing to three sides, will risk discomfort glare.

Another characteristic difference between diffuse and concentrated light is quite important to the accurate and pleasant viewing of color. Diffuse light

a

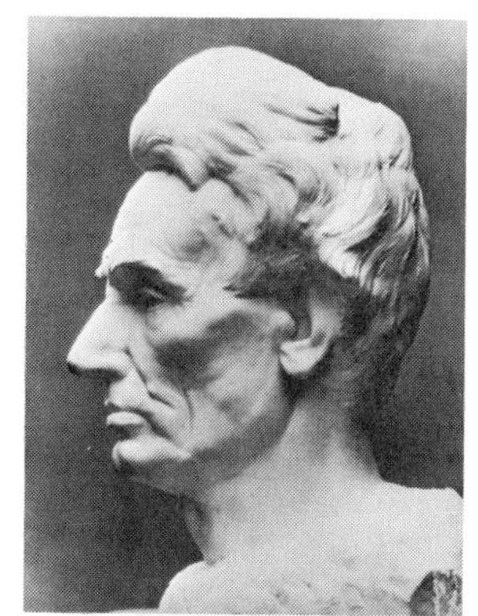

The Metropolitan Museum of Art, Morris K. Jessup Fund, 1929.

b

The Metropolitan Museum of Art, Rogers Fund, 1912.

c

The Metropolitan Museum of Art, Rogers Fund, 1917.

Fig. 12–22. Lighting of sculpture. a. Concentrating sources, alone, from front left. b. Total overhead diffuse lighting conforms to expression of features. c. Low diffuse lighting and strong concentrated accent adds to stern expression. d. Light concentration from upper right. Strong overhead diffuse lighting aids in viewing the details of this complex sculpture.

alone, tends to "desaturate" colors and imparts a dullness to them. Concentrated (directional) light strongly renders saturation in colors.[15] Paintings, such as those of Van Gogh, which have very irregular surface textures in addition to the interspersing of saturated color were apparently intended to produce a specific *post-retinal* relation between the colors and the texture. Both concentrated and diffuse light together are desirable for full appreciation of such surface characteristics. This post-retinal phenomenon is often experienced in judging relative brightness contrasts. This difference between diffuse and concentrated light shows up markedly on paintings having more than one coat of varnish.[16] The latter are usually kept cleaned and conditioned by experienced restorers in larger museums.

Color of Light Source

For two or three-dimensional lighting requirements (wall-hanging and freestanding), aspects of both color and texture must be specifically visualized. In interior spaces, a lower color temperature or (warmer) source than daylight is usually preferred. The color temperature of daylight can range from 5000 K to 50,000 K. Electric sources available today generally range from 2500 K to 7500 K.

The three major categories of electric light sources, incandescent, fluorescent, and high-intensity discharge, all now have lamp varieties that can offer good to excellent color rendering. Incandescent light sources, including the many tungsten-halogen types, are high in color rendering ability although they tend toward the warmer tones, *e.g.*, yellow-orange. Standard white fluorescent lamps such as warm white, cool white and daylight, are limited in their color rendering capabilities. However, the deluxe cool white lamp and the higher color temperature daylight-matching fluorescent sources do an excellent color rendering job. High-intensity discharge sources are compact, efficient, and produce a great deal of light but present sources have limited color rendering capability. As these sources are developed they will become more suitable for museum and art gallery lighting.

Tinting of standard incandescent projector lamp output may often produce outstanding results by the use of pale, blue-white, "daylight" color roundels. Paintings which have the delicate blues or other colors of unsaturated quality, such as French impressionist, particularly respond to "daylight" tonality in accent lighting. Tinting, other than with "daylight" tones should be done only under curatorial judgment in the lighting of fine arts.

Entrance Foyer

An entrance foyer generally should be regarded as a vision-conditioning space in which the visitors' eyes are permitted to adjust from daylight levels to lower lighting levels in most art galleries, textile exhibits, costume collections, and other displays susceptible to deterioration in exposure to strong light. It is desirable to have some control over the general illumination level in entrance foyers to adjust the lighting level to the changing need. In the evening the adaptation process is reversed; the visitor must accommodate to the low level of exterior lighting when he leaves the building.

Gallery Design Principles

Presentation of concentrated light to wall displays (tapestries, paintings, etc.) should be at an incident angle of 60 degrees with the horizontal, centered at an adult sight-line height of 5 feet 6 inches from the floor as shown in Fig. 12–23.* This angle provides a good balance between frame shadow, specular reflections from protection glass or varnishes on paintings, the "raking" of surface textures (see Fig. 12–24), and maintaining a practical width of viewing zones. These, of course, will change with the height of hanging displays. (See Fig. 12–25).

A nominal level of illumination of 30 footcandles maintained (being total flux from all sources), on both horizontal and vertical planes, is recommended to meet all normal visitor functions of gallery viewing, copying, and studying. Vertical footcandles

* These points control the over-all geometry of viewing.

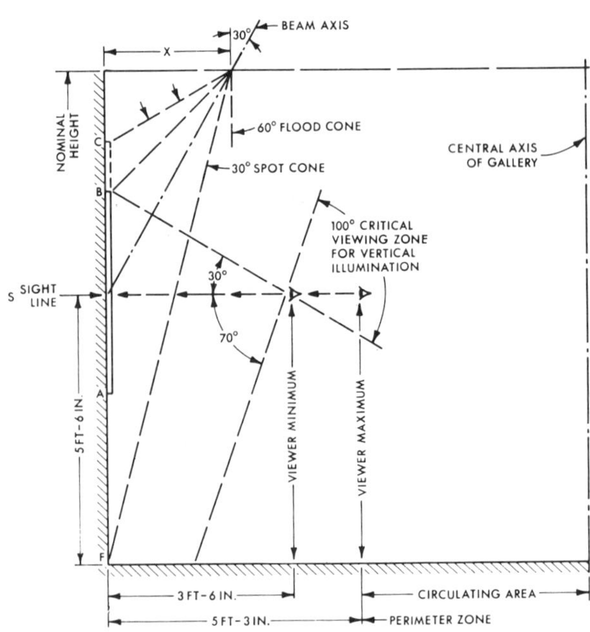

Fig. 12–23. Model perimeter (viewing) zones at nominal ceiling height. Model based on: (1) primary diffuse component (of vertical footcandles) at approximately 40 per cent of horizontal footcandles at S, (2) height of wall-hung display, (3) ideal utilization of beam cones, and (4) minimum effective viewing distance relative to a nominal height of object (A to B = 52 inches for a 30-degree cone, A to C = 65 inches for a 60-degree cone). To calculate viewing zones for higher objects, increase horizonal dimensions 1.5 inches for each 1.0 inch increase in height of object.

Fig. 12–24. Raking of weave of Renaissance tapestry by high angle directional light (left). Same tapestry with concentrated light at 60 degrees incident angle (right).

The Metropolitan Museum of Art, Bequest of George Blumenthal, 1941.

Fig. 12–25. Range of perimeter zone width. A—with fully controlled wall illumination. B—with full width luminous ceiling.

should be figured for these purposes on a 60 per cent full wall coverage basis, but, due to variable factors, may not, in practice, average that figure.

Sight-Lines, Ceiling Heights, and Viewing Zones. A sight-line height of 5 feet 6 inches from the floor has been found most responsive to adult seeing habits. Sight-lines in rooms set up for school classes may be lowered, but not more than one foot, because of wide age differences between grades. High schools invariably use the regular galleries.

For purposes of lighting design, the determination of the 60 degree incident angle for concentrated, vertical illumination, is of the first order of procedure. At gallery height H in inches, specified by the museum, the distance in inches, x, from the walls to the line of light sources is:

$$x = (H - 66)(.577)$$

This provides a frame of reference around the gallery ceiling for the respective designs for lighting the perimeter (viewing) zone and central area. However, such calculation neither defines the viewing zone itself, nor its basic relation *to objects being viewed.* The control of ceiling heights should be based on the anticipated maximum height of wall-hung displays. In a small museum, the higher ceilings may be needed in only one or two rooms.

Unless ceiling heights are arbitrarily established, the dimensions of perimeter (viewing) zones should be related, for the purpose of electric illumination, to three pertinent factors (see Fig. 12–23):

1. The height of paintings or tapestries, etc.
2. Ideal utilization of concentrated light-beam cones.
3. Minimum effective viewing distances for the maximum permissible height of exhibits.

At the specified 60-degree incident angle of presentation, a projected light-beam cone becomes oval. Because the maximum candlepower at the beam axis is the one critical point governing (a) relative brightness as a function of good seeing, and (b) any over-exposure to photochemical damage hazards (see below), measurement of vertical illumination at the sight-line becomes one of the most trustworthy, yet simplest, of criteria for the management of lighting in a fine arts museum. It will also provide a guide for the number of horizontal spacing of emphasis lighting units needed for any single display setup.

A study of Fig. 12–23 with performance data on projector and reflector lamps will show that for good visual performance and maximum safety of fine arts collections, the following approximate, sight-line values of *concentrated* illumination (in "maintained" footcandles) should *not* be exceeded:

1. 60 footcandles for short-term, or temporary exhibits.
2. 20 footcandles for fixed or permanent exhibits.

Combined with the described diffuse component, the total, vertical footcandles will, according to the best available information, conform to both seeing performance and museum conservation limits.

The above specified values may be selected according to standard, concentrating lamp ratings of maximum candlepower within 10-degree beam cones, as follows:

$$E = \frac{.214\ I}{h^2}$$

where:
E = maintained illumination in footcandles,
I = the (manufacturer's) listed, initial candlepower, and
h = ceiling height *above* sight-line.

Since such lamps are handled much more frequently, they are cleaned more often. The above equation is, therefore, based on an 80 per cent light loss factor, or 95 per cent median candlepower, with a 10-degree cone, over the life of the lamp, the length of beam throw squared, and the cosine of the specified incident angle.

It will be seen at once that any real balance between horizontal and vertical illumination contradicts the idea of flexibility. However, it is equally plain that the cost of vertical lighting in terms of good seeing favors diffuse light. The over-employment of concentrated light may be shown to *decrease* visual efficiency and *increase* photochemical damage hazards. The over-employment of concentrated light has become a commonly demonstrated tendency in the absence or lack of sufficient diffuse light. It is a potentially harmful trend. To the extent that the two are made complementary, and kept under control, unnecessarily high electric current costs, and the impaired visual performance of visitors are avoided.

There are, of course, a large number of ways to meet these needs. They depend upon the architectural and structural character of the building, whether or not lighting auxiliary to daylight is involved, availability of ceiling cavity space, the incidence of arches and doors, dimensions of rooms, the physical features of the material for exhibition, and available operating funds. Ceilings may be fully luminous, or with luminaires distributed in the central area.

For the best seeing a viewer should be unaware of where the light is coming from, but in many museums the ceiling and the lighting appear in the view of approaching visitors. Viewers then quickly become aware of any over-busy, or cluttered ceiling appearance. It is recommended that luminance ratios between adjacent luminaires or surfaces be reduced to a ratio of 3 to 1. If fully luminous ceilings are preferred, they should extend to within one foot of walls for galleries having ceiling heights of 15 feet or less, and within two feet of walls for those over 16 feet in height, in order to insure that at least 40 per cent of the calculated diffuse horizontal illumination is available as vertical footcandles at the wall, measured along the sight-line height. Such a ceiling should have its control divided between the perimeter and the central areas; the latter may be subject to dimming—no closer than a 50 to 1 range being necessary.

Floor luminance is important in all rooms devoted to wall hung displays. It is very distracting at or near the base of walls in the normal position of viewing. This can be clearly seen by reference to Fig. 3–6, page 3–4, showing that the angle of visual sensitivity for both eyes, increases to 70 degrees below the horizontal, at the bottom of the field of view (see Fig. 12–23). Floor reflectance in galleries should be held below 10 per cent (see Fig. 12–21), being totally unlike the ideal conditions for task lighting where the higher floor reflectances often aid the available working flux. Floors of this reflectance appear lighter under diffuse light when waxed, as is frequently the case, but this effect disappears when directly viewed at the walls.

Lighting for Displayed Materials

Display areas are devoted to case or vitrine exhibits, dioramas, habitat groups, etc., where viewers stand or move in relatively low levels of illumination. Unlighted display cases or table cases are now rarely used in modern museums of fine arts. They are employed there largely for study-storage. In natural history and science museums, however, they are invaluable for large cases requiring the exhibition and protection of varieties of specimen and large objects of archeological, ethnological, or geological interest, etc. These cases can be well illuminated by general lighting through the top glass without significant problems of specular reflection.

The principal consideration in case lighting, for the comfortable visual performance of visitors, is the brightness of any exterior object, when seen reflected from enclosing glass fronts, should be substantially less than that of the objects on display within the case. Competing images, including brightness of "viewers" faces and those of the exterior surround, detract greatly from visual efficiency. These interferences may be minimized if the reflectances of the interior backgrounds, decks, and mounts within cases strongly favor the objects exhibited. This also means *as low a luminance as possible of the illuminating source within the case.*

The image brightness of viewers' faces is almost solely due to unshielded, translucent diffusing panels in the light openings. Particular care should be exercised in laying out case galleries, to avoid image interference from the brightness of the room ceiling luminaires, and of lighted cases on opposite walls. Frequent practice is to provide wall cases which have their glass fronts tilted forward slightly at the top, to throw reflections downward below normal sight-lines. The frequent practice of mounting free-standing cases *endwise (normal) to one wall* of a room, their exposed ends being paneled, also minimizes this problem.

It is regular practice to conceal the interior light

sources within enclosures called "attics", mounted at the top of wall or free-standing cases. If displays are sensitive to temperature, humidity, irradiation from light sources, or because their rarity or intrinsic value must be safeguarded, attics are placed *over* the glass top enclosure to control opening of the case. Illumination in either event is best presented through relatively deep (60-degree cut-off) low-luminance louvering. The etched aluminum type with small, hexagonal cells has very satisfactorily met these requirements. Matte black or gray finished square cells, or plain longitudinal louvering, with the deeper cut-off and the same finish, have also been acceptably used.

Daylight fluorescent color enhances silverware; other tonalities of the family of white fluorescent lamps are sometimes used over the great variety of decorative arts, antique jewelry, folk artifacts, and archeological material, when color rendering is not important.

Lighted case interiors should have clear, transparent shelves, or the well-planned use of fabric-covered block mounts of varying heights. The latter scheme lends itself well to labeling, and very striking effects are achieved with them. Case interior illumination levels very rarely should exceed 30 footcandles, measured at the usual waist-high deck, or three feet from the floor in full-length cases. In measuring results, the factor of internal reflections from the inside surface of glass fronts should be noted. These add considerably to the flux available in the lower half of a case. Footcandle measurement may properly be made only with the front closed.

Consistent with safety and room-cleaning maintenance, the level of general illumination in case-lighted galleries may pleasantly range between 5 and 10 footcandles. Substantial spill-out of light from cases will always add considerably to lighting the circulating areas of the room. Floors should be lighter than elsewhere at 40 per cent (maximum) reflectance, and ceiling finishes should be at the same level. Low-luminance downlights, recessed or surface-mounted and finished to blend with ceiling luminance, will conform well to the foregoing. General illumination at these low levels should have the warmer tones of incandescent lamps.

Period and Historic Room

The over-riding limitations in these applications is the faithful maintenance of the orginal aspects of the room, and this is uncompromising in rooms of historical significance. The problem of integrating the instruction of visitors and school classes with the original illumination levels was never anticipated by the first owners or designers. Rarely, does the original lighting means serve to permit critical examination of the interior.

If the period room is not specifically historic, but rather an example of period design, its elements usually are chiefly original, and therefore, irreplaceable. Visitors are rarely allowed to enter and circulate at will around the room, but may be limited to a restricted area near the doorway, or "in" and "out" another nearby. Any treatment must be specially created if a museum chooses to provide modern illumination levels in the room. Various means have been used, such as employing optical beam projection, under curatorial supervision and responsibility, but all methods of this nature must involve penetration of walls or ceiling, for which no suggestion here is possible.

Assuming the existence of a barrier behind which viewers stand, it is usually feasible to mount out of sight, two or three swivel type reflectors in one cluster over an outlet above, or at the end of the entrance doorway, fully exposed, but above and behind the group. Not more than two 75-watt reflector floods at sides, and one 75-watt reflector spot in the middle, are needed if the cluster is near the corner of the room. Two blue-white, 100-watt, PAR-38 flood lamps are usually ideal for the average room. This unit is connected to control switch at the door jamb, where the guide stands to demonstrate. The room is first seen in its original aspect for the full verbal explanation. Then, the examination light is turned on to show detail, and extinguished when the group leaves. This is a forthright alternative to either a flashlight, or a portable indirect torchiere, plugged in to base outlets of the room, and left lighted.

The treatment of *false window lighting* is of major importance in this class of room. When this is required, it should be kept in mind that originally such window light presented all the characteristics of outside daylighting, the sources of which were not only sun and sky, but also very largely from *ground reflections*. It is not necessary, nor desirable, that false-window cavity reflectors reproduce landscaping, nor city streets, nor any noticeable added indication of outdoors. But excessive direct glare from the lower half of a false window is unnatural and a direct burden on occular accommodation. It can be minimized if a plaster-smooth surface is on the reflector; the cavity floor returned as far below the sill as possible—the return to have a darker gray paint finish (Munsell GY 5/2). The lower half of the reflector is then finished in three stages of graduated, matte, low chroma (green-yellow-gray), beginning about Munsell value GY 7/2, and growing lighter in two more stages of G 8/2 and G 9/2, (Gray-Green) respectively, to about half-way up the window opening, or where it may be reasonably judged as the area of the horizon. Above this, the finish would be as usual, a high reflectance matte white. If this is done, the effect will be that of a light, early morning fog, obscuring the surround. This is not too difficult a job for an experienced painter. Appropriate glazing of sash is desirable.[17]

The light source should be standard cool white fluorescent lamps in strips mounted vertically on both sides, concealed behind the outside reveals, and located in the upper part of the opening. With a shallow, curved reflector, the total flux of these lamps should be at approximately 300 lumens per square foot of window opening, on a maintained basis. If there is sufficient cavity depth available, projector lamps may be beamed obliquely through the sash from an upper corner of the cavity, across the sill and down onto the floor, simulating high angle sunlight. For a window height of six feet or less, not more than two 150-watt R-40 lamps are adequate for this purpose. Ventilation is necessary.

Museum Conservation

The great art and other collections of this country are part of our national wealth, and today museum conservation is a scientifically organized technology in the restoration and preservation of historic and artistic works. The extent of deterioration depends primarily upon (a) the level of the radiation, (b) the time of exposure, (c) the spectral characteristics of the radiation, and (d) the capacity of the individual materials to absorb and be affected by the radiant energy. Certain external factors may increase the rate of deterioration: high temperature, high humidity, and active gases in the atmosphere.

The so-called reciprocity law, while important to consider, is far less an uncertainty than the character of the shorter wave spectral distribution in daylight and electric light sources, together with the wide variation in resistances of materials themselves to deterioration. The first definitive advance was made by the U.S. National Bureau of Standards in 1950, for the preservation of the Declaration of Independence, and the Constitution of the United States.[18] Extensive research is continuing.* [19] Any exposure under the relatively high levels of common merchandising practices is usually of little significance when goods are continually sold and delivered. Title to museum objects rarely passes.

Data now available[20] show that the commonly employed light sources in museums provide an order of photochemical damage hazard, as shown in Fig. 12-26. These values are expressed in the Probable Damage per Footcandle (D/fc), of six different light sources, compared with clear zenith sky light taken as 100 per cent, both under direct exposure and through a tested available commercial acrylic filter material.

It can be noted that the most hazardous radiations are those having spectral distributions which are dominant in the shorter wavelengths. The relative damage factor increases logarithmically in inverse ratio to wavelength. Thus, ultraviolet is far more hazardous than visible light. If, however, footcandle levels are sufficiently high, such shorter wavelength visible radiation may become significantly hazardous, even if all ultraviolet is filtered out.[21]

The foregoing should not be taken to infer that damage must occur regardless of the fastness or light stability of any substance, but the possible hazards of exposure to irreplaceable materials must be carefully considered in museums. Exhibition case lighting, due to the levels at which objects are mounted in proximity to light sources, such as on high shelves, provides the point of the most critical exposure to fading or structural damage. The filters widely used are those which test equally with the performances of types UF-1 and UF-3 acrylics. Type UF-1 has a zero cut-off at 360 nanometers, a cut-off of 60 per cent at 400; and of 89 per cent at 420, giving "water-white" transmittance of the visible spectrum.

Type UF-3, slightly yellowish in appearance, has a zero cut-off at 395; of 56 per cent at 400; and 79 per cent at 420 nanometers. Since, at least, 79 per cent of blues are thus transmitted, color rendering is not noticeably affected, and the material further reduces the hazard in zenith sky light to 50 per cent below that achieved with UF-1.[22] This filter material comes in two forms, one in ⅛″ thick cast sheets, which would be cut to size and laid on top of the vitrine beneath the lamp attic. The second form comes in ⅛″ thick, extruded tubing, made to slip over the fluorescent lamp itself. The tubing should not be used over tubular incandescent lamps because excessive heat will deteriorate both lamp and the acrylic. Further, if any incandescent lamps are used without the use of an adequate filter, for case lighting of susceptible material, the maximum flux density *at the point of closest* proximity to exposed material should not exceed 12 footcandles.

* Particularly by the Mellon Institute, Pittsburgh, Pa., in cooperation with The National Gallery of Art (Smithsonian Institution, Washington, D.C.)

Fig. 12-26. Probable Damage per Footcandle (D/fc) Inherent in Museum Light Sources

Light Source	Color Temperature (degree K)	D/fc (Per Cent)	
		Unfiltered	Filtered*
Zenith Sky Light	11000	4.800 (100.0)	.407 (8.5)
Overcast Sky Light	6400	1.520 (31.7)	.243 (5.1)
Sunlight	5300	.790 (16.5)	.192 (4.0)
Fluorescent (CWX)	4300	.554 (11.5)	.147 (3.1)
Fluorescent (WWX)	2900	.444 (9.2)	.086 (1.8)
Fluorescent (Daylight)	6500	.402 (8.4)	.245 (5.1)
Incandescent	2854	.136 (2.8)	.062 (1.3)

*A special ultraviolet absorbing formulation of polymethylmethacrylate (UF-1 Rohm & Haas) producing clear transparency to visible radiation.

REFERENCES

1. Rambusch, E.: "Public Buildings," *Illum. Eng.*, Vol. LI, p. 108, January, 1956.
2. Church Lighting Subcommittee of the Institutions Committee of the IES: "Church Lighting," *Illum. Eng.*, Vol. LVII, p. 67, February, 1962.
3. Subcommittee on Kitchen Lighting of the Institutions Committee of the IES: "Lighting for Commercial Kitchens," *Illum. Eng.*, Vol. LI, p. 553, July, 1956.
4. Subcommittee on Hospital Lighting of the Institutions Committee of

the IES: "Lighting for Hospitals," *Illum. Eng.*, Vol. LXI, p. 417, June, 1966.

5. Griffin, N. L.: "The Hospital Electroencephalographic Suite," Public Health Publication No. 930-D-13, U. S. Government Printing Office.
6. "Report of the Surgeon General's Ad Hoc Committee on Planning for Mental Health Facilities," U. S. Department of Health, Education, and Welfare, January, 1961.
7. "Nursing Homes Environmental Health Factors—Volume 13, Lighting," U. S. Department of Health, Education, and Welfare, March, 1963.
8. "Lighting for Hospital Patient Rooms," PHS Publication No. 930-D-3, U. S. Department of Health, Education, and Welfare, Public Health Service, Division of Hospital and Medical Facilities, October, 1962.
9. Antes, J.: "Bibliography of Hospital Illumination," *Guthrie Clin. Bull.*, Vol. 40, p. 96, 1970.
10. Subcommittee on Hotel Lighting of the Institutions Committee of the IES: "Lighting for Hotels," *Illum. Eng.*, Vol. LIII, p. 359, July, 1958.
11. Committee on Library Lighting of the IES: "Recommended Practice of Library Lighting." *to be published.*
12. Harrison, L.: "Auxiliary Lighting for Museum Galleries," *Illum. Eng.*, Vol. XLV, p. 676, November, 1950.
13. Logan, H. L.: "Modeling with Light," *Illum. Eng.*, Vol. XXXVI, p. 216, February, 1941.
14. Helson, H., Judd, D. B. and Wilson M.: "Color Rendition with Fluorescent Sources of Illumination," *Illum. Eng.*, Vol. LI, p. 329, April, 1956.
15. Evans, R. M.: *An Introduction to Color,* Vol. V, Plate II, John Wiley & Sons, Inc., N. Y., 1948.
16. Harrison, L.: "Visual Experience and Museum Lighting," Paper, CIE Zurich Congress, 1955.
17. Harrison, L.: "Museum Lighting," *Illum. Eng.* Vol. LV, p. 65, February, 1960.
18. "Preservation of the Declaration of Independence and the Constitution of the United States," U. S. Nat. Bur. of Stand., *Circular 505,* 1951.
19. Feller, R. L.: "Control of Deteriorating Effects of Light Upon Museum Objects," *Museum (ICOM)*, Vol. XVII, No. 2, p. 57, 1964.
20. Harrison, L.: "An Investigation of the Damage Hazard in Spectral Energy," *Illum. Eng.*, Vol. LIX, p. 253, May, 1954.
21. Harrison, L.: "Recent Advances in Conservation," Butterworths, p. 1, London, 1963.

SECTION

13

LIGHTING FOR MERCHANDISING

Merchandising and Architectural Trends 13–1 *Lighting and Merchandising Relationships 13–2*
Merchandise Emphasis 13–4 *Show Windows 13–7*
Exteriors 13–9 *Parking Areas 13–12*

Perhaps no other application area presents as great a challenge to the Illuminating Engineer as the merchandising area. Current trends in merchandising and architecture have established basic considerations effecting lighting design.[1]

TRENDS IN MERCHANDISING

Surburban Growth. The trend in decentralization of population from urban to suburban areas has had a profound effect on marketing and merchandising patterns for retail stores. New suburban shopping centers have become an established means of retail distribution, influencing every phase of store design and construction.

Modernization. In direct challenge to the building of new shopping centers, existing urban stores have modernized at an increasing frequency. Remodeling offers vitality to the downtown shopping area.

Changes in Store Character. Radical changes in merchandising philosophy have resulted in the introduction of new techniques and patterns in store design. There has been an increasing emphasis on diversification, with many shops, such as drug, hardware, and food stores expanding the types of merchandise carried.

Concurrent with these shifts in merchandising trends there have been changes in the methods of presenting and displaying the merchandise itself. Self-service and self-selection, both designed to reduce sales personnel, have become increasingly popular. Lighting can assist in presenting the merchandise to the customer in a manner which virtually makes the merchandise sell itself.

Circulation and Merchandise. The lighting system often must be suitable for the display and appraisal of many different types of merchandise. Distinctive yet coordinated changes in the lighting may also serve as a means of identifying or setting off various departments.

The lighting pattern, together with supplementary accents, can be employed to attract customers into the store, influence the traffic flow of customers once there, and assist in emphasizing different types of merchandise.

NOTE: References are listed at the end of each section.

TRENDS IN ARCHITECTURE

Influence of Merchandising Techniques. Changing trends in merchandising have influenced the architectural and engineering design of new shopping center construction and to a somewhat lesser extent, the modernization of existing buildings.

Since the shopping center units are basically horizontal they lend themselves to a merchandising layout influenced by horizontal circulation. Large horizontal floor areas are frequently divided into individual "shops" of various sizes, each requiring differentiated lighting.

Influence of Types of Stores. The type and size of store may greatly influence the lighting plan and technique employed. They range from small shops to large department stores, with low to high ceilings and regular to irregular shapes.

Whether the lighting is to be designed in a high or low key may depend on the style of architecture and the class of merchandise to be displayed. A bare-lamp installation, or even a merely functional installation, has often conveyed the implication of economy-priced merchandise. On the other hand, a neat, well-shielded, unobtrusive lighting system will help create a distinctive atmosphere and the tone generally associated with higher-grade merchandise.

Basic Changes. The use of large glass areas, the introduction of long-span bays, the increasing use of air conditioning, and the integration of store fixture design with the architectural structure all are important influences on the lighting design as an integral part of the structure.

Although the trend to large glass areas has provided the opportunity to utilize an effective and appealing interior to attract passers-by, it has also created a visual problem by providing a surface for veiling reflections. To make the store interior and merchandise visible from the outside in the daytime, a sufficiently high level of lighting must be provided to overcome, as much as possible, the effect of these veiling reflections. It is particularly important that vertical surfaces in such stores be well illuminated. See Fig. 13–1.

Structural features of newer buildings, such as larger bays and clear spans, may give the lighting designer much more latitude in his layout pattern by producing clear ceiling areas. However, recessing space may be more restricted due to other mechanical equipment concealed in the plenum.

The relationship between the lighting and air-conditioning system has become increasingly important. Not only must the lighting be designed with the heat load in mind, but the actual layout and type of luminaire used is often affected by the location and size of air-conditioning ducts and diffusers, and by the location of sprinkler heads when used. There are luminaires also available which incorporate air distribution and air return. Additional information concerning the relationship of lighting and air conditioning will be found in Section 16.

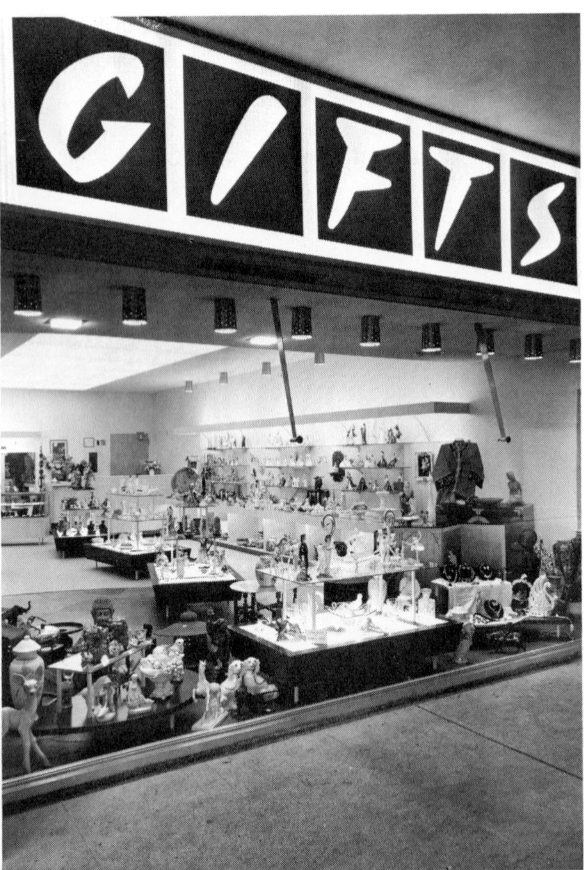

Fig. 13–1. This gift shop illustrates the fact that in open-front stores the interior illumination is also "show-window" lighting.

Interior design and the planning of store display and storage fixtures have become so closely related to store architecture that the lighting of these elements should be carefully integrated. Lighted valances, cornices, shadow boxes, illuminated counter tops and other devices, can contribute greatly to the over-all lighting effect. In addition, showcase and display lighting offer the designer a varied selection of equipment to give the store interior added sparkle and interest.

RELATIONSHIP BETWEEN LIGHTING AND MERCHANDISING

Physical Characteristics of Merchandise

Along with the consideration of the over-all architectural effect, it may be desirable to design the lighting specifically for the type of merchandise displayed. Color, pattern, texture, size, shape, and finish of the merchandise will influence the type of lighting in a given sales area. The directional quality of point sources such as incandescent and high-intensity discharge sources enhance some merchandise qualities that the diffuse quality of a fluorescent linear source does not.

To show facets of jewelry, contours of silver, weave in fabrics, pile in rugs or carving in furniture, directional light is needed. Diffuse light is suitable for hard goods and massed packaged merchandise. The emphasis will shift between point and linear depending on the merchandise and the type of store.

Horizontal and Vertical Display

A store lighting system should be designed to provide the desired quantity and quality of light in the planes in which the merchandise is displayed and examined by the shopper. General lighting is usually designed in terms of the horizontal illumination which is satisfactory for merchandise displayed on counters and tables. However, many types of merchandise, such as draperies, clothing, and shelf goods, are displayed vertically and, therefore, careful consideration of the illumination in the vertical plane is required. See Fig. 13–2.

Color

Lighting which does not show merchandise as it will appear where it is to be used is often responsible for customer dissatisfaction and the return of goods. It is generally advisable to use incandescent lighting on merchandise such as: home furnishings; wearing apparel; jewelry; cosmetics; and meats, fresh fruits, and vegetables. The availability today of fluorescent, incandescent, or even high-intensity

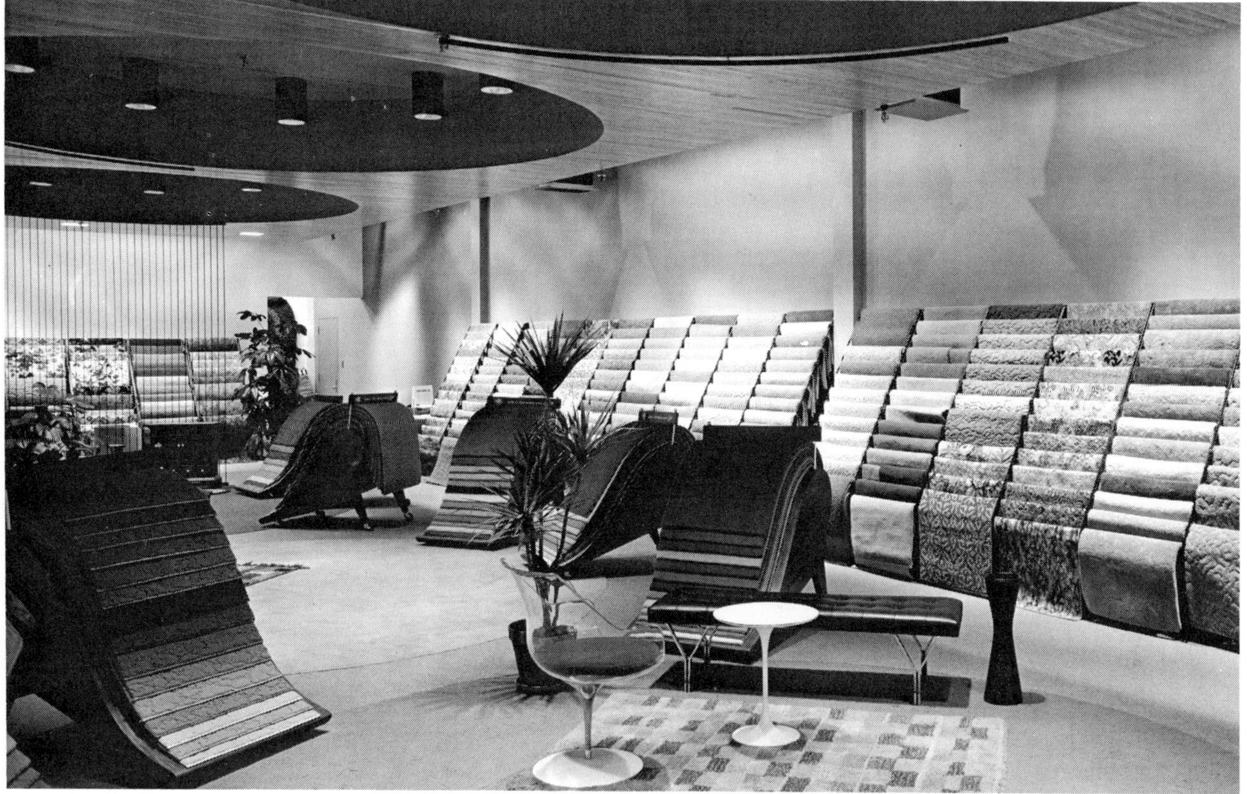

Fig. 13-2. Careful consideration of the illumination in the vertical plane is illustrated in this display of carpets.

discharge sources offers a variety of colors by direct or mixed use to provide the spectral distribution of light under which the specified merchandise is most frequently used. It is generally advisable to use a cool quality of light on furs, outdoor sporting goods, and garden supplies.

If used with great caution and taste, color tints are sometimes suitable. For example, straw tint may be used to dramatize beach and southern vacation equipment and apparel, or pale pinks can be used to glamorize lingerie and romantic evening wear.

Color is often the determining factor in selecting the lamp type to use. The following effects should be considered in making a choice:

1. Incandescent filament lamps produce a warm atmosphere in the store by emphasizing reds, oranges, and yellows. They give a familiar appearance to colors of merchandise, room finishes, and customer complexions.

2. Improved-color cool-white fluorescent lamps make colors appear almost as they would under daylight. Their use creates a neutral or cool atmosphere in the store, but they are still flattering to skin tones.

3. Improved-color warm-white fluorescent lamps give an effect on colors similar to that of incandescent lamps, creating a warmer atmosphere in the store.

4. Cool-white and warm-white fluorescent lamps which have a higher efficacy lack sufficient red to be flattering.

5. Daylight, and soft-white fluorescent lamps are not usually recommended for general lighting except for special applications.

6. High-intensity discharge lamps with improved spectral characteristics may be considered where long life and high efficacy are required. Careful consideration should be given to the type and characteristics of the merchandise and the spectral distribution of the light under which the merchandise will eventually be used.

Factors in Seeing Merchandise

Buying decisions are based largely on seeing tempered by a subjective response to environment. Package design is a science in itself, involving choice of texture, color, and shape in order to capture attention and add visual appeal, even though the contents may fulfill desires in terms of the other senses. Hence, providing conditions under which the customer can see effectively has a direct bearing on buying. Three of the factors which determine the customers' seeing performance are: inherent visibility of the merchandise, time, and level of illumination.

Inherent Visibility of the Merchandise.

Contrast. High contrasts, such as a dark monogram on a white handkerchief, are easy to see. Low contrast visual tasks, such as subtle brown and black suit fabric patterns, are difficult to see.

Visual Size. Often the marks of quality are in small details, such as stitching on apparel or an intricate design on jewelry. The factor of distance is just as important as actual size. A boldly printed package label that is easy to read at arm's length becomes difficult to read at 20 feet because its visual size is greatly decreased. Particularly in self-service and self-selection, ability of customers to see effectively at a distance of several feet is vital to creating buying impressions of impulse items.

Reflectance. Dark-colored merchandise is more difficult to see than light-colored merchandise even though size of details and contrasts in patterns, stitching, etc., may be similar.

Time. Time is a major key to profitable store functions. In clerk-service operation, there may appear to be plenty of time to see. However, easier seeing helps speed buying decisions, enabling the customer to consider more purchases while in the store, thus enabling the clerks to handle more customers. In self-service and self-selection higher levels of illumination increase the speed of seeing, assisting the customer in making more rapid buying impressions and selections.

Level of Illumination. One important factor which the store designer should consider is to provide a lighting system which will produce the desired level of illumination for adequate seeing. See discussion below on recommended levels.

Practical Considerations

The above relates only to the shopper's eyes and the seeing tasks involved in appraising merchandise. This is a key consideration, but the design of store lighting must also consider:

Minimum Distraction from the Merchandise. In general as levels of illumination are increased, more attention needs to be given to control of luminaire brightness and its relation to the environment.

Control of Heat. Lighting heat can be an asset in cold weather, but should be disposed of in warm weather. The coordination of lighting, heating, and cooling may effect savings in initial and operating costs.

Fading. Fading of merchandise may be caused by exposure to higher levels of illumination for extended periods of time. Not all products have the same color stability, and not all products fade or change chemical composition in the same environment. Exposure to higher levels of illumination affects some merchandise more than others. Attention should be given to the environmental conditions that cause fading so that steps can be taken to correct the problem. See Section 25 for a more detailed discussion.

Economics

The precise relationship between sales and lighting is difficult to establish, since sales also depend on so many other factors. Based on experience and on cost relationships, improved lighting does offer a reasonable expectation of a profitable return.

Recommended Levels

Recommendations should always be based on current knowledge. Fig. 9–80 lists current recommended illumination levels for merchandising areas and other service facilities. These values do not necessarily refer to average levels throughout the store, since in many types of stores it is desirable and practical to provide non-uniform lighting between aisles and merchandise locations, as well as for accent lighting, creation of contrast, etc.

Where merchandise is displayed in one location and appraised in another, it is desirable not to exceed a 3-to-1 ratio in illumination levels between the two locations.

During the day, reflections in the glass of show windows determine the levels needed to enable customers to see through them effectively. Higher or lower values than recommended may be desirable, depending on such considerations as severity of reflections, reflectance of merchandise and backgrounds, and illumination levels in adjacent store windows.

LIGHTING FOR MERCHANDISE EMPHASIS

Merchandise lighting should be both exciting and unobtrusive. Total finesse should be observed by the skilled designer to show the merchandise to best advantage, without glare or other disquieting influence intruding on the consciousness of the customer. The totality of the interior then becomes a display and product information center, and the impulse to buy of paramount concern for the display lighting designer. See Fig. 13–3.

Creation of Interest

Once the type of store, class of merchandise to be handled, and clientele to be served are determined, the lighting should be designed in keeping with its character. Store lighting design should consider all surfaces in the customer's field of view.

While merchandise should dominate the scene by having the highest brightness, the creation of a pleasing environment is vital. Environmental lighting involves balance and variation in brightness of architectural forms, including ceilings, luminaires, floors, and space dividers, and particularly walls.

To attract and retain customer attention, featured merchandise should be on the order of five times as bright as its surroundings. Compelling, brightly lighted displays strategically placed throughout the store can conduct shoppers on a buying tour. Even poorly located areas can be made sales productive.

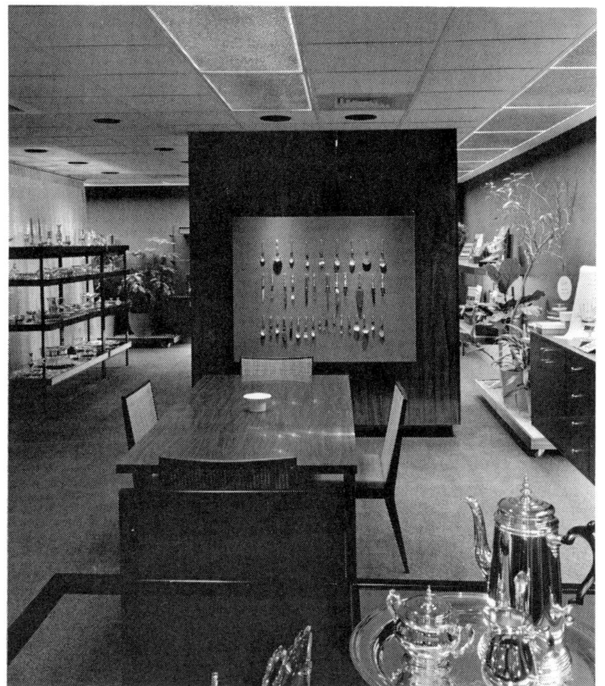

Fig. 13-3. This store uses a combination of fluorescent and incandescent luminaires to create an unobtrusive but attractive atmosphere that emphasizes merchandise display.

Luminaires

The direction of light from luminaires may be totally down (direct-below horizontal), totally up (indirect-above horizontal) or a combination of these two. In direct-lighting luminaires, the distribution of light may be concentrating, medium, or wide. Concentrating distribution is used for predominantly horizontal illumination, or to create sharp and dramatic high-lights and shadows, whereas medium or wide distribution is used for vertical as well as horizontal illumination. Totally indirect systems produce a softer, general atmosphere of lighting, where shadows are almost nonexistent and merchandise textures and shapes are less prominent. See also Section 10.

Lighting Techniques

There are two basic approaches to lighting merchandising areas in stores. One is to design a specific lighting plan to conform to the layout of the merchandise fixtures, showcases, gondolas, etc., resulting in a luminaire layout which may not conform with the building architecture. The other technique is to employ a pattern of luminaires to provide uniform illumination throughout the sales area. Both systems may be supplemented by various forms of accent lighting for merchandise emphasis.

Specific Method. In the specific method, luminaires are placed in relation to merchandise locations to emphasize the merchandise itself. Higher

Fig. 13-4. The specific method of lighting where recessed downlights follow the irregular layout of the showcases and illuminate tops and interiors of the cases. Chandeliers provide glitter for atmosphere. Wall cases are internally illuminated by vertical fluorescent lamps and reflectors.

levels of illumination are provided at appraisal areas and lower levels in circulation areas. This system may be enhanced by uplighting from tops of store fixtures, by curtain wall lighting, and where the character of store permits, by chandeliers or other decorative luminaires. See Fig. 13-4.

Incandescent luminaires are frequently used in the specific method because of their ability to project and control light toward the point of sale in the appraisal areas. Units lighting the showcase and counter tops generally employ reflector contours and light-control devices for rectangular light patterns.

General Pattern Method. Luminaires are basically located symmetrical with the building architecture in the general pattern method. Store fixture layout may be planned and rearranged without regard to the luminaire layout. General pattern lighting may be incandescent, fluorescent, high-intensity discharge, or a combination of any. As environmental lighting, it should help create a pleasant, attractive atmosphere that emphasizes the store's activity and makes it a desirable place in which to shop. See Fig. 13-5.

Fig. 13-5. Merchandise appraisal is made easy in this store with a combination fluorescent-incandescent general pattern system.

Many variations in store layout are possible with the general pattern method. However, if only general area lighting is used, the uniformity of the lighting plan and illumination level may tend to produce monotony and lack of focal points. Merchandise displayed in showcases and along walls is usually not adequately lighted by general area lighting and, therefore, supplementary lighting should be provided.

Supplementary Lighting

It is important to add supplementary lighting to attract attention to merchandise, to influence traffic circulation, and to create interest. The proper balance of general and supplementary lighting is dependent on the type of merchandise, methods of presentation, and type of store. As a rule, two to five times the general illumination level is required for the merchandise to be effectively displayed. The resultant brightness patterns created play a major part in achieving the complete selling environment for a store.

Downlights. Luminaires producing direct lighting from a single source are called downlights. Downlights can be fixed or adjustable and are used to provide a higher level of controlled illumination and emphasis at the merchandising level.

General-service incandescent and high-intensity discharge lamps in reflectors, as well as projector or reflector lamps, are employed as sources. The light pattern may be circular, rectangular, triangular, or spread. Where radiant heat from lamps would damage merchandise, reflector lamps with special dichroic filters on the reflector, which deflect unwanted heat rays away from the sales area, can be used.

Fig. 13–6. Downlighting units should be mounted above the front side of the counter so that the reflected image of the source will be directed away from the customer's eyes. This location also provides the maximum illumination at the point of greatest need—on the merchandise as it is examined closely by the shopper.

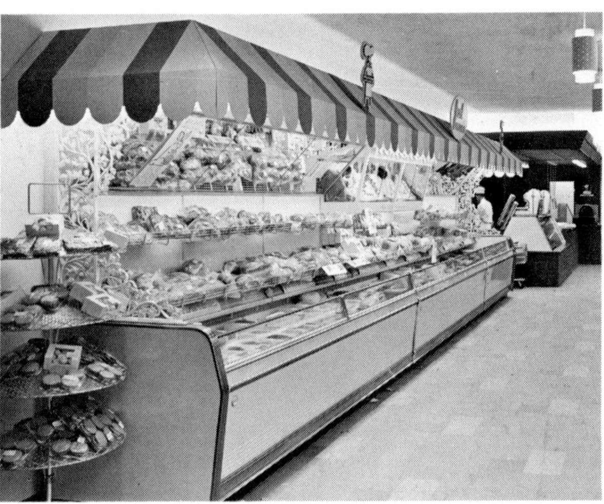

Fig. 13–7. It is often possible to "departmentalize" an otherwise uniform lighting layout through the use of custom-designed shielding media, such as in this valance above bakery cases.

The general rule is to make luminaires inconspicuous both by design and location. Very low surface brightness in the normal line of vision is a most desirable objective.

When used to light merchandise in glass-top store fixtures, it is best to locate luminaires above the front edge of the fixture so that reflections are minimized from the customer's standpoint. See Fig. 13–6. For lighting long counter tops, fluorescent luminaires, affording good control through reflectors and louvers or prismatic lenses, can be used.

Valances and Cornices. Valances and cornices are simple, direct ways to achieve vertical illumination. See Fig. 13–7. Valances and cornices may take many forms; opaque, perforated, or luminous, and may carry lettering. Lamps in the valance should be located far enough in front of the merchandise to light vertical surfaces effectively. Very often valances also are used to illuminate walls above the merchandise. See Fig. 13–8.

Fig. 13–8. Valance lighting provides effective vertical illumination on merchandise below and on the wall area above.

Luminous Soffits and Canopies. Vertical surfaces can also be illuminated by means of luminous soffits and canopies. These are usually large horizontal elements utilizing continuous lighting and are shielded by lenses, louvers, or diffusers. To heighten the dramatic effect on some types of merchandise, incandescent spotlights are sometimes added.

Bullet and Hooded Units. Pendant and wall-mounted lamp housings such as spot lights are also used for supplementary lighting. Integral reflector and general service incandescent lamps with suitable reflectors and/or lenses or louvers can be used. Fluorescent equipment is also available. Once the effect to be achieved is determined, units should be spaced according to the distribution pattern of the luminaire. For best results, it is important to consider the distance from merchandise, the angle of incidence of the light, and how heat and color rendition will affect the displays.

Vertical Reflectors. Lamps placed in vertical strips can be mounted in individual shadow boxes or niches, or in continuous runs of shelving as in Fig. 13-4. In the latter case particularly, it is important to space the shielded light strips close enough together and far enough out to avoid dark areas. Incandescent lamps may be used where it is desirable to achieve sparkle and a warmer color characteristic.

Showcase Lighting. To make merchandise attractive to the buyer and to minimize reflected images from glass enclosures, the level of illumination in the case should be at least twice that of the general system. Sheen, texture, and heat affect the choice of a system. Luminaries using high intensity, low-voltage, showcase and standard incandescent lamps, fluorescent and cold cathode are all available for the purpose. All of these types produce differing results and should be evaluated as they may affect the saleability of the merchandise on display.

Fig. 13-9. These displays of hollow ware are illuminated effectively by internally lighted shelves with diffusing glass at top and bottom.

Integral Shelf Lighting. There are various techniques of lighting merchandise on continuous shelves. For open-shelf presentation, as for glassware and small appliances, and to avoid shadows, individual shelf lighting is most effective. The simplest method is to provide a narrow shielded light strip under the front edge of each shelf.

Another method is to use a double translucent shelf. See Fig. 13-9. A narrow opaque front should be used to shield a thin lighting strip contained within the double shelf. Adequate space should be left around the objects illuminated to permit the object to receive the light from the sources and to be readily seen. Care should be used that the shelf brightness does not distract from the merchandise.

Display Lighting. Feature display areas can be treated with more dramatic impact than the general merchandise displays. Higher levels of illumination, two to five times merchandising area levels, can be used and color introduced both on the merchandise and backgrounds.

Mirror Lighting. At the mirror, shoppers finally appraise hats, dresses, shoes, cosmetics, and hairdos in terms of their color, fit, and how well they complement the buyer's personality.

Three things are important in lighting mirrors: the face should be softly lighted with a color that flatters the skin tone; the sales item should be adequately lighted over its entirety; and the lighting should be of a quality consistent with the illumination under which the merchandise will be worn.

SHOW WINDOW LIGHTING

The show window can be a powerful advertising medium—the link between the merchandise inside the store and the potential customers who walk or ride past. Adequate lighting of show-window displays has these objectives:

1. Attract attention.
2. Hold attention despite competing distraction from street traffic, other displays, and reflections from the glass.
3. Leave a favorable, lasting, and accurate impression of the goods or idea displayed.
4. Invite the shopper into the store by suggesting a bright, cheerful interior.

Daytime Effectiveness

The selling force of a show window is very valuable during the daylight shopping hours, but the daylight often creates severe competition for the shopper's attention. This competition can be minimized with modern lighting equipment and latest techniques.

With windows on the shady side of the street, the glass tends to act as a mirror, reflecting an image of the sky, buildings across the street, pave-

Fig. 13-10. Under certain conditions, slanted or curved glass windows can help to eliminate veiling glare. Glass that slants inward from top to bottom (left) reflects the sidewalk. If this is in the shade, or a canopy is used, reflections are reduced. Curved glass (center) can almost completely eliminate reflections, by reflecting a dark well in front of the window. Glass that slants outward from top to bottom (right) can be arranged to reflect the dark underside of a canopy.

ment, street traffic, etc. Measurements show that these objects, in direct sunlight, have a typical luminance (photometric brightness) of 1000 footlamberts. The window reflects about 10 per cent of this luminance, or 100 footlamberts. These reflected images are seen superimposed on the window display. Without adequate lighting they will completely veil the display itself.

Studies have shown that in order to "see through" reflected images in the window glass, surfaces inside the window must be at least as bright as the reflected images; preferably several times as bright. Thus, when reflections have a luminance of 100 footlamberts, vertical display surfaces should also have a luminance of at least 100 footlamberts. With light colored displays and background (averaging 50 per cent reflectance), 200 footcandles of general illumination on vertical surfaces will produce this luminance, while for darker colored displays, or for greater effectiveness, higher levels are needed.

On the sunny side of the street, building and pavement surfaces adjacent to the windows are very bright. By comparison, inadequately lighted windows appear to be dark recesses in the store front. Therefore, it is necessary to light show windows on the sunny side to a higher illumination level in the daytime than is needed at night. Window reflection can be partially controlled by an awning, marquee, or canopy. Dark paving of the sidewalk also helps reduce window reflections.

Nighttime Effectiveness

With the increase in evening shopping hours, nighttime lighting of show windows is becoming more important. Special effects with color and contrast can be particularly potent since daylight is not present to dilute the color or shadow effect. However, reflections from the glass may also be present at night from signs, lighted buildings, and street lighting. Therefore, medium to higher illumination levels may be required to overcome the reflections, and gain competitive advantage.

Lighting Requirements

Under typical conditions, effective show-window lighting (using the components in Fig. 13-11) would provide:

1. Large, bright areas to attract attention both day and night, to help overcome reflections from the glass, and to allow accurate appraisal of merchandise.
2. Areas of very high brightness on important parts of displays for maximum attention value; for dramatic, effective composition; and to compete with surrounding distractions and reflection from the glass.
3. Strong contrast, modeling, highlights, and color to enhance the display.
4. Good color rendering for accurate customer impressions.
5. Flexibility to allow for changing displays and use of extra equipment for special effects.

Lighting Application

General Lighting. General lighting may be accomplished with incandescent, fluorescent, or high-intensity discharge luminaires. However, because of their high luminous efficacy and low heat radiation per footcandle, fluorescent lamps are desirable as part of the general lighting. Where the lighting is mostly fluorescent, or high-intensity discharge, and good color effect is essential, improved-color lamps should be used.

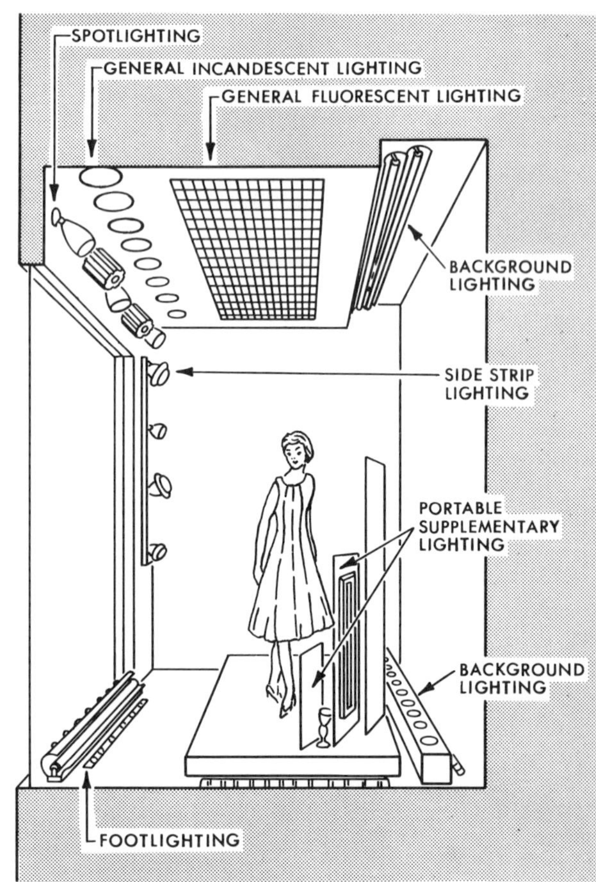

Fig. 13-11. Components of show-window lighting.

The prime purpose of general illumination is to reduce reflections on the glass by lighting the vertical planes seen in the window. General illumination is usually used in conjunction with accent or supplementary lighting.

For lighting massed displays of small items that are too small to be accented effectively, general illumination may be used alone. Some directional component of light may be desirable for modeling.

Since merchandise is the most important element in the window, the light sources should be well shielded to prevent distracting brightness from the luminaires and light source.

Accent Lighting. Individual items can be emphasized with directional light. It is possible to obtain higher level vertical illumination, good definition of form and texture, and eye catching highlights from specular or faceted merchandise by using accent lighting. Compact light sources in combination with light-control devices reduce spill light and direct the light beam toward the display target. Although directional units are commonly mounted at the top front of the window, dramatic effects may be obtained by angling them from the sides or below. Flexibility in aiming is an important mounting consideration.

Spotlighting. Spotlights make it easy and practical to provide 1000 or more footcandles on the important areas of a display. Spotlighting equipment includes theatrical-type spots and a variety of reflectorized lamp types. Units are commonly mounted at the top front of the window, and often at the sides as well. See Fig. 13-12. Flexibility in aiming is an important consideration. The use of reflectorized lamps with dichroic filters, when properly ventilated, permits a reduction in the heat in the window area.

Supplementary Lighting. Supplementary lighting offers opportunities to create a more dramatic display, as well as to increase the brightness and attracting power of the window. By using lamps close to display surfaces and background, built into display fixtures or concealed within the display, the high brightness necessary for daytime visibility can be efficiently obtained. Background lighting is especially desirable to silhouette dark-colored merchandise.

Small-diameter fluorescent lamps, and general-service, PAR and R-type incandescent lamps all lend themselves to easy concealment within the display. In some cases, lighted floor and table lamps, or wall brackets may form part of the setting. Colored light used on background or within the display can create compelling and dramatic effects.

Luminous panels are also a means of creating background lighting and of overcoming window reflections.

Colored Light. White light (or a very pale tint) is generally desirable on the merchandise to avoid distortion of colors. However, colored light presents an opportunity to heighten the dramatic quality of the windows. For example, modeling light from the sides or from below can be used to fill in shadows with color, or backgrounds lighted with color can provide eye-catching contrast with the merchandise displayed.

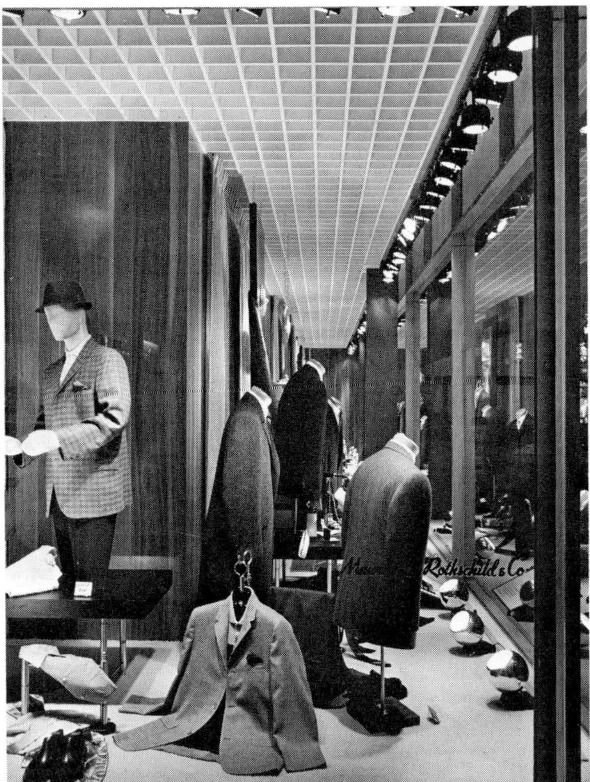

Fig. 13-12. Massed displays of relatively large items lend themselves well to "general spotlighting." One or two beams of appropriate size cover each item, with spill light and supplementary lighting from the floor to brighten display backgrounds.

Fluorescent lamps in tints and saturated colors are useful for floodlighting large areas. Colored light may be produced with incandescent lamps by the use of glass filters or spread lenses over spot and flood lamps, lighting strips, and theatrical type spotlights. Reflector color lamps are a convenient and effective source of colored light.

Separately switched circuits of red, green, and blue incandescent lamps or pink, green, and blue fluorescent lamps provide a high degree of flexibility in tinting backgrounds or the whole window. Practical equipment is now available for dimming fluorescent, as well as incandescent, lamps. If the color circuits are on dimmer controls, a nearly infinite variety of hues and tints can be created.

EXTERIOR LIGHTING

Over-All Objectives

The function of most exterior lighting (see also Section 17) is to attract favorable attention to places of business or related objects and areas. The brightness of a light source itself seldom is desirable for attraction purposes. Exceptions are signs and Christmas decorations where patterns of light sources convey a message or create an impression.

Location. Lighting can help pull traffic to a store in an unfavorable location. Bold brightnesses and patterns are needed rather than soft, subtle effects. Stores in high-traffic locations can use lighting to stop those who are intent on going some place else.

Attraction and Impact. A distinction is drawn between diverting one's attention (attraction power) and conveying a message or impression once attention is captured (impact). In this sense, attraction power does not depend on what a sign says, for example, or what merchandise is in a window. It depends on brightness, area, and contrast with surroundings.

Design Coordination. Coordination in design is vital to good appearance and hence customer appeal. The effectiveness of the exterior of a store depends on handling of the facade, the sign, the windows, and the interior. Transitional effects as customers approach, enter, and leave the store should be visually comfortable and pleasing. In groups of stores, some degree of unity in design is desirable to overcome the hodgepodge that has marred so many new and old commercial areas. See Fig. 13–13.

Approaches to Design

Solo Stores. Solo stores give the greatest opportunity for individualized architectural and lighting design freedom. During busy periods, when the view of the windows tends to be obscured by parked cars, attraction power should be built up by the lighting of other areas. Generally a solo store needs particular attention given to identification at a distance, at the street, and on building surfaces above car level.

In solo stores there is a maximum opportunity to control visual adaptation of customers. At night, brightnesses can be stepped up from the parking lot, to the entrance and then to the interior. Conversely, during the day, brightness need to be stepped down from the bright outside.

Shopping Centers. Identification of the center at a distance can be accomplished by signs, insignia, or other devices mounted on high poles. Safety and convenience are served if the name of the center is legible at a sufficient distance to allow motorists to change lanes and slow down, well before they come to entrance points. Because of the low silhouette of single-story buildings, it may be desirable to create higher architectural features such as pylons or fins to which lighting can be applied for attraction. Upper floors of multi-story buildings are key surfaces.

From the standpoint of attraction and attractiveness of the entire center, lighting of landscaping, fountains, sculpture, etc., are obvious considerations. Best results accrue from planning and relating all areas to each other with regard to balance of brightnesses, harmony of color, and the build-up of primary and secondary focal points.

Street Stores. Street stores present, in some ways, the most complex design problem of all. While attraction power can be built into a store on an individual basis, lighting for attractiveness of a block or business district should follow planning for architectural harmony. Lacking this, improvements can be made by the cooperation of stores in adopting compatible lighting techniques for signs and fronts.

Lighting of the front is primarily of institutional value since it is seen most effectively from across the street. Lighting of planes perpendicular to the sidewalk is of prime importance in getting attention.

Lighting and Building Materials

Brightness requirements vary depending on store location. Illumination to produce a given brightness depends on how much light is reflected by building surfaces. Since brighter areas look lighter in terms of weight than dark ones, it is usually desirable to have brighter areas above to give the building a feeling of stability.

Texture of a building surface can be emphasized

Fig. 13–13. In shopping centers fluorescent lighting can be used effectively for store identification and to emphasize entire pedestrian arcades. Key design principles are spacings that assure adequate spread of illumination, satisfactory shielding of sources from view, and selection of lamps and equipment that will maintain adequate output at lowest normal operating temperatures.

Fig. 13–14. Floodlighting building facades with fluorescent lamps is often practical and efficient. Techniques usually involve lamp and luminaire designs that operate effectively at lower temperatures. Here the linear lighting element ties in with the building lines and produces a uniformly illuminated building surface.

by lighting from a grazing angle. However, this technique is also critical of construction workmanship. This can be minimized by designing so that variations form a deliberate pattern rather than emphasize mistakes.

Shiny surfaces such as pebbles, exposed aggregate concrete panels, flecks of mica, or configurated metal are best revealed by incandescent or high-intensity discharge sources. Fluorescent lighting is desirable for soft effects.

Color, both variation in "whites" and strong colors, is too often overlooked in building floodlighting. The tendency is to choose the lowest cost system without regard to the appearance of the building material. In general, warm colors are enhanced by incandescent, cool by mercury, while fluorescent is comparatively well balanced.

Lighting and Building Forms

Lighting properly applied can bring out the best features of a building and enhance the architect's design.

Line. Small cross-section, continuous lighting elements, for example, can be fitted to existing cornices to maintain original lines; or can be designed into a new building to avoid a heavy line. In general, it is preferable to have linear lighting elements tie in with building lines rather than contrast with them. See Fig. 13-14.

Form. The use of large glass areas results in emphasis on skeletal structure and the space which is enclosed, rather than the monolithic effect of opaque walls. Both exterior and interior lighting are involved in making buildings effective as viewed from outdoors. Ideally, the balance of brightness and color between exterior and interior should be such that the interior looks brighter and warmer or cooler depending on the season.

Pattern. Where surfaces are designed with patterns in relief, grazing-angle lighting is indicated. Where pattern is created by color on a flat surface, uniform, flat illumination is logical except that variations are sometimes desirable to accent key areas. Pattern can be created by light on a plain surface and this technique has the advantage of creating a different appearance at night than during the day.

Sparkle. Small points of high brightness further enrich the appearance of a building. Such points of brightness may be light-source images reflected in shiny surfaces, or small lamps exposed to direct view. In scope they may be 6-watt clear bulbs to be seen from a few yards, or spotlights mounted high on a building to be seen from a mile or more.

Sign-Lighting Design

In general, the simpler the message and the cleaner the letter design, the greater the legibility. Embellished letters and script have to be larger than block letters for the same effectiveness. See Section **23** for a discussion of sign-lighting design technique.

Sidewalks and Entrances

One of the most intriguing new techniques is the use of infrared lamps in the soffit of marquees and entrances in combination with conventional luminaires to provide warmth as well as additional light in cold weather. These installations not only help attract customers, but can keep sidewalks dry, thereby reducing slush and dirt tracked into the store. See Section 25 for a discussion of "people-heating."

PARKING AREA LIGHTING[2]

Objectives

From the standpoints of traffic safety; protection against assault, theft, and vandalism; convenience and comfort to the user; and in many instances, for business attraction, adequate parking area lighting is as vital as roadway lighting in our motorized society. A well lighted parking area spells security particularly to the female motorists, who constitute a major portion of the shopping public.

Illumination Requirements

The illumination requirements of a parking area (see Fig. 9-80, Exterior Lighting portion) depend on the type of usage the area receives. Reasonably uniform illumination is desirable, that is, the lowest footcandle value at any point on the pavement should not be less than one-fourth the recommended average. The small suburban shopping area design may result in illumination which is decorative, as well as functional.

Exits, entrances, loading zones, and collector lanes should be given special consideration to permit ready identification, to facilitate the handling of packages, and to aid in providing a safe installation. The illumination level should not be less than twice the illumination of the adjacent parking area or the adjoining street, which ever is greater.

For competitive reasons, many stores and shopping plazas use higher illumination levels to attract attention.

Control of Brightness

In an area where parking is close to residential property, there may be a problem of annoyance to residents from stray light from parking area lighting equipment. Careful consideration should be given in such areas, not only to the level of illumination, but to the type and aiming of luminaires employed to keep objectionable spill light to a minimum.

The type of source used for parking area lighting will depend upon the color of light desired, light control, maintenance, and economics.

In larger parking areas, exceeding two or three lanes, the parking lanes should be considered in the selection of pole locations. Poles should be spaced to permit sufficient overlapping of adjacent beams to minimize harsh shadows and avoid dark pockets throughout the area.

REFERENCES

1. Committee on Merchandising Areas of the IES: "Recommended Practice For Lighting Merchandise Areas," *Illum. Eng.*, Vol. LVIII, p. 397, June, 1963.
2. Subcommittee on Lighting Service Stations and Parking Areas of the Store Committee of the IES: "Recommended Practice of Outdoor Parking Area Lighting," *Illum. Eng.*, Vol. LV, p. 307, May, 1960.

SECTION

14

INDUSTRIES*

General Lighting 14–1 Supplementary Lighting 14–6 Emergency Lighting 14–10
Steel Mills 14–12 Foundries 14–15 Machining Metal Parts 14–17
Sheet Metal Shops 14–19 Automotive Assembly 14–20 Rubber Tires 14–22
Graphic Arts—Printing 14–24 Cotton Gins 14–26 Textiles 14–27
Cleaning and Pressing 14–31 Shoe Manufacturing 14–33 Men's Clothing 14–34
Fluid Milk Industry 14–35 Flour Mills 14–36 Bakeries 14–37
Candy Manufacturing 14–39 Fruit and Vegetable Packaging 14–41
Petrochemicals 14–43 Electrical Generating Stations 14–43
Sawmills 14–46 Railroad Yards 14–47 Farm Lighting 14–49

Industrial lighting encompasses seeing tasks, operating conditions, and economic considerations of a wide range.[1] Visual tasks may be extremely small or very large; dark or light; opaque, transparent, or translucent; on specular or diffuse surfaces; and may involve flat or contoured shapes. With each of the various task conditions, lighting must be suitable for adequate visibility in developing raw materials into finished products. Physical hazards exist in manufacturing processes and, therefore, lighting must contribute to the utmost as a safety factor in preventing accidents. The speed of operations may be such as to allow only minimum time for visual perception and, therefore, lighting must be a compensating factor to increase the speed of vision (see Section 3).

Lighting must serve not only as a production tool and as a safety factor but should also contribute to the over-all environmental conditions of the work space. The lighting system should be a part of an over-all planned environment.

The design of a lighting system and selection of equipment may be influenced by many economic factors. Economic decisions in regard to the lighting system should not only be based on the initial and operating costs of the lighting, but also on the relationship of lighting costs to other plant producing facilities and costs of labor.

* The information in this Section serves as a guide in designing lighting for various industries. Additional information can be obtained from the *American National Standard Practice for Industrial Lighting,* A11.1-1965 (R1970), and IES study reports on specific industries. See references at the end of this section.

NOTE: references are listed at the end of each section.

FACTORS OF GOOD INDUSTRIAL LIGHTING

Quantity of Illumination

The illumination recommended for an installation depends upon the seeing task. Variations in contrast, time for seeing the task, and reflecting characteristics of the task all contribute to the severity of the seeing task and, therefore, determine the amount of illumination required for optimum visibility. As illumination on a task is increased, luminance (photometric brightness) increases, and the ease, speed, and accuracy of vision are improved.

Specific recommended values of maintained illumination levels for industrial tasks and areas are given in Fig. 9–80, page 9–81. These values are considered as being minimum on a task at any time and higher values will generally produce increased benefits of visual performance. If it is desired to determine the level produced by an existing installation, the measurement procedure outlined in Section 4 should be followed.

A general lighting system should be designed to provide a uniform distribution of light over the entire work area. Where work areas are close to walls, such as work benches, the first row of luminaires should be located closer to the wall or additional lighting should be provided over the particular work spaces.

To insure that a given illumination level will be maintained, it is necessary to design a system to give initially more light than the required minimum.

In locations where dirt will collect very rapidly on luminaire surfaces and where adequate maintenance is not provided, the initial value should be even higher. For typical light loss data and a further discussion see pages 9–2 through 9–5.

Where safety goggles are worn, the light reaching the eye is likely to be materially reduced and the general level of illumination should therefore be increased in accordance with the loss of transmission.

Quality of Illumination

Quality of illumination pertains to the distribution of brightness in the visual environment. The term is used in a positive sense and implies that all brightnesses contribute favorably to visual performance, visual comfort, ease of seeing, safety, and esthetics for the specific visual task involved. Glare, diffusion, direction, uniformity, color, luminance, and luminance ratios all have a significant effect on visibility and the ability to see easily, accurately, and quickly. See Section **3**. Certain seeing tasks, such as discernment of fine details, require much more careful analysis and higher quality illumination than others. Areas where the seeing tasks are severe and performed over long periods of time require much higher quality than where seeing tasks are casual or of relatively short duration. See Fig. 14–1.

Industrial installations of very poor quality are easily recognized as uncomfortable and are possibly hazardous. Unfortunately, moderate deficiencies are not readily detected, although the cumulative effect of even slightly glaring conditions can result in material loss of seeing efficiency and undue fatigue. See Section 3 for detailed information.

Direct Glare. When glare is caused by the source of lighting within the field of view, whether daylight or electric, it is described as direct glare.

To reduce direct glare in industrial areas, the following steps may be taken: (1) decrease the luminance of light sources or lighting equipment, or both; (2) reduce the area of high luminance causing the glare condition; (3) increase the angle between the glare source and the line of vision; (4) increase the luminance of the area surrounding the glare source and against which it is seen.

Unshaded factory windows are frequent causes of direct glare. They may permit direct view of the sun, bright portions of the sky, or bright adjacent buildings. These often constitute large areas of very high luminance in the normal field of view.

Luminaires which are too bright for the environment in which they are located produce direct glare. This glare may be in the form of discomfort glare and disability glare. Discomfort glare produces discomfort but does not necessarily interfere with visual performance or visibility. On the other hand, disability glare reduces visual performance and visibility and often is accompanied by discomfort. To reduce direct glare, lighting equipment should be mounted as far above the normal line of sight as possible. The luminance and quantity of light emitted from luminaires should be limited and controlled in the visual field from horizontal down to 45 degrees.

There is such a wide divergence of tasks and environmental conditions in industry that it may not be economically feasible to recommend a degree of quality which will satsify all cases. The luminance control required depends on the task, length of time in which the task is performed, and the mounting height of the luminaires. In production areas, luminaires in the normal field of view should be shielded to at least 25 degrees from the horizontal but preferably down to 45 degrees.

Luminance and Luminance Ratios. The ability to see detail depends upon the contrast between the detail and its background. The greater the contrast, difference in luminance, the more readily the seeing task is performed. However, the eyes function most comfortably and more efficiently when the luminances within the remainder of the environment are relatively uniform. Therefore, all luminances in the

Fig. 14–1. Factors of good industrial lighting are well illustrated here. A 100-footcandle grid layout provides good visibility on machine tool scales and verniers; 30 per cent upward component and cross louver shielding combined with a light, cheerful color scheme make a pleasant and comfortable environment.

field of view should be carefully controlled. In manufacturing there are many areas where it is not practical to achieve the same luminance relationships as are easily achieved in areas such as offices. But, between the extremes of heavy manufacturing and office spaces lie the bulk of industrial areas. Therefore, Fig. 14–2 has been developed as a practical guide of recommended maximum luminance ratios for industrial areas.

To achieve the recommended luminance relationships, it is necessary to select the reflectances of all the finishes of the room surfaces and equipment as well as control the luminance distribution of the lighting equipment. Fig. 14–3 lists the recommended reflectance values for industrial interiors and equipment. High reflectance surfaces are generally desirable to provide the recommended luminance relationships and high utilization of light. They also improve the appearance of the work space. See Fig. 14–4.

In many industries machines are painted such that they present a completely harmonious environment from the standpoint of color. It is desirable that the background be slightly darker than the seeing task. It appears desirable to paint stationary and moving parts of machines with contrasting colors to reduce accident hazard by aiding identification.

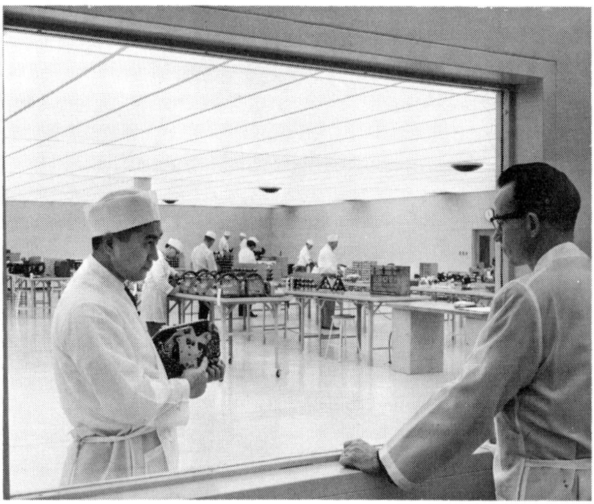

Fig. 14–4. A super-clean white room for the manufacturing of precision aerial-photography systems. The luminous ceiling provides an average level of 250 footcandles in this area. Recommended room-surface reflectances help achieve good luminance relationships.

Fig. 14–2. Recommended Maximum Luminance Ratios

	Environmental Classification*		
	A	B	C
1. Between tasks and adjacent darker surroundings	3 to 1	3 to 1	5 to 1
2. Between tasks and adjacent lighter surroundings	1 to 3	1 to 3	1 to 5
3. Between tasks and more remote darker surfaces	10 to 1	20 to 1	†
4. Between tasks and more remote lighter surfaces	1 to 10	1 to 20	†
5. Between luminaires (or windows, skylights, etc.) and surfaces adjacent to them	20 to 1	†	†
6. Anywhere within normal field of view	40 to 1	†	†

* Classifications are:

A—Interior areas where reflectances of entire space can be controlled in line with recommendations for optimum seeing conditions.

B—Areas where reflectances of immediate work area can be controlled, but control of remote surround is limited.

C—Areas (indoor and outdoor) where it is completely impractical to control reflectances and difficult to alter environmental conditions.

† Luminance ratio control not practical.

Fig. 14–3. Recommended Reflectance Values (Applying to Enviromental Classifications A and B in Fig. 14–2)

Surfaces	Reflectance* (per cent)
Ceiling	80 to 90
Walls	40 to 60
Desk and bench tops, machines and equipment	25 to 45
Floors	not less than 20

* Reflectance should be maintained as near as practical to recommended values.

Veiling Reflections. Where seeing task details are specular, care should be taken to minimize veiling reflections which will decrease task visibility. See Section 3.

Reflected Glare. Reflected glare is caused by the reflection of high luminance light sources from shiny surfaces. In manufacturing processes this may be a particulary serious problem where critical seeing is involved with highly polished surfaces such as polished sheet metal, vernier scales, and critically machined metal surfaces.

Reflected glare can be minimized or eliminated by using light sources of low luminance or by orienting the work so reflections are not directed in the normal line of vision. Often it is desirable to use reflections from a large-area, low-luminance luminaire located over the work. The section that follows on supplementary lighting covers in detail the solutions to such problems.

Distribution, Diffusion, and Shadows. The general lighting system for a factory should be designed for uniformly distributed illumination. Uniform illumination is defined as a distribution of light with a maximum and minimum illumination at any point which is not more than $1/6$ above or below the average level in the area.

Harsh shadows should be avoided, but some shadow effect may be desirable from the general lighting system to accentuate the depth and form of objects.

There are a few specific visual tasks where clearly defined shadows improve visibility and such effects should be provided by supplementary lighting equipment arranged for the particular task.

Color Quality of Light. For general seeing tasks in industrial areas there appears to be no effect upon visual acuity by variations in color of light. However, where color discrimination or color matching are a part of the work process, the color of light should be very carefully selected. One example is in the printing industry, covered in detail on page 14-24. Color, of course, has an effect upon the appearance of the work space and upon the complexions of personnel. Therefore, the selection of the lighting system and the decorative scheme should be carefully coordinated.

General Considerations of Design for Lighting Industrial Areas

The designer of an industrial lighting system should consider the following factors as the first and all-important requirements of good planning:

1. Determine the quantity and quality of illumination desirable for the manufacturing processes involved.
2. Select lighting equipment that will provide the quantity and quality requirements by examining photometric characteristics, and mechanical performance that will meet installation, operating, and actual maintenance conditions.
3. Select and arrange equipment so that it will be easy and practical to maintain.
4. Balance all of the economic factors including initial, operating, and maintenance costs, versus the quantity and quality requirements for optimum visual performance. The choice of the electric distribution system may affect over-all economics.

Types of Lighting Equipment. The manner in which the light from the lamps is controlled by the lighting equipment governs to a large extent the important effects of glare, shadows, distribution, and diffusion. Luminaires are classified in accordance with the way in which they control the light. Fig. 10-10, page 10-11 gives the standard CIE classifications for interior lighting equipment.

Most industrial applications call for either the direct or semi-direct types. Luminaires with upward components of light are preferred for most areas because an illuminated ceiling or upper structure reduces luminance ratios between luminaires and the background. The upward light reduces the "dungeon" effect of totally direct lighting and creates a more comfortable and more cheerful environment as shown in Fig. 14-5. Industrial luminaires for fluorescent, high-intensity discharge, and incandescent filament lamps are available with upward components. See Fig. 14-6. Good environmental luminance relationships can also often be achieved with totally direct lighting if the illumination level and room surface reflectances are high.

In selecting industrial lighting equipment, it will be noted that other factors leading to more comfortable installations include:

1. Light-colored finishes on the outside of luminaires to reduce luminance ratios between the outside of the luminaire and the inner reflecting surface and light source.
2. Higher mounting heights to raise luminaires out of the normal field of view.
3. Better shielding of the light source by deeper reflectors, cross baffles, or louvers. This is particularly important with high-wattage incandescent filament or high-intensity discharge sources and the higher output fluorescent lamps.
4. Selecting light control material, such as specular or nonspecular aluminum or prismatic configurated glass or plastic that can limit the luminaire luminance in the shielded zone.

Top openings in luminaires generally minimize dirt collection on the reflector and lamp by allowing an air draft path to move dirt particles upward and

Fig. 14-5. Comfort and visibility were both incorporated in the lighting design of this paper mill trimmer room. The use of mercury luminaires with an upward component, mounted at 21 feet above the floor, plus high reflectance room finishes, create an effective seeing environment.

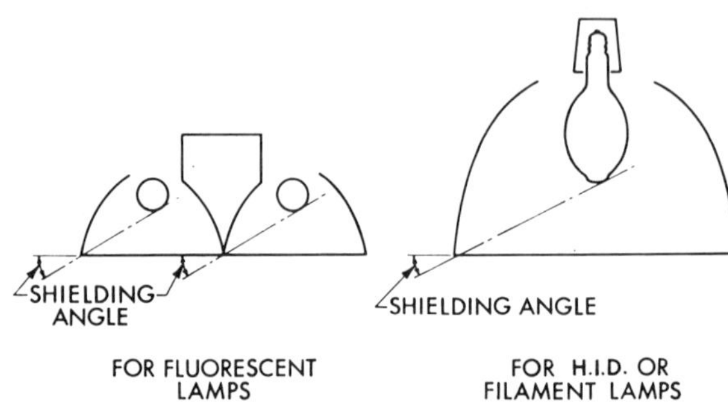

Fig. 14-6. Luminaires require adequate shielding for visual comfort. This is particularly important for higher luminance sources. An upward component also contributes to visual comfort by balance of luminances between luminaires and their backgrounds. Top openings help minimize dirt accumulation.

through the luminaire to the outer air. Therefore, ventilated types of luminaires have proven their ability to minimize maintenance of fluorescent, high-intensity discharge, and incandescent filament types of luminaires. Gasketed dust-tight luminaires are also effective in preventing dirt collection on reflector surfaces.

Direct Lighting Equipment. Direct industrial lighting equipment is usually classified according to the distribution of its direct component of light.[2] These vary form Highly Concentrating to Wide Spread. The classification of luminaires by distribution is expressed in terms of permissible spacing-to-mounting-height ratios, as shown in Fig. 14–7. See also Fig. 9–14.

The spread types are comprised of porcelain-enameled reflectors and various other types of diffuse white reflecting surfaces. Aluminum, mirrored-glass, prismatic glass, and other similar materials may also be used to provide a spread distribution when the reflector is designed with the proper contour. This type of light distribution is advantageous in industrial applications where a large proportion of the seeing tasks are vertical or nearly vertical.

Concentrating distributions are obtained with prismatic-glass, mirrored-glass, and aluminum reflectors. This type of light distribution is useful where the mounting height is approximately equal to or greater than the width of the room or where high machinery and processing equipment necessitate directional control for efficient illumination between the equipment.

In making a choice between spread and concentrating equipment on the basis of horizontal illumination levels, a comparison of coefficients of utilization for the actual room conditions involved will serve as a guide in selecting the most effective distribution. See Section 9. Care should be taken to use such coefficients based as close as practical to actual ceiling, wall, and floor reflectance as well as actual room proportions.

If, however, it is desired to determine illumination at specific points, then a point calculation method should be used to obtain accurate results (see Section 9). This is particularly true for high mounting heights with either spread or directional lighting equipment.

Other Types of Direct Lighting Equipment. Where low reflected luminance is a necessity, large-area types of low-luminance luminaires should be used. Such a luminaire may consist of a diffusing panel on a standard type of fluorescent reflector, an indirect light hood, or a completely luminous ceiling.

Semi-Direct Lighting Equipment. This classification of distribution is useful in industrial areas because the upward component (10 to 40 per cent) is particularly effective in creating more comfortable seeing conditions. A variety of fluorescent and high-intensity discharge luminaires of this distribution are available and designed specifically for industrial application. See Figs. 14–8 and 14–9.

While the semi-direct type of distribution has a sufficient upward component to illuminate the ceiling, the downward component of 90 to 60 per cent of the output contributes to good illumination efficiency, particularly where ceiling obstructions may minimize the effectiveness of the indirect component.

Industrial Applications of Other Distribution Classifications. The general diffuse, semi-indirect, and indirect systems are suitable for industrial applications where a superior quality of diffused, low-luminance illumination is required and where environmental conditions make such systems practical. An example of such applications includes the precision industries where a completely controlled environment is important, including lighting, air conditioning, and carefully planned decoration.

Fig. 14–7. Classification of Luminaires by Distribution

Spacing-to-Mounting Height Ratio (Above Work Plane)	Luminaire Classification
Up to 0.5	Highly Concentrating
0.5 to 0.7	Concentrating
0.7 to 1.0	Medium Spread
1.0 to 1.5	Spread
Over 1.5	Wide Spread

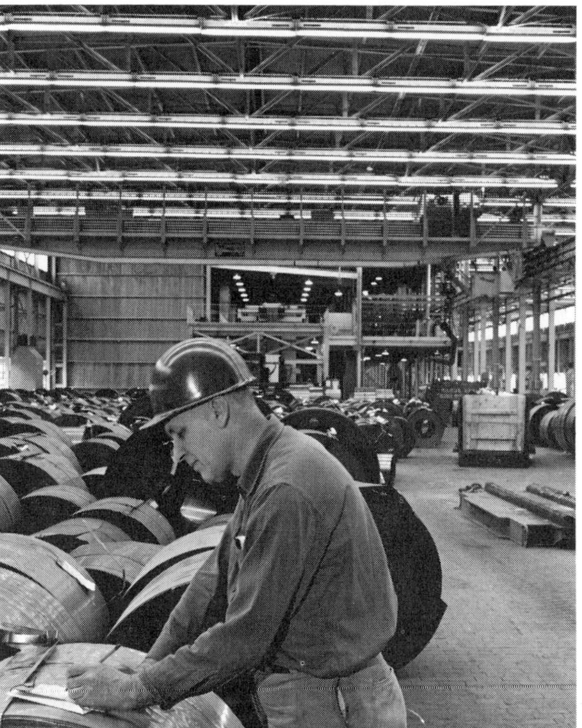

Fig. 14–8. Uniform lighting of nearly 100 footcandles is provided in this shipping area for steel coils. Continuous rows of fluorescent luminaires, with 25 per cent uplight and 25-degree shielding of extra-high-output lamps, are mounted on 20-foot centers at the bottom of the ceiling trusses, 38 feet above the floor.

Fig. 14-9. This 200-footcandle installation utilizes 1000-watt improved color mercury lamps in high bay luminaires which also allow for uplight. A few fluorescent units are also provided for safety lighting while the mercury lamps are restarting following power interruptions.

Factors of Special Consideration

Lighting and Space Conditioning. With the use of higher illumination levels, it is often practical to combine the lighting, heating, cooling, and atmospheric control requirements in an integrated system. The lighting system can often provide most of the energy during the heating period. When cooling is required, much of the lighting heat can be removed by the air exhaust system. See Section 16 for further details.

High Humidity or Corrosive Atmosphere, and Hazardous Location Lighting. Enclosed gasketed luminaries are used in non-hazardous areas where atmospheres contain non-inflammable dusts and vapors, or excessive dust. Enclosures protect the interior of the luminaire from conditions prevailing in the area. Steam processing, plating areas, wash and shower rooms, and other areas of unusually high humidity are typical areas that require enclosed luminaires. Severe corrosive conditions necessitate knowledge of the atmospheric content to permit selection of proper material for the luminaire.

Hazardous locations are areas where atmospheres contain inflammable dusts, vapors, or gases in explosive concentrations. They are grouped by the National Electrical Code on the basis of their hazardous characteristics, and all electrical equipment must be approved for use in specific classes and groups. Luminaires are available specifically designed to operate in these areas, which are noted in Article 500 of the National Electrical Code as Class I, Class II, and Class III locations.

For definitions of luminaires used in these areas, such as, *explosion-proof, dust-tight, dust-proof,* and *enclosed and gasketed,* see Section 1.

Abnormal Temperature Conditions. Low ambient temperatures must be recognized as existing in such areas as unheated, heavy industrial plants, frozen food plants, and cold storage warehouses. Equipment should be selected to operate under such conditions and particular attention should be given to lamp starting and light output characteristics if fluorescent equipment is considered. With high-intensity discharge equipment, temperature variation has practically no effect on light output, but the proper starting characteristics must be provided. With incandescent filament lamp equipment, neither the starting nor the operation is a problem at low temperature.

Abnormally high temperatures may be common at truss height in foundries, steel mills, forge shops, etc. Caution should be observed in selecting lighting equipment for mounting in such locations. It is particularly important to consider the temperature limitations of fluorescent and high-intensity discharge ballasts under such conditions. Often ballasts should be remotely located at a lower and cooler level or special high temperature equipment should be used. The reduction in fluorescent lamp output at high operating temperatures should be recognized. See Section 8.

Maintenance. Regular cleaning and prompt replacement of lamp outages is essential in any well-operated industrial lighting system. It is important for the lighting designer to analyze luminaire construction and reflector finish and also to make provisions for maintenance access so the system can be properly serviced. Another point that should be considered is that it may often be necessary to do the servicing during the plant operating hours. Further details on maintenance, access methods, and servicing suggestions are found in Section 16.

SUPPLEMENTARY LIGHTING IN INDUSTRY

Difficult seeing tasks often require a specific amount or quality of lighting which cannot readily be obtained by standard general lighting methods.

To solve such problems supplementary luminaires often are used to provide higher illumination levels for small or restricted areas. Also, they are used to furnish a certain luminance, or color, or to permit special aiming or positioning of light sources to produce or avoid highlights or shadows to best portray the details of the task.

Before supplementary lighting can be specified it is necessary to recognize the exact nature of the visual task and to understand its light reflecting or transmitting characteristics. An improvement in the visibility of the task will depend upon one or more of the four fundamental visibility factors—luminance, contrast, size, and time. Thus, in analyzing the problem, the engineer may find that seeing difficulty is caused by insufficient luminance, poor contrast (veiling reflections), small size, or that task motion is too fast for existing seeing conditions.

The planning of supplementary lighting also entails consideration of the visual comfort of both those workers who benefit directly and those who are in the immediate area. Supplementary equipment must be carefully shielded to prevent glare for the user and his associates. Luminance ratios should be carefully controlled. Ratios between task and immediate surroundings should be limited as recommended in Fig. 14–2. To attain these limits it is necessary to coordinate the design of supplementary and general lighting.

Luminaires for Supplementary Lighting

Supplementary lighting units can be divided into five major types according to candlepower distribution and luminance. These are:

Type S-I. The directional type includes all concentrating units. Examples are a reflector spot lamp or units employed concentrating reflectors or lenses. Also included in the group are concentrating longitudinal units such as a well-shielded fluorescent lamp in a concentrating reflector.

Type S-II. The spread type includes high-luminance, small-area sources, such as incandescent or mercury. A deep-bowl diffusing reflector with an incandescent filament lamp and without diffusing cover is an example of this type.

Type S-III. The spread, moderate-luminance type includes all fluorescent units having a variation in luminance of more than two to one.

Type S-IV. The uniform-luminance type includes all units having less than two-to-one variation of luminance. Usually this luminance is less than 2000 footlamberts. An example of this type is an arrangement of lamps behind a diffusing panel.

Type S-V. The uniform-luminance type with a pattern is a luminaire similar to Type S-IV except that a pattern of stripes or lines is superimposed.

Portable Luminaires

Wherever possible, supplementary luminaires should be permanently mounted in the location to produce the best lighting effect. See Fig. 14–10. Adjustable arms and swivels will often adapt the luminaires to required flexibility. Portable equipment, however, can be used to good advantage where it must be moved in and around movable machines or objects such as in airplane assembly, garages, or where internal surfaces must be viewed. The luminaires must be mechanically and electrically rugged to withstand possible rough handling. Lamps should be guarded and of the rough-service type. Guards or other means should protect the user from excessive heat. All possible precautions should be taken to prevent electrical shock to the user.

Classification of Visual Tasks and Lighting Techniques

Visual tasks are unlimited in number, but can be

Fig. 14–10. Supplementary lighting for inspection on automobile assembly line consists of closed-end industrial reflectors (two 40-watt) with diffusing covers. General lighting from two-lamp 96-inch slimline units with 25 per cent upward component.

Fig. 14-11. Classification of Visual Tasks and Lighting Techniques

Part I—Flat Surfaces

Classification of Visual Task	Example		Lighting Technique	
General Characteristics	Description	Lighting Requirements	Luminaire Type	Locate Luminaire
A. Opaque Materials				
1. Diffuse detail & background				
a. Unbroken surface	Newspaper proofreading	High visibility with comfort	S-III or S-II	To prevent direct glare & shadows (Fig. 14-12a)
b. Broken surface	Scratch on unglazed tile	To emphasize surface break	S-I	To direct light obliquely to surface (Fig. 14-12c)
2. Specular detail & background				
a. Unbroken surface	Dent, warps, uneven surface	Emphasize unevenness	S-V	So that image of source & pattern is reflected to eye (Fig. 14-12d)
b. Broken surface	Scratch, scribe, engraving, punch marks	Create contrast of cut against specular surface	S-III or	So detail appears bright against a dark background
			S-IV or S-V when not practical to orient task	So that image of source is reflected to eye & break appears dark (Fig. 14-12d)
c. Specular coating over specular background	Inspection of finish plating over underplating	To show up uncovered spots	S-IV with color of source selected to create maximum color contrast between two coatings	For reflection of source image toward the eye (Fig. 14-12d)
3. Combined specular & diffuse surfaces				
a. Specular detail on diffuse, light background	Shiny ink or pencil marks on dull paper	To produce maximum contrast without veiling reflections	S-III or S-IV	So direction of reflected light does not coincide with angle of view (Fig. 14-12a)
b. Specular detail on diffuse, dark background	Punch or scribe marks on dull metal	To create bright reflection from detail	S-II or S-III	So direction of reflected light from detail coincides with angle of view (Fig. 14-12b)
c. Diffuse detail on specular, light background	Graduations on a steel scale	To create a uniform, low-brightness reflection from specular background	S-IV or S-III	So reflected image of source coincides with angle of view (Fig. 14-12b or d)
d. Diffuse detail on specular, dark background	Wax marks on auto body	To produce high brightness of detail against dark background	S-III or S-II	So direction of reflected light does not coincide with angle of view (Fig. 14-12a)
B. Translucent Materials				
1. With diffuse surface	Frosted or etched glass or plastic, lightweight fabrics, hosiery	Maximum visibility of surface detail	Treat as opaque, diffuse surface—See A-1	
		Maximum visibility of detail within material	Transilluminate behind material with S-II, S-III, or S-IV (Fig. 14-12e)	
2. With specular surface	Scratch on opal glass or plastic	Maximum visibility of surface detail	Treat as opaque, specular surface—See A-2	
		Maximum visibility of detail within material	Transilluminate behind material with S-II, S-III, or S-IV (Fig. 14-12e)	
C. Transparent Materials				
Clear material with specular surface	Plate glass	To produce visibility of details within material such as bubbles & details on surface such as scratches	S-V and S-I	Transparent material should move in front of Type S-V, then in front of black background with Type S-I directed obliquely. Type S-I should be directed to prevent reflected glare

D. Transparent over Opaque Materials				
1. Transparent material over diffuse background	Instrument panel	Maximum visibility of scale & pointer without veiling reflections	S-I	So reflection of source does not coincide with angle of view (Fig. 14–12a)
	Varnished desk top	Maximum visibility of detail on or in transparent coating or on diffuse background		
		Emphasis of uneven surface	S-V	So that image of source & pattern is reflected to the eye (Fig. 14–12d)
2. Transparent material over a specular background	Glass mirror	Maximum visibility of detail on or in transparent material	S-I	So reflection of source does not coincide with angle of view. Mirror should reflect a black background (Fig. 14–12a)
		Maximum visibility of detail on specular background	S-V	So that image of source & pattern is reflected to the eye (Fig. 14-12d)
Part II—Three-Dimensional Objects				
A. Opaque Materials				
1. Diffuse detail & background	Dirt on a casting or blow holes in a casting	To emphasize detail with a poor contrast	S-III or S-II or	To prevent direct glare & shadows (Fig. 14–12a)
			S-I or	In relation to task to emphasize detail by means of highlight & shadow (Fig. 14–12 b or c)
			S-III or S-II as a black light source when object has a fluorescent coating	To direct ultraviolet radiation to all points to be checked
2. Specular detail & background				
a. Detail on the surface	Dent on silverware	To emphasize surface unevenness	S-V	To reflect image of source to eye (Fig. 14–12d)
	Inspection of finish plating over underplating	To show up areas not properly plated	S-IV plus proper color	To reflect image of source to eye (Fig. 14–12d)
b. Detail in the surface	Scratch on a watch case	To emphasize surface break	S-IV	To reflect image of source to eye (Fig. 14–12d)
3. Combination specular & diffuse				
a. Specular detail on diffuse background	Scribe mark on casting	To make line glitter against dull background	S-III or S-II	In relation to task for best visibility. Adjustable equipment often helpful
				Overhead to reflect image of source to eye (Fig. 14–12b or d)
b. Diffuse detail on specular background	Micrometer scale	To create luminous background against which scale markings can be seen in high contrast	S-IV or S-III	With axis normal to axis of micrometer
	Coal picking	To make coal glitter in contrast to dull impurities	S-I, S-II	To prevent direct glare (Fig. 14–12b)
B. Translucent Materials				
1. Diffuse surface	Lamp shade	To show imperfections in material	S-II	Behind or within for transillumination (Fig. 14–12e)
2. Specular surface	Glass enclosing globe	To emphasize surface irregularities	S-V	Overhead to reflect image of source to eye (Fig. 14–12d)
		To check homogeneity	S-II	Behind or within for transillumination
C. Transparent Materials				
Clear material with specular surface	Bottles, glassware—empty or filled with clear liquid	To emphasize surface irregularities	S-I	To be directed obliquely to objects
		To emphasize cracks, chips, and foreign particles	S-IV or S-V	Behind for transillumination. Motion of objects is helpful (Fig. 14–12e)

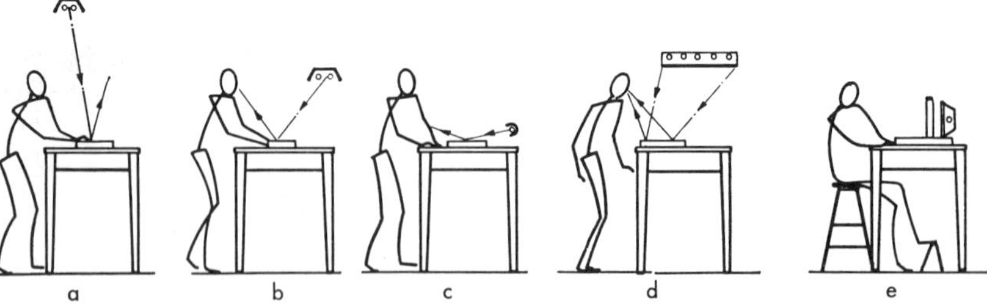

Fig. 14-12. Examples of placement of supplementary luminaires: a. Luminaire located to prevent veiling reflections and reflected glare; reflected light does not coincide with angle of view. b. Reflected light coincides with angle of view. c. Low-angle lighting to emphasize surface irregularities. d. Large-area surface source and pattern are reflected toward the eye. e. Transillumination from diffuse source.

classified according to certain common characteristics. The detail to be seen in each group can be emphasized by an application of certain lighting fundamentals. Fig. 14-11 classifies tasks according to their physical and light controlling characteristics and suggests lighting techniques for good visual perception.

It should be noted when using Fig. 14-11 that the classification of visual tasks is based on the prime and fundamental visual task characteristics and not on the general application. For example, on a drill press the visual task would be the discernment of a punch mark on metal. This could be specular detail with a diffuse, dark background, classification A-3(b) in Fig. 14-11. Luminaire Types S-II or S-III are recommended. S-II on an adjustable arm bracket is a practical recommendation due to space limitations. Several or all of the luminaire types are applicable for many visual task classifications and the best luminaire for a particular job will depend upon physical limitations, possible placements of luminaires, and the size of the task to be illuminated.

Special Effects and Techniques

Color as a part of the seeing task can be very effectively used to improve contrast. While black and white are the most desirable combinations for continual tasks such as the reading of a book, it has been found that certain color combinations have a greater attention value. Black on yellow is most legible and the next combinations in order of preference are green on white, red on white, blue on white, white on blue, and finally black on white.

The color of light can be used to increase contrast by either intensifying or subduing certain colors inherent in the seeing task. To intensify a color, the light source should be strong in this color; to subdue a color the source should have relatively low output in the particular color. For example, it has been found that chromium plating imperfections over nickel plating can be emphasized by using a bluish color of light such as the daylight fluorescent lamp.

Three-dimensional objects are seen in their apparent shapes because of the shadows and highlights resulting from certain directional components of light. This directional effect is particularly useful in emphasizing texture and defects on uneven surfaces.

Silhouette is an effective means of checking contour with a standard template. Illumination behind the template will show brightness where there is a difference between the contour of the standard and the object to be checked.

Fluorescence under ultraviolet radiation is often useful in creating contrast. Surface flaws in metal and non-porous plastic and ceramic parts can be detected by the use of fluorescent materials.

The detection of internal strains in glass, mounted lenses, lamp bulbs, radio tubes, transparent plastics, etc., may be facilitated by transmitted polarized light. The nonuniform spectral transmittance of strained areas causes the formation of color fringes that are visible to an inspector. With transparent models of structures and machine parts, it is possible to analyze strains under operating conditions.

Inspection of very small objects may be greatly simplified by viewing them through lenses. For production work the magnified image may be projected on a screen. Because the projected silhouette is many times the actual size of the object, any irregular shapes or improper spacings can be detected readily. Similar devices are employed for the inspection of machine parts where accurate dimensions and contours are essential. One typical device now in common use projects an enlarged silhouette of gear teeth on a profile chart. The meshing of these production gears with a perfectly cut standard is examined on the chart.

It is sometimes necessary to inspect and study moving parts while they are operating. This can be done with stroboscopic illumination which can be adjusted to "stop" or "slow up" the motion of constant speed rotating and reciprocating machinery. Stroboscopic lamps give flashes of light at controllable intervals (frequencies). Their flashing can be so timed that when the flash occurs an object with rotating or reciprocating motion is always in exactly the same position and appears to stand still.

EMERGENCY LIGHTING

Emergency lighting is defined as the lighting designed to supply illumination essential to safety of life and property in the event of failure of the normal lighting.

Objectives

Generally, emergency lighting is designed into an industrial building system to fulfill two basic objectives, as follows:

1. Provide minimal illumination required for personnel safety and evacuation purposes. Minimal illumination is defined as sufficient to demark, at least by silhouette, any obstructions or safety hazards in the direct travel path of exiting personnel. The Life Safety Code[4] requires that "every exit and the necesary ways of exit access thereto shall be illuminated . . . continuously during the time that the conditions of occupancy require that the means of egress be available for use. . . . The floors of exits and of ways to exit access shall be illuminated at all points such as angles and intersections of corridors and passageways, stairways, landings of stairs, and exit doors. . . . Emergency lighting facilities shall be arranged to maintain the specified degree of illumination in the event of failure of the normal lighting for a period of at least ½ hour. . . ." Such systems should be so designed and installed that the failure of any individual lighting element will not leave an occupied area in total darkness. Exit signs and directional signs are required to be illuminated either internally or externally and incorporated into the emergency lighting system or provided with emergency lighting facilities. The illumination contributed by the exit sign can be included in providing the required illumination.
2. Provide sufficient illumination for longer durations, in areas where required for security purposes, hazardous processes, or continuity of critical operations. Typical of such areas are guard headquarters, gate houses, fire headquarters, fire pump rooms, first aid areas, central control rooms, boiler operating areas, emergency generator rooms, telephone switchboard rooms, electrical switchgear rooms, and areas where hazardous processes must be controlled or shut down in orderly sequence. Toilet and washrooms used by key personnel who will remain in the building also should be provided with emergency lighting. Battery powered lighting should be considered for the emergency generator room to provide lighting for trouble shooting if the generator fails to start.

Types of Systems

Although the National Electrical Code[5] recognizes six methods of supplying emergency lighting, three of the methods rely on one or more services from off-premises sources and, hence, are vulnerable to large area disruptions of public utility power. The other three recognized methods, which are inherently more reliable, are (1) individual battery powered lighting units, (2) centralized storage battery powered systems, and (3) mechanically driven generator powered systems. Regardless of the source of emergency power, the National Electrical Code requires that luminaires which are designated as emergency lighting units be wired independently of the general lighting and power system in separate raceways, boxes, and cabinets, "except in transfer switches or in exit or emergency lighting fixtures supplied from two separate sources."

Individual Battery Units. Individual battery powered lighting units consist of a dry-cell or wet-cell type battery, a battery charger when the battery is of the wet-cell type, one or more lamps mounted in suitable lamp heads, and a relaying device arranged to energize the lamps automatically upon failure of the power supply to the general lighting in the area. Wet-cell type batteries usually have a visual indicator of state-of-charge and the charger should be of the automatic high-rate and trickle-charge-rate type. The unit should have a test switch to simulate power failure and may have pilot lights to indicate normal power on, fast charge, and trickle charge. The National Electrical Code requires that these units be permanently fixed in place (not portable) and permanently connected to the power supply (not connected by a cord and plug). The battery capacity, the number of lamps, and their wattages must be so correlated that the light output is maintained for at least the ½ hour required by the codes. The battery powered lighting units are installed and the lamp heads are directed in such manner as to provide the illumination required to fulfill objective (1) above. Both dry-cell and wet-cell type battery units require frequent inspection, tests, and maintenance if they are to perform their intended function.

Centralized Battery Systems. To provide the illumination required to fulfill objective (2), it is usually necessary to install one or more centralized storage battery systems or mechanically driven generator powered systems. With the centralized storage battery system, the emergency luminaires usually employ incandescent filament lamps, since they must operate on direct current when energized by the battery. Many different arrangements are acceptable. The emergency luminaires may be a part of the general lighting layout, wired from the emergency distribution system but powered from the normal supply. When the normal source fails, an automatic transfer switch transfers the emergency distribution system to the battery. When the normal supply is re-established, the automatic transfer switch retransfers the system to the normal source. The battery is maintained in a charged condition by a charger of the motor-generator or rectifier type. If the capacity of the charger is sufficient to carry the emergency lighting load and keep the battery charged, the charger can provide the normal source with the battery "floating" on the system. Upon failure of the supply to the charger, the battery automatically becomes the source of supply. A transfer switch is not required with this scheme.

The emergency luminaires may be supplemental to the general lighting layout and energized continuously by either of the schemes just described. Since the illumination provided by such emergency luminaires is not required during normal operation, this arrangement is wasteful of electrical energy and has the further disadvantage that there will be a tendency to neglect lamp replacement in the emergency luminaires, particularly if the general lighting employs lamp types with inherently longer life. These disadvantages can be overcome by designing the emergency system so that it is normally de-energized. Upon failure of the normal supply, the emergency system is connected to the battery by an automatic switch actuated by a relay.

Mechanically Driven Generator Systems. The extent of the areas requiring emergency illumination, and particularly the duration of time available, are severely limited by the capacity of the battery in centralized storage battery systems. Systems utilizing mechanically driven alternating current generators offer many advantages if properly designed. The prime mover may be an internal combustion engine, or gas, or steam turbine, utilizing self-contained or immediately available fuel. A diesel engine drive is preferable over one utilizing gasoline because of the relatively high flammability hazard of gasoline and the deterioration of gasoline with age. If gas fuel from an off-site supply is used, the contract with the supplying utility should not include permissible interruptable clauses. Pumping facilities of the supplying utility should be powered by gas, or self-contained fuel units, or equivalent. Steam driven generators should be used only if the steam supply is always immediately available. Electrical facilities used in the production of steam should be provided with back-up emergency electrical power.

With the mechanically driven generator powered system, the emergency luminaires may employ incandescent filament or fluorescent lamps. Because of the time required for high-intensity discharge lamps to achieve acceptable light output after power interruption, these lamps should not be used as the sole source of emergency lighting unless the emergency system is continuously powered from the generator, without transfer to and from another source. Even a momentary outage or severe voltage dip will result in these lamps being inoperative until they have cooled to a point where the arc can restrike after restoration of normal voltage to the ballast. The time interval required for cooling, restrike, and return to significant light output may be from three to ten minutes. Luminaires are available which employ high-intensity discharge lamps in combination with incandescent filament, or tungsten-halogen lamps and photocells, or timing relays. When these luminaires are re-energized, the auxiliary emergency lamps light immediately. When the high-intensity discharge lamp reaches a predetermined light output (as monitored by the photocell or timing relay) the auxiliary lamp is automatically disconnected from the circuit. The long-term emergency illumination is thus furnished by the more efficient high-intensity discharge lamps.

The generator powered system may be designed to operate in any of three different modes: (1) generator always running and continuously supplying the emergency lighting system, (2) generator running idle and automatically connected to the emergency lighting system when the normal source fails, or (3) generator normally shut down and starts automatically upon failure of the normal power supply. The last arrangement requires that the generator prime mover be of a type which starts and achieves rated speed and output in a few seconds and be provided with all necessary automatic starting facilities and controls. The emergency luminaires may be a part of the general lighting layout or may be supplemental and the emergency lighting system may be designed to function similarly to the schemes described for the centralized battery system.

Other Considerations

Either the centralized storage battery powered system or the mechanically driven generator powered system can be utilized to accomplish both objectives as previously outlined. Emergency luminaires, whether or not part of the general lighting layout, can usually be located to provide illumination for off-hour security guard tours when the balance of the general lighting system is switched off. If these guard-tour emergency luminaires are switched off during regular working hours, provision must be included to automatically bypass this switching upon failure of the general lighting supply.

Regardless of the type of emergency lighting system employed, all components must be periodically tested and maintained in order that the system will operate satisfactorily when called upon. In addition to the national codes mentioned, the installation must comply with applicable state and local codes. Requirements for certain components of the emergency lighting system have been established by the Underwriters' Laboratories, Inc.[6]

STEEL MILLS

The manufacture of steel is an integrated process involving three major steps: the blast furnaces refine the iron ore into pig iron; the pig iron is transferred to either an open hearth furnace, basic oxygen furnace, or electric furnace where it is converted into steel; and the steel ingots are sent to various rolling mills where they are formed into sheets, bars, rails, structural forms, etc. See Fig. 14-13. Each one of these processes can be considered as a separate operation. They may be located anywhere from a few hundred yards to many miles apart.

Regardless of any other consideration, enough lighting equipment should be provided to satisfy minimum safety requirements. Also, there are certain areas of critical seeing where higher levels and proper quality are important. See Fig. 9–80, page 9–86, for the recommended levels of illumination. Beyond these minimums there are some examples of far-reaching management policies which also recognize good lighting as a useful tool to aid the production of increased outputs of higher quality material. Since considerable capital expenditures are authorized for lighting systems, the selection of the most practical and trouble-free lighting system pays off when preliminary planning is complete in every detail.

First costs of lighting systems are not a true measure of the systems' effectiveness, especially when steel mills operate at full capacity, 21 shifts per week or 8760 hours per year. Prime study should be given to the annual operating costs of these systems. In virtually all cases, lighting equipment which performs efficiently with dependable and trouble-free operation will prove economically sound for steel mill application.

The following characteristics are common to steel mill building construction as well as the operation of the industry and each has its influence on the design of practical lighting systems:

1. All production areas have overhead cranes which must be used to move raw materials and finished products.

2. The raw materials and finished products are of such composition and size as to require considerable head room, with the result that lighting units must be installed 30 feet or higher above the floor.

3. Arrangement of equipment on floor plans influences the sizes and spans of overhead cranes, and the flow of products from raw material to finished form dictates the locations of lighting outlets to a degree seldom, if ever, found in any other industry.

4. The range of ambient temperatures where lighting equipment must operate runs from zero degrees Fahrenheit in unheated rolling and beam mills during winter months, to soaking pit and hot top areas where readings of 140° F are common above crane cabs in parts of June, July, and August.

5. Atmospheric conditions are poor at best even in the most modern mills where well-planned ventilating systems are installed. The basic production methods which employ a succession of heating and cooling cycles make for conditions which induce fast rising air currents which carry metal dust, products of combustion, and oil vapors upward to locations where lighting equipment must be installed.

6. Instead of finished ceilings with high reflectances of 60 to 80 per cent, there are open truss frameworks below metal roof decks; instead of smooth, light colored walls with 30 to 50 per cent reflectances, there are corrugated metal sidings and a succession of columns marking the boundaries between bays; and finally, instead of floors with smooth finishes rated 10 to 30 per cent reflectance, there are pieces or stacks of dark colored material and equipment in most locations, or black dirt and unswept concrete where parts of floor space are visible.

In many areas, the high mounting heights favor the use of 400-watt and 1000-watt high-intensity discharge lamps. Special consideration should be given to maintaining lighting continuity in case of momentary or long-term power outages. See Emergency Lighting, page 14–10. The choice of light source will depend upon the power supply available. High-intensity discharge lamps installed on cranes served by direct current should use appropriate ballasts.

Open Hearth[7]

Stock Yard. In this area scrap metal, ore, and limestone are loaded by overhead cranes. Generally, overhead lighting should be provided. In outdoor areas in which no overhead framing is available, directional lighting units having wide distribution can be installed on the underside of crane girders.

Charging Floor. Charging boxes are emptied into the furnaces by machine and alloy materials added

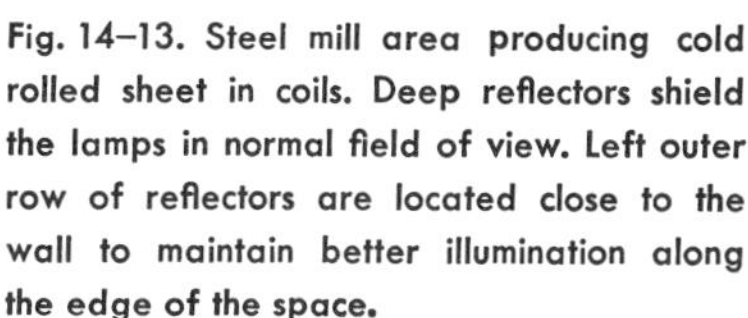

Fig. 14–13. Steel mill area producing cold rolled sheet in coils. Deep reflectors shield the lamps in normal field of view. Left outer row of reflectors are located close to the wall to maintain better illumination along the edge of the space.

by shovel or machine. In addition to general lighting, localized lighting should be provided on the instrument panel which controls the furnace operation. Additional lighting should be provided over the small material storage bins. This usually can be best accomplished by mounting directional lighting equipment on the steel columns or on the underside of the crane girders.

Pouring Side. The molten metal is an area of very concentrated high luminance. When the furnaces are tapped and the ladles emptied into molds there are clouds of fumes and vapor. General overhead lighting is recommended. Auxiliary lighting may be provided for the slag pits in either of two ways: directional lighting equipment can be mounted permanently on the steel columns above the platform level which is part of the charging floor area, or, if this not feasible, portable units mounted on a standard having either a tripod or flat platform with castors can be set on the platform above the slag pit area. At the ingot pouring platform supplementary lighting is necessary and can be obtained by mounting directional lighting equipment on the columns back of the platform. Portable lighting is not satisfactory due to low mounting height and glare from pour and rear directional light sources.

Mold Yard. Here the ingot molds are cleaned, coated, sprayed, and stored. This is a very dirty area with a high concentration of fumes from the spraying operation. General overhead lighting with supplementary lighting on the underside of the crane is recommended. Consideration should be given to suitable maintenance features such as dirt-resistant design, dust-tight covers, etc.

Hot Top Relining and Repairing. General overhead lighting should be provided. Supplemental lighting at the point of hot top repairs and mold relining is also necessary. Directional lighting equipment can be mounted on the overhead beams; for close inspection of the lining it is necessary to use extension cords.

Hot Top Storage. The operator must be able to see in order to place a thin steel gasket accurately between the hot top and mold. General overhead lighting plus lighting units under the crane are required.

Checker Cellar. Brick repair work is sometimes done in this low-ceiling area. Overhead lights should be located wherever possible and portable work lights provided.

Buggy and Door Repair. This is a repair shop for charging boxes, buggies, and furnace doors. Operations include rough machine and bench work, cutting, and welding. General lighting is required.

Stripping Yard. The molds are removed from the ingots by a stripping machine. For indoor locations high-bay general lighting is required. Outdoors, directional lighting equipment should be mounted under crane girders; overhead units mounted on messenger cable are also satisfactory.

Scrap Stock Yard. In this outdoor area with overhead cranes, scrap is cut by torch and shears and then stored. Standard industrial lighting equipment can be mounted on messenger cable strung between the crane-supporting steel structure with floodlights on towers or poles erected beside the crane runway.

Mixer Building. In this area molten metal from blast furnaces is poured into a tilting vessel which acts as a reservoir. The vessel, called a mixer, has a capacity of 300 to 1500 tons. Owing to the very large mixers and the bins containing raw material mounted above them, it is almost impossible to provide general illumination over the entire area. Lighting equipment should be located where possible and convenient. Supplementary lighting should be provided for important areas such as at the top of mixers, pouring level, walkways, and stairwells.

Calcining Building. This is a very dirty high-bay area in which limestone is stored and crushed. General overhead lighting is not advisable due to the height of crushing machinery and bins for feeding them. Industrial lighting units should be provided at convenient locations on all working platforms or levels.

Skull Cracker. In this outdoor area old ingot molds and solid material, which freezes on the bottom or sides of ladles, are broken up. Directional lighting equipment should be mounted on the underside of crane girders. Lighting is also necessary under the cranes.

Cinder Dump. Furnace cinders are stored outside. General lighting from overhead or directional lighting equipment and additional units under the cranes are necessary.

Basic Oxygen Furnace Plant

The lighting techniques are much the same as with the open hearth. Mounting heights are very high with servicing accomplished from cranes or disconnect and lowering hangers. High concentration of graphite particles, dust, etc., dictates use of ventilated luminaires, gasketed enclosed luminaires or reflector lamps for this facility.

Continuous Casting

The continuous casting process takes molten steel and produces the required cross sections of slabs or blooms and, therefore, bypasses the soaking pits and blooming or slabbing mills. The tundish area can be lighted similar to other hot metal handling facilities. Vertical type machines will usually require directional lighting from the sides. These luminaires can be mounted on building columns or structural supporting members of the casting units.

FOUNDRIES

Improvements in the design, construction, and operation of modern foundries are most notable in the fields of materials handling and dust control, so that today, it is not uncommon to find semi-automatic production processes and work conducted in a comparatively clear atmosphere.

Although atmospheric conditions and materials handling facilities have been improved, there is still a requirement for careful selection of: (1) proper light sources, (2) luminaires with acceptable performance characteristics, and (3) locations for luminaires so that installation and maintenance can be handled easily.

Ferrous and non-ferrous castings are made in a vast variety of sizes and shapes. They may vary from a few ounces to a hundred tons. Some castings must be held to very close tolerances, while with others, no close tolerances are required. The lighting requirements for various foundry operations must, therefore, vary in accordance with the desired accuracy and severity of the seeing task. See Fig. 9–80, page 9–84, for recommended levels of illumination. Where workers wear goggles or helmets, additional lighting should be provided to compensate for the loss of transmission.

Many operations such as molding and core making involve non-specular-surface seeing tasks and equipment, thus permitting a wide choice of luminaires, even concentrating types, since reflected glare will not be present. Smoke and steam and dirt cause severe maintenance problems that are minimized through the use of reflector lamps, gasketed enclosed luminaires or luminaires with ventilation openings at the top of reflectors.

Core Making. There are three general methods used to form sand cores—hand ramming, machine ramming, and blowing. Hand and machine rammed cores are formed by packing the sand in the core boxes by hand or by machine. Core blowing machines inject the sand into the core box and pack it tightly by means of compressed air.

The most critical seeing tasks in core making are (1) inspection of empty core boxes for foreign material or sand, and (2) inspection of the cores for flaws such as missing sand or heavy parting lines.

The severity of the seeing task varies with the size of the cores and degree of perfection required. Core making is a rapid and continuous operation requiring an almost instantaneous inspection at frequent intervals.

The contrast between the sand and the metal boxes is fairly good, but it is extremely poor between brown sand and orange shellacked boxes or black sand and black boxes. Contrast and seeing conditions may be improved considerably by finishing the inner surfaces of wooden boxes with white paint. Metal boxes may cause annoying specular reflections if small high-luminance light sources are employed. Bench tops having a light, natural wood finish are both practical and very desirable, for they can be easily kept clean and will provide a comfortable visual environment with low luminance ratios.

Ventilated industrial fluorescent luminaires with a large-area low-luminance source provide excellent seeing conditions in a localized general lighting system. The luminaires should be centered over the worker's side of each bench running parallel to the bench and at right angles to the normal line of sight. This minimizes reflected glare and shadows.

Molding Area. The seeing tasks involved in forming molds from treated sand are: (1) inspecting the pattern for foreign material; (2) setting the pattern in the flask and packing sand around it; (3) removing the pattern and inspecting the mold for loose sand and for accuracy of mold contour; (4) inserting core supports and cores (operator must be able to see the core supports); (5) smoothing mold surfaces, checking core position, and checking clearance between parts. The critical seeing tasks are: (1) inspection of the mold, and (2) placing of the cores (and chaplet supports, if employed).

The size and detail of the seeing tasks may vary through a wide range. The smallest size has a visual angle of about 10 minutes, corresponding to the size of separate grains of sand. A defect involving the presence of only 5 or 6 grains of sand out of place will cause imperfections in small castings. The more exacting seeing tasks are repetitive, and interrupted-short-time rather than prolonged in duration.

The lighting requirements are for intermittent tasks involving critical seeing of materials which have low reflectances and luminances and contrasts of the most unfavorable nature. Mold cavities of various depths require good vertical illumination without harsh shadows.

The typical lighting arrangement in Fig. 14–14 represents good practice for molding machine areas where small castings up to 18 inches maximum dimension are made.

Deep pit molds require additional consideration in planning proper lighting. The walls of the pit may block some of the light from the general lighting system and result in shadows and lower illumination levels, especially on the vertical surfaces of the molds. The pit areas will benefit by the installation of additional general lighting luminaires, located so as to avoid conflict with materials-handling equipment.

Where sand is supplied from overhead ducts and conveyors, localized general lighting is recommended (see Fig. 14–15). However, installation of a general lighting system which satisfies local lighting requirements is preferred in most foundry applications.

Charging Floor. The weighing and handling of metal for charging furnaces is a simple seeing task. This is one of several areas where the seeing task is

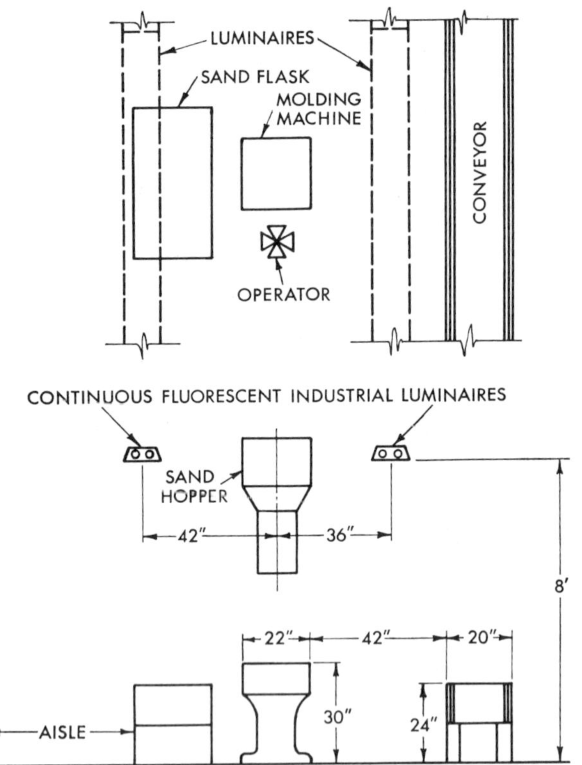

Fig. 14–14. Lighting layout for molding machines used for small castings.

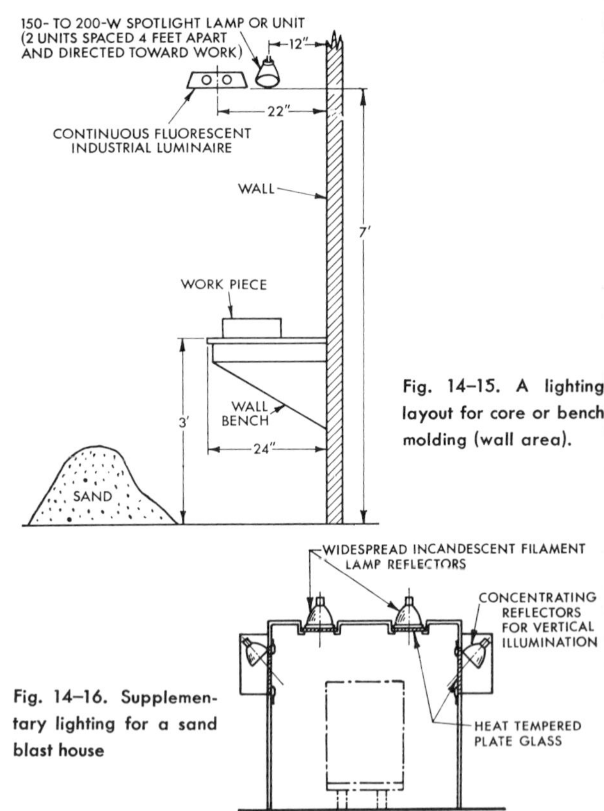

Fig. 14–15. A lighting layout for core or bench molding (wall area).

Fig. 14–16. Supplementary lighting for a sand blast house

not exacting, thus the general lighting system will satisfy the illumination requirements.

Pouring Area. The most exacting seeing task is one of accurately directing the molten metal into the pouring basin. If the pouring basin is not hit accurately, splashing results. This splashing may push sand into the mold causing rejects, as well as causing possible burns for the workers.

The eyes of the worker have to become adapted to the contrast of the bright molten metal with the dark surroundings. This adaptation affects the safety of the worker if there are obstructions to be seen on the floor. Adequate general lighting should be provided to prevent dark surroundings.

The low reflectance of the sand sometimes is offset by placing white parting sand about the opening in order to increase contrast and improve visibility. When weights are used, the opening in the weight indicates the general location of the pouring basin.

Shake-Out Area. Castings are removed from the molds and freed of sand in the shake-out area. The most critical seeing task is the removal of gates and risers. If a ventilation hood is employed over the grate, supplementary lighting is required on the grate.

Heat Treating Area. Small castings are prepared in annealing ovens in malleable iron foundries. The main considerations of the seeing tasks are safety and the packing of castings for annealing so that they will not warp.

Sand-Blasting or Cleaning Area. Three methods are used for cleaning castings: (1) sand blasting (in a blast room), (2) sand blasting in a cabinet or on a rotary table, and (3) friction in a tumbling barrel. The principal visual tasks are: (1) handling castings, (2) directing the blast stream (when manual), and (3) inspecting the castings to see that they are clean.

Where blast cleaning cabinets or houses are used, the general room lighting should be supplemented by lighting units mounted outside the cabinet and projecting their light inside through tempered glass windows. See Fig. 14–16.

Grinding and Chipping Areas. Removal of excess metal is usually accomplished by (1) breaking off the greater part of fins, sprues, risers, and gates with a hammer; (2) chipping remaining projections down further with hand or power chisel; and (3) grinding to a finished condition. With small castings, tumbling in the cleaning operation may have removed all or most of the excess metal in fins, etc.

In grinding, the operator removes excess metal and fins from castings, grinding to contour, to a mark, or to a gauge. Protective glasses worn by the operators often become fogged. The seeing task is fairly severe. For stationary grinders, a combination general and supplementary lighting system should be used. Good practice for stationary grinders is to locate the center line of the luminaires approximately six inches from the edge of the wheel on the side toward the operator. See Fig. 14–17.

Inspection Area. Quality control in the foundry is largely a seeing task. A casting meets the tolerance specified in the drawing because: (1) patterns are carefully checked against the drawings; (2) flasks are inspected for fit; (3) cores and molds are inspected for size, accuracy, and alignment; (4) core clearances are gauged prior to mold closing; (5) castings are checked against templates and gauges; and (6) surfaces are inspected and defective castings are sorted out.

In a foundry, as in many high-production industries, inspections are generally conducted at many intermediate stages in the production of the complete product. In foundries the inspections at some stages are either combined with the particular operation or at least performed in the same immediate area. Where this is the case, the inspection task will dictate the necessary quality and quantity of illumination.

Typical of these conditions is the inspection of cores by the core-maker prior to baking. Similarly, pieces may be scrapped by the shake-out handlers or by grinder operators, thus saving subsequent wasted labor on defective parts. Sorting is frequently combined with inspection of small castings.

In sorting areas, a general lighting system may be mounted four or more feet above the sorting table or conveyor. Seeing into deep cavities and tubular areas may require the use of small, shielded portable luminaires.

For "medium" inspections, fluorescent luminaires may be used for reduced reflected glare and improved diffusion of light. "Fine" inspection sometimes requires special lighting techniques. See Supplementary Lighting, page 14–6.

MACHINING METAL PARTS[9]

The major operations that are performed in the machining of metal parts consist of: the setting up and operation of machines such as lathes, grinders (internal, external, and surface), millers (universal and vertical), shapers, drill presses; bench work; and inspection of metal surfaces. The precision of such machine operation usually depends upon the accuracy of the setup and the careful use of the graduated feed indicating dials rather than the observation of the cutting tool or its path. The work is usually checked by portable measuring instruments and only in rare cases is a precision cut made to

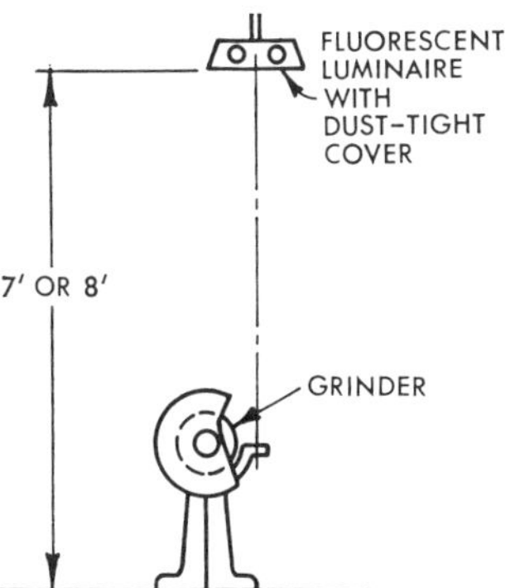

Fig. 14–17. Lighting for a stand grinder.

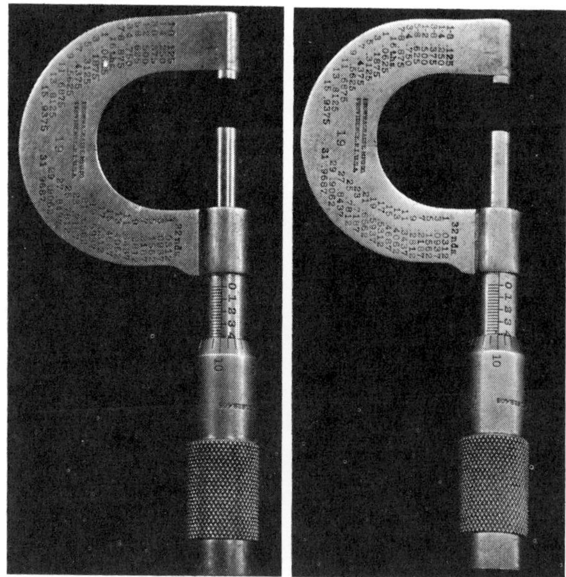

Fig. 14–18. (Left) Micrometer illuminated with a system of small, bright sources is seen with bright streak reflections against a dark background. (Right) When illuminated with a large area, low luminance source, the micrometer graduations are seen in excellent contrast against a luminous background.

scribed line. The fundamental seeing problem is the discrimination of detail on plane or curved metallic surfaces.

Visibility of Specific Seeing Tasks*

Convex Surfaces. The discrimination of detail on a convex surface, such as reading a convex scale on a micrometer caliper, is typical of many seeing tasks encountered. The reflected image of a large-area low-luminance source on the scale provides excellent contrast between the dark figures and divisions and the bright background without producing reflected glare. The use of a nearly point source for such applications results in a narrow brilliant (glaring) band that obscures the remainder of the scale because of the harsh specular reflection and loss of contrast between the figures or divisions and the background. See Fig. 14–18.

Flat Surfaces. In viewing a flat surface, such as a flat scale, the seeing task is similar to that in reading a convex scale. With a flat scale, however, it is possible, depending on the size, location, and shape of the source, to reflect the image of the source either on the entire scale, or only on a small part of it. If the reflected image of the source is restricted to too small a part of the scale the reflection is likely to be glaring.

Scribed Marks. The visibility of scribed marks depends on the characteristics of the surface and frequently on the orientation of the scribed mark as well as on the nature of the light source. When light of a directional characteristic is used, good visibility of scribed marks may be obtained on untreated

* See also Fig. 14-11.

cold rolled steel if the marks are oriented for maximum visibility. This position is such that the brightness of the source is reflected from the side of the scribed mark to the observer's eye. Unfortunately the visibility of other scribed marks is poor when this lighting system is used, and better average results are obtained if a large-area low-luminance source is used. If the surface to be scribed is treated with a low reflectance dye, the process of scribing will remove the dye, exposing the surface of the metal. Such scribing can be made to appear bright against a dark background by reflecting the image of a large-area low-luminance source from the scribed marks to the observer's eye. In lighting specular or diffuse aluminum, excellent visibility may be obtained if the image of a large-area low-luminance source is reflected to the observer's eyes. In this case the scribed marks will appear dark against a bright background.

When there is considerable difference in the contrast between the scribed line and the background, the large-area low-luminance source accentuates this contrast and increases the visibility. When there is little difference in the contrast between the scribed line and the background, directional lighting will heighten contrast either by utilizing the difference in specular reflection between the scribed line and background or by the introduction of shadows.

Center-Punch Marks. A visual task quite similar to scribing is that of seeing typical center-punch marks. The maximum visibility is obtained when the side of the punch opposite the observer reflects the brightness of a light source to the observer's eye. A directional source located between the observer and the task provides excellent results when the light makes an angle of about 45 degrees with the horizontal.

Concave Specular Surfaces. Because of reflections from most light sources in the surrounding area the inspection of concave specular surfaces is a difficult visual task. Large-area, low-luminance sources have been found to provide the best visibility. Scratches, blow holes, and other irregularities are visible because of the luminance contrast.

Lighting for Specific Visual Tasks

For the seeing tasks involved in the machining of small metal parts a luminance of 500 footlamberts for the low-luminance source has been found to be desirable for most applications. The size of the source required depends on the shape of the surface being viewed and on the area on which it is desired to reflect the source brightness. The procedures outlined are based on specular reflections, but they can also be applied to surfaces that are semi-specular.

Flat Specular Surfaces. The procedure for determining the size of source required for flat areas is illustrated in Fig. 14-19. The first step is to draw

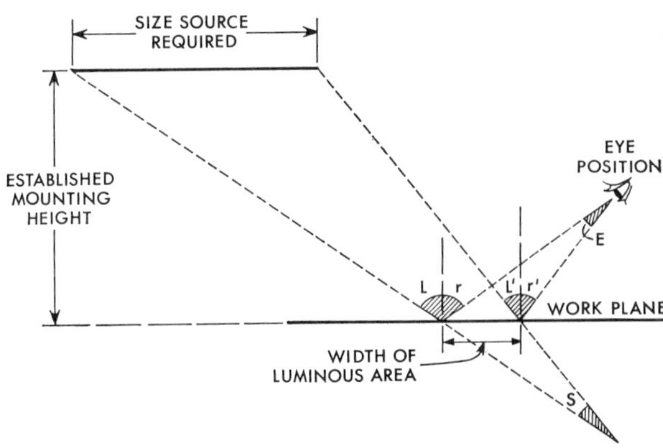

Fig. 14-19. Procedure to be used in establishing the size of source necessary to obtain the desired area of reflected brightness when applied to a flat specular surface:

$$S = E$$

where S — angle subtended by source on work, and
E = angle subtended by luminous area at eye

lines from the extremities of the area that is to be luminous, to the position where the observer's eye is to be located. At the points where these lines intersect the plane of the luminous area, erect normals to the luminous area and construct lines so that the angle of reflection is equal to the angle of incidence. These lines form angle S (equal to angle E) and this angle defines the solid angle that the source must subtend.

Convex Specular Surfaces. After determining the width of the luminous area needed, construct lines from the eye to the margins of the luminous area as shown in Fig. 14-20. At these points erect normals to the surface and construct lines from the surface so that the angle of reflection is equal to the angle of incidence. The angle S that is formed when these lines are projected below the cylindrical surface can be shown to equal twice the angle C at the center of the cylinder subtended by the reflection of the source, plus the angle E at the eye that is subtended by the reflection of the source. Angle S defines the size of the source required. The same general procedure can be applied to concave surfaces.

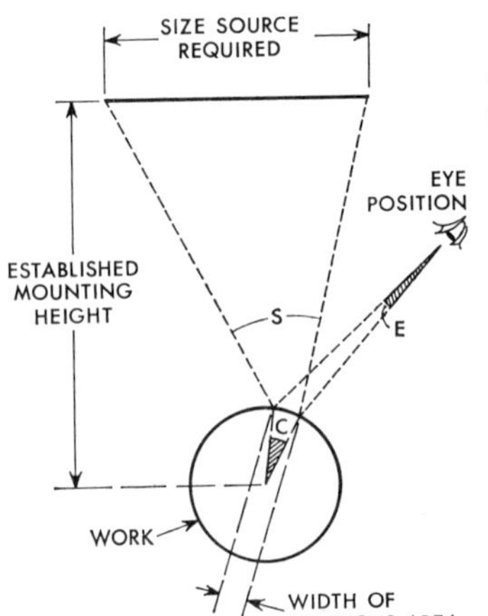

Fig. 14-20. Procedure to be used in establishing the size of source necessary to obtain the desired area of reflected brightness when applied to a convex specular surface:

$$S = 2C + E$$

where S = angle subtende by source on work, C = angle subtended by luminous area on work, and E = angle subtended by luminous area at eye

General Lighting

The desirability of the large-area low-luminance source for most of the visual tasks in the machining of metal parts is evident. The ideal general lighting system would be one having a large indirect component. While both fluorescent and high-intensity discharge sources are used for general lighting, fluorescent luminaires, particularly in a grid pattern, are preferred where servicing is not difficult from the floor. High-reflectance room surfaces are recommended to improve utilization of illumination and seeing performance.

SHEET METAL SHOPS

The problem of seeing encountered in the sheet metal shop is made difficult by one or more of the following factors: the sheet metal after pickling and oiling has a reflectance similar to the working surface of the machine, resulting in poor contrasts between the machine and work; the low reflectance of the metal results in a low task luminance; high speed operation of small presses reduces the available time for seeing; bulky machinery obstructs the distribution of light from general lighting luminaires; and noise contributes to fatigue.

Punch Press

The seeing task is essentially the same for the large press as it is for the small press except that in the case of the small press less time is available for seeing. The shadow problem, however, is much greater with the large press. In both cases the operator must have adequate illumination to move the stock into the press, inspect the die for scrap after the operating cycle is completed, and inspect the product. Where an automatic feed is employed the speed of operation is so great that the operator has time only to inspect the die for scrap clearance.

The general lighting system in the areas where the presses are located should provide the level of illumination listed in Fig. 9–80, page 9–89. This is necessary for the safe and rapid handling of stock in the form of unprocessed metal, scrap, or the finished products. In areas where large presses are located, such as those shown in Fig. 14–21, the illumination should be furnished by high-bay lighting equipment, or a combination of high-bay and supplementary lighting as shown in Fig. 14–22. For low-bay areas as shown in Fig. 14–23, the illumination should be supplied by luminaires having a widespread distribution to provide uniform illumination for the bay and die surface area. In such areas where the mounting height of fluorescent lighting systems exceeds 20 feet, careful consideration should be given to maintenance costs.

The operator's ability to inspect the die is more directly related to the reflected brightness of the die surface than to the amount of light striking it (incident footcandles). For example, a concentrated light placed on the operators' side of the press and directed toward the die may produce results much less satisfactory than a large-area source of low luminance placed at the back or side of the press. See Fig. 14–24. The luminance required for optimum visibility of the die has not as yet been established, but it appears that 500 footlamberts would be satisfactory.

Paint applied to both the exterior and the throat surfaces of the press plays a very important part in the operator's ability to see. The reflectance of the paint selected for the exterior of the press should not be less than 40 per cent. This treatment of vertical surfaces on the exterior provides for maximum utilization of light from the general lighting system. Similarly, the paint selected for throat surfaces should have a reflectance of 60 per cent or higher.

Fig. 14–21. High bay large press shop lighted with 1000-watt improved color mercury lamp units.

Fig. 14–22. Large press with built-in supplementary lighting to facilitate set-up and operation.

Fig. 14–23. Low bay press shop with fluorescent lighting. Note: upward component and center shield for better comfort.

Fig. 14–24. Highlights and shadows on the die of a small press are produced by supplementary or general lighting at the rear of the press. The operator can quickly inspect the open die for loose pieces of scrap.

Shear

The operator must be able to see a measuring scale in order to set the stops for gauging the size of "cut." When the sheet has to be trimmed, either to square the sides or to cut off scrap from the edges, the operator must be able to see where the cut will be made in order to hold the amount of scrap to a minimum.

The general lighting system should provide the level of illumination listed in Fig. 9–80 in the area around the shear for the safe feeding of the sheets at the front, collecting the scrap at the back, and stacking the finished pieces in preparation for removal.

Localized lighting, as shown in Fig. 14–25, produces a line of light to indicate where the cut will be made and consequently the amount of scrap that will be trimmed off. It also provides lighting for the oil-soaked gloves on the hands of the operators. This enables the operator who is responsible for pressing the foot-release bar to see quickly that all hands are clear of the guard.

AUTOMOTIVE ASSEMBLY

The automotive assembly process is continuous and requires both general and supplementary lighting along the major portion of the assembly line. Supplementary lighting is needed to illuminate the many difficult seeing tasks, since the maze of ducts, piping, supporting steel, and conveyors around the assembly line reduce the effectiveness of the general lighting system.

The flow chart shown in Fig. 14–26 identifies the major steps in the assembly operation. Descriptions of the major operations, related seeing tasks, and typical lighting systems are as follows:

Body Framing Area. Metal parts are placed in large jigs or fixtures and automatically or manually welded.

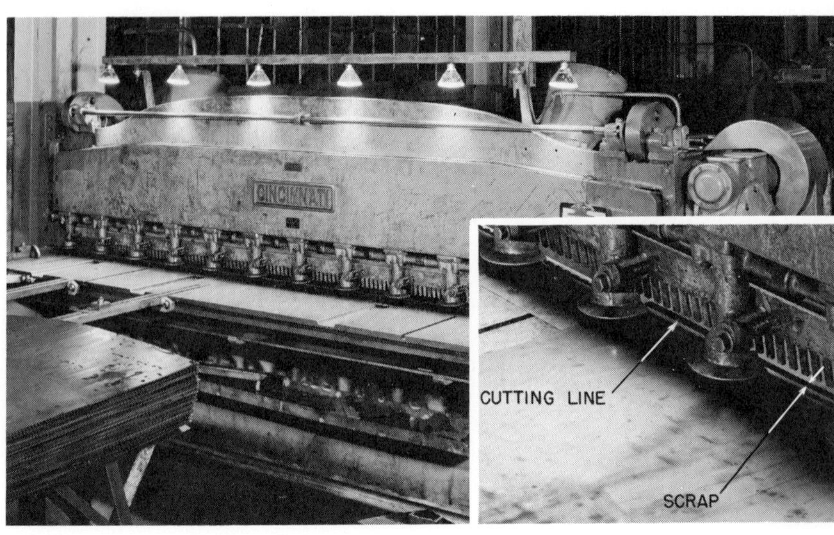

Fig. 14–25. Shear. Six 150-watt floodlamps provide a narrow line of light on the metal sheet to indicate where the cut will be made. Sockets are mounted in channel supported by springs. A 2-lamp fluorescent ventilated industrial luminaire will also provide good illumination and is less subject to short lamp life due to vibration.

Seeing Task—Locating (lining-up) parts in jigs, welding, and inspection of welding tips.

Lighting—Localized general lighting with luminaires positioned slightly behind the operator.

Body Soldering Area. Soldering joints where body parts have been welded to give a smooth appearance. Also, filling dents and scratches with solder.

Seeing Task—To see all welds, joints, dents, and scratches and cover with solder so metal finishers can grind and polish to exact contour of surface.

Lighting—Luminaires are positioned on both sides of line and oriented above and behind worker. Usually, luminaires are built into exhaust canopy which extends over this area.

Metal Finishing Area. Raw metal and solder are ground and polished to the desired surface contour.

Seeing Task—To see all welds and soldered areas and grind and polish to desired contour and smoothness.

Lighting—Luminaires are located along both sides of line and usually mounted parallel or perpendicular to assembly line.

Body Inspection Area. The body is inspected to locate all metal defects for correction so that when paint is applied, the surface will be uniform.

Seeing Task—To locate and mark all metal body defects (dents, scratches, high spots, etc.) so that they can be repaired before painting.

Lighting—Illumination is provided along both sides of the assembly line. Fluorescent luminaires are mounted at 45-degree angle with the horizontal to provide illumination for the hood, roof, and upper vertical surfaces. Vertical mounted fluorescent luminaires are positioned along the line to illuminate the lower vertical surfaces and quarter panels.

Body Repair Area. Defects that have been marked by inspection personnel are repaired.

Seeing Task—To see and distinguish the various marks that identify the location and nature of the defect. Also, to see that surface of body is smooth and uniform after grinding or polishing.

Lighting—Same quantity and quality of illumination to that of Body Inspection Area to assure that defect has been properly repaired.

Painting Area. Paint is applied to all body surfaces.

Seeing Task—To see identification marks so that the proper color paint is applied. Also, to see that paint is applied completely and evenly over the entire surface to eliminate running and insufficient coverage.

Lighting—Fluorescent luminaires are installed in spray booth walls and ceiling so that illumination is provided over all horizontal and vertical body surfaces.

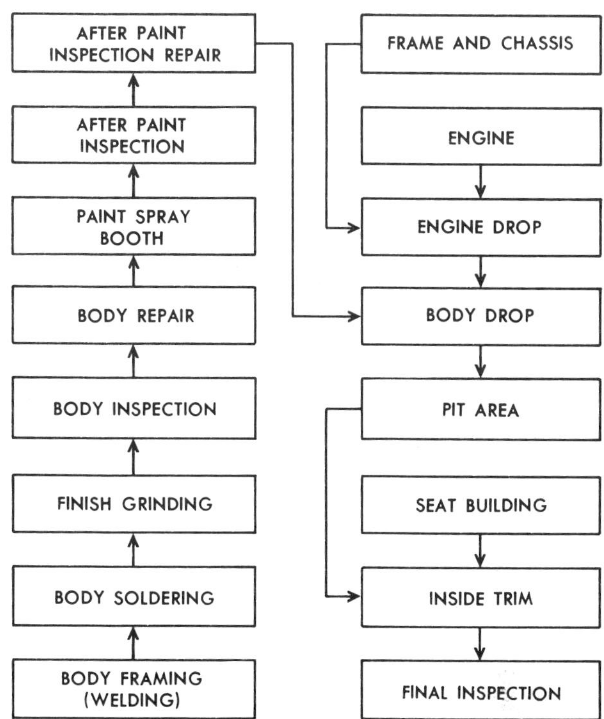

Fig. 14-26. Flow chart showing areas on automotive assembly lines.

After Paint Inspection Area. Inspect for proper paint coverage of body as well as for defects not identified and corrected prior to painting.

Seeing Task—To detect any non-uniformity in body surface, color, and insufficient coverage.

Lighting—Fluorescent luminaires usually mounted on 45-degree angle, along both sides of body and perpendicular to assembly line. Luminaires should be positioned to enable inspectors to see defects by observing the reflected light source image in the specular body surface.

Frame and Chassis Area. The frame of the automobile is started down the assembly line. The gas tank, exhaust system, and suspension are all assembled on the frame during this phase of the assembly operation.

Seeing Task—There are no critical seeing tasks in this area. Parts are generally large and alignment is usually all that is required.

Lighting—General or localized lighting.

Engine Drop Area. The engine is placed on the frame, fuel lines connected, and miscellaneous other parts bolted into their proper position.

Seeing Task—Relatively large parts are bolted in place with hand tools.

Lighting—Continuous row fluorescent luminaires on both sides of line angled to provide horizontal and vertical illumination.

Body Drop Area. The body is lowered onto the chassis and fastened into position.

Seeing Task—The alignment of the body is the most severe seeing task.

Lighting—General and vertical illumination is required. The illumination must be sufficient to ensure proper positioning of the body and to bolt the body in place with hand tools.

Pit Area. Bolting of body to frame along with other operations to undercarriage is accomplished by workers from pits with hand tools.

Seeing Task—To see that bolts are properly installed and tightened.

Lighting—Luminaires are usually recessed in pit wall on both sides of pit to provide light over worker's shoulder.

Inside Trim Area. Seats, head liner, instrument panel, and other interior accessories are installed.

Seeing Task—To position and fasten seats and instrument panel. Also, see proper location for fastening head liner to interior roof of car.

Lighting—Fluorescent luminaires are located along both sides of assembly line. General illumination in this area will insure the type of illumination to satisfy seeing requirements inside the automobile.

Final Inspection Area. Automobile is inspected for all defects, *i.e.*, body, trim, etc., so that repairs can be effected.

Seeing Task—Most critical task is identification of body defects and surface damage which may have occurred during the subsequent assembly process.

Lighting—Fluorescent luminaires are installed above and to the side of the automobile so that all surfaces can be inspected. Inspection personnel spot body defects by viewing reflected lamp image in specular surface of automobile.

RUBBER TIRES[10]

Recommended levels of illumination for the tasks in manufacturing rubber tires are listed in Fig. 9–80.

Processes and Flow Chart

The following descriptions are used to clarify the processes through which the raw materials must pass to become a finished tire. Fig. 14–27 is a flow chart showing the major areas.

Banbury Area (1). Bales of synthetic and natural rubber are removed from storage and split into small pieces. Then various grades of rubber are blended to meet specific compound requirements, and are softened together in a plasticator. The resulting blend of rubber and exact amounts of carbon black and pigments are emptied from a conveyor into a Banbury machine. After mixing, this "Master Batch" drops from the Banbury into a Pelletizer where it is cut into small marble-like pellets of uniform size and shape to facilitate cooling, handling and processing. Sulphur and accelerators are added to the pellets in a final mixing process; and, after a trip through the Banbury mixer, the "Final Batch" drops onto an automatic mill where it is rolled into a thick continuous sheet and conveyed to a "wig-wag" loader which places it on a skid, ready for further processing.

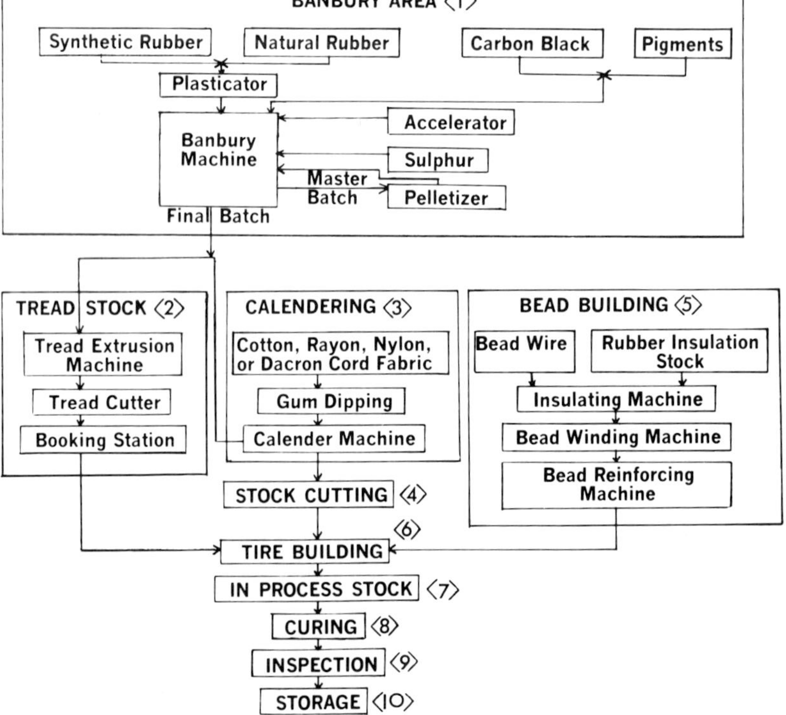

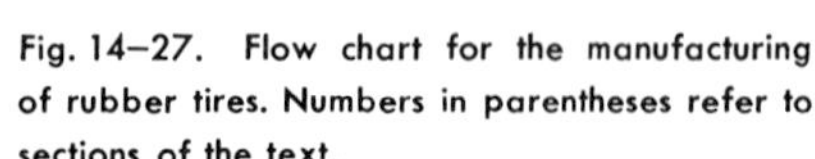

Fig. 14–27. Flow chart for the manufacturing of rubber tires. Numbers in parentheses refer to sections of the text.

When needed for the preparation of treads, sidewalls, plies, or beads, a batch of compounded rubber stock is conveyed to a warming mill where it is kneaded and heated to make it more workable. It is then removed in continuous strips on conveyor belts to the machines which turn out the components.

Tread Stock (2). Rubber stocks are conveyed from warming mills and are fed into two opposing feed hoppers at a tread extrusion machine. The tread and sidewall are usually extruded through one die in a continuous strip which is then stamped with an identifying code and check weighed. The entire underside of the tread and sidewall unit is then coated with cement.

The continuous unit is conveyed over a series of belts for cooling and on to a tread cutter where it is cut to specific lengths for use in different sizes and types of tires. The individual tread and sidewall units are then conveyed to booking stations where they are inspected, placed on racks, and trucked to storage for aging before going to the tire assembly operation.

Calendering (3). Huge rolls containing cotton, rayon, nylon, dacron, or other cord fabric are taken to a gum-dipping operation where the cord fabric is dipped in a special compound, heat treated, and dried under tension.

The fabric passes on to calendering machines where rubber stock from warm-up mills, fed automatically into the calender rolls, is pressed onto both sides and between the cords of the fabric as it passes between the rolls. The fabric goes over cooling drums and is rolled into cloth liners to prevent sticking.

Square yard weights and widths are recorded at the control board. Many quality checks, including use of the Beta Gauge and statistical quality control charts, aid the calender operations. Accuracy of these measurements is very important to insure a quality product.

Stock Cutting (4). From calendering, the rolls of fabric are trucked to the stock cutting operation where the fabric is cut on the bias, diagonally across the cords. Cut to specified widths and bias angles, these pieces, known as plies, pass from the cutting table, and are turned end to end, and precision-spliced into continuous strips, ready for tire assembly.

Bead Building (5). Copper plated high tensile strength steel wire is brought in on large reels. A specified number of strands are brought together from the reels and guided through a die in the head of a small tube machine. Here a special rubber insulation stock is squeezed around and between the separate strands of wire.

The rubber covered wire strip is then run into a bead winding machine, where it is wound on a chuck or collapsible ring. The machine automatically winds the proper number of turns, cuts the wire and ejects the wound bead on to a storage arm.

Fig. 14-28. Forming and curing presses where "green" tires are fed in to be formed and vulcanized.

In the next operation, bead reinforcing, the wound bead wire is covered with a strip of calendered fabric called a reinforce. The reinforce not only ties the bead into the tire but also plays a vital part in giving the tire added strength and stability.

Tire Building (6). All the parts of a tire are brought together in the tire assembly operation and are combined on a semi-automatic precision machine with a collapsible drum.

In Process Stock (7). Green tires are accumulated along with other parts still in process, such as fabrics and calendered stock.

Curing (8). Green tires are sprayed automatically with lubricants, inside and out, to aid in molding. They are then conveyed to long batteries of automatic forming and curing presses where they are transformed into tires. See Fig. 14-28.

Inspection (9). In final inspection, tires are subjected to a series of quality checks. Passenger tires arriving from the curing room pass along a conveyor to automatic trimmers and on to spray painting machines for identification symbols. The tires are then conveyed to inspectors who check each tire inside and out.

The entire tread is examined to see that the tread angles are sharply molded so that they will grip the road as designed. The sidewalls of the tire are inspected for gouges, pinvents, and whitewall adhesion. The bead is examined for wrinkles or any defects which might prevent the tire from seating tightly and securely when mounted on a wheel or in any other way affect the service of the tire. Tire inner liners are inspected for any separations.

Finished Tire Storage (10). Tires are stored for later shipment.

Lighting Design for Specific Tasks

Banbury Area (1). When the Banbury machine is manually charged, the ingredients for the batch are made up, weighed, and inserted into the charging hopper. The operator should be able to read temperature recorders, time clocks, and scale dials during each machine cycle.

General lighting should be provided with particular attention given to locations where the operators perform their task. Also, it should be noted that the dirt accumulation on luminaires and room surfaces will be high due to large quantities of carbon black in the air.

Tread Stock (2). General lighting is required in this area, with the higher level given in Fig. 9–80 at the extruder, over the check weighing and width measuring station, and over the booking and inspection station, where the tread is inspected for physical defects, weight, length and width.

Calendering (3). General light should be provided in the calendering area with the higher level as listed in Fig. 9–80 at the letoff and windup, for the crew to handle the fabric, and at the instrumentation panel and over the measurement and inspection area, where the operator looks for such imperfections as bare cords.

Stock Cutting (4). This area should have both general and localized general lighting, with the higher recommended level from the localized general units over the splicing table. The operator should be able to inspect fabric for bare cords and to make manual splices or to set up an automatic splicer machine.

Bead Building (5). General lighting is needed in the bead room to allow for proper machine setup, seeing coatings on the wire, and for reinforcing applications.

Tire Building (6). General lighting should be provided throughout the entire area with supplementary lighting at the tire building machines to enable the operator to properly feed the materials onto the drum and to inspect laps.

In Process Stock (7). General lighting is needed for reading tags or color codes on items in this active storage area.

Curing (8). In addition to general lighting, supplementary lighting is required at the molds. Supplementary luminaires should be positioned so that molds may be readily inspected during cleaning and coating.

Inspection (9). This area should be provided with general and supplementary illumination or localized general at each inspection stand. Luminaires should be positioned on both sides of the tires to illuminate both the inside and outside of the tire casings. Luminaires should be shielded and installed in locations to eliminate as much glare as possible.

Finished Tire Storage (10). See In Process Stock.

GRAPHIC ARTS—PRINTING

The graphic arts industry is one of the oldest of our industries. Much of its history spans a period when daylighting was a major source of interior illumination.

Experience in the industry has shown, however, that a modern, well designed lighting installation, excluding daylighting, has a marked beneficial effect on quantity and quality of work. Recommended levels of illumination are listed in Fig. 9–80.

Receiving Area

The most difficult tasks in this area are those of reading markings on shipments, labels, and bills of lading.

General illumination will provide sufficient light for the above tasks and for the operation of hand operated and automatic fork lift trucks as well as for general traffic in the area.

Supplementary illumination may be necessary to provide illumination for the interior of transport carriers bringing material to the plant. Angle or projector type luminaires may be utilized to provide this supplementary illumination, but care must be taken to prevent glare from these sources. If the conveyances are deep, reel type or other suitable portable equipment may be necessary.

Yard or loading dock lighting should be installed for night operation.

Stock Room Area

Large bulky materials, rolls of paper, and crates or boxes with identification on the side need special attention to assure vertical illumination on the labels or side markings.

Where stock is piled quite high and cuts off general lighting, additional illumination should be provided over the aisles.

If volatile materials are stored in a separate area, local building code requirements should be checked as to permissible luminaires.

Copy Preparation Area

This is the area where first preparations for actual printing take place. The area may include paste up, layout, art, and design work. The point size of type, the style of type, size, and placement of cuts (pictures, drawings, charts, etc.), arrangement of text, and page size are decided here.

If color is to be used this is the area where its exact specification is made. It should be understood that general illumination is not suitable for accurate color appraisal. Therefore, the equipment used for this purpose (see color appraisal area below) should be located near this area.

General illumination in this area should be provided by a well diffused glare-free system.

Composing Room Area

In modern printing plants, type composition is accomplished by machines such as Ludlow, Monotype, and Linotype. The operator sits at a keyboard similar to a typewriter to perform his task. In this area general illumination from well shielded luminaires with good upward component is recommended.

Where hand tasks are performed by compositors, special lighting problems are encountered. In inspecting type, the visibility varies with the number of times it is run (impressions made), reflectance of the type, contrast between the face and shoulder of the type, spacing, and type size. Lighting in this area should be provided by large-area, high-luminance sources. See Fig. 14–29. The high luminance is reflected from the shoulder of the type and gives the necessary contrast between the mirror-like shoulder and the type face. Research has indicated that the downward luminance of the large-area source should be in excess of 3000 footlamberts. Care must be taken to select and locate the luminaires to provide a well diffused glare-free system throughout the composing room.

Proofreading is also done in this area. The task involved here is proofing a printed page. Well shielded diffuse illumination is recommended.

Color Appraisal Area

In 1959 a joint report was prepared by the Illuminating Engineering Society and the Research and Engineering Council of the Graphic Arts Industry[11] to define a standard light source for the color appraisal of reflection type material. In 1968 a more comprehensive report, incorporating the CIE color rendering system, was issued.* That report calls for a light source (luminaire) with a correlated color temperature of 5000 K for appraisal of color quality and 7500 K for appraisal of color uniformity, both sources with a color rendering index of 90 or higher.

Fig. 14–29. Composing room lighted for high visibility of type. It was determined by visibility tests that optimum conditions would result from well-shielded, large-area fluorescent luminaires with fairly high downward luminance and good upward component (40 per cent in this case).

Plate Preparation Area

The visual task is severe and prolonged in this area.

The galley (camera) area needs only enough general illumination for traffic in the area. Care must be taken to keep light source glare from the camera lens. Packaged lighting units for picture taking are usually furnished by the supplier of the camera.

Stripping and opaquing are done on a luminous top table. Care should be taken in the table design to provide good diffusion with low luminance to assure visual comfort. Any part of the luminous area not covered by the negative should be masked out.

General overhead, low-level lighting for traffic in this area should be located so as to eliminate reflections on the table tops.

Plate making requires low levels of illumination. Plates are processed over a considerable length of time and higher levels are injurious to them. Colored lamps are frequently utilized in this area.

Pressroom Area

The pressroom is usually a large area with a high ceiling, necessitated by the dimensions of the equipment used in this area. The utilization of light will be reduced due to the size of this equipment.

In the pressroom, the tasks can be divided into the following three groups:

1. A group involving tasks with type for make-ready, register, and correction of errors in both the proof press and the production press. The movement of semi-finished products from one press to another, the movement of finished sheets from presses to other departments, and the movement of raw materials to the presses also fall in this group.

2. A group involving mechanical functions such as adjustment of the presses, installing of frames and cylinders on the presses, and adjustment of ink fountains for the inking rollers and the feeding of the paper.

3. A group involving the inspection of semi-finished and finished products. If this involves color appraisal, see above on color appraisal areas.

General illumination is recommended, using a luminaire with good shielding and with a minimum of 10 per cent upward component.

* *Draft USA Standard Viewing Conditions for the Appraisal of Color Quality and Color Uniformity in the Graphic Arts Industry*, American National Standards Institute, 1430 Broadway, New York, N.Y. 10018, 1968.

Ink and drying compounds in the atmosphere create a difficult maintenance condition. Therefore, ease of maintenance is an important item in luminaire selection. Supplementary lighting should be used wherever required and this can only be determined by careful examination of the equipment being used. Low mounted luminaires tilted on an angle to penetrate recesses in the presses may be necessary. See Fig. 14-30. Pressmen need a large-area high-luminance light source, as recommended above for inspecting type.

Bindery Area

After the actual printing operations are completed, if binding or one of the operations of the bindery is required, the finished product of the printing plant becomes the raw product of the bindery.

Production in this area involves hand labor in practically all operations.

Collating, folding, stapling, stitching, gluing, backing, and trimming pages all involve some hand operations. Although not much critical seeing is involved in many operations, speed here is essential. Diffuse general lighting should be provided by well shielded luminaires allowing an upward component.

In book binding, additional operations, which are more tedious and exacting, are quite often performed. These include corner rounding, indexing, cover imprinting, and gold leaf or gold ink application. In these areas additional luminaires or closer spacing is necessary to provide a higher level for these critical seeing tasks.

Shipping Area

The tasks involved in this location are similar to those in the receiving area, with the addition of wrapping, packaging, labeling, typing, and weighing.

In addition, this is a more critical area than receiving because the finished product is being sent to the customer. It is important that the product is not damaged, that it is sent to the proper address, and that the shipping be accomplished without delay.

COTTON GINS[12]

Areas included in the operation of a cotton gin contain overhead equipment, gin stands, lint cleaners, and bale press. These areas should be illuminated with a well planned general lighting system to provide the illumination levels recommended in Fig. 9-80.

The function of a cotton gin is to separate seed and trash from cotton lint, gather the seed, dispose of the trash, and place the lint in a form that may be easily handled and transported.

Fig. 14-30. In the newspaper press room, supplementary fluorescent luminaires should be mounted along the edge of the catwalk as shown. Equipment should be enclosed and gasketed to minimize maintenance (ink mist is in suspension in the atmosphere).

Overhead Equipment

The overhead equipment in a cotton gin includes separators, driers, cleaners, stick machines, and feeders. These machines are automatic and do not require an operator. Catwalks are usually provided around the overhead equipment for maintenance purposes. General lighting should be provided for safety and maintenance.

Gin Stand and Lint Cleaners

The gin stand separates cotton seed from the cotton lint. The gin stand and the adjacent control console are operated by a ginner, who starts the gin and adjusts the suction conveyor to regulate the flow of cotton into the gin stands. He examines ginned cotton for foreign material and adjusts spiked drums and saw blades within the gin stands. The seeing tasks for the lint cleaner are similar.

The seeing tasks at the gin stand, lint cleaner, and control panel are on a vertical plane and, therefore, the general lighting system should provide a high vertical component. It is suggested that supplementary lighting be mounted in the seed channel inside the gin stand, to help the ginner to observe the rate of flow of the cotton seed.

Bale Press

The bale press is a hydraulic press that compresses the cotton into a bale that is about 60 by 42 by 24 inches and weighs about 500 pounds. Most bale presses are operated by two pressmen, who spread a piece of burlap or bale covering on the bed of the chamber of the press. Then they start the equipment that dumps and tamps bulk cotton into the chamber. They observe the level of cotton in the chamber and stop the loading equipment when the chamber contains the required amount of cotton. The chamber is then positioned under the press ram. The pressmen then start the press to compress the cotton. One pressman inserts metal bands through the channel in the upper ram of the press to the pressman on the opposite side of the compressed bale. The second pressman returns the band through the channels in the lower ram to encircle the bale. The pressman then buckles the ends of the bands together. General lighting should be provided in this area. Seeing tasks are on both the horizontal and vertical plane.

TEXTILES

The seeing tasks in the textile field include both simple ones and some of the most severe found in industry. Improvements in various kinds of textile machinery and methods have increased productivity but at the same time have increased the severity of many visual tasks. For example, in some weaving sheds one weaver now may operate as many as 50 to 100 looms, especially in the weaving of some cotton or synthetic fabrics, or only a single loom on complicated patterns or fabrics. It is necessary for him to see quickly and accurately. The lighting requirements are determined by the color, weave, and fineness of the material being fabricated, as well as by the specific operation under consideration. Textile operations can be classified into three groups according to the type of fabric involved: (1) cotton, (2) synthetic, and (3) wool. Many of the operations and lighting requirements in these three groups are similar or comparable.

One difference in these similar operations is the name for the operation. Each group has its own nomenclature.

Fig. 14–31. This lighting system, with spectral quality similar to daylight, provides constant quality and quantity of illumination for cotton classing. Each 2-foot by 4-foot luminaire contains three 40-watt daylight fluorescent, two 20-watt blue fluorescent, and four 25-watt incandescent lamps above diffusing glass.

Cotton Mill Lighting

Areas including the operations of *opening, mixing, picking, carding,* and *drawing* should be illuminated with a well-planned general lighting system to provide the illumination levels recommended in Fig. 9–80. See Fig. 14–31.

Operations such as *roving* and *slubbing, spinning, spooling,* and *warping* present more severe seeing tasks. The basic seeing task in all of these operations is to detect broken ends as soon as the break occurs and to make immediate repairs. Loss in production is a result of stopping an entire machine while repairs are being made on one thread. General lighting is needed to minimize high luminance ratios, since most of the light is concentrated on the working area. Most of the work areas of these machines are relatively long and narrow. See Fig. 14–32. A linear source aids in the elimination of shadows and has the desired light distribution characteristics. The lighting design should be coordinated with the air conditioning and cleaning system.

Drawing In. This is probably the most difficult seeing task in the mill because of the small size of the details to be seen and the unrelenting visual concentration required. In this operation, the warp ends are drawn by hand through drop wires, harnesses, and reeds with a thin instrument called a reed hook. At any one time the operator's attention, as he moves from one side of the warp to the other, is confined to a space about 4 inches square. This task requires well-diffused illumination, such as would

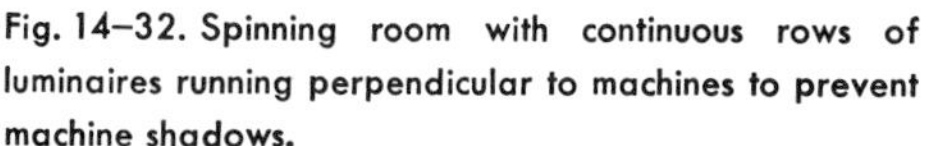

Fig. 14–32. Spinning room with continuous rows of luminaires running perpendicular to machines to prevent machine shadows.

be provided by industrial fluorescent luminaires suspended over the operator's head and aimed at the work. Another satisfactory solution of this problem is to use a 60- or 100-watt incandescent lamp in an industrial reflector of parabolic shape, designed to be moved from one side of the frame to the other as the work progresses.

Automatic Tying In. The ends of a full loom beam are tied to the ends of a loom beam which is nearly exhausted, whenever possible, in order to eliminate the drawing-in operation. The work lies primarily on a horizontal plane. Prolonged visual effort is involved, and localized general illumination should be provided. A diffusing luminaire similar to the industrial fluorescent type or a special local incandescent type should be supplied for each operator.

Weaving. Weaving involves visual tasks of various degrees of difficulty. The warp strands which run lengthwise of the cloth are drawn through the eyes of heddle wires which create the warp shed. Illumination has to be furnished for the "fixer" to repair and oil the loom, for the "cleaner" to brush away lint, for the "creeler" to fill the bobbin creel, for other operators to install the full loom beams with accessories, and for still others to remove the full cloth roller. Broken ends must be located and "pulled in" (repaired), defects in the cloth must be "picked out" (removed by picking out the yarn from the filling bobbin), and the cloth must be inspected as it is woven. The most difficult of these tasks in the manufacture of gray goods is to see the detail of the finished cloth well enough to determine whether or not all of the specifications for perfect material are being met. More difficult tasks are met when weaving dark materials.

The looms are designed to stop automatically when an end breaks; however, there are defects which are not the result of a broken end. It is possible also for ends to break and cause a defect without the drop wires falling and stopping the loom. These more obvious flaws must be noted in addition to the smaller ones, such as a bent reed, too many ends through one opening between the reeds, lint on the back of the loom which will in time cause a break, etc.

Shadows are a problem in a weave shed. Light sources in the back aisle cast machine shadows on the work; those centered on the loom over the work aisle may cast shadows of the weaver's hands. The ideal form of illumination for weave sheds, other than those using Jacquard looms, is obtained from a general lighting system consisting of continuous rows of lighting units. The lighting system should be designed to provide minimum glare from any angle and to give good diffusion. Practice has shown that lighting rows running perpendicular to the long axis of the looms, as shown in Fig. 14–33, give best results.

Inspection. Inspection is a specialized task peculiar to each mill. Recommended levels of illumination are listed in Fig. 9–80.

Synthetic Fabric Plants

Most areas in the synthetic fabric plant can be best illuminated by a general lighting system. Illumination levels as given in Fig. 9–80 should be used as a guide in the designing of the lighting.

Soaking and Fugitive Tinting. Preparatory to twisting or throwing, yarns which are received in the form of skeins are soaked or lubricated. Also, they may be fugitive-tinted during the same operation. Tints are used to identify the direction and amount of twist and occasionally to distinguish different lots of yarn.

The soaking operation may be carried out in a number of different ways; the simplest is to submerge a number of skeins in a tank of soaking solution. It is not necessary to see the individual threads, and the visual effort required during the process is not great. Uniform illumination should be provided throughout the entire working area.

Winding or Spooling. Each skein is mounted on a light wooden or wire wheel-shaped frame, known as a swift, from which the thread is wound

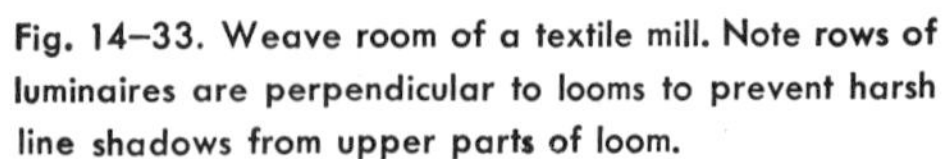
Fig. 14–33. Weave room of a textile mill. Note rows of luminaires are perpendicular to looms to prevent harsh line shadows from upper parts of loom.

onto a horizontal friction-driven spool. The machines, which may be either single- or double-deck construction, normally are arranged in rows.

When the thread is broken during the winding operation, the ends must be found against a background consisting of the rest of the thread on the skein and the spool. Since the contrast is extremely low, the visual task is very severe and touch is called upon to assist the repair of the mend.

The recommended illumination for the spool and the portion of the swift which normally is in the field of view can be provided by semi-direct fluorescent luminaires.

Doubling and Twisting. The seeing problem is similar to that in winding. Good illumination is needed throughout the entire length of the threads from their origin on the spools of untwisted thread to their terminations on the receiving bobbin. The machines ordinarily are arranged in rows, but are higher than winding machines. The lighting requirements, which are similar to those for winding, although in many cases considerably more severe, may best be met with a semi-direct fluorescent unit. Most twisted yarns are fugitive-tinted.

Quilling. Light should be supplied on the spool or cone from which the thread is being wound, on the quill on which it is being received, and on the entire length of the thread between them. The thread must be seen against a low-reflectance background, consisting of various parts of the machine.

Warping. The spools necessary to supply all the warp ends required for a single section are mounted on the creel and are threaded through the appropriate spacing devices (reeds) and tension-control apparatus. See Fig. 14–34. All the ends in one section are gathered together in a single knot and hooked to a pin on the reel. As the reel rotates, the yardage is indicated on a large dial. After a section has been completed, knotted, and tied, the next section is placed on the reel alongside the first by exactly the same process until the required number of warp ends has been obtained.

Through the action of drop wires at the top of the creel, the machine will stop automatically if an end breaks or the tension falls. If the break occurs at the creel, it usually is possible to locate both ends and splice them directly, but if it occurs at the reeds, the location of the end on the creel is somewhat more difficult.

Illumination on the creel is necessary to enable the operator to locate and repair the broken ends and to place new cones, the threads of which must be tied to the ends of the threads on the cones in use.

Drawing-In or Entering. Throughout the entire process it is necessary to see the threads against a low-reflectance background comprising the mass of heddles. The seeing task is very severe for both operations. General illumination supplemented by concentrating luminaires may be used. The concentrating luminaires should be fixed in position and should illuminate the entire working area without repeated adjustments by the worker.

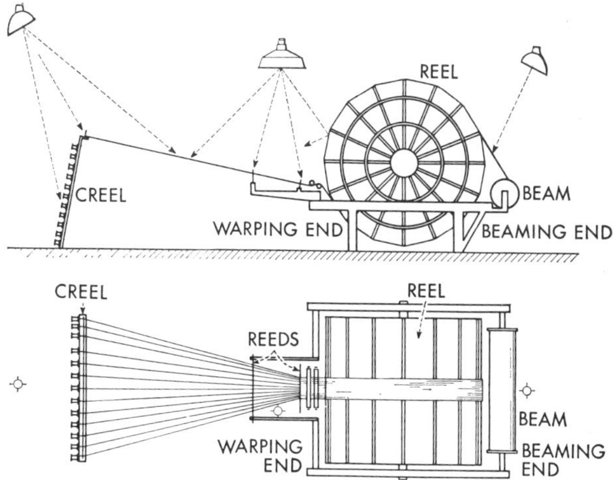

Fig. 14–34. Plan and elevation of inclined-creel warper.

Weaving. Weavers are constantly on the alert to see that looms are producing according to specifications. A loom may continue to operate after an end has been broken if the drop wire fails to fall. Other causes, such as a bent reed, produce defects which can be determined only by an inspection of the cloth. These defects should be located as soon as they occur so that corrections can be made and high shrinkage losses avoided.

Burling and Mending. The object of burling and mending is to locate and, when practicable, remove any defects in woven cloth prior to the final finishing process. Several types of defects may exist: (1) broken filaments and knots, (2) loose filling, (3) slack or tight ends, (4) pulled warp, (5) temple cut, and (6) stretched yarn.

Each of these six defects can be observed best in silhouette against a flashed-opal glass plate, lighted from beneath. The optimum luminance of the plate is a function of the transmittance of the sample; high for low-transmittance (opaque) materials, and low for high-transmittance (sheer) fabrics.

For the best silhouette vision the illumination on the cloth should be low. A luminance of 100 to 400 footlamberts for the flashed-opal glass plate has been found satisfactory for some purposes. The opal glass plate should be set at an angle of 45 degrees. Cloth is drawn over it at a fairly rapid rate. See Fig. 14–35.

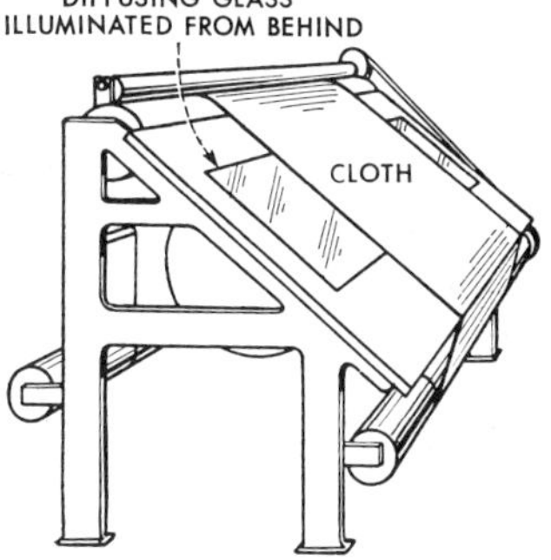

Fig. 14–35. A burling and mending table.

Synthetic Fiber Plants

The synthetic fiber plant is the beginning of the fibers which are used in the manufacture of man-made fabrics. Generally, the illumination requirements are the same for nylon, polyester, acrylics, and polypropylene. Areas of plants which require good general illumination are: polymerization, spinning, drawing and twisting, and warping and beaming.

Polymerization. This is the process whereby crude petrochemical monomers are combined by the use of controlled heat and pressure to form polymers. Most of this process is one of chemical reaction in enclosed vessels, and is primarily automated. Good quality general illumination is usually all that is required. With some polymerization processes, equipment may have to be suitable for use in hazardous locations.

Spinning. This process is the production of fibers by extruding the molten polymer through small capillaries called spinnerets. The polymer is then solidified and conditioned before being wound onto a bobbin. Good quality general illumination is required throughout the spinning area. Supplementary lighting is needed on the machinery to see fibers smaller than human hair, and is usually achieved by mounting continuous rows of fluorescent luminaires the entire length of the machine to illuminate the spinneret and the wind-up sections.

Drawing and Twisting. After the fiber is spun it is placed on a draw-twister or draw-winder and stretched and twisted, or stretched and wound. Continuous rows of fluorescent luminaires are usually located parallel to the machinery to provide illumination for the operator to thread a very fine fiber into a machine operating at a high rate of speed.

Warping and Beaming. The most difficult seeing task involved in the manufacture of synthetic fibers is that of warping and beaming. The pirns of fiber are placed on creels and rewound together under slight tension on large rolls called beams. The entire length of the creel should be uniformly illuminated to enable the operators to thread the fibers and supplementary illumination should be provided so that an operator will be able to see a single broken fiber among thousands.

Knitting

Manufacture of knit goods is done with either flat or circular knitting machines. On flat machines, the needles are arranged in either a single plane or two planes arranged at right angles to each other. Circular machines have one or two sets of needles arranged in a circle. Fabrics produced on either type machine vary from coarse to fine goods depending on yarn size and needle gauge. Typical knitted products include shirts, hosiery, seat cover materials, industrial filter cloth, and suitings.

Yarn from cones or beams, located above or below the knitting head, is threaded through a series of guides to the needles, which latch or pick the yarn forming it into a series of loops or interlocking latches making the fabric. The knitted cloth is then wound onto a take-up roll, which may be above or below the knitting head.

Threading the yarn through the guides has varying difficulty depending on yarn count and gauge of the needles. Dark colored yarn increases the difficulty of the visual task.

Woolen and Worsted Mills[14]

The seeing problems encountered in woolen and worsted mills are in many respects comparable to those of silk or synthetic fiber textile mills and more severe than those of cotton mills. The lighting levels given in Fig. 9–80 are practicable maintained-in-service levels of illumination required by the processes or operations involved. Fluorescent light sources facilitate reduction of shadows and provide diffusion of light required for seeing threads and machine operations. Fluorescent open-top industrial units should be mounted not less than 8 feet above the floor.

Opening and Grading. Raw wool arriving at the mill is separated according to grade with consideration of length, strength, degree of fineness, and color. Illumination should be constant in color and approximate average daylight conditions.

Carding. Fibers are blended and aligned by feeding them unbroken through rollers equipped with continuous wire combs. General lighting is required, particularly on the sides of the machines, for cleaning, adjusting, and repairing. Color is unimportant, but localized lighting is required at the take-off end of the cards where the roping comes from the machine to the roping spools. Two portable vertical fluorescent floor-mounted units are best suited for this localized lighting requirement.

Spinning Mules. This operation involves the final twisting and elongation of roving or roping to produce thread. The area involved is relatively large and general lighting is suitable.

Spinning Frames. Both ring spinners and cap spinners require well diffused general or localized general lighting with rows of luminaires parallel to the working aisles. Continuous-row fluorescent units are desirable to eliminate shadows of the spindle barriers from falling on the spindles.

Weaving. The weaver must inspect the finished cloth as it is woven, detect broken threads, and repair them. In addition, lighting must be of suitable quantity and distribution for inserting the loom beam and cleaning and maintaining the loom. Good results are obtained from continuous rows of fluorescent industrial luminaires mounted perpendicular to the axis of the weave alley.

Perching. This operation involves the final examination of finished worsted and woolen textiles. Proper quality and color of light are very important here. The operation consists of the inspector pulling cloth over the top of a near vertical framework (perch board). The inspector examines for imperfections such as wrong color of yarns, off-shades, spots, shadiness of goods, and mistakes in pattern. The most desirable illumination is provided by fluorescent lamps mounted in banks of specular reflectors, adjusted so that the angle of reflection from the cloth is not troublesome. Constancy and higher level of illumination are extremely important here. Daylight should be eliminated. See Fig. 14–36.

CLEANING AND PRESSING

Recommended levels of illumination for the tasks involved in the operations tabulated below are listed in Fig. 9–80.

Operations. The operations in dry-cleaning plants are functionally divided as follows:

1. Receiving.
2. Checking and sorting.
3. Dry cleaning.
   a. Naphtha-solvent process
   b. Synthetic-solvent process
4. Steaming.
5. Examining and spotting.
6. Laundry or wet cleaning.
7. Repair and alteration.
8. Machine finishing.
9. Hand finishing.
10. Final inspection.
11. Shipping.

Receiving. Cleaning and pressing establishments receive soiled items from their pick-up trucks at a receiving platform at which garments are transferred from motor truck to hand truck. The garments are then wheeled to the checking and sorting tables.

Checking and Sorting. For special instructions a checker reads a driver's ticket (written in pencil) attached to incoming garments and pins tags numbered in indelible ink to each garment for matching with the original ticket after the completed operation. Pockets are searched for matches or articles of value and finally the sorter divides the garments into synthetics, cottons, silks, and woolens, dark and light colored, or other classifications necessitated by the cleaning operation. The penciled notations are difficult to read. Contrast, as well as the handwriting, often is poor.

Dry Cleaning by the Naphtha-Solvent Process. The solvent used in this process is inflammable and, under certain conditions, explosive. For this reason the cleaning operation is carried on in a separate building or in a section of the plant divided off by a firewall. *Explosion-proof lighting equipment is mandatory.*

Fig. 14–36. Perch board illuminated by 48-inch daylight fluorescent lamps in specular reflectors.

No attempt is made in the washing and drying room to determine whether the cleaning has removed all spots, and no other difficult seeing problems are involved. The explosion-proof luminaires must be located so the washer, extractor, and drying-tumbler interiors are well illuminated when the covers are thrown back. In addition, distribution must be such as to light properly pressure and flow gauges on the filters and in the piping. The time of washing is largely determined by the clarity of the naphtha coming from the washer. This dirt can best be seen in silhouette against a white background while it is passing through the filter gauge.

Dry Cleaning by the Synthetic-Solvent Process. This process differs from the naphtha type in using a nonexplosive solvent and a closed system. The seeing tasks are related to loading and unloading the cylinder and reading the various temperature, pressure, and flow gauges. Light should be directed into the cylinder and toward the gauges from locations such that an image of the source will not be formed in the field of view.

Examining and Spotting. The dry-cleaning process takes practically all of the oil and grease out of stains of various types unless it is ground into the fabric very firmly. Nearly always, however, some spots remain to be taken out by chemicals or steam. Many stains have characteristic colors by which they may be identified by a skilled "spotter."

Through long experience this workman is trained to detect, classify as to type, and remove all types of spots after choosing the proper chemicals. The critical seeing task lies in detecting the spot and its type.

After the washing process, spots present a very

subdued appearance with little contrast between themselves and the material. Also, the reflectance of many materials has a strong specular component. Low contrast stains are particularly difficult to see on such garment materials as taffeta, gabardine, crepe, and many of the synthetic varieties.

The manner in which the eye responds in examination of garments for stains is affected by the kind of light to which it is adapted. The spectral characteristics of the light source directly affect the appearance of the garment. A blue garment (dress or suit) appears blue simply because it reflects more blue light than any other color. Similarly, a red garment reflects more red light than any other color. It is apparent, therefore, that the color of the garment and the spectral characteristics of the light source play a major part in the color appearance of the garment.

It is practical to examine individual colors of garments for low contrast stains by increasing the contrast of the color so that such stains become more easily seen. This can be accomplished effectively by employing a complementary color light source. This is illustrated by the use of deluxe warm white fluorescent lamps for inspection of low contrast stains on blue and green garments and daylight fluorescent lamps for the inspection of yellow and red garments. The combination of deluxe warm white and daylight lamps can be used effectively on black and brown garments.

A low-luminance, large-area luminaire is recommended for use over spotting tables. This could be a two-foot by four-foot luminaire equipped with an acrylic or glass diffuser (opal or polarized) housing two deluxe warm white fluorescent lamps controlled by one toggle switch and two daylight fluorescent lamps controlled by a second toggle switch. Toggle switches should be located so as to be convenient for the spotter's use. Separate or combined switching can be applied as needed for the color of the garment being examined.

Fig. 14-37 illustrates recommended mounting locations for these luminaires over hand spotting boards and steam spotting tables.

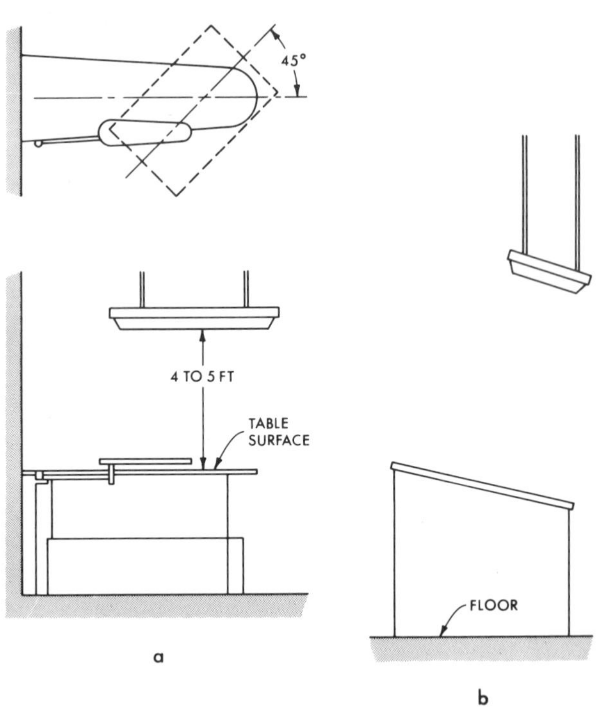

Fig. 14-37. Plan and elevation of installation for inspection lighting for (a) spotting boards and (b) steam tables.

Laundry or Wet Cleaning. On some garments the spots are so numerous, large, and widely distributed that it is uneconomical to use the spotting method for removing them. Each cleaning plant maintains a small laundry for such garments. A conventional cylindrical washer, centrifugal extractor, and drying tumbler are used. The lighting problem is very similar to that of the dry-cleaning operation. Vapor-proof luminaires are recommended and should be located so as to light the interior of the machines.

Repair and Alteration. Both hand and machine sewing is done in this section, often with dark thread on dark material. For hand work a well-diffused light is recommended. Supplementary illumination at the needle point of the machines is recommended.

Machine Finishing. Machine presses usually are lined up in a row for convenience and for minimum cost in the steam-piping installation. The operator combines speed with good workmanship. Each garment is moved several times as a small section is finished and another moved onto the buck (working surface). The workman watches to see that all wrinkles are eliminated. The buck of the press should be uniformly illuminated without shadows from the head of the press or the workman's body. A continuous row of fluorescent luminaires overlapping the ends of the presses is recommended. This method takes care of the working area on the buck and in addition illuminates the clothes racks, aisles, and machine space.

One of the most difficult tasks is to prevent double creases in trouser legs. A concentrating reflector at the rear of the buck causes a crease to cast a shadow, making it more easily discernible.

Hand Finishing. Hand finishing (ironing) boards usually are installed in rows spaced 3½ to 5 feet apart. The volume of handwork is decreasing gradually because of improvements in machines. However, the hand iron still is used to achieve the best results on lightweight materials. The hand finisher watches to see that wrinkles are eliminated, watches that the garment is completely pressed and that it is not scorched, and corrects minor defects.

The seeing task is moderately critical because careful handling of the iron is required for pleats, shirring, ruffles, and trimming.

Final Inspection. Garments on individual hangers are delivered to the final inspector on portable racks or by a power-chain conveyor. Each garment in turn is hung on an overhead support in such a way that it will rotate easily. The inspector examines the garment carefully, watching for inferior finishing, for spots, for damage done to the material during the cleaning process, and for the completion of any customer ordered repairs or alterations.

Most of the critical visual work is done with the garment at approximately a 45-degree angle with the vertical and at short range. The lighting requirements are about the same as for spotting and similar luminaires with two colors of lamps and separate switching are recommended. To increase the vertical plane illumination the luminaire should be tilted parallel to the usual garment plane.

Shipping. The shipping section covers the garments with protective bags, attaches the original ticket, and loads the delivery trucks. The identifying tags attached to the garments often are difficult to read because the ink is partially washed out during the cleaning process.

SHOE MANUFACTURING[15]

Recommended levels of illumination for typical shoe manufacturing operations are given in Fig. 9-80, page 9-89.

Seeing Task Groups. Leather and rubber shoe manufacturing processes may be separated into three groups according to the type of seeing tasks involved in each:

1. Simple seeing tasks include:

Leather: storage, staying, sole laying, beveling, nailing, heel scouring, burnishing, spraying, box making, dinking, last racks, lasting, pulling over, trimming, channeling, heel breasting, edge setting.

Rubber: washing, compounding, calendering.

2. Seeing tasks of average difficulty include the mechanized operations:

Leather: skiving and splitting, treeing, welting, rough rounding, perforating, buttonholing, eyeletting on both light and dark materials, certain types of bench work.

Rubber: sole rolling milling, completed stages of compounding.

3. Seeing tasks of considerable difficulty include:

Leather: cutting, bench work, stitching, inspection, rounding, sole stitching, fine edge trimming on both light and dark materials.

Rubber: cutting, making, calendering.

Leather-Shoe Manufacturing

Sole Department. In a sole department, leather, sorted for grain and thickness, is stored in 5- to 6-foot piles on low platforms which are arranged with passageways between them.

For grading according to color, some advantage may be gained by the use of illumination of spectral characteristics similar to daylight.

Beam Dinkers. A beam dinker stamps soles and insoles out of hides by means of dies. It consists of a heavy cast-iron frame and a large beam that exerts pressure through a vertical motion on a cutting die. There is some hazard of finger injury in operating the machine. Therefore, special attention must be given to this location with an adequate, well diffused general lighting system plus supplementary lighting to prevent shadows.

To avoid casting shadows of the beam on the platform, all luminaires in the area occupied by the beam dinkers should be placed at the operator's side of the machine.

Last Storage. Last storage bins usually are located in a segregated section of the sole department. Generally there is a 3-foot aisle between the bins. Luminaires should be mounted over the aisles in a manner to illuminate the inside of the bins so that lasts may be selected.

Upper Department. An upper department generally is divided into the following sections: (1) sorting; (2) trimming, cutting, and staying; (3) lining; (4) upper cutting; (5) marking and skiving; and (6) assembling.

When an order is received for a certain grade of shoes, the sorting department grades the leather as to color and quality. For this work a well diffused, general lighting system is best. A daylight color quality is desirable.

Skilled workers then split each piece of leather into as many sheets as possible and cut out individual parts for uppers. This work generally is done on tables 30 to 36 inches above the floor.

The various pieces go to the counting department where they are counted and marked with job numbers. Skiving, which consists of the mechanical thinning of edges of the uppers so that they can be turned over to present a finished appearance, is the next step. The work of assembling consists of bringing together the various parts which make up the uppers such as lining, stay, vamp, counter, toe, tip, etc. For these operations a uniform illumination level is recommended.

Stitching Department. In the stitching department the following operations are typical: (1) lining; (2) tip; (3) closing and staying; (4) boxing; (5) top stitching; (6) buttonholing and stamping; and (7) toe closing.

These operations present difficult seeing tasks. A high level of uniform general illumination should be provided plus supplementary lighting on the machines.

Making Department. The making department in the average plant is subdivided according to operations as follows: (1) vamping; (2) welt-bottoming; (3) bottoming; (4) heeling; (5) turning; and (6) standard, screw, nail, or pegged shoe making.

In some plants this department is called the *gang room* and occupies an entire floor. A general lighting system plus local lighting should be installed to mitigate shadows of overhanging machine parts.

Lasters, sole layers, levelers, and nailing machines may be illuminated by diffused light. The source location is not critical. Other machines in the making department should be illuminated from the rear and to the right of the operator. The vertical as well as horizontal illumination level is most important.

Finishing Room. In the finishing room shoes are inspected and faults are corrected. Treeing machines are used in this area for ironing out wrinkles. From here the shoe goes to the final inspection and then to the packers and shippers.

Rubber-Shoe Manufacturing

In rubber-shoe manufacturing plants typical operations include the following: (1) washing; (2) compounding and milling; (3) cutting and calendering; (4) drying; (5) sole rolling and cutting; (6) making; (7) varnishing and vulcanizing; and (8) packing and shipping.

In the washing department crude rubber is cut up by band saws. General illumination is recommended for washing and cutting and also for the compound and mill area except where hoods are placed over the compounding machines. In such cases, local lighting should be provided by luminaires installed under the hood, with a reflector directed at the point of work.

As materials pass over the cutting and calendering machines, care must be taken to see that the coating is applied correctly. Calenders, especially the three- or four-roller type, should be lighted by luminaires on both sides of the machines. The light should be well diffused to avoid sharp shadows and glare.

After cutting or gumming, rolls go to the drying room where they are dried by steam heat. Where this department is confined to the center of the building and has no direct general ventilation, there is an explosion hazard. Where such is the case, explosion-proof lighting units are recommended. Supplementary lighting equipment provides light at both front and rear ends of the sole rolling machine.

In the sole and upper cutting department, operators work rapidly with sharp knives. A uniform illumination level throughout the area is recommended. Luminaires should be mounted as high as possible. In some plants beam dinkers are used, and these should be lighted in the manner described above.

The making department is the most important in this type of plant. All parts are supplied, cut to shape, to bench workers who use cement to attach and complete a shoe. In some cases there is a shelf or rack over the center of the bench, extending its entire length. The lasts are placed on this shelf and if luminaires are placed over this shelf and hung low, the shelf causes a sharp shadow on the working areas of the bench. A general lighting installation is recommended.

In the varnishing and vulcanizing areas, a uniform illumination level is recommended.

MEN'S CLOTHING[16]

Recommended levels of illumination for the general tasks in manufacturing men's clothing are listed in Fig. 9–80.

Definitions and Flow Chart

The descriptions that follow are used to clarify the processes through which the cloth must pass to become a finished garment. The processes are also shown in Fig. 14–38 along with a letter key on the left (the key letters are used after the process name in the following descriptions):

Preparation of Cloth (A to H).

Receiving (A). Shipment of cloth received from mills.

Opening (B). Bolts of cloth are opened.

Examining (C). Checking and marking of all material for imperfections in texture or color.

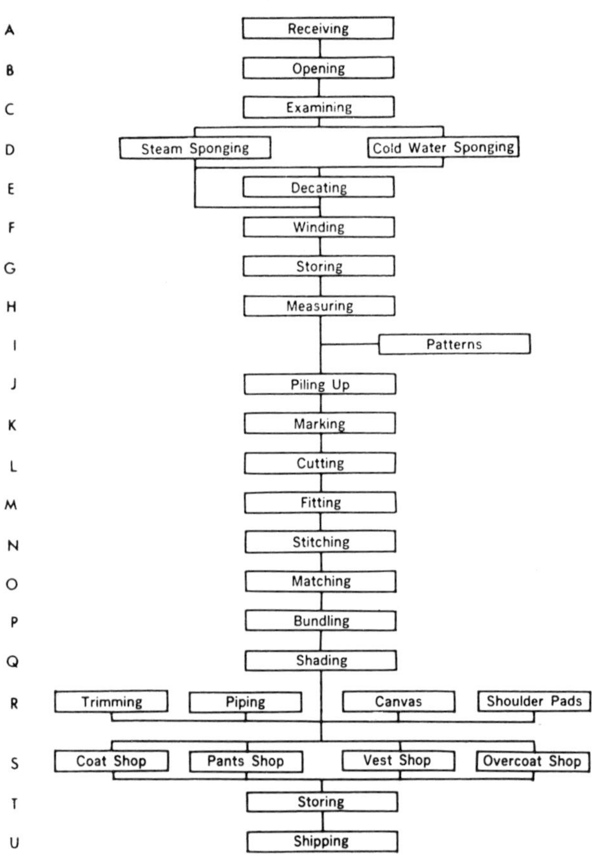

Fig. 14–38. Flow chart and footcandle recommendations for the manufacturing of men's clothing.

***Sponging* (*D*).** Shrinking of cloth by applying moisture and drying. There are two types: Steam Sponging and Cold Water (London) Sponging.

***Decating* (*E*).** Process of setting the cloth by means of steaming. (This is not done to all materials.)

***Winding* (*F*).** Rerolling the cloth.

***Storing* (*G*).** Holding cloth until needed in cutting room.

***Measuring* (*H*).** An accurate record of the cloth sent to cutting room.

Patterns (I). Making of patterns for different styles.

Cutting (J to Q).

***Piling Up* (*J*).** Placing the layers of cloth one on top of the other to obtain the desired number.

***Marking* (*K*).** Marking the patterns on the cloth with chalk.

***Cutting* (*L*).** Operating the cutting machine.

***Fitting* (*M*).** Bundling of similar pieces of cloth.

***Stitching* (*N*).** Marking cloth for pockets, etc.

***Matching* (*O*).** Selecting pockets, etc., of same color and pattern.

***Bundling* (*P*).** Tying up bundles and moving to trimming room.

***Shading* (*Q*).** Marking of lot number on cloth.

Preparation (R). Trimming, piping, canvas, shoulder pads, etc., for different garments.

Sewing (S). Manufacture of different garments. The operations in the shops can be broken down into the following operations:

1. Machine sewing.
2. Hand finishing and sewing (standing or sitting).
3. Hand pressing.
4. Jumper pressing.
5. Machine pressing.
6. Inspection.

There are over 100 operations in making a coat, all of which can be classified in the above categories.

Handling of Finished Garments (T and U).

***Storing* (*T*).** Storing finished garments.

***Shipping* (*U*).** Shipping finished garments.

Lighting Design for Specific Tasks

Preparation of Cloth (A to H), Examining. Of the eight steps in this processing division, only the examining requires special lighting.

Historically in the clothing industry, all of the examining has been done by daylight, preferably north sky light. However, with need for increased and continuous production, it has been necessary to use electric lighting for this important task. Daylight is not always dependable and it is difficult to obtain the necessary illumination at the perch in many buildings.

Since the task is the same as that of perching in the woolen textile mills, the system shown in the report on that industry is recommended (see page 14-31).

The cloth is examined in a vertical position and is inspected for shading and defects. Lighting equipment should provide uniform vertical illumination.

Patterns (I). There is no special lighting required.

Cutting (J to Q). General lighting or localized general lighting should provide the base lighting for the different operations in this section. If more than one operation is performed in the same area, the higher footcandle level required will govern the design. Supplementary lighting on the cutting machine should be used to provide the necessary increase in illumination levels over the general lighting. If the cutting is done by hand the localized general lighting will have to be increased.

Preparation (R). There is no special lighting required.

Sewing (S). General lighting should provide the base lighting for the general operation in this section. Supplementary or localized general lighting should be used to obtain the required footcandles for the different tasks.

Handling of Finished Garments (T and U). There is no special lighting required.

FLUID MILK INDUSTRY

Recommended minimum footcandle values to be used for guidance in the solution of lighting problems in the fluid milk industry are listed in Fig. 9-80.

Loading and Unloading Platforms

Entrance driveways, truck ramps, and platforms used during dark winter mornings and evenings and at night for loading and unloading should be well lighted. The interior areas used by checkers who make a detailed count of kinds and numbers of milk bottles being loaded or unloaded should be illuminated by a general lighting system.

Return Bottle Sorting Area

A general lighting system will enable employees to pick out foreign articles, remove caps, sort out very dirty and foreign bottles, and to sort and segregate various types of bottles, such as retail and store bottles.

Bottle Washers

One of the most difficult problems in a modern dairy is to light milk bottles as they are discharged from a bottle-washing machine so that any remaining foreign matter, fractures, etc., are quickly and easily seen.

Method of Operation. The complete operating cycle of a typical bottle-washing machine lasts 6 seconds. The clean bottles stand still (up-ended) for 1½ seconds in full view of the operator, at a distance of 40 inches. During this time (0.2 second per bottle) he inspects them while picking up a load of dirty bottles. If the dirty bottles do not require his attention, he may have a fraction of the remaining 4½ seconds to inspect some of the clean ones as they move out.

Inspection. Bottles are inspected both while being placed in the washer and after discharge from the washer. The purpose of the inspection is to discard foreign or store bottles, bottles stained beyond recovery, bottles with chipped necks or cracks, and incompletely washed bottles. Foreign bodies such as caps, wire, and nails must be removed.

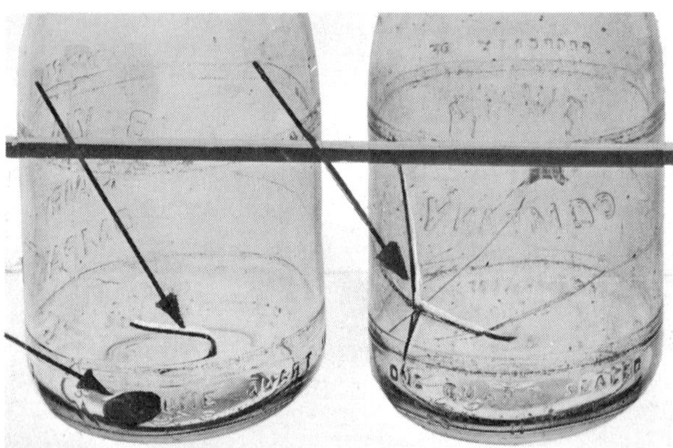

Fig. 14-39. Close-up of bottles in front of luminous inspection panel. Bottles can be inspected efficiently for foreign matter, and fractures are made easy to see.

Fig. 14-40. Illumination of bottle washers which cannot be modified to accommodate a luminous panel may be improved by installing a large-area luminaire directly above the inspection end.

It was found experimentally that inspection of bottles silhouetted against a low-luminance luminous surface, as in Fig. 14-39, is the most efficient method. A luminous inspection panel may be incorporated in the unloading mechanism in the form of an inspection light-box. A typical box consists of a sheet-metal enclosure containing lamps and auxiliary equipment, the open front of which is covered with a translucent plastic sheet. The assembly is placed in the unloading mechanism in such a way that when the bottles are pushed out of the washer they are silhoutted against the luminous panel. Incomplete washing, cracks, chips, foreign matter, etc., can be detected readily. See also Fig. 14-40.

A maximum luminance of 500 footlamberts is permissible with 250 footlamberts the preferred value on the luminous panel.

Manufacturing Areas

A general lighting system should be provided for in the manufacturing areas of a fluid milk plant. Rooms should be finished with ceiling reflectance of 75 per cent or more and wall reflectance of from 50 to 60 per cent.

The lighting should be such that there will be a minimum number of specular images of the light sources formed on the surface of a bottle, whether empty or full, that will interfere with the proper inspection of the bottle or the finished product. The distribution characteristics and spacing of luminaires should be such that no sharp shadows will be cast, and there will be no dark corners.

FLOUR MILLS

The inspection of the material in process is done by a combination of visual examination plus sense of feel.[13] The material to be judged is usually examined in the hands of the operator. Illumination installed in sufficient quantity (see Fig. 9-80) and of daylight quality makes inspection and judgment of the material more accurate under visual examination. The numerous obstructions, such as spouts, shafting, etc., encountered above the equipment as on the roll floor, need not interfere with installations of luminaires. See Fig. 14-41.

All lighting equipment must meet the Underwriters' approval for use in Class II, Group G locations. Such equipment will also reduce the lighting maintenance problem caused by accumulation of dust, dirt, and grease. If possible, luminaires should deliver some upward lighting component for illumination of overhead line shafting, belting, etc.

Roll, Sifter, and Purifier Floors

Illumination from daylight fluorescent lamps provides satisfactory lighting for inspection purposes. The contrast between the flour and the impurities is higher under daylight quality of illumination. Fluorescent luminaires can produce uniform illumi-

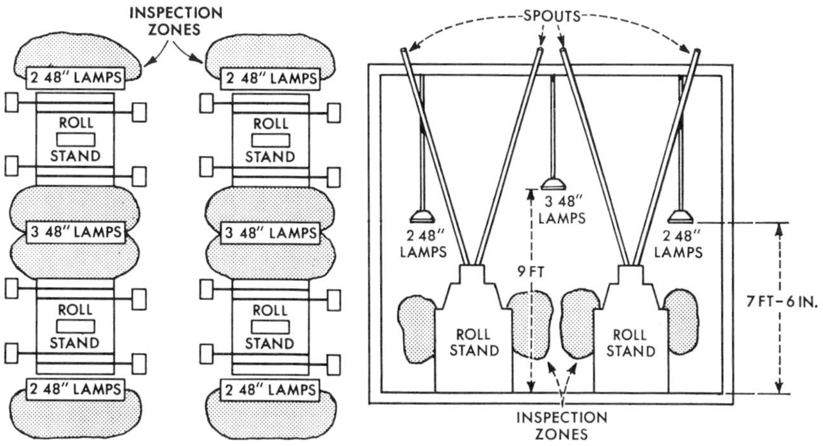

Fig. 14–41. Recommended arrangement of luminaires for roll floor: (left) plan view and (right) elevation.

nation without dark shadows and high luminance differences. Fig. 14–41 shows a suggested arrangement of dust-tight fluorescent lighting equipment for roll floors.

Parking Areas

Color quality is not important. Supplementary illumination may be necessary for the scale dials and sewing heads.

Product Control

The illumination requirements here are similar to those for inspection on the roll, sifter, and purifier floors.

Miscellaneous Areas

Lighting is required for use in the *bins* to inspect material level. The necessary type of lighting unit should be Class II, Group G, lightweight, portable, and provide adequate illumination.

Floodlighting by the use of permanently located units is recommended around the *cleaning screens* and *man lift* areas.

For most *non-critical seeing areas* such as stairs, aisles, and overhead runways, see Fig. 9–80.

BAKERIES

Visual tasks in bakeries are not severe. Most hand operations are done almost automatically with no close attention to detail. General lighting systems will provide adequate illumination for most operations; a few will need supplementary lighting. Fig. 9–80 lists the recommended levels of illumination.

Besides illuminating the task, lighting is an aid to sanitation, safety, and morale.[19] Cleanliness, of great importance in all food-producing establishments, is far easier to maintain in well-lighted interiors. In bakeries a large part of production areas is taken up by trucks, racks, mixing bowls, etc. Adequate illumination can help prevent injuries due to congestion in such areas and can also enhance the safety of employees operating moving machinery, working adjacent to hot surfaces, and handling hot pans.

Mixing Room

Flour is normally stored in bins directly above the mixers so that it can be weighed and sifted directly into the mixing bowl. Sometimes it is conveyed to scale hoppers located above the mixers. Standard dial or beam scales are used to weigh the flour and meters to measure the liquid ingredients. Some mixing rooms have a side bench where additional ingredients are mixed and weighed. General illumination with vertical illumination on the face of the shelves is required in this area. For vertical mixers supplementary lighting should be provided to light the inside of the mixing bowl.

Fermentation Room

From the mixing room, bread dough or sponges are taken in large troughs (holding 1200 to 1500 pounds) to the fermentation room which is maintained at about 80 degrees Farenheit and 80 per cent humidity. Since little attention is paid to the dough during fermentation, only sufficient light is required to insure safe handling of the equipment. Due to the high humidity, enclosed and gasketed luminaires should be used in this area. Experiments have shown that ultraviolet radiation can control wild molds in the Fermentation Room.

Make-Up Room

In the make-up room the dough is divided, shaped, and panned preparatory to baking. In modern bakeries production line procedures are used, and the operations are largely mechanical, requiring very little handling.

Proofing Room

During proofing the panned dough rises, to acquire its final shape. Localized general illumination is required in the proofing room directly in front of the tray racks so that a quick inspection of the panned dough can be made. Due to the low ceiling of most proofing rooms, symmetrical, angular directional luminaires or luminaires with asymmetric distribution can be used. Due to high humidity in the proofing room, enclosed and gasketed luminaires are recommended.

Oven Room

Ovens present a special lighting problem because (1) many of them are finished black and white and (2) they are lighted on the interior by lamps with high temperature bases in glass enclosures. The front of the oven should be illuminated to balance the luminances of the lighted oven opening to keep luminance ratios within 10 to 1. Luminaires must be mounted so that tray trucks placed near the oven mouth do not cause shadows when the oven is being filled or emptied.

Fillings and Other Ingredients

Preparation of fillings involves weighing and mixing various ingredients. Accuracy is required to maintain the proper standards of quality of product. Many ingredients are of a perishable nature and must be carefully inspected before use. Luminaires should be placed to provide good illumination for cleaning the inside of the mixing bowl.

For cooking fruit fillings, an exhaust canopy and fan are usually provided above the kettles. Auxiliary luminaires can be mounted inside the canopy, which provides good shielding and properly directs the light. Enclosed and gasketed luminaires protect the lamps.

Tables for fruit washing, cutting, and mixing are normally placed near the windows. Care should be taken that the workers do not face the windows, unless provision is made for the control of glare by means of Venetian blinds, shades, etc.

Decorating and Icing

In many large operations, icing is applied to bakery products mechanically. The lighting problem is similar to that in the make-up room.

Hand decorating and icing, however, require greater skill and care on the part of the operator if uniformly good results are to be obtained. Because of the detail involved in hand decorating and the necessity of working rapidly if satisfactory rates of production are to be maintained, quick and accurate seeing is required. Lighting on the decorating benches can be best provided by industrial type luminaires. Properly shielded fluorescent luminaires will minimize reflected brightness from the glossy icings toward the decorator.

If the operators are stationed on only one side of the decorating benches, luminaires may be placed above the heads of the operators parallel to the benches. This location is recommended because:

1. Illumination of the vertical surfaces of the product will be improved and the slight shadows produced will bring out the detail of the decoration.
2. Reflections from the highly reflective surfaces of the icing will be directed away from the operator.

The luminaires may be tilted slightly toward the decorating benches if the arrangement of the benches is such that a particular lighting unit is not expected to light more than one bench.

If the operators are located on both sides of the benches, the luminaires should be placed across the benches between the positions of the operators.

The color of the light sources in the decorating and icing department should match that under which the product will be purchased and consumed.

Illumination of Scales and Thermometers

For quick, accurate reading of the graduations, instruments should be illuminated. However, special care should be taken to prevent direct or reflected glare from luminaires or windows. See Fig. 14-42. Where local luminaires cannot be used, suitable shielded spotlights are recommended.

Wrapping Room

Wrapping is done on automatic machines. Interruptions may be caused by torn wrappers, clogging of the mechanism, etc. General lighting is required to expedite location and removal of obstructions and for cleaning, oiling, and adjusting the machines.

To control mold in packaged bakery goods, bactericidal lamps can be used to advantage.

Storage

Proper illumination of the storage room is important. Correct level of illumination promotes good housekeeping and thus tends to decrease storage losses.

Since low ceilings are frequently found in storage rooms, special attention should be paid to proper shielding of the lamps. Correct placing of the units with regard to the bins, platforms, etc., is essential to prevent blocking of light from the working areas.

Shipping Room

The finished product is delivered to the shipping room by tray trucks or conveyors. Shipping room tasks include keeping shipping records, loading trucks, and in some places maintenance and cleaning of the trucks.

Truck Platform

Since the truck bodies are enclosed and are loaded from the rear door, it becomes necessary to use

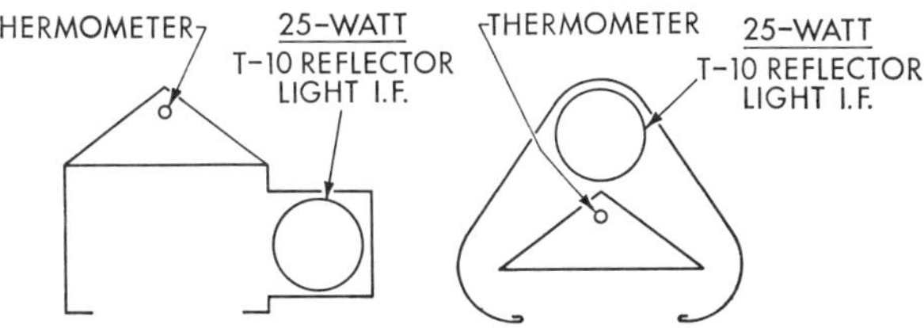

Fig. 14-42. Two suggested methods for illuminating thermometers.

angle or projector equipment permanently attached to wall, column, or other facilities appropriately located to provide adequate seeing conditions within the truck body. Sufficient illumination is necessary in the garage proper to insure proper cleaning of the trucks.

CANDY MANUFACTURING

In compliance with stringently enforced pure food laws and to foster good will, progressive candy manufacturers utilize every means for promoting cleanliness and efficient plant operation. Recommended levels of illumination for candy manufacturing are listed in Fig. 9-80.

Chocolate Making

In the manufacture of chocolate, the cacao beans first are toasted and then are passed through shell-removing machines. The bean then is conveyed by gravity feed to the crushers which press out liquid cacao butter. After milling and mixing with powdered milk and confectionery sugar, the pulverized beans are pressed through a series of rollers and then mixed with the cacao butter in a conche.

Many of these operations are gravity fed and utilize portions of two or three floors in a large plant with conveyors or chutes passing through the floors. There is very little handwork because practically all processes described are confined to the inside of hoppers, refiners, conches, and other machines. Consequently, no difficult seeing tasks are encountered in chocolate manufacture. However, the five rollers of the roller mill, where a careful setting of the rollers must be made periodically, is a difficult task. Supplementary lighting, having a predominant vertical component, should be used at this point. See Fig. 14-43.

Chocolate Dipping

Dipping is carried on in various sections of large plants, because this arrangement facilitates the manufacture and minimizes the conveyance of the different fillings. Dipping tables generally are located symmetrically in the area provided, with the operator sitting beside a depressed section of the table. Drippings from the operator's fingers are set in a design on top of the candy for decoration. The dipper must see the relative position of the drippings from the hand over the confection in order to make a neat and orderly design. General illumination should be provided in each dipping room.

Cream Making

Glucose, which is the base for most creams and fillings, is cooked, beaten by paddles, then remelted and recooked to increase its viscosity. It is then flavored, beaten again, and finally pressure-formed in plaster-of-Paris molds. The seeing task in cream making is of moderate severity. General illumination is recommended.

Kiss Wrapping

A kiss-wrapping production line consists of many individual kiss-wrapping machines, arranged on both sides of a belt conveyor. General illumination should be provided over the entire area, with supplementary lighting at the critical seeing points. These vary in location with the type of wrapping machine.

Gumdrop and Jellied Form Making

In this process plaster-of-Paris patterns are used to make smooth molds of fine-milled cornstarch. The molds are arranged symmetrically in shallow wooden trays which then are moved by a belt conveyor into such a position that one row of molds is placed directly under a series of injectors which automatically place the proper quantity of syrup in each. This operation is repeated until all the

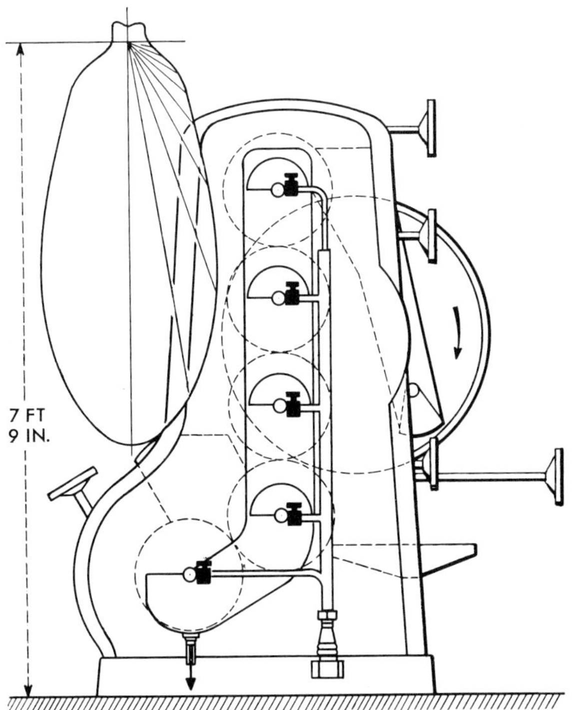

Fig. 14-43. Typical five-roller refiner. Periodic adjustments are made at the five rollers. Light should be distributed so as to illuminate the entire refining area.

molds in the tray are filled. In mold-fillers for gumdrops and similar candies, the automatic injectors which press the fluid candy into the molds are kept clean by an attendant.

Illumination should be supplied by luminaires with concentrating distributions hung above the equipment and directed toward the molds.

Hard Candy Making

In the manufacture of hard candy, sugar is cooked, flavored, and placed in a semi-solid state on water-cooled tables where a batch is kneaded into an oblong shape. Fillings are added at this stage. The batch then is worked into a cylinder about 10 inches in diameter and 6 feet long. After tapering in a heated canvas hammock the point is fed through a die-casting machine (see Fig. 14-44), which automatically shapes and cuts the candy.

General illumination should be provided for ingredient mixing and cooking, and the level increased by supplementary lighting at the die-casting machine. Supplementary luminaires should be located between the operator and the die-cutting machine. To minimize specular reflections in hard candy, luminaires with a large low-luminance luminous surface should be centered 4 feet above each hand-mixing table. Continuous fluorescent-lamp luminaires also may be used.

Assorted Candy Packing

There are three methods of packing candy:

1. *Progressive method.* In the progressive method candy is placed in simple containers in front of operators who sit on each side of a long table with a centrally-located conveyor belt.

2. *Stationary method.* In the stationary method long flat tables, 36 inches high and 36 inches wide, are used. Directly over the center of the table a stock rack, 18 inches wide, is suspended from the ceiling or fastened to the table so that its bottom is 18 inches above the top of the table. The operator removes eight or ten different types of candy from the rack and packs them in a box in front of her.

3. *Circular method.* In the circular method, which is not used as much as the other two, a ring table, 36 inches high and 18 inches wide, is used. The outside diameter is about 6 feet, the inside diameter about 3 feet. The operator sits on a swivel stool in the center. Candy for packing is placed on the circular table, one kind to a container. By rotating her stool 360 degrees an operator is able to pick a complete assortment.

Special Holiday Mold Candy Making

Holiday candy usually is made on the north side of the building where daylight is available to aid in the hand artistry generally required. At window tables operators with small artist's brushes decorate molded candy with a thin, colored mixture of cream filling. Because of the intricate positions in which decorations must be placed on the confection, and the fine details of the decorations themselves, the seeing task is severe.

Box Making and Scoring

In many candy factories, containers and boxes are made on the premises in a department divided into two main sections, one devoted to making standard boxes, the other to special boxes.

Scoring, the first operation in making boxes, is mechanical. Care should be taken that the frame which holds the knives in position does not cast a shadow on the flat cardboard surface. All light sources should be located between the operator and the frame of the scorer, thus avoiding shadows under the frame holding the scorers.

Flat cardboard usually is fed over rollers at the front of the machine and the first set of scorings is made by circular knives. In manufacturing these boxes, scorings must be made also at right angles to the original scorings.

After the cardboard has been scored, it is conveyed to a box-forming machine. This machine bends the cardboard at the scorings, applies the gummed corner supports, and automatically shapes the container. The machine is pedal-controlled, and all work is accomplished on a horizontal plane, with the tool and forming-die completing the work.

Most box cover papers have high reflectances compared with machine backgrounds. The luminance ratio may be as high as 5 to 1.

General illumination is recommended for box manufacturing.

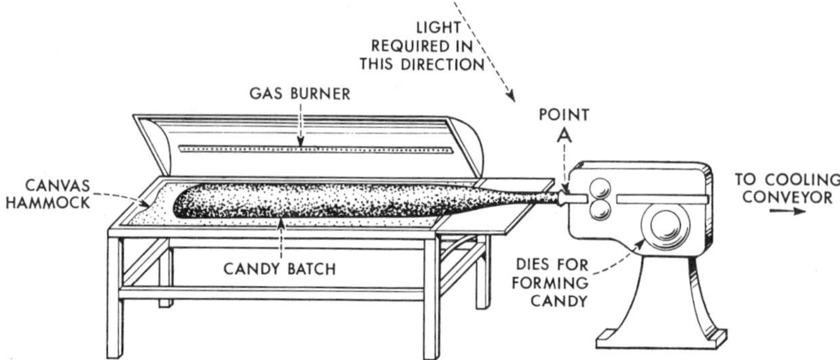

Fig. 14-44. Hard-candy-forming machine. The batch is revolved slowly in the canvas hammock. Heat is applied for surface glazing. The operator tapers one end to enter the dicing machine at point *A*, which cuts and forms in one operation and delivers the pieces to a cooling conveyor.

FRUIT AND VEGETABLE PACKAGING[20]

Lighting recommendations should be based on full knowledge and consideration of the tasks which will be performed in the specific packaging plant. Each food has special characteristics of origin, composition, size, and color.

Most seeing tasks in a cannery can be illuminated by a general lighting system. The lighting of sorting and canning belts and other critical seeing tasks should be from a system of supplementary or localized general lighting designed specifically for each individual task. The supplementary lighting recommendations given below for apricots and peaches may be used as a guide for other light-colored products. The illumination levels for tomatoes are for medium-dark products and those for olives represent values for most dark-colored products. For footcandle levels see Fig. 9-80.

Color Grading

Investigations of seeing conditions have disclosed that one of the major problems in food processing is the lighting of sorting and inspection belts for the separation of quality grades where maturity is an important factor. Maturity of most products is judged primarily by color. Thus, selection between maturities presents problems in detecting characteristic colors and small differences between these colors.

An extensive investigation of colored lights to find the spectral quality of illumination most suitable for increasing color differences was made at Stanford University.[21] This study has indicated that a continuous distribution of energy in the light source, in other words a "white" light, is essential. For most sorting and canning operations, the improved-color cool white fluorescent lamp is adequate. The illuminant best suited for extremely critical color-difference problems is one similar to daylight from the north sky; that is, one having approximately equal distribution throughout the visible spectrum. For such cases, the "daylight" fluorescent lamp is usually satisfactory, but where a closer approximation is needed, light sources used in the graphic arts industry should be used. (See page 14-24.) There may be some specific types of blemishes or fruit conditions that the applications of specific colors of light will accentuate.

Beginning stages of some types of decay or fungi can be detected by exposing the incoming fruit or vegetable to black light. This can be determined by field experimentation.

Tests using colored backgrounds and surrounds have indicated that the colors of all surfaces within the field of view are of first importance in seeing color differences. Small color differences between similarly colored objects can best be detected if the objects are viewed against a background that has a reflectance and color equal to the average of those of the objects.

Container Inspection

The inspection of empty glass jars or metal cans presents a difficult seeing task when the containers pass in single file on a conveyor belt. Since it takes an appreciable amount of time for the eyes to focus on a container and to move from one container to another, the time necessary for proper inspection, as well as the lighting, is an important factor. Glass containers are inspected to detect defects such as chips, spikes, cracks, and air bubbles. Cans are inspected for such defects as cracks, seams, solder pellets, droops, cutovers, and scratches in the protective inside coatings.

Fig. 14-45 shows an inspection booth for glass jar inspection. The luminous horizontal and vertical panels should be illuminated by fluorescent lamps from behind so that they will have a luminance of from 300 to 400 footlamberts. Above the containers is placed an inclined mirror for viewing the interior of the jars.

Since metal cans are opaque and specular, two mirrors instead of luminous panels are used, one behind the cans and one at an angle of about 45 degrees as shown in Fig. 14-46. The whole is enclosed in a booth with interior finished in a flat white and indirectly illuminated to have a luminance of about 400 footlamberts. All exterior light should be excluded to prevent specular images of outside light sources.

Cutting Rooms

Since the final check on the quality of grading being done in the cannery is usually made in the cutting rooms it is important that cutting rooms be conditioned so that accurate visual inspection for form and color is possible. The ceiling should be

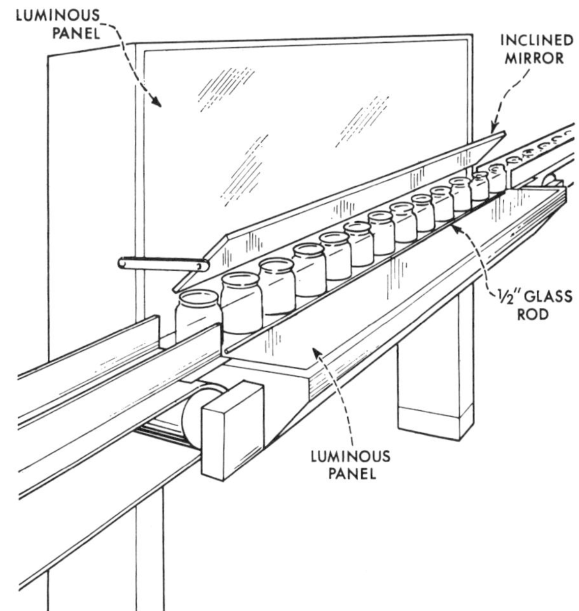

Fig. 14-45. Inspection booth for glass jars; horizontal and vertical translucent glass panels illuminated from behind by fluorescent lamps.

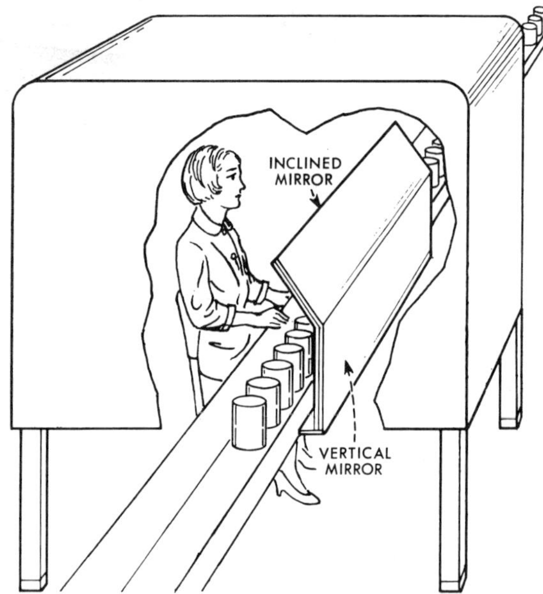

Fig. 14-46. Can inspection booth.

painted white and the walls a very light neutral gray with an eggshell finish and with a reflectance of 65 per cent or better. This can be produced by adding a trace of black only to white paint.

Direct sunlight should be excluded and the room illuminated with luminaires using daylight fluorescent lamps. Critical inspection for color should be made under one controlled source of illumination. This can be best obtained by using a light hood painted white with a reflectance of 75 per cent or higher and illuminated by light sources used in the graphic arts industry. (See page 14-24.) All extraneous light should be excluded from the hood.

Daylight should not be used because of the extreme changes in spectral quality during the day and from one day to another.

Cutting and Pitting Areas. Fruit is cut and pitted by hand or by machine. Operators must sort fruit and discard obviously defective or immature fruit.

Cut apricots or peaches are inspected rapidly on belts, usually wet, and reflected glare becomes a problem. Low luminaire luminance and careful location of lighting units with respect to sorting belts are important. The seeing task requires maximum speed of vision. The detecting of pieces of broken pits in fruits is very critical, especially in dark colored fruits such as prunes or olives.

Sorting and Canning Tables. Arrangements vary but, typically, a short sorting belt for removing off grade or defective fruit is followed by a row of canning sinks or tables. Operators select and separate grades of fruit in packing, and must observe small defects at high speed. Fluorescent luminaires, mounted end-to-end on the center line of the belt, meet these requirements but introduce a possibility of excessive reflected glare. This last can be minimized by locating the units 12 inches back from the edge of the belt and by use of a suitable baffle over the center line.

Tomatoes

Receiving Department. Tomatoes are unloaded and either held temporarily or transferred at once to the washers. Samples drawn from deliveries may be graded for quality here. Inspection for quality requires illumination suitable for color grading.

Preliminary Sorting and Trimming. Washed tomatoes on conveyors are inspected and sorted to remove defects such as sunscald, worms, rot, green tomatoes, etc. These are critical seeing tasks requiring very high acuity.

Sorting and Canning Tables. Operators select tomatoes by size and color for whole pack on sorting and canning belts or table. Glare-free illumination of spectral quality suitable for color grading is required.

Olives

Sorting, Grading, and Inspection Belts. Olives are sorted as received to remove immature, overripe, or damaged fruit and to remove sticks, stems, or other foreign material. A second sorting and inspection operation takes place when cured olives are made ready for canning. This involves, in addition to the above, the segregation of fruit because of difference in uniformity or darkness of color. A higher level of illumination from luminaires having low surface luminance is required.

Pitting and Chopping. Pitting and chopping of olives are performed mechanically. The operations involve, primarily, control of the machines and examination of the products as they emerge, to check on the proper operation of the machines.

Hand Packing Olives in Jars. Certain types of olives are hand packed in glass jars in symmetrical arrangements, each olive being picked up with tongs and placed carefully in its proper position. Localized general illumination is recommended from properly positioned, shielded luminaires.

Carton Lighting

In general, no special lighting is required for empty carton inspection as the cartons are received flat in bulk and are shaped by machines prior to filling. Plastic bag containers come in rolls and are expanded prior to filling. Broken or torn bags can be detected by normal inspection lighting as described under Supplementary Lighting (see page 14-6).

Frozen Food, Processing, Packing, and Storage

This branch of the industry has all the problems of inspection and sorting plus the additional problems of temperature, air movement, and moisture.

Careful consideration of lamp temperature and lumen output should be considered in the basic application. In addition, the luminaire should be selected to afford maximum protection from moisture and corrosion. Fluorescent lamps operated at cold temperatures suffer from low lumen output as discussed in Section 8. Enclosed luminaires or jacketed lamps can be utilized to solve both the temperature problem as well as sanitation. High-intensity discharge lamps generally are not affected by temperature and should be considered for cold room lighting when all other factors such as brightness, distribution, and effects of voltage interruptions are considered.

PETROCHEMICALS

Hazardous vs Non-Hazardous Locations

Many areas in petrochemical manufacturing plants involve what are defined by the National Electrical Code as hazardous locations. Article 500 has classified and grouped the common dangerous materials found in most manufacturing processes. Within a given group are those materials which have similar properties and therefore give rise to similar degrees of hazard. There are three classes and seven groups. Each of the three classes is subdivided for the purpose of further defining the type of equipment required into Division 1 or Division 2, the essential difference between the two being whether the hazardous material is present in dangerous quantities under normal conditions (Division 1), or whether it is only present under abnormal conditions (Division 2). Most processes found in the petroleum industry are Class I Group D, Hazardous Locations. The industry has developed standards which for common processes define the type and extent of these Division 1 and Division 2 areas.

Three general types of luminaires may be used: namely those specifically designed for use in Division 1 Hazardous Locations, those which may be used in Division 2 locations, and those for use in ordinary locations. Therefore, to properly specify equipment for hazardous locations, it is necessary to know the applicable Class, Group, and Division.

Consideration must also be given to the other environmental conditions of moisture, corrosive atmospheres, etc., that may be involved. The possibility of floodlighting outdoor Division 1 Hazardous Locations by locating non-explosion-proof floodlights beyond the hazard boundaries should not be overlooked.

Once the environmental conditions of hazardous location, corrosive vapors, and other ambient atmospheric conditions of moisture, temperature, etc., have been considered, the lighting task follows accepted industrial practices.

Process Equipment Buildings

Luminaires within reach, or where exposed to breakage, should always be equipped with strong metal guards.

Special Equipment

Special lighting equipment often is required, such as that for illuminating the insides of filters or other equipment whose operation must be inspected through observation ports. If the equipment does not include built-in luminaires, concentrating-type reflector luminaires should be mounted at ports in the equipment housing.

Portable luminaires are utilized where manholes are provided for inside cleaning and maintenance of tanks and towers. Explosion-proof types (where hazardous conditions may exist) with 50-foot portable cables are connected at industrial receptacles (either explosion-proof or standard) provided near manholes on towers and at other locations.

Outdoor Tower Platforms, Stairways, Ladders, Etc.

Luminaires should provide uniform illumination and be shielded from direct view of persons using these facilities. Special luminaires often are required at gauges. Enclosed and gasketed or weather-proof luminaires equipped with refractors or clear gasketed covers may be used. Luminaires above top platforms or ladder tops should be equipped with refractors or reflectors. Reflectors may well be omitted on intermediate platforms around towers so that the sides of the towers will receive some illumination and the reflected light therefrom will mitigate sharp shadows.

ELECTRICAL GENERATING STATIONS

It is convenient to classify electric power generating stations by the type of energy utilized: fossil fuel, hydroelectric, and nuclear. Such a classification is convenient because, while many areas, such as turbine buildings, transformer yards, and switchyards, have similar lighting problems, there are few similarities in the problems encountered in lighting the various types of energy handling areas.

For recommended levels of illumination see Fig. 9-80.

Switchyards, Substations and Transformer Yards[22]

These areas are occupied by station transformers, oil or air circuit breakers, disconnect switches, and high voltage buses together with their auxiliary equipment. Outdoor substations present a peculiar

Fig. 14-47. Substation lighted with bracket-mounted substation luminaires.

lighting problem since in many cases the seeing tasks are a considerable distance above eye level, and there is a high degree of contrast variation.

Especially designed stanchion or bracket mounted substation luminaires are available to provide the required upward lighting component (see Fig. 14-47). In addition, floodlights and spotlights are often utilized. The use of high-intensity discharge lamps, because of their long life, helps to reduce the maintenance problem caused by the inaccessibility of this area.

Two aspects of lighting are required: general illumination for movement in the area; and directed illumination for viewing such seeing tasks as fans, bushings, oil gauges, panels, and disconnect switch jaws.

Control Rooms and Load Dispatch Rooms[23]

Because the control room is the nerve center of the power plant and must be monitored continuously, its lighting must be designed with special attention given to the comfort of the operator. To do this direct and reflected glare and veiling reflections must be minimized and luminance ratios must be kept low.

Along with ordinary office type seeing tasks, meters must be read from long distances, often 10 to 15 feet. Care must be taken, especially with curved face meters, to eliminate reflected glare and veiling reflections from the meter face.

While the practice is by no means standardized, most control room lighting of recent construction falls into one of two general categories: diffuse lighting and directional lighting. Diffuse lighting may be low-luminance luminous indirect lighting, solid luminous plastic ceilings, or louverall ceilings (see Fig. 14-48). Directional lighting (see Fig. 14-49) may take the form of troffer units recessed into the ceiling and following the general contour of the control board; these luminaires must be accurately located to keep reflected light away from the glare zone. Illumination for the balance of the room may utilize any type of low-luminance general lighting equipment.

Giving operators full control of illumination levels through the use of multiple switching or dimming systems increases the flexibility of an installation. Load dispatch rooms resemble control rooms in many respects and may be illuminated similarly.

Turbine Buildings[24]

Most indoor areas of turbine buildings are characterized by medium and high ceilings. Seeing tasks include general inspection, meter and gauge reading, and pedestrian passage. Where detailed maintenance or inspection is to be done, higher level portable lighting equipment is recommended.

In low and medium bay applications high-intensity discharge or fluorescent industrial luminaires suspended below major obstructions are appropriate for general illumination in the turbine building (see Fig. 14-50). Where structural limitations exist, floodlights mounted on walls or platforms may be required. Supplementary lighting is recommended to provide vertical illumination on such equipment as control panels, switchgear, and motor control centers. The use of luminaires with an upward light component helps to improve visual comfort.

Fig. 14-48. Control room of a large generating station illuminated with a louverall ceiling.

High bay areas (generally 25 feet or higher) are common in turbine buildings. For these areas, either high-intensity discharge or fluorescent luminaires can be used. Lighting continuity during momentary or extended power interruptions must be considered (see Emergency Lighting, page 14–10).

Often there are small enclosed rooms such as relay rooms, dispatch rooms, laboratories, electronic equipment rooms, and computer rooms.

Fossil Fuel Areas

Fossil fuel plants, including those which use gas, coal, oil, or lignite, require facilities to receive, store, and transport the fuel. Coal handling usually involves extensive outdoor storage areas. High-intensity discharge floodlights will provide the illumination required for stacker, reclaimer, and bulldozer operations. Conveyor platforms are often lighted by post top type luminaires, the light required for both pedestrian traffic and conveyor inspections. Crusher houses, transfer towers, thaw sheds, unloading hoppers, and other enclosures generally require indoor industrial type luminaires. Many of these areas are classified as hazardous and, where required, they should be lighted with equipment which meets the applicable provisions of the National Electrical Code.

Oil fired stations must have large oil storage tanks, pumps which must be lighted for inspection, and may have barge or tanker unloading facilities. Gas fired stations usually have gas metering areas. General area illumination should be provided with floodlights. Care must be taken to assure compliance with the requirements of the National Electrical Code.

Boilers on fossil fuel stations have a series of platforms, some indoors and some outdoors. These platforms, and their associated stairs and landings, should be illuminated for safe pedestrian passage and for inspection of pumps, valves, gauges, soot blowers, etc. Where no overhead structure is available for the support of general purpose industrial luminaires, stanchion type luminaires may be used.

Fig. 14–50. View of condenser well lighted from general lighting system in turbine room ceiling. Lighting layout keyed to special acoustical treatment of ceiling and light tile walls. Note visitors' gallery.

Extreme heat is associated with certain indoor boiler areas and care should be exercised to select luminaires for such areas that are designed to operate in high ambient temperature conditions.

Hydroelectric Stations

The turbine building of a hydroelectric station is commonly called the power house and its lighting requirements are generally the same as those previously outlined under Turbine Buildings.

Lighting of the intake and discharge areas, where applicable, is best accomplished with floodlighting.

Nuclear Stations

The selection of lighting equipment for nuclear stations is often limited by factors other than economy or efficiency. There are extensive station areas, especially around the reactor and fuel storage facilities, where there may be a restriction on the kinds of lamps and metals used in the luminaires. There-

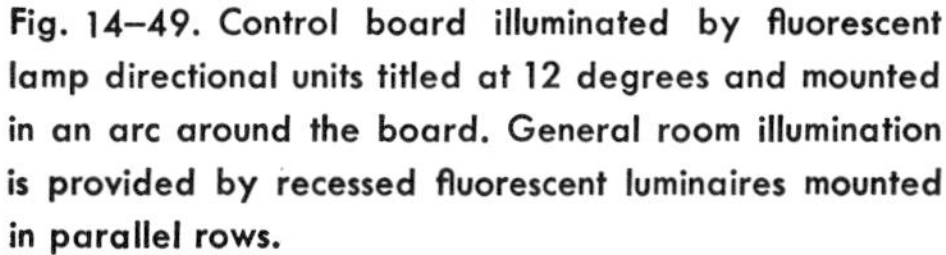

Fig. 14–49. Control board illuminated by fluorescent lamp directional units titled at 12 degrees and mounted in an arc around the board. General room illumination is provided by recessed fluorescent luminaires mounted in parallel rows.

fore, matters concerning luminaire and lamp selection should be coordinated with the appropriate authorities.

Emergency Lighting

In the event of failure of the normal lighting source, standby lighting will be required in certain areas, to enable operating personnel to continue the performance of critical functions and for safe building egress. Areas where emergency lighting should be considered are control rooms, first aid rooms, turbine rooms, exit stairs and passages, battery rooms, and standby generator rooms.

In nuclear stations it is recommended that emergency lighting be provided in all areas subject to contamination, especially the decontamination room and lab area.

SAWMILLS

Lighting the Redwood Green Chain[26]

The severity of the seeing task makes proper lighting and environment necessary so that maximum speed and uniformity in grading may be achieved.

Fig. 14–51. Perspective view of inspector and boards.

As shown in Fig. 14–51, the grader normally views up to twenty feet of board moving along past him on a conveyer. The chains in mills travel approximately 25 to 39 feet per minute, and the boards are usually spaced two inches to two feet apart. Both the speed of the chain and spacing of the lumber vary a great deal from mill to mill. The grading is based on the many types of defects that can be found in boards.*

For most satisfactory results, it is important to provide light over the full length of the longest board which might be graded on the chain. The maximum illumination should be at the far end of the board.

In no case should the bright surfaces of luminaires or of a bare lamp be left exposed to the grader's eyes.

Directional Lighting. Directional lighting produces the best results in increasing the visibility of shake, checks, splits, and knots.

It appears from observations made in the laboratory and in the field that lighting units should be mounted in rows parallel to the length of the board and tilted to project the light at a fairly high angle (45 degrees or more) to the normal to the plane of the board and in the direction and flow of the board.

Incandescent reflectorized lamps can be used with specially constructed shielding as shown in Fig. 14–52.

Diffuse Lighting (Non-Directional). Diffuse lighting was found to be generally satisfactory for grading tables, as a second choice.

Existing fluorescent equipment in mills can be adapted with diffusing panels so as to give improved results.

Consideration should be given to the use of fluorescent units with good maintenance features. The general area is normally very dusty.

* For illustrations of board defects see *Lighting for Sawmills: Redwood Green Chain*, Illuminating Engineering Society, New York, 1957.

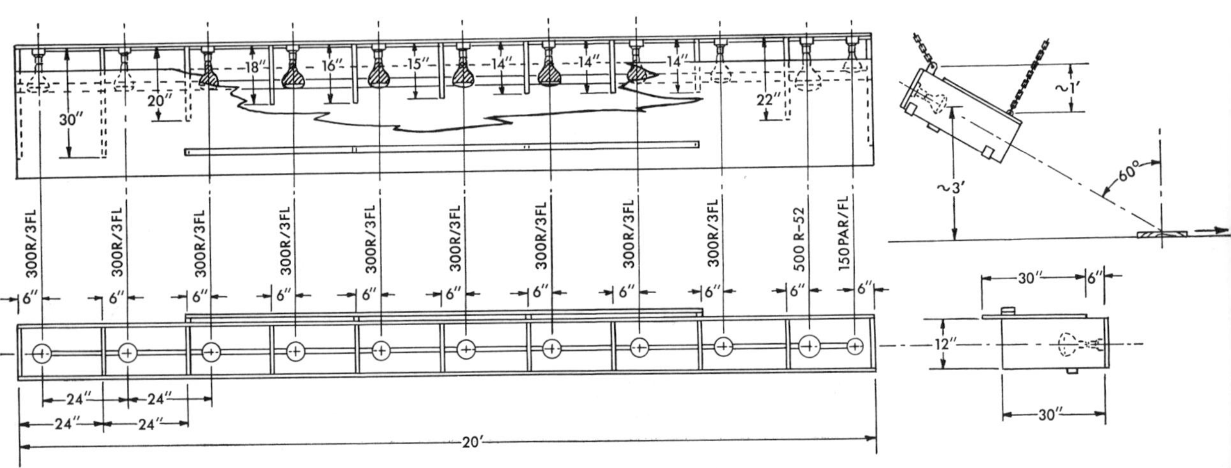

Fig. 14–52. Directional lighting using incandescent reflectorized lamps.

Illumination Level. From data and field observation it is shown that lighting levels of 300 footcandles or more are needed for the grading of redwood lumber.

Color. This variable is not of primary importance in grading of redwood. Conclusive tests for color preferences have not been made but accuracy of grading does not appear to be greatly affected by minor color variations.

Environment. Field conditions emphasize the need for shielding glare from bright sky areas by enclosing the grading area. Painting of background areas is strongly indicated. A flat gray with a reflectance of 45 per cent is satisfactory.

RAILROAD YARDS[27]

The lighting of railroad yard work and storage areas and platforms is essential to promote safety to personnel, expedite operations, and reduce pilferage and damage to equipment. Recommended levels of illumination for the tasks encountered in the railroad yard are listed in Fig. 9-80 on page 9-91.

Because light is absorbed by moisture, smoke, and dust particles, even in an apparently clean atmosphere, when luminaires are located at a distance from task and/or the task is viewed from a distance, the absorption of light must be taken into consideration. For example, if the atmospheric transmittance were 80 per cent per 10-foot distance and the task were viewed at 100 feet, the illumination on the task would have to be increased by a factor of 10 to obtain the same visibility as at 100 per cent atmospheric transmittance.

Seeing Tasks

Railroad yards can be divided into general areas which have different seeing tasks, as covered below.

Retarder Classification Yard. The large and often highly automated retarder classification yard, with its supporting yards and servicing facilities, presents a number of different seeing tasks that are considered under the following categories:

1. *Receiving Yard.* Seeing tasks throughout the area consist of walking between cars, bleeding air systems, opening journal box covers, and observing air hoses, safety appliances, etc.

2. *Hump Area.* Seeing tasks in the hump area are diversified. The scale operator and hump conductor are usually required to check each car number visually. Car inspectors should have illumination on the underneath surfaces of the cars and on the running gear to permit ready and precise inspection of a car that is in motion. There also should be enough light at the top of cars to permit judgment of height. Car uncouplers should be able to see the coupling mechanism. The operator of the wedge inserter, if one is used, must be able to see the coupler accurately in order to apply the wedge, again with the car in motion. The hump conductor, car inspector, car uncoupler, and wedge operator should have specifically directed lighting of a higher level than that provided by the general lighting in other parts of the hump area.

3. *Control Tower and Retarder Area.* In some yards the operator may be required to visually check the extent of track occupancy in the yard, gauge the speed of the car coming from the hump, and manually set the amount of retardation to be applied to the car. Even in automated yards, the operator may also be required to do this manually in event of failure of one or more of the automatic features. In many yards, the control tower operator is expected to check the car number against a switching list and see that the car goes to the correct track. Under clear atmospheric conditions, it is important that there be no *direct* light projected toward the operator. However, under adverse atmospheric conditions of dense fog, for example, it is general practice to utilize auxiliary lighting equipment on the side of the tracks opposite the retarder control tower to reveal the outlines of cars in silhouette.

4. *Head End of Classification Yards.* The operator should be able to see that cars entering the classification yard actually clear switch points and clearance points so that following cars will not be impeded or perhaps wrecked.

5. *Body of Classification Yard.* In many yards, the operator must be able to see the body of the yard sufficiently well to determine the extent of track occupancy. On some railroads, men are required to move along cars in the body of the classification yard to couple air hoses, pack journal boxes, close journal box covers, etc. At the leaving end of the body of the classification yard skatemen generally are required to place the track skates for stopping moving cars at the desired location and to remove them for switching.

6. *Pull-Out End of Classification Yard.* In this area, switchmen are required to walk along tracks to determine switch positions and operate them if necessary. There is very little requirement for illumination other than to provide safe walking conditions along the switch tracks.

7. *Dispatch or Forwarding Yard.* As in the receiving yard, the main seeing task in the dispatch yard consists of walking between cars.

Hump and Car Rider Classification Yards. The seeing tasks in the classification yard, and around the hump, are considerably different in the rider-type yard than in the retarder yard. Around the hump area, a yard clerk should be able to read car numbers, cars must be uncoupled, and car riders must be able to see grab irons, ladders, etc., to climb safely onto the cars. Switchmen operating along the lead track must have safe seeing conditions to enable them to walk along the lead track and operate switches. Car riders on the cars rolling into the yard should be able to see cars on the track ahead so that they can brake adequately to reduce impact and prevent consequent damage to lading. The rider must then be able to see to get off the car and walk along yard tracks to the hump.

Flat Switching Yards. The only seeing requirements in most yard areas of this type are for safe walking conditions for switchmen around the head end and pull-out end switches and for pulling pins and throwing switches. A yard supervisor may also be required to read car numbers at the head end of the yard in order to assign cars to their proper tracks. A locomotive pushes cars into the body of the yard, and in most cases, the locomotive headlight furnishes sufficient light to provide adequate seeing for the locomotive engineer. Sufficient illumination is needed at clearance points to be sure of no interference of cars going to adjoining track. General lighting is recommended in the area of the switches at the head end and pull-out end of the yard. In addition, if a supervisor must read car numbers, localized lighting should be provided at his location.

Trailer-on-Flatcar Yards. Seeing tasks involved in yards where highway-type trailers are loaded on special railroad flatcars require the tractor operator to be able to back up or drive along the tops of the flatcars, uncouple the tractor, and pull off. Men must then tie down the trailers to the flatcars. To do this they must be able to see beneath the trailers.

Container-on-Flatcar Yards. The principal seeing tasks require crane operators to pick up containers (a) from any part of the trailer parking yard and place them in precise locations on the flatcars, or (b) from the flatcars and place them in precise locations on the trailers. Another seeing task is concerned with the job performed by those who tie down or release containers. Lighting should come from all sides of vehicles.

Lighting Systems

Two different systems of lighting are generally used in illuminating railroad yards: the Projected (Long Throw) Lighting System and the Distributed Lighting System. Each has its advantages under specific yard situations and the characteristics as listed below for each should be thoroughly weighed.

In general, the principles in lighting for railroad yards are the same as for other outdoor locations; however, railroad regulations should be observed with respect to the location of any lighting equipment above or adjacent to the tracks.

Projected Lighting System. It is the function of this system to provide illumination throughout the various work areas of the yard from a minimum of locations. See Fig. 14–53.

The basic characteristics of a projected lighting system are:

1. A minimum amount of ground space is required.
2. Light distribution is flexible. Both general and localized lighting are readily achieved.
3. The units employed are effective over long ranges.
4. Maintenance problems are restricted to a few concentrated areas.
5. High mounting heights are generally used.
6. Aiming of individual projectors is required.
7. Physical and visual obstructions are minimized.
8. Electrical distribution system serves a small number of concentrated loads.

Distributed Lighting System. Distributed lighting differs from the projected technique in that luminaires are at many locations throughout the area rather than at relatively few locations. See Fig. 14–54.

Some characteristics of the distributed lighting system are:

1. Good uniformity of illumination on the ground.
2. Good utilization of light.
3. Reduction of undesirable shadows.
4. Less critical aiming.
5. Lower mounting heights.
6. Reduced losses due to atmospheric absorption and scattering.
7. Electrical distribution system serves a large number of small distributed loads.

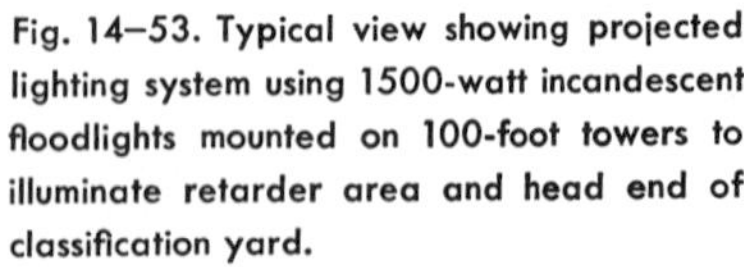

Fig. 14–53. Typical view showing projected lighting system using 1500-watt incandescent floodlights mounted on 100-foot towers to illuminate retarder area and head end of classification yard.

Fig. 14-54. View of a distributed lighting system along switch lead tracks at pull-out end of classification yard.

FARM LIGHTING

The drastic changes that have taken place in agriculture in the last few years have greatly affected the tasks to be performed in and around farm buildings. The tasks have not only changed in nature, but the time available to perform them has been reduced, in most cases. New lighting concepts to keep pace with these developments have necessitated research that will provide the basic data on lighting requirements. Because of their universal application, Dairy Farm Lighting and Poultry Lighting have been selected by the IES-ASAE (Joint Farm Lighting) Committee to be the first to receive special attention. The recommendations under Dairy Farm Operations and Lighting for the Poultry Industry are based on the results of these studies. Lighting for other types of farming will be the same where the tasks are similar.

Quantity of Illumination. The illumination levels required to perform visual tasks on the dairy farm vary through wide extremes. For example, many hundreds of footcandles may be required in the farm shop, or where inspections are made for cleanliness and disease control, such as in milking operations. But relatively low levels of illumination are adequate on walkways between buildings, or in storage areas where seeing tasks may be much less critical. Recommended levels of illumination for the various dairy and poultry farm operations are listed in Fig. 9-80.

Quality of Illumination. Specifying the quantity of light in no way establishes the quality of a lighting installation. Some important factors of the lighting installation are uniformity, glare, color, and environment (see page 14-2).

Uniformity of illumination may be expressed as a ratio of the maximum footcandle value to the minimum footcandle value over an area. Based on this, satisfactory uniformity ratios for farms vary from 1.5 to 1 for critical seeing task areas, to 5 to 1 for less critical areas. In general, with greater mounting heights and closer spacing of luminaires, better uniformity can be achieved.

The quality of the lighting installation can be greatly influenced by certain environmental factors. Room surfaces should have matte finishes of high reflectance to help prevent excessive luminance (photometric brightness) ratios. The ceilings, walls, and floors can increase the utilization of the light within the room by acting as a secondary light source of large area. Luminaires which have some light directed upward toward a ceiling having a relatively high reflectance will help create a more comfortable visual environment. Recommended reflectance values for farm interiors are consistent with those of other industrial areas. See Fig. 14-3.

Codes. The use of lighting equipment on the farm is governed by many federal, state, and local codes. Some public health codes specify minimum illumination levels required in the milking and milk handling areas for maintenance of health standards. These public health lighting requirements are normally below those levels shown in Fig. 9-80, since they are concerned only with sanitation. The levels recommended in Fig. 9-80 were selected for efficient performance of the visual task in addition to the concern for sanitation.

The modern farm includes many types of occupancies which may be wet, damp, corrosive, dirty, surrounded by combustible materials, or saturated with gasoline fumes. Therefore, it is important to follow the National Electrical Code and any local regulation which may be in effect when installing lighting equipment. See also Reference 30.

Dairy Farm Operations[28]

Dairy farms vary in size, location, amount of investment, and degree of mechanization; however, there are some areas common to every dairy facility. These are the housing areas, including community, maternity, and calf pens; the feeding areas; the milking operation areas; the milk handling equipment and storage areas; and the feed storage areas.

Livestock Housing. The two basic housing systems are stall barn and loose housing. The stall barn is a structure for sheltering dairy cattle and/or young stock, where the adult animals are confined to stalls by means of stanchions, straps, halters, or chains during part of the year, and usually for milking purposes. There may be one or more rows of stalls and pens. Roughages and concentrates may be fed in mangers at the individual stalls. All, part, or none of the feeds and bedding may be stored in the structure. The stall barn is sometimes referred to as a stanchion barn or warm-housing.

Loose housing, sometimes referred to as cold-housing, is a management system for dairy cattle wherein the adult animals are given access to a feeding area, a resting area, and an adjoining open lot. At milking time, the lactating animals are passed through a milking parlor. Other dairy animals may be in separate pens, lots, or buildings.

Adequate lighting is required to observe the condition of the animals and to detect hazards to the livestock and operator. Portable supplementary lighting units can be used to examine or treat individual animals when required.

Milking Operation. Both systems, stall barn and loose housing, are quite different in layout, but each contains similar basic areas of operation which require lighting. In the stall barn, the milking operation is performed in the stall. In the loose housing system, the animals are milked in a milking parlor, sometimes called a milking room. There are various milking parlor layouts, but all contain milking stalls, an alley for cow entrance and exit, and an operator's area for use by the personnel performing the milking operation.

Adequate lighting is required to determine cleanliness of cow, to detect undesirable milk, to handle milking equipment readily, and to detect dirt and foreign objects on the floor. Recommended levels should be provided at cow-edge of gutter and on the floor. Supplementary lighting should be available to determine cleanliness of the udder and to clean and examine the udder.

Milk Room. The milk room is a room with one or more sections for handling raw milk, wholly or partly enclosed by the structure in which the cows are milked. The milk house is the same as a milk room except that it is not a part of, but may or may not be connected with, any other structure. The milk room or milk house would be the same for stall barn and loose housing systems.

Illumination is required for the operator to move about readily and safely and to determine floor cleanliness. In the washing area it is necessary to detect dirt and other impurities on the milk handling equipment. Supplementary portable ultraviolet luminaires should be available in this area to aid in the detection of milkstone on the equipment.

Lighting is also necessary to adequately inspect the bulk tank interior for cleanliness. Additional lighting may be required to illuminate dip-stick or scale.

Feeding Operation. The feeding area is a part of a barn, shed, or open lot where cows are fed roughages and water, and sometimes concentrates. This area may or may not include feed storage. Feed storage and processing areas include haymow, silo, silo room, grain bins, concentrate storage areas, and feed mill and mixing areas.

Adequate lighting is required in the feeding area for the detection of foreign objects in the grain, hay, or silage.

In forage storage areas, lighting is required for safety of the operator in moving about, but adequate lighting is required for the detection of foreign objects in the grain, hay, or silage.

Luminaires should be mounted at the top of the silo near the ladder chute for ease in luminaire cleaning and lamp replacement.

In grain and concentrate storage areas adequate lighting is required to inspect amount and condition of grain. When grain is suspected of being moldy, containing foreign objects, etc., samples should be inspected under higher illumination levels. Adequate lighting is required to read concentrate labels and again higher illumination levels are required for critical inspection for impurities and spoilage.

Illumination is required in the processing area to see to move about readily, and to read labels, scales, and equipment dials. If machine repairs are necessary, additional light should be supplied by portable luminaires or daylighting.

Miscellaneous Areas. Other buildings and/or areas, common not only to dairy farms but to all types of farms, include: the machine storage areas, such as the garage and machine sheds; the farm shop; the farm office; restrooms; and the pump house. (See earlier portions of this Section for industrial type areas and Section 11 for Offices.) Many exterior areas on the farm require lighting such as paths, steps, outdoor storage areas, feed lots, shop aprons, and building entrances. Protective lighting is also required in many exterior areas (see Section 18).

Poultry Industry Operations[29]

Poultry farms and processing plants vary in size, layout, and degree of mechanization; however, there are some areas common to most facilities. The following are brief discussions describing the operations in the major areas.

Laying Houses. There are basically two types of laying houses used on most poultry farms, the floor type house and the cage type house. See Fig. 14-55. In the floor type houses, the hens may move about freely over the floor area. In the cage type houses, hens are confined in cages with three to four hens per cage. A common size cage for housing four birds is 12 inches wide, 18 inches deep, and 16 inches high. Both the floor and cage system may range in size from a few thousand to several hundred thousand birds. Depending on the degree of mechanization, feed is mechanically conveyed to the birds in both type systems. Water is also provided continuously to both systems. Eggs are normally collected twice per day using a hand push cart, mechanical self-propelled cart, or automatic egg collectors.

General lighting provided by a lighting circuit separate from the circuit used to stimulate production and growth is needed for feeding, inspection and cleaning. Localized lighting is also needed for reading charts and records and to read thermometers, thermostats, and time clocks.

Egg Handling, Packing, and Shipping. The egg handling area may be located in a separate or

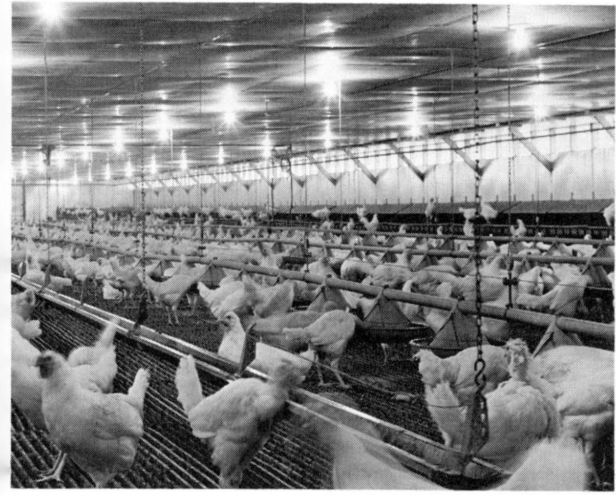

Fig. 14–55. Typical laying houses of the floor type (left) and cage type (right).

adjoined building. When the eggs arrive at the egg processing area, they are either stored in a refrigerated area or loaded directly onto a washer. After the eggs have been washed, they are sorted and graded. Rough shells, cracked shells, and dirty or stained eggs are removed at the grading table. Candling equipment is used inside an enclosed booth to sort eggs that have internal defects such as blood spots or meat spots. If the more modern equipment is used, the eggs are sorted according to size by a machine and placed in an egg carton for shipment. See Fig. 14–56. The cartons are held in refrigerated storage until they are ready for shipment to retail outlets.

General lighting is needed to keep the area clean, detect any unsanitary conditions, and to examine and grade eggs. In the loading platform and egg storage areas, general lighting is needed for safe operation of mechanical and loading equipment.

Raw Egg Processing. Eggs which are to be marketed as liquid, frozen or powdered are processed in this area. The area must meet the sanitary requirements of a public food preparation area as set up by the health department. Cracked eggs, eggs with shell defects and shell stains, plus grade A eggs, are utilized in the processing of liquid eggs. The eggs are broken out of their shells and pumped into a holding tank. The liquid eggs for interstate shipment are pasteurized and packaged. Some states also require that all broken-out eggs be pasteurized for intrastate shipment. Also, liquid eggs may be frozen at the processing area and shipped in a frozen state.

General lighting is needed to provide for cleanliness for food preparation.

Hatchery. Fertile eggs are brought to the hatchery and loaded onto trays to be placed in incubators. The incubators maintain a temperature required to develop the embryo and the chick is hatched in twenty-one days. When hatched, baby chicks that are to be grown for layers are sorted according to sex. The male chicks are disposed of and only the females are marketed. Most of the female chicks have their combs dubbed, a process of clipping off the top of the comb which results in the mature bird having a smooth comb. This helps to prevent the adult bird from catching her comb in the wires of the laying cages. Broiler chicks are handled in much the same manner except they are not sex-sorted or dubbed. Broilers are not normally reared in cages due to the increased tendency of bruising from the cages.

General lighting is needed for operators to move about readily and safely and for cleaning. Supplementary lighting is required for inspecting and cleaning inside incubators, at dubbing stations to prevent cuts and injuries to chicks, and for sex sorting (should be provided in a closed area to prevent excessive luminance ratios between task area and the immediate surrounding area).

Poultry Processing Plant. The birds are brought to the plant in crates of about 20 per crate. Crates

Fig. 14–56. Eggs being sorted according to size by machine and placed in cartons for shipment.

are unloaded and the birds hung by the feet on a continuously revolving overhead carriage. They pass by an area where they receive a slight electrical shock which stuns them just before killing. They move through a bleeding area and then into the scalding tanks. Feathers are then removed by machine and the birds move on to the processing area. All birds are government inspected for wholesomeness after eviscerating, are thoroughly washed, inspected again, and sorted according to grade. Generally, the processed birds are then packed in ice and shipped to retail outlets.

General lighting is needed for cleanliness, inspection, and sanitation. Supplementary lighting is required to detect diseases and blemishes (vertical illumination if the birds are hanging).

Farm Home

The farm home is essentially the same as the urban or suburban home and, therefore, has the same basic lighting requirements for family activities which involve close vision. See Section 15, Residential Lighting.

Outside Lighting

All farmsteads should have a good outdoor lighting system. Workers must frequently visit farm buildings during periods of darkness; therefore, it is essential that outdoor areas be adequately lighted for protection of people, livestock, and equipment, and for safety and convenience. Building entrances should be well-lighted to allow for good visibility of doorsills and steps. This will also permit safe passage from a lighted area to a darker one. Outdoor incandescent filament or high-intensity discharge luminaires can be either pole or building mounted to provide good area lighting. All equipment used out-of-doors should be designed to withstand the elements.

REFERENCES

1. *American National Standard Practice for Industrial Lighting*, A11.1-1965 (R1970), American National Standards Institute, New York (Sponsored by the Illuminating Engineering Society).
2. Committee on Illumination Performance Recommendations of the IES: "Classification of Luminaires by Distribution," *Illum. Eng.*, Vol. XLIX, p. 545, November, 1954.
3. Subcommittee on Supplementary Lighting of the Committee on Lighting Study Projects in Industry of the IES: "Recommended Practice for Supplementary Lighting," *Illum. Eng.*, Vol. XLVII, p. 623, November, 1952.
4. "Code for Safety to Life from Fire in Buildings and Structures," No. 101-1967, Sections 5-10 and 5-11, National Fire Protection Association, 60 Batterymarch Street, Boston, Massachusetts.
5. *National Electrical Code* (NFPA 70), Article 700, National Fire Protection Association, 60 Batterymarch Street, Boston, Massachusetts, 1968.
6. "Emergency Lighting Equipment," *Standards for Safety*, UL924-1958 (Revised October, 1969), Underwriters' Laboratories, Inc., 207 East Ohio Street, Chicago, Illinois.
7. Subcommittee on Lighting in Steel Mills of the Committee on Lighting Study Projects in Industry of the IES: "Lighting for Steel Mills—Part 1: Open Hearth," *Illum. Eng.*, Vol. XLVII, p. 165, March, 1952.
8. Subcommittee on Lighting for Foundries of the Committee on Lighting Study Projects in Industry of the IES: "Lighting for Foundries," *Illum. Eng.*, Vol. XLVIII, p. 279, May, 1953.
9. Committee on Lighting for the Machining of Small Metal Parts of the IES: "Lighting for the Machining of Small Metal Parts," *Illum. Eng.*, Vol. XLIV, p. 615, October, 1949.
10. Subcommittee on Lighting for the Rubber Industry of the Industrial Committee of the IES: "Lighting for Manufacturing Rubber Tires," *Illum. Eng.*, Vol. 64, p. 112, February, 1969.
11. Color Committee of the Graphic Arts Subcommittee of the Industrial Committee of the IES: "Lighting for the Color Appraisal of Reflection-Type Materials in Graphic Arts," *Illum. Eng.*, Vol. LII, p. 493, September, 1957.
12. Subcommittee on Cotton Gin Lighting of the Industrial Committee of the IES: "Lighting for Cotton Gins," *J. Illum. Eng. Soc.*, Vol. 1, October, 1971.
13. Committee on Industrial and School Lighting of the IES: "Lighting for Silk and Rayon Throwing and Wide Goods Weaving," *Trans. Illum. Eng. Soc.*, Vol. XXXIII, p. 17, January, 1938.
14. Joint IES-ASME Committee on Lighting in the Textile Industry, "Lighting for Woolen and Worsted Textile Mills," *Illum. Eng.*, Vol. XLIV, p. 364, June, 1949.
15. Committee on Industrial and School Lighting of the IES: "Report on Lighting in the Shoe Manufacturing Industry," *Trans. Illum. Eng. Soc.*, Vol. XXXII, p. 289, March, 1937.
16. Subcommittee on Lighting in the Clothing Industry of the Industrial Lighting Committee of the IES: "Lighting for the Manufacturing of Men's Clothing," *Illum. Eng.*, Vol. LVII, p. 379, May, 1962.
17. Committee on Lighting Practice of the IES: "Report on Lighting in the Fluid Milk Division of the Dairy Industry," *Illum. Eng.*, Vol. XXXVII, p. 561, November, 1942.
18. Subcommittee on Lighting in Flour Mills of the IES: "Lighting for Flour Mills," *Illum. Eng.*, Vol. XLIV, p. 691, November, 1949.
19. Subcommittee on Lighting in Bakeries of the Committee on Lighting Study Projects in Industry of the IES: "Lighting in Bakeries," *Illum. Eng.*, Vol. XLV, p. 387, June, 1950.
20. Subcommittee on Lighting in the Canning Industry of the Committee on Lighting Study Projects in Industry of the IES: "Lighting for Canneries," *Illum. Eng.*, Vol. XLV, p. 45, January, 1950.
21. White, David C.: "Maximizing Color Differences," a doctoral dissertation submitted to Stanford University in 1949.
22. Subcommittee on Lighting Outdoor Locations of the Committee on Lighting Central Station Properties of the IES: "Lighting Outdoor Locations of the Central Station Properties," *Illum. Eng.*, Vol. LII, p. 105, February, 1957.
23. Committee on Lighting of Central Station Properties of the IES: *Lighting of Central Station Properties; Part I: Lighting of Control Rooms; Part II: Lighting of Load Dispatch Rooms*, Illuminating Engineering Society, New York, 1951.
24. Subcommittee on High Bay Lighting of the Committee on Lighting of Central Station Properties of the IES: "Lighting of Central Station High Bay Areas," *Illum. Eng.*, Vol. L, p. 395, August, 1955.
25. Subcommittee on Lighting of Indoor Locations of the Committee on Lighting of Central Station Properties of the IES: "Lighting Indoor Locations of Central Station Properties," *Illum. Eng.*, Vol. LII, p. 423, August, 1957.
26. Subcommittee on Sawmill Lighting of the Industrial Lighting Committee of the IES: "Lighting for Sawmills: Redwood Green Chain," *Illum. Eng.*, Vol. LII, p. 381, July, 1957.
27. Subcommittee on Outdoor Productive Areas of the Industrial Lighting Committee of the IES: "Railroad Yard Lighting," *Illum. Eng.*, Vol. LVII, p. 239, March, 1962.
28. IES-ASAE (Joint Farm Lighting) Committee: "Lighting for Dairy Farms," *Illum. Eng.*, Vol. LXII, p. 441, July, 1967.
29. Joint Farm Lighting Committee of the IES and ASAE: "Lighting for the Poultry Industry," *Illum. Eng.*, Vol. 65, p. 440, July, 1970.
30. *Agricultural Wiring Handbook*, Edison Electric Institute, 90 Park Avenue, New York, N.Y. 10016, 1971.

SECTION

15

RESIDENTIAL LIGHTING[1]

Light as an Element of Design 15–1 *Environment Objectives 15–1*
Lighting Equipment 15–3 *Design Considerations for Specific Visual Tasks 15–12*
Decorative Accents 15–26 *Exteriors and Grounds 15–26*

LIGHT AS AN ELEMENT OF RESIDENTIAL DESIGN

Light is an element of design which should be used not only for visual comfort but also to achieve predetermined emotional responses from the lighted environment. Light has certain characteristics that affect the mood and atmosphere of the space—influencing the emotional responses of the people who occupy the space. See Section 10. The definition and character of space is greatly dependent on the distribution and pattern of illumination. Luminaires themselves have dimensional qualities that may be used to strengthen or minimize architectural line, form, color, pattern, and texture.

As a general statement of principles: higher levels of controlled general lighting are usually cheerful, and stimulate people to alertness and activity. Lower levels tend to create an atmosphere of relaxation, intimacy, and restfulness. Control of the levels of illumination is generally desirable and often necessary. Such control makes it possible to change the mood or tone of the room to suit its various uses. It provides psychological change.

Lighting can be described as being "soft" or "hard." Soft or diffused light minimizes harsh shadows, and provides a more relaxing and less visually compelling atmosphere. When used alone the effect can be lacking in interest, as for example, an overcast day. The artful use of hard or directional light can provide highlights and shadows that emphasize texture and add beauty to form, as, for example, a shaft of sunlight.

Another characteristic is the brilliance or sparkle obtained from small unshielded sources such as a bare lamp or candle flame. Usually such sources are not used as the prime origin of illumination, but are generally decorative, and therefore need other illumination to supplement it. The glitter of crystal, jewels, polished brass, and the luster and sheen from table settings and some types of surface materials create a sense of aliveness and gaiety.

Color is an additional dimension of lighting design. Human beings are emotionally responsive to their surroundings, and color is one of the chief factors that determines how these surroundings appear. Since color of light intensifies surface colors of the same color and grays opposites, it can greatly affect this emotional response. See Section 5.

Association of color with objects, experiences, locale, or cultures is a powerful factor in the personal reaction to color. It is usually desirable that colored light should enhance the color identification of the object or area. However, colored light as a general light source should be low in saturation, since the use of saturated or special colors can destroy the accepted appearance of materials and people.

The importance of lighting as a design element can hardly be overemphasized and in his design the lighting designer should consider the esthetic and psychological effects as well as the fulfillment of seeing needs.

ENVIRONMENT OBJECTIVES

Residence lighting practice differs from other lighting fields in that there are no set "lighting systems" for most areas. Individual seeing tasks may be performed in almost any room. Therefore, designers often consider the lighting of the individual task areas first—then tie these lighting elements together with some form of general or environmental lighting system. This "tying-together" process is important in achieving a comfortable visual environment. Individually-lighted task areas can run the risk of uncomfortable excessive differences between the actual task lighting and the lighting in the visual surroundings.

NOTE: References are listed at the end of each section.

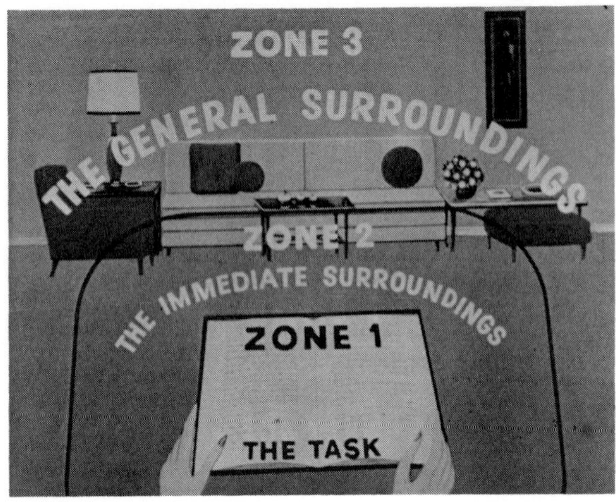

Zone	Luminance (footlamberts)
2—Area adjacent to the visual task	
a) Desirable ratio	⅓ to equal to task†
b) Minimum acceptable ratio*	⅕ to equal to task†
3—General surrounding	
a) Desirable ratio	⅕ to 5 times task†
b) Minimum acceptable ratio	⅒ to 10 times task†

* Not to apply to reading in a chair, see page 15-15.
† Typical task luminance range is 12 to 35 footlamberts (seldom exceeds 60 footlamberts).

Fig. 15-1. Seeing zones and luminance ratios for visual tasks.

Quality

Seeing Zones. In general, an individual's visual field is considered to consist of three major zones (see Fig. 15-1). First is the task itself (zone 1 of Fig. 15-1), second is the area immediately surrounding the task (zone 2), and third is the general surroundings (zone 3). For visual comfort, there should be a harmonious relationship among these seeing zones.

Luminance (Photometric Brightness) Ratios. In the immediate surround (zone 2 of Fig. 15-1), luminances that are noticeably greater than the task luminance or which are less than ⅓ the task luminance contribute to visual discomfort and are apt to be distracting. The luminance of substantial areas in the general surround should not exceed ten times the task luminance or be much less than ⅒ as bright. These luminance relationships become more important as illumination levels increase and as the duration of tasks is lengthened.

Luminance Balance. There are a number of ways in which luminances in the visual field may be brought into balance. In general, they involve the following:

1. Reduce unpleasantly high luminances of windows and luminaires by careful shielding and low-transmission shading material.
2. Help to bring large areas of room surfaces and furnishings into comfortable balance by the use of favorable reflectances in finishes. See below.
3. Provide additional lighting in the visual surround from other light sources.

Room Surface Reflectance. Every colored room surface reflects some portion of the light it receives, and this portion, expressed in percentage of the total light, is its reflectance.

To attain recommended luminance ratios of the visual task to the immediate surround and the more remote surround as listed in Fig. 15-1, use of matte finish surfaces with reflectances shown in Fig. 15-2, are recommended. In addition to listing reflectances, Fig. 15-2 also shows approximate corresponding Munsell values. For a discussion of Munsell values and of reflectances and the relationship of colored surfaces and light, see Section 5, pages 5-7 and 5-16.

Surface Colors and Color Schemes. Colors of objects often appear to change with surface finish. Specular or mirror reflection from glossy surfaces may, in extreme cases, increase the chroma and darkness of the sample at one angle and wipe out all color at other angles, as well as cause distracting glare. A matte finish reflects light diffusely, and thus gives an object the appearance of its "natural" color. Deeply textured finishes, such as velvet and deep-pile carpeting, cause shadows that make materials appear darker than smooth-surfaced materials, such as satin, silk, or plastic laminates, of the same inherent color.

The quality of light reaching a surface is the average of the light that reaches it from all points in the environment. If daylight from a bright blue sky is reflected from the green leaves of a tree into a room with a red rug and pink walls, the color of the light falling on an object in the room is not that of the daylight, but is the result of the daylight as modified by these several reflections. As light is reflected or bounced around a room, inter-reflections from large areas of a single color can cause the "amplification" of chroma or saturation of the wall color and cause it to appear more vivid than the original small sample from which it was selected.

There are no "rules" for color harmony. Certain generalities have been arrived at, however, through experience. These include the principle of order, in which there is a recognizable pattern in the color relationship; the principle of similarity, in which the same hue family or same value group is used; and the principle of familiarity, such as national identification (red, white, and blue), seasonal identification (red and green), or personal identification (pink and blue). These principles merely indicate the general direction in which esthetic judgement may be applied with some safety. More sophisticated color schemes tend toward subtle relationships

Fig. 15-2. Reflectances and Approximate Munsell Values for Interior Surfaces of Residences

Surface	Reflectance (per cent)	Approximate Munsell Values
Ceiling	60 to 90	8 and above
Curtain and drapery treatment on large wall areas	35 to 60	6.5 to 8
Walls	35 to 60*	6.5 to 8
Floors	15 to 35*	4.0 to 6.5

* In areas where lighting for specific visual tasks takes precedence over lighting for environment, minimum reflectances should be 40% for walls, 25% for floors.

of color and cannot be analyzed to conform to a given axiomatic pattern.

A well designed lighting system cannot be planned independently of the color and character of the surfaces it is to illuminate. The effect attained depends upon the way surfaces within an area reflect and absorb light. Thus ceilings, walls, floors, and major furnishings become an integral part of the lighting design.

Task Surfaces. Visual comfort and ease of seeing are also influenced by a task which has glossy or specular surfaces. Some areas of a glossy task may contain veiling reflections from light sources or luminaires. Other areas may contain reflected glare. See pages 3–27 and 3–28. Veiling reflections and reflected glare cannot always be completely eliminated. Reduction of the light source or luminaire luminances or more careful placement of the source (or the task) tends to alleviate the condition. Reflected images are further "softened" if part of the light on the task comes "indirectly" from walls, ceilings, or other luminaires.

Distribution of Light. General lighting can run the gamut from the perfectly uniform illumination of a luminous ceiling to the distinctly non-uniform effect of a grouping of open-top portable lamps. As a rule, the function of general lighting is to provide brightness-balancing illumination and to "tie the room together." (However, in the case of some kitchen lighting, the general illumination may be providing the task lighting, or vice versa.) Many luminaires and virtually all structural lighting elements are primarily for general or environmental lighting.

Livable Environment. Fundamental also in the design plan is the over-all impression created by the environment. It must "feel" in harmony with those basic esthetic and emotional sentiments associated with the home. Brightness patterns should avoid monotonous "even" light and permit interesting interplay of light and shadow. For areas used for activities which do not involve close vision, illumination levels are less critical and substantially higher luminance ratios may be employed. Scintillating, exciting effects, or more subdued, intimate ones, may be geared to the desired mood. There are many cases where the usefulness of the lighting system can be enlarged through the use of switches or dimmers to change the lighting effects to fit the use of the space. Occasionally it is even desirable to have more than one "lighting system" in a given space so that distinct changes in atmosphere and mood can be created. Fig. 15–3 illustrates a few of the many design possibilities.

Quantity

Recommended illumination levels for all specific visual tasks and general lighting levels within designated areas are listed in Fig. 9–80, page 9–88. These should be regarded as minimums.

All illumination and luminance measurements should be made in accordance with the methods given in Section 4, Measurement of Light.

LIGHTING EQUIPMENT FOR RESIDENTIAL APPLICATION

There are literally hundreds of manufacturers making lighting equipment for the residential field—ranging from novelty plug-in lamps to completely engineered package luminous ceilings. Key categories of luminaire types are listed below by their lighting distribution characteristics. Also, for each type, some indication is given of their good features and of suitable applications.

Recessed Lighting

Recessed luminaires are built into the ceiling and, therefore, distribute all of their light downward. They have what is known as a *direct* distribution. However by the use of various lens, reflector, and louver configurations, and by employing different types of lamp bulbs, recessed luminaires can be made to distribute their light in a wide variety of patterns (see Fig. 15–3 upper left). Candlepower distribution curves offer a good means of evaluation. See Fig. 15–4.

Asymmetric Distribution. While recessed luminaires commonly used in home lighting have symmetrical distributions, there is an increasing demand for units that are adjustable and which will direct light to walls and other vertical surfaces. See Fig. 15–4. A popular type of asymmetric luminaire is the "wall washer." This is designed to direct a broad spread of light at a nearby vertical surface to create nearly uniform brightness patterns on that surface. While a distribution curve of such a unit could enable the engineer to calculate footcandles on the wall by the point-by-point method, manufacturers usually provide footcandle charts to simplify this design procedure.

Applications. Fig. 15–5 shows features, applica-

Fig. 15-3. The use of a variety of lighting systems in one room provides flexibility for the desired mood or for the visual tasks to be performed. Recessed luminaires with a symmetrical distribution dramatically accent the chairs and coffee table in the view at the upper left. In the upper right, light for reading is provided by portable lamps at the couch and one chair, and by a pendant direct-indirect luminaire at the other chair. Asymmetric recessed luminaires accentuate the texture of the brick fireplace wall while a lighted cornice highlights the vertical louvered window treatment in the view at the lower left. Together, as in the view at the lower right, all lighting systems provide an interesting, livable environment.

tions, and a sketch of recessed luminaires with wide, medium, narrow, and asymmetric profiles (candlepower distributions).

Ceiling-Mounted Luminaires

General Diffusing. General diffusing ceiling-mounted luminaires (see Fig. 15-6a and 15-6b) direct their light in a very wide pattern and are normally used for general illumination because they light walls and ceilings. Their luminance should be carefully considered in relation to room reflectances. Light sources, either incandescent or fluorescent, should not be visible either through or over the luminaire. Although fluorescent lamps are lower in luminance than incandescent, they should always be shielded from direct view. The close-to-ceiling position of luminaires of this type allows them to have higher luminance than similar ones hung lower into the line of sight. For typical residential sizes the average luminance of the total luminous area should not exceed 500 footlamberts, except in utility areas. For equipment used in strictly utility spaces luminances as high as 800 footlamberts are acceptable. Within the diffusing element, luminance

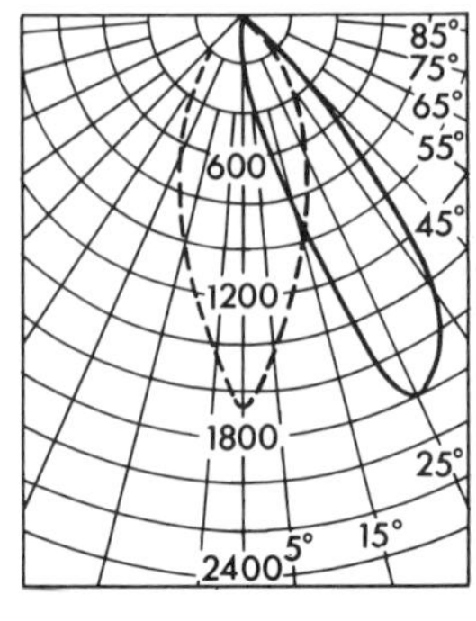

Fig. 15-4. A typical distribution curve for a recessed adjustable luminaire using a 150-watt, R-40 spot lamp.

Fig. 15-5. Recessed Lighting Equipment

Section Through Typical Units	Applications
Wide Profile	
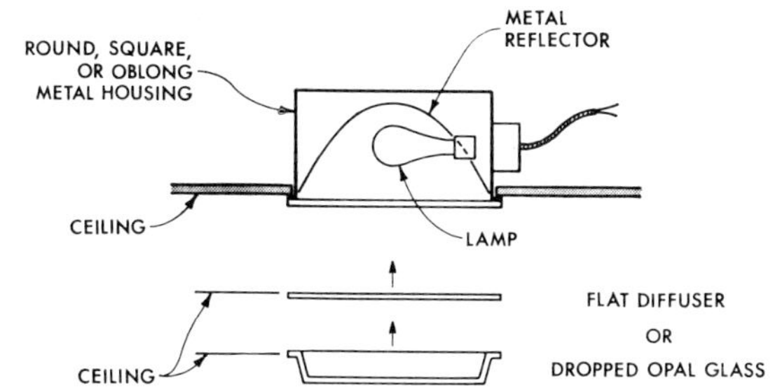 ROUND, SQUARE, OR OBLONG METAL HOUSING METAL REFLECTOR CEILING LAMP FLAT DIFFUSER OR CEILING DROPPED OPAL GLASS	For general illumination (almost always used in multiple) in recreation rooms, kitchens, laundries, halls (service). Used singly or in small groups for small areas, such as: walk-in closets, garages, entry doors, overhangs in porches. Because of high luminance of diffuser, seldom used in living or social areas.
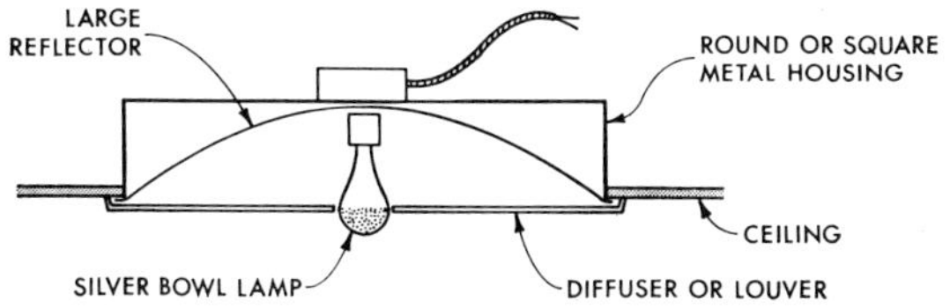 LARGE REFLECTOR ROUND OR SQUARE METAL HOUSING CEILING SILVER BOWL LAMP DIFFUSER OR LOUVER	Used in multiple for general lighting in kitchens, baths, laundries, recreation rooms, and family rooms. Used singly or in groups over game tables. Lower luminance allows this type to be used in living areas if styling permits.
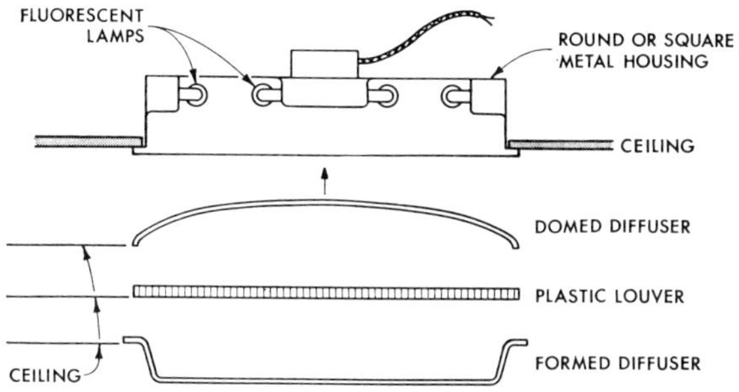 FLUORESCENT LAMPS ROUND OR SQUARE METAL HOUSING CEILING DOMED DIFFUSER PLASTIC LOUVER CEILING FORMED DIFFUSER	Very wide distribution excellent as general lighting for kitchens, laundries, recreation rooms, baths. Large size and high lumen output mean fewer luminaires required. Often used singly or in pairs for entry halls and foyers and for skylight effect in interior halls and stairways.
Medium Profile	
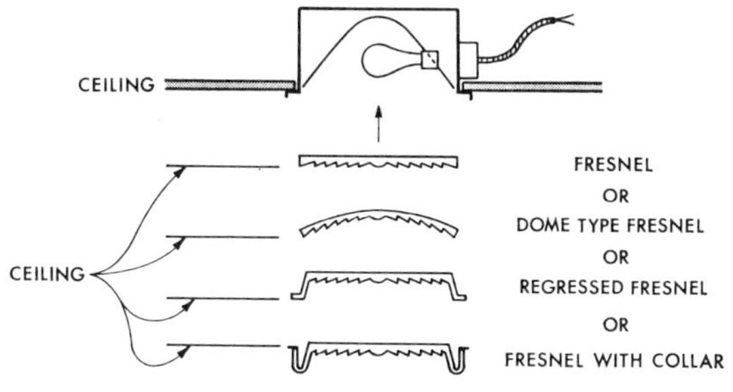 CEILING FRESNEL OR DOME TYPE FRESNEL OR CEILING REGRESSED FRESNEL OR FRESNEL WITH COLLAR	Used for specific task lighting where task area is large, such as: kitchen sink, kitchen island counter or range, laundry tubs and ironing, game tables, workbenches, hobby areas. Used for general lighting in restricted areas, such as: halls, entries, baths. Multiple groupings are satisfactory for the general lighting of kitchens and recreation rooms. If weather-proof, outdoor uses, including overhangs, porches, and entries.
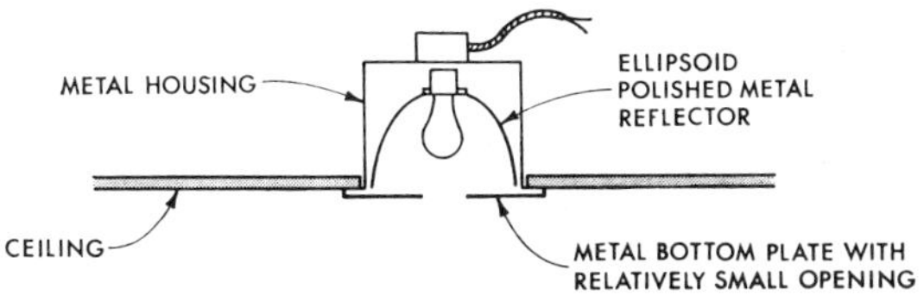 METAL HOUSING ELLIPSOID POLISHED METAL REFLECTOR CEILING METAL BOTTOM PLATE WITH RELATIVELY SMALL OPENING	Uses are basically the same as the Fresnel unit listed above, except that the lower luminance makes this type of unit more usable in living and dining areas.
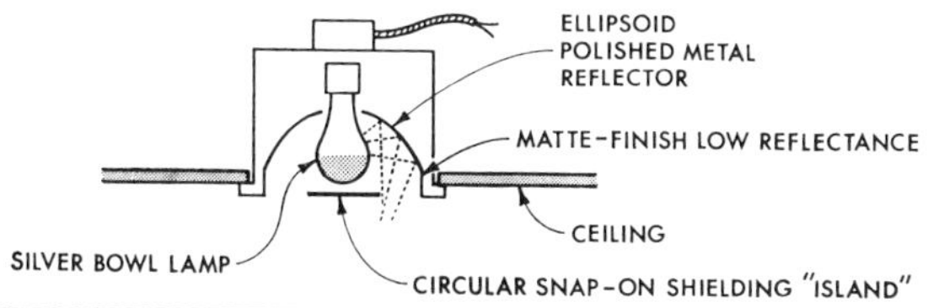 ELLIPSOID POLISHED METAL REFLECTOR MATTE-FINISH LOW REFLECTANCE CEILING SILVER BOWL LAMP CIRCULAR SNAP-ON SHIELDING "ISLAND"	Same uses as open ellipsoidal units listed above but has a far better control of lamp luminance.

Fig. 15-5. *Continued*

Section Through Typical Units	Applications
Narrow Profile	
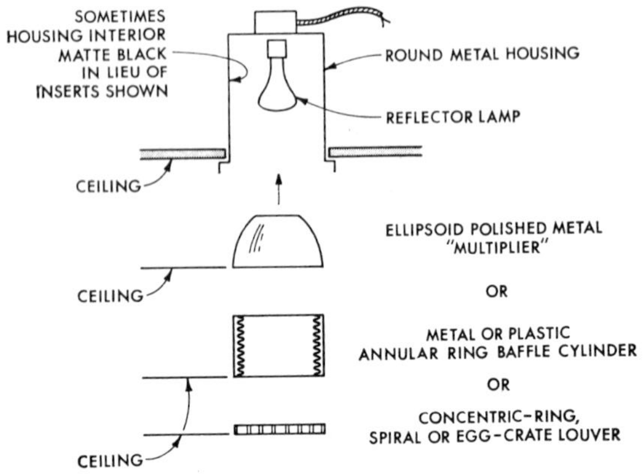 SOMETIMES HOUSING INTERIOR MATTE BLACK IN LIEU OF INSERTS SHOWN; ROUND METAL HOUSING; REFLECTOR LAMP; CEILING; ELLIPSOID POLISHED METAL "MULTIPLIER"; CEILING; OR; METAL OR PLASTIC ANNULAR RING BAFFLE CYLINDER; OR; CONCENTRIC-RING, SPIRAL OR EGG-CRATE LOUVER; CEILING	Accent lighting over plants, cocktail tables, etc. Wall lighting—mounted close to textured surfaces, such as: brick, stone, rough wood, fabrics. Task lighting—food preparation areas (avoid specular reflections). In multiple on quite close spacing for general lighting. Most effective when used near perimeter of room so that some light spills onto wall. Dramatic effects for family rooms and formal living areas. Supplementary stair lighting—shadow patterns define treads and risers. Dining tables—provide actual light on dining table to supplement decorative effect from hanging luminaire.
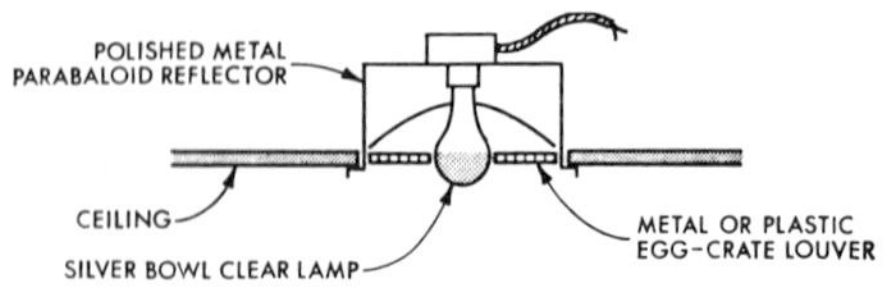 POLISHED METAL PARABALOID REFLECTOR; CEILING; SILVER BOWL CLEAR LAMP; METAL OR PLASTIC EGG-CRATE LOUVER	Same uses as type luminaire above. Low luminance of luminaires very desirable, but larger size sometimes prohibits use in highly styled interiors.
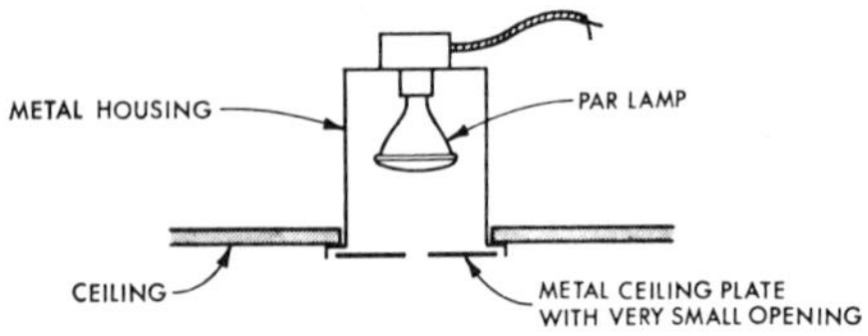 METAL HOUSING; PAR LAMP; CEILING; METAL CEILING PLATE WITH VERY SMALL OPENING	Unit does not effectively utilize lumens from lamp and much energy is trapped in can with possible overheating. A 2-inch bottom opening recommended as minimum. Can be used for low level accent lighting where luminaire cost is vital. Suitable for use over coffee and cocktail tables and for "starlight" effect in recreation rooms.
Special Asymmetric Profile	
SPOTLIGHT LAMP; ENTIRE LAMP, SHUTTER AND LENS ASSEMBLY ADJUSTABLE IN POSITION; FRAMING SHUTTERS; CEILING; METAL CEILING PLATE WITH VERY SMALL APERTURE; ADJUSTABLE CONDENSER LENS ASSEMBLY	Adjustable beam can be framed precisely to outline paintings, pictures, maps, and niches. When aimed directly down, shutters can frame dining, cocktail, or coffee tables, or other horizontal elements. Special high lumen output lamps with relatively short life are often used in these units. Easy access required for frequent relamping. May come equipped with top access openings for relamping from above.
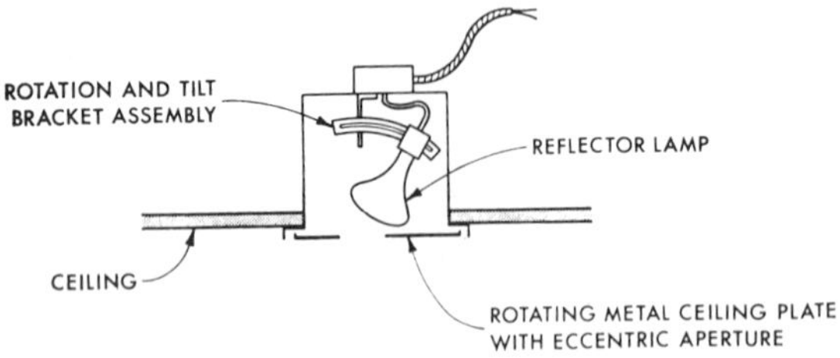 ROTATION AND TILT BRACKET ASSEMBLY; REFLECTOR LAMP; CEILING; ROTATING METAL CEILING PLATE WITH ECCENTRIC APERTURE	Useful for gallery or picture lighting and to light sculpture. If scalloping effect is acceptable it can be used for wall lighting and to accent fireplace surfaces. Large size of bottom aperture sometimes makes this unacceptable for stylized interiors. Also used for lighting piano music and sewing machines.
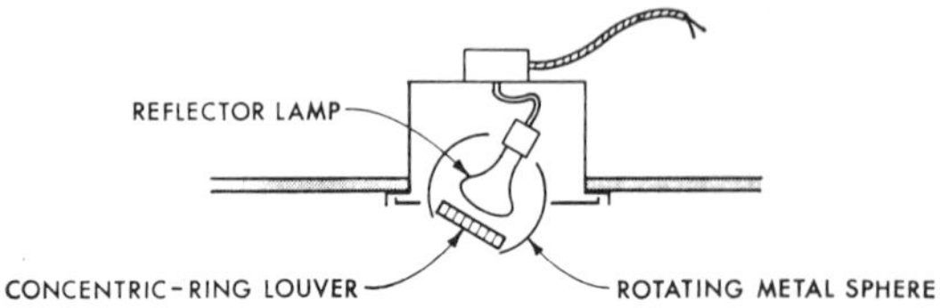 REFLECTOR LAMP; CONCENTRIC-RING LOUVER; ROTATING METAL SPHERE	Same uses as recessed adjustable luminaire as listed above.

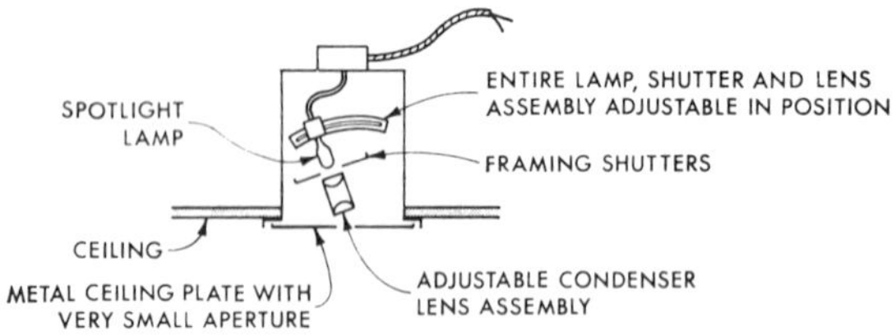

Fig. 15–5. *Continued*

Section Through Typical Units	Applications
Special Asymmetric Profile *Continued*	
REFLECTOR LAMP; METAL INTERIOR REFLECTOR; CEILING; METAL CEILING PLATE WITH DROPPED SCOOP REFLECTOR; WALL	For uniform illumination of plane wall surfaces. Extremely effective for lighting murals and for minimizing wall imperfections. *Not* generally to be used for lighting textured wall surfaces because it directs no grazing light at wall. Spacing of units is critical—follow recommendations closely.

of the brightest square inch should not materially exceed twice the average. Luminance ratios between the luminaire and the ceiling should not be more than 20 to 1. Even with glass or plastic of excellent diffusing qualities, spottiness will occur if the lamps are widely spaced or too close to the diffuser.

Diffuse Downlighting. There are many variations of the basic luminaire type, but they are usually characterized by an opaque or partly opaque side or border. See Fig. 15–6c. This restricts the distribution somewhat and limits the effectiveness as a general lighting device. Such luminaires seldom put light on the ceiling, and in order to light walls effectively they must be located rather close to them. If they are used for general room lighting, they should be limited to no more than 100 square feet per luminaire and the room finishes should be very light in color.

Downlighting. Downlighting or directional luminaires (see Fig. 15–6d) are designed to be used for accent lighting or supplementary illumination for critical visual tasks such as sewing. If used in

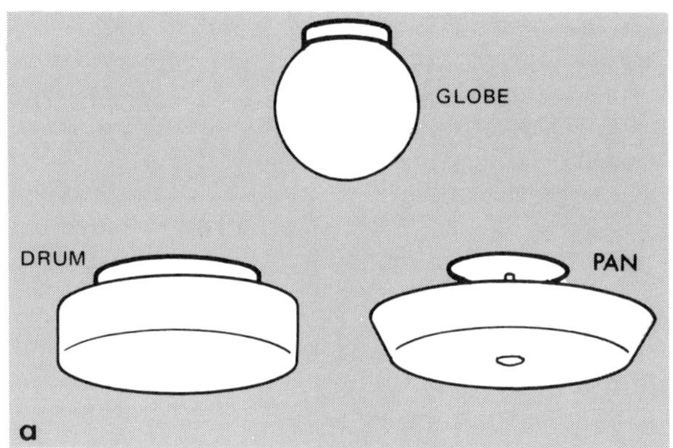

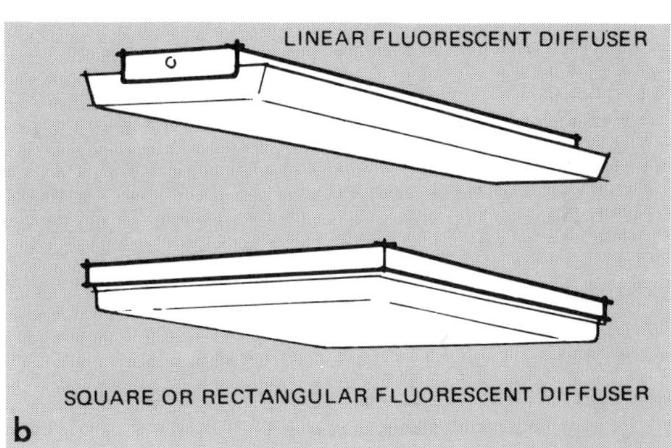

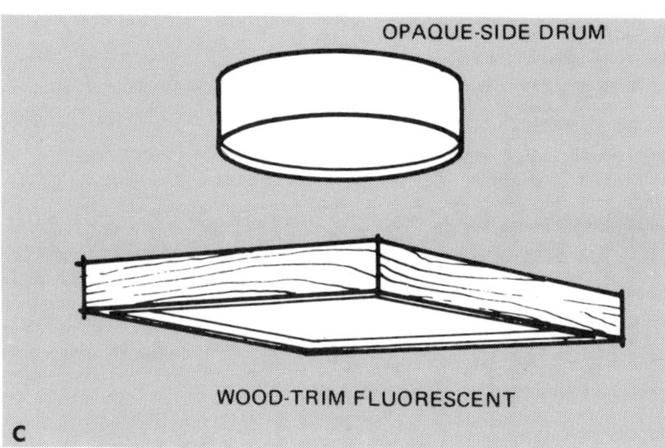

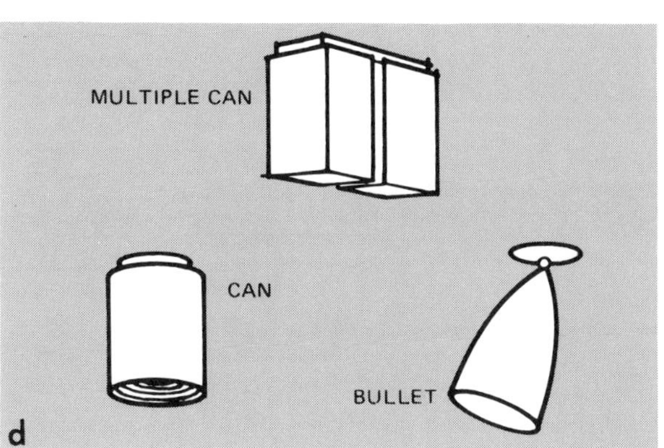

Fig. 15–6. Ceiling mounted luminaires: group a, small general diffusing type; group b, large general diffusing type; group c, diffuse downlighting type; group d, downlighting type.

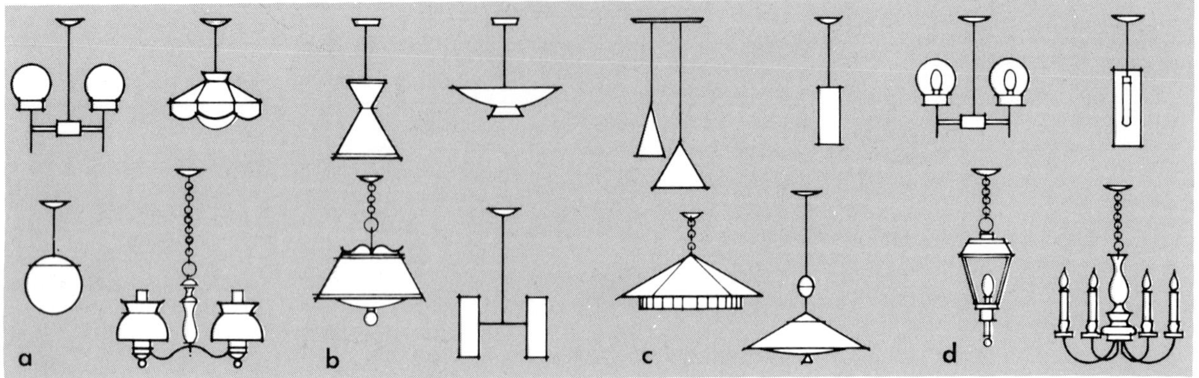

Fig. 15–7. Pendant luminaires: group a, general diffusing; group b, direct-indirect; group c, direct downlighting; and group d, exposed lamp.

multiple for general illumination of larger areas they must be spaced quite close together because of their narrow distribution. At typical room heights one luminaire cannot effectively light an area of more than 25 square feet. Sharp shadows are inherent in the lighting result.

Pendant Luminaires

General Diffusing. Pendant luminaires with general diffusing characteristics are used for both functional and decorative purposes. See Fig. 15–7a. For the latter they are lamped with low-wattage or tinted light bulbs, or provided with dimmer control to keep the luminance low, and hung as points of decorative interest or as "lighted ornaments." When intended for functional lighting and located well above the normal line of sight, they are subject to the same luminance limitations as general diffusing ceiling-mounted luminaires. If hung low enough to be in the line of sight they should be restricted to substantially lower luminances. Fig. 15–8 gives lamping recommendations for this type of equipment. Even relatively low luminances may be annoyingly bright if viewed against dark walls or dark night-time windows. When luminaires of this kind are used in dining areas the luminance should be rigidly controlled, and additional lighting with a directional component should be installed to illuminate the table top.

Fig. 15–8. Suggested Lamp Wattages for Visual Comfort for Spherical Uniformly Luminous General Diffusing Luminaires*

Sphere Diameter (inches)	Mounting Height of Luminaire		
	84 Inches or Above		40 to 72 Inches
	Utility Spaces	Living Areas	Living Areas
6 or less	60W	40W	25W
8 to 10	75W	60W	40W
12 to 16	150W	100W	60W
18 to 22	200W	150W	75W
24 to 28	300W	200W	100W

* Assuming typical diffusing material of 50 per cent transmittance and ceilings and walls of recommended reflectances. See Fig. 15–2.

Direct-Indirect. Direct-indirect luminaires (see Fig. 15–7b) provide generous upward light which contributes to room lighting, as well as functional downlighting. They are characterized by opaque or slightly luminous shades or shields, which often make them suitable for low hanging where they will be in the line of sight (see Fig. 15–3 upper right), and they are commonly used to light dining tables, game tables, snack bars, etc. When they are hung quite low over coffee tables or planters the downward component can provide accent highlights and in this case top shielding should be considered. Such equipment is often pulley-mounted, and can have added flexibility when suspended from a track. Glass or plastic diffusers should not be visible below the shading element. Bottom shielding or louvering is important when luminaires are mounted in a higher position and viewed from below.

Downlighting. The main function of pendant downlights (see Fig. 15–7c) is to provide accent illumination. They are generally hung at or below eye level. When mounted above eye level, as in stairways, halls, entryways, etc., they require some form of bottom shielding. The lighting distribution is characterized by sharp beam definition and shadow patterns. In general, the larger the luminaire the less sharply defined are the shadows. Almost without exception luminaires of this type need additional general room lighting for visual comfort.

Exposed Lamp. The main function of exposed-lamp luminaires (see Fig. 15–7d) is to provide decorative highlights, sparkle and accent. They are not suitable for general room illumination or for the lighting of specific tasks. Since their use is primarily decorative, lamps should be kept low in wattage or be used on dimmer control. Other room lighting is necessary to insure visual comfort. Because of the high luminances of all incandescent filament lamps, even low-wattage ones and particularly those with

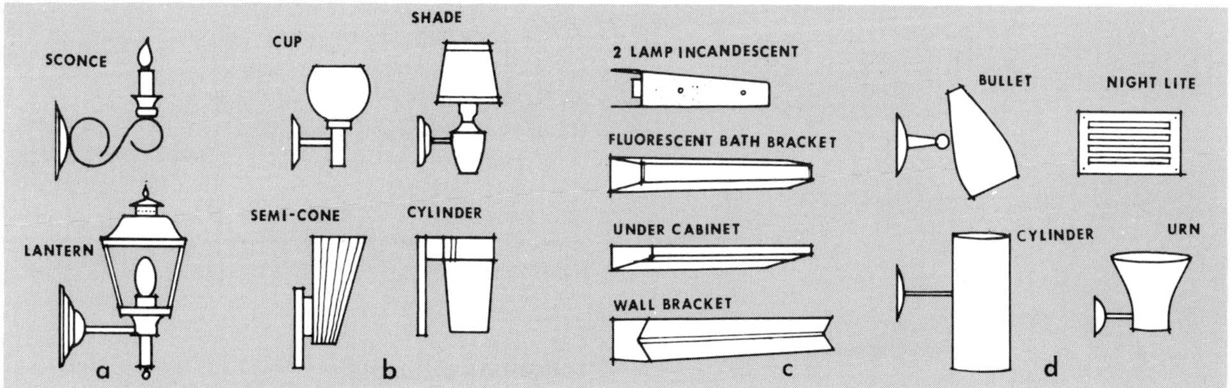

Fig. 15-9. Wall luminaires: group a, exposed lamp; group b, small diffuser; group c, linear diffuser; group d, directional (adjustable and fixed).

clear bulbs, mounting above the normal line of sight is recommended. When luminaires must be suspended low, lamps should be of very low luminance and walls against which they are viewed should be light in color.

Wall-Mounted Luminaires

Basically, wall-mounted luminaires (see Fig. 15-9) are junior versions of similar luminaires mounted on the ceiling. They have the same distribution characteristics and qualities described in previous sections. However, their low mounting places them much more in the field of view, and therefore the designer must exercise more strict control over luminance. Below is a brief summary of some of the more popular types of wall luminaires and their uses.

Exposed Lamp. Quite often used for decorative purposes—especially with clear light bulbs. See Fig. 15-9a. Other room lighting is almost invariably necessary. One type of exposed-lamp unit used for functional purposes is a mirror strip utilizing several low-wattage light bulbs with diffusing white coating. The comments on the luminance of pendant exposed-lamp luminaires apply also to wall-mounted equipment.

Small Diffuser. Quite often used in pairs, in groups, or in lines for lighting hallways, stairways, doorways, and mirrors. See Fig. 15-9b. In areas like living rooms and family rooms where the luminaires will be in view for long periods of time, the luminance should be very low. Where they will be seen only briefly in passing, as in hallways or doorways or providing functional lighting at a mirror, higher luminances can be tolerated, but other lighting in the room is usually necessary.

Linear. Linear wall brackets are commonly placed over mirrors, under cabinets, and in other locations where functional lighting is required for close-up tasks. See Fig. 15-9c. They are occasionally used for general lighting in hallways, passages, etc. Good diffusing quality of glass or plastic, or satisfactory opaque shielding is important because of the high light output of the lamps employed. Higher luminances can be tolerated when the luminaire is seen briefly, or when it is used for a specific task, as at a bathroom mirror. Well-designed fluorescent brackets, nearly always with opaque shielding, may be used in living areas to light walls, provide general room lighting, or even to supply functional light at desks or seating arrangements placed against the wall. They may be extended to cover the entire length of a wall. In these applications they perform exactly like structural wall brackets.

Directional (Adjustable and Fixed). Directional luminaires having strong downward components are often mounted on the wall for accent and display lighting, over planters, room dividers, and sculpture. See Fig. 15-9d. Directional luminaires providing upward lighting can be used in groups for general illumination. Both types are most effective when shielding is opaque and the light source cannot be seen from any point in the room.

Portable Lamps

Although all portable lamps are considered by homemakers and designers alike as decorative elements in the furnishing scheme, there are many cases where they must be functional as well as good-looking. See Fig. 15-3, upper right. In many interior-living spaces, portable lamps are relied upon to make a major contribution to the general illumination as well as provide illumination for the most difficult visual tasks. Such lamps must be chosen not only for their decorative suitability, but also for the amount and quality of the light they deliver on the task area and their emission of upward light to add to the general illumination. Moreover, a well-designed lamp can perform effectively only if it is placed correctly in relation to the task and to the eyes of the user. The following paragraphs list some check points that will serve as criteria in the selection of lamps and their proper placement.

Under-the-Shade Components. If the lighting in a room is well balanced and the room has favorable

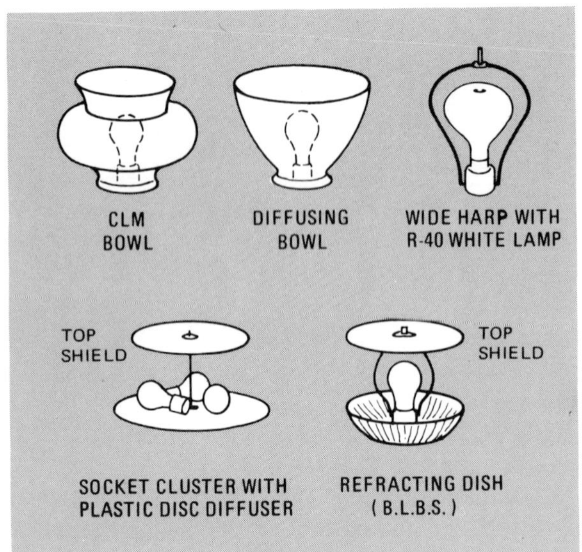

Fig. 15–10. Typical under the shade components for portable lamps.

reflectances it is not essential that each portable lamp have a light-control device under the shade; the ordinary inside-frosted or white light bulb will provide a reasonable amount of diffusion. However, the addition of diffusing, reflecting, refracting, or shielding elements results in improved quality that is highly desirable, particularly if the visual task is to be difficult or prolonged. Lamps for the study desk *must* make provision for excellent lighting quality.[3] Fig. 15–10 illustrates some of the constructions to be found in portable lamps on the market today, taking into account the sizes and types of light bulbs currently available. As new, more powerful, light sources are introduced in portable lamps, the need for control and quality-improving elements will increase.

Provision for Varying Light Output. Some means of varying the light output of portable lamps is a desirable design feature, because they are often used for purposes other than specific task illumination. For television viewing, entertaining, or conversation a lower lighting level is often preferred. The utmost in control is obtained when a lamp has a built-in full-range dimmer, or when it is used with a separate table-top dimmer. However, high-low and three-way switches for three-way light bulbs or clusters of low-wattage bulbs are suitable means of achieving flexibility.

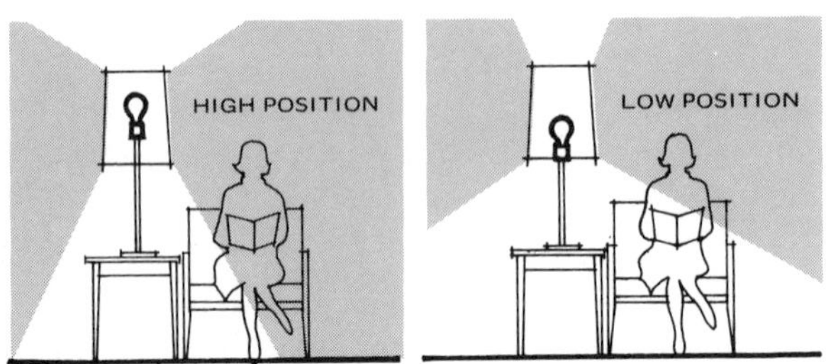

Fig. 15–11. Influences of light source position on light distribution of portable lamps.

Fig. 15–12. Influence of the shade dimensions on light distribution of portable lamps.

Position of Light Source. If the light bulb is close to the bottom edge of the shade, more light will reach the reading plane. See Fig. 15–11. However, the present practice is to center the bulb so that the luminance of the shade is more nearly uniform from top to bottom. A better solution would be to mount one bulb high in the shade to provide upward light, while placing bulbs responsible for the task-plane illumination near the lower edge.

Characteristics of Shade. Deep, narrow shades should be avoided for lamps intended to provide useful illumination; they are double-ended bottlenecks, restricting the spread of both the downward and upward light. Fig. 15–12 illustrates the difference in performance between a 14-inch and a 16-inch shade. The inner surface of the shade should have a very high reflectance (otherwise much light is lost by absorption), and the shade material should have good diffusing quality and suitable transmittance.

A pleasing effect is achieved when the shades blend smoothly rather than contrast sharply with their backgrounds, in both color and luminance. White or light shades usually should be used against light walls, and darker shades of lower luminance against darker walls. Translucent shades provide horizontal or cross-lighting, and if all shades have approximately the same luminance the effect is more harmonious. Opaque shades create pools of light above and below, and little light on vertical surfaces. Unless there are other light sources in the room, such as lighted valances or brackets, the effect can be visually uncomfortable. On the other hand, shades of too high transmittance and too little diffusion, such as some made of white fibered glass or laminated plastic, show "hot spots" that are unattractive and distracting. The degree of translucence is vital to visual comfort, and especially important in a room lighted exclusively by portable lamps.

A light meter, preferably one that reads to 500 footcandles, will help to determine suitable shade luminances. The light-sensitive cell is placed in direct contact with the shade, and moved from top to bottom to get an average. See Fig. 15–13. If the shade is to be seen against walls having 40 to 60 per cent reflectance, the meter should read 150 or less. With wood-paneled and other walls having reflectances substantially lower than 40 per cent, visual comfort demands much denser shades, and the meter

Fig. 15–13. Translucency of shade materials is measured easily by slipping a piece of material into a square opening in the side of a standard 16-inch shade. The meter cell is held against the material as shown, and the reading is checked against published luminance data. This simple procedure can be followed before materials are purchased in quantity.

should read no more than 50, preferably less. (Meter reading equals luminance of shade in footlamberts.)

Typical Lamp Placement. The sketches in Fig. 15–14 indicate the correct positioning for lamps that are to be used at eye level and also for those to be used just above eye level. There must be a very definite correlation between the height of the lamp to the bottom of the shade and the height of the table on which it is placed.

Dresser and Dressing Table Lamps. Lamps used on dresser and dressing tables are notable exceptions to the general rules for portable lamps. Here, instead of light being directed from the bottom of the shade to a seeing-task, the light useful for make-up must be transmitted through the shade onto the face, and the outside of the shade becomes the light source. White fibered glass, cellulose acetate, and high-transmittance silks or plastics are suitable materials. For a shade with a lower diameter of 9 to 11 inches a footcandle meter in contact with the surface should show an average reading of 250 to 400.

Centers of the shades should be approximately at cheek level. This usually means that dresser lamps must measure about 22 inches from the bottom of the base to the center of the shade, and dressing table lamps 15 inches to the center of the shade. In general, dressing table lamps cannot be placed closer than 36 inches apart or they will be bothersome visually and will also tend to obscure the mirror.

Lamps for General Room Lighting. Many portable lamps are designed not to provide local illumination but to contribute to general room lighting (Fig. 15–15). Some are of the completely indirect type, sending all their light to the ceiling to be redirected over the room. Others are luminous forms, with or without an indirect component; their soft glow can be a most effective element in the lighting pattern of a room. Light sources should never be visible from any seated or standing position within the room, and large luminous surfaces at eye level should not give a reading in excess of 50 on a footcandle meter placed against them (a luminance of 50 footlamberts). The lighting effectiveness of some of the indirect types can be increased by equipping them with reflectorized light bulbs.

Lamps for Directional Lighting. Portable lamps providing directional light are made in a wide variety of forms, a few of which are illustrated in Fig. 15–16. Some are used only to supply decorative accent light on a restricted area. Others have excellent

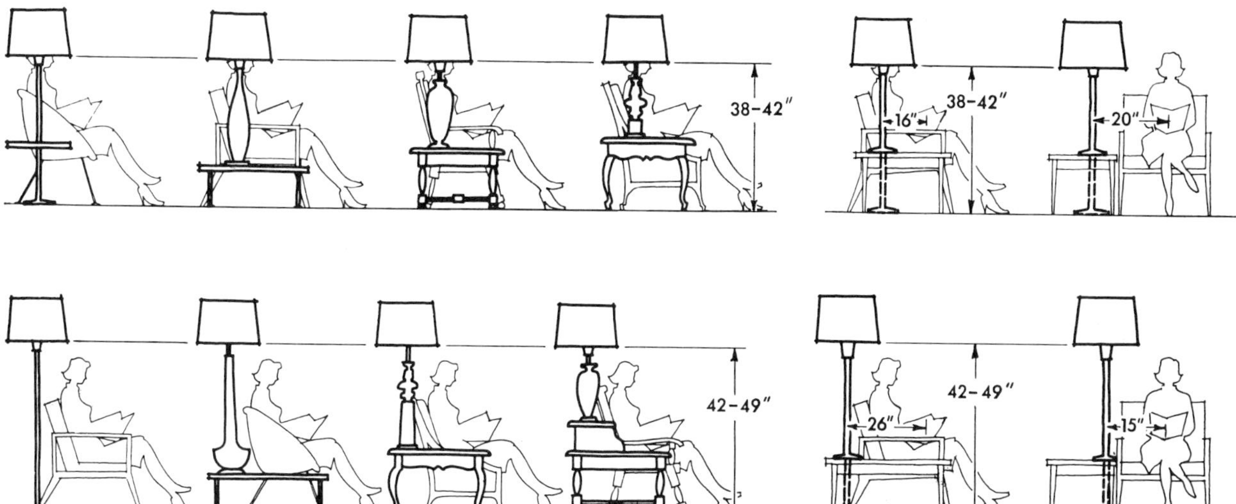

Fig. 15–14. Average seated eye level is 38 to 42 inches above the floor. Lower edge of floor or table lamp shades should be at eye level when lamp is beside user. This is the correct placement for most table lamps, and for floor lamps serving furniture placed against a wall. Floor lamps with built-in tables should have shades no higher than eye level. For user comfort—when floor-lamp height to lower edge of shade or lamp-base-plus-table height is above eye level (42 to 49 inches), placement should be close to right or left rear corner of chair. This placement is possible only when chairs or sofas are at least 10 to 12 inches from wall.

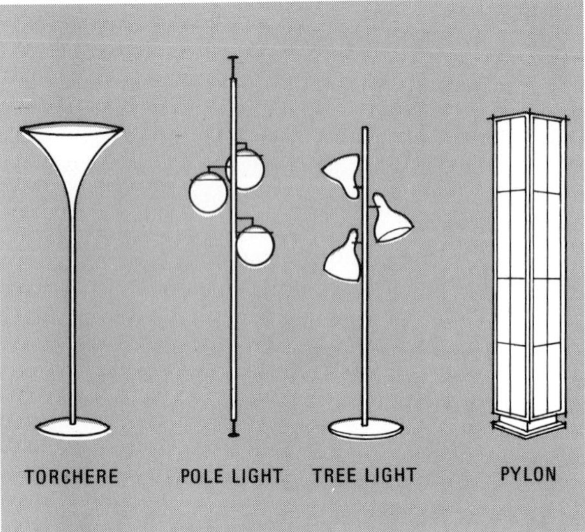

Fig. 15-15. Typical portable lamps for general room illumination.

application as sources of supplementary illumination for difficult visual tasks. For this purpose they should never be used without other lighting in the room; otherwise the strongly directional light will create sharp shadows, and luminance ratios between the task and the surrounding area will be excessively high.

Structural Lighting Elements

This lighting form is usually built in as part of the structure or uses structural elements (such as the joist space) as a part of the unit luminaire.

Structural lighting not only includes those widely used types shown in Fig. 15-17, but also includes: bookshelves with integral concealed lighting, lighted niches, lighted coffers, artificial skylights, extended soffits, lighted mantels, lighted handrails, lighted china cabinets, luminous ceilings, and many others.

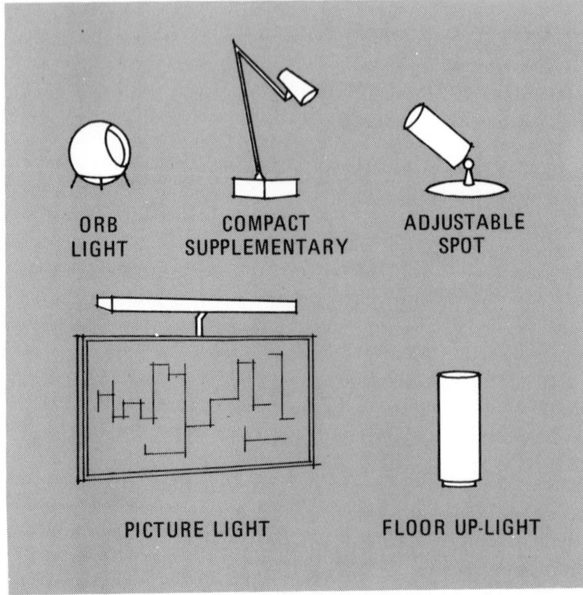

Fig. 15-16. Typical portable lamps producing directional lighting.

The principal advantage of structural lighting is its close correlation with the architecture of the room, offering a very unobtrusive method of providing light. See Fig. 15-3, lower right. Perhaps its principal disadvantage is the fact that the designer has such great freedom. He must make sure that the installation will be architecturally correct, will be simple to construct, and will provide optimum lighting quality. The same basic considerations of lamp concealment, diffuser luminance, direction of light, and light-source color that apply to all types of luminaires apply here also. Structural lighting is usually designed for linear sources, most commonly fluorescent. Dimming devices or alternate switching arrangements may be included to provide more than one lighting effect from the same luminaire. Many manufacturers of light source equipment have prepared excellent technical guides for simplifying the design of good structural lighting. Some manufacturers now offer completely assembled structural lighting "packages."

DESIGN CONSIDERATIONS FOR THE LIGHTING OF SPECIFIC VISUAL TASKS

General Considerations

Certain general design considerations apply to virtually all of the specific visual tasks covered in this section. In addition to these general considerations there are rather specific characteristics of each of the individual seeing tasks that call for special attention.

1. There are some children and adults, particularly the elderly, who have reduced vision. Among them are those frequently referred to as partially seeing, many of whom may need increased amounts of good quality illumination, not only for sewing, but for general reading, reading in bed, and household activities requiring critical seeing over extended periods. Beyond certain minimum recommendations, therefore, their needs should be considered on an individual basis.

2. If an individual faces a window during a performance of any critical seeing task during *daylight hours,* extreme care should be exercised in the control of:

a. Daylight glare by use of blinds, shades, and draperies.

b. Comfort within the visual field by the use of high reflectance wall colors adjacent to the fenestration.

3. If an individual performing a critical visual task faces a window during *nighttime hours,* the darkness of the window should be controlled with light-colored window covering.

4. Good brightness balance within the visual field should be a major design goal:

a. Wall surfaces in the visual field should be of

Fig. 15-17. Structural Lighting Elements

LIGHTED CORNICES

Cornices direct all their light downward to give dramatic interest to wall coverings, draperies, murals, etc. May also be used over windows where space above window does not permit valance lighting. Good for low-ceilinged rooms.

VALANCES

Valances are always used at windows, usually with draperies. They provide up-light which reflects off ceiling for general room lighting and down-light for drapery accent. When closer to ceiling than 10 inches use closed top to eliminate annoying ceiling brightness.

WALL BRACKETS (HIGH TYPE)

High wall brackets provide both up and down light for general room lighting. Used on interior walls to balance window valance both architecturally and in lighting distribution. Mounting height determined by window or door height.

WALL BRACKETS (LOW TYPE)

Low brackets are used for special wall emphasis or for lighting specific tasks such as sink, range, reading in bed, etc. Mounting height is determined by eye height of users, from both seated and standing positions. Length should relate to nearby furniture groupings and room scale.

LIGHTED SOFFITS

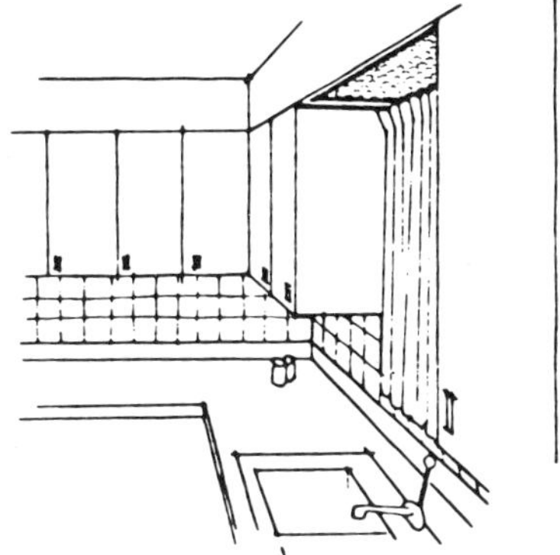

Soffits over work areas are designed to provide higher level of light directly below. Usually they are easily installed in furred-down area over sink in kitchen. Also are excellent for niches over sofas, pianos, built-in desks, etc.

LIGHTED SOFFITS

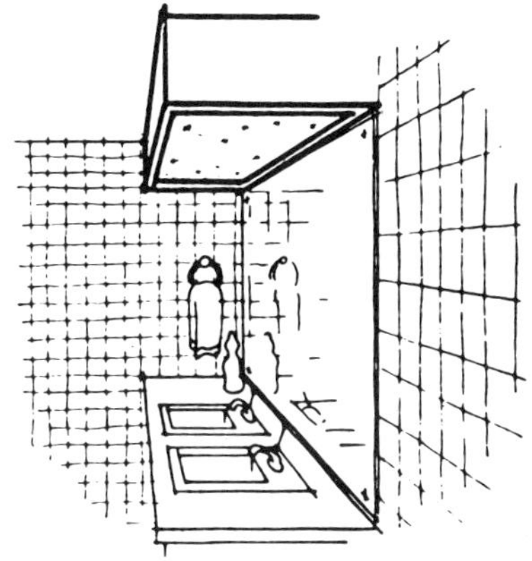

Bath or dressing room soffits are designed to light user's face. They are almost always used with large mirrors and counter-top lavatories. Length usually tied to size of mirror. Add luxury touch with attractively decorated bottom diffuser.

Fig. 15-17. *Continued*

LUMINOUS WALL PANELS

Luminous wall panels create pleasant vistas; are comfortable background for seeing tasks; add luxury touch in dining areas, family rooms and as room dividers. Wide variety of decorative materials available for diffusing covers.

LIGHTED CANOPIES

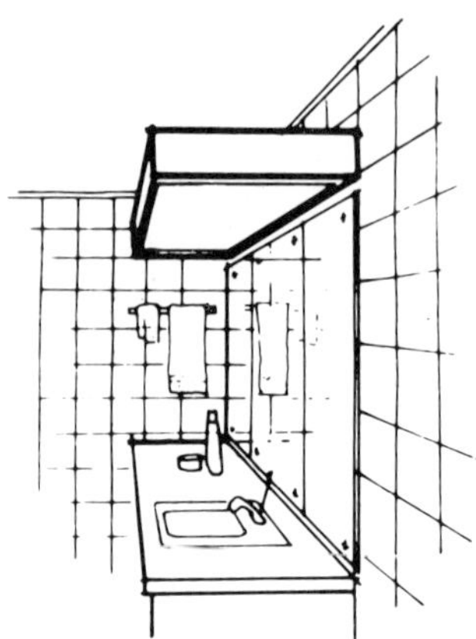

The canopy overhang is most applicable to bath or dressing room. It provides excellent general room illumination as well as light to the user's face.

COVES

Coves direct all light to the ceiling. Should be used only with white or near-white ceilings. Cove lighting is soft and uniform but lacks punch or emphasis. Best used to supplement other lighting. Suitable for high-ceilinged rooms—and for places where ceiling heights abruptly change.

LUMINOUS CEILINGS

Totally luminous ceilings provide skylight effect very suitable for interior rooms or utility spaces, such as kitchens, baths, laundries. With attractive diffuser patterns, more decorative supports, and color accents they become acceptable for many other living spaces such as family rooms, dens, etc. Dimming controls desirable.

"TYPICAL" DIMENSIONAL DRAWING

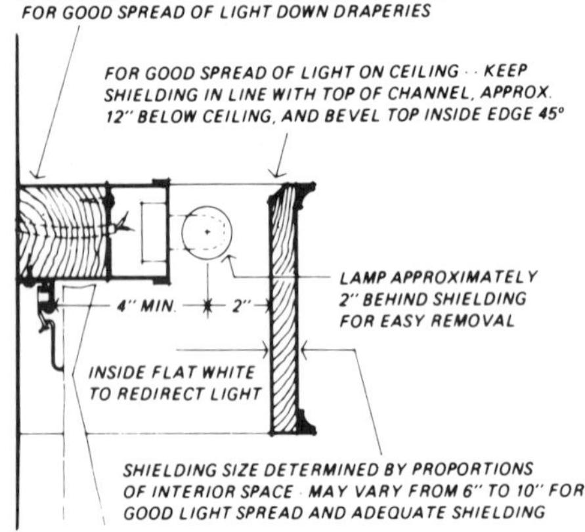
WOOD BLOCKING LOCATES LAMP OUT FROM WALL TO MINIMIZE UPPER WALL BRIGHTNESS, AND APPROXIMATELY 4" IN FRONT OF DRAPERY TRACK FOR GOOD SPREAD OF LIGHT DOWN DRAPERIES
FOR GOOD SPREAD OF LIGHT ON CEILING -- KEEP SHIELDING IN LINE WITH TOP OF CHANNEL, APPROX. 12" BELOW CEILING, AND BEVEL TOP INSIDE EDGE 45°
LAMP APPROXIMATELY 2" BEHIND SHIELDING FOR EASY REMOVAL
4" MIN.
2"
INSIDE FLAT WHITE TO REDIRECT LIGHT
SHIELDING SIZE DETERMINED BY PROPORTIONS OF INTERIOR SPACE - MAY VARY FROM 6" TO 10" FOR GOOD LIGHT SPREAD AND ADEQUATE SHIELDING

This "typical" dimensional drawing applies only to commonly-encountered window valance situations. Obviously, other window treatments could necessitate modifications in these critical dimensions; *i.e.*, vertical blinds, double track situations, curved bay windows, etc.

The same "job-tailored" variations can occur in the design of any type of structural lighting device. Therefore no other dimensional drawings have been included here.

medium to high reflectance (40 per cent or more) without a strong pattern.

b. Floor surfaces should be of as high reflectance as possible (above 25 per cent).

c. If the individual faces into the room while performing the visual task, additional room illumination should be provided to bring brightnesses into balance in the visual surround.

5. Equipment should be shielded either by design or position so that neither the light source nor any luminous luminaire part exceeding 200 footlamberts is seen by the user's eye in a normal position while performing the task.

6. The light source should be positioned so that the reflected image of the light source in glossy work materials is not seen by the user from a normal eye position.

Specific Visual Tasks

The recommendations which follow in Figs. 15-18 through 15-36 are organized for each specific task on the basis of (1) statement of the task, (2) location of the task plane, (3) recommended illumination levels, (4) special considerations and requirements of the task, and (5) typical equipment.

General Lighting for Task Surroundings

Areas Where Visual Work is Performed. Residence lighting is planned on the basis of activities, not on the basis of rooms. Many seeing tasks can be, and are, performed in almost any part of the home. The designer must determine what activities are to be carried on in each room, and provide light for them first. The individual task areas are then tied together by means of a general or environmental lighting system. This is essential in order to prevent a spotty effect, to maintain recommended luminance ratios in the field of view, to provide light throughout the interior for safety in moving about and for general housekeeping activities, and to avoid the necessity for rapid eye adaptation required by excessive differences in illumination levels between adjoining rooms.

Fig. 15-18. Reading in a chair.

1. THE TASK: Typical reading tasks in a home encompass a wide range of seeing difficulty, from short-time casual reading of material with good visibility (large print on white paper) to prolonged reading of poor material (small type on low-contrast paper).

2. DESCRIPTION OF THE TASK PLANE: The task plane measures 14 inches (36 centimeters) wide by 12 inches (31 centimeters) high with the center of the plane approximately 26 inches (66 centimeters) above the floor. The plane is tilted at 45 degrees from the vertical. The reader's eyes are approximately 40 inches (102 centimeters) above the floor.

3. ILLUMINATION RECOMMENDATIONS (MINIMUM ON THE TASK PLANE AT ANY TIME): a) Normal reading of books, magazines, papers—30 footcandles (32 dekalux). b) Prolonged difficult reading of handwriting, reproductions, poor copies—70 footcandles (75 dekalux).

4. SPECIAL DESIGN CONSIDERATIONS: Normal seated eye level [38 to 42 inches (97 to 107 centimeters) above the floor] is a critical consideration when the light source is to be positioned beside the user. The lower edge of the shielding device should not be materially above or below eye height. This will prevent discomfort from bright sources in the periphery of the visual field and yet permit adequate distribution of light over the task area. Variations in chair and table heights necessitate selection and placement of equipment to achieve this relationship for each individual case.

5. TYPICAL EQUIPMENT LOCATIONS:
Table-mounted alongside or behind the user.
Floor-mounted alongside or behind the user.
Wall-mounted alongside or behind the user.
Ceiling-mounted
a) Suspended beside or behind the user
b) Surface—large-area
c) Recessed—large-area
d) Large luminous element or luminous ceiling.
Directional small-area luminaires may be used (wall, ceiling, pole-mounted) in combination with other larger-area diffusing luminaires.

Fig. 15-19. Study desk.

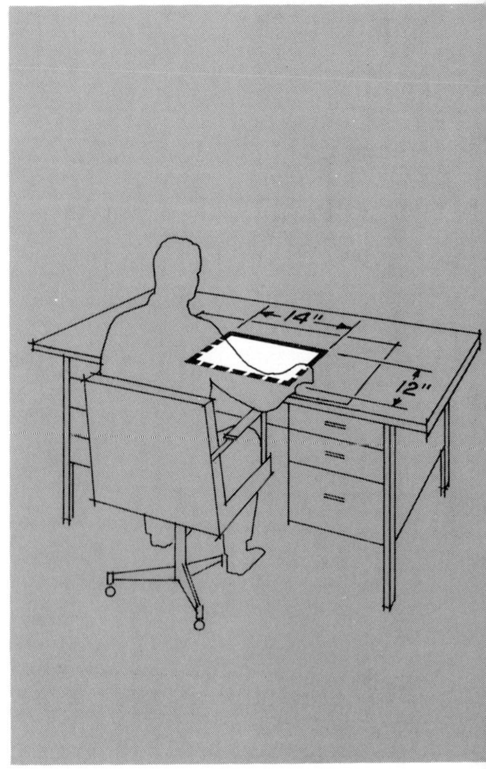

1. *THE TASK:* The task is prolonged, including difficult reading, handwriting, typing and drawing. It involves fine print and close detail.

2. *DESCRIPTION OF THE TASK PLANES:* The primary task is a plane 14 by 12 inches (36 by 31 centimeters) parallel with the desk top. The bottom edge of the task plane is 3 inches (7.6 centimeters) from the front edge of the desk. A secondary task plane for reference books, large drawings, etc., measures 24 inches (61 centimeters) deep by 36 inches (91 centimeters) wide with the front edge at the front of the desk top.

3. *ILLUMINATION RECOMMENDATIONS (MINIMUM ON THE TASK PLANE AT ANY TIME):*
a) Primary task plane—70 footcandles (75 dekalux).
b) Maximum illumination on the primary plane should not exceed the minimum by more than 3 to 1.
c) Secondary task plane—30 footcandles (32 dekalux).

4. *SPECIAL DESIGN CONSIDERATIONS :* The lighting equipment should be located so that shadows are not cast on the task area by the user's hand. The surface of the desk top should be nonglossy and light in color (30 to 50 per cent reflectance). The luminance of any luminaire visible from a normal seated position should be no more than 150 footlamberts (510 candelas per square meter), no less than 50 (170).

5. *TYPICAL EQUIPMENT LOCATIONS:*
Desk-mounted (1 or more luminaires).
Wall-mounted (1 or more luminaires).
Ceiling-mounted (1 or more luminaires).
Floor mounted.

Fig. 15-20. Reading in bed.

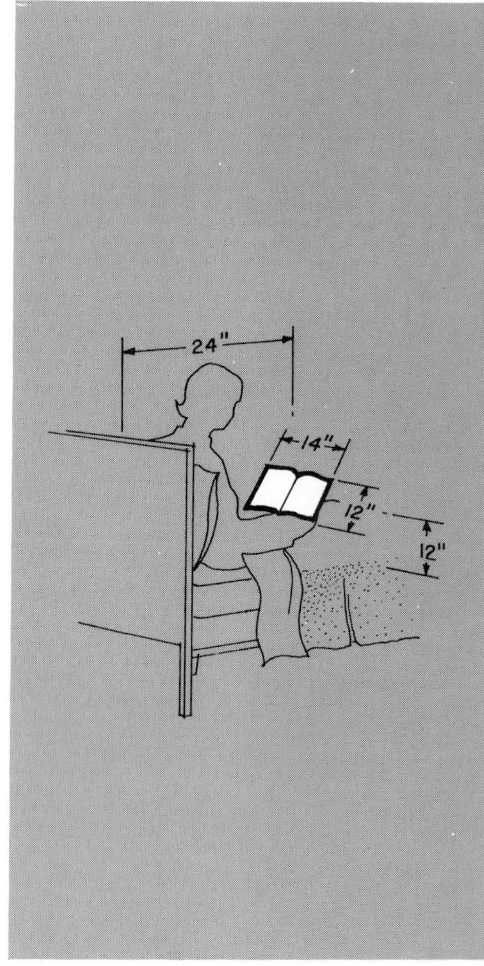

1. *THE TASK:* The majority of people who read in bed are only casual readers, perhaps reading for a few minutes before going to sleep. They are often interested in closely confining the light distribution so as not to disturb another occupant of the room. Such lighting arrangements are not satisfactory for comfortable reading over a long period. The following recommendations are for the individual who reads for a more extended period, or for the person who performs very critical eye tasks while confined to bed. The normal materials vary from books and magazines to pocket editions and to newspaper print. Occasionally other eye work is performed in bed (usually by people confined to bed) including hobbies, crafts, sewing, embroidery, etc.

2. *DESCRIPTION OF THE TASK PLANE:* The task plane is 12 by 14 inches (31 by 36 centimeters) tilted at an angle of 45 degrees from the vertical. The center of the task plane is 24 inches (61 centimeters) out from the headboard or wall and 12 inches (31 centimeters) above the mattress top. There are no customary reading positions or habits. These recommendations are based on the assumption that the reader is in an upright or semireclined position.

3. *ILLUMINATION RECOMMENDATIONS (MINIMUM ON THE TASK PLANE AT ANY TIME):* Normal reading—30 footcandles (32 dekalux). Serious prolonged reading or critical eye work—70 footcandles (75 dekalux).

4. *SPECIAL DESIGN CONSIDERATIONS:* Equipment should be located so that no shadows are cast on the reading plane by the head or body, and so that the luminaire does not interfere with a comfortable position for the reader.

5. *TYPICAL EQUIPMENT LOCATIONS:* Wall-mounted directly in back of or to one side of the user (both linear and non-linear designs).
Luminaire on bedside table or storage headboard.
Pole-type luminaires (floor-to-ceiling or table-to-ceiling).
Ceiling-mounted
a) Suspended: adjustable or stationary
b) Surface-mounted: directional or non-directional
c) Luminous-area panel: surface mounted or recessed
d) Recessed: directional or nondirectional
Luminaire incorporated into furniture design.

Fig. 15-21. Kitchen counter.

1. THE TASK: Typical seeing tasks at the kitchen counter include reading fine print on packages and cookbooks, handwritten recipes in pencil and ink, and numbers for speeds and temperatures on small appliances. Other tasks include measuring and mixing, color and texture evaluation of foods, safe operation of small appliances, and cleanup.

2. DESCRIPTION OF THE TASK PLANE: The task area is a plane 20 inches (51 centimeters) deep (starting at the front edge of the counter) and the length of the counter used.

3. ILLUMINATION RECOMMENDATIONS (MINIMUM ON THE TASK PLANE AT ANY TIME): Food preparation and cleaning involving difficult seeing tasks—150 footcandles (160 dekalux). Other noncritical tasks—50 footcandles (54 dekalux). Variable intensity controls, such as multiple position switching and dimming equipment, can be utilized to lower the level of illumination to meet the needs of the noncritical tasks.

4. SPECIAL DESIGN CONSIDERATIONS: Although there are a great many ways in which counter surfaces may be lighted, luminaires are commonly mounted under the wall cabinets, above the counter. These may be well shielded because of the cabinet structure itself, but if not, there is need for added shielding. Also, care should be exercised to see that the luminances of the luminaires are comfortable to other users of the room particularly in seated positions.

5. TYPICAL EQUIPMENT LOCATIONS:
Under-side of wall cabinets.
Ceiling—recessed, surface-mounted or pendant.
Structural—lighted soffits, wall brackets, canopies.

Fig. 15-22. Kitchen range.

1. THE TASK: Typical seeing tasks at the kitchen range are the determination of the condition of foods in all stages of the cooking process (color and texture evaluation) and reading range controls, instructions, and recipes.

2. DESCRIPTION OF THE TASK PLANE: The task area is a range top. Generally this is located 36 inches (91 centimeters) above the floor.

3. ILLUMINATION RECOMMENDATIONS (MINIMUM ON THE TASK PLANE AT ANY TIME):
Food preparation and cleaning involving difficult seeing tasks—150 footcandles (160 dekalux).
Other noncritical tasks—50 footcandles (54 dekalux).
Variable intensity controls, such as multiple position switching and dimming equipment, can be utilized to lower the level of illumination to meet the needs of the noncritical tasks.

4. SPECIAL DESIGN CONSIDERATIONS: Reflected glare is inherent in the shiny finish of utensils and range tops. Some reduction in the luminance of reflected images may be obtained by the use of diffuse luminaires and/or sources. Color-rendering qualities of light source are especially important in the kitchen. Lighting equipment installed on the range by the manufacturer seldom is located high enough to direct light into cooking utensils.

5. TYPICAL EQUIPMENT LOCATIONS:
Range hood.
Ceiling—recessed, surface-mounted or pendant.
Structural—lighted soffits, wall brackets, canopies.
Underside of wall cabinets.

Fig. 15-23. Kitchen sink.

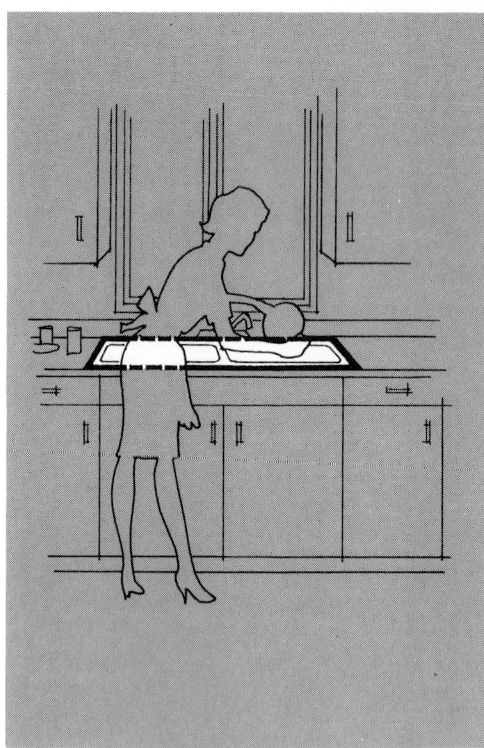

1. *THE TASK:* Typical eye tasks at the kitchen sink involve cleaning and inspection of dishes and utensils, evaluation of color and texture of foods in preparation, reading and measuring.

2. *DESCRIPTION OF THE TASK PLANE:* The task area measures 20 by 30 inches (51 by 76 centimeters). It is usually at a height of 36 inches (91 centimeters) above the floor.

3. *ILLUMINATION RECOMMENDATIONS (MINIMUM ON THE TASK PLANE AT ANY TIME):*
Food preparation and cleaning involving difficult seeing tasks—150 footcandles (160 dekalux).
Other noncritical tasks—50 footcandles (54 dekalux).
Variable intensity controls, such as multiple position switching and dimming equipment, can be utilized to lower the level of illumination to meet the needs of the noncritical tasks.

4. *SPECIAL DESIGN CONSIDERATIONS:* Color-rendering qualities of the light source are particularly important in kitchen illumination. The limited space available for luminaire mounting at the sink location increases the possibility of shadows being cast on the work plane by the head or body of the user.

5. *TYPICAL EQUIPMENT LOCATIONS:*
Ceiling—recessed, surface-mounted or pendant.
Structural—lighted soffits, wall brackets, canopies.
Underside of wall cabinets.

Fig. 15-24. Laundry

1. *THE TASK:* Preparation area—the tasks are sorting of fabrics by color and type, determination of location and type of soil, prewash treatment, tinting, bleaching, and starching.
Tub area—the tasks are soaking, handwashing, tinting, rinsing, bleaching and starching.
Washing machine and dryer area—the tasks are loading, setting of dials and controls, removal of clothes.

2. *DESCRIPTION OF THE TASK PLANES:* Preparation area—the general task area is 20 by 24 inches (51 by 61 centimeters) with a critical seeing area 12 by 12 inches (31 by 31 centimeters).
Tub area—the general task area is 20 by 24 inches (51 by 61 centimeters) on a single laundry tub with a critical seeing area 12 by 12 inches (31 by 31 centimeters) in the center.
Washing machine and dryer area—this task area has no definable boundaries and must be illuminated by the general room lighting.

3. *ILLUMINATION RECOMMENDATIONS (MINIMUM ON THE TASK PLANE AT ANY TIME):* General task planes at both preparation and tub areas—30 footcandles (32 dekalux).
Critical seeing areas at both preparation and tub areas—50 footcandles (54 dekalux).
Washer and dryer area—30 footcandles (32 dekalux).

4. *SPECIAL DESIGN CONSIDERATIONS:* Totally direct and not highly diffused light sources can contribute to the visibility of certain laundry tasks. In most laundry locations task lighting equipment also must provide the general room illumination. Luminaires should in this case be designed and positioned to illuminate the ceiling and side walls for comfortable luminance relationships.

5. *TYPICAL EQUIPMENT LOCATIONS:* Ceiling-mounted (pendant, surface or recessed) linear or non-linear luminaires centered over the front edge of the laundry equipment. As above, supplemented by wall-mounted or cabinet-mounted linear luminaires.
Large-area luminous panels or complete luminous ceilings.

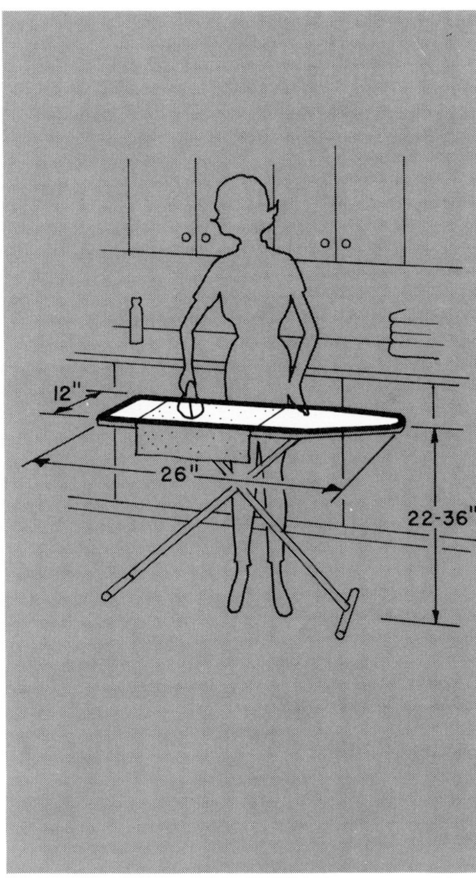

Fig. 15-25. Hand ironing.

1. *THE TASK:* The basic visual task in hand ironing is the detection and removal of wrinkles from garments and the detection of possible scorch.

2. *DESCRIPTION OF THE TASK PLANE:* The task plane is 12 by 26 inches (31 by 66 centimeters) and varies in its height, depending upon the ironing board. In general such boards are adjustable from 22 to 36 inches (56 to 92 centimeters) in height, to be used either standing or sitting. The seat of the average stool is located 24 inches (61 centimeters) above the floor, which places the average seated individual's eye level at 53 inches (135 centimeters). In a standing position the eye level is 61 inches (155 centimeters).

3. *ILLUMINATION RECOMMENDATIONS (MINIMUM ON THE TASK PLANE AT ANY TIME):*
50 footcandles (54 dekalux).

4. *SPECIAL DESIGN CONSIDERATIONS:* A light source with directional quality may frequently reveal shadows cast by small wrinkles, creases, etc., to the advantage of the user. Ironing and television viewing are often done at the same time. Under these circumstances it is important to ensure a good balance among the luminances of task, television screen and other room surfaces in the line of sight.

5. *TYPICAL EQUIPMENT LOCATIONS:*
Ceiling-mounted
a) Suspended, adjustable
b) Fixed, directional
c) Fixed, nondirectional
d) Combination of luminaires (general diffusing plus directional component).

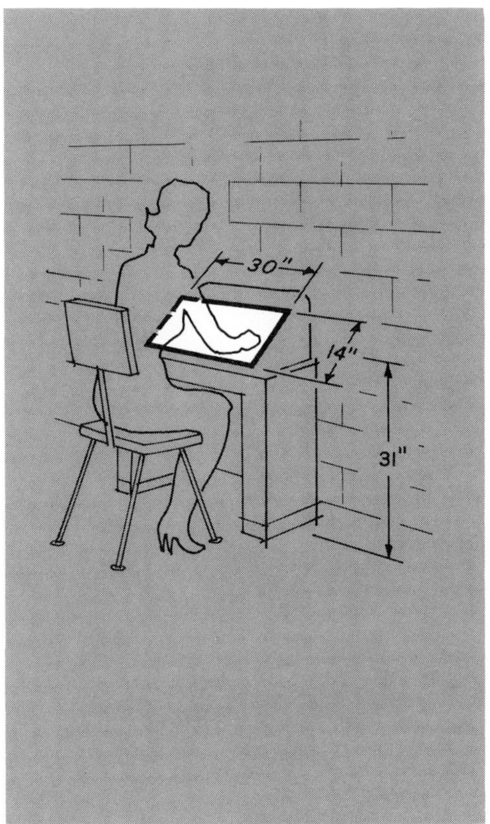

Fig. 15-26. Machine ironing.

1. *THE TASK:* The task is the positioning and inspection of fabrics for scorches and wrinkles from a seated position.

2. *DESCRIPTION OF THE TASK PLANE:* The task plane is fixed: 14 inches (36 centimeters) deep by 30 inches (76 centimeters) wide inclined at a 45-degree angle, with its center 31 inches (79 centimeters) above the floor. The regulation chair with seat 18 inches (46 centimeters) above the floor places the average individual eye level at 46 inches (117 centimeters).

3. *ILLUMINATION RECOMMENDATIONS (MINIMUM ON THE TASK PLANE AT ANY TIME):*
50 footcandles (54 dekalux).

4. *SPECIAL DESIGN CONSIDERATIONS:* A light source with directional quality reveals shadows cast by small wrinkles, creases, etc., to the advantage of the user.

5. *TYPICAL EQUIPMENT LOCATIONS:*
Floor-mounted or pole-mounted.
Ceiling-mounted
a) Suspended, adjustable
b) Fixed, directional
c) Fixed, nondirectional
d) Combination of luminaires (general diffusing plus directional component).

Fig. 15-27. Personal grooming.

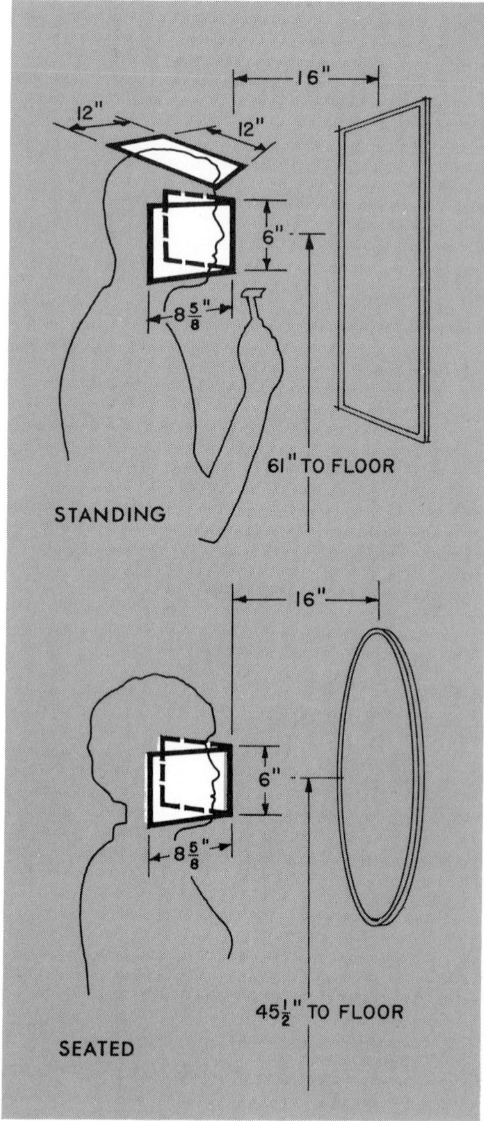

1. THE TASK: The chief tasks are shaving and makeup. Because the apparent distance of the face or figure as viewed in the mirror is twice its actual distance from the mirror, and because the details to be seen in shaving and critical inspection of dress or makeup are usually small and of low contrast with their background, the visual task may be a severe one.

2. DESCRIPTION OF THE TASK PLANES: Standing position—the task area consists of two 6 by 8⅝ inch (15 by 22 centimeters) planes at right angles with each other, converging at a point 16 inches (41 centimeters) out from the mirror, and centered vertically 61 inches (155 centimeters) above the floor. They represent the front and sides of the face. A third plane 12 inches (31 centimeters) square, its front edge also 16 inches (41 centimeters) out from the mirror, is tilted up 25 degrees above the horizontal and represents the top of the head. Seated position—the two facial planes are identical in size and position to those mentioned above, except that the center of the planes is 45½ inches (116 centimeters) above the floor. No top-of-the-head plane is considered for seated grooming.

3. ILLUMINATION RECOMMENDATIONS (MINIMUM ON THE TASK PLANE AT ANY TIME): 50 footcandles (54 dekalux).

4. SPECIAL DESIGN CONSIDERATIONS: Lighting equipment at the mirror should direct light *toward the person* and not onto the mirror. The luminances of surfaces reflected in the mirror and seen adjacent to the face reflection should not be distracting.

a) Adjacent walls should have a 50 per cent or higher reflectance.
b) Luminaires should be mounted outside the 60-degree visual cone, the center line of which coincides with the line of sight.
c) No luminaire should exceed 600 footlamberts (2100 candelas per square meter) in luminance. [*i.e.*, an illumination meter held against it should not read more than 600 (6500)].

5. TYPICAL EQUIPMENT LOCATIONS: Wall-mounted linear or nonlinear luminaires over the mirror. Wall-mounted linear or nonlinear luminaires over and at the sides of the mirror.
Combination of wall- and ceiling-mounted luminaires flanking the mirror and over the head of the user.
Structural devices (such as soffits) over the mirror, extending the length of the mirror.
Portable luminaires with luminous shades flanking the mirror.
Ceiling-pendant luminaires with luminous sides flanking the mirror.
NOTE: If grooming is performed in a seated position, the relationship of the luminaires to the face should remain as specified above for the standing position.

Fig. 15-28. Full length mirror.

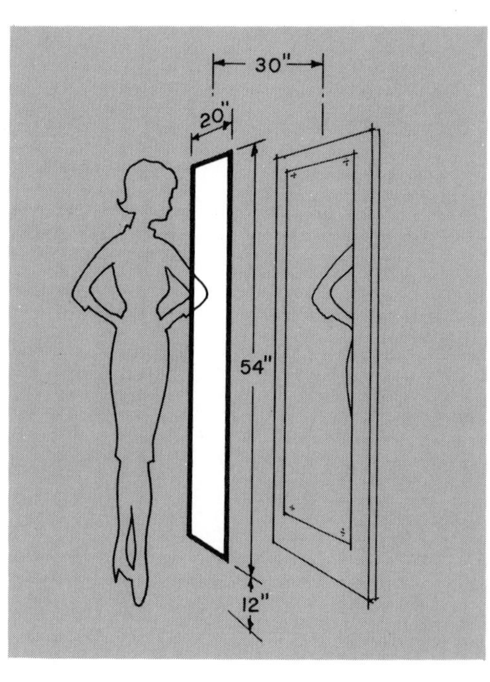

1. THE TASK: The task is the alignment of clothing and casual over-all appraisal.

2. DESCRIPTION OF THE TASK PLANE: The task area is a plane 20 inches (51 centimeters) wide by 54 inches (137 centimeters) high with the lower edge 12 inches (31 centimeters) above the floor; it is centered on and parallel with the mirror, 30 inches (76 centimeters) from the mirror surface.

3. ILLUMINATION RECOMMENDATIONS (MINIMUM ON THE TASK PLANE AT ANY TIME): 30 footcandles (32 dekalux).

4. SPECIAL DESIGN CONSIDERATIONS: Lighting equipment at the mirror should direct light *toward the person* and not onto the mirror. The luminances of surfaces reflected in the mirror and seen adjacent to the face reflection should not be distracting.

a) Luminaires should be mounted outside the 60-degree visual cone, the center line of which coincides with the line of sight.
b) No luminaire should exceed 600 footlamberts (2100 candelas per square meter) in luminance (*i.e.*, a footcandle meter in contact with the surface should not read more than 600).

5. TYPICAL EQUIPMENT LOCATIONS:
Vertical linear luminaires wall-mounted beside the mirror.
As above, supplemented by wall-mounted or ceiling-mounted over-mirror luminaires.

Fig. 15-29. Workbench hobbies.

1. THE TASK: Activities carried on at the workbench include woodworking (sawing, hammering, vise operation, planing, assembling parts, drilling, etc.) and craft hobbies. Lighting for power-tool operation is not considered here.

2. DESCRIPTION OF THE TASK PLANE: The task plane area is 20 inches (51 centimeters) wide and 48 inches (122 centimeters) long, 36 inches (91 centimeters) above the floor. (Home workbenches vary in length. The task plane extends the full length of the bench.)

3. ILLUMINATION RECOMMENDATIONS (MINIMUM ON THE TASK PLANE AT ANY TIME): 70 footcandles (75 dekalux).
Hobbies vary greatly in visual difficulty, and often require additional illumination and consideration of directional quality. The table below will provide some guidance for the more difficult tasks.

Difficult Seeing Tasks — 100 footcandles (110 dekalux)
*leather work, ceramic enamelling, pottery, mosaic, wood carving, block cutting (linoleum-wood)

Very Difficult Seeing Tasks — 150 footcandles (160 dekalux)
*model assembly, electrical and electronic assembly, fly tying

Critical Seeing Tasks — 200 footcandles (220 dekalux)
*Metal engraving, embossing, lapidary—gem polishing, jewelry making

4. SPECIAL DESIGN CONSIDERATIONS: Luminance balance within the visual field

a) The wall immediately behind the workbench should have a reflectance above 40 per cent. This reflected light is necessary to provide good illumination on the task plane as well as eye comfort from the standpoint of luminance differences.
b) Additional room illumination should be provided to contribute to the luminance balance in the visual surrounding when the worker faces into the room.
c) If the individual is facing a window while working:
   1) daylight glare should be controlled by blinds or shades.
   2) light-colored window coverings should be used at night.

The light source should be so positioned that its image reflected in glossy materials is not visible to the user in normal position, unless desired for some special application.

5. TYPICAL EQUIPMENT LOCATIONS:
Ceiling-mounted or suspended linear luminaire running parallel with the task plane.
Ceiling-mounted or suspended non-linear luminaires in symmetrical arrangement over work area to provide uniform distribution of light.
Wall- or shelf-mounted luminaire or luminaires directly in front of user.
Luminous ceiling area.
Portable equipment to provide added directional lighting for special conditions as indicated in the table above.

* These tasks require large-area low-luminance reflections in order to see fine detail which shows up as an interruption in a surface sheen.

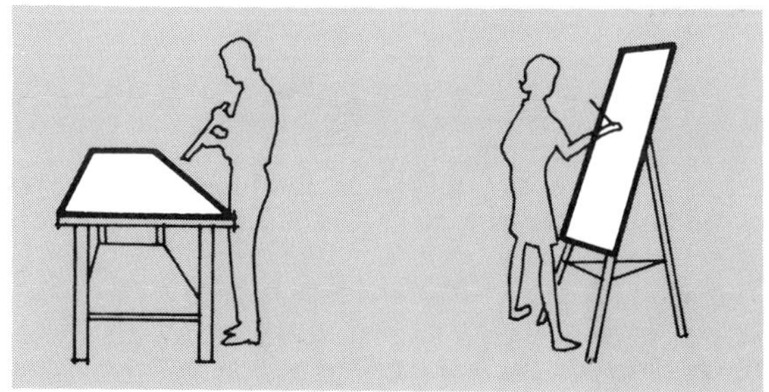

Fig. 15-30. Easel hobbies.

1. THE TASK: Easel hobbies include painting, sketching, collage.*

2. DESCRIPTION OF THE TASK PLANE: The task area is a surface up to 3 feet (0.9 meter) square. The plane of the task is inclined from the vertical to suit the user and the task.
Average eye height—48 inches (122 centimeters) sitting, 62 inches (158 centimeters) standing.
There is also a palette and often an object being copied. The locations of these are not fixed.

3. ILLUMINATION RECOMMENDATIONS (MINIMUM ON THE TASK PLANE AT ANY TIME): 70 footcandles (75 dekalux)**

4. SPECIAL DESIGN CONSIDERATIONS: Luminance balance within the visual field

a) Reflectance of wall surfaces should be above 35 per cent.
b) Additional room illumination should be provided to contribute to the luminance balance in the visual surroundings when the painter looks away from his work.
c) If the person faces a window while working
   1) Glare should be controlled by blinds or shades.
   2) Light-colored window coverings should be used.

The light source should be so positioned that its image reflected in glossy materials is not visible to the artist in normal working position, unless specifically desired. For instance, large low-luminance reflection may be required in order to see fine detail in a glossy paint surface; the detail shows up as an interruption on a surface sheen.
a) A general recommendation would be to paint under the light source by which the painting will ultimately be seen. This is usually incandescent lighting.
b) Many painters prefer to work under "North Skylight" conditions. An approximation of this, as far as color rendering is concerned, can most simply be obtained with deluxe cool white fluorescent lamps.***

5. TYPICAL EQUIPMENT LOCATIONS:
Ceiling-mounted or suspended linear luminaires running parallel with the task plane.
Ceiling-mounted or suspended non-linear luminaires in a line to provide uniform distribution of light.
Portable equipment to provide added lighting for fine detail.

* Unlike most tasks which can be described in relatively exact terms, easel hobbies include widely divergent activities.

** The definition of a "standard painting" would be absurd. If it is assumed that the artist wishes to see small applications of nearly identical colors, a standard task may be described as the application of a spot of color one-quarter-inch (.64 centimeters) in diameter on a background of the next nearest color in a scale comprising 1800 different colors (Munsell).

*** There are sources currently available or under development with a high color rendering index rating designed to simulate daylight.

Fig. 15-31. Formal dining area.

1. *THE TASK:* The task is essentially one of creating the desired mood or atmosphere for dining. Therefore, the illumination level, the luminances in the room, and the choice of luminaire are largely matters of personal taste.

2. *DESCRIPTION OF THE TASK PLANE:* The entire tabletop should be considered as the task plane. If carving, serving, etc. is done at a separate location, this becomes another task area and can generally be equated with the task plane for Kitchen Counter.

3. *ILLUMINATION RECOMMENDATIONS:*
The illumination is usually 15 footcandles (16 dekalux) on the table, or more, but this is essentially a question of personal preference.
Variable-intensity controls, such as multiposition switching and dimming equipment, can often add to the enjoyment of dining-area lighting by adapting the level of illumination to the particular occasion.

4. *SPECIAL DESIGN CONSIDERATIONS:* Lighting with a strong downward component will accent the table setting, creating attractive focal highlights. However, this type of distribution, if used alone, will render faces poorly, causing harsh shadows. Strong downward lights should be kept away from people's faces (*i.e.*, confined within the perimeter of the table itself), or should be well balanced by indirect light from the table top, walls, or ceiling. The nature of the table top may also influence the choice of lighting distribution: downlighting may cause annoying specular reflections from glossy table tops such as glass or marble, and if all the light is directed on the table a colored tablecloth may appreciably tint the light by reflection.
Exposed sources such as unshielded low-wattage bulbs can often be tolerated, especially if some general lighting is provided and the background is not too dark. The darker the walls, the more general lighting is required to keep luminance relationships in the room within a comfortable range.
In situations where the dining table is moved from time to time, a flexible means of mounting the luminaire is desirable. One such method is a traverse-track suspension.

5. *TYPICAL EQUIPMENT LOCATIONS:*
Task Area: Over-center-of-table luminaires:
a) Recessed (a group of recessed units is generally required)
b) Surface-mounted
c) Pendant, generally mounted so that the bottom of the luminaire is 30 to 36 inches (76 to 91 centimeters) above the table top.

Area Surrounding the Task:
a) Luminous ceiling or large luminous area
b) Luminous wall
c) Cornice
d) Valance
e) Cove
f) Brackets, *e.g.*, linear fluorescent or decorative incandescent
g) Recessed luminaires, *e.g.*, incandescent downlight or wall washers
h) Ceiling-mounted luminaires, *e.g.*, shallow large-area types
i) Pendant luminaires, *e.g.*, small pendants
j) Table lamp
k) Floor lamp or torchere

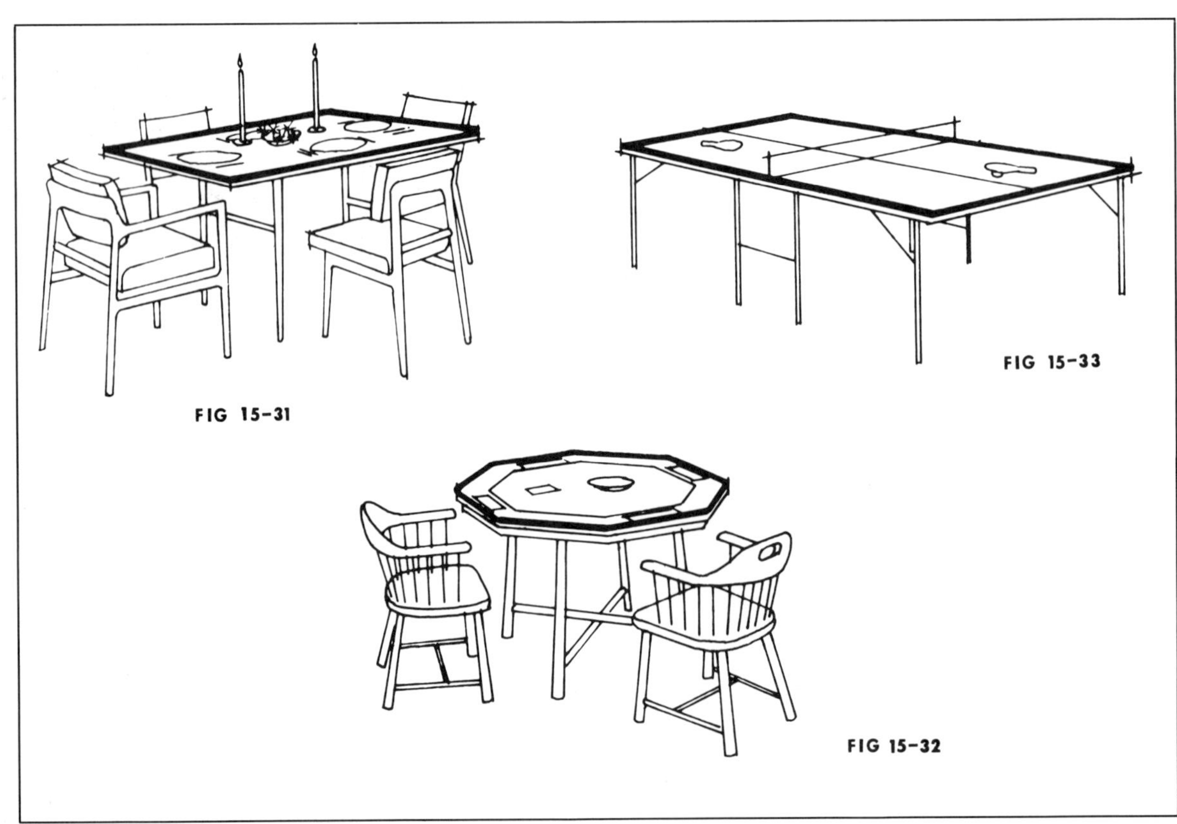
FIG 15-31
FIG 15-33
FIG 15-32

Fig. 15-32. Multipurpose dining area.

1. *THE TASK:* The task includes both the creation of the desired mood or atmosphere for dining (see Fig. for Formal Dining Area) and provision for other general eye tasks such as sewing activities, reading, hobbies and playing table games.

2. *DESCRIPTION OF THE TASK PLANE:* The entire table top must be considered as the task plane. If carving, serving, etc. is done at a separate location, this becomes another task area and can generally be equated with the task plane for Kitchen Counter.

3. *ILLUMINATION RECOMMENDATIONS:* Dining requires generally about 15 footcandles (16 dekalux) on the table, but this is essentially a question of personal preference. Variable-intensity controls, such as multiposition switching and dimming equipment, can often add to the enjoyment of dining-area lighting by adapting the level of illumination to the particular occasion. For casual or relatively easy work or play activities such as card or table games, simple hobbies or crafts, sewing-pattern layout and cutting, casual reading, a minimum of 30 footcandles (32 dekalux) should be provided over the entire table top. For prolonged or difficult seeing tasks performed at the dining table, see recommendations listed for the particular task.

4. *SPECIAL DESIGN CONSIDERATIONS:*
a) For dining: Lighting with a strong downward component will accent the table setting, creating attractive focal highlights. However, this type of distribution, if used alone, will render faces poorly, causing harsh shadows. Strong downward lights should be kept away from people's faces (*i.e.*, confined within the perimeter of the table itself), or should be well balanced by indirect light from the table top, walls or ceiling. The nature of the table top may also influence the choice of lighting distribution: downlighting may cause annoying specular reflections from glossy table tops such as glass or marble, and if all the light is directed on the table a colored tablecloth may appreciably tint the light by reflection. Exposed sources such as unshielded low-wattage bulbs can often be tolerated, especially if some general lighting is provided and the background is not too dark. The darker the walls, the more general lighting is required to keep luminance relationships in the room within a comfortable range.
In situations where the dining table is moved from time to time, a flexible means of mounting the luminaire is desirable. Two such methods are the traverse-track suspension and the swag.

b) For other activities: For activities other than dining, a broad distribution pattern of light is required to illuminate the entire table top rather uniformly. To minimize veiling reflections, the light sources should have a high degree of diffusion and/or substantial indirect components.

c) Flexibility of distribution and level: It is difficult for a single static light source to provide drama and atmosphere for dining and widespread, diffused lighting (at twice the footcandle level) for other table activities. Therefore, a multipurpose table usually requires more than one lighting system or a means of switching from one effect to another.

5. *TYPICAL EQUIPMENT LOCATIONS:*
Task Area: Over-center-of-table luminaires:
a) Recessed (a group of recessed units is generally required)
b) Surface-mounted
c) Pendant, generally mounted so that the bottom of the luminaire is 36 inches (91 centimeters) above the tabletop.

Area Surrounding the Task:
a) Luminous ceiling or large luminous area
b) Luminous wall
c) Cornice
d) Valance
e) Cove
f) Brackets, *e.g.*, linear fluorescent or decorative incandescent
g) Recessed luminaires, *e.g.*, incandescent downlights or wall washers
h) Ceiling luminaires, *e.g.*, shallow large-area types
i) Pendant luminaires, *e.g.*, small pendants
j) Table lamp
k) Floor lamp or torchere

Fig. 15-33. Table tennis.

1. *THE TASK:* The basic game of table tennis, although it always has the same rules and table size, varies greatly in its visual difficulty according to the skill of the players. Lighting recommendations here are for "recreational play" where the speed of play is not high and the players' skill is minimal. At this level the game is often called "ping-pong."

2. *DESCRIPTION OF THE TASK PLANE:* In "recreational play" the task-plane area is to be considered as the 5- by 9-foot (1.5- by 2.7-meter) table only.

3. *ILLUMINATION RECOMMENDATIONS (MINIMUM ON THE TASK PLANE AT ANY TIME):* 20 footcandles (22 dekalux).

4. *SPECIAL DESIGN CONSIDERATIONS:* Although the general guides for luminance balance within the visual field still hold true in table tennis, the background surfaces seen by the player should not be too light or they will not provide sufficient contrast with the white ball for easy visibility. Wall and ceiling surfaces must not have strong distracting patterns. Spottiness or uneven distribution of light can cause seeing difficulty as the ball could appear to speed up or slow down as it passes in and out the areas of greater illumination. Because the ceiling plane is the major part of the visual field, the luminance ratios at the ceiling become more important than usual.

5. *TYPICAL EQUIPMENT LOCATIONS:* Ceiling-mounted (surface or recessed) linear sources with the center lines of the luminaires crosswise of the table, located approximately one foot (.3 meter) in from each end of the table plus one or more in each runback area.
Ceiling-mounted (surface or recessed) linear sources lengthwise of the table, centered over the outer edges of the table and extending into runback areas.
Ceiling-mounted (surface or recessed non-linear sources arranged in a symmetrical pattern over the entire task area).
Large-area luminaires of *low luminance* symmetrically located to cover the task area, *i.e.*, equipped with louvers or other material providing a minimum of 45-degree shielding.

Fig. 15-34. Hand sewing.

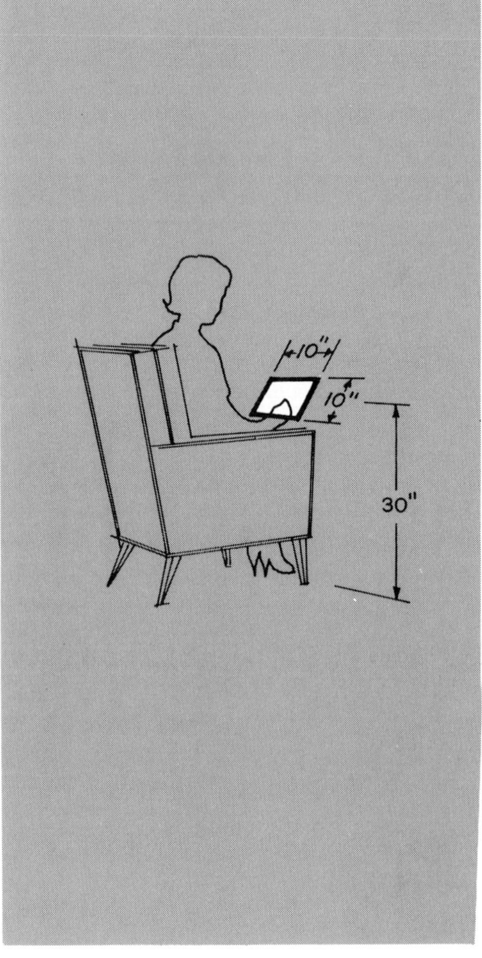

1. THE TASK: The seeing task encompasses a wide range of difficulty from coarse threads to fine; from light materials to very dark; from high contrast to virtually no contrast at all.

2. DESCRIPTION OF THE TASK PLANE: The task area is a plane 10 inches (25 centimeters) square tilted at 45 degrees toward the eye. The plane is centered at 30 inches (76 centimeters) from the floor. Eye position is approximately 42 inches (107 centimeters) from floor.

3. ILLUMINATION RECOMMENDATIONS (MINIMUM ON THE TASK PLANE AT ANY TIME):
a) Dark fabrics (fine detail, low contrasts)—200 footcandles (220 dekalux).
b) Prolonged periods (light to medium fabrics)—100 footcandles (110 dekalux).
c) Occasional periods (light fabrics)—50 footcandles (54 dekalux).
d) Occasional periods (coarse thread, large stitches, high contrast thread to fabric)—30 footcandles (32 dekalux).
e) Maximum illumination level on the task plane should not exceed the minimum by more than 3 to 1.

4. SPECIAL DESIGN CONSIDERATIONS: Equipment should be located opposite the hand being used, so that shadows are not cast on the task area.

5. TYPICAL EQUIPMENT LOCATIONS:
Floor-mounted or pole-mounted.
Ceiling-mounted
a) Suspended, adjustable
b) Fixed, directional (surface or recessed)
c) Fixed, luminous-area
d) Combination luminaire (general diffusing plus directional component).
Wall-mounted (both linear and non-linear) sources located beside or behind the user.

Fig. 15-35. Music study at piano or organ.

1. THE TASK: The task is the reading of musical scores ranging from very simple ones with large notes and staff lines to very difficult substandard-size scores with notations printed on the lines.

2. DESCRIPTION OF THE TASK PLANE: The primary task plane is on the music rack in an area 12 by 18 inches (31 by 46 centimeters); it is tilted back from the viewer about 17 degrees. The lower edge is 32 to 35 inches (81 to 89 centimeters) from the floor. The secondary plane includes an additional 9 by 12 inches (23 by 31 centimeters) on each side of the primary plane. The average piano keyboard, 48 inches (122 centimeters) long and 28 inches (71 centimeters) above floor, is also a secondary plane. NOTE: These dimensions will vary greatly with electric organs and miniature pianos. The eye is approximately 47 inches (119 centimeters) above the floor.

3. ILLUMINATION RECOMMENDATIONS (MINIMUM ON THE TASK PLANE AT ANY TIME):
a) Simple scores—30 footcandles (32 dekalux).
b) Advanced scores—70 footcandles (75 dekalux).
c) When score is substandard size, and notations are printed on the line—150 footcandles (160 dekalux).
The maximum footcandle level should not exceed the minimum by more than 3 to 1. The minimum illumination level on the secondary task plane should not be less than one-third of the minimum on the primary task plane.

4. SPECIAL DESIGN CONSIDERATIONS: The instrument is in the best position for control of luminance values if the player faces a wall.

5. TYPICAL EQUIPMENT LOCATIONS:
Ceiling-mounted or recessed above ceiling:
1) Directional source
   a) Should be adjustable to strike the plane of the task at about 90 degrees.
   b) Should be located above the user's head to avoid a shadow of his body.

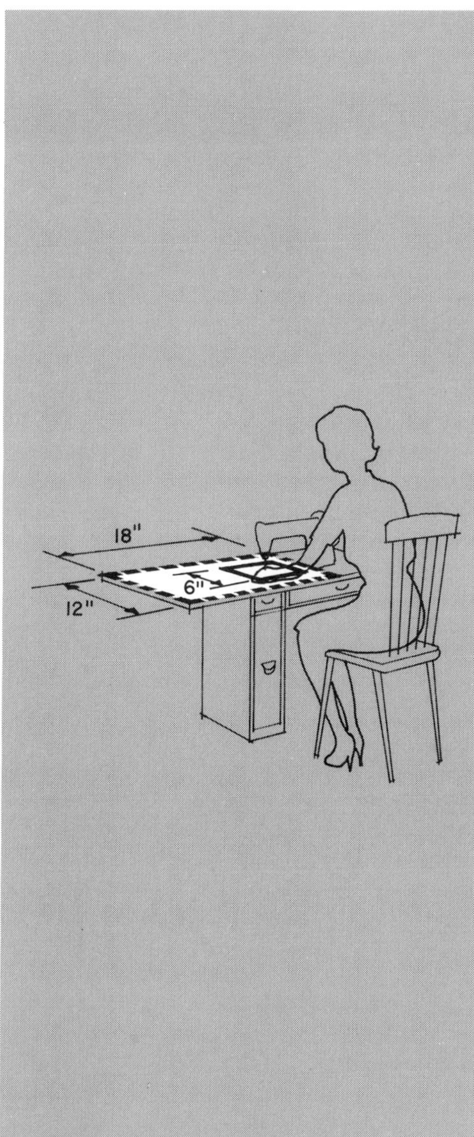

Fig. 15-36. Machine sewing

1. *THE TASK:* The small detail and low contrast between thread and material usually involved in machine sewing make it a visually difficult task. The degree of difficulty varies with thread and stitch size, reflectance of materials and contrast between thread and fabric.

2. *DESCRIPTION OF THE TASK PLANE:* The primary task area is a plane 6 inches (15 centimeters) square located so that the needle point is 2 inches (5 centimeters) forward from the center of the back edge.
The secondary task area of less critical seeing measures 12 by 18 inches (31 by 46 centimeters) with the needle point centered on the shorter dimension and 6 inches (15 centimeters) in from the right-hand edge.

3. *ILLUMINATION RECOMMENDATIONS (MINIMUM ON THE PRIMARY TASK PLANE AT ANY TIME):*
a) Dark fabrics (fine detail, low contrast)—200 footcandles (220 dekalux).
b) Prolonged periods (light to medium fabrics)—100 footcandles (110 dekalux).
c) Occasional periods (light fabrics)—50 footcandles (54 dekalux).
d) Occasional periods (coarse thread, large stitches, high contrast thread to fabric)—30 footcandles (32 dekalux).
e) Maximum level on the primary task plane should not exceed the minimum by more than 3 to 1.
f) The minimum level on the secondary task plane should not be less than 1/3 of the minimum on the primary task plane.

4. *SPECIAL DESIGN CONSIDERATIONS:* Equipment should be located so that shadows are not cast on the task area by the user's hand. Most sewing tasks involve low-sheen materials which minimize veiling reflections and allow the use of light with a moderate directional component to increase the visibility of threads by casting slight shadows.

5. *TYPICAL EQUIPMENT LOCATIONS* (does not take into account the light built into the machine):
Wall-mounted directly in front of the user. (Both linear and nonlinear sources.)
Ceiling-mounted (location of luminaire and/or machine should avoid the possibility of the user's head blocking out light or casting a shadow on the task).
1) Suspended-adjustable
2) Fixed, directional (surface or recessed)
3) Fixed, nondirectional
4) Luminous area.
Floor-mounted or pole-mounted.

---

c) Should be located and shielded to prevent glare to other persons occupying or passing through the area.
d) Downlights are not desirable; distribution is not good, and reflected glare and veiling reflections may be a problem.

2) Large-area nondirectional source
a) The luminance should be within the comfort range and esthetic considerations of the room.

Mounted on the instrument:
1) Uniformity of distribution over the task plane may be difficult to achieve.
2) There should be no luminous part within the user's field of view having a luminance of more than 50 footlamberts (170 candelas per square meter).

Pole-type luminaires:
Because of the directional quality of this type of light source, it is possible to get acceptable illumination levels on the task as well as general surround light. Care should be exercised to avoid glare to other occupants in the room as well as veiling reflections on the task area.

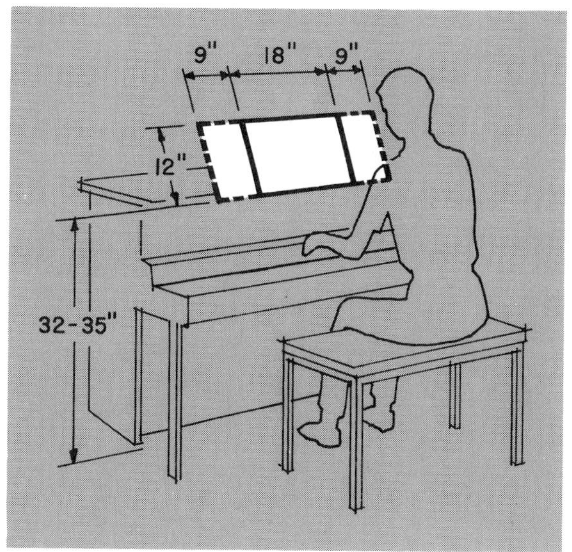

In some cases, notably in utilitarian areas such as kitchens and laundries, a general lighting system can be designed to supply all, or nearly all, of the illumination needed for visual activities. However, the relatively high level and uniform distribution of general lighting necessary in a kitchen would be unacceptable, practically and esthetically, in a living area. The equipment most commonly used to light room surfaces and create a satisfactory background for visual work includes general diffusing ceiling luminaires, wall lighting elements, indirect portable lamps, and coves. In small rooms general illumination may even be supplied by the luminaires employed for specific task lighting—the mirror lighting in a bathroom, for example, or a number of well-designed open-top floor and table lamps used to supply reading light in a living room.

Areas Devoted to Relaxation. For periods when no close visual work is being done in a living area, or for the occasional room designed solely for relaxation, a low level of general lighting creates a pleasant atmosphere for conversation, watching television, or listening to music. Uniformity of illumination need not be an objective here. On the contrary, moderate variations create a more attractive pattern of light. Any of the methods mentioned above, plus a wide range of possibilities limited only by the imagination and ingenuity of the designer, are applicable. The primary considerations are comfort and esthetic satisfaction.

Halls and Stairways. Changes in eye adaptation determine the proper level of illumination here. If the hall or stairway adjoins an interior area with a higher level, the illumination should be one-fifth that of adjacent areas. Wall luminances are most important in creating a sensation of lightness, and in affecting adaptation. Therefore luminaires should direct light to the walls, and wall and floor finishes should have high reflectance values. Lighting in the entry hall should be flexible, so that adjustment can be made for eye adaptation in the daytime and at night. On stairs, good lighting of the top and bottom steps is especially critical for safety. Differences in color and directional quality of light can help to emphasize treads. Under no circumstances should a luminaire or open-top portable lamp be located where a person descending the stairway can look directly at the light source.

Garages. The major areas where light is needed in a garage are on each side of the car or between cars, especially over the front and rear. Luminaires are usually located slightly to the rear of the trunk area, and approximately in line with the front wheels. A portable trouble light should be provided for repair work.

Closets. Light sources in closets should be located out of the normal view, generally to the front of the closet and above the door. In walk-in closets a ceiling luminaire should be mounted in the center of the traffic area so that shelves do not block the lighting of the garments.

DECORATIVE ACCENT POSSIBILITIES

Variation and areas of interest created by interplay of light and shadows are essential components of esthetically pleasing residence lighting design. Lighting can create decorative accents to enhance the appearance of art objects and/or create a focus of interest in a particular area. Fig. 15-37 discusses design considerations and suggests techniques for a variety of decorative possibilities.

EXTERIORS AND GROUNDS

The lighting design of residential exteriors should be developed to attain three major goals: quick, sure seeing for safe passage in and out of the home and for discouraging trespassing; smooth integration between indoor and outdoor living areas for increased hours of use and enjoyment; and nighttime beauty of garden and grounds.

Entrances and Walkways

The designer should bear in mind that against the darkness of night even light sources of relatively low luminance can cause annoying contrasts. Dark-adapted eyes can see with very little light, and therefore low levels of illumination are sufficient, provided that careful shielding is employed to prevent direct and disability glare. Where it is necessary or desirable to use highly decorative or traditional luminaires having unshielded light sources, it is usually necessary to keep lamp wattages extremely low, and to make provision for carefully controlled lighting for safety. Light-colored reflecting surfaces can help to guide pedestrians along walks, drives, and steps.

Walkways. In general, lighting equipment for walkways should be located high up in trees, on poles, or on buildings, so that the luminous parts are above the normal line of sight; or luminaires should be located quite low above the walkway so that luminous parts are below the line of sight. Low-mounted luminaires usually should be rather closely spaced for adequate coverage. Intermediate-height luminaires (in the line of sight) are the poorest choice for walkway lighting, but often are required for appearance or other reasons. Extreme care should be exercised in the selection and placement of such luminaires:

1. Units in which bare lamps may be seen are unacceptable.
2. Luminaires should be designed to reduce the luminance toward the eye and to direct light toward the walkway.
3. Where luminaires have luminous parts in the line of sight it may be necessary to use lower-wattage lamps and place the units closer together.

Steps. All considerations on the height of luminaires for lighting walkways apply to the lighting of steps as well. It is important to call attention to both treads and risers. This can be done by:

Fig. 15-37. Decorative Accent Lighting Design Considerations and Techniques

Design Considerations	Typical Techniques
Paintings, Tapestries, and Murals	
Avoid 1. Shadows of frames. 2. Specular reflections from picture, frame, or glass. 3. Excessive difference between lighted object and surrounding areas. 4. Higher levels, for extended periods may cause deterioration of the paint surface. Consider 1. Sight-lines of people seated and standing. 2. Above all, the artist's intent. 3. Lighting equipment placement—should be placed so that light rays reach the center of the painting at an angle of 30 degrees to avoid surface reflection of the light source and any disturbing shadows of frame, heavy paint texture, etc.	1. Entire picture wall lighted by cornice or wall-wash equipment. 2. Individual frame-mounted luminaires. 3. Individual framing spots. 4. Individual spot or flood lamps not confined to picture area. 5. Lighting from below by luminaires concealed in decorative urns, planters, mantels, etc.
Sculpture	
1. A sculpture is a three-dimensional object in space—to obtain a good three-dimensional effect with transparent shadows, use a luminance ratio between 2 and 6.[5] If the modeling ratio is reduced below 2, the lighting becomes too "flat" and solid objects appear two-dimensional. If the ratio is above approximately 6, the contrast tends to become unpleasant, with loss of detail in the shadows or in the highlights. 2. Amount of pleasurable specular reflection. 3. Experimentation with diffused and/or a directional source or sources—will help to determine the most acceptable solution.	Adjustable spots, floods, individual framing spots, back lighting.
Planters	
1. Lighting that provides for plant enhancement may not necessarily be suitable for plant growth. See Section 25. Plants may have to be rotated to a plant growth area from time to time to keep their beauty. 2. Plants tend to grow toward the light. This should be kept in mind when locating equipment. 3. Light sources, particularly incandescent, should not be located too close to plant material because the heat may be detrimental. 4. Backlighting of translucent leaves often will reveal leaf structure, color, and texture. 5. Front lighting of opaque leaves often will reveal leaf structure, color, and texture. 6. Front lighting of opaque leaves often will add sheen, texture, and accentuate form. 7. Silhouetting of foliage can be an additional dimension. 8. Take the necessary fire safety precautions if artificial plants are used. 9. Bear in mind, live plants are not static, they change in height and bulk.	1. Incandescent downlights, recessed, surface mounted, or suspended above planting area. 2. Luminous panel or soffit using fluorescent and/or incandescent sources. 3. Silhouette lighting with luminous wall panel, lighted walls, concealed up-lights. 4. Low-level incandescent stake units. 5. Special planting racks containing fluorescent and/or incandescent lamps especially selected for plant growth. 6. Luminaires recessed in earth to provide uplight.
Niches	
1. Lighting method is determined by what is to be displayed in the niche. 2. Brightness of shielding media and interior niche surfaces should be carefully controlled to avoid extreme difference to surrounding areas.	1. Incandescent spotlights from one or more directions. 2. Incandescent or fluorescent sources concealed vertically or horizontally at edges. 3. Luminous sides, bottom, top, back, or combinations. 4. Lamps concealed in back of object to create silhouette effect.

Fig. 15-37. *Continued*

Design Considerations	Typical Techniques
Bookcases	
1. Distribution of equipment to light faces of book covers and other objects on all shelves. 2. Luminance ratios of bookcase area and surrounding wall surfaces.	1. Adjustable spot or flood lamps or wall washers aimed into shelves from ceiling in front of bookcase. 2. Lighted bracket, cornice, or soffit extending in front of shelves. 3. Tubular sources concealed vertically at sides of bookcase or horizontally at front edge of individual shelves. 4. Lighting concealed at back of bookcase for silhouette effects. 5. Some bookshelves may require modification to allow ample spread of light over entire bookcase. For example: cut shelves back, or use plastic or glass shelves.
Fireplaces	
1. Brightness relations of surrounding area to facing of brick, stone, wood, etc. 2. The mantel as a decorative wall treatment and as the focal point of interest. 3. Orientation in the room.	1. Cornice and wall brackets. 2. Recessed or surface-mounted downlights close to fireplace surface for grazing light. 3. Adjustable spots or floods (ceiling mounted) aimed at decorative objects, mantel—or grouped together to light entire fireplace wall. 4. Recessed or surface-mounted wall washers for even distribution of light over entire fireplace surface. 5. Lighting equipment concealed in mantel to light picture or objects above and/or surfaces below.
Textured Walls and Draperies	
1. Grazing light to emphasize textured surfaces. 2. Light pattern created by luminaires close to walls.	1. Recessed or surface-mounted downlights located close to the vertical surface to direct grazing light over the surface. 2. Cornice, valance, wall brackets, or wall washers. 3. Individual adjustable incandescent luminaires. 4. Concealed up-lighting equipment.
Luminous Panels and Walls	
1. Luminous elements larger than about 16 square feet should not exceed 50 footlamberts average luminance when they will be viewed by people seated in a room. 2. Luminous elements of smaller size (or larger elements seen only in passing) may approach 200 footlamberts average luminance. 3. Variable controls are valuable in adjusting luminance of panel to desired level for room activities. 4. Uniform luminance of a luminous surface may or may not be desirable and is dependent on spacing of the light source. 5. Over-all pattern, shielding devices, structural members, or other design techniques to add interest and relieve over dominance and monotony of large, evenly illuminated panels.	1. Fluorescent lamps uniformly spaced vertically or horizontally in a white cavity. 2. Fluorescent lamps placed top and bottom or on each side of a white cavity. 3. Random placement of incandescent lamps (perhaps with color) in a white cavity. 4. Uniform pattern of incandescent lamps in a white cavity. 5. Incandescent reflectorized lamps concealed at top or bottom of a white cavity to reflect light from back surface of cavity. 6. For outdoor luminous walls, floodlamps directing light to diffusing screen from behind.
Other Architectural Elements	
Many residences have especially distinctive architectural features that require special lighting. Because of the variety of problems it is difficult to list design considerations that apply to all. In general, however, the following guides apply to all:	a. Conceal the light source from normal view. b. Avoid "unnatural" lighting effects. c. Do not introduce lighting patterns that conflict or distract from architectural appearance of the element being lighted.

Fig. 15-37. *Continued*

Design Considerations	Typical Techniques
Luminaires as Decorative Accents	
1. Luminance of luminous parts of luminaires can be critical. 2. Luminaires should not introduce distracting or annoying light patterns. 3. Where exposed lamps are used for a desired decorative effect, dimmer controls to provide higher or lower luminance should be considered.	1. Decorative crystal chandeliers, wall sconces, lanterns, and brackets. 2. Decorative portable lamps. 3. Pendant luminaires.
Dining Areas	
1. Type of illumination for a formal dining area is largely a matter of personal taste. The desired mood or atmosphere, choice of luminaire, and brightness in the room are all to be considered. 2. Footcandle level is also a matter of personal preference; 15 footcandles or more is generally desirable. Variable controls provide flexibility to adapt a level of illumination to fit the particular occasion. 3. Strong downward component will accent the table setting, creating attractive focal highlights; but if unbalanced, will render faces poorly and create harsh shadows. 4. Unshielded sources in low wattage can often be tolerated, especially if some general lighting is present and the background is not too dark. 5. In situations where the dining table is moved from time to time, a flexible means of mounting the luminaire or luminaires is desirable.	1. Decorative chandelier or pendant cluster suspended over center of table and supplemented by two or more recessed louvered ceiling downlights. Both chandelier and downlights on dimmer control. 2. Same as above, but augmented by some form of structural or built-in lighting (valance, cornice, luminous panel, etc.) to enhance and bring out character of rest of the room. 3. Pattern of dimmer-controlled ceiling downlights, designed to accent the table settings, but to avoid harsh downlighting directly on people at table. Carefully used tinted lighting can be employed here. 4. Pull-down pendant luminaire with strong downward component plus generous upward light located over center of table. Best supplemented by other structural room lighting.

1. Using light reflecting surfaces.
2. Building lighting into risers or side walls.
3. Using low or high-level equipment placed to light both vertical and horizontal surfaces.

Doorways. Light is needed at doorways for safety of passage and also for identifying callers. Therefore, it is important in entranceways to get light on a caller's face, by either direct or reflecting techniques. Lighted house numbers, visible from the street, are thoughtful courtesies. Numerals should be three inches high with a one-half inch stroke to be visible at 75 feet.

Outdoor Living Spaces

Lighting planned to fit the range of activities on porches and terraces expands their usefulness and the living space of the house. The extent of wall enclosure, whether there are roofs, overhead trees, or no covering modifies the method and degree of needed weatherability of equipment. Illumination levels in keeping with those specified for similar indoor activities generally require the use of about 3 to 5 watts per square foot. Comfortable brightness patterns are essential. Control of lighting is desirable for adjustment to mood and visual need.

Grounds

In garden scenes, the choice of what to light goes to the best in intrinsic beauty, colorfulness, form, and composition. Garden sculpture, pools, or special landscape features influence choice of area as does its location with relation to points of viewing.

Guiding Principles. Non-uniform floodlighting should be provided to form pleasing patterns in light, shade, shadow, and color. The over-all may be subdued, stimulating, or dramatic, depending on the effect desired and manner of light application. Maximum luminance should not in general exceed 10 footlamberts (for the focal center of interest), unless the scene is viewed from a fully lighted terrace or interior. The highest luminance should usually be 2 to 5 times more than other features—avoiding black "holes" in the scene. When recommended illumination levels exist, it is wise to provide low (1 to 2 footcandles) over-all floodlighting to unify the composition. The eyes of viewers, neighbors, and passers-by must be protected from the glare of lamps; even low wattage ones, if bare or poorly shielded, are offensive. Heavy ground cover, shrubs, hedges, and foliage may provide "natural" shielding and daytime concealment.

Techniques. Subject matter should be analyzed for texture, form, line, reflectance, and background to determine what and how to give emphasis. Head-on floodlighting tends to make objects flat; modeling is obtained by lighting from both sides with more light from one side than the other. Light, striking a surface at a very narrow angle, emphasizes texture. Translucency, depth, form, and pattern are best accentuated by silhouetting, with very little light on the front side.

Equipment. Weatherability of equipment is vital for continued outdoor use. Aluminum, brass, copper, or stainless steel are generally applicable. When equipment is not concealed, it should be in keeping with the surround in color, form, or concept.

Illumination Levels. Levels recommended in Fig. 9–80 are a dependable guide. They may be reduced to one-tenth where viewers are dark-adapted, fantasy effects are sought, or no street lights or houses are adjacent.

Color. In addition to tinted, colored bulbs, color is obtained by tinted cover glass. Tinted colored light of the same hue as the object lighted heightens it. Yellow or low-wattage bulbs tend to deaden grass and foliage which is enhanced by blue-green or blue-white sources. Cool colors add depth, and pale blues simulate moonlight.

REFERENCES

1. Residence Committee Lighting of the IES: "Design Criteria for Lighting Interior Living Spaces," *Illum. Eng.*, Vol. 64, p. 511, August, 1969.
2. Committee on Residence Lighting of the IES: "Lighting Keyed to Today's Homes," *Illum. Eng.*, Vol. LV, pp. 527–559, 594–613, 638–654, October, November, December, 1960.
3. "IES Lighting Performance Requirements for Table Study Lamps," *Illum. Eng.*, Vol. LX, p. 463, July, 1965.
4. Crouch, C. L. and Kaufman, J. E.: "Illumination Performance for Residential Study Tasks," *Illum. Eng.*, Vol. LX, p. 591, October, 1965.
5. Moon, P. and Spencer, D. E.: "Modeling with Light," *J. Franklin Inst.*, Vol. 25–252, p. 453, April, 1951.
6. Color Committee of the IES: "Color and the Use of Color by the Illuminating Engineer," *Illum. Eng.*, Vol. 57, p. 764, December, 1962.

SECTION

16

LIGHTING SYSTEM DESIGN FACTORS

Lighting and Air Conditioning 16–1 *Wiring for Lighting 16–7*
Dimming Devices 16–9 *Maintenance of Lighting 16–11*
Economic Analysis 16–20

As an introduction to the subject of lighting design, Section 10 discusses design approaches and lighting systems. Sections 11 through 15 cover the illumination recommendations (levels, visual comfort, luminance ratios, etc.) for specific types of interiors. This section completes the subject of interior lighting design by presenting other system design factors common to all applications—coordination of lighting and air conditioning, wiring and dimming for lighting, and maintenance—and ends by providing an economic analysis procedure for comparing designs.

LIGHTING AND AIR CONDITIONING[1]

The Total Environment

Today's buildings represent an evolution in design—the total environment concept. Modern structures, in addition to providing useful space for efficient operation, are designed to satisfy basic human needs. These physical and psychological requirements include such diverse elements as sight, sound, thermal comfort, and esthetics. The integration of these factors requires a more complex technology in design than has been the practice.

In the past, the design approach to building has featured individual efforts of engineers who separately designed lighting, heating, cooling, and acoustics. This technique ignored the essential interdependence of these factors in achieving the balance necessary to provide total indoor environment.[2]

As technological advances reduced lighting costs, designers found it both practical and desirable to utilize more light for better seeing. The increased lighting provides, as a by-product, more heat. This heat is useful in replacing building heat losses during the heating season. Since the building's thermal needs vary not only seasonally but from area to area within the structure, control of lighting heat and integration with the heating and cooling system are essential. This is particularly evident when the thermal factors dictate that the interior zones be on a virtually continuous cooling cycle. Good design today requires provision for efficient utilization or dissipation of the lighting heat.

Comfort Parameters

Temperature. Any light source adds heat to the interior in which it operates. This creates a relationship between lighting and room temperature and, in turn, human thermal comfort. However, comfort depends not only on the absolute room temperature, but other factors, as well.

Heat gains and losses in a room result from heat transfer through walls and ceilings, heat transfer with air changes, solar gain from sunshine through windows, electric lighting, heat emitted by people, and heat from occupational processes in the space—production equipment, computers, electric typewriters, etc. The proportion of heat gain from each factor varies widely, depending on building design, use of the space, climate, etc. Procedures for determining the magnitude of various loads are outlined in the Guide of the American Society of Heating, Refrigerating and Air-Conditioning Engineers.*

Any source of energy which changes the temperature is termed *sensible* heat. Convection energy from lamps is sensible heat which raises the air temperature. Lamps also produce radiant energy which does not heat air directly, but is converted to sensible heat when intercepted and absorbed by a person or surface.

Use of lamps of higher efficacy, such as fluorescent or mercury, will provide a given level of illumination with less heat than with incandescent

* *ASHRAE Guide and Data Book*, American Society of Heating, Refrigerating and Air-Conditioning Engineers, Inc., New York.

NOTE: References are listed at the end of each section.

lamps. However, incandescent lamps frequently have characteristics which make them a superior choice for an application regardless of efficacy.

Humidity. Temperature is measured by a thermometer which, unless specified otherwise, is of the dry bulb type. However, since the human body regulates its temperature to maintain 98.6° F by evaporation of moisture from the skin as well as by radiation and convection, humidity is important. Air has the property of sharing space with water vapor up to a specific amount. The variation of water vapor below this maximum, expressed in per cent, is called *relative humidity.*

Relative humidity is measured by comparing wet bulb and dry bulb thermometer readings. The wet bulb type has its bulb area covered with a wet cloth. Evaporation will lower the wet bulb reading below the dry bulb, to a degree depending on the amount of water vapor already in the air. The relation between the two temperatures is a measure of relative humidity.

Relative humidity values over 90 per cent are called extreme, between 60 and 90 per cent humid, between 40 and 60 per cent normal, and under 40 per cent dry. These are generalizations since ambient temperature, air motion, and activity make appreciable differences in the apparent sensation experienced.

Comfort Limits. The relationship between comfort and temperature and humidity has been determined by a study based on human experiences and voiced reactions. A summary of such observations is shown in Fig. 16–1 for the air velocities as stated. The chart indicates that the *effective* temperature is a few degrees below the dry bulb temperature, the amount below being indicative of the dryness of the air. For example, a room having 75° F dry bulb temperature and 62.5° F wet bulb has a relative humidity of 50 per cent and an effective temperature of 70° F, according to the chart. In summer, approximately 93 per cent of people, on the average, would feel comfortable under these conditions and in winter, 85 per cent.

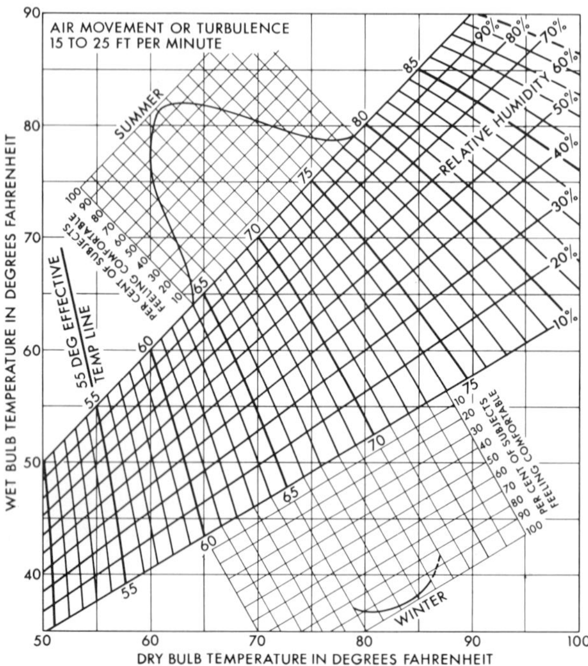

Fig. 16–1. Still-air comfort chart of the American Society of Heating, Refrigerating and Air-Conditioning Engineers.

Radiation from occupants to room surfaces (or the reverse) and between occupants has an important bearing on the feeling of warmth and may affect to a measurable degree the optimum conditions for comfort. The rate of heat loss by radiation depends on the exposed surface of the body and the difference between its temperature and the mean surface temperature of the surrounding walls and/ or objects, called *mean radiant temperature* (MRT). The mean radiant temperature is affected by cold walls and windows as well as by conventional heating units or luminaires, the temperature of which may be significantly different from that of the people in the space.

Tests by the American Society of Heating, Refrigerating and Air-Conditioning Engineers indicate that each degree of difference between the room dry bulb and the mean radiant temperature requires a compensating change in the opposite direction of ½-degree in effective temperature. This relation holds approximately for radiant panel heating which makes possible lower room effective temperature by increasing the mean radiant temperature.

Lighting Load on Air Conditioning

Fig. 16–2 gives an example of interior load distribution in a particular office building. The amount of exterior masonry is small enough to neglect its heat gain. Fluorescent lighting is provided consuming 5 watts per square-foot. It was assumed that occupants lowered the Venetian blinds on the sunny side of the building.

In typical buildings, factors such as solar radiation through windows, heat transmitted through walls and roof, and cooling of ventilation air, people, and machines comprise 50 to 70 per cent of the total cooling load. The refrigeration and/or air handling capacity required for lighting may be reduced if some of the principles described later in this section can be applied.

For the building of Fig. 16–2 the relative magnitude of all heat sources at the time of the total building peak cooling load is shown in Column A of Fig. 16–3. For comparison, analyses of total cooling load for other installations are shown in Columns B, C, and D. The lighting load in Columns B and C is 3 watts per square-foot.

Electric Lamps as Heat Sources. Electric lamps are efficient converters of electric power to heat energy. Each watt of electric power consumed by a lamp generates 3.4 British thermal units (Btu) per hour, just as any electric heating device. The energy takes two principal forms: (1) conduction-convection energy, and (2) radiant energy (including infrared, light, and ultraviolet). From this it is obvious that only a part of the energy generated by electric lamps is light. However, light itself produces heat. It does not heat air as convection sources do, but it raises the temperature of any surface which absorbs it.

A knowledge of the relative amount of each type of energy emanating from an electric lamp can be helpful in analyzing its performance and the effect it may have on thermal considerations. Figs. 16-4, 16-5, and 16-6 show approximate data for some representative fluorescent, incandescent, and high-intensity discharge lamps, respectively. These values are for lamps suspended in space under specific operating conditions.[3] Energy output for an individual luminaire in space, or for a system of luminaires installed in a room is likely to vary considerably from that for lamps alone.

Luminaires as Heat Sources. Performance characteristics of luminaires are well documented in terms of luminous efficiency, light control and candlepower distribution because equipment designers have been concerned primarily with the purposeful distribution of visible light. Now it is necessary to consider the total energy distribution of any luminaire destined to become a component of a building.[4] Performance studies should range from near ultraviolet through the far infrared wavelengths[5, 6]

From Figs. 16-4, 16-5, and 16-6 it can be seen that the greatest percentage of energy converted by electric lamps is radiation lying predominantly in the near infrared or far infrared regions—the proportions depend on the light source. Because the properties of lighting materials are different in the range from visible to invisible radiation, it is important to consider the underlying physics.

Fig. 16-2. Example of Cooling Load Distribution in a Modern Office Building

Heat Source	Exterior Offices*				Interior Offices
	North	East	South	West	
Glass	27%	56%	51%	56%	—
Lighting	61	37	41	36	84%
Occupants	12	7	8	8	16
	100%	100%	100%	100%	100%

* Per cent at time of maximum load in each office.

Fig. 16-3. Example of Cooling Load for Several Types of Installations

	Office Building A	Loan Office B	Chain Store C	Clinic D
Glass	11.8%	27.0%*	5.7%	14.8%
Lighting	42.3	12.9	23.3	16.9†
Roof and walls	0.7	16.8	12.0	33.4
Occupants	15.4	12.2	33.4	15.5
Ventilation	26.6	31.1	21.4	19.4
System power	3.2	—	4.2	—
	100.0%	100.0%	100.0%	100.0%
Glass + lighting	54.1%	39.9%	29.0%	31.7%

* Large expanse of window exposed west.
† Assumes 50% of lighting in use.

Fig. 16-4. Energy Output for Some Fluorescent Lamps of Cool White Color (Lamps Operated at Rated Watts on High Power Factor, 120-Volt, 2-Lamp Ballasts; Ambient Temperature 77° F, Still Air)

Type of Energy	40WT12	96 Inch T12 (800 mA)	PG17† (1500 mA)	T12 (1500 mA)
Light	19.0%	19.4%	17.5%	17.5%
Infrared (est.)*	30.7	30.2	41.9	29.5
Ultraviolet	0.4	0.5	0.5	0.5
Conduction-convection (est.)	36.1	36.1	27.9	40.3
Ballast	13.8	13.8	12.2	12.2
Approximate average bulb wall temperature	106° F	113° F	140° F	

* Principally far infrared (wavelengths beyond 5000 nanometers).
† Grooves sideways.

Fig. 16-5. Energy Output for Some Incandescent Lamps

Type of Energy	100-Watt* (750-hour life)	300-Watt (1000-hour life)	500-Watt (1000-hour life)	400-Watt‡ (2000-hour life)
Light	10.0%	11.1%	12.0%	13.7%
Infrared†	72.0	68.7	70.3	67.2
Conduction-convection	18.0	20.2	17.7	19.1

* Coiled-coil filament.
† Principally near infrared (wavelengths from 700 to 5000 nanometers).
‡ Tungsten-halogen lamp.

Fig. 16-6. Energy Output for Some High-Intensity Discharge Lamps

Type of Energy	400-Watt Mercury	400-Watt Metal Halide	400-Watt High Pressure Sodium
Light	14.6%	20.6%	25.5%
Infrared	46.4	31.9	37.2
Ultraviolet	1.9	2.7	0.2
Conduction-convection	27.0	31.1	22.2
Ballast	10.1	13.7	14.9

Fig. 16–7 shows that some materials used in luminaires can be good reflectors of light and good absorbers of far infrared. Several materials used to transmit light show significant differences in the far infrared reflected. Reference to these data will be helpful in understanding the transfers by which luminaires absorb low temperature radiation from fluorescent lamps and become radiating sources of heat energy.

Any quantitative analysis of luminaires as heat sources should assume conditions of temperature stabilization, constant voltage, and service position. In this state, total energy may not follow the distribution of light energy. However, it will be helpful to compare total energy distribution with the general classifications assigned to candlepower distribution curves.

Thermal distribution characteristics would narrow the CIE classifications (see Fig. 10–10) to (1) semi-direct, (2) direct-indirect, and (3) semi-indirect, as illustrated in Fig. 16–8. Totally direct or indirect lighting classifications are unlikely in total energy distribution curves. Of the two general diffuse lighting classifications, direct-indirect would be more appropriate.

Several test methods have been employed to assess the *total energy* distribution from a particular luminaire. One involves an adaptation of photometric techniques. Two others involve calorimetry, including a continuous water flow calorimeter[7] and continuous air flow calorimeters.[8, 9] Though procedures and equipment varied widely, test results were of the same order of magnitude.[1]

Testing guides have been published by the IES,[10] NEMA,[11] and ADC.[12]

Lighting Systems as Heat Sources. Visual and thermal conditions are two of the most important considerations in a planned interior environment. Visual comfort is partly due to quantity and quality of illumination. Thermal comfort is the result of a proper balance in temperature, relative humidity, and air motion. If these factors are to be fully evaluated, consideration should be given to the total energy distribution of luminaires, their relationship to room surfaces, and to the type of conditioning system contemplated.

While total energy should ultimately be considered within the building cube, comfort conditions will be primarily affected by that portion of the energy distributed into the occupied space. Thus, actual lighting and heating characteristics will be influenced by luminaire performance, ambient temperatures, surrounding materials, and surface reflectances.

Section 9 contains a formula for calculating the number of luminaires required to deliver a desired illumination level. The *ASHRAE Handbook of Fundamentals* contains a formula for calculating the lighting heat load from a number of luminaires.[13] Because the two are associated, it will be helpful to calculate the instantaneous rate of heat gain using a single relationship:

$$q_{el} = 3.413 \left[\frac{A \times E}{\phi \times CU \times LLF} \right] W \times UF \times EDF$$

where

q_{el} = sensible lighting heat load (Btu/h)
A = total area in square feet
E = average illumination in footcandles
ϕ = total lumens per luminaire
CU = coefficient of utilization
LLF = light loss factor
W = actual wattage per luminaire in service in watts

Fig. 16–7. Properties of Lighting Materials (Per Cent Reflectance (R) and Transmittance (T) at Selected Wavelengths)

Material	Visible Wavelengths						Near Infrared Wavelengths						Far Infrared Wavelengths							
	400 nm		500 nm		600 nm		1000 nm		2000 nm		4000 nm		7000 nm		10,000 nm		12,000 nm		15,000 nm	
	R	T	R	T	R	T	R	T	R	T	R	T	R	T	R	T	R	T	R	T
Specular aluminum	87	0	82	0	86	0	97	0	94	0	88	0	84	0	27	0	16	0	14	0
Diffuse aluminum	79	0	75	0	84	0	86	0	95	0	88	0	81	0	68	0	49	0	44	0
White synthetic enamel	48	0	85	0	84	0	90	0	45	0	8	0	4	0	4	0	2	0	9	0
White porcelain enamel	56	0	84	0	83	0	76	0	38	0	4	0	2	0	22	0	8	0	9	0
Clear glass (.125 inch)	8	91	8	92	7	92	5	92	23	90	2	0	0	0	24	0	6	0	5	0
Opal glass (.155 inch)	28	36	26	39	24	42	12	59	16	71	2	0	0	0	24	0	6	0	5	0
Clear acrylic (.120 inch)	7	92	7	92	7	92	4	90	8	53	3	0	2	0	2	0	3	0	3	0
Clear polystyrene (.120 inch)	9	87	9	89	8	90	6	90	11	61	4	0	4	0	4	0	4	0	5	0
White acrylic (.125 inch)	18	15	34	32	30	34	13	59	6	40	2	0	3	0	3	0	3	0	3	0
White polystyrene (.120 inch)	26	18	32	29	30	30	22	48	9	35	3	0	3	0	3	0	3	0	4	0
White vinyl (.030 inch)	8	72	8	78	8	76	6	85	17	75	3	0	2	0	3	0	3	0	3	0

Note: (a) Measurements in visible range made with General Electric Recording Spectrophotometer. Reflectance with black velvet backing for samples (b) Measurments at 1000 nm and 2000 nm made with Beckman DK2-R Spectrophotometer. (c) Measurements at wavelengths greater than 2000 nm made with Perkin-Elmer Spectrophotometer. (d) Reflectances in infrared relative to evaporative aluminum on glass.

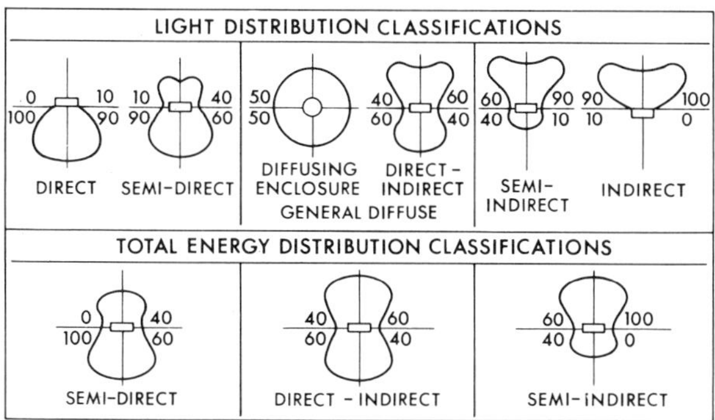

Fig. 16-8. Light distribution curves by CIE classifications of luminaires, compared with typical total energy distribution types. Upward and downward components are in percentage ranges.

UF = use factor, ratio of wattage in use for the conditions under which the load estimate is being made to the total installed wattage.

EDF = energy distribution factor, to account for fluorescent and other luminaires which are either ventilated or installed so that only part of their heat goes to the conditioned space.

A ceiling has an important function in a lighting system. It can have an equally important role in the distribution of thermal energy. Fig. 16-9 illustrates typical heat flows for various types of ceiling-to-luminaire relationships. Total energy distribution involves all three mechanisms of heat transfer. Assuming a 77°F room ambient, all luminaires in Fig. 16-9 approximate the direct-indirect energy distribution classification. The input of the suspended luminaire in Fig. 16-9(A) would be convected and radiated in all directions to be reflected or absorbed and reradiated. Essentially, all of the input energy would remain within the occupied space, and the energy distribution factor (EDF) would be unity.

Heat transfers from the surface mounted semi-direct luminaire in Fig. 16-9(B) involve radiation, conduction and convection. Assuming good contact with the ceiling, upper surfaces of the luminaire will transfer energy to or from the ceiling by conduction. Since many acoustical ceiling materials are also good thermal insulators, it may be assumed that temperatures within the luminaire will be elevated. Thus, lower luminaire surfaces will tend to radiate and convect to the space below at a somewhat higher rate. Unless the ceiling material is a good heat conductor and can reradiate above, essentially all of the input energy will remain in the space.

A different situation exists when components of the system are separated from the space. The recessed luminaire in Fig. 16-9(C) distributes some portion of input wattage above the suspended ceiling.

Lighting systems of luminous and louvered ceiling types are illustrated in Fig. 16-9(D). A similarity with the heat transfers of Fig. 16-9(A) is noted. Although luminaires are separated from the occupied space, plastics and glass used in luminous ceilings are good absorbers of infrared. White synthetic enameled louvers are also good absorbers, whereas aluminum louvers reflect a high percentage of infrared.

Perhaps the greatest variable among heating characteristics of a lighting system can be introduced by the air conditioning system itself.[14] Figs. 16-10 and 16-11 show how a combined assembly of lighting, ceiling, and conditioning systems becomes the final determinant of energy distribution factors. Two basic methods are illustrated for conditioning spaces from the ceiling plane. All-air systems, as the name implies, distribute conditioned air to heat or cool by convection. Air-water systems use circulating water to heat or cool by radiation and air to ventilate the room.

Methods of Controlling Lighting Heat

The methods of controlling lighting energy require the use of materials as heat transfer mechanisms. The materials which may be used in the transfer of energy are gaseous, liquid, or solid. These may be supplemented by filters designed to selectively reflect or transmit visible light and infrared for the purpose of separating undesired heat from the light flux.[15]

The fluids finding most application today are air and water. Air is necessary to the proper ventilation of spaces and is useful as a heat transfer mecha-

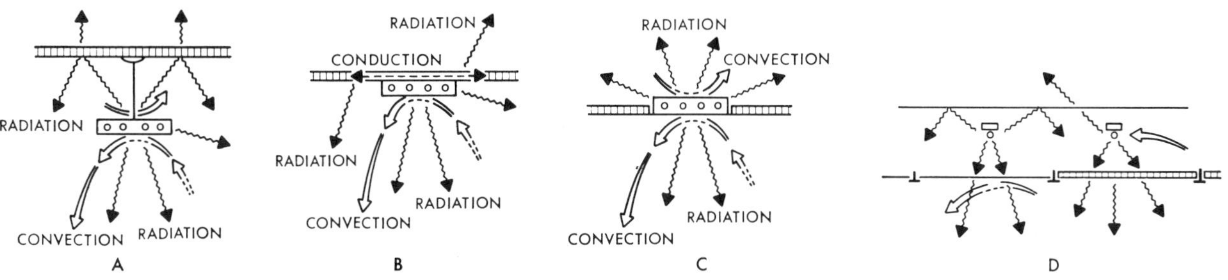

Fig. 16-9. Effect of ceiling-to-luminaire relationship upon lighting system heat transfer. Fluorescent luminaire has direct-indirect total energy distribution classification. (A) Suspension mounting. (B) Surface mounting. (C) Recessed. (D) Luminous and louvered ceiling.

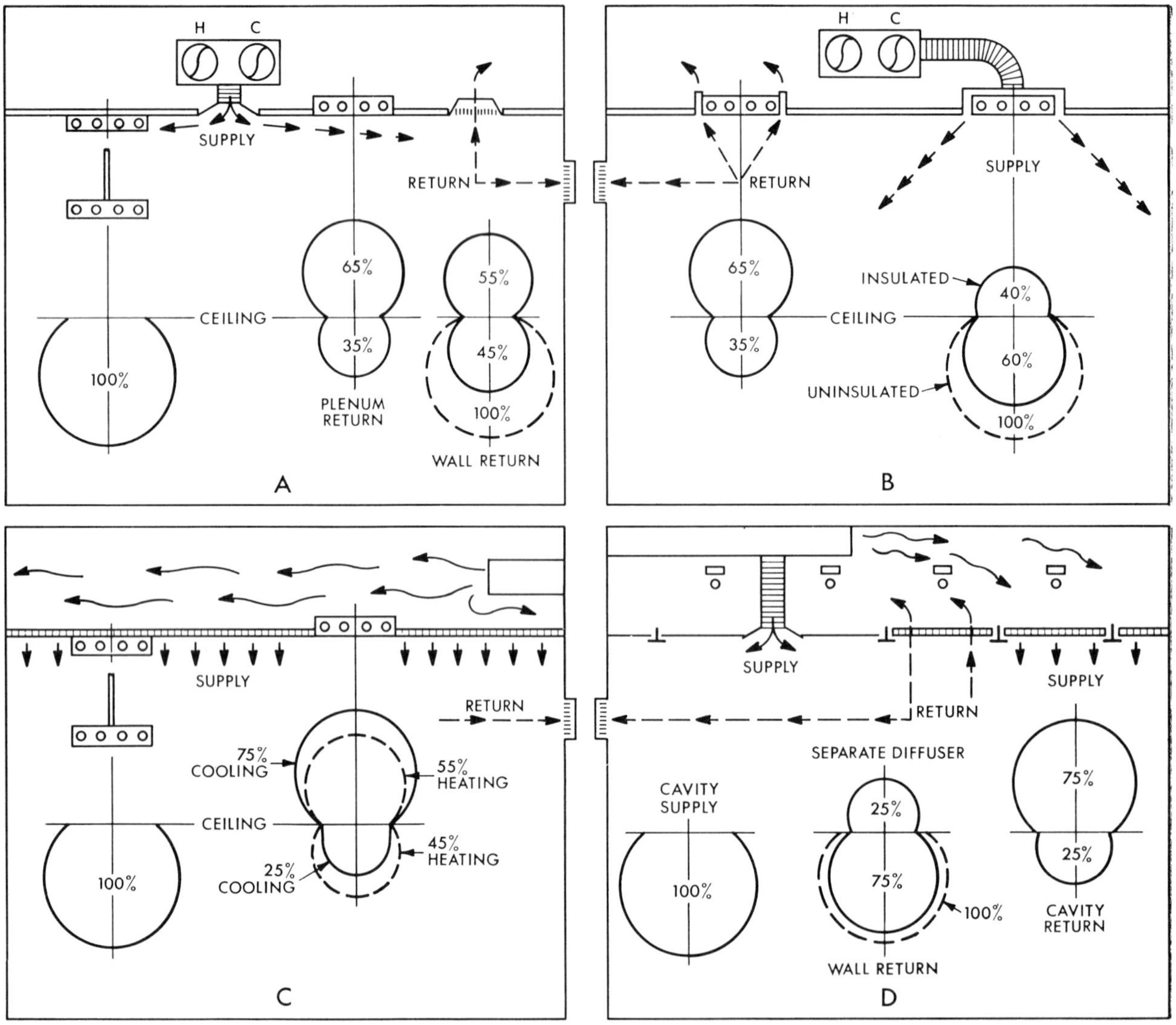

Fig. 16–10. Relative effects upon energy distribution factors for various lighting systems, as influenced by typical all-air systems, thermal properties of structures, and usage. (A) Separate diffusers and grilles. (B) Air-handling luminaires. (C) Ventilating acoustical ceiling. (D) Luminous and louvered ceilings.

nism when properly applied.[16] Fig. 16–12 illustrates several methods of employing air flow in and around luminaires and lighting systems.

The suspended luminaire in Fig. 16–12(A) will readily transfer energy to an air stream flowing over the lamps and heated surfaces. The same principle may be applied to surface mounted luminaires as shown in Fig. 16–12(B). In each instance, control should be maintained over the path of the heated air stream if any benefit is to be realized in the occupied space. Air returns located in the upper part of a room would accomplish this. This technique could prevent stratification of warm air in the upper part of a room, reduce ceiling temperatures (and radiant effects), and carry off much lighting heat before it enters the occupied space.

Fig. 16–12(C) illustrates the transfer of energy from heated surfaces to air flowing over the face or directly through the interior of a recessed luminaire. The amount of energy removed through in-

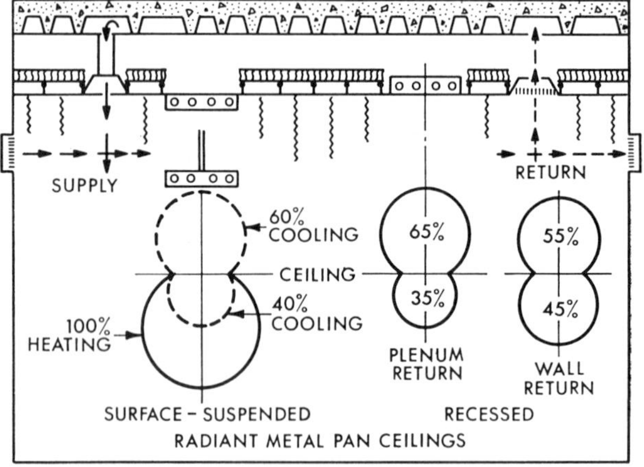

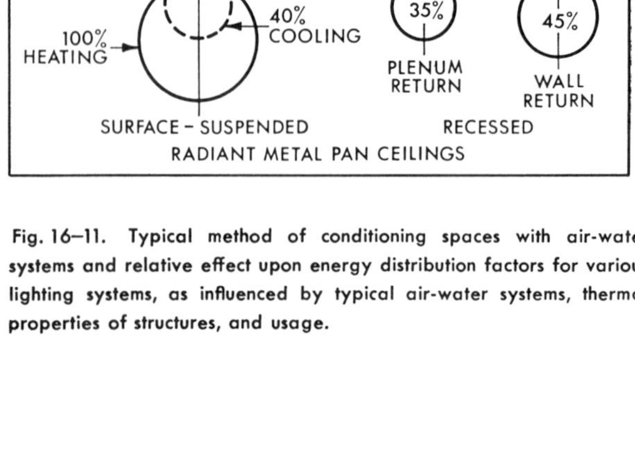

Fig. 16–11. Typical method of conditioning spaces with air-water systems and relative effect upon energy distribution factors for various lighting systems, as influenced by typical air-water systems, thermal properties of structures, and usage.

ternal flows can be substantial if the heated air is carried into a plenum space and returned to a mechanical room. The use of air is also applicable to luminous and louvered lighting systems as shown in Fig. 16–12(D).

The use of circulating water as a control mechanism for lighting heat[17, 18] is shown in Fig. 16–13. Water has an advantage compared with air of requiring substantially less volume for the same heat capacity. In most applications where water is a principal cooling (and/or heating) source, air is also needed to satisfy humidity control, air motion, and minimum ventilation requirements. However, the volume of air needed may be substantially reduced compared with all-air systems. (It is not suggested that air change be reduced below the minimum representing good practice.)

Fig. 16–14 shows the addition of infrared reflecting filters to a lens. This results in infrared energy being intercepted and reflected, while light is transmitted. The reflected infrared is largely absorbed in the luminaire, converted to conduction-convection energy, and becomes available for transfer out of the luminaire by means of air or water. The use of infrared filters in luminaires requires that heat removal methods be employed to prevent excessive heat build-up which may adversely affect lamp and ballast performance.

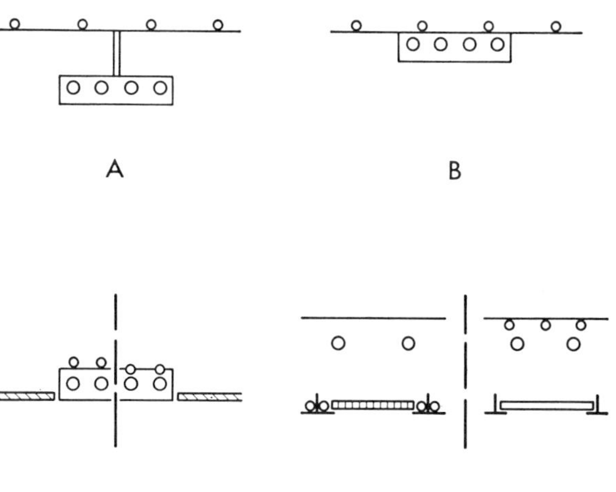

Fig. 16–13. Use of water as a heat control mechanism.

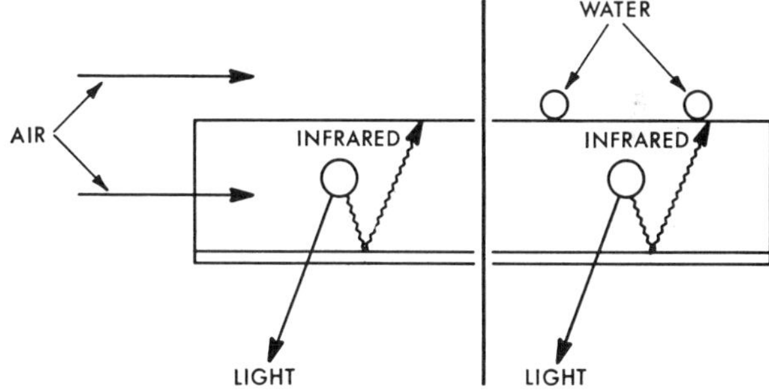

Fig. 16–14. An infrared reflecting filter can be employed on a luminaire to reduce the total energy in the light coming from the unit. Heat removal techniques should be used to prevent luminaire overheating.

Benefits of Integrated Designs

The benefits of integrating building heat in lighting design are: (1) improved performance of the air conditioning system, (2) more efficient handling of lighting heat, and (3) more efficient lamp performance if fluorescent lamps are the light source.[19]

The control and removal of lighting heat can reduce heat in the occupied space, reduce air changes and fan horsepower, lower temperature differentials required in the space, enable a more economical cooling coil selection because of the higher temperature differential across the coil, and reduce luminaire and ceiling temperature thereby minimizing radiant effects.

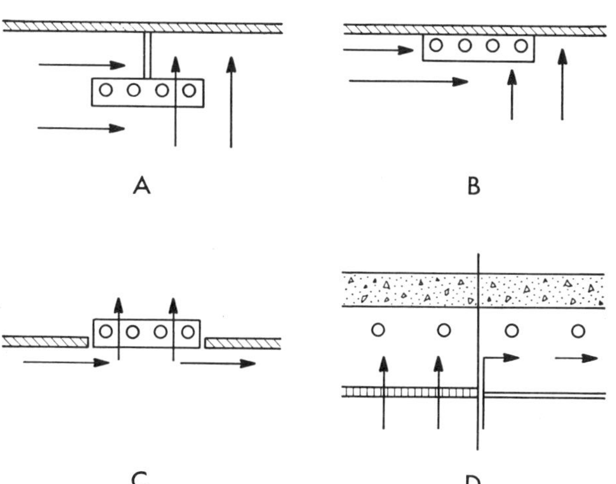

Fig. 16–12. Use of air as a heat control mechanism.

The degree to which any of these benefits may be obtained depends on many variables such as the quantity of energy involved, the type of heat transfer mechanism, the temperature difference between source and sink, and the velocity and quantity of fluids and/or air available for heat transfer. However, in most applications luminaire temperature will be higher than room temperature so fluids at room temperature can be effective in heat transfer. Any unwanted heat that can be removed at room temperature or above can be removed much more economically than at lower temperatures.

WIRING FOR LIGHTING

Every electric lighting system, regardless of its size, scope, simplicity, or complexity should have a well designed, trouble free electrical wiring system. Its size and capacity, its electrical characteristics (voltages, frequencies, phases, etc.), feeders, branch circuit layouts, and switch and dimmer controls, must all be specifically selected and designed to conform to the layout and design of the lighting system which it is to operate and control.

The lighting designer needs to know the basic fundamentals of electrical wiring system design for lighting. He needs this information to insure that he can get the maximum lighting flexibility and lighting desired or required for his lighting system results. Quite often the lighting designer is the consulting electrical engineer on the project. In this case, he usually is qualified to design and specify the electrical wiring system for the lighting. However, if the lighting designer, is not qualified to design the electrical wiring system, he should cooperate with an electrical consultant when designing the lighting system, to insure that the wiring system and controls will provide all the lighting variations and flexibility that is desired and intended. All electrical systems must be designed and installed in accordance with the provisions and requirements of the *National Electrical Code* and any other local or state code requirements, and the electrical consultant is qualified to include these provisions and requirements in his design and specifications.

The first step in the design of the electrical wiring system is to determine the total electrical load for the lighting system. On large projects, it may be desirable to break down the lighting load into logical sub-loads, for serving individually from separate load centers. These sub-loads can then be further broken down for individual panelboard control. On smaller projects, this usually resolves itself into selection and location of one or more lighting panels, each conveniently located near the center of the lighting load it serves.

When the lighting loads have been determined and load centers located with respect to the building structure, the next step should be to determine the type and rating of the electric power which is available from the electric utility. For relatively small projects, most utilities supply 120/208-volt, three-phase four-wire electric service. For larger projects, utilities may be able to supply 277/480-volt, three-phase, four-wire service. Lighting loads from these two systems are supplied from single-phase 120-volt and single-phase 277-volt circuits. Low-voltage light sources which may be incorporated in the lighting system may be operated at their rated voltages by using normal step-down transformers operating from any of the higher standard voltages.

In the United States 60-hertz power is almost universal for lighting systems. However, some lighting engineers and electrical consultants sometimes consider the use of high-frequency electric power for the operation of fluorescent lighting systems, especially when the advantages of high-frequency operation are important (see Section 8). One such advantage might be the reduction of the excessive weight of normal ballasts used for 60-hertz power. Another might be to obtain the maximum light output of the fluorescent lamps being considered for the project. In earlier installations, frequencies of 360, 420, and 840-hertz were used. More recently, a 3000-hertz system has been used, which reportedly converts power from 60- to 3000-hertz at about 93 per cent efficiency, compared with overall conversion efficiencies of only 80 to 85 per cent for the earlier model converters operating at the lower frequencies.

The total electrical load required for the lighting system is first calculated in total watts, or kilowatts. For purposes of selection of transformers, main switchboards, circuit breakers, and other similar electrical distribution considerations, kilowatts (kW) or kilovolt amperes (kVA), are used. However, when converted to "watts per square foot," the term becomes more meaningful and useful not only to the lighting designer, from the standpoint of an economical lighting system, but also to the air-conditioning and heating engineers. The chart shown in Fig. 16–15 shows maintained illumination levels for each of a variety of luminaires based on their wiring capacity expressed in watts per square foot. The data in this chart are only *approximate,* and are based on a room of average shape with a Room Cavity Ratio of 2.5, and for high reflectances of 80 per cent for the ceiling cavity, 50 per cent for the walls, and 20 per cent for the floor cavity. For more accurate results, data for specific luminaires, and for specific sized rooms and reflectances should be used. However, the chart is useful for quick appraisals when one type of luminaire is being considered versus other types, especially in early stages of the lighting design procedure.

Rooms with lower Room Cavity Ratios will require higher loadings (watts per square foot); with higher Room Cavity Ratios, the loadings will be correspondingly reduced. With other parameters constant, the required loading for any luminaire in an area with another Room Cavity Ratio is inversely proportional to the Coefficients of Utilization associated with the Room Cavity Ratios of the two areas. Similarly, changes in reflectances will also affect the required loading in inverse proportion to the Coefficients of Utilization associated with the reflectance conditions.

The wiring system should be designed to provide maximum flexibility with adequate capacity for present and anticipated future needs. Overloading or excessive extensions of circuits, in addition to the hazards involved, results in a lowering of the light output of both incandescent filament and fluorescent lamps.

The light output of an incandescent filament lamp is about three per cent less for each volt the lamp is operated under its rated voltage. Fluorescent lamp light output is also affected by undervoltage operation but not to so great an extent; starting difficulties and reduced lamp life may also result. For further information, see Section 8.

With gaseous discharge sources, the wattage consumed and the current carrying capacity required

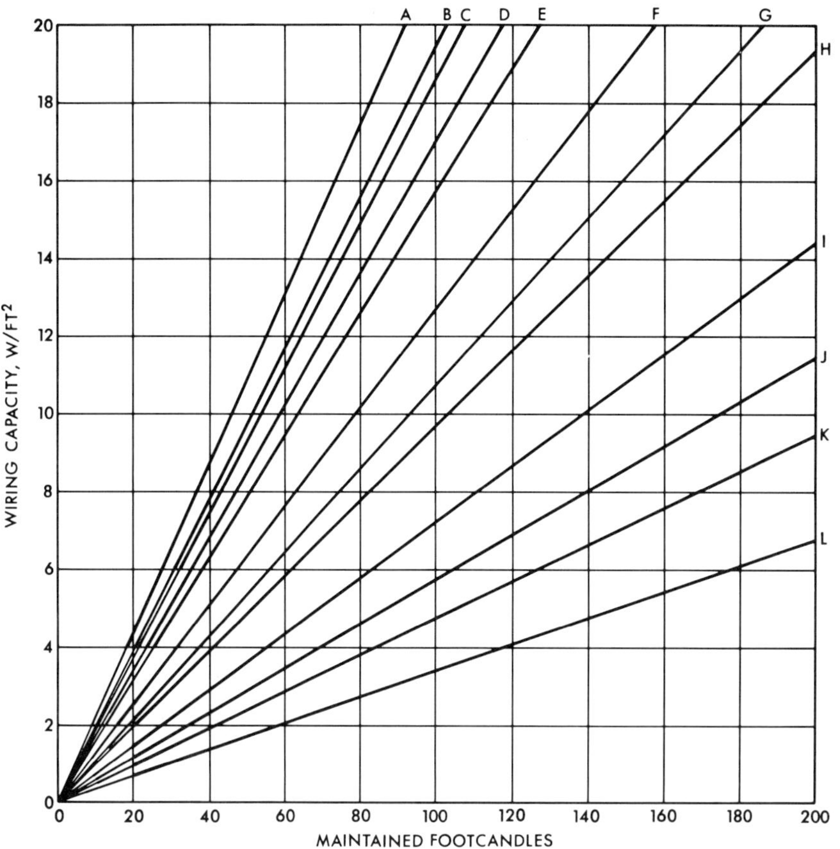

Fig. 16-15. Approximate wiring capacity to provide a given maintained level of illumination in a room of 2.5 Room Cavity Ratio by means of the following:

A—Indirect, incandescent filament (silvered bowl)
B—Direct, incandescent filament (with diffuser)
C—Direct, incandescent filament (downlight)
D—General diffuse, incandescent filament
E—Direct, incandescent filament (lens)
F—Direct, incandescent filament (industrial) Indirect, fluorescent (cove)
G—Indirect, fluorescent (extra high output)
H—Direct, fluorescent (extra high output) (louvered)
I—Direct, fluorescent (louvered)
J—Luminous ceiling, fluorescent
K—Direct, fluorescent (lens) Direct, HID (mercury)
L—Direct/Semi-direct, fluorescent (industrial)

will, in general, be greater than the rated wattage of the light sources. Provision for auxiliary wattage losses and for power factors lower than unity must be made in the wiring system. Mercury lamps require a warm-up period of several minutes. During this time, with certain types of auxiliaries, the current may be as much as twice the operating current. Wiring capacity must be provided for the starting current where it exceeds the operating current.

Minimum wiring capacity requirements are given in the National Electrical Code. The purpose of the Code is the practical safeguarding of persons and buildings and their contents from hazards arising from the use of electricity for light, heat, power, radio, signaling, and other purposes. The Code is not a design manual; compliance with its requirements does not insure adequacy either for present use or future growth.

Wiring design information based on current good practice for specific applications may be found in the following publications:

Electric Systems for Commerical Buildings—1964, IEEE* No. 241

Agricultural Wiring Handbook—1971, EEI†

Electric Power Distribution for Industrial Plants—1969, IEEE* No. 141

Electrical Systems for Power and Light—1964, McGraw-Hill Book Company, New York.

General wiring information may be found in:

Standard Handbook for Electrical Engineers—McGraw-Hill Book Company, New York.

Electrical Engineers' Handbook—John Wiley & Sons, Inc., New York

American Electricians' Handbook—McGraw-Hill Book Company, New York.

DIMMING DEVICES

The most practical method of controlling light output for nearly every purpose is to control the electrical input to the light source. There are several methods by which this may be accomplished, based on one of two principles: the input current can be varied by changing the amplitude of the current, or it can be varied by changing the amount of time during a cycle that it is permitted to flow.

Change in amplitude can be accomplished by resistance dimmers. The metallic rheostatic dimmer, one of the earliest controls developed, is still made and offers features which prescribe its use under certain circumstances. A resistance dimmer is connected in series in the circuit and the voltage which appears across the lamp is equal to the line voltage less the voltage drop (IR) across the resistance dimmer. To vary the amount of light, adjustment

* Institute of Electrical and Electronics Engineers, 345 East 47th Street, New York.

† Edison Electric Institute, 90 Park Avenue, New York.

to the rheostat is made by moving a contact which adds or subtracts resistance to the circuit. The dimmer capacity must be selected to match the lamp load fairly closely if dimming control to blackout is to be achieved. The resistance loss in the dimmer is an appreciable percentage of the lamp load at low luminance levels, liberating large amounts of heat which must be disposed of and consuming considerable power.

Change in amplitude can also be accomplished by continuously adjustable autotransformers. A single layer of copper magnet wire is wound over an iron core to form a toroid. Each wire turn is "bared" to form a commutator for the carbon brush. When a 120-volt ac line is applied to this winding, sliding the brush from turn to turn allows tapping off the desired ac output voltage. Wire size, number of turns, and brush dimensions have been carefully designed so brush contact is made with the next turn before leaving the previous conductor, assuring perfectly smooth, flickerless type of light control of almost infinite fineness from zero to full luminance. A dimmer of such design has excellent regulation—there is no visual change in light intensity as lamps are added or removed from the circuit, and it will also dim equally well all sizes of lamps at the same time. Equally important, the dimmer operates at very high efficiency so that there is little heat loss and one pays for only the energy consumed by the lamp.

Autotransformer dimmers are available for wall mounting in capacities from 200 watts to 1800 watts, and for heavier loads from 2000 watts to 5000 watts. These larger sizes can be operated manually or by motor drive, in which case there is a choice of motor speeds from 5 seconds to 60 seconds, from full brightness to blackout.

The saturable reactor was the first of the electronic type or phase angle control dimmers. It consists of a magnetic core associated with dc and ac windings. By adjusting the current of the dc winding, the inductive reactance of the ac winding may be varied smoothly from approximately zero to its maximum value. Saturable reactors are not in much use today in lighting applications because of their weight, size, and slowness of response; however, because they are good in the control of very heavy loads, they do fill some requirements.

Thyratron tube control is another of the earlier electronic methods. A gating type of control is obtained from a thyratron tube when ac voltage is applied to its plate and an ac voltage of adjustable phase is applied to its grid. As the phase angle between the voltage on the plates and grids is adjusted, the tube can be made to conduct various length portions of the cycle. Two thyratrons are connected back to back so that each one may conduct during each half cycle. While response is very rapid, all filaments in this tube system must be preheated approximately ½ minute or more. Should a power failure occur during use, the need to preheat becomes a great inconvenience.

Within the past twenty years the magnetic amplifier or self saturating reactor has been developed. It was a development of the saturable reactor through the use of solid-state power rectifiers and improved metals. Today, further improvements—smaller size, less weight, rugged construction—make this a leader in the field of dimming devices. The performance of the magnetic amplifier gives excellent regulation with a long maintenance free life. It will control from full intensity to blackout any load from 3 per cent to 100 per cent of its capacity rating. It is capable of handling an overload of 50 per cent for 15 minutes without damage, and there are no surge or "hot patched" limitations. The speed of response of the magnetic amplifier dimmer can be considered instantaneous. These units are designed in 3-, 6-, and 12-kilowatt sizes for theatre type switchboard installations. They can be either permanently connected or are available for plug-in.

The silicon controlled rectifier (SCR) dimmer is a relatively recent development in the field of light control. In its simplified form it consists of two silicon controlled rectifiers connected inverse-parallel or back-to-back. Each element thus controls the amount of current to pass during its half cycle.

They are available in small capacities (typically 500 watts) for mounting in a standard switchbox. They provide infinitely fine control from 100 per cent light output to blackout of incandescent lamp load. Because of the small size very little protection against current surge or heat can be incorporated, and they are therefore somewhat fragile. They can also introduce radio interference when installed near radio, television, or high fidelity systems. Some types of lamp bulbs will hum when controlled by SCR devices.

Large silicon controlled rectifier units for theatre-type switchboard installations have advantages similar to the magnetic amplifier—excellent regulation, long life, and minimum maintenance. They are light in weight and have approximately the same over-all dimensions as the magnetic amplifier. They are highly sensitive to momentary overload (inrush current) and voltage transients, and therefore, built-in protective devices as well as filters to prevent radio frequency interference account for much of the cost of heavy duty SCR equipment. These controls are only as good as the built-in protective devices. They are available in capacities from 2.5 kilowatts to 12 kilowatts and can be permanently connected or are available for plug-in.

For dimming fluorescent lamps, a number of arrangements are available using autotransformers or SCR devices. Depending on the dimming system, 50 to 90 per cent of full lamp luminance can be achieved, with dimming control down to 1/300 of full luminance. Autotransformer systems are available to dim two or more lamps, while up to 1200 lamps may be controlled on an SCR system. The lamps best suited to this application are the 40-watt and 30-watt T-12 rapid start. The two wattages should not be used on the same circuit. For smooth operation, the lamps on any circuit should be made by the same manufacturer, at the same time, in the same color, and of the same age in use. Group replacement is the most satisfactory procedure. Lamps should be operated free from drafts at 50° to 80° F and should be seasoned 100 hours at full output prior to dimming. Special dimming ballasts are required and components must be securely grounded. See Section 8.

MAINTENANCE OF LIGHTING

Physical maintenance is the only way to continue the effectiveness of any lighting installation. Planned maintenance includes all the means that can be used to keep the output of a lighting system as near to its initial level as is practical.

The use of *light loss factors* in planning installations is a necessary admission that no amount of physical maintenance can keep the output of a system up to its initial level. The value of the light loss factor used indicates the amount of the expectedly uncontrollable depreciation and the amount of effort it is expected will be devoted to try to overcome this depreciation.

Trained maintenance people are needed to implement systematic maintenance plans. Large plants, commercial buildings, etc., can afford to equip, train, and supervise such personnel, although many find it advantageous to hire outside specialists. In medium and small installations, the lighting maintenance contractor can supply planning services, equipment, and manpower, and by more thorough work, may justify his existence with a resulting lower cost per unit of light delivered.

The following paragraphs enumerate and discuss causes of light loss; advantages of planned relamping and cleaning; and operating programs, methods, material, and equipment. There also is a discussion of the remedies for mechanical and electrical difficulties that can develop in lighting systems.

Causes of Light Loss

It is important to recognize that the following factors contribute to the over-all loss of light (see Fig. 16–16, and Section 9, page 9–2:[20, 25]

1. Luminaire ambient temperature.
2. Voltage to luminaire.
3. Ballast factor.
4. Luminaire surface depreciation.
5. Room surface dirt depreciation.
6. Burnouts.
7. Lamp lumen depreciation.
8. Luminaire dirt depreciation.

The individual effect of each factor varies with the kind of work performed and the atmospheric location of the building. Air is dirtier in a foundry than in an air-conditioned office; the amount and type of dirt found in office air is different for an industrial area compared to suburbs; and black steel mill dirt is unlike some of the light-colored dust of a woodworking shop.

Lamp Lumen Depreciation. Light output of lamps decreases as the lamps progress through life. This decrease is called lumen depreciation and is an inherent characteristic of all lamps. (See Section 8.) Losses due to this effect can be reduced by lamp replacement programs such as planned relamping.

Luminaire Dirt Depreciation. A significant amount of light loss can generally be attributed to dirt accumulation on luminaire surfaces. In addition to the kind and amount of dirt in the area, the amount of light loss depends on luminaire design, lamp type and shapes, and luminaire finish.

Ventilated lighting units tend to collect dirt less rapidly than those with closed tops.[20, 21] This requires proper luminaire design and placement of the opening. The temperature difference between the lamp and surrounding air causes convection currents that help carry much of the dust and dirt through the luminaire rather than allowing it to accumulate on its reflector.

Dirt accumulation on a reflecting surface can be minimized if the reflector is sealed from the air, as in a dust-tight luminaire or a reflectorized lamp. A suitable protective shield should be used with reflectorized lamps to reduce glare and, if they are subject to moisture dropping on the bulb, to reduce lamp breakage.

Burnouts. Lamp burnouts contribute to loss of light. If lamps are not replaced promptly after burnout, the average illumination level will be decreased proportionately. In some instances, more than just the faulty lamp may be lost, *i.e.*, when series sequence fluorescent ballasts are used and one lamp fails, both lamps go out.

Luminaire Surface Depreciation. Materials used in luminaire construction differ in their resistance to deterioration. Processed aluminum finishes, while not as high in initial reflectance as the enamels, tend to have a slower rate of depreciation. Enamels, on the other hand, are usually easier to clean. In addition to the absorption of light, accumulation of dirt in certain luminaires may change the light distribution. For example, dirt accumulation on a specular aluminum reflector in a high bay luminaire can change the beam shape from narrow to wide. The loss in utilization here is much greater than the loss in efficiency.

The use of plastics in luminaire construction has increased markedly in the past decade. The types most widely used for light transmission and control are acrylics and polystyrenes followed by cellulosics, polycarbonates and vinyls. Over a period of time, the transmittance and color of all will change upon exposure to ultraviolet radiation and to heat. Acrylics are the most resistant to these changes. The other types of plastics are less resistant to varying degrees.

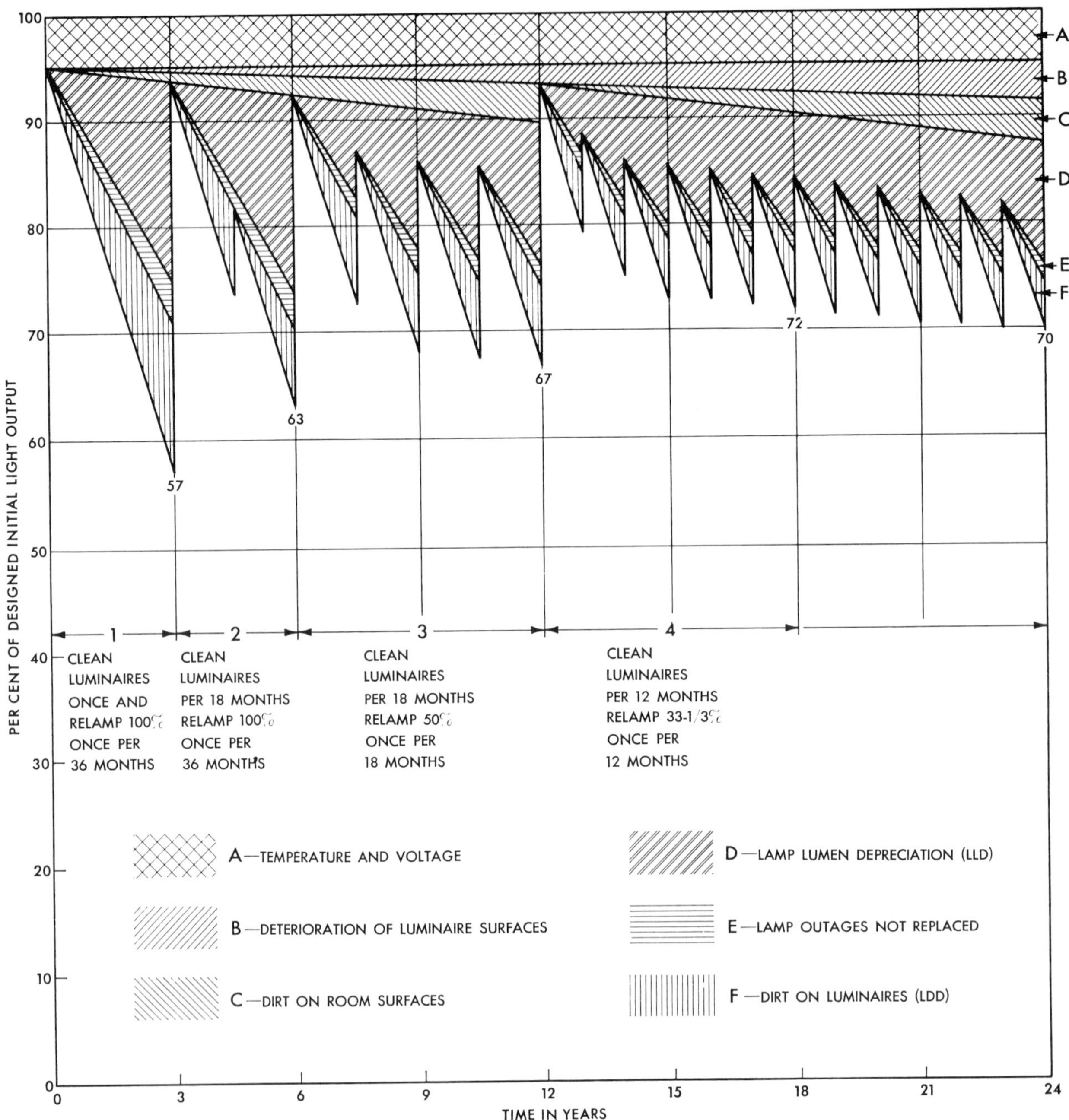

Fig. 16-16. Effect of light loss on illumination level. Example above uses 40-watt T-12 cool white rapid start lamps in enclosed surface mounted units, operated 10 hours per day, 5 days per week, 2600 hours per year. All four maintenance systems are shown on the same graph for convenience. For a relative comparison of the four systems, each should begin at the same time and cover the same period of time.

The rate of change in transmittance and color also depends upon the specific application: type of lamp used, distance of the plastic from the lamp, and ambient temperature of the plastic during the operating period of the luminaire. Use of improper cleaning materials and/or techniques can cause added changes in transmittance due to chemical action and/or scratching of the surface. Therefore, hard surfaced, chemically resistant materials are most resistant to changes from these causes.

Exactly how much change occurs in transmittance, reflectance, absorption, and color of the plastics from application to application is not known. For further information on plastics, see Section 6.

Room Surface Dirt Depreciation. General practice in all lighting fields is to use finishes with high reflectance to balance brightnesses and to utilize light most efficiently. Room proportion and distribution of light from the lighting units determine the amount of light which strikes the walls and ceilings. Dirt collection on room surfaces tends to reduce the amount of reflected light. While periodic cleaning and painting of walls and ceilings is necessary in all installations, it should be done more frequently in those areas where a larger per cent of light is reflected by these surfaces.

It should be remembered that there may be instances of luminaire and room appreciation, *i.e.*, the bleaching out of wall paints, curtains, or anodized aluminum parts, or where light color dust may have higher reflectance than the room surface on which it collects.

Temperature and Voltage. There are other factors which may cause a variation in light output from day to day in addition to the causes discussed above. One factor is the variation in voltage. Another factor is ambient temperature. Fluorescent lamps are particularly affected by changes in temperature. See Section 8 for data on the effect of voltage and temperature on lamps. For specific information about a specific luminaire design, it is best to consult the manufacturer.

Planned Relamping and Cleaning

Periodic Planned Relamping. A properly planned relamping program will arrest lumen depreciation and prevent many burnouts, thereby improving and maintaining illumination levels. Reduction of burnouts gives an added advantage in saving the time and expense otherwise involved in burnout replacement.

Periodic Planned Cleaning. Clean luminaires and room surfaces produce these results:
(a) *More light delivered per dollar.* Cleaning improves the light output of a system when the dirt on luminaires and room surfaces has a lower reflectance than do the surfaces themselves.
(b) *Pride in ownership.* Clean luminaires and room surfaces give noticeable evidence of pride of ownership and pride of good housekeeping. The appearance of the installation is improved. Only by cleaning may such improvement be gained.
(c) *Improved morale.* Clean luminaires and room surfaces can improve appearance, thereby directly improving morale and indirectly improving production and sales.
(d) *Reduced capital investment.* When a lighting system is periodically relamped and cleaned on a correctly planned schedule, it provides more light than when lamps are replaced only after burnout and when luminaires remain dirty. If the lighting designer knows that a planned maintenance program will be performed, he can design for a given average level of illumination using fewer luminaires. This will, of course, result in reduced capital investment as well as reduced operating cost.

Operations—Programs and Methods

Planning. The timing of relamping and cleaning should be in accordance with the plans of the lighting system designer. If intervals between operations are too long, excessive loss of light results. If intervals are too short, labor, equipment, and lamps are wasted.

Lighting systems are becoming more complex. As a result, the requirements are increasing for labor and equipment to relamp and clean them. The size of a lighting installation is the greatest single factor in determining how efficiently it can be cleaned. For servicing a few luminaires, the purchase or use of specialized equipment is impractical, and the training of labor in special skills is hardly necessary. As the number of luminaires increases, the value of the savings which result from mechanization and employee training also increases.

Lighting systems are installed in so many different kinds of locations and atmospheres that luminaire accessibility and rate of dirt accumulation vary in almost every circumstance. Thus, programs of relamping and cleaning should be planned to fit each set of circumstances.

Performing. The sequence of cleaning steps will vary with types of luminaires and locations. For example, one man with a ladder, sponge, and pail can handle, unaided, open strip units ten feet from the floor. On the other hand, a sizeable crew with an elaborate scaffolding assembly may be required for a transilluminated ceiling 30 feet above the floor. In general, the operations of a two-man team represent typical methods and are as follows:

(1) *Remove shielding material and lamps.* Louvers, plastic or glass panels, etc., and lamps are removed from luminaire and passed to floor man.

(2) *Make luminaire shock-free.* Care should be taken to prevent shock when working around electric sockets. Electrical circuit can be turned off or sockets can be covered with tape, dummy lamp bases, etc.

(3) *Clean basic unit.* If required, heavy deposits of dirt can be removed first from top surfaces of channel, reflector, etc., by vacuuming, wiping, or brushing. The entire unit then can be washed with a suitable solution, using brushes, sponges, or cloths. The unit then should be rinsed to remove any residue of solution and dirt.

(4) *Clean shielding material and lamps.* While ladder man works on luminaire as in (3), floor man takes shielding material and lamps to a cleaning station or cleans them at the ladder. Plastic materials should be allowed to drip dry after rinsing or be damp dried with toweling or some other material. Dry wiping can cause the formation of electrostatic charges. New lamps should be dry wiped before installation.

(5) *Replace lamps and shielding material.* When ladder man finishes cleaning, he installs clean shielding and new or cleaned lamps, passed to him by the floor man.

Incandescent and high-intensity discharge luminaires usually do not require as many cleaning steps as fluorescent units but the same general method applies to cleaning all lighting equipment.

Cleaning Compounds

A knowledge of cleaning compounds is needed to determine the ones best suited to any particular cleaning application. Correct cleaning compounds, properly used, save time and money. The following information has been found to apply in most cases to the commonly used luminaire finishes:

Aluminum. Very mild soaps and cleaners can be used on aluminum and will not affect the finish, if the material is thoroughly rinsed with clean water immediately after cleaning. Strong alkaline cleaners should never be used.

Porcelain Enamel. This finish is not injured by non-abrasive cleaners. Detergents and most automobile and glass cleaners do a good job under average conditions.

Synthetic Enamel. Some strong cleaners may injure this finish; particularly in cases where the enamel is left to soak in the solution. Alcohol or abrasive cleaners should not be used. Detergents produce no harmful effects.

Glass. As with porcelain enamel, most non-abrasive cleaners can be used satisfactorily on glass. Dry cleaners are usually preferred on clear glass panels, but not on etched or sand blasted surfaces. Most detergents will work well under average conditions.

Plastics. Very often dust is attracted by a static charge developing on plastic. Most common detergents do not provide a high degree of permanence in their anti-static protection. In most areas, however, if the plastic is cleaned at least twice a year with a detergent, a satisfactory relief from static dirt collection is obtained. Destaticizers are available which have greater permanence than common detergents in this respect. Plastic should not be wiped dry after the application of a rinse solution.

Cleaning Equipment

Time, labor, and expense of maintaining a lighting system can be greatly reduced by choosing maintenance equipment with features most suitable to the requirements of each system. Many different kinds of maintenance devices are available to facilitate the cleaning task. The choice of equipment will depend on several factors such as: mounting height, size of area, accessibility of lighting units, and obstacles in the area. Some available maintenance equipment are:

Ladders. Ladders are often used in lighting maintenance because their low weight, low cost, and simplicity make them desirable for simple maintenance tasks. See Fig. 16–17. However, safety and mobility restrictions limit their use in many cases.

Scaffolding. Portable scaffolding generally has greater safety and mobility than ladders. See Fig. 16–17. More equipment can be carried and the maintenance man has a firm platform from which to work. In general, scaffolds should be light, sturdy, adjustable, mobile, and easy to assemble and dismantle. Special requirements often dictate the type of scaffolding which can be used, for example, for mounting on uneven surfaces or for clearance of obstacles such as tables or machines. See Fig. 16–18.

Fig. 16–17. Maintenance from a ladder and a portable scaffold.

Telescoping Scaffolding. The telescoping scaffold provides a quick means for reaching lighting equipment at a variety of mounting heights. This equipment comes in various sizes that have platforms which can be raised or lowered either manually or electrically. See Fig. 16–19.

Lift Truck. Often the quickest and most efficient maintenance device is the lift truck. See Fig. 16–20. Although there are different types available, the method of operation is basically the same. The platform can be raised or lowered automatically and, in some types, the truck can be driven from the platform. While the initial investment for such equipment often is high, the maintenance savings often are large enough to make this device economical.

Disconnecting Hangers. Disconnecting hangers lower lighting units to a convenient work level, enabling the worker to maintain them with a minimum of equipment. See Fig. 16–21. When a lighting unit is raised into place, the hanger positions the unit and makes the proper electrical circuit connection automatically. An additional safety feature of this type of device is that the electrical circuit is disconnected when the luminaire is lowered.

Lamp Changers. Lamp replacement can often be simplified by the use of lamp changers. By gripping the lamps either mechanically or with air pressure, as in a vacuum type, the lamp changer can be used to remove and replace lamps. See Fig. 16–22.

Catwalks, Cranes, Cages, etc. Lighting maintenance can be incorporated as an integral part of the

Fig. 16–18. Portable scaffolding for uneven surfaces (left) and for high mounting heights (right).

Fig. 16–19 (left). Telescoping scaffold.

Fig. 16–20 (below.) Lift truck.

lighting system. This can be accomplished in many ways. Luminaires can be maintained from catwalks, cranes, or maintenance cages (see Fig. 16–23). The catwalks and maintenance cages can be installed alongside each row of lighting units so that maintenance can be performed from them with safety, speed, and efficiency.

Vacuum Cleaners and Blowers. A blower or vacuum cleaner is sometimes used to remove dust from lighting units. While some of the dirt can be removed in this way, the units still have to be washed for a thorough job. But, the periodic use of a vacuum cleaner or blower can prolong the cleaning interval.

Fig. 16–21. Disconnecting and lowering hangers.

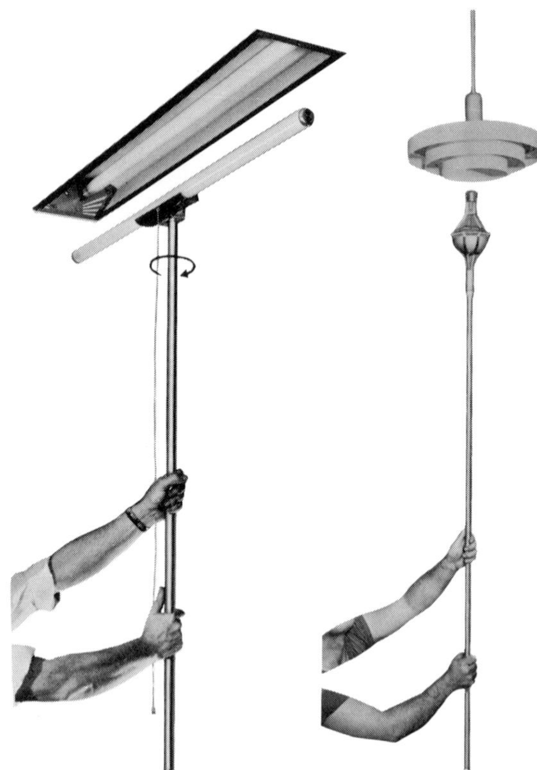

Fig. 16–22. Pole lamp changers.

Fig. 16–23. Maintenance cage.

Wash Tanks. It is desirable to have a wash tank specifically designed for lighting maintenance. Tanks should have both wash and rinse sections and be the proper size for the luminaire parts to be washed. Heating units, mounted in each section, are generally desirable to keep the cleaning solution hot. Louvers or reflectors can be set on a rack to drip dry after washing and rinsing while another unit is being cleaned. Special cleaning tanks have been designed for fluorescent luminaire parts and for flexible types of ceiling panels. See Fig. 16–24.

Ultrasonic Cleaning. This method removes foreign matter from metals, plastics, glass, etc. by the use of high frequency sound waves. Basic equipment consists of a generator, a transducer, and a suitable tank. See Fig. 16–25. The generator produces high frequency electrical energy which the tank mounted transducer converts to high frequency sound waves that travel through the cleaning solution. These waves cause the "cavitation" effect—formation of countless microscopic bubbles which grow in size, then violently collapse—thus creating a scrubbing action that forcibly and rapidly removes dirt from the material immersed in the solution.

Trouble Shooting and Maintenance Hints

Lighting maintenance incorporates lamp replacement and cleaning, planned maintenance to prevent trouble, as well as repairs to the lighting components. While the operation of fluorescent and high-intensity discharge lamps is more complex than incandescent filament lamps, trouble can generally be diagnosed and corrected quickly with simple test equipment.

Preheat Starting Fluorescent Lamp Circuits.

Trouble Shooting. See Section 8 for diagram and explanation of the circuit.

1. Replace existing lamps with lamps known to be operative.

Fig. 16-24. Hand wash-tank (upper) and machine wash-tank (lower).

Fig. 16-25. Ultrasonic cleaner.

2. Replace existing starters with starters known to be operative. No blink type starters are recommended for replacement. Refer to Section 8 for description of various types and the features of each.

3. Check luminaire wiring for incorrect connections, loose connections, or broken wires. Refer to the wiring diagram printed on the ballast.

4. Replace the ballast.

Maintenance Hints.

1. Deactivated lamps should be replaced as quickly as possible. Blinking lamps cause abnormal currents to flow in the ballast which will cause ballast heating and thereby reduce ballast life.

2. Blinking lamps will also reduce starter life.

Rapid Starting Fluorescent Lamp Circuits.

Trouble Shooting. Refer to Section 8 for the diagram and explanation of the circuit. Constant heater current is essential for proper starting of all rapid start type lamps. For 800 mA and 1500 mA types, the constant heater current is also essential for proper lamp operation.

1. If lamp requires 5 or 6 seconds to start, one cathode is probably not receiving the cathode heating current. This usually results in excessive end darkening of lamps after a short period of operation. With lamps out of the sockets, check heater voltages. This can be done with available testers which have a flashlight lamp mounted on a fluorescent lamp base. If a voltmeter is used, a 10-ohm, 10-watt resistor should be inserted in parallel with the meter. If proper voltage is available, check for poor contact between lampholder and base pins or contacts on the lamp. If no voltage is measured, check for open circuit (poor or improper connections, broken or grounded wires, open heater circuit on the ballast). Check for proper spacing of lampholders.

2. If one lamp is out and the other lamp is operating at low brightness or if both lamps are out, only one lamp may be deactivated. Refer to circuit diagram in Section 8 and note that 2 lamp circuit is a sequence starting, series operating design.

3. Replace the ballast.

Maintenance Hints.

1. Deactivated lamps should be replaced as quickly as possible. The 800 mA and 1500 mA lamps require both heater current and operating current for proper operation. If either is missing, poor starting or short lamp life will result. In the 2 lamp series circuit, one lamp can fail and the second lamp will operate at reduced current. This condition will reduce the life of the second lamp.

2. Lamps should be kept reasonably clean. The bulbs of all rapid start type lamps are coated with a silicone to provide reliable starting in conditions of high humidity. Dirt can collect on the lamp surface which could absorb moisture in high humidity atmospheres, thus nullifying the silicone coating, and prevent starting or cause erratic starting.

Instant Starting Fluorescent Lamp Circuits.

Trouble Shooting. Refer to Section 8 for the diagram and explanation of the circuit. Note that 2-lamp circuits can be either lead-lag or series-sequence design.

1. Replace existing lamps with lamps known to be operative.
2. Check lampholders for broken or burned contacts. Check circuit for improper or broken wires. Refer to wiring diagram on the ballast.
3. If ballast is suspected of being defective, replace with ballast known to be operative. Measurement of ballast voltages in the luminaire is difficult because the primary circuit of the ballast is open when a lamp is removed. Refer to circuit diagram in Section 8.

Maintenance Hints. Deactivated lamps should be replaced as soon as possible. In the 2-lamp series circuit, one lamp can fail and the second lamp will operate at low brightness. This condition will reduce the life of the second lamp and also will cause an abnormal current to flow in the ballast, giving rise to ballast heating and a reduction in ballast life.

Incandescent Lamps.

Maintenance Hints. Troubles with incandescent lamps are usually the result of misapplication, improper operating conditions, or poor maintenance practice. Most problems can be avoided by applying the following maintenance hints:

1. *Over-voltage or over-current operation.* Rating of lamps should correspond with actual circuit operating conditions. Over-voltage or over-current operation may shorten lamp life drastically. For example, a 120-volt lamp operated on a 125-volt circuit suffers a 40 per cent loss in life. Refer to Section 8.
2. *Shock and vibration conditions.* Under such conditions, the use of vibration service or rough service lamps is recommended. The use of general service lamps under these conditions results in short life.
3. *Sockets.* Lamps of higher wattage should not be operated in sockets designed for a specified wattage or excessive lamp and socket temperatures may result. Excessive temperatures may affect lamp performance or may shorten the life of insulated wire, sockets, etc. Refer to Section 8.
4. *Luminaires.* Only the proper lamps for which the luminaire was designed should be used. Contact of any metal part of a luminaire with a hot lamp may result in violent failure of the lamp.
5. *Cleaning lamps.* A wet cloth should not be used to clean a hot lamp. A violent failure may occur with a wet cloth.
6. *Proper burning position.* Lamps should be operated in their proper burning position as specified by the lamp manufacturer. Operation of the lamps in the wrong position may cause a lamp to fail immediately or after short life.
7. *Replacing lamps.* Whenever possible, lamps should be replaced with power "off." Replacing lamps with power "on," particularly high voltage types, can result in the drawing of an arc between the lamp base and the socket.
8. *Tungsten-halogen lamps.* Lamps should be installed with the power "off." It is also recommended that the bulb be held with a clean cloth or tissue to avoid fingerprints on the bulb. Fingerprints may result in bulb discoloration and a subsequent reduction in light output. Follow lamp manufacturers' instructions on the carton of each lamp.
9. *Dichroic reflector lamps.* Certain lamps utilize a dichroic reflector designed to radiate heat back through the reflector portion of the bulb. Luminaires using these lamps should be ventilated, or otherwise designed to provide adequate cooling of the socket and wiring adjacent to the bulb.

Mercury Lamps.

Trouble Shooting.

1. Replace lamp with lamp known to be operative. Be sure operative lamp is cool as hot lamps will not restart immediately.
2. Check lampholder for proper seating of the lamp and for proper contact.
3. Check ballast name plate reading, particularly if low temperatures are involved.
4. Check ballast wiring. If a tapped ballast is used, be sure ballast tap matches supply voltage at the ballast.
5. Check circuit wiring for open circuit or incorrect connections.
6. Replace ballast.
7. If lamps fail prematurely, check for the following: *a. Lamp breakage.* Check lamps for cracks or scratches in the outer bulb. These can be caused by rough handling, by contact with metal surfaces in bulb changer or luminaire, or by moisture falling on overheated bulb. *b.* Bulb touching the luminaire, the lampholder, or other hard surfaced object.
8. If arc tube is cracked, blackened or swollen early in life, or if connecting leads inside outer bulb are burned-up, check for the following: *a. Overwattage operation.* Check the ballast rating and the voltage at the ballast and if proper tap on ballast is used. *b. Excessive current.* Check if ballast is shorted. Check for possible voltage surges or transients on the supply line.

Caution: Do not replace with new bulb until circuit is checked and cause of trouble has been corrected.

Maintenance Hints.

1. If tapped ballasts are used, check should be made to be sure tap matches supply voltage at ballast. Low voltage will cause low light output, poor

Fig. 16-26. Lighting Cost Comparison

	Lighting Method #1	Lighting Method #2
Installation Data		
Type of installation (office, industrial, etc.)	....	....
Luminaires per row	....	....
Number of rows	....	....
Total luminaires	....	....
Lamps per luminaire	....	....
Lamp type	....	....
Lumens per lamp	....	....
Watts per luminaire (including accessories)	....	....
Hours per start	....	....
Burning hours per year	....	....
Group relamping interval or rated life	....	....
Light loss factor	....	....
Coefficient of utilization	....	....
Footcandles maintained	....	....
Capital Expenses		
Net cost per luminaire	....	....
Installation labor and wiring cost per luminaire	....	....
Cost per luminaire (luminaire plus labor and wiring)	....	....
Total cost of luminiares	....	....
Assumed years of luminaire life	....	....
Total cost per year of life	....	....
Interest on investment (per year)	....	....
Taxes (per year)	....	....
Insurance (per year)	....	....
Total capital expense per year	....	....
Operating and Maintenance Expenses		
Energy expense		
Total watts	....	....
Average cost per kWh	....	....
Total energy cost per year*	....	..
Lamp renewal expense		
Net cost per lamp	....	....
Labor cost each individual relamp	....	....
Labor cost each group relamp	....	....
Per cent lamps that fail before group relamp	....	....
Renewal cost per lamp socket per year**	....	....
Total number of lamps	....	....
Total lamp renewal expense per year	....	....
Cleaning expense		
Number of washings per year	....	....
Man-hours each (est.)	....	....
Man-hours for washing	....	....
Number of dustings per year	....	....
Man-hours per dusting each	....	....
Man-hours for dustings	....	....
Total man-hours	....	....
Expense per man-hour	....	....
Total cleaning expense per year	....	....
Repair expenses		
Repairs (based on experience, repairman's time, etc.)	....	....
Estimated total repair expense per year	....	....
Total operating and maintenance expense per year	....	....
Recapitulation		
Total capital expense per year	....	....
Total operating and maintenance expense per year	....	....
Total lighting expense per year	....	....

$$\text{* Total energy cost per year} = \frac{\text{Total watts} \times \text{burning hours per year} \times \text{cost per kWh}}{1000}$$

** The following formulas give the annual cost per socket for lamps and replacement, and can be used for determining the most economical replacement method.

$$\text{Individual replacement} = \frac{B}{R}(c + i) \text{ dollars/socket/year.}$$

$$\text{Group replacement (early burnouts replaced)} = \frac{B}{A}(c + g + KL + Ki) \text{ dollars/socket/year.}$$

$$\text{Group replacement (no replacement of early burnouts)} = \frac{B}{A}(c + g) \text{ dollars/socket/year.}$$

where B = burning hours per year
R = rated average lamp life, hours
A = burning time between replacements, hours
c = net cost of lamps, dollars
i = cost per lamp for replacing lamps individually, dollars
g = cost per lamp for replacing lamps in a group, dollars
K = proportion of lamps failing before group replacement (from mortality curve)
L = net cost of *replacement* lamps, dollars

No general rule can be given for the use of group replacements; each installation should be considered separately. In general, group replacement should be given consideration when individual replacement cost i is greater than half the lamp cost c and when group replacement cost g is small compared to i.

lumen maintenance, and reduced lamp life. High voltage will cause short lamp life.

2. The circuit should be reasonably free from voltage fluctuations. Replacement ballasts should match the particular voltage, frequency, and lamp type.

3. The proper lamp type should be used for the ballast in the installation. Certain lamps with the same wattage rating are available in two distinct lamp types, each designed for operation on specific ballasts having completely different electrical characteristics. Incorrect matching of lamp and ballast may result in short lamp life or lamps going on and off repeatedly.

4. Lamps should be handled carefully. Rough handling can cause cracks and scratches in outer bulb which will result in short lamp life.

Metal Halide Lamps. Recommendations given for mercury lamps also apply to metal halide lamps; however, the following additional information is pertinent:

1. Until there is industry standardization of metal halide lamps, it is important to be sure that the ballast will operate a given variety and brand of metal halide lamp satisfactorily.

2. Time to restrike after an outage may be much longer than for a mercury lamp.

3. Metal halide lamps may have a vacuum in outer bulb area, as a result all metal halide lamps should be handled carefully. In some cases enclosed luminaires are recommended.

4. When first turned on, metal halide lamps may require up to two days to reach characteristic color and light output. Also, a change in burning position may cause a change in color and light output.

High Pressure Sodium Lamps. Recommendations for mercury lamp trouble shooting and maintenance apply. In addition, it is important to replace lamps with power "off" since ballast may have a high voltage spike. Handle vacuum-type bulbs carefully. Restriking time for hot lamps is generally very short—a minute or two.

Self-Ballasted Mercury Lamps. Items 1, 2, and 5 of mercury lamp trouble shooting section and item 4 of maintenance hints apply. In addition, supply voltage should match voltage rating for the lamp within manufacturer's approved tolerances. Self-ballasted mercury lamps take longer than mercury lamps to restrike following a current interruption.

ECONOMIC ANALYSIS

The total cost of a lighting system is the sum of owning and operating charges. While initial investment may in some cases be a dominant factor in selecting specific luminaires or lamp types, there are capital expenses (amortization, interest, taxes, and insurance) which also should be considered.

Fig. 16-26 shows a typical analysis form, which lists the various elements normally included in making economic comparisons of two or more methods being studied for a given illumination system. Lighting cost calculations by different persons should be checked to insure similarity of energy rates, burning hours, depreciation basis, etc.[26, 27]

REFERENCES

1. Committee on Lighting and Air Conditioning of the IES: "Lighting and Air Conditioning," *Illum. Eng.*, Vol. LXI, p. 123, March, 1966.
2. Fisher, W. S. and Flynn, J. E.: "Integrated Lighting—Air Conditioning Systems," *Illum. Eng.*, Vol. LIV, p. 615, October, 1959.
3. Committee on Light Sources of the IES: "The Effect of Temperature on Fluorescent Lamps," *Illum. Eng.*, Vol. LVIII, p. 101, February, 1963.
4. Quin, M. L.: "Overcoming the Lighting Thermal Barrier," *Progr. Architect.*, Vol. XLII, p. 165, October, 1961.
5. Stewart, H. S.: "The New Optics," *Int. Sci. Technol.*, April, 1962.
6. Benedict, R. P.: "Temperature and Its Measurement," *Electro-Technol.*, Vol. 72, p. 71, July, 1963.
7. Bonvallet, G. G.: "Method of Determining Energy Distribution Characteristics of Fluorescent Luminaires," *Illum. Eng.*, Vol. LVIII, p. 69, February, 1963.
8. Mueller, T. and Benson, B. S.: "Testing and Performance of Heat-Removal Troffers," *Illum. Eng.*, Vol. LVII, p. 793, December, 1962.
9. Ballman, T. L., Bradley, R. D., and Hoelscher, E. C.: "Calorimetry of Fluorescent Luminaires," *Illum. Eng.*, Vol. LIX, p. 779, December, 1964.
10. Committee on Testing Procedures of the IES: "IES Approved Method for the Photometric and Thermal Testing of Air and Liquid Cooled Heat Transfer Luminaires," *Illum. Eng.*, Vol. 63, p. 485, September, 1968.
11. "Fluorescent Luminaires," NEMA Standards Publication, Pub. No. LE 1-1968, National Electrical Manufacturers Association, 155 East 44th Street, New York, N.Y.
12. *ADC Test Code TC-66*, Air Diffusion Council, 435 North Michigan Avenue, Chicago, Illinois.
13. *ASHRAE Handbook of Fundamentals*, Chapter 28, p. 496, American Society of Heating, Refrigerating and Air-Conditioning Engineers, Inc., New York, 1967.
14. Dally, G. R. and Ballman, T. L.: "Higher Lighting Levels and Interior Zone Air-Distribution Systems: Action-Reaction," *Illum. Eng.*, Vol. LX, p. 43, January, 1965.
15. Beesley, E. M., Makulec, A., and Schroeder, H. H.: "Lamps with Multilayer Interference-Film Reflectors," *Illum. Eng.*, Vol. LVIII. p. 380, May, 1963.
16. Hamel, J. S.: "Integrated Ceilings for Illumination," *Illum. Eng.*, Vol. LIV, p. 229, April, 1959.
17. Leopold, C. S.: "Design Factors in Panel and Air Cooling Systems," *Heating, Piping and Air Conditioning*, May, 1951.
18. Humphreville, T. N., Meckler, G., and Folsom, W. E.: "Installation and Operation of Integrated All-Electric Total Energy Environmental Systems," *Illum. Eng.*, Vol. LX, p. 453, July, 1965.
19. Fisher, W. S., Roehr, L. A., and Smith, J. M.: "Some Factors Affecting Heat Transfer in Luminaires," *Illum. Eng.*, Vol. LX, p. 51, January, 1965.
20. Steiner, J. W.: "Practical Reduction of Dirt Accumulation in High Wattage Luminaires," *Illum. Eng.*, Vol. XLVIII, p. 184, April, 1953.
21. Sell, F. W.: "Ventilation—The Key to Self Maintenance," *Illum. Eng.*, Vol. XLVIII, p. 500, September, 1953.
22. Edwards, R. N., Jr.: "Maintenance Factors for Interior Lighting," *Illum. Eng.*, Vol. LVI, p. 355, May, 1961.
23. Clark, F.: "Accurate Maintenance Factors," *Illum. Eng.*, Vol. LVIII, p. 124, March, 1963.
24. Clark, Francis: "Accurate Maintenance Factors—Part Two (Luminaire Dirt Depreciation)," *Illum. Eng.*, Vol. LXI, p. 37, January, 1966.
25. Clark, Francis: "Light Loss Factor in the Design Process," *Illum. Eng.*, Vol. 63, p. 575, November, 1968.
26. Barr, A. C. and Amick, C. L.: "Fundamentals of Lighting Analysis," *Illum. Eng.*, Vol. XLVII, p. 260, May, 1952.
27. Amick, C. L.: *Fluorescent Lighting Manual*, McGraw-Hill Book Co., New York, 1961.

SECTION

17

OUTDOOR LIGHTING APPLICATIONS

Design Procedure 17–1 Building Floodlighting 17–1
Monuments, Expositions, and Public Gardens 17–6 Fountains 17–7
Christmas Lighting 17–8 Service Stations 17–8

Building exteriors and surrounds are lighted for both utilitarian and decorative purposes. This section is concerned with the decorative aspects of floodlighting buildings, monuments, fountains, exhibitions and gardens, and with the attraction and utilitarian aspects of service station lighting. For other utilitarian outdoor lighting applications see Section 18 for Protective Lighting, Section 19 for Sports Lighting, Section 20 for Roadway Lighting, Section 21 for Aviation Lighting, and Section 23 for Lighting for Advertising.

Floodlighting Design Procedure

The following procedure may be helpful in designing a floodlighting installation:

1. Determine the effect desired.
2. Determine the location of the floodlights.
3. Determine the illumination level required. (See Fig. 17–1 and Fig. 9–80.)
4. Select the appropriate equipment.
5. Determine the number of units required and the wattage of the lamps to be used.
6. Check the uniformity and coverage of lighting. (Formulas and tabular data are given on pages 9–68 and 9–69.)

Building Floodlighting

The floodlighting of stores, shopping centers, offices, and other places of business is intended to attract attention to these buildings and to create a favorable impression with passersby. In this sense, the floodlighting of these buildings is often a subtle and dignified, yet highly effective, form of advertising.

Public buildings, churches, and monuments are generally lighted as an expression of civic pride, although here, too, the advertising aspect is present if the end result is to attract new people, business, and industry to the community.

Decorative floodlighting is essentially an art rather than a science. While calculations of luminance (photometric brightness) will generally be necessary, successful floodlighting depends to a large extent on the designer's ability to manipulate brightness relationships, textures, and colors. Thus, floodlighting is part of the architectural vocabulary and as such can be utilized to help create a nighttime image of a structure, sculpture, or garden, thereby extending the hours of their usefulness.

Principles of Floodlighting Design. The first step in floodlighting design, as listed in the above design procedure, is to establish the effect desired, or more accurately, to investigate the effects possible. The daylighted appearance may be helpful. Day-

Fig. 17–1. Recommended Illumination Levels for Floodlighting*

Surface Material	Reflectance (per cent)	Surround: Bright — Recommended Level (footcandles)	Surround: Dark — Recommended Level (footcandles)
Light marble, white or cream terra cotta, white plaster	70–85	15	5
Concrete, tinted stucco, light gray and buff limestone, buff face brick	45–70	20	10
Medium gray limestone, common tan brick, sandstone	20–45	30	15
Common red brick, brownstone, stained wood shingles, dark gray brick	10–20†	50	20

* See also Fig. 9-80. For poster panels, see Fig. 23-13.
† Buildings or areas of materials having a reflectance of less than 20 per cent usually cannot be floodlighted economically, unless they carry a large amount of high-reflectance trim.

NOTE: References are listed at the end of each section.

Fig. 17-2. Unobtrusive interior floodlighting transforms the Lincoln Memorial in Washington, D.C. into an inspiring hall of reverence at night. Ground mounted luminaires outside are aimed at columns to soften the stark silhouette effect.

lighting is usually a combination of strongly directional sunlight and diffuse sky light. The color of the former is warm, while that of the latter is cool. Shadows, therefore, are never "black", but simply less bright and bluer. Daylighting varies continually with the time of day and year and with the weather. Floodlighting, by contrast, is highly controllable and can therefore be utilized to present the building in a continuously favorable aspect, or in a character that is not seen during the day. See Fig. 17-2.

The major viewing locations or directions may help determine a floodlighting approach. If the building is to be seen primarily from moving automobiles, these viewing locations will most often be at some considerable distance and generally not head-on. The over-all impression created by the major elements of the building, especially the upper areas, may be the most important consideration. At close viewing distance the main concern may be with the ground level facade, sidewalk, and landscaping and with the effect of light on the building material and construction details.

It is usually desirable to locate the main, or key, lighting so that there will be some modeling effect. See Fig. 17-3. If floodlights are aimed on the line of sight from the viewer to the building, the effect will tend to be flat and often uninteresting.

The key lighting need not be a single source or several sources located at a single point. Modeling can be achieved if a series or line of floodlights are aimed in the same direction.

Deep shadows may be softened by low levels of

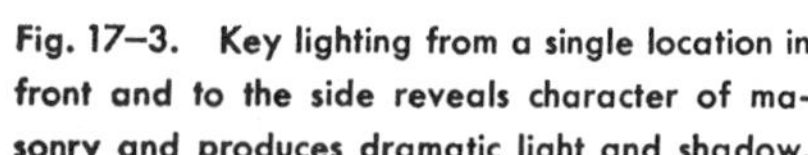

Fig. 17-3. Key lighting from a single location in front and to the side reveals character of masonry and produces dramatic light and shadow.

Fig. 17–4. Two examples of ground-mounted floodlighting. High pressure sodium floodlighting, on the left, produces essentially a uniform brightness pattern over the entire surface of a large building. On the right, architectural treatment is re-enforced with a single floodlight per bay, producing a regular pattern which predominates when seen from a distance. Masonry texture becomes important for close viewing.

diffuse floodlighting at an angle relative to the key lighting. A cool color may help recall the daylighted appearance.

Where it is desirable to have key lighting from two directions, modeling effects may be achieved by using contrasting tints or saturated colors.

Location of Floodlighting Equipment. Physical limitations imposed by either the relationship of the building to its surrounding or by local regulations may drastically limit the designer in the number of solutions or effects available. In general, four locations may be considered: on the building itself (see Fig. 13–14, page 13–11); on adjacent ground (see Fig. 17–4); on poles or ornamental standards (see Fig. 17–5); and on adjacent buildings (see Fig. 17–6). To completely express the structure in light, several, or all, of these locations may be required.

Floodlighting located close to the surface to be lighted and aimed at a grazing angle will tend to emphasize the texture of the surface (see Fig. 17–7), especially when viewed from nearby. However, defects in the surface—ripples, dents, dinges, misalignment of building panels, etc.—will also tend to be emphasized. Depending on the spacing and nearness to the surface, "scallops" from individual units may also be visible although these are not necessarily objectionable. See Fig. 17–4. Mounting the floodlights farther and farther from the surface de-emphasizes or flattens the texture, and usually improves brightness uniformity.

Setbacks generally offer ideal opportunities to mount and conceal floodlighting. See Fig. 17–8. It is desirable that each of the setback areas be lighted to a different level or should vary in brightness vertically so as to retain the setback appearance. Increasing the average luminances of the upper setback areas 2 to 4 times over that of the lower areas is generally considered to create an apparent equality of brightness of all the areas and to increase the impression of height.

Direct or reflected glare from floodlights can detract from even the most interesting floodlighting installation and can be a source of annoyance to neighbors. The lamp and reflector should be shielded or louvered so that brightness cannot be seen from any normal viewing locations. Special attention should be given to entrance areas of buildings used at night so as to avoid direct glare which could make steps and curbing hazardous to pedestrians.

Floodlights should be located or shielded so that units do not light adjacent units thereby revealing their presence.

Landscaping, walls, or wells are highly useful in minimizing direct glare. In addition, they help conceal the equipment during the day, a most important aspect of the total design.

Fig. 17–5. The cool color effect of metal halide lamps, pole-mounted to light the front and ground-mounted to light the sides, is contrasted with the warmth of tungsten-halogen lamps lighting the portico from the ceiling behind the columns.

Fig. 17-6. An example of "flat" floodlighting with equipment mounted on buildings across the street.

Fig. 17-7. Grazing floodlighting from units concealed in moat emphasizes the texture of the sculptured facade.

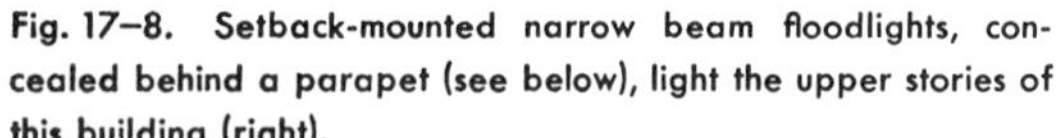

Fig. 17-8. Setback-mounted narrow beam floodlights, concealed behind a parapet (see below), light the upper stories of this building (right).

Reflected glare from specular surfaces such as polished marble, tile, glass, and metal trim should also be considered. Sight lines should be checked to insure that reflected images of lamps or reflectors will not detract from the over-all effect. This is especially important in areas where there is considerable pedestrian traffic and the surface is seen from nearby.

Fig. 17-9. Major brightness pattern of this building is created by deliberately operating the interior lighting systems at night.

Interior Floodlighting. Modern construction, utilizing large expanses of glass, as windows or curtain walls, can often be expressed most effectively at night in terms of interior brightnesses. The simplest technique is to operate all or parts of the interior lighting of spaces next to windows. See Fig. 17-9. In spaces which are electrically space conditioned this technique may in some cases also provide a measure of the nighttime heating required in cold weather.

Where there is no control of the interior lighting, such as in tenanted office buildings, separate interior lighting may be provided. This could take the form of supplementary wall and ceiling lighting such as coves and valances. Lamps can often be concealed in soffits, in window headers, or in sills. These would light the window reveal and blinds or curtains which are kept closed at night.

A major advantage of the interior approach to floodlighting is that indoor lighting equipment can be utilized and maintenance can be simplified.

Fig. 17-10. Approximate Factors by Which Clear Bulb Incandescent Filament Lamp Wattage Must Be Multiplied to Compensate for the Absorption of Various Color Filters

Desired Effect Relative to Clear Bulb	Approximate Multiplying Factor			
	Amber	Red	Green	Blue
Equal illumination	2	6	15	25
Equal advertising or decoration	1.5	2	4	6

Illumination Level. To serve as a design and calculation guide, recommended illumination levels for building floodlighting are given in Fig. 17-1. Because of the decorative and advertising nature of building floodlighting, these should be considered as guides only. Variation from these levels is to be expected depending on the type of building, its location, and the ultimate purpose for floodlighting.

Color in Floodlighting. The levels recommended in Fig. 17-1 are based on "white" light. Where saturated colors are desired, color filters can be applied. Since these produce the wanted color by absorbing or reflecting the unwanted colors, the amount of light transmitted by the filter is sharply reduced. The transmittance of saturated color filters usually falls within the following ranges: amber, 40 to 60 per cent; red, 15 to 20 per cent; green, 5 to 10 per cent; and blue, 3 to 5 per cent. Fig. 17-10 indicates the factors by which incandescent filament lamp wattage must be increased when it is desired to provide equal illumination in white and color. Relatively less colored light than white light is needed for equal advertising or decorative effect. The second line of Fig. 17-10 gives factors by which clear-bulb incandescent filament lamp wattage should be multiplied in order to achieve an advertising or decorative effect in color comparable to that obtained with a given wattage lamp emitting white light.

Actually even "white" light sources produce a color of light that is characteristic with the source. Incandescent lamps, including the tungsten-halogen types, produce light which is warm in appearance, rich in red, and weak in blues and greens. Most "white" fluorescent lamp colors used in floodlighting appear cooler or bluer than incandescent. Fluorescent lamps are also efficient producers of colored light. The high-intensity discharge sources, such as mercury, metal-halide, and high pressure sodium, have individual color characteristics. For example, clear mercury emphasizes blues and greens, deluxe white mercury is similar to cool-white fluorescent, and high pressure sodium is warmer in appearance than incandescent, but is weak in the deep reds and blues. See Section 8, page 8-20. These color differ-

Fig. 17–11. The major program area at Expo '70 in Osaka, Japan featured the "Grand Roof," a steel lattice structure lighted from below by 800 metal halide floodlights.

ences can be used to visually separate or accentuate elements of building design, or a combination of sources can be employed to create the desired color.

Maintenance. Although maintenance is important in most lighting systems, the ultimate success of a floodlighting installation often depends upon just how well it is maintained. To this end the functions of relamping, cleaning, and reaiming should be considered as a fundamental part of the total design. See also Section 16, Maintenance of Lighting.

Monument and Statue Floodlighting

The design of floodlighting for monuments and statues aims at the achievement of a natural lighted appearance. See Fig. 12–22. The relationship of shadows and brightness is of utmost importance. This is particularly true where the human form is concerned. When obelisk-type structures are lighted, depth can be maintained by adjusting the brightness of each face so that there will be a contrast between each face in the field of view of an observer at any one time.

Exposition and Public Garden Floodlighting

In the lighting of spectaculars, such as expositions and public gardens, the primary considerations are to create beauty and to enhance form and design of surface. Lighting is utilized to reveal or conceal, to accent or subdue, to create moods, to control traffic, to provide heat (infrared radiation), and to illuminate and create animation. See Fig. 17–11.

Light can be used as a tool to transform a daytime environment to a completely new setting at night. Buildings and grounds become a fairyland under the spell of properly applied lighting schemes. For example, the ordinary daytime appearance of a group of buildings and exhibits, can at night, under the creative genius of the architect and illu-

Fig. 17–12. Translucent glass screen walls highlight traditional Japanese architecture at Expo '70 in Osaka. Fluorescent lamps inside the walls are spaced to provide maximum brightness at the front and gradually decreasing brightness down the sides.

minating engineer, become a delight to the visitor, designed to transport him into an environment of specific moods. Rather than creating areas of high brightness for attraction and resultant gaudiness, often the goal is to achieve an elegant atmosphere of mystery. See Fig. 17–12.

When lighting outdoor areas for decoration, it is not necessary to maintain the natural daytime appearance. Originality and novelty are the dominant factors. Beautiful and unusual effects may be achieved by imaginative application of the almost infinite list of light sources and equipment available today. The engineer and architect who may be involved in the over-all planning should maintain extremely close cooperation because the ideas and equipment used are limited only by their mutual imagination.

General Principles. Decorative effect is the primary objective rather than efficacy in terms of lumens per watt as is usually the case in forms of functional lighting.

Probably the most important single rule that should be followed in decorative as well as functional type lighting installations is to conceal the light source. (See Fig. 17–13 and Location of Floodlighting Equipment on page 17–3.) Lighting equipment may be shielded from view by trees, shrubs, rocks, buildings, and structures, or shubbery planted expressly for this purpose. Conventional equipment may be placed in a suitable cowling, which takes on an appearance in keeping with that of the area. For example, an optical assembly might be mounted inside of a hollowed stone or the cowling may be made to simulate a toadstool. In addition, all equipment might be painted as an aid in camouflaging.

Sometimes bright spots appearing immediately in front of lighting equipment mounted on the ground can be virtually as objectionable as viewing the light source itself. Also, spill light on surrounding branches and foliage may create undesirable effects. Therefore, easily adjusted equipment is preferred for garden and architectural lighting. Sometimes special louvers and shields will be required.

Illumination Level. Every scene at an exposition or public garden has a center of attraction. See Fig. 17–11. The lighting installation should be designed to create a higher level of illumination at that location with adequate surrounding light to prevent this attraction from appearing detached from the immediate environs.

Color. Considerable caution should be exercised when selecting colors of light sources. For example, clear mercury sources should not be used to illuminate a reddish brick facade. However, clear mercury sources are excellent for enhancing the green of foliage. During the autumn when leaves become yellow, red, and brown, incandescent filament sources, which are rich in red and yellow, are a better choice than the clear mercury lamp. At the same time, the imaginative designer should not overlook the possibility of using other high-intensity discharge sources which efficiently produce a wide range of less subtle colors including yellow, red, green, and blue. Furthermore, the entire rainbow of colors is available from fluorescent sources.

In choosing the light sources to illuminate flower beds which contain a variety of delicate natural colors care should be taken and, in some cases, it may be found necessary to experiment before the final choice is made. Generally, daylight fluorescent lamps provide a good "white" for most cases. Pale blue light on some structures will simulate moonlight; yellow or amber light may be used effectively to light the surrounding area.

Fountains

The use of water has been an important architectural element for centuries and certainly has been one of the major attractions at world's fairs.[8] Still water acts as a mirror and will reflect clear images of lighted objects. When churned into spray and foam, water is an excellent diffuse reflector and will absorb incident light and will appear to change color to match that of the light.

Fig. 17–13. Integrated landscaping and building lighting with light sources hidden from view.

A single jet may be made attractive by installing directly beneath it, in a suitable water-tight enclosure, a narrow beam of light projecting along the stream. In the case of more complex displays, the effects which may be obtained by varying water flow, number of in-service jets, and colors of light are unlimited.

Floodlight Types. Attempts have been made to illuminate sprays and jets of water from floodlights operating in air as well as under water. Although the initial investment is usually more, the results are generally more desirable when underwater locations are used.

When wide area water effects are to be lighted such as water curtains, sprays, and jet rings, wider beam spread floodlights should be used. Narrow beams are needed to adequately illuminate high projected streams.

The cascade formed by water spilling over the lip from an inner basin can also be effectively lighted by installing colored lamps under the lip. The lip should be designed to churn the curtain of water thereby making it turbulent and diffuse in appearance. Lamp spacing around the basin should not exceed the distance back from the waterfall. For optimum uniformity in apparent brightness, fluorescent strips may be preferred in this application.

Size and Location of Fountains. As a practical point, the radius of the outer pool is usually not less than the maximum water projection height. In extremely windy locations the radius may be increased to two times the maximum water projection, so as to not continually wet the landscape and/or spectators.

If a fountain is to be illuminated the actual location is best selected where there is a minimum of ambient light, especially when colored light is to be used. Any environmental light tends to wash out the relatively low brightness produced by underwater color floodlighting either by reducing its contrast with the surroundings or by adapting the spectator's eyes to a higher level.

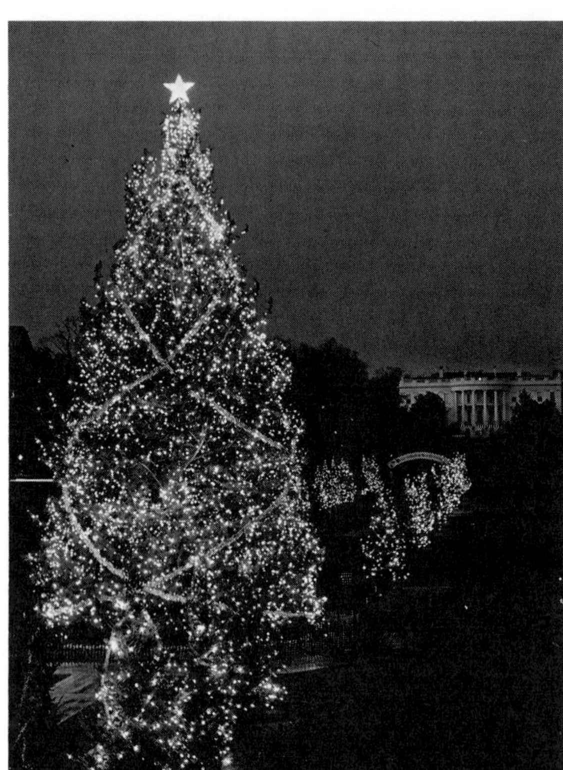

Fig. 17–15. Christmas trees on the Ellipse in back of the White House—lighted with D27 clear flasher lamps and with 10S11 sign lamps.

Fig. 17–14. An attractive entrance is lighted with 7-watt C-7½ lamps in foil covered stylized tree on door and 10-watt C-9½ lamps in festooning and hard board tree in front of entrance. Two 150-watt PAR lamps are used, a green one behind the tree and a clear spot out front. This produces a shadow tree in green.

Christmas Lighting

Christmas is the one season of the year when everyone is especially conscious of decorative lighting. See Figs. 17–14, and 17–15. Several publications, appealing especially to the homemaker, offer helpful hints on the many uses of miniature lamps at Christmas time. Other applications include the use of miniature lamps for other festive occasions and for colorful patio and garden lighting. For the latter application low-voltage miniature equipment is available. It should be noted that not all string sets and devices have Underwriters' Laboratories, Inc. approval.

Service Station Lighting

The design of service stations tends toward a specific architectural style, such as colonial, ranch-type structures, etc., with landscaping, fencing, walkways, and other treatments utilizing outdoor lighting. Light sources for such installations should be considered in the initial design. See also Exposition and Public Garden Floodlighting, page 17–6, and Section 8.

The basic objectives of service station lighting are:

1. To aid the rapid identification of the station and its product.
2. To facilitate safe entrance and exit into and out of the station.
3. To provide adequate light on the pump island and its adjacent area to permit the attendent to easily perform his tasks (see Fig. 9–80).
4. To provide a well lighted building interior and exterior.
5. To provide an over-all installation that is attractive to the prospective user.

Additional lighting objectives are:

A. To provide driveway lighting between the approach and island to complete the traffic pattern.
B. To illuminate the auxiliary service areas.
C. To illuminate additional driveway areas.
D. To illuminate parking areas.

See Fig. 17–16 for a correlation of service station areas and the above objectives.

Identification. The first objective to be considered in designing service station lighting is the identification of the product, services rendered, and facilities available. This is achieved by adequately illuminating the product identification sign and the principal vertical surfaces of the station.

Signs. Location, legibility, and brightness are the primary factors to be considered in sign lighting. Signs should be located to provide maximum advertising and identification value from all possible traffic angles. Recommended illumination levels for both internally and externally lighted signs are given in Section 23, on outdoor signs.

When utilizing a sign pole for the mounting of approach or driveway lighting units, caution should be taken so that the lighting units will not distract from the advertising appeal of the sign. The use of a minimum pole height of 15 feet should permit the mounting of a properly designed approach light below the sign where there is less tendency to distract from the sign itself. When a sign pole is located 10 feet or more from the approach curb, use of a separate light pole should be considered.

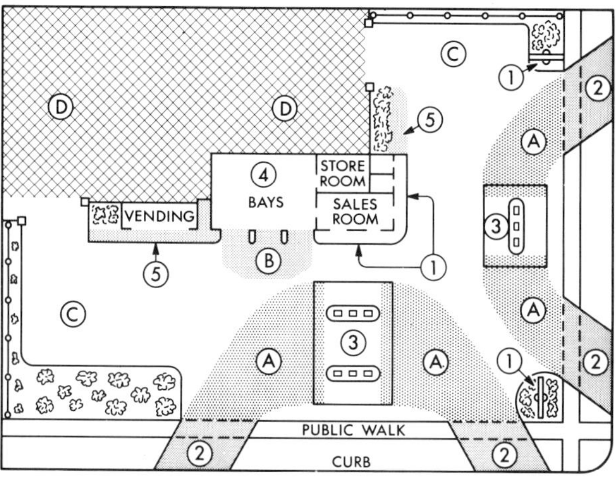

Fig. 17–16. Service station plot plan defining areas for lighting. Numbers and letters refer to text above.

Building Lighting. Lighting the prominent faces of the station building serves to advertise the station's presence and indicate that the station is open and ready for business. Soffit type lighting is a commonly used method for the lighting of the building face. The internal lighting of spires, cupolas, etc., as well as the accenting of the building's architectural features with appropriate floodlighting also is used. See Building Floodlighting, page 17–1.

Approach Lighting. Adequate lighting of the approaches from the edge of the highway or street curb to the station property is a vital part of good service station lighting. The limits of the approaches should be readily identified to permit the motorist to see them well in advance. While some luminaire brightness may be desirable from an attraction viewpoint, extreme care should be taken to insure that the type of luminaire or the aiming does not cause glare that will interfere with the vision of passing motorists or those entering or leaving the station. A minimum clearance of 14 feet should be contemplated where luminaires are designed to extend out over the approaches.

Pump Island Area. The pump island is a primary point of sale, and its lighting should reflect its commercial importance. Sufficient illumination should be provided in this area not only to enable the operator to perform his tasks quickly and accurately, but also to prominently illuminate the product dispenser and the merchandise displays in the area. As part of its evaluation, the types of pumps to be used (lighted or unlighted) should be considered.

Building Interior. The building interior is the office and major work area of the station. The salesroom, which is used for merchandise displays, limited stock supplies, and sales transactions should be illuminated in accordance with the requirements of a general office area. Areas for automatically vended foods, beverages, etc., also may be involved. Lighting in the service bay areas should be adequate to permit efficient performance of the various services rendered. Auxiliary lighting for merchandise displays also may be required. Well lighted rest rooms are essential to encourage cleanliness and acceptance by the public. Special lighting consideration should be given to the walkways and approaches to the rest rooms as well.

Driveway. Limited driveway lighting (sections marked *A* in Fig. 17–16) completes the path-of-light from entrance to island to exit and adds to the active business like appearance of the station. Depending on the magnitude of the operation and services available, extended driveway lighting (C) plus auxiliary lighting (B) may be required to provide sufficient visibility for outdoor service operations such as replacing headlamps, servicing batteries, changing tires, or installing chains. Where there is a patio or housing for vending machines within the driveway area, there should be adequate lighting to promote cleanliness and instill public confidence.

The Surround. The over-all objective in lighting a service station is to make it stand out by contrast from the areas around it, to attract the attention of the motorist in need of its services. In achieving these objectives, care must be taken that the type of equipment used, mounting heights, etc. do not create an annoyance, through glare and spill light, to surrounding neighbors. The requirements of applicable codes and ordinances must be considered in the selection of equipment as well as the economic factors of maintenance, operating costs, and flexibility for future expansion.

REFERENCES

1. Benson, B. S., Jr.: "The Application of Floodlighting to Buildings," *Illum. Eng.*, Vol. XLIX, p. 367, August, 1954.
2. Eshelby, R. A.: "Outdoor Decorative Lighting Techniques," *Illum. Eng.*, Vol. L, p. 155, April, 1955.
3. Nightingale, F. B.: *Lighting as an Art*, Knight Publishing Co., Skyforest, California, 1962.
4. Schuler, S.: *Outdoor Lighting*, D. Van Nostrand Co., New York, 1962.
5. Nightingale, F. B.: *Garden Lighting*, Knight Publishing Co., Skyforest, California, 1958.
6. Gladstone, B.: *The Complete Book of Garden & Outdoor Lighting*, Harthside Press, Inc., New York, 1956.
7. Travis, B. A. and Faucett, R. E.: "Outdoor Lighting at Seattle World's Fair—Century 21," *Illum. Eng.*, Vol. LVII, p. 651, October, 1962.
8. Hamel, J. S., Langer, R. A., and Culter, C. M.: "New York World's Fair Lighting," *Illum. Eng.*, Vol. LIX, p. 644, October, 1964.
9. "Expo 67 and Its Lighting," *Illum. Eng.*, Vol. LXII, P. 453, August, 1967.
10. "Expo 70," *Illum. Eng.*, Vol. 65, p. 458, August, 1970.
11. Faucett, R. E.: "Engineering of Lighted Fountains," *Arch. Rec.*, Vol. 126, p. 218, July, 1959.

SECTION

18

LIGHT PROJECTION EQUIPMENT AND PROTECTIVE LIGHTING

Light Projection Equipment 18–1 Protective Lighting 18–5

LIGHT PROJECTION EQUIPMENT

Light projection equipment pertains to luminaires which produce a beam of light using optical elements. These luminaires include searchlights, beacons and solar simulators.

Aids to Marine Navigation

Most lighted aids to marine navigation utilize incandescent filament light sources; however, increasing use is being made of gaseous discharge light sources in new and in modernized aids to navigation.

The light beams projected by lighthouse beacons may be produced by one of the following types of apparatus: (a) flashing incandescent filament lamps, flash tubes, or flashing xenon short-arcs mounted in fixed cylindrical Fresnel lens assemblies; and (b) steady burning incandescent filament lamps, mercury lamps, or mercury-xenon short-arcs mounted in rotating assemblies of Fresnel flash panels or in rotating assemblies of reflectors. The effect of both types of apparatus is to produce a distinctive flashing characteristic which permits ready sighting and identification of the lighthouse by reference to appropriate navigational charts on a Light List.

Some lighthouses and other lighted aids to marine navigation (minor battery operated shore lights) are characterized by steady burning or flashing lights projecting distinctive colors over various sectors of azimuth. This effect is accomplished by use of one of the two basic types of apparatus described above together with suitable opaque screens and color filters.

A rotating "high intensity" lighthouse beacon of a "dual intensity" installation is shown in Fig. 18–1. It consists of two 36-inch parabolic mirrors mounted back to back and two 2500-watt mercury-xenon short-arc light sources. This combination produces a peak beam candlepower of 10^8 candelas. When rotated at 6 revolutions per minute it produces a flash every five seconds. Such equipment is used during periods of low visibility (haze or fog). A similar two-mirror "low intensity" beacon using incandescent filament lamps and mounted near the "high intensity" beacon is used during periods of good visibility. Such combinations of equipment are designed to permit sighting of the lighthouse beacon at its "operational range" for conditions of atmospheric transmissivity prevailing for from 82 to 90 per cent of the nights in the particular geographic location. Equipment of recent Coast Guard design

Fig. 18–1. High intensity lighthouse beacon consisting of two 36-inch mirrors mounted back to back and two 2500-watt mercury-xenon short-arc light sources. Peak beam candlepower is 100 million candelas.

NOTE: References are listed at the end of each section.

Fig. 18–2. Lightweight aircraft searchlight for anti-submarine warfare.

uses up to four 2500-watt mercury-xenon short-arc lamps burning simultaneously.

Major lighthouses comprise but one of several classes of lighted aids to marine navigation. By far, the bulk of money, maintenance time and material is applied by the United States Coast Guard to more than 15,000 minor and secondary lighted aids to navigation, consisting of battery powered lighted buoys and shore mounted marine navigational beacons.

Complete details of design and application of lighting equipment for aids to marine navigation are contained in United States Coast Guard, *Civil Engineering Report No. 37* (*CG-250-37*), "Visual Signalling: Theory and Application to Aids to Navigation." [6]

Military Searchlights

Until World War II, searchlights were in general use in military operations primarily for coastal defense, anti-aircraft, and miscellaneous signaling purposes. Subsequently, however, development has been concentrated on searchlights for new tactical applications. One of these developments is a lightweight carbon-arc searchlight mounted in naval aircraft for anti-submarine warfare, as shown in Fig. 18–2.

Another development is the 30-inch general purpose searchlight used for battlefield illumination. See Fig. 18–3. This unit, designated A/N TVS-3, utilizes a 20-kilowatt liquid cooled xenon short-arc lamp and was primarily designed to replace the more cumbersome 60-inch unit. The maximum beam candlepower is 800 million candelas. Illumination levels of up to 140 footcandles are achieved to provide camera and television coverage of the various pre-flight operations for Apollo launches.[9] A battery of up to 50 have been used, beginning with Apollo 8 (Fig. 18–4), and operate unattended for periods up to 90 hours continuous.

Fig. 18–3. Army A/N TVS 30-inch mobile 10-15-20-kilowatt searchlight with liquid cooled xenon short-arc lamp.

Fig. 18–4. A/N TVS-3 searchlights at Kennedy Space Center illuminate Apollo-8 space vehicle for pre-flight checkout.

Fig. 18–5. Xenon short-arc 2200-watt searchlight for combat use.

Fig. 18–6. Airborne application of gimbal mounted 30-kilowatt xenon searchlight. Provides 1 million candelas and 35° beam spread.

A 2200-watt xenon short-arc lamp is the light source for the searchlight shown mounted on the M-60 tank in Fig. 18–5. The searchlight has five modes of operation and is fully controlled from within the tank. The modes are: (a) blackout with lamp operating; (b) visible wide beam; (c) visible narrow beam; (d) infrared wide beam; and (e) infrared narrow beam. Intensities of 140 million candelas are achieved and the circuitry is arranged to permit a lamp current overload of 50 per cent for short periods of time.

Xenon lamps are also utilized in air-borne illumination such as the 30-kilowatt helicopter-mounted unit shown in Fig. 18–6. Some have self-contained gas turbine generators; others operate from the helicopter generating system at power ratings less than 30 kilowatts.

Searchlights equipped with 1000-watt mercury-xenon and xenon short-arc lamps and mechanical shutters are used for visual signaling on naval ships. See Fig. 18–7. Recent developments utilizing solid state circuitry in conjunction with dc xenon lamps makes practical the transmission of coded messages without the need for a mechanical shutter.

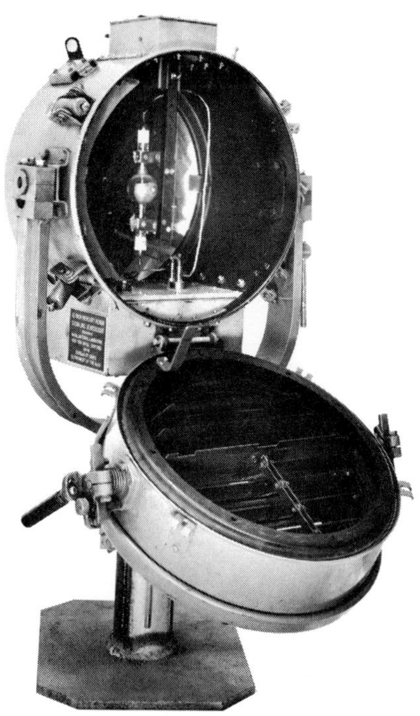

Fig. 18–7. 12-inch signaling searchlight with 1000-watt short-arc mercury-xenon lamp.

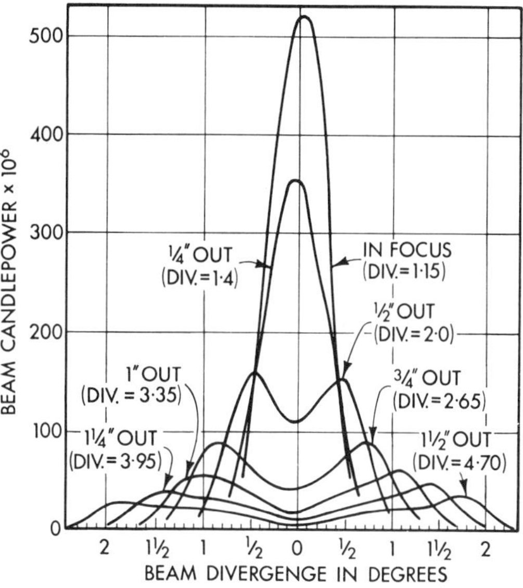

Fig. 18–8. Candlepower distributions of 60-inch searchlight with the lamp in focus and out of focus by varying amounts. The lamp is a carbon arc with a 16 mm positive operated at 150 amps. The focal length of the reflector is 26 inches.

Fig. 18–9. Characteristics of Typical Searchlights and Beacons

Type	Optic	Light Source	Electrical Characteristics	Peak Candlepower (approx. candelas)	Beam Divergence (degrees)	
					Horizontal	Vertical
Navigation	24-inch parabolic reflector	incandescent filament lamp	500 watts, 110 volts	2,000,000	4¼	3½
Beacons	24-inch parabolic reflector	incandescent filament lamp	1000 watts, 30 volts	3,000,000	3¾	3¾
		mercury-xenon short-arc lamp	2500 watts	40,000,000	2	2
	36-inch double-ended doublet lens (white and green)	incandescent filament lamp	1000 watts, 110 volts	2,000,000	4½	4½
			1000 watts, 30 volts	3,000,000	4½	4¼
	36-inch parabolic reflector	incandescent filament lamp	1000 watts, 30 volts	8,000,000	2	2½
		mercury-xenon short-arc lamp	2500 watts	100,000,000	1	1¼
Aircraft anti-submarine	19-inch parabolic reflector	carbon-arc	70 volts, 120 amps	75,000,000	3½	4
Target or battlefield illumination	23 × 16⁵⁄₃₂-inch parabolic reflector	xenon short-arc lamp	2200 watts	140,000,000	2	.8
	30-inch parabolic reflector	xenon short-arc lamp	20,000 watts	800,000,000	1¾	1¾
	60-inch parabolic reflector	carbon-arc	80 volts, 150 amps	500,000,000	1¼	1½
Air-borne	24-inch elipsoidal	xenon short-arc	30,000 watts	1,000,000	35	35
Signaling	12-inch parabolic reflector	short-arc mercury-xenon	1000 watts	3,000,000	2	2¾
	8-inch parabolic sealed-reflector PAR-64 bulb	short-arc xenon	150 watts	200,000	4	6
Handheld	15-inch parabolic reflector	xenon short-arc	150 watts	1,000,000	3	3

Searchlight Characteristics

Because of the high luminances of the carbon arc, it has been used extensively where maximum beam intensity is required (mainly in high intensity searchlights and motion picture equipment). The carbon arc lamp mechanism is complex and difficult to maintain and the carbon electrodes are consumed rapidly. Automatic magazine fed lamp mechanisms have been developed to provide operation over longer periods of time; however, the total unattended operating time is still relatively short.

Other high intensity sources such as mercury, mercury-xenon, and xenon short-arc lamps are being used increasingly in a wide variety of searchlight applications. These sources have become available in ratings up to 20 kilowatts. Lamps of this type can operate unattended for long periods of time and do not require the complex feed mechanisms common to carbon arcs; however, they have disadvantages, such as high voltage ignition circuitry for starting (up to 50 kilovolts), and average luminances less than those of the carbon arc.

Searchlights may be defocused to increase the beam spread (divergence), but at considerable sacrifice of intensity. Fig. 18–8 shows the effect of defocusing in a typical searchlight. Fig. 18–9 gives important characteristics of several representative searchlights. For Searchlight Calculations see Section 9.

Solar Simulators

With the advent of space exploration, systems and components must be tested under conditions that approximate outer space. The difficult task of projecting light, of a quality that approximates solar energy in outer space, into a high vacuum space simulator requires either a carbon arc, a mercury-xenon, or a xenon lamp projection system. If the item to be irradiated is relatively small in diameter, then a single lamp system operating outside of the vacuum chamber can be used satisfactorily.

At a point in space just outside the earth's atmos-

Fig. 18-10. Typical direct illumination solar simulation system including evacuated chamber.

phere, the sun subtends an angle of about ½ degree and the irradiance is 1393 watts per square meter. The characteristics of solar simulator systems should approach these values. Fig. 18-10 shows a typical direct illumination solar simulation system.

PROTECTIVE LIGHTING

Protective lighting[10] is necessary for nighttime policing of outdoor areas to discourage would-be intruders and to render them visible to plant guards should they attempt entry. It may also reduce fire risk. Illumination for policing, however, is not usually adequate for efficient plant operation; therefore, protective lighting is generally treated as an auxiliary to productive lighting.

Principles of Protective Lighting

Protective seeing is achieved by adequate light upon bordering areas of buildings and by producing glaring light in the eyes of the tresspasser, with no light on the guard. There should be a high brightness difference between tresspasser and background. This may be accomplished by adequate light and by proper use of paint and ground cover materials.

Two basic systems or a combination of both may be used to provide practical and effective protective lighting. The first method is to light the boundaries and approaches; the second, to light the area and structures within the general boundaries of the property.

To be effective, protective lighting should:

1. Discourage or deter attempts at entry by intruders. Proper illumination may lead a potential intruder to believe detection inevitable. However, lighting should not be used alone. It should be employed with other measures such as fixed or patrolling guards, fences, alarms, etc.
2. Make detection certain should entry be effected.
3. Avoid glare that handicaps the guards and annoys roadway traffic, marine or railroad operations, workers on neighboring projects, or occupants of adjacent buildings. Glare directed at the intruder, if not in violation of the above, is effective in handicapping the intruder and preventing the discovery of the guard.
4. Provide complete reliability. Failure of a single lamp should not leave a dark spot vulnerable to entry.
5. Provide special treatment for such situations as railroad sidings, alleys, roofs of abutting buildings, wooded or water approaches, entrances, and exits.
6. Provide adequate illumination. See Fig. 18-11. The amount of light required depends upon the accessibility and vulnerability of the property.
7. Provide convenient control and maintenance. Controls for blackout purposes may be necessary.
8. Provide supplementary searchlights controllable by the guards to permit exploration outside of property, or to augment fixed lighting systems for emergencies.
9. Provide for poles, lighting equipment, and electrical auxiliaries to be located inside the property fence, or where they are not readily accessible to malicious damage.

Typical Protective Lighting Applications

Typical protective lighting application data together with recommended minimum levels of illumination are shown in Fig. 18-11. A typical plant protective lighting layout is shown in Fig. 18-12.

For the use of television for surveillance, see Section 24.

Protective Lighting Equipment

In general there are four broad forms of lighting units used in protective lighting systems. These classifications are: floodlights, street lights, Fresnel lens units, and searchlights. The choice in type of equipment depends upon the pattern of light distribution, considerations of glare, convenience in servicing, and the mechanical construction.

Floodlighting Luminaires. These are used to emphasize definite areas or where it is desired to locate the lighting unit in an inaccessible location, and project the light to the critical area. Floodlighting may also be used for the illumination of boundaries, fences, buildings, and for local emphasis of vital areas or buildings. Reflectorized lamps with suitable holders are applicable for lighting small areas and irregular spaces such as around building setbacks, piles of materials, and tanks and for boundary lighting where the light must be confined to the immediate fence area.

Fig. 18–11. Recommended Levels of Illumination and Typical Protective Lighting Applications

Application Reference	Plant[a] Classification	Illumination[b] (footcandles)	Luminaire	Lamp Size W = watts L = lumens	Location of Luminaire Inside Boundary	Minimum Mounting Height (feet)	Spacing (feet)
A. Isolated Fenced Boundary (lighted strip 210 feet wide)	Critical	0.15[c]	Fresnel	500W	10 feet	20	165
	Critical		Fresnel	300W	10 feet	20	110
	Critical		Floodlight Type 2 Spread Lens	500W	80 feet, 2 per location	25	225
	Critical		Floodlight Type 2 Spread Lens	300W	80 feet, 2 per location	25	180
	Average	0.07[c]	Fresnel	500W	10 feet	20	270
	Average		Fresnel	500W	10 feet	20	165
	Average		Floodlight Type 2 Spread Lens	300W	80 feet, 2 per location	25	225
B. Isolated Semi-Isolated Fenced Boundary (lighted strip 80 feet wide)	Critical	0.04	Streetlight Type III	10,000L	10 feet	30	190
	Critical		Streetlight Type I	10,000L	10 feet tilted 25°	30	170
	Critical		Streetlight Type III	6,000L	10 feet tilted 10°	25	145
	Critical		Streetlight Type I	6,000L	10 feet tilted 15°	25	90
	Critical		Floodlight Type 2 Spread Lens	500W	80 feet, 2 per location	40	225
	Critical		Floodlight Type 2 Spread Lens	300W	80 feet, 2 per location	40	180
	Average	0.02	Streetlight Type III	10,000L	10 feet	30	205
	Average		Streetlight Type I	10,000L	10 feet tilted 25°	30	190
	Average		Streetlight Type III	6,000L	10 feet	25	160
	Average		Streetlight Type I	6,000L	10 feet tilted 15°	25	130
	Average		Floodlight Type 2 Spread Lens	300W	80 feet, 2 per location	40	180
	Average		Floodlight Type 2	200W	10 feet, 2 per location	25	150
	Average		Floodlight Type 3	150W	40 feet, 3 per location	40	150
C. Non-Isolated Fenced Boundary (lighted strip 20 feet inside & 40 feet outside fence)	Critical	0.08	Streetlight Type III	10,000L	20 feet	30	190
	Critical		Streetlight Type III	6,000L	20 feet	25	150
	Critical		Streetlight Type III	4,000L	20 feet	25	110
	Average	0.04	Streetlight Type III	10,000L	20 feet	30	215
	Average		Streetlight Type III	6,000L	20 feet	25	185
	Average		Streetlight Type III	4,000L	20 feet	25	165
D. Non-Isolated Fenced Boundary (lighted strip 30 feet inside & 30 feet outside fence)	Critical	0.10	Streetlight Type III	10,000L	30 feet	30	170
	Critical		Streetlight Type III	6,000L	30 feet	25	130
	Critical		Streetlight Type III	4,000L	30 feet	25	110
	Average	0.05	Streetlight Type III	10,000L	30 feet	30	215
	Average		Streetlight Type III	6,000L	30 feet	25	175
	Average		Streetlight Type III	4,000L	30 feet	25	150
E. Building Face Boundary (lighted strip 50 feet wide)	Critical	0.10	Streetlight Type III	10,000L	4 feet outside	30	150
	Critical		Streetlight Type III	6,000L	4 feet outside	25	125
	Critical		Streetlight Type III	4,000L	4 feet outside	25	110
	Average	0.05	Streetlight Type III	10,000L	4 feet outside	30	205
	Average		Streetlight Type III	6,000L	4 feet outside	25	185
	Average		Streetlight Type III	4,000L	4 feet outside	25	155
F. Unfenced Boundary (lighted strip 80 feet wide)	Average	0.04	Streetlight Type III	10,000L	4 feet from building face	30	190
	Average		Streetlight Type III	6,000L	4 feet from building face tilted 10°	25	145
	Average		Floodlight Type 2 Spread Lens	500W	2 per location at building face	40	225
	Average		Floodlight Type 2 Spread Lens	300W	2 per location at building face	40	155
G. Waterfront Boundary (lighted strip 60 feet wide)	Critical	0.10	Streetlight Type III	10,000L	10 feet†	30	170
	Critical		Streetlight Type III	6,000L	10 feet†	25	115
	Critical		Streetlight Type III	4,000L	10 feet†	25	90
	Critical		Floodlight Type 5	500W	50 feet,† 2 per location	60	190
	Critical		Floodlight Type 5	300W	50 feet,† 2 per location	60	125
	Average	0.05	Streetlight Type III	10,000L	10 feet†	30	200
	Average		Streetlight Type III	6,000L	10 feet†	25	160
	Average		Streetlight Type III	4,000L	10 feet†	25	155
	Average		Floodlight Type 5	500W	50 feet,† 2 per location	60	260
	Average		Floodlight Type 5	300W	50 feet,† 2 per location	60	180
H. Pedestrian Entrance	Any	2.0	Streetlight Type IV	15,000L	1 at each of 4 locations	30	12½
	Any		Floodlight Type 6	750W	1 at each of 3 locations	25	25
	Any		Floodlight Type 6	500W	1 at each of 4 locations	25	12½
	Any		Floodlight Type 4	150W	5 at each of 3 locations	25	50
J. Vehicular Entrance	Any	1.0	Streetlight Type IV	15,000L	1 at each of 4 locations	30	25
	Any		Floodlight Type 6	750W	1 at each of 3 locations	25	50
	Any		Floodlight Type 6	500W	1 at each of 4 locations	25	25

The requirements of applications C to G may be satisfied by other types of streetlighting distribution if appropriate mounting height, spacing, and tilting are selected.

[a] Plant classification should be determined by Federal Government Agency.

[b] Minimum horizontal illumination at ground level, except where otherwise stated.

[c] Footcandle values are on a vertical plane 3 feet above ground and parallel to fence. In addition there should be sufficient illumination on the ground from 10 feet inside to 25 feet outside the fence to assure ready detection of persons.

† Location of waterfront units should be such that no appreciable shadow is shown on the water by sea wall or bank. U.S. Coast Guard should be consulted for approval of any proposed lighting adjacent to navigable waters.

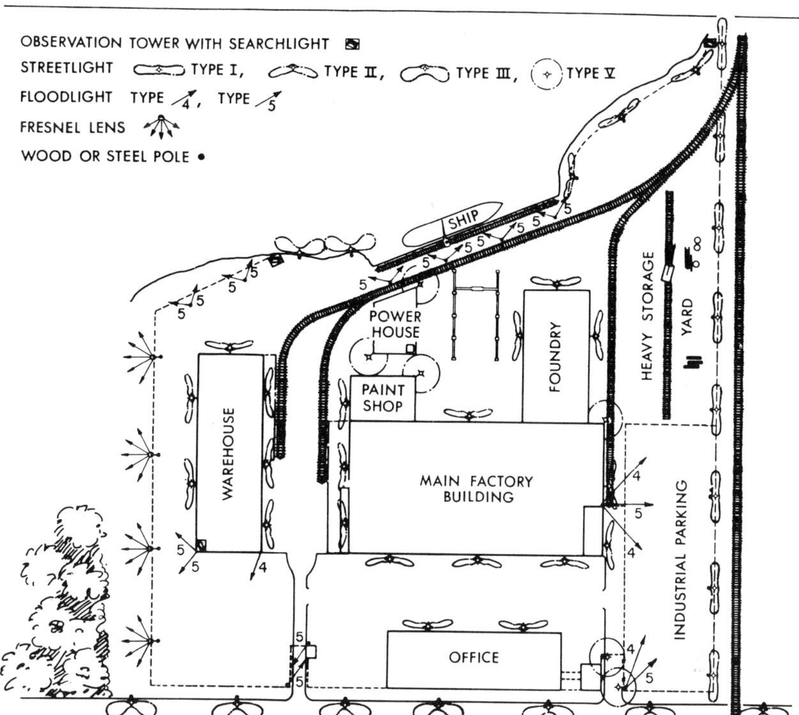

Fig. 18-12. Typical plant protective lighting layout.

Floodlighting luminaires are commonly operated from a multiple circuit. An isolating transformer may be used for operating floodlights from a series circuit, if circumstances make this necessary.

The classification of floodlights used in Fig. 18-11 is that established by IES and NEMA. See Section 9, Fig. 9-64, for luminaire designations.

Street Lighting Luminaires. Luminaires in this classification are operated from series or multiple circuits, and it is common practice to use up to 10,000-lumen lamps in series luminaires and up to 500-watt lamps in multiple luminaires for protective lighting. The distribution of light may be symmetrical or asymmetrical.

Luminaires with symmetrical light distributions find application in lighting large areas where the luminaires may be located centrally with respect to the area to be lighted; also at entrances, exits, and for special boundary conditions.

Luminaires with asymmetrical light distribution find application at situations where the position of the lighting unit is restricted as to location with respect to the area lighted, under which conditions a symmetrical distribution would be wasteful or ineffective. An example of the use of the asymmetric distribution is for boundaries where it is advisable to locate the lighting unit inside the property and deliver the light largely outside the fence; also for roadways where the pole must be placed outside the limits of the roadway but the effective light is that reaching the road surface.

For a discussion of street lighting luminaire types, as used in Fig. 18-11, see Section 20, Roadway Lighting.

Fresnel Lens Luminaires. Fresnel lens units used for protective lighting deliver a fan-shaped beam of light approximately 180 degrees in the horizontal and from 15 to 30 degrees in the vertical. It is intended to afford protection to a property by directing the light outward to illuminate the approaches and inflict glare on the would-be intruder while affording the guard comparative concealment in darkness. The 300-watt multiple lamp or 6000-lumen series lamp is commonly used with this unit. Its application is limited to locations where advantage may be taken of its unique characteristics without causing objectionable glare to neighboring activities.

Searchlights. Where searchlights are employed in protective lighting systems they are usually of the incandescent type because of the small amount of attention required for their operation and because of their simplicity and dependability. Searchlights are rated in diameter of reflector and lamp wattage, and for application referred to herein fall within the range of 12 to 24 inches and 250 to 3000 watts. The beam angles of these searchlights range between 3 and 8 degrees.

The general types of mounting in use are the pilot house control type and the trunnion and pedestal types. The pilot house control type finds application on top of guard houses and places the control of the searchlight in the hands of the guard inside the house to permit exploration of areas within or without the property or to supplement, when and where required, the fixed lighting within the property. The trunnion and pedestal types may be directed at will by the guard, or used with a fixed setting to illuminate local areas.

The general application of searchlights is to supplement the fixed lighting at the plant; size and wattage will depend upon local requirements.

REFERENCES

1. *Symposium on Searchlights.* Illuminating Engineering Society, London, 1948.
2. Benford, F.: "Studies on the Projection of Light," *Gen. Elec. Rev.* No. 3, 1923 to No. 12, 1926.
3. Blackwell, H. R.: "Battlefield Illumination by Visible Light," U. S. Army Corps of Engineers, January, 1955.
4. Hulbert, E. O.: "Optics of Searchlight Illumination," *J. Opt. Soc. Am.*, August, 1946.
5. Segal, S. M.: "Study of High Intensity Light Sources," *Illum. Eng.*, Vol. L, p. 259, May, 1955.
6. U. S. Coast Guard: "Visual Signalling: Theory and Application to Aids to Navigation," *Civil Engineering Report No. 37*, Civil Engineering Division, U. S. Coast Guard Headquarters, Washington, D. C., September, 1964.
7. U. S. Dept. of the Air Force, Air Force Cambridge Research Center: *Handbook of Geophysics*, The MacMillan Co., New York, 1960.
8. Committee on Testing Procedures of the IES: "IES Guide for Photometric Testing of Searchlights," *Illum. Eng.*, Vol. LII, p. 155, March, 1958.
9. Ayling, R. J. and Freeman, R. S.: "Xenon-Arc Searchlight Illumination of the Apollo 8 Launch Area," *SMPTE J.*, April, 1970.
10. *American National Standard Practice for Protective Lighting*, Sponsored by the Illuminating Engineering Society, *Illum. Eng.*, Vol. 53, p. 174, March, 1957; ANSI A85.1-1956, American National Standards Institute, New York, Reaffirmed June, 1970.

SECTION

19

SPORTS LIGHTING[1]

Seeing Problems 19–1 *Quality and Quantity of Illumination* 19–2
Light Sources 19–3 *Indoor Sports* 19–4 *Outdoor Sports* 19–10

SEEING PROBLEMS IN SPORTS

The objective of a sports lighting installation is to control the brightness of the object and the background to the extent that the object will be visible regardless of its size, location, path, and velocity, for any normal viewing position of spectator or player. In a majority of sports, this objective is achieved by illumination of vertical rather than horizontal surfaces.

Objects To Be Seen

Dimensions and reflectances of typical objects to be seen in sports activities vary over a wide range resulting in different lighting requirements. For example, objects may vary in size from a man to a golf ball, and in reflectance from 1 per cent (a hockey puck) to 80 per cent (a golf ball). The combination of these sizes, reflectances, and how they are viewed in contrast with their backgrounds will influence the lighting design.

Background Brightness

In many sports, the normal background against which an object must be viewed by a player comprises all surfaces or space above, below, and on all sides of the player's position.

Because a ball or other object may move rapidly through the field of view, the background brightness, which is seldom uniform, will vary rapidly. For example, outdoors in daylight a baseball may be viewed against the relatively dark shaded grandstand area at one instant and in the next be silhouetted against the sky or sun. A football may be viewed against dark green grass, white snow, clear sky, a mottled pattern of spectators, or a player's jersey.

With illumination from a few high candlepower sources concentrated on an outdoor playing field and filling the space above to a limited altitude only, most of the background area is relatively dark. Great care should be taken to be sure that, in addition to providing relatively uniform illumination, the sources are placed so that a ball will seldom be viewed against any bright portions of the sources.

One of the most effective ways to reduce the effects of glare is to keep the luminance (photometric brightness) of the surround to a reasonable level. This can be done very effectively for indoor sports by the proper selection of reflectances and finishes for walls and ceilings. Control of the surround luminance is much more difficult in outdoor locations; however, a great deal can be done in this regard. Adequate light in the stands for safety considerations, light colored fences, together with provisions for providing some illumination on the ground immediately around the playing field, aid considerably in improving the surround conditions.

Observer Location

In designing lighting for sports, careful consideration should be given to the requirements and comfort of each of the three observer groups; players, officials, and spectators, whose orientation with respect to the object of play differs. The normal fields of view of each group differ also, and in the case of player and official, there may be no fixed location. The probable variation in location and field of view will be different for each sport.

In providing adequate illumination of proper quality for one group, no glare should be introduced into the field of view of the other two groups, if at all possible.

NOTE: References are listed at the end of each section.

QUALITY AND QUANTITY OF ILLUMINATION

Diffuse illumination, such as that provided by an overcast sky on an outdoor playing field during the day, or that provided by an indirect lighting system in an interior with high reflectance ceilings, walls, and floors, is considered to be of excellent quality for sports. Indoors, the design problems are quite similar to those encountered in other interior areas. Outdoors, the problem is more difficult, and it has been necessary to develop locations and mounting heights for concentrated sources which will provide satisfactory visibility of the object. See Fig. 19–1.

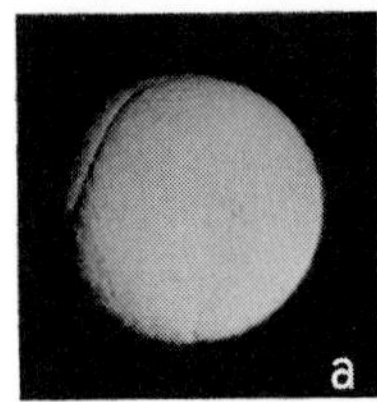

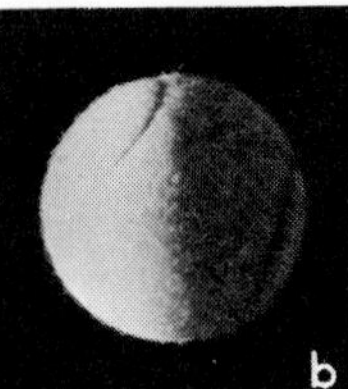

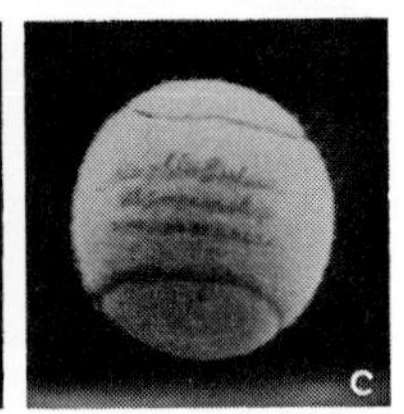

Fig. 19–1. Appearance of a tennis ball lighted in different ways: (a) by single source 30 degrees to right of line of sight; (b) by single source 90 degrees to left of line of sight; (c) by light from two sources above and at 85 degrees from line of sight, and by light reflected by the ground.

Number and Location of Sources

The shape and surface characteristics of the object to be seen and its probable orientation with respect to the observer are important factors in establishing satisfactory luminaire mounting heights and locations. Fortunately, balls with diffuse surface reflectance are the most common objects to be viewed. A point source located in such a position that its central axis forms an angle of not more than 30 degrees with the observer's line of sight (apex at the ball) will for practical purposes illuminate the entire side of the ball facing the observer. See Fig. 19–1a. If the angle is increased to 90 degrees, the ball will remain lighted over half its visible surface, as shown in Fig. 19–1b. Fig. 19–1c shows the same tennis ball lighted from above by two point sources that form angles of 85 degrees with the line of sight and from the ground by reflected light. It appears that the arrangement shown in Fig. 19–1c will provide illumination satisfactory for more observer locations than will the others. However, in the practical case, the specific source locations selected will be those which offer the best compromise between desired illumination diffusion and minimum glare in the majority of observer locations.

It is necessary that illumination be fairly uniform (with no sharp changes in level) at points throughout the entire space above the playing area through which the object may travel, since a fast-moving object passing quickly from a light to a dark space will appear to accelerate. This occurs when there is inadequate overlap of floodlight beams. Such a condition distorts the players' judgment of the object's trajectory. In terms of horizontal illumination, for acceptable uniformity (for sports in which play is skillful, the visual task is severe, or there are likely to be spectators) the ratio of maximum to minimum illumination should not exceed three to one within a given area.

Glare

The reduction or elimination of objectionable glare is one of the principal quality objectives in sports lighting. A floodlight is inherently a glare source, and whenever possible, it is imperative to diminish the effects of glare by locating the luminaires away from the normal lines of sight. The angle between the luminaire and the normal line of sight is affected both by the luminaire location in a horizontal plane and by the mounting height.

To remove all of the luminaires in an installation from the normal lines of sight of each observer group, however, is not always feasible. For many sports, the luminaire locations must reflect a compromise between reducing glare and providing light from the proper direction. Physical obstructions may also require changes from recommended luminaire locations. In all cases, a careful evaluation should be made of the possible lines of sight of both players and spectators.

In those instances where glare cannot be avoided in positioning the luminaires, consideration should be given to the use of glare shields or some form of louvering to reduce the luminaire luminance. These means can also be used to reduce the amount of spill light into surrounding areas, if the sports installation, for example, is located in the heart of a residential zone. When shields are used, particularly the louver type, additional floodlights may be required to compensate for the reduced efficiency of the floodlights.

For unidirectional sports, such as bowling, racing, handball, archery, golf, etc., it is desirable and possible to provide much higher vertical footcandles from one direction as well as to locate the luminaires so that they are almost completely removed or shielded from the normal field of view.

The aiming of floodlights, even with correct luminaire locations and mounting heights, determines, to a large extent, whether the uniformity, direction, and candlepower toward the eye are satisfactory. For information concerning the correct fundamentals of aiming floodlights, see page 19–14.

Illumination Level

It is important that levels of illumination be sufficient for comfortable and accurate seeing, first, to enable the players to perform their visual task, and second, to enable the spectators and the audience viewing the event on television[2] to follow the course of the play. See also Section 24.

In those sports where large numbers of spectators are expected, such as in large football and baseball stadiums, the illumination level is determined by the amount required for the spectators in the row of seats farthest removed from the playing area to follow the course of the play. This condition may require several times the amount of light found satisfactory to the players. The illumination levels suggested for sports (see Fig. 9–80, page 9–91) are those which are currently considered values of good practice, taking into consideration both players and spectators.

The illumination values in Fig. 9–80 are in most cases stated as horizontal footcandles in service. It is recognized that the vertical component of the illumination on the playing area is important in most sports. This is particularly true in the "aerial" games, where both players and spectators rely, to a considerable degree, on the vertical illumination on or near the playing area, and in some cases well above the playing area. The vertical components of illumination have usually been found adequate where the horizontal illumination meets the values in the table, except where noted otherwise, and the lighting equipment is positioned at mounting heights and locations conforming to accepted good practice, as covered later.

Daylighting for Sports

Daylight usually provides adequate illumination to permit satisfactory participation in morning and afternoon outdoor contests, even on cloudy and overcast days. In the design of gymnasiums and other interiors used for daytime athletics, the daylighting principles set forth in Section 7 should be applied.

Windows. Windows in a gymnasium are not considered desirable, particularly behind the baskets of the basketball court, as they are unsatisfactory in the amount of illumination that they provide, and are a potential blinding glare source. Skylights should be screened to prevent breakage. Since screening may have a very low transmittance, the utilization factor for screened apertures will be low (15 to 60 per cent, depending on screen transmittance characteristics).

LIGHT SOURCES

For sports lighting applications, there are three basic types of light sources in use today: incandescent (including tungsten-halogen), fluorescent, and high-intensity discharge. Each type has certain advantages and disadvantages, and the selection from among these sources will depend upon the particular requirements of the installation being considered, the economics, and perhaps some personal preference of the system designer or owner.

Incandescent Lamps

The chief advantages of incandescent lighting are its low initial cost, good color rendering properties, and optical control capabilities. Disadvantages are shorter lamp life and lower efficacy (lumens per watt) as compared to the other light sources.

Over-voltage operation often can be used to economic advantage in sports lighting. This is especially important if a lighting system is used for only a few hundred hours or less each year. With over-voltage operation, the lamps will deliver more lumens per watt but their average life will be reduced. In general, operation at 10 per cent above rated lamp voltage is recommended if the lamps are in use for less than 200 hours per year, and 5 per cent over-voltage if annual use is 200 to 500 hours per year. If annual use exceeds 500 hours, lamp operation at rated voltage is recommended. The approximate increases in light output, lumens per watt, watts, and the reduction in average lamp life are shown in Fig. 8–12.

Fluorescent Lamps

Advantages of fluorescent lighting are its high luminous efficacy, long lamp life, and good color rendering properties. Its inherently lower luminance offers a further advantage for some applications, especially those in which mounting heights are relatively low and short projection distances are acceptable. Such applications can include tennis, bowling, curling, trampolines, spectator stands, and a variety of indoor sports.

In a manner somewhat similar to the over-wattage operation that is permissible for incandescent and mercury lamps, some fluorescent lamps are given dual ratings (such as 0.8 and 1.0 ampere) with performance characteristics published at each rating.

High-Intensity Discharge Lamps

The family of high-intensity discharge lamps include mercury lamps, metal halide lamps, and high pressure sodium lamps. Although each of these lamp types has its own specific characteristics, they have the following characteristics in common: long lamp life and high luminous efficacy when compared to incandescent lamps, a time delay and slow build up of light output when the lighting system is first energized or when there is a power interruption. Because of this time delay characteristic, it may be desirable to include an incandescent lighting system to provide emergency stand-by illumination, particularly over spectator areas. Properly designed, a high intensity discharge lighting system may require fewer luminaires for a given lighting requirement, but these luminaires are usually higher in initial cost than incandescent luminaires.

In those sports where color rendition is of some importance, fluorescent-mercury lamps are usually recommended rather than clear mercury lamps. It should be noted, however, that phosphor-coated lamps provide inherently wide beam spreads. Overwattage operation of mercury lamps is feasible, within limits, resulting in an increase in light output proportional to the increase in lamp wattage with a major reduction in mercury lamp life.

Special Design Considerations

In selecting either a high-intensity discharge or a fluorescent lighting system for use in a sports or multi-purpose area, some important factors should be given special consideration:

1. Both types of light sources, when operated singly on alternating circuits, may produce a flicker on rapidly-moving objects. This condition, stroboscopic effect, can be minimized by connecting lamps or luminaires on alternate phases of a three-phase supply, or by employing two-lamp lead-lag or series sequence start ballasts where available.
2. If a quiet surround is an important factor in multi-purpose areas, consideration should be given to the possibly objectionable disturbance resulting from ballast "hum". Remote mounting of ballasting equipment may be desirable.
3. When fluorescent lamps are used outdoors, they may need protection from the wind and low temperature in order to maintain light output and starting ability. Special lamp designs are available for operation under such conditions. See Section 8, page 8-25.

LIGHTING FOR INDOOR SPORTS

The walls and ceilings of interiors used for sports provide a means for controlling background brightnesses, assist in diffusing the available light, and make possible a variety of convenient lighting equipment arrangements. The design and calculation procedures for interiors used for sports are outlined in Section 9. However, in addition to luminaire mounting height, spacing, and lumen output, and illumination uniformity on a horizontal reference plane, which are important factors in all installation plans, it is necessary in designing sports lighting to consider the following factors:

1. Observers have no fixed visual axis or field of view. During the course of the game, the ceiling and luminaires may frequently be included in the visual field.
2. The object of regard will have no fixed location, and may be viewed at floor level, near the ceiling, or at any level in between.
3. It is particularly important for observers to be able to estimate accurately object velocity and trajectory.

The location of sport play can be divided into general areas used for more than one sport and areas designed for a particular sport. The lighting system must meet the varied or particular requirements for the sport, or sports, played in the given area.

The sports, themselves, may be generally divided into two classes: sports which are aerial in part or whole, and sports which are at or close to ground level.

Aerial Sports

Badminton, basketball, handball, jai-alai, squash, tennis, and volleyball are considered aerial sports. The type of action encountered during normal participation in these activities is such that the ceiling may be in the observer's field of view during a large portion of the playing period. In planning general lighting installations for these sports, therefore, every effort should be made to select, locate, and shield the light sources to avoid introducing glare into the observer's field of view. For these sports in particular, adequate overlapping of the luminaire beam patterns is imperative to insure proper vertical illumination over the entire height of the playing area.

Low Level Sports

Archery, billiards, bowling, fencing, curling, hockey, shuffleboard, skating, rifle and pistol ranges, swimming, boxing, and wrestling, and other sports in which observers in the normal course of play do not look upward are called low level sports. General lighting may be planned more easily for these sports than for aerial sports, since luminaire luminance is less critical.

General Areas

General areas used for sports would be field houses, gymnasiums, community center halls, and other multi-purpose areas. The sports normal to such areas are badminton, basketball, volleyball, fencing, shuffleboard, and other similar sports. The criteria used for designing the lighting system for such areas can best be demonstrated by the design criteria for a gymnasium.

Gymnasiums

The modern gymnasium has become a multi-purpose as well as a multi-sport area, serving a variety of needs of the student body and community in general, including such activities as assemblies, concerts, and dances. Where the creation of a mood or atmosphere is the objective, this may be done by the use of portable or temporary auxiliary lighting equipment. It is possible, however, and at times desirable, to incorporate dimming, switching, or other means of illumination control into the over-all scheme of design.

For most high school applications, a general lighting system providing a level of 30 footcandles will answer the majority of requirements. If, however, exhibitions and matches are played, it is recommended that this level be increased to a minimum of 50 footcandles. Principles of good design would indicate that this could probably be most easily achieved by designing for the higher level and providing switching means for the reduction in level. Fig. 19-2 is a layout for the basketball area lighting for a gymnasium. Fig. 19-3 shows installations meeting current recommendations.

To prevent breakage, it may be necessary to protect luminaires with wire guards or grids. Any reduction of the light output of the unit by the addition of such a device should be taken into consideration and compensated for in the initial system design. Ceiling heights of gymnasiums may vary, but minimum recommended mounting height for luminaire mount is 22 feet.

The position of luminaires and windows in a gymnasium can present serious problems. Fig. 19-4 demonstrates the hazards of improperly located luminaires and unshielded fenestration.

Field Houses

The field house and the gymnasium closely resemble each other as far as sports activities are concerned. The field house may, however, be larger in dimension and serve a somewhat wider range of sports. Among these are indoor track and field events, skating, and such outdoor sports as may be driven indoors by inclement weather. Portable floors and seating facilities are in common use. General lighting levels and methods dictated by particular sports will meet the needs for the participants, but

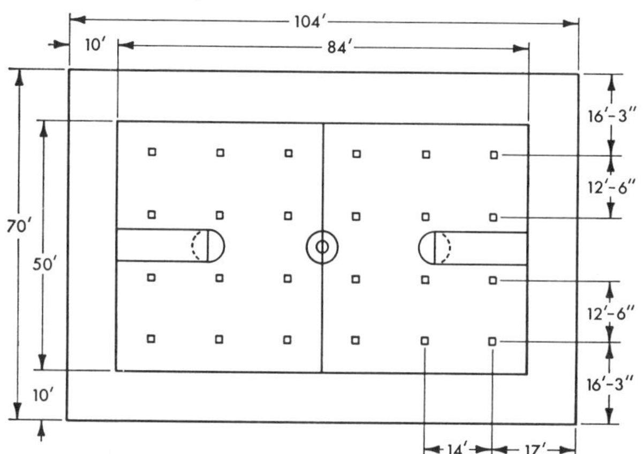

Fig. 19-2. Lighting layout for an indoor basketball court. A distance of 10 feet between court boundary and wall is recommended. Minimum luminaire mounting height should be 22 feet.

a

b

c

Fig. 19-3. (a) A gymnasium illuminated to 100 footcandles by the use of 1500 mA fluorescent lamps in suspended semi-direct, porcelain enamel reflector units with prismatic plastic shielding. (b) A gymnasium illuminated by continuous rows of louvered fluorescent units. (c) A gymnasium illuminated by individual direct incandescent units in layout as shown in Fig. 19-2.

may require considerable increases to meet the needs of the spectators. The resultant lighting system design should therefore meet the requirements for the anticipated activities in the field house as well as provide for the spectators. This could include consideration for aerial and low level sports, versatile control or individual systems for the various sports, and increased illumination levels and beam control to meet the needs of the spectators.

Specialized Areas

Lighting layouts which illustrate the adaption of the previously stated principles to certain specialized indoor sports areas are shown in Figs. 19–5 through 19–14. It is important to recognize that these layouts are not the only acceptable method which can be used for lighting a particular sports area. Other types of luminaires, light sources, and in some instances, luminaire locations may satisfactorily be used. These layouts merely show one or more ways in which the lighting objective has been accomplished. For recommended levels of illumination see Fig. 9–80, page 9–91.

Badminton. Badminton is an aerial sport, and requires ceiling heights of 25 feet minimum and upwards to 40 feet desirable. A brown or green color is recommended for the walls and ceiling to provide good contrast for the white shuttle. A dark finish is also recommended for the floor. To minimize glare, a well controlled lighting system mounted along the sideline, or an indirect system, is recommended. See Fig. 19–5.

Billiards. It is desirable to have a layout of the location of the tables before establishing the location of luminaires, so that they can be placed over the tables, to provide the best lighting possible and create the fewest number of shadows. Luminous ceilings or other general lighting systems can be utilized. Billiard tables are approximately 5 by 9 feet in size and are usually located so that they are five feet apart, side by side and at least six feet from the adjoining wall. The minimum recommended mounting height of the ceiling and light source is seven and one-half feet. The preferred height is ten to twelve feet. The ceiling should be a light color with a reflectance of 75 to 85 per cent. The recommended

Fig. 19–4. Caution should be exercised in positioning luminaires relative to critical surfaces such as glass basketball backboards (left), to avoid blinding, reflected glare. Windows behind glass backboards in gymnasiums (lower left) can produce direct glare, and unshielded windows (below) are a potential source of both direct and reflected glare. Note how reflections on the floor veil the floor markings.

HAZARDS OF IMPROPERLY LOCATED LUMINAIRES AND UNSHIELDED FENESTRATION—

GLARE SITUATIONS TO BE AVOIDED

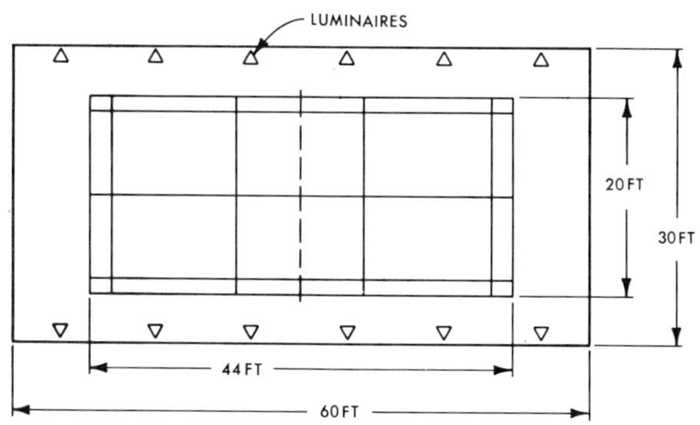

Fig. 19–5. Recommended lighting layout for badminton. Lamp size and luminaire quantities for each class of play are dependent upon the specific room characteristics and luminaires used. Luminaires should be semi-direct and carefully shielded.

illumination levels might be substantially increased for public attraction or business considerations.

Bowling. Lighting for bowling is often governed more by public attraction and increased business considerations than any other factor. Bowling is considered a low-level sport which is divided into three areas—the approaches, the lanes, and the pins. General illumination methods are utilized in the approach area. This area often includes seating for spectators as well as participants with lighting utilized to create a pleasant atmosphere. The lighting of the lanes should be well shielded for the bowler and the shielding is often an architectural element of the structure. This ceiling area should be finished with a high reflectance, non-gloss, light paint which maintains a 70 to 85 per cent reflectance. The illumination of the pins is so directed as to provide high vertical footcandle levels as seen by the bowler. A typical layout is shown in Fig. 19–6.

Boxing and Wrestling. The recommendations for illumination are governed by the requirements of the spectators which completely outweigh the requirements of the participants. A recommended layout is shown in Fig. 19–7.

Curling. The indoor curling rink is classified as a low-level installation system. Direct or semi-direct luminaires, with wide spread distribution, mounted between rinks provide the best method of illumination. The minimum mounting height of the luminaires is twelve feet and the ceiling and wall finishes should have a reflectance of over 60 per cent to provide good luminance ratios. Fig. 19–8 shows a recommended layout for curling.

Handball and Squash. The handball and squash court with its white walls and ceiling presents definite luminaire beam control problems for this aerial sport. The wall and ceiling finish should be a white, non-glossy paint with a reflectance of 75 to 85 per cent. The luminaires should be recess-mounted in the ceiling with a carefully shielded spread distribution. In these areas, adequate protection of the luminaires from possible breakage through the use of guards or impact-resistant covers is vitally important. See Fig. 19–9.

Fig. 19–6. Typical lighting arrangements for bowling. The ceiling luminaires should be completely shielded from the view of the bowler. To avoid severe luminance differences and to make maximum use of reflected light, the ceiling should be maintained at a reflectance of 70 per cent or better.

PLAN VIEW

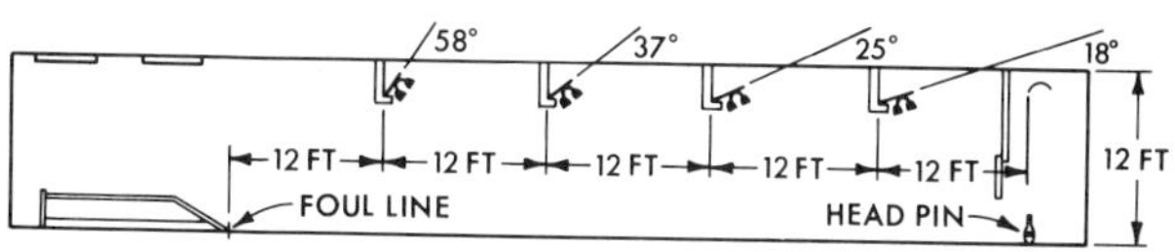

ELEVATION

DIRECT FLUORESCENT TILTED 15° TO 60° FROM HORIZONTAL AND CONCEALED

DIRECT FLUORESCENT OR EQUIVALENT INCANDESCENT

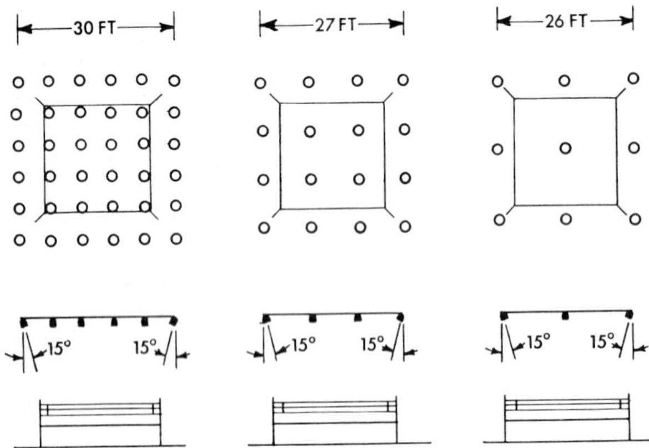

Fig. 19–7. Recommended layouts for indoor boxing or wrestling rings. Luminaires are direct with concentrating distribution from 20-foot mounting height.

Hockey Rink. Lighting for indoor hockey rinks requires extreme care in the selection and location of luminaires. Not only should direct glare from the luminaires be considered, but the possible loss of visibility due to reflected glare from the ice is of equal importance. Care should be exercised so that no shadows from the boards and nets cause difficulty in following the course of play. All luminaires should be mounted above the line of sight of the spectator in the most elevated seat at the greatest distance from the playing area. This provides an uninterrupted view of the playing area, minimizes possible direct glare to the spectators, and improves general appearance. See Fig. 19–10.

Jai-Alai Court. Due to extreme speed of the ball in play (over 150 miles per hour), careful consideration must be given to level of illumination, shielding of luminaires, and surface texture of paint on the court walls and floor. Glare is to be avoided at all costs. Play is fast and serious accidents are not uncommon. Colors recommended are: grass green for the frontis, lateral, and rebote; off-white for the floor; and dark red for the foul stripes. Luminaires should be mounted above the top screen for physical protection. Viewing is done from the open side of the court, again through a protective screen. It is good practice to provide for dimming in the audience area at the start of play. See Fig. 19–11.

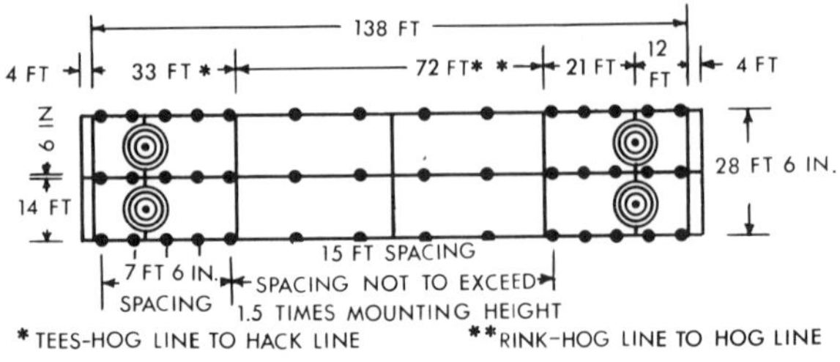

Fig. 19–8. Recommended lighting layout for curling. Lamp size and luminaire quantities for each class of play are dependent upon the specific room characteristics and luminaires used.

Shooting (Archery, Pistol, and Rifle Ranges). Indoor archery, pistol and rifle ranges present similar illumination problems. Major emphasis is placed upon the illumination at the target and the distance from the firing line to the target. In the case of the indoor pistol and rifle range, which has a 50 foot distance, the recommended vertical illumination on the target meets the requirements for the distance from the firing line to the targets and the size of the targets. In the case of archery, distances of 60 to 150 feet are normal between the firing line and target. The recommended vertical illumination level on the target again considers the distance and the size of

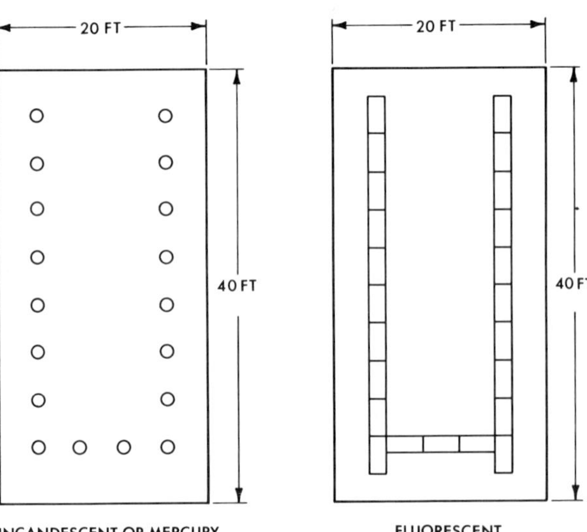

Fig. 19–9. Recommended lighting layouts for squash and four-wall handball. Lamp size and luminaire quantities for each class of play are dependent upon the specific room characteristics and luminaires employed. Tournament class of play is illustrated here.

the target. The typical layout for shooting ranges is shown by the example in Fig. 19–12 which illustrates the standard pistol and rifle range.

Swimming. The lighting of swimming pools is multi-fold. It is to light: (1) the water surface; (2) the floor of the pool; and (3) the deck area around the pool adequately, and for the safety of the persons using the pool. Underwater luminaires should be so located to give complete illumination to all underwater areas. Refer to the National Electrical Code and applicable local codes for specific placement of luminaires.

For underwater lighting, luminaires should be properly located in the pool walls to provide adequate illumination throughout the pool, but should not be placed in line with a swimming lane where competitive swimmers would make a turn and possibly kick the light during the turn. It is therefore quite important that the luminaires should be located between the lanes so that it would not in any way interfere with competitive swimming.

The overhead lighting of the indoor pools can be executed in a way similar to lighting any indoor space with proper spacing and location throughout the ceiling. In the event there is crawl space above the ceiling, it is desirable to select luminaires that can be relamped from above. In the event the luminaires must be relamped from below, it would seem desirable to locate them over the deck rather than over the water and aim some of them toward the water. This will eliminate the need for servicing the overhead luminaires from a pool location. Fig. 19–13 illustrates the recommended practice for swimming pools.

Indoor Tennis Court. The area under consideration for indoor play should approximate that recommended for outdoor, *i.e.*, 120 feet by 48 feet. Suggested interior finishes are: ceiling and upper walls, light, non-glossy, 80 to 85 per cent reflectance; floors, natural hardwood, clay or concrete, non-glossy, 15 to 30 per cent reflectance; walls, lower 8 feet, gray, non-glossy with a maximum reflectance of 60 per cent. Luminaires should be provided with vertical baffles, louvers, or other shielding techniques to reduce the possibility of glare distracting the players. These shielding elements should provide cut-off at 45 degrees in the direction of play. This shielding design should be such as to allow adequate illumination to reach high balls. Luminaires should be mounted toward the side of the courts or between courts as shown in the recommended layout in Fig. 19–14.

Indirect Lighting

Many aerial sports require the upper walls and ceiling to be finished with a high reflectance semi-gloss white paint. This area is illuminated by the upward component of the semi-direct type luminaires and becomes an added factor in the overall

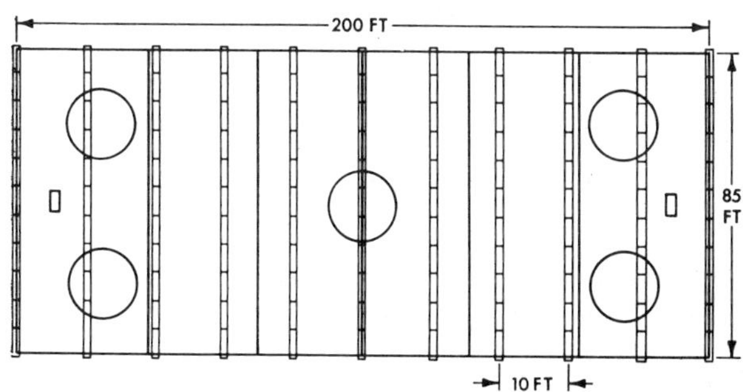

Fig. 19–10. Typical lighting layout of semi-direct fluorescent luminaires on an indoor hockey arena.

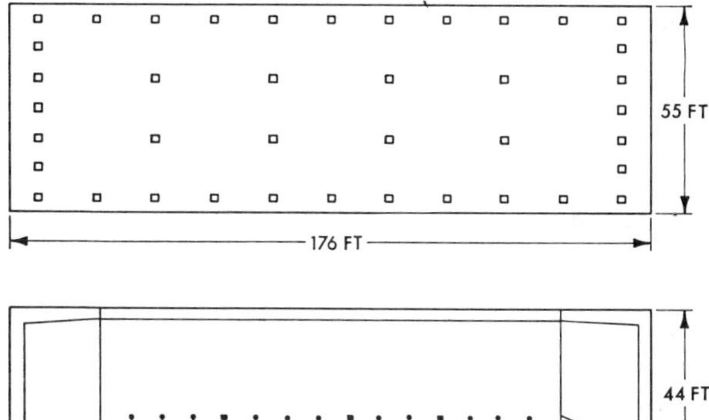

Fig. 19–11. Typical lighting layout for jai-alai court with basic dimensions of 176 × 44 × 55 feet.

quality of illumination. One method of increasing the quality of illumination is to utilize a totally indirect lighting system. Such systems are less efficient than a semi-direct system, but often provide other benefits and at times present the only adequate method of obtaining satisfactory results.

Inflatable structures are finding a wide usage in the sports field. This is especially true for skating rinks, swimming pools and tennis courts. These inflatable structures normally mount on a foundation that varies between ground level and a wall of up to 8-foot height. The structures cannot support overhead items such as luminaires and, therefore, indirect lighting answers the lighting need. The design of such structures normally employs the use of materials which provide a high reflectance matte surface. It is very important that such material is utilized to eliminate glare caused by specular reflections and to prevent increasing the lighting load as the surface reflectance decreases.

A major consideration in the design of an indirect lighting system is the uniformity of illumination over the entire surface. Hot spots around the luminaire's location can be as distracting as a direct view of the luminaire or light source by the participant. The number of luminaire locations need only be governed by the uniformity which can be achieved

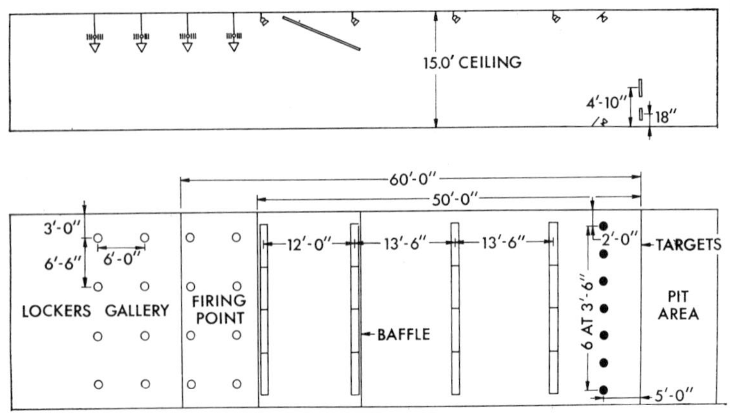

o SEMI-DIRECT FLUORESCENT OR INCANDESCENT
▯ DIRECT FLUORESCENT OR INCANDESCENT (MAIN BEAM AIMED 30 TO 40 DEGREES FROM HORIZONTAL)
o TYPE 5 FLOODLIGHT

Fig. 19–12. Recommended layout for an indoor rifle and pistol range. Lamp size and luminaire quantities for each class of play are dependent on the specific room characteristics and luminaires employed.

and architectural or surface elements which could create deep shadows on the surface being illuminated. These could be as distracting as a black cloud might be in an otherwise clear blue sky. An example of the results obtained in an indirect lighting system utilized in an inflatable structure is shown in Fig. 19–15.

LIGHTING FOR OUTDOOR SPORTS

In the following discussion, where various "classes" of sports are indicated, the classifications follow league ratings where they exist. In general, these ratings are indicative of the skill and speed of play to be expected, and correlate closely with the relative number of spectators regularly accommodated. This latter factor determines the maximum distance at which a spectator may be observing the playing area, and consequently has a direct bearing on the angular size of the object to be seen and therefore on the quantity of the light required. Figs. 19–16 through 19–22 present data for layouts considered good practice.* For recommended levels of illumination see Fig. 9–80, page 9–91.

Baseball

Baseball presents a severe, though not prolonged, seeing task. The ball is small, moves rapidly, and is viewed at varying distances against variable background brightness. The necessity for concentration is intermittent. The large number of possible observer locations and the movement of the players also introduce difficulties. See layout shown in Fig. 19–16.

In providing adequate and uniform illumination for baseball, it is standard practice to consider the infield as including a 30-foot strip outside all baselines and to consider the outfield as including a 30-foot strip outside each foul line.

The floodlights should be aimed so that the beam overlap will provide lighting from two directions at almost every outfield point and from four directions over most of the infield.

Junior League Baseball

This classification of baseball includes such leagues as Pony, Colt, Khoury, Little, Teen-Age, etc. In general, the standard baseball principles apply here also. However, an auxiliary strip outside the baselines and foul lines equal to one-third the length of the baseline is recommended in each instance to be lighted to the same level of illumination as the adjacent playing area. See Fig. 19–17.

Combination Sports Field

The combination layout is never as satisfactory as two individual lighting systems. Nevertheless, athletic fields are laid out for daytime seasonable playing of several sports, usually for a two- or three-game combination of baseball, softball, and football. Lighting such a combination field for night play requires special attention, since the lighting requirements for each individual sport must be considered in developing the final lighting plan. The final design will be largely affected by the relative

* For other sports see reference 1. For floodlight types, see Fig. 9–64, page 9–69. For floodlight classes: HD = heavy duty, GP = general purpose or enclosed ground-area, O = ground-area open, and OI = ground-area with reflector insert (see Section 1).

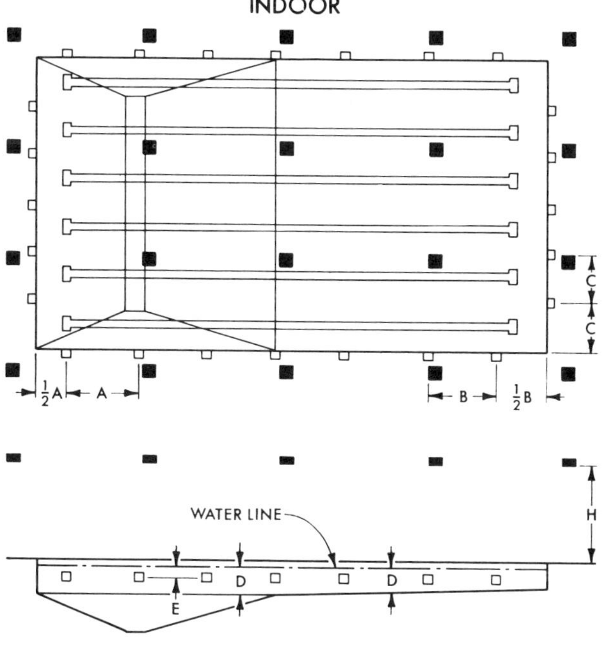

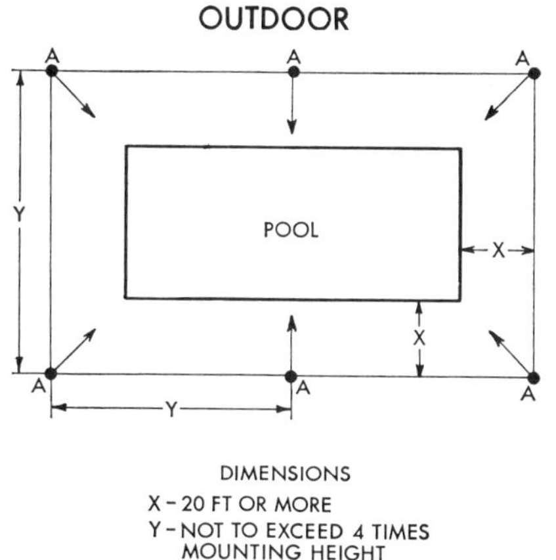

UNDERWATER*

Location of Pool	Lamp Lumens Per Square Foot of Pool Surface (width × length)
Outdoors	60
Indoors	100

Dimensions

Lamp Lumens	A Maximum (feet) (where D is over 5 feet)	B Maximum (feet) (where D is under 5 feet)	E (inches below water line) Minimum	Maximum
3750 to 8000	8	10	12	15
9900 to 33,000	12	15	18	24

* C dimension is equal to the swimming lane width to minimize glare and accidental damage.

Above lighting uses especially designed floodlights not covered by IES Classification or Type. Two systems are used—wet niche and dry niche. The former uses submersible units, while in the latter the casings or niche linings are cast in the pool walls with the floodlights behind them. Use minimum number of floodlights that will satisfy distribution and lumens per square foot. At the ends of the pool, the C dimension can be doubled or units eliminated especially at the shallow end or for narrow pools.

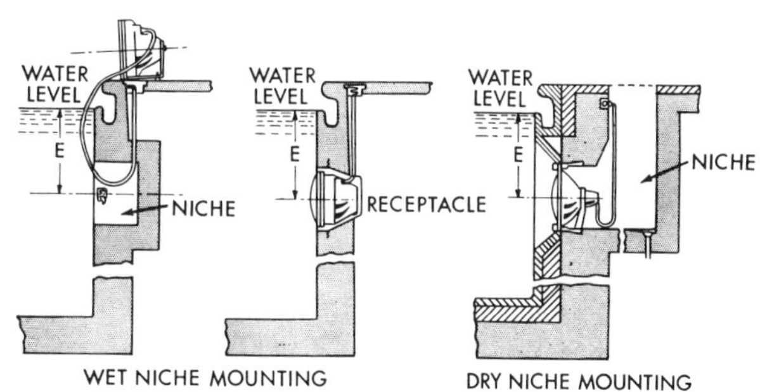

Fig. 19-13. Lighting recommendations for swimming pools. Locate lighting equipment or life guards' positions so as to minimize direct and reflected glare. For overhead lighting in indoor and outdoor pools mounting height should be 20 feet minimum. For outdoor pools use floodlights type 5 or 6, class GP, and underwater lighting and layout shown.

location of the several fields, and the limiting restrictions which each specific arrangement may impose.

Sometimes baseball and softball are played with the same home plate and foul line locations. In these cases, baseball pole locations and mounting heights can be made entirely satisfactory for softball lighting by means of a system (switching or other) that will permit lighting only as many floodlights as are necessary and properly aimed to cover the softball area.

A great number of equipment locations is possible when overlapping baseball or softball and football fields. It is sometimes necessary either to reaim floodlights on certain poles between seasons or mount additional floodlights on those poles otherwise requiring reaiming. Mounting height on a pole should be the greatest height recommended for any sport served by that particular pole. Portable poles may be desirable for certain locations in order to avoid too great a departure from the standard layout for each sport. See Fig. 19-18.

Football, Soccer, and Rugby

Football, soccer, and rugby are a combination of aerial and ground play requiring adequate lighting from ground level to approximately 50 feet above ground. The symmetrical field utilized for these sports affords easy provision for good lighting. Pole locations may prove the biggest problem because of existing facilities. See layout Fig. 19-19.

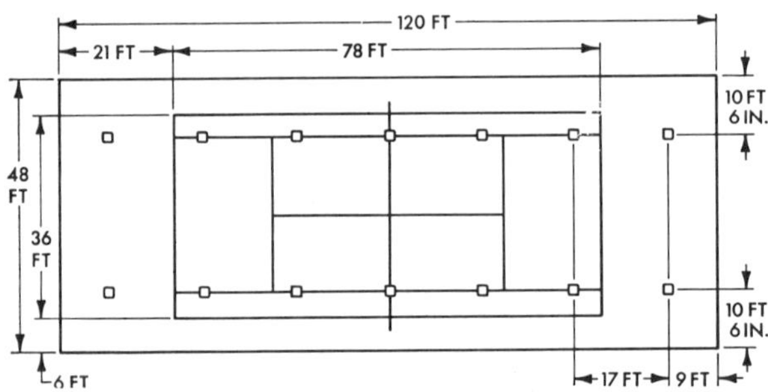

Fig. 19–14. Lighting layout for indoor tennis court. Minimum luminaire mounting height is 22 feet.

Fig. 19–15. Indirect lighting in an inflatable structure.

Golf

The lighting of a golf course for night play involves problems not generally encountered in other sports. The area involved is many times larger than the average sports area. Although the sport is basically unidirectional in nature, the frequent orientation of fairways in direct opposition to each other, and the extreme variations in terrain make the selection of pole locations, beam types, and luminaire orientation a much more critical problem than in most sports areas.

The tee should be lighted so that neither a right- nor left-hand player will shadow his ball. High vertical values of illumination down the fairway should be provided to permit the player to follow the small sphere for the full length of that area while it is traveling at a speed of 100 miles per hour or more and to locate it after it has come to rest.

Each green should be lighted from at least two directions to minimize harsh shadows. Care should be taken in the selection and aiming of the floodlights so that glare from the units does not handicap either the player or those on adjacent fairways. See Fig. 19–20. Special consideration must be given areas not covered by the general lighting system. Some will present physical hazards or require special accent. Examples: sand traps, water hazards, bridges, steep grades, roughs, areas adjacent to greens, pathways, etc.

Softball

Softball also follows the same general principles as baseball. Fields may vary in outfield distance from 160 to 280 feet. See layout Fig. 19–21. Dimensions for slow-pitch softball are essentially the same as standard softball. See layout and footnotes relative to slow-pitch softball.

Tennis

Tennis is a fast, aerial sport, confined to a smaller area than are baseball, football, and softball. Consequently, less equipment is required to provide the recommended illumination. In order to maintain the recommended quality, however, particular care should be employed in designing for play behind the baselines. See layout Fig. 19–22.

Quantity and Location of Poles

Positions of poles are generally dictated by the existing facilities. Care should be taken to insure that luminaires are displaced from the normal lines of sight of players and spectators. Recommended pole locations should be used when designing new installations.

Class of Baseball (regulation)	Floodlights		Minimum Mounting Height to Bottom Crossarm (feet)
	Type	Class	
Major league	3, 4 or 5	GP	120
AAA or AA	3, 4 or 5	GP	110
A and B	3, 4 or 5	GP	90
C and D	3, 4 or 5	GP	70
Semi-professional and municipal	3, 4 or 5	GP	70
	4, 5 or 6	OI	
Recreational	3, 4 or 5	GP	70
	4, 5 or 6	OI	

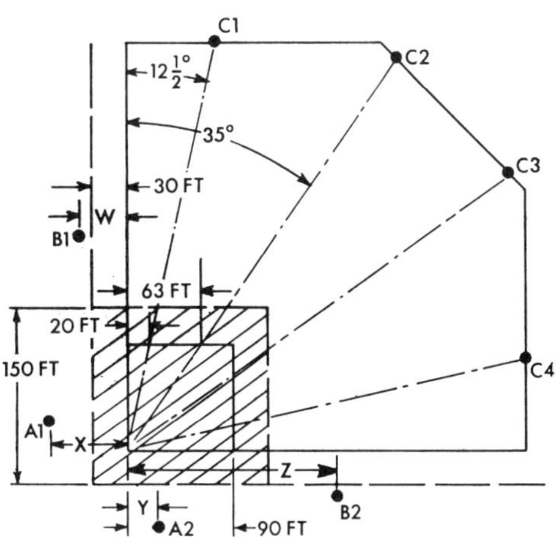

Fig. 19–16. Recommended lighting layouts for baseball fields. These layouts are based on the following total playing area including a strip 30 feet wide outside each foul line—132,500 square feet. Infield Area (shaded)—22,500 square feet. Outfield Area—110,000 square feet. Dimensions: W = 30 feet to 60 feet; X = 40 feet to 80 feet; Y = 20 feet to 30 feet; Z = 130 feet to 180 feet.

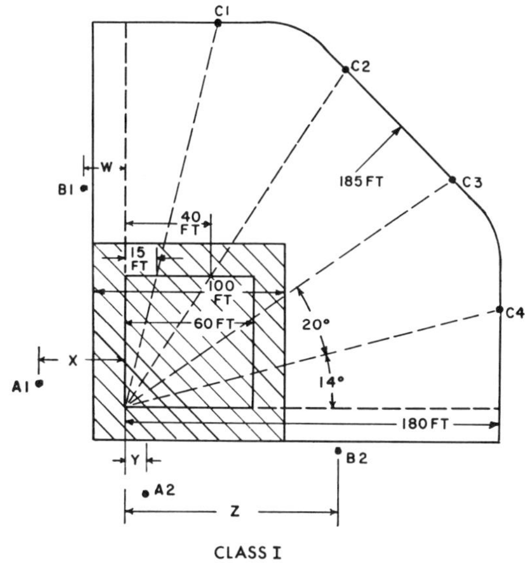

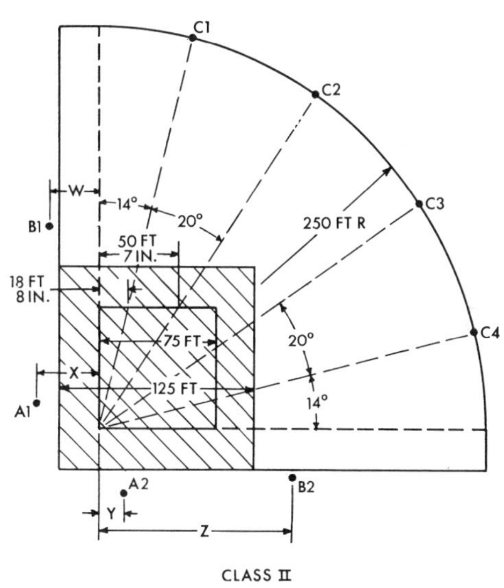

Class of Baseball	Floodlights		Dimension (feet)				Area (square feet)		Minimum Mounting Height to Bottom Floodlight (feet)	
	Type	Class	W	X	Y	Z	Infield	Outfield	Poles A and B	Poles C
I	3, 4 or 5	GP	20–30	30–50	5–15	90–110	10,000	24,700	40	50
	4, 5 or 6	O or OI								
II	3, 4 or 5	GP	24–45	35–65	10–25	110–145	15,625	46,600	50	60
	4, 5 or 6	OI								

Fig. 19–17. Lighting recommendations for Junior League Baseball. (a) Class I—baselines 60 feet or less. (b) Class II—baselines 60 feet and up to 75 feet.

Sport (class)	Floodlights		Minimum Mounting Height to Bottom Floodlight Crossarm (feet)	
	Type	Class	Poles A1, A2, B1, B2, C1 and C4	Poles C2 and C3
Baseball (semiprofessional and municipal)	3, 4 or 5	GP	70	90
Football	4, 5 or 6	OI		

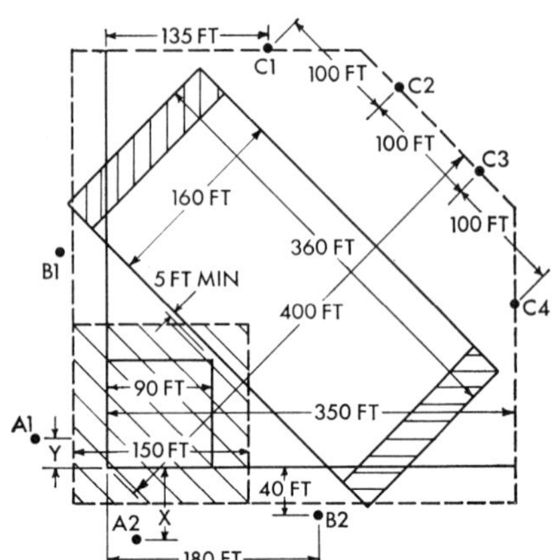

Fig. 19–18. Recommended lighting layout for a combination baseball and football field where baseball is the primary sport. The combination layout is not as satisfactory for either sport as individual layouts. Re-aiming for each sport will increase the effectiveness. The layouts are based on the following: Total playing area, including a strip 30 feet outside each foul line—132,500 square feet. Infield Area (baseball)—22,500 square feet. Outfield Area—110,000 square feet. Dimensions: X = 40 feet to 80 feet; Y = 20 feet to 30 feet.

Mounting Height

The angle between the horizontal playing surface and a line drawn through the lowest mounted floodlight and a point one-third the distance across the playing field should not be less than 30 degrees. At the same time, a minimum of 20 feet for ground sports and 30 feet for aerial sports should be observed. This can be expressed in a formula described in Fig. 19–23. To minimize calculations, the mounting height chart of Fig. 19–24 allows simple pole height selection.

Aiming of Floodlights

In any sports lighting project, proper aiming of the luminaires upon installation is vitally important in order that the user may secure the full benefits of the quality that the manufacturer has built into his equipment, and of the layout that the engineer has provided. Each luminaire must be carefully directed to its appropriate point on the playing field if the lighting system is to provide both the horizontal and vertical uniformity and the freedom from objectionable glare for which the installation was designed. To facilitate the actual aiming process, an aiming or "spotting" diagram, prepared in advance, is generally employed.

Definite calculation methods make it possible to predetermine accurately the footcandle distribution provided by any given aiming pattern. Because however, such calculations are long and tedious, it is general practice to base spotting or aiming diagrams for certain sports employing a symmetric field (such as football, or for minor sports where relatively few floodlights are employed) on scale plots of the floodlight beam spread and the area to be lighted, previous calculations, and practical ex-

Fig. 19–19. Recommended lighting layouts for football fields. Any of the above six pole plans or any intermediate longitudinal spacings are considered good practice with local field conditions dictating exact pole locations.

Classification*	Distance—Nearest Sideline to Floodlight Poles (feet)	Spectator Seating Capacity	Number of Poles	Floodlights	
				Type	Class
I	Over 140 100–140	Over 30,000	6 6	1 or 2 2 or 3	GP GP
II	75–100 50–75	10,000–30,000	6 8	3 3, 4	GP GP
III	30–50	5000–10,000	8	4	GP
IV	15–30	5000	10	5 6 6	GP OI O
V	15–30	No fixed seating facilities	10	5 6 6	GP OI O

* It is generally conceded that distance between the spectators and the play is the first consideration in determining the class and lighting requirements. However, the potential seating capacity of the stands should also be considered. For minimum mounting height see Fig. 19-24.

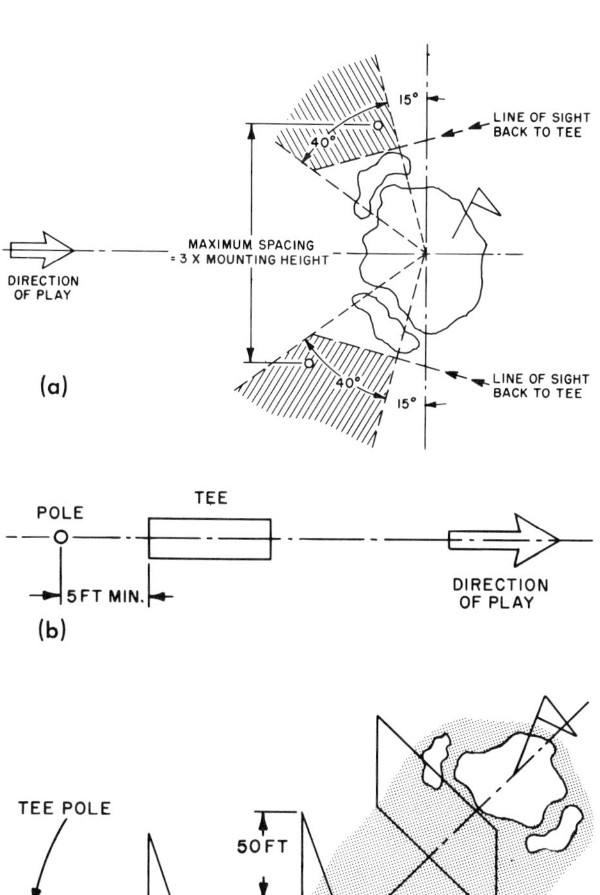

1. Each green should be lighted from at least two directions to minimize harsh shadows.
2. Pole locations should be confined to the 40° cross-hatched zone indicated in front of the green.
3. Pole spacing should be equal to or less than 3 times mounting height.
4. The maximum horizontal illumination measured at any place on the green area should not be greater than 3 times the minimum illumination measured at any other place on the green area.
5. Care should be taken in placement of luminaire poles around green so as to neither obstruct the approaching drive nor create objectionable glare in the eyes of the approaching golfer.

1. Use one pole located a minimum of 5 feet behind back edge of tee. Extremely wide tees may require more than one floodlight location.
2. Floodlight mounting height above tee should be equal to or greater than one half the width of the tee but in no case less than 30 feet. Good practice indicates higher mounting heights for deep tees.

1. Vertical planes should be considered to:
   (a) Extend the full width of the fairway at the point in question,
   (b) Be perpendicular to the centerline of the fairway,
   (c) Extend from fairway centerline elevation to a point 50 feet above the fairway centerline.
2. Vertical planes should be considered to be at points midway between fairway poles.
3. The first vertical plane should be considered to be no less than 100 feet from the tee pole.
4. The ratio of average to minimum illumination at any point in the plane under consideration should be no more than 7 to 1.
5. Minimum mounting height should be 35 feet above the pole base; however, it may be necessary to adjust this if unusual terrain features exist.
6. Spacing between poles should be coordinated with photometric characteristics of floodlight employed, terrain existing at site, and other lighting design criteria.

Fig. 19–20. Recommended lighting layouts for (a) golf course greens, (b) tees, and (c) fairways.

perience with similar installations. The procedure is as follows:

From an end elevation view, similar to that shown for a football field in Fig. 19–25, the vertical aiming of the floodlight beam axes can be determined to obtain approximately uniform horizontal illumination across the field together with sufficient "spill," "direct filament," or "beam-edge" light in the space above to provide adequate illumination to a height of approximately 50 feet above the field. In this connection, care should be taken to minimize the amount of light from the upper portion of the floodlight beams falling in the opposite stands. A limited number of point calculations (see page 9–44) based on a single group of floodlights aimed along a line perpendicular to the axis of the pole crossarms can be made without excessive difficulty to check the graphical appraisal of the proper vertical aiming angle, and such calculations will increase the accuracy of the aiming diagram; particularly if more than one row of floodlights from a single pole is required. The plan view of the field makes it possible to plan horizontal aiming of the floodlights to provide approximately uniform horizontal illumination in the longitudinal direction of the field.

It will be noted in Fig. 19–25b that relatively wide beam floodlights (Type 5) are used because the poles are close to the playing field. It will also be noted that the upper parts of the beams of the two sets of floodlights indicated fall in the opposite stands. However, since these are the wide beam type, the candlepower in the upper portions of the beam (more than 16 degrees from the beam center) will be low, and the spill brightness from them will be well within comfortable limits when evaluated with respect to the relatively high brightness of the field itself.

For installations differing appreciably from the standard recommendations or involving a large number of floodlights, it is desirable to obtain an aiming diagram prepared by the manufacturer of the lighting equipment.

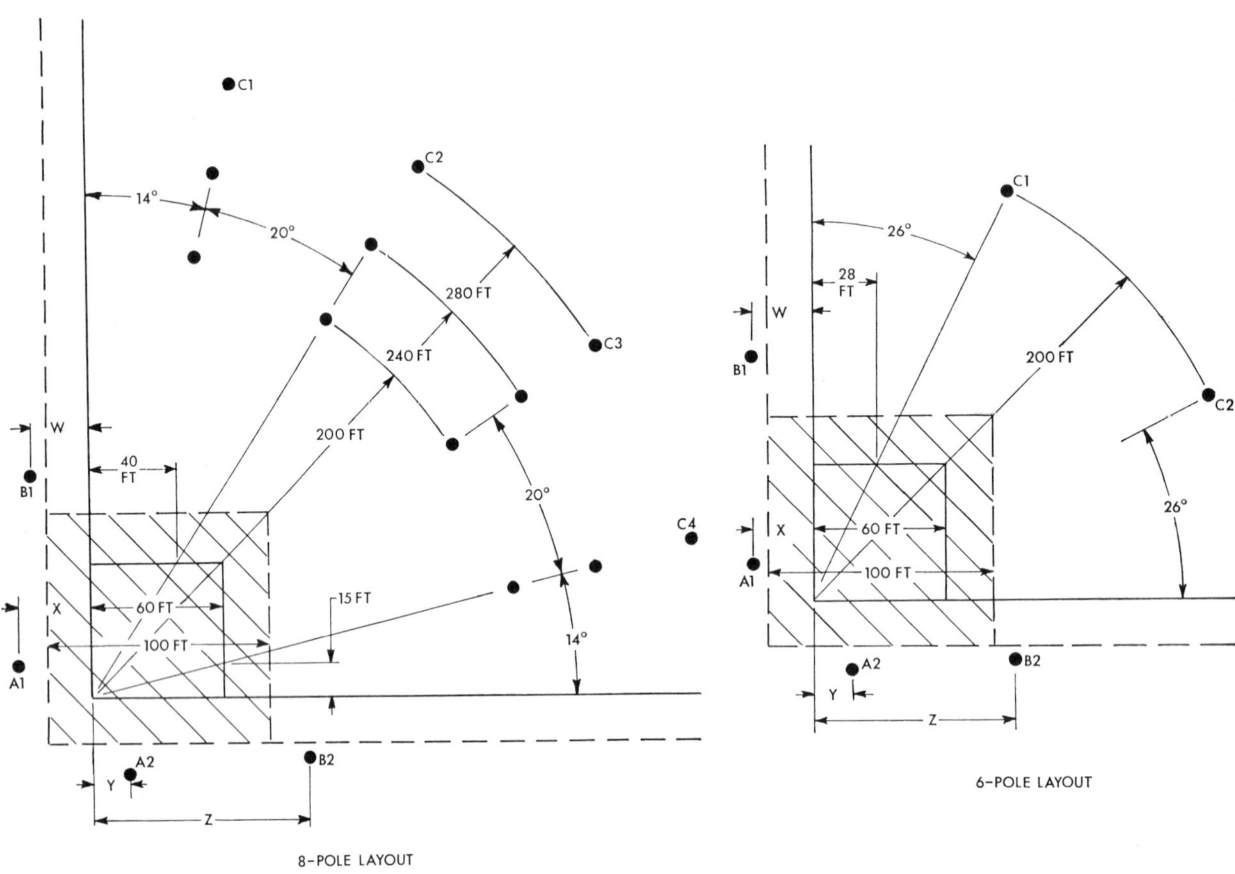

Class	Outfield (feet)	Floodlights		Minimum Mounting Height to Bottom Floodlight Crossarm (feet)	
		Type	Class	A and B Poles	C Poles
8-Pole Layout					
Professional and championship	280 240	3, 4, or 5	GP	50 50	60 55
Semi-professional	280 240	3, 4, or 5 4, 5, or 6	GP OI	40 40	55 50
Industrial league	280 240 200	3, 4, or 5 4, 5, or 6 6	GP OI O	35 35 35	50 45 40
6-Pole Layout					
Recreational	200	5 4, 5, or 6 6	GP OI O	35	40

Poles: 6 for recreational, 8 for other classes.
Note: Supplementary corner poles may be installed to carry overhead wire around boundary rather than across playing area.
For slow-pitch softball—tournament class is same as industrial league, recreational class same as recreational above.

Fig. 19–21. Recommended lighting layouts for softball. These layouts are based on the following total playing area including a strip 20 feet wide outside each foul line: Infield Area—10,000 square feet; Outfield Area (200 feet)—29,815 square feet; Outfield Area (240 feet)—45,240 square feet; Outfield Area (280 feet)—63,200 square feet. Dimensions: W = 20 feet to 30 feet; X = 25 feet to 50 feet; Y = 5 feet to 15 feet; Z = 90 feet to 110 feet.

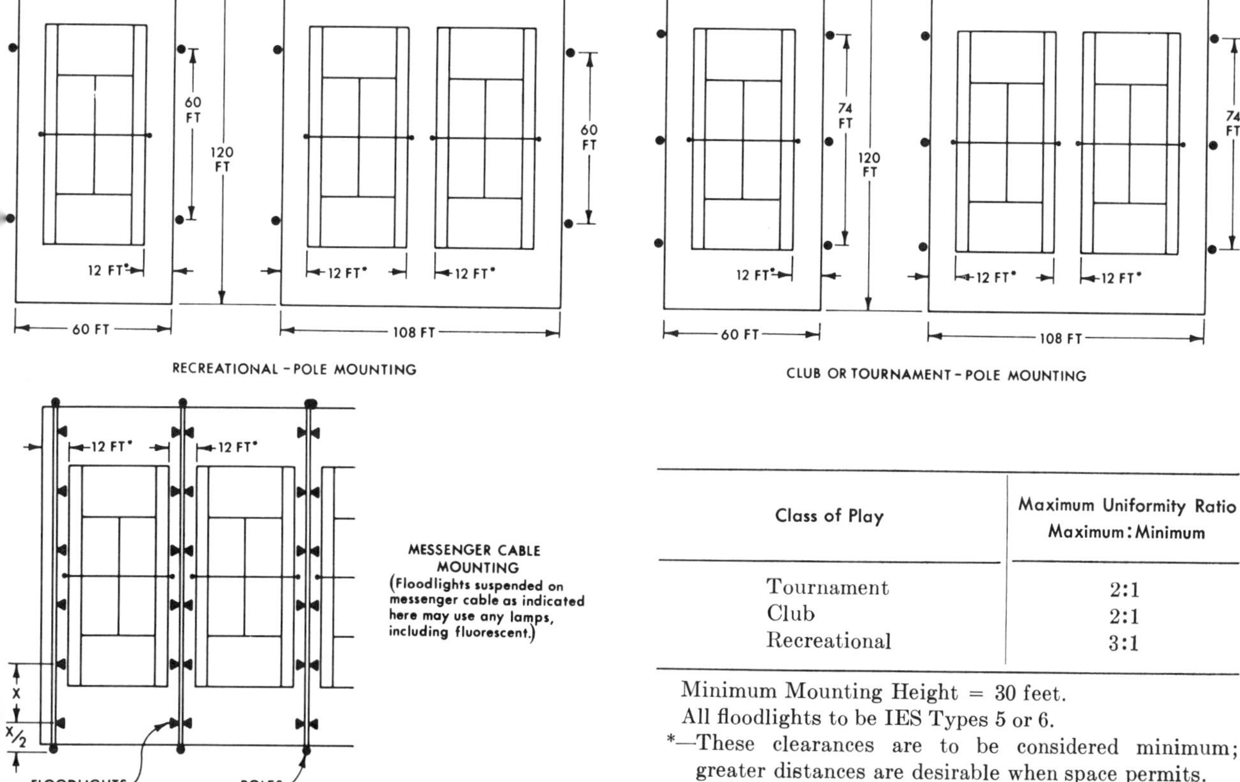

Class of Play	Maximum Uniformity Ratio Maximum:Minimum
Tournament	2:1
Club	2:1
Recreational	3:1

Minimum Mounting Height = 30 feet.
All floodlights to be IES Types 5 or 6.
*—These clearances are to be considered minimum; greater distances are desirable when space permits.

Fig. 19–22. Recommended lighting layouts for single and double outdoor tennis courts.

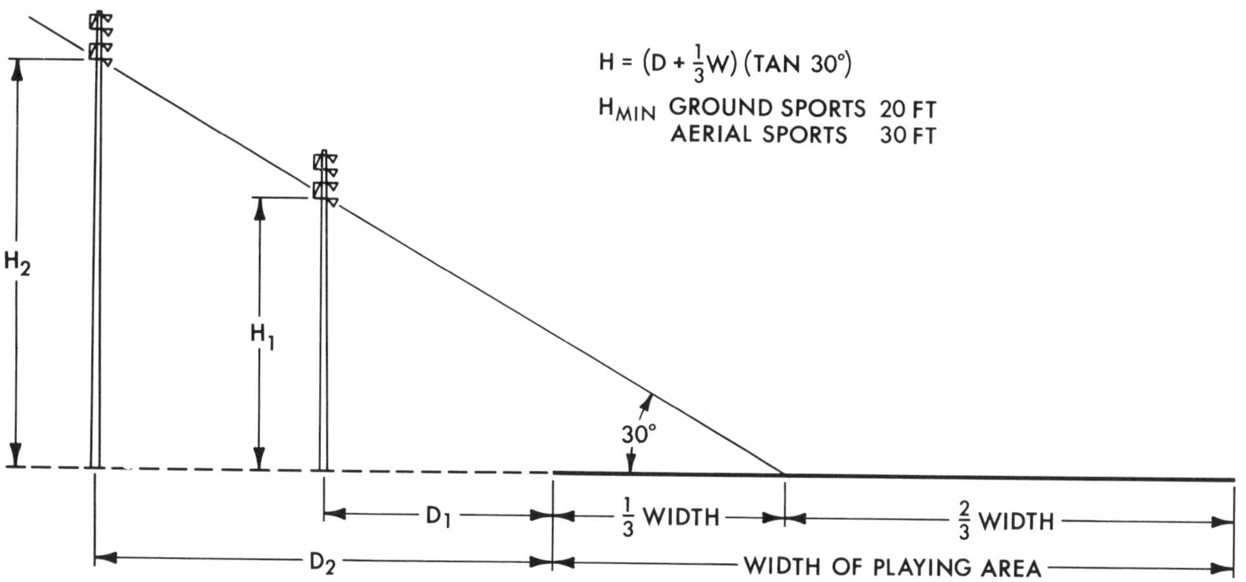

Fig. 19–23. For adequate mounting heights, a line drawn from a point one third the distance across the playing field to the lowest mounted floodlight should form an angle with the horizontal of not less than 30 degrees. In addition, minimum height for ground sports should not be less than 20 feet; for aerial sports not less than 30 feet.

Field Methods of Adjusting Floodlights

There are several ways to put spotting or aiming information to use in making an installation. First, and most accurate, is manual aiming of the floodlight beam centers at predetermined spots on the playing field area. This may be accomplished by using built-in beam sights, or by placing necessary beam sights against the floodlights parallel to their optical axes. Then by gridding both the aiming diagram and the playing area into sections, perhaps 30 feet square, markers can be placed at the aiming

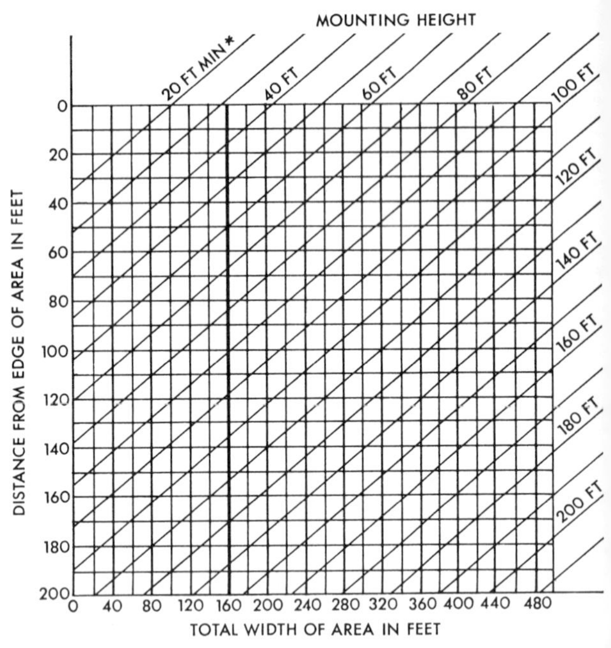

Fig. 19–24. (Right) Mounting height chart for all sports areas—minimum height to bottom floodlight crossarm. Read mounting height along diagonal at intersection of appropriate horizontal and vertical lines. For example where Area Width = 160 feet and Pole Setback = 50 feet, minimum height of 60 feet is indicated by diagonal at intersection of 50 feet and 160 feet.

* For ground area sports. 30 feet is minimum for aerial sports.

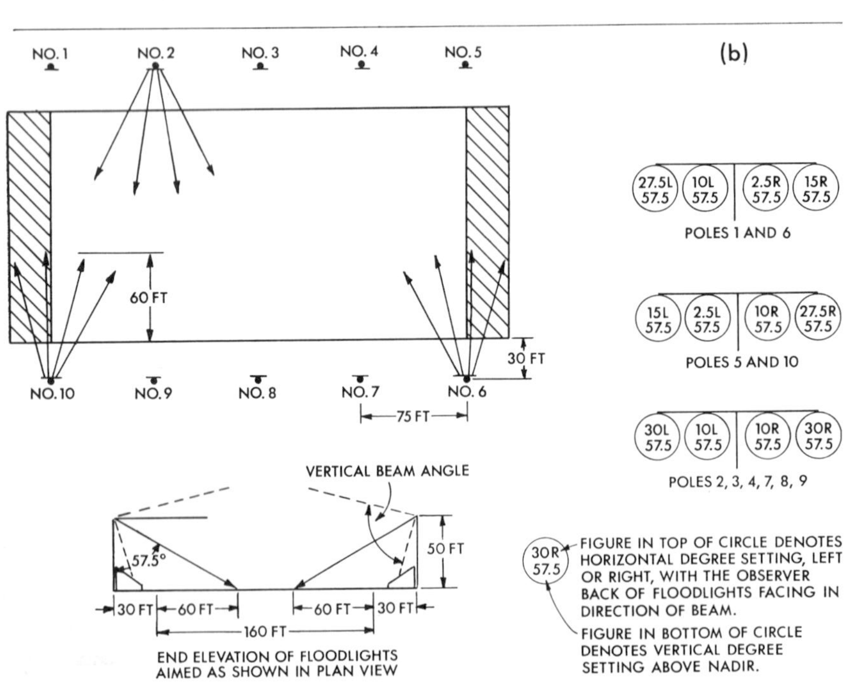

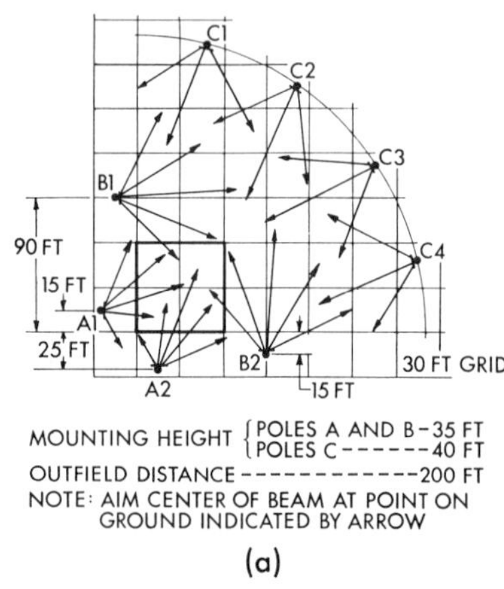

Fig. 19–25. Floodlight spotting diagrams (a—above) Manual aiming diagram for a 32-unit softball field for use with a floodlight sight. (b—left) Angular aiming diagram for a 40-unit football field.

points on the field designated on the drawing, and the sights aimed at those points.

A second aiming method for directing the floodlights is to calculate or determine graphically from the aiming diagram, the vertical and horizontal angular settings of each floodlight. Most floodlights are equipped with degree scales which may then be set to those angles. The accuracy of this system may be less than that of the one described above unless the poles are set accurately and the crossarms carefully leveled and aligned. A difference of only a few degrees may move the beam center 20 feet or more on the field.

A third aiming method which may be used successfully with practice is to stand an observer on the field, short of the aiming point (so the line from the floodlight to the aiming point passes approximately through the observer's eyes), and observe the floodlight, preferably through binoculars. As the floodlight is moved by an assistant, the observer then estimates the position in which the lamp filament (or concentric reflector rings) appears exactly centered in the floodlight aperture. An alternate observation method that may be used with the narrow-beam type (specular reflector) floodlight is to light the lamp and, with smoked glasses on (preferably with binoculars), estimate when the entire reflector appears uniformly bright and at a maximum brightness. The latter methods are inherently less accurate than the first method but can be satisfactory when relatively large numbers of medium or wide beam floodlights are directed into the same general area.

DETAILED DESIGN FACTORS

Details of floodlighting calculations pertinent to sports lighting can be found in Section 9 and in reference 1.

REFERENCES

1. Committee on Sports and Recreational Areas of the IES: "Current Recommended Practice for Sports Lighting," *Illum. Eng.*, Vol. 64, p. 457, July 1969.*
2. Committee on Sports and Recreational Areas and the Committee on Theatre, Television and Film Lighting of the IES: "Interim Report—Design Criteria for Lighting of Sports Events for Color Television Broadcasting," *Illum. Eng.*, Vol. 64, p. 191, March, 1969.

* Contains a very comprehensive list of sports lighting references.

SECTION

20

ROADWAY LIGHTING

Visibility, Discernment, Glare 20–1 Roadway, Walkway, and Area Classifications 20–2 Luminaire Distribution Classifications 20–3 Design of Roadway and Walkway Lighting 20–5

The principal purpose of roadway lighting is to produce quick, accurate, and comfortable seeing at night. These qualities of seeing combine to safeguard, facilitate, and encourage vehicular and pedestrian traffic. Every designer should keep in mind the basic fact that the facility he produces should provide all the inherent qualities required by the user. The driver must be able to see distinctly, and locate with certainty and in time, all significant details; notably the alignment of the road (its direction and its surrounds) and any possible obstacles in or about to enter the roadway. The pedestrian must be able to see distinctly the edges of the walkways, vehicles, and obstacles; dark patches should not occur. Where good seeing is provided, efficient night use is made of the large investment in roadway construction and motor vehicles. Thus, the proper use of roadway lighting as an operative tool provides economic and social benefits to the public, including:

(a) Reduction in night accidents and attendant human misery and economic loss.
(b) Prevention of crime and aid to police protection.
(c) Facilitation of flow of traffic.
(d) Promotion of business and industry during night hours.
(e) Inspiration for community spirit and growth.

Factors Which Influence Seeing and Visibility

Most aspects of traffic safety involve visibility. The fundamental factors which directly influence visibility are:

1. The brightness of an object on or near the roadway.
2. The general brightness of the background of the roadway.
3. The size of an object and its identifying detail.
4. The contrast between an object and its surroundings.
5. The ratio of pavement luminance (photometric brightness) to the surroundings as seen by the observer.
6. The time available for seeing the object.
7. Glare.

Good visibility on roadways at night results from lighting which provides adequate pavement brightness with good uniformity and appropriate illumination of adjacent areas, together with reasonable freedom from glare.

Method of Discernment

Discernment by Silhouette. An object is discerned by silhouette when the general luminance level of all or a substantial part of the object is lower or higher than the luminance of its background. This method of discernment predominates in the observation of distant objects on lighted roadways. Silhouette discernment depends upon the pavement surface reflectance.

Discernment by Surface Detail. When an object is seen by virtue of variations in brightness or color over its own surface, without regard to its contrast with its background, it is discerned by surface detail.

Glare in Roadway Lighting

The common term "glare", as it affects human vision, is subdivided into two components which are not completely independent but which are discussed separately below. These are:

(1) *Disability Glare* (which may not be apparent to the observer). It acts to reduce the ability to see or spot an object. It is sometimes also referred to as "blinding glare" or "veiling glare."

NOTE: References are listed at end of each section.

(2) *Discomfort Glare.* It produces a sensation of ocular discomfort but does not affect the visual acuity or the ability to discern an object.

While both forms of glare reactions are caused by the same light flux, the many factors involved in roadway lighting such as source size, displacement angle of the source, illumination at the eye, adaptation level, surround luminance, exposure time, and motion do not affect both forms of glare in the same manner, nor to the same degree. The only two factors common to both forms of glare are illumination at the eye and the angle of flux entrance into the eye. Even these factors have varying effects on the two forms of glare. It is generally true that when Disability Glare is reduced, it follows that there will also be a reduction in Discomfort Glare, but not necessarily in the same relative amount. On the contrary, it is entirely possible to reduce the Discomfort Glare of a system but at the same time increase the Disability Glare.

It is impossible to eliminate Disability Glare completely since the pavement, surrounding buildings, and the objects which are viewed have a definite luminance which projects some light flux into the eye.

The amount of Disability Glare can be calculated and measured readily. Unfortunately, such is not the case with Discomfort Glare, because it must be evaluated subjectively. Different people vary considerably in their appraisals of the borderline between comfort and discomfort.

Too often judgment is passed on the anticipated effectiveness of a roadway lighting system as a result of casual visual static observations. It should be recognized that in such instances an appraisal can only be made of the Discomfort Glare effect since the eyes may not be conscious of and cannot evaluate Disability Glare as such. Often this becomes misleading, especially when observing a single luminaire at close range rather than a system of several luminaires at different distances. It should also be recognized that comparing a single large area luminaire with a smaller one, especially at short distance, can be misleading as to the anticipated glare effects of an entire system.

ROADWAY, WALKWAY, AND AREA CLASSIFICATIONS

Roadway and Walkway Classifications

Major. The part of the roadway system that serves as the principal network for through traffic flow. The routes connect areas of principal traffic generation and important rural highways entering the city.

Collector. The distributor and collector roadways serving traffic between major and local roadways. These are roadways used mainly for traffic movements within residential, commercial, and industrial areas.

Local. Roadways used primarily for direct access to residential, commercial, industrial, or other abutting property. They do not include roadways carrying through traffic. Long local roadways will generally be divided into short sections by collector roadway systems.

Expressway. A divided major arterial highway for through traffic with full or partial control of access and generally with interchanges at major crossroads. Expressways for non-commercial traffic within parks and park-like areas are generally known as parkways.

Freeway. A divided major highway with full control of access and with no crossings at grade.

Alleys. A narrow public way within a block, generally used for vehicular access to the rear of abutting properties.

Sidewalks. Paved or otherwise improved areas for pedestrian use, located within public street rights-of-way which also contain roadways for vehicular traffic.

Pedestrian Ways. Public sidewalks for pedestrian traffic generally not within rights-of-way for vehicular traffic roadways. Included are skywalks (pedestrian overpasses), subwalks (pedestrian tunnels), walkways giving access to park or block interiors and crossings near centers of long blocks.

Area Classifications

Commercial. That portion of a municipality in a business development where ordinarily there are large numbers of pedestrians and a heavy demand for parking space during periods of peak traffic or a sustained high pedestrian volume and a continuously heavy demand for off-street parking space during business hours. This definition applies to densely developed business areas outside of, as well as those that are within, the central part of a municipality.

Intermediate. That portion of a municipality which is outside of a downtown area but generally within the zone of influence of a business or industrial development, characterized often by a moderately heavy nighttime pedestrian traffic and a somewhat lower parking turnover than is found in a commercial area. This definition includes densely developed apartment areas, hospitals, public libraries, and neighborhood recreational centers.

Residential. A residential development, or a mixture of residential and commercial establishments, characterized by few pedestrians and a low parking demand or turnover at night. This definition includes areas with single family homes, townhouses, and small apartments. Regional parks, cemeteries, and vacant lands are also included.

CLASSIFICATION OF LUMINAIRE LIGHT DISTRIBUTIONS

Proper distribution of the light flux from luminaires is one of the essential factors in efficient roadway lighting. The light emanating from the luminaires is directionally controlled and proportioned in accordance with the requirements for seeing and visibility. Light distributions are generally designed for a typical range of conditions which include luminaire mounting height, transverse (overhang) location of the luminaires, longitudinal spacing of luminaires, widths of roadway to be effectively lighted, arrangement of luminaires, percentage of lamp light directed toward the pavement and adjacent areas, and maintained efficiency of the system.

For practical operating reasons the range in luminaire mounting heights may be kept constant. Therefore, it becomes necessary to have several different light distributions in order to light effectively different roadway widths, using various luminaire spacing distances at a particular luminaire mounting height. All luminaires can be classified according to their *Lateral* and *Vertical* distribution patterns and their control of light distribution above maximum candlepower. Different Vertical distributions are available for different *Spacing-to-Mounting Height Ratios.* Distributions with higher vertical angles of maximum candlepower emission are necessary to obtain the required uniformity of illumination where longer luminaire spacings are used (as on residential and light traffic roadways). These higher vertical emission angles produce a more favorable pavement luminance which may be desired for silhouette seeing, but due to higher glare they should generally be used only where traffic volume is light. Distributions with lower vertical angles of maximum candlepower emission are used in order to reduce system glare. This becomes more important when using high lumen output lamps. The lower the emission angle, the closer the luminaire spacing must be to obtain required illumination uniformity. Therefore, to achieve specific illumination results, it becomes necessary as a part of any lighting system design to consider and to check the uniformity of illumination by checking ratios of average footcandles to minimum footcandles as set forth on page 20-7.

Classification of light distribution should be made on the basis of an isocandela diagram which, on its rectangular coordinate grid, has superimposed a series of *Longitudinal Roadway Lines* (LRL) and a series of *Transverse Roadway Lines* (TRL) both in multiples of *Mounting Height.* The relationship of LRL and TRL to an actual street and the representation of such a web are shown in Figs. 20-1 and 20-2.

Vertical Light Distributions. Vertical light distributions of luminaires are divided into three groups. When the maximum candlepower point lies between 1.0 and 2.25 MH TRL the luminaire is classified as having a *Short* (S) distribution; between 2.25 and 3.75 MH TRL a *Medium* (M) distribution; and between 3.75 and 6.0 MH TRL a *Long* (L) distribution.

Lateral Light Distributions. Lateral light distributions are divided into two groups. One group is based on the location of the luminaire at or near the center of the roadway, and the other group at or near the side of the roadway. The center located luminaires have similar light distributions on both the *House Side* and *Street Side* of the *Reference Line.* The second group, side located luminaires, is subdivided into divisions with regard to width of the "Street Side" roadway area in terms of the MH ratio. The Street Side segment of the half maximum candlepower isocandela trace within the longitudinal range in which the point of maximum candlepower falls (Short, Medium, or Long) may or may not cross the reference line and it is preferable that it remains near the reference line.

Only the segments of the half maximum candlepower isocandela trace which fall within the longi-

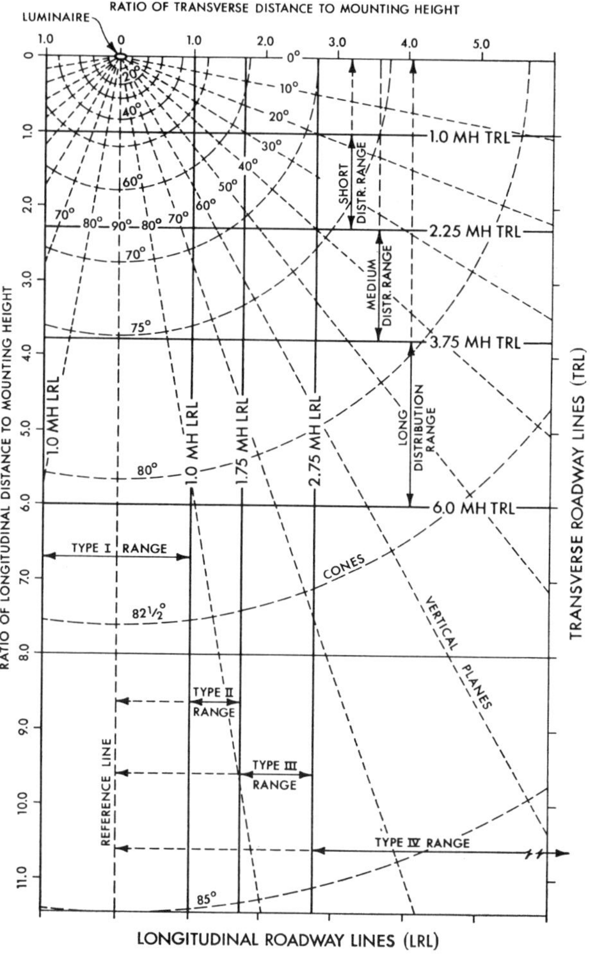

Fig. 20-1. Plan view of roadway coverage for different types of luminaires.

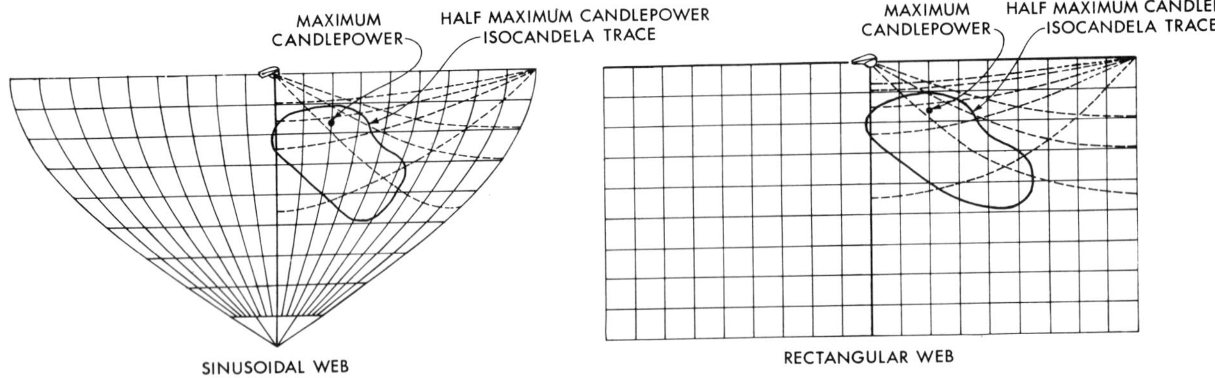

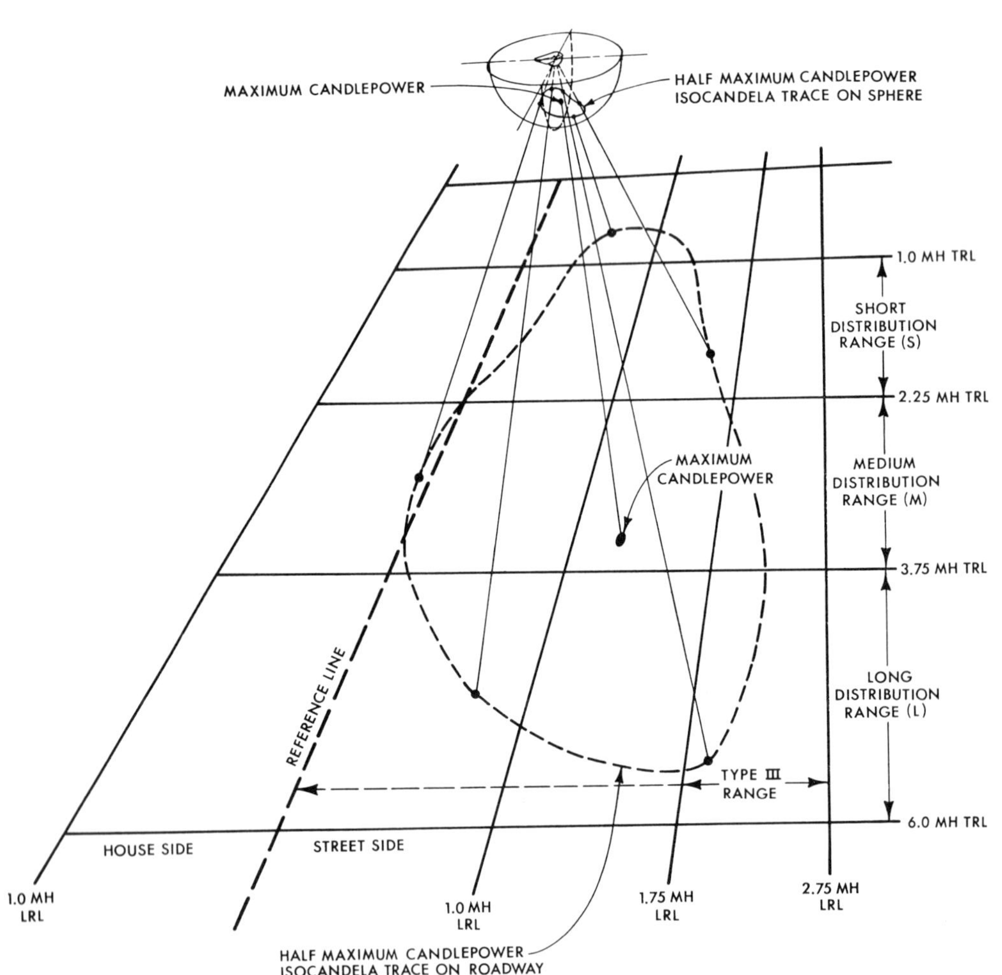

Fig. 20–2. (Center) Diagram showing projection of maximum candlepower and half maximum candlepower isocandela trace from a luminaire having a Type III—Medium distribution, on the imaginary sphere and the roadway. Sinusoidal web (upper left) and rectangular web (upper right) representation of sphere are also shown, with maximum candlepower and half maximum candlepower isocandela line. Rectangular web (lower right) showing Longitudinal Roadway Lines (LRL) and Transverse Roadway Lines (TRL) as solid lines.

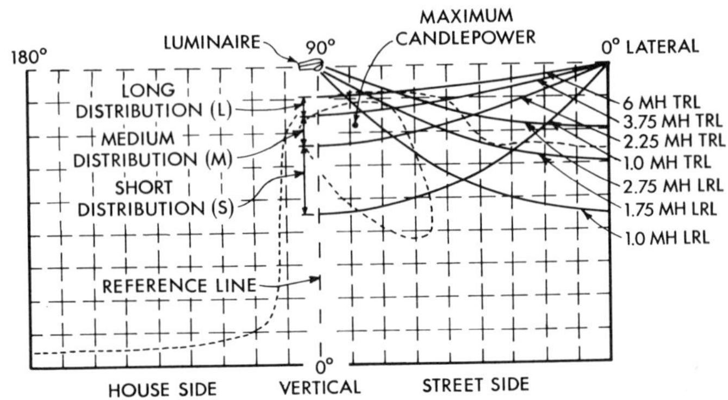

tudinal distribution range as determined by the point of maximum candlepower (Short, Medium, or Long) are used for the purpose of establishing the luminaire width classifications. Fig. 20-3 lists the *Standard Lateral Classifications*.

Control of Distribution Above Maximum Candlepower. Although the pavement luminance generally increases with a higher vertical angle of light flux emission, it should be emphasized that the disability and discomfort glare also increase. Since the respective rates of increase and decrease of these factors are not the same, design compromises become necessary in order to achieve balanced performance. Therefore, varying degrees of control of candlepower in the upper portion of the beam above maximum candlepower are required. This control of the candlepower distribution is divided into three categories.

Cutoff. A luminaire light distribution is designated as cutoff when the candlepower per 1000 lamp lumens does not numerically exceed 25 (2½ per cent) at an angle of 90 degrees above nadir (horizontal); and 100 (10 per cent) at a vertical angle of 80 degrees above nadir. This applies to any lateral angle around the luminaire.

Semicutoff. A luminaire light distribution is designated as semicutoff when the candlepower per 1000 lamp lumens does not numerically exceed 50 (5 per cent) at an angle of 90 degrees above nadir (horizontal); and 200 (20 per cent) at a vertical angle of 80 degrees above nadir. This applies to any lateral angle around the luminaire.

Noncutoff. A luminaire light distribution is designated as noncutoff when there is no candlepower limitation in the zone above maximum candlepower.

Variations and Comments. With the variations in roadway width, type of surface, luminaire mounting height, and spacing which may be found in actual practice, there could be a large number of "ideal" lateral distributions. For practical applications, however, a few types of lateral distribution patterns may be preferable to many complex arrangements. This simplification of distribution types will be more easily understood and consequently there will be greater assurance of proper installation and more reliable maintenance.

When luminaires are tilted upward the angle of the Street Side light distribution is raised. Features such as cutoff or width classification may be changed appreciably. When the tilt is planned the luminaire should be photometered and classified in the position in which it will be installed.

For typical roadway conditions it is desirable to approach very closely the light distributions prescribed. Purposeful variations from these distributions are permissible when such variations become necessary. Several examples of these purposeful variations are:

Fig. 20-3 Lateral Light Distribution Classification of Luminaires

Type	Luminaire Location	Lateral Width of Half Maximum Candlepower Isocandela Trace in Terms of MH Ratio to LRL
I	At or near center of roadway	Between 1 MH LRL House Side and 1 MH LRL Street Side
I-4-WAY	At or near center of intersection	Same as Type I except it has four beams
II	At or near side of roadway	Within 1.75 MH Street Side LRL
II-4-WAY	At or near corner of intersection	Same as Type II except it has four beams
III	At or near side of roadway	Partly or entirely within 1.75 MH Street Side and 2.75 MH Street Side LRL
IV	At or near side of roadway	Partly or entirely beyond the 2.75 MH Street Side LRL

(1) Fluorescent luminaires which provide broad Type I or Type II distributions and which project the maximum candlepower lower than specified.
(2) Directional lighting for one-way streets and divided highways, where the light projected in the direction of traffic is substantially reduced in the high vertical angle.
(3) Fluorescent luminaires parallel to the street to obtain reduced glare and increased utilization.
(4) Luminaires mounted at low mounting heights.
(5) Types IV and V luminaire distributions with extra upward light for illuminating building fronts. (A distribution is classified as Type V when the distribution has a circular symmetry of candlepower distribution which is essentially the same at all lateral angles around the luminaire.) Other purposeful variations from the distributions specified may be found advantageous from time to time for special applications.

DESIGN OF ROADWAY AND PEDESTRIAN WALKWAY LIGHTING

The design of a lighting system involves many variables including economics, esthetics, and visibility related factors. The design process follows these major steps:
(1) Determination of the level of illumination. See Fig. 20-4 and Roadway and Area Classifications.
(2) Formulation of a tentative concept as to luminaire location, light source, and mounting height

relative to the area to be lighted. Lamps in general use today in roadway lighting are shown in Figs. 8–92, 8–93, 8–100, 8–116, and 8–118.
(3) Selection of a luminaire distribution type classification to be used in trial calculations. See Fig. 20–5.
(4) Calculations using several tentative light source types and sizes, luminaires, mounting heights, and maintenance conditions to determine spacings, luminaire locations, and footcandle levels achieved (average and minimum).
(5) Comparative calculations on several possible systems to determine relative factors of uniformity, economics, disability glare and pavement luminance.
(6) Selection of final design or re-entry of design process at any step above.

It is important that roadway lighting be planned on the basis of traffic information which includes the factors necessary to provide traffic safety and pedestrian security. Some of the factors applicable to the specific problem which are to be carefully evaluated are:
(a) Type of lane use development (Area Classification) abutting the roadway or walkway.
(b) Type of route (Roadway or Walkway Classification).
(c) Traffic accident experience.
(d) Street crime experience and security requirements.
(e) Roadway construction features.

Fig. 20–4. Recommendation for Average Maintained Horizontal Illumination

Roadway and Walkway Classification*	Area Classification**					
	Commercial		Intermediate		Residential	
	Foot-candle	Lux	Foot-candle	Lux	Foot-candle	Lux
Vehicular Roadways						
Freeway†	0.6	6	0.6	6	0.6	6
Major and Expressway†	2.0	22	1.4	15	1.0	11
Collector	1.2	13	0.9	10	0.6	6
Local	0.9	10	0.6	6	0.4	4
Alleys	0.6	6	0.4	4	0.2	2
Pedestrian Walkways						
Sidewalks	0.9	10	0.6	6	0.2	2
Pedestrian Ways	2.0	22	1.0	11	0.5	5

* See page 20–2.
** See page 20–2.
† Both mainline and ramps.
Note: The recommended illumination values shown are meaningful only when designed in conjunction with other elements. The most critical elements, as described in Design of Roadway Lighting and Pedestrian Walkway Lighting, are as follows:
(a) Illumination Depreciation
(b) Quality
(c) Uniformity
(d) Luminaire Mounting Heights
(e) Spacing
(f) Transverse Location of Luminaires
(g) Luminaire Selection
(h) Traffic Conflict Areas
(i) Border Areas
(j) Transition Lighting
(k) Alleys
(l) Roadway Lighting Layouts

Illumination Requirements

Recommended illumination levels are given in Fig. 20–4. They represent the lowest average levels which are currently considered appropriate for the kinds of roadways or walkways in various areas. Numerous installations have been made at higher values. Furthermore, the recommendations assume design of proper uniformity and use of applicable types of luminaire distributions, lamp sizes, mounting heights, spacings, and transverse locations. These values do not represent initial illumination, but should be in-service values of systems designed with proper light loss factors. These elements are reviewed below.

Illumination Depreciation

The recommended levels in Fig. 20–4 represent average illumination when the luminaires are at their lowest output. This condition occurs just prior to lamp replacement and luminaire washing. It is impossible to attempt the design of a lighting system without knowing in advance the light losses to be expected. Even when light loss factors are considered and allowance for them is incorporated in an operating service plan, lighting levels may still be reduced to less than 60 per cent of initial at terminal points in the servicing schedule. In the absence of group lamp replacement and luminaire washing schedules appropriate to local conditions, the average system illumination can fall below 50 per cent of the initial value. See Section 9 for the use of light loss factors in calculations.

Quality

Quality in roadway lighting relates to the ability of the available light to provide a visual scene in such a manner that people may make quick, accurate, and comfortable recognition of the cues required for the seeing task. The quality of lighting of installation "A" is higher than that of installation "B" if, with the same average illumination level, visual recognition of typical tasks is faster, easier and/or done more comfortably under installation "A."

Many factors are interrelated to produce improved quality of lighting. The following factors are involved, but quantitative values and relative importance cannot be given:
(1) Disability glare.
(2) Reflected glare.
(3) Pavement luminance.
(4) Light on vertical surfaces.
(5) Uniformity of horizontal and vertical illumina-

Fig. 20-5. Guide for Luminaire Lateral Light Distribution Type Selection and Placement

Rectangular Roadway Area					
Side of the Roadway Mounting			Center of the Roadway Mounting		
One Side or Staggered	Staggered or Opposite	*Grade Intersections	Single Roadway	Twin Roadways (Median Mtg.)	*Grade Intersections
Width up to 1.5 MH	Width beyond 1.5 MH	Width up to 1.5 MH	Width up to 2.0 MH	Width up to 1.5 MH (each pavement)	Width up to 2.0 MH
Types II–III–IV	Types III & IV	Type II 4-Way	Type I	Types II & III	Types I 4-Way & V

Note: In all cases suggested *maximum* longitudinal spacings and associated vertical distribution classifications are:
Short Distribution = 4.5 MH
Medium Distribution = 7.5 MH
Long Distribution = 12.0 MH
* Local Category Intersection.

tion, as well as uniformity of pavement luminance and other background areas.

It should be recognized that in many instances, changes intended to optimize one factor relating to quality will adversely affect another and the resultant total quality of the installation may be degraded.

Uniformity

The illumination values in Fig. 20-4 are minimum and provide effective visibility only when combined with uniformity, or even illumination spread on pavement and sidewalks.

Uniformity may be expressed using the Average Level-to-Minimum Point method where the average illumination of the roadway design area between two adjacent luminaires is compared to (divided by) the lowest value at any point in the area. Under this method, the average-to-minimum ratio should not exceed 3 to 1 for any roadway in Fig. 20-4, except for Local Residential streets, which should have a ratio not exceeding 6 to 1.

Luminaire Mounting Height

Mounting heights of luminaires have in general increased substantially during the past decade. The advent of modern more efficient and larger size (lumen output) lamps has been the basic reason. Engineers have increased mounting heights in order to obtain economic and esthetic gains in addition to increased illumination when utilizing newer lamps. Examples are the 40, 50, 60-foot and higher mounting heights used along roadways and the cluster mounting of luminaires at interchanges. The advent of suitable servicing equipment and means has made this practical.

During this same period there has been a trend to lower mounting heights in some cases. In general, this has been due to esthetic considerations. An example is the use of pole top mounted luminaires in residential areas.

When designing a system, mounting height should be considered in conjunction with spacing and lateral positioning of the luminaires as well as the luminaire type and distribution. See Fig. 20-6. *Uniformity and levels of illumination should be maintained as recommended* regardless of the mounting height selected.

Increasing mounting height may, but will not necessarily, reduce direct glare, reflected glare or disability glare. Higher mounting increases the angle between a luminaire and the line of sight; however, luminaire light distribution and candlepower also are significant factors. See Section 3.

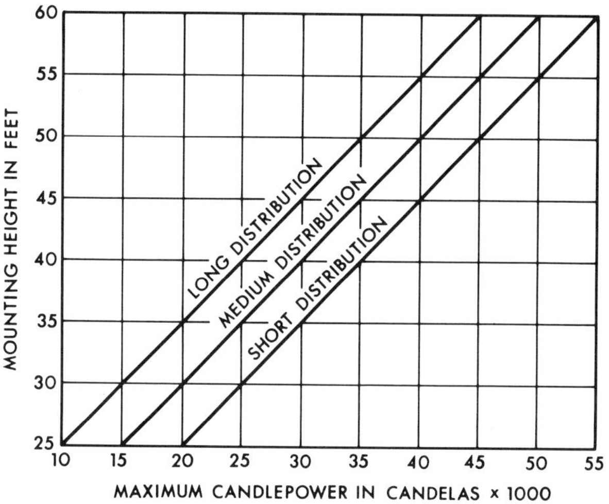

Fig. 20-6. Minimum luminaire mounting heights based on current practice (*for roadway lighting only*).

Luminaire Spacing

The spacing of luminaires is often influenced by the location of utility poles, block lengths, property lines, and the geometric configurations of the terrain features. It is generally more economical to use larger lamps at reasonable spacings and mounting heights than to use small lamps at more frequent intervals with lower mounting heights. This is usually in the interest of good lighting, provided the spacing-to-mounting height ratio is within the range of light distribution for which the luminaire is designed. The desired ratio of lowest illumination at any point on the pavement to the average illumination should be maintained. Disregarding luminaire distribution characteristics and exceeding maximum spacing-to-mounting height ratios may cause loss of visibility of objects between luminaires.

Transverse Location of Luminaires

Types II, III and IV luminaires are intended to be mounted over or near the edge of the roadway. Type I is an exception that is designed to be mounted over or near the center of the roadway. Generally, luminaire overhang exceeding 0.25 mounting height does not contribute to visibility and often increases system glare and cost.

Optimum luminaire location is best determined by reference to the photometric data showing illumination distribution and utilization. Other factors that should be considered are:

(a) Access to luminaires for servicing.
(b) Vehicle-pole collision probabilities.
(c) System glare aspects.
(d) The visibility (both day and night) of traffic signs and signals.
(e) Esthetic appearance.

Luminaire Selection

Luminaire light distribution classifications are intended to serve as a means of selecting a luminaire which is a good candidate for further calculations to determine if it is optically and economically suitable for lighting a particular roadway from the proposed mounting height and mounting location. Fig. 20-5 tabulates preferred lateral light distributions and maximum recommended longitudinal spacings for various geometric factors encountered in common practice. The fact that a luminaire is assigned a particular classification does not assure the designer that it will produce the recommended quantity and quality of illumination for the roadway configuration and mountings shown in Fig. 20-5. The relative amount and control of light in areas other than the cone of maximum candlepower are equally important in producing good visibility in the final system and are not considered in the classification system.

No hesitation should be attached to using a luminaire of a particular classification on roadway configurations other than those recommended in Fig. 20-5 if adequate calculations confirm that it meets performance requirements.

Traffic Conflict Areas

The levels in Fig. 20-4 are for roadway sections which are approximately straight and nearly level. Intersecting, converging or diverging roadway areas require higher illumination. The illumination within these areas should be at least equal to the *sum* of the values recommended for each roadway which forms the intersection. Such areas include ramp divergences or connections with streets or freeway mainlines. They also include very high volume driveway connections to public streets and mid-block pedestrian crosswalks.

Border Areas

There is value in illuminating areas beyond the roadway proper provided it is appropriate to the environment and not objectionable to the adjacent property use. It is desirable to widen the narrow visual field into the peripheral zone in order to reveal objects and enhance eye adaptation. It also improves depth perception and perspective thus facilitating the judgment of speed, distance, etc. Such illumination should diminish gradually and uniformly away from the road.

Transition Lighting

It is good practice to gradually decrease luminance in the driver's field of view when emerging from an adequately lighted section of roadway. This may be accomplished by extending the lighting system in each exit direction using approximately the same spacing and mounting height but graduating the size of the lamp used. A recommended procedure to achieve this graduation is to sector the extension of the best lighted portion of the principal roadway using the designed value of this section as the calculation base. Using the design speed of the roadway, the lowered level sectors should be illuminated for a 15-second continuous exposure to the sector illumination level of one-half of the preceding higher lighted sector, but the terminal illumination in the lowest sector should not be less than 0.25 footcandle (2.7 lux) nor more than 0.5 footcandle (5.4 lux).

Alleys

Experience has proven that well-lighted alleys remove the criminal's opportunity to operate and hide under cover of darkness. Alleys should be adequately lighted to facilitate police patrolling from sidewalks and cross streets, especially in commercial areas. Generally, such lighting also meets vehicular traffic needs.

Pedestrian Walkways

All sidewalk lighting provided as incidental to roadway lighting should be evaluated for adequacy independently of the level of illumination on the associated roadway. The photometric data provided by the supplier of roadway luminaires can be used for the checking and design of sidewalk illumination as well as for roadway illumination.

Open margins of walkways should be illuminated to not less than one-half the levels specified in Fig. 20–4 for at least 6 feet on either side, and to not less than one-tenth of the specified levels for at least 15 feet on either side. Levels should be verified in service to include reflections from building fronts, fences and walls, which can often contribute significantly to the illumination of the walkway.

The uniformity ratio should not exceed 4 to 1 for walks or pedestrian ways, except that in Residential areas a ratio of 10 to 1 is acceptable.

Security Problem Locations. For all walkways in areas with high crime experience, the recommended illumination levels should be doubled, but the average should not in any case be less than 0.5 footcandles (5.4 lux). Features closely adjacent to any walkway which offer unusual hazards should be well illuminated to the same level as the walkway, or should be eliminated.

Roadway Complexities and Special Situations

Vertical and horizontal curves as well as interchanges and intersections produce at least three factors which are fundamentally different from straight, level roadways:

(1) Motor vehicle operators are burdened with increased visual and mental tasks upon approaching and negotiating these areas.

(2) Silhouette seeing cannot be provided in many cases due to the locations of vehicles, pedestrians, obstructions, and the general geometry of the roadway.

(3) Adequate headlight illumination often cannot be provided.

The apparent complications of the lighting problem can be reduced to basic situations or a combination of these situations.

Curves and Hills. (See Fig. 20–7.) Headlight illumination is not effective on abrupt curves and hills and silhouette seeing cannot always be provided. Luminaires should be located to provide ample illumination on vehicles, road curbings, berms, and guard rails. Many drivers are strange to these areas and illumination on the surround is helpful.

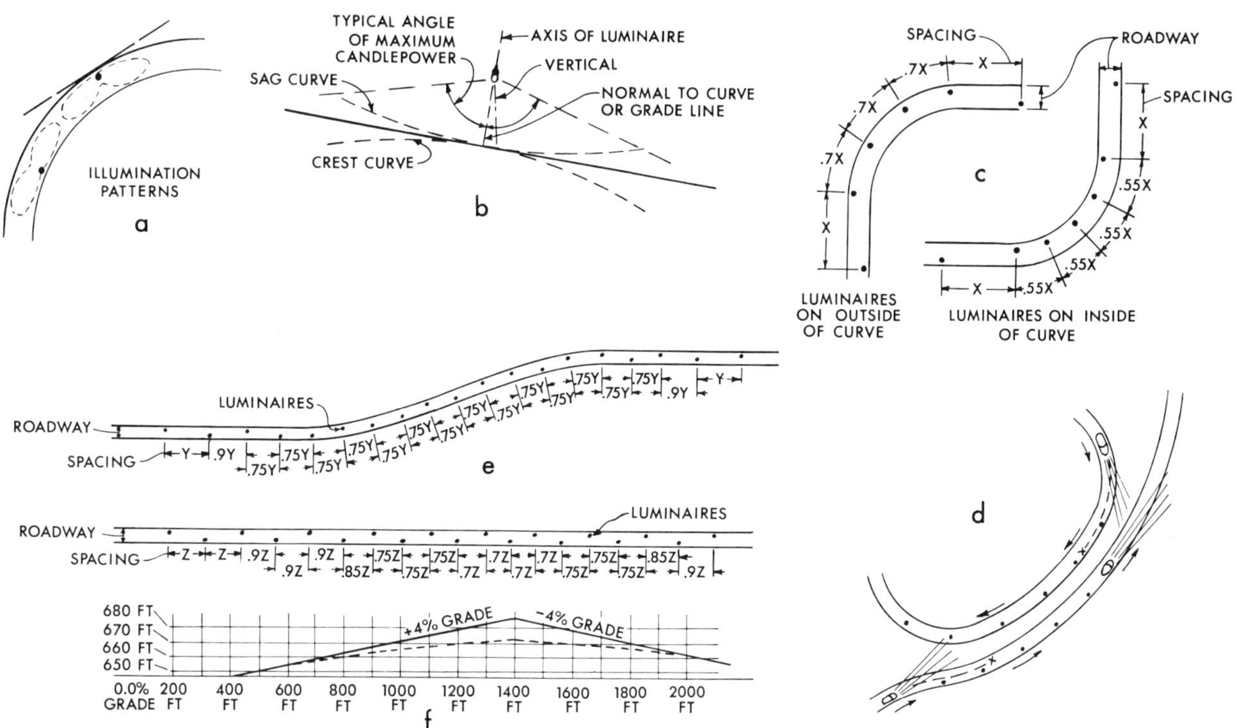

Fig. 20–7. Typical lighting layouts for horizontal curves and vertical curves. (a) Luminaires oriented to place reference plane perpendicular to radius of curvature. (b) Luminaire mounting on hill (vertical curves and grade). (c) Short radius curves (horizontal). (d) Vehicle illumination limitations. (e) Horizontal curve, radius 1000 feet, super elevation 0.06 feet. (f) 1250-foot vertical curve with four per cent grade and 750-foot sight distance.

Converging Traffic Lanes. (See Fig. 20–8.) Frequently this lighting requires the same treatment as abrupt curves. Also, direct illumination is needed on the sides of vehicles entering the traffic lanes.

Diverging Traffic Lanes. (See Fig. 20–8.) Such lanes are often confusing to the driver. Luminaires should be located to illuminate curbs, abutments, guard rails, and vehicles, starting at the entrance to the deceleration zone.

Interchanges. (See Fig. 20–8.) It is important to reveal all the complexes and features of the entire scene, allowing the driver to know with certainty at all times his position and where he is going. An inadequately lighted interchange with too few luminaires randomly placed may lead to confusion by giving misleading information. The critical sections, including points of access and egress, curves, ramps, etc., need higher levels of illumination than the approaches and departure zones to which transition lighting should be applied.

Railroad Grade Crossings. (See Fig. 20–9.) The direction and level of illumination should permit visual recognition of identification signs and pavement markings. Minor variations of the basic layouts shown in Fig. 20–9 may be desirable depending on the exact locations of such signs and markings. The footcandle level within 100 feet on both sides of the track should be in accordance with Fig. 20–4 or twice the level of the adjacent area of the same roadway, whichever is higher, but never less than 1.0 footcandle (11 lux).

Trees. (See Figs. 20–10 to 20–12.) Roadway lighting and tree foliage need not conflict. Judicious pruning will permit effective lighting, reduce system glare, and generally improve the appearance of the street. Design compromises, involving deviations from preferred system layouts with respect to luminaire spacings, mounting height, and transverse locations, may be necessary. Any such deviation can generally be compensated for by resorting to center suspension or lower mounting heights and closer spacings, with smaller lamp sizes and, if necessary, lower angle of maximum candlepower. Also, irregular spacing of individual luminaires up to 20 per cent of average spacing can be tolerated, providing no two consecutive luminaire locations are involved. Transverse deviation of an individual luminaire should only be made where there is no other reasonable compromise.

Although foliage interference mostly affects roadway illumination, there may be instances on local traffic residential streets where it can also affect important sidewalk illumination. Generally this problem can be solved by either altering the luminaire positions, by pruning, or a combination of both methods.

Tunnels[2]

The basic objective of tunnel illumination is adequate and comfortable visibility both by day and at night for the users of the tunnel. Some of the main factors involved in the illumination design are:

(1) A minimum of "black-hole effect" at the tunnel entrance. (In daylight hours when the motorist approaches the tunnel entrance, an object in the tunnel cannot be seen if its luminance and that of its immediate surroundings is much lower than the luminance to which the eye of the motorist is still adapted at that moment.)
(2) The design of an adequate number of tunnel zones of sufficient length and different illumination levels to provide eye adaptation from bright sunlight to the minimum level of tunnel lighting used.
(3) The location and alignment of light sources.
(4) A minimum of glare.
(5) A minimum of flicker effect of alternate light and dark areas.
(6) Adequate reflectance in tunnel linings.
(7) Pleasing color contrast in tunnel linings.
(8) Proper signing and signal lighting.

For lighting purposes every artificial or natural covering of a road, irrespective of the length and nature of the covering, is considered a tunnel. Although the term underpass is in common usage, there is no agreement on the definition of an underpass. Hence, the term tunnel is used exclusively.

Recommended illumination levels for tunnels are given in average horizontal values on the roadway at minimum conditions. With wall reflectance of 70 per cent or higher, and the use of luminaires which illuminate the roadway and walls, the horizontal illumination recommended will usually produce satisfactory visibility. It is considered that under these conditions, adequate horizontal illumination will also provide adequate vertical illumination.

In vehicular tunnels, the design speed may be as high as 70 miles per hour, and in heavy traffic, the vehicular spacings may be as close as 50 feet or less. Much of the driver's attention is directed to the pavement markings and the curbing defining the traffic lanes, as well as to other vehicles ahead in his lane and to the vehicles in adjacent lanes.

Uniform illumination on horizontal and vertical surfaces is necessary for good visibility within tunnels. In the lanes of tunnels which are adjacent to the walls, luminance of the vertical surface is of major importance. At the tunnel entrances a high wall luminance is particularly valuable in reducing the "black-hole effect."

Short Tunnels. A short tunnel is one where, in the absence of traffic, the exit is clearly visible from a point ahead of the entrance portal. For lighting purposes, the length of a short tunnel is usually limited to approximately 150 feet. Some tunnels up to 400 feet long may be classified as short if they are straight, level and have a high width and/or height to length ratio.

In most cases no lighting system is required inside short tunnels for adequate driver visibility. Daytime penetration from each end, plus the silhouette effect of the opposite end brightness, gener-

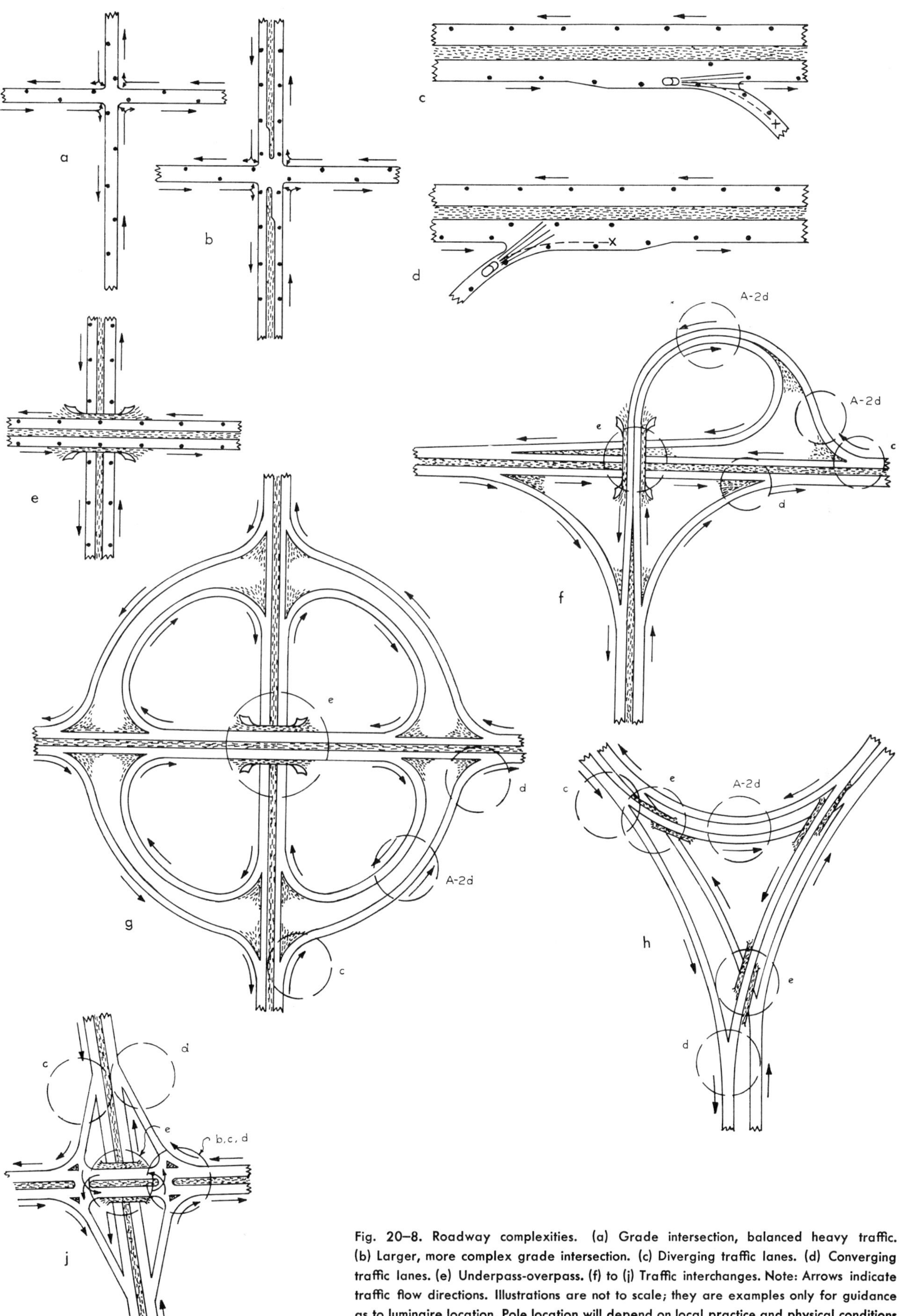

Fig. 20–8. Roadway complexities. (a) Grade intersection, balanced heavy traffic. (b) Larger, more complex grade intersection. (c) Diverging traffic lanes. (d) Converging traffic lanes. (e) Underpass-overpass. (f) to (j) Traffic interchanges. Note: Arrows indicate traffic flow directions. Illustrations are not to scale; they are examples only for guidance as to luminaire location. Pole location will depend on local practice and physical conditions of the area.

ally assures satisfactory visibility. Tunnels between 75 and 150 feet in length may require lighting if daylighting is restricted due to roadway depression or the proximity of tall buildings in urban areas. In this case the tunnels should be treated as long tunnels and entrance lighting provided.

For tunnels approximately 75 feet or less in length, the proper positioning and mounting height of pole-mounted street lights adjacent to each end of these tunnels, usually provides satisfactory nighttime lighting. When higher mounting heights are used for these street lights, the poles must be located farther away from the tunnel to get the light into the tunnel entrance. This may mean that the uniformity ratio is exceeded and some tunnel lighting may be necessary.

On roadways over 50 feet in width, or where the tunnel has a center row of columns, it may be desirable to position pole-mounted units on opposite sides of the roadways at each end. Wherever a shoulder-mounted pole can be eliminated by attaching luminaires to a necessary bridge structure, it should be done for the sake of safety.

For tunnels from approximately 75 to 150 feet in length, a nighttime lighting system is usually required to provide a maintained level of approximately two times, but not over three times, that recommended in Fig. 20–4 for the connecting open roadway.

Long Tunnels. In straight tunnels, generally over 150 feet in length, where the exit brightness takes up too small a part of the driver's field of vision to serve as an effective background for silhouette discernment, the structure is classified as long for lighting purposes. All tunnels where the exits cannot be seen from points ahead of the entrance portals, are also classified as long.

After the Open Road Zone the following main zones are considered:

(a) For day conditions—
 (1) the Grid Zone, which is optional.
 (2) the Main Tunnel Entrance Zone.
 (3) the Central Tunnel Zone.

If the Grid Zone is not used, two or more illumination levels are required in the Main Tunnel Entrance Zone, with the first section at a high level of illumination.

(b) For night conditions—the tunnel is considered as a single zone for the entire tunnel length.

Grids used may be of many different designs but are usually constructed to permit passage of snow and rain but not direct sunlight. The average grid is designed to reduce the daylight on the roadway by about 70 per cent. This can be reduced further at the end of the Grid Zone by the use of black paint on the grid, and/or by decreasing the width of the grids and increasing the width of the solid section on each side of the grids.

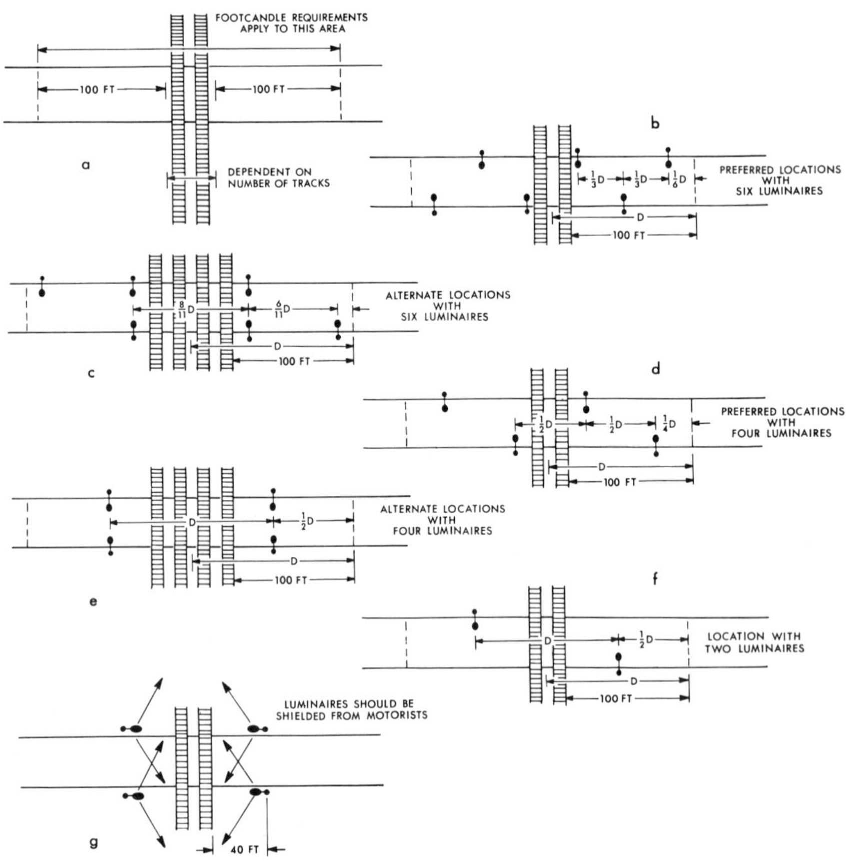

Fig. 20–9. Railroad grade crossings.

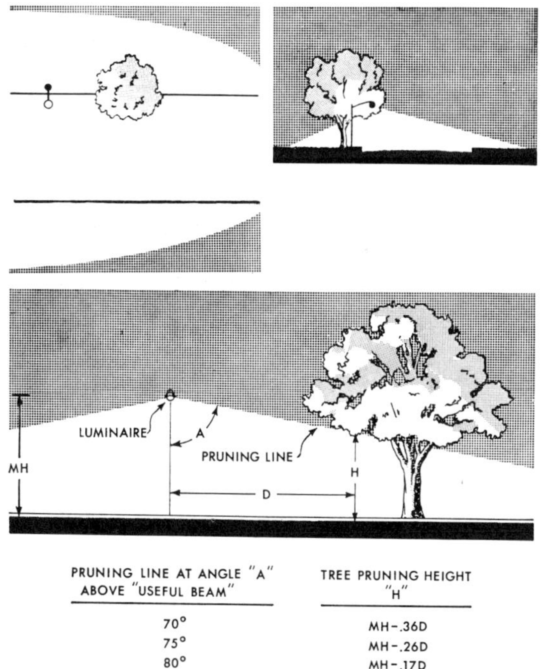

PRUNING LINE AT ANGLE "A" ABOVE "USEFUL BEAM"	TREE PRUNING HEIGHT "H"
70°	MH-.36D
75°	MH-.26D
80°	MH-.17D

Fig. 20-10. Recommended tree pruning to minimize conflict with roadway lighting.

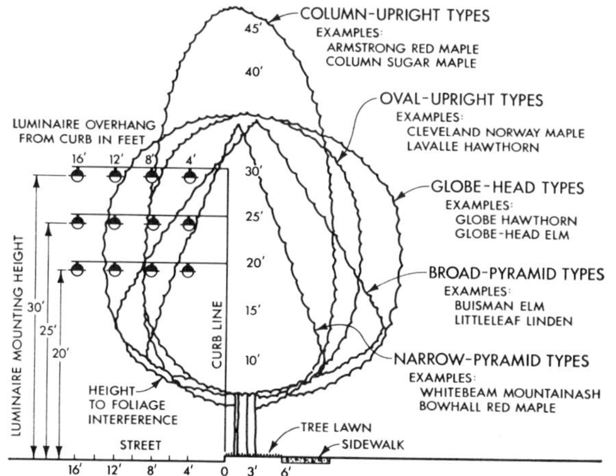

Fig. 20-11. Height to foliage interference for different types of trees and luminaire overhang from curb. Tree examples by E. H. Scanlon.

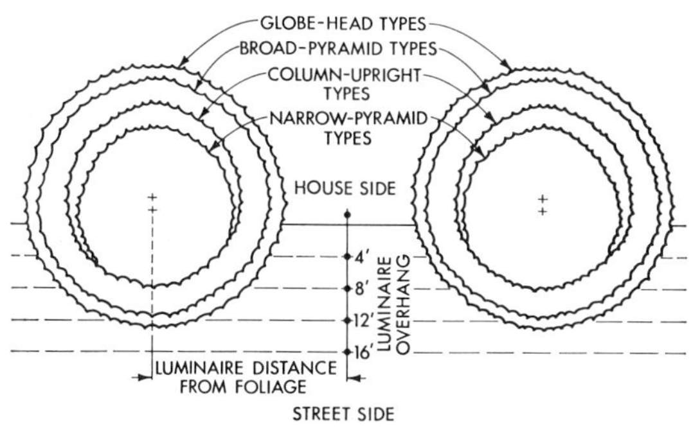

Fig. 20-12. Longitudinal and transverse location of luminaires as related to different types of trees.

Daytime Lighting. For daytime lighting two simple principles are recommended to produce comfort and safety for the motorist:

(1) Illumination levels under daytime conditions in adjacent zones of long tunnels should not exceed a reduction ratio of 10 to 1.

(2) The time that the motorist requires in passing from the highest to the lowest illumination, should be 4 seconds or more.

The lighting in the Main Tunnel Entrance Zone under daytime conditions is a major lighting design problem. It is generally assumed that the luminance near a tunnel entrance can be regarded as the luminance to which the driver is adapted. This adaptation luminance is determined by the average luminance of the visual field at this location. As the driver approaches the tunnel entrance, the central part of the driver's visual field is mostly focused on the darker tunnel entrance. As this becomes larger on the approach to the tunnel, the adaptation luminance of the driver is greatly reduced.

Considering as an example a case where there is no grid and the driver's adaptation level is conservatively assumed to be that produced by an average of 5000 footcandles, the illumination level in the Main Tunnel Entrance Zone just after the Main Tunnel Portal should therefore be a minimum of 500 footcandles horizontal maintained, according to the 10 to 1 maximum reduction ratio for adjacent zones. If the Main Tunnel Portal has a large opening which gradually slopes down to the main tunnel cross section, this will help in the transition from full daylight to the tunnel lighting.

With no grid used, the Main Tunnel Entrance Zone is divided into two or more zones for lighting. If only two zones are used in the example considered the second zone should have at least a 50 footcandle, horizontal maintained level and each of the zones would have a minimum travel time of 2 seconds.

The illumination level for the Central Tunnel Zone of all long tunnels should be at least 5 footcandles horizontal maintained, which is between two and three times the recommended level for *nighttime* lighting of the open road at each end of the long tunnel. For tunnels up to 2000 feet in length it may be desirable to use up to 10 footcandles in that zone.

To facilitate the transition from high daylight levels to practical levels of lighting at the Main Tunnel Entrance, a grid is sometimes installed over the roadway ahead of the Main Tunnel Portal. This use of a Grid Zone aids adaptation to high and low values of daylight and appears to be a good method of combating the black-hole effect. The Grid Zone, if used, becomes part of the tunnel.

Since the average grid is designed to reduce the illumination on the road at a point soon after the entrance to the Grid Zone by 70 per cent, in the example above, the 5000 footcandle level on the Open Road would be reduced to 1500 footcandles in the Grid Zone. By use of solid sections of increasing width at the side of the grid, the illumination on the road can be further reduced to 500 footcandles at the end of the Grid Zone. Black paint may also be used at the last part of the Grid to cut down on the light transmission.

With a reduction to 500 footcandles at the end of the Grid Zone, the Main Tunnel Entrance Zone should have a level of 50 footcandles and the Central Zone 5 footcandles minimum maintained. In the example given, the length of the Grid Zone and the Main Tunnel Entrance Zone should each have a minimum travel time of 2 seconds.

The length of the Grid Section should be *increased* as compared to the example given, if there is considerable direct sunlight penetration into the entrance of the grids. The Grid Section may be *decreased:*

(1) For long straight tunnels from 150 to 1000 feet in length and made in proportion to the tunnel length.

(2) For entrance from a depressed roadway with close vertical side walls since this results in a reduced lighting level.

(3) When one or more short tunnels for cross streets over the depressed roadway are comparatively near the long tunnel.

(4) Where large buildings or trees shade the tunnel entrance.

Nighttime Lighting. At nighttime the full length of long tunnels is considered as a single zone for lighting purposes, and the illumination level of the entire zone should be between two and three times the level of lighting on the open roadway approaches as specified in Fig. 20-4.

Adjacent Areas. Special lighting is normally not used at tunnel exits as the motorist's eyes adjust quickly to the increasing luminance as the exit is approached; however, when a tunnel has two-way traffic, both portals must be treated as entrances.

All tunnels, urban and non-urban, which have nighttime lighting, should have at least 500 feet of roadway lighting before each tunnel entrance and after each tunnel exit in accordance with the recommended levels in Fig. 20-4.

REFERENCES

1. "American National Standard Practice for Roadway Lighting," Sponsor: Illuminating Engineering Society, to be published.
2. Subcommittee on Lighting of Tunnels and Underpasses of the Roadway Lighting Committee of the IES: "Lighting of Tunnels," to be published.

SECTION

21

AVIATION LIGHTING

Aircraft Lighting 21–1 Ground Lighting to Aid Navigation 21–6 Heliport and VTOL Lighting 21–12 STOL Port Lighting 21–12 Obstruction Lights 21–13 Ground Operational Lighting 21–13

Much of the lighting installed on aircraft and on the landing areas for aircraft is signal lighting. This lighting conveys information to the pilot by means of color, location, flash characteristic, and pattern of the lights rather than by illumination of areas or objects. Some exceptions to this generalization are the lighting of instruments in the aircraft cockpit, the lighting of passenger cabins and cargo holds, the floodlighting of runways by fixed lights or by lights on the aircraft, and the lighting of the service areas of the airport.

Standardization

In no field of lighting is standardization more important than in aviation lighting. The interstate and international scope of air operations make imperative the establishment of standards of color,[1, 2] flash characteristics, and pattern of the lighting systems essential to the safe operation of aircraft. The Federal Aviation Administration (FAA) is the agency responsible for the establishment of domestic standards for civil operation. Military aviation lighting in the U. S. is regulated by the Departments of the Army, Navy, and Air Force.

In many cases the FAA, the Army, the Navy, the Air Force, and other government and industry organizations have collaborated in reaching joint standards.

Ground lighting aids at Canadian airports are quite similar to systems defined for US airports. In Canada, the Ministry of Transport, through the Director of Civil Aviation, regulates the use of illuminated visual aids to air navigation at all Canadian Airports.

International standards and practices pertaining to civil aviation are formulated by the International Civil Aviation Organization (ICAO), an agency of the United Nations, which is composed of representatives of nearly all nations interested in air transport. Standards adopted by this body are usually accepted by the member nations and are made mandatory minimum requirements.

The requirements for aviation lighting are constantly changing. For this reason, the applicable agencies should be consulted before planning an aviation lighting project in order to obtain the latest requirements.

AIRCRAFT LIGHTING

Aircraft lighting refers to the use of luminaires and lighted equipment attached to the airframe and can be broadly assigned to three categories: (1) *exterior*—to provide a signaling or identification function, (2) *crew station*—to aid the crew in operating the aircraft and (3) *passenger and service areas*—special forms of illumination for passengers or for service areas. Luminaires are normally designed to perform a specific lighting function and to fit a specific position in a specific aircraft model. Therefore, it is very difficult to achieve standardization of hardware. However, standardization of visual elements of displays, such as, color, intensity, flash characteristics, is attempted for some aircraft lights with varying degrees of success. Small single-engine aircraft are equipped most commonly with 14-volt dc power systems. Power systems of 28 volts dc are common for light twin engined aircraft and most turbine-powered aircraft have 120/208-volt, 400 hertz primary electrical systems with transformation to the lower voltages required by the light sources used—5- to 75-volt range for large military and commercial aircraft.

Exterior Lighting[3, 4]

In a primary sense the purpose of exterior lighting of an aircraft is to provide warning of its presence in a spherical environment. This requirement is sometimes difficult to meet with conventional

NOTE: References listed at the end of each section.

placement of light sources because of the design of today's airframes, which employ swept wings, variable position wings, high mounted horizontal stabilizers, tail mounted engines, and strict aerodynamic requirements for a "clean" structure. All of these factors should be considered by both the designer of the lighting hardware and the aircraft structure engineers in early stages of design if optimum lighting in accordance with military and FAA requirements is to be achieved. The high speed of today's aircraft has created very high vibration and temperature environments for the exterior lighting equipment mounted on wing tips and tail structures with excessive rates of lamp failure resulting. Ambient temperatures of 500°F and vibration environments of 30 g's acceleration from 10 to 2000 hertz are not uncommon. Here again, it is very important to know the vibration characteristics of the specific airframe at the lighting equipment mounting positions, to insure the selection of proper lamps or the design of suitable vibration resistant mounts.

Navigation Lights (Position Lights). Navigation lights are signal lights displayed on the aircraft to provide aspect information in three azimuthal sectors to pilots of other aircraft and to ground observers. These lights, normally located on the wing tips and tail of an airplane and on the fuselage of a helicopter, are arranged to give a red signal throughout the 110-degree sector to the left of the forward axis of the aircraft, a green signal in the 110-degree sector to the right of that axis, and a white signal in the 70-degree sector on each side of the aft axis of the aircraft. In the simplest form, navigational lights for a small private aircraft consist of red and green wing tip lights and a white tail light. On larger aircraft of the transport category, it's likely that two lights will be used at each position to provide the required coverage. It is becoming quite common for the white tail lights to be mounted on the trailing edges of each wing tip, thus providing easier maintenance and also providing more information about the attitude of the aircraft to an observer. See Fig. 21–1.

The above coverages are also required for all other aircraft structures such as vertical take-off or landing (VTOL) and lighter than air designs. In the mounting of position lights on helicopters having large transparent bubble cockpit enclosures, any illumination of the transparent enclosures by the red and green position lights should be avoided or minimized.

Primary light sources used are of the reflector-type incandescent and tungsten-halogen types. In addition, Xenon white flashing lights are being used on many aircraft to supplement the existing navigational lights.

Anti-collision Lights. Anti-collision lights are red, high-intensity, flashing lights designed to provide a long range signal in the more critical areas around the plane of flight of the aircraft, for the purpose of making it very conspicuous to oncoming aircraft. These lights are normally installed on the top of the vertical fin; or on the top, or top and bottom of the fuselage, to be visible in all directions in the plane of flight and for 30 degrees above and below that plane. Anti-collision lights are required on all large civil aircraft (more than 12,500 pounds maximum certificated weight), on all small aircraft for which application for type certificate was made after March 31, 1957, and on military aircraft. It is common on larger aircraft to use supplementary Xenon white flashing lights to supplement the red anti-collision lights.

On military aerial tankers a special type of anti-collision light or rendezvous light is often used, designed so that white, red, green or blue or alternating combinations of these colors can be displayed.

Landing Lights. Landing lights are high-intensity spotlights designed to aid the pilot in judging the height and attitude of the aircraft, and to illuminate obstacles that might constitute hazards during landing or taking off. There are two standard types of landing light installations: the leading edge type, shown in Fig. 21–2(a), and the retractable type, shown in Figs. 21–2(b) and 21–3.

Another type of retractable landing light, which is controllable in both elevation and azimuth, is also sometimes used. This type of light is useful for taxiing, since it can be remotely controlled to look from side to side as well as directly ahead.

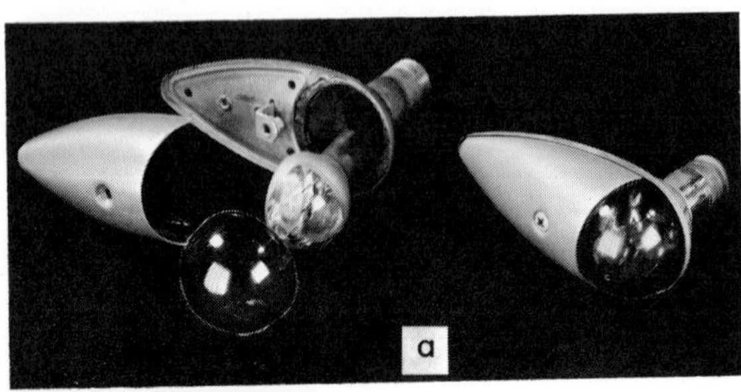

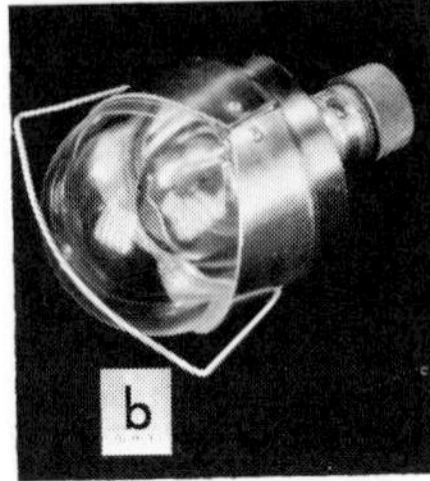

Fig. 21–1. Navigation Lights: (a) wingtip (forward); (b) tail (rear).

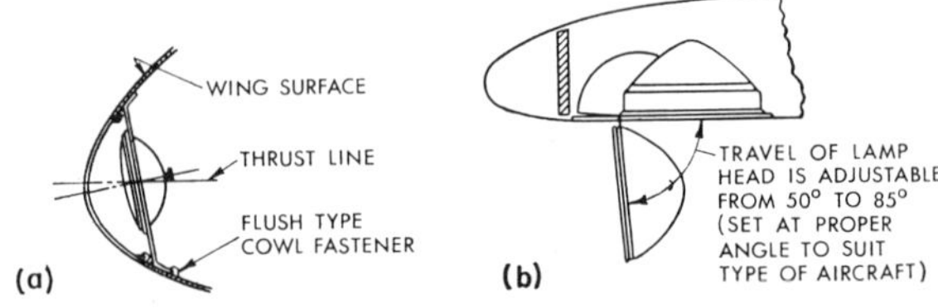

Fig. 21–2. (a) Leading edge installation of landing light. (b) Retractable type landing light.

Fig. 21-3. Retractable landing light.

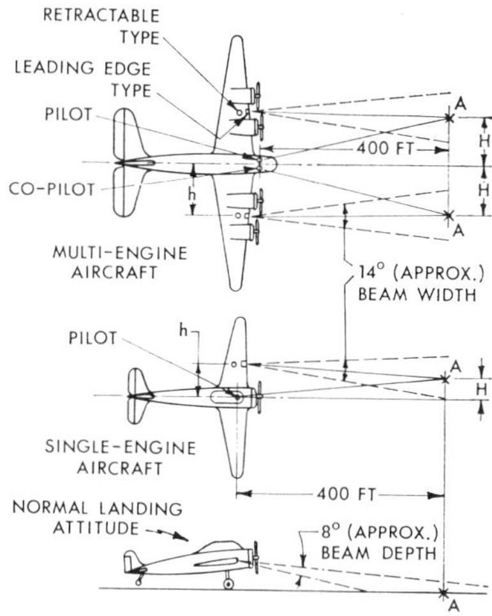

Fig. 21-4. Location and aiming of landing lights; point *A* lies on the middle of the beam 400 feet ahead of the aircraft. *H* is the distance from *A* to the centerline of the aircraft, *h* is the distance from the light source to the centerline of the aircraft.

In some aircraft a landing light is also located in the nose of the fuselage. In this position it may also serve as a taxi light to illuminate the ground while taxiing, or as a signaling light useful when the aircraft is flying at speeds which prohibit the extension of the retractable landing lights.

Two landing lights are normally required on an aircraft equipped for night flying, although in some single engine aircraft, one landing light in the left wing is permissible. Landing lights may be omitted altogether on carrier-based aircraft. Usually one light is placed in each wing as far outboard as practicable. See Fig. 21-4. In leading-edge installations two lamps may be located side-by-side to provide more light during normal use and to guard against complete outage in the case of lamp filament failure. Lamps for landing lights are usually of the sealed-reflector type.

Auxiliary Lights:

Taxi Lights. Taxi lights are provided to furnish illumination in front of taxiing aircraft to permit obstacles or curves in the taxiway to be seen by the aircraft pilot in sufficient time to stop or change course in order to avoid damage to the aircraft. They also are the light source for retroreflector guidance systems. See Fig. 21-5. Landing and taxi lights may be incorporated into one system suitable for both applications. Possible installation points for taxi lights are on the steerable nosewheel strut as shown in Fig. 21-6 and on the wing root.

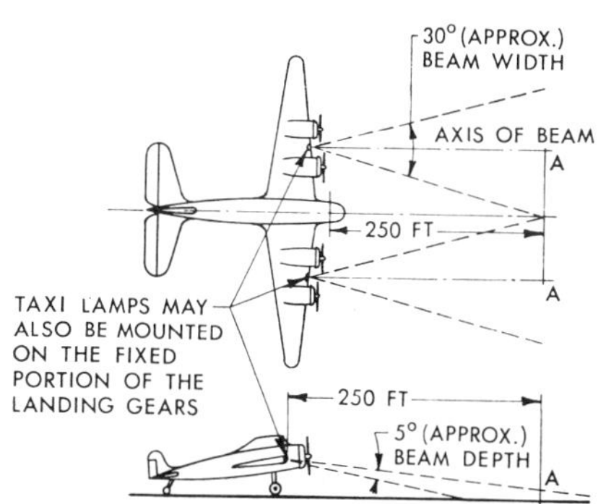

Fig. 21-5. Location and aim of taxiing lights.

Fig. 21-6. Taxiing light mounted on nosewheel strut.

Formation Lights. Formation lights are special purpose exterior lights common to military types of aircraft and not normally used on civil aircraft. These include lights and systems for inflight refueling and formation flying. Formation lights in some cases are floodlights which light up vertical areas on the aircraft. In other cases they are small one by five-inch incandescent or electroluminescent lighted panels, mounted on the side of the fuselage. On newer aircraft, larger areas of electroluminescent lamps are becoming common for formation lighting. Troop-carrying helicopters are equipped with rotor tip lights as part of the formation-flying system.

Aerial tankers are commonly equipped with rendezvous lights which are a special type of anti-collision lights which can display white, red, green or blue color or alternating combinations.

Emergency Lighting. Exterior emergency lights are now required on all airline aircraft. These lights are used to light up escape slides and emergency egress areas over the wing and on the ground. These lights usually operate on a power-failure basis and have their own rechargeable battery power supplies.

Ice Detection Lights and Engine Check Lights. These are installed on the fuselage to illuminate the leading edge of the wings and the engine cowling to reveal icing and engine operating conditions. They are mounted flush on either side of the fuselage or, in the case of four engine aircraft, on the outboard side of the engine nacelle.

Approach Lights. Approach lights are used primarily on Naval aircraft but have recently been applied to USAF aircraft. These are three color coded signal lights (red, yellow, green) which receive signals from attitude sensing equipment within the aircraft and indicate, by the color of the activated light, the correct or incorrect attitude for landing of the approaching aircraft to the Landing Signal Officer aboard an aircraft carrier. These lights must have sufficient intensity to be seen in daylight during the final stage of approach to the carrier and must be greatly reduced in intensity for night use.

Clearance Lights. Clearance lights are lights of relatively low intensity, mounted on or near the extremities of the aircraft structure to provide collision avoidance information during ground operations.

Hover Lights. Hover lights are usually fixed, high intensity, wide beam lights and are used by helicopters for illuminating the area below in low altitude search operations or for landing. These lights also have application on VTOL aircraft for landing operations.

Interior Lighting

Interior lighting of aircraft falls into three broad categories: (1) *crew station lighting,* (2) *cabin lighting,* and (3) *emergency lighting.* Crew station lighting represents, perhaps, the most complex problem in illumination of any aircraft. This is particularly true for certain types of military aircraft where the rigid requirements of pilot dark adaptation are in conflict with other requirements for various colored signal indicator lights and for high brightness cathode ray tube displays. These factors, coupled with the necessity for eliminating reflections in the windshield, a problem common to all aircraft, and the need for crowding many illuminated components into a small area, make it essential to consider the crew station lighting task at the earliest stage of the cockpit design.

Lighting of the passenger cabin space is complicated by low ceilings, space, heat and weight limitations, and the desire of some passengers to read while others wish to rest or sleep. Emergency lighting has its problems in finding areas for the proper location of lighting hardware and in providing a system of completely reliable emergency power not connected to the normal electrical system of the aircraft.

Crew Station Lighting.[5, 6, 7] Three distinct colors for instrument and control panel lighting are in common use today.

1. Naval aircraft use red light because the pilot must remain dark adapted. The red color must have a minimum cone-to-rod[8] ratio of 18 to meet dark adaptation requirements.
2. Commercial aircraft specifying white light utilize unfiltered incandescent lamps.
3. USAF aircraft use colored illumination matching the chromaticity coordinates of a 2854 K source with specified tolerances for blue, yellow, green and magenta departures from the blackbody locus.

Dimming of lighting systems is controlled as a group, not individually, thereby preserving brightness uniformity within the group.

Cockpit Lighting:

Instrument Lighting. For new design, all instrument lighting is of the integral type, *i.e.,* all lamps for the illumination of the instrument dials, pointers, counters, etc., are contained within the instrument case and may or may not be replaceable. Most instrument lighting design is predicated on the use of the high reliability T-1, 5.0-volt, subminiature lamps having very long life and great resistance to vibration failure. See Fig. 21-7. Electroluminescent light sources also offer possibilities of usefulness in hermetically sealed instrument displays because of their uniform brightness and reliability under con-

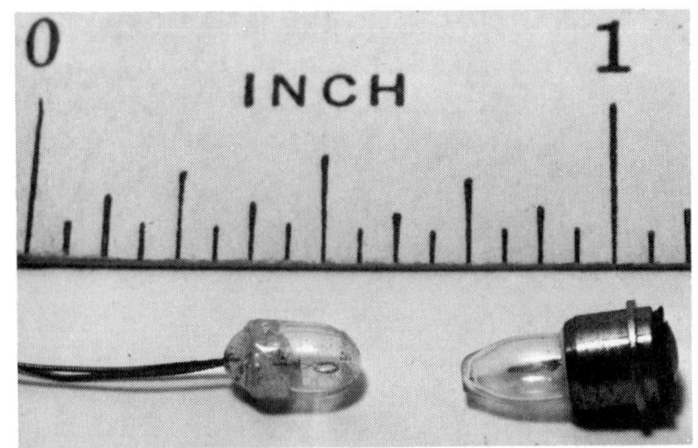

Fig. 21-7. Subminiature lamps (T-1, 5.0 volts) for instrument and control panel lighting.

trolled environmental conditions. Nearly all aircraft require some form of secondary lighting to be used in the event of failure of the primary system. In most cases this consists of a group of individual floodlights spaced in such a manner as to provide relatively uniform lighting of at least all essential instruments. Power for this secondary system must be taken from a source other than that supplying the primary system.

Control Panel Lighting. Control panel lighting of switches, knobs, and other controls for most commercial and all military aircraft has been standardized around the plastic lighting panel utilizing the T-1, 5.0-volt lamp and circuitry either embedded within the thickness of the plastic plate or contained on a separate printed circuit board mounted against the rear of the control panel. Secondary lighting of the plastic lighting panels is by floodlighting from optimally placed equipment. Electroluminescent lighting of these control areas is also coming into use, utilizing techniques for encapsulating electroluminescent lamps within the plastic panels. See Fig. 21-8.

Warning and Indicator Lighting. The design trend today is to categorize indicator lights as Warning Lights, Caution Lights, and Advisory Lights and to use legend type lights grouped together as to classification. Colors normally used are red for warning, yellow for caution, and green for advisory. Fig. 21-9 shows a grouping of the legend type indicators.

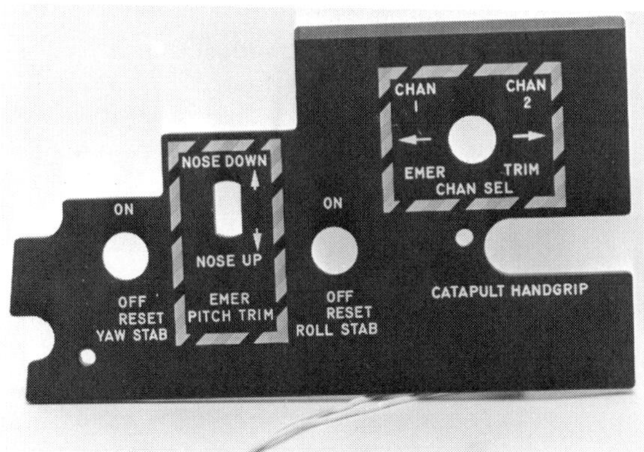

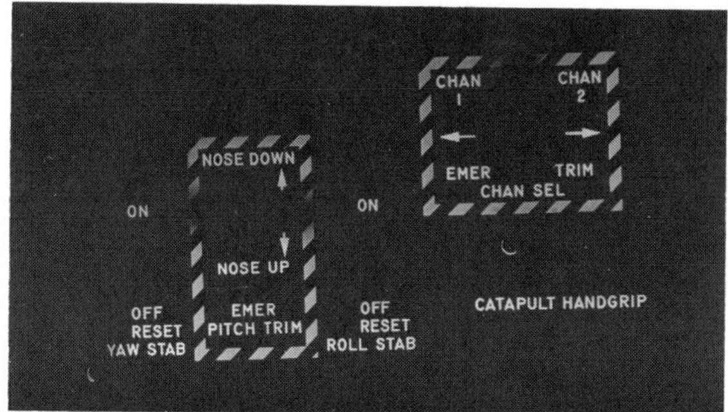

Fig. 21-8. Electroluminescent control panel. Top is unlighted daylight view. Bottom is lighted night view.

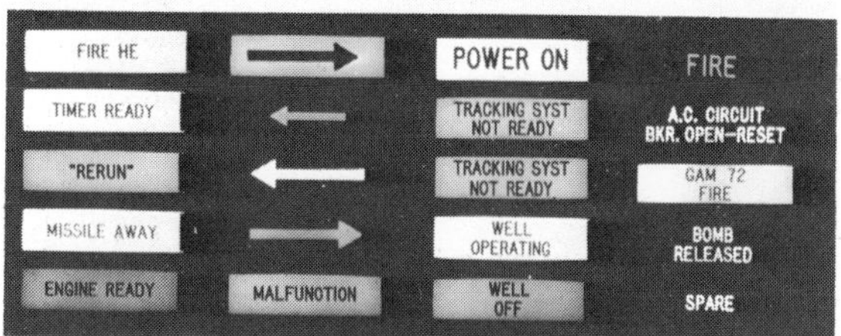

Fig. 21-9. Grouped legend type indicator lights. Top is unlighted daylight view. Bottom is lighted night view.

Floodlighting and General Lighting. Special floodlighting is provided for map reading or other detailed visual tasks. Individual luminaires, usually removable from the airframe structure and adjustable as to intensity and beam distribution, are used. There also is higher level lighting for thunderstorm conditions. In these latter conditions, white floodlighting of the instrument panel and console areas is required. It is desirable to obtain a minimum of 100 footcandles of white light on the instrument panel containing the essential flight instruments.

Cabin Lighting. In the passenger cabin considerable attention should be given to providing general illumination, comfortable luminance ratios, and good seeing conditions for reading and writing. For general illumination, a relatively uniform level of 5 footcandles, measured on a horizontal plane over the entire cabin, should be provided; this general lighting supplemented by 20 footcandles over a small reading or writing area for individual passengers.

Either fluorescent or incandescent filament lamps are used for general lighting. Fluorescent lamps are operated from 400-hertz, 120-volt (or 115–208 volt, 3-phase) power supplies. Inductive ballasts for 400-cycle operation are available up to 40-watt sizes and are appreciably lighter than the 60-cycle equivalent ballasts. Capacitor type ballasts for 400-cycle operation are much lighter even than the inductive ballasts. A leading (negative) power factor of much less than unity should be avoided so that the aircraft rotary inverter will not reach an unstable operating condition.

Each passenger on the aircraft is provided with individual reading lights (see Fig. 21-10). The intent is to produce a relatively uniform level of illumination of 20 footcandles over about an 18-inch circular area on the normal reading plane. For greater seeing comfort it is preferable to have the

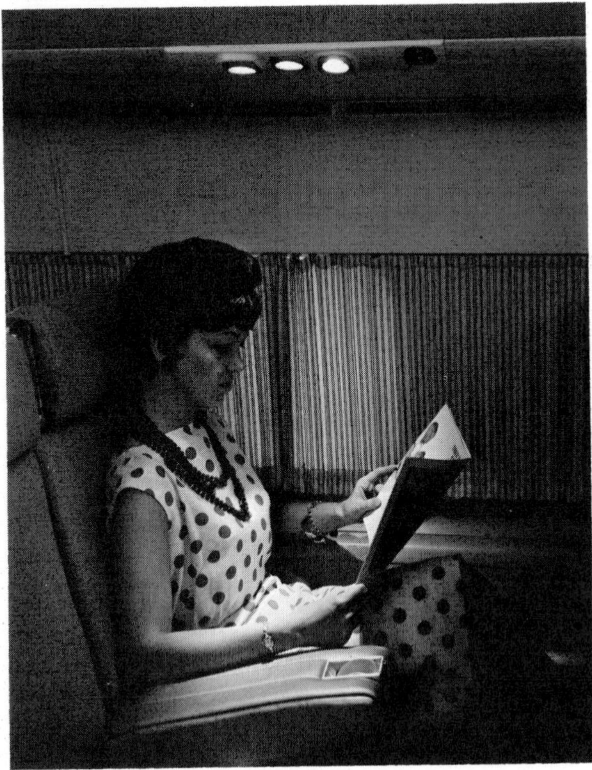

Fig. 21–10. Reading lights.

beam taper off gradually rather than have a sharp cutoff.

GROUND LIGHTING TO AID NAVIGATION

There are various lighting aids used by a pilot for the visual problems encountered in making a visual approach to an airport and in landing. These visual problems include: location and identification of the airport; location and identification of the runway; location of and distance from the runway threshold; determination, from visual reference, of altitude, attitude, alignment, and progress along runway and taxiway to the airport service area.

Location and Identification of the Airport

Airport beacons[9] are installed to provide airport location and identification information under VFR conditions. See Fig. 21–11. Such beacons are high intensity (approximately 250,000 candelas effective intensity in white)[10] displaying alternate green-white beams for land airports or yellow-white for seadromes. Military airports display a double peaked white flash and single peaked green or yellow flashes. The flash rate is 12 to 15 flashes per minute. Airport beacons required to be located more than 5000 feet from the airport in order to provide 360 degree azimuth coverage are supplemented with flashing green beacons which may be coded to identify the specific airport.

Fig. 21–11. Glossary of Abbreviations Used in Text

ALS	Approach Lighting System
Cat. I	Aircraft operations down to a 200-foot decision height and with a runway visual range not less than 2400 feet.
Cat. II	Aircraft operations down to a 100-foot decision height and with a runway visual range not less than 1200 feet.
FLOLS	Fresnel Lens Optical Landing System
GCA	Ground Controlled Approach
IFR	Instrument Flight Rules
ILS	Instrument Landing System
MALS	Medium (Intensity) Approach Lighting System
MALSF	Medium (Intensity) Approach Lighting System with (sequenced) Flashes
MALSR	Medium (Intensity) Approach Lighting System with RAILS thereafter
PAR	Precision Approach Radar—previously referred to as GCA
RAILS	Runway Alignment Indicator Lighting System—7 sequenced flashing lights spaced 200 feet apart
REILS	Runway End Identifier Lighting System
RVR	Runway Visual Range
VASI	Visual Approach Slope Indicator
VFR	Visual Flight Rules

In Canada, airport beacons are also used. They are similar in physical characteristics to those used in the United States but provide only yellow beams of light. Authorization for the use of a beam must be obtained from the Ministry of Transport.

Airport beacons are classified as "true lights," however, lawful authority for their operation is no longer required. Severe penalties are provided for interfering with the operation of a true light, or exhibiting a light with the intent to interfere with air navigation within the United States.

True lights should be operated whenever the sun's disc is 6 or more degrees below the horizon and may be operated during the day in periods of restricted visibility. Any changes in the hours of operation shall first be published in "Notices to Airman" through proper channels.

Fig. 21–12. Runway end identifier light with condenser discharge source.

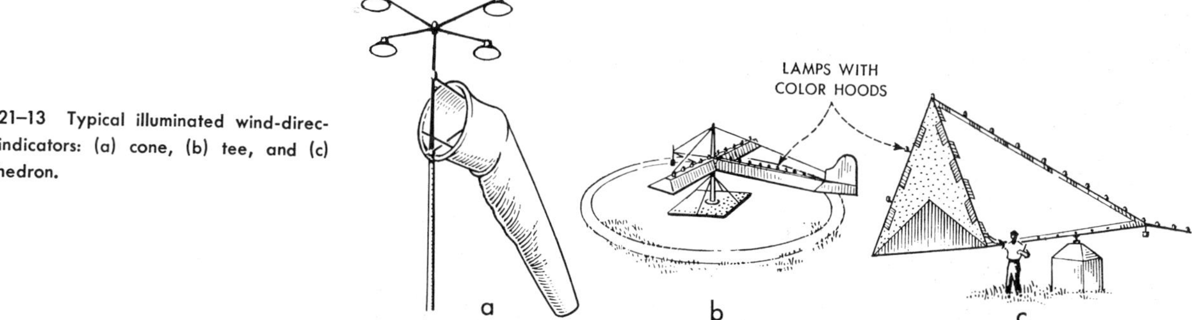

Fig. 21-13 Typical illuminated wind-direction indicators: (a) cone, (b) tee, and (c) tetrahedron.

Runway Visual Approach Guidance. After the pilot has visually located and identified the airport, it is necessary for him to locate and identify the runway to which an approach and landing is to be made. Under VFR conditions this identification is obtained from visual clues provided by runway markings, natural terrain features, man-made objects by day and runway lights, runway identifiers and various types of approach lights by night. VASI systems, where installed, provide a visual glide slope path that is obstruction free. Runway End Identifier Light Systems (REILS), see Fig. 21-12, consist of two identifiers that flash simultaneously, twice per second, and are located approximately 40 feet outboard of each edge of the runway in line with the runway threshold lights. The REILS is usually installed where there is no ALS, but a requirement exists to provide early identification of the runway. The Navy uses the REILS for guidance on the downwind leg of a circling approach in conjunction with the ALS.

In Canada, the REILS are also used but are located 100 feet from the runway threshold lights, and are angled 20 degrees away from the longitudinal axes and 7½ degrees horizontally.

Wind and Landing Direction Indicators. The illuminated *wind cone* (see Fig. 21-13) is used by the pilot to determine the direction of the wind by day or night. It is a large cloth "sock" free to swing about a vertical shaft and is illuminated from above for night use. Since the greater part of its length is not rigidly supported, it gives a rough indication of wind velocity as well as wind direction. *Wind tees* or *tetrahedrons* (see Fig. 21-13) are used to indicate the preferred landing direction and may or may not be free swinging. The tee is a large T-shaped wind vane outlined by rows of lamps with colored hoods. The tetrahedron is a triangular pyramid with the long axis horizontal with all edges outlined by rows of lamps covered with colored hoods. Both units are mounted on vertical shafts and, when not free

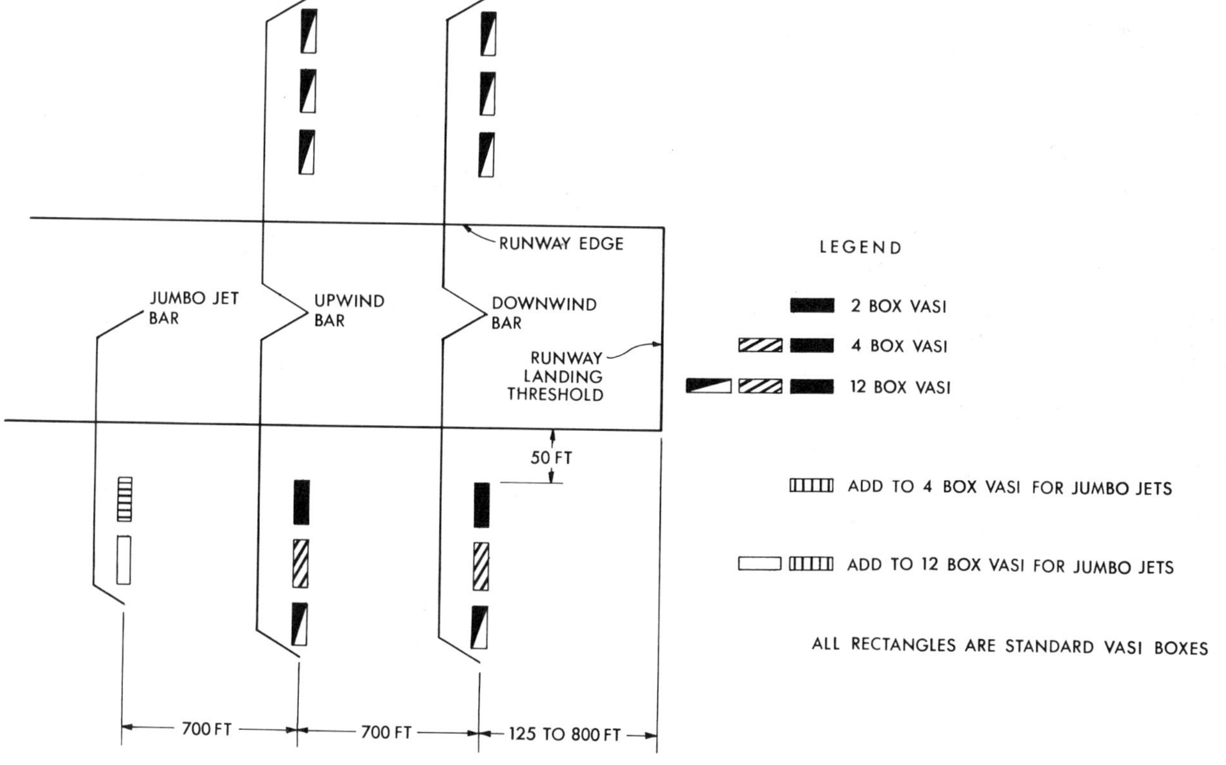

Fig. 21-14. Three versions of visual approach slope indicator (VASI) system.

Fig. 21-15. Visual Approach Slope Indicator (VASI) system.

swinging, may be adjusted to the proper direction either by a remote controlled motor or manually moved and locked in the desired landing direction. Wind tees or tetrahedrons are not used in Canada.

Visual Approach Slope Indicators. VASI is normally installed on runways having non-precision instrument approach aids, or on runways where VFR approaches only are authorized and approach slope guidance is necessary. VASI is not normally installed where approach slope guidance is provided electronically.

Three versions of the VASI are shown in Fig. 21-14. The 2 box system (VASI-2) is usable during daytime approaches at distances up to 3 nautical miles, the 4 box VASI up to 4 nautical miles and the 12 box VASI up to 5 nautical miles. The individual VASI box shown in Fig. 21-15 contains three halogen cycle PAR-64 lamps and emits a two colored light beam; the upper being white, the lower being red. Approach slope information provided by all VASI systems is based on a color change of the upwind and downwind bars. When on the proper approach slope the upwind bar is red, the downwind bar white, above approach slope both bars are white and below approach slope both bars are red. Of the three versions of VASI mentioned, only the 4 and 12 box systems are used in Canada.

A low cost 2 box VASI version, containing one halogen-cycle lamp per box, is available for use where visual ranges of not over 1½ miles are required such as needed for utility airports supporting aircraft operations where additional visual range is not required.

Modifications of the standard VASI configurations are being made so that it can be used by long bodied jumbo aircraft. This modification provides a second upwind bar which is located upwind of the standard upwind bar. Long bodied aircraft would use the two upwind bars which provide an approach slope permitting proper wheel clearance over the runway threshold. Conventional aircraft would use the two downwind bars. Thus this configuration actually provides two approach slopes to meet requirements of any aircraft likely to use the runway.

Non-Precision Approach Lighting Systems. Two non-precision approach lighting systems, MALS and MALSF, indicated in Fig. 21-16, provide alignment, roll guidance and distance to threshold information. They are basic 1400-foot systems and are normally used in conjunction with non-precision electronic approach aids.

The MALS consists of only steady burning lighted bars containing five PAR-38 type, 150-watt incandescent lamps. Intensity control of these lamps is provided to meet operational requirements.

The MALSF is the MALS to which has been added three sequenced flashing lights, one at each of the three outer bars. This system is installed

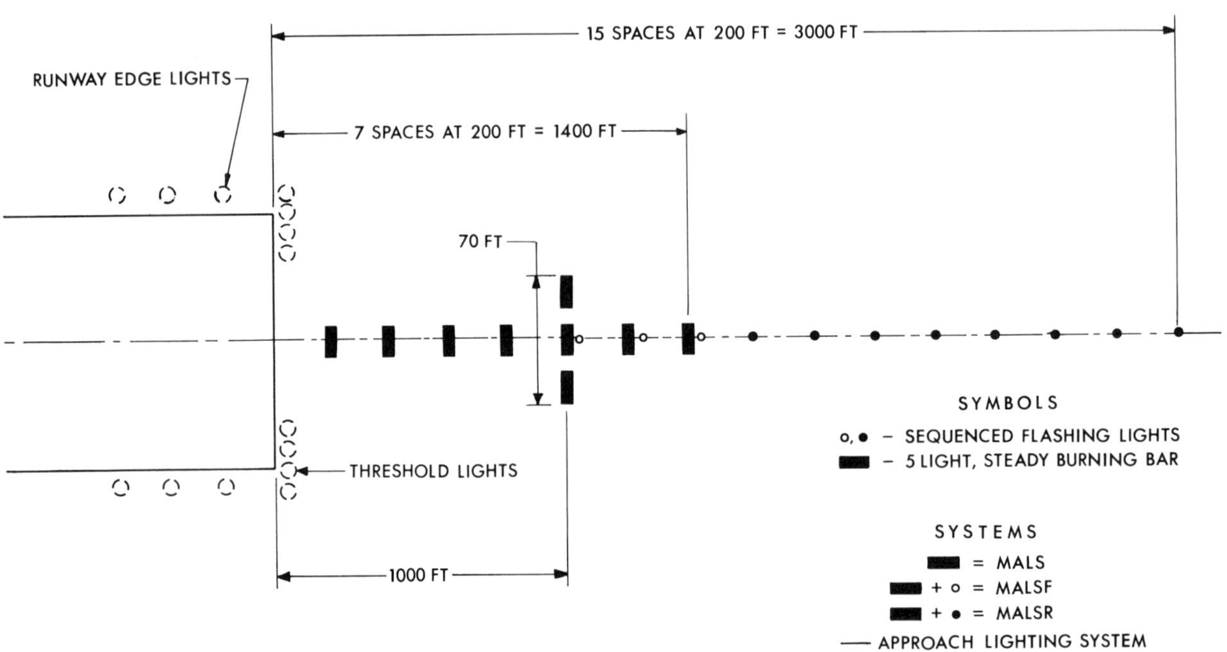

Fig. 21-16. MALS, MALSF, and MALSR non-precision approach lighting systems.

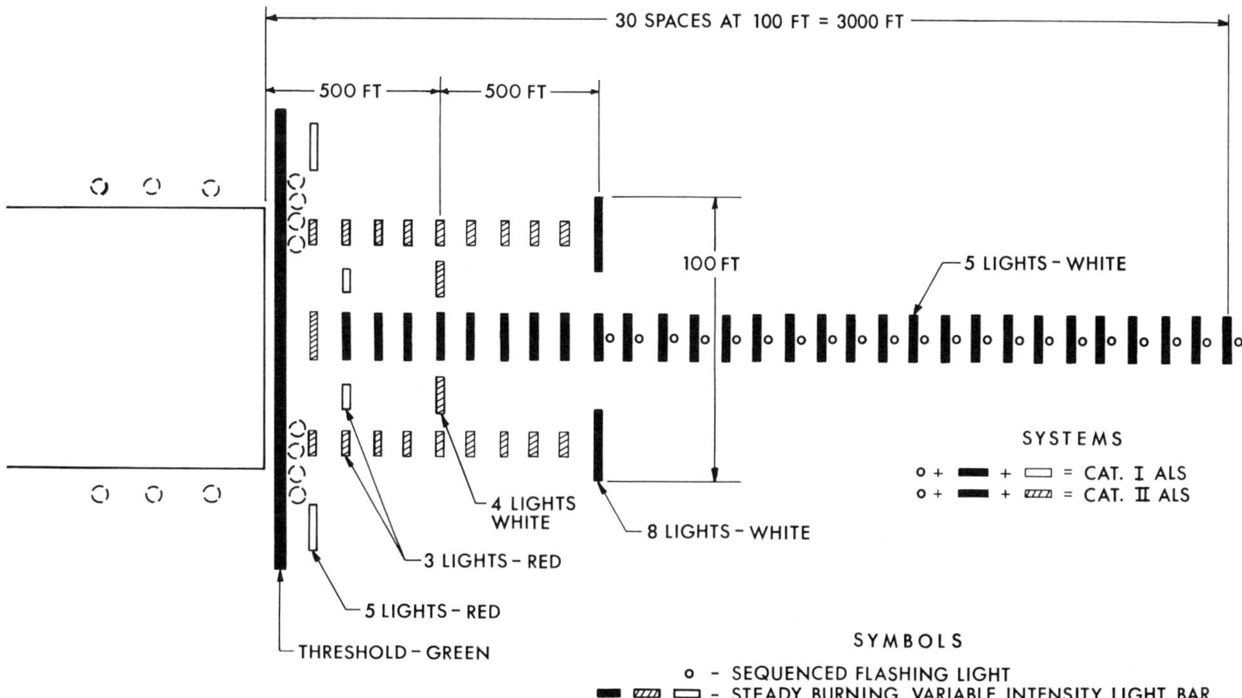

Fig. 21-17. Precision approach lighting systems—Cat. I ALS and Cat. II ALS. Note: Navy System Cat. I and Cat. II varies slightly from these configurations.

when early identification of the system is required.

In Canada, a low intensity approach lighting system is used. This system consists of a line of 2 light bars extending 3000 feet from the end of the runway on 200-foot centers. The system centerline is offset 5 feet from the centerline of the runway. Lights of 100-watt size are used.

Precision Approach Lighting Systems. These systems used in conjunction with complete ILS or PAR electronic approach systems form a portion of the ground based aids required for aircraft landings under IFR conditions. Three configurations of these systems are used to provide adequate visual guidance for all types of aircraft operating down to the lowest weather conditions expected at a specific airport.

MALSR (see Fig. 21-16) can be used by aircraft having approach speeds of less than 160 knots in Cat. I weather conditions (see Fig. 21-11). Aircraft with higher approach speeds are limited to 300-foot ceiling and ¾-mile visibility conditions.

Cat. I ALS (see Fig. 21-17) can be used by all types of aircraft in Cat. I (2400-foot RVR) weather conditions. The Canadian system is very similar in configuration to the US system.

Cat. II ALS (see Fig. 21-17) can be used by all types of aircraft, in conjunction with other required landing aids in Cat. II (1200-foot RVR) weather conditions. This configuration differs from the Cat. I ALS in the inner 1000 feet of the system. The Canadian CAT II is very similar in configuration.

All three systems have centerline bars consisting of five each PAR-type incandescent lamps. See Fig. 21-18. The centerline bars provide alignment and roll guidance. The 1000-foot crossbar (and 500-foot crossbar Cat. II configuration only) provide distance to threshold information, and the sequence flashers provide early recognition. All systems have intensity control of the incandescent lamps. MALSR systems use 150-watt PAR-38 lamps. Cat. I and CAT. II ALS use 300- and 500-watt PAR-56 lamps respectively. In Canada, CAT. I and CAT. II High Intensity Systems use 250-watt PAR-56 lamps.

Fresnel Lens Optical Landing System. FLOLS is a visual glide slope guidance system used by the Navy on aircraft carriers and land based installations. The FLOLS is normally located on the port

Fig. 21-18. High intensity approach light bar with condenser discharge source.

Fig. 21-19. Fresnel lens optical landing system. Lighting configuration is for pilot too high on his glide path.

side of the angled deck of the carrier, as shown in Fig. 21-19. The system consists of green datum bars located and centered on either side of five stacked sharp cutoff optical cells. When the pilot is on glide path the center cell appears lighted and in line with the green datum bar. When the pilot is high or low on the glide path, the upper or lower cells appear lighted thus indicating to the pilot he is off the glide path and providing information for correction. Figure 21-19 shows the lighting configuration viewed by a pilot that is high on his glide path. The vertical lines of light sources shown are red "wave-off" lights that flash at 90 flashes per minute to indicate wave-off. In addition, a horizontal row of green lights are positioned above the cells. These lights are turned on to indicate a "cut" to the pilot of propeller-driven aircraft.

Runway Lighting

Runway lighting is installed to give a visual reference to the limits of the runway during periods of darkness or low visibility conditions. Runway edge lighting systems are classified as low (not less than 10 candelas), medium (not less than 1500 candelas), or high (more than 10,000 candelas) intensity. The minimum system for use in good weather conditions consists of threshold, edge and runway end lights of low intensity. At locations where operations are conducted under restricted visibility conditions, either medium or high intensity lights are recommended. The choice between the installation of high or medium intensity runway edge lights is normally made on the basis of the operating minimums assigned to the runway. Where Cat. II operating minimums are assigned to a runway, high intensity edge lights and in-pavement runway centerline and touchdown zone lights are required. See Fig. 21-20 for a typical Cat. II runway lighting configuration.

Edge, Threshold and Runway End Lights. Edge lights are located in parallel straight lines along the longitudinal sides of the runway within ten feet of the full strength pavement. Maximum light spacing, in each line, is 200 feet. Threshold lights are located at the approach end of the runway in groups located symmetrically about the runway centerline. Runway end lights are located across

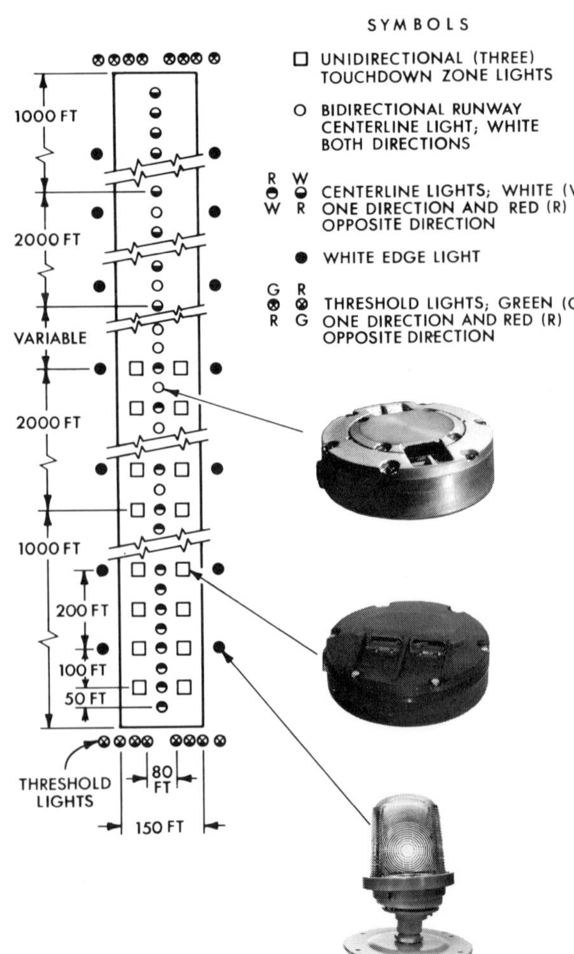

Fig. 21-20. Typical Cat. II runway lighting system.

the rollout end of the runway (see Fig. 21–20). The lines of threshold and runway end lights are not less than two feet, nor more than ten feet from the designated threshold or end of the runway. Edge lights emit white light, threshold lights emit green light and runway end lights emit red light, unless otherwise required for special applications. The low intensity edge, threshold and runway end lights have a symmetrical lens. The medium intensity and high intensity edge lights have asymmetrical lenses. The medium and high intensity threshold and runway end lights have asymmetrical lenses except for Navy facilities where symmetrical lenses are used.

Canadian airport runway lighting is similar to that used in the United States, except that the light units are placed not more than 5 feet from the designated edge or end of runway.

Displaced Thresholds. When the threshold is displaced from the extremity of the runway, the threshold lighting should consist of a line of lights outboarded on each side of the runway. These lights extend at least 40 feet outward from and at right angles to the line of runway lights. If the area created by the displaced threshold is useable for specific operations (take-off only, taxiing) and denied for others, it should be lighted to indicate the correct signal to the pilot for the operation. Denied and unuseable areas may be lighted by adding split colored filters and/or blank portions of luminaires. The luminaires indicate a green signal at the threshold area, a red signal for denied areas, a white signal for useable runway areas, and a blue signal for useable taxiing areas. The same principles are employed in Canada in defining a displace threshold area, except that a blue signal indicates either a usable runway or taxiway area.

Centerline and Touchdown Zone Lights. Runway centerline lights are bidirectional in-pavement lights installed in the runway pavement in the configuration shown in Figure 21–20. The touchdown lights shown in this configuration are unidirectional. Centerline and touchdown zone lights have intensities of 5000 and 7500 candelas respectively.

Taxiway Lighting

Taxiing guidance is provided by a system of lights which may include: taxiway edge lights, entrance-exit and destination signs, and centerline lights.

Taxiway Edge Lights. Taxiway edge lights are usually elevated type units, with a symmetrical blue cover lens having an intensity of 2 candelas (in blue). The lights are located not more than 10 feet from the pavement. The maximum spacing is 200 feet. Closer spacing is provided on short straight sections and curves. Canadian systems are equivalent to US except Canadian lights are located not more than 5 feet from pavement edge.

Taxiway Centerline Lights. Taxiway centerline lights emitting green light may be installed in lieu of elevated blue lights (1) where operations are authorized in low visibilities, (2) in ramp and apron areas where other lighting may cause confusion to aircraft taxiing or parking operations, and (3) in complex taxiway systems. The maximum longitudinal spacing of taxiway centerline lights on straight sections is 100 feet. Closer light spacing is provided on curved sections of taxiways. The intensity of these lights is 20 candelas (in green).

Taxiway Guidance Signs. Taxiway guidance signs are internally lighted, externally lighted or provided with retroreflective markers. Internally lighted type signs emit yellow light. Taxiway guidance signs are normally used at intersections of taxiways with other taxiways or runways, and to indicate destinations. Canadian systems are equivalent to US except Canada normally uses retroreflective markers.

Circuitry and Intensity Control

Series and multiple circuits are used to supply power to airport luminaires. The series circuits are usually designed for either 6.6 or 20 amperes. Isolating transformers are used at each light to isolate the unit and to change the lamp current when required. In Canada only 6.6 ampere series circuits are used.

Control of glare from these systems is achieved by reducing the current in clear weather to obtain intensities as low as 0.2 per cent of full-current intensity. Additional control is obtained through the design of the intensity distribution of the lights. However, the effectiveness of glare control by means of intensity distribution is limited due to required beam spreads, which must be made great enough to cover the expected tracks of the aircraft.

Pavement Markings

Runways. White markings are used on the runway surface to provide runway end identification and to mark serviceable portions of the landing area. Unserviceable areas are identified by yellow markings.

Taxiways. Yellow markings are provided on the taxiways to indicate serviceable and unserviceable portions of the taxiway, and holding positions for aircraft.

Runway Distance Markers. These markers were developed by and for the use of the military. The markers provide information as to the number of thousand feet remaining during takeoff and landing operations. Markers are located near both sides of the runway and are externally floodlighted or internally illuminated for night operations.

Retroreflectors. Retroreflectors, of the same type as used for highway lane markers, are coming into use as runway and taxiway centerline markers on runways where the installation of centerline lights is not warranted. Also retroreflector delineators are installed along the edges of taxiways lighted only with centerline lights.

HELIPORT AND VTOL LIGHTING

Nighttime operation under VFR (see Fig. 21-11) requires landing area perimeter lights, a lighted wind indicator and obstruction lights. When the heliport is not at an airport, a beacon is recommended. The perimeter lights are yellow with an odd number no less than five on each side and with a maximum space between lights of 40 feet. The lighted wind indicator and obstruction lights are standard airport equipment. The beacon flashes the sequence green, yellow and white at a total 30 to 60 flashes per minute.

Daytime operation requires the standard day surface and perimeter surface marking (see Fig. 21-21). White "paint" is used for these markings. When the pavement is a light color, the markings should be outlined with black "paint" to increase conspicuity. Daytime marking in Canada is a 30-foot diameter circle, 3 feet wide, with an H in the circle oriented on true North.

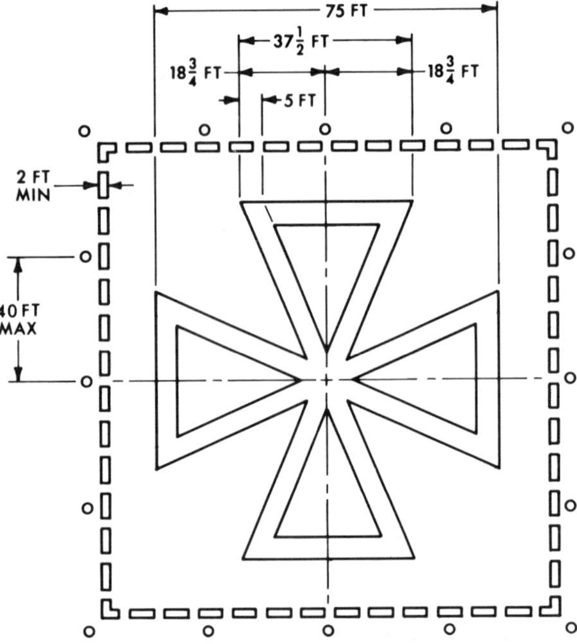

Fig. 21-21. Standard heliport marking and lighting. Circles in above drawing represent yellow perimeter lights, odd number on each side, not less than 5 lights on each side. Standard pattern marking dimensioned for pad size 90 feet or larger. For smaller sized pads, scale pattern proportionally smaller.

STOL PORT LIGHTING

STOL (Short Takeoff or Landing) aircraft approach and takeoff on steep glide slopes (6° to 8°) and can operate into and from a runway 1500 feet in length. The visual aids they require are shown in Fig. 21-22.

Nightime Operation:

Runway Edge Lights. Runway edge lights are alternate white and yellow elevated units with asymmetric lenses and are equally spaced along the runway edge at spacings no greater than 200 feet.

Threshold Lights. Threshold lights consist of two groups of lights forming wing bars symmetrically located about the runway centerline at the threshold. The lights are alternate yellow and green in the approach direction and red in the opposite direction. The innermost light is yellow and located 10 feet outside the runway edge.

Runway End Lights. Runway end lights consist of a bar of seven red semi-flush unidirection in-pavement lights on line with the threshold lights, symmetrically disposed about the centerline, and spaced five feet apart at the end of the runway. The center light is on the extended runway centerline. In the event of bidirectional operation, bidirectional lights are used with red lights to mark the end of the runway and alternate yellow and green lights in the approach direction. The center light in the approach direction is yellow.

Runway Distance Remaining Lights. Four, red unidirectional in-pavement lights are located along the centerline beginning 50 feet from the end of the runway, spaced 50 feet apart and show red light toward the approaching aircraft.

Visual Approach Slope Indicator (VASI-2). The (two-box) VASI-2 is located with the downwind box 250 feet from the threshold and the upwind box 425 feet from the threshold.

Runway End Identifier Light System (REILS). The two flashing lights are located 75 feet from the runway edge and are in line with the threshold lights.

Other STOL Port Lighting. A beacon is required for STOL Ports remote from an airfield. Its identifying characteristics have not been determined. In addition, a lighted wind indicator and standard obstruction lights together with conventional taxiway, apron and parking lights will be installed as necessary. Approach lighting for instrument operations is under study.

Daytime Operation:

Runway Marking. Runway marking consists of a threshold bar, the letters "STOL" 60 feet high at the threshold, runway direction number, runway edge marking, and touchdown aim point markings as shown.

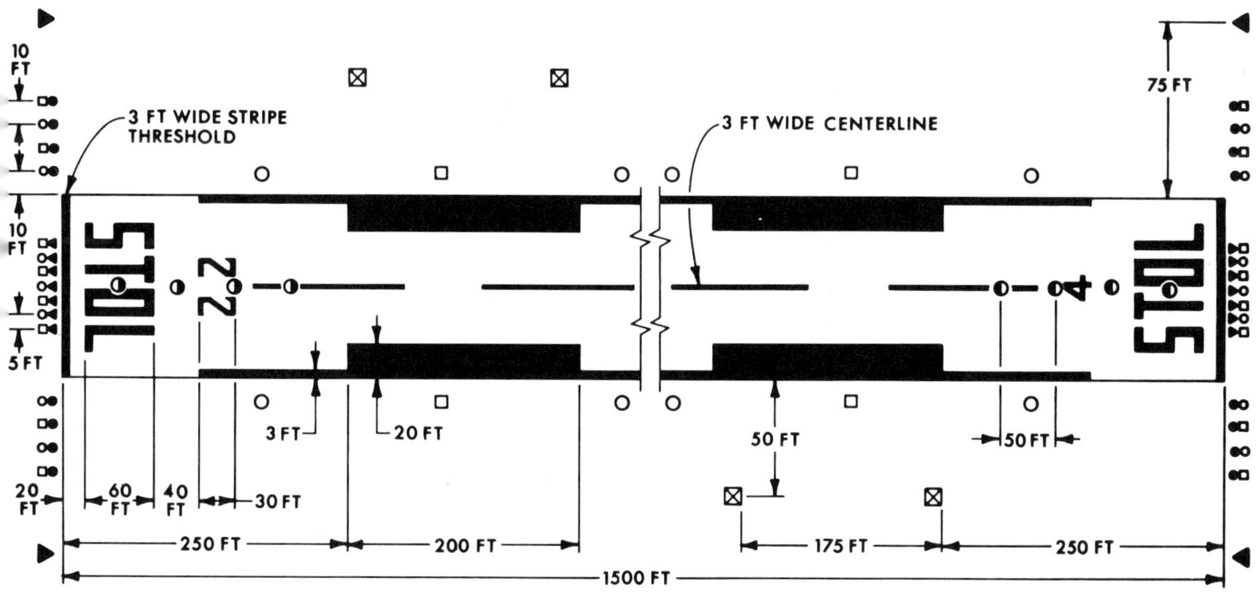

Fig. 21–22. STOL port lighting and marking.

SYMBOL	DESCRIPTION OF LIGHT
○	White, 360° Bidirectional Runway Edge Light
□	Yellow, 360° Bidirectional Runway Edge Light
◑	Blanked Out-Red Unidirectional Semi-flush Inset Light
○◀	Yellow-Red Bidirectional Semi-flush Inset Light
□◀	Green-Red Bidirectional Semi-flush Inset Light
○●	180° Yellow-180° Red Bidirectional Above Ground Light
□●	180° Green-180° Red Bidirectional Above Ground Lgiht
▶	REILS (Runway End Identifier Light System)
⊠	VASI-2 (Two Box Visual Approach Slope Indicator)

OBSTRUCTION LIGHTS

Nighttime Lighting. Obstruction lighting defines the vertical and horizontal limits of natural and man-made objects which are hazards to navigation. The arrangement of the lights and the characteristics of the light signal are dependent upon the kind of object to be defined. Several types of obstruction lights are used. Low intensity obstruction lights with a peak intensity of approximately 80 candelas (in red) are used to mark obstructions less than 150 feet in height and are used at intervals on higher obstructions. Flashing hazard beacons having a peak effective intensity[10] of approximately 2000 candelas (in red) are used to mark the top and at certain levels of obstructions more than 150 feet in height. Rotating 24-inch beacons producing 12 flashes per minute with a peak effective intensity of approximately 50,000 candelas (in red) are used to mark areas in which a visible or invisible hazard exists. Obstruction lighting is similar in Canada except that the use of 24 inch rotating beacons for obstruction markings are not authorized.

Daytime Lighting for Tall Obstructions. Standards now being considered for tall thin structures more than 500 feet in height require installation of three or more (depending on the height) flashing lights producing an effective intensity of at least 200,000 candelas of white light in order to be seen in the day and a lower intensity at night. The lights will flash simultaneously at about 40 flashes per minute. For power lines which are aviation hazards, a system of three white flashing lights on the supporting structures at either end of the span are under consideration. These lights will flash at about 50 flashes per minute, in a specified sequential arrangement and will have a daytime effective intensity of no less than 40,000 candelas (in white) in the specified directions and a lower night intensity.

GROUND OPERATIONAL LIGHTING

Airport Service Area[11, 12, 13]

The objectives of a service area lighting installation are: (1) to enable the pilot to guide (taxi) his aircraft into final position for loading and servicing; and (2) to provide lighting suitable for personnel to perform the functions of loading and unloading passengers and cargo, loading fuel, and performing other apron service functions. This lighting installation should be capable of providing

Fig. 21-23. Jet aircraft hangar illuminated by 400-watt improved-color mercury lamps in twin high-bay luminaires. Emergency lighting is provided by 1000-watt incandescent luminaires.

an average level of illumination of 0.5 footcandle, horizontal (maintained in service), over the surface of the entire area, with a uniformity range within 4 to 1 (average to minimum). It should also provide a level of 2.0 footcandles, average vertical, at the near side of the aircraft for visual tasks performed in connection with servicing the aircraft.

Service areas generally extend out over 200 feet from the building with plane positioning circles of 150 to 200 feet in diameter. These large areas are often illuminated using high brightness incandescent or mercury luminaires mounted on buildings or poles. Mounting heights of these units should be as great as possible, the minimum not less than 50 feet for floodlights, and not less than 30 feet for street lighting types. Lower brightness luminaires, such as fluorescent, may be mounted at heights as low as 15 feet, if proper shielding and diffusion are used. In determining the height of all equipment, obstruction height limitations must be adhered to; however, reduction of glare may often be more important than restrictions on mounting height.

Aircraft Hangar Lighting[14, 15, 16, 17]

Aircraft hangars are large, high bay areas used for storage, inspection, and maintenance. Direct lighting with concentrating type industrial luminaires is generally recommended in these high bay areas. Where highly specular finishes are to be worked upon and reflections interfere with the task, low brightness luminaires should be installed. For the lighting of repair shops where motor and parts repairs are performed, whether in the hangar or an adjacent building, see Section 14 on industrial lighting.

Recommended levels of illumination should be provided by a general lighting system, but when internal work or shadowed parts around the aircraft prevent this, a supplementary lighting system should be employed.

For ease in maintaining the high bay lighting equipment, lowering hangars, catwalks, or traveling monorail cranes can be used. Regardless of the system selected, luminaires should be accessible even when a hangar is full.

Hangar aprons should be lighted to an average level of one footcandle for a distance of about 50 feet out from the hangar.

Airport Parking Area[18, 19]

The lighting of automobile parking areas and roadways in and around an airport must provide visibility for control tower operators and pilots and promote safe and efficient movement of motor vehicles and pedestrians.

The lighting must not interfere with nighttime visibility of the control tower operators and incoming pilots. At night, control tower operators, work in semi-darkness. Their eyes must be dark-adapted to enable them to see aircraft maneuvering in the air and on the ground. Any appreciable amount of brightness in their fields of vision will greatly reduce their ability to see.

The same is true for the incoming pilot. But, in addition to his ability to see under dark adaptation, the nearby roadway and parking area luminaires should not be visible above the horizontal to avoid confusion between the pattern they may form and the pattern of runway marker lights.

For the parking area, where the public park their own cars, the major considerations in providing illumination are to eliminate accidents, make it easier to locate parking spaces and to locate cars on return, and to discourage petty larceny and criminal assault.

Illumination Required. It is considered good practice to provide a minimum of at least one footcandle with a maximum of two footcandles average maintained in service, with a uniformity ratio of not greater than 3 to 1 between the average and minimum point anywhere on the roadway, and 4 to 1 in the parking area.

Focal points in parking areas, such as entrances and exits, should have an average level of illumination at least twice that of the general parking area.

Also, points of heavy pedestrian crossings should have that same increased illumination.

Special Requirements. It is desirable to limit the ambient light on the control tower windows to no more than 0.1 footcandle of ambient light. Tests at 0.8 footcandle and 0.3 footcandle proved to be objectionable. Lowering this to 0.1 footcandle did satisfy the control tower operators. The illumination (vertical) on the control tower windows is ambient light accumulated from all sources, including reflected light from paved surfaces and direct high-angle light from luminaires. In areas that are subject to snowfall, the reflected light may become a significant factor.

Although it is difficult to control reflected light and glare, some factors that contribute to reflected light can be controlled to some degree, such as: the location of lighting equipment with relation to the location of the control tower, and the candlepower and angle at which the main beam is directed in relation to the control tower. The reflected angle should not be in the direction of the tower.

Direct light or stray light falling on the tower may be controlled in several ways: by using luminaires or floodlights which have positive optical control such that no direct or stray light is emitted above the horizontal, and by proper location of these sources with relation to the tower and the height of the tower.

The view of the runways and runway approaches, taxiways, and ramp areas should be considered when locating luminaires, so that during darkness they provide minimum obstruction as glare sources. The lighting poles used must be designed so that the obstruction clearance requirements specified in Federal Aviation Regulations Part 77, "Objects Affecting Navigable Airspace" * are met.

* Copies may be obtained from the Superintendent of Documents, U. S. Government Printing Office, Washington, D. C. 20402.

REFERENCES

1. "Colors, Aeronautical Lights and Lighting Equipment, General Requirements for," MIL-C-25050, Naval Publications and Forms Center, 501 Taber Avenue, Philadelphia, Pa., December 2, 1963.
2. "Covers, Light-Transmitting, for Aeronautical Lights, General Specifications for," MIL-C-7989A, Naval Publications and Forms Center, 501 Taber Avenue, Philadelphia, Pa., January 6, 1959.
3. A20 Aircraft Lighting Committee: Aircraft Lighting Standards, Recommended Practices and Information Bulletins, Society of Automotive Engineers.
4. Projector, Theodore H.: "The Role of Exterior Lights in Mid-Air Collision Prevention," Final Report No. 4, prepared for the Federal Aviation Administration under Contract No. FAA/BRD-127, July, 1962.
5. "Panel, Information, Integrally Illuminated," MIL-P-7788D, Naval Publications and Forms Center, 501 Taber Avenue, Philadelphia, Pa., April 14, 1967.
6. "Lighting, Integral, Aircraft Instrument, General Specifications for," MIL-L-25467C, Naval Publications and Forms Center, 501 Taber Avenue, Philadelphia, Pa., May 29, 1964.
7. "Lighting, Instrument, Integral, White, General Specifications for," MIL-L-27160B, Naval Publications and Forms Center, 501 Taber Avenue, Philadelphia, Pa., December 4, 1969.
8. Projector, T. H. and Hardesty, G. K. C.: "The Computation and Use of Cone-to-Rod Ratio Specifications," Naval Ship Research and Development Laboratory Report ELECLAB 25/69, September, 1969.
9. "Extraneous Lighting in the Vicinity of Airfields, Aerial Beacons and Other Light Beacons," *Light and Lighting*, August, 1945.
10. Aviation Committee of the IES: "IES Guide for Calculating the Effective Intensity of Flashing Signal Lights," *Illum. Eng.*, Vol. LIX, p. 747, November, 1964.
11. Aviation and Signal Committee of the IES: "Recommended Practice for Airport Service Area Lighting," *Illum. Eng.*, Vol. LV, p. 407, July, 1960.
12. Karns, E. B.: "Lighting of Airport Service Areas," *Illum. Eng.*, Vol. LV, p. 565, October, 1960.
13. Magee, A. E. and Magee, D. G.: "A New Luminaire Design for Airport Finger Aprons," *Illum. Eng.*, Vol. LX, p. 66, February, 1965.
14. O'Neill, D. J.: "American Airlines Specifies. New Lighting for Idlewild Hangar," *Illum. Eng.*, Vol. LV, p. 363, July, 1960.
15. Faktor, W., Fishman, I., and Wenson, Jr., H. W.: "Hangar Lighting—Design, Techniques, Economics," *Illum. Eng.*, Vol. LVII, p. 39, January, 1962.
16. Adams, J. C.: "Lighting for Aviation—Part Five: Lighting of Aircraft Works and Hangars," *Light and Lighting*, June, 1962.
17. Jones, E. M.: "A Paint Hangar Lighting Installation," *Illum. Eng.*, Vol. 65, p. 326, April, 1970.
18. Aviation Committee of the IES: "Recommended Practice for Airport Parking Area Lighting," *Illum. Eng.*, Vol. 63, p. 590, November, 1968.
19. Karns, E. B.: "Airport Parking Areas," *Illum. Eng.*, Vol. LX, p. 308, April, 1965.
20. "Circular Checklist and Status of Federal Aviation Regulations," Department of Transportation, Federal Aviation Administration, Federal Register Vol. 35, No. 79, Part II, (issued triannually).
21. Stiles, W. S., Bennett, M. G. and Green, H. K.: "Visibility of Light Signals with Special Reference to Aviation Lights," Aeronautical Research Committee Reports and Memoranda No. 1793, British Air Ministry, May, 1937.
22. Langmuir, I. and Westendorp, W. F.: "A Study of Light Signals in Aviation and Navigation," *Physics*, November, 1931.
23. Breckenridge, F. C.: "Trends in Aviation Lighting," *Trans. Illum. Eng.*, Vol. XXXIII, p. 262, March, 1938.
24. Breckenridge, F. C.: "Airport Lighting in the United States and Europe," *Illum. Eng.*, Vol. XXXVI, p. 1157, December, 1941.
25. Calvert, E. S.: "Visual Judgements in Motion," *J. Inst. Navigation*, No. 3, 1954.
26. Gibson, J. J.: "The Optical Expansion—Pattern in Aerial Location," *Amer. J. Psychol.*, September, 1955.
27. Breckenridge, F. C.: "Fifty Years of Signal Lighting," *Illum. Eng.*, Vol. LIII, p. 311, June, 1958.
28. Finch, D. M.: "Recent Developments in Airport Lighting," *J. Air Transp. Div., Proc. ASCE*, No. AT2, August, 1961.
29. Brady, F. B.: "All-Weather Aircraft Landing," *Sci. Amer.*, March, 1964.

SECTION

22

TRANSPORTATION LIGHTING

Automobile Lighting 22–1 *Specifications 22–3* *Reflex Devices 22–3*
Public Conveyance Lighting—Road and Rail 22–5 *Lamps and Power Systems 22–11*
Marine Lighting 22–12 *Railway Guidance Systems 22–15*

The general principles established for good interior and exterior lighting apply also in the transportation field. However, limited supply of electric power and the special basic characteristics of this power in automobiles, buses, and railway cars often make it a more difficult and expensive problem to provide interior illumination of recommended quantity and quality. Also, the tremendous length of roadways and railways, the fact that they are used intermittently, and their exposure to a wide variety of weather conditions are important factors which complicate the lighting problem.

The following requirements are of special importance in the transportation field. They apply to the lighting of practically all types of vehicles but are, of course, more essential in some than in others.

1. Adequate level of illumination on the reading plane. Where the reading matter tends to vibrate, a relatively higher level of illumination is needed than for a fixed seeing task.
2. Adequate illumination at special locations to permit safe entrance, exit, and movement of passengers.
3. Lighting for fare and ticket collection.
4. Adequate vision of operator, particularly freedom from reflections or glare spots in his field of vision.
5. Minimum glare and luminance (photometric brightness) ratios for passengers' eye comfort.
6. Cheerful and attractive appearance of the vehicle inside and also as seen from without.

Most vehicles are long and narrow and have low ceilings and large window areas. Also, the desires of the passengers and the loading of the conveyance vary. These factors make the lighting problems difficult. On the other hand, the position of the occupants and the operator is generally known and fixed so that good lighting results are obtainable through a careful study of all phases of the problem.

In the transportation field, direct current power supplies are predominant, a wide variety of voltages are involved, and power is often limited. For these reasons, one of the principal problems is to adapt lamps and lighting equipment to the power available and particularly to provide suitable power for fluorescent lighting, which is discussed beginning on page 22–11.

NOTE: References are listed at the end of each section.

AUTOMOBILE LIGHTING

Most automobiles depend on a 12-volt wet storage battery kept charged by a generator (dc or rectified ac) driven by the car engine. In addition to the lighting, the electrical system must supply power for starting, heater blowers, radios, air conditioners, power seats, power windows, and other special equipment. A single wire grounded wiring system is commonly used.

Exterior Lighting

Two categories of exterior lighting equipment are commonly used on motor vehicles—lighting units to see by and lighting units to be seen. The first group comprises headlights and back up lights; the second includes a variety of signal lights whose function is to convey information between drivers of vehicles.

Headlighting. The most difficult and important illuminating engineering problem in the automobile field is headlamps. Because of the speed at which modern cars operate, because most roads are used for two-way traffic, and because a few feet above the road surface is the most convenient headlamp location, it is not easy to provide good road lighting without creating glare for an approaching driver.

One headlighting system in limited use comprises two two-filament sealed-beam units, each of which provides an upper and a lower beam. A pedal switch permits the use of the lower beam only or both beams together.

Fig. 22-1. Lamps Used in a Dual Headlighting System

Lamp Type	Filaments (watts)	Filament Position	Provides
1 (Inner or Lower)	1 (37½)	At focus	Primary part of upper beam
2 (Outer or Upper)	1 (50)	At focus	Lower beam
	1 (37½)	Below focus	Part of upper beam

The dual headlighting system, now widely used, consists of four 5¾-inch diameter sealed-beam units of two types arranged in pairs. The details of this system are summarized in Fig. 22-1.

A pedal switch permits the use of the two 50-watt filaments only in traffic and in passing or all six filaments (250 watts) for country driving. Each lamp in the dual headlighting system has three bosses on the lens face which are used in conjunction with a mechanical aiming device to provide correct beam alignment, even in broad daylight.

Current practice in headlamps is uniform and is serving its purpose satisfactorily. However, as the number of vehicles using highways during the hours of darkness increases, it may become necessary to use new techniques and make further improvements. There is also the possibility that new sources, improvement in existing sources, and cheaper electric power may make it economically feasible to use fixed roadway lighting on the portions of the highways carrying large volumes of traffic.

Signal Lighting. Signal lamps include tail lights, stop lights, turn signals, parking lights, and reflex reflectors. Signal indications must be unmistakable. The number of indications or messages conveyed must be kept to a minimum to avoid confusion and to be easily understood by the public.[8] The following indications will cover most situations:

1. Indication of the presence of a vehicle proceeding in a normal manner and its direction of travel.
2. Indication that the brakes are being applied and the vehicle is stopping or is stopped.
3. Indication that the vehicle is disabled or obstructing a traffic lane.
4. Indication that the driver intends to change the direction of travel.

Effective signal indications may be accomplished by the following means:

Color of Signal Indication. For automotive use the colors available are uncolored, yellow, red, and blue. The usually accepted meaning of these colors is: uncolored—indication of the presence of the vehicle, commonly considered as approaching; yellow—indication that caution is needed in approaching the vehicle displaying the yellow light; and red—indication that the vehicle displaying the red light is a traffic hazard and that the approaching vehicle may need to stop. Volunteer firemen and police are permitted in many states to use a blue light as an identification.

Intensity of Signal Light. Intensity of the signal light has a definite bearing on its conspicuity. In giving signal indications, different intensities may be used, provided the ratio of the intensities is sufficiently great. In the contemplated use on automobiles, the signal indication must be recognized unmistakably when first seen. Consider the situation where a car comes over the brow of a hill or around a sharp curve, and suddenly the driver sees a vehicle ahead of his car. The driver must know at once whether this vehicle is proceeding normally or is stopping or stopped. A change of intensity, as used on many vehicles today, is useless, because the ratio of the two intensities is too small. For example, a driver not in sight of the car ahead when the intensity changed and therefore not seeing the change in intensity occur, has no sure way of knowing whether he is looking at a bright tail light or a dim stop light—information which may be of vital importance.

Difference in intensity, to be a reliable means of conveying different indications, must be from units having a ratio of intensities much larger than the current ratio of 5 to 1.

Pattern of Display. Patterns of display, such as a single unit for one indication, two units, one above the other, for another indication, or three units in a triangular pattern for still another indication, are good signal patterns.

Method of Display. The signal light may be displayed as a steady burning light or as a flashing light. Two units may be displayed by alternate or simultaneous flashing. Thus there are definite unmistakable indications conveyed by different methods of display.

Interior Illumination

The average person does not expect to read or write continuously in a passenger automobile either while driving or while the vehicle is parked. Standards of luminance and illumination have not been established for interior illumination for passenger cars; however, installations should be planned to provide illumination for casual inspection of road maps and other printed matter, and for safety in getting out. The installation should be in harmony with the style motive of the car interior. Lamps employed range from 1½ to 21 candelas and are shielded to prevent direct glare. In addition, all lamps should be located and/or shielded to prevent reflections in the windshield from obscuring the driver's view of the road, if the lamps should accidentally be turned on while the car is in motion.

Panel-board or instrument lighting for automobiles should be designed for utilitarian requirements with decorative considerations given second place.

The average driver uses the various meters for reference rather than continuous viewing; nevertheless they should be easily and quickly readable. It is essential that lighting units and meter faces be so placed that they are not reflected from the windshield or shiny trim surfaces into the driver's eyes. Provision for dimming or other control of panel and turn signal indicator lights is desirable to avoid excessive brightness and interference with the driver's view of the road at night. Illumination is provided by: small lamps recessed behind glass, plastic, or other light-transmitting materials; similar lamps used for edge lighting of recessed or raised numerals; lamps located at the top or bottom or in front of the panel faces (direct illumination); or ultraviolet excitation of fluorescent numerals and pointers. Recently electroluminescent lamps have been used to advantage.

SPECIFICATIONS FOR EXTERIOR LIGHTING OF MOTOR VEHICLES

The mass production methods characteristic of the automotive industry encourage extensive standardization and through the cooperation of the Society of Automotive Engineers, the Illuminating Engineering Society, safety engineers, and state motor vehicle administrators, standards have been developed over a period of years covering the characteristics and procedures for testing automotive lighting equipment. The Department of Transportation may issue standards for automotive lighting equipment. The American Association of Motor Vehicle Administrators has an *equipment approval program* with participation by 35 states and 5 Canadian provinces, for not only lighting devices, but other safety equipment as well. The standards most commonly used are those published in the *SAE Handbook* (also available in reprint) which are reviewed and continued or revised annually. The standards outline specifications and tests for the various lighting devices covering such details as photometry, color, vibration, moisture, dust, and corrosion. These are referenced in Fig. 22-2.[9]

REFLEX DEVICES IN TRANSPORTATION LIGHTING

Retro-reflecting devices or reflex reflectors are important in transportation lighting, signaling, and for directional guides. They have been standard equipment on the rear of automobiles since about 1935. Other applications include—reflector flares for highway emergency markers, railroad switch signals, clearance markers for commercial vehicles, luminous warnings and direction signals, delineators of highway contours, marine buoys, contact markers for airplane landing strips, bicycle front and rear markers, belts and markers for traffic officers,

Fig. 22-2. SAE Standardized Basic Tests, Test Methods, and Requirements Applicable to Motor Vehicle Lighting Devices and Components (Indication Given of Applicable Test Procedures for Color and Photometry)

Device	SAE Report No.	SAE Identification Code Letters	Color	Photometry
Sealed Beam Headlamp Units for Motor Vehicles	J579a		×	×
Sealed Beam Headlamps	J580a	H		
Electric Supplementary Driving Lamps	J581	Y	×	×
Electric Supplementary Passing Lamps	J582	Z	×	×
Fog Lamps	J583b	F	×	×
Motorcycle and Motor Driven Cycle Headlamps	J584a	M	×	×
Motorcycle and Motor Driven Cycle Turn Signal Lamps	J131		×	×
Tail Lamps	J585d	T	×	×
Stop Lamps	J586c	S	×	×
License Plate Lamps	J587d	L	×	×
Turn Signal Lamps	J588e	I D	×	×
Turn Signal Switch	J589a	Q QB		
Turn Signal Flashers	J590c			
Spot Lamps	J591a	O	×	
Identification or Parking Lamps	J592c	P	×	×
Clearance or Side Marker Lamps	"	P1	×	×
Combination Clearance and Side Marker Lamps	"	PC	×	×
Back Up Lamps	J593c	R	×	×
Reflex Reflectors	J594e	A B	×	×
Flashing Warning Lamps for Authorized Emergency, Maintenance and Service Vehicles	J595b	W1	×	×
Electric Emergency Lanterns	J596	X	×	×
Liquid Burning Emergency Flares	J597	V		×
Sealed Lighting Units for Construction and Industrial Machinery	J598a			×
Emergency Reflex Reflectors	J774b		×	×
360 Deg Emergency Warning Lamps	J845	W2	×	×
Cornering Lamps	J852b	K	×	×
School Bus Red Signal Lamps	J887	W3	×	×
Hazard Warning Signal Switch	J910a	QC		
Side Turn Signal Lamps	J914	E	×	×
Vehicle Hazard Warning Signal Flasher	J945			
Supplemental High Mounted Stop and Turn Signal Lamps	J186		×	×

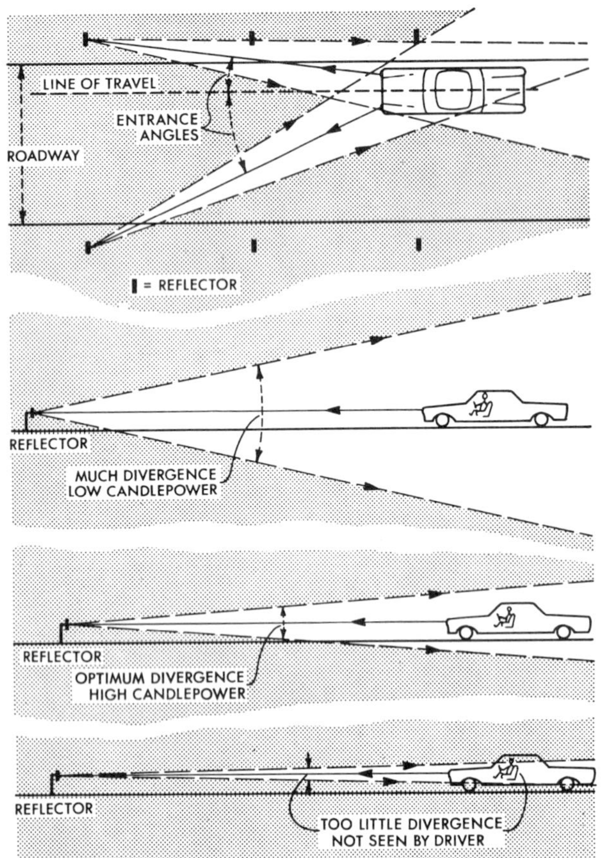

Fig. 22-3. Effect of the divergence of reflex devices on their angular coverage and intensity.

luminous paving strips, and luminous advertising display signs.

Principles of Operation

A reflex reflector is a device that turns light back toward its source. There are several specific types of these devices; however, the principle of operation, namely the production of brightness in the direction of the source, as shown in Fig. 22-3, is the same for all types. The greater the accuracy of the design of the reflex reflector, the narrower will be the cone of reflected visual brightness. The narrower the cone, the brighter will be the signal. Two optical systems in general use are the triple reflector (corner of a cube) and the lens-mirror device.

Triple Reflectors

The triple reflector makes use of the principle of reflection from plane surfaces where the angle of incidence is such that total reflection takes place. Three optically flat surfaces arranged mutually at right angles, as the inside corner of a cube, form a system such that any ray of light which has been successively internally reflected from the three surfaces will be reflected back upon the source.

A plaque of transparent glass or plastic with a continuous pattern of small adjacent cube corners molded into the back as shown in Fig. 22-4 is a commonly used form of reflex reflector. Acrylic plastic is more adaptable than glass to accurate shaping of the prisms, resulting in greater intensity of the return beam. Economy of manufacture, lightness, and shatter resistance are other advantages. A three-inch plastic reflex can be readily seen from up to 1000 feet from an automobile with headlights set for country driving. Emergency signals made up of four of these plaques arranged back-to-back in pairs have been approved by the Interstate Commerce Commission.

Lens-Mirror Reflex

The lens-mirror button consists of a short focal length lens and mirror combination designed to some extent with respect to chromatic and spherical aberration such that the lens focuses the light source upon the mirror, the mirror and lens returning the reflection in the direction of the source. See Fig. 22-5. An aggregation of small lenses pressed into a plaque with a mirrored backing formed into con-

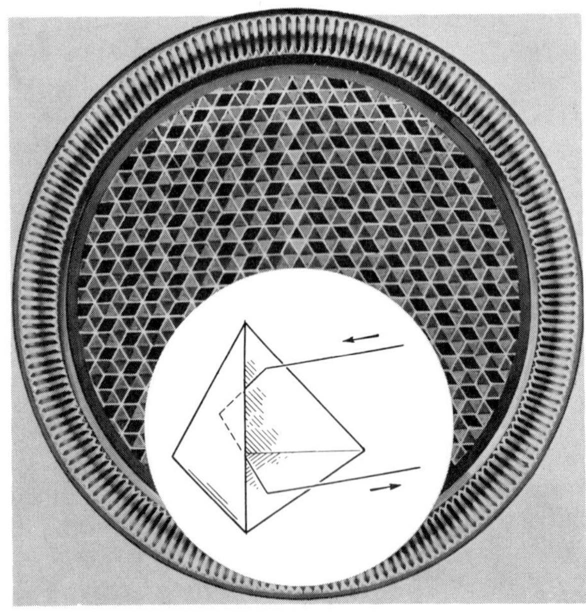

Fig. 22-4. Triple mirror reflectors comprise aggregates of concave cube corners.

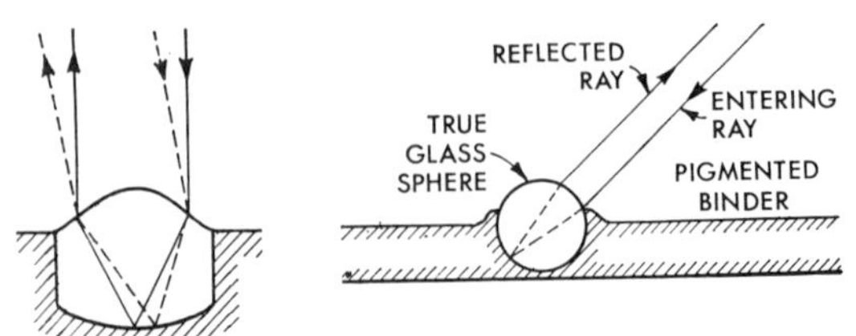

Fig. 22-5. Light paths in button and spherical ball type lens-mirror reflexes.

cave surfaces properly designed will produce a very satisfactory reflex reflector.

Another device that produces a wide spread of light but lower luminance is the spherical transparent glass bead embedded in a diffuse reflecting material such as white or aluminum paint. This type of reflex is used largely in signboards, center stripes on highway pavements, etc.

Maintenance and Construction

For good maintenance, all of the reflecting and transmitting surfaces should be kept clean and where possible free from moisture. The construction should be such that the rear surfaces are sealed and waterproofed. Moisture on a totally reflecting surface will lessen the reflection. If moisture should accumulate on a dusty surface, optical contact takes place and light would pass through the surface rather than be reflected. Any roughening or etching of the transmitting surface will tend to reduce the efficiency.

PUBLIC CONVEYANCE LIGHTING—ROAD AND RAIL[11]

The material that follows covers the lighting of interiors of public passenger road and rail vehicles. The general principles established for lighting of fixed interiors as covered in earlier sections apply to the lighting of public conveyance interiors. This has been made feasible through the development of solid state devices and the availability of new light sources and luminaires.

Illumination Levels. In modern road and rail conveyances, passenger seeing tasks vary widely; from seeing to board or exit, deposit fares or have fares collected, find seat accommodations, read and write, view advertising cards, to residential type tasks on inter-city trains such as found in dining cars, lounges, and washrooms.

Illumination recommendations for the specific tasks found in road and rail conveyances are shown in Fig. 9–80. In general, they are consistent with similar tasks in other land applications and have been tempered only to recognize the factors of adaptation and comfort where the exterior surround is in darkness and when passengers must move from lighted to unlighted areas.

The illumination values in Fig. 9–80 are minimum values which should be provided on the visual tasks regardless of their location or plane. To insure the availability of these values, the original lighting design should make allowance for the decrease in light output caused by luminaire dirt accumulations and depreciation of other interior surfaces such as walls, ceilings, floors, and upholstery and by depreciation in light source output. See Section 9.

Quality of Illumination. In addition to providing the recommended quantity of illumination, the quality of illumination should also be considered as essential to the creation of a comfortable, livable environment and for visual efficiency. High luminaire and window luminances and high luminance ratios will produce uncomfortable seeing conditions (see Fig. 22–6), and prolonged exposure will generally result in eye fatigue. To avoid discomfort from high luminances, which may reach the eyes directly or indirectly through reflections from shiny surfaces or from improperly shielded light sources, lighting equipment should be located as far from the line of sight as possible. Glossy reflecting surfaces should be covered or modified to reduce glare, and luminaires and windows should have proper brightness control. See Sections 6 and 7.

In passenger spaces, the average luminance of the total luminous area of a luminaire should not exceed 500 footlamberts. Within a diffuser area, the luminance of the brightest square inch should not materially exceed *twice* the average. In strictly utility spaces luminances as high as 800 footlamberts are acceptable. In either case, the luminance ratio between the luminaire and the ceiling should not be more than 20 to 1 as shown in Fig. 22–6.

Luminances within the remainder of the environment also should be balanced in accordance with Fig. 22–6. The use of matte finishes of reflectances recommended in Fig. 22–7 will help to achieve this balance.

Fig. 22–6. Recommended Luminance Ratios for Public Conveyances

To achieve a comfortable balance of luminances, it is desirable to limit luminance ratios between areas of appreciable size from normal viewing points as follows:

Areas Involved	Acceptable Limit
Between task and adjacent surroundings	1 to ⅕
Between task and more remote darker surfaces	1 to 1/10
Between task and more remote lighter surfaces	1 to 10
Between luminaire or windows and surfaces adjacent to them	20 to 1
Anywhere within the normal visual field	40 to 1

Fig. 22–7. Suggested Reflectance Values for Surfaces in Public Conveyance Interiors

Surface	Reflectance
Ceilings	60–90%
Walls	
Upper	35–90%
Lower	35–60%
Floors	15–35%
Upholstery	15–35%
Furniture	25–45%

Colors of objects appear to change with the surface finish of the object. It should be noted that matte finishes will reflect diffuse light and give an object a consistent color appearance while glossy surfaces may lose their color when viewed at a direction near the specular angle. Finishes such as velvet and deep pile carpeting appear darker than smooth surface materials such as vinyl or a plastic laminate of the same color.

Choice of Light Source. Incandescent and fluorescent lamps are normally used in the lighting of public conveyance interiors. Incandescent lamps are usually considered where (1) operating hours are short, (2) a high degree of light control is necessary, (3) interior surfaces are warm in color, and (4) general illumination levels are low. Fluorescent lamps are considered where (1) operating hours are long, (2) general illumination levels are higher, (3) the linear shape is desired, and (4) surfaces are cool in color.

Road Conveyances—City, Inter-City, and School

General lighting is provided for passenger movement along aisles and for seat selection. City buses leave general lighting on at all times because of the frequent movement of passengers when boarding and exiting the bus. Inter-city buses usually turn general lighting off when on the road to allow the driver's eyes to adjust to the outside luminances. For school bus lighting, the general principles and requirements are the same as those for city buses and inter-city buses; however, higher levels of general lighting are provided for surveillance of active children, whether the bus is moving or stopped, being boarded or exited. Lighting should be directed or shielded in such a way that the driver's vision is not impaired by reflections or glare spots in the field of vision.

Boarding and Exiting. The seeing tasks are the steps, ground, or platform area. The steps and/or an area extending 4 feet outside the door should be illuminated to allow passengers to see the base of steps, curbs, or platform area. The plane of the task is horizontal and on steps, curbs, or platforms, with the viewing distance varied according to the height of the individual.

Recommended illumination levels should be provided at the center of each step and 18 inches from the bottom step on the ground, centered on the doorway.

Fare Collection. The seeing task is one of identifying money or tickets and depositing fares, and, therefore, the fare box and the immediate area around the box should be illuminated. The plane of the task is horizontal and at the top of the fare box.

Recommended illumination levels should be provided at the top of the fare box on a horizontal plane.

Aisle Lighting. The task is one of observing the floor area for obstacles and the seat area for accommodation, generally while the vehicle is moving. The plane of the task for walking is at floor level and for seat selection at the back of seats.

Recommended illumination levels should be provided on a horizontal plane at the center line of the aisle floor and for seat selection, on a horizontal plane at the top center of each seat back.

Advertising Cards. The seeing task is viewing opaque or back-lighted translucent advertising cards placed at the top of side walls. The plane of the task is vertical to 45 degrees.

Recommended illumination levels on opaque cards should be provided on the face of the card. Luminance of back-lighted advertising cards should be measured at the face of the luminaire without the card over the diffuser.

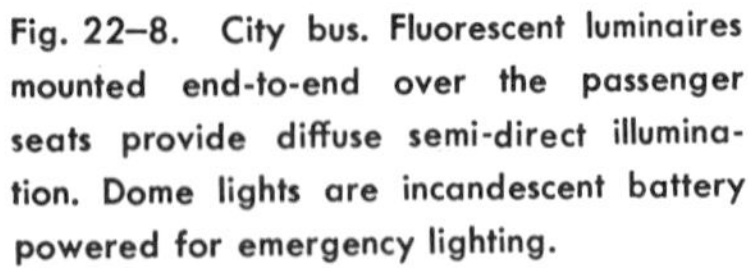

Fig. 22–8. City bus. Fluorescent luminaires mounted end-to-end over the passenger seats provide diffuse semi-direct illumination. Dome lights are incandescent battery powered for emergency lighting.

Reading. The seeing task is generally one of reading newspapers, magazines, and books and is generally the most difficult seeing task encountered in conveyances. The task plane, which is at 45 degrees, should be as free from reflection as possible.

For a seated passenger, the recommended illumination level should be provided on a plane 17 inches above the front edge of the seat on a 45 degree angle. For the standing passenger, measurement should be taken 56 inches above the floor on a 45 degree plane at the edge of the aisle seat as if a passenger were reading facing the seat.

Typical Lighting Methods.

City Buses. For boarding and exiting, luminaires should be designed, located, and arranged so as to minimize the casting of shadows, prevent glare for both passenger and driver, maintain a uniform level of illumination, and remain permanently in adjustment. Typical luminaire locations are the ceiling, over the steps, or at the side of the step well.

For fare collection luminaires can be ceiling mounted or a local light may be used.

Fig. 22-8 illustrates one approach for general lighting for aisle seat selection, advertising cards, and reading.

Inter-City Buses. For boarding and exiting, lighting of step wells and the outside ground area is the same as for city buses.

Fig. 22-9 is an example of luminaires located for reading. Illumination for fare collection, aisle lights and seat selection is provided by an indirect system. A low-level night light system is often desirable so that the driver may clearly see throughout the entire bus. At minor stops, a special aisle lighting system of moderate level is turned on to facilitate passenger movement without disturbing those who are sleeping.

Special luminaires are often provided for the baggage compartment above the seats.

Rail Conveyances—Rapid Transit

Rapid transit is used here to indicate intra-city rail service such as subway or surface railways. It is generally characterized by fast travel conditions ranging from a few minutes to an hour. Standing passengers are a normal condition of this service. The seeing tasks and their lighting are the same as those described for Road Conveyances, except that the recommended illumination level for boarding and exiting should be provided on the tread of the entrance and platform areas.

Typical Lighting Methods. For boarding and exiting, provisions should be made for illumination of the threshold and the steps of the car. In addition, car illumination should supplement platform illumination for at least 4 feet from the car body at the location of the doors. In general, fare collection is made before entering the boarding platform; however, where rapid transit cars have fare collection facilities, it is necessary to provide illumination either by an internal light in the fare box and/or a ceiling light.

The use of incandescent sources for aisle, seat selection, advertising cards, and reading has been discarded in favor of fluorescent luminaires. This is due, primarily, to the efficiency, long life and general diffused type of illumination which is available

Fig. 22-9. Inter-city bus. Individually controlled beam-type incandescent lights are mounted in the base of the baggage rack to provide each passenger with a reading light. An indirect lighting system is built into the edge of the baggage rack for general interior illumination.

Fig. 22-10. Rapid transit. Continuous rows of 40-watt fluorescent luminaires along each side provide diffuse illumination. Lamps are operated on 32-volt inverter ballasts.

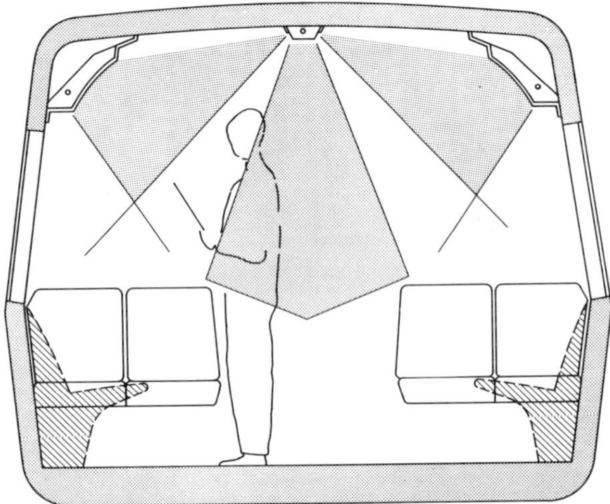

Fig. 22-11. Back-lighted car card luminaires with supplementary center strip lighting increases the level of illumination.

from fluorescent luminaires. The normal type installations use a row of luminaires down the center of a car, or a row of luminaires on each side of a car center as shown in Fig. 22-10, or as a cornice on each side of the car. In some cars a third row of fluorescent luminaires is added between the outside rows, often with an air distributor combined for an integrated ceiling. Where transverse seats are installed in the car, the lighting can be accomplished by use of transverse luminaires. They can be mounted above each seat provided there are no difficulties in connection with the car structural elements and air distribution system.

Some back-lighted advertising card luminaires are designed to provide adequate lighting for both standees and seated passengers (Fig. 22-11). Supplementary center strip luminaires are also needed to provide illumination. In cars using opaque cards, illumination can be supplied from two rows of continuous fluorescent luminaires over the passenger seats as in Fig. 22-10.

Rail Conveyances—Inter-City and Commuter

Seeing tasks on inter-city and commuter trains are much the same as on other public conveyances. The only differences are those additional tasks in the facilities provided for passenger comfort during long distance travel, such as food preparation areas, diners and washrooms. Because of the shorter travel distances and less travel time involved, some tasks described here do not apply to the commuter train.

Boarding and Exiting. The seeing tasks are the platform, steps, and vestibule floor, and all are horizontal at floor level. The viewing angle is practically vertical for the individual to observe conditions of tread surface and platform alignment with the car floor.

Recommended illumination levels are provided at the center of the vestibule floor and at the longitudinal center line of the steps.

Fare or Ticket Collection. There are no special lighting requirements for fare or ticket collection since this normally takes place while passengers are seated.

Aisle Lighting and Seat Selection. The seeing task is one of observing the floor area for obstacles and the seat area for accommodations. The plane of the task is at floor level for walking and at the top center of each seat back for seat selection.

The recommended illumination level should be provided on a horizontal plane at the center line at floor level. For seat selection, illumination should be measured on a horizontal plane at the top center of each seat back.

Advertising cards. The seeing task is the same as for road conveyances except that some cards may be placed on bulkheads at the end of passenger compartments.

Reading. The seeing task is one of reading magazines, newspapers, books or business correspondence for an extended period of time. For the plane of the task and measurement of illumination levels, see Road Conveyances.

Food Preparation. The seeing task is one of preparing foods and beverages and for cleaning the area. The plane of the task is horizontal on work counters 36 inches from the floor.

Recommended illumination levels should be provided on the work counters from the front to the back edge.

Dining. The seeing tasks are eating and drinking, and menu and check reading. The plane of the task is horizontal to 45 degrees from the horizontal 8 inches in from the front edge of the table at each seat position.

Recommended illumination levels should be provided on a plane 45 degrees up facing the seated diner, 8-inches from the front edge of the table or counter.

Lounge. In addition to relaxing and conversing, there are seeing tasks such as reading and card playing. For reading, the task plane is the same as for Road Conveyances. For card playing the task is horizontal at the playing surface.

General illumination should be provided on a horizontal plane 30 inches from the floor.

Washrooms. The chief tasks are shaving or makeup. Because the apparent distance of the face or figure as viewed in the mirror is twice its actual distance from the mirror, and because the details to be seen in shaving and critical inspection are usually small and of low contrast with background, the visual task may be a severe one. The task area in standing position consists of two 6- by 8⅝-inch planes at right angles with each other, converging at a point 16 inches out from the mirror, and centered vertically 61 inches above the floor. See page 15–20. They represent the front and sides of the face. A third plane 12 inches square, its front edge also 16 inches out from the mirror, is tilted up 25 degrees above the horizontal and represents the top of the head.

Recommended illumination levels should be provided 61 inches above the floor and 16 inches from the mirror with the plane facing in the direction of the light source.

Typical Lighting Methods. The design of the interior lighting system of inter-city and commuter coaches are much the same. The main differences are the additional facilities on inter-city trains such as washrooms and lounges that provide passenger comfort during long distance travel.

All of the general requirements of good lighting such as elimination of direct and reflected glare or deep shadows and excessive luminance ratios that will interfere with good vision apply to railroad cars. A railway passenger car has certain fundamental limitations which affect the design of lighting systems. The length, height, and width of the car are fixed by track gauge and clearances. The inherent physical characteristics of the car with its maximum utilization of space, air ducts, wire ways, structural members and limitations of power supply, also may complicate the lighting design.

Boarding and Exiting. Vestibule and platform lighting should be designed so that passengers board and detrain safely. Both track and vestibule levels are generally provided with a luminaire over each trap door, illuminating vestibule and step areas. Supplementary lighting may be provided by step lights located in the step well to increase illumination levels over steps, and/or a leading light located adjacent to steps, beamed to give increased illumination in front of the steps.

Fare or Ticket Collection. See Inter-City Buses.

Aisle and Seat Selection. General lighting can be provided indirectly where the luminaire is mounted in the extreme corner of the baggage rack and outer wall of the car, by ceiling mounted luminaires which direct the light to the desired areas as in Figs. 22–12 and 22–13, and by luminous ceilings. Individual reading lights are desirable as shown in Fig. 22–12 since the baggage rack casts shadows on the window seats.

Reading. Illumination for reading and writing can be accomplished by individually controlled reading lights mounted either in the aisle edge of the baggage rack, or at the joint line of the baggage rack and the outer wall of the car. In the case of the closed type of baggage rack, these lights are mounted directly above the seats.

Food Preparation. Higher levels of comfortable illumination are required in these areas for the utmost efficiency of operating personnel and to insure good appearance of food and its proper inspection. Quality of lighting is important, especially in regard to color. In addition to general lighting, usually supplied by ceiling mounted luminaires, supplementary lighting should be provided over all work areas. The service window between the kitchen and pantry should be well lighted to facilitate food inspection. Refrigeration cabinets should have at least one lamp per compartment operated by automatic door switches.

Fig. 22–12. Inter-city railway passenger coach. A single row of fluorescent luminaires mounted at the center of the ceiling provides general lighting. Individual reading lights are located under the baggage rack.

Fig. 22–13. Commuter railway passenger coach. Two rows of fluorescent luminaires are ceiling mounted over the seats to provide general lighting for reading.

Dining Area Lighting. The functions of the lighting system in dining areas are to enhance the appeal of the interior decorations, table settings and food, and to assist in providing a comfortable, pleasant atmosphere for the diners. Quality and color of light under these circumstances are of more importance than quantity which should, however, be adequate for safety and convenience. The design of the lighting system should be governed by the over-all decor of the car and the effects desired. Direct incandescent downlighting on a table provides sparkle to the silverware and glassware that cannot be obtained from diffuse illumination, and is an important aid in the stimulation of eye appeal.

Lounge. General illumination should be such as to meet the requirements for relaxing or card playing and to provide sufficient illumination to the upper walls and ceiling for the elimination of high luminance ratios. Supplementary lighting units may be required to furnish a higher level for prolonged reading and typing or processing business forms. Fig. 22-14 shows a typical lighting design for a lounge car with supplementary individual downlights at each chair.

Sleeping Car (Pullman). The bedroom compartment for all intents and purposes can be considered the passenger's travelling apartment and should afford some home environments. It should lend to the comfort, convenience and beauty of the accommodations. As in any good lighting installation, the ability to perform the visual task is the primary consideration. The visual tasks in any sleeping accommodation are similar to those in the home.

The arrangement of berths and provisions for upper storage in the daytime severely limits the ceiling area available for luminaires. Structural members, air ducts and conduit runs also have a limiting effect on useable ceiling area. In general, the location and maximum size of a luminaire in the ceiling is fixed by the conditions outlined above.

Berth lighting units should be designed to provide suitable illumination for reading in bed. Here again, as in the case of the ceiling luminaires, freedom of design and optimum use of materials are limited by the physical characteristics of the application. Berth lighting units should be so designed as to provide a concentrated beam of light for the reading task and a component for general illumination to relieve excessive luminance ratios.

Washrooms and Toilet Sections. The lighting design should provide general illumination and mirror lighting from luminaires on ceiling, side walls or in combination. A luminaire generally located in the ceiling should be provided in the toilet section.

Fig. 22-14. Railroad parlor car. Two continuous rows of 40-watt fluorescent lamps are concealed by the parcel rack on each side of the car. Supplementary lighting is provided by locally controlled fluorescent lamps centered above each chair.

LAMPS AND ELECTRIC POWER SYSTEMS FOR TRANSPORTATION LIGHTING

Lamps

Both incandescent filament and fluorescent lamps are used for public conveyance lighting, but there is a trend toward use of more fluorescent lighting in the newer units. The high efficacy, long tubular shape, and low luminance of the fluorescent lamp make it well suited to vehicular lighting. However, few public conveyances have the normal 60-hertz 118-volt alternating current for which most fluorescent lamps and accessories are designed; consequently, special electrical systems have been devised to facilitate the use of fluorescent lamps in this field.

Multiple incandescent filament lamps suitable for transportation are described in Fig. 8–88, series lamps in 8–81, and fluorescent in 8–110 through 8–113.

Electric Systems for Operating Fluorescent Lamps in Public Conveyances

Fluorescent lamps were primarily developed for ac operation and are generally more efficient and satisfactory when operated on ac, but a dc power supply for lighting has been the accepted standard in the transportation field until recent years. Certain sizes of fluorescent lamps may be operated directly on the dc available on the vehicle or ac may be generated from the available dc supply by means of various types of conversion units.

With dc operation of fluorescent lamps, a resistance type ballast is used to control the current.[12] Ballast loss should be included when determining the total power required for the lighting. Also, in lamps over 24 inches the direction of the current flow through the lamps should be periodically reversed to prevent the reduction in light output at the positive end of the lamp caused by the gradual drift of the mercury to the negative end. The useful lamp life on dc burning is reduced to approximately 80 per cent of that on ac burning. Also, special provisions are needed to assure dependable lamp starting at low line voltage and at lower ranges of ambient temperature.

The use of power conversion equipment to convert from dc to ac is now common practice. See Fig. 22–15. The conversion may be to 60 hertz; however, the trend is toward higher frequencies in order to gain over-all efficiency and reduced weight in the auxiliary equipment.

There are two common methods of conversion in use to produce ac from the basic dc power source available:

Rotary Machines. Such devices as rotary converters, motor alternators, booster inverters, etc., are being used on dc voltages to generate various ac output voltages. Gasoline-electric or diesel electric[13]

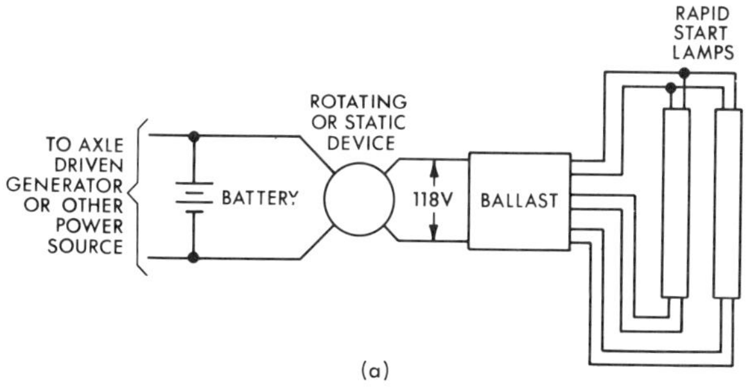

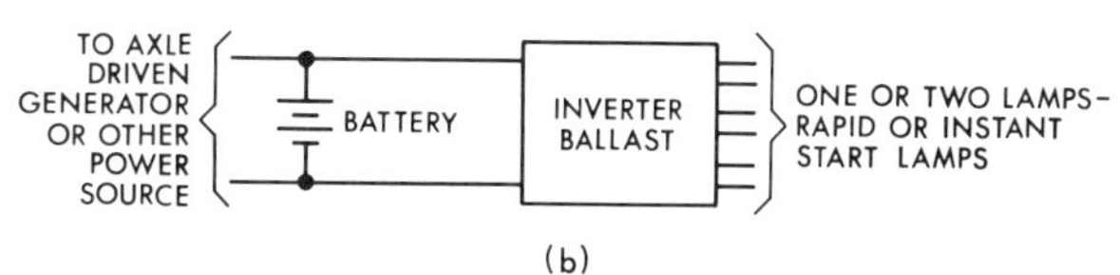

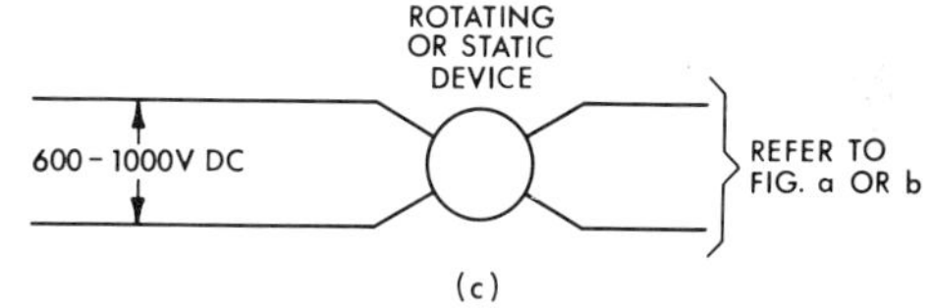

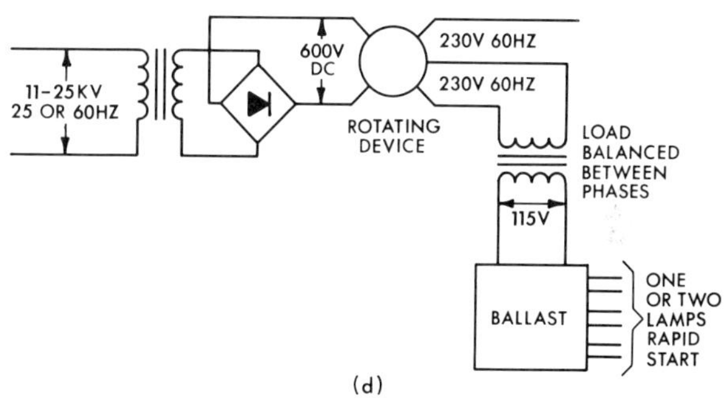

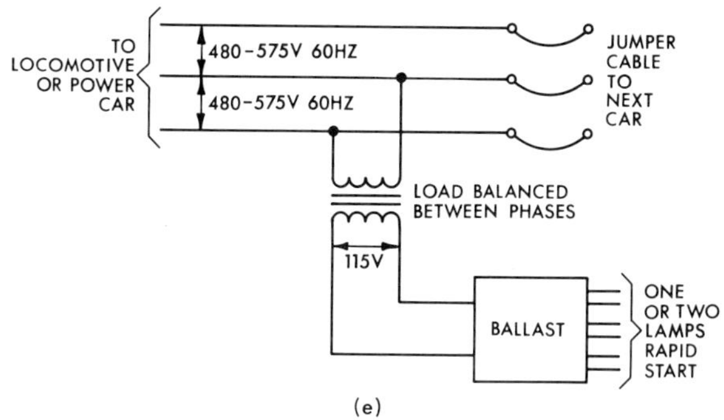

Fig. 22–15. Typical circuits for operating fluorescent lamps in the transportation field: (a) circuit for converting available dc to 60 Hz ac; (b) circuit using available dc; (c) circuit for converting high voltage dc; (d) circuit for converting high voltage ac; (e) circuit used with power source from locomotive.

equipment is available for mounting beneath a car and for a "head-end" ac power system.

Inverter Systems. These systems produce ac energy usually of a high frequency ranging from 400 hertz to 25 kilohertz. The lower frequencies are sometimes used in applications where the noise of the vehicle overcomes the audible frequency hum. Optimum frequencies are those above the hearing range. These frequencies also provide minimum size of equipment and maximum over-all system efficiency.

Any standard fluorescent lamp can be used to advantage on high frequency power; however, the ballast and all component equipment should be designed for good lamp performance, taking into consideration ample open circuit voltage and correct operating watts for the fluorescent lamp involved.

MARINE LIGHTING[14]

The objectives of shipboard lighting are to provide for the safety and well being of the passengers and crew, to provide adequate illumination for the various tasks encountered aboard ship, and to provide a home-like environment with comfortable, well-lighted staterooms and public spaces.

Interior Lighting

Lighting in the interior of a ship involves spaces similar to many of those encountered on land. However, due to low ceiling heights, space limitations, and the requirements for watertight, guarded, or explosion-proof luminaires in certain locations, there are some differences necessary in the design and application of luminaires from those encountered on land.

The principles set forth in such Sections as 10, 11, 12, 14, 19, and 24, having to do with lighting practices in land installations, will be applicable aboard ship. Recommended minimum illumination levels for the tasks in the spaces normally encountered on board ship are found in Fig. 9-80, page 9-94. The values listed represent minimum values, measured without daylight or light supplied by supplementary units such as mirror lights, berth lights, gauge lights, etc., but include all units normally contributing to the general illumination such as wall brackets, floor lamps, and table lamps. They are based on the safety, utilitarian, and decorative characteristics of the areas concerned.

Regulatory Body Requirements and Other Standards

The requirements for safety to life for ships of United States registry are established by the US Coast Guard. Regulations regarding ships' electrical plant and lighting are contained in *US Coast Guard Electrical Engineering Regulations.*[15] These regulations reflect the lighting requirements published by the 1960 International Conference on Safety of Life at Sea. Detailed requirements for the construction of marine-type luminaires are contained in Underwriters' Laboratories' "Standard for Marine-Type Electric Lighting Fixtures"[16] and "Standards for Electrical Lighting Fixtures for use in Hazardous Locations."[17]

Luminaires which have been examined and found to comply with the Underwriters' Laboratories' standards are identified in two different ways:

1. *Underwriters' Laboratories' service label*—provided for most luminaires, the exception being general utilitarian deck and bulkhead luminaires. Labeled luminaires are identified by Underwriters' Laboratories' label service symbol together with the designation "Marine-Type Electric Fixture" or "Marine-Type Recessed Electric Fixture." The labels also include other applicable luminaire information, such as "Inside-Type," "Inside Drip-proof-Type," "Outside-Type (Fresh Water)," or "Outside-Type (Salt Water)."
2. *Underwriters' Laboratories' re-examination service*—used for unwired general utilitarian type luminaires for deck and bulkhead mounting where exposed to the weather or other wet or damp locations. These luminaires are listed by catalog number in Underwriters' Laboratories' "Electrical Construction Materials List,"[18] but are not labeled. Listed luminaires are characterized by having junction boxes, globes, and frequently guards. Special luminaires that are not within the scope of Underwriters' Laboratories' "Standard for Marine-Type Electric Lighting Fixtures" or luminaires of such limited use that inspection by Underwriters' Laboratories would be uneconomical may be given special consideration by the US Coast Guard.

The American Bureau of Shipping establishes requirements for construction of vessels for certifying them as being eligible for insurance issued by members of marine insurance underwriters.[19] The Committee on Marine Transportation of the Institute of Electrical and Electronics Engineers (IEEE) has published a "Recommended Practice for Electric Installations on Shipboard"[20] which serves as a guide for the equipment of merchant vessels with an electric plant system.

Seeing Tasks in Marine Lighting

Living and Public Spaces. Stateroom, dining, recreation, office, and medical spaces for passengers, officers and crew should have general illumination as well as local lighting where seeing tasks are involved. Direct- or indirect-type luminaires can be used to give the recommended illumination levels without glare, and to allow safe movement throughout these areas, as well as sufficient light for cleaning and maintenance purposes. Supplementary lighting should be provided in all areas where reading, serving, dressing and make-up are intended. A

Fig. 22–16. Stateroom. General illumination is from fluorescent valance and incandescent downlights; supplementary lighting is provided by incandescent portable lamps.

Fig. 22–17. Officers' dining room. General lighting is provided by fluorescent lighting from 2-lamp units on approximately 6-foot centers. A warm tone is provided by incandescent lamps in the bracket above the wide mirror.

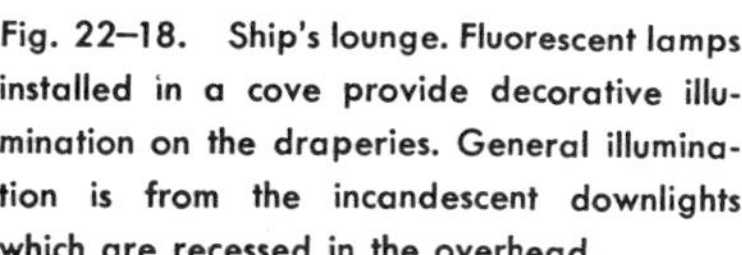

Fig. 22–18. Ship's lounge. Fluorescent lamps installed in a cove provide decorative illumination on the draperies. General illumination is from the incandescent downlights which are recessed in the overhead.

means for controlling the general lighting should be conveniently located at each main entry, as well as local control for supplementary lighting. Figs. 22-16, 22-17, and 22-18 show typical living and public spaces.

Passageways, stair foyers, and stairs should be uniformly illuminated by luminaires installed overhead, in the corner between the bulkhead and the overhead, or in other suitable locations. Care should be exercised in locating the luminaires to illuminate cross passages, stairs, room numbers, etc. Some of the lamps in these areas should be supplied from the emergency lighting system(s) to provide for safety and escape in an emergency.

Passenger entrances are the first spaces on board the ship seen by the passenger, and therefore, particular attention should be paid to the quality of illumination provided. Emergency lighting should be provided to enable escape.

Passengers' and officers' dining room, passengers' lounges, libraries, smoking rooms, cocktail lounges, bars, and ballrooms are all public spaces and the principles of lighting involved are the same as for similar land installations (see Figs. 22-17 and 22-18), but emergency lighting should be provided to enable escape.

Ship's Offices and Navigation Spaces. During nighttime operation of the ship, the *wheelhouse* must be in complete darkness, except for necessary instrument lights. General lighting should be provided for cleaning and maintenance purposes only. Passageway lights, which would interfere with navigation when a door to the wheelhouse is opened, should be controlled by a door-operated switch. In order not to interfere with dark adaptation of the eyes of the personnel on duty, instruments should be illuminated by red light of wavelengths not less than 590 nanometers and luminance not greater than 0.1 footlambert.

General lighting in the *chart room* should be provided by ceiling luminaires. Adjustable chart table lamps, fitted with red filters, and a lamp for illumination of the chronometers should be installed and controlled by a momentary contact switch. The chart table lamp should be supplied from the emergency lighting system.

Service and Operating Spaces. Careful attention should be paid to the quality of illumination in *galleys, shops,* and *other service spaces.* Reflected glare and high luminance ratios should be avoided. The general illumination should be provided by ceiling luminaires with a concentration over benches and working surfaces; supplementary detail lighting should be provided at griddles and range tops, sewing machines, and other points where specific tasks are performed. See Fig. 22-19.

Machinery spaces, steering gear rooms, and other ship operating spaces should be provided with general illumination by means of luminaires installed so that piping and other interferences do not obstruct light. The general lighting should be supplemented by detail lighting of gauges, switchboards, and other points where meter reading or specific functions require higher levels of illumination. See Fig. 22-20.

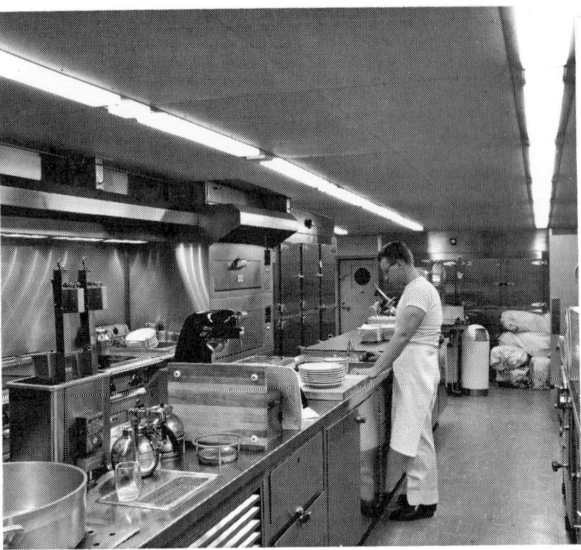

Fig. 22-19. Two rows of two-lamp fluorescent luminaires alternating with 60-watt incandescent units provide general illumination in this galley.

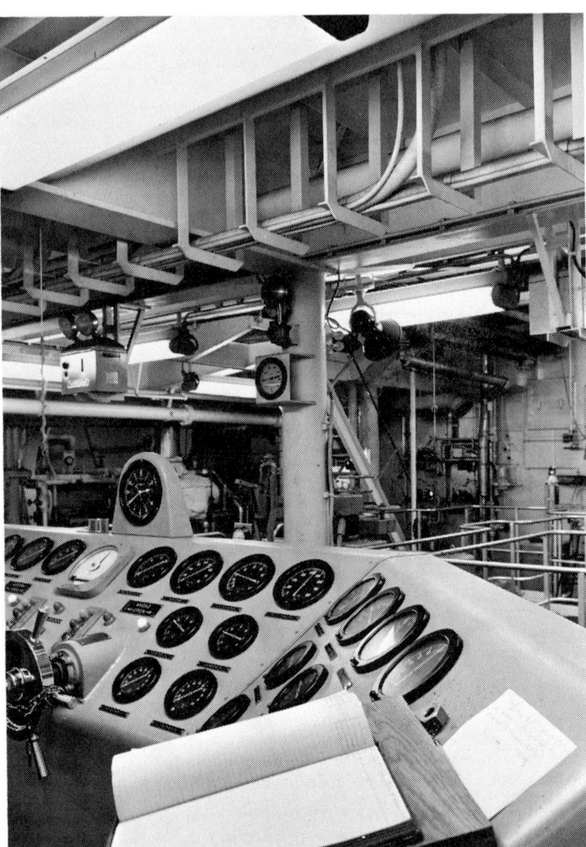

Fig. 22-20. Operating deck console and control panel lighting is provided by a fluorescent general lighting system in this installation. Supplementary lighting is also used at the control panel.

Exterior Lighting

General Deck Lighting. Outside lighting of decks should be by means of watertight deck and bulkhead luminaires. Protective guards should be provided except in certain passenger areas such as passengers' promenades where it is desirable to achieve a certain decorative effect. In certain areas, such as those below, levels of illumination in excess of that required for safe night passage should be provided.

Cargo Handling. Modern ships are usually fitted with permanent lighting installations in the cargo holds, and receptacles are normally located at the hatches for use with portable cargo clusters. In addition to the foregoing, floodlighting of the deck area in the vicinity of cargo hatches is recommended. This may be accomplished by the use of floodlights suitably located on the ship's structure. At least two such floodlights should be provided at each hatch. See Fig. 22-21.

Recreation Areas. Where illumination is desired for nighttime sports activities, floodlights suitably located to illuminate the game areas are recommended.

Lifeboat and Life Raft Launching. Incandescent floodlights should be provided at each lifeboat and life raft for launching. The location of the light should be such that it can be trained on the boat or the boat gear during launching operations. These floodlights should be permanently connected to the emergency system.

Stack Lighting. Where stack floodlighting is desired, floodlights equipped with focusing adjustment are recommended. They should be provided in sufficient quantity and located to insure an even light distribution on the entire stack, except when special effects are desired.

Fig. 22-21. Tanker with high-tower deck lighting using floodlights housing two high-intensity discharge 400-watt lamps.

Navigation Lights. Requirements for running, anchor and signal lights should conform to the "Rules of the Road" and should be installed to conform to the governing rules for the waters in which the ship is to navigate.[21] A running light indicator panel, equipped with both visible and audible indication of light failure, should be provided for control of the side, masthead, range, and stern lights. In general, all navigation and other special lights used for navigational purposes should be energized from the emergency lighting system.

Searchlights. Where a navigational searchlight is desired, it should be located on top of the wheelhouse and operable from within the wheelhouse. A portable signaling searchlight should be provided for operation from either bridge wing. The light should be either battery operated or energized from the emergency lighting system. Signaling may be effected by keying the lamp current or by movement of shutters fitted to the searchlight.

Naval Vessels

Naval vessels are designed for specific military purposes to a set of military characteristics. One of these is "Habitability", of which the lighting installation is a very important factor. The US Navy Bureau of Ships has made an exhaustive study of naval shipboard lighting and has established and standardized on a very efficient group of luminaires to insure meeting the exacting tasks encountered on board and to provide as pleasant an environment as is possible.

RAILWAY GUIDANCE SYSTEMS

Railway train operating personnel receive guidance through lighting units of three main categories:

1. Exterior lights on the train, including headlights and marker lights,
2. Interior cab signals, and
3. Wayside signals.

Voice communications by wire and radio serve various crew coordinating functions and yard movements; however, actual train movement into any main line segment of the rail system is always directed by a signal light indication.

Exterior Lights on Trains

Locomotive headlights are classified either as road service giving 800-foot object visibility, or as switching service for 300-foot object visibility, as governed by the regulations of the Interstate Commerce Commission. New headlight equipment consists of two all glass sealed beam lamps mounted in a single housing, each lamp projecting a beam about

Fig. 22-22. In this type of rapid transit cab signal, lighted segments on the speedometer show the motorman the highest speed permitted. The six windows below the speedometer light to show yellow, green, and red aspects and speed limits.

5 degrees wide of 300,000 candelas. The pair of lamps exceeds the performance and reliability of the old single reflector headlight. Shielding is provided to minimize veiling glare from stray light illuminating atmospheric particles in the line of sight. In addition to lighting possible obstructions on the right of way the headlight also activates colored relectorized markers at switch locations. The rear lights on trains consist of two red markers spaced horizontally apart.

Locomotive Cab Signals

By suitable track circuits and electric receiving equipment, automatic signal lights inside the cab can be made to show signal aspects corresponding to those of the wayside signals. This is useful in times of poor visibility due to atmospheric conditions or other obstructions. Changes of wayside signal aspects in advance of the train may be displayed promptly in the cab thus expediting response to the change.

Rapid Transit Cab Signals

In rapid transit systems speed commands are continuously transmitted, through the rails, precisely and exclusively to the train intended. Onboard, the cab signal displays the commands, and the overspeed control system compares the actual train speed with the maximum speed allowed by the cab signal. See Fig. 22-22. If the actual speed exceeds the limit displayed, the system warns the motorman, audibly, that a brake application is required. If he fails to take action immediately, the control system automatically stops the train.

Wayside Signals

The movement and speed of trains into each segment of trackage is permitted only by the adjacent signal indication, with some advance information provided by the range of the signal beam and the proceeding signal.

Wayside Signal Range. The beam intensity and range considerations for a lighted signal are based upon the estimated safe visual range by day in clear weather. For red and green signals it is common to use the formula: $D = \sqrt{2000\ I}$, where D is the range in feet and I the intensity in candelas of the same type signal when equipped with colorless optical parts.

Yellow and lunar white lenses provide a somewhat longer range, but blue only about one-third the distance D.

By the use of this formula and the candlepower distribution curve of a signal beam, it is possible to lay out a chart or plan that shows the ground area over which a particular signal will be within visible range in clear weather. This signal range plan can be superimposed over a track plan to see whether the signal will have visibility over a particular track approach. See Fig. 22-23. Signal manufacturing companies have prepared range charts for their various signal units embodying the large variety of horizontal beam spreading and deflecting auxiliary lenses available.

A horizontal deflecting or spreading prismatic element may be chosen to provide visibility along a curved track approach. A vertical deflecting prismatic element is necessary to enable an engineer at very close range to see a signal high overhead or

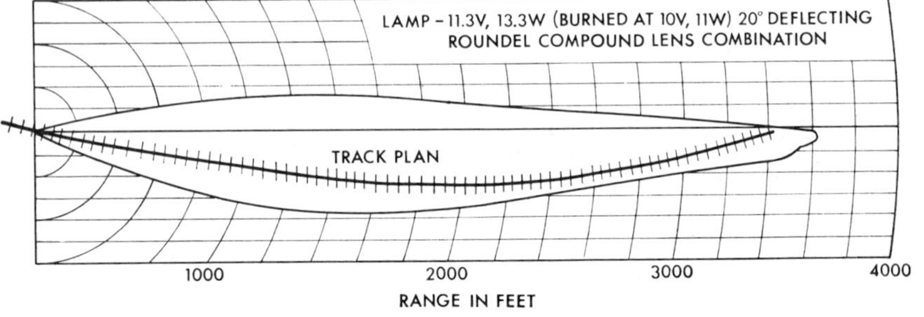

Fig. 22-23. Range chart for searchlight-type signal-unit with part of a track plan superimposed to show range of useful coverage.

to see a dwarf signal close to the ground.

External Light Interference. By making the front surface of lenses and roundels convex, rather than flat, it is possible to scatter most of the external light reflected from the front surface. Also, so that daylight will produce negligible interference with the signal under most conditions, hoods or visors are always used. Occasionally flat auxiliary roundels are inclined at selected angles. The incorporation of reflectors in the optics of a signal involves particularly careful analysis to prevent reflected external light. A typical deep parabolic reflector, as used in ordinary spotlights, could flash false indications from external light if used in signals.

Color. Train operating personnel are selected as having normal color vision. The colors used by railroads in North America are with very few exceptions governed by the Association of American Railroads Specifications 59 and 69.[23] See Fig. 22-24. These color specifications contain basic definitions for the colors to be displayed in service and the tolerances for color-limit filters to be used to inspect signal glassware. The primary standard filters controlling these inspection filters are deposited with the National Bureau of Standards in Washington, D.C.

The Association of American Railroads currently specifies five colors: red, yellow, green, blue, and lunar white; however, only red, yellow, and green are used for long range color signals. Both the red and yellow are somewhat deeper colors than those used in street traffic signals. Blue performance is limited by the very low blue emission of incandescent signal lamps and consequent transmission of about 2 per cent by blue glass. Lunar white is the term applied to white light as filtered by a bluish glass which raises the apparent color temperature or whiteness. The lunar white aspect from an incandescent signal lamp will appear about 4000 to 5000 K and from a kerosene unit about 3000 to 4000 K. Purple is no longer recommended as a signal color because of low filter transmission and because of the variable impression made upon different observers.

As is commonly understood, red is associated with the most restrictive signal indications, green with the least restrictive, and yellow intermediate. For the specific meaning of the signal aspects, many of which involve two or more lights shown together, see *American Railway Signalling Principles and Practices,* Chapter II, published by the Association of American Railroads.

Wayside Signal Types. Modern signal units and their arrangement on a tower all have some feature suggestive of the early semaphore unit which had a blade for day viewing and associated color discs that swing in front of a lamp for night viewing. The three types of signals which depend entirely upon light are described below as: position-light signal, color-light signal, and color-position-light signal. See Fig. 22-25. All utilize large targets or black backgrounds to permit relatively low wattage lamps to show up without blending into the sky. Lamps are usually of 5 to 18 watts with very small filaments that must be precisely located at the focal point by a prefocus base. Each signal unit is accurately aimed using a sighting device during installation or is adjusted by radio instructions from a viewer down the track.

Color-Light Signal. A color-light signal stand may involve separate lights with colored lenses for each color, or the signal lighting units may have internal mechanism and moveable filters to change the color within each unit. This latter moveable filter unit is called a searchlight signal and permits three units on a tower to display the widest variety of color combinations, for instance, "red over yellow over yellow," or "green over yellow over red," etc.

Position-Light Signal. The position-light signal is a type of wayside signal which does not depend upon color discrimination by the engineer. In this type, a number of lamps (maximum nine) are mounted on a circular target—eight lights arranged in a circle, one in the center. By operating three lamps at a time, the aspect of the signal may be a vertical, a horizontal, or a diagonal row. Each of the target lights is aligned by its own projector system in the direction of the approaching train. Yellow lenses are normally used to achieve distinctiveness from other non-signal lights.

Color-Position-Light Signal. The color-position-light signal is a type which utilizes a combination of the principles of the color-light and the position-light systems. Here also there are several lights on a target. These may be lighted in pairs: vertical pair (green), horizontal pair (red), right and left diagonal pairs (yellow and lunar white, respectively).

Power Sources for Signals. The lighted aspect displayed is controlled by relays at the signal, actuated by coded impulses in the track circuit; however, power for the lamp comes from storage batteries at the location for complete dependability. The batteries may be used alone or as standby for ac service. To provide long lamp life and reduce the probability of a dark signal, lamps are usually burned at 90 per cent of rated voltage.

Control Panels for Signal System

The movement of trains and sections of track system thus occupied are represented on control panels at "interlocking" or at "centralized traffic control" centers. On such a panel (see Fig. 22-26) the operator has before him push buttons that operate relays for switches and signals along a portion of the rail line which may be a local yard or several hundred miles of track. Lights indicate the response of the switches and signals.

A track diagram for the territory is studded with

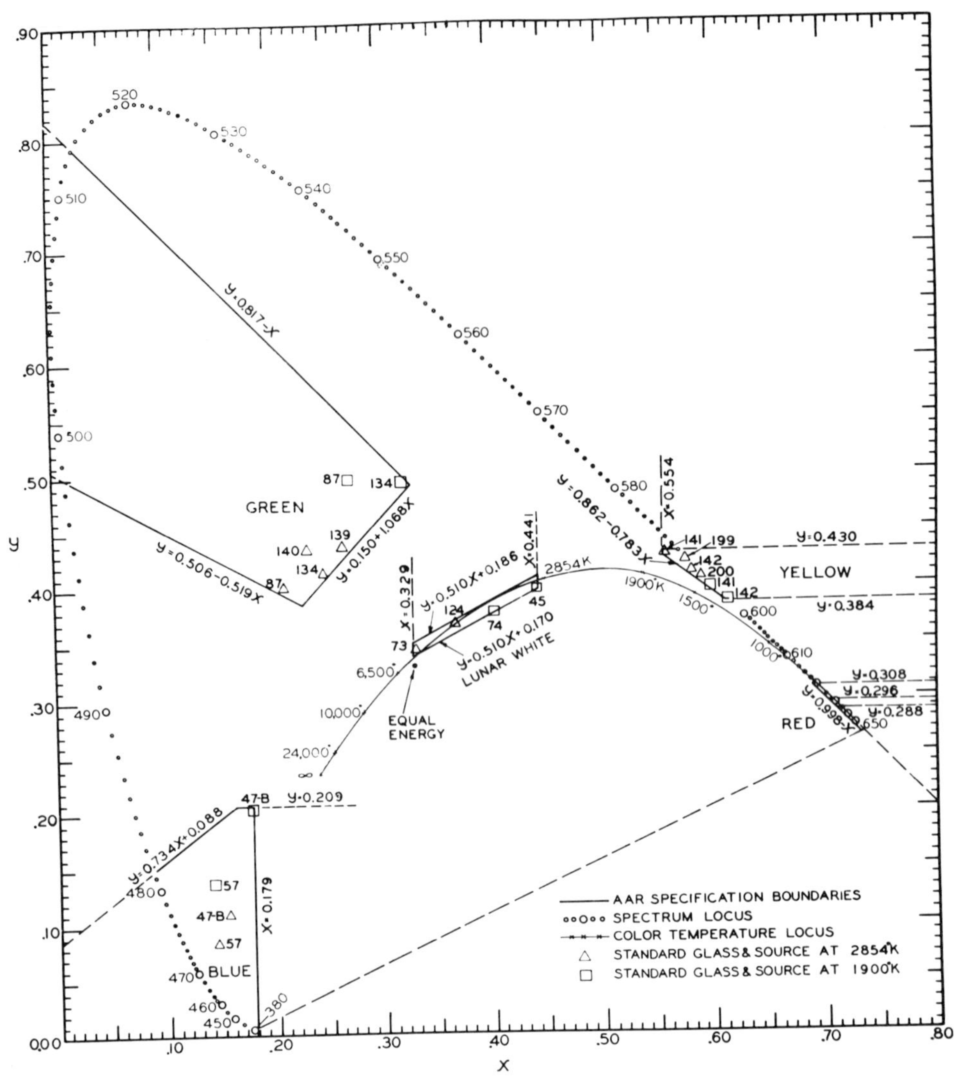

Fig. 22–24. Railway signal color specifications plotted on CIE chromaticity diagram.

Fig. 22–25. *a*. Position-light signal. *b*. Color-light signal. *c*. Color-position light signal. *d*. Search-light-type of color-light signals.

Fig. 22-26. Traffic master push button control console for centralized traffic control.

indicator lights which show when a train occupies certain sections of track along the line or what route has been established.

Railroad-Highway Grade Crossing Lights

The warning for automotive traffic at grade crossings is accomplished by railroad owned equipment. Its functioning condition is checked by train operating personnel in passing. These lights are two horizontally spaced alternate flashing red lights controlled by track circuits which sense the presence of the approaching train. The light beam width is usually 30 degrees and must provide a safe visibility of 1000 feet on straight highway approaches.

REFERENCES

1. Roper, V. J. and Howard, E. A.: "Seeing with Motor Car Headlamps," *Trans. Illum. Eng. Soc.*, Vol. XXXIII, p. 417, May, 1938. Roper, V. J., and Scott, K. D.: "Silhouette Seeing with Motor Car Headlamps," *Trans. Illum. Eng. Soc.*, Vol. XXXIV, p. 1073, November, 1939. Roper, V. J. and Meese, G. E.: "Seeing Against Headlamp Glare," *Illum. Eng.*, Vol. XLVII, p. 129, March, 1952.
2. Land, E. H., Hunt, J. H., and Roper, V. J.: "The Polarized Headlight System," *Highway Research Board Bulletin No. 11*, National Research Council, Washington, D. C., 1948.
3. Davis, D. D., Ryder, F. A., and Boelter, L. M. K.: "Measurements of Highway Illumination by Automobile Headlamps under Actual Operating Conditions," *Trans. Illum. Eng. Soc.*, Vol. XXXIV, p. 761, July, 1939.
4. de Boer, J. B. and Vermeulen, D.: "On Measuring the Visibility with Motorcar Headlighting," *Proc. Int. Comm. Illum.*, 1951; de Boer, J. B. and Vermeulen, D.: "Motorcar Headlights," *Philips Tech. Rev.*, Vol. 12, p. 305, May, 1951.
5. Bone, E. P.: "Automobile Glare and Highway Visibility Measurements," *Highway Research Board Bulletin No. 34*, National Research Council, Washington, D. C., 1951.
6. Boelter, L. M. K. and Ryder, F. A.: "Notes on the Behavior of a Beam of Light in Fog," *Illum. Eng.*, Vol. XXXV, p. 223, March, 1940.
7. Finch, D. M.: "Lighting Design for Night Driving," *Illum. Eng.*, Vol. XLV, p. 371, June, 1950.
8. Committee on Motor Vehicle (Exterior) Lighting of the IES: "Lighting Study Project Report on Motor Vehicle (Exterior) Lighting," *Illum. Eng.*, Vol. LIX, p. 660, October, 1964.
9. "Tests for Motor Vehicle Lighting Devices and Components," *SAE Handbook*, p. 271, 1971.
10. Spencer, D. E. and Levin, R. E.: "Guidance in Fog on Turnpikes," *Illum. Eng.*, Vol. LXI, p. 251, April, 1966.
11. Committee on Interior Lighting for Public Conveyances of the IES: "Interior Lighting of Public Conveyances—Road and Rail," to be published.
12. Brady, C. I., Slauer, R. G., and Wylie, R. R.: "Fluorescent Lamps for High Voltage Direct Current Operation," *Illum. Eng.*, Vol. XLIII, p. 50, January, 1948.
13. "Uses of Electric Power," *Westinghouse Eng.*, Vol. 8, p. 2, January, 1948.
14. Subcommittee on Marine Transportation of the Committee on Interior Lighting for Public Conveyances of the IES: "Recommended Practice for Marine Lighting," to be published.
15. *Electrical Engineering Regulations*, Publication No. CG-259, 46CFR, Parts 110 to 113, US Coast Guard, Washington, D.C., latest issue.
16. "Standard for Marine-Type Electric Lighting Fixtures," Publication No. UL-595, Underwriters' Laboratories, Inc., Chicago, Illinois, latest issue.
17. "Standard for Electric Lighting Fixtures for Use in Hazardous Locations," Publication No. UL-844, Underwriters' Laboratories, Inc., Chicago, Illinois, latest issue.
18. "Electrical Construction Materials List," Underwriters' Laboratories, Inc., Chicago, Illinois, latest issue.
19. *Rules for Building and Classing Steel Vessels for Service on Rivers and Intracoastal Waterways*, American Bureau of Shipping, R. R. Donnelley and Sons, Co., Lakeside Press, Chicago, 1963.
20. "Recommended Practice for Electrical Installation on Shipboard," Publication No. 45, Institute of Electrical and Electronics Engineers, New York, New York, latest issue.
21. *Rules of the Road:* "International—Inland," Publication No. CG-169, latest issue; "Great Lakes," Publication No. CG-172, latest issue; "Western Rivers," Publication No. CG-184, latest issue, US Coast Guard, Washington, D.C.
22. Campbell, J. H.: "New Parameters for High Frequency Lighting Systems," *Illum. Eng.*, Vol. LV, p. 247, May, 1960.
23. *Signal Manual, Specification No. 69*, Part 136, April, 1960. Association of American Railroads, Washington, D.C. 20036.
24. "Equivalent Indications for Semaphore, Color Light Position and Light and Color Position Light Signal Aspects," *Proc. of Signal Section*, Vol. XLII, No. 2. Association of American Railroads, Washington, D.C. 20036.
25. Gage, H. P.: "Practical Considerations in the Selection of Standards for Signal Glass in the United States," *Proc. Int. Congr. on Illum.*, p. 834, 1928.
26. Gibson, K. S. and Haupt, G. W.: "Standardization of the Luminous-Transmission Scale Used in the Specification of Railroad Signal Glasses," *J. Res. Nat. Bur. Stand.*, Research Paper RP 1688, January, 1946.

SECTION

23

LIGHTING FOR ADVERTISING

Exposed Lamp Signs 23–1 *Luminous Element Signs 23–4*

Floodlighted Signs 23–7

Through the knowledge and use of new materials, new light sources, and advanced techniques a wide variety of signs can be produced for today's needs. Illuminated advertising signs—exposed lamp, luminous tube, luminous element, etc.—while differing in several respects from other forms of advertising, definitely tie in with any over-all promotional activity. Signs, whether small identification or large spectacular, mass viewed signs (see Fig. 23–1) can quickly gain the observers' attention through the combined use of size, color, and motion.

Sign Characteristics

Electric signs may be classified by illumination method:

1. Luminous letter signs (illuminated letters, non-illuminated background) such as exposed lamp signs, exposed luminous tube signs, raised glass or plastic letter signs, etc.
2. Luminous background signs (illuminated background, silhouette) such as translucent plastic or glass faced signs with interior light sources such as fluorescent lamps, incandescent lamps, or luminous tubing.
3. Floodlighted signs, such as painted bulletins and poster panels.

Signs may also be classified by the application function, such as: single-faced luminous elements, window or roof; double-faced projecting, horizontal or vertical, etc.

Size. Physical location, desired legibility range, and brightness determine the minimum letter height required for legibility. However, to attain advertising effectiveness, letter heights of twice minimum height for legibility generally are employed. Vertical columns of letters, though usually an aid in increasing the apparent size of a sign, are more difficult to read than horizontal arrangements.

Brightness. Letter or background brightness and contrast between letter and background are factors influencing the legibility of a letter and the rapidity with which it is recognized. Contrast between the average sign brightness and that of its surround determines, in a large measure, the manner in which the sign stands out. Brightness and contrast attract attention.

Location and Position. The advertising value of a sign depends on the greatest possible number of persons seeing it. This is a function of its location.

Distinctiveness. One of the elements of a good electric sign is that it possesses distinctiveness and individuality. It should create a pleasing, favorable impression, should have public appeal, and should be remembered easily.

Motion. Motion increases the attracting power and memory value of a sign. It capitalizes on the instinctive trait of people to be aware of and to give heed to moving things.

Color. Color is an important factor in legibility. Often color is incorporated in a sign because it provides contrast. It may aid in attracting attention. It may add distinctiveness.

EXPOSED LAMP SIGNS

Exposed Incandescent Filament Lamp Signs

These signs are constructed so that the lamps are exposed to direct view. This type is well suited for application where long viewing distances are involved, as well as for small, high brightness signs. Motion and color can be incorporated very easily in such signs.

Legibility. Legibility is primarily a function of letter size and form or design, lamp spacing, and brightness, and also contrast between letter or design and background.

Block letters possess greater legibility than ornamental styles, script, and special forms, although

NOTE: References are listed at the end of each section.

Fig. 23-1. A 180-foot tall spectacular sign utilizing exposed yellow lamps on a blue neon tubing background for the word "Dunes," clear scintillating lamps for the diamond, red luminous tubing for the outline, and fluorescent lamp transilluminated plastic panels for luminous elements near the base.

the latter types may be used to gain distinctiveness. Wide, extended letters are more legible than tall, thin letters.

Reflectors. When viewing at wide angles is relatively unimportant, reflectors for increasing the directional candlepower or reflectorized lamps may be used with advantage. In this way, the brightness at the viewed angle can be greatly increased, even to the point of being effective during daylight.

Letter Size. The letter height employed on an exposed lamp sign usually is greater than the minimum height necessary to gain recognition. For purposes of advertising and quick reading, it is common practice to provide exposed lamp signs with letter heights that are 50 to 100 percent greater than those necessary for legibility.

For simple block letters (width equal to three-fifths the height) with a *single row of lamps,* in typical locations, the minimum height for legibility is given by the formula:

$$H_r = \frac{D}{500}$$

where H_r = minimum vertical height of letter, for recognition, from top lamp to bottom lamp (feet).

D = maximum distance at which letter is legible to a majority of people (feet).

For letters with *multiple rows of lamps,* the height should be increased by three times W, the distance between outside rows of lamps in a stroke:

$$H_r = \frac{D}{500} + 3W$$

Letter width, height, and stroke and lamp spacing are illustrated in Fig. 23-2.

Lamp Spacing. The proper spacing between lamps to obtain an apparently continuous line of light is determined by the minimum viewing distance. Spacing may be estimated by the following formula:

$$s = \frac{D_{min}}{1500}$$

where s = spacing between centerlines of lamps (feet).

D_{min} = minimum viewing distance (feet).

In very bright locations the above spacing should be decreased by 25 to 35 percent.

To produce a "smooth" line of light the above spacing should be decreased by 50 percent. At viewing distances of less than 500 feet, a smooth line of light will generally not be possible because low wattage, medium based lamps (11- or 15-watt S-14) require spacings of 2 to 2½ inches to permit easy maintenance.

Lamp Wattage Rating. The incandescent lamp wattage employed depends upon the general brightness of surroundings, and background, as the sign is viewed. Consequently, a roof sign, even if located in a brightly lighted area in the business center of a city, might at night always be viewed against a a dark sky. Such a sign would require the same lamps called for in low brightness areas.

Fig. 23-3 indicates the typical lamp wattages found in signs in various areas, classified according to brightness.

If incandescent lamps with colored bulbs or clear

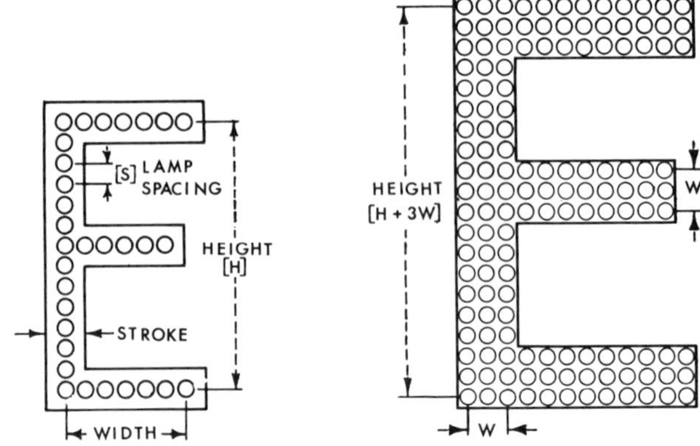

Fig. 23-2. Important dimensions in the design of exposed lamp letters.

bulbs with colored accessories are employed, lower letter brightness will result than when equal wattage lamps with clear bulbs are used alone.

Usually, however, less colored light is necessary to create equal advertising effectiveness. It is, therefore, not necessary to increase wattage in direct proportion to the output of the color lamps. This is taken into account in Fig. **23-4.**

Both transparent and ceramic coatings are used to color bulbs. In general, the transparent coatings have higher transmittances than the ceramic, thus appear brighter. In addition, the filament is visible for added "glitter" at near viewing distances.

Note that signs lighted in cool colors, blue and green for example, will generally be less legible than those with clear or warm colored lamps, because cool colors appear to swell or "irradiate" more than warm.

Lamp Types. For exposed lamp signs located where rain or snow could fall on relatively hot glass, vacuum-type incandescent lamps are recommended. They are available in 6-, 10-, 11-, 15-, 25-, and 40-watt ratings in both clear and colored bulbs.

For high speed motion effects, a 20-watt gas-filled clear lamp is available. The filament heats and cools very rapidly producing a clean, sharp, on-off action. It is used for scintillation effects, running borders and traveling message signs, and wherever afterglow is undesirable.

Channels. Incandescent lamps are often set into channels. This improves legibility of the sign when viewed at an angle, and increases contrast by reducing background spill light. It does not prevent the strokes of the letters from merging together when viewed at a distance since this phenomenon occurs in the eye.

It is very desirable to employ electrically grounded metal channels to separate incandescent filament lamps and luminous tubing when combined in a sign. Without the channel the electric field generated by the tubing causes the filament to vibrate, thereby reducing the life of the lamp.

Fig. 23-3. Lamp Wattages for Various District Brightnesses

District Brightness	Typical Sign Lamp Wattage
Low	6, 10, 11
Medium	10, 11, 15, 25
Bright	25, 40

Fig. 23-4. Relative Wattage of Clear and Transparent Colored Incandescent Filament Lamps Required for Approximately Equal Advertising Value

Color	clear	yellow	orange	red	green	blue
Wattage	10-11	10-11	15	25	25	40

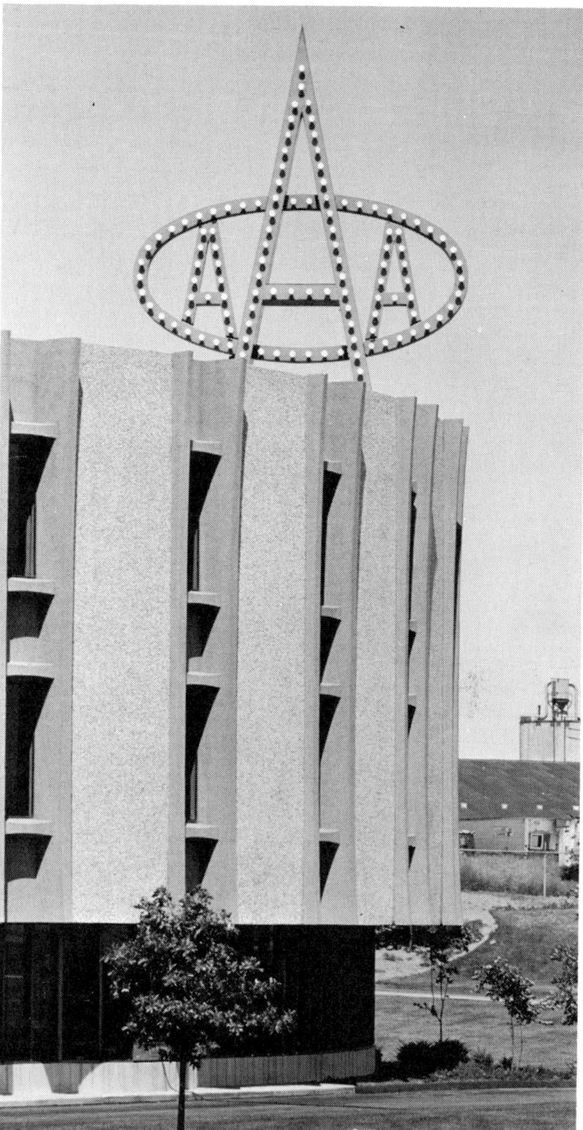

Fig. 23-5. Example of a high candlepower exposed lamp sign.

Daytime Effective Exposed Lamp Signs

Exposed high-candlepower light sources can be used to create electric signs that have as much or more advertising value during the day as do conventional exposed lamp signs at night. Since traffic is generally greater during daylight hours, greatly increased readership usually results with the cost per advertising impression remaining comparable to night viewed lamp signs. The technique is adaptable to signs ranging in size from small store signs to major spectaculars. See Fig. **23-5.**

Letter height and lamp spacing in a daytime lamp sign depend primarily on lamp candlepower and on maximum and minimum viewing distances.[1]

For the greatest majority of daytime sign applications, 75-watt PAR-38 floodlamps on 6-inch centers will adequately meet advertising and identification needs. Higher candlepower sources should be

Fig. 23-6. Daytime Attraction Power of Lamp Types (Viewed Perpendicular to Plane of Sign)

Lamp	Spacing (inches)	Distance (feet)		
		1100	2500	4300
25-watt PAR-38 Flood	6	F	P	NR
	12	P	NR	NR
	18	—	—	NR
75-watt PAR-38 Flood	6	E	E	G
	12	G	G	F
	18	—	—	NR
150-watt PAR-38 Flood 75-watt PAR-38 Spot	6	*	E	E
	12	*	E	G
	18	—	—	F
150-watt PAR-38 Spot	6	*	E	E
	12	*	E	E
	18	—	—	E

— —Spacing inadequate.
* —Brighter than normally necessary.
E —Excellent.
G —Good.
F —Fair.
P —Poor.
NR —Not recommended.

used with discretion since there is a real possibility of making the sign too bright for comfort. A guide for choosing applicable lamp size is given in Fig. 23-6.

Minimum letter height for legibility is the same as for nighttime exposed lamp signs.

The designer should recognize that a daytime sign utilizing PAR-type lamps is a highly directional display, that is the sign's luminance is a function of the candlepower distribution of the lamp. With a PAR flood, for example, the luminance is reduced to 10 per cent of maximum when viewed 30 degrees off axis. This characteristic also provides automatic dimming of the sign as the motorist drives toward the sign, and under the beam.

Nighttime Viewing. A sign of sufficient brightness to compete successfully with daylight will, in most cases, require dimming at night in order to prevent loss of legibility due to irradiation and the possibility of excessive glare. The need for dimming appears to occur for the 75-watt PAR-38 flood at about 25 footcandles daylight illumination, vertically, on the back of the sign. Except for the highest candlepower lamps, dimming by reducing line voltage 50 per cent has generally proved satisfactory. This can be accomplished by placing the primary windings of the supply transformers in series. Greater dimming, especially for very high candlepower sources, through multiple tap or variable transformers may be required. Continuously variable brightness depending on sky brightness may be accomplished by regulating a dimming system with a photocell.

Luminous Tube Signs

Luminous tube signs are constructed of gas-filled glass tubing which, when subjected to high voltage, becomes luminescent in a color characteristic of the particular gas used, the gas and the color of the tubing combined, or of the fluorescent phosphors coating the inner wall.

Color. Fluorescent tubing may be made to emit almost any desired color by mixing different phosphors. Most colors have a higher lumen output per watt rating than the gaseous tubing without a fluorescent coating. Color produced by any one of the gases may be modified by using colored glass tubing, which will transmit only certain colors.

Effective Range. The range of effectiveness for advertising purposes of tube signs is approximately that of exposed incandescent lamp signs of the same size, color, and luminance—250 feet to several miles.

Legibility. For block letters of a width equal to three-fifths of their height, the minimum letter height that will be legible to most people is approximately the same as that for exposed lamp signs.

Tubing Sizes. Standard sizes of tubing for signs range from 9 to 15 millimeters, outside diameter, but larger tubing is available.

Transformers. Several forms of high-leakage-reactance type transformers are manufactured to supply the high voltage necessary to start and operate sign tubing. This voltage is of the order of 5000 to 15,000 volts. After a tube sign is lighted, one-third to one-half of the starting voltage is necessary to keep it operating. The usual range of operating current for tube signs is between 10 and 50 milliamperes.

LUMINOUS ELEMENT SIGNS

A luminous element sign can be created by transilluminating or backlighting a plastic or glass panel which may be either integrally pigmented, externally painted, or opaqued. The pigmentation or paint film diffuses the light, providing uniform brightness over the desired portion of the sign face. See Fig. 23-7. A wide variety of colors is possible with both the integrally pigmented media and the available translucent lacquers.

Design Data for Luminous Elements

Proper lighting is important in assuring the best attraction value and readability in these signs, but other points must be considered along with the lighting.

Contrast. High contrast, either in color or brightness, between message and background panels should be provided. Opaque letters on a light back-

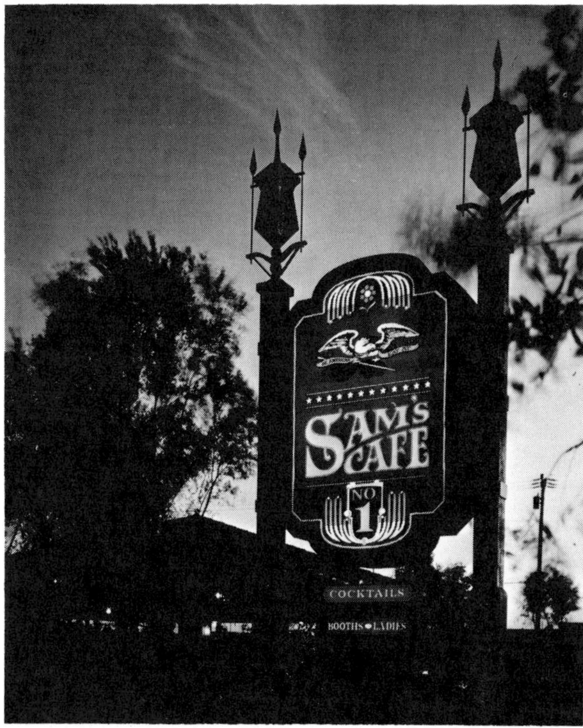

Fig. 23-7. Luminous sign for attraction both day (left) and night (right).

ground are generally preferred for commercial signs because of the attraction value. Where communication is of prime importance, such as in illuminated traffic signs, very light letters on dark backgrounds are generally specified.

Light Sources. Selection of the light source is based on the brightness required, size and shape of the sign, desired color effects, flashing or dimming requirements, environmental temperature conditions, and service access requirements. Line sources of light such as fluorescent lamps, luminous tubing, or custom sign tubing provide efficient illumination. Special diffusing screens should be used with spot sources, such as incandescent or mercury lamps, to prevent appearance of "hot spots" of brightness on sign faces.

Sign Brightness. Adequate sign brightness should be provided but it is important that it should not be "overdone." This depends primarily upon the desired sign brightness, use, and its environment (see Fig. 23-8).

Calculating Number of Lamps Needed. When white, yellow, or ivory backgrounds are used, a formula for estimating the spacing of the lighting elements is as follows:

$$\text{Spacing Between Lamps (Equal on Centers)} = \frac{9.4 \times \tau \times \phi/\text{ft}}{L}$$

where 9.4 = a constant for the combined interreflectance characteristics of the sign enclosure.

L = luminance required (from Fig. 23-8).

τ = transmittance of the media (from manufacturer's literature or measurement).

ϕ/ft = the lumens per foot of lamp to be used. This is obtained by dividing the manufacturer's initial lumen output for the given lamp by the number of feet of length.

This spacing is based on providing a clearance between the lighting elements and the sign face material equal to the center-to-center spacing figure just obtained. However, with both internally pigmented media and with lacquer coatings, it is possible to obtain satisfactory diffusion in many cases

Fig. 23-8. Recommended Luminous Background Sign Luminances

Range of Sign Luminance (footlamberts)	Potential Areas of Application
20 to 100	Lighted facades and fascia signs
75 to 150	Bright fascia signs as in shopping centers
125 to 200	"Low" brightness areas where signs are relatively isolated or have dark surrounds
200 to 300	Average commercial sign such as for gas station identification
300 to 400	High rise signs and signs in areas of high sign competition
400 to 500	For emergency traffic control conditions where communication is critical

where the clearance between the lighting elements and the sign face is less than the center-to-center spacing of the lighting elements. There are minimums, in all cases, where light element streaking will appear on the sign face if the clearance-to-spacing ratio is too low. Actual clearance between lamps and sign face media can be determined in a test mock-up where experience or published data are lacking.

Adaptation of Formulas for Point Light Sources. The formulas above may be used for point sources where the initial lumen output of the lamp is substituted for the ϕ/ft value. With all point sources of light, the clearance between the surface of the lamps and the sign face media should be not less than that derived from the following formula:

Minimum Clearance (in inches)

$$= \tfrac{1}{2}\sqrt{\text{Wattage of Lamp}}$$

This is to prevent excessive overheating of the sign face media by direct radiation from the lamp. Having determined the necessary spacing, the number of lamps can be readily calculated.

Obscuring Lamp Sockets. Where fluorescent lamps are to be used, the dimensions of the sign should be such that the lamp sockets are located just beyond the translucent face area. This will prevent shadows at the edges directly over the lamp sockets. Series arrangement of tubes in large signs using more than one tube per row requires an overlapping of the tubes of at least 3 inches to prevent shadows similar to those at the sockets.

All internal sign components such as structural framing, sheet metal backgrounds, and ballasts should be coated with at least two coats of high reflectance white paint to derive maximum lighting efficiency and prevent shadows on the sign faces.

Venting. Provision for venting and air circulation may be necessary depending upon the environmental temperature conditions of the sign. This is primarily to maintain the efficiency and prolong the life of the lighting elements and supporting equipment (ballasts, etc). In signs with lamps on very close centers, forced ventilation may be necessary to prevent overheating of the sign face media.

Readability. Readability of a luminous panel sign depends to a great extent on four factors:

Fig. 23-9. Maximum Spacing of Three Dimensional Letters

Observation Angle (degrees from plane of sign)	Minimum Spacing*
5	12 D
10	6 D
15	4 D
20	3 D

* D = depth of formed or fabricated letter.

1. The size and proportions of letters as well as letter design configuration.
2. Letter spacing.
3. Color and brightness contrast between letter and background.
4. Brightness of a sign face.

Size and Proportions of Letters. With dark letters and light background colors, the following formula may be used to determine the minimum letter height:

$$H = \frac{D}{500}$$

where H = letter height (feet), and
D = maximum distance of legibility (feet).

The proportions of a letter designed for maximum readability are—width = 60 per cent of height, and stroke = 15 per cent of the height in a sans serif block Gothic style.

Spacing of Letters. The above width and stroke proportions are effective in preventing blending of letter lines where maximum distance of legibility is required. Considerable license is possible with spacing of painted letters; however, for maximum distance legibility, the spacing between letters should be balanced between 15 per cent of letter height and visual equalization of "white masses such as between a "W" and an "A", etc.

Letters and illustrations, or insignia, of bold silhouette rather than fine detail are preferable for long distance legibility. Fine detail and stylized script letters find their greatest use where they will normally be observed from relatively close quarters, such as in downtown shopping areas and in shopping malls.

Three dimensional formed and fabricated letters should be spaced on the basis of their depth of forming and the minimum acuteness of angle of observation. See Fig. 23-9.

The ratios given in Fig. 23-9 are also useful in the design of letters to be formed or fabricated where their principal legibility will be a function of the acuteness of angle of observation. For example, based on a minimum observation angle of 10 degrees and a proposed depth of forming of 2 inches, the minimum opening in a letter like an "O" should be 6 x 2 = 12 inches. Hence, letter designs for acute observation angles will be "extended" and the average letter width will be as much as the letter height, or more, to meet the requirements of acute observation angles.

Luminance and Readability of Sign Face. The brightness of the sign face has a significant influence on the readability of the sign. A sign which is too bright can suffer loss of readability from the halo effect around the letters. Insufficient lighting will reduce distance of legibility. The recommendations for sign luminance in given districts, shown in Fig. 23-8, are within safe limits. In some cases, where high background brightness is required, elimination of this halo effect is achieved by applying a ½-inch wide stripe of black opaque paint around the outline of the letter. This applies particularly to those signs employing flat cut-out, formed, or fabricated letters attached to the light colored backgrounds. In signs using dark backgrounds with light letters, "debossing," or forming depressed areas, rather than the conventional raised letter areas, eliminates the causes of halation.

Luminous Building Fronts or Facades

The same basic data for design of luminous elements applies, in general, to luminous portions of building fronts. However, the illumination levels need not be designed for greater than 100 footlamberts of surface luminance. In an area of low level environmental lighting, 25 footlamberts of surface luminance will be found adequate.

Building Fascia or Belt Signs

Service access requirements in relatively long fascia or "belt" signs often call for use of a single row of lamps along the top or bottom edges of the enclosures. This low level light input system produces a surface luminance in the 20 to 100 footlambert range. A double row raises the surface luminance to the 75 to 175 footlambert range.

Matters of major design importance in systems for obtaining uniform light distribution or even lighting of the fascia surface, are:

1. The depth of the sign cabinet is the major factor in obtaining uniform light distribution.
2. The use of specially shaped sign enclosures with sloping, parabolic or elliptical contoured backs does not improve the light distribution over the straight back sign cabinet.
3. It is possible to improve the lighting uniformity with special reflectors at the light source. (A parabolic reflector troffer will permit shallower depth of sign cabinets for given heights.)
4. The luminance uniformity ratio of the sign face media, as shown in the following formula:

$$\frac{\text{Footlamberts Near Light Source}}{\text{Footlamberts Farthest Away}} = \text{Uniformity ratio}$$

(A ratio of 1 would be optimum. A ratio of 2 may be tolerated in some installations but should be considered the maximum allowable. Ratios of 1.3 to 1.5 will be satisfactory for a majority of installations.)

5. Very high output fluorescent lamps are as efficient in obtaining uniform light distribution as aperture lamps.

Examples of Fascia Signs.

Up to Four Feet in Height: Satisfactory luminance uniformity, depending upon the depth of the sign cabinet can be produced with a straight backed sign box, interior painted with high reflectance white paint, equipped with one or two lines of fluorescent lamps (1500 mA) shielded with a baffle plate and light output controlled with a thin gauge sheet metal egg crate grid. See Fig. 23–10*a*. Surface luminance values versus uniformity ratios are tabulated in Fig. 23–11.

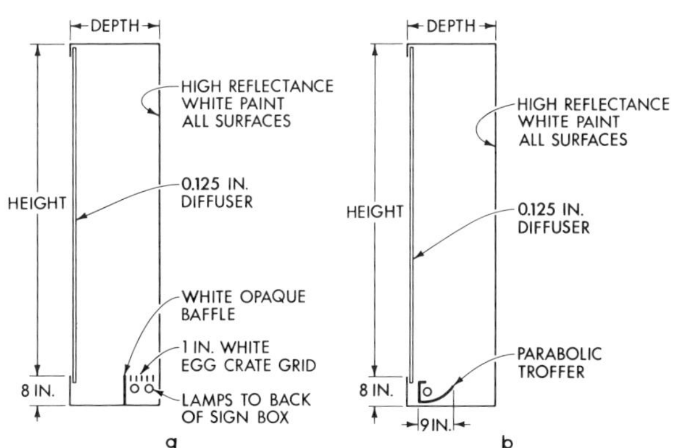

Fig. 23–10. Typical construction of fascia signs (a) 4 or 6 feet in height, (b) 8 feet in height.

Up to Six Feet in Height: A straight sign cabinet with either baffles and egg crate grid (see Fig. 23–10*a*), or with a parabolic reflector lighting troffer (see Fig. 23–10*b*), may be used depending upon the required luminance and available depth of the cabinet.

In Fig. 23–11, two 1500 mA fluorescent lamps produce fair luminances for a fascia sign, but at the expense of a uniformity ratio close to the maximum. By using the parabolic troffer system, a more satisfactory uniformity ratio can be achieved as long as the depth of the sign cabinet is from 18 to 24 inches deep. The lower luminances with the troffer are due to the use of a single tube.

Up to Eight Feet in Height: A straight sign cabinet with a parabolic troffer (see Fig. 23–10*b*) is required to obtain reasonably uniform lighting distribution. The data supplied in Fig. 23–11 will be helpful in both the design and lighting of "belt" or fascia signs.

Luminous Fascia Colors Other than White. The information above is confined to the results using .125 inch thick integrally pigmented sign face media having a transmittance of 40 per cent. Other whites and other colors with lower transmittance values will produce surface luminance values below those shown. When using colors other than white, it is necessary to apply a spray coating of white paint to the inside surface in order to obtain comparable light distribution qualities.

FLOODLIGHTED SIGNS

Lighting Poster Panels, Bulletin Boards, and Vertical Surface Signs

The most important factors which contribute to the conspicuity of an illuminated sign are its area

Fig. 23-11. Luminance of Most Distant Area in Footlamberts Versus Uniformity Ratios[a]

Type of Lights	Depth of Sign Box (inches)					
	8	9	12	18	20	24
a. For Facia Signs up to 4 Feet in Height						
One row of lamps	65/3.0	—	70/1.9	65/1.3	—	—
Two rows of lamps	100/3.5	—	130/2.1	130/1.4	—	—
b. For Facia Signs Up to 6 Feet in Height						
Two 1500 mA lamps[b]	—	—	60/4.0	80/1.9	90/1.8	95/1.6
Parabolic troffer[c]	—	45/2.8	48/2.0	50/1.4	55/1.3	60/1.2
c. For Facia Signs Up to 8 Feet in Height						
Two 1500 mA lamps[b]	—	50/9.0	—	75/3.2	—	50/1.8
Parabolic troffer[c]	—	35/2.9	—	40/1.1	—	20/1.5

[a] All data based on 40 per cent transmittance for sign face media.
[b] See Fig. 23-10*a*.
[c] See Fig. 23-10*b*.

and its brightness. However, there exist several relatively complex factors affecting legibility of signs, many of which are psychological as well as physical. See page 23-1 under Sign Characteristics.

Standardized sign dimensions are shown in Fig. 23-12.

General Guides for Floodlighting Signs. The following is a list of recommendations to be considered in designing the floodlighting of signs.

1. The brightness of the sign panel should be sufficient to stand out in contrast with its surroundings. Fig. 23-13 lists recommended levels of illumination.
2. The luminance should be sufficiently uniform to provide equal legibility over the message area. A maximum to minimum ratio of 4 to 1 is desirable.
3. The lighting should cause neither direct nor objectionable reflected glare at the normal viewing positions.
4. The lighting equipment should not obstruct the reading of the sign from normal viewing position, nor produce daytime shadows objectionably reducing the legibility of the sign.
5. The lighting equipment should require a minimum of maintenance and provide low cost annual operation.
6. The maintenance of the system should be adequate to achieve the designed illumination level.

Fig. 23-12. Standard Structure Dimensions of the Outdoor Advertising Association of America, Inc.

Display	Outside Dimensions (Uniformly Lighted Area)
Association Loewy Poster Panel	12′3″ x 24′6″
Association Standard Poster Panel	12′ x 25′
Standard Streamliner	15′ x 46′
Deluxe Urban Bulletin	13′4″ x 46′10″
Standard Highway Bulletin	13′ x 41′8″
Junior Highway Bulletin	12′4″ x 24′8″

Fig. 23-13. Recommended Minimum Illumination for Poster Panels, Bulletin Boards, and Other Advertising Signs

Average Reflectance of Advertising Copy	Recommended Illumination Levels (footcandles)	
	Bright Surrounds	Dark Surrounds
Low	100	50
High	50	20

Location of Lighting Equipment. Some of the factors to be considered when determining whether luminaires should be mounted across the top or bottom of a floodlighted sign are:

A. *For top mounted units*
   1. Cover may collect less dirt, snow, etc.
   2. Luminaires will not hide message.
   3. Reflected glare more apparent.
   4. Luminaires may produce daytime shadows.
   5. Sign itself usually shields direct view of lamps from opposing traffic.
   6. Luminaires may be more difficult to service.

B. *For bottom mounted units*
   1. Cover may collect more dirt, snow, etc.
   2. Luminaires may hide message from some viewing angles.

Fig. 23-14. Typical floodlighted sign installations and illumination results when using equipment with (a) metal halide lamps, (b) fluorescent-mercury lamps, and (c) tungsten-halogen lamps.

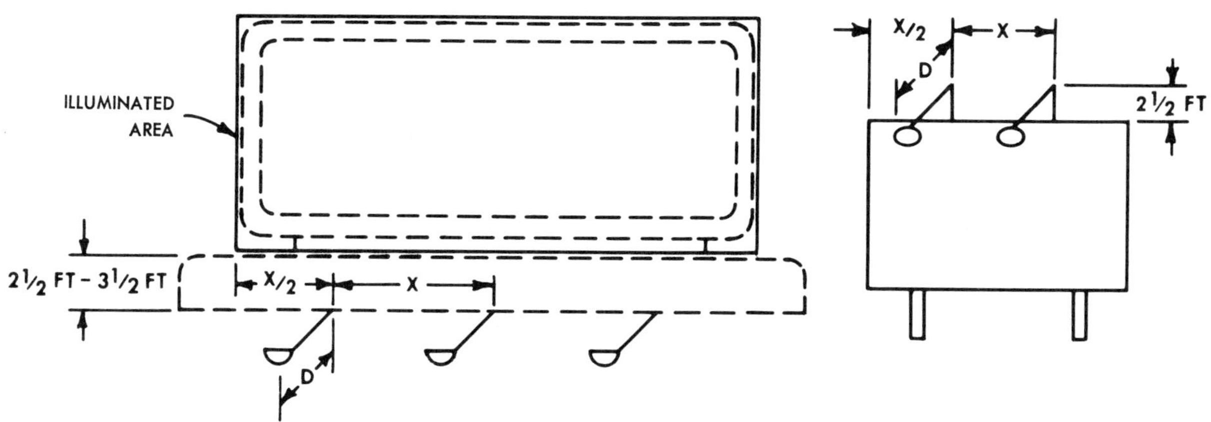

d = Length of 1¼-inch conduit (measured from copy face to luminaire hub).

X = Space between luminaires $\left(\frac{\text{Length of illuminated area}}{\text{Number of luminaires}}\right)$

X (max.) = 2½d

(a) For Metal Halide Luminaires

Display	Number of Luminaires	(d) Conduit Length	(x) Spacing	Maintained Average Footcandles*
Association Loewy Poster Panel	3	4′0″	8′2″	55
Association Standard Poster Panel	3	4′0″	8′4″	55
Standard Streamliner	6	4′9″	7′8″	50
Deluxe Urban Bulletin	5	4′9″	9′4″	50
Standard Highway Bulletin	4	4′2″	10′5″	50
Junior Highway Bulletin	3	4′0″	8′2″	55

* Based on 400-watt metal halide lamp, Light Loss factor = .71.

(b) For Fluorescent-Mercury Luminaires

Display	Number of Luminaires	(d) Conduit Length	(x) Spacing	Maintained Average Footcandles*
Association Loewy Poster Panel	4	4′0″	6′1½″	55
Association Standard Poster Panel	4	4′0″	6′3″	55
Standard Streamliner	8	4′9″	5′9″	50
Deluxe Urban Bulletin	7	4′9″	6′8½″	50
Standard Highway Bulletin	6	4′2″	6′11″	50
Junior Highway Bulletin	4	4′0″	6′2″	55

* Based on 400-watt fluorescent-mercury lamp, Light Loss factor = .77. If same quantities of 250-watt lamps are used, multiply footcandles by .52. If same quantities of 175-watt lamps are used, multiply footcandles by .37.

(c) For Tungsten-Halogen Luminaires

Display	1500-Watt			500-Watt		
	Number of Luminaires	(x) spacing	Maintained Average Footcandles*	Number of Luminaires	(x) Spacing	Maintained Average Footcandles*
Association Loewy Poster Panel	3	8′2″	115	4	6′1½″	50
Association Standard Poster Panel	3	8′4″	115	4	6′3″	50
Standard Streamliner	6	7′8″	100	8	5′9″	55
Deluxe Urban Bulletin	5	9′4″	95	8	5′10″	50
Standard Highway Bulletin	5	8′4″	105	7	5′11″	50
Junior Highway Bulletin	3	8′2″	115	4	6′2″	50

* Light loss factor = .89. Distance from sign to luminaire = 8-feet, aiming point is at a distance down from top of sign equal to ⅓ height of sign.

3. Reflected glare is minimized.
4. No daytime shadows.
5. Shielding may be necessary to hide direct view of lamps from opposing traffic.
6. Luminaires may be easier to service—as from a catwalk.

Light Sources for Floodlighted Signs

There is no single type of source that can be described as "best" for sign floodlighting. Most lamps, regardless of type, can be used with different reflector and lens combinations to realize various beam patterns which may be required. Therefore, choices of light source for a given sign are generally made for reasons of initial cost, operating cost (including maintenance cost), end result desired, color, or novelty. Section 8 contains a detailed discussion of available light sources, but there are some general guidelines which should be followed, based on the inherent characteristics of the various lamps. Below are characteristics to be considered in selecting the light source. Implied comparisons are relative to the other sources and are based on equal illumination on the sign.

Incandescent. Advantages include good color rendering, small size permits accurate beam control, and very good operation in cold weather. Disadvantages are relatively low output (lumens per watt), short lamp life (usually 1000 hours), and higher operating costs.

Tungsten-Halogen. Advantages include good color rendering, high lumen maintenance (output remains practically constant throughout life), and very good operation in cold weather. Disadvantages are relatively low output (lumens per watt), medium lamp life (usually 2000 hours), and higher operating costs.

Mercury. Advantages include long lamp life (as much as 24,000 hours), high output (lumens per watt), and low operating cost. Disadvantages are high initial cost, fair beam control (due to larger source), and color rendering capability below that of incandescent or tungsten halogen—(although usually acceptable).

Metal Halide. Advantages are good lamp life (usually 10,000 hours), exceptionally high output (lumens per watt), good color rendering capability, and low operating cost. Disadvantage is high initial cost.

Fluorescent. Advantages are moderately long life (at least 7500 hours), high output (lumens per watt), and low operating cost. Disadvantages are high initial cost, very little beam control due to large source, and variable output due to changing temperatures.

Lighting Systems for Floodlighted Signs

Concurrent with consideration of a particular light source, there should be an evaluation of the other elements such as lamp housing, mounting arrangements, and auxiliary equipment if required. Primarily, due to improved lamp performance, the metal halide and mercury lamp systems have made significant inroads in this area. The majority of existing signs, however, are still lighted by fluorescent equipment. Incandescent lamps, with their associated housings are, in general, less expensive to install initially, since they require no auxiliaries (such as ballasts), but their use has dwindled considerably, in favor of the more efficient sources.

Application Data. Regardless of which system is used, the most economical floodlighting system is one which utilizes the fewest number of floodlights containing the highest wattage lamps. It is easier to install, control, and maintain. It also uses less power for the same illumination level than a system using more but smaller units. However, uniformity of illumination and appearance may require the selection of a system utilizing a larger number of smaller units. It may be necessary to draw a careful balance between the two extremes. Regardless of which type of source is selected, the beam patterns should be overlapped so that any given area receives light from at least two units. This requirement is usually satisfied if the acceptable uniformity ratio is achieved.

Metal Halide Systems. The 400-watt metal halide lamp is the size most frequently used for sign lighting. Fig. 23–14 includes information on positioning a typical luminaire for both top and bottom mounting and shows the maintained illumination obtained with specific luminaire location data. For a view of a typical sign, see Fig. 23–15.

Mercury Systems. Fluorescent-mercury lamps in the 175-, 250-, and 400-watt sizes have many appli-

Fig. 23–15. View of lighted sign using metal halide lamps.

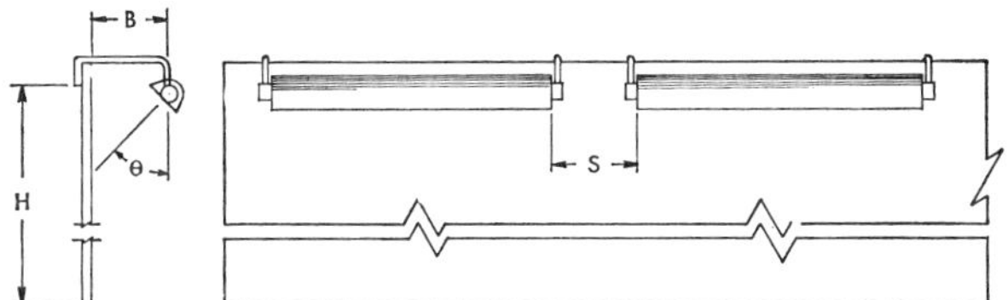

Fig. 23-16. General rules for applying fluorescent floodlights.*
1. *S* should not exceed *B*.
2. Overhang *B* should not be less than approximately 0.4*H*.
3. If *H* exceeds 15 feet, floodlights are recommended across both bottom and top of area in order to assure an acceptable uniformity ratio.
4. *B* should be as large as practical.

* These "Rules of Thumb" are based upon the use of a white enameled reflector which produces a symmetrical distribution in a plane perpendicular to the lamp. *B* and Θ may vary with specific reflector types.

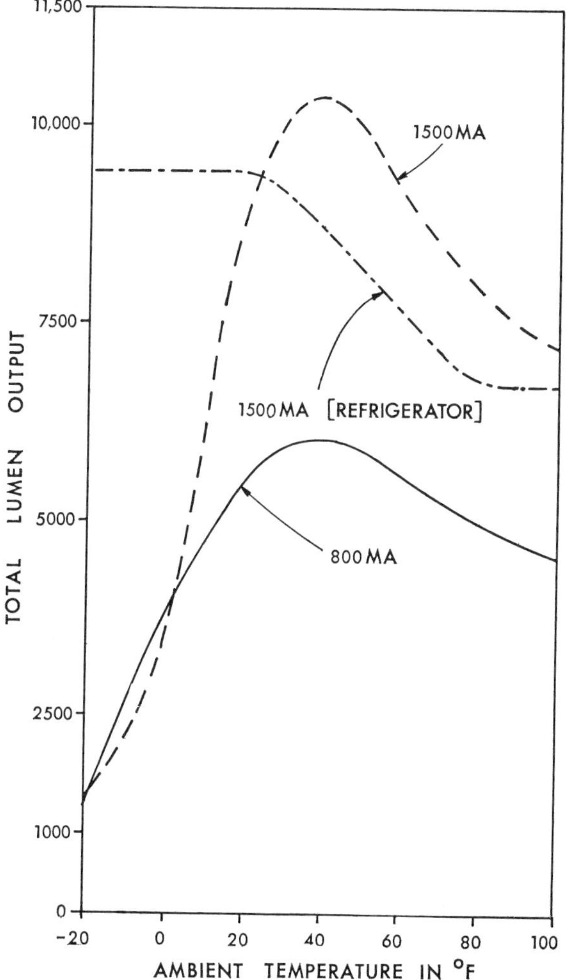

Fig. 23-17. Lumen output versus temperature characteristics for typical fluorescent floodlights.

Fig. 23-18. Illumination Obtained on Vertical Surfaces From Enclosed Fluorescent Floodlights[a] (See Drawing in Fig. 23-16)

Height (feet) H	Number of 8-foot Lamps[b]	Average Level of Illumination Maintained in Service (footcandles)[c] at 50° F (96-inch Fluorescent Lamps Approximate MA—Rating)	
		800	1500
Standard Poster Panel	2-units	13	21
	3- "	19	32
Deluxe Urban Bulletin	3-units	11	18
	4- "	14	23
	5- "	17	29
Standard Highway Bulletin	3-units	12	19
	4- "	15	25
	5- "	19	33
Loewy Poster Panel	2-units	13	23
	3- "	20	33
Junior Highway Bulletin	2-units	13	23
	3- "	20	34
Standard Streamliner	3-units	11	18
	4- "	13	23
	5- "	17	29

[a] Calculations of illumination based upon $B = 5$ feet.
[b] See Fig. 23-16 for general overhangs *B* and aiming angles θ.
[c] Assuming a light loss factor of 0.65. Uniformity ratio max/min should be less than 4/1; all values are expressed to the nearest footcandle.

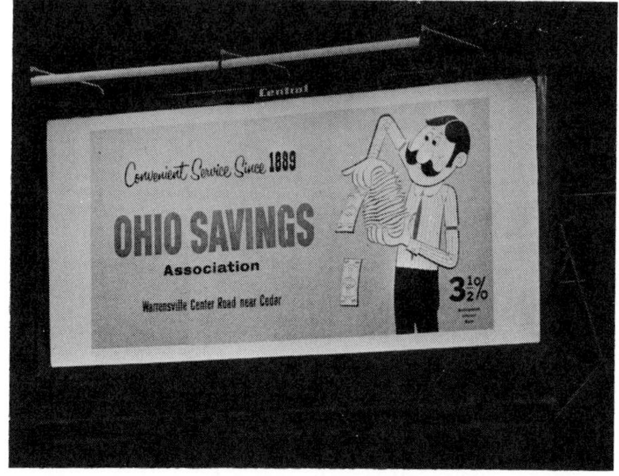

Fig. 23-19. Typical fluorescent floodlights mounted in tandem to illuminate advertising signs.

cations in sign lighting where their color rendering capability is acceptable. The life of these lamps is unusually long, and they are well-suited to locations where maintenance is infrequent, or where the signs are relatively inaccesible. Luminaire location is usually determined in the same manner as for metal-halide systems. Maintained illumination values obtained on typical signs are shown in Fig. 23-14.

Tungsten-Halogen Systems. Both 500- and 1500-watt tungsten-halogen lamps are used for sign lighting. See Fig. 23-14 for typical location details for bottom or top mounted units and the maintained illumination obtained on typical signs.

Fluorescent Systems. General rules for applying fluorescent units are shown in Fig. 23-16. Aiming angle θ is usually 45 degrees. One important consideration is that fluorescent lamps are temperature and wind velocity sensitive, and most photometric data are obtained at an ambient temperature of 77°F. See Fig. 23-17 for typical lumen output data for fluorescent floodlights in varying ambient temperatures. Illumination values obtained on typical signs are shown in Fig. 23-18. For higher values of illumination, twin units are used or units are added at the bottom of the sign and aimed upward. See Fig. 23-19 for typical fluorescent floodlights mounted in tandem.

REFERENCES

1. Hart, A. L.: "Some Factors that Influence the Design of Daytime Effective Exposed Lamp Signs," *Illum. Eng.*, Vol. LI, p. 677, October, 1956.
2. O'Day, R.: "Fluorescent Lighting," *Outdoor Advertising Association News*, February, 1956.
3. Peek, S. C. and Keenan, J. P.: "Outdoor Applications of New Reflector Contour Designs for Higher Output Fluorescent Lamps," *Illum. Eng.*, Vol. LIV, p. 77, February, 1959.

SECTION

24

THEATRE, TELEVISION, AND PHOTOGRAPHIC LIGHTING

Luminaires, Lamps, and Control Systems 24-1 *Lighting for Theatres 24-6*
Lighting for Television 24-12 *Photographic Lighting 24-19*
Picture Projection Lighting 24-25 *Photochemical Reproduction Processes 24-35*

Lighting for theatre and television have many similarities. The design of the lighting system, as well as the choice of luminaires and light source, is based on production plans and the purpose for which the facility is being constructed. Some installations are planned as part of the training facility required by a high school, college, or university—others are planned as part of a commercial operation. In general, the lighting system will be governed by the allocated budget of the facility, but planning considerations should allow for future expansion or modification without creating expensive obsolescence.

Theatre design requires information concerning types of programs (opera, ballet, drama, variety) which will be produced by resident groups or touring companies. Television design requires information concerning types of productions (variety shows, dramas, news, panel shows) which will be produced for local broadcasting, closed circuit, syndication, or network release. However, the required flexibility in illumination level and in illumination distribution over a wide range necessitates the use of many high-intensity sources in television. The color requirements depend on the type of television camera tube used.

The primary function of lighting in photography is to produce photochemical change in a photosensitive material such that subsequent processing will result in a satisfactory permanent image. It is desirable in most cases that the change be effected with a minimum expenditure of energy, and often in the shortest possible time. Since photosensitive materials vary widely in their spectral and their absolute sensitivity, these factors influence the photographic applications of radiant energy sources.

NOTE: References are listed at the end of each section.

Infrared, ultraviolet, and x-ray radiation, as well as light, can be used to create a latent image.

Light sources, optical systems, and screens used for picture projection are planned in combination for a particular range of viewing distances and viewing angles and for a given range of surrounding luminances (photometric brightnesses). The lighting design objective is to provide a capacity for creating realistic contrasts between highlights and shadows on the screen at a satisfactory average luminance level.

LUMINAIRES, LAMPS, AND CONTROL SYSTEMS

Luminaires for Theatre, Television, and Photographic Lighting

In theatre, television, and photographic lighting different types of luminaires are used to produce a quality of light which falls generally within three basic categories:

1. *Soft light*—diffuse illumination with indefinite margins. It produces poorly defined shadows.
2. *Key (intermediate) light*—illumination with defined margins. It produces defined, but soft edged shadows.
3. *Hard light*—illumination which produces sharply defined geometrically precise shadows.

The basic types of luminaires used in theatre, television, and film production are described as follows (see Fig. 24–1):

Fresnel Spotlight. The Fresnel spotlight is a luminaire which embodies a lamp, a Fresnel lens, and is generally equipped with a spherical reflector

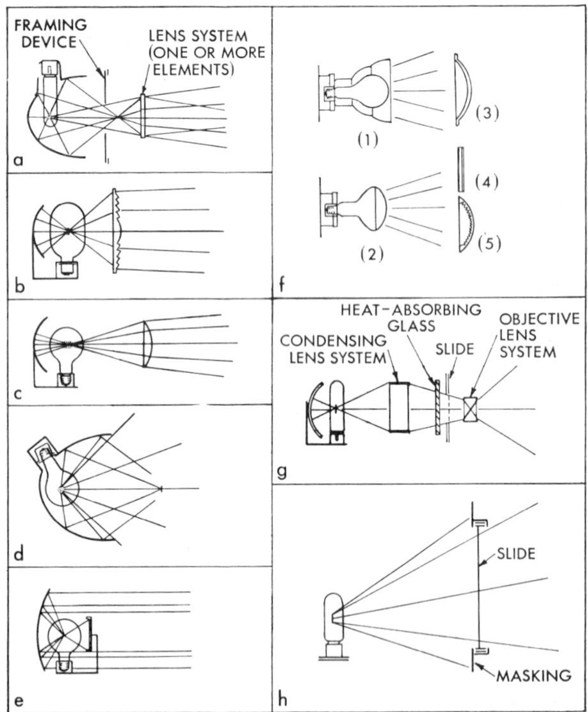

Fig. 24-1.. Optical characteristics of stage lighting equipment. a. Ellipsoidal-reflector spotlight. b. Fresnel-lens spotlight. c. Plano-convex-lens spotlight. d. Scoop-type floodlight. e. Parabolic-reflector floodlight. f. Striplight: (1) reflector with general service lamp; (2) reflectorized lamp; (3) glass roundel, plain or colored. (4) sheet color medium; (5) spread lens roundel, plain or colored. g. Lens-type scenic slide projector. h. Non-lens type scenic slide projector (Linnebach type).

behind the lamp. The field and beam angles* can be varied by changing the distance between the lamp and the lens. This action is termed focusing. The distance between the lamp and reflector is set by optical design and does not change during focus.

The quality of illumination produced by a Fresnel spotlight tends to be intermediate or hard, and the beam is soft edged. This illumination varies considerably, depending on the optics of the luminaire. Typical luminaires have a beam angle of 10 to 40 degrees, depending on the relative position of the lamp and lens.

Fresnel spotlights are equipped generally with incandescent lamps, including tungsten-halogen lamps, with C-13 or C-13D filaments. They may be provided with barn-doors† to control the shape of the light beam, funnels‡ to control the stray light, and color frames to hold color or diffusion material.

Fresnel spotlights are manufactured in a number of configurations related to wattage, size of lens, and housing. Lens sizes vary from 3 to 24 inches and wattages range from 75 to 10,000 watts. Some types of Fresnel spotlights are now equipped for dual filament lamps and switches which permit a unit to be, for example, either 2½ or 5 kilowatts. Mechanical features of Fresnel spotlights (pan, tilt, focus, etc.) may be manually set on the luminaire or pole-operated.

Scoop. The scoop is a floodlight consisting primarily of a lampholder, lamp, and ellipsoidal matte-finished reflector. Lamp and reflector may have a fixed or variable relationship. Scoops are equipped with runners for frames containing either diffuser or color media. The scoop produces illumination having a field angle of less than 100 degrees and a cut-off angle of less than 180 degrees. The quality of illumination is considered soft, and shadows are dependant primarily on the texture of the reflector. Scoops are manufactured in various sizes from approximately 12 to 18 inches in diameter, and may be equipped with incandescent lamps, including tungsten-halogen, in wattages from 300 to 2000 watts. Lamps used in scoops are usually frosted. If the lamp bulb is relatively small in diameter, a shield is often used to eliminate the direct light which modifies the soft nature of the illumination.

Non-Lens Luminaire. The non-lens luminaire embodies a lamp, reflector, and frequently a focus mechanism which changes the relationship between the lamp and reflector, thus varying the field and beam angles. The quality of illumination produced by a non-lens luminaire tends to be intermediate or hard depending on lamp type and reflector. The beam may be partly controlled by the focus mechanism, external accessary holders equipped with barn-doors, or funnels. In addition the luminaires may be equipped with scrims, heat filters, daylight correction filters, and color frames for diffusing material. There are many different types of non-lens luminaires and some can be considered as floodlights or as spotlights. The majority use tungsten-halogen lamps in the range of 650 to 2000 watts.

PAR-Luminaire. PAR-luminaires embody a PAR lamp, lampholder, and housing; the performance of the luminaire depends upon the type of PAR lamp selected.

As the beam pattern of most PAR lamps tends to be oval, the luminaire is designed so that the lamp may be rotated to cover the desired area. The beam of a PAR lamp can be shaped and glare reduced by barn-doors, and the intensity can be increased by an *intensifier skirt*—an external reflector that continues the parabolic contour of the internal lamp reflector.

PAR luminaires using 650- or 1000-watt lamps are designed to accommodate either single lamps or groups of lamps in clusters of 3, 6, 9, 12, etc. Luminaires using PAR-38 or PAR-56 lamps may be arranged in lineal strips defined as *borderlights*.

* Those points of the candlepower distribution curve where the candlepower is 10 per cent of the maximum candlepower defines the field of the lighting unit. The included angle is the field angle.

Those points of the candlepower distribution curve where the candlepower is 50 per cent of maximum candlepower defines the beam of the lighting unit. The included angle is the beam angle.

† A set of swinging flaps, usually two or four (two-way or four-way) which may be attached to the front of a luminaire in order to control the shape and spread of the light beam.

‡ A metal tube that can be mounted on the front of a spotlight to control stray light. Also known as a *snoot, top hat*, or *high hat*.

Striplight. A striplight is a compartmentalized luminaire. Each compartment contains a reflector lamp, or a lamp and reflector, and a color frame. Compartments are arranged in line, and wired on 2, 3, or 4 alternate circuits, each circuit being colored with a suitable color media.

Striplights provide an overall wash of illumination over an area of a stage and provide cyclorama lighting (borderlights). They are located on the floor approximately 4 feet from the cyclorama to provide lighting for the lower portion of the cyclorama. Also striplights may be suspended and used to light the top portion of the cyclorama. Striplights may also be located at the front of a stage to provide over-all low level of illumination (termed *footlight*).

Ellipsoidal Spotlight. The ellipsoidal spotlight, or pattern spotlight, consists of a lamp and ellipsoidal reflector, mounted in a fixed relationship. Light is directed into an aperture where the beam may be shaped by an internal shutter or iris. This beam then passes through a single or compound lens system which focuses the beam as required.

The ellipsoidal spotlight produces a sharply defined beam of light with controlled and variable margins. This beam may be shaped and colored by a pattern inserted in the aperture of the beam and focussed to produce a hard or sharp image by adjustment of the lens system. Some ellipsoidal spotlights are equipped with dual sets of shutters, one set has hard edges, the other feathered edges. Correctly adjusted, these dual shutters can produce a light beam with some hard and some soft edges. The soft edges can better blend with overlapping spotlight patterns.

Follow Spot. A follow spot is generally a special form of ellipsoidal spotlight mounted on a stand with iris, shutter, and other mechanical controls brought out to external handles. Some follow spots utilize PAR lamps and lenses in place of the ellipsoidal reflector. Follow spots are designed to be hand operated—that is, rotated and tilted as required to follow the movement of performers. They produce a hard, even beam of illumination, usually circular in shape. The beam can also be shaped by gate shutters and iris, softened by an internal spread lens, and focused by the objective lens.

A follow spot can be equipped with a "color boom", containing several color filters, which may be placed in front of the objective lens as required, or an external iris placed in front of the lens to reduce the intensity of the beam.

Light sources for follow spots may be carbon, xenon, or other concentrated arc lamps, and incandescent filament lamps.

Parabolic Spotlight. The parabolic spotlight consists of a lamp and parabolic specular reflector. Some luminaires have a reflector in front of the lamp to redirect light into the main reflector. Other luminaires are equipped with spill-rings to minimize spill light and glare. In most types of parabolic spotlights the lamp and reflector are adjustable to produce a wide or a narrow beam of light—the closer the lamp to the parabolic reflector the wider the beam. This luminaire produces a hard beam of illumination which cannot be easily controlled, except in part by spill-rings. (A parabolic spotlight is also known as a *sun spot* or *beam projector*.)

Long Range Scoop. The long range scoop consists of a lamp with combination matte and specular reflector. Relationship between lamp and reflector is adjustable to permit some beam adjustment. This luminaire produces a relatively hard but soft edged beam of light which may be directed into areas as required. The beam angle is smaller and the effective range of this luminaire is considerably greater than that of the regular scoop.

Soft Light. A soft light luminaire is a well diffused light source which produces soft or poorly defined shadows when an object is placed between the luminaire and a light background. The luminaire consists of one or more lamps mounted in sockets, with close-fitting reflectors that direct the illumination to a large matte-surfaced reflector which, in turn, directs illumination to the subject. Thus, the final source of the illumination is a large diffused surface, and the resultant illumination is very soft.

Arc Light. An arc light luminaire uses a carbon arc as a source of illumination. Such luminaires are manufactured in two basic forms: (1) equipped with a Fresnel lens for general illumination, and (2) equipped with a reflector and optical system for follow spot operation. Carbons available for these luminaires are of various types producing corresponding intensities and color temperatures (see Figs. 24–2 and 24–3). Carbons and ballast equipment are available for either dc or ac operation.

Lamps for Theatre, Television, and Photographic Lighting

Data for lamps most frequently used in lighting for theatre stages, television and motion picture production, and professional still photography are found in Section 8, Figs. 8–90, 8–91, 8–106, and 8–108. In many cases lamps may appear to be mechanically interchangeable with each other, *i.e.*, they have the same base and light source location. However, caution should be exercised. For example, in the newer compact luminaires, designed specifically for tungsten-halogen, the bulbs of some incandescent lamps may be too large to fit. Furthermore, there may be differences in filament configuration that could affect the luminaire optical performance.

Another possible limitation on lamp interchangeability is that some luminaires may not provide adequate heat dissipation for higher-wattage lamps. Lighting equipment manufacturers should be consulted for maximum allowable wattage.

Fig. 24-2. Arc Lamps Used in Motion Picture Set Lighting

Unit	Beam Divergence (degrees)		Amperes	Arc Volts	Positive Carbons	Negative Carbons
	Minimum	Maximum				
Type 40 Duarc	90	90	40	36	8 mm x 12 inch CC MP studio	7 mm x 9 inch CC MP studio
Type 90	8	44	120	58	13.6 mm x 12 inch MP studio	7/16 x 8½ inch CC MP studio
Type 170	8	48	150	68	16 mm x 20 inch HI MP studio	½ x 8½ inch CC MP studio
Type 450 Brute	12	48	225	73	16 mm x 22 inch Super HI MP studio (white flame) or 16 mm x 22 inch Super HI YF MP studio (yellow flame)	17/32 x 9 inch H.D. Orotip
Titan	12	48	350 white flame 300 yellow flame	78 white flame 68 yellow flame	16 mm x 25 inch ULTREX HI WF studio (white flame) or 16 mm x 25 inch ULTREX HI YF 300 special studio (yellow flame)	11/16 x 9 inch CC MP studio

ANSI Codes. The American National Standards Institute (ANSI) assigns three-letter designations for incandescent lamps (including tungsten-halogen lamps) used in photographic, theatre, and television applications. The letters are arbitrarily chosen and do not describe the lamps. When an ANSI code is assigned to a lamp type at the request of a lamp manufacturer, it may be used by all manufacturers as a commercial ordering code. The assignment of an ANSI code defines some basic parameters of the lamp such as wattage, type of base, size of bulb, light center length, color temperature, etc., so that physical interchangeability of lamps is assured. However, if lighting equipment performance is particularly sensitive to filament size, coil spacing, etc., the ANSI code does not guarantee equal performance for lamps of different manufacture. Furthermore, lamp manufacturers sometimes make improvements in lamp performance characteristics (life, lumens, etc.) beyond the values that are stated on the original ANSI code.

Manufacturer's Ordering Codes. The manufacturer's ordering code is another type of lamp designation. Often two or more manufacturers will agree on the code designation for a lamp type. Manufacturer's codes are usually descriptive of some of the lamp characteristics, and often include lamp wattage, bulb size, and shape. Some manufacturers of lighting equipment supply lamps marked with their own private codes to identify the proper lamps for their equipment.

Low-Noise Construction. All tungsten-halogen lamps for theater, television, and film application, as well as many other types of incandescent lamps, have special *low-noise* construction to minimize audible noise generation when operated on ac circuits. The lamp manufacturer should be consulted for information on which incandescent lamps have low-noise construction. Lamps, sockets, wiring, etc., tend to generate more audible noise when used with dimmers that distort the normal ac sine wave than with ordinary autotransformer or resistance dimmers (See page 16–9). Generally noise is not generated on dc circuits.

Caution Notices. Caution notices, are generally included with most lamps for stage and studio service. All tungsten-halogen lamps operate with internal pressure above that of the atmosphere; therefore, protection from lamp abrasion and avoidance of over-voltage operation is advised. The use of screening techniques is advised where appropriate to protect people and surroundings in the remote possibility that a lamp shatters.

Fig. 24-3. Beam Characteristics of Typical Motion Picture Studio Arc Lighting Units

Unit	Lamp	Beam Width (degrees)	Lumens	Approximate Peak Candlepower (candelas)
Type 40—Duarc	115V dc arc (40 amp, two 36V arcs in series)	150	76,500	32,000
Type 90—High-intensity arc spot	115V dc arc (120 amp, 58 arc V)	10	18,700	2,100,000
		18	26,000	550,000
		44	62,500	150,000
Type 170—High-intensity arc spot	115V dc arc (150 amp, 68 arc V)	10	47,000	5,700,000
		18	75,000	2,300,000
		48	130,000	300,000
Type 450—Brute high-intensity arc spot	115V dc arc (225 amp, 73 arc V)	12	117,000	10,000,000
		16	159,000	7,300,000
		20	172,000	5,000,000
		30	228,000	2,500,000
		48	260,000	1,000,000

Control Systems for Theater and Television Lighting

The lighting system designer as well as the rigging consultant should be involved in the design of a lighting control system at the earliest development. Consideration should be given to maintenance routines and availability of knowledgeable maintenance personnel in the planning of all lighting control systems.

The simplest lighting control systems consist of a few manually operated autotransformer dimmers with luminaires plugged into the dimmer load sockets. Where budget and simplicity requirements are paramount, the autotransformer dimmer is still useful in small capacity systems.

The more complex systems may have (a) a dimmer for each outlet, (b) dimmers permanently connected to some loads and other dimmers connected through a power programming system to luminaires, and (c) all dimmers connected to a power programming system and thus to luminaires. The control consoles are generally multi-preset manually operated systems. Electronic control systems employing magnetic data storage are being introduced in the larger theatre installations to provide flexibility, large cue storage, and repeat accuracy necessary for the more complex productions.

Dimmers have been constructed using autotransformers (manual and motor driven), thyratrons, saturable reactors, magnetic amplifiers, and silicon controlled rectifiers. The more recent installations have been designed around semi-conductor dimmers which use the silicon controlled rectifier and provide the stability, ease of control, flexibility, and operating economy which is now required.

Control Equipment for Stage Lighting. Some dimming facilities must be available, even for the smallest stage. The dimming facilities must at least be capable of reducing the luminance of all the stage lighting that may be in use at any one time. Thus if the total lighting load is 36 kW, the dimmer controls should collectively have a capacity of at least 36 kW (for example six 6 kW dimmers) and should be provided with means for connecting them to the lighting so that they control all the lighting circuits in use at any one time. However, it is desirable to have dimmer facilities that also will enable any required lighting instrument or instruments to be independently dimmed. Various types of cross connect systems, which enable lighting circuits to be connected to individual dimmers, are available including (a) cord and jack, (b) slider patch, (c) rotary selector, and (d) push button selector systems.

There should be a minimum number of six dimmers, even for the smallest stage. While these dimmers may provide the total capacity required by the stage lighting load, a more workable arrangement would be to have a minimum of twelve dimmers, so that some degree of flexibility is achieved. Dimmer systems of this size are suitable for use with from two to four times the number of load circuits. When the dimmer system includes 24 or more dimmers, the number of load circuits per dimmer may be increased when necessary from four to six per dimmer. Thus, a dimmer system of 50 dimmers can be employed in conjunction with a cross connect system having as many as 200 to 300 lighting circuits.

Dimmer systems may be grouped into three broad categories: (a) manual direct control, (b) remote control, (c) remote control with preset facilities.

Manual Direct Control usually employs autotransformer dimmers, except when a dc power supply is involved, in which case, resistance-type dimmers are necessary.

Remote Control Systems for stage lighting employ electronic-type dimmers, as for example, silicon controlled rectifier, magnetic amplifier, or thyratron tube-type dimmers.

Remote Control with Preset Facilities also employ electronic type dimmers as above. They allow the operator to work from a position where a better view of the stage is available, compared with the position where manual direct type dimmers would, of necessity, be placed.

Presetting facilities range from two-scene preset to fully automatic, so-called infinite preset systems. Some degree of presetting facility is desirable for any remote control system and where budget problems exist there should at least be means for presetting one scene ahead of the one in use, *i.e.*, two-scene preset.

Control Equipment for Television Production. Switchboards are equipped with various types of dimmers, the most common of which are the autotransformer, magnetic amplifier, and silicon controlled rectifier (solid state).

The latter two types are generally used in conjunction with preset boards. Preset boards are designed with from two to ten scenes as desired. There are two methods commonly used to cross connect loads or outlets to dimmer boards—the Patch system and the Selector Switch system.

Patch System. This system consists of a counterweighted single conductor cord with a male plug on one end. There is a plug and cord for each outlet in the studio. These in turn plug into a jack field. The jack field consists of a panel of female receptacles in groups of five or six and each group is connected to a dimmer or non-dim circuit. This enables the operator to connect any light or group of lights to any dimmer or non-dim circuit.

Selector Switch System. This system consists of a switch with the same number of positions as there are dimmers. In some cases positions on the selector switch are assigned to non-dim circuits, enabling the operator to group his lights either in dim or non-dim groups.

Memory Devices. In very large studios both presetting and cross connecting may be controlled by some means of automatic memory device, such as punched cards and magnetic tape. Lighting consoles are now available which are controlled by computer. This innovation has increased the storage capacity, provided the operator with real time status information of every control channel, and affected great savings in setup and rehearsal time.

LIGHTING FOR THEATRES

The two basic categories of theatres are the live-production type and the film or motion picture theatre. The former can be further classified as legitimate, community, and school theatres. In the case of motion picture theatres, the two basic classifications are the indoor auditorium-type and the outdoor drive-in theatre.

Lighting requirements for the marquee, lobby, and foyer are similar for both live and film auditorium-type theatres. But, separate considerations must be given to drive-in theatres. For auditorium-type theatres, it is a case of reducing the lighting level as a patron proceeds from the brightly lighted marquee and street area to the lobby, the foyer, and eventually to the darkened auditorium. Basically, the same type of planning is required for the approach to drive-in theatres.

At this point, however, the lighting requirements for the various types of theatres change. The so-called houselights, which illuminate the seating or parking area, are quite different in nature. Lighting within the film theatre is minor in scope as compared to house and stage lighting required for live productions. Further, a large portion of the on-stage luminaires are designed to be portable so as to accommodate a variety of needs as productions change.

Marquee. Attracting attention is one of the most important factors in the design of theatre exteriors. Flashing signs, running borders, color-changing effects, floodlighting, and architectural elements are but a few of the many techniques employed. Many present marquee attraction panels are lighted with incandescent filament lamps, fluorescent sign tubing, or fluorescent lamps behind diffusing glass or plastic. Opaque or colored letters on a lighted field are generally more effective than luminous letters on a dark field. The principal requirement is uniformity of luminance (photometric brightness), because spottiness interferes with legibility of letters. The actual luminance is usually determined by the general district luminance. The following values are a guide: low luminance district, 90 to 150 footlamberts; medium, 120 to 200; and high, 150 to 350.

A studded pattern of small sign lamps is a typical soffit treatment for many theatres. Reflectorized lamps are being employed for this puropse in many theatres to provide illumination on the sidewalk and some degree of luminance at the soffit, as well as effective ceiling pattern. A range of 30 to 50 footcandles is typical. Infrared units for heating and snow melting should be considered in localities where required by climatic conditions. See Section 25 for further discussion

Lobby. A minimum illumination level of 20 footcandles is desirable in theatre lobbies. The ceiling treatment is often integrated with the marquee soffit. Many lighting treatments are applicable here; some requirements are easy maintenance, designs that retain appearance and architectural suitability, and a pattern of luminances that attracts as well as influences the flow of traffic.

Built-in lighting with fluorescent lamps, spotlighting, or transillumination are successful techniques for poster panels. Luminances range from 50 to 200 footlamberts, depending on surroundings and brightness competition. An important consideration is to allow sufficient depth so fairly uniform illumination may be obtained.

Proceeds from refreshment stand sales may be a large share of the total gross income. Built-in case lighting, spotlighting, luminous elements, and signs (often in color) help gain attention. Lamps are used in hot food dispensers both for attraction purposes and heat.

Foyer. Usually a restful, subdued atmosphere is desirable in the foyer. Illumination from large, low luminance elements, such as coves, is one good method. Wall lighting and accents on statuary, paintings, posters, and plants are important in developing atmosphere. Light must not spill into the auditorium. General illumination levels of from 5 footcandles for motion picture theatres, to 15 footcandles for live-production theatres are recommended.

Live-Production Theatres

Included among the live-production theatres are the legitimate, community, and educational theatres. The term live-production refers to the presence of live actors on-stage as opposed to the presentation of a film show. Although there are many varieties of indoor and outdoor live-production theatres, such as amphitheatre, music tent, arena, open stage, etc., the most common are the traditional proscenium and the open stage or "thrust-type." See Fig. 24-4.

The proscenium-type of theatre is composed, typically, of a seating area and a stage area. It may serve not only as a theatre, but also as an assembly and lecture hall, study room, concert or audio-visual aids (including television) area, and it may also house numerous other activities. Considerable attention is being given to the development of speech and theatre arts programs not only in schools but also as a community activity among adult groups. The many uses of the theatre require well planned and properly equipped lighting facilities.

Seating Area. The seating area should be provided with well diffused comfortable illumination. Because

Fig. 24–4. Two of the most common types of live-production theatre stages: (left) the traditional proscenium stage in the New York State Theatre in Lincoln Center, and (right) the open stage or "thrust-type."

of the many purposes the seating area often serves, different levels of illumination are necessary. Basic illumination, well diffused and of the order of 10 footcandles should always be provided in the seating area. It should be under dimmer control, preferably from several stations (stage lighting control board, projection booth, etc.). However, there should be transfer capabilities so that the lighting will not be accidentally turned on during performances. Lighting equipment for the basic illumination may include general downlighting units, coves, sidewall urns, curtain and mural lights, etc. Supplementary illumination is required where the seating area is also used for visual tasks to provide an additional 30 footcandles over the basic illumination level. This supplementary illumination should be directed evenly over the seats and should be controlled separately from the basic illumination. The supplementary illumination should preferably be achieved by a downlighting system, properly shielded to prevent direct rays from becoming distracting. Selected circuits of the supplementary lighting system may also be used as work lights for cleaning, rehearsals, etc. Separate control of certain basic or supplementary circuits for emergency use should be provided at the rear of the seating area and other locations as may be required. This control must be independent of dimmer or switch settings. In accordance with local and national codes, a system of emergency lighting which would include panic lights, exit lights, shielded aisle and step lights, and other required lighting must be provided.

Stage Area.[1] Proper lighting for dramatic presentations extends beyond visibility to the achievement of artistic composition, production of mood effects, and the revelation of forms as three-dimensional. These functions of stage lighting result from the manipulation of various qualities, quantities, colors, and directions of light and these vary from one performance to the next and even continually throughout a single performance. The layout is affected by the amount and kind of use planned for the theatre.

Stage areas to be lighted include stage apron or forestage, acting areas upstage from the proscenium arch, including areas above and below the stage floor, extension stage areas, auxiliary acting areas in the auditorium, foreground scenery and properties, and background scenery and properties.

Basic locations for lighting equipment (see Fig. 24–5) may be divided into two groups as follows: locations in front of proscenium opening (which include the auditorium ceiling, side walls of auditorium or proscenium, or both, balcony front, follow spot booth, and edge of the stage apron) and locations behind the proscenium opening (which include relatively fixed positions, such as tormentor pipes for side lights, overhead cyclorama pipes for top lights, cyclorama pit for base lights, and locations employed as needed, such as overhead light pipes, free space at sides and rear of the stage main acting area for floor mounted or hanging equipment, and below stage areas).

Locations of Luminaires in Front of the Proscenium Opening.

Luminaires in the Auditorium Ceiling. Stage lighting luminaires in the auditorium ceiling are generally used for the basic purpose of lighting down-stage and apron acting areas. Also, they are sometimes used for the illumination of the proscenium curtains when the angles of throw permit reasonable coverage of the curtains. Each luminaire should produce a clearly defined light beam, which can provide an average illumination of 35 to 50 footcandles of white light on a vertical plane, with adjustable means for a controlled cut-off. so that the beam can be varied in shape to cover a desired area with little or no spill onto adjacent areas. The ellipsoidal reflector spotlight incorporating framing

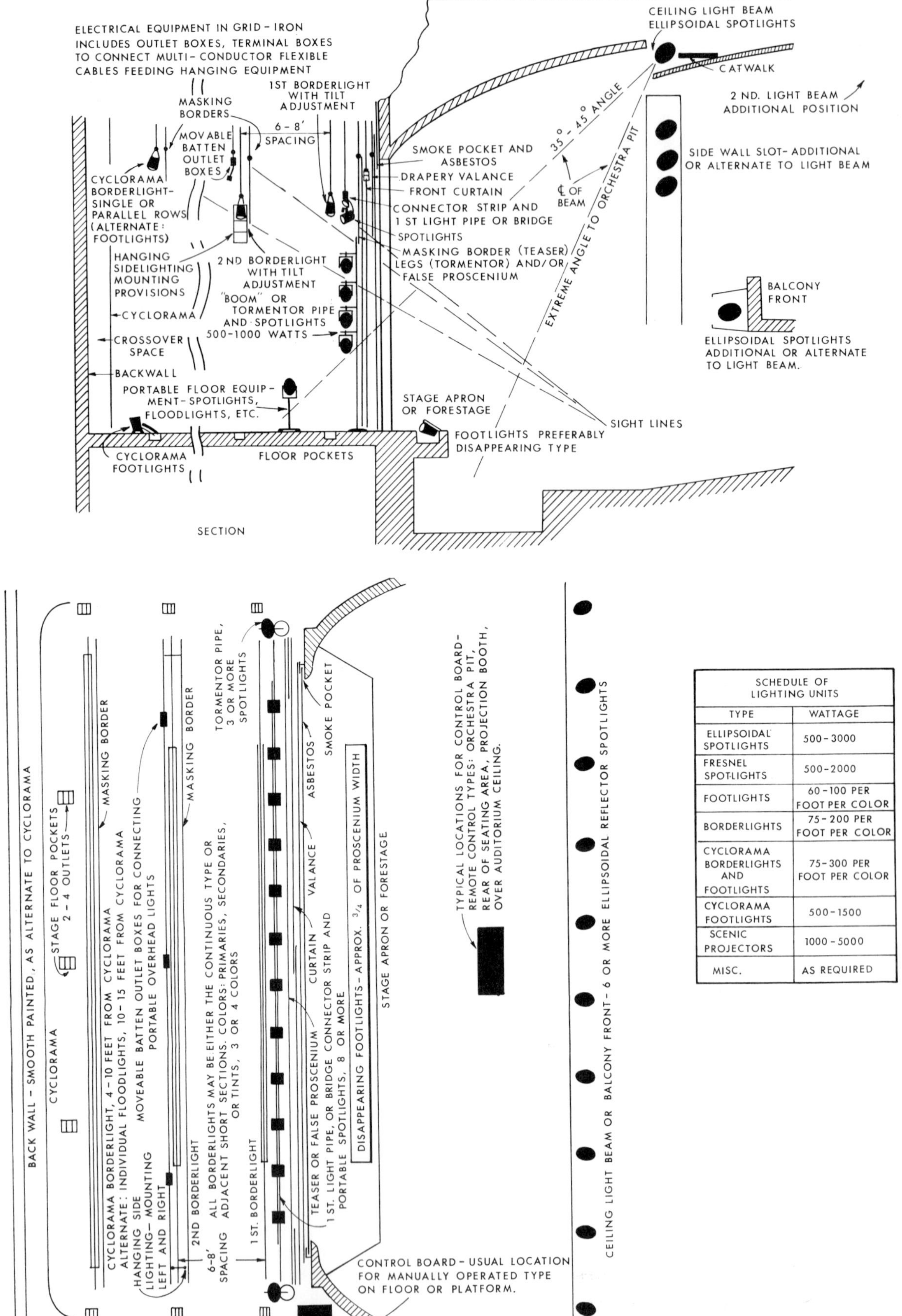

SCHEDULE OF LIGHTING UNITS

TYPE	WATTAGE
ELLIPSOIDAL SPOTLIGHTS	500 - 3000
FRESNEL SPOTLIGHTS	500 - 2000
FOOTLIGHTS	60 - 100 PER FOOT PER COLOR
BORDERLIGHTS	75 - 200 PER FOOT PER COLOR
CYCLORAMA BORDERLIGHTS AND FOOTLIGHTS	75 - 300 PER FOOT PER COLOR
CYCLORAMA FOOTLIGHTS	500 - 1500
SCENIC PROJECTORS	1000 - 5000
MISC.	AS REQUIRED

Fig. 24–5. Typical plan and section for an average size theatre.

shutters is an example of such an instrument, and is the type usually used.* See Fig. 24-1. The spotlights are best located behind slots in the ceiling surface and are ideally mounted in a continuous slot stretching across the ceiling from side wall to side wall. Whether in one or more rows, spotlights should be angled to project so that beam centers come within 30 to 60 degrees with the horizontal when seen in side elevations, and up to 45 degrees with the stage axis when viewed in plan.

Luminaires in Auditorium and Proscenium Side Walls. Luminaires located on or in the side walls, although not absolutely required, are recommended. They are used mainly as a supplement to the ceiling spotlights, and are of a similar type. Preferably, these luminaires should be recessed in wall slots. They provide lower angles of throw than the ceiling units and offer an excellent opportunity for a combination of side and top lighting with a wide range of different angles of throw.

Luminaires on Balcony Front. Angle lighting from spotlights in the side walls and/or ceiling will provide effective front lighting for most purposes, but there are occasions when the balcony position affords desirable low angles of throw or a soft wash of directional front lighting.

Follow Spot Booth. Follow spots are used to highlight selected performers. A follow spot should be capable of providing a level of at least 200 footcandles in a 8-foot diameter area, and should embody means to enable the size and shape of the light beam to be varied so that it may be reduced to cover the head of a person only, or be widened to flood a considerable portion of the stage. In addition to the iris diaphragm, spread lens, and horizontal paired shutters—accessories usually required for this purpose—follow spot equipment often includes either a color wheel or a multi-slide color boomerang (or tormentor pipe) that can be operated either from the side or rear of the spotlight, in order to change the color of the light.

Luminaires at Edge of Stage Platform or Apron. Space for footlight equipment should be provided at the edge of the stage platform. Footlights are a desirable part of a modern stage system, in spite of the trend to eliminate them in the belief that other front lighting facilities make them unnecessary. They should normally be used for supplementary lighting only, rather than as sources of primary illumination.

Locations Behind The Proscenium Opening.

Overhead Locations. The greatest number of luminaires in any one location upstage of the proscenium will be mounted on the first pipe, or a bridge immediately upstage of the proscenium. The instrumentation for the first pipe or bridge may include spotlights, borderlights, and scenic projectors. The majority of the spotlight units are required to produce a soft-edged beam that is widely variable in focus, as for example, is obtained from a Fresnel lens type spotlight. A number of ellipsoidal reflector spotlights are usually mounted in the same row. There should be provisions for mounting additional rows of lights on pipes parallel with the proscenium opening every 6 to 8 feet of stage depth.

Connector Strips for Stage Light Pipes. The basic purpose of a connector strip is to provide a simple and quick method of electrically connecting a number of lighting units, wherever mounted on an associated pipe, by means of a series of outlets. The length of a connector strip mounted on a pipe or bridge should approximate the width of the stage proscenium opening. Outlets should not be spaced closer than 12 inches apart and each outlet or group of outlets should be on an individual circuit with a separate neutral. So that any available pipe may be used for lighting instruments, a more flexible alternative to connector strips fixed to specified light pipes is a number of multi-conductor cables, permanently attached to the gridiron, with a box of minimum proportions housing an appropriate number of outlets at the other end. Each cable is long enough so its outlet box can be lowered and attached to either end of a stage pipe.

Borderlights for Stage Light Pipes. A borderlight provides general downlighting across an area of the stage and provides overhead illumination of hanging curtains and scenery. It contributes tonal quality to the over-all lighting effect. Separate control of different sections of a borderlight enables parts of the stage width to be variously accented in brightness and color. Borderlights (other than for cyclorama illumination) should be wired on three or four separate color circuits and should be capable of providing illumination of the whole width of a curtain or flat scenic drop mounted 4 feet or more upstage of the borderlight equipment. The level of illumination provided by a borderlight in the center of the vertical surface should be not less than 25 footcandles of white light when measured at a point 6 feet from the stage floor.

Cyclorama Top Lighting. Cyclorama borderlights must be long enough to illuminate the whole of the visible width of the background, independent of illumination from cyclorama bottom lighting. Cyclorama lighting requires at least double the footcandle level provided by other borderlights, and when the cyclorama is an important feature and is to be illuminated with 3 or 4 color lighting in conjunction with deep color filters such as red, green, and blue, then the wattage of the associated borderlight equipment may be from 2 to 4 times that of a regular borderlight—sometimes even more. The required lighting level may necessitate two parallel rows of borderlights, using, for example, 300-watt R-40 lamps on 6-inch centers or 500-watt R-40 lamps on 8-inch centers in each strip.

Backlighting from an Upstage Pipe. It is desirable to have a row of high intensity narrow beam luminaires, as for example, parabolic reflector

* For lighting the curtain *only*, Fresnel lens spotlights may be used.

floodlights or ellipsoidal reflector spotlights, suspended on an upstage pipe, but directed downstage to provide backlighting of artists in the main acting area. For example, there may be one 500- to 750-watt luminaire for every 4 to 6 feet of effective stage width.

Mounting for Stage Side Lights. Although other mounting methods may be used where conditions permit, there are two general methods of providing side stage lighting, either on suspended three or four rung ladders or from vertical, floor mounted boomerangs or tormentor pipes.

Special Theatrical Effects. Fluorescent paints, fabrics, or other materials responding to long-wave ultraviolet are often used for special theatrical effects. Sources for exciting the fluorescing materials include mercury lamps with filters for absorbing visible radiation, fluorescent "black light" lamps (which also require an auxiliary filter), and integral-filtered fluorescent "black light" lamps. Carbon arc follow spots are sometimes filtered for black light effects. Fluorescent lamps are the most efficient "black light" sources, but the energy cannot be readily focused for long throws or to cover restricted areas.

For greatest dramatic effect, visible light in the scene should be held to a minimum.

Scenic Projection. Increased understanding of the techniques of projection by theatre personnel, has refined the basic optical design for lens projection in the live theatre. Presently, the principal refinements involve: increased wattages (up to 10 kW) to provide additional scene illumination and increased area coverage; improved slide-making and handling techniques; programmed remote slide changing; extension of wide-angle projection to screen widths up to 1½ times the projection distance; standardization of units to permit easy interchangeability; and the availability of relatively inexpensive and easily mastered remote-control 35mm projectors.

The general availability of the 35mm or 2- x 2-inch slide projectors, as well as the increased availability of larger units, has led to much experiment. Multi-sources, wide ranges in the message, wide ranges in composition and build-up of the total picture, multi-screens and mosaic designs, fractured pictures, time and dimming variants, programmed variation, and adjustment of the controllable properties of auxiliary light sources—all combine to make projection an exciting light tool in all theatrical presentations.

Meeting, Convention, and Industrial Show Facilities[2]

Lighting for meetings and conventions requires comfortable illumination and accent lighting for non-theatrical participants. Where free discussion takes place between speakers and audience, the lighting should be free of glare so that prompt recognition of each speaker occurs. Lighting for industrial shows, new product presentations, etc. require the dramatic lighting of the theatre.

Stage locations may vary considerably from meeting to meeting. A show that uses rear projection may move the stage area 15 to 20 feet forward. Another meeting may require a simple platform with maximum space for an audience seated in school room or conference style. Many meetings use a center area, arena, or theatre-in-the-round arrangement. Other producers find a projected stage along a wall more satisfactory for their presentation.

Many meetings are conducted in multiple-purpose spaces that are used for food service, fashion shows, motion pictures, social events and meetings. The ease and speed in which these areas can be changed from one arrangement to another is an important economic operating factor.

Lighting must be coordinated with many other elements. These include wall surface brightness, projection screen locations, communications, sound systems, etc. Projection from audio-visual equipment and followspots require unobstructed views of screens, stages, and acting areas. Chandeliers must not be placed in locations that will interfere.

Theatre-Restaurants and Lounges

Stage lighting design criteria for theatres and auditoriums is generally applicable for theatre-restaurants, night clubs, and lounges. However, theatrical lighting in small areas such as lounges will utilize more compact luminaires. For low ceilings, a basic luminaire is an "inky" with a 3-inch Fresnel lens, or adapter accessory with individually adjustable framing shutters, and a 100- to 150-watt lamp. In larger spaces with longer throws, a Fresnel-lens spotlight, or a floodlight for 250-watt to 400-watt lamp and a beamshaper accessory may be used. An alternative would be a small ellipsoidal-reflector framing spot of 650 watts or less, available with wide, medium and narrow beam lens system.

Luminaire Locations. Lights can normally be used at sharper angles, closer to the 30-degree rather than the 45-degree limits. Downlights are used to produce "pools of lights" on dancers and set pieces. Uplights recessed in the stage floor are also used. Side-front, side, and side-back lighting is essential for three-dimensional effects particularly in dance and production numbers. Floor-mounted linear strips are used for horizon effects on cycloramas. These may be of the disappearing type or recessed with expanded metal covers to permit performers to walk over them.

Followspots. Locations for several followspots should be provided for front-side as well as front spotlighting. One or more followspots should be able to cover audience areas. Some performers enter from the audience, and runways are frequently used to bring the chorus closer to the viewers. Side-stages on each side of the main stage are frequently used for bands and stage action and provision should be made for adequate lighting of these areas.

Transparencies. Scrims are frequently used to *hide* the band when playing for a show. The band and performers are frequently *revealed* by bringing up lighting behind the scrim and keeping light off the front of the scrim. These changes may occur on the side or principal center stage.

Special Effects. Mounting devices and no-dim control circuits and receptacles are required for "black lights," projectors, electronic flash, motor-driven color wheels, and dissolves, fog and smoke machines, mirrorballs, etc. Color-organs are used to pulsate lights with music. Plastic covered floors for dancing and entertainment should provide selectable color and pattern effects. In minimum spaces, fluorescent lamps and dimming ballasts are used. If ventilation can be provided, incandescent lamps can be used for special effects.

Controls. Single lights are frequently used. Receptacles should be on individual cross-connect circuits. Permanent grouping should be avoided or minimized if necessary. For more flexibility a greater proportion of dimmers to receptacle circuits are used than in a conventional theatre. Control cross-connects may be used to supplement or in the place of load patching. No-dim controls should be integrated with the dimmer controls, *i.e.*, voltage sensitive relays controlled by the same type potentiometers as used for the dimmer controls.

Motion Picture Theatres

Auditorium-Type Theatres. The objectives of auditorium lighting in the motion picture theatre may be outlined as follows: (a) to create a pleasing distinctive environment; (b) to retain brightness and color contrasts inherent in the picture; (c) to create adequate visibility for safe and convenient circulation at all times; and (d) to provide comfortable viewing conditions.

For general lighting during intermission, five footcandles is considered minimum. During the picture, illumination is necessary for safe and convenient circulation of patrons. Illumination values of 0.1 to 0.2 footcandles represent good practice. Screen luminance with picture running is in the range of 5 to 6 footlamberts to less than 1. The need to eliminate stray light on the screen dictates controlled lighting for at least the front section of the auditorium. Downlighting is one of the most effective methods for this purpose. In general, diffusing elements, such as coves, allow too much light to fall on the screen if they provide adequate illumination in the seating area. Diffusing wall brackets, semi-direct luminaires, and luminous elements are generally too bright to be used for supplying illumination during the picture presentation.

Contrast between the screen and its black border is very high (sometimes more than a thousand to one), creating uncomfortable viewing conditions. Areas around the screen can be raised in luminance if they are not treated with distracting decorations. Light for this purpose may be reflected from the screen under special conditions, supplied by supplementary projectors, or by elements behind the screen.

Curtains may be lighted in color with a projector border during intermissions. Adequate spotlighting on the stage is desirable for announcements and special occasions.

Aisle lights should be low in luminance and spaced close enough to give fairly uniform illumination in the aisle.

Houselights should be dimmer controlled, rather than controlled merely by on/off switches.

Drive-In Theatres. Lighting for safety is a major concern of the drive-in theatre manager because of the interference of pedestrian traffic going from cars to the refreshment stand, and cars entering and leaving without benefit of headlighting. Screen luminance is much lower than in conventional theatres, which imposes greater limitations on tolerable extraneous luminances. Screen orientation with respect to moonlight, twilight, and sky haze is a major factor. Unique opportunities for spectacular and decorative effects include lighting flower beds, plants, pools, and fountains, which are not within view of the audience watching the screen.

Entrance signs must be legible at long distances to allow enough time for patrons traveling at high speed to slow down and maneuver to enter. Conventional sign lighting methods for regular theatres are used. The entrance drive can be lighted by street lighting equipment, luminous pylons, or low-level units. Particular care is needed to control any luminance so that it will not be distracting from the screen area. Light reflected from trees, surrounding structures, or even haze or mist can be disturbing. Floodlighting, luminous elements, lighted murals, and patterns outlined in tubing are popular means of drawing attention to the screen tower.

At least 30 footcandles is recommended in the money exchange area at the box office (generally some distance from the entrance). Spotlighting for cashiers in costume and decorative lighting at this location offer opportunities for showmanship. Lighted ramp markers, lighted ushers' wands, and illuminated fences facilitate movement after car headlights are turned out. Speaker stands are available equipped with lights for ground area and for signaling the refreshment vendor.

Intermission lighting should be at least 0.5 footcandle, and can be satisfactorily provided from units mounted on poles. Experience has shown that floodlights for this purpose located at top of the screen tower are glaring and lead to complaints by patrons. One to two footcandles is recommended for exit lanes for safety of fast-moving, highly congested traffic. Floodlights along the fence aimed with the flow of traffic plus floodlights on a pole at the rear of the area is one solution. At least 2 footcandles is recommended for convergence of theatre traffic with highway traffic.

Refreshment stand interior light must be shielded from the outside. High luminance luminaires are objectionable because patrons are dark-adapted when they come in. Lighting the counter area is usually sufficient. In stands with a viewing room, vertical glass windows require highly-shielded, low-luminance luminaires to prevent reflections. Windows splayed out at the front reflect patrons instead of the ceiling pattern, but the former method can be better if properly handled.

LIGHTING FOR TELEVISION

Television broadcasting requires extensive pre-planning of the lighting arrangements, switching, dimming, etc., due to the necessity of continuous dramatic action. An exceptionally high degree of mechanical and electrical flexibility is mandatory. Multiple scenes are arranged for continuous camera switching. Consequently, the lighting for each scene is pre-set without interference with adjoining sets. The quantity and quality of the lighting needed for television production depends upon the type of camera pick-up tube used (Image Orthicon, Vidicon, Plumbicon) and the reflectance of the subject. The choice of lenses (size and focal length) is governed by the same optical principles commonly applied to all photographic devices.

Studio Lighting for Monochrome Television

There are two aspects of lighting for television production. The first has to do with controlling the light in terms of quantity, color, and distribution to produce a technically good picture. The second has to do with the design of lighting to produce a dramatic and artistic visual effect for broadcasting.

Camera Pickup Tubes. The 5820 Image Orthicon, the Vidicon (SB_2S_3), and the lead oxide Target Vidicon (PbO) of the Plumbicon* type are the pickup tubes most commonly used in studio telecasting. Their spectral sensitivity characteristics are shown in Fig. 24–6 and the normal lighting level required is 90 footcandles. The tubes do not lend themselves to a contrast range exceeding 40 to 1. Lens stops are adjusted for proper exposure, usually between *f*4 and *f*11, depending on the lighting used. For example, with an illumination level of 100 to 125 footcandles of incident light a lens stop of *f*8 normally results in a broadcast quality picture (provided the average reflectance of the subject matter equals that of flesh tones, *i.e.*, a 40 to 50 per cent reflectance). White and some dark areas can be used in each picture to aid camera level adjustment. These should be of limited area and

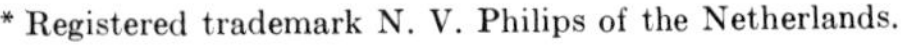

* Registered trademark N. V. Philips of the Netherlands.

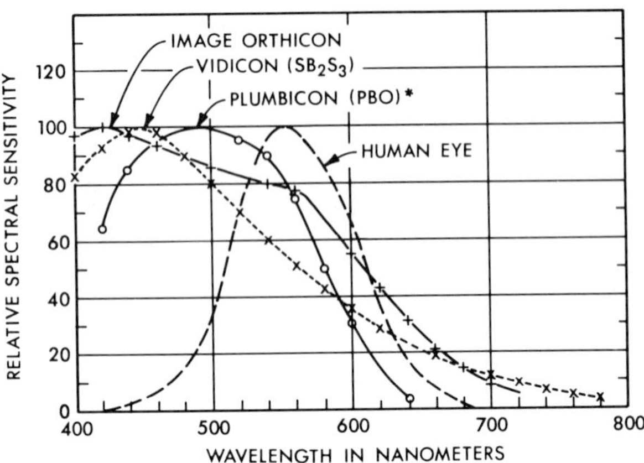

Fig. 24–6. An equal energy relative spectral response curve of an Image Orthicon, an SB_2S_3 Vidicon, a Plumbicon,* and the human eye.

have enough reflectance to guarantee the contrast range mentioned above.

Using the Vidicon Camera. Before giving the recommended illumination level at which to operate the Vidicon camera, it is necessary to know some of the interlocking facets of high- and low-level lighting. Lower levels (10 to 40 footcandles) are usually found in existing room lighting since most modern offices and classrooms have a lighting level of over 40 footcandles. The higher levels (100 footcandles and up), when not available under the existing room lighting, may almost always be obtained by using standard television spotlights and floodlights.

To determine whether to use ordinary room lighting (which may be a combination of incandescent and fluorescent permanently mounted luminaires) or to use spotlights and floodlights, the telecaster should consider the quality of the resulting picture and the standard he sets as acceptable for transmission. For instance, if there is no motion in the scene either on the part of objects in the scene or on the part of the camera itself (panning, tilting) and if it does not matter whether the picture is somewhat grainy, then lower level lighting is adequate. If, however, the scene contains moving objects, or the camera makes a fast pan, and the picture must approach broadcast quality, then a higher level should be used.

The Vidicon tube, however, has special characteristics that under certain conditions will deliver an "acceptable" picture with ordinary room lighting as the only illumination. In this case, the picture is far too noisy and otherwise unacceptable for minimum broadcast transmission. If, however, the system is only used to observe and count people entering and leaving a store, for instance, detail and picture quality may not be necessary. For this purpose an acceptable picture may be accomplished with as low a level as 20 footcandles.†

The Vidicon phenomenon called image retention is also given other names, some of which are stick-

† For further information on light levels and camera stops on 106 special and industrial applications, see reference 8.

ing, trailing, and lag. It occurs when the actor moves quickly or the camera pans or tilts rapidly to a new position. It is especially noticeable in low-level lighting and becomes less of a problem as the lighting level is increased.

Studio Lighting for Color Television

Response characteristics of color cameras are critical and experience has shown that the spectral power distributions of the light sources within a scene should be essentially uniform. Where filament lamps are used, current practice indicates that they should remain within a 200 K range for accurate color rendering. While a camera chain and system can be adjusted to operate with almost any continuous spectrum light source by adjusting the gains of the red, blue, and green channels, a wide variation in the spectral characteristics of the light sources within the picture area will cause unpredictable changes in color reproduction.

Incandescent filament lamps have been favored for color studio lighting due to their continuous spectrum, their relatively low cost and long life, and the fact that they are available in a wide range of sizes with approximately the same Kelvin temperature. Incandescent filament light source color temperatures of 2900 K to 3200 K are currently used. These lamps are favored because their housing equipment is generally less bulky and of lighter weight, and the smaller filament light sources allow for better optical control of the light beam. Furthermore, they readily permit use of color filters at the light source.

A camera initially balanced for incandescent filament lighting may be used to take scenes outdoors in daylight with the color temperature variations corrected through the use of suitable filters on the camera lens. Variations in the amount of light can be controlled by the camera iris.

Mixed light sources, such as incandescent, high-intensity gas discharge, and fluorescent ordinarily should not be used to light a scene simultaneously because the camera will not transmit a good color picture where the spectral power distribution of the light sources vary greatly. However, new developments in fluorescent and high-intensity gas discharge light sources are now providing greater flexibility for mixing sources. This can be attributed to discharge lamps being manufactured with much greater and, in some cases, continuous spectral power distributions. Some sports arenas have mixed 75 per cent high-intensity gas discharge and 25 per cent incandescent sources successfully.

For televising color pictures, lighting levels of 175 to 350 footcandles incident are found satisfactory. Many studios, especially those producing color spectaculars, have found it desirable to provide up to 500 footcandles. This is done to compensate for older model color cameras or to provide greater depth of field.

Videotape and Kinescope Recordings

The same requirements as given in all the above apply to video and kinescope recording. In addition, the contrast range should average 20 to 1, but under controlled conditions the contrast range can be extended to 30 to 1.

Television Film Production

The significant element in lighting for television film production is the control of subject contrast range. Most television film reproduction systems require a picture luminance (photometric brightness) range not exceeding 25 to 1. This contrasts with the current theatre projection range of about 100 to 1. It is important, therefore, to limit the luminance range of filmed subjects. A fair amount of such control can be obtained by the introduction of "fill" light to raise the luminance of significant shadow areas. This does not mean that flat lighting is desirable, but rather that the lighting ratios used for modeling should be lower than is normally used when making films for theatrical projection.

In shooting color film for television, the same general rules apply. Additionally, it is necessary to bear in mind that color contrast is not enough to insure an acceptable signal. Brightness contrasts essentially constitute the signal seen on a monochrome receiver tuned to a color broadcast. Consequently, it may be necessary to light in such a way that areas having hue differences only, are placed at different lighting levels. A poor monochrome picture is particularly conspicuous when scenery is painted in several pastel shades, all having similar reflectances.

Projected Backgrounds

Rear projection is the projection of a scene onto a translucent screen from behind. It is used to simulate background scenery, which may take the form of stationary objects as produced by a slide; moving effects such as clouds, water, etc., as produced by an effects machine; or continuous motion simulating moving trains, motion from an automobile, etc., as produced by a motion picture film.

For realism, projected highlight levels should be between ½ to 1 times live highlight levels. It is desirable to have a projected highlight luminance of 70 footlamberts measured facing the screen from the camera side of the screen when the lighting level in the acting area is 100 footcandles.

Chroma Key

The production technique known as *chroma key* is a special effect that enables any background material to be matted into a scene. In the studio a color camera views the subject against a backdrop of a primary color that has sufficient saturation to produce a full output level in the corresponding channel of the camera. This signal output is used to key a special effects generator so that all information except the wanted subject is matted out of the original studio scene. Information from any other source such as a film chain or video tape recorder can then be inserted in the matted portions of the signal.

The primary color used in the backdrop is chosen on the basis that it is not present in the color of the wanted subject. When human subjects are used, blue is usually the optimum background color because of its absence in flesh tones.

Additional precaution must be taken to avoid the use of the background color in costumes or stage props.

The lighting level on the backdrop must be high enough to produce a full output signal from the camera without excessive noise. However, the same lighting must not be so high that light reflected from the background will appear on the subject and thereby create spurious keying signals.

The color camera used for the chroma key technique must have sufficient video bandwidth in the color channels so that key pulses with good transient edges can be developed.

Field Pickups of Sports Events[12]

The lighting recommendations for sports (see Section 19) are based upon player and spectator visibility; however, when telecast in color additional lighting may be required in certain sports events.

It is important that the lighting system (1) provide the illumination necessary to permit use of close-up shots; (2) maintain sharp focus over the area of critical action; and (3) yield good color and monochrome picture quality.

Figs. 24–7 and 24–8 indicate the recommended illumination levels for baseball and football respectively. It must be emphasized that the indicated values are footcandles at a point 36 inches above the ground and with the photocell oriented at an angle of 15 degrees with a vertical plane (see Fig. 24–9). If camera positions are fixed, the illumination values with the photocell facing the camera are of major concern. If camera positions are not fixed or are unknown, the value shown should be provided with the photocell facing in all four directions as indicated. All values are maintained in service values defined as the amount of light available when the lighting system is at its lowest output.

The illumination values shown in Figs. 24–7 and 24–8 refer to the readings in the four near-vertical planes. The horizontal plane readings may be significantly different from those found in the near vertical planes. The purpose of recording the horizontal data is to determine resultant uniformity of illumination incident on the ground. It is important that the brightness pattern be uniform as seen by

Footcandles Readings at Test Positions						
1	2	3	4	5	6	
—	150	150	150	—	—	A
—	—	—	—	150	—	B
200	200	200	200	—	150	C
200	300	300	200	—	150	D
—	300	300	200	—	150	E
200	—	200	200	—	—	F

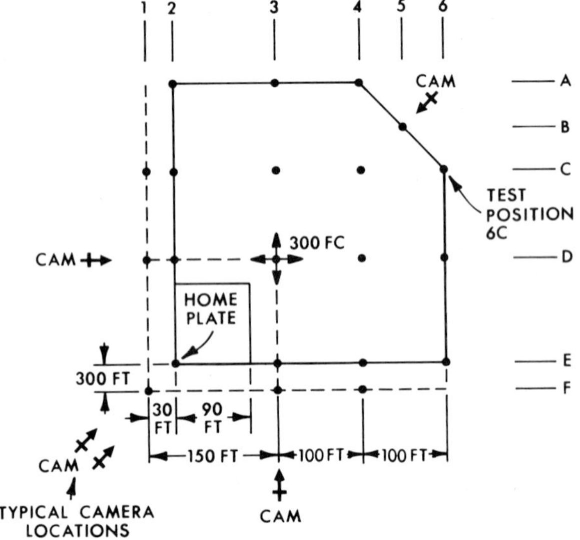

Fig. 24–7. Recommended illumination levels and survey test positions for color television pickup of baseball.

the camera from either a narrow (close-up) or wide-angle shot.

Illumination also should be provided in the spectator area adjacent to the playing field to permit camera pickup of celebrities; a minimum of 150 footcandles should be provided for the area in which it is expected to have camera pickup.

Evaluation of uniformity ratios resulting from the five specified readings should be between values in similarly oriented planes or the values in planes which are perpendicular to the line of sight of the camera locations.

For good picture quality, any variation in uniformity must be gradual. The ratio of maximum to minimum illumination values over the designated area should not exceed 2 to 1.

Luminaire or Pole Location. Pole locations as shown for baseball in Section 19 are applicable when designing a lighting installation for color television. However, for football, pole locations will be required in accordance with Fig. 24–10 to provide the illumination for the camera located at the end of the field. Extreme caution should be exercised in the selection and aiming of floodlights so as not to create objectionable glare for players, spectators, or side-line cameras.

Footcandles Readings at Test Positions

	1	2	3	4	5	6	7	8	9
A	160	160	160	160	160	160	160	160	160
B	180	210	230	230	230	230	230	210	180
C	200	275	300	300	300	300	300	275	200
D	180	210	230	230	230	230	230	210	180
E	160	160	160	160	160	160	160	160	160

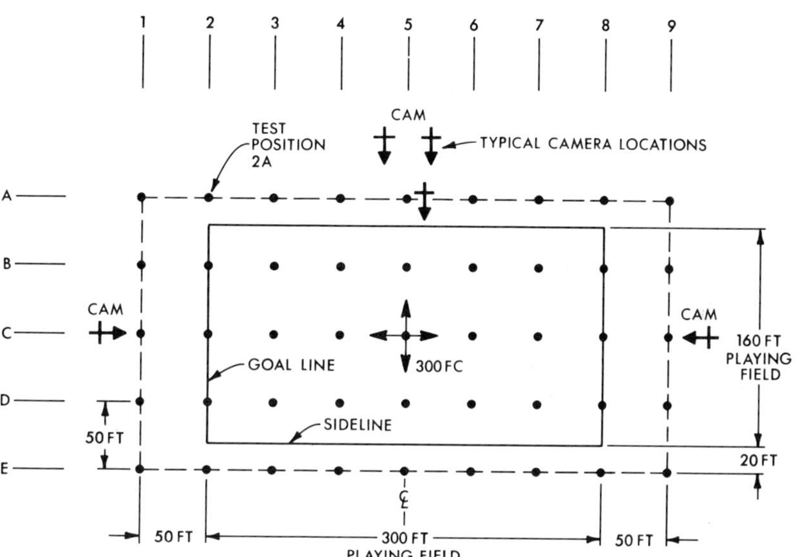

Fig. 24-8. Recommended illumination levels and survey test positions for color television pickup of football.

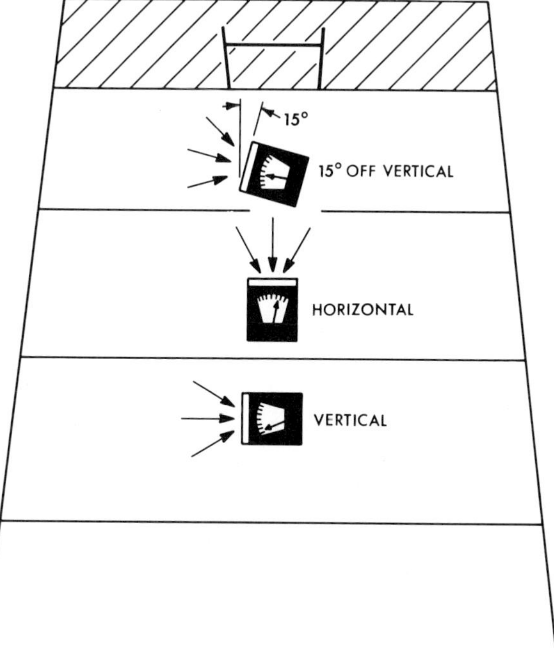

Fig. 24-9. Meter orientations to measure illumination in 15 degrees off vertical, horizontal, and vertical planes.

Measurement of Illumination

Meters generally used in television production measure incident light and consequently do not take into account the reflectances of the scene components. Since measurements by such meters are not indicative of the scene luminance variations, the operator must keep this constantly in mind when planning the lighting for different sets. However, an incident light meter in the hands of an experienced person can be a very useful tool. The quantity of light is measured with a photocell meter holding the photosensitive element in the scene at approximately camera lens height and at the angle of and facing the lens. Extreme light contrasts can be eliminated and exposures can be based upon such measurements. The practical scene contrast range for operational use is 30 to 1.* Color contrast should not be confused with brightness since two different colors can have the same luminance. Unless careful consideration is given to tonal rendering of colors, a scene that appears beautiful to the eye can appear on a receiver as a flat monotone picture.

* *See* Videotape and Kinescope Recordings.

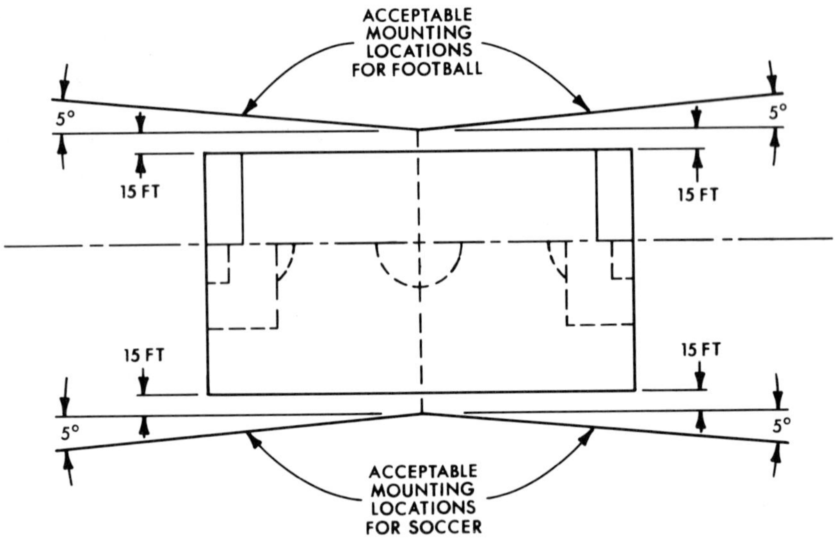

ACCEPTABLE MOUNTING LOCATIONS FOR FOOTBALL

5° 15 FT 15 FT 5°

15 FT 15 FT

5° 5°

ACCEPTABLE MOUNTING LOCATIONS FOR SOCCER

Fig. 24-10. Recommended pole locations for color television pickup of football and soccer. Mounting locations to be in areas defined in Fig. 24-8; minimum mounting height to be in accordance with Fig. 19-23. (See Section 19.)

Types of Illumination

Base Light or Fill Light. Base light or fill light is usually supplied by floodlights using incandescent lamps, including etched tungsten-halogen which supply a broad source of soft illumination. It is desirable to aim base lights at a 12- to 15-degree angle above a horizontal through the eye level height of the subject.

Key or Modeling Light. Key or modeling light is usually supplied by Fresnel lens spotlights ranging in lamp size from 500 to 5000 watts. The Fresnel lens unit has a specular spherical reflector and generally the lens risers are opaque to reduce spill light. The unit can be focused from spot to flood, with the principal characteristic of the beam being a soft edge rather than a well defined hard edge beam. Fresnels are equipped to hold supplementing masking devices, such as barn-doors. The barn-door fits in front of the lens and is used to limit the bottom, top, or sides of the light beam. These units can either be hung or used on floor stands and are generally aimed at a 20- to 40-degree angle above a horizontal through eye level height of the subject.

Back Light. Back light is used for separation. Back lights are hung behind a subject and are aimed at a 45-degree angle to light the back of the head and shoulders, and to separate the subject from the background. Back light can be from one-half to equal that of the front light, depending upon the reflectance of the hair and the costume.

Set Light. Set light is used to decorate or help give dimension to scenery. The amount of light necessary is totally dependent on the reflectance of the scenery. It must be kept in mind that flesh (a person's face) reflects 40 to 45 per cent of the light falling upon it. Therefore, the major part of the background must be kept below the reflected light level of the face.

A graphic representation of the above types of lighting is shown in Fig. 24-11.

There are many other luminaires that can be used to help dramatize a show. A few are: sun spots, ellipsoidal follow spots, pattern projectors, and strip lights.

Balancing for Correct Contrast. It is important to have the proper balance among the four types of light discussed above. If, for instance, the set is painted in a color that reflects more light than the flesh of the actor, skin tones may appear darker than desired in the picture. This means that the set light should be reduced. A quick way to do this is to use spun glass diffuser material on the set-lighting units. This material is available in 3-foot by 12-foot rolls and is cut to fit the unit. One .015-inch thickness cuts about 20 per cent of the light. Additional thicknesses can be used until the correct contrast is obtained. Another medium that is sometimes used to balance the lighting is one or more layers of ordinary house window screening. Likewise, if the front, or fill light, is brighter than the key light the use of spun glass will help in solving the problem.

If the installation includes dimmers and a cross-connecting system, the various lighting units can be grouped and then dimmed until the desired contrast is obtained. This method is highly recommended because it saves much setup time and effort and is standard procedure in a commercial broadcast studio.

Lighting Equipment Installation

The method of supporting the luminaires depends to a great extent on the ceiling height and the intended use of the studio. Where the height is low (12 to 16 feet) a permanent pipe or track grid is usually installed from which the units are hung directly or through available pantographs which permit individual vertical setting. See Fig. 24-12. In any case, the units are capable of complete rotation and tilting. In high-ceiling studios and in television theatres, the lighting units are supported either from

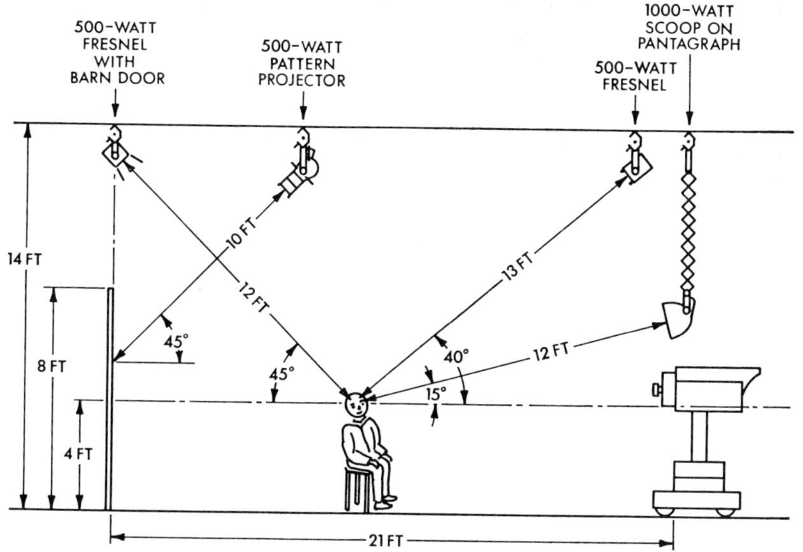

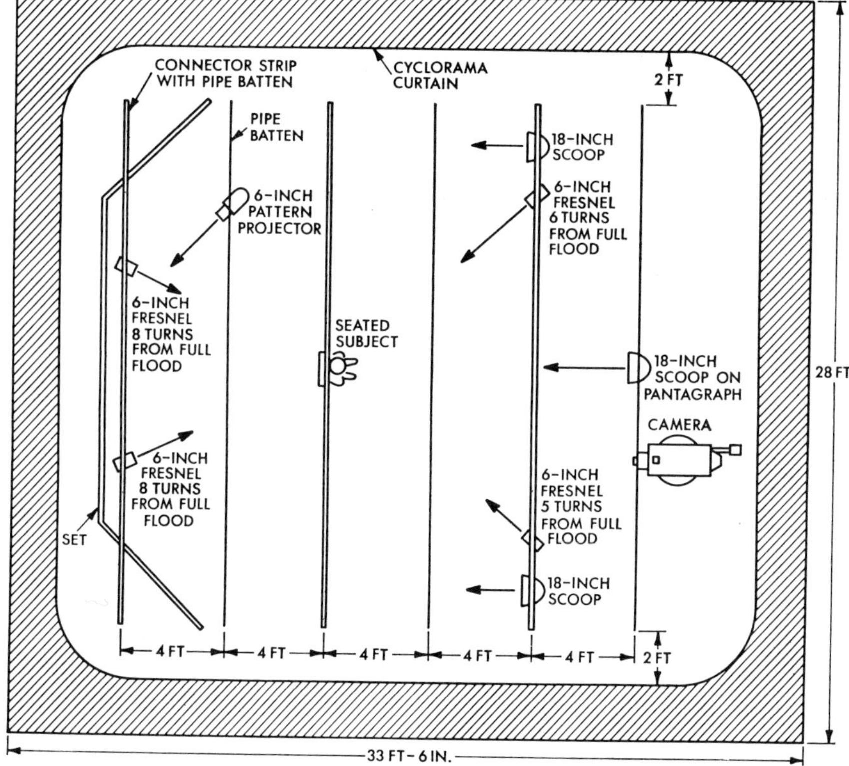

Fig. 24-11. Diagram showing good practice in luminaire location and aiming angles. Area enclosed by shaded portion is the net production area (NPA). Shaded area is not included when making watt per square-foot calculations.

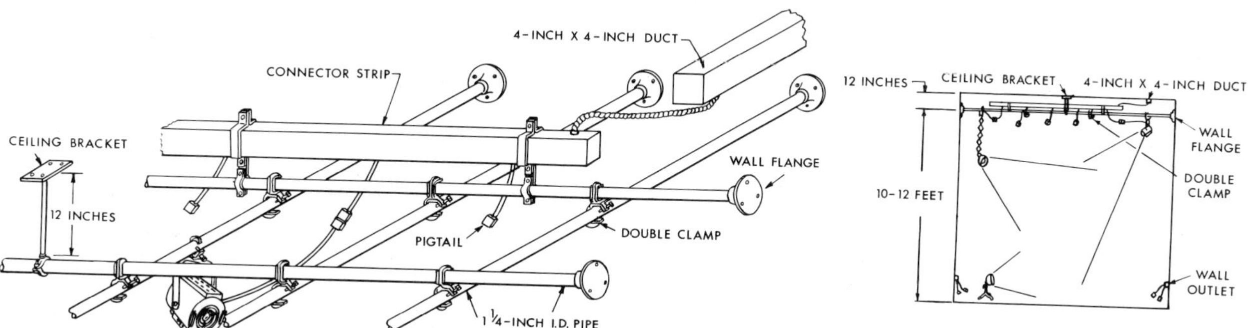

Fig. 24-12. Diagram showing a typical overhead grid system for mounting lighting equipment in a small low-ceiling studio.

fixed pipe or track grids or on counterweighted pipe or track battens (see Fig. 24-13).

Educational Television

In addition to commercial television channels, there are certain channels which have been allocated for educational broadcasting exclusively. Most of these are owned and operated by universities or other educational bodies. Their operations closely parallel the commercial stations in that the same type cameras and lighting are used, even though the budget they have to work with may be only a small fraction of that of their commercial counterparts.

Closed Circuit Television (Cable Television)

A recent trend, which may in time become the largest television field, is the use of television on a closed circuit basis. Such telecasts cannot be picked up by the conventional home receiver but can be by those who have a set wired-in to the originating circuit.

In closed circuit educational television a person giving a lecture or classroom demonstration in one room can be viewed in other rooms in the same building or in other buildings on the campus. One lecturer can thus contact hundreds or thousands of students at one time. Such installations have been installed in many elementary schools, high schools, and colleges.

The principal difference between the usual lighting practices of commercial and closed circuit television is that most closed circuit systems, due to financial limitations, utilize Vidicon cameras. The techniques for lighting closed circuit television are identical to those of the commercial counterpart of the particular camera type being utilized as explained in previous sections.

Lighting in Screen Viewing Rooms (Radar)

The primary concern in radar rooms is the detection and interpretation of signals on the cathode ray tube (CRT) screen. Room illumination may interfere with the detection of signals on the CRT screen by: (1) reflection from the scope face, (2) changing the adaptation state of the eye, and (3) exciting the screen phosphor and thus reducing contrast. Of the three, glare due to reflected light is the most frequently encountered source of interference. On intensity modulated scopes, signal detection is based upon the luminance and shape difference between the signal and its background. Accordingly, stimulus conditions which influence luminance and shape discrimination (for example, signal size or duration) play their normal role. Ambient light reflected from the scope face effectively increases background luminance, and thus reduces the contrast of the signal to the background.

Tolerance for Ambient Illumination. The amount of reflected ambient illumination that can be tolerated is directly related to the screen luminance. It has been calculated[13] that if reflected light equals screen luminance, a 1- to 4-decibel loss in signal visibility occurs. If reflected light is more than the screen luminance, additional losses in visibility occur.

As noted earlier, ambient light, whether reflected from the scope or impinging directly upon the eyes of the observer, also influences signal detection by changing the light adaptation level of the eyes.

Fig. 24-13. Typical large studio grid system with raising and lowering capability

However, provided it is diffuse, a small amount of illumination may improve signal visibility. It has been found[14] that typical screens operated with a surround illumination of about 0.1 footcandle give better signal visibility than when operated in complete darkness. These findings also indicated that the eye may be previously adapted to a light level approximately two log units above screen luminance without interfering seriously with signal visibility.

Another effect of ambient illumination is the apparent flicker of the display. Normal television standards call for repetition rates of 30 cycles per second, and while this rate may vary among radar systems, it is a commonly employed standard for cathode ray tube displays. As the ambient illumination on such displays increases, it is normal for the operator to increase the luminance of the display. Should the resultant luminance be too high, the images will no longer appear to be of a steady intensity, and most observers will report major dissatisfaction with the display. It is critical that ambient illumination on the scope face does not reduce the image contrast to an extent that would require the observer to adjust image luminance beyond that for suitable apparent steady-state luminance.

Selective Lighting Methods. In rooms that require high levels of illumination for tasks other than CRT screen observing, polarizing filters and selective spectrum lighting have been used. In one polarizing method, reflected light is screened from the CRT display by placing a polarizing filter over the light source and a crossed polarizing filter over the scope face. In another method, a linear polarizing filter and a quarter wave retardation filter are placed over the scope face. Ambient light, after plane polarization by the linear filter, passes twice through the quarter wave retardation filter and emerges as light polarized in a plane perpendicular to that of the linear filter. The linear filter then effectively prevents such reflected light from reaching the observer.

In a selective spectrum lighting system, the spectral characteristics of the light are obtained by using a special light source or a special filter. Light from the source is then absorbed by a scope filter, which at the same time transmits most of the light emitted by the CRT. Characteristic of the type is the Broad Band Blue (BBB) light system,[15] in which a blue filter is placed over a fluorescent luminaire and an orange filter is placed over the scope. The blue filter transmits and the orange filter absorbs energy below about 540 nanometers.

Placing a filter over the scope face reduces the luminance of both signal and background by absorption and reflection. The extent of the loss depends upon the particular combination of screen phosphor and filter used, and may be calculated by superimposing the relative spectrophotometric curves for phosphor emission and filter transmission.[16] The introduction of a filter does not influence the contrast between signal and background. However, if the filter interposed between the CRT screen and the eye reduces the luminance level to the range where an increase in contrast is required for discrimination, a reduction in signal visibility occurs. Weak signals will disappear from view. Nevertheless, if the screen luminance is sufficiently high, a small filter loss can be tolerated without significant effect upon visibility.

Recommendations. Lighting practices are dependent upon the purpose and methods of operating the radars, and the following suggestions are for conditions that require higher levels of ambient illumination: (1) glare should be minimized by keeping the light for other visual tasks that must be performed in radar rooms at the lowest effective level for those tasks, and by avoiding the location of light sources or their reflections directly in the visual field (full advantage should be taken of louvers and shields to decrease the level of illumination on the CRT); (2) to avoid adverse light adaptation, diffuse illumination around the radar screen should be held below the luminance on the CRT screen; and (3) selective spectrum lighting and filtering systems can be utilized when room illumination must be high.

PHOTOGRAPHIC LIGHTING

Commonly used photosensitive films and plates include the following:

Panchromatic (sensitive to all color—black and white image)

Orthochromatic (sensitive to all color except orange and red—black and white image)

Color (sensitive to all colors—color image)

Infrared (sensitive to red and infrared—black and white image)

Their spectral sensitivity curves are given in Fig. 24-14.

For photography light sources must emit energy in the spectral region in which the photographic material is sensitive. Even in black-and-white photography, color delineation in the form of faithful gray values is required. In black-and-white photography, the photographer endeavors to secure in his negative a scale of grays corresponding to the various brightnesses of the subject. thus, it is necessary that the film and the light source complement each other. Where this is not possible, it is general practice to employ a filter at the camera lens.

For photography in color, the spectral quality of the illumination is even more critical. Color emulsions are "balanced" for use with a particular quality of light. Because most color photography materials are based on three emulsion layers, each sensitive to a relatively narrow spectral band, adjustment by filtering to a light source other than the one for which the material was originally intended calls for precise filter selection.

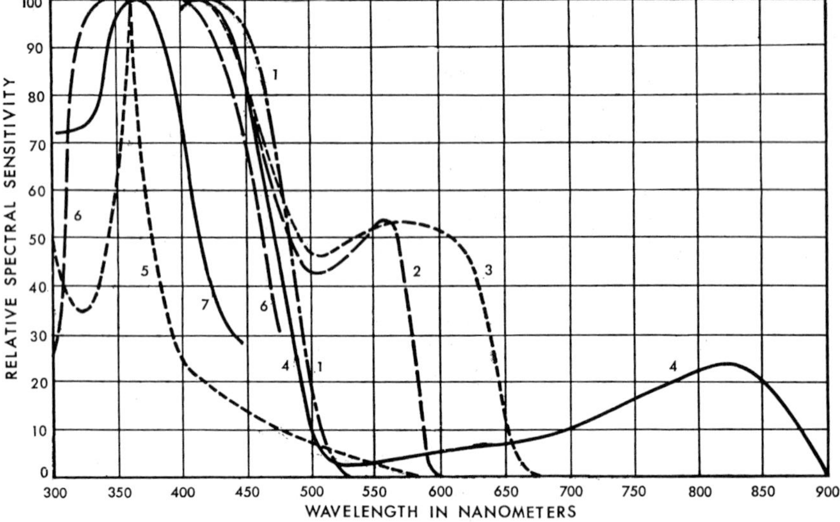

Fig. 24-14. Spectral sensitivity curves for common types of photographic and photo-process materials
1. Blue Sensitive
2. Orthochromatic
3. Panchromatic
4. Infrared Sensitive
5. Bichromate Coating
6. Diazotype Paper
7. Blueprint Paper

The quantity of light that a film receives is a function of object luminance, of exposure time, and of lens aperture. "Exposure" equals *illumination at the film* × *time*. This relationship frequently is referred to as the reciprocity law. Longer or shorter exposure than a particular film is designed for will result in "reciprocity failure." The film manufacturer will suggest adjustments to compensate. Exposure time may be governed by factors such as the necessity for stopping motion, flashlamp and flashtube characteristics, and subject reflectance.

The most common system of expressing the light-gathering ability of a lens is the *f*-system in which the *f*-value of a lens is given as the *focal length* ÷ *the diameter of the lens opening*. For adjustable-aperture lenses, any particular aperture setting is referred to as the "*f*-stop." For a given scene luminance, illumination at the film is proportional to the inverse of the square of the *f*-value.

There are a number of systems for evaluating the sensitivity (speed) of film and plates.[17] The American National Standards Institute (formerly ASA) has standardized a procedure for determining film speed. For example, a particular film may be said to have an ASA rating of 64.

The following formula, embodying the reciprocity law and the factors of lens aperture and film rating, gives the relationship of the several elements affecting exposure for objects of average reflectance:

$$E = \frac{K \times f^2}{T \times S}$$

where E = illumination in line with camera on subject being photographed (in footcandles).

f = f-value at which the lens aperture is set.

S = speed of film, according to the ASA system.

T = time of exposure (seconds).

K = a constant based on the various elements used. 20 is a satisfactory value for negatives of average density (ASA).

Instead of basing the exposure on incident illumination, the average luminance L may be substituted for E if a corresponding change is made in K.

Exposure Meters

The formula above is the basis of exposure meter design and operation, since all exposure meters of the photoelectric-cell type are essentially luminance-measuring devices. However, some may be used also as illuminometers to measure the illumination on the subject. (See Section 4.) The meter consists of a photovoltaic cell, an ammeter of high sensitivity, and a calculator. A hood or louver is provided in front of the sensitive cell to limit the acceptance angle to approximately 30 degrees, a rough average of the angle intercepted by the lenses of both still and movie cameras.

The customary method of using a photoelectric exposure meter (luminance type) is to hold it near the camera and point it toward the subject, thereby assuming that the meter "sees" the area being photographed, much as does the camera lens. Frequently a scene may include large areas, such as an open sky or a dark surrounding doorway, that may result in a luminance indication on the meter scale having little relation to the brightness of the subject. An underexposure or overexposure of the subject will result unless the proper precautions are taken. These include holding the meter at such a distance from the subject as to include only the subject.

The design of some meters permits the removal of this hood so that the cell will respond to illumination from an almost 180-degree solid angle when making illumination measurements. When using a meter of this type, a different method (often called the incident-light method) is used. The meter is held close to the subject but pointed in the general direction of the camera. The meter reading indicates the illumination on the subject. Meters of this type usually include a provision in the calculator for arriving at the correct shutter speed and lens aperture. If not, the formula given on page 24–20 can be applied.

In motion-picture photography the lens aperture forms the only variable, inasmuch as the exposure time is fixed by picture frequency. Exposure meters designed for this work give *f*-numbers for a specific film speed.

Guide Number System. Since it is not practical to employ exposure meters in connection with the use of flash lamps, there has come into general use a system of guide numbers which greatly simplifies the statement and use of exposure information in connection with these sources.

Flash Lamp Photography

The five important elements affecting exposure in flash photography are:

1. Luminance of the subject (affected by light output of flash source used, reflector used, reflectance of subject).
2. Film speed rating.
3. Shutter timing.
4. Distance from the light source to the subject.
5. Lens aperture.

A photographer usually has a particular size or sizes of photographic flash lamp and reflector, a particular type of film, and an established practice as to the shutter speed he prefers to use. Thus, items 1, 2, and 3 are fixed and it is possible to combine them empirically to provide a guide number that is the product of the aperture (*f*-number) and the distance (feet) from subject to lamp. Since these are both second power functions and in inverse relationship, it remains merely to divide the guide number by the distance from lamp to subject to obtain the aperture setting. It becomes a simple matter to remember the guide number applicable to a particular lamp, film, and shutter speed. The guide-number system has been found useful also in conjunction with other light sources, when an exposure meter is not available.

The American National Standards Institute, through its Photographic committees, has worked out a standard procedure for obtaining guide numbers. This standard sets up the following formula:

$$\text{Guide Number} = \sqrt{0.004Lt \times M \times S}$$

where Lt = light output of the flash source, expressed in lumen-seconds
M = reflector factor
S = film speed rating (ASA system).

For "time," "bulb," and exposures to $\frac{1}{30}$ second the full lumen-second rating of the flash lamp may be used in the formula as the value of Lt. For higher shutter speeds the rated lumen-seconds of the lamp must be reduced since not all of the light will pass through the shutter during the time it is open. The following are approximate values for typical foil-filled lamps:

Shutter Speed (seconds)	Percentage of Rated Lumen-Seconds of Lamp
Open, $\frac{1}{25}$, $\frac{1}{30}$	100
$\frac{1}{50}$, $\frac{1}{60}$	70–80
$\frac{1}{100}$, $\frac{1}{125}$	55–65
$\frac{1}{200}$, $\frac{1}{250}$	35–40
$\frac{1}{400}$, $\frac{1}{500}$	18–23

Reflector factors are controlled by the contour, size, and surface finish of the reflector and by the physical size of the flash lamp.

Typical values are as follows:

Reflector Diameter (inches)	Finish	Flash Lamp Bulb Diameter (inches)	M
2	Polished*	$\frac{15}{32}$	9
3	Polished	$\frac{27}{32}$	8.5
4–6	Polished	$1\frac{1}{2}$	8
4–6	Satin	$1\frac{1}{2}$	5
6–7	Polished	$1\frac{7}{8}$	6
6–7	Polished	$2\frac{3}{8}$	4
Large studio	Polished	$2\frac{5}{8}$	6

The film speed ratings are usually supplied with the film or obtainable from the film manufacturers.

Exposures with flashtubes (see Section 8) depend principally on the watt-seconds output of the power supply unit, but are also influenced by the design of the reflector and the flashtube used. A chart applicable to flashtubes (see Fig. 24–15) gives recommended guide numbers for loadings up to 500 watt-seconds.

* Due to the wide range of reflectors in use for this small diameter flash lamp, this information is intended only as a guide for a good 2-inch polished parabolic reflector.

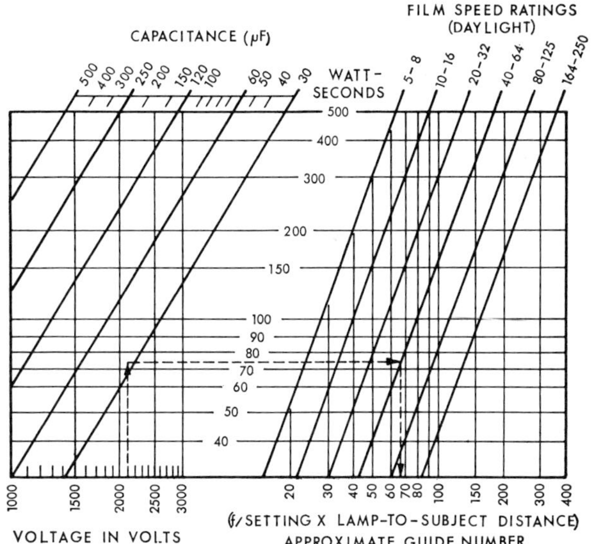

Fig. 24–15. Approximate guide numbers for flashtubes operated at various loadings.

Flash Lamp Synchronizers. For most flash lamp photography, it is desirable to operate the camera shutter at speeds of 1/125 to 1/250 of a second in order to minimize the effect of any ambient illumination and to stop motion. Since approximately 5 milliseconds is required for a shutter of the pre-set type to reach full opening and 17.5 to 20 milliseconds for the lamp to reach peak light output, the synchronizing device must first apply the current to the lamps and then 12.5 to 15 milliseconds later trip the shutter.

Two or three flashlight-type dry cells customarily are used to supply the igniting power. Within two or three milliseconds (0.002 to 0.003 second) a flash-lamp filament is heated to a sufficient temperature to ignite the priming material. In the case of lamps filled with shredded aluminum or zirconium wire, the burning primer sends a shower of sparks through this material, initiating its combustion at about 10 milliseconds. The burning metal reaches peak light output at about 17.5 to 20 milliseconds for the smaller and average-size lamps and at 30 milliseconds for the largest size. See Figs. 24–16 and 8–106.

A recent flash lamp development provides self-contained ignition energy in the form of a cocked spring which ignites the primer when triggered by the camera synchronizer, thus permitting simple batteryless flash lamp systems.

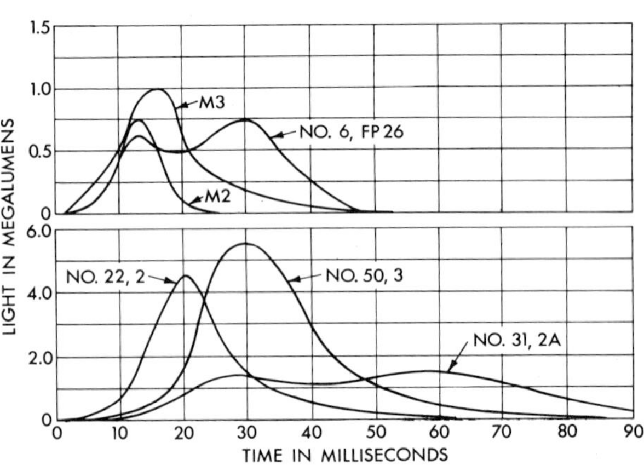

Fig. 24–16. Approximate time-light curves for several photoflash lamps; for further data see Fig. 8–106.

Photographic Lighting Equipment

Reflectors. Still camera lenses ordinarily cover an angle of about 45 degrees; therefore, for lighting equipment placed at or near the camera, reflector beam patterns for complete light utilization should fill an angle of about 45 degrees. However, difficulties caused by inaccurate aiming of the reflector and other variables are minimized by filling a 60-degree cone with reasonably uniform lighting levels. A luminaire with a 60-degree beam angle, of course, provides lower illumination toward the edges of the scene (slightly less than ½ of the center-beam value), but this is seldom objectionable since the point of interest in a picture is generally in the middle and a lower exposure at the edges is not serious.

The shadows and contrasts that help to light a person as normally seen are usually "soft," such as those produced by a light source of appreciable angular size. Large reflectors (16 to 24 inches in diameter) produce more natural modeling and are commonly used in portrait studios or other applications where size is not a handicap. Flash lamp equipment used with hand-carried cameras necessarily have smaller reflectors which produce somewhat sharper shadows (less natural).

Lens Spots. The lens spotlight frequently is used to provide a higher level of illumination over a limited and well-defined area. It is employed by professional motion picture, commercial, and portrait photographers and (in small sizes) by amateurs. In its usual form, it consists of a lens of either plano-convex or Fresnel type behind which is placed a concentrated source such as an arc or an incandescent lamp.

For incandescent-lamp spots, a spherical mirrored reflector is used behind the lamp to redirect back through the source light that otherwise would be wasted. Control of spot size is obtained by movement of the source to and from the lens.

Photographic Lighting Techniques

Successful photography requires a rather narrow range of illumination so that both the brightest parts (highlights) and the darkest parts (shadows) will be fully rendered in the final print or transparency. This range is much narrower than can be used for vision, particularly in the case of color photography. With typical subject reflectance ranges, the recommended maximum illumination range within a scene is 10 to 1 for black and white, and 4 to 1 for color film.

A general requirement peculiar to color photography is that the apparent color temperature of all of the light sources used must be the same. The eye readily accepts illumination of mixed color temperature; photographic film does not.

Another general requirement of photographic lighting arises from the monocular vision of the camera. To compensate for the lack of stereo depth, the best lighting on photographic subjects emphasizes their roundness, form, and spatial relationship. This is largely a matter of lighting direction, such as lighting from the side or the back.

In photography, two types of illumination are needed to produce a likeness of a subject:

1. *General illumination,* if used alone, produces a negative that is flat and without modeling. Such illumination does not produce prominent shadows, and density differences in the negative are caused for the most part by differences in the reflectance of various portions of the subject. This general, overall illumination goes by several names, among which

are front light, broad light, flat light, camera light, and basic light.

2. *Modeling light*, if used alone, produces a negative in which the highlights can be well exposed but the shadows are clear and show no detail at all. Modeling lights are usually highly directional and are used for the express purpose of casting shadows and forming highlights.

Background Luminance. A factor closely related to lighting is background luminance. For ordinary subjects, the background should not be very dark, very light, or too close behind the subject; neither should it be of exactly the same luminance as important parts of the subject, because such a condition would have the effect of merging the subject with the background. The less detail and the fewer the distracting spots in the background, the better.

Lighting Installation Photography

The photography of "existing light" installations does not require special photographic materials or equipment. Attention to focus, proper exposure, and composition are necessary; the use of a tripod is recommended.

The finished picture should represent what the actual installation looked like. Because most installations have a luminance range that exceeds the acceptance of the film, various techniques in the taking and finishing are used to compensate or "compress" the luminance of the scene to produce a satisfactory photograph.

In the taking of the picture, exposures may be split, "fill light" may be introduced, lamp shades may be lined, bulbs may be substituted. When fill light is used, great care must be taken not to introduce unnatural shadows into the scene. This may mean bouncing light off of walls or ceiling. In black and white photography, a 10 to 1 ratio of scene illumination (maximum) is desirable; in color, a 4 to 1 ratio (maximum).

The old rule of exposing for shadows and developing for highlights is still valid for black and white photography. Through experience and testing, the photographer finds the best combination of lighting technique, film, and processing that yields an easily printed negative. The printer can also use various methods of compensation such as dodging, flashing, and burning-in to overcome the deficiencies of a poorly executed negative.

When using color film, exposure should be for the highlight areas where detail is desired. It is also imperative that all the light sources be similar in color. Cool white fluorescent lamps cannot be combined with incandescent fill light, for example, without a noticeable color mismatch in the photograph. One combination of lamps that has proved generally compatible is incandescent lamps and warm white deluxe fluorescent lamps, as often used in residential photography.

Installations of discharge lamps photographed with color positive material present balancing problems. Because all films are designed to respond to a continuous spectrum and high-intensity discharge lamps have discontinuous spectrums, filtration in some degree is required to produce a realistic color balance. The exception is clear mercury which has no red and, therefore, can never be corrected properly.

Lamps and film manufacturers can generally furnish recommendations of suitable filters for balancing the various color-positive films to specific discharge lamps—both fluorescent and high-intensity types.

A much simpler procedure is to photograph the installation on unfiltered *color negative* material and work with a professional color laboratory to produce prints or slides with the correct color—the necessary filtration being done in the laboratory.

Professional Motion Picture Photography

Motion picture set lighting is both a science and a creative art. The objectives in the entertainment type of motion picture photography from the standpoint of light are twofold. First, it is necessary that sufficient light be used to properly expose the film, and second, that the types and sizes of units are available which will give the director of photography or cameraman a maximum of control over the illumination level, distribution of light, and color temperature.

The entire illusion created by motion pictures is done with light. The director of photography uses light as the painter uses pigment. In order to obtain depth, roundness, smooth skin texture, color and tonal separation, sharp shadows, no shadows, streak-light (as in sunlight coming through a window or doorway), or the effect of a single source when using a hundred or more lights, he must have not only a wide assortment of light sources but all practical devices for the control of such sources.

The basis of motion picture set lighting is the "key-light" which simulates an actual indicated source appearing in the scene (a window or a table lamp, for example), or arbitrarily establishes a source if none is indicated. It is almost always the brightest area of light in the scene and is the reference point which the cinematographer uses to calculate basic exposure.

Customarily, key-light is established at a level which will permit a negative of the desired density for optimum print quality. All other illumination on the set (such as "fill" or back lighting) is arranged for position and level in relation to the key-light. The balance to be achieved between the key-light and the secondary illumination is largely dependent upon the psychological, emotional and dramatic mood required in the scene to be photographed. These are artistic considerations dictated by the script and/or the director's style.

When it is realized that the director of photography is actually making a series of snapshots of approximately 1/48-second exposure; that the characters themselves are moving about; and that the camera is often traveling on a dolly, or swinging through the air on a camera crane, it becomes apparent that motion picture photography is really an art as well as a science, and all lighting equipment and accessories should be designed for maximum operational latitude and versatility even if sacrifices in engineering efficiency are to be made.

Since the cinematographer must deal with mechanical problems as well as artistic intangibles, the palate is somewhat more limited than that of the painter who can take great liberties in the use of basic materials. The cinematographer is restricted by the speed of the film stock, the technical characteristics of lenses, and the shifting spatial relationships of the elements within a given scene. For example, the slower the film, the more the aperture of the lens must be opened—with a proportional decrease in the depth of field. Thus, when using a film stock rated as ASA 64, an illumination level of 160 footcandles must be available to permit shooting at *f*/2.8. This *f*-stop has its own sharply defined depth of field; however, if the available level is only 40 footcandles, the lens aperature must be opened to *f*/1.4, with a resultant decrease in depth of field. Conversely, if the artistic requirement of the scene demands a greater depth of field (*e.g.*, that which is available at *f*/5.6), then the level will have to be raised to 640 footcandles.

Black and white production is complex since it requires that visual effect be achieved by *chiarescuro*, or black and white shadow patterns. This permits the use of dimmer control of individual luminaires without concern for the effect of variations in color temperature. In color motion picture production, the cinematographer has the additional tool of color, but with the restriction that color temperature be maintained so that it is consistent within each scene. As a result, dimmers can only be used in a restricted fashion.

Professional color film is manufactured to meet either of two basic lighting conditions. "Daylight" balanced emulsions are created for exposure under conditions approximating 5600 K. "Tungsten" balanced emulsions are for exposure with approximately 3200 K. However, it should be understood that there is a reasonable degree of latitude in the color temperature range that each emulsion will reasonably expose.

The acceptable ranges for the color temperature of the exposure lighting cannot be precisely specified. With most professional film, variations may be readily corrected in the laboratory so long as the following limits are not exceeded:

Daylight 5600 ± 600 K
Tungsten 3200 ± 200 K

Since the only "objective" standard is one that must conform to reality, the subjective test of the "skintone" of a human face is usually chosen as the standard when final print corrections are determined.

Light Sources. The use of lighting in "naturally" illuminated situations, requires great care in order to be certain that severe color distortions are not introduced. These may be uncorrectable in the laboratory due to the distorted character of the mixture. Examples might be (1) an interior location "naturally" lighted with fluorescent light and supplemented with incandescent sources, and (2) a natural exterior (daylight) where electric lighting is used as supplemental fill to reduce shadow contrast.

Incandescent filament, including tungsten-halogen types, are available in a wide variety of forms. For motion picture production lighting, see Fig. 8–90. Among the most interesting types developed for film use are the dichroic coated PAR-36 and PAR-64 types which emit a "daylight corrected" light. Carbon arc sources can be used for mixing with 3200 K incandescent light sources, with the use of a yellow-flame carbon and YF-101 filter. See Fig. 24–17.

Daylight sources are used under quite different conditions from those involving the 3200 K sources. When the 3200 K sources are used they usually provide all of the illumination that the cinematographer may require on the stage. Daylight sources are usually used where the predominant light is daylight itself. "Daylight" is a broad term; its color temperature varies widely depending upon the time of day, the proportion of sunlight to skylight, the existence and character of clouds, etc. "Daylight" sources are basically used as supplemental lighting to modify the effect of real daylight with respect to varying color temperature and density of shadows. Daylight sources include:

1. High-wattage carbon arcs. These operate on dc current and require a resistance "grid" (ballast) to provide the proper arc voltage when operating from 120-volt systems. The sizes range from 120-amp arcs to 350-amp arcs.

2. Short-arc, high-pressure xenon sources have recently been used for this application. They are available in sizes up to 4.5 kilowatts and require a somewhat complex power supply and high-voltage start system.

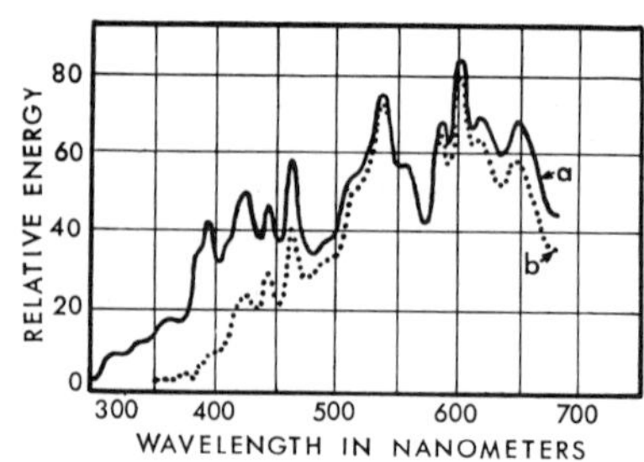

Fig. 24–17. Spectral energy distribution curves for yellow flame carbon. a. Without filter. b. With YF-101 filter.

In the past several years, because of increasing studio costs, plus a greater emphasis upon dramatic realism, there has been an increasing trend toward shooting on locations. This trend includes not only the filming of exterior scenes, but those which take place within actual interiors as well. This modification of production procedure, which will surely become more prevalent, demands the development of lighter weight, more highly efficient filming equipment in all categories—most definitely including lighting units.

Luminaires. The selection of luminaires is dictated more by the dramatic requirements of the script than by the size of a particular set. In general, motion picture production requires what is called "controlled" lighting and because of this the Fresnel lens spotlight is the most popular type of luminaire. Fresnel lens luminaires offer a controllable beam ranging from an 8- to 10-degree "spot" to a 30- to 45-degree "flood." Because they can be successfully "barn-doored," they offer the kind of control that permits the cinematographer to compose his lighting.

The introduction of the compact and the coiled-coil filaments of tungsten-halogen lamps has resulted in the design of a new type of luminaire which produces "semi-controlled" illumination. These luminaires are "focusing" but their "barn-door" control is very limited. They do not use a lens to achieve their focusing ability. This is done by moving the filament with respect to the reflector or vice versa. Since there is direct illumination from the filament itself, in addition to the varying illumination from the functioning of the reflector, it is possible to change the level and coverage of the light, but the shadow tends to be double. This restricts predictable control of the "barn-dooring" function. The effective lumen output is quite good, since very high collection efficiencies are possible and there is no loss through a lens. In general, these luminaires are very light in weight and very compact. They are available in wattages up to 2kW.

Another important luminaire in film production is the "shadowless" or "soft" light. Since the cinematographer must always control the "scene" it is desirable to be able to illuminate the subject and reduce the shadows (wrinkles on a face, etc.) on the subject and behind the subject. This can be achieved with the uses of large, even sources. Large incandescent lamps have been used in luminaires such as the bulky, generally awkward and inefficient "cone" light. The long straight filaments of certain of the tungsten-halogen lamps have permitted development of more manageable and efficient soft lights, since the filaments are easily hidden.

The compactness of tungsten-halogen lamps has led to the redesign of other luminaires and the creation of new and more convenient special luminaires. The long filament of some of the tungsten-halogen lamps has permitted a positive redesign of the family of luminaires called "broads". These luminaires are excellent for broad, general base illumination and provide a very flat pattern with high efficiency. A miniature version of this type of luminaire has been developed. It provides very flat illumination and can be barn-doored to a very sharp cut-off in the horizontal direction. These small luminaires can be placed in tight areas and permit the cinematographer greater latitude in developing his desired effects.

The development of tungsten-halogen PAR lamps with compact filaments, plus the development of dichroic coatings which can survive the high temperatures, have resulted in an entire line of luminaires for daylight fill use on location.[24] The most popular luminaires of this type utilize either PAR-36 lamps, of about 650 watts each, or PAR 64, 1000-watt units. Since the reflector and the precise positioning of the filament result in illumination of considerable "punch," they can provide powerful daylight-type illumination when used in clusters. The loss of light due to the dichroic filter is offset by raising the color temperature of the tungsten source to approximately 5600 K. The weight and size per lumen output of these luminaires is very favorable, as compared to the daylight arcs, so that they have become exceptionally popular on location.

A relatively new luminaire for daylight "fill" is the "focusing" xenon compact arc light. This features a minimum amount of focusing control, and a 4kW xenon arc provides a light-weight luminaire with good output.

PICTURE PROJECTION LIGHTING

Satisfactory picture projection requires not only careful selection of light source and optical elements for projecting the picture but also of the screen and its surroundings in relation to the seating area from which it is to be viewed.[25] The basic requirement is that the picture luminance be of a value such that the proper contrasts of highlights and shadows are achieved at a satisfactory over-all level.[26] Some illumination in the seating area is essential for the convenience of the audience, safety, discipline, etc.[27] See Theatres, page 24-6. However, if light from the seating area is allowed to fall on the screen the desired contrasts are reduced, and the over-all luminance must be increased to restore the proper relations.[28]

The logical place to start in planning projection is the area in which the pictures will be shown. This establishes the luminance level, the type of screen to be used, and the amount of light needed from the projector.

Luminance Levels for Motion Pictures

Luminance levels recommended by the Society of Motion Picture and Television Engineers (SMPTE) are predicated upon the presence of minimum stray light on the screen itself and a practicable balance between characteristics of photographic materials and available light. The weight given these factors, and therefore the determination of the most desirable projection conditions, differs somewhat among the five typical applications of projected pictures: (1) review rooms, (2) theatres, (3) drive-in theatres, (4) auditoriums, and (5) classrooms.

Definitions. Screen measurements in motion picture projection are made with the projection light source adjusted to normal operating conditions, with the projector shutter running, and with no film in the gate. This screen luminance level is approximately 10 times the average luminance of the pictures projected from normal films.[28, 29] Generally, the incident light falling upon the projection screen decreases from the center toward the edges; the incident light measured at points on the horizontal center line which are at a distance in from the screen edge equal to 5 per cent of the screen width, is normally 60 to 80 per cent of the illumination at the center.[30] The luminance of the screen from the audience position may parallel the variation of the incident illumination, or may vary from it in a more complex pattern as indicated in the description of the properties of screen surfaces.[31, 32, 33]

Review Rooms. During the preparation of motion pictures, the producer, the motion picture film laboratory personnel, and others examine the film many times from the original test shots through many stages to the final release print. The films are projected in a specialized theatre known as a "review room." These installations are designed specifically for the inspection of motion pictures and are built to accommodate a small reviewing group of usually 10 to 20 people. The actual picture size may be small or large depending upon the space available, but the viewing conditions are chosen to duplicate as nearly as possible actual theatre viewing from the most desirable seating locations. All of the viewing conditions are capable of precise control and it is generally practical in review rooms to hold these variables to a minimum tolerance.

For 35mm review rooms, luminance levels are generally based upon the standard[34] for screen luminance and viewing conditions for 35mm review rooms. This provides for an aim luminance of 16 ± 2 footlamberts at a color temperature of 5400 K ± 400 K, at the center of the screen.

For 16mm review rooms there is a specific standard[35] specifying 16 ± 2 footlamberts at the center of the screen.

Theatres. Facilities constructed primarily for the presentation of projected pictures to an audience are designed for a minimum of stray light, adequate and permanent projection facilities, optimum viewing angles, screen sizes and surfaces adequate for the facility in which they are installed, etc. Under these optimum conditions the audience receives the best in projected pictures.

For 35mm theatre projection the standard[36] specifies 16 ±2 footlamberts at the center of the screen. The Projection Practice Committee of the SMPTE is currently working on the inclusion of other important characteristics in this specification; it is suggested that current reports of the Committee be consulted.

For 16mm projection in theatre installations, the same luminance aims are applied as for 35mm.

Drive-In Theatres. Viewing conditions are quite different from those of the indoor theatres with perhaps the most significant change being the much smaller angle which the screen subtends at the observer's eye; this in turn changes several psychophysical factors. Permissible screen luminances have been shown by experience to be lower than those for comparable quality projected images in indoor theatres, even though the drive-in projects the same identical standard prints. An applicable luminance standard has not yet been devised. As a guide, the SMPTE has issued a recommended practice[37] in which the minimum screen luminance in drive-ins is 4.5 footlamberts.

Auditoriums. These are facilities where provision is made for projecting motion pictures to an audience although this is not the sole use of the facility and compromises may be made in the interest of other functions. Stray light may not be as well controlled, projection facilities are more limited, screen sizes may be smaller, etc. It is suggested, nevertheless, that the theatre conditions be used as a design aim.

Classrooms. Frequently projection in classrooms is handicapped by an irreducibly high stray light level, generally temporary projection equipment and screens, and frequently little control of seating arrangements. Under these conditions optimum pictorial quality can seldom be achieved, although a satisfactory presentation of information to the classroom group is frequently possible. Acceptable luminance levels are 16 ± 2 footlamberts[38]; consideration should be given to many other factors in the selection of the particular room, its equipment, etc.[39, 40]

Luminance Levels for Slide Projection, Etc.

There has been less study of optimum projection conditions for slides, slide films, opaque projectors, etc., but all proposed standards have agreed on 10 footlamberts as a desirable aim luminance. The projection of continuous-tone pictures seems to be governed by the same factors as affect projection of conventional motion pictures; the projection of line drawings, tables, and other high contrast images is less critical of luminance level and less critical of stray light, as discussed in the following section.

Surround Illumination

Illumination levels on surfaces other than the projection screen are controlled for both esthetic and practical reasons. In review rooms illumination levels must be adequate for limited movement among a small group; in theatres and auditoriums provision must be made for complete seating of the audience in semi-darkness, and for reasonable safety, etc.; in drive-in theatres illumination must permit both automobile and pedestrian traffic; in classrooms illumination should permit note-taking—although the usual problem is the adequate reduction of ambient illumination. In all applications visual comfort is desired.

In the design of ambient lighting the two prime requirements are avoidance of all high luminance sources visible to the audience, and direction of the light so that as little as possible falls onto the screen surface.[28] (With directional screen materials the disadvantages of stray light will depend upon both its level and its angle of incidence upon the screen.) Good pictorial reproduction demands an adequate tone scale in the projected picture; maximum highlight luminance is limited by and cannot exceed screen luminance; minimum shadow luminance must be at least equal to the stray light luminance. The ambient illumination must be so designed and directed as to hold stray light luminance of the screen to as low a value as possible; suggested luminance ratios are listed in Fig. 24-18.

Screen Types

Screens can be classified generally as reflective or translucent, depending upon whether the projected picture is viewed from the same side as the projector or from the opposite side. Reflective types may be either directional or nondirectional depending upon whether or not the luminance changes with viewing angle. Translucent screens generally are of the directional type. All reflective theatre screens are perforated to permit sound transmission from speakers located behind the screen; these perforations represent about 10 per cent of the total surface area.

The directional characteristics of a screen surface result from the nature of the reflection or transmission, and produce a variation in luminance as a function of both incidence and viewing angles.[33] For screens showing significant directionality, it is customary to curve the screen surface into a portion of a cylindrical surface, with axis vertical and radius equal to the distance from projector to screen; this minimizes the variations in incidence angles.

The general nature of such reflective patterns is suggested in Fig. 24-19. Off-axis response introduces many complexities which should be studied in detail.[31, 32, 33, 41] The number of special screen surfaces is increasing rapidly but they may be classified among the following general types:

Fig. 24-18. Permissible Luminance Ratios for Projected Pictures: Conventional Screen Luminance[a] vs Screen Luminance Resulting from All Non-Image Luminance[b]

Class of Projection	Projection Facility[c] Type 1	2	3	4	5	Type of Material	Recommended Ratio[d]
AA	×	×	×	×		Motion pictures at optimum.	500:1
A			×	×	×	Full scale black and white and color, where pictorial values are important and color differences must be discriminated.	100:1
B					×	Color diagrams and continuous tone black and white in high key.	25:1
C					×	Simple line material such as text, tables, diagrams, and graphs.	5:1

[a] Measured with no film in the aperture; therefore maximum image highlight luminance will normally be 25 to 60 per cent of the screen luminance.[29]
[b] Measured with the projection lens capped; therefore minimum shadow luminance will approach this screen luminance resulting from all non-projected illumination.
[c] Type of projection facilities: 1—Review Room
2—Theatre
3—Drive-In Theatre
4—Auditorium
5—Classroom.
[d] Reference 28.

Matte Surface. Such surfaces are practically non-directional. In other words their luminance is substantially the same at all viewing angles. Practical reflective matte screens have surfaces of high reflectance but since the light is distributed throughout a complete hemisphere the maximum attainable luminance is limited. On typical clean, new screens an incident illumination of 1 footcandle produces a luminance of 0.75 to 0.90 footlamberts. Matte surfaces are recommended where viewers will be distributed over a wide angle in the viewing area.

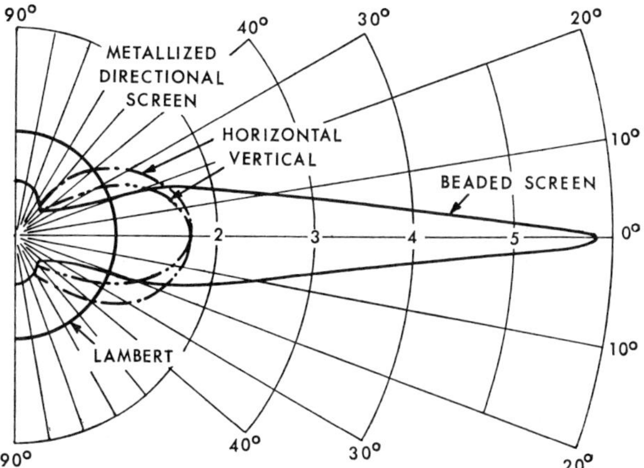

Fig. 24-19. Luminance patterns for typical screens. Goniometric distribution for a non-absorbing perfect diffuser (Lambert), a beaded screen, and a metallized screen showing different horizontal and vertical patterns. The radial scale is graduated in footlamberts observed under illumination of one footcandle.

They are applied in theatres, auditoriums, and classrooms whenever the screen sizes and available light permit a satisfactorily bright picture. Translucent matte screens have such a low transmittance that they are not of practical importance. Whenever a higher luminance than that available from matte screens is desired, one of the following types of directional screens may be specified if their more restrictive installation requirements can be met.

Semi-Matte Surface. In this type of surface a material is incorporated in the paint or surface finish to give it a slight gloss. This can be done in such a way that it improves the reflectance or "gain" over a certain distribution angle. When carefully done, with central luminance gain held to values not in excess of 1.5 to 2, an over-all improvement in picture luminance can be obtained without excessive "hot-spotting" or glare appearing near that point on the screen where an image of the projector lens would be reflected by a specular reflector.

Metallized Surface. A metallic pigment (usually aluminum) may be incorporated in a paint or otherwise applied to the surface of a suitable supporting material to make a useful reflective screen. Such a surface is mostly very directional but can be made to vary in its reflection characteristics. Because of this versatility these screens can be "tailor-made" to give maximum luminance consistent with acceptable viewing in any given situation. Directional screens of this type will show a "hot-spot". Furthermore, plane screens will appear brighter on the near than on the far side when viewed from side positions. Such effects can be minimized by curving the screens approximately to the projection radius. Therefore, installations of screens of this type should be made only after careful consideration has been given to all viewing factors, and then in accordance with specifications carefully drawn with the special characteristics of the screen in mind. A metallized screen must be provided whenever projection of 3-D pictures or slides with polarized light is anticipated.

Lenticulated Surface. Practical reflective screens have been developed with small uniformly-shaped and spaced lens-elements impressed in the surface. These control the direction of light reflection so that the maximum luminance will be obtained within certain specified viewing angles. Within these angles, moreover, the luminance is generally more uniform than for non-lenticulated types of directional screens. The highest luminance for a given incident illumination is obtained with screens having lenticulated surfaces. Since these are even more specialized in their characteristics than are the metallized surfaces, the recommendations of the manufacturers must be followed in their selection and installation if satisfactory results are to be achieved.

Practical lenticulated translucent screens have been produced so far only in limited sizes; this has restricted their use to specialized applications such as projection under conditions of high ambient light, where the controlled luminance of a small lenticulated screen permits the presentation of acceptable pictures.

Beaded Surface. This may be either a reflective or translucent screen; the surface is covered with small glass or plastic spheres which reflect or transmit most of the light in the general direction of the projection axis. Such a screen will appear the brightest when viewed along the axis of projection and will darken quite rapidly as the viewing angle increases away from this axis. The property of reflecting a high proportion of the light back to its source offers some control of stray light from sources that are not located in line with the audience. Such screens are useful for homes and small classrooms where the viewing angle requires limited horizontal distribution.

Maximum and Minimum Viewing Distances

Viewing distances are determined by a number of factors including resolution of picture detail,[42] limitations of graininess and sharpness in the projected image, psycho-physical impressions of presence and identity within the projected action, limitations of comfortable viewing posture, structural limitations of viewing angle, etc.

Over the past decade there has been intensive effort in the motion picture industry to increase the horizontal viewing angles so that the viewer can more readily identify himself with the projected story, and simultaneously to reduce the imperfections in the projected image in order to produce a more complete illusion of reality. At the same time the vertical viewing angles have remained relatively constant since they are more effectively limited by design of theatres and auditoriums (especially balcony overhangs), etc. For design purposes, therefore, Fig. 24–20 may be taken as an approximate guide. Improvements in picture sharpness, graininess, and resolution eventually may make closer viewing positions possible, but further studies on the psycho-physical response are still not conclusive concerning the so-called wide-screen presentation methods.

Limitation of Viewing Angle

The angle (measured from the projection axis) within which an audience may see a presentable picture depends upon both geometric distortions and luminance variations on the screens. Objectionable geometric distortions of the image on a flat screen become apparent at angles beyond approximately 30 degrees from the normal to the screen. Since screen surfaces may be curved cylindrically to obtain better luminance distribution from direc-

Fig. 24-20. Limitations of Viewing Distances and Angles[a]

	Review Rooms[b]	Theatres and Auditoriums[c]	Drive-in Theatres[d]	Classrooms[e]	
				Matte Screens	Beaded Screens[f]
Front row of seats		1.2	2.8		
Minimum viewing distance	3.0			2.7	3.3
Maximum viewing distance	5.0	8.0	14	8.0	8.0
Maximum viewing angle (degrees)	10	30	30	30	20

[a] Distances are measured from the screen surface in multiples of picture height; angles are measured from the screen normal in degrees.

[b] Suggested values; no current standard.

[c] Recommendations: Society of Motion Picture and Television Engineers Theatre Engineering Committee.

[d] Values typical of present practice; no formal recommendations or standards available.

[e] Recommendations: Society of Motion Picture and Television Engineers Committee on Non-Theatrical Equipment.[38]

[f] These values should also be used for metallized and lenticular screens unless the manufacturer's data indicate that performance at greater viewing angles is satisfactory.

tional screens or to increase audience participation, geometrically desirable viewing angles may thereby be further limited. In addition, since all of the directional screens show some luminance fall-off as the viewing angle is increased, acceptable maximum values are also set by the point at which screen luminance falls below the minimum tolerable.

Practical viewing angles are determined by both the geometry of the theatre and the goniometric reflection characteristics of the screen surface. It is suggested that these two factors be so adjusted that at the maximum useful viewing angle, the luminance of the center of the screen be no less than half its maximum luminance.

Projected Screen Image Dimensions

The projected screen image size is determined by three factors, namely: projector aperture dimensions; focal length of projection lens; and throw, *i.e.*, distance from lens to screen. In motion picture projection, be it 8, 16, 35, or 70mm, the projector aperture is always smaller than the available printed image on the film to allow for printer aperture variations, height unsteadiness, and side-weave. The projected image area through the aperture is further masked on the screen to permit a clean-cut picture outline with a minute "spill" being absorbed by the black velvet masking. In still, slide, or filmstrip projection, the screen image borders are the same as determined by the original mounts.

The ratio of width to height of the projected or finally masked screen image is referred to as aspect ratio and should be considered as an esthetic rather than a technical measurement.

For many years motion picture projection was limited to 1.33:1 aspect ratio which still prevails for the 8mm format and 16mm and 35mm films for television. Theatrical wide-screen systems from a single piece of film have, in search for the optimum audience appeal, greatly departed from that aspect ratio. In 35mm, two methods are generally adopted. Using linear, non-anamorphic lenses, pictures are presented as "wide screen" productions in 1.66:1, 1.75:1, or 1.85:1 aspect ratios, the last of these being the most commonly used.

Introducing an anamorphic "squeeze" ratio of 2:1 led to establishing an aspect ratio of 2.35:1 for productions justifying more "scope" for stories so told with 35mm film. There is of course an increasing amount of 16mm prints now available with 2:1 "squeezed" image information which will result in presentation aspect ratios of 2.35:1 up to 2.66:1.

The actual screens used in any projection system must of course be oversized but will have approximately the aspect ratio of the system involved. In theatrical installations it is customary to use the largest screen that can be accommodated on the stage. All kinds of aspect ratios are presented with variable screen masking, mostly remote controlled.

Since filmstrips and all professional slides are designed for aspect ratios of 1.33:1 or greater, and the 2 x 2 slides can be positioned either horizontally or vertically, it is necessary to employ a square screen for such showings.

The basic factors to determine the requirements for any linear, *non-anamorphic* projection set-up are related as follows:

$$\text{Focal Length} = \frac{\text{Aperture Width} \times \text{Throw}}{\text{Picture Width}}$$

$$\text{Picture Width} = \frac{\text{Aperture Width} \times \text{Throw}}{\text{Focal Length}}$$

$$\text{Throw} = \frac{\text{Focal Length} \times \text{Picture Width}}{\text{Aperture Width}}$$

$$\text{Aspect Ratio} = \frac{\text{Picture Width}}{\text{Picture Height}} \text{ or } \frac{\text{Aperture Width}}{\text{Aperture Height}}$$

$$\text{Picture Height} = \frac{\text{Picture Width}}{\text{Aspect Ratio}}$$

In linear projection the aspect ratio of the projected image is equal to the aspect ratio of the projector aperture or slide mount. This is not the case involving projection of *anamorphic* or "squeezed" pictures. Their requirements for a projection set-up are as follows:

$$\text{Focal Length} = \frac{\text{Aperture Width} \times \text{Squeeze Ratio} \times \text{Throw}}{\text{Picture Width}}$$

$$\text{Picture Width} = \frac{\text{Aperture Width} \times \text{Squeeze Ratio} \times \text{Throw}}{\text{Focal Length}}$$

$$\text{Throw} = \frac{\text{Focal Length} \times \text{Picture Width}}{\text{Aperture Width} \times \text{Squeeze Ratio}}$$

$$\text{Screen Image Aspect Ratio} = \frac{\text{Picture Width}}{\text{Picture Height}} \text{ or } \frac{\text{Aperture Width} \times \text{Squeeze Ratio}}{\text{Aperture Height}}$$

Due to changing projection and wide screen systems, as well as standards, most tables compiling such data as picture sizes, aspect ratios, and available focal length of lenses become easily obsolete or misleading. The preceding formulas permit the calculation of all data required for each individual projection condition. The presently used aperture dimensions are listed in Fig. 24–21. Fig. 24–22 lists picture widths.

Wide Screen and Special Processes

With the expansion of the screen in theatres different production and presentation methods were devised.

These processes can be classified into two categories according to film formats used,[43, 44, 45] namely:

1. Standard 35mm film plus optical devices for wide screen presentation.
2. 65mm camera negative and 70mm release print methods.

There were several special processes in use utilizing multiple negative and multiple print methods such as in Cinerama with three 35mm prints or Circarama whereby nine camera originals and prints are used for a 360-degree picture presentation method.

The aim of these systems is to obtain a "wrap-around" or "audience participation effect" to enhance the story telling impact. Due to inherent defects in "lacing" projection methods such as Cinerama it is now attempted to utilize a single 70mm print and project it into the deeply curved screen. A recent addition to these special processes is called "Dimension 150," meaning that a 150-degree field of view can be used and projected to the viewer. Numerous other systems have been published, most of which did not reach commercial application for entertainment purposes.

Projection Booths[46]

For detailed information on the design and facilities recommended for the projection room and the screen presentation in theatres it is suggested that the reports of the Projection Practice Committee of the Society of Motion Picture and Television Engineers be studied in the *Journal of the Society of Motion Picture and Television Engineers.*

Required Light Output of Projectors

In order to determine the required light ouput of a projector, it is necessary to know the picture size that satisfies the viewing conditions and the average reflectance at the applicable viewing angles of the screen to be used. With this information, the lumens required to meet the luminance recommendations can be calculated by the formula:

$$\phi_s = \frac{L_c \times K_d \times A}{L_f}$$

where ϕ_s = Luminous flux reaching the screen, in lumens.

L_c = Luminance of the screen at its center in footlamberts.

Fig. 24–21. Projector Aperture Dimensions

Projection Mode	Film Size	Aperture		Projection		Applicable ANSI Standard[b]
		Width (inches)	Height (inches)	Optics	Aspect Ratio[a]	
Motion Picture Film						
Standard	8mm	0.172	0.129	Normal	1.33/1	PH22.20—1969
Standard	16mm	0.380	0.284	Normal	1.33/1	PH22.8—1969
Anamorphic Print	16mm	0.380	0.284	Anamorphic[c]	2.66/1	
Standard (TV)	35mm	0.825	0.600	Normal	1.33/1	PH22.58—1954
"Wide Screen"	35mm	0.825	0.447	Normal	1.85/1	PH22.58—1954
Anamorphic Print	35mm	0.839	0.715	Anamorphic[c]	2.35/1	PH22.106—1965 (R 1969)
"Todd AO"						
"Super Panavision"	70mm	1.913	0.868	Normal	2.21/1	
"Ultra Panavision"	70mm	1.913	0.868	Anamorphic[d]	2.35/1 plus	
Slides, etc.						
2 x 2 Slides[e]	35mm	0.906	1.344	Normal	0.67/1	
Filmstrips	35mm	0.906	0.680	Normal	1.33/1	PH1.24—1955
2 x 2 Slides[e]	35mm	1.344	0.906	Normal	1.48/1	PH3.43—1969
Lantern Slides	4 x 3¼ inches	3.00	2.25	Normal	1.33	PH3.43—1969

[a] Aspect ratio is the quotient of projected width divided by projected height. This ratio will be reduced if the vertical projection angle is not 0 degrees.
[b] Present standards or latest revisions thereof as issued by the American National Standards Institute, New York, N.Y.
[c] These anamorphic processes require projection lenses with a horizontal image magnification greater than the vertical magnification by a ratio of 2:1.
[d] In present 70 mm anamorphic processes the horizontal magnification is 1.25:1 greater than the vertical.
[e] Slides in the 2 x 2 format will fit in the projector with the 1.344 inch frame dimension oriented either horizontally or vertically. Horizontal orientation is preferred for professional use.

K_d = Distribution weighting factor. This factor will be 1.00 for a screen which is uniformly illuminated; for the more general situation of side-to-center distributions of 60 to 80 per cent in illumination, factors of 0.72 to 0.86 are suggested.

A = Area of screen in square feet.

L_f = Luminance factor. The ratio of the luminance of the screen to the luminance of a nonabsorbing perfect diffuser receiving the same illumination. This factor may be a function of the angles of illumination and viewing.

Classroom Projection. Lumens-at-screen values to satisfy the recommended luminance values for classroom projection are given in Fig. 24–23 for several screen sizes. Only one set of values is given for

Fig. 24–22. Picture Sizes Obtained with Various Lenses and Projection Distances

Focal Length of Lens (inches)	Type of Projector	Lens to Screen Distance (feet)																		
		5	10	15	20	25	30	35	40	50	70	80	100	150	200	250	300	400	500	600
		Width of Picture (feet)																		
0.5	8mm	1.9	3.8	5.7																
1.0	8mm	1.0	1.9	2.8	3.8	4.8	5.7													
	16mm	1.8	3.8	5.6	7.5	9.4	11.3	13.2	15.0	18.8	26.3	30.0	37.6							
2.0	16mm	1.0	1.9	2.8	3.7	4.7	5.6	6.5	7.4	9.3	13.0	14.9	18.6	27.9	37.2					
	35mm			6.1	8.2	10.3	12.3	14.4	16.4	20.6	28.8	32.9	41.2	61.8	82.4	103.0	123.6	164.8		
	35mm Anamorphic			12.6	16.8	20.9	25.2	29.4	33.5	41.9	58.7	67.1	83.9	125.9	167.8					
2.5	16mm		1.5	2.3	3.0	3.8	4.5	5.3	6.0	7.5	10.4	11.9	14.9	22.4	29.9	37.3				
	35mm				6.6	8.2	9.9	11.5	13.2	16.5	23.1	26.4	33.0	49.5	66.0	82.5	99.0	132.0	165.0	
	35mm Anamorphic				13.4	16.8	20.1	23.5	26.8	33.5	47.0	53.7	67.1	100.7	134.2	167.8				
3.0	16mm		1.3	1.9	2.5	3.2	3.8	4.4	5.0	6.2	8.8	10.0	12.5	18.8	25.0	31.2	37.5			
	35mm					6.9	8.2	9.6	11.0	13.7	19.2	22.0	27.5	41.3	55.0	68.8	82.6	110.0	137.6	165.2
	35mm Anamorphic					13.9	16.8	19.5	22.3	27.9	39.1	44.7	55.9	83.9	111.9	139.8	167.8			
	2 x 2 Slide		4.7	7.2	9.3	11.6	14.0	16.3	18.6	23.2	32.5	37.1	46.4	69.6						
3.5	35mm					5.9	7.1	8.2	9.4	11.8	16.5	18.8	23.6	35.4	47.1	58.9	70.8	94.3	117.8	141.6
	35mm Anamorphic					11.9	14.4	16.8	19.2	23.9	33.5	38.3	47.9	71.9	95.9	119.8	143.8	191.8		
4.0	16mm			1.4	1.9	2.4	2.9	3.3	3.8	4.8	6.7	7.6	9.5	14.2	19.0	23.7	28.4	37.9		
	35mm						6.2	7.2	8.2	10.3	14.4	16.5	20.6	30.9	41.2	51.5	61.8	82.5	103.0	123.6
	35mm Anamorphic						12.6	14.6	16.8	20.9	29.3	33.5	42.0	62.9	83.9	104.8	125.8	177.8		
	2 x 2 Slide		3.6	5.4	7.1	8.8	10.6	12.4	14.1	17.6	24.6	28.1	35.0	52.5	70.0					
	Lantern Slide		9.1	14.2	18.8	23.5	28.2	32.9	37.5	46.8	65.4	74.8	93.3	140.0						
5.0	35mm							5.8	6.6	8.2	11.5	13.2	16.5	24.8	33.0	41.2	49.6	66.0	82.4	99.2
	35mm Anamorphic							11.8	13.4	16.7	23.5	26.8	33.6	50.3	67.1	84.0	100.6	134.2	168.0	
	2 x 2 Slide		2.9	4.2	5.6	7.0	8.4	9.7	11.1	13.8	19.8	22.1	27.6	41.2	56.0	68.8				
	Lantern Slide																			
6.0	35mm								5.5	6.9	9.6	11.0	13.8	20.6	27.5	34.5	41.2	55.0	69.0	82.4
	35mm Anamorphic								11.2	13.9	19.6	22.4	28.0	42.0	55.9	70.0	84.0	111.9	140.0	168.0
	2 x 2 Slide		2.4	3.5	4.7	5.8	7.0	8.1	9.3	11.6	16.0	18.3	23.0	34.5	46.0	57.5	69.0			
	Lantern Slide		6.5	9.6	12.7	15.8	18.9	21.9	25.0	31.2	43.6	49.8	62.0	93.1	124.0					
8.0	35mm									5.1	7.2	8.2	10.3	15.5	20.6	25.5	31.0	41.2	51.0	62.0
	35mm Anamorphic									10.5	14.7	16.8	21.0	31.5	42.0	52.4	63.0	83.9	104.8	126.0
	2 x 2 Slide		1.9	2.8	3.6	4.5	5.4	6.3	7.1	8.9	12.3	14.0	17.8	26.5	35.2	44.0	52.7	70.3		
	Lantern Slide		4.9	7.2	9.4	11.8	14.1	16.5	18.8	23.4	32.6	37.2	46.4	69.4	92.4	115.0	138.0			
10.0	35mm										5.8	6.6	8.2	12.4	16.5	20.6	24.8	33.0	41.2	49.6
	35mm Anamorphic										11.8	13.4	16.8	25.1	33.6	42.0	50.3	67.2	84.0	100.6
	Lantern Slide		4.0	5.9	7.7	9.5	11.4	13.2	15.1	18.8	26.1	29.8	37.2	55.3	74.1	92.6	111.0	148.0		
20.0	Lantern Slide		2.1	3.0	3.9	4.8	5.7	6.6	7.5	9.3	12.9	14.7	18.4	27.4	36.4	45.4	54.4	72.4	90.2	108.0

Fig. 24-23. Screen Requirements for Classroom Projection

1. Screen Lumen Requirements

Screen Size (feet)	Matte Screen		Beaded Screen
	Lumens for 5 Footlamberts	Lumens for 20 Footlamberts	Lumens for 5-20 Footlamberts
2.5 x 3.33	55	210	45
3 x 4	75	305	65
3.75 x 5	120	475	105
4.5 x 6	170	690	150
5.25 x 7	235	940	205
6 x 8	305	1,225	265
6.75 x 9	385	1,540	340
7.5 x 10	480	1,915	415
9 x 12	690	2,750	600
10.5 x 14	935	3,745	815
12 x 16	1,230	4,920	1,070

2. Screen Size Requirements[28]

Picture Width		Maximum Audience Size	
(inches)	(feet)	Matte Screen	Beaded Screen
40	3.3	35	20
50	4.2	50	30
60	5.0	75	45
70	5.8	100	60
84	7.0	150	90

beaded screens because the luminance differences encountered over the range of viewing positions embrace the recommended luminance range.*

Motion Picture Theatre Projection

In order to ensure sufficient screen luminance for proper viewing conditions[34-37] it is necessary to employ light sources of very high luminance to conform with these standards. In addition, the light source must be of a color quality permitting the faithful rendition of color motion pictures.[47, 48] For these reasons, carbon arc lamps are almost universally used for 35mm and 70mm projectors. Professional 16mm projectors are also operated with arc lamps, however of lower power.[49]

Carbon arcs ranging in luminance up to 100,000 candelas per square centimeter and higher are available. See Figs. 8-125 and 24-2 for characteristics of various arcs and carbons.

For efficient operation of an optical projection system all components must be compatible optically as well as mechanically.

In motion picture projectors, reflector and condenser type optical systems as shown in Fig. 24-24 are used. These focus an image of the positive carbon crater upon the aperture to illuminate the film which in turn is imaged by the projection lens on the screen. Fig. 24-25 shows the spectral energy distribution of the light on the screen with two typical high-intensity carbon arc combinations of which more data are given in Fig. 24-26 for both 35mm

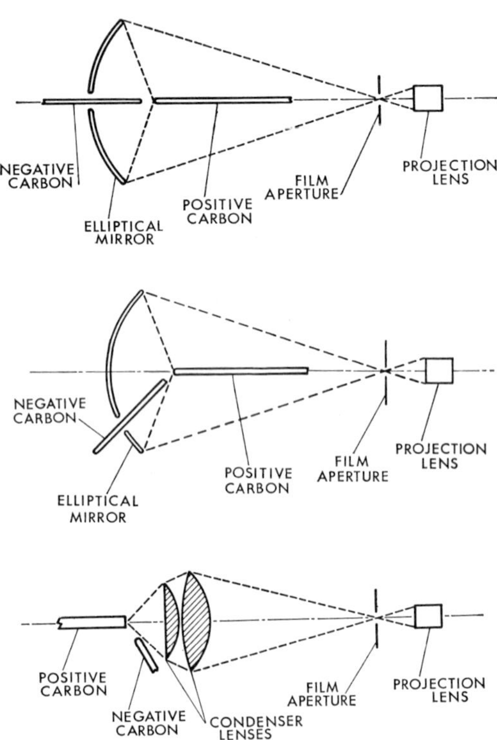

Fig. 24-24. Reflector and condenser optical systems for motion picture projection.

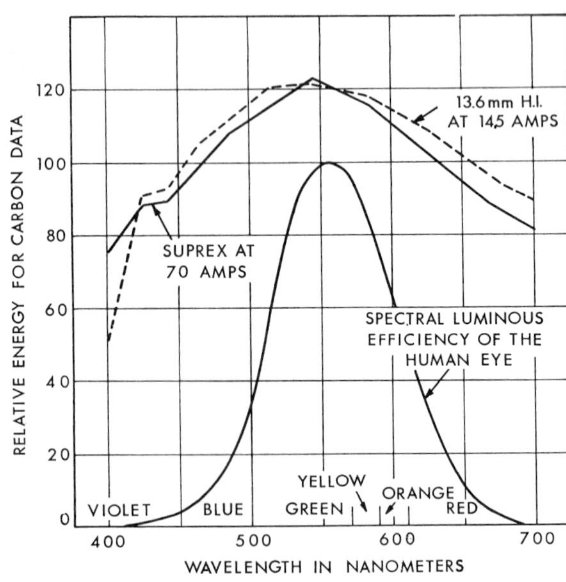

Fig. 24-25. Spectral energy distribution data for light at center of projection screen with Suprex-type lamp at 70 amperes, and condenser-type lamp at 145 amperes; heights of curves are adjusted to give equal footcandles.

* For comprehensive information see: "Foundation for Effective Audio Visual Projection," Kodak Pamphlet No. S-3, Eastman Kodak Co., Rochester 4, N. Y. and reference 54.

Fig. 24-26. Screen Illumination Provided by Typical High Intensity Carbon Arc Motion Picture Projection Systems

Carbons				Arc		Lamp Optical System	Screen at Maximum Center Light[f] Standard Aperture[a]								Screen at Maximum Center Light[f] Projection Aperture for Anamorphic Prints[b]							
Positive		Negative					Screen Lumens[d]	Distribution Ratio (per cent)	Footlamberts at Screen Center[e]						Screen Lumens[d]	Distribution Ratio (per cent)	Footlamberts at Screen Center[e]					
									Screen Width (feet)								Screen Width (feet)					
Diameter (mm)	Length (inches)	Diameter (mm)	Length (inches)	Amps	Volts				20	30	40	65	100	150			20	30	40	65	100	150
35mm Film Projection Systems																						
7	12 or 14	6	9	50	38	14″ OD $f/2.3$ Mirror[g]	11,700	60	20.3	9.0	5.1				14,000	60	20.9	9.3	5.2			
8	12 or 14	7	9	70	40	14″ OD $f/2.3$ Mirror[g]	15,700	60		12.1	6.8				18,800	60		12.5	7.0			
9	12 or 14	8	9	80	45	14″ OD $f/2.3$ Mirror[g]	16,600	65		12.3	6.9				19,900	65		12.7	7.1			
9	20	5/16″	9	85	55–60	16″ OD $f/1.9$ Mirror[h]	19,500	55		15.7	8.8				23,000	55		15.9	8.9			
10	20	11/32″	9	110	59–60	18″ OD $f/1.7$ Mirror[l]	25,200	60		19.4	10.9	4.1	1.8	0.8	31,600	60		21.0	11.8	4.5	1.9	0.8
10	20	1/2″	9	135	66–70	18″ OD $f/1.7$ Mirror[l]	30,300	65		22.5	12.6	4.8	2.0	0.9	36,200	65		23.1	13.0	4.9	2.1	0.9
11	20	3/8″	9	120	59–68	18″ OD $f/1.7$ Mirror[j]	27,600	60		21.3	12.0	4.5	1.9	0.8	34,700	60		23.0	12.9	4.9	2.1	0.9
11	20	5/16″	9	185	55	18″ OD $f/1.7$ Mirror[i]	14,900	84		9.3	5.2	2.0	0.84	0.37	17,900	84		9.7	5.4	2.05	0.87	0.39
11	20	11/32″	9	105	61	18″ OD $f/1.7$ Mirror[i]	18,600	84		11.6	6.5	2.5	1.05	0.46	22,300	84		12.0	6.8	2.6	1.08	0.48
13.6	22	1/2″	9	160	77	$f/2.0$ Condenser Lenses[g]	20,500	60		15.8	8.9	3.4	1.4	0.6	24,600	60		16.3	9.2	3.5	1.5	0.65
13.6	22	1/2″	9	180	74	$f/2.0$ Condenser Lenses[g]	24,000	60		18.5	10.4	3.9	1.7	0.75	29,800	60		19.8	11.1	4.2	1.8	0.8
13.6	20	3/8″	9	125	60	18″ OD $f/1.7$ Mirror[i]	22,000	65		16.2	9.1	3.5	1.5	0.65	26,400	65		16.9	9.5	3.6	1.5	0.66
13.6	20	7/16″	9	135	64	18″ OD $f/1.7$ Mirror[i]	23,500	65		17.3	9.75	3.7	1.6	0.693	28,200	65						
13.6	20	1/2″	9	160	74	18″ OD $f/1.7$ Mirror[i]	29,000	65		21.3	12.0	4.5	1.96	0.85	34,800	65						
10	25	7/16″	12	155	74[k]	21″ OD $f/1.7$ Mirror[j]	46,000	75		31.0	17.4	6.6	2.8	1.2	55,000	75		32.0	18.0	6.8	2.9	1.3
16mm Film Projection Systems									10	15	20	30										
6	8½	5.5	6	30	28	10¼″ OD $f/1.6$ Mirror[l]	2,900	70	18.1	8.1	4.3											
7	12	6	9	50	37	10¼″ OD $f/1.6$ Mirror[l]	5,800	70		16.1	9.1	4.0										
7	12	6	9	46	35	11⅜″ OD Mirror—Aperture field lens $f/1.5$ combination	6,300	70		17.5	9.8	4.4										

[a] Aperture—0.600 x 0.825 inch.
[b] Aperture—0.715 x 0.839 inch.
[c] The lamp system can be adjusted to vary the brightness between the side and center of the screen. The per cent ratio of the side to center screen light is the light distribution ratio. There will be a loss in total light whenever the per cent distribution exceeds the values listed.
[d] Screen lumen figure is for systems with no shutter, film, or filters of any kind.
[e] Footlambert figure at center of screen assumes 50 per cent shutter transmission, a perfectly diffusing screen with 75 per cent reflectance, and no film or filters of any sort. Directional screens provide luminance 2 to 3 times as great over restricted viewing angles.
[f] For aspect ratios between 1.33 and 2.35 total light is directly proportional to the area of the aperture.
[g] 5-inch focal length $f/2.0$ coated projection lens.
[h] 5-inch focal length $f/1.9$ coated projection lens.
[i] 4-½-inch focal length $f/1.7$ mirror interference coated.
[j] Mirror interference coased.
[k] Data refer to "blown-arc"; see text.
[l] 0.284- x 0.380-inch aperture, 2-inch focal length f 1.5-1.6 projection lens.

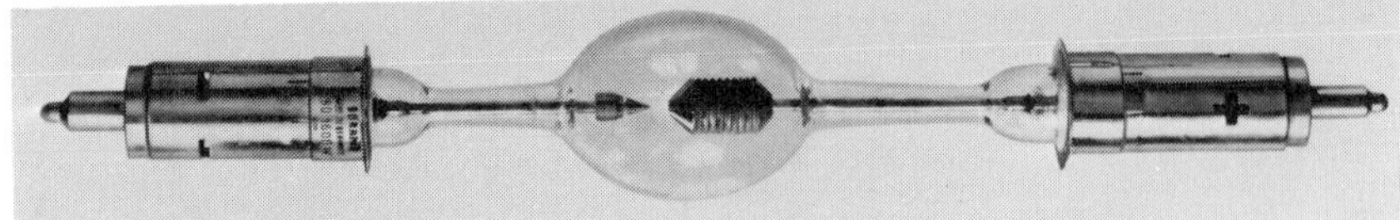

Fig. 24–27. Typical xenon short arc bulb (1600 watts) as used in motion picture projector lamp houses, showing connectors, quartz envelope, and tungsten electrodes.

and 16mm film. For highest luminance requirements, such as special 70mm operations and 35/70mm drive-in projection, a completely different arc lamp known as "blown-arc" or "jet-arc" is used.[50]

In order to avoid permanent damage to film, this arc, as most high-intensity carbon arc lamps, uses so-called interference type reflectors, mostly called "cold-mirrors," that do not reflect the infrared part of the spectrum, thus reducing heat on the film.[51, 52]

Motion Picture Projection with Xenon Lamps

A recent addition to the range of light sources for 35mm and 16mm motion picture projection, as well as for many still-projectors, is the xenon short-arc lamp.[53]

The structure of this lamp is quite simple. Two tungsten electrodes are enclosed in a fused quartz envelope, and the envelope is filled with xenon gas under pressure of 8 to 10 atmospheres in the cold stage, which rises to approximately 30 atmospheres when the lamp is in operation. A momentary high voltage, supplied by a special ignition unit or starter, is required to establish a conducting path across the gap between the electrodes, ionizing the gas, after which the lamp operates on low voltage direct current.

The ionized gas emits a light similar to daylight—approximately 6050 K and intensely bright. All of this light comes from the arc and not from the electrodes. It does not change in color temperature over its entire operating and/or luminance output range.

Lamps suitable for 35mm projection are presently rated at 900, 1600, and 2500 watts. Fig. 24–27 shows a xenon lamp of 1600-watt capacity which is extensively used in motion picture projection today.

The optical system conventionally used in conjunction with xenon lamps is illustrated in Fig. 24–28. Since the "arc" is somewhat bell-shaped an auxiliary condenser or field flattener is often inserted in 35mm projectors as specified at *D* in the schematic diagram. Light distribution ratios as high as 95 per cent can thus be obtained.

A complete lamphouse for a 35mm projector with a 1600-watt xenon bulb, main mirror, auxiliary mirror, flexible optical alignment shafts, and starter switches is shown in Fig. 24–29. Fig. 24–30 indicates lumen output for 10-footlambert screen luminance, including different picture widths for a matte white screen and a screen having a "gain" of 1.5.

A larger lamp of 6500-watt rating is available, the light intensity of which is approximately 75 per cent higher than that of 1600-watt lamp as shown.

Recently 450, 900 as well as 1600-watt xenon lamps have been successfully adapted to 16mm projectors. Many projectors, so far equipped with tungsten filament projection lamps, have been converted or, as it is called, "xenonized" to these new lamps which have a usable burning rate of between 1500 and 2000 hours and more.

The instantaneously available full light output,

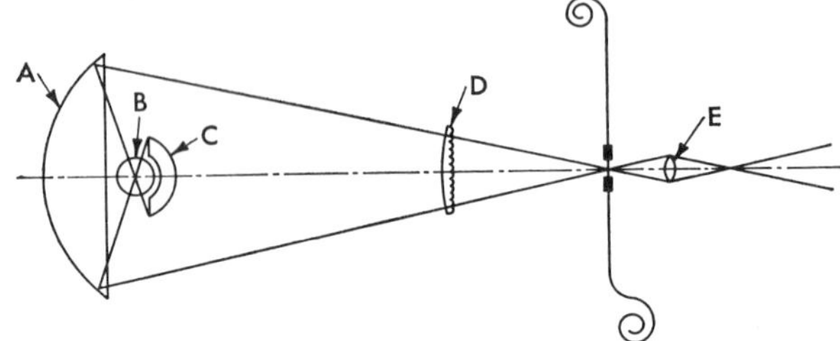

Fig. 24–28. Schematic of optical system for use of xenon lamp in motion picture projection. A—reflector. B—lamp. C—auxiliary mirror. D—auxiliary condenser. E—lens.

Fig. 24–29. 35mm motion picture projector lamphouse with 1.6 kW xenon lamp. Door removed to show main mirror, xenon bulb, and auxiliary mirror arrangement.

Fig. 24–30. Light Output Rating of Xenon Lamps Used for Motion Picture Projection

Wattage	Lumens for 10 Footlamberts	Screen Width (feet)	
		85 Per Cent Matte Reflectance	150 Per Cent Directional
900	4500	16	21
1600	9000	23	30
2500	12,500	27	35

constant burning characteristic, and clean, virtually maintenance free operation of these "captivated" xenon arcs, have paved the way for great advancements toward projection automation. Wherever the range of light output of the xenon lamp satisfies the requirements on hand it should therefore be given preferred consideration prior to planning new projection set-ups.

PHOTOCHEMICAL REPRODUCTION PROCESSES

Projection printing, photocopying by contact and optical methods, diazo printing, blueprinting, vandyke printing, and the graphic arts processes of photoengraving, lithography, and photogravure are photochemical reproduction processes. See Fig. 24-31.

Darkroom Lighting

In general, any type of darkroom safelight filter must transmit light which will have the least effect on the photographic material and yet will provide most illumination for the eye. Any photographic material will fog if left long enough under safelight illumination.

The placement as well as the size and type of the safelight lamp will depend on the purpose which the light is to serve. The two types of darkroom illumination are (1) general, to supply subdued illumination over the whole room without concentration at any one point and (2) local, to supply higher illumination on some particular point or object. These are combined, dependent upon the size of the room and the type of work.

Because of the varying sensitivities of the different classes of photographic materials, several safelight filters are available, differing in both color and intensity. These have been scientifically prepared by the manufacturers and it is, therefore, never safe to use substitutes. Other materials may appear to the eye to have the same color as a tested safelight filter, but they will frequently have a much greater photographic action. The use of makeshift safelight substitutes is a very fertile source of darkroom troubles.

The following listing indicates available types of safelight filters:

Color	Material Used With
Clear yellow	Contact printing papers
Bright orange	Bromide and other fast papers and lantern slide plates
Greenish-yellow	Better than orange for judging print quality
Orange-red	Ordinary films and plates
Deep red	Orthochromatic films and plates
Green	Panchromatic films and plates
Yellowish-green	X-ray film
Special green	Infrared films and plates only

Fig. 24-31. Photochemical Reproduction Process Data for Flat Copy Lighted with Incandescent Filament Lamps, Electric Discharge Lamps, or Arcs

Process	Area to be Lighted (inches)	Photosensitive Material	Wavelength Limits of Useful Sensitivity (nanometers)*	Type of Light Source** (watts per square foot of exposure area)	Exposure Time With Indicated Light Source (seconds)
Photocopying by contact printing	11 x 14 Sheets up to 40 wide. Paper from rolls up to 42 wide moving 1 to 32 ft per min	Chloride paper	350-500	50-500 of general service incandescent or tungsten-halogen lamps 50-150 of daylight or black-light fluorescent lamps	3-5
Photocopying by optical printing	Copyboard 11 x 14 to 40 x 50‡	Bromide paper		150-300 of 3200 K incandescent or tungsten-halogen lamps 100-250 of green fluorescent lamps	3-15
Projection enlarging	Up to 8 x 10 negatives	Bromide paper Chlor-bromide paper	350-510	Diffuse Lighting Photographic enlarger lamps Circline fluorescent	10-30
				Specular Lighting 1000-W, coiled-coil filament, incandescent or tungsten-halogen lamp	3-5

Fig. 24-31. *Continued*

Process	Area to be Lighted (inches)	Photosensitive Material	Wavelength Limits of Useful Sensitivity (nanometers)*	Type of Light Source** (watts per square foot of exposure area)	Exposure Time With Indicated Light Source (seconds)
Phototemplates	12 x 24	Orthochromatic	350–550	3000–6000 of carbon arc or mercury lamps	30–60
Enlarging on metal plate	5 ft x 10 ft (approx. max.)	Silver bromide	350–500	10–15 of blacklight fluorescent	5
Contact printing on metal plate	Up to 6 ft x 10 ft	Diazo coating	350–500	200–250 of blacklight fluorescent	60–75
Photoengraving	Copyboard 20 x 30 34 x 48	Process film and dry plates	350–700 (color) 350–550 (B&W)	1500–2000 3200 K tungsten-halogen 50–150 green fluorescent	3–5 min
Making negative	Prevailing size 30 x 40	Wet plates	350–420	Carbon arc	
Contact printing	Vacuum frame 18 x 24 to 30 x 40	Bichromate coating on zinc or copper	300–420	Carbon arc; 200–500 of mercury and 100–150 blacklight fluorescent	3–7 min
Photolithography **Making negative**	Copyboard sizes to 40 x 48	Process film and dry plates	350–700 (color) 350–550 (B&W)	1500–2000 3200 K tungsten-halogen 50–150 green fluorescent 300–400 of mercury; carbon arc	3–5 min
Contact printing	Vacuum frame, including photocomposing printer 20 x 24 to 50 x 70	Bichromate coating on thin zinc plates	300–550	Carbon arc; 200–500 of mercury; 100–150 of blacklight fluorescent	3–5 min
Photogravure (making negative)	Copyboard 30 x 40	Orthochromatic and panchromatic film or plates	350–700 (color) 350–550 (B&W)	1500–2000 of 3200 K tungsten-halogen lamps; carbon arc 50–150 green fluorescent 300–400 of mercury; carbon arc	2–4
Making positive transparency by contact printing	Vacuum frame	Process film or plates	300–550	50–150 of daylight or blacklight fluorescent lamps	5–20
Making carbon tissue image	Vacuum frame 20 x 30	Bichromate sensitized carbon tissue	300–500	Carbon arc; 100–150 of blacklight fluorescent	3–4
Blueprinting Contact printing Continuous single print	Paper from rolls up to 42 wide, moving 2 to 24 ft per min; also cut sizes	Blueprint paper	300–440	400–1500 of mercury lamps	6–20
Diazo printing Continuous single print	Same as blueprinting (1 to 40 ft per min)	Diazo paper (B&W paper)	350–480 Different colors vary within this range	400–1500 of mercury lamps	2–20
Vandyke printing Contact printing Continuous single print	Same as blueprinting	Vandyke paper	300–440	400–1500 of mercury lamps	6–20

* In all photoprocesses glass or plastic is interposed between the lamp and the light-sensitive material as a lens or as a contact plate. Glass lenses do not transmit appreciable energy of wavelengths shorter than 350 nanometers and glass contact plates do not transmit below 300 nanometers, hence these values are given as the lower limit of the useful range, although many materials are sensitive well below 300 nanometers.

** Where incandescent lamps are suggested, it is, in general, advantageous to use tungsten-iodine types. They will provide longer life and significantly better maintenance than ordinary incandescent lamps; this is especially true for lamps operating at 3200

High pressure mercury lamps, where recommended, are usually 400-watt or 1000-watt clear general lighting types or linear tubular types (T-3 and T-9) having lighted lengths from 12 to 48 inches.

Fluorescent lamps 36 inches or longer, in general, should be of rapid start, high output design, as they provide the higher output per foot than either preheat or instant start types. Where rapid start lamps will be started frequently, a flashing type ballast is recommended; this will permit innumerable starts without materially reducing lamp life.

‡ Photocopying by paper cameras: maximum area to be lighted is 36 x 48 inches. Photocopying by microfilming: maximum area to be lighted is 48 x 54 inches. usually 36 x 48 inches.

REFERENCES

1. Committee on Theatre, Television, and Film Lighting of the IES: "Lighting for Theatrical Presentations on Educational and Community Proscenium-Type Stages," *Illum. Eng.*, Vol. 63, p. 327, June, 1968. Bentham, F.: *The Art of Stage Lighting*, Taplinger Publishing Company, Inc., New York, 1969.
2. "Making Available Light Available," *J. Soc. Motion Pict. Telev. Eng.*, March, 1966. "Elements of a Successful Meeting Area," *Film and Audio-Visual Annu.*, 1966.
3. "The History of Stage Lighting," *Illum. Eng.*, Vol. LI, p. 113, January, 1956.
4. Parker, W. O. and Smith, H. K.: *Scene Design & Stage Lighting*, Holt, Reinhart & Winston, New York, 1963.
5. Rubin, J. E. and Watson, L. H.: *Theatrical Lighting Practice*, Theatre Arts Books, New York, 1954.
6. Rubin, J. E.: "Stage Lighting in the School Theatre," *Amer. Sch. Univ.*, (year book) 1957–58.
7. Williams, R. G.: *The Technique of Stage Lighting*, Sir Isaac Pitman & Sons, Ltd., London, 2nd Edition, 1952.
8. Subcommittee on Lighting for the Vidicon Camera of the Theatre-Television Committee of the IES: "Lighting for the Vidicon Camera," *Illum. Eng.*, Vol. LVIII, p. 387, May, 1963.
9. Committee on Television Lighting of the IES: "Current Lighting Practice for Television Production," *Illum. Eng.*, Vol. XLVI, p. 494, September, 1951.
10. Committee on Television Production Lighting of the IES: "Progress in Television Studio Lighting," *Illum. Eng.*, Vol. XLIX, p. 313, June, 1954.
11. Committee on Lighting for Television Production of the IES: "Current Practices for Color Television Production," *Illum. Eng.*, Vol. L, p. 624, December, 1955.
12. Committee on Sports and Recreational Areas and the Committee on Theatre, Television, and Film Lighting of the IES: "Interim Report—Design Criteria for Lighting of Sports Events for Color Television Broadcasting, *Illum. Eng.*, Vol. 64, p. 191, March, 1969.
13. Morgan, C. T.: *Theory and Problems of Radar Visibility*, NRL Report 3965, Naval Research Laboratory, Washington, D. C., April, 1952.
14. Hanes, R. M. and Williams, S. B.: "Visibility on Cathode Ray Tube Screens: The Effects of Light Adaptation," *J. Opt. Soc. Amer.*, Vol. 38, p. 363, 1938.
15. Kraft, C. L.: *A Broad Band Blue Lighting System for Radar Approach Control Centers: Evaluation and Refinements Based on Three Years of Operational Use*, WADC Technical Report 56-71, Wright Air Development Center, Wright-Patterson Air Force Base, Ohio, August, 1956.
16. Baker, C. A. and Grether, W. F.: *Visual Presentation of Information*, WADC Technical Report 54-160, Wright Air Development Center, Wright-Patterson Air Force Base. Ohio. August, 1954.
17. Jones, L. A.: "Measurement of Radiant Energy with Photographic Materials," *Measurement of Radiant Energy*, Forsythe, W. E., Editor, McGraw-Hill Book Co., Inc., New York, 1937.
18. Taylor, F. C.: "35mm Kodachrome Film Used in Photographing Lighting Installations in Color," *Illum. Eng.*, Vol. XXXV, p. 869, December, 1940.
19. Meyers, Jr., G. J. and Mooney, V. J.: "Measuring the Brightness of Street by Means of Photography," *Illum. Eng.*, Vol. XXXVI, p. 643, June, 1941. Hopkinson, R. G.: "The Photographic Representation of Street Lighting Installations," *Illum. Eng.*, Vol. XLI, p. 169, February, 1946. Marsh, C. and Marsh, E.: "A Photographic Method for Brightness Recording," Vol. LIII, p. 355, June, 1958.
20. Knowles, T.: "Photographing Lighting Installations," *Illum. Eng.*, Vol. XLVII, p. 374, July, 1952. Knowles, T.: "More Tips on Lighting Photography," *Illum. Eng.*, Vol. XLIX, p. 271, June, 1954. Knowles, T.: "How to Capture Realism in Lighting Photographs," *Illum. Eng.*, Vol. LVII, p. 29, January, 1962.
21. Knowles, T. and Toenjes, D. A.: "Techniques for Photography of Lighted Streets," *Illum. Eng.*, Vol. XLVIII, p. 49, January, 1953.
22. Allphin, W.: "Photographing Fluorescent Lighting Installations in Color," *Illum. Eng.*, Vol. XLVIII, p. 639, December, 1953. Allphin, W.: "Color Photography Under Fluorescent and Mercury Lighting," *Illum. Eng.*, Vol. LVII, p. 211, March, 1962.
23. Jones, B. F.: "Good Color Slides without Gadgetry," *Illum. Eng.*, Vol. LVIII, p. 116, March, 1963.
24. Levin, R. E. and Lemons, T. M., "Application of Tungsten Halogen Lamps in Theatrical Luminaires," *Illum. Eng.*, Vol. 64, p. 47, January, 1969.
25. Stote, H. M. (Editor): *The Motion Picture Theatre*, Society of Motion Picture and Television Engineers, New York, 1948.
26. Guth, S. K., Logan, H. L., Lowry, E. M., MacAdam, D. L., Schlanger, B., Hoffberg, W. A., and Spragg, S. D. S.: "Screen Viewing Factors Symposium," *J. Soc. Motion Pict. Telev. Eng.*, Vol. 57, p. 185, September, 1951.
27. Allen, C. J.: "Lighting the School Auditorium and Stage," *Illum. Eng.*, Vol. XLVI, p. 131, March, 1951.
28. Estes, R. L.: "Effects of Stray Light on the Quality of Projected Pictures at Various Levels of Screen Brightness," *J. Soc. Motion Pict. Telev. Eng.*, Vol. 61, p. 257, August, 1953. Eastman Kodak Company: "Legibility Standards for Projected Material," *Kodak Sales Service Pamphlet No. S-4*, 1956. "The Foundation for Effective Audio-Visual Projection," *Kodak Sales Service Pamphlet*, Rochester, New York.
29. Tuttle, C. M.: "Density Measurements of Release Prints," *J. Soc. Motion Pict. Eng.*, Vol. 26, p. 548, May, 1936.
30. Lozier, W. W.: "Reports on Screen Brightness Committee Theatre Survey," I. *J. Soc. Motion Pict. Telev. Eng.*, Vol. 57, p. 238, September, 1951; and II. *J. Soc. Motion Pict. Telev. Eng.*, Vol. 57, p. 489, November, 1951.
31. Berger, F. B.: "Characteristics of Motion Picture and Television Screens," *J. Soc. Motion Pict. Telev. Eng.*, Vol. 55, p. 131, August, 1950.
32. D'Arcy, E. W. and Lessman, G.: "Objective Evaluation of Projection Screens," *J. Soc. Motion Pict. Telev. Eng.*, Vol. 61, p. 702, December, 1953.
33. Hill, Armin J.: "A First Order Theory of Diffuse Reflecting and Transmitting Surfaces," *J. Soc. Motion Pict. Telev. Eng.*, Vol. 61, p. 19, July, 1953.
34. "Screen Luminance and Viewing Conditions for 35mm Review Rooms," PH22.133—1963 (R1969), American National Standards Institute, New York.
35. "Screen Luminance and Viewing Conditions for 16mm Review Rooms," PH22.100—1967, American National Standards Institute, New York.
36. "Specifications for Screen Luminance for Indoor Motion-Picture Theatres," PH22.124—1970, American National Standards Institute, New York.
37. "Minimum Screen Luminance for Drive-in Theatres," SMPTE Recommended Practice RP12, *J. Soc. Motion Pict. Telev. Eng.*, Vol. 71, p. 514, July, 1962.
38. Committee on Nontheatrical Equipment of the SMPE, "Recommended Procedure and Equipment Specifications for Educational 16mm Projection," *J. Soc. Motion Pict. Eng.*, Vol. 37, p. 22, July, 1941.
39. "Planning Schools for Use of Audio-Visual Materials": No. 1, "Classrooms," July, 1952; No. 2, "Auditoriums," February, 1953; and No. 3, "Audio Visual Instructional Materials Center," January, 1954, National Education Association, Dept. of Audio-Visual Instruction, Washington, D. C.
40. Will, Jr., P.: "Eyes and Ears in School," *Architect. Rec.*, Vol. 99, p. 66, February, 1946.
41. Vlahos, P.: "Selection and Specification of Rear-Projection Screens," *J. Soc. Motion Pict. Telev. Eng.*, Vol. 70, p. 89, February, 1961.
42. Lowry, E. M.: "Screen Brightness and the Visual Functions," *J Soc. Motion Pict. Eng.*, Vol. 26, p. 490, May, 1936.
43. Beyer, W.: "Wide Screen Systems," *Amer. Cinematographer*, p. 44, October, 1960.
44. Beyer, W.: "Wide Screen Production Systems," *Amer. Cinematographer*, p. 296, May, 1962.
45. *Wide Screen Motion Pictures*, Society of Motion Picture and Television Engineers, New York, New York.
46. Beyer, W.: "The Research Council Developments for Better Theatre Projection," *J. Soc. Motion Pict. Telev. Eng.*, Vol. 69, p. 792, November, 1960.
47. Harrington, R. E. and Bowditch, F. T.: "Color Measurement of Motion Picture Screen Illumination," *J. Soc. Motion Pict. Telev. Eng.*, Vol. 54, p. 63, January, 1950.
48. Null, M. R., Lozier, W. W., and Joy, D. B.: "The Color of Light on the Projection Screen," *J. Soc. Motion Pict. Eng.*, Vol. 38, p. 219, March, 1942.
49. Heppberger, C. E. and Bowen, E. A.: "Carbon Arcs for 16mm Film Projection," *J. Soc. Motion Pict. Telev. Eng.*, Vol. 73, p. 862, October, 1964.
50. Ayling, R. J. and Hatch, A. J.: "Improvements in the Blown-arc for Projection," *J. Soc. Motion Pict. Telev. Eng.*, Vol. 67, p. 693, October, 1958.
51. Dimmick, G. L. and Widdop, M. E.: "Heat Transmitting Mirror," *J. Soc. Motion Pict. Telev. Eng.*, Vol. 58, p. 36, January, 1952.
52. Balzers Laboratories: "Cold Mirrors for Projection Heat Control," *J. Soc. Motion Pict. Telev. Eng.*, Vol. 67, p. 175, March, 1958.
53. Ulffers, H.: "Xenon High Pressure Lamps in Motion Picture Theatres," *J. Soc. Motion Pict. Telev. Eng.*, Vol. 67, p. 389, June, 1958. Reese, W. B.: "The Xenon Arc Projection Lamp," *J. Soc. Motion Pict. Telev. Eng.*, Vol. 67, p. 392, June, 1958. Seeger, B. and Jaedicke, W.: "The Xenon Short Arc Lamp in Motion Picture Projection," *J. Soc. Motion Pict. Telev. Eng.*, Vol. 69, p. 474, July, 1960. Kleopfel, D. V.: "Xenon Projection Lamps: A Resume," *J. Soc. Motion Pict. Telev. Eng.*, Vol. 73, p. 479, June, 1964.
54. School and College Committee of the IES: "Guide for Lighting Audiovisual Areas in Schools," *Illum. Eng.*, Vol. LXI, p. 477, July, 1966.

SECTION

25

MISCELLANEOUS APPLICATIONS OF RADIANT ENERGY

Light 25-1 *Ultraviolet Energy 25-14* *Infrared Energy 25-19*

Earlier sections of this Handbook cover the uses of light as an aid to seeing; however, there are many applications of radiant energy (light, ultraviolet, and infrared) that do not involve vision. These include:

Light
- Photoelectric control
- Plant growth—aiding photosynthesis and production of chlorophyll
- Fading of colored materials
- Insect attraction and trapping

Ultraviolet
- Development of erythema
- Production of vitamin D
- Prevention and cure of rickets
- Photochemical actions
- Catalysis of chemical reactions
- Microorganism growth control, as in air and liquid sterilization

Infrared
- Radiant heating and heat therapy
- Production drying, softening, heating
- Dehydration
- Comfort heating

Radiant-Energy Sources

Many of the light sources described in Section 8 produce small quantities of ultraviolet energy (wavelength less than 380 nanometers) and infrared energy (wavelength more than 780 nanometers) as well as light energy. In most cases, the amount of ultraviolet energy emitted by sources used for general lighting is not of practical importance. However, 75 per cent or more of the output of standard incandescent filament lamps, including those with high ratings of 20 to 30 lumens per watt, is emitted in the infrared spectral region. Incandescent filament lamps designed as infrared emitters may produce 90 per cent or more of their output in the infrared wavelengths.

The production of ultraviolet and infrared energy may be accomplished in much the same manner as the production of light, as explained in Section 2. The principles of light control described in Section 6 are equally valid in most cases for infrared and ultraviolet energy as well.

Fig. 25–1 shows the characteristics of solar energy at the earth's surface.

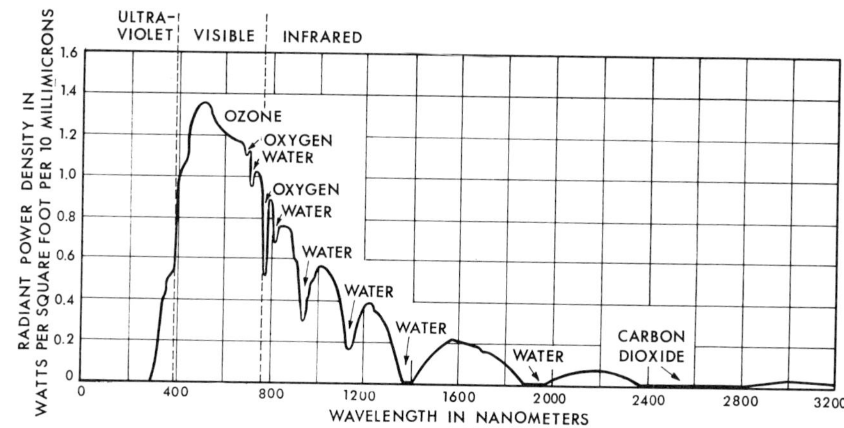

Fig. 25–1. Spectral distribution of solar radiant power density at sea level, showing the ozone, oxygen, water, and carbon dioxide absorption bands.

NOTE: References are listed at the end of each section.

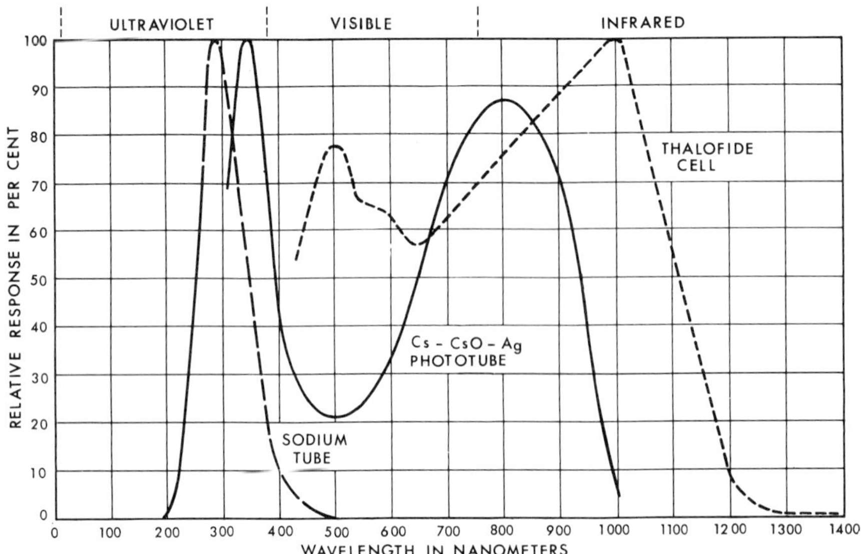

Fig. 25–2. Relative response of several sensitive elements to energy of different wavelengths.[1]

Reflection, Transmission, and Absorption of Radiant Energy

Ultraviolet and infrared radiant energy are similar in all respects, except wavelength, to light. As indicated in Section 3, the normal human eye, though blind (from the standpoint of ordinary seeing tasks) to radiation of wavelengths shorter than 380 nanometers and longer than 780 nanometers, does react slightly to these "extra-visual" wavelengths. Other radiant-energy-sensitive receptors, such as photoelectric cells and some chemical compounds, exhibit response curves which may have peaks in the ultraviolet, visible, or infrared regions. See Fig. 25–2. The reflection and the transmission characteristics of materials vary with wavelength also. See Figs. 25–3 and 25–29.

Fig. 25–3. Reflectance of Various Materials for Energy of Wavelengths in the Region of 253.7 Nanometers

Material	Reflectance (per cent)
Aluminum	
Untreated surface	40–60
Treated surface	60–89
Sputtered on glass	75–85
Paints	55–75
Stainless steel	25–30
Tin plate	25–30
Magnesium oxide	75–88
Calcium carbonate	70–80
New plaster	55–60
White baked enamels	5–10
White oil paints	5–10
White water paints	10–35
Zinc oxide paints	4–5

MISCELLANEOUS APPLICATIONS OF LIGHT

Automatic Control with Photoelectric-Cell-Operated Relays

A number of energy-beam, photocell-operated relay applications have been developed.[2] Typical of the common uses are door opening and closing, burglar alarm operation, safety shutoff of hazardous machinery, conveyor control, production-unit counting, and production quality control. In schoolrooms photocell control of electrical illumination has been used.

A light, ultraviolet, or infrared energy beam and photocell combination may be arranged in such a manner that action takes place either upon incidence of the beam on the cell or upon interception of the beam and darkening of the cell. Typical installations are shown in Fig. 25–4.

Horticultural Lighting[3–36]

The purpose of light in horticulture is to duplicate the growth responses induced in plants by sunlight and daylight, but not necessarily to duplicate the sun in its light energy or spectral emission. Plant responses can be accomplished by the use of lighting techniques which provide the irradiance and period and wavelengths of light most effective for the response desired. Because plants utilize only certain portions of the sunlight spectrum and because both level and spectral distribution of sunlight can change extensively from dawn to dusk, from cloud to cloud, and from day to day, there has been no effort to design or combine light sources to stimulate these variations for horticultural lighting. However, the natural photoperiod is easily and frequently simulated with electric light sources.

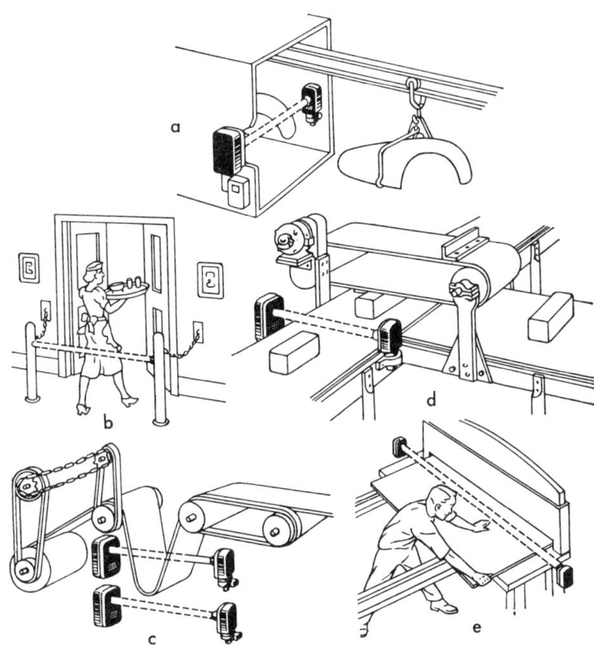

Fig. 25-4. Typical applications of photoelectric-cell-operated relays: (a) Light source and photocell-operated relay arranged as an automatic counter. (b) Photocell-operated relay controls door-opening mechanism. Photocell-operated relays may be used to operate limit switches (c), to sort objects of different sizes (d), or to operate safety shutoff switches (e).

All plants and animals depend upon *photosynthesis* for food. The process of photosynthesis may be very simply defined as the manufacture of carbohydrates from carbon dioxide and water by the photochemistry of the photoreceptor, chlorophyll. In addition to photosynthesis, there are three other major photoresponses of plants including *chlorophyll synthesis, phototropism* and *photomorphogenesis*. Chlorophyll synthesis is the synthesis of chlorophyll by light, which causes the reduction of its precurser and photoreceptor, protochlorophyll. Phototropism is a light induced growth movement of a plant organ, generally controlled by carotenoid photoreceptors. Photomorphogenesis is light controlled growth, development and differentiation of a plant from responses initiated largely by the photoreceptor, phytochrome. The action spectra of these photoresponses are shown in comparison to that of photopic vision in Fig. 25-5.

The wavelength region from about 300 to 800 nanometers is utilized in the photosynthesis of plants. This region is somewhat broader than that used in photopic vision (380 to 780 nanometers). Even though these plant responses utilize a broader spectral region and have spectral absorption characteristics and action spectra very different from that of vision, the radiant energy utilized in the photochemistry of plants is referred to as light. Because of the differences between the photoresponses of plants and the eyes, plant response lighting is not the same as illumination for vision. However, many of the basic lighting principles in the use and applications of light sources in illumination may be applied to plant lighting.

Light Sources. Electric light sources which emit sufficient energy over the entire 300 to 800-nanometer spectral region are effective in photosynthesis and other photoresponses of plants. Light sources which have been made to emit light in limited spectral regions by filtering or other means have been used for special photoresponse purposes. In experimental work, for example, lamps which emit light in

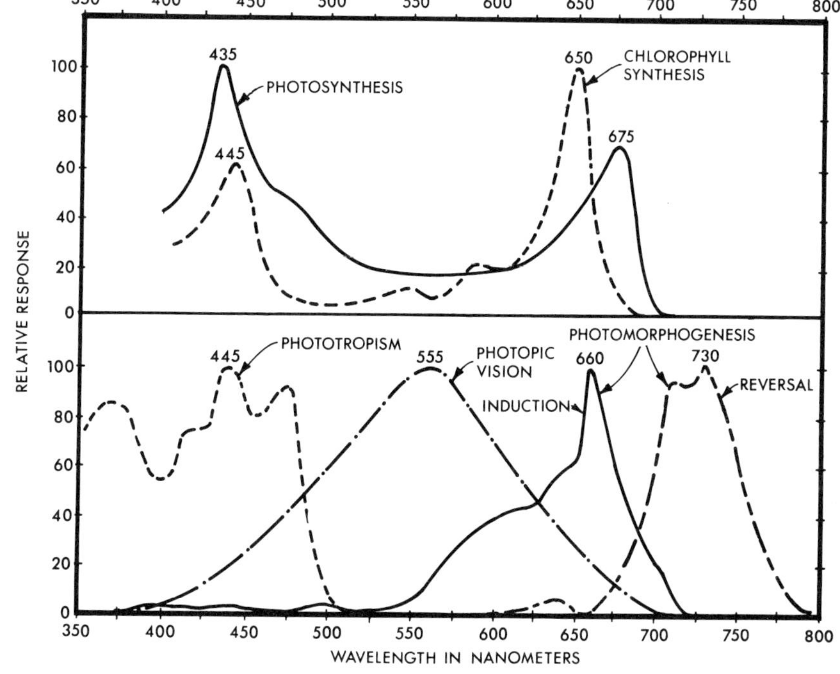

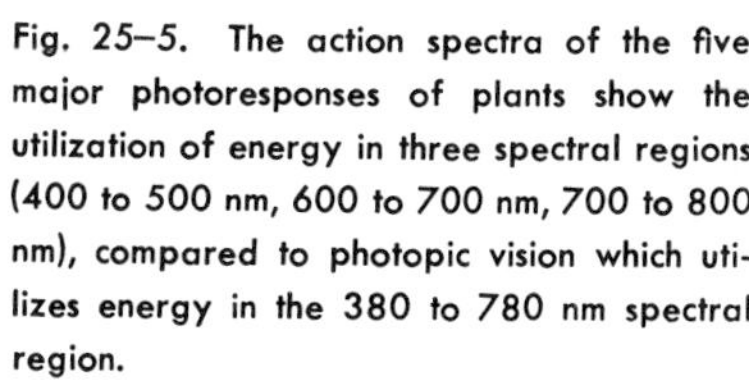
Fig. 25-5. The action spectra of the five major photoresponses of plants show the utilization of energy in three spectral regions (400 to 500 nm, 600 to 700 nm, 700 to 800 nm), compared to photopic vision which utilizes energy in the 380 to 780 nm spectral region.

the 500 to 580-nanometer region have been used as "safe lights" because this energy is in a low plant response wavelength region. In contrast, lamps emitting energy in the blue (400 to 500 nanometers), the red (600 to 700 nanometers) or far-red (700 to 800 nanometers) are used in studying various photoresponses which absorb strongly the energy emitted in these spectral regions.

Experimental work in horticultural science uses many types of light sources including carbon arc, incandescent (including tungsten-halogen), fluorescent, mercury, xenon, sodium, and metal halide lamps. Various combinations of lamp types are sometimes used; the most common combination being fluorescent and incandescent. It has been generally determined that the type of light source, or combination of types, is more efficient in producing plant material (fresh and dry weight) when the greater portion of the energy, emitted throughout the 300 to 800-nanometer spectral region, is in the 580 to 800-nanometer region. For example, mercury lamps with a phosphor coating which emits in the red (600 to 700 nanometers) spectral region, have been found to be more effective in producing plant material than clear mercury lamps or lamps with a phosphor coating which emits throughout the 400 to 700-nanometer region. Of the various "white" fluorescent lamps, the warm white lamp has been found to be most efficient in the synthesis of dry matter in plants. Generally the ability of fluorescent lamps to synthesize dry matter in plants is enhanced with the addition of energy from incandescent lamps. This phenomenon is due to additional energy from the incandescent lamps in the red (600 to 700 nanometers) and especially the energy in the far-red (700 to 800 nanometers) where most "white" fluorescent lamps are relatively deficient. Incandescent lamps which emit a greater percentage of their output in these regions have found wide usage in the control of flowering of horticultural crops. However, many variations in plant responses to light sources apparently caused by species differences do occur.

Plant growth lamps are presently fluorescent lamps which have been designed exclusively for plant growth and responses. The emission spectra of plant growth lamps are tailored so that nearly all of the energy emitted is that capable of being absorbed by plant photoreceptors for utilization in plant growth. Two types of lamps are manufactured. One type produces emission maxima at 450 and 660 nanometers, closely matching the chlorophyll synthesis curve, but with little emission beyond 700 nanometers. This lamp has generally found wide usage in residential lighting for the growth and color enhancement of house plants requiring relatively low energy levels, especially African violets and gloxinias. The other lamp type has high emission in the blue and red spectral regions but differs from the previously described plant growth lamp and other fluorescent lamps because of its relatively high emission in the 700 to 800 nanometer spectral region. With its far-red emission, the relative growth of several plant species is enhanced and its use generally precludes the use of supplementary incandescent light. This lamp has found use in the growth and flowering of plants which are normally grown in full sunlight, as are many commercial florist and vegetable crops. The spectral energy distribution of these two lamps is compared in Fig. 25-6.

Radiant Energy Measurement. As mentioned above horticultural lighting is not illumination because it does not conform to the definition, as given in Section 2, i.e., visually evaluated radiant energy. Horticultural lighting is plant evaluated radiant energy as illustrated in Fig. 25-5. Because horticultural lighting is not illumination, the footcandle as a unit of radiant energy measurement is limited in its value, especially in comparing plant responses

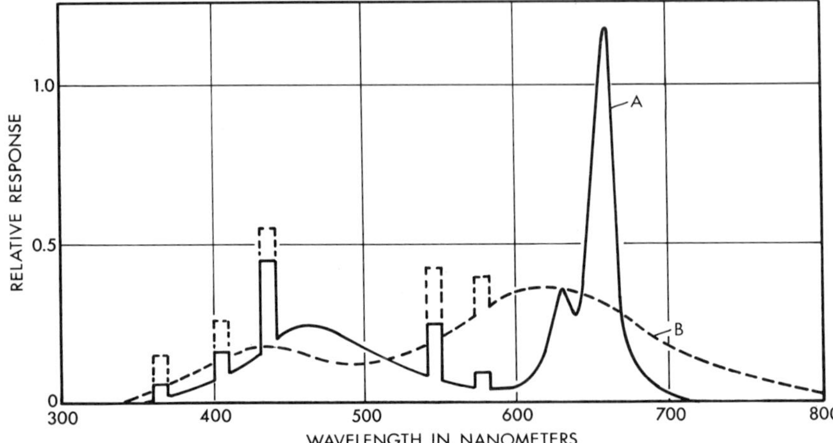

Fig. 25-6. Fluorescent plant growth lamps have been designed with emission spectra tailored so that nearly all of the energy emitted is capable of being absorbed and utilized by the plant photoreceptors for utilization in plant growth. Lamp A has emission maxima at 450 nm and 660 nm with little emission beyond 700 nm, compared to Lamp B which has its major emission also in the blue and red regions as well as emission beyond 700 nm.

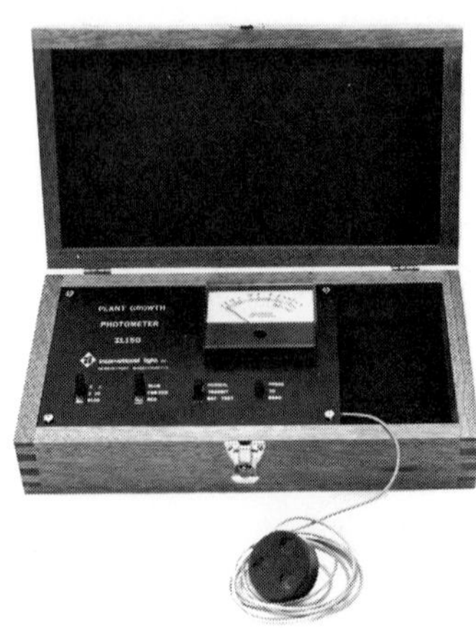

Fig. 25-7. The plant growth photometer is designed to measure radiant energy in absolute units ($\mu W/cm^2$) in the three wavelength regions of maximum plant responses (400 to 500 nm, 600 to 700 nm, 700 to 800 nm).

with light from sources having different spectral energy distribution characteristics. In horticultural science, footcandle measurements are either discussed in terms of the spectral emission of the light source or are converted to absolute units by calibrating the footcandle meter for the amount of energy ($\mu W/cm^2$) per footcandle for each light source with a different spectral emission. A better alternative is to use a thermopile, bolometer or spectroradiometer which is linear with wavelength and energy level. A plant growth photometer has been developed to measure plant growth energy in absolute units ($\mu W/cm^2$) in the three spectral regions of major plant response (400 to 500, 600 to 700, and 700 to 800 nanometers). See Fig. 25-7.

For practical considerations, it may be more convenient to express horticultural lighting levels in terms of lamp watts per square foot of growing area with the light source at a distance from the plants compatible with the plants and the light source. This enables the grower to determine the number of lamps needed for a particular area and the power required. The formula for calculating the number of lamps for an installation is:

$$\text{Number of Lamps} = \frac{\text{Growing Area} \times \text{Lamp Watts per Square Foot Required}}{\text{Individual Lamp Watts}}$$

In horticulture there are two general uses for lighting: Photosynthesis and Photoperiodism.

In lighting for photosynthesis, light is applied to plants to sustain in part or in total the photosynthetic processes necessary for desired growth. In lighting for photoperiodism, light from various sources is applied to plants to sustain in part or in total the photoperiod necessary to produce a desired flowering response. For many plants, the quantity of light required for photosynthesis can range from 10 to 100 times greater than that required for photoperiodic lighting.

Photosynthetic Lighting. Photosynthetic lighting is used in the greenhouse to supplement daylight during periods of diminished sunlight in winter months, for the growth of out-of-season crops. See Figs. **25-8** and **25-9.** This supplementary lighting can be much less than full sunlight—the level being determined by the requirements of the particular plant species.

The different applications of lighting for photosynthesis are as follows:

1. Day Length Extension—Lighting in the greenhouse before sunrise or after sundown to extend the light period.
2. Dark Day—Lighting in the greenhouse on dark, overcast days for the total light period.
3. Night—Lighting in the middle of the dark period when natural carbon dioxide concentrations are high; an application for low-cost off-peak power.
4. Underbench—Lighting under benches can double growing area within a greenhouse.

Fig. 25-8. Over-bench photosynthetic lighting in the greenhouse is a means of increasing crop production efficiency and of timing crops for a market advantage. Luminaires should provide maximum incident light on plants, minimum shading of sunlight, and minimum interference with greenhouse routine. Such a luminaire using high output reflector fluorescent lamp is shown.

Fig. 25-9. Under-bench photosynthetic lighting enables the grower to double the growing area within a greenhouse by "stacking" plants.

5. Growth Room—All lighting is provided by lighting equipment. Such rooms may range from controlled environmental rooms for research to nonproductive areas where seedlings, cuttings, and plants with a low-light requirement can be economically grown. See Figs. **25-13** and **25-17.**

General lighting requirements for the above lighting applications are shown in Fig. 25-10. Specific

Fig. 25-10. The Requirements for Photosynthetic and Photoperiodic Lighting

Object of Lighting	Applications	Time Applied	Total Effective Light Period (hours)	Lamp (W/ft²)	Light Sources	Luminaires
I. Photosynthetic						
A. Supplementary						
1. Daylength extension	a. Seed germination, seedlings, cuttings, bulb forcing b. Mature plants	4 to 10 hours before sunrise and/or after sunset	a. 12 to continuous b. 10 to continuous	a. 5 to 20 b. 10 to 40	Fluorescent, mercury and fluorescent-mercury lamps of various wattages with and without internal reflectors and used with or without 10 to 30 per cent of installed watts of incandescent	Moisture-resistant luminaires of industrial or custom made designs with mountings fixed or adjustable providing minimum interference with greenhouse routine and uniform light distribution
2. Dark day	as above	Total light period	as above	as above		
3. Night	as above	4 to 6 hours in middle of dark period	16	as above		
4. Underbench	as above	Total light period	10 to continuous	as above	Fluorescent lamps	Moisture-resistant, direct reflector units with mounting for uniform light distribution
B. Growth room						
1. Professional horticulture	Seed germination, seedlings, cuttings, bulb forcing	Total light period	12 to continuous	5 to 30	Fluorescent lamps with or without 10 to 30 per cent incandescent or combination of plant growth lamps	Industrial direct reflector luminaires which are moisture resistant and are mounted in a shelf arrangement
2. Amateur horticulture	Seed germination, seedlings, cuttings, bulb forcing, mature plants, etc.	Total light period	10 to continuous	5 to 30	Fluorescent lamps (plant growth lamps) with and without incandescent	as above
3. Experimental horticulture	All types of plant responses	Total light period	0 to continuous	0 to 140 and higher	Many types used to fit the requirements of tests. Generally fluorescent with 10 to 30 per cent incandescent lamps or combination of plant growth and wide spectrum lamps	Custom built with minimum spacing for maximum light output of lamps with uniform light distribution
II. Photoperiodic						
A. Supplementary						
1. Daylength extension	Long day effect to prevent flowering of short day plants and induce flowering of long day plants	4 to 8 hours before sunrise and/or after sunset	14 to 16	.5 to 5	Fluorescent, fluorescent-mercury and incandescent lamps	As for photosynthetic supplementary lighting
2. Night break	as above	2 to 5 hours in middle of dark period	14 to 16	.5 to 5	as above	as above
3. Cyclic	as above	1 to 4 seconds per minute, 1 to 4 or 10 to 30 minutes per hour	14 to 15	1 to 5	Mostly incandescent, or fluorescent lamps with flashing ballasts	as above

requirements depend upon plant species and varietal requirements. Particular consideration should be given to photoperiodic plants.

Although definite advantages can be obtained in the greenhouse with photosynthetic lighting alone, it is evident from experimental work that the benefits in the regulation of plant growth are obtained when other interacting factors such as temperature, water, nutrients, and carbon dioxide are considered. The greatest growth response is obtained when these factors are present at optimum levels. In research, such optimum levels can be achieved without great concern for costs, but the commercial grower is constantly faced with supplying practical levels of these factors which the particular crop can economically support.

Photoperiodic Lighting. In the U. S. today the greatest use of lighting in horticulture is in the use of photoperiodism to control the out-of-season flowering of certain species of economic plants which require specific ratios of light to dark periods for flowering. Such plants will remain vegetative rather than flower until these requirements are met. Therefore, plants are classified as to the relative length of light period to dark period needed to set flower buds and to bloom. This knowledge is used to bring plants into bloom when there is a particular market advantage. In Florida and California the flowering of several hundred acres of chrysanthemums in the field is controlled by this type of lighting. See Fig. 25-11.

During winter months it is essential to extend the day length to promote the flowering of long-day (short-night) plants and to inhibit the flowering of short-day (long-night) plants. It is also essential that the grower be able to shorten the day length to promote the flowering of short-day plants and to inhibit the flowering of long-day plants. During summer months, the grower must apply an opaque cloth or plastic covering over the plants for part of the day to simulate a short day. It is essential that the material used be opaque and that the plants be exposed to no light at this time because very low levels are effective in this response. The use of lighting and opaque covering permits the growth and flowering of both long-day and short-day plants the year round.

Long-day responses for both short-day and long-day plants are usually obtained by lighting plants 4 to 8 hours before sunrise or after sunset or by a 2- to 5-hour light period in the middle of the dark period (called "night break"). See Fig. 25-10.

The most important economic group of photoperiodic plants are short-day plants, chrysanthemums and poinsettias. Such short-day plants remain vegetative with a continuous light period of greater than 12 hours or by a "night break". When flowering is desired, the photoperiod is shortened to about 10 hours and the "night break" is discontinued. By providing long-day plants such as China aster and Shasta daisy with a continuous 16- to 18-hour day with supplementary light, they can be brought into flower, while continuous short days will cause them to remain in the non-flowering or vegetative state.

Fig. 25-11. Photoperiodic lighting enables the grower in southern climates to control the flowering of chrysanthemums in the field. This installation utilized 100-watt incandescent lamps on 12-foot by 14-foot centers.

Fig. 25-12. Over-bed or over-bench photoperiodic lighting in the greenhouse used in conjunction with opaque coverings enables the grower to grow both short-day and long-day plants the year round and to time flowering for the best market period.

Incandescent and fluorescent lighting are used for photoperiodic lighting. Clear incandescent in industrial reflector luminaires or reflector incandescent lamps are commonly used in the field (see Fig. 25-11) or in the greenhouse (see Fig. 25-12). However, incandescent lamps are not always satisfactory because the internodal elongation produced in some plants may be undesirable. For either fluorescent or incandescent light, the "night break" type of day-length control is also an application for low-cost, off-peak power.

It is often desired to use the higher levels of illumination required for photosynthetic lighting for photoperiodic lighting; however, these distinctly different lighting applications should not be confused.

Plant Growth Chambers. Plant growth climatology chambers are now used extensively in agricultural experiment stations, educational institutions, and by industrial research laboratories for the growth of plants under controlled environmental conditions (see Fig. 25–13). The environmental conditions which are controlled and monitored include light, temperature, nutrients, carbon dioxide, and oxygen. A research facility which consists of several plant growth chambers (controlled environment rooms) is called a phytotron. For most scientific work, the light level is measured in absolute units. However, growth chamber manufacturers rate the light output within growth chambers in footcandles.

For illumination levels up to 5000 footcandles, 1500-mA 8-foot T-12 or T-17 fluorescent (white or cool white) lamps are closely spaced and mounted to a white perforated ceiling through which lamp heat is exhausted. Spaced at uniform intervals between the fluorescent lamps are 100-watt, 2500-hour-life incandescent lamps which provide the infrared component necessary for cuttings and seedlings. For higher footcandle levels, reflector-type T-10 1500-mA fluorescent lamps are used to permit even closer spacing.

Recent studies have shown that the combination of plant growth lamps (one type A and one type B of Fig. 25–6) has produced more foliage and earlier

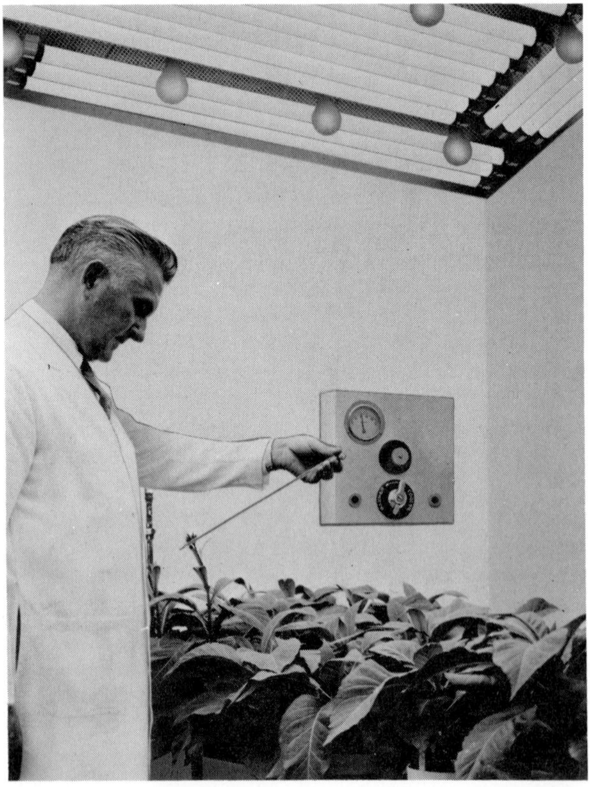

Fig. 25–13. Plant growth chamber (5 × 10 × 8 feet) equipped with closely-spaced 1500 mA fluorescent lamps mounted on a perforated white ceiling through which lamp heat is exhausted. Also, 100-watt, 2500-hour life incandescent lamps at uniform spacings provide infrared component. Glossy-white thermal-fabricated walls absorb radiant heat from lamps and reflect light downward to plant area. Initial illumination level was approximately 5100 footcandles at bench. Circuitry of lamp groups provides steps in illumination levels.

fruit set than the combination of cool white fluorescent and incandescent lamps.

For closer climatology control, the growth chamber walls are made of glossy-white, thermal fabricated material which absorbs the infrared energy received from the lamps and reflects a high percentage of the light downward to plant growth areas. Uniformly spaced groups of lamps are separately circuited to provide several steps of illumination levels, and at the same time maintain uniform light distribution. To compensate for early light output depreciation of fluorescent lamps, the system should be designed to produce initially 25 per cent more footcandles than the maximum level. Circuitry control is used to regulate levels as lamps age. Light meters designed to measure the higher footcandles are used periodically to check illumination levels.

Depending on the layout and circuitry desired, most growth chamber fluorescent systems are operated from 2-lamp or 3-lamp ballasts which are located remotely outside of the chamber. This arrangement divorces ballast heat from the chamber and reduces the air-conditioning load factor.

In chambers where experiments are performed with isotopes to study plant mutation, a shielding medium of lead is built into the chamber walls to prevent transmission of radiation.

Home, Hobby, and Office Horticulture. During the past decade, much research has been conducted to determine minimum levels of electric lighting under which many new kinds of ornamental plants can be maintained in healthy condition. In Fig. 25–14, names of ornamental foliage plants are tabulated under minimum footcandle ranges that will provide sufficient light for maintaining growth. The footcandle level, which would provide equivalent energy to the plant would be about one-third of that given in Fig. 25–14 for lamp A of Fig. 25–6 and about two-thirds of that value for lamp B.

With the aid of lighting and the available lighting equipment for indoor plant culture (see Figs. 25–15 and 25–16), flowering and foliage plants can be taken off the window sill to a place in the room where they can be grown and displayed to the best advantage. Some luminaires are equipped with trays to hold moisture to raise humidity about plants and with timers to turn lights on and off automatically.

Some amateurs, unsatisfied with the number and types of plants which can be grown for decorative purposes, have set up basement gardens of varying sizes in which plants are grown from seed, cuttings, and bulbs as shown in Fig. 25–17. A wide variety of flowering and foliage plants, including all plants of the "house plant" category, have been successfully grown under lights.

Fluorescent lamps in T-8 and T-12 sizes and with typical loadings have been accepted for this type of horticulture because incandescent lighting tends to elongate most plants. For some plant species, however, it is often considered desirable to include some incandescent light along with fluorescent light to enhance growth and flowering.

Fig. 25-14. Foliage Plants Grouped According to Level of Electric Illumination Needed to Maintain Them for at Least 12 Months*

Low-Light Requirement (15-25 footcandles)	Medium-Light Requirement (25-50 footcandles)	High-Light Requirement (50-100 footcandles)
Aglaonema commutatum	All Plants Listed in Previous Column Do Better in This Light Range	All Plants Listed in Previous Columns Do Better in This Light Range
" pictum	Aglaonema marantifolium	Fatshedra lizei
Araucaria excelsa	" roebelini	Ficus doescheri
Aspidistra lurida	Anthurium hybrids	" exotica
Aucuba japonica	Begonia metallica	" pandurata
Dieffenbachia amoena	Bromeliads—various	Hedra helix var. Eali
" picta	Chlorophytum species	" " " Maple Queen
Dracaena deremensis	Cyrtomium rockfordianum	Pothos (Scindapsus)
" rothiana	Dieffenbachia Roehrs var.	Spathiphyllum kochii
" sanderiana	Kentia fosteriana	Tolmiea menziesi
Philodendron cordatum (oxycardium)	Nephrolepsis bostoniensis	Vitis (cissus) rhombifolia
Philodendron hastatum	Peperomia obtusifolia	
" panduriforme	" " var.	
" pertusum	" sandersi	
Sansevieria species	Pilea cadieri	
Schefflera actinophylla	Philodendron dubium	
Syngonium podophyllum	" sodiroi	
	Pothos (scindapsus) Aureus	

* Based on daily photoperiod of 16 hours from cool white, white, or warm white lamps.

With the illumination levels being used in modern office buildings, an office design technique called "office landscaping" becomes practical. It is essentially the elimination of conventional partitions and the use of a combination of large foliage plants, planters, screens and baffles to achieve audio and visual privacy, yet provide a degree of informality to occupants and visitors. "Office landscaping" may cost less, provide greater flexibility and have greater appeal than conventional partitions.

Aquarium and Terrarium Lighting. Aquaria and terraria are also found in the home, office, and school for hobby, decorative, and educational purposes.

Aquarium lighting serves both a functional and an ornamental purpose when plants are part of the

Fig. 25-15. Attractive equipment for the home have changed the culture and types of plants grown. Such lighting has replaced "window sill" culture and enables the grower to display and grow plants anywhere in the house.

Fig. 25-16. Portable carts made of tubing, with two 40-watt fluorescent lamps in a special reflector mounted over each tray, makes a convenient rack for growth of African violets, gloxinias, and similar house plants.

Fig. 25-17. Both amateur and professional growers have found value in basement gardens. They allow the amateur to increase the size of his hobby and enable the professional grower to utilize unproductive space for rooting of cuttings and growth of seedlings.

Fig. 25-18. Lighting the aquarium provides illumination for the fish and the viewer as well as providing the energy which enables aquarium plants to carry on photosynthesis, improving the environment for the fish.

aquarium environment. See Fig. **25-18.** Through the process of photosynthesis, lighted aquarium plants increase the oxygen level essential for fish respiration and at the same time reduce the carbon dioxide level, preventing the buildup of carbonic acid which can be harmful to fish. The light also illuminates both the fish and the aquarium. Dramatic colors of both fish and plants are observed when special fluorescent plant growth lamps are used because of the high red and blue emission of such lamps.

Both fluorescent (T-5, T-8, T-12) and incandescent lamps are used, with a preference for fluorescent because they produce more light and less heat per watt. Lighting requirements for aquaria usually range from one to two lamp watts per gallon of tank capacity.

Terrarium lighting usually requires both fluorescent and incandescent lighting. Flourescent light is applied at about 20 watts per square foot for the plant life while an incandescent lamp is used to light a portion of the terrarium to simulate the infrared of sunlight for the animal life (lizards, frogs, etc.) usually found in such environments.

Fading and Bleaching

Fading and bleaching of colored textiles and other materials upon exposure to light and other radiant energy is of special interest because of the higher illumination levels now employed in merchandising. Consequently, a knowledge of some of the factors involved is important. Some of these (not necessarily arranged in order of importance) are as follows:

1. Level of illumination.
2. Duration of exposure.
3. Spectral distribution of the radiation.
4. Moisture.
5. Temperature of the material.
6. Chemical composition of dye or other colorant.
7. Saturation of dye (tints versus saturated colors).
8. Composition and weave of fabric.
9. Intermittency of exposure.
10. Chemical fumes in the atmosphere.

While many researches on fading and color-fastness have been carried out and results published, especially in textile journals, most of them are deficient in data on illumination levels involved. In general, the primary purpose has been improvements in dyes and dyeing methods. Such tests have involved exposures to daylight in various geographical regions and to standardized types of arc lamps, so-called "fading lamps." The National Bureau of Standards has developed standardized methods of conducting such tests and is prepared to standardize carbon arcs in terms of their standards.

One publication[37] presenting a lengthy review, with extensive bibliography, of researches in this field summarizes the subject in this way:

The rate at which a dye fades is governed by seven factors:

1. The photochemistry of the dye molecule.
2. The physical state of the dye.

3. The chemistry of the substrate.
4. The fine structure of the substrate.
5. The presence of foreign substances.
6. The atmosphere.
7. The illumination.

From publication in technical journals over the past forty years certain generalized conclusions have been derived.[38-40] However, in view of the fact that the tests have been limited to an infinitesimal percentage of the dyes and textiles in general use, it must be realized that such conclusions have limited application and that many exceptions will be found. With this reservation in mind, the *average* results for several hundred specimens of colored textiles will be discussed.

Level of illumination and duration of exposure to any particular light source are obviously the most important factors. Two researches[41, 42] indicate an approximate reciprocity relationship between time and level of illumination in the production of fading, *i.e.*, the fading is dependent upon the product of these two factors and is substantially unaffected by variations in both as long as the product is unchanged. A third research[43] disagrees with this conclusion, indicating that the relationship varies from direct reciprocity at higher illumination levels.

The spectral distribution of the radiant energy used affects the rate of fading. It has been found that ultraviolet energy of wavelengths shorter than 300 nanometers[39] (not present in energy radiated by most light sources) may cause very rapid fading—and other forms of product deterioration—in some cases. Energy in the region from 300 to 400 nanometers is present in the radiation from most electric light sources in common use but to a much smaller amount per lumen than in daylight. This spectral region apparently produces more fading per unit of energy than an equal amount in the visible spectrum. Absorption of the near ultraviolet in sunlight by filters which absorb very little energy in the visible spectrum has been found to reduce the fading somewhat,[40] but not by as large an amount as is sometimes suggested. In sunlight it has been found that fading is produced by energy through the whole region shorter than the orange-red (approximately 600 nanometers).

Because of the greater amount of energy in the near ultraviolet and blue, per lumen of daylight, than in the light from tungsten and fluorescent lamps, it is logical to expect a somewhat higher rate of fading in daylight, and this is found to be the case. Fig. 25-19[38] shows the results of comparative tests of 108 colored ribbons under daylight, tungsten filament, and daylight fluorescent lamps. While daylight fluorescent lamps with since-obsoleted phosphors were used in that test, all evidence at hand would indicate that results would not be greatly different if lamp colors most commonly used at present had been used. The specimens were exposed to sunlight and skylight through ⅛-inch window glass between 9:00 a.m. and 3:30 p.m. on clear days in midsummer. The exposures in footcandle-hours were automatically recorded throughout the test.

Fig. 25-20 shows spectral reflectance curves for new and slightly faded specimens of pink silk cloth. It will be noted that the spectral changes indicate bleaching in regions of maximum absorption and darkening in regions of minimum absorption. These changes are typical of many specimens tested.

Fig. 25-19. Distribution of 108 Colored Textiles with Respect to Exposure-Ratios (Footcandle-Hours) Required to Produce Equal Amounts of Fading with Three Light Sources

Exposure Relative to Daylight	Number of Specimens Equally Faded	
	Incandescent	Daylight Fluorescent
Below 0.5	4	4
0.51 to 0.75	12	9
0.76 to 1.00	12	12
1.01 to 1.25	10	14
1.26 to 1.50	15	13
1.51 to 1.75	14	17
1.76 to 2.00	8	10
2.01 to 2.25	14	14
2.26 to 2.50	2	1
2.51 to 3.00	2	3
3.00 to 4.00	9	6
Greater than 4	6	5
	108	108

Average ratio
Incandescent filament light 1.81
Daylight fluorescent 1.68
For 110 specimens the *average* exposure ratio, daylight fluorescent to incandescent-filament light, was 0.99.

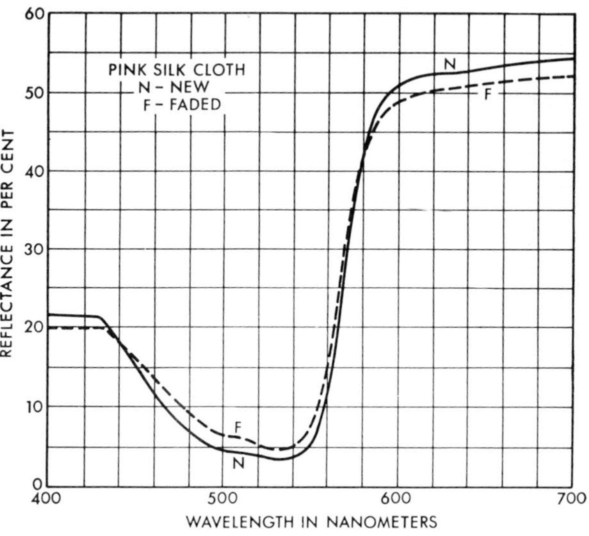

Fig. 25-20. Spectral reflectance of a specimen of pink silk before and after exposure sufficient to cause moderate fading.

Fading appears to be a photochemical process requiring oxygen and is inhibited or greatly reduced in a vacuum.

An increase in moisture content can cause a very large increase in fading, on cellulose in particular, but has less effect on wool.[37]

Temperature appears to have little effect on fading rate of silk and cotton at temperatures below 120° F, but the rate is approximately twice as great at 150° F as at 85° F.

It is often found that a light tint is more fugitive than a higher concentration of the same dye.

The permissible exposure before some fading may be expected is of major importance in the merchandising field. Tests of approximately 100 textile specimens performed about 1940 showed that half of them showed some fading after exposures of approximately 50,000 footcandle-hours to incandescent lamps. A more recent research[43] showed that it required approximately ten times that exposure to produce a minimum perceptible fading with incandescent and fluorescent lamps on more than 100 commercial fabrics. Obviously, improvements in dyes have greatly improved their light-fastness, and fading of merchandise in display cases is not the critical problem which it was when the fluorescent lamp first became an important factor in lighting, leading to usage of greatly increased illumination levels.

Fading of merchandise is most readily apparent under conditions where an area receiving high illumination is adjacent to areas not exposed. Typical examples are folded neckties and socks stacked on shelves in display cases, the folded edges approaching the lamp closely. In order to reduce the hazard of fading, goods displayed on shelves near the lamps might be rotated to lower shelves on a 1-week to 10-day cycle.

In many modern grocery stores, especially those of the self-service type, packaged meats are displayed in refrigerated cases with relatively high illumination. Fresh meats show no appreciable color change due to light within any reasonable display period, although unwrapped meats may show changes due to dehydration. Many of the processed meats (veal and pork loaves, bologna, etc.) receive their red color from a curing process using salt or sodium nitrate. Through some reaction of light and air, the processed meats will return to their original or grayed color, and this "fading" takes place very rapidly with some meats. Some, especially veal loaf and bologna, will show perceptible color change in 150 to 200 footcandle-hours. Since the illumination in some of these cases may be as high as 100 footcandles, it is found that undesirable changes may occur in 1½ to 2 hours. The most susceptible meats should be placed as far away from the lamps as possible.

Research has shown that lighting levels from 50 footcandles to 200 footcandles have approximately the same effect on shelf life of frozen meats.[44] Depending on the degree of original muscle pigmentation, frozen meat was considered saleable for three to six days; however, above 200 footcandles shelf life was considerably reduced. Differences in spectral energy distribution of the light sources resulted in no apparent or statistically significant differences in rate of color degradation.

Cigars displayed in cases illuminated by fluorescent lamps within the cases may be somewhat bleached by the light, but the exposures required are very long. In a test using seven brands of cigars it was found that an exposure of approximately 40,000 footcandle-hours produced a just-noticeable change, and that this exposure could be doubled before the color change reached an objectionable degree. Exposures of this magnitude are much greater than any to be expected in normal merchandising of cigars.

The bactericidal lamp, producing high energy at 253.7 nanometers, has been found to be a potent source for accelerated fading.[39] It might be used for accelerated fading tests, but there is no simple relationship to fading by sunlight.

Light for Insect Trapping[45–49]

Light sources are used in agriculture and many other fields as lures for phototropic insects. They are of particular value in determining the periods and intensities of moth flight to estimate the time of egg laying and the need for insecticide treatment. They are used as indicating devices for European corn borers, coddling moths, fruit flies, night-feeding beetles, and many other insects. General conclusions have been reached as follows:

1. The closer light wavelengths approach the blue end of the spectrum, the more insects they attract.
2. The closer light wavelengths approach the red end of the spectrum, the fewer insects they attract.
3. For a given spectral distribution, the brighter the source the greater its attractive power.
4. The substitution of yellow lamps for white lamps of equal candlepower reduces the number of insects attracted.
5. Because bare lamps attract insects from all directions and only a small percentage of the light emitted by a bare lamp falls on the area it is desired to light, they attract more insects than lamps in reflectors.
6. The use of projector or reflector spotlight-type bulbs results in the maximum reduction of insects when the sources are located 20 or more feet away from the area it is desired to illuminate.
7. The addition of opal diffusing globes or other means of reducing bare lamp luminance will reduce the numbers of insects attracted.
8. There is a variety of electric insect traps available, using black light fluorescent lamps, which have been found to be effective in attracting night-flying insects.

In outdoor lighting, the probable luminance of areas illuminated to a level of less than 100 footcandles will not attract phototropic insects to any degree comparable to the attraction of an exposed source. Bright sources should be placed at considerable distances from lighted areas if insects are a nuisance near the area, or such sources should be well shielded with reflectors and louvers.

In applying these conclusions, the principles of good lighting for vision described in the preceding sections of this Handbook should be followed.

Luminescence and Luminescent Materials[50]

The emission of light resulting from causes other than thermal stimulation (incandescence) is known as luminescence. Some of the more important luminescence terminology are discussed in Section 2 and defined in Section 1.

Photoluminescence occurs to a practical degree in many hundreds of materials when they are exposed to radiation from long wavelengths in the visible through the ultraviolet to the x-ray and gamma-ray regions. The most important practical application of photoluminescent materials (lamp phosphors) is in light sources where mercury ultraviolet is the exciting radiation (see pages 2–7 through 2–10). These phosphors are oxygen dominated inorganic crystalline materials. Other materials such as the zinc and cadmium sulfides and a wide variety of organic compounds excited by the near ultraviolet (approximately 360 nanometers) are used extensively to achieve spectacular theatrical effects and in various signs and instrument dials. So-called "optical bleaches" are fluorescent organics used as whiteners in laundered items such as shirts, sheets, etc.; they are excited by the near ultraviolet and radiation in daylight to fluoresce a bright blue, thus compensating for the natural yellow-white appearance of the unimpregnated cloth. Super bright orange and red organic dyes which are fluorescent under near ultraviolet excitation are widely used as identification and warning markers—for example in high speed aircraft to aid in rapid visual acquisition to avoid collision. Fluorescent paint, ink, and dyed fabrics are available in many colors, including red, orange, yellow, blue, and a white that appears blue under ultraviolet. Because these materials transform ultraviolet, violet, and even blue energy into light, as well as reflect incident light, their brightness under daylight is striking. This is true because of the ultraviolet energy in daylight, which, after striking the materials, returns to the eye as light in addition to the daylight reflected by the materials and gives some fluorescent materials an apparent reflectance (under daylight) of 110 per cent or more; that is, they send back more light of a given color than strikes them.

These colored fluorescent materials are especially useful on signal flags and signal panels since they can be identified at greater distances than those with nonfluorescent surfaces. The increased range over which the fluorescent flags can be identified is most apparent during the half-light conditions of dawn and twilight. Organic fluorescent dyed materials are at times used to produce spectacular signs, such as used on streetcars or buses. This use of fluorescent paints and dyes is commercially applicable wherever the long-distance identification of objects is important throughout the hours of daylight. Such materials are also used to produce very colorful clothing.

Other photoluminescence applications include x-ray and γ-ray stimulable crystals which find extensive use in scintillation counters—used for detecting the exciting radiation itself. Chemical analyses are often based upon the use of the characteristic luminescence of certain activator ions in known host media.

Cathodoluminescent materials find their most important application in television screens and in scientific instrumentation such as oscilloscopes, electron microscopes, image intensifiers, and radar screens. Here zinc and cadmium sulfides and oxygen dominated phosphors such as the silicates, phosphates, and tungstates are used. A recent improvement in color television screens has resulted from the development of a rare earth (europium) activated deep red phosphate phosphor. This permits color fidelity in the receiving screen unattainable heretofore.

Luminescence attending chemical reactions has been observed widely in both organic and inorganic systems. One of the most interesting is the reaction between the naturally occurring chemicals luciferin, luciferase, and adenosine triphosphate (ATP) as it takes place in the firefly. A space-age application of this reaction invloves the detection of extra-terrestrial life. A mixture of luciferin and luciferase will be rocketed into space where the slightest bit of extra-terrestrial ATP (present in every life form) coming into contact will cause a burst of light which will be detected by multiplier phototubes and telemetered to the earth receiving station thus signaling the existence of life outside the earth.

Ion, sound, friction, and electric field excitation of phosphors remain essentially phenomena of relatively little practical application except that the latter has found use in read-out devices (see pages 2–10 and 2–11).

Phosphorescent Materials. Phosphorescent materials, excited by ultraviolet energy, daylight, or light from electric lamps, have been shown to have a high brightness of afterglow for periods of from 6 to 9 hours, and some of these have a noticeable brightness for as long as 24 hours after the exciting source has been removed.

Phosphorescent materials, generally combinations of zinc, calcium, cadmium, and strontium sulfides, can be incorporated into adhesive tapes (plastic over-coatings), paints, and certain molded plastics. Because of the tendency of many plastics either to transmit moisture—which decomposes the sulfide—or to react directly with the phosphor, care should be exercised in the choice of a plastic to carry the phosphorescent powders. Both vinyl and polystyrene plastics have been found well suited to this application.

Phosphorescent materials are suitable only for applications where exposure to light prior to use is possible. While some can be used in spots where a visible brightness is necessary for from 6 to 9 hours, only a few of the many phosphorescent compounds have this degree of persistence. Those manufactured from zinc sulfide have high initial brightness after the light source has been removed, but their useful brightness period does not extend beyond 20 to 30 minutes. Before refinements in the processing of calcium and strontium phosphors were made in 1944, the useful brightness of these types did not extend beyond from 2 to 3 hours after activation. However, now that long-persistence phosphors are available phosphorescent materials are, in many applications, suitable for night-long use.

Brightness reduction (decay) rates are hastened by high temperatures. At very low temperatures (60 kelvin) luminescence may be completely arrested to be released later upon warming.

Radioactive Excitation. This is simply excitation by electrons, ions (atoms, nuclear fragments), or gamma rays—singly or in combination—resulting from the fission or radioactive decay of certain elements. For example, radium emits alpha particles which can excite luminescence when they strike a suitable phosphor. Krypton-85 excites by emission of beta rays or high energy electrons. The sulfide phosphors not only emit light when exposed to ultraviolet energy or light, but also exhibit this property under bombardment by the rays from radioactive materials. Thus, by compounding a mixture of such a radioluminescent material, *e.g.*, zinc sulfide, and a small amount of radioactive material, a self-luminous mixture can be produced. Such a radioactive luminous compound will continue to emit light without the help of external excitation for a very long time (several years) in practical applications. Radioactive-luminous materials have been used for many years on watch and clock dials, and on the faces of other instruments that must be read in the dark. They are the only type of commercially available luminous materials that maintain self-luminosity over long periods of time. The power source was formerly some salt of radium or more frequently the lower-priced mesothorium. Lately strontium-90, and more recently polonium, which has certain advantages with regard to cost and safety, are also coming into use. The bombardment of the fluorescent materials by the radiations from the exciting radioactive materials causes a decomposition of the fluorescent materials, which, of course, limits their life. A good-quality material will be useful for a few years and will maintain a relatively constant brightness during this period. The actual life of a radioactive-luminous paint is controlled to a great extent by its initial brightness, which is varied by changing the concentration of radioactive material in the mixture. Increased brightness means increased radioactive content, and more rapid decomposition of the glowing salt.

Because of the expense of the radioactive substances used to activate this material, radium-luminous paint seldom is used in large quantities or to cover large areas.

MISCELLANEOUS APPLICATIONS OF ULTRAVIOLET ENERGY

Ultraviolet Energy

Located directly below the blue wavelengths of the visible spectrum is the invisible but very active ultraviolet region. This region contains four main bands. These bands are the "black light", the erythemal, the bactericidal, and the ozone producing wavelengths. See page 1–21.

The near ultraviolet or "black light" band has the property of causing certain materials to fluoresce. It produces photochemical reactions in other materials. The second ultraviolet region is the erythemal which is noted for its ability to provide our sunburn and summer tan and to generate vitamin D in the body. The far ultraviolet or bactericidal region is highly lethal to all microorganisms. Secondary properties are the ability to fluoresce certain materials and to produce a superficial erythema of the skin. Ultraviolet energy below the bactericidal wavelengths generates a polymer of oxygen known as ozone (O_3).

Sources of Ultraviolet Energy

The ultraviolet energy emitted by a general service incandescent filament lamp (color temperature 3000 K) is equal to only about 1 per cent of the energy in the visible spectrum. Therefore, it is not of practical consequence. Tungsten-halogen lamps operated at a color temperature of 3100 K or over and with ultraviolet transmitting tubes can cause erythema and eye injury.[51] Fluorescent lamps also emit some ultraviolet, in particular the 365.2-nanometer band of the mercury discharge. This radiant energy likewise is only a fraction of the light energy emitted. Some fluorescent lamps have been made with phosphors to provide radiation favorable to plant growth and animal health. Generally, illumination sources are not considered useful producers of ultraviolet radiation. Such ultraviolet as they do emit normally is composed of longer wavelengths, near the visible spectrum.

Mercury arcs enclosed within ultraviolet transmitting glass or fused quartz emit ultraviolet energy in addition to light. The ultraviolet component of the energy emitted by a high-pressure mercury arc may be equal to or nearly twice as large as that of the visible component, depending upon lamp design.

Low-pressure quartz mercury lamps produce about 85 per cent of their total (light and ultraviolet energy) output in the ultraviolet spectrum. Over 90 per cent of their ultraviolet energy output is emitted in one band at 253.7 nanometers. When an ultraviolet transmitting glass such as 96 per cent silica glass is substituted for fused quartz, the ultraviolet transmission of 253.7-nanometer energy is reduced by about 8 per cent. The transmission of the longer wavelengths is approximately the same for 96 per cent silica glass and fused quartz.

Most ultraviolet sources require special circuits and external ballasts for operation. A few have their ballast built into the bulb, either in the form of an incandescent filament or a low-temperature resistor, and may be attached to and operated from a standard electrical outlet.

Several types of fluorescent lamps with ultraviolet emitting phosphors are available. See Figs. 8–114, 8–119, 25–21, and 25–22.

Carbon arcs are also used for the production of ultravolet radiation. See Fig. 8–126. Employment of different coring materials in the electrodes alters the distribution of energy through the spectrum. Sources are available in sizes ranging from a few hundred watts to several kilowatts, with high concentration due to the small physical volume of the arc.

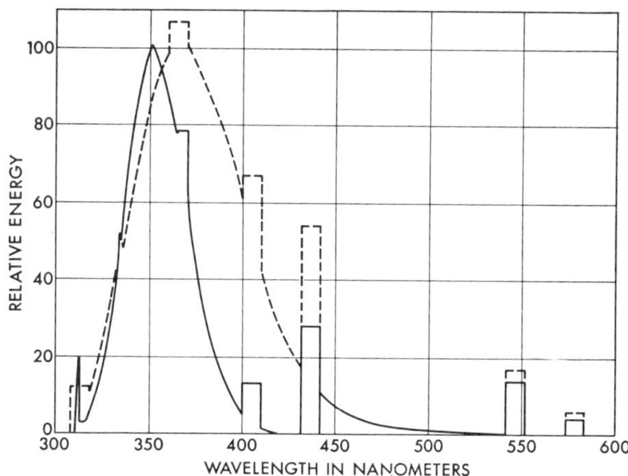

Fig. 25–21. Spectral distribution of typical BL fluorescent lamps.

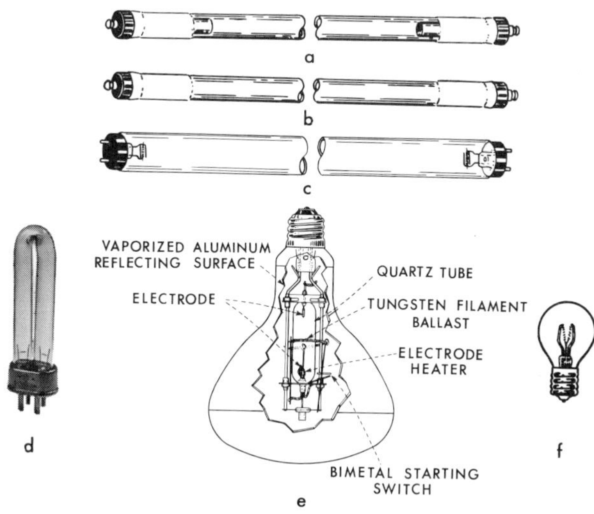

Fig. 25–22. Typical sources of ultraviolet energy: a. cold cathod bactericidal; b. Slimline bactericidal; c. 360 BL fluorescent—15-watt T-8, 20-watt T-12, or 40-watt T-12; also 20- or 40-watt T-12 fluorescent sunlamps; d. hot cathode bactericidal lamp; e. RS sunlamp; f. 3.5- to 4.0-watt ozone lamp.

Eye Protection

Eye protection is essential for all who are exposed to the direct or reflected radiation from lamps emitting ultraviolet, especially shortwave U.V. Ordinary window or plate glass or goggles that exclude radiations of wavelength shorter than 340 nanometers usually are sufficient protection. However, if the radiation is intense, or is to be stared at for some time, special goggles should be used. Failure to protect the eyes can result in temporary but painful inflammation of the conjunctiva, cornea, and iris; photophobia; blepharospasm; and ciliary neuralgia.

Erythemal and Biological Ultraviolet

Generators of ultraviolet energy used as sunlamps are designed to have erythemal effectiveness in the region of 290 to 300 nanometers.

Production of Vitamin D. Ultraviolet energy of 296.7 nanometers wavelength has the greatest erythemal effectiveness. Energy in the 253.7-nanometer wavelength is 55 per cent as effective, watt for watt.

Lamps designed to produce erythema also increase the lime, phosphorus, and carbohydrate metabolism and develop antirachitic vitamins (especially D), since the absorption of ergosterol in the human skin is maximum in the region between 250 and 300 nanometers.

Tan, differing from erythema but likely to follow it, may result from somewhat longer wavelengths. When incident radiation exceeds wavelengths of 390 nanometers up to approximately 1400 nanometers, the result is a skin reddening and a dilation of the capillaries.

Figs. 25–23 and 25–24 record the reflectance and transmittance of the human skin for different wavelengths. Fig. 25–25 indicates a secondary peak of erythemal efficiency at wavelengths in the neighborhood of 250 nanometers. This ultraviolet wavelength is not found in daylight or in the output of commercial sunlamps.

Sunlamps. The most common sources of erythemal ultraviolet energy are mercury-vapor-discharge lamps. Certain fluorescent-lamp phosphors emit considerable ultraviolet energy in the erythemal region. (See Fig. 8–114.)

Sunlamps of the RS type are rated at 100 watts in the arc, and 175 watts in the ballasting filament.

Fig. 25-23. Reflectance at Various Wavelengths of Average Untanned Human Skin (Caucasian)[52, 53]

Wavelength (nanometers)	Reflectance (per cent)	Wavelength (nanometers)	Reflectance (per cent)
240	3	400	28
260	4	450	35
280	4	500	42
300	5	550	48
320	11	600	54
340	16	650	65
360	21	700	68

Fig. 25-24. Transmittance at Various Wavelengths of Different Thicknesses of Human Skin[54]

Wavelength (nanometers)	Transmittance (per cent)		
	0.1 mm	0.5 mm	1.0 mm
253.7	0.00	0.00	0.00
289.4	0.01	—	—
296.7	2	—	—
302.4	8	—	—
313.2	30	0.3	0.008
334.2	42	1.3	0.02
366.3	49	3	0.08
405.0	55	5	0.3
435.9	59	7	0.5

Fig. 25-25. Erythemal and Bactericidal Efficiency of Ultraviolet Radiation

Wavelength (nanometers)	Erythemal Efficiency	Tentative Bactericidal Efficiency
*235.3	....	0.35
240.0	0.56	....
*244.6	0.57	0.58
*248.2	0.57	0.70
250.0	0.57	....
*253.7	0.55	0.85
*257.6	0.49	0.94
260.0	0.42	....
265.0	....	1.00
*265.4	0.25	0.99
*267.5	0.20	0.98
*270.0	0.14	0.95
*275.3	0.07	0.81
*280.4	0.06	0.68
285.0	0.09	....
*285.7	0.10	0.55
*289.4	0.25	0.46
290.0	0.31	....
*292.5	0.70	0.38
295.0	0.98	....
*296.7	1.00	0.27
300.0	0.83	....
*302.2	0.55	0.13
305.0	0.33	....
310.0	0.11	....
*313.0	0.03	0.01
315.0	0.01	....
320.0	0.005	....
325.0	0.003	....
330.0	0.000	....

* Emission lines in the mercury spectrum; other values interpolated.

The small quartz envelope is enclosed in an outer bulb that prevents emission of any energy of a wavelength less than 280 nanometers. The emitted ultraviolet energy of a wavelength longer than 280 nanometers is similar in nature to that from a therapeutic quartz high pressure mercury arc, but under comparable conditions is less intense.

Photochemical Lamps and Their Uses

Many radiant energy applications in the photochemical field require radiators of near-visible ultraviolet energy. The uses include such diverse operations as production of uranium, of "smoke gases" for military concealment, of synthetic rubbers and some plastic preparations, and processes in photography, blueprinting and other photo-copying, and photolithograhy (see Section 24). Photochemical applications merge with laundry bleaching, the treatment of wood-pulp and textile fibers, and the fixation of hydrocarbons.

Bactericidal Ultraviolet[55-65]

Ultraviolet radiation between 180 and 300 nanometers destroys bacteria, mold, yeast, and virus. The relative effectiveness of these radiations is shown in Fig. 25-25. Because of absorption in the ozone layer of the upper atmosphere, practically none of this short-wave ultraviolet radiation reaches the earth's surface from the sun.

The bactericidal effectiveness of ultraviolet radiation results from the absorption of the radiation by nuclear protein, the vital part of bacteria, after transmission through air, water, and the ordinary protein of the bacteria. In contrast, theoretically more lethal higher frequency energy, such as x-rays and gamma radiation, passes through the organism with little absorption and little killing action. Radiation of 253.7 nanometers in wavelength produced efficiently by bactericidal lamps, as shown in Fig. 25-26, is unique in its bactericidal action without objectionable heating or photochemical effects.

Bactericidal Lamps. The most practical method of generating bactericidal radiations is by passage of an electric discharge through low pressure mer-

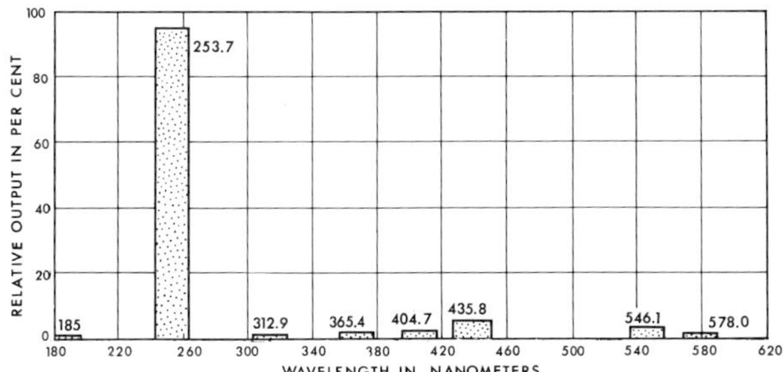

Fig. 25-26. Relative spectral distribution of energy emitted by ozone producing bactericidal lamps.

cury vapor enclosed in a special glass tube which transmits these short ultraviolet radiations. About 95 per cent of the energy is radiated in the 253.7-nanometer band which is very near to the greatest lethal effectiveness. The bactericidal output of various commercial lamps is given in Fig. 8-119.

Hot cathode bactericidal lamps are similar in physical dimensions and electrical characteristics to the standard preheat 8-watt, 15-watt, and 30-watt fluorescent lamps and they operate on the same auxiliaries. Slimline bactericidal lamps are instant-start lamps capable of operating at several current densities within their design range (120 to 420 milliamperes), depending on the ballast with which they are used. Cold cathode bactericidal lamps are instant-start lamps with the cylindrical cold cathode type of electrode. They are made in many sizes and operate from a transformer.

The life of the hot cathode and slimline bactericidal lamps is governed by the electrode life and frequency of starts. (The effective life of hot cathode bactericidal lamps is sometimes limited by the transmission of the bulb. This is particularly true when operated at cold temperatures.) The electrodes of cold cathode lamps are not affected by number of starts, and useful life is determined entirely by the transmission of the bulb. All bactericidal lamps should be checked periodically for ultraviolet output to make sure that their bactericidal effectiveness is being maintained.

The majority of bactericidal lamps operate most efficiently in still air at room temperature, and ultraviolet output is measured at an ambient temperature of 77° F. Temperatures either higher or lower than this optimum value decrease the output of the lamp. Lamps operating in a room at 40° F produce only about two-thirds to three-fourths as much ultraviolet as at 80° F. Cooling the lamp by passing air currents over it or by submerging it in liquid likewise lowers its output.

Slimline bactericidal lamps operated at from 300 to 420 milliamperes and certain preheat bactericidal lamps operated at 600 milliamperes are designed exceptions to this general rule. At these high-current loadings the lamp temperature is above the normal value for optimum operation; therefore cooling of the bulb does not have the same adverse effect as with other lamps. Thus these lamps are well suited for air-conditioning duct applications.

In addition to the energy at 253.7 nanometers, some bactericidal lamps generate a controlled amount of 184.9-nanometer radiation which pro-

Fig. 25-27. Incident Energy at 253.7 Nanometers Necessary to Inhibit Colony Formation in 90 Per Cent of the Organisms

Organism		Energy (μW-sec/cm^2)
Bacillus anthracis		4520
S. enteritidis		4000
B. megatherium sp. (veg.)		1300
B. megatherium sp. (spores)		2730
B. paratyphosus		3200
B. subtilis		7100
B. subtilis spores		12000
Corynebacterium diphtheriae		3370
Eberthella typhosa		2140
Escherichia coli		3000
Micrococcus candidus		6050
Micrococcus sphaeroides		10000
Neisseria catarrhalis		4400
Phytomonas tumefaciens		4400
Proteus vulgaris		2640
Pseudomonas aeruginosa		5500
Pseudomonas fluorescens		3500
S. typhimurium		8000
Sarcina lutea		19700
Seratia marcescens		2420
Dysentery bacilli		2200
Shigella paradysenteriae		1680
Spirillum rubrum		4400
Staphylococcus albus		1840
Staphylococcus aureus		2600
Streptococcus hemolyticus		2160
Streptococcus lactis		6150
Streptococcus viridans		2000
Yeast		
Saccharomyces ellipsoideus		6000
Saccharomyces sp.		8000
Saccharomyces cerevisiae		6000
Brewers' yeast		3300
Bakers' yeast		3900
Common yeast cake		6000
Mold Spores	**Color**	
Penicillium roqueforti	Green	13000
Penicillium expansum	Olive	13000
Penicillium digitatum	Olive	44000
Aspergillus glaucus	Bluish green	44000
Aspergillus flavus	Yellowish green	60000
Aspergillus niger	Black	132000
Rhizopus nigricans	Black	111000
Mucor racemosus A	White gray	17000
Mucor racemosus B	White gray	17000
Oospora lactis	White	5000

duces ozone in the air. See Fig. 25–26. Ozone is a deodorant and in the presence of water vapor is bactericidal and fungicidal.

Exposure Time. A lethal exposure of an organism is determined by its susceptibility, the wavelength of radiation, the density of the radiant flux (watts per unit area) and the time of exposure. Fig. 25–27 gives the amount of 253.7-nanometer energy in microwatt-seconds per square centimeter to destroy 90 per cent of various common microorganisms. The bactericidal effectiveness is proportional to the product of intensity times time from one microsecond to a few hours.

A non linear relationship exists between ultraviolet exposure and bactericidal efficacy. If a specific ultraviolet exposure produces a 90 per cent kill of bacteria, doubling the exposure can only produce a 90 per cent kill of the residual 10 per cent for a final 99 per cent. In reverse a 50 per cent decrease in intensity or exposure results only in a decrease in bactericidal efficacy from 99 to 90 per cent.

Precautions. Bactericidal ultraviolet irradiation can produce erythema. The American Medical Association has set a limit of 0.1 microwatt per square centimeter for continuous exposure and 0.5 for 7 hours per day. This conservative limitation can be extrapolated to 22 microwatts per square centimeter for 10 minutes, and a limit of 2.5 minutes exposure to 90 microwatts per square centimeter at 1 meter, from a G30T8 lamp, and to 6 to 7 seconds exposure at the distance required to read the caution label on the same lamp while in operation. See Fig. 8–119.

Applications

Air Disinfection in Rooms. In occupied rooms, irradiation should be confined to the area above the heads of occupants as shown in Fig. 25–28. Louvered equipment should be used where ceilings are less than 9 feet high to avoid localized high concentration of flux which may be reflected down onto occupants. An average irradiation of 20 to 25 microwatts per square centimeter is effective for slow circulation of the upper air, and will maintain a freedom from respiratory disease organisms comparable with outdoor air.

The upper air disinfection as practiced in hospitals, schools, and offices is effective at the breathing level of room occupants. Personnel movement, body heat, and winter heating methods create convection currents through the whole cross section of a room sufficient to provide "sanitary ventilation" of 1 to 2 air changes per minute. This has been compared to a removal of the room ceiling for access to outdoor air. There may be "ultraviolet barriers" down across doors to provide irradiation sufficient to disinfect air at usual movement through them. In all cases the ceilings and upper walls should have a bactericidal ultraviolet reflectance as low as the 5 per cent characteristic of most oil and some water-base paints. See Fig. 25–3. "White coat" plaster or gypsum product surfaced wallboard and acoustical tile may have higher germicidal reflectances and should always be painted. See Fig. 25–29. These precautions are especially important in the hospital infants' ward and somewhat less so in the schoolroom.

In hospital operating rooms, especially where such prolonged surgery as heart, brain, and lung operations is done, a combination of upper air and vertical barriers at 25 microwatts per square centimeter is used with head, eye, and ear protection in addition to the usual face mask

Air Duct Installations. It is possible to provide a sufficiently high level of ultraviolet for a 90 to 99 per cent kill of most bacteria in the very short exposure times of duct air at usual air velocities. The limitation of the method is that it can only make the duct air equivalent to good outdoor air and its value is in the treatment of recirculated air and contaminated outdoor air in hospitals, pharmaceutical, and food processing plants. Duct installations are especially valuable where central air heating and ventilating systems recirculate air through all of the otherwise isolated areas of an institution. Bactericidal lamps, often especially designed for the cooling effect of high velocity duct air, are installed on doors in the sides of ducts, or inserted across their axis, depending upon the size and shape of the duct and access for servicing. Where possible the best location is across the duct to secure longer travel of the energy before absorption on the duct walls and to promote turbulence to offset the variation in ultraviolet level throughout the irradiated part of the duct.

Liquid Disinfection. Ultraviolet disinfection of

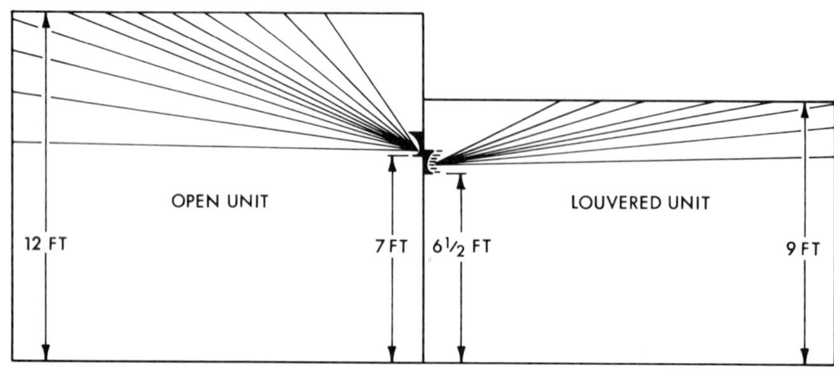

Fig. 25–28. Bactericidal lamps for air disinfection in occupied rooms: (a) open unit used in rooms over 9 feet; (b) louvered unit used where ceilings are less than 9 feet.

water is used where it is essential that there is no residual substance or taste.

Water disinfection methods are similar to those for air except that allowance should be made for some ultraviolet absorption by traces of such natural chemical contaminants as iron compounds and for an 8 to 10 fold greater exposure for wet than for dry bacteria. The product of these two factors calls for water disinfection ultraviolet exposures of 700 to 1000 microwatt-minutes per cubic centimeter. Such exposures are secured by slow gravity flow of water through shallow tanks under banks of lamps, or by lamps in isolating jackets of ultraviolet transmitting glass, immersed directly in the water.

Liquids of high absorptance, such as fruit juices, milk, blood, serums, and vaccines, are disinfected with "film spreaders" ranging from high speed centrifugal devices and films adhering to the surfaces of rotating cylinders to gravity flow down screens and inclined planes. The centrifugal devices spread a film of a liquid down to a thinness of the order of its molecular size.

Granular Material Disinfection. The surfaces of granular materials such as sugar are disinfected on traveling belts of vibrating conveyors designed to agitate the material during travel under banks of closely spaced bactericidal lamps. During a transit which defines the exposure time, all the particles are brought to the surface frequently and long enough to provide an effective exposure of their surfaces to the irradiation. In the case of sugar, thermoduric bacteria survive the vacuum evaporator temperatures of sugar-syrup concentration and, rejected by the sugar crystals during formation, remain in the final film of dilute syrup left on the crystal surfaces. Ordinarily harmless, they may cause serious spoilage in canned foods and beverages.

Product Protection and Sanitation. Product protection and sanitation is by both air disinfection and surface irradiation as with sugar. In this field, however, the usefulness of bactericidal ultraviolet is generally limited to the prevention of contamination during processing instead of a disinfection of an otherwise final product. Bactericidal lamps in concentrating reflectors are used to disinfect any air which might contaminate a product during processing and packaging, as during the travel of bottles from washing through filling to capping. They serve to replace or supplement heat in processes where sterilization by heat might be destructive. Where sufficient irradiation to kill mold spores may be impractical, the vegetative growth of mold itself can be prevented by continuous irradiation at the levels lethal to ordinary bacteria as in bakeries and breweries, as on the surfaces of liquid sugar syrup, fruit juices, and beverages during tank storage. Intensive irradiation of container surfaces can supplement or replace washing between usages.

MISCELLANEOUS APPLICATIONS OF INFRARED ENERGY[66-76]

Heat may be transferred from one body to another by conduction, convection, or radiation, or by a combination of these processes. Infrared heating involves energy transfer primarily by radiation, although some convection heating may exist simultaneously due to natural or forced air movement.

Transfer of energy or heat occurs whenever any amount of radiant energy emitted by one body is

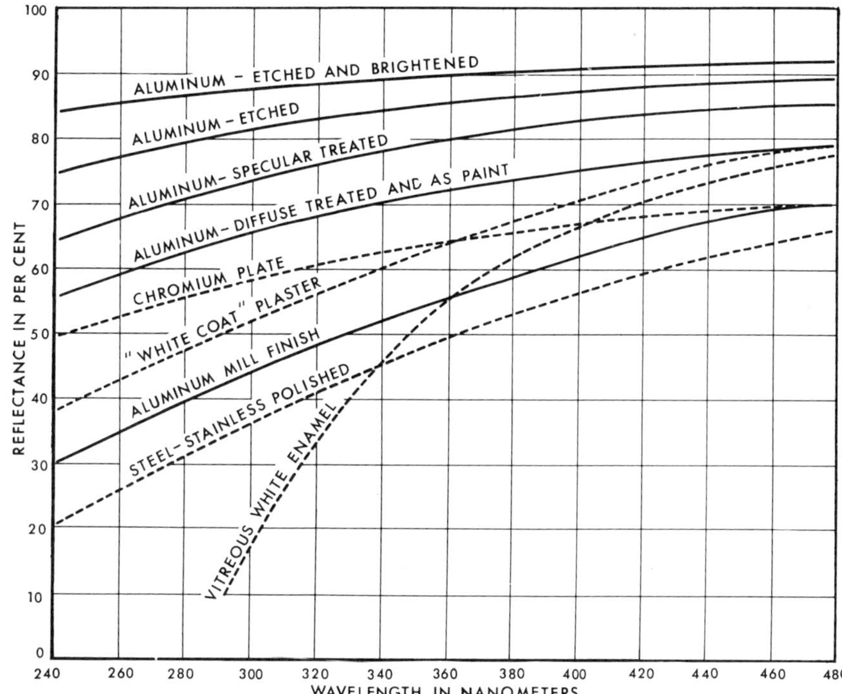

Fig. 25-29. Spectral reflectance characteristics of various materials in the blue, violet, and ultraviolet spectral regions.

absorbed by another. However, it is the electromagnetic spectrum wavelengths longer than those of visible energy (780 nanometers) and shorter than those of radar waves that are utilized for radiant heating. Best energy absorption is obtained with white, many pastel colored, and translucent products by using wavelength emissions longer than 2500 nanometers whereas the majority of dark pigmented and oxide-coated materials will readily absorb the full range of wavelength emissions from 760 to 9000 nanometers. Water vapor and visible aerosals, such as steam, readily absorb the longer infrared wavelengths.

Applications of Infrared Energy

Infrared radiant energy may be used for any heating application where the principal product surfaces can be arranged for exposure to the heat sources.[66] Modern conveyorized methods of material handling have greatly accelerated use, with heat sources arranged in banks or tunnels as illustrated in Fig. 25–30. Typical applications include:

1. Drying and baking of paints, varnishes, enamels, adhesives, printers' ink, and other protective coatings.
2. Preheating of thermoplastic materials for forming and tacking operation, molds.
3. Heating of metal parts for shrink fit assembly, forming, thermal aging, brazing, radiation testing, and conditioning surfaces for application of adhesives and welding.
4. Dehydrating of textiles, paper, leather, meat, vegetables, pottery ware, and sand molds.
5. Spot and localized heating for any desired objective.

Rapid rates of heating can be provided in relatively cold surroundings by controlling the intensity of radiant energy, absorption characteristics of the exposed surfaces, and rate of heat loss to the surroundings.[67] Highly reflective enclosures, with or without thermal insulation, are commonly employed to assure maximum energy utilization. Limited amounts of air movement are often essential in portions or all of the heating cycle, to avoid temperature stratification and assure removal of water or solvent vapors.[70] Product temperature control is normally provided by the exposure time to infrared energy, or by the wattage of heaters employed per square foot of facing tunnel area. With modern linear heaters, this is readily designed for energy densities from 0.5 to 12.0 kilowatts per square foot to accommodate high automation speeds.

Where precise temperatures are needed, the design condition may then be modified by voltage or current input controls to add flexibility for a variety of product conditions, handling speeds, chemical formulations or other, as the process may require. The temperature of moving parts can be accurately measured by scanning with a radiation pyrometer to provide indication or full automatic control of the heating cycle. Where temperatures can vary by 3 to 5 per cent, and meet quality standards desired, an initial installation test may be made with portable instrumentation and thereafter, the cycle will repeat itself with a degree of reliability consistent with the power supply voltage, thus avoiding need for the usual mandatory controls required for other types of process heating.

Heat Sources

Many sources for producing infrared energy are now available. These can be classified generally as point, line and area sources. Their temperatures, spectral energy distribution, and life characteristics vary widely, although selection is not critical unless the products to be heated are highly selective as to wavelength penetration or absorption as in the case of many translucent plastics.

Maximum design flexibility and economy for industrial installations are generally obtained by using the tungsten filament quartz lamps, alloy resistor quartz tubes, or rod type metal sheath heaters in air or water cooled external reflectors. These are available in energy ratings from 40 to 200 watts per inch of length, and in sizes from ⅜-inch diameter by 8 inches long to ¾-inch diameter by 63 inches long. Additionally, a variety of screw base lamps, with and without internal reflectors, are available for special applications. All are listed in Section 8, Fig. 8–109. Precise voltage ratings as used for lighting service lamps need *not* be followed in using these infrared sources, often employed at voltages as low as 50 per cent of manufacturers' rating. Most of the tungsten filament heaters are designed for a color temperature of 3800 K and a life rating of 5000 hours barring over-voltage use. Metal rod heaters and quartz tubes using coiled alloy resistors are usually designed for 1450° F operating temperature and a life span of approximately 10,000 hours.

Tungsten filament heaters alone provide instant on-off response from a power source, and their radiant energy efficiency at 86 per cent of power input makes them a preferred infrared source. Other heaters have thermal inertia varying from about 1 minute for quartz tubes to 4 or 5 minutes for metal sheath heaters. Operating efficiencies are substantially influenced by the design and maintenance of external reflector systems and to a lesser extent by air temperature and velocity within the heating zone. Pay load efficiencies of 35 to 60 per cent are readily obtained in well designed systems where long holding time at designed product temperature is not required. All of the quartz heat sources can accept high thermal shock. However, metal heaters are best qualified for applications subject to mechanical shock and vibration. A variety of porcelain holders and terminals are available for these sources. Specular reflectors of anodized aluminum, gold or rhodium are recommended to direct the radiant energy to product surfaces as desired. Typical installations are shown in Fig. 25–30.

a

d

b

c

Fig. 25–30. Typical applications of infrared radiant heating. (a) Refinishing automobile bodies. (b) Speed cooking meats for airline catering service. (c) Curing wood finishes on television and musical appliance cabinets. (d) Baking plastisol coatings on wire baskets.

By comparison, gas infrared systems require far heavier and more costly construction to comply with insurance safety standards. Also their operating efficiencies must take into account a 50 per cent energy loss in the combustion flue products plus other design factors as to eventual energy utilization.

Advantages

The ease with which electric infrared heating can be controlled quickly and reliably by modern radiation pyrometers and solid-state voltage devices has advanced its acceptance and range of use greatly over the past decade. Processes that once required hours in convection ovens, and periods up to 10 minutes in early infrared systems, are often handled in seconds with present day coordination of materials, chemistry and infrared application. The absence of combustion fuel hazard with need for handling large volumes of air, and the space saving afforded for high volume, clean, quality results, keep this form of electric heat in the foreground of manufacturing technology.

Comfort Heating Using High Temperature Infrared Sources

In the past decade, the use of infrared radiation for personnel heating in commercial and industrial areas has become quite popular. The T-3 quartz lamp as a semi-luminous infrared source has distinguished itself for a wide variety of applications in commercial buildings, marquee areas, industrial plants, warehouses, stadiums, pavilions, and other public areas. Units are usually of the pendant or recessed type, with reflector control for the combined visible and infrared radiation, using the quartz tube and quartz lamp sources listed in Section 8, Fig. 8-109. In contrast, residential use except for bathroom areas is mostly confined to low temperature sources such as electric base boards and plastered radiant ceilings.

Applications. Radiant comfort heating applications fall into two broad classifications—general heating and spot heating.[74] General heating installations irradiate complete room areas and usually appear in high, poorly insulated industrial spaces such as shops, warehouses, storage sheds, or garages. The installation in this case often consists of a uniform radiant wattage density in the range of 10 to 30 watts per square foot incident to the floor surface. However, some system designers prefer equipment layouts using asymmetric units to provide a somewhat higher density in the areas adjacent to outside walls to help off-set the wall thermal loss. To date, like convection heating systems, over-all radiant systems have an installed Btu capacity sufficient to hold the desired indoor temperature and overcome the building heat loss at the specified outdoor design temperature; but performance data on some installations indicate a capacity 70 to 90 per cent of the building thermal loss is adequate.[72] This reduction is probably due to the direct personnel heating and an improved mean radiant temperature in the space.

Infrared energy passes through air with little absorption and this is particularly true of near infrared. Therefore, installations involving quartz infrared lamps may be mounted at much greater heights than those with far infrared sources. By selecting equipment of narrower beam spread so the radiation can be confined primarily to the floor where it will be most beneficial with limited losses through the walls, the mounting height may be increased without requiring a greater installed capacity. It is good practice to keep the radiation from striking the walls at heights more than 8 feet above the floor to limit wall losses.

Although infrared heating of the air is minimal, air in radiation heated areas is warmed from energy absorption by the floor and other solid surfaces. This causes upward gravitational movement, permitting conventional control with air thermostats[72] which are shielded from the infrared sources.

In the case of quartz tubes, metal sheath heaters, and gas-fired infrared units, on-off cycling of the equipment is permissible. When lamps are used, unless the visible energy (about 7 to 8 lumens per watt) is reduced by filtering, the cycling should be from full to half-voltage to prevent a severe change in illumination. Operation of this sort not only gives some lighting but also provides some amount of radiant heat at all times.

Infrared heating systems have an advantage over convection air heating systems for spaces that are subject to high air changes per hour (for example, where overhead doors are opened frequently). In these areas, the warm air is lost immediately and air temperature recovery can be lengthy with convection heating. With a radiant system, most objects are warmer than the air so the air in the space recovers temperature faster.

Spot Heating. The greatest potential use for high intensity radiant heating lies in spot or zone heating of swimming pools, marquees, passenger waiting platforms, loading docks, and infrequently used areas as stadiums, arenas, race track pavilions, etc.

The radiation intensity needed for spot heating varies with a number of factors. The major ones are:

1. The degree of body activity as dictated by the task. The more physical effort expended by the worker, the lower the temperature at which heat is needed. Type of clothing also influences this temperature.

2. The minimum temperature which is apt to exist in the space (or the lowest temperature at which the owner wants to provide comfort).

3. The amount of air movement at the location. Indoor drafts and slight air movements outdoors can be overcome by higher energy densities, but compensation for wind velocities of more than 5 to

Fig. 25-31. Quartz lamp installations at store fronts, entrances, and marquees warm shoppers, melt snow, dry sidewalks, and provide illumination.

10 miles per hour at temperatures below 30° F will not be sufficiently rewarding. Wind screens are far more beneficial than increased radiation levels.

For spot heating, units should be positioned to supply radiation from at least two directions,[74] preferably above and to the side of the area to be heated. Care should be taken to avoid locating equipment directly over a person's head.[72] In practice, levels for spot heating vary from 10 watts per square foot (at waist level) for an indoor installation supplementing an inadequate convection system to more than 100 watts per square foot for a marquee or sidewalk people heating system.

Snow melting by overhead infrared installations (see Fig. 25-31) is a by-product of outdoor spot heating.[75] Such installations should be energized before the snow starts to fall to avoid snow accumulation and the consequent high reflection to infrared energy. Installations of 60 to 90 watts per square foot are often capable of melting 90 per cent of the winter snow falling in most areas of the United States.

REFERENCES

General

1. Koller, L. R.: "Measurement of Spectral Radiation by Means of the Photoelectric Tube," *Measurement of Radiant Energy,* Forsythe, W. E., Editor, McGraw-Hill Book Company, New York, 1937.
2. Knowlton, A. E.: *Standard Handbook for Electrical Engineers,* 9th Edition, McGraw-Hill Book Company, New York, 1957.

Horticultural Lighting

3. Davidson, O. W.: "Footcandles and Green Leaves," *Circular 558,* New Jersey Agricultural Experiment Station, Rutgers University, New Brunswick, N.J., March, 1954.
4. Zahour, R. L.: "Indoor Footcandles for Maintaining Plants," *Illum. Eng.,* Vol. XLIX, p. 383, August, 1954.
5. Schulz, P.: *Growing Plants Under Artificial Light,* M. Barrows & Co., New York, 1955.
6. Went, F. W.: *The Experimental Control of Plant Growth,* Chron. Bot. Co., Waltham, Mass., 1957.
7. Downs, R. J., Borthwick, H. A., and Piringer, A. A.: "Comparison of Incandescent and Fluorescent Lamps for Lengthening Photoperiod," *Proc. Amer. Soc. Hort. Sci.,* Vol. 71, p. 568, 1958.
8. Van Der Veen, R. and Meijer, G.: *Light and Plant Growth,* The Macmillan Company, New York, 1959.
9. Kofranek, A. M.: "Artificial Light for Controlling the Flowering of Asters and Daisies," *Trans. Amer. Soc. Agr. Eng.,* Vol. 2, p. 106, 1959.
10. Dunn, S. and Went, F. W.: "Influence of Fluorescent Light Quality on Growth and Photosynthesis of Tomato," *Lloydia,* Vol. 22, p. 302, 1959.
11. Harrison, M. R. and Thames, G.: "Foliage Plants for Interiors," *New Jersey Bulletin 327,* College of Agriculture, Rutgers University, New Brunswick, N. J., March, 1960.
12. Mpelkas, C. C.: "Radiant Energy Sources for Plant Growth," Sylvania Lighting Products, Salem, Mass.
13. Cathey, H. M. and Borthwick, H. A.: "Cyclic Lighting for Controlling Flowering of Chrysanthemums," *Proc. Amer. Soc. Hort. Sci.,* Vol. 78, p. 545, 1961.
14. Downs, R. J.: "Phytotrons—Controlled Environments for Plant Studies," *Res./Develop.,* Vol. 20, p. 30, November, 1969.
15. Austin, R. B.: "The Effectiveness of Light from Two Artificial Sources for Promoting Plant Growth," *J. Agr. Eng. Res.,* Vol. 10, p. 15, 1965.
16. Helson, V. A.: "Growth and Flowering of African Violets under Artificial Lights," *Greenhouse—Garden—Grass,* Vol. 7, p. 4, 1968.
17. Butterfield, N. W., Hemerick, G. A. and Kinoshita, H.: "Effect of Fluorescent Lighting Beyond the Normal Day on the Flowering of Carnation," *Hort. Sci.,* Vol. 3, p. 78, 1968.
18. Bickford, E. D.: "Effect of Supplemental Lighting on Growth and Flowering of Roses," *Roses Inc. Bul.,* p. 17, December, 1968.
19. Withrow, R. B. (Editor): "Photoperiodism and Related Phenomena in Plants and Animals," *Amer. Assoc. Adv. Sci.,* Publ. 55, Washington, D. C., 1959.
20. Helson, V. A.: "Comparison of Gro-Lux and Cool White Fluorescent Lamps With and Without Incandescent as Light Sources used in Growth Rooms for Growth and Development of Tomato Plants," *Can. J. Plant Sci.,* Vol. 45, p. 461, 1965.
21. Lammerts, W. E.: "Comparative Effects of Gro-Lux and Incandescent Light on Growth of Camellias," *Amer. Camellia Yearbook,* p. 158, 1965.
22. Thomas, A. S., Jr.: "Plant Growth and Reproduction with New Fluorescent Lamps," *Planta,* Vol. 72, p. 198 and 208, 1964.
23. Wittwer, S. H.: "Carbon Dioxide and its Role in Plant Growth," *XVII Int. Hort. Congr. Proc.,* Vol. 3, p. 311, 1967.
24. Rabinowitch, E. and Govinjee, *Photosynthesis,* John Wiley & Sons, Inc., New York, 1969.
25. Hendricks, S. B.: "How Light Interacts with Living Matter," *Sci. Amer.,* Vol. 219, p. 174, September, 1968.
26. Bickford, E. D.: "Modern Greenhouse Lighting," *Illum. Eng.,* Vol. LXII, No. 5, May, 1967.
27. Canham, A. E.: *Artificial Light in Horticulture,* Centrex Publishing Co., Eindhoven, Holland, 1966.
28. Kleischnin, A. F.: "Die Pflanze and das Licht," *Akademie-Verlag,* Berlin, 1960.
29. Cherry, E. C.: *Fluorescent Light Gardening,* D. van Nostrand Co., Princeton, N. J., 1965.
30. Halpin, J. E. and Farrar, M. D.: "The Effect of Four Different Fluorescent Light Sources on the Growth of Orchid Seedlings," *Amer. Orchid Soc. Bul.,* Vol. 34, p. 416, May, 1965.
31. Halpin, J. E.: "Exotic Plants for the Light Room," *Indoor Light Gard. News,* Special Supplement, 1966.
32. Lindstrom, R. S.: "Supplemental Light and Carbon Dioxide on Flowering of Floricultural Plants," *Florist Rev.,* Vol. 144, p. 21, May, 1969.
33. Arditti, J.: "Factors Affecting the Germination of Orchid Seeds," *Bot. Rev.,* Vol. 33, January-February, 1967.
34. Mastalerz, J. W.: "Environmental Factors: Light, Temperature and Carbon Dioxide," *Roses, Penn Flower Growers,* N. Y. State Flower Growers Assn., Inc. and Roses Inc., 1969.
35. Mpelkas, C. C.: "Programmed Supplemental Lighting System For Increasing Carnation Blooms," *Carnation Science Symposium Proceedings,* American Carnation Society Conference, Boston, Mass. p. 29, March, 1970.
36. Crouch, C. L.: "Burölanclschaft—A New Concept for Office Environment," *Illum. Eng.,* Vol. 64, Sect. II, April, 1969.

Fading and Bleaching

37. "The Light-Fastness of Dyes, a Review," *Textile Res. J.,* p. 528, July, 1963.
38. Luckiesh, M. and Taylor, A. H.: "Fading of Dyed Textiles by Radiant Energy," *Amer. Dyestuff Reporter,* October 14, 1940.
39. Taylor, A. H.: "Fading of Colored Textiles," *Illum. Eng.,* Vol. XLI, p. 35, January, 1946.

40. Pracejus, W. G. and Taylor, A. H.: "Fading of Colored Materials by Light and Radiant Energy," *Illum. Eng.*, Vol. XLV, p. 149, March, 1950.
41. Luckiesh, M. and Taylor, A. H.: "Fading of Colored Materials by Daylight and Artificial Light," *Trans. Illum. Eng. Soc.*, Vol. XX, p. 1078, December, 1925.
42. AATCC Committee on Color-Fastness to Light: "A Study of the Variables in Natural Light Fading," *Amer. Dyestuff Reporter*, p. 861, November 18, 1957.
43. Delaney, W. B. and Makulec, A.: "A Review of the Fading Effects of Modern Light Sources on Modern Fabrics," *Illum. Eng.*, Vol. LVIII, p. 676, November, 1963.
44. Hansen, L. J. and Sereika, H. E.: "Factors Affecting Color Stability of Prepackaged Frozen Beef in Display Cases," *Illum. Eng.*, Vol. 64, p. 620, October, 1969.

Insect Trapping

45. Gui, H. L., Porter, L. C., and Prideaux, G. F.: "Response of Insects to Color, Intensity and Distribution of Light," *Agr. Eng.*, Vol. 23, p. 51, February, 1942. Taylor, J. G. and Deay, H. O.: "Use of Electric Lamps and Traps in Corn Borer Control," *Agr. Eng.*, Vol. 31, p. 503, October, 1950.
46. Weiss, H. B., Soraci, F. A., and McCoy, E. E., Jr.: "Insect Behavior to Various Wavelengths of Light," *J. of N. Y. Entomol. Soc.*, June, 1943.
47. Taylor, J. G., Deay, H. O., and Orem, M. T.: "Some Engineering Aspects of Electric Traps for Insects," *Agr. Eng.*, Vol. 32, p. 496, September, 1951.
48. Taylor, J. G., Altman, L. B., Hollingsworth, J. P., and Stanley, J. M.: "Electric Insect Traps for Survey Purposes," *USDA Agricultural Research Service* (ARS-42-3), May, 1956.
49. Ditchman, J. P. and Staley, K. A.: "Fighting Insects with Light Bulbs," *Amer. Motel Mag.*, October, 1953.

Luminescent Materials

50. Pringsheim, P. and Vogel, M.: *Luminescence of Liquids and Solids*, Interscience Publishers, New York, 1943. Fonda, G. R. and Seitz, F.: *Solid Luminescent Materials*, John Wiley and Sons, New York, 1948. Kroger, F. A.: *Aspects of the Luminescence of Solids*, Elsevier Publishing Co., New York, 1948. Pringsheim, P.: *Fluorescence and Phosphorescence*, Interscience Publishers, New York, 1949. Garlick, G. F.: *Luminescent Materials*, Clarendon Press, Oxford, 1949. Leverenz, H. W.: *Luminescence of Solids*, John Wiley and Sons, New York, 1950.

Infrared and Ultraviolet Energy

51. "Questions and Answers on Light Sources," *Illum. Eng.*, Vol. LXII, p. 139, March, 1967.
52. Luckiesh, M.: *Applications of Germicidal, Erythemal, and Infrared Energy*, D. Van Nostrand Company, New York, 1946.
53. Luckiesh, M. and Taylor, A. H.: "Erythemal and Tanning Effectiveness of Ultraviolet Energy," *G. E. Rev.*, Vol. 42, p. 274, June, 1939.
54. Hasselbalch, K. A.: "Chemische und Biologische Wirkung der Lichtstrahlen," *Strahlentherapie*, p. 403, 1913.
55. Wells, W. F.: "Bactericidal Irradiation of Air, Physical Factors," *J. Franklin Inst.*, Vol. 229, p. 347, 1940.
56. Buchbinder, L., Solowey, M. and Phelps, E. B.: "Studies on Microorganisms in the Simulated Room Environments, Part III—The Survival Rates of Streptococci in the Presence of Natural, Daylight and Sunlight, and Artificial Illumination," *J. Bacteriol.*, Vol. 42, p. 353, 1941.
57. Hollaender, A.: "Aerobiology," *Publication No. 17*, American Association for the Advancement of Science, F. R. Moulton, Editor, 1942.
58. Rentschler, H. C., Nagy, R., and Mouromseff, G.: "Bactericidal Effects of Ultraviolet Radiations," *J. of Bacteriol.*, June, 1941. Rentschler, H. C. and Nagy, R.: "Bactericidal Action of Ultraviolet Radiations on Airborne Organisms," *J. of Bacteriol.*, July, 1942.
59. Zelle, M. R. and Hollaender, Alexander: "Effects of Radiation on Bacteria," *Radiat. Biol.*, Vol. II, p. 365, McGraw-Hill Co., New York, 1955.
60. Buttolph, L. J.: "Practical Applications and Sources of Ultraviolet Energy," *Radiat. Biol.*, Vol. II, p. 11, McGraw-Hill Co., New York, 1955.
61. Oppenheimer, Franz, Benesi, Egon and Taylor, Alton R.: "The Ultraviolet Irradiation of Biological Films in Thin Flowing Films," *Amer. J. Public Health*, Vol. 49, p. 903, 1959.
62. Hart, D. and Nicks, J.: "Ultraviolet Radiation in Operating Room," *Arch. of Surg.*, Vol. 82, p. 449, 1961. Hart, D.: "Bactericidal Ultraviolet Radiation in Operating Room," *J. of the Amer. Med. Ass.*, Vol. 172, p. 1019, 1960.
63. Koller, L. R.: *Ultraviolet Radiation*, 2nd Edit., John Wiley & Sons, Inc., New York, 1965.
64. Summer, W.: *Ultraviolet and Infrared Engineering*, Interscience Publishers, New York, New York, 1962.
65. Smith, K. C. and Hanawalt, P. C.: *Molecular Photobiology*, Academic Press, New York, 1969.
66. Hall, J. D.: *Industrial Applications of Infrared*, McGraw-Hill Book Company, New York, 1947.
67. Tiller, F. M. and Garber, H. J.: "Infrared Radiant Heating," *Ind. Eng. Chem.*, July, 1942, and March, 1950.
68. Goodell, P. H.: "Radiant Heating—A full-Fledged Industrial Tool," *Trans. Amer. Inst. of Elec. Eng.*, Vol. 60, p. 464, January, 1941.
69. Goodell, P. H.: "Faster Baking and Drying with Infrared Heat," *Electrical World*, Vols. 127, 128, pp. 58, 72, June 21 and July 5, 1947.
70. "Standards for Class A Ovens and Furnaces (Including Industrial Infrared Heating Systems)," *Bulletin No. 86A*, National Fire Protection Association, 60 Batterymarch Street, Boston, Mass., 1969.
71. Christensen, M. and Foote, A. G.: "Development and Application of Quartz Infrared Lamps," *Illum. Eng.*, Vol. LI, p. 377, May, 1956.
72. "Heating with Infrared," *Elec. Constr. and Maintenance*, Vol. 61, pp. 92, 133, August and October, 1962.
73. "How to Apply High Intensity Infrared Heaters for Comfort Heating," *Heating, Piping & Air Conditioning*, January, 1961.
74. Frier, J. P. and Stephens, W. R.: "Design Fundamentals for Space Heating with Infrared Lamps," *Illum. Eng.*, Vol. LVII, p. 779, December, 1962.
75. Frier, J. P.: "Design Requirements for Infrared Snow Melting Systems," *Illum. Eng.*, Vol. LIX, p. 686, October, 1964.
76. Jeffery, R. W.: "Economics of Radiant Process Heating Methods," *Automation*, September, 1969.

SECTION

26

UNDERWATER LIGHTING*

Terms and Definitions 26-1 *Filtering Properties of Sea Water 26-2*
Seeing Distance in Water 26-2 *Light Sources 26-4*
Sensor Characteristics 26-4 *Underwater Lighting Calculations 26-5*
Measurement Techniques and Instrumentation 26-7

Divers, manned submersible vehicles, unmanned instrument platforms, and permanent bottom installations use underwater light for a variety of tasks. Underwater television as well as motion picture and still photography are dependent on light.

Seeing ranges in water vary from a few inches in harbors and bays to a few hundred feet in unusually clear parts of the oceans. The underwater world is blue-green due to the color filtering effects of water. The great external pressure of water at depth and the corrosive effects of sea water also provide unusual challenges to the designer of lighting equipment or lighting systems for underwater applications.

Terms and Definitions

The terms and definitions given here are peculiar to underwater lighting. The definitions here generally follow the recommendations of the Committee on Radiant Energy in the Sea of the International Association of Physical Oceanography (IAPO). Neither the notation nor the units are universal, but they are widely used and are essentially those recommended by IAPO. Additional terms and definitions useful in underwater lighting calculations are found in Section 1.

It should be noted that in the past some authors have used terms such as "absorption" or "extinction" to mean "beam attenuation" as it is defined here, and that sometimes decade values for the beam attenuation, absorption, or scattering coefficients have been given, *i.e.*, coefficients smaller by the factor $1/\log_e 10$. To further confuse matters either the term "extinction coefficient" or "vertical extinction coefficient" has sometimes been used in lieu of "diffuse attenuation coefficient."

Absorption Coefficient: the ratio of radiant flux lost through absorption (dF_a) (in an infinitesmally thin layer of medium normal to the beam) to the incident flux (F)*, divided by the thickness of the layer (dx).

$$a = -(1/F)\ dF_a/dx \qquad \text{unit: meter}^{-1}$$

Volume Scattering Function: the radiant intensity ($dI(\theta)$) from a volume element (dV) in a given direction (θ) per unit of irradiance (E) of a beam incident on the volume per unit volume. (θ is ordinarily measured from the forward direction.)

$$\beta(\theta) = (1/E)\ dI(\theta)/dV \qquad \text{unit: meter}^{-1}$$

(Total) Scattering Coefficient: the ratio of radiant flux lost through scattering (in an infinitesimally thin layer of the medium normal to the beam) (dF_s) to the incident flux (F), divided by the thickness of the layer (dx); equivalent to the integral of the volume scattering function over all directions.

$$b = -(1/F)\ dF_s/dx = \int_0^{4\pi} \beta(\theta)\ d\omega$$

$$= 2\pi \int_0^{\pi} \beta(\theta) \sin\theta\ d\,\theta \qquad \text{unit: meter}^{-1}$$

Beam Attenuation Coefficient: the sum of the absorption coefficient (a) and the scattering coefficient (b).

$$c = a + b \qquad \text{unit: meter}^{-1}$$

Note: Sometimes the symbol α is used instead of c. Hence, a transmissometer is often called an "alphameter".

Diffuse Attenuation Coefficient for Irradiance: the ratio of irradiance lost through absorption and scattering (in an infinitesimally thin horizontal layer of the medium) (dE) to the incident irra-

* Swimming pool lighting is discussed in Section 19.
Note: References are listed at the end of each section.

* IAPO symbol. IES symbol is ϕ.

diance (E), divided by the thickness of the layer (dz).

$$K = -(1/E)\ dE/dz \qquad \text{unit: meter}^{-1}$$

Filtering Properties of Sea Water

Different wavelengths of light are absorbed by different amounts in water. Sea water that is both deep and clear has a spectrum that closely resembles distilled water with peak transmission in the blue near 480 nanometers.

The absorption curves for other bodies of water differ greatly from this theoretical or ideal distribution due to the presence of silt, decomposition of plant and animal material, pollution, and living organisms. A common source of change is plankton in the water; plankton absorb short wavelengths of energy much more than long wavelengths. Therefore, the peak of the transmission curve in a body of water moves to the yellow-green (510- to 570-nanometer) portion of the spectrum. The amount of shift depends on the density of the plankton.

Some typical spectral transmittance curves for different types of water are shown in Fig. 26–1.[1] The shapes of the spectral transmittance curves result primarily from absorption, but the curves include losses from both absorption and scattering processes. Since scattering in natural water is nearly independent of wavelength, it affects the level of the curves and not the shape.

The curve for Morrison Springs in Florida is essentially the same as distilled water and has a maximum transmittance of over 90 per cent at 480 nanometers. The only difference between the samples from Morrison Springs and the Gulf of Mexico is a lower transmittance in the violet and blue portions, presumably due largely to plankton. The Long Island Sound water shows less transmittance throughout the spectrum, with the greatest loss in the blue and blue-green portions. Therefore, it can be said, that typical transmission curves are more symmetrical than that of distilled water. For coastal or moderately shallow water the spectrum of interest lies between about 480 to 500 nanometers.

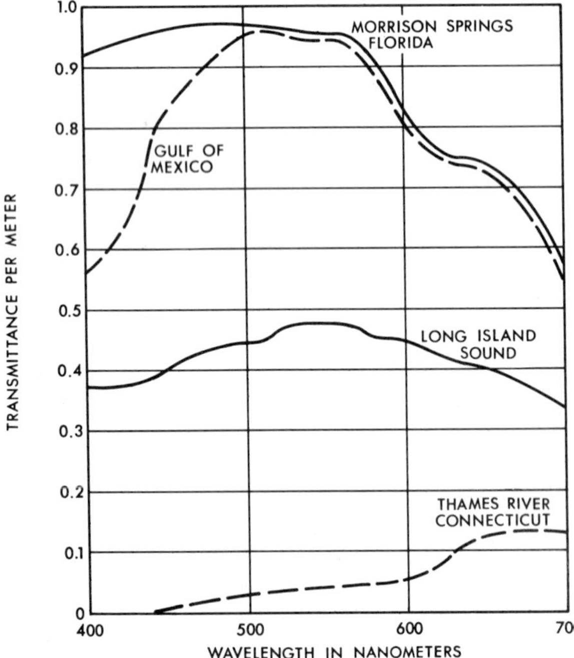

Fig. 26–1. Spectral transmittance of one meter of various bodies of water.

Waters in heavily polluted areas such as the Thames River in Connecticut,[1] very little light is transmitted at all and the shape of the curve has been completely transformed, with the greatest transmittance in the long wavelengths.

These measured differences are for a one meter water distance. Since transmittance is related to distance in water by a power function, the wavelength absorption becomes extreme as the distance the light travels increases. Therefore, the relative visibility of different colors can be expected to vary considerably in different types of water and at different distances.

Seeing Distance in Water

Assuming sufficient light, seeing distance in water is limited by scattered light. The loss of contrast and limited seeing distance in water is like that encountered in fog, that is, a diver using artificial lighting has a seeing problem similar to that of a person driving a car at night in dense fog.

At short distances man or camera can see acceptably well in water with almost any lighting arrangement. At longer underwater distances the quality of underwater images made with the sensor (eye or camera) located near to the light source is seriously degraded by light scattered within the volume of water which is common to both the illuminating and sensing systems; *i.e.*, that part of the line of sight which is lighted by the source. Successful long range underwater viewing requires elimination of scattering within the common volume.

Three simple techniques are used to suppress common volume scattering.

1. A light shield (septum) can be used to shield most of the line of sight from direct light from the lamp.
2. The light can be offset to the side. No part of the line of sight is close to the lamps; therefore, the inverse-square law and attenuation by absorption and scattering operate to avoid intense lighting of the water close to the sensor. Offset of the lamp may be to the rear or to the front as well as to the side. Offset to the rear offers good uniformity of illumination for large fields of view, but requires increased light source intensity to compensate for the longer distance between light source and object. Offset to the front (source between sensor and object) is effective and efficient, but requires more complex light control to achieve uniform illumination.

Fig. 26-2. Values of One Attenuation Length for Several Bodies of Water[2]

Location	Attenuation Length (meters)
Caribbean	8
Pacific N. Equatorial Current	12
Pacific Countercurrent	12
Pacific Equatorial Divergence	10
Pacific S. Equatorial Current	9
Gulf of Panama	6
Galapagos Islands	4

3. A third method for reducing common volume scattering is provided by crossed polarizers. Both the lamp and the sensor must have polarizers and the object must depolarize in the process of reflection in order to be visible. Absorption of light by the polarizers requires an increase in camera integration time, lens speed, or sensitivity, or increased lamp output.

To simplify quantitative discussion of seeing distance in water, the concept of the *attenuation length* is often used. One attenuation length is defined as the reciprocal of the beam attenuation coefficient and is the distance in which the radiant flux is reduced to 1/e or about 37 per cent of its initial value. The attenuation length is analogous to the mean free path in physics and the time constant in electronics. Typical values of one attenuation length are shown in Fig. 26-2.[2]

Most underwater photography and viewing is done at distances of less than one attenuation length, because images can be seen with acceptable contrast through such distances in water even with the light source very close to the camera. With natural lighting and horizontal viewing a swimmer can detect a dark object at a maximum of about 4 attenuation lengths and a light object at about 4 to 5 attenuation lengths.[2] At 3 to 4 attenuation lengths good images can be obained with the aid of the polarizers and septum techniques. Good images can be obtained at 6 attenuation lengths with the light moved near the object in combination with the septum and polarizers. With an ideally arranged combination of these aids and high contrast film developing, usable photographs have been obtained at a distance of 12 attenuation lengths. This distance corresponds to 120 meters (394 feet) in average clear ocean water (one attenuation length equaling 10 meters).

There is a limit to the underwater viewing range which becomes apparent at slightly greater distances. Even with unlimited sensor integration time, self-luminous objects, and silhouetted objects in otherwise unlighted water, objects disappear in their own forward scattered light beyond approximately 15 attenuation lengths. Contrast enhancement extends this limit only slightly.

It should be noted that extremely high attenuation of light exists at such long distances. The laboratory experiments from which the above numbers were derived were performed with stationary objects, moderate sensitivity film (ASA 160), and $f/2$ lens, and a DVY lamp (see Fig. 8-90). At distances of 1 or 2 attenuation lengths the exposure times were small fractions of a second, but at 5 or 6 attenuation lengths several minutes were required. A time exposure for several hours was needed in order to obtain a picture at 12 attenuation lengths. Such sensor integration times are not usually feasible, but they give an idea of the amount of light (or sensor sensitivity) required for viewing at scatter limited ranges in water.

Two other techniques to suppress scatter can be used where their complexity and cost is justifiable.

1. A synchronously scanned narrow beam source and sensor system can reduce the common volume to a minimum.
2. A system using a pulse of light of a few nanoseconds duration (a few feet distance in water) with a sensor which is gated to see only the pulse can also minimize the common volume. Lasers are capable of such pulse lengths.

Such systems are described in articles referenced by a recent bibliography.[3]

Fig. 26-3. Currently Available Underwater Lighting Systems
(For additional light source data, see Figs. 8-90, 8-116, and 8-121.)

Light Source	Power (watts)	Spectral Power Distribution	Recommended Uses
Incandescent (Tungsten-Halogen)	75-1000	2800-3400 K Color Temperature @ 1000-25 hours life. Fig. 8-2	Vision, photography, television
Mercury	175 250	Fig. 8-23	Television, vision
Metal Halide Thallium	150 250 400	Fig. 8-23	Television, vision
Metal Halide Dysprosium Thallium	400	Fig. 8-23	Vision, photography, television
High Pressure Sodium	400	Fig. 8-23	Vision, photography, television
Xenon (Pulsed)	100-1000 watt-seconds	Fig. 8-57	Photography, slow scan television

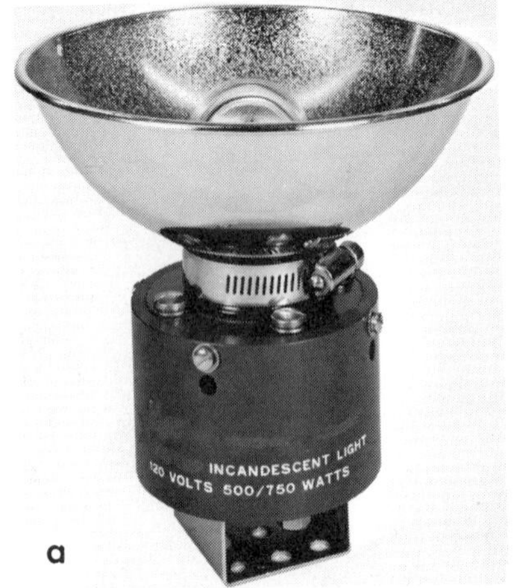

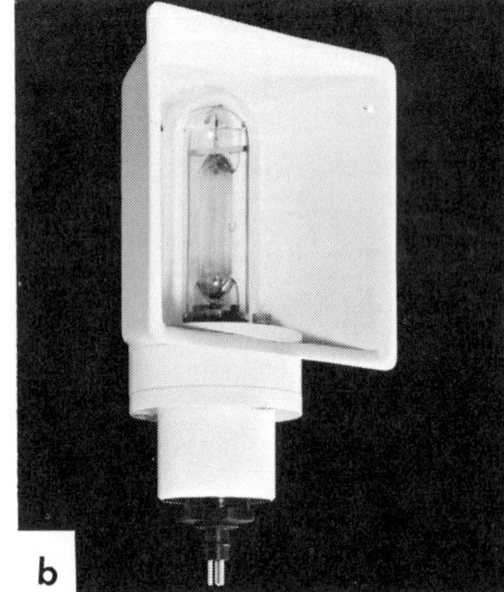

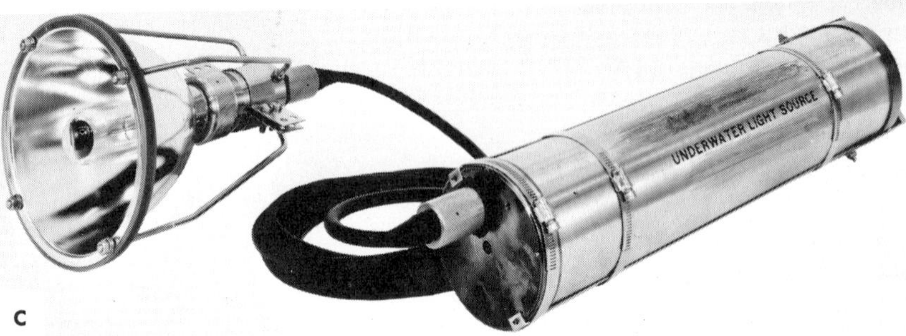

Fig. 26–4. Typical underwater lighting units: (a) incandescent, (b) arc discharge, and (c) pulsed xenon.

Light Sources for Underwater Use

Light sources used in currently available underwater lighting systems are shown in Fig. 26–3. Incandescent sources are used where instantaneous starting, simplicity, and small size are requirements. The arc discharge (the others) are used where their higher efficacy in converting electrical power into radiant flux and longer life are requirements. Typical underwater lighting units are shown in Fig. 26–4.

In choosing light sources for underwater applications, one should consider the filtering properties of the water. Red light is rapidly absorbed by the water. Only at short ranges is it practical to attempt the careful color balance achieved routinely by color photography in air.

Other light sources applicable to underwater systems, but not yet available in underwater lighting equipment are listed in Fig. 26–5.

Electrical characteristics and power supplies for the various light sources are discussed in Section 8, Light Sources. Also, there are auxiliary circuits available from underwater lighting equipment manufacturers which are specially designed for the unusual requirements of various underwater lighting applications. Information peculiar to the design of pressure housings for underwater lights has been presented.[4]

Lasers in underwater systems are discussed in several articles listed in a recent bibliography.[3]

Fig. 26–5. Currently Available Light Sources Applicable to Underwater Systems
(For additional light source data, see Figs. 8–116 and 8–121.)

Light Source	Power (watts)	Spectral Power Distribution	Recommended Uses
Scandium Sodium Metal Halide	175 400 1000	Fig. 8–23	Vision, photography, television
Mercury Short Arc	75 100 200	Fig. 8–56	Vision, photography, television
Xenon Short Arc	75 150	Fig. 8–57	Vision, photography, television

Sensor Characteristics

A wide variety of sensors is available for underwater use. The choice of sensor type will depend on the over-all system requirements and will have a bearing on the characteristics required of the associated underwater lighting equipment.

These sensors may be divided into two general categories: imaging, as in the case of television or film; and non-imaging, for use in instrumentation applications such as photometry and attenuation or scattering measurement.

Photographic Films. Black and white films used in underwater photography have ASA ratings ranging from 12 to 400, and with special development, ASA ratings up to 3200 can be obtained. Negative and positive color films are available with ASA ratings from 25 to 500. The spectral response of a typical panchromatic negative film and of a color film are shown in Fig. 26-6.

Television Sensors. Typical characteristics of some television imaging sensors of interest to the underwater lighting system designer are shown in Figs. 26-7 through 26-10.

The Vidicon is the least sensitive as indicated by Fig. 26-7, but is small and requires relatively simple camera circuitry. The range of spectral response is shown in Fig. 26-8. These devices exhibit lag on the order of 20 to 25 per cent (3rd field) and can be damaged by overexposure.

Solid state Vidicons employing lead oxide or silicon diode array targets are similar to the Vidicon in size, are generally more sensitive, and provide higher signal to noise ratios. Residual image, or lag is less. The silicon diode target type is highly resistant to effects of overexposure. Both types have

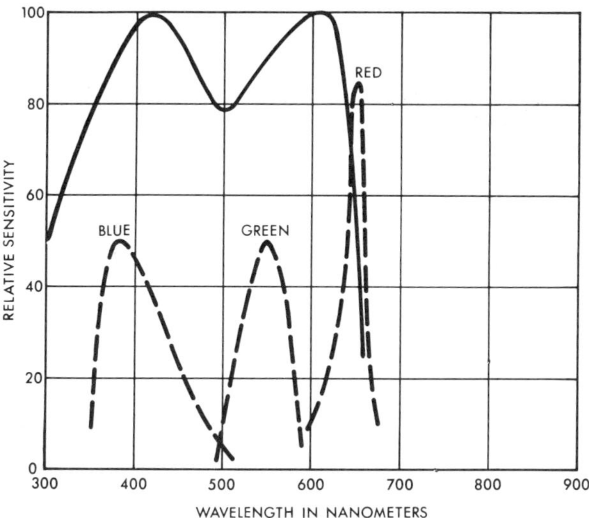

Fig. 26-6. Spectral response of black and white panchromatic negative film (solid line) and of a typical color reversal film (broken line).

a limited dynamic range of light input. Typical absolute spectral response curves of these devices are shown in Fig. 26-9, together with curves for Vidicons used in underwater applications.

The Image Orthicon and Image Isocon are much larger than the Vidicons. These devices are sensitive and capable of high resolution. They exhibit appreciable loss in sensitivity with relative scene motion and require more complex circuitry than the Vidicon types.

The Secondary Electron Conduction or SEC Camera Tube is intermediate in size and sensitivity. It has low lag (10 per cent or less) and is least affected by relative scene motion. Newly developed versions of this type of sensor incorporate burn resistant targets or highly burn resistant silicon diode array targets. The latter type is also more sensitive as shown in Fig. 26-7 and provides a higher signal to noise ratio.

The Orthicon, Isocon, and the SEC Camera Tubes use semi-transparent photocathodes having absolute spectral response curves as shown in Fig. 26-10. Any of these devices may be fiber-optics coupled to an intensifier, with increased sensitivity and a change in absolute spectral sensitivity to that of the intensifier used.

Non-Imaging Sensors. Photodiodes and photomultipliers available for use in underwater instrumentation generally employ semi-transparent photocathodes with a spectral response similar to those shown in Fig. 26-10. Characteristics of other types of photosensitive devices are described in Section 4. Because of the extreme variations possible in the spectral transmittance of water, it is often necessary to match the sensor spectral response to the system application or to make a number of narrowband measurements.

Underwater Lighting Calculations

The amount of light needed for an underwater task is difficult to predict with any accuracy. Even

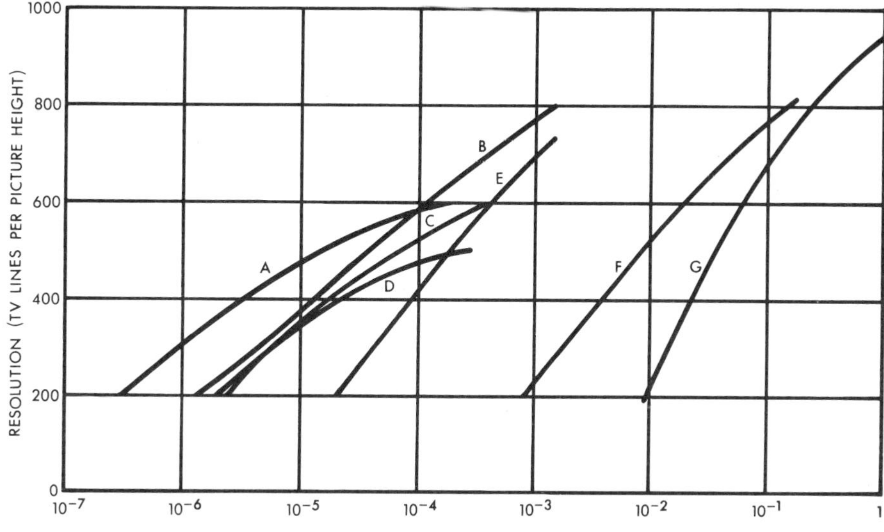

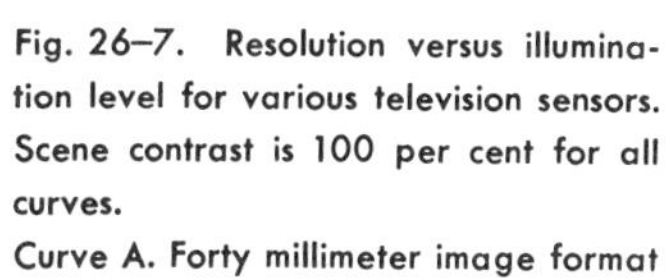

Fig. 26-7. Resolution versus illumination level for various television sensors. Scene contrast is 100 per cent for all curves.

Curve A. Forty millimeter image format Secondary Electron Conduction (SEC) tube with 40 millimeter intensifier.

Curve B. Image Orthicon.

Curve C. Twenty-five millimeter Silicon Target SEC tube.

Curve D. Twenty-five millimeter SEC tube with 25 millimeter intensifier.

Curve E. Forty millimeter SEC tube.

Curve F. Broadcast Vidicon.

Curve G. Vidicon.

if underwater light were accurately predictable, wide variations exist in the color filtering and contrast reducing properties of water from place to place, and from day to night and season to season in the same place. For these reasons, the calculations described here are simple approximations, but they are often sufficient for engineering purposes.

Daylighting Calculations. The illumination due to daylighting on a surface parallel to the water surface can be calculated by

$$E = E_0 e^{-Kd}$$

where E_0 is the illumination at the surface of the water, K is the diffuse attenuation coefficient for irradiance, and d is the depth from the water surface to the illuminated surface. An observer looking at this surface would see an apparent luminance

$$L = E\rho e^{-cr_2}$$

where ρ is the reflectance of the illuminated surface, c is the beam attenuation coefficient and r_2 is the distance from the illuminated surface to the underwater observer.

Extensive studies of daylight in water have been made.[2]

Artificial Lighting Calculations. The luminance observed by an underwater viewer or camera is

$$L = I\rho\tau/r_1^2$$

where I is the source luminous intensity, ρ is object reflectance, τ is water transmittance (two way path) and r_1 is the source to object distance. The water transmittance can be approximated by

$$\tau = e^{-c(r_1+r_2)}$$

where c is the beam attenuation coefficient of the water, and r_2 is the object to sensor distance. If $r_1 + r_2$ equals one meter, this equation also defines the relation between c and τ for Fig. 26–1. Two factors must be considered in evaluating the validity of using these equations in this simple form: spectral effects (*i.e.*, c, I, ρ, and sensor response are functions of wavelength) and scattering.

A semi-empirical formula has been developed[2] to allow approximate calculation of the amount of scattered light that illuminates the object. This forward scattered component is not included in the

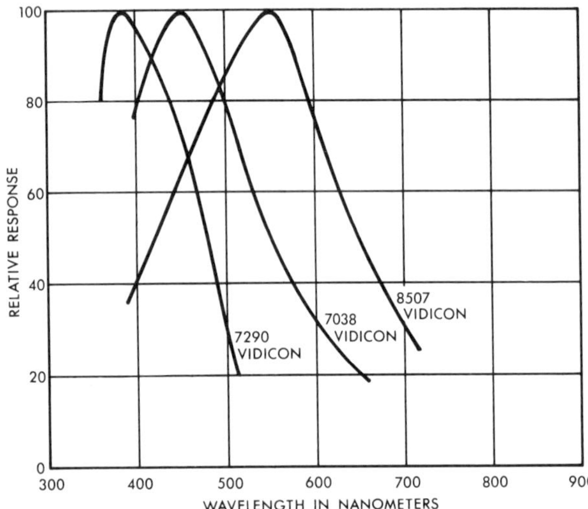

Fig. 26–8. Relative spectral response of Vidicons.

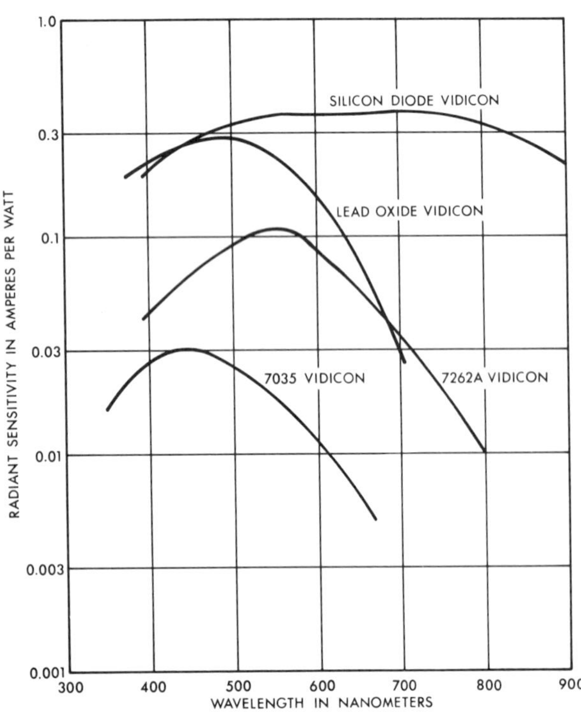

Fig. 26–9. Absolute spectral response of Vidicons and solid state Vidicons.

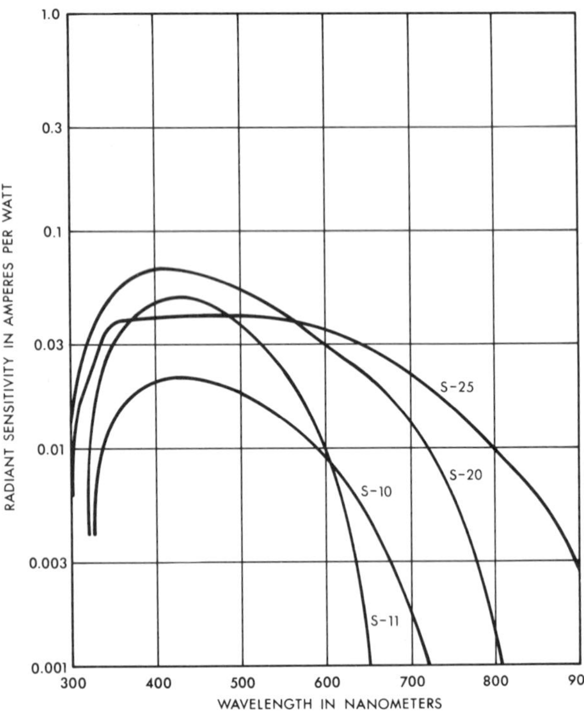

Fig. 26–10. Absolute spectral response of some semi-transparent photocathodes.

simple expression given above for τ, but it can become quite large. The proportion of scattered light in the object illumination increases with range. At one attenuation length, the unscattered object illumination (calculated by the simple expression given above) is nearly equal to the illumination due to scattering. At four attenuation lengths, the scattered illumination is about ten times unscattered illumination. The increase in calculated object luminance due to scattering is partially offset by losses due to spectral filtering (see Fig. 26–1).

The value of the beam attenuation coefficient, c, used in approximate calculations is usually the minimum value with respect to wavelength (maximum value in the transmittance curves of Fig. 26–1). For photometric calculations where the photopic eye is the sensor, Fig. 26–1 shows that this choice is a reasonable approximation, because the curves are nearly flat in the region of greatest eye response. When the sensor has a spectral response much different from that of the eye or when more accurate predictions are needed, the radiance observed by the underwater sensor at each wavelength can be calculated as

$$L_\lambda = \rho_\lambda \tau_\lambda I_\lambda / r_1^2$$

where I_λ is the spectral radiant intensity of the source. The expression for τ_λ is the same as for τ except that the attenuation coefficient c is now c_λ, a function of wavelength. The response of the underwater sensor is

$$S = \int_{\lambda_1}^{\lambda_2} S_\lambda E_\lambda d\lambda$$

where S_λ is the spectral response of the sensor (see Figs. 26–6, 26–8, 26–9 and 26–10) and E_λ is the irradiance of the sensor due to L_λ. E_λ and L_λ are related by the usual imaging equation:

$$E_\lambda = \pi t_\lambda L_\lambda / 4(f\text{-number})^2$$

where t_λ is the lens spectral transmittance. Further discussion of spectral calculations can be found in the references.[5, 6, 7]

As one might expect from inspection of Fig. 26–1, the light loss due to greater absorption in the blue and red in water is usually smaller than the gain due to scattered light. This means that the simple expressions for L and τ will usually give conservative predictions; *i.e.*, they will predict the need for more light from the source than is actually required. Additional discussion of underwater lighting calculations can be found in a book by Mertens.[8]

Measurement Techniques and Instrumentation

Performance of underwater light sources has been tested according to standard measuring techniques for light sources used in air. These data are usually supplied on the information sheets, but can be misleading because, as pointed out above, light absorption is relative to the wavelength. The photometric units often given do not apply when these light sources are used in water.

There are no established methods of instrumentation to evaluate a light source in the water environment. A number of research programs by light source manufacturers are underway but standardization is distant.

REFERENCES

1. Luria, S. M. and Kinney, Jo Ann S.: "Underwater Vision", *Sci.*, Vol. 167, March 13, 1970.
2. Duntley, S. Q.: "Light in the Sea", *J. Opt. Soc. Amer.*, Vol. 53, February, 1963.
3. *Bibliography on Underwater Photography and Photogrammetry*, Kodak Pamphlet No. P-124, Eastman Kodak Company, Rochester, New York, 1968.
4. Stachiw, J. D. and Gray, K. O.: *Light Housings for Deep Submergence Applications, Parts I and II*, Naval Civil Engineering Laboratory, Report TR-532, 1967, and TR-559, 1968.
5. *Source-Detector Spectral Matching Factors*, Technical Note No. 100, October, 1966, ITT Industrial Laboratories, Fort Wayne, Indiana.
6. Biberman, L. M.: "Apples, Oranges and Unlumens", Opt. Soc. Amer., Long Abstract 1967, Spring Meeting ThG 11-1.
7. Moon, Parry: *The Scientific Basis of Illuminating Engineering*, McGraw-Hill Book Co., New York, 1936.
8. Mertens, L. E.: *In-Water Photography*, John Wiley & Sons, Inc., New York, 1970.

CREDITS FOR ILLUSTRATIONS AND TABLES

The Illuminating Engineering Society is indebted to the many individuals, committees, and organizations which contributed the multitude of illustrations and tables published in this Handbook. Many of the illustrations and tables omitted from the following listing appeared in previous publications issued by the Society or were supplied by committees of the Society especially for use in this volume. Most other omissions comprise material published in previous editions without indication of origin.

Contributors

1. Acme Lite Products Co., Inc., Congers, N. Y.
2. The American Institute of Physics, *Journal of the Optical Society of America,* New York, N. Y.
3. American Society of Heating, Refrigerating and Air-Conditioning Engineers, New York, N. Y.
4. Association of American Railroads, New York, N. Y.
5. Bausch & Lomb Optical Co., Rochester, N. Y.
6. H. R. Blackwell, Ohio State University, Columbus, O.
7. Carlisle & Finch, Cincinnati, O.
8. Canadian Standards Association, Rexdale, Ontario, Canada.
9. Consolidated Vultee Aircraft Corporation, San Diego, Calif.
10. Corning Glass Works, Corning, N. Y.
11. Crouse-Hinds Co., Syracuse, N. Y.
12. Day-Brite Lighting, Inc., Division of Emerson Electric Co., St. Louis, Mo.
13. Duquesne Light Co., Pittsburgh, Pa.
14. Duro-Test Corporation, North Bergen, N. J.
15. Eastman Kodak Co., Rochester, N. Y.
16. EG & G, Environmental Equipment Division, Waltham, Mass.
17. J. R. Fairweather, Lansdale, Pa.
18. Genarco, Inc., Jamaica, N. Y.
19. General Electric Co., Cleveland, O.; Schenectady, N. Y.
20. General Railway Signal Company, Rochester, N. Y.
21. A. Girard & Associates, New York, N. Y.
22. P. H. Goodell, Akron, O.
23. Gotham Lighting Corp., Long Island City, N. Y.
24. Hanovia Lamp Div., Engelhard Hanovia, Inc., Newark, N. J.
25. D. B. Harmon, *The Co-ordinated Classroom,* American Seating Co., Grand Rapids, Mich., and F. W. Wakefield Brass Co., Vermilion, O.
26. L. S. Harrison, Coral Gables, Fla.
27. Holophane Co. Inc., New York, N. Y.
28. Indiana University, Bloomington, Ind.
29. International Light, Inc., Newburyport, Mass.
30. Inter-Society Color Council, Washington, D. C.
31. K. L. Kelly, Washington, D. C.
32. Leeds & Northrup Co., Phila., Pa.
33. *Lighting & Lamps,* New York, N. Y.
34. Lincoln Center for the Performing Arts, Inc., New York, N. Y.
35. McGraw-Hill Book Co., Inc., New York, N. Y.
  a. *Measurement of Radiant Energy,* Forsythe.
  b. *The Principles of Optics,* Hardy & Perrin.
  c. *Scientific Basis of Illuminating Engineering,* Moon.
36. The Metropolitan Museum of Art, New York, N. Y.
37. W. E. K. Middleton, Ottawa, Canada.
38. Monsanto Chemical Co., St. Louis, Mo.
39. Munsell Color Co., Baltimore, Md.
40. National Bureau of Standards, U. S. Dept. of Commerce, Washington, D. C.
41. National Carbon Co., Cleveland, O.
42. National Electrical Manufacturer's Assn., New York, N. Y.
43. Newport News Shipbuilding and Dry Dock Company, Newport News, Va.
44. Owens-Illinois Glass Co., Toledo, O., *Daylight in School Classrooms,* Paul.
45. Philadelphia Electric Co., Philadelphia, Pa.
46. Photo Research Corp., Hollywood, Calif.
47. Polaroid Corp., Cambridge, Mass.
48. Prismo Safety Corp., Huntingdon, Pa.
49. T. H. Projector, Washington, D. C.
50. Public Service Company of Indiana, Plainfield, Ind.
51. Pullman-Standard, Chicago, Ill.
52. Rohm & Haas Co., Philadelphia, Pa.
53. F. K. Sampson, Los Angeles, Calif.
54. Sylvan R. Shemitz, West Haven, Conn.
55. Shinichi Ishizaka, Matsushita Electric Works, Ltd., Japan.
56. Society of Automotive Engineers, Inc., New York, N. Y.
57. Society of Motion Picture and Television Engineers, New York, N. Y.
58. Stimsonite Plastics, Chicago, Ill.
59. Sylvania Electric Products, Inc., New York, N. Y.; Salem, Mass.
60. Uitg. Stichting "Prometheus," N. Z. Voorburgwal 27⅓, Amsterdam, Netherlands.
61. Union Switch & Signal, Division of Westinghouse Air Brake Co., Pittsburgh, Pa.
62. United Air Lines, Chicago, Ill.
63. U. S. Army, Washington, D. C.
64. U. S. Department of Agriculture, Washington, D. C.
65. U. S. Navy Department, Washington, D. C.
66. D. Van Nostrand Co. Inc., New York, N. Y.
  a. *Light, Vision & Seeing,* Luckiesh.
  b. *The Science of Seeing,* Luckiesh & Moss.
67. Western Cataphote Corp., Toledo, O.
68. Westinghouse Electric Corp., Bloomfield, N. J.; Cleveland, O.
69. Weston Electrical Instrument Corp., Newark, N. J.
70. Wide-Lite Corp., Houston, Tex.
71. John Wiley & Sons, New York, N. Y., *Color Science,* G. Wyszecki and W. S. Stiles.
72. R. L. Zahour, Bloomfield, N. J.

Credits

Section 2.

2–4: **35b.** 2–5: **35c.** 2–6: **38.** 2–8. **68.** 2–10: **68.** 2–22: **69.**

Section 3.

3–4: **66b.** 3–5: **2.** 3–6: **19.** 3–18: **66b.** 3–19: **6.** 3–20: **6.** 3–21: **6.** 3–22: **25.** 3–23: **66a.** 3–24: **6.** 3–25: **6.** 3–26: **6.** 3–28: **6.** 3–29: **6.** 3–41: **53.** 3–42: **53.** 3–43: **53.** 3–47: **40.** 3–49: **37.** 3–51: **2.**

Section 4.

4–3a: **32.** 4–3b: **19.** 4–3c: **46.** 4–3d: **46.** 4–4a: **19.** 4–4b: **69.** 4–4c: **19.** 4–4d: **46.** 4–6: **19.** 4–7: **8.** 4–8: **8.** 4–9: **10.** 4–12: **11.** 4–17: **40.**

Section 5.

5–1a: **71.** 5–1b: **71.** 5–4: **39.** 5–5: **39.** 5–6: **71.** 5–7: **71.** 5–12: **31.** 5–14: **71.** 5–17A: **45.** 5–17C: **45.** 5–18: **15.** 5–22: **30.** 5–23: **64.**

Section 6.

6–3: **35b.** 6–10a: **27.** 6–10c: **10.** 6–10d: **5.** 6–13: **5.** 6–14: **68.** 6–15: **19.** 6–18: **47.**

Section 7.

7–7: **44.**

Section 8.

8–6: **19.** 8–7: **19.** 8–9: **19.** 8–17: **19.** 8–19: **19.** 8–25: **59.** 8–45: **19.** 8–55: **59, 68.** 8–69: **41.**

Section 9.

9–30: **19.** 9–34: **68.** 9–35: **68.** 9–58: **19.** 9–64: **42.** 9–67: **19.**

Section 10.

10–1: **19.** 10–2: **19.** 10–3: **19.** 10–4: **19.** 10–5: **19.** 10–6: **19.** 10–7: **19.** 10–8: **19.** 10–9: **33.**

Section 11.

11–3. **68.** 11–4: **68.** 11–6 upper: **19.** 11–8: **12.** 11–9: **19.** 11–10: **19.** 11–11: **19.**

Section 12.

12–1 left: **13.** 12–1 right: **59.** 12–7 left: **21.** 12–7 upper right: **54.** 12–8: **19.** 12–9: **50.** 12–11: **50.** 12–15: **19.** 12–16: **19.** 12–17: **19.** 12–21: **26.** 12–22a: **26:** 12–22b: **36.** 12–22c: **36.** 12–22d: **36:** 12–23: **26.** 12–24: **36.** 12–25: **26.** 12–26: **26.**

Section 15.

15–3: **19.** 15–13: **19.**

Section 16.

16–1: **3.**

Section 17.

17–2: **11.** 17–3: **19.** 17–4: **19.** 17–5: **19.** 17–6: **19.** 17–7: **19:** 17–8: **19.** 17–11: **55:** 17–12: **55.** 17–13: **19.** 17–14: **19.** 17–15: **19.** 17–16: **17.**

Section 18.

18–1: **7.** 18–2: **65.** 18–4: **14.** 18–5: **63.** 18–6: **63.** 18–7: **65.** 18–8: **49.** 18–10: **18.**

Section 19.

19–1: **68.** 19–3b: **59.** 19–3c: **70.** 19–6 photo: **68.** 19–15: **59.** 19–25 photos: **68.**

Section 21.

21–6: **9.** 21–10: **62.** 21–19: **65.**

Section 22.

22–2: **56.** 22–3: **48.** 22–4: **58.** 22–5: **67.** 22–10: **51.** 22–16: **43.** 22–17: **19.** 22–18: **43.** 22–19: **19.** 22–20: **19.** 22–21: **60.** 22–22: **20.** 22–24: **4.** 22–25: **61.** 22–26: **20.**

Section 23.

23–5: **19.** 23–19: **19.**

Section 24.

24–2: **57.** 24–3: **57.** 24–4 left: **34.** 24–5: **57.** 24–13: **28.** 24–15: **19.** 24–16: **68, 19.** 24–17: **57.** 24–18: **57.** 24–19: **57.** 24–20: **57.** 24–21: **57.** 24–22: **57.** 24–23: **57.** 24–24: **57.** 24–25: **57.** 24–26: **57.** 24–27: **57.** 24–28: **57.** 24–29: **57.** 24–30: **57.** 24–31: **19, 15, 68.**

Section 25.

25–1: **19:** 25–2: **35a.** 25–3: **19.** 25–7: **29.** 25–8: **59.** 25–9: **59.** 25–10: **59.** 25–11: **19.** 25–12: **19.** 25–13: **72.** 25–14: **72.** 25–15: **59.** 25–16: **72.** 25–17: **59.** 25–18: **1.** 25–19: **19.** 25–20: **19.** 25–21: **19.** 25–22: **19, 24, 59, 68.** 25–26: **68.** 25–27: **19.** 25–29: **19.** 25–30: **22:** 25–31: **19.**

Section 26.

26–4(a): **16.** 26–4(c): **16.**

INDEX*

Pages are numbered consecutively within each section

A

ABSORPTION COEFFICIENT, of water: 26-1

ABBERRATIONS, LENS: 6-5, 6-7

ABBOT-GIBSON DAYLIGHT: 5-21, 5-22

ABBREVIATIONS
for use in text: 1-23
in aviation lighting: 21-6
photometric quantities: 1-2
radiometric quantities: 1-2
scientific and engineering: 1-23

ABSORPTANCE
definition: 1-1
symbol and unit: 1-2

ABSORPTION
atmospheric: 4-24, 9-71
definition: 1-1
light: 6-9

ABSORPTION BANDS, in solar spectrum: 25-1

ACCENT LIGHTING
definition: 1-1
in residences: 15-26 to 15-29
in show windows: 13-9

ACCOMMODATION
definition: 1-1
effection of age: 3-11
in visual process: 3-1, 3-2, 3-3

ACOUSTICS
in lighting design: 10-2, 10-3
lighting equipment: 6-23, 6-24

ACRONYMS, in aviation lighting: 21-6

ACTIVATORS, in lamp phosphors: 2-9, 2-11

ACUITY (see **VISUAL ACUITY**)

ADAPTATION
color: 5-11, 5-12
dark, definition: 1-7
definition: 1-1
eye: 3-3
light, definition: 1-12
state of chromatic, definition: 1-19

ADAPTIVE COLOR SHIFT
definition: 1-1
in color rendering: 5-19

ADVERSE-WEATHER LAMPS (see **FOG LAMPS**)

ADVERTISING LIGHTING (see also **FLOODLIGHTING** and **SIGNS**)
exposed lamp signs: 23-1 to 23-4
luminous element signs: 23-4 to 23-7
luminous tube signs: 23-4
poster panels, bulletin boards, vertical surface signs: 23-7 to 23-12

AERODROME LIGHTING (see **AIRPORT LIGHTING**)

AERONAUTICAL LIGHTING (see **AIRCRAFT LIGHTING** and **AIRPORT LIGHTING**)

AERONAUTICAL LIGHTING TERMS
aeronautical beacon: 1-1
aeronautical ground light: 1-1
aeronautical light: 1-1
airway beacon: 1-1
approach-light beacon: 1-2
approach lights: 1-2
bar, of lights: 1-4
barette: 1-4
beacon: 1-4
boundary lights: 1-4
channel lights: 1-5
circling guidance lights: 1-5
course light: 1-6
fuselage lights: 1-9
taxi-channel lights, definition: 1-20
taxi-light, definition: 1-20
taxiway-centerline lights, definition: 1-20
taxiway-edge lights, definition: 1-20
taxiway holding-post light, definition: 1-20
taxiway lights, definition: 1-20
touchdown zone lights, definition: 1-20
VASIS, definition: 1-21

AGE
effect on accommodation: 3-11
effect on contrast sensitivity: 3-12, 3-13
effect on disability glare susceptibility: 3-12
effect on pupil size: 3-11, 3-12
effect on visual acuity: 3-12

AIR CONDITIONING
air-comfort chart: 16-2
air-water systems: 16-5 to 16-7
all-air systems: 16-5 to 16-7
comfort limits: 16-2
comfort parameters: 16-1, 16-2
controlling lighting heat: 16-5 to 16-7
cooling load distributions, typical: 16-3
coordination with lighting system: 10-3
lighting load effect: 16-2 to 16-5
methods of: 16-5 to 16-7
offices: 11-6, 11-7
total environment concept: 16-1

AIRCRAFT HANGARS
illumination level: 9-81
lighting of: 21-14

AIRCRAFT LIGHTING (see also **AIRPORT LIGHTING**)
anti-collision lights: 21-2
approach lights: 21-4
auxiliary lights: 21-3
aviation agencies: 21-1
cabin lighting: 21-5, 21-6
clearance lights: 21-4
cockpit lighting: 21-4, 21-5
control panel lighting: 21-5
crew station lighting: 21-4
emergency lighting: 21-4
engine check lights: 21-4
escape slide lighting: 21-4
exterior: 21-1 to 21-4
formation lights: 21-3, 21-4
helicopters: 21-2
hover lights: 21-4
ice detection lights: 21-4
illumination levels: 9-94
instrument lighting: 21-4, 21-5
interior lighting: 21-4 to 21-6
lamps for: 8-13, 8-78, 8-79
landing lights: 21-2, 21-3
navigation lights: 21-2
passenger cabin lighting: 21-5, 21-6
position lights: 21-2
power supply for: 21-1, 21-5
purpose of: 21-1
reading lights: 21-5, 21-6
taxi lights: 21-3
warning and indicator lights: 21-5

AIRCRAFT LIGHTING TERMS
aircraft aeronautical light: 1-1
anchor light: 1-2
anti-collision light: 1-2
formation light, definition: 1-9
fuselage lights: 1-9
wing clearance lights, definition: 1-22

AIRCRAFT MANUFACTURING, illumination levels: 9-81

AIR MOVEMENT, in equipment design: 6-23, 6-25

AIRPLANE LIGHTING (See **AIRCRAFT LIGHTING**)

AIRPORT LIGHTING (see also **AIRCRAFT LIGHTING**)
airport beacon: 21-6
approach lighting: 21-7
beacons: 21-6
centerline lights: 21-11, 21-12
circuitry: 21-11
edge lights: 21-10, 21-11
final approach guidance: 21-7 to 21-10
Fresnel lens optical landings system: 21-9, 21-10
ground lighting to aid navigation: 21-6 to 21-12
ground operational lighting: 21-13, 21-14
hangar lighting: 21-14
heliport and VTOL lighting: 21-12
identification of airport: 21-6
illumination levels: 9-94
intensity control of lights: 21-11
lamps for: 8-13, 8-76, 8-77
non-precision approach lighting: 21-8
obstruction lights: 21-13
parking area: 21-14, 21-15
pavement markings: 21-11, 21-12
precision approach lighting: 21-9
runway distance markers: 21-11
runway identification: 21-11
runway lights: 21-10, 21-11
runway visual approach guidance: 21-7 to 21-12
service area: 21-13, 21-14
standardization: 21-1
STOL port lighting: 21-12
take-off guidance: 21-11
taxiway lights: 21-11
threshold lights: 21-11
touchdown lights: 21-11
visual approach slope indicator: 21-8
visual problems encountered: 21-6
wind direction indicators: 21-7

AIRPORT LIGHTING TERMS
aerodrome beacon: 1-1
angle-of-approach lights: 1-2
bar, of lights: 1-4
beacon: 1-4
boundary lights: 1-4
channel lights: 1-5
circling guidance lights: 1-5

AIRPORT SERVICE AREA LIGHTING: 21-13, 21-14

ALABASTER, characteristic of: 6-6

ALLARD'S LAW: 3-33

ALTITUDE
definition: 1-1
solar: 7-1, 7-3, 7-4, 7-7

ALUMINUM
characteristics of: 6-6
finishes: 6-17
maintenance of: 16-14
physical properties of: 6-17, 6-19
types and uses in lighting: 6-17
ultraviolet reflectance of: 25-2, 25-19

AMBIENT TEMPERATURE (see **TEMPERATURE**)

AMERICAN NATIONAL STANDARDS INSTITUTE (see **ANSI**)

AMERICAN SOCIETY FOR TESTING MATERIALS (see **ASTM**)

AMERICAN STANDARDS ASSOCIATION (see **ANSI**)

ANGLE OF INCIDENCE
Brewster's angle: 6-2, 6-8
critical angle: 6-3
effect on photovoltaic cell readings: 4-6
reflectance at various angles: 6-2
specular reflection: 6-1

ANGLE OF REFLECTION: 6-1

ANGSTROM:
conversion factors: 1-26
definition: 1-2
wavelength: 1-22, 2-2

ANGULAR COORDINATE-DIC METHOD, of illumination calculations: 9-49, 9-51

ANGULAR MEASURE, conversion factor: 1-27

ANSI
equipment design standards: 6-22, 6-25
method of measuring and specifying color: 5-10, 5-11
nomenclature and definitions for Illuminating Engineering: 1-1
standard definitions of electrical terms: 1-1

ANTI-COLLISION LIGHTS: 1-2, 21-2

APERTURE FLUORESCENT LAMPS: 8-21

APOSTILB
conversion factors: 1-26
definition: 1-2

APPARENT REFLECTANCE: 25-13

APPROACH LIGHTS: 1-2, 21-4

AQUARIUM LIGHTING: 25-9, 25-10

ARC CRATER: 2-7

ARC LIGHTS, motion picture: 24-4

* For terms not found listed in the index, see Section 1, Dictionary of Lighting Terms.

ARC SOURCES (see **CARBON ARC SOURCES**)

ARCHERY
illumination levels: 9–91
lighting for: 19–4, 19–8, 19–9

ARCHITECTURAL LIGHTING
architectural study of systems: 10–8
brightness in: 10–8
color in: 10–8
comparison with engineering study: 10–8
context with luminaires: 10–8, 10–9
detailing in: 10–8
environmental systems: 10–8
form in: 10–8
light and architecture: 10–2
"period" designs in: 10–8
scale in: 10–8
space requirements: 10–8
styles of architectural periods: 10–9, 10–10
visually-subordinate lighting systems: 10–8
visually-prominent lighting systems: 10–10

ARCHITECTURAL PERIODS AND STYLES: 10–9, 10–10

AREA SOURCE CALCULATIONS: 9–58 to 9–61

ARGON
for incandescent lamp fill: 8–6
for mercury lamp starting: 8–36, 8–41
in glow lamps: 8–44

ARGON GLOW LAMPS (see **GLOW LAMPS**)

ARMORIES, illumination levels: 9–81

ART GALLERIES (see **MUSEUMS**)

ART WINDOWS, in churches and synagogues: 12–2 to 12–4

ARTIFICIAL DAYLIGHT, for color inspection: 5–22

ASPECT-FACTOR METHOD, in illumination calculations: 9–52 to 9–56

ASSEMBLY WORK, illumination levels: 9–81

ASSOCIATION OF AMERICAN RAILROADS, color specifications: 22–17, 22–18

ASTM
equipment design standards: 6–22
method of measuring and specifying color: 5–10, 5–11

ATHLETIC FIELDS (see **SPORTS LIGHTING**)

ATMOSPHERIC ABSORPTION: 4–24, 9–71

ATMOSPHERIC TRANSMISSIVITY: 1–3, 3–34, 3–35, 4–24

ATOMIC STRUCTURE: 2–4, 2–5

ATTENUATION LENGTH, of water: 26–3

AUDITORIUMS
convention facilities: 24–10
hotel: 12–14
illumination levels: 9–81
industrial show facilities: 24–10
meeting facilities: 24–10
school: 11–13, 11–14
theatre: 24–6, 24–7, 24–11

AURORA AUSTALIS: 2–6

AURORA BOREALIS: 2–6

AUTOMOBILE LIGHTING
exterior: 22–1, 22–2
headlighting: 22–1, 22–2
illumination levels: 9–94
interior: 22–2, 22–3
lamps: 8–14, 8–81, 8–82, 8–83
panel boards: 22–2, 22–3
signal lighting: 22–2
meaning of signal colors: 22–2

AUTOMOBILE LIGHTING TERMS
backup lamp: 1–4
clearance lamp, definition: 1–5
dual headlighting system: 1–7
intensity of signal light: 22–2
lower (passing) beams, definition: 1–12
multiple-beam headlamps, definition: 1–13
parking lamp, definition: 1–14
pattern of display of signal light: 22–2
stop lamp, definition: 1–20
tail lamp, definition: 1–20
turn signal operating unit, definition: 1–21
upper driving beams, definition: 1–21

AUTOMOBILE MANUFACTURING
illumination levels: 9–81
lighting for assembly: 14–20 to 14–22

AUTOMOBILE SHOWROOMS (see **STORE LIGHTING**)

AUTOPSY AND MORGUE, lighting of: 12–6, 12–7

AUXILIARY EQUIPMENT
for high-intensity discharge lamps: 8–38 to 8–40
for short-arc lamps: 8–43

AVERAGE ILLUMINATION CALCULATIONS: 9–6 to 9–36

AVERAGE LUMINANCE
of luminaires: 1–3, 3–24, 3–25
of surfaces: 1–3

AVERAGE LUMINANCE CALCULATIONS: 9–33

AVIATION AGENCIES: 21–1

AVIATION LAMPS: 8–13, 8–76 to 8–79

AVIATION LIGHTING (see **AIRCRAFT LIGHTING** and **AIRPORT LIGHTING**)

AZIMUTH
definition: 1–3
solar: 7–1, 7–3, 7–4

B

BACK LIGHT
definition: 1–3
in television: 24–16

BACKGROUND, projected in television: 24–13

BACTERICIDAL LAMPS
list of: 8–110
characteristics of: 25–15 to 25–17

BACTERICIDAL ULTRAVIOLET RADIATION (see also **ULTRAVIOLET RADIANT ENERGY**)
air conditioning: 25–18
applications of: 25–18, 25–19
bactericidal effectiveness: 1–4
efficiency, bactericidal: 1–4, 25–16
exposure, bactericidal: 1–4
exposure time for: 25–18
flux, bactericidal: 1–4
flux density, bactericidal: 1–4
inhibition of colony formation: 25–17
lamps for: 8–110, 25–16, 25–17
liquid sterilization: 25–18, 25–19
personal protection: 25–18
product protection: 25–19
spectral energy distribution from ozone producing lamp: 25–17
region of spectrum: 1–21

BADMINTON: 9–91, 19–4, 19–6, 19–7

BAFFLES: 1–4, 6–9, 6–10

BAKERY INDUSTRY
illumination levels: 9–81
lighting for: 14–37 to 14–39

BALLAST, FLUORESCENT
class P: 8–30
considerations in equipment design: 6–25
construction: 8–30
definition: 1–4, 8–17
life: 8–30
operation: 8–28 to 8–30
power factor: 8–30
power requirements: 8–101
reference ballast: 1–16, 4–13, 4–14
sound from: 8–30
testing procedures: 4–11 to 4–14
trigger start: 8–29

BALLAST, HIGH-INTENSITY DISCHARGE LAMP
considerations in equipment design: 6–25, 6–26
definition: 1–4
reference ballast: 4–13, 4–14
testing procedures: 4–11 to 4–14
types of: 8–38 to 8–40

BALLAST FACTOR, effect on Light Loss Factor: 9–4

BALLROOMS AND FUNCTION ROOMS, hotels and motels: 12–14

BANKS
illumination levels: 9–81
lighting of: 12–1, 12–2

BAR PHOTOMETER: 4–9

BARBER SHOPS AND BEAUTY PARLORS, illumination levels: 9–81

BARN LIGHTING (see **DAIRY FARMS**)

BARNES COLORIMETER: 5–14

BARNES FILTER: 4–7

BARRIER-LAYER CELLS (see **PHOTOVOLTAIC CELLS**)

BASE LIGHT
definition: 1–4
in television: 24–16

BASEBALL: 9–91, 19–10 to 19–14

BASES, LAMP
fluorescent: 8–18, 8–21
incandescent: 8–5

BASKETBALL: 9–92, 19–3, 19–4, 19–5, 19–6

BATHING BEACHES: 9–92

BATHROOM LIGHTING
hotel: 9–85, 12–14
residences: 9–88, 15–13, 15–14, 15–20

BATTERY POWERED LIGHTING SYSTEMS: 14–11, 14–12

BAUMGARTNER REFLECTOMETER: 4–7, 4–8

BEACON
aerodrome: 1–1
airport: 21–6
airway: 1–1
approach-light: 1–2
beam divergence: 18–4
candlepower: 18–4
characteristics: 18–4
definition: 1–4
hazard or obstruction: 1–10
identification, definition: 1–10
landmark, definition: 1–11
lighthouse: 18–1, 18–2

BEAM ATTENUATION COEFFICIENT, of water: 26–1

BEAM CHARACTERISTICS, of arc lighting units: 24–4

BEAM DIVERGENCE, searchlight: 9–70

BEAM EFFICIENCY, floodlight: 9–68, 9–69

BEAM-LUMEN METHOD, floodlighting: 9–68 to 9–69

BEAM SPREAD
definition: 1–4
floodlight: 9–68
searchlight: 9–70

BEAUTY PARLORS, illumination level: 9–81

BICYCLE LAMPS: 8–14, 8–62

BILLIARDS: 9–92, 19–4, 19–6, 19–7

BINOCULAR VISION: 3–4, 3–6

BIOLUMINESCENCE, definition of: 2–6

"BLACK LIGHT"
applications of: 25–13, 25–14
definition: 1–4
fluorescent lamps: 8–109, 25–15
flux: 1–4
flux density: 1–4
ultraviolet region: 1–21

BLACKBOARDS (see **CHALKBOARDS**)

BLACKBODY
definition: 1–4, 2–3
locus: 1–4, 5–5, 5–11
luminance: 8–55
radiation curves: 2–3

BLACKBODY RADIATION: 2–2, 2–3

BLACKWELL VISUAL TASK EVALUATOR: 3–17

BLEACHING: 25–10 to 25–12

BLONDEL, conversion factor: 1–26

BLUEPRINTING: 24–36

BOLOMETERS: 4–8

BOOK MANUFACTURING: 9–82, 14–24 to 14–26

BORDERLIGHTS, in stage lighting: 24–8, 24–9

BOUGUER'S LAW: 6–7

BOWL REFLECTOR LAMPS: 8–11, 8–60

BOWLING: 9–92, 19–4, 19–7

BOWLING ON THE GREEN: 9–92

BOXING OR WRESTLING (RING): 9–92, 19–4, 19–7, 19–8

BREWERIES: 9–82

BREWSTER'S ANGLE: 6–2, 6–8

BRIGHTNESS (see **LUMINANCE** for Photometric Brightness)
definition: 1–4
in interior lighting design: 10–8
subjective, definition: 1–20
veiling brightness, definition: 1–21

BRIGHTNESS CALCULATIONS (see **LUMINANCE CALCULATIONS**)

BRIGHTNESS COEFFICIENTS (see **LUMINANCE COEFFICIENTS**)

BRIGHTNESS CONTRAST (see **LUMINANCE CONTRAST**)

BRIGHTNESS METERS (see **LUMINANCE METERS**)

BUILDINGS AND MONUMENTS, illumination levels: 9–90

BUILDING (CONSTRUCTION), illumination levels: 9–90

BUILDING EXTERIORS
floodlighting effect from interior: 17–5
floodlighting of: 17–1 to 17–6
illumination levels: 9–90, 17–1, 17–5, 17–7
location of equipment for floodlighting: 17–3, 17–4
reflectances: 17–1
service stations: 17–9
setback construction: 17–3, 17–4

BUILDING OFFICIALS CONFERENCE OF AMERICA, local codes of: 6–21

BUILDING SECTIONS: 7–5, 7–6, 7–9

BUILT-IN LIGHTING (see **ARCHITECTURAL LIGHTING** and **STRUCTURAL LIGHTING ELEMENTS**)

BULB
fluorescent lamp: 8–17, 8–18
incandescent lamp: 8–3 to 8–5, 8–6, 8–7

BULLETIN BOARDS AND POSTER PANELS
illumination levels: 9–90
lighting of: 23–7 to 23–12

BUNSEN "GREASE SPOT" PHOTOMETER: 4–4

BUREAU OF SHIPS, industry standards: 6–22

BUS LIGHTING, city and inter-city buses: 22–6 to 22–7

C

CAFETERIAS (see **FOODSERVICE FACILITIES**)

CAGES, MAINTENANCE: 16–14, 16–15, 16–16

CALCIUM CARBONATE, ultraviolet reflectance: 25–2

CALCULATIONS
area sources: 9–58 to 9–61
average illumination: 9–6
average luminance: 9–14 to 9–28, 9–33
beam-lumen method for floodlights: 9–68, 9–69
cost: 16–19
daylighting: 9–74 to 9–80
floodlighting: 9–68, 9–69
footcandles from survey: 4–27
for underwater illumination: 26–5 to 26–7
illumination, form for: 9–7
illumination at a point: 9–44 to 9–61
interior illumination: 9–1 to 9–33
lighting levels: 9–81 to 9–95
linear sources: 9–56 to 9–58
lumen method: 9–1 to 9–33
lumens from candlepower data: 4–18 to 4–21
luminance: 9–14 to 9–28, 9–33
point-by-point method, definition: 1–15
point illumination: 9–44 to 9–61
point methods: 9–44 to 9–61
point sources: 9–44 to 9–61
procedure: 9–1, 9–2
roadway illumination: 9–71 to 9–74
room cavity ratio, definition: 1–17
searchlight: 9–69 to 9–71
shelf and garment case: 9–63
show windows: 9–64 to 9–67
showcase illumination: 9–67 to 9–68
spacing-to-mounting height ratio, definition: 1–18
specific application methods: 9–63
vertical surface illumination: 9–68
zonal-cavity inter-reflectance method, definition: 1–22
zonal-cavity method: 9–8
zonal factor interflection method, definition: 1–22
zonal lumens: 4–18

CAMERA PICKUP TUBES: 24–12

CANADIAN STANDARDS ASSOCIATION, codes and standards: 6–21, 6–26, 6–27

CANDELA: 1–4, 1–5

CANDELAS PER UNIT AREA
conversion factors: 1–26
conversion to SI Unit: 1–24

CANDLE (see **CANDELA**)

CANDLEPOWER
airport beacons: 21–6
apparent: 1–2
definition: 1–5
locomotive headlights: 22–15, 22–16
reference standard: 2–3
searchlights: 9–70

CANDLEPOWER DISTRIBUTION CURVE
definition: 1–5
general lighting luminaires: 4–17, 4–18
luminaires, typical: 9–12 to 9–30
measurement: 4–9, 4–14, 4–15, 4–16, 4–17
width line, definition: 1–22
reflector and PAR Lamps: 8–12
searchlight: 18–3

CANDLEPOWER RATIOS FOR FLUORESCENT LAMPS: 4–25

CANDOLUMINESCENSE: 2–7

CANDY MANUFACTURING
illumination levels: 9–82
lighting for: 14–39, 14–40

CANNERIES
illumination levels: 9–82
lighting of: 14–41 to 14–43

CANOPY LIGHTING, store lighting: 13–7

CAR LIGHTING (see **AUTOMOBILE LIGHTING**)

CARBON ARC SOURCES
chromaticity data: 5–22
flame arcs: 8–50, 8–51, 8–119
high-intensity arcs: 8–51, 8–118
lamp, definition: 1–5
low intensity arcs: 8–50, 8–118
luminance measurements: 4–26
luminance of: 8–51, 8–52, 8–55
motion picture photography: 24–4, 24–23, 24–24
motion picture projection: 24–32, 24–33, 24–34
operation of: 2–7, 8–50
power sources for: 8–51
spectral distribution: 8–52
tables of: 8–118, 8–119
types of arcs: 8–50, 8–51

CARRELS, library: 12–16

CASTING, illumination levels: 9–92

CATHODOLUMINESCENCE: 2–11, 25–13

CATWALKS, maintenance: 16–14, 16–15

CAUSES OF LIGHT LOSS
interior installations: 16–11 to 16–13
roadway: 9–71, 9–72

CAVITY RATIO
definition: 1–5
formula for: 9–8
table of: 9–9

CEILING CAVITY RATIO
definition: 1–5
formula for: 9–8
table of: 9–9

CEILINGS
louvered: 1–12, 9–61 to 9–63
luminous: 1–13, 9–61 to 9–63

CENTRAL STATION PROPERTIES
illumination levels: 9–82, 9–90
lighting of: 14–43 to 14–46

CENTRAL VISION: 1–5, 3–4, 3–6

CERTIFIED BALLAST MANUFACTURERS, codes and standard practices: 6–22

CHALKBOARDS: 11–11, 11–14

CHEMICAL WORKS, illumination levels: 9–82

CHEMILUMINESCENCE: 2–12

CHLOROPHILL SYNTHESIS: 25–3

CHRISTMAS LIGHTING: 17–8

CHROMA, MUNSELL: 1–13, 5–6, 5–7

CHROMATICITY
calculation of CIE: 5–4, 5–7, 5–8, 5–9
color, of a: 1–5
coordinates (of a light): 1–5, 8–19, 8–27
diagrams: 1–5, 5–5, 8–19, 8–27
fluorescent lamp: 8–19, 8–24, 8–26, 8–27
railway signals: 22–18
u,v-diagram: 5–19

CHROMIUM, characteristics of: 6–6

CHURCH and SYNAGOGUE LIGHTING
art windows: 12–3, 12–4
church architecture and lighting: 12–3, 12–4
entrances: 12–2
illumination levels: 9–82
main worship areas: 12–2

CIE (COMMISSION INTERNATIONALE de l'ECLAIRAGE)
chromaticity coordinates: 5–7
chromaticity diagram: 5–5
color specification system: 5–4, 5–5, 5–7
luminaire classification: 10–11, 10–12
method of obtaining color specification: 5–4, 5–7
spectral luminous efficiency curve: 2–5
standard chromaticity diagram: 1–5
standard sources: 5–8
standard sources, spectral power: 5–7
test color method for color rendering: 5–19 to 5–21
tristimulus computational data: 5–8
uniform chromaticity diagram: 5–19

CIGARS, bleaching and fading: 25–12

CIRCLINE FLUORESCENT LAMPS: 8–17, 8–94

CIRCUITS
dimming: 8–32
mercury lamp operation: 8–39
flashtubes: 8–46
fluorescent lamps: 8–29, 8–32, 8–33, 22–11, 22–12

CIRCULAR SOURCE CALCULATIONS: 9–59

CIVIL AERONAUTICS BOARD, federal standards: 6–22

CLASS P BALLASTS: 6–25

CLASSES OF STANDARDS
laboratory reference: 1–11
national: 1–14
national standard of length: 1–14
primary: 1–15
reference: 1–16
secondary: 1–17
working: 1–22

CLASSING ROOMS for COLOR GRADING: 5–23

CLASSROOM LIGHTING (see **SCHOOL LIGHTING**)

CLAY PRODUCTS AND CEMENTS, illumination levels: 9–82

CLEANING (see also **MAINTENANCE**)
compounds for: 16–13, 16–14
equipment for: 16–14 to 16–16
in interior lighting design: 10–8
periodic planned: 16–33

CLEANING AND PRESSING INDUSTRY
illumiation levels: 9–82
lighting for: 14–31 to 14–33

CLOTH PRODUCTS, illumination levels: 9–82

CLOTHING MANUFACTURE (MEN'S)
illumination levels: 9–82
lighting for: 14–34, 14–35

CLUBS AND LODGE ROOMS, illumination levels: 9–82

COAL TIPPLES AND CLEANING PLANTS, illumination levels: 9–82

COAL YARDS, illumination level: 9–90

COCKTAIL LOUNGES: 12–5, 12–6

CODES AND STANDARDS PRACTICES
federal standards: 6–22
industry standards: 6–22
international standards: 6–21
local codes: 6–21
national codes: 6–21
national standards: 6–21
manufacturers' standards: 6–22
thermal test limits: 6–27

COEFFICIENT OF BEAM UTILIZATION: 1–6, 9–68

COEFFICIENT OF UTILIZATION
definition: 1–6
multipliers for various effective floor cavity reflectances: 9–32
roadway: 9–71
sidelighting: 9–77 to 9–79
tables of: 9–12 to 9–30
toplighting: 9–76, 9–77
use of: 9–6 to 9–8

COHERENT LIGHT: 2–10

COLD-CATHODE LAMPS
bactericidal: 25–15, 25–17
definition: 1–6
operation of: 8–18
table of: 8–100

COLOR (see also **CHROMATICITY, COLOR RENDERING, COLOR SPECIFICATION, COLORIMETRY**)
adaptation: 5–11, 5–12
appearance: 5–14, 5–15, 5–16
appearance of light source: 5–14
ANSI colorimetric standards for fluorescent lamps: 5–15
ANSI methods of measuring and specifying: 5–10, 5–11
artificial daylighting: 5–22
ASTM methods of measuring and specifying: 5–10, 5–11
automobile signals: 22–2
Barnes colorimeter: 5–14
basic concepts of: 5–1 to 5–14
characteristics of phosphors: 2–9
chromatic adaptation, state of, definition: 1–19

chromaticity coordinates of CIE standard sources: 5–7
chromaticity diagram: 5–5, 5–12, 5–19
chromaticity of a, definition: 1–5
CIE standard observer: 5–3
CIE standard sources: 5–7
CIE supplementary observer: 5–3
CIE system of color specification: 5–4 to 5–7
CIE test color method: 5–19 to 5–21
color matching terms: 5–4
color rendering: 5–2, 5–18 to 5–21
color rendering index, basis of: 5–19
color temperature: 5–11, 5–15
color terms: 5–1, 5–2
color vision: 3–4 to 3–6
colorimetric calculations: 5–4
colorimetry: 5–14 to 5–16
comparison, or color grading, definition: 1–6
complementary wavelength, definition: 1–5
conditional color match: 5–13, 5–14
consistency of: 5–11, 5–12
contrast: 5–12, 5–13
control of: 5–24, 5–25
correction, definition: 1–6
correlation of specification methods: 5–11
daylight standards: 5–21, 5–22
definition of: 1–6
determination of CIE coordinates: 5–9
discrimination, definition: 1–6
documents on color: 5–10
dominant wavelength: 5–7
effect of transmission through various bodies of water: 26–2
effect on apparent size of objects: 5–18
effect on light utilization: 5–17, 5–18
excitation purity: 5–7
floodlighting: 17–5
fluorescent lamp: 5–15, 8–24, 8–26
fluorescent materials: 25–13
garden lighting: 15–30
grading: 5–23, 5–24
graphic arts: 5–23, 5–24, 14–25
guide to selection: 5–18
Hardy-Rand-Rittler test: 5–15
Hering opponent-color theory: 3–4, 3–5
hue names and abbreviations: 5–10
hue of a perceived light source: 1–10
hue of a perceived object, definition: 1–10
in hospitals: 12–6, 12–11, 12–12
in industry: 14–4, 14–10
in insect attraction: 25–12
in interior lighting design: 10–6, 10–7, 10–8
in lighting design: 10–6
in offices: 11–6
ISCC-NBS method of designating colors: 5–9, 5–10
light source: 5–14
light source, for merchandising: 13–2, 13–3
light source, in schools: 11–12, 11–13
light sources for special purpose applications: 5–21 to 5–25
lightness index: 5–4
lightness of a perceived object, definition: 1–12
luminance factor: 5–4, 5–10
luminous reflectance: 5–7, 5–16, 5–17
luminous signals: 3–36
MacAdam ellipses: 5–3
matching: 1–6, 5–24
measurement of light source color: 5–14
measurements, basis of: 5–2, 5–3
metamerism: 5–13, 5–14
Munsell spacing: 5–3
Munsell system: 1–13, 5–7, 5–9
museum and art galleries: 12–20
NBS method of designating colors: 5–9, 5–10
object color: 1–14, 5–2, 5–11
object color, perceived: 5–1
Ostwald system: 5–11
perceived object color: 5–1
photoelectric colorimeters: 5–14
photography: 5–22, 5–23
Planckian distribution curves: 5–13
preference: 5–25
purity: 5–7
railway signal specifications: 22–17, 22–18
reflectance vs light source: 5–16, 5–17
relative spectral distributions: 5–13
residence surfaces: 15–2, 15–3
SAE standardized tests: 22–3
schemes: 5–17, 5–18
selection: 5–24
shift in, due to interreflections: 5–16
shift in, due to light source: 5–19
show window lighting: 13–9
signs: 23–1, 23–3, 23–4
specification of: 5–4 to 5–11
spectral distributions of daylight: 5–13
spectral tristimulus values for equal power source: 5–3
spectroradiometry: 5–15
surgical light: 12–12
textiles: 5–22
tristimulus computational data: 5–8
tristimulus source data: 5–8
use of color: 5–16 to 5–18
visual comparisons, direct: 5–15, 5–16
visual evaluation: 5–2, 5–3
Young-Helmholtz theory: 3–4, 3–5

COLOR DESIGNATION (see **COLOR SPECIFICATION**)

COLOR FILTERS
railway signal: 22–17, 22–18
transmittance of: 17–5

COLOR GRADING: 5–23, 5–24, 14–41

COLOR MATCHING: 1–6, 5–24

COLOR PHOTOGRAPHY: 5–22, 5–23, 8–15, 8–16

COLOR PREFERENCE: 5–25

COLOR RENDERING
characteristics of common light sources: 5–20, 5–21
CIE test color method: 5–19 to 5–21
concepts of: 5–2, 5–18 to 5–21
definition: 1–6
index: 5–19 to 5–21
in interior design: 10–6

COLOR SCHEMES: 5–17, 5–18, 10–6

COLOR SHIFT
adaptive color shitfs: 1–1, 5–19
colorimetric shift: 1–6, 5–19
from interreflections: 5–16, 5–19
resultant color shift: 1–17

COLOR SPECIFICATION
ANSI methods: 5–10, 5–11
ASTM methods: 5–10, 5–11
by color temperature: 5–15
chromaticity coordinates: 5–7
CIE system: 5–4, 5–5, 5–7
correlation among methods: 5–11
dominant wavelength: 5–7
excitation purity: 5–7
ISCC-NBS method: 5–9, 5–10
lightness index: 5–4
luminance factor: 5–4
luminous reflectance: 5–7
Munsell system: 1–13, 5–6, 5–7, 5–9
Ostwald system: 5–11
purity: 5–7

COLOR TEMPERATURE
aviation cockpit lighting: 21–4, 21–5
basic concepts: 5–11
color specification with: 5–15
correction factors for uncorrected cells: 4–7
correlated: 1–6, 5–11
definition: 1–6, 2–4
diagram for determining: 5–12
measurement of: 4–2
photographic lamps: 8–15, 8–16, 8–85
Planckian locus: 5–11
specification: 5–15
sunlight: 2–6
tungsten filaments: 8–2

COLOR TERMS (see also **SECTION 1** under specific term): 5–1, 5–2

COLOR VISION: 3–4 to 3–6

COLORED LAMPS
incandescent: 8–4, 8–5
used in signs: 23–3, 23–4

COLORED LIGHT SIGNALS: 3–36

COLORIMETERS: 5–14

COLORIMETRIC SHIFT: 1–6, 5–19

COLORIMETRY
Barnes colorimeter: 5–14
of light source: 5–14 to 5–16
photoelectric colorimeters: 5–14
standard source: 1–19
tristimulus: 5–14
visual colorimeters: 5–15

COMBINATION SPORTS FIELD: 9–92, 19–10, 19–11

COMFORT HEATING
applications of: 25–22
general heating: 25–22
sources for: 25–22
spot heating: 25–22, 25–23

COMFORT PARAMETERS, Environment: 16–1, 16–2

COMMERCIAL BUILDING LIGHTING (see **PUBLIC BUILDING LIGHTING**)

COMMERCIAL KITCHENS: 11–14, 12–4, 12–5

COMPACT ARC LAMPS (see **SHORT-ARC LAMPS**)

COMPLETE RADIATOR (see **BLACK-BODY**)

COMPOUND REFLECTION: 6–2

COMPUTER ROOM: 11–9, 11–10

CONDITIONAL COLOR MATCHES: 5–13, 5–14

CONES, of eye: 1–6, 3–2, 3–3, 3–4, 3–5

CONFERENCE ROOMS: 11–9, 12–1

CONSTANTS
solar: 1–18
zonal: 1–22, 4–18, 4–19

CONTACT PRINTING: 24–35, 24–36

CONTRAST (see also **LUMINANCE CONTRAST**)
color: 5–12, 5–13
definition: 1–6
office visual tasks: 11–1

CONTRAST BORDERS: 3–7

CONTRAST, EQUIVALENT: 3–17, 3–18

CONTRAST RENDITION FACTOR
definition: 1–6
determination of: 3–18
use of: 3–18, 3–23, 3–33

CONTRAST SENSITIVTY
definition: 1–6
variation with luminance: 3–7

CONTROL OF LIGHT (see **LIGHT CONTROL**)

CONTROL PANEL LIGHTING, aircraft: 21–5

CONTROL ROOMS AND DISPATCH ROOMS
illumination levels: 9–83
lighting of: 14–44, 14–45

CONTROL SYSTEMS, for theatre and television: 24–5, 24–6

CONVERSION FACTORS
angular measure: 1–27
illumination: 1–27
length: 1–26
luminance: 1–26
metric (or SI) units: 1–24
unit prefixes 1–27

CORNICE LIGHTING: 1–6, 13–6, 15–13

CORPS OF ENGINEERS, federal standards: 6–22

CORPUSULAR THEORY: 2-1

CORRIDORS
hospital: 12–7
hotel: 12–14
office: 11–9
school: 11–14

CORRISIVE LOCATIONS, in **Equipment Design**: 6–24

COSINE-CUBED LAW: 1–6, 4–2, 4–3, 9–45

COSINE ERROR, in illumination meters: 4–6, 4–7

COSINE LAW: 1–6, 4–2, 4–3, 9–44

COST (see **ECONOMIC COMPARISON**)

COTTON CLASSING ROOMS: 5–23

COTTON GINS
illumination levels: 9–83
lighting of: 14–26, 14–27

COTTON MILLS
illumination levels: 9–89
lighting for: 14–27, 14–28

COURT ROOMS, illumination levels: 9–83

COVE LIGHTING
definition: 1–6
design of: 9–62, 9–63
residence: 15–14

CRANES, Maintenance: 16–14, 16–15

CRF (see **CONTRAST RENDITION FACTOR**)

CROQUET: illumination levels: 9–92

CRYSTALLOLUMINESCENCE: 2–12

CU (see **COEFFICIENT OF UTILIZATION**)

CURLING
illumination levels: 9–92
lighting for: 19–4, 19–7, 19–8

CURVES, ROADWAY: 20–9

CUTOFF, ROADWAY LUMINAIRE CATEGORY: 20–5

CUTOUT STARTER: 8–31

CYSTOSCOPIC ROOM: 12–7

D

DAIRY FARMS
illumination levels: 9–83
lighting for: 14–49, 14–50

DAIRY PRODUCTS INDUSTRY
illumination levels: 9–83
lighting for: 14–35, 14–36

DANCE HALLS, illumination level: 9–83

DARKROOM LIGHTING: 24–35, 24–36

DAYLIGHT
Abbot-Gibson: 5–21, 5–22
artificial: 5–22
availability of: 7–1, 7–2, 7–3
chromaticity data for: 5–22
fading from: 25–10 to 25–12
fluorescence under: 25–13
for underwater illumination: 26–6
ground as source: 7–3, 7–5
illumination level: 2–6
relative spectral distributions: 5–13
sky as source: 7–1, 7–3
standards for color: 5–21, 5–22
sun as source: 7–1
variability of: 7–1, 7–2, 7–3

DAYLIGHT INCANDESCENT LAMPS: 8–4

DAYLIGHTING
architectural design: 7–5, 7–6, 7–8, 7–9
bilateral building design: 7–5
building sections: 7–5, 7–9
calculations: 9–74 to 9–80
clerestories: 7–5
controls: 7–6, 7–8, 7–9
daylight: 7–1 to 7–5, 7–6, 7–7, 7–8
design factors: 7–1
effect of interior reflectances: 7–9
illumination, distribution from: 7–10, 7–11
industrial buildings: 7–10, 7–11
light loss factors for daylighting design: 7–12
louvers: 7–8
materials: 7–6, 7–8, 7–9
measurements: 7–12
offices: 7–10
overhangs: 7–8
roof monitors: 7–5
sawtooth construction: 7–5
schools: 7–10
shades: 7–8
sky luminance: 7–3, 7–6, 7–7
skylight illumination: 7–7, 7–8
skylights: 7–5, 7–6
solar altitude: 7–1, 7–3
solar azimuth: 7–1, 7–3
solar illumination: 7–5
sports: 19–3
unilateral building design: 7–5
window orientation: 7–6

DAYLIGHTING CALCULATIONS
illumination computations: 9–74, 9–75
lumen method: 9–76, to 9–79
luminance: 9–79, 9–80
non-uniform overcast sky: 9–76
overhangs: 9–80
sidelighting: 9–77, 9–79
toplighting: 9–76, 9–77

DC SERIES INCANDESCENT LAMPS: 8–11, 8–58

DECORATIVE ACCENT LIGHTING, in residences: 15–26 to 15–29

DECORATIVE INCANDESCENT LAMPS: 8–13

DEFINING EQUATIONS, Photometric and Radiometric: 1–2, 1–3

DEFINITIONS (see **SECTION 1, DICTIONARY OF LIGHTING TERMS**)

DEKALUX LEVELS (see **ILLUMINATION LEVELS**)

DENSITY, OPTICAL
equation for: 6–7
measurement: 4–2

DENTAL SUITES
illumination levels: 9–84
lighting for: 12–7

DEPOTS, TERMINALS, AND STATIONS, illumination levels: 9–83

DEPRECIATION (see also **MAINTENANCE**)
factors for roadway: 9–72
incandescent lamp output: 8–9
lamp lumen factor: 1–11, 9–5
luminaire dirt factor: 1–12, 9–5

DESIGN APPROACHES (see **INTERIOR LIGHTING DESIGN APPROACHES**)

DESIGN DATA
for equipment design: 6–21 to 6–29
for installations: 9–1
for luminaires: 6–10 to 6–14
for luminous elements: 23–4 to 23–6
for reflectors: 6–10 to 6–13
for refractors: 6–13, 6–14

DESIGN OF LIGHTING INSTALLATIONS, practical guides: 3–33

DGF (see **DISABILITY GLARE FACTOR**)

DIAMOND GRADING, light source for: 5–21

DIFFRACTION: 6–9

DIFFUSE LIGHTING, on appearance of objects: 10–6

DIFFUSE REFLECTION: 1–7, 6–2

DIFFUSE TRANSMISSION: 1–7, 6–7

DIFFUSING MATERIALS: 6–6, 6–7, 7–6

DIFFUSION
in office lighting: 11–5
light: 6–9
on appearance of objects: 10–6

DIMMING
devices for: 16–9 to 16–11
fluorescent lamp: 8–32
television studio: 24–5, 24–6
theatre stage: 24–5

DINING AREAS
illumination levels: 9–84, 9–88
lighting for: 12–5, 12–6, 12–14, 15–22, 15–23

DIOPTERS, definition of: 3–3

DIRECT GLARE (see also **DISCOMFORT GLARE**)
definition: 1–7
from various lighting systems: 10–11, 10–12
in industrial lighting: 14–2
in offices: 11–2 to 11–4
in roadway lighting: 20–1, 20–2
in schools: 11–11
VCP: 3–24 to 3–27

DIRECT LIGHTING: 10–11, 10–12

DIRECT LUMINAIRES: 10–11

DIRECT PHOTOMETRY: 4–3

DIRECT RADIATION COMPONENT, in illumination calculations: 9–44 to 9–60

DIRECT-INDIRECT LIGHTING: 10–11, 10–12

DIRECT-INDIRECT LUMINAIRES: 10–11

DISABILITY GLARE
allowance for in design: 3–18
allowance for in evaluation of ESI: 3–23
cause of: 3–10
definition: 1–7
effect of age: 3–12
effect on contrast: 3–10
in industrial lighting: 14–2
in office lighting: 11–3, 11–4
in roadway lighting: 20–1, 20–2
in school lighting: 11–11, 11–12
veiling luminance: 3–10

DISABILITY GLARE FACTOR, use of: 3–18

DISCHARGE LAMPS (see also specific type)
electrical measurements: 4–11 to 4–14
photometric measurements: 4–14

DISCOMFORT GLARE
causes of: 3–24
definition: 1–7
evaluation of: 3–24, 3–26, 3–27, 3–28
in industrial lighting: 14–2
in offices: 11–2 to 11–4
in roadway lighting: 20–2
in schools: 11–11
visual comfort criteria: 3–24 to 3–27

DISCONNECTING HANGERS, for maintenance: 16–14, 16–16

DISPATCH BOARDS
illumination level: 9–83
lighting of: 14–44, 14–45

DISPERSION: 6–4, 6–5

DISPLAY LIGHTING
in museums and art galleries: 12–22, 12–23
in stores: 13–6, 13–7

DISTRIBUTION, CANDLEPOWER (see **CANDLEPOWER DISTRIBUTION CURVE**)

DISTRIBUTION (GONIO) PHOTOMETER: 4–9, 4–14, 4–17, 4–24

DIVERGENCE, BEAM: 9–70

DOMINANT WAVELENGTH: 1–7, 5–7

DORMITORY ROOMS: 11–14

DOWNLIGHTING
in residences: 15–4 to 15–8
in stores: 13–6

DRAFTING ROOMS
illumination levels: 9–87
in offices: 11–7, 11–8
in schools: 11–14, 11–15

DREDGING, illumination level: 9–90

DRIVE-IN THEATRES: 24–11, 24–12, 24–26

DRY CELL SIZES: 8–14

DRY-CLEANING INDUSTRY
illumination levels: 9–82
lighting for: 14–31 to 14–33

E

ECONOMIC ANALYSIS: 16–19, 16–20

ECONOMIC CONSIDERATIONS
in equipment design: 6–28, 6–29
in interior lighting design: 10–2

EDUCATIONAL TELEVISION: 24–18

EFFECTIVE CAVITY REFLECTANCES: 9–8, 9–10, 9–11

EFFICACY, LUMINOUS
definition: 1–13, 2–6
fluorescent lamps: 8–22, 8–23
high-intensity discharge lamps: 8–35, 8–36
maximum obtainable: 2–6
sodium: 8–44
symbol and unit: 1–3

EFFICIENCY
bactericidal, of ultraviolet: 25–16
erythemal, of ultraviolet: 25–16
luminaire: 1–12

EFFICIENCY, LAMP (see **EFFICACY, LUMINOUS**)

EFFICIENCY, LUMINOUS, of Radiant Energy: 1–3, 2–5, 2–6

EFFICIENCY, SPECTRAL LUMINOUS
curve of: 2–5
values: 3–5, 3–6

ELECTRIC DISCHARGE LAMPS (see also specific type)
definition: 1–7
electrical measurements of: 4–11
life performance testing: 4–16
photometric measurements of: 4–14

ELECTRIC POWER SYSTEMS, most common: 16–8

ELECTRIC SIGNS (see **SIGNS**)

ELECTRIC EQUIPMENT MANUFACTURING, illumination levels: 9–83

ELECTRICAL GENERATING STATIONS
illumination levels: 9–83
lighting of: 14–43 to 14–46

ELECTRICAL MEASUREMENTS
ballasts: 4–13
electric discharge lamps: 4–11 to 4–14
fluorescent lamps: 4–11 to 4–14
high-intensity discharge lamps: 4–11 to 4–14
incandescent filament lamps: 4–11
typical circuit for: 4–12

ELECTRODE POTENTIAL SERIES: 2–13

ELECTROENCEPHALOGRAPHIC SUITE: 12–7

ELECTROLUMINESCENCE: 2–10 to 2–12

ELECTROLUMINESCENT LAMPS
characteristics of: 8–47, 8–48
color of light: 2–11
construction: 2–11
phosphors: 2–11
uses: 21–4, 21–5

ELECTROMAGNETIC SPECTRUM: 1–8, 1–16, 2–1, 2–2

ELECTROMAGNETIC THEORY: 2–1

ELECTROMYOGRAPHIC SUITES: 12–7

ELEVATORS
illumination level: 9–83
lighting of: 11–9

ELLIPSOIDAL REFLECTORS: 6–10, 6–11

EMERGENCY LIGHTING
aircraft: 21–4
definition: 1–8
hospital: 12–13
industrial: 14–10 to 14–12, 14–46
schools: 11–16

EMMISSIVITY
definition: 1–8
spectral, definition: 1–18
symbol and unit: 1–2
total: 2–4

EMITTANCE (see **EXITANCE**)

ENAMEL, ULTRAVIOLET REFLECTANCE: 25–2, 25–19

ENCLOSED AND GASKETED (see **VAPOR-TIGHT**)

ENERGY DISTRIBUTION CLASSIFICATION, Luminaire: 16–5

ENERGY SPECTRUM: 2–2

ENGRAVING (WAX) illumination level: 9–83

ENLARGER LAMPS: 8–86

ENTRANCE FOYER
hotels and motels: 12–14
museum and art galleries: 12–20

EQUATIONS
average illumination from survey: 4–28
Bouguer's law: 6–7
color specification: 5–4
cosine law: 4–2, 4–3, 9–44
cosine-cubed law: 4–2, 4–3
ellipse: 6–10
Holladay and Stiles: 3–10
hyperbola: 6–10
illumination for photography: 24–20, 24–21
incandescent lamp characteristics: 8–8
inverse-square law: 4–2, 4–3, 9–44
Lambert's cosine law: 4–2, 4–3
Lambert's law of transmission: 6–7
luminous signal visibility: 3–33 to 3–36
Planck's, for blackbody radiation: 2–3, 2–4
radiation: 2–4
Rayleigh: 5–22
sky luminance: 9–76
Snell's law: 6–4, 6–5
Stefan-Boltzmann: 2–3
veiling luminance: 3–10
velocity of radiant energy: 2–2
Wien displacement and radiation: 2–3

EQUIPMENT DESIGN
codes and standard practices: 6–21, 6–22
economic considerations: 6–28, 6–29
electrical considerations: 6–25, 6–26, 6–27
environmental considerations: 6–23 to 6–25
luminaire characteristics: 6–22, 6–23
mechanical considerations: 6–26, 6–27
safety considerations: 6–27, 6–28
standard practices: 6–21, 6–22
thermal considerations: 6–27

EQUIVALENT CONTRAST: 3–17, 3–18

EQUIVALENT SPHERE ILLUMINATION
definition: 1–8
evaluation of: 3–20, 3–23
in lighting design: 3–33
recommended values: 9–80

ERYTHEMA (see **ERYTHEMAL ULTRAVIOLET RADIATION**)

ERYTHEMAL ULTRAVIOLET RADIATION
efficiency, erythemal: 25–16
production of vitamin D: 25–15
reflectance of skin for: 25–16
sources of: 8–100, 25–15
sunlamps: 25–15
transmittance of skin for: 25–16

ESCALATORS: 11–9

ESI (see **EQUIVALENT SPHERE ILLUMINATION**)

EXAMINING ROOMS, hospital: 12–8

EXCITATION PURITY: 5–7

EXPENSE ANALYSIS (see **ECONOMIC ANALYSIS**)

EXPLOSIVES MANUFACTURE, illumination levels: 9–83

EXPOSITIONS, lighting of: 17–6, 17–7

EXPOSURE, PHOTOGRAPHIC: 24–19 to 24–22

EXPOSURE METERS: 24–20, 24–21

EXTENDED SERVICE INCANDESCENT LAMPS: 8–11, 8–59

EXTERIOR LIGHTING (see also **ADVERTISING LIGHTING, FLOODLIGHTING, SPORTS LIGHTING**)
design data: 17–1 to 17–10
illumination levels: 9–90, 9–91
of hotels and motels: 12–14
of residences: 15–26, 15–29, 15–30
of stores: 13–9 to 13–12
relationship with interior: 10–3

EYE
accommodation. 3–1, 3–2, 3–3
adaptation: 3–3
age effects: 3–11
bipolar nerve cells: 3–2
central and peripheral vision: 3–4
ciliary muscle: 3–1, 3–2
color vision: 3–4 to 3–6
cones: 3–2, 3–3, 3–4, 3–5
conjunctiva: 3–3
cornea: 3–1, 3–2, 3–3
field of view: 3–4
fixation: 3–6
fovea: 3–2, 3–5
ganglion cells: 3–2
iris: 3–1, 3–2, 3–3
lens: 3–1, 3–2, 3–3
macular pigment: 3–5
Maxwell's spot: 3–6
mesopic vision: 3–4
opsin: 3–2
optic nerve fibers: 3–2
peripheral vision: 3–4
photochemical processes: 3–2, 3–3
photopic vision: 3–4, 4–1
photoreceptors: 3–2
pupil: 3–1, 3–3, 3–4
retina: 3–2
retinal: 3–2
retinene: 3–2
rhodopsin: 3–2, 3–3
rods: 3–2, 3–3, 3–4, 3–5
sclera: 3–1, 3–2
scotopic vision: 3–4, 4–1
spectral luminous efficiency data: 3–5, 3–6, 4–1
spectral response characteristics: 2–5
structure of: 3–1, 3–2
ultraviolet protection: 25–15
visual purple: 3–2

EYE SENSITIVITY CURVE (see **EFFICIENCY, SPECTRAL LUMINOUS**)

F

FACTORY LIGHTING (see **INDUSTRIAL LIGHTING**)

FADING
accelerated testing: 25–12
cigars: 25–12
colored textiles: 25–11
effect of temperature on: 25–10, 25–12
factors involved in: 25–10
museum conservation: 12–24
processed meats: 25–12
relative effect of different sources: 25–11
store lighting: 13–4

FARMS
illumination levels: 9–83, 9–87, 9–90
lighting of: 14–49 to 14–52

FEDERAL AVIATION ADMINISTRATION, federal standards: 6–22

FEDERAL STANDARDS, in **Equipment Design:** 6–22

FENCING: 9–92, 19–4

FENESTRATION
definition: 1–8
materials and control elements: 7–6, 7–8, 7–9
purposes of: 7–9

FIBER OPTICS: 6–3

FIELD HOUSES (see **GYMNASIUMS**)

FIELD LUMINANCE, in glare evaluation: 3–24

FIELD MEASUREMENTS: 4–26 to 4–30

FILAMENT LAMPS (see **INCANDESCENT FILAMENT LAMPS**)

FILAMENT SOURCES: 2–7

FILAMENTS, for incandescent lamps: 8–1 to 8–3, 8–8

FILES, lighting of: 11–8

FILL LIGHT
definition: 1–8
motion picture photography: 24–23
television: 24–16

FILM SPEED, ASA: 24–19, 24–20

FILMS
low-reflectance: 6–9
photographic: 24–19, 24–20

FILTERING PROPERTIES OF SEA WATER: 26–2

FILTERS
Barnes: 4–7
darkroom: 24–35
definition: 1–8
infrared reflecting: 16–7
neutral: 4–3, 4–4
photographic: 24–19, 24–20
spectral transmittance curves: 4–7
transmittance of saturated color: 17–5

FINISHES, LUMINAIRE: 6–19 to 6–21, 6–28, 6–29

FIRE HALL (see **MUNICIPAL BUILDINGS**)

FIXTURES (see **LUMINAIRES**)

FLAGS, FLOODLIGHTED (see **BULLETIN BOARDS AND POSTER PANELS**)

FLAME LUMINESCENCE: 2–7

FLAME TYPE CARBON ARCS: 8–50, 8–51, 8–119

FLASH PHOTOGRAPHY: 24–21, 24–22

FLASHER LAMPS: 8–15, 8–80

FLASHLAMPS (see **FLASHTUBES** and **PHOTOFLASH LAMPS**)

FLASHLIGHT LAMPS: 8–14, 8–62

FLASHTUBES
energy input limits: 8–47
guide numbers for: 24–21
list of: 8–117
power supply for: 8–46
spectral distribution: 8–45
types of: 8–45

FLICKER
fluorescent lamp: 8–27, 8–28
high-intensity discharge lamp: 8–40

FLICKER INDEX
fluorescent lamp: 8–27, 8–28
high-intensity discharge lamp: 8–40

FLICKER PHOTOMETER: 4–4

FLOODLIGHT INCANDESCENT LAMPS: 8–11

FLOODLIGHTING (see also **FLOODLIGHTS, GARDENS, PROTECTIVE LIGHTING, SPORTS LIGHTING**)
building exteriors: 17–1 to 17–6
calculations: 9–68, 9–69
color filters for: 17–5
color in: 17–5, 17–6, 17–7
design procedure: 17–1
expositions: 17–6, 17–7
fountains: 17–7, 17–8
illumination levels for: 17–1, 17–5, 17–7
interior floodlighting of buildings: 17–5
light sources for: 17–5, 17–7
location of equipment: 17–3, 17–4
maintenance: 17–6
monuments and statues: 12–19, 17–6
principles of: 17–1 to 17–3
protective lighting: 18–5
public gardens: 17–6, 17–7
setback construction: 17–3, 17–4
signs: 23–7 to 23–12

FLOODLIGHTING CALCULATIONS: 9–68, 9–69

FLOODLIGHTS
aiming for sports: 19–14, 19–15, 19–18, 19–19
beam spread: 9–68
coefficient of beam utilization: 9–68
color filters for: 17–5
definition: 1–8
fluorescent: 23–10 to 23–12
for fountains: 17–7, 17–8
for signs: 23–7 to 23–12
incandescent: 23–10
light loss factors: 9–69
location for building lighting: 17–3, 17–4
location for setback construction: 17–3, 17–4
mercury: 23–10 to 23–11
metal halide: 23–10
minimum efficiencies: 9–69
NEMA types: 9–69
photometry of: 4–22 to 4–25
protective lighting: 18–5, 18–7
shielding in gardens: 17–7
tungsten-halogen: 23–10, 23–11, 23–12

FLOOR CAVITY RATIOS (FCR)
formula for: 9–8
table of: 9–9

FLOUR MILLS
illumination levels: 9–84
lighting of: 14–36, 14–37

FLUID MILK INDUSTRY
illumination levels: 9–83
lighting for: 14–35, 14–36

FLUORESCENCE
atomic activity in: 2–8, 2–9
definition of: 1–9, 2–8
fluorescent materials: 25–13, 25–14
steps in luminescent process: 2–7, 2–8
Stokes law: 2–9

FLUORESCENT LAMPS
activators: 2–9
air movement, effect on: 8–26
ANSI colorimetric standards for: 5–15
aperture lamps: 8–21
aquaria: 25–10
arc length: 8–24, 8–25
auxiliary circuits for: 8–28, 8–29, 8–32
aviation: 8–77
ballasts: 8–28 to 8–30, 8–101
bases for: 8–18, 8–21
"black light": 8–109
bulb designation: 8–17, 8–18
candlepower ratios: 4–25
chromaticity: 8–19, 8–24, 8–26, 8–27
circline: 8–17, 8–94
circuits for operation in transportation lighting: 22–11, 22–12
cold cathode: 8–18, 8–100
color: 8–24, 8–26, 8–27
color measurement: 5–14, 5–15, 5–16
color rendering characteristics: 5–20, 5–21
color temperature, correlated: 5–14, 5–15
construction: 8–17, 8–18, 8–21
definition: 1–9, 8–17
dimming of: 8–32
effect on air conditioning: 16–2, 16–3
efficacy: 8–22, 8–23
efficacy vs frequency: 8–33
electrical measurements: 4–11 to 4–14
electrodes: 8–18
energy conversion efficacy: 8–22
energy distribution: 8–24, 8–25
energy output: 16–3
flashing of: 8–32
flicker: 8–27, 8–28
floodlighting with: 23–11, 23–12
gas fill: 8–18
high frequency operation: 8–32, 8–33
hot cathode: 8–18, 8–94 to 8–99
in equipment design: 6–24, 6–26
instant start lamps: 8–29, 8–98
inverters for: 8–33
lamp circuits: 8–28, 8–29, 8–32
lampholders: 8–31, 8–32

life: 8–23, 8–24
life performance testing: 4–16
light output: 8–22, 8–23, 8–24, 8–25, 8–94 to 8–100
light production process: 2–7, 2–8
LLD factors: 8–94 to 8–99
lumen maintenance: 8–24, 8–25
luminance: 8–55, 8–94 to 8–100
luminance formula: 4–25
luminescent process: 2–7, 2–8
luminous efficacy: 8–22, 8–23
performance parameters: 8–22 to 8–25
phosphors: 2–8, 2–9, 8–18
photochemical reproduction: 24–35, 24–36
photometric measurements: 4–14
plant growth: 25–3 to 25–10
power requirements: 8–101
preheat lamps: 8–28, 8–29, 8–96
public conveyance: 22–11, 22–12
radio interference: 8–28
rapid start lamps: 8–29, 8–30, 8–94
reflector lamps: 8–21
signs, use in: 23–10, 23–11, 23–12
slimline: 8–98
spectral power distribution: 8–19, 8–24
sports lighting: 19–3, 19–4
starters: 8–29, 8–30, 8–31
starting: 8–28 to 8–30
stroboscopic effect: 8–27
sunlamps: 8–100
temperature effects on: 8–25, 8–26, 8–28
ultraviolet producing: 25–14, 25–15
U-shaped: 8–18, 8–94
voltage vs length: 8–23, 8–24

FLUORESCENT LUMINAIRES
coefficients of utilization: 9–14 to 9–30
effect of temperature on: 23–11
maintenance of: 16–17, 16–18
trouble shooting: 16–16, 16–17, 16–18

FLUORESCENT MATERIALS: 25–13

FLUORESCENT PLANT GROWTH LAMPS: 25–4

FLUX TRANSFER THEORY: 9–36

FOG
daylighted luminance of: 3–36
effect on searchlight range: 9–71

FOODSERVICE FACILITIES
commercial kitchens: 12–4, 12–5
dining spaces: 12–5, 12–6
illumination levels: 9–84
school cafeterias: 11–14

FOOTBALL
illumination levels: 9–92
lighting for: 19–11, 19–14

FOOTCANDLE (see also **CALCULATIONS**)
conversion factors: 1–27
conversion to SI Unit: 1–24
definition: 1–9
meters: 4–5, 4–7
survey procedures: 4–26 to 4–30

FOOTCANDLE LEVELS (see **ILLUMINATION LEVELS**)

FOOTCANDLE METER (see **ILLUMINATION METER**)

FOOTLAMBERT
conversion factors: 1–24, 1–26
definition: 1–9

FORGE SHOPS, illumination level: 9–84

FORM FACTORS, for **Zonal-Cavity System**: 9–36

FORMULAS
Allard's law: 3–33
apparent contrast: 3–10
area sources: 9–59, 9–60
attenuation of light in water: 26–1
cavity ratios: 9–8
contrast: 3–7
dimension of sign letters: 23–2, 23–5, 23–6, 23–9
discomfort glare: 3–24, 3–25
flashtube characteristics: 8–46
flicker index: 8–27
floodlighting: 9–68
fluorescent lamp luminance: 4–25
illumination from daylighting: 9–76, 9–77
line sources: 9–56 to 9–58
lumen method: 9–6, 9–8
luminance, illumination, reflectance: 3–7
luminance of fluorescent lamp: 4–25
luminance, walls and ceiling: 9–33
luminous ceilings: 9–62
minimum distance, inverse-square law for searchlights: 4–23
mountng height of floodlights, sports: 19–17
museum and art gallery design: 12–21, 12–22
overhangs: 9–80
per cent flicker: 8–27
point method: 9–44
roadway illumination: 9–72
searchlight: 9–70
sensible lighting heat load: 16–4
show window illumination: 9–64 to 9–67
showcase illumination: 9–67 to 9–68
signal lighting: 3–33 to 3–36
transmittance of toplights: 9–76
underwater lighting: 26–6, 26–7
veiling luminance: 3–10

FOUNDRIES
illumination levels: 9–84
lighting of: 14–15 to 14–17

FOUNTAINS, lighting of: 17–7, 17–8

FRESNEL LENS
luminaires for, protective lighting: 18–7
refractive properties: 6–4, 6–5

FREUND LUMINANCE SPOT METER: 4–5, 4–8

FRUIT AND VEGETABLE PACKAGING, lighting for: 14–41 to 14–43

***f*-SYSTEMS, LENS**: 24–20

FUNDUS REFLECTOMETRY, definition of: 3–3

G

GALVANOLUMINESCENCE: 2–12

GARAGES, illuminaton levels: 9–84

GARDENS
color in: 17–7
general principles: 17–6, 17–7
illumination levels: 9–91
public: 17–6, 17–7
residential: 15–29, 15–30

GAS LASER: 2–11, 2–12

GAS MANTLE: 2–7

GASEOUS DISCHARGE, phenomena in producing light: 2–7

GASEOUS DISCHARGE LAMPS (see specific type)

GASLIGHTING: 8–52 to 8–54

GASLIGHTS
high-intensity type: 8–54
isofootcandle diagram of typical: 8–54
life: 8–54
light output: 8–53
luminance: 8–55
mantle construction: 8–52
mantle performance: 8–53
open flame type: 8–54

GASOLINE STATIONS (see **SERVICE STATIONS**)

GENERAL DIFFUSE LIGHTING: 1–9, 10–11, 10–12

GENERAL DIFFUSE LUMINAIRES: 10–11

GENERAL LIGHTING: 1–9, 10–11, 10–12

GENERAL SERVICES ADMINISTRATIONS, federal standards: 6–22

GENERAL SERVICE INCANDESCENT LAMPS: 8–10, 8–57

GENERATION OF LIGHT: 2–6, 2–7

GERMICIDAL (see **BACTERICIDAL**)

GLARE (see also **DISABILITY GLARE, DISCOMFORT GLARE, REFLECTED GLARE, VEILING REFLECTIONS**)
blinding, definition: 1–4
definition: 1–9
in industrial lighting: 14–2, 14–3
in office lighting: 11–2 to 11–5
in roadway lighting: 20–1, 20–2
in school lighting: 11–11, 11–12
in sports lighting: 19–2, 19–5, 19–6

GLASS
for fenestration: 7–6
infrared reflectance: 16–4
infrared transmittance: 16–4
lamp bulb: 8–4
maintenance of: 16–14
properties of lighting glass: 6–16
reflectance: 6–6, 16–4
spectral transmittance for various: 4–10
transmittance: 6–6, 7–9, 16–4
used in light control: 6–14, 6–15, 6–16

GLASS BLOCK, for fenestration: 7–6, 7–9

GLASS BOTTLE SORTING AND WASHING, lighting for: 14–35, 14–36

GLASS WORKS, illumination levels: 9–84

GLOVE MANUFACTURING: illumination levels: 9–84

GLOW LAMPS: 8–44, 8–80, 8–116

GLOW MODULATOR LAMP: 8–47

GLOW-SWITCH STARTERS: 8–31

GOLF
illumination levels: 9–92
lighting for: 19–12, 19–15

GONIOMETER: 4–9, 4–17, 4–24

GONIOPHOTOMETER: 4–9, 4–14

GOODEVE'S DATA: 2–6, 3–5, 3–6

GRAPHIC ARTS INDUSTRY
carbon arcs for: 8–52
color work: 5–23, 5–24
illumination levels: 9–87
lighting for: 14–24 to 14–26

GRAYBODY: 1–9, 2–4

GREASE SPOT PHOTOMETER: 4–4

GREEK ALPHABET: 1–27

GROUP REPLACEMENT, of lamps: 16–19

GROWTH OF PLANTS (see **HORTICULTURAL LIGHTING**)

GUIDE NUMBER SYSTEM, in photography: 24–21

GYMNASIUMS (see also **SPORTS LIGHTING**)
illumination levels: 9–92
installations, typical: 19–4, 19–5
lighting layout, typical: 19–5
windows in: 19–3, 19–6

H

HALLWAYS (see **CORRIDORS**)

HANDBALL
illumination levels: 9–92
lighting for: 19–4, 19–7, 19–8

HANDLANTERN LAMPS: 8–14, 8–62

HANGARS (see **AIRCRAFT HANGARS**)

HANGERS, DISCONNECTING, for maintenance: 16–14, 16–16

HARDY-RAND-RITTLER TEST: 5–15

HAT MANUFACTURING, illumination levels: 9–84

HAZARDOUS LOCATION LIGHTING: 14–6, 14–43

HEADLAMPS
automobile: 22–1, 22–2
definition: 1–10
lamps: 8–14, 8–81, 8–82, 8–83

HEADLIGHTING
automotive: 22–1, 22–2
lamps: 8–14, 8–81, 8–82, 8–83
locomotive: 22–15, 22–16

HEALTH CARE FACILITIES (see **HOSPITAL LIGHTING** and **NURSING HOMES**)

HEAT (see also **INFRARED RADIANT ENERGY** and **INFRARED HEATING**)
lamps as source of: 16–2, 16–3
lighting systems as source of: 16–4, 16–5
luminaires as source of: 16–3, 16–4
methods of control: 16–5 to 16–7
sensible heat: 16–1 to 16–5
sources in interiors: 16–1
thermal energy distribution: 16–4, 16–5
transfer, in lighting systems: 16–4, 16–5

HEAT LAMPS: 8–92, 8–93

HEATING (see **COMFORT HEATING, INFRARED HEATING**)

HEFNER, conversion factor: 1–26

HELICOPTER LIGHTING (see **AIRCRAFT LIGHTING**)

HELIPORT LIGHTING: 21–12

HERING COLOR THEORY: 3–4, 3–5

HID LAMPS (see **HIGH-INTENSITY DISCHARGE LAMPS**)

HIGH FREQUENCY LIGHTING
aircraft: 21–1, 21–5
electroluminescent lamp: 8–47, 8–48
fluorescent lamp: 6–25, 8–32, 8–33
power for: 16–8

HIGH-INTENSITY CARBON ARCS: 8–51, 8–118

HIGH PRESSURE SODIUM LAMPS
auxiliary equipment for: 8–38, 8–40
ballast for: 8–38, 8–40
construction of: 8–35, 8–36
definition: 1–10
designation: 8–36
effect of operating position: 8–37
effect of temperature: 8–38
efficacy: 8–36
electrical measurements: 4–11 to 4–13
energy output: 16–3
gas fill: 8–35
life: 8–37, 8–107
life performance testing: 4–16
lumen maintenance: 8–37
operation of: 8–35, 8–36
photometric measurements: 4–14
spectral power distribution: 8–20, 8–36
starting: 8–36, 8–37
table of: 8–107
trouble shooting and maintenance: 16–20

HIGH-INTENSITY DISCHARGE LAMPS (see also **MERCURY LAMPS, METAL HALIDE LAMPS, HIGH PRESSURE SODIUM LAMPS**)
color rendering characteristics: 5–20, 5–21
definition: 1–10, 8–33
effect on air conditioning: 16–2, 16–3
energy output: 16–3
in equipment design: 6–24
sports lighting: 19–3, 19–4
trouble shooting and maintenance: 16–18, 16–20

HIGH TEMPERATURE APPLIANCE LAMPS: 8–12

HIGH VOLTAGE INCANDESCENT LAMPS: 8–10, 8–11, 8–58

HIGHWAY LIGHTING (see **ROADWAY LIGHTING**)

HOCKEY
illumination levels: 9–92
lighting for: 19–4, 19–8, 19–9

HOME LIGHTING (see **RESIDENCE LIGHTING**)

HORSE SHOES, illumination levels: 9–93

HORSE SHOWS, illumination levels: 9–93

HORTICULTURAL LIGHTING
action spectra for: 25–3
aquarium lighting: 25–9, 25–10
design data for: 25–5, 25–6
footcandle levels for ornamental foliage: 25–9
light sources for: 25–3, 25–4
luminaires for: 25–8 to 25–10
photoperiodic lighting: 25–7
photoperiodism: 25–3, 25–5, 25–7
photosynthesis: 25–2, 25–3, 25–4, 25–5
photosynthetic lighting: 25–5
plant growth chambers: 25–7 to 25–8
terrarium lighting: 25–9, 25–10
wattage requirements for: 25–6

HOSPITAL LIGHTING
anesthesia rooms: 12–11, 12–12
autopsy and morgues: 12–6, 12–7
corridors: 12–7
cystoscopic rooms: 12–7
delivery rooms: 12–10
dental suites: 12–7
electroencephalographic suites: 12–7
electromyographic suites: 12–7
emergency lighting: 12–13
emergency rooms: 12–8
eye, ear, nose, and throat suites: 12–8
fluoroscopy rooms: 12–8
illumination levels: 9–84, 9–85
intensive care rooms, 12–8, 12–9
laboratories: 12–9
mental health facilities: 12–9
night-lighting: 12–10
nuclear medicine facilities: 12–9
nurseries: 12–9
nurses stations: 12–9, 12–10
nursing homes: 12–10
operating rooms: 12–11 to 12–13
patient holding areas: 12–13
patients' room: 12–10, 12–11
pharmacy: 12–11
radiation, diagnostic: 12–11
radiation therapy: 12–11
reflectances recommended: 12–7, 12–11
scrub room: 12–13
surgical recovery: 12–13
surgical suite: 12–11 to 12–13

HOT CATHODE LAMPS
operation of: 8–18
tables of: 8–94 to 8–99

HOT-COLD RESISTANCE, effect on current in incandescent lamps: 8–56

HOTEL LIGHTING
ballrooms: 12–14
corridors: 12–14
dining room: 12–14
entrance foyer: 12–14
exteriors: 12–14
guest room: 12–14
illumination levels: 9–85
lobbies: 12–14
lounges: 12–14
principles: 12–13
reception desk: 12–14

HUE
abbreviations: 5–10
definition: 1–10
Munsell: 1–13, 1–14, 5–6, 5–7
names: 5–10
perceived object color: 1–14

HUMIDITY, relation to comfort: 16–2

HYDROGEN, for incandescent lamp fill: 8–6

HYPERBOLIC REFLECTORS: 6–10

I

ICE HOCKEY (see **HOCKEY**)

ICE MAKING, illumination levels: 9–86

ICI (see **CIE**)

IERI (see **ILLUMINATING ENGINEERING RESEARCH INSTITUTE**)

ILLUMINANCE (see **ILLUMINATION**)

ILLUMINANTS (see **LIGHT SOURCES**)

ILLUMINATING ENGINEER, responsibility of: 3–1

ILLUMINATING ENGINEERING RESEARCH INSTITUTE: 3–1

ILLUMINATION (see also **CALCULATIONS, ILLUMINATION LEVELS,** and **METHOD FOR PRESCRIBING ILLUMINATION**)
calculation sheet: 9–7
conversion factors for units: 1–27
definition: 1–10
effect on photoelectric detection: 2–12
equivalent sphere illumination: 1–8, 3–33
measurement: 2–13, 4–26, to 4–29
meters: 4–6
sky: 7–8
solar: 7–5
sphere: 1–19
symbols and units: 1–3

ILLUMINATION CALCULATIONS (see **CALCULATIONS**)

ILLUMINATION LEVELS (see also under specific application listing)
for growth of plants: 25–9
from moon: 2–6
from overcast sky: 2–6
from sun: 2–6
protective lighting: 18–6
roadway lighting: 20–6
table of footcandle and dekalux values: 9–81 to 9–95
wiring capacity for: 16–9

ILLUMINATION METERS
Macbeth illuminometer: 4–4, 4–5
photovoltaic cells: 2–13, 4–5 to 4–7
portable photoelectric: 4–5 to 4–7

INCANDESCENCE
forms of: 2–6, 2–7
radiation equations: 2–4

INCANDESCENT FILAMENT LAMPS
aircraft: 8–13, 8–78, 8–79
airport: 8–13, 8–76, 8–77
assembly of: 8–2, 8–3
automotive: 8–14, 8–81, 8–82, 8–83
aviation: 8–13, 8–76 to 8–79
bases for: 8–5
bicycle: 8–14, 8–62
bowl reflectors: 8–13, 8–60
bulb finishes and colors: 8–4, 8–5
bulb shapes and sizes: 8–3, 8–4
bulb temperatures: 8–6, 8–7, 8–8
classes of: 8–10 to 8–13
color rendering characteristics: 5–20, 5–21
color temperature: 5–14, 5–15
color temperature vs efficacy: 8–3
colored lamps: 8–5
construction of: 8–2, 8–3
dc series: 8–11, 8–58
decorative: 8–13
definition: 1–10
depreciation of: 8–9
effect on air conditioning: 16–2, 16–3
efficacy: 8–3
efficacy-voltage relationship: 8–8
electrical measurements of: 4–11
energy characteristics: 8–6
energy output: 16–3
extended service: 8–11, 8–59
filament notching: 8–8
filaments: 8–1 to 8–3, 8–4
fill gas: 8–5, 8–6
flasher lamps: 8–15, 8–80
flashlight: 8–14, 8–62
floodlight: 8–11
general service: 8–10, 8–57
glass for: 8–4
handlantern: 8–14, 8–62
high temperature appliance: 8–12
high voltage: 8–10, 8–11, 8–58
in equipment design: 6–24, 6–26
indicator: 8–15, 8–84
infrared: 8–16, 8–17, 8–92, 8–93, 25–20, 25–23
life performance testing: 4–16
life-voltage relationships: 8–8
LLD factors: 8–57
locomotive: 8–11, 8–62
lumen-voltage relationship: 8–8
lumiline: 8–12, 8–80
luminance: 8–55
luminous characteristics: 8–78
maintenance of: 16–18
miniature: 8–13 to 8–15, 8–62, 8–80
mortality rate: 8–9, 8–10
motion picture studio: 8–11, 8–64 to 8–72
multiple street lighting: 8–13, 8–75
oven lamps: 8–12
PAR lamps: 8–11, 8–12, 8–73
photo-enlarger: 8–16, 8–86
photoflash: 8–16, 8–86
photoflood: 8–16, 8–85
photography: 8–15, 8–16
photometric measurements of: 4–14
projection: 8–16, 8–87 to 8–91
R lamps: 8–11, 8–12, 8–73
radio panel: 8–15, 8–84
reflectorized: 8–11, 8–12, 8–73
renewal rate: 8–9, 8–10
RF (radio frequency) lamp: 8–17
ribbon filament lamps: 8–13, 8–14
rough service: 8–12, 8–63
sealed beam: 8–14, 8–15, 8–83
seals for: 8–2, 8–3
series street lighting: 8–13, 8–74
showcase: 8–13, 8–59
sign: 8–13
signs, use in: 23–1 to 23–4, 23–10
socket temperatures: 8–6, 8–7
spotlight: 8–11
street lighting: 8–13, 8–74, 8–75
television studio: 8–11, 8–64 to 8–72
theatre stages: 8–11, 8–64 to 8–72
thermal characteristics: 8–78
thermal radiation from: 8–7, 8–8
three light: 8–13, 8–61
traffic signal: 8–13, 8–75
train and locomotive: 8–11, 8–62
tungsten-halogen: 8–11, 8–60, 8–61, 8–64 to 8–72
tungsten-halogen cycle: 8–6
vibration service: 8–12, 8–81
wattage-voltage relationship: 8–8

INDEX OF REFRACTION: 6–3, 6–4

INDICATOR LAMPS: 8–15, 8–84

INDIRECT LIGHTING: 10–11, 10–12

INDIRECT LUMINAIRES: 10–11

INDUSTRIAL LIGHTING (see also specific industry)
automotive assembly: 14–20 to 14–22
bakeries: 14–37 to 14–39
candy manufacturing: 14–39, 14–40
canneries: 14–41 to 14–43
central station properties: 14–43 to 14–46
cleaning and pressing: 14–31 to 14–33
color appraisal: 14–25
color quality of light: 14–4, 14–25
corrosive atmosphere: 14–6
cotton gins: 14–26, 14–27
daylighting in: 7–10
design considerations: 14–4, 14–5
direct glare: 14–2
distribution and diffusion of light: 14–3
electrical generating stations: 14–43 to 14–46
emergency lighting: 14–10 to 14–12, 14–46
equipment, types of: 14–4, 14–5
factors of good lighting: 14–1 to 14–7
farms: 14–49 to 14–52
flour mills: 14–36, 14–37
fluid milk industry: 14–35, 14–36
foundries: 14–15 to 14–17
fruit and vegetable packaging: 14–41 to 14–43
graphic arts: 14–24 to 14–26
grinders: 14–16, 14–17
hazardous location: 14–6
high humidity atmosphere: 14–6
illumination levels: 9–81 to 9–91
lighting techniques: 14–7 to 14–10
luminaire locations for various visual tasks, (see also specific industries): 14–8, 14–9
luminaire shielding: 14–2, 14–4
luminaires for supplementary lighting: 14–7
luminaires, types of: 14–4, 14–5
luminance ratios: 14–2, 14–3
machining metal parts: 14–17 to 14–19
maintenance: 14–6
men's clothing manufacture: 14–34, 14–35
petrochemicals: 14–43
petroleum and petroleum products: 14–43
printing: 14–24 to 14–26
quality of illumination: 14–2 to 14–4
quantity of illumination: 14–1, 14–2
railroad yards: 14–47, 14–48
reflectances recommended: 14–3
reflected glare: 14–3
rubber tires: 14–22 to 14–24
sawmills: 14–46, 14–47
selection of equipment: 14–4
shadow effect: 14–3, 14–4
sheet metal shops: 14–19, 14–20
shoe manufacturing: 14–33, 14–34
space conditioning: 14–6
spacing-to-mounting height ratios: 14–5
steel mills: 14–12 to 14–14
supplementary lighting: 14–6 to 14–10
temperature conditions: 14–6
textile mills: 14–27 to 14–31

INDUSTRY STANDARDS, Equipment Design: 6–22

INFRARED HEATING (see also **INFRARED RADIANT ENERGY**)
advantages of: 25–22
applications: 25–20 to 25–23
snow melting: 25–23
sources: 8–16, 8–17, 8–92, 8–93, 25–20, 25–22

INFRARED LAMPS: 8–16, 8–17, 8–92, 8–93, 25–20, 25–22

INFRARED RADIANT ENERGY (see also **INFRARED HEATING**)
ability to see: 2–6
applications of: 25–20 to 25–23
comfort heating, use for: 25–22, 25–23
reflectance of various materials: 16–4
sources of: 8–16, 8–17, 8–92, 8–93, 25–20, 25–22
transmittance of various materials: 16–4
wavelengths of, used for heating: 25–20

INFRARED REFLECTING FILTERS: 16–7

INSECT TRAPPING: 25-12

INSIDE FROSTED LAMPS: 8-4

INSPECTION LIGHTING
color in: 5-24, 14-10
foundry: 14-16, 14-17
glass containers: 14-36, 14-41
illumination levels: 9-86
metal containers: 14-41, 14-42
silhouette lighting: 14-10
stroboscopic illumination: 14-10
use of fluorescence: 14-10
use of polarized light: 14-10

INSTANT START FLUORESCENT LAMPS: 8-29, 8-98

INSTITUTE OF ELECTRICAL AND ELECTRONIC ENGINEERS, industry standards: 6-22

INSTITUTIONS (see **PUBLIC BUILDINGS** and alphabetical listings under specific type)

INSTRUMENT LIGHTING, AIRCRAFT: 21-4, 21-5

INSTRUMENTS (see also **PHOTOMETERS**)
bar photometers: 4-9
Barnes colorimeter: 5-14
Baumgartner reflectometer: 4-7, 4-8
Blackwell visual task evaluator: 3-17
Bunsen "grease spot" photometer: 4-4
colorimeters: 5-14, 5-15
distribution (gonio) photometer: 4-9, 4-14, 4-17, 4-24
field: 4-2
flicker photometer: 4-4
Freund luminance spot meter: 4-5, 4-8
goniometer: 4-9, 4-17, 4-24
integrating sphere (Ulbricht) photometers: 4-9
laboratory: 4-2
Luckiesh-Taylor luminance meter: 4-5
Lummer-Brodhun photometer: 4-4
Macbeth illuminometer: 4-4, 4-5
photoelectric tube meters: 4-8
photovoltaic cell meters: 4-5 to 4-7
portable visibility meters: 3-17
Pritchard photometer: 4-5, 4-8
reflectometers: 4-5, 4-7, 4-8
sector disk: 4-3
sensors for underwater instrumentation: 26-5
spectrophotometer: 4-10
Taylor reflectometer: 4-5
visual task evaluators: 3-17

INTEGRATING SPHERE (ULBRICHT) PHOTOMETER: 4-9, 4-14

INTENSITY, LUMINOUS (see also **CANDLEPOWER**)
definition: 1-13
symbol and unit: 1-3

INTENSITY, RADIANT
definition: 1-16
symbol and unit: 1-2

INTERFERENCE
light control: 6-8, 6-9
radiation: 6-23

INTERIOR ILLUMINATION CALCULATIONS (see also **CALCULATIONS**)
cavity ratios: 9-8, 9-9
coefficient of utilization: 9-6 to 9-8, 9-12 to 9-30
cove lighting: 9-62, 9-63
direct radiation component: 9-44 to 9-60
louvered (louverall) ceiling systems: 9-30, 9-61, 9-62
lumen method: 9-1 to 9-33
luminaire spacing: 9-12 to 9-28, 9-33 to 9-35
luminous ceiling systems: 9-30, 9-61 to 9-63
point methods: 9-44 to 9-61
reflected component: 9-60
shelf and garment case: 9-63
show window: 9-64 to 9-67
showcase: 9-67, 9-68
spacing ratios, maximum: 9-12 to 9-28, 9-33 to 9-35
uniformity of illumination: 9-33 to 9-35
vertical surfaces: 9-63, 9-68
zonal-cavity method: 9-8

INTERIOR LIGHTING DESIGN
acoustical systems: 10-2, 10-3
air conditioning and lighting: 16-1 to 16-7
appearance of objects: 10-6, 10-8
architectural detailing: 10-8
architectural periods and styles: 10-9, 10-10
architectural study of: 10-8
building function, effect on: 10-2
color in: 5-17, 5-18, 10-6, 10-7, 10-8
components and modules: 10-3
diffusion: 10-6
dimensional characteristics and form: 10-8
direct-indirect lighting: 10-11, 10-12
direct lighting: 10-11, 10-12
economics in: 10-2
engineering, study of: 10-8
environmental systems: 10-2, 10-3, 10-8
focal centers: 10-4, 10-5, 10-7
form: 10-8
general diffuse lighting: 10-11, 10-12
general lighting: 10-11, 10-12
indirect lighting: 10-11, 10-12
interior-exterior relationships: 10-3
light and architecture: 10-2
light source choice: 10-2
lighting systems: 10-1, 10-8, 10-11, 10-12
local lighting: 10-11
localized general lighting: 10-11
luminaire classification: 10-11, 10-12
luminaire selection: 10-8, 10-10
luminaires, architectural context of: 10-8, 10-10
luminance level: 10-6
luminance ratio: 10-1
luminous environment design approach: 10-3 to 10-10
maintenance: 10-8, 16-11 to 16-20
mechanical systems: 10-2, 10-3
modules: 10-3
occupied zone: 10-6, 10-7
operating costs: 10-8
overhead zone: 10-4, 10-5, 10-7
perimeter zone: 10-4, 10-5, 10-6, 10-7
"period" designs: 10-8
scale, architectural: 10-8
semi-direct lighting: 10-11, 10-12
semi-indirect lighting: 10-11, 10-12
shade in: 10-6
space requirements: 10-8
sparkle in: 10-6
stimulation, level of: 10-6
transitional considerations: 10-6
visual composition: 10-3 to 10-6, 10-8, 10-10
visual task oriented approach: 10-10, 10-11
visually-prominent lighting systems: 10-10
visually-subordinate lighting systems: 10-8
wiring: 16-7 to 16-9

INTERIOR LIGHTING MEASUREMENTS: 4-26 to 4-29

INTERNATIONAL COMMISSION ON ILLUMINATION (see **CIE**)

INTER-SOCIETY COLOR COUNCIL (see **ISCC**)

INVERSE-SQUARE LAW: 1-11, 4-1, 4-2, 9-44

INVERTERS: 8-33

IRON AND STEEL MANUFACTURING, illumination levels: 9-86

IRRADIANCE
definition: 1-11
of sun outside atmosphere: 18-5

ISCC-NBS COLOR DESIGNATION: 5-9, 5-10

ISOCANDELA DIAGRAM, in roadway lighting: 20-3, 20-4

ISOFOOTCANDLE DIAGRAM
use in interior calculations: 9-56, 9-57
use in roadway calculations: 9-73

J

JAI-ALAI
illumination levels: 9-93
lighting for: 19-4, 19-8, 19-9

JEWELRY AND WATCH MANUFACTURING, illumination level: 9-86

JUNIOR LEAGUE BASEBALL
illumination levels: 9-92
lighting for: 19-10

K

KEY LIGHT
definition: 1-11
in floodlighting buildings: 17-2
in motion picture photography: 24-23, 24-24
in television: 24-16

KITCHENS
commercial: 9-84, 12-4, 12-5
residential: 9-88, 15-17 to 15-19
school: 11-14

KOSCHMIEDER'S LAW: 3-34

KRYPTON, for lamp fill gas: 8-6

L

LABORATORIES
hospital: 12-9
school: 11-15

LABORATORY MEASUREMENTS, of light: 4-10 to 4-26

LACROSSE, illumination level: 9-93

LADDERS, for maintenance: 16-14

LAMBERT
definition: 1-11
conversion factor: 1-26

LAMBERT'S COSINE LAW: 1-11, 4-2, 4-3

LAMBERT'S LAW OF TRANSMISSION: 6-7

LAMP BURNOUTS: 9-4, 16-11, 16-12

LAMP CHANGERS, for maintenance: 16-14, 16-16

LAMP LUMEN DEPRECIATION: 8-9, 9-5, 16-11, 16-12

LAMP LUMEN DEPRECIATION FACTORS (LLD), for specific lamps: 8-57 to 8-107

LAMP SHADES: 15-10, 15-11

LAMP SHIELDING: 6-9, 6-10

LAMP SPACING
in coves: 9-62
in luminous and louvered ceilings: 9-61, 9-62
in signs: 23-2, 23-5

LAMP STARTING
fluorescent: 8-28 to 8-30
high-intensity discharge: 8-36, 8-37

LAMPHOLDERS: 6-26, 8-31, 8-32

LAMPS (see specific type)

LANDING LIGHTS, aircraft: 21-2, 21-3

LASERS: 2-10, 2-11, 26-3, 26-4

LAUNDRIES
commercial: 9-86, 14-32
home: 9-88, 15-18

LAYOUTS, LIGHTING
airport approach system: 21-7
protective lighting: 18-7
sports: 19-5, 19-7 to 19-18
stage: 24-8

LDD FACTOR (see **LUMINAIRE DIRT DEPRECIATION FACTOR)**

LEATHER MANUFACTURING, illumination levels: 9-86

LEATHER SHOE MANUFACTURING
illumination level: 9-89
lighting for: 14-33, 14-34

LEATHER WORKING, illumination levels: 9-86

LEF (see **LIGHTING EFFECTIVENESS FACTOR)**

LENGTH, UNITS OF, conversion factor: 1-26

LENSES
aberrations: 6-5, 6-7
f-value: 24-20
Fresnel: 6-4, 6-5
prismatic: 6-4, 6-5
T-value: 24-20

LETTERS
Greek: 1-27
proportions for signs: 23-6
size for signs: 23-2, 23-6
spacing for signs: 23-6

LEVELS OF ILLUMINATION, Recommended Master Table: 9-81 to 9-95

LIBRARIES
lighting of: 12-15 to 12-18
illumination levels: 9-86

LIFE, LAMP
electroluminescent: 8-48
fluorescent: 8-23, 8-24
high-intensity discharge: 8-37
incandescent: 8-8
testing: 4-16

LIFT TRUCK, for maintenance: 16-14, 16-15

LIGHT (see also **LIGHT PRODUCTION**)
bleaching: 25-10 to 25-12
concepts of: 2-1 to 2-6
control of: 6-1 to 6-14
definition of: 1-11, 2-1
detection of: 2-12
detectors: 2-12
diffraction: 6-9
diffusion: 6-9
dispersion: 6-4, 6-5
fading: 25-10 to 25-12
frequency: 2-2
generation of: 2-6
horticultural uses of: 25-2 to 25-10
insect trapping: 25-12
luminescent materials: 25-13 to 25-14
measurement of: 2-12, 4-1 to 4-32
miscellaneous applications of: 25-2 to 25-14
polarization: 6-7, 6-8
response of human eye: 2-5
spectral quality for inspection and grading: 5-21 to 5-25
spectrum: 2-2
speed: 1-19, 2-2
theories of: 2-1
wavelength limits: 2-1

LIGHT AND ARCHITECTURE (see **ARCHITECTURAL LIGHTING)**

LIGHT AND VISION: 3-1 to 3-39

LIGHT BULBS (see **INCANDESCENT FILAMENT LAMPS)**

LIGHT CONTROL
absorption: 6-9
diffraction: 6-9
diffusion: 6-9
equipment design: 6-21 to 6-29
finishes used in luminaires: 6-19 to 6-21
interference: 6-8, 6-9
lamp shielding: 6-9, 6-10
lenses: 6-4, 6-5, 6-7
materials used in: 6-14 to 6-19
physical principles of: 6-1 to 6-9
polarization: 6-7, 6-8
practical concepts: 6-21 to 6-29
reflection: 6-1 to 6-3
reflector design: 6-10 to 6-12
refraction: 6-3 to 6-5
refractor design: 6-13, 6-14
refractors: 6-5
transmission: 6-7

LIGHT DETECTION: 2-12, 2-13

LIGHT EMITTING DIODES: 2-11, 8-48, 8-49

LIGHT LOSS, causes of: 9-2 to 9-5

LIGHT LOSS FACTOR (LLF)
components of: 9-3 to 9-5
definition: 1-12
for floodlights: 9-69

LIGHT METERS (see **INSTRUMENTS**)

LIGHT OUTPUT
lamps (see specific type)
measurement of: 4-14 to 4-23

LIGHT PRODUCTION
atomic activity, relation to: 2-4, 2-5
effect of activators on: 2-8, 2-9
electroluminescence: 2-10
fluorescence: 2-8, 2-9
gaseous discharge: 2-7
incandescence: 2-7
luminescence: 2-7 to 2-10
methods of: 2-6, 2-7
natural phenomena: 2-6
phosphorescence: 2-9

LIGHT PROJECTION, equipment for: 18-1 to 18-5

LIGHT-SENSITIVE CELLS (see **PHOTOVOLTAIC CELLS** and **PHOTOELECTRIC CELLS**)

LIGHT SOURCES (see also specific types)
as heat sources: 16-2, 16-3
color rendering characteristics: 5-20, 5-21
color temperature of, definition: 1-6
colorimetry of light sources: 5-14 to 5-16
daylight: 7-1 to 7-12
distribution temperature of, definition: 1-7
effect on fading and bleaching: 25-10 to 25-12
electric: 8-1 to 8-52, 8-55 to 8-119
energy output: 16-3
for attracting phototropic insects: 25-12
for aquarium and terrarium lighting: 25-9, 25-10
for home, hobby and office horticulture: 25-8, 25-9
for office landscaping: 25-8, 25-9
for photoperiodic lighting: 25-7
for plant growth chambers: 25-8
for plant photoresponses: 25-3, 25-4
for underwater lighting: 26-3, 26-4
gas: 8-52 to 8-54
heat sources: 25-20, 25-22
in motion picture photography: 24-24, 24-25
luminous efficacy of a source of light, definition: 1-13
luminance of various: 8-55
"man-made": 2-6
museums and art galleries: 12-19, 12-20
natural phenomena: 2-6
radiant energy sources: 25-1
roadway: 20-6
sources of ultraviolet: 25-14 to 25-17
special purpose color application: 5-21 to 5-25
sports lighting: 19-3, 19-4

LIGHT TERMS (see **SECTION 1, DICTIONARY OF LIGHTING TERMS**)

LIGHT WELL EFFECT: 9-76, 9-77

LIGHTHOUSE BEACON: 18-1, 18-2

LIGHTING AND AIR CONDITIONING (see **AIR CONDITIONING**)

LIGHTING CALCULATIONS (see **CALCULATIONS**)

LIGHTING DESIGN (see **INTERIOR LIGHTING DESIGN** and **CALCULATIONS**)

LIGHTING EFFECTIVENESS FACTOR
definition: 1-12
use of: 3-19, 3-20

LIGHTING EFFECTS, of architectural periods: 10-9, 10-10

LIGHTING EQUIPMENT (see **LUMINAIRES**)

LIGHTING EQUIPMENT MANUFACTURERS' ASSOCIATION, industry standards: 6-22

LIGHTING INSTALLATION PHOTOGRAPHY: 24-23

LIGHTING LAYOUTS (see **LAYOUTS, LIGHTING**)

LIGHTING SURVEYS
daylighting: 7-12
interior: 4-26 to 4-29
outdoor: 4-29, 4-30

LIGHTING SYSTEMS
architectural detailing of: 10-8
architectural study of: 10-8
as heat sources: 16-4, 16-5
components and modules: 10-3
coordination with mechanical and acoustical: 10-2, 10-3
diffusion: 10-6
direct: 10-11, 10-12
direct-indirect: 10-11, 10-12
engineering study of: 10-8
environmental systems: 10-8
general: 10-11
general diffuse: 10-11
heat transfer in: 16-4, 16-5
indirect: 10-11, 10-12
layout, classifications by: 10-11
local: 10-11
localized general: 10-11
location, classification by: 10-11
modules: 10-3
"period" designs: 10-8
semi-direct: 10-11, 10-12
semi-indirect: 10-11, 10-12
supplementary lighting: 10-11
visually-prominent: 10-10
visually-subordinate: 10-8

LIGHTING TERMS (see **SECTION 1, DICTIONARY OF LIGHTING TERMS**)

LIGHTNESS INDEX: 5-4

LIGHTNING: 2-6, 8-55

LINEAR SOURCES, illumination calculations: 9-56 to 9-58

LIVE-PRODUCTION THEATRES: 24-6 to 24-10

LLD FACTOR (see **LAMP LUMEN DEPRECIATION FACTOR**)

LOADING AND UNLOADING PLATFORMS, illumination levels: 9-91

LOBBY LIGHTING
in hotels: 12-14
in libraries: 12-18
in theatres: 24-6

LOCAL CODES, in **Equipment Design:** 6-21

LOCAL LIGHTING (see also **SUPPLEMENTARY LIGHTING**): 1-12, 10-11

LOCALIZED GENERAL LIGHTING: 1-12, 10-11

LOCKER ROOMS, illumination level: 9-86

LOCOMOTIVE AND TRAIN LAMPS: 8-11, 8-62

LOCOMOTIVE CAB SIGNALS: 22-16

LOCOMOTIVE HEADLIGHTS: 22-15, 22-16

LOCUS
blackbody, definition: 1-4
Planckian: 5-11
spectrum: 1-18, 5-5

LOUVER
as daylight control: 7-8
definition: 1-12
lamp concealment: 6-9, 6-10
shielding angle, definition: 1-12

LOUVERED CEILING
definition: 1-12
design of: 9-61, 9-62
typical coefficients of utilization: 9-30

LOW-INTENSITY CARBON ARCS: 8-50, 8-118

LOW PRESSURE SODIUM LAMPS: 8-83, 8-84, 8-109

LOW-REFLECTANCE FILMS: 6-9

LUCKIESH-TAYLOR LUMINANCE METER: 4-5

LUMBER YARDS, illumination level: 9-91

LUMEN MAINTENANCE
electroluminescent lamps: 8-48
fluorescent lamps: 8-22, 8-23, 8-24, 8-25
high-intensity discharge lamps: 8-37
incandescent filament lamps: 8-9

LUMEN METHOD
calculation sheet for: 9-7
cavity ratios: 9-8
coefficients of utilization: 9-6 to 9-8, 9-12 to 9-30
daylighting: 9-74 to 9-80
effective cavity reflectances: 9-8
formulas for: 9-6, 9-8
lamp lumen depreciation factor (LLD): 8-57 to 8-107, 9-5
limitations of: 9-33
luminaire dirt depreciation factor (LDD): 9-3, 9-5
luminaire spacing: 9-12 to 9-28, 9-33 to 9-35
maintenance categories: 9-5, 9-12 to 9-28
reflectances, effective cavity: 9-8, 9-10, 9-32
sidelighting: 9-77 to 9-79
top lighting: 9-76
uniformity of illumination: 9-12 to 9-28, 9-33 to 9-35
Zonal-Cavity Method: 9-8

LUMILINE LAMPS: 8-12, 8-80

LUMINAIRE BRIGHTNESS LIMITATIONS (see **LUMINAIRE LUMINANCE LIMITATIONS**)

LUMINAIRE CHARACTERICTICS: 6-22, 6-23

LUMINAIRE CLASSIFICATIONS: 10-11, 10-12

LUMINAIRE DIRT DEPRECIATION: 9-3, 9-5, 16-11, 16-12

LUMINAIRE DIRT DEPRECIATION FACTOR (LDD): 9-3, 9-5, 9-6, 9-12 to 9-28

LUMINAIRE-LAMP-BALLAST OPERATING FACTOR: 4-23

LUMINAIRE LUMINANCE LIMITATIONS: 3-26, 11-3, 11-4

LUMINAIRE SPACING RATIOS: 9-12 to 9-28, 9-33 to 9-35

LUMINAIRE SURFACE DEPRECIATION: 9-4, 16-11, 16-12

LUMINAIRES
appearance of: 6-22, 10-8
applications of residential recessed: 15-3 to 15-8
architectural context of: 10-8, 10-10
as decorative accents: 15-26, 15-29
as heat source: 16-3, 16-4
average luminance of: 3-24, 3-25, 11-3
beam spread, floodlights: 9-69
bullet and hooded units: 13-7
candlepower distributions, typical: 9-12 to 9-28, 9-31
characteristics: 6-22, 6-23
CIE classification: 10-11, 10-12
classification by distribution: 14-5
cleaning of: 16-13 to 16-16
ceiling mounted residential: 15-4, 15-7, 15-8
coefficients of utilization: 9-12 to 9-30
components: 6-27
cornices: 15-13
coves: 15-14
cut-off angle: definition: 1-6
definition: 1-12
design of: 6-21 to 6-29
direct: 10-11, 10-12
direct-indirect: 10-11, 10-12
dirt depreciation: 9-3, 9-5, 9-6, 9-12 to 9-28
downlighting: 15-4 to 15-8
downlights: 13-6
dust-proof, definition: 1-7
dust-tight, definition: 1-7
efficiency of: 6-22
explosion-proof, definition: 1-8
exposed lamp, residential: 15-8, 15-9
finishes used for: 6-19 to 6-21
floodlight designations: 9-69
general diffuse: 10-11, 10-12
hazardous locations: 14-6
heat transfer in: 16-4, 16-5
indirect: 10-11, 10-12
industrial, type of: 14-4, 14-5
infrared reflecting filters for: 16-7
lampholders for: 6-26
life: 6-23
light distribution, roadway: 20-3 to 20-5
lighted soffits: 15-13
luminance coefficients for typical: 9-12 to 9-30
luminous ceiling: 15-14
luminous walls: 15-14, 15-28
maintenance of: 6-23, 16-11 to 16-20
maintenance categories for: 9-5, 9-12 to 9-28
materials used in: 6-14 to 6-19
maximum luminance of: 3-26, 11-3, 11-4
measurements of: 4-16 to 4-25
motion picture photography: 24-1 to 24-3, 24-25
night-lights, hospital: 12-10
pendant, residential: 15-8, 15-9
"period" design: 10-8
photography: 24-1 to 24-3, 24-22
photometry of: 4-16, to 4-25
plant growth: 25-7 to 25-10
plastics used in, deterioration: 16-11, 16-12
portable lamps: 15-9 to 15-12
recessed residential: 15-3 to 15-8
residence lighting: 15-3 to 15-14
roadway classifications for: 20-2
selection in industrial lighting: 14-4
selection in interior lighting design: 10-8, 10-10
semi-direct: 10-11

semi-indirect: 10-11, 10-12
shielding: 6-9, 6-10, 14-2, 14-4
sockets and lampholders for: 6-26
spacing ratios for: 9-12 to 9-28, 9-33 to 9-35, 14-5
stage lighting: 24-1 to 24-3, 24-7 to 24-10
standards vs. specials: 6-28
store lighting: 13-5
structural lighting elements: 15-12 to 15-14
supplementary: 14-7
surface depreciation: 16-11, 16-12
surgical: 12-12
television lighting: 24-1 to 24-3
thermal testing of: 4-26
total energy distribution classification: 16-5
trouble shooting and maintenance of: 16-16 to 16-20
valances: 15-13, 15-14
veiling reflections from: 3-31, 3-32
ventilation: 6-23
vibration: 6-23
walkway classifications for: 20-2

LUMINANCE
average, of a surface: 1-3
average, of luminaires: 1-3, 3-24, 3-25
blackbody: 8-55
calculations: 9-14 to 9-30, 9-33, 9-79, 9-80
carbon arcs: 8-55, 8-118
combustion sources: 8-55
conversion factors: 1-26
daylighting calculations: 9-79, 9-80
definition: 1-12
electroluminescent lamps: 8-55
fenestration: 9-79, 9-80
field, in glare evaluation: 3-24
fluorescent lamps: 8-55, 8-94 to 8-100
fog: 3-36
for picture projection: 24-25 to 24-27
gaslights: 8-55
ground: 3-36
hospital, luminaire limits: 12-11
in visual comfort evaluation: 3-24, 3-26, 3-27
incandescent lamps: 8-55
industrial locations: 14-2, 14-3
lamp: 8-55
lamp and luminaire measurement: 4-25, 4-29
light sources, various: 8-55
maximum, of luminaires: 3-26
measurement of lamp and luminaire: 4-25, 4-29
meters: 4-4 to 4-8
moon: 2-6, 8-55
nuclear: 8-55
of surfaces: 11-3
sky: 2-6, 3-36, 8-55
sun: 2-6, 8-55
survey measurements: 4-29
symbol and units: 1-3, 1-26
veiling: 3-10

LUMINANCE CALCULATIONS
fluorescent lamps: 4-25
room surfaces: 9-14 to 9-30, 9-33

LUMINANCE COEFFICIENTS: 9-14 to 9-30

LUMINANCE CONTRAST
definition: 1-12
equation for: 3-7
minimum perceptible: 3-7

LUMINANCE FACTOR
definition: 1-13
relation to Munsell value: 5-6

LUMINANCE MEASUREMENTS: 4-25, 4-29

LUMINANCE METERS
calibration of: 4-25
Freund luminance spot meter: 4-5, 4-8
Luckiesh-Taylor: 4-5
Macbeth illuminometer: 4-4, 4-5
Pritchard photometer: 4-5, 4-8

LUMINANCE RATIOS
commercial kitchens: 12-4, 12-5
in lighting design: 10-1
industrial locations: 14-3
museums and art galleries: 12-22
offices: 11-3
residences: 15-2, 15-7
schools: 11-12

LUMINESCENCE (see also **FLUORESCENCE** and **PHOSPHORESCENCE**)
definition of: 1-13, 2-7
forms of: 2-6, 2-7 to 2-10
materials: 25-13, 25-14
radioactive excitation: 25-14
types of: 25-13

LUMINOSITY (see **EFFICIENCY, SPECTRAL LUMINOUS**)

LUMINOUS CEILINGS
coefficients of utilization for: 9-30
definition: 1-13
design of: 9-61 to 9-63
in residence lighting: 15-14

LUMINOUS EFFICACY (see **EFFICACY, LUMINOUS**)

LUMINOUS EFFICIENCY (see **EFFICIENCY, LUMINOUS**)

LUMINOUS ELEMENT SIGNS: 23-4 to 23-7

LUMINOUS ENVIRONMENT DESIGN APPROACH: 10-3 to 10-10

LUMINOUS FLUX
constants for calculating: 4-18 to 4-20
definition of: 1-13, 2-1
measurement of: 4-1 to 4-32
symbol and units: 1-3

LUMINOUS INTENSITY (see also **CANDLEPOWER**)
definition: 1-13
measurement of: 4-25, 4-29
symbols and units: 1-3

LUMINOUS REFLECTANCE: 5-7, 5-16, 5-17

LUMINOUS SIGNALS (see **SIGNALS, LUMINOUS**)

LUMINOUS SOFFITS: 13-7

LUMINOUS WALLS, in residence lighting: 15-14, 15-28

LUMMER-BRODHUN PHOTOMETER: 4-4

LUX
conversion factors: 1-25, 1-27
definition: 1-13

LUX LEVELS (see **ILLUMINATION LEVELS**)

M

MACADAM ELLIPSES: 5-3, 8-27

MACBETH ILLUMINOMETER: 4-4, 4-5

MACHINE SHOPS
illumination levels: 9-86
lighting of: 14-17 to 14-19

MACHINING METAL PARTS, lighting for: 14-17 to 14-19

MACULAR PIGMENT: 3-5

MAGNESIUM OXIDE, ultraviolet reflectance: 25-2

MAGNETIC AMPLIFIER DIMMERS: 16-10

MAINTENANCE
causes of light loss: 9-2 to 9-5, 16-11 to 16-13
cleaning compounds: 16-13, 16-14
dirt on room surfaces: 16-12, 16-13
equipment for: 16-14 to 16-16
in interior lighting design: 10-8
in luminaire design: 6-23, 6-27
industrial lighting: 14-6
lamp lumen depreciation: 16-11
lamp outages: 16-11
luminaire dirt depreciation: 16-11
luminaire surface depreciation: 16-11, 16-12
method of: 16-13
need for: 16-11
planned relamping and cleaning: 16-13
temperature and voltage effect: 16-13
trouble shooting: 16-16 to 16-20

MAINTENANCE FACTOR (see **LIGHT LOSS FACTOR**)

MANUFACTURERS' STANDARDS, Equipment Design: 6-22

MANUFACTURING (see specific industry)

MARINE LIGHTING
exterior: 22-15
illumination levels: 9-94, 9-95
interior: 22-12
naval vessels: 22-15
seeing tasks: 22-12 to 22-14

MARINE NAVIGATION AIDS: 18-1, 18-2

MARQUEE LIGHTING: 24-6

MATERIALS HANDLING, illumination levels: 9-86

MATERIALS, Light Control
aluminum: 6-6, 6-17, 6-19
copper: 6-19
glass: 6-14, 6-15, 6-16
non-ferrous alloys: 6-19
plastics: 6-15, 6-18
stainless steel: 6-17, 6-19
steel: 6-15, 6-17

MAXWELL'S SPOT, in eye: 3-6

MEAN RADIANT TEMPERATURE (MRT): 16-2

MEASUREMENT, FIELD (see also **INSTRUMENTS** and **PHOTOMETERS**)
daylighting surveys: 7-12
interior illumination: 4-26 to 4-29
interior luminance: 4-29
luminance: 4-29
measurable quantities: 4-2
outdoor illumination: 4-29 to 4-30
projection screen luminance: 24-25, 24-26
shade luminance: 15-11
spot illumination: 4-28, 4-29
television studio: 24-15

MEASUREMENT, LABORATORY (see also **ELECTRICAL MEASUREMENT, INSTRUMENTS, PHOTOMETRIC TESTS, PHOTOMETERS, PHOTOMETRY**)
ballasts: 4-11 to 4-14
candlepower: 4-9, 4-14, 4-16
electrical measurements: 4-11 to 4-14
floodlights: 4-22
lamps: 4-10 to 4-16
life performance testing: 4-16
light output: 4-9, 4-14, 4-16
luminaire photometry: 4-16 to 4-23
luminance: 4-25
measurable quantities: 4-2
photometric: 4-14 to 4-26
precision of: 4-10
projector photometry: 4-23, 4-24
searchlights: 4-23, 4-24
spectral emittance: 4-10
spectral reflectance: 4-10
spectral transmittance: 4-10

MEASUREMENT OF LIGHT (see also **MEASUREMENT, LABORATORY; MEASUREMENT, FIELD; PHOTOMETRIC MEASUREMENT AND TESTS; PHOTOMETRY; and PHOTOMETERS**)
basic units: 1-2, 1-3
color: 5-2 to 5-4, 5-14
spectroradiometric: 5-2

MEASUREMENT OF RADIANT ENERGY: 4-8, 4-9, 25-4

MEASUREMENT SURVEY PROCEDURES
indoor: 4-26
outdoor: 4-29

MEAT PACKING, illumination levels: 9-86

MEATS, fading of: 25-12

MEETING, CONVENTION, AND INDUSTRIAL SHOW FACILITIES: 24-10

MEGA, unit prefix: 1-27

MEN'S CLOTHING MANUFACTURING
illumination levels: 9-82
lighting for: 14-34, 14-35

MENTAL HEALTH FACILITIES: 12-9

MERCHANDISING, Lighting for (see **SHOWCASE LIGHTING, STORE LIGHTING,** and **SHOW WINDOW LIGHTING**)

MERCURY LAMPS
argon in: 8-36, 8-41
auxiliary equipment for: 8-38 to 8-40
ballasts for: 8-38 to 8-40
circuits, typical: 8-39
color rendering index: 5-20
construction of: 8-33, 8-34
definition: 1-13
designations: 8-36, 8-102 to 8-106
effect of temperature: 8-38
effect of operating position: 8-37
electrical measurements: 4-11 to 4-13
electrodes: 8-33
energy output: 16-3
flicker index: 8-40
for underwater illumination: 26-4
life: 8-37, 8-102 to 8-106
life performance testing: 4-16
lumen maintenance: 8-37
maintenance of: 16-18, 16-20
operation of: 8-33, 8-34
phosphors: 2-9, 8-33
photochemical reproduction: 24-35, 24-36
photometric measurements: 4-14
self-ballasted: 8-40, 8-108
short-arc: 8-40, 8-41, 8-114
signs, use in: 23-10
spectral power distribution: 8-20, 8-34
starting: 8-36
stroboscopic effect: 8-40
sunlamps: 8-100
table of: 8-102 to 8-106
ultraviolet sources: 8-109 to 8-113
vapor pressure: 8-33
warm-up characteristics: 8-36

MERCURY-ARGON LAMPS: 8-114, 8-115

MERCURY-XENON LAMPS: 8-41, 8-42, 8-114, 8-115

MESOPIC VISION, definition: 1-13, 3-4

METAL HALIDE LAMPS
auxiliary equipment for: 8-38, 8-40
ballasts for: 8-38, 8-40
construction of: 8-34, 8-35
definition: 1-13
designations: 8-36
effect of operating position: 8-37
effect of temperature: 8-38
efficacy: 8-35
electrical measurements: 4-11 to 4-13
energy output: 16-3
halides used: 8-35
lamp lumen depreciation factors: 8-107
life: 8-37, 8-107
life performance testing: 4-16
light output: 8-107
lumen maintenance: 8-37
maintenance: 16-20
operation of: 8-34, 8-35
photometric measurements: 4-14
signs, use in: 23-10

spectral power distribution: 8-20, 8-35
starting: 8-36
table of: 8-107
trouble shooting: 16-20

METALLIC IODIDE VAPOR ARC LAMP: 8-47

METAMERISM: 5-13, 5-14

METEOROLOGICAL RANGE: 3-35

METERS (see **INSTRUMENTS**)

METHOD FOR PRESCRIBING ILLUMINATION
allowance for disability glare: 3-18, 3-23
allowance for transient adaptation: 3-18, 3-19, 3-23
allowance for veiling reflections: 3-17, 3-18, 3-23
determining values of design illumination: 3-17 to 3-20
determining values of required illumination: 3-16, 3-17
determining visibility level: 3-23
evaluation of equivalent sphere illumination: 3-20
practical guides for design: 3-33
procedure for achieving design values: 3-19
reflected glare in: 3-27
use of RCS function: 3-20, 3-23
veiling reflections in: 3-28 to 3-33
visual comfort criteria in: 3-24 to 3-27
visual performance criteria: 3-14
visual performance potential, selection of level: 3-14 to 3-16

METRIC (OR SI) UNITS, conversion factors: 1-24, 1-25, 1-26, 1-27

MICRO, unit prefix: 1-27

MICRON: 1-26, 2-2

MILITARY SEARCHLIGHTS: 18-2, 18-3

MILK INDUSTRY
illumination levels: 9-83
lighting for: 14-35, 14-36

MILLI, unit prefix: 1-27

MILLILAMBERT, conversion factors: 1-26

MILLIMICRON (see **NANOMETER**)

MILLIPHOT, conversion factors: 1-27

MINIATURE LAMPS (see also **INCANDESCENT FILAMENT LAMPS**)
definition: 8-13, 8-14
flasher lamps: 8-15, 8-80
glow lamps: 8-44, 8-80, 8-116
power sources for: 8-14
uses of: 8-15
tables of: 8-62, 8-80
types of: 8-13 to 8-15

MINIMUM PECEPTIBLE CONTRAST, variation with background luminance: 3-7

MIRED, definition: 5-11

MIRROR LIGHTING
residences: 15-20
stores: 13-7

MIRRORS
ellipsoidal: 6-10
parabolic: 6-10
reflection characteristics: 6-1, 6-2, 6-3

MIXED TRANSMISSION: 6-7

MODELING LIGHT
in photography: 24-23
in television: 24-28

MODULES, in interior lighting design: 10-3

MONOCHROMATOR: 4-10

MONUMENTS, floodlighting of: 17-6

MOON, luminance of: 2-6, 8-55

MOONLIGHT: 2-6

MORTALITY CURVE, incandescent lamp: 8-10

MOTEL LIGHTING (see **HOTEL LIGHTING**)

MOTION PICTURE PHOTOGRAPHY
carbon arc lamps: 8-118, 24-4
lamps for: 8-64 to 8-72, 8-118, 8-119, 24-3
light sources for: 24-24
luminaires for: 24-1, 24-25
set lighting: 24-4
stage lighting equipment for: 24-2
studio equipment for: 24-23 to 24-25

MOTION PICTURE PROJECTION
screen illumination: 24-33
xenon lamps, use of: 24-34

MOTION PICTURE STUDIO LIGHTING, lamps for: 8-11, 8-64 to 8-72

MOTION PICTURE THEATRES
auditorium-type: 24-11, 24-26
drive-in: 24-11, 24-12, 24-26

MOTOR VEHICLES LIGHTING (see also specific term)
automobile: 22-1 to 22-3
circuits for operating fluorescent lamps: 22-11
public conveyances: 22-5 to 22-7
specifications for exterior lighting of motor vehicles: 22-3

MOUNTING METHODS (see under specific type)

MULTIPLE STREET LIGHTING INCANDESCENT LAMPS: 8-13, 8-75

MUNICIPAL BUILDINGS, Fire and Police, illumination levels: 9-86

MUNSELL COLOR SYSTEM
basic concepts: 5-7 to 5-9
color solid: 5-6
definitions: 1-13, 1-14
value-luminance relationship: 5-6, 5-10, 5-16
value scales: 5-16

MUSEUMS
illumination levels: 9-81
lighting of: 12-18 to 12-24

MUSIC STUDY, HOME, lighting for: 15-24, 15-25

N

NANOMETER: 1-14, 1-22, 2-2

NATIONAL CODES, in **Equipment Design:** 6-21

NATIONAL ELECTRICAL CODE: 6-21, 6-24, 6-27, 14-6, 14-11, 14-43, 16-8, 16-9

NATIONAL ELECTRICAL MANUFACTURERS' ASSOCIATION, industry standards: 6-22

NATURAL LIGHT (see **DAYLIGHT** and **DAYLIGHTING**)

NAVIGATION LIGHTS
aircraft: 21-2
marine: 22-15

NEON GLOW LAMPS (see **GLOW LAMPS**)

NEUTRAL FILTERS: 4-3, 4-4

NICOL PRISM: 4-2

NIGHT VISION (see **SCOTOPIC VISION**)

NIT, conversion factors: 1-26

NITROGEN, for lamp fill gas: 8-5

NOMENCLATURE: 1-1 to 1-22

NOMOGRAMS
for lateral angles: 9-70
for obtaining zonal lumens from candle-power data: 4-21
for vertical angles: 9-46

NUCLEAR LIGHT SOURCES: 8-49, 8-50, 8-55

NURSING HOMES
extended care and long term care facilities: 12-10
illumination levels: 9-86, 9-87

O

OBJECT COLOR: 1-14, 5-2, 5-11

OBSTRUCTION LIGHTS, aviation: 1-14, 21-13

OFFICE LIGHTING
air conditioning: 11-7
business machines: 11-9, 11-10
color of office surfaces: 11-6
computer rooms: 11-9, 11-10
conference rooms: 11-9
corridors and hallways: 11-9
daylighting: 7-10, 11-6
data processing: 11-9, 11-10
diffusion: 11-5
discomfort glare: 11-3, 11-4
direct discomfort glare: 11-3, 11-4
drafting rooms: 11-7, 11-8
effect of time on visual tasks: 11-1
elevators: 11-9
environmental factors: 11-5
escalators: 11-9
filing facilities: 11-8
furniture and machine finishes: 11-5, 11-6
general offices: 11-7
glare: 11-2 to 11-5
horticulture: 25-8, 25-9
illumination levels: 9-87
illumination uniformity: 11-4, 11-5
influence of lighting: 11-2
interior lighting design: 10-7
luminance: 11-1, 11-2, 11-3
luminance ratios: 11-3
luminaire selection for: 11-6, 11-7
machine finishes: 11-6
mail rooms: 11-8, 11-9
maintenance: 11-10
modular construction: 11-6, 11-7
personal service: 11-9
private offices: 11-7
quality of illumination: 11-2 to 11-5
quantity of illumination: 11-2
readability of office tasks: 11-2
reception rooms: 11-9, 11-10
reflectances recommended: 11-5
reflected glare: 11-4
room finishes: 11-5
seeing tasks: 11-1, 11-2
shadows: 11-8
size of visual tasks: 11-1
stairways: 11-9
supplementary lighting: 11-9
transillumination for drafting: 11-8
uniformity of illumination: 11-4
veiling reflections: 11-4
visual comfort probability: 11-3, 11-4
visual discomfort, evaluation of: 11-3, 11-4
visual tasks: 11-1, 11-2

OPAL GLASS
characteristics of: 6-6
properties of: 6-16

OPERATING ROOMS
illumination levels: 9-85
lighting for: 12-11 to 12-13

OPTICAL DENSITY
formula for: 6-7
measurement of: 4-2

OPTICAL ELEMENTS, design of: 6-10 to 6-14

ORIENTATION
definition: 1-14
window for daylighting: 7-6

OSTWALD COLOR SYSTEM: 5-11

OUTDOOR ILLUMINATION MEASUREMENTS: 4-29, 4-30

OUTDOOR LIGHTING (see specific application)

OVEN LAMPS: 8-12

OVERHANGS, Daylighting: 9-80

OVER-VOLTAGE OPERATION, of lamps: 8-8, 19-3

OZONE-PRODUCING LAMPS: 25-15, 25-17

OZONE-PRODUCING RADIATION: 1-21

P

PACKING AND BOXING (see **MATERIALS HANDLING**)

PAINT, ultraviolet reflectance: 25-2

PAINT MANUFACTURING, illumination levels: 9-87

PAINT SHOPS, illumination levels: 9-87

PAINTINGS, lighting of: 12-20 to 12-22, 15-27

PANEL-BOARD LIGHTING, automobile: 22-2, 22-3

PAPER MANUFACTURING, illumination levels: 9-87

PAPER-BOX MANUFACTURING, illumination level: 9-87

PAR-LAMPS: 8-11, 8-12, 8-73

PARABOLIC REFLECTORS: 6-10

PARKING AREAS
at airports: 21-14, 21-15
in business areas: 13-12

PATCH SYSTEMS DIMMER CONTROL: 24-5

PATIENTS' ROOMS, hospital: 12-10, 12-11

PEDESTRIAN WALKWAYS: 20-5, 20-6, 20-9

PENDANT LUMINAIRES, for residences: 15-8, 15-9

PEOPLE HEATING (see **COMFORT HEATING**)

PER CENT FLICKER
fluorescent lamps: 8-27
high-intensity discharge lamps: 8-40

PERCEIVED OBJECT COLOR: 1-12, 5-1

PERIMETER ZONE, in interior lighting design: 10-4, 10-5, 10-6, 10-7

PERIPHERAL VISION: 1-4, 3-4, 3-6

"PERIOD" DESIGNS, in interior lighting design: 10-8

PETROCHEMICAL INDUSTRY, lighting for: 14-43

PETROLEUM AND PETROLEUM PRODUCTS MANUFACTURING, lighting for: 14-43

PHARMACY, hospital: 12–11

PHOSPHORESCENCE
definition: 1–14
phosphorescent materials: 25–13, 25–14
physics of: 2–9

PHOSPHORS
activators: 2–9, 2–11
color characteristics: 2–9, 2–11
electroluminescent lamp: 2–11
energy diagram: 2–8
excitation of: 2–7 to 2–9
fluorescent lamp: 2–8, 2–9, 8–18
high-intensity discharge lamp: 8–33
mercury lamp: 2–9, 8–33

PHOT: 1–14, 1–27

PHOTOCELLS (see **PHOTOELECTRIC CELLS**)

PHOTOCHEMICAL LAMPS: 25–16

PHOTOCHEMICAL PROCESS IN EYE: 3–2, 3–3

PHOTOCHEMICAL REPRODUCTION PROCESSES: 24–35, 24–36

PHOTOCONDUCTOR CELLS: 2–13

PHOTOELECTRIC CELLS (see also **PHOTOVOLTAIC CELLS** and **PHOTOELECTRIC PHOTOMETERS**)
illumination meters: 4–5 to 4–7
physics of: 2–12, 2–13
relay devices using: 25–2, 25–3
spectral response of: 4–7, 25–2

PHOTOELECTRIC COLORIMETERS: 5–14

PHOTOELECTRIC DETECTION: 2–12

PHOTOELECTRIC EFFECT: 2–12

PHOTOELECTRIC PHOTOMETERS
Baumgartner reflectometer: 4–7, 4–8
exposure meters: 24–20, 24–21
Freund luminous spot meter: 4–5, 4–8
photoelectric tube meters: 4–8
photovoltaic cell meters: 4–5 to 4–7
Pritchard photometer: 4–5, 4–8

PHOTOELECTRIC TUBE METERS: 4–8, 4–9

PHOTOELECTRIC TUBES, physics of: 2–12, 2–13

PHOTOENGRAVING: 24–36

PHOTO-ENLARGER LAMPS: 8–16, 8–86

PHOTOFLASH LAMPS
definition: 1–14
description of: 8–16
exposure: 24–20
guide number system: 24–21
synchronizers for: 24–22
table of: 8–86
time-light curves: 24–22

PHOTOFLOOD LAMPS: 8–16, 8–85

PHOTOGRAPHIC LAMPS: 8–15, 8–16, 8–85 to 8–91

PHOTOGRAPHIC LIGHTING (see **PHOTOGRAPHY**)

PHOTOGRAPHY
ANSI designation for lamps: 24–4
application of light and equipment: 24–19 to 24–25
background luminance for: 24–23
color photography: 5–22 to 5–23
color sensitivity of films: 24–19, 24–20
darkroom lighting: 24–35, 24–36
equipment for: 24–1 to 24–4, 24–22
exposure: 24–19 to 24–22
exposure meters: 24–20
f-value of lens: 24–20
film speed: 24–20, 24–21
films for: 24–19
flash lamp photography: 24–21, 24–22
flash lamp synchronizers: 24–22
general illumination for: 24–22
guide number systems: 24–21
hard light: 24–1
illumination formula: 24–20
key light: 1–11, 24–1
key light/*f*-stop, for motion pictures: 24–23, 24–24
lamps for: 8–15, 8–16, 8–85, 8–86, 24–3
lighting equipment for: 24–1 to 24–3, 24–22, 24–25
lighting installation: 24–23
low-noise construction lamps for: 24–4
luminaires in installation: 24–23
manufacturers' ordering codes for lamps: 24–4
meters, exposure: 24–20, 24–21
modeling light for: 24–23
motion picture photography: 24–23 to 24–25
reciprocity law: 24–20
set lighting, motion picture: 24–4, 24–23, 24–24
soft light: 24–1
spectral sensitivity of films: 24–19
studio lighting equipment: 24–1 to 24–4, 24–22
techniques: 24–22, 24–23
T-value, of lens: 24–20
underwater photography: 26–3, 26–4, 26–5

PHOTOGRAVURE: 24–36

PHOTOLITHOGRAPHY: 24–36

PHOTOLUMINESCENCE: 2–7, 2–8, 25–13

PHOTOMETERS
bar: 4–9
Baumgartner reflectometer: 4–7, 4–8
Bunsen "grease spot" disk: 4–4
definitions: 1–14, 4–1
distribution (gonio): 4–9
flicker photometer: 4–4
for electric discharge lamp measurements: 4–14
Freund luminous spot meter: 4–5, 4–8
integrating (Ulbricht) sphere: 4–9
Luckiesh-Taylor luminance meter: 4–5
Lummer-Brodhun cube: 4–4
Macbeth illuminometer: 4–4, 4–5
photoelectric: 4–5 to 4–8
photoelectric tube meters: 4–8
photovoltaic cell: 4–5 to 4–7
Pritchard photometer: 4–5, 4–8
radiometric: 4–8
spectrophotometers: 4–10
Taylor reflectometer: 4–5
visual photometers: 4–4, 4–5
visual task photometer: 4–8

PHOTOMETRIC BRIGHTNESS (see **LUMINANCE**)

PHOTOMETRIC MEASUREMENTS AND TESTS (see also **PHOTOMETRY**)
automobile headlamps: 22–1, 22–2, 22–3
color appearance: 4–14, 5–14 to 5–16
electric discharge lamps: 4–14
floodlamp-type luminaires: 4–22
fluorescent lamps: 4–14 to 4–15
general lighting luminaires: 4–16 to 4–22
high-intensity discharge lamps: 4–14
incandescent filament lamps: 4–14, 4–15
luminaires: 4–16, 4–22
mercury lamps: 4–14
reflector-type lamps: 4–15
searchlights: 4–23

PHOTOMETRIC UNITS, SYMBOLS, AND EQUATIONS: 1–2, 1–3

PHOTOMETRY (see also **PHOTOMETERS; MEASUREMENT, FIELD; MEASUREMENT, LABORATORY;** and **PHOTOMETRIC MEASUREMENT AND TESTS**)
attenuation means: 4–3
basis of: 4–1
direct photometry: 4–3
general methods: 4–3
luminaire: 4–16, 4–22
projector: 4–23
relative photometry: 4–3
substitution photometry: 4–3

PHOTOMORPHOGENESIS: 25–3

PHOTOMULTIPLIER TUBE METERS: 4–8

PHOTOPERIODIC LIGHTING: 25–7

PHOTOPERIODISM: 25–3, 25–5, 25–7

PHOTOPIC VISION: 3–4, 4–1

PHOTOSYNTHESIS: 25–2, 25–3, 25–4, 25–5

PHOTOSYNTHETIC LIGHTING: 25–5

PHOTOTROPISM: 25–3

PHOTOTUBES, spectral response: 25–2

PHOTOVOLTAIC CELL METERS
accuracy of: 4–7
color connection: 4–6, 4–7
cosine effect: 4–6, 4–7
effect of frequency: 4–7
effect of temperature: 4–7
fatigue of: 4–6
physics of: 2–13
spectral response: 4–6

PHYSICAL PHOTOMETERS (see **PHOTOELECTRIC PHOTOMETERS**)

PHYSICS OF LIGHT: 2–1 to 2–13

PICTURE PROJECTION
aperture dimensions: 24–30
auditoriums: 24–26
classrooms: 24–26, 24–31, 24–32
drive-in theatre: 24–26
image dimensions: 24–29, 24–30, 24–31
lamps for: 8–87 to 8–91
light output of projectors: 24–30
luminance levels: 24–25, 24–26
measurement of luminance: 24–25, 24–26
motion picture theatre: 24–32
picture dimensions: 24–29, 24–30, 24–31
projection booths: 24–30
review rooms: 24–26
screen illumination: 24–33
screen requirements for classrooms: 24–32
screens: 24–27, 24–28
slide projection: 24–26
surround illumination: 24–27
theatre: 24–25, 24–26, 24–32, 24–33, 24–34
viewing angles: 24–28, 24–29
viewing distances: 24–28
wide screen: 24–30
xenon sources, use of: 24–34

PIERS, illumination levels: 9–91

PIEZOLUMENESCENCE: 2–12

PISTOL RANGE
illumination levels: 9–93
lighting for: 19–4, 19–8, 19–9, 19–10

PLANCKIAN LOCUS: 5–11

PLANCKIAN RADIATOR, spectral energy distribution: 5–13

PLANCK'S CONSTANT: 2–1

PLANCK'S EQUATION: 2–3, 2–4

PLAN-SCALE METHOD, of point calculation: 9–49, 9–50

PLANT GROWTH (see **HORTICULTURAL LIGHTING**)

PLANT GROWTH CHAMBERS: 25–7, 25–8

PLANT GROWTH LAMPS: 25–4

PLASTER
characteristics of: 6–6
ultraviolet reflectance of: 25–2, 25–18

PLASTICS
for fenestration: 7–9
maintenance of: 16–14
properties of those used in lighting: 6–18
transmittance of various: 7–9
used in light control: 6–15
uses of important resins in lighting: 6–17

PLATING, illumination level: 9–87

PLAYGROUNDS (see also **SPORTS LIGHTING**), illumination level: 9–93

POINT CALCULATION METHODS, for illumination: 9–44 to 9–61

POINT SOURCE CALCULATIONS: 9–44 to 9–49, 9–56

POLARIZATION
Brewster's angle: 6–2, 6–8
definition: 1–15
dichroic polarizers: 6–8
effect on photoelectric detection: 2–12
in reducing veiling reflections: 3–32
in sky: 2–6
measurement of: 4–2
multilayer polarizers: 6–8
per cent polarization: 6–7

POLARIZED LIGHT
for improving underwater vision: 26–3
graphic representation of: 6–8
industrial inspection with: 14–10
per cent vertical polarization: 6–7
production of: 6–7, 6–8

POLARIZERS: 3–32, 6–8

POLICE (see **MUNICIPAL BUILDINGS**)

POLISHING AND BURNISHING, illumination levels: 9–87

POOLS, GARDEN: 17–7, 17–8

POOLS, SWIMMING (see **SWIMMING POOLS**)

PORCELAIN ENAMEL, maintenance of: 16–14

PORTABLE LAMPS, residential type: 15–9 to 15–12

POSITION INDEX, in glare evaluation: 1–15, 3–24

POST OFFICES, illumination levels: 9–87

POSTER PANELS
illumination levels: 9–90
lighting for: 23–7 to 23–12

POULTRY INDUSTRY
illumination levels: 9–87
lighting for: 14–50 to 14–52

POWER
conversion equations: 1–27
requirements for fluorescent lamps: 8–101
sources for miniature lamps: 8–14
sources for railroad signals: 22–17
sources for transportation lighting: 22–11
systems for aircraft: 21–1, 21–5
systems for airports: 21–11

POWER PLANTS (see **CENTRAL STATION PROPERTIES**)

PREHEAT FLUORESCENT LAMPS: 8–28, 8–29, 8–96, 8–97

PRESCRIBING ILLUMINATION (see **METHOD FOR PRESCRIBING ILLUMINATION**)

PRESSING AND CLEANING
illumination level: 9–82
lighting for: 14–31 to 14–33

PRIMARY LINE OF SIGHT: 1–15, 3–6

PRIMARY STANDARDS: 1–15, 4–2

PRINTING INDUSTRY
illumination levels: 9–87
lighting for: 5–23, 5–24, 14–24 to 14–26

PRISMS: 6–4, 6–5, 6–7

PRISON YARDS, illumination level: 9–91

PRITCHARD PHOTOMETER: 4–5, 4–8

PROFESSIONAL OFFICES (see **HOSPITAL LIGHTING** and **OFFICE LIGHTING**)

PROJECTION (see **PICTURE PROJECTION**)

PROJECTION BOOTHS: 24–30

PROJECTION LAMPS: 8–16, 8–87 to 8–91

PROJECTION SCREENS
dimensions of: 24–29, 24–30
types of: 24–27, 24–28

PROJECTOR PHOTOMETRY: 4–23

PROJECTORS, PICTURE
aperture dimensions: 24–30
aspect ratio: 24–30
lamps for: 8–87 to 8–91
required light output: 24–30 to 24–32
theatre: 24–32 to 24–35

PROTECTIVE LIGHTING: 18–5 to 18–8

PUBLIC BUILDING LIGHTING
art galleries: 12–18 to 12–24
banks: 12–1, 12–2
churches and synagogues: 12–2 to 12–4
commercial kitchens: 12–4, 12–5
dining areas: 12–5, 12–6
health care facilities: 12–6 to 12–13
hospitals: 12–6 to 12–13
hotels and motels: 12–13, 12–14
libraries: 12–15 to 12–18
museums: 12–18 to 12–24
theatres: 24–6 to 24–12

PUBLIC CONVEYANCE LIGHTING (see also **TRANSPORTATION LIGHTING**)
aircraft: 21–1 to 21–6
electric power systems for: 22–11, 22–12
illumination levels: 9–94, 9–95, 22–5
marine lighting: 22–12 to 22–15
rail conveyances—inter-city and commuter: 22–8 to 22–10
rail conveyances—rapid transit: 22–7, 22–8
road conveyances—city, inter-city, and school: 22–6, 22–7

PUBLIC GARDENS, lighting of: 17–6, 17–7

PULSED XENON ARC LAMPS: 8–45

PUNCH PRESS, lighting of: 14–19, 14–20

PUPIL SIZE, of eye: 3–3, 3–4, 3–11, 3–12

PURITY, EXCITATION: 5–7

PURKINJE PHENOMENON: 1–15, 3–6

PXA LAMPS: 8–45

PYROLUMINESCENCE: 2–7

Q

QUALITY OF ILLUMINATION
disability glare: 3–10
discomfort glare: 3–24 to 3–27
industrial lighting: 14–2 to 14–4
office lighting: 11–2 to 11–5
residential lighting: 15–2, 15–3
school lighting: 11–11, 11–12
shadows: 3–13
sports lighting: 19–2
veiling reflections: 3–28 to 3–33

QUANTITY OF ILLUMINATION (see **ILLUMINATION LEVELS**)

QUANTUM THEORY: 2–1, 2–13

QUARRIES, illumination level: 9–91

QUARTZ-IODINE LAMPS (see **TUNGSTEN-HALOGEN LAMPS**)

QUOITS, illumination level: 9–93

R

R LAMPS, incandescent: 8–11, 8–12, 8–73

RACING, illumination levels: 9–93

RADAR (SCREEN VIEWING ROOMS): 24–18, 24–19

RADIAN, value in degrees: 1–27

RADIANT ENERGY (see also **BACTERICIDAL ULTRAVIOLET RADIATION, ERYTHEMAL ULTRAVIOLET RADIATION, INFRARED RADIANT ENERGY, SPECTRAL ENERGY DISTRIBUTION,** and **ULTRAVIOLET RADIANT ENERGY**)
absorption of: 25–2
applications of: 25–1 to 25–23
definition: 1–16
effect on luminaires: 6–22
evaluation of: 2–1
luminous efficiency of: 2–5
measurement: 25–4, 25–5
reflection of: 25–2
response of photoelectric cells: 25–2
solar radiant power density: 25–1
sources of: 25–1
spectrum: 2–2
symbol and units: 1–2
transmission of: 25–2

RADIANT FLUX: 1–2, 1–16, 2–1

RADIATION EQUATIONS: 2–4

RADIATORS, SELECTIVE: 2–4

RADIO FREQUENCY (RF) LAMP: 8–17

RADIO INTERFERENCE, from fluorescent lamps: 8–28

RADIO PANEL LAMPS: 8–15, 8–84

RADIOACTIVE LUMINESCENCE: 25–14

RADIOLUMINESCENCE: 2–12, 25–14

RADIOMETERS (see **RADIOMETRIC PHOTOMETERS**)

RADIOMETRIC PHOTOMETERS: 4–8

RAIL CONVEYANCES
illumination levels: 9–94
lighting for: 22–7 to 22–10
power supplies for: 22–11, 22–12

RAILROAD CROSSINGS: 20–10, 20–12

RAILROAD LIGHTING (see **RAIL CONVEYANCES** and **RAILWAY GUIDANCE SYSTEMS**)

RAILROAD YARDS
illumination levels: 9–91
lighting of: 14–47, 14–48

RAILWAY GUIDANCE SYSTEMS
control panels for signal system: 22–17
exterior lights on trains: 22–15
grade crossing lights: 22–19
locomotive cab signals: 22–16
locomotive headlights: 22–15, 22–16
rapid transit cab signals: 22–16
traffic control panels: 22–17, 22–18
wayside signals: 22–15 to 22–19

RAILWAY LAMPS: 8–11, 8–62

RAILWAY MAIL CARS, illumination levels: 9–94

RAILWAY PASSENGER CAR LIGHTING (see **RAIL CONVEYANCES**)

RAPID START FLUORESCENT LAMPS: 1–16, 8–29, 8–30, 8–94, 8–95

RAPID TRANSIT CARS
illumination levels: 9–94
lighting for: 22–7, 22–8
power supply for: 22–11, 22–12

RAYLEIGH SCATTERING EFFECT: 2–6, 5–22

RCS (see **RELATIVE CONTRAST SENSITIVITY**)

REACTOR, ballast type: 6–25

READABILITY, of office tasks: 11–2

READING, HOME, lighting for: 15–15, 15–16

READING AREAS, of libraries: 12–16

READING LIGHTS, aircraft: 21–5, 21–6

RECEIVING AND SHIPPING (see **MATERIALS HANDLING**)

RECEPTION ROOM LIGHTING: 11–9, 11–10

RECESSED LIGHTING, for residences: 15–3 to 15–8

RECIPROCITY LAW, Photographic: 24–20

RECOMMENDED ILLUMINATION LEVELS, master table: 9–81 to 9–95

REFERENCE STANDARDS: 1–16, 4–2

REFLECTANCE
aluminum, of light and infrared: 6–6, 6–17, 16–4
aluminum, ultraviolet: 25–2, 25–19
apparent, of fluorescent materials: 25–13
building materials: 6–6, 7–9, 7–10
calcium carbonate: 25–2
characteristics of materials: 6–6
chromium: 6–6, 25–19
effect on daylight illumination: 7–10
effect on utilization of light in interiors: 5–16 to 5–18
effective ceiling and floor cavity: 9–8, 9–10, 9–11
enamel, ultraviolet: 25–2, 25–19
exterior surfaces: 7–10
finishes, luminaire: 6–20
glass, of light and infrared: 6–6, 16–4
human skin: 25–16
illumination level, relation to: 5–17
infrared, of various materials: 16–4
interiors of public conveyances: 22–5
limestone: 6–6
luminous: 5–7
magnesium oxide: 25–2
measurement of: 4–2, 4–5, 4–7, 4–8
paints, various: 6–6, 25–2
plaster: 6–6, 25–2, 25–19
plastics, of light and infrared: 16–4
porcelain enamel: 6–6, 16–4
spectral, of human skin: 25–16
spectral, of various materials: 25–19
steel: 6–6, 25–2, 25–19
surfaces in dental operatories: 12–7
surfaces in surgical suites: 12–11
surfaces for offices: 11–5
surfaces in industries: 14–3
surfaces in residences: 15–2, 15–3
surfaces in schools: 11–13
symbol and unit: 1–2
test plate: 25–2
white enamel, of light and infrared: 16–4

REFLECTED COMPONENT in illumination calculations: 9–60

REFLECTED GLARE (see also **VEILING REFLECTIONS**)
cause of: 3–27
definition: 1–16
from various lighting systems: 10–11, 10–12
in industrial lighting: 14–3
in offices: 11–4
in schools: 11–12
in sports lighting: 19–6
reduction of: 3–27

REFLECTION
air-glass surface: 6–2
characteristics of finishes: 6–20
compound reflection: 6–2
curved surfaces: 6–2
definition: 1–16
diffuse reflection: 1–7, 6–2
fiber optics: 6–3
low reflectance films: 6–9
reflex reflectors: 22–3 to 22–5
regular, definition: 1–16
specular reflection: 6–1, 6–2
spread reflection: 6–2
total reflection: 6–3

REFLECTION FACTOR (see **REFLECTANCE**)

REFLECTIONS, in store windows: 13–2, 13–7, 13–8

REFLECTOMETERS
Baumgartner: 4–7, 4–8
Taylor: 4–5

REFLECTORIZED LAMPS
fluorescent: 8–21
incandescent: 8–11, 8–12, 8–73
photometric measurements of: 4–15

REFLECTORS
basic contours: 6–10, 6–11
contour determination: 6–11, 6–12
design of: 6–10 to 6–12, 6–22
diffuse: 6–2, 6–11
diffuse-specular: 6–2
ellipsoidal: 6–10
general contours: 6–11, 6–12
hyperbolic: 6–10, 6–11
parabolic: 6–10
photographic lighting: 24–22
reflex: 22–3 to 22–5
specular: 6–1, 6–11, 6–12
spherical: 6–11
spread: 6–2

REFLEX REFLECTORS: 22–3 to 22–5

REFRACTION
definition: 1–16
dispersion: 6–4, 6–5
examples of: 6–4
index of: 6–3, 6–4
law of: 6–4
prisms: 6–4
Snell's law: 6–4

REFRACTION-REFLECTION SYSTEMS: 6–13, 6–14

REFRACTORS
definition: 1–16
design of: 6–13, 6–14
Fresnel lenses: 6–4, 6–5
prisms: 6–4, 6–5, 6–7
ribbed surfaces: 6–5

RELAMPING, PLANNED (see also **GROUP REPLACEMENT**): 16–13

RELATIVE CONTRAST SENSITIVITY
as function of luminance: 3–20, 3–21, 3–22
determination of: 3–20, 3–23
table of values: 3–21, 3–22

RELATIVE ENERGY DISTRIBUTION OF SOURCES A, B, C: 1–19

RELATIVE PHOTOMETRY: 4–3

RELATIVE SPECTRAL LUMINOUS EFFICIENCY CURVES: 3–6

REPRODUCTION PROCESS LIGHTING: 24–35, 24–36

RESEARCH AGENCIES, Light and Vision: 3–1

RESIDENCE LIGHTING
application of recessed lighting: 15–5 to 15–7
art objects: 15–27
bathroom: 15–13, 15–14, 15–20
bookcases: 15–28
ceiling-mounted luminaires: 15–4, 15–7, 15–8
color of surfaces: 15–2
cornices: 15–13
coves: 15–14
decorative accent: 15–26 to 15–29
dining areas: 15–29
distribution of light: 15–3
downlighting: 15–4 to 15–8
dressing table lamps: 15–11
entrance, exterior: 15–26, 15–29
environment objectives: 15–1 to 15–3
equipment: 15–3 to 15–14
exteriors and grounds: 15–26, 15–29, 15–30
fireplaces: 15–28
floor and table lamps: 15–9 to 15–12
general lighting: 15–15, 15–26
grooming: 15–20
grounds: 15–29, 15–30
illumination levels: 9–88, 15–15 to 15–25
ironing: 15–19
kitchen activities: 15–17, 15–18
laundry: 15–18
light as element of design: 15–1
luminaires: 15–3 to 15–14
luminance ratios: 15–2, 15–7
luminous ceilings: 15–14
luminous walls: 15–14, 15–28
makeup: 15–20
measuring shade luminance: 15–11
mirror lighting: 15–20
music study: 15–24, 15–25
niches: 15–27
outdoor living spaces: 15–29
painting, tapestries, murals: 15–27
pendant luminaires: 15–8, 15–9
planters: 15–27, 25–8, 25–9
portable lamps: 15–9 to 15–12
quality of illumination: 15–2, 15–3
quantity of light: 9–88, 15–3
reading: 15–15, 15–16
recessed lighting: 15–3 to 15–8
reflectances of surfaces: 15–2, 15–3
reflected glare: 15–3
seeing zones in: 15–2
sewing: 15–24, 15–25
sculpture: 15–27
shaving: 15–20
soffit lighting: 15–13
structural lighting elements: 15–12 to 15–14
study desk: 15–16
study lamps: 15–9, 15–10
surface mounted luminaires: 15–4, 15–7, 15–8
table tennis: 15–23
valances: 15–13, 15–14
veiling reflections in tasks: 15–3
visual tasks: 15–12, 15–15 to 15–26
wall brackets: 15–13
wall-mounted luminaires: 15–9
workbench: 15–21
writing: 15–16

RESISTANCE DIMMERS: 16–9

RESTAURANT-THEATRES AND LOUNGES: 24–10, 24–11

RESTAURANTS, LUNCH ROOMS, CAFETERIAS
illumination levels: 9–84
lighting of: 12–4 to 12–6

RESULTANT COLOR SHIFT: 1–17, 5–19

RETRO-REFLECTING DEVICES (see **REFLEX REFLECTORS**)

RF (RADIO FREQUENCY) LAMP: 8–17

RIBBON FILAMENT LAMPS: 8–13, 8–14

RIFLE AND PISTOL RANGES
illumination levels: 9–93
lighting for: 19–4, 19–8, 19–9, 19–10

ROAD CONVEYANCES—CITY, INTER-CITY AND SCHOOL
advertising card lights: 22–6
aisle lighting: 22–6
boarding and exiting lighting: 22–6
fare collection lighting: 22–6
illumination levels: 9–94
reading lighting: 22–7
typical lighting methods: 22–7

ROADWAY ILLUMINATION CALCULATIONS: 9–71 to 9–74

ROADWAY LIGHTING
alleys: 20–8
area classification: 20–2
benefit of: 20–1
calculations: 9–71 to 9–74
classification of areas and roadway: 20–2
classification of luminaire light distributions: 20–3
control of luminaire distribution: 20–5
converging traffic lanes: 20–10, 20–11
curves: 20–9
daytime lighting of tunnels: 20–13, 20–14
design of: 20–7 to 20–14
disability glare: 20–1, 20–2
discernment of roadway objects: 20–1
discomfort glare: 20–1, 20–2
diverging traffic lanes: 20–10, 20–11
effect of trees on: 20–10
factors influencing visibility: 20–1
fluorescent lamps for: 8–94, 8–95
glare in: 20–1, 20–2
high-intensity discharge lamps for: 8–102 to 8–107
hills: 20–9
illumination calculations: 9–71 to 9–74
illumination levels: 20–6, 20–13, 20–14
illumination measurement: 4–29, 4–30
incandescent lamps for: 8–13, 8–92, 8–93
interchanges: 20–10, 20–11
isocandela diagrams, use of: 20–3, 20–4
lateral light distribution: 20–3
light sources for: 8–13, 8–92 to 8–95, 8–102 to 8–107, 8–109, 20–6
luminaire light distributions: 20–3 to 20–5
luminaire mounting heights: 20–7
luminaire spacing: 20–8
nighttime lighting of tunnels: 20–14
pedestrian walkways: 20–9
railroad grade crossings: 20–10, 20–12
roadway classifications: 20–2
silhouette discernment: 20–1
sodium lamps for: 8–109
surface detail discernment: 20–1
transition lighting: 20–8
tree pruning: 20–10, 20–13
tunnels and underpasses: 20–10 to 20–14
vertical light distribution: 20–3
visibility: 20–1

ROADWAY LIGHTING LAMPS
fluorescent: 8–94, 8–95
high-intensity discharge: 8–102 to 8–107
incandescent: 8–92, 8–93
sodium: 8–109

RODEO, illumination levels: 9–93

RODS, of the eye: 1–17, 3–2, 3–3, 3–4, 3–5

ROOM CAVITY RATIO
definition: 1–17
formula for: 9–8
table of: 9–9

ROOM INDEX (see **CAVITY RATIO** and **ROOM CAVITY RATIO**)

ROOM POSITION MULTIPLIERS (RPM): 9–60

ROOM RATIO (see also **ROOM CAVITY RATIO**), for toplighting calculations: 9–76

ROOM SURFACE CHARACTERISTICS
effect of reflectance on illumination: 5–17, 5–18
effect on daylighting: 7–9, 7–10
industrial: 14–3
office: 11–5, 11–6
school: 11–5, 11–13

ROOM SURFACE DIRT DEPRECIATION (RSDD)
in maintenance: 16–12, 16–13
interior calculations: 9–4

ROQUE, illumination levels: 9–92

ROUGH SERVICE INCANDESCENT LAMPS: 8–12, 8–63

RUBBER GOODS—MECHANICAL, illumination levels: 9–88

RUBBER SHOE MANUFACTURING
illumination levels: 9–88
lighting for: 14–34

RUBBER TIRE MANUFACTURING
illumination levels: 9–88
lighting for: 14–22 to 14–24

RUBY LASER: 2–10

S

SAE SPECIFICATIONS, motor vehicle exteriors: 22–3

SAFETY
considerations in equipment design: 6–27, 6–28
emergency lighting: 14–10 to 14–12, 14–46
protective lighting: 18–5
roadway lighting: 20–1, 20–6
school lighting: 11–16

SAWMILLS
illumination level: 9–88
lighting of: 14–46, 14–47

SCAFFOLDING, for maintenance: 16–14, 16–15

SCATTERING COEFFICIENT, water: 26–1

SCHOOL BUS LIGHTING: 22–6

SCHOOL LIGHTING
art rooms: 11–13
auditoriums: 11–13, 11–14
cafeterias: 11–14
chalkboards: 11–14
corridors: 11–14
daylighting in: 7–10
direct glare: 11–11
dormitory rooms: 11–14
drafting rooms: 11–14, 11–15
emergency lighting: 11–16
for partially seeing and hearing: 11–14
for safety: 11–16
general purpose rooms: 11–18, 11–19
illumination levels recommended: 9–88
improvement in visual tasks: 11–11
kitchens: 11–14
laboratories: 11–15
lamps for: 11–12, 11–13
lecture rooms: 11–15
locker rooms: 11–15
luminaire luminance: 11–4
luminance ratios: 11–12
quality of illumination: 11–11, 11–12
quantity of illumination: 9–88, 11–2
reflectances recommended: 11–13
reflected glare: 11–12
sewing rooms: 11–15
shops: 11–15
stairs: 11–15
supplementary lighting: 11–9, 11–11, 11–14
veiling reflections: 11–11
visual aids: 11–15, 11–16
visual environment: 11–10, 11–11
visual tasks: 11–11

SCOTOPIC VISION: 1–17, 3–4, 4–1

SCREEN VIEWING ROOMS (RADAR): 24–18, 24–19

SCREENS, PICTURE PROJECTION
dimensions: 24–29 to 24–32
types: 24–27, 24–28

SCULPTURE, lighting of
in art galleries and museums: 12–19, 12–20
in residences: 5–27
outdoor floodlighting: 17–6

SEA WATER, filtering properties: 26–2

SEALED BEAM LAMPS: 8–14, 8–15, 8–83, 22–1, 22–2

SEARCHLIGHT CALCULATIONS: 9–69 to 9–71

SEARCHLIGHTS
aircraft: 18–2
atmospheric light absorption: 9–71
beam divergence: 9–70, 18–3, 18–4
beam spread: 9–70
calculations: 9–69 to 9–71
candlepower: 9–70
characteristics: 18–4
marine navigation: 18–1, 18–2
military: 18–2, 18–3
photometry of: 4–23 to 4–25
protective lighting: 18–6 to 18–8
range: 9–71
signaling: 18–1, 18–2, 18–4

SECONDARY STANDARDS: 1–17, 4–2

SECTOR DISK PHOTOMETRY: 4–3

SEEING DISTANCE IN WATER: 26–2

SEEING TASKS (see **VISUAL TASKS**)

SELECTIVE RADIATORS: 2–4

SELENIUM CELLS (see **PHOTOVOLTAIC CELLS**)

SELF-BALLASTED MERCURY LAMPS: 8–40, 8–108, 16–20

SELF-LUMINOUS MATERIALS: 25–13, 25–14

SEMI-DIRECT LIGHTING: 1–17, 10–11, 10–12

SEMI-DIRECT LUMINAIRES: 10–11, 10–12

SEMI-INDIRECT LIGHTING: 1–17, 10–11, 10–12

SEMI-INDIRECT LUMINAIRES: 10–11, 10–12

SENSIBLE HEAT: 16–1, 16–4

SERIES STREET LIGHTING INCANDESCENT LAMPS: 8–13, 8–74

SERVICE SPACE, illumination levels: 9–88

SERVICE STATIONS
illumination levels: 9–88, 9–91
lighting of: 17–8 to 17–10

SET LIGHT: 1–17, 24–16

SEWING
home: 15–24, 15–25
school: 11–15

SHADOW BOXES, store lighting: 13–2

SHADOWS
effect on visual ability: 3–13
in floodlighting: 17–2, 17–3
in office lighting: 11–8
under various lighting systems: 10–11, 10–12
use in industrial lighting: 14–3, 14–10

SHEET METAL SHOPS
illumination levels: 9–89
lighting of: 14–19, 14–20

SHELF AND GARMENT CASES: 9–63, 13–7

SHIELDING ANGLE: 1–17, 6–9, 6–10

SHIP LIGHTING (see **MARINE LIGHTING**)

SHIP YARDS, illumination levels: 9–91

SHOE MANUFACTURING
illumination levels: 9–89
lighting for: 14–33, 14–34

SHORT-ARC LAMPS
application of: 8–43
auxiliary equipment: 8–43
efficacy: 8–42
mercury and mercury-xenon lamps: 8–41, 8–42, 8–114, 8–115
mercury-argon lamps: 8–41, 8–114, 8–115
operating enclosures: 8–42
spectral distribution curves: 8–41, 8–42
xenon lamps: 8–41, 8–42, 8–114, 8–115

SHOW WINDOW LIGHTING
accent lighting: 13–9
calculations: 9–64 to 9–67
colored light: 13–9
daytime effectiveness: 13–7, 13–8
equipment: 13–8, 13–9
illumination levels: 9–89, 13–4
nighttime effectiveness: 13–8
open-front stores: 13–1
requirements: 13–8

SHOWCASE INCANDESCENT LAMPS: 8–13, 8–59

SHOWCASE LIGHTING: 9–67 to 9–68, 13–7

SHUFFLEBOARD, illumination levels: 9–93

SI (or METRIC) UNITS, conversion factors: 1–24, 1–25, 1–27

SIDELIGHTING, daylighting: 9–77 to 9–79

SIGN LAMPS
comparison colored-uncolored: 23–3
for exposed lamp signs: 23–3
for luminous elements: 23–4, 23–5
types of: 8–13
wattages for various locations: 23–3

SIGNAL LIGHTS
automobile: 22–2, 22–3
marine: 22–15
railroad: 22–15 to 22–19

SIGNALS, LUMINOUS
Allard's law: 3–33
atmospheric effects on: 3–34
automobile: 22–2, 22–3
colored lights: 3–36
flashing lights: 3–36, 3–37
luminances of backgrounds: 3–36
railroad: 22–15 to 22–19
size of: 3–36
threshold illumination: 3–35
visual range of: 3–33 to 3–37

SIGNS
attraction power of exposed lamp signs: 23–4
brightness of: 23–1, 23–5
bulletin boards: 23–7 to 23–12
characteristics of: 23–1
color in: 23–1
colored lamps in: 23–3, 23–4
daytime effective exposed lamp signs: 23–3, 23–4
dimensions of standard poster panels and bulletin boards: 23–8
exposed incandescent lamp signs: 23–1
floodlighted signs: 23–7 to 23–12
floodlighting of signs and other vertical surfaces: 23–7 to 23–12
illumination levels for poster panels and bulletin boards: 9–90, 23–8
illumination on vertical surfaces from fluorescent floodlights: 23–11
incandescent (tungsten-halogen) floodlighting of signs: 23–10, 23–11, 23–12
lamp spacing in: 23–2, 23–5
lamps for: 8–13
legibility: 23–1, 23–4
letter size: 23–1, 23–2, 23–6
letter spacing: 23–6
light sources for floodlighted signs: 23–10
location of floodlighting equipment: 23–8
luminance for luminous background signs: 23–5, 23–6
luminous element signs: 23–4 to 23–7
luminous building fronts: 23–7
luminous tube signs: 23–4
poster panels: 23–7 to 23–12
temperature effect on fluorescent floodlights: 23–10, 23–11
transformers for luminous tubing: 23–4
wattage of lamps used in: 23–3

SILHOUETTE
discernment by, on roadways: 20–1
signs: 23–4 to 23–7
use in inspection: 14–10

SILICON CONTROLLED RECTIFIER (SCR) DIMMERS: 16–10, 16–11

SILK TEXTILE MILLS, illumination levels: 9–89

SILVERED-BOWL LAMPS: 8–60

SKATING, illumination levels: 9–93

SKEET, illumination levels: 9–93

SKI SLOPE, illuminatin levels: 9–93

SKIN, HUMAN
reflectances at various wavelengths: 25–16
transmittance at various wavelengths: 25–16

SKY
as a light source: 7–1, 7–2
equivalent luminance of: 7–6, 7–7
illumination from: 7–8
luminance: 3–36, 8–55

SKY LIGHT: 2–6, 7–7, 7–8

SKYLIGHT
building section: 7–5, 7–6
calculation of illumination: 9–76

SLIDE PROJECTION: 24–26

SLIMLINE LAMPS: 8–98, 8–99, 25–15, 25–17

SNELL'S LAW: 6–4

SOAP MANUFACTURING, illumination levels: 9–89

SOCCER, illumination levels: 9–92

SOCIETY OF AUTOMOTIVE ENGINEERS (see **SAE SPECIFICATIONS**)

SODIUM LAMPS (see **HIGH PRESSURE SODIUM LAMPS** and **LOW PRESSURE SODIUM LAMPS**)

SODIUM PHOTOTUBE, spectral response: 25–2

SOFFIT LIGHTING: 15–13

SOFTBALL
illumination levels: 9–93
lighting for: 19–12, 19–16

SOLAR ALTITUDE: 7–1, 7–3, 7–4

SOLAR AZIMUTH: 7–1, 7–3, 7–4

SOLAR ILLUMINATION
as function of altitude: 7–5
footcandle level: 2–6

SOLAR SIMULATORS
carbon arcs for: 8–41
characteristics of: 18–4, 18–5

SOLAR TIME: 7–4

SOLID LASERS: 2–10

SONOLUMINESCENCE: 2–12

SOURCE A
relative energy distribution of: 1–19
spectral power distributions: 5–7
standard, definition: 1–19
tristimulus computational data: 5–8

SOURCE B
relative energy distribution of: 1–19
spectral power distributions: 5–7
standard, definition: 1–19
tristimulus computational data: 5–8

SOURCE C
relative energy distribution of: 1–19
spectral power distribution: 5–7
standard, definition: 1–19
tristimulus computational data: 5–8

SOURCE D_{55}
spectral power distributions: 5–7
tristimulus computational data: 5–8

SOURCE D_{65}
spectral power distributions: 5–7
standard, definition: 1–19
tristimulus computational data: 5–8

SOURCE D_{75}
spectral power distributions: 5–7
tristimulus computational data: 5–8

SOURCES (see also **LIGHT SOURCES**)
bactericidal ultraviolet: 8–110, 25–14 to 25–17
erythemal ultraviolet: 8–100, 25–14, 25–15
infrared: 8–16, 8–17, 8–92, 8–93, 25–20, 25–22
ozone-producing ultraviolet: 25–15, 25–17
relative energy distribution: 1–19
ultraviolet energy: 8–100, 8–109 to 8–113, 25–14 to 25–17

SOUTHERN BUILDING OFFICIALS' CONFERENCE OF AMERICA, local codes of: 6–21

SPACING-TO-MOUNTING HEIGHT RATIOS
classification of luminaires by: 14–5
determination of: 9–34
of typical luminaires: 9–12 to 9–28
use of: 9–33, 9–35

SPECTRAL DISTRIBUTIONS
BL fluorescent lamp: 25–15
blackbody at various temperatures: 2–3, 2–4
carbon arc sources: 8–52
CIE standard sources: 5–7
fluorescent lamps: 8–19, 8–24
graybody: 2–4
high-intensity discharge lamps: 8–20, 8–34, 8–35, 8–36
incandescent lamps: 8–2
light emitting diodes: 8–49
mercury-xenon lamps: 8–41
motion picture sources: 24–24, 24–32
ozone-producing bactericidal lamp: 25–15, 25–17
Planckian: 5–13
short arc lamps: 8–41, 8–42
solar radiation: 25–1
sources A, B, C: 1–19
tungsten: 2–4
xenon flashtube: 8–45
xenon lamps: 8–42

SPECTRAL EMISSIVITY: 1–18, 2–3

SPECTRAL ENERGY DISTRIBUTION (see **SPECTRAL DISTRIBUTIONS**)

SPECTRAL LUMINOUS EFFICIENCY (see **EFFICIENCY, SPECTRAL LUMINOUS**)

SPECTRAL POWER DISTRIBUTION (see **SPECTRAL DISTRIBUTIONS**)

SPECTRAL REFLECTANCE (see **REFLECTANCE**)

SPECTRAL SENSITIVITY
human eye: 3–5, 3–6
photographic materials: 24–20
photovoltaic cells: 4–5 to 4–7

SPECTRAL TRANSMITTANCE (see **TRANSMITTANCE**)

SPECTRAL TRISTIMULUS VALUES, for equal power source: 5–3, 5–5

SPECTROPHOTOMETERS: 4–10

SPECTRORADIOMETERS: 4–10

SPECTRORADIOMETRY, color data from: 5–15

SPECTROSCOPE: 4–10

SPECTRUM
electromagnetic: 1–8, 2–2
radiant energy: 2–2
region attracting insects: 25–12
region effective in fading: 25–11
regions effective in major photoresponses of plants: 25–3
ultraviolet regions: 25–14
visible: 2–1, 2–2

SPECTRUM LOCUS: 1–18, 5–5

SPECULAR REFLECTION (see also **VEILING REFLECTIONS**): 6–1, 6–2

SPEED OF LIGHT: 1–19, 2–2

SPHERE ILLUMINATION: 1–19, 3–17, 3–18

SPHERE PHOTOMETER: 4–9, 4–10

SPHERICAL REFLECTORS: 6–11

SPORTS LIGHTING (see also under specific sport)
aerial sports: 19–4
aiming of floodlights: 19–14, 19–15, 19–18, 19–19
archery: 19–4, 19–8, 19–9
background brightness in: 19–1
badminton: 19–4, 19–6, 19–7
baseball: 19–10 to 19–14
basketball: 19–3, 19–4, 19–5, 19–6
billiards: 19–4, 19–6, 19–7
bowling: 19–4, 19–7
boxing: 19–4, 19–7, 19–8
combination sports field: 19–10, 19–11
curling: 19–4, 19–7, 19–8
daylighting for: 19–3
design factors: 19–4, 19–19
direction of light: 19–2
fencing: 19–4
field houses: 19–4, 19–5, 19–6
football: 19–11, 19–14
glare in: 19–2, 19–6
golf: 19–12, 19–15
gymnasiums: 19–3, 19–4, 19–5, 19–6
handball: 19–4, 19–7, 19–8
hockey: 19–4, 19–8, 19–9
illumination levels for: 9–91 to 9–93, 19–2, 19–3
illumination measurement procedure: 4–29, 4–30
indirect lighting: 19–9, 19–10
indoor sports: 19–4 to 19–10
jai alai: 19–4, 10 8, 19–9
junior league baseball: 19–10
light sources for: 19–3, 19–4
location of sources for: 19–2
low level sports: 19–4
luminaire location: 19–2, 19–6
measurement of: 4–29, 4–30
mounting height of floodlights: 19–14, 19–17, 19–18
objectives of: 19–1
objects to be seen: 19–1
observer location in: 19–1
outdoor sports: 19–10 to 19–19
over-voltage lamp operation: 19–3
over-wattage lamp operation: 19–4
pistol ranges: 19–4, 19–8, 19–9, 19–10
pole quantity and locations: 19–12
quality of illumination: 19–2

reflected glare: 19–6
rifle ranges: 19–4, 19–8, 19–9, 19–10
rugby: 19–11
shuffleboard: 19–4
skating: 19–4
soccer: 19–11
softball: 19–12, 19–16
squash: 19–4, 19–7, 19–8
stroboscopic effect: 19–4
swimming pools: 19–4, 19–9, 19–11
television field pick-ups: 24–14
tennis: 19–4, 19–9, 19–12, 19–17
uniformity of illumination: 19–2
volleyball: 19–4
windows in gymnasiums: 19–3, 19–6
wrestling: 19–4, 19–7, 19–8

SPOTLIGHT LAMPS: 8–11, 8–64 to 8–72

SPOTLIGHTS
motion picture and television: 24–1 to 24–3, 24–22
photography: 24–1 to 24–3, 24–22
stage lighting: 24–1 to 24–3, 24–7 to 24–19

SPREAD REFLECTION: 6–2

SPREAD TRANSMISSION: 6–7

SQUASH
illumination levels: 9–92
lighting for: 19–4, 19–7, 19–8

STACK LIGHTING
library: 12–16, 12–17
ship: 22–15

STAGE LIGHTING
control equipment: 24–5
lamps for: 8–64 to 8–72, 24–3, 24–4
location and function of equipment: 24–7 to 24–10
luminaires for: 24–1 to 24–3, 24–7 to 24–10
plan and section of average size theatre: 24–8

STAINED GLASS WINDOWS (see **ART WINDOWS**)

STAINLESS STEEL
for luminaires: 6–17, 6–19
reflectance: 25–2, 25–19

STAIRWAY LIGHTING
in offices: 11–9
in schools: 11–15

STANDARD OBSERVER
tristimulus computational data: 5–3, 5–8

STANDARD PRACTICES (see **CODES AND STANDARD PRACTICES**)

STANDARDS
AAR for color of railroad signals: 22–17, 22–18
aviation lighting: 21–1
color designation systems: 5–4 to 5–11
daylight: 5–21, 5–22
laboratory reference, definition: 1–11
national, definition: 1–14
primary standards: 1–15, 4–2
reference standards: 1–16, 4–2
SAE for motor vehicle lighting devices: 22–3
secondary standards: 1–17, 4–2
wiring: 16–9
working standards: 1–22, 4–3

STANDARD UNITS, SYMBOLS, AND EQUATIONS: 1–2, 1–3

STARTERS, LAMP: 6–26, 8–29, 8–30, 8–31

STATIONS AND TERMINALS, illumination levels: 9–83

STATUES (see also **SCULPTURE**), floodlighting of: 17–6

STEEL, for luminaires: 6–15, 6–17

STEEL MILLS
illumination levels: 9–86
lighting of: 14–12 to 14–14

STEPHAN-BOLTZMANN LAW: 2–3, 2–4

STILB: 1–19, 1–26

STILL-AIR COMFORT CHART: 16–2

STOKES LAW: 2–9

STONE CRUSHING AND SCREENING, illumination levels: 9–89

STORAGE BATTERY MANUFACTURING, illumination level: 9–89

STORAGE ROOMS OR WAREHOUSES, illumination levels: 9–89

STORAGE YARDS, illumination levels: 9–91

STORE LIGHTING
air conditioning and: 13–1, 13–2
architectural trends: 13–1, 13–2
canopies: 13–7
color: 13–2, 13–3
cornices: 13–2, 13–6
counter-top reflection: 13–6
creation of interest: 13–4
displays: 13–6, 13–7
economics: 13–4
exteriors: 13–9 to 13–12
factors in seeing merchandise: 13–3, 13–4
fading: 13–4, 25–10 to 25–12
fluorescent luminaires in: 13–5
general pattern method: 13–5, 13–6
illumination levels: 9–89, 13–4
incandescent luminaires in: 13–6
integral shelf lighting: 13–7
interior: 13–1 to 13–9
interior lighting design: 10–7
layout techniques: 13–4 to 13–6
light source selection: 13–2, 13–3
luminaires for: 13–5
merchandise emphasis: 13–4 to 13–7
merchandising techniques: 13–1
merchandising trends: 13–1, 13–2
mirror lighting: 13–7
parking area lighting: 13–12
relationship to merchandising: 13–2 to 13–4
self-service: 13–1
shadow boxes: 13–2
show windows: 9–64, 13–7 to 13–9
showcase lighting: 9–67, 13–6, 13–7
specific method: 13–5
supplementary lighting: 13–1, 13–6, 13–7
valances: 13–2, 13–6
veiling reflections in store windows: 13–2, 13–7, 13–8
vertical plane illumination: 13–3

STRAY LIGHT, effect on vision: 3–10

STREET LIGHTING (see **ROADWAY LIGHTING**)

STROBOSCOPIC EFFECT
in inspection lighting: 14–10
with fluorescent lamps: 8–27
with high-intensity discharge lamps: 8–40

STRUCTURAL LIGHTING ELEMENTS: 15–12 to 15–14

STRUCTURAL STEEL FABRICATION, illumination level: 9–89

STUDY DESK, Home, lighting for: 15–16

STYLES OF ARCHITECTURAL PERIODS: 10–9, 10–10

SUBSTITUTION PHOTOMETRY: 4–3

SUBWAY CAR LIGHTING (see **RAPID TRANSIT CARS**)

SUGAR REFINING, illumination levels: 9–89

SUN
as light source: 7–1
average luminance: 2–6
hours of sunshine: 7–3
illumination as function of altitude: 7–5
irradiance outside atmosphere: 18–4, 18–5
luminance of: 8–55

SUNLAMPS: 8–100, 25–15, 25–16

SUNLIGHT
illumination level at sea level: 2–6
spectral distribution: 25–1

SUPPLEMENTARY LIGHTING
classification of visual tasks for: 14–7 to 14–10
definition: 1–20
in interior lighting design: 10–11
industrial lighting: 14–6 to 14–10
luminaire classification: 14–7
offices: 11–9
schools: 11–11, 11–14
store lighting: 13–1, 13–6, 13–7
techniques for: 14–7 to 14–10

SURFACE MOUNTED LUMINAIRES, for residences: 15–4, 15–7, 15–8

SURGERY
illumination levels: 9–85
lighting for: 12–11 to 12–13

SURVEY PROCEDURES (see **LIGHTING SURVEYS**)

SWIMMING POOLS
illumination levels: 9–93
lighting of: 19–4, 19–9, 19–11

SYMBOLS, Photometric and Radiometric: 1–2, 1–3

SYNAGOGUE LIGHTING (see **CHURCH LIGHTING**)

SYNTHETIC ENAMEL, maintenance of: 16–14

SYNTHETIC FABRIC MILLS
illumination levels: 9–89
lighting for: 14–28, 14–29

SYNTHETIC FIBER PLANTS: 14–30

T

TABLE OF RECOMMENDED FOOTCANDLE LEVELS: 9–81 to 9–95

TABLE TENNIS, Home, lighting for: 15–23

TAF (see **TRANSIENT ADAPTATION FACTOR**)

TALBOT'S LAW: 4–3, 4–7

TASK (see **VISUAL TASK**)

TAYLOR REFLECTOMETER: 4–5

TELEVISION LIGHTING
aiming angles: 24–17
ANSI codes for lamps: 24–4
arc light for: 24–3, 24–4
back light: 1–3, 1–4, 24–16
base light: 1–4, 24–16
cable television: 24–18
camera pickup tubes: 24–12
carbon arcs for: 8–118, 8–119
chroma key: 24–13
closed circuit: 24–18
color: 24–13
contrast, balancing for: 24–16
control equipment: 24–5, 24–6
dimmers, types of: 24–5, 24–6
educational: 24–18
ellipsoidal spotlight for: 24–3
equipment installation: 24–16 to 24–18
field pickups, sports events: 24–14
fill light: 1–8, 24–16
film production: 24–13
follow spot for: 24–3
hard light: 24–1
illumination levels: 24–14, 24–15
illumination, types of: 24–16
key light: 1–11, 24–1, 24–16
kinescope recordings: 24–13
lamps for: 8–11, 8–64 to 8–72, 24–3, 24–4
luminaire installation: 24–16, 24–17, 24–18
luminaires for: 24–1 to 24–3
measurement of illumination: 24–15
memory devices for: 24–6
modeling light: 24–16
monochrome: 24–12, 24–13
net production area: 24–17
patch system for: 24–5
Plumbicon pickup tube: 24–12
projected backgrounds: 24–13
selector switch system for: 24–5
set light: 1–17, 24–16
soft light: 24–1
studio lighting: 24–12 to 24–19
underwater use: 26–5
videotape recordings: 24–13
Vidicon camera: 24–12, 24–13

TEMPERATURE (see also **COLOR TEMPERATURE**)
as comfort parameter: 16–1, 16–2
bulb, incandescent lamp: 8–6, 8–7, 8–8
conversion equations: 1–27
effect of photovoltaic cells: 4–7
effect on equipment design: 6–23, 6–24, 6–25
effect on fluorescent floodlamps: 23–11
effect on fluorescent lamps: 8–25, 8–26, 8–28
effect on high-intensity discharge lamps: 8–38
effect on light output: 16–13
mean radiant: 16–2
melting point of tungsten: 8–1
socket, incandescent lamp: 8–6, 8–7

TENNIS
illumination level: 9–93
lighting for: 19–4, 19–9, 19–12, 19–17

TERRARIUM LIGHTING: 25–9, 25–10

TERMS AND TERMINOLOGY (see **Section 1, DICTIONARY OF LIGHTING TERMS**)

TESTING, illumination levels: 9–89

TESTING PROCEDURES
electrical measurements: 4–11 to 4–14
life performance: 4–16
photometric measurements: 4–14 to 4–16
SAE for motor vehicles: 22–3

TEXTILE MILLS (see also specific types)
illumination levels: 9–89, 9–90
lighting for: 14–27 to 14–31

TEXTILES, light source for color matching: 5–22

THALOFIDE CELL, spectral response: 25–2

THEATRE LIGHTING
ANSI codes for lamps: 24–4
arc light for: 24–3
community theatres: 24–6 to 24–10
control equipment for: 24–5
convention facilities: 24–10
dimmerboards: 24–5
drive-in theatres: 24–11, 24–12
foyer: 24–6
hard light: 24–1
illumination levels: 9–90
industrial show facilities: 24–10
key light: 1–11, 24–1, 24–16
lamps for stage: 8–11, 8–64 to 8–72, 24–3, 24–4
legitimate theatres: 24–6 to 24–10
live-production theatres: 24–6 to 24–10
lobby: 24–6
luminaires for: 24–1 to 24–3, 24–7 to 24–10
marquees: 24–6
meeting facilities: 24–10
motion picture theatres: 24–11
scenic projection: 24–10
school theatres: 24–6 to 24–10

seating areas: 24–6, 24–7
soft light: 24–1
special effects: 24–11
stage area: 24–7 to 24–10
theatre-restaurants and lounges: 24–10
theatrical effects: 24–10

THERMAL SWITCH STARTER: 8–31

THERMAL TESTING OF LUMINAIRES: 4–26

THERMOLUMINESCENCE: 2–12

THREE-LITE LAMPS: 8–13, 8–61

THRESHOLD ILLUMINATION, for signal visibility: 3–35, 3–36

TIRE AND TUBE MANUFACTURING, illumination levels: 9–88

TOBACCO PRODUCTS, illumination levels: 9–90

TOILETS AND WASH ROOMS, illumination levels: 9–90

TOPLIGHTING, Daylighting: 9–76

TORRS, definition of: 8–41

TOTAL REFLECTION: 6–3

TRAFFIC SIGNAL LAMPS: 8–13, 8–75

TRAIN AND LOCOMOTIVE LAMPS: 8–11, 8–62

TRAIN LIGHTING (see **RAIL CONVEYANCE** and **RAILWAY GUIDANCE SYSTEMS**)

TRANSFORMERS
for high-intensity discharge lamps: 8–38 to 8–40
for sign tubing: 23–4

TRANSIENT ADAPTATION
effects of: 3–10, 3–11
estimation of effect in design: 3–18, 3–19
in visual sensory process: 3–3
nature of: 3–3

TRANSIENT ADAPTATION FACTOR, use of: 3–18, 3–19, 3–23

TRANSILLUMINATION
for drafting: 11–8
industrial: 14–8, 14–9, 14–10

TRANSMISSION: 6–7

TRANSMISSION FACTOR (see **TRANSMITTANCE**)

TRANSMISSIVITY, ATMOSPHERIC (see **ATMOSPHERIC TRANSMISSIVITY**)

TRANSMITTANCE
alabaster: 6–6
characteristics of various materials: 6–6, 7–9, 16–4
glass: 6–6, 6–14, 6–15, 7–9, 16–4
human skin: 25–16
infrared, of various materials: 16–4
marble: 6–6
measurement of: 4–2, 4–7, 4–8
of light through water: 26–1 to 26–3
plastics: 6–6, 7–9, 16–4
spectral curves of filters: 4–7
spectral curves of various glasses: 4–10
spectral, of human skin: 25–16

TRANSPORTATION LIGHTING (see also specific mode of travel)
automobile lighting: 22–1 to 22–3
aviation: 21–1 to 21–6
illumination levels: 9–94
lamps for (see also specific use): 22–11
marine lighting: 22–12 to 22–15
motor vehicle lighting: 22–1 to 22–7
power systems for: 22–11, 22–12
public conveyance interior: 22–5 to 22–10
railway guidance systems: 22–15 to 22–19
railway passenger car lighting: 22–8 to 22–10
rapid transit car lighting: 22–7, 22–8
reflex devices: 22–3 to 22–5
requirements of: 22–1
SAE specifications: 22–3
ship lighting: 22–12 to 22–15

TRAP SHOOTING, illumination levels: 9–93

TRIBOLUMINESCENCE: 2–12

TRICHROMATIC COEFFICIENTS
for railroad signals: 22–18
for surgical lights: 12–12

TRIGONOMETRIC FUNCTIONS: 9–45

TRISTIMULUS VALUES
computational data for standard sources: 5–8
for equal power source: 5–3

TROUBLE SHOOTING, lamp circuits: 16–16 to 16–18, 16–20

TUNGSTEN FILAMENT LAMPS (see **INCANDESCENT FILAMENT LAMPS**)

TUNGSTEN FILAMENTS
color temperature of: 8–2
designations: 8–3, 8–4
form: 8–3, 8–4
properties of: 8–1, 8–2
resistance characteristics: 8–2

TUNGSTEN-HALOGEN CYCLE: 8–6

TUNGSTEN-HALOGEN LAMPS
for general lighting: 8–11, 8–60, 8–61
photoflood: 8–16, 8–85
projection lamps: 8–87 to 8–91
signs, use in: 23–10, 23–11
theatre, motion picture, television: 8–64 to 8–72

TUNNELS, lighting of: 20–10 to 20–14

TYPES OF LIGHTING SYSTEMS: 10–11, 10–12

U

ULBRICHT SPHERE: 4–9, 4–10

ULTRASONIC CLEANING: 16–16, 16–17

ULTRAVIOLET LAMPS: 8–100, 8–109 to 8–113, 25–14, 25–15

ULTRAVIOLET RADIANT ENERGY (see also **BACTERICIDAL ULTRAVIOLET RADIATION** and **ERYTHEMAL ULTRAVIOLET RADIATION**)
bactericidal: 25–14
bactericidal lamps: 25–16, 25–17
"black light": 25–14, 25–15
effect on plastics: 6–20
efficiency of, bactericidal: 25–16
efficiency of, erythemal: 25–16
erythemal: 25–15, 25–16
eye protection for: 25–15
fluorescence and phosphorescence: 25–13
generation of: 2–7, 2–8
industrial inspection: 14–10, 14–50
ozone producing: 25–15, 25–17
photochemical lamps: 25–16
reflectance of skin for: 25–16
reflectance of various materials to: 25–2
response of eye: 2–6, 25–3
sources of: 25–14 to 25–17
spectral distribution of BL fluorescent lamp: 25–15
spectral reflectance of various materials: 25–19
sunlamps: 25–15
transmittance of skin for: 25–16

UNDERPASSES, lighting of: 20–10 to 20–14

UNDERWATER LIGHTING
filtering properties of sea water: 26–2
light sources: 26–4
measurement techniques and instrumentation: 26–7
seeing distance in water: 26–2
sensor characteristics: 26–4
terms and definitions: 26–1
underwater lighting calculations: 26–5

UNDERWRITERS' LABORATORIES, INC., safety standards of: 6–21

UNIFIED THEORY, of light: 2–1

UNIFORMITY OF ILLUMINATION
luminaire spacing for: 9–33
office lighting: 11–4, 11–5
roadway lighting: 20–7

UNIT PREFIXES, conversion factor: 1–27

UNITED STATES COAST GUARD, federal standards: 6–22

UNITS
IES units to SI units: 1–24
length: 1–26
luminance: 1–21
prefixes: 1–27
radiometric and photometric: 1–2, 1–3
SI units to IES units: 1–25

UPHOLSTERING, illumination level: 9–90

UTILIZATION, COEFFICIENT OF (see **COEFFICIENT OF UTILIZATION**)

V

VALANCE LIGHTING: 1–21, 13–2, 13–6, 15–13, 15–14

VALUE, MUNSELL: 1–14, 5–7, 5–10

VCP (see **VISUAL COMFORT PROBABILITY**)

VEILING LUMINANCE: 3–10, 3–18

VEILING REFLECTIONS
allowance for in design: 3–17, 3–18
allowance for in evaluating equivalent sphere illumination: 3–23
causes of: 3–28
contrast loss from: 3–28
definition: 1–21
examples of: 3–29, 3–30
guides for reducing: 3–32, 3–33
methods of evaluating: 3–32
offending zone in: 3–30
office tasks: 11–4
polarization relation: 3–32
school tasks: 11–11
store windows: 13–7, 13–8
visual significance: 3–32

VELOCITY OF LIGHT: 2–2

VERTICAL SURFACE ILLUMINATION: 9–68, 23–7 to 23–12

VIBRATION, in **Equipment Design:** 6–24

VIBRATION SERVICE INCANDESCENT LAMPS: 8–12, 8–81

VIEWING ANGLES
for projected pictures: 24–28, 24–29
for visual tasks: 3–30, 3–31

VIEWING DISTANCE, for projected pictures: 24–28, 24–29

VISIBLE SPECTRUM: 2–2

VISIBILITY
definition: 1–21
luminous signals: 3–33 to 3–37
meters: 3–17
reference function: 3–14, 3–15
roadway: 20–1

VISIBILITY LEVEL: 3–14, 3–23

VISIBILITY METERS: 3–17

VISIBILITY REFERENCE FUNCTION: 3–14, 3–15

VISION (see also **VISUAL SENSORY PROCESSES** and **VISUAL ABILITY**)
binocular: 3–4, 3–6
central: 1–5, 3–4, 3–6
color: 3–4, 3–6
contrast borders: 3–7
contrast sensitivity: 3–7
effect of age on: 3–11, 3–12, 3–13
effect of stray light: 3–10
effects of surrounds on: 3–9 to 3–11
mesopic: 1–13, 3–4
peripheral: 1–14, 3–4, 3–6
photopic: 1–14, 3–4, 4–1
scotopic: 1–17, 3–4, 4–1

VISUAL ABILITY
effect of age: 3–11 to 3–13
effect of disability glare: 3–10
effect of shadows: 3–13, 3–14
effect of surroundings: 3–9 to 3–11
effect of transient adaptation: 3–10
physical factors: 3–7 to 3–9
physiological factors: 3–6

VISUAL ACUITY
definition: 1–21
effect of age on: 3–12
test objects for determining: 3–8
vs background luminance: 3–8

VISUAL ANGLE, definition of: 1–22, 3–8

VISUAL COLORIMETERS: 5–15

VISUAL COMFORT CRITERIA: 3–24 to 3–27

VISUAL COMFORT PROBABILITY (VCP)
discomfort glare rating: 3–26
evaluation procedure: 3–24
in offices: 11–3, 11–4
in schools: 11–11
simplified procedures: 3–26, 3–27, 3–28
typical table of VCP values: 3–27
use of VCP value: 3–26

VISUAL FIELD: 1–22, 3–4

VISUAL PERFORMANCE CAPABILITY
definition of: 3–14
studies of: 3–15, 3–16

VISUAL PERFORMANCE CRITERIA, types of: 3–14

VISUAL PERFORMANCE CRITERION FUNCTION: 3–14, 3–18

VISUAL PERFORMANCE POTENTIAL
definition of: 3–14
selection of a criterion level: 3–14 to 3–16

VISUAL PHOTOMETERS: 4–1, 4–4

VISUAL PURPLE (RHODOPSIN): 3–2

VISUAL RANGE OF LUMINOUS SIGNALS: 3–33 to 3–37

VISUAL SENSORY PROCESSES: 3–1 to 3–6

VISUAL SIZE: 3–8

VISUAL TASK
effect of luminance on: 11–2
evaluation in design: 10–10
evaluation of illumination for: 3–23
fundamental factors of: 3–7
in visual ability: 3–7

library: 12–15
lighting for residential: 15–12, 15–15 to 15–26
lighting of various industrial: 14–7 to 14–10
office: 11–1, 11–2
oriented design approach: 10–10, 10–11
school: 11–11
veiling reflections in: 3–28 to 3–30
viewing angles 3–31
visibility of: 3–14

VISUAL TASK EVALUATOR: 3–17

VISUAL TASK PHOTOMETER: 4–8

VITAMIN D, production of by ultraviolet: 25–15

VOLLEYBALL, illumination levels: 9–93

VOLTAGE
effect on incandescent lamps: 8–8
to luminaires, effect on light loss factor: 9–3, 16–13

VOLUME SCATTERING FUNCTION, water: 26–1

VTE (see **VISUAL TASK EVALUATOR**)

VTOL LIGHTING, aviation: 21–12

W

WALKWAYS: 20–5, 20–6, 20–9

WALL LUMINANCE COEFFICIENTS
for typical luminaires: 9–13 to 9–29
work sheet for computing: 9–42

WAREHOUSE (see **STORAGE ROOMS**)

WASH TANKS, for maintenance: 16–16, 16–17

WATTS PER SQUARE FOOT, required for various lighting systems: 16–8, 16–9

WAVE THEORY, of light: 2–1

WAVELENGTH
effect on photoelectric detection: 2–12
measurement of: 4–2
of light: 2–1, 2–2
relation to velocity: 2–2

WELDING, illumination levels: 9–90

WELSBACH MANTLE: 2–7

WIEN DISPLACEMENT LAW: 1–22, 2–3

WIEN RADIATION LAW: 1–22, 2–3, 2–4

WINDOWS
art windows: 12–3, 12–4
daylighting controls for: 7–6, 7–8, 7–9
false, in museums: 12–23
gymnasium: 19–3, 19–6
light loss factors for: 7–12

WIRING, for interior lighting: 16–7 to 16–9

WOODWORKING, illumination levels: 9–90

WOOLEN AND WORSTED MILLS
illumination levels: 9–89, 9–90
lighting of: 14–30, 14–31

WORKING STANDARDS: 1–22, 4–3, 4–14

WORSHIP AREAS, churches and synagogues: 12–2 to 12–4

WRESTLING
illumination levels: 9–92
lighting for: 19–4, 19–7, 19–8

X

XENON LAMPS
characteristics of: 8–42
color rendering characteristics: 5–20, 5–21
for underwater illumination: 26–4
picture projection: 24–34, 24–35
table of: 8–114, 8–115

Y

YOUNG-HELMHOLTZ COLOR THEORY: 3–4, 3–5

Z

ZIRCONIUM CONCENTRATED ARC LAMPS: 8–44, 8–117

ZONAL-CAVITY METHOD, OF ILLUMINATION CALCULATION (see **LUMEN METHOD**)

ZONAL CONSTANTS
for computing lumens from candlepower data: 4–18
for projector type luminaires: 4–19, 4–20

ZONAL LUMENS
constants for computing: 4–18
nomogram for obtaining: 4–21

ILLUMINATING ENGINEERING SOCIETY
345 EAST 47TH STREET • NEW YORK, N.Y. 10017 PLAZA 2 • 6800

Section/Chapter Use Only
(Signature of Chairman, Section/Chapter Board of Examiners)
Section/Chapter
Headquarters/Board of Examiners Use Only
Date ______ Action ______
Signature ______

APPLICATION FOR MEMBERSHIP

All Applicants are required to complete the entire form.

I HEREBY make application for ☐ **admission** ☐ **transfer** ☐ **reinstatement—date of original election** ______ **to the grade of** ☐ **MEMBER**** ☐ **ASSOCIATE** ☐ **AFFILIATE** ☐ **STUDENT**

I wish to be affiliated with Section/Chapter closest to my business address ☐ **to my home** ☐

1. Full Name .... (Print last name first)
2. Present Occupation .... (Title and Department)
3. Name of Company .... Telephone .... (Area Code)

Mail to √ one
4. ☐ Business Address .... (Street) (City) (State) (Zip Code)
5. ☐ Home Address .... (Street) (City) (State) (Zip Code)
6. Application Obtained from ....
7. Other Society Affiliations .... (Give grade of membership)
8. Date and Place of Birth ....
9. Type of Lighting of Interest to Applicant? ....
10. College .... Degree .... Year .... Curriculum ....
11. Business Experience *(Please begin with most recent experience):*

DATES From Mo./Yr.	To Mo./Yr.	Detail of qualifying professional experience (Attach separate sheet if needed for complete statement)	COMPANY	TITLE

**MEMBER grade applicants are urged to supply five references or endorsements in MEMBER or FELLOW grade (*minimum of three required*). ASSOCIATE, AFFILIATE, and STUDENT grades require one sponsor in FELLOW, MEMBER, or ASSOCIATE grade.

TO BE COMPLETED BY REFERENCES

I hereby certify that I am qualified** to act as a reference for this applicant and that the applicant is eligible for ______ grade, in accordance with the By-laws I have listed below (see back of application for By-laws).

	SIGNATURES OF REFERENCES	GRADE	COMPANY	QUALIFYING BY-LAWS
1				
2				
3				
4				
5				

I hereby certify that the above statements are correct, and agree, if elected by the Society, to be classified in the membership grade designated by the General Board of Examiners, and to be governed by its Constitution and By-laws for the duration of my membership.

(Over)

.... (Signature of Applicant) ..../..../.... Date

ILLUMINATING ENGINEERING SOCIETY

QUALIFICATIONS FOR MEMBERSHIP

CONSTITUTION and BY-LAWS, Article II

EDUCATIONAL QUALIFICATIONS, *Section 2*

a. In determining the qualifications of an applicant (in either MEMBER or Associate grade) under the requirement "shall be a graduate from a 4 year or longer accredited college curriculum related to the science or art of illumination,"
Credit shall be given but not limited to—Graduates of colleges who have received a degree in engineering, architecture, industrial design, physics, science, psychology, ophthalmology, optometry, interior decoration or design.

MEMBER Grade, *Section 2*

Applicants for admission or transfer to MEMBER grade may qualify under any of the following provisions:

a. A MEMBER shall be a graduate from a 4 year or longer accredited college curriculum related to the science or art of illumination and shall have been for at least 5 years actively engaged in a professional or other responsible capacity in the practice or teaching of illumination, or in other fields or activities directly related thereto;

In determining the qualifications of an applicant for MEMBER grade under the requirements "actively engaged in a professional or other responsible capacity in the practice or teaching of illumination, or in other fields or activities directly related thereto," consideration shall be given to experience as:

(i) An engineer, architect, designer or decorator, who designs lighting systems or lighting installations based on the recommended standard practices of the Society.

(ii) A lighting application engineer employed by a manufacturer, public utility, contractor or distributor, who plans or designs lighting systems or lighting installations based on the recommended and standard practices of the Society.

(iii) An engineer or designer who designs or who supervises the manufacture or the testing of lighting equipment, lamps, luminaires, control or other equipment associated with lighting.

(iv) An executive who is in immediate responsible charge of activities of others qualified for MEMBER grade.

(v) An engineer, physicist, scientist, ophthalmologist, or optometrist engaged in research or application in the fields of light, vision, or optics.

(vi) A teacher of illuminating engineering or related subjects. As such, under general direction, he shall have taken responsibility for instructions in the subject in a school of recognized standing.

b. A MEMBER shall be a graduate from a 4 year or longer accredited college curriculum related to the science or art of illumination and shall have made some valuable contribution to the science or art of illuminating engineering or to the literature pertaining thereto;

c. A MEMBER shall have been for at least 10 years actively engaged in the practice or teaching of illumination, or in other fields or activities directly related thereto, of which 5 years are in a professional or other responsible capacity;

d. A MEMBER shall have made some exceptionally valuable contribution to the science or art of illuminating engineering or to the literature pertaining thereto.

In determining the qualifications of an applicant for MEMBER grade under the requirements "and shall have made some exceptionally valuable contribution to the science or art of illuminating engineering or to the literature pertaining thereto," consideration shall be given to:

(i) The development of new basic knowledge, techniques, or instrumentation in the field of light and vision, or fields related thereto.

(ii) Invention or perfection of materials, devices or systems which advance the science or art of illumination.

(iii) Papers or treatises on original work on lighting or related subjects, particularly those accepted for publication in recognized journals.

ASSOCIATE Grade, *Section 3*

Applicants for admission or transfer to Associate Member grade may qualify under either of the following provisions:

a. An Associate Member shall be a graduate from a 4 year or longer accredited college curriculum related to the science or art of illumination;

b. An Associate Member shall have been for at least 6 years actively engaged in the practice or teaching of illumination, or in other fields or activities directly related thereto.

AFFILIATE Grade, *Section 5*

An Affiliate may be anyone interested in the aims of the Society. At the time of his election, he shall be at least 18 years of age. He shall receive the Society's monthly journal and have the privilege of participating in meetings. He shall have the right to vote only in Chapter, Section and Regional affairs. He may not serve as a national officer or as an officer of a national committee.

STUDENT Grade, *Section 4*

A Student Member shall be one registered in a College or University for study toward a degree in an engineering, science, fine arts or other related curriculum acceptable to the Board of Directors. A Student Member may have the privilege of participating in meetings, but shall not have the right to vote or hold elective office except in a Student Branch.

a. A Student Member shall be eligible to continue in that grade through the fiscal year following that in which his status as a student ceases, and in no case for a period of more than four years.

PROCEDURE FOR MAKING APPLICATION

Submit your application to the local officers of the Section or Chapter where you are located. These officers will approve and forward your application to the Headquarters of the Society. If you are not located near a Section or Chapter, send the application to the Illuminating Engineering Society, 345 East 47th Street, New York, N. Y. 10017.

Your application should be accompanied by your check or money order covering entrance fee and proportion of the dues below. *Complete* applications received at Headquarters by the 15th of any month will, in general, be considered for election in the following month. Make check payable to Illuminating Engineering Society.

ENTRANCE FEES are in addition to Annual Dues and must accompany the Application.

Application received for consideration	Entrance Fee	MEMBER GRADE Dues to June 30th End of Fiscal Year	ASSOCIATE GRADE Dues to June 30th End of Fiscal Year	AFFILIATE GRADE Dues to June 30th End of Fiscal Year	*STUDENT GRADE Dues to June 30th End of Fiscal Year
July 1—December 31	$2.50	$30.00	$25.00	$20.00	$5.00
January 1—June 30	2.50	15.00	12.50	10.00	5.00 *No Entrance Fee

USE OF COLOR CHART BELOW

The facing page displays a series of neutral grays and chromatic colors, with their reflectances, selected for use in a variety of interiors. These represent a sampling of the top half of the Munsell Color Solid (see Section 5) and will permit the visualization of other colors in the same region, with their reflectances. The chart displays ten hues, one for each of the ten hue families, on values 6/, 7/ and 8/, with one yellow color on 8.50/ value. A selection of neutral grays also is included. Munsell notations are shown below each color.

All colors in a vertical column are of the same hue and are therefore harmonious. With the exception of Row 6, all colors in a horizontal row have the same value notation, and therefore the same reflectance.

All colors within a row are in the same chroma range. The weak chroma colors in Rows 3, 4, and 5 represent the colors usually employed in large areas of walls and floors. There are many more than these from which to choose, produced by intermixing adjacent hues and varying slightly values and chromas.

Ceilings are usually white or an off-white in the direction of the wall color hue. One can visualize such colors readily by imagining the Row 3 colors mixed with white.

The top two rows (1 and 2) display colors of higher chroma, used for smaller wall and floor areas and as colors for furniture and office equipment. Such colors may appear as elements in patterns used for large walls and floors, and are occasionally used as principal colors in large areas. Row 6 shows a popular high chroma color for each hue, regardless of value, for use as accent colors. When using these colors, the usual rule is the higher the chroma the smaller the area.

The hue families are listed across the bottom of the chart. Note that the hues have been grouped into the warm, cool, and essentially neutral categories. The selection and grouping of colors has been done to facilitate the selection of color schemes, to include those using colors intermediate to those on the chart. This last factor is very important. With the hues arranged in their normal hue relationship the user will have little trouble visualizing colors falling between those displayed. In this manner a limited number of colors permits the effective employment of many times their number.